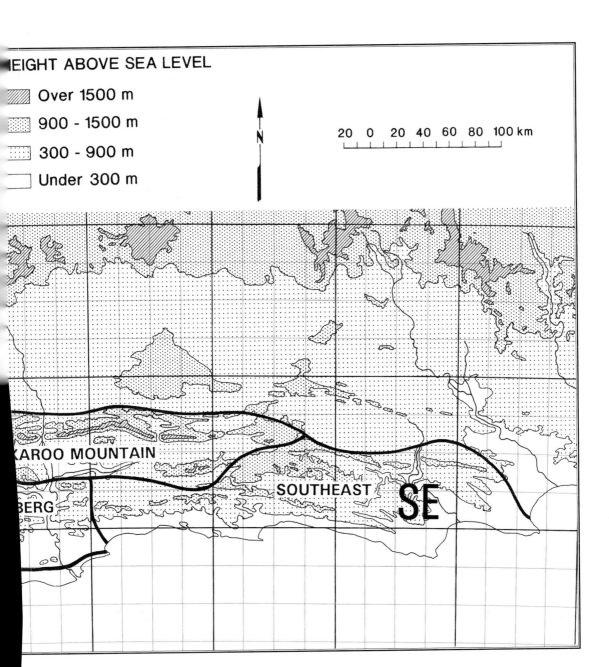

STRELITZIA 9

CAPE PLANTS
A Conspectus of the Cape Flora of South Africa

Peter Goldblatt and John Manning

STRELITZIA

This series has replaced *Memoirs of the Botanical Survey of South Africa* and *Annals of Kirstenbosch Botanic Gardens* which the NBI inherited from predecessor organizations.

The genus *Strelitzia* occurs naturally in eastern southern Africa. It comprises three arborescent species, known as wild bananas, and two acaulescent species, known as crane flowers or bird-of-paradise flowers. The logo of the National Botanical Institute is based on the striking inflorescence of *Strelitzia reginae*, a native of the Eastern Cape and KwaZulu-Natal that has become a garden favourite worldwide. It symbolizes the commitment of the National Botanical Institute to promote the sustainable use, conservation, appreciation and enjoyment of the exceptionally rich plant life of South Africa, for the benefit of all its people.

Authors
Peter Goldblatt, B.A. Krukoff Curator of African Botany,
Missouri Botanical Garden

John Manning, Compton Herbarium,
National Botanical Institute of South Africa

Technical Editors: E. du Plessis and G. Germishuizen

ISBN 0-620-26236-2
Cape Town
September 2000

Copyright © text and photographs 2000 by the authors.

Published jointly by

National Botanical Institute of South Africa
Private Bag X101, Pretoria 0001, South Africa

MBG Press, Missouri Botanical Garden
P.O. Box 299, St. Louis, Missouri 63166, U.S.A.

Design: Red Roof Design, Cape Town, South Africa
Reproduction by Fairstep, Cape Town, South Africa
Printed by ABC Press, Epping, South Africa

Contributors

Clare Archer (Cyperaceae)
Robert H. Archer (Celastraceae in part)
Kevin Balkwill (Acanthaceae)
Bruce Bayer (*Haworthia, Oxalis*)
P.A. Bean (*Agathosma*)
Angela Beaumont (*Gnidia*)
Josephine P.B. Beyers (*Arctotheca, Arctotis, Eriocephalus, Haplocarpha, Lachnaea, Struthiola*)
Christien Bredenkamp (*Passerina*)
Peter V. Bruyns (asclepioid Apocynaceae, *Euphorbia*)
Matt H. Buys (*Lobostemon*)
G.J. Campbell (*Rafnia*)
Graham D. Duncan (*Lachenalia*)
Trevor J. Edwards with A.L. Schutte (*Argyrolobium*)
Anna C. Fellingham (*Cliffortia*)
Gerrit Germishuizen with A.L. Schutte (*Rhynchosia*)
Paul Herman (*Gymnostephium*)
O.M. Hilliard (*Globulariopsis, Microdon, Selago*)
Marie Jordaan (Celastraceae in part)
Cornelia Klak (Aizoaceae: Ruschioideae and Mesembryanthemoideae)
Marinda Koekemoer (*Amphiglossa, Elytropappus, Stoebe*)
Hubert Kurzweil (Orchidaceae)
H. Peter Linder (Poaceae in part, Restionaceae)
Nicole L. Meyer & G.F. Smith (*Astroloba*)
Rodney O. Moffett (Anacardiaceae)
E.G.H. Oliver & I.M. Oliver (Ericaceae)
Peter B. Phillipson (Campanulaceae: Lobelioideae)
Elizabeth Retief (Boraginaceae except *Lobostemon*, Vitaceae)
J.P. Roux (Pteridophytes)
Brian D. Schrire (*Indigofera, Tephrosia* with A.L. Schutte)
Anne Lise Schutte (Fabaceae in part)
Gideon F. Smith (*Aloe, Astroloba* with N.L. Meyer)
Deidre A. Snijman (Agapanthaceae, Alliaceae, Amaryllidaceae, Hyacinthaceae in part, Hypoxidaceae)
Kim E. Steiner (*Colpias, Diascia, Hemimeris*)
C.H. Stirton with A.L. Schutte (*Otholobium, Psoralea*)
Ben-Erik van Wyk (*Annesorhiza* with P. Tilney, *Centella* with M. Schubert, *Chamarea*)
L. van Zyl (*Zygophyllum*)
A.V. Verboom (*Ehrharta*)
P. Vorster (*Pelargonium*)
W.G. Welman (*Solanum*)

Sponsor
Elizabeth Parker

Preface

It is now some 15 years since *Plants of the Cape flora* was completed by Pauline Bond and one of us (Peter Goldblatt), after more than five years in preparation. The volume, published as a Supplement to the *Journal of South African Botany*, enjoyed unexpected success and was used by a surprisingly broad range of people. For biologists of many fields it was a vital reference while naturalists and plant people of all kinds found the book a valuable resource. We were often surprised and gratified to see an annotated, and much thumbed copy on the bookshelf of a hiker, farmer, or wild-flower enthusiast. That volume certainly had its shortcomings, but was something that Pauline and I had worked hard to bring to completion and we were proud of it. We acknowledged then, and we are glad to acknowledge again, the help that botanical colleagues gave us, making the volume a truly collaborative effort. *Plants of the Cape flora* had less than 100 copies left for sale in 1992 when Barrie Low, then responsible for publications of the National Botanical Institute, asked John Rourke, Curator of the Compton Herbarium at the National Botanical Institute, and one of us (Peter Goldblatt) if we might undertake the preparation of a new edition.

We agreed that there was a need for a new volume, not only because the present one would soon be unavailable, but because advances in botanical knowledge had rendered the first volume very much out of date. Names of plants are subject to change in the light of new information and, at a time of unusual activity in systematic botany associated with the growth of molecular systematics, the circumscription of families too was changing. Administrative responsibilities and other endeavours took up all of John Rourke's time and John Manning willingly filled his place as coauthor-cum-editor. As in *Plants of the Cape flora*, we have depended very much on all the botanical expertise available on Cape plants and we have had the benefit of generous help from colleagues at the National Botanical Institute, the University of Cape Town, and several other institutions in South Africa and abroad.

While *Plants of the Cape flora* was a milestone and valuable baseline, we felt a need to advance the quality of the content. We felt particularly that geographical information needed to be as accurate as knowledge allowed. In collaboration with Steve Johnson, then at the University of Cape Town, and by proxy with Richard Cowling, Director of the Institute of Plant Conservation, also at the University of Cape Town, we adopted a system of citing ranges of species by phytogeographic centre, of which there are six within the Cape flora region. Botanical knowledge of the Cape flora is still imperfect and, with many genera not revised for over 50 years and some not within the last century, the ranges of species are not always completely documented. Nevertheless, we now have a baseline for the number of species for each centre and that list can be enlarged with relative ease.

One of the most important uses of *Cape plants* is as an aid to plant identification. To that end we have adopted another innovation, the addition of keys to the families and genera in the flora. We hope they prove usable and useful. To assist with identification of species, especially among the larger genera, we have expanded the diagnoses, attempting within space limitations to provide critical descriptive information, and have arranged species in clusters by major characters. It should prove possible, with care and effort, to identify most species of the flora now by using the generic keys, species arrangement, and the diagnoses in combination with distributional information. Larger genera remain a problem with no easy solution. We have further expanded the scope of the treatment by including the pteridophytes, a group that was not included in *Plants of the Cape flora*.

The preservation of species and ecosystems is of concern to all biologists, and the provision of accurate information of all kinds about organisms is essential to the successful outcome of conservation efforts. We need to know how many species there are, where they are, and what they look like before we can effectively direct conservation resources. With this in mind we have tried to provide within the scope of a single volume much of this basic information about the plants of the Cape flora. There is always a small gap between nomenclatural correctness and biological reality and whenever the two have conflicted we have favoured biological reality. We include, as was the case in *Plants of the Cape flora,* undescribed species, either with a manuscript name if one is available, or simply a number. Conversely, species that seem to us to be identical are treated under the earliest name with the later name included in parenthesis. In this way the present treatment is far more critical, and we trust, reflects more accurately biological reality. The only formal synonyms listed are those species that were recognized in *Plants of the Cape flora*.

The past ten years have been a time of considerable advance in the understanding of the evolution and radiation of plants. One of the most striking results has been the realization that some genera are misplaced in a particular family. In some instances, families have been found to have no foundation, but are specialized lineages of other families. Examples are legion. We favour recognizing only those families that are natural (monophyletic) entities. Consequently we have adopted the system of family and ordinal classification proposed by the Angiosperm Phylogeny Group published in 1998 that incorporates the new molecular data into a classification of the orders and families of the flowering plants. Readers who follow plant family classification will note especially the inclusion of Asclepiadaceae in Apocynaceae, Capparidaceae in Brassicaceae, Sterculiaceae and Tiliaceae in Malvaceae, and Eriospermaceae in Convallariaceae. Conversely, several other families are now recognized. These include Menyanthaceae (with *Nymphoides* and *Villarsia* removed from Gentianaceae), Celtidaceae (with *Celtis* removed from Ulmaceae), Behniaceae and Hemerocallidaceae separated from Anthericaceae, and Prioniaceae separated from Juncaceae. Particularly noteworthy, Mesembryanthemaceae, recognized in *Plants of the Cape flora*, are included in Aizoaceae, but several genera until now included in the latter family, have been removed to Molluginaceae. We have felt compelled to adopt all the new changes proposed by the Angiosperm Phylogeny Group in the interest of a better treatment of the flora. We sympathize with readers who may initially be confused and hope the cross-references in the text will help them find their way.

This volume ends with a chapter of formal taxonomy. Where we felt confident that nomenclatural changes were called for we have taken the opportunity to make them. Such changes include the validation of new genera, sinking of others, the reduction of species that we have found through field study to have no foundation, and the correction of various nomenclatural problems. It seemed to us better to make these changes forthwith, rather than publish them piecemeal in scattered journals whenever we found the time to write small, completely mechanical papers.

Despite our best endeavours, we realize that this volume has shortcomings. Plant systematics remains an unending synthesis. Species continue to be discovered and new discoveries may show that others lack reality. We are, however, confident that this volume will prove even more useful than *Plants of the Cape flora*. Again, we express our thanks to all who collaborated with us, sharing their knowledge, providing written accounts of families or genera, giving advice, or simply encouraging us in this undertaking. In particular we must express our gratitude to Ingrid Nänni for her review of the manuscript, Susette Foster, for help with many aspects of manuscript preparation, Emsie du Plessis and Gerrit Germishuizen for their thorough editing, and Brian Huntley, John Rourke and Peter Raven for their support through the several years that it took to bring this project to completion.

Peter Goldblatt and John Manning   Cape Town, August 2000

Contents

Introduction to the Cape Flora — 7

    Physical characteristics of the Cape Region — 8
    Floristic composition — 10
    Diversity — 15
    Plates — 20
    Reasons for Cape floristic diversity — 32
    How to use this account — 35

The Cape Flora — 37

    Pteridophytes: Ferns and fern allies — 37
    Gymnosperms — 50
    Angiosperms: Flowering plants — 51
    Palaeodicotyledons — 51
    Monocotyledons — 52
    Eudicotyledons — 219

Keys to the Families — 691

Taxonomic Notes — 703

    Families recognized and generic realignments — 703
    Taxonomic and nomenclatural changes — 704

Bibliography — 717

Appendix — 724

    Numbers of genera and species — 724

Index of Families and Genera — 731

Introduction to the Cape Flora

Situated at the southwestern tip of the African continent between latitudes 31° and 34°30' S, the area that has come to be called by biologists the Cape Region has a flora, and to a lesser extent fauna, that is so sharply distinct from that of the lands immediately surrounding it that it has impressed naturalists from the time of its discovery by European explorers in the sixteenth century. Indeed, the floristic characteristics of the Cape Region are so unusual that it is often regarded as one of the world's six floral kingdoms (e.g. Good 1974; Takhtajan 1986). There are no accepted criteria for distinguishing 'floral kingdoms' and recognition of a Cape Floral Kingdom is not universal. We use the neutral appellation 'floristic region' here simply for convenience. Whatever the terminology used, however, the flora of the Cape Region is always accorded special status in classifications of the world's floras.

Comprising a land area of about 90 000 km^2, less than 4 % of the total area of the southern African subcontinent (Goldblatt 1978), the Cape Floristic Region is one of the world's richest regions in terms of botanical diversity. An estimated 9 000 species of vascular plants (ferns and fern allies, gymnosperms, and flowering plants) are native to this area, almost 69 % of which are endemic. The great majority of these species, some 8 888 in total, are flowering plants. Thus, the flora of the Cape Region comprises almost 44 % of the approximately 20 500 species that occur in all of southern Africa (Arnold & De Wet 1993; C. de Wet pers. comm.). In fact, the species richness of the Cape Region is notable not only in an African context, but is remarkable for the temperate zone. Species richness of the Cape Region compares favourably with that areas of comparable size in the wet tropics rather than to any temperate region. Only over the past 50 years as our knowledge of plant species diversity has increased, has it become evident just how species-rich the Cape Region actually is. Despite claims that the Cape Region has, for its size, one of the richest floras in the world, this is not strictly true. Although species richness in the Cape Region is comparable to many areas of similar size in the moist tropics, which are undoubtedly the most species-rich habitats on earth, it is substantially lower than in some Neotropical areas, including Costa Rica, Ecuador or Guatemala. The species richness of the Cape Region is, however, remarkable when compared to other parts of Africa, in which the tropical flora is relatively depauperate in terms of overall species numbers.

Compared with floras of other parts of the world, including other parts of Africa, the composition of the Cape flora is also extremely unusual (Bond & Goldblatt 1984; Goldblatt 1997). Unexceptional for a region of fairly dry climate, the largest families are the Asteraceae and Fabaceae, together comprising some 20 % of the total species. The families next in size, however, are remarkable for any flora. Nowhere else in the world do Iridaceae, Aizoaceae, Ericaceae, Proteaceae and Restionaceae assume numerical significance, except for parts of Australia where Proteaceae and Restionaceae are also well represented.

Other floristic peculiarities of the flora are the dominance of fine-leaved sclerophyllous shrubs, the very few trees, and a remarkably large number of geophytes, here defined as seasonal herbaceous perennials with bulbs, corms, or prominent rhizomes (thus excluding shrubs and subshrubs that resprout from a woody caudex, usually after fire). Such plants, especially numerous among the monocots (notably Iridaceae, Hyacinthaceae and Amaryllidaceae), also include many species of *Oxalis* and *Pelargonium* as well as species of several other genera among the dicots. Geophytes as so defined comprise somewhat more than 17 % of the species in the Cape flora. Conversely, the Cape flora has a surprisingly low proportion of annuals for an area of largely semi-arid climate. Approximately 6.8 % of the species are annuals, which is a striking contrast to California (30 % annuals) or central Chile (nearly 16 % annuals) (Arroyo *et al.* 1994), areas of comparable latitude and climate. Both geophytes and annuals are primarily adapted to seasonally dry climates and escape the time of year unfavourable for growth by retreating to underground storage organs or by ensuring continued survival only by production of seeds.

Geologically, the Cape Region consists of a mosaic of sandstone and shale substrates that give rise to soils of quite different types. In addition, local areas of limestone contribute significantly to the edaphic diversity. Climates across the region are extremely variable and the predominant orographic rainfall ranges locally from 2 000 mm to less than 100 mm per year, with extremely steep gradients as a result of the mountainous landscape. These sharp local gradients in precipitation compound the edaphic diversity resulting from the mosaic of different soils and create an unusual number of local habitats. A feature of many soils in the Cape Region is a low nutrient status and many plants on such soils have poor seed dispersal capabilities, an important factor in explaining the high levels of local endemism. Species richness in the Cape Region then, seems to be the result of an interplay between a complex mosaic of diverse habitats and steep ecological gradients against a background of *relatively stable* climate and geology after the establishment of a Mediterranean climate there sometime after the beginning of the Pliocene, c. 5 mya.

Physical characteristics of the Cape Region

Geology and soils. Over most of its surface the Cape Region is covered by soils derived from rocks of pre-Carboniferous age, thus more than 400 mya (Figure 1). Most of these rocks comprise part of the Cape System, an ancient Devonian–Ordovician series of sedimentary strata, consisting of alternating series of quartzitic sandstones (the Table Mountain and Witteberg Groups) and fine-grained shales (Bokkeveld Group). During the Jurrassic the land surface was folded and warped as Antarctica separated from the south coast of southern Africa and South America rifted away from the west coast. The folds consistently run parallel to the coasts, resulting in a series of east-west trending mountain ranges in the southern half of the Cape Region and north-south trending ranges in the west.

Differential weathering of the components of the Cape System yielded two fundamentally different soil types, coarse-grained sandy soils, poor in essential plant nutrients (Groves *et al*. 1983), and richer, clay soils of nutrient-intermediate status. At the low precipitation levels that are usual in the Cape Region, these factors become so limiting that the different soils support sharply different vegetation with different suites of species and different facies. Apart from differences in nutrient status, the soils differ significantly in their structure and water-retention properties. Erosional patterns differ on the two rock types and the result is that the mountains consist primarily of sandstone rocks and the valleys of shale. Where folding or faulting have been severe, more ancient rocks of the Precambrian Malmesbury Group are exposed. These are largely shales that give rise to clay soils of the same type as the shales of the Cape System. Granitic schists are locally exposed in deep valleys and along the west coast, and limestones, mainly of Tertiary age, are exposed near the coast where they are extensive only along the southern coast from the Agulhas Peninsula to Mossel Bay. The coastal plain includes areas with aeolian sandy soils derived from reworking of Cape Sandstones. Moving from the coastal plain to the interior, the resultant landscape is a mosaic of coastal limestones and deep sands, or valleys with clay soils alternating with mountain ranges of nutrient-poor sands. Local faulting has added a secondary component of islands of one rock type embedded in another. Both the nutrient-poor and nutrient-intermediate soils favour the development of a fairly uniform, sclerophyllous, shrubby vegetation that is fire-adapted (see discussion under Diversity).

The climate is largely Mediterranean, and strictly so in the west, but the eastern half of the Cape Region receives substantial summer precipitation. Rain thus falls mainly in the winter months and while summers are hot and dry, they are relatively less so in the east. In areas of low total rainfall, the average precipitation may be higher in the summer months, notably in the Little Karoo, although effective rainfall is still mainly in the winter. South-facing mountain slopes benefit from summer moisture in the form of rain or fog from the southeast trade winds. The narrow coastal plain in the Knysna area has a particularly equable climate and high rainfall and it supports an evergreen, broad-leaved forest. Likewise, sheltered valleys and locally wet sites throughout the region are covered in evergreen forest vegetation.

Local variation in rainfall is particularly pronounced in mountainous areas, and this is particularly important when precipitation is orographic. Mountain slopes facing prevailing winds receive considerably more precipitation than those that face away from these winds. Rainfall patterns in the Cape Region show dramatic variation in quantity, dropping from 2 000 mm per year on the high mountains of the ranges immediately facing the coast, to less than 200 mm on the leeward slopes of the interior ranges. Mosaic effects of soils across the entire region are thus complicated by variation in precipitation from the coast to the interior, as well as changes in seasonality from the west of the region to the east. In addition, elevation and aspect affect precipitation depending on the direction of moisture-bearing winds.

The number of ecological niches available to plant life is multiplied by soil differences, and this is particularly pronounced as precipitation levels drop. With ample rain the effect of soil on vegetation composition is less prominent. Rainfall is limiting almost throughout the region, however, and vegetation varies conspicuously with soil and available moisture. Climatic gradients are steep, although perhaps not more so than in most other areas of Mediterranean climate, but the effect may be compounded in the Cape Region by soil diversity. Different soil types in the Cape Region support characteristic vegetation types depending on associated levels of precipitation. Forest vegetation is typical of deeper soils where preciptation is high and fairly evenly spread throughout the year. As soil qualities change and precipitation becomes lower or more seasonal, forest gives way to shrubby or herbaceous vegetation types. On sandy soils, forest gives way to a sclerophyllous vegetation (fynbos) and the species diversity of fynbos decreases and composition changes until rainfall minimums reach about 300–250 mm per year when a succulent shrubland becomes dominant. On clay soils, forest gives way to fynbos, and then to the characteristic renosterveld, a shrubland dominated by shrubby, microphyllous Asteraceae. At precipitation levels below 100 mm per year, renosterveld is replaced by a vegetation increasingly dominated by succulent perennials. The dissected nature of the landscape ensures that broad sweeps of one vegetation type are isolated from one another by habitats that will not support their growth. The mosaic of different soil types alone contributes to increasing diversity, but the peculiar nature of nutrient-poor soils may result in higher than expected effects on plant diversity. As we elaborate below, such soils appear to have a significant impact on plant dispersal and hence on gene flow.

Landscape and climate. Mountain belts of the Cape Region are not particularly high, generally 1 000–2 000 m in elevation, and although the peaks are well below a truly alpine zone they are high enough for winter freezing to be a factor affecting the vegetation. The mountain slopes are broken and dissected, with cliffs and exposed rock evident everywhere. The rugged topography and varied landscape amplify the effects of climatic variation, with the result that the mountains offer a greater diversity of habitats than are present in the lowlands.

Winter-rainfall and climatic stability. How important is the current Mediterranean climate that prevails over most of the region to the flora? The vegetation of the Cape Region prior to the establishment of a winter-rainfall climate was very different from that now found here. Evergreen forest has been decreasing since the middle of the Tertiary and its diversity has dropped dramatically since the mid-Miocene as well (Coetzee 1993). Families such as Casuarinaceae, Chloranthaceae, Sarcolaenaceae and Winteraceae no longer found on the African mainland but still extant in Madagascar, were present in the Cape Region until the mid-Miocene (Coetzee & Praglowski 1984; Coetzee & Muller 1984). In addition, early to mid-Miocene deposits on the Cape west coast indicate a fauna adapted to forest and woodland (Hendey 1982).

The establishment of the cold Benguela Current along the west coast of southern Africa in the Miocene, with its cooling and drying effects on the west coast, was probably the single most important factor for vegetation change on the subcontinent. Summer drought was likely to have become increasingly severe in the west as this current was strengthened as a result of the spread of the Antarctic ice sheet at the end of the Miocene, c. 5 mya. Even in the late mid-Miocene, however, there was a fairly rich subtropical flora replete with palms (Coetzee & Rogers 1982) near Saldanha Bay on the west coast of the Cape Region that today supports a largely treeless, succulent or sclerophyllous shrubland. No palm species occur in the Cape Flora Region today. Faunal remains from this period suggest that by the late Miocene the widespread forest and woodland were being replaced by more open savanna (Hendey 1982). It was probably not until after the beginning of the Pliocene, i.e. less than 5 mya, that the present Cape flora could be distinguished. Elements of that flora are recognized in Oligocene pollen cores taken within the Cape Region and nearby (Scholz 1985). It seems clear that a climatic change that included increasing summer drought and lower overall rainfall was the driving force for vegetational change in southern Africa into the Pliocene.

Although post-Pliocene changes in climate of the Cape Region are poorly documented, the climate of the Cape Region appears to have been relatively stable. In comparison with southern Europe, North America, and southern South America, all of which experienced cycles of extreme climatic fluctuations with periods of mild climate alternating with extreme cold and dryness (Villagrán 1994), the climate of the Cape Region appears to have been relatively stable throughout the Quaternary. Whereas in Central Chile, southern Europe and North America nearby mountain glaciers were developed and winter temperatures must have fallen to levels that their floras could not tolerate, the Cape appears to have merely suffered cycles of drier and cooler or wetter and warmer conditions. The ameliorating affects of large oceans to the south and west would have prevented the extreme conditions that result in major extinction events. Although no glaciers developed, there is evidence of colder climates in the past (Deacon 1979) consistent with a temperature depression of the order of 5 °C at the latitude of the Cape Region.

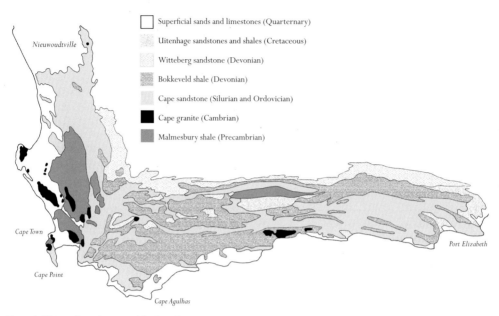

Figure 1. Main geological groups of the Cape Region.

The data of Meadows & Sugden (1991) are among few studies of the history of the Cape flora over the past 20 000 years. Pollen profiles from the Cedarberg Mountains in the northwestern part of the region show no vegetational changes comparable to those known for Chile (Villagrán 1994) or California (Raven & Axelrod 1978). Instead, there seems to have been a series of subtle shifts in conditions that favoured one community type over another in the 14 600 years covered in their sampling. The Cedarberg data are especially notable because that range lies at the northern, more arid end of the Cape Region, and would therefore have been particularly sensitive to climatic change. In the southern Cape the lowering of sea levels as much as 120 m at times during the Pleistocene resulted in the extension of the coastal plain off the Bredasdorp coast. The vegetation along the coast at this time was probably a grass-dominated flora that supported the dominant alcephaline and equid fauna (Klein 1977). Even today, the clay soils of this area have a large grass component. Vegetational changes at higher taxonomic level are thus not evident.

Floristic composition

Major families and genera. The Asteraceae, usually the largest family in floras of arid to semi-arid regions, is also the most speciose family in the Cape flora (Table 1). Additions to several genera of Fabaceae since the Cape flora was last analyzed (Bond & Goldblatt 1984), now makes this family the second-largest in the flora (previously believed to be fourth largest). This is also unexceptional, as Fabaceae are well developed in most parts of the world. However, the huge contribution made by Aizoaceae, Ericaceae and Iridaceae, next in size (Table 1), is a unique aspect of the Cape flora (and consequently of the southern African flora as a whole). Scrophulariaceae, Proteaceae and Restionaceae follow in size, showing a pattern that is somewhat different from that described in the now outdated analysis of the Cape flora (Bond & Goldblatt 1982). This is due to the publication of numerous critical taxonomic revisions, especially of genera of Aizoaceae, Fabaceae and Scrophulariaceae. In this earlier analysis, Mesembryanthemaceae was the third largest family, but critical taxonomic revisions of its constituent genera unfailingly recognize substantially fewer species. With the union in this account of Mesembryanthemaceae with Aizoaceae, the family remains among the largest five in the flora, and with just one species less than the Iridaceae.

The large numbers of species of Proteaceae and Restionaceae, seventh- and eighth-largest families respectively, is another striking feature of the flora. The importance of Ericaceae, Proteaceae and Restionaceae both in terms of biomass and species diversity is widely appreciated, but the remarkable number of species of Iridaceae, predominantly a family of herbaceous, seasonal geophytes, is especially striking. Nowhere else in the world does this family provide more than a small proportion of the total species. Indeed, the adaptive radiation of Ericaceae and Iridaceae in the Cape flora is one of its most unusual aspects. The diversification of the Ericaceae, Proteaceae, Restionaceae and even Cyperaceae is closely associated with the impoverished sandstone soils of the Cape mountain ranges and these families are often poorly represented on other soils. The massive radiation of the Fabaceae and Iridaceae shows no such correlation and must have other explanations.

Table 1. Ranking of the 20 largest families in the Cape flora by size.

	Species	Endemic (%)	Genera (endemic)	Species : genus
1. Asteraceae	1036	655 (63.2)	121 (30)	8.6
2. Fabaceae	760	627 (82.5)	37 (6)	20.5
3. Iridaceae	661	520 (79.1)	32 (6)	20.7
4. Aizoaceae	660	525 (79.6)	76 (18)	8.7
5. Ericaceae	658	635 (96.8)	1 (0)	655
6. Scrophulariaceae	418	297 (71.1)	33 (8)	12.7
7. Proteaceae	330	319 (96.7)	14 (9)	23.6
8. Restionaceae	318	294 (92.5)	19 (10)	16.8
9. Rutaceae	273	258 (94.5)	15 (6)	18.2
10. Orchidaceae	227	138 (60.8)	25 (2)	9.1
11. Poaceae	207	80 (38.7)	61 (3)	3.4
12. Cyperaceae	206	101 (49.0)	29 (3)	7.3
13. Hyacinthaceae	192	87 (45.3)	14 (0)	13.7
14. Campanulaceae	184	140 (76.1)	13 (6)	14.2
15. Asphodelaceae	158	81 (51.3)	8 (0)	19.8
16. Geraniaceae	155	91 (58.7)	3 (0)	51.7
17. Polygalaceae	141	122 (86.5)	3 (0)	47.0
18. Rhamnaceae	137	126 (92.0)	5 (1)	27.4
19. Thymelaeaceae	124	94 (76.4)	4 (1)	31.0
20. Crassulaceae	123	35 (28.5)	5 (0)	24.6
Total species	6966	(= 77.4 % of the total flora.)		

Although the wealth of Scrophulariaceae, sixth-largest family in the Cape flora, seems remarkable in a world context, the family is well represented across Africa, especially in the floras of drier areas (Maggs *et al.* 1998). In the Cape flora Scrophulariaceae contribute 166 species to the annual flora, far more than the Asteraceae, with 138 annuals.

Poaceae are comparatively poorly represented in the Cape flora. Although third-largest in number of genera, the family is eleventh in size in total species, with fewer species than Restionaceae and barely more than Cyperaceae, the two families that grow in habitats normally occupied by grasses. This situation is paralleled only in southwestern Australia but contrasts sharply with surrounding southern Africa where Poaceae are a prominent family. Poaceae are the largest family in the flora of Namibia (Maggs *et al.* 1998) and one of the five largest families in southern Africa excluding the Cape Flora Region.

Although 150 families of seed plants and another 23 families of ferns and fern allies are represented in the flora, remarkably few account for the bulk of the species. While 23 families in the flora have over 100 species, only 12 have over 200 species. In contrast, 38 families have only one species each. The largest 10 families account for 5 360 species, over half the flora, and the largest 20 families account for 6 987 species, more than 77 % of the flora (Table 1).

Endemic families. The unique floristic composition of the Cape Region with its high representation of Ericaceae, Iridaceae, Proteaceae and Restionaceae (Table 1), is emphasized by the presence of several families that are endemic or nearly so. The endemic families are all dicotyledons of diverse affinity and relatively low evolutionary specialization (Table 2). The largest is Penaeaceae (Myrtales), followed by Stilbaceae (including Retziaceae) (Lamiales), Grubbiaceae (Cornales), Roridulaceae (Ericales), and Geissolomataceae (of uncertain ordinal position) (ordinal classification following the Angiosperm Phylogeny Group 1998).

Bruniaceae, one of the distinctive families of the Cape flora, is almost endemic. It has an estimated 64 species in 11 genera and only three species in two genera extend outside the confines of the Cape Region, one as far east as southern KwaZulu-Natal. The relationships of the Bruniaceae are uncertain but recent phylogenetic study places the family close to Eucommiales and Icacinaceae (Nandi *et al.* 1998), the latter consisting largely of forest trees, including four species in the Cape flora, notably *Apodytes*. Retziaceae (1 genus : 1 species) has often been considered an endemic Cape family (e.g. Bond & Goldblatt 1984), although its affinities have long been in dispute (Dahlgren *et al.* 1979). Ribosomal DNA sequence analysis indicates that it is nested in Stilbaceae (Bremer *et al.* 1994), and has floral specialization for bird pollination, the source of most of its peculiarities.

In contrast to the Cape Region which alone has five endemic families, all of southern Africa has only 12 endemic families. In addition to the five absolutely restricted to the Cape Region, they are the eudicot families Bruniaceae (ordinal position uncertain, 11 genera: 64 species), Achariaceae (Flacourtiales, 3 genera : 3 species), two species of which also occur in the Cape Region, Greyiaceae (Geraniales, 1 genus: 3 species), Rhynchocalycaceae (Myrtales, 1 genus : 1 species); two monotypic genera of the monocots, *Lanaria* and *Prionium*, which are also regarded as comprising their own families, Lanariaceae (Asparagales) and Prioniaceae (Poales) respectively; and the cycad family Stangeriaceae (1 genus : 1 species). Both *Lanaria* and *Prionium* are widespread in the Cape Region but extend outside its confines a short distance to the east. Aitoniaceae, Curtisiaceae and Oftiaceae have at times been recognized but they are no longer thought to be separate families. They are readily referable to Meliaceae (Pennington & Styles 1975), Cornaceae (Xiang *et al.* 1993) or Scrophulariaceae (Goldblatt 1979) respectively.

Table 2. Endemic families of the Cape flora.

Family	Genera	Species
Penaeaceae	7	23
Stilbaceae	5	14
Grubbiaceae	1	3
Roridulaceae	1	2
Geissolomataceae	1	1

Genera. Some 942 native genera of seed plants (or 988 genera of vascular plants), about half of those occurring in southern Africa, are represented in the flora, of which some 160 or 16 % are endemic. The level of generic endemism is modest, but gives only a partial indication of the unusual nature of the flora. The number of near-endemic genera (those of moderate size with just one or two species extending locally outside the Cape Region) is unusually high. The largest genus in the flora by far is *Erica* with some 656 species (Table 3). Changes in the circumscription of *Erica* (Oliver, in prep.) have now resulted in the inclusion of all the minor genera of southern African Ericaceae: Ericoideae in *Erica*, leaving this one genus with over 7 % of the species in the Cape flora. It is unclear whether this remarkable pattern of radiation without generic diversification is associated with the relatively recent arrival of ancestral ericaceous stock in the Cape Region, or with rapid evolution following the establishment of a Mediterranean climate at the Cape. By comparison, the smaller families Proteaceae and Restionaceae appear to belong to old African (or even Gondwanan) groups, now poorly represented elsewhere in Africa, and they show the highest levels of endemism at the generic level. These two families plus Bruniaceae are the only nonendemic families that show greater than 50 % generic endemism.

The second-largest genus is *Aspalathus* (Fabaceae), with 272 species, followed by *Pelargonium* (Geraniaceae), *Agathosma* (Rutaceae), *Phylica* (Rhamnaceae), *Lampranthus* (Aizoaceae), and *Oxalis* (Oxalidaceae), each with between 118 and 148 species (Table 3). There are 36 genera with over 50 species and 12 with over 100 species. The 10 largest genera contribute over 21 %, or 1 932 species, to the flora. The next 10 largest genera contribute an additional 900 species. The 20 largest genera in the Cape flora therefore contain over 30 % of the total species.

There is no unifying ecological pattern evident in the species-rich genera. The genera *Erica*, *Aspalathus*, *Phylica*, *Agathosma*, *Cliffortia*, and the two largest genera of Proteaceae, *Leucadendron* and *Protea*, among others, are best developed on sandy soils and are most diverse in montane habitats. In contrast, species of *Lampranthus*, *Moraea*, *Pelargonium*, *Oxalis*, *Gladiolus*, and *Crassula* appear to occur with equal frequency on nutrient-poor, nutrient-intermediate, or comparatively rich soils and favour lowland habitats. *Lampranthus* and *Crassula* are succulents, and *Disa, Oxalis, Gladiolus* and *Moraea* are seasonal geophytes, as are some species of *Pelargonium*. The remaining genera among the largest 20 comprise mostly shrubs or small trees. A few species of *Pelargonium*, *Senecio*, *Crassula* and *Helichrysum* are annuals. Species of *Thesium* are hemiparasitic shrubs.

The most obvious shared factors in successful genera in the Cape flora seem to be either a shrubby habit or seasonal geophytism. Over 17 % of the total species in the flora are geophytes with bulbs, corms, tubers or rhizomes. The number of species with underground perennating buds would be even higher if plants with woody caudexes were regarded as geophytes.

Table 3. Ranking of the 20 largest genera in the Cape flora. (with number of endemic species in parentheses).

Erica	658	(635)	Muraltia	106	(100)
Aspalathus	272	(257)	Gladiolus	105	(86)
Pelargonium	148	(79)	Selago	101	(79)
Agathosma	143	(138)	Crassula	95	(26)
Phylica	133	(126)	Disa	92	(78)
Lampranthus	124	(118)	Ruschia	88	(79)
Oxalis	118	(94)	Restio	85	(82)
Moraea	115	(79)	Leucadendron	82	(79)
Cliffortia	114	(104)	Helichrysum	81	(35)
Senecio	110	(58)	Thesium	81	(35)
Total	**1 935**	(= 21.5 %)		**2 849**	(=31.7 %)

The ratio of species per genus, 9.1 (9.3 excluding ferns) in the Cape flora (Table 4) is particularly high compared to other floras, and is one of the highest in the world (Fenner *et al.* 1997), although southern Africa, including the Cape Region, has a comparable ratio (9.6, fide Goldblatt 1998). A ratio of three dicots to one monocot species in the Cape flora is close to the average for floras across the world. Although the proportion of monocots does not seem unusual, the monocot families that are represented in the Cape flora are most unexpected. Some half of the species in the monocot families are geophytic and belong in the petaloid monocot families, notably Iridaceae, Orchidaceae, Hyacinthaceae and Amaryllidaceae. The proportion of monocot to dicot species matching that elsewhere in the world is therefore no more than coincidence.

Palaeoendemic genera. The endemic and near-endemic families of the Cape Region are all small and contribute relatively few genera and species to the flora. These families are all taxonomically isolated and comprise part of the small proportion of species in the Cape Region that are best regarded as palaeoendemics. With the exception of the rhizomatous perennial, *Lanaria*, members of the endemic and near-endemic families are all evergreen, sclerophyllous shrubs. They are often summer-flowering, have small flowers and grow mostly on sandstone-derived soils. These plants thus appear to be relicts of an ancient temperate southern African flora adapted to nutrient-poor soils.

Among nonendemic families, the endemic or near-endemic genera that are taxonomically isolated and with one or few species, and thus palaeoendemic, are genera of the geologically oldest communities, tropical thicket and evergreen forest. Especially notable are *Platylophus*, a monotypic genus of Cunoniaceae and one of two continental African members of this family. The other genus, *Cunonia*, is common in the Cape Region but extends into eastern southern Africa. Other monotypic genera such as *Laurophyllus* and *Heeria* (Anacardiaceae), *Hartogiella* and *Maurocenia* (Celastraceae), *Lachnostylis* (Euphorbiaceae), and *Smelophyllum* (Sapindaceae) also exemplify the palaeoendemic component of the depauperate tree element of the flora and mostly have ranges restricted to the southern or eastern portion of the Cape Region.

There are few palaeoendemic genera or species in unforested habitats, apart from members of the endemic (and near-endemic) families of the Cape Region. These include the monotypic shrublets, *Empleuridium* (Celastraceae) and *Ixianthes* (Scrophulariaceae). The small tree *Hyenanche* (Euphorbiaceae), also monotypic, and *Metrosideros angustifolia* (Myrtaceae), a member of an otherwise Australasian genus, show an odd pattern for the tree flora. Their ranges are restricted to the western half of the Cape Region where there are few tree species. *Metrosideros angustifolia*, the only African member of the otherwise Australasian Myrtoideae: Metrosiderinae, seems as geographically isolated as *Cunonia* and *Platylophus*. Like *Metrosideros*, *Bulbinella* is also a Cape–Australasian disjunct, but in this case the radiation within the genus has occurred largely in the Cape Region. The small number of palaeoendemics emphasizes the huge contribution that recent speciation in a narrow range of families and genera has made to the total species diversity in the Cape Region.

Table 4. Selected statistics for the Cape flora and various comparable regions. The figures represent the percentage of the total flora.

Region	Species per genus	Ten largest genera	Monocots %	Asteraceae %	Annuals %
Cape flora	9.1	21.5	24.4	11.5	6.8
Southern Africa	9.6	15.1	23.0	11.0	7.0
Cape Peninsula	4.2	17.5	34.6	11.5	9.6
KwaZulu-Natal	3.9	17.0	27.1	11.4	c. 6.5
Eastern North America	5.2	21.8	28.2	12.7	8.7
Europe	7.8	14.0	18.0	12.0	?
California flora	5.3	15.2	19.2	13.6	27.4
Sonoran Desert	3.3	12.8	12.1	15.0	21.4
Texas	3.9	10.2	24.4	13.4	20.4
Hawaii	4.4	81.0	11.8	15.9	0.04
New Zealand	5.1	26.3	27.3	12.5	6.0

Life forms. Like all Mediterranean floras, the Cape flora has relatively few trees, and this life form accounts for some 220 species, less than 2.5 % of the flora. The remaining c. 65 % of the flora are shrubs and perennial herbs. The shrubby habit is the most common life form in the Cape Region, accounting for an estimated 4 797 species, 53.3 % of the flora. Shrubs are diverse in form, but typically include species with sclerophyllous and mostly microphyllous leaves, the characteristic that has given rise to the word *fynbos*, an Afrikaans word describing fine-leaved vegetation. Shrubs also include large numbers of species with succulent leaves (especially Aizoaceae). Stem-succulents include species of Apocynaceae and Euphorbiaceae, some of which are so reduced in size that the term shrublet hardly seems deserved. The Cape flora stands out when compared to both California and Chile in the overwhelming proportion of shrubs (Table 5) and this is largely explained by the nutrient-poor soils which favour a shrubby life form.

Table 5. Comparison of life forms in the California flora, central Chile and the Cape flora. Figures are percentage of total species; perennials below includes geophytes and graminoids. Data for California and Chile are from Arroyo et al. (1994).

	Perennials %	Shrubs %	Trees %	Annuals %
Cape Region	37.4	53.3	2.5	6.8
California	56.2	11.0	4.6	30.2
Central Chile	63.4	17.8	2.9	15.8

The Cape flora shows some striking differences with other Mediterranean floras. One of these is a surprisingly low proportion of annuals; only some 609 species, about 7 % of the total flora, are annuals (Table 6). This is strikingly different from the floras of both the Mediterranean basin and those of the Mediterranean climate zones of California and Chile, which have 30 % and 15 % annuals respectively (Arroyo *et al.* 1994; Cowling *et al.* 1996). Although the proportion of annuals in the Cape flora is low, the annual flora is quite rich. The total number of annual species is actually considerably higher than the 378 annuals in Chile, an area of comparable size. Although the Cape flora has less than half the 1 279 annuals of the California Floristic Province, it has an area only about one-fourth the size of the California Floristic Province. For its geographic area then, the Cape flora is not depauperate in annuals, but its wealth of other life forms makes the annual habit appear under-represented. No comparable figures are available for the Mediterranean basin. A small annual flora is also characteristic of southwestern Australia, an area which has a recent geological history comparable to that of the Cape Region and a similar pattern of nutrient-poor sandstone soils and richer clays. The low proportion of annuals in the Cape flora has remained without a satisfactory explanation since it was first noted by Bond & Goldblatt (1984). Part of the explanation may simply lie in the disproportionate numbers of other life forms, especially microphyllous shrubs, which are particularly well adapted to the nutrient-poor soils.

Two families, Scrophulariaceae and Asteraceae, are particularly important in their contribution to the annual flora. Scrophulariaceae, with 166 species, contributes the largest number of annuals, and not as might be expected the Asteraceae, which has some 138 annual species (Table 6). The Aizoaceae, Brassicaceae, Campanulaceae, Crassulaceae, Cyperaceae, Fabaceae and Poaceae, each contribute between 20 and 35 species to the annual flora. Campanulaceae, in particular, need taxonomic study and our estimation of the number of species in the family, including its annual component, is subject to significant revision.

In contrast to the low proportion of annuals, the Cape Region has perhaps the highest proportion of geophytes of any part of the world, and is four to five times richer than other Mediterranean floras (Esler & Rundel 1998; Esler *et al*. 1999). At least 1 550 species, over 17 % of the total, have specialized underground organs including bulbs, corms, rhizomes or tubers and are seasonally dormant (Table 7). The overwhelming number of geophytes are monocots, with over 1 300 geophytic species, 635 of which are species of Iridaceae. Most of these species are seasonal and lie dormant underground in the dry season, but we have included the few more or less evergreen species (e.g. *Agapanthus*, *Kniphofia*) with similar underground organs in the geophyte category. The other main category of the monocots is the graminoids and the perennial species of Cyperaceae, Juncaceae, Poaceae, Restionaceae and a few other families account for 795 species.

Table 6. Families with the highest numbers of annuals in the Cape flora.

Scrophulariaceae	166	Asteraceae	138
Campanulaceae	135	Brassicaceae	33
Poaceae	31	Aizoaceae	25
Fabaceae	25	Crassulaceae	23
Cyperaceae	21	Gentianaceae	18
Molluginaceae	16		
Total: 609 species (6.8 % of entire flora)			

Table 7. Total species (percentage in parenthesis) of the different life forms in the Cape flora.

Trees	220	(2.5)
Shrubs	4800	(53.3)
Perennials	1025	(11.4)
Geophytes	1551	(17.2)
Graminoids	795	(8.8)
Annuals	609	(6.8)
Total	**9000**	**100 %**

Diversity

Vegetation. Far from having uniform vegetation, the Cape Region encompasses four biomes (or five by some estimates) and several distinctive vegetation types (Rutherford & Westfall 1994; Cowling & Holmes 1992a), each with their own suites of species and physical characteristics. The most common and distinctive is heathland, locally called fynbos, an analogue of California chaparral and Mediterranean maquis. Shrubs with ericoid or short, narrow, often needle-like leaves predominate, but most species of Proteaceae, a family common in this vegetation, have broad sclerophyllous leaves. Fynbos typically occurs on sandstone soils. A second distinctive vegetation type, renosterveld, is usually restricted to richer fine-grained soils. It shares few species with fynbos although the two vegetation types often grow adjacent to one another. Microphyllous Asteraceae are common in renosterveld which consists of a dense shrubland with a rich herbaceous understorey that becomes evident after fire or clearing but is often suppressed under a mature shrub cover. Dry sites with rainfall less than 200 mm per year support a vegetation of small succulent-leaved shrubs, including many Aizoaceae and Asteraceae in a biome called karoo steppe or succulent shrubland. Forest thicket and evergreen forest make up the remaining biomes.

Fire is an integral part of the ecology of the Cape Region and accounts for several aspects of the flora. Growth form in mature fynbos and renosterveld is a relatively uniform, closed, low canopy of twiggy and microphyllous to sclerophyllous shrubs. The vegetation is highly prone to periodic fire. Fire itself has a disruptive effect on the vegetation. It has obviously been a feature of the ecology for so long that there is a large flora of ephemerals, geophytes, other perennials and short-lived shrubs that appear in the years following a fire, and subsequently disappear as they are succeeded by longer-lived shrubs. The long-term ecological consequence of the fire component of the flora is the existence of a niche for species that grow rapidly after fire and bloom and reproduce in the immediate post-fire years. This suite of species contributes substantially to the overall diversity in the flora. Mature vegetation is affected by fire in more subtle ways, but fire may cause local fluctuations in species composition and the elimination of some taxa. This can create opportunities for diversification and speciation.

Regional diversity. The patterns of endemism within the Cape Region are fairly consistent, and an analysis of these patterns in selected genera that have diversified largely on sandstone substrates has resulted in the recognition of several regional centres of endemism. Weimarck (1941) pioneered this field which has now been refined by Cowling and his co-workers (Cowling 1992). The presence of these centres suggests that exchange between them is limited because of effective geographic isolation or because different microclimates in each centre favour local species at the expense of migrants.

One of the innovations in this account of the Cape flora is the recognition of phytogeographic subcentres in the flora and, as far as available information allows, distributional data for species are arranged according to phytogeography. Thus, we have had to define each centre geographically (Front Endpaper). The result is a first estimation of floristic biodiversity for each subcentre. We realize that the results are merely approximations because some centres are under-collected (or under-cited in taxonomic accounts). We suspect that the Karoo Mountain Centre (KM) and the Agulhas Plain Centre (AP) have more species (and thus lower levels of endemism) than our data suggest. However, it is unlikely that the broad patterns of species numbers and endemicity will change to the extent that our conclusions need more than minor revision.

At the geographic centre of the Cape Region, the Southwestern Centre (SW) has the largest flora (4 651 species) and the highest level of endemism (31.9 %). The Northwestern Centre (NW) follows it in size (4 058 species) and endemicity (26.1 %). The KM and SE Centres have much smaller floras, and substantially lower levels of endemism. Much smaller in extent, the Langeberg (LB) and AP Centres understandably have smaller floras. The SE Centre, of almost the same physical size as the NW and SW Centres, has a markedly smaller flora (2 830 species), and only 9.7 % endemism. This may be explained by its more equable and apparently less diverse climate which translates to fewer local microhabitats available for species diversification. Differences in levels of endemism across the Centres are striking. The SW and NW Centres each have about twice the proportion of endemic species that the others have, a reflection not only of their greater climatic diversity, but perhaps of their greater seasonality.

Table 8. Comparison of species richness, endemism and the proportion of life forms in the floras of the six phytogeographic centres of the Cape flora (n/a = not available).

Region	Area 10^3 km^2	Total species	Endemism	Trees	Annuals (%)	Geophytes (%)
Northwestern Centre	22	4062	26.1 %	69	415 (10.3)	855 (21.2)
Southwestern Centre	23	4654	31.9 %	95	312 (6.8)	846 (18.2)
Agulhas Plain Centre	3	1374	14.9 %	24	92 (6.7)	202 (14.7)
Karoo Mountain Centre	19	2148	15.4 %	47	130 (6.1)	330 (15.5)
Langeberg Centre	7	2365	11.7 %	100	127 (5.4)	389 (16.4)
Southeastern Centre	18	2832	9.7 %	163	156 (5.5)	427 (15.1)
Cape Peninsula	4.7	2250	7.5 %	n/a	n/a	n/a
Cape Region	90	9000	68.8 %	220	609 (6.8)	1552 (17.2)

The different life forms are unevenly distributed across the Cape Region and the numbers of species of the two most distinctive life forms, annuals and geophytes, drop dramatically from west to east. The summer-dry NW and SW Centres have the largest numbers of geophytes and the highest proportion of geophyte species out of the total species for the Centre, 21.2 % and 18.7 % respectively. These two Centres each have over 50 % of all the geophytes in the entire Cape flora. In comparison, the other Centres have between 14.7 % and 16.4 % geophytes. Annuals are more commonn in the west and the NW Centre has 10.3 % annual species, while the remaining Centres each have no more than 6.8 % annuals. The NW Centre alone has 65 % of the total annual species in the Cape flora. The distribution of trees shows the converse, with relatively few tree species in the NW Centre and the highest numbers by far in the SE Centre.

These patterns seem directly related to climate. Both geophytes and annuals seem best adapted to a seasonally extreme climate with a wet winter and dry summer. A climate with higher, seasonally more evenly distributed rainfall, characteristic of the LB and SE Centres, favours a tree flora and fewer geophytes and annuals. Like annuals, geophytes are adapted for survival in semi-arid, seasonal habitats. This is reflected in the greater representation of annuals and geophytes in the western half of the Cape Region where summer precipitation is lowest. The comparatively high numbers of trees in the SW Centre reflect the dissected landscape with the presence of sheltered valleys and the regular occurrence of rainfall in the summer along its southern coast and interior.

Comparisons with other floras. An aspect of the Cape flora that is of particular interest is the high level of species diversity, both regional and local. For its size (c. 90 000 km^2), the number of species of vascular plants, c. 9 000 (8 888 seed plants plus 112 pteridophytes), is comparable with areas of the wet Neotropics (Table 9). Thus Panama (75 000 km^2) has 7 300 seed plant species and Costa Rica (54 000 km^2) may have over 9 000 species. In fact, southern Africa as a whole has a particularly rich and diverse flora for a predominantly temperate region. The area customarily treated for floristic purposes as Southern Africa (Botswana, Lesotho, Namibia, South Africa and Swaziland) has about 20 400 native vascular plant species in an area of 2 674 000 km^2, and South Africa alone may have some 18 500 species (c. 18 250 seed plants plus 245 pteridophytes) (C. de Wet pers. comm.). This is striking compared with an estimated 19 000 species in the whole of North America north of Mexico, or the estimated 16 500 native vascular plant species (15 800 seed plants plus c. 700 pteridophytes) currently recognized for Peru, an area of 1 285 000 km^2 (Brako & Zarucchi 1993; J.L. Zarucchi pers. comm.). To put this in a regional context, all of tropical Africa may have about 26 500 species (Lebrun & Stork 1997), in an area nearly ten times larger than that of southern Africa and about 250 times as large as the Cape Region. Southern Africa and North Africa have approximately 21 500 additional species, making a total of c. 47 000 species for the entire African continent. The tiny Cape Region, less than 0.5 % of the total area of Africa, then has almost 20 % of all the species on the continent. Subtropical southern Africa, excluding the Cape Region, has only about 14 300 species, a figure comparable with that for Tropical East Africa. For the African continent then, not only is the Cape Region remarkably rich in species, but southern Africa has a higher species diversity than would normally be predicted on the basis of species diversity increasing toward the equator.

The diversity of species in the Cape flora is, by any measure, remarkable. Moreover, some 6 190 or about 68.8 % of the species are endemic there (Table 9). The high degree of species endemism in the Cape flora compared to the California Floristic Province (Table 9), for example, emphasizes the peculiarities of the Cape flora. Such levels of endemism are usually associated with islands that have been isolated for long periods of geological time or have very sharp boundaries that limit direct plant migration. In a biological sense, the Cape Region is virtually an island, not surrounded by ocean, but by a zone of dry climate or sharply different soils, or seasonal rainfall distribution. The flora of southwestern Australia shares with the Cape Region an unusually high endemicity for a local continental flora, and so does that of southern Africa as a whole (Table 9). Why the last-named region should have such a high level of endemism is not at all clear.

The high level of diversity and local endemism of the Cape Region is starkly emphasized by comparison of the subcentres within the region with other areas. The SW and NW Centres both have over 4 000 species and over 31 % or 26 % endemism respectively, compared with about 2 400 species and an estimated 22–23 % endemism (depending on the geographical definition of the region) for the entire Mediterranean flora of Chile (Table 9), an area over five times greater than either of these subcentres of the Cape Flora Region. Likewise, important centres of local endemism (hot spots) within the

Mediterranean Basin, including the islands of Sicily, Sardinia and Crete, or the Peloponnese Peninsula, all of more or less comparable geographic size to the NW or SW Centres, have approximately half or less than half the number of species and between 6 % and 12.5 % (Médail & Quézel 1997).

Table 9. Comparison of endemism of native vascular plants in selected regions of the world.
References: 1. C. de Wet, pers. comm.; 2. Brako & Zarucchi 1993, Zarucchi, pers. comm.; 3. Raven & Axelrod 1978; 4. Arroyo et al. 1994; 5. Arroyo & Cavieres 1997; 6. Beard 1970; 7. Hopper 1992; 8. Hamel & Grayum, pers. comm.; 9. d'Arcy 1987; 10. Wagner 1991; Wagner et al. 1990; 11. Schatz et al. 1996; 12. Médail & Quézel 1997.

	Area 10^3 km^2	Genera	Endemic %	Species	Endemic %	Reference
Continental areas						
Southern Africa	2,674	2,130	20.0	20,367	80.3	1
Peru	1,285	2,210	2.1	16,500	31.2	2
Areas of mediterranean climate						
Cape Region	90	988	16.2	9,000	68.8	
California FP	324	806	6.5	4,240	47.7	3
Central Chile	104	591	-	2,395	c. 22.5	4
or	155	-	-	2,537	c. 23.4	5
SW Australia	270	462	c. 20	3,650	68	6
or				c. 8,000	c. 75	7
Moist to wet tropics						
Costa Rica	54	1,877	-	c. 9 000	-	8
Panama	75	1,800	-	7 300	c. 15	9
Tropical or Temperate Islands						
Hawaii	16.6	26,715	1,138	86	11	10
New Zealand	268	393	10	1,996	81	3
Madagascar	594	1,000	-	c. 11,500	c. 80	11
Mediterranean Islands						
Crete	c. 9	-	-	c. 1,706	c. 10	12
Peloponnese	21	-	-	2,400	c. 12.5	12
Sardinia	24	-	-	2,054	c. 6	12
Sicily	26	-	-	c. 2,700	c. 10	12

Alpha diversity. Aspects of plant species diversity have been addressed on several levels and it has been shown that at the local level selected areas within the Cape Region are not unusual on a world scale (Cowling 1992; Goldblatt 1997) and are less species-rich than many areas sampled in the New and Old World lowland tropics (Gentry 1988a; 1988b). Patterns of alpha diversity (the number of species in a homogenous community) in a range of vegetation types in the Cape Region, including fynbos, renosterveld, forest thicket and evergreen forest, are surprising. Fynbos sites (with seasonal species not included in species counts) have a mean alpha diversity of 68 species per 1 000 m^2, with 121 the highest number of species recorded at any site (Cowling & Holmes 1992a). Fynbos diversity is by no means uniform and there is ample evidence that diversity is higher in the west than in the south and in more mesic than dry sites (Bond 1983).

Non-fynbos sites have been less well studied. Figures in the literature for renosterveld include means of 66 (Tilman et al. 1983) and 84 (Cowling & Holmes 1992b) species per 1 000 m^2. Forest thicket sites have a mean of 59 species per 1 000 m^2, forest sites have ranges of 44 to 52 species, and succulent karoo shrublands a mean of 43 species in the same area (Tilman et al. 1983).

By comparison, California chaparral communities have alpha diversity levels around 34 species per 1 000 m^2, but the more comparable southwestern Australian region has an average of 69 species per 1 000 m^2 in heathland (kwongan), a vegetation type similar to fynbos. While these figures are consistent with higher total species richness in the Cape Region and southwestern Australia compared with the California FP, they do not explain the comparable regional diversity of the Cape Region and the lowland wet tropics. In the wet tropics mean alpha diversity of trees (including woody lianas) alone has been found

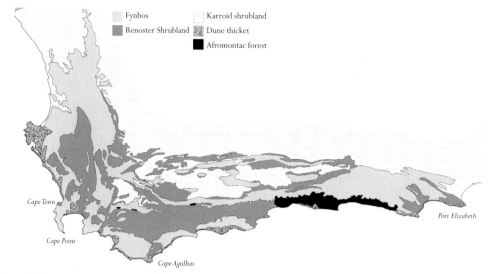
Figure 2. Major vegetation communities of the Cape Region (adapted from Low & Rebelo 1996).

to range from 129 species (Africa) to 140 species (Neotropics), to 193 species (Asia) per 1 000 m^2 (Gentry 1988b), about twice the alpha diversity found in Mediterranean communities on nutrient-poor soils. Inclusion of epiphytes and other herbaceous plants raises alpha diversity in some tropical sites. Gentry & Dodson (1987) have shown that a major component of the plant species diversity in tropical forests actually lies with the epiphytes. Similarly, the inclusion of seasonal geophytes would increase the figures for the Cape Region and until more comprehensive surveys are made any comparisons can only be tentative.

Beta and gamma diversity. Beta diversity (species turnover along a habitat or environmental gradient) is relatively high in the Cape Region. Cowling (1990), for example, has reported nearly complete replacement in sites along the Agulhas Plain that differed in soil features but were climatically and topographically similar. Differences between communities on sandstone versus clay soils are nearly complete, so that the plants on these two soils types are treated as belonging to different vegetation types.

Gamma diversity (species turnover in equivalent habitats along geographic gradients, also called delta diversity) is likewise extremely high in the Cape Region, and is reflected in the high levels of regional endemism. Species replacement values of 46–70 % have been reported by Kruger & Taylor (1979) for sites 25 km apart and Linder (1985) has suggested that geographic replacement may account for 30 % of the differences between species composition along geographic gradients in similar habitats. These figures are, however, lower than some estimates for lowland Neotropical sites (B. Boyle pers. comm.).

Plate 1. Endemic families of the Cape Flora: Penaeaceae and Stilbaceae. 1. **Penaea mucronata** (Penaeaceae).
2. **Endonema retzioides** (Penaeaceae). 3. **Brachysiphon fucatus** (Penaeaceae). 4. **Stilbe ericoides** (Stilbaceae).
5. **Stilbe vestita** (Stilbaceae). 6. **Retzia capensis** (Stilbaceae).

Plate 2. Endemic and near-endemic families of the Cape Flora. 1. **Roridula gorgonias** (Roridulaceae). 2. **Grubbia rosmarinifolia** (Grubbiaceae). 3. **Geissoloma marginata** (Geissolomataceae). 4. **Raspalia variabilis** (Bruniaceae). 5. **Berzelia lanuginosa** (Bruniaceae). 6. **Staavia glutinosa** (Bruniaceae).

Plate 3. Forest and some characteristic forest species. 1. Relict forest in sheltered gorges in Du Toitskloof. The grass **Pentameris thourii** covers the rocky butresses. 2. **Metrosideros angustifolia** (Myrtaceae). 3. **Cunonia capensis** (Cunoniaceae). 4. **Olea capensis** (Oleaceae). 5. **Virgilia divaricata** (Fabaceae).

Plate 4. Ericoid fynbos and associated species. 1. **Erica mammosa** (Ericaceae). 2. **Erica coccinea** on upper slopes of the Riviersonderend Mts. 3. **Erica junonii**. 4. **Syncarpha eximea** (Asteraceae). 5. **Phylica plumosa** (Rhamnaceae).

Plate 5. Proteoid fynbos and some characteristic Proteaceae. 1. **Leucadendron daphnoides** on DuToitskloof Pass. 2. **Protea repens**. 3. **Leucospermum conocarpodendron**. 4. **Serruria trilopha**. 5. **Protea cynaroides**.

Plate 6. Some characteristic plants of fynbos vegetation. 1. **Mimetes cucullatus** (Proteaceae) on steep, Restionaceae-covered slopes on the Riviersonderend Mts. 2. **Witsenia maura** (Iridaceae) in a seep on Kogelberg summit. 3. **Klattia stokoei** (Iridaceae). 4. **Disa ferruginea** (Orchidaceae). 5. **Roella ciliata** (Campanulaceae).

Plate 7. Landforms, high mountain fynbos vegetation and some typical fynbos plants. 1. Summit of Hansiesberg in the Skurweberg Range near Ceres, c. 2000 m. 2. Summit of the Witteberg near Touwsrivier, c. 1500 m. 3. The high altitude cushion plant, **Bryomorphe lycopodioides** (Asteraceae). 4. **Liparia splendens** (Fabaceae). 5. **Aspalathus callosa** (Fabaceae), endemic to the Cape Peninsula.

Plate 8. Restiod fynbos and characteristic fynbos plants. 1. **Syncarpha vestita** (Asteraceae) growing among clumps of Restionaceae on sandy flats near Cape Point. 2. **Adenandra uniflora** (Rutaceae). 3. **Agathosma thymifolia** (Rutaceae), a local endemic of limestone enriched sands on the Cape west coast. 4. **Pelargonium cucullatum** (Geraniaceae). 5. Stands of **Elegia** (Restionaceae) on sandstone slopes in the southern Cape Peninsula.

Plate 9. Karroid vegetation and some plants of dry habitats. 1. **Nymania capensis** (Meliaceae). 2. Little Karoo near Garcia's Pass. 3. **Astroloba rubriflora**, a narrow endemic of the Robertson Karoo. 4. Quartzite pebble patches with **Astroloba corrugata** (Asphodelaceae). 5. **Carpobrotus acinaciformis** (Aizoaceae).

Plate 10. Cape geophytes. 1. Geophytes flowering profusely in recently burned renosterveld near Villiersdorp, with **Moraea comptonii** and **Romulea rosea** (Iridaceae) dominant. 2. **Lachenalia viridiflora** (Hyacinthaceae). 3. **Oxalis glabra** (Oxalidaceae) 4. **Moraea villosa** (Iridaceae). 5. **Amaryllis belladonna** (Amaryllidaceae), a fire-adapted species.

Plate 11. Cape geophytes and some specialized habitats. 1. Seasonal pools on the west coast which support several aquatic geophytes, including **Oxalis disticha** (Oxalidaceae), **Romulea aquatica** (Iridaceae) and **Onixotis stricta** (Colchicaceae). 2. **Brunsvigia orientalis** (Amaryllidaceae) in coastal sands. 3. **Romulea tortuosa** (Iridaceae), a winter-flowering species of moist clay depressions. 4. **Gladiolus sempervirens** (Iridaceae) a summer-flowering species dependent on summer moisture from coastal fog. 5. **Ornithogalum dubium** (Hyacinthaceae).

Plate 12. Annuals. 1. **Diascia longicornis** (Scrophulariaceae). 2. **Nemesia cheiranthus** (Scrophulariaceae). 3. **Dorotheanthus bellidiformis** (Aizoaceae). 4. **Heliophila arenaria** (Brassicaceae). 5. Annual Asteraceae including **Ursinia cakilefolia** (radiate yellow), **Cotula barbata** (disciform yellow), **Dimorphotheca sinuata** (white), **Felicia australis** (blue) and **Senecio cakilefolius** (pink).

Reasons for Cape floristic diversity

Richer in species than any other temperate flora and most tropical ones of comparable geographic area (Table 9), the Cape flora is also highly distinctive. One of five regions in the world with a Mediterranean climate, the Cape has substantially more species than either California or central Chile, which are substantially larger in area. Although southwestern Australia has a flora that may have about the same number of species as the Cape Region (Table 9), it is at least three times larger in area. Only the Mediterranean Basin, approximately 25 times larger in area, has a flora that is larger than the Cape flora, but with about 2.5 times as many species. The reasons for the substantially higher species diversity of the Cape Region compared with these floras are several, and include a range of factors, both physical and historic.

The diversity of soils, a rugged landscape, and extremely variable and complex rainfall patterns in the Cape Region have combined to produce a mosaic of sharply different habitats that lie in close proximity to one another in a repeated pattern across its entire area. This high physical heterogeneity, although striking, is not unique to the region and may even be greater in other regions. The California Floristic Province, too, has a wide range of soils, including serpentine substrates not present in the Cape Region, diverse climates, a rugged topography with higher mountains than those in the Cape, plus a wider latitudinal extension. Likewise, the Mediterranean Basin, orders of magnitude larger than the Cape Region, has a wide diversity of soils and a rugged landscape, with the mountains also higher than those found in the Cape Region. Both areas are often cited as being species-rich compared to neighbouring regions, yet both are substantially poorer in species than the Cape Region, California in absolute terms and the Mediterranean substantially less diverse for its huge area. In contrast, southwestern Australia, which has a flora approaching that of the Cape in size, noticeably lacks the rugged topography of other regions, although it exhibits some of the ecological features of the Cape Region. Therefore physical heterogeneity alone cannot account for the richness of the Cape flora and edaphic factors may be more significant.

Of the five Mediterranean regions of the world only the Cape and southwestern Australia have soils that include large areas of nutrient-poor quartzitic sands, and at least part of the explanation for the higher species numbers here has been thought to relate directly to the peculiar effects of this substrate on plant life. High levels of local diversity have been considered characteristic of nutrient-poor soils (Tilman 1982, 1983). If this is correct, then the mere existence of large areas of such soils should account for the comparatively high alpha diversity of heathland vegetation in both South Africa and southwestern Australia compared with that in California or Chile. However, this hypothesis is not supported either in the Cape or southwestern Australia where alpha diversity levels on nutrient-poor and nutrient-intermediate soils differ only minimally (Goldblatt 1997). Although they may not support a flora significantly richer than that occurring on soils of other nutrient status, nutrient-poor soils in the Cape are able to maintain particularly high beta diversity in the associated fynbos vegetation, both on geographical gradients and on slightly different soils under the same climatic conditions. Comparable data for nutrient-intermediate and nutrient-rich sites are not yet available. The high frequency of fire is another aspect that contributes to diversity in the Cape Region. Fire is also significant in southwestern Australia and California but not in the Mediterranean zone of Chile.

The combination of edaphic and topographic diversity, steep local climatic gradients, peculiar nutrient-poor soils, and frequent fires, is undoubtedly important in promoting species diversity in the Cape Region. But it is still an inadequate explanation for the presence of the unusually rich flora in the Cape Region, particularly when compared to southwestern Australia. A notable and perhaps crucial difference between the Cape and other areas of Mediterranean climate, possibly excluding southwestern Australia, lies in their Pliocene–Pleistocene climatic history. Available data indicate a history of southern Africa very different from that experienced in the northern hemisphere and to a lesser extent in Chile. Cycles of extreme cold and aridity alternating with warm wet phases made these areas largely uninhabitable by their current floras, elements of which either became extinct or were confined to local sites of relatively mild climate. A similarly dynamic history of the Cape flora hypothesizes that Pleistocene glacial cycles caused a northward shift in the winter-rainfall zone, which in turn caused a northward expansion of the flora during the pluvial periods but local extinction and its restriction to refugia during dry periods (Axelrod & Raven 1978). The little evidence available for the Cape Region, however, indicates a more moderate climate without changes of such cataclysmic dimensions. Pollen cores reflect comparatively modest shifts in the flora (Meadows & Sugden 1991), even in the semi-arid and ecologically sensitive Cedarberg Mountains. Changes there might be expected to have been more severely influenced by the drier and colder climatic conditions that are postulated to have occurred during glacial periods when belts of vegetation adapted to Mediterranean climates contracted away from the dry zones that lay towards the equator. The absence of any evidence of major changes in the vegetation of the Cape Region makes it seem likely that the Pleistocene glacial cycles did not have the catastrophic effects on plant life in southern Africa that they did in the northern hemisphere or Chile. In southern Africa relatively drier and wetter cycles may simply have changed the local composition of vegetation, perhaps causing some local extinction, which in turn created opportunities for speciation.

The unusually high species richness of the Cape flora is therefore in all likelihood a consequence of sustained climatic stability and reliability, enabling a more or less uninterrupted evolution of the flora to occur in a region of high physical complexity. The history of this evolution can be traced to some extent by considering the modes of speciation evident in the flora. Although there are few such studies for plant groups centred in the Cape Region, the available evidence suggests that parapatric speciation linked to substrate or microclimatic differences is an important mode of speciation in some families. In this model, vicariant species are either more or less allopatric or even sympatric, and exhibit differences in ecology such as edaphic, microclimatic, seasonal or pollination characteristics. For example, vicariant species of *Rhodocoma* (Restionaceae), a genus restricted to nutrient-poor sandstone soils in montane habitats, favour different habitats and are not significantly isolated geographically (Linder & Vlok 1991). Parapatric speciation also appears to have been more important than geographic isolation in the genus *Lapeirousia* (Iridaceae) in the Cape Region and adjacent parts of the southern African west coast (Goldblatt & Manning 1996). In addition, high levels of both beta and gamma diversity support the hypothesis that microgeographic speciation has played a major role in speciation in the Cape flora. Nearly adjacent habitats under the same climatic and topographical conditions which differ only in their substrates (coarse sand or fine sand or limestone), can support plant communities that differ radically in their species composition while still broadly similar in family and generic composition (Cowling & Holmes 1992a).

In plants, parapatric or microgeographic speciation may actually be the rule, not the exception (Levin 1993), and is favoured by reduced gene flow across strong selection differentials. The mosaic of strikingly different substrates which characterize the Cape Region appears to provide such a strong selective differential. Although the Cape flora may not differ from other floras in mean pollen dispersability (Linder 1985), many of its most characteristic elements have low seed dispersal distances. Most species in the Cape Region show no evident adaptations for dispersal and are regarded as passively dispersed, with seed dispersal distances under 5 m (and most likely much less than this). Dispersal in most Aizoaceae is by rain drops falling on hydrochastic capsules, and this mechanism, although an active one, results in very small dispersal distances. Active seed dispersal by ants is disproportionately well represented in both the Cape Region (Bond & Slingsby 1983) and Australia (Berg 1975). Some 1 000 Cape species, notably in the families Fabaceae, Proteaceae, Restionaceae, Rhamnaceae and Rutaceae, produce seeds with fatty bodies (elaiosomes) that are attractive to ants, and an undetermined additional number are transported to underground nests by harvester ants. In vegetation types prone to frequent fires, such as fynbos, the burial of seeds is a valuable adaptation. However, dispersal distances for ant-dispersed seeds is also short, up to 6 m. More effective dispersal strategies are relatively restricted in their occurrence. Wind dispersal is characteristic of many Asteraceae and Orchidaceae (which have comparatively low levels of endemism in the Cape flora) while long-distance dispersal involving flying vertebrates (birds and bats) is least common, and is especially rare in plants on nutrient-poor substrates. There is an assumption that plants on such soils cannot afford to allocate resources to protein-rich berry or capsular fruits (Bond & Slingsby 1983). The low frequency of fruits with burs and spines, adapted for exochory, suggests that dispersal by nonflying vertebrates has always been unimportant in the flora, perhaps because the fauna has historically been a small one in terms of numbers of species and individuals. This is largely the result of the unpalatability, low nutrient status and low productivity of the flora as a whole.

Indirect evidence of the importance of reduced gene flow distances in stimulating local species diversity comes from a comparison of the number of species and their level of endemism between taxa with widely dispersed seeds and those with reduced dispersal distances. Genera with fleshy diaspores or those which are well adapted for wind dispersal tend to have wide ranges, few species per genus and low levels of local endemism. Compare the berry-fruited *Nylandtia* (Polygalaceae), which has one or two species, with its relative *Muraltia*, which has dry fruits and over 100 species, most with narrow ranges. Similarly, *Chasmanthe*

(Iridaceae), which has fleshy or deceptive (brightly coloured) seeds, has two widespread species and one localized one, whereas its relative *Tritonia*, which has dry seeds, is speciose and most of the species have narrow ranges. Another striking example is *Chrysanthemoides* (Asteraceae) of which the seeds are enclosed in a fleshy pericarp. The two species extend throughout the Cape Region and one far beyond it into tropical Africa. The numerous species of the closely related *Osteospermum* and *Tripteris* have dry seeds and have small geographic ranges. This comparison is also instructive at the family level. Low seed dispersability is typical of many of the larger and most characteristic families in the flora, many of which have both unusually high ratios of species to genus and high levels of local endemism. Ericaceae, Iridaceae and Fabaceae, which largely lack highly developed mechanisms for long-distance seed dispersal, have ratios of above 20 species per genus and higher than 80 % endemism. Asteraceae and Orchidaceae, with wind-dispersed seeds, and Poaceae and Anacardiaceae, with fruits adapted to animal dispersal, show levels of local and regional endemism below the mean for the flora.

The massive speciation in the Cape flora is, we suggest, most likely explained by a model of local speciation in the absence of severe climatic or topographic perturbations. It appears that a relatively stable climate prevailed in the Cape during the Pleistocene and that local parapatric speciation across steep environmental gradients may account for a considerable proportion of the speciation events that occurred in the Region. Because of the relative stability it seems reasonable to postulate that extinction rates in the main vegetation zones, fynbos, renosterveld and succulent shrubland, were low and more than compensated for by local speciation events. The nutrient-poor soils scattered in a mosaic across the Region must have raised local speciation rates because of the characteristic low vagility of the seeds in the great majority of the plants adapted to these soils. Likewise, the low vagility of many of the species in succulent shrubland, although presumably a consequence of different selective forces, has the same consequences, with high levels of local speciation, and thus high levels of diversity across geographic, environmental and edaphic gradients.

The unusually high diversity of the Cape flora is matched by its extraordinary composition of families and genera. Instead of a balanced flora with relatively small numbers of species per genus there has been massive local radiation in a series of unrelated genera. This is so pronounced that almost 22 % of the total species in the Cape Region fall in just 10 genera while the 20 largest genera account for over 30 % of the total species (Table 3). Typical examples of these genera are *Erica* (over 650 species), *Aspalathus* (272 species), *Agathosma* (143 species), *Phylica* (133 species) and *Cliffortia* (114 species). Significantly, none of the genera that display a pattern of massive local radiation in the Cape Region are endemic here, but extend northwards into southern KwaZulu-Natal or further north into tropical Africa. They are, however, primarily restricted to nutrient-poor soils wherever they occur. It is likely, therefore, that their radiation in the Cape Region has been relatively recent and largely post-Pliocene.

In sharp contrast to these examples are the endemic families of the Cape flora. These are without exception depauperate in species, although they may contain several genera. These families are typically restricted to montane habitats in acidic sandstone soils and many of their constituent species are highly local endemics of particular mountain chains or peaks. They display the characteristics of palaeoendemism and probably represent elements of a previously more widespread southern temperate flora adapted to nutrient-poor soils in a summer-rainfall regime. It is probably no coincidence that most of these species flower in summer, that is out of phase with the spring-flowering peak of most members of the flora. With the development of a winter-rainfall climate in the late Pliocene, it is reasonable to infer that these pre-Cape elements were gradually restricted to mesic sites in which some moisture was present over the summer months. Concomitantly it appears that other elements of the flora were able to radiate into emerging niches, thereby establishing the huge neo-endemic element of the flora. The highly sclerophyllous or microphyllous habit developed by taxa adapted to nutrient-poor substrates can thus be regarded as an important pre-adaptation to the Mediterranean climate. The highly seasonal nature of the Mediterranean climate, as well as its reliability, would also have favoured families or genera with a geophytic habit, for example Amaryllidaceae, Hyacinthaceae and Iridaceae.

The rapid and extensive radiation of plants in the Cape Region must have been promoted by both the emergence of new habitats through climatic change and exposure of the coastal plain as sea levels fell at times during the Pleistocene, as well as by the stability of the climate. At the same time the flora was increasingly isolated by the winter-wet and summer-dry climate regime from recruitment from the summer-rainfall-adapted flora of adjacent parts of southern Africa which largely lack the nutrient-poor or nutrient-intermediate soils so characteristic of the Cape Region.

A final characteristic of the flora is the great, and often extreme diversity of flower form that is a feature of many of the genera. This is linked to a diversity of pollination strategies, many of which are poorly exploited outside the region. In particular, pollination strategies using sunbirds, long-proboscid flies, monkey-beetles, rodents, and the butterfly *Aeropetes* are more extensively developed in the Cape flora than elsewhere in Africa. All these strategies favour differently shaped, large, brightly coloured flowers. Both Iridaceae and Ericaceae, for example, have adopted a range of pollination strategies not evident or only weakly expressed elsewhere across their range (Vogel 1954; Goldblatt & Manning 1998), and are often striking when in bloom. The generally low insect diversity in the Cape Region, combined with a short season favourable for both plant and insect growth, is probably the major factor responsible for this diversification in pollination strategies and, more than anything else, has made the Cape flora so extremely appealing to human sensibilities.

How to use this account

Keys: Keys to the families are included at the end of the account. Keys to the genera within each family precede each family treatment which deals with more than one genus.

Arrangement of taxa: The family circumscriptions and classification system used here are those recommended by the Angiosperm Phylogeny Group (1998). Plant families are arranged alphabetically within the major categories Pteridophytes, Gymnosperms and Angiosperms: Palaeodicotyledons, Monocotyledons and Eudicotyledons. This departure from a traditional dichotomy in the angiosperms more accurately reflects the current understanding of relationships among the families. Genera are arranged alphabetically within each family. Species within each genus are generally also listed alphabetically but in many instances have been grouped in various ways as an aid to identification at the species level. Where possible, the larger genera have been subdivided using more or less readily visible characters, but in two instances (*Drimia* and *Moraea*) the generic circumscriptions have been so radically altered from those currently in use elsewhere that the species are grouped informally in their previous arrangements as an aid to locating them.

The grouping of species in the larger genera is usually indicated in the form of a dichotomous key. Each lead is labelled alphabetically in turn (e.g. A, A; B, B). In rare instances a trichotomy occurs and in these cases the second and third options are distinguished by single and double apostrophes respectively after the initial letter (e.g. A, A', A"). In a few instances a dichotomous (or trichotomous) division using strictly contrasting character states is not practical (e.g. *Centella* and *Pelargonium*) and here the final division is into a number of *Groups* distinguished by a combination of characters.

Recognized species are listed in **bold**. Species that we consider conspecific but that are currently recognized nomenclaturally are included in parenthesis in (**bold**). Synonyms are indicated in *italics*. Common names that apply to particular species are provided after the species name and synonyms.

Species entries: Each species entry contains information on habit and morphology, flowering time, habitat, occurrence in phytogeographical region and geographical distribution. In some instances information on flowering time or habitat may be lacking and this in indicated by a ? in the appropriate place.

Habit and morphology. The brief descriptions are intended to provide a working picture of the species, including, where appropriate or available, diagnostic characters useful in distinguishing the species. Most entries include the following information: habit (tree, shrub, subshrub, perennial, geophyte or annual) and a measure of size, usually height, at flowering; leaf arrangement if otherwise than alternate, shape, vesture and other features appropriate to the group; and flower arrangement, colour, shape if variable within a genus, and various other critical features; fruit characters are included if striking or diagnostic.

Flowering time. The flowering period is indicated in months; in species that occur outside the Cape Region flowering time applies specifically to populations occurring within our area. Some species have very extended periods of flowering and in these instances only the peak flowering periods during the year are indicated. Several geophytic species in particular flower only in the season immediately after a fire has cleared the surrounding vegetation and this is indicated where known.

Habitat. Information on the habitat of the species is given when available, including slope and soil or vegetation type. Many species in the Cape flora are more or less specific to sandy, clay or limestone substrates.

Phytogeography and distribution. The six phytogeographical subcentres recognized in the Cape Region are shown on the Front Endpaper. Occurrence of a species in each of these subcentres is indicated by the abbreviations **NW** (Northwest Centre), **SW** (Southwest Centre), **AP** (Agulhas Plain), **KM** (Karoo Mountain Centre), **LB** (Langeberg Centre) and **SE** (Southeast Centre). A more complete indication of the range, from north to south and west to east, follows in parenthesis. This is particularly precise for species endemic to the Cape flora but less so for species extending beyond its boundaries. Note that in this account the term E Cape refers to ranges east of Port Elizabeth and not the Eastern Cape provincial boundaries. The major mountain ranges most commonly referred to in the text are indicated on the Back Endpaper.

Endemic taxa: Families, genera and species endemic to the Cape flora are marked with the symbol • after each entry.

Introduced taxa: Families, genera and species that are not native to the Cape flora are marked with the symbol * before each entry. Introduced species have been kept to a minimum and include only those which have become naturalized and might be mistaken for part of the native flora.

The Cape Flora

PTERIDOPHYTES Ferns & Fern-allies
by J.P. Roux

ANEMIACEAE

MOHRIA SCENTED FERN 7 spp., Africa, Madagascar and Réunion

caffrorum (L.) Desv. Rhizomatous perennial. Fronds erect, the fertile usually longer than the sterile, stipe terete, paleate, to 16 cm long, lamina 2-pinnatifid to 3-pinnate, to 27 cm long, sparsely hairy above, densely hairy and paleate beneath. Sporangia near-marginal. Fynbos and renosterveld, to 500 m, NW, SW, KM, LB, SE (Bokkeveld Mts to E Cape).

saxatilis J.P.Roux Rhizomatous perennial. Fronds caespitose, erect, the fertile marginally longer than the sterile, stipe terete, paleate, to 60 mm long, lamina 2-pinnatifid, to 12 cm long, sparsely hairy above, densely paleate beneath. Sporangia near-marginal. Rock ledges, 900–1200 m, NW (Pakhuis to Hex River Mts).•

ASPLENIACEAE

1. Lamina abaxially sparsely set with small paleae; sori indusiate ....**Asplenium**
1.' Lamina abaxially densely set with large paleae obscuring the sori; indusium obsolete ....**Ceterach**

ASPLENIUM SPLEENWORT c. 650 spp., nearly cosmopolitan

adiantum-nigrum L. BLACK SPLEENWORT, BLACK MAIDENHAIR Perennial with short rhizome. Fronds erect to arching, stipe initially paleate, lamina soft, narrowly deltoid, 2- or 3-pinnate, pinnae unequally cuneate, pinnules oblong, shallowly lobed to deeply pinnatifid, serrate to dentate, minutely paleate beneath. Sori linear, indusium entire. Cliffs and boulders, to 1500 m, NW, SW, KM, LB, SE (Gifberg to Eurasia, Mexico).

aethiopicum (Burm.f.) Bech. AFRICAN SPLEENWORT Perennial with short rhizome. Fronds arching, soft, stipe paleate, lamina lanceolate, 2-pinnate to 4-pinnatifid, pinnae divided into narrowly obovate to narrowly oblong segments, irregularly incised, subglabrous above, paleate beneath. Sori linear, of irregular lengths, indusium subentire. Forest floors, to 400 m, NW, SW, KM, LB, SE (Grootwinterhoek Mts to W Africa, Madagascar).

erectum Bory ex Willd. Perennial with short rhizome. Fronds erect to arching, stipe subglabrous, lamina pinnate to 2-pinnatifid, narrowly elliptic, pinnae falcate, serrate. Sori to 15 per pinna, indusium entire to erose. Forest floors, 300–500 m, SW, LB, SE (Cape Peninsula to W Africa and India).

flexuosum Schrad. Perennial with short rhizome. Fronds caespitose, some gemmiferous, coriaceous, stipe sparsely paleate, lamina ovate-lanceolate, pinnate to 2-pinnatifid, pinnae auriculate acroscopically, more deeply incised above into 2- or 3-fid oblong lobes. Sori to 9 mm long, indusium entire. Shaded forest floors, 150–400 m, LB, SE (George to N Province).

NONSTANDARD SYMBOLS OR ABBREVIATIONS USED	
•	Endemic to the Cape Flora Region.
*	Not native to the Cape Flora Region.
=	Indicates that a species has been formally included in another.
?	Information is lacking or doubtful.
incl.	Applied to a species that we consider to be the same as another but which has not yet been formally included in it.
NW, SW, AP, KM, LB, SE	Occurring in these phytogeographical centres (see *Phytogeography and distribution* above).

gemmiferum Schrad.   MOTHER FERN   Perennial with erect rhizome. Fronds arching, stipe initially densely paleate, lamina oblong-lanceolate, pinnate, soft, gemmiferous, pinnae ovate-lanceolate, serrate, sparsely paleate beneath. Sori linear, indusium entire. Forest floors and mossy rocks, 150–500 m, LB, SE (Heidelberg to W Africa).

lobatum Pappe & Rawson   Perennial with short rhizome. Fronds arching, soft, stipe subglabrous, lamina narrowly elliptic, 2-pinnate to 4-pinnatifid, pinnae narrowly oblong, pinnules rhombic, coarsely serrate or lobed, sparsely paleate beneath. Sori 3–6 per pinnule, indusium entire. Moist deeply shaded forests, to 800 m, SE (Knysna to E Africa and Madagascar).

lunulatum Sw.   Perennial with short rhizome. Fronds arching, gemmiferous, stipe glabrous at maturity, lamina linear-lanceolate, soft, pinnate, pinnae oblong-obtuse, falcate, auriculate acroscopically, crenate-serrate, sparsely minutely paleate beneath. Sori linear, indusium entire. Moist forest floors and boulders, 100–800 m, SE (George to E Africa).

monanthes L.   SINGLE SORUS SPLEENWORT   Perennial with erect rhizome. Fronds suberect, stipe paleate initially, often gemmiferous, lamina linear, pinnate, pinnae oblong-falcate, broadly cuneate-flabellate at base, outer margins crenate-dentate, sparsely paleate beneath. Sori solitary, parallel to basiscopic margin, indusium subentire. Forests and ravines, 350–600 m, SW, LB, SE (Cape Peninsula to W Africa, Madagascar and tropical America).

platyneuron (L.) Oakes   EBONY SPLEENWORT   Rhizome suberect. Fronds erect, tufted, stipe initially paleate, lamina linear, pinnate, pinnae to 50 pairs, soft, linear, auriculate acroscopically, margins irregularly crenate-serrate, sparsely paleate beneath. Sori to 14 pairs per pinna, indusium erose. Forest floors and streambanks, 400–800 m, LB, SE (Langeberg to Mpumalanga and N America).

protensum Schrad.   Perennial with suberect rhizome. Fronds arching, soft, gemmiferous, stipe paleate, lamina narrowly elliptic, pinnate to 2-pinnatifid, pinnae to 55 pairs, pinnatifid into linear lobes, sparsely paleate beneath. Sori linear, indusium entire. Mossy boulders in forests, 400–600 m. SE (George to W Africa and Madagascar).

rutifolium (P.J.Bergius) Kunze   CARROT FERN   Perennial with erect rhizome. Fronds caespitose, coriaceous, stipe glabrous at maturity, lamina narrowly elliptic, 3- or 4-pinnatifid, pinnae to 19 pairs, ultimate lobes oblanceolate, sparsely paleate beneath. Sori 1 per lobe, indusium entire. Low-level epiphyte or lithophyte in forests, to 400 m, KM, LB, SE (Swellendam to tropical Africa).

simii A.F.Braithw. & Schelpe   sim spleenwort   Perennial with erect rhizome. Fronds tufted, coriaceous, stipe glabrous at maturity, lamina narrowly oblong, 2-pinnatifid to 2-pinnate, pinnae to 10 pairs, trapeziform, incised into 3–5 obcuneate lobes, margins dentate, paleate beneath. Sori of irregular lengths, indusium entire. Low-level epiphyte and on forest floors, 200–500 m, SE (Knysna to E Africa).

theciferum (Kunth) Mett.   CARROT FERN   Perennial with erect rhizome. Fronds caespitose, coriaceous, stipe sparsely paleate, lamina narrowly ovate-lanceolate, 2- or 3-pinnatifid, pinnae to 10 pairs, oblong to trapeziform, lobes spathulate, sparsely paleate beneath. Sori cupuliform, 1 per lobe, indusium entire. Forest epiphyte or lithophyte, 200–850 m, SE (George to Knysna and S to W Africa).

trichomanes L.   MAIDENHAIR SPLEENWORT   Perennial with erect rhizome. Fronds tufted, soft, gemmiferous, stipe glabrous, lamina linear, pinnate, to 32 pairs, broadly oblong-circular, minutely paleate beneath. Sori 2–7 per pinna, indusium erose. Cliffs and boulders, 800–1400 m, SW (Tulbagh to Stellenbosch and South Africa to Eurasia, N America, Australasia).

CETERACH   RESURRECTION FERN   c. 6 spp., Africa and Europe

cordatum (Thunb.) Desv.   Perennial with short rhizome. Fronds caespitose, suberect, thinly coriaceous, involute when dry, stipe densely paleate, lamina elliptic to narrowly elliptic, pinnatifid to 2-pinnate, pinnae narrowly oblong, weakly undulate, densely paleate beneath. Sori linear. Rock crevices and boulder bases in drier situations, 200–400 m, NW, SW, AP, KM, LB, SE (Bokkeveld Mts to E Africa and Madagascar).

AZOLLACEAE

*AZOLLA   MOSQUITO FERN   7 spp., pantemperate and tropical

*filiculoides Lam.   Perennial with horizontal, minutely papillate rhizome, roots borne singly. Upper leaf lobe ovate, lower leaf lobe hyaline. Megasporocarps with prominent dark apex, microsporocarps borne singly or subtended by a megasporocarp. Dams, vleis and rivers, SW (S American weed, Cape Flats).

BLECHNACEAE

BLECHNUM DEER FERN   c. 150 spp., nearly cosmopolitan

australe L. SOUTHERN DEER FERN   Perennial with branched rhizome. Fronds erect or arching, soft, stipe sparsely paleate, lamina narrowly elliptic, to 50 cm long, pinnate, sterile pinnae oblong-acute, mucronate, base somewhat auriculate, fertile pinnae linear, falcate. Sori extending most of the length of fertile pinnae, indusium lacerate. Rock crevices and forest floors, 250–1200 m, NW, SW, KM, LB, SE (Gifberg to E Africa, Madagascar and S Atlantic islands).

capense Burm.f. CAPE DEER FERN   Perennial with massive rhizome. Fronds arching, soft, stipe paleate, sterile lamina oblong-acute, pinnae shortly petiolate, minutely serrate, fertile lamina narrowly oblong-acute, pinnae undulate, linear. Sori extending most of the length of the fertile pinnae, indusium lacerate. Moist streambanks, 170–1400 m, NW, SW, LB, SE (Citrusdal to E Africa).

giganteum (Kaulf.) Schltdl. LARGE DEER FERN   Rhizomatous perennial. Fronds arching, coriaceous, stipe paleate, sterile lamina elliptic, pinnate, proximal pinnae decrescent, narrowly oblong-attenuate, entire, sessile, fertile lamina elliptic, pinnate, pinnae linear, sessile. Sori extending most of the length of the fertile pinnae, indusium erose. Wet shady sites, streams and in forests, 100–800 m, SW, LB, SE (Cape Peninsula to E Africa and Madagascar).

inflexum (Kunze) Kuhn   Perennial with erect rhizome. Fronds erect, soft, stipe paleate, sterile lamina oblong-elliptic, pinnatifid to pinnate, pinnae narrowly oblong, entire, sessile, fertile lamina shorter, narrowly oblong, pinnae linear, subsessile. Sori extending most of the length of fertile pinnae, indusium lacerate. Moist streambanks and forest margins, 600–1200 m, NW, SW, KM, SE (Clanwilliam to Knysna to Mpumalanga).

punctulatum Sw.   Rhizomatous perennial. Fronds arching, soft, stipe proximally paleate, lamina oblong-elliptic, to 75 cm long, pinnate, sterile pinnae oblong-attenuate from an auriculate base, fertile pinnae linear, base somewhat auriculate acroscopically. Sori extending most of the length of fertile pinnae, indusium erose. Moist forests, terrestrial or lithophytic, 250–800 m, NW, SW, KM, LB, SE (Pakhuis Mts to Zimbabwe and Madagascar).

tabulare (Thunb.) Kuhn MOUNTAIN DEER FERN   Perennial with erect rhizome. Fronds erect to suberect, coriaceous, stipe proximally paleate, lamina narrowly oblong, to 1.4 m, pinnate, sterile pinnae subsessile, entire, fertile pinnae subsessile to shortly petiolate. Sori extending most of the length of fertile pinnae, indusium lacerate. Exposed moist slopes, 300–1200 m, NW, SW, KM, LB, SE (Pakhuis Mts to W Africa and Madagascar).

CYATHEACEAE

CYATHEA TREE FERN   c. 40 spp., pantropical

capensis (L.f.) Sm. CAPE TREE FERN   Shrub or tree with caudex to 4.5 m high, slender. Fronds arching, stipe paleate, lamina to 3-pinnatifid, to 2 m long, pinnule segments narrowly oblong, dentate, with bullate paleae along costules beneath. Sori 1 or 2 at base of each pinnule segment, indusium asymmetric. Moist forests in ravines, 350–1200 m, SW, KM, LB, SE (Cape Peninsula to Knysna and E Africa, Brazil).

dregei Kunze COMMON TREE FERN, BOOMVARING   Shrub or tree with caudex to 5 m high, stout. Fronds arching, stipe paleate, lamina 3-pinnate, to 3 m long, pinnule lobes narrowly oblong, entire to crenate, glabrous to densely tomentose beneath. Sori to 12 per pinnule lobe, indusium cupuliform. Streambanks in grassland, 900–1200 m, LB (Heidelberg to Zimbabwe).

DENNSTAEDTIACEAE

1. Sori with paraphyses; venation always with some anastomoses:
    2. Lamina glabrous ....**Histiopteris**
    2.' Lamina set with acicular hairs ....**Blotiella**
1.' Sori without paraphyses; venation free:
    3. Sporangia borne along a marginal vascular commisure protected by a well-differentiated reflexed marginal indusium...**Pteridium**
    3.' Sporangia borne terminally on anterior branch of a vein protected by a differentiated lobe margin ...**Hypolepis**

BLOTIELLA   c. 15 spp., tropical America, Africa and Madagascar

glabra (Bory) A.F.Tryon   Rhizomatous perennial. Fronds to 1.5 m long, arching, stipe pubescent, lamina elliptic, 2- or 3-pinnatifid, pinna costa winged for at least three-quarters of its length, pinna lobes oblong, adnate, crenate to pinnatifid into sinuate lobes, pubescent. Sori semicircular in small sinuses of frond segments, indusium membranous. Moist shaded forest floors, 100–600 m, SE (George to Knysna and S and E Africa, and Madagascar).

natalensis (Hook.) A.F.Tryon   NATAL LANCE FERN   Rhizomatous perennial. Fronds soft, to 2 m long, arching, stipe thinly pubescent, lamina 2-pinnatifid to 2-pinnate, lanceolate, pinnae oblong-acuminate, incised into adnate to sinuate lobes, thinly pubescent. Sori subcircular, in marginal sinuses of pinna segments, indusium membranous. Wet forest floors and streambanks, 150–400 m, SE (Knysna to E Africa and Madagascar).

HISTIOPTERIS   c. 1 sp., pantropical and temperate

incisa (Thunb.) J.Sm.   Rhizomatous perennial. Fronds arching, to 3 m long, stipe glabrous, lamina pinnatifid to 2-pinnatifid, broadly lanceolate, proximal pinnae segments auriculate and developed close to the rachis, ultimate lobes oblong, entire to sinuate. Sori linear along margins of ultimate lobes, indusium entire. Moist streambanks, 180–800 m, NW, SW, LB, SE (Gifberg to W Africa, Madagascar).

HYPOLEPIS   c. 40 spp., pantropical

sparsisora (Schrad.) Kuhn   Rhizomatous perennial. Fronds to 3 m long, stipe glabrous at maturity, lamina 3- to 5-pinnatifid. ultimate segments oblong, crenate to pinnatifid. Sori singly on acroscopic margin of lobes of ultimate segments, pseudo-indusium subentire. Along streams on forest margins, 200–900 m, SW, LB, SE (Cape Peninsula to E and W Africa, Madagascar).

PTERIDIUM   BRACKEN FERN, ADELAARSVARING   1 sp., cosmopolitan

aquilinum (L.) Kuhn   Perennial with subterranean rhizome. Fronds erect, to 1.5 m long, stipe glabrous at maturity, lamina deltate, 3- or 4-pinnate, pinnule segments deeply pinnatifid into obtuse, oblong lobes, glabrous or thinly pubescent above, densely pubescent-tomentose beneath. Sori elongate, submarginal, pseudo-indusium ciliate. Fynbos and forest margins, to 1000 m, NW, SW, KM, LB, SE (Gifberg to Europe).

DRYOPTERIDACEAE

1. Indusium inferior ....Cystopteris
1.' Indusium superior:
   2. Basal pinnae never basiscopically developed ....Polystichum
   2.' Basal pinnae conspicuously basiscopically developed:
      3. Indusium peltate ....Rumohra
      3.' Indusium reniform ....Dryopteris

CYSTOPTERIS   c. 6 spp., nearly cosmopolitan

fragilis (L.) Bernh.   BRITTLE FERN   Rhizomatous perennial. Fronds erect, stipe sparsely paleate, lamina oblong-lanceolate, 3- or 4-pinnatifid, to 15 cm long, ultimate segments obtusely dentate. Sori circular, indusium peltate, opening towards margin, lacerate. Moist cliffs and rocks, 1300–1500 m, NW, KM (Worcester to Ladismith and S to E Africa, pan-boreal).

DRYOPTERIS   WOOD FERN   c. 150 spp., nearly cosmopolitan

antartica (Baker) C.Chr.   Rhizomatous perennial. Fronds arching, stipe paleate, lamina ovate-deltate, 3-pinnatifid, proximal pinnae basiscopically developed, pinnules narrowly deltate, lobes dentate, with filiform paleae above and clathrate paleae beneath. Sori on ultimate segments, indusium orange. Moist shady gullies and cliff bases, 1200–1600 m, NW, SW, KM, LB, SE (Cold Bokkeveld Mts to E Africa).

inaequalis (Schltdl.) Kuntze   Rhizomatous perennial. Fronds arching, stipe paleate, lamina ovate to broadly deltate, 3- or 4-pinnatifid, proximal pinna pair basiscopically developed, pinnules oblong, crenate-serrate, sparsely paleate beneath. Sori circular, indusium subentire. Moist forest floors and streambanks, 250–600 m, SW, LB, SE (Cape Peninsula to E and W Africa).

POLYSTICHUM SHIELDFERN c. 200 spp., nearly cosmopolitan

incongruum J.P.Roux Perennial with decumbent, branched rhizome. Fronds suberect to arching, firmly herbaceous to coriaceous, stipe densely paleate, lamina ovate-truncate to broadly ovate-truncate, 2- or 3-pinnate, pinnules narrowly trullate to somewhat falcate, pungent, subglabrous above, sparsely paleate beneath. Sori circular, indusium peltate, repand to erose, often black-centred. Forest floors, 50–1000 m, LB (Swellendam to Hogsback).

monticola N.C.Anthony & Schelpe MOUNTAIN SHIELDFERN Perennial with short rhizome. Fronds suberect or arching, soft, stipe densely paleate, lamina ovate-truncate, 2-pinnate to 3-pinnatifid, pinnules often lunate, aristate, subglabrous above, sparsely paleate beneath. Sori circular, indusium peltate, erose. Forest floors, shaded cliffs and rocks, 600–1200 m, NW, SW, KM, LB, SE (Kamiesberg and Cape Peninsula to Free State).

pungens (Kaulf.) C.Presl PRICKLY SHIELDFERN Rhizomatous perennial. Fronds suberect to arching, firm, stipe moderately paleate, lamina broadly ovate, proximal acroscopic pinnule largest, pinnules ovate to rhomboid, punctate, subglabrous above, sparsely paleate beneath, large bullate palea often subtending each pinnule. Sori circular, indusium peltate, minute, amorphous. Shaded forest floors and gullies, 250–800 m, SW, LB, SE (Cape Peninsula to Mpumalanga).

transvaalense N.C.Anthony TRANSVAAL SHIELDFERN Perennial with short, suberect rhizome. Fronds suberect, caespitose, stipe paleate, lamina 2-pinnate, narrowly ovate-attenuate, pinnules auriculate, aristate, subglabrous above, with twisted paleae beneath. Sori circular, indusium peltate, fimbriate. Streamsides in forest, 500 m, SW (Riviersonderend Mts and S to W Africa).

wilsonii Christ WILSON SHIELDFERN Perennial with short, suberect rhizome. Fronds suberect to arching, caespitose, stipe densely paleate, lamina 2-pinnate, ovate-lanceolate, pinnules aristate, moderately paleate above, densely paleate beneath. Sori circular, indusium peltate, erose. Moist rock shelters, 1300–1500 m, NW, KM (Worcester to Oudtshoorn and tropical Africa to E Asia).

RUMOHRA c. 6 spp., circum-austral

adiantiformis (G.Forst.) Ching SEVEN WEEKS FERN, SEWEWEEKSVARING Rhizomatous perennial. Fronds arching, coriaceous, stipe paleate, lamina pentagonal, 3-pinnate to 4-pinnatifid, proximal pinnae basiscopically developed, ultimate segments ovate, margins crenate-dentate, sparsely paleate beneath. Sori circular, indusium peltate. Forests and rocky ravines, 50–1200 m, SW, LB, SE (Cape Peninsula to Mpumalanga, Madagascar, S Oceanic islands, Australia, Polynesia, Mesomerica).

EQUISETACEAE

EQUISETUM HORSETAIL 15 spp., temperate regions of the world, except Australasia

ramosissimum Desf. AFRICAN HORSETAIL, PERDESTERT, LIDJIESGRAS Perennial with underground rhizome. Aerial stems erect, hollow, ribbed, leaves reduced to a short, toothed sheet at each node. Strobili terminal on branches. Wet sites in fynbos, 100–1200 m, AP, KM, SE (Cape Infanta and George to W Africa, Madagascar).

GLEICHENIACEAE

GLEICHENIA c. 110 spp., southern Africa, Malaysia, Australasia, Madagascar

polypodioides (L.) Sm. CORAL FERN, CREEPING FERN Rhizomatous perennial. Fronds widely spaced, falsely dichotomously branched with a terminal bud in each false dichotomy, all branches bearing foliar segments, pinnules linear, lobed. Sori partially immersed in lamina bearing 2–4 sporangia. Shady moist rocks, 500–1600 m, NW, SW, KM, LB, SE (Cedarberg to E Africa, Madagascar).

GRAMMITIDACEAE

GRAMMITIS c. 400 spp., pantropical and austral

poeppigiana (Mett.) Pic.Serm. Rhizomatous perennial. Fronds closely spaced, sessile, lamina rounded-oblong, to 4 cm long, sparsely paleate beneath. Sori elongate, to 2.5 mm long, set in a line on either side of midrib at angles of 20° to midrib. Rock crevices on peaks above 1700 m, SW, LB (Stellenbosch to Swellendam and circum-Antarctic).

HYMENOPHYLLACEAE

1. Venation catadromous ....Trichomanes
1.' Venation anadromous:
    2. Fronds closely set with stellate hairs ....Sphaerocionium
    2.' Fronds not bearing stellate hairs ....Hymenophyllum

HYMENOPHYLLUM   FILMY FERN   c. 200 spp., pantemperate and tropical

capense Schrad.   CAPE FILMY FERN   Perennial with branching rhizome. Fronds 2-pinnatifid, stipe partially narrowly winged, lamina to 10 cm long, bifurcate to pinnatifid into 3–7 entire lobes. Sori on distal acroscopic segments with entire indusial valves. Wet shaded rocks, 800–1500 m, SW, LB, SE (Cape Peninsula to E Africa and Madagascar).

peltatum (Poir.) Desv.   Perennial with branched rhizome. Fronds 2-pinnatifid to 2-pinnate, stipe partially narrowly winged, lamina to 18 cm long, unequally pinnatifid acroscopically, lobes serrate. Sori on proximal acroscopic segments with entire indusial valves. Wet cliffs and rocks in forest, 200–1200 m, NW, SW, LB, SE (Gifberg to Europe, S America, Australasia).

tunbridgense (L.) Sm.   TUNBRIDGE FILMY FERN   Perennial with branched rhizome. Fronds 2- or 3-pinnatifid, stipe partially narrowly winged, lamina to 10 cm long, bilaterally pinnatifid to 2-pinnatifid, lobes linear, serrate. Sori on proximal acroscopic lobes with serrate indusial valves. Wet shaded rocks and tree trunks in forest, 200–1200 m, NW, SW, KM, LB, SE (Gifberg to S Europe, Madagascar).

SPHAEROCIONIUM   c. 25 spp., pantemperate and tropical

aeruginosum (Poir.) Pic.Serm. (= *Sphaerocionium marlothii* (Brause) Pic.Serm.)   Perennial with branched rhizome. Fronds 2-pinnatifid to 2-pinnate, stipe set with stellate hairs, lamina to 8 cm long, set with stalked stellate hairs on margins and veins. Sori at lobe apices with small shallow indusial valves set with stellate hairs. Wet rocks in gullies, 850–1500 m, NW, SW, LB, SE (Grootwinterhoek Mts and Cape Peninsula to Knysna, S Atlantic islands).

TRICHOMANES   BRISTLE FERN   c. 300 spp., pantemperate and tropical

melanotrichum Schltdl.   Perennial with branched rhizome set with black hairs. Fronds pinnatifid or 2-pinnatifid, stipe narrowly winged distally, lamina to 7 cm long, divided into rounded entire lobes. Sori conical, partly narrowly winged. On wet rocks in forest, 250–1000 m, LB, SE (George to Knysna and E and W Africa, Madagascar).

ISOETACEAE

ISOETES   QUILLWORT   c. 150 spp., nearly cosmopolitan

capensis Duthie   CAPE QUILLWORT   Tufted geophyte with horny deltate scales. Sporophylls 5–35, slender, to 20 cm long, ligule cordate. Sporangium with complete velum. Clay soils in submerged vleis, to 300 m, SW (Darling to Stellenbosch).•

stellenbossiensis Duthie   STELLENBOSCH QUILLWORT   Tufted geophyte with tricuspidate scales. Sporophylls 5–33, to 12 cm long, ligule small, elongate-deltate. Sporangium lacking velum. Waterlogged soils, to 300 m, SW (Cape Flats).•

LOMARIOPSIDACEAE

ELAPHOGLOSSUM   TONGUE FERN   c. 500 spp., pantemperate and tropical

acrostichoides (Hook. & Grev.) Schelpe   Rhizomatous perennial. Fronds erect or arching, stipe sparsely paleate, lamina simple, coriaceous, narrowly elliptic, to 35 cm long, minutely stellate-paleate above, moderately substellate-paleate beneath. Sporangia acrostichoid, exindusiate. Moist protected or exposed rocks, 300–1200 m, NW, SW, LB, SE (Grootwinterhoek Mts and Cape Peninsula to E and W Africa, Mascarenes).

angustatum (Schrad.) Hieron.   BUCK TONGUE FERN   Rhizomatous perennial. Fronds erect or arching, stipe moderately paleate, lamina simple, coriaceous, oblanceolate, to 25 cm long, with minute stellate paleae above, moderately stellate-paleate beneath. Sporangia acrostichoid, exindusiate. Moist boulders along streams, 280–700 m, SW, LB, SE (Cape Peninsula to KwaZulu-Natal).

conforme (Sw.) Schott ex J.Sm.   Rhizomatous perennial. Fronds erect, stipe paleate, lamina simple, entire, coriaceous, viscid when young, sterile elliptic, fertile oblong-elliptic, sterile smaller than fertile, paleate along midrib above, minutely substellate-paleate beneath. Sporangia acrostichoid, exindusiate. Moist rock ledges in fynbos, 600–1000 m, SW (Cape Peninsula to Stellenbosch and E Africa and St Helena).

LYCOPODIACEAE

1. Stems isotomous, without elongate indeterminate main stems; sporophylls and vegetative leaves alike, or if smaller then not ephemeral ....**Huperzia**
1.' Stems anisotomous, differentiated into long indeterminate rhizomatous or trailing main stems and determinate aerial branchlet systems; sporophylls strongly modified, ephemeral:
   2. Strobili erect, pedunculate, borne on branchlet systems which arise dorsolaterally on main stem ....**Lycopodium**
   2.' Strobili pendulous and sessile or terminating in simple erect branches which arise dorsally on creeping stems ....**Lycopodiella**

HUPERZIA   c. 300 spp., pantemperate and tropical

gnidioides (L.f.) Trevis.   Perennial with erect, arching or pendulous, dichotomously branched aerial stems from a compact branching horizontal stem. Leaves narrowly oblong, sporophylls shorter than foliage leaves. Montane forests and ravines, 300–2000 m, SW, KM, LB (Cape Peninsula to N Province, Madagascar).

saururus (Lam.) Trevis.   Perennial with erect, unbranched aerial stems from a compact branching horizontal stem. Leaves closely imbricate. Sporangia hidden, sporophylls indistinguishable from foliage leaves. Rock ledges and cliffs above 1700 m, SW, KM, LB (Riviersonderend Mts to George and E and W Africa, Madagascar).

LYCOPODIELLA   club moss   c. 40 spp., pantemperate and tropical

caroliniana (L.) Pic.Serm.   SLENDER CLUB MOSS   Perennial with prostrate main stem, somewhat dorsiventrally flattened. Leaves lanceolate, lateral leaves spreading horizontally, dorsal leaves smaller than lateral leaves, adpressed. Stobili solitary, erect. Marshes and seeps in fynbos, to 1500 m, NW, SW, KM, LB, SE (Pakhuis Mts to Knysna and E southern Africa, temperate America).

cernua (L.) Pic.Serm.   NODDING CLUB MOSS   Perennial with prostrate main stem and erect, much-branched leafy stems, the lateral branches bearing strobili. Leaves subulate. Strobili solitary at branch apices. Moist streambanks and road cuttings, to 200 m, NW, SW, LB, SE (Gifberg to tropical Africa, Madagascar).

LYCOPODIUM   CLUB MOSS   c. 40 spp., pantemperate and tropical

clavatum L.   RUNNING CLUB MOSS   Perennial with prostrate main stem and dichotomously branched, erect stems. Leaves with a translucent hair point. Strobili pedunculate, in groups of 2–5. Forest margins and road cuttings, 300–1500 m, NW, SW, LB, KM, SE (Grootwinterhoek Mts to Knysna, E and W Africa, Mascarenes).

complanatum L.   RUNNING PINE   Perennial with prostrate or underground main stem and repeatedly dichotomously branched aerial stems. Leaves subulate. Strobili pedunculate, in groups of 2–8. Sheltered rocky slopes, 1000–2000 m, NW, SW, KM, SE (Grootwinterhoek Mts to Port Elizabeth, and Madagascar).

MARATTIACEAE

MARATTIA   c. 40 spp., pantropical

fraxinea J.Sm. ex J.F.Gmel.   Perennial with massive rhizome. Fronds arching, stipe to 1.5 m long, with a pair of carnose stipules, lamina to 2 m long, 2-pinnate, pinnules linear-attenuate, to 16 cm long. Synangia submarginal. Forests, 500–1000 m, SE (George to E Africa).

MARSILEACEAE

MARSILEA WATER CLOVER c. 50 spp., nearly cosmopolitan

burchellii (Kunze) A.Br. BURCHELL SE WATERKLAWER Aquatic perennial. Fronds with slender, glabrous stipes, pinnae variable, pedicels free, straight, slender, pilose at first. Sporocarps crowded, small, subcircular, pyriform, lateral ribs not apparent, superior tooth developed, conical. Seasonal vleis and seeps, 250–400 m, NW, SW (Namaqualand and Karoo, Vanrhynsdorp to Malmesbury).

capensis A.Br. CAPE WATER CLOVER Aquatic perennial. Fronds with slender, usually crowded stipes, pinnae narrowly obdeltate, flanks slightly convex, pedicels solitary, free, slender. Sporocarps obliquely broad-oblong or irregularly rhombic in lateral view, lateral ribs absent, superior tooth distinct, conical, erect or recurved. Seasonal vleis, 250–400 m, NW, SW, KM, AP, SE (Namaqualand and Karoo to Port Elizabeth, and N Africa).

macrocarpa C.Presl Aquatic perennial. Fronds with stipes usually glabrous, pinnae with brown interstitial streaks beneath, flanks concave, pedicels free. Sporocarps narrowly rectangular in dorsiventral cross section, lateral ribs distinct, superior tooth present, short. Seasonal vleis, 250–400 m, SW (Cape Peninsula and S to E Africa).

schelpeana Launert Aquatic perennial. Fronds with stipes crowded, slender, pinnae narrowly obdeltoid, flanks slightly convex, outer margins bilobate, appressed pilose at first, pedicels from base of stipe, erect. Sporocarps solitary or crowded, obliquely ovate-oblong to elliptic in lateral view, lateral ribs absent, superior tooth distinct. Seasonal vleis, 150–500 m, AP, SE (Bredasdorp to E Cape).

*NEPHROLEPIDACEAE

***NEPHROLEPIS** c. 20 spp., pantropical and subtropical

***exaltata** (L.) Schott BOSTON FERN Perennial with erect rhizome. Fronds erect, stipe paleate, lamina ovate-lanceolate, pinnate, pinnae subsessile, oblong-acute, auriculate, serrate-dentate, subglabrous. Sori semilunate, indusium reniform. Disturbed areas, to 400 m, SW, SE (tropical American weed, Cape Peninsula to Knysna).

OPHIOGLOSSACEAE

OPHIOGLOSSUM ADDER'S-TONGUE c. 30 spp., cosmopolitan

bergianum Schltdl. BERGIUS'S ADDER'S TONGUE Perennial with fusiform rhizome, roots proliferous. Leaves 2–4, tropophore narrowly oblanceolate, to 5 cm long. Sporophore arising from leaf base, apparently independently, with 4–7 pairs of sporangia. Seasonally wet sites in fynbos, to 650 m, NW, SW (Gifberg to Cape Peninsula).•

gracile Pocock ex J.E.Burrows Perennial with linear rhizome, roots proliferous. Leaves 1 or 2, tropophore narrowly elliptic to narrowly ovate. Sporophore to 14 cm long, with 13–28 pairs of sporangia. Seasonally moist soils in fynbos, 250–900 m, NW, SW, SE (Pakhuis Mts to George).•

nudicaule L.f. SLENDER ADDER'S TONGUE Perennial with linear rhizome, roots proliferous. Leaves 2–5, tropophore elliptic to narrowly ovate. Sporophore to 4 cm long, with 5–9 pairs of sporangia. Seasonally moist clay soils in karroid scrub, 500–1200 m, NW, SW, KM, LB, SE (Clanwilliam to Oudtshoorn).•

polyphyllum A.Br. Perennial with fusiform rhizome, roots proliferous. Leaves 1 or 2, tropophore elliptic to narrowly ovate, to 6 cm long. Sporophore to 10 cm long, with 11–36 pairs of sporangia. Seasonally moist soils in fynbos and renosterveld, 250–1200 m, AP, KM, LB (Bredasdorp to Oudtshoorn, tropical and N Africa to India).

OSMUNDACEAE

1. Sporangia on contracted apical fertile segments..Osmunda
1.' Sporangia on noncontracted basal fertile segments...Todea

OSMUNDA c. 6 spp., pantemperate and tropical mountains

regalis L. ROYAL FERN Perennial with erect rhizome. Fronds erect, stipe glabrous at maturity, lamina to 1 m, 2-pinnate, fertile pinnae borne in apical portion, sterile pinnules naFrrowly oblong, to

6 cm long, fertile pinnules with groups of sporangia. Streambanks in fynbos, 700–1500 m, NW, SW, SE (Cedarberg to N Province, pantropical).

TODEA   2 spp., South Africa and Australasia

barbara (L.) T.Moore   Perennial with massive rhizome. Fronds erect, stipe with loose tomentum, lamina to 1 m, 2-pinnatifid, pinnules oblong with broadened adnate bases, sporangia on distal third of pinnae. Sori confluent at maturity. Streambanks in fynbos, to 1500 m, NW, SW, KM, LB, SE (Cedarberg to E southern Africa).

POLYPODIACEAE

1. Lamina simple, entire; venation reticulate:
    2. Areoles without included veinlets or if these present then few and excurrent .... Pleopeltis
    2.' Areoles with simple and forked excurrent and recurrent included veinlets .... Lepisorus
1.' Lamina unevenly lobed or pinnatifid; venation free or reticulate:
    3. Receptacle with conspicuous peltate clathrate paraphyses .... × Pleopodium
    3.' Receptacle with inconspicuous simple hair-like paraphyses .... Polypodium

LEPISORUS   c. 40 spp., Africa and Asia

schraderi (Mett.) Ching   Rhizomatous perennial. Fronds spaced, erect, simple, entire, stipe glabrous, lamina narrowly elliptic, to 30 cm long. Sori round, in a line on either side of midrib in upper half of lamina, to 6 mm in diam. Low-level epiphyte in forests, 200–600 m, LB, SE (Langeberg at Heidelberg to George, and to E Africa, Madagascar).

PLEOPELTIS   SHIELD SORUS POLYPODY   c. 10 spp., America, Africa, Madagascar to India

macrocarpa (Bory ex Willd.) Kaulf.   Perennial with widely spreading rhizome. Fronds spaced, coriaceous, stipe sparsely paleate, lamina simple, entire, narrowly elliptic, to 20 cm long, minutely lacerate-paleate beneath. Sori oval, to 4 mm in diam. Forest epiphyte or lithophyte, to 800 m, SW, LB, SE (Cape Peninsula to tropical Africa, Madagascar, tropical and S. America).

× PLEOPODIUM   c. 15 spp., Africa and tropical America

simianum Schelpe & N.C.Anthony   Rhizomatous perennial. Fronds spaced, stipe paleate, lamina sinuate to deeply pinnatifid below, irregular, segments unequally deltate, sinuate around sori, paleate beneath. Sori in 2 rows, one on either side of midrib in upper half of lamina, oval. Forest epiphyte or lithophyte, 400–1200 m, SE (Knysna to E southern Africa).

POLYPODIUM   POLYPODY   c. 150 spp., mostly extratropical regions of the world

ensiforme Thunb.   Perennial with short rhizome. Fronds spaced, erect, coriaceous, stipe sparsely paleate, lamina oblong to lanceolate, deeply pinnatifid almost to midrib into linear, shallowly crenate lobes. Sori round, to 2 mm in diam., in a line on either side of the costa. Low-level epiphyte in forests, 100–1300 m, SW, KM, LB, SE (Betty's Bay to E Cape).

vulgare L.   COMMON POLYPODY   Rhizomatous perennial. Fronds spaced, erect, soft, stipe glabrous, lamina ovate-oblong, pinnatifid to near midrib into oblong, entire lobes. Sori round, to 22 per lobe, to 3 mm in diam. Damp shaded rocks, 800–1400 m, NW, SW, LB, SE (Grootwinterhoek Mts to Mpumalanga and Europe).

PTERIDACEAE

1. Sori exindusiate; sporangia borne along veins; small ephemeral plants .... Anogramma
1.' Sori marginal or submarginal:
    2. Sori borne on reflexed marginal flaps .... Adiantum
    2.' Sori borne on lamina surface near margin and often covered by it:
        3. Ultimate fertile segments with a distinct serrate sterile apex .... Pteris
        3.' Ultimate fertile segments without a distinct sterile apex:
            4. Ultimate segments articulated .... Pellaea
            4.' Ultimate segments not articulated .... Cheilanthes

ADIANTUM MAIDENHAIR FERN, VROUEHAAR   c. 150 spp., nearly cosmopolitan

aethiopicum L.   Perennial with widely spreading rhizome. Fronds arching, stipe terete, lamina ovate, 3-pinnate, pinnules cuneate to rhomboid, petiolate, minutely dentate on outer margins with veins ending in teeth. Sori 1 or 2 per pinnule on outer margin, indusial flaps glabrous, orbicular to deeply reniform. Moist banks in forest, 100–400 m, NW, SW, LB, SE (Grootwinterhoek Mts to George and N America).

capillus-veneris L.   BLACK MAIDENHAIR, SWART VROUEHAAR   Rhizomatous perennial. Fronds arching, soft, stipe glabrous, lamina ovate-deltate, 3-pinnate, pinnules cuneate, shallowly to deeply lobed, petiolulate, minutely crenate-dentate, veins ending in teeth. Sori along outer margins of pinnules, indusial flaps lunate to oblong. Shaded moist rock faces, to 800 m, NW, SW, KM, SE (Gifberg to tropical Africa, Madagascar).

*****hispidulum** Sw.   ROSY MAIDENHAIR   Perennial with short rhizome. Fronds erect, pedate, stipe hispid, lamina deltate, repeatedly dichotomously divided with up to 8 linear pinnae, pinnules rhombic, hispid beneath. Sori on acroscopic outer margin of pinnules, indusial flaps pilose, rotund to oblong. Moist banks in forest, SW (?Asian weed, Cape Peninsula).

poiretii Wikstr.   Perennial with widely spreading rhizome. Fronds arching, soft, stipe glabrous, lamina ovate-deltate, 3- or 4-pinnate, pinnules obcuneate to shallowly reniform, articulated at apex of filiform petiolules. Sori along outer margins of pinnules, indusial flaps lunate. Moist forest floors, 250–800 m, SW (Cape Peninsula and S to E Africa to India, C and S America).

*****raddianum** C.Presl   Perennial with widely spreading rhizome. Fronds arching, stipe glabrous, lamina ovate-deltate, 3- or 4 pinnate, ultimate segments obcuneate, outer margins minutely cuneate-serrate, veins ending in sinuses. Sori on outer margins of ultimate segments, indusial flaps reniform. Naturalised on moist streambanks, SW, SE (S American weed, Cape Peninsula to George).

ANOGRAMMA   c. 5 spp., pantemperate and tropical

leptophylla (L.) Link   ANNUAL FERN   Annual with minute rhizome. Fronds few, tufted, erect, stipe glabrous, lamina 2- or 3-pinnatifid, to 7 cm long, ultimate segments broadly cuneate, shallowly lobed. Sori at vein endings. Damp, shady, moist earth slopes, 150–1000 m, NW, SW (Cedarberg to Worcester, nearly worldwide).

CHEILANTHES LIP FERN   c. 150 spp., subcosmopolitan

bergiana Schltdl.   BERGIUS'S LIP FERN   Perennial with short, erect rhizome. Fronds soft, stipe densely pilose, lamina pentagonal, 3-pinnate to 5-pinnatifid, proximal pinnae basiscopically developed, rounded lobes set with hairs along costae and veins on both surfaces. Sori at margins of lobes, indusium subentire. Deep shade in forests, 250–1200 m, SE (George to E Africa, Madagascar).

capensis (Thunb.) Sw.   CAPE LIP FERN   Rhizomatous perennial. Fronds soft, stipe nitid, usually glabrous, lamina ovate-deltate, to 12 cm long, 3-pinnatifid, proximal pinna pair basiscopically developed, ultimate segments serrate-crenate, with few paleae along costa and costules. Sori at vein endings at margin, indusium erose. Rock outcrops in fynbos and scrub, to 1200 m, NW, SW, AP, KM, LB, SE (Bokkeveld Mts to Port Elizabeth, southern Africa).

concolor (Langsd. & Fisch.) R.M. & A.F.Tryon   GERANIUM FERN, OAK LEAF FERN   Perennial with short, procumbent rhizome. Fronds suberect, stipe paleate, lamina hastate-pentagonal, 3-pinnatifid, proximal pinnae basiscopically developed. Sori marginal, indusium continuous or discontinuous. Shaded forest floors, 300–600 m, SE (George to tropical Africa, Asia, Australasia, C and S America).

contracta (Kunze) Mett. ex Kuhn   Rhizomatous perennial. Fronds erect, stipe with hairs and paleae, lamina lanceolate, 2-pinnate to 3-pinnatifid, to 30 cm long, pinnules to 7-lobed, hairy on both surfaces. Sori on ultimate lobes, protected by revolute margin, indusium absent. Shady rocks, 200–1200 m, NW, SW, AP, KM, SE (Namaqualand to Port Elizabeth and E southern Africa).

deltoidea Kunze   Perennial with short rhizome. Fronds erect, soft, stipe glabrous, lamina broadly deltate, to 10 cm long, 2- or 3-pinnatifid, proximal pinnae basiscopically developed, rachis winged. Sori linear, indusium continuous, erose. Rock outcrops and open ground in karroid scrub, 150–800 m, NW (Namibia to Citrusdal).

depauperata Baker   Rhizomatous perennial. Fronds erect, coriaceous, stipe paleate, lamina linear, to 17 cm long, 2-pinnate to 3-pinnatifid, pinnule lobes oblong-lanceolate with crenate recurved margins, with yellowish hairs beneath. Sori beneath revolute margins, exindusiate. Rocks and cliffs, 500–1000 m, SW, KM, LB, SE (Endemic to SW Cape).

eckloniana (Kunze) Mett.   ECKLON'S LIP FERN, RESURRECTION FERN   Rhizomatous perennial. Fronds erect, coriaceous, stipe paleate, lamina narrowly oblong, to 11 cm long, 2-pinnate to 3-pinnatifid, pinna segments pinnatifid into obtuse lobes, white-hairy above, densely tomentose beneath. Sori marginal, forming an interrupted soral line, indusium small. Rock outcrops, 1000–1200 m, KM (Witteberg and Karoo, throughout southern Africa and Zimbabwe).

hastata (L.f.) Kunze   Perennial with short creeping rhizome. Fronds erect, soft, stipe glabrous at maturity, lamina linear-lanceolate, to 30 cm long, ultimate segments acute or obtuse, crenate. Sori marginal, linear, indusium membranous. Sheltered rocks and under bushes, 120–1200 m, NW, SW, KM, LB, SE (Namibia to George).

hirta Sw.   PARSLEY FERN   Perennial with short creeping rhizome. Fronds erect, soft, stipe pilose, lamina linear-lanceolate, 2-pinnate to 3-pinnatifid, pinnae sparsely pubescent. Sori discrete on margins of incurved pinnule lobes, indusium minute. Rock crevices and boulder bases, 250–1400 m, KM, SE (George to N Province and E Africa, Madagascar).

induta Kunze   Perennial with spreading rhizome. Fronds erect, coriaceous, stipe paleate, lamina broadly lanceolate, 3-pinnate to 4-pinnatifid, ultimate segments deeply pinnatifid into rounded-oblong, crenate segments, tomentose beneath. Sori at margins of ultimate segments, indusium discontinuous. Rock outcrops, 600–1200 m, NW, SW, LB (Cedarberg to Heidelberg and South Africa).

involuta (Sw.) Schelpe & N.C.Anthony   Perennial with short rhizome. Fronds erect, soft, stipe with hairs and paleae, lamina linear to lanceolate, 2-pinnate to 3-pinnatifid, ultimate segments unequally oblong-hastate, paleate along costules beneath. Sori along margins of ultimate segments, linear, indusium continuous. Dry rocky slopes in karroid scrub, 500–800 m, KM (Witteberg to E Africa).

multifida (Sw.) Sw.   Perennial with short creeping rhizome. Fronds erect, coriaceous, stipe subglabrous, lamina oblong-ovate, 4- or 5-pinnatifid, proximal pinnae basiscopically developed, pinnules pinnatifid into ovate-oblong lobes. Sori discrete, around margins of ultimate segments, indusium entire to lacerate. Rock outcrops, 150–1200 m, NW, SW, KM, LB, SE (Vanrhynsdorp to E Africa).

namaquensis (Baker) Schelpe & N.C.Anthony   NAMAQUA LIP FERN   Rhizomatous perennial. Fronds erect, thinly coriaceous, to 20 cm long, stipe distally shallowly sulcate, lamina oblong-ovate, 3-pinnatifid, proximal pinnae largest, ultimate segments oblong-obtuse to hastate. Sori on margins of ultimate segments, indusium erose. Rock crevices, 300–600 m, NW, SW, KM (Namibia to Cape Peninsula and Witteberg).

parviloba (Sw.) Sw.   Rhizomatous perennial. Fronds erect, soft, stipe pilose, lamina ovate-lanceolate, 2-pinnate to 3-pinnatifid, pinnae pinnatifid into oblong segments, viscid beneath. Sori at apices of ultimate segments, exindusiate. Dry rocky slopes and rocks, 300–1400 m, NW, SW, KM (Grootwinterhoek Mts to E southern Africa and Namibia).

robusta (Kunze) R.M.Tryon   ROBUST LIP FERN   Rhizomatous perennial. Fronds erect, coriaceous, to 10 cm long, stipe glabrous, lamina oblong-ovate, 3-pinnatifid, ultimate segments narrowly hastate to ovate. Sori marginal, indusium linear, irregularly minutely erose. Exposed gravelly soils among low scrub, 200–400 m, SW (Namibia to Moorreesburg).

viridis (Forssk.) Sw.   Rhizomatous perennial. Fronds arching, stipe initially paleate, lamina soft, deltate-pentagonal, 2- to 4-pinnate, proximal pinnae basiscopically developed, ultimate segments narrowly hastate. Sori marginal, confluent, indusium membranous, subentire. Forest margins and scrub, 200–800 m, SW, LB, SE (Cape Peninsula to Arabia and India).

PELLAEA   CLIFF BRAKE   c. 35 spp., nearly pantropical

calomelanos (Sw.) Link   HARD FERN   Perennial with a short rhizome. Fronds erect, coriaceous, stipe terete, lamina narrowly ovate, 3-pinnate, pinnules cordate to broadly hastate, entire, glaucous, articulated to petiolules. Sori marginal, continuous, entire. Rock crevices and boulder bases, 300–1500 m, SW, KM, LB, SE (Cape Peninsula to E Africa, Madagascar, Spain, India).

leucomelas (Mett. ex Kuhn) Baker   Perennial with a short rhizome. Fronds erect, coriaceous, stipe proximally paleate, lamina narrowly ovate-deltate, 2- or 3-pinnate, pinnules oblong-ovate to broadly hastate, articulated to short petiolules, glabrous or thinly hairy. Sori marginal on revolute margin, indusium entire. Rock outcrops in karroid scrub, 600–1000 m, KM, LB, SE (Montagu to Port Elizabeth).•

pteroides (L.) Prantl MYRTLE FERN Perennial with a short rhizome. Fronds erect, coriaceous, stipe glabrous with age, lamina ovate-deltate, 2- or 3-pinnate, proximal pinnae basiscopically developed, ultimate segments oblong to ovate, minutely crenate, articulated to short petiolules. Sori marginal, discrete, indusium broadly oblong. Fynbos and forest, 300–1700 m, NW, SW, KM (Clanwilliam to Ladismith).

rufa A.F.Tryon Rhizomatous perennial. Fronds erect, coriaceous, stipe proximally paleate, lamina narrowly oblong-elliptic, 2- or 3-pinnate, pinnules broadly elliptic, entire, articulated to short petiolules. Sori linear, indusium subentire. Rock outcrops, 900–1100 m, KM (Ladismith to Willowmore and S Karoo).

PTERIS BRAKE c. 200 spp., pantemperate and tropical

buchananii Baker ex Sim BUCHANAN'S BRAKE Perennial with widely spreading rhizome. Fronds arching, soft, to 1.5 m, stipe glabrous, lamina broadly deltate, 4-pinnatifid, proximal pinnae basiscopically developed, ultimate lobes falcate, adnate, apices serrate, costules spinulose beneath. Sori along lobe margins, indusium erose. Moist soil in forests, 150–800 m, LB, SE (Heidelberg to E Africa).

cretica L. CRETAN BRAKE, AVERY FERN Rhizomatous perennial. Fronds erect, soft, stipe glabrous, lamina deltate, pinnate, pinnae linear, margins serrate-dentate. Sori in marginal lines, indusium linear, subentire. Forests, to 1000 m, SW, LB, SE (Heidelberg to S Europe, Madagascar; adventive on the Cape Peninsula).

dentata Forssk. Rhizomatous perennial. Fronds arching, to 1.5 m long, stipe glabrous, lamina ovate, 3-pinnatifid, proximal pinnae basiscopically developed, deeply pinnatifid, lobe apices serrate. Sori in a marginal line, indusium erose. Forest floors and ravines, 100–1200 m, NW, SW, LB, SE (Grootwinterhoek Mts and Cape Peninsula to tropical Africa, Madagascar).

*****tremula** R.Br. SHAKING BRAKE Perennial with short, erect rhizome. Stipe glabrous, lamina ovate, proximal pinnae basiscopically developed, 3- or 4-pinnate, venation free, ultimate segments linear, apices obtuse, dentate. Sori in marginal lines. Forests and clearings, SW (Australasian weed, Cape Peninsula to Stellenbosch).

vittata L. CHINESE LADDER BRAKE Rhizomatous perennial. Fronds arching, stipe proximally paleate, lamina elliptic-oblong, pinnate, pinnae linear-attenuate, margins minutely crenate. Sori in submarginal lines extending for most of pinna length, indusium subentire. Shaded rock crevices and forest margins, to 500 m, SW, LB, SE (Cape Peninsula to Port Elizabeth, pantemperate and tropical).

*SALVINIACEAE

***SALVINIA** c. 10 spp., temperate and tropical

*****molesta** D.S.Mitch. KARIBA WEED Floating perennial with horizontal rhizome. Floating fronds in pairs, emarginate, base cordate, folded along midrib, with multicellular papillae with 4 separate curved segments adjoining apically above, submerged fronds dissected. Sporocarps spherical, hairy. Standing inland waters, to 250 m, SW, SE (tropical African weed, Cape Peninsula to Knysna).

SCHIZAEACEAE

SCHIZAEA GRASS FERN, TOOTHBRUSH FERN c. 30 spp., pantropical

pectinata (L.) Sw. CURLY GRASS FERN Perennial with underground rhizome. Fronds erect, linear, to 20 cm long, fertile portion with a recurved rachis, pinnae to 9 mm long. Sporangia in 2 rows on either side of costa. Dry mountain slopes, 100–600 m, NW, SW, AP, KM, LB, SE (Clanwilliam to E Africa and Madagascar).

tenella Kaulf. Perennial with underground rhizome. Fronds erect, linear, to 25 cm long, fertile portion with a straight ascending rachis, pinnae to 4 mm long. Sporangia borne in 2 rows on either side of costa. Moist streambanks and cliff ledges, 300–1500 m, fynbos, NW, SW, KM, LB, SE (Cold Bokkeveld to N Province).

SELAGINELLACEAE

SELAGINELLA SPIKE MOSS   c. 700 spp., nearly cosmopolitan

caffrorum (Milde) Hieron.   RESURRECTION PLANT   Mat-forming perennial with branched prostrate stems, leafy stems radially symmetrical or occasionally slightly dorsiventral. Fronds narrowly deltate, margins ciliate. Strobili with sporophylls in 4 ranks. Rock faces and crevices, 800–1200 m, KM (Little Karoo to tropical Africa).

kraussiana (Kunze) A.Br.   KRAUSS' SPIKE MOSS   Perennial with prostrate, spreading stems, branched, 1- to 3-furcate. Leaves heteromorphic, median leaves sparsely toothed, lateral leaves serrate. Strobili at apex of ultimate branches, sporophylls undifferentiated. Forest floors, to 500 m, SE (George to E Africa and Macaronesia).

pygmaea (Kaulf.) Alston   Annual with erect, slender stems, simple or sparsely branched, terminating in 1 or 2 strobili. Leaves distant, ovate. Sporophylls in 4 ranks. Seasonally moist ground, 150–400 m, NW, SW, AP, SE (Clanwilliam to Port Elizabeth).•

THELYPTERIDACEAE

1. At least basal pair of veins regularly anastomosing with excurrent vein running to sinus .... **Cyclosorus**
1.' Veins all free, or basal veins sometimes touching below sinus between pinna lobes:
    2. Pinna lobe veins mostly once-forked .... **Thelypteris**
    2.' Pinna lobe veins simple:
        3. Sori exindusiate .... **Stegnogramma**
        3.' Sori indusiate:
            4. Indusium with unicellular acicular hairs and unicellular oblong glands .... **Amauropelta**
            4.' Indusium closely set with unicellular acicular hairs only .... **Christella**

AMAUROPELTA   c. 200 spp., America, Africa and Madagascar

bergiana (Schltdl.) Holttum   Perennial with erect rhizome. Fronds arching, stipe sparsely pubescent, lamina narrowly elliptic, 2-pinnatifid, proximal 4 or 5 pinna pairs decrescent and deflexed, pinnae deeply pinnatifid, with straight hairs on costa and costules above, with short hooked hairs beneath. Sori medial, to 14 per lobe, indusium minute. Moist streambanks in forests, 100–400 m, SW, LB, SE (Cape Peninsula to tropical Africa).

knysnaensis (N.C.Anthony & Schelpe) B.S.Parris   Perennial with erect rhizome. Fronds arching, stipe thinly pubescent, lamina elliptic, pinnate, 5 proximal pinna pairs decurrent and deflexed, pinnae pinnatifid into oblong pinnules, proximal acroscopic pinnule largest, lamina hairy beneath. Sori medial, indusium with minute stalked yellow glands and hairs. Moist forests, to 400 m, SE (Knysna).•

CHRISTELLA   c. 70 spp., pantropical

gueinziana (Mett.) Holttum   Perennial with erect rhizome. Fronds arching, soft, stipe thinly pubescent, lamina elliptic, 2-pinnatifid, the proximal 2 pinnae pairs decurrent and deflexed, pinnae deeply pinnatifid, pubescent, basal pair of veins anastomosing at sinus. Sori medial, to 16 per lobe, indusium orange, pilose. Streambanks in forest, 200–400 m, SE (George to S and E Africa).

CYCLOSORUS   c. 70 spp., pantropical

interruptus (Willd.) H.Itô   Perennial with widely spreading rhizome. Fronds erect, soft, stipe glabrous, lamina oblong-lanceolate, pinnae narrowly oblong, shallowly incised, glabrescent, basal pair of veins anastomosing well below sinus. Sori circular, to 18 per lobe, indusium densely pilose. Riverbanks and other marshy areas, 100–600 m, SW, SE (Worcester to W and E Africa and Mauritius).

STEGNOGRAMMA   c. 12 spp., pantropical

pozoi (Lag.) K.Iwats.   Perennial with erect rhizome. Fronds arching, stipe thinly pubescent, lamina lanceolate, pinnate, proximal pinnae decurrent and deflexed, pinnae adnate, narrowly oblong, incised about midway to costa into rounded undulate lobes, pilose. Sori linear along veins, exindusiate. Shaded forest floors and streambanks, 100–400 m, SW, SE (Cape Peninsula to W and E Africa and Spain).

THELYPTERIS   2 spp., pantropical

confluens (Thunb.) C.V.Morton   SCALY LADY FERN, BOG FERN   Perennial with widely spreading rhizome. Fronds erect, soft, 2-pinnatifid, stipe glabrous, lamina lanceolate, pinnae deeply pinnatifid, costa paleate or thinly pilose. Sori circular, to 18 per lobe, medial, indusium glabrous. Marshes and streambanks, 100–400 m, SW, LB, SE (Cape Peninsula to W and E Africa, Madagascar and Australasia).

VITTARIACEAE

VITTARIA   SHOESTRING FERN   c. 50 spp., pantropical

isoetifolia Bory   Perennial with short rhizome. Fronds simple, sessile, pendent, to 20 cm long, lamina linear. Sori in 2 inframarginal grooves with paraphyses. Moist forests or at higher elevations in sheltered rock crevices, to 1000 m, SW, SE (Cape Peninsula to N Province and Madagascar).

GYMNOSPERMS   CONE-BEARING PLANTS

CUPRESSACEAE

WIDDRINGTONIA   CEDAR   3 spp., W Cape to S tropical Africa

cedarbergensis Marsh   CLANWILLIAM CEDAR, SEDERBOOM   Monoecious, often gnarled tree to 10(–20 m). Adult leaves scale-like, ovate, adpressed. Female cones with rough valves, tubercled on margins. Sandstone slopes and rocks above 1000 m, NW (Cedarberg Mts).•

nodiflora (L.) Powrie (= *Widdringtonia cupressoides* (L.) Endl.)   MOUNTAIN CYPRESS, BERGSIPRES   Monoecious shrub or tree to 10 m (to 50 m in tropics). Adult leaves scale-like, narrowly oblong, keeled, adpressed. Female cones with smooth to wrinkled valves. Rocky sandstone and clay slopes, SW, LB, SE (Cape Peninsula to E Cape and to S tropical Africa).

schwarzii (Marloth) Mast.   BAVIAANSKLOOFSEDER   Like **W. cedarbergensis** but 17–26 m, sometimes to 40 m, and seeds broadly winged. Rocky ravines, SE (Baviaanskloof and Kouga Mts).•

*PINACEAE

***PINUS**   PINE   c. 90 spp., north temperate and subtropical

***pinaster** Aiton   CLUSTER PINE   Monoecious, pyramidal tree to 40 m. Leaves needle-like, in bundles of 2, grey-green, sparse and rigid. Cones large. Sandy and rocky mountain slopes, NW, SW, LB, SE (Ceres, Cape Peninsula to E Cape, widespread Mediterranean weed).

[**P. pinea** L. and **P. radiata** D. Don. occur in plantations but are not naturalised. Volunteers occur only in the immediate vicinity of original plantings.]

PODOCARPACEAE

1. Fruit borne on scaly or leafy axillary branchlets, the receptacle not swollen ....**Afrocarpus**
1.' Fruit borne on naked axillary branchlets, the receptacle leathery or swollen and fleshy ....**Podocarpus**

AFROCARPUS   YELLOWWOOD   c. 3 spp., southern and tropical Africa

falcatus (Thunb.) C.N.Page (= *Podocarpus falcatus* (Thunb.) R.Br. ex Mirb.)   OUTENIQUA YELLOWWOOD   Dioecious, tall tree, 20–60 m. Leaves narrowly lanceolate to linear-elliptic, twisted at base, 20–40 × 2–6 mm. Seeds borne on scaly or leafy stalks, yellow to brown, without a fleshy arillode. Mostly coastal forests, LB, SE (Swellendam to tropical Africa).

PODOCARPUS   YELLOWWOOD   c. 100 spp., mostly S hemisphere, temperate and tropical

elongatus (Aiton) L'Hér. ex Pers.   BREËRIVIERGEELHOUT   Like **P. latifolius** but sometimes a spreading shrub 3–6 m, leaves narrower, 20–70 × 3–5 mm and seeds dark green. Mainly sandstone slopes along streams, NW, SW, KM, LB (Bokkeveld Mts to Swellendam).•

latifolius (Thunb.) R.Br. ex Mirb.   OPREGTE GEELHOUT   Dioecious tree to 30 m. Leaves linear-elliptic, 35–60 × 6–10 mm. Seeds grey to purple, borne on nude stalks swollen above to form a fleshy reddish arillode. Forests and open mountain slopes, SW, LB, SE (Cape Peninsula to tropical Africa).

ZAMIACEAE

ENCEPHALARTOS   CYCAD   c. 50 spp., southern and tropical Africa

caffer (Thunb.) Lehm.   DWARF CYCAD   Dioecious tree with subterranean stem. Leaves pinnate, leaflets many, linear-lanceolate, mostly entire, twisted at various angles. Female cones lime-green, to 30 cm. Jan.–Mar. Coastal grassland, SE (Humansdorp to E Cape).

horridus (Jacq.) Lehm.   BLUE CYCAD   Dioecious stemless or short, single-stemmed tree to 1 m. Leaves pinnate, glaucous, leaflets distant below, ovate to lanceolate, lobed and sharply toothed, pungent. Female cones bluish green, to 40 cm. Oct.–Feb. Rocky grassland, woodland or karroid scrub, SE (Baviaanskloof Mts to Port Elizabeth).•

lehmannii Lehm.   KAROO CYCAD   Dioecious tree to 2 m, often branched from base. Leaves pinnate, dark green but glaucous when young, leaflets linear-lanceolate, pungent, sometimes 1- or 2-toothed. Female cones green, to 50 cm. Feb.–May. Karroid scrub, SE (Baviaanskloof Mts to E Cape).

longifolius (Jacq.) Lehm.   THUNBERG'S CYCAD   Dioecious tree to 3 m, often branched below. Leaves pinnate, dark green, leaflets finely pubescent, oblong-lanceolate, pungent, usually entire. Female cones olive-green, to 60 cm. May–June. Stony sandstone slopes, SE (Uniondale to E Cape).

ANGIOSPERMS   Flowering Plants

PALAEODICOTYLEDONS

LAURACEAE

1. Parasitic, twining herbs with yellowish, thread-like stems and scale-like leaves ....**Cassytha**
1.' Trees or shrubs with well-developed leaves:
    2. Flowers usually unisexual; anthers 4-valved; leaves usually with pits on underside in axils of lowest 1 or 2 pairs of veins ....**Ocotea**
    2.' Flowers bisexual; anthers 2-valved; leaves without pits ....**Cryptocarya**

CASSYTHA   FALSE DODDER, DEVIL'S TRESSES, NOOIENSHAAR   c. 16 spp., Old World tropics, mostly Australia

ciliolata Nees   Yellowish, achlorophyllous perennial vine parasitic on trees and shrubs. Leaves rudimentary, scale-like. Flowers in crowded clusters, yellowish. Fruits fleshy, red. Mainly Sept.–Jan. Various trees and shrubs, NW, SW, AP, KM, LB, SE (Clanwilliam to E Cape).

CRYPTOCARYA c. 200 spp., Old World tropics

angustifolia E.Mey. ex Meisn.   BLUE LAUREL   Shrub or small tree to 3 m, finely hairy on young twigs. Leaves linear-lanceolate. Flowers in axillary panicles, whitish. Berries black. Nov. Rocky river valleys, NW, SW, LB (Gifberg to Langeberg Mts: Swellendam).•

OCOTEA   STINKWOOD, STINKHOUT   c. 200 spp., mostly tropical America, 4 spp. in Africa

bullata (Burch.) Baill.   Monoecious, evergreen tree, 8–30 m, with bark pale when young. Leaves elliptic-oblong, glossy, with pits in vein axils. Flowers in axillary cymes, polygamous, yellowish. Mainly Dec.–Feb. Coastal and mistbelt forests, SW, LB, SE (Cape Peninsula to Mpumalanga).

NYMPHAEACEAE

NYMPHAEA   WATERLILY, WATERLELIE   c. 60 spp., pantropical and warm temperate

nouchali Burm.f. (= *Nymphaea capensis* Thunb.)   BLUE WATER LILY   Aquatic perennial with floating leaves, woolly below. Leaves orbicular, deeply notched at base, margins undulate or scalloped, petioles long or short. Flowers blue, scented, closing at night. Dec.–Mar. Pools, NW, SW, AP, LB, SE (Elandsbaai, Cape Peninsula to tropical Africa).

PIPERACEAE

1. Shrubs; stamens 2–6; anther thecae usually distinct; stigmas 2–4..**Piper**
1.' Herbs; stamens 2; anther thecae usually confluent; stigma 1 ....**Peperomia**

PEPEROMIA   1000+ spp., pantropical and subtropical, mainly America

retusa (L.f.) A.Dietr.   Soft, creeping perennial to 20 cm, rooting from the nodes. Leaves lightly succulent, obovate, glossy, minutely ciliolate along apex. Flowers in thread-like, terminal spikes, green. Jan.–Mar. Forest floors and humus on sheltered rocks, SW, KM, LB, SE (Cape Peninsula to Mpumalanga).

tetraphylla (G.Forst.) Hook. & Arn.   Creeping perennial to 15 cm, with jointed, divaricately branched stems rooting from nodes. Leaves in whorls of 4, rhomboid, leathery, 3-veined from base. Flowers in narrowly club-shaped, terminal spikes, green. Jan.–Apr. Lithophytic or epiphytic in forests, LB, SE (Swellendam to tropical Africa, pantropical).

PIPER   PEPPER   1000+ spp., pantropical and subtropical

capense L.f.   BOSPEPER, WILDEPEPER   Soft, straggling shrub to 3 m. Leaves ovate, acuminate, 5–7-veined from base, glossy above, paler and hairy on veins beneath. Flowers in spadix-like spikes, greenish. Oct.–Nov. Forests and sheltered cliffs, LB, SE (Swellendam to tropical Africa).

MONOCOTYLEDONS

AGAPANTHACEAE
by D.A. Snijman

AGAPANTHUS   AGAPANTHUS, BLOULELIE   10 spp., southern Africa, absent from arid areas

africanus (L.) Hoffmanns.   Rhizomatous evergreen geophyte, 25–70 cm. Leaves suberect, strap-shaped. Flowers broadly funnel-shaped, deep blue, thick in texture. Dec.–Apr., mainly after fire. Rocky sandstone slopes, SW, LB, AP (Cape Peninsula to Swellendam).•

praecox Willd.   Rhizomatous evergreen geophyte, 50–100 cm. Leaves suberect, strap-shaped. Flowers broadly funnel-shaped, blue to white, thin in texture. Dec.–Feb. SE (Knysna to E Cape).

walshii L.Bolus   Rhizomatous evergreen geophyte, 60–70 cm. Leaves suberect, strap-shaped. Flowers nodding, tubular, deep blue, thick in texture. Jan.–Feb., mainly after fire. Rocky sandstone slopes, SW (Steenbras Mts).•

ALLIACEAE
by D.A. Snijman

1. Rootstock a bulb; flowers campanulate to urceolate, white to pink, tepals connate at base, without a corona; inner filaments tricuspidate..**Allium**
1.' Rootstock a rhizome; flowers more or less tubular, mauve, cream-coloured, green or brown, tepals fused below into a tube, with a corona; anthers sessile ....**Tulbaghia**

ALLIUM   WILD ONION   c. 550 spp., mainly N hemisphere

dregeanum Kunth   AJUIN, WILDEUI   Bulbous geophyte, 35–80 cm, strongly aromatic. Leaves linear, dry at flowering time, suberect. Flowers in a compact spherical head, white to pink. Oct.–Dec. Dry stony slopes and flats, NW, LB, SE (Namaqualand, Clanwilliam to Langkloof, dry areas throughout southern Africa).

TULBAGHIA   WILD GARLIC   20 spp., South Africa to S tropical Africa

alliacea L.f.   Similar to **T. capensis** but the corona lobes completely fused into a fleshy collar 6–8 mm long and the upper anthers inserted on it. Mar.–May. Widespread, NW, SW, LB, SE (Clanwilliam to Cape Peninsula to Port Elizabeth, throughout southern Africa).

capensis L.   Bulbous geophyte, 15–35 cm, strongly aromatic. Leaves spreading, linear. Flowers brownish to purplish and green with orange corona, corona lobes 6, free or fused below, to 5 mm long. Apr.–Oct. Rocky slopes, SW, AP, LB, SE (Cape Peninsula to Langkloof).•

dregeana Kunth   Bulbous geophyte, 15–25 cm, strongly aromatic. Leaves spreading, linear. Flowers cream to greenish yellow, corona lobes completely fused into a fleshy ring 2 mm long. May–Oct. Stony soils, NW, SW (Namaqualand and W Karoo to Worcester, Cape Peninsula to Stanford).

violacea Harv.   ICINZINI   Bulbous geophyte, 20–35 cm, strongly aromatic. Leaves suberect, linear, glaucous. Flowers mauve, corona lobes 3, free, to 2 mm long. Nov.–Apr. Forest margins and streambanks, KM, SE (Ladismith, Knysna to KwaZulu-Natal).

AMARYLLIDACEAE
by D.A. Snijman

1. Scape hollow or rarely solid; fruit dehiscent and papery; seeds dry, black and flattened ....**Cyrtanthus**
1.' Scape solid; fruit dehiscent and papery or indehiscent and papery to fleshy; seeds moist, cream-coloured, pink, reddish or green, ovoid or subglobose and slightly angled by compression; embryo usually green:
    2. Leaves often spirally twisted; inflorescence acaulescent, 1-flowered; fruit clavate, cylindrical or occasionally ellipsoid, much longer than wide, not ribbed or conspicuously veined:
        3. Leaves few–many, often conspicuously sheathed and pubescent; stamens in a single series; anthers 6–many ....**Gethyllis**
        3.' Leaves 1–few, not conspicuously sheathed, glabrous; stamens in 2 series; anthers 6 ....**Apodolirion**
    2.' Leaves rarely spirally twisted; inflorescence well developed above ground, (1–)many-flowered; fruit subfusiform to globose or trigonous, rarely much longer than wide and then ribbed or conspicuously veined:
        4. Plants rhizomatous or bulbous; tunics not producing extensible threads when torn; leaves often speckled with red; spathe valves 4 or more, often conspicuous; fruit indehiscent, fleshy or pulpy:
            5. Plants with large, fleshy bulbs; leaves sword- or tongue-shaped, without a midrib, succulent and often pubescent ....**Haemanthus**
            5.' Plants rhizomatous; leaves with a petiole-like base and prominent midrib, thin-textured and glabrous ....**Scadoxus**
        4.' Plants bulbous; tunics producing extensible threads when torn; leaves immaculate, or if speckled then fringed with long bristles; spathe valves 2, inconspicuous; fruit dehiscent or indehiscent, papery or membranous:
            6. Leaf margins hyaline, more or less fringed with short, branched cilia; pedicels usually shorter than flowers at anthesis; filaments free to base; fruit indehiscent, often beaked; seeds thinly cork-covered:
                7. Leaves annual, closely abutting each other to form an erect fan; all leaf tips subacute to obtuse; fruiting head detaching from scape apex; fruit trigonous, prominently 3-ribbed ....**Boophone**
                7.' Leaves perennial, suberect or prostrate; tips of all but the youngest leaves truncate; fruiting head not detaching from scape apex; fruit irregularly shaped, smooth or 6-ribbed:
                    8. Leaves evenly spreading, channelled, suberect to recurved; perianth tube curved, narrowly cylindrical in lower half (in Cape species) ....**Crinum**
                    8.' Leaves biflabellate, flat, prostrate; perianth tube straight, not narrowly cylindrical below:
                        9. Flowers actinomorphic; pedicels not elongating after anthesis; fruiting head drooping; fruit membranous, smooth..**Ammocharis**
                        9.' Flowers weakly zygomorphic; pedicels elongating and radiating after anthesis; fruiting head detaching at ground level; fruit papery, 6-ribbed ....**Cybistetes**
            6.' Leaf margins smooth, softly pubescent or raised and fringed with long bristles or short, branched, red cilia; pedicels rarely shorter than flowers at anthesis; filaments connate at base, or if free then fused to style base; fruit dehiscent, beakless; seeds cutinous, not corky:
                10. Leaves with a prominent midrib; perianth longer than 8 cm; seeds slightly angled by compression, pink to colourless, only embryo green ....**Amaryllis**
                10.' Leaves without a midrib; flowers shorter than 8 cm; seeds ovoid, usually reddish, integument and embryo green:
                    11. Flowers zygomorphic (sometimes only by the deflexed style), rarely actinomorphic but then bright red:
                        12. Leaves narrow, usually narrower than 2.5 cm, subsucculent; pedicels slender, rarely longer than flowers; tepal margins often undulate; capsules subglobose, membranous, without conspicuous transversal veins ....**Nerine**
                        12.' Leaves broad, usually wider than 2.5 cm, leathery; pedicels stout, usually much longer than flowers; tepal margins rarely crisped; capsules trigonous to fusiform, with conspicuous transversal veins:
                            13. Leaves immaculate, margins smooth or with short, branched cilia; pedicels obscurely 3-angled in cross section; flowers longer than 25 mm, tepals more or less spreading, filaments tightly clustered, not bulbous at base; capsules prominently ribbed, tardily dehiscent ......**Brunsvigia**

13.' Leaves speckled with red, margins with long bristles; pedicels sharply triangular in cross section; flowers shorter than 20 mm, tepals reflexed, filaments more or less separate, bulbous at base; capsules not ribbed, readily dehiscent ....**Crossyne**
11.' Flowers actinomorphic:
    14. Leaves 2(3), glabrous; flowers persisting after anthesis; filaments connate into a short to long tube, free from style; anthers centrifixed to subcentrifixed; style slender ....**Hessea**
    14.' Leaves 2–6, glabrous or hairy; flowers withering after anthesis; filaments free or if fused then filament tube trilocular; at least 1 filament whorl adnate to style base; anthers subcentrifixed to dorsifixed:
        15. Leaves pubescent or glabrous, ovate to filiform; tepals free; outer filaments adnate to style base; style winged or swollen basally ....**Strumaria**
        15.' Leaves glabrous, filiform; tepals connate into a distinct tube; inner filaments adnate to style; style columnar throughout...**Carpolyza**

AMARYLLIS MARCH LILY, BELLADONNA LILY   2 spp., Namaqualand and W Cape

belladonna L.   Bulbous geophyte to 90 cm. Leaves dry or absent at flowering, several, spirally arranged, suberect, channelled, midrib prominent. Flowers large, trumpet-shaped, pink to white, fragrant. Feb.–Apr., mainly after fire. Loamy soils on lowlands, NW, SW, AP, LB, SE (Olifants River valley to George).•

AMMOCHARIS SEEROOGBLOM   5 spp., southern Africa to tropical E Africa

coranica (Ker Gawl.) Herb.   Bulbous geophyte, 25–35 cm. Leaves several, distichous, trailing, falcate with truncate tips. Flowers in a rounded cluster, widely funnel-shaped, pink to red. Nov.–Feb. Seasonally damp karroid flats, KM, SE (Oudtshoorn to S Angola and Zimbabwe).

APODOLIRION GROUND LILY   6 spp., W Cape to NE southern Africa

cedarbergense D.Müll.-Doblies   Acaulescent bulbous geophyte. Leaves dry at flowering, suberect, 2 or 3, linear. Flower solitary, white or pale pink, fragrant. Fruit a purple berry. Jan. Sandy flats and slopes, NW (Cedarberg Mts).•

lanceolatum (L.f.) Benth.   Acaulescent bulbous geophyte to 3 cm. Leaf dry at flowering, solitary, spreading to prostrate, strap-shaped, margins sometimes crisped. Flower solitary, white. Fruit an ovoid, yellow berry. Dec.–Feb. Karroid lowlands in rocks, KM, LB (Ladismith, Oudtshoorn, Swellendam).•

macowanii Baker   Acaulescent bulbous geophyte to 8 cm. Leaves dry at flowering, 2 or 3, suberect, coiled. Flower solitary, white. Fruit a long, yellow berry. Dec.–Apr. SE (Uitenhage to E Cape).

BOOPHONE OXBANE, KOPSEERBLOM   2 spp., southern Africa to tropical E Africa

disticha (L.f.) Herb.   Bulbous geophyte to 25 cm. Leaves usually dry at flowering, many, in an upright fan, grey. Flowers in a dense, spreading cluster, pink to red, pedicels elongating in fruit. Sept.–Mar. Rocky slopes and flats, SW, AP, LB, SE (Robertson and Bredasdorp to tropical E Africa).

haemanthoides F.M.Leight.   Bulbous geophyte, 30–50 cm. Leaves dry at flowering, many, in an upright fan. Flowers in a compact, brush-like cluster, cream with reddish centres, fragrant, pedicels elongating in fruit. Mainly Nov.–Feb. Coastal sands, limestone or dolerite rocks, NW, SW (Namaqualand and W Karoo to Saldanha).

BRUNSVIGIA CANDELABRA LILY, KANDELAAR   c. 18 spp., southern Africa

A. Flowers pink

bosmaniae F.M.Leight. (incl. **B. appendiculata** F.M.Leight.)   Bulbous geophyte to 20 cm. Leaves dry at flowering, 5 or 6, prostrate. Flowers c. 20 in a large, rounded head, tepals broadly oblong, pink with darker veins, outer stamens about half as long as inner, scented of narcissus. Capsules sharply 3-angled, heavily ribbed. Mar.–May. Clay, coastal sand, clay and granite soils, NW, SW (Namaqualand and W Karoo to Tygerberg).

gregaria R.A.Dyer   Bulbous geophyte, 12–25 cm. Leaves dry at flowering, 4–6, prostrate. Flowers 20–40 in a dense, hemispherical head, tepals flaring, tube well-developed, pink to crimson. Capsules rounded above, heavily ribbed. Mar.–Apr. Clay soils, SE (Humansdorp, Baviaanskloof, Port Elizabeth to E Cape).

striata (Jacq.) Aiton (incl. **B. minor** Lindl.)   Bulbous geophyte, 15–25 cm. Leaves dry at flowering, 3–5, prostrate. Flowers c. 20 in a rounded cluster, tepals flaring, recurved above, pink with dark midrib. Capsules rounded above, not strongly ribbed. Mar.–Apr. Stony slopes and flats, NW, AP, KM, LB, SE (Bokkeveld Mts to Montagu, Oudtshoorn and Cape Infanta).•

A. Flowers red

josephinae (Redouté) Ker Gawl. (incl. **B. gydobergensis** D. & U.Müll.-Doblies)   LANTANTER   Bulbous geophyte to 65 cm, bulbs often exposed. Leaves dry at flowering, 8–20, suberect. Flowers 20–60 in a large, open hemispherical head, red, tepals fused below for 10–20 mm, unequally rolled back. Capsules narrowly ovoid. Feb.–Mar. Rocky slopes and clay flats in renosterveld, NW, SW, AP, KM, SE (W Karoo to Worcester, Malgas to Baviaanskloof).

litoralis R.A.Dyer   Like **B. josephinae** but bulbs subterranean, leaves smaller, to 40 cm long and flowers mostly to 20. Feb.–Mar. Coastal sands, SE (Tsitsikamma to Port Elizabeth).•

marginata (Jacq.) Aiton (= *Nerine marginata* (Jacq.) Herb.)   Bulbous geophyte to 20 cm. Leaves dry at flowering, 4, prostrate. Flowers 10–20 in a compact, hemispherical head, actinomorphic, bright scarlet, tepals widely flared, stamens prominent, central. Capsules sharply 3-angled. Mar.–June. Rocky sandstone slopes in shale bands, NW, SW (Citrusdal to Worcester and Paarl).•

orientalis (L.) Aiton ex Eckl.   KONINGSKANDELAAR   Bulbous geophyte, 40–50 cm. Leaves dry at flowering, usually 6, prostrate. Flowers 20–40 in a dense, spherical head, red, tepals fused below for 2–5 mm, unequally rolled back. Capsules sharply 3-angled, heavily ribbed. Feb.–Apr. Sandy, mainly coastal lowlands, NW, SW, AP, SE (S Namaqualand to Cape Peninsula and Plettenberg Bay).

CARPOLYZA•   WIRE LILY   1 sp., W Cape

spiralis (L'Hér.) Salisb.   Bulbous geophyte, 5–15 cm, scape wiry, spirally twisted below. Leaves 4–6, filiform, spreading. Flowers small, erect, funnel-shaped, white to pink. May–Aug. Seasonally wet flats and rock crevices, SW, AP, KM, LB (Cape Peninsula to Heidelberg and Oudtshoorn).•

CRINUM   MARSH LILY   c. 65 spp., pantropical but mainly sub-Saharan Africa

lineare L.f.   Bulbous geophyte, 18–60 cm. Leaves several, suberect, channelled, to 20 mm wide. Flowers trumpet-shaped, white, with pale to dark pink keels, pedicels to 25 mm long, anthers black. Jan.–Mar. Sandy soils, SE (Port Elizabeth to E Cape).

variabile (Jacq.) Herb.   Bulbous geophyte, 45–60 cm. Leaves several, spreading, to 50 mm wide. Flowers large, trumpet-shaped, pale to deep pink, pedicels to 50 mm long, heavily scented, anthers yellow. Jan.–May. Streambeds, NW (Namaqualand and W Karoo to Biedouw River).

CROSSYNE   PARASOL LILY, SAMBREELBLOM   2 spp., Namaqualand, W Karoo and W Cape

flava (W.F.Barker ex Snijman) D. & U.Müll.-Doblies (= *Boophone flava* W.F.Barker ex Snijman)   GEELSAMBREELBLOM   Like **C. guttata** but flowers pale yellow and stamens declinate. Mar.–May. Shale flats and rocky slopes, NW (Namaqualand to Cedarberg Mts).

guttata (L.) D. & U.Müll.-Doblies (= *Boophone guttata* (L.) Herb.)   Bulbous geophyte to 45 cm. Leaves dry at flowering, 4–6, prostrate, margins bristly. Flowers many in a large rounded head, small, with tepals reflexed, maroon to dusky pink, stamens spreading. Feb.–Apr., usually after fire. Shale and granite flats and lower slopes, NW, SW, AP, LB (Piketberg to Swellendam).•

CYBISTETES   MALGAS LILY, MALGASLELIE   1 sp., N and W Cape

longifolia (L.) Milne-Redh. & Schweick.   Bulbous geophyte, 25–35 cm. Leaves dry or emergent at flowering, 9–14, prostrate, falcate with truncate tips. Flowers widely funnel-shaped, cream to pink, lily-scented. Dec.–Apr. Sandy flats, NW, SW (Namaqualand to Cape Peninsula and Bonnievale).

CYRTANTHUS   FIRE LILY, BRANDLELIE   c. 50 spp., southern and tropical Africa

*A. Flowers cream or yellow to salmon, sometimes with pink to red stripes (see also **C. obliquus**)*

leptosiphon Snijman   Bulbous geophyte to 30 cm. Leaves usually green at flowering. Flowers 1–4, suberect, narrowly funnel-shaped, somewhat bilabiate, segments pale apricot with darker keels, tube pale pink. Feb.–Mar. Gravelly clay, LB (Swellendam).•

leucanthus Schltr.   WITBRANDLELIE   Bulbous geophyte, 15–25 cm. Leaves usually dry at flowering. Flowers 1–4, suberect, tubular, throat narrow, segments broadly elliptic, cream, sweetly scented. Jan.–Mar. Sandstone or limestone slopes or flats, SW, AP (Betty's Bay to Potberg).•

loddigesianus (Herb.) R.A.Dyer (incl. **C. speciosus** R.A.Dyer)   GRASVELDLELIE   Bulbous geophyte, 8–20 cm. Leaves sometimes dry at flowering. Flowers 1–4, suberect, funnel-shaped, delicate, cream with green or pink keels. Oct.–Mar., after rain. Grassland or grassy fynbos in sandy soils, SE (Humansdorp to E Cape).

ochroleucus (Herb.) Burch. ex Steud.   Bulbous geophyte, 15–20 cm. Leaves dry at flowering. Flowers 2–4, suberect, tubular, dull yellow, segments narrow, oblong, acrid scented. Nov.–Feb. Dry, sandy or stony middle slopes or flats, AP, LB (Langeberg Mts: Garcia's Pass and Stilbaai).•

sp. 1   Bulbous geophyte to 50 cm. Leaves dry at flowering. Flowers 2–3, spreading, narrowly funnel-shaped, peach to salmon, with red stripes in the tube from the tepal sinuses, filaments incurved, style declinate, 3-lobed. Mar. Rocky S-facing slopes in fynbos, SE (Baviaanskloof).•

A. Flowers red, bright pink or purple
B. Flowers widely funnel-shaped

elatus (Jacq.) Traub (= *Cyrtanthus purpureus* (Aiton) Traub)   GEORGE LILY, BERGLELIE   Bulbous geophyte to 45 cm. Leaves several, broad, green at flowering. Flowers 2–9, suberect, large, widely funnel-shaped, scarlet, rarely pink, style capitate. Nov.–Mar., often after fire. Forest margins and moist mountain slopes, SE (George to Humansdorp).•

flammosus Snijman & Van Jaarsv.   Bulbous geophyte, 20 cm. Leaves green at flowering, greyish, somewhat leathery. Flowers solitary or rarely paired, erect, large, widely funnel-shaped, scarlet, style 3-branched, pedicel stout. Mar. Shaded rock crevices on cliffs, SE (Baviaanskloof Mts).•

guthrieae L.Bolus   BREDASDORP LILY   Bulbous geophyte, 10–12 cm. Leaves dry at flowering, few, linear. Flowers solitary or rarely paired, erect, large, widely funnel-shaped, bright red, smaller than in **C. flammosus**, style capitate, pedicel very short. Mar.–Apr., after fire. Lower sandstone slopes, SW (Bredasdorp).•

montanus R.A.Dyer   Bulbous geophyte to 10 cm, bulbilliferous. Leaves green at flowering. Flowers 5–10, erect, widely funnel-shaped from a narrow tube, red, stamens well-exserted. Jan.–Mar. Rocky crevices on upper slopes, SE (Baviaanskloof Mts).•

B. Flowers tubular or narrowly funnel-shaped
C. Perianth weakly bilabiate, the upper 3 segments hooded over the stamens

fergusoniae L.Bolus   Leaves often dry at flowering, narrow, somewhat fleshy, minutely papillate-ribbed beneath. Flowers 4–8, spreading, tubular, bright red, segments oblong, about half as long as the tube, 3 upper projecting forward, 2 lateral spreading, lowest curved downwards, style subcapitate, lying over upper stamens. Dec.–Jan. Loam or sand on limestone ridges, AP (Bredasdorp to Stilbaai).•

inaequalis O'Brien   Bulbous geophyte to 30 cm. Leaves green at flowering, linear. Flowers 4–9, suberect, tubular, coral-red, segments narrowly oblong, about a third as long as the tube, 3 upper hooded, 2 lateral spreading, lower decurved, irregular form becoming more pronounced with age, style tricuspidate, not exserted. Jan. Rocky slopes in gravelly soil, KM, LB, SE (Barrydale, Ladismith, George).•

labiatus R.A.Dyer   Bulbous geophyte to 30 cm, forming bulbils. Leaves green at flowering, strap-shaped. Flowers c. 7, suberect, tubular, coral-red, segments narrow, short, 4 upper overlapping and forming a hood, 2 lower decurved, style subcapitate, strongly decurved. Dec.–Jan. Shaded sandstone rock faces, SE (Kouga and Baviaanskloof Mts).•

C. Perianth segments equally spreading

angustifolius (L.f.) Aiton   FIRE LILY, BRANDLELIE   Bulbous geophyte to 45 cm. Leaves several, broad, rarely dry at flowering. Flowers 4–10, nodding, long, tubular, scarlet, somewhat fleshy, filaments inserted in throat, style exserted. Oct.–Feb., after fire. Mountain slopes and flats in seasonal streams and vleis, NW, SW, AP, KM, LB, SE (Cedarberg Mts to Port Elizabeth).•

carneus Lindl.   Bulbous geophyte to 1 m, robust. Leaves broad, twisted, usually green at flowering. Flowers 15–25, nodding, long, tubular, crimson or pink. Dec.–Feb. Coastal sand, SW, AP (Caledon to Potberg).•

collinus Ker Gawl.   Bulbous geophyte, 15–25 cm. Leaves narrow, greyish, sometimes dry at flowering. Flowers 4–10, nodding, narrowly funnel-shaped, bright red, filaments inserted in throat, style shortly included and arching against upper tepal. Dec.–May. Rocky sandstone slopes, SW, KM (Genadendal to Oudtshoorn).•

obliquus (L.f.) Aiton   KNYSNA LILY, JUSTAFINA   Bulbous geophyte to 60 cm, robust. Leaves broad, twisted, usually green at flowering. Flowers 6–12, nodding, large, funnel-shaped, yellow or orange with green tips. Aug.–Feb. Grassland and grassy fynbos, SE (Knysna to KwaZulu-Natal).

odorus Ker Gawl.   Bulbous geophyte to 20 cm. Leaves present or absent at flowering. Flowers 2–5, suberect to spreading, tubular, dark red to maroon, sweetly scented. Feb.–Apr., after fire. Lower to middle slopes, LB (Langeberg Mts: Swellendam).•

spiralis Burch. ex Ker Gawl.   VARKSTERTLELIE   Bulbous geophyte, 15–30 cm. Leaves usually dry at flowering, narrowly strap-shaped, firm, coiled. Flowers 4–7, nodding, narrowly trumpet-shaped, flame-red. Dec.–Feb. Flats and lower slopes in semi-arid habitats, SE (Uitenhage to Port Elizabeth).•

staadensis Schönland   Bulbous geophyte to 40 cm. Leaves narrow, greyish. Flowers 3–6, spreading horizontally, tube narrow and curved below, then abruptly inflated and urn-shaped, vermilion, style shortly included and deflexed. Feb.–Mar. Grassy fynbos on moist slopes, SE (Uitenhage to E Cape).

ventricosus (Jacq.) Willd. (incl. **C. pallidus** Sims)   BRANDLELIE   Bulbous geophyte, 10–20 cm. Leaves usually dry at flowering. Flowers 2–12, nodding, tubular but slightly swollen above, vermilion to bright red, segments sometimes pink, filaments attached to the base of the tube, arching under upper segment. Dec.–May, after fire. S-facing sandstone slopes in fynbos, SW, AP, SE (Cape Peninsula to Baviaanskloof Mts).•

wellandii Snijman.   Bulbous geophyte to 40 cm. Leaves usually dry at flowering. Flowers 4–7, spreading horizontally, small, funnel-shaped, vermilion to scarlet, style much shorter than the filaments. Feb. Stony slopes in grassy renosterveld, SE (Humansdorp).•

sp. 2   Bulbous geophyte, 10–15 cm. Leaves dry at flowering, linear. Flowers 1–3, suberect, trumpet-shaped, bright pink with darker pink midveins. Nov.–Apr., after fire. Stony, sandy soil on rocky slopes, SE (Outeniqua Mts).•

GETHYLLIS   KUKUMAKRANKA   c. 32 spp., S Namibia, N and W Cape

A. Anthers more than 6

afra L.   Bulbous geophyte, 10–14 cm. Leaves dry at flowering, erect, spiralled, glabrous or shortly ciliate. Flowers white, keeled with red on reverse, anthers 9–18 in 6 clusters. Dec.–Jan. Sandy flats, NW, AP, LB (Clanwilliam to Heidelberg).•

britteniana Baker   Bulbous geophyte to 15 cm. Leaves dry at flowering, erect, spiralled, glabrous, basal sheaths paired, spotted. Flowers large, white to pink, anthers 35–60 in more than 6 clusters. Oct.–Mar. Sandveld, NW (Namaqualand to Clanwilliam).

campanulata L.Bolus (incl. **G. multifolia** L.Bolus)   Bulbous geophyte to 12 cm. Leaves dry at flowering, suberect, scarcely twisted, lightly hairy. Flowers large, white to cream, anthers 12 in 6 pairs. Nov.–Jan. Stony clay flats, NW (Bokkeveld Escarpment, Worcester, Montagu).•

ciliaris (Thunb.) Thunb. (incl. **G. undulata** Herb.)   Bulbous geophyte, 10–20 cm. Leaves dry at flowering, suberect, spiralled, fringed with long, firm, upturned hairs, sometimes undulate, basal sheath minutely spotted. Flowers white or pink, anthers 15–30 in 6 clusters. Dec.–Feb. Flats, NW, SW (Namaqualand to Cape Peninsula).

A. Anthers 6
*B. Leaves glabrous (see also **G. gregoriana**)*

kaapensis D.Müll.-Doblies   Bulbous geophyte. Leaves few, green at flowering, suberect, weakly spiralled. Flowers white with pink undersurface. Nov.–Dec., after fire. Sandy flats, SW (Cape Peninsula).•

spiralis (Thunb.) Thunb.   Bulbous geophyte to 6 cm. Leaves dry at flowering, suberect, curled, glabrous. Flowers white with pink reverse. Nov.–Jan. Habitat? SW, SE (Worcester, Long Kloof).•

transkarooica D.Müll.-Doblies   Bulbous geophyte. Leaves dry at flowering, spiralled, mostly smooth. Flowers rose-pink, anthers tailed. Nov.–Dec. Slopes or flats in sand or clay, NW (Ceres and W Karoo, Free State).

verticillata R.Br. ex Herb.   Bulbous geophyte to 20 cm. Leaves dry at flowering, suberect, coiled, glabrous, basal sheaths paired, spotted, fringed. Flowers large, white. Nov.–Feb. Rock outcrops on middle slopes, NW (Namaqualand to Piketberg).

B. Leaves hairy or scaly

barkerae D.Müll.-Doblies   Bulbous geophyte. Leaves dry at flowering, rosulate, elliptic, stellate-pubescent. Flowers pale pink. ?Nov.–Dec. Rocky slopes, NW (S Namaqualand to Nardouwsberg).

gregoriana D.Müll.-Doblies   Bulbous geophyte to 15 cm. Leaves few, dry at flowering, erect, not twisted, strap-shaped, grey, glabrous or pubescent. Flowers large, white. Dec. NW (Namaqualand to Cedarberg Mts).

lanuginosa Marloth   Bulbous geophyte. Leaves dry at flowering, erect, loosely to tightly coiled, softly hairy. Flowers white. Nov. NW, SW (Namaqualand and W Karoo to Hopefield).

verrucosa Marloth   Bulbous geophyte to 5 cm. Leaves dry at flowering, suberect, with silvery scales. Flowers white. Oct.–Dec. Lower slopes and flats in heavy soil, AP (Bredasdorp, W Karoo).

villosa (Thunb.) Thunb.   Bulbous geophyte to 10 cm. Leaves dry at flowering, spreading and spiralled, pubescent with medifixed hairs. Flowers pink or white. Oct.–Dec. Flats and S-slopes in sand or clay, NW, SW, AP (Bokkeveld Mts and W Karoo to Mossel Bay).

[**Species excluded**   No authentic material found: **G. marginata** D.Müll.-Doblies, **G. oligophylla** D.Müll.-Doblies, **G. oliverorum** D.Müll.-Doblies]

HAEMANTHUS   PAINTBRUSH, POWDERPUFF   22 spp., southern Africa, mainly Namaqualand

A. Flowers pink or white (see also H. canaliculatus, H. crispus and H. sanguineus)

albiflos Jacq.   Bulbous geophyte, 6–30 cm. Leaves 2–6, prostrate to spreading, usually pubescent. Flowers in a compact head, white, bracts thin-textured. Apr.–Aug. Coastal and riverine scrub, AP, KM, SE (Stilbaai and Oudtshoorn to KwaZulu-Natal).

amarylloides Jacq.   Bulbous geophyte to 25 cm. Leaves dry at flowering, 2, prostrate to erect, glabrous, plane. Flowers spreading, pink, bracts thin-textured. Feb.–Apr. Seasonally moist sites, NW (Namaqualand, Bokkeveld Mts to Clanwilliam).

pumilio Jacq.   Bulbous geophyte, 5–14 cm. Leaves dry at flowering, suberect, 2, channelled, narrow and twisted, barred with red. Flowers slightly spreading, pink, bracts thin-textured. Mar.–Apr. Seasonally wet clay flats, SW (Hermon, Paarl, Stellenbosch).•

A. Flowers red

canaliculatus Levyns   Bulbous geophyte to 20 cm. Leaves dry at flowering, 2, erect, glabrous, canaliculate, basally speckled with red. Flowers slightly spreading, scarlet, rarely pink, bracts somewhat leathery. Dec.–Mar., after fire. Swampy coastal flats, SW (Rooi Els to Betty's Bay).•

coccineus L.   APRIL FOOL   Bulbous geophyte, 6–20 cm. Leaves dry at flowering, 2, spreading, fleshy, often ciliate, usually speckled. Flowers in a compact head, scarlet, bracts many, stiff and leathery. Feb.–Apr. Coastal scrub and rocky slopes, NW, SW, AP, KM, LB, SE (S Namibia to Port Elizabeth).

crispus Snijman (= *Haemanthus undulatus* Herb.)   Bulbous geophyte, 4–10 cm. Leaves dry at flowering, 2, spreading to suberect, channelled, with wavy edges, speckled basally. Flowers in a small, compact head, red, sometimes pink, bracts 4 or 5, blunt, waxy. Mar.–Apr. Stony lower slopes, NW (Namaqualand to Olifants River valley).

nortieri Isaac   Bulbous geophyte to 20 cm. Leaf dry at flowering, solitary, erect, rough, leathery, often sticky, scape hispid. Flowers in a compact head, deep red, bracts leathery, acuminate. Feb.–Mar. Seasonal washes, NW (Nardouw Mts).•

pubescens L.f.   POEIERKWAS   Bulbous geophyte, 7–12 cm. Leaves dry at flowering, 2, prostrate, pubescent. Flowers in a tight head, red, bracts usually 4, large, acute, fleshy. Feb.–Apr. Sandy flats, NW, SW (Namaqualand to Cape Peninsula and Worcester).

sanguineus Jacq.   VELSKOENBLAAR   Bulbous geophyte, 5–30 cm. Leaves dry at flowering, 2, prostrate, leathery, often outlined with red. Flowers crowded in a dense head, red or pink, bracts leathery. Jan.–Apr., especially after fire. Lower slopes, NW, SW, AP, KM, LB, SE (Nardouw Mts to Port Elizabeth).•

HESSEA   SAMBREELTJIE   14 spp., S Namibia, N and W Cape

breviflora Herb.   Bulbous geophyte to 15 cm. Leaves dry at flowering, 2, spreading, basal sheath conspicuous and red. Flowers in a small cluster, pink. Apr.–May. Sandy pockets between rocks, lower slopes, NW, SW (Namaqualand, Olifants River valley to Hopefield).

cinnamomea (L'Hér.) T.Durand & Schinz   Bulbous geophyte, 6–15 cm. Leaves dry at flowering, 2, spreading. Flowers north-facing, tepals crisped, pale pink with claret centre, with spicy scent. May–June, after fire. Peaty lowlands, SW (Cape Peninsula).•

mathewsii W.F.Barker (= *Dewinterella mathewsii* (W.F.Barker) D. & U.Müll.-Doblies)   Bulbous geophyte to 15 cm. Leaves dry at flowering, 2, spreading, shallowly channelled. Flowers pale pink with dark pink to crimson centre, filaments with basal, incurved hooks. May. Limestone flats, SW (Saldanha Bay).•

monticola Snijman   Bulbous geophyte to 20 cm. Leaves dry at flowering, 2, spreading, shallowly channelled. Flowers in widely spreading clusters, tepals crisped, white to pale pink with red lines leading into throat. Mar.–May, usually after fire. Rocky slopes or seasonally wet valleys, NW, SW (Piketberg and Cedarberg to Franschhoek Mts).•

pulcherrima (D. & U.Müll.-Doblies) Snijman (= *Dewinterella pulcherrima* (D. & U.Müll.-Doblies) D. & U.Müll.-Doblies)   Bulbous geophyte to 10 cm. Leaves dry at flowering, 2, spreading, spiralled apically. Flowers delicate, in a small head, white, centre usually crimson, filaments basally hooked. May. Clay flats, NW (Bokkeveld Escarpment and W Karoo).

pusilla Snijman   Bulbous geophyte to 15 cm. Leaves dry at flowering, 2, spreading, narrow. Flowers in a small head, pale pink. Apr.–May. Sandy sandstone plateaus, NW (Bokkeveld Mts).•

stellaris (Jacq.) Herb.   Bulbous geophyte, 7–15 cm. Leaves dry at flowering, 2, spreading, narrow. Flowers in an umbel-like cluster, pale to deep pink, occasionally with a dark star-shaped centre. Apr.–June. Sandy or clay flats, NW, KM (W Karoo, Karoopoort to Oudtshoorn).

undosa Snijman   Bulbous geophyte to 10 cm. Leaves dry at flowering, 2, spreading, shallowly channelled. Flowers in a dense cluster, tepals crisped, pink with reddish centre. June–July. Seasonally waterlogged, sandstone rock pockets, NW (Gifberg).•

NERINE   NERINE   c. 23 spp., southern Africa

humilis (Jacq.) Herb. (incl. **N. breachiae** W.F.Barker, **N. peersii** W.F.Barker, **N. tulbaghensis** W.F.Barker)   Bulbous geophyte, 15–35 cm. Leaves dry or emergent at flowering, several, spreading to prostrate, 4–10 mm wide. Flowers flared upwards, tepals c. 4 mm wide, undulate, pink. Apr.–June. Loamy soils among rocks, NW, SW, AP, KM, LB, SE (Clanwilliam to Worcester, Bredasdorp, Montagu to Baviaanskloof Mts).•

pudica Hook.f.   Bulbous geophyte, 25–35 cm. Leaves several, emergent at flowering, suberect, 2–5 mm wide. Flowers few, trumpet-shaped, tepals c. 6 mm wide, almost flat, pale pink with dark keels. Mar.–May, usually after fire. Steep rocky slopes, SW (Paarl to Caledon Mts).•

ridleyi E.Phillips   Like **N. humilis** but leaves broader, 15–25 mm wide. Feb.–Apr. Steep south-facing sandstone ledges, NW (Cold Bokkeveld to Hex River Mts).•

sarniensis (L.) Herb.   GUERNSEY LILY   Bulbous geophyte, 25–45 cm. Leaves dry at flowering, several, spreading, 8–20 mm wide. Flowers actinomorphic, tepals scarlet with golden sheen, rarely pink or white, 5–8 mm wide, filaments elongate. Mar.–May. Rocky slopes, NW, SW (Citrusdal to Caledon).•

SCADOXUS   BLOOD LILY   9 spp., W Cape to tropical Africa

membranaceus (Baker) Friis & Nordal   Bulbous geophyte, 12–30 cm. Leaves erect, not sheathing to form a false stem. Flowers shades of red, bracts 4, uniform in size. Jan.–Mar. Coastal sand, SE (Port Elizabeth to KwaZulu-Natal).

puniceus (L.) Friis & Nordal   Bulbous geophyte, 30–45 cm. Leaves several, often dry at flowering, erect, basally sheathing to form a false stem. Flowers red to salmon-pink, bracts usually more than 4, variable in size. Sept.–Jan. Coastal bush, AP, SE (Stilbaai to tropical Africa).

STRUMARIA   CAPE SNOWFLAKE, TOLBOL   23 spp., S Namibia to W Cape and Karoo

A. Flowers bell- or funnel-shaped

picta W.F.Barker   Bulbous geophyte, 6–13 cm. Leaves dry at flowering, 2, spreading, broadly strap-shaped, leathery, greyish, minutely ciliate. Flowers widely bell-shaped, white with broad reddish brown central bands on reverse, scented. May–June. Clay flats, NW (Bokkeveld Escarpment).•

salteri W.F.Barker   Bulbous geophyte to 20 cm. Leaves dry at flowering, 2, prostrate, elliptic, dark green, minutely ciliate. Flowers widely funnel-shaped, rose-pink, keels dark reddish pink. May. Sandstone rock crevices, NW (Nardouw and Pakhuis Mts).•

truncata Jacq. NAMAQUALAND SNOWFLAKE Bulbous geophyte, 20–35 cm. Leaves usually dry at flowering, 2–6, erect, spreading into a fan. Flowers nodding, funnel-shaped, white to pink. Apr.–June. Sandy or stony flats, NW (Namaqualand to W Karoo and Bokkeveld Escarpment).

watermeyeri L.Bolus Bulbous geophyte, 10–15 cm. Leaves dry at flowering, 2, prostrate, elliptic, with glutinous margins. Flowers funnel-shaped, tepals spreading above, pale pink. Apr.–May. Shallow soil on sandstone pavement, NW (Bokkeveld Mts).•

A. Flowers star-shaped

chaplinii (W.F.Barker) Snijman Bulbous geophyte to 10 cm. Leaves dry at flowering, 2, prostrate, ovate, softly hairy. Flowers star-shaped, white. Mar.–Apr. Granite outcrops, SW (Saldanha) •

discifera Marloth ex Snijman Bulbous geophyte to 15 cm. Leaves dry at flowering, 2, spreading strap-shaped, softly hairy. Flowers star-shaped, white, thinly striped with green or pink, often scented. Mar.–May. Heavy soils, NW (Knersvlakte and Bokkeveld Plateau to W Karoo).

gemmata Ker Gawl. Bulbous geophyte to 40 cm. Leaves dry at flowering, 2, spreading to prostrate, usually pubescent. Flowers star-shaped, pale lemon, rarely cream, tepals crisped. Feb.–May. Stony slopes or flats, clay or limestone, AP, KM, LB, SE (Bredasdorp to Riversdale, Ladismith to Port Elizabeth, Karoo).

leipoldtii L.Bolus Bulbous geophyte to 15 cm. Leaves dry at flowering, 2, prostrate, oval, with a dense white fringe. Flowers star-shaped, white. Mar.–Apr. Sandstone rock ledges, NW (Lambert's Bay to Olifants River valley).•

tenella (L.f.) Oberm. (= *Hessea tenella* (L.f.) Oberm.) Bulbous geophyte, 10–20 cm, slender. Leaves usually present at flowering, up to 6, spreading, filiform. Flowers star-shaped, white, sometimes flushed with pink. Apr.–July. Seasonally damp, loamy flats, NW, SW (Bokkeveld Escarpment and W Karoo to Cape Peninsula and Montagu, Free State).

unguiculata (W.F.Barker) Snijman Bulbous geophyte to 35 cm. Leaves usually dry at flowering, 2, suberect, broadly elliptic, glabrous, light green. Flowers star-shaped, white with wine-red stripes on reverse, strongly scented. May. Loamy, stony soils, NW (Botterkloof valley).•

ANTHERICACEAE

CHLOROPHYTUM (= *ANTHERICUM* in part) GRASS LILY c. 150 spp., mainly tropical Africa, also Asia

A. Inflorescence divaricately branched; pedicels articulated near the base; capsule transversely ridged

monophyllum Oberm. Rhizomatous geophyte to 50 cm. Leaf single, linear. Flowers in much-branched racemes, white, filaments rough. Capsules 3-winged, pedicels articulated near the base. Dec. Sandy flats, NW (Cold Bokkeveld).•

rigidum Kunth Rhizomatous geophyte to 50 cm, roots fleshy. Leaves distichous, lanceolate, stiff. Flowers in few-branched racemes, white with brown keels, filaments rough. Capsules 3-winged, pedicels articulated near the base. May–Oct. Stony slopes and flats, NW, SW (Ceres to Villiersdorp).•

viscosum Kunth Rhizomatous geophyte to 60 cm, roots slightly swollen, pinkish. Leaves more or less linear, stiff, closely ribbed, glandular. Flowers in simple or laxly branched, glandular racemes, white with dark keels, filaments rough. Capsules 3-winged, pedicels articulated near the base. June–Oct. Rocky sandstone slopes, NW (S Namibia to Piketberg).

A. Inflorescence simple or branched; pedicels articulated near the middle; capsules smooth

capense (L.) Voss Rhizomatous geophyte with compressed stems to 1 m, roots long, fleshy. Leaves linear to lanceolate, glaucous, closely ribbed. Flowers in much-branched racemes, white, filaments smooth. Capsules 3-winged, pedicels articulated above the middle. Oct.–Apr. Shaded bush, often near sea, SE (Humansdorp to King William's Town).

crispum (Thunb.) Baker Rhizomatous geophyte to 50 cm, roots unevenly swollen. Leaves in a flat rosette, lanceolate, margins crisped and fringed. Flowers in much-branched racemes, white with green keels, filaments rough. Capsules 3-winged, pedicels articulated in the middle. Sept.–Apr. Stony flats, LB, SE (Riversdale to Alexandria).

comosum (Thunb.) Jacq. HEN AND CHICKENS, SPIDER PLANT Rhizomatous geophyte to 80 cm, roots swollen toward tips. Leaves soft, channelled. Flowers in long few-branched racemes often with leafy tufts at nodes, white, filaments smooth. Capsules 3-winged, pedicels articulated near the middle. Oct.–Apr. Forests and bush in shade, LB, SE (Swellendam to Mpumalanga).

lewisiae Oberm.   Rhizomatous geophyte, 15–30 cm, roots slender. Leaves few, linear, densely hairy, with blunt dark tips. Flowers in unbranched hairy racemes, white with green keels, filaments rough. Capsules 3-winged, pedicels articulated below the middle. Sept. Sandstone slopes, NW (Botterkloof).•

pauciphyllum Oberm.   Rhizomatous geophyte, 15–30 cm, roots slender, swollen toward tips. Leaf single, linear, rigid, margins ciliate, sheath purple-spotted. Flowers in unbranched racemes, white with green keels, filaments rough. Capsules 3-winged, pedicels articulated near the middle. Apr.–Sept. Rocky sandstone slopes, NW (Nardouw to Cedarberg Mts).•

rangei (Engl. & Krause) Nordal (= *Anthericum rangei* Engl. & Krause)   Rhizomatous geophyte to 40 cm, roots wiry with scattered tubers. Leaves grass-like. Flowers in lax branched racemes, white, filaments smooth. Capsules ovoid, pedicels articulated below the middle. Nov.–Mar. Sand or stony slopes, often granite, NW, SW (S Namibia to Cape Peninsula).

triflorum (Aiton) Kunth   Rhizomatous geophyte to 1 m, roots hard, dark and tapering. Leaves lanceolate, margins ciliate. Flowers in unbranched racemes, white, reddish on reverse, filaments rough. Capsules 3-winged, pedicels articulated near the middle. July–Oct. Sandy slopes and flats, NW, SW (Elandsbaai to Cape Peninsula).•

undulatum (Jacq.) Oberm.   Rhizomatous geophyte to 50 cm, roots slender and stiff, sometimes with short tubers. Leaves lanceolate, margins ciliate. Flowers in unbranched racemes, white with red keels, filaments rough. Capsules 3-winged, pedicels articulated near the middle. July–Oct. Stony flats and slopes, NW, SW (Namaqualand and W Karoo to Somerset West).

APONOGETONACEAE

APONOGETON   WATERBLOMMETJIE   c. 30 spp., palaeotropics and southern Africa

angustifolius Aiton   Rhizomatous aquatic. Leaves floating, small, elliptic to lanceolate. Flowers distichous, white, with 2 tepals. June–Sept. Pools and ditches, SW (Malmesbury to Worcester).•

desertorum Zeyh. ex A.Spreng.   Rhizomatous aquatic. Leaves floating, oblong, cordate. Flowers spirally arranged and crowded, whitish, with 2 tepals. Nov.–May. Rock pools, SE (Uitenhage to Botswana and Namibia).

distachyos L.f.   Rhizomatous aquatic. Leaves floating, oblong. Flowers distichous, white, with 1 tepal, scented. July–Dec. Pools and ditches, NW, SW, AP, LB, SE (Bokkeveld Mts to Knysna).•

junceus Lehm. ex Schltdl.   RAMSHORN   Rhizomatous aquatic. Leaves floating, lanceolate to linear. Flowers distichous, white to lilac, with 1–3 tepals. Oct.–Dec. Ponds, marshes and rivers, AP, LB, SE (Riversdale to Zimbabwe and N Namibia).

ranunculiflorus Jacot Guill. & Marais   Rhizomatous aquatic. Leaves submerged, subterete. Flowers few, white, with 1 or 2 large tepals. Sept. Seasonal pools, NW (Bokkeveld Mts and Lesotho).

ARACEAE (= LEMNACEAE)

1. Terrestrial plants; leaves present, well developed; flowers in a spike-like spadix subtended by a large, petaloid spathe..**Zantedeschia**
1.' Minute floating aquatic herbs; leaves absent:
   2. Roots absent; plants globular...**Wolffia**
   2' Roots present; plants discoid:
      3. Roots several, usually short...**Spirodela**
      3.' Root solitary, usually long ....**Lemna**

LEMNA   DUCKWEED, DAMSLYK   9 spp., nearly cosmopolitan

gibba L.   Minute, floating aquatic forming large colonies. Leaves spongy. Flowers minute. Nov.–Dec. Freshwater pools and dams, SW, LB, SE (Cape Peninsula to Uitenhage, cosmopolitan).

minor L.   Minute, floating aquatic forming colonies. Leaves thin. Flowers minute. Nov.–Dec. Pools, SE (Uitenhage, cosmopolitan).

SPIRODELA   4 spp., nearly cosmopolitan

punctata (G.Mey.) Thomps.   Minute, floating aquatic forming dense colonies. Leaves purple beneath. Oct.–Dec. Freshwater pools, SW (Cape Peninsula, cosmopolitan).

WOLFFIA 7 spp., nearly cosmopolitan

arrhiza (L.) Horkel ex Wimm.   Minute, floating aquatic forming colonies, without roots. Oct.–Dec. Freshwater pools, SW, SE (Cape Peninsula to Knysna, widespread in Old World).

ZANTEDESCHIA   CALLA LILY, VARKBLOM   8 spp., southern Africa

aethiopica (L.) Spreng.   Rhizomatous geophyte, 60–100 cm. Leaves sagittate, on long spongy petioles. Flowers with large white spathe and yellow spadix. June–Dec. Sandy or rocky places, usually seasonally damp, NW, SW, AP, LB, SE (Richtersveld, Kamiesberg, Bokkeveld Mts to N Province).

ASPARAGACEAE

ASPARAGUS (= *MYRSIPHYLLUM, PROTASPARAGUS*)   KATDORING, KRULKRANSIE, WILDE ASPERSIE   c. 120 spp., mainly Africa, also Asia

A. Tepals fused below; cladodes usually broad and leaf-like and solitary in the axils

alopecurus (Oberm.) Malcomber & Sebsebe   Erect, brush-like shrublet to 80 cm. Cladodes linear. Flowers nodding, solitary in axils, tepals fused below, filaments straight. Apr.–Oct. Rocky sandstone slopes and flats, NW (Namaqualand to Clanwilliam).

asparagoides (L.) W.Wight   Scrambler to 3 m. Cladodes ovate. Flowers nodding, solitary in axils, tepals fused below, filaments straight. July–Sept. Widespread in bush, NW, SW, AP, KM, LB, SE (Gifberg to Port Elizabeth to tropical Africa).

declinatus L.   Much-branched scrambler to 1 m. Cladodes in threes, linear. Flowers nodding, solitary in axils, tepals fused below, filaments straight. June–Oct. Mostly rock outcrops, fynbos and coastal scrub, NW, SW, AP, KM, LB (S Namibia to Riversdale).

fasciculatus Thunb.   Shrublet to 1 m, sprawling or scrambling in bush. Cladodes fascicled, filiform. Flowers nodding, 1–3 in axils, tepals fused below, filaments straight. Mar.–June. In bush, NW, SW (Namaqualand and W Karoo to Saldanha).

kraussianus (Kunth) J.F.Macbr.   Sprawling shrublet to 1 m. Cladodes lanceolate with a distinct midrib. Flowers nodding, solitary in axils, tepals fused below, filaments straight. Sept.–Oct. Strandveld and fynbos, NW, SW, AP (Clanwilliam to Stilbaai).•

multituberosus R.A.Dyer   Scandent to erect shrublet to 45 cm. Cladodes ovate to heart-shaped, many-veined. Flowers nodding, 1–3 in axils, tepals fused below, filaments straight, styles 3. July–Sept. Rocky lower slopes and flats, NW (Namaqualand and W Karoo to Karoopoort).

ovatus T.M.Salter   Scandent shrublet to 1 m. Cladodes ovate, many-veined. Flowers nodding, 1–3 in axils, tepals fused below, filaments straight. July. Mostly coastal, SW, AP, SE (Saldanha to East London).

scandens Thunb.   Scandent shrublet to 2 m. Cladodes in one plane, in threes with a solitary smaller one opposing a larger pair, narrow, lightly sigmoid. Flowers nodding, 1–3 in axils, tepals spreading, filaments straight. Sept.–Jan. Forest and bush in shade, NW, SW, AP, LB, SE (Gifberg to Tsitsikamma Mts).•

undulatus (L.f.) Thunb.   Erect, branched shrublet to 40 cm. Cladodes lanceolate, ribbed. Flowers nodding, 1–3 in axils, tepals fused below, filaments straight. July–Oct. Sandy slopes, often in shade, NW, SW (Namibia to Hottentots Holland Mts).

volubilis Thunb.   Scandent shrublet to 1 m. Cladodes narrowly elliptic, several-veined. Flowers nodding, solitary in axils, tepals fused below, filaments straight. June–Oct. Coastal scrub or forest, NW, SW, AP, LB, SE (Citrusdal to E Cape).

A. Tepals free; cladodes linear to needle-like, usually in fascicles
B. Flowers 1–3(–many) on an apical disc

burchellii Baker   Erect to sprawling spiny shrub to 1.5 m, spines spreading. Cladodes 2 or 3 in fascicles, terete. Flowers 1–3 on an apical disc, tepals and filaments spreading. Feb.–May. Mainly dry bush, SW, AP, KM, LB, SE (Stellenbosch to Queenstown).

capensis L. KATDORING Erect spiny shrub to 1 m, stems brush-like, with spines in threes. Cladodes mostly 5 in clusters, sublinear, hairy. Flowers 1 or 2 on an apical disc, tepals and filaments spreading. Mainly Apr.–Aug. Rocky slopes, NW, SW, AP, KM, LB, SE (S Namibia to Transkei).

exsertus (Oberm.) Fellingham & N.L.Mey. Erect spiny shrublet to 80 cm, stems thin, spines single or paired. Cladodes 1–3 in fascicles, terete, pubescent. Flowers 1 or 2 on an apical disc, tepals and filaments spreading. Nov.–Dec. Shale slopes in renosterveld, NW (Worcester district).•

mariae (Oberm.) Fellingham & N.L.Mey. Erect spiny shrub to 1 m, stems densely puberulous, pale, spines in threes. Cladodes 1–4 in fascicles, filiform. Flowers 1–3 on an apical disc, tepals and filaments spreading. Mar.–May. Stony flats, SW, LB (Bredasdorp to Mossel Bay, also Grahamstown).

setaceus (Kunth) Jessop FEATHERY ASPARAGUS Spiny climbing shrub to 2 m, stems spiny only at base. Cladodes numerous in fascicles, filiform. Flower solitary on an apical disc, tepals and filaments spreading. Sept.–Apr. Mainly forests, AP, LB, SE (Port Beaufort to S tropical Africa).

stipulaceus Lam. Erect, brush-like shrublet to 50 cm, stems minutely ribbed, pubescent, spines in threes. Cladodes 3 in fascicles, terete. Flowers 1 or 2 on an apical disc, tepals and filaments spreading. Apr.–July. Coastal dunes, SW, AP (Cape Peninsula to Bredasdorp).•

striatus (L.f.) Thunb. Erect shrublet to 60 cm, stems minutely striate with weak spines. Cladodes single, linear, hard and striate. Flowers few to many on an apical disc, tepals and filaments spreading. Aug.–Jan. AP, KM, LB, SE (Agulhas to E Cape and Free State).

suaveolens Burch. Erect spiny shrub to 1 m, stems sometimes brush-like, spines in twos or threes. Cladodes 1–6 in fascicles, terete. Flowers 1–3 on an apical disc, tepals and filaments spreading. Mainly Apr.–Sept. NW, KM, AP, LB, SE (widespread in southern Africa).

subulatus Thunb. Scandent or erect shrub to 2 m, stems minutely grooved, with adpressed spines below. Cladodes 3–6 in fascicles, filiform. Flowers on an apical disc, tepals and filaments spreading. Sept.–Dec. Dry rocky sites, SE (Humansdorp to Komga).

B. Flowers few in the axils or several in racemes

aethiopicus L. Spiny climber to 3 m, stems pale and ribbed, spines hooked. Cladodes 4–6 in fascicles, terete. Flowers in racemes. Jan.–June. Mainly dry bush, NW, SW, AP, KM, LB, SE (Namaqualand to Transkei).

africanus Lam. Spiny shrub to 1 m or climber to 3 m, stems with straight or spreading brownish spines. Cladodes c. 12 in fascicles. Flowers up to 6 in axils, tepals and filaments spreading. Aug.–Dec. Usually moist places, SW, KM, LB, SE (Saldanha to N KwaZulu-Natal).

confertus K.Krause Erect spiny shrub to 1 m, stems ribbed, papillate, with orange-brown spines. Cladodes 1–3 in fascicles, linear. Flowers 1–3 in axillary racemes, tepals and filaments spreading. Flowering time? Stony flats and slopes, NW, KM, LB (Roberston to S Karoo).

crassicladus Jessop Spiny scandent or sprawling shrub to 2 m, branches short, flexuose, pubescent, spines hooked. Cladodes in fascicles, succulent, curved. Flowers in fascicles on short branches, tepals and filaments spreading. Nov.–Dec. SE (Plettenberg Bay to E Cape).

densiflorus (Kunth) Jessop Erect or spreading spiny shrub to 60 cm, stems striate with hooked spines. Cladodes 1–few in fascicles, linear. Flowers in racemes, tepals and filaments spreading. Oct.–Mar. Mainly coastal, SE (Uitenhage to S Mozambique).

exuvialis Burch. Erect or scrambling shrub, 0.5–2 m, stems with white membranous bark, sometimes with minute spines. Cladodes in fascicles, filiform. Flowers 2–6 in axils, tepals and filaments spreading. Oct.–Apr. Mainly dry areas, NW, SW, KM, LB, SE (drier parts of South Africa, Namibia, Botswana).

filicladus (Oberm.) Fellingham & N.L.Mey. Erect spiny shrub to 60 cm, stems pale grey with hooked spines. Cladodes 15–25 in fascicles, falcate. Flowers in racemes, tepals and filaments spreading. Flowering time? Stony slopes, SE (Knysna to Queenstown).

lignosus Burm.f. Spiny shrublet to 80 cm, stems spreading to erect, pale, striate. Cladodes terete, in fascicles. Flowers 1–4 in axils. Oct.–May. Rocky sandstone slopes and marshy flats, NW, SW, AP, KM, LB (Clanwilliam to Mossel Bay).•

macowanii Baker Erect or scandent spiny shrub to 2 m, stems with hooked spines below. Cladodes to 50 in fascicles. Flowers many in fascicles on young branches, tepals and filaments spreading. Mainly Sept.–Nov. Moist sites often near rivers, SE (Knysna to Mozambique).

mucronatus Jessop Erect spiny shrub to 1 m, branches pubescent with spreading spines. Cladodes grey, 1–4 in fascicles. Flowers c. 2 in axils, tepals and filaments spreading. Oct.–Dec. Stony flats, SW, KM, LB, SE (Paarl to Queenstown and Kimberley).

multiflorus Baker   Coarse scrambling or tangled shrub, stems striate and velvety, bearing blunt triangular knobs. Cladodes 7 in fascicles. Flowers few in axils, tepals and filaments spreading. Dec.–Jan. Stony slopes, SW, LB, SE (Swellendam to E Cape).

natalensis (Baker) Fellingham & N.L.Mey.   Spiny climber or shrub, stems pale grey, with small spines. Cladodes 1–4 in fascicles, linear. Flowers in racemes, tepals and filaments spreading. Aug.–Sept. Scrub, SE (Plettenberg Bay to tropical Africa).

oliveri (Oberm.) Fellingham & N.L.Mey.   Erect, spiny, puberulous shrublet to 20 cm, stems brush-like, striate. Cladodes solitary, terete and spiny. Flowers 1–few in axils, succulent, ?cleistogamous. Nov. Rocky sandstone slopes, KM (Rooiberg Mts).•

ramosissimus Baker   Scandent shrublet 1–2.2 m. Cladodes 3 in fascicles, linear, keeled. Flowers nodding, solitary in axils, tepals spreading, filaments straight. Sept.–Feb. Moist shady places, LB (Potberg to Mpumalanga).

recurvispinus (Oberm.) Fellingham & N.L.Mey.   Erect spiny shrub, stems dark, bearing hooked spines. Cladodes 1–3 in fascicles, terete. Flowers paired in axils, tepals and filaments spreading. Nov. Succulent karoo, KM (Ladismith to Oudtshoorn).

retrofractus L.   Scrambling spiny shrub to 3 m, stems grey and ribbed when young, with spreading spines. Cladodes in feathery fascicles, filiform. Flowers 2–7 in axils, tepals and filaments spreading. Apr.–June. NW, SW, KM, LB (S Namibia to E Cape).

rubicundus P.J.Bergius   Erect spiny shrubs to 1.5 m, stems dark brown, glossy, with spreading spines. Cladodes c. 10 in fascicles, terete. Flowers 1 or 2 in axils, tepals and filaments spreading. Mainly Mar.–June. Sandy and granite slopes, NW, SW, AP, KM, LB, SE (Kamiesberg, Gifberg to Uitenhage).

ASPHODELACEAE
Aloe by G.F. Smith, **Astroloba** by N.L. Meyer & G.F. Smith, **Haworthia** by B. Bayer

1. Tepals free or connate at base, spreading or campanulate:
    2. Flowers long-lived; perianth persistent; filaments smooth; seeds 1 or 2 per locule, shield-shaped .....**Bulbinella**
    2.' Flowers lasting less than a day; perianth caducous; filaments scabrid or bearded; seeds usually many, angled:
        3. Filaments retrorsely scabrid; flower white ....**Trachyandra**
        3.' Filaments densely bearded; flowers yellow or orange ....**Bulbine**
1.' Tepals fused below into a tube, erect or shortly spreading above:
    4. Inflorescence terminal, usually a simple dense spike; leaves basal, soft, immaculate, usually keeled....**Kniphofia**
    4.' Inflorescence apparently axillary, sometimes branched; leaves various but usually succulent, hard, thick, prickly, maculate or immaculate, in basal or apical rosettes or cauline, rounded on back:
        5. Stamens as long as or longer than perianth:
            6. Perianth tube distinctly curved near the middle and more or less inflated below ....**Gasteria**
            6.' Perianth more or less straight, at most upturned near tip but then not inflated below ....**Aloe**
        5.' Stamens shorter than perianth and included in it:
            7. Perianth distinctly bilabiate...**Haworthia**
            7.' Perianth regular or at most weakly bilabiate ....**Astroloba**

ALOE   ALOE, AALWYN   c. 350 spp., Africa, Arabia, Madagascar, Socotra

A. Stem short or absent
B. Racemes mostly branched

buhrii Lavranos   Acaulescent succulent forming dense clumps. Leaves lanceolate, erect, yellowish green, copiously marked with H-shaped spots, margins at most minutely toothed. Flowers in capitate, rounded panicles, nodding, inflated below, red, orange or yellow. Aug.–Oct. Rocky slopes, NW (Bokkeveld Escarpment).•

falcata Baker   Acaulescent or short-stemmed succulent. Leaves slightly asymmetrically lanceolate, incurved, surfaces granulate, margins coarsely toothed. Flowers in conical, branched racemes, nodding, red or yellow with green tips, anthers exserted. Dec. Rocky slopes, NW (Richtersveld to Klawer).

maculata All. (= *Asparagus saponaria* (Aiton) Haw.)   BONTAALWYN, SEEPAALWYN   Acaulescent or short-stemmed succulent, sometimes freely suckering from base. Leaves deltoid-lanceolate, spotted dull white, margins coarsely toothed. Flowers in capitate, branched and flat-topped racemes, nodding, inflated below, pink to orange or yellow. Nov.–Dec. Rocky slopes, SW, LB, SE (Cape Peninsula to Zimbabwe).

striata Haw.   MAKAALWYN, CORAL ALOE   Acaulescent or short-stemmed succulent. Leaves grey-green, broadly deltoid, spreading, unarmed. Flowers in capitate, rounded panicles, nodding, inflated below, coral with pink tips. Aug.–Sept. Rocky slopes, NW, KM, SE (Worcester Karoo to Port Elizabeth and E Karoo).

B. Racemes simple

bowiea Schult. & J.H.Schult.   Dwarf, acaulescent, mat-forming succulent. Leaves linear-subulate, minutely spotted below, margins finely toothed. Flowers in a lax raceme, horizontal, greenish white, anthers well exserted. Dec.–Mar. Karroid thicket, SE (Uitenhage to Port Elizabeth).•

brevifolia Mill.   KLEINDUINE AALWYN   Dwarf, acaulescent succulent. Leaves lanceolate, margins and sometimes underside with coarse white prickles. Flowers in conical racemes, nodding, red. Oct.–Nov. Coastal dunes, AP (Agulhas to Riversdale).•

claviflora Burch.   AANTEELAALWYN, KANONAALWYN   Acaulescent or shortly sprawling succulent. Leaves asymmetric, narrowly lanceolate, glaucous, margins and underside with coarse prickles. Flowers in dense, sprawling, cylindrical racemes, red, yellowing with age, club-shaped, anthers well exserted. Aug.–Sept. KM (Namibia to Oudtshoorn and S Free State).

framesii L.Bolus   Acaulescent succulent forming dense colonies. Leaves lanceolate-attenuate, usually copiously white-spotted, margins toothed. Flowers in conical racemes, nodding, dull scarlet with green tips. June–July. Coastal rock outcrops, NW (Port Nolloth to Lambert's Bay).

glauca Mill.   BLOU AALWYN   Acaulescent or short-stemmed succulent to 1 m. Leaves glaucous, lanceolate, margins coarsely toothed. Flowers in stout, conical racemes, nodding, red with green tips. July–Oct. Mountain slopes, NW, SW, LB (Namaqualand and W Karoo to Swellendam).

haemanthifolia A.Berger & Marloth   Acaulescent succulent. Leaves in a distichous fan, lorate, greenish grey, spineless, margins reddish. Flowers in capitate racemes, nodding, scarlet. Oct.–Nov. Sandstone ledges and cliffs, NW, SW (Hex River Mts to Jonkershoek).•

humilis (L.) Mill.   DWARF HEDGEHOG ALOE   Dwarf, acaulescent succulent growing in dense colonies. Leaves suberect, narrowly lanceolate, margins and both surfaces with soft, white prickles. Flowers few, in short racemes, scarlet, sometimes orange. Sept.–Oct. Coastal hills in renosterveld or dry thicket, LB, SE (Mossel Bay to E Cape).

longistyla Baker   RAMENAS   Dwarf, acaulescent succulent. Leaves narrowly lanceolate, margins and both surfaces with soft white prickles. Flowers succulent in dense, capitate racemes, curving upward, salmon to red, anthers well exserted. July–Aug. Stony flats and lower slopes, KM (Calitzdorp through Little Karoo to Grahamstown).

micracantha Haw.   WATERAALWYN   Tufted grass-like perennial to 50 cm, with thick fleshy fusiform roots. Leaves linear, slender, scarcely succulent, rigid, copiously white-spotted, unarmed. Flowers in capitate racemes, nodding, salmon-pink. Dec.–Jan. Stony grassland, SE (Langkloof to E Cape).

microstigma Salm-Dyck   Acaulescent or short-stemmed succulent to 50 cm. Leaves lanceolate to deltoid, copiously white-spotted, margins coarsely toothed. Flowers in conical racemes, nodding, orange to yellow. May–July. Dry karroid slopes, NW, KM (Ceres and Little Karoo to S Karoo and E Cape).

pictifolia D.S.Hardy   Succulent with short, creeping or hanging stems to 30 cm. Leaves narrowly lanceolate, ascending, spreading above, copiously white-spotted, margins toothed. Flowers in conical racemes, nodding, red, greenish at tips. June–Aug. Rock outcrops and cliffs, SE (Humansdorp district).•

variegata L.   BONTAALWYN, KANNIEDOOD   Acaulescent succulent to 50 cm. Leaves in compact, 3-ranked rosettes, keeled, green to brown, boldly mottled white, margins white and horny, minutely toothed. Flowers in conical racemes, nodding, pink to red. July–Sept. Partial shade, stony ground, NW, KM (S Namaqualand and W Karoo to Ladismith and Uniondale).

A. Stem well developed, erect or sprawling
C. Racemes branched

africana Mill.   UITENHAAGSE AALWYN   Single-stemmed or rarely branched succulent shrub to 4 m, covered below with dry leaves. Leaves dull green, lanceolate, spreading, margins coarsely toothed. Flowers in dense, conical, 2–4-branched racemes, curving upward, yellow-orange, anthers well exserted. July–Sept. Thick bushveld, SE (Humansdorp to E Cape).

comptonii Reynolds   Stems erect or creeping, succulent, to 1 m. Leaves ascending, lanceolate, margins coarsely toothed. Flowers in capitate, branched racemes, nodding, scarlet. Aug.–Dec. Stony flats and slopes, KM, SE (Montagu to Uniondale and S Karoo).

distans Haw.   Like **A. mitriformis** but smaller, with leaves to 15 × 7 cm, copiously spotted with white. Nov.–Dec. Coastal granite outcrops, SW (St Helena Bay to Saldanha Bay).•

ferox Mill.   TAPAALWYN, OPREGTE AALWYN   Single-stemmed succulent shrub to 3 m, covered below with dry leaves. Leaves broadly lanceolate, margins and often lower surface coarsely toothed. Flowers in dense, cylindrical, branched racemes, nodding and slightly upcurved, orange to reddish, anthers well exserted. May–Nov. Rocky slopes, SW, AP, KM, LB, SE (Stormsvlei to KwaZulu-Natal).

mitriformis Mill.   KRANSAALWYN   Sprawling, often branched succulent with stems 1–2 m. Leaves narrowly ovate, dark green at most sparsely speckled with white, margins coarsely toothed. Flowers in capitate, branched racemes, nodding, scarlet. Dec.–Feb. Rocky slopes and cliffs, NW, SW (Bokkeveld Mts to Kleinmond).•

pluridens Haw.   FRANSAALWYN, GARAA   Single- or several-stemmed succulent shrub, 2–3 m. Leaves spreading above, coarsely toothed. Flowers in conical, few-branched racemes, nodding, salmon to dull scarlet. May–July. Succulent thicket, SE (Humansdorp to KwaZulu-Natal).

C. Racemes mostly simple

arborescens Mill.   KRANSAALWYN   Much-branched succulent shrub to 3 m. Leaves spreading, deflexed above, margins toothed. Flowers in conical, mostly unbranched racemes, nodding, scarlet to orange, sometimes yellow. May–June. Bush and forest, SW, AP, LB, SE (Caledon to S tropical Africa).

arenicola Reynolds   Succulent with creeping, decumbent, simple or branched stems to 1 m. Leaves triangular to lanceolate, margins finely toothed, blue-green with white spots. Flowers in capitate racemes, rarely branched, nodding, red. July–Jan. Coastal sandveld, NW (Namaqualand to Lambert's Bay).

ciliaris Haw.   Much-branched, sprawling succulent shrub with slender stems to 5 m. Leaves lanceolate, basally sheathing, usually auriculate, margins distinctly ciliate. Flowers in oblong racemes, nodding, red with yellowish tips. Nov.–Apr. Dry thicket, SE (Uitenhage district to E Cape).

commixta A.Berger   Succulent with creeping or suberect branched stems to 1 m. Leaves lanceolate, ascending, margins finely toothed. Flowers in stout, oblong racemes, nodding, yellowish to orange. Aug.–Sept. Rocky slopes, SW (Cape Peninsula).•

comosa Marloth & A.Berger   Single-stemmed succulent shrub to 2 m, covered below with dry leaves. Leaves erect, spreading above, margins finely toothed. Flowers nodding and adpressed in very long, slender racemes, ovoid, buds dull pink opening ivory. Dec.–Jan. Dry rocky slopes, NW (Botterkloof to Clanwilliam).•

gracilis Haw.   RANKAALWYN   Shrublet to 2 m, much branched from below. Leaves narrowly lanceolate, margins finely toothed. Flowers nodding in lax, 1- or 2-branched racemes, red with greenish yellow tips. May–Aug. Partly shaded slopes, LB, SE (Garcia's Pass to Port Elizabeth).•

lineata (Aiton) Haw.   Single-stemmed shrub to 1.5 m, covered below with dry leaves. Leaves lanceolate, striate, green or yellowish green, margins coarsely toothed. Flowers nodding in conical racemes, salmon. Feb.–Mar. Karroid scrub, KM, LB, SE (Ladismith to Grahamstown).

plicatilis (L.) Mill.   BERGAALWYN   Stout, dichotomously branched shrub or small tree to 5 m. Leaves oblong, in tight distichous fans, unarmed. Flowers nodding, in short lax racemes, scarlet. Aug.–Oct. Rocky sandstone slopes, NW, SW (Tulbagh to Jonkershoek).•

speciosa Baker   SPAANSAALWYN, SLAPHORINGAALWYN   Single-stemmed or branched shrub to 4 m, covered below with dry leaves. Leaves in oblique rosettes, twisted and decurved, glaucous, margins reddish, scarcely toothed. Flowers spreading in dense, cylindrical racemes, buds reddish opening greenish white. Aug.–Sept. Succulent thicket, KM, SE (Montagu to E Cape and S Karoo).

succotrina Lam.   BERGAALWYN   Simple or dichotomously branched shrub to 2 m, sometimes in dense clumps. Leaves dark green, lanceolate, margins with coarse white teeth. Flowers nodding in conical racemes, red with green tips. June–Sept. Sandstone rocks and cliffs, SW (Cape Peninsula to Hermanus).•

ASTROLOBA (= *POELLNITZIA*)   8 spp., W and E Cape, Karoo

bullulata (Jacq.) Uitewaal   Caulescent succulent to 30 cm. Leaves dark green, pungent, usually with fairly prominent tubercles, scattered or in rows. Flowers in lax racemes, erect, greenish brown with yellow tepals. Nov.–June. Karroid scrub, KM (W Karoo to Witteberg).

corrugata N.L.Mey. & G.F.Smith (= *Astroloba rugosa* Roberts Reinecke ined.)   Caulescent succulent to 20 cm. Leaves not pungent, with tubercles in longitudinal rows. Flowers in lax racemes, erect, white or green tinged pink, midribs green, outer tepals sometimes slightly inflated. Oct.–Feb. Clay flats, NW, KM (Worcester to Ladismith).•

foliolosa (Haw.) Uitewaal   Caulescent succulent to 20 cm. Leaves smooth and shiny green, not pungent. Flowers in lax racemes, erect, greenish white or pale cream with white or cream tepals, midribs green. June–Mar. Karroid flats, KM, SE (Ladismith to Uitenhage and Karoo).

herrei Uitewaal   Caulescent succulent to 20 cm. Leaves light green, not pungent, smooth and finely striate. Flowers in lax racemes, erect, white with yellow tepals, midribs pale green, outer tepals sometimes slightly inflated below. June–Oct. Karroid flats and slopes, KM, SE (Prince Albert to Uniondale).•

rubriflora (L.Bolus) G.F.Smith & J.C.Manning (= *Poellnitzia rubriflora* (L.Bolus) Uitewaal)   Sprawling caulescent succulent to 45 cm. Leaves smooth, pungent. Flowers secund in horizontal racemes, erect, orange-red with greenish tepals. Dec.–Apr. Rocky karroid flats and low hills, NW, SW (Robertson to Bonnievale).•

spiralis (L.) Uitewaal   Caulescent succulent to 20 cm. Leaves light to dark green, not pungent, smooth sometimes striate. Flowers in lax racemes, erect, white with yellow tepals, midribs pale green, outer tepals inflated below. Dec.–May. Karroid flats and lower slopes, KM (Little Karoo and E Cape).

sp. 1 (= *Astroloba hallii* Roberts Reinecke ined.)   Caulescent succulent to 15 cm. Leaves light green with dark veins, pungent, with inconspicuous whitish tubercles. Flowers in lax racemes, erect, greenish white with bright to creamy yellow tepals. Nov.–May. Shaly ridges, KM (Laingsburg and Prince Albert).

sp. 2 (= *Astroloba smutsiana* Roberts Reinecke ined.)   Caulescent succulent to 30 cm. Leaves light green with reddish brown tinge, not pungent, smooth and finely striate, sometimes with elongate, raised shiny patches. Flowers in lax racemes, erect, greenish cream. Jan.–Mar. Shaly ridges, KM (Little Karoo).•

BULBINE   BULBINE, KOPIEVA   c. 50 spp., mainly southern and tropical Africa, also Australia

A. Rootstock a rhizome with wiry roots

abyssinica A.Rich. (= *Bulbine asphodeloides* (L.) Willd.)   Geophyte, 40–60 cm, forming large tufts, rootstock a rhizome. Leaves linear. Flowers in a dense raceme on long pedicels, yellow. Capsules globose, spreading. Mainly Aug.–Nov. Stony flats and slopes, KM, SE (W Karoo and Worcester to tropical Africa).

annua (L.) Willd.   Annual, 15–40 cm, roots wiry. Leaves many in a basal cluster, terete. Flowers in a dense raceme on long pedicels, yellow. Capsules globose, spreading. Sept.–Dec. Sandy soils, SW, LB (Saldanha to Riversdale).•

frutescens (L.) Willd.   RANKKOPIEVA   Shrublet with wiry roots, 20–60 cm. Leaves subterete. Flowers in a dense, elongate raceme, yellow, orange or white. Capsules subglobose, spreading-upcurved. Sept.–Apr. AP, KM, SE (dry areas throughout southern Africa).

lagopus (Thunb.) N.E.Br.   Geophyte to 40 cm, forming tufts, roots wiry. Leaves linear. Flowers in a dense, elongate raceme, yellow. Capsules subglobose, erect. July–Dec. Rocky slopes, NW, SW, AP, KM, LB (Gifberg to Lesotho).

latifolia (L.f.) Roem. & Schult.   Geophyte, 30–60 cm, rootstock a thick rhizome with thin wiry roots. Leaves in a rosette, broadly lanceolate. Flowers in a dense, elongate raceme, yellow. Capsules globose, erect. Mainly Aug.–Nov. Rocky slopes, SE (Uniondale to Mpumalanga).

A. Rootstock a tuber
B. Leaves dry or emergent at flowering

alooides (L.) Willd.   Geophyte, 20–40 cm, often clumped, rootstock a tuber with swollen roots. Leaves emergent at flowering, lanceolate, margins often ciliate. Flowers in a crowded raceme, yellow. Capsules globose, spreading. Mar.–May. Rocky slopes, NW, SW (Namaqualand to Darling).

cepacea (Burm.f.) Wijnands (= *Bulbine tuberosa* (Mill.) Oberm., *B. pugioniformis* (Jacq.) Link; incl. **B. bachmanniana** Schinz)   Geophyte, 20–40 cm, rootstock a large flat-based tuber. Leaves often dry at flowering, narrowly lanceolate, surrounded at base by fibrous sheaths. Flowers in a dense raceme, yellow. Capsules oblong, suberect. Mar.–May. Stony flats, NW, SW, LB (Clanwilliam to Riversdale).•

favosa (Thunb.) Roem. & Schult. (= *Bulbine dubia* Schult. & Schult.f.)   Geophyte, 15–50 cm, rootstock a tuber. Leaves dry at flowering, few, linear to filiform. Flowers in a lax raceme, yellow, fragrant.

Capsules globose, erect. Mainly Mar.–May. Sandy and limestone flats and slopes, NW, SW (Piketberg to Riviersonderend Mts).•

flexuosa Schltr.   Geophyte with wiry peduncle, 10–20 cm, rootstock a tuber. Leafs often dry at flowering, 1 or 2, filiform. Flowers in a lax, flexuose raceme, yellow, pedicels elongating in fruit and persistent. Capsules obovoid, erect. Apr. Dry rocky slopes, NW (S Namaqualand to Pakhuis Mts).

foleyi E.Phillips (= *Bulbine tenuifolia* (Baker ex Kuntze) Baijnath ms)   Geophyte, 8–40 cm, rootstock a tuber. Leaves, usually dry at flowering, linear to filiform, enclosed below by a long, softly fibrous neck. Flowers in a long, dense raceme, usually yellow, fragrant. Capsules globose, ascending. Oct.–Feb. Mainly shale flats and slopes, NW, SW, AP, KM, SE (W Karoo and Clanwilliam to Albertinia).

B. Leaves fully developed at flowering

diphylla Schltr. ex Poelln.   Geophyte, 10–15 cm, rootstock a tuber. Leaves 2 or 3, unequal, ovoid, fleshy, distinctly cauline. Flowers in a lax raceme, yellow. June–Sept. Sandstone and granite outcrops, NW (S Namaqualand to Pakhuis Mts).

esterhuyseniae Baijnath   Geophyte to 5 cm, rootstock a tuber. Leaves filiform. Flowers few in an umbel-like raceme, yellow. Capsules globose, spreading. Apr. Sandstone outcrops, NW (Cedarberg to Cold Bokkeveld Mts).•

longifolia Schinz   Similar to **B. praemorsa** but much smaller, 15–20 cm. Leaves few, narrowly lanceolate. Flowers in a lax raceme. Capsules ovoid, suberect. Aug.–Oct. Rocky slopes, NW, SW, AP, KM (Elandskloof to Swartberg Mts and S Karoo).

mesembryanthemoides Haw.   WATERGLAS, WATERKANNETJIES   Geophyte, 8–20 cm, rootstock a tuber. Leaves 2, short, erect, succulent, truncate and transparent across apex. Flowers few in a lax raceme, yellow, tepals reflexed. Capsules ovoid, spreading. Aug.–Nov. Rocky slopes and flats, clay or sandstone, succulent karoo and fynbos, NW, KM, SE (Namaqualand to Uitenhage and E Cape).

monophylla Poelln.   Geophyte, 15–25 cm, rootstock a tuber. Leaf solitary, fleshy and terete. Flowers in a slender raceme, yellow. Capsules globose, suberect. Aug.–Sept. Sandy flats, NW (near Porterville).•

praemorsa (Jacq.) Roem. & Schult.   Geophyte, 40–60 cm, rootstock a small tuber. Leaves thick and fleshy, narrowly channelled, surrounded at base by a short fibrous neck. Flowers in a lax raceme, yellow to salmon. Capsules oblong, erect. June–Sept. Mostly rocky sandstone slopes, NW, SW, AP, KM (Namaqualand to Bredasdorp).

succulenta Compton   Geophyte, 10–20 cm, rootstock a woody tuber. Leaves several short and fleshy, with a basal collar of stiff, stout fibres. Flowers in a lax raceme, yellow. Capsules ovoid, spreading. July–Sept. Succulent karoo, NW, KM (W Karoo and Bokkeveld Mts to Witteberg).

torta N.E.Br.   Geophyte, 8–25 cm, rootstock a tuber. Leaves several, coiled. Flowers in a lax or dense raceme, yellow, tepals reflexed. Capsules globose, suberect. July–Sept. Sandstone outcrops, NW (Namaqualand and W Karoo to Cedarberg Mts).

BULBINELLA   BULBINELLA, KATSTERT   c. 23 spp., winter-rainfall southern Africa, New Zealand

*A. Leaves all subequal; plants mostly less than 50 cm (see also **B. caudafelis**)*

chartacea P.L.Perry   Rhizomatous geophyte to 40 cm, with papery cataphylls. Leaves few, filiform. Flowers in a narrowly conical raceme, yellow. Feb.–Apr. Sandstone in fynbos, NW, SW (Olifants River Mts to McGregor).•

divaginata P.L.Perry   Rhizomatous geophyte to 45 cm. Leaves filiform. Flowers in a narrowly cylindrical raceme, yellow. Mar.–June.   Mainly clay soils in renosterveld, NW, SW (Namaqualand to False Bay).

elegans Schltr. ex P.L.Perry   Rhizomatous geophyte to 60 cm, cataphylls netted. Leaves filiform, margins finely and irregularly toothed. Flowers in a compact, cylindrical raceme, yellow to white with pink tinge. Mainly Aug.–Sept. Various soils, NW, KM (Bokkeveld Escarpment and W Karoo to Witteberg).

trinervis (Baker) P.L.Perry   Rhizomatous geophyte to 40 cm. Leaves filiform. Flowers in a narrowly cylindrical raceme, white to pinkish. Mainly Mar.–Apr. Rocky sandstone slopes, SW, KM, LB, SE (Malmesbury to Baviaanskloof Mts).•

triquetra (L.f.) Kunth   Rhizomatous geophyte to 35 cm. Leaves filiform, margins finely toothed. Flowers in a subcorymbose to narrowly conical raceme, yellow. Mainly Sept.–Nov. Damp sand and granite, NW, SW (Bokkeveld Mts and W Karoo to Bredasdorp).

A. Leaves solitary or inner smaller; plants usually more than 50 cm

barkerae P.L.Perry   Rhizomatous geophyte to 60 cm. Leaves linear, channelled, margins ciliate. Flowers in a narrow, cylindrical raceme, white, with musty odour. Sept.–Oct. Shale flats in renosterveld, SW, AP, LB (Botrivier to Riversdale).•

caudafelis (L.f.) T.Durand & Schinz   Rhizomatous geophyte to 80 cm. Leaves linear, channelled, margins sometimes finely toothed. Flowers in a narrowly conical raceme, white with pink tinge. Aug.–Dec. Sandstone, granite or clay, NW, SW, AP, KM, LB, SE (Namaqualand to Avontuur).

eburniflora P.L.Perry   Rhizomatous geophyte to 75 cm. Leaves linear, channelled, margins finely toothed. Flowers in a cylindrical raceme, ivory or pale straw-coloured, with musty odour. Aug.–Sept. Clay and sand, NW (Bokkeveld Escarpment).•

elata P.L.Perry   Rhizomatous geophyte to 1 m. Leaves linear, flat. Flowers in a long, slender raceme, cream. July–Aug. Clay and granite soils, NW, SW (Pakhuis Mts to Mamre).•

graminifolia P.L.Perry   Rhizomatous geophyte to 65 cm. Leaves linear, channelled, margins finely toothed. Flowers in a narrowly cylindrical raceme, white. July–Aug. Clay in renosterveld, NW (Namaqualand to Citrudsal).

latifolia Kunth   Like **B. nutans** but leaves broader, to 65 mm wide and raceme narrower, to 45 mm wide. Flowers deep yellow or orange. Aug.–Oct. Seasonally damp sandstone or granite, rarely dolerite, NW (Namaqualand to Cedarberg Mts).

nutans (Thunb.) T.Durand & Schinz   Rhizomatous geophyte to 1 m. Leaves linear, channelled. Flowers in a conical raceme, yellow or cream. July–Oct. Damp peaty soils, NW, SW, LB (Loeriesfontein and W Karoo to Swellendam).

potbergensis P.L.Perry   Rhizomatous geophyte to 60 cm, cataphylls netted. Leaf solitary, linear, leathery. Flowers in a narrowly conical raceme, yellow. Sept. Silcrete, SW (Potberg).•

punctulata A.Zahlbr.   Rhizomatous geophyte to 1 m, cataphylls netted. Leaves few, linear, channelled. Flowers in a cylindrical raceme, yellow. Aug.–Oct. Rocky sandstone, often in wet places, NW (Gifberg to Porterville Mts).•

GASTERIA   GASTERIA, BONTAALWYN   c. 16 spp., dry areas of southern Africa

*A. Leaves distichous (see also **G. carinata**, **G. nitida**)*

brachyphylla (Salm-Dyck) Van Jaarsv.   Acaulescent succulent to 1 m. Leaves distichous, lanceolate to oblong, smooth, margins crenulate. Flowers in an inclined, usually simple raceme, nodding, pink and green, inflated below. July–Feb. Succulent karoo, KM, SE (Barrydale to Willowmore).•

disticha (L.) Haw.   BEESTONGBLAAR   Acaulescent succulent to 90 cm. Leaves distichous, oblong, surface rough, margins toothed. Flowers in an inclined, usually simple raceme, nodding, pink and green, inflated below. July–Feb. Shale soils, NW, KM (Worcester to Great Karoo).

glomerata Van Jaarsv.   Acaulescent succulent to 20 cm. Leaves distichous, oblong, rough to tuberculate, margins smooth. Flowers in a suberect raceme, nodding, red and green, inflated at base. Sept. Vertical sandstone cliffs, SE (Patensie).•

pillansii Kensit   Acaulescent succulent to 1.5 m. Leaves distichous, oblong, surface rough to tuberculate, margins toothed. Flowers in an inclined raceme, nodding, pink and pale green, barely inflated below. Nov.–Apr. Rocky sandstone outcrops, NW (Namaqualand to Clanwilliam).

rawlinsonii Oberm.   Caulescent succulent with long, nodding leafy stems to 1 m. Leaves distichous, rough, margins sparsely toothed. Flowers in a suberect raceme, nodding, pink or white and green, much inflated below. Aug.–Oct. Rock outcrops, SE (Baviaanskloof Mts).•

A. Leaves in a rosette

acinacifolia (Jacq.) Haw.   Acaulescent succulent to 1 m. Leaves in a spiral rosette, smooth, margins and keels toothed. Flowers in an inclined, usually branched raceme, nodding, pink and green, elongate, not inflated below. Sept.–Dec. Coastal dune thicket, SE (Knysna to Port Alfred).

carinata (Mill.) Duval   Acaulescent succulent to 90 cm. Leaves in a rosette or distichous, triangular, often channelled, smooth to tuberculate, margins toothed. Flowers in an inclined raceme, nodding, pink and green, not inflated below. July–Nov. Renosterveld on clay and limestone, NW, SW, AP, LB (Worcester to Mossel Bay).•

ellaphieae Van Jaarsv.   Acaulescent succulent to 40 cm. Leaves in a rosette, triangular, tuberculate. Flowers in an inclined, simple or branched raceme, nodding, reddish, inflated below. Jan.–Feb. Sandstone outcrops, SE (Patensie).•

glauca Van Jaarsv.   Acaulescent succulent to 70 cm, clump-forming. Leaves in a rosette, linear-triangular, tuberculate-asperulous. Flowers in an inclined raceme, nodding, reddish, inflated below. Dec.–Jan. Sandstone cliffs, SE (Kouga Mts: Guerna Kop).•

nitida (Salm-Dyck) Haw.   Acaulescent succulent to 1 m. Leaves in a spiral or distichous rosette, smooth, triangular, margins and keels smooth. Flowers in an inclined, simple or branched raceme, nodding, reddish, inflated below. Dec.–Feb. Grassy fynbos and renosterveld, often on sandstone soils, SE (Uniondale to Great Fish River Mouth).

pulchra (Aiton) Haw.   Acaulescent succulent to 1.5 m. Leaves smooth, linear-triangular, in a spiral rosette, margins and keels toothed. Flowers in an inclined, usually branched raceme, nodding, reddish, inflated below. July–Nov. Valley bushveld, SE (Hankey to Humansdorp).•

vlokii Van Jaarsv.   Acaulescent succulent to 60 cm. Leaves in a rosette, lanceolate to triangular, surfaces rough, margins toothed. Flowers in an inclined raceme, nodding, reddish and green, hardly inflated below. Jan.–Feb. Sandstone soils in fynbos, KM (Witteberg and Swartberg Mts).•

HAWORTHIA   HAWORTHIA   c. 70 spp., dry parts of southern Africa

A. Peduncle robust, freely branching; perianth tube straight, hexangular or rounded-hexangular at base, abruptly joined to pedicel; capsules rounded

kingiana Poelln.   Acaulescent succulent to 15 cm diam. Leaves erect, pungent, bright green with poorly defined whitish tubercles. Flowers white. Nov.–Dec. Sparse grassland in valley bushveld and renosterveld, LB, SE (Herbertsdale to Mossel Bay).•

marginata (Lam.) Stearn   Acaulescent succulent. Leaves silver-green to yellowish, deltoid, pungent, smooth. Flowers white. Nov.–Dec. Shale or sandstone flats in renosterveld, SW, LB (Robertson to Riversdale).•

pumila (L.) M.B.Bayer (= *Haworthia maxima* (Haw.) Duval)   VRATJIESAALWEE   Acaulescent succulent to 15 cm diam., rosettes solitary. Leaves brown to olive-green with large whitish to brown tubercles. Flowers yellow with green veins, tipped with brown. Nov.–Dec. Karroid scrub, NW, KM (Worcester to Montagu and W Karoo).

minima (Aiton) Haw. (= *Haworthia poellnitziana* Uitewaal)   Acaulescent succulent, forming clumps to 25 cm diam. Leaves erect, to 10 cm, blue-green with white tubercles. Flowers white with pinkish tips. Nov.–Dec. Coastal renosterveld, SW, AP, LB (Bredasdorp to Hartenbos).•

A. Peduncle slender, lax, usually unbranched; perianth tube curved; capsules oblong
B. Perianth tube obcapitate, hexangular or rounded-hexangular at base, substipitate

attenuata Haw. (= *Haworthia radula* (Jacq.) Haw.)   Acaulescent succulent to 15 cm diam., forming clumps. Leaves erect, attenuate, green to brownish, with small to large banded tubercles. Flowers white. Nov.–Dec. Clearings in valley bushveld, SE (Patensie to Kei River).

bruynsii M.B.Bayer   Acaulescent succulent with few-leaved rosettes. Leaves retuse-truncate, opaque, dark green, scabrid. Flowers white. Jan.–Feb. Dry flats in karroid veld, KM, SE (Steytlerville).

fasciata (Willd.) Haw.   Acaulescent or caulescent succulent, forming dense rosettes, often in clumps. Leaves green, fairly broad, incurved, with white, banded tubercles. Flowers white. Oct.–Nov. Grassy fynbos, SE (Humansdorp to Port Elizabeth).•

longiana Poelln.   Acaulescent succulent, forming robust rosettes. Leaves often curving sideways from base, slender, rigid, light green, nearly smooth. Flowers white. Nov.–Dec. Grassy fynbos, steep rocky conglomerate slopes in grassy fynbos, SE (Humansdorp to Uitenhage).•

scabra Haw. (= *Haworthia starkiana* Poelln.)   Acaulescent succulent with forming few-leaved rosettes. Leaves dark brownish green, smooth or scabrid and tubercled, thick and short. Flowers white. Nov.–Dec. Quartzite patches in fynbos and bush thicket, KM, LB, SE (Ladismith to Uniondale).•

sordida Haw   Acaulescent succulent, forming large dark green rosettes. Leaves dark green, finely tuberculate, slightly viscous. Flowers white, on wiry peduncles. Nov.–Dec. Rocky karroid slopes, SE (Kirkwood to Steytlerville).

venosa (Lam.) Haw.   KLEINKANNIEDOOD   Acaulescent or sometimes caulescent succulent. Leaves dark green, short and spreading, reticulate patterned. Flowers white. Nov.–Dec. Rocky slopes among grass in renosterveld. SW, LB (Namibia and Karoo to Breede River valley).

viscosa (L.) Haw.   KOEDOEKOS   Caulescent or acaulescent succulent, forming clumps. Leaves brownish, trifarious, recurved and pungent. Flowers white. Oct.–Nov. Rocky karroid slopes, KM, LB (Little Karoo to Baviaanskloof Mts, E Cape and Karoo).

B. Perianth tube obclavate, triangular or rounded-triangular at base

arachnoidea (L.) Duval (= *Haworthia aranea* (Berger) M.B.Bayer, *H. aristata* auct., *H. habdomadis* Poelln., *H. unicolor* Poelln.)   SPINNEKOPBOLLETJIE   Acaulescent succulent, forming compact rosettes to 10 cm diam. Leaves uniformly green with long white spines. Flowers white. Nov.–Dec. Rocky slopes and under bushes, NW, SW, KM, LB (Namaqualand and W Karoo to Worcester and Little Karoo).

bayeri Hammer & Venter   Solitary, acaulescent succulent. Leaves abruptly recurved, dark green, translucent along veins, often scabrid. Flowers white. Sept.-Oct. Dry mountain fynbos or renosterveld, KM, SE (Little Karoo to Uniondale).•

blackburniae W.F.Barker (= *Haworthia graminifolia* G.G.Sm.)   Acaulescent succulent forming clumps, roots fusiform, stem fibrous. Leaves slender, elongate, 5–30 cm long, canaliculate, grey-green, smooth or lightly toothed. Flowers white. Nov.–Dec. Steep, rocky, southern slopes in transitional karroid veld, KM (Swartberg and Little Karoo Mts).•

chloracantha Haw.   Acaulescent succulent, rosettes usually clustered. Leaves pale to deep green, erect, firm, 3–6 cm long, with short open spaced teeth. Flowers white. Sept.–Oct. Valley bushveld, rocky grassy patches, LB (Gourits valley to Great Brak River).•

cymbiformis (Haw.) Duval   Acaulescent succulent, forming clumps. Leaves soft, fleshy, spineless. Flowers white. Nov.–Dec. Cliffs and rock faces, SE (Baviaanskloof Mts to Port Elizabeth).•

emelyae Poelln. (= *Haworthia comptoniana* G.G.Sm.)   Acaulescent succulent t0 12 cm diam., rosettes withdrawn to ground level. Leaves few, abruptly recurved, opaque, dark green with pinkish flecks, scabrid above to spinescent. Flowers white. Aug.-Sept. Under low shrubs in karroid veld, KM, LB, SE (W Karoo to Little Karoo and Uniondale).

floribunda Poelln.   Acaulescent succulent with small erect rosettes to 3 cm diam. Leaves few, lanceolate, twisted, dark green, margins denticulate. Flowers white. Sept.–Oct. Grassy coastal fynbos, LB (Heidelberg to Gourits River).•

gracilis Poelln. (= *Haworthia translucens* (Haw.) Haw.)   Acaulescent succulent with small, proliferous rosettes. Leaves grey-green, translucent, toothed. Flowers white. Oct.–Nov. Karroid valley bushveld to fynbos, SE (Humansdorp to Grahamstown).

heidelbergensis G.G.Sm.   Acaulescent succulent with tiny rosettes. Leaves fleshy, tapering to slender points, brownish green, pellucid on upper surface, lightly serrate. Flowers white. Aug.–Sept. Rocky slopes under shrubs in renosterveld, SW, LB (Bredasdorp to Heidelberg).•

herbacea (Mill.) Stearn   Acaulescent succulent to 7 cm diam., rosettes usually solitary. Leaves firm, incurved, mottled yellow-green, with white teeth. Flowers beige, bud-tips curved upward. Sept.–Oct. Karroid broken veld, under shrubs, SW, LB (Worcester to McGregor).•

maculata (Poelln.) M.B.Bayer   Acaulescent succulent to 8 cm diam., rosettes solitary. Leaves purplish, speckled, lightly dentate. Flowers narrow, yellowish. Sept.–Oct. Dry rocky slopes in mountain fynbos, NW, SW (Hex River valley to Worcester).•

magnifica Poelln.   Acaulescent succulent to 8 cm diam., rosettes dark. Leaves dark green, opaque, flat above and rough, apices usually abruptly recurved and tapering to point, lightly tubercled. Flowers white, green on tepal keels. Apr.–May. Rock outcrops in renosterveld, KM, LB (Heidelberg to Gourits River valley).•

maraisii Poelln.   Acaulescent succulent to 5 cm diam., rosettes dark. Leaves dark green, opaque, flat above and rough, apices usually abruptly recurved and triangular, lightly dentate. Flowers white, green on tepal keels. Apr.–May. Rock outcrops in renosterveld, SW, KM, LB (Worcester to Heidelberg and Bredasdorp).•

mirabilis Haw.   Acaulescent succulent to 10 cm diam. Leaves translucent, brownish, triangular, abruptly recurved, with pellucid tips above, often lightly dentate and spotted beneath. Flowers white. Feb.–Mar. Rocky slopes in renosterveld and coastal fynbos, SW, AP (Caledon to Bredasdorp).•

monticola Fourcade (= *Haworthia divergens* M.B.Bayer)   Acaulescent succulent, forming clumps. Leaves small, slender, erect to incurved at tips, dark green to reddish, with white teeth. Flowers white. Aug.–Sept. Rocky southern slopes in fynbos, KM, SE (Outeniqua foothills, and Oudtshoorn to Uniondale).•

mucronata Haw. (= *Haworthia rycroftiana* M.B.Bayer)   Acaulescent succulent to 15 cm diam. Leaves pale to bright emerald-green, fleshy, smooth or spinescent, often solitary. Flowers white. Aug.–Sept. Karroid or marginal fynbos vegetation, KM, LB (Little Karoo, Anysberg to Uniondale).•

mutica Haw.   Acaulescent succulent, forming solitary rosettes. Leaves short, abruptly recurved, round at tips, purple-brown, glaucous. Flowers white. Aug.–Sept. Shale rocks in coastal renosterveld, SW, LB (Bredasdorp to Heidelberg).•

nortieri G.G.Sm.   Acaulescent succulent with small rosettes to 5 cm diam. Leaves light green, mottled. Flowers greyish. Sept.–Oct. Rock outcrops in dry mountain fynbos, NW (S Namaqualand to Ceres).

outeniquensis M.B.Bayer   Acaulescent succulent with dense, many-leaved rosettes to 4 cm diam. Leaves elongate, slender, margins whitish cartilaginous and densely spiny. Flowers white. Oct.–Nov. Rocky north slopes in fynbos, SE (Outeniqua Mts).•

parksiana Poelln.   Acaulescent succulent with small compact rosettes to 3 cm diam. Leaves abruptly recurved, dark green, rough. Flowers white. Oct.–Nov. Under shrubs, among lichens and leaf debris in renosterveld, LB (Great Brak River).•

pubescens M.B.Bayer   Acaulescent succulent with small solitary rosettes to 4 cm diam. Leaves dark green, finely pubescent, erect and incurved. Flowers white. Sept.–Oct. Quartzite rocks in dry mountain fynbos, NW (Hex River valley).•

pulchella M.B.Bayer   Acaulescent succulent, usually with solitary, compact rosettes. Leaves bright green, reticulate, sharply spiny, erect. Flowers white. Oct.–Nov. Rock cracks and crevices in dry mountain fynbos, KM (W Little Karoo).•

pygmaea Poelln.   Acaulescent succulent with solitary rosettes to 10 cm diam. Leaves grey-green, abruptly recurved, often scabrid to papillate, flat above, obtuse. Flowers white. Sept.–Oct. Sparse grassland in coastal renosterveld, LB (Mossel Bay).•

reticulata Haw.   Acaulescent succulent, forming clumps. Leaves usually glabrous, turgid below, yellowish green with pink flush, mottled reticulate. Flowers pinkish, bud tips upturned. Aug.–Sept. Karroid broken veld, rocky slopes, NW, SW (Worcester to Robertson karoo).•

retusa (L.) Duval   Acaulescent succulent, usually with solitary rosettes to 15 cm diam. Leaves yellowish green to brown, somewhat translucent, smooth, abruptly recurved, flat above. Flowers white. Aug.–Sept. Sparse grassland and rock outcrops in coastal renosterveld, LB (Swellendam to Mossel Bay).•

serrata M.B.Bayer   Acaulescent succulent with solitary rosettes. Leaves narrowly lanceolate, erect, incurved, white-toothed. Flowers white. Sept.–Oct. Sparsely grassed rocky slopes in coastal renosterveld, LB (Heidelberg).•

truncata Schönland (= *Haworthia maugahnii* Poelln.)   PERDETANDE   Acaulescent succulent. Leaves distichous or multifarious, withdrawn to ground-level, grey-green, truncate, windowed, spineless, rarely piliferous. Flowers white. Dec.–Jan. Rocky sites in karroid veld, KM, LB (Calitzdorp to Oudtshoorn).•

turgida Haw.   Acaulescent succulent with proliferous rosettes borne at ground level. Leaves turgid, pale yellowish green, mottled translucent, spreading and flat above, smooth. Flowers white. Aug.–Sept. Rock crevices in sandstone or shale in fynbos, renosterveld, or karroid veld, SW, LB, AP (Bredasdorp to Mossel Bay).•

variegata L.Bolus   Acaulescent succulent, forming dense clumps. Leaves dark green and variegated, fairly slender, toothed. Flowers white. Sept.–Oct. Limestone slopes in coastal fynbos, SW, AP (Bredasdorp to Riversdale).•

vlokii M.B.Bayer   Acaulescent succulent with small dense rosettes. Leaves opaque, with small rounded white dots, margins spinescent. Flowers white. Sept.-Oct. Rock cracks in fynbos, KM (De Rust).•

wittebergensis W.F.Barker   Acaulescent succulent. Leaves erect, subulate, dark green to purple with small white spines. Flowers white. Nov.–Dec. Rock crevices in light shade in dry fynbos, KM (Witteberg and W Karoo).

zantneriana Poelln.   Acaulescent succulent with soft green rosettes. Leaves with pale mottled markings. Flowers white. Oct.–Nov. Rocky slopes in dry mountain fynbos, KM, SE (Little Karoo to Baviaanskloof).•

KNIPHOFIA   RED HOT POKER, VUURPYL   65 spp., sub-Saharan Africa, S Arabia

citrina Baker   Rhizomatous perennial, 40–65 cm, forming clumps. Leaves strap-shaped, coarsely fibrotic. Flowers in globose racemes, yellow. Mar.–May. Grassland, SE (Humansdorp to Grahamstown).

praecox Baker   Rhizomatous perennial, 1.5–2 m, forming clumps. Leaves strap-shaped, margins finely serrate. Flowers in cylindrical racemes, reddish, opening yellow to yellow-green. Nov.–Jan. Streambanks and wet hollows, SE (George to Komga).

sarmentosa (Andrews) Kunth   Rhizomatous perennial to 60 cm. Leaves greyish, strap-shaped. Flowers in ovoid to cylindrical racemes, reddish, opening buff. June–Oct. Mountain streams and moist hollows, NW, KM (Hex River Mts to W Karoo).

tabularis Marloth   Rhizomatous perennial, 60–120 cm. Leaves strap-shaped, somewhat fleshy. Flowers in laxly cylindrical racemes, red to orange, blackish at tips. Dec.–Jan. Wet sandstone cliffs, NW, SW (Tulbagh to Hottentots Holland Mts).•

uvaria (L.) Oken   Rhizomatous perennial, 50–120 cm, in small clumps. Leaves strap-shaped, fibrotic. Flowers in oblong to globose racemes, orange to greenish yellow. Mostly Oct.–Dec. Seeps, marshes and streams on sandstone slopes, NW, SW, KM, LB, SE (Namaqualand to Barkly East).

TRACHYANDRA   CAPE SPINACH, WILDEBLOMKOOL   52 spp., southern and tropical E Africa and Madagascar, mainly W Cape

A. Stem base without membranous sheaths

adamsonii (Compton) Oberm.   Sparsely branched, subwoody shrub to 1 m, roots many, slender. Leaves lanceolate, succulent. Flowers in an unbranched raceme, white. Aug.–Sept. Quartzite slopes in karroid scrub, NW (Olifants River valley and Richtersveld).

affinis Kunth   Rhizomatous perennial to 80 cm, roots many, wiry. Leaves lanceolate to linear, usually pubescent. Flowers in a sparsely branched raceme with pubescent peduncle, often developing axillary plantlets, white. Mostly Aug.–Dec. Grassy coastal flats and lower slopes, AP, KM, SE (Stilbaai to S KwaZulu-Natal).

brachypoda (Baker) Oberm.   Rhizomatous perennial to 60 cm, often on cliffs, roots many, wiry. Leaves linear to filiform. Flowers in a slender, sparsely branched raceme, white. Fruits 5 mm long, on short pedicels. Nov.–Apr. Seasonally marshy sandy flats and lower slopes, SW (Mamre to Cape Peninsula).•

esterhuysenae Oberm.   Rhizomatous perennial to 50 cm, roots slender. Leaves 1 or 2, stiff, subtcrete. Flowers in a slender, congested raceme, white. Fruits on short pedicels, 1-seeded. Dec.–Feb. Sandstone seeps, 600–1000 m, NW, SW (Porterville Mts to Kogelberg and Riviersonderend Mts).•

gerrardii (Baker) Oberm.   Rhizomatous perennial to 60 cm, roots many, wiry but thickened near tips. Leaves many, linear, triquetrous, soft, roughly glandular-hairy. Flowers in a much-branched, roughly glandular-hairy raceme, white. Fruits on long, spreading pedicels, globose, 10 mm long, densely covered with glandular protuberances. Oct.–Feb. Grassy slopes, SE (Port Elizabeth to Mpumalanga).

gracilenta Oberm.   Rhizomatous perennial to 40 cm, roots many, slender. Leaves tufted, filiform, velvety. Flowers in an unbranched raceme, white. Sept. Shale slopes in karroid scrub, NW (Gifberg to Cedarberg Mts).•

hirsuta (Thunb.) Kunth   Rhizomatous perennial to 60 cm, roots wiry. Leaves in a fan, lanceolate, soft, usually velvety. Fowers in a sparsely branched raceme, peduncle usually pubescent, white. Sept.–Dec. Mostly shale slopes and flats in renosterveld, NW, SW, AP (Piketberg to Agulhas).•

sabulosa (Adamson) Oberm.   Rhizomatous perennial to 60 cm, roots many, wiry. Leaves many, linear, triquetrous, soft and usually roughly hairy. Flowers in a roughly glandular-hairy, trailing, unbranched raceme, white. Fruits obscurely stalked, 7–9 mm long, densely covered with branched protuberances, on long, spreading pedicels. Sept.–Oct. Coastal sand flats, SW, AP (Hopefield to Cape Agulhas).•

scabra (L.f.) Kunth   Like **T. sabulosa** but fruits conspicuously stalked, 5 mm long, sparsely covered with simple protuberences. Sept.–Dec. Coastal sand flats, NW, SW (Klawer to Cape Peninsula).•

tabularis (Baker) Oberm.   Rhizomatous perennial to 1.2 m, roots wiry. Leaves lanceolate to linear, fibrotic. Flowers congested in a sparsely branched, ascending raceme, white. Fruits 9 mm long. Sept.–Feb. Sandy flats and lower slopes, strandveld and lowland fynbos, SW (Yzerfontein to Kleinrivier Mts).•

A. *Stem base with membranous sheaths*
B. *Raceme simple or few-branched*

chlamydophylla (Baker) Oberm.   Rhizomatous perennial to 70 cm, often in clumps, roots many, thick. Leaves clumped, linear, erect. Flowers congested in an unbranched raceme, white. Fruits on very long, deflexed pedicels. Aug.–Oct. Clay flats and slopes in renosterveld, SW (Darling to Somerset West and Worcester).•

ciliata (L.f.) Kunth   Rhizomatous perennial to 50 cm, roots many, rather fleshy and swollen. Leaves straggling, channelled, soft and spongy, usually hairy. Flowers in a straggling, elongate, usually hairy raceme with conspicuous bracts, white. Fruits on long, deflexed pedicels. June–Sept. Damp sandy coastal flats, NW, SW, LB, SE (Namibia to Grahamstown).

falcata (L.f.) Kunth   Rhizomatous perennial to 1 m, roots many, thickened. Leaves few, lanceolate and falcate, leathery. Flowers in a stout, unbranched or sparsely branched raceme, imbricate, white, bracts conspicuous, usually hairy. July–Oct. Sandy or clay flats and slopes, karroid scrub, NW, SW, KM (Namibia to Worcester and W Karoo).

filiformis (Aiton) Oberm.   Rhizomatous perennial to 60 cm, slender, roots swollen. Leaves few, linear, often mottled near base. Flowers in a lax, usually unbranched raceme, white. Fruits on very long, spreading pedicels. Aug.–Oct. Damp sandy flats near sea level, NW, SW, AP (Elandsbaai to Bredasdorp).•

hirsutiflora (Adamson) Oberm.   Rhizomatous perennial to 60 cm, roots many, more or less fleshy and swollen. Leaves linear, tough, scabrid. Flowers congested in a hairy, subcorymbose, unbranched raceme on long hairy pedicels, white to grey. Fruits densely hairy. Mostly Sept.–Oct. Sandy flats and lower slopes, NW, SW (Piketberg to Caledon).•

hispida (L.) Kunth   Rhizomatous perennial to 30 cm, roots few, fleshy and swollen. Leaves few, lanceolate, pubescent. Flowers in a subcorymbose, unbranched raceme, white, peduncle pubescent. Fruits on very long, spreading, pubescent pedicels. June–Sept. Shale slopes in renosterveld, NW, SW, AP, LB (Piketberg to Albertinia).•

B. *Raceme divaricately much-branched*

divaricata (Jacq.) Kunth   Similar to **T. revoluta** but stouter and glabrous, leaves somewhat fleshy and fruits larger, c. 1 cm long. July–Sept. Littoral dunes and sand flats, NW, SW, AP, SE (Namaqualand to Port Alfred).

flexifolia (L.f.) Kunth (incl. **T. oligotricha** (Baker) Oberm.)   Rhizomatous perennial to 30 cm, roots few, thickened. Leaves few to many, linear, often undulate, usually hairy. Flowers in a branched raceme, nodding, white, base of peduncle pubescent. Fruits on short, erect pedicels. May–Sept. Sandy and shale flats and slopes, NW, SW (Namaqualand to Bredasdorp).

jacquiniana (Roem. & Schult.) Oberm.   ANYSBLOM   Rhizomatous perennial to 50 cm, roots few, bulbous. Leaves many, linear, undulate when young, often softly hairy. Flowers in a panicle with many ascending branches, nodding, white. Fruits narrowly ovoid. July–Sept. Stony clay slopes in karroid scrub, NW, KM (Namaqualand to Montagu).

muricata (L.f.) Kunth   Like to **T. revoluta** but leaves few to several, lanceolate and not sheathed individually by cataphylls. July–Oct. Stony clay slopes in karroid scrub and renosterveld, NW, SW (S Namibia to Caledon).

paniculata Oberm.   Rhizomatous perennial to 30 cm, roots few, bulbous. Leaves few, lanceolate, usually hairy. Flowers in a panicle, nodding, white. Fruits globose. Sept.–Oct. Stony clay slopes, karroid scrub, NW (Richtersveld to Olifants River valley).

patens Oberm.   Rhizomatous perennial to 30 cm, roots few, thickened. Leaves few to many, filiform, sometimes hairy. Flowers in a branched raceme, white, base of peduncle hairy. Fruits on long, spreading pedicels. Aug.–Oct. Stony clay slopes, karroid scrub, NW (Springbok to Worcester).

revoluta (L.) Kunth   Rhizomatous perennial to 90 cm, roots many, more or less swollen. Leaves many, linear, scabrid, sheathed individually by cataphylls. Flowers in a divaricate panicle, nodding, white, tepals recurved, base of peduncle scabrid. Aug.–Nov. Sandy flats, NW, SW, AP, SE (Richtersveld to Port Alfred).

tortilis (Baker) Oberm.   Rhizomatous perennial to 20 cm, roots few, bulbous. Leaves few, lanceolate or linear-lanceolate, undulate. Flowers in a compact, branched raceme, nodding, white. Fruits on spreading pedicels. July–Sept. Sandy and clay flats, karroid scrub, NW (Springbok to Clanwilliam).

BEHNIACEAE

BEHNIA   AFRICAN SOLOMON'S SEAL   1 sp., eastern southern Africa

reticulata (Thunb.) Didr.   Climbing perennial. Leaves ovate, glossy, with reticulate venation. Flowers in axillary cymes, nodding, urn-shaped, cream to green. Berries cream or greenish. Mainly Sept.–Dec. Forests and scrub, SE (Knysna to Zimbabwe).

COLCHICACEAE

1. Flowers all bracteate:
    2. Flowers few, erect on short pedicels, usually in a head overtopped by green or petaloid bracts; styles more or less erect ....**Androcymbium**
    2.' Flowers several, racemose, nodding on long pedicels; styles slender and spreading...............**Ornithoglossum**
1.' Flowers all or only the upper ebracteate:
    3. Flowers pedicellate, the lower subtended by slender bracts, orange or yellow ....**Baeometra**
    3.' Flowers sessile, all ebracteate, white, cream-coloured or pink to maroon:
        4. Tepals fused at base, not conspicuously auriculate ....**Wurmbea**
        4.' Tepals free, conspicuously auriculate above claw:
            5. Styles short and hooked, arising laterally on truncate ovary lobes; tepals attenuate, greenish ...**Neodregea**
            5.' Styles slender, arising terminally on ovary; tepals obovate, white or pink..............................**Onixotis**

ANDROCYMBIUM   CUP AND SAUCER, PATRYSBLOM   c. 40 spp., mainly southern Africa, and tropical Africa to Eurasia

A. Leaves grading into the bracts; bracts green, leaf-like

cuspidatum Baker   Acaulescent cormous geophyte. Leaves tristichous, lanceolate, prostrate. Flowers 1 or 2, surrounded by inconspicuous green, leaf-like bracts, exposed, tepals auriculate below. July–Aug. Stony flats, NW, KM, SE (W Karoo and Cedarberg Mts to Uniondale).

dregei C.Presl   Short-stemmed, cormous geophyte. Leaves sublinear. Flowers 1 or 2, often with aerial pedicels, enclosed by green, leaf-like linear bracts, tepals plane. June–Aug. Sheltered rocky slopes, NW, KM (Namaqualand to Montagu).

eucomoides (Jacq.) Willd. (incl. **A. austrocapense** U. & D.Müll.-Doblies, **A. eghimocymbion** U. & D.Müll.-Doblies, **A. undulatum** U. & D.Müll.-Doblies)   Acaulescent cormous geophyte. Leaves lanceolate-attenuate, prostrate. Flowers enclosed in greenish leaf-like bracts, tepals auriculate below. July–Aug. Clay flats and slopes, NW, SW, AP, KM, LB, SE (Namaqualand to E Cape).

hughocymbion U. & D.Müll.-Doblies   Plants acaulescent. Leaves tristichous, prostrate, lanceolate-attenuate, aristate. Flowers surrounded by inconspicuous green, leaf-like bracts, exposed, tepals auriculate below. June–July. Stony flats and slopes, NW, SW (Worcester to Potberg).•

longipes Baker   Acaulescent, cormous geophyte. Leaves spreading, narrowly lanceolate, channelled. Flowers enclosed in leaf-like, green bracts, tepals with very long claws, auriculate below. Apr.–Jan. Moist slopes and stony grasslands, KM, SE (Ladismith to E Cape).

A. Leaves abruptly differentiated from the bracts; bracts often coloured

burchellii Baker   Acaulescent, cormous geophyte. Leaves ovate, prostrate, sometimes ciliate. Flowers enclosed in green and white bracts, tepals with very short limbs, auriculate below, stamens twice as long as tepal limbs. June–Aug. Stony flats, NW, KM (W Karoo and Bokkeveld Mts to Little Karoo).

capense (L.) Krause (= *Androcymbium fenestratum* Schltr. & Krause; incl. **A. ciliolatum** Schltr. & Krause, **A. hantamense** Engler, **A. irroratum** Schltr. & Krause)   Acaulescent, cormous geophyte. Leaves lanceolate, margins ciliate, sometimes undulate. Flowers enclosed in large white bracts, sometimes striped green, tepals auriculate below. June–Aug. Clay or loam flats, NW, SW, LB (Namaqualand and W Karoo to Swellendam).

latifolium Schinz (= *Androcymbium pulchrum* Schltr. & Krause)   ROOIPATRYSBLOM   Acaulescent, cormous geophyte. Leaves ovate, prostrate, ciliate. Flowers enclosed in large, red to purple bracts, tepals with very short limbs, auriculate below, stamens twice as long as tepal limbs. July–Aug. NW (W Karoo and Bokkeveld Plateau).

orienticapense U. & D.Müll.-Doblies (= *Androcymbium melanthioides* auct. in part)   BOBBEJAANSKOEN   Short-stemmed, cormous geophyte. Leaves linear, channelled. Flowers on an aerial stem, enclosed by large white bracts veined with green, tepals auriculate below. Mainly June–Aug. Grassland, SE (Knysna to E Cape).

volutare Burch.   Acaulescent, cormous geophyte. Leaves linear, coiled at tips. Flowers concealed in large, pale green bracts, tepals auriculate below. Aug.–Sept. Clay soils, KM (W Karoo to Little Karoo).

[**Species excluded**   No authentic material found and possibly conspecific with one of the above: **A. worsonense** U. & D.Müll.-Doblies]

BAEOMETRA•   BEETLE LILY   1 sp., W Cape

uniflora (Jacq.) G.J.Lewis   Cormous geophyte to 25 cm. Leaves lanceolate and channelled, spirally arranged. Flowers usually pedicellate, orange, styles free, hooked. Aug.–Oct. Mainly rocky sandstone and granite slopes, SW, AP, LB (Malmesbury to Riversdale).•

NEODREGEA   MOSQUITO LILY   1 sp., W and E Cape

glassii C.H.Wright   Cormous geophyte to 4 cm. Leaves 3, lanceolate, the uppermost small and subtending the lowest flower. Flowers minute, sessile in short spikes, tepals attenuate, yellow, styles free, hooked. May–June. Mainly clay slopes in renosterveld, SW, LB, SE (Somerset West to E Cape).

ONIXOTIS   WATER PHLOX   2 spp., W Cape and Namaqualand

punctata (L.) Mabberley   HANEKAMMETJIE   Cormous geophyte, 10–20 cm. Leaves lanceolate, margins ciliate, uppermost about halfway up the stem. Flowers sessile in short spikes, white to pink, styles free, filiform. July–Sept. Rocky and clay slopes, NW, SW, AP, KM, LB (Bokkeveld Mts to Swellendam and Agulhas).•

stricta (Burm.f.) Wijnands (= *Onixotis triquetra* (L.f.) Mabberley)   RYSBLOMMETJIE   Cormous geophyte, 20–50 cm. Leaves subterete, triangular in section, upper 2 leaves set just below spike. Flowers sessile in elongate spikes, pink, styles shortly united, filiform. Aug.–Oct. Marshes and pools, NW, SW (Namaqualand to Cape Peninsula and Worcester).

ORNITHOGLOSSUM   SNAKE LILY   8 spp., W Cape to S tropical Africa

gracile B.Nord.   Cormous geophyte, 2–10 cm. Leaves short, lanceolate, undulate to crisped. Flowers nodding, actinomorphic, dull greenish, filaments often with a hump-like swelling in the middle, 10–12 mm long. Apr.–May. Rocky sandstone slopes, NW (Vanrhyn's Pass to Botterkloof).•

parviflorum B.Nord.   SLANGKOP   Cormous geophyte, 6–30 cm. Leaves linear-lanceolate. Flowers nodding, actinomorphic, green with maroon margins or red to brown, filaments thickened below, 2–5 mm long. June–Oct. Stony slopes, NW, KM (Namaqualand to Worcester, W Karoo).

undulatum Sweet   Cormous geophyte, 5–20 cm. Leaves lanceolate, margins sometimes crisped. Flowers nodding, zygomorphic-asymmetric, white to pink with purple or maroon tips, filaments filiform, 15–25 mm long. Apr.–July. Rocky sandstone slopes, NW, KM, LB, SE (S Namibia to Somerset East).

viride (L.f.) Aiton   EENDJIES, SLANGKOP   Like **O. parviflorum** but nectary a small round pocket or mouth-like flap much narrower than tepal claw. July–Oct. Mostly deep sandy soils, NW, SW, AP, LB (Clanwilliam to Riversdale).•

vulgare B.Nord.   Like **O. parviflorum** but flowers larger, filaments 5–13 mm long. Aug.–Oct. Stony slopes, KM (Little Karoo to tropical Africa).

WURMBEA   SPIKE LILY   37 spp., Africa, Australia

A. Perianth tube longer than the tepals

capensis Thunb.   Cormous geophyte, 5–10 cm. Leaves linear. Flowers few, cream and brown, tepals erect, tube slightly longer than the tepals, styles hooked. Aug.–Sept. Stony slopes, NW, SW (Clanwilliam to Stellenbosch).•

dolichantha B.Nord.   Cormous geophyte, 10–30 cm. Leaves lanceolate, channelled. Flowers white or cream with purple marks, fragrant, tube much longer than the tepals. Sept.–Oct. Mostly sand or clay, NW (Bokkeveld Escarpment to Piketberg).•

inusta (Baker) B.Nord.   Cormous geophyte, 5–20 cm. Leaves linear. Flowers greenish or cream with purple margins and median spot, tube longer than the tepals, fragrant. Sept.–Nov. Damp flats, NW, SW (Tulbagh to Bredasdorp).•

robusta B.Nord.   Cormous geophyte, 15–25 cm. Leaves lanceolate. Flowers white, tepals with purple margins, tube slightly longer than the tepals, filaments about as long as the tepals. July–Sept. Clay and granite slopes in renosterveld, NW, SW (Moorreesburg to Malmesbury).•

A. Perianth tube as long as or shorter than the tepals

compacta B.Nord.   Cormous geophyte, 5–18 cm. Leaves lanceolate, falcate. Flowers in dense spikes, pink, tube as long as the tepals, filaments as long as the tepals. June–July. KM (Montagu).•

elongata B.Nord.   Cormous geophyte, 7–20 cm. Leaves linear. Flowers greenish white to cream with dark margins, lightly fragrant, tube shorter than the tepals. Sept.–Oct. Rocky sandstone slopes, NW (Cedarberg Mts to Piketberg).•

hiemalis B.Nord.   Cormous geophyte, 4–15 cm. Leaves linear. Flowers white with dark margins, tube about as long as the tepals. May–Aug. Damp sandy slopes, SW (Cape Peninsula).•

marginata (Desr.) B.Nord.   SWARTKOPPIE   Cormous geophyte, 6–22 cm. Leaves lanceolate. Flowers rotate, red or purple with darker margins, foul-scented, tube shorter than the tepals, filaments very short. Sept.–Oct. Mostly clay or loam, SW, AP, LB (Hopefield to Albertinia).•

minima B.Nord.   Cormous geophyte, 2–5 cm. Leaves lanceolate. Flowers white, rotate, tube vestigial. Oct. Rocky slopes, NW (Cedarberg to Porterville Mts).•

monopetala (L.f.) B.Nord.   Cormous geophyte, 5–25 cm. Leaves narrowly lanceolate, channelled. Flowers greenish or cream with dark margins in upper two-thirds, tube slightly shorter than the tepals, filaments very short. Aug.–Nov. Sandstone and granite slopes, NW, SW (Pikeberg to Caledon).•

recurva B.Nord.   Cormous geophyte, 5–20 cm. Leaves narrowly lanceolate. Flowers red to purplish brown, tepals recurved, tube shorter than the tepals, filaments very short. Sept.–Oct. NW, SW (Tulbagh to Somerset West).•

spicata (Burm.f.) T.Durand & Schinz (= *Wurmbea ustulata* B.Nord.)   WITKOPPIE   Cormous geophyte, 5–20 cm. Leaves narrowly lanceolate. Flowers white to cream, sometimes with dark margins, tube shorter than the tepals. Aug.–Nov. Mostly clay and granite slopes in renosterveld, NW, SW, KM, LB (Bokkeveld Mts to Swellendam).•

variabilis B.Nord.   Cormous geophyte, 5–20 cm. Leaves ovate. Flowers cream, tepals with median spot, foul-scented, tube shorter than the tepals. Aug.–Oct. NW, SW, KM, LB, SE (Clanwilliam to Port Elizabeth, W Karoo).

COMMELINACEAE

1. Flowers actinomorphic; inflorescences not enclosed in spathes; fertile stamens 6; filaments densely bearded ....**Cyanotis**
1.' Flowers zygomorphic; inflorescences consisting of 1 or 2 cymes enclosed in folded or obliquely funnel-shaped spathes; fertile stamens (2)3; filaments smooth...**Commelina**

COMMELINA   WANDERING JEW   c. 230 spp., cosmopolitan

africana L.   Scrambling perennial to 50 cm from a woody rootstock. Leaves oblong to linear, sometimes hairy. Flowers yellow, spathes folded, dry inside. Nov.–June. Near streams or lower slopes, SW, AP, LB, SE (Cape Peninsula to Arabia and Madagascar).

*****benghalensis** L.   BLOUSELBLOMMETJIE   Spreading, hairy annual to 30 cm. Leaves ovate. Flowers blue, spathes obliquely fused, mucilaginous inside. Nov.–Apr. Shady and damp sites, SE (cosmopolitan weed).

eckloniana Kunth   Sparsely white-hairy perennial with annual stems to 35 cm from a knobbly rootstock. Leaves linear, hairy. Flowers blue, spathes obliquely fused, mucilaginous inside. Nov.–Feb. SE (Humansdorp to Ethiopia).

CYANOTIS   DOLLS' POWDER PUFF   c. 50 spp., palaeotropics

speciosa (L.f.) Hassk.   BLOUPOEIERKWASSIE   Tufted, spreading, hairy perennial, 15–50 cm, rhizome geniculate. Leaves lanceolate, hairy beneath. Flowers blue to mauve. Nov.–May. Grassland, LB, SE (Riversdale to S Tanzania, Madagascar).

CONVALLARIACEAE (= DRACAENACEAE, ERIOSPERMACEAE)

1. Rootstock a tuber; flowers shorter than 10 mm; fruit a capsule with woolly seeds; plants usually leafless at flowering ....**Eriospermum**
1.' Rootstock a rhizome; flowers longer than 20 mm; fruit a fleshy berry; plants leafy at flowering:
   2. Trees or shrubs with cauline leaves; inflorescence paniculate ....**Dracaena**
   2.' Rhizomatous herbs with basal leaves; inflorescence racemose ....**Sansevieria**

DRACAENA   PALM LILY   c. 60 spp., pantropical, mostly tropical Africa

aletriformis (Haw.) Bos (= *Dracaena hookeriana* K.Koch)   Single-stemmed or sparsely branched tree to 5 m. Leaves crowded apically, lanceolate, with white cartilaginous margins. Flowers tufted, in dense panicles, greenish white. Jan.–Feb. Forest and coastal bush, SE (Van Staden's Mts to Kenya).

ERIOSPERMUM   COTTONSEED   c. 102 spp., sub-Saharan Africa, especially W Cape to Namaqualand

A. Tepals equal or subequal
B. Filaments oblong-ovate, erect around ovary

bayeri P.L.Perry   Tuberous geophyte to 45 cm. Leaf erect, sword-shaped, margins wavy. Flowers star-shaped, nearly sessile, pale greenish, filaments erect, oblong. Mar.–May. Shale slopes in renosterveld, NW, SW (W Karoo to Robertson Karoo).

bifidum R.A.Dyer   Tuberous geophyte to 30 cm. Leaf erect, lanceolate to heart-shaped, leathery. Flowers star-shaped, yellowish, on long pedicels, filaments erect, oblong-bifid. Jan.–Apr. Shaly flats, KM, SE (Namaqualand and W Karoo to Grahamstown).

brevipes Baker   Tuberous geophyte to 30 cm. Leaf spreading, heart-shaped, sparsely hairy. Flowers cup-shaped, white, fragrant, filaments erect, oblong. Jan.–Mar. Sandstone slopes, grassland or fynbos, SE (Plettenberg Bay to Transkei).

breviscapum Marloth ex P.L.Perry   Tuberous geophyte to 12 cm. Leaf prostrate, heart-shaped, fleshy. Flowers crowded, star-shaped, white, fragrant, filaments erect, oblong. Feb.–Mar. Shale slopes in renosterveld and succulent karoo, SW, AP (Robertson to Stilbaai).•

cernuum Baker   Tuberous geophyte to 35 cm. Leaf erect, lanceolate to heart-shaped, margins sometimes red. Flowers cup-shaped, white, filaments erect, oblong. Feb.–Apr. Damp sites on sandstone soils, NW, SW, KM (Clanwilliam to Bredasdorp).•

crispum P.L.Perry   Tuberous geophyte to 35 cm. Leaf erect, sword-shaped, leathery, margins crisped. Flowers star-shaped, white, filaments erect, oblong. Mar.–Apr. Habitat ?, KM (Calitzdorp).•

porphyrium Archibald   Tuberous geophyte to 70 cm. Leaf prostrate, heart-shaped. Flowers star-shaped, cream to greenish, fragrant. Dec.–May. Clay soils in grassland, SE (Kouga Mts to N Province).

rhizomatum P.L.Perry   Tuberous geophyte to 50 cm, spreading by rhizomes to form tufts. Leaf spreading, heart-shaped, leathery. Flowers cup-shaped, nearly sessile, tepals suberect, white, filaments erect, oblong. Feb.–Mar. Sandstone rocks in shade, KM (Calitzdorp to Oudtshoorn).•

zeyheri R.A.Dyer   Tuberous geophyte to 50 cm. Leaf prostrate, heart-shaped. Flowers star-shaped, cream to greenish, fragrant, filaments erect, oblong. Dec.–Mar. Clay soils in renosterveld, SW, KM, LB, SE (McGregor to Grahamstown).

B. Filaments subulate, spreading

aequilibre Poelln.   Tuberous geophyte to 45 cm. Leaf erect, sword-shaped, rugose with prominent veins. Flowers star-shaped, nearly sessile, light green. Mar.–Apr. Rocky succulent karoo, KM (De Rust to Kammanassie Mts).•

aphyllum Marloth   Tuberous geophyte to 8 cm. Leaf erect, filiform. Flowers star-shaped, whitish to pink, on long persistent pedicels. Mar.–Apr. Hard stony clay, NW (Namaqualand and W Karoo to Nardouw Mts).

arenosum P.L.Perry   Tuberous geophyte to 20 cm. Leaf erect, heart-shaped. Flowers star-shaped, white, on long pedicels. Mar.–Apr. Coastal sands, NW (Namaqualand to Aurora).

bruynsii P.L.Perry   Tuberous geophyte to 25 cm. Leaf erect, sword-shaped, petiole hairy. Flowers star-shaped, nearly sessile, pale green. Mar.–Apr. Habitat? KM (Calitzdorp).•

ciliatum P.L.Perry   Tuberous geophyte to 24 cm. Leaf prostrate, ovate to heart-shaped with ciliate margins. Flowers star-shaped, bright yellow. Feb.–Apr. Sandstone slopes, fynbos or grassland, SE (Humansdorp to Port Elizabeth).•

dielsianum Schltr. ex Poelln.   Tuberous geophyte to 25 cm. Leaf erect, lanceolate to heart-shaped, petiole and blade hairy. Flowers star-shaped, white. Jan.–Apr. Mostly sandstone soils, NW, SW, AP, KM, LB, SE (Cold Bokkeveld to Port Elizabeth and W Karoo).

flavum P.L.Perry   Tuberous geophyte to 6 cm. Leaf erect, terete, stem wiry and coiled. Flowers star-shaped, bright yellow, on long persistent pedicels. May. Sandstone outcrops NW (Nardouw Mts to Cold Bokkeveld).•

inconspicuum P.L.Perry   Tuberous geophyte to 8 cm. Leaf erect, elliptic to ovate, rugose with prominent veins. Flowers star-shaped, white. Apr. Sandstone soils in fynbos, SW, LB, SE (Caledon to Outeniqua Mts).•

paradoxum (Jacq.) Ker Gawl.   Tuberous geophyte to 10 cm. Leaf blade small, bearing a woolly, plumose appendage. Flowers crowded, star-shaped, white, fragrant, Apr.–May. Sandy and clay soils, NW, SW, AP, KM, LB, SE (Namaqualand to Grahamstown).

parvifolium Jacq.   Tuberous geophyte to 30 cm. Leaf erect, elliptic-ovate. Flowers star-shaped, white. Mar.–Apr. Stony clay soils, NW (Namaqualand to Bokkeveld Escarpment).

patentiflorum Schltr.   Tuberous geophyte to 40 cm. Leaf erect, ovate to sword-shaped, minutely white-pilose, petiole abruptly swollen, red, persisting as loosely sheathing collars. Flowers star-shaped, white, on long pedicels. Mar. Rocky slopes in arid fynbos, NW (S Namaqualand to Olifants River valley).

pumilum T.M.Salter   Tuberous geophyte c. 10 cm. Leaf erect, elliptic, petiole hairy, margins red. Flowers star-shaped, white. Mar.–Apr. Sandstone slopes, NW, SW (Namaqualand and W Karoo to False Bay).

schlechteri Baker   Tuberous geophyte to 23 cm. Leaf erect, elliptic, surface ribbed. Flowers on long pedicels, star-shaped, bright yellow. Mar.–Apr. Sandstone slopes in fynbos, SW (Kogelberg to Shaw's Mt).•

spirale C.H.Bergius ex Schult.   Tuberous geophyte to 6 cm, peduncle wiry and coiled. Leaf terete. Flowers star-shaped, white to yellow, on long persistent pedicels. Apr.–June. Sandstone flats and granite outcrops, NW, SW (Gifberg to False Bay).•

A. Tepals dimorphic
*C. Leaves with enations (see also **E. paradoxum**)*

alcicorne Baker   Tuberous geophyte to 8 cm. Leaf ovate, sometimes hairy, often with several enations. Flowers crowded, inner tepals erect, white. Jan.–Apr. Clay and sandstone soils, NW, KM (Namaqualand to Willowmore).

bowieanum Baker   Tuberous geophyte to 80 cm. Leaf blade reduced, bearing simple or branched, terete enations. Flowers crowded, subsessile, inner tepals erect, white. Feb.–Mar. Clay soils, NW (Worcester to Ashton).•

dregei Schönland   Tuberous geophyte to 14 cm. Leaf reduced, bearing simple or branched, terete, hairy enations. Flowers with inner tepals erect, white. Mar. Grassland, KM, SE (Montagu to Grahamstown).

erinum P.L.Perry   Tuberous geophyte to 20 cm. Leaf suberect, ovate-cordate, bearing short, cylindrical enations each with an apical tuft. Flowers on long pedicels, inner tepals erect, whitish. Feb.–Apr. Tillite flats in renosterveld, NW (Bokkeveld Mts).•

flabellatum P.L.Perry   Tuberous geophyte to 80 cm. Leaf blade reduced, bearing branched terete enations. Flowers crowded, inner tepals erect, white. Mar.–Apr. Shale slopes, KM (Montagu to Barrydale and W Karoo).

proliferum Baker   Tuberous geophyte to 30 cm, sometimes in clumps. Leaf small, with thread-like processes, petioles hairy. Flowers on long pedicels, inner tepals erect, white. Feb.–Mar. Clay and sand, NW, SW, AP, KM, SE (Namaqualand to Baviaanskloof Mts).

C. Leaves without enations

capense (L.) Thunb.   Tuberous geophyte to 50 cm, sometimes clumped. Leaf spreading, heart-shaped, often with red ridges, margins sometime ciliate. Flowers with inner tepals erect, on long pedicels, yellowish. Nov.–Mar. Mainly clay soils, NW, SW, AP, KM, LB, SE (Namaqualand to Grahamstown).

cordiforme T.M.Salter   Tuberous geophyte to 40 cm. Leaf spreading, heart-shaped, surface wrinkled. Flowers crowded, pedicels short, inner tepals erect, cream. Jan.–Feb. Sandstone and granite soils, SW, LB, SE (Darling to Alexandria).

dissitiflorum Schltr. Tuberous geophyte to 40 cm. Leaf erect, elliptic to lanceolate. Flowers with inner tepals erect, white. Jan.–Apr. Clay and sandstone soils, fynbos and renosterveld, LB, SE (Riversdale to Transkei).

exigium P.L.Perry Tuberous geophyte to 15 cm. Leaf erect, sword-shaped, petiole wiry. Flowers with inner tepals erect, white. Mar. Sandstone rocks, NW (Bokkeveld Mts to Gifberg).•

exile P.L.Perry Tuberous geophyte to 30 cm. Leaf erect, sword-shaped. Flowers nearly sessile, inner tepals erect with attenuate tips, white to pale yellow. Jan.–Mar. Quartzite and shale in shade, NW, KM (Worcester to De Rust).•

glaciale P.L.Perry Tuberous geophyte to 5 cm. Leaf erect, elliptic-lanceolate, white-woolly beneath. Flowers nearly sessile, few, inner tepals erect, white. Apr. Clay flats in renosterveld, NW (Bokkeveld Mts).•

graminifolium A.V.Duthie Tuberous geophyte to 30 cm. Leaf long and sword-shaped, margins and petiole hairy. Flowers on long pedicels, inner tepals erect, white. Feb.–Apr. Sand and clay, NW, SW, AP, KM, LB, SE (Bokkeveld Mts to George).•

lanceifolium Jacq. Tuberous geophyte to 40 cm. Leaf erect, sword-shaped, leathery, bluish, margins wavy, sometimes hairy. Flowers sometimes nearly sessile, inner tepals erect, white. Mar.–Apr. Sandstone or granite soil, NW, SW, KM, LB (Olifants River Mts to Albertinia).•

lanuginosum Jacq. Tuberous geophyte to 38 cm. Leaf spreading, heart-shaped, white-woolly. Flowers with erect inner tepals, cream. Feb.–Mar. Sandstone slopes in fynbos, NW, SW (Bokkeveld Mts to Gouda).•

laxiracemosum P.L.Perry Tuberous geophyte to 30 cm. Leaf erect, sword-shaped. Flowers on long pedicels, inner tepals erect, white. Feb.–Apr. Sandstone rocks, NW (Gifberg to Pakhuis Mts).•

marginatum Marloth ex P.L.Perry Tuberous geophyte to 25 cm. Leaves prostrate, heart-shaped, leathery, margins hairy. Flowers on long pedicels, inner tepals erect, white. Jan.–Mar. Stony slopes in karroid scrub, NW, KM (Namaqualand and W Karoo to Barrydale).

minutipustulatum P.L.Perry Tuberous geophyte to 12 cm. Leaf small, prostrate, ovate-cordate, sparsely pustulate-hairy. Flowers with erect inner tepals, white. March. Clay slopes, NW (Kobee Pass).•

nanum Marloth Tuberous geophyte to 30 cm. Leaf spreading, heart-shaped. Flowers on long pedicels, inner tepals erect, cream. Feb.–May. Sandstone soils, NW, SW, AP (Pakhuis Mts to De Hoop).•

orthophyllum P.L.Perry Tuberous geophyte to 40 cm. Leaf erect, lanceolate to elliptic, leathery, margins yellow or purple. Flowers on long pedicels, inner tepals erect, white. Jan.–Feb. Sandy soils, SE (Port Elizabeth to Transkei).

pubescens Jacq. Tuberous geophyte to 30 cm, sometimes in clumps. Leaf prostrate, heart-shaped, with adpressed straight hairs. Flowers on long pedicels, inner tepals erect, white. Feb.–Apr. Mainly clay soil in renosterveld, NW, SW, KM, LB, SE (Ceres to Somerset West to Knysna).•

pustulatum Marloth ex A.V.Duthie Tuberous geophyte to 40 cm. Leaf prostrate, heart-shaped, silvery white, hairy and pustulate. Flowers on long pedicels, inner tepals erect, white. Nov.–Dec. Clay in succulent karoo, NW (Nardouw Mts to Karoo).

subincanum P.L.Perry Tuberous geophyte to 25 cm. Leaf prostrate, heart-shaped, shiny green above, densely white-woolly beneath, margins undulate and red. Flowers on long pedicels, inner tepals erect, yellowish green. Feb.–Mar. Rocky outcrops, NW (Gifberg to Biedouw).•

subtile P.L.Perry Tuberous geophyte to 20 cm. Leaf erect, ovate. Flowers on long pedicels, inner tepals erect with attenuate tips, white. Mar.–Apr. Shale in renosterveld, NW, KM (Bokkeveld Mts to Koo).•

vermiforme Marloth ex P.L.Perry Tuberous geophyte to 15 cm, forming clumps from stolons. Leaf heart–shaped, margins ciliate. Flowers with inner tepals erect, white. Feb.–Mar. Sandstone flats, LB (Mossel Bay).•

villosum Baker Tuberous geophyte to 20 cm. Leaf erect, lanceolate, grey, densely hairy. Flowers on long pedicels, inner tepals erect, white. Dec.–Feb. Granite and shale, NW (Namaqualand to Piketberg).

SANSEVIERIA MOTHER IN LAW'S TONGUE c. 12 spp., dry palaeotropics and subtropics, mainly Africa

hyacinthoides (L.) Druce WILDEDATEL Acaulescent succulent to 50 cm, spreading from branched rhizomes. Leaves sword-shaped, leathery, irregularly banded with grey, margins red and white. Flowers in tufts on elongate racemes. Nov.–Mar. Dry bush and scrub, SE (Uitenhage to S tropical Africa).

CYPERACEAE
by C.A. Archer

1. Plants functionally dioecious .... **Scirpoides dioecus**
1.' Plants hermaphroditic to monoecious:
   2. Spikelets unisexual or bisexual; florets all unisexual or functionally unisexual; some empty glumes and/or spikelet bracts present:
      3. Spikelets solitary, pseudolateral; composed of numerous male florets and 1 terminal female floret .... **Chrysitrix**
      3.' Not as above:
         4. Female spikelet not enclosed by a perigynium (modified bract):
            5. Bisexual, male and female spikelets clustered together; nutlet borne on a stalk or gynophore .... **Scleria**
            5.' Male spikelets borne at upper nodes of inflorescence, female spikelets borne at lower nodes; nutlet not borne on a stalk or gynophore .... **Tetraria crinifolia**
         4.' Female spikelet partially or wholly enclosed by perigynium:
            6. Perigynium entire or split unevenly in one or two places; rachilla present within perigynium, usually bearing male florets, which are exserted .... **Schoenoxiphium**
            6.' Perigynium entire, bottle-shaped, frequently apex rostrate, 2-toothed; rudimentary rachilla occasionally present within perigynium .... **Carex**
   2.' Spikelets, at least aerial ones, bisexual; florets all bisexual, or bisexual and unisexual or functionally unisexual together; some empty glumes and/or spikelet bracts sometimes present:
      7. Glumes distichous:
         8. Spikelets with 0–2 empty glumes and/or bracts at base; nutlets not accompanied by bristles:
            9. Stigmas 3 (sometimes very short), or if 2, nutlet dorsiventrally compressed:
               10. Glumes of spikelet disarticulating in age sequence from base, rachilla persistent .... **Cyperus**
               10.' Entire spikelet with rachilla disarticulating from cushion above spikelet bract .... **Mariscus**
            9.' Stigmas 2; nutlet laterally compressed:
               11. Inflorescence digitate or anthelate to compound-anthelate .... **Pycreus**
               11.' Inflorescence capitate (occasionally with accessory capitula) .... **Kyllinga**
         8.' Spikelets with 3 or more empty glumes and/or bracts at base; nutlets sometimes accompanied by bristles:
            12. Bristles present:
               13. Spikelets of 2 kinds, subterranean female and aerial bisexual; bristles 3 .... **Trianoptiles**
               13.' Spikelets all aerial, bisexual; bristles 6 .... **Carpha**
            12.' Bristles absent, or if present, minute:
               14. Rachilla of spikelets between bisexual florets elongated, thickened and curved .... **Schoenus**
               14.' Rachilla of spikelet elongated and sometimes curved above uppermost bisexual floret.. **Epischoenus**
      7.' Glumes subdistichous to spirally arranged:
         15. Spikelets with 3 or more empty glumes and/or bracts at base:
            16. Stigmas 2 .... **Rhynchospora**
            16.' Stigmas 3 or more:
               17. Leaves mostly reduced to sheaths, uppermost leaf cylindrical and similar to culms .... **Neesenbeckia**
               17.' Leaves not as above:
                  18. Nutlet long-clawed .... **Capeobolus**
                  18.' Nutlet sessile or shortly clawed:
                     19. Plants large, robust; culm nodose, rounded and hollow; inflorescence a panicle .... **Cladium**
                     19.' Plants variable in height and robustness; culm variable, if nodose then triangular in cross section and solid; inflorescence variable .... **Tetraria**
         15.' Spikelets with 0–2 empty glumes and/or bracts at base:
            20. Stamens 6 .... **Cyathocoma**
            20.' Stamens 1–3:
               21. Style base enlarged:
                  22. Inflorescence a single spikelet (occasionally with accessory spikelets); inflorescence bracts scale-like; style base persisting .... **Eleocharis**
                  22.' Inflorescence seldom a single spikelet, if rarely so (in depauperate specimens), then inflorescence bracts not scale-like and style base not persisting:
                     23. Mouth of leaf sheath glabrous; style base not persisting on nutlet .... **Fimbristylis**
                     23.' Mouth of leaf sheath pilose; style base persisting as a button-like structure on nutlet.. **Bulbostylis**
               21.' Style base not enlarged:
                  24. Nutlet borne on gynophore:
                     25. Ligule usually well developed, papyraceous .... **Ficinia**
                     25.' Ligule not well developed .... **Scirpoides nodosus, Isolepis marginata**
                  24.' Nutlet not borne on gynophore:

26. Leaves cauline; culms usually markedly 3(–5)-angled:
   27. Inflorescence apparently umbellate or compound-umbellate, occasionally capitate or reduced to a single spikelet, but if so with at least 2 leaf-like involucral bracts ....**Bolboschoenus**
   27.' Inflorescence paniculate or occasionally capitate, but if so with only 1 involucral bract ....**Fuirena**
26.' Leaves ± basal; culms terete or elliptical in section:
   28. Leaf blades well developed ....**Isolepis**
   28.' Leaf blades reduced to sheaths (sometimes a very short blade developed):
      29. Inflorescence pseudolateral:
         30. Inflorescence either a lateral sessile cluster of relatively few spikelets, or if clusters stalked then clusters ± digitate..**Schoenoplectus**
         30.' Inflorescence either a dense ± spherical sessile cluster of spikelets, or if stalked then clusters ± spherical..**Scirpoides**
      29.' Inflorescence terminal:
         31. Inflorescence paniculate, of many spikelets............................**Pseudoschoenus**
         31.' Inflorescence capitate, of (1–)several spikelets:
            32. Dwarf annuals or slender perennials up to 200 mm tall (usually less); spikelets less than 3 mm diam. ....**Isolepis**
            32.' Densely tufted robust perennials 400–800 mm tall; spikelets c. 7 mm diam. ....**Hellmuthia**

BOLBOSCHOENUS   c. 16 spp., cosmopolitan

maritimus (L.) Palla (= *Scirpus maritimus* L.)   SNYGRAS, SNYRUIGTE   Robust or slender perennial to 1.2 m. Spikelets golden to dark brown. Oct.–Mar. Marshy flats near water, mainly coastal, below 700 m, NW, SW, AP, LB, SE (Clanwilliam to tropical Africa, pantropical).

BULBOSTYLIS   c. 100 spp., cosmopolitan

contexta (Nees) Bodard   Tufted perennial, 10–25 cm. Spikelets brown. July–Jan. Flats and lower sandstone slopes below 700 m, SE (Uitenhage to tropical Africa).
humilis (Kunth) C.B.Clarke (incl. **B. breviculmis** Kunth, **B. striatella** C.B.Clarke)   Densely tufted annual, 5–10 cm. Spikelets pale brown. July–Oct. Rock ledges inland and littoral sand below 500 m, NW, SW, SE (Ceres to Cape Peninsula to northern South Africa).

CAPEOBOLUS•   1 sp., southern Africa

brevicaulis (C.B.Clarke) J.Browning (= *Costularia brevicaulis* C.B. Clarke)   Densely tufted perennial, 8–20 cm. Spikelets hidden among leaves, greenish. Oct.–Mar. Rocky slopes in mountain fynbos below 1500 m, SW, LB, SE (Cape Peninsula to Tsitsikamma Mts).•

CAREX   c. 1 700 spp., cosmopolitan

***acutiformis** Ehrh.   Tufted perennial to 1 m. Spikelets blackish. Sept.–Feb. Marshes, LB, SE (widespread and sporadic in South Africa, indigenous to Europe).
aethiopica Schkuhr   Tufted perennial to 1.2 m. Spikelets greenish brown to reddish brown. Sept.–Dec. Shady areas near water in forest, SW, AP, LB, SE (Cape Peninsula to E Cape).
clavata Thunb.   Tufted perennial to 1.7 m. Spikelets shiny brown. Sept.–Nov. Marshy flats and lower slopes below 100 m, SW, AP, LB, SE (Malmesbury to E Cape).
cognata Kunth   Tufted perennial to 80 cm. Spikelets green. Nov.–Dec. Flats and slopes, SW, SE (sporadic from Cape Peninsula to northern South Africa, Namibia, Botswana).
ecklonii Nees   Tufted perennial to 70 cm. Spikelets greenish brown. Oct.–Apr. Coastal flats, SW, AP, SE (sporadic from Cape Peninsula to E Cape, worldwide).
vulpina L. (incl. **C. glomerata** Thunb.)   FOXTAIL SEDGE   Perennial to 70 cm. Spikelets brown. Aug.–May. Marshy flats generally at low altitude, SW, AP, LB, SE (Ceres and Cape Peninsula to northern South Africa).

CARPHA   13 spp., nearly cosmopolitan

capitellata (Nees) Boeck. (= *Asterochaete capitellata* Nees, *A. bracteosa* C.B.Clarke)   Perennial to 50 cm. Spikelets straw-coloured. Oct.–Dec. Marshy flats or lower slopes, NW, SW, ?AP, LB, SE (Ceres to Cape Peninsula to Mpumalanga).

glomerata (Thunb.) Nees   VLEIBIESIE, VLEIRIET   Robust tufted perennial to nearly 2 m. Spikelets golden-brown. Aug.–Jan. Marshy flats, lower slopes and watercourses, NW, SW, LB, SE (Citrusdal to Cape Peninsula to S KwaZulu-Natal).

schlechteri C.B.Clarke   Tufted perennial to 80 cm. Spikelets golden-brown. Dec.–Jan. Slopes near watercourses, NW (Skurweberg).•

CHRYSITRIX   4 spp., W Cape and Australia

capensis L.   KWASBIESIE   Sparsely tufted perennial with flattened stems, 16–40 cm. Spikelets rusty brown. Apr.–Nov. NW, SW, LB, SE (Cape Peninsula to Ceres and to Humansdorp).•

dodii C.B.Clarke   Densely tufted perennial, 30–40 cm. Spikelets brownish. Oct.–Feb. Mid to upper slopes, SW, SE (Cape Peninsula to Humansdorp).•

junciformis Nees   Perennial to 30 cm. Spikelets brownish. Apr.–Dec. Upper slopes, NW, SW, ?LB (Cedarberg Mts to Riversdale).•

CLADIUM   c. 60 spp., nearly cosmopolitan but especially Australia

mariscus (L.) Pohl   SAW GRASS   Stout perennial to nearly 3 m. Spikelets brown. Dec.–Mar. Marshy flats and watercourses, SW, AP, SE (sporadic throughout South Africa and nearly cosmopolitan).

CYATHOCOMA (= *MACROCHAETIUM*)   3 spp., W Cape to KwaZulu-Natal

ecklonii Nees (= *Macrochaetium ecklonii* (Nees) Levyns)   Robust perennial to 1.25 m. Spikelets pale yellow-brown. Nov.–Apr. Seeps on mountain slopes below 1000 m, NW, SW (Cedarberg Mts to Tulbagh).•

hexandra (Nees) J.Browning (= *Macrochaetium hexandrum*(Nees) Pfeiffer)   Robust perennial, 50–150 cm. Spikelets dark red. Aug.–Mar. Marshes and watercourses on mountain slopes below 800 m, SW, AP, LB, SE (Cape Peninsula to Humansdorp).•

CYPERUS   MATJIESGOED   c. 550 spp., cosmopolitan

albostriatus Schrad.   Tufted perennial, 20–50 cm. Spikelets pale brown. Jan.–Apr. Forest floor below 300 m, SE (Uitenhage to northern South Africa).

brevis Boeck.   Rhizomatous perennial to 45 cm. Spikelets brown and red. Oct.–May. Coastal dunes, SE (Riversdale to KwaZulu-Natal).

denudatus L.f.   Leafless perennial to 60 cm, rhizome woody. Spikelets greenish brown. June–Jan. Streambanks, SW, AP, SE (W Cape to tropical Africa).

dives Delile (= *Cyperus immensus* C.B.Clarke)   Robust tufted perennial to 1.2 m. Glumes golden. Nov.–Mar. Marshy coastal flats below 50 m, SE (Port Elizabeth to tropical Africa).

*****esculentus** L.   Tuberous perennial. Glumes yellow-brown. Dec.–Feb. Disturbed places, NW, SW, SE (widespread weed of cultivation).

fastigiatus Rottb.   Stout stems to 1.2 m. Spikelets reddish to golden-brown. Oct.–Feb. Marshy flats or lower slopes along watercourses, NW, SW, AP, LB, SE (Piketberg to Uitenhage, E Cape to northern South Africa).

laevigatus L. (= *Juncellus laevigatus* (L.) C.B.Clarke)   RIVIERKWEEK   Rhizomatous to tufted perennial, 10–60 cm. Spikelets light brown or purplish. Oct.–Apr. Damp brackish areas, sea level to 1400 m, NW, SW, AP, KM, LB, SE (Clanwilliam to Humansdorp, widespread in southern Africa and worldwide).

longus L.   WATERBIESIE, DOOIWORTEL, WATERKWEEK   Stout perennial to 1 m. Spikelets dark reddish brown with pale margins. Oct.–Apr. Damp flats and watercourses, NW, SW, AP, KM, LB, SE (Clanwilliam to Avontuur, widespread in southern and tropical Africa).

marginatus Thunb.   MATJIESGOED   Leafless perennial to 60 cm. Spikelets chestnut-brown with pale margins. Oct.–Feb. Rock crevices in seasonal watercourses, NW, SW (Ceres and Robertson, widespread in southern and tropical Africa).

pulcher Thunb.   Perennial to 40 cm. Spikelets reddish brown. Oct.–Feb. Marshes and watercourses, SE (Uitenhage to KwaZulu-Natal).

*****rotundus** L.   Tuberous perennial to 15 cm. Spikelets red. Dec.–Mar. Disturbed places, SW, KM (widespread weed of cultivation).

rubicundus Vahl   Tufted perennial. Spikelets red and straw-coloured. Dec.–Apr. Limestone flats, SE (Port Elizabeth, also widespread in Old World).

rupestris Kunth   Tufted perennial to 15 cm. Spikelets dark reddish brown. Feb. SW (Cape Peninsula, also E Cape to northern South Africa).

semitrifidus Schrad.   Low perennial, 8–25 cm. Spikelets reddish brown. Dec.–Apr. Damp flats or lower slopes, SE (Uitenhage to northern South Africa).

sphaerospermus Schrad.   Tufted perennial to 60 cm. Spikelets golden- to reddish brown. Oct.–Mar. Marshes and watercourses below 600 m, NW, SW, LB, SE (Clanwilliam to Uitenhage, widespread in southern Africa).

tenellus L.f.   Small annual to 15 cm. Spikelets green or reddish. Sept.–Jan. Lower slopes, NW, SW, AP, LB, SE (Namaqualand to E Cape).

textilis Thunb.   UMBRELLA SEDGE, MAT SEDGE, MATJIESGOED   Stout, tufted, leafless perennial to 1.2 m. Spikelets pale rust-red. Oct.–Mar. Marshes and watercourses below 150 m, NW, SW, AP, KM, LB, SE (Piketberg to S KwaZulu-Natal).

usitatus Burch.   INDIAN GRASS, BOESMANUINTJIE, HOENDERGRAS   Tuberous perennial to 25 cm. Spikelets dark red. Dec.–Apr. Mid to upper slopes below 250 m, SW, LB, SE (Paarl to Port Elizabeth, also widespread in central southern Africa).

ELEOCHARIS   c. 200 spp., nearly cosmopolitan

lepta C.B.Clarke (= *Scirpus leptus* (C.B.Clarke) Levyns)   Aquatic with thread-like stems to 45 cm. Spikelets brown. Nov. Pools, SW (Cape Peninsula).•

limosa (Schrad.) Schult.   Stems 20–60 cm. Spikelets brown. Aug.–Dec. Pools or marshes, SW, AP, SE (Namibia to Cape Peninsula to KwaZulu-Natal, Madagascar).

schlechteri C.B.Clarke   Stems to 6 cm. Spikelets pale brown. Nov.–Dec. Marshes, SW (Cape Peninsula to Caledon).•

EPISCHOENUS•   8 spp., W Cape

adnatus Levyns   Densely tufted perennial to over 1 m. Spikelets brown. Feb.–Aug. Damp mountain slopes, SW, AP, LB, SE (Worcester to Humansdorp).•

cernuus Levyns   Slender tufted perennial to 50 cm. Spikelets brown. Oct.–Nov. Mountain slopes, NW, SW (Clanwilliam to Caledon).•

complanatus Levyns   Tufted perennial, 20–65 cm. Spikelets pale brown. Dec.–Mar. Upper slopes, NW, SW (Tulbagh to Caledon).•

dregeanus (Boeck.) Levyns (= *Tetraria dregeana* (Boeck.) C.B.Clarke)   Tufted perennial, 20–70 cm. Spikelets pale (golden?) brown. Dec.–Apr. Middle to upper slopes, NW, SW (Cedarberg to Franschhoek Mts).•

gracilis Levyns   Tufted perennial, 20–90 cm. Spikelets pale brown. Dec.–Mar. Mountain slopes, NW, SW, LB, SE (Cedarberg Mts to George).•

lucidus (C.B.Clarke) Levyns   Tufted perennial to 60 cm. Spikelets brown and white. Dec. Upper slopes, NW, SW (Cold Bokkeveld to Bainskloof Mts and Worcester).•

quadrangularis (Boeck.) C.B.Clarke   Tufted perennial, 30–75 cm. Spikelets dark brown. Nov.–May. Mountain slopes, SW, LB, SE (Cape Peninsula to Port Elizabeth).•

villosus Levyns   Tufted perennial to 1 m. Spikelets brown. Jan.–Feb. Coastal flats to upper slopes, SW (Cape Peninsula to Caledon).•

FICINIA   c. 60 spp., tropical and southern Africa, mainly W Cape

acuminata (Nees) Nees (incl. **F. involuta** Nees, **F. elongata** Boeck.)   Tufted perennial, 20–70 cm. Spikelets reddish. Aug.–Dec. Rock crevices and shade of boulders near watercourses on mountain slopes, below 1700 m, NW, SW, AP, KM, SE (Cedarberg Mts to Uniondale).•

anceps Nees   Fairly robust perennial, 10–25 cm. Spikelets dull red. Apr.–Nov. Rock crevices, below 400 m, SW (Cape Peninsula).•

angustifolia (Schrad.) Levyns (= *Ficinia longifolia* C.B.Clarke)   Mat-forming perennial to 45 cm. Spikelets chestnut-brown or grey. Oct.–Feb. Moist rock ledges below 1500 m, NW, SW, SE (Cedarberg Mts to Humansdorp, also Mpumalanga).

arenicola T.H.Arnold & Gordon-Gray   Loosely tufted, erect perennial to 30 cm. Spikelets green, tinged red. Aug.–Nov. Coastal sand or clay flats, SE (Mossel Bay to E Cape).

argyropa Nees   Compact, tufted perennial to 25 cm. Spikelets reddish. July–Oct. Sandy flats, mostly near coast, NW, SW, AP, LB (Namaqualand to Riversdale).

bergiana Kunth   Stiff-leaved perennial to 25 cm. Spikelets brown. June–July. Lower to middle slopes, SW, LB (Cape Peninsula to E Cape).

brevifolia Nees ex Kunth (= *Ficinia composita* (Nees) Nees)   Densely tufted, robust perennial, 30–60 cm. Spikelets dark brown. Aug.–Dec. Watercourses on mountain slopes below 1200 m, NW, SW, AP, KM, LB (Namaquland to Riversdale).

bulbosa (L.) Nees   Tufted perennial to 30 cm. Spikelets chestnut-brown. Feb.–Sept. Sandy flats and slopes below 1000 m, strandveld, coastal and mountain fynbos, NW, SW, AP, LB, SE (Cedarberg Mts to E Cape).

capillifolia (Schrad.) C.B.Clarke   Loosely tufted, sprawling perennial, 20–30 cm. Spikelets chestnut-brown. Sept.–Nov. Upper slopes, NW, SW, SE (Cedarberg Mts to Humansdorp).•

capitella (Thunb.) Nees   Compact, tufted perennial, 5–25 cm. Spikelets reddish. July–Nov. Flats and slopes below 1700 m, NW, SW (W Karoo, Ceres to Caledon).

cedarbergensis T.H.Arnold & Gordon-Gray   Relatively robust perennial to 60 cm. Spikelets brown. Sept.–Apr. Upper slopes below 1100 m, NW (Cedarberg Mts).•

compacta (C.B.Clarke) T.H.Arnold   Stiff, tufted perennial to 30 cm. Spikelets dark brown. Oct.–Dec. Upper slopes, SW (Worcester).•

deusta (P.J.Bergius) Levyns   Relatively robust perennial to 40 cm. Spikelets large, dark. Mar.–Aug. Mountain slopes below 1700 m, NW, SW, AP, KM, LB, SE (Namaqualand to Knysna).

distans C.B.Clarke   Tufted perennial to 60 cm. Spikelets dark red. Dec.–Apr. Watercourses below 60 m, NW, SW (Ceres to Hermanus Mts).•

dunensis Levyns   Sparsely tufted perennial to 20 cm. Spikelets reddish. Aug.–Oct. Coastal dunes or mountain slopes, NW, SW, AP, SE (Cedarberg Mts to Port Elizabeth).•

dura Turrill   Perennial to 27 cm. Spikelets green to brown. Apr. Lower slopes below 100 m, SW, AP (Stanford to Bredasdorp).•

ecklonea (Steud.) Nees (= *Scirpus eckloneus* Steud.)   Stiff perennial, 30–45 cm. Spikelets in dense brown heads. Oct.–Dec. Sandy slopes below 1000 m, coastal fynbos, SW, AP (Malmesbury to Albertinia).•

elatior Levyns   Perennial to 40 cm. Spikelets reddish. Aug.–Dec. Flats near streams and pools below 100 m, SW, AP (Cape Peninsula to Bredasdorp).•

fascicularis Nees   Perennial with slender stems, 30–60 cm. Spikelets dull brown. Oct.–Nov. Middle to upper slopes below 2000 m, damp places in mountain fynbos and forest, SW, KM, SE (Caledon to KwaZulu-Natal).

fastigiata (Thunb.) Nees   Slender tufted perennial, 25–40 cm. Spikelets grey-green. Apr.–Oct. Flats and lower slopes, SW (Cape Peninsula).•

grandiflora T.H.Arnold & Gordon-Gray   Perennial, 35–70 cm. Spikelets maroon and white. July–Oct. Sandstone or granite slopes below 800 m, SW (Du Toitskloof to Hottentots Holland Mts).•

gydomontana T.H.Arnold   Perennial to 40 cm. Spikelets light to dark reddish. Oct.–Jan. Upper slopes below 2000 m, NW (Cold Bokkeveld to Hex River Mts).•

indica (Lam.) Pfeiffer (= *Ficinia striata* Kunth)   Perennial, 10–40 cm. Spikelets chestnut-brown. July–Nov. Flats and lower slopes, NW, SW, AP, SE (Namaqualand to E Cape).

ixioides Nees   Tufted perennial, 5–25 cm. Spikelets chestnut-brown. Aug.–Jan. Upper slopes, NW, SW, SE (Cedarberg to Tsitsikamma Mts).•

laciniata (Thunb.) Nees   Perennial to 20 cm. Spikelets pale brown. July–Oct. Coastal flats below 300 m, LB, SE (Albertinia to KwaZulu-Natal).

laevis Nees   Relatively robust perennial to 25 cm. Spikelets brown. Oct. Habitat? SW (Namaqualand to E Cape).

lateralis (Vahl) Kunth   Relatively robust, tufted perennial, 15–25 cm. Spikelets brown. Mar.–Aug. Coastal sands, SW, AP, SE (Cape Peninsula to E Cape).

latifolia T.H.Arnold & Gordon-Gray   Relatively robust perennial to 15 cm. Spikelets greenish. Apr. Sand over coastal limestone below 100 m, AP (Bredasdorp).•

levynsiae T.H.Arnold & Gordon-Gray   Relatively robust perennial to 15 cm. Spikelets dark brown. Oct.–Jan. Middle to upper slopes, SW (Ceres to Franschhoek Mountains).•

macowanii C.B.Clarke   Slender perennial, 30–60 cm. Spikelets white-edged. Oct.–May. Lower slopes, SW, LB (Caledon to Swellendam).•

micrantha C.B.Clarke   Slender tufted perennial to 12 cm. Spikelets white. Oct.–Jan. Middle slopes, SW (Cape Peninsula).•

minutiflora C.B.Clarke   Very slender perennial, 5–15 cm. Spikelets pale brown and white. Oct.–Mar. Middle slopes, SW (Hottentots Holland Mts and Kogelberg).•

monticola Kunth   Relatively robust perennial, 15–40 cm. Spikelets dark maroon and white. June–Nov. Middle to upper slopes below 1500 m, SW, LB (Worcester to Caledon and Langeberg Mts).•

mucronata C.B.Clarke   Slender tufted perennial to 15 cm. Spikelets light brown. Sept.–Nov. Rocky upper slopes below 1500 m, NW (Cedarberg Mts).•

nigrescens (Schrad.) J.Raynal (= *Ficinia bracteata* Boeck.)   Tufted perennial, 7–40 cm. Spikelets brown. May–Oct. Flats to upper slopes below 2000 m, NW, SW, AP, KM, LB, SE (S Namibia to E Cape).

oligantha (Steud.) J.Raynal (= *Ficinia filiformis* auct.; incl. **F. capillaris** (Nees) Levyns)   Slender, tufted perennial to 25 cm. Spikelets brown. Sept.–Jan. Lower slopes, NW, SW, AP, LB, SE (Clanwilliam to Knysna).•

pallens (Schrad.) Nees   Relatively robust tufted perennial, 15–30 cm. Spikelets yellow-green tinged with red. Mar.–July. Flats and lower slopes below 300 m, SW (Cape Peninsula to Caledon).•

paradoxa (Schrad.) Nees   Densely tufted perennial, 10–35 cm. Spikelets light brown. May–Nov. Flats to upper slopes, SW, AP, LB (Cape Peninsula to Albertinia).•

petrophila T.H.Arnold & Gordon-Gray   Stiff-leaved, tufted perennial to 30 cm. Spikelets yellow and reddish. Oct.–Jan. Rock crevices at high altitude, KM (Anysberg).•

pinguior C.B.Clarke   Relatively robust perennial to 60 cm. Spikelets brown. Mar.–June. Flats and lower slopes below 700 m, SW (Cape Peninsula to Caledon).•

polystachya Levyns (= *Ficinia angustifolia* C.B.Clarke)   Perennial to 30 cm. Spikelets dark brown. Oct.–Jan. Moist rock ledges above 1000 m, SW (Ceres to Jonkershoek Mts).•

praemorsa Nees   Tufted perennial to 40 cm. Spikelets yellow to brown. Jan.–Dec. Limestone flats below 600 m, AP (Stanford to Mossel Bay).•

pygmaea Boeck. (= *Ficinia limosa* Levyns)   Tufted perennial, 3–40 cm. Spikelets light brown. Aug.–Nov. Coastal dunes below 100 m, NW, SW, AP (Lambert's Bay to Bredasdorp).•

quinquangularis Boeck.   Perennial to 25 cm. Spikelets chestnut-red. Mar.–May. Rocky mountain slopes and plateaus below 1500 m, LB, SE (Swellendam to Humansdorp).•

radiata (L.f.) Kunth (= *Sickmannia radiata* (L.f.) Nees)   STERGRAS   Robust tufted perennial, 5–25 cm. Spikelets yellow, in heads with broad, yellow radiate bracts. Sept.–Nov. Flats and slopes, SW, AP (Ceres to Stilbaai).•

ramosissima Kunth   Erect, branched perennial to 20 cm. Spikelets light brown. July–Sept. Lower slopes and rock crevices in shade, SW, AP, KM, SE (Cape Peninsula to E Cape).

repens (Nees) Kunth   Rhizomatous perennial to 15 cm. Spikelets brown. Oct.–Feb. Salt flats, AP, LB, SE (Bredasdorp to E Cape).

rigida Levyns   Stiff perennial to 35 cm. Spikelets brown. Jan.–Mar. Sandy lower slopes, SW (Cape Peninsula to Kleinmond).•

secunda (Vahl) Kunth   Perennial to 60 cm. Spikelets chestnut-red. Mar.–Oct. Sandy flats below 1000 m, NW, SW, AP, LB (Cedarberg Mts to Mossel Bay).•

stolonifera Boeck. (incl. **F. contorta** (Nees) Pfeiffer, **F. pusilla** C.B.Clarke, **F. thyrsoidea** Pfeiffer)   Perennial, 7–15 cm. Spikelets brown. Sept.–Dec. Flats to upper mountain slopes, NW, SW, LB, SE (Cape Peninsula to tropical Africa).

tenuifolia Kunth   Perennial, 25–45 cm. Spikelets light brown. Apr.–Nov. Lower slopes, SW, KM (Cape Peninsula to Ladismith).•

trichodes (Schrad.) Benth. & Hook.f.   Much-branched, sprawling perennial to 45 cm long. Spikelets light brown. June–Sept. Rocky lower to middle slopes, SW, SE (Cape Peninsula to E Cape).

trispicata (L.f.) Druce (= *Ficinia leiocarpa* Nees, *F. sylvatica* Kunth)   Tufted perennial, 20–60 cm. Spikelets white to light brown. Sept.–Nov. Lower slopes below 1000 m in forest, LB, SE (Swellendam to E Cape).

tristachya (Rottb.) Nees   Tufted perennial to 20 cm. Spikelets brown. Mar.–May. Flats and lower slopes, SW, AP, LB, SE (Cape Peninsula to E Cape).

truncata (Thunb.) Schrad.   Robust, grey-green tufted perennial. Spikelets chestnut-brown. June–Oct. Limestone hills below 200 m, AP, SE (Bredasdorp to E Cape).

zeyheri Boeck. Slender tufted perennial, 15–30 cm. Spikelets brown. Aug.–Oct., after fire. Sandy soil in mountain seeps below 1700 m, in fynbos, SW, LB, SE (Cape Peninsula to Uniondale).•

[**Species excluded** **Scirpus lucida** C.B.Clarke, a putative interspecific hybrid between **F. cedarbergensis** & **F. ixioides** subsp. **glabra**]

FIMBRISTYLIS   c. 250 spp., cosmopolitan

bisumbellata (Forssk.) Bub.   Annual, 10–30 cm. Spikelets small, brown. Aug.–Dec. Lower slopes, ?NW (Namibia, Clanwilliam to tropical Africa).

complanata (Retz.) Link   Perennial, 15–60 cm. Spikelets reddish. Oct.–Nov. Lower slopes below 100 m, SE (George to northern South Africa and nearly worldwide).

squarrosa Vahl   Annual, 5–20 cm. Spikelets greyish. Jan.–June. Lower slopes, NW (Clanwilliam and southern Africa, also worldwide in tropics and subtropics).

FUIRENA   c. 40 spp., nearly cosmopolitan

coerulescens Steud.   Perennial, 20–40 cm. Spikelets pubescent, blue-green. Sept.–Mar. Marshy flats and lower slopes below 100 m, SW, AP, SE (Cape Peninsula to northern South Africa).

hirsuta (P.J.Bergius) P.L.Forbes (= *Fuirena hottentotta* (L.) Druce)   Perennial with pubescent stems, 30–50 cm. Spikelets pubescent, blue-green. Oct.–Feb. Marshy flats and watercourses on lower slopes to 1000 m, NW, SW, AP, LB, SE (Namaqualand to Mpumalanga).

HELLMUTHIA•   1 sp., W Cape

membranacea (Thunb.) R.Haines & K.Lye (= *Scirpus membranaceus* Thunb.)   BIESIE   Reed-like, tufted perennial to 80 cm. Spikelets dark brown, clustered. May–Oct. Coastal sands below 500 m, SW, AP, SE (Saldanha to Knysna).•

ISOLEPIS   c. 100 spp., cosmopolitan

antarctica (L.) Roem. & Schult.   Annual, 8–20 cm. Spikelets straw-coloured and dark red. Oct.–Nov. Damp flats and slopes to 800 m, SW, AP, LB (Cape Peninsula to Langeberg Mts).•

brevicaulis (Levyns) J.Raynal   Minute annual to 2 cm. Spikelets green. Sept.–Oct. Near pools on gravel flats, NW, SW (Namaqualand to Bredasdorp and Robertson).

cernua (Vahl) Roem. & Schult.   Densely tufted annual to 14 cm. Spikelets reddish. Nov.–Mar. Marshes and watercourses, SW, AP, SE (Cape Peninsula to Port Elizabeth, cosmopolitan).

diabolica (Steud.) Schrad.   Stoloniferous perennial to 30 cm. Spikelets straw-coloured and dark red. Sept.–Nov. Marshes and watercourses below 300 m, NW, SW, SE (Ceres to KwaZulu-Natal and Free State).

digitata Schrad.   Flaccid, tufted perennial, 15–35 cm. Spikelets green. Sept.–Jan. Attached to rocks in streams below 1000 m, NW, SW, LB (Clanwilliam to Riversdale).•

fluitans (L.) R.Br.   WATERBIESIE, WATERGRAS   Aquatic branching perennial, 15–30 cm. Spikelets green. Jan.–June. Pools and watercourses on flats or slopes, LB, SE (Riversdale to Mossel Bay, widespread in the Old World).

hystrix (Thunb.) Nees   BIESIE   Small, densely tufted annual, 2–10 cm. Spikelets green. Aug.–Nov. Damp flats, sometimes to 1600 m, NW, SW (Namaqualand to Cape Peninsula).

incomptula Nees   Tufted annual, 2–8 cm. Spikelets green with white or red edges. Aug.–Oct. Muddy flats or sandy pockets in sandstone, NW, SW, SE (Namaqualand to Port Elizabeth).

inconspicua (Levyns) J.Raynal   Minute, tufted, turf-forming annual, 2–3 cm. Spikelets reddish. Aug.–Sept. Low-lying wet sites, SW (Cape Peninsula and Cape Flats).•

karroica (C.B.Clarke) J.Raynal   Small, tufted annual to 3 cm. Spikelets straw-coloured. Oct.–Feb. Watercourses, KM (Namibia and W Karoo to Witteberg).

leucoloma (Nees) C.Archer (= *Cyperus leucoloma* Nees)   Tufted annual, 2–5 cm. Spikelets dark red and white. Oct.–Nov. Mountain slopes to 1000 m, NW, SW (Cedarberg Mts to Cape Peninsula).•

ludwigii (Steud.) Kunth   Stoloniferous, turf-forming perennial to 10 cm. Spikelets straw-coloured to brown-red. Sept.–Dec. Damp, often disturbed flats near water, mostly below 1000 m, SW, LB, SE (Cape Peninsula to E Cape).

marginata (Thunb.) A.Dietr. (= *Isolepis cartilaginea* R.Br.)   Annual, 5–15 cm. Spikelets straw-coloured and dark red. Sept.–Dec. Dunes, flats and slopes in seasonally damp sandy soil, to 1200 m, NW, SW, AP, LB, SE (Namaqualand to E Cape, also Australia).

minuta (Turrill) J.Raynal   Annual to 3 cm. Spikelets green-keeled. Nov. Moist sandstone soil, NW (Ceres).•

natans (Thunb.) A.Dietr. (= *Scirpus rivularis* (Schrad.) Boeck.)   Tufted annual with flattened stems, 5–25 cm. Spikelets dark brown. Oct.–Nov. Marshes and pools below 400 m, NW, SW, LB, SE (Cedarberg Mts to Cape Peninsula to Mpumalanga).

prolifer R.Br.   Moderately robust perennial, stems to 50 cm. Spikelets light brown. Oct.–Mar. Streamsides and seeps below 1000 m, NW, SW, AP, KM, LB, SE (Namaqualand to KwaZulu-Natal, Australasia, St Helena).

pusilla Kunth (= *Scirpus nanodes* Levyns)   Slender annual to 5 cm. Spikelets reddish with green keels. Oct.–Dec. Damp mountain slopes, SW (Cape Peninsula).•

rubicunda Kunth   (= *Scirpus globiceps* C.B.Clarke)   Often submerged perennial with branching stems to 30 cm. Spikelets red. May–Oct. Seasonal pools on flats or lower slopes, SW (Langebaan to Cape Peninsula).•

sepulcralis Steud. (incl. **Scirpus chlorostachyus** Levyns)   Annual to 6 cm. Spikelets dark red. Aug.–Feb. Watercourses, SW, SE (Namaqualand to KwaZulu-Natal).

setacea (L.) R.Br.   Annual to 15 cm. Spikelets pale, often with red markings. Jan. Habitat? SE (Humansdorp, nearly cosmopolitan).

striata (Nees) Kunth   Branching aquatic perennial. Spikelets green or reddish. Aug.–Dec. Pools at lower altitudes, NW, SW, SE (Namaqualand to Riviersonderend and Uitenhage).

tenuissima (Nees) Kunth   Slender, branching aquatic to 60 cm. Spikelets minute, dark red and green. Dec.–Apr. Rivers or pools, NW, SW, SE (Ceres to Cape Peninsula to KwaZulu-Natal, Madagascar).

venustula Kunth (= *Scirpus venustulus* (Kunth) Boeck.)   Rhizomatous perennial, 5–25 cm. Spikelets dark red. Sept.–Nov. Coastal flats in damp soil or shallow pools, SW (Cape Peninsula to Caledon).•

verrucosula (Steud.) Nees   Low, densely tufted annual to 5 cm. Spikelets straw-coloured or reddish, with green keels. Sept.–Dec. Marshy flats, NW, SW, SE (Namaqualand to Uitenhage).

sp. 1 (**Scirpus bulbiferus** Boeck.; incl. **S. delicatulus** Levyns)   Submerged aquatic to 20 cm. Spikelets green and red. Nov.–Jan. Seasonal pools, SW (Cape Peninsula).•

sp. 2 (**Scirpus burchellii** C.B.Clarke)   Small tufted annual to 5 cm. Spikelets pale. Nov. Flats, LB (Riversdale).•

sp. 3 (**Scirpus dregeanus** C.B.Clarke)   Like **I. marginata** but only 2–4 cm. Spikelets red to brown. Sept.–Oct. Lower slopes, SW (Bainskloof to Riviersonderend Mts).•

[**Species excluded**   No authentic material found and probably conspecific with one of the above: **I. leptostachya** Kunth, **I. trachysperma** Nees]

KYLLINGA   c. 60 spp., nearly cosmopolitan in warm, moist regions

alata Nees   Tufted perennial, 9–25 cm. Spikelets golden. Oct.–Mar. Grassy flats and slopes below 100 m, LB, SE (Swellendam to tropical Africa).

erecta Schumach.   Rhizomatous perennial, 10–30 cm. Spikelets golden-brown. Jan.–Mar. Grassy flats and lower slopes below 100 m, LB, SE (Swellendam to tropical Africa, Mascarene Islands).

melanosperma Nees   Slender perennial to 45 cm. Spikelets olive-green. Nov.–Jan. Lower slopes, SE (Humansdorp to E Cape, widespread in palaeotropics).

MARISCUS   c. 200 spp., nearly cosmopolitan

capensis (Steud.) Schrad.   Perennial to 20 cm. Spikelets green. Dec.–Apr. Grassy slopes, SE (Port Elizabeth to northern South Africa).

congestus (Vahl) C.B.Clarke   HEDGEHOG SEDGE   Sparsely tufted perennial, 20–90 cm. Spikelets reddish. Dec.–Apr. Damp flats and watercourses on slopes to 600 m, NW, SW, AP, KM, SE (widespread in southern Africa, Mediterranean, Australia, S Atlantic islands).

durus (Kunth) C.B.Clarke   Tufted perennial to 95 cm. Spikelets dark red. Oct. Marshy coastal flats, SE (Knysna to E Cape and S tropical Africa).

solidus (Kunth) P.J.Vorster (incl. **M. involutus** C.B.Clarke)   Robust perennial. Spikelets dark red. Oct.–Jan. Habitat? SE (Uitenhage to KwaZulu-Natal).

tabularis (Schrad.) C.B.Clarke   Perennial to 60 cm. Spikelets dark red. Sept.–Dec. Damp flats and watercourses below 200 m, NW, SW, SE (Namaqualand to E Cape).

thunbergii (Vahl) Schrad. (= *Mariscus riparius* Schrad.)   Robust perennial to 1.5 m. Spikelets dull red-brown. June–Dec. Near water below 500 m, NW, SW, AP, LB, SE (Namaqualand to E Cape).

uitenhagensis Steud.   Perennial, 5–25 cm. Spikelets green. Oct.–Mar. Grassy slopes below 50 m, SE (Uitenhage, widespread in southern Africa).

NEESENBECKIA•   1 sp., W Cape

punctoria (Vahl) Levyns (= *Tetraria punctoria* (Vahl) C.B.Clarke)   Rigid, robust, tufted perennial to 2 m. Spikelets light brown. Mar.–Apr. Streamsides on lower slopes to 800 m, SW (Cape Peninsula to Caledon).•

PSEUDOSCHOENUS   1 sp., southern Africa

inanis (Thunb.) Oteng-Yeboah (= *Scirpus inanis* (Thunb.) Steud., *S. spathaceus* Hochst.)   Rhizomatous perennial with stout, reed-like stems to 1 m. Spikelets pale brown. Nov.–Dec. Streamsides at high altitudes to 1500 m, KM (Namibia to Swartberg Mts, and Karoo to Lesotho).

PYCREUS   c. 70 spp., nearly cosmopolitan

mundii Nees   Creeping perennial, 20–60 cm. Spikelets brown. Oct.–Apr. Edges of pools and watercourses on flats and lower slopes, SW, AP, SE (Cape Peninsula to Port Elizabeth, S to tropical Africa, Spain, Mascarenes).

nitidus (Lam.) J.Raynal (= *Pycreus lanceus* (Thunb.) Turrill) WATERBIESIE   Tufted perennial, 20–50 cm. Spikelets dark red. Oct.–Apr. Marshy flats and lower slopes, NW, SW, LB, SE (Clanwilliam to E Cape and to northern South Africa, widespread).

polystachyos (Rottb.) P.Beauv.   Tufted perennial, 30–60 cm. Spikelets yellow-green. Dec.–Apr. Damp lower slopes, NW, SW, AP, LB, SE (Clanwilliam to Cape Peninsula to Port Elizabeth, pantropical).

RHYNCHOSPORA   c. 200 spp., nearly cosmopolitan

brownii Roem. & Schult.   Tufted perennial, 30–70 cm. Spikelets brown. Dec.–Jan. Seeps at lower altitudes, SW, SE (Cape Peninsula to Humansdorp, also widespread in eastern southern Africa and nearly worldwide in warmer regions).

SCHOENOPLECTUS   c. 60 spp., southern and tropical Africa

decipiens (Nees) J.Raynal   Rhizomatous perennial to 30 cm. Spikelets black and straw-coloured. Oct.–Apr. Marshes below 100 m, LB, SE (Riversdale to E Cape and widespread in southern Africa).

leucanthus (Boeck.) J.Raynal (= *Scirpus leucanthus* Boeck., *S. supinus* auct.)   Small tufted annual, 2–6 cm. Spikelets straw-coloured. Nov. Riverbanks, NW (Namibia to Clanwilliam, Karoo to Botswana).

paludicola (Kunth) Palla ex J.Raynal (= *Scirpus paludicola* Kunth) STEEKBIESIE, STEEKRIETJIE   Tufted perennial, 30–50 cm. Spikelets chestnut-brown. Oct.–Feb. Marshes below 100 m, LB, SE (Swellendam to Mpumalanga, Madagascar).

scirpoideus (Schrad.) J.Browning (= *Scirpus litoralis* auct.) PAPGRAS, STEEKBIESIE   Reed-like perennial with soft stems to 150 cm. Spikelets brown. Nov.–Jan. Marshes and riverbanks, mostly coastal areas below 200 m, SW, AP, SE (Namibia and Cape Peninsula to Mozambique).

*****triqueter** (L.) Palla (= *Scirpus triqueter* L.)   Perennial to 90 cm. Spikelets brown. Nov.–Jan. Edges of pools near the coast, SW, AP (Langebaan to Bredasdorp, also sporadic in rest of South Africa, introduced from Eurasia).

SCHOENOXIPHIUM   c. 8 spp., tropical and southern Africa

altum Kukkonen   Tufted perennial, 50–70 cm. Spikelets green. Mar.–Apr. Forests, SE (George).•

ecklonii Nees (= *Carex zeyheri* C.B.Clarke)   Tufted perennial to 30 cm. Spikelets gold. June–Oct. Open, bushy slopes, SW, SE (Cape Peninsula to Uitenhage).•

lanceum (Thunb.) Kük. (= *Scirpus sickmannianum* Kunth, *S. capense* Nees)   Tufted perennial to 60 cm. Spikelets dull gold. June–Nov. Shady lower slopes, SW, SE (Cape Peninsula to Humansdorp).•

lehmannii (Nees) Steud.   Tufted perennial, 20–40 cm. Spikelets green to brownish. Oct.–Jan. Shady forested slopes, SW, SE (Cape Peninsula to E Cape to Tanzania).

sparteum (Wahlenb.) C.B.Clarke   Tufted perennial to 40 cm. Spikelets green. May–Nov. Damp shady slopes, SW, SE (Cape Peninsula to Humansdorp, widespread in southern Africa).

thunbergii Nees   Tufted perennial, 20–40 cm. Spikelets greenish. July–Oct. Grassy or bushy slopes, SW (Cape Peninsula and Stellenbosch).•

SCHOENUS   c. 100 spp., cosmopolitan

nigricans L.   Tufted perennial, 20–50 cm. Spikelets brown. May–Oct. Marshes and watercourses on flats and lower slopes below 200 m, SW, AP, SE (Cape Peninsula to E Cape, more or less worldwide).

SCIRPOIDES   c. 5 spp., nearly cosmopolitan

dioecus (Kunth) J.Browning (= *Scirpus dioecus* (Kunth) Boeck.)   BIESIE   Reed-like perennial with robust stems to 50 cm. Spikelets light brown in dense heads. Sept.–Dec. Riverbanks and seasonal pans to 1000 m, NW, SW, AP, KM, LB (Namibia to Bredasdorp, widespread in South Africa).

nodosus (Rottb.) Soják (= *Scirpus nodosus* Rottb.)   VLEIBIESIE   Tufted perennial with rigid stems to 100 cm. Spikelets in dense, brown heads. Dec.–Mar. Damp sandy flats in coastal areas to 250 m, SW, AP, LB, SE (Namaqualand to KwaZulu-Natal and widespread in S hemisphere).

thunbergii (Schrad.) Soják (= *Scirpus thunbergianus* (Nees) Levyns)   Tufted perennial with flattened stems, 15–70 cm. Spikelets in dense clusters, brown. Oct.–May. Damp flats near the coast to 300 m, NW, SW, SE (Cape Peninsula to E Cape).

SCIRPUS see **Bolboschoenus, Isolepis, Pseudoschoenus, Schoenoplectus** and **Scirpoides**.

SCLERIA   c. 200 spp., pantropical

natalensis C.B.Clarke   SNYGRAS   Tufted perennial to 85 cm. Spikelets light brown. May–June. Damp places in coastal forest below 300 m, SE (Humansdorp to KwaZulu-Natal).

TETRARIA   c. 40 spp., Africa and Australasia, mainly South Africa

autumnalis Levyns   Tufted perennial to 40 cm. Spikelets dull brown. Apr.–June. Damp lower slopes to 1000 m, SW (Cape Peninsula).•

bolusii C.B.Clarke   Slender stems to 30 cm. Spikelets chestnut-red. July–Aug. Below 1200 m, SW, ?LB (Cape Peninsula and ?Langeberg Mts).•

brachyphylla Levyns   Densely tufted perennial to 60 cm. Spikelets reddish. July–Nov. Sandy coastal dunes and lower slopes below 200 m, SW, AP, SE (Cape Peninsula to Plettenberg Bay).•

bromoides (Lam.) Pfeiffer (= *Tetraria rottboellii* (Schrad.) C.B.Clarke)   BERGPALMIET   Robust tufted perennial to over 1.5 m. Spikelets pinkish to dark brown. Oct.–Feb. Dry mountain fynbos up to 1500 m, NW, SW, AP, LB, SE (Porterville to Cape Peninsula to Uitenhage).•

burmanii (Schrad.) C.B.Clarke   Tufted perennial, 10–30 cm. Spikelets rusty brown. Nov.–Apr. Mountain slopes, SW, AP, SE (Bainskloof to Potberg to Uitenhage).•

capillacea (Thunb.) C.B.Clarke   Tufted perennial, 30–80 cm. Spikelets dark brown. Mainly Oct.–Nov. Mountain slopes to 1500 m, dry to moist mountain fynbos, SW, AP, LB, SE (Cape Peninsula to E Cape).

compacta Levyns   Densely tufted perennial to 60 cm. Spikelets yellow-brown. Aug.–Nov. Lower slopes, SW (Villiersdorp to Kleinmond).•

compar (L.) Lestib.   Perennial, 30–90 cm. Spikelets viscid, light to dark brown. Apr.–June. Sandy lower slopes and coastal fynbos, SW, AP (Cape Peninsula to Albertinia).•

compressa Turrill   Robust perennial to over 1 m. Spikelets brown. Feb. Mountain slopes, SE (Knysna to KwaZulu-Natal).

crassa Levyns   Robust tufted perennial to 60 cm. Spikelets brown. Apr.–June. Lower mountain slopes, SW (Hottentots Holland Mts to Hangklip).•

crinifolia (Nees) C.B.Clarke   Tufted perennial to 30 cm. Spikelets dark red-brown. Aug.–Oct. Sandstone slopes to 1600 m, NW, SW (Clanwilliam to Caledon Swartberg).•

cuspidata (Rottb.) C.B.Clarke   Tufted perennial, 30–60 cm. Spikelets dark brown. Aug.–Nov. Mountain slopes, NW, SW, AP, SE (Cedarberg Mts to Cape Peninsula to Mpumalanga).

exilis Levyns   Tufted perennial, 20–30 cm. Spikelets brown. Apr.–June. Flats and slopes, SW (Du Toit's Kloof Mts to Kleinmond).•

eximia C.B.Clarke   Relatively robust perennial, 30–90 cm. Spikelets rusty brown. May–Aug. Flats to 1200 m, SW (Cape Peninsula to Caledon).•

fasciata (Rottb.) C.B.Clarke (incl. **T. pleosticha** C.B.Clarke)   Tufted perennial, 30–80 cm. Spikelets light grey-brown. Nov.–May. Sandy lower or middle slopes to 1400 m, SW, LB, SE (Cape Peninsula to Humansdorp).•

ferruginea C.B.Clarke   Tufted perennial to 60 cm. Spikelets rusty brown. Mar.–July. Mountain slopes above 1200 m, NW, SW (Cedarberg to Hex River Mts and Bainskloof Mts).•

fimbriolata (Nees) C.B.Clarke   Tufted perennial, 30–45 cm. Spikelets brown. Jan.–Apr. Sandy flats and lower slopes and mountain fynbos to 1200 m, SW, AP, KM, LB, SE (Cape Peninsula to Uitenhage).•

flexuosa (Thunb.) C.B.Clarke   Tufted perennial, 45–75 cm. Spikelets dark brown. Jan.–May. Flats to middle slopes, NW, SW, LB (Ceres to Cape Peninsula to Riversdale).•

fourcadei Turrill & Schönland   Tufted perennial to 80 cm. Spikelets reddish to dark brown. Jan.–May. Mountain slopes up to 1800 m, SW, KM, LB, SE (Hex River Mts to Great Winterhoek Mts).•

gracilis Turrill   Slender tufted perennial to 60 cm. Spikelets brown. Dec.–Jan. Middle to upper slopes, SW, LB, SE (Paarl to Uniondale).•

graminifolia Levyns   Densely tufted perennial to 40 cm. Spikelets light brown. July–Nov. Lower slopes, SW (Cape Peninsula).•

involucrata (Rottb.) C.B.Clarke   Robust reed-like perennial to 2 m. Spikelets light brown. Jan.–Apr. Moist sandstone slopes to 2000 m, NW, SW, AP, KM, LB, SE (Ceres to Humansdorp).•

maculata Schönland & Turrill   Tufted perennial, 30–45 cm. Spikelets dark brown with pale margins. Dec.–Feb. Seeps in fynbos on high peaks, above 1200 m, NW, SW, KM (Cedarberg to Kammanassie Mts).•

microstachys (Vahl) Pfeiffer   Tufted perennial to 25 cm. Spikelets brown or pale. Dec.–Feb. Sandy flats and slopes to 600 m, SW, LB, SE (Tulbagh to Cape Peninsula to Humansdorp).•

nigrovaginata (Nees) C.B.Clarke   Tufted perennial, 20–45 cm. Spikelets brown. Jan.–Apr. Sandy mountain slopes and plateaus to 1200 m, NW, SW, KM, LB (Cedarberg Mts to Cape Peninsula to Witteberg).•

paludosa Levyns   Densely tufted perennial to 100 cm. Spikelets reddish. Aug.–Nov. Marshy lower slopes, SW (Cape Peninsula).•

picta (Boeck.) C.B.Clarke   Tufted perennial, 20–40 cm. Spikelets shining, brown. Dec.–May, after fire. Moist sands above 1200 m, NW, SW, KM (Cedarberg Mts to Caledon Swartberg and Anysberg).•

pillansii Levyns   Slender tufted perennial to 25 cm. Spikelets greenish. Jan.–Feb. Lower slopes, SW, AP, LB, SE (Worcester to Humansdorp).•

pubescens Schönland & Turrill   Densely tufted perennial to 12 cm, flowering stems shorter than leaves. Spikelets light brown. Oct.–Apr. Rocky sandstone slopes above 1200 m, NW, SW (Cedarberg to Du Toitskloof Mts).

pygmaea Levyns   Tufted perennial to 15 cm. Spikelets yellow-brown. Feb.–Apr. Sandstone slopes below 1200 m, SW, AP, LB, SE (Cape Peninsula to Langkloof).•

robusta (Kunth) C.B.Clarke   Robust perennial to 1 m. Spikelets dull brown. Mar. Mountain slopes, SE (Humansdorp).•

secans C.B.Clarke   BERGKLAPPER   Robust tufted perennial to 2.5 m. Spikelets dusky brown. Apr.–May. Marshy forest margins below 200 m, LB, SE (Riversdale to Humansdorp).•

sylvatica (Nees) C.B.Clarke   Tufted perennial, 15–40 cm. Spikelets shining brown. May–Dec. Flats and lower slopes, SW, AP, SE (Cape Peninsula to George).•

thermalis (L.) C.B.Clarke   BERGPALMIET   Robust tufted perennial to over 2 m. Spikelets dusky brown. June–Oct. Rocky flats and slopes below 1000 m, SW, AP, LB (Cape Peninsula to Riversdale).•

triangularis (Boeck.) C.B.Clarke (incl. **T. macowaniana** B.L.Burtt)   Tufted perennial to 60 cm. Spikelets rusty brown. Feb.–Apr. High altitudes, NW, SW, LB, SE (Cedarberg Mts to Cape Peninsula to KwaZulu-Natal).

ustulata (L.) C.B.Clarke   Tufted perennial, 45–90 cm. Spikelets shiny brown. Jan.–May. Sandy flats, lower slopes and plateaus to 1200 m, NW, SW, AP, LB, SE (Namaqualand to Cape Peninsula to Outeniqua Mountains).

vaginata Schönland & Turrill   Tufted perennial, 10–30 cm. Spikelets reddish with papery margins. Sept.–Apr. Rocky upper slopes to 2000 m, NW, SW (Cedarberg and Hex River Mts).•

variabilis Levyns   Tufted perennial, 20–40 cm. Spikelets dull brown. Apr.–June. Flats, SW (Cape Peninsula).•

TRIANOPTILES   3 spp., W Cape and Namaqualand

capensis (Steud.) Harv.   Grass-like annual, 6–25 cm. Spikelets greenish. Aug.–Nov. Damp flats or lower slopes, NW, SW, LB, SE (Ceres to Cape Peninsula to Knysna).•

solitaria (C.B.Clarke) Levyns   Grass-like annual, 10–30 cm. Spikelets greenish. Aug.–Oct. Damp flats, SW (Cape Peninsula).•

stipitata Levyns   Like **T. capensis** but ovary stalked. Aug.–Oct. Flats, NW, SW (Namaqualand to Caledon).

DIOSCOREACEAE

DIOSCOREA   YAM   400 spp., pantropical and warm temperate

burchellii Baker   Perennial vine with branched tuber, stems twisting to the left, ribbed and warty, to 2 m. Leaves lanceolate to sagittate. Male flowers in nodding racemes, female flowers in nodding spikes, cream. Apr.–June. Damp sandstone slopes, SE (George to Humansdorp).•

cotinifolia Kunth   Perennial with tubers on slender roots from small crown, stems twisting to the right. Leaves ovate to heart-shaped. Male flowers in erect racemes, female flowers in nodding spikes, white. Oct.–Dec. Coastal bush, SE (Van Staden's River to Mpumalanga).

elephantipes (L'Hér.) Engl.   ELEPHANT'S FOOT, HOTTENTOTSBROOD   Perennial with exposed, armour-plated tuber, stems twisting to the left, to 90 cm. Leaves heart-shaped. Male flowers in erect, spiny racemes, female flowers in spinescent, nodding to spreading spikes, yellowish green. Nov.–Feb. Dry rocky slopes, NW, SE (Clanwilliam and Uniondale to Graaff-Reinet and S Karoo).

hemicrypta Burkill   Perennial with half-exposed and armour-plated tuber, base broadly lobed, stems erect or twisting to the left, to 1.2 m. Leaves broadly ovate. Male flowers in erect racemes, female flowers in speading spikes, yellowish. Jan.–Apr. Dry stony slopes, KM (Richtersveld and W Karoo to Little Karoo).

mundii Baker   Perennial vine with branched, gnarled tuber, stems twisting to the left, to 5 m. Leaves heart-shaped. Male flowers in erect racemes, female flowers in nodding spikes, greenish. Apr.–Aug. Coastal bush or forest, SE (George to Nature's Valley).•

sylvatica (Kunth) Eckl.   Perennial with large, lobed tuber sometimes partly exposed, with corky armour, stems twisting to the left, to 6 m. Leaves heart-shaped to sagittate. Male flowers in more or less erect racemes, female flowers in nodding spikes, yellowish green. Nov.–Mar. Bush or forest, SE (Plettenberg Bay to tropical Africa).

ERIOSPERMACEAE see **CONVALLARIACEAE**

HAEMODORACEAE

1. Ovary superior; stamens equal; leaves plicate ....**Wachendorfia**
1.' Ovary inferior; 1 stamen shorter and with a larger anther; leaves flat or ribbed....................................**Dilatris**

DILATRIS•   BLOODROOT, ROOIWORTEL   4 spp., W Cape

corymbosa P.J.Bergius   Rhizomatous geophyte, 40–60 cm, stems grey-hairy. Leaves narrowly sword-shaped. Flowers in a corymbose panicle, enantiostylous, mauve, tepals ovate, long stamens about as long as tepals, with anthers half as long as anther of short stamen. Aug.–Jan. Sandstone slopes and flats, NW, SW (Tulbagh to Hottentots Holland Mts).•

ixioides Lam.   Rhizomatous geophyte, 20–40 cm, stems grey-hairy. Leaves linear. Flowers in a corymbose panicle, enantiostylous, mauve, tepals ovate, long stamens twice as long as tepals, with anthers one-fourth as long as anther of short stamen. Sept.–Feb. Rocky sandstone slopes, NW, SW, LB, SE (Bokkeveld Mts to George).•

pillansii W.F.Barker   Rhizomatous geophyte, 20–45 cm, stems grey-hairy. Leaves linear. Flowers in a corymbose panicle, enantiostylous, mauve, tepals ovate, long stamens shorter than tepals, with anthers two-thirds as long as anther of short stamen. Aug.–Jan. Rocky sandstone slopes, NW, SW, AP, LB (Cedarberg Mts to Agulhas).•

viscosa L.f. (= *Dilatris paniculata* L.f.)   Rhizomatous geophyte, 45–60 cm, stems reddish glandular-hairy. Leaves sword-shaped. Flowers in a corymbose panicle, enantiostylous, dull orange or yellow, tepals linear-oblanceolate, long stamens slightly longer than tepals, with anthers less than half as long as anther of short stamen. Aug.–Dec. Marshy places on mountain slopes and plateaus, NW, SW, LB (Ceres to Cape Peninsula to Riversdale).•

WACHENDORFIA BUTTERFLY LILY, ROOIKANOL   4 spp., winter-rainfall South Africa

brachyandra W.F.Barker   Rhizomatous geophyte, 20–65 cm. Leaves narrow, glabrous. Flowers enantiostylous, in a lax panicle, apricot-yellow, stamens and style half as long as tepals, bracts scarious. Aug.–Dec. Damp sandstone or granite, SW (Hermon to Cape Peninsula and Franschhoek).•

multiflora (Klatt) J.C. Manning & Goldblatt (= *W. parviflora* W.F.Barker)   Rhizomatous geophyte, 10–30 cm. Leaves narrow, softly hairy. Flowers enantiostylous, in a dense, rounded panicle, dull yellow to browish purple, bracts green. Aug.–Sept. Sandstone and granitic soils, NW, SW, KM (Namaqualand to Cape Peninsula and Robertson).

paniculata Burm.   KOFFIEPIT   Rhizomatous geophyte, mostly 20–70 cm. Leaves narrow, usually hairy. Flowers enantiostyous, in a lax to dense panicle, apricot to yellow, bracts scarious. Mostly Aug.–Nov. Mainly sandstone soils, NW, SW, AP, KM, LB, SE (Bokkeveld Mts to Port Elizabeth).•

thyrsiflora Burm.   Rhizomatous geophyte, 1–2 m. Leaves broad and glabrous. Flowers enantiostylous, in a crowded, cylindrical panicle, golden-yellow, bracts scarious. Sept.–Dec. Permanent marshes and streams. NW, SW, AP, LB, SE (Clanwilliam to Cape Peninsula to Humansdorp).•

HEMEROCALLIDACEAE *(= ANTHERICACEAE in part)*

CAESIA   BLUE GRASS LILY   c. 12 spp., South Africa, Madagascar, Australia

capensis (Bolus) Oberm.   Rhizomatous geophyte to 8 cm, forming compact cushions. Leaves linear. Flowers hidden in leaves, raised on long pedicels, blue, filaments smooth. Mainly Oct.–Jan. Sandstone rocks at high elevations, NW, SW, KM (Grootwinterhoek Mts and Du Toit's Peak to Seweweekspoort).•

contorta (L.f.) T.Durand & Schinz   Rhizomatous geophyte, 15–30 cm, stems sprawling, becoming diffuse. Leaves strap-shaped to linear, with flattened pseudopetiole. Flowers in lax racemes, blue, nodding, filaments scabrid and striped blue and white. Nov.–Mar. Mainly sandstone slopes, NW, SW, AP, KM, LB, SE (Namaqualand to Stutterheim).

sp. 1   Like **C. contorta** but more robust, 30–40 cm, flowers large, cream to mauve. Nov.–Feb. Coastal sands, SW (Saldanha to Blouberg).•

HYACINTHACEAE
with D.A. Snijman, **Lachenalia** by G.D. Duncan

1. Bracts (at least the lower) spurred; leaves often dry at flowering; seeds with a loose testa:
    2. Flowers long-lived, tepals persistent in fruit, free; inflorescence branched and voluble, somewhat fleshy ....**Bowiea**
    2.' Flowers short-lived, tepals caducous, cohering above when faded, circumscissile below, often connate below; inflorescence simple or rarely branched but then wiry ....**Drimia**
        a. Anthers subsessile or filaments to 2.5 mm long, often included in perianth; flowers moodding, often campanulate:
            b. Flowers 1 or 2, cylindric; anthers subsessile, inserted in middle of perianth tube, opening by longitudinal slits ....***Litanthus* group**
            b.' Flowers several, urn-shaped or cup-shaped; anthers connivent over ovary, inserted near base of perianth tube, opening by pores or by longitudinal slits ....***Rhadamanthus* group**
        a.' Anthers exserted and filaments longer than 1 mm:
            c. Raceme erect, firm; perianth tubular below, tepals united for about a third of their length, recurved above; stamens usually connivent around style; capsules large, c. 10 mm long, conspicuously 3-angled or -winged...***Drimia* group**
            c.' Raceme usually flexuose or wiry; perianth campanulate or stellate, tepals free or shortly united below, spreading or recurved; stamens erect or spreading, rarely ent around style; capsules smaller, obscurely 3-lobed:
                d. Leaf bases enclosed in an elongated, transversely ridged and banded sheath; tepals free; stamens clustered around ovary, with basifixed anthers; style declinate ....***Tenicroa* group**
                d.' Leaf bases without an elongate, banded sheath; tepals free or united below; anthers usually dorsifixed; style straight or declinate:
                    e. Raceme simple, deciduous; pedicels erect or spreading in fruit ....***Urginea* group**
                    e.' Raceme branched, persistent; pedicels erect or deflexed in fruit ....***Schizobasis* group**
1.' Bracts not spurred; leaves mostly green at flowering; seeds with an adherent testa:

3. Ovary stipitate, abruptly expanded above base and mushroom-shaped; ovules 2 in each locule, apparently basal; leaves often spotted; inflorescences often several and apparently axillary ....**Ledebouria**
3.' Ovary usually sessile, subglobose or ovoid; ovules 2 to several in each locule; axillary; inflorescence solitary and obviously central:
    4. Filaments either inserted in 2 series at different levels or inserted obliquely on perianth; bracts often vestigial:
        5. Plants robust with several oblong-lanceolate leaves: perianth tube much longer than the tepals, the filaments inserted in a single series in the middle of the tube; ovules 2 per locule ....**Veltheimia**
        5.' Plants not as above:
            6. Tepals fused for more than 1/4 their length, the inner and outer similar ....**Polyxena**
            6.' Tepals fused for less than 1/4 their length, or only the outer fused for more, the inner and outer often dissimilar...**Lachenalia**
    4.' Filaments inserted in 1 series at the same level (rarely obliquely in the lower flowers only, which are then highly zygomorphic); bracts always well developed:
        7. Leaves 2, spreading or prostrate; inflorescence not distinctly pedunculate; tepals fused below into a cup or tube:
            8. Flowers yellow or red; bracts shorter or longer than flowers..**Daubenya**
            8.' Flowers white to pink; bracts longer than flowers:
                9. Inflorescence spicate and erect above leaves; bracts all green and fleshy..................**Whiteheadia**
                9.' Inflorescence corymbose or capitate and sessile at ground level between leaves, rarely somewhat racemose; bracts mostly membranous ....**Massonia**
        7.' Leaves several, usually ascending; inflorescence distinctly pedunculate; tepals more or less free or fused at base, rarely more:
            10. Inflorescence topped by a coma of leafy bracts longer than flowers.................................**Eucomis**
            10.' Inflorescence not topped by a coma of leafy bracts:
                11. Flowers usually secund on an inclined raceme; tepals fused below into an elongate tube at least half as long as lobes; stamens included; filaments closely adpressed to inner tepals and anthers appearing sessile ....**Dipcadi**
                11.' Flowers never secund; tepals free or fused at base; stamens more or less exserted and free from tepals above:
                    12. Flowers blue; tepals recurved above..**Scilla**
                    12.' Flowers white, yellow, orange or green; tepals erect or spreading:
                        13. Inner tepals erect, cucullate, enclosing stamens and ovary; inner stamen filaments at least pinched below, outer whorl sometimes sterile; style usually prismatic and papillate ....**Albuca**
                        13.' Tepals all spreading, subequal; filaments sometimes expanded below but never pinched; style terete and smooth:
                            14. Stigma bluntly 3-lobed; seeds angular to discoid ....**Ornithogalum**
                            14.' Stigma with 3 divergent branches; seeds pear-shaped or subglobose ....**Neopatersonia**

ALBUCA   SLIME LILY, SLYMLELIE, TAMARAK   c. 60 spp., W Cape to Arabia

A. Inner tepals with a hinged fleshy flap at apex, flowers nodding

acuminata Baker (= *Albuca convoluta* E.Phillips)   Bulbous geophyte, 20–30 cm, bulb scales becoming fibrous above. Leaves several, channelled, slender, clasping below. Flowers yellow to green with green keels, outer stamens sterile. Sept.–Oct. Deep sands, NW, SW, KM, LB (Namaqualand to Mossel Bay).

ciliaris U.Müll.-Doblies   Bulbous geophyte to 20 cm, bulb depressed-globose. Leaves several, short, not clasping, flat, narrowly oblong, often twisted, margins ciliate. Flowers few, dull greenish, outer stamens sterile. Sept.–Oct. Rocky sandstone slopes, NW (Namaqualand to Clanwilliam).

cooperi Baker (= *Albuca karooica* U.Müll.-Doblies)   Bulbous geophyte, 35–60 cm, bulb tunics fibrous above. Leaves 2 or 3, slender, channelled, conspicuously clasping and warty beneath. Flowers yellow with green keels, outer stamens sterile, septa sometimes lightly crested. Sept.–Nov. Stony, mostly sandy slopes and flats, sometimes limestone, NW, SW, AP, KM, LB, SE (Richtersveld and W Karoo to Cape Peninsula to Willowmore).

echinosperma U.Müll.-Doblies   Like **A. flaccida** but more delicate with leaves fewer, 1–3, and seeds distinctly papillate. Aug.–Oct. Rocky sandstone slopes, NW, SW (Piketberg to Hermanus).•

flaccida Jacq. (= *Albuca canadensis* auct., *A. materfamilias* U.Müll.-Doblies)   Bulbous geophyte, 40–100 cm, bulb tunics membranous. Leaves channelled, fleshy, clasping below. Flowers mostly yellow, sometimes with green keels, lightly fragrant, outer stamens sterile. Aug.–Oct. Mostly coastal in deep sandy soils, NW, SW, AP (S Namaqualand to Stilbaai).

hallii U.Müll.-Doblies (= *Albuca brucebayeri* U.Müll.-Doblies) Bulbous geophyte, 10–15 cm. Leaves few, linear, glandular, corkscrewed above, obtuse. Flowers yellow with green keels, outer stamens sterile. Mar.–May. Stony slopes, NW, KM (Vredendal, Little Karoo).•

juncifolia Baker (= *Albuca imbricata* F.M.Leight.) Bulbous geophyte, 15–30 cm. Leaves several, slender and stiff, channelled below but often terete above, not clasping below. Flowers in drooping racemes, yellow with green keels, outer stamens sterile. Sept.–Oct. Sandy and calcareous flats, NW, SW, AP, LB (Ceres to Cape Peninsula to Mossel Bay).•

massonii Baker Like **A. acuminata** but flowers smaller, c. 1 cm long and style short, about half as long as ovary. Sept.–Oct. Sandstone slopes, NW (Gifberg to Pakhuis Mts).•

maxima Burm.f. (= *Albuca altissima* Dryand.) WITTAMARAK Bulbous geophyte, 40–150 cm, bulb tunics slightly fibrous above. Leaves lanceolate, channelled, fleshy, clasping below. Flowers white with green keels, outer stamens sterile, filaments oblanceolate, septa crested. Aug.–Oct. Rocky sandstone or granitic soils, NW, SW, KM, LB (Namaqualand to Riversdale).

namaquensis Baker GROWWETAMARAK Bulbous geophyte to 30 cm, bulb tunics membranous. Leaves many, channelled or rolled, usually coiled above, not clasping below, scabrid or hairy. Flowers yellow with green keels, outer stamens sterile. Sept.–Oct. Stony sandstone slopes, NW, SW, KM, SE (Namibia E Cape).

navicula U.Müll.-Doblies Bulbous geophyte to 20 cm, bulb depressed-globose. Leaves several, short, oblong, closely ciliate, clasping below, longitudinally folded and boat-shaped above. Flowers few, dull greenish, outer stamens sterile. Sept.–Oct. Red sandy flats, NW (Namaqualand to Clanwilliam).

paradoxa Dinter Like **A. flaccida** but bulb conspicuously depressed and fragmenting into segments. July–Aug. Sandy flats and slopes, NW (S Namibia to Clanwilliam).

spiralis L.f. Bulbous geophyte, 20–40 cm. Leaves few to many, linear, channelled or rolled, often spirally twisted above, clasping below, glandular-hairy and peduncle also glandular-hairy below. Flowers green, outer tepals with cream to yellow margins, sweetly fragrant, outer stamens sterile. Aug.–Oct. Sandy and stony slopes, NW, SW (Namaqualand to Cape Peninsula).

A. Inner tepals hooded or cowled at apex
B. Flowers nodding

clanwilliamigloria U.Müll.-Doblies Bulbous geophyte to 2 m, bulb depressed-globose, producing bulbils. Leaves narrow, channelled, fleshy, dry at flowering. Flowers in slender, elongate racemes, scarcely flaring, large, dull yellow with dark yellow or green keels, outer anthers slightly smaller. Oct.–Nov. Deep sandy soils in restioid fynbos, NW (Olifants River valley to Piketberg).•

foetida U.Müll.-Doblies Bulbous geophyte, 20–40 cm. Leaves few, channelled, clasping below, glandular-hairy, peduncle also glandular-hairy. Flowers yellow with green keels, outer stamens sterile. Sept.–Oct. Stony slopes, NW, SW, KM (Namaqualand and W Karoo to Tulbagh, Oudtshoorn).

fragrans Jacq. Slender, sometimes stout, bulbous geophyte to 1 m, bulb sometimes bulbilliferous. Leaves linear, channelled, shortly clasping below. Flowers in drooping racemes, yellow with green keels, outer anthers slightly smaller. Sept.–Oct. Sandy slopes and flats, often coastal, NW, SW (Bokkeveld Mts to Hermanus).•

goswinii U.Müll.-Doblies Bulbous geophyte, 15–60 cm. Leaves 2 or 3, linear, channelled, clasping below. Flowers in drooping racemes, yellow with green keels, outer stamens sterile. Sept.–Oct. Stony slopes, SW, LB (Houwhoek to Riversdale).•

papyracea J.C.Manning & Goldblatt Slender, bulbous geophyte, 50–80 cm, bulb tunics papery and fibrous above. Leaves 2, linear, clasping, sometimes warty below. Flowers in drooping racemes, yellow with green keels, inner anthers curved, outer slightly smaller, ovary with diverging septal ridge. Sept.–Nov. Stony shale slopes, KM (Little Karoo).•

viscosa L.f. (= *Albuca aspera* U.Müll.-Doblies, *A. bontebokensis* U.Müll.-Doblies, *A. jacquinii* U.Müll.-Doblies, *A. viscosella* U.Müll.-Doblies) TAAITAMARAK Bulbous geophyte, 20–40 cm, bulbs ovoid, often pink, tunics dry and wrinkled above. Leaves semiterete, often spirally twisted above, glandular-hairy, not clasping below, peduncle base also glandular-hairy. Flowers yellow with green keels, outer anthers smaller, fragrant. Aug.–Oct. Rocky flats, NW, SW, AP, KM, LB (Namibia and Karoo to Riversdale).

B. Flowers erect

aurea Jacq.   Bulbous geophyte to 50 cm, bulbs often bluish. Leaves few, lanceolate, channelled, clasping below. Flowers erect, often subsecund on an inclined peduncle, whitish or yellow with green keels, inner tepals cowled and yellowish, outer anthers smaller, ovary septa crested. Mainly Sept.–Dec. Stony clay slopes, NW, SW, KM (Clanwilliam to Worcester and Little Karoo).

batteniana Hilliard & B.L.Burtt   Bulbous geophyte to 60 cm, bulb epigeal, green, scales firm, truncate and fibrous above. Leaves lanceolate, firm. Flowers erect on long pedicels, subsecund on an inclined peduncle, white with pale greenish keels, inner tepals cowled, outer anthers smaller, ovary septa crested. July–Oct. Coastal cliffs, SE (Knysna and Kei Mouth).

decipiens U.Müll.-Doblies   Bulbous geophyte to 80 cm. Leaves few, channelled, clasping below. Flowers erect, white with green keels, outer stamens sterile, ovary septa crested. Aug.–Sept. Rocky slopes, NW (Namaqualand to Olifants River valley).

exuviata Baker   Bulbous geophyte to 30 cm, bulb tunics coarsely fibrous above, forming a thick neck with conspicuous, woody rings. Leaves linear-filiform, not clasping. Flowers erect, subcorymbose, white to yellow with green keels, inner tepals cowled, all stamens fertile, ovary septa crested. Mainly Aug.–Sept. Clay soils, KM, SE (Little Karoo to E Cape).

glandulosa Baker   Bulbous geophyte to 35 cm, glandular hairy, bulb tunics membranous with rings around neck. Leaves narrow, channelled, clasping below. Flowers erect, yellow with greenish keels, inner tepals cowled, outer stamens sterile, ovary septa crested. Aug.–Sept. Dry stony shale slopes, KM (Namaqualand and W Karoo to Little Karoo).

longipes Baker   Bulbous geophyte to 30 cm, outer bulb tunics often dry and wrinkled above. Leaves few, linear, channelled, not clasping below, dry at flowering. Flowers often corymbose, erect on long pedicels, white with green keels, inner tepals cowled and bright yellow at tips, outer anthers smaller, ovary septa crested. Sept.–Nov. Clay or lime slopes, NW, SW, AP, LB, SE (Richtersveld and W Karoo to Cape Peninsula to Willowmore).

schonlandii Baker   Bulbous geophyte to 30 cm, bulb tunics dry and firm. Leaves oblong, flat, margins hyaline. Flowers erect on long pedicels, subcorymbose, white and green, inner tepals cowled, outer anthers smaller, ovary septa crested. Sept.–Nov. Dry sandstone slopes, KM, SE (Oudtshoorn to E Cape).

setosa Jacq.   DIKTAMARAK   Bulbous geophyte, 15–60 cm, bulb large, basal sheaths fibrous. Leaves several, narrow, fleshy, not clasping below, margins hyaline and often minutely ciliolate. Flowers erect on long pedicels, yellow with green keels, inner tepals cowled, outer anthers slightly smaller, ovary septa crested. Oct.–Nov. Rocky clay flats and slopes, NW, SW, KM, SE (S Namaqualand to Swaziland).

[Species excluded   No authentic material found and possibly conspecific with one of the above: **A. hesquasportensis** U.Müll.-Doblies, **A. robertsoniana** U.Müll.-Doblies, **A. weberlingiorum** U.Müll.-Doblies]

BOWIEA   KNOLKLIMOP   1 sp., southern and tropical Africa

volubilis Harv. ex Hook.f.   Bulbous geophyte, bulb partly exposed, green. Leaves filiform, dry at flowering. Flowers in a trailing, fleshy, diffusely branched raceme, green, tepals reflexed. Jan.–Feb. Rock outcrops and bush margins, SE (S Namibia, Baviaanskloof Mts to tropical Africa).

DAUBENYA (= *ANDROSIPHON, AMPHISIPHON*)   PINCUSHION LILY   5 spp., N and W Cape

angustifolia (L.f.) A.M.van der Merwe & J.C.Manning (= *Massonia zeyheri* Kunth)   Bulbous geophyte to 10 cm. Leaves prostrate, shiny green, bracts small to 10 mm. Flowers clustered between leaves, tubular below, yellow or red, filaments orange. May–June. Sandy and clay flats, NW, SW (Bokkeveld Mts and W Karoo, Saldanha).

capensis (Schltr.) A.M.van der Merwe & J.C.Manning   (= *Androsiphon capense* Schltr.)   Bulbous geophyte to 5 cm, bracts large. Leaves prostrate, ovate, bracts large. Flowers clustered between the leaves, tubular below, yellow and red, stamens fused below into a slender tube c. 10 mm long. June–Sept. Doleritic clays, NW (Bokkeveld Escarpment and W Karoo).

stylosa (W.F.Barker) A.M.van der Merwe & J.C.Manning   (= *Amphisiphon stylosus* W.F.Barker)   Bulbous geophyte to 5 cm. Leaves prostrate, ovate, bracts small to 10 mm. Flowers clustered between leaves, tubular below, tepals fused above into a tube with minute lobes, stamens fused below into a slender tube c. 10 mm long, yellow, scented. May–June. Clay soils, NW (Bokkeveld Escarpment).•

DIPCADI SLANGUI   30 spp., Africa, Mediterranean and India

brevifolium (Thunb.) Fourc.   Bulbous geophyte, 20–40 cm. Leaves 2–4, sometimes dry at flowering, linear to filiform, straight or coiled. Flowers brown, green or cream. Aug.–Apr. Stony flats or slopes, NW, SW, AP, KM, LB, SE (Namaqualand and W Karoo to E Cape).

ciliare (Zeyh. ex Harv.) Baker   Bulbous geophyte to 40 cm. Leaves c. 6, linear-lanceolate, often coiled and usually roughly hairy beneath, margins ciliate and usually crisped or undulate. Flowers brown, green, or yellowish. Nov.–May. Stony flats or slopes, KM, SE (Oudtshoorn to Port Elizabeth, Karoo to northern South Africa).

crispum Baker   KRULUI   Bulbous geophyte to 30 cm. Leaves c. 4, linear-lanceolate, coiled, grey, usually softly hairy, margins usually crisped. Flowers brown to grey-green, outer tepals with short, caudate appendages. Apr.–Dec. Stony flats or slopes, NW (Namibia and W Karoo to Clanwilliam).

viride (L.) Moench   SKAAMBLOMMETJIE   Bulbous geophyte, 15–120 cm. Leaves 1–4, long, linear to lanceolate. Flowers green to brown, outer tepals with short or long, filiform appendages. Sept.–Feb. Stony flats or slopes, KM, LB, SE (Ladismith and Riversdale to Port Elizabeth, widespread in eastern southern Africa and N Namibia to Ethiopia).

DRIMIA (= *LITANTHUS, RHADAMANTHUS, SCHIZOBASIS, TENICROA, URGINEA*)   POISON SQUILL   c. 60 spp., Africa, Mediterranean, Asia

Drimia-group   BRANDUI, JEUKBOL

capensis (Burm.f.) Wijnands (= *Drimia altissima* (L.f.) Ker Gawl., *D. forsteri* (Baker) Oberm.)   MAERMAN   Bulbous geophyte, 1–2 m. Leaves dry at flowering, spreading, oblong to lanceolate. Flowers in an elongate densely whorled raceme, white or cream, often subsessile, tepals reflexed, stamens connivent, anthers green, 6–8 mm long. Dec.–Mar. Clay and lime soils, NW, SW, AP, KM, LB, SE (S Namaqualand to Port Elizabeth).

elata Jacq. ex Willd.   Bulbous geophyte to 100 cm. Leaves dry at flowering, erect to spreading, linear-lanceolate, often undulate, sometimes hairy, margins ciliate. Flowers in erect racemes, silvery white, green or purple, tepals reflexed, stamens connivent, anthers c. 2 mm long, blue to purple. Mainly Dec.–Apr. Sandy and clay soils, NW, SW, AP, KM, LB, SE (S Namaqualand to Cape Peninsula to East Africa).

haworthioides Baker   Bulbous geophyte, 20–40 cm, bulbs exposed, with loose scales. Leaves dry at flowering, spreading, lanceolate, margins often ciliate. Flowers like **D. elata**. Nov.–Feb. Dry karroid areas, NW, KM, SE (Worcester to E Cape and Karoo).

media Jacq. ex Willd.   Bulbous geophyte, 30–55 cm, usually evergreen. Leaves suberect, subterete, firm. Flowers like **D. elata**. Jan.–Mar. Sandy coastal flats and slopes, SW, AP, LB, SE (Saldanha, Cape Peninsula to Knysna).•

pusilla Jacq.   Like **D. elata** but much smaller, 6–15 cm, few-flowered, capsules depressed-oblong, at least as broad as long and seeds large, discoid. Mainly Dec.–Apr. Clay soils, NW, SW (S Namaqualand to Cape Peninsula).

Litanthus-group   FAIRY SNOWDROP

uniflora J.C.Manning & Goldblatt (= *Litanthus pusillus* Harv.)   Bulbous geophyte, 2–8 cm. Leaves dry at flowering, suberect, filiform. Flower solitary, sometimes 2, nodding, tubular, white to pale pink. Dec.–Mar. Rock outcrops and flushes, NW, KM, SE (Namaqualand to Zimbabwe).

Rhadamanthus-group   AFRICAN SNOWDROP

A. Anthers apiculate or tailed below

arenicola (B.Nord.) J.C.Manning & Goldblatt (= *Rhadamanthus arenicola* B.Nord.)   Bulbous geophyte, 10–15 cm, with bulb scales loosely overlapping. Leaves suberect, linear. Flowers in racemes, nodding, urn-shaped, whitish or light brown with brown keels, anthers apiculate below, porose. Oct.–Nov. Sandy habitats, NW (Namaqualand and W Karoo to Clanwilliam).

platyphylla (B.Nord.) J.C.Manning & Goldblatt (= *Rhadamanthus platyphyllus* B.Nord.)   Bulbous geophyte, 3–15 cm. Leaves dry at flowering, usually 2, prostrate, elliptic, velvety. Flowers in racemes, nodding, bell-shaped, reddish brown to creamy pink, anthers barbellate below, porose. Nov.–Jan. Widespread, NW, SW, AP, KM, SE (Namibia, Namaqualand and Karoo to Bredasdorp, Uniondale).

uranthera (R.A.Dyer) J.C.Manning & Goldblatt (= *Rhadamanthus urantherus* R.A.Dyer)   Bulbous geophyte, 12–20 cm. Leaves dry at flowering, suberect, linear. Flowers in racemes, nodding, urn-shaped, light brown, anthers tailed, porose. Mar.–Apr. Stony slopes, KM (Oudtshoorn).•

A. Anthers rounded below

albiflora (B.Nord.) J.C.Manning & Goldblatt (= *Rhadamanthus albiflorus* B.Nord.)   Bulbous geophyte, 12–24 cm. Leaves unknown. Flowers in racemes, nodding, dish-shaped, white with brown keels, anthers porose. Dec. Mountain slopes, SW, KM (Montagu to Stormsvlei).•

convallarioides (L.f.) J.C.Manning & Goldblatt (= *Rhadamanthus convallarioides* (L.f.) Baker, *R. montanus* B.Nord.)   Bulbous geophyte, 5–30 cm. Leaves dry at flowering, suberect, subterete, sometimes with a barred sheath below. Flowers in racemes, nodding, bell-shaped, creamy pink to light brown, anthers porose. Oct.–Feb. Sandy soils, NW, SW, KM (Namaqualand and W Karoo to Jonkershoek Mts and Little Karoo).

involuta (J.C.Manning & Snijman) J.C.Manning & Goldblatt (= *Rhadamanthus involutus* J.C.Manning & Snijman)   Bulbous geophyte, 10–15 cm. Leaves dry at flowering, suberect, filiform. Flowers in racemes, cupped, white with green basal markings, outer tepals spreading apically, inner tepals conduplicate above, anthers porose. Nov.–Dec. Loamy soil on exposed sandstone pavement, NW (Bokkeveld Mts).•

karrooica (Oberm.) J.C.Manning & Goldblatt (= *Rhadamanthus karooicus* Oberm.)   Bulbous geophyte to 20 cm. Leaves dry at flowering, 4–6, spreading, oblong. Flowers in racemes, nodding, urn-shaped, pale lilac to green, anthers longitudinal. Jan.–Feb. KM (W and Little Karoo).

Schizobasis-group VOLSTRUISKOS

intricata (Baker) J.C.Manning & Goldblatt (= *Schizobasis intricata* (Baker) Baker)   Bulbous geophyte, 10–50 cm. Leaf ephemeral, dry at flowering. Flowers in densely branched panicles, small, white, pale yellow or pink. Jan.–Mar. Stony slopes, NW, KM, SE (Botterkloof to Port Elizabeth, dry areas of southern and tropical Africa).

Tenicroa-group GIFBOL

exuviata (Jacq.) Jessop (= *Tenicroa exuviata* (Jacq.) Speta)   Bulbous geophyte to 1 m. Leaves about as long as raceme, few, erect, leathery, greyish, 3–4 mm diam., enclosed below in a banded sheath. Flowers rotate, white with green keels, often flushed purple, fragrant, tepals more or less free. Sept.–Oct. NW, SW, AP, KM, LB, SE (Namaqualand to Port Elizabeth, E Cape).

filifolia (Jacq.) J.C.Manning & Goldblatt (= *Tenicroa filifolia* (Jacq.) Oberm.)   Like **D. exuviata** but to 30 cm, leaves few to many, c. 1 mm diam. Sept.–Dec. Clay soils, NW, SW, AP, KM, LB (Bokkeveld Escarpment to Swellendam).•

fragrans (Jacq.) J.C.Manning & Goldblatt (= *Tenicroa fragrans* (Jacq.) Raf.)   Bulbous geophyte, 30–80 cm. Leaves subsucculent, shorter than raceme, many, lightly flexuose, c. 1 mm diam., enclosed below in a banded sheath. Flowers rotate, white with green or purple keels, fragrant, tepals more or less free. Sept.–Nov. Sandy flats, NW (Bokkeveld Mts to Hex River Valley).•

multifolia (G.J.Lewis) Jessop (= *Tenicroa multifolia* (G.J.Lewis) Oberm.)   Like **D. fragrans** but to 20 cm, leaves filiform, coiled, c. 0.5 mm diam. Sept.–Oct. Poorly drained soils, NW, SW (S Namibia, Namaqualand and W Karoo to Breede River valley).

Urginea-group SLANGKOP

A. Inflorescence corymbose

marginata (Thunb.) Jessop   Bulbous geophyte to 20 cm. Leaves dry at flowering, 2 or 3, prostrate, leathery, shiny, oblong, broadly obtuse, margins thickened and cartilaginous. Flowers like **D. minor**. Oct.–Jan. Sandy or stony slopes, NW (W Karoo to Hex River Mts).

minor (A.V.Duthie) Jessop   Bulbous geophyte, 8–20 cm. Leaves dry at flowering, erect, terete. Flowers in a head-like corymbose raceme, stellate, white to brownish, open in the evening. Feb.–Apr. Sandy flats, NW, SW, KM, LB (Bokkeveld Mts to Swartberg and Langeberg Mts).•

physodes (Jacq.) Jessop (incl. **Urginea pusilla** (Jacq.) Baker)   BERGSLANGKOP   Bulbous geophyte to 10 cm, bulbs usually large. Leaves dry at flowering, erect, twisted, lanceolate. Flowers in a subcorymbose raceme on long, spreading pedicels, white, open in the afternoon. Oct.–Apr. Stony flats, NW, KM (Namaqualand and W Karoo to Worcester and Little Karoo).

virens (Schltr.) J.C.Manning & Goldblatt   Bulbous geophyte to 20 cm. Leaves 1 or 2, falcate, fleshy. Flowers like **D. minor**. Sept.–Feb. Stony flats, NW, AP, KM (W Karoo to Little Karoo and Stilbaai).•

sp. 1 (= *Urginea barkerae* Oberm. ms)   Bulbous geophyte to 20 cm, scape finely hairy below. Leaves elliptic, coarsely hairy. Flowers like **D. minor**. Oct.–Dec. Shale slopes, NW (Piketberg).•

A. Inflorescence racemose

ciliata (L.f.) Baker (= *Urginea rosulata* Oberm. ms)   Bulbous geophyte, 7–15 cm. Leaves rosulate, spreading, elliptic to oblanceolate, leathery, margins hyaline and papillate-ciliolate. Flowers like **D. sclerophylla**. Dec.–Feb. Rocky outcrops, AP, SE (De Hoop to Grahamstown).

dregei (Baker) J.C.Manning & Goldblatt (= *Urginea dregei* Baker; incl. **U. gracilis** A.V.Duthie)   Bulbous geophyte to 30 cm. Leaf dry at flowering, solitary, erect, terete. Flowers in slender, crowded racemes, mostly white with dark keels, tepal margins downturned, open during the morning. Nov.–Mar. Damp sandy flats and slopes, SW, LB (Cape Peninsula to Swellendam).•

revoluta (A.V.Duthie) J.C.Manning & Goldblatt (= *Urginea revoluta* A.V.Duthie)   Bulbous geophyte, 10–30 cm, scape scabrid below. Leaves dry at flowering, single, sometimes 2, terete, firm, erect. Flowers in flexuose racemes, brown, tepals more or less free, reflexed, stamens connivent, lightly scented, open at night. Jan.–Feb., mostly after fire. Rocky slopes and flats, SW, AP (Du Toitskloof to De Hoop).•

salteri (Compton) J.C.Manning & Goldblatt   Bulbous geophyte to 25 cm. Leaves dry or emergent at flowering, 2–several, erect or falcate, terete. Flowers crowded in slender racemes, whitish to brown or maroon, open at night to midmorning. Oct.–Feb. Sandy or stony soils, SW (Bainskloof to Hermanus).•

sclerophylla J.C.Manning & Goldblatt (= *Urginea rigidifolia* Baker)   Bulbous geophyte, 10–30 cm, scape scabrid below. Leaves dry at flowering, 2 or 3, erect, terete, sclerotic-striate. Flowers in lax racemes on wiry pedicels, white, stellate, tepals more or less free, sweetly scented, opening in late afternoon. Dec.–Jan. Rocky slopes, AP, KM, LB, SE (De Hoop to Port Elizabeth).•

sp. 2 (= *Urginea cataphyllata* Oberm. ms)   Bulbous geophyte to 50 cm, with banded, papery cataphylls, stems scabrid below. Leaves usually dry at flowering, single, terete, firm, erect. Flowers in a narrow raceme, whitish, tepals lighly reflexed. Oct.–Feb. Sandy or stony soils, SE (Baviaanskloof and S Karoo).

[Species excluded   No authentic material found and probably conspecific with one of the above. **U. ecklonii** Baker]

EUCOMIS   PINEAPPLE LILY   10 spp., N and W Cape to S tropical Africa

autumnalis (Mill.) Chitt.   Bulbous geophyte, 6–30 cm. Leaves suberect, oblong-lanceolate, uniformly green, margins undulate. Flowers white to greenish, spike leafy above, pedicels 3–9 mm long. Dec.–Feb. Rocky, grassy slopes, SE (Knysna northwards throughout E southern Africa).

comosa (Houtt.) H.R.Wehrh.   Bulbous geophyte, 17–100 cm. Leaves suberect, oblong-lanceolate, speckled purple beneath. Flowers greenish with dark centre, spike leafy above, pedicels 15–30 mm long. Dec.–Feb. Grassland and vleis, SE (Port Elizabeth to KwaZulu-Natal).

regia (L.) L'Hér.   Bulbous geophyte, 8–15 cm. Leaves usually prostrate, oblanceolate-spathulate, uniformly green. Flowers cream to greenish, spike leafy above, pedicels to 2 mm long. July–Sept. Mostly cooler S-facing clay slopes, NW, SW, AP, KM, LB (Namaqualand and W Karoo to Bredasdorp, Little Karoo).

LACHENALIA   LACHENALIA, VIOOLTJIE   c. 110 spp., Namibia to E Cape

*A. Anthers well exserted (see also **L. contaminata**, **L. hirta**)*

anguinea Sweet   Bulbous geophyte, 10–35 cm. Leaf solitary, narrowly lanceolate, banded green and maroon. Flowers on long pedicels, campanulate, cream with green markings, anthers well exserted. July–Sept. Deep coastal sands, NW (Richtersveld to Klawer).

gillettii W.F.Barker   Bulbous geophyte, 12–22 cm. Leaves 2, lorate. Flowers shortly pedicellate, oblong-campanulate, white and lilac with green markings, anthers exserted. Aug.–Sept. Clay soils, often in large colonies, NW (Piketberg to Citrusdal).•

haarlemensis Fourc.   Bulbous geophyte, 12–22 cm. Leaves 1 or 2, linear-lanceolate, erect, banded with maroon at base. Flowers shortly pedicellate, campanulate, greenish grey, anthers exserted and mauve. Sept.–Oct. Habitat? SE (Kammanassie Mts to Langkloof).•

juncifolia Baker (incl. **L. esterhuysenae** W.F.Barker) Bulbous geophyte, 7–40 cm. Leaves 2, linear, filiform, terete or subterete, with maroon bands. Flowers on long pedicels, oblong-campanulate, white or pink, tinged darker pink or blue, with purple or green markings, anthers exserted. Aug.–Nov. Often in sand in large colonies, NW, SW, AP (Cedarberg Mts to Stilbaai).•

karooica W.F.Barker ex G.Duncan Bulbous geophyte, 4–22 cm. Leaves 1 or 2, lanceolate, with blotches on upper surface. Flowers sessile or shortly pedicellate, oblong-campanulate, greenish white and pale blue with maroon or brown markings, anthers exserted. June–Sept. Rocky outcrops, NW (Worcester, Great Karoo to W Free State).

lactosa G.D.Duncan Bulbous geophyte, 10–25 cm. Leaves 1 or 2, oblong, maroon beneath. Flowers pedicellate, urceolate, bluish or greenish white, anthers shortly exserted, peduncle heavily blotched. Sept.–Oct. Sandy coastal flats, SW (Botrivier to Elim).•

leipoldtii G.D.Duncan Bulbous geophyte, 10–28 cm. Leaf 1, lanceolate, spotted. Flowers subsessile, campanulate, cream to greenish yellow, anthers well-exserted, peduncle sometimes inflated. Aug.–Sept. Sandstone slopes, NW, KM (Biedouw to Waboomsberg).•

latimerae W.F.Barker Bulbous geophyte, 15–28 cm. Leaves 1 or 2, linear-lanceolate. Flowers on long pedicels, campanulate, pale pink with greenish brown markings, anthers exserted. July–Aug. Sand in large colonies, KM, SE (Swartberg and Kouga Mts).•

mathewsii W.F.Barker Bulbous geophyte, 10–20 cm. Leaves 2, narrowly lanceolate tapering to a long terete apex. Flowers shortly pedicellate, yellow with green markings, anthers exserted. Sept. Moist lower slopes, SW (Vredenburg).•

moniliformis W.F.Barker Bulbous geophyte, 12–17 cm. Leaves several, terete with circular, raised, fleshy bands along upper two-thirds. Flowers on long pedicels, pale blue and pink with reddish brown markings, anthers well exserted. Sept. Sandy flats, NW (Worcester district).•

montana Schltr. ex W.F.Barker Bulbous geophyte, 10–33 cm. Leaves 2, linear, conduplicate, unmarked. Flowers on long magenta pedicels, campanulate, nodding, cream or pink with large brownish green markings, anthers exserted. Oct.–Dec., only after fire. Sandy mountain slopes, SW (Franschhoek to Hermanus).•

nervosa Ker Gawl (= *Lachenalia latifolia* Tratt.) Bulbous geophyte, 15–30 cm. Leaves 2, ovate, prostrate, plain or with pustules on upper surface. Flowers shortly pedicellate, campanulate, white with reddish pink markings, anthers exserted. Sept.–Nov. Habitat? AP, LB, SE (Swellendam to George).•

physocaulos W.F.Barker Bulbous geophyte, 13–30 cm. Leaf solitary, linear, widening abruptly into a white clasping base. Flowers shortly pedicellate, campanulate, pale magenta with brownish green markings, anthers exserted, peduncle heavily spotted, swollen. Aug.–Sept. Sandy flats and slopes, NW, LB (Robertson to Swellendam).•

polyphylla Baker Bulbous geophyte, 6–18 cm. Leaves several, erect, terete,. Flowers on long pedicels, narrowly campanulate, pale blue and pink with brownish markings, anthers exserted. Sept.–Oct. Gravel flats, NW (Piketberg to Tulbagh).•

purpureocaerulea Jacq. Bulbous geophyte, 10–28 cm. Leaves 2, lanceolate or lorate, densely pustulate. Flowers campanulate, white and purplish blue with greenish brown markings, anthers exserted. Oct.–Nov. Gravel flats, SW (Darling and Mamre).•

pusilla Jacq. Bulbous geophyte, 1–4 cm. Leaves in a rosette, prostrate, linear to lanceolate, plain or spotted. Flowers cylindrical, erect, borne at ground level, in a congested raceme, white, heavily scented, anthers exserted. Apr.–June. Common on sandy flats and slopes, NW, SW, AP, LB (Bokkeveld Mts to Swellendam).•

pustulata Jacq. Bulbous geophyte, 15–35 cm. Leaves 1 or 2, lanceolate or lorate, smooth or densely pustulate, unmarked. Flowers on long pedicels, oblong-campanulate, shades of cream, blue or pink with green or brownish markings. Aug.–Oct. Often in large colonies in heavy soil, SW (St Helena Bay to Cape Peninsula).•

salteri W.F.Barker Bulbous geophyte, 15–35 cm. Leaves 2, lanceolate, leathery, plain or with large brown blotches. Flowers shortly pedicellate, oblong-campanulate, cream, reddish purple or a combination of pale blue and pink, anthers exserted. Oct.–Dec. Marshy areas around seasonal pools, SW, AP (Cape Peninsula to Bredasdorp).•

splendida Diels (= *Lachenalia roodiae* E.Phillips) Bulbous geophyte, 6–25 cm. Leaves 2, lanceolate, unmarked, peduncle swollen just below base of inflorescence. Flowers sessile, oblong-campanulate, pale blue and bright lilac, with greenish brown markings, anthers exserted. July–Aug. Usually on quartzite flats, NW (Garies to Klawer).

stayneri W.F.Barker   Bulbous geophyte, 12–30 cm. Leaves 2, lanceolate or lorate, prostrate, with large pustules on upper surface. Flowers on long pedicels, campanulate, pale blue and cream with reddish markings, anthers exserted. Aug.–Sept. Habitat? NW (Worcester to Robertson).•

thomasiae W.F.Barker ex G.Duncan   Bulbous geophyte, 12–38 cm. Leaves 2, lanceolate or lorate, yellowish green. Flowers on long pedicels, oblong-campanulate, white with green or brown markings, anthers well exserted. Sept.–Oct. Habitat? NW (Clanwilliam district).•

unicolor Jacq.   Bulbous geophyte, 8–30 cm. Leaves 2, lanceolate or lorate, densely pustulate on upper surface. Flowers on long pedicels, cream, lilac, pink, magenta, blue or purple, with green or purplish markings, anthers exserted. Sept.–Oct. Heavy soil in large colonies, NW, SW (Bokkeveld Mts to Somerset West).•

ventricosa Schltr. ex W.F.Barker   Bulbous geophyte, 20–48 cm. Leaf solitary, lanceolate to lorate, margins undulate. Flowers sessile, oblong urn-shaped, pale yellow with white tips, anthers exserted. Aug.–Sept. Usually in sand in large colonies, NW (Nardouw and Pakhuis Mts).•

violacea Jacq.   Bulbous geophyte, 10–35 cm. Leaves 1 or 2, lanceolate, plain or heavily spotted. Flowers on long pedicels, campanulate, bluish green at base, with magenta or purple tips, anthers mauve, magenta or white, exserted. July–Sept. Very variable, usually in rocky places, NW (Namaqualand to Clanwilliam).

A. Anthers included or barely protruding
*B. Flowers on long pedicels (see also **L. trichophylla**)*

aloides (L.f.) Engl. (= *Lachenalia tricolor* Thunb.)   CAPE COWSLIP, VIERKLEURTJIE   Bulbous geophyte, 5–31 cm. Leaves 1 or 2, lanceolate or lorate, plain or densely spotted with green or purple. Flowers on long pedicels, nodding, cylindrical, combinations of orange, red, yellow or greenish blue, with greenish markings, anthers included. May–Oct. Granite and sandstone outcrops, NW, SW (Lambert's Bay to Bredasdorp).•

bolusii W.F.Barker   Bulbous geophyte, 10–35 cm. Leaf solitary, ovate-lanceolate or lorate, banded maroon below. Flowers on long pedicels, campanulate, nodding, pale blue and white with brownish markings, anthers included. Aug.–Sept. Rocky outcrops, NW (Richtersveld to Clanwilliam).

bulbifera (Cirillo) Engl. (= *Lachenalia pendula* Aiton)   ROOINAELTJIE   Bulbous geophyte, 8–30 cm. Leaves 1 or 2, lanceolate, lorate or ovate, plain or blotched. Flowers on fairly long pedicels, cylindrical, nodding, orange to red with darker red or brown markings and green tips, anthers included. April–Sept. Sandy slopes and flats, mainly coastal, NW, SW, AP (Klawer to Mossel Bay).•

hirta (Thunb.) Thunb.   Bulbous geophyte, 10–30 cm. Leaf solitary, linear, with stiff hairs on margins and below, banded with maroon. Flowers on long pedicels, oblong-campanulate, blue to blue-grey with brown markings and pale yellow tips, anthers usually included, sometimes exserted. Aug.–Sept. Often in sandy soil in large colonies, NW, SW (Namaqualand to Malmesbury).

leomontana W.F.Barker   Bulbous geophyte, 10–30 cm. Leaf solitary, lanceolate or lorate, plain or with purple spots on upper surface. Flowers on long pedicels, oblong-campanulate, white with pale green markings, anthers included. Oct.–Nov. LB (Langeberg Mts: Swellendam).•

peersii Marloth ex W.F.Barker   Bulbous geophyte, 15–30 cm. Leaves 1 or 2, lorate, green or purpish, unmarked. Flowers on fairly long pedicels, urn-shaped, cream or white with greenish brown markings, strongly carnation-scented. Oct.–Nov. Often in partial shade, sandy soil, SW (Cape Hangklip to Hermanus).•

rosea Andrews   Bulbous geophyte, 8–30 cm. Leaves 1 or 2, lanceolate, plain or blotched with maroon or brown. Flowers shortly pedicellate, oblong-campanulate, pink or combinations of pink and blue, with darker pink markings, anthers included. Aug.–Dec. Mainly coastal, on moist flats, SW, AP, KM, LB, SE (Cape Peninsula to Knysna, Ladismith and Montagu).•

sargeantii W.F.Barker   Bulbous geophyte, 20–30 cm. Leaves 2, linear-lanceolate. Flowers on long pedicels, nodding, cream or pale green with green or brown markings, anthers included. Nov., only after fire. Sandstone slopes, SW (Bredasdorp Mts).•

unifolia Jacq.   Bulbous geophyte, 10–35 cm. Leaf solitary, linear, banded with green and maroon. Flowers variable, on short or long pedicels, oblong-campanulate, blue, pink or pale yellow, with white tips, anthers included. Aug.–Oct. Sandy granitic of sandstone soils, NW, SW, AP (Namaqualand to Bredasdorp).

variegata W.F.Barker   Bulbous geophyte, 10–40 cm. Leaf solitary, lanceolate or lorate, margins thickened, undulate. Flowers shortly pedicellate, oblong-campanulate, greenish grey with darker green, blue, purple or brown markings and white tips, anthers included. Aug.–Oct. Mainly coastal, in deep sand, NW, SW (Clanwilliam to Cape Peninsula).•

youngii Baker   Bulbous geophyte, 7–30 cm. Leaves 2, narrowly lanceolate. Flowers on fairly long pedicels, campanulate, pale blue and pink with darker purplish pink markings, anthers included. July–Nov. Coastal areas, SE (Mossel Bay to Humansdorp).•

B. Flowers sessile or on pedicels to 2 mm long

algoensis Schönland   Bulbous geophyte, 6–30 cm. Leaves 1 or 2, linear, lanceolate or lorate. Flowers shortly pedicellate, facing upward, yellow to greenish yellow, fading to dull red, anthers included. July–Aug. Habitat? SE (Knysna to E Cape).

ameliae W.F.Barker   Dwarf bulbous geophyte, 4–11 cm. Leaves 1 or 2, broadly lanceolate, smooth or hairy. Flowers sessile, urn-shaped or oblong, greenish yellow, with or without purple tips, anthers included. Aug.–Sept. Clay flats, NW, KM (Ceres to Montagu and Touwsrivier).•

arbuthnotiae W.F.Barker   Bulbous geophyte, 18–40 cm. Leaves 1 or 2, lanceolate, plain or densely spotted. Flowers sessile, oblong, yellow with pale green markings, anthers included. Aug.–Oct. Marshy flats, SW (Cape Flats).•

aurioliae G.D.Duncan   Like **L. obscura** but leaf bases not heavily barred and flower oblong-urceolate with inner tepals without magenta tips. July–Aug. Stony and sandy slopes, KM, SE (S Karoo and Montagu to Kammanassie Mts).

bachmannii Baker   Bulbous geophyte, 15–30 cm. Leaves 2, linear. Flowers shortly pedicellate, campanulate, white with brownish markings, anthers included. Aug.–Sept. Edges of seasonal pools, SW (Piketberg to Stellenbosch).•

bowkeri Baker (= *Lachenalia subspicata* Fourc.)   Bulbous geophyte, 10–26 cm. Leaves 1 or 2, lanceolate, blade unmarked. Flowers shortly pedicellate, oblong-campanulate, pale blue and white with purple markings, anthers included. Aug. Sandy soil, LB, SE (Riversdale to Riebeek East).

capensis W.F.Barker   Bulbous geophyte, 15–25 cm. Leaves 1 or 2, lanceolate or lorate, with or without brown blotches. Flowers sessile or shortly pedicellate, oblong-cylindrical, white or cream, anthers included. Sept.–Oct. Sandstone slopes, SW (Cape Peninsula).•

contaminata Aiton   Bulbous geophyte, 6–25 cm. Leaves several, subterete, erect. Flowers shortly pedicellate, campanulate, white with brown or reddish markings, anthers included or exserted. Aug.–Oct. Wet places, often common, NW, SW, AP (Citrusdal to Bredasdorp).•

dehoopensis W.F.Barker   Bulbous geophyte, 8–16 cm. Leaves 2, linear, banded with green and maroon. Flowers shortly pedicellate, oblong-campanulate, pale blue and cream with reddish markings, anthers included. Aug.–Sept. Sandy flats, AP (De Hoop, Bredasdorp).•

elegans W.F.Barker   Bulbous geophyte, 10–30 cm. Leaves 1 or 2, lanceolate to ovate-lanceolate, with or without green or maroon spots. Flowers sessile, urn-shaped, in shades of yellow, blue, mauve or purple, with white tips, anthers included. July–Oct. Sandy, mostly moist slopes, often in large colonies, NW (Bokkeveld Mts and W Karoo to Clanwilliam).

fistulosa Baker (= *Lachenalia convallariodora* Stapf)   Bulbous geophyte, 8–30 cm. Leaves 2, lorate, plain or spotted with brown. Flowers sessile, oblong-campanulate, cream, yellow, blue, lilac or violet, with pale brown markings, heavily scented, anthers included. Sept.–Oct. Rocky mountain slopes, NW, SW (Piketberg to Caledon).•

liliiflora Jacq.   Bulbous geophyte, 10–20 cm. Leaves 2, lanceolate, usually densely pustulate on upper surface. Flowers shortly pedicellate, oblong-campanulate, white with brownish markings and dark magenta tips, anthers included. Sept.–Oct. Hilly slopes in renosterveld, SW (Tygerberg to Paarl).•

longibracteata E.Phillips   Bulbous geophyte, 7–35 cm. Leaves 1 or 2, lanceolate, leathery, plain or spotted. Flowers sessile or shortly pedicellate, oblong-campanulate, each flower with a long bract at base, pale blue, or yellow with a blue base, with brown or green markings, anthers included. July–Sept. Clay flats and slopes, NW, SW (Piketberg to Malmesbury).•

margaretae W.F.Barker   Dwarf bulbous geophyte, 3–12 cm. Leaves 1 or 2, lorate, sometimes spotted. Flowers shortly pedicellate, campanulate, white with large brown or green markings, anthers very shortly exserted. Oct.–Dec. Rock ledges in partial shade, NW (Cedarberg Mts).•

marginata W.F.Barker.   Bulbous geophyte, 11–30 cm. Leaf solitary, ovate to lanceolate, glaucous, marked brown or green, margins thickened. Flowers sessile, oblong-cylindrical, greenish yellow with large dark brown markings, anthers included. July–Aug. Sandy flats or slopes, NW (Bokkeveld Mts to Clanwilliam).•

martinae W.F.Barker   Bulbous geophyte, 10–25 cm. Leaf solitary, ovate-lanceolate with maroon bands on clasping base, margins undulate. Flowers shortly pedicellate, oblong-campanulate, dull white and grey, with greenish brown markings, anthers included. July–Aug. Sandstone outcrops, NW (Olifants River valley and Mts).•

maximiliani Schltr. ex W.F.Barker   Bulbous geophyte, 10–20 cm. Leaf solitary, lanceolate, unmarked. Flowers shortly pedicellate, oblong-campanulate, pale blue and white with magenta tips, anthers included. July–Aug. Sandy slopes, often in large colonies, NW (Cedarberg Mts).•

mediana Jacq.   Bulbous geophyte, 20–40 cm. Leaves 1 or 2, lanceolate, unmarked. Flowers shortly pedicellate, oblong-campanulate, pale blue and white, or in shades of pinkish blue, with green or purplish markings, anthers included. Aug.–Sept. Clay soil, often in large colonies, NW, SW (Porterville to Cape Peninsula and Caledon).•

muirii W.F.Barker   Bulbous geophyte, 10–25 cm. Leaves 1 or 2, linear, withered at flowering time. Flowers sessile, urn-shaped to oblong, pale blue and white with brown or maroon markings, anthers included. Oct.–Dec. Limestone hills and flats, AP (Bredasdorp to Stilbaai).•

mutabilis Sweet   BONTVIOOLTJIE   Bulbous geophyte, 10–45 cm. Leaf solitary, lanceolate, erect, with crisped margins. Flowers oblong to urn-shaped, pale blue and white with yellow tips, or yellowish green, with brown markings, anthers included. July–Sept. Sandy and stony slopes, NW, SW (Namaqualand to Langebaan, Worcester and Riviersonderend).

obscura Schltr. ex G.D.Duncan   Bulbous geophyte, 6–30 cm, with hard tunic. Leaves mostly 2, lanceolate, often banded purple beneath. Flowers sessile or subsessile, often ternate below, oblong-campanulate, cream or yellowish to brownish, fading dull purple, anthers scarcely exserted. June–Oct. Stony karroid flats, NW, KM (Namaqualand and W Karoo to Little Karoo).

orchioides (L.) Aiton   Bulbous geophyte, 10–40 cm. Leaves 1 or 2, lanceolate or lorate, plain or densely spotted. Flowers oblong-cylindrical, greenish yellow or pale to dark blue, anthers included. Aug.–Oct. In heavy soil, often in partial shade, in large colonies, NW, SW, KM, AP (Gifberg to Albertinia and Little Karoo).•

orthopetala Jacq.   Bulbous geophyte, 9–27 cm. Leaves several, grass-like, plain or with brown spots. Flowers shortly pedicellate, oblong-campanulate, upward-facing, white with maroon markings, anthers included. Sept.–Oct. Clay soils in large colonies, NW, SW (Piketberg to Durbanville).•

pallida Aiton   Bulbous geophyte, 12–30 cm. Leaves 1 or 2, lanceolate, sometimes with pustules above. Flowers shortly pedicellate, oblong-campanulate, cream to dark yellow with brown or green markings. Aug.–Oct. In clay in large colonies, NW, SW (Piketberg to Stellenbosch).•

perryae G.Duncan   Bulbous geophyte, 12–32 cm. Leaf solitary, narrowly lanceolate, banded maroon below. Flowers shortly pedicellate, very pale blue and white with green or brown markings. July–Sept. Clay or sandy soil among succulents, NW, SW, LB (Worcester to Albertinia and E Cape).

reflexa Thunb.   Bulbous geophyte, 3–19 cm. Leaves 1 or 2, lanceolate or lorate, plain or densely spotted on upper surface. Flowers shortly pedicellate, cylindrical-ventricose, erect, bright yellow, anthers included. June–Aug. Wet sandy flats in large colonies, SW (Malmesbury to Cape Peninsula and Franschhoek).•

rubida Jacq.   SANDVIOOLTJIE   Bulbous geophyte, 6–25 cm. Leaves 1 or 2, lanceolate or lorate, plain green or spotted with darker green or purple. Flowers shortly pedicellate, nodding, cylindrical, plain or densely spotted with pink or red, anthers included. Mar.–July. Sandy flats and slopes, NW, SW, AP, SE (Hondeklipbaai to Cape Peninsula to George).

trichophylla Baker   Bulbous geophyte, 8–20 cm. Leaf solitary, heart-shaped, prostrate with stellate hairs on upper surface and margin. Flowers sessile or with short to long pedicels, oblong-cylindrical, shades of yellow, or yellow flushed with pink, with green markings, anthers included. Aug.–Sept. Sandy slopes, NW (Namaqualand to Citrusdal).

viridiflora W.F.Barker   Bulbous geophyte, 8–20 cm. Leaves 2, lanceolate, plain or spotted. Flowers shortly pedicellate, cylindrical-ventricose, turquoise, anthers included. May–July. Sandy slopes, SW (Saldanha district).•

zeyheri Baker   Bulbous geophyte, 6–20 cm. Leaves 1 or 2, subterete. Flowers shortly pedicellate, campanulate, white with reddish brown or green markings, anthers included. Sept.–Oct. Marshes and seeps, NW (Elandskloof to Ceres).•

LEDEBOURIA   AFRICAN SQUILL, UNTLOKWANA   c. 30 spp., India, Madagascar and sub-Saharan Africa, mainly southern Africa

ovalifolia (Schrad.) Jessop   Bulbous geophyte, 7–12 cm. Leaves ascending, soft, lanceolate to ovate, narrowed below, reddish beneath. Flowers in broad racemes, purple or pink and white, pedicels longer than the flowers. Dec.–Apr. Rocky places, often coastal limestone, NW, SW, AP, SE (Hex River valley to Humansdorp).•

ovatifolia (Baker) Jessop   Like **L. revoluta** but bulb scales loosely arranged, very fleshy and producing copious threads when torn. Sept.–Oct. Stony slopes, SE (Port Elizabeth to tropical Africa to Sri Lanka).

revoluta (L.f.) Jessop   Bulbous geophyte to 15 cm, scales producing sparse threads when torn. Leaves ascending, lanceolate, firm, spotted with red, margin hyaline, often slightly crisped and ciliolate. Flowers in broad racemes, purple and greenish, pedicels longer than the flowers. Mainly Oct.–Dec. Stony slopes, LB (Swellendam to Riversdale, E southern Africa to India).

undulata (Jacq.) Jessop   Bulbous geophyte, 10–15 cm. Leaves dry at flowering, ascending, narrowly lanceolate, mottled with dark green, margins undulate. Flowers in dense racemes, whitish and purple-pink or greenish, pedicels about as long as the flowers. Nov.–Jan. Rocky places, including sandstone and limestone, NW, SW (Namaqualand and W Karoo to darling).

sp. 1 (*Scilla ensifolia* (Eckl.) Britten)   Like **L. revoluta** but bulb cylindrical with a papery or leathery neck, scales not producing threads when torn and flowers in narrow racemes with pedicels as long as the flowers. Aug.–Dec. Stony slopes, SE (Humansdorp to E Cape).

MASSONIA   HEDGEHOG LILY, BOBBEJAANBOEK, KRIMPVARKIE   c. 6 spp., widespread in dry areas, S Namibia and W Cape to Lesotho

A. Anthers large, 2.5–3 mm long; perianth tube with a wide mouth

depressa Houtt.   Bulbous geophyte to 5 cm. Leaves prostrate, bracts large. Flowers clustered between the leaves, deeply cup-shaped below, green or yellowish to white or pink, anthers large, c. 2.5 mm long. May–July. Sandy and clay flats, NW, SW, AP, KM, SE (Namaqualand to Langkloof, E Cape, Karoo).

grandiflora Lindl.   Bulbous geophyte to 5 cm. Leaves prostrate, bracts large. Flowers clustered between the leaves, shallowly cup-shaped, greenish to white, anthers large. c. 2.5 mm long. May–July. Clay flats, NW, KM (W and Great Karoo to Bonteberg).

A. Anthers small, to 2 mm long; perianth tube narrow

echinata L.f. (= *Massonia hirsuta* Link & Otto, *M. setulosa* Baker, *M. tenella* Soland. ex Baker)   Bulbous geophyte to 5 cm. Leaves prostrate, glabrous or papillate-hairy above, bracts large. Flowers clustered between the leaves, cylindrical below, cream or white, fading pink, tubular below. May–July. Sandy and clay flats, NW, SW, AP, KM, LB, SE (Bokkeveld Mts to Port Elizabeth, Karoo).

pustulata Jacq.   Bulbous geophyte to 5 cm. Leaves large, prostrate, pustulate, bracts large. Flowers clustered between the leaves, deeply cup-shaped below, pink, white or cream. June–Sept. Sandy and clay flats, NW, SW, AP, KM, SE (Namaqualand to Port Elizabeth and Karoo).

pygmaea Kunth (= *Massonia heterandra* (Isaac) Jessop)   Bulbous geophyte to 5 cm. Leaves small, prostrate, pustulate-papillate, bracts large. Flowers clustered between the leaves, tubular below, pink or white, filaments usually of two lengths. Apr.–June. Mountains, NW, SW (Kamiesberg, Cedarberg Mts to Villiersdorp).

NEOPATERSONIA   FLY LILY   3 spp., S Namibia to E Cape

uitenhagensis Schönland   Bulbous geophyte, 15–20 cm. Leaves spreading, lanceolate. Flowers greenish with white stamens, filaments fleshy and lobed below. Sept.–Oct. Stony, mostly limestone slopes, NW, AP, LB, SE (Robertson to Addo).

ORNITHOGALUM   CHINCHERINCHEE, TJIENK   c. 120 spp., Africa and Eurasia

A. Style longer than ovary, deflexed; flowers yellow (or cream to white) with broad green keels

concordianum (Baker) U. & D.Müll.-Doblies (= *Ornithogalum apertum* (I.Verd.) Oberm.)   Bulbous geophyte, 10–20 cm. Leaves many, narrow, grey, coiled. Flowers firm, yellow with broad green keels. Aug.–Sept. Stony flats, NW, KM, SE (S Namibia and W Karoo to Clanwilliam through Little Karoo to Uniondale).

diluculum Oberm. Bulbous geophyte to 25 cm. Leaf dry at flowering, solitary, spreading, margin thickened. Flowers firm, yellow with greyish green keels, slightly nodding, opening in the early morning. Sept. Habitat? KM (W Karoo to Montagu: Bloutoring).

pentheri Zahlbr. Like **O. suaveolens** but plants sparsely glandular-hairy. Aug.–Oct. Stony and sandy slopes, NW (Calvinia to Citrusdal).

polyphyllum Jacq. (= *Ornithogalum semipedale* (Baker) U. & D.Müll.-Doblies) Bulbous geophyte, 25–60 cm. Leaves many, erect, narrow, loosely coiled apically on drying. Flowers firm, white or rarely yellow with green keels, fragrant. Aug.–Sept. Stony slopes, NW (Namaqualand to Cedarberg Mts).

sabulosum U. & D.Müll.-Doblies Like **O. suaveolens** but leaves often strap-shaped, glandular-papillate and covered in sand. Sept.–Oct. Deep red sands, NW (Namaqualand to Clanwilliam).

secundum Jacq. Bulbous geophyte, c. 35 cm. Leaves dry at flowering, many, oblong, margin often hyaline or fimbriate. Flowers firm, yellow with green keels. Aug.–Nov. Habitat? NW, SW (Namaqualand and W Karoo to Saldanha).

suaveolens Jacq. (= *Ornithogalum albucoides* (Aiton) Thunb., *O. namaquanum* U. & D.Müll.-Doblies; incl. **O. vittatum** (Ker Gawl.) Kunth) BONTTJIENK Bulbous geophyte, 10–50 cm. Leaves several, sometimes dry at flowering, linear, clasping basally. Flowers firm, yellow with green keels. Sept.–Nov. Dry slopes and flats, NW, SW, AP, KM, LB, SE (Namibia, Namaqualand and W Karoo to Humansdorp).

unifolium Retz. (= *Ornithogalum ovatum* Thunb.) Bulbous geophyte, 6–30 cm. Leaf usually dry at flowering, usually solitary, sometimes 2 or 3, broadly ovate, prostrate or spreading. Flowers pale yellow to cream or buff with a broad dark central band. Sept.–Nov. Dry karroid places, NW (Namibia, Namaqualand and W Karoo to Robertson).

A. Style erect, mostly shorter than ovary; flowers white, yellow or orange, mostly without broad green keels
  B. Bracts large, boat-shaped, entire

conicum Jacq. CHINCHERINCHEE Similar to **O. thyrsoides** but flowers never with a dark centre or ovary and inner filaments filiform or at most with an ovate expansion below. Nov.–Dec. Clay or loam flats, often moist, NW, SW, KM (Bokkeveld Plateau and W Karoo to Cape Peninsula, and Lesotho).

constrictum F.M.Leight. Bulbous geophyte, 20–40 cm. Leaves dry at flowering, prostrate, oblong, margins minutely fimbriate. Flowers on stout pedicels, white with green midrib, stamens unequal, dimorphic. Nov.–Feb. Clay soils in renosterveld, AP, KM, SE (Worcester, Little Karoo, Bredasdorp to E Cape).

dubium Houtt. (= *Ornithogalum citrinum* Schltr. ex Poelln.) GEELTJIENK Bulbous geophyte, 10–50 cm, outer bulb tunics often black. Leaves sometimes dry at flowering, ascending, margins minutely ciliolate. Flowers with large bracts, yellow to orange or rarely white, often with a green or brown centre, style very short. Aug.–Dec. Mountains and flats, NW, SW, AP, KM, LB, SE (Clanwilliam to Paarl, Caledon to Port Elizabeth, W Karoo, E Cape).

fimbrimarginatum F.M.Leight. Bulbous geophyte, 20–45 cm, bulbs with hard dark tunics. Leaves usually dry at flowering, spreading, oblong, margins usually fimbriate. Flowers with large bracts, white, centre sometimes dark. Sept.–Jan. Widespread, NW, SW, KM, SE (Gifberg to Port Elizabeth, E Cape).

inclusum F.M.Leight. Bulbous geophyte to 30 cm. Leaves dry at flowering, suberect, margins minutely fimbriate. Flowers white with green midribs, stamens unequal, dimorphic. Aug.–Sept. Karroid flats, NW (Botterkloof valley: Doornbosch).•

maculatum Jacq. ORANJETJIENK Bulbous geophyte, 8–50 cm. Leaves usually suberect, glabrous, somewhat fleshy. Flowers with large bracts, orange to orange-red or yellow, outer tepal tips often with a dark or transparent blotch, style very short. Sept.–Oct. Usually sandy soils, often on rocks, NW, SW (Namaqualand to Paarl, W Karoo).

multifolium Baker KLIPTJIENK Bulbous geophyte, 3–25 cm. Leaves spreading, terete, succulent. Flowers with large bracts, orange-yellow, unmarked, style very short. Sept.–Oct. Shallow soil on rocks, NW, SW (Namaqualand, Bokkeveld Mts to Mamre, W Karoo).

rupestre L.f. Bulbous geophyte, 2–10 cm. Leaves 1–3, spreading, linear, succulent. Flowers with large bracts, few, white, often flushed pink, stigma sessile. Sept.–Dec. Sandy pockets on granite boulders, NW, SW (Bokkeveld Mts to Saldanha).•

subcoriaceum L.Bolus   Bulbous geophyte to 20 cm, bulb with hard, dark tunics. Leaves few, spreading to erect, broad. Flowers with large bracts, white. Sept.–Oct. Sandy middle to upper slopes and plateaus, NW (Bokkeveld Plateau to Gydo Pass).•

thyrsoides Jacq.   CHINCHERINCHEE   Bulbous geophyte, 20–80 cm, bulb tunics soft, whitish. Leaves sometimes dry at flowering, suberect, long. Flowers with large bracts, white, shiny, centre often dark, inner filaments with broad membranous wings below. Oct.–Dec. Sandy flats and lower slopes, often in vleis, NW, SW, AP (Namaqualand to Pearly Beach).

B. Bracts usually small, deltoid, denticulate, rarely larger but then alternate

bicornutum F.M.Leight.   Bulbous geophyte, 10–16 cm. Leaves dry at flowering, spreading, with sheathing membranous base. Flowers white, nocturnal, filament appendages horn-like. Oct.–Dec. Dry stony situations, NW (Botterkloof and W Karoo).

ciliiferum U. & D.Müll.-Doblies (= *Ornithogalum gifbergense* U. & D.Müll.-Doblies)   Bulbous geophyte, 10–20 cm. Leaves dry at flowering, clasping stem, margins long-ciliate, basal sheaths long, tubular, bracts small, long-ciliate. Flowers white. Nov.–Dec. Stony slopes, NW (Namaqualand to Clanwilliam).

comptonii F.M.Leight.   Dwarf bulbous geophyte, 5–10 cm. Leaves sometimes dry at flowering, short, arcuate, linear and keeled, sclerotic, margins retrorse-ciliate. Flowers densely subcorymbose, white with brown midrib. Oct.–July. Shale flats, KM (W and Little Karoo).

dregeanum Kunth   Like **O. juncifolium** but sheaths usually orange-spotted, ageing to form a long fibrous neck. Dec.–Jan. Sandy, often wetter sites, NW, SW (Tulbagh to Kleinrivier Mts).•

esterhuyseniae Oberm.   Bulbous geophyte, 50–70 cm. Leaves erect, linear to oblong channelled. Flowers subcorymbose, small, white, shortly tubular below, ovary stipitate. Dec.–Feb., usually after fire. Wet places at high alt., NW, SW (Hex River to Hottentots Holland Mts).•

flexuosum (Thunb.) U. & D.Müll.-Doblies (= *Ornithogalum ornithogaloides* (Kunth) Oberm.)   Like **O. paludosum** but flowers in open, elongate racemes with pedicels to 40 mm in fruit. Nov.–Mar. Vleis and riverbanks, LB, SE (Riversdale to Port Elizabeth to Malawi).

graminifolium Thunb.   Bulbous geophyte, 10–30 cm. Leaves often dry at flowering, linear, sometimes hairy, reddish below and forming a papery neck. Flowers usually in a narrow, spike-like raceme, white, dull yellow or pale pink, only a few open at a time. Dec.–Mar. Stony clay flats and slopes, often moist sites, NW, SW, AP, KM, LB, SE (W Karoo and Bokkeveld Mts to KwaZulu-Natal).

hispidum Hornem.   GROWWETJIENK   Bulbous geophyte, 10–40 cm. Leaves dry at flowering, clasping stem, often pubescent, basal sheaths long, tubular, often spotted, bracts small. Flowers white. Aug.–Dec. Clay flats or rock outcrops, NW, SW, KM (Namaqualand to Cape Peninsula to Little Karoo).

juncifolium Jacq. (= *Ornithogalum comptum* Baker)   Bulbous geophyte, 10–40 cm. Leaves suberect, slender, usually strongly ribbed. Flowers usually in a narrow, spike-like raceme, small, white, only a few open at a time. Nov.–Mar. Dry flats or exposed rocky slopes, SW, AP, KM, LB, SE (Little Karoo and Caledon to eastern southern Africa).

longibracteatum Jacq.   PREGNANT ONION   Bulbous geophyte, 1–1.5 m, bulbilliferous. Leaves spreading, often flaccid. Flowers many, in a dense raceme, white with a broad, green central band, with long-attenuate bracts. Aug.–May. Shaded slopes and forest margins, SE (Mossel Bay to tropical East Africa).

nannodes F.M.Leight. (= *Ornithogalum hesperanthum* U. & D.Müll.-Doblies)   Bulbous geophyte to 15 cm, with a neck. Leaves dry at flowering, suberect, filiform, tufted. Flowers usually corymbose, whitish with brown midribs. Oct.–Dec. Hard stony soil, NW, SW (Namaqualand and W Karoo to Stellenbosch).

paludosum Baker   Stiffly erect bulbous geophyte, 20–50 cm, bulb elongate and poorly developed, somewhat rhizomatous below. Leaves linear, erect, sclerotic. Flowers in a subspicate raceme, white, tubular below, ovary stipitate. Oct.–Jan. Wet grassy slopes, SW, KM, SE (Caledon to Mpumalanga).

pilosum L.f. (= *Ornithogalum perparvum* Poelln.)   Bulbous geophyte, 15–30 cm, bulbs with hard black tunics, often with more than 1 raceme. Leaves sometimes dry at flowering, erect, sheathing below. Flowers white with pinkish reverse. Oct.–Dec. Clay flats and lower slopes, NW, SW, KM, LB (Pakhuis Mts to Riversdale).•

sardienii Van Jaarsv.   Dwarf evergreen, bulbous geophyte to 4 cm, with epigeal bulb. Leaves numerous, evergreen, *Haworthia*-like, triangular in section, with 6 rows of white cilia. Flowers white. Jan.–Mar. Enon conglomerate hills, KM (Oudtshoorn).•

schlechterianum Schinz. (= *Ornithogalum niveum* auct., *O. oreogenes* Poelln., *O. vallisgratae* Schltr. ex Poelln; incl. **O. rogersii** Baker)   Bulbous geophyte, 10–30 cm, straggling. Leaves spreading, soft, linear to linear-oblanceolate. Flowers on a flaccid peduncle, often subcorymbose, translucent white. Dec.–Feb. Rock ledges at middle to upper alt., SW, KM, LB, SE (Cape Peninsula to Outeniqua and Swartberg Mts).•

synadelphicum U. & D.Müll.-Doblies   Slender bulbous geophyte, 10–20 cm. Leaves filiform, minutely hairy. Flowers spreading on short pedicels, white, filaments widened below and fused into a cup. Sept.–Dec. Shale flats, KM (Ladismith to Oudtshoorn).•

tenuifolium F.Delaroche   BOSUI   Bulbous geophyte, 10–60 cm. Leaves c. 5, suberect. Flowers whitish with green keels, bracts long, narrow. Nov.–Mar. Grassland, SE (Humansdorp to tropical Africa).

thermophilum F.M.Leight.   Like **O. hispidum** but flowers in a dense, narrow raceme and tepals linear, white with orange midribs. Dec. Rocky slopes, NW (Namaqualand to Clanwilliam).

tortuosum Baker (= *Ornithogalum thunbergianulum* U. & D.Müll.-Doblies)   Bulbous geophyte, 10–20 cm. Leaves filiform, tortuose, minutely hairy. Flowers white with green keels. Aug.–Feb. Clay flats, NW, KM, LB, SE (W Karoo and Worcester to Steytlerville).

zebrinellum U. & D.Müll.-Doblies   Bulbous geophyte, 4–6 cm. Leaves dry at flowering, spreading, with striped basal sheaths. Flowers delicate, white. Feb.–Mar. NW (Namaqualand and W Karoo to Hex River Mts).

[**Species excluded**   No authentic material found and possibly conspecific with one of the above: **O. nathoanum** U. & D.Müll.-Doblies]; **O. adseptentrionesvergentulum** U. & D.Müll.-Doblies

POLYXENA (= *PERIBOEA*)   CAPE HYACINTH   5 spp., winter-rainfall South Africa

corymbosa (L.) Jessop   Short-stemmed bulbous geophyte to 15 cm. Leaves 2–6, suberect, linear-channelled, 1–5 mm wide. Flowers shortly racemose, pale lilac with dark keels, perianth tube shorter than tepals, 3.5–6 mm long, filaments 4–5 mm long, anthers exserted. Apr.–June. Loamy flats, NW, SW (Citrusdal to Gordon's Bay).•

ensifolia (Thunb.) Schönland (= *Polyxena pygmaea* (Jacq.) Kunth)   Short-stemmed bulbous geophyte to 5 cm. Leaves 2, spreading to prostrate, lanceolate to ovate, 10–25 mm wide. Flowers corymbose, clustered between the leaves, white to mauve or pale blue, perianth tube slender, longer than tepals, 10–25 mm long, filaments 4–6.5 mm long, anthers exserted. Apr.–June. Clay or granite flats, NW, SW, AP, KM, LB, SE (Namaqualand and W Karoo to Port Elizabeth).

maughanii W.F.Barker   Like **P. ensifolia** but filaments 1–2 mm long, inner anthers at least included and style shorter, about half as long as tube. May–June. Dolerite or rarely sandstone flats, NW (W Karoo and Bokkeveld Mts).

paucifolia (W.F.Barker) A.M.van der Merwe & J.C.Manning (= *Periboea oliveri* U. & D.Müll.-Doblies)   Like **P. corymbosa** but leaves broader, 5–10 mm wide, flowers darker purple, filaments 1–2 mm long, anthers included and style very short, 1–2.5 mm long. Apr.–June. Coastal granite and limestone outcrops, SW (Paternoster to Langebaan).•

[**Species excluded**   No authentic material available and identity uncertain: **P. calcicola** U. & D.Müll-Doblies]

SCILLA   SQUILL   c. 40 spp., W Cape to Eurasia

plumbea Lindl.   Bulbous geophyte, 20–40 cm. Leaves suberect, fleshy, deeply channelled. Flowers purple-blue. Dec.–Jan. Sandstone slopes on wetter sites, SW (Bainskloof Mts).•

VELTHEIMIA   VELTHEIMIA, SANDLELIE   2 spp., W and E Cape, Namaqualand

bracteata Harv. ex Baker   Bulbous geophyte, 20–40 cm, bulb tunics fleshy. Leaves suberect, glossy green, margins undulate, seldom all deciduous. Flowers tubular, spreading to nodding, pink or pale yellow, finely speckled with red. Aug.–Sept. Coastal scrub, SE (Humansdorp to E Cape).

capensis (L.) DC.   Bulbous geophyte, 20–40 cm, outer bulb tunics papery. Leaves suberect, greyish, margins undulate. Flowers as in **V. bracteata**. Apr.–July. Rocky slopes, NW, SW, AP, KM (Namaqualand and W Karoo to Darling, Potberg and Little Karoo).

WHITEHEADIA PAGODA LILY 2 spp., S Namibia to W Cape

bifolia (Jacq.) Baker   Bulbous geophyte, 8–12 cm. Leaves prostrate, fleshy, fragile, bracts large and green. Flowers in a dense, conical spike, white. June–Aug. Mostly in lee of rocks, NW (Namaqualand to Pakhuis Pass).

HYDROCHARITACEAE

1. Marine plants; leaves linear to lanceolate-ovate, in pairs from a creeping rhizome; flowers apetalous, submerged ....**Halophila**
1.' Fresh-water plants; leaves linear to linear lanceolate, widely spaced below, dense above, attenuate, subopposite or whorled from long leafy submerged stems arising from a perennial rhizome; flowers with petals, exserted above water during anthesis..**Lagarosiphon**

HALOPHILA c. 9 spp., tropical coasts of Indian and Pacific Oceans

ovalis (R.Br.) Hook.f.   Monoecious or dioecious, creeping marine aquatic exposed only at very low tides. Leaves paired, linear to ovate. Flowers minute, solitary in axils of secondary branches. Jan. Marine, SE (Knysna to Natal, tropical coasts of Indian and Pacific Oceans).

LAGAROSIPHON c. 16 spp., mainly Africa and Madagascar

muscoides Harv.   Monoecious or dioecious, submerged aquatic perennial, rootstock a rhizome, stems long and leafy. Leaves linear, toothed. Flowers tiny, axillary, white or pink. Flowering time? Fresh water, SE (Humansdorp to tropical Africa).

HYPOXIDACEAE
by D.A. Snijman

1. Plants densely pubescent, especially on reverse of flowers; hairs branched, stellate or tufted ....**Hypoxis**
1.' Plants glabrous or if sparsely pubescent then hairs simple and absent from flowers:
   2. Stamens 3 ....**Pauridia**
   2.' Stamens 6:
      3. Leaves often pleated; flowers with a conspicuous solid neck between ovary and tepals; ovary usually hidden in leaf sheaths during flowering, unilocular; fruit indehiscent, subsucculent ....**Empodium**
      3.' Leaves not pleated; flowers with tepals usually free to base, rarely with a solid neck between ovary and tepals; ovary usually well exposed during flowering, trilocular or occasionally unilocular; fruit dehiscent, thin-walled ....**Spiloxene**

EMPODIUM AUTUMN STAR c. 9 spp., W Cape to KwaZulu-Natal

flexile (Nel) M.F.Thomps. ex Snijman (= *Forbesia flexilis* Nel)   Like **E. plicatum** but corm with thick fibrous neck, flowers pale yellow and sweetly scented and anthers with elongate, fleshy appendages. Feb.–June. Stony flats, NW, KM (Namaqualand to Oudtshoorn).

gloriosum (Nel) B.L.Burtt   Cormous geophyte to 20 cm. Leaves many, with long narrow tips, softly pleated. Flowers yellow, ovary at ground level with glabrous beak to 70 mm long, style longer than stigma. Mar.–June. Sandy or loamy lower slopes, AP, KM, LB, SE (Montagu, Bredasdorp to E Cape).

namaquensis (Baker) M.F.Thomps.   Cormous geophyte to 30 cm, basal sheaths pale. Leaves tufted, thin-textured, broad and pleated. Flowers yellow, ovary exserted from leaf sheaths, with beak to 20 mm long. Apr.–May. Rocky outcrops, NW (Namaqualand to Cedarberg Mts).

plicatum (Thunb.) Garside PLOEGTYDBLOMMETJIE   Cormous geophyte, 10–20 cm, basal sheaths pale. Leaves 1–4, narrow and emergent at flowering, widening later, suberect to spreading, deeply pleated, ribs more or less hispid beneath. Flowers yellow, ovary at ground level with somewhat villous beak to 100 mm long. Apr.–June. Clay and granite flats and lower slopes, NW, SW, AP (Bokkeveld Mts to Bredasdorp).•

veratrifolium (Willd.) M.F.Thomps.   Cormous geophyte to 30 cm, basal sheaths dark brown. Leaves spreading, thin-textured, broad and pleated. Flowers yellow, ovary exserted on long pedicels, with beak to 10 mm long. May–June. Granite rocks along the coast, NW, SW (Lambert's Bay to Saldanha Bay).•

HYPOXIS STAR GRASS, INKBOL c. 80 spp., pantropical and subtropical, mainly Africa

angustifolia Lam.   Cormous geophyte, 8–25 cm. Leaves falcate to spreading, with long, soft hairs. Flowers yellow. Sept.–Feb. Grassland, LB, SE (Mossel Bay to tropical Africa, Madagascar).

argentea Harv. ex Baker   Cormous geophyte, 15–25 cm, with fibrous neck. Leaves suberect, grey-green, linear, with adpressed hairs. Flowers yellow. Oct.–Feb. Grassland, AP, LB, SE (Swellendam to Gauteng).

floccosa Baker   Cormous geophyte, mostly 5–10 cm, with a blackish papery neck. Leaves suberect to spreading, with soft rufous hairs. Flowers often exceeding leaves, solitary or few, yellow. Jan.–Apr. SW, AP, SE (Caledon to E Cape).

longifolia Baker   Cormous geophyte, 25–40 cm, with a fibrous neck. Leaves suberect, much exceeding flowers, narrow, fibrotic, with short hairs on margins. Flowers yellow. Nov.–Feb. Habitat? SE (Knysna to E Cape).

setosa Baker   Cormous geophyte, 15–25 cm, often with a soft fibrous neck. Leaves spreading, lanceolate, with adpressed hairs, margins and midribs thickened. Flowers usually exceeding the leaves, yellow. Jan.–Apr. Habitat? LB, SE (Swellendam to E Cape).

stellipilis Ker Gawl.   Cormous geophyte, 20–30 cm, with softly fibrous neck. Leaves falcate, discolorous, silvery-felted beneath, margins thickened. Flowers yellow. Nov.–Apr. Grassy fynbos, SE (Humansdorp to Port Elizabeth).•

villosa L.f. (incl. **H. sobolifera** Jacq.)   GOLDEN WINTER STAR   Like **H. setosa** but leaves with rufous hairs in distinct tufts. Mar.–Apr. Habitat? AP, LB, SE (Swellendam to KwaZulu-Natal).

PAURIDIA•   KLIPSTERRETJIE   2 spp., W Cape

longituba M.F.Thomps.   Cormous geophyte to 10 cm. Leaves suberect, linear. Flowers white, tube much longer than tepals, fragrant; pedicels short, erect in fruit. May–June. Granite outcrops, SW (St Helena Bay to Saldanha).•

minuta (L.f.) T.Durand & Schinz   Cormous geophyte, 3–5 cm. Leaves several, suberect, lanceolate. Flowers white or pale pink, tube shorter than tepals; pedicels long, recurved in fruit. Apr.–June. Damp flats and rock pavement, NW, SW, AP, LB (Langebaan to Riversdale).•

SPILOXENE   CAPE STAR, STERRETJIE   c. 25 spp., S Namibia, W to E Cape; 4 spp. in Australia and New Zealand

*A. Flowers more than 1 per scape (see also **S. schlechteri**)*

alba (Thunb.) Fourc.   WITSTERRETJIE   Cormous geophyte to 10 cm, corm non-fibrous. Leaves suberect, firm, subterete, usually shorter than inflorescence. Flowers (1)2 per scape, with a short solid perianth tube, white with maroon reverse; bracts 1 or 2, sheathing. Apr.–June. Marshes or damp places, NW, SW (Cold Bokkeveld to Hermanus and Breede River valley).•

aquatica (L.f.) Fourc.   WATERSTERRETJIE   Cormous geophyte, 10–30 cm, corm non-fibrous. Leaves erect, firm, subterete, exceeding inflorescence. Flowers 2–7 per scape, white with green reverse, tepals free to base; bracts 2 or more, broad, spreading. June–Nov. Pools or marshes, NW, SW, AP, LB (Namaqualand to Bredasdorp and Swellendam).

flaccida (Nel) Garside   Cormous geophyte, 6–25 cm, corm fibres spreading and pungent. Leaves spreading, channelled. Flowers 2 per scape, yellow; bracts 2, leaf-like with translucent margins, spreading. July–Sept. Damp flats and southern slopes, SW, AP, LB, SE (Paarl, Worcester to Humansdorp).•

minuta (L.) Fourc.   Cormous geophyte, 2–5 cm, minute, corm flat-based. Leaves spreading. Flowers 2 per scape, white with green reverse; bracts 2, filiform, spreading. Apr.–June. Damp flats, NW, SW (Clanwilliam to Cape Peninsula).•

trifurcillata (Nel) Fourc.   Cormous geophyte to 12 cm, corm softly fibrous. Leaves spreading, V-shaped in section. Flowers 2 per scape, small, yellow or white; bracts filiform. Apr.–Aug. Damp slopes and rock ledges, KM, SE (Swartberg Mts to E Cape).

umbraticola (Schltr.) Garside (incl. **S. maximiliani** (Schltr.) Garside)   Cormous geophyte to 15 cm, corm small, softly fibrous. Leaves spreading, strap-shaped, soft. Flowers 2 per scape, yellow or white, ovary elongated; bracts 2, broad, spreading. Aug. Damp rock crevices, NW (Bokkeveld Mts to Citrusdal).•

sp. 1   Cormous geophyte, 10–30 cm, corm small. Leaves several, lax, delicate. Flowers 1 or 2 per scape, small, yellow or white, sometimes tetramerous; bracts 2, leaf-like, spreading. Aug.–Oct. Moist sand in lee of rocks, NW (Bokkeveld Escarpment to Cedarberg Mts).•

*A. Flower 1 per scape (see also **S. alba**)*

canaliculata Garside   GEELPOUBLOM   Like **S. capensis** but leaves channelled, flowers firm-textured, yellow or orange with dark, non-iridescent centre and seeds J-shaped. July–Nov. Wet flats, SW (Darling to Cape Peninsula).•

capensis (L.) Garside   PEACOCK FLOWER, POUBLOM   Cormous geophyte, 10–30 cm. Leaves several, spreading, keeled and V-shaped in section. Flower variable in size, 1 per scape, yellow or white, sometimes pink, centre usually iridescent blue or green, sometimes dark and non-iridescent, banded on reverse; bract solitary, long and leaf-like, sheathing. July–Oct. Seasonally wet flats, NW, SW, AP, LB, KM (Clanwilliam to Oudtshoorn).•

curculigoides (Bolus) Garside (incl. **S. declinata** (Nel) Garside)   Cormous geophyte to 15 cm, corm neck usually of long, hard, straight light brown fibres. Leaves few, suberect to spreading, thin-textured, surrounded by a ribbed basal sheath. Flower 1 per scape, lemon-yellow; bract solitary, long, sheathing. Apr.–Aug. Damp rocky areas, SW, AP (Tulbagh, Cape Peninsula to Bredasdorp)•

monophylla (Schltr.) Garside   Cormous geophyte, 4–9 cm, corm small. Leaves spreading, firm, subterete. Flower 1 per scape, yellow, stamens very short; bract solitary, leaf-like, sheathing. Feb.–Mar., especially after fire. Sandstone slopes, SW (Kogelberg to Napier).•

ovata (L.f.) Garside (incl. **S. cuspidata** (Nel)Garside, **S. gracilipes** (Schltr.) Garside)   Cormous geophyte, 6–20 cm, corm covered with hard roots. Leaves recurved, narrow to broad, V-shaped above, many veined. Flower 1 per scape, yellow or white, rarely orange, with reddish reverse, tepals broadly elliptical; bract solitary, linear. June–Oct. Seasonally wet rocks and depressions, clay or sandy soils, NW, SW, AP (Namaqualand to Riviersonderend Mts).•

schlechteri (Bolus) Garside   Cormous geophyte, 6–9 cm, corm with dark woody tunics. Leaves suberect, firm, hemiterete to flattened. Flowers 1 or 2 per scape, yellow with green or reddish reverse; bracts leaf-like, sheathing. June–Aug. Marshy flats, NW, SW (Bokkeveld Mts to Cape Peninsula, Worcester).•

serrata (Thunb.) Garside (incl. **S. linearis** (Andrews) Garside)   Cormous geophyte, 6–20 cm. Leaves several, suberect, channelled, margin minutely toothed. Flower 1 per scape, yellow, orange or white with green reverse; bracts 2, linear. May–Oct. Flats and lower slopes, NW, SW (Namaqualand to Cape Peninsula and Worcester).

sp. 2 (*Empodium occidentale* (Nel) B.L.Burtt) Cormous geophyte, 1–3 cm. Leaves suberect, hemiterete to channelled, firm. Flower 1 on subterranean scape, with a long, solid perianth tube, white with yellow centre and pink reverse, ovary at ground level; bract solitary, sheathing. June–Sept. Damp depressions, NW (Ceres and W Karoo).

sp. 3   Cormous geophyte, 3–10 cm, corm tunics hard and ribbed. Leaves suberect, linear-channelled, firm. Flower 1 per scape, yellow; bract 1, linear, minute. May–June. Sandstone/clay interface, LB (Montagu to Barrydale).•

[**Species excluded**   No authentic material found and probably conspecific with one of the above: **S. acida** (Nel) Garside, **S. aemulans** (Nel) Garside]

IRIDACEAE

1. Flowers in umbellate clusters (rhipidia) enclosed by a pair of opposed leafy bracts (spathes), rarely solitary on peduncles or plants acaulescent but then style either dividing below anthers into tangentially compressed, petal-like branches or dividing below or above base of anthers and obscurely 3-lobed apically, lobes entire or fringed; individual flowers sessile or pedicellate; rootstock a woody caudex, a rhizome or a corm; tepals free, connate below or united in an extended tube:
    2. Plants evergreen shrubs with woody aerial stems; rootstock a woody caudex; individual flowers sessile and tepals always united in a well-developed tube and lasting at least two days:
        3. Inflorescence compound, forming a compressed capitulum enclosed by enlarged green or coloured leaves; perianth tube shorter than the linear-spathulate tepals ....**Klattia**
        3.' Inflorescence either compound, forming branched panicles or corymbs, or flowers borne in isolated pairs, enclosed in green or brown spathes; perianth tube shorter or longer than the oblong to ovate tepals:
            4. Flowers blue; tepals patent, not villous on reverse; stamens and/or style well exserted from flower ....**Nivenia**
            4.' Flowers green to blackish and yellow; tepals remaining closed during flowering, densely villous on reverse; stamens included in flower ....**Witsenia**

2.' Plants perennial, either with aerial stems which are not woody or acaulescent, sometimes evergreen; rootstock a rhizome or corm; individual flowers stalked or sessile, tepals free, connate below or united in a tube; lasting one to several days:
    5. Style eccentric, apically notched or lobed, lobes sometimes fringed; flowers usually deep blue (occasionally lilac, white or pale blue); tepals shortly connate basally..**Aristea**
    5.' Style central, usually dividing near base of anthers into distinct branches, these either extending between anthers or appressed against them, sometimes style exceeding anthers; flowers variously coloured; tepals free, connate basally or united in a tube:
       6. Rootstock a creeping or erect rhizome; plants evergreen; pedicels hairy above:
          7. Tepals with well-defined ascending claws and spreading limbs, outer larger than inner; style branches broad and petal-like, compressed tangentially, terminating in paired erect crests; anthers adpressed to abaxial side of style branches ....**Dietes**
          7.' Tepals not clawed, inner and outer whorls subequal; style branches filiform, spreading horizontally; anthers ascending, alternating with style branches ....**Bobartia**
       6.' Rootstock a corm; plants deciduous; pedicels without hairs near apices or flowers sessile:
          8. Leaves unifacial, oriented edgewise to stem; corm persisting for some years, tunics membranous or absent; tepals with crisped edges; style branches terminating in a feathery, plumose tuft ....**Ferraria**
          8.' Leaves bifacial or terete; corm usually resorbed annually, tunics fibrous; tepals with plane to undulate edges; style branches rarely feathery and plumose but then other characters not as above.........**Moraea**
             a. Flowers with a perianth tube; style lobed apically, lobes entire or fringed ....*Galaxia* group
             a.' Flowers usually without a perianth tube; style dividing below or opposite anthers, compressed tangentially, adpressed to anthers, terminating in paired erect crests:
                b. Style branches filiform and each divided to base, thus with 6 branches, branches extending below, between or above anthers ....*Hexaglottis* group
                b.' Style branches flattened tangentially, as wide as or much wider than anthers, branches ascending to upright and opposite anthers, sometimes concealed by them:
                   c. Plants acaulescent with flowers crowded basally; flowers either with ovary borne below or close to ground level or raised above ground on contractile pedicels...............*Moraea* group
                   c.' Plants with aerial stems; flowers with ovary borne well above ground level:
                      d. Ovary more or less sessile and extended distally in an elongate tubular sterile beak ....*Gynandriris* group
                      d.' Ovary borne on long pedicels or occasionally subsessile but then tepals united in a tube, without an elongate sterile beak:
                         e. Flowers with prominent style branches wider than anthers and terminating in paired erect crests; outer tepals larger than inner and with long ascending claws..*Moraea* group
                         e.' Flowers with style branches as wide as or narrower than anthers, often hidden by them, with a short bilobed apex opposite the stigmatic lobe(s); outer tepals only slightly larger than inner, with long or short claws:
                             f. Flowers yellow or salmon to pink, stems never sticky ....*Homeria* group
                             f.' Flowers either blue to purple or yellow but then stems sticky ....*Moraea* group
1.' Flowers in spikes or solitary on peduncles, sometimes in pseudopanicles, or acaulescent, individual flowers always sessile and style branches filiform, simple or deeply divided; rootstock a corm; tepals united below in a tube:
    9. Outer and inner bracts membranous to scarious, usually translucent to transparent with veins often darkly coloured, outer occasionally solid below but then margins lacerate:
       10. Plants acaulescent; leaves mostly entirely bifacial, usually channelled to adaxially grooved, sometimes terete but never grooved; bracts tubular below; corm tunics woody; flowers blue to purple...............**Syringodea**
       10.' Plants acaulescent or with aerial stems; leaves unifacial, rarely bifacial, plane or terete but then grooved; bracts with margins free to base or united below; corm tunics woody or papery to fibrous; flowers variously coloured:
          11. Bracts pale, dry, papery and crinkled or solid, irregularly streaked with dark flecks or veins, not 3-toothed:
             12. lowers nodding, borne on wiry stems; leaves linear, narrow, tough and fibrotic without a prominent midrib; E Cape, Humansdorp to Uitenhage ....**Dierama**
             12.' Flowers upright or facing to the side, borne on firm, somewhat fleshy stems; leaves lanceolate, relatively soft, usually with a prominent midrib; Nieuwoudtville to Riversdale ....**Sparaxis**
          11.' Bracts pale or rust-coloured, membranous or dry but not papery and crinkled, sometimes streaked with dark flecks or veins, often 3-toothed:
             13. Perianth zygomorphic with stamens unilateral, rarely actinomorphic but then stamens irregularly spreading and style eccentric or tepals orange with conspicuous brown veining; stems firm and relatively thick, never wiry ....**Tritonia**
             13.' Perianth actinomorphic with stamens either symmetrically disposed around a central style or unilateral with anthers drooping and porose; stems often more or less wiry ....**Ixia**

9.' Outer and inner bracts firm to soft-textured, green or leathery and dry, never lacerate, sometimes inner bracts with broad membranous to scarious margins or rarely almost entirely membranous but then leaves terete and 4-grooved:
  14. Style branches deeply divided, occasionally multifid:
    15. Leaves terete with narrow longitudinal grooves, or 4-winged; corm tunics brittle and woody; flowers actinomorphic, solitary on branches, not arranged in spikes; plants often acaulescent .... **Romulea**
    15.' Leaves usually plane but if round in transverse section then without longitudinal grooves; corm tunics leathery to fibrous; flowers zygomorphic or if actinomorphic then arranged in spikes; plants never acualescent:
      16. Corms bell-shaped, with a flat base; leaves plane or corrugate .... **Lapeirousia**
      16.' Corms globose to obconic, round or pointed at base:
        17. Flowers solitary on branches; leaves prostrate .... **Xenoscapa**
        17.' Flowers in spikes; leaves usually erect:
          18. Spikes inclined to horizontal, flowers borne on upper side; bracts green or dry above, then often dark brown at apices .... **Freesia**
          18.' Spikes erect, flowers distichous or spirally arranged; bracts green or partly to entirely dry but then never dark brown at apices:
            19. Flowers small, shorter than 12 mm, crowded in dense distichous spikes; bracts solid below with broad membranous margins.. **Micranthus**
            19.' Flowers medium to large, usually at least 20 mm long, in distichous or spiral spikes; bracts without broad membranous margins:
              20. Spikes distichous; leaf blades plane, relatively broad, margins moderately to strongly thickened; flowers never blue or purple; perianth tube always curved so that flowers face to the side .... **Watsonia**
              20.' Spikes spiral; leaf blades rounded in section or plane but then narrow and without thickened margins; flowers purple or blue to nearly white; perianth tube straight, flowers facing upward .... **Thereianthus**
  14.' Style branches undivided or at most notched apically:
    21. Inflorescence a panicle, individual flowers always pedunculate; flowers actinomorphic, bright orange .... **Pillansia**
    21.' Inflorescence a simple or branched spike, flowers sessile, rarely inflorescence reduced to a single flower; flowers actinomorphic or zygomorphic, variously coloured:
      22. Floral bracts fairly short, coriaceous and often partly to entirely dry at flowering, inner bracts always substantially longer than outer; leaves sometimes with a long pseudopetiole, without a pseudomidrib and usually with more than 1 prominent vein....................................... **Tritoniopsis**
      22.' Characters not combined as above:
        23. Floral bracts short, about twice as long as ovary, firm and leathery, green or dry, about as long as bracteoles; flowers orange to scarlet:
          24. Flowers actinomorphic, nodding, perianth tube narrow throughout; seeds dark brown to blackish... **Crocosmia**
          24.' Flowers zygomorphic, perianth tube narrow and cylindric below, abruptly expanded into a broad upper cylindrical part; seeds bright orange ....**Chasmanthe**
        23.' Floral bracts usually more than twice as long as the ovary, softer, green or dry but never firm and leathery:
          25. Leaf blades pleated, sometimes more or less linear and striate; stems, leaves or bracts hairy; seeds smooth and glossy .... **Babiana**
          25.' Leaf blades various but never pleated; plants sometimes hairy; seeds never glossy:
            26' Flowers usually zygomorphic with tepals unequal and stamens arcuate, rarely actinomorphic with stamens central or arcuate:
              27. Corms bell-shaped with a flat base; inflorescence wiry, much-branched; ovary deeply 3-lobed above; seeds subglobose.. **Melasphaerula**
              27.' Corms subglobose or obconic, rounded at the base; inflorescence never wiry nor much-branched; ovary ovoid; seeds broadly winged... **Gladiolus**
            26.' Flowers usually actinomorphic, if zygomorphic then tepals subequal and stamens declinate:
              28. Style usually dividing well above mouth of perianth tube, branches relatively short and recurved... **Geissorhiza**
              28.' Style dividing at apex of perianth tube or within tube, branches long and laxly spreading .... **Hesperantha**

ARISTEA ARISTEA, BLOUSUURKANOL   50 spp., sub-Saharan Africa and Madagascar

A. Capsules oblong to cylindrical and 3-lobed
B. Stems unbranched

biflora Weim.   Rhizomatous perennial, 20–40 cm, forming clumps, stems unbranched, lateral flower clusters sessile or lacking. Leaves linear, narrow. Flowers large, lilac to purple, inner tepals with transparent to translucent bronze windows on lower margins. Capsules elongate. Aug. Oct. Loamy clay in renosterveld, SW (Caledon to Drayton).•

cantharophila Goldblatt & J.C.Manning   Rhizomatous perennial, 20–40 cm, forming clumps, stems unbranched, lateral flower clusters sessile. Leaves linear, narrow. Flowers lilac to cream with a dark centre. Capsules elongate. Aug.–Sept. Clay and granite slopes in fynbos or renosterveld, SW (Kuilsrivier to Greyton).•

lugens (L.f.) Steud.   Rhizomatous perennial, 30–40 cm, stem compressed below, unbranched, lateral flower clusters sessile. Leaves broad, short, falcate, spathes greenish. Flowers pale blue to whitish, inner tepals small and dark blue-black. Capsules elongate. Sept.–Oct. Low granitic hills in renosterveld, SW (Riebeek-Kasteel to Stellenbosch).•

pauciflora Wolley-Dod   Rhizomatous perennial, 20–40 cm, forming clumps, stems elliptic, unbranched, lateral flower clusters sessile. Leaves narrow, linear, spathes greenish. Flowers deep blue. Capsules elongate, locules acute. Oct.–Dec. Mainly clay and granite slopes, SW (Cape Peninsula).•

pusilla (Thunb.) Ker Gawl.   Rhizomatous perennial to 20 cm, stem flattened and 2-winged. Leaves sword-shaped, fairly soft. Flowers blue, spathes with hyaline margins. Capsules cylindric-trigonous, elongate. Sept.–Nov. Mainly clay flats and lower slopes in renosterveld, AP, KM, LB, SE (Swartberg Mts and Swellendam to E Cape).

spiralis (L.f.) Ker Gawl.   Rhizomatous perennial, 20–50 cm, often in small clumps, stems flattened and 2-winged, usually unbranched, lateral flower clusters sessile. Leaves fairly broad, soft. Flowers secund, white or pale blue, stamens and style long, spathes green with hylaine margins. Capsules elongate. Mainly Sept.–Nov. Rocky sandstone and granite slopes, to 600 m, SW, AP, LB, SE (Cape Peninsula to Knysna).•

simplex Weim.   Rhizomatous perennial to 50 cm, forming clumps, stems elliptic, unbranched, lateral flower clusters sessile. Leaves narrow. Flowers secund, pale blue, spathes green. Capsules elongate. Sept.–Oct. Clay flats and lower slopes in renosterveld, SW, KM, LB, SE (Stellenbosch to George, and Swartberg Mts).•

teretifolia Goldblatt & J.C.Manning   Rhizomatous perennial, 20–40 cm, stem rounded, unbranched, lateral flower clusters sessile. Leaves linear often terete. Flowers large, lilac to cream, inner tepals with a large dark mark below, spathes green. Capsules elongate. Aug.–Sept. Low clay hills in renosterveld, SW (Shaw's Pass to Napier and Elim).•

B. Stems usually branched (if unbranched then broadly 2-winged and with a short leaf under inflorescence)

abyssinica Pax (= *Aristea cognata* N.E.Br. ex Weim.)   Slender rhizomatous perennial, 10–15 cm, stem compressed and 2-winged, consisting of 1 long internode, with a short, subterminal leaf. Leaves linear, narrow. Flowers blue, spathes dry-membranous, lacerate with age. Capsules ovoid. Nov.–Jan. Coastal grassland and forest margins, SE (Humansdorp to Ethiopia and Cameroon).

anceps Eckl. ex Klatt   Rhizomatous perennial, 10–30 cm, stem flattened and 2-winged, consisting of 1 long internode with a short, subterminal leaf. Leaves sword-shaped. Flowers blue, spathes rusty, dry, lacerate with age. Capsules oblong. Sept.–Jan. Mainly coastal slopes and flats, SE (Humansdorp to Transkei).

ecklonii Baker   Rhizomatous perennial to 50 cm, stem 2-winged, usually much-branched. Leaves broad, sword-shaped, soft. Flowers deep blue, spathes scarious, brownish. Capsules oblong, 3-lobed. Sept.–Dec. Coastal and montane, mostly forest margins, SE (Humansdorp to Uganda and Cameroon).

ensifolia Muir   Rhizomatous perennial, to 50 cm, stem compressed and 2-winged, few- to many-branched. Leaves sword-shaped, soft. Flowers blue, spathes dry-membranous. Capsules elongate, cylindric-trigonous, indehiscent, decaying with age. Sept.–Nov. Coastal forests, shade or clearings, LB, SE (Riversdale to Uitenhage).•

latifolia G.J.Lewis   Rhizomatous perennial to 1 m, stem lightly compressed, usually much-branched above. Leaves very broad, soft. Flowers deep blue, spathes pale, scarious. Capsules ovoid, with 3 shallow, rounded lobes. Nov.–Jan. Shady kloofs and gullies, 500–1500 m, SW (Bainskloof to Hottentots Holland Mts).•

schizolaena Harv. ex Baker   Rhizomatous perennial to 80 cm, stem elliptic, usually few-branched and with sessile lateral flower clusters. Leaves sword-shaped. Flowers blue, spathes scarious, rust-brown, lacerate with age. Capsules ovoid. Dec.–June. Mainly coastal grassland, SE (Plettenberg Bay to Mpumalanga).

A. Capsules with 3 narrow wings
C. Style 3-lobed apically and fringed; stems dichotomously branched (sometimes simple)

africana (L.) Hoffmanns.   Rhizomatous perennial, mostly 10–15 cm, stem compressed, dichotomously branched, sometimes simple. Leaves linear. Flowers blue, narrow, spathes hyaline with dark keels, closely and deeply fringed and curled, sometimes rusty above. Capsules short, 3-winged. Mainly Oct.–Jan. Sandy flats and mountain slopes, NW, SW, AP, LB (Gifberg to Bredasdorp and Riversdale).•

dichotoma (Thunb.) Ker Gawl.   VENSTERVRUG   Rhizomatous perennial, 15–30 cm, forming cushions, stems flattened below, dichotomously 3–5-branched. Leaves narrow, glaucous. Flowers blue, spathes translucent with dark keels. Capsules short, 3-winged. Dec.–Mar. Sandy flats and lower slopes, NW, SW (Namaqualand to Cape Peninsula).

glauca Klatt   Rhizomatous perennial, 10–15 cm, forming diffuse low cushions, spreading by stolons, stems strongly compressed and 2-winged, dichotomously branched or simple. Leaves linear, narrow, glaucous. Flowers blue, spathes translucent with dark keels. Capsules short, 3-winged. Oct.–Dec. Coastal and lower slopes, NW, SW, AP, LB (Ceres and Cape Peninsula to Riversdale).•

oligocephala Baker   Rhizomatous perennial 15–25 cm, stems subterete, dichotomously branched. Leaves linear, narrow, rigid, elliptic in section. Flowers blue, spathes translucent with dark keels. Capsules short, 3-winged. Nov.–Jan. Sandstone slopes, NW, SW, AP (Hottentots Holland Mts to Bredasdorp).•

palustris Schltr.   Rhizomatous perennial to 1 m, stem rounded, dichotomously branched. Leaves sword-shaped, soft. Flowers blue, spathes translucent with dark keels. Capsules short, 3-winged. Nov.–Jan. Coastal and lower slopes in wet sites, SW, AP (Bredasdorp).•

recisa Weim.   Rhizomatous perennial, 15–35 cm, forming small tufts, stems dichotomously branched, flattened, slender. Leaves sword-shaped, soft. Flowers blue, spathes translucent with dark keels, margins rusty, closely and shallowly fringed. Capsules short, 3-winged. Mainly Sept.–Oct. Sandstone slopes in wet sandy places, 600–1200 m, SW, LB (Hottentots Holland Mts to Hermanus and Swellendam).•

singularis Weim.   Rhizomatous perennial to 40 cm, forming diffuse tufts, stems slightly flattened, dichotomously branched and rooting at nodes. Leaves sword-shaped. Flowers blue, nodding, soft, spathes greenish translucent and lightly lacerate. Capsules short, 3-lobed. July–Aug. Sandstone slopes near streams and in shade, NW (Pakhuis Mts).•

C. Style undivided to minutely 3-fid; stem variously branched, but seldom dichotomously, or simple with lateral flower clusters sessile

bakeri Klatt (incl. **A. confusa** Goldblatt)   BLOUSUURKANOL   Rhizomatous perennial to 1 m, stems rounded, usually much branched. Leaves linear-sword-shaped, broad, fibrotic. Flowers blue, style undivided, spathes ovate, usually dry-rusty with transparent margins. Capsules oblong, 3-winged. Mainly Oct.–Dec., mainly after fire. Stony sandstone slopes, NW, SW, LB, SE (Piketberg to Port Elizabeth and E Cape).

cuspidata Schinz   Rhizomatous perennial, 20–60 cm, stems lightly compressed. Leaves linear, fairly narrow. Flowers blue, style undivided, spathes rust-brown, transparent on margins. Capsules 3-winged. Oct.–Dec. SW, KM, LB, SE (Cape Peninsula to Knysna, and Swartberg Mts).•

fimbriata Goldblatt & J.C.Manning   Rhizomatous perennial, 20–30 cm, forming cushions, stems rounded. Leaves linear to sword-shaped, narrow, firm. Flowers blue, style undivided, spathes translucent, margins closely fringed and rust-brown. Capsules oblong, 3-winged. Dec.–Jan. Rocky sandstone slopes in fynbos, 500–800 m, NW (Piketberg).•

inaequalis Goldblatt & J.C.Manning   Rhizomatous perennial to 1.5 m, stems rounded, with long spreading branches, lateral flower clusters sessile, widely spaced. Leaves linear, narrow, grey with reddish margins, fibrotic. Flowers blue, secund, stamens unequal, style undivided, spathes dry-membranous, translucent with dark keels. Capsules short, 3-winged. Oct.–Nov. Sandstone rocks, 500–900 m, NW (Bokkeveld Mts to Pakhuis Pass).•

juncifolia Baker   Rhizomatous perennial, 12–30 cm, with rounded stems, usually unbranched, lateral flower clusters sessile. Leaves subterete to ellipsoid, tough and fibrotic. Flowers with prominently keeled bracts, blue, style undivided, spathes rust-brown, transparent on the edges. Capsules oblong, 3-winged. Nov.–Dec. Coastal and lower mountain slopes, SW (Cape Peninsula to Kleinrivier Mts).•

macrocarpa G.J.Lewis   Rhizomatous perennial to 1 m, stem subterete, usually with short ascending branches below. Leaves broad, fibrotic, spathes rust-brown, hairy below, transparent on margins. Flowers blue, style undivided. Capsules oblong, 3-winged. Nov.–Dec. Rocky mountain soils, 200–1500 m, SW (Cape Peninsula to Caledon).•

major Andrews (= *Aristea thyrsiflora* D.Delaroche)   BLOUVUURPYL   Rhizomatous perennial to 1.5 m, stems rounded, bearing short branches above, flower clusters crowded and overlapping. Leaves linear-sword-shaped, broad, fibrotic. Flowers blue, style undivided, spathes lanceolate, dry-membranous, translucent with dark keels. Capsules short, 3-winged. Oct.–Dec. Mountain slopes, 100–900 m, NW, SW, LB, SE (Piketberg to George).•

monticola Goldblatt (= *Aristea coerulea* (Thunb.) Vahl)   Rhizomatous perennial to 1.3 m, stem subterete, often much branched, branches short. Leaves linear to sword-shaped, often broad, fibrotic. Flowers blue, style undivided, spathes large, dry-membranous, rusty, translucent on edges, lightly hairy or scabrid. Capsules short, 3-winged. Aug.–Oct. Sandstone slopes, 300–1000 m, NW, SW (Cedarberg to Du Toitskloof Mts).•

racemosa Baker   Rhizomatous perennial to 40 cm, stems rounded, usually unbranched, lateral flower clusters sessile. Leaves subterete, fibrotic. Flowers blue, style undivided, spathes dry-membranous, rusty, translucent on edges. Capsules oblong, 3-winged. Oct.–Dec. Rocky sandstone slopes, 200–1000 m, SW, AP, KM, LB, SE (Paarl to George, Swartberg Mts).•

rigidifolia G.J.Lewis   Rhizomatous perennial to 1.5 m, stems rounded, with short branches or lateral flower clusters sessile. Leaves elliptic in section, fibrotic, narrow. Flowers blue, style undivided, spathes ovate, rusty, becoming lacerate. Capsules oblong, 3-winged. Oct.–Nov. Sandy flats, SW (Cape Peninsula to Hermanus Mts).•

rupicola Goldblatt & J.C.Manning   Rhizomatous perennial to 40 cm, stems rounded, divaricately branched. Leaves linear, narrow, fibrotic. Flowers light blue, style undivided, spathes ovate, rusty, margins translucent flecked with brown, hairy to scabrid on inside. Capsules short, 3-winged. Dec.–Mar. Sandstone outcrops c. 1000 m, NW (N Cedarberg Mts).•

zeyheri Baker   Rhizomatous perennial, 15–30 cm, stem terete, slender, unbranched with 1 or 2 sessile lateral flower clusters. Leaves terete, filiform. Flowers blue, style somewhat lobed, spathes greenish with hyaline margins. Capsules elongate, 3-winged. Nov.–Dec. Sandstone slopes, usually damp sites, SW (Cape Peninsula to Hermanus Mts).•

BABIANA (= *ANTHOLYZA*)   BABIANA, BOBBEJAANTJIE   c. 65 spp., southern Africa and Socotra

A. Flowers actinomorphic

blanda (L.Bolus) G.J.Lewis   Cormous geophyte to 6 cm. Leaves lanceolate, hairy. Flowers rosy pink, actinomorphic, tepals obovate or suborbicular, bracteoles usually divided. Aug.–Sept. Sandy flats, SW (Darling to Paarl).•

foliosa G.J.Lewis   Cormous geophyte to 4 cm. Leaves many, lanceolate, hairy. Flowers actinomorphic, mauve with cream markings, bracts entirely green, bracteoles divided. Aug. Habitat? SW (Riviersonderend).•

leipoldtii G.J.Lewis   Cormous geophyte, 6–15 cm. Leaves lanceolate, hairy. Flowers actinomorphic, blue-violet with a dark centre, bracteoles divided. Aug.–Sept. Damp sandy flats, SW (Darling to Klipheuwel).•

purpurea (Jacq.) Ker Gawl.   Like **B. stricta** but flowers pink to purple, with lobes mostly shorter than tube and stigmas large and orbicular. Aug.–Sept. Clay flats and slopes in renosterveld, SW (Robertson to Bredasdorp).•

pygmaea (Burm.f.) N.E.Br.   GEELBOBBEJAANTJIE   Cormous geophyte, 4–10 cm. Leaves lanceolate, hairy. Flowers large, actinomorphic, yellow with a dark centre, tepals obovate. Aug.–Sept. Gravelly flats, SW (Hopefield to Darling).•

rubrocyanea (Jacq.) Ker Gawl.   ROOIBLOUBOBBEJAANTJIE   Cormous geophyte, 5–15 cm. Leaves lanceolate, hairy. Flowers actinomorphic, bright blue with red centre, tepals broadly clawed, bracteoles divided, stigmas large and flattened. Aug.–Sept. Granitic sands in renosterveld, SW (Darling to Mamre).•

stricta (Aiton) Ker Gawl.   Cormous geophyte, 10–20 cm. Leaves lanceolate, hairy. Flowers almost actinomorphic, purple to blue, white or yellow, bracteoles divided, anthers sagittate, broad. Aug.–Oct. Clay soils in renosterveld, NW, SW (Piketberg to Riviersonderend).•

villosa (Aiton) Ker Gawl.   ROOIBOBBEJAANTJIE   Like **B. stricta** but flowers red to red-purple or pink, with tube filiform to near tip and then abruptly expanded. Sept. Clay flats and slopes in renosterveld, NW, SW (Tulbagh to Malmesbury).•

villosula (Gmelin) Ker Gawl. ex Steud.   Cormous geophyte, 3–8 cm, stem underground. Leaves lanceolate, hairy. Flowers actinomorphic, pale blue to mauve, white in centre, tube filiform, bracteoles divided. May–July. Sandy flats and lower slopes in fynbos, SW (Malmesbury to Gordon's Bay).•

*A. Flowers zygomorphic (see also **B. purpurea** and **B. stricta**)*
B. Perianth tube longer than the tepals or flowers red

ecklonii Klatt   Cormous geophyte, 10–30 cm. Leaves lanceolate, hairy. Flowers zygomorphic, violet with dark blue and cream markings, tube elongate, geniculate, bracteoles divided, ovary hairy. Aug.–Sept. Sandstone crevices, mountain slopes and flats, 200–500 m, NW (Gifberg to Elandskloof and Piketberg).•

framesii L.Bolus   Cormous geophyte to 10 cm, stem shorter than leaves. Leaves lanceolate, hairy. Flowers zygomorphic, dark blue to purple with white markings, tube elongate, slightly curved. Aug.–Sept. Rock outcrops in karroid scrub, NW (Namaqualand to Bokkeveld Escarpment and W Karoo).

geniculata G.J.Lewis   Cormous geophyte, 2–6 cm, stem underground. Leaves lanceolate, hairy. Flowers zygomorphic, purple with white markings, tube elongate, geniculate, style branches 10 mm long, ovary hairy. Aug. Rocky sandstone in dry fynbos, NW (Pakhuis Mts).•

patersoniae L.Bolus   Cormous geophyte, 15–25 cm. Leaves lanceolate, hairy. Flowers zygomorphic, white to pale blue to mauve with yellow markings, fragrant, bracteoles divided, ovary hairy. Aug.–Oct. Clay slopes in renosterveld, SW, LB, SE (Caledon to E Cape).

pauciflora G.J.Lewis   Cormous geophyte to 10 cm, stems extending shortly above ground. Leaves obliquely oblong-lanceolate, pleated below. Flowers zygomorphic, violet with yellow markings, tube elongate, fragrant. June. Rocky flats, NW (Bokkeveld Escarpment).•

ringens (L.) Ker Gawl. (= *Antholyza ringens* L.)   ANTHOLYZA, ROTSTERT   Cormous geophyte, 15–40 cm, main spike axis sterile. Leaves linear-lanceolate, glabrous. Flowers on a side branch, highly zygomorphic, upper tepal tubular below. Aug.–Oct. Sandy flats in fynbos, 30–500 m, NW, SW, AP (Bokkeveld Mts to Bredasdorp).•

sambucina (Jacq.) Ker Gawl.   Cormous geophyte, 5–14 cm, stem usually underground, shorter than leaves. Leaves lanceolate, hairy. Flowers zygomorphic, blue to purple with white markings, fragrant, tube elongate, straight. July–Sept. Rocky slopes and flats, fynbos and renosterveld, NW, SW, AP, KM, LB, SE (Bokkeveld Mts to Port Elizabeth, Karoo, E Cape).

thunbergii Ker Gawl. (= *Antholyza plicata* (L.f.) Goldblatt)   ROOIHANEKAM   Cormous geophyte, 40–70 cm, with short, horizontal branches. Leaves lanceolate, minutely velvety-hairy. Flowers zygomorphic, bright red. July–Oct. Sandy flats and dunes, coastal, NW, SW (Orange River Mouth to Saldanha).

truncata G.J.Lewis   Cormous geophyte to 8 cm, stem underground. Leaves cuneate, abruptly truncate, hairy. Flowers zygomorphic, mauve marked with pale yellow. July–Aug. Clay soils, renosterveld, NW (Namaqualand to Bokkeveld Plateau).

tubulosa (Burm.f.) Ker Gawl.   Cormous geophyte, 7–15 cm, stems shorter than leaves. Leaves linear, hairy. Flowers zygomorphic, white to cream sometimes with red markings, tube very long, lobes somewhat clawed. Sept.–Oct. Sandy flats and lower slopes, NW, SW, AP, LB (Elandsbaai to Riversdale).•

B. Perianth tube as long as or shorter than the tepals

ambigua (Roem. & Schult.) G.J.Lewis   Cormous geophyte, 5–8 cm. Leaves narrow, longer than stem, hairy. Flowers zygomorphic, blue to mauve with white to cream markings, fragrant, inner bracts usually free. Aug.–Sept. Sandy flats and lower slopes, NW, SW, AP, LB (Gifberg to Riversdale).•

angustifolia Sweet (= *Babiana pulchra* (Salisb.) G.J.Lewis)   Cormous geophyte, 10–20 cm. Leaves lanceolate, hairy. Flowers often inverted, dark blue to violet, lower tepals with black markings, bracteoles divided. Aug.–Sept. Damp clay flats and lower slopes, renosterveld, NW, SW (Piketberg to Somerset West).•

auriculata G.J.Lewis   Cormous geophyte, 4–10 cm. Leaves lanceolate, hairy. Flowers zygomorphic, facing stem apex, mauve and yellow, lower tepals auriculate at base. Sept. Rock crevices, NW (Pakhuis Mts).•

cedarbergensis G.J.Lewis   Cormous geophyte, 4–6 cm, stem very short. Leaves rigid, almost pungent, velvety hairy. Flowers zygomorphic, mauve with yellow markings. Sept. Rocky sandstone soils, NW (Cedarberg Mts).•

crispa G.J.Lewis   Cormous geophyte to 13 cm, stem underground. Leaves crisped and undulate, lanceolate, hairy. Flowers zygomorphic, mauve with yellow markings. July. Hard clay and shale, NW (Kamiesberg to Botterkloof and Biedouw valley).

disticha Ker Gawl. (= *Babiana plicata* Ker Gawl.)   Cormous geophyte, 7–20 cm. Leaves lanceolate, hairy. Flowers zygomorphic, violet to pale blue, marked with yellow, fragrant, bracteoles divided, ovary hairy. July–Sept. Sandstone slopes, NW, SW (Ceres to Cape Peninsula).•

fourcadei G.J.Lewis   Cormous geophyte, 7–15 cm. Leaves lanceolate, hairy. Flowers zygomorphic, mauve with yellow and violet markings, bracteoles divided. Sept.–Oct. Mountain slopes, KM, LB, SE (Riversdale to George and Little Karoo Mts).•

fimbriata(Klatt) Baker   Cormous geophyte, 15–20 cm. Leaves linear, spirally coiled above. Flowers zygomorphic, windowed, dorsal tepal narrow, arcuate, blue or mauve with yellow markings, stamens arcuate. Aug.–Sept. Sandy flats and slopes, NW (Namaqualand to Klawer).

klaverensis G.J.Lewis   Cormous geophyte, 6–10 cm, stems shorter than leaves. Leaves rigid, lanceolate, hairy, with thickened margins. Flowers zygomorphic, lower tepals clawed and united with upper laterals below, blue to mauve, bracteoles divided, ovary hairy. June–July. Rocky sandstone slopes, NW (Bokkeveld Mts to Gifberg).•

lineolata Klatt   Cormous geophyte, 13–25 cm. Leaves lanceolate, hairy. Flowers small to 30 mm, zygomorphic, pale blue to mauve with yellow markings, bracteoles divided, bracts c. 10 mm, ovary hairy. Sept. Sandy flats, NW (Piketberg to Cold Bokkeveld).•

minuta G.J.Lewis   Cormous geophyte, 7–13 cm, stem shorter than leaves. Leaves lanceolate, hairy. Flowers zygomorphic, pale reddish mauve with yellow markings, fragrant. July–Sept. Shale and rocky sandstone soils, NW (S Namaqualand to Karoopoort and W Karoo).

montana G.J.Lewis   Cormous geophyte, 6–7 cm. Leaves lanceolate, hairy. Flowers zygomorphic, mauve with yellow and purple markings, bracteoles divided, stamens unequal, 1 longer, stigmas flattened and orbicular. June–Aug. Sandstone and limestone slopes, SW, AP (Caledon to Bredasdorp).•

mucronata (Jacq.) Ker Gawl.   Cormous geophyte, 5–15 cm. Leaves lanceolate, hairy. Flowers zygomorphic, lower tepals somewhat clawed, contiguous below and diverging obliquely above, pale blue with yellow lower tepals, scented, bracteoles divided, ovary hairy. July–Sept. Rocky sandstone slopes and flats, NW (Bokkeveld Mts to Tulbagh).•

nana (Andrews) Spreng.   Cormous geophyte, 3–10 cm. Leaves often very broad, hairy. Flowers zygomorphic, blue or purple with white markings, fragrant. Aug.–Sept. Coastal flats and dunes, NW, SW, AP, LB (Lambert's Bay to Mossel Bay).•

obliqua E.Phillips   Cormous geophyte to 16 cm. Leaves very obliquely lanceolate, not plicate, hairy. Flowers zygomorphic, lilac with yellow markings, bracteoles divided. July. Sandy coastal soils, NW, SW (Lambert's Bay to Mamre).•

odorata L.Bolus   Like **B. mucronata** but flowers yellow and fragrant with tube to 14 mm long. July–Sept. Clay soils in renosterveld, NW, SW (Porterville to Tygerberg).•

patula N.E.Br.   Cormous geophyte, 2–8 cm. Leaves longer than stem, lanceolate, hairy. Flowers zygomorphic, mauve to blue with yellow markings or entirely dull yellow, fragrant, bracteoles divided. Aug.–Sept. Clay flats and lower slopes, SW, LB, AP (Tulbagh to Albertinia).•

scabrifolia Brehm. ex Klatt   Cormous geophyte, 5–9 cm, stems shorter than leaves. Leaves lanceolate, hairy, narrow and twisted when young. Flowers zygomorphic, blue to lilac with yellow and purple markings, fragrant, bracteoles divided, ovary hairy. June–Aug. Sandy soils in dry fynbos, NW (Olifants River valley).•

scariosa G.J.Lewis   Cormous geophyte, 10–40 cm. Leaves lanceolate, hairy. Flowers zygomorphic, mauve with pale yellow markings, bracts papery, bracteoles divided. Aug.–Sept. Dry sandstone or clay, in fynbos or karroid scrub, NW, KM (Bokkeveld Mts and W Little Karoo).•

secunda (Thunb.) Ker Gawl.   Cormous geophyte, 15–35 cm. Leaves lanceolate, hairy. Flowers zygomorphic, inverted, blue with yellow or white markings, bracteoles divided, outer bracts mostly brown and lacerate. Oct.–Nov. Clay flats and lower slopes in renosterveld, SW (Hopefield to Paarl).•

sinuata G.J.Lewis   Cormous geophyte, 10–25 cm, stems much-branched. Leaves narrow, undulate and twisted, margins crisped and long-hairy. Flowers zygomorphic, windowed, dorsal tepal narrow, arcuate, blue with yellow markings, stamens arcuate, anthers connate. Aug.–Sept. Rocky shale slopes, NW (Namaqualand to Clanwilliam).

unguiculata G.J.Lewis   Cormous geophyte, 10–20 cm. Leaves linear, hairy. Flowers zygomorphic, windowed, dorsal tepal arcuate, yellow, stamens arcuate, ovary hairy. Aug.–Sept. Lower mountain slopes, NW (Nardouw Mts).•

vanzyliae L.Bolus   Cormous geophyte, 4–12 cm. Leaves lanceolate, hairy, longer than stem. Flowers zygomorphic, yellow to mauve, fragrant. Aug.–Sept. Rocky sandstone soils, fynbos, NW (Bokkeveld Mts to Pakhuis Pass).•

BOBARTIA   RUSH LILY, BLOMBIESIE   15 spp., W to E Cape

*A. Leaves plane or elliptic in section (see also **B. macrospatha**)*

filiformis (L.f.) Ker Gawl.   Slender rhizomatous perennial, 15–55 cm. Leaves subterete to linear. Flowers in a much-reduced inflorescence of 1–few cymes, yellow. Sept.–Dec., mostly after fire. Sandy lower to middle slopes in fynbos, SW, AP (Paardeberg to Michell's Pass to Agulhas).•

gladiata (L.f.) Ker Gawl.   Rhizomatous perennial, 20–80 cm, stems usually flat. Leaves linear, elliptic in section. Flowers in a flattened inflorescence of 3–12 cymes, yellow. Sept.–Dec. Mountain slopes and coastal flats, in fynbos, SW (Cape Peninsula and Bainskloof to Hermanus).•

lilacina G.J.Lewis   Rhizomatous perennial, 30–70 cm, stem laxly branched and sticky below nodes. Leaves narrowly sword-shaped. Flowers in solitary cymes, purple. Jan.–Mar. Mountain slopes on shale, 600–1500 m, SW (Bainskloof to Franschhoek).•

paniculata G.J.Lewis   Rhizomatous perennial, 40–100 cm, stem branched above and sticky below nodes. Leaves linear. Flowers in solitary cymes loosely clustered near top of stem, yellow. Jan.–Feb. Middle to upper slopes in dry fynbos, KM (Kammanassie Mts).•

*A. Leaves terete (see also **B. filiformis**)*
B. Spathes fibrotic, brown or grey

fasciculata J.B.Gillett ex Strid   Rhizomatous perennial, 80–150 cm. Leaves terete, spathes fibrotic, dilated. Flowers in a dense, distinctly fasciculate inflorescence of 16–60 cymes, yellow. Aug.–Nov. Mountain slopes, NW, SW (Porterville and Olifants River Mts).•

macrospatha Baker   BERGBLOMBIESIE   Rhizomatous perennial, 50–110 cm. Leaves terete to linear, spathes fibrotic, dilated. Flowers in a dense inflorescence of (3–)8–30 cymes, yellow, with a perianth tube. Mainly Aug.–Nov. Mountain slopes in fynbos, AP, LB, SE (Swellendam to Humansdorp).•

parva J.B.Gillett   Slender rhizomatous perennial, 15–40 cm. Leaves terete, spathes fibrotic, dilated. Flowers in a small inflorescence of 3–7 cymes, yellow. Nov.–Feb. Moist mountain slopes, LB (Langeberg Mts: Swellendam to Lemoenshoek).•

robusta Baker   Rhizomatous perennial, 70–210 cm. Leaves terete, spathes fibrotic, dilated. Flowers in a dense inflorescence of 8–40 cymes, yellow, ovary and capsules tuberculate. Mainly Aug.–Oct. Coastal and lower mountain slopes, LB, SE (Riversdale to George).•

rufa Strid   50–90 cm. Rhizomatous perennial. Leaves terete, spathes fibrotic, dilated, reddish. Flowers in a compact, indistinctly fasciculate inflorescence of c. 50 cymes, yellow. Sept.–Nov. Mountain fynbos, NW (Cedarberg to Hex River Mts).•

B. Spathes green and leaf-like

aphylla (L.f.) Ker Gawl.   BIESROEI   Slender rhizomatous perennial, 30–60 cm. Leaves terete. Flowers in a narrow inflorescence of 3–10 cymes, yellow, ovary and capsules tuberculate. Mainly Nov.–Mar. Grassy mainly coastal slopes, LB, SE (Mossel Bay to Plettenberg Bay).•

indica L.   BIESIESRIET   Rhizomatous perennial, 50–100 cm. Leaves terete, longer than stems and trailing. Flowers in a dense inflorescence of 6–40 cymes, yellow. Mainly Oct.–Mar. Sandy flats and slopes, SW, AP (Mamre to Caledon).•

longicyma J.B.Gillett   Robust rhizomatous perennial to 1.8 m. Leaves terete. Flowers in an inflorescence of 2–20 long cymes, yellow. Capsule obovoid, 10–20 mm. Aug.–Dec. Sandy flats and lower slopes, SW, AP (Kuilsrivier to Potberg).•

macrocarpa Strid   Robust rhizomatous perennial to 1 m. Leaves terete. Flowers in a slender, shortly pedunculate inflorescence of 2–6 cymes, yellow. Aug.–Feb. Grassy slopes, SE (Humansdorp to E Cape).

orientalis J.B.Gillett   Rhizomatous perennial, 40–130 cm. Leaves terete. Flowers in a dense inflorescence of 10–100 short cymes, yellow. Capsule subglobose, 4–8 mm. Mainly Aug.–Nov. Mainly dry, stony sandstone slopes, NW, SW, LB, SE (Piketberg and Riviersonderend Mts to Transkei).

CHASMANTHE   COBRA LILY, KAPELPYPIE   3 spp., NW to E Cape

aethiopica (L.) N.E.Br.   Cormous geophyte, 40–65 cm, corm tunics papery, stems unbranched. Leaves sword-shaped. Flowers secund, tube flaring abruptly and almost pouched below, orange. Apr.–July. Hills and flats on granite, sandstone, or shale, mainly coastal in bush or forest margins, SW, AP, LB, SE (Darling to E Cape).

bicolor (Gasp. ex Tenore) N.E.Br.   Cormous geophyte, 70–90 cm, corm tunics papery, stems usually branched. Leaves sword-shaped. Flowers secund, tube flaring gradually, orange-red with green markings. July–Aug. Sheltered ravines, probably on shale, SW (Robertson district).•

floribunda (Salisb.) N.E.Br.   Cormous geophyte, 45–100 cm, corm tunics firm-papery, becoming fibrous with age, stems branched. Flowers distichous, tube flaring gradually, orange-red (rarely yellow). July–Sept. Coastal and montane on sandstone and granite, NW, SW (Bokkeveld Mts to Hermanus).•

*CROCOSMIA   MONTBRETIA   7 spp., southern and tropical Africa

*aurea (Pappe ex Hook.) Planch.   Cormous geophyte, 40–130 cm. Leaves linear-lanceolate. Flowers actinomorphic, nodding, orange, filaments elongate. Mainly Feb.–Mar. Forest margins, SE (E Cape to tropical Africa, ?naturalised near George).

DIERAMA   HAIRBELL, GRASKLOKKIE   44 spp., S Cape to Ethiopia

pendulum (L.f.) Baker   Cormous perennial, 70–110 cm, stem slender and wiry, laxly branched. Leaves linear, fibrotic. Flowers nodding, in drooping spikes, pink. Oct.–Dec. Rocky sandstone slopes and flats, SE (Knysna to Grahamstown).

DIETES   WOOD IRIS   6 spp., S Cape to E Africa, Lord Howe Island (1 sp.)

iridioides (L.) Sweet ex Klatt   Evergreen rhizomatous perennial, 30–60 cm. Leaves sword-shaped. Flowers white with violet style arms, fugacious. Mainly Aug.–Dec. Evergreen forest and margins, SW, LB, SE (Riviersonderend Mts to Ethiopia).

[Species excluded: **D. flavida** Oberm. was probably recorded from the Baviaanskloof Mts in error]

FERRARIA   SPIDER LILY, SPINNEKOPBLOM   10 spp., W Cape through Namibia to tropical Africa

crispa Burm.   KRULLETJIE   Cormous geophyte, 40–100 cm, leafy and much-branched. Leaves sword-shaped. Flowers brown and speckled, anther thecae parallel, . Aug.–Oct. Mainly coastal, sandstone or granite rocks, NW, SW, AP, KM, LB (Lambert's Bay to Mossel Bay, Little Karoo).•

densepunctulata M.P.de Vos   Slender, cormous geophyte, 12–35 cm. Lower leaves linear. Flowers greenish to grey, anther thecae parallel. May–July. Rocky sites, mostly coastal, NW, SW (Lambert's Bay to Langebaan).•

divaricata Sweet   GEEL SPINNEKOPBLOM   Cormous geophyte, 6–20 cm, often much-branched. Leaves sword-shaped, crowded basally, margins thickened. Flowers yellowish to blue with dark spots and margins, anther thecae divergent, ovary with a sterile beak. Aug.–Nov. Sandy and shale flats and rock outcrops, NW, SW, AP, KM (S Namibia to Clanwilliam and W Karoo to Oudtshoorn).

ferrariola (Jacq.) Willd.   Slender cormous geophyte, 15–60 cm, stems partly exposed, spotted with red below. Lower leaves with long linear blades. Flowers greenish, blue or yellow, anther thecae parallel, later diverging, ovary beaked. June–Aug. Granite and sandstone slopes, NW (N Namaqualand to Clanwilliam).

foliosa G.J.Lewis   Like **F. crispa** but leaves without midribs and flowers maroon, purple or dark brown. Aug.–Oct. Deep sand, west coast, NW (Namaqualand to Elandsbaai).

uncinata Sweet   Cormous geophyte, 10–40 cm. Leaves lanceolate, obtuse, in a tight fan, margins thickened and crisped. Flowers yellow, blue or purple, tepal tips attenuate and coiled, anther thecae divergent, ovary beaked. Aug.–Oct. Rock outcrops, NW (Namaqualand to Tulbagh).

FREESIA (= *ANOMATHECA* in part)   FREESIA, KAMMETJIE   14 spp., W Cape and Karoo to tropical Africa

A. Bracts green and leathery

alba (G.L.Mey.) Gumbleton   Cormous geophyte, 12–40 cm. Leaves sword-shaped. Flowers white, often mauve on reverse, sweetly scented. July–Oct. Sandy or stony soils, mainly coastal, SW, AP, LB, SE (Hermanus to Plettenberg Bay).•

caryophyllacea (Burm.f.) N.E.Br. (= *Freesia elimensis* L.Bolus)   Cormous geophyte, 5–10 cm. Leaves sword-shaped, obtuse, often prostrate. Flowers white with yellow markings, sweetly scented. Apr.–June. Clay soils and limestones, renosterveld and coastal bush, SW, AP, LB (Villiersdorp to Swellendam and Bredasdorp).•

fergusoniae L.Bolus   RIVERSDALE FREESIA   Cormous geophyte, 10–20 cm. Leaves sword-shaped, obtuse, prostrate. Flowers yellow with orange markings, sweetly scented. Aug.–Sept. Clay soils, renosterveld, LB (Heidelberg to Mossel Bay).•

leichtlinii Klatt   DUINE FREESIA   Cormous geophyte, 8–20 cm. Leaves sword-shaped. Flowers cream with broad yellow markings, sweetly scented. Aug.–Sept. Deep sands, coastal fynbos, AP, LB (Cape Agulhas to Mossel Bay).•

sparrmannii (Thunb.) N.E.Br.   Cormous geophyte, 12–18 cm. Leaves sword-shaped, often prostrate. Flowers narrowly funnel-shaped with lower part of tube 12–15 mm, white, flushed purple on reverse, faintly scented. Sept. Forest margins in loam, LB (Langeberg foothills: Swellendam to Heidelberg).•

A. Bracts soft-textured, often with brown tips

corymbosa (Burm.f.) N.E.Br.   FLISSIE, KAMMETJIE,   Cormous geophyte, 25–50 cm. Leaves sword-shaped, acute. Flowers yellow, sometimes pink, lightly scented. Mainly Aug.–Nov. Mainly stony sandstone slopes, SE (Langkloof to E Cape).

occidentalis L.Bolus   Cormous geophyte, 9–50 cm. Leaves sword-shaped, obtuse. Flowers creamy white and yellow, lightly scented. July–Sept. Stony soils, NW, KM (Cedarberg Mts to Touwsrivier and W Karoo).

refracta (Jacq.) Klatt   FLISSIE, KAMMETJIE   Cormous geophyte, 18–45 cm. Leaves sword-shaped, bracts uniformly pale. Flowers white, greenish or purple with orange markings, spice-scented. July–Sept. Dry stony karoo and arid fynbos, SW, KM (Worcester to Oudtshoorn).•

speciosa L.Bolus   Cormous geophyte, 8–20 cm. Leaves sword-shaped, obtuse, inclined to prostrate. Flowers large, 50–70 mm long, white and deep yellow, sweetly scented. Aug–Sept. Stony karroid flats, KM (Montagu to Calitzdorp).•

verrucosa (Vogel) Goldblatt & J.C.Manning (= *Anomatheca verrucosa* (Vogel) Goldblatt)   Cormous geophyte, 8–20 cm. Leaves sword-shaped. Flowers pink, with narrow tube and spreading tepals, faintly sweet-scented. Aug.–Oct. Clay soils, renosterveld, KM, SE (Oudtshoorn to Humansdorp).•

viridis (Aiton) Goldblatt & J.C.Manning (= *Anomatheca viridis* (Aiton) Goldblatt)   GROENAGRETJIE   Cormous geophyte, 10–35 cm. Leaves sword-shaped, sometimes crisped and glaucous. Flowers small, green to dull reddish, tepals attenuate, tube slender and curved, faintly clove-scented. July–Sept. Stony clay and limestone or sometimes sandstone slopes, NW, SW (S Namibia to Mamre).•

GEISSORHIZA   SATIN FLOWER, SYSIE   c. 85 spp., mainly W Cape, also Namaqualand and W Karoo

*A. Flowers blue to violet, sometimes with red centre (see also **G. inconspicua**, **G. inflexa** and **G. intermedia**)*

alticola Goldblatt   Cormous geophyte, 20–30 cm, tunics fibrous. Leaves linear. Flowers blue-violet, tepals spreading. Dec.–Feb. Mountain peaks, 1500–2000 m, SW (Bainskloof to Wemmershoek Mts).•

arenicola Goldblatt   Cormous geophyte, 12–20 cm, tunics woody, imbricate, stem velvety. Leaves linear, with thick margins. Flowers blue, tepals spreading. Aug.–Sept. Sandy mountain soils in fynbos, NW (Bokkeveld Mts to Gifberg).•

aspera Goldblatt   BLOU SYSIE   Cormous geophyte, 10–35 cm, tunics woody, imbricate, stem velvety. Leaves sword-shaped, margins and midrib lightly thickened. Flowers blue-violet, sometimes white, tepals spreading, stamens usually slightly unequal. Aug.–Sept. Mostly sandy soils, flats and slopes, 10–700 m, NW, SW, AP (Gifberg to Agulhas).•

burchellii R.C.Foster   Cormous geophyte, 12–20 cm, tunics papery, imbricate. Leaves linear. Flowers dark purple, tepals spreading, stamens unequal and unilateral. Dec.–Jan., mostly after fire. Rocky sandstone slopes, 200–400 m, SW, LB (Bainskloof to Langeberg Mts).•

cataractarum Goldblatt   Cormous geophyte, 10–30 cm, tunics soft. Leaves slender, trailing. Flowers pale blue, tepals spreading. Nov.–Jan. Waterfalls and damp cliffs, SW (Betty's Bay to Hermanus).•

eurystigma L.Bolus   WINE CUP, KELKIEWYN   Cormous geophyte, 8–20 cm, tunics woody, imbricate. Leaves ribbed. Flowers deep blue with a red centre, tepals cupped, stigmas broad and villous. Sept.–Oct. Granitic soils in renosterveld, SW (Darling to Malmesbury).•

hesperanthoides Schltr.   Cormous geophyte, 15–30 cm, tunics fibrous. Leaves linear, margins and midribs thickened. Flowers blue to violet, tepals spreading. Nov.–Jan. Damp and marshy mountain slopes, 800–1500 m, SW (Bainskloof to Bredasdorp).•

heterostyla L.Bolus   Cormous geophyte, 12–45 cm, tunics woody, imbricate, stem velvety. Leaves linear to sword-shaped, margins winged and ciliate. Flowers blue to purple, sometimes dark in centre, tepals spreading, stamens unequal. Aug.–Oct. Mainly on clay slopes in renosterveld, NW, SW, KM, LB, SE (Bokkeveld Mts to Port Elizabeth and W Karoo).

inaequalis L.Bolus   Cormous geophyte, 8–15 cm, tunics woody, imbricate, stem velvety. Leaves sword-shaped. Flowers blue, tepals spreading, stamens unequal. Aug.–Oct. Heavy clay soils, NW (Bokkeveld Escarpment and W Karoo).

lithicola Goldblatt   Cormous geophyte, 15–30 cm, tunics woody concentric. Leaves terete. Flowers violet to purple, tepals spreading. Oct. Lower rocky slopes, 50–200 m, SW (Kogelberg).•

mathewsii L.Bolus   WINE CUP, KELKIEWYN   Cormous geophyte, 8–18 cm, tunics woody, imbricate. Leaves ribbed. Flowers violet with red centre, tepals cupped, stigmas broad and villous. Aug.–Sept. Wet sandy flats, SW (Darling district).•

monanthos Eckl.   Cormous geophyte, 6–20 cm, tunics woody, imbricate, stem velvety. Leaves sword-shaped. Flowers glossy dark blue, often pale in centre with a darker ring, tepals cupped, stamens usually unequal. Aug.–Oct. Sandy slopes, granite outcrops, NW, SW (Citrusdal, Saldanha to Somerset West).•

nigromontana Goldblatt   Cormous geophyte, 10–16 cm, tunics woody, concentric. Leaves lanceolate, trailing. Flowers 1–few per spike, blue, tepals spreading. Jan.–Feb. Along streams, KM (Swartberg Mts).•

pseudinaequalis Goldblatt   Cormous geophyte, 9–30 cm, tunics papery. Leaves linear. Flowers blue to violet, tepals spreading, stamens unequal. Oct.–Jan. Mountain slopes and cliffs, SW (Bainskloof to Simonsberg).•

purpurascens Goldblatt   Cormous geophyte, 18–30 cm, tunics woody, concentric. Leaves linear. Flowers mauve, tepals spreading. Sept.–Oct. Sandy flats, NW, SW (Piketberg to Stellenbosch).•

pusilla (Andrews) Klatt   Cormous geophyte, 7–25 cm, tunics woody, concentric. Leaves sword-shaped, lightly villous. Flowers blue to mauve, tepals spreading. Aug.–Oct. Damp shady places, SW (Cape Peninsula to Paarl).•

radians (Thunb.) Goldblatt   WINE CUP, KELKIEWYN   Cormous geophyte, 8–16 cm, tunics woody, imbricate. Leaves linear. Flowers deep blue with red centre and white ring, tepals cupped, stamens and style unilateral. Sept.–Oct. Damp sandy soils, SW (Malmesbury to Gordon's Bay).•

ramosa Ker Gawl. ex Klatt   Cormous geophyte, 20–45 cm, tunics woody, imbricate, stem often several-branched. Leaves linear. Flowers small, blue to purple, tepals spreading, stamens unequal. Oct.–Dec. Rocky mountain slopes, NW, SW, LB (Tulbagh to Swellendam).•

scopulosa Goldblatt   Cormous geophyte, 6–20 cm, tunics soft, stem velvety. Leaves linear. Flowers blue-violet, tepals spreading, stamens unequal. Nov. Rocky sandstone soils, 500–1500 m, NW (Hex River Mts).•

splendidissima Diels   BOKKEVELD PRIDE   Cormous geophyte, 8–20 cm, tunics woody, imbricate, stem velvety. Leaves linear. Flowers glossy blue-violet, tepals cupped, stamens unilateral, anthers red-brown. Aug.–Sept. Clay soils in renosterveld, NW (Bokkeveld Plateau).•

subrigida L.Bolus   Cormous geophyte, 12–30 cm, tunics woody, imbricate. Leaves linear to sword-shaped, pilose. Flowers blue-violet, tepals spreading. Aug.–Sept. Rocky sandstone soils, NW (Bokkeveld Mts).•

tabularis Goldblatt   Cormous geophyte, 25–35 cm, tunics woody, imbricate, stem often branched. Leaves sword-shaped. Flowers mauve, rarely white, tepals spreading, stamens unequal. Oct.–Dec. Cool, damp sandstone slopes, SW (Cape Peninsula).•

A. Flowers white to pink or yellow (see also **G. aspera** *and* **G. tabularis**)

barkerae Goldblatt   Cormous geophyte, 10–20 cm, tunics woody, imbricate. Leaves linear, ribbed. Flowers yellow with a purple centre, tepals cupped, stamens and style unilateral. Sept.–Oct. Marshes and seeps at foot of mountains, NW (Piketberg to Citrusdal).•

bolusii Baker   Cormous geophyte, 3–10 cm, tunics woody, concentric. Leaves lanceolate to ovate, often prostrate. Flowers small, white, tepals spreading, buds often aborted and replaced by cormlets, never producing capsules. Oct.–Jan. Damp shady sandstone slopes in moss, 400–1500 m, NW, SW (Cedarberg to Cape Peninsula to Worcester).•

bonaspei Goldblatt   Cormous geophyte, 12–20 cm, tunics woody, concentric. Leaves linear, narrow, H-shaped in section, viscid. Flowers pink, darker on veins, tepals spreading, tube elongate, stamens and style unilateral. Sept.–Nov., mainly after fire. Rocky sandstone slopes in fynbos, SW (Cape Peninsula).•

bracteata Klatt   Cormous geophyte, 6–18 cm, tunics woody, concentric. Leaves sword-shaped. Flowers 1–few per spike, white, tepals spreading. Sept.–Oct. Mostly clay slopes in renosterveld, LB, SE (Albertinia to Grahamstown and Somerset East).•

brehmii Eckl. ex Klatt   Cormous geophyte, 20–30 cm, tunics woody, imbricate. Leaves terete. Flowers white to cream, tepals spreading. Aug.–Oct. Seasonal pools, mainly lowlands, SW, AP (Malmesbury and Cape Peninsula to Bredasdorp).•

brevituba (G.J.Lewis) Goldblatt   Cormous geophyte, 12–20 cm, tunics woody, concentric. Leaves sword-shaped, margins and midrib thickened and viscid. Flowers deep pink, darker on veins, tepals cupped, stamens and style unilateral. Sept., mainly after fire. Rocky sandstone slopes, NW (Piketberg).•

bryicola Goldblatt   Cormous geophyte, 15–30 cm, tunics woody, imbricate. Leaves narrowly sword-shaped, soft and trailing. Flowers white, stamens unequal, tepals spreading. Sept.–Nov. Wet rocks, waterfalls, stream edges, SW (Hermanus Mts).•

callista Goldblatt   Cormous geophyte, 15–25 cm, tunics papery. Leaves linear, margins thickened. Flowers bright pink with dark purple centre, tepals cupped, stamens and style unilateral. Oct.–Nov., mainly after fire. Wet rocks, SW, LB (Riviersonderend and Langeberg Mts).•

cedarmontana Goldblatt   Cormous geophyte, 7–35 cm, tunics woody, concentric. Leaves linear. Flowers pale pink, dark red at tepal bases, tepals spreading, tube elongate, stamens and style included in tube. Oct.–Nov. Damp south slopes and rocks, NW (Cedarberg Mts).•

ciliatula Goldblatt   Cormous geophyte, 6–12 cm, tunics woody, imbricate, stem velvety. Leaves linear and soft. Flowers white fading mauve, tepals spreading. Oct.–Nov. Moist rocky sandstone slopes, 600–1000 m, NW (Cedarberg Mts).•

confusa Goldblatt   Cormous geophyte, 12–30 cm, tunics woody, concentric. Leaves linear to sword-shaped, margins and midribs thickened and viscid. Flowers creamy beige fading pinkish, darker on veins, tepals spreading, stamens and style unilateral. Mainly Oct.–Nov., mainly after fire. Rocky sandstone slopes in fynbos, 400–1200 m, NW, SW (Gifberg to Caledon).•

darlingensis Goldblatt   Cormous geophyte, 7–15 cm, tunics woody, concentric. Leaves linear, margins and midribs thickened and viscid. Flowers yellow with a dark brown centre, tepals cupped. Sept.–Oct. Damp flats, SW (Darling).•

delicatula Goldblatt   Cormous geophyte, 3–12 cm, tunics soft. Leaves linear to sword-shaped, soft, often prostrate. Flowers lilac, tepals spreading. Aug.–Dec. Sandstone outcrops, KM (Swartberg Mts).•

divaricata Goldblatt   Cormous geophyte, 20–45 cm, tunics woody, imbricate, stems divaricately branched. Leaves sword-shaped, margins and veins thickened and ciliate. Flowers small, white, rarely pale mauve, dark mauve on reverse, tepals spreading. Sept.–Oct. Sandstone rocks, NW (Bokkeveld Mts to Gifberg).•

elsiae Goldblatt   Cormous geophyte, 14–25 cm, tunics soft. Leaves sword-shaped to linear, soft. Flowers pink. Oct.–Nov. Damp southern slopes, KM (Kammanassie Mts).•

erubescens Goldblatt   Cormous geophyte, 8–15 cm, tunics woody, imbricate, stem velvety. Leaves narrowly sword-shaped, margins and midribs thickened. Flowers small, cream, bright red on reverse, tepals spreading. Sept. Shale and loam slopes, NW (Pakhuis Mts).•

esterhuyseniae Goldblatt   Cormous geophyte, 7–8 cm, tunics woody, concentric. Leaves sword-shaped, margins and midrib lightly thickened. Flowers white, tepals spreading, stamens and style included in tube. Oct. Rocky S-facing sandstone slopes, c. 1200 m, NW (Grootwinterhoek Mts).•

exscapa (Thunb.) Goldblatt   LANGPYP SYSIE   Cormous geophyte, 18–30 cm, tunics woody, concentric, stem shorter than leaves. Leaves linear, H-shaped in section, viscid on margins. Flowers creamy beige, fading pink, darker on veins, tube elongate, stamens and style unilateral. Oct.–Nov. Sandy soils, coastal and montane, 120–400 m, NW, SW (Namaqualand to Melkbos).

foliosa Klatt   Cormous geophyte, 8–20 cm, tunics woody, concentric, stems usually branched. Leaves short, lanceolate, soft. Flowers lilac to mauve-pink with white anthers. Sept.–Nov. Clay slopes and flats in renosterveld, LB (Swellendam to Riversdale).•

fourcadei (L.Bolus) G.J.Lewis   Cormous geophyte, 12–30 cm, tunics woody, concentric, stems branched. Leaves terete. Flowers single on branches, large, pink to mauve, tepals spreading, stamens and style unilateral. Mainly Mar.–May. Sandstone rocks, KM, SE (Swartberg and Outeniqua Mts to Humansdorp).•

furva Ker Gawl. ex Baker   Cormous geophyte, 8–14 cm, tunics woody, concentric. Leaves terete. Flowers large, golden-yellow, tepals cupped. Sept.–Oct. Stony flats and lower slopes, NW, SW (Piketberg to Paarl).•

geminata E.Mey. ex Baker   Cormous geophyte, 12–30 cm, tunics woody, imbricate. Leaves linear. Flowers white to cream, tepals cupped. Sept.–Nov. Marshes and pools, NW (Cold Bokkeveld to Worcester).•

grandiflora Goldblatt   Cormous geophyte, 16–35 cm, tunics soft. Leaves linear. Flowers large, pink, tepals cupped, stamens unequal and unilateral. Nov.–Dec. Rocky sandstone slopes, NW, SW (Grootwinterhoek Mts to Villiersdorp).•

hispidula (R.C.Foster) Goldblatt   Cormous geophyte, 7–25 cm, tunics woody, concentric. Leaves linear, margins and midribs thickened and sticky. Flowers small, cream or white, tepals spreading. Aug.–Sept., mainly after fire. Sandy flats and mountain slopes, SW, AP, LB (Cape Peninsula to Albertinia).•

humilis (Thunb.) Ker Gawl.   Cormous geophyte, 8–14 cm, tunics woody, concentric. Leaves linear, margins thickened and sticky. Flowers bright yellow, tepals lightly cupped. Aug.–Oct., mainly after fire. Sandy soils in fynbos, 50–300 m, SW (Cape Peninsula to Paarl).•

imbricata (D.Delaroche) Ker Gawl.   Cormous geophyte, 8–25 cm, tunics woody, imbricate. Leaves ribbed. Flowers white to yellow, tepals spreading. Aug.–Nov. Wet sandy flats, marshes, streamsides, NW, SW, AP (Gifberg to Bredasdorp).•

inconspicua Baker   Cormous geophyte, 10–30 cm, tunics woody, concentric. Leaves sword-shaped to linear. Flowers white or purple, tepals spreading. Oct.–Feb. Mountains and flats, LB, SE (Swellendam to Uitenhage).•

inflexa (D.Delaroche) Ker Gawl.   WITSYBLOM   Cormous geophyte, 12–35 cm, tunics woody, imbricate. Leaves sword-shaped to linear, margins winged and ciliate on veins margin edges. Flowers white, sometimes red or purple. Aug.–Sept. Clay flats and slopes in renosterveld, NW, SW, AP (Piketberg to Bredasdorp).•

intermedia Goldblatt   Cormous geophyte, 5–12 cm, tunics woody, imbricate, stem velvety. Leaves sword-shaped, sticky below. Flowers white. Sept.–Oct. Sandstone soils in fynbos, NW, SW (Porterville to Stellenbosch).•

juncea (Link) A.Dietr.   Cormous geophyte, 20–40 cm, tunics woody, concentric. Leaves terete. Flowers small, cream to yellow, tepals spreading. Aug.–Nov. Sandy flats and slopes, to 400 m, NW, SW (Cedarberg Mts to Bredasdorp).•

leipoldtii R.C.Foster   Cormous geophyte, 12–30 cm, tunics woody, imbricate, stem velvety. Leaves sword-shaped, lightly pilose. Flowers white, pink or mauve, tepals spreading, stamens often unequal. Aug.–Sept. Mostly S-facing shale slopes, NW (Pakhuis Mts to Citrusdal).•

longifolia (G.J.Lewis) Goldblatt   Cormous geophyte, 12–20 cm, tunics woody, concentric. Leaves linear, margins and midribs heavily thickened, sticky. Flowers white, fading pink with darker veins, tepals spreading, tube elongate. Sept.–Nov. Stony, often shale soils, NW (Gifberg to Piketberg and Ceres).•

louisabolusiae R.C.Foster   Cormous geophyte, 15–20 cm, tunics woody, imbricate. Leaves linear to terete. Flowers pale yellow, stamens usually unequal. Aug.–Sept. Wet sandy flats, NW (Olifants River valley).•

malmesburiensis R.C.Foster   Cormous geophyte, 5–8 cm, corms with flat base, tunics woody, concentric. Leaves linear-filiform. Flowers 1–few per spike, yellow. Sept.–Oct. Granitic sand in renosterveld, SW (Malmesbury).•

minuta Goldblatt   Cormous geophyte, 3–12 cm, tunics woody, imbricate. Leaves linear. Flowers small, white, tepals spreading. Sept.–Oct. Wet sites, NW (Matsikamma to Pakhuis Mts).•

nana Klatt   Cormous geophyte, 5–7 cm, tunics woody, concentric. Leaves linear. Flowers mostly 1 per spike, tiny, white, tepals spreading. Sept.–Oct. Clay slopes and flats in renosterveld, SW, AP, LB (Caledon to Riversdale).•

nubigena Goldblatt   Cormous geophyte, 15–30 cm, tunics fibrous. Leaves linear. Flowers rose to mauve, tepals spreading, tube fairly long, stamens and style unilateral. Dec.–Jan. High rocky slopes, 1000–1800 m, NW, SW (Grootwinterhoek Mts to Kogelberg).•

ornithogaloides Klatt   Cormous geophyte, 4–10 cm, corms often with a flat base, tunics woody, concentric. Leaves linear. Flowers 1–few per spike, bright yellow, tepals spreading. Aug.–Oct. Mostly clay flats and lower slopes, NW, SW, LB, SE (Cedarberg Mts to Humansdorp).•

outeniquensis Goldblatt   Cormous geophyte, 20–50 cm, tunics soft, stems often trailing. Leaves trailing. Flowers pink to purple. Oct.–Feb. Streambanks, waterfalls and wet rocks in shade, SE (Outeniqua Mts).•

ovalifolia R.C.Foster   Cormous geophyte, 3–9 cm, tunics woody, concentric. Leaves oblong, often prostrate. Flowers white. Oct.–Nov. Damp places, NW, SW (Cold Bokkeveld to Franschhoek Mts).•

ovata (Burm.f.) Aschers. & Graebn.   Cormous geophyte, 6–15 cm, tunics woody, concentric. Leaves ovate, leathery, prostrate. Flowers few per spike, white, pink on reverse, tube fairly long. Aug.–Oct., mainly after fire. Sandstone slopes and flats, NW, SW, AP (Olifants River Mts to Riversdale).•

pappei Baker   Cormous geophyte, 5–10 cm, tunics woody, concentric. Leaves linear, margins and midribs thickened and sticky. Flowers small, white to cream, tepals spreading. Sept.–Oct. Sandy mountain soils, NW, SW (Ceres to Caledon).•

parva Baker   Cormous geophyte, 4–12 cm, tunics woody, concentric. Leaves sword-shaped to oblong, leathery, often prostrate. Flowers few per spike, small, cream to yellow. Aug.–Nov. Sandy soils, mountains and flats, NW, SW, LB (Cedarberg Mts to Swellendam).•

purpureolutea Baker   Cormous geophyte to 15 cm, tunics woody, imbricate. Leaves linear. Flowers yellowish, usually with dark centre, tepals spreading. Aug.–Sept. Wet sandy flats, NW, SW (Piketberg to Paarl).•

roseoalba (G.J.Lewis) Goldblatt   Cormous geophyte, 15–20 cm, tunics woody, concentric. Leaves sword-shaped, margins and midribs thickened. Flowers pale pink with red markings, pink on reverse, tepals loosely cupped, stamens and style unilateral. Aug.–Sept., mostly after fire. Sandstone soils in dry fynbos, KM, LB, SE (Little Karoo to Uitenhage).•

rupicola Goldblatt & J.C.Manning   Cormous geophyte, 15–20 cm, tunics fibrous. Leaves linear, margins thickened. Flowers pale pink, tepals loosely cupped. Nov.–Dec. Wet sandstone cliffs, NW (Ceres).•

schinzii (Baker) Goldblatt   Cormous geophyte, 10–20 cm, tunics woody, imbricate. Leaves linear, margins and midribs thickened and sticky. Flowers large, pink with darked veins, tepals spreading, stamens and style unilateral. Aug.–Oct., after fire. Stony sandstone slopes, SW (Houwhoek to Bredasdorp Mts).•

scillaris A.Dietr.   Cormous geophyte, 12–35 cm, tunics woody, imbricate. Leaves linear, margins heavily thickened, second leaf entirely sheathing. Flowers small, white to pale blue or mauve, tepals spreading. Aug.–Nov. Rocky sandstone slopes and flats, NW, SW (Cedarberg Mts to Caledon).•

setacea (Thunb.) Ker Gawl.   Cormous geophyte, 4–8 cm, tunics woody, concentric. Leaves linear to sword- shaped. Flowers small, 1–few per spike, white or cream, tepals spreading. June–Sept. Damp sandy and stony flats, SW (Gouda to Gordon's Bay).•

silenoides Goldblatt & J.C.Manning   Cormous geophyte, 20–30 cm, tunics woody, imbricate, stem velvety. Leaves linear, margins thickened. Flowers pale pink, tube long, stamens unequal. Sept.–Oct. Shale slopes in fynbos, NW (Ceres).•

similis Goldblatt   Cormous geophyte, 12–35 cm, tunics woody, imbricate. Leaves linear. Flowers white. Aug.–Oct. Sandy slopes and flats, SW (Bainskloof to Cape Peninsula).•

stenosiphon Goldblatt   Cormous geophyte, 15–30 cm, tunics woody, concentric. Leaves terete, 4-grooved. Flowers 1 or 2, tubular, white with pink reverse. Nov.–Dec. Rocky sandstone slopes above 400 m, NW (Cold Bokkeveld Mts).•

sulphurascens Schltr. ex R.C.Foster   Cormous geophyte, 12–20 cm, tunics woody, imbricate. Leaves linear. Flowers cream. Aug.–Sept. Wet sandy soils, NW (Bokkeveld Mts).•

tenella Goldblatt   Cormous geophyte, 10–30 cm, tunics woody, concentric. Leaves narrow, H-shaped in section, viscid. Flowers white to pink, zygomorphic. Oct.–Dec. Sandy flats and dunes, SW (Darling to Bredasdorp).•

tulbaghensis F.Bolus   Cormous geophyte, 8–15 cm, tunics woody, imbricate, stem velvety. Leaves linear. Flowers white with dark centre, zygomorphic. Aug.–Sept. Clay flats, NW, SW (Porterville to Wellington).•

uliginosa Goldblatt & J.C.Manning   Cormous geophyte, 15–30 cm, tunics soft. Leaves linear. Flowers dark pink, zygomorphic. Dec.–Jan. Waterfalls and wet cliffs, KM (Swartberg Mts).•

umbrosa G.J.Lewis   Cormous geophyte, 12–30 cm, tunics mostly fibrous. Leaves linear. Flowers white to cream. Oct.–Dec. Sandstone slopes in damp sites mostly above 400 m, NW, SW (Cedarberg Mts to Cape Peninsula and Riviersonderend Mts).•

unifolia Goldblatt   Cormous geophyte, 5–10 cm, tunics papery, concentric. Leaves mostly 1, linear. Flowers white, small. Oct. High elevations, NW (Cedarberg Mts).•

GLADIOLUS (= *ANOMALESIA, HOMOGLOSSUM*)   GLADIOLUS, AFRIKANER, PYPIE   c. 250 spp., Africa and Madagascar, Eurasia

*A. Flowers red to orange (see also **G. meliusculus**)*

abbreviatus (Andrews) Goldblatt & M.P.de Vos (= *Homoglossum abbreviatum* (Andrews) Goldblatt)   Cormous geophyte, 30–65 cm, tunics woody. Leaves cross-shaped in section. Flowers long-tubed, tube cylindrical, orange to reddish, lower tepals very short, green to blackish. June–Sept. Clay soils in renosterveld, SW, LB (Botrivier to Riversdale).•

alatus L.   KALKOENTJIE, KIPKIPPIE   Cormous geophyte, 8–25 cm, tunics papery. Leaves falcate, ribbed. Flowers bilabiate, upper tepal erect, orange marked yellow to greenish, scented. Aug.–Sept. Flats, slopes and plateaus, mainly in sand, NW, SW, AP (Bokkeveld Mts to Bredasdorp).•

bonaspei Goldblatt & M.P.de Vos (= *Homoglossum merianellum* (Thunb.) Baker)   VLAMME   Cormous geophyte, 30–50 cm, tunics fibrous. Leaves sword-shaped, villous. Flowers long-tubed, tube cylindrical, tepals equal, orange to yellow. Apr.–Aug. Sandy flats and slopes to 250 m, SW (Cape Peninsula).•

brevitubus G.J.Lewis   Cormous geophyte, 12–35 cm, tunics fibrous. Leaves linear. Flowers rotate, almost actinomorphic, tube obsolete, orange, faintly scented. Sept.–Nov. Rocky sandstone slopes, SW (Somerset West to Riviersonderend Mts and Hermanus).•

cardinalis Curtis   NEW YEAR LILY, NUWEJAARSBLOM   Cormous geophyte, 60–120 cm, tunics papery. Leaves sword-shaped, stem inclined to drooping. Flowers large, funnel-shaped, red with white splashes on the lower tepals. Dec.–Jan. Waterfalls and wet cliffs, SW (Bainskloof to Riviersonderend Mts).•

cunonius (L.) Gaertn. (= *Anomalesia cunonia* (L.) N.E.Br.)   SUIKERKANNETJIE, LEPELBLOM   Cormous geophyte, 20–45 cm, stoloniferous, tunics papery. Leaves sword-shaped, soft. Flowers tubular, upper tepal elongate and spooned, bright red, lower tepals small, green. Sept.–Oct. Coastal in sandy soils, SW, AP, SE (Saldanha to Knysna).•

huttonii (N.E.Br.) Goldblatt & M.P.de Vos   EASTERN CAPE FLAME   Cormous geophyte, 30–60 cm, tunics coarsely fibrous. Leaves cross-shaped in section. Flowers long-tubed, red to orange with cylindrical tube streaked maroon, lower tepals smaller and sometimes yellow. June–Sept. Sandstone slopes, SE (Plettenberg Bay to Grahamstown).

insolens Goldblatt & J.C.Manning   Cormous geophyte, 40–65 cm, tunics papery. Leaves linear, slightly fleshy. Flowers long-tubed, tulip-shaped, tepals subequal, scarlet. Dec.–Jan. Wet sandstone cliffs and rocks, c. 1200 m, NW (Piketberg: Zebra Kop).•

miniatus Eckl.   Cormous geophyte, 15–40 cm, tunics papery. Leaves sword-shaped. Flowers long-tubed, salmon, tube cylindrical. Oct.–Nov. Coastal limestone outcrops, SW, AP (Hermanus to Agulhas).•

nerineoides G.J.Lewis   Cormous geophyte with reduced leaves, 30–40 cm, tunics fibrous. Leaf produced after flowering on a separate shoot, linear, solitary, hairy. Flowers clustered, long-tubed, narrowly funnel-shaped, scarlet. Jan.–Mar. Rocky sandstone slopes and cliffs, 500–1500 m, SW (Bainskloof to Somerset West).•

overbergensis Goldblatt & M.P.de Vos (= *Homoglossum guthriei* (L.Bolus) L.Bolus)   Cormous geophyte, 35–55 cm, tunics fibrous. Leaves sword-shaped, scabrid. Flowers tubular, lower tepals smaller, red to orange. July–Sept. Sandstone slopes, SW, AP (Hermanus to Agulhas).•

priorii (N.E.Br.) Goldblatt & M.P.de Vos (= *Homoglossum priorii* (N.E.Br.) N.E.Br.)   Cormous geophyte, 30–40 cm, tunics fibrous. Leaves linear with short blades. Flowers tubular, tepals subequal, red with yellow throat. Apr.–June. Sandstone and granite slopes, SW (Saldanha to Hermanus).•

pulcherrimus (G.J.Lewis) Goldblatt & J.C.Manning   KALKOENTJIE   Like **G. alatus** but plants often taller, 20–50 cm, leaves not ribbed, dorsal tepal erect. Sept.–Oct. Sandstone slopes, NW (Trawal to Piketberg).•

quadrangularis (Burm.f.) Ker Gawl. (= *Homoglossum quadrangulare* (Burm.f.) N.E.Br.)   Cormous geophyte, 50–90 cm, tunics fibrous. Leaves cross-shaped in section. Flowers like **G. watsonius**, red. Aug.–Oct. Rocky sandstone slopes, NW (Cold Bokkeveld to Koo).•

saccatus (Klatt) Goldblatt & M.P.de Vos (= *Anomalesia saccata* (Klatt) Goldblatt)   ROGGEVELD SUIKER KANNETJIE   Cormous geophyte, 25–80 cm, tunics fibrous. Leaves sword-shaped. Flowers tubular, dorsal tepal elongate and spooned, lower tepals reduced to tiny scales, bright red. June–Aug. Dry shale slopes, NW (Namibia to Pakhuis Mts).•

sempervirens G.J.Lewis   GEORGE FLAME   Evergreen cormous geophyte, 40–60 cm, corm reduced and rhizomatous. Leaves sword-shaped. Flowers large, funnel-shaped, pinkish red with white streaks on the lower tepals. Mar.–May. Seeps on sandstone slopes, 300–1500 m, SE (George to Kareedouw).•

speciosus Thunb.   KALKOENTJIE   Cormous geophyte, 10–20 cm, stoloniferous, tunics papery. Leaves sword-shaped. Flowers like **G. alatus** but upper tepal hooded. Sept.–Oct. Deep sandy soils in fynbos, NW, SW (Bokkeveld Escarpment to Mamre).•

stefaniae Oberm.   Cormous geophyte, 40–60 cm, tunics papery. Leaves reduced. Flowers large, funnel-shaped, red with white streaks on the lower tepals. Mar.–Apr. Rocky sandstone slopes, 100–800 m, NW, SW (Montagu and Potberg).•

stokoei G.J.Lewis   Cormous geophyte with reduced leaves, 30–45 cm, tunics fibrous. Leaf produced after flowering on a separate shoot, linear, solitary, hairy. Flowers tulip-shaped, scarlet. Mar.–Apr. Marshy sandstone slopes, 500–1000 m, SW (Riviersonderend Mts).•

teretifolius Goldblatt & M.P.de Vos (= *Homoglossum muirii* (L.Bolus) L.Bolus)   Cormous geophyte, 30–60 cm, tunics woody. Leaves slender, oval to terete in section and 4-grooved. Flowers tubular, tepals subequal, red. May–Aug. Clay slopes in renosterveld, SW, LB (Caledon to Mossel Bay).•

vandermerwei (L.Bolus) Goldblatt & M.P.de Vos (= *Homoglossum vandermerwei* (L.Bolus) L.Bolus)   Cormous geophyte, 30–60 cm, stoloniferous, tunics papery. Leaves sword-shaped. Flowers tubular, bright red, lower tepals linear, marked with green. Aug.–Sept. Shale slopes in renosterveld, SW, LB (Botrivier to Heidelberg).•

watsonius Thunb. (= *Homoglossum watsonium* (Thunb.) N.E.Br.)   ROOI AFRIKANER   Cormous geophyte, 30–50 cm, tunics woody. Leaves linear with thickened margins and midribs. Flowers tubular, tepals subequal, red to orange. Aug.–Sept. Clay and granite slopes in renosterveld, NW, SW (Piketberg to Stellenbosch).•

A. Flowers white, yellow, pink, shades of blue to mauve, brown or green
B. Leaves terete and 4-grooved or cross-shaped in transverse section (see also **G. virescens**)

ceresianus L.Bolus   Cormous geophyte, 8–15 cm, tunics fibrous. Leaves imbricate, terete with 4 grooves. Flowers bilabiate, dorsal tepal erect, dull purple to brownish with dark veining, fragrant. Aug.–Oct. Sandstone slopes, NW, KM (Cold Bokkeveld to Witteberg and W Karoo).

cylindraceus G.J.Lewis   Cormous geophyte, 30–50 cm, tunics fibrous. Leaves cross-shaped in section. Flowers long-tubed, pale pink with dark marks on the lower tepals. Dec.–Jan. Sandstone slopes, NW (Cold Bokkeveld and Tulbagh Mts)•

engysiphon G.J.Lewis   Cormous geophyte with reduced leaves, 35–50 cm, tunics fibrous. Leaf blades lacking, foliage leaves terete and 4-grooved. Flowers long-tubed, cream with red streaks on the lower tepals. Mar.–Apr. Clay and granitic loam in renosterveld and grassland, LB (Swellendam to Mossel Bay).•

fourcadei (L.Bolus) Goldblatt & M.P.de Vos (= *Homoglossum fourcadei* (L.Bolus) N.E.Br.)   Cormous geophyte, 40–60 cm, tunics fibrous. Leaves cross-shaped in section. Flowers tubular, tube cylindrical, red or yellow-green. Sept.–Oct. Clay soils in renosterveld, KM, SE (George to Humansdorp).•

inflatus Thunb. BLOUKLOKKIE, TULBAGH BELL   Cormous geophyte, 25–60 cm, tunics woody. Leaves terete with 4 grooves. Flowers bell-like, white, mauve or pink with yellow markings. Aug.–Nov. Rocky sandstone slopes, NW, SW, LB (Cedarberg Mts to Swellendam).•

jonquilliodorus Eckl. ex G.J.Lewis   Cormous geophyte with reduced leaves, 30–45 cm, tunics papery. Leaves produced after flowering on a separate shoot, 2 or 3, terete with 4 grooves. Flowers bilabiate, cream to pale yellow, fragrant. Nov.–Dec. Sandy coastal flats, SW (Darling to Cape Peninsula).•

longicollis Baker   Cormous geophyte, 35–60 cm, tunics woody. Leaves terete and 4-grooved. Flowers long-tubed, brown to cream with brownish speckling, fragrant in the evening. Sept.–Oct. Sandstone slopes, KM, SE (Oudtshoorn to N Province).

martleyi L.Bolus (= *Gladiolus pillansii* G.J.Lewis)   Cormous geophyte with reduced leaves, 20–35 cm, tunics papery. Leaves 1 or 2, produced after flowering on a separate shoot, terete and 4-grooved. Flowers bilabiate, white to mauve with dark pink markings, fragrant. Feb.–May. Sandy and rocky flats and lower slopes to 200 m, NW, SW, AP, LB (Bokkeveld Mts to Riversdale).•

nigromontanus Goldblatt   Cormous geophyte, 30–40 cm, tunics woody. Leaf blades short, terete with 4 grooves. Flowers long-tubed, white with red streaks on the lower tepals. Mar. Wet sandstone slopes, KM (Swartberg Pass).•

patersoniae F.Bolus   Cormous geophyte, 30–50 cm, tunics coarsely fibrous. Leaves terete with 4 grooves. Flowers bell-like, blue to pearly grey with yellow markings on the lower tepals, fragrant. Aug.–Oct. Rocky sandstone slopes, SW, KM, SE (Worcester to Great Winterhoek Mts).•

subcaeruleus G.J.Lewis   Cormous geophyte with reduced leaves, 20–30 cm, tunics fibrous. Leaf produced after flowering on a separate shoot, solitary, terete and 4-grooved, thinly hairy. Flowers bilabiate, pale blue to mauve with yellow markings on the lower tepals. Mar.–May. Sandy loam and clay slopes in renosterveld, SW (Caledon to Bredasdorp).•

sufflavus (G.J.Lewis) Goldblatt & J.C.Manning   Cormous geophyte, 45–70 cm, tunics fibrous. Leaves slender, cross-shaped in section, sheaths hairy. Flowers bell-like, greenish yellow, fragrant. Aug.–Sept. Marshy sandstone soils, NW (Bokkeveld Mts).•

trichonemifolius Ker Gawl. (= *Gladiolus tenellus* auct., *G. citrinus* Klatt)   Cormous geophyte, 10–25 cm, tunics woody. Leaves terete and 4-grooved. Flowers funnel-shaped, sometimes actinomorphic, yellow to whitish, occasionally with a dark centre, usually fragrant. July–Oct. Wet sandy flats, 50–1000 m, NW, SW, AP (Hopefield and Ceres to Bredasdorp).•

tristis L.   MARSH AFRIKANER, TROMPETTERS   Cormous geophyte, 40–150 cm, tunics fibrous. Leaves slender, cross-shaped in section. Flowers long-tubed, cream with brown shading, fragrant in evening. Aug.–Dec. Usually marshy sites on sandstone, clay, or limestone soils, NW, SW, AP, KM, LB, SE (Bokkeveld Mts to Port Elizabeth).•

B. Leaves plane or ribbed

acuminatus F.Bolus   Cormous geophyte, 20–30 cm, tunics fibrous. Leaves linear. Flowers long-tubed, tepals attenuate, greenish yellow, fragrant. Aug.–Sept. Stony clay soils, SW (Caledon to Bredasdorp).•

albens Goldblatt & J.C.Manning   Cormous geophyte, 30–40 cm, tunics fibrous. Leaves linear, blades often short. Flowers long-tubed, white to cream, fragrant. Mar.–May. Grassy slopes, SE (George to Alexandria).

angustus L.   Like **G. carneus** but flowers with tube much longer than tepals, white to cream with red markings on the lower tepals. Oct.–Nov. Streams and marshes on sandstone soils, NW, SW (Cedarberg Mts to Cape Peninsula).•

aquamontanus Goldblatt   Cormous geophyte, 40–100 cm, tunics papery. Leaves sword-shaped, stems inclined to drooping. Flowers long-tubed, mauve-pink with purple markings on the lower tepals. Nov.–Dec. Streams and wet cliffs, KM (Swartberg Mts).•

arcuatus Klatt   Cormous geophyte, 8–30 cm, tunics papery. Leaves narrowly sword-shaped, velvety below. Flowers bilabiate, greyish mauve with yellow lower tepals, windowed, fragrant. June–Aug. Gravelly flats and slopes, NW (Namaqualand to Klawer).

atropictus Goldblatt & J.C.Manning   Cormous geophyte to 40 cm, tunics fibrous. Leaves linear, midribs and margins lightly raised. Flowers bilabiate, blue with reddish streaking on the lower tepals, fragrant. July–Aug. Rocky sandstone slopes, SW (Riviersonderend Mts).•

aureus Baker   Cormous geophyte, 30–50 cm, tunics fibrous. Leaves linear, hairy, blades often very short. Flowers slender-tubed, tepals subequal, bright yellow. Aug.–Sept. Seeps on rocky sandstone slopes, SW (Cape Peninsula).•

bilineatus G.J.Lewis   Cormous geophyte, 20–40 cm, tunics papery. Leaves sword-shaped, blades often short. Flowers long-tubed, pink. Mar.–Apr. Clay and loamy sand, fynbos, LB (Swellendam to Albertinia).•

blommesteinii L.Bolus   Cormous geophyte, 30–60 cm, tunics woody. Leaves linear with raised margins. Flowers bilabiate, mauve or pink with dark streaks on the lower tepals, bracts ridged. Aug.–Oct. Sandstone slopes in fynbos, SW (Du Toitskloof to Riviersonderend Mts).•

brevifolius Jacq.   HERFSPYPIE   Cormous geophyte with reduced leaves, 15–50 cm, tunics fibrous. Leaf produced after flowering on a separate shoot, solitary, linear, hairy. Flowers bilabiate, pink, rarely brownish or grey, with yellow markings. Mar.–May. Sandstone and shale slopes, NW, SW (Clanwilliam to Riviersonderend and Bredasdorp).•

buckerveldii (L.Bolus) Goldblatt   Cormous geophyte, 80–120 cm, tunics papery. Leaves sword-shaped, stem inclined to drooping. Flowers long-tubed, cream with red markings on the lower tepals, bracts large. Dec.–Jan. Rocky streambanks and waterfalls, NW (Cedarberg Mts).•

bullatus Thunb. ex G.J.Lewis   CALEDON BLUEBELL   Cormous geophyte, 50–80 cm, tunics woody. Leaves linear, blades short. Flowers bell-like, blue marked with yellow on the lower tepals, bracts ridged. Aug.–Oct. Sandstone slopes in fynbos, SW, AP (Kogelberg to Potberg).•

caeruleus Goldblatt & J.C.Manning   Cormous geophyte, 40–60 cm, tunics woody. Leaves narrow, with winged margins (H-shaped), soft. Flowers bilabiate, blue with dark spotting on the lower tepals, fragrant. Aug.–Sept. Limestone outcrops, calcareous sands, SW (Saldanha to Yzerfontein).•

carinatus Aiton   BLOU AFRIKANER   Cormous geophyte, 30–60 cm, tunics fibrous, stem base mottled. Leaves linear, midrib prominent. Flowers bilabiate, blue to violet or yellow, occasionally pink, fragrant. Aug.–Sept. Sandstone slopes or deep coastal sands, NW, SW, AP, LB, SE (Namaqualand to Knysna).

carmineus C.H.Wright   CLIFF GLADIOLUS   Cormous geophyte to 35 cm, tunics papery. Leaf blades reduced, sword-shaped. Flowers funnel-shaped, deep pink with white streaks on the lower tepals. Feb.–Apr. Coastal sandstone cliffs and rocks, SW, AP (Cape Hangklip to Cape Infanta).•

carneus D.Delaroche   PAINTED LADY, BERGPYPIE   Cormous geophyte, 25–60 cm, tunics papery. Leaves sword-shaped. Flowers funnel-shaped, pink or white, often with dark pink markings on the lower tepals. Mainly Oct.–Nov. Sandstone slopes, often wet sites, to 500 m, NW, SW, LB, SE (S Cold Bokkeveld to Knysna).•

caryophyllaceus (Burm.f.) Poir.   SANDPYPIE, PIENK AFRIKANER   Cormous geophyte, 18–75 cm, tunics fibrous. Leaves sword-shaped, pubescent. Flowers large, funnel-shaped, pink to mauve, speckled, fragrant. Aug.–Oct. Sandstone flats and slopes, NW, SW, KM (S Namaqualand to Mamre and Swartberg Mts).

comptonii G.J.Lewis   Cormous geophyte, 45–60 cm, tunics fibrous. Leaves linear. Flowers bilabiate, tepals attenuate, yellow with brown streaks on the lower tepals. June–July. Rocky sandstone slopes, NW (Heerenlogement Mt).•

crispulatus L.Bolus   Cormous geophyte, 35–45 cm, tunics fibrous. Leaves linear, 2-veined. Flowers bilabiate, tepal margins crisped, dark pink, lower tepals with triangular median streaks and dark spots in the throat. Nov.–Dec., especially after fire. Rocky sandstone slopes, LB (Langeberg Mts: Swellendam to Riversdale).•

debilis Ker Gawl.   LITTLE PAINTED LADY   Cormous geophyte, 35–50 cm, tunics woody. Leaves linear with raised margins. Flowers long-tubed, white with red markings on the lower tepals, bracts ridged. Sept.–Oct. Rocky sandstone slopes, SW (Cape Peninsula to Bredasdorp).•

delpierrei Goldblatt   Cormous geophyte, 40–45 cm, tunics fibrous. Leaves linear, 2-veined. Flowers bilabiate, yellowish cream with yellow and red marking on the lower tepals. Dec.–Jan. Marshy sandstone slopes, 1200 m, NW (Cedarberg Mts: Sneeuberg).•

emiliae L.Bolus   Cormous geophyte, 20–50 cm, tunics fibrous. Leaf blades reduced. Flowers long-tubed, yellowish to light brown with brown or purplish speckles, fragrant. Mar.–Apr. Rocky loam, SW, KM, LB, SE (Riviersonderend to George and Gamkaberg).•

exilis G.J.Lewis   Cormous geophyte, 25–45 cm, tunics fibrous. Leaves linear, blades reduced. Flowers bilabiate, white to pale blue with dark streaks on the lower tepals, fragrant. Apr.–May. Clay loam in fynbos, NW, SW (Porterville to Du Toitskloof).•

floribundus Jacq. Cormous geophyte, 15–45 cm, tunics papery. Leaves sword-shaped. Flowers long-tubed, lower tepals smaller, white to pink with dark median streaks. Sept.–Nov. Dry clay, sandy or limestone flats and slopes, NW, SW, AP, KM, LB, SE (Cedarberg Mts to Alexandria).

geardii L.Bolus (= *Gladiolus robustus* Goldblatt) Cormous geophyte, 80–150 cm, tunics papery, stems branched. Leaves sword-shaped. Flowers funnel-shaped, pink with darker markings on the lower tepals. Nov.–Jan. Moist sandstone slopes, SE (Humansdorp to Uitenhage).•

gracilis Jacq. BLOUPYPIE Cormous geophyte, 30–60 cm, tunics woody. Leaves linear with winged margins. Flowers bilabiate, blue to grey, occasionally pink or yellow, with dark streaks on the lower tepals, fragrant. June–Aug. Mostly clay slopes, sometimes on granite, NW, SW, AP, LB (Aurora to Albertinia).•

grandiflorus Andrews (= *Gladiolus floribundus* subsp. *milleri* (Ker Gawl.) Oberm.) Cormous geophyte, 25–50 cm, tunics papery. Leaves sword-shaped. Flowers funnel-shaped, cream to greenish, sometimes with darker median streaks on lower or all the tepals. Sept.–Oct. Clay slopes in renosterveld, SW, KM, LB, SE (Botrivier to Port Elizabeth).•

griseus Goldblatt & J.C.Manning Like **G. carinatus** but flowers smaller, greyish with pale yellow markings on the lower tepals. May–July. Calcareous coastal soils in fynbos and strandveld, SW (Saldanha to Milnerton).•

gueinzii Kuntze Cormous geophyte, 25–50 cm, producing aerial cormlets, tunics papery. Leaves linear, leathery. Flowers nearly actinomorphic, funnel-shaped, mauve with purple and white markings. Mainly Oct.–Dec. Coastal sand dunes, AP, LB, SE (Agulhas and Mossel Bay to KwaZulu-Natal).

guthriei F.Bolus (= *Gladiolus odoratus* L.Bolus) KANEELPYPIE Cormous geophyte, 30–50 cm, tunics fibrous. Leaf blades short, sword-shaped, sometimes hairy. Flowers long-tubed, pink to red or browish with dark speckles, fragrant. Apr.–June. Sandstone outcrops, 100–800 m, NW, SW, KM (Bokkeveld Mts to Elim).•

hirsutus Jacq. (= *Gladiolus punctulatus* Schrank) LAPMUIS Cormous geophyte, 30–60 cm, tunics fibrous. Leaves sword-shaped, blades usually short, hairy. Flowers bilabiate, pink to purple or white, lower tepals streaked with dark colour on a pale background. Mainly June–Oct. Rocky sandstone slopes, NW, SW, LB (Citrusdal to Mossel Bay).•

hyalinus Jacq. SMALL BROWN AFRIKANER Cormous geophyte, 25–50 cm, tunics woody. Leaves linear with thickened margins and midribs. Flowers long-tubed, brownish to cream with dark speckles, rarely fragrant. June–Sept. Shale, granite and sandstone slopes, fynbos or renosterveld, NW, SW, SE (Namaqualand to Port Alfred).

inflexus Goldblatt & J.C.Manning Cormous geophyte, 15–25 cm, tunics woody. Leaves linear. Flowers bilabiate, pale blue to mauve with dark speckling on the lower tepals, fragrant. July–Aug. Rocky sandstone or limestone flats in fynbos, SW (Worcester and Bredasdorp).•

involutus D.Delaroche Cormous geophyte, 30–50 cm, stoloniferous, tunics papery. Leaves linear. Flowers bilabiate, upper lateral tepals attenuate, lower tepals involute, white with yellow-green markings, often fading pink. Aug.–Oct. Clay slopes in renosterveld and grassland, LB, SE (Swellendam to East London).

leptosiphon F.Bolus Cormous geophyte, 30–50 cm, tunics fibrous. Leaves linear and whip-like. Flowers long-tubed, upper tepals attenuate, cream with purple streaks on the lower tepals. Oct.–Nov. Dry, stony sandstone slopes, KM, SE (Ladismith to Uitenhage).•

liliaceus Houtt. LARGE BROWN AFRIKANER, AANDPYPIE Cormous geophyte, 35–70 cm, tunics woody. Leaves linear with thickened margins and midribs, bracts attenuate. Flowers long-tubed, funnel-shaped, brown to russet or beige, turning mauve in the evening and then fragrant. Aug.–Nov. Clay slopes, mainly in renosterveld, NW, SW, AP, KM, LB, SE (Cedarberg Mts to Port Elizabeth).•

maculatus Sweet BRUINAFRIKANER Cormous geophyte, 30–60 cm, tunics fibrous. Leaves linear, blades often short. Flowers long-tubed, funnel-shaped, brownish with dark speckling, fragrant. Mar.–July. Mainly clay slopes, SW, AP, KM, LB, SE (Cape Peninsula to Grahamstown).

meliusculus (G.J.Lewis) Goldblatt & J.C.Manning KALKOENTJIE Cormous geophyte, 12–25 cm, tunics papery. Leaves falcate, ribbed. Flowers bilabiate, dorsal tepal erect, pink to orange with black and greenish markings on the lower tepals, fragrant. Sept.–Oct. Damp sandstone and granite slopes and flats, SW (Hopefield to Cape Peninsula).•

meridionalis (G.J.Lewis) Goldblatt & J.C.Manning   Cormous geophyte, 35–45 cm, tunics fibrous. Leaves linear, blades usually short. Flowers long-tubed, tube cylindrical, salmon-pink to yellowish cream, lightly scented. May–July. Sandstone slopes, SW, AP, SE (Pearly Beach to Port Elizabeth).•

monticola G.J.Lewis   Like **G. brevifolius** but flowers long-tubed with tube longer than tepals, pink with red markings on the lower tepals. Dec.–Mar. Rocky sandstone slopes, SW (Cape Peninsula).•

mostertiae L.Bolus   Cormous geophyte, 15–30 cm, tunics fibrous. Leaves linear, pubescent. Flowers bilabiate, pale pink with yellow-green markings on the lower tepals. Nov.–Dec. Wet sandy soils, NW (Bokkeveld Mts).•

mutabilis G.J.Lewis   BROWNIES   Cormous geophyte, 25–50 cm, tunics fibrous. Leaf blades short, linear. Flowers bilabiate, purple to pink or brown with purple streaks on the lower tepals, fragrant. July–Aug. Sandstone slopes, SW, LB, SE (Albertinia to Grahamstown).•

orchidiflorus Andrews   GROENKALKOENTJIE   Cormous geophyte, 30–80 cm, tunics papery to fibrous. Leaves linear to sword-shaped. Flowers bilabiate, windowed in profile, greenish to purple with dark purple markings, fragrant. Aug.–Oct. Clay and sandstone soils, NW, SW, KM (Namibia to Cape Flats, to Free State).

oreocharis Schltr.   Cormous geophyte, 30–60 cm, tunics papery. Leaves linear, 2-veined. Flowers funnel-shaped, dark pink with red and white markings on the lower tepals. Dec.–Jan., mostly after fire. Wet sandstone slopes, 1000–2000 m, NW, SW, LB (Cedarberg to Langeberg Mts).•

ornatus Klatt   Cormous geophyte, 40–60 cm, tunics papery. Leaves linear, margins thickened. Flowers funnel-shaped, pink with white and red markings on the lower tepals. Aug.–Nov. Marshy sandstone and granite slopes, SW (Mamre to Cape Flats).•

pappei Baker   Cormous geophyte, 20–35 cm, tunics papery. Leaves linear. Flowers long-tubed, funnel-shaped, dark pink with red and white markings on the lower tepals. Oct.–Nov., mainly after fire. Marshes on sandstone slopes, SW (Cape Peninsula and Jonkershoek Mts).•

permeabilis D.Delaroche   Cormous geophyte, 30–50 cm, tunics fibrous. Leaves linear, whip-like. Flowers bilabiate, windowed in profile, cream to brownish or mauve usually with yellowish markings. Aug.–Oct. Shale slopes in renosterveld, SW, LB, KM, SE (Caledon eastwards throughout southern Africa).

phoenix Goldblatt & J.C.Manning   Cormous geophyte, 50–75 cm, tunics papery, stems branched. Leaves sword-shaped. Flowers funnel-shaped, deep pink with red and white markings on the lower tepals. Nov.–Dec., only after fire. Wet sandstone slopes above 600 m, SW (Bainskloof Mts).•

pritzelii Diels   Cormous geophyte, 30–50 cm, tunics fibrous. Leaves linear, scabrid to pilose. Flowers bell-like, dull yellow with brown marking on the lower tepals, fragrant. Aug.–Oct. Rocky sandstone slopes, 800–2000 m, NW (W Karoo, Cedarberg Mts to Cold Bokkeveld).

quadrangulus (D.Delaroche) Barnard   Cormous geophyte to 30 cm, tunics fibrous. Leaves linear with prominent midribs. Flowers rotate, actinomorphic, lilac to pink. Aug.–Oct. Wet sandy flats, SW (Darling to Cape Flats).•

recurvus L.   VOORLOPERTJIE   Like **G. gracilis** but flowers long-tubed with attenuate tepals, grey to cream or pinkish, fragrant. June–Oct. Clay flats and lower slopes, NW, SW (Ceres to Somerset West).•

rhodanthus J.C.Manning & Goldblatt   Cormous geophyte to 50 cm, tunics fibrous. Leaves linear, hairy. Flowers funnel-shaped with a slender tube, pink with red markings on the lower tepals. Dec.–Jan. Rocky sandstone slopes at high alt., SW (Stettynskloof Mts).•

rogersii Baker   RIVERSDALE BLUEBELL   Cormous geophyte, 30–60 cm, tunics fibrous to woody. Leaves linear or solidly terete. Flowers bell-like or bilabiate, blue to purple with yellow or white markings on the lower tepals, fragrant. Mainly Sept.–Oct. Sandstone and limestone slopes to 1000 m, AP, KM, LB, SE (Pearly Beach to Humansdorp).•

roseovenosus Goldblatt & J.C.Manning   Cormous geophyte, 20–40 cm, tunics fibrous. Leaf blades reduced, linear. Flowers long-tubed, funnel-shaped, pink with darker streaks. Mar.–Apr. Sandstone slopes, SE (Outeniqua Mts).•

rudis Lichtenst. ex Roem. & Schult.   Like **G. grandiflorus** but leaf bases conspicuously marked with textured white speckles, perianth tube to 20 mm and flowers cream with spear-shaped markings on the lower tepals. Sept.–Oct. Sandstone slopes in fynbos, SW (Grabouw to Elim).•

stellatus G.J.Lewis   Cormous geophyte, 20–50 cm, tunics fibrous. Leaves linear, whip-like. Flowers rotate, actinomorphic, white to lilac, fragrant. Sept.–Nov. Clay slopes in renosterveld, AP, KM, LB, SE (Swellendam to Port Elizabeth).•

taubertianus Schltr.   Cormous geophyte, 18–25 cm, tunics woody. Leaves linear. Flowers bilabiate, purple with dark streaks, lower tepals with yellow markings. Aug.–Sept. Rocky sandstone soils, NW (Pakhuis Pass to Cold Bokkeveld).•

uitenhagensis Goldblatt & Vlok   Like **G. permeabilis** but perianth tube longer, 22–28 mm. Flowers grey-blue. Sept.–Oct., only after fire. Rocky sandstone slopes, SE (Uitenhage).•

undulatus L.   VLEIPYPIE   Like **G. carneus** but flowers longer-tubed with tepals attenuate and undulate to crisped, greenish white or cream, often with red markings on the lower tepals. Nov.–Dec. Marshy sandstone slopes, NW, SW (Kamiesberg, Bokkeveld Mts to Stellenbosch).

uysiae L.Bolus ex G.J.Lewis   Cormous geophyte, 7–20 cm, corm depressed-globose with papery tunics, producing long stolons. Leaves falcate. Flowers bilabiate, dorsal tepal erect, brownish purple with conspicuous dark veining, fragrant. Aug.–Sept. Clay slopes in renosterveld, 600–1000 m, NW (Bokkeveld Escarpment and W Karoo to Ceres).

vaginatus F.Bolus   Cormous geophyte, 20–70 cm, tunics fibrous. Leaves 2, entirely sheathing. Flowers bilabiate, blue to grey with dark streaks on the lower tepals, fragrant. Feb.–Apr. Limestone and clay-loam slopes, fynbos and renosterveld, SW, LB, SE (Cape Peninsula, Caledon to Knysna).•

variegatus (G.J.Lewis) Goldblatt & J.C.Manning   Cormous geophyte, 20–40 cm, tunics woody. Leaves linear. Flowers funnel-shaped, white to pale pink, lower tepals irregularly spotted with dark red. Sept.–Oct. Limestone outcrops, AP (Gansbaai to Cape Agulhas).•

venustus G.J.Lewis   Cormous geophyte, 20–60 cm, tunics woody. Leaves linear to falcate. Flowers bilabiate, lower tepals pinched and geniculate below, purple to pink with yellow lower tepals, fragrant. Aug.–Oct. Clay and sandstone slopes, NW, SW, KM, LB (Bokkeveld Escarpment to Swellendam).•

vigilans Barnard   Cormous geophyte, 30–40 cm, tunics woody. Leaves linear, margins and midribs thickened. Flowers long-tubed, pink with darker markings on the lower tepals. Oct.–Nov. Sandstone slopes, SW (Cape Peninsula, ?Kogelberg).•

violaceolineatus G.J.Lewis   Cormous geophyte, 35–60 cm, tunics fibrous. Leaves linear, midrib winged on one side. Flowers bilabiate, tepals attenuate, blue with violet veins, fragrant. July–Aug. Rocky sandstone slopes, 500–1000 m, NW (Gifberg to Cedarberg Mts).•

virescens Thunb.   Cormous geophyte, 10–25 cm, tunics papery. Leaves linear and ribbed to terete. Flowers bilabiate, dorsal tepals erect, yellow to pink with dark veins, fragrant. Aug.–Sept. Sandstone or clay slopes, NW, SW, KM, LB, SE (Ceres to Port Elizabeth).•

virgatus Goldblatt & J.C.Manning   Like **G. blommesteinii** but flowers long-tubed, funnel-shaped, pale to deep pink with red blotches on the lower tepals. Sept.–Nov. Rocky sandstone slopes, SW (Du Toitskloof to Somerset West).•

viridiflorus G.J.Lewis   Cormous geophyte, 10–20 cm, stem base speckled, tunics woody. Leaves sword-shaped, twisted. Flowers bilabiate, greenish with purple markings, fragrant. May–July. Rocky sandstone slopes, NW (Orange River to Clanwilliam).

watermeyeri L.Bolus   Cormous geophyte, 10–30 cm, tunics papery. Leaves narrow, ribbed. Flowers bilabiate, windowed in profile, pearly grey with dark veins, lower tepals green, fragrant. July–Sept. Rocky sandstone slopes, NW (Bokkeveld Mts to Wuppertal).•

HESPERANTHA   HESPERANTHA, AANDBLOM   c. 55 spp., sub-Saharan Africa

*A. Flowers pink, yellow or purple (see also **H. acuta**, **H. falcata** and **H. pilosa**)*

elsiae Goldblatt   Cormous geophyte, 25–30 cm. Leaves linear. Flowers long-tubed, bright pink, facing to the side, stamens and style branches included in tube, bract margins united below around spike axis. Dec. Rocky sandstone slopes, NW (Cedarberg Mts).•

fibrosa Baker   Cormous geophyte, 8–30 cm, corm rounded, tunics drawn into long fibres above. Leaves sword-shaped, fleshy with thickened margins. Flowers mauve or purple, sometimes white. Aug.–Sept. Clay slopes in renosterveld, SW, LB (Caledon to Heidelberg).•

humilis Baker   Acaulescent cormous geophyte, 3–8 cm, corm rounded. Leaves falcate, glaucous. Flowers long-tubed, pink to reddish. July–Sept. Sandstone and shale slopes in renosterveld, NW, KM (W Karoo and Hex River Mts to Witteberg).

pallescens Goldblatt   Cormous geophyte, 10–20 cm, corm rounded, tunics imbricate. Leaves linear, fairly soft. Flowers long-tubed, pale yellow, fragrant. Aug.–Sept. Clay slopes in renosterveld, NW (Olifants River Mts).•

pauciflora G.J.Lewis   Cormous geophyte, 8–24 cm, corm with a flat base. Leaves sword-shaped. Flowers pink to purple, occasionally yellow. Aug.–Sept. Sandy and clay soils, NW (Namaqualand to Bokkeveld Plateau).

truncatula Goldblatt   Cormous geophyte, 8–12 cm, corm rounded. Leaves short, oblong. Flowers pale blue. Aug.–Sept. Dry N-facing, shale slopes in renosterveld, KM (N foothills of the Klein Swartberg).•

vaginata (Sweet) Goldblatt   PERDEBLOM   Cormous geophyte, 12–18 cm, corm rounded. Leaves sword-shaped. Flowers large, cup-shaped, yellow often marked with dark brown. Aug.–Sept. Heavy clay soil, NW (Bokkeveld Escarpment and W Karoo).

A. Flowers white or cream (see also **H. fibrosa***)*

acuta (Licht. ex Roem. & Schult.) Ker Gawl.   Cormous geophyte, 10–30 cm, corm rounded. Leaves linear to sword-shaped, upper leaf sheathing the stem. Flowers yellow or white, fragrant at night. July–Sept. Clay slopes in renosterveld and succulent karoo, NW, KM, SE (Worcester to George, S Karoo).

bachmannii Baker   WITROKKIE   Cormous geophyte, 15–30 cm, corm rounded, tunics imbricate. Leaves linear to sword-shaped. Flowers white, nodding on recurved tube, tepals reflexed, sweetly scented. July–Sept. Mostly clay slopes in renosterveld, NW, SW, KM, LB, SE (widespread, Namaqualand to East London).

brevifolia Goldblatt   Cormous geophyte, 15–40 cm, corm with oblique flat base. Leaves short, sword-shaped. Flowers white, red on reverse, tube curved outward, bract margins united below around spike axis. Sept.–Nov. Shale and sandstone slopes, NW, SW (Nardouw Mts to Bainskloof).•

cedarmontana Goldblatt   Cormous geophyte, 12–25 cm, corm with a flat base. Leaves sword-shaped. Flowers long-tubed, white, stamens included in tube, fragrant. Sept.–Oct. Sandstone outcrops, NW (Pakhuis Mts to Piketberg).•

cucullata Klatt   Cormous geophyte, 15–30 cm, corm rounded. Leaves sword-shaped. Flowers white, red to brown on reverse, fragrant. July–Sept. Sandy and shale slopes, mostly renosterveld, NW (W Karoo and Bokkeveld Mts to Biedouw valley).

erecta (Baker) Benth. ex Baker   Cormous geophyte, 10–22 cm, corm rounded. Leaves narrowly sword-shaped. Flowers white to cream, fragrant. Aug.–Sept. Granite outcrops and granitic sands, sandveld and renosterveld, NW, SW (Klawer to Mamre).•

falcata (L.f.) Ker Gawl.   BONTROKKIE   Cormous geophyte, 6–30 cm, corm with a flat base. Leaves sword-shaped to falcate. Flowers white or yellow, red to brown on reverse, fragrant. July–Oct. Sandstone and shale slopes, widespread, NW, SW, AP, LB, SE (Bokkeveld Mts to Port Elizabeth).•

juncifolia Goldblatt   Cormous geophyte, 18–20 cm, corm with a flat base. Leaves terete. Flowers white, pink on outside, bract margins united below around spike axis. Sept.–Oct. Limestone flats, AP (Agulhas Peninsula).•

marlothii R.C.Foster   Cormous geophyte, 4–15 cm, corm with a flat base. Leaves linear, slightly fleshy. Flowers nodding on a recurved tube, cream, brown or red on outside, bracts more or less fused below around spike axis. July–Sept. Sandstone rocks, NW (Bokkeveld Mts to Cold Bokkeveld and W Karoo).

montigena Goldblatt   Cormous geophyte, 5–15 cm, corm rounded. Leaves oblong, often prostrate, with thickened margins. Flowers white, red on reverse. Oct.–Nov. Sandstone outcrops, 1000–1600 m, NW, SW (Worcester Mts to Jonkershoek).•

muirii (L.Bolus) G.J.Lewis   Cormous geophyte, 10–20 cm, corm with a flat base. Leaves linear. Flowers large, nodding, with a recurved tube, large, cream with pink veins, bract margins fused below around spike axis. Oct.–Nov. Clay slopes in renosterveld, SW, LB (Bredasdorp to Albertinia).•

pilosa (L.f.) Ker Gawl.   Cormous geophyte, 10–30 cm, corm rounded. Leaves pilose, the lower 2 linear or sword-shaped, the third leaf sheathing stem. Flowers white, open in the evening or blue to purple and open in the day. Aug.–Oct. Sandstone and clay slopes, NW, SW, AP (Bokkeveld Mts to Bredasdorp, and W Karoo).

pseudopilosa Goldblatt   Like **H. pilosa** but lower 2 leaves lanceolate to oblong, bearing a short, scale-like leaf below spike, flowers always white, greenish to brown on reverse. Aug.–Sept. Clay soils in renosterveld, NW, KM (Bokkeveld Mts and W Karoo to Klein Swartberg).

radiata (Jacq.) Ker Gawl. Cormous geophyte, 20–40 cm, corm obliquely flattened below. Leaves short, linear, fleshy. Flowers nodding, tube recurved, white to cream, red to brown on reverse, bract margins fused below around spike axis. Aug.–Oct. Sandstone granite and clay soils, fynbos and renosterveld, NW, SW, KM, LB, SE (widespread, Namaqualand to Swaziland).

rivulicola Goldblatt Cormous geophyte, 15–30 cm, corm rounded. Leaves subterete, inflated and hollow. Flowers white, brown on reverse. Sept. Along streams, NW (Bokkeveld Mts to Calvinia).

saldanhae Goldblatt Cormous geophyte, 15–25 cm, corm obliquely flattened below. Leaves sword-shaped. Flowers small, white, stamens and style branches included in tube. Aug. Granite outcrops, SW (Vredenburg).•

spicata (Burm.f.) N.E.Br. KANEELTJIE Cormous geophyte, 12–35 cm, corm with flat base. Leaves terete and hollow or lanceolate to falcate and margins sometimes crisped. Flowers small, white, dark on reverse, sweetly cinnamon-scented especially in the evening. Aug.–Sept. Clay and sandy soils, NW, SW (Piketberg to Cape Peninsula).•

IXIA IXIA, KALOSSIE 50 spp., Namaqualand to E Cape

A. Perianth tube filiform; filaments inserted at tepal bases
B. Anthers short, bilobed, often dehiscing incompletely from base and pollen not exposed

collina Goldblatt & Snijman Cormous geophyte, 50–90 cm. Leaves sword-shaped. Flowers pale pink, tube filiform, stamens unilateral, anthers broadly oblong with a right-angled bend near base. Aug.–Sept. Shale hills in renosterveld, SW (Breede River valley near Worcester).•

erubescens Goldblatt KLEINAGRETJIE Cormous geophyte, 12–30 cm. Leaves lanceolate, margins undulate to crisped. Flowers pink, tube short, anthers oblong. Aug.–Sept. Clay flats and slopes in renosterveld, NW, SW (Piketberg to Caledon).•

micrandra Baker Slender cormous geophyte, 25–50 cm, usually unbranched. Leaves linear. Flowers pink, tube filiform, anthers oblong. July–Sept. Sandstone slopes, SW, AP, KM, LB (Paarl to Oudtshoorn).•

scillaris L. AGRETJIE Cormous geophyte, 25–50 cm. Leaves sword-shaped, margins sometimes undulate. Flowers pink, tube filiform, anthers oblong, unilateral and nodding. Sept.–Nov. Sand and clay flats and slopes, NW, SW (Namaqualand to Caledon).

stricta (Eckl. ex Klatt) G.J.Lewis Cormous geophyte, 35–55 cm. Leaves sword-shaped. Flowers pale to deep pink, tube filiform, anthers oblong. Nov.–Dec. Lower sandy loam slopes in renosterveld, SW (Caledon to Bredasdorp).•

trinervata (Baker) G.J.Lewis Cormous geophyte, 20–40 cm. Leaves lanceolate, becoming dry at flowering. Flowers pink to mauve, tube filiform, anthers oblong. Sept. Rocky slopes, SW (Elgin to Riviersonderend).•

B. Anthers oblong to linear, not bilobed, dehiscing longitudinally and pollen exposed
C. Flowers yellow to orange (see also I. lutea)

curta Andrews Cormous geophyte, 15–40 cm. Leaves sword-shaped. Flowers orange with a brownish centre, tube filiform, filaments united. Sept.–Oct. Sandy flats and slopes, SW (Hopefield to Darling).•

dubia Vent. (incl. **I. frederickii** M.P.de Vos) Cormous geophyte, 25–60 cm. Leaves sword-shaped to nearly linear. Flowers orange to yellow, often dark in centre, tube filiform, bracts translucent pink. Oct.–Dec. Sandstone and granite flats and slopes, NW, SW, AP (Piketberg to Caledon).•

maculata L. Cormous geophyte, 20–50 cm, corms with stolons. Leaves sword-shaped, bracts large, truncate and rusty. Flowers orange to yellow with dark star-like centre, tube filiform, filaments united below. Sept.–Oct. Granite and sandstone flats and slopes, mostly fynbos, NW, SW (Clanwilliam to Melkbos, extinct on Cape Peninsula).•

C. Flowers white to mauve, green or red

campanulata Houtt. ROOIKALOSSIE Cormous geophyte, 10–35 cm. Leaves linear. Flowers white or bright red, tube short and filiform. Oct.–Nov. Damp sandstone slopes, NW, SW (Tulbagh to Villiersdorp).•

flexuosa L. WAAIKALOSSIE Cormous geophyte, 35–65 cm, stems wiry, often unbranched. Leaves linear. Flowers in congested spikes, pink, mauve or white, lightly fragrant, tube filiform. Mainly Aug.–Sept. Mostly clay flats and slopes, SW, AP, LB (Cape Peninsula to Riversdale).•

gloriosa G.J.Lewis   Cormous geophyte, 35–65 cm, stems wiry. Leaves linear. Flowers deep pink with a purple-black centre, tube filiform. Aug.–Sept. Clay slopes in renosterveld, KM (Barrydale district).•

lutea Eckl. (= *Ixia conferta* R.C.Foster)   Cormous geophyte, 15–35 cm. Leaves sword-shaped. Flowers red to purple or cream to yellow, with dark centre, tube filiform. Aug.–Oct. Clay flats and slopes in renosterveld, NW, SW (Citrusdal to Paarl).•

metelerkampiae L.Bolus   Cormous geophyte to 70 cm. Leaves linear. Flowers pink to lilac with a purple centre outlined white, tube filiform. Nov.–Dec. Sandstone slopes, SW (Bainskloof to Paarl).•

monadelpha D.Delaroche   BONTKALOSSIE   Cormous geophyte, 15–40 cm. Leaves sword-shaped. Flowers shades of turquoise to purple, rarely whitish, with blackish centre, tube filiform, filaments united, black. Sept.–Oct. Wet sandy flats and lower slopes, SW (Darling to Cape Peninsula).•

mostertii M.P.de Vos   Cormous geophyte, 35–45 cm, stem with a fibrous neck. Leaves narrow. Flowers pink, mauve or white with a dark centre, tube filiform. Sept.–Oct. Clay slopes in renosterveld, NW (Romans River to Worcester).•

patens Aiton   Cormous geophyte, 20–50 cm. Leaves sword-shaped. Flowers red, pink or white with a pale or dark centre, tube filiform. Sept.–Oct. Clay slopes in renosterveld, NW, SW (Clanwilliam to Riviersonderend).•

polystachya L.   Cormous geophyte, 40–80 cm, corms with stolons. Leaves linear to sword-shaped. Flowers in dense spikes, white to pink or mauve, often with darker centre, sometimes yellow, tube filiform, rarely lightly fragrant. Oct.–Dec. Granitic and sandstone slopes and flats, NW, SW (Cedarberg to Caledon).•

purpureorosea G.J.Lewis   Cormous geophyte, 20–60 cm. Leaves sword-shaped. Flowers mauve-pink with blackish centre, tube filiform. Sept.–Oct. Limestone and calcareous sands in strandveld, SW (Saldanha Bay).•

rouxii G.J.Lewis   Cormous geophyte, 35–50 cm. Leaves linear. Flowers white, pink, bluish, green or yellow with dark centre, tube filiform. Oct.–Nov. Clay flats, NW (Porterville to Saron).•

stolonifera G.J.Lewis   Cormous geophyte, 20–50 cm, corms with stolons. Leaves sword-shaped. Flowers mauve with a purple centre, tube filiform-shaped, bracts setaceous. Sept. Sandstone slopes, c. 1000 m, KM (Montagu: Kiesiesberg).•

vanzijliae L.Bolus   Cormous geophyte, 18–40 cm. Leaves sword-shaped, twisted with undulate margins. Flowers pinkish with darker centre, tube filiform. Aug.–Sept. Clay flats, LB (Bonnievale).•

versicolor G.J.Lewis   Cormous geophyte, 15–35 cm. Leaves linear, twisted. Flowers white or purple with dark centre, tube filiform. Oct. Sandy flats, SW (Simondium to Gordon's Bay).•

vinacea G.J.Lewis   Cormous geophyte, 40–45 cm, stem with a fibrous neck. Leaves linear. Flowers red with dark centre, tube filiform. Aug.–Sept. Stony clay flats, NW (Tulbagh).•

viridiflora Lam.   GROENKALOSSIE   Cormous geophyte, 50–100 cm. Leaves narrow. Flowers in elongate, lax spikes, green with dark purple-black centre, tube filiform. Sept.–Oct. Rocky, mostly clay slopes, NW, SW (Tulbagh to Wolseley, Paardeberg).•

A. Perianth tube not filiform; filaments inserted within tube
D. Perianth tube cylindrical, usually longer than 20 mm

fucata Ker Gawl.   Cormous geophyte, 15–40 cm. Leaves linear. Flowers white to pale pink, tube elongate, cylindric. Sept.–Nov. Sandstone mountain slopes, NW, SW, KM (Ceres to Caledon and Montagu).•

longituba N.E.Br. (= *Ixia bellendenii* R.C.Foster)   Cormous geophyte, 35–70 cm. Leaves sword-shaped. Flowers white or pink, tube long, cylindric. Sept.–Oct. Shale slopes, SW, AP, LB (Caledon to Swellendam).•

paniculata D.Delaroche   PYPKALOSSIE   Cormous geophyte, 40–100 cm. Leaves sword-shaped. Flowers cream to biscuit, tube elongate, stamens and style branches often included in tube. Oct.–Dec. Wet sandy slopes and flats, NW, SW (Bokkeveld Mts to False Bay).•

paucifolia G.J.Lewis   Cormous geophyte, 12–55 cm. Leaves linear to lanceolate, sometimes falcate. Flowers white to pink, tube elongate, cylindrical, filaments shortly exserted from the tube. Sept.–Nov. Stony middle to upper mountain slopes, NW, KM (Cedarberg Mts to Ladismith).•

splendida G.J.Lewis   Cormous geophyte, 30–60 cm. Leaves linear. Flowers pale pink, tube elongate, cylindrical, stamens included in tube. Oct.–Nov. Mountain slopes, SW (Piketberg: Zebra Kop).•

D. Perianth tube funnel-shaped, to 20 mm long

aurea J.C.Manning & Goldblatt   Cormous geophyte, 15–40 cm. Leaves sword-shaped, loosely coiled above. Flowers yellowish orange, tube narrowly funnel-shaped, bracts translucent. Sept.–Oct. Granite slopes, SW (Darling).•

brunneobracteata G.J.Lewis   Cormous geophyte, 25–45 cm. Leaves linear, narrow. Flowers cream, tube funnel-shaped, bracts glossy, dark brown. Sept.–Oct. Sandstone soils in marshes, NW (Bokkeveld Mts).•

capillaris L.f.   Cormous geophyte, 20–45 cm. Leaves linear, very narrow. Flowers white to mauve, tube funnel-shaped. July–Sept. Sandy or clay slopes, NW, SW, KM, LB (Citrusdal to Ladismith).•

cochlearis G.J.Lewis   Cormous geophyte, 20–40 cm. Leaves linear. Flowers rose to salmon-pink with dark veins, tube narrowly funnel-shaped. Nov. Lower mountain slopes, SW (Stellenbosch Mts).•

esterhuyseniae M.P.de Vos   Cormous geophyte, 10–15 cm. Leaves linear. Flowers yellow, reddish on reverse, tube funnel-shaped. Dec.–Jan. Sandstone rocks, SW (Jonkershoek and Hottentots Holland Mts).•

latifolia D.Delaroche   Cormous geophyte, 20–50 cm. Leaves broad, often falcate. Flowers pink to purple, tube narrowly funnel-shaped. Sept.–Nov. Mostly clay soils in renosterveld, NW, SW, KM (Namaqualand to Paarl, Montagu and W Karoo).

leipoldtii G.J.Lewis   Cormous geophyte, 11–25 cm. Leaves narrow. Flowers white with dark purple centre, tube funnel-shaped. Sept. Clay soils in renosterveld, KM (Barrydale to Prince Albert).•

odorata Ker Gawl.   SOETKALOSSIE   Cormous geophyte, 20–50 cm. Leaves linear to lanceolate, often coiled above. Flowers small, in dense spikes, pale yellow, fragrant, tube funnel-shaped. Sept.–Nov. Sandstone and granite slopes, NW, SW (Citrusdal to Hermanus).•

orientalis L.Bolus   Cormous geophyte, 25–70 cm. Leaves linear. Flowers cream to mauve-pink, tube narrowly funnel-shaped. Sept.–Oct. Flats and slopes, SW, AP, KM, LB, SE (Villiersdorp to Port Alfred).

pauciflora G.J.Lewis   Cormous geophyte, 20–40 cm. Leaves linear. Flowers cream or pink to violet, usually fragrant, tube elongate and wide. Aug.–Sept. Sandstone soils in fynbos, NW (Cedarberg Mts to Ceres).•

pumilio Goldblatt & Snijman   Cormous geophyte, 12–20 cm. Leaves linear. Flowers brick-red, tube funnel-shaped. Aug.–Sept. Sandy alluvial flats, SW (Breede River valley near Worcester).•

rapunculoides Delile   BLOUKALOSSIE   Cormous geophyte, 15–70 cm, stem usually branched. Leaves sword-shaped. Flowers blue, mauve or pink, tube funnel-shaped. Aug.–Sept. Mostly clay soils in renosterveld, NW, KM (Namaqualand to Oudtshoorn and W Karoo).

stohriae L.Bolus   Cormous geophyte, 15–30 cm. Leaves linear. Flowers pink or white, stamens included in long, narrowly funnel-shaped tube. Sept.–Oct. Mountain slopes, 300–800 m, LB (Langeberg Mts: Swellendam).•

tenuifolia Vent. (= *Ixia framesii* L.Bolus)   Cormous geophyte, 15–40 cm. Leaves linear. Flowers orange to red with dark centre, tube cylindrical, slightly wider above. Sept.–Oct. Sandy flats, SW (Darling and Kalbaskraal).•

KLATTIA• KWASBOS   3 spp., W Cape

flava (G.J.Lewis) Goldblatt   Evergreen shrub, 80–130 cm. Leaves lanceolate. Flowers in congested, brush-like heads, bright yellow, style shortly exceeding tepals, inflorescence bracts pale green. Nov.–Dec. Mountain slopes in fynbos, 400–1250 m, SW (Bainskloof to Mt Lebanon).•

partita Baker   Evergreen shrub, 60–120 cm. Leaves narrowly lanceolate. Flowers in congested, brush-like heads, dark purple-black, styles not reaching tepal apices, inflorescence bracts green. Oct.–Dec. Mountain slopes, 600–1250 m, SW, LB (Cape Peninsula, Hottentots Holland and Langeberg Mts).•

stokoei L.Guthrie   Like **K. partita** but leaves lanceolate and flowers and inflorescence bracts bright red, styles exceeding tepals. Dec.–Feb. Mountains slopes in seeps, SW (Kogelberg and Palmiet River Mts).•

LAPEIROUSIA   LAPEIROUSIA, CABONG   40 spp., sub-Saharan Africa, mainly SW southern Africa

A. Leaves plane; floral bracts subequal

azurea (Eckl. ex Baker) Goldblatt   Cormous geophyte, 6–12 cm. Leaf blades plane, falcate, broad, margins often undulate. Flowers in corymbose panicles, deep blue with blackish markings, stamens unilateral, pollen dark blue to brown. Sept.–Oct. Granitic soils in renosterveld, SW (Gouda to Paarl).•

corymbosa (L.) Ker Gawl.   BLOUCABONG   Cormous geophyte, 5–15 cm. Leaf blades plane, falcate, undulate. Flowers in corymbose panicles, actinomorphic, pale to deep blue with white central star. Mainly Sept.–Nov. Sandy and granitic slopes, NW, SW, AP (Piketberg to Agulhas).•

falcata (L.f.) Ker Gawl.   Cormous geophyte to 10 cm. Leaf blades plane. Flowers in short spikes, pink to mauve with red markings, tubular, stamens unilateral. Sept.–Oct., mainly after fire. Rocky sandstone slopes in shallow sand, NW (Cold Bokkeveld to Worcester).•

fastigiata (Lam.) Ker Gawl.   Cormous geophyte, 6–9 cm. Leaf blades plane, falcate and often loosely crisped. Flowers in corymbose panicles, actinomorphic, pale yellow with brown markings. Sept.–Oct. Clay slopes and flats in renosterveld, to 300 m, NW, SW (Piketberg to Malmesbury).•

micrantha (E.Mey. ex Klatt) Baker   Cormous geophyte, 15–35 cm. Leaf blades plane. Flowers in corymbose panicles, actinomorphic, small, cream to maroon, fragrant. Oct.–Nov., only after fire. Rocky sandstone soils in fynbos, NW, SW, LB (Gifberg to Riversdale).•

neglecta Goldblatt & J.C.Manning   Cormous geophyte, 30–80 cm. Leaf blades plane. Flowers in open panicles, white or blue, stamens unilateral. Nov.–Dec., only after fire. Rocky sandstone slopes above 800 m, SW (Bainskloof to Hottentots Holland Mts).•

A. Leaves ribbed; inner floral bracts small and 2-keeled

anceps (L.f.) Ker Gawl.   Cormous geophyte, 10–30 cm. Leaf blades ribbed. Flowers in short spikes, cream to pink with red markings on lower tepals, tube elongate. Sept.–Nov. Deep sand or stony slopes in fynbos, NW, SW, AP, LB (S Namaqualand to Mossel Bay).

divaricata Baker   Cormous geophyte, 7–25 cm. Leaf blades ribbed. Flowers bilabiate and short-tubed, white to pale pink, fragrant. Aug.–Oct. Damp sandy places, NW (Bokkeveld Mts to Citrusdal).•

fabricii (D.Delaroche) Ker Gawl.   CABONG, CHABI   Cormous geophyte, 15–25 cm. Leaf blades ribbed. Flowers long-tubed, large, cream to pink with red markings on lower tepals, lower tepals with claw-like appendages. Sept.–Oct. Stony sandstone slopes in fynbos, NW, SW (Namaqualand to Malmesbury).

jacquinii N.E.Br.   Cormous geophyte, 8–12 cm. Leaf blades ribbed. Flowers long-tubed, dark purple with cream and reddish streaks on lower tepals, bracts 2-keeled below, broadly obtuse. Aug.–Sept. Sandstone soils, NW, SW (S Namaqualand to Worcester).

oreogena Schltr. ex Goldblatt   Acaulescent cormous geophyte, 5–10 cm. Leaf blades ribbed. Flowers actinomorphic, long-tubed, violet with cream and blackish markings. Aug.–Sept. Clay soils, NW (Bokkeveld Escarpment and W Karoo).

plicata (Jacq.) Diels   HAASCABONG   Acaulescent cormous geophyte, 3–5 cm. Leaf blades ribbed. Flowers actinomorphic, long-tubed, small, blue to white. July–Sept. Dry shale flats, succulent karoo or renosterveld, NW, KM (Worcester to Riversdale, Namibia, Bushmanland, Karoo).

pyramidalis (Lam.) Goldblatt   NAELTJIE   Cormous geophyte, 5–10 cm. Leaf blades ribbed. Flowers long-tubed, cream to bluish and fragrant, or dark purplish to magenta and scentless, bracts spreading, broad and retuse above. July–Sept. Shale and sandstone soils, fynbos to renosterveld, NW, SW, AP, LB, KM (S Namaqualand to Oudtshoorn).

violacea Goldblatt   Cormous geophyte to 10 cm. Leaf blades ribbed. Flowers long-tubed, violet with dark red markings on lower tepals, bracts large and inflated. Aug.–Sept. Sandstone soils in renosterveld and arid fynbos in sand, NW (Bokkeveld Escarpment to Biedouw valley).•

MELASPHAERULA   FAIRY BELL, BAARDMANNETJIE, FEEKLOKKIE   1 sp., S Namibia to W Cape

ramosa (L.) N.E.Br.   Cormous geophyte, 30–60 cm, diffusely branched. Flowers lax, wiry spikes, small, bilabiate, cream to pale yellow, lower tepals streaked with red-brown in midline, sour-smelling. July–Sept. Mostly sheltered sites on sandstone or limestone slopes, NW, SW, AP, KM (S Namibia to De Hoop and Swartberg Mts).

MICRANTHUS•   COMB FLOWER, VLEIBLOMMETJIE   3 spp., W Cape

alopecuroides (L.) Rothm.   Cormous geophyte, 25–40 cm. Leaves sword-shaped, plane with distinct midvein. Flowers pale to deep blue. Oct.–Dec. Sandstone soils, SW (Bainskloof to Elgin and Cape Peninsula).•

junceus (Baker) N.E.Br.   Cormous geophyte, 25–45 cm. Leaves terete, slender and hollow. Flowers usually dark blue, occasionally white. Nov.–Jan. Wet sites on granite or sandstone soils, NW, SW, AP, LB (Bokkeveld Mts to Riversdale).•

tubulosus (Burm.) N.E.Br.   Cormous geophyte, 25–45 cm. Leaves tubular, hollow, tips truncate-apiculate. Flowers blue to mauve, fragrant. Nov.–Dec. Mainly clay soils in renosterveld, NW, SW (Gifberg to Cape Peninsula).•

MORAEA (= *GALAXIA, GYNANDRIRIS, HEXAGLOTTIS, HOMERIA, RHEOME, ROGGEVELDIA, SESSILISTIGMA*)   UINTJIE   c. 195 spp., sub-Saharan Africa and Mediterranean to Middle East

Galaxia-group   CLOCK FLOWER, HORLOSIEBLOM

A. Flowers pink, mauve or purple, sometimes reddish

barnardiella Goldblatt (= *Galaxia barnardii* Goldblatt)   Acaulescent cormous geophyte, 3–5 cm. Leaves broadly lanceolate, margins undulate, often prostrate. Flowers pink to purple with blackish centre, tepals spreading horizontally, style reaching middle of anthers, stigmas lobed. Aug.–Sept. Clay flats and hills, renosterveld, SW (Villiersdorp to Caledon).•

melanops Goldblatt & J.C.Manning   Acaulescent cormous geophyte, 2–4 cm. Leaves broadly lanceolate, prostrate. Flowers purple with blackish centre, tepals shallowly cupped to laxly spreading, stamens free in upper half or entirely, style exceeding anthers, stigmas lobed. Aug.–Sept. Clay slopes in renosterveld, SW (Shaw's Pass to Potberg).•

variabilis (G.J.Lewis) Goldblatt (= *Galaxia variabilis* G.J.Lewis)   Acaulescent cormous geophyte, 2–5 cm. Leaves ovate to oblong, channelled, prostrate. Flowers purple to mauve with yellow centre, tepals cupped, anthers sessile, style exceeding anthers, stigmas lobed. Sept.–Oct. Mainly clay soils, 400–1000 m, NW (Cold Bokkeveld to Ceres).•

versicolor (Salisb. ex Klatt) Goldblatt (= *Galaxia versicolor* Salisb. ex Klatt)   Acaulescent cormous geophyte, 2–5 cm. Leaves ovate to lanceolate, prostrate, margins undulate. Flowers pink to purple, often with yellow centre, tepals cupped, style exceeding anthers, stigmas lobed. Aug.–Sept. Clay and granite flats and slopes, to 300 m, SW (Tulbagh to Cape Peninsula and Houwhoek).•

A. Flowers yellow or white

albiflora (G.J.Lewis) Goldblatt (= *Galaxia albiflora* G.J.Lewis)   Acaulescent cormous geophyte, 2–3 cm. Leaves linear, channelled. Flowers white with yellow centre, style shorter than stamens, stigmas fringed. May–Aug. Coastal, on sand or granite or limestone outcrops, SW, AP (St Helena Bay to Agulhas).•

angulata Goldblatt (= *Galaxia alata* Goldblatt)   Acaulescent cormous geophyte, 2–4 cm, tunics woody and vertically winged. Leaves terete. Flowers yellow or white, tepals cupped, style exceeding anthers, stigmas fringed. June–Aug. Wet sandy flats, SW (Malmesbury to Gordon's Bay).•

citrina (G.J.Lewis) Goldblatt (= *Galaxia citrina* G.J.Lewis)   Acaulescent cormous geophyte, 3–6 cm. Leaves oblong, channelled. Flowers yellow, tepals cupped, anthers sessile, style exceeding anthers, stigmas lobed. July–Oct. Shallow sandy or stony soils, 500–1000 m, NW (Bokkeveld Mts to Gydo Pass).•

fugacissima (L.f.) Goldblatt (= *Galaxia fugacissima* (L.f.) Druce)   Acaulescent cormous geophyte, 3–6 cm. Leaves linear to terete. Flowers yellow, tepals cupped, style exceeding anthers, stigmas fringed. July–Sept. Wet sand and clay flats, NW, SW, AP, LB, SE (Namaqualand to Humansdorp).

galaxia (L.f.) Goldblatt (= *Galaxia ovata* Thunb.)   Acaulescent cormous geophyte, 2–4 cm. Leaves ovate, prostrate, margins thickened and ciliate. Flowers yellow, tepals cupped, style exceeding anthers, stigmas fringed. July–Sept. Flats and plateaus, mainly on sandstone soils, NW, SW, AP (Cedarberg Mts to Bredasdorp).•

luteoalba (Goldblatt) Goldblatt (= *Galaxia luteoalba* Goldblatt)   Acaulescent cormous geophyte, 2–5 cm. Leaves ovate and prostrate, conspicuously ciliate. Flowers yellow fading to white at edges, tepals cupped, stamens free, style exceeding anthers, stigmas fringed. July–Sept. Sandstone outcrops, 400–1000 m, NW (Matsikamma to N Cedarberg Mts).•

minutiflora Goldblatt (= *Galaxia parva* Goldblatt)   Acaulescent cormous geophyte, 2–3 cm. Leaves lanceolate, spreading, margins undulate. Flowers tiny, white marked with green, tepals cupped, style reaching middle of anthers, stigmas lobed. July–Aug. Clay flats in renosterveld, SW (Bredasdorp).•

pilifolia Goldblatt (= *Galaxia ciliata* Pers.)   Like **M. galaxia** but leaf margins with cilia longer than the thickened margins. June–July. Stony slopes, NW (Namaqualand to Gifberg).

stagnalis (Goldblatt) Goldblatt (= *Galaxia stagnalis* Goldblatt)   Acaulescent cormous geophyte, 2–4 cm. Leaves linear to lanceolate. Flowers yellow, tepals cupped, style reaching anther apices, stigmas fringed. June–Aug. Wet sites on sandstone soils, 400–800 m, NW (Namaqualand to Pakhuis Pass).

Gynandriris-group   PAPIERUINTJIE

australis (Goldblatt) Goldblatt (= *Gynandriris australis* Goldblatt)   Cormous geophyte, 8–25 cm. Leaf solitary, linear, channelled and trailing. Flowers large, pale blue, nectar guides cream, inner tepals erect. Sept.–Nov. Coastal sand dunes and rocky flats, SE (George to Humansdorp).•

cedarmontana (Goldblatt) Goldblatt (= *Gynandriris cedarmontana* Goldblatt)   Cormous geophyte, 10–30 cm. Leaves 2, linear, channelled and trailing. Flowers white, nectar guides yellow, inner tepals erect, sweetly scented. Sept.–Oct. Streambanks in sandy soils, NW (Pakhuis Mts to Citrusdal).•

hesperantha (Goldblatt) Goldblatt (= *Gynandriris hesperantha* Goldblatt)   Cormous geophyte, 40–60 cm. Leaves 2, linear, trailing, margins inrolled. Flowers dark blue, inner tepals reflexed, opening late afternoon. Oct.–Nov. Heavy clay slopes in renosterveld, NW (Bokkeveld Plateau and W Karoo).

pritzeliana Diels (= *Gynandrisris pritzeliana* (Diels) Goldblatt)   KURKTREKKERUINTJIE   Cormous geophyte, 10–25 cm. Leaves 2, surface plane, translucent along midline, entire blade helically coiled. Flowers dark blue, nectar guides cream, inner tepals reflexed. Sept.–Oct. Sandstone and clay soils, mainly in renosterveld, NW (Bokkeveld Plateau and W Karoo).

setifolia (L.f.) Druce (= *Gynandriris setifolia* (L.f.) R.C.Foster)   BOKUINTJIE, PAPIERUINTJIE   Cormous geophyte, 5–20 cm. Leaves 1 or 2, linear, chanelled and trailing. Flowers small, pale mauve, nectar guides orange and white, inner tepals reflexed. Sept.–Nov. Sandy and gravelly flats and slopes, NW, SW, AP, KM, LB (Namaqualand to Grahamstown).

Hexaglottis-group   THREAD STAR, VOLSTRUISUINTJIE

lewisiae (Goldblatt) Goldblatt (= *Hexaglottis lewisiae* Goldblatt)   Cormous geophyte, 20–90 cm. Leaves 1–3, linear, channelled and trailing. Flowers deep yellow. Capsules oblong to ellipsoid, exserted from the spathes. Oct.–Dec. Various soils and habitats, mostly dry sites, NW, SW, KM, LB, SE (Namaqualand to Humansdorp).

longifolia (Jacq.) Pers. (= *Hexaglottis longifolia* (Jacq.) Salisb.)   Cormous geophyte, 60–150 cm. Leaves 3–5, linear, channelled and trailing. Flowers pale yellow. Capsules club-shaped, exserted from the spathes, 12–20 mm. Oct.–Nov. Shady, moist sites on sandstone, SW (Cape Peninsula to Du Toitskloof).•

nana (L.Bolus) Goldblatt & J.C.Manning (= *Hexaglottis nana* L.Bolus)   Cormous geophyte, 10–30 cm, stem usually branched above the leaves and branches crowded. Leaves 2–several, clustered above ground. Flowers pale yellow to salmon, style arms 6, filiform and extending between filaments. Sept.–Nov. Rocky sandstone slopes, NW (Namaqualand to Citrusdal).

riparia (Goldblatt) Goldblatt (= *Hexaglottis riparia* Goldblatt)   Cormous geophyte, 45–90 cm. Leaves 2 or 3, linear, channelled and trailing. Flowers yellow. Capsules club-shaped, exserted from the spathes, 6–12 mm. Oct.–Nov. Along streams and rivers in rocky sandstone, NW (Clanwilliam to Tulbagh).•

virgata Jacq. (= *Hexaglottis virgata* (Jacq.) Sweet)   PYPIEVOLSTRUISUINTJIE   Cormous geophyte, 20–85 cm. Leaves 2 or 3, linear, channelled and trailing. Flowers pale to deep yellow, tepals forming a tube below and ovary subsessile. Capsules narrowly ellipsoid, enclosed in the spathes. Mainly Sept.–Nov. Shale and granite soils, rarely sandstone, NW, SW, AP, LB, SE (S Namaqualand and W Karoo to Port Elizabeth).

Homeria-group   CAPE TULIP, TULIP

*A. Foliage leaves more than 1 (see also **M. karooica** and **M. ochroleuca**)*

aspera Goldblatt (= *Homeria spiralis* L.Bolus)   Cormous geophyte, 10–30 cm. Leaves 3, linear, loosely coiled, stem minutely hairy. Flowers stellate, with short tepal claws, salmon, anthers exserted on a slightly bulbous filament column. Aug.–Sept. Clay slopes in renosterveld, NW (Bokkeveld Escarpment).•

cedarmonticola Goldblatt (= *Homeria cedarmontana* Goldblatt)   Cormous geophyte, 70–100 cm. Leaves 2 or 3, linear-lanceolate, trailing above, glaucous. Flowers yellow, stamens included in the narrow cup. Aug.–Oct., only after fire. Rocky sandstone slopes, 1000–1500 m, NW (Cedarberg Mts).•

miniata Andr. (= *Homeria miniata* (Andrews) Sweet)   PRONKTULP   Cormous geophyte, 15–60 cm. Leaves 2 or 3, linear, trailing. Flowers stellate, with short tepal claws, usually salmon, sometimes yellow or white, minutely speckled in centre, anthers exserted on a bulbous filament column. Aug.–Sept. Mainly clay slopes, renosterveld and karroid scrub, NW, SW, LB (Namaqualand to Riversdale, and Karoo).

A. Foliage leaf solitary
B. Tepal claws short and erect, often clasping filament base; anthers on a long, exserted filament column

bifida (L.Bolus) Goldblatt (= *Homeria bifida* L.Bolus)   Like **M. miniata** but leaf solitary, fairly broad below and clasping lower half of stem. Aug.–Sept. Clay soils in renosterveld, NW (Bokkeveld Plateau and W Karoo to Pakhuis Pass).

brachygyne (Schltr.) Goldblatt (= *Homeria brachygyne* Schltr.)   Cormous geophyte, 8–25 cm, stem flexed outward above leaf sheath. Leaf solitary, channelled and trailing. Flowers small, stellate, with short tepal claws, pink with yellow, speckled centre, anthers exserted on a cylindrical, pilose filament column. July–Sept. Rocky sandstone slopes, NW (Bokkeveld Mts to Clanwilliam).•

bulbillifera (G.J.Lewis) Goldblatt (= *Homeria bulbillifera* G.J.Lewis)   UINTJIESTULP   Cormous geophyte, 30–50 cm, stems often with clusters of cormlets at nodes. Leaf solitary, linear, trailing. Flowers stellate, tepals with short claws forming a shallow cup, yellow to salmon, anthers exserted on a slender filament column. Aug.–Sept. Sandstone and limestone soils, mainly coastal, SW, AP, LB, SE (Cape Peninsula to Alexandria).

cookii (L.Bolus) Goldblatt (= *Homeria cookii* L.Bolus)   Cormous geophyte, 30–60 cm. Leaf solitary, sheathing lower stem, channelled and trailing above. Flowers yellow with a darker yellow-speckled centre, tepal claws short, anthers exserted on a thick filament column. Aug.–Sept. Rocky sandstone slopes, NW, KM, SE (Cedarberg Mts to Karoo and Lesotho).

flavescens (Goldblatt) Goldblatt (= *Homeria flavescens* Goldblatt)   Cormous geophyte, 12–30 cm, stem flexed outward above leaf sheath. Leaf solitary, linear, trailing. Flowers yellow with a greenish speckled centre, tepal claws forming a short cup, anthers exserted on a slender filament column. Sept. Rocky sandstone slopes, NW (Bokkeveld to Cedarberg Mts).•

fuscomontana (Goldblatt) Goldblatt (= *Homeria fuscomontana* Goldblatt)   Cormous geophyte, 10–25 cm, stem flexed outward above leaf sheath. Leaf solitary, linear, channelled. Flowers stellate, tepal claws short, clasping, pale yellow with a green, speckled centre, anthers exserted on a slender filament column. Sept. Rocky sandstone slopes in dry fynbos, NW (Swartruggens).•

louisabolusiae Goldblatt (= *Homeria bolusiae* Goldblatt)   Cormous geophyte, 15–40 cm. Leaf solitary, stem flexed outward above leaf. Flowers stellate, with short tepal claws, yellow or salmon-pink, stamens exserted from a short cup. Aug.–Sept. Rock outcrops, NW (Namaqualand to Nardouw Mts).

marlothii (L.Bolus) Goldblatt (= *Homeria marlothii* L.Bolus)   Cormous geophyte, 50–75 cm. Leaf solitary, clasping lower half of stem, channelled and trailing above. Flowers yellow or salmon-pink, tepal claws short and clasping, anthers exserted on a thick filament column. Aug.–Oct. Sandstone rocks or heavy clay soils, NW (Bokkeveld Mts and W Karoo).

patens (Goldblatt) Goldblatt (= *Homeria patens* Goldblatt)   Cormous geophyte, 25–45 cm, stem flexed above leaf sheath. Leaf solitary, channelled and trailing. Flowers yellow or salmon-pink, tepal claws short, forming a short cup, anthers exserted on a slender filament column. Aug.–Sept. Sandstone soils, NW (Nardouw and Pakhuis Mts).•

radians (Goldblatt) Goldblatt (= *Sessilistigma radians* Goldblatt)   Cormous geophyte, 12–25 cm, stem flexed outward above leaf sheath. Leaf solitary, linear, channelled, glaucous. Flowers cream with a yellow centre, tepals spreading from base, stamens free, style branches subsessile. Aug.–Sept. Clay soils in renosterveld, SW (near McGregor).•

B. Tepal claws fairly long, erect or ascending, forming a cup including the filaments and sometimes the anthers

autumnalis (Goldblatt) Goldblatt (= *Homeria autumnalis* Goldblatt)   Cormous geophyte, 20–30 cm. Leaf solitary, linear, trailing. Flowers yellow, tepal claws froming a narrow cup, anthers partly exserted. Apr.–July. Sandstone slopes, NW (Cold Bokkeveld:Elandskloof).•

britteniae (L.Bolus) Goldblatt (= *Homeria britteniae* L.Bolus)   Cormous geophyte, 20–45 cm. Leaf solitary, clasping lower half of stem, channelled, trailing above. Flowers pale yellow to cream, tepals claws forming a narrow cup, anthers on a slender filament column, included. Sept.–Oct. Sandy slopes, SE (Knysna to Grahamstown).

collina Thunb. (= *Homeria collina* (Thunb.) Salisb.)   AASUINTJIE, GEELTULP   Cormous geophyte, 20–50 cm, stem flexed outward above leaf sheath. Leaf solitary, linear, channelled. Flowers yellow or salmon, tepal claws froming a cup including the stamens, lightly scented. July–Sept., common after fire. Lower mountain slopes and flats on sand or clay, SW (Bainskloof to Caledon).•

comptonii (L.Bolus) Goldblatt (= *Homeria comptonii* L.Bolus)   RUIKTULP   Cormous geophyte, 18–40 cm. Leaf solitary, linear, channelled, glaucous, clasping stem below. Flowers yellow or salmon with a yellow centre, tepals often with a large green mark, tepal claws forming a wide, shallow cup, filaments included, intensely fragrant. Aug.–Sept. Clay slopes in renosterveld, SW (Villiersdorp to Stanford).•

demissa Goldblatt (= *Homeria tenuis* Schltr.)   Cormous geophyte, 7–20 cm, stem flexed outward above leaf sheath. Leaf solitary, linear, channelled. Flowers small, pale yellow, tepal claws forming a shallow cup, anthers exserted on a thick filament column. Aug.–Sept. Rocky sandstone slopes, 400–1400 m, NW (Gifberg to Citrusdal).•

elegans Jacq. (= *Homeria elegans* (Jacq.) Sweet)   POUTULP   Like **M. comptonii** but tepals widest below apex and flowers yellow with outer tepals orange, often with large green blotches. Aug.–Sept. Clay slopes in renosterveld, SW (Teslaarsdal to Bredasdorp).•

flaccida Sweet (= *Homeria flaccida* (Sweet) Steud.)   GROOTTULP, ROOITULP   Cormous geophyte, 35–60 cm, stem flexed outward above leaf sheath. Leaf solitary, linear, channelled. Flowers salmon with a yellow centre or entirely yellow, tepal claws forming a wide, shallow cup, anthers exserted on a thick filament column. Aug.–Oct. Wet sandstone and granitic soils, NW, SW (Bokkeveld Mts to Caledon).•

fragrans Goldblatt (= *Homeria odorata* L.Bolus)   Like **M. bifida** but flowers pale yellow and tepal claws forming a deep cup including the stamens, sweetly fragrant. Aug.–Sept. Mainly clay soils in renosterveld, NW (Bokkeveld Plateau and W Karoo).

karooica Goldblatt (= *Homeria tricolor* G.J.Lewis)   Cormous geophyte, 15–30 cm. Leaf usually solitary, occasionally 2, usually clasping stem below. Flowers salmon with a yellow centre, tepal claws forming a shallow cup, anthers exserted on a slender filament column. Sept.–Oct. Clay soils in renosterveld, NW, KM (Ceres to Barrydale and W Karoo).

longistyla (Goldblatt) Goldblatt (= *Homeria longistyla* Goldblatt)   Cormous geophyte, 15–30 cm, stem flexed outward above leaf sheath. Leaf solitary, linear, channelled. Flowers yellow or salmon, tepal claws forming a deep cup including the stamens, style branches usually united in lower half. Aug.–Oct. Mainly clay soils, renosterveld or arid fynbos, NW, SW, KM (Ceres to Montagu and Caledon).•

minor Eckl. (= *Homeria minor* (Eckl.) Goldblatt)   KLEINTULP   Like **M. longistyla** but ovary and capsules elongate and cylindrical and style branches free to base. Aug.–Sept. Sandstone and granitic soils, to 800 m, NW, SW (Gifberg to Cape Peninsula).•

ochroleuca (Salisb.) Drapiez (= *Homeria ochroleuca* Salisb.)   APRICOT TULP   Cormous geophyte, 35–75 cm. Leaf solitary, occasionally 2, linear, channelled. Flowers yellow to orange or bicoloured, tepal claws forming a wide cup, filament column slender, anthers partly exserted. Mainly Aug.–Nov., mainly after fire. Rocky sandstone slopes, NW, SW (Citrusdal to Caledon).•

pyrophila Goldblatt (= *Homeria galpinii* L.Bolus)   Cormous geophyte, 15–30 cm, stem flexed outward above leaf sheath. Leaf solitary, linear, channelled, apex often flattened. Flowers yellow, tepal claws forming a narrow cup including the stamens. Mainly Mar.–June, only after fire. Rocky sandstone slopes, NW, SW (Piketberg to Bredasdorp).•

vallisbelli (Goldblatt) Goldblatt (= *Homeria vallisbelli* Goldblatt)   SLAPTULP   Cormous geophyte, 15–30 cm, stem flexed outward above leaf sheath. Leaf solitary, linear, channelled. Flowers yellow or pink with a yellow nectar guide outlined in dark colour, tepal claws froming a narrow cup, anthers partly exserted on a slender filament column. July–Sept. Rocky sandstone soils, NW (Bokkeveld Mts to Botterkloof).•

Moraea-group MORAEA, UINTJIE

*A. Plants acaulescent (but see **M. nubigena**)*

ciliata (L.f.) Ker Gawl.   Acaulescent cormous geophyte, 5–10(–20) cm. Leaves 3–5, sparsely to densely hairy, usually grey. Flowers blue or yellow, rarely white, spicy-fragrant. July–Sept. Sandy and clay slopes, NW, SW, AP, KM, LB (Namaqualand to Riversdale and S Karoo).

falcifolia Klatt   Acaulescent cormous geophyte to 5 cm. Leaves several, spreading, channelled and somewhat twisted. Flowers white with yellow nectar guides on outer tepals and purple on inner tepals. May–Aug. NW (Namaqualand to Clanwilliam and W Karoo to Alexandria).

macrocarpa Goldblatt   Acaulescent cormous geophyte, 8–12 cm. Leaf solitary, linear, channelled. Flowers violet, outer tepals with white marks, ovary and capsules elongate, beaked. Sept. Deep sand in arid fynbos, NW (Lambert's Bay to Worcester).•

macronyx G.J.Lewis   Acaulescent cormous geophyte, mostly 9–15 cm. Leaves 3–5, lightly pilose. Flowers yellow and white, tepal claws longer than limbs, fragrant. Sept.–Oct. Rocky sandstone slopes, NW, LB, SE (Cold Bokkeveld to Avontuur).•

tricolor Andrews   Acaulescent cormous geophyte, 5–15 cm. Leaves 3–5, pilose or glabrous. Flowers yellow, red or purple, style crests broad, fragrant. July–Sept. Wet sandy flats, SW (Darling to Caledon).•

A. Plants with aerial stems
*B. Foliage leaves more than 1 (see also **M. crispa**, **M. fugax**, **M. gracilenta**)*

bipartita L.Bolus   Cormous geophyte, 15–45 cm. Leaves 2–4, linear, channelled., stems much branched. Flowers blue. June–Nov. Clay flats, KM, LB, SE (Ladismith to E Cape and S Karoo).

bituminosa (L.f.) Ker Gawl.   TEERUINTJIE   Cormous geophyte, 20–50 cm, stems sticky. Leaves 2 or 3, linear, channelled and trailing. Flowers fairly large, yellow (rarely mauve). Oct.–Dec. Granitic and sandstone slopes and flats, SW (Darling to Agulhas).•

bubalina Goldblatt   Cormous geophyte, 30–45 cm, stems sticky. Leaves 2–5, linear, channelled and trailing. Flowers brownish with green markings. Oct.–Nov. Rocky sandstone slopes, NW (Gifberg to Botterkloof and W Karoo).

cooperi Baker   Cormous geophyte, 20–35 cm, stem much branched. Leaves 2 or 3, linear, channelled and trailing. Flowers yellow, tubular below, inner tepals absent, ovary subsessile and capsules concealed in spathes. Sept.–Oct. Rocky sandstone slopes and flats, often near water, NW, SW (Tulbagh to Stanford).•

elsiae Goldblatt   Cormous geophyte, 20–40 cm, stems sticky. Leaves 2 or 3, linear, channelled and trailing. Flowers yellow, style crests vestigial. Nov.–Dec. Deep sandy soils, SW (Cape Peninsula to Bredasdorp).•

fergusoniae L.Bolus   Cormous geophyte, 10–20 cm, only branching near ground. Leaves several, linear-lanceolate, channelled, margins undulate, sometimes crisped. Flowers white, sometimes blue, inner tepals often tricuspidate. July–Aug. Clay slopes in renosterveld, SW, LB (Botrivier to Mossel Bay).•

gawleri Spreng.   Cormous geophyte, 15–45 cm. Leaves 2 or 3, margins often crisped. Flowers yellow, cream or brick-red, sometimes bicoloured. July–Oct. Sandy or clay slopes, usually in renosterveld, NW, SW, AP, KM, LB (Namaqualand to Humansdorp).

inconspicua Goldblatt   TAAI UINTJIE   Cormous geophyte, 20–45 cm, stems sticky. Leaves 2 or 3, linear, channelled and trailing or loosely coiled. Flowers tiny, yellow to brown or cream, tepal limbs often strongly reflexed. Sept.–Nov. Sandy and clay slopes, NW, SW, AP, KM, LB, SE (Namaqualand to Port Elizabeth).

linderi Goldblatt   Cormous geophyte, 35–45 cm, stem branching above and branches short and crowded. Leaves 2 or 3, linear, channelled. Flowers yellow. Oct.–Dec. Sandstone soils in fynbos, NW (Piketberg and Cold Bokkeveld).•

lugubris (Salisb.) Goldblatt   KERSBLAKERTJIE   Cormous geophyte, 6–16 cm. Leaves 2 or 3, linear, channelled. Flowers deep blue, style branches feathery. Aug.–Nov., mainly after fire. Mostly damp sandstone soils, NW, SW, AP (Bokkeveld Mts to Bredasdorp).•

maximiliani (Schltr.) Goldblatt & J.C.Manning (= *Rheome maximiliani* (Schltr.) Goldblatt)   Cormous geophyte to 15 cm, stem branched above the leaves and branches crowded. Leaves 2–several, clustered well above ground. Flowers buff, filaments united and anthers concealing minute style arms, crests vestigial. Aug.–Sept. Sandstone slopes and flats, NW (Pakhuis Mts).•

papilionacea (L.f.) Ker Gawl.   Cormous geophyte, 10–20 cm, branching only from base. Leaves 3 or 4, often pilose. Flowers yellow or salmon. Aug.–Oct. Mostly sandstone soils, sometimes clay, renosterveld and transitional fynbos, to 500 m, NW, SW (Cedarberg Mts to Bredasdorp).•

polyanthos L.f.   BLOUTULP   Cormous geophyte, 15–60 cm, stem often much branched. Leaves several, linear, channelled. Flowers blue or white, tepals cupped below, style crests vestigial. Aug.–Sept. Flats and lower slopes, mainly clay, NW, KM, LB, SE (Worcester to E Cape).

polystachya (Thunb.) Ker Gawl.   Cormous geophyte, 50–80 cm, much-branched. Leaves several. Flowers blue. Mainly Mar.–June. Dry karroid slopes, KM (Ladismith to Oudtshoorn, Karoo to S Namibia).

ramosissima (L.f.) Druce   VLEIUINTJIE   Cormous geophyte, 50–120 cm, much-branched, roots spiny. Leaves several in a distichous fan, linear, channelled. Flowers yellow. Oct.–Dec., mainly after fire. Damp sandstone flats and slopes, NW, SW, LB, SE (Gifberg to E Cape).

serpentina Baker   SLANGUINTJIE   Cormous geophyte, 4–15 cm, branching mainly from base. Leaves 2 or 3, linear, margins inrolled, twisted or coiled. Flowers white and yellow, style branches and crests often flushed violet. Sept.–Oct. Dry stony flats, NW (Namaqualand to Olifants River valley).

umbellata Thunb. (= *Rheome umbellata* (Thunb.) Goldblatt)   Cormous geophyte, 15–45 cm, stems branching above the leaves and branches crowded. Leaves 2–several, clustered above ground. Flowers pale yellow, style branches minute, appressed to anthers, crests vestigial. Sept.–Nov. Seasonally wet sandstone flats and plateaus, NW, SW (Piketberg to Caledon).•

vegeta L.   BRUINUINTJIE   Cormous geophyte, 15–30 cm, stem minutely hairy. Leaves 3 or 4, linear-lanceolate, channelled. Flowers buff to dull purple-brown. Sept.–Oct. Damp clay or granite slopes and flats, SW (Darling to Caledon).•

viscaria (L.f.) Ker Gawl.   Cormous geophyte, 20–45 cm, stems sticky. Leaves 2 or 3, linear, channelled and trailing. Flowers white, fragrant. Sept.–Dec. Sandy flats, SW, AP (Saldanha to Cape Agulhas).•

vlokii Goldblatt   Cormous geophyte, 15–20 cm. Leaves 1 or 2, linear, channelled. Flowers tiny, yellow, style branches reduced, concealed by anthers, crests vestigial. Oct. Rocky sandstone slopes in arid fynbos, KM (Montagu to Swartberg Mts).•

B. Foliage leaf solitary
C. Inner tepals lacking, filiform or trifid

amissa Goldblatt   Cormous geophyte, 20–30 cm. Leaf solitary, linear, channelled. Flowers violet with a dark eye, inner tepals trifid with a long, straight central cusp. Oct. Stony granitic slopes, SW (Malmesbury).•

aristata (D.Delaroche) Asch. & Graebn.   BLOUOOGUINTJIE   Cormous geophyte, 25–35 cm. Leaf solitary, linear, channelled. Flowers white with a blue eye, inner tepals trifid with a long, straight central cusp. Sept. Clay soils, SW (Cape Peninsula: Observatory).•

atropunctata Goldblatt   Cormous geophyte, 15–20 cm. Leaf solitary, linear, channelled, hairy below. Flowers grey-white with dark speckles, browish on the reverse, inner tepals trifid with a long central cusp. Sept. Clay slopes, SW (Caledon: Eseljacht Mts).•

barnardii L.Bolus   Cormous geophyte, 15–30 cm. Leaf solitary, linear, narrowly channelled. Flowers white, speckled with blue, inner tepals absent. Sept.–Oct. Rocky sandstone slopes, SW (Caledon: Shaw's Mts).•

bellendenii (Sweet) N.E.Br.   PATRYSUINTJIE   Cormous geophyte, 50–100 cm, stem willowy. Leaf solitary, linear, channelled and trailing. Flowers yellow, inner tepals trifid with a short, obliquely twisted central cusp. Oct.–Nov. Granitic, sandy or clay slopes, SW, LB, AP, SE (Darling to Plettenberg Bay).•

caeca Barnard ex Goldblatt   Cormous geophyte, 20–40 cm. Leaf solitary, linear, channelled. Flowers mauve with a small dark or sometimes yellow eye, inner tepals trifid with a long, straight central cusp. Sept.–Oct. Rocky sandstone slopes in fynbos, NW (Piketberg to Porterville Mts).•

calcicola Goldblatt   Cormous geophyte, 30–40 cm. Leaf solitary, linear, channelled. Flowers mauve to purple with dark markings, inner tepals trifid with a long, straight central cusp. Sept. Limestone hills, SW (St Helena Bay to Saldanha).•

debilis Goldblatt   Cormous geophyte, 15–40 cm. Leaf solitary, linear, channelled, hairy below. Flowers mauve, becoming mottled with age, inner tepals trifid with a straight, filiform central tricusp. Sept.–Oct. Clay slopes in renosterveld, SW, LB (Botrivier to Swellendam).•

derustensis Goldblatt & J.C.Manning   Cormous geophyte, 18–25 cm. Leaf solitary, linear, channelled. Flowers deep violet, inner tepals trifid with a long, incurved central cusp. Aug.–Sept. Rocky slopes, KM (De Rust).•

gigandra L.Bolus   GROOTFLAPPIE   Cormous geophyte, 20–40 cm. Leaf solitary, linear, channelled, hairy below. Flowers blue, rarely white or orange, inner tepals trifid with a long, straight central cusp. Sept.–Oct. Clay soils, NW (Piketberg to Porterville).•

longiaristata Goldblatt   Cormous geophyte, 15–30 cm. Leaf solitary, linear, channelled. Flowers white, speckled blue, inner tepals filiform, straight and erect. Sept.–Oct. Rocky sandstone lower slopes, SW (Caledon Swartberg).•

loubseri Goldblatt   SPINNEKOPFLAPPIE   Cormous geophyte, 15–20 cm. Leaf solitary, linear, channelled, hairy below. Flowers deep blue to purple, with purple hairs on outer tepals, inner tepals trifid with a long, laxly spreading central cusp. Aug.–Sept. Limestone on granite hills, SW (Saldanha Bay).•

tricuspidata (L.f.) G.J.Lewis   RIETUINTJIE   Cormous geophyte, 25–60 cm. Leaf solitary, linear, channelled. Flowers white to cream, inner tepals trifid with a short, obliquely twisted central cusp. Mainly Sept.–Oct. Clay or granitic slopes, NW, SW, LB, SE (Cedarberg Mts to Grahamstown).

tripetala (L.f.) Ker Gawl.   BLOU UINTJIE   Cormous geophyte, 20–45 cm. Leaf solitary, linear, channelled, occasionally hairy below. Flowers blue to violet, rarely white, inner tepals vestigial or absent, sometimes trifid with filiform cusps. Aug.–Sept. Rocky sandstone and clay soils, to 1200 m, NW, SW, AP, KM, LB (Bokkeveld Mts and W Karoo to Riversdale and Swartberg Mts).

tulbaghensis L.Bolus   ROOIFLAPPIE   Cormous geophyte, 25–35 cm. Leaf solitary, linear, channelled, hairy below. Flowers orange to reddish, marked green, inner tepals trifid with a long, straight central cusp. Sept. Clay flats in renosterveld, NW, SW (Tulbagh to Wellington).•

unguiculata Ker Gawl.   Cormous geophyte, 20–50 cm. Leaf solitary. Flowers white to cream or brownish, rarely violet, inner tepals tricuspidate with inrolled central cusp. Sept.–Nov. Mostly shale slopes in renosterveld, NW, SW, AP, LB, SE (Namaqualand to Port Elizabeth, Karoo Mts).

villosa (Ker Gawl.) Ker Gawl.   BLOUFLAPPIE, UILTJIE, PEACOCK MORAEA   Cormous geophyte, 30–40 cm. Leaf solitary, pilose. Flowers purple, blue or orange, nectar guides large and dark, inner tepals tricuspidate. Aug.–Sept. Stony granite and clay slopes and flats, NW, SW (Piketberg to Gordon's Bay and Ceres).•

sp. 1 (= *Moraea caryophyllacea* Goldblatt & J.C.Manning ms)   Cormous geophyte, 20–35 cm. Leaf solitary, linear, channelled. Flowers pale pink with a bright yellow mark on each outer tepal, inner tepals trifid with a long, incurving central cusp. Aug.–Sept. Clay slopes, LB (Attakwas Mts).•

C. Inner tepals lanceolate or oblanceolate to linear

algoensis Goldblatt   Cormous geophyte, 20–40 cm. Leaf solitary, linear, channelled. Flowers purple, inner tepals oblanceolate sometimes 3-lobed. July–Sept. Clay slopes in renosterveld, NW, KM, LB, SE (Worcester to Port Elizabeth).•

angusta (Thunb.) Ker Gawl.   Cormous geophyte, 20–40 cm, stem unbranched, sticky on nodes. Leaf solitary, terete. Flowers yellow, ovary 3-angled. Aug.–Nov. Rocky sandstone flats and slopes, NW, SW, LB, SE (Cedarberg Mts to Knysna).•

anomala G.J.Lewis   Cormous geophyte, 20–40 cm, stem unbranched, sticky on nodes. Leaf solitary, terete. Flowers yellow, ovary 3-angled. Sept.–Nov. Mountains and flats, often on clay, NW, SW (Pakhuis Pass to Cape Peninsula, Caledon).•

barkerae Goldblatt   Cormous geophyte, 15–40 cm. Leaf solitary, linear, channelled. Flowers pale salmon to pink with purple markings, inner tepals attenuate. Oct.–Nov. Rocky sandstone slopes, NW (Cedarberg and Cold Bokkeveld Mts).•

crispa Thunb.   Cormous geophyte, 8–20 cm. Leaf usually solitary, linear, channelled, often lightly twisted, margins rarely crisped. Flowers blue-mauve with yellow to orange markings, tepals subequal, style branches reduced, crests vestigial. Oct.–Nov. Mainly clay slopes in renosterveld, NW, KM, SE (Cedarberg Mts to Baviaanskloof, S and W Karoo).

deltoidea Goldblatt & J.C.Manning   Like **M. unguiculata** but inner tepals oblanceolate and style crests obsolete. Oct.–Nov. Seeps on sandstone, SW (Kleinmond to Kleinrivier Mts).•

elliotii Baker   Cormous geophyte, 15–50 cm. Leaf solitary, linear, channelled. Flowers blue-violet with yellow markings. Aug.–Mar. Grassy sandstone slopes, LB, SE (Mossel Bay to Malawi).

exiliflora Goldblatt   Cormous geophyte, 15–25 cm. Leaf solitary, linear, channelled. Flowers small, pale blue-mauve with yellow markings. Sept. Sandstone outcrops in fynbos, KM (Swartberg Mts).•

fugax (D.Delaroche) Jacq.   SOETUINTJIE, HOTTENTOTUINTJIE   Cormous geophyte, 12–80 cm, branches often crowded. Leaves 1 or 2, inserted well above ground, linear, channelled, often trailing. Flowers blue, white or yellow, fragrant, ovary and capsules beaked. Aug.–Nov. Deep sands and rocky sandstone and granitic soils, NW, SW, AP, LB (Namaqualand to Swellendam).

gracilenta Goldblatt   Like **M. fugax** but stem laxly branched, flowers small, pale blue, opening late afternoon. Sept.–Oct. Sandy soils, NW (Clanwilliam to Tulbagh).•

incurva G.J.Lewis   Cormous geophyte, 35–40 cm. Leaf solitary, linear, channelled. Flowers deep blue with yellow or white markings. Oct. Clay soils in renosterveld, NW (Tulbagh valley).•

insolens Goldblatt   Cormous geophyte, 20–35 cm. Leaf solitary, linear, channelled. Flowers deep orange or cream with brown centre, style branches narrow and crests short. Sept. Clay slopes in renosterveld, SW (Caledon Swartberg).•

lurida Ker Gawl.   AASUINTJIE   Cormous geophyte, 20–30 cm. Leaf solitary, linear, channelled. Flowers maroon, sometimes marked with yellow, or entirely cream, usually foetid-smelling, style crests short. Aug.–Oct., mainly after fire. Rocky sandstone and granitic slopes in fynbos, SW (Sir Lowry's Pass to Bredasdorp).•

neglecta G.J.Lewis   Cormous geophyte, 20–50 cm, stem unbranched, nodes sticky. Leaf solitary, terete. Flowers yellow with darkly stippled markings, ovary 3-angled. Sept.–Nov. Usually deep sandy soils, NW, SW, AP (Bokkeveld Mts to Agulhas coast).•

neopavonia R.C.Foster   PEACOCK MORAEA   Cormous geophyte, 25–50 cm. Leaf solitary, linear, channelled, hairy below. Flowers orange with speckled or iridescent blue markings. Sept. Clay soils in renosterveld, NW, SW (Piketberg to Paarl).•

nubigena Goldblatt   Dwarf cormous geophyte, 3–5 cm. Leaf solitary, linear-lanceolate, shallowly channelled. Flowers small, blue-mauve, style crests short. Sept.–Oct. Rock seeps at 1200 m, NW (Worcester: Fonteintjiesberg).•

obscura Goldblatt (= *Roggeveldia montana* Goldblatt)   Cormous geophyte to 40 cm. Leaf solitary, terete. Flowers blue or white, stamens free, extending between filiform style arms. Nov.–Dec. Rocky lower slopes, KM (Swartberg Mts and W Karoo).

spathulata (L.f.) Klatt   Cormous geophyte to 1 m, unbranched. Leaf solitary, linear, flat or channelled, often fairly broad. Flowers large, yellow, inner tepals erect. June–Sept. Sandstone and peaty slopes and flats, KM, SE (Kammanassie Mts and George to Zimbabwe).

thomasiae Goldblatt   Cormous geophyte, 15–30 cm. Leaf solitary, linear, channelled. Flowers pale yellow with dark veins, inner tepals erect. Aug.–Sept. S-facing clay slopes in renosterveld, NW, KM (Worcester to Barrydale).•

vallisavium Goldblatt   Cormous geophyte, 10–35 cm, unbranched. Leaf solitary, terete. Flowers yellow with brown spotted markings, ovary 3-angled. Dec.–Jan. Sandstone rocks, 500–1000 m, SW, LB (Kleinrivier and Langeberg Mts).•

verecunda Goldblatt   Cormous geophyte, 15–25 cm. Leaf solitary, linear, channelled. Flowers tiny, violet, tepals subequal, style branches narrow and crests vestigial. Oct.–Nov. Rocky sandstone in fynbos, NW (Bokkeveld Mts).•

worcesterensis Goldblatt   Cormous geophyte to 15 cm. Leaf solitary, linear, channelled. Flowers stellate, tepals spreading from base, purple with diamond-shaped nectar guides, style branches narrow, crests vestigial. Sept.–Oct. Rocky flats in fynbos, NW (Worcester).•

NIVENIA•   BUSH IRIS   10 spp., W Cape

A. Inflorescence a few-flowered pseudoraceme or single flower pair

argentea Goldblatt   (= *Nivenia capitata* (Klatt) Weim.)   Shrub to 80 cm. Leaves sword-shaped. Flowers paired, heterostylous, blue, tube 30–40 mm, bracts conspicuous, silvery. Nov. Rocky sandstone slopes in fynbos, KM, LB (Rooiberg and Langeberg Mts: Riversdale).•

concinna N.E.Br.   Shrublet to 40 cm. Leaves sword-shaped. Flowers dark blue, tube c. 15 mm. Jan.–Feb. Rocky sandstone slopes in fynbos, SW (Groenland Mts).•

fruticosa (L.f.) Baker   Cushion-like shrublet, 9–20 cm. Leaves sword-shaped. Flowers paired, pale blue, tube 20–30 mm. Nov.–Jan. Rocky sandstone slopes, LB (Langeberg Mts).•

levynsiae Weim.   Shrublet to 25 cm. Leaves sword-shaped. Flowers pale blue, tepals cupped, tube short, c. 7 mm. Dec.–Feb. Rocky sandstone outcrops, SW (Kogelberg to Kleinmond).•

A. Inflorescence a lax corymbose pseudopanicle

binata Klatt   Shrub to 50 cm. Leaves sword-shaped. Flowers paired, heterostylous, deep blue, anthers blue, tube 9–15 mm. Mainly Aug.–Oct. Rocky sandstone slopes in fynbos, KM (Swartberg Mts: Ladismith to Meiringspoort).•

corymbosa (Ker Gawl.) Baker   Shrub 50–200 cm. Leaves sword-shaped. Flowers solitary, heterostylous, deep blue, tube 11–13 mm. Feb.–Mar. Rocky sandstone slopes near water, SW (Bainskloof to Tulbagh Kloof).•

dispar N.E.Br.   Shrub to 80 cm. Leaves sword-shaped. Flowers solitary, heterostylous, light blue, tube 16–20 mm. Feb.–Apr. Sandstone cliffs and rocks, 200–400 m in river valleys, SW (Riviersonderend Mts).•

parviflora Goldblatt   Shrub 50–80 cm. Leaves sword-shaped. Flowers paired, heterostylous, tiny, pale to deep blue, anthers blue, tube c. 5.5 mm, shorter than lobes. Mar.–Apr. Sandstone slopes c. 300 m, KM (Swartberg Mts: near Ladismith).•

stenosiphon Goldblatt   Shrub 40–100 cm. Leaves sword-shaped. Flowers paired, heterostylous, blue, anthers blue, tube 20–25 mm. Apr. Rocky sandstone slopes in fynbos, 500–800 m, KM (Touwsberg and Swartberg Mts: near Ladismith).•

stokoei (L.Guthrie) N.E.Br.   Shrub 40–60 cm. Leaves sword-shaped. Flowers solitary, large, pale to deep blue, tube 27–37 mm. Feb.–Mar. Rocky sandstone in fynbos, SW (Kogelberg to Kleinmond).•

PILLANSIA•   1 sp., W Cape

templemannii (Baker) L.Bolus   Cormous evergreen perennial, 60–90 cm. Leaves linear, strap-like, loosely twisted, fibrotic, without a midrib. Flowers in flat-topped panicles, rotate, orange. Oct.–Nov. Sandstone slopes after fire, SW (Kogelberg to Kleinrivier Mts).•

ROMULEA   ROMULEA, KNIKKERTJIE, FROETANG   c. 85 spp., South Africa to S Europe and Middle East, mainly W Karoo and W Cape

*A. Corms symmetrical and bell-shaped (see also **R. sanguinalis**)*

amoena Schltr. ex Bég.   Cormous geophyte, 8–12 cm, corms symmetrical, bell-shaped. Basal leaves 2–4. Flowers pink to red with black blotches at edge of cream cup. Aug.–Sept. Sandstone soils, NW (Bokkeveld Mts).•

discifera J.C.Manning & Goldblatt   Cormous geophyte, 8–15 cm, corms symmetrical, depressed-discoid. Basal leaves 2 or 3. Flowers yellow with darker cup, with or without black markings. July. Sandy flats, NW (Bokkeveld Mts).•

gracillima Baker   Like **R. hirsuta** but flowers smaller, pale pink with yellow cup. Aug.–Sept. Sandstone slopes, SW, AP (Cape Peninsula to Agulhas flats).•

hirsuta (Steud. ex Klatt) Baker   Cormous geophyte, 6–10 cm, corms symmetrical, bell-shaped, stem branching above ground. Basal leaves 2, sometimes solitary. Flowers pink to rose or coppery orange with dark marks at edge of yellow cup. Aug.–Sept. Sandstone or clay slopes and flats, NW, SW (Clanwilliam to Elim).•

sladenii M.P.de Vos   Cormous geophyte, 7–9 cm, corms symmetrical, bell-shaped, basal margins finely lacerate, stems branched above ground. Basal leaves 2. Flowers white with yellow cup. Aug.–Sept. Rocky sandstone flats, NW (Gifberg).•

tortilis Baker   Like **R. hirsuta** but leaves spirally twisted. July–Sept. Sandstone slopes, NW (Clanwilliam to Piketberg).•

triflora (Burm.f.) N.E.Br.   Cormous geophyte to 20 cm, corms symmetrical, bell-shaped, stem usually branched above ground. Basal leaves usually 2. Flowers yellow, sometimes white, with yellow cup. Aug.–Oct. Sandstone slopes, NW, SW (Citrusdal to Stanford).•

A.' Corms rounded or pointed at base

atrandra G.J.Lewis   Cormous geophyte, 8–15 cm, corms rounded at base. Basal leaves several. Flowers magenta to pale pink or white with dark veins and dark blotches at edge of yellow cup, bracts with broad membranous margins. July–Oct. Clay soils, NW, SW, KM, LB, SE (Gifberg to E Cape and W Karoo).

cedarbergensis M.P.de Vos   Like **R. obscura** but leaves 1–3. Flowers white to pale pink with yellow cup, filaments longer than anthers. July–Sept. Sandstone rocks, NW (Cedarberg Mts).•

cruciata (Jacq.) Baker   Cormous geophyte, 5–12 cm, corm pointed at base. Basal leaves 2–several. Flowers magenta to lilac with dark blotches at edges of yellow cup. July–Sept. Sandstone and granite slopes and rocks, NW, SW, AP, LB (Bokkeveld Mts to Gourits River).•

eximia M.P.de Vos   Like **R. cruciata** but flowers larger, longer than 35 mm, old rose. Aug.–Sept. Sandy flats, SW (Yzerfontein to Melkbos).•

hirta Schltr.   Cormous geophyte, 5–10 cm, corms rounded at base. Leaves several, 4-winged, ciliate. Flowers dull yellow. July–Sept. Sandstone soils, NW (W Karoo and Bokkeveld to Cedarberg Mts).

luteoflora (M.P.de Vos) M.P.de Vos   Like **R. atrandra** but flowers yellow with brown streaks or blotches. July–Sept. NW, KM, LB (Namaqualand, Cedarberg Mts to Riversdale, W Karoo and Lesotho).

malaniae M.P.de Vos   Cormous geophyte, 8–12 cm, corms rounded at base. Basal leaves 1–3. Flowers on geniculate peduncles, small, pale yellow, bracts with broad membranous margins. Aug. Sandstone outcrops, NW (Matroosberg to Koo).•

membranacea M.P.de Vos   Cormous geophyte to 12 cm, corm ovoid, pointed at base. Basal leaves several. Flowers uniformly deep yellow with yellowish cup, inner and outer bracts mostly membranous and spotted with brown. July–Aug. Sandy flats, NW (Lokenberg and W Karoo).

monticola M.P.de Vos   Like **R. obscura** but flowers yellow with dark streaks. June–Sept. Sandy loam in fynbos, NW (Bokkeveld Mts to Gifberg).•

obscura Klatt   Cormous geophyte, 5–8 cm, corms rounded at base. Basal leaves 1–several. Flowers apricot or rose, often with dark blotches at edge of greenish to yellow cup. Aug.–Oct. Sandy flats, NW, SW, AP (Clanwilliam to Agulhas).•

rosea (L.) Eckl.   ROOIKNIKKERTJIE   Cormous geophyte, 10–40 cm, corms rounded at base. Basal leaves several. Flowers pink to purple, sometimes white, with yellow cup. Ripe capsules erect. July–Oct. Sandy and clay slopes and flats, NW, SW, AP, KM, LB, SE (Bokkeveld Mts to Port Elizabeth).•

sabulosa Schltr. ex Bég.   SATYNBLOM   Cormous geophyte, 6–10 cm, corms rounded at base. Basal leaves several. Flowers dark red, rarely pink with black blotches at edge of creamy green cup, inner bracts 2-keeled. July–Sept. Clay slopes in renosterveld, NW (Bokkeveld Escarpment).•

viridibracteata M.P.de Vos   Cormous geophyte, 4–8 cm, corms rounded at base. Basal leaves several. Flowers yellow with dark brown markings, inner bracts 2-keeled. Aug.–Sept. Sandstone slopes, NW (Bokkeveld Mts to Pakhuis Pass).•

vlokii M.P.de Vos   Like **R. cruciata** but bracts conspicuously veined. Flowers pink with orange cup. July–Aug. Wet sands, KM (Kammanassie Mts).•

A". Corms obliquely ridged at base or laterally compressed

albomarginata M.P.de Vos   Cormous geophyte to 6 cm, corm with an oblique basal ridge. Basal leaves 2. Flowers magenta to pink with orange-yellow cup. Aug.–Oct. Sandstone flats, NW (Cold Bokkeveld).•

aquatica G.J.Lewis   WATERFROETANG   Like **R. multisulcata** but basal leaf solitary. Flowers white with yellow cup, fragrant. Fruiting peduncles erect. Aug.–Sept. Seasonal pools, NW, SW (Pools to Hopefield).•

austinii E.Phillips   Cormous geophyte to 20 cm, corm with a broad, fan-like oblique base. Basal leaves several, channelled to apex. Flowers yellow usually with brown markings. May–July. Damp stony flats, KM (Montagu to Uniondale, S and W Karoo).

barkerae M.P.de Vos   Cormous geophyte, 4–6 cm, corm with an oblique basal ridge, stem branching above ground. Basal leaf solitary, with 2 grooves. Flowers white with black cup edged in yellow, inner bracts papery. July–Aug. Limestone rocks, SW (Paternoster to Saldanha).•

biflora (Bég.) M.P.de Vos   Cormous geophyte, 10–15 cm, corm with an oblique basal ridge, stem branched above ground. Basal leaves 2. Flowers pink to rose with yellow cup edged with dark blotches. July–Sept. Clay foothills, NW (Gifberg and Biedouw valley).•

dichotoma (Thunb.) Baker   Cormous geophyte to 35 cm, corm with an oblique basal ridge, stem branching divaricately above ground. Basal leaf usually solitary. Flowers pink to salmon with yellowish cup. Sept.–Oct. Sandy flats and slopes, SW, AP, LB, SE (Stanford to Humansdorp).•

elliptica M.P.de Vos   Cormous geophyte, 15–20 cm, corm with an oblique basal ridge, stem branching above ground. Basal leaves 2. Flowers yellow with dark streaks. Aug. Sandy flats, SW (Vredenburg).•

fibrosa M.P.de Vos   Cormous geophyte to 35 cm, corm with an oblique basal ridge, stem with a fibrous neck, branching above ground. Basal leaves 2. Flowers magenta to pink with yellow cup. Oct.–Dec. High altitudes, LB, KM, SE (Langeberg to Great Winterhoek Mts).•

flava (Lam.) M.P.de Vos   GEELFROETANG, GEELKNIKKERTJIE   Cormous geophyte, 5–30 cm, corm with an oblique basal ridge, stem short or branching just above ground. Basal leaves 1 or 2, clasping below. Flowers white or yellow, rarely blue or pinkish, with yellow cup, inner bracts papery. June–Sept. Sandy and clay soils, fynbos or renosterveld, NW, SW, AP, LB, SE (Bokkeveld Mts to Humansdorp).•

flexuosa Klatt   Like **R. schlechteri** but flowers white with buff to brown cup, anthers with long pointed appendages. May–July. Sandstone rocks, in fynbos, NW, SW (Lokenberg to Hottentots Holland Mts).•

gigantea Bég.   Like **R. schlechteri** but flowers smaller, white, lilac or blue with greenish yellow cup. Sept.–Oct. Moist places, SW, LB, SE (Kleinmond to Port Alfred).

jugicola M.P.de Vos   Cormous geophyte to 30 cm, corm with an oblique basal ridge, stems with fibrous neck, branching above ground. Basal leaf solitary, hairy. Flowers orange with yellow cup. Aug. Clay soils in renosterveld, SW, KM, SE (Potberg, Outeniqua and Little Karoo Mts).•

leipoldtii Marais   Like **R. tabularis** but basal leaves usually 2. Flowers cream with yellow cup, not blotched on reverse, inner bracts greenish in the centre. Sept.–Oct. Damp sandy sites, NW, SW (Bokkeveld Mts to Piketberg).•

longipes Schltr.   Cormous geophyte, 4–35 cm, corm with an oblique basal ridge, stem branched above ground. Basal leaves 2. Flowers cream to apricot with yellow cup. July–Nov. Sandy flats, SE (Port Elizabeth to East London).

minutiflora Klatt   Cormous geophyte, 3–6 cm, corm with a spathulate basal projection. Basal leaves several. Flowers tiny, to 15 mm long, pale mauve with yellowish cup, anthers c. 2 mm, inner bracts conspicuously spotted with brown. July–Sept. NW, SW, AP, KM, LB, SE (widespread, Bokkeveld Mts to Grahamstown).

montana Schltr. ex Bég.   Cormous geophyte, 5–15 cm, corm with a wide basal ridge, stem often branching above ground. Basal leaves 2. Flowers yellow with dark streaks or blotches. July–Sept. Sandstone outcrops, NW (Bokkeveld to Cedarberg Mts).•

multisulcata M.P.de Vos   WATERFROETANG   Cormous geophyte, 6–15 cm, corm with an oblique basal ridge, stem branching above ground. Basal leaves 2, blades 6–8-grooved. Flowers white or pale yellow with darker cup, fruits globose on sharply spreading peduncles. Aug.–Sept. Seasonal pools, NW (Bokkeveld Mts and Gifberg flats).•

papyracea Wolley-Dod   Like **R. schlechteri** but outer bracts longer than the inner and keeled, flowers lilac-pink with yellow cup. Oct. Sandstone soils, SW (Cape Peninsula).•

pratensis M.P.de Vos   Cormous geophyte to 10 cm, corm with an oblique basal ridge. Basal leaves several. Flowers rose to white with greenish yellow cup. July–Sept. Grassland, SE (Avontuur to Alexandria).

saldanhensis M.P.de Vos   Similar to **R. flava** but flowers orange-yellow with dark lines in cup. Aug.–Sept. Wet sand or clay flats, SW (Vredenburg to Darling).•

sanguinalis M.P.de Vos   Cormous geophyte to 35 cm, corm almost bell-shaped with incomplete basal ridge, stem branched above ground. Basal leaves 2. Flowers uniformly red, filaments smooth. Aug. Rocky flats, NW (Bokkeveld Mts near Botterkloof).•

saxatilis M.P.de Vos   Similar to **R. schlechteri** but with slender, elongate stem. Flowers magenta-pink with yellow cup, filaments swollen in middle. Sept.–Oct. Sandstone rocks, NW (Cedarberg Mts to Ceres).•

schlechteri Bég.   Cormous geophyte, 8–30 cm, corm with an oblique basal ridge, stem usually branching above ground. Basal leaves 2. Flowers pink to cream with deep yellow cup. July–Sept. Streams and vleis on sandy soils, NW, SW (Bokkeveld Mts to Caledon).•

setifolia N.E.Br.   Cormous geophyte, 4–12 cm, corm with an oblique basal ridge. Basal leaves 2. Flowers yellow or apricot, sometimes with dark blotches at edge of cup. July–Sept. Sandstone slopes and flats, NW, SW, AP, KM, LB, SE (Bokkeveld Mts to Port Elizabeth, and W Karoo).

sinispinosensis M.P.de Vos   Like **R. minutiflora** but flowers 15–20 mm long, white with yellow cup, anthers 3.5–4 mm long, inner bracts with inconspicuous spots. Aug. Sandy slopes, NW (Doringbaai to Sauer).•

sphaerocarpa M.P.de Vos   Cormorous geophyte to 20 cm. Corm with a broad fan-like oblique base. Leaf mostly solitary, sticky. Flower usually 1, yellow with brown streaks. June. Sandstone outcrops, NW, KM (Swartruggens to Waboomsberg: Ouberg Pass).•

stellata M.P.de Vos   Cormous geophyte, 3–5 cm, corms with an oblique basal ridge. Leaves 1 or 2. Flowers hypocrateriform, violet, rarely white, with yellow in throat, tube elongate. May–July. Sandstone pavement, NW (Gifberg to N Cedarberg Mts).•

sulphurea Bég.   Cormous geophyte, 4–6 cm, corm with an oblique basal ridge, stem branching just above ground. Basal leaves 2. Flowers yellow with dark marks in cup, filaments twice as long as anthers and hairy. Aug. Sandstone rocks, NW (Pakhuis Mts).•

tabularis Eckl. ex Bég.   Cormous geophyte to 10 cm, corm with an oblique basal ridge, stem branching above ground. Basal leaves 1 or 2. Flowers blue to white with yellow cup, often blotched on reverse, sometimes fragrant. July–Oct. Moist sandy or limestone flats, NW, SW, AP (S Namaqualand to Agulhas).

tetragona M.P.de Vos   Cormous geophyte, 4–12 cm, corms with an oblique basal ridge. Basal leaves several, 4-winged, ciliate. Flowers rose to lilac or pink with yellowish cup. Aug.–Sept. Clay soils, NW, KM (W Karoo and Cold Bokkeveld to Tweedside).

tortuosa (Licht. ex Roem. & Schult.) Baker   Cormous geophyte to 20 cm. Corm laterally compressed and fan-like. Leaves several, channelled, twisted to coiled. Flowers yellow, often with brown markings June–Sept. Sandstone and clay soils, NW, KM (Kamiesberg and W Karoo to Gydo Pass).

toximontana M.P.de Vos   Cormous geophyte to 10 cm, corm with a wide, obliquely flattened basal ridge, stem usually branching above ground. Basal leaves 2. Flowers cream with orange cup. Aug. Sandy soils, NW (Bokkeveld Mts to Gifberg).•

vinacea M.P.de Vos   Like **R. schlechteri** but flowers blue-violet with cream and yellow cup, outer tepals shiny and wine-coloured on reverse. Aug. Sandy soils, NW (Pakhuis Pass).•

SPARAXIS (= *SYNNOTIA*)   CAPE BUTTERCUP, SPARAXIS, FLUWEELTJIE   15 spp., W Cape to W Karoo

A. Flowers actinomorphic or almost so with erect stamens

bulbifera (L.) Ker Gawl.   BOTTERBLOM   Cormous geophyte, 15–45 cm, stem branched and with axillary cormlets after blooming. Leaves sword-shaped. Flowers nearly actinomorphic, with unilateral stamens and style, white to cream, often purplish on reverse. Sept.–Oct. Wet sandy or clay flats, SW, AP (Darling to Agulhas).•

elegans (Sweet) Goldblatt   Cormous geophyte, 9–20 cm. Leaves lanceolate. Flowers actinomorphic, stamens and style central, salmon, rarely white, marked purple and yellow, anthers coiled. Aug.–Sept. Clay soils, NW (Bokkeveld Escarpment and W Karoo).

fragrans (Jacq.) Ker Gawl.   Cormous geophyte, 10–25 cm. Leaves linear. Flowers actinomorphic, stamens and style central, yellow to buff, scented. Aug.–Sept. Clay slopes in renosterveld, SW (Botrivier to Bredasdorp).•

grandiflora (D.Delaroche) Ker Gawl.   BOTTERBLOM   Cormous geophyte, 10–25 cm. Leaves lanceolate, sometimes prostrate. Flowers almost actinomorphic, with unilateral stamens and style, white or yellow to plum-red. Aug.–Sept. Clay flats and slopes in renosterveld, NW, SW (Clanwilliam to Bredasdorp).•

maculosa Goldblatt   Cormous geophyte, 10–20 cm. Leaves broad, falcate. Flowers actinomorphic, stamens and style central, yellow with black centre. Sept. Clay slopes in renosterveld, SW (Villiersdorp).•

tricolor (Schneev.) Ker Gawl.   HARLEQUIN FLOWER   Cormous geophyte to 30 cm. Leaves lanceolate. Flowers actinomorphic, stamens and style central, orange with black and yellow centres. Sept.–Oct. Damp clay and sandstone soils in renosterveld, NW (Bokkeveld Escarpment).•

A. Flowers zygomorphic with unilateral arcuate stamens
B. Corms globose, tunic fibres coarse and clawed below

caryophyllacea Goldblatt   Cormous geophyte, 8–20 cm, tunics clawed below. Leaves oblong, obtuse, prostrate, darkly speckled on sheaths. Flowers zygomorphic, yellow and white with speckled throat, dorsal tepal arcuate, fragrant. Aug. Rocky sandstone slopes, NW (Nardouw Mts).•

metelerkampiae L.Bolus   Cormous geophyte, 15–30 cm, tunics clawed below. Leaves lanceolate, obtuse, sometimes prostrate. Flowers zygomorphic, tube elongate, sharply bent, violet marked with white, style branches short, not overtopping anthers. Aug.–Sept. Rocky sandstone slopes, NW (Bokkeveld Mts to Eendekuil).•

variegata (Sweet) Goldblatt   Cormous geophyte, 25–40 cm, tunics clawed below. Leaves lanceolate, obtuse, often inclined. Flowers zygomorphic, tube elongate, sharply bent, yellow and violet, throat streaked with purple, style branches long, overtopping anthers. Aug.–Sept. Rocky sandstone slopes, NW (Olifants River valley).•

villosa (Burm.f.) Goldblatt   Cormous geophyte, 12–35 cm, tunics clawed below. Leaves lanceolate, often obtuse. Flowers zygomorphic, small, yellow, upper tepals purple. Aug.–Sept. Clay and granite slopes, NW, SW (Citrusdal to Cape Peninsula).•

B. Corms globose or conic, tunic fibres medium to fine

auriculata Goldblatt & J.C.Manning   Cormous geophyte, 15–50 cm. Leaves lanceolate, darkly speckled on sheaths. Flowers zygomorphic, mauve and yellow, dorsal tepal erect, lightly scented. Aug. Rocky sandstone slopes, NW (Gifberg).•

galeata (Jacq.) Sweet   Cormous geophyte, 12–35 cm. Leaves lanceolate, obtuse, often prostrate, speckled on sheaths. Flowers zygomorphic, cream and yellow with purplish markings, dorsal tepals erect or reflexed, usually fragrant. July–Sept. Dry clay slopes in renosterveld, NW (W Karoo and Vanrhyn's Pass to Clanwilliam).

parviflora (G.J.Lewis) Goldblatt   Cormous geophyte, 15–30 cm. Leaves linear to sword-shaped. Flowers zygomorphic, small, cream sometimes flushed purple, tube very slender, lightly scented. Aug.–Sept. Granite slopes and rocks, SW (Saldanha to Darling).•

roxburghii (Baker) Goldblatt   Cormous geophyte, 25–40 cm. Leaves sword-shaped, speckled on sheaths. Flowers zygomorphic, mauve to lilac with yellow on lower tepals, dorsal tepal erect, tube fairly long. Aug.–Sept. Rocky sandstone slopes, NW (rare and local S of Clanwilliam).•

SYRINGODEA   CAPE CROCUS   8 spp., W Cape to Vaal River

derustensis M.P.de Vos   Cormous geophyte, 5–8 cm, corm turbinate. Leaves few, lanceolate, flattened and prostrate. Flowers violet, rarely white. May–June. Stony hills, KM (Little Karoo: Dysselsdorp to De Rust).•

longituba (Klatt) Kuntze   Cormous geophyte, 4–10 cm, corm turbinate. Leaves several, filiform, often twisted. Flowers blue to violet with white or yellow centre. Apr.–June. Sandstone and shale soils, NW, SW, KM, LB (Namaqualand and W Karoo to Caledon, Little Karoo to Mossel Bay).

saxatilis M.P.de Vos   Cormous geophyte, 4–8 cm, corm turbinate. Leaf solitary, terete and fleshy. Flowers lilac to violet. May–June. Low rocky hills, KM (Ladismith).•

unifolia Goldblatt   Cormous geophyte, 5–12 cm, corm flattened, fan-shaped. Leaf solitary, terete. Flowers violet or white with yellow centre. May–June. Clay flats, NW, KM (Hex River Mts to Touwsrivier and W Karoo).

THEREIANTHUS•   SOMERPYPIE   8 spp., W Cape

A. Perianth tube longer than tepals

juncifolius (Baker) G.J.Lewis   Cormous geophyte, 30–35 cm. Leaves terete, slender. Flowers in a lax spike, blue to purple, white in throat, tube slender. Nov.–Mar. Mountain seeps, marshes and streams in sandstone, NW, SW (Cold Bokkeveld to Riviersonderend Mts).•

longicollis (Schltr.) G.J.Lewis   Like **T. spicatus** but flowers pale lilac or white with purple markings and tube at least 3 times as long as tepals. Nov.–Jan. Sandstone slopes, NW, SW (Porterville Mts to Tulbagh Falls).•

minutus (Klatt) G.J.Lewis (= *Thereianthus lapeyrousioides* (Baker) G.J.Lewis)   Cormous geophyte, 10–20 cm. Leaves sword-shaped with definite midrib. Flowers purple, tube elongate. Nov.–Jan. Seeps and wet sandstone rocks, NW, SW (Cold Bokkeveld to Kogelberg).•

sp. 1.   Cormous geophyte to 35 cm. Leaves linear with prominent veins. Flowers purple, tube elongate. Jan.–Feb. Sandstone slopes, SW (Riviersonderend Mts).•

A. Perianth tube shorter than to as long as tepals

bracteolatus (Lam.) G.J.Lewis   Cormous geophyte, 15–25 cm. Leaves terete without prominent veins. Flowers deep blue to white or purple, tepals prominently veined. Nov.–Jan. Dry sandstone slopes, NW, SW (Citrusdal to Bredasdorp).•

ixioides G.J.Lewis   Cormous geophyte, 25–45 cm. Leaves terete without prominent veins. Flowers clustered in head-like racemes, actinomorphic, white with violet markings, tube filiform, the mouth closed by the filament bases. Oct.–Nov. Damp sandstone slopes, SW (Bainskloof to Du Toitskloof).•

racemosus (Klatt) G.J.Lewis   Cormous geophyte, 15–25 cm. Leaves linear to sword-shaped with definite midrib. Flowers pale blue with dark markings on lower tepals, tube very short and included in the bracts. Oct.–Dec. Rocky sandstone slopes, NW (Piketberg and Porterville Mts).•

spicatus (L.) G.J.Lewis   Cormous geophyte, 18–30 cm. Leaves linear with prominent veins or terete. Flowers pale blue to mauve. Oct.–Nov. Sandstone slopes, SW (Elandskloof Mts to Kleinmond).•

TRITONIA   TRITONIA, BASTERKALKOENTJIE   28 spp., S Cape and Namaqualand to S tropical Africa

A. Flowers or at least perianth actinomorphic, stamens unilateral or central

crocata (L.) Ker Gawl.   MOSSELBAAI BASTERKALKOENTJIE   Cormous geophyte, 25–50 cm. Leaves sword-shaped. Flowers almost actinomorphic, rotate, with unilateral stamens and style, orange to reddish, tepal margins nearly tranparent below. Sept.–Nov. Clay slopes in renosterveld, AP, LB, SE (Swellendam to Humansdorp).•

deusta (Aiton) Ker Gawl.   Cormous geophyte, 15–25 cm. Leaves sword-shaped. Flowers almost actinomorphic, rotate, with unilateral stamens and style, orange, often with dark marks on outer tepals. Sept.–Oct. Clay or granite slopes in renosterveld, SW, AP, LB (Cape Peninsula to Riversdale).•

dubia Eckl. ex Klatt   Cormous geophyte, 12–25 cm. Leaves falcate. Flowers actinomorphic, pink to orange with dark veins, stamens central around erect style. Aug.–Sept. Clay slopes in renostersveld and bush, SE (Humansdorp to Port Elizabeth).•

squalida (Aiton) Ker Gawl.   Cormous geophyte, 25–40 cm. Leaves sword-shaped. Flowers almost actinomorphic, rotate, with unilateral stamens and style, pink to purple, tepals with lower margins transparent. Sept.–Oct. Limestone outcrops and calcareous sands, AP, LB (Riversdale to Albertinia).•

A. Flowers zygomorphic
B. Perianth tube shorter than tepals

laxifolia Benth. ex Baker   Like **T. securigera** but autumn-flowering, calluses blunt and capsules larger, 10–15 mm long. Mar.–May. Grassy slopes, SE (Uitenhage to Tanzania).

lineata (Salisb.) Ker Gawl.   Cormous geophyte to 60 cm. Leaves sword-shaped with a prominent vein near margins. Flowers weakly bilabiate, cream or yellow with dark veins. Aug.–Nov. Grassland, KM, SE (Swartberg Mts and Jeffreys Bay to Mpumalanga).

parvula N.E.Br.   Like **T. securigera** but leaves linear, 1–4 mm wide and lateral calluses sometimes reduced. Sept.–Nov. Stony sandstone soils, KM, LB, SE (Riversdale to Willowmore).•

securigera (Aiton) Ker Gawl.   Cormous geophyte, 15–40 cm. Leaves sword-shaped. Flowers bilabiate, reddish to orange, lower tepals each with a large, tooth-like yellow callus. Sept.–Nov. Clay slopes, KM, LB, SE (Riversdale to Katberg).

watermeyeri L.Bolus   Like **T. securigera** but leaves twisted and tightly crisped along margins. Aug.–Sept. Clay flats in succulent karoo, KM (Montagu to Barrydale).•

B. Perianth tube longer than tepals

bakeri Klatt   Cormous geophyte, 40–70 cm. Leaves linear, subterete. Flowers long-tubed, perianth nearly actinomorphic but stamens unilateral, cream to pale mauve. Oct.–Dec. Rocky slopes, KM, LB, SE (Ladismith to Avontuur).•

chrysantha Fourc.   Cormous geophyte, 15–30 cm. Leaves sword-shaped. Flowers long-tubed, bright yellow, lower tepals each with a large tooth-like callus. Aug.–Oct. Dry karroid slopes, KM, SE (Oudtshoorn to Grahamstown).

cooperi (Baker) Klatt   Cormous geophyte, 40–60 cm. Leaves narrowly sword-shaped with flanged margins, H-shaped in section. Flowers long-tubed, white to cream, fading pink, with red markings, bracts small, obtuse. Nov.–Dec. Rocky sandstone soils, SW, LB (Dutoitskloof to Riversdale).•

crispa (L.f.) Ker Gawl.   Cormous geophyte, 18–35 cm. Leaves sword-shaped with undulate and crisped margins. Flowers long-tubed, cream with red markings, pink on outside, bracts small, obtuse. Oct.–Dec. Rocky sandstone slopes in fynbos, NW, SW (Bokkeveld Mts to Grabouw).•

flabellifolia (D.Delaroche) G.J.Lewis   Cormous geophyte, 25–40 cm. Leaves sword-shaped. Flowers long-tubed, cream with red streaks, bracts large, rust-coloured above and attenuate. Mainly Oct.–Nov. Rocky sandstone and shale slopes in renosterveld or fynbos, SW, KM, LB (Villiersdorp to Albertina).•

lancea (Thunb.) N.E.Br.   Cormous geophyte, 10–30 cm. Leaves narrow with inrolled margins and petiole-like base. Flowers long-tubed, white with red markings, bracts small, obtuse. Aug.–Sept. Sandstone slopes, NW (Piketberg).•

pallida Ker Gawl.   Cormous geophyte, 20–40 cm. Leaves sword-shaped. Flowers long-tubed, cream or pink to pale lilac, lower tepals each usually with a yellowish green tooth, bracts fairly short, acute. Sept.–Oct. Sandstone and clay slopes, NW, KM, LB (Robertson to Oudtshoorn).•

TRITONIOPSIS (= *ANAPALINA*)   SNAKE FLOWER   21 spp., W to E Cape

A. Flowers red to salmon

antholyza (Poir.) Goldblatt (= *Anapalina nervosa* (Thunb.) G.J.Lewis, *Anapalina longituba* Fourc.)   KARKARBLOM, BERGPYPIE   Cormous geophyte to 90 cm. Leaves lanceolate to linear, 3–6-veined. Flowers tubular, yellowish pink to red, dorsal tepal largest, arising below lower, lower tepals often with dark streaks, tube 25–30 cm. Nov.–Apr. Rocky sandstone slopes, NW, SW, AP, KM, LB, SE (Bokkeveld Mts to Port Elizabeth).•

burchellii (N.E.Br.) Goldblatt (= *Anapalina burchellii* (N.E.Br.) N.E.Br.)   Cormous geophyte, 50–90 cm. Leaves linear, 3-veined, narrowed below into a petiole, upper leaves filiform, brown. Flowers tubular, scarlet, upper tepals larger, tube 30–40 mm. Feb.–Apr. Rocky sandstone slopes, 200–600 m, NW, SW, AP, LB (Riebeek-Kasteel to Albertinia).•

caffra (Ker Gawl. ex Baker) Goldblatt (= *Anapalina caffra* (Ker Gawl. ex Baker) G.J.Lewis)   Cormous geophyte, 20–80 cm. Leaves sword-shaped to linear, 2–4-veined. Flowers tubular, red, dorsal tepal largest, elongate-spathulate, arising 6–8 mm beyond lower, tube 20–30 mm. Sept.–Dec. Sandstone slopes, LB, SE (Heidelberg to East London).

intermedia (Baker) Goldblatt (= *Anapalina intermedia* (Baker) G.J.Lewis)   Like **T. caffra** but leaves shorter than stem, lower tepals with purple-black markings at sinuses and dorsal tepal arising 3–5 mm beyond lower. Sept.–Dec. Rocky sandstone slopes, SE (Uniondale to Port Elizabeth).•

lesliei L.Bolus   Cormous geophyte, 35–65 cm. Leaves linear, 1- or 2-veined, stems usually branched. Flowers actinomorphic, bright red, tube elongate, 20–25 mm, anthers short-apiculate. Feb.–Apr. Sandstone rocks along streams, NW (Skurfdeberg: Ceres).•

pulchra (Baker) Goldblatt (= *Anapalina pulchra* (Baker) N.E.Br.)   Cormous geophyte, 25–50 cm. Leaves linear to lanceolate, 2- or 3-veined. Flowers dark rose, tepals nearly equal, tube elongate, tube 30–33 mm,. Feb.–June. Rocky sandstone slopes, SW, AP (Hottentots Holland to Agulhas coast).•

triticea (Burm.f.) Goldblatt (= *Anapalina triticea* (Burm.f.) N.E.Br.)   Like **T. burchellii** but flowers smaller, tube 25–30 mm and tepals subequal, to 12 mm. Mainly Feb.–Apr. Rocky granite and sandstone slopes, NW, SW, LB (Cape Peninsula and Porterville Mts to Mossel Bay).•

williamsiana Goldblatt   Like **T. caffra** but leaves and tunics softer and not forming a neck, dorsal tepal arising 2–3 mm beyond lower, capsules small and verrucose, to 10 mm and seeds flattened. Dec.–Jan. Marshes on peaty sandstone soil, SW (Hermanus Mts).•

*A. Flowers pink, cream or yellow (see also **T. antholyza**)*
B. Perianth tube 20–70 mm long

flexuosa (L.f.) G.J.Lewis   Cormous geophyte, 15–30 cm. Leaves ovate, 2-veined, pseudopetiolate, margins undulate. Flowers pink, tube elongate, 35–40 mm, anthers long-apiculate, bracts 25–35 mm, dry and brown. Jan.–Feb. Clay slopes in renosterveld, SW, AP (Shaw's Mts to Agulhas flats).•

nervosa (Baker) G.J.Lewis   Cormous geophyte, 50–100 cm. Leaves sword-shaped, 2- or 3-veined. Flowers pale yellow to cream or white, tube elongate, 30–40 mm, fragrant. Dec.–Jan. Rocky mountain slopes, NW (Pakhuis Mts to Ceres).•

revoluta (Burm.f.) Goldblatt (= *Tritoniopsis apiculata* (F.Bolus) G.J.Lewis)   Cormous geophyte, 25–40 cm. Leaves sword-shaped, 3–5-veined. Flowers pink with darker markings, tube elongate, 40–70 mm, anthers long-apiculate, bracts large and brown, 20–50 mm. Mar.–May. Stony soils, coastal to 1000 m, AP, KM, LB (Potberg, Langeberg and Swartberg Mts).•

B. Perianth tube to 15 mm long

caledonensis (R.C.Foster) G.J.Lewis   Like **T. unguicularis** but leaves narrower, linear, 1-veined, 1–2 mm wide and filaments shorter, 3–5 mm. Nov. Rocky sandstone slopes, SW (Houwhoek to Shaw's Mts).•

dodii (G.J.Lewis) G.J.Lewis   Like **T. ramosa** but never branched and bracts obtuse, rigid when dry. Feb.–Apr. Sandstone slopes, SW, AP (Cape Peninsula to Agulhas flats).•

elongata (L.Bolus) G.J.Lewis   Like **T. dodii** and **T. ramosa** but flower spike very long and elongate and bracts conspicuously apiculate. Mar.–Apr. Sandy loam at low altitudes, SW (Paarl, Caledon).•

lata (L.Bolus) G.J.Lewis   Like **T. ramosa** but more robust and rarely branched, dorsal tepal larger, 7–12 mm wide and longer than upper laterals. Mar.–May. Sandstone mountain slopes, SW (Bainskloof to Hermanus).•

latifolia G.J.Lewis   Like **T. ramosa** but stouter, flowers pale pink to cream and leaves lanceolate, 4–6-veined. Dec.–Jan. Rocky sandstone slopes, NW (N Cedarberg Mts).•

nemorosa (E.Mey. ex Klatt) G.J.Lewis   Cormous geophyte, 1–1.5 m, stems often branched. Leaves sword-shaped, 5–7-veined. Flowers yellow with brown markings, tube c. 8 mm, anthers shortly apiculate. Nov.–Jan. Rocky slopes in fynbos, NW (Cedarberg Mts to Elandskloof).•

parviflora (Jacq.) G.J.Lewis   Cormous geophyte, 15–40 cm. Leaves linear, 1- or 2-veined. Flowers yellow with brown to maroon markings, tube c. 4 mm. Nov.–Jan. Rocky sandstone slopes, NW, SW, AP (Citrusdal to Agulhas).•

pulchella G.J.Lewis   Like **T. ramosa** but tube c. 15(–20) mm, tepals more or less truncate and upper 3 subequal in width and wider than the 3 lower. Dec.–Feb. Sandstone slopes, SW (Bainskloof Mts).•

ramosa (Eckl. ex Klatt) G.J.Lewis   Cormous geophyte, 15–50 cm, stems usually with slender branches. Leaves linear to lanceolate, 1–3-veined. Flowers deep pink with dark markings, tube 7–10 mm, anthers shortly apiculate. Jan.–Apr. Sandstone slopes, NW, SW, LB, SE (Gifberg to Humansdorp).•

unguicularis (Lam.) G.J.Lewis   Cormous geophyte, 20–55 cm. Leaves linear-lanceolate, 2- or 3-veined. Flowers small, 12–15 mm, tube c. 3 mm, cream to yellowish with mauve markings, fragrant. Dec.–Mar. Rocky sandstone slopes, SW (Cape Peninsula to Elim).•

WATSONIA   WATSONIA, KANOLPYPIE   51 spp., Namaqualand to N Province, mainly W Cape

A. Flower tube short or long, gradually flaring above
*B. Stamens declinate (see also **W. borbonica**)*

amabilis Goldblatt   Cormous geophyte, 15–50 cm. Leaves sword-shaped. Flowers pink with dark markings in throat, stamens declinate. Oct.–Nov. Granite slopes in renosterveld, SW (Paarl to Sir Lowry's Pass).•

dubia Eckl. ex Klatt   Cormous geophyte, 25–40 cm. Leaves sword-shaped. Flowers pink, tube elongate, slender, stamens declinate, bracts long, becoming dry and lacerated. Capsules oblong. Oct.–Nov. Granite and clay slopes in renosterveld, NW, SW (Citrusdal to Wellington).•

laccata (Jacq.) Ker Gawl.   Dwarf cormous geophyte, 30–40 cm. Leaves sword-shaped. Flowers pink to orange or white, sometimes with pink keels, stamens declinate. Sept.–Nov. Sandstone slopes in fynbos, NW, SW, AP, SE (Kleinmond to Humansdorp).•

strictiflora Ker Gawl.   Cormous geophyte, 25–45 cm. Leaves sword-shaped. Flowers pink with dark markings in throat, tube elongate, stamens declinate. Capsules ovoid-oblong. Nov.–Dec. Mainly sandstone outcrops, SW (Durbanville to Klapmuts).•

B. Stamens arcuate or erect

borbonica (Pourr.) Goldblatt (= *Watsonia ardernei* J.W.Mathews & L.Bolus, *W. pyramidata* (Andrews) Klatt; incl. **W. rogersii** L.Bolus)   Robust cormous geophyte, 50–200 cm, stems often purple. Leaves sword-shaped. Flowers purple-pink, rarely white, tube short, stamens declinate or arcuate. Capsules ovoid-oblong. Oct.–Jan. Mainly rocky sandstone slopes, also granite and clay, 100–1500 m, NW, SW (Tulbagh to Bredasdorp).•

distans L.Bolus   Cormous geophyte, 15–35 cm, slender. Leaves sword-shaped. Flowers deep pink, short-tubed. Nov.–Dec. Mountain marshes, SW (Franschhoek to Kleinmond).•

elsiae Goldblatt   Cormous geophyte, 40–65 cm. Leaves sword-shaped, margins thickened. Flowers scarlet, stamens included in tube. Capsules ovoid. Mainly Nov.–Dec. Stony sandstone slopes, SE (Uniondale to Joubertina).•

emiliae L.Bolus   Cormous geophyte, 40–60 cm. Leaves sword-shaped. Flowers pale pink. Capsules ovoid. Nov.–Dec. Rocky sandstone slopes, KM, LB (Swartberg and Langeberg Mts: Riversdale).•

galpinii L.Bolus   Cormous geophyte to 1.5 m, spikes dense. Leaves sword-shaped. Flowers dark red. Capsules obovoid. Dec.–May. Streamsides, KM, SE (Swartberg and Outeniqua Mts).•

marginata (L.f.) Ker Gawl. Cormous geophyte, 50–200 cm. Leaves broad with thick hyaline margins and midribs. Flowers actinomorphic, pink, occasionally white or purple. Sept.–Dec. Sandy and granitic soils, often damper sites, NW, SW (Bokkeveld Mts to Hermanus).•

stenosiphon L.Bolus Cormous geophyte, mostly 20–35 cm, slender. Leaves sword-shaped. Flowers bright orange, tube elongate, slender. Capsules cylindric. Sept.–Oct. Sandy coastal flats, SW (Hermanus to Potberg).•

A. Flower tube long, slender below but widening abruptly into a broadly cylindrical upper part
C. Floral bracts short, 5–14 mm long (see also W. aletroides)

angusta Ker Gawl. ROOIKANOLPYPIE Cormous geophyte to 1.2 m. Leaves sword-shaped. Flowers scarlet, tube elongate. Capsules attenuate. Mainly Nov.–Jan. Montane marshes and streambanks in fynbos, NW, SW, AP, LB, SE (Cedarberg Mts to S KwaZulu-Natal).

fourcadei J.W.Mathews & L.Bolus SUURKANOL Cormous geophyte to 2 m. Leaves sword-shaped. Flowers mostly orange to red, sometimes pink or purple, red or orange, tube elongate, bracts short. Capsules tapering. Nov.–Jan. Rocky sandstone slopes, NW, SW, LB, SE (Cedarberg Mts to Port Elizabeth).•

knysnana L.Bolus Cormous geophyte to 1.6 m, robust. Leaves sword-shaped. Flowers mostly pink to purple, tube elongate. Capsules obovoid. Nov.–Jan. Sandstone slopes in fynbos, LB, SE (Mossel Bay to East London).

marlothii L.Bolus Cormous geophyte, 60–120 cm. Leaves sword-shaped, with thick hyaline margins and midrib. Flowers mainly red to pink, tubes elongate. Capsules globose. Mainly Nov.–Jan. Rocky sandstone slopes, KM (Swartberg Mts).•

minima Goldblatt Cormous geophyte, 10–45 cm. Leaves few, sword-shaped. Flowers orange-red, tube elongate. Capsules obovoid-truncate. Nov.–Dec. Seasonally waterlogged rocky sandstone slopes, SW (Riviersonderend Mts).•

paucifolia Goldblatt Cormous geophyte, 20–45 cm. Leaves few, narrow. Flowers pink, tubes elongate. Capsules ovoid. Nov.–Jan. Rocky sandstone slopes, 500–1200 m, NW, SW (Tulbagh Mts to Greyton).•

stokoei L.Bolus Cormous geophyte to 1 m. Leaves sword-shaped. Flowers in a lax spike, small, red, tube elongate. Capsules obovoid. Nov.–Jan. Sandstone soils in seeps and marshes, NW (Gifberg to Cold Bokkeveld).•

tabularis J.W.Mathews & L.Bolus Cormous geophyte to 1.5 m. Leaves sword-shaped, cauline leaves inflated. Flowers orange or pink, tube elongate. Capsules oblong-truncate. Nov.–Dec. Rocky sandstone soils, 50–400 m, SW (Cape Peninsula).•

wilmaniae J.W.Mathews & L.Bolus Cormous geophyte, 80–150 cm. Leaves sword-shaped. Flowers orange, red or purple, tube elongate, slender. Capsules obovoid. Nov.–Jan. Rocky sandstone soils, often wet sites, KM, SE (Ladismith to Knysna).•

zeyheri L.Bolus Cormous geophyte, 50–120 cm. Leaves sword-shaped. Flowers bright orange, tube elongate. Capsules fusiform. Nov.–Jan. Marshes on sandstone, coastal to 100 m, SW, AP (Cape Peninsula to Agulhas coast).•

C. Floral bracts longer, 15–45 mm long (but see also W. fourcadei, W. knysnana, W. zeyheri)

aletroides (Burm.f.) Ker Gawl. Cormous geophyte to 45 cm. Leaves sword-shaped. Flowers red, sometimes pinkish, nodding on a recurved tube, tepals short, barely spreading. Capsules fusiform-attenuate. Sept.–Oct. Clay slopes, mainly renosterveld, SW, AP, LB, SE (Caledon to Knysna).•

coccinea Herb. ex Baker Cormous geophyte, 14–40 cm. Leaves sword-shaped. Flowers purple, pink or scarlet, tubes elongate. Capsules oblong. Aug.–Nov. Sandstone flats and plateaus, SW, AP (Malmesbury to Bredasdorp).•

fergusoniae L.Bolus Cormous geophyte to 80 cm. Leaves sword-shaped. Flowers bright orange-red, tube elongate. Capsules oblong-cylindric. Oct.–Nov. Limestone outcrops, AP (Agulhas to Albertinia).•

humilis Mill. (= *Watsonia roseoalba* Ker Gawl.) Dwarf cormous geophyte, 15–40 cm. Leaves sword-shaped. Flowers pale pink to white, tube elongate, bracts imbricate. Sept.–Nov. Sandstone or granite flats, SW (Malmesbury to Gordon's Bay).•

hysterantha J.W.Mathews & L.Bolus Cormous geophyte, 50–90 cm, with slender stems. Leaves dry at flowering, linear with prominent margins. Flowers red, tube elongate. Capsules ovoid. Apr.–July. Coastal granite outcrops, SW (Saldanha to Langebaan).•

meriana (L.) Mill. (= *Watsonia bulbillifera* J.W.Mathews & L.Bolus) LAKPYPIE, WASPYPIE Cormous geophyte, 60–200 cm, sometimes with cormlets at nodes. Leaves sword-shaped. Flowers red, pink or mauve, tube elongate. Capsules oblong. Sept.–Nov. Sandy or granitic soils, often vleis and streambanks, NW, SW, AP (Namaqualand to Bredasdorp).

pillansii L.Bolus (= *Watsonia beatricis* J.W.Mathews & L.Bolus) Cormous geophyte, 50–120 cm. Leaves sword-shaped. Flowers scarlet, tube elongate. Capsules obovoid. Nov.–Jan. Sandy soils mostly at low elevations, SE (George to central KwaZulu-Natal).

schlechteri L.Bolus Cormous geophyte, 40–100 cm. Leaves sword-shaped, with very thick margins. Flowers scarlet, tube elongate. Capsules obovoid. Nov.–Feb., mainly after fire. Rocky sandstone slopes in fynbos, NW, SW, AP, KM, LB, SE (Citrusdal to Kouga Mts).•

spectabilis Schinz Cormous geophyte, 25–50 cm. Leaves sword-shaped. Flowers large, scarlet, tube elongate. Capsules obovoid to globose. Aug.–Nov. Sandy flats and plateaus, often near water, NW, SW, AP (Bokkeveld Mts to Potberg).•

vanderspuyiae L.Bolus Cormous geophyte, 1–2 m, robust. Leaves broad, with thick margins. Flowers dark red, tube elongate. Capsules oblong. Sept.–Nov. Sandstone outcrops, NW (Cedarberg Mts to Ceres and Piketberg).•

versfeldii J.W.Mathews & L.Bolus Cormous geophyte, 1–2 m, robust. Leaves broad. Flowers pink purple, tube elongate. Capsules obconic. Oct.–Nov. Sandstone slopes, NW (Piketberg and Porterville Mts).•

WITSENIA• BOKMAKIERIESTERT, WAAIERTJIE 1 sp., W Cape

maura Thunb. Slender woody shrub to 2 m. Leaves narrowly lanceolate. Flowers paired in pseudoracemes, tubular, tepals connivent, dark green with velvety yellow tips. Mainly Apr.–Aug. Marshy coastal flats and mountain slopes, to 1250 m, SW, LB (Cape Peninsula to Swellendam).•

XENOSCAPA FEEPYPIE 2 spp., W Cape to Namaqualand, W Karoo

fistulosa (Spreng. ex Klatt) Goldblatt & J.C.Manning (= *Anomatheca fistulosa* (Spreng. ex Klatt) Goldblatt) Cormous geophyte, 3–20 cm. Leaves ovate, prostrate. Flowers small, white, long-tubed. Aug.–Oct. Damp clay soil, sometimes on granite, NW, SW, KM (S Namibia and W Karoo to Cape Peninsula and Little Karoo).

JUNCACEAE

JUNCUS c. 250 spp., cosmopolitan

A. Annuals

*****bufonius** L. TOADRUSH Soft annual, 15–30 cm. Leaves linear. Flowers in leafy, branched spikes, green. Oct.–Jan. Disturbed areas, NW, SW, AP, LB, SE (cosmopolitan).

cephalotes Thunb. Tufted annual, 6–15 cm. Leaves linear. Flowers in solitary capitula, streaked with black or brown. Sept.–Nov. Damp flats, NW, SW (Namaqualand to Swellendam).

obliquus Adamson Similar to **J. rupestris** but capitula usually solitary, stamens 3 and style short. Sept.–Oct. Streamsides, NW (Cold Bokkeveld Mts).•

rupestris Kunth Small, delicate, tufted annual, 6–16 cm. Leaves filiform. Flowers in branched capitula, yellow. Oct.–Nov. Damp sandy places, NW, SW (Gifberg to Cape Peninsula).•

scabriusculus Kunth Low, tufted, scabrid annual, 6–12 cm. Leaves linear. Flowers in branched capitula, yellow-brown. Nov.–Dec. Marshes, NW, SW, AP (Cape Peninsula to Bredasdorp).•

A. Perennials

acutus L. Hard, tufted perennial to 2 m. Leaves cylindric. Flowers in pseudolateral panicles, red to brown. Sept.–Jan. Brackish marshes, NW, SW, KM, LB, SE (pantemperate).

capensis Thunb. Tufted perennial, 6–60 cm. Leaves filiform to linear. Flowers in cymes, pale with dark keels. Nov.–Apr. Damp flats and lower slopes, NW, SW, AP, KM, SE (Clanwilliam to E Cape).

*****capillaceus** Lam. Tufted perennial to 30 cm. Leaves filiform, ribbed. Flowers in pseudolateral spikes, green. Mar.–Apr. Disturbed areas, SW (South American weed).

dregeanus Kunth Similar to **J. capensis** but style vestigial. Flowers dark brown. Nov.–Dec. Marshes, SW, LB, SE (Darling to East Africa).

effusus L.   Tufted perennial to 1 m. Leaves cylindrical, stem-like. Flowers in tight pseudolateral clusters, green. Dec.–Mar. Streambeds or marshes, SW, LB, SE (Cape Peninsula to E Cape, cosmopolitan).

exsertus Buchenau   BIESIE   Similar to **J. oxycarpus** but capsule cylindrical and exserted from perianth. Nov. Streamsides or marshes, SW, KM, LB, SE (Malmesbury to S tropical Africa).

*imbricatus Laharpe   Similar to **J. capillaceus** but to 60 cm and more sclerotic. Nov.–Mar. Disturbed areas, SW (South American weed).

kraussii Hochst.   RUSH, BIESIE   Hard dark green tufted perennial to 1.2 m. Leaves cylindrical, stem-like. Flowers in pseudolateral panicles, brown. Sept.–Jan. Saline marshes, SW, AP, LB, SE (Cape Peninsula to Mozambique, Australia, S America).

lomatophyllus Spreng.   Tufted perennial, 30–80 cm. Leaves strap-shaped, soft. Flowers in cymes, dark brown. Oct.–Apr. Streamsides, marshes and seeps, NW, SW, AP, LB, SE (Clanwilliam to Zimbabwe).

oxycarpus E.Mey. ex Kunth   Tufted perennial with soft stems, 40–80 cm. Leaves cylindrical, septate. Flowers in round capitula in pseudolateral panicles, brown. Oct.–Jan. Streamsides and marshes, NW, SW, LB, SE (Gifberg to Eritrea).

punctorius L.f.   Hard, tufted, pale green perennial to 1.5 m. Leaf solitary per flowering stem, cylindrical, septate. Flowers in round capitula in pseudolateral panicles, brown. Oct.–Feb. Freshwater marshes, NW, SW, AP, LB, SE (Namibia to Cape Peninsula to N Province, also N Africa, Eurasia).

stenopetalus Adamson   Similar to **J. cephalotes** but capsule with a long beak. Oct. Damp sandy lower slopes, NW, SW (Tulbagh to Cape Peninsula).•

JUNCAGINACEAE

TRIGLOCHIN   c. 14 spp., cosmopolitan

bulbosa L.   Tufted, rhizomatous perennial, 5–50 cm, sometimes with fibrous tunics. Leaves green or emergent at flowering, terete to filiform. Flowers green. Fruits tapering-conical, 5–10 mm long. mostly July–Nov. damp or marshy places, NW, SW, AP, KM, LB, SE (Bokkeveld Mts to tropical Africa and Mediterranean).

striata Ruiz & Pav.   Like **T. bulbosa** but leaves more distichous, flowers in dense spiral whorls and fruits globose, 2 mm long. Mostly July–Dec. Marshes and seeps, NW, SW, LB, SE (Clanwilliam to E Cape and more or less worldwide).

LANARIACEAE

LANARIA   CAPE EDELWEISS, PERDEKAPOK   1 sp., W to E Cape

lanata (L.) T.Durand & Schinz   Evergreen rhizomatous perennial, 30–80 cm. Leaves tufted, narrow, channelled, fibrotic, margins serrate. Flowers in white-woolly panicles, small, mauve. Nov.–Jan. clay and sandstone slopes, SW, LB, AP, SE (Bainskloof to E Cape).

NAJADACEAE

NAJAS   35–50 spp., cosmopolitan

marina L.   submerged aquatic annual to 60 cm, stems prickly. Leaves narrow, serrate. Flowers minute. Flowering time? Brackish water, SE (George to tropical Africa, Eurasia, Australia).

ORCHIDACEAE
by H. Kurzweil

1. Plants epiphytic or lithophytic with slender, smooth white or grey roots on surface of substrate; leaves distichous, leathery, jointed basally; pollinia hard and waxy:
   2. Plants sympodial, with pseudobulbs of several internodes; inflorescence terminal; flowers not resupinate ....**Polystachya**
   2.' Plants monopodial, without pseudobulbs; inflorescence axillary; flowers resupinate:
      3. Rostellum deeply notched ....**Angraecum**
      3.' Rostellum beaked:
         4. Viscidia 2 ....**Mystacidium**
         4.' Viscidia 1:
            5. Lip simple; pollinia attached to separate stalks (stipes) ....**Cyrtorchis**
            5.' Lip 3-lobed; pollinia attached to a single stalk (stipes) ....**Tridactyle**

1.' Plants normally terrestrial with roots subterranean and often tuberous; leaves various; pollinia various:
  6. Pollinia under a deciduous anther cap:
    7. Saprophytic plants with reduced, scale-like leaves; flowers nodding ....**Gastrodia**
    7.' Eutrophic plants with well-developed green leaves:
      8. Inflorescence terminal:
        9. Leaves few to numerous, linear, distichous, leathery, scattered on an unthickened stem .....**Acrolophia**
        9.' Leaves 2 or 3, ovate to elliptic, flat and often leathery, on underground or above-ground pseudobulbs ....**Liparis**
      8.' Inflorescence lateral:
        10. Spur thread-like, 20–40 mm long; pollinia 8, rather soft; lip and gynostemium extensively united basally ....**Calanthe**
        10.' Spur absent or short and broad, to 9 mm long; pollinia 2, hard; lip free from gynostemium ...**Eulophia**
  6.' Pollinia in 2 persistent thecae with longitudinal slits:
    11. Lip with a single basal spur; sepals never spurred; anther usually erect:
      12. Stigma concave:
        13. Leaves several, lanceolate, cauline; plants glabrous ....**Brachycorythis**
        13.' Leaves 1 or 2, orbicular, adpressed to the ground; plants mostly variously pubescent:
          14. Leaves 2; flowers few to many, small, with lip entire or lacerated ....**Holothrix**
          14.' Leaf 1; flowers normally solitary, large, with lip deeply lacerated..............................**Bartholina**
      12.' Stigma convex, with 2 lobes or arms:
        15. Flowers pink; petals free, simple, concealing gynostemium; lip spotted ....**Stenoglottis**
        15.' Flowers green or green and white; petals adhering to median sepal, often bilobed; lip unmarked:
          16. Lip united with base of stigmatic arms; central rostellum lobe deeply concave, helmet shaped ....**Bonatea**
          16.' Lip free from stigma; central rostellum lobe flat or folded.....................................**Habenaria**
    11.' Lip with paired spurs or without spurs; sepals sometimes with a single spur; anther usually reflexed:
      17. Gynostemium slender and elongated; petals and sepals similar:
        18. Lip scarcely different from petals and never spurred ....**Pachites**
        18.' Lip hooded and with 2 spurs (sometimes reduced to sacs or rarely absent) ....**Satyrium**
      17.' Gynostemium short and squat; petals and sepals usually dissimilar:
        19. Petals free from median sepal; lip not adpressed to gynostemium, usually spreading, never with an appendage:
          20. Lip complex with an ascending basal part, flat middle part, and tooth-like apex; stem slender, wiry and often flexuose..**Schizodium**
          20.' Lip simple, rarely fimbriate or spathulate; stem seldom wiry..................................**Disa**
            a. Viscidium 1; lip fleshy; rostellum unlobed; spur pendent ....*Monadenia* group
            a.' Viscidia 2, sometimes connate; lip not fleshy; rostellum 3-lobed; spur various:
              b. Lip spathulate, elliptic tic-ovate, often fringed; leaves grass-like, usually dry at flowering; petals falcate..*Herschelia* group
              b.' Lip mostly linear or lorate; leaves various, mostly green at flowering..................*Disa* group
        19.' Petals normally connate with median sepal to form a hood; lip erect at base and adpressed to gynostemium, mostly fused with it, a lip appendage mostly present:
          21. Lateral sepals spurred or saccate...**Disperis**
          21.' Lateral sepals never spurred or saccate:
            22. Median sepal spurred; lip erect, minute (about a third of gynostemium length) ....**Brownleea**
            22.' Median sepal normally not spurred; lip never erect and minute:
              23. Lip blade anchor-shaped, rarely rhomboid or kidney-shaped; leaves numerous, cauline, linear or narrowly lanceolate; lip appendage present or absent:
                24. Lip appendage absent or 2-lobed ....**Ceratandra**
                24.' Lip appendage entire and elongate..**Evotella**
              23.' Lip blade not anchor-shaped; leaves 1-many, elliptic to lanceolate, rarely linear; lip appendage generally present:
                25. Median sepal and petals forming a deeply globose hood; lip appendage a shield on top of gynostemium, often with lateral processes ....**Corycium**
                25.' Median sepal and petals forming a shallow to flat hood; lip appendage elongate, solid or funnel-shaped ....**Pterygodium**

ACROLOPHIA   CINDERELLA ORCHID   7 spp., W and E Cape

A. Anther cap horned

capensis (P.J.Bergius) Fourc.   Robust geophyte to 80 cm, with short rhizome. Leaves linear, channelled, leathery. Flowers in a lax, branched raceme, c. 10–15 mm long, green to purplish brown, lip white with purple callus and side lobes, narrowly obovate, 3-lobed, with 5–7 rows of papillae, spur 1.5–2.5 mm long, anther cap with 2 prominent lateral horns. Dec.–Jan. Rocky outcrops and slopes, NW, SW, AP, KM, LB, SE (Ceres to Grahamstown).

lamellata (Lindl.) Schltr. & Bolus   Like **A. capensis** but flowers larger, sepals 13–20 mm long. Mainly Oct. Sandy coastal flats, SW (Darling to Bredasdorp).•

A. Anther cap not horned

barbata (Thunb.) H.P.Linder (= *Acrolophia lunata* (Schltr.) Schltr. & Bolus)   Robust rhizomatous geophyte to 1 m. Leaves linear, channelled, leathery. Flowers few to many in a dense raceme, white or pale rose, c. 12 mm long, lip oblanceolate, 3-lobed, with 5 rows of papillae, spur 3 mm long. Nov.–Dec. Sandstone slopes, LB, SE (Swellendam to Humansdorp).•

bolusii Rolfe   Robust rhizomatous geophyte to 80 cm. Leaves linear, channelled, leathery. Flowers many in a lax, branched raceme, c. 10 mm diam., not resupinate, brownish, lip dull yellow, broadly obovate, often recurved, disc with slender tall papillae, spur conical, 1 mm long. Oct.–Dec. Sandy coastal flats, SW, AP (Hopefield to Bredasdorp).•

cochlearis (Lindl.) Schltr. & Bolus   Robust rhizomatous geophyte to 1 m. Leaves linear, channelled, leathery. Flowers many in a branched raceme, c. 7 mm long, not resupinate, brownish, lip yellow, obovate, obscurely 3-lobed, disc densely papillate, spur saccate, 1.5 mm long. Sept.–Dec. Rocky grassland, SW, AP, LB, SE (Bredasdorp to KwaZulu-Natal).

micrantha (Lindl.) Schltr. & Bolus   Slender rhizomatous geophyte to 60 cm. Leaves linear, channelled, leathery. Flowers many in a lax, branched raceme, c. 7 mm diam., purplish green or brown, lip white, distinctly 3-lobed, disc with 3 rows of papillae, spur 1.5 mm long. Sept.–Dec. Coastal forelands, SW, AP, SE (Cape Peninsula to East London).

ustulata (Bolus) Schltr. & Bolus   BLACK ORCHID   Slender rhizomatous geophyte to 10 cm. Leaves linear-lanceolate, lightly channelled, leathery. Flowers few in a lax raceme, dark maroon or greenish yellow, 7–8.5 mm long. Nov.–Dec., after fire. Fynbos, SW, AP, LB, SE (Cape Peninsula to George).•

ANGRAECUM   SHELL ORCHID   c. 200 spp., mainly tropical Africa and Madagascar

conchiferum Lindl.   Monopodial epiphyte to 30 cm, often branched. Leaves 8–16, narrowly lanceolate. Flowers solitary in axils, white, sepals 25–30 mm long, lip broadly ovate-acuminate, spur 30–45 mm long. Sept.–Jan. Montane forests, SE (George to tropical Africa).

pusillum Lindl.   Monopodial epiphyte to 2.5 cm. Leaves 5–10, grass-like, linear. Flowers few in a lax axillary raceme, to 5 mm diam., white, sepals 1.5 mm long, spur saccate, 1 mm long. Throughout the year. Temperate forests, LB, SE (Swellendam to N Province).

sacciferum Lindl.   Monopodial epiphyte to 4 cm. Leaves 4–8, linear-lorate. Flowers few in a lax axillary raceme, lime-green, sepals ovate to oblong, 3–4 mm long, lip hooded, spur 2 mm long. Dec.–Mar. Montane forests near the coast, SE (Knysna to tropical Africa).

BARTHOLINA   SPIDER ORCHID   2 spp., S Namibia to E Cape

burmaniana (L.) Ker Gawl.   Slender tuberous geophyte to 23 cm, scape hairy. Leaf single, basal, ovate, prostrate, hairy. Flower usually solitary, white, lip much-divided with segments acute, spur 8–12 mm long. Aug.–Oct., usually after fire. Clay slopes and flats, NW, SW, AP, LB, SE (Clanwilliam to E Cape).

etheliae Bolus   Like **B. burmaniana** but lip segments clavate at tips. Mainly Oct.–Dec. Sandy slopes and flats, NW, SW, KM, LB, SE (S Namibia to E Cape).

BONATEA   PHANTOM ORCHID   20 spp., Africa and Arabia

cassidea Sond.   Tuberous geophyte to 50 cm. Leaves scattered, linear-lanceolate. Flowers few to many in a lax raceme, green and white, to 40 mm long, petals bilobed, lip 3-fid, spur 11–25 mm long, stigmas c. 6 mm long. Aug.–Oct. Bush and scrub, SE (Uitenhage to Zimbabwe).

speciosa (L.f.) Willd.   Robust tuberous geophyte to 1 m. Leaves oblong to broadly lanceolate. Flowers usually many in a lax to dense raceme, usually many-flowered, green and white, petals bilobed, lip 3-fid, spur 25–47 mm long, stigmas c. 25 mm long. June–Feb. Coastal scrub and forest margins, SW, AP, LB, SE (Yzerfontein to Zimbabwe).

BRACHYCORYTHIS   HELMET ORCHID   c. 32 spp., Africa and Asia

macowaniana Rchb.f.   Tuberous geophyte to 20 cm. Leaves narrowly lanceolate, imbricate. Flowers many in a dense raceme, brownish and green flowers, lateral sepals c. 4 mm long, lip with a spur c. 2 mm long. Nov.–Dec., after fire. Grassland or restio veld, LB, SE (Swellendam to E Cape).

BROWNLEEA FALSE DISA   7 spp., southern and tropical Africa, Madagascar

recurvata Sond.   Slender tuberous geophyte to 50 cm. Leaves 2 or 3, linear to narrowly lanceolate. Flowers few in a lax raceme, c. 10 mm diam., white to pale pink with darker spots, lip c. 1 mm long. Feb.–Apr. Mainly stony slopes or flats, SE (Knysna to Mpumalanga).

CALANTHE FAIRY ORCHID   c. 200 spp., mainly Asia

sylvatica (Thouars) Lindl.   Robust rhizomatous perennial to 65 cm, with pseudobulbs largely obscured by leaves. Leaves lanceolate, pleated. Flowers many in a shortly hairy raceme, white to pink, lip 3-lobed with a callus of 3 small ridges, spur slender, 15–30 mm long. Dec.–Mar. Temperate or submontane forests, SE (Knysna to tropical Asia).

CERATANDRA SHIELD ORCHID   6 spp., W and E Cape

A. Lip appendage absent

atrata (L.) T.Durand & Schinz   Tuberous geophyte to 35 cm, with thickened roots and small tubers. Leaves linear, cauline leaves longer. Flowers many in a lax raceme, greenish yellow, not resupinate, lateral sepals 11–12 mm long, lip anchor-shaped, appendage absent, rostellum arms horn-like, projecting, 7–8 mm long. Oct.–Dec., after fire. Marshes or rock flushes and streambanks, NW, SW, KM, LB, SE (Cold Bokkeveld to Knysna).•

globosa Lindl.   Like **C. atrata** but flowers many in a corymbose raceme, sepals pale purplish green to pink, petals and lip white, lateral sepals c. 5–7 mm long, lip spathulate, appendage absent, rostellum arms c. 4 mm long. Nov.–Jan., mostly after fire. Sandstone slopes, NW, SW, KM, LB, SE (Cedarberg to Baviaanskloof Mts).•

grandiflora Lindl.   Like **C. atrata** but flowers many in a corymbose raceme, orange-yellow, not resupinate, lateral sepals 10–12 mm, lip spathulate-cordate, appendage absent, rostellum arms c. 6 mm long. Mainly Oct.–Dec., after fire. Damp sandstone flats, AP, SE (Gourits River to Grahamstown).

A. Lip appendage present

bicolor Sond. ex Bolus (= *Evota bicolor* (Sond. ex Bolus) Rolfe)   Slender tuberous geophyte to 35 cm, with thickened roots. Leaves linear-lanceolate, cauline leaves lax. Flowers 1–9 in a lax raceme, sepals green, petals and lip yellow, lateral sepals 11–13 mm long, lip spathulate-lunate, appendage a pair of horns 6–9 mm long. Nov.–Dec., after fire. Sandstone flats, NW, SW (Cedarberg to Riviersonderend Mts).•

harveyana Lindl. (= *Evota harveyana* (Lindl.) Rolfe)   Like **C. bicolor** but flowers 2–many in a lax to dense raceme, sepals pale green, petals and lip yellow, lateral sepals 9–12 mm long, lip anchor-shaped, appendage a pair of fleshy wings. Nov.–Dec., after fire. Sandstone slopes, SW (Cape Peninsula to Kleinrivier Mts).•

venosa (Lindl.) Schltr. (= *Evota venosa* (Lindl.) Schelpe)   Tuberous geophyte to 26 cm. Leaves linear-lanceolate. Flowers many in a dense raceme, sepals green, petals and lip white with pink veins, crenulate-erose, lateral sepals c. 7 mm long, lip subrotund, crenulate, appendage a pair of erect, laterally flattened, rounded horns. Oct.–Nov., after fire. Sandstone slopes, SW (Cold Bokkeveld to Hottentots Holland Mts).•

CORYCIUM MONKSHOOD   14 spp., southern Africa, extending into tropical Africa

bicolorum (Thunb.) Sw.   Robust to slender tuberous geophyte to 40 cm. Leaves narrowly lanceolate. Flowers many in a dense raceme, c. 10 × 4 mm, greenish yellow, lateral sepals fused, lip 2-lobed, appendage forming a shield over gynostemium, with 2 small lateral lobes. Oct.–Nov., after fire. Sandy flats, SW, AP, LB, SE (Cold Bokkeveld to Mossel Bay).•

bifidum Sond.   Slender tuberous geophyte to 15 cm. Leaves short, narrowly triangular. Flowers many in a dense raceme, c. 10 × 4 mm, sepals and petals yellowish green, lip green, lateral sepals fused for most of their length, lip elliptic-deltate, appendage arching forward. Mainly Nov., after fire. Sandy flats, SW (Pringle Bay and Botrivier).•

carnosum (Lindl.) Rolfe   Slender to fairly robust tuberous geophyte to 55 cm, turning black on drying. Leaves narrowly lanceolate. Flowers many in a dense raceme, c. 18 × 7 mm, sepals green, petals pink and lip whitish pink, lateral sepals free, lip with 2 broad diverging lobes, appendage a beaked hood over gynostemium. Oct.–Jan., after fire. Mostly sandstone seeps, SW, AP, KM, LB, SE (Cape Peninsula to Port Elizabeth).•

crispum (Thunb.) Sw.   Robust tuberous geophyte to 40 cm. Leaves many, lanceolate, margins crisped. Flowers many in a fairly dense raceme, c. 20 × 5 mm, yellow with a green lip appendage, lateral sepals fused for three-quarters of their length, lip with 2 broad apical lobes, appendage with 2 deflexed lobes. Sept.–Oct. Sandy flats, NW, SW, AP, KM, LB (Namaqualand to Albertinia).

deflexum (Bolus) Rolfe   Robust tuberous geophyte to 25 cm. Leaves many, lanceolate, sometimes withered at flowering. Flowers 6–many in a lax to dense raceme, c. 12 × 6 mm, yellow with green lip appendage, lateral sepals fused to about halfway, lip broadened and 2-lobed apically, appendage shield-like, with 2 strongly deflexed lobes. Mainly Oct. Dry scrub, NW (W Karoo and Cedarberg Mts).

dracomontanum Parkman & Schelpe   Slender tuberous geophyte to 30 cm. Leaves lanceolate to linear. Flowers many in a dense raceme, c. 14 × 6 mm, sepals dry and black at flowering, petals and lip green tinged purple, lip rotund, lateral lobes of appendage oblong and not recurved. Nov.–Mar. Grassland, SE (Knysna to Mpumalanga, Malawi).

excisum Lindl.   Slender to fairly robust tuberous geophyte to 26 cm, turning black on drying. Leaves many, linear. Flowers many in a dense raceme, c. 8 × 5 mm, lime-green, lateral sepals fused except apically, lip subobdeltate, appendage shield-like with 2 obscure horizontal lobes. Nov.–Dec., after fire. Sandy flats, NW, SW, AP (Piketberg to Agulhas).•

ingeanum E.G.H.Oliv.   Fairly robust tuberous geophyte to 20 cm. Leaves c. 7, erect-spreading, lanceolate. Flowers many in a fairly dense raceme, greenish yellow but sepals membranous-brown, petals tipped reddish to black and lip bright green, lateral sepals fused, lip with 2 spreading lobes apically, appendage shield-like, with 2 lobes arching backwards. Sept. Sandy or clay flats in renosterveld, NW (Bokkeveld Plateau).•

microglossum Lindl.   Robust tuberous geophyte to 30 cm. Leaves many, linear. Flowers many in a dense raceme, c. 24 × 7 mm, sepals and petals dry, greyish brown, lip appendage green, lateral sepals fused in lower portion, lip linear-attenuate, appendage broadly elongate-ovate, arching over gynostemium. Mainly Nov., after fire. Sandy flats and slopes, SW (Cold Bokkeveld to Cape Peninsula).•

orobanchoides (L.f.) Sw. (= *Corycium vestitum* Sw.)   Slender or robust tuberous geophyte to 42 cm. Leaves many, lanceolate, barred with red below. Flowers many in a dense raceme, c. 5 × 15 mm, yellow-green with purple petal apices, lateral sepals fused, lip with 2 diverging lobes, appendage shield-like, with long horizontal lobes pointing backwards. Sept.–Oct. Sandy flats, NW, SW, AP, KM, LB (Klawer to Albertinia).•

CYRTORCHIS   BIRD ORCHID   20 spp., Africa

arcuata (Lindl.) Schltr.   Monopodial epiphyte or lithophyte with stems to 40 cm. Leaves strap-shaped, channelled. Flowers few to many in a dense axillary raceme, white, tepals subequal, recurved, narrowly lanceolate, apiculate, 20–30 mm long, spur 25–40 mm long. Jan.–May. Montane and lowland forests, SE (Knysna to tropical Africa).

DISA (= *HERSCHELIA, MONADENIA*)   DISA   162 spp., sub-Saharan Africa, Madagascar and Reunion

Disa-group

*A. Spur obsolete (see also **D. cylindrica, D. obtusa** and **D. tenuis**)*

atricapilla (Harv. ex Lindl.) Bolus   Slender or robust tuberous geophyte to 30 cm. Leaves cauline, subequal, linear-lanceolate. Flowers few to many in a corymbose raceme, median sepal white, lateral sepals red, black and white, petals and lip maroon, lateral sepals 10–17 mm long, keeled above. Nov.–Dec., mostly after fire. Seeps and moist sandstone slopes, NW, SW, KM (Cedarberg Mts to Cape Peninsula to Swartberg Mts).•

begleyi L.Bolus   Slender tuberous geophyte to 10 cm. Leaves cauline, more or less equal, lanceolate. Flowers few in a subcorymbose raceme, pale pink with darker speckles, lateral sepals oblong, 7–9 mm long. Dec.–Jan., after fire. Stony sandstone slopes at high altitude, SW (Jonkershoek to Elgin).•

bivalvata (L.f.) T.Durand & Schinz   Like **D. atricapilla** but sepals white, not keeled above and petals and lip pale to dark red. Sept.–Jan., usually after fire. Sandstone slopes and seeps, NW, SW, LB, SE (Cape Peninsula to Humansdorp).•

bodkinii Bolus   Robust tuberous geophyte, 3–25 cm. Leaves lanceolate. Flowers 1–15 in a corymbose raceme, purplish brown and black, lateral sepals narrowly ovate, c. 15 mm long. Nov., usually after fire. Slight seeps on sandstone slopes, SW, LB (Cape Peninsula to Riversdale).•

brachyceras Lindl.   Slender tuberous geophyte to 7 cm. Leaves many, spiralling, linear, reaching to raceme base. Flowers many in a cylindrical raceme, white, lateral sepals oblong, c. 2.5 mm long. Aug.–Sept. Damp sandy slopes, NW, SW (Hex River to Riviersonderend Mts).•

elegans Sond. ex Rchb.f.   Robust tuberous geophyte to 60 cm. Leaves cauline, linear-lanceolate. Flowers few in a corymbose raceme, white, petals tipped yellow and maroon, lateral sepals broadly oblong, 15-22 mm long. Nov.–Dec., usually after fire. Mountain marshes and seeps, NW, SW, KM (Cedarberg Mts to Cape Peninsula to Swartberg Mts).•

filicornis (L.f.) Thunb.   Slender or robust tuberous geophyte. Leaves dimorphic: radical leaves linear, cauline leaves lanceolate. Flowers few in a lax raceme, white to bright pink, lateral sepals oblong, 10–20 mm long. Oct.–Dec. Mostly dry sandstone slopes, NW, SW, AP, KM, LB, SE (Cedarberg Mts to Humansdorp).•

introrsa Kurzweil, Liltved & H.P.Linder   Slender tuberous geophyte to 15 cm. Leaves 3–5, cauline, linear. Flowers few in a lax raceme, not resupinate, facing inward, dull carmine-red and cream-coloured, median sepal 3 mm long. Nov.–Dec., after fire. Sandstone slopes, NW (Ceres).•

lineata Bolus   Slender tuberous geophyte to 40 cm. Leaves subimbricate, narrowly lanceolate. Flowers few in a lax raceme, pale brown, lip purple with yellow base and apex, petals with purple tips, lateral sepals oblong, c. 6 mm long. Sept.–Nov., mostly after fire. Sandstone slopes, NW, SW (Cedarberg to Kleinrivier Mts).•

maculata L.f.   Slender, subflexuose tuberous geophyte to 30 cm. Leaves in a basal rosette, elliptic to oblanceolate. Flower solitary, blue, petals with purple markings, lateral sepals obovate, 15–18 mm long. Oct.–Nov. Mossy sandstone ledges, NW, SW, KM (Piketberg to Cape Peninsula to Swartberg Mts).•

minor (Sond.) Rchb.f.   Slender tuberous geophyte to 10 cm. Leaves basal, narrowly elliptic. Flowers few in a corymbose raceme, pink, lateral sepals obovate, c. 7 mm long. Nov.–Dec. Montane marshes, NW (Grootwinterhoek to Hex River Mts).•

neglecta Sond.   Slender tuberous geophyte to 18 cm. Leaves narrowly lanceolate. Flowers many in a dense raceme, dull yellow and purple, petals strongly reduced and 1 mm long. Nov.–Dec., after fire. Sandstone slopes, NW, SW (Grootwinterhoek and Hex River Mts).•

oligantha Rchb.f.   Slender tuberous geophyte to 15 cm. Leaves narrowly oblanceolate. Flowers few in a corymbose raceme, sepals cream, petals and lip yellow from white bases, lateral sepals oblong, 6–7 mm long. Dec.–Jan. Sandstone slopes at high alt., SW (Cape Peninsula, Jonkershoek Mts).•

pillansii L.Bolus   Slender tuberous geophyte to 30 cm. Leaves mainly basal, elliptic. Flowers few in a corymbose or subcorymbose raceme, purplish pink, lateral sepals obovate, 10–14 mm long. Oct.–Dec. Streambanks and seeps, SW (Hottentots Holland Mts).•

racemosa L.f.   Slender tuberous geophyte to 1 m. Leaves basal, narrowly lanceolate. Flowers few in a lax raceme, pale pink with darker veins, median sepal dish-shaped, lateral sepals oblong, 15–25 mm long. Nov.–Dec., after fire. Sandstone seeps and marshes, NW, SW, KM, LB, SE (Cold Bokkeveld and Cape Peninsula to Grahamstown).

richardiana Lehm. ex Bolus   Slender tuberous geophyte to 30 cm. Leaves mainly basal, elliptic. Flowers few in a corymbose raceme, white, petals and lip yellow, lateral sepals obovate, 9–12 mm. Sept.–Nov. Mossy sandstone ledges, SW (Cape Peninsula to Riviersonderend Mts).•

rosea Lindl.   Flexuose tuberous geophyte to 20 cm. Leaves 2 or 3, basal, broadly ovate, purple beneath. Flowers few in a lax, subcorymbose raceme, white to rose, lateral sepals elliptic-oblong, 8–13 mm long. Nov.–Dec. Mossy sandstone ledges, NW, SW (Grootwinterhoek Mts to Cape Peninsula to Riviersonderend Mts).•

schizodioides Sond.   Slender tuberous geophyte to 25 cm. Leaves narrowly elliptic. Flower usually solitary, white to violet-mauve, lateral sepals obovate, 13–20 mm long. Dec.–Jan., often after fire. Seeps or sandstone ledges, NW, SW, KM, LB, SE (Hex River Mts to Langkloof).•

tenuifolia Sw. (= *Disa patens* (L.f.) Thunb.)   Slender tuberous geophyte to 30 cm. Leaves dimorphic: radical leaves linear, cauline leaves lanceolate. Flowers few in a lax raceme, bright yellow, lateral sepals lanceolate, 12–18 mm long. Nov.–Jan., often after fire. Mountain seeps, NW, SW, KM, LB, SE (Porterville to Riversdale).•

venosa Sw.   Like **D. racemosa** but plants smaller and median sepal narrower. Nov., mainly after fire. Seeps on sandstone slopes, NW, SW, KM, SE (Cedarberg and Cape Peninsula to Outeniqua Mts).•

virginalis H.P.Linder   Like **D. maculata** but flower not resupinate, white. Oct. Sandstone slopes and outcrops, NW, SW (Piketberg to Bainskloof).•

A. Spur well developed
B. Leaves in a basal tuft, sometimes dry at flowering (see also **D. glandulosa**)

arida Vlok   Like **D. gladioliflora** but flowers mauve-purple with blue iridescence or pink and petals oblong, minutely tridentate. Mar.–Apr. North-facing, inland sandstone slopes, KM, SE (Rooiberg and Outeniqua Mts).•

aurata (Bolus) L.Parker & H.Koopowitz (= *Disa tripetaloides* (L.f.) N.E.Br. subsp. *aurata* (Bolus) H.P.Linder)   Like **D. tripetaloides** but flowers yellow, lateral sepals 14–16 mm long and spurs 0.5–2.5 mm long. Dec.–Jan. Streambanks on sandstone, LB (Langeberg Mts: Swellendam).•

cardinalis H.P.Linder   Like **D. tripetaloides** but flowers red and lateral sepals elliptic, 18–28 mm long. Oct.–Dec. Streambanks on north-facing, inland sandstone slopes, LB, SE (Langeberg Mts: Barrydale to Robinson Pass).•

cochlearis S.D.Johnson & Liltved   Slender tuberous geophyte to 45 cm. Leaves dry at flowering, basal, linear. Flowers few in a lax raceme, white with pale mauve tinge, spur ascending, c. 19 mm long, lateral sepals lanceolate, 10 mm long, lip linear, 8 mm long, with warty tubercle near apex. Jan.–Feb. Clay slopes in renosterveld, KM (Elandsberg).•

draconis (L.f.) Sw.   WHITE DISA   Slender or robust tuberous geophyte. Leaves dry at flowering, basal, lorate to linear. Flowers few to many in a lax raceme, cream or white with purple markings, spur 35–45 mm long, lateral sepals oblong, 15–22 mm long, petals lorate, included in median sepal galea. Oct.–Nov. Sandy coastal flats, SW (Yzerfontein to Cape Peninsula).•

esterhuyseniae Schelpe ex H.P.Linder   Slender, grass-like, tuberous geophyte to 30 cm. Leaves dry at flowering, basal, lorate to lanceolate. Flowers in a lax raceme, yellow-green suffused with brown, spur cylindrical, horizontal, 6 mm long, lateral sepals lanceolate, 6 mm long. Jan.–Feb. Sandstone slopes, NW (Cedarberg to Hex River Mts).•

ferruginea (Thunb.) Sw.   CLUSTER DISA   Slender tuberous geophyte to 45 cm. Leaves dry at flowering, basal, linear, bracts dry. Flowers many in a condensed raceme, bright red to orange, spur slender, acute, grading imperceptably into the galea, 7–20 mm long, lateral sepals elliptic, with apiculi up to 4 mm long, lip narrowly ovate to linear-lanceolate, 10–12 mm long. Feb.–Mar. Sandstone slopes, NW, SW, LB (Cape Peninsula to Albertinia).•

gladioliflora Burch. ex Lindl.   Slender tuberous geophyte to 50 cm. Leaves dry at flowering, basal, linear. Flowers few in a lax raceme, white to pink, often with darker markings, spur slender, 6–14 mm long, lateral sepals oblong, 8–16 mm long, petals acute. Feb.–May. Sandstone slopes, KM, LB, SE (Swartberg to Tsitsikamma Mts).•

harveiana Lindl.   LILAC DISA   Like **D. draconis** but flowers cream or mauve with purple or red streakings, spur slender, 20–90 mm long, petals exserted from median sepal. Oct.–Jan. Sandstone slopes, NW, SW, KM (Bokkeveld Mts to Cape Peninsula, Swartberg and Outeniqua Mts).•

longicornu L.f.   DRIP DISA   Slender, subflexuose tuberous geophyte to 20 cm. Leaves basal, narrowly elliptic. Flower solitary, pale greyish blue, spur cylindrical, 20–35 mm long, lateral sepals ovate, 20–30 mm long. Dec.–Jan. Wet cliffs, SW (Cape Peninsula and Jonkershoek Mts).•

marlothii Bolus   Slender tuberous geophyte to 35 cm. Leaves mostly basal, narrowly elliptic. Flowers few in a corymbose raceme, purplish red, spur very slender, 15–20 mm long, lateral sepals patent, oblong, 10–15 mm long. Dec.–Jan. Streambanks, NW, SW, SE (Cedarberg to Hex River Mts, and Tsitsikamma Mts).•

porrecta Sw.   Grass-like tuberous geophyte to 60 cm. Leaves dry at flowering, basal, linear, bracts dry. Flowers many in a dense raceme, bright red to scarlet with yellow petals and lip, spur cylindrical, ascending, 20–40 mm long, lateral sepals ovate to obovate, 6–8 mm long, apiculi up to 3 mm long. Mainly Mar. Grassland or stony slopes, KM, SE (Kammanassie Mts to Lesotho).

sagittalis (L.f.) Sw.   Slender tuberous herb to 30 cm. Leaves basal, lorate to narrowly elliptic. Flowers few to many in a lax to dense raceme, white to mauve, median sepal shallowly galeate with lateral extensions, spur 2–3 mm long, lateral sepals oblong, c. 7–8 mm long. Sept.–Nov. Rock pavements or ledges, LB, SE (Swellendam to E Cape).

salteri G.J.Lewis   Slender, grass-like, tuberous geophyte to 60 cm. Leaves dry at flowering, basal, linear. Flowers few to many in a lax raceme, brown, spur slender, usually ascending, 20–30 mm long, lateral sepals lanceolate, 2–4 mm long. Apr., usually after fire. Sandstone or shale, NW, SW, KM, SE (Cedarberg Mts to Cape Peninsula to George).•

tenuis Lindl.   Slender, grass-like, tuberous geophyte to 50 cm. Leaves dry at flowering, basal, linear. Flowers many in a lax raceme, greenish brown to white, spur to 1 mm long, subconical, lateral sepals elliptic-oblong, 3–6 mm long. Mainly Mar.–May. Sandstone slopes, SW (Cape Peninsula to Kleinmond).•

triloba Lindl.   Slender tuberous geophyte to 20 cm. Leaves lorate or elliptic. Flowers few in a subcorymbose raceme, mauve with purple markings, spur slender, c. 5 mm long, lateral sepals narrowly obovate, 8–13 mm long. Dec. Sandstone slopes, NW, SW, LB (Cedarberg to Outeniqua Mts).•

tripetaloides (L.f.) N.E.Br.   Slender tuberous geophyte to 60 cm, stoloniferous. Leaves basal, narrowly oblanceolate. Flowers few in a lax raceme, white to pink, spur conical to cylindrical, 2–3 mm long, lateral sepals obovate, elliptic or oblong, 8–15 mm long. Nov.–Jan. Streambanks and mountain seeps, SW, LB, SE (Hottentots Holland Mts to KwaZulu-Natal).

uniflora Bergius   RED DISA   Flexuose to erect tuberous geophyte to 60 cm, stoloniferous. Leaves clustered below, narrowly lanceolate. Flowers 1 to few in a lax raceme, carmine-red to orange, spur 10–15 mm long, lateral sepals narrowly ovate, 35–65 mm long. Jan.–Mar. Wet cliffs, streamsides and seeps, NW, SW (Cedarberg Mts to Betty's Bay and Riviersonderend Mts).•

vaginata Harv. ex Lindl.   Slender tuberous geophyte to 20 cm. Leaves clustered below, lanceolate to lorate. Flowers few in a dense, usually subcorymbose raceme, pink with red spots, spur slender, 5–6 mm long, lateral sepals oblong, 5–6 mm long. Nov.–Dec. Rock outcrops, NW, SW, LB, SE (Cedarberg to Tsitsikamma Mts).•

vasselotii Bolus ex Schltr.   Slender tuberous geophyte to 25 cm. Leaves 4–10, elliptic. Flowers 1 to few in a dense, subcorymbose to corymbose raceme, white, petals with pink and yellow horizontal bars, spur slender, 2–4 mm long, lateral sepals obovate, 9–15 mm long. Nov.–Dec., usually after fire. Mountain seeps LB, SE (Swellendam to Humansdorp).•

B. Leaves cauline, always green at flowering

aconitoides Sond.   Slender tuberous geophyte to 60 cm. Leaves cauline, lanceolate. Flowers many, mauve, spur massive, conical, laterally flattened, usually with rounded apex, 5–8 mm long, lateral sepals narrowly oblong, 5–7 mm long. Nov.–Dec. Grassland, SE (Knysna to tropical Africa).

brevipetala H.P.Linder   Slender tuberous geophyte to 20 cm. Leaves cauline, lax, linear. Flowers 5–20 in a subdense raceme, pink to reddish purple, spur 3–4 mm long, lateral sepals 6–7 mm long, petals with a large semi-orbicular basal anticous lobe. Oct.–Nov. Sandstone slopes, SW (Kleinmond).•

caulescens Lindl.   Slender tuberous geophyte to 40 cm. Leaves narrowly lanceolate. Flowers few in a lax raceme, white, petals barred maroon, spur slender, 2–3 mm long, lateral sepals obovate to elliptic, 7.5–11 mm long. Nov.–Jan. Streambanks on sandstone, NW, SW (Cedarberg to Riviersonderend Mts).•

cedarbergensis H.P.Linder   Tuberous geophyte to 18 cm. Leaves cauline, few, linear-lorate. Flowers c. 15 in a subdense raceme, white with purplish galea and spur, spur 3.5 mm long, lateral sepals 6 mm long. Dec. Peaty streamsides at high alt., NW (Cedarberg Mts).•

chrysostachya Sw.   Robust tuberous geophyte to 1 m. Leaves cauline, many, narrowly elliptic. Flowers many in a dense and very narrow raceme, c. 8 mm diam., orange, spur pendent and clavate, 5–11 mm long, lateral sepals 5.5–9 mm long, anther erect. Nov.–Jan. Damp or marshy grassland, SE (Knysna to N Province).

cornuta (L.) Sw.   Robust tuberous geophyte to 1 m. Leaves cauline, imbricate, lanceolate. Flowers many in a dense raceme, purple and silvery green, spur 10–20 mm long, lateral sepals oblong, 12–16 mm long. Sept.–Feb. Sandstone slopes and grassland, NW, SW, AP, LB, SE (Elandsbaai to Zimbabwe).

cylindrica (Thunb.) Sw.   Slender tuberous geophyte to 35 cm. Leaves cauline, lorate. Flowers many in a dense raceme, green, spur subobsolete to 1 mm long, lateral sepals oblong, 4–5 mm long. Oct.–Dec., after fire. Seeps or damp sandstone slopes, NW, SW, AP, LB, SE (Tulbagh to Cape Peninsula to Humansdorp).•

fasciata Lindl.   Tuberous geophyte to 25 cm. Leaves cauline, ovate, sheath usually barred with purple. Flowers few in a corymbose raceme, white with some purplish spots, pseudoactinomorphic with sepals and lip borne in the same plane, spur slender, 3–5 mm long, lateral sepals broadly oblong, 8–13 mm. long. Mainly Oct., after fire. Sandstone slopes, SW, AP, LB (Cape Peninsula to Riversdale).•

glandulosa Burch. ex Lindl.   Slender, glandular-hairy, tuberous geophyte to 20 cm. Leaves dimorphic: basal leaves spreading, elliptic, cauline leaves sheathing. Flowers few in a corymbose raceme, pink with red spots, spur subconical to rarely cylindrical, 2–3 mm long, lateral sepals oblong, 5–6 mm long. Dec.–Jan. Wet moss or turf on rock pavement, SW, LB, SE (Cape Peninsula to Humansdorp).•

hallackii Rolfe   Robust tuberous geophyte to 50 cm. Leaves cauline, imbricate, lanceolate. Flowers with bracts prominent, horizontal, sepals green, petals purplish, spur semipendent, subclavate, 5 mm long, lateral sepals oblong, 10–16 mm long. Oct.–Nov. Sandstone slopes, SW, AP, SE (Cape Peninsula to Port Elizabeth).•

longifolia Lindl.   Tuberous geophyte to 40 cm. Leaves cauline, linear. Flowers few to many in a dense raceme, pink or white, spur pendent, 2–4 mm long, lateral sepals oblong, 5–7 mm long, lip finely hairy. Oct., usually after fire. Peaty sandstone slopes, NW, SW (Piketberg to Riviersonderend Mts).•

micropetala Schltr.   Slender tuberous geophyte to 30 cm. Leaves cauline, many, linear. Flowers many in a dense raceme, pink to purple with white base, spur filiform, 1 mm long, lateral sepals 2–4 mm long. Nov.–Dec., usually after fire. Peaty sandstone slopes and ledges, SW, LB, SE (Cape Peninsula to Tsitsikamma Mts).•

obtusa Lindl.   Slender tuberous geophyte to 40 cm. Leaves cauline, linear. Flowers many in a dense raceme, white to brown or purple, often spotted, sepals 3–8 mm long, spur with a deep dorsal groove, 0.2–2.2 mm long. Sept.–Dec., after fire. Sandstone seeps, SW, KM, LB, SE (Cape Peninsula to Baviaanskloof and Tsitsikamma Mts).•

ocellata Bolus   Slender tuberous geophyte to 30 cm. Leaves cauline, linear. Flowers few in a lax raceme, white with dull purple-brown markings, spur 2 mm long, horizontal, clavate, lateral sepals c. 5 mm long. Dec. Sandstone slopes, SW, KM (Cape Peninsula to Swartberg Mts).•

ovalifolia Sond.   Tuberous geophyte to 20 cm. Leaves cauline, imbricate. Flowers few in a dense raceme, sepals green, lip brown, spur ascending, 10–15 mm long, lateral sepals oblong, c. 8 mm long. Sept.–Oct. Sandstone slopes, NW, SW (Cedarberg to Cold Bokkeveld Mts).•

polygonoides Lindl.   Slender or robust tuberous geophyte to 70 cm. Leaves cauline, few, linear to lorate. Flowers many in a dense raceme, red, spur pendent, lateral sepals 6–7 mm long, anther erect. Oct.–Mar. Marshy grassland, SE (Humansdorp to S Mozambique).

subtenuicornis H.P.Linder   Robust tuberous geophyte to 25 cm. Leaves linear, densely imbricate, clustered below. Flowers many in a dense raceme, white with purple spots, spur pendent, 3 mm long, lateral sepals c. 10 mm long. Dec., after fire. Sandstone seeps at high alt., LB (Langeberg Mts: Riversdale).•

telipogonis Rchb.f.   Slender tuberous geophyte, usually to 6 cm. Leaves linear, as tall as or overtopping raceme. Flowers few in a corymbose raceme, yellow, spur 2–4 mm long, lateral sepals narrowly ovate, 5–6 mm long, petals with long apiculi. Nov. Sandstone outcrops, NW, SW (Cedarberg Mts to Cape Peninsula).•

tenella (L.f.) Sw.   Slender tuberous geophyte to 15 cm. Leaves cauline, linear, spiralling, reaching raceme base. Flowers few to many in a dense raceme, white, pink or red, often with darker mottling, spur slender, to 5 mm long, lateral sepals oblong, 4–6.5 mm long. Aug.–Oct. Sandy flats and slopes, NW, SW (Gifberg to Hottentots Holland Mts).•

tenuicornis Bolus   Robust tuberous geophyte to 50 cm. Leaves cauline, linear-acuminate, imbricate. Flowers many in a dense raceme, white with red lines and spots, spur pendent, 3–4 mm long, lateral sepals 8–10 mm long. Oct., usually after fire. Sandstone slopes, SW (Cape Peninsula to Babilonstoring).•

tysonii Bolus   Robust tuberous geophyte to 60 cm. Leaves cauline, imbricate, narrowly ovate. Flowers many in a dense raceme, cream-yellow with green veins, lip with purple spots, spur cylindrical to subclavate, 4–6 mm long, lateral sepals oblong-elliptic, 6–8 mm long. Dec.–Jan. Damp, rocky grassland, SE (Knysna to KwaZulu-Natal).

uncinata Bolus   Slender, flexuose, tuberous geophyte to 50 cm. Leaves few, narrowly lanceolate. Flowers few in a lax raceme, white or cream, spur horizontal and constricted basally, 2–4 mm long, lateral sepals oblong, 5–8 mm long. Oct.–Jan. Mountain streams and wet cliffs, NW, SW, KM, LB, SE (Tulbagh to Knysna).•

Herschelia-group

*A. Lip entire (see also **D. hians**)*

forcipata Schltr. (= *Herschelia forcipata* (Schltr.) Kraenzl.)   Slender tuberous geophyte to 60 cm. Leaves? Flowers few in a lax raceme, greenish yellow, median sepal c. 13 mm long, spur cylindrical, 3–4 mm long, bifid, lip lanceolate. Flowering time? Habitat? SE (?Knysna).•

forficaria Bolus (= *Herschelia forficaria* (Bolus) H.P.Linder)   Tuberous geophyte to 50 cm. Leaves basal, linear or narrowly ovate. Flowers few in a lax raceme, hyperresupinate, sepals greenish red, petals and lip maroon, lateral sepals narrowly ovate, 8–12 mm long, petals acicular, finely hairy, lip reniform, apex finely hairy. Jan.–Feb., after fire. Dry, stony slopes, SW (Wellington, Cape Peninsula to Grabouw).•

graminifolia Ker Gawl. ex Spreng. (= *Herschelia graminifolia* (Ker Gawl ex Spreng.) T.Durand & Schinz)   BLOUMOEDERKAPPIE   Slender tuberous geophyte to 1 m. Leaves dry at flowering, basal, linear. Flowers few in a lax raceme, blue to violet-purple, petal apices green, lip dark purple, elliptic, spur clavate, 2–4 mm long, lateral sepals oblong, 13–18 mm, down-curved. Jan.–Mar. Sandstone slopes, SW, LB, SE (Cape Peninsula to Port Elizabeth).•

newdigateae L.Bolus (= *Herschelia newdigateae* (L.Bolus) H.P.Linder)   Like **D. forficaria** but petals flattened and obscurely bilobed. Mar.–Apr. Stony slopes, SE (Outeniqua Mts: Knysna).•

purpurascens Bolus (= *Herschelia purpurascens* (Bolus) Kraenzl.)   BLOUMOEDERKAPPIE   Like **D. graminifolia** but spur conical and lip margin upcurved. Oct.–Nov. Sandstone slopes and outcrops, SW, AP (Cape Peninsula to Agulhas).•

schlechteriana Bolus (= *Herschelia schlechteriana* (Bolus) H.P.Linder)   Tuberous geophyte to 60 cm. Leaves basal, linear. Flowers few in a lax raceme, cream with mauve veins, spur slender cylindrical, 30–50 mm long, lateral sepals oblong, 20–25 mm long, lip oblong. Dec. North-facing sandstone slopes, LB (Langeberg Mts: near Riversdale).•

A. Lip lacerate or spathulate with a long claw

barbata (L.f.) Sw. (= *Herschelia barbata* (L.f.) Bolus)   OUMAN MET SY BAARD   Slender tuberous geophyte to 50 cm. Leaves dry at flowering, basal, linear. Flowers few in a lax raceme, white to very pale blue, spur conical, 1–5 mm long, lateral sepals narrowly oblong, 15–25 mm long, lip ovate-lacerate. Oct.–Nov. Marshy coastal sands, SW (Darling, extinct on Cape Peninsula).•

hians (L.f.) Spreng. (= *Herschelia hians* (L.f.) A.V.Hall)   Slender tuberous geophyte to 60 cm. Leaves often dry at flowering, basal, linear. Flowers few in a lax raceme, pale to purplish blue, spur conical, 4–6 mm long, lateral sepals oblong, 8–12 mm long, lip oblong-elliptic, crenulate-lacerate, occasionally entire, 7–12 mm long. Dec.–Jan., often after fire. Sandstone slopes, AP, LB, SE (Riversdale to Port Elizabeth).•

lugens Bolus (= *Herschelia lugens* (Bolus) Kraenzl.)   BLOUMOEDERKAPPIE   Slender tuberous geophyte to 1 m. Leaves dry at flowering, basal, linear. Flowers few in a lax raceme, cream-green, sometimes almost black, lateral sepals mauve, lip green to grey-green, spur cylindrical, 1–5 mm long, lateral sepals oblong, 8–13 mm long, lip ovate-lacerate. Oct.–Nov. Sandstone slopes, SW, AP, LB, SE (Cape Peninsula to Grahamstown).•

multifida Lindl. (= *Herschelia multifida* (Lindl.) Rolfe)   OUPA MET SY PYP   Slender tuberous geophyte to 60 cm. Leaves dry at flowering, basal, linear. Flowers few in a lax raceme, blue with green lip, spur conical, 1–6 mm long, lateral sepals narrowly ovate, 10–15 mm long, lip spathulate with slender claw 30–100 mm long, blade deeply lacerate. Nov.–Dec. Sandstone slopes, NW, SW, AP, KM (Cedarberg to Swartberg Mts).•

procera H.P.Linder (= *Herschelia excelsa* sensu Rolfe)   Slender tuberous geophyte to 60 cm. Leaves linear, green at flowering. Flowers few in a lax raceme, deep cerise-red, spur conical, 2–3 mm long, lateral sepals oblong, 8–10 mm long, lip crenulate to lacerate, rarely entire, 8–10 mm long. Oct. Habitat? SE (Sedgefield).•

spathulata (L.f.) Sw. (= *Herschelia spathulata* (L.f.) Rolfe)   OUPA MET SY PYP   Slender tuberous geophyte to 30 cm. Leaves green at flowering, narrow. Flowers few in a lax raceme, maroon to pale lime or green and blue, spur clavate, 1.5–3 mm long, lateral sepals ovate, 6–16 mm long, lip spathulate with slender claw, limb sagittate to trisect. Sept.–Oct. Sandstone and shale, NW, SW, SE (Namaqualand to Caledon, Uniondale).•

venusta Bolus (= *Herschelia venusta* (Bolus) Kraenzl.)   Slender tuberous geophyte to 60 cm. Leaves dry at flowering, basal, linear, stiff. Flowers few in a lax raceme, blue, spur conical, 1.5–3 mm long, lateral sepals oblong, 12–16 mm long, lip ovate, variably lacerate, shorter than lateral sepals. Jan. Coastal sands, SW, AP (Cape Peninsula to Hermanus, Hogsback Mts).

Monadenia-group

A. Spur longer than median sepal

atrorubens Schltr. (= *Monadenia atrorubens* (Schltr.) Rolfe)   Slender, purple-flushed, tuberous geophyte to 40 cm. Leaves lanceolate. Flowers few to many, sepals and spur dull red, lip and petals almost black, spur cylindrical, 15–30 mm long, lateral sepals narrowly oblong, 6–8 mm long. Aug.–Oct., after fire. Sandy flats, NW, SW (Bokkeveld Mts to Hermanus).•

bolusiana Schltr. (= *Monadenia bolusiana* (Schltr.) Rolfe)   Slender or robust tuberous geophyte to 30 cm. Leaves 2–5, basal, narrowly oblong. Flowers many in a dense raceme, lime-green, sometimes tinted red, spur slender, 16–22 mm long, lateral sepals oblong, 6–8 mm long. Nov.–Dec. Gravelly ridges, SW, KM, LB (Cape Peninsula to Riversdale).•

brevicornis (Lindl.) Bolus (= *Monadenia brevicornis* Lindl.)   Slender or robust tuberous geophyte to 50 cm. Leaves usually narrowly lanceolate. Flowers many in a lax raceme, lime-green and maroon, spur cylindrical, 7–11 mm long, lateral sepals oblong, 5–9 mm long. Nov.–Feb. Montane grassland, SE (Knysna to tropical Africa and Madagascar).

cernua (Thunb.) Sw. (= *Monadenia cernua* (Thunb.) T.Durand & Schinz)   Robust tuberous geophyte to 60 cm. Leaves basal, narrowly lanceolate. Flowers many in a dense raceme, sepals cream-green mottled maroon, petals and lip lime-green, spur clavate, rounded, 11–17 mm long, lateral sepals oblong, 9–13 mm long. Mainly Oct. Sandy coastal flats, SW, AP, SE (Cape Peninsula to Humansdorp).•

comosa (Rchb.f.) Schltr. (= *Monadenia comosa* Rchb.f.)   Erect or subflexuose tuberous geophyte to 30 cm. Leaves basal, elliptic. Flowers few in a lax raceme, lime-green, sometimes tinted red, spur cylindrical, 17–24 mm long, lateral sepals oblong-ovate, 6–7 mm long. Sept.–Nov. Sandstone slopes and outcrops, NW, SW, AP, KM, LB, SE (Clanwilliam to Kouga Mts).•

ophrydea (Lindl.) Bolus (= *Monadenia ophrydea* Lindl.)   Slender or robust, red-flushed, tuberous geophyte to 40 cm. Leaves narrowly lanceolate. Flowers few in a lax raceme, purple-red, lateral sepals paler and often yellow or white, spur slender, 20–24 mm long, lateral sepals ovate, 8–10 mm long. Oct.–Nov., after fire. Mountain seeps, NW, SW, LB, SE (Cape Peninsula to Van Staden's Mts).•

reticulata Bolus (= *Monadenia reticulata* (Bolus) T.Durand & Schinz)   Like **D. rufescens** but flowers many in a dense raceme, lime-green, sometimes tinted or mottled maroon, spur 10–20 mm long, lateral sepals oblong, 6–7 mm long, bracts with prominent reticulate venation. Nov.–Dec., often after fire. Seasonal seeps, SW, LB, SE (Cape Peninsula to Knysna).•

rufescens (Thunb.) Sw. (= *Monadenia rufescens* (Thunb.) Lindl.)   Tuberous geophyte to 40 cm. Leaves linear-lanceolate. Flowers few in a slender raceme, lime-green with purple petals and lip, spur slender, cylindrical, 10–16 mm long, lateral sepals oblong, 6–8 mm long. Sept.–Nov. Seasonal seeps, NW, SW, KM, LB, SE (Cedarberg Mts to Cape Peninsula).•

A. Spur shorter than to as long as median sepal

bracteata Sw. (= *Monadenia bracteata* (Sw.) T.Durand & Schinz)   Slender or robust tuberous geophyte to 30 cm. Leaves linear-lanceolate. Flowers many in a dense raceme, green with sepals tinted maroon, spur pendent, 3–4.5 mm long, lateral sepals 2.5–3.5 mm long. Sept.–Nov. Fynbos, especially roadsides, NW, SW, AP, KM, LB, SE (Vredendal to E Cape).

conferta Bolus (= *Monadenia conferta* (Bolus) Kraenzl.)   Slender, dark red-flushed, tuberous geophyte to 22 cm. Leaves linear, imbricate. Flowers many in a dense raceme, c. 10 mm diam., lime-green at anthesis, sepals soon turning deep red, lateral sepals oblong to narrowly ovate, 2.5 mm long, spur obsolete. Sept.–Dec., usually after fire. Sandy slopes in fynbos, SW, AP (Cape Peninsula to Elim).•

densiflora (Lindl.) Bolus (= *Monadenia densiflora* Lindl.)   Slender tuberous geophyte to 20 cm. Leaves linear-lanceolate, imbricate. Flowers many in a dense, slender raceme, petals and lip dull green, sepals rusty red to green, spur constricted at base, 2–3 mm long, lateral sepals oblong-ovate, 3–5 mm long. Mainly Oct. Sandy soils, SW, AP, SE (Cape Peninsula to Bredasdorp and Storms River mouth).•

ecalcarata (G.J.Lewis) H.P.Linder (= *Monadenia ecalcarata* G.J.Lewis)   Slender tuberous geophyte to 13 cm. Leaves linear-lanceolate, imbricate. Flowers many in a dense raceme, not resupinate, lime-green, lateral sepals 4 mm long, spur obsolete. Oct. Fynbos, SW (Cape Peninsula).•

nubigena H.P.Linder Slender tuberous geophyte to 24 cm. Leaves lanceolate. Flowers few to many in a dense raceme, 10–12 mm diam., sepals white, petals and lip yellow-brown, spur 2 mm long. Oct. Restioid vegetation, SW (Cape Peninsula).•

physodes Sw. (= *Monadenia physodes* (Sw.) Rchb.f.) Robust tuberous geophyte to 60 cm. Leaves linear-lanceolate. Flowers many in a cylindrical raceme, lime-green and often mottled or suffused with maroon, spur clavate, rounded, 7–9 mm long, lateral sepals oblong, 7–10 mm long. Sept.–Oct. Habitat? NW, SW (Piketberg to Cape Peninsula).•

pygmaea Bolus (= *Monadenia pygmaea* (Bolus) T.Durand & Schinz) Slender tuberous geophyte to 15 cm. Leaves narrowly ovate. Flowers many in a dense raceme, lime-green, spur c. 2.5 mm long, cylindrical, lateral sepals oblong, 4 mm long. Oct.–Nov., often after fire. Sandy lower slopes, SW (Cape Peninsula to Genadendal).•

sabulosa Bolus (= *Monadenia sabulosa* (Bolus) Kraenzl.) Slender tuberous geophyte to 10 cm. Leaves lanceolate, imbricate. Flowers many in a dense, obovate raceme, lime-green and yellow, spur pendent, 10–15 mm long, lateral sepals 7–9 mm long. Oct.–Nov. Sandy coastal flats, SW (Cape Peninsula to Hangklip).•

DISPERIS WITCH ORCHID c. 80 spp., mostly S and tropical Africa and Madagascar, also Australasia

A. Stem hairless

circumflexa (L.) T.Durand & Schinz. Tuberous geophyte to 20 cm. Leaves 2, suberect, linear-lanceolate. Flowers 2–10, greenish and white, lip limb horizontal-canaliculate with tip inflexed, warty within, rostellum arms curled outwards. Aug.–Sept. Sand, clay or granite, NW, SW, LB (Kamiesberg to Riversdale).

disaeformis Schltr. Slender tuberous geophyte to 15 cm. Leaves 2, cauline, alternate, ovate. Flowers 1 or few in a lax raceme, whitish or tinged with lilac, lip blade triangular and reflexed, appendage oblong, apical part with papillae. July–Oct. Dry and open woodland, LB, SE (Riversdale, E Cape to KwaZulu-Natal, and N Province).

lindleyana Rchb.f. Tuberous geophyte to 30 cm. Leaf 1, spreading, cordate. Flowers 1–4, white, lip limb deflexed, convex-ovate, rostellum arms upcurved. Dec.–Feb. Montane forest, SE (George to Mpumalanga).

paludosa Harv. ex Lindl. Tuberous geophyte to 50 cm. Leaves c. 3, suberect, lanceolate. Flowers 2–7, magenta and green, lip limb horizontal, lanceolate-conduplicate, rostellum arms porrect with outflexed tips. Nov.–Jan., especially after fire. Marshy sandstone slopes, SW, AP, LB, SE (Hex River Mts to Cape Peninsula to S KwaZulu-Natal).

thorncroftii Schltr. Tuberous geophyte to 25 cm. Leaves 2, spreading, cordate. Flowers 1–3, white or lilac, lip limb deflexed, ovate, rostellum arms porrect with inflexed tips. Nov.–Dec. Montane forest, SE (Knysna and E Cape to Zimbabwe).

A. Stem hairy

bodkinii Bolus Like **D. macowanii** but flowers smaller and often cleistogamous with green sepals and lip appendage inflexed at tip. Aug.–Sept. Clay flats under scrub, NW, SW, KM (Tulbagh to Cape Peninsula and Anysberg).•

bolusiana Schltr. ex Bolus Tuberous geophyte with shortly hairy stem to 20 cm. Leaves 2 or 3, spreading, ovate. Flower 1, yellowish green to white, lip limb cup-shaped, rostellum arms obsolete with deflexed tip. Sept.–Oct. Clay flats, NW, SW, KM, LB (Bokkeveld Escarpment to Riversdale).•

capensis (L.f.) Sw. Tuberous geophyte with softly hairy stem to 50 cm. Leaves 2, suberect, lanceolate. Flower 1, green and magenta or cream, sepals conspicuously tailed, lip limb curled downwards, lanceolate, warty, rostellum arms tightly coiled. July–Sept. Sandstone seeps, NW, SW, AP, KM, LB, SE (Pakhuis Mts to Grahamstown).

cucullata Sw. WITCH ORCHID Tuberous geophyte with shortly hairy stem to 20 cm. Leaves 2, suberect and clasping below, ovate. Flower 1, pale green, lip limb goblet-shaped and apiculate, rostellum arms curled over. Aug.–Oct. Seasonally damp, often sandy flats, NW, SW (Clanwilliam to Riviersonderend Mts).•

macowanii Bolus Tuberous geophyte with shortly hairy stem to 20 cm. Leaves 2, spreading, ovate. Flower 1, white or mauve, lip limb minute, upcurved, rostellum arms incurled. Feb.–Mar. Damp grassland, SE (Mossel Bay and Port Elizabeth to Free State).

purpurata Rchb.f.   Like **D. bolusiana** but flowers magenta to white with rostellum arms exserted. Aug.–Sept. Clay and granite slopes, NW, KM, SE (Richtersveld and W Karoo to Humansdorp).

villosa (L.f.) Sw.   Tuberous geophyte with shortly hairy stem to 20 cm. Leaves 2, lower petiolate, elliptic. Flowers 1–4, yellowish green, lip limb slipper-shaped, rostellum arms corkscrew-twisted. Aug.–Sept. Clay and granite slopes, NW, SW, AP, LB, SE (Clanwilliam to Mossel Bay and Port Elizabeth).•

EULOPHIA   HARLEQUIN ORCHID   c. 250 spp., pantropical and subtropical

A. Lip crests of low fleshy ridges tapering gradually to lip apex

parviflora (Lindl.) A.V.Hall   Rhizomatous geophyte to 50 cm. Leaves emergent at flowering, leathery. Flowers few to many in a dense, elongate raceme, sepals brownish green, petals and lip yellow, lip side lobes tinged purple, median sepal 10–14 mm long, lip crests of low warty ridges, spur conic to subcylindrical, 2–5 mm long. Aug.–Nov. Grassy fynbos and coastal thornveld, SE (Langkloof to N Province).

speciosa (R.Br. ex Lindl.) Bolus   Geophyte mostly to 60 cm, with subterranean or partly exposed corms/pseudobulbs. Leaves more or less succulent, fully developed at flowering. Flowers few to many, yellow, median sepal 11–13 mm long, lip crests of broad fleshy ridges, spur conical, 1–3 mm long. Oct.–Jan. Mostly coastal bushveld, SE (Wilderness to tropical Africa).

streptopetala Lindl.   Geophyte to 2 m, with partly exposed pseudobulbs. Leaves lanceolate, pleated. Flowers few to many in a lax, elongate raceme, sepals green mottled with brown, petals and lip midlobe yellow, lip side lobes purple, median sepal 11–18 mm long, lip crests of low ridges, spur subcylindrical, 1.5–2.2 mm long. Oct.–Jan. Bush, SE (Uniondale to tropical Africa).

A. Lip crests of papillae or lamellae abruptly terminating at or lacking on outer third

aculeata (L.f.) Spreng.   Rhizomatous geophyte to 25 cm. Leaves partly or fully developed at flowering, linear-lanceolate, pleated. Flowers few to many in a dense raceme, dull ivory to white, tepals scarcely spreading, median sepal oblong, 6–16 mm long, lip crests of ridges and lamellate papillae, spur absent. Nov.–Jan. Fynbos and grassland, SW, AP, LB, SE (Cape Peninsula to Mpumalanga).

clavicornis Lindl.   Rhizomatous geophyte to 73 cm. Leaves partly to fully developed at flowering, linear-lanceolate. Flowers few to many in a lax raceme, sepals purplish, petals white tinged pale pink to pale blue or yellow, lip crests pale pink to yellow, median sepal 8–18 mm long, lip crests of pubescent ridges and lamellae, spur slender, 1.4–8.9 mm long. Aug.–Feb. Grassland, SE (George to tropical Africa and Madagascar).

hereroensis Schltr.   Rhizomatous geophyte to 55 cm, often with aerial pseudobulbs. Leaves partly to fully developed at flowering, linear-lanceolate. Flowers few to many in a lax raceme, sepals yellowish green, petals and lip pale yellowish green, lip crests pale yellow, median sepal 12–-19 mm long, lip crests of lamellae, spur 3–4 mm long. Sept.–Jan. Arid vegetation, SE (Humansdorp to N Province, Namibia).

litoralis Schltr.   Rhizomatous geophyte to 66 cm. Leaves usually absent to vestigial. Flowers few to many in a rather lax raceme, sepals yellowish green tinged olive, petals and lip yellow, lip side lobes tinged purple, crests yellow, median sepal lanceolate, 17–24 mm long, lip crests of fleshy ridges and papillae, spur slender, 2.5–4.5 mm long. Nov.–Jan. Coastal dunes, SW, LB, SE (Cape Flats to Plettenberg Bay).•

ovalis Lindl.   Rhizomatous geophyte to 65 cm. Leaves leathery, fully developed at anthesis. Flowers few in a lax raceme, sepals purple, petals and lip white and tinged purple, lip crests yellow, median sepal oblong to lanceolate-oblong, 14–36 mm long, lip crests of ridges and papillae, spur conical to cylindrical, 1–5.6 mm long. Nov.–Feb. Grassland, SE (Humansdorp to N Province).

platypetala Lindl.   Rhizomatous geophyte to 45 cm. Leaves partly developed at flowering, somewhat leathery. Flowers few in a lax raceme, sepals purple brown, petals and lip greenish yellow, median sepal 15–17 mm long, lip crests of 3–5 lamellae, spur conical, 2–3 mm long. Oct.–Nov. Coastal renosterveld and forest clearings, SW, LB, SE (Swellendam to E Cape).

tabularis (L.f.) Bolus   Rhizomatous geophyte to 40 cm. Leaves absent or partly developed at flowering, linear-lanceolate, pleated. Flowers 2–few in a dense raceme, yellow, median sepal narrowly elliptic, 20–22 mm long, lip crest of a single ridge, spur absent. Nov.–Jan., often after fire. Sandy flats and slopes, NW, SW, LB, SE (Biedouw to E Cape).

tenella Rchb.f.   Rhizomatous geophyte to 60 cm. Leaves fully developed at anthesis, linear-lanceolate. Flowers few to many in a dense raceme, sepals dark green to purple, petals yellow and brown, lip brown, lip crest bright yellow, median sepal 6–8 mm long, lip crests of fleshy ridges and lamellae, spur stout, cylindrical, 2–3 mm long. Nov.–Jan. Coastal bushveld, SE (Uitenhage to Zimbabwe).

tuberculata Bolus   Rhizomatous geophyte to 40 cm. Leaves absent to partly developed at flowering, leathery. Flowers few to many in a lax raceme, sepals yellowish green and purple, petals and lip white tinged maroon, crests yellow, median sepal 6–7 mm long, lip crests of ridges and lamellae terminating abruptly near midlobe apex, spur vestigial. Sept.–Dec. Grassland and karroid scrub, SE (Knysna to tropical Africa).

EVOTELLA•   LIVER ORCHID   1 sp., W Cape

rubiginosa (Sond. ex Bolus) Kurzweil & H.P.Linder (= *Corycium rubiginosum* (Sond. ex Bolus) Rolfe)   Slender tuberous geophyte to 30 cm. Leaves linear-lanceolate. Flowers many in a dense raceme, c. 15 × 15 mm, sepals green tinged maroon, petals rusty red to maroon, lip white tinged with maroon, lip appendage green, lip blade broadly deltate-hastate, appendage erect. Oct.–Nov., after fire. Sandstone flats and seeps, NW, SW (Grootwinterhoek to Hermanus).•

*GASTRODIA   POTATO ORCHID   c. 17 spp., Australasia

*sesamoides R.Br.   Slender rhizomatous saprophyte to 1 m. Leaves reduced, scale-like, membranous. Flowers in a lax raceme, bell-shaped, not resupinate, patent or pendulous, light brown and cream, sepals and petals fused, lip free, 3-lobed with 2 crests. Nov.–Dec. Forest and plantations, SW (Cape Peninsula, introduced from Australia).

HABENARIA   GHOST ORCHID   c. 800 spp., pantropical and subtropical

anguiceps Bolus   Stout tuberous geophyte to 40 cm. Leaves many, cauline, lanceolate. Flowers in a dense raceme, green, petals obliquely lanceolate, lip oblong with bulbous spur c. 8 mm long, stigmas c. 1 mm long. Nov.–Feb. Grassy slopes, SE (Humansdorp to Mpumalanga).
arenaria Lindl.   Slender tuberous geophyte to 40 cm. Leaves few, mostly basal, spreading, elliptic, often mottled with grey. Flowers in a lax raceme, green, petals lanceolate, lip tripartite with spur 15–20 mm long, stigmas 0.5 mm long. Apr.–July. Coastal scrub and forest, LB, SE (Riversdale to Mpumalanga).
epipactidea Rchb.f.   Robust tuberous geophyte to 50 cm, usually densely leafy. Leaves cauline, lanceolate. Flowers many in a dense raceme, green, with greenish white petals and white lip, petals elliptic to suborbicular, lip oblong with basal whiskers and a spur 20–30 mm long, anther reflexed, stigmas c. 1.5 mm long, central rostellum lobe broadly triangular and covering anther. Nov.–Apr. Savanna and grassland, SE (E Cape to tropical Africa).
falcicornis (Burch. ex Lindl.) Bolus   Robust tuberous geophyte to 70 cm. Leaves many, cauline, lanceolate. Flowers in a fairly dense raceme, green and white, petals bipartite with anterior lobe elliptic, lip tripartite with spur 20–40 mm long, stigmas 2–4 mm long. Feb.–Apr. Seasonally damp grassland, LB, SE (Swellendam to N Province).
laevigata Lindl.   Slender tuberous geophyte to 40 cm. Leaves many, cauline, elliptic, with cartilaginous margins. Flowers in a dense raceme, green, petals obliquely ovate, lip tripartite with filiform spur 16–35 mm long, stigmas c. 1 mm long. Nov.–Mar. Stony grassland, LB (George, Grahamstown to Mpumalanga).
lithophila Schltr.   Slender tuberous geophyte to 30 cm. Leaves 2, basal, adpessed, suborbicular. Flowers in a dense raceme, yellowish green, petals bipartite, puberulous, anterior lobe linear, lip tripartite with spur 8–11 mm long, stigmas 1–2 mm long. Nov.–Mar. Stony grassland, SE (Mossel Bay to tropical Africa).

HOLOTHRIX   THREAD ORCHID, TRYPHIA   c. 55 spp., Africa, Arabia and Madagascar

A. Scape with bracts

burchellii (Lindl.) Rchb.f.   Slender tuberous geophyte to 50 cm, scape thinly hairy. Leaves 2, basal, ovate. Flowers many in a dense raceme, dimorphic, upper flowers with much longer petals and lip lobes, sepals green, petals and lip cream, petals with 5–10 filiform lobes, lip spurred and divided into 5–13 filiform lobes. Aug.–Nov. Grassland, NW, AP, LB, SE (Piketberg to E Cape).
grandiflora (Sond.) Rchb.f.   Slender tuberous geophyte to 25 cm, scape thinly hairy. Leaves 2, basal, ovate, often withered at flowering. Flowers many in a dense raceme, sepals and petals green, lip white with pale green or lilac base, petals divided into 5–9 filiform lobes, lip spurred and divided into 13–26 filiform lobes. Mar. Rock crevices, NW, LB, SE (Vredendal to Port Elizabeth).•

schlechteriana Schltr. ex Kraenzl.   Slender tuberous geophyte to 27 cm, scape thinly hairy. Leaves 2, basal, ovate. Flowers many in a dense raceme, sepals green, petals and lip green or yellow, petals with 4–9 filiform lobes, lip spurred and divided into 5–11 filiform lobes. Oct.–Feb. Among rocks and shrubs, SW, KM, SE (Namaqualand to E Cape).

A. Scape without bracts

aspera (Lindl.) Rchb.f.   Slender tuberous geophyte to 25 cm, scape thinly hairy. Leaves 2, basal, ovate. Flowers many in a dense raceme, sepals green, petals and lip white with maroon or purple stripes, petals entire, lip spurred and with 7 unequal, rounded lobes. June–Oct. Sandy slopes or rock pavements, NW, SW, LB (Namaqualand and W Karoo to Swellendam).

brevipetala Immelman & Schelpe   Slender tuberous geophyte to 31 cm, scape stiffly deflexed-hairy. Leaves 2, basal, ovate, stiffly hairy or scaly. Flowers many in a dense raceme, green, petals entire, lip shortly spurred and with 3–5 short, broad lobes. Usually Aug.–Jan. Rock pavements, NW, SW, SE (Cape Peninsula to Port Elizabeth).

cernua (Burm. f.) Schelpe   Slender tuberous geophyte to 24 cm, scape stiffly deflexed-hairy. Leaves 2, basal, ovate, stiffly hairy or scaly. Flowers many in a dense raceme, green, petals entire, lip spurred and with 3–5(–7) narrow lobes. July–Jan., mostly after fire. Sandstone slopes and flats, NW, SW, AP, LB, SE (Namaqualand to Grahamstown).

exilis Lindl.   Slender tuberous geophyte to 29 cm, scape thinly hairy. Leaves 2, basal, ovate, glabrescent. Flowers few to many in a dense raceme, green, petals entire, lip entire and shortly spurred. Oct.–Mar. Coastal regions, SW, AP, SE (Cape Peninsula to Kei Mouth).

longicornu G.J.Lewis   Slender tuberous geophyte to 15 cm, scape deflexed-hairy. Leaves 2, basal, ovate, hairy. Flowers with basally connate sepals, lip spurred and with 3–5 short, oblong lobes. Oct. Habitat? SE (Port Elizabeth).•

mundii Sond.   Slender tuberous geophyte to 16 cm, scape thinly hairy. Leaves 2, basal, ovate, fleshy. Flowers in a condensed raceme, sepals green, petals and lip white, petals entire, lip spurred and with 7 broad lobes. Sept.–Nov. Grassland, SW, AP, LB, SE (Cape Peninsula to Port Elizabeth).•

parviflora (Lindl.) Rchb.f.   Slender tuberous geophyte to 24 cm, scape glabrous. Leaves 2, basal, ovate, fleshy. Flowers in a dense raceme, sepals green, petals and lip white or marked with purple, petals entire, broadly linear and membranous, lip spurred and with 5 narrow lobes. July–Nov. Damp places and on mossy boulders, SW, LB, SE (Cape Peninsula to KwaZulu-Natal).

pilosa (Burch. ex Lindl.) Rchb.f.   Slender tuberous geophyte to 55 cm, scape densely hairy below. Leaves 2, basal, ovate, densely hairy on reverse. Flowers many in a dense raceme, sepals green, petals and lip white or cream, petals entire, linear, lip spurred and with 3–7 short, broad fleshy lobes. Nov.–Mar. Dry, stony places, SW, LB, SE (Riviersonderend to Port Elizabeth).•

secunda (Thunb.) Rchb.f.   Slender tuberous geophyte to 30 cm, scape thinly hairy. Leaves 2, basal, ovate, fleshy. Flowers many in a lax raceme, sepals green, petals and lip cream, petals entire and fleshy, lip spurred and with 5 linear lobes. June–Oct. Dry rocky slopes and pavements, NW, SW, AP, KM, LB (Richtersveld and W Karoo to Breede River valley and Little Karoo).

villosa Lindl.   Slender tuberous geophyte to 36 cm, scape silky hairy. Leaves 2, basal, ovate, hairy. Flowers many in a dense raceme, green, petals entire, lip spurred and with 3–5 lobes. Oct.–Jan. Sandstone and granite slopes and pavements, NW, SW, AP, KM, LB, SE (Richtersveld and W Karoo to E Cape).

LIPARIS   FLY ORCHID   c. 250 spp., worldwide

capensis Lindl.   Terrestrial herb to 10 cm, with pseudobulbs largely subterranean. Leaves 2 or 3, ovate, spreading. Flowers many in a dense raceme, green, lateral sepals oblong, 4 mm long, lip oblong with obscure side lobes, emarginate. Apr.–July. Sandstone slopes, SW, AP, SE (Cape Peninsula to Algoa Bay).•

remota J.L.Stewart & Schelpe   Terrestrial, rarely epiphytic herb to 20 cm, pseudobulbs separated by creeping stems with membranous cataphylls. Leaves 3, elliptic. Flowers 3 in a lax raceme, green, lateral sepals broadly elliptic, 5–6 mm long, largely united, lip spathulate, rhomboid to lunate with a small basal callus. Nov.–Mar. Forests, SE (Knysna to Swaziland).

MYSTACIDIUM   TREE ORCHID   9 spp., southern and tropical Africa

capense (L.f.) Schltr.   Monopodial epiphyte with stems to 2.5 cm. Leaves oblong-oblanceolate, channelled, apically notched. Flowers several in a lax raceme, white, 25–30 mm diam., tepals lanceolate, lip with short side lobes and a filiform spur 35–60 mm long. Sept.–Jan. Montane and lowland forests, SE (George to Swaziland).

PACHITES•   2 spp., W Cape

appressa Lindl.   Slender geophyte to 40 cm. Leaves 5–12, cauline, linear. Flowers few in a lax raceme, pink but lip pale greenish yellow below, tepals ovate-oblong, 5–7 mm long. Jan.–Feb., after fire. Sandstone slopes, LB (Langeberg Mts: Swellendam to Riversdale).•

bodkinii Bolus   Slender or robust geophyte to 20 cm. Leaves 3–7, cauline, linear. Flowers in a dense raceme, dull purplish pink, tepals linear-lanceolate, 12–15 mm long, lip with minute side lobes, anther loculi dark purple. Nov.–Dec., after fire. Sandstone seeps, SW, LB (Cape Peninsula to Mossel Bay).•

POLYSTACHYA   SNOUT ORCHID   c. 200 spp., pantropical, but mostly African

ottoniana Rchb.f.   Epiphyte to 15 cm, pseudobulbs often forming chains, 10–25 mm long. Leaves 2 or 3, oblong-lanceolate. Flowers few in a lax raceme, not resupinate, cream or white, lip with a median yellow streak, sepals 7.5–13 mm long, lip hooded and recurved above, with a central ridge-like callus. Aug.–Dec. Forest, LB, SE (Swellendam to N Province).

pubescens (Lindl.) Rchb.f.   Epiphyte or lithophyte to 20 cm, pseudobulbs clustered, 20–60 mm long. Leaves 1–3, oblong-elliptic. Flowers few in a lax raceme, not resupinate, deep yellow, lip and lateral sepals often with reddish markings, sepals 8–11 mm long, lip 3-lobed, flat, ecallose. Oct.–Dec. Forests or forest margins, SE (Port Elizabeth to Zimbabwe).

PTERYGODIUM (= *ANOCHILUS*)   BONNET ORCHID, MOEDERKAPPIE   18 spp., mostly southern Africa

A. Lip uppermost (flowers hyper-resupinate)

hallii (Schelpe) Kurzweil & H.P.Linder (= *Anochilus hallii* Schelpe)   Like **P. inversum** but lip appendage undivided. Sept.–Oct. Short dry scrub, NW, SW (Namaqualand and W Karoo to Langebaan).

inversum (Thunb.) Sw. (= *Anochilus inversus* (Thunb.) Rolfe)   Robust tuberous geophyte to 55 cm. Leaves cauline, imbricate, lanceolate. Flowers many in a dense raceme, c. 25 × 12 mm, pale green, hyper-resupinate, lip appendage ovate-oblong, distinctly bilobed apically. Sept.–Oct. Mostly coastal forelands, NW, SW (Piketberg to Kleinmond).•

A. Lip lowermost (flowers resupinate)
B. Lip appendage with 2 broad lobes spanning flower

alatum (Thunb.) Sw.   Slender tuberous geophyte to 20 cm. Leaves clustered basally, narrowly elliptic. Flowers 2–many in a fairly dense raceme, c. 24 × 15 mm, pale greenish yellow, lip with 2 broad lobes and a small pointed midlobe, appendage spathulate, 6–8 × 3 mm. Aug.–Oct. Sandy flats and slopes, NW, SW, AP, LB, SE (Cedarberg Mts to Humansdorp).•

caffrum (L.) Sw.   Slender to fairly robust tuberous geophyte to 30 cm. Leaves cauline, ovate-lanceolate. Flowers many in a fairly dense raceme, c. 28 × 15 mm, sepals pale greenish, petals and lip yellow, lip with 2 broad lobes, appendage erect and rather stout. Sept.–Nov. Sandstone slopes, NW, SW, LB, SE (Piketberg to Port Elizabeth).•

pentherianum Schltr.   Like **P. caffrum** but flowers smaller and pale green. Aug.–Sept. Clay flats in renosterbos, NW (Clanwilliam to Cold Bokkeveld).•

B. Lip appendage with lateral lobes small or lacking

acutifolium Lindl.   Slender tuberous geophyte to 35 cm. Leaves cauline, oblong. Flowers few in a lax to subdense raceme, c. 17 × 15 mm, sepals lime-green, petals and lip yellow, lip deltate, appendage triangular, apex smooth and down-curved, 7–10 × 4 mm. Mainly Sept.–Dec., usually after fire. Marshes, rock flushes and seeps in fynbos, NW, SW, AP, LB, SE (Piketberg to Port Elizabeth).•

catholicum (L.) Sw.   Like **P. acutifolium** but flowers with yellowish green sepals and lip, often flushed red and apex of lip appendage denticulate. Sept.–Nov., often after fire. Renosterveld, NW, SW, AP, KM, LB, SE (Bokkeveld Mts to Port Elizabeth).•

connivens Schelpe   Slender tuberous geophyte to 25 cm. Leaves 1 or 2, linear. Flowers few in a lax raceme, c. 25 × 10 mm, yellowish green, sepals connivent, lip appendage cruciform, c. 8 mm tall. Oct. Marshes in fynbos, SW (Cape Peninsula).•

cruciferum Sond.   Slender to fairly robust tuberous geophyte to 40 cm. Leaves 2 or 3, linear to elliptic. Flowers few in a lax raceme, c. 35 × 30 mm, yellowish green, lip very narrowly linear, appendage cruciform, 8–14 mm tall. Sept.–Nov. Sandy flats, SW, AP (Mamre to Pearly Beach).•

newdigateae Bolus   Slender tuberous geophyte to 35 cm. Leaves cauline, 2–4, elliptic-oblong. Flowers few in a lax raceme, c. 30 × 15 mm, greenish yellow, mostly cleistogamous. Mainly Nov. Stony slopes, SE (Knysna to E Cape).

platypetalum Lindl.   Slender, often flexuose, tuberous geophyte to 20 cm. Leaves usually 2, narrowly elliptic. Flowers 1 or 2, c. 18 × 15 mm, pale yellowish green, lip deltate, appendage triangular, central lobe reflexed, lateral lobes erect, 4–5 mm long. Sept.–Oct., often after fire. Sandstone slopes, often on ledges, NW, SW, LB (Piketberg to Swellendam).•

schelpei H.P.Linder   Like **P. volucris** but flowers fewer, white. Sept.–Oct. Damp clay flats and slopes, NW, SW, KM (Richtersveld and W Karoo to Stellenbosch).

volucris (L.f.) Sw.   Slender tuberous geophyte to 50 cm. Leaves cauline, ovate-oblong. Flowers many in a lax to dense raceme with deflexed bracts, sepals green, petals and lip lime, c. 15 × 12 mm, lip deltate, appendage cup-like, anther erect. Sept.–Oct. In sandy or clayey soils among bushes, NW, SW, AP, KM, LB, SE (Bokkeveld Mts to Humansdorp).•

SATYRIUM (= *SATYRIDIUM*)   SATYR ORCHID, TREWWA   c. 88 spp., Africa, Madagascar, extending to Asia

A. Spurs absent or saccate

bicallosum Thunb.   Slender to stout tuberous geophyte to 20 cm. Leaves 1–3 near soil surface, partly spreading, ovate, bracts spreading. Flowers many in a dense raceme, dull white, faintly greenish and with a pale purple patch above entrance to spurs, sepals 2–4 mm long, spurs saccate. Oct.–Nov., after fire. Sandy flats and slopes, NW, SW, LB, SE (Cedarberg Mts to Knysna).•

bracteatum (L.f.) Thunb.   Slender to stout geophyte to 30 cm. Leaves 3–8, cauline, lanceolate, bracts spreading. Flowers few to many in a lax to dense raceme, dull white or yellow with dark reddish brown markings, sepals 4–6 mm long, lip hemispherical, spurs saccate. Sept.–Nov. Sandy flats or peaty ledges, NW, SW, LB, SE (Cedarberg Mts to Cape Peninsula to N Province).

muticum Lindl.   Stout tuberous geophyte to 35 cm. Leaves 1 or 2, adpressed to the ground, broadly ovate to orbicular, bracts deflexed. Flowers few in a dense raceme, pink, petals ivory with purple speckles, sepals 14–16 mm long, spurs minute, saccate or lacking. Aug.–Sept. Moist slopes, LB, SE (Heidelberg to Plettenberg Bay).•

pumilum Thunb.   Tuberous, mostly acaulescent geophyte to 5 cm. Leaves 3–5, cauline, lanceolate. Flowers usually 2–4, relatively large, dull green outside, lip inside dull greenish yellow marked with transverse bars of dark maroon, sepals and petals fused for most of their length, 8–11 mm long, spurs saccate. Sept.–Oct. Damp flats and rock ledges, NW, SW, LB (Namaqualand to Riversdale).

retusum Lindl.   Slender tuberous geophyte to 15 cm. Leaves basal, 1 or 2, broadly ovate, bracts spreading. Flowers many in an elongate raceme, greenish to yellowish white with deep red lines inside lip, sepals 1–4 mm long, spurs saccate. Sept.–Nov., often after fire. Seeps and peaty soils, NW, SW, KM, LB, SE (Cold Bokkeveld to Cape Peninsula to Humansdsorp).•

striatum Thunb.   Slender tuberous geophyte to 10 cm, viscous on most surfaces. Leaves basal, 1 or 2, spreading, broadly ovate, bracts erect. Flowers 1–8 in a lax raceme, dull yellow to greenish white, with dark purple, striped markings, sepals 4–6 mm long, spurs saccate. Sept.–Oct. Moist flats and slopes, NW, SW (Cold Bokkeveld to Cape Peninsula).•

A. Spurs slender
*B. Spurs shorter than ovary (see also **S. stenopetalum**)*

coriifolium Sw.   EWWA TREWWA   Stout tuberous geophyte to 80 cm. Leaves 2–4, basal and spreading, purple-spotted at base, leathery, elliptic to ovate, bracts deflexed. Flowers 7–20 in a dense raceme, bright yellow to bright orange, sepals 7–13 mm long, spurs 9–12 mm long. Aug.–Oct. Moist clay and sand, NW, SW, AP, KM, LB, SE (Cedarberg Mts to Cape Peninsula to Port Elizabeth).•

erectum Sw.   GEEL TREWWA   Stout tuberous geophyte to 60 cm. Leaves 2, prostrate, ovate-elliptic, bracts deflexed. Flowers 11–37 in a dense raceme, pale to deep pink with darker tinges and spots on petals, spurs 5–11 mm long. July–Oct. Dry sandstone and clay flats, NW, SW, KM, LB, SE (Namaqualand and W Karoo to Baviaanskloof).

ligulatum Lindl.   Slender or stout tuberous geophyte to 55 cm. Leaves 1–4, cauline but clustered near stem base, ovate to narrowly ovate, bracts deflexed. Flowers many in a slender raceme, yellowish green to dull creamy white, tepals attenuate and soon drying above, sepals 5–10 mm long, spurs 5–10 mm long. Sept.–Dec. Scrub, forest and grassland, NW, SW, AP, KM, LB, SE (Namaqualand to KwaZulu-Natal).

rhynchanthum Bolus   (= *Satyridium rostratum* Lindl.)   Robust tuberous geophyte to 40 cm. Leaves 2–6, cauline, spreading to partly erect, linear-lanceolate, bracts deflexed. Flowers 10–40 in a rather dense raceme, white to pale pink with dark purple spots, sepals 5–7 mm long, spurs 5–6 mm long. Nov.–Dec., after fire. Marshy sandstone slopes, NW, SW (Cold Bokkeveld to Bredasdorp).•

rupestre Schltr. ex Bolus   Slender tuberous geophyte to 40 cm. Leaves 2–4, borne near ground level, spreading, ovate, bracts deflexed. Flowers few to many in a rather dense raceme, white with pale green tinges, sepals 2–4 mm long, lip with 2–4 mm long spurs. Nov.–Dec. Moist forest and rock ledges, NW, KM, SE (Worcester to Kareedouw).•

B. Spurs longer than ovary

acuminatum Lindl.   Slender to stout tuberous geophyte to 60 cm, usually much less. Leaves 2, prostrate, ovate to subrotund, bracts deflexed. Flowers many in a dense raceme, white to pale pink, sepals c. 4 mm long, spurs 12–22 mm long. Sept.–Dec., especially after fire. Damp sandstone slopes, KM, LB, SE (Swellendam to E Cape).

bicorne (L.) Thunb.   Stout tuberous geophyte to 60 cm, stem with conspicuous tubular sheaths. Leaves usually 2, prostrate, ovate to rotund, bracts deflexed. Flowers 4–40 in a dense to lax raceme, pale greenish yellow, sepals 6–9 mm long, spurs 10–22 mm long. Sept.–Oct., especially after fire. Scrub, NW, SW, AP, LB, SE (Namaqualand to Knysna).

candidum Lindl.   WIT TREWWA   Stout tuberous geophyte to 60 cm. Leaves 2, prostrate, broadly ovate to subrotund, bracts deflexed. Flowers few in a lax raceme, white to ivory, faintly pink-tinged, sepals 9–12 mm long, spurs 10–20 mm long. Sept.–Oct., often after fire. Sandy flats, NW, SW (Citrusdal to Hermanus).•

carneum (Dryand.) Sims   ROOI TREWWA   Stout tuberous geophyte to 80 cm. Leaves 2–4, thick and fleshy, lowest 2 partly prostrate, bracts deflexed. Flowers many in a dense raceme, pale pink to rose, sometimes white, sepals 13–18 mm long, spurs 14–20 mm long. Sept.–Nov. Coastal flats, SW, AP (Cape Peninsula to Riversdale).•

foliosum Sw.   Slender or robust tuberous geophyte to 30 cm. Leaves 2 or 3, near base, ovate, bracts deflexed. Flowers 4–30 in a dense raceme, pale yellowish green, tinged purple-brown on lateral sepals and lip, sepals 5–6 mm long, spurs 20–25 mm. Jan.–Feb. Peaty soil on sandstone slopes, SW (Cape Peninsula and Kogelberg).•

hallackii Bolus   Stout tuberous geophyte to 60 cm. Leaves 4–6, cauline, partly erect, narrowly oblong to elliptic-oblong, bracts partly deflexed. Flowers many in a dense raceme, pink, sepals 6.5–10.5 mm long, spurs 7–31 mm long. Dec.–Jan. Coastal flats and inland marshes, SW, SE (Cape Peninsula to Betty's Bay, Port Elizabeth to N Province).

humile Lindl.   Slender tuberous geophyte to 40 cm. Leaves usually 2, spreading near the ground, ovate, bracts deflexed. Flowers many in a dense raceme, pale cream tinged green, pink or brownish, sepals 6–9 mm long, spurs 12–26 mm long and diverging from ovary. Mainly Oct., especially after fire. Stony sandstone slopes, NW, SW, AP, KM, LB (Citrusdal to Riversdale).•

longicauda Lindl.   Stout tuberous geophyte to 80 cm. Leaves 1 or 2, ovate, on a separate shoot next to flowering stem, bracts deflexed. Flowers many in a lax raceme, white to pink and usually partly with pink tinge, spurs filiform, 15–46 mm long. Dec.–Feb. Moist grassland, SE (Plettenberg Bay to tropical Africa).

longicolle Lindl.   Slender to stout tuberous geophyte to 40 cm. Leaves 2, prostrate, broadly ovate to rotund, bracts deflexed. Flowers 10–26 in a lax raceme, pale pink to ivory, petals and sepals with dark purple markings, spurs 19–30 mm long. Oct.–Dec., mainly after fire. Damp slopes, LB, SE (Riversdale to Grahamstown).

lupulinum Lindl.   Slender to stout tuberous geophyte to 30 cm. Leaves 1–3, basal, suberect, often purple on reverse, bracts deflexed. Flowers 12–20, in a rather dense raceme, dull yellowish green and usually tinged dark purple, sepals 6–8 mm long, spurs 15–18 mm long. Sept.–Oct., often after fire. Moist, sandy slopes or flats, SW, AP, LB, SE (Cape Peninsula to Port Elizabeth).•

membranaceum Sw.   Stout tuberous geophyte to 60 cm. Leaves 2, prostrate, broadly ovate-elliptic, leaf sheaths on the stem dry and membranous, bracts deflexed, dry. Flowers 8–22, pale to deep pink, sepals 9–11 mm long, petal margins lacerate, spurs 20–27 mm long. Sept.–Nov. Grassy slopes, LB, SE (Swellendam to E Cape and Lesotho).

odorum Sond.   SOET TREWWA   Robust tuberous geophyte to 55 cm. Leaves 2–6, cauline, ovate to narrowly ovate, bracts deflexed. Flowers few to many in a moderately dense raceme, yellow with a purple tinge, sepals 3–7 mm long, lip with narrow entrance and spurs 13–18 mm long. Aug.–Oct. Scrub, NW, SW, AP, SE (Saldanha Bay to Riversdale).•

outeniquense Schltr.   Slender tuberous geophyte to 30 cm. Leaves 2–4, basal, bracts deflexed. Flowers few in a lax raceme, pale yellow, lip with 2 spurs 12–16 mm long. Oct.–Nov., mostly after fire. Sandstone slopes, LB, SE (Swellendam to Humansdorp).•

pallens S.D.Johnson & Kurzweil   Stout tuberous geophyte to 50 cm. Leaves 2, prostrate, fleshy, ovate-elliptic, bracts deflexed. Flowers few to many in a dense raceme, white or cream, sepals 10–15 mm long, spurs 19–22 mm. Oct. Dry rocky sandstone slopes and karroid scrub, NW, KM (Karoo Poort to Outeniqua Mts: Perdepoort).•

parviflorum Sw.   Slender or robust tuberous geophyte to 30 cm, sometimes leafless. Leaves (if present) 2–4, on a separate shoot, elliptic to broadly ovate, bracts deflexed. Flowers many in a lax to dense raceme, yellowish green to maroon, petals drying shortly after anthesis, sepals 2–4 mm long, lip with small entrance and spurs 5–15 mm long. Dec.–Feb. Sandy flats and marshy grassland, SE (Mossel Bay to N Province).

princeps Bolus   Stout tuberous geophyte to 70 cm. Leaves 2, very broadly ovate-elliptic, prostrate, leaf sheaths dry and membranous, bracts deflexed, dry. Flowers many in a dense raceme, rose-pink to carmine-red, petals with crisped margins, sepals 11–14 mm long, spurs 16–20 mm long. Mainly Oct. Coastal dunes, SE (Wilderness to Port Elizabeth).•

pygmaeum Sond.   Slender tuberous geophyte to 15 cm. Leaves 1–3, spreading near the ground, ovate-acute, bracts deflexed. Flowers many in a moderately dense raceme, dull yellowish green tinged pale to dark purple, sepals 3–4 mm long, spurs 6–10 mm long. Oct.–Nov. Wet mossy ledges, NW, SW, KM, LB, SE (Citrusdal to Plettenberg Bay).•

stenopetalum Lindl.   Slender or robust tuberous geophyte to 60 cm. Leaves stiff-textured, basal, suberect, elliptic, bracts erect. Flowers rather few in a lax raceme, white, sepals 5–13 mm long, spurs 9–30 mm long. Sept.–Dec. Sandstone slopes and rock flushes, NW, SW, AP, KM, LB, SE (Clanwilliam to Humansdorp).•

SCHIZODIUM   WIRE ORCHID, KAPOTJIE   6 spp., mainly W Cape

A. Sepals linear-lanceolate, acuminate

cornutum (L.) Schltr.   Slender tuberous geophyte to 30 cm, with wire-like, flexuose stem. Leaves rosulate, obovate-oblanceolate. Flowers few in a lax raceme, white to rose-red, lip pale green, sepals recurved, narrowly lanceolate, median sepal 7–12 mm long and spur 5–10 mm long. Aug.–Sept. Sandy flats, NW, SW (W Karoo to Kleinrivier Mts).

longipetalum Lindl.   Like **S. cornutum** but to 20 cm, flowers greenish yellow, with recurved lateral sepals, 7–12 mm tall, spur 2–3 mm long, petals and lip with a long apical mucro. Aug.–Sept. Proteoid shrubland, NW, SW (Piketberg to Paarl).•

A. Sepals ovate-lanceolate, acute

bifidum (Thunb.) Rchb.f.   Slender tuberous geophyte to 30 cm, with wiry flexuose stem. Leaves basal, obovate-oblanceolate. Flowers few in a lax raceme, pink, lip apex greenish to purple, sepals lanceolate, 10–15 mm long, spur 10–13 mm long, upcurved. Mainly Sept.–Oct. Sandy flats, NW, SW, AP, KM, LB, SE (Cedarberg Mts and S Karoo to Port Elizabeth).

flexuosum (L.) Lindl.   BACON AND EGGS, SPEK EN EIERS   Like **S. bifidum** but flowers with white sepals and yellow petals and lip, the latter with black spots, sepals ovate, 7–11 mm long, spur 2–4 mm long. Sept.–Oct. Seasonally moist, sandy flats, NW, SW (Bokkeveld Mts to Caledon Swartberg).

inflexum Lindl.   Like **S. bifidum** but flowers smaller, sepals 8–10 (–12) mm long and spur straight or decurved. Oct.–Nov. Sandstone slopes, NW, SW, KM, LB, SE (Bokkeveld Mts to E Cape).

obliquum Lindl.   Like **S. inflexum** but spur constricted at base. Mainly Aug.–Sept. Sandstone slopes, NW, SW, LB (Ceres to Cape Peninsula to Riversdale).•

STENOGLOTTIS   MADONNA ORCHID   4 spp., southern and tropical Africa

woodii Schltr.   Slender tuberous geophyte to 20 cm. Leaves basal, narrowly elliptic. Flowers few to many in a lax raceme, white to pink, lip 3-lobed, 5–9 mm long, spur 2–3 mm long. Dec.–Mar. Sandstone cliffs, SE (Wilderness and E Cape to Zimbabwe).

TRIDACTYLE   TRIDENT ORCHID   c. 30 spp., southern and tropical Africa

bicaudata (Lindl.) Schltr.   Stout monopodial epiphyte or lithophyte to 35 cm. Leaves leathery, somewhat succulent, oblong, notched apically. Flowers few to many in a dense, axillary raceme, c. 10 mm diam., pale yellowish brown, lip tridentate with side lobes fringed apically, spur 9–13 mm long. Oct.–Feb. Dry forests, SE (George to tropical Africa).

POACEAE
with H.P. Linder, **Ehrharta** by A.V. Verboom

1. Spikelets 2-flowered, lower floret male or sterile (exceptionally bisexual) and upper floret bisexual; lower lemma not dorsally awned; spikelets often falling entire at maturity but if breaking up usually lower floret and glumes persistent (reduction or suppression of lower floret or rarely the upper sometimes results in spikelets being apparently 1-flowered, such spikelets may usually be distinguished from truly 1-flowered spikelets by being dorsally compressed or not laterally compressed and by falling entire at maturity) :
   2. Glumes usually membranous, rarely one or both indurated, often unequal (the lower usually shorter); lower lemma like glumes in texture, upper lemma usually firmer, harder and more rigid; spikelets solitary, in pairs, threes or clusters, and more or less alike in form:
      3. Upper lemma awned from a 2-lobed apex, awn bent and twisted; spikelets laterally compressed, breaking up readily between lower and upper floret; upper floret not conspicuously harder and firmer than the lower:
         4. Spikelets sessile or subsessile, arranged in clusters of 3 ....**Tristachya**
         4.' Spikelets pedicelled, solitary or paired ....**Arundinella**
      3.' Upper lemma awnless or with a straight awn from entire apex; spikelet dorsally compressed or not compressed (rarely laterally compressed), usually falling entire; upper floret usually conspicuously harder and firmer than the lower:
         5. Spikelets surrounded by bristles that fall with them at maturity; inflorescence a spike-like panicle:
            6. Bristles flattened below, often spiny, some or all joined together below ....**Cenchrus**
            6.' Bristles not flattened or spiny, free ....**Pennisetum**
         5.' Spikelets not subtended by bristles or if so these persistent, spikelets falling without them at maturity:
            7. Upper glume and lower lemma lobed or notched at apex, shortly awned from between lobes; lower glume almost equally wide throughout; spikelets densely hairy with long silky hairs ....**Melinis**
            7.' Upper glume and lower lemma entire at apex, awnless or with a terminal awn; lower glume, if developed, wider at base and more or less surrounding spikelet there or if not then inflorescence not an open or contracted panicle:
               8. Glumes conspicuously awned, awns viscid; racemes often reduced to small clusters of spikelets..**Oplismenus**
               8.' Glumes awnless or minutely awned:
                  9. Spikelets, or some of them, subtended by 1 to many bristles; inflorescence frequently spike-like, dense and cylindrical..**Setaria**
                  9.' Spikelets not subtended by bristles, pedicels or rachis sometimes hairy but inflorescence not spike-like, dense and cylindrical:
                     10. Inflorescence an open or contracted panicle, sometimes spike-like and cylindrical, branches sometimes reduced to racemes but then not dense and conspicuously one-sided ......**Panicum**
                     10.' Inflorescence a one-sided true or false spike or spike-like raceme, usually 2 to many such units variously arranged:
                        11. Upper lemma with flat, translucent margins that are thinner in texture than rest of lemma, covering much or most of the palea ....**Digitaria**
                        11.' Upper lemma with narrowly inrolled margins of same texture as rest of lemma, exposing much of palea:
                           12. Inflorescence a spike, or of few to many spikes sunk in cavities on central axis...**Stenotaphrum**
                           12.' Inflorescence usually of spikes or spike-like racemes variously arranged, not sunk in cavities nor pressed against central axis; rachis of raceme terminated by a spikelet:
                              13. Lower glume absent or turned away from rachis................................**Paspalum**
                              13.' Lower glume always present and adjacent to rachis of raceme ....**Brachiaria**

2.' Glumes of bisexual or all spikelets usually indurated and equal or almost equal, enclosing florets; lower lemma like upper in texture, both thinly membranous and usually transparent or the upper reduced to a stalk-like scale at base of a stout awn; spikelets often in pairs with 1 of each pair sessile, the other pedicelled, those of each pair often dissimilar in sex, shape and form, rarely pedicelled spikelet completely suppressed and represented by an empty pedicel:
  14. Inflorescence a panicle, open, contracted or spike-like, with racemes shorter than central axis; racemes usually numerous, not supported by spathes:
    15. Panicle plumose with white or cream-coloured hairs, small, narrow and usually spike-like; spikelets awnless; stamens 1 or 2; lodicules absent...**Imperata**
    15.' Panicle light or bright brown or purplish, large and usually contracted; spikelets usually awned; stamens 3; lodicules 2 ....**Miscanthidium**
  14.' Inflorescence various, if paniculate then with racemes either longer than central axis or else supported by spathes individually, in pairs or in groups:
    16. Inflorescence of few to many spathe-supported, more or less triangular units on slender, usually drooping branches, a unit consisting of 1–5 short racemes, each with an involucre of spikelets at base and supported by a spathe..**Themeda**
    16.' Inflorescence not as above:
      17. Racemes solitary at end of culm and each flowering branch; branches, if present, sometimes numerous so that inflorescence consists of many solitary racemes; spikelets awned or awnless:
        18. Spikelets awnless or with straight or curved awns from glumes, lemmas awnless
          19. Racemes conspicuously white-hairy; pedicels not fused to rachis ....**Elionurus**
          19.' Racemes glabrous; sessile spikelets more or less sunk in depressions on rachis; pedicels flattened, wholly or partly fused to rachis ....**Hemarthria**
        18.' Bisexual sessile spikelets with a bent and twisted awn from upper lemma:
          20. Spikelets with 1 of each pair sessile, the other pedicelled; racemes with lowest 1 to many sessile spikelets awnless and like the pedicelled in shape and sex ....**Heteropogon**
          20.' Spikelets all pedicelled; racemes with spikelets of all pairs consisting of 1 awned and 1 awnless spikelet ....**Trachypogon**
      17.' Racemes paired, or 3 to many arranged subdigitately or on a central axis shorter than racemes, inflorescence consisting of 1 to many such units; sessile or all spikelets with a bent and twisted awn:
        21. Spikelets all with a bent and twisteexuous awn from upper lemma:
          22. Racemes almost glabrous; spikelet pairs not crowded; awns flexuous, hair-like; lower glume with a narrow longitudinal median groove ....**Microstegium**
          22.' Racemes conspicuously hairy; spikelet pairs crowded; awns twisted and bent; lower glume shallowly or deeply concave ....**Eulalia**
        21.' Sessile (or short-pedicelled) spikelets with a bent and twisted awn, pedicelled spikelets awnless or suppressed.
          23. Raceme pairs or groups not supported by spathes, or spathes inconspicuous and racemes far exserted from them:
            24. Sessile spikelets with upper lemma 2-lobed, awned from between lobes ....**Andropogon**
            24.' Spikelets all pedicelled, short-pedicelled spikelets with upper lemma gradually passing into awn ....**Trachypogon**
          23.' Raceme pairs or groups supported by and often partly enclosed in spathes:
            25. Lower glume of awned sessile spikelets rounded, or keeled only towards apex; awn from upper lemma hairy on lower part; plants not aromatic ....**Hyparrhenia**
            25.' Lower glume of awned sessile spikelets sharply keeled near margins throughout; awn from upper lemma glabrous; plants usually aromatic...**Cymbopogon**
1'. Spikelets 1–many-flowered, when more than 1-flowered the lowest floret usually bisexual and more rarely male, sterile or greatly reduced, and then spikelet either more than 2-flowered or lower lemma awned from low down on back; spikelets often laterally compressed, breaking up at maturity (floret/s deciduous, glumes persistent) :
  26. Spikelets strictly 1-flowered, without reduced florets or rudiments above or below:
    27. Glumes minute or absent; palea 3–5-nerved; stamens 6 ....**Leersia**
    27.' Both glumes well developed, or the lower greatly reduced or absent; palea 2-nerved or nerveless; stamens 1–3:
      28. Inflorescence a spike or spike-like raceme or of 2 to many one-sided spikes or spike-like racemes variously arranged:
        29. Spikelets arranged in 1 or 2 rows on one side of rachis of 1 to many variously arranged spikes or spike-like racemes:
          30. Fertile lemma 1-nerved; spikelets falling entire at maturity......................................**Spartina**
          30.' Fertile lemma 3-nerved, occasionally with 2–4 additional short nerves (rarely nerveless); spikelets usually breaking up at maturity with glumes often persistent; inflorescence of 2 to many spikes arranged digitately or subdigitately ....**Cynodon**
        29.' Spikelets arranged on opposite sides or all round rachis of a solitary spike or spike-like raceme (rarely a panicle):

31. Lemma 5-nerved; spikelets in clusters of 3, of which usually only the middle one has bisexual floret, lateral ones male, sterile and greatly reduced, sometimes to awns ....**Hordeum**
31.' Lemma 1–3-nerved; spikelets pressed flat against rachis of spike, awnless, sessile, glabrous:
   32. Both glumes present, placed side by side ....**Parapholis**
   32.' Lower glume absent except in terminal spikelet..**Hainardia**
28.' Inflorescence a panicle, open, contracted, or spike-like and sometimes dense:
  33. Lemma hardened and rigid at maturity, usually cylindrical and tightly rolled, with 1 or 3 awns from entire or minutely 2-lobed apex:
    34. Lemma 3-awned ....**Aristida**
    34.' Lemma 1-awned:
      35. Awn placed off-centre; female-fertile florets swollen ....**Nassella**
      35. Awn terminal; female-fertile floret not swollen ....**Stipa**
  33.' Lemma membranous, frequently translucent, often laterally compressed, awnless or awned, awn when present usually dorsal, more rarely terminal or from between 2 short lobes:
    36. Spikelets falling entire and in clusters at maturity, burr-like; glumes with hooked hairs ....**Tragus**
    36.' Spikelets usually breaking up at maturity, not burr-like:
      37. Lemma 1–3-nerved, awnless; glumes often shorter than floret; similar to lemma in texture ....**Sporobolus**
      37.' Lemma 3–7-nerved or almost nerveless, often awned; glumes as long as or longer than floret:
        38. Ligule a fringe of hairs or a short membrane fringed with hairs ....**Pentaschistis**
        38.' Ligule a membrane:
          39. Glumes at least 10 mm long, firmer than lemmas; leaves mostly basal ....**Ammophila**
          39.' Glumes 1–7 mm long, similar in texture to lemmas or thinner; leaves not distinctly basally aggregated:
            40. Plants softly and densely hairy all over; panicle spike-like, dense, ovoid, woolly from plumose glumes and with long, fine awns from lemmas radiating from it.........**Lagurus**
            40.' Plants glabrous or scantily hairy; panicle open, contracted or spike-like, but not ovoid or woolly; glumes glabrous or shortly hairy:
              41. Glumes swollen at base; lemmas densely hairy ....**Gastridium**
              41.' Glumes not swollen below; lemmas glabrous or hairy:
                42. Spikelets disarticulating above the glumes; glumes awnless or minutely awned, awn much shorter than glumes, gaping in fruit.................................**Agrostis**
                42.' Spikelets falling entire; glumes usually awned, awn 2 to many times the length of glumes, spike-like panicle bristly from awns ....**Polypogon**
26.' Spikelets 2- to many-flowered, sometimes with only 1 bisexual floret and 1 or more male or sterile florets or rudiments above or below it:
  43. Spikelets 3-flowered, 2 lower florets usually sterile (more rarely male), either well developed or reduced to inconspicuous lemmas, uppermost floret bisexual:
    44. Lower florets represented by 2 well-developed lemmas (rarely the lower small and inconspicuous), 1 or both often transversely ridged, awnless or with a terminal awn, upper sometimes with appendages at base; inflorescence various, but not a cylindrical, spike-like, dense panicle of numerous spikelets ....**Ehrharta**
    44.' Lower florets represented either by 2 small inconspicuous awnless lemmas (both shorter than bisexual floret) or well developed and longer than bisexual floret (male or sterile) but then awned from back or from between 2 lobes, not transversely ridged and not with appendages; inflorescence a spike-like, usually dense more or less cylindrical panicle of numerous spikelets, rarely reduced to a scanty raceme:
      45. Glumes equal; lower 2 lemmas awnless, reduced to scales or up to half as long as uppermost one (rarely absent); stamens 3..**Phalaris**
      45.' Glumes unequal; lower 2 lemmas awned, larger than uppermost one; stamens 2 in bisexual floret ....**Anthoxanthum**
  43.' Spikelets 2- to many-flowered, lowest floret typically bisexual but if not then spikelet with more than 1 bisexual floret or with 1 or more reduced florets above bisexual floret or with only 1 male floret below it:
    46. Florets enveloped by long, silky hairs from rachilla or lower part of lemmas; lemmas awnless or with terminal awn; tall reed-like grasses with showy, plumose panicles:
      47. Plants dioecious, tussock-forming; leaves basal; spikelets unisexual; glumes 1-nerved ....**Cortaderia**
      47.' Plants bisexual, rhizomatous; leaves cauline; spikelets bisexual; glumes 3–5-nerved:
        48. Spikelets with lowest floret male or sterile; fertile lemmas glabrous but enveloped by long hairs from callus ....**Phragmites**
        48.' Spikelets with lowest floret bisexual; fertile lemmas with long hairs on back in lower part, callus shortly hairy ....**Arundo**
    46.' Florets not usually enveloped by long, silky hairs; lemmas sometimes conspicuously hairy but then plants either not reed-like with showy, plumose panicles or lemmas with a bent and twisted awn from between 2 lobes:
      49. Lemmas 9-lobed and 9-awned, awns scabrid or plumose ....**Enneapogon**

49.' Lemmas awnless or 1–3-awned:
  50. Lemmas 1–3-nerved (occasionally with 1–4 additional shorter nerves)
    51. Spikelets with 2 to many bisexual florets (rarely 1 but then spikelets not arranged in one-sided spikes or spike-like racemes and sterile florets smaller but not different in form to bisexual ones):
      52. Lemmas glabrous or with tubercle-based hairs on side nerves:
        53. Inflorescence of one-sided spikes arranged digitately ....**Eleusine**
        53.' Inflorescence paniculate, sometimes contracted:
          54. Leaves rigid and pungent; rachis spinescent......................................**Cladoraphis**
          54.' Leaves softer; rachis slender ....**Eragrostis**
      52.' Lemmas hairy on side or all nerves, hairs not tubercle-based:
        55. Lemmas 3-awned, lobed; awns straight ....**Triraphis**
        55.' Lemmas 1-awned or awnless:
          56. Leaves rigid, pungent, cauline ....**Odyssea**
          56.' Leaves not pungent and cauline ....**Diplachne**
    51.' Spikelets with 1 bisexual floret, exceptionally with 2; spikelets arranged in 1 or 2 rows on one side of rachis of 1 to many spikes or spike-like racemes (except in **Fingerhuthia**); reduced florets often different from the bisexual in size and form:
      57. Inflorescence a spike-like panicle, spikelets densely clustered on short branches and falling entire at maturity...**Fingerhuthia**
      57.' Inflorescence of 1 to many one-sided spikes or spike-like racemes variously arranged on a central axis; glumes persistent at maturity:
        58. Inflorescence a solitary, terminal, one-sided spike (rarely 2 spikes but then rachis hairy or woolly)..**Harpochloa**
        58.' Inflorescence of 2 to many spikes arranged digitately or subdigitately in 1 or more whorls; spikelets 2-flowered:
          59. Lower glume keeled, upper flat or rounded on back, 2-lobed and shortly awned..**Eustachys**
          59.' Both glumes keeled, awnless ....**Cynodon**
  50.' Lemmas 5–11-nerved (rarely 3-nerved):
    60. Ligule a fringe of hairs or a short membrane fringed with hairs:
      61. Glumes shorter than spikelet (rarely longer); lemmas awnless or awned from apex, awn not bent or twisted...**Tribolium**
      61.' Glumes as long as or longer than rest of spikelet, rarely slightly shorter; lemmas often with a bent and twisted awn from between 2 lobes:
        62. Spikelets 3–10-flowered, sometimes only 2 lowest florets bisexual, the upper or uppermost sterile and reduced:
          63. Lemmas notched or with 2 minute awnless lobes, with or without a short, straight, rounded awn from between them; spikelets 4–7 mm long; ....**Schismus**
          63.' Lemmas with 2 awned or awnless lobes, central awn usually bent and twisted and flattened in lower part; spikelets 7–55 mm long excluding awns:
            64. Pedicels articulated, bearded with long hairs at and above joint; lowest lemma usually differing from the others; hairs on lemma, if present, not in tufts or fringes ....**Chaetobromus**
            64.' Pedicels not articulated; lowest lemma like the others; hairs on lemma often in tufts or fringes:
              65. Spikelets 8–25 mm long; inflorescences longer than 6 cm............**Merxmuellera**
              65.' Spikelets 4–7 mm long; inflorescence to 6 cm long ....**Karoochloa**
        62.' Spikelets (1)2-flowered, both florets bisexual (rachilla often produced and occasionally bearing rudiment of another floret):
          66. Inflorescence a spike or spike-like raceme; spikelets awnless; plants annual..**Prionanthium**
          66.' Inflorescence usually a panicle, rarely a raceme but then spikelets awned and plants perennial:
            67.' Ovary apex hairy; glumes 1–3-nerved ....**Pentameris**
            67.' Ovary apex glabrous:
              68. Female-fertile lemmas and glumes of similar texture; lower glume 1–3-nerved..**Pentaschistis**
              68.' Female-fertile lemmas decidedly firmer than glumes; lower glume 5–7-nerved, 18–60 mm long ....**Pseudopentameris**
    60.' Ligule a membrane not fringed with hairs:
      69' Inflorescence a spike or spike-like raceme; both glumes developed, sometimes almost reduced to bristles:

70. Lemmas awnless, conspicuously hairy with long, silky hairs; uppermost 2 or 3 florets reduced to lemmas, usually tightly enclosing each other and forming a club-shaped or oblong body; inflorescence a spike-like, often one-sided raceme, spikelets solitary, frequently drooping; ovary glabrous ....**Melica**
70.' Lemmas awnless or awned, not conspicuously hairy with long, silky hairs; uppermost 2 or 3 florets not reduced to lemmas; inflorescence a spike or spike-like raceme, spikelets solitary or in clusters, not drooping; ovary minutely hairy at top or with a hairy appendage:
    71. Inflorescence a false spike with clusters of spikelets on reduced axes; spikelets of sexually distinct forms on same plant, female-fertile spikelets in triplets, the triplets shed together ....**Hordeum**
    71.' Inflorescence a single spike or raceme; spikelets solitary, all alike:
        72. Inflorescence a spike; lemmas 5-nerved, awnless; shoots aromatic ....**Thinopyrum**
        72.' Inflorescence a raceme; lemmas 7-nerved, awned; not aromatic ......**Brachypodium**
69.' Inflorescence a panicle, open, contracted or spike-like (rarely a spike but then with lower glume absent except in terminal spikelet):
    73. Glumes shorter than spikelet (upper glume sometimes longer); lemmas awnless, or awned from apex or from between 2 short lobes, awn not bent and twisted:
        74. Ovary with a conspicuous, hairy terminal appendage, styles arising laterally below it ....**Bromus**
        74.' Ovary glabrous or hairy above but without a hairy terminal appendage, styles terminal but sometimes rather far apart on top of ovary:
            75. Spikelets mixed, fertile and sterile, rigidly awned ....**Cynosurus**
            75.' Spikelets alike in sex and structure:
                76. Lemmas about as wide as long, cordate at base, closely imbricate, spreading horizontally...**Briza**
                76.' Lemmas longer than wide, erect or spreading obliquely:
                    77. Spikelets sessile and solitary on opposite sides of rachis of a spike (rarely inflorescence scantily branched in lower half); lower glume absent in all but terminal spikelet...**Lolium**
                    77.' Spikelets pedicelled in open or contracted panicles; all spikelets with 2 glumes (rarely not):
                        78. Lemmas tapering into a straight awn longer than body of lemma; plants annual..**Vulpia**
                      78.' Lemmas awnless, or awn shorter than body of lemma and plants perennial:
                            79. Plants perennial:
                                80. Lemmas acute or awned ....**Festuca**
                                80.' Lemmas obtuse, awnless ....**Puccinellia**
                        79.' Plants annual; lemmas awnless:
                            81. Upper glume 1-nerved, lower 0-nerved ....**Sphenopus**
                            81.' Upper glume 2- or 3-nerved, lower 1–3-nerved:
                                82. Lemmas rounded on back; panicle branches short and rigid, bearing spikelets almost from base; leaf blades attenuate ....**Catapodium**
                                82.' Lemmas keeled; panicle branches filiform, spikelets mostly near the ends; leaf blades hooded at tips ....**Poa**
73.' Glumes (at least the upper) usually as long as or longer than lowest floret, often longer than rest of spikelet; lemmas awnless or awned from back (rarely from apex or from between 2 lobes), awn often bent and twisted:
    83. Lemmas awnless, conspicuously hairy with long, silky hairs; uppermost 2 or 3 florets reduced to lemmas, usually tightly enclosing each other and forming a club-shaped or oblong body ....**Melica**
    83.' Lemmas awned (rarely awnless and then glabrous or minutely hairy); uppermost 2 or 3 florets not reduced to lemmas:
        84. Spikelets falling entire at maturity, 2-flowered; glumes not shining; lower lemma awnless, upper with a short, hooked awn from back near apex ....**Holcus**
        84.' Spikelets breaking up at maturity, 2–6-flowered; glumes shining, at least towards margins; lemmas usually awned but not as above, or awnless:
            85. Panicle contracted, usually spike-like and dense; lemmas awned from uppermost quarter or third of back or from between 2 short lobes; awn short, straight or slightly bent:
                86. Plants perennial; lemmas 3-nerved, entire, awnless or minutely awned from apex ....**Koeleria**
                86.' Plants annual; lemmas 5-nerved, awned from apex or from back just below apex, this splitting readily into 2 short lobes as far as awn insertion ....**Lophochloa**

85.' Panicle open, rarely contracted; lemmas awned from near base or middle of back, awn usually bent and twisted:
    87. Spikelets 7–35 mm long; lemmas 5–11-nerved; ovary hairy:
        88. Plants perennial; spikelets (2–)3–5(–6)-flowered; glumes 1–3-nerved, shorter than rest of spikelet ....**Helictotrichon**
        88.' Plants annual; spikelets 2- or 3-flowered; glumes 7–11-nerved, usually longer than rest of spikelet..**Avena**
    87.' Spikelets 2–8 mm long; lemmas 3–5-nerved; ovary glabrous:
        89. Plants perennial; spikelets 4–8 mm long; rachilla hairy, produced ....**Deschampsia**
        89.' Plants annual; spikelets 2–4 mm long; rachilla glabrous, not or minutely produced:
            90. Awn clavate, with a ring of minute hairs in outer part ....**Corynephorus**
            90.' Awn tapering, without a ring of hairs ....**Aira**

AGROSTIS 150–200 spp., mainly temperate and tropical mountains

***avenacea** C.C.Gmel. Annual to 60 cm. Leaves linear. Spikelets in an open panicle, 2.5–4 mm long. July–Mar. Disturbed areas, SW, AP, SE (Australian weed).

bergiana Trin. Annual or perennial to 75 cm. Leaves linear. Spikelets in a diffuse panicle, 1.5–2.5 mm long, pedicels 5–10 mm long. Oct.–Feb. Damp areas, NW, SW, AP, LB, SE (Saldanha to Mpumalanga and St Helena).

lachnantha Nees VINKAGROSTIS Annual or short-lived perennial to 90 cm. Leaves linear. Spikelets in a narrow, sinuous panicle, 1.5–3 mm long, branches suberect. Oct.–Mar. Damp sites, NW, SW, KM, LB, SE (S and tropical Africa).

***montevidensis** Spreng. ex Nees FOG GRASS Annual to 60 cm. Leaves linear. Spikelets in a diffuse panicle with hair-like branches, 1.5–2.5 mm long, pedicels longer than 20 mm. Nov.–Apr. Montane grassland, SW, SE (S American weed).

polygonoides Stapf Perennial to 90 cm. Leaves linear. Spikelets in a panicle, 4–4.5 mm long, glumes awned. Oct.–Jan. Wet places, NW, SW (Bokkeveld Mts to Stellenbosch).•

schlechteri Rendle Annual to 65 cm. Leaves linear. Spikelets in an open panicle, spikelets green and purple, 3–3.5 mm long, branches ascending. Jan.–Apr. Near water, NW, SW (Tulbagh to Hottentots Holland Mts).•

***AIRA** c. 8 spp., Old World, mainly Mediterranean

***cupaniana** Guss. Annual to 30 cm. Leaves linear. Spikelets in a delicate, open panicle, 2–3 mm long, silvery, awned. Sept.–Dec. Flats and slopes, NW, SW, AP, KM, LB, SE (European weed).

***AMMOPHILA** MARRAM GRASS 2 spp., N America and Eurasia

***arenaria** (L.) Link Robust rhizomatous perennial to 1 m. Leaves rolled, rigid. Spikelets in a narrow, spike-like panicle, straw-coloured. Oct.–Dec. Coastal dunes, NW, SW, AP, SE (European weed).

ANDROPOGON c. 113 spp., pantropical and subtropical, especially Africa and America

appendiculatus Nees BLUE GRASS Perennial to 1.2 m. Leaves linear. Spikelets in 4–20 narrow racemes per spathe, purple, shortly hairy, paired, sessile spikelet awned. Oct.–Apr. Moist sandstone slopes, NW, SW, AP, LB, SE (Olifants River Mts to tropical Africa).

eucomus Nees SILVER THREAD GRASS Perennial to 1 m. Leaves linear. Spikelets in 2–5 racemes per spathe, with white-silky hairs twice as long as spikelets, pedicellate spikelet vestigial, sessile spikelet awned. Nov.–May. Vleis and wet places, NW, SW, LB, SE (Olifants River Mts to tropical Africa and Madagascar).

ANTHOXANTHUM VERNAL GRASS 20 spp., N temperate and tropical mountains, Africa, Europe, Asia

dregeanum (Nees ex Trin.) Stapf SWEET VERNAL GRASS Aromatic perennial to 60 cm. Leaves rigid, linear-lanceolate, often folded, to 9 mm wide. Spikelets in spike-like panicle, 6–7 mm long, lower glume 3-nerved. Oct.–Jan. Upper mountain slopes, NW, SW, LB (Bokkeveld Mts to Swellendam).•

ecklonii (Nees ex Trin.) Stapf   Perennial to 80 cm, culms often bulbous below. Leaves linear-lanceolate, 4–9 mm wide. Spikelets in a spike-like panicle, 6–8 mm long, lower glume 1-nerved. Dec.–Apr. Moist sandstone slopes, LB, SE (Langeberg Mts: Swellendam to tropical Africa).

tongo (Trin.) Stapf   Like **A. dregeanum** but diffuse, leaves filiform, to 2 mm wide and panicle scanty. Sept.–Feb. Moist shady slopes, often rock ledges, NW, SW, AP (Grootwinterhoek to Agulhas).•

ARISTIDA   STEEKGRAS   300 spp., mainly tropical and subtropical

adscensionis L.   Annual to 1 m. Leaves linear. Spikelets in a narrow, often spike-like panicle, 10–40 mm long, lower glume mucronate. Dec.–Sept. Stony slopes and flats, often disturbed, KM (widespread in tropical Africa).

congesta Roem. & Schult.   KATSTERTSTEEKGRAS   Slender perennial or annual to 90 cm, densely tufted. Leaves linear. Spikelets in a congested panicle, 25–30 mm long, lower glume awned, lemmas articulated above. Dec.–May. Stony, mostly clay slopes, NW, SW, KM (widespread in tropical Africa).

diffusa Trin.   BESEMSTEEKGRAS   Perennial to 1 m. Leaves linear. Spikelets in an open panicle, 25–45 mm long, lower glume obtuse, lemmas sometimes articulated above. Oct.–Apr. Rocky slopes, NW, SW, AP, KM, SE (widespread in southern Africa).

junciformis Trin. & Rupr.   WIRE GRASS   Perennial to 90 cm. Leaves linear. Spikelets in a congested panicle, 20–30 mm long, lower glume awned, lemmas without articulation. Nov.–May. Mountain slopes, NW, SW, AP, KM, LB, SE (Cedarberg Mts to tropical E Africa).

*ARUNDINELLA   55 spp., pantropical

*nepalensis Trin.   BEESGRAS, RIETGRAS   Robust rhizomatous perennial to 1.5 m. Leaves linear, densely hairy. Spikelets in a dense panicle, 4–6 mm long, awned. Dec.–Mar. Marshes, NW, SE (Uitenhage through Africa to Asia, introduced weed in the Olifants River valley).

*ARUNDO   SPANISH REED   3 spp., Mediterranean and Taiwan

*donax L.   SPAANSRIET   Robust bamboo-like perennial to 3 m. Leaves lanceolate, deciduous. Spikelets 12–18 mm long, in a large, compact, plumose panicle 30–60 cm long. Jan.–Mar. Riverbanks and marshes, SW, SE (Mediterranean weed).

*AVENA   OATS   27 spp., Mediterranean to central Asia, widely introduced as weeds and crop plants

*barbata Brot.   WILD OATS, WILDEBAARDHAWER   Like **A. byzantina** but lemmas densely hairy. Sept.–Nov. Common in disturbed areas, NW, SW, AP, SE (Asian weed).

*byzantina K.Koch   Annual to 1 m. Leaves linear-lanceolate. Spikelets drooping, in an open panicle, 17–35 mm long, lemmas glabrous or sparsely hairy, awned. Sept.–Dec. Disturbed areas, NW, SW, SE (European weed).

*fatua L.   COMMON WILD OATS, GEWONE WILDEHAWER   Like **A. barbata** but lemma awnless and lemma teeth 1–1.5 mm long. Sept.–Nov. Widespread in disturbed areas, NW, SW, SE (European weed).

*sativa L.   OATS, HAWER   Like **A. byzantina** but awn with a distinct column, sometimes awnless. Sept.–Nov. Widespread in disturbed areas, NW, SW, LB, SE (cultivated cereal introduced from Europe).

*sterilis L.   Like **A. fatua** but spikelets to 46 mm long (versus 32 mm), lemma teeth to 0.5 mm long. Sept.–Nov. Disturbed areas, SW (European weed).

BRACHIARIA   c. 90 spp., worldwide, tropics and subtropics

serrata (Thunb.) Stapf   RED TOPPED SIGNAL GRASS, FERWEELGRAS   Perennial to 75 cm. Leaves linear to lanceolate. Spikelets in short, spike-like branches, with long, silky white or purple hairs. Oct.–May. Grassland, AP, LB, SE (Bredasdorp to tropical Africa).

BRACHYPODIUM   FALSE BROME   c. 17 spp., temperate and tropical mountains

*distachyon (L.) P.Beauv.   Annual to 45 cm. Leaves linear. Spikelets sessile, 2–6 in distichous racemes, lemma awn 10–25 mm long. Sept.–Jan. Disturbed areas, NW, SW, AP, KM (Mediterranean weed).

flexum Nees   VALSBROMUS   Straggling perennial to 90 cm. Leaves linear, soft. Spikelets 5–9 in flexuose racemes, lemma awns 4–8 mm long. Oct.–Apr. Shady slopes, bush and thicket, SW, LB, SE (Saldanha to Sudan).

*BRIZA   QUAKING GRASS   20 spp., N temperate Old World and S America

*maxima L.   GROOTBEWERTJIE   Annual to 40 cm. Leaves linear. Spikelets 3–12 in open panicles, drooping, longer than wide, to 25 × 15 mm. July–Dec. Widespread in disturbed areas, NW, SW, AP, LB, SE (Mediterranean weed).

*minor L.   KLEINBEWERTJIE   Like **B. maxima** but spikelets more than 20 per panicle, about as wide as long, to 5 × 6 mm. Sept.–Dec. Widespread in disturbed areas, NW, SW, AP, LB, SE (European weed).

BROMUS   BROME   50 spp., pantemperate

*catharticus Vahl   (= *Bromus unioloides* Humb., Bonpl. & Kunth, *B. willdenowii* Kunth)   Like **B. commutatus** but spikelets laterally compressed, glumes and lemmas sharply keeld, lemma awns to 3 mm long. Oct.–Apr. Moist or disturbed areas, NW, SW, AP, KM, LB, SE (S American weed).

commutatus Schrad.   Like **B. pectinatus** but awns not exceeding lemmas, 3–8 mm long. Sept.–Dec. Disturbed, wet places, SW, LB, SE (Cape Peninsula to Drakensberg).

*diandrus Roth   LANGNAALDBROMUS   Annual to 1 m. Leaves linear, hairy. Spikelets in a lax panicle, 30–90 mm long, lower glume 1-nerved, upper glume 3-nerved, lemma awn stiff and straight, 30–70 mm long. Sept.–Jan. Roadsides, NW, SW, AP, LB, SE (Mediterranean weed).

*hordeaceus L.   SOFT BROME   Like **B. commutatus** but panicle dense, contracted with all pedicels shorter than spikelets. Oct.–Feb. Disturbed areas. SW, SE (European weed).

leptoclados Nees   MOUNTAIN BROME GRASS   Perennial to 1.5 m. Leaves linear-lanceolate. Spikelets in an open panicle, 10–30 mm long, lower glume 3–5-nerved, upper glume 3–5-nerved, lemma awns to 5 mm long. Oct.–Feb. Moist shady places along rivers, SE (Knysna to tropical Africa).

*pectinatus Thunb. (= *Bromus japonicus* Thunb.)   JAPANESE BROME, HOOIGRAS   Annual to 80 cm. Leaves linear, hairy. Spikelets in an open panicle, green and purple, 10–30 mm long, drooping, lower glume 1–3-nerved, upper glume 5–7-nerved, lemma awns 6–18 mm long. July–Nov. Disturbed areas on richer soils, NW, SW, AP, KM, LB, SE (Asian weed).

*rigidus Roth.   RIPGUT BROME   Like **B. diandrus** and often hybridising with it, but panicle usually dense, lemma callus pointed. Sept.–Oct. Disturbed areas, NW, SW (European weed).

*CATAPODIUM   2 spp., Europe

*rigidum (L.) C.E.Hubb.   FERN GRASS   Annual to 35 cm. Leaves linear. Spikelets in a narrow rigid panicle, 5–7 mm long, awnless. Oct.–Dec. Waste land, shady areas, SW, AP, SE (European weed).

CENCHRUS   SANDBURR   22 spp., tropical and warm temperate

ciliaris L.   BUFFELSGRAS   Perennial to 1 m. Leaves linear. Spikelets in a bristly, false spike, bristles slender, 5–10 mm long, joined below spikelet, straw-coloured or purple. Aug.–Apr. Sandy flats, SW, KM, SE (Saldanha to India).

*incertis M.A.Curtis   Annual to 40 cm. Leaves linear. Spikelets in a spiny, open or compact, false spike, bristles/spines 2–5 mm long, joined below to form a hard spiny involucre. Jan.–Mar. Sandy flats, SW, SE (pantropical weed).

CHAETOBROMUS   GHAGRAS   1 sp., S Namibia to W Cape

dregeanus Nees   Stoloniferous or tufted perennial to 40 cm. Leaves linear-lanceolate, sometimes sparsely hairy. Spikelets in a panicle, 12–17 mm long, awned, with a tuft of hair at base of each spikelet. Sept.–Nov. Sandy slopes, NW, SW, KM (S Namibia and Karoo to Cape Peninsula).

*CHLORIS   55 spp., pantropical

*gayana Kunth   RHODES GRASS   Stoloniferous perennial to 1.2 m. Leaves linear. Spikelets in 1-sided, digitate spike-like racemes, 3–5 mm long, awn to as long as lemma, 1–10 mm. Nov.–May. Disturbed places, SW, SE (Indian weed).

*truncata R.Br.   Stoloniferous perennial to 45 cm. Leaves narrowly oblong, obtuse. Spikelets in 1-sided spike-like racemes, awns longer than lemma, 6–12 mm long. June–July. Disturbed places in cultivated paddocks, SW (Australian weed).

*virgata Sw.   FEATHERED CHLORIS, KLOSSIEGRAS   Like **C. gayana** but lemma with apical tuft of stiff hairs, awns longer than lemma, 2–15 mm long. Dec.–June. Disturbed places, NW, SE (cosmopolitan weed).

CLADORAPHIS (= *ERAGROSTIS* in part)   2 spp., coastal, W southern Africa

cyperoides (Thunb.) S.M.Phillips (= *Eragrostis cyperoides* (Thunb.) P.Beauv.)   STEEKRIET   Like **C. spinosa** but primary branches of panicle widely spaced, to more than twice their length apart and spikelets usually clustered and appressed, 4–8 mm long. Aug.–May. Coastal dunes, NW, SW (Angola to Cape Peninsula).

spinosa (L.f.) S.M.Phillips (= *Eragrostis spinosa* (L.f.) Trin.)   VOLSTRUISDORING   Spiny, bushy perennial to 60 cm. Leaves lanceolate, rolled, rigid, pungent. Spikelets in rigid panicles, primary branches persistent, spiny, less than their own length apart, 6–18 mm long, perpendicular to branchlets. Aug.–May. Sandy flats, mostly coastal, NW, AP (Namibia to Agulhas).

***CORTADERIA**   PAMPAS GRASS   c. 16 spp., South America and New Zealand

***selloana** (Schult.) Aschers. & Graebn.   Robust perennial to 3 m, in large tussocks. Leaves linear, keeled, serrulate. Spikelets in dense feathery panicles to 60 cm long, awned. Feb.–Apr. Sandy slopes and flats, SW, SE (Argentinian weed).

***CORYNEPHORUS**   5 spp., Europe

***fasciculatus** Boiss. & Reut.   Annual to 55 cm. Leaves linear, rolled. Spikelets in a panicle, c. 3 mm long, lemma awn basal. Oct.–Nov. Disturbed sandy soils, SW (European weed).

CYMBOPOGON   TURPENTINE GRASS   c. 40 spp., tropical and subtropical Africa, Asia

marginatus (Steud.) Stapf ex Burtt Davy   MOTWORTELTERPENTYNGRAS   Aromatic perennial to 80 cm. Leaves linear, 3–6 mm wide. Spikelets in paired racemes clustered in leafy spathes, rachis and pedicels long-hairy, spikelets 5–6.5 mm long, lower glume of sessile spikelet winged. Oct.–May. Rocky lower slopes, NW, SW, AP, KM, LB, SE (Namaqualand to E Cape).

plurinodis (Stapf) Stapf ex Burtt Davy   BITTER TURPENTINE GRASS   Perennial to 1 m. Leaves linear, usually folded, 2–4 mm wide. Spikelets as in **C. marginatus** but lower glume of sessile spikelets deeply concave, wingless. Oct.–May. Habitat? SW, AP, LB, SE (throughout S and tropical Africa).

prolixus (Stapf) E.Phillips   TAMBOEKIEGRAS   Like **C. marginatus** but racemes appearing glabrous, rachis and pedicels hairy only on sides. Oct.–Apr. Habitat? LB, SE (Swellendam to northern South Africa).

validus (Stapf) Stapf ex Burtt Davy (= *Cymbopogon nardus* (L.) Rendle)   GIANT TURPENTINE GRASS, TAMBOEKIEGRAS   Like **C. prolixus** but culms robust, 1.2–2.4 m tall and leaves broader, 5–10 mm wide. Oct.–Nov. Damp areas, SE (southern Africa to India and Burma).

CYNODON   QUICK GRASS   8 spp., pantropics and subtropics

dactylon (L.) Pers.   Mat-forming perennial to 40 cm. Leaves linear. Spikelets secund, mostly 4 or 5 in digitately branched spikes, 2–2.5 mm, lemma keel wingless, glumes to three-quarters the length of spikelet. Sept.–May. Mountains and flats, NW, SW, AP, KM, LB, SE (throughout Africa).

incompletus Nees   KAROO QUICK GRASS, SOETKWEEK   Stoloniferous perennial to 30 cm. Leaves linear. Spikelets secund in digitate spikes, 2.5–3 mm long, lemma keel winged, glumes mostly to half as long as spikelet. Nov.–May. Habitat? NW, KM, SE (Clanwilliam to Free State).

***transvaalensis** Burtt Davy   TRANSVAAL QUICK GRASS   Like **C. dactylon** but spikes usually 2 and reflexed at maturity. Nov.–May. Habitat? SW (N African weed).

***CYNOSURUS**   8 spp., Europe

***coloratus** Lehm. ex Nees   Like **C. echinatus** but awns purple below, fertile spikelets 1-flowered, anthers 0.4–0.6 mm long. Mar.–Apr. Habitat? SW (Mediterranean weed).

***echinatus** L.   DOG'S TAIL   Annual to 60 cm. Leaves linear. Spikelets in a congested, softly bristly panicle, sexually dimorphic, 2- or 3-flowered, anthers 3–4 mm long. July–Jan. Habitat? SW, KM (European weed).

DIGITARIA   FINGER GRASS   c. 200 spp., pantropical and subtropical

***abyssinica** (A.Rich.) Stapf   Mat-forming perennial to 35 cm. Leaves linear. Spikelets in 3–11 digitately arranged racemes, lemmas glabrous. Nov.–June. Habitat? SW, SE (tropical African weed).

argyrograpta (Nees) Stapf SILVER FINGER GRASS Perennial 60 cm. Leaves linear. Spikelets in paired racemes adhering by tangled hairs, lower lemma hairy, intervein area narrowest next to midrib. Nov.–Mar. Flats and slopes, SW, KM, LB, SE (Riviersonderend to Mozambique).

*****debilis** (Desf.) Willd. Like **D. sanguinalis** but upper and lower glumes separated by an internode. Nov.–June. Habitat? SW (Mediterranean weed).

eriantha Steud. WOOLLY FINGER GRASS Perennial to 1 m. Leaves linear to lanceolate. Spikelets in 3–15 digitately arranged racemes, lower lemma hairy, intervein area equal or broadest next to midrib. Jan.–Apr. Lower to middle slopes, NW, SW, AP, LB, KM, SE (throughout South Africa).

natalensis Stent Like **D. eriantha** but lower leaf sheaths rusty brown, ligule longer than 4 mm. Dec.–June. Sandy slopes, SW, SE (Cape Peninsula and George to Mpumalanga).

*****sanguinalis** (L.) Scop. CRAB GRASS, KRUISGRAS Spreading annual to 50 cm. Leaves linear. Spikelets in 3–12 digitately arranged racemes, lemmas scabrid. Nov.–May. Disturbed areas, SW, NW, LB, SE (European weed).

DIPLACHNE 18 spp., pantropics and subtropics

fusca (L.) P.Beauv. ex Roem. & Schult. SWAMP GRASS Perennial to 1.5 m. Leaves linear. Spikelets in a panicle, olive, without awns. Oct.–May. Damp to wet areas, NW, SW, AP, KM, LB, SE (throughout Africa and Old World tropics).

*****ECHINOCHLOA** MILLET 30–40 spp., pantropics and subtropics

*****crusgallii** (L.) P.Beauv. BARNYARD MILLET Stoloniferous annual to 1 m. Leaves lanceolate, ligule absent. Spikelets in panicles of short, 2–several-rowed racemes. Jan.–Apr. Marshes, seasonally damp places, SW, LB, SE (Eurasian weed).

EHRHARTA c. 36 spp., Africa, Mascarene Islands, SE Asia, Australasia

A. Annuals or weak (facultative) perennials

brevifolia Schrad. Annual to 50 cm. Leaf blades flat. Spikelets in a contracted panicle, glumes equal or longer than spikelet, sterile lemmas smooth, glabrous, truncate to mucronate, the second with a pair of ear-like, basal appendages. Aug.–Nov. Sandy coastal flats, NW, SW, AP (Namaqualand to Agulhas).

delicatula (Nees) Stapf Annual to 45 cm. Leaf blades flat, with wavy margins. Spikelets in an open panicle, small, green or green and purple, glumes shorter than spikelet, sterile lemmas corrugate, glabrous, blunt, the second with a pair of basal granular appendages. July–Nov. Shady habitats, often under bushes, NW, SW, LB (S Namibia to Swellendam).

erecta Lam. Sprawling or tufted perennial or annual to 80 cm. Leaf blades flat, with wavy margins. Spikelets in an open panicle, green, glumes shorter than spikelet, sterile lemmas corrugate, glabrous, blunt to mucronate, the second slightly constricted basally. Mainly Sept.–Jan. Shady habitats, often weedy, SW, AP, LB, SE (Cape Peninsula to E Africa).

longiflora J.E.Sm. Annual to 90 cm. Leaves linear-lanceolate, margins undulate. Spikelets in a verticillate panicle, spikelets green, sometimes with purple, glumes shorter than spikelet, sterile lemmas smooth to corrugate, glabrous, tapering into a straight awn, the second lightly constricted at base. July–Nov. Damp or shady habitats, sometimes weedy, NW, SW, AP, LB (Namaqualand to Mossel Bay).

triandra Nees ex Trin. BOKHORINGGRAS Annual to 45 cm. Leaves linear-lanceolate, margins undulate. Spikelets in a verticillate panicle, green sometimes with purple, glumes shorter than spikelet, sterile lemmas corrugate, glabrous, apically tapering and bending outwards to resembling springbok horns. July–Oct. Shady habitats, often among rocks, NW (S Namibia to Cedarberg Mts).

A. Perennials
B. Lowest culm internode(s) swollen and bulbous; leaves rolled or flat, often with wavy margins

bulbosa J.E.Sm. Spreading perennial to 70 cm, culm base swollen, whitish, matt, globose. Leaves linear-lanceolate, margins undulate. Spikelets in a verticillate panicle, straw-coloured and purple, glumes shorter than spikelet, sterile lemmas corrugate, glabrous, keel sometimes bristly, bilobed, with a straight awn, the second constricted at base. Oct.–Dec., mostly after fire. Flats and lower slopes, NW, SW, AP, LB, SE (Cape Peninsula to George).•

capensis Thunb. KNOL EHRHARTA Spreading perennial to 1 m, culm base swollen, yellow, shiny, elongate. Leaves linear-lanceolate, margins thickened, undulate. Spikelets in a verticillate panicle, straw-coloured and purple, glumes shorter than spikelet, sterile lemmas corrugate, glabrous, mucronate to aristate, the second constricted at base. Sept.–Dec., especially after fire. Rocky sandstone slopes, NW, SW, AP, KM, LB, SE (Cedarberg Mts to George).•

eburnea Gibbs-Russ.   Spreading perennial to 15 cm, culm base swollen, white, shiny, elongate. Leaves linear-lanceolate, margins undulate. Spikelets in a verticillate, raceme-like panicle, green and purple, glumes shorter than spikelet, sterile lemmas smooth, glabrous to shortly villous, keel long-bristly, mucronate. Sept.–Nov. Rocky slopes, KM (W Karoo and Bokkeveld Mts to Witteberg).

longifolia Schrad.   Perennial, to 1.2 m. Leaf blades rolled to tightly rolled, culm base swollen, whitish, matt, spherical. Spikelets in a verticillate panicle, straw-coloured and purple, glumes to as long as spikelet, sterile lemmas corrugate, glabrous, truncate to mucronate, the second strongly constricted basally. Nov.–Jan., ?mostly after fire. Mountain slopes, NW, SW (Cedarberg Mts to Cape Peninsula).•

ottonis Kunth ex Nees   Perennial, to 1.2 m, culm base swollen, orange, shiny, elongate. Spikelets in a verticillate panicle, straw-coloured or straw-coloured and purple or pink, glumes shorter than spikelet, sterile lemmas corrugate, glabrous, mucronate, the second constricted at base. Sept.–Nov., ?mostly after fire. Rocky slopes, SW, LB, SE (Cape Peninsula to Uitenhage).•

B. Lowest culm internode(s) not swollen and bulbous
C. Tufted perennials
D. Spikelets shorter than 8.5 mm; lemmas awnless to aristate

calycina J.E.Sm.   ROOIGRAS, POLGRAS   Perennial (sometimes annual) to 70 cm. Leaf blades linear-lanceolate, sometimes rolled, margins often undulate. Spikelets in an open panicle, green and pink, glumes to as long as spikelet, sterile lemmas villous, mucronate to aristate, the second with a pair of ear-like, basal appendages. Mainly July–Dec. Flats and slopes, NW, SW, AP, KM, LB, SE (Namaqualand to KwaZulu-Natal).

melicoides Thunb.   Perennial to 80 cm. Leaves linear-lanceolate, sometimes rolled. Spikelets in an open panicle, small, green or straw-coloured, glumes as long as to slightly longer than spikelet, sterile lemmas smooth, glabrous, blunt, the second with a pair of ear-like, basal appendages. Sept.–Nov. Rocky slopes, NW, SW, LB (Namaqualand to Swellendam).

D. Spikelets longer than 9 mm; sterile lemmas straight-awned

dura Nees ex Trin.   BRANDGRAS   Perennial to 80 cm. Leaves linear-lanceolate, flat to rolled. Spikelets in a verticillate panicle, green, glumes shorter than spikelet, sterile lemmas smooth to scabrid, glabrous, tapering into a long straight awn, 2–16 mm long. Oct.–Feb., only after fire. Middle to upper slopes, NW, SW, LB, SE (Tulbagh to Uitenhage).•

microlaena Nees ex Trin.   Tufted perennial, to 1 m. Leaf blades setaceous. Spikelets in a verticillate panicle, green,. glumes shorter than spikelet, sterile lemmas smooth to scabrid, glabrous, tapering into a long straight awn (13–25 mm long). Nov.–Feb. Damp upper slopes after fire, NW, SW (Tulbagh to Stellenbosch).•

C. Branching perennials
E. Leaves evergreen; lemmas blunt with canoe-shaped tips

rupestris Nees ex Trin.   Perennial to 45 cm. Leaves in 2 ranks, linear-lanceolate, sometimes rolled. Spikelets in a raceme, green, glumes less than half as long as spikelet, sterile lemmas smooth to granular, blunt, tips canoe-shaped. Nov.–Feb. High rocky slopes in damp areas, SW, KM, LB, SE (Cape Peninsula to Tsitsikamma Mts).•

setacea Nees   Perennial to 60 mm. Leaves in 2 ranks, linear-lanceolate, rolled. Spikelets in a raceme, green to purple, glumes half to as long as spikelets, sterile lemmas smooth to granular, glabrous, tips canoe-shaped. Sept.–Jan. High rocky slopes in damp areas, SW, LB (Cape Peninsula to Riversdale).•

E. Leaves often summer-deciduous, lemmas truncate to aristate

ramosa (Thunb.) Thunb.   Rigid, branching perennial to 1 m. Leaves linear-lanceolate, flat, short-lived or absent. Spikelets in a raceme-like panicle or raceme, green or whitish, glumes shorter to longer than spikelet, sterile lemmas smooth, glabrous, truncate to mucronate, the second with a pair of ear-like, basal appendages. Sept.–Jan. Mountain slopes, NW, SW, KM, LB, SE (Cedarberg Mts to Uitenhage).•

rehmannii Stapf   Loosely tufted or spreading perennial to 1 m. Leaves linear-lanceolate, flat to lightly rolled. Spikelets in a raceme-like panicle or raceme, green, glumes equal to or longer than spikelet, sterile lemmas weakly corrugate, glabrous, truncate to mucronate, the second with a pair of ear-like basal appendages. Aug.–Jan. Mountain slopes, damp areas and forest margins, SW, LB, AP, SE (Cape Peninsula to Tsitsikamma Mts).•

thunbergii Gibbs-Russ.   Rigid, branching perennial to 1.5 m, with well-developed hairy rhizomes. Leaves linear-lanceolate, short-lived. Spikelets in a contracted or verticillate panicle, straw-coloured and purple, glumes shorter than to as long as spikelet, sterile lemmas smooth, villous, mucronate to aristate. Sept.–Dec. Flats and slopes. NW, SW, KM, LB (Vredendal to Mossel Bay).•

villosa Schult.f.   PYPGRAS   Perennial to 1.5 m, with well-developed, naked rhizomes with elongate internodes. Leaves linear, fairly short-lived, margins undulate. Spikelets in contracted or verticillate panicle, straw-coloured, glumes to as long as spikelet, sterile lemmas villous, truncate to mucronate. Sept.–Mar. Coastal dunes, SW, AP, SE (St Helena Bay to Port Elizabeth).•

*ELEUSINE   9 spp., mostly tropical Africa

*coracana (L.) Gaertn.   AFRICAN FINER MILLET   Annual to 60 cm. Leaves linear, often folded. Spikelets in 3–13 digitately arranged secund spikes, 5–8 mm long. Oct.–May. Disturbed places, NW, SW, SE (tropical African weed).

ELIONURUS   14 spp., tropical Africa and America, 1 in Australia

muticus (Spreng.) Kunth   WIRE GRASS, DRAADGRAS   Perennial to 1 m. Leaves aromatic, linear, rolled. Spikelets crowded in a dense, narrow, white-silky spike. Sept.–May. Grassland, SE (Plettenberg Bay to tropical Africa, Arabia).

ENNEAPOGON   SUURGRAS   30 spp., pantropics and subtropics

desvauxii P.Beauv.   Glandular-hairy perennial or annual to 30 cm. Leaves linear to filiform. Spikelets in a dense spike-like panicle. Lemma awns plumose. Mostly Nov.–Mar. Habitat? KM (Little Karoo to tropical Africa).

scaber Lehm.   KLIPGRAS   Perennial to 30 cm. Leaves linear. Spikelets in an open panicle, 3 mm long, lemma 9-awned, awns glabous to scabrid. Sept.–Mar. Stony upper slopes, NW, SW, KM, LB, SE (Namibia to Knysna).

ERAGROSTIS   LOVE GRASS   300 spp., cosmopolitan

A. Spikelets wider than 2 mm

bergiana (Kunth) Trin.   KALKKWEEK   Creeping perennial to 10 cm. Leaves linear, basal sheaths densely woolly. Spikelets in a lax panicle, densely clustered on side branches, straw-coloured and pink, 4–8 × 2–4 mm, lowest lemma obtuse, lateral nerves conspicuous. Sept.–Feb. Flats, LB, SE (Little and Great Karoo, Namibia).

capensis (Thunb.) Trin.   HARTJIEGRAS   Perennial to 90 cm. Leaves linear. Spikelets in a raceme or sparsely branched panicle, 4–15 × 3–7 mm, plump, straw-coloured and purplish, lowest lemmas obtuse, lateral nerves distinct. Sept.–Apr. Flats and slopes, NW, SW, AP, KM, LB, SE (Clanwilliam to S tropical Africa and Madagascar).

cilianensis (All.) F.T.Hubb.   STINKGRAS   Annual to 90 cm. Leaves linear-lanceolate, basal sheaths sometimes hairy. Spikelets in a narrow panicle with side branches longer than 40 mm, spikelets 3–20 × 2–4 mm, lemma obtuse. Oct.–June. Habitat? LB, SE (Swellendam to palaeotropics).

elatior Stapf   Like **E. capensis** but spikelets 5–8 × c. 2 mm, lowest lemmas acute, with lateral nerves distinct. Dec.–Mar. Rocky streambanks, SW (Cape Peninsula to Riviersonderend).•

obtusa Munro ex Ficalho & Hiern   KWAGGAKWEEK   Perennial to 45 cm. Leaves linear. Spikelets in a lax panicle, 3–5 × 3–4 mm, lowest lemma obtuse, lateral nerves distinct. July–May. Mostly near streams, KM, LB, SE (southern Africa).

*racemosa (Thunb.) Steud.   SMALHARTJIEGRAS   Perennial to 80 mm. Leaves linear, glabrous or thinly silky hairy. Spikelets in a lax panicle, 3–10 × 2–5 mm, olive, lowest lemma with lateral nerves indistinct. Aug.–May. Grassy lower slopes, SW (SE African weed).

A. Spikelets to 2 mm wide

chloromelas Steud.   Perennial to 80 cm. Leaves filiform, curly, basal sheaths glabrescent. Spikelets in an open much-branched panicle with long hairs in axils, 4–6 × 1–1.5 mm. Dec.–May. SE (Humansdorp to northern South Africa).

curvula (Schrad.) Nees   BERG SOETGRAS, BLOUSAADGRAS   Perennial to 1 m. Leaves linear, often rolled and filiform, basal sheaths densely hairy. Spikelets in open, much-branched panicles, 4–10 × 1–1.5 mm, dark green. Jan.–Dec. Disturbed sites and grassland, NW, SW, AP, KM, LB, SE (throughout S and tropical Africa).

homomalla Nees   REËNGRASSIE   Annual to 5 cm. Leaves linear. Spikelets in a rigid, sparsely branched panicle, irregularly and densely condensed on primary branches, 2–7 × c. 1 mm. Jan.–May. Habitat? NW, AP, KM, SE (Namibia to Karoo and Stilbaai).

*****pilosa** (L.) P.Beauv.   Annual to 70 cm. Leaves linear. Spikelets distant in a delicate open panicle, axils bearded, 3–7 × c. 1 mm. Oct.–May. Sandy soils, NW, SW (Eurasian weed).

plana Nees   TAAIPOL ERAGROSTIS   Perennial to 1 m. Leaves linear, basal sheaths strongly compressed, smooth and shiny. Spikelets in a panicle, appressed, 6–10 × c. 1–2 mm. Nov.–May. Habitat? SE (George to tropical Africa).

planiculmis Nees   Perennial to 1.2 m. Leaves linear, rolled, culm bases strongly compacted. Spikelets in open much-branched panicles, yellow and pink, c. 8 × 2 mm. Nov.–Apr. Damp slopes, SW, LB, SE (Saldanha to Mpumalanga).

sabulosa (Steud.) Schweick.   Creeping, rhizomatous perennial to 15 cm. Leaves short, linear, mainly cauline. Spikelets crowded-appressed in narrow, compact panicles, branches appressed, olive, 7 × 2 mm, lower lemma 1.6–2.5 mm. Mainly Oct.–Nov. Coastal dunes, SW (Saldanha to False Bay).•

sarmentosa (Thunb.) Trin.   Like **E. sabulosa** but spikelet groups often distant on thick pedicels, lower lemma 1.5 mm long. July–May. Winter-wet sand, NW, SW, AP, LB, SE (Namaqualand to George and tropical Africa).

*****virescens** C.Presl   CHILEAN LOVE GRASS   Annual to 70 cm. Leaves linear-lanceolate, margins scabrid. Spikelets in an open panicle, 3–4.5 × c. 1 mm. Dec.–Mar. Disturbed sandy soils, NW, SW, KM, SE (S American weed).

EULALIA   30 spp., palaeotropical and subtropical

villosa (Thunb.) Nees   Perennial to 1 m. Leaves linear, basal sheaths hairy. Spikelets in digitately arranged spike-like branches, white-hairy, 5–7 mm long, awns 15–20 mm. Sept.–May. Grassy slopes, SE (Humansdorp to India).

EUSTACHYS   HOENDERSPOOR   12 spp., tropical and subtropical America, South Africa

paspaloides (Vahl) Lanza & Mattei   BRUINHOENDERSPOOR   Perennial to 1 m. Leaves linear-lanceolate, folded. Spikelets secund, in digitately arranged spike-like branches, brown, 1.5–2.5 mm long. Oct.–May. Grassy slopes, NW, AP, LB, SE (Worcester to Arabia).

FESTUCA   FESCUE   80 spp., pantemperate and subtropical, tropical mountains

*****arundinacea** Schreb.   MEADOW FESCUE   Perennial to 2 m. Leaves linear-lanceolate, dark green, auricles well developed. Spikelets in a nodding panicle, 10–18 mm long, awn to 4 mm. Sept.–Apr. Damp places, SW, SE (Eurasian weed).

scabra Vahl   MUNNIKSGRAS   Perennial to 90 cm, culms with swollen bases. Leaves linear, often rolled, velvety below, fibrous with age. Spikelets in narrow, contracted panicle, green and straw-coloured, 7–15 mm long, minutely awned. Sept.–Dec. Dry flats and slopes, NW, SW, AP, KM, LB, SE (widespread in southern Africa).

FINGERHUTHIA   VINGERHOEDGRAS   2 spp., tropical and southern Africa

africana Lehm.   KALKVINGERHOEDGRAS   Perennial, sometimes annual to 80 cm. Leaves linear. Spikelets in a dense spike-like raceme or panicle, silvery and purple, glumes densely hairy, lemmas obtuse-awned. Mostly Sept.–Dec. Middle to upper slopes, NW, SW, KM, SE (widespread in southern and S tropical Africa).

sesleriiformis Nees   Like **F. africana** but rhizomes and plant more robust, glumes sparsely hairy and lemmas acute-awned. Nov.–Apr. Marshes and along streams, SE (Langkloof to Mpumalanga).

*****GASTRIDIUM**   2 spp., Mediterranean

*****phleoides** (Nees & Meyen) C.E.Hubb.   Annual to 60 cm. Leaves linear-lanceolate. Spikelets in a narrow, spike-like panicle, awned, glumes swollen at base. Oct.–Dec. Disturbed flats, SW (Mediterranean weed).

*HAINARDIA   1 sp., Mediterranean and Middle East

*cylindrica (Willd.) Greuter (= *Monerma cylindrica* (Willd.) Coss. & T.Durand)   Annual to 30 cm. Leaves linear. Spikelets in a simple spike, green, 5–8 mm long, alternate and embedded in an articulated rachis, awnless. Nov.–Dec. Moist places, SW, LB, SE (Mediterranean weed).

HARPOCHLOA   CATERPILLAR GRASS   1 sp., South Africa

falx (L.f.) Kuntze   Rhizomatous perennial 60 cm. Leaves linear, often rolled. Spikelets secund in an inflexed simple spike, 6–7 mm long. Sept.–Apr. Stony grassland, SW, LB, SE (Tulbagh to northern South Africa).

HELICTOTRICHON   90 spp., pantemperate, including tropical mountains

capense Schweick.   Like **H. hirtulum** but spikelets c. 15 mm long. Oct.–Nov. Damp lower slopes, SW, AP, LB, SE (Cape Peninsula to KwaZulu-Natal).

dodii (Stapf) Schweick.   Like **H. turgidulum** but spikelets 12–15 mm long, yellowish, lemma lobes above awn longer, 6–8 mm. Oct.–Dec. Damp lower slopes, SW, LB, SE (Cape Peninsula to Lesotho).

hirtulum (Steud.) Schweick.   Perennial to 1 m. Leaves linear. Spikelets 8–11 mm long, in a narrow panicle to 20 cm long, upper glume half to two-thirds as long as spikelet, lemmas scabrid. Nov.–Mar. Clay slopes and flats, NW, SW, SE (Bokkeveld Mts to N Province).

leoninum (Steud.) Schweick.   Perennial to 50 cm. Leaves linear. Spikelets 12–14 mm long, in a narrow panicle 6–10 cm long, upper glumes about half as long as spikelet, lemmas dense papillose, rachilla internodes glabrous. Aug.–Nov. Mountain slopes, SW (Cape Peninsula to Hottentots Holland Mts).•

longum (Stapf) Schweick.   HAWERGRAS   Rhizomatous perennial to 1.5 m. Leaves linear-lanceolate. Spikelets 15–30 mm long, in a narrow panicle 15–30 cm long, green, loosely flowered, upper glume to half as long as spikelet, lemmas minutely granular. Sept.–Oct. Coastal sandy flats, NW, SW (Namaqualand to False Bay).

turgidulum (Stapf) Schweick.   Perennial to 1 m. Leaves linear. Spikelets green, 10–12 mm long, in a panicle 7–30 cm long, upper glume two-thirds as long as spikelet, lemmas smooth or papillate, often purple. Oct.–Apr. Wet flats and slopes, SW, LB, SE (Saldanha to northern South Africa).

HEMARTHRIA   c. 10 spp., Africa, Madagascar, tropical Asia

altissima (Poir.) Stapf & C.E.Hubb.   PERDEGRAS   Erect or sprawling perennial to over 1 m. Leaves linear. Spikelets in a simple spike, embedded in jointed axis. Oct.–June. Marshes and streambanks, NW, SW, AP, SE (Clanwilliam to tropical Africa).

HETEROPOGON   PYLGRAS   6 spp., Africa, S Europe

contortus (L.) P.Beauv. ex Roem. & Schult.   Rhizomatous perennial to 70 cm. Leaves linear. Spikelets in a spike-like raceme, the upper awned, awns twisted together above spikelets. Oct.–June. Flats and slopes, SW, LB, SE (widespread through Africa).

HOLCUS   8 spp., Mediterranean, Asia, South Africa

*lanatus L.   VELVET GRASS, FLUWEELGRAS   Perennial to 1 m. Leaves linear-lanceolate, softly hairy. Spikelets in a panicle, pink and grey, upper glume awn to 1 mm. Oct.–Dec. Damp areas, SW, LB, SE (Eurasian weed).

setiger Nees   Annual to 30 cm. Leaves linear, shortly hairy. Spikelets in a panicle, pale green, upper glume awn 2–6 mm. Oct.–Dec. Damp lower slopes, NW, SW, SE (Namaqualand to George).

HORDEUM   BARLEY   20 spp., temperate

capense Thunb.   CAPE WILD BARLEY   Perennial to 70 cm. Leaves linear, rigid, without auricles, becoming fibrous below. Spikelets in a 2-ranked false spike, green and purple, long-awned. Nov.–Dec. Wet areas or roadsides, SW, KM, LB, SE (Namaqualand to northern South Africa).

*marinum Huds.   Like **H. murinum** but glumes of lateral spikelets smooth or scabrid. Sept.–Nov. Disturbed places, NW, SW (Mediterranean weed).

*murinum L.   FALSE BARLEY   Annual to 50 cm. Leaves linear, soft, auricled. Spikelets in a 2-ranked false spike, green and purple, long-awned, glumes of lateral spikelets long-ciliate. Oct.–Dec. Disturbed places, NW, SW, AP, KM, LB, SE (European weed).

HYPARRHENIA   53 spp., mainly Africa and Mediterranean, also tropical America and Asia

*****anamesa** Clayton   Like **H. hirta** but leaves broader, c. 4 mm wide, racemes 15–25 mm long and raceme pairs with 4–7 awns. Oct.–May. Dry soils, in open places, SW, AP, LB, SE (SE African weed).

hirta (L.) Stapf   THATCH GRASS   Perennial to 2 m. Leaves linear, 1–2 mm wide. Spikelets in paired racemes 20–40 mm long, subtended by spathes, hairy, raceme pairs with 8–14 awns. Nov.–May. Disturbed areas and grassland, NW, SW, AP, KM, LB, SE (widespread through Africa and Mediterranean).

*****IMPERATA**   10 spp., pantropical and subtropical

*****cylindrica** (L.) Raeuschel   SILWERAARGRAS, SYGRAS   Strongly rhizomatous perennial to 1 m. Leaves linear-lanceolate, Spikelets in a dense, cylindrical, white-silky panicle, awnless. Jan.–Dec. Wet habitats, SW, SE (tropical Africa weed).

KARROOCHLOA   4 spp., South Africa

curva (Nees) Conert & Tuerpe   OULANDEGRAS   Perennial to 40 cm. Leaves linear, sometimes folded, glabrous, sometimes thinly hairy. Spikelets in a contracted panicle, green, straw-coloured and purple, lemmas densely hairy below, awn 4–6 mm long. Oct.–Dec. Shady lower slopes, NW, SW, AP, KM, LB, SE (Lambert's Bay to KwaZulu-Natal).

purpurea (L.f.) Conert & Tuerpe   QUAGGAGRAS   Perennial to 20 cm. Leaves linear, rolled, sparsely hispid. Spikelets in a contracted panicle, straw-coloured and purple, lemmas with hairs in tufts c. 2 mm long, awn 3–4 mm long. Sept.–Dec. Upper slopes, NW, SW, KM (Namaqualand and Karoo to De Hoop).

tenella (Nees) Conert & Tuerpe   Dwarf annual to 15 cm. Leaves linear, rolled, sparsely hispid. Spikelets in a contracted panicle, green and purple, lemmas with hairs in tufts 0.5–1.2 mm long, awn 3–4 mm long. Aug.–Oct. Rocky sandstone slopes, NW, KM (Namaqualand and Karoo to Witteberg).

KOELERIA   c. 60 spp., N and S temperate

capensis (Steud.) Nees   POLGRAS, STRANDGRAS   Perennial to 70 cm. Leaves linear. Spikelets 3.5–4 mm long, in a dense cylindrical spike-like panicle, sometimes interrupted. Nov.–Feb. Coastal sands and mountain slopes, NW, SW, AP, KM, LB, SE (Namaqualand to KwaZulu-Natal).

*****LAGURUS**   HARE'S TAIL, HAASSTERT   1 sp., Mediterranean

*****ovatus** L.   Velvety annual to 40 cm. Leaves lanceolate. Spikelets in a dense, ovoid, silky white, bristly panicle. Oct.–Nov. Disturbed areas, SW, AP, SE (European weed).

LEERSIA   15 spp., tropical and warm temperate

hexandra Sw.   MOERASGRAS, WATERKWEEK   Slender, long-rhizomatous perennial to 1 m, hairy at nodes. Leaves linear-lanceolate, scabrid. Spikelets in an open panicle, purple. Jan.–Feb. Shady places near water, SW, SE (Cape Peninsula to Old World tropics).

*****LOLIUM**   RYE   12 spp., temperate Eurasia

*****multiflorum** Lam.   ITALIAN RYE GRASS, DRABOK   Like **L. perenne** and hybridising with it but sterile shoots absent, leaves rolled when young and lemmas awned. Oct.–Dec. Disturbed coastal areas, SW, AP, KM, LB, SE (Eurasian weed).

*****perenne** L.   PERENNIAL RYE   Perennial to 50 cm, with sterile shoots. Leaves linear. Spikelets in a simple spike, 10–15 × 3–10 mm, awnless. Oct.–Dec. Disturbed areas, NW, SW, AP, KM, LB, SE (European weed).

*****rigidum** Gaudin   Annual to 30 cm. Leaves linear. Spikelets in a simple spike, embedded in rachis, 1–3 mm wide, partly concealed by appressed glumes, awned or awnless. Sept.–Jan. Common in dry, disturbed areas, NW, SW, AP, KM, SE (Mediterranean weed).

*****temulentum** L.   Annual to 1 m. Leaves linear. Spikelets in a simple spike, green, 8–28 mm long, lemmas swollen at maturity. Sept.–Feb. Disturbed areas, NW, SW, LB, SE (European weed).

*LOPHOCHLOA   c. 85 spp., temperate areas

*cristata (L.) Hyl.   Annual to 40 cm. Leaves linear. Spikelets in a contracted panicle, 3–5 mm long, upper glume glabrous. Oct.–Dec. Dry exposed areas, NW, SW, KM, LB, SE (Mediterranean weed).

*pumila (Desf.) Bor.   Like **L. cristata** but upper glume densely pubescent. Sept.–Jan. Dry or rocky places, NW, SW, KM (European weed).

MELICA   70 spp., N and S temperate, excluding Australia

racemosa Thunb.   HAAKGRAS, DRONKGRAS   Perennial to 1 m. Leaves linear, often scabrous. Spikelets in silvery racemes or panicles, lemmas hairy on margins. Oct.–Mar. Slopes and flats, NW, SW, AP, KM, LB, SE (Namaqualand to Lesotho).

MELINIS (= *RHYNCHELYTRUM*)   c. 26 spp., mostly circum-Indian Ocean, 1 in South America

nerviglumis (Franch.) Zizka (= *Rhynchelytrum setifolium* (Stapf) Chiov.)   FERWEELGRAS   Like **M. repens** but leaves rolled, sheaths strongly overlapping. Nov.–Dec. Road verges and disturbed areas, SW, LB, SE (Tulbagh to tropical Africa).

*repens (Willd.) Zizka (= *Rhynchelytrum repens* (Willd.) C.E.Hubb.)   NATAL RED TOP, BERGROOIGRAS   Annual or perennial to 80 cm. Leaves linear, leaf sheaths not overlapping. Spikelets shining red, silver and mauve. Oct.–Feb. Disturbed areas, widespread, NW, SW, LB, SE (SE African weed).

MERXMUELLERA   14 spp., southern and S tropical Africa, Madagascar

A. Leaf sheaths densely woolly

decora (Nees) Conert   Perennial to 60 cm, culm bases swollen. Leaves linear, rolled, sheaths densely woolly. Spikelets in a loosly contracted, interrupted panicle, straw-coloured and purple, glumes 5–7-nerved, lemmas 9–15 mm long, hairy below with transverse median row of white hairs, awn 12–20 mm long, geniculate. Sept.–Nov. Damp lower slopes, SW, LB, SE (Cape Peninsula to George).•

lupulina (Thunb.) Conert   Like **M. rufa** but glumes usually 1-nerved, lemmas 6–8 mm long, awn 4–8 mm long, mostly straight. Oct.–Jan. Sandstone slopes, SW (Tulbagh to Bredasdorp).•

rufa (Nees) Conert   BRANDGRAS   Perennial to 45 cm, culm bases swollen. Leaves linear, rolled, sheaths densely woolly. Spikelets in a contracted panicle, straw-coloured and purple, glumes 3–5-nerved, lemmas 7–12 mm long, hairy below with transverse median row of white hairs, awn 6–16 mm long, geniculate. Sept.–Nov. Sandstone slopes, NW, SW (Bokkeveld Mts to Hermanus).•

A. Leaf sheaths not woolly

arundinacea (P.J.Bergius) Conert   OLIFANTSGRAS   Reed-like perennial to 1 m. Leaves linear, rolled. Spikelets in a dense panicle, straw-coloured, glumes 1-nerved, lemma 6–8 mm long, backs pubescent, not tufted, awn 9–13 mm long. Sept.–Nov. Dry sandstone slopes, NW, SW, KM, LB, SE (Bokkeveld Mts to Port Elizabeth).•

cincta (Nees) Conert   Like **M. arundinacea** but to 2 m, lemma glabrous below, with a transverse median band of tufted, white hairs 5–12 mm long. Oct.–Apr. Streamsides, NW, SW, AP, LB, SE (Olifants River Mts to E Cape).

disticha (Nees) Conert   KOPERDRAAD   Perennial to 40 cm. Leaves linear, rolled. Spikelets distichous in a spike-like panicle, straw-coloured and green, glumes 1–3-nerved, lemmas 10–15 mm long, fringed on margins, with a basal tuft of white hairs on each side, awns 10–16 mm long, geniculate. Nov.–Dec. Streams and dry stony slopes, SW, LB, SE (Riviersonderend Mts to Free State).

papposa (Nees) Conert   Perennial to 50 cm. Leaves linear, rolled. Spikelets in a dense panicle, silvery and straw-coloured, glumes 3-nerved, lemmas 9–10 mm long, sparsely hairy below, with 3 tufts of long white hairs c. 6 mm, awn 15–18 mm long, geniculate. Nov.–Jan. Sandstone slopes, SE (Uitenhage).•

setacea N.P.Barker   Perennial to 1 m, bulbous below. Leaves linear, rolled, short and pungent. Spikelets in a contracted panicle, glumes 1-nerved, lemmas 8–11 mm long, with a tuft of hairs on each margin, awn 11–15 mm long, geniculate. Nov. Seeps on sandstone slopes, NW (Cold Bokkeveld to Ceres).•

stricta (Schrader) Conert   BOKBAARDGRAS   Perennial to 80 cm. Leaves linear. Spikelets in an interrupted panicle, straw-coloured and purple, glumes 3–7-nerved, lemmas 6–9 mm long, with 4 or more tufts of marginal hairs, awn 6–7 mm. Sept.–Dec. Middle to upper slopes, NW, SW, AP, KM, LB, SE (Cold Bokkeveld Mts to Karoo).

MICROSTEGIUM   30 spp., mainly tropical and subtropical Asia, also Africa

nudum (Trin.) A.Camus (= *Microstegium capense* (Hochst.) A.Camus)   Trailing, tangled annual to 60 cm. Leaves lanceolate, ligule membranous. Spikelets on 3 or 4 slender racemes, green, 3.5–4.5 mm long, long-awned. Jan.–May. Forests, SE (Knysna to tropical Africa).

MISCANTHUS (= *MISCANTHIDIUM*)   RUIGTEGRAS   6 or 7 spp., pantropical and subtropical

capensis (Nees) Andersson (= *Miscanthidium capensis* (Nees) Stapf, *Miscanthidium sorghum* (Nees) Stapf)   Robust perennial to 2.4 m. Leaves linear-lanceolate. Spikelets in a large silky panicle, reddish, 4–6 mm wide, awned. Dec.–May. Riverbanks and forest margins, SE (Humansdorp to Swaziland).

*****NASSELLA**   15 spp., S America

*****trichotoma** (Nees) Hack. ex Arech.   NASSELLA TUSSOCK   Densely tufted perennial to 60 cm. Leaves linear, rolled, tough. Spikelets 6–8.5 mm long, in an open panicle, awns slender, 35 mm long. Aug.–Jan. Mountain grassland, SW, LB (South American invasive weed).

ODYSSEA   2 spp., Africa and Arabia

paucinervis (Nees) Stapf   Mat-forming perennial to 75 cm, often spinescent. Leaves needle-like, woody, pungent. Spikelets in a contracted panicle, awnless. Oct.–May. Salt marshes, NW (tropical Africa to Rocher Pan).

OPLISMENUS   5 spp., pantropical

hirtellus (L.) P.Beauv.   Trailing perennial, stems to 80 cm. Leaves lanceolate to ovate. Spikelets 6–20 in racemes, awns smooth, sticky. Jan.–June. Forest shade, SE (Knysna to tropical Africa).

undulatifolius (Ard.) Roem. & Schult.   Like **O. hirtellus** but spikelets 2–6 in fascicles. Jan.–July. Forest shade, SE (Knysna to tropical Afica).

PANICUM   PANIC GRASS   600 spp., pantropical and warm temperate

coloratum L./**stapfianum** Fourc. complex   WITBUFFELGRAS   Perennial to 1 m. Leaves linear, sometimes tubercled-hairy. Spikelets distant in an open panicle with spreading branches, 2.5–3 mm long, lower lemma 7–9-nerved. Oct.–May. Damp places, SW, SE (Cape Peninsula to tropical Africa).

deustum Thunb.   RIETBUFFELGRAS   Like **P. coloratum** but spikelets 3.5–5.5 mm long, panicle branches with clavellate hairs. Sept.–Apr. Coastal forest or grassland, LB, SE (Riversdale to tropical Africa).

ecklonii Nees   Perennial to 80 cm. Leaves linear, usually densely tubercled-hairy. Spikelets in an open panicle, c. 3 mm long, glumes and lemma pectinate. Sept.–Apr. Habitat? SE (Langkloof to tropical Africa).

maximum Jacq.   GUINEA GRASS   Perennial, sometimes annual to 2 m. Leaves lanceolate. Spikelets c. 3 mm long, in an open panicle, purple, secondary branches flexuose, fertile lemma transversely wrinkled. Jan.–May. Shady places, SW, SE (Cape Peninsula to tropical Africa).

repens L.   COUCH PANICUM, KRUIPGRAS   Rhizomatous perennial to 1 m. Leaves mostly cauline, often pungent. Spikelets in an open panicle, narrowly oblong, 2–3 mm long, lower lemma 7–9-nerved. Oct.–June. Wet sandy soils, NW, SW, SE (Namaqualand to Cape Peninsula, George to tropical Africa).

schinzii Hack.   BLOUSAADGRAS   Sprawling annual to 1 m. Leaves linear-lanceolate. Spikelets in an open panicle, 2–3 mm long, lower lemma 9–11-nerved. Nov.–May. Moist sites, NW, SW, SE (Lambert's Bay to N Province).

*****PARAPHOLIS**   6 spp., Mediterranean and Middle East

*****incurva** (L.) C.E. Hubb.   Annual to 30 cm. Leaves reduced. Spikelets in a simple spike, 4–7 mm long, laterally compressed, alternate on a straight rachis. Aug.–Oct. Infrequent roadside weed, NW, SW, AP, LB, SE (European weed).

*****PASPALUM**   250 spp., mainly neotropics and subtropics

*****dilatatum** Poir.   DALLIS GRASS   Rhizomatous perennial to 2 m. Leaves lanceolate. Spikelets in 4–9 subdigitate, spike-like branches, secund, in 4 rows, 3–4 mm long, fringed with white hairs. Oct.–May. Disturbed damp places, NW, SW, LB, SE (S American weed).

*__distichum__ L. BANKROTKWEEK   Perennial hydrophyte to 30 cm. Leaves linear. Spikelets in 2 or 3 digitate, spike-like branches, secund in 2 rows, 2.5–3.5 mm long, upper glume minutely hairy. Nov.–May. Near fresh or brackish water, NW, SW, SE (S American weed).

*__urvillei__ Steud. LANGBEEN PASPALUM   Like **P. dilatatum** but often taller, spikelets 1.6–2.8 mm long, in 10–30 spikes. Oct.–Apr. Near water, SW, SE (S American weed).

*__vaginatum__ Sw. BRAK PASPALUM   Like **P. distichum** but upper glume glabrous, spikelets 3–4.5 mm long, lanceolate. Dec.–Apr. Vlei margins. SW, SE (pantropical weed).

PENNISETUM   70 spp., warm regions

*__clandestinum__ Chiov. KIKUYU GRASS   Rhizomatous, mat-forming perennial, sometimes to 30 cm. Leaves linear-oblong. Spikelets 10–20 mm long, enclosed in an upper leaf sheath. Aug.–Apr. Clay and loamy soils, SW, LB, SE (E African weed).

__macrourum__ Trin. BEDDINGGRAS   Perennial to 1.7 m. Leaves linear. Spikelets 4–6 mm long, in a congested, cylindrical panicle 12–25 cm long, straw-coloured, involucral bristles as long as spikelets. Nov.–May. Marshes, NW, SW, AP, KM, LB, SE (Bokkeveld Mts to tropical Africa).

*__setaceum__ (Forssk.) Chiov. FOUNTAIN GRASS   Like **P. macrourum** but bristles 4–5 times as long as spikelets, the inner plumose, spikelets 4–6.5 mm long. Nov.–July. Stony slopes, NW, SW, SE (N African weed).

__thunbergii__ Kunth NAPIER MILLET   Like **P. macrourum** but most bristles to twice as long as spikelet, panicle 30–50 mm long, spikelets 3 mm long. Oct.–June. Vlei margins, NW, SW, LB, SE (Tulbagh to tropical Africa and Arabia).

PENTAMERIS•   9 spp., W and E Cape

A. Glumes 14–24 mm long, leaves usually needle-like or rigid

__hirtiglumis__ N.P.Barker   Like **P. macrocalycina** but leaves rolled, falcate, sheaths pubescent, glumes hairy. Sept.–Nov. Shale bands, SW (Hottentots Holland Mts).•

__longiglumis__ (Nees) Stapf   Robust perennial to 1.5 m, culms sometimes branched at base. Leaves rigid and rolled, sheath sometimes shortly hairy, often with green auricles. Spikelets in a lax, globose panicle 17–30 cm long, glumes 15–24 mm long. Sept.–Dec. High sandstone slopes, SW (Cape Peninsula to Kogelberg).•

__macrocalycina__ (Steud.) Schweick.   Branched perennial to 1 m. Leaves terete, filiform, cauline. Spikelets 2-flowered, in a narrow panicle 6–12 cm long, spikelets 16–24 mm long, glumes 6–24 mm long. Sept.–Dec. Middle to upper sandstone slopes, NW, SW, KM, LB, SE (Pakhuis Mts to Uitenhage).•

__oreophila__ N.P.Barker   Like **P. macrocalycina** but sometimes cushion-like, leaves rolled, falcate, pungent, sheaths pubescent. Sept.–Dec. Shale bands at high alt., NW, SW (Hex River and Riviersonderend Mts).•

__thuarii__ P.Beauv.   Robust, erect or decumbent, branched perennial, culms to 1.7 m. Leaves linear, flat or folded, cauline, sheaths pubescent, with purple auricles. Spikelets 2-flowered, in a lax globose panicle 7–22 cm long, glumes 16–22 mm long. Sept.–Dec. Lower sandstone slopes, SW, LB, SE (Du Toitskloof to Outeniqua Mts).•

A. Glumes to 14 mm, leaves usually softer

__distichophylla__ (Lehm.) Nees (= *Pentaschistis dregeana* Stapf)   Bushy perennial with culms to 1.2 m. Leaves largely basal, linear, flat or rolled, hairy, sheaths densely woolly. Spikelets 2-flowered, in dense, ovoid panicle 5–11 cm long, glabrous, 12–15 mm long, glumes 12–15 mm long. Sept.–Dec. Rocky sandstone slopes, NW, SW, KM, LB, SE (N Cedarberg to Kouga Mts).•

__glacialis__ N.P.Barker   Like **P. swartbergensis** but leaves shorter, to 10 cm, panicle to 70 mm long with fewer than 15 spikelets. Oct.–Dec. High S-facing gulleys, KM (Waboomsberg).•

__swartbergensis__ N.P.Barker   Like **P. distichophylla** but leaf sheaths glabrous or pubescent only at margins and lemma lateral bristles shorter, 2.3–5.5 mm. Sept.–Dec. Rocky summit ridges, KM (Klein Swartberg).•

__uniflora__ N.P.Barker   Sprawling perennial to 65 cm, with thin flexuose culms. Leaves linear, mostly cauline. Spikelets 1-flowered, 5–20 in a small panicle 4–6 cm long, glumes 11–12 mm long. Sept.–Dec. Shady, S-facing cliffs on wetter mountains, SW, LB (Riviersonderend Mts to Robinson Pass).•

PENTASCHISTIS   c. 68 spp., temperate and montane Africa and Madagascar

A. Lemmas awnless, plants with or without stalked glands

ampla (Nees) McClean   Perennial to 70 cm, with woody base. Leaves linear, hairy, glandular. Spikelets 3.3–4.6 mm long, in an open, hemispherical panicle, pedicels glandular, longer than spikelets. Dec.–Mar. Sandstone slopes at low to middle alt., NW, SW, LB, SE (Clanwilliam to Somerset East).

aurea (Steud.) McClean   Perennial to 45 cm. Leaves linear, sometimes hairy, sheath mouth villous. Spikelets 4–5 mm long, in an open panicle, pedicels with obscure linear glands, lemma lobe setae obscure, c. 0.1 mm long. Jan.–Mar. Cool damp rocky ledges, NW, SW, LB, SE (Namaqualand to S Cape and to Drakensberg).

capillaris (Thunb.) McClean   Annual to 40 cm, with stalked glands on leaves and pedicels. Leaves linear, villous beneath. Spikelets c. 3 mm long, in an open panicle, glumes obtuse. Sept.–Oct. Sandy flats and lower slopes, NW, SW (Namaqualand to Saldanha).

ecklonii (Nees) McClean   Perennial to 30 cm. Leaves cauline, linear, usually rolled. Spikelets 3–4 mm long, in spike-like panicles, pedicels with stalked glands. Jan.–Mar. Seasonally waterlogged soils, often over laterite, SW (Hopefield to Somerset West).•

malouinensis (Steud.) Clayton   Perennial to 30 cm. Leaves linear, often rolled. Spikelets 3.4–4.5 mm long, in a lax open or narrow panicle, glumes rounded, lemmas without awns. Nov.–Jan. Dry sandstone slopes, often on rocks and ledges, NW, SW, AP, KM, LB, SE (Cedarberg to Humansdorp).•

pusilla (Nees) H.P.Linder (= *Poagrostis pusilla* (Nees) Stapf)   Cushion-forming perennial to 12 cm. Leaves linear, rigid, sometimes rolled, sparsely hairy above, recurving with age. Spikelets 2.5–3 mm long, single-flowered, in an open panicle. Dec.–Feb. Cool shady or damp habitats, SW (Bainskloof to Cape Peninsula and Jonkershoek).•

reflexa H.P.Linder   Stoloniferous biennial to 35 cm. Leaves cauline, linear, often rolled. Spikelets 3–4 mm long, reflexed, in an open panicle, pedicels and glumes with stalked glands. Oct.–Dec., after fire. Dry sandstone slopes in arid fynbos, NW (Cedarberg Mts to Piketberg).•

A. Lemmas awned
B. Plants with stalked glands

airoides (Nees) Stapf   Annual to 35 cm. Leaves linear, villous above. Spikelets 2.5–5 mm long, in an open, hemispherical panicle to 6 cm long, awns much exserted, anthers 0.3–1 mm long. Aug.–Oct. Light shade on lower slopes, NW, SW, KM (Namaqualand and Karoo and Drakensberg).

aspera (Thunb.) Stapf   Perennial to 60 cm. Leaves linear, ridged above, pseudoserrate, with prominent stalked glands. Spikelets 5–7 mm long, in an open panicle, pedicels and glumes also with stalked glands. Sept.–Dec. Stony coastal slopes, often disturbed, SW (Cape Peninsula to Kleinmond).•

barbata (Nees) H.P.Linder   Perennial to 60 cm, with stalked glands on leaf blades, pedicels and glumes. Leaves linear, usually villous, flaccid, to 12 mm wide. Spikelets 5–6 mm long, in large panicles, open, soon contracting, lemmas awned. Sept.–Nov. Mainly coastal sands, NW, SW, SE (Elandsbaai to Betty's Bay and Knysna).•

cirrhulosa (Nees) H.P.Linder   Perennial to 1 m, with glands on leaf sheaths, pedicels, and glumes. Leaves linear, rolled or margins upcurved, truncate. Spikelets 5–9 mm long in an open panicle. Oct. Sandstone gravels, SW, LB, KM, SE (Villiersdorp to Humansdorp).•

densifolia (Nees) Stapf   Cushion-forming perennial, culms to 25 cm. Leaves linear, villous above, with stalked glands common on margins, often turning pink. Spikelets 3.5–4.5 mm long in an open panicle. Dec.–Jan. Sandstone ledges and in rock cracks at middle elevation, NW, SW, KM, LB (Simonsberg and Cedarberg to Swartberg Mts).•

glandulosa (Schrad.) H.P.Linder   Perennial to 35 cm. Leaves linear, grooved, with margins incurved, with scattered raised glands, often white-spotted. Spikelets 4–5.5 mm long, many in large open panicles. Oct.–Dec. Mostly heavy soils, NW, SW, LB, SE (Gifberg to Queenstown).

pallida (Thunb.) H.P.Linder (= *Pentaschistis thunbergii* (Kunth) Stapf)   Loosely tufted to almost matted perennial to 40 cm. Leaves linear, sometimes rolled, often with stalked glands. Spikelets 6–7 mm long, in contracted panicles, awned. Sept.–Oct. Slopes and flats, NW, SW, AP, KM, LB, SE (Namaqualand to E Cape).

papillosa (Steud.) H.P.Linder   Tangled perennial to 40 cm. Leaves linear, sometimes hairy below. Spikelets 5–7 mm long, in a loose panicle, pedicels with stalked glands, lemmas awned. Oct.–Nov. Lower sandstone slopes, SW (Cape Peninsula to Hermanus).•

patula (Nees) Stapf   Annual to 30 cm, often with stalked glands. Leaves linear, usually hairy, sheath mouth with a ring of bristles. Spikelets 3.3–5 mm long, in an open panicle, lemmas awned. Sept.–Oct. Sandstone slopes and flats, NW, SW (Namaqualand to Elandskloof Mts and Worcester).

rupestris (Nees) Stapf   Perennial to 1 m, often with stalked glands on leaves and pedicels. Leaves linear, sheaths villous, sheath mouth with a ring of bristles. Spikelets 7–8 mm long, numerous in a large open panicle, awned. Oct. Rocky sandstone slopes, NW, KM (Cedarberg Mts to Cold Bokkeveld and Witteberg).•

tomentella Stapf   Perennial to 30 cm, with stalked glands on leaf sheaths, pedicels and glumes. Leaves linear, puberulous or villous. Spikelets 4–5 mm long, in a contracted panicle, lemmas awned. Sept.–Oct. Dry slopes and flats, NW (Namaqualand to Clanwilliam).

veneta H.P.Linder   Perennial to 40 cm. Leaves linear, flaccid, villous, margins with stalked glands. Spikelets 5–6.5 mm long, in an open panicle, lemmas 2.8–3 mm long, awned. Dec.–Jan. Seeps and below cliffs on sandstone slopes, NW, SW (Bokkeveld Mts to Du Toitskloof).•

B. Plants without stalked glands

acinosa Stapf   Erect or cushion-like perennial, culms to 30 cm. Leaves cauline, distichous in a herringbone fashion, lanceolate. Spikelets 9–10 mm long, few in a small, lax panicle, lemmas awned, 4–4.5 mm long. Oct.–Jan. Rock ledges, usually damp S-facing slopes, NW, SW, LB (Cold Bokkeveld to Langeberg Mts).•

alticola H.P.Linder   Biennial to 30 cm. Leaves linear, rolled, rigid or flaccid, curly. Spikelets 4–6 mm long. Nov.–Jan. High rocky slopes, NW (Cedarberg to Hex River Mts).•

argentea Stapf   Like **P. viscidula** but stoloniferous, culm not swollen at base, leaf blades glabrous. Oct.–Nov. Dry sandstone slopes, SW, AP, LB (Cape Peninsula to Langeberg Mts).•

aristidoides (Thunb.) Stapf   Perennial with short compact rhizomes, culms to 1 m. Leaves radical. Spikelets 12–15 mm long, in wide lax panicles 6–11 cm wide, pedicels with obscure linear glands. Sept.–Nov., mostly after fire. Rocky sandstone slopes, NW, SW, AP (Bokkeveld Mts to Agulhas).•

calcicola H.P.Linder   Perennial to 30 cm. Leaves linear, rolled and rigid, hairy beneath and on sheaths. Spikelets 5–6 mm long, in a compact panicle, awned. Oct. Limestone pavement, AP (Gansbaai to Cape Infanta).•

capensis (Nees) Stapf   Tangled perennial, culms to 35 cm, drooping, leaf sheaths white, shiny, persistent. Leaves linear, flaccid. Spikelets 6–9 mm long, in a lax panicle, lemmas with long, soft, curving awns. Dec.–Jan. Rocky streams, often along waterfalls, SW (Cape Peninsula and Du Toitskloof to Hermanus).•

caulescens H.P.Linder   Like **P. acinosa** but leaves rigid, margins incurved. Sept.–Oct. Shale band on dry stony slopes, NW (Hex River Mts).•

colorata (Steud.) Stapf   Tangled or caespitose perennial to 60 cm. Leaves linear, rolled, curly, sheath mouth sometimes with a few bristles. Spikelets 8–13 mm long, in a lax open panicle. Aug.–Dec. Sandstone slopes, sometimes in seeps, SW, LB (Cape Peninsula to Swellendam).•

curvifolia (Schrad.) Stapf   Perennial to 50 cm, with shiny white persistent leaf sheaths. Leaves basal, linear, rigid, sometimes rolled, margins thickened. Spikelets 8–12 mm long, in a dense head-like panicle, ivory, lemmas awned. Oct.–Nov. Sandstone slopes, especially shallow soils and paths, NW, SW, KM, LB, SE (Bokkeveld Mts to Grahamstown).

elegans (Nees) Stapf   Cushion-forming perennial, culms to 30 cm. Leaves linear, villous beneath and on sheaths, sheath mouth with a ring of bristles. Spikelets 7–9 mm long, few in small panicles, lemmas awned. Sept. Calcareous sands, AP (Gansbaai to Agulhas).•

eriostoma (Nees) Stapf   Perennial to 90 cm, in dense tussocks. Leaves basal, linear, rigid, usually rolled, sheaths and sheath mouth woolly. Spikelets 8–12 mm long, in a lax open panicle, lemmas awned. Sept.–Nov. Sandstone and clay slopes and flats, NW, SW, KM, LB, SE (Namaqualand to E Cape).

holciformis (Nees) H.P.Linder   Perennial to 60 cm. Leaves linear, rolled rigid, margins hairy. Spikelets 6–7 mm long, in a lax pancile, lemmas awned, setae 0.3 mm long. Mar.–Apr. Sandstone slopes and plateaus, SW (Jonkershoek to Hermanus).•

montana H.P.Linder   Perennial to 20 cm, with very short stolons. Leaves basal, linear, rolled, rigid. Spikelets 4.5–5 mm long, in an open panicle, lemma awns 5–6 mm long. Nov. Rocky sandstone slopes, NW, KM (Hex River Mts to Witteberg).•

pallescens (Schrad.) Stapf (= *Pentaschistis silvatica* Adamson)   Perennial to 1.2 m. Leaves linear, usually villous above, to 60 cm broad. Spikelets 10–12 mm long, c. 200 in an expanded panicle, pedicels with obscure linear glands, awned. Nov.–Dec. Sandstone slopes, SW (Cape Peninsula to Bainskloof and Grabouw).•

pseudopallescens H.P.Linder   Biennial to 80 cm, with obscure linear glands. Leaves aggregated basally, linear, densely villous above. Spikelets 10–12 mm long, in a lax panicle, lemmas awned. Oct.–Dec., especially second year after fire. Sandstone slopes at middle alt., NW (Cold Bokkeveld to Hex River Mts).•

pungens H.P.Linder   Perennial to 50 cm. Leaves linear, rigid, sometimes pungent, becoming recurved with age. Spikelets 11–15 mm long, in a contracted panicle, lemmas awned. Sept.–Nov. Sandstone slopes at higher alt., NW (Cedarberg to Grootwinterhoek Mts).•

pyrophila H.P.Linder   Perennial to 60 cm. Leaves linear, rigid and rolled, usually pungent. Spikelets 10–12 mm long, in a lax panicle, awned. Oct.–Jan. Dry stony slopes at high alt., NW, SW, KM, SE (Cedarberg to Great Winterhoek Mts).•

rigidissima Pilg. ex H.P.Linder   Cushion-forming perennial, culms to 30 cm, sometimes in dense mounds. Leaves linear, rolled, rigid, sometimes pungent. Spikelets 7–8 mm long, in a contracted panicle, lemmas awned. Oct.–Feb. Dry slopes, rock crevices, NW, SW, KM, LB, SE (Cedarberg to Great Winterhoek Mts).•

rosea H.P.Linder   ROOIGRAS   Biennial to 35 cm. Leaves cauline, linear, with linear glands. Spikelets 8–12 mm long, in an open panicle, awned, glumes tuberculate hairy, pedicels with obscure, linear glands. Oct.–Dec., second year after fire. High rocky slopes, NW, SW (Cedarberg and Hex River Mts to Hottentots Holland).•

scandens H.P.Linder   Tangled perennial, culms to 50 cm, spreading. Leaves cauline, linear, rigid, truncate. Spikelets 10–11 mm long, few in a small open panicle, lemmas awned. Aug. Sandy flats, SW, AP (Bredasdorp to Elim).•

tortuosa (Trin.) Stapf   Perennial to 1 m. Leaves linear, often rolled and curly, to 50 cm long. Spikelets 7–11 mm long, in contracted, often pendent panicles, lemmas awned. Oct.–Dec. Damp sandstone slopes, SW, LB, SE (Tulbagh to Langkloof).•

triseta (Thunb.) Stapf   Annual to 60 cm, with 1 to several culms, with obscure linear glands. Leaves linear, villous. Spikelets 15–18 mm long, 3–10 in a lax panicle, reddish, awned. Sept.–Oct., after fire. Sandstone slopes, NW, SW (Bokkeveld Mts to Cape Peninsula).•

velutina H.P.Linder   Perennial to 60 cm, culms 1–few, with a swollen, woolly base. Leaves linear, with hairs on adaxial surface directly above ligule often as a web of interlocking hairs. Spikelets 12–15 mm long, in a wide, lax panicle, awned, pedicels with obscure linear glands. Oct.–Nov. Gravel plateaus, NW (Cedarberg to Porterville Mts).•

viscidula (Nees) Stapf   Perennial to 50 cm, culms with a swollen, woolly base. Leaves radical, sheaths hairy. Spikelets 7–10 mm long, in an open panicle, awned, pedicels with obscure linear glands. Oct.–Nov. Mountain plateaus, NW, SW, LB, SE (Cedarberg Mts to Langkloof).•

*PHALARIS   15 spp., N and S temperate

*aquatica L.   TOWOOMBA CANARY GRASS   Perennial to 1.5 m. Leaves lanceolate. Spikelets 4–8 mm long, in a cylindrical, spike-like panicle 2–11 cm long, glumes broadly winged, sterile floret 1, sometimes 2. Nov.–Apr. Wet disturbed areas, SW (Mediterranean weed).

*canariensis L.   Annual to 60 cm. Leaves linear-lanceolate. Spikelets 7–8 mm long, in a cylindrical spike-like panicle 2–4 cm long, glumes prominently winged, sterile florets 2, more or less equal. Oct.–Dec. Disturbed areas, SW, SE (NW Africa and Canary Islands weed).

*minor Retz.   SMALL CANARY GRASS, KANARIEGRAS   Like **P. canariensis** but glume narrowly winged, spikelets 4–6 mm long, sterile floret 1. Sept.–Jan. Disturbed ground, NW, SW, AP, KM, LB, SE (Mediterranean weed).

PHRAGMITES   COMMON REED, FLUITJIESRIET   3 spp., cosmopolitan

australis (Cav.) Trin. ex Steud.   Robust, reed-like perennial to 3 m. Leaves cauline, lanceolate. Spikelets in a plumose panicle, brown and white, awned. Feb.–May. Marshes, streams and seeps, NW, SW, AP, KM, LB, SE (worldwide).

POA c. 300 spp., cosmopolitan

*****annua** L.   ANNUAL WINTER GRASS   Annual, sometimes biennial to 30 cm. Leaves linear-oblong, flaccid. Spikelets in a pyramidal panicle, anthers to 0.8 mm. Jan.–Dec. Disturbed areas, NW, SW, AP, KM, LB, SE (European weed).

binata Nees   Rhizomatous perennial to 60 cm. Leaves linear, thick, hooded, old leaf sheaths split into fibres. Spikelets in an ovoid-pyramidal panicle, aggregated on periphery, anthers c. 1 mm long. Sept.–May. Moist, cool grasslands, SE (George to Drakensberg and E Zimbabwe).

bulbosa (= *Poa vivipara* (L.) Willd.)   Perennial to 30 cm, culms swollen below ground. Leaves linear-filiform, mostly basal. Spikelets in panicles, producing plantlets directly, rarely making functional flowers. Aug.–Oct. Sandstone slopes in richer soils, often in renosterveld, NW, SW (Namaqualand and W Karoo to Stellenbosch).

*****pratensis** L.   KENTUCKY BLUEGRASS   Like **P. binata** but lowest panicle branches whorled, basal sheaths not fibrous and lemmas woolly below. Sept.–Jan. Moist shady areas, SW, SE (European weed).

POLYPOGON   BAARDGRAS   15 spp., pantropical and warm temperate

*****monspeliensis** (L.) Desf.   BRAKBAARDGRAS   Annual to 50 cm. Leaves linear. Spikelets in an ovoid, contracted, bristly panicle, pale green, glume awns 4–10 mm long, lemma awns to 2.5 mm. Sept.–Apr. Damp lower slopes, NW, SW, AP, KM, LB, SE (cosmopolitan weed).

strictus Nees   Like **P. monspeliensis** but glume awns 10–25 mm long, lemma awns 5–10 mm long. Oct.–Apr. Wet places, SW, AP, SE (Saldanha to E Cape).

*****viridis** (Gouan) Breistr.   Annual or perennial to 60 cm. Leaves linear-lanceolate. Spikelets in an open panicle, glumes and lemmas awnless. Sept.–Jan. Damp lower slopes, NW, SW, KM, LB, SE (European weed).

PRIONANTHIUM•   3 spp., W Cape

dentatum (L.f.) Henr.   Like **P. pholiuroides** but spikelets densely aggregated, not secund, glumes with prominently stalked glands, lemmas pubescent. Sept. Heavy soils, NW (Bokkeveld Mts).•

ecklonii (Nees) Stapf   Like **P. pholiuroides** but spikelets arranged alternately in pairs, 4.4–6.1 mm long, glumes with glands prominently stalked. Sept.–Oct. Slopes and flats, NW (Olifants River Mts and Piketberg).•

pholiuroides Stapf   Annual to 30 cm. Leaves linear. Spikelets 2-ranked and alternate in a simple, secund spike to 60 mm long, spikelets arranged singly, glumes with glands sessile or subsessile, lemmas glabrous. Oct.–Dec. Local in seasonally wet sandy depressions, SW (Cape Peninsula to Stellenbosch).•

PSEUDOPENTAMERIS•   4 spp., W Cape

brachyphylla (Stapf) Conert   Tangled perennial with woody base, culms to 80 cm. Leaves cauline, linear, flat or folded, to 15 cm. Spikelets in a contracted panicle, glumes 27–37 mm long, lemmas entirely pubescent, 5.5–6.5 mm long. Sept.–Oct. Lower sandstone slopes, SW (Hottentots Holland to Kleinrivier Mts).•

caespitosa N.P.Barker   Like **P. macrantha** but culms unbranched, leaves basal, spikelets and panicle purple, lemmas larger, 8–10 mm long. Oct.–Nov. Lower sandstone slopes, NW, SW (Piketberg to Bredasdorp).•

macrantha (Schrad.) Conert·   Rhizomatous perennial to 1.5 m, often forming large clumps. Leaves cauline, linear, flat or rolled, to 35 cm. Spikelets in a contracted panicle, glumes 22–50 cm long, lemmas basally glabrous, 5–8 mm long. Aug.–Sept. Sandstone slopes, SW, AP (Cape Peninsula to Stilbaai).•

obtusifolia (Hochst.) N.P.Barker   Perennial with branched culms from woody base, sometimes to 3 m. Leaves cauline, linear, firm and rolled, to 12 cm. Spikelets in a contracted, ovoid panicle, lower glumes 18–25 mm long, lemmas 4.5–6 mm long. Nov.–Jan. Sandstone slopes in fynbos, SW (Hottentots Holland to Kleinrivier Mts).•

PUCCINELLIA   BRAKGRAS   100 spp., N temperate and South Africa

angusta (Nees) C.A.Sm. & C.E.Hubb.   VINKBRAKGRAS   Perennial to 60 cm. Leaves linear. Spikelets distant or sometimes shortly overlapping in a narrow panicle 5–15 mm diam., branches appressed or ascending. Aug.–Dec. Disturbed, saline soils, NW, SE (scattered throughout southern Africa).

*distans (L.) Parl.   Perennial to 65 cm. Leaves linear. Spikelets in a pyramidal to elongate panicle more than 15 mm diam., closely overlapping, at least some branches naked in lower half and spreading. Apr.–Oct. Rivers, canals and furrows, SE (European weed)

*fasciculata (Torr.) Bickn.   Like **P. distans** but panicle more compact with stiff branches bearing spikelets to the base. Sept.–Jan. Wet habitats, often saline and disturbed, NW, SW, AP (European weed).

SCHISMUS   5 spp., Africa, Mediterranean to India

barbatus (Loefl. ex L.) Thell.   WINTERHAASGRAS   Annual to 25 cm. Leaves linear, rolled. Spikelets narrowly lanceolate, 5–10-flowered, in a compact panicle, lemma backs pubescent with club-shaped hairs, mucro to 1 mm. June–Dec. Sandy slopes, NW, SW, AP, KM (Namibia and Karoo to E Cape).

inermis (Stapf) C.E.Hubb.   Perennial to 40 cm. Leaves linear. Spikelets 4–6-flowered, in a compact panicle, green or purple, lemmas densely pubescent, with a short mucro to 1 mm. June–Feb. Dry slopes, NW, SW, AP, KM, LB, SE (S Namibia to E Cape).

pleuropogon Stapf   Like **S. inermis** but lemmas fringed in lower third and mucro 1–1.5 mm. Nov. Damp places, LB (Langeberg Mts: Swellendam to Riversdale).•

scaberrimus Nees   Like **S. inermis** but lemmas sparsely hairy, hairs often in a row across lemma back. Sept.–Oct. Sandy areas, NW, SW (Namaqualand and W Karoo to Cape Peninsula).

SETARIA   BRISTLE GRASS   140 spp., pantropical and warm temperate

***geniculata** (Lam.) P.Beauv.   KNOTROOT   Perennial with a knotty, slender and branching rhizome, to 80 cm. Leaves linear. Spikelets 2–3 mm long, in a cylindrical, spike-like panicle, bristles 2 or 3 per cluster, lower glume 2- or 3-nerved. Dec.–Jan. Adventive, SW (American weed).

incrassata (Hochst.) Hack.   Rhizomatous perennial to 2 m, culm nodes pubescent. Leaves linear. Spikelets 2.5–3 mm long, in a contracted spike-like panicle often interrupted below, bristles 4–10 per cluster, lower glume 3-nerved. Oct.–May. Habitat? SW, LB, SE (Cape Peninsula to tropical Africa).

sphacelata (Schum.) Stapf & C.E.Hubb. ex Moss   GOLDEN TIMOTHY   Rhizomatous perennial to 1 m. Leaves linear, mostly rolled. Spikelets 2–3 mm long, in a spike-like panicle, golden-yellow, bristles 7–15 per cluster, lower glume 3-nerved. Sept.–Mar. Disturbed areas, SW, AP, LB, SE (Cape Peninsula to tropical Africa).

***verticillata** (L.) P.Beauv.   BUR BRISTLE GRASS, KLITSSETARIA   Annual, often sprawling, to 1 m. Leaves linear-lanceolate. Spikelets 1.5–2.5 mm, in a spike-like panicle often shortly branched and interrrupted below, bristles 1–4 per cluster, retrorsely barbed and often tangled, lower glume 1-nerved. Dec.–May. Disturbed places, NW, SW, SE (Old World tropical weed).

***SPARTINA**   16 spp., mostly America, Europe, Africa

***maritima** (Curtis) Fern.   CORD GRASS, STRANDKWEEK   Creeping perennial to 80 cm. Leaves linear, rolled. Spikelets secund, in a spike-like raceme, lower glume three-quarters as long as spikelet, upper glume as long as spikelet. Nov.–Mar. Intertidal mud flats, SW, SE (N Atlantic coastal weed).

***SPHENOPUS**   2 spp., Mediterranean to Western Asia

***divaricatus** (Gouan) Reichb.   Slender annual to 20 cm. Leaves linear, rolled, setaceous. Spikelets 2–3 mm long, in a delicate, open panicle. Aug.–Oct. Coastal mud flats and dune hollows, NW (European weed).

SPOROBOLUS   DROPSEED   150 spp., pantropical and warm temperate

africanus (Poir.) Robyns & Tournay   TAAIPOL   Perennial to 1 m. Leaves linear. Spikelets 2–2.8 mm long, in a dense, narrow, more or less spike-like panicle, branches short and rigid, not obscuring central axis, glume up to half as long as spikelet. Oct.–Apr. Disturbed soils, SW, AP, LB, SE (Saldanha to Ethiopia).

fimbriatus (Trin.) Nees   Like **S. africanus** but to 1.6 m, spikelets 1.4–2.2 mm long, upper glume two-thirds as long , panicle open with branches spreading at c. 60 degrees. Dec.–May. Disturbed areas, SW, LB, SE (Cape Peninsula to tropical Africa).

fourcadei Stent   Like **S. africanus** but panicle open and branches almost horizontal at maturity. Nov.–Mar. Flood plain and forest margins, SE (Mossel Bay to KwaZulu-Natal).

*****virginicus** (L.) Kunth   BRAKGRAS, SEA RUSH GRASS   Creeping stoloniferous and rhizomatous perennial to 30 cm. Leaves linear, rolled, pungent. Spikelets in a dense, spike-like panicle. Oct.–Apr. Dunes, beaches and coastal marshes, NW, SW, AP, SE (worldwide).

STENOTAPHRUM   7 spp., pantropical and subtropical

secundatum (Walter) Kuntze   BUFFELSGRAS   Mat-forming prostrate perennial, sometimes to 30 cm. Leaves oblong, folded. Spikelets partly embedded in a secund, 2-ranked spike, swollen and rounded on one side. Oct.–Jan. Sandy coastal slopes and flats, SW, AP, SE (Cape Peninsula to pantropical).

STIPA   BOESMANGRAS   300 spp., mainly temperate to subtropical

capensis Thunb.   Annual to 50 cm. Leaves linear, often rolled. Spikelets 12–15 mm long, in an open panicle, glumes c. 15 mm long, awn bent and twisted, 50–80 mm long. Aug.–Nov. Open slopes, often disturbed areas, NW, SW, SE (Namaqualand to Mossel Bay and Eurasia).

dregeana Steud.   Perennial to over 1 m. Leaves linear, broad. Spikelets 5–7 mm long, in an open or contracted panicle, glumes 5–10 mm long, awn to 18 mm long. Aug.–May. Forest margins, SW, AP, LB, SE (Cape Peninsula to E tropical Africa).

STIPAGROSTIS   BOESMANGRAS   c. 50 spp., Eurasia and Africa

anomala De Winter   Weak perennial or annual to 60 cm. Leaves linear, rolled, setaceus, often curved, scabrid. Spikelets 9–12 mm long, in a narrow, interrupted panicle, awn solitary, spikelet callus pungent. Jan.–June. Sandy soils and gravel flats, KM (Namibia to Little Karoo).

ciliata (Desf.) De Winter   LANGBEENBOESMANGRAS   Densely or laxly tufted perennial to 1 m, culm nodes with collar of stiff hairs. Leaves mainly basal, linear, rolled. Spikelets 6.2–15 mm long in an open or contracted panicle, central awn plumose, spikelet callus pungent. Aug.–Oct. Sandstone slopes, KM (Namibia and Karoo to Swartberg Mts and Free State).

obtusa (Del.) Nees   KORTBEENBOESMANGRAS   Compact, densely tufted perennial to 60 cm. Leaves basal, linear, rolled, often curved. Spikelets 11–12 mm long, in an interrupted panicle, central awn plumose, spikelet callus usually bifid. July–May. Dry sandy slopes and flats, KM (Namaqualand and Karoo to Little Karoo).

zeyheri (Nees) De Winter   CAPE BUSHMAN GRASS   Robust perennial to 1 m. Leaves linear, rolled, setaceus. Spikelets in an open panicle, white and purple, all 3 awns distinctly plumose, spikelet callus pungent. Oct.–Apr. Sandy flats, NW, SW, AP, LB, SE (Namaqualand to Mpumalanga).

THEMEDA   10 spp., tropical and temperate Africa and Asia

triandra Forssk.   RED GRASS, ROOIGRAS   Perennial to 80 cm. Leaves linear. Spikelets long-awned, in triangular clusters subtended by spathes, reddish, brown and yellow. Oct.–July. Widespread in grassland, NW, SW, AP, KM, LB, SE (throughout tropical Africa and Asia).

THINOPYRUM   5 spp., Europe

distichum (Thunb.) Loeve (= *Agropyron distichum* (Thunb.) P.Beauv.)   COASTAL WHEATGRASS, SEA WHEAT   Hard, robust perennial to 80 cm, with strong, thick, creeping rhizomes. Leaves linear, rolled, rigid, pungent. Spikelets in a distichous spike, awnless. Oct.–Jan. Coastal dunes, NW, SW, AP, SE (S Namaqualand to Transkei).

TRACHYPOGON   13 spp., tropical America, Africa, Madagascar

spicatus (L.f.) Kuntze   GIANT SPEAR GRASS   Perennial to 1.2 m, culm nodes with ring of hairs. Leaves linear, folded, ligule membranous, 3-lobed. Spikelets in a spike-like raceme, awns velvety. Oct.–May. Clay and sandstone slopes, SW, LB, SE (Cape Peninsula to tropical Africa).

*TRAGUS   KOUSKLITS   6 spp., tropical Africa, 1 cosmopolitan

*****berteronianus** Schult.   SMALL CARROT SEED GRASS   Loosely tufted annual to 60 cm. Leaves lanceolate, margins roughly ciliate. Spikelets crowded in a narrow false spike, awnless, burr-like with hooked glume hairs. Jan.–Dec. Disturbed places, KM, SE (American weed).

TRIBOLIUM (= *LASIOCHLOA, PLAGIOCHLOA, UROCHLAENA*)   10 spp., Namibia to SW Cape and Drakensberg

A. Spikelets distichous

brachystachyum (Nees) Renvoize (= *Plagiochloa brachystachya* (Nees) Adamson & Sprague)   Like **T. uniolae** but mostly to 30 cm, spikelets mostly 4–5 mm long and glumes glassy hairy. Nov.–Dec. Sandstone slopes, NW, SW (Cedarberg Mts to Houwhoek).•

uniolae (L.f.) Renvoize (= *Plagiochloa alternans* (Nees) Adamson & Sprague, *Tribolium amplexum* Renvoize, *Plagiochloa uniolae* (L.f.) Adamson & Sprague)   KORINGGRAS   Perennial to 60 cm. Leaves linear, with innovation buds extravaginal. Spikelets to 6 mm long, in a distichous spike, lemmas with club-shaped hairs below, awnless. Oct.–Dec. Mostly clay and granite flats, NW, SW, AP, KM, LB, SE (Bokkeveld Mts to Port Elizabeth).•

A. Spikelets not distichous
B. Annuals

ciliare (Stapf) Renvoize (= *Plagiochloa echinatum* (Stapf) Adamson & Sprague)   Like **T. hispidum** but to 10 cm, spikelets 2–2.5 mm long, in a lax panicle. Sept. Limestone slopes and flats, AP (Agulhas Peninsula).•

echinatum (Thunb.) Renvoize (= *Lasiochloa echinata* (Thunb.) Adamson)   Annual to 30 cm. Leaves linear, hairy. Spikelets 3–5.5 mm long, in a contracted panicle partly enclosed by upper leaf, glumes glassy-hairy, lemma hairy on lower margins. Sept.–Oct. Flats and lower slopes, often on richer soils, NW, SW, AP (Bokkeveld Mts to Mossel Bay).•

pusillum (Nees) H.P.Linder & Davidse (= *Urochlaena pusilla* Nees)   Annual to 25 cm. Leaves linear. Spikelets 5–6 mm long, in a dense, subglobose panicle subtended by an enclosing, auriculate leaf sheath, disarticulating below sheath, lemmas glassy-hairy, long-awned. Aug.–Sept. Dry slopes, NW (S Namaqualand to Botterkloof and Nardouw Mts).

utriculosum (Nees) Renvoize (= *Lasiochloa utriculosa* Nees)   Like **T. echinatum** but spikelets shortly awned and glume and lemma hairs club-shaped. Sept.–Oct. Stony slopes, NW (S Namibia to Piketberg and Hex River Mts).

B. Perennials

acutiflorum (Nees) Renvoize (= *Plagiochloa acutiflora* (Nees) Adamson & Sprague)   Like **T. obliterum** but spikelets 4–6, in terminal and axillary panicles, glume hairs clavate. Sept.–Dec. Mostly clay slopes, NW, SW (Namaqualand to Cape Peninsula).

hispidum (Thunb.) Renvoize (= *Lasiochloa longifolia* (Schrad.) Kunth)   Like **T. obtusifolium** but glumes densely long-bristly, microscabridulous. Sept.–Dec. Flats and slopes, NW, SW, AP, KM, LB, SE (Namaqualand to E Cape).

obliterum (Hemsl.) Renvoize (= *Plagiochloa oblitera* (Hemsl.) Adamson & Sprague)   Perennial to 40 cm, often stoloniferous. Leaves linear. Spikelets 3.5–4.5 mm long, in a lax panicle, 5–10-flowered, florets exserted from glumes. Sept.–Nov. Flats and plateaus on loamy soils, NW, SW, AP, LB (Grootwinterhoek to Langeberg Mts).•

obtusifolium (Nees) Renvoize (= *Lasiochloa obtusifolia* Nees)   Like **T. obliterum** but spikelets 3–6-flowered, florets largely included, glumes glassy hairy, lemmas stiffly ciliate. Oct.–Dec. Clay flats, NW, SW, AP, KM, LB, SE (Olifants River Mts to E Cape).

TRIRAPHIS   10 spp., Africa, Australia

andropogonoides (Steud.) E.Phillips   KOPERDRAADGRAS   Perennial to 1.2 m, with long creeping rhizomes. Leaves linear, rolled. Spikelets in a dense panicle to 30 cm long, 5–15-flowered, shortly awned. Oct.–May. Stony riverbanks, SW, LB, SE (Potberg to northern South Africa).

TRISTACHYA   TRIDENT GRASS   c. 20 spp., Africa, Madagascar, tropical America

leucothrix Nees (= *Tristachya hispida* (L.f.) Schum.)   ROOISAADGRAS   Perennial to 90 cm. Leaves linear, roughly hairy, basal leaf sheaths with dense brown hairs at base. Spikelets in threes, long-awned, glumes with tuberculate hairs. Oct.–Mar. Marshy grassland and sandstone slopes, LB, SE (Swellendam to Mpumalanga).

*VULPIA   c. 25 spp., pantemperate

*bromoides (L.) S.F.Gray   SQUIRREL TAIL FESCUE   Annual to 70 cm. Leaves linear. Spikelets usually secund, in a panicle exserted from leaf sheaths, glume awns shorter than 3 mm, calluses rounded. Aug.–Jan. Disturbed areas, SW, SE (worldwide weed).

*fasciculata (Forssk.) Samp.   Like **V. bromoides** but upper glume awn 10–20 mm long, calluses pointed. Oct.–Nov. Disturbed areas, SW (European weed).

*muralis (Kunth) Nees   Like **V. bromoides**, but lower glume to half as long as upper glume. Sept.–Dec. Disturbed areas, NW, SW (European weed).

*myuros (L.) C.C.Gmel.   RATS TAIL FESCUE, LANGBAARD SWENKGRAS   Like **V. bromoides** but panicle partly enclosed in upper leaf sheath, lower glume always less than half a long as upper glume. Sept.–Nov. Disturbed areas, NW, SW, LB, SE (European weed).

POTAMOGETONACEAE

POTAMOGETON   FONTEINGRAS, FONTEINKRUID   c. 100 spp., cosmopolitan

pectinatus L.   Submerged aquatic. Leaves linear to filiform, often clustered above, adnate to stipules. Flowers green, detached and floating, inconspicuous. Oct.–Jan. Fresh or brackish water, NW, SW, AP, LB, SE (Namibia to Port Elizabeth, cosmopolitan).

pusillus L.   Submerged aquatic. Leaves linear. Flowers green, inconspicuous. Dec.–Mar. Fresh water, NW, SW, SE (Africa and N hemisphere).

schweinfurthii A.W.Benn.   Large submerged aquatic. Leaves sometimes floating, linear-lanceolate. Flowers on swollen peduncles, green. Nov.–Dec. Fresh water pools and streams, SE (Uitenhage to tropical Africa).

thunbergii Cham. & Schltdl.   Aquatic forming large mats. Leaves submerged and floating, petiolate, floating blades ovate. Flowers on slightly swollen peduncles, inconspicuous. Dec.–Mar. Fresh water or mud banks, KM, SE (Little Karoo to E tropical Africa).

trichoides Cham. & Schltdl.   Submerged aquatic forming dense mats. Leaves filiform. Flowers green, inconspicuous. Flowering time? Ponds and streams, SE (Uitenhage, Africa and Eurasia).

PRIONIACEAE

PRIONIUM   PALMIET   1 sp., W Cape to S KwaZulu-Natal

serratum (L.f.) Drège ex E.Mey.   Robust shrub to 2 m. Leaves lanceolate, glaucous, crowded, margins sharply serrate. Flowers in large panicles, minute, brown. Sept.–Feb. Streams and rivers, often forming dense stands, NW, SW, AP, LB, SE (Gifberg to S KwaZulu-Natal).

RESTIONACEAE
by H.P. Linder

1. Style solitary:
    2. Fruit dehiscent..**Ischyrolepis**
    2.' Fruit indehiscent:
        3. Perianth chartaceous; a rare plant of the Groot Swartberg...**Staberoha**
        3.' Perianth cartilaginous:
            4. Male spikelets erect...**Calopsis**
            4.' Male spikelets pendent...**Thamnochortus**
1.' Styles 2 or 3:
    5. Styles 2:
        6. Ovary dehiscent:
            7. Styles fused at base, or if free then bracts longer than flowers ....**Ischyrolepis**
            7.' Styles free; bracts shorter than flowers..**Chondropetalum**
        6.' Ovary indehiscent:
            8. Culms simple:
                9. Male flowers in racemes:
                    10. Tepals of male flowers valvate, lanceolate ....**Ceratocaryum**
                    10.' Tepals of male flowers linear..**Willdenowia**
                9.' Male flowers in spikelets:

11. Male spikelets pendent; nut enclosed by perianth ....**Staberoha**
11.' Male spikelets erect; perianth shorter than nut, or if as large then hyaline:
    12. Nut flattened on one side; female spikelets usually with several flowers....................**Cannomois**
    12.' Nut round in cross section:
        13. Male bracts acute; female flowers always with a single flower ....**Hypodiscus**
        13.' Male bracts rounded; female spikelets with several flowers ....**Nevillea**
8.' Culms branching:
    14. Female spikelets with several flowers:
        15. Bracts of female flowers hyaline-chartaceous ....**Hydrophilus**
        15.' Bracts of female flowers coriaceous to osseous:
            16. Bracts shorter than perianth ....**Chondropetalum**
            16.' Bracts longer than perianth:
                17. Nut shorter than 5 mm ....**Calopsis**
                17.' Nut longer than 10 mm...**Cannomois**
    14.' Female spikelets with a solitary flower:
        18. Bracts shorter than perianth and nut ....**Elegia**
        18.' Bracts overtopping perianth and nut:
            19. Bracts of male flowers like perianth lobes, linear, hyaline; male flowers in racemes.....**Willdenowia**
            19.' Bracts of male flowers larger than perianth, not linear; male flowers in spikelets:
                20. Bracts of female flowers osseous; male flowers in cone-like spikelets ....**Mastersiella**
                20.' Bracts of female flowers chartaceous; male flowers in few-flowered spikelets ......**Anthochortus**
5.' Styles 3:
    21. Sheaths usually deciduous; bracts shorter than flowers, or if longer then hyaline and lacerated:
        22. Ovary unilocular, indehiscent ....**Elegia**
        22.' Ovary 3-locular, usually at least some locules dehiscent:
            23. Culms branching..**Dovea**
            23.' Culms simple:
                24. Bracts longer than flowers...**Askidiosperma**
                24.' Bracts shorter than flowers, or if longer then not hyaline ....**Chondropetalum**
    21.' Sheaths persistent; bracts usually cartilaginous, often overtopping flowers:
        25. Culms flattened; sheaths green with a stout mucro; seeds white with a fragile surface...........**Platycaulos**
        25.' Culms terete, or if flattened then not with the above characte:
            26. Culms with a simple central axis and with sterile and/or fertile branches clustered at nodes; male spikelets pendent ....**Rhodocoma**
            26.' Culms usually branching, or if simple then without branches clustered at nodes; male spikelets erect:
                27. Fruit a unilocular nut:
                    28. Culms branching..**Calopsis**
                    28.' Culms simple ....**Staberoha**
                27.' Fruit a capsule, or if a nut then bracts narrowly lanceolate, concolorous:
                    29. Culms branching ....**Restio**
                    29.' Culms simple:
                        30. Bracts shorter than flowers; spikelets numerous at several nodes ....**Chondropetalum**
                        30.' Bracts at least as long as flowers; spikelets few to several ....**Restio**

ANTHOCHORTUS• 7 spp., W Cape

capensis Esterh.   Dioecious, tangled perennial to 50 cm, culms branched, slender, spathes and bracts setaceous. Flowering time? Along streams, SW (Cape Peninsula).•

crinalis (Mast.) H.P.Linder   Dioecious, tangled perennial to 80 cm, culms much branched, densely tuberculate. Female flowers single or paired in spikelets, spathes and bracts setaceous. May. Damp slopes and seeps, SW, LB (Cape Peninsula to Worcester, Riversdale).•

ecklonii Nees   Dioecious, tangled perennial to 50 cm, producing stolons, culms sparsely branched, striate. Flowering time? Damp slopes and seeps, NW, SW, LB, SE (Ceres to Oudtshoorn and George).•

graminifolius (Kunth) H.P.Linder   Dioecious, tangled perennial, producing stolons, culms branched, compressed, sheaths green, often with large free blades. Flowering time? Damp slopes and seeps, SW (Paarl to Caledon).•

insignis (Mast.) H.P.Linder   Dioecious, rhizomatous perennial to 70 cm, culms sparsely branched, sheaths with awns to 10 mm. Female flowers in capitate spikelets, bracts acuminate. May. Rocky streambanks, NW (Cold Bokkeveld).•

laxiflorus (Nees) H.P.Linder   Dioecious, caespitose perennial, 15–50 cm, culms branched, basal sheaths often with leaf-like blade, spathe margins hyaline. Flowering time? Damp slopes and seeps, SW (Cape Peninsula to Caledon).•

singularis Esterh.   Dioecious, caespitose perennial to 30 cm, with slender rhizomes and sparsely branched, densely rugulose culms. Flowering time? Dry stony slopes, NW (Bokkeveld Mts).•

ASKIDIOSPERMA•   10 spp., W Cape

albo-aristatum (Pillans) H.P.Linder   Dioecious, caespitose perennial to 1 m, sheaths deciduous. Flowers with hyaline bracts, female spikelets many-flowered. Sept.–Oct. Habitat? NW (Pakhuis and Cedarberg Mts).•

andreaeanum (Pillans) H.P.Linder   Dioecious, caespitose perennial to 80 mm. Flowers of female spikelets few, obscured by hyaline bracts. Jan. Habitat? SW (Du Toitskloof Mts).•

capitatum Steud.   Dioecious, caespitose perennial to 80 cm. Female flowers in spikelets aggregated into capitate heads. Flowering time? Dry stony slopes and sandy flats, NW (Cedarberg to Ceres).•

chartaceum (Pillans) H.P.Linder   Dioecious, caespitose perennial to 1 m. Flowers of female spikelets many. Aug.–Sept. Marshy mountain slopes and plateaus, NW, SW (Tulbagh to Kogelberg and Riviersonderend Mts).•

esterhuyseniae (Pillans) H.P.Linder   Dioecious, caespitose perennial to 1 m. Flowers of female spikelets obscured by spathes. May. Higher mountains, SW (Jonkershoek to Sir Lowry's Pass).•

insigne (Pillans) Linder   Dioecious, caespitose perennial, 1–2 m. Sept.–Oct. Marshes above 1800 m, NW (Cold Bokkeveld Mts to Ceres).•

longiflorum (Pillans) H.P.Linder   Dioecious, caespitose perennial to 80 mm. Female flowers longer than 6 mm. Oct.–Nov. Sandstone slopes, NW (Cold Bokkeveld and Hex River Mts).•

nitidum (Mast.) H.P.Linder   Dioecious, caespitose perennial to 1.2 m. Flowers of female spikelets many. Aug.–Sept. Sandstone slopes, NW, SW (Cedarberg to Jonkershoek Mts).•

paniculatum (Mast.) H.P.Linder   Dioecious, caespitose perennial to 1 m. Flowers of female spikelets few, obscured by spathes. Apr. Sandstone mountain slopes, NW, SW, LB (Cold Bokkeveld to Langeberg Mts).•

rugosum Esterh.   Dioecious, caespitose perennial to 1 m. Aug.–Sept. Sandstone slopes, SW (Hottentots Holland Mts).•

CALOPSIS   c. 25 spp., Namaqualand to KwaZulu-Natal

A. Lateral sepals winged or keeled

esterhuyseniae (Pillans) H.P.Linder   Dioecious, caespitose perennial to 40 cm, culms simple, with flat-topped tubercles, sheaths closely convolute. Flowers with spikelets 5–8 mm long, ovate to rotund, styles 2. Flowering time? Sandstone slopes, NW, SW, KM, (Cold Bokkeveld to Stellenbosch, and Little Karoo).•

hyalina (Mast.) H.P.Linder   Dioecious, caespitose, rhizomatous perennial to 50 cm, culms branched, with white tubercles, spathes with hyaline apices. Flowers 2 or 3 on distichously branched spikelets 6–8 mm long, styles 3. Nov. Flats, SW, LB (Elandskloof Mts to Riversdale).•

impolita (Kunth) H.P.Linder   Dioecious, caespitose perennial, culms branched, with white tubercles. Flowers 2 or 3 on distichously branched spikelets 10–15 mm long, styles 3. Flowering time? Sandy coastal flats, NW, SW (Lambert's Bay to Cape Peninsula).•

levynsiae (Pillans) H.P.Linder   Dioecious, caespitose perennial to 50 cm, culms branched, sheaths closely convolute. Female flowers many on spikelets 10–20 mm long, styles 3. June. Sandstone slopes, NW (Cold Bokkeveld).•

monostylis (Pillans) H.P.Linder   Dioecious, caespitose perennial to 40 cm, culms simple, coarsely tubercled, sheaths closely convolute. Female flowers on spikelets to 8 mm long, narrowly ovate, style solitary. Feb. Rocky slopes, LB (Langeberg Mts: Montagu to Riversdale).•

A. Lateral sepals not winged or keeled
B. Sheaths broadly membranous above and soon decaying

aspera (Mast.) H.P.Linder   Dioecious, caespitose perennial to 50 cm, culms branched, coarsely tubercled, spathes with wide hyaline apices. Flowers in spikelets 6–8 mm long. Flowering time? Habitat? SW (Hottentots Holland to Shaw's Mts).•

clandestina Esterh.   Dioecious, caespitose perennial to 40 cm, culms branched, densely tuberculate, sheaths with long hyaline apices. Flowers in spikelets with elongate internodes, perianth glabrous. Flowering time? Coastal slopes, SW (Kogelberg to Kleinrivier Mts).•

membranacea (Pillans) H.P.Linder   Dioecious, rhizomatous perennial to 30 cm, with culms branched, spreading through the vegetation, spathes membranous above. Flowers exceeding bracts, sepals acute. Flowering time? Habitat? NW, SW (Tulbagh and Cape Peninsula to Riviersonderend Mts).•

nudiflora (Pillans) H.P.Linder   Dioecious, caespitose perennial to 20 cm, culms branched and tangled, roughly tubercled. Flowers exceeding bracts, sepals obtuse. Flowering time? Habitat? SW (Jonkershoek to Kogelberg).•

paniculata (Rottb.) Desv.   Dioecious perennial to 2 m, with long rhizomes, culms branched. Flowers in compound inflorescences with numerous spikelets on main branches. Flowering time? Mainly along streams, NW, SW, KM, LB, SE (Cedarberg Mts to KwaZulu-Natal).

pulchra Esterh.   Dioecious, caespitose perennial to 50 cm, culms sparsely branched, densely tuberculate. Flowers in racemes with spikelet internodes much elongated, perianth villous. Sept. Limestone and sandstone flats, SW, AP (Pearly Beach to Struisbaai).•

rigida (Mast.) H.P.Linder   Dioecious, rhizomatous perennial with trailing, branched culms, sheaths lightly convolute. Flowers obscured by cartilaginous bracts. Apr. Habitat? NW, KM (Ceres to Witteberg and Swartberg Mts).•

sparsa Esterh.   Dioecious, caespitose perennial to 50 cm, culms branched, rugose, spathes with wide hyaline apices. Flowers in spikelets 10–15 mm long. Flowering time? Rocky sandstone slopes, SW (Kogelberg).•

B. Sheaths of uniform colour and texture

adpressa Esterh.   Dioecious, caespitose perennial, culms branched, spreading, sheaths loosely convolute, but not flat, awn often recurved. Flowering time? Coastal slopes and flats, often on limestone, SW, AP (Cape Hangklip to Stilbaai).•

andreaeana (Pillans) H.P.Linder   Dioecious, caespitose perennial to 60 cm, culms branched, sheaths closely convolute. Flowers on spikelets 8–12 mm long. Flowering time? Sandstone slopes, KM (Rooiberg and Swartberg Mts).•

burchellii (Mast.) H.P.Linder   Dioecious, caespitose perennial to 45 cm, culms branched, densely white-tuberculate, sheaths loosely convolute. June. Sandstone slopes, LB, SE (Riversdale to Humansdorp).•

dura Esterh.   Dioecious, caespitose perennial to 60 cm, culms branched, obscurely tubercled or smooth, spreading, sheaths loosely convolute. Female flowers solitary on spikelets. Flowering time? Habitat? NW (Cedarberg Mts to Ceres).•

filiformis (Mast.) H.P.Linder   Dioecious, caespitose perennial to 60 cm, culms branched, sheaths tightly convolute, awn straight. Flowers on spikelets to 5 mm. Flowering time? Coastal sandstone slopes, SW, AP (Sir Lowry's Pass to Agulhas).•

fruticosa (Mast.) H.P.Linder   Dioecious, caespitose perennial to 60 cm, culms spreading, branched, sheaths loosely convolute, with awns to 5 mm, about a third as long as sheath. Flowers on spikelets 5–10 mm long. Aug.–Nov. Mainly coastal limestone, SW, AP (Cape Peninsula to Stilbaai).•

gracilis (Mast.) H.P.Linder   Dioecious, caespitose perennial to 60 cm, culms slender, much branched, white-tubercled, sheaths tightly convolute. Flowers on solitary spikelets to 5 mm long. Nov. SW (Cape Peninsula).•

marlothii (Pillans) H.P.Linder   Dioecious, caespitose perennial to 1 m, culms sparsely branched, sheaths free, awns less than a third as long as sheath. Female flowers on compound inflorescences with spikelets well spaced. May–Aug. Habitat? NW, KM, LB (Namaqualand and Cedarberg to Little Karoo Mts and Langeberg).•

muirii (Pillans) H.P.Linder   Dioecious, loosely caespitose perennial to 1 m, culms spreading, branched, sheaths loosely convolute, margins and apices hyaline, awn less than a third as long as sheath. Female flowers in crowded spikelets. Flowering time? Lower sandstone slopes, LB (Langeberg Mts: Riversdale).•

viminea (Rottb.) H.P.Linder   Dioecious, caespitose perennial to 1 m, culms spreading, rugulose, branched, sheaths free, with awns to 5 mm long. Female flowers several on spikelets. Flowering time? Habitat? NW, SW, AP, KM, LB, SE (Namaqualand to Port Elizabeth).

CANNOMOIS•   7 spp., W Cape

aristata Mast.   Dioecious, caespitose perennial to 1 m, culms simple, sheaths with awns to 10 mm, bracts acuminate. Aug. Habitat? NW (Cedarberg Mts).•

congesta Mast.   Dioecious, caespitose perennial to 1 m, culms usually drooping, simple. Flowers with tepals almost as long as nut. Flowering time? Gravelly mountain slopes, often above shale bands, NW, SW (Bokkeveld Mts to Bredasdorp Mts).•

nitida (Mast.) Pillans   Dioecious, caespitose perennial to 70 cm, culms simple. Female flowers solitary on spikelets, bracts acuminate, erect or reflexed with awns to 10 mm. Flowering time? Sandstone slopes above 1500 m, NW, SW, KM (Cold Bokkeveld to Swartberg Mts).•

parviflora (Thunb.) Pillans   Dioecious, caespitose perennial to 1 m, with rhizomes often long and creeping, culms simple, spathes and bracts dark, shiny, coriaceous, acute. Flowers with perianth about a tenth as long as nut. Aug. Lower slopes and coastal plains, NW, SW, AP, SE (Bokkeveld Mts to Humansdorp).•

scirpoides (Kunth) Mast. (= *Cannomois dregei* Pillans)   Dioecious, caespitose perennial to 1 m, culms usually simple, spathes shiny. Flowers with perianth about a tenth as long as nut. Flowering time? Habitat? KM, SE (Oudtshoorn to Humansdorp).•

taylorii H.P.Linder   Dioecious, caespitose perennial to 2 m, with spreading rhizomes, often forming large stands. Flowering time? Dry sandstone slopes in arid fynbos and *Passerina* scrub, NW (Karoopoort to Moedverloor).•

virgata (Rottb.) Steud.   Dioecious, caespitose perennial to 3 m, culms branched above. Female flowers solitary, sometimes paired, acute, to 30 mm. Flowering time? Along streams and seeps, sometimes in dense bamboo-like stands, NW, SW, KM, LB, SE (Bokkeveld Mts to Humansdorp).•

CERATOCARYUM•   6 spp., W Cape

argenteum Nees ex Kunth   Dioecious, caespitose perennial to 1.5 m, culms simple. Female flowers on spikelets to 30 mm. Nuts to 10 mm diam. Feb.–Apr. Mostly sandy flats, SW, AP, LB (Jonkershoek to Albertinia).•

decipiens (N.E.Br.) H.P.Linder   Dioecious, caespitose perennial to 2 m, culms simple. Female flowers on spikelets to 15 mm, spathes nitid, dark brown. Nuts pedicellate. Feb.–Mar. Sandstone slopes, SW, LB (Du Toitskloof to Riversdale).•

fimbriatum (Kunth) H.P.Linder   Dioecious, caespitose perennial to 60 cm, spathes shiny, dark brown. Nuts stipitate, rugulose. June. Sandstone slopes, NW, SW (Cold Bokkeveld and Hex River Mts to Du Toitskloof).•

fistulosum Mast.   Dioecious, caespitose perennial to 2 m, culms simple, hollow. Sept.–Oct. Sandstone slopes at high alt., LB (Langeberg Mts: Swellendam to Riversdale).•

pulchrum H.P.Linder   Dioecious, caespitose perennial to 70 cm. Nuts sessile, dark brown, surface rugulose. Oct. Coastal sandstone slopes, SW (near Elim).•

xerophilum (Pillans) H.P.Linder   Dioecious, caespitose perennial to 60 cm, culms simple. Nuts stipitate, smooth. Sept.–Oct. Subalpine sandstone slopes, NW, KM (Hex River Mts to Klein Swartberg).•

CHONDROPETALUM•   10 spp., W Cape

acockii Pillans   Dioecious, caespitose perennial to 1 m, culms simple, sheaths persistent, several per culm. Flowering time? Sandy flats, SW (Hopefield to Cape Peninsula).•

aggregatum (Mast.) Pillans   Dioecious, caespitose perennial to 80 cm, sheaths deciduous. Female flowers 3 or 4 per spikelet. Flowering time? Sandstone slopes, SW (Riviersonderend Mts).•

decipiens Esterh.   Dioecious, caespitose, rhizomatous perennial to 1 m, culms 1- or 2-noded, sheaths deciduous. Female flowers 1 or 2 per spikelet, aggregated into a subcapitate head. Flowering time? Sandstone slopes, SW (Kleinrivier Mts).•

deustum Rottb.   Dioecious, caespitose perennial to 70 cm, sheaths persistent, solitary per culm. Flowering time? Seasonally wet places on sandstone slopes, SW (Cape Peninsula to Bredasdorp).•

ebracteatum (Kunth) Pillans   Dioecious, caespitose perennial to 80 cm, sheaths deciduous. Female flowers 1 or 2 per spikelet. Oct.–Dec. Sandstone slopes, SW, LB (Cape Peninsula to Langeberg Mts: Riversdale).•

microcarpum (Kunth) Pillans   Dioecious, caespitose perennial, culms much branched, sheaths free, persistent. Apr. Coastal sands and limestones, SW, AP, SE (Melkbos to Port Elizabeth).•

mucronatum (Nees) Pillans   Dioecious, caespitose perennial to 2 m. Flowers on large inflorescences with spathes 5–10 cm. Oct.–Nov. Marshes and seeps, mainly montane, SW, LB (Bainskloof to Langeberg Mts: Riversdale).•

nudum Rottb.   Like **C. tectorum** but culms shorter and more slender and petals scabrid on outside. Apr.–June. Sandy flats, SW, AP, LB (Darling to Albertinia).•

rectum (Mast.) Pillans   Dioecious, caespitose perennial to 60 cm, with long slender rhizomes, culms slender, sheaths deciduous. Flowering time? Damp flats, often on clay, SW, AP (Cape Peninsula to Agulhas).•

tectorum (L.f.) Raf.   DAKRIET   Dioecious, caespitose perennial to 1 m, sheaths deciduous. Flowers shorter than 3 mm, petals scabrid only in upper half, or smooth. Flowering time? Marshes and seeps on deep sands, NW, SW, AP, LB, SE (Clanwilliam to Port Elizabeth).•

DOVEA•   1 sp., W Cape

macrocarpa Kunth   Dioecious, caespitose perennial to 1 m, culms branched, sheaths deciduous. Female flowers large. Flowering time? Sandstone slopes, NW (Cedarberg Mts to Piketberg).•

ELEGIA   34 spp., W to E Cape

A. Culms branched

capensis (Burm.f.) Schelpe   Dioecious, caespitose perennial to 2 m, culms with dense sterile verticillate branches. Flowering time? Mainly sandstone slopes along streams, NW, SW, AP, KM, LB, SE (Clanwilliam to Port Elizabeth).•

equisetacea (Mast.) Mast.   Dioecious, caespitose perennial to 1 m, often with a few verticillate branches. Sept. Habitat? LB, SE (Langeberg Mts: Swellendam to Van Staden's Mts).•

grandis (Nees) Kunth   Dioecious, caespitose perennial to 1 m, culms branched, spathes large, persistent, green. May. Dry sandstone slopes, SW (Bainskloof to Riviersonderend Mts).•

hutchinsonii Pillans   Dioecious, caespitose perennial with persistent sheaths. Flowers with bracts c. 2 mm long, styles 3. Flowering time? Sandstone slopes, NW, SW (Cold Bokkeveld to Du Toitskloof).•

muirii Pillans   Dioecious, caespitose perennial to 60 cm, sheaths persistent, free, spathes larger than sheaths. Flowers with bracts 7–8 mm long, styles 3. Apr. Coastal, mainly on limestone, AP (Agulhas to Gouritsmond).•

neesii Mast.   Dioecious, caespitose perennial to 70 cm, culms sparsely branched, sheaths green, closely convolute. Oct. Sandstone slopes, NW, SW, LB, SE (Cedarberg to Humansdorp).•

prominens Pillans   Dioecious, caespitose perennial with culms much branched. Female flowers with ovary exceeding perianth. Flowering time? Coastal flats, SW, AP (Yzerfontein to Bredasdorp).•

stipularis Mast.   Dioecious, caespitose perennial to 50 cm, culms much branched, sheaths persistent, like the spathes. Female flowers with styles slender, 2. June–July. Habitat? SW, AP, LB (Cape Peninsula to Mossel Bay).•

stokoei Pillans   Dioecious, caespitose perennial with culms sparsely branched, sheaths deciduous. Styles slender, 2. Flowering time? Sandstone slopes, SW (Elandskloof Mts to Villiersdorp).•

A. Culms simple

altigena Pillans   Like **E. juncea** but styles 2. Flowering time? Rocky sandstone slopes, SE (Uniondale).•

amoena Pillans   Dioecious, caespitose perennial to 1.5 m. Female inflorescence cylindrical, flowers obscured by spathes. Feb. Sandstone slopes above 1500 m, SW (Slanghoek Mts).•

asperiflora (Nees) Kunth   Dioecious, caespitose perennial to 70 cm, culms lightly compressed, sheaths persistent. Feb.–Mar. Seeps on sandstone slopes, NW, SW, KM, LB, SE (Cedarberg Mts to E Cape).•

atratiflora Esterh.   Dioecious, caespitose perennial to 70 cm, sheaths dark brown, nitid, deciduous. Flowers papillose, reddish. Sept. Habitat? SW (Hottentots Holland Mts to Kleinrivier Mts).•

caespitosa Esterh.   Dioecious, caespitose perennial to 40 cm, culms compressed, sheaths persistent or deciduous. Dec. Habitat? SW, LB (Bainskloof to Langeberg Mts).•

coleura Nees ex Mast.   Dioecious, caespitose perennial to 60 cm, culms lightly compressed, sheaths deciduous. Flowering time? Damp sandy flats, SW, AP, LB, SE (Cape Peninsula to Humansdorp).•

cuspidata Mast.   Dioecious, caespitose perennial to 1 m. Flowers with bracts and spathellae setaceous, chestnut-brown. Flowering time? Coastal sands, SW (Bokbaai to Kleinmond).•

esterhuyseniae Pillans   Dioecious, caespitose perennial to 30 cm, sheaths deciduous. Flowers in linear inflorescences, spathes variegated, obscuring female flowers. Flowering time? High sandstone slopes, NW, SW (Cedarberg to Riviersonderend Mts).•

extensa Pillans   Dioecious, caespitose perennial to 50 cm. Flowering time? Habitat? SW, AP (Wolseley to Bredasdorp).•

fenestrata Pillans   Dioecious, caespitose perennial to 60 cm, sheaths deciduous. Flowers on non-linear inflorescences, bracts shorter than flowers. Seeps on coastal flats, SW, AP (Cape Peninsula to Bredasdorp).•

filacea Mast.   Dioecious, caespitose perennial to 50 cm, culms slender, sheaths deciduous. Flowers on exsposed inflorescences. Feb. Damp flats and sandy slopes to 2100 m, NW, SW, AP, KM, LB, SE (Cedarberg to Port Elizabeth).•

fistulosa Kunth   Dioecious, caespitose perennial to 70 cm, culms fistulose, often spreading through the vegetation, sheaths deciduous. Apr.–Nov. Seeps, SW, AP, LB, SE (Malmesbury to Van Staden's Mts).•

galpinii N.E.Br.   Dioecious, caespitose perennial to 1 m, with culms clustered from creeping rhizomes, sheaths and female spathes deciduous, almost black. Aug. Dry sandsone slopes, LB (Langeberg Mts: Montagu to Attaquaskloof).•

grandispicata Linder   Dioecious, caespitose or diffuse perennial to 1.5 m, forming large stands, culms rough. Flowers with bracts acuminate. May–Aug. Seeps and along streams above 1000 m, NW, SW, LB (Cedarberg to Langeberg Mts).•

intermedia (Steud.) Pillans   Dioecious, caespitose perennial to 1 m, culms rough to the touch, sheaths deciduous. Flowers on crowded inflorescences, spathes papery. Dec. Moist sites, SW (Cape Peninsula).•

juncea L.   Dioecious, caespitose perennial to 80 cm, with creeping rhizomes, sheaths deciduous. Flowers on exposed inflorescences, styles 3. Aug. NW, SW, AP, KM, LB (Grootwinterhoek Mts and Cape Peninsula to Swartberg Mts).•

persistens Pillans   Dioecious, caespitose perennial to 1 m, with spreading rhizomes, spathes variegated, completely obscuring female flowers. Flowering time? Sandstone slopes, SW (Houwhoek to Bredasdorp Mts).•

racemosa (Poir.) Pers.   Dioecious, caespitose perennial to 1 m, with spreading rhizomes, culms simple, sheaths deciduous. Flowers concealed by the large spathes, spathe margins pale. Sept. SW, LB, SE (Du Toitskloof Mts to Uitenhage).•

rigida Mast.   Dioecious, caespitose perennial to 30 cm. Flowers with female spathes overlapping, apices reflexed. Jan. Habitat? SW (Slanghoek Mts).•

spathacea Mast.   Dioecious, caespitose perennial to 60 cm. Female flowers obscured by spathes. Flowering time? Sandstone slopes, SW (Jonkershoek to Kleinrivier Mts).•

squamosa Mast.   Dioecious, caespitose perennial to 20 cm, with spreading rhizomes, sheaths deciduous. Flowering time? Seasonally wet flats, SW (Cape Peninsula and Paarl to Caledon).•

thyrsifera (Rottb.) Pers.   Dioecious, caespitose perennial to 1 m, sheaths deciduous. Flowers with large spathellae, bracts shorter than flowers. Flowering time? Damp mountain slopes, SW, AP, LB, SE (Cape Peninsula to Knysna).•

thyrsoidea (Mast.) Pillans   Dioecious, caespitose perennial to 80 cm, with spreading rhizomes, spathes without pale margins. Flowers obscured by the overlapping spathes. Flowering time? Sandy mountain slopes, SE (Outeniqua and Tsitsikamma Mts).•

vaginulata Mast.   Dioecious perennial to 40 cm, with culms arising from spreading rhizomes. Flowering time? Sandstone slopes, NW, SW, AP, KM, LB, SE (Cedarberg to E Cape).

verreauxii Mast.   Dioecious perennial to 30 cm, with slender spreading rhizomes, culms in small tufts, sheaths deciduous. Flowering time? Damp flats, SW, AP (Malmesbury to Agulhas).•

HYDROPHILUS   1 sp., W to E Cape

rattrayi (Pillans) H.P.Linder   Dioecious pernnial to 70 cm, forming spreading clumps, often with stolons, culms sparsely branched. Flowers with bracts chartaceous, pale, spikelets 15–25 mm. Flowering time? Streams on dry slopes, NW, KM (Cedarberg to Swartberg Mts and Hogsback).

HYPODISCUS   15 spp., Namaqualand to SW Cape

A. Culms more or less striate

argenteus (Thunb.) Mast.   Dioecious, caespitose perennial to 1 m, culms simple, round. Male spikelets many, to 5 mm, bracts hyaline, female spikelets to 20 mm. Nuts without perianth. Mar.–July. Habitat? NW, SW, LB (Clanwilliam to Riversdale).•

neesii Mast.   Dioecious, caespitose perennial to 50 cm, culms simple, round. Male and female spikelets dissimilar, male spikelets several, bracts hyaline. Nuts without elaiosome. June. Dry lower mountain slopes, NW, KM (Cedarberg to Witteberg).•

procurrens Esterh.   Dioecious perennial to 20 cm, with long, creeping rhizomes, culms simple, round. Floral bracts acuminate. Flowering time? Habitat? AP (Agulhas Peninsula).•

rigidus Mast.   Dioecious, caespitose perennial to 60 cm, culms simple, round. Floral bracts acuminate. Nuts 3–5 mm long, with elaiosome. Flowering time? Coastal limestones, AP (Soetanysberg to Witsand).•

striatus (Kunth) Mast.   Dioecious, caespitose perennial, 20–60 cm, culms simple, round. Male and female spikelets dissimilar, female spikelet solitary, rarely paired. Apr.–June. NW, SW, AP, LB, SE (Namaqualand to Port Elizabeth).•

sulcatus Pillans   Dioecious, caespitose perennial to 40 cm, culms flexuose, stout, sulcate, round. Female spikelets solitary. Nuts c. 10 mm long, with elaiosome. Flowering time? Rocky sandstone soils, KM (Witteberg and Bonteberg).•

willdenowia (Nees) Mast.   Dioecious, rhizomatous perennial to 40 cm, culms simple, compressed. Spikelets solitary. June.–Aug. Sandy slopes and flats, NW, SW, AP, LB, SE (Cold Bokkeveld to Humansdorp).•

A. Culms smooth

albo-aristatus (Nees) Mast.   Dioecious, caespitose perennial to 70 cm, culms simple, round, with a single node. Male and female spikelets similar, to 10 mm, bracts long-acuminate. May–Sept. Habitat? NW, SW, AP, KM, LB, SE (Grootwinterhoek to Swartberg Mts and Humansdorp).•

alternans Pillans   Dioecious, caespitose perennial to 55 cm, culms simple, round. Male spikelets 5–10 mm, bracts reddish, acute. Flowering time? Habitat? SW (Caledon).•

aristatus (Thunb.) Mast.   Dioecious, caespitose perennial to 80 cm, culms simple, round. Male and female spikelets similar, to 20 mm, floral bracts long-acuminate. Nuts smooth, often without a perianth. May. Mostly sandstone soils, NW, SW, AP, KM, LB, SE (Clanwilliam to Baviaanskloof Mts).•

laevigatus (Kunth) H.P.Linder   Dioecious, caespitose perennial to 80 cm, culms simple, often slightly compressed. Male and female spikelets similar, 5–10 mm, bracts acute. Flowering time? Stony lower slopes, NW, SW, LB (Cedarberg Mts to Riversdale).•

montanus Esterh.   Dioecious, caespitose perennial to 40 cm, culms simple, obscurely sulcate. Spikelets entirely obscured by spathes. Feb. Rocky summits, LB (Langeberg Mts: Misty Point).•

rugosus Masters   Dioecious, caespitose perennial to 25 cm, culms often somewhat compressed. Male and female spikelets similar, to 7 mm, bracts acuminate. Flowering time? Coastal flats, SW, AP (Malmesbury to Agulhas flats).•

squamosus Esterh.   Dioecious, caespitose perennial to 60 cm, culms simple, compressed. Feb. SW (Riviersonderend Mts).•

synchroolepis (Steud.) Mast.   Dioecious, caespitose perennial to 40 cm, culms simple. Male and female spikelets similar, 5–10 mm, bracts acuminate. Nuts with linear fleshy lobes. Feb. Sandstone slopes, KM, LB, SE (Swartberg Mts to Riversdale and Humansdorp).•

ISCHYROLEPIS   c. 47 spp., Namaqualand and Karoo to E Cape

Group 1: Plants erect tussocks, culms branching dichotomously; spikelets many-flowered; bract awn longer than bract body; lateral sepals villous-carinate; fertile locules 2

capensis (L.) H.P.Linder   Dioecious, tangled-caespitose perennial to 50 cm, culms branched, often sterile, tubercled. Floral bracts long-acuminate, recurved, ovaries often smutted. Oct.–Nov. Mostly clay slopes, NW, SW, AP, KM, LB, SE (Clanwilliam to Port Elizabeth).•

curvibracteata Esterh.   Dioecious, caespitose perennial to 40 cm, culms simple. Floral bracts acuminate, recurved. Flowering time? Sandstone slopes, SW (Du Toitskloof to Hottentots Holland Mts).•

fuscidula (Pillans) H.P.Linder   Dioecious, caespitose perennial to 50 cm, culms simple. Spikelets subglobose. Flowering time? Habitat? NW (Worcester: Audensberg).•

hystrix (Mast.) H.P.Linder   Dioecious, caspitose perennial to 1.5 m, culms branched. Floral bracts long-acuminate, erect. Flowering time? Sandstone slopes, NW, KM, LB, SE (Ceres to Langkloof).•

longiaristata Pillans ex H.P.Linder   Dioecious, caespitose perennial to 50 cm, sheaths loosely convolute, culms branched, tubercled. Spikelets several-flowered. Oct. Rocky sandstone slopes, NW (Bokkeveld Mts to Gifberg).•

marlothii (Pillans) H.P.Linder   Dioecious, caespitose perennial to 50 cm, culms branched, tubercled. Female spikelets solitary, floral bracts acuminate. Flowering time? Rocky lower slopes, NW, KM (Gifberg to Witteberg).•

ocreata (Kunth) H.P.Linder   Dioecious, caespitose perennial to 80 cm, culms branched, sheaths spreading. Female spikelets usually solitary, male spikelets densely clustered, floral bracts long-acuminate, recurved. Flowering time? Dry rocky slopes, NW, SW, KM (Namaqualand to Jonkershoek and Swartberg Mts).•

setiger (Kunth) H.P.Linder   Dioecious, caespitose perennial to 60 cm, culms sparsely branched, leaf sheaths closely convolute. Floral bracts long-acuminate, recurved. Flowering time? NW (Cedarberg Mts).•

virgea (Mast.) H.P.Linder   Dioecious, caespitose perennial to 60 cm, culms branched. Spikelets densely clustered, floral bracts acuminate, erect. Dec. Sandstone slopes, NW, SW (Cedarberg to Riviersonderend Mts).•

Group 2: Plants with erect culms, branching often verticillate; coriaceous portion of sheaths acute; spikelets usually few-flowered

leptoclados (Mast.) H.P.Linder   Dioecious, mat-forming perennial to 60 cm, culms erect, with numerous verticillate branches. Spikelets solitary, terminal. Flowering time? Coastal sands, SW, AP, SE (Betty's Bay to Knysna).•

rivulus Esterh.   Dioecious, tangled, mat-forming, stoloniferous perennial to 80 cm, culms branched. Spikelets usually solitary. Flowering time? Seeps and streambanks, NW (S Cedarberg and Cold Bokkeveld Mts).•

subverticillata Steud.   Dioecious, caespitose perennial to 1.5 m, branches numerous, verticillate, mostly fertile. Feb.–June. Streambanks and riverine bush, SW (Bainskloof to Riviersonderend).•

tenuissima (Kunth) H.P.Linder   Dioecious, tangled-caespitose, stoloniferous perennial, culms slender, branched. Spikelets usually solitary. Flowering time? Habitat? NW, SW, LB (Cold Bokkeveld to Langeberg Mts).•

Group 3: Plants generally small, tangled; culms tuberculate; coriaceous portion of sheath truncate; female spikelets 1-flowered; ovary often with 1 fertile locule

caespitosa Esterh.   Dioecious, caespitose perennial to 20 cm, culms much branched, tubercled, usually flexuose. Flowering time? Habitat? SW (Kogelberg and Riviersonderend Mts to Bredasdorp).•

cincinnata (Mast.) H.P.Linder   Dioecious, caespitose perennial to 50 cm, culms much branched, tubercled, flexuose. Nov. Habitat? SW (Cape Peninsula to Riviersonderend Mts).•

curviramis (Kunth) H.P.Linder   Dioecious, tangled-caespitose, often mat-forming perennial to 50 cm, leaf sheaths with coriaceous section truncate. Aug. Habitat? NW, SW, LB, SE (Cedarberg Mts to George).•

duthieae (Pillans) H.P.Linder   Dioecious, caespitose perennial 50 cm, stoloniferous, culms branched, finely tuberculate. Flowering time? Flats, often on granitic soils, SW (Paarl to False Bay).•

eleocharis (Nees ex Mast.) H.P.Linder   Dioecious, caespitose perennial to 40 cm, culms much branched. Spikelets solitary, terminal. Flowering time? Coastal slopes, mainly limestone, SW, AP, SE (Cape Peninsula to Port Elizabeth).•

macer (Kunth) H.P.Linder   Dioecious, caespitose perennial to 40 cm, culms branched, densely tubercled. Spikelets many-flowered. Aug. Habitat? NW, SW (Bokkeveld Mts to Bredasdorp).•

nana Esterh. Dioecious, stoloniferous perennial to 20 cm, forming small cushions, culms branched, subflexuose. Sept.–Nov. Rocky sandstone slopes above 1200 m, NW, SW (Cedarberg to Goudini).•

pratensis Esterh. Dioecious, stoloniferous perennial to 20 cm, culms branched. Male spikelets 10–15 mm long. Flowering time? Habitat? NW, SW (Worcester to Cape Peninsula).•

pygmaea (Pillans) H.P.Linder Dioecious, caespitose perennial to 20 cm, culms branched, sheaths loosely convolute. Flowering time? High rocky slopes, NW (Witsenberg and Hex River Mts).•

rottboellioides (Kunth) H.P.Linder Dioecious, caespitose perennial, culms branched, finely tuberculate. Female spikelets closely adpressed to scape. Flowering time? Stony slopes, NW (Namaqualand to Piketberg).

sabulosa (Pillans) H.P.Linder Dioecious, rhizomatous perennial to 30 cm, culms branched. Flowering time? Coastal flats and slopes, SW, AP (Cape Flats to Bredasdorp).•

saxatilis Esterh. Dioecious, caespitose perennial to 20 mm, forming small tangled cushions. Flowering time? Sandstone slopes, SW (Hottentots Holland Mts).•

Group 4: Plants with erect culms; sheaths tightly convolute; female spikelets ovate, bracts finely acute

coactilis (Mast.) H.P.Linder Dioecious, caespitose perennial to 30 cm, culms branched, leaf sheaths with a woolly bract. Flowering time? Sandstone slopes, SW (Tulbagh to Wellington).•

esterhuyseniae (Pillans) H.P.Linder Dioecious, caespitose, often tangled perennial to 40 cm, culms branched, prominently tubercled. Stylopodium glabrous. Flowering time? Rocky sandstone slopes, NW (Karoopoort).•

karooica Esterh. Dioecious, caespitose perennial to 30 cm, culms more or less erect, branched, prominently tubercled. Stylopodium pilose. Oct. Sandstone slopes, NW, KM (Karoopoort to Touwsrivier).•

Group 5: Plants various; culms slender-ascending; sheaths acute-acuminate; spikelets slender spindle-shaped

affinis Esterh. Dioecious, caespitose perennial to 60 cm, culms branched. Floral bracts acuminate. Flowering time? Moist slopes, c. 1000 m alt., LB (Langeberg Mts: Swellendam).•

arida (Pillans) H.P.Linder Dioecious, caespitose perennial to 30 cm, culms finely tuberculate, branched, sheaths often with an exserted woolly bract. Flowering time? Sandstone slopes, NW, KM (Namaqualand to Witteberg).•

distracta (Mast.) H.P.Linder Dioecious, caespitose perennial to 50 cm, culms branched and curly, finely and densely tuberculate, sheaths with exserted woolly bract. Flowering time? High rocky slopes, NW, KM, LB (Cedarberg to Willowmore, Karoo and E Cape).

feminea Esterh. Dioecious, rhizomatous perennial to 30 cm, culms branched. Style solitary. Flowering time? Limestone slopes, AP (W Agulhas coast).•

gaudichaudiana (Kunth) H.P.Linder Dioecious, caespitose, often tangled perennial to 1 m, culms branched, obscurely tubercled, sheaths spreading, often golden-speckled, free from axis. Flowering time? Dry rocky slopes, NW, SW, AP, KM, LB (Clanwilliam to Uniondale).•

gossypina (Mast.) H.P.Linder Dioecious, caespitose perennial to 50 cm, culms slender, branched, with woolly bracts in sheath axes. Style base conspicuously villous. Flowering time? Light seeps and moist slopes, NW, SW, KM (Namaqualand and W Karoo to Hottentots Holland and Swartberg Mts).

helenae (Mast.) H.P.Linder Dioecious, caespitose perennial to 60 cm, culms branched, sparsely punctate, sheaths obtuse. Spikelets subulate to cylindrical, acute. Flowering time? Habitat? NW, SW (Cold Bokkeveld to Riviersonderend Mts and E Cape to KwaZulu-Natal).

laniger (Kunth) H.P.Linder Dioecious, caespitose perennial to 30 cm, culms branched, often flexuose, smooth, leaf axils with woolly scales. Flowering time? Rocky sandstone slopes, mostly above 1500 m, NW, SW, KM (Gifberg to Swartberg Mts).•

monanthos (Mast.) H.P.Linder Dioecious, caespitose perennial to 60 cm, culms branched. Flowering time? Coastal flats and slopes, NW, SW (Kamiesberg to Bredasdorp Mts).

nubigena Esterh. Like **I. sieberi** but forming compact cushions and seeds shiny. Flowering time? Sandstone slopes above 1650 m, NW, SW (Hex River and Riviersonderend Mts).•

paludosa (Pillans) H.P.Linder Dioecious, caespitose perennial to 30 cm, culms branched, bases swollen, reddish, sheaths setaceous. Flowering time? Seasonally wet sands, NW, SW (Citrusdal to Bredasdorp).•

papillosa Esterh. Dioecious, caespitose perennial to 50 cm, culms branched, often swollen at base, reddish. Sept.–Dec. Coastal flats and slopes, SW, AP, SE (Malmesbury to Langkloof).•

schoenoides (Kunth) H.P.Linder   Dioecious, caespitose, sometimes stoloniferous perennial to 60 cm, culms branched. Flowering time? Rocky slopes, NW, KM, SE (Cedarberg and Swartberg Mts to N Province).

sieberi (Kunth) H.P.Linder   Dioecious, caespitose, sometimes tangled perennial, culms branched, frequently with shorter sterile culms, sheaths always golden-speckled, sometimes with exserted woolly bracts in axils. Jan.–Mar. Rocky slopes and flats, NW, SW, AP, KM, LB, SE (Namaqualand to E Cape).

sporadica Esterh.   Dioecious, tangled perennial to 20 cm, culms branched, flexuose, finely tuberculate, spreading. Flowering time? Seasonally wet, sandy coastal flats, SW, AP (Saldanha to Bredasdorp).•

triflora (Rottb.) H.P.Linder   Dioecious, caespitose perennial to 60 cm, culms finely punctate, branched, branches often sterile, young sheaths with hyaline shoulders. Spikelets narrowly elliptic. Flowering time? Mostly silcrete flats, SW, AP, LB, SE (Paarl to E Cape).

unispicata H.P.Linder   Dioecious, caespitose perennial to 60 cm, culms branched, bluish, densely and finely tubercled. Female spikelets solitary. Flowering time? Seasonally dry sandy plateaus, NW, SW, KM (Cedarberg to Swartberg Mts).•

wallichii (Mast.) H.P.Linder   Dioecious, rhizomatous-caespitose perennial to 30 cm, culms branched, finely spotted, sheaths setaceous. Spikelets subulate to cylindrical, acute. Flowering time? Mostly along streamlines and riverbeds, NW, SW (Bokkeveld Mts to Riviersonderend).•

wittebergensis Esterh.   Dioecious, caespitose perennial to 50 cm, culms branched, densely and obsoletely tubercled. Spikelets densely aggregated. Flowering time? Habitat? KM (Witteberg).•

MASTERSIELLA•   3 spp., W Cape

digitata (Thunb.) Gilg-Ben.   Dioecious, caespitose perennial to 1 m, culms branched, spreading to erect. Male spikelets 5–15 mm, much exceeding bracts. Mar.–Sept. Habitat? SW (Cape Peninsula to Potberg).•

purpurea (Pillans) H.P.Linder   Dioecious, caespitose perennial to 1 m, culms branched, spreading to rarely erect. Male spikelets to 5 mm, as long as bracts. Feb. Habitat? LB, SE (Langeberg Mts: Riversdale to Humansdorp).•

spathulata (Pillans) H.P.Linder   Dioecious, caespitose perennial to 1 m, culms branched, somewhat compressed near apices, spreading to erect. Male spikelets and bracts to 5 mm. Sept. Habitat? SW, AP, LB, SE (Bredasdorp to Uitenhage).•

NEVILLEA•   2 spp., W Cape

obtusissimus (Steud.) H.P.Linder   Dioecious, caespitose perennial to 80 cm, sheaths rounded apically. Male spikelets 10–20 mm, obtuse, bracts rounded, to 4 mm. Mar. Marshy slopes, SW (Elandskloof Mts and Cape Peninsula to Kleinrivier Mts).•

singularis Esterh.   Dioecious, caespitose perennial to 60 cm, sheaths obtuse. Flowering time? Habitat? SW (Riviersonderend Mts).•

PLATYCAULOS•   8 spp., W Cape

acutus Esterh.   Dioecious, caespitose perennial to 30 cm, culms compressed, branched. Female spikelets solitary, ovary unilocular. Feb.–Mar. Montane marshes, LB (Langeberg Mts: Swellendam).•

anceps (Mast.) H.P.Linder   Dioecious, mat-forming perennial, culms more or less compressed, branched. Female spikelets numerous per branch, ovary unilocular. Flowering time? Habitat? SW, LB, SE (Kleinrivier to Tsitsikamma Mts).•

callistachyus (Kunth) H.P.Linder   Dioecious, caespitose perennial to 2,5 m, culms branched, only ultimate branches compressed. Female spikelets solitary, ovary bilocular. Apr. Habitat? NW, SW, AP, KM, LB, SE (Ceres to Uitenhage).•

cascadensis (Pillans) H.P.Linder   Dioecious, caespitose to sprawling perennial, culms compressed, thickened at margins, branched, usually trailing through vegetation. Female spikelets solitary or paired, ovary bilocular. Flowering time? Streams and waterfalls, SW (Kogelberg to Betty's Bay).•

compressus (Rottb.) H.P.Linder   Dioecious, matted to caespitose perennial to 1 m, culms branched, compressed. Female spikelets several, 10–20 mm, ovary bilocular. Aug. Habitat? NW, SW, LB, SE (Ceres and Mamre to Tsitsikamma Mts).•

depauperatus (Kunth) H.P.Linder   Dioecious, caespitose perennial to 40 cm, culms branched, compressed, flexuose. Female spikelets solitary, ovary unilocular. Mar. Habitat? SW (Tulbagh Kloof to Stellenbosch).•

major (Mast.) H.P.Linder   Dioecious, sprawling perennial to 1 m, culms trailing through vegetation, compressed. Female spikelets usually several, 10–40 mm. Flowering time? Habitat? SW (Cape Peninsula to Caledon Swartberg).•

subcompressus (Pillans) H.P.Linder   Dioecious, caespitose perennial to 20 cm, culms branched, compressed. Female spikelets solitary, ovary bilocular. Flowering time? SW (Bainskloof to Elgin).•

RESTIO   c. 88 spp., S and S tropical Africa, Madagascar

Odd species

ambiguus Mast.   Dioecious, caespitose perennial to 50 cm, culms branched, sheaths red-brown with white speckling, acuminate, decaying to truncate. Nov. SW (Cape Peninsula to Kleinrivier Mts).•

Group 1: Culms square

quadratus Mast.   Dioecious perennial to 2 m, culms square, branched, spreading through vegetation, inflorescence much branched. Sept. Habitat? NW, SW (Worcester and Cape Peninsula to Riviersonderend).•

tetragonus Thunb.   Dioecious, caespitose perennial to 1 m, culms square, sometimes branched, inflorescence compound. Flowering time? Sandstone slopes, SW, AP, LB, SE (Cape Peninsula to Tsitsikamma Mts).•

Group 2: Plants with slender, erect, rarely branched culms; spikelets small, globular, male and female similar; seeds colliculate

confusus Pillans   Dioecious, caespitose perennial to 60 cm, culms simple, sheath mucro about half sheath length. Spikelets 2–8 per branch. Flowering time? Habitat? NW, SW (Cedarberg Mts to Kogelberg).•

debilis Nees   Dioecious, caespitose perennial to 40 cm, culms branched, sheaths with awns to 4 mm. Flowering time? Habitat? NW, SW (Cold Bokkeveld to Caledon Swartberg).•

filicaulis Pillans   Dioecious, caespitose perennial to 30 cm, culms slender, simple. Spikelets solitary. Flowering time? Sandstone slopes at high alt., NW (Ceres).•

miser Kunth   Dioecious, caespitose perennial to 50 cm, culms slender, simple. Spikelets several, racemose. Flowering time? Mostly perennially wet, mountain seeps, NW, SW (Clanwilliam to Hottentots Holland Mts).•

pedicellatus Mast.   Dioecious, caespitose perennial to 40 cm, generally forming dense tufts, culms simple, slender, bases thickened. Spikelets often black. Flowering time? Habitat? NW, SW (Cedarberg Mts to Houwhoek).•

similis Pillans   Dioecious, caespitose perennial to 60 cm, culms slender, sparsely branched, sheaths with awns to 5 mm. Spikelets solitary or paired. Flowering time? Habitat? NW, SW, AP (Ceres to Agulhas).•

stereocaulis Mast.   Dioecious, caespitose perennial to 40 cm, culms compressed, simple or sparsely branched, sheath with awns to 5 mm. Flowering time? Habitat? SW (Franschhoek Mts).•

subtilis Nees ex Mast.   Dioecious, caespitose perennial to 50 cm, culms simple, rarely branched, basal sheaths chartaceous. Spikelets several, to 4 m. Flowering time? Habitat? SW (Bainskloof to Hottentots Holland Mts).•

Group 3: Plants with erect culms; male and female spikelets similar; bracts large, flat, margins undulate; seed triangular, beaded or colliculate

bifarius Mast.   Dioecious, caespitose perennial to 80 cm, culms sparsely branched. Female spikelets solitary or paired, male flowers numerous, pendulous, floral bracts narrowly ovate, reddish, margins undulate. Apr.–May. Habitat? SW (Jonkershoek and Hottentots Holland Mts).•

bifidus Thunb.   Dioecious, caespitose perennial to 60 mm, culms sparsely branched, bracts narrowly ovate, brown, concolorous, to 10 mm. Flowering time? Habitat? SW (Cape Peninsula to Kleinrivier Mts).•

exilis Mast.   Dioecious, caespitose perennial to 50 cm, culms sparsely branched. Female spikelets solitary or paired, bracts with purple margin or apex. Flowering time? Habitat? SW (Riviersonderend Mts).•

papyraceous Pillans   Dioecious, caespitose perennial to 60 cm, culms simple, sheaths with apices decaying, loosely convolute. Spikelets globose, 15–20 mm. Flowering time? Rocky summits, KM (Klein Swartberg).•

Group 4: Plants frequently tangled, culms usually fine, hair-like, often obscurely tubercled; spikelets with 1 or a few flowers, often single, tepals often glabrous

arcuatus Mast.   Dioecious, tangled perennial to 60 cm, culms branched, with ultimate branches flexuose, sheaths loosely convolute. Spikelets solitary, 5 mm. Flowering time? Habitat? LB (Langeberg Mts: Swellendam to Riversdale).•

capillaris Kunth   Dioecious, tangled perennial to 60 cm, culms slender, branched. Spikelets single or several, bracts chartaceous. Flowering time? Sandstone slopes, NW, SW, LB (Paarl to Swellendam).•

colliculospermus H.P.Linder   Dioecious, tangled perennial to 30 cm, culms slender, branched, obscurely rugulose. Flowering time? Habitat? SW (Riviersonderend Mts).•

distans Pillans   Dioecious, caespitose perennial to 50 cm, culms wrinkled, much branched, spikelets 4–8, to 5 mm long. Flowering time? Habitat? SW (Jonkershoek to Hottentots Holland Mts).•

fourcadei Pillans   Dioecious, tangled perennial to 1 m, culms much branched near apices, tubercled, most branchlets fertile. Spikelets 1 or 2. Flowering time? Mostly in riverine forest, LB, SE (Swellendam to Humansdorp).•

fragilis Esterh.   Dioecious, tangled perennial to 30 cm, culms slender, rugulose, branched. Flowering time? Habitat? LB (Langeberg Mts: Swellendam).•

harveyi Mast.   Dioecious, caespitose to mat-forming, stoloniferous perennial to 15 cm, culms slender, much branched, tuberculate. Flowering time? Clay and gravel flats, SW, AP (Cape Peninsula to Agulhas).•

implicatus Esterh.   Dioecious, tangled perennial to 60 cm, culms slender, rugulose, sheaths loosely convolute. Spikelets solitary or paired. Flowering time? Habitat? LB (Langeberg Mts: Swellendam).•

intermedius (Pillans) H.P.Linder   Dioecious, tangled perennial, culms much branched, sheaths with large hyaline shoulders. Spikelets 1 to several, to 8 mm. Flowering time? Rocky slopes, SW (Jonkershoek to Caledon).•

patens Masters   Dioecious, tangled perennial to 50 cm, culms branched. Spikelets 1 to several per branch, bracts with hyaline margins. Flowering time? Rocky sandstone slopes, NW (Cedarberg Mts to Gydouberg).•

perplexus Kunth   Dioecious, tangled perennial to 60 cm, culms slender, much branched, tuberculate. Spikelets single, bracts with hyaline margins. Aug. Habitat? NW, SW (Cold Bokkeveld and Cape Peninsula to Langeberg Mts).•

pumilus Esterh.   Dioecious, caespitose perennial to 20 cm, culms branched, with flat-topped tubercles, sheaths loosely convolute with wide hyaline margins. Flowering time? Habitat? SW (Hottentots Holland to Kleinrivier Mts).•

scaberulus N.E.Br.   Dioecious, caespitose perennial, 30–80 cm, culms roughly and densely tuberculate, subverticillately branched. Spikelets solitary or paired. Flowering time? Habitat? LB, SE (Riversdale to Humansdorp).•

singularis Esterh.   Dioecious, spreading, tangled perennial to 50 cm, culms branched, densely tuberculate, sheaths loosely convolute, with large hyaline shoulders. Flowering time? Sandstone slopes, SW (Du Toitskloof).•

stokoei Pillans   Dioecious, caespitose perennial to 1 m, culms branched, tubercled. Spikelets racemose. Flowering time? Habitat? SW, LB (Stellenbosch to Swellendam).•

verrucosus Esterh.   Dioecious, caespitose, tangled perennial to 40 cm, culms slender, much branched, roughly and unevenly tuberculate. Aug. Habitat? SW (Houwhoek Mts).•

versatilis H.P.Linder   Dioecious, tangled, caespitose perennial to 40 cm, culms much branched, obscurely tubercled. Spikelets 1 to several. Flowering time? Habitat? SW (Cape Peninsula and Du Toitskloof to Houwhoek).•

zwartbergensis Pillans   Dioecious, stoloniferous, tangled perennial to 30 cm, culms much branched, with flat-topped tubercles, sheaths loosely convolute. Flowering time? Habitat? SW (Riviersonderend Mts and Caledon Swartberg).•

Group 5: Plants caespitose; spikelets very numerous, often with few flowers per spikelet; bract apices slightly discoloured; seed shiny, orange

brachiatus (Mast.) Pillans   Dioecious, caespitose perennial to 80 cm, culms branched. Female spikelets in a compound raceme, bracts with hyaline margins. Oct. Habitat? NW, SW, KM (Tulbagh and Paarl to Witteberg).•

cymosus (Mast.) Pillans   Dioecious, caespitose perennial to 1.5 m, culms sparsely branched. Female spikelets numerous, several-flowered, bract margins hyaline. Flowering time? Habitat? NW, SW (Namaqualand to Paarl).

fusiformis Pillans   Dioecious, caespitose perennial to 40 cm, culms sparsely branched. Male and female spikelets ellipsoid, acute, 10–15 mm. Flowering time? Habitat? SW (Jonkershoek to Hangklip).•

occultus (Mast.) Pillans   Dioecious, caespitose perennial to 1 m, culms branched. Spikelets numerous, male spikelets pendent. Flowering time? Habitat? NW, SW, KM (Cedarberg to Hottentots Holland and Witteberg).•

Group 6: Plants with erect culms usually roughly tuberculate, especially near base; sheaths with a tough hyaline margin; seed shiny, orange

communis Pillans   Dioecious, caespitose perennial to 70 cm, culms branched, conspicuously tuberculate. Female spikelets solitary or paired, bracts and spathes membranous to chartaceous, often speckled. Flowering time? SW (Cape Peninsula).•

ingens Esterh.   Dioecious, caespitose perennial to 1 m, culms often much branched towards apex, roughly tuberculate. Spikelets solitary, bracts and spathes chartaceous. Flowering time? Sandstone slopes, SW (Riviersonderend Mts).•

involutus Pillans   Dioecious perennial to 60 cm, culms stout below, much branched above, usually with large tubercles. Spikelets 7–15 mm, racemose. Flowering time? Habitat? SW (Hottentots Holland Mts).•

montanus Esterh.   Dioecious, caespitose to sprawling perennial to 60 cm, culms much branched towards apices, tangled, roughly tubercled, sheaths loosely convolute. Flowering time? Habitat? SW (Slanghoek Mts).•

scaber Mast.   Dioecious perennial to 40 cm, culms sparsely branched, roughly tubercled. Spikelets solitary. Flowering time? Rocky sandstone slopes, SW (Caledon Swartberg).•

tuberculatus Pillans   Dioecious, caespitose perennial to 50 cm, culms stout, erect to decumbent, much branched above, roughly tuberculate. Female spikelets exceeding bracts. Flowering time? Habitat? NW (Gifberg to N Cedarberg Mts).•

Group 7: Plants with erect, often much-branched and tuberculate culms; inflorescence of many-flowered spikelets, bracts without hyaline margins, cartilaginous, about as tall as the flowers

inconspicuus Esterh.   Dioecious, stoloniferous perennial to 1 m, culms spreading, branched. Spikelets 4–8, imbricate. Flowering time? Habitat? LB, SE (Swellendam to George).•

multiflorus Spreng.   Dioecious, caespitose to spreading perennial to 1 m, culms sparsely branched. Flowers slightly longer than bracts. Flowering time? Habitat? NW, SW (Piketberg to Bredasdorp).•

peculiaris Esterh.   Dioecious, caespitose perennial to 40 cm, culms branched, stout, densely tuberculate, bracts and sheaths with wide hyaline margins. Jan. Rocky summits, LB (Langeberg Mts: Swellendam).•

quinquefarius Nees   Dioecious, caespitose perennial to 80 cm, culms branched. Female spikelets solitary, 10–20 mm, bracts c. 5 mm. Aug. Sandy coastal flats, NW, SW (Lambert's Bay to Cape Peninsula).•

secundus (Pillans) H.P.Linder   Dioecious, sprawling perennial to 50 cm, culms branched, tuberculate. Spikelets several, obscured by spathes with wide hyaline apices. Flowering time? Habitat? LB (Langeberg Mts: Swellendam to Riversdale).•

sejunctus Mast.   Dioecious, caspitose to spreading perennial to 60 cm, culms much branched. Flowers and bracts equal. Apr. Habitat? SW, KM, LB, SE (Worcester to KwaZulu-Natal and Karoo).

strictus N.E.Br.   Dioecious, caespitose perennial to 40 cm, culms sparsely branched, sheaths membranous in upper half. Spikelets 1–3, to 15 mm. Flowering time? Sandstone slopes, SW, LB, SE (Riviersonderend Mts to George).•

triticeus Rottb.   Dioecious, caespitose perennial to 80 cm, culms often with sterile branches below, white-tubercled. Feb.–Apr. Habitat? SW, KM, LB, SE (Malmesbury to E Cape).

Group 8: Plants caespitose, with erect culms; bracts narrowly lanceolate, much taller than flowers, often pale brown or yellow; spikelets elongate, with 1–3 flowers, or with internodes elongate

alticolus Pillans   Dioecious, caespitose perennial to 60 cm, culms much branched, sheaths free. Spikelets solitary. Flowering time? Habitat? SW (Bainskloof to Paarl).•

corneolus Esterh.   Dioecious, caespitose perennial, 15–90 cm, culms sparsely branched, obscurely rugulose. Spikelets 1 to several. Flowering time? Habitat? SW (Jonkershoek to Betty's Bay).•

decipiens (N.E.Br.) H.P.Linder   Dioecious, caespitose, culms ascending, much branched. Female spikelets solitary, shorter than spathes. Flowering time? Habitat? LB (Swellendam to Riversdale).•

dispar Mast.   Dioecious, caespitose perennial to 1 m, culms branched. Spikelets 20–50 mm, spathes reddish with yellow speckling, exceeding spikelets. Mar.–Apr. Flowering time? Habitat? SW (Cape Peninsula and Worcester to Caledon).•

distichus Rottb.   Dioecious, caespitose perennial to 50 cm, culms branched. Female spikelets several per branch, bracts cartilaginous, narrowly lanceolate, apices hyaline. Nov. Habitat? NW, SW, KM, LB (Cold Bokkeveld to Cape Peninsula, Swartberg and Langeberg Mts).•

dodii Pillans   Dioecious, caespitose perennial to 1 m, culms sparsely branched, upper half of sheaths and bracts abruptly hyaline. Flowering time? Habitat? SW (Cape Peninsula to Bredasdorp).•

ejuncidus Mast.   Dioecious, caespitose perennial to 50 cm, culms erect, branched, with white tubercles. Spikelets with long acute chartaceous bracts. Apr. Sandstone slopes, NW, SW, LB (Olifants River to Langeberg Mts: Swellendam).•

festuciformis Nees ex Mast.   Dioecious, caespitose perennial to 40 cm, culms sparsely branched. Spikelets numerous, lax, bracts narrowly lanceolate, chartaceous, acute. Flowering time? Habitat? SW (Hottentots Holland to Bredasdorp Mts).•

leptostachyus Kunth   Dioecious, tangled perennial, culms branched. Female spikelets solitary, bracts acute, setaceous. May. Habitat? SW (Paarl to Caledon).•

micans Nees   Dioecious, caespitose, perennial to 50 cm, culms sparsely branched, often compressed below. Female spikelets narrow-elliptic, to 20 mm. Flowering time? Wet coastal flats, SW (Malmesbury to False Bay).•

purpurascens Nees ex Mast.   Dioecious, caespitose perennial to 1 m, culms branched, somewhat compressed. Spikelets 15–25 mm, bracts purplish. Flowering time? Habitat? SW (Wellington to Cape Peninsula and Caledon).•

rarus Esterh.   Dioecious, caespitose perennial to 30 cm, culms branched, tuberculate-rugulose. Spikelets solitary. Flowering time? Rocky sandstone slopes, KM (Klein Swartberg).•

sarocladus Mast.   Dioecious, caespitose perennial to 50 cm, culms obscurely rugulose, branched, sheaths with awns 5–10 mm. Floral bracts narrowly lanceolate, chartaceous, concolorous, acute. Flowering time? Habitat? SW (Cape Peninsula and Paarl to Kleinrivier Mts).•

vallis-simius H.P.Linder   Dioecious, stoloniferous perennial to 1 m, culms branched and spreading. Flowers with cartilaginous bracts without hyaline margins. Flowering time? Sandstone slopes, SE (Baviaanskloof).•

Group 9: Plants caespitose; spikelets compact, usually many-flowered, often few per inflorescence; floral bracts with distinct, usually darker-coloured upper margins

acockii Pillans   Dioecious, caespitose perennial to 80 cm, culms compressed, sparsely branched. Spikelets narrowly elliptic, bracts with purple apices. Flowering time? Seasonally wet sands, SW (Saldanha to Cape Peninsula).•

aureolus Pillans   Dioecious, caespitose perennial to 20 mm. Female spikelets solitary or paired, elliptic, to 8 mm, bracts with hyaline margins. Apr. Habitat? NW (Grootwinterhoek and Hex River Mts).•

bifurcus Nees ex Mast.   Dioecious, caespitose perennial to 1 m, culms sparsely branched. Inflorescences with c. 10 spikelets. Aug. Rocky slopes, SW, KM (Cape Peninsula to Caledon, Witteberg).•

bolusii Pillans   Dioecious, caespitose perennial to 60 cm, culms sparsely branched. Female spikelets 10–20 mm, ellipsoid, acute, fewer than 6 per spike. Flowering time? Habitat? NW, SW (Worcester to Bredasdorp).•

brunneus Pillans   Dioecious, caespitose perennial to 80 cm, culms sparsely branched. Spikelets 10–20 mm, bracts acuminate with spreading apices. Aug. Rocky slopes, NW (Cedarberg Mts).•

burchellii Pillans   Dioecious, caespitose perennial to 50 cm, culms sparsely branched. Spikelets ellipsoid, acute, to 10 mm. Flowering time? Habitat? SW (Hottentots Holland to Kleinrivier Mts).•

echinatus Kunth   Dioecious, caespitose perennial to 50 cm, culms scarcely branched. Floral bracts acuminate, reflexed. Flowering time? Sandstone slopes, NW, SW (Tulbagh to Wellington).•

egregius Hochst.   Dioecious, caespitose perennial to 1 m, culms sparsely branched. Female spikelets solitary, 20–40 mm, male spikelets paniculate, pendent. Mar. Habitat? SW (Cape Peninsula to Villiersdorp and Bredasdorp).•

filiformis Poir.   Dioecious, caespitose perennial to 60 cm, culms erect, sparsely branched. Female spikelets 1–6, flowers and bracts subequal. Aug. Sandstone slopes, NW, SW, LB (Gifberg to Riversdale).•

insignis Pillans   Dioecious, caespitose perennial to 60 cm, culms sparsely branched. Spikelets globose to ovate, 20–30 mm long, solitary or paired. Flowering time? Rocky slopes, NW (Clanwilliam to Ceres).•

inveteratus Esterh.   Dioecious, caespitose perennial to 60 cm, culms sparsely branched. Female spikelets laxly racemose, elliptic. Flowering time? Sandstone slopes, SW (Paarl to Worcester).•

nodosus Pillans   Dioecious, caespitose perennial to 40 cm, culms branched. Spikelets solitary, globose, to 10 mm, bract apices crisped, black. Flowering time? Sandstone slopes, NW (Worcester).•

nuwebergensis Esterh.   Dioecious, caespitose perennial to 30 cm, culms simple. Spikelets 10–20 mm diam., with numerous recurved bracts. Flowering time? Rocky slopes, SW (Hottentots Holland Mts).•

obscurus Pillans   Dioecious, caespitose perennial, 20–80 cm, culms branched. Spikelets subglobose, c. 5 mm diam., bract apices stout, reflexed, black. Sept. Sandstone slopes, SW (Bainskloof to Stellenbosch).•

pachystachyus Kunth   Dioecious, caespitose perennial to 80 cm, culms sparsely branched. Female spikelets 10–20 mm, bracts acuminate. Flowering time? Sandstone slopes, NW, SW (Ceres to Caledon).•

perseverans Esterh.   Dioecious, caespitose perennial to 40 cm, culms branched. Female spikelets solitary, to 15 mm. Feb. Sandstone slopes, LB (Langeberg Mts: Swellendam to Riversdale).•

praeacutus Mast.   Dioecious, caespitose perennial to 1 m, culms flattened or grooved, sparsely branched. Female spikelets several, racemose. Flowering time? Sandstone slopes, NW, SW (Clanwilliam to Paarl).•

pulvinatus Esterh.   Dioecious, caespitose-tangled, stoloniferous perennial to 20 cm, culms branched. Spikelets solitary or paired. Flowering time? Habitat? SW (Worcester, Paarl to Caledon).•

rupicola Esterh.   Dioecious, caespitose perennial to 1 m, culms branched. Spikelets numerous, elliptic, 7–10 mm, 1- or 2-flowered. Flowering time? Sandstone slopes, NW, SW (Ceres to Worcester).•

strobilifer Kunth   Dioecious, caespitose perennial to 80 cm, culms sparsely branched. Spikelets 10–20 mm, ellipsoid, acute, solitary or paired. Aug.–Oct. Habitat? NW, KM (Cedarberg Mts to Klein Swartberg).•

RHODOCOMA   6 spp., W Cape to KwaZulu-Natal

alpina H.P.Linder & Vlok   Dioecious, spreading perennial to 50 cm, with elongate rhizomes, culms simple. Flowers with spikelets obscured by large spathes. Jan.–Feb. Rocky slopes above 1600 m, KM, LB (Langeberg and Swartberg Mts).•

arida H.P.Linder & Vlok   Dioecious, caespitose perennial to 2 m, culms simple, without basal sterile branches. May–July. Quartz outcrops, KM (Little Karoo).•

capensis Nees ex Steud.   Dioecious, caespitose perennial to 2 m, culms with numerous verticillate branches, the upper terminating in spikelets. Oct.–Nov. Habitat? NW, KM, SE (Ceres to E Cape).

fruticosa (Thunb.) H.P.Linder   Rhizomatous, spreading perennial to 60 cm, culms simple. Male flowers in drooping, paniculate spikelets sometimes with branched sterile culms. May–July. Habitat? NW, SW, AP, KM, LB, SE (Ceres to KwaZulu-Natal).

gigantea (Kunth) H.P.Linder   Dioecious, caespitose perennial with fertile culms to 3 m. Male spikelets paniculate, pendent, sterile branches verticillate. June–July. Moist coastal slopes, LB, SE (Langeberg to Tsitsikamma Mts).•

gracilis H.P.Linder & Vlok   Dioecious, caespitose perennial to 70 cm, culms simple slender, more than 1 mm diam. Mar.–July. Moist habitats, LB (Langeberg Mts).•

STABEROHA•   9 spp., W Cape

aemula (Kunth) Pillans   Dioecious, caespitose perennial to 60 cm. Female flowers winged, tepals obtuse, male spikelets globose. June–Sept. Habitat? NW, SW, KM (Cedarberg Mts to Prince Albert).•

banksii Pillans   Dioecious, caespitose perennial to 60 cm, culms simple. Female flowers 1–3 in spikelets 2.5–5 cm long, bracts 1.2–1.5 cm long. Mar.–Apr. Sandstone slopes at low alt., NW, SW (Worcester and Cape Peninsula to Bredasdorp).•

cernua (L.f.) T.Durand & Schinz   Dioecious, caespitose perennial to 60 cm, culms simple. Female flowers with finely lacerate wings, male spikelets numerous, 5–7 mm wide. Mar., Aug. Habitat? NW, SW, KM (Ceres to Cape Peninsula, Swartberg Mts).•

distachya (Rottb.) Kunth   Dioecious, caespitose perennial to 60 cm, with well developed rhizomes, culms simple. Female spikelets solitary, about half as wide as long. Feb.–Dec. Habitat? NW, SW, AP, LB, SE (Gifberg to Mossel Bay).•

multispicula Pillans   Dioecious, caespitose perennial to 60 cm, culms simple. Female spikelets several. Flowering time? Coastal sands, SW (Betty's Bay to Bredasdorp).•

ornata Esterh.   Dioecious, caespitose perennial to 30 cm, culms simple. Inflorescence spathes reddish. Aug.–Oct. Sandy mountain plateaus, NW (Cedarberg to Hex River Mts).•

remota Pillans   Dioecious, caespitose perennial to 50 cm, rhizomatous, culms simple, female spikelets solitary, obtuse, female flowers with keeled lateral sepals. Flowering time? Habitat? NW, SW (Ceres to Caledon).•

stokoei Pillans   Dioecious, caespitose perennial to 30 cm, with well-developed rhizomes, culms simple. Female spikelet solitary, male spikelets erect. Flowering time? High rocky slopes, KM (Swartberg Mts: Prince Albert to Meiringspoort).•

vaginata (Thunb.) Pillans   Dioecious, caespitose perennial to 60 cm, culms simple. Female flowers with keeled lateral sepals, bracts 0.6–1 cm long. Flowering time? Habitat? SW (Cape Peninsula to Caledon).•

THAMNOCHORTUS   c. 31 spp., Namaqualand to KwaZulu-Natal

A. Culms velvety pubescent

acuminatus Pillans   Dioecious, caespitose perennial to 30 cm, culms obscurely velvety pubescent, sheaths often subimbricate and obscuring the culms, spathes hyaline with reddish keels. Oct. Dry sandstone slopes, NW (Cedarberg Mts to Worcester).•

cinereus H.P.Linder   Dioecious, caespitose perennial to 80 cm, fertile culms simple, with sterile branches clustered at nodes, long and soft. May–Sept. Habitat? SW, AP, KM, LB, SE (Malmesbury to Humansdorp).•

fruticosus P.J.Bergius   Dioecious, caespitose perennial to 60 cm, with well-developed rhizomes, often with tufts of sterile branches on the simple culms, velvety pubescent. July–Nov. Habitat? NW, SW, AP, KM, SE (Tulbagh to KwaZulu-Natal).

rigidus Esterh.   Dioecious, caespitose perennial to 80 cm, fertile culms usually with sterile branches clustered at nodes, stout and firm. July.–Nov. Dry sandstone slopes, NW, KM, SE (Clanwilliam to Baviaanskloof Mts).•

A. Culms glabrous
B. Fruits shorter than to as long as wide

amoena H.P.Linder   Dioecious, caespitose perennial to 1 m, culms with dense clusters of sterile branches at nodes, bracts silvery membranous. Female spikelets 3 or 4, 40–60 mm long. June. Rocky N-facing slopes, LB (Langeberg Mts).•

arenarius Esterh.   Dioecious, rhizomatous perennial to 60 cm, culms sometimes present, simple, sterile, much branched. May–June. Habitat? SW (Cape Peninsula to Hermanus).•

bachmannii Mast.   Dioecious, caespitose perennial to 80 cm, culms simple, often sulcate. Female spikelets to 10 mm. June. Sandy flats, NW, SW (Namaqualand to Malmesbury and Worcester).

erectus (Thunb.) Mast.   WYFIERIET   Dioecious, caespitose perennial to 1,5 m, with prominent rhizomes, culms simple. Spikelets subglobose, to 10 mm. Sept.–Oct. Habitat? SW, AP, LB, SE (Malmesbury to Knysna).•

gracilis Mast.   Dioecious, caespitose perennial to 70 cm, fertile culms branched, spreading. Female spikelets to 10 mm, bracts acuminate, margins hyaline above. Apr. Habitat? SW (Cape Peninsula to Caledon).•

insignis Mast.   MANNETJIESRIET   Dioecious, caespitose perennial to 2 m, fertile culms simple. Female spikelets 15–25 mm. Mar.–Apr. Loamy soils between dunes, AP, LB (Agulhas to Gouritsmond).•

karooica H.P.Linder   Dioecious, caespitose perennial to 1 m, fertile culms simple. Female spikelets 5–27, 15–25 mm long. May. Low sandstone slopes, LB (N foothills of Langeberg Mts).•

lucens (Poir.) H.P.Linder   Dioecious, caespitose perennial to 60 cm, sometimes with sterile much-branched culms. Female spikelets 10–20 mm, bracts cartilaginous. Mar.–July. Habitat? NW, SW, LB (Malmesbury and Ceres to Riversdale).•

platypteris Kunth   Dioecious, caespitose perennial to 60 cm, culms simple. Flowers with bracts entirely or largely hyaline, female flowers with wide wings. Aug.–Sept. Sandstone slopes, NW, KM, (Cedarberg to Witteberg, Little Karoo Mts to Barrydale).•

punctatus Pillans   Dioecious, caespitose perennial to 60 cm, with well-developed rhizomes, culms simple, with sunken brown spots. Female spikelets to 20 mm. Mar.–May. Sandy flats and slopes, NW, SW (Bokkeveld Mts to Cape Peninsula).•

spicigerus (Thunb.) Spreng.   Dioecious, caespitose perennial to 1.5 m, with well-developed rhizomes, culms slightly flexuose, simple. Flowers wider than bracts, female spikelets to 20 mm. Apr.–May. Coastal sands in strandveld, SW (Langebaan to Cape Peninsula).•

sporadicus Pillans   Dioecious, rhizomatous perennial to 50 cm, fertile culms simple, sometimes with sterile branched culms, bracts cartilaginous. Female spikelets 10–15 mm. Oct.–Nov. Sandstone slopes, NW, SW (Piketberg to Cape Peninsula).•

B. Fruits longer than wide

dumosus Mast.   Dioecious, caespitose perennial to 30 cm, fertile culms simple, sterile branched culms to 10 cm. Female floral bracts hyaline and acuminate. May. Habitat? SW (Caledon to Bredasdorp).•

ellipticus Pillans   Dioecious, rhizomatous perennial to 50 cm, fertile culms simple, sometimes with sterile branches in second year. Apr. Lower, gravelly slopes, LB (Langeberg Mts: Riversdale).•

fraternus Pillans   Dioecious, caespitose perennial to 1 m, culms simple, all fertile. Female spikelets to 20 mm, bracts narrowly lanceolate, erect, acute, female tepals decurrent on stipe, leaving narrow strips of stipe exposed. Apr.–May. Usually on limestone, SW, AP (Cape Peninsula to Bredasdorp).•

glaber (Mast.) Pillans   Dioecious, caespitose perennial to 60 cm, with well-developed rhizomes. Female spikelets to 30 mm. Aug.–Sept. Habitat? SE (Knysna to E Cape).

guthrieae Pillans   Dioecious, caespitose perennial to 50 cm, culms simple, sometimes with sterile branches. Female spikelets 10–15 mm, bracts cartilaginous. Jan.–June. Habitat? SW (Malmesbury to Bredasdorp).•

levynsiae Pillans   Dioecious, rhizomatous perennial to 1 m, culms branched, spathes as tall as or taller than spikelets. Mar. Rock ledges, SW (Cape Peninsula).•

muirii Pillans   Dioecious, caespitose perennial mostly to 60 cm, culms simple. Female spikelets several, to 15 mm. July–Sept. Coastal flats, AP, LB (Bredasdorp to Mossel Bay).•

nutans (Thunb.) Pillans   Dioecious, caespitose perennial to 60 cm, culms simple, sometimes with sterile much-branched culms. Female spikelets to 20 mm, bracts cartilaginous. Aug.–Sept. Rocky summits, SW (Cape Peninsula).•

obtusus Pillans   Dioecious, rhizomatous perennial to 30 cm, culms simple, sterile branched culms produced from rhizomes. Female spikelets to 15 mm, bracts cartilaginous. Jan.–Apr. Coastal flats, SW, AP (Saldanha to Agulhas).•

paniculatus Mast.   Dioecious, rhizomatous perennial to 1 m, culms simple. Female spikelets 20–30 mm, bract apices somewhat reflexed, female flowers with tepals decurrent on stipe, leaving narrow strips of stipe exposed. June.–Nov. Mainly limestone slopes and flats, AP (Agulhas).•

papyraceous Pillans   Dioecious, caespitose perennial to 50 cm, culms simple. Floral bracts entirely hyaline. Flowering time? Habitat? KM (Swartberg Mts: Ladismith to Prince Albert).•

pellucidus Pillans   Dioecious, caespitose perennial to 60 cm, fertile culms simple, sterile culms branched, to 20 cm. Female spikelets to 20 mm, several aggregated into a head. Flowering time? Habitat? SW (Caledon to Bredasdorp).•

pluristachyus Mast.   Dioecious, caespitose perennial to 1 m, culms simple, sheaths tightly convolute. Female spikelets narrowly elliptic, to 10 mm. June. Limestone flats and slopes, AP (Agulhas coast).•

pulcher Pillans   Dioecious, caespitose perennial to 40 cm, fertile culms simple, sterile branched culms produced from base. Female spikelets to 20 mm, bracts slender-acuminate. Aug.–Dec. Habitat? SW (Stellenbosch to Bredasdorp).•

scabridus Pillans   Dioecious, caespitose perennial to 40 cm, culms simple, densely and finely tubercled. Female spikelets solitary or paired, spathes often hyaline. Aug.–Sept. Dry sandstone slopes, NW, KM (Worcester to Witteberg Mts).•

schlechteri Pillans   Dioecious, caespitose perennial to 30 cm, with well-developed rhizomes, culms simple. Female spikelets solitary, bracts acuminate, chartaceous, with upper margins hyaline. Oct.–Nov. Sandstone slopes mostly above 1200 m, NW (Cedarberg Mts to Ceres).•

stokoei Pillans   Dioecious, caespitose perennial to 40 cm, culms simple. Female spikelets 10–25 mm, bracts with wide hyaline margins. Sept.–Nov. Sandstone slopes, SW (Paarl to Caledon).•

WILLDENOWIA   11 spp., Namaqualand to W Cape

A. Perianth stipitate; nuts with an elaiosome

arescens Kunth   Dioecious, caespitose perennial to 1 m, culms branched, smooth, sheath apices hyaline. Female flowers in spikelets to 15 mm, tepals narrowing towards base. Nuts pitted. Aug.–Nov. Habitat? NW, SW (Namaqualand to Malmesbury and Worcester).

bolusii Pillans   Dioecious, caespitose perennial to 1 m, culms branched, striate, spathes greenish with red margins. Feb. Lower sandstone slopes, KM (Swartberg and Little Karoo Mts).•

glomerata (Thunb.) H.P.Linder   Dioecious, caespitose perennial to 1 m, culms branched, smooth, sheath apices hyaline. Flowers with tepals minute or lost. Nuts smooth, black. Apr.–Aug. Habitat? NW, SW, AP, KM, SE (Clanwilliam to Uniondale).•

sulcata Mast.   Dioecious, caespitose perennial to 50 cm, culms spreading, branched, striate. Nuts pitted. Sept. Habitat? NW, SW (Clanwilliam to Caledon).•

teres Thunb.   Dioecious perennial to 1 m, forming erect to spreading tussocks, culms branched, smooth. Flowers with tepals widening below. Nuts smooth. Sept. Habitat? NW, SW, AP, KM, LB, SE (Ceres to Uniondale).•

A. Perianth sessile; nuts without an elaiosome

affinis Pillans   Dioecious, caespitose perennial with culms branched. Female spikelets to 20 mm. Flowering time? Habitat? SW (Cape Peninsula).•

humilis Mast.   Dioecious, caespitose perennial to 30 cm, with spreading rhizomes, culms simple. Female spikelets closely adpressed to scape. Oct.–Nov. Habitat? NW, SW (Clanwilliam to Cape Peninsula).•

incurvata (Thunb.) H.P.Linder   Dioecious, caespitose perennial to 2 m, culms striate and branched. Nuts pitted. June. Sandy coastal flats, NW, SW (Namaqualand to Cape Peninsula).

purpurea Pillans   Dioecious, caespitose perennial to 30 cm, culms branched, smooth, stoloniferous, sheaths with awns to 7 mm. Flowers with female spikelets enclosed by spathes. Flowering time? Marshy sandstone flats, SW (Franschhoek Mts to Viljoen's Pass).•

rugosa Esterh.   Dioecious, caespitose perennial to 30 cm, culms simple, with sheaths only at base. Nut ornamented. ?May. Habitat? SW (Hottentots Holland Mts).•

stokoei Pillans   Dioecious, caespitose perennial to 60 cm, culms simple or rarely branched. Female spikelets clustered. Nuts pitted. Oct. Habitat? NW, KM (Clanwilliam to Swartberg Mts).•

RUPPIACEAE

RUPPIA   2 spp., temperate and subtropical

maritima L.   Submerged grass-like aquatic perennial. Leaves linear to filiform, acute. Flowers on short, straight peduncles. Sept.–Mar. Salt pans or brackish streams, SW, SE (Namibia to Port Elizabeth, cosmopolitan).

spiralis L. ex Dumort.   Similar to **R. maritima** but flowers on long, spirally coiled peduncles. Mostly Dec.–Feb. Brackish coastal pools, NW, SW, AP, SE (Namaqualand to Port Elizabeth, cosmopolitan).

STRELITZIACEAE

STRELITZIA CRANE FLOWER c. 5 spp., E Cape to Zimbabwe

alba (L.f.) Skeels kaapse wildepiesang large single-stemmed tree to 10 m. Leaves with large oblong blades. Flowers white. July–Dec. forests, SE (George to Humansdorp).•

juncea Link Like **S. reginae** but leaf blades absent or very reduced. Nov.–July. Bush, SE (Humansdorp to Uitenhage).•

reginae Banks Evergreen perennial to 1.5 m. Leaves long-petiolate with large, ovate blades. Flowers orange with blue styles. Jan.–May. Riverbanks, in coastal bush and thicket, SE (Humansdorp to Transkei).

TECOPHILAEACEAE

1. Stem erect or scandent, sometimes prickly, with cauline leaves; flowers axillary; rootstock a tuber ...... **Walleria**
1.' Stem erect with basal, usually rosulate leaves; flowers in racemes or rarely solitary; rootstock a corm .. **Cyanella**

CYANELLA LADY'S HAND, RAAPTOL 8 spp., southern Africa

alba L.f. TOE TOE UINTJIE Cormous geophyte, 12–25 cm. Leaves filiform-terete. Flowers enantiostylous, 1 or 2 on long peduncles, white to pink, or yellow, fragrant, with 5 upper and 1 larger lower stamen, anthers cohering. Aug.–Oct. Stony clay and sandstone soils, NW (Bokkeveld Plateau to Ceres and W Karoo).

hyacinthoides L. BLOURAAPTOL Cormous geophyte, 25–40 cm. Leaves linear-lanceolate, glabrous to finely pubescent. Flowers in branched racemes, mauve, sometimes white, stamens bright yellow, fragrant, with 5 upper and 1 larger lower stamen, filaments fused. Aug.–Nov. Mostly clay and granite slopes, often in renosterveld, NW, SW, KM, LB (Namaqualand to Riversdale).

lutea L.f. GEELRAAPTOL Cormous geophyte, 12–25 cm. Leaves lanceolate. Flowers in branched racemes, yellow, sometimes pink, fragrant, pedicels suberect, with 5 upper and 1 larger lower stamen, filaments free to base. Sept.–Oct. Mostly clay, or limestone flats, NW, SW, AP, KM, LB, SE (S Namibia to Lesotho, Botswana).

orchidiformis Jacq. Cormous geophyte, 30–40 cm. Leaves lanceolate, soft, often undulate. Flowers in branched racemes, mauve with darker centre, fragrant, pedicels suberect, with 3 upper and 3 larger lower stamens. July–Sept. Rocky flats and lower slopes, often wet sites, NW (S Namibia to Clanwilliam).

WALLERIA POTATO LILY 4 spp., W Cape to tropical Africa

gracilis (Salisb.) S.Carter Slender scrambling tuberous geophyte to 30 cm, with prickly stems. Leaves linear, ending in tendrils, midribs prickly beneath. Flowers axillary, nodding, white with purple centre, fragrant, anthers yellow with purple tips. June–July. Low sandstone outcrops, NW (Richtersveld, Gifberg to Pakhuis Mts).

TYPHACEAE

TYPHA BULRUSH, PAPKUIL 10 spp., cosmopolitan

capensis (Rohrb.) N.E.Br. Monoecious, tufted perennial to 2 m. Leaves strap-like, twisted, spongy. Flowers small, in cylindrical spikes, male flowers above and females below, brown. Dec.–Mar. Streambanks and marshes, NW, SW, AP, KM, LB, SE (southern and tropical Africa).

XYRIDACEAE

XYRIS YELLOW EYED GRASS c. 250 spp., pantropical and subtropical

capensis Thunb. Grass-like perennial, 8–50 cm. Leaves linear. Flowers in tight clusters among brown bracts, yellow, lasting a few hours. Oct.–Apr. Marshes and seeps, NW, SW, LB, SE (Clanwilliam to tropical Africa, India, Brazil).

ZANNICHELLIACEAE

1. Anther 2–4-locular; leaves opposite or alternate, linear; fruits curved, pectinate .... **Zannichellia**
1.' Anther 1-locular; leaves crowded above, setaceous; fruits straight, smooth .... **Althenia**

ALTHENIA   1 sp., Cape to Namibia, S Europe

filiformis Petit   Small, creeping, submerged aquatic annual to 15 cm. Leaves in tufts, filiform. Flowers green. Oct. Brackish pools near the sea, NW, SW, SE (Namibia to Port Elizabeth, southern Europe).

ZANNICHELLIA   2 spp., cosmopolitan

aschersoniana Graebn.   Slender, submerged aquatic annual to 60 cm. Leaves alternate, linear. Flowers minute. Oct.–Nov. Brackish pools near coast, NW, SW, AP (Lambert's Bay to Stilbaai).•
palustris L.   Submerged mat-forming annual to 50 cm. Leaves opposite or ternate, linear. Flowers inconspicuous. Nov. Fresh or brackish pools, SW, KM, SE (Cape Peninsula to Port Elizabeth, cosmopolitan except Australia).

ZOSTERACEAE

ZOSTERA   SEEGRAS   11 spp., cosmopolitan in temperate regions

capensis Setch.   Submerged, mat-forming marine perennial to 12 cm. Leaves 2–4 in fascicles, linear. Flowers inconspicuous. Mostly Oct.–Mar. Brackish or sea water, SW, SE (Saldanha to Mozambique and Madagascar).

EUDICOTYLEDONS

ACANTHACEAE
by K. Balkwill

1. Corolla 1-lipped with adaxial suture slit; stamens 4; anthers all monothecous; bracts spiny or pungent:
    2. Anticous (outer) filaments with a short process near apex; stigma lobes short, subequal ....**Blepharis**
    2.' Anticous (outer) filaments without a process; stigma lobes unequal, anticous lobe enlarged and flattened and posticous lobe minute ....**Acanthopsis**
1.' Corolla actinomorphic or 2-lipped, adaxial suture not slit; stamens usually 2, rarely 4 but then at least 2 with bithecous anthers; bracts mostly unarmed:
    3. Style carried in a distinct median channel in upper lip:
        4. Pollen spheroidal, biporate and spinose; anthers attached at right angles to filaments; flowers in narrow spikes with bracts and calyx glandular-hairy ....**Isoglossa**
        4.' Pollen prolate, bi- or tricolporate, without spines; anthers attached parallel to filaments; flowers 1 or few in axils with bracts and calyx glabrous or hairy but not glandular:
            5. Seeds 1 per locule, smooth and shiny...**Monechma**
            5.' Seeds 2 per locule, rough and dull:
                6. Corolla tube shorter than lips ....**Justicia**
                6.' Corolla tube longer than lips..**Siphonoglossa**
    3.' Style not carried in a median channel in upper lip:
        7. Corolla not resupinate, usually with contorted aestivation (if imbricate not ascending); seeds with hygroscopic hairs:
            8. Stamens not didynamous nor attached to a flange of tissue; calyx 4-fid, 2 outer lobes large; bracts sometimes spiny and pungent ....**Barleria**
            8.' Stamens didynamous (unequal in length with 2 short and 2 long), often attached to a flange of tissue decurrent on corolla; calyx 5-fid; bracts never spiny:
                9. Stamens 2 (sometimes with 2 staminodes); anthers tailed; ovules 2 per locule; corolla 2-lipped ....**Chaetacanthus**
                9.' Stamens 4; anthers not tailed; ovules more than 2 per locule; corolla nearly actinomorphic..**Ruellia**
        7.' Corolla resupinate, with ascending imbricate aestivation; seeds without hygroscopic hairs:
            10. Anthers monothecous ....**Hypoestes**
            10.' Anthers bithecous:
                11. Capsules with inelastic placentae not breaking free at dehiscence; bracts narrower than 5 mm..**Peristrophe**
                11.' Capsules with elastic placentae which break free at dehiscence; bracts wider than 5 mm..**Dicliptera**

ACANTHOPSIS 7 spp., southern Africa

disperma Nees   Spiny, acaulescent perennial to 10 cm. Leaves tufted, oblanceolate, incised and spinescent. Flowers in dense spikes among spiny bracts, blue, sometimes white; bracts reflexing in fruit. Mar.–Oct. Stony slopes and flats, KM (Namibia and Karoo to Little Karoo).

BARLERIA 250 spp., pantropical, but only 1 sp. in Americas

irritans Nees   Stiffly branched, shortly hairy, spiny shrublet to 20 cm. Leaves opposite, ovate-elliptic, pungent, margins white. Flowers in dense axillary clusters, blue; outer bracts forming long spreading spines. Dec.–Apr. Karroid scrub and dry bush, KM, SE (Little Karoo to E Cape and Karoo).

obtusa Nees   Sprawling, shortly hairy shrub to 2 m. Leaves opposite, ovate, often obtuse. Flowers in lax axillary cymes, mauve to blue; bracts subspathulate, glandular-hairy. Feb.–Dec. Bush and forest margins, SE (Humansdorp to Zimbabwe).

pungens L.f.   Stiffly branched, thinly felted, spiny shrublet to 40 cm. Leaves opposite, subsessile, sharply toothed and pungent. Flowers in axillary clusters, mauve, pink or white; bracts ovate, spiny and pungent. Oct.–May. Stony flats, LB, SE (Swellendam to E Cape).

BLEPHARIS 80 spp., Africa to East Indies and Mediterranean

capensis (L.f.) Pers.   Stiffly erect, spiny shrub, mostly to 1.5 m. Leaves appearing dimorphic, oblanceolate, margins revolute and spiny, tufted in axils of hard, spreading spines. Flowers clustered in terminal spikes, whitish; bracts stiff, oblanceolate, sharply toothed and spine-tipped, white on the nerves. Mainly Dec.–Apr. Dry slopes on sand or clay, KM, LB, SE (Riversdale to E Cape and Karoo).

hirtinervia (Nees) T.Anderson   Like **B. capensis** but outer bracts obovate, with few small teeth. Jan.–Dec. Karroid bush, LB, SE (Riversdale to Bathurst).

inermis (Nees) C.B.Clarke   Twiggy, dichotomously branched, spiny shrublet to 30 cm. Leaves in axillary tufts, oblanceolate, margins revolute. Flowers in dense clusters at branch tips, white or cream-coloured, outer bracts spiny, inner bracts oblanceolate, pungent. Oct.–Dec. Karroid areas, KM (Ladismith to Garcia's Pass).•

integrifolia (L.f.) E.Mey. ex Schinz.   RANKKLITS   Prostrate, shortly hairy shrublet to 10 cm. Leaves opposite, elliptic. Flowers solitary in axils, blue or pink; bracts oblanceolate, shortly hairy, with short, barbed spines. Oct.–Apr. Grassy slopes, LB, SE (Riversdale to tropical Africa and India).

procumbens (L.f.) Pers.   Prostrate, glabrescent subshrub to 5 cm. Leaves narrowly oblanceolate, margins slightly revolute, toothed and softly prickly. Flowers in dense terminal spikes, blue; bracts obovate, shortly hairy, with thin marginal setae. July–Mar. Scrub and coastal bush, SE (Jeffreys Bay to Port Elizabeth).•

CHAETACANTHUS c. 5 spp., southern Africa

etiger (Pers.) Lindl.   Sprawling, shortly hairy, twiggy shrublet to 30 cm. Leaves opposite, obovate, obtuse, axillary leaves sparsely glandular-hairy. Flowers few in axils, white, sometimes blue. Aug.–Jan. Grassland and renosterveld, LB, SE (Mossel Bay to Mpumalanga).

sp. 1   Like **C. setiger** but axillary leaves associated with inflorescences glabrous or sparsely eglandular-hairy. Sept.–Dec. Grassland, LB, SE (Riversdale to E Cape).

DICLIPTERA 150 spp., cosmopolitan in warm and tropical areas

capensis Nees   Shortly hairy, lax to dense shrublet to 50 cm. Leaves opposite, ovate, 10–20 mm long. Flowers in short axillary clusters to 10 mm long, pink or white. Mar.–Dec. Valley bushveld, SE (Port Elizabeth to E Cape).

extenta S.Moore (= *Dicliptera zeylanica* auct.)   Lax, stoloniferous perennial to 60 cm. Leaves opposite, ovate, acute to attenuate, 30–80 mm long. Flowers in axillary clusters, 9–50 mm long, white. Apr.–Sept. Forest, SE (Knysna to Zimbabwe).

HYPOESTES over 150 spp., Africa, Asia, and Australia

aristata (Vahl) Sol. ex Roem. & Schult.   SEEROOGBLOMMETJIE   Erect, shortly hairy subshrub to 1.5 m. Leaves opposite, ovate; bracts aristate. Flowers in axillary clusters, appearing verticillate, mauve, calyx 4-fid. Jan.–May. Coastal forest margins, AP, SE (De Hoop to tropical Africa).

forskaolii (Vahl) R.Br.   Like **H. aristata** but bracts subspathulate, flowers white, sometimes mauve, in slender axillary clusters, not appearing verticillate and calyx 5-fid. Jan.–Apr. Forest margins, LB, SE (Mossel Bay to NE Africa).

ISOGLOSSA   c. 50 spp., Old World tropics

ciliata (Nees) Lindau (incl. **I. sylvatica** C.B.Clarke)   Sprawling, glabrescent subshrub sometimes to 1 m. Leaves opposite, ovate. Flowers in axillary spikes arranged in panicles, pink or white with spotted throat; bracts glandular-hairy but fringed with eglandular hairs. Feb.–Nov. Forest and forest margins, SE (Knysna to KwaZulu-Natal).

origanoides (Nees) Oerst.   Like **I. ciliata** but bracts not fringed with eglandular hairs. Mainly Apr.–Nov. Forest and forest margins, SE (E Cape to KwaZulu-Natal).

[Species excluded   Not definitely recorded from the area: I. eckloniana (Nees) Lindau]

JUSTICIA   c. 420 spp., cosmopolitan in warm and tropical areas

capensis Thunb.   Sprawling, shortly hairy shrub to 2.5 m. Leaves opposite, ovate to obovate. Flowers few in axillary cymes, white or magenta. Sept.–July. Bush and forest margins, SE (Humansdorp to KwaZulu-Natal).

cuneata Vahl   BLOUBOS   Twiggy shrublet to 60 cm, with stiffly erect, whitish stems. Leaves opposite, small, sessile, ascending, elliptic. Flowers solitary in axils, pedunculate, cream-coloured with pink markings. July–Sept. Sandy and rocky flats, NW, KM, LB, SE (Namaqualand to Clanwilliam, Little Karoo to Port Elizabeth and S Karoo).

orchioides L.f.   Like **J. cuneata** but sometimes thorny and flowers smaller, shorter than 10 mm. Sept.–Mar. Dry stony slopes, KM, SE (Little Karoo to E Cape and NW Province).

MONECHMA   c. 40 spp., Africa and India

robustum Bond   Grey, stiffly branched shrublet to 40 cm. Leaves opposite, grey-mealy, leathery, shortly petiolate, often with axillary tufts. Flowers solitary in axils, subsessile, white. Mainly July–Nov. Dry stony slopes, KM (W Little Karoo).•

spartioides (T.Anderson) C.B.Clarke   Glabrescent shrub to 2 m, often with horizontal branching. Leaves opposite, linear to ovate, usually glabrous, yellowish green. Flowers solitary in axils, white, cream-coloured, pale mauve or blue. Feb.–Dec. Dry stony slopes, KM (Namibia and Karoo to Little Karoo).

PERISTROPHE   15 spp., Africa to East Indies

cernua Nees   Sprawling, shortly hairy shrublet to 50 cm. Leaves opposite, ovate. Flowers in axillary cymes, magenta. Mainly Mar.–Sept. Valley bushveld, SE (Gamtoos River to KwaZulu-Natal).

RUELLIA   150 spp., pantropical

pilosa L.f.   Sprawling, shortly hairy, twiggy shrublet to 8 cm. Leaves opposite, obovate, obtuse, minutely ciliate. Flowers solitary in axils, blue to mauve. Sept.–Mar. Clay slopes in renosterveld, LB (Swellendam to Mossel Bay).•

SIPHONOGLOSSA   8 spp., Africa and America

leptantha (Nees) Immelman   Sprawling, shortly hairy shrub to 50 cm. Flowers solitary in uppermost axils, white to lilac with pink spots in throat. Dec.–June. Coastal bush and forest margins, SE (Knysna to KwaZulu-Natal).

THUNBERGIA   100 spp., Old World tropics

capensis Retz.   Softly hairy, sprawling perennial to 40 cm. Leaves opposite, subrotund. Flowers 1 or 2 in axils on hairy pedicels, creamy to yellow. Oct.–Apr. Grassland and coastal bush, SE (Knysna to E Cape).

dregeana Nees   Glabrescent vine. Leaves broadly sagittate, obscurely toothed. Flowers solitary in axils on glabrescent pedicels, dull orange-yellow. Nov.–Feb. Forest margins and streamsides in bush, SE (Humansdorp to KwaZulu-Natal).

ACHARIACEAE

1. Male flowers in short racemes; corolla not winged, lobes ciliate .... **Ceratiosicyos**
1.' Male flowers solitary or paired; corolla usually winged .... **Acharia**

ACHARIA   1 sp., South Africa

tragodes Thunb.   Monoecious thinly hairy perennial to 40 cm. Leaves palmatisect, toothed. Flowers 1–few in axils, nodding, greenish. Sept.–Mar. Scrub and forest margins, SE (Uitenhage to KwaZulu-Natal).

CERATIOSICYOS   1 sp., South Africa

laevis (Thunb.) A.Meeuse   Monoecious herbaceous climber. Leaves palmately 5–7-lobed, toothed. Flowers axillary, male flowers in racemes, female flowers solitary, greenish. Fruits cylindrical. Apr.–Aug. Forest margins, SE (Knysna to N Province).

AIZOACEAE
by C. Klak, except **Acrosanthes**, **Aizoon**, **Galenia** and **Tetragonia**

1. Petaloid staminodes absent:
   2. Fruit indehiscent, a winged or horned nut; ovary inferior or half-inferior (subfamily TETRAGONIOIDEAE) .... **Tetragonia**
   2.' Fruit a loculicidal or septicidal capsule; ovary superior (subfamily AIZOOIDEAE):
      3. Ovary incompletely 2-locular; ovules basal, solitary in each locule .... **Acrosanthes**
      3.' Ovary completely 2–5-locular; ovules mainly pendulous, 1–few in each locule:
         4. Ovary 4- or 5-locular with 2–many ovules in each locule; capsules with distinct expanding keels .... **Aizoon**
         4.' Ovary 2–5-locular with 1 ovule in each locule; capsules with expanding tissue lacking or rudimentary .... **Galenia**
1.' Petaloid staminodes present:
   5. Placentation axile; keels reaching centre of fruit; nectaries a series of separate grooves (subfamily MESEMBRYANTHEMOIDEAE):
      6. Plants with articulate assimilating stems with persistent succulent green internodes:
         7. Bladder cells similar on stems and leaves, more or less reduced; filamentous staminodes gathered into a cone .... **Psilocaulon**
         7.' Bladder cells dissimilar on stems and leaves; stems with densely arranged, tall, cylindrical, xeromorphic bladder cells, firm to the touch and visible in cross section as a conspicuous white layer; leaves with scattered, mesomorphic bladder cells, soft to the touch; filamentous staminodes not gathered into a cone:
            8. Sepals remaining erect during anthesis; stems cylindrical .... **Brownanthus**
            8.' Sepals bending down during anthesis; stems mostly 4-angled, rarely cylindrical .... **Aptenia**
      6.' Plants with a continuous stem, only youngest stem growth green, soon becoming woody:
         9. Plants with a smooth epidermis, i.e. bladder cells much flattened:
            10. Stems herbaceous or only a little woody; lower part of capsule longer than upper part .... **Prenia**
            10.' Stems woody; lower part of capsule as long as upper part .... **Aridaria**
         9.' Plants with a papillate epidermis:
            11. Plants annual or biennial .... **Mesembryanthemum**
            11.' Plants perennial:
               12. Old leaves withering to a skeleton and persisting, enclosing young leaves .... **Sceletium**
               12.' Old leaves never withering to a skeleton .... **Phyllobolus**
   5.' Placentation basal or parietal, rarely axile; keels never reaching centre of fruit; nectaries a series of small teeth in a continuous ring or in groups (subfamily RUSCHIOIDEAE):
      13. Fruit an indehiscent, fleshy berry .... **Carpobrotus**
      13.' Fruit dry:
         14. Fruit a nut, finally disintegrating into nutlets, or breaking into separate units:
            15. Perennial herb; fruit woody .... **Caryotophora**
            15.' Annual herb; fruit leathery .... **Hymenogyne**
         14.' Fruit a loculicidal capsule:
            16. Capsule xerochastic, expanding tissue absent:
               17. Locules more than 8 .... **Conicosia**
               17.' Locules 5–7:
                  18. Perennial; leaves c. 50 × 8 mm, with bladder cells .... **Saphesia**
                  18.' Annual or biennial; leaves c. 80 × 20 mm, without bladder cells .... **Skiatophytum**
            16.' Capsule hygrochastic, expanding tissue present:
               19. Expanding tissue reduced, consisting of either expanding sheets or expanding keels:

20. Expanding keels only present:
    21. Locules open, seeds visible; leaf margins more or less smooth ....**Apatesia**
    21.' Locules covered by septa, seeds invisible; leaf margins with long hairs ....**Carpanthea**
20.' Expanding sheets only present:
    22. Leaves lyrate...**Aethephylum**
    22.' Leaves entire ....**Dorotheanthus**
19.' Expanding tissue consisting of both expanding sheets and keels:
    23. Capsule opening once only, expanding keels breaking after opening:
        24. Locules 8...**Stayneria**
        24.' Locules 5:
            25. Leaves very soft; flowers yellow ....**Scopelogena**
            25.' Leaves hard; flowers white to pale pink...**Stoeberia**
    23.' Capsule opening repeatedly, expanding keels remaining functional:
        26. Leaves with a diadem or terminal whorl of spines ....**Trichodiadema**
        26.' Leaves without a diadem:
            27. Seeds echinate:
                28. Leaves more or less finger-like ....**Antegibbaeum**
                28.' Leaves triquetrous-apiculate, with serrate and cartilaginous keel and
                    margins ....**Braunsia**
            27.' Seeds smooth or papillate but not echinate:
                29. Fruits more than 6-locular:
                    30. Fruits without a closing body:
                        31. Plants compact, rarely shrubby..**Gibbaeum**
                        31.' Plants low shrubs:
                            32. Leaves united at base and continuous with stem ....**Smicrostigma**
                            32.' Leaves free or shortly fused towards base but never continuous with stem:
                                33. Straggling or procumbent to creeping shrublets....................**Malephora**
                                33.' Erect shrubs:
                                    34. Shrubs 30–100 cm tall; flowers large, to 5 cm diam.................**Erepsia**
                                    34.' Shrubs to 10 cm tall; flowers to 3 cm diam................**Zeuktophyllum**
                    30.' Fruits with a closing body:
                        35. Erect shrubs:
                            36. Capsules c. 10-locular ....**Leipoldtia**
                            36.' Capsules to 8-locular:
                                37. Shrub with spines developing after capsules have been shed........**Arenifera**
                                37.' Shrubs not spinescent ....**Octopoma**
                        35.' Clump-forming or creeping shrublets:
                            38. Plants creeping:
                                39. Capsules with a small closing body ....**Jordaaniella**
                                39.' Capsules with a large closing body ....**Cephalophyllum**
                            38.' Plants clump-forming or with a central clump and long straggling branches:
                                40. Plants with a central clump and long straggling branches:
                                    41. Leaves heterophyllous..**Vanzijlia**
                                    41.' Leaves homophyllous:
                                        42. Covering membrane with closing rodlet at distal end .....**Hallianthus**
                                        42.' Covering membrane with thick closing bulge at distal
                                            end ....**Cephalophyllum**
                                40.' Plants clump-forming:
                                    43. Leaves soft...**Glottiphyllum**
                                    43.' Leaves hard:
                                        44. Leaves with conspicuous dark dots:
                                            45. Fruit dark brown, more than 10 mm diam. ....**Pliospilos**
                                            45.' Fruit light brown, less than 10 mm diam. ....**Tanquana**
                                        44.' Leaves without dark dots:
                                            46. Leaves cylindrical or more or less terete ....**Cylindrophyllum**
                                            46.' Leaves sabre-shaped or triangular:
                                                47. Leaves sabre-shaped; flowers mostly 3 ....**Machairophyllum**
                                                47.' Leaves triangular, rarely flattened or hood-shaped; flowers
                                                    solitary ....**Cheiridopsis**
                29.' Fruits to 6-locular:
                    48. Plants annual ....**Cleretum**
                    48.' Plants perennial:
                        49. Capsules with a closing body:
                            50. Flowers yellow:

51. Leaves laterally compressed in upper part, smooth ....**Hereroa**
51.' Leaves trigonous, hunched in lower part, margins with some teeth, keel not dentate ....**Carruanthus**
50.' Flowers white, purple or pink:
  52. Closing body very large:
    53. Closing body lens-shaped; plants compact; leaf margins with scattered cartilaginous teeth ....**Acrodon**
    53.' Closing body stalked; plants compact or shrubby; leaf margins smooth:
      54. Plants creeping, without a compact centre; leaves slightly longer than broad ....**Lampranthus**
      54.' Plants shrubby, clump-forming or creeping but then leaves much longer than broad...**Antimima**
  52.' Closing body small to medium sized:
    55. Fruits scarcely woody; closing body bipartite ....**Disphyma**
    55.' Fruits woody; closing body rod-shaped or occasionally reduced:
      56. Valve wings absent ....**Ruschia**
      56.' Valve wings present...**Amphibolia**
49.' Capsules without a closing body:
  57. Plants short-stemmed or stemless:
    58. Leaves united for most of their length, forming a corpuscle:
      59. Leaves very fleshy and soft, velvety from long epidermal hairs ....**Gibbaeum**
      59.' Leaves hard, smooth or papillose to hairy......................**Conophytum**
    58.' Leaves free or shortly fused towards base:
      60. Leaves very soft, pulpy, deciduous ....**Diplosoma**
      60.' Leaves hard:
        61. Flowers pink, purple or white:
          62. Leaves dark green to blackish; flowers 1–3, bracteolate ....**Cerochlamys**
          62.' Leaves green to whitish grey; flowers solitary, ebracteolate ....**Gibbaeum**
        61.' Flowers yellow:
          63. Flowers stalked ....**Rhinephyllum**
          63.' Flowers sessile or subsessile:
            64. Leaves usually with long awn-shaped teeth but if not then with pronounced cartilaginous margins ....**Faucaria**
            64.' Leaves with small white teeth ....**Stomatium**
  57.' Plants shrubby:
    65. Surface of ovary (flower) concave, forming a hypanthium ....**Erepsia**
    65.' Surface of ovary (flower) flat to raised:
      66. Fruits without wings ....**Esterhuysenia**
      66.' Fruits with wings:
        67. Young stems hirsute or rough............................**Drosanthemum**
        67.' Young stems smooth:
          68. Covering membranes reduced to narrow rims..........**Delosperma**
          68.' Covering membranes more or less complete:
            69. Plant with a large, tuberous rootstock:
              70. Leaves smooth, perennial ....**Mestoklema**
              70.' Leaves scabrid, deciduous.........................**Lampranthus**
            69.' Plants not tuberous:
              71. Leaves sharply triquetrous with serrulate margins...**Circandra**
              71.' Leaves trigonous, semicylindrical to subcylindrical:
                72. Fruits scarcely woody; leaves grey; nectaries consisting of a series of small teeth arranged in 5 groups ......**Oscularia**
                72.' Fruits woody; leaves grey or various shades of green; opening of locules blocked by sterile funicles; nectaries consisting of a series of small teeth in a continuous ring...**Lampranthus**

ACRODON•   5 spp., W Cape

bellidiflorus (L.) N.E.Br. (= *Ruschia constricta* L.Bolus, *R. duplessiae* L.Bolus, *R. longifolia* L.Bolus, *R. macrophylla* L.Bolus)   Tufted, succulent perennial. Leaves trigonous, keel and margins smooth or with 3 or 4 broad-based, flexible teeth. Flowers white to pale pink, c. 35 mm diam. Fruits longer than

20 mm. Apr.–July. Renosterveld, grassland, or coastal fynbos, SW, KM, LB (Bredasdorp and McGregor to Little Karoo).•

parvifolius Du Plessis   Prostrate succulent perennial with visible internodes. Leaves trigonous, keel smooth or with up to 4 broad-based flexible teeth, margins mostly smooth. Flowers white to pink, c. 20 mm diam. Fruits shorter than 20 mm. Aug.–Sept. Clay streambeds, SW (Hermanus coast).•

purpureostylus (L.Bolus) P.M.Burgoyne (= *Cerochlamys purpureostyla* (L.Bolus) H.E.K.Hartmann, *Ruschia purpureostyla* (L.Bolus) Bruyns)   Tufted succulent to 15 cm, sometimes with horizontal runners. Leaves trigonous, very fleshy, 25–35 × 4–8 mm, margins and keel sparsely toothed. Flowers on pedicels 20–40 mm long, pale pink to white with a central pink stripe. Capsule 4–6-locular. June. Stony slopes, SW (McGregor to Stormsvleikloof).•

quarcicola H.E.K.Hartmann   Compact, tufted, succulent perennial. Leaves scimitar-shaped, keels and margins minutely toothed, smooth with age, 15–20 mm long. Flowers pink, petals with dark lateral and central stripes. Fruits shorter than 20 mm. Flowering time? Quartzitic slopes, SW (Agulhas Peninsula).•

subulatus (Mill.) N.E.Br. (= *Ruschia leptophylla* L.Bolus)   Compact, tufted, succulent perennial. Leaves subulate, keel and margin often with few broad-based flexible teeth. Flowers white or pale pink, petals with darker central stripe and margins. Fruits longer than 20 mm. Nov.–Apr. Lower slopes in renosterveld and coastal fynbos, SW (Caledon to Stanford).•

ACROSANTHES•   SPEKVYGIE   5 spp., W Cape

anceps (Thunb.) Sond.   Sprawling shrublet to 10 cm, branches often trailing. Leaves oblanceolate, mucronate, fleshy, 1–2 cm. Flowers solitary at nodes, white, stamens 15–20. Sept.–Nov. Stony sandstone slopes, NW, SW (Piketberg and Clanwilliam to Goudini).•

angustifolia Eckl. & Zeyh.   Like **A. anceps** but leaves to 1 cm. Nov.–Dec. Sandstone slopes, NW (Pakhuis Mts to Worcester).•

humifusa Sond.   Like **A. angustifolia** but leaves yellow-green, flowers subsessile, stamens c. 10. Dec.–Jan. Rocky ridges and sandy plateaus, NW, SW (Matsikamma to Riviersonderend Mts).•

microphylla Adamson   Like **A. angustifolia** but leaves blackish green, elliptic-ovate, pseudopetiolate, 4–6 mm, stamens 8–10. Oct.–Dec. Rocky sandstone ridges and summits, NW, KM (Cedarberg to Witteberg.)•

teretifolia Eckl. & Zeyh.   Sprawling, diffuse perennial to 80 cm, with woody base. Leaves opposite, subterete, 1–2 cm, fleshy, mucronate. Flowers solitary at nodes, white, stamens more than 25. Oct.–Feb. Stony flats and slopes, NW, SW, AP (Tulbagh to Agulhas).•

AETHEPHYLLUM•   1 sp., W Cape

pinnatifidum (L.f.) N.E.Br. Prostrate annual. Leaves flat, lyrate, papillate. Flowers solitary, small, yellow. Fruits 5-locular with funnel-shaped base. Sept.–Oct. Rocky slopes, NW, SW (Gifberg to Elandskloof Mts).•

AIZOON   SPEKVYGIE   20 spp., Africa, Mediterranean, Australia

A. Flowers terminal

paniculatum L. Prostrate or sprawling subshrub to 50 cm, adpressed-hairy when young. Leaves mostly basal and opposite, oblanceolate. Flowers sessile in flattened, dichotomous paniculate cymes, magenta, sometimes cream-coloured. July–Oct. Dry sandstone and clay slopes and flats, NW, SW (Bokkeveld Mts to Cape Peninsula).•

sarmentosum L.f   Like **A. paniculatum** but leaves subterete, flowers 1–3(–5) at branch tips, white. June–Oct. Dry flats and lower slopes, NW, SW (Namaqualand to Montagu).

A. Flowers apparently axillary

canariense L.   Prostrate annual or short-lived perennial to 30 cm, shortly hairy when young. Leaves pseudopetiolate, oblanceolate-spathulate, adpressed-hairy. Flowers sessile in flattened, axillary clusters, yellow-green, sepals small, upcurving. July–Sept. Dry stony flats, often disturbed ground, NW, SE (Namaqualand to Clanwilliam, Karoo to N Africa and Arabia).

glinoides L.f.   Like **A. rigidum** but with silky, spreading hairs, and capsules smooth above. Oct.–Jan. Dry stony flats, SE (Langkloof to KwaZulu-Natal).

karooicum Compton   Gnarled shrublet to 20 cm. Leaves elliptic-lanceolate, often folded, retrorsely adpressed-hairy. Flowers sessile, apparently axillary, yellow. Mar.–June. Stony flats, often on quartz outcrops, NW, KM (Worcester to Little Karoo).•

rigidum L.f.   Grey, adpressed-hairy, prostrate subshrub to 15 cm. Leaves obovate-spathulate, pseudopetiolate. Flowers sessile, solitary in axils, yellow. Capsules sharply angled above. Sept.–Oct. Dry stony slopes, often near the sea, SW, AP, KM, SE (Gansbaai and Little Karoo to East London).

AMPHIBOLIA   c. 5 spp., S Namibia and Karoo to W Cape

hutchinsonii (L.Bolus) H.E.K.Hartmann (= *Ruschia hutchinsonii* L.Bolus)   Sprawling shrublet to 20 cm. Leaves pale green to grey, club-shaped, c. 15 × 6 mm. Flowers solitary, pink, to 20 mm diam. Fruits 5-locular, base funnel-shaped, valve wings present. July. Coastal plains, NW, SW (Vredendal to Melkbosstrand).•

ANTEGIBBAEUM•   BASTERVOLSTRUISTONE   1 sp., W Cape

fissoides (Haw.) Schwantes ex C.Weber   Perennial to 8 cm, branched from base into clumps. Leaves in unequal pairs, curved, blunt. Flowers on short pedicels with 2 pairs of bracts below flower, violet-red, to 60 mm diam. Fruits 6- or 7-locular, closing bodies absent. Seeds short and spiny. Aug. Quartz patches, often with *Gibbaeum*, KM (W Little Karoo).•

ANTIMIMA (= *RUSCHIA* in part)   KUSSINGVYGIE   c. 100 spp., S Namibia to E Cape

biformis (N.E.Br.) H.E.K.Hartmann (= *Ruschia biformis* (N.E.Br.) Schwantes)   Branching, tufted shrublet to 2.5 cm. Leaves almost free, 2–7 × 2–3 mm, trigonous, acute to somewhat acuminate. Flowers and fruits unknown. Flowering time? Habitat?, LB (Swellendam).•

bina (L.Bolus) H.E.K.Hartmann (= *Ruschia bina* L.Bolus)   Erect shrublet to 9 cm. Leaves glaucous, fused below, to 5 mm long. Flowers pink, c. 16 mm diam. Aug.–Sept. Habitat?, NW, SW (Clanwilliam to Malmesbury).•

brevicarpa (L.Bolus) H.E.K.Hartmann (= *Ruschia brevicarpa* L.Bolus)   Tufted shrublet to 20 cm, with branches covered with old leaf remains. Leaves velvety, 4–12 mm long, sheath 2–4 mm long. Flowers pink, c. 15 mm diam. Oct. Rock outcrops at high alt., NW (Clanwilliam).•

brevicollis (N.E.Br.) H.E.K.Hartmann (= *Ruschia brevicollis* (N.E.Br.) Schwantes)   Low, tufted shrublet. Leaves free, shortly fused below, abaxial surface sharply keeled, pale grey-green. Flowers short-stalked, pink, to 12 mm diam. Flowering time? Habitat?, KM (Ladismith).•

concava (L.Bolus) H.E.K.Hartmann (= *Ruschia concava* L.Bolus)   Spreading shrublet to 12 cm, with branches to 16 cm. Leaves erect, abaxial surface keeled, with ciliate edges, to 3 mm long, sheath 1.5 mm long. Flowers solitary, magenta, 15 mm diam. Aug.–Sept. Habitat?, NW, KM (W Karoo and Bokkeveld Mts to Ladismith).

concinna (L.Bolus) H.E.K.Hartmann (= *Ruschia concinna* L.Bolus)   Erect shrublet, to 7 cm, with short shoots in axils. Leaf pairs dimorphic, erect, to 4 mm, shiny green, with a bristle at apex, margins finely ciliate. Flowers pale pink or magenta, to 15 mm diam. June–July. Habitat?, NW, SW (Bokkeveld Mts to Malmesbury).•

condensa (N.E.Br.) H.E.K.Hartmann (= *Ruschia condensa* (N.E.Br.) Schwantes)   Like **A. propinqua** but leaves 6–8 mm long, not hairy and flowers short-stalked, 10 mm diam., pink. Flowering time? Habitat?, LB (Montagu).•

dasyphylla (Schltr.) H.E.K.Hartmann (= *Ruschia dasyphylla* (Schltr.) Schwantes)   Compact shrublet to 20 cm, with branches at first 2-angled. Leaves fused basally, 3–5 mm long, trigonous, keeled, keel and margins finely ciliate. Flowers solitary, magenta, 12 mm diam. Aug. Upper slopes and plateaus, NW (Cedarberg Mts).•

distans (L.Bolus) H.E.K.Hartmann (= *Ruschia distans* (L.Bolus) L.Bolus)   Erect shrublet to 15 cm, stems angled and narrowly winged. Leaves in pairs at branch tips, 7–10 mm long. Flowers solitary, magenta, 15 mm diam. May. Habitat?, NW (Clanwilliam).•

fergusoniae (L.Bolus) H.E.K.Hartmann (= *Ruschia fergusoniae* L.Bolus)   Tufted shrublet, with branches to 15 mm long. Leaves thick, adaxial surface keeled, 2–4 mm long. Flowers solitary, sessile, pink with white centre, 10 mm diam. Sept. Habitat?, KM (Barrydale).•

gracillima (L.Bolus) H.E.K.Hartmann (= *Ruschia gracillima* L.Bolus)   Slender shrublet to 25 cm. Leaves slender, erect, with a recurved bristly tip, margins and keel finely hairy, green, 13 mm long. Flowers on pedicels to 20 mm long, purplish, 10 mm diam. Sept. Habitat?, NW (Nieuwoudtville).•

granitica (L.Bolus) H.E.K.Hartmann (= *Ruschia granitica* (L.Bolus) L.Bolus)   Prostrate shrublet with slender, spreading branches. Leaves erect, mucronate, back surface keeled, glaucous. Flowers solitary, pink, 12 mm diam. Sept. Granite rocks, NW, SW (Hondeklipbaai to Saldanha Bay).

hamatilis (L.Bolus) H.E.K.Hartmann (= *Ruschia hamatilis* L.Bolus)   Prostrate shrublet, with grey branches to 8–12 cm long. Leaves ascending, with flat adaxial surface, tip with small bristle, velvety. Flowers magenta, 14 mm diam. June–Sept. Karroid plains, NW (Worcester).•

leipoldtii (L.Bolus) H.E.K.Hartmann (= *Ruschia leipoldtii* L.Bolus)   Like **A. peersii** but with longer, more slender branchlets, leaves erect, mucronate, sheath 2–3 mm long, and flowers 16 mm diam. Aug.–Oct. Habitat?, NW, SW (Worcester to Robertson).•

limbata (N.E.Br.) H.E.K.Hartmann (= *Ruschia limbata* (N.E.Br.) Schwantes)   Low tufted shrublet, to 30 mm. Leaves fused basally, 5–10 mm long, trigonous. Flowers on short pedicels, magenta, to 16 mm diam. Flowering time? Habitat?, SW (Saldanha).•

lokenbergensis (L.Bolus) H.E.K.Hartmann (= *Ruschia lokenbergensis* L.Bolus)   Erect shrublet to 15 cm. Leaves ascending, adaxial surface flat, 5–15 mm long. Flowers 15–19 mm diam. May. S-facing, rocky slopes in arid fynbos, NW (Bokkeveld Mts).•

maxwellii (L.Bolus) H.E.K.Hartmann (= *Ruschia maxwellii* L.Bolus)   Robust shrublet to 6 cm, with branches to 9 mm diam. Leaves enclosing internodes, ascending, mucronate, 17 mm long, sheath 3 mm long. Flowers solitary, stalked, pink, 20 mm diam. Flowering time? Habitat?, KM, SE (Oudtshoorn and Willowmore to Knysna).•

menniei (L.Bolus) H.E.K.Hartmann (= *Ruschia menniei* L.Bolus)   Low shrublet, with stems to 4 mm diam. Leaves linear, acute, with small cilia along margins and keel, to 15 mm long. Flowers pink, sometimes with purplish reverse, petals 4–6 mm long. July. Habitat?, SW (Malmesbury).•

microphylla (Haw.) Dehn (= *Ruschia microphylla* (Haw.) Schwantes)   Wiry stemmed shrublet to 10 cm. Leaves crowded on short shoots, enclosed by dried older leaves, 4–5 mm long, set with large transparent dots. Flowers solitary on short pedicels, white with pink tips, c. 20 mm diam. Aug. Habitat?, NW, SW (Piketberg to Paarl and Worcester).•

minutifolia (L.Bolus) H.E.K.Hartmann (= *Ruschia minutifolia* L.Bolus)   Compact shrublet to 8 cm. Leaves apiculate, with ciliate margins, dissimilar, lower ones ovate, 3 mm long, upper lanceolate, to 4 mm long. Flowers magenta, 12–15 mm diam. July. Upper rocky slopes, NW (Clanwilliam).•

mucronata (Haw.) H.E.K.Hartmann (= *Ruschia mathewsii* L.Bolus, *R. mucronata* (Haw.) Schwantes)   Rigid shrublet to 20 cm. Leaves laterally compressed, keeled adaxially, acute, bristle-tipped, velvety, 10–15 mm long, old leaves dry and persisting. Flowers solitary, pink, 16 mm diam. Apr.–June. Habitat?, SW (Malmesbury).•

paucifolia (L.Bolus) H.E.K.Hartmann (= *Ruschia paucifolia* L.Bolus)   Erect shrublet to 20 cm. Leaves small, 4–5 mm long. Flowers solitary, magenta, c. 10 mm diam. May. Rocky sandstone slopes in dry fynbos, NW (Bokkeveld Mts).•

peersii (L.Bolus) H.E.K.Hartmann (= *Ruschia peersii* L.Bolus)   Compact shrublet to 6 cm. Leaves fused below, sheath 8 mm long, keeled abaxially, to 2 cm long, slender, old leaves persisting. Flowers solitary, magenta with stripe, c. 22 mm diam. Nov. Habitat?, LB (Riversdale).•

persistens (L.Bolus) H.E.K.Hartmann (= *Ruschia persistens* L.Bolus)   Compact shrublet to 7 cm. Leaves shortly fused below, acute, green with translucent dots, older leaves drying and persisting. Flowers pink, to 15 mm diam. Aug. Habitat?, LB, KM (Montagu, Ladismith).•

propinqua (N.E.Br.) H.E.K.Hartmann (= *Ruschia propinqua* (N.E.Br.) Schwantes)   Tufted shrublet. Leaves 5–6 mm long, glaucous. Flowers subsessile, pink with dull stripe, 8–16 mm diam. Flowering time? Habitat?, LB (Montagu).•

prostrata (L.Bolus) H.E.K.Hartmann (= *Ruschia prostrata* L.Bolus)   Robust mat-forming shrublet, with spreading branches to 30 cm long. Leaves fused below, c. 3 mm long, sheath 2 mm long. Flowers subsessile. Flowering time? Habitat?, NW (S Namaqualand to Clanwilliam).

pygmaea (Haw.) H.E.K.Hartmann (= *Ruschia pygmaea* (Haw.) Schwantes)   Small mat-forming shrublet. Leaves dimorphic, upper pair fused almost to tips, 4–5 mm long, drying and enveloping a later pair of free, widely spreading leaves. Flowers ?magenta, 18 mm diam. Flowering time? Habitat?, NW, KM (Worcester to Witteberg).•

roseola (N.E.Br.) H.E.K.Hartmann (= *Ruschia roseola* (N.E.Br.) Schwantes)   Sprawling shrublet. Leaves fused below, 6–8 mm long, blunt, set with transparent dots. Flowers pale pink, c. 15 mm diam. Flowering time? Habitat?, KM (Montagu).•

saturata (L.Bolus) H.E.K.Hartmann (= *Ruschia saturata* L.Bolus)   Erect shrublet to 10 cm. Leaves ascending, glossy, tapering, bristle-tipped, margins ciliate. Flowers pink with purple reverse. c. 18 mm diam. July. Habitat?, NW (Tulbagh).•

sobrina (N.E.Br.) H.E.K.Hartmann (= *Ruschia sobrina* (N.E.Br.) Schwantes)   Tufted shrublet. Leaves free or fused basally, 3–5 mm long. Flowers magenta with stripe, petals c. 8 mm long. Flowering time? Habitat?, LB (Riversdale).•

stokoei (L.Bolus) H.E.K.Hartmann (= *Ruschia stokoei* L.Bolus)   Compact shrublet to 5 cm, with stiff spreading branches. Leaves united below, sheath 5–11 mm, free part of leaf 9 mm long, mucronate. Flowers magenta with stripe, c. 22 mm diam. July–Aug. Upper slopes, NW, SW (Worcester, Caledon).•

subtruncata (L.Bolus) H.E.K.Hartmann (= *Ruschia subtruncata* L.Bolus)   Creeping shrublet, with branches to 25 cm long. Leaves dimorphic, shorter ones to 6 mm long, longer ones 6–10 mm long, fused below. Flowers stalked, pink, 25 mm diam. Sept. Habitat?, NW (Nieuwoudtville to Sutherland).•

ventricosa (L.Bolus) H.E.K.Hartmann   Compact, rounded perennial to 10 cm. Leaves dimorphic, free parts trigonous, papillate, 40–80(–120) × c. 13 mm diam. Flowers solitary, purple, 54 mm diam. Fruits 4–6-locular, without valve wings. June–Sept. Rock outcrops, NW, SW (Vanrhynsdorp to Pakhuis Pass, Saldanha).•

verruculosa (L.Bolus) H.E.K.Hartmann (= *Ruschia verruculosa* L.Bolus)   Compact shrublet to 7 cm, with red branches later fading grey, minutely papillate when young. Leaves shortly fused basally, bristle-tipped, margins glabrous, to 4 mm long. Flowers short-stalked, magenta, 12 mm diam. Feb.–Sept. Habitat?, NW (Hex River Mts).•

APATESIA•   3 spp., W Cape

helianthoides (Aiton) N.E.Br.   Annual to 15 cm. Leaves flat, slightly succulent, margins with idioblasts. Flowers yellow. Fruits flat at base, 5-angular, conical above, valve wings lacking. Seeds smooth, light brown. Sept.–Oct. Habitat?, NW (Vanrhynsdorp to Tulbagh).•

pillansii N.E.Br.   Annual to 15 cm. Leaves flat, slightly succulent, margins with idioblasts. Flowers yellow with whitish centre. Fruits with a shallow base, slightly domed above, valve wings represented by very narrow membranous rims. Seeds slightly papillate, light brown. Sept. Coastal dunes, SW (Velddrif to Cape Peninsula).•

sabulosa (Thunb.) L.Bolus (= *Apatesia maughanii* N.E.Br.)   Annual to 20 cm. Leaves flat, slightly succulent, margins with idioblasts. Flowers yellow. Fruits with a shallow base, flat to slightly domed above, tips of valves forming a cone, valve wings narrow. Seeds papillate, brown. Aug.–Sept. Habitat?, NW (Clanwilliam to Tulbagh).•

APTENIA (= *PLATYTHYRA*)   4 spp., southern Africa (after Gerbaulet, in prep.)

*****cordifolia** (L.f.) Schwantes   Prostrate, mat-forming perennial with 4-angled stems. Leaves flat, heart-shaped. Flowers magenta, 20 mm diam. Fruits without valve wings. Dec.–Apr. Shady places, widely grown in gardens (E Cape weed).

geniculiflora (L.) Bittrich ex Gerbaulet (= *Sphalmanthus geniculiflorus* (L.) L.Bolus)   Decumbent to erect, often scrambling, deciduous perennial with terete or slightly 4-angled stems. Leaves channelled or cylindrical, free. Flowers white to pale yellow, 10–30 mm diam. Fruits with valve wings. Flowering time?. Dry flats, often scrambling in bush, SE (Namibia to E Cape).

haeckeliana (A.Berger) Bittrich ex Gerbaulet (= *Platythyra haeckeliana* (A.Berger) N.E.Br.)   Prostrate to decumbent perennial with 4-angled stems. Leaves flat, ovate to lanceolate. Flowers pale yellow, 20 mm diam. Fruits with valve wings. Dec. Karroid flats, SE (Port Elizabeth to Alexandria).

ARENIFERA   PLAKKERTJIE   4 spp., Richtersveld to W Cape and Karoo

spinescens (L.Bolus) H.E.K.Hartmann (= *Ruschia spinescens* L.Bolus)   Shrublet to 40 cm, branches becoming spiny after capsules have been shed. Leaves sticky, trigonous. Flowers in cymes, pink, filamentous staminodes white with pink tips. Fruits 8-locular, complete covering membranes, closing bodies present. Flowering time? Gravel and loamy flats, NW (S Namaqualand and W Karoo to Ceres).

ARIDARIA   4 spp., W Cape to S Namibia

brevicarpa L.Bolus   Shrub to 1 m. Leaves to 30 × 4 mm. Flowers white to pale pink, closing at dusk. Fruits hemispherical below and above, remaining open when dry. Aug.–Sept. Sandy soils, NW (S Namibia to Clanwilliam).

noctiflora (L.) Schwantes (= *Aridaria leipoldtii* L.Bolus)   Shrub to 1 m. Leaves to 50 × 7 mm. Flowers white, opening at dusk, closing near dawn. Fruits slender below, conical to hemispherical above. Sept.–Nov. Sandy places, sometimes among rocks, NW, KM, LB, SE (S Namibia to E Cape).

serotina L.Bolus   Shrub to 30 cm. Leaves to 25 × 10 mm. Flowers white, opening at dusk, closing during the night. Fruits slender below, hemispherical to cubical above, hygrochastic. Sept.–Nov. Rocky outcrops, sometimes on sand, NW (S Namibia to Clanwilliam).

BIJLIA   PRINCE ALBERT VYGIE   2 spp., N and W Cape

tugwelliae (L.Bolus) S.A. Hammer   Densely leafy, clump-forming shrublet. Leaves suberect, sabre-shaped, trigonous, to 60 mm long. Flowers more or less sessile, yellow, 40–50 mm diam. Fruits 5-locular. Aug. Sandstone outcrops, KM (Bosluiskloof to Prince Albert).•

BRAUNSIA   c. 5 spp., W Cape and Karoo

apiculata (Kensit) L.Bolus   Like **B. geminata** but leaves velvety hairy. July–Sept. Mainly sandstone, NW, KM (Ceres Karoo and Cedarberg Mts to Little Karoo).

geminata (Haw.) L.Bolus   Dwarf, woody perennial, with ascending branches to 15 cm long. Leaves erect, trigonous, united in lower half, scarcely gaping above, smooth. Flowers solitary, white, to 40 mm diam. Fruits woody, 5-locular. Apr.–June. Shale slopes, KM, LB (Ceres Karoo to Little Karoo).•

maximiliani (Schltr. & A.Berger) Schwantes (= *Lampranthus maximiliani* (Schltr. & A.Berger) L.Bolus)   Prostrate shrublet with branches trailing and rooting at nodes, to 17 cm long, fertile branches erect, densely 4-leaved. Leaves velvety, swollen, trigonous, boat-shaped, grey-green. Flowers solitary, pink, filaments reddish at tips, c. 20 mm diam. Fruits small, woody, 5-locular. July–Sept. Rock pavements, NW (Bokkeveld Mts to Clanwilliam).•

[**Species excluded**   Genus uncertain, see end of family: **B. vanrensburgii**]

BROWNANTHUS   12 spp., dry parts of southern Africa

ciliatus (Aiton) Schwantes   Erect shrublet to 25 cm. Leaves with a conspicuous ring of white cilia at base. Flowers in few-flowered dichasia. Fruits 5-locular, with or without seed bags. Oct.–Nov. Dry flats, usually a pioneer, NW, KM (S Namibia to Ceres and Willowmore).

fraternus Klak   Erect or decumbent shrublet to 35 cm. Leaves free below, deciduous. Flowers solitary. Fruits 5-locular. Nov.–Dec. Quartzite patches on clay, SW, LB (Bredasdorp to Swellendam).•

CARPANTHEA•   VETKOUSIE   1 sp., W Cape

pomeridiana (L.) N.E.Br. (= *Carpanthea calendulacea* (Haw.) L.Bolus, *C. pilosa* (Haw.) L.Bolus)   Annual to 20 cm, with softly hairy-papillate stems. Leaves flat, slightly succulent, margins with hair-like papillae. Flowers yellow, calyx with hair-like papillae. Fruits hygrochastic, expanding keels well developed, breaking into mericarps, seed bags incomplete. Sept.–Nov. Sandy flats, NW, SW (Bokkeveld Mts to Cape Peninsula).•

CARPOBROTUS   HOTTENTOT FIG, SUURVY   13 spp., W Cape to KwaZulu-Natal, Chile, California, Australia

A. Receptacle top-shaped, tapering into pedicel

edulis (L.) L.Bolus   Succulent perennial with trailing stems to 2 m long. Leaves straight or slightly curved. Flowers yellow, fading to pink with age, stigmas about as long as stamens, receptacle top-shaped, tapering into pedicel. Fruits clavate to subglobose, yellowish. Aug.–Oct. Coastal and inland slopes, NW, SW, AP, KM, LB, SE (Namaqualand to E Cape).

mellei (L.Bolus) L.Bolus (= *Carpobrotus pageae* L.Bolus, *C. pillansii* L.Bolus)   Succulent perennial with trailing stems to 1.5 m long. Leaves more or less straight, 8–18 mm diam. Flowers pink or purple, 50–80 mm diam., stigmas overtopping stamens, receptacle top-shaped, tapering into pedicel. Fruits clavate, bicarinate when still green. Sept.–Jan. Sandstone slopes, SW, KM, LB (W Cape).•

A. Receptacle subclavate to globose, not tapering into pedicel

acinaciformis (L.) L.Bolus (= *Carpobrotus concavus* L.Bolus, *C. vanzijliae* L.Bolus)   Succulent perennial with trailing stems. Leaves robust, sabre-shaped, 15–25 mm diam. Flowers purple, 70–100 mm diam., top of ovary flattish or depressed towards centre, receptacle bicarinate, oblong or subglobose, not tapering into pedicel. Fruits oval. Aug.–Dec. Coastal sands, SW, AP (Saldanha to Mossel Bay).•

deliciosus (L.Bolus) L.Bolus (= *Carpobrotus dulcis* L.Bolus, *C. fourcadei* L.Bolus)   Robust succulent perennial with trailing stems. Leaves almost straight, 12–18 mm diam. Flowers purple, pink or white, 60–80 mm diam., top of ovary raised towards centre, receptacle bicarinate, subglobose, not tapering into pedicel. Fruits globose to subglobose. June–Oct. Sand dunes to rocky grassland, AP, SE (Riversdale to KwaZulu-Natal).

muirii (L.Bolus) L.Bolus   Like **C. deliciosus** but leaves narrow, 5–7 mm diam., and flowers 60–90 mm diam., top of ovary depressed in centre. Sept.–Nov. Coastal areas, AP (De Hoop to Stilbaai).•

quadrifidus L.Bolus (= *Carpobrotus sauerae* Schwantes)   Succulent perennial with trailing stems to 2.5 m long. Leaves straight, 18–25 mm diam. Flowers white, pale pink or purplish, 120–150 mm diam., receptacle bicarinate, subclavate, not tapering into pedicel, top of ovary convex or flat. Flowering time? Sandy lowlands, NW, SW (S Namaqualand to Saldanha).

CARRUANTHUS   TIERBEKVYGIE   c. 2 spp., W Cape and Karoo

ringens (L.) Boom (= *Carruanthus caninus* (Haw.) Schwantes)   Short-stemmed branching perennial with fleshy roots. Leaves crowded, lanceolate, trigonous, finely toothed along margins and towards apex, to 60 × 18 mm. Flowers solitary or in dichotomous cymes on pedicels to 10 cm long, yellow, 40–50 mm diam. Fruits 5-locular. Aug.–Oct. Rock crevices, KM, SE (Vondeling to Willowmore).•

CARYOTOPHORA•   1 sp., W Cape

skiatophytoides Leistner   Weakly succulent perennial, with spreading branches. Leaves flat, fleshy, lanceolate-spathulate. Flowers terminal on pedicels 3–10 cm long, white, 40–60 mm diam. Fruits with nut-like, woody mericarps. Oct.–Nov. Sandy flats, AP (Agulhas Peninsula: Brandfontein).•

CEPHALOPHYLLUM   30 spp., S Namibia to W Cape

A. Leaves trigonous

curtophyllum (L.Bolus) Schwantes (= *Cephalophyllum rhodandrum* L.Bolus)   Compact or sprawling succulent. Leaves trigonous, pointed, dark green, 30–55 mm long. Flowers cream-coloured with purple filaments or purple with white filaments and yellow anthers. Fruits with narrow valve wings, covering membranes narrowly recurved. Aug. Open slopes or in karroid scrub, NW (Cedarberg Mts and W Karoo).

diversiphyllum (Haw.) H.E.K.Hartmann (= *Cephalophyllum bredasdorpense* L.Bolus, *C. caledonicum* L.Bolus, *C. vandermerwei* L.Bolus)   Like **C. subulatoides** but flowering stems annual. Sept. Coastal renosterveld or fynbos, SW, AP, LB (Bredasdorp to Mossel Bay).•

subulatoides (Haw.) N.E.Br. (= *Cephalophyllum diminutum* (Haw.) L.Bolus)   Compact succulent with short, thickened stems. Leaves trigonous, dark green. Flowers magenta, filamentous staminodes white. Fruits usually with rounded base, 12-locular, with persistent stalks. July–Aug. Humus-rich soils under shrubs, KM, LB (Little Karoo).•

A. Leaves terete

alstonii Marloth ex L.Bolus (= *Cephalophyllum franciscii* L.Bolus)   Clump-forming succulent. Leaves 60–120 mm long, whitish grey, erect. Flowers in many-flowered cymes, magenta. Fruits 18–20-locular, stalks persistent. June–Sept. Dry sandstone slopes, NW (W Karoo and Clanwilliam to Ceres).

loreum (L.) Schwantes (= *Cephalophyllum cedrimontanum* L.Bolus, *C. decipiens* (Haw.) L.Bolus)   Compact to sprawling succulent. Leaves terete. Flowers yellow with whitish stamens. Fruits 15–17-locular, to 16 mm diam. June–Aug. Sandstone slopes in loam above 600 m alt., NW (Gifberg to Swartruggens).•

parviflorum L.Bolus   Prostrate succulent? Leaves to 50 mm. Flowers less than 25 mm diam., yellow. Fruits unknown. July. Sandy coastal flats, SW (Saldanha to Milnerton).•

parvulum (Schltr.) H.E.K.Hartmann (= *Cheiridopsis parvula* (Schltr.) N.E.Br.   Compact succulent. Leaves rounded, short, 10–17 mm long. Flowers pink. Fruits unknown. Flowering time? Habitat?, NW (Clanwilliam).•

purpureo-album (Haw.) Schwantes (= *Cephalophyllum gracile* L.Bolus, *C. littlewoodii* L.Bolus, *C. middlemostii* L.Bolus, *C. worcesterense* L.Bolus)   Clump-forming succulent. Leaves terete, dark green. Flowers yellow with white or light yellow stamens. Fruits 13–17-locular, stalks slender, ephemeral. May–Sept. Gravel flats, SW, KM, LB (Little Karoo, Robertson Karoo).•

rostellum (L.Bolus) H.E.K.Hartmann (= *Vanzijlia rostellum* L.Bolus)   Prostrate perennial with trailing branches with long internodes. Leaf pairs adpressed like a beak, to 20 mm long, purple-tinged. Flowers magenta. Fruits small (7–9 mm diam.). June. Humus-rich soils near the sea, SW (Saldanha).•

tricolorum (Haw.) Schwantes (= *Cephalophyllum crassum* L.Bolus, *C. tenuifolium* L.Bolus)   Prostrate succulent. Leaves terete, usually dark green. Flowers yellow with orange-purple filaments and purple anthers. Fruits without persisting stalks, 15–17-locular. July–Sept. In low karroid bush, NW (S Namaqualand to Clanwilliam).

CEROCHLAMYS   pronkvingertjies   3 spp., W Cape

pachyphylla (L.Bolus) L.Bolus   Stemless perennial, tufted with age. Leaves trigonously clavate, smooth, to 40 mm long. Flowers 1–3, terminal, subsessile, pink, to 30 mm diam. Fruits 5-locular, closing bodies absent or obscure. May–July. Sandstone rocks, KM (Barrydale to Oudtshoorn and Laingsburg).•

trigona N.E.Br.   Compact perennial. Leaves trigonous, keeled, mostly as broad as thick. Flowers purple with white centre, 30–40 mm diam., staminodes in a central cone. Fruits 5-locular. Flowering time? Stony flats, LB, SE (Swellendam, Oudtshoorn).•

CHEIRIDOPSIS   23 spp., Namibia to W Cape

cigarettifera (A.Berger) N.E.Br. (= *Cheiridopsis duplessii* L.Bolus)   Like **C. rostrata** but leaves with serrated keels and fruits more than 10-locular, with narrow flag-shaped valve wings. July–Sept. Shale slopes and flats, 300–900 m, NW, SW, KM, LB (Namaqualand and W Karoo to Little Karoo).

rostrata (L.) N.E.Br. (= *Cheiridopsis carnea* N.E.Br., *C. inspersa* (N.E.Br.) N.E.Br., *C. purpurascens* (Salm-Dyck) N.E.Br., *C. velutina* L.Bolus)   Compact, densely branched succulent. Leaves weakly heterophyllous, drying to form a cylindrical papery sheath. Flowers yellow fading red. Fruits less than 10-locular, valve wings awn-shaped. July–Sept. Granite outcrops, NW, SW (Lambert's Bay to Yzerfontein).•

CIRCANDRA•   1 sp., W Cape

serrata (L.) N.E.Br. (= *Erepsia serrata* (L.) L.Bolus)   Sparsely branched shrublet to 60 cm. Leaves to 35 mm long, trigonous, acute, free at bases, margins toothed. Flowers solitary, yellow, to 50 mm diam., filamentous staminodes absent. Fruits 5-locular, c. 12 mm diam. Nov.–Jan. Habitat?, NW, SW (Tulbagh to Villiersdorp).•

CLERETUM (= *MICROPTERUM*)   3 spp., W Cape, Karoo, Namaqualand

herrei (Schwantes) Ihlenf. & Struck (= *Micropterum herrei* Schwantes)   Prostrate papillose annual. Leaves lyrate. Flowers small, sessile, white. Fruits with large valve wings. Flowering time? Habitat?, SW (Cape Peninsula to Kleinmond).•

papulosum (L.f.) L.Bolus (= *Micropterum papulosum* (L.f.) Schwantes)   Trailing papillose succulent. Leaves ligulate to spathulate. Flowers large (to 40 mm diam.) on a long pedicel or small on a short pedicel, yellow. Fruits with high rims. Sept.–Oct. Habitat?, NW, SW, KM, LB (Namaqualand to Mossel Bay).

CONICOSIA   gansies, snotwortel, varkslaai   2 spp., Namaqualand to E Cape

elongata (Haw.) N.E.Br.   Prostrate perennial with tuberous rootstock. Leaves cylindrical to half-cylindrical in section. Flowers white or yellow, styles free. Fruits xerochastic, shed at maturity. Sept. Habitat?, NW, SW (Richtersveld and Karoo to Touwsrivier).

pugioniformis (L.) N.E.Br. (= *Conicosia australis* L.Bolus)   Tufted perennial to 40 cm, with thick tap root. Leaves trigonous. Flowers yellow, styles free. Fruits xerochastic. Sept.–Nov. Sandy flats, mostly coastal, NW, SW, AP, LB, SE (Richtersveld to Port Elizabeth).

CONOPHYTUM   TOONTJIES, CONE PLANT   86 spp., S Namibia to W Cape

A. Flowers diurnal

albiflorum (Rawe) S.A.Hammer   Succulent perennial forming dense flat mats. Leaf bodies 8–15 mm long, lobes acutely keeled, glabrous, spotted and streaked, sheath whitish or reddish brown, persistent. Flowers diurnal, white. Fruits 4- or 5-locular. Mar.–June. Depressions in granite outcrops, NW, SW (S Namaqualand to St Helena Bay).

bicarinatum L.Bolus   Like **C. luckhoffii** but leaf bodies much larger, 12–30 × 8–15 × 5–10 mm, and lighter green. Flowering time? Sandstone kloofs in shaded crevices or ledges, NW (Clanwilliam to Porterville).•

luckhoffii Lavis   Densely tufted perennial with very short internodes. Leaf bodies bilobed, keeled, spotted and streaked, mostly 8–15 × 4–8 × 2–5 mm, sheath papery, pale brown, spotted, persistent. Flowers diurnal, outer petals magenta, inner white. Fruits 4- or 5-locular. Mar.–June. Habitat?, NW (Bokkeveld Mts to Citrusdal).•

minusculum (N.E.Br.) N.E.Br.   Succulent perennial forming low mats. Leaf bodies 5–15 mm long, often depressed or slightly keeled, spotted and streaked, glabrous or papillate, sheath light brown or white, spotted and streaked, persistent. Flowers diurnal, tube yellow, petals magenta, sometimes white, filamentous staminodes white to gold. Fruits 4- or 5-locular. Mainly Mar.–June. Sandstone slopes and pavement, NW (Bokkeveld Mts to Gifberg).•

minutum (Haw.) N.E.Br.   Succulent perennial forming mats or cushions. Leaf bodies 15–20 mm long, obconic, truncate, sometimes concave, glabrous or minutely papillate, spotted, sheath papery, whitish or pale brownish, persistent and forming collars. Flowers diurnal, magenta to pink, sometimes white. Capsule 4- or 5-locular. Mar.–June. Quartzite, calcrete, shale or granite, NW (Namaqualand to Olifants River mouth).

rubrolineatum Rawe   Tufted perennial forming low mats. Leaf bodies 5–8 × 4–8 × 4–8 mm, cylindrical to obconic, top convex or depressed, papillate, greyish green or purplish brown, with radiate streaks, sheath white, persisting. Flowers diurnal, yellow. Fruit 4–6-locular. Mar.–May. In moss on sandstone, NW (Kobee Mts).•

swanepoelianum Rawe   Succulent perennial forming mats or domes, 10–30 mm diam. Leaf bodies flat to convex, keeled or concave, 2–7 mm diam., with raised spots or streaks, sheath whitish, spotted, persisting in fragments. Flowers diurnal, magenta. Fruits 4-locular. Mar. Sandstone pavement, NW (Bokkeveld Mts).•

turrigerum (N.E.Br.) N.E.Br.   Succulent perennial forming loose tufts. Leaf bodies 5–25 mm long, clavate and bilobed at 'turreted' or faceted apex, usually spotted and lined, sheath whitish, spotted, not persisting. Flowers diurnal, pale pink, lilac or white. Fruits (3)4- or 5-locular. Apr. Granite outcrops, SW (Darling to Paarl).•

A. Flowers nocturnal

comptonii N.E.Br.   Succulent perennial forming cushions to 20 mm. Leaf bodies 3–5 mm long, obconic, glabrous, markings radiate, sheath dark brown, persistent. Flowers nocturnal, yellowish pink to pale orange or brownish. Fruits 4- or 5-locular. Mar.–June. Sandstone pavement, NW (Bokkeveld Mts to Matsikamma).•

ficiforme (Haw.) N.E.Br.   Succulent perennial forming compact mats or domes. Leaf bodies 20–30 mm long, fig-shaped, often 2-lobed, glabrous or slightly pubescent, spotted or streaked, sheath yellowish to white, wrinkled, marked with tannin spots, persistent. Flowers initially vespertine later diurnal, occasionally paired, pale yellow, pink or whitish. Fruits 4–6-locular. Mar.–June. Laminated, upturned shale or tillite, NW, SW (Worcester, Robertson, Montagu).•

joubertii Lavis   Succulent perennial forming tight domes to 30 mm. Leaf bodies 5–9 mm long, elongate-pyriform or top-shaped, sometimes slightly bilobed and keeled, glabrous, often shiny, faintly streaked, sheath pale tan, spotted, persisting as a collar. Flowers initially nocturnal, tube to 4 mm long, petals white to pale yellow. Fruits 4-locular. Mar.–June. Shale, quartzite, KM (Barrydale to Vanwyksdorp).•

minimum (Haw.) N.E.Br. (= *Conophytum pictum* (N.E.Br.) N.E.Br., *C. wittebergense* De Boer)   Succulent perennial forming loose mats or domes. Leaf bodies 8–15 mm long, elliptic in outline, glabrous to finely papillate, finely spotted and streaked, sheath white, often striate, persistent. Flowers nocturnal, white, pale yellow or pale pink. Fruits 4- or 5-locular. May–June. Shale or sandstone rocks, KM (Witteberg and Little Karoo).•

obcordellum (Haw.) N.E.Br. (= *Conophytum parviflorum* N.E.Br., *C. spectabile* Lavis) Succulent perennial, forming dense mats or cushions, sometimes single or double. Leaf bodies 8–20 mm diam., round, elliptic, reniform or roughly hexagonal in outline, spotted and streaked, sheath papery, whitish to rusty brown, densely spotted, persistent. Flowers nocturnal, white to yellowish or pink. Fruits 4–6-locular. Mar.–June. Sandstone, NW (Namaqualand and W Karoo to Ceres).

piluliforme (N.E.Br.) N.E.Br. (= *Conophytum archeri* Lavis, *C. subconfusum* Tischer) Succulent perennial forming mats or domes. Leaf bodies pill-shaped, 4–8 mm long, finely spotted and streaked, sheath white or brownish, often spotted, semipersistent. Flowers nocturnal, maroon, pink, salmon or purplish. Fruits 4-locular. May. Exposed shale bands, with or without quartz, KM, LB (Little Karoo).•

truncatum (Thunb.) N.E.Br. (= *Conophytum muirii* N.E.Br., *C. multipunctatum* Tischer, *C. pisinnum* (N.E.Br.) N.E.Br.) Succulent perennial forming domes to 15 cm diam. Leaf bodies 6–30 mm long, obconic to cylindric, truncate, marked with tannin spots, sheath whitish, spotted, papery, persistent. Flowers nocturnal, tube white or yellow, petals white, grey or amber. Fruits 4–7-locular. Mar.–Apr. In between stones, in crevices or under bushes, NW, LB, SE (Robertson to Steytlerville).

uviforme (Haw.) N.E.Br. Succulent perennial forming mats or domes. Leaf bodies 8–25 mm long, subglobose, sometimes flattened, glabrous or slightly papillate, often streaked or lined, sheath brownish grey, persistent. Flowers nocturnal, white to pink or yellow. Fruits 5- or 6-locular. Mar. Quartzite patches on shale, granite or calcrete, NW (Namaqualand to W Karoo and Bokkeveld Mts).

CORPUSCULARIA   c. 8 spp., Uitenhage to E Cape

lehmannii (Eckl. & Zeyh.) Schwantes (= *Delosperma lehmannii* (Eckl. & Zeyh.) Schwantes) Compact, densely leafy perennial with some long shoots and later with numerous short shoots. Leaves trigonous, hard to the touch, glaucous. Flowers orange or yellow. Fruits 6(–10)-locular. Nov.–Apr. Karroid slopes, SE (Port Elizabeth to Uitenhage).•

CYLINDROPHYLLUM   c. 5 spp., W Cape and Karoo

comptonii L.Bolus Cushion-like, branched perennial 13 × 25 cm. Leaves erect to spreading, subcylindrical, acuminate or acute, to 90 × 10 mm. Flowers solitary, terminal, silvery white, 75 mm diam. Fruits 5–8-locular, with small closing bodies. Sept.–Dec. Karroid veld, KM (Witteberg and Swartberg Mts to Barrydale).•

DEILANTHE   3 spp., Little and Great Karoo to Free State

peersii (L.Bolus) N.E.Br. Perennial with fleshy, turnip-like roots. Leaves thick, triangular, smooth, punctate. Flowers nocturnal, solitary, yellow, to 2.5 cm diam. Fruits 8–12-locular, with small closing bodies. Sept. Shale gravel, silt or quartz pebbles, KM (Anysberg to Uniondale and Karoo).•

DELOSPERMA   skaapvygie   c. 158 spp., W Cape and Namibia to E Africa

asperulum (Salm-Dyck) L.Bolus (= *Drosanthemum bredai* L.Bolus) Shrublet to 50 cm, with stems brownish grey, papillate when young. Leaves subcylindrical to trigonous, spreading, apically recurved. Flowers solitary on pedicels to 10 mm long, pale pink. Fruits 4-locular. Aug.–Oct. Pioneer, NW, SW, LB (Namaqualand to Riversdale).

burtoniae L.Bolus Erect, densely branched shrublet to 15 cm, stems slightly papillate. Leaves acute, recurved, to 25 mm long, narrow. Flowers yellow or coppery red, to 30 mm diam. Fruits 5-locular. Jan. Habitat?, NW (?Hex River Mts).•

calitzdorpense L.Bolus Densely branched shrublet to 10 cm, with grey stems. Leaves spreading, keeled. Flowers 1–3, white. Fruits unknown. Dec. Habitat?, KM (Calitzdorp).•

ecklonis (Salm-Dyck) Schwantes Diffuse shrublet with slender branches. Leaves trigonous, recurved, papillate, to 25 mm long, narrow. Flowers in cymes, white. Fruits 5-locular. July–Oct. Habitat?, SE (Humansdorp to Uitenhage).•

esterhuyseniae L.Bolus Compact to cushion-like shrublet, with stems covered by old leaves. Leaves subclavate, erect, apex rounded. Flowers solitary, white, 35 mm diam. Fruits with high rims on valves. Nov.–Mar. Rock crevices and cliffs, SE (Uniondale and Langkloof).•

expersum (N.E.Br.) L.Bolus   Slender shrublet to 30 cm, stems finely papillate when young. Leaves terete, obtuse, to 9 mm long, narrow. Flowers solitary, pink to red. Fruits unknown. Nov.–Jan. Rock ledges at high alt., NW, LB (Calvinia to Worcester).

fredericii Lavis   Shrublet to 25 cm. Leaves trigonous, acute, 10–15 mm long, narrow. Flowers in many-flowered cymes on pedicels 6–13 mm long, pale brick fading darker. Fruits unknown. Feb.–Apr. Habitat?, SE (Uitenhage).•

gratiae L.Bolus (= *Delosperma longii* L.Bolus)   Erect shrublet to 60 cm. Leaves suberect, S-shaped. Flowers in lax cymes on pedicels 5–10 mm long, magenta. Fruits with valves often gaping. Dec.–June. Habitat?, SE (Port Elizabeth).•

guthriei Lavis   Sprawling succulent perennial. Leaves ascending, triangular, apiculate. Flowers solitary on pedicels to 20 mm long, white, to 26 mm diam., filamentous staminodes recurved. Fruits unknown. Feb.–Mar. Coastal sand or rocks, SW (Caledon).•

hollandii L.Bolus   Creeping succulent perennial, with 2-angled stems to 30 cm. Leaves suberect, apiculate, acute. Flowers in threes on pedicels to 15 mm long, white, to 25 mm diam. Fruits 5-locular, 5–6 mm diam. Feb.–June. Habitat?, SE (Uitenhage to Port Elizabeth).•

inaequale L.Bolus   Erect shrublet to 20 cm. Leaves suberect. Flowers several on pedicels to 20 mm long, magenta, c. 40 mm diam. Fruits ?-locular, 6 mm diam., rims erect. Sept. Habitat?, SW (Bredasdorp).•

inconspicuum L.Bolus   Prostrate perennial with branches to 40 cm from a woody caudex. Leaves suberect, acuminate, recurved. Flowers in many-flowered cymes, white, small. Fruits 5-locular. Feb.–Apr. Habitat?, LB, SE (Swellendam to Knysna).•

litorale (Kensit) L.Bolus   KALKKLIPVYGIE   Decumbent shrublet to 15 cm, with trailing stems to 35 cm long. Leaves spreading, subfalcate with white margins. Flowers in threes on pedicels to 20 mm long, white. Fruits unknown. Apr. Coastal limestone flats, NW, SW, AP, SE (Lambert's Bay to Humansdorp).•

longipes L.Bolus (= *Delosperma grandiflorum* L.Bolus)   Erect, deciduous shrublet to 30 cm, with tuberous rootstock, branches reddish brown and papillate. Leaves deciduous, flat, papillate. Flowers solitary, magenta to reddish golden, to 80 mm diam. Fruits unknown. Aug.–Oct. Stony slopes, NW (Botterkloof to N Cedarberg Mts).•

macrostigma L.Bolus   Tufted shrub with slender branches. Leaves spreading, subterete, papillate. Flowers 1–3 in groups, white or pale pink, to 20 mm diam. Fruits ?-locular, c. 7 mm diam., with rims of valves elevated. Jan.–Apr. Habitat?, LB (Swellendam).•

mariae L.Bolus   Cushion-like shrublet to 30 cm. Leaves ascending, subterete, acute. Flowers in cymes, purplish, to 10 mm diam. Fruits ?-locular, with narrow rims from septa. Jan.–Feb. Limestone flats, AP (Agulhas Peninsula).•

multiflorum L.Bolus Prostrate to ascending shrublet, with finely papillate branches to 25 cm long. Leaves spreading, subterete to terete. Flowers in many-flowered cymes, pink, to 12 mm diam. Fruits unknown. Mar.–July. Habitat?, SE (George to Hankey).•

neethlingiae (L.Bolus) Schwantes   Slender, tuberous-rooted shrublet to 15 cm, with papillate stems. Leaves subterete, apically recurved, with apical papillae resembling a diadem. Flowers 1–3 in groups, magenta, to 13 mm diam. Fruits 5-locular. Apr.–Sept. Habitat?, SW, LB (Caledon to Mossel Bay).•

pageanum (L.Bolus) Schwantes   Erect shrublet to 30 cm. Leaves spreading, papillate. Flowers 1–3 in groups on papillate pedicels, purple, to 16 mm diam. Fruits c. 4 mm diam., valve wings broad, septa without rims. May–June. Karroid hills, NW, KM (Worcester, Montagu).•

papillatum (L.Bolus) L.Bolus   Sprawling deciduous shrublet to 20 cm, stems brown with white papillae. Leaves spreading, subterete, obtuse, long and narrow. Flowers solitary, white, to 60 mm diam. Fruits 5-locular. Nov.–Feb. Habitat?, NW (Worcester).•

parviflorum L.Bolus   Tufted shrublet to 6 cm. Leaves spreading, to 14 mm long. Flowers solitary, subsessile, bright yellow to golden, to 8 mm diam. Fruits ?-locular, c. 5 mm diam. Jan.–Mar. Habitat?, SE (Uitenhage).•

patersoniae (L.Bolus) L.Bolus   Dwarf shrublet with compressed stems. Leaves trigonous, recurved, margins ciliate. Flowers 1–3 in groups on pedicels to 15 mm long, white. Fruits unknown. Nov.–Mar. Habitat?, SE (Port Elizabeth).•

pubipetalum L.Bolus   Erect, lax deciduous shrublet to 30 cm. Stems papillate. Leaves more or less erect, subterete, papillate. Flowers solitary, pink. Fruits ?-locular, c. 7 mm diam. Dec. Habitat?, KM (Montagu).•

saxicola Lavis   Tufted succulent to 4 cm. Leaves spreading, linear, acute. Flowers solitary on pedicels to 6 mm long, pink. Fruits unknown. Oct. Coastal cliffs, SE (Humansdorp).•

subincanum (Haw.) Schwantes   Erect to prostrate shrublet to 7 cm. Leaves finely papillate, trigonous, apically recurved. Flowers in a few-flowered inflorescence, white. Fruits 5-locular. Aug.–Dec. Habitat?, SE (Humansdorp).•

uitenhagense L.Bolus   Densely branched, prostrate to decumbent shrublet, stems to 30 cm long. Leaves spreading, subterete. Flowers in cymes on pedicels 2–5 mm long, white, to 20 mm diam. Fruits ?-locular, c. 5 mm diam. Mar.–Apr. Habitat?, SE (Uitenhage).•

uncinatum L.Bolus   Sprawling succulent perennial, with grey stems to 45 cm long. Flowers in cymes on pedicels to 15 mm long, white. Fruits 5-locular, c. 6 mm diam. Mar.–Apr. Habitat?, SE (Uitenhage).•

virens L.Bolus   Erect shrublet to 50 cm. Leaves falcate, bright green. Flowers solitary, magenta, to 35 mm diam. Fruits ?-locular, with narrow valve wings. Jan.–May. Limestone hills, LB (Mossel Bay).•

DICROCAULON   c. 9 spp., Namaqualand to Klawer

microstigma (L.Bolus) Ihlenf.   Robust cushion forming shrubs, 10 × 30 cm. Leaves heterophyllous, deciduous, papillate. Flowers solitary on pedicels to 40 mm long, magenta. Fruit mostly 6-locular. July. Shallow soil, mostly in quartz pebble patches, NW (S Namaqualand to Klawer).

DIPLOSOMA   EENDEVOETVYGIE   2 spp., S Namaqualand to W Cape

retroversum (Kensit) Schwantes (= *Diplosoma leipoldtii* L.Bolus)   Tufted, stemless, deciduous perennial to 3 cm, with fibrous roots. Leaves paired, to 25 mm long, fused basally or up to half way on one side, soft, pulpy, with some hyaline dots. Flowers solitary, purplish, c. 18 mm diam. Fruits 6- or 7-locular, without a tubercle. Sept. Quartzitic gravel over clay, NW (Eendekuil to Piekenierskloof).•

DISPHYMA   c. 4 spp., W Cape coast, Australia and New Zealand

crassifolium (L.) L.Bolus   Mat-forming perennial, rooting at nodes. Leaves 25–35 mm long, trigonous, shortly acuminate, smooth, dark green, with translucent dots. Flowers 1–3, terminal, white to rose-red, 40 mm diam. Fruits 5-locular, light yellow, sponge-like. July–Oct. Coastal rocks just above high-tide mark, NW, SW, AP, SE (Lambert's Bay to Port Elizabeth, S Australia).

dunsdonii L.Bolus   Like **D. crassifolium** but leaves 5–9 mm long and flowers 20–25 mm diam. Sept.–Oct. Salt pans, SW (Bredasdorp).•

DOROTHEANTHUS   7 spp., Namaqualand to W Cape

apetalus (L.f.) N.E.Br. (= *Dorotheanthus gramineus* (Haw.) Schwantes)   Small tufted annual. Leaves linear to ligulate. Flowers subsessile, white, petals shorter than sepals. Fruits with small, half-circular closing bulge. July–Sept. Coastal flats, SW, AP (Yzerfontein to Cape Agulhas).•

bellidiformis (Burm.f.) N.E.Br. (= *Dorotheanthus acuminatus* L.Bolus, *D. bidouwensis* L.Bolus, *D. flos-solis* (A.Berger) L.Bolus, *D. hallii* L.Bolus, *D. littlewoodii* L.Bolus, *D. martinii* L.Bolus, *D. muirii* N.E.Br., *D. oculatus* N.E.Br.)   BOKBAAIVYGIE   Tufted annual. Leaves ligulate to spathulate. Flowers on pedicels to 25 mm long, red, yellow, salmon or white, 20–30 mm diam., petals longer than sepals. Fruits with large or small closing bulge. Aug.–Sept. Mostly on sandy flats, NW, SW, AP, KM, LB (Namaqualand to Stilbaai).

clavatus (Haw.) Struck (= *Dorotheanthus gramineus* auct.)   Tufted annual. Leaves linear. Flowers stalked, orange, white or pink, petals much longer than sepals. Fruits with small, half-circular closing bulge. Aug.–Sept. Saline places, SW (Hopefield to Darling).•

ululariis Brusse   Like **D. bellidiformis** but seed surface rough (vs smooth). July–Sept. Calcareous sands on limestone, AP (Gansbaai).•

DROSANTHEMUM   DOUVYGIE   110 spp., Namibia to E Cape

acuminatum L.Bolus   Slender, papillate shrublet to 20 cm. Leaves suberect, terete. Flowers 1–3 in groups, golden. Fruits 5-locular. Sept.–Oct. Sheltered sandstone slopes, KM (Montagu).

acutifolium (L.Bolus) L.Bolus   Tufted, papillate shrublet to 7 cm, with red stems. Leaves acute with dark apical bristle. Flowers in many-flowered cymes, magenta. Fruits 5-locular. Flowering time? Shale with quartz gravel in crevices, KM, LB (Montagu to Riversdale).•

albiflorum (L.Bolus) Schwantes   Slender, much-branched, papillate shrublet to 15 cm. Leaves terete, obtuse. Flowers solitary at branch tips, white. Fruits 5-locular. Oct. Habitat?, LB (Riversdale).•

ambiguum L.Bolus   Mat-forming, papillate shrublet, with woody stems to 50 cm. Leaves terete, obtuse. Flowers in cymes, magenta. Fruits 5-locular, false septum present. Nov.–Jan. Pioneer, often in disturbed sites, SW (Malmesbury to Bredasdorp).•

anomalum L.Bolus   Tufted, papillate shrublet to 2 cm, from a tuberous caudex. Leaves subterete. Flowers in cymes, golden yellow. Fruits 5-locular. Aug.–Oct. Habitat?, KM (Montagu).•

attenuatum (Haw.) Schwantes   Like **D. striatum** but prostrate with slender stems, petals white with red stripes. Nov. Habitat?, SW, LB (Caledon to Riversdale).•

aureopurpureum L.Bolus   Slender, papillate shrublet to 20 cm, stems red with white papillae. Leaves trigonous. Flowers in cymes, golden yellow with purple reverse. Fruits 5-locular. Aug.–Oct. Habitat?, LB (Swellendam).•

austricola L.Bolus   Laxly branched, decumbent, papillate shrublet, stems to 15 cm. Leaves terete, obtuse. Flowers solitary, magenta. Fruits 5-locular. Sept. Limestone flats, AP (Agulhas Peninsula).•

autumnale L.Bolus   Densely branched, papillate shrublet to 45 cm, with red stems. Leaves terete, obtuse. Flowers 3 or 4 in groups, purple. Fruits 5-locular, deciduous, pedicel spinescent. Mar.–Apr. Coastal, LB (Mossel Bay).•

barkerae L.Bolus   Erect, papillate shrublet to 60 cm. Leaves terete, obtuse. Flowers solitary on short shoots, 22 mm diam., purple. Fruits unknown. Flowering time? Habitat?, KM (Oudtshoorn).•

barwickii L.Bolus   Erect shrub to 70 cm, stems papillate when young, blackish with age, with short spines between bases of leaf pairs. Leaves trigonous, falcate, papillate. Flowers solitary, magenta. Fruits 6-locular, top with low rims. Sept. Lower sandstone slopes, LB (Langeberg Mts: Swellendam).•

bellum L.Bolus   Laxly branched, papillate shrublet to 30 cm, with red stems. Leaves subterete, apically recurved. Flowers solitary, to 50 mm diam., pink, cream-coloured or white. Fruits 5-locular. Sept.–Dec. Habitat?, NW (Ceres to Worcester).•

bicolor L.Bolus   Erect, rigid shrub to 1 m, stems with small papillae when young. Leaves subterete, erect, with small papillae. Flowers 1–3, golden yellow with maroon tips. Fruits 5-locular, with very high rims. Oct.–Dec. Stony sandstone slopes, KM (Little Karoo).•

calycinum (Haw.) Schwantes   Sprawling, papillate shrublet to 50 cm. Leaves cylindrical, obtuse, with shiny papillae. Flowers in cymes, white, filamentous staminodes in a cone, later recurving. Fruits 5-locular, false septum present. Sept.–Nov. Stony slopes, NW, SW (Clanwilliam to Ceres and Malmesbury).•

candens (Haw.) Schwantes   Mat-forming, papillate shrublet. Leaves cylindrical, incurved, obtuse, grey, glittering. Flowers in cymes, pale pink or white. Fruits unknown. Oct.–Jan. Coastal rocks, SW, AP (Cape Peninsula to Bredasdorp).•

cereale L.Bolus   Erect, papillate shrublet to 15 cm, with slender, red stems. Leaves more or less trigonous, tips recurved. Flowers solitary, yellow, margins red at tips. Fruits 5-locular, with high rims. Dec.–Feb. Habitat?, NW, SW (Ceres, Caledon).•

collinum (Sond.) Schwantes   Erect shrub to 30 cm, with slender, rough branches. Leaves cylindrical to trigonous, erect, blunt, papillate. Flowers 12 mm diam., yellow. Fruits unknown. Dec. Habitat?, NW (Worcester).•

crassum L.Bolus   Stiff shrublet to 25 cm, with yellow-brown stems papillate when young, becoming twisted with age. Leaves ovate, papillate. Flowers solitary. Fruits 5-locular, with low rims, deciduous, pedicels bluntly spinescent. Dec. Karroid scrub in loamy sand, KM (Little Karoo).•

croceum L.Bolus   Low, intricately branched shrublet to 15 cm, with rough stems. Leaves subcylindrical, tapering, papillate. Flowers 25 mm diam., saffron. Fruits 5-locular. Oct.–Nov. Cliffs, LB (Swellendam).•

cymiferum L.Bolus   Erect shrublet to 10 cm, stems spreading at right angles, papillate when young. Leaves terete, obtuse, papillate. Flowers 3–5 in cymes, lateral flowers shed leaving a spiny pedicel. Fruits 5-locular, only on central pedicel of cyme, false septum present, deciduous, bundles persisting as stars. Flowering time? Dwyka tillite, NW (W Karoo to Ceres).

delicatulum (L.Bolus) Schwantes   Stiffly branched shrublet to 10 cm, with red, sparsely papillate stems. Leaves subglobose, papillate. Flowers solitary, subsessile, to 10 mm diam., pale pink. Fruits 5-locular, with low rims. Oct.–Nov. Shale slopes, KM, LB (Montagu, Swellendam).•

duplessiae L.Bolus   Erect, papillate shrublet to 20 cm, with ochre to reddish stems. Leaves terete, obtuse. Flowers pink. Fruits 5-locular, with low rims, without false septum. Jan. Habitat?, KM (Oudtshoorn).•

edwardsiae L.Bolus   Lax shrublet to 15 cm, with sparsely papillate stems. Leaves subterete, acuminate, densely papillate. Flowers 1–3 in cymes, orange. Fruits 5-locular, with high rims. Aug.–Nov. Stony slopes, SE (George).•

flammeum L.Bolus   Erect, papillate shrublet to 25 cm. Leaves trigonous. Flowers 1–3 in cymes, orange. Fruits 5-locular, without false septum. Oct. Habitat?, NW (Robertson).•

flavum (Haw.) Schwantes   Low shrublet to 15 cm, with slender, roughly scabrous stems. Leaves subterete to cylindrical, papillate, with purple tips. Flowers solitary, few petals, yellow. Fruits 5-locular. Sept.–Oct. Habitat?, SW (Caledon, Bredasdorp).•

floribundum (Haw.) Schwantes   Sprawling, papillate shrublet to 30 cm. Leaves subterete, shiny. Flowers solitary on short shoots, purple with white centre. Fruits 5-locular. Sept.–Dec. Habitat?, NW, SW, KM, LB, SE (Namaqualand to E Cape).

fourcadei (L.Bolus) Schwantes   Erect shrublet to 25 cm, with red, scarcely papillate branches Leaves subcymbiform, densely papillate. Flowers solitary, to 15 mm diam., reddish. Fruits 6-locular. Oct.–Dec. Stony slopes, SE (Hankey).•

framesii L.Bolus   Cushion-like shrublet to 10 cm, with yellow, hispid branches. Leaves terete, obtuse, densely papillate. Flowers solitary on erect short shoots, to 30 mm diam., cream-coloured. Fruits 5-locular, false septum present. Flowering time? Flats on Dwyka tillite, NW (Ceres).•

giffenii (L.Bolus) Schwantes   Low sprawling shrublet, with red, slightly papillate branches. Leaves trigonous, densely papillate. Flowers solitary, pink. Fruits 5-locular, false septum absent. Oct.–Nov. Brackish places, KM, LB (Little Karoo).•

glabrescens L.Bolus   Rigid erect shrublet to 20 cm, with brown, scarcely papillate branches. Leaves subterete, obtuse, densely papillate. Flowers 20–30 mm diam., magenta. Fruits 5-locular, with high rims, false septum present. Sept. Loam, NW (W Karoo to Bokkeveld Mts).

globosum L.Bolus   Erect, much-branched shrublet to 45 cm, with brown, minutely papillate branches. Leaves globose, densely papillate. Flowers to 17 mm diam., magenta. Fruits 5-locular, almost star-shaped. June–July. Habitat?, KM (Montagu).•

gracillimum L.Bolus   Sprawling shrublet to 20 cm, with slender, red, scarcely papillate branches. Leaves terete to subterete, obtuse, sparsely papillate. Flowers to 15 mm diam., magenta. Fruits 5-locular, with high rims, valve wings rather narrow. May–Aug. Mountain slopes, in grass or climbing in bushes, SE (Uniondale).•

hallii L.Bolus   Papillate shrublet to 25 cm. Leaves terete, densely papillate. Flowers solitary, yellow, filamentous staminodes black. Fruits 5-locular. Sept.–Oct. Stony slopes, NW (Worcester).•

hispidum (L.) Schwantes   Erect or spreading shrublet to 60 cm, with red, mostly hispid branches. Leaves terete, obtuse, bending downwards, densely papillate. Flowers solitary, magenta. Fruits 5-locular. Sept.–Jan. Pioneer, NW, SW, AP, KM, LB, SE (dry parts of southern Africa).

hispifolium (Haw.) Schwantes   Low shrublet with hispid branches. Leaves hispid. Flowers 20–30 mm diam., white with purple midvein. Fruits with high, undulate rims, endocarpal closing bodies present, without false septa. Flowering time? Slopes after fire, SW (Malmesbury).•

insolitum L.Bolus   Slender shrublet to 15 cm, with hispid branches. Leaves apically recurved, slightly papillate. Flowers yellow, filamentous staminodes black. Fruits unknown. Nov. Habitat?, SW (Caledon).•

intermedium (L.Bolus) L.Bolus   Sprawling shrublet with slender, hispid branches. Leaves subterete, densely papillate. Flowers solitary on erect short shoots, to 25 mm diam., magenta. Fruits 5-locular, endocarpal closing bodies, false septa occasionally present. Aug.–Sept. Coastal rocks, SW, AP, LB (Simonstown to Mossel Bay).•

lavisii L.Bolus   Erect, densely branched, papillate shrublet, with red branches. Leaves S-shaped, terete, pointed. Flowers solitary on long shoots, reddish, coppery on reverse. Fruits 5-locular. Sept. Shale outcrops, SW (Greyton to Bredasdorp).•

laxum L.Bolus   Low shrublet to 10 cm, with red, sparsely papillate branches. Leaves terete, obtuse densely papillate. Flowers in spreading cymes, magenta. Fruits 5-locular. Feb. Habitat?, NW (Robertson).•

leipoldtii L.Bolus   Stiffly erect shrublet to 30 cm, with hispid branches. Leaves terete, densely papillate. Flowers to 17 mm diam., magenta. Fruits 5-locular, with low rims. Oct. Habitat?, NW (Clanwilliam to Piketberg).•

leptum L.Bolus   Erect shrublet to 35 cm, branches with scattered, flat papillae. Leaves trigonous, slightly recurved, densely papillate. Flowers solitary, white. Fruits 5-locular. Oct. Rocky ravines, NW (Robertson).•

lignosum L.Bolus   Tufted shrublet to 15 cm, with yellow branches with scattered, large papillae. Leaves terete, obtuse, densely papillate. Flowers magenta. Fruits with low rims. Oct. Habitat?, NW (Ceres).•

lique (N.E.Br.) Schwantes   Erect shrublet to 45 cm, with rough branches. Leaves spreading widely, cylindrical, obtuse, papillate, to c. 11 mm long. Flowers stalked, solitary, c. 22 mm diam., purple, filamentous staminodes in a cone around stamens. Fruits 5-locular. Oct.–Feb. Loamy soils, often a pioneer, SE (Humansdorp to Uitenhage).•

macrocalyx L.Bolus   Robust shrublet to 30 cm, with red, rough branches with short spines at bases of leaf pairs. Leaves subcymbiform, obtuse, papillate. Flowers to 35 mm diam., pink. Fruits with high valves and a central dip, false septa present, covering membranes reduced. Flowering time? Rocky sandstone slopes, KM (Ladismith, Montagu).•

marinum L.Bolus   Mat-forming shrublet, with yellowish, hispid branches. Leaves terete, obtuse, densely papillate. Flowers in cymes, pink or magenta. Fruits with low rims, false septa present. Aug.–Sept. Seashore, SW (Langebaan to Yzerfontein).•

micans (L.) Schwantes   Densely branched shrublet to 60 cm, with scabrid branches. Leaves subulate to subcylindrical, papillate. Flowers solitary, pedicels to 40 mm, petals red, filamentous staminodes black. Fruits 5-locular. Oct. Habitat?, NW, LB (Worcester, Swellendam).•

montaguense L.Bolus   Like **D. macrocalyx** but fruits smaller and less papillate, tops of capsules lower, with covering membranes broad, and false septa hardly developed. Oct.–Nov. Loamy soils with quartz pebbles, KM, LB (Montagu, Swellendam).•

muirii L.Bolus   Erect shrublet to 40 cm, with yellowish, hispid branches. Leaves terete, obtuse, densely papillate. Flowers pink or red. Fruits 5-locular. Oct. Rocky habitats, LB (Riversdale).•

opacum L.Bolus   Sprawling shrublet to 30 cm, with branches to 50 cm long, papillate when young. Leaves terete, obtuse, densely papillate. Flowers on short shoots, to 22 mm diam., pink. Fruits 5-locular. Oct. W-facing shale slopes, NW (Tulbagh).•

parvifolium (Haw.) Schwantes   Sprawling shrublet to 15 cm, with rough, slender, reddish branches to 80 cm long. Leaves trigonous, spreading, densely papillate, to 3 mm long. Flowers solitary or in small cymes on short branches, pedicels to 14 mm, magenta. Fruits 5-locular, false septa absent, bracteoles persisting as 2 spines below fruit. Oct.–Dec. Disturbed places, sandstone or quartzitic slopes, SW (Bredasdorp).•

pickhardii L.Bolus   Slender, papillate shrublet to 15 cm, with reddish branches bearing short spines at base of leaf pairs. Leaves terete, obtuse, densely papillate. Flowers solitary, yellow-orange. Fruits 5-locular. Oct. Habitat?, KM (Montagu).•

praecultum (N.E.Br.) Schwantes   Erect shrublet to 30 cm, branches papillate when young. Leaves cylindrical to subglobose, papillate. Flowers often solitary, on lateral shoots, petals 6–7 mm long. Fruits 5-locular. Flowering time? Habitat?, KM (Montagu).•

prostratum L.Bolus   Prostrate shrublet, with yellowish, hispid branches to 35 cm. Leaves terete, obtuse, densely papillate. Flowers to 30 mm diam., reddish. Fruits 5-locular, false septa present. Aug. Habitat?, NW (Clanwilliam).•

pulchellum L.Bolus   Prostrate mat-forming shrublet, branches hispid when young. Leaves subterete, obtuse, densely papillate. Flowers to 25 mm diam., reddish. Fruits 5-locular, false septa present. July–Sept. Habitat?, NW (Clanwilliam).•

pulchrum L.Bolus   Lax, papillate shrublet to 25 cm. Leaves trigonous, acuminate. Flowers with pedicels to 50 mm, reddish gold with a yellow and greenish centre, filamentous staminodes black. Fruits 5-locular, false septum absent, rims very low. Oct. Habitat?, NW (Worcester).•

salicola L.Bolus   Compact, densely branched, papillate shrublet to 25 cm. Leaves globose to subglobose. Flowers solitary, pale pink. Fruits 5-locular. Nov. Saline flats, NW (Piketberg).•

semiglobosum L.Bolus   Slender, papillate shrublet to 40 cm, with reddish branches. Leaves subterete, recurved apically. Flowers solitary, golden with red tips, stigmas red. Fruits 5-locular. Nov. Habitat?, NW (Worcester).•

speciosum (Haw.) Schwantes   Erect, papillate shrublet to 60 cm. Leaves subcylindrical, obtuse, glistening when young. Flowers red to orange. Fruits 5-locular. May–Oct. Shale slopes, NW, KM, LB (Worcester to Barrydale).•

splendens L.Bolus   Erect papillate shrublet to 50 cm, with reddish branches. Leaves subterete. Flowers golden with maroon tips, filamentous staminodes black. Fruits unknown. Sept.–Dec. Habitat?, KM (Montagu).•

stokoei L.Bolus   Mat-forming, papillate shrublet with beige branches. Leaves crowded, terete, obtuse. Flowers to 17 mm diam., pink, filamentous staminodes in a cone, later spreading. Fruits 5-locular. Aug.–Jan. Coastal rocks, SW (Simonstown to Kleinmond).•

striatum (Haw.) Schwantes   PORSELEINBOS   Like **D. hispifolium** but leaves covered with scattered, round papillae. Aug.–Sept. Loamy soils exposed to occasional fires, NW, SW, KM (Tulbagh to Montagu).•

strictifolium L.Bolus   Erect, papillate shrublet to 30 cm, with reddish branches. Leaves subterete, pointed. Flowers in small cymes, to 50 mm diam., golden yellow with red tips, pedicels 30–60 mm. Fruits unknown. Oct. Habitat?, LB (Riversdale).•

subcompressum (Haw.) Schwantes (= *Drosanthemum roridum* L.Bolus)   Densely branched shrublet to 30 cm, with hispid branches. Leaves trigonous. Flowers solitary, magenta. Fruits 6-locular. Sept. Habitat?, KM (Montagu).•

thudichumii L.Bolus   Woody shrub to 1 m, branches papillate on young parts. Leaves subterete, pointed, densely papillate. Flowers 1–3, white, orange or yellow. Fruits 5-locular, shed when mature leaving a blunt spine, false septa absent. Oct.–Nov. Habitat?, NW (Worcester).•

tuberculiferum L.Bolus   Mat-forming, papillate shrublet with branches to 15 cm long. Leaves terete. Flowers solitary on short shoots, to 25 mm diam., magenta. Fruits 5-locular, base flat, rims low, false septa present. Sept.–Oct. Open karroid veld on gravelly loam, NW (Worcester).•

vandermerwei L.Bolus   Prostrate, papillate shrublet, with reddish branches to 40 cm long. Leaves subterete to terete, obtuse. Flowers to 27 mm diam., pink. Fruits 5-locular. Sept.–Oct. Habitat?, LB (Swellendam).•

vespertinum L.Bolus   Laxly branched, papillate shrublet to 30 cm. Leaves subglobose to subcylindrical. Flowers to 14 mm diam., white or pale pink, opening in late afternoon. Fruits 5-locular, false septa only partly developed. Oct.–Apr. Habitat?, NW, KM (Ceres, Prince Albert).•

wittebergense L.Bolus   Erect, papillate shrublet to 25 cm. Leaves subterete to subulate. Flowers 1 or 2, magenta or white. Fruits 5-locular. Nov. Habitat?, KM (Witteberg).•

worcesterense L.Bolus   Densely branched, erect shrublet to 30 cm, with yellow, papillate branches. Leaves terete, obtuse, densely papillate. Flowers in cymes, pale pink. Fruits 5-locular. Nov. Habitat?, NW (Worcester).•

zygophylloides (L.Bolus) L.Bolus   Sprawling shrublet to 14 cm, branches red, with small papillae. Leaves subterete, obtuse, more or less smooth, very soft. Flowers in many-flowered cymes, mustard-yellow. Fruits 5-locular. Sept.–Oct. Habitat?, NW (Clanwilliam).•

[**Species excluded**: Insufficiently known and possibly conspecific with one of the above or outside our area: **D. maculatum** (Haw.) Schwantes]

EREPSIA• (= *KENSITIA, SEMNANTHE*)   ALTYDVYGIE   27 spp., W Cape

A. Leaves broader, usually 5–10 mm diam, walls of hypanthium sloped (except **E. inclaudens***)*

aristata (L.Bolus) Liede & H.E.K.Hartmann (= *Ruschia aristata* L.Bolus)   Robust stiff erect shrublet to 60 cm. Leaves erect, laterally compressed, sharply keeled, glaucous, covered with crowded dots. Flowers red or yellow, to 30 mm diam. Sept.–Nov. Conglomerate outcrops, SE (Uitenhage).•

babiloniae Liede   Dwarf shrublet to 20 cm, branched from base. Leaves 3–5mm diam. Flowers? Fruits 6–9-locular, 10–15 mm diam., base funnel-shaped. Flowering time? Rocky sandstone slopes, after fire, 700–1000 m, SW (Babilonstoring).•

dunensis (Sond.) Klak (= *Lampranthus dunensis* (Sond.) L.Bolus)   Compact sprawling shrublet, branches decumbent or creeping. Leaves laterally compressed, sharply keeled, 25–45 × 3–5 mm. Flowers on erect branchlets 3–6 cm high, white, to 35 mm 6diam. Sept.–Oct. Sand dunes on limestone, SW (Cape Peninsula and Cape Flats).•

forficata (L.) Schwantes (= *Erepsia mutabilis* (Haw.) Schwantes, *Ruschia filamentosa* (L.) L.Bolus, *Ruschia forficata* (L.) L.Bolus)   Shrublet to 30 cm, branching from base, branches erect, becoming decumbent. Leaves 5–8 mm diam. Flowers mostly solitary, whitish to pink, petals very numerous, reproductive parts partly hidden. Fruits 9–11 mm diam. May–June. Rock crevices, 300–900 m, SW (Cape Peninsula).•

heteropetala (Haw.) Schwantes   Sprawling shrublet to 20 cm, stems to 4.5 mm diam. Leaves more than 5 mm diam. Flowers mostly solitary, whitish to pink, petals very numerous, reproductive parts partly hidden. Fruits 10–13 mm diam. Oct.–Feb. Rock ledges at 400–1300 m, SW (Du Toitskloof and Jonkershoek Mts).•

inclaudens (Haw.) Schwantes   Densely branched shrublet to 40 cm. Leaves with red tips, 6–8 mm diam. Flowers in cymes, deep mauve, petals spathulate, reproductive parts hidden. Fruits 5-locular, base funnel-shaped, 8–10 mm diam. Oct.–Dec. Rock crevices at 300–1300 m, SW (Du Toitskloof to Caledon).•

lacera (Haw.) Liede (= *Semnanthe lacera* (Haw.) N.E.Br.)   Erect, sparsely branched shrub to 80 cm, branches to 10 mm diam. Leaves more than 5 mm diam., toothed. Flowers in dichasia, magenta, petals very numerous, reproductive parts partly hidden. Fruits 9–11-locular, 13–15 mm diam. Oct.–Nov. Granitic slopes at 300–700 m, SW (Malmesbury to Stellenbosch).•

pillansii (Kensit) Liede (= *Kensitia pillansii* (Kensit) Fedde)   Erect, sparsely branched shrublet to 80 cm, with branches 4–8 mm diam. Leaves falcate, more than 5 mm diam. Flowers in dichasia, magenta, filamentous staminodes white, 30–60 mm diam., reproductive parts hidden. Fruits (9)10(–13)-locular, c. 13 mm diam,. Aug.–Sept. Sandstone slopes, 500–1000 m, NW (Piketberg).•

steytlerae L.Bolus   Shrublet to 20 cm, branches to 4 mm diam. Leaves 4–7 mm diam. Flowers mostly solitary, magenta, filamentous staminodes yellow, petals numerous, reproductive parts hidden. Fruits c. 10 mm diam. Jan.–Feb. In rocks along the coast, SW (Kleinmond).•

A. Leaves narrower, usually less than 5 mm diam, walls of hypanthium erect

anceps (Haw.) Schwantes   Slender, erect shrublet to 30 cm, sparsely branched from base. Leaves less than 5 mm diam. Flowers in monochasia, pink or magenta, remaining open, petals spathulate, filamentous staminodes papillate, bright yellow, reproductive parts hidden. Fruits 4–8 mm diam. Dec.–Mar., mainly after fire. Sandstone slopes to 1000 m, NW, SW (Tulbagh to Cape Peninsula and Caledon).•

aperta L.Bolus   Mat-forming shrublet to 15 cm, branching from base. Leaves less than 5 mm diam. Flowers 1–few at branch tips, pink, petals spathulate, filamentous staminodes pale yellow, reproductive parts only partly hidden. Fruits with reduced covering membranes. Dec.–Jan., after fire. Rock ledges at high alt., NW (Grootwinterhoek and Hex River Mts).•

aspera (Haw.) L.Bolus (= *Erepsia tuberculata* N.E.Br.)   Shrublet to 40 cm, branching from base, stems slender, 0.7–1.5 mm diam. Leaves less than 5 mm diam. Flowers in monochasia, magenta, remaining open, filamentous staminodes whitish or yellow, papillate at least at base, reproductive parts hidden. Fruits 5–6 mm diam. Dec.–Mar., often after fires. Rocky slopes to 2000 m, NW, SW, LB (Cedarberg Mts to Riversdale).•

bracteata (Aiton) Schwantes (= *Erepsia caledonica* L.Bolus)   Erect shrublet to 50 cm, sparsely branched from base, stems 2–4 mm diam. Leaves to 5 mm diam. Flowers in monochasia, magenta, remaining open, filamentous staminodes pale yellow, papillate at least at base, reproductive parts hidden. Fruits 6–8 mm diam. Jan.–Apr. Rocky slopes to 500 m, NW, SW (Cedarberg to Langeberg Mts).•

brevipetala L.Bolus   Robust shrublet to 20 cm, densely branched from base. Leaves less than 5 mm diam. Flowers in reduced dichasia, magenta, remaining open, filamentous staminodes whitish, reproductive parts hidden. Fruits 6–8 mm diam. Nov.–Feb. Open strandveld, salt-tolerant, . SW (Malmesbury).•

distans L.Bolus   Erect shrublet to 70 cm, branching from base, stems to 2.5 mm diam. Leaves less than 5 mm diam. Flowers in monochasia, magenta, remaining open, petals linear, filamentous staminodes whitish, reproductive parts hidden. Fruits 4–6 mm diam. Nov.–Mar., after fire. At 700–1000 m, NW (Cedarberg Mts).•

dubia Liede   Erect shrublet to 40 cm, densely branched from base, stems slender, less than 1 mm diam. Leaves less than 3 mm diam. Flowers in many-flowered cymes on short pedicels, pink, filamentous staminodes whitish, reproductive parts only partly6 hidden. Fruits 6–7 mm diam. Flowering time? Pioneer, at 100 m, LB (Swellendam).•

esterhuyseniae L.Bolus   Mat-forming shrublet, branched from base, to 15 cm. Leaves less than 5 mm diam. Flowers 1–few, magenta, petals spathulate, filamentous staminodes whitish, reproductive parts only partly hidden. Fruits 4–6 mm diam., with reduced covering membranes. Dec.–Feb., after fire. High sandstone slopes, SW (Bainskloof to Riviersonderend Mts).•

gracilis (Haw.) L.Bolus (= *Erepsia carterae* L.Bolus)   Slender shrublet to 30 cm, branching from base, branches to 2.5 mm diam. Leaves less than 5 mm diam. Flowers in monochasia, magenta, remaining open, filamentous staminodes yellow, reproductive parts hidden. Fruits 4–6 mm diam. Feb.–Apr., after fires. Sandstone slopes at 300–2000 m, NW, SW (Cedarberg to Riviersonderend Mts).•

hallii L.Bolus   Erect, regularly branched shrublet to 30 cm, stems slender, 1.5–2.5 mm diam. Leaves less than 5 mm diam. Flowers in monochasia, sessile, small (less than 5 mm diam.), white, closing at night, filamentous staminodes yellow with red tips, reproductive parts hidden. Fruits 4–6 mm diam. Feb.–May. Granite hills, SW (Darling).•

insignis (Schltr.) Schwantes   Dwarf shrublet to 15 cm, branching from base, stems 1–2 mm diam. Leaves less than 5 mm diam. Flowers solitary, magenta, filamentous staminodes whitish, reproductive parts partly hidden. Fruits 6–7 mm diam., with reduced covering membranes. Oct. Nov. Rocky slopes at 1200–1700 m, SW (Bainskloof Mts).•

oxysepala (Schltr.) L.Bolus (= *Erepsia stokoei* L.Bolus)   Sparsely branched shrublet to 30 cm. Leaves 4–6 mm diam. Flowers in cymes, 15–25 mm diam., pink, filamentous staminodes white, reproductive parts hidden. Fruits 5-locular, 5–6 mm diam., base funnel-shaped. Dec.–Jan., after fires. Sandstone slopes at 200–1400 m, SW (Caledon).•

patula (Haw.) Schwantes (= *Erepsia pageae* L.Bolus)   Like **E. hallii** but flowers remaining open and petals with pink tips. Jan.–May. Loamy soils, to 500 m, in slightly disturbed places, SW (Cape Peninsula).•

pentagona (L.Bolus) L.Bolus   Erect shrublet to 40 cm, with flattened stems. Leaves 4–6 mm diam. Flowers in monochasia, magenta, outer filamentous staminodes deep red, 25–35 mm diam., reproductive parts hidden. Fruits 5-locular, pentagonal. Oct.–Feb., after fires. S-facing sandstone slopes, 500–1000 m, LB (Langeberg Mts).•

polita (L.Bolus) L.Bolus   Like **E. pentagona** but flowers with outer filamentous staminodes not deep red and fruits indistinctly pentagonal. June. Dry sandstone slopes, c. 350 m, LB (Langeberg Mts).•

polypetala (A.Berger & Schltr.) L.Bolus   Densely branched shrublet to 40 cm, with slender branches 2–3 mm diam. Leaves less than 5 mm diam. Flowers in many-flowered cymes, pink, filamentous staminodes pale yellow, reproductive parts partly hidden. Fruits c. 5 mm diam. Apr. Coastal fynbos, AP (Bredasdorp).•

promontorii L.Bolus   Erect, densely branched shrublet to 20 cm, branches becoming decumbent. Leaves less than 5 mm diam. Flowers in many-flowered cymes, magenta, filamentous staminodes whitish, reproductive parts partly hidden. Fruits less than 5 mm diam. Jan. Coastal lowlands, SW (Cape Peninsula to Hermanus).•

ramosa L.Bolus (= *Erepsia roseo-alba* L.Bolus)   Like **E. patula** but flowers stalked. Mar.–June. Mostly on shale, to 750 m, NW, SW (Piketberg to Cape Flats).•

saturata L.Bolus (= *Erepsia marlothii* N.E.Br.)   Erect shrublet to 40 cm, with slender branches from base. Leaves less than 5 mm diam. Flowers in monochasia, magenta, remaining open, outer filamentous staminodes red, inner yellow with red tips, reproductive parts hidden. Fruits 5–7 mm diam. Feb.–May. Sandstone slopes at 700–1000 m, NW (Gifberg to Piketberg).•

villiersii L.Bolus   Erect densely branched shrublet to 70 cm, with slender branches to 1.5 mm diam., branching from base. Leaves less than 3 mm diam. Flowers in monochasia, white with magenta tips, remaining open, petals linear, filamentous staminodes white with red tips, reproductive parts hidden. Fruits 4–6 mm diam. Dec. Gravel slopes, c. 400 m, SW (Villiersdorp).•

ESTERHUYSENIA•   4 spp., W Cape

alpina L.Bolus   Compact shrublet to 30 cm. Leaves ascending or spreading, obscurely keeled, apiculate. Flowers solitary on short pedicels, magenta, to 17 mm diam. Fruits 5-locular, keels parallel at base but diverging strongly above, valve wings absent. Dec. Rock ledges at high alt., NW (Hex River Mts).•

drepanophylla (Schltr. & A.Berger) H.E.K.Hartmann (= *Ruschia drepanophylla* (Schltr. & A.Berger) L.Bolus)   Low shrublet to 15 cm, blackish at base, branches reddish when young. Leaves free, trigonous, falcately incurved to lunate, blunt, mucronate, 12 × 3–4 mm. Flowers subsessile, magenta, c. 15 mm diam. Fruits 5-locular, without valve wings. Aug.–Oct. Upper slopes, NW (Cedarberg Mts).•

inclaudens (L.Bolus) H.E.K.Hartmann (= *Ruschia inclaudens* L.Bolus)   Compact shrublet to 25 cm. Leaves oblong, sharply tapered, apiculate, to 20 × 5–7 mm, fused below into a sheath to 4 mm long with an impressed line. Flowers solitary, white to pinkish purple, 25–30 mm diam. Fruits 5-locular. Sept. Upper slopes, NW (Worcester).•

stokoei (L.Bolus) H.E.K.Hartmann (= *Ruschia thomae* L.Bolus)   Compact low shrublet, with slender, trailing branches. Leaves sabre-shaped, glaucous, c. 14 × 8 mm. Flowers solitary, purplish, c. 23 mm diam. Fruits 5-locular. Nov.–Jan. Cliff faces at high alt., NW, SW (Hex River to Du Toitskloof Mts).•

FAUCARIA   TIERBEKVYGIE   6 spp., W and E Cape and Karoo

felina (L.) Schwantes ex Jacobsen (= *Faucaria duncanii* L.Bolus, *F. lupina* (Haw.) Schwantes, *F. uniondalensis* L.Bolus)   Tufted perennial, short-stemmed with age. Leaves crowded, 4–8, ovate-rhomboid to linear-lanceolate, keeled, toothed. Flowers 1 or 2, yellow, sometimes white, to 50 mm diam. Fruits without covering membranes or closing body. Mar.–Aug. Dry grassland and karroid bush or thicket, SE (Uniondale to E Cape and Karoo).

GALENIA   BRAKBOS   27 spp., southern and S tropical Africa

A. Flowers in terminal, symmetrically branched cymes; styles 2

africana L.   KRAALBOS, GEELBRAKBOS   Yellow-green, softly woody shrublet, sometimes to 1 m. Leaves opposite, linear-oblanceolate. Flowers in terminal panicle, inconspicuous, c. 1.5 mm, styles 2. Oct.–Dec. Dry flats and lower slopes, often on disturbed ground, NW, SW, AP, KM, LB (Namaqualand to Uniondale, Karoo and E Cape).

ecklonis Walp.   ROOILOODJIE   Sprawling to prostrate subshrub to 10 cm. Leaves opposite, linear, channelled, united around stem at base. Flowers in terminal, dichotomous cymes, white or red, 3–4 mm, styles 2. Sept.–Oct. Sandy flats, NW, SW (Gouda to Worcester).

hispidissima Fenzl   Bristly haired sprawling shrublet or perennial to 20 cm. Leaves opposite, linear-oblanceolate, bristly. Flowers crowded at branch tips, 2–3 mm, styles 2. Sept.–Oct. Sandy lower slopes, NW (Cold Bokkeveld to Piketberg).•

procumbens L.f.   Stiffly branched, yellowish shrublet to 1 m. Leaves opposite, small, oblanceolate, recurved. Flowers in small, dichotomous cymes, whitish, c. 2 mm, styles 2. Aug.–Oct. Stony flats in karroid scrub, KM (Namaqualand to S Karoo and Free State).

rigida Adamson   Like **G. africana** but to 50 cm, leaves, rigid, not drooping when dry, flowers in flat-topped umbels. Oct.–Jan. Rocky slopes, NW (W Karoo to Clanwilliam).

A. Flowers in secund cymes; styles 2–5
B. Styles 2

crystallina (Eckl. & Zeyh.) Fenzl   Sprawling, grey-mealy shrublet to 15 cm, with branches to 40 cm. Leaves oblanceolate, folded. Flowers in secund cymes, yellow or pink, c. 2 mm, styles 2. July–Dec. Sandy flats, NW, SW, KM, SE (Namaqualand to Worcester, Little Karoo to Uitenhage).

fruticosa (L.f.) Sond.   Twiggy, grey-felted shrublet to 50 cm. Leaves opposite, obovate, folded. Flowers in secund cymes on stiff branchlets often becoming spiny, red or yellow, c. 1 mm, styles 2. Aug.–Dec. Shale flats in karroid scrub, NW, SW, KM (Namibia and Karoo to Tygerberg and Little Karoo).

pruinosa Sond.   Like **G. fruticosa** but leaves glabrous, flowers mauve, styles 2, sometimes 3. Aug.–Sept. Rocky flats, NW (Namibia to Piketberg).

B. Styles 3–5

affinis Sond.   Like **G. filiformis** but always glabrous, plants twiggy and styles usually 3, sometimes 5. Sept.–Oct. Dry rocky slopes, KM, LB (Namaqualand and Karoo to Montagu and Langeberg Mts).

collina (Eckl. & Zeyh.) Walp.   Like **G. filiformis** but leaves often sticky, flowers larger, 2.5–3.0 mm, styles 3, sometimes 5. Mainly Sept.–Nov. Rocky slopes, NW, SW, KM, LB, SE (Namibia to George).

cymosa Adamson   Like **G. filiformis** but flowers in terminal, forked inflorescences, styles mostly 5 or 4. Oct.–Nov. Dry stony slopes and flats, KM (Karoo, Little Karoo and E Cape).

filiformis (Thunb.) N.E.Br.   Prostrate subshrub with branches to 80 cm, glabrescent. Leaves oblanceolate. Flowers crowded in secund cymes on short, lateral branchlets, pink or white, c. 2 mm, styles 5, sometimes 3. Mainly Aug.–Oct. Sandy slopes, especially dry riverbeds, NW, SW, AP, KM, LB (W Karoo and Bokkeveld Mts to King William's Town).

herniariifolia (C.Presl) Fenzl   Like **G. filiformis** but sepals and styles 4 or 3 (not 5). Nov.–Jan. Coastal sand or gravel, SW, AP, SE (Cape Peninsula to George).•

papulosa (Eckl. & Zeyh.) Sond.   Like **G. pubescens** but leaves succulent. Aug.–Oct. Dry slopes and watercourses, NW, KM (Namibia to Piketberg, Ceres to Uniondale).

portulacacea Fenzl   Like **G. pubescens** but leaves grey-mealy and sparsely long-hairy. Aug.–Sept. Dry sandy areas, NW, LB (Namaqualand and Karoo to Mossel Bay).

pubescens (Eckl. & Zeyh.) Druce   BLOUBRAKBOSSIE   Like **G. filiformis** but erect and twiggy or prostrate, styles 3, sometimes 5. Oct.–Dec. Rocky or disturbed flats, NW, SW, AP, KM, LB, SE (Karoo and Clanwilliam to Port Elizabeth).

sarcophylla Fenzl   JOUBERTSBRAKBOSSIE, VANWYKSBRAKBOSSIE   Like **G. pubescens** but leaves grey-mealy, softly hairy and succulent. July–Oct. Habitat? NW, KM, SE (Namaqualand and Karoo to Clanwilliam and Uniondale).

secunda (L.f.) Sond.   VANWYKSBRAKBOSSIE   Like **G. filiformis** but leaves obovate and densely hairy, flowers white or pink, styles 3, sometimes 5. July–Dec. Dry stony slopes, NW, KM, LB, SE (Karoo to Piketberg, Ceres to KwaZulu-Natal).

GIBBAEUM (= *MUIRIA*)   VOLSTRUISTONE   c. 16 spp., W Karoo to W Cape

A. Leaves about as long as broad

album N.E.Br.   Whitish acaulescent succulent. Leaves united, forming an oblique-ovate body 60–120 × 60–80 mm, covered with fine white hairs. Flowers solitary, white or pink, to 25 mm diam. Fruits 6-locular. Dec. Dry slopes and flats in quartz pebbles on shale, LB (N slopes of Langeberg Mts: Muiskraal).•

cryptopodium (Kensit) L.Bolus   Like **G. pilosulum** but plants white-pubescent to glabrous. July. Shale and quartz, KM (W Little Karoo).•

dispar N.E.Br.   DUIMPIE-SNUIF, PAPPEGAAIBEK   Clumped aulescent succulent. Leaves in pairs of unequal size forming ovoid bodies 10–15 × 10–14 mm, with a definite keel, thick, deeply fissured, finely velvety-hairy. Flowers pink or mauve to red, 8–10 mm diam. Fruits 6-locular. Apr. Shale slopes at c. 400 m, KM (Little Karoo: Vanwyksdorp).•

heathii (N.E.Br.) L.Bolus (= *Gibbaeum blackburnii* L.Bolus, *G. comptonii* (L.Bolus) L.Bolus, *G. luckhoffii* L.Bolus)   HONDEBAL, VOLSTRUISWATER   Compact succulent forming subglobose clumps to 30 cm diam., from woody rootstock. Leaves subequal to equal, closely pressed together, forming subglobose bodies 20–30 × 15–20 mm, with definite fissure, green to whitish green. Flowers solitary on pedicels 10–12 mm long, white or red, 10–30 mm diam. Fruits 8- or 9-locular. June–Oct. Habitat?, KM (Little Karoo).•

hortenseae (N.E.Br.) Thiede & Klak (= *Muiria hortenseae* N.E.Br.)   MUISKOPVYGIE   Stemless, clump-forming dwarf succulent. Leaves fused, forming an ovate-conical body, velvety, enclosed in membranous sheaths during the resting period. Flowers solitary, white to mauve, to 20 mm diam. Fruits 6- or 7-locular, without covering membranes or closing bodies. Jan. On quartz, KM (Little Karoo: Springfontein).•

nebrownii Tischer (= *Imitaria muirii* N.E.Br.; incl. **G. johnstonii** Van Jaarsv. & S.A.Hammer)   Like **G. dispar** but imbedded in the ground, usually solitary, sometimes in clumps of 6–8, leaf bodies with a flat or conical top, without a keel. May–July. Shale slopes and flats, KM (Little Karoo: S of Ladismith).•

petrense (N.E.Br.) Tischer   MIMICRY PLANT   Aaulescent succulent perennial to 5 cm, forming tufts 5–10 cm diam. Leaves in annual increments of 1 or 2 pairs, with a definite fissure, deltoid-ovate or triangular in outline, with sharp edges and acute or obtuse at apex, sharply keeled, grey-green. Flowers solitary, magenta, to 15 mm diam. Fruits 6- or 7-locular. Sept.–Oct. Quartz pebbles on clay, KM (Little Karoo: Springfontein).•

pilosulum (N.E.Br.) N.E.Br.   Succulent perennial forming dense mats to 10 cm diam. embedded in the ground. Leaf pairs united, forming obovate bodies c. 2.5 mm long, white-pubescent, the fissure off-centre, 3–4 mm deep. Flowers bright pink, 6–7 mm diam. Fruits 7-locular. July–Aug. Habitat?, KM (Touwsberg).•

A. Leaves very unequal and usually 2–5 times as long as broad

angulipes (L.Bolus) N.E.Br.   Perennial succulent with prostrate stems to 6 cm long, flowering branches erect, branches with 4 leaves. Leaves spreading, fused for c. 8 mm at base, 23–26 × 9 mm at base, top flat, glaucous, velvety. Flowers solitary on pedicels 20–24 mm long, purplish, 25 mm diam. Fruits 6- or 7-locular. Oct.–Nov. N-facing slopes, LB (Langeberg Mts: Muiskraal).•

esterhuyseniae L.Bolus   Like **G. haagei** but plants glabrous. Oct.–Nov. Quartz patches, SW (McGregor to Stormsvleikloof).•

geminum N.E.Br.   Cushion-like perennial with short, prostrate branches. Leaves in annual increments of 2 or 3 decussate pairs, fused, spreading, unequal, the larger to 15 × 6 mm, slightly compressed, the smaller only a quarter as long. Flowers red or magenta, 12–15 mm diam. Fruits 6- or 7-locular. June–Sept. Sandy, more or less alluvial flats, KM (W Little Karoo: N of Warmwaterberg).•

gibbosum (Haw.) N.E.Br.   Deep green, acaulescent succulent to 10 cm, forming compact clumps 6–10 cm diam., with a woody rootstock. Leaves unequal, larger leaf slightly incurved, subcylindrical towards apex, abaxial surface bluntly 2-keeled above, smooth. Flowers 20–30 mm diam., pink to purple. Fruits 6- or 7-locular. Aug. Pebbly shale or quartz patches, KM (W Karoo to Matjiesfontein and Little Karoo).•

haagei Schwantes   Like **G. velutinum** but with distinctive white epidermis and striped reddish petals. Oct. Rocky sandstone slopes overlying clayey sand with or without quartz pebbles, LB (Swellendam to Heidelberg).•

pachypodium (Kensit) L.Bolus   Rounded succulent perennial, 25–40 cm diam. Leaves erect, unequal, minutely pubescent, greenish to greenish grey, trigonous to subterete, slightly keeled. Flowers solitary on elongate pedicels 40–75 mm long, pink to reddish, 20 mm diam. Fruits 6-locular. Nov.–Dec. Rocky riverbanks, KM (Little Karoo: Muiskraal to Ladismith).•

pubescens (Haw.) N.E.Br.   VISBEKVYGIE   Succulent perennial forming compact cushions, with a woody, shortly branched rootstock raised well above the ground. Leaves unequal, larger one to 3 cm long, cylindric-ovoid, the smaller semilunate, about one-third as long as the larger, silvery white-hairy. Flowers solitary, pale to deep purple, to 15 mm diam. Fruits 6-locular. June–Aug. White quartz flats, KM (Barrydale to Ladismith).•

shandii N.E.Br.   Like **G. pubescens** but plants mostly less branched, leaves shorter and thicker and epidermal hairs branched. June–Aug. On quartz or shale, KM (Little Karoo: N of Warmwaterberg).•

velutinum (L.Bolus) Schwantes   Tufted succulent perennial with a strongly developed rootstock. Leaves spreading, prostrate, unequal, longer leaf to 6 cm, 3 cm wide at base, tapering to a point, velvety papillose, grey-green. Flowers solitary on pedicels to 25 mm long, borne in angle between old and new leaves, white or pink, 40–50 mm diam. Fruits 6-locular. Oct.–Nov. Habitat?, KM (Little Karoo: Barrydale to Muiskraal).•

GLOTTIPHYLLUM   SKILPADKOS   16 spp., Karoo, W and E Cape

A. Leaves decussate

cruciatum (Haw.) N.E.Br. (= *Gibbaeum apiculatum* N.E.Br.)   Mat-forming perennial with short, decumbent stems. Leaves decussate, suberect, basally dilated, subequal. Fruits with very high tops and disintegrating stalks. May–July. Under shrubs on sandy loam, KM, SE (central Little Karoo).•

fergusoniae L.Bolus   Compact succulent perennial. Leaves erect, decussate, pointed, tapering from a broad pustulate base, slightly unequal, grey. Fruits persisting, valves with awns. Mar.–Apr. Shale or sandstone rock crevices, KM, LB (Little Karoo).•

regium N.E.Br. (= *Gibbaeum compressum* L.Bolus)   Succulent perennial forming compact mats. Leaves decussate, longer than 65 mm, slightly unequal, narrowed below, bright green. Fruits on persistent stalks, without valve wings or awns, base spongy, valves with high rims. Flowering time? Sandstone slopes or sandy loam, KM, LB (Little Karoo).•

salmii (Haw.) N.E.Br.   Compact succulent with 3–5 branches. Leaves decussate, with several pairs per branch, erect to ascending, slightly unequal, falcate, laterally compressed, larger with an impression from the opposite, shorter leaf, grey. Fruits tumbling. Mar.–Apr. Sandstone slopes, KM (E Little Karoo).•

surrectum (Haw.) L.Bolus (= *Gibbaeum concavum* N.E.Br., *G. parviflorum* L.Bolus)   Like **G. regium** but leaves shorter than 60 mm and fruits somewhat woody, rounded, valves with low rims. Apr.–June. Weathered shaly sandstone, KM, LB (Little Karoo).•

A. Leaves distichous

carnosum N.E.Br. (= *Gibbaeum jordaanianum* Schwantes)   Compact succulent perennial, with 3–5 decumbent branches. Leaves distichous, suberect, oval, apically hunched, with a thick waxy cuticle, margins distinct. Seeds with long papillae in dorsal region. Flowering time? Mostly sandstone slopes, often with **G. regium** and **G. cruciatum**, KM, SE (Little Karoo to Willowmore).•

depressum (Haw.) N.E.Br. (= *Gibbaeum barrydalense* Schwantes, *G. marlothii* Schwantes)   Branched succulent perennial. Leaves decumbent or ascending, distichous or nearly so, green. Fruits with high, spongy tops, valves thickened and with low rims. July–Aug. Under shrubs on loamy soils, NW, KM, LB, SE (W Karoo, Ceres and Little Karoo to Humansdorp).

linguiforme (L.) N.E.Br. (= *Gibbaeum cilliersiae* Schwantes, *G. latifolium* N.E.Br.)   TONGBLAARVYGIE   Spreading, few-branched succulent perennial. Leaves ascending to erect, distichous, broadly tongue-shaped, rather flat, narrowing to pustulate base, grey. Fruits tumbling. Mar.–June. Under shrubs on loamy sand with quartzite, KM (Little Karoo).•

longum (Haw.) N.E.Br. (= *Gibbaeum latum* (Salm-Dyck) N.E.Br.)   Succulent perennial with 3–5, later many branches. Leaves decumbent or ascending, distichous, flat, strap-shaped with rounded margins, green, several to many pairs per branch. Fruits on persistent stalks, tops rather flat. Oct.–Dec. Karroid scrub on sandy loam, NW, SW, LB, SE (Ceres to S Little Karoo to E Cape).

suave N.E.Br. (= *Gibbaeum herrei* L.Bolus)   Like **G. linguiforme** but leaves thickened. Apr.–May. On quartz or shaly sandstone, KM, LB (W Little Karoo).•

HALLIANTHUS   1 sp., N and W Cape

planus (L.Bolus) H.E.K.Hartmann (= *Leipoldtia compressa* L.Bolus)   Compact succulent with prostrate shoots. Leaves trigonous to falcate. Flowers solitary or in 3-flowered cymes, petals white or pink, filamentous staminodes numerous. Fruits 10-locular, top taller than base, valve wings rectangular, large closing bodies present. Flowering time? Crevices in granite, sandstone or quartz ridges, NW (Namaqualand to Clanwilliam).

HEREROA   CLOCK PLANT, SLAAPVYGIE   c. 30 spp., Namibia to E Cape

acuminata L.Bolus   Low tufted shrublet to 5 cm. Leaves erect, tapering, mostly green, rough, obliquely blunt, margins and keel serrated. Flowers solitary, yellow with reddish reverse, to 30 mm diam. Oct. Habitat?, KM (Ladismith).•

aspera L.Bolus   Erect, branching shrublet to 8 cm. Leaves inclined, subcylindrical, rough, glaucous, to 25 mm long, narrow. Flowers solitary on pedicels c. 7 mm long, yellow with reddish tips, to 36 mm diam. Jan.–Feb. Habitat?, KM, LB (Little Karoo, Swellendam).•

gracilis L.Bolus   Cushion-forming perennial, with flowering branches to 25 mm. Leaves slender, erect to spreading, 4 per branch, roughened with dots, to 35 mm long. Flowers on pedicels c. 13 mm long, yellow, to 25 mm diam. Mar.–Apr. Habitat?, KM (Little Karoo).•

muirii L.Bolus   Low branched perennial to 10 cm. Leaves spreading, subcylindrical, slender, rough, to 50 mm long. Flowers on pedicels c. 20 mm long, small, yellow. Feb. Habitat?, KM (Little Karoo: Barrydale).•

odorata (L.Bolus) L.Bolus   Small sprawling shrublet with erect flowering branches. Leaves spreading to inclined, mostly subcylindrical, glaucous. Flowers 3 per branch on pedicels to 10 mm long, yellow with red tips, to 22 mm diam. Sept.–Oct. KM, LB (S Karoo to Swellendam).

tenuifolia L.Bolus   Erect, laxly branched shrublet to 25 cm. Leaves spreading to erect, semicylindrical, glaucous, rough. Flowers 1–3 per branch on pedicels 25–30 mm, yellow, to 30 mm diam. Feb.–Mar. Habitat?, NW, KM (Robertson, Barrydale).•

HYMENOGYNE   2 spp., N and W Cape

conica L.Bolus   Erect or prostrate annual to 3 cm. Leaves flat, margins papillate, sometimes with tannin cells. Flowers yellow, styles fused into a cone. Fruits breaking into mericarps, seed bags filling locule, seeds 2 per locule. Aug.–Oct. Habitat?, NW (S Namaqualand to Biedouw).

glabra (Aiton) Haw.   Prostrate annual. Leaves flat, margins papillate, some papillae tanniniferous. Flowers yellow or straw-coloured, styles fused into a funnel. Fruits breaking into mericarps, seed bags filling locule, seeds 2 per locule. Sept.–Oct. Habitat? NW, SW (Clanwilliam to Cape Peninsula).•

JORDAANIELLA STRANDVYGIE   4 spp., N and W Cape

dubia (Haw.) H.E.K.Hartmann (= *Cephalophyllum procumbens* (Haw.) Schwantes, *C. maritimum* (L.Bolus) Schwantes, *C. vanputtenii* L.Bolus)   Prostrate succulent, stems with internodes longer than 20 mm. Leaves slender, terete, without papillae. Flowers usually yellow, sometimes white or magenta. Fruits 10–15-locular, subglobose. May–Sept. Coastal sands, NW, SW, AP (Elandsbaai to Mossel Bay).•

LAMPRANTHUS VYGIE   c. 155 spp., S Namibia to E Cape, 1 in Australia

acrosepalus (L.Bolus) L.Bolus (?incl. **L. rubroluteus** (L.Bolus) L.Bolus)   Erect shrublet to 25 cm, branches willowy, 3 mm thick. Leaves ascending, slender, compressed and tapering, sharply keeled, 17–30 mm long. Flowers solitary, orange or white, to 60 mm diam. Fruits 10–13 mm diam. Oct. Rocky sandstone slopes at c. 1100 m, NW (Piketberg).•

acutifolius (L.Bolus) N.E.Br.   Dwarf shrublet to 20 cm. Leaves erect, subfalcate, set with fairly large dots, 10 × 2 mm. Flowers 1–3 on pedicels to 20 mm long, pink, to 30 mm diam. Fruits c. 4.5 mm diam. Jan. Stony slopes, SW (Malmesbury).•

aduncus (Haw.) N.E.Br. (?incl. **L. curvifolius** (Haw.) N.E.Br.   Low shrublet with erect branches. Leaves crowded at branch tips, with short shoots in axils, fused below, tapering, spreading and recurved at tip, 15–20 × c. 2 mm. Flowers solitary, terminal, magenta-red, to 18 mm diam., stamens and staminodes in a central cone, stigmas slender. Fruits with large valve wings. May–Aug. Stony slopes, NW, SW (Clanwilliam to Cape Peninsula).•

aestivus (L.Bolus) L.Bolus (?incl. **L. austricola** (L.Bolus) L.Bolus)   Erect, stiffly branched shrublet to 30 cm. Leaves erect, glaucous, punctate, 10–15 mm long, slender, laterally compressed and widened above. Flowers numerous on pedicels to 20 mm long, pink or purplish, 28 mm diam. Fruits unknown. Dec.–Mar. Sandy flats, SW (Cape Peninsula to Kleinmond).•

algoensis L.Bolus   Erect shrublet to 35 cm. Leaves acuminate, tip recurved, 10–14 mm long. Flowers 1–3, scarlet, to 35 mm diam. Fruits c. 8 mm diam. Oct. Habitat?, SE (Port Elizabeth: Algoa Park).•

altistylus N.E.Br. (incl. **L. maturus** N.E.Br.)   Delicate, decumbent shrublet to 10 cm, sometimes rooting at nodes. Leaves ascending to spreading, somewhat falcate, minutely papillose, to 4 mm long. Flowers solitary, pink, c. 20 mm diam. Fruits less than 5 mm diam. Aug. Sandy flats, SW (Kalbaskraal to Cape Peninsula).•

amabilis L.Bolus   Laxly branched shrublet to 30 cm. Leaves erect, tapering above, to 25 mm long. Flowers solitary, white, rarely pale pink, 35–52 mm diam. Fruits c. 8 mm diam. Sept. Stony lower slopes in low bush, SW (Potberg).•

amoenus (Salm-Dyck ex DC.) N.E.Br.   Shrublet to 40 cm. Leaves weakly spreading, cylindrical to trigonous, shortly mucronate, to 40 mm long. Flowers ternate, white to purple, 35–40 mm diam. July–Oct. Sandy flats, SW (Malmesbury to Cape Peninsula).•

antemeridianus (L.Bolus) L.Bolus   Shrublet with prostrate to ascending branches to 20 cm long. Leaves erect, with minute dots, to 25 mm long. Flowers solitary on pedicels 30–35 mm long, yellow to salmon-red to pink, to 45 mm diam. Aug.–Sept. Sand dunes, AP (Riversdale: Welgevonden).•

antonii L.Bolus   Mat-forming perennial with trailing stems to 12 cm long. Leaves linear, acute, apiculate, 2–5 cm long. Flowers solitary, white to pale pink, 50–65 mm diam. Fruits c. 12 mm diam. Sept. On moist rocks, NW (Agterwitzenberg).•

arbuthnotiae (L.Bolus) L.Bolus (?incl. **L. monticola** (L.Bolus) L.Bolus)   Like **L. tegens** but leaves more slender and shorter, and fruits fairly shallow (vs funnel-shaped), c. 4 mm diam. Aug.–Oct. Stony calcareous sands, AP (Agulhas Peninsula).•

arenarius H.E.K.Hartmann (= *Ruschia arenosa* L.Bolus)   Erect, stiffly branched shrublet to 30 cm. Leaves fused, sheath to 2 mm long, erect, upper side flat, back surface keeled, glaucous, to 23 × 4 mm. Flowers stalked, pink, to 15 mm diam. Fruits without closing bodies. Sept. Habitat?, NW (Nieuwoudtville).•

arenosus L.Bolus   Laxly branched, decumbent shrublet to 25 cm. Leaves erect, sides widened towards apex, pale glaucous to yellow-green, 15–20 mm long. Flowers 1–3, pink, to 45 mm diam. Fruits c. 10 mm diam. Oct.–Nov. Sandy flats, SW (Saldanha).•

aurantiacus (DC.) Schwantes   Sparsely branched shrublet to 45 cm. Leaves bluntly trigonous, pruinose, 20–30 mm long. Flowers solitary, stalked, orange, 40–50 mm diam. Sept.–Oct. Sandy flats, SW (Malmesbury to Cape Flats).•

aureus (L.) N.E.Br.   Erect perennial to 40 cm. Leaves fused basally, pruinose, to 5 cm long. Flowers stalked, shiny orange, to 60 mm diam. Aug. Granite outcrops, SW (Vredenburg to Saldanha).•

baylissii L.Bolus   Low spreading shrublet to 15 cm, 50 cm diam. Leaves falcate, rough, bluish green, c. 2 cm long. Flowers solitary, bright yellow, to 20 mm diam. Nov. Habitat?, SE (Knysna).•

bicolor (L.) N.E.Br. (?incl. **L. inaequalis** (Haw.) N.E.Br.)   Stiffly branched, erect shrublet to 30 cm. Leaves almost trigonous, green, rough, 12–25 mm long. Flowers 1–3, yellow with scarlet or copper reverse. Oct.–Jan. Sandy flats or slopes, NW, SW, AP (Tulbagh and Cape Peninsula to Bredasdorp).•

brevistamineus (L.Bolus) L.Bolus   Slender shrublet to 14 cm. Leaves cuneate to falcate, sharply tapering, to 7 mm long. Flowers solitary, purplish, c. 30 mm diam. Nov. Habitat?, NW (foot of Matroosberg).•

brownii (Hook.f.) N.E.Br. Slender shrublet to 30 cm. Leaves shortly fused, subcylindrical, prominently dotted, to 10 mm long. Flowers mostly ternate, orange-red with yellow reverse, fading light red, c. 20 mm diam. Fruits unknown. Flowering time? Habitat?, (exact range unknown).

calcaratus (Wolley-Dod) N.E.Br. (?incl. **L. tenuis** L.Bolus)   Slender shrublet to 25 cm. Leaves erect, acute, 6–11 mm long. Flowers solitary on pedicels 5–6 mm long, pink, 16–20 mm diam. Fruits 5–6 mm diam., base funnel-shaped. June–Sept. Marshy, sand, SW (Malmesbury to Cape Peninsula).•

candidus L.Bolus   Loosely branched, slender shrublet to 30 cm. Leaves spreading, subfalcate, rough, 6 mm long. Flowers 1–3, white, to 20 mm diam. Fruits c. 4 mm diam. Oct. ?, NW (Piketberg).•

capillaceus (L.Bolus) N.E.Br.   Succulent perennial to 14 cm, with slightly hairy branches. Leaves spreading to ascending, subfalcate, bluish, c. 5 mm long. Flowers stalked, purplish, to 20 mm diam. Jan. Habitat?, NW, SW (Tulbagh to Bainskloof).•

caudatus L.Bolus   Rigid shrublet to 30 cm, with slender, densely leafy branches. Leaves spreading to ascending, recurved at tip, rough, glaucous, 9–14 mm long. Flowers 1–3 on slender pedicels 10–20 mm long, pink to pale pink, to 20 mm diam. Fruits 5–7 mm diam. Apr. ?, AP (Agulhas).•

ceriseus (L.Bolus) L.Bolus   Erect shrublet to 15 cm. Leaves fused basally, ascending, sharply keeled, laterally compressed, glaucous, c. 10 mm long. Flowers solitary, pink, to 35 mm diam. Fruits c. 8 mm diam. Jan. Limestone hills, AP (Cape Agulhas to Stilbaai).•

coccineus (Haw.) N.E.Br.   Erect shrublet to 90 cm. Leaves crowded on short shoots, spreading, trigonous compressed, grey-green, 15–25 mm long. Flowers 1–3, red, to 40 mm diam. Sept.–Dec. Sandy flats, SW (Saldanha to False Bay).•

creber L.Bolus   Robust shrublet to 35 cm. Leaves densely crowded, spreading, bluntly keeled, rough, to 10 mm long. Flowers pink, to 25 mm diam. July. ?, LB (Albertinia).•

cyathiformis (L.Bolus) N.E.Br. (?incl. **L. vernalis** (L.Bolus) L.Bolus)   Erect shrublet to 20 cm. Leaves erect to ascending, slightly recurved above, acuminate, green, c. 30 mm long. Flowers crimson, c. 45 mm diam. Sept.–Oct. Habitat?, NW (Clanwilliam).•

densifolius (L.Bolus) L.Bolus   Erect shrub with stiff branches. Leaves erect, bluish green, shortly tapered, 17–27 × 2–3 mm. Flowers solitary, pink to purple, c. 60 mm diam. Jan.–Feb. Habitat?, SW (Vredenburg).•

dependens (L.Bolus) L.Bolus (?incl. **L. pauciflorus** (L.Bolus) N.E.Br.)   Sprawling shrublet with trailing branches to 30 cm. Leaves spreading to recurved, green, 35–55 × 3–4 mm. Flowers stalked, pale mauve, c. 60 mm diam. Oct.–Jan. Moist cliffs, NW, LB, SE (Worcester to Humansdorp).•

diffusus (L.Bolus) N.E.Br. (incl. **L. microstigma** (L.Bolus) N.E.Br., **L. occultans** L.Bolus, **L. rabiesbergensis** (L.Bolus) L.Bolus)   Spreading, loosely branched shrublet to 30 cm. Leaves narrow, rough, 10–25 mm long. Flowers solitary on pedicels 35 mm long, white to purplish, to 30 mm diam. Fruits unknown. Sept.–Nov. Stony slopes, NW, SW, LB, SE (Ceres to Bredasdorp and to Humansdorp).•

dilutus N.E.Br.   Sprawling shrublet to 25 cm long. Leaves spreading, subfalcate, acute, tip recurved, rough, glaucous, 5–7 mm long. Flowers pale pink to white, 23 mm diam. June–Aug. Habitat?, NW (Porterville).•

dissimilis (G.D.Rowley) H.E.K.Hartmann (= *Ruschia dissimilis* G.D.Rowley)   Erect or sprawling shrublet with maroon internodes, branches to 12 cm long; roots thickened and turnip-shaped. Leaves subulate, recurved, scabrid, to 12 × 2 mm. Flowers solitary, purplish pink. Fruits with large rectangular valve wings. Sept. Stony slopes, NW (Hex River Mts: Warm Bokkeveld).•

diutinus (L.Bolus) N.E.Br.   Like **L. multiseriatus** but pedicels to 30 mm long and flowers larger, 40–80 mm diam., purplish. July–Nov. Sand, LB (Riversdale).•

dregeanus (Sond.) N.E.Br.   Erect shrublet to 60 cm, with branches compressed and 2-angled, red, finely pruinose. Leaves compressed-trigonous, finely denticulate, to 16 × 6 mm. Flowers subsessile, 1–3, white. Sept. Stony slopes, NW, KM (Tulbagh to Montagu).•

dulcis (L.Bolus) L.Bolus   Erect, loosely branched shrublet to 45 cm. Leaves ascending, falcately incurved, acute or tapering, glaucous, to 35 mm long. Flowers solitary, pale pink, 40 mm diam. Oct. Rocky slopes, NW (Olifants River Mts).•

edwardsiae (L.Bolus) L.Bolus (?incl. **L. argenteus** (L.Bolus) L.Bolus, **L. hiemalis** (L.Bolus) L.Bolus, **L. montaguensis** (L.Bolus) L.Bolus, **L. nardouwensis** (L.Bolus) L.Bolus, **L. pleniflorus** L.Bolus)   Stiffly erect shrublet to 30 cm. Leaves trigonous, glaucous, to 20 × 5 mm. Flowers solitary or ternate, pinkish, staminodes in a central cone. Fruits bell-shaped below, valve wings broad and rectangular. June–Aug. Mostly sandstone slopes, NW, SW, KM (Tanqua Karoo to Montagu).

egregius (L.Bolus) L.Bolus   Erect shrublet to 40 cm. Leaves ascending to spreading, dull green, to 25 × 3–4 mm. Flowers 1–3 in groups, purplish, to 60 mm diam. Sept. Habitat?, KM (Montagu).•

elegans (Jacq.) Schwantes (?incl. **L. paarlensis** L.Bolus)   Sprawling, much-branched shrublet to 30 cm, branches prostrate, erect or spreading. Leaves rough, slender, 10–25 mm long. Flowers ternate, pink or purple-rose, 20–30 mm diam., filamentous staminodes and stamens in a central cone. Fruits 5–8 mm diam. Sept.–Nov. In clay and rocky slopes and flats, NW, SW, LB?, SE (Cape Peninsula to Worcester, Uitenhage).•

emarginatus (L.) N.E.Br.   Like **L. elegans** but without filamentous staminodes. Sept.–Dec. Habitat?, SW (Langebaan to Gordon's Bay).•

eximius L.Bolus   Sprawling shrublet, with prostrate branches to 5 cm long. Leaves ascending to erect, acute, 15–22 mm long. Flowers on erect branches 15–30 cm long, purplish, c. 50 mm diam. Oct.–Nov. Habitat?, SW (Langebaan to Yzerfontein).•

explanatus (L.Bolus) N.E.Br. (incl. **L. sternens** L.Bolus)   Like **L. reptans** but plants not mat-forming, branches slender, creeping and leaves narrower, to 2 mm diam. Aug. Sandy flats, SW, AP (Cape Flats to Cape Agulhas).•

falcatus (L.) N.E.Br.   Small, freely branching shrublet with very slender branches. Leaves subfalcate, grey-green, with round papillae, to 6 mm long. Flowers ternate, pink, to 15 mm diam. Nov.–Dec. Habitat?, NW, SW (Tulbagh to Bredasdorp).•

falciformis (Haw.) N.E.Br. (= *Oscularia falciformis* (Haw.) H.E.K.Hartmann)   Decumbent to erect shrublet to 20 cm with reddish stems. Leaves sabre-shaped to falcate, with large dots, glaucous. Flowers in terminal, crowded cymes, pink, filaments white. Fruits turbinate, 5-locular. Nov.–Feb. Rock outcrops and ledges, NW (Piketberg).•

fergusoniae (L.Bolus) L.Bolus   Prostrate to decumbent shrublet with slender branches. Leaves erect, compressed, bluntly keeled, acute, rough, to 17 mm long, slender. Flowers solitary, yellow or orange-red, c. 30 mm diam. Nov. Limestone dunes, AP (Riversdale).•

filicaulis (Haw.) N.E.Br.   Mat-forming succulent to 10 cm. Leaves crowded at branch tips, incurving, apices recurved, to 25 × 3 mm. Flowers solitary, rose to magenta, to 20 mm diam., filamentous staminodes and stamens in a central cone. Fruits c. 6 mm diam. June–July. Moist depressions on laterite, SW (Cape Peninsula and Cape Flats).•

foliosus L.Bolus   Sprawling shrublet, with trailing branches rooting at nodes. Leaves crowded, erect to spreading, acute, 15–30 mm long. Flowers stalked, solitary, purplish, to 40 mm diam. Aug.–Sept. Habitat?, SW (Gansbaai).•

framesii (L.Bolus) N.E.Br.   Diffuse, erect shrublet to 20 cm. Leaves shortly acuminate, rough, to 10 mm long, slender. Flowers numerous, pink to white, to 25 mm diam. Fruits c. 6 mm diam. Jan. Habitat?, SW (Hermanus to Bredasdorp).•

furvus (L.Bolus) N.E.Br.   Erect shrublet to 20 cm. Leaves somewhat falcate, to 10 mm long, slender. Flowers 1–3 in groups, magenta to purple, to 25 mm diam., stigmas dark purple. Fruits c. 6 mm diam. Nov.–Feb. Stony slopes, NW, SW (Clanwilliam to Caledon).•

galpiniae (L.Bolus) L.Bolus   Erect shrublet to 10 cm, with slender branches. Leaves erect, rarely falcate, subcylindrical, narrow, acute, to 22 mm long. Flowers solitary, white with pink centre, to 60 mm diam. Fruits 5-locular, 8 mm diam. Oct.–Nov. Limestone hills, AP (Agulhas coast).•

glaucus (L.) N.E.Br. (incl. **L. citrinus** (L.Bolus) L.Bolus, **L. longistamineus** (L.Bolus) N.E.Br., **L. palustris** (L.Bolus) L.Bolus)   Erect shrublet to 30 cm. Leaves fused basally, compressed-trigonous, rough, grey pruinose, 15–30 mm long. Flowers stalked, solitary, golden yellow to lemon-yellow, rarely cream-coloured, to 30 mm diam. July–Nov. Habitat?, NW, SW (Gifberg to Cape Peninsula).•

glomeratus (L.) N.E.Br.   Erect shrublet to 30 cm. Leaves fused basally, acute, 12–18 mm long, slender, with prominent dots. Flowers numerous, violet or rose-purple, to 25 mm diam. Fruits c. 6 mm diam. Dec.–Jan. Sandy slopes, NW, SW (Tulbagh to Cape Peninsula).•

gracilipes (L.Bolus) N.E.Br.   Erect shrublet to 20 cm, branches 4-angled. Leaves spreading, subfalcate, mucronate, 7 mm long. Flowers pink, c. 32 mm diam. Oct. Rocky slopes, NW (Olifants River Mts and Piketberg).•

gydouwensis (L.Bolus) H.E.K.Hartmann (= *Amphibolia gydouwensis* (L.Bolus) L.Bolus)   Shrublet to 40 cm, stems to 1 cm diam. Leaves fused, sheath swollen and with an impressed line, apiculate, to 35 mm long, 4 mm diam. Flowers in a many-flowered inflorescence, purple, 18 mm diam. Fruits funnel-shaped. Oct. Habitat?, NW (Cedarberg Mts to Ceres Karoo).

hallii L.Bolus   Erect shrublet to 35 cm, becoming prostrate with age. Leaves fused basally, ascending to erect, acutely to bluntly keeled, laterally compressed, 20–80 mm long. Flowers stalked, pink, to 35 mm diam. Sept. Habitat?, LB (Langeberg Mts: Cloete's Pass).•

haworthii (Donn ex Haw.) N.E.Br. (?incl. **L. coralliflorus** (Salm-Dyck) N.E.Br.)   Erect, freely branched shrublet to 60 cm. Leaves fused basally, spreading, densely pruinose, 25–40 mm long. Flowers solitary, light purple, to 70 mm diam. Sept.–Oct. Habitat?, NW (Clanwilliam).•

henricii (L.Bolus) N.E.Br. (?incl. **L. mariae** (L.Bolus) L.Bolus)   Erect, stiffly branched shrublet to 20 cm. Leaves inclined, falcate, blunt to acute, glaucous, c. 7 × 1.5 mm. Flowers solitary, purplish, 14 mm diam. Sept.–Oct. Sandy slopes and flats, NW (W Karoo and Cedarberg Mts).

hurlingii (L.Bolus) L.Bolus   Erect, loosely branched shrublet to 30 cm. Leaves erect, trigonous, acute, glaucous, set with dots, 22–50 mm long. Flowers stalked, c. 25 mm diam., golden yellow. Fruits c. 10 mm diam. Sept. Habitat?, NW (Robertson).•

immelmaniae (L.Bolus) N.E.Br. (?incl. **L. stenopetalus** (L.Bolus) N.E.Br., **L. subtruncatus** L.Bolus) Erect shrublet to 25 cm. Leaves ascending, tapering, set with round papillae, 10–20 × 2 mm. Flowers pale pink, to 20 mm diam. Oct. Habitat?, NW, SW (Graafwater to Mamre).•

intervallaris L.Bolus   Laxly branched shrub. Leaves tapered, obtuse to acute, green, 15–20 mm long, slender. Flowers pale pink, 22–28 mm diam. Aug. Habitat?, NW (Clanwilliam).•

laetus (L.Bolus) L.Bolus   Erect, tangled shrublet to 20 cm. Leaves ascending to erect, slightly rough, dull green, 7–13 mm long, slender. Flowers purplish, 20–23 mm diam. Oct. Rocky sandstone slopes, KM (Montagu).•

laxifolius (L.Bolus) N.E.Br.   Erect shrublet to 30 cm. Leaves ascending to erect, falcately incurved, keeled, acute to acuminate, 18–26 × c. 3 mm. Flowers pink to magenta, 20–23 mm diam. Feb.–Mar. Habitat?, LB (Langeberg Mts: Riversdale).•

leightoniae (L.Bolus) L.Bolus (?incl. **L. berghiae** (L.Bolus) L.Bolus)   Erect shrublet to 50 cm. Leaves ascending, trigonous, glaucous, to 13 × c. 3 mm. Flowers in few-flowered dichasia, pink to pale pink, to 40 mm diam. Fruits 9 mm diam. Aug.–Sept. Rocky sandstone slopes, often among restios, NW, SW (Clanwilliam to Kalbaskraal).•

leipoldtii (L.Bolus) L.Bolus   Erect shrublet to 20 cm, with slender branches. Leaves erect, slender, dull green, set with round papillae, 10–15 mm long. Flowers 1–3 in groups, pale pink to pink, c. 35 mm diam. Oct. Habitat?, SW (Worcester to Villiersdorp).•

leptaleon (Haw.) N.E.Br.   Slender, lax shrublet, 12–15 cm. Leaves trigonous, tapering and recurved above, minutely papillate, bluish, 5–10 × 1–2 mm. Flowers 1–3, pink, to 15 mm diam., staminodes in a central cone. Fruits 5-locular. Oct. Stony flats, SW (Wellington to Somerset West).•

leptosepalus (L.Bolus) L.Bolus (?incl. **L. walgateae** L.Bolus)   Slender, erect shrublet to 15 cm. Leaves erect, narrowed above, dull green, set with round papillae, 10–15 mm long. Flowers 1–3 in groups, magenta, c. 30 mm diam. Dec.–Mar. Rocky sandstone slopes, NW, LB (Cedarberg to Langeberg Mts at Tradouw Pass).•

lewisiae (L.Bolus) L.Bolus   Erect shrublet to 10 cm. Leaves erect, narrowed above, rough, to 14 × 2 mm. Flowers stalked, solitary, pink, 30 mm diam. Sept. Habitat?, NW (Gydouw Pass).•

lunulatus (A.Berger) L.Bolus (?incl. **L. convexus** (L.Bolus) L.Bolus, **L. pakpassensis** H.E.K.Hartmann, = *Ruschia pakhuisensis* L.Bolus)   Like **L. edwardsiae** but to 60 cm, leaves 8–10 ¥ 3–4 mm and green and flowers numerous. Sept. Sandy slopes, NW (Clanwilliam).•

macrocarpus (A.Berger) N.E.Br.   Robust, spreading shrublet, with short shoots from leaf axils. Leaves fused basally, inclined to recurved, 10–25 × 2–4 mm. Flowers 1–3 in groups, rose-purple, to 50 mm diam. Fruits to 18 mm diam., seeds brown. Oct.–Dec. Habitat?, SE (Uitenhage to Redhouse).•

macrosepalus (L.Bolus) L.Bolus   Sprawling to prostrate shrublet, with branches to 20 cm long. Leaves fused basally, spreading to erect, blunt to shortly acuminate, 3–7 cm long. Flowers 1–3 in groups, pink, to 60 mm diam. Sept. Habitat?, SE (Assegaaibos).•

macrostigma L.Bolus   Erect shrublet to 20 cm. Leaves erect, rough, glaucous, 7–14 × 1.5 mm. Flowers mostly solitary, white, c. 25 mm diam. Dec. Sandy montane valleys, NW (Cedarberg Mts).•

magnificus (L.Bolus) N.E.Br.   Sprawling to prostrate shrublet, fertile branches slender, ascending or erect. Leaves fused basally into a short sheath, erect, subcylindrical, acute, 25–45 mm long. Flowers solitary, crimson. c. 50 mm diam. Oct. Habitat?, SW (Malmesbury).•

marcidulus N.E.Br.   Shrublet to 20 cm, with flaccid branches. Leaves ascending to spreading, fused basally, trigonous, 20–40 × 4–5 mm. Flowers solitary on long pedicels, solitary, golden yellow, c. 35 mm diam. Flowering time? Habitat?, KM (Montagu).•

martleyi (L.Bolus) L.Bolus   Erect shrublet, branches angled below, shiny brown. Leaves erect, trigonous, glaucous, to 25 × 3 mm. Flowers silvery pink, to 40 mm diam. Nov. Habitat?, NW (Piketberg).•

matutinus (L.Bolus) N.E.Br.   Prostrate shrublet, branches rooting at nodes, fertile branches to 6 cm long. Leaves ascending, fused basally, subcylindrical, set with prominent dots, to 20 × 2 mm. Flowers solitary on short pedicels, golden yellow with red tips, c. 35 mm diam. Aug. Habitat?, NW (Piketberg).•

microsepalus L.Bolus   Low shrublet to 20 cm. Leaves bluntly keeled, recurved and oblique at tip, c. 9 × 1.5 mm. Flowers solitary, pink, c. 22 mm diam. July. Habitat?, NW (Kleinberg, near Tulbagh).•

middlemostii (L.Bolus) L.Bolus   Loosely branched shrublet, branches to 30 cm, fertile branches to 6 cm. Leaves fused basally, falcate, tapering above, to 8 mm long. Flowers mostly solitary on short pedicels, white to pale pink. Dec.–Jan. Habitat?, SW (Houwhoek).•

mucronatus L.Bolus   Compact shrublet to 15 cm. Leaves fused below into a sheath to 8 mm long, apiculate, 20–35 × 5 × 4–6 mm. Flowers solitary, white, 25–30 cm diam. Sept. Dry sandstone slopes, NW (Swartruggens).•

multiradiatus (Jacq.) N.E.Br. (incl. **L. incurvus** (Haw.) Schwantes, **L. roseus** (Willd.) Schwantes)   Spreading shrublet to 60 cm. Leaves compressed-trigonous, 15–25 × 2–4 mm. Flowers 1–3 in groups, pale pink, to 50 mm diam. Aug.–Nov. Stony slopes, SW (Cape Peninsula).•

multiseriatus (L.Bolus) N.E.Br. (?incl. **L. perreptans** L.Bolus)   Prostrate shrublet, branches slender and trailing, rooting at nodes. Leaves subfalcate, acute above, set with fine dots, to 15 mm long, narrow. Flowers solitary on pedicels 6–23 mm long, to 43 mm diam. Aug.–Oct. Sandy flats, LB (Swellendam to Riversdale).•

mutans (L.Bolus) N.E.Br. (?incl. **L. lavisii** (L.Bolus) L.Bolus)   Erect shrublet to 20 cm, with virgate branches. Leaves fused basally, narrow, erect, subcylindrical, tip somewhat recurved. Flowers on long pedicels, pink with yellow centre, c. 35 mm diam. Oct.–Nov. Habitat?, SE (Humansdorp to Port Elizabeth).•

neostayneri L.Bolus Erect shrublet to 30 cm, branches ascending. Leaves spreading to recurved, rough, green, to 15 mm long. Flowers 1–3 in groups, purplish pink, to 45 mm diam. Aug. Habitat?, NW (Piketberg).•

obconicus (L.Bolus) L.Bolus   Erect shrublet to 30 cm, branches shiny brown. Leaves ascending, trigonous, slightly rough, to 10 mm long. Flowers pink, c. 25 mm diam. Sept. Sandy flats, NW (Bokkeveld Plateau).•

pakhuisensis (L.Bolus) L.Bolus   Erect shrublet to 25 cm. Leaves erect, shortly acuminate, rough, to 15 × 2 mm. Flowers solitary on short pedicels, pink. Sept.–Oct. Sandstone slopes, NW (Pakhuis Mts).•

parcus N.E.Br.   Loosely branched, slender shrublet to 20 cm. Leaves obscurely keeled, rough, 4–9 × c. 1.5 mm. Flowers on short pedicels, petals pink with deeper pink margins, to 20 mm diam. Aug.–Sept. Habitat?, SW (Caledon).•

paucifolius (L.Bolus) N.E.Br.   Erect perennial to 25 cm. Leaves indistinctly keeled, acute, to 17 × 1–2 mm. Flowers on short pedicels, purplish, c. 25 mm diam. Sept. Habitat?, NW (Bokkeveld Plateau).•

peacockiae (L.Bolus) L.Bolus   Like **L. filicaulis** but leaves much shorter, more slender, and minutely papillate. Aug.–Sept. Clay hill slopes or flats, SW (Darling to Cape Peninsula).•

peersii (L.Bolus) N.E.Br.   Erect shrublet to 35 cm. Leaves inclined or ascending, trigonous, acute, glaucous, to 20 mm long. Flowers stalked, golden orange fading to red then rose-purple, 35–44 mm diam. July. Habitat?, NW (Graafwater).•

persistens (L.Bolus) L.Bolus   Erect shrublet to 25 cm, with slender, stiffly erect branches. Leaves recurved, to 25 mm long. Flowers in many-flowered dichasia on pedicels 25–40 mm long, purplish, to 14 mm diam. Fruits c. 5 mm diam. Oct. Habitat?, NW (Bokkeveld Mts).•

plenus (L.Bolus) L.Bolus   Spreading shrublet to 30 cm. Leaves ascending to erect, shortly acuminate, green, 10–15 × 2–3 mm. Flowers on long pedicels, pink, 35 mm diam. Aug.–Nov. Habitat?, NW (Olifants River valley).•

pocockiae (L.Bolus) N.E.Br.   Delicate, tangled shrublet, to 9 cm. Leaves erect or spreading, incurved at apex, rough, c. 11 × 2 mm. Flowers on short pedicels, solitary, pink to magenta, to 20 mm diam. Oct.–Dec. Habitat?, KM (Swartberg Mts).•

profundus (L.Bolus) H.E.K.Hartmann (= *Ruschia profunda* L.Bolus)   Erect shrublet to 36 cm, with old branches cylindrical and young ones compressed-winged. Leaves fused, sheath 2–3 mm long, ascending, back surface keeled, keel serrate, 10–17 × 5 mm. Flowers short-stalked, pink to purple, to 17 mm diam. Sept. Habitat?, NW (Piketberg).•

prominulus (L.Bolus) L.Bolus   Erect shrublet to 35 cm, with rigid branches. Leaves erect, with recurved tips, scabrid, 12–20 × c. 2 mm. Flowers stalked, purplish, c. 25 mm diam. Sept.–Oct. Habitat?, SE (George).•

promontorii (L.Bolus) N.E.Br.   Erect shrublet to 20 cm, with ascending branches. Leaves erect or spreading, fused basally, compressed-trigonous, shortly acuminate, 10–23 × 3 mm. Flowers on short pedicels, pale yellow, 14–23 mm diam. Dec.–Jan. Habitat?, SW (Cape Peninsula).•

purpureus L.Bolus   Loosely branched shrublet to 40 cm. Leaves erect, rough, glaucous, 15–35 mm long, slender. Flowers 1–3 on pedicels 35–45 mm long, purplish, to 30 mm diam. Fruits c. 8 mm diam. Nov. Rocky slopes, NW (Olifants River Mts: Keerom).•

recurvus (L.Bolus) Schwantes (= *Lampranthus tulbaghensis* (L.Bolus) N.E.Br., Habitat?*L. argillosus* L.Bolus)   Erect shrublet to 20 cm. Leaves spreading, adaxial surface convex, laterally compressed, 8–13 mm long. Flowers purplish, 18 mm diam. Sept. Habitat?, NW (Clanwilliam).•

reptans (Aiton) N.E.Br. (incl. **L. serpens** (L.Bolus) L.Bolus, **L. woodburniae** (L.Bolus) N.E.Br.)   Mat-forming perennial with trailing branches. Leaves in tufts, erect, 15–25 × 5–6 mm. Flowers on long pedicels, white or yellow. Aug.–Oct. Sandy flats, SW (Kalbaskraal to Cape Flats).•

rupestris (L.Bolus) N.E.Br.   Slender shrublet, to 15 cm, branches spreading, prostrate to creeping. Leaves fused basally, grey, rough, to 12 mm long. Flowers 1–3 on pedicels c. 20 mm long, white, to 25 mm diam. Oct. Shallow depressions on granite outcrops, SW (Paardeberg).•

rustii (A.Berger) N.E.Br.   Sprawling shrublet to 30 cm, branches slender, trailing, with numerous short shoots from leaf axils. Leaves trigonous, recurved with mucro, to 14 mm long, smaller on branchlets. Flowers many per inflorescence on pedicels 10–30 mm long, to 20 mm diam. Flowering time? Habitat?, LB (Riversdale).•

salicola (L.Bolus) L.Bolus Erect shrub, 20–30 cm. Leaves shortly fused, ascending, semicylindrical, shortly acuminate, glaucous, to 15 mm long, 2–3 mm wide and thick. Flowers stalked, pink, to 40 mm diam. Oct. Saline flats, NW (Rocher Pan to Velddrif).•

salteri (L.Bolus) L.Bolus   Sprawling shrublet to 15 cm, branches to 20 cm long. Leaves erect, acute or shortly acuminate, 15–30 × 1.5–3.0 mm. Flowers solitary on short pedicels, reddish with purple reverse, c. 30 mm diam. Dec. Habitat?, AP (Cape Agulhas).•

saturatus (L.Bolus) N.E.Br. (incl. **L. vanputtenii** L.Bolus)   Erect shrublet branching from base, to 50 cm. Leaves ascending to spreading, shortly fused below, to 40 mm long. Flowers solitary, magenta, without staminodes, to 60 mm diam. Fruits 5-locular. June–Aug. Sandstone outcrops, NW (Olifants River valley).•

sauerae (L.Bolus) L.Bolus   Like **L. stanfordiae** but leaves mucronate, flowers with few petaloid staminodes, petals red within, coppery on reverse. Sept. Habitat?, SW (Yzerfontein).•

scaber (L.) N.E.Br.   Richly and irregularly branched, deciduous shrublet to 30 cm, with tuberous rootstock to 15 cm long. Leaves sessile, trigonous, scabrid, 10–20 × 2–4 mm. Flowers mostly solitary, pale shiny lilac, to 30 mm diam. Aug. Granite hills in disturbed renosterveld, SW (Darling).•

schlechteri (Zahlbr.) L.Bolus   Shrublet with suberect to ascending branches, 13–20 cm. Leaves shortly fused, inclined, acute, 10–20 × 1 mm. Flowers 1–3 on short pedicels, salmon, to 50 mm diam. Nov. Sandy flats, SW (near Franschhoek).•

sociorum (L.Bolus) N.E.Br.   Prostrate perennial, branches rooting at nodes, internodes 2–3 mm long. Leaves ascending, trigonous to half-cylindrical, acute, glaucous, 10–25 × 3 mm. Flowers solitary on erect pedicels c. 4 mm long., pink, 30–35 mm diam., staminodes in a cone. Oct. Sandy slopes, NW, SW (Citrusdal to Koeberg).•

spiniformis (Haw.) N.E.Br. (incl. **L. nelii** L.Bolus)   Like **L. aduncus** but leaves more slender, stigmas thick and short and valve wings narrow. May–June. Stony slopes, NW, SW (Saron to Genadendal).•

staminodiosus (L.Bolus) Schwantes   Erect shrublet to 15 cm. Leaves slender, shortly acuminate, slightly scabrid, glaucous with purple tips, 8 mm long, slender. Flowers solitary on pedicels 10–18 mm long, dark purple, 12 mm diam. Fruits c. 7 mm diam. Nov. Sandstone slopes, NW (Elandskloof).•

stanfordiae L.Bolus   Slender erect shrublet to 25 cm. Leaves acute, slightly scabrid, to 17 × 2.5 mm. Flowers 1–3 in groups on pedicels to 30 mm long, orange, to 37 mm diam., without petaloid staminodes. Sept. Habitat?, SW (Yzerfontein).•

stenus (Haw.) N.E.Br.   Slender, much-branched shrublet to 50 cm, with arched, spreading branches. Leaves fused basally, subcylindrical, tapering, 12–30 × 2 mm. Flowers 1–3 in groups, pale mauve, to 30 mm diam. Nov.–Feb. Sandy flats, SW (Cape Peninsula and Cape Flats).•

subaequalis (L.Bolus) L.Bolus   Erect shrublet to 28 cm. Leaves erect, rough, mucronate, 8–12 mm long, slender. Flowers 1–3 in groups on pedicels 20–35 mm long, pale pink, c. 25 mm diam., sepals unequal. Oct.–Nov. Habitat?, SE (Uniondale).•

swartbergensis (L.Bolus) N.E.Br.   Erect shrublet to 35 cm, with angled branches. Leaves somewhat rough, 18 × 3 mm. Flowers solitary, purplish with a yellow centre, to 30 mm diam. Fruits c. 6 mm diam. Jan. Rocky slopes at c. 1600 m, KM (Swartberg Mts).•

tegens (F.Muell.) N.E.Br. (incl. **L. caespitosus** (L.Bolus) N.E.Br., **L. ernestii** (L.Bolus) L.Bolus) RANKVYGIE   Mat-forming perennial, branches slender, rooting at nodes, fertile branches erect, 10–15 mm long. Leaves ascending, trigonous, acute, to 8 mm long, rough. Flowers solitary on pedicels 6–12 mm long, purplish with paler centre, to 16 mm diam. Aug.–Oct. Flats, SW, LB (Darling to Swellendam, sometimes a weed in Australia).•

tenuifolius (L.) N.E.Br.   Sprawling shrublet to 20 cm, with slender branches often rooting at nodes. Leaves fused basally, subcylindrical, with translucent dots, c. 40 × 2 mm. Flowers solitary on long pedicels, deep pink, to 40 mm diam. Oct.–Nov. In sand, SW (Cape Peninsula).•

tulbaghensis (A.Berger) L.Bolus   Erect shrublet to 30 cm, with angled branches. Leaves trigonous, with rough dots, 5–10 mm long, slender, mucronate. Flowers ternate, white to magenta, c. 20 mm diam. Flowering time? Rocky slopes, often in moist sites, NW (Tulbagh to Worcester).•

uncus (L.Bolus) H.E.K.Hartmann (= *Ruschia marginata* L.Bolus, *Ruschia unca* (L.Bolus) L.Bolus; ? incl. **Lampranthus franceseae** H.E.K.Hartmann, **R. dubitans** (L.Bolus) L.Bolus, **R. marginata** L.Bolus, **R. leightoniae** L.Bolus)   Erect, stiff shrublet to 16 cm. Leaves fused below into a sheath, spreading, tip recurved, mucronate, 2–15 mm long. Flowers solitary, pink, 12 mm diam. Aug. Stony slopes and flats, NW (Matroosberg).•

uniflorus (L.Bolus) L.Bolus (= *Drosanthemum uniflorum* (L.Bolus) Friedr.   Cushion-forming shrublet to 1 × 0.8 m, stems whitish grey. Leaves cylindrical, obtuse, soft-fleshy, green, to 25 mm long. Flowers solitary, pink, stamens and staminodes in a central cone. Fruits funnel-shaped below. July–Sept. Stony flats, KM (W and Little Karoo to Great Karoo).

vallisgratiae (Schltr. & A.Berger) N.E.Br.   Slender shrublet to 20 cm. Leaves rough, 4–6 mm long, blunt. Flowers in small terminal clusters on pedicels 10–20 mm long, white to magenta, c. 15 mm diam. Nov.–Apr. Sandstone slopes, SW (Caledon).•

vanzijliae (L.Bolus) N.E.Br.   Like **L. reptans** but leaves longer (to 55 mm) and flowers larger (to 60 mm diam.), red. Aug.–Nov. Habitat?, NW (near Worcester).•

variabilis (Haw.) N.E.Br.   Shrublet with branches prostrate or spreading. Leaves fused basally, crowded on short shoots, trigonous, to 25 mm long, slender. Flowers 1–3, petals salmon-pink to pale mauve, c. 35 mm diam. Oct. Habitat?, SW (Saldanha).•

verecundus (L.Bolus) L.Bolus   Slender, laxly branched shrublet to 20 cm. Leaves slightly falcate, rough, 10–20 mm long, slender. Flowers solitary, pale pink, c. 25 mm diam. Fruits c. 7 mm diam. Dec. Sandstone slopes at c. 400 m, LB (Langeberg Mts: Riversdale).•

villiersii (L.Bolus) L.Bolus   Erect shrublet to 12 cm. Leaves subcylindrical, scabrid, tapering, to 15 × 1.5 mm. Flowers on short pedicels, ternate, purple, to 25 mm diam. Nov. Habitat?, SW (Mts near Villiersdorp).•

virgatus L.Bolus   Erect shrublet to 45 cm. Leaves linear, rough, 20–25 mm long. Flowers ternate, bright pink, to 25 mm diam. Sept. Habitat?, NW (Cedarberg Mts).•

watermeyeri (L.Bolus) N.E.Br.   Erect shrublet to 30 cm. Leaves fused basally, subcylindrical, 20–25 × 6 mm. Flowers mostly solitary on long pedicels, white, rarely purple, to 70 mm diam. July–Oct. NW (Namaqualand to Gifberg).

wordsworthiae (L.Bolus) N.E.Br.   Densely leafy, sprawling shrublet to 3 cm. Leaves fused basally, apex recurved or spreading, subcylindrical, mucronate, 15–28 × 3–5 mm. Flowers solitary on short pedicels, c. 25 mm diam. May–July. Sandstone slopes at c. 500 m, SW (Hottentots Holland Mts).•

[**Species excluded**   Incompletely known and possibly conspecific with one of the above: **L. curviflorus** (Haw.) H.E.K.Hartmann, **L. deflexus** (Aiton) N.E.Br., **L. inconspicuus** (Haw.) Schwantes, **L. versicolor** (Haw.) L.Bolus)]

LEIPOLDTIA   8 spp., S Namibia to E Cape

schultzei (Schltr. & Diels) Friedr. (= *Leipoldtia amplexicaulis* (L.Bolus) L.Bolus, *L. britteniae* (L.Bolus) L.Bolus, *L. constricta* (L.Bolus) L.Bolus, *L. jacobseniana* Schwantes)   Sprawling shrublet, branches to 70 cm, red, sometimes ochre. Leaves trigonous. Flowers 6–30 mm diam., petals purple. Fruits mostly 10-locular, base greyish, bell- to trumpet-shaped. Apr.–Sept. Sandy flats and slopes, NW, KM, LB, SE (Namaqualand to Humansdorp).

LITHOPS   36 spp., dry parts of southern Africa

localis (N.E.Br.) N.E.Br. (= *Lithops terricolor* N.E.Br.)   Acaulescent succulent. Leaf bodies compact, obconic, truncate, slightly convex, speckled. Flowers 20–30 mm diam., yellow. Fruits mostly 5-locular. Apr.–May. Black shale and gravel, KM, SE (W Karoo and Witteberg to E Cape).

MACHAIROPHYLLUM   DOLKVYGIE   10 spp., W to E Cape

acuminatum L.Bolus   Tufted perennial to 4 cm. Leaves pale green, trigonous. Flowers solitary on pedicels 40–60 mm long, yellow, to 50 mm diam. Nov. In crevices on rocky crests, SE (Uniondale to Humansdorp).•

albidum (L.) Schwantes   Tufted perennial to 10 cm, mat-forming with age. Leaves crowded, whitish. Flowers ternate, on long, 2-angled pedicels, yellow with reddish reverse, to 60 mm diam. Nov.–Dec. Rocky slopes, LB (Langeberg Mts: Robinson Pass).•

baxteri L.Bolus   Tufted perennial to 7 cm. Leaves bluish green, subtrigonous. Flowers solitary on pedicels to 75 mm long, red, 50–60 mm diam. Aug. Habitat?, SE (George).•

bijlii (N.E.Br.) L.Bolus   Tufted perennial to 20 cm. Leaves pale green, trigonous, angles acute. Flowers solitary on pedicels to 10 mm long, golden yellow with red reverse, to 60 mm diam. Oct. Rocky slopes, KM (Swartberg Pass).•

brevifolium L.Bolus Tufted perennial to 2 cm. Leaves grey-green. Flowers solitary on pedicels c. 15 mm long, yellow, to 35 m diam. Aug.–Sept. Red cliffs, KM (Little Karoo: Oudtshoorn).•

cookii (L.Bolus) Schwantes   Tufted perennial to 8 cm. Leaves bluish green, spreading, trigonous, fairly narrow, c. 9 mm diam. Flowers on pedicels to 40 mm long, greenish yellow, to 45 mm diam. Oct.–Nov. Stony slopes, KM (Barrydale).•

latifolium L.Bolus   Tufted perennial to 3 cm. Leaves 4–6 per branch, pale bluish, almost as wide as long. Flowers on pedicels c. 12 mm long, yellow, to 30 mm diam. Feb. Stony slopes, KM (Little Karoo: Oudtshoorn).•

stayneri L.Bolus   Like **M. albidum** but leaves blue-green. Aug.–Sept. Habitat?, SE (Uitenhage).•

MALEPHORA   VINGERKANNA   c. 15 spp., Namibia to E Cape

crassa (L.Bolus) H.Jacobsen & Schwantes   Prostrate perennial, branches to 5 mm diam., internodes 25–40 mm long. Leaves erect, fused below, almost terete, mucronate, to 40 mm long, 14 mm diam. Flowers solitary, golden yellow, to 60 mm diam. Aug.–Oct. Stony or gravelly slopes, NW (Clanwilliam to Karoopoort).•

crocea (Jacq.) Schwantes   Erect shrublet to 20 cm. Leaves crowded on short shoots, erect, older ones spreading, 25–45 × 6 mm. Flowers solitary on pedicels 25–45 mm long, terminal, red, to 30 mm diam. May–Sept. Sandy flats, NW, SW (Namaqualand to Saldanha).•

framesii (L.Bolus) H.Jacobsen & Schwantes   Prostrate succulent with white branches to 30 cm long. Leaves flat to slightly channelled, to 40 × 14 mm. Flowers solitary on pedicels to 10 mm long, yellow, opening in the evening. Aug.–Oct. Habitat?, NW, SW (S Namaqualand to Saldanha).

lutea (Haw.) Schwantes   Sprawling shrublet, branches with brown internodes. Leaves spreading, compressed-trigonous, to 45 × 4 mm. Flowers orange and yellow, to 25 mm diam. June–Dec. Habitat?, KM, LB, SE (Karoo and Little Karoo to Port Elizabeth).

luteola (Haw.) Schwantes   Like **M. lutea** but stems more slender, profusely branched and flowers numerous, small, yellow. Aug.–Sept. Habitat?, KM (Little Karoo).•

thunbergii (Haw.) Schwantes   Prostrate perennial with trailing branches to 40 cm long. Leaves crowded, to 50 × 8 mm, subcylindrical. Flowers solitary, terminal, yellow, 30–40 mm diam., opening at noon. Aug.–Nov. Habitat?, NW, KM, LB, SE (Vanrhynsdorp to Uitenhage, Karoo and E Cape).

uitenhagensis (L.Bolus) H.Jacobsen & Schwantes   Mat-forming succulent with trailing stems to 40 cm long. Leaves terete, to 25 × 5 mm, smooth, green. Flowers solitary on pedicels 15–25 mm long, yellowish or copper-red, to 35 mm diam. Apr.–Sept. KM, SE (Little Karoo to Uitenhage).•

MARLOTHISTELLA   1 sp., W and E Cape

stenophylla (L.Bolus) S.A.Hammer   Tufted perennial, root turnip-like, fleshy. Leaves acute, more or less half-cylindrical, smooth, punctate. Flowers solitary, purplish, to 35 mm diam. Fruits 5-locular, with large closing body. July–Aug. Grassland or quartz flats, KM, LB, SE (Karoo and Little Karoo to Uniondale).

MESEMBRYANTHEMUM   BRAKSLAAI, OLIFANTSLAAI   16 spp., dry parts of southern Africa (after Gerbaulet 1999, in prep.)

A. Basal leaves forming a rosette

crystallinum L.   Prostrate annual. Leaves at base forming a small rosette, flat, ovate-spathulate, bladder cells large. Flowers white or pinkish, 15–30 mm diam. Fruits with valve wings inflexed over valves. Nov.–Dec. Coastal sands, NW, SW, AP, SE (Lambert's Bay to E Cape).

guerichianum Pax (= *Mesembryanthemum alatum* (L.Bolus) L.Bolus, *M. intransparens* L.Bolus, *M. macrostigma* L.Bolus, *M. magniflorum* L.Bolus, *M. perlatum* Dinter, *M. purpureoroseum* L.Bolus, *M. quinangulatum* L.Bolus)   Like **M. crystallinum** but bladder cells much smaller and flowers larger, 25–55 mm diam. Sept.–Dec. Sandy flats, roadsides, NW, SW, LB, KM, SE (dry parts of southern Africa).

A. Basal leaves not forming a rosette

aitonis Jacq. (= *Mesembryanthemum louiseae* (L.Bolus) L.Bolus, *M. paulum* (N.E.Br.) L.Bolus)   Prostrate or ascending annual. Leaves flat, ovate- to lanceolate-spathulate, bladder cells distinct. Flowers white or pinkish, 5–20 mm diam. Fruits with valve wings reflexed and fused in pairs. Sept.–Jan. Pioneer, SW, LB, KM, SE (Caledon to E Cape and Karoo).

excavatum (L.Bolus) L.Bolus   Prostrate or decumbent annual. Leaves narrowly obovate, bladder cells distinct. Flowers yellow, c. 15 mm diam. Fruits with valve wings reflexed and fused in pairs. Oct. Disturbed sites, NW, KM (Ceres to Beaufort West).

longistylum DC. (= *Mesembryanthemum cryocalyx* L.Bolus)   Decumbent to erect annual to 40 cm. Leaves narrowly oblong, bladder cells inconspicuous. Flowers white, 10–20(–25) mm diam. Fruits with valve wings reflexed and fused in pairs. Sept.–Oct. Disturbed sites, NW, SW, KM, LB (Vanrhynsdorp to Malmesbury, Worcester to Witteberg and Riversdale).•

nodiflorum L. (= *Mesembryanthemum paucandrum* L.Bolus)   Like **M. excavatum** but leaves linear and flowers white or pinkish. Oct.–Nov. Disturbed places, roadsides, NW, SW, LB, SE (Namaqualand to E Cape).

stenandrum (L.Bolus) L.Bolus (= *Mesembryanthemum galpinii* (L.Bolus) L.Bolus)   Like **M. nodiflorum** but flowers larger, 15–30 mm diam., and petals more numerous. Nov.–Dec. Disturbed sites, NW (Namaqualand to Clanwilliam and Karoo).

subtruncatum L.Bolus   Prostrate annual. Leaves subcylindrical, almost truncate, bladder cells inconspicuous. Flowers pink, 20–25 mm diam. Oct. Disturbed places in quartz, NW, KM (Ceres, Montagu).•

MESTOKLEMA   DONKIEVYGIE   c. 6 spp., S Namibia to E Cape

tuberosum (L.) N.E.Br. ex Glen   Much-branched shrublet to 70 cm, with subglobose, tuberous roots. Leaves spreading, slender, trigonous, recurved at tips, finely papillate, 10–15 mm long. Flowers in terminal cymes, reddish yellow c. 8 mm diam. Fruits 5-locular, without closing bodies. Feb.–June. Dense scrub in alluvial soils, KM, SE (S Namibia to Little Karoo and E Cape).

MONILARIA   ERTJIEVYGIE   5 spp., N and W Cape

moniliformis (Thunb.) Ihlenf. & Jörgens.   Succulent shrublet to 12 cm, internodes barrel-shaped, soft, 5–12 × 6–12 mm. Leaves papillate, 10–15 mm long. Flowers on pedicels to 50 mm long, white, to 40 mm diam. Aug. Quartz patches, NW (Namaqualand to Klawer).

pisiformis (Haw.) Schwantes   Like **M. moniliformis** but internodes hard, 4–9 × 4–8 mm and flowers pink to white. Aug. Quartz patches, NW (S Namaqualand to Olifants River mouth).

OCTOPOMA   c. 8 spp., N and W Cape

abruptum (A.Berger) N.E.Br.   Stiffly erect shrublet to 20 cm. Leaves spreading, trigonous, acute, edges finely dentate and velvety. Flowers 1–3 in groups on short pedicels, c. 25 mm diam. Fruits 8-locular, with small closing bodies, with or without narrow wings. Aug. Dry stony slopes at c. 500 m, NW (W Karoo to Clanwilliam).

octojuge (L.Bolus) N.E.Br.   Erect, much-branched shrublet to 10 cm. Leaves spreading, somewhat trigonous above, keels finely denticulate, to 7 mm × 2–4 mm. Flowers solitary on pedicels to 10 mm long, white, 20–30 mm diam. Fruits 8-locular, with large, stalked closing bodies, expanding keels with awn-like extensions. Nov. Quartzite patches, KM (Little Karoo: Laingsburg, Ladismith).•

quadrisepala (L.Bolus) H.E.K.Hartmann (= *Ruschia quadrisepala* L.Bolus)   Like **O. octojuge** but leaves subglobose above, keels smooth and flowers pale pink. Flowering time? Slopes among quartzite pebbles, KM (Little Karoo: Laingsburg, Ladismith).•

rupigena (L.Bolus) L.Bolus   Compact shrublet to 15 cm, with tangled, contorted branches. Leaves spreading, c. 12 × 3–4 mm. Flowers solitary, bright pink, c. 25 mm diam. Fruits 8-locular, with small closing bodies. Sept. Among rocks, NW (Clanwilliam).•

OSCULARIA (= *LAMPRANTHUS* in part)   SANDSTEENVYGIE   c. 25 spp., N and W Cape

alba (L.Bolus) H.E.K.Hartmann (= *Lampranthus albus* (L.Bolus) L.Bolus)   Shrublet to 20 cm. Leaves lunate, pale glaucous, apiculate. Flowers in cymes, white. Fruits funnel-shaped below, 5-locular. Oct. Rock crevices, NW (W Karoo and Bokkeveld Mts).

cedarbergensis (L.Bolus) H.E.K.Hartmann (= *Lampranthus cedarbergensis* (L.Bolus) L.Bolus)   Erect shrublet to 15 cm. Leaves trigonous. Flowers 3–7 per cyme, pink. Fruits 5-locular. Flowering time? High rocky slopes, NW (Cedarberg Mts).•

compressus (L.Bolus) H.E.K.Hartmann (= *Lampranthus compressus* L.Bolus)   Diffusely branched shrublet to 20 cm. Leaves sublunate, truncate. Flowers in many-flowered cymes, pale pink, filaments purplish pink at tips. Fruits 5-locular. Flowering time? Habitat?, NW (Clanwilliam).•

deltoides (L.) Schwantes (= *Lampranthus deltoides* (L.) Glen)   Sprawling or rounded shrublet to 20 cm, with reddish, shining branches. Leaves deltoid, trigonous, glaucous, keel and margins toothed. Flowers in crowded cymes, pink. Fruits 5-locular. Oct.–Dec. Sandstone rocks, SW (Bainskloof to Jonkershoek).•

ebracteata (L.Bolus) H.E.K.Hartmann (= *Lampranthus ebracteatus* L.Bolus)   Erect shrublet to 25 cm. Leaves lunate, keeled. Flowers in cymes, pale pink. Fruits 5-locular. Nov. Habitat?, NW (Olifants River mouth).•

excedens (L.Bolus) H.E.K.Hartmann (= *Lampranthus excedens* (L.Bolus) L.Bolus)   Erect shrublet to 16 cm. Leaves laterally compressed, falcate. Flowers in 3–5-flowered cymes, bright pink, filaments paler. Fruits 5-locular. Oct. Habitat?, NW (Clanwilliam to Piketberg).•

guthrieae (L.Bolus) H.E.K.Hartmann (= *Lampranthus guthrieae* (L.Bolus) N.E.Br.   Shrublet to 15 cm, with stiff, tangled branches covered with persistent leaves. Leaves spreading, subfalcate, glaucous. Flowers 1–3 in groups, pink, to 20 mm diam. Fruits 5- or 6-locular. Oct. Rocky slopes, NW (Ceres).•

lunata (Willd.) H.E.K.Hartmann (= *Lampranthus lunatus* (Willd.) N.E.Br.) Sprawling shrublet. Leaves sabre-shaped to lunate, without dots. Flowers pink to purple. Flowering time? Rock crevices in light shade, NW (Bokkeveld Mts to Clanwilliam).•

ornata (L.Bolus) H.E.K.Hartmann (= *Lampranthus ornatus* L.Bolus) Erect shrublet to 25 cm. Leaves falcate, margins and keel reddish near apex, pale glaucous green. Flowers in terminal cymes, pink. Fruits c. 5 mm diam., top low. Sept. Rock crevices, NW (Clanwilliam).•

paardebergensis (L.Bolus) H.E.K.Hartmann (= *Lampranthus paardebergensis* (L.Bolus) L.Bolus) Compact shrublet to 20 cm, branches decumbent. Leaves lunate, red at apex. Flowers in many-flowered cymes, pink. Fruits 5-locular. Nov.–Dec. Rock crevices, SW (Malmesbury).•

piquetbergensis (L.Bolus) H.E.K.Hartmann (= *Lampranthus piquetbergensis* (L.Bolus) L.Bolus) Diffuse shrublet to 10 cm. Leaves subfalcate, apiculate, glaucous. Flowers in small cymes, pink. Flowering time? High rocky slopes, NW (Piketberg).•

prasinus (L.Bolus) H.E.K.Hartmann (= *Lampranthus prasinus* L.Bolus) Much-branched shrublet to 30 cm. Leaves falcate to almost globose, apiculate, finely papillate. Flowers in small cymes, purplish pink. Fruits 5-locular. Oct. High rocky slopes, NW (Piketberg).•

primivernus (L.Bolus) H.E.K.Hartmann (= *Lampranthus primivernus* (L.Bolus) L.Bolus) Like **O. ornata** but leaves lunate. July–Oct. Rock crevices, NW (Piketberg).•

steenbergensis (L.Bolus) H.E.K.Hartmann (= *Lampranthus steenbergensis* (L.Bolus) L.Bolus) Decumbent shrublet to 10 cm. Leaves falcate. Flowers in cymes, deep pink, filamentous staminodes covering stamens. Fruits 5-locular. June–July. Granite outcrops, SW (St Helena Bay to Langebaan).•

superans (L.Bolus) H.E.K.Hartmann (= *Lampranthus superans* (L.Bolus) L.Bolus) Erect, diffusely branched shrublet to 25 cm. Leaves falcate, glaucous. Flowers in cymes, bright pink, filamentous staminodes white with pink tips. Fruits 5-locular. Oct. Habitat?, NW (Vanrhynsdorp to Clanwilliam).•

thermara (L.Bolus) H.E.K.Hartmann (= *Lampranthus thermarum* (L.Bolus) L.Bolus) Erect shrublet to 20 cm. Leaves falcate, glaucous. Flowers in compact cymes, petals pink, filamentous staminodes apically recurved. Fruits 5-locular. Oct. Sandstone slopes, NW (Olifants River Mts).•

vernicolor (L.Bolus) H.E.K.Hartmann (= *Lampranthus vernicolor* (L.Bolus) L.Bolus) Diffusely branched shrublet to 10 cm. Leaves falcate, apiculate. Flowers in cymes, petals bright pink, filamentous staminodes recurved, filaments white with pink tips. Fruits 5-locular. Sept.–Oct. Rocky slopes, NW (Tulbagh).•

vredenburgensis (L.Bolus) H.E.K.Hartmann (= *Lampranthus vredenburgensis* L.Bolus) Diffusely branched shrublet to 25 mm. Leaves lunate. Flowers in small cymes, bright pink, filamentous staminodes white with pink tips. Fruits 5-locular. Aug. Granite rocks, SW (Vredenburg).•

PHYLLOBOLUS (= *SPHALMANTHUS*) VINGERKANNA 32 spp., southern Africa

A. Shrublets with fibrous roots

nitidus (Haw.) Gerbaulet (= *Sphalmanthus littlewoodii* L.Bolus) Decumbent to erect shrub to 30 cm (sometimes cushion-like or scrambling), stems weakly woody and with a conspicuous cork layer. Leaves decussate, with conspicuous bladder cells. Flowers 20–30 mm diam., reproductive parts exposed. Fruits 4- or 5-locular. Sept.–Oct. Dry flats, NW, LB, KM (W and S Karoo to Swellendam).

saturatus (L.Bolus) Gerbaulet Sprawling to erect shrublet with a thickened base and woody stems, roots fibrous. Leaves decussate, with small bladder cells. Flowers c. 30 mm diam., reproductive parts exposed. Fruits 5-locular. Sept.–Oct. Karroid bush, NW, LB (Namaqualand to Ceres).

splendens (L.) Gerbaulet (= *Aridaria brevifolia* L.Bolus, *Sphalmanthus splendens* (L.) L.Bolus) Like **P. saturatus** but taller, to 1 m, and leaves becoming alternate in inflorescence. Oct.–Dec. Dry flats and lower slopes, LB, SE (Worcester to E Cape and Karoo).

suffruticosus (L.Bolus) Gerbaulet Erect to sprawling shrublet with woody stems to 70 cm, roots fibrous. Leaves decussate, alternate in inflorescence, with small bladder cells. Flowers 20–30 mm diam., reproductive parts exposed. Fruits 5-locular. Flowering time? Rocky slopes, SW (Malmesbury).•

trichotomus (Thunb.) Gerbaulet (= *Sphalmanthus trichotomus* (Thunb.) L.Bolus) Erect shrublet with woody stems to 20 cm, roots fibrous. Leaves almost cylindrical, narrowly ovate, decussate, with small, xeromorphic bladder cells. Flowers 10–20 mm diam., reproductive parts exposed. Fruits 4-locular. Oct.–Nov. Rocky flats, sometimes on quartz gravel, NW (Namaqualand to Piketberg).

A. Geophytes with tuberous roots

canaliculatus (Haw.) Gerbaulet (= *Sphalmanthus canaliculatus* (Haw.) N.E.Br.)   Geophyte with long, creeping branches, softly woody and rooting at nodes, roots tuberous. Leaves decussate, alternate in inflorescence, with prominent bladder cells. Flowers 20–30 mm diam., reproductive parts exposed. Fruits 5-locular. Nov. Coastal dunes, SW, AP, SE (Cape Peninsula to Port Elizabeth).•

caudatus (L.Bolus) Gerbaulet (= *Sphalmanthus caudatus* (L.Bolus) N.E.Br.)   Prostrate to decumbent or scrambling geophyte with softly woody stems, roots tuberous. Leaves decussate, alternate in inflorescence, with prominent bladder cells. Flowers to 40 mm diam., reproductive parts exposed. Fruits 5-locular. Sept. Rocky flats, NW (Worcester Karoo).•

grossus (Aiton) Gerbaulet (= *Sphalmanthus grossus* (Aiton) N.E.Br., *S. subpetiolatus* (L.Bolus) L.Bolus)   Prostrate to scrambling geophyte with stems becoming woody at base, roots tuberous. Leaves decussate, alternate in inflorescence, with prominent bladder cells. Flowers 20–30 mm diam., reproductive parts exposed. Fruits 5-locular. Sept.–Oct. Slopes and flats, often scrambling in bush, NW, LB, SE (Ceres to Grahamstown).

pumilus (L.Bolus) Gerbaulet   Tufted geophyte with short, softly woody stems, roots tuberous. Leaves decussate, alternate in inflorescence, with large bladder cells. Flowers c. 30 mm diam., reproductive parts exposed. Fruits 5-locular. Flowering time? Rocky areas, LB, KM (Montagu to Sutherland).

resurgens (Kensit) Schwantes (= *Sphalmanthus micans* L.Bolus, *S. resurgens* (Kensit) L.Bolus)   Tufted geophyte with a thick caudex and short stems, corky below. Leaves alternate, tufted, with large bladder cells. Flowers often solitary, c. 40 mm diam., reproductive parts hidden. Fruits 5-locular. June–Sept. Shallow soil, NW, KM (Namaqualand to Ceres and Witteberg).

viridiflorus (Aiton) Gerbaulet   Sprawling geophyte with weakly lignified stems with a thick cork layer at base. Leaves decussate, alternate in inflorescence, with large bladder cells. Flowers c. 40 mm diam., reproductive parts hidden, blue to greenish white. Fruits 5-locular. Flowering time? Quartzite flats or slopes, NW (Clanwilliam).•

PLEIOSPILOS   KWAGGAVY   4 spp., W Cape and Karoo to Free State

compactus (Aiton) Schwantes (= *Pleiospilos brevisepalus* L.Bolus, *P. grandiflorus* L.Bolus, *P. kingiae* L.Bolus)   Stemless, much-branched succulent perennial. Leaves more than twice as long as broad, becoming broader towards apex. Flowers solitary, pale yellow. Fruits with large closing bodies. Apr.–June. Stony, NE-facing slopes, KM, LB (Karoo and Little Karoo).

PRENIA   6 spp., S Namibia to E Cape

englishiae (L.Bolus) Gerbaulet   Prostrate perennial with whitish stems. Leaves obtusely trigonous, linear. Flowers whitish to yellow. Fruits 4-locular. Oct.–Nov. Disturbed places, NW, LB (Worcester-Robertson Karoo).•

pallens (Aiton) N.E.Br. (= *Prenia relaxata* (Willd.) N.E.Br.)   Like **P. englishiae** but leaves depressed-trigonous, ovate or narrowly ovate. Fruits 4- or 5-locular. Sept.–Dec. Mainly clay and granite slopes, NW, SW (Namaqualand to Cape Peninsula).

radicans (L.Bolus) Gerbaulet   Prostrate perennial rooting at nodes. Leaves subcylindrical, linear. Flowers whitish to pale yellow. Fruits 4-locular. Oct.–Mar. Karroid flats and disturbed areas, SE (Uniondale to E Cape).

tetragona (Thunb.) Gerbaulet   Erect perennial. Leaves subcylindrical, slender. Flowers white to yellowish or pinkish. Fruits 4-locular. Sept.–Mar. Karroid flats or disturbed areas, NW, KM, LB, SE (S Namibia and Free State to S Cape).

vanrensburgii L.Bolus   Mat-forming perennial. Leaves flat, ovate. Flowers whitish to slightly yellowish. Fruits 5-locular. Oct. Along seashore, SW, AP (Hermanus to Bredasdorp).•

PSILOCAULON   ASBOS   13 spp., southern Africa

articulatum (Thunb.) N.E.Br. (= *Psilocaulon hirtellum* L.Bolus)   Decumbent to erect annual or perennial to 30 cm, branches with hair-like bladder cells. Flowers several, pink or white. Fruits 5-locular. Nov.–Dec. Pioneer, mostly disturbed sites, NW, SW, KM, LB, SE (S Namibia to E Cape and Free State).

bicorne (Sond.) Schwantes   Perennial, prostrate with cluster of short branches in centre and long trailing flowering shoots. Flowers few, white. Fruits 5-locular. Oct.–Nov. Pioneer, mostly disturbed sites, NW, KM, LB, SE (Worcester to E Cape and Karoo).

coriarium (Burch. ex N.E.Br.) N.E.Br. (= *Psilocaulon absimile* N.E.Br., *P. mentiens* (A.Berger) N.E.Br.) Decumbent to erect shrub to 1.5 m. Flowers few to many, white, 10–13 mm diam. Fruits 4- or 5-locular. Oct.–Jan. Pioneer of disturbed sites, KM, LB, SE (central Namibia and Karoo to Little Karoo and E Cape).

dinteri (Engler) Schwantes   Spreading to prostrate shrublet to 45 cm, branch internodes becoming barrel-shaped in saline soils, smooth or with domed bladder cells. Flowers few to many, purple or pink. Fruits 5-locular. Oct.–Nov. Pioneer, often in quartz gravel or saline soils, NW, SW, KM (S Namibia to W Cape and Little Karoo).

granulicaule (Haw.) Schwantes   Decumbent to erect annual or biennial to 25 cm, branches with dome-shaped bladder cells. Flowers 1–few, pale yellow or cream-coloured. Fruits 5-locular. Oct.–Nov. Pioneer, KM, LB, SE (Namibia to Little Karoo, E Cape and Free State).

junceum (Haw.) Schwantes (= *Psilocaulon acutisepalum* (A.Berger) N.E.Br., *P. utile* L.Bolus)   Like *P. coriarium* but only 15–55 cm and flowers usually larger, to 15 mm diam., white or pink, petals usually broad, to 2 mm. Oct.–Nov. Pioneer, NW, SW, KM, LB, SE (W Cape to E Cape and Karoo).

parviflorum (Jacq.) Schwantes   Prostrate, mat-forming perennial, with stems smooth, slender, 1–2 mm diam., when young. Flowers 1–few, white. Fruits 4-locular. Oct.–Dec. Pioneer, on clay flats, NW, SW, AP, LB (Clanwilliam to Mossel Bay).•

RHINEPHYLLUM   c. 14 spp., W Cape and Karoo

muirii N.E.Br.   Succulent perennial with a fleshy caudex, forming clumps of many shoots. Leaves inclined, 10–25 × 5–10 mm, with small whitish tubercles in upper half, margins and keel with a white cartilaginous edge. Flowers solitary on pedicels to 12 mm long, yellowish white, to 14 mm diam. Fruits 5-locular, without tubercle or covering membranes. Oct.–Nov. Quartz fields and shales, KM, LB (Little Karoo).•

RHOMBOPHYLLUM   5 spp., E Cape

dolabriforme (L.) Schwantes   Tufted perennial forming small, densely branched solitary tufts. Leaves sickle-shaped, with a tooth-like projecting tip, dull green. Flowers to 4 cm diam., yellow. June–Aug. On rocky, shale hills, SE (Uitenhage to Graaff-Reinet).

rhomboideum (Salm-Dyck) Schwantes   Compact, little-branched perennial. Leaves hatchet-shaped above, broadly channelled below, margins often twisted. Flowers in tall cymes, yellow with pink reverse. Fruits 5-locular, with 2-lobed closing bodies. Nov. Gravelly patches in bush, SE (Port Elizabeth to Graaff-Reinet).

RUSCHIA   VYGIE   c. 400 spp., dry parts of southern Africa

A. Leaves free to base or very shortly fused

capornii (L.Bolus) L.Bolus (= *Lampranthus capornii* (L.Bolus) L.Bolus)   Erect shrublet to 30 cm. Leaves trigonous, easily shed. Flowers in axillary cymes, magenta, to 10 mm diam. Fruits unknown. Mar.–Apr. Habitat?, NW (Namaqualand to Tulbagh).

cradockensis (O.Kuntze) H.E.K.Hartmann & Stüber (= *Eberlanzia horrescens* (L.Bolus) L.Bolus, *E. triticiformis* (L.Bolus) L.Bolus)   Erect shrublet to 30 cm, with dominant long shoots. Leaves to 10 mm long. Flowers few per dichasia, 7–14 mm diam. Fruits mostly on short shoots, sometimes in terminal spiny dichasia, valve wings absent. July–Aug. Stony flats in karroid bush, NW (Namaqualand to Ceres to Cradock).

decurrens L.Bolus   Erect, stiffly branched shrublet to 15 cm. Leaves decurrent at base, trigonous, finely velvety. Flowers in ternate cymes, sometimes solitary, to 16 mm diam. Fruits 5-locular. Flowering time? Habitat?, NW (Vanrhynsdorp to Clanwilliam).•

divaricata L.Bolus (= *Eberlanzia divaricata* (L.Bolus) L.Bolus)   Shrublet to 35 cm, often with thickened fusiform roots, branches whitish beige. Leaves 10–15 mm long, with small prominent dots. Flowers in dense, spiny, terminal cymes, pink, c. 15 mm diam. Fruits with narrow valve wings. Flowering time? Sandy flats, NW (S Namibia and Beaufort West to Ceres).

esterhuyseniae L.Bolus   Erect shrublet to 20 cm, branches virgate, reddish brown. Leaves triquetrous, smooth, with recurved tips, 7–11 mm long. Flowers solitary, terminal, purplish, 36 mm diam. Fruits 5-locular, with funnel-shaped base. Nov. Lower slopes, SE (Uniondale).•

fourcadei L.Bolus   Erect, rigidly branched shrublet to 15 cm. Leaves trigonous with recurved mucro. Flowers solitary or in ternate cymes, pink, to 26 mm diam. Fruits grey. Dec.–Jan. In loam in karroid scrub, KM, SE (Swartberg Mts to Uitenhage).•

fredericii (L.Bolus) L.Bolus   Spreading shrublet to 50 cm, with ascending to erect branches. Leaves subcylindrical, glaucous, c. 20 × 3 mm. Flowers 1–3 in groups, purple, c. 16 mm diam. Fruits unknown. Sept.–Oct. Habitat?, NW (Clanwilliam to Worcester).•

intricata (N.E.Br.) H.E.K.Hartmann & Stüber (= *Eberlanzia persistens* (L.Bolus) L.Bolus)   Shrublet to 50 cm, branches turning grey with age. Leaves obtuse, smooth with convex epidermal cells. Flowers in much-branched, spiny cymes. Fruits without valve wings. Sept.–Oct. Habitat?, NW (Clanwilliam).•

intrusa (Kensit) L.Bolus   Low shrublet to 15 cm, branching from base, branches enclosed by remains of old leaves. Leaves swollen-trigonous, obtuse, entire, 55 × 9 mm. Flowers solitary, to 40 mm diam., pinkish-purple. Fruits 5-locular. June–July. Habitat?, NW, LB (Worcester to Swellendam).•

knysnana (L.Bolus) L.Bolus   Erect shrublet to 40 cm. Leaves trigonous, apiculate, light shining green, shortly fused below, 33 × 5 mm. Flowers in cymes on pedicels to 25 mm long, pink, 25 mm diam. Fruits 6-locular. Jan.–Mar. Habitat?, SE (Knysna to Humansdorp).•

lapidicola L.Bolus   Erect, many branched, deciduous shrublet to 30 cm. Leaves subfalcate, acute, green, 9–15 × 3–6 mm. Flowers in cymes, magenta, c. 15 mm diam. Fruits 5-locular. July–Oct. Rocky upper slopes, NW (Clanwilliam).•

leptocalyx L.Bolus   Prostrate shrublet to 10 cm. Leaves trigonous, with sharp keels, to 20 × c. 5 mm. Flowers solitary or in cymes, red, petals c. 8 mm long. Fruits 5-locular, grey. July–Aug. Coastal slopes, AP (Stilbaai).•

neovirens Schwantes ex H.E.K.Hartmann   Spreading shrublet to 60 cm, with slender, compressed branches, greenish or brownish, becoming grey and rounded. Leaves compressed trigonous, covered with dots. Flowers in cymes on pedicels 20–30 mm long, red, to 40 mm diam. Fruits unknown. Flowering time? Sandy flats, SE (Bethelsdorp to Port Elizabeth).•

orientalis L.Bolus   Stiffly erect, lax shrublet to 35 cm. Leaves trigonous, to 20 × 4 mm. Flowers solitary, purplish, 20–30 mm diam. Fruits 5-locular, c. 8 mm diam. Flowering time? In open bushland, SE (Port Elizabeth to Alexandria).

spinosa (L.) Dehn (= *Eberlanzia mucronifera* (Haw.) Schwantes)   Shrublet to 70 cm, with whitish grey branches turning dark with age. Leaves 10–20 mm long, epidermal cells convex to papillate. Flowers in spiny, branched cymes, magenta, to 28 mm diam. Fruits mostly with narrow valve wing. Sept.–Oct. Dry karroid flats, NW (S Namibia to Clanwilliam).

staminodiosa L.Bolus   Erect, much-branched shrublet to 15 cm. Leaves trigonous, tip recurved, glaucous, to 15 × 5 mm. Flowers solitary, purplish, c. 25 mm diam. Fruits 5-locular. Oct. Habitat?, KM, LB (Ladismith, Riversdale).•

vanbredai L.Bolus   Slender tangled shrublet to 15 cm. Leaves falcate to subfalcate, 10–17 × 3–4 mm. Flowers in cymes, magenta, petals 3–8 mm long. Fruits 5-locular. June. Habitat?, SE (Uniondale).•

virgata (Haw.) L.Bolus   Erect shrublet, with slender, ascending to prostrate branches, at first compressed, later cylindrical, grey. Leaves shorter than internodes, trigonous, 15–20 mm long, glabrous. Flowers solitary, terminal, stalked, red, 20–25 mm diam. Fruits 5-locular. Oct. Gravelly soil, SE (Uniondale).•

A. Leaves connate below
B. Leaf blades usually much shorter than connate sheath, rarely as long

altigena (L.Bolus) L.Bolus   Decumbent shrublet with branches to 15 cm. Leaves subterete. Flowers solitary, magenta, to 23 mm diam. Fruits 5-locular. Sept.–Dec. Rocky sandstone slopes at c. 1700 m, KM (Witteberg).•

approximata (L.Bolus) Schwantes   Laxly branched shrublet to 15 cm. Leaves trigonous, keel 1–3-denticulate. Flowers solitary on pedicels enclosed in bracteoles, pink. Fruits 5-locular. Nov. Habitat?, KM (Montagu).•

archeri L.Bolus   Succulent shrublet to 30 cm. Leaves trigonous, basally fused. Flowers solitary, magenta. Fruits 5(6)-locular. July. Habitat?, KM, LB (Little Karoo: Ladismith to Swellendam).•

grisea (L.Bolus) Schwantes   Lax shrublet to 30 cm, with stiff, spreading, grey branches. Leaves fused with stem in lower 5 mm, free parts 2–4 mm long, with (1–)2 small apical teeth. Flowers solitary, subsessile, pale pink, c. 14 mm diam. Fruits 5-locular, with high rims. Oct.–Nov. Habitat?, KM (Little Karoo: Montagu).•

impressa L.Bolus   Compact shrublet to 6 cm, with spreading branches. Leaves fused with stem in lower 10 mm, free parts trigonous, 2–5 mm long, pale glaucous green. Flowers solitary, pink, c. 10 mm diam. Fruits 5-locular, top with high rims. Nov.–Dec. Habitat?, KM (Ladismith).•

vanniekerkiae L.Bolus   Low shrublet, with branchlets 1–2 mm diam. Leaves trigonous, recurved, glaucous, 4–7 × 2–3 mm. Flowers solitary, subsessile, pink. c. 18 mm diam. Fruits 5-locular. Oct.–Nov. Habitat?, KM (Ladismith).•

B. Leaf blades longer than connate sheath

amicorum (L.Bolus) Schwantes   Robust shrublet to 40 cm. Leaves subterete. Flowers in ternate cymes, magenta, c. 15 mm diam. Fruits 5-locular. Apr.–Sept. Habitat?, KM (Montagu).•

aristulata (Sond.) Schwantes   Prostrate succulent, rooting from nodes, branches compressed. Leaves trigonous, acute, margins finely ciliate. Flowers enclosed by bracteoles, ?pink. Fruits 5-locular. Flowering time? Rock outcrops, SW (Cape Peninsula).•

bolusiae Schwantes   Erect shrublet to 15 cm. Leaves glaucous green, almost smooth. Flowers in cymes, pink, c. 20 mm diam. Fruits 5-locular. May–Sept. Habitat?, NW (Clanwilliam).•

burtoniae L.Bolus   Erect shrublet to 60 cm, branches shiny dark brown, with spongy pith. Leaves subterete, basal sheath slightly swollen. Flowers in cymes with thin stalks, pink, to 20 mm diam. Fruits 5-locular, with low rims, closing bodies absent. May–June. Rocky flats, NW (Bokkeveld Mts and Gifberg).•

calcicola (L.Bolus) L.Bolus   Sprawling shrublet, branches trailing, becoming white with age. Leaves trigonous, apiculate, keel serrulate, slightly scabrid. Flowers solitary, enclosed by bracteoles, magenta, to 18 mm diam. Fruits 5-locular, top and rims low. July. Limestone slopes and flats, AP (Agulhas to Riversdale).•

caroli (L.Bolus) Schwantes   Spreading shrub with decumbent, grey to reddish branches. Leaves trigonous, with green dots. Flowers in cymes, magenta. Fruits unknown. Aug.–Sept. Habitat?, NW, KM (Clanwilliam to Montagu and Robertson).•

cedarbergensis L.Bolus   Erect shrublet to 25 cm. Leaves of a pair almost U-shaped, separated at sheath by a line 4–6 mm long, subterete, slightly scabrid. Flowers in many-flowered cymes, magenta. Fruits 5-locular, top low, base funnel-shaped, closing bodies small. Sept. Habitat?, NW (Clanwilliam).•

ceresiana L.Bolus   Erect shrub to 1 m. Leaves subterete, sheath a little swollen, leaves completely separated by a deep line along sheath. Flowers in many-flowered terminal cymes on pedicels to 30 mm long, magenta. Fruits 5-locular, top low, base shortly funnel-shaped, closing bodies moderately broad hoods. June–Nov. Habitat?, NW (Clanwilliam to Ceres).•

cincta (L.Bolus) L.Bolus   Sprawling shrublet to 25 cm. Leaves trigonous, with recurved mucro, keel and margins minutely toothed. Flowers mostly solitary, on pedicels to 12 mm long, pink with purple margins, to 22 mm diam. Fruits 5-locular, top raised, base funnel-shaped, closing bodies small hooks. Aug. Habitat?, LB (Riversdale).•

copiosa L.Bolus   Erect shrublet, 30–50 cm, densely branched, stems maroon. Leaves semiterete, sheath basally tumid, smooth. Flowers in many-flowered terminal cymes with additional lower storeys, magenta. Fruits 5-locular. Sept. Habitat?, NW (Clanwilliam).•

costata L.Bolus   Prostrate perennial with trailing branches with internodes 25–35 mm long. Leaves ascending to erect, subterete, 25–35 × 4 mm. Flowers in cymes on pedicels 30–50 mm long, magenta, to 20 mm diam. Fruits 5-locular. Aug. Habitat?, LB (Montagu).•

cupulata (L.Bolus) Schwantes   Erect shrublet to 20 cm, with rigid, grey, ascending branches. Leaves erect, with short shoots from axils, trigonous, keel finely papillate, 15–30 × 3–4 mm. Flowers in cymes, to 28 mm diam. Fruits 5-locular, base funnel-shaped. Oct.–Dec. Coastal areas, SW (Vredenburg to Bokbaai).•

cymbifolia (Haw.) L.Bolus   Shrublet to 30 cm, branches filiform. Leaves spreading, apically recurved, boat-shaped, blunt, grey-green, punctate, to 15 mm long. Flowers solitary or few in cymes, purplish. Fruits 5-locular. May–July. Habitat?, LB (Riversdale to Mossel Bay).•

decumbens L.Bolus   Prostrate shrublet with branches 4–6-leaved, old dry leaves persistent. Leaves erect, glaucous to purple, scabrid, 10–18 × 6 mm. Flowers solitary, purplish, to 35 mm diam. Fruits 5-locular, base bell-shaped. Dec. Habitat?, KM (Ladismith).•

decurvans L.Bolus   Erect shrublet to 45 cm, branches with grey bark. Leaves erect to spreading, glaucous, to 50 × 5 mm. Flowers in many-flowered terminal cymes raised above leaves, purplish, 20–26 mm diam. Fruits unknown. Sept.–Oct. Habitat? NW (S Namaqualand to Clanwilliam).

densiflora L.Bolus   Rounded shrublet to 45 cm. Leaves subterete, scabrid. Flowers in compact cymes raised above leaves, purple, c. 14 mm diam. Fruits 5-locular, funnel-shaped, with high rims. Nov. Coastal flats, NW (Lambert's Bay).•

dichroa (Rolfe) L.Bolus   Decumbent shrub, with short branches to 10 cm. Leaves trigonous, glaucous, minutely dotted, margins denticulate. Flowers subsessile, white or red with white centre, 35–40 mm diam. Fruits 6-locular. May–July. Sandstone sheets, NW (Bokkeveld to Pakhuis Mts).•

diversifolia L.Bolus   Sprawling shrublet to 14 cm, branches trailing, stiff, reddish turning grey, to 25 cm long. Leaves trigonous, edges cartilaginous, slightly serrate, to 70 mm long on long shoots, shorter on short shoots. Flowers in cymes, purplish, to 28 mm diam. Fruits unknown. May–June. Habitat?, NW (Tulbagh).•

duthiae (L.Bolus) Schwantes   Prostrate perennial with trailing branches, flowering branches erect, 7–10 cm long. Leaves trigonous, 20–80 × 7 mm. Flowers solitary or in cymes, purplish with darker streaks, c. 24 mm diam. Fruits 5-locular. Sept. Habitat?, SE (Knysna).•

exigua L.Bolus   Compact shrublet. Leaves subglobose, c. 10 × 3 mm. Flowers solitary, pink, 15–20 mm diam. Fruits unknown. July. Habitat?, LB (Montagu).•

festiva (N.E.Br.) Schwantes   Erect, deciduous shrublet to 30 cm, branches papillose on young parts. Leaves erect or suberect, c. 30 mm long, subcylindrical, glaucous, with recurved tips. Flowers in cymes, to 25 mm diam. Fruits 5–7-locular. Sept.–Nov. Habitat?, NW, SW (Vanrhynsdorp to Langebaan).•

filipetala L.Bolus   Erect shrublet to 30 cm, branches with cylindrical internodes 25–35 mm long. Leaves trigonous, erect, c. 23 × 3.5 mm. Flowers in much-branched inflorescences, mauve, petals filiform, c. 15 mm diam. Fruits 5-locular. Sept.–Oct. Habitat?, NW (Clanwilliam).•

geminiflora (Haw.) Schwantes   Prostrate shrublet with slender, trailing, grey branches to 90 cm long. Leaves trigonous, apically recurved, keel and margins slightly dentate. Flowers in cymes or paired, purple, striate. Fruits 5-locular. Sept.–Oct. In sand, SW (Kalbaskraal to Cape Peninsula).•

gracilis L.Bolus   Sprawling shrublet to 15 cm, with trailing branches to 60 cm. Leaves trigonous, hamate at tip, 15–40 × 2–3 mm diam. Flowers in small clusters, pink to pale pink, slightly striate. Fruits 5-locular. Sep.–Oct. Coastal slopes, often on limestone, NW, SW, AP (Clanwilliam to Stilbaai).•

incumbens L.Bolus   Erect shrublet to 45 cm. Leaves trigonous to subterete, mucronate, glaucous, 25–40 × 4 mm. Flowers in cymes, pink, c. 15 mm diam. Fruits 5-locular. Sept.–Oct. Habitat?, NW (Klawer to Clanwilliam).•

incurvata L.Bolus   Erect shrublet to 50 cm. Leaves subterete, shortly acuminate and mucronate, c. 40 × 5 mm, sheath 6 mm long. Flowers in many-flowered cymes, pale pink with an obscure pink stripe, to 20 mm diam. Fruits 5-locular. Sept.–Oct. Habitat?, NW (S Namaqualand to Clanwilliam).•

indecora (L.Bolus) Schwantes   Erect shrublet to 40 cm. Leaves recurved at tips, swollen at sheaths, glaucous, c. 21 × 3 mm at base. Flowers in cymes, silvery white with staminodes tipped magenta, 8–10 mm diam. Fruits 5-locular. Sept.–Oct. Coastal sands, SW (Melkbosstrand to Cape Peninsula).•

insidens L.Bolus   Densely branched shrublet to 15 cm, with grey stems c. 2 mm thick. Leaves trigonous, to 11 × 3 mm, papillate. Flowers pink, 18 mm diam. Fruits unknown. Apr. Habitat?, NW (Bokkeveld Mts to Gifberg).•

intermedia L.Bolus   Erect shrublet to 35 cm. Leaves trigonous to subterete and subfalcate, 20–30 × 3 mm. Flowers in cymes, purplish, petals striate, 10–12 mm long. Fruits 5-locular. Oct. Rocky sandstone flats, NW (Porterville Mts).•

klipbergensis L.Bolus   Erect, laxly branched shrublet to 40 cm, with ascending, purple stems. Leaves subterete, apically recurved, 25–35 × 2–4 mm. Flowers in cymes, petals pink, c. 10 mm long. Fruits 5-locular, base funnel-shaped. Mar.–July. Habitat?, SW (Darling).•

langebaanensis L.Bolus   Robust, prostrate shrublet, with stiff branches c. 6 mm diam. Leaves erect, mucronate, green to light blue, 35 × 4 mm. Flowers 3–5 in cymes, magenta, petals 5–10 mm long. Fruits 5-locular. Aug.–Sept. Rock crevices on granite, SW (Saldanha).•

lavisii L.Bolus   Erect shrublet to 40 cm. Leaves trigonous, acuminate, sheaths swollen, pale glaucous, 20–25 × 5 mm. Flowers 3–5 in cymes, magenta, c. 22 mm diam. Fruits c. 5 mm diam., with conspicuous closing bodies. Aug. Rock crevices, SW (Gordon's Bay).•

lineolata (Haw.) Schwantes   Sprawling, much-branched shrublet to 15 cm. Leaves trigonous, acuminate, with short sheath clasping stem, with line on sheath, c. 10 × 4 mm. Flowers solitary, white or purple, to 20 mm diam. Fruits unknown. Aug.–Sept. Habitat?, SW, LB, SE (Caledon to Humansdorp).•

littlewoodii L.Bolus   Compact succulent to 5 cm, with thickened rootstock to 9 cm long. Leaves trigonous, subulate, 10–18 × 4 mm at base. Flowers solitary, magenta. Fruits unknown. Aug. Dry fynbos or karroid bush, NW (Ceres).•

macowanii (L.Bolus) Schwantes   Sprawling shrublet to 20 cm, with prostrate branches. Leaves subterete, 20–35 × 4 mm, sheath swollen, c. 5 mm long. Flowers in cymes, pink, c. 22 mm diam. Fruits 5-locular. Aug.–Oct. Coastal rocks, SW, AP (Yzerfontein to Agulhas).•

maxima (Haw.) L.Bolus   Shrub to 1.5 m. Leaves lunate, trigonous, laterally compressed with acute keel, 70 mm long, flanks 20 mm wide, grey to whitish grey. Flowers in many-flowered cymes, pink, to 20 mm diam. Fruits 5-locular. July. Rocky places, NW (Clanwilliam).•

misera (L.Bolus) L.Bolus   Stiffly branched shrublet to 40 cm. Leaves subfalcate, fused below, 17 × 2.5 mm, sheath c. 2 mm long, covered with fine velvety papillae. Flowers in many-flowered cymes, petals absent, filaments white, c. 7 mm diam. Fruits 5-locular. Oct.–Nov. Habitat?, NW (Clanwilliam).•

montaguensis L.Bolus   Erect, stiffly branched shrublet to 30 cm. Leaves erect, subterete, glaucous, 13–20 × 4 mm, fused below, sheath with a distinct line. Flowers in cymes, magenta, petals c. 7 mm long, with an indistinct stripe. Fruits 5-locular. Sept.–Oct. Habitat?, LB (Montagu).•

muiriana (L.Bolus) Schwantes   Erect shrublet to 20 cm. Leaves ascending, subulate with recurved tips, surface rough from elevated dots, c. 25 × 3 mm, sheath c. 6 mm long. Flowers solitary, purplish, c. 13 mm diam. Fruits 5-locular. Aug.–Oct. Habitat?, KM (Ladismith).•

multiflora (Haw.) Schwantes   Shrub to 1 × 2 m, with erect branches. Leaves fused below into a sheath, trigonous, shortly mucronate. Flowers in many-flowered cymes, subsessile, white, to 30 mm diam. Fruits unknown. Oct.–Dec. Habitat?, KM, LB (W Karoo to Swellendam).

patens L.Bolus   Erect shrublet to 25 cm, with reddish branches, later turning grey. Leaves spreading, recurved, later withering and persisting, keel and margins serrate. Flowers 1–3 per cyme, purplish, to 24 mm diam. Fruits 5-locular. May–June. Habitat?, NW (? near Piketberg).•

pauciflora L.Bolus   Shrublet with creeping branches, to 40 cm, branches compressed, winged, light brown, shining. Leaves spreading, trigonous, mucronate, c. 25 × 4 mm. Flowers solitary, purplish, 12 mm diam. Fruits 5-locular. May. Habitat?, NW (probably Piketberg district).•

pulchella (Haw.) Schwantes   Prostrate shrublet with trailing branches covered with old leaves. Leaves crowded, trigonous, fused below, leaf tip ending in a fine point, 7–10 × 3 mm, margins ciliate-pubescent. Flowers solitary, rose-pink, c. 20 mm diam. Fruits 5-locular. July–Sept. Clay flats and slopes, SW (Cape Peninsula to Stellenbosch).•

pungens (A.Berger) H.Jacobsen (= *Ruschia cymosa* (L.Bolus) Schwantes)   Erect shrublet with grey-brown branches. Leaves fused below and forming a swollen sheath, cylindrical to trigonous, mucronate, 20–30 mm long. Flowers in compound cymes, magenta, to 15 mm diam. Fruits 5-locular, Sept.–Dec. Habitat?, KM, SE (Montagu to E Cape).

radicans L.Bolus   Creeping shrublet, branches to 35 cm, rooting at nodes. Leaves trigonous, finely velvety, 10–15 × 6 mm at base. Flowers solitary, pink, 18 mm diam. Fruits 5-locular. May–June. Habitat?, NW (Clanwilliam).•

rariflora L.Bolus   Sprawling shrublet with branches spreading to prostrate. Leaves subcylindrical, recurved apically, glaucous, 40–80 × 5 mm diam. sheath 8 mm long, with a line. Flowers in a compound cyme, pink, 24 mm diam. Fruits 5-locular. Aug.–Sept. Habitat?, NW (Clanwilliam).•

rigida (Haw.) Schwantes   Rounded shrublet to 40 cm. Leaves trigonous, acuminate, sheath 5–10 mm long, free parts usually slightly shorter, rough on keel. Flowers white, to 10 mm diam. Fruits unknown. Jan.–Feb. Habitat?, SW, LB (Caledon to Swellendam).•

rigidicaulis (Haw.) Schwantes   Sprawling shrublet with rigid branches. Leaves trigonous, dotted. Flowers 3 or 4 in clusters, pink with central red stripes, c. 30 mm diam. Fruits unknown. Aug.–Sept. Habitat?, NW (Clanwilliam).•

rostella (Haw.) Schwantes   Shrublet with slender prostrate branches. Leaves curved upwards when young, each pair forming a beak, later forming a basal persisting sheath around younger leaves, subterete, grey-green. Flowers solitary, terminal, whitish, to 25 mm diam. Fruits 5-locular. July–Aug. Habitat?, KM (Ladismith).•

rubricaulis (Haw.) L.Bolus   Small shrublet with angular branches, reddish when young. Leaves trigonous, margins and keel serrulate, 25–40 × 4–6 mm. Flowers, solitary, rose-pink, to 15 mm diam. Fruits 5-locular. May–Aug. Coastal slopes, SW (Cape Peninsula).•

sarmentosa (Haw.) Schwantes   Prostrate shrublet, branches to 45 cm, trailing and rooting at nodes. Leaves trigonous, slightly rough, to 50 × c. 6 mm. Flowers 1–3 in terminal clusters, reddish with a dark stripe, petals 8 mm long. Fruits 5-locular. July–Aug. Sandy flats, SW (Malmesbury to Cape Flats).•

schollii (Salm-Dyck) Schwantes   Low shrublet, with erect, grey branches. Leaves crowded, trigonous, recurved, 15–20 × 2 mm, dotted. Flowers 1–3 in terminal clusters, magenta, c. 18 mm diam. Fruits base funnel-shaped, top with high rims. Aug.–Sept. Habitat?, SW (Stellenbosch to Caledon).•

strubeniae (L.Bolus) Schwantes   Erect shrublet to 75 cm, with compressed, red branches. Leaves shortly fused below, more or less falcate, margins serrate, 40–60 × 5–6 mm. Flowers in cymes, pink with purple stripe, 32 mm diam. Fruits unknown. Aug.–Nov. Habitat?, NW (Piketberg).•

suaveolens L.Bolus   Erect shrublet to 50 cm. Leaves trigonous, apically recurved, velvety from small papillae, 30–60 × 3–4 mm. Flowers in many branched inflorescences, pink, to 15 mm diam. Fruits 5-locular. Oct.–Jan. Habitat?, NW (Vanrhynsdorp to Clanwilliam).•

subpaniculata L.Bolus   Erect shrub to 30 cm. Leaves subterete, to 30 × 3 mm. Flowers in many-flowered cymes, pink, petals 6 mm long. Fruits 5-locular. Aug.–Sept. Coastal sands, NW (Lambert's Bay).•

subteres L.Bolus   Erect shrublet to 25 cm. Leaves inclined, subcylindrical, more or less smooth, glaucous, 25–30 × 5 mm. Flowers 5–15 per inflorescence, magenta, petals 6 mm long. Fruits 5-locular. Oct. Habitat?, NW (Worcester).•

tardissima L.Bolus   Compact shrub with prostrate, compressed branches to 40 cm. Leaves trigonous, fused below forming a sheath, the lower 50–70 mm long, the upper shorter, c. 5 mm broad. Flowers ternate, pink with purplish stripe, petals 14–17 mm long. Fruits 5-locular. June–July. Habitat?, NW (W Karoo to Clanwilliam and Ceres).

tecta L.Bolus   Erect shrub to 1 m. Leaves erect, S-shaped, fused below into a sheath, sheath slightly swollen. Flowers in many-flowered cymes, purplish with white centre, petals 17 mm long. Fruits 6-locular. Sept.–Oct. Sandy coastal flats, SW (Saldanha to Melkbosstrand).•

tenella (Haw.) Schwantes   Much-branched slender shrublet. Leaves fused below, free parts 6–12 mm long, sharply trigonous, recurved. Flowers 1 or 2, small, white. Fruits 5-locular. Nov.–May. Habitat?, LB, SE (Riversdale to Uniondale).•

triflora L.Bolus   Compact shrublet to 10 cm, internodes grey to whitish, enclosed by the dry and persistent leaf sheaths. Leaves erect, subcylindrical, c. 30 × 3 mm. Flowers ternate, pink, 16 mm diam. Fruits 5-locular. Mar.–Apr. Habitat?, NW (Clanwilliam).•

tumidula (Haw.) Schwantes (= *Ruschia caudata* L.Bolus)   Erect shrublet to 40 cm with reddish branches. Leaves subterete, slightly rough. Flowers in many-flowered cymes, magenta. Fruits 5–8-locular. Nov.–Dec. Sandy flats, SW, LB (Malmesbury to Swellendam).•

uitenhagensis (L.Bolus) Schwantes   Erect, densely branched shrublet to 30 cm. Leaves recurved, shortly fused below, glaucous dotted, 5–9 × 3 mm at base. Flowers solitary, pink, 14 mm diam. Fruits 5-locular. Flowering time? Karroid scrub, SE (Uitenhage).•

victoris (L.Bolus) L.Bolus   Much-branched shrublet with branches to 6 mm diam. Leaves fused at base into a sheath for 5 mm, velvety, 20–65 × 6 mm. Flowers in cymes, pinkish with purple stripe, 15 mm diam. Fruits 5-locular. June. Flats, NW (Clanwilliam).•

virens L.Bolus   Erect shrublet to 30 cm, with robust branches to 12 mm diam. at base. Leaves falcate, truncate above, 20–28 × 6 mm. Flowers 1–3 in clusters, pink, c. 28 mm diam. Fruits 4–6-locular. Sept.–Oct. Sandstone slopes at c. 360 m alt., LB (Riversdale to Mossel Bay).•

[**Species excluded**   Genus uncertain, see end of family: **R. promontorii**. Poorly known and probably conspecific with one of the above: **R. umbellata** (L.) Schwantes]

SAPHESIA•   1 sp., W Cape

flaccida (Jacq.) N.E.Br.   Deciduous, woody perennial to 1.5 m, with long black caudex and long branches. Leaves alternate, sessile, flat, flaccid, 50 × 8 mm. Flowers terminal on long pedicels, yellow and white, to 40 mm diam. Fruits 5–7-locular, xerochastic. Oct.–Nov. Sandy flats, NW, SW (Clanwilliam to Malmesbury).•

SCELETIUM   KANNA, KOUGOED   8 spp., W Cape, Karoo, and Namaqualand

crassicaule (Haw.) L.Bolus   Prostrate to decumbent, sometimes scrambling perennial. Leaves not imbricate, tips recurved, to 40 mm long, bladder cells flattened. Flowers stalked, white to pale yellow or pale pink, 20–30 mm diam. Fruits 5- or 6-locular, with valve wings. Oct.–Nov. Dry karroid slopes, KM, SE (Willowmore to Grahamstown).

expansum (L.) L.Bolus (= *Sceletium regium* L.Bolus)   Prostrate to sprawling perennial. Leaves not imbricate, tips recurved, to 65 mm long, bladder cells small. Flowers stalked, yellow, c. 40 mm diam., filamentous staminodes concealing reproductive parts. Fruits 5-locular, with valve wings. Sept.–Nov. Stony slopes, NW, SW (Clanwilliam to Malmesbury).•

strictum L.Bolus   Prostrate to decumbent, sometimes scrambling perennial. Leaves not imbricate, tips recurved, to 40 mm long, bladder cells flattened. Flowers stalked, white to pale yellow, 20–30 mm diam. Fruits 4- or 5-locular, with valve wings. May–Sept. Rock outcrops, KM (Ladismith).•

tortuosum (L.) N.E.Br. (= *Sceletium compactum* L.Bolus, *S. framesii* L.Bolus, *S. joubertii* L.Bolus) Prostrate to scrambling perennial. Leaves imbricate, tips incurved, to 40 mm long, bladder cells large. Flowers subsessile, white to pale yellow, pale salmon or pale pink, c. 20–30 mm diam. Fruits 4- or 5-locular, with valve wings. July–Sept. Often on quartzite, NW, KM, LB, SE (Namaqualand to Montagu and Aberdeen).

varians (Haw.) Gerbaulet (= *Sceletium subvelutinum* L.Bolus)   Erect or decumbent and scrambling perennial. Leaves not imbricate, tips recurved, to 20 mm long, bladder cells small. Flowers stalked, white to pale yellow, c. 40 mm diam., filamentous staminodes concealing reproductive parts. Fruits (4)5-locular, with valve wings. Aug.–Oct. Stony slopes, NW, SW (Worcester, Robertson).•

SCOPELOGENA   2 spp., N and W Cape

bruynsii Klak   Like **S. veruculata** but flowers yellow, pink or salmon and fruits smaller, opening and closing repeatedly. Sept.–Oct. Low sandstone cliffs, NW (Namaqualand and Tanqua Karoo to Clanwilliam).

veruculata (L.) L.Bolus (incl. **S. gracilis** L.Bolus)   Woody perennial to 25 cm, forming cushions when old. Leaves crowded, fused basally, erect, trigonous to cylindrical, glaucous, very soft. Flowers in cymes, yellow, rarely white, to 15 mm diam. Fruits 5-locular, remaining open, valves without wings. Oct.–Dec. Rocky sandstone slopes, often below low cliffs or rocky ledges, SW, LB (Cape Peninsula to Riversdale).•

SKIATOPHYTUM•   PLATBLAARVYGIE   1 sp., W Cape

tripolium (L.) L.Bolus   Annual or biennial to 20 cm, with stout, fleshy, branching roots. Leaves opposite or alternate, flat, shining, margins wavy when young, c. 80 × 20 mm. Flowers (1)2–5 in terminal cymes on pedicels to 60 mm long, white, to 30 mm diam. Fruits 5-locular, xerochastic. Sept. Shady slopes, NW, SW (Clanwilliam to Bredasdorp).•

SMICROSTIGMA•   1 sp., W Cape

viride (Haw.) N.E.Br.   Low shrubby perennial to 40 cm, with erect branches succulent when young, woody with age. Leaves fused below into a laterally compressed sheath 20–25 mm long, blades shorter, slightly trigonous, with recurved tip, smooth. Flowers subsessile, solitary, terminal, pink, to 30 mm diam. Fruits 7–10-locular, without closing bodies. Aug.–Nov. Karroid scrub, NW, LB, SE (Worcester to Uniondale).•

STAYNERIA•   1 sp., W Cape

neilii (L.Bolus) L.Bolus   Stout, erect, loosely branched shrub with long woody branches to 60 cm. Leaves ascending, clasping stem, 70 × 8 mm. Flowers ternate, white to pink, c. 40 mm diam. Fruits 6–9-locular, woody, without closing bodies. June. Clay slopes, NW (Nuy to Robertson).•

STOEBERIA   BOOMVYGIE   5 spp., S Namibia to W Cape

utilis (L.Bolus) Van Jaarsv. (= *Ruschia utilis* (L.Bolus) L.Bolus)   Erect, densely branched shrub to 2 m. Leaves blunt, 10–15 × 3–5 mm, velvety. Flowers white, c. 10 mm diam. Fruits without closing bodies, with stiff persistent funicles. Oct. Strandveld, NW (Alexander Bay to Elandsbaai).

STOMATIUM   40 spp., Namaqualand to W Cape and Karoo

braunsii L.Bolus   Low tufted succulent, with fertile branches 6-leaved. Leaves of a pair unequal, adaxial surface flat, densely spotted, abaxial surface convex, margins 3- or 4-dentate, teeth tipped with bristles. Flowers sessile, yellow, c. 24 mm diam., noc6turnal. July–Aug. Habitat? NW (Vanrhynsdorp to Clanwilliam).•

suricatinum L.Bolus   Like **S. braunsii** but leaves obscurely dotted, flowers on short pedicels, lemon-yellow, c. 20 mm diam. May–Oct. Habitat? KM (Laingsburg to Little Karoo).

TANQUANA   TANKWABEESKLOU   3 spp., W Karoo and Little Karoo

hilmarii (L.Bolus) H.E.K.Hartmann & Liede (= *Pleiospilos hilmarii* L.Bolus)   Succulent perennial sunken in the ground, unbranched. Leaves in equal pairs, shorter than 20 mm. Flowers deep yellow. Fruits with small closing bodies. Mainly Mar.–July. Shale banks, KM (Little Karoo S of Laingsburg).•

TETRAGONIA   KINKELBOS, KLAPPERBRAK   60 spp., Africa, S America, Australia

A. Stamens as many as the sepals

caesia Adamson   Like **T. echinata** but flowers solitary in axils and fruits barrel-shaped with 3 or 4 vertical ridges. July–Oct. Sandy flats, SW (False Bay to Gansbaai).•

echinata Aiton (incl. **T. microptera** Fenzl)   Prostrate, succulent annual to 30 cm. Leaves ovate to orbicular. Flowers subsessile, 2–4 in axils, greenish, stamens as many as the sepals. Fruits globose, with spiny ridges and horns. June–Sept. Sandy slopes and disturbed ground, NW, SW, KM (Namibia to Stellenbosch, Ladismith to Grahamstown).

A. Stamens more than twice as many as the sepals
B. Plants entirely herbaceous, annual or tuberous
C. Flowers sessile or subsessile

chenopodioides Eckl. & Zeyh.   Tuberous perennial with sprawling branches to 40 cm. Leaves rhomboid, fleshy, reddish beneath. Flowers in axillary and terminal clusters, subsessile, yellow-green, styles twice as long as perianth. Fruits 8-ridged. July–Sept. Coastal sand, NW, SW (Clanwilliam to Mamre).•

galenioides Fenzl   Like **T. chenopodioides** but leaves ovate to oblong, stamens 5–8 (vs numerous), style as long as perianth and fruits 4-winged. Flowering time? Sandstone slopes, NW (Cedarberg Mts).•

C. Flowers pedicellate

halimoides Fenzl   Like **T. nigrescens** but fruits larger, 2.0–2.5 cm long (vs 0.6–1.7 mm long), flowers 4–5 mm. Aug.–Oct. Sand or clay among shrubs, NW, SW (Bokkeveld Mts to Paarl).

herbacea L.   Tuberous perennial with sprawling stems to 50 cm. Leaves obovate to oblanceolate. Flowers long-pedicellate, in terminal umbels and solitary in upper axils, bright yellow, 4–6 mm. Fruits pear-shaped, smooth, ridged when dry. Mainly June–Aug. Mostly clay and granite slopes, NW, SW (Klawer to Bredasdorp).•

nigrescens Eckl. & Zeyh. (incl. **T. portulacoides** Fenzl)   Tuberous perennial with sprawling stems to 50 cm. Leaves obovate to suborbicular, often red beneath. Flowers in axillary and terminal umbels mostly on slender pedicels, yellow or cream-coloured, sometimes orange, reverse often purple, 3–4 mm. Fruits winged. July–Oct. Sandy and clay slopes and flats, NW, SW, AP, LB, SE (Namaqualand and W Karoo to Humansdorp).

sphaerocarpa Adamson   Like **T. nigrescens** but fruits rounded, covered with rigid, spreading scales. Sept.–Oct. Sandy flats, NW (Piketberg).•

B. Plants softly woody at least below, from a woody base
D. Sepals connate basally

hirsuta L.f.   Coarsely hairy subshrub with sturdy erect stems to 50 cm. Leaves rhomboid-oblanceolate, often crowded below. Flowers subsessile, in axillary and terminal spikes or tufts, yellow, large, 5–8 mm, sepals connate and constricted below. Fruits 4-winged, with knobs between wings, coarsely hairy, 15–25 mm. Aug.–Oct. Rocky sandstone slopes, NW (Bokkeveld Mts to Porterville, ? Du Toitskloof).•

rosea Schltr.   Glabrescent subshrub with sprawling branches to 60 cm. Leaves rhomboid-oblanceolate. Flowers subsessile or pedicellate, 1–few in upper axils or terminal, magenta, sepals fused basally. Fruits 4-winged with knobs between wings, 15–20 mm. Mainly Aug.–Sept. Sandstone slopes, NW (Bokkeveld Mts to Piketberg).•

D. Sepals free

arbuscula Fenzl   Like **T. spicata** but leaves smaller, flowers all axillary, fruits 4-winged. Mainly June–Sept. Dry stony slopes NW, KM, LB (Karoo and Clanwilliam to Mossel Bay).

decumbens Mill.   Sprawling perennial with branches to 1 m. Leaves papillose-hirsute, obovate-oblong, fleshy. Flowers in branched axillary clusters shorter than leaves, yellow. Fruits with rigid wings. Mainly Aug.–Nov. Coastal dunes, NW, SW, AP, SE (S Namibia to E Cape).

fruticosa L.   KLIMOPKINKELBOSSIE   Like **T. spicata** but branches often long and trailing through bush, leaves with margins recurving. Mainly Sept.–Nov. Granite and sandstone slopes, especially along the coast, NW, SW, AP, KM, SE (Namaqualand to Clanwilliam to Port Elizabeth .

glauca Fenzl   Softly woody shrublet to 50 cm. Leaves elliptic-lanceolate, glaucous, margins narrowly revolute. Flowers solitary in axils, pedicels slender, yellow. Fruits pendent, 4-winged with narrow ridges between. July–Sept. Karroid areas, NW, KM (Namibia and W Karoo to Little Karoo).

namaquensis Schltr.   Subshrub with ascending branches to 30 cm. Leaves elliptic, papillose hirsute, margins slightly revolute. Flowers on axillary clusters on short pedicels, yellow, 3–4 mm. Fruits softly winged, to 10 mm. June–Oct. Shale rocks, NW, SW (Namaqualand and W Karoo to Hermanus).

saligna Fenzl (incl. **T. erecta** Adamson, **T. lasiantha** Adamson)   Sprawling shrublet to 60 cm. Leaves linear-oblanceolate, leathery. Flowers subsessile, in axillary and terminal clusters, yellow, styles 1 or 2(–4). Fruits ovoid, ridged to subglobose. Sept.–Nov. Rocky slopes, NW, SW, KM (Namaqualand and Karoo to Caledon and Oudtshoorn).

sarcophylla Fenzl (incl. **T. distorta** Fenzl, **T. robusta** Fenzl)   Compact, twiggy shrub to 40 cm. Leaves small, oblanceolate, margins revolute, midribs prominent beneath. Flowers 1–few in upper axils and in terminal racemes, yellow, 2–3 mm. Fruits 4-winged, 4–15 mm. June–Sept. Dry stony slopes, NW, KM, SE (Namaqualand to Grahamstown).

spicata L.f. (incl. **T. calycina** Fenzl)   Erect or sprawling shrub to 1 m. Leaves rhomboid-lanceolate. Flowers mostly in terminal racemes, or 1–few in upper axils, yellow, 3–4 mm. Fruits broadly winged, with knobs between wings. July–Oct. Granite and sandy slopes, NW, SW, KM, LB, SE (Namaqualand and Karoo to Grahamstown).

verrucosa Fenzl   Glistening subshrub with sprawling branches to 30 cm. Leaves oblanceolate, margins slightly revolute, often incurved-ascending. Flowers subsessile in upper axils and on terminal racemes, yellow. Fruits winged, 15–20 mm. Aug.–Sept. Stony karroid slopes, NW, KM (Namibia to W Little Karoo).

virgata Schltr.   Like **T. spicata** but flowers mostly axillary and fruits small. July–Aug. Sandy slopes, NW (Namaqualand to Clanwilliam).

[**Species excluded**   Poorly known and probably conspecific with one of the above: **T. haworthii** Fenzl, **T. macroptera** Pax]

TRICHODIADEMA   DIADEM VYGIE   34 spp., S Namibia to W Cape and Free State

A. Leaves with apical hairs but no diadem of coloured, sclerotinized bristles

attonsum (L.Bolus) Schwantes   Erect shrublet to 15 cm, branches shortly papillate. Leaves with 1–3 apical papillae, 10–13 × 3–4 mm. Flowers solitary on pedicels to 12 mm long, white, with filamentous staminodes. Fruits (4)5-locular. Mar.–June. Quartz outcrops, LB (Riversdale district).•

calvatum L.Bolus   Shrublet to 15 cm, with thickened rootstock, branches papillate. Leaves without a diadem, sparsely bristly when young, glaucous, 10–20 × 3 mm. Flowers on pedicels to 15 mm long, pink, without filamentous staminodes. Fruits 5-locular. Sept. Habitat? NW (Robertson).•

fergusoniae L.Bolus   Compact, closely leafy shrublet to 30 mm, with fibrous roots. Leaves soft, densely papillate, to 9 × 2.5 mm, with 1–few apical hairs. Flowers on pedicels to 8.5 mm long, salmon with red midline, with filamentous staminodes. Fruits 5-locular, with 2-lobed closing bodies. Aug. Habitat?, LB (Riversdale).•

occidentale L.Bolus   Closely leafy, succulent perennial to 20 mm, with tuberous roots. Leaves imbricate, erect, glaucous, papillate, with apical papillae. Flowers solitary on pedicels to 8 mm long, salmon, with filamentous staminodes. Fruits 5-locular, with 2-lobed closing bodies. June–July. Habitat? SW (Bredasdorp).•

pygmaeum L.Bolus   Compact shrublet to 3 cm, with fibrous roots. Leaves imbricate, with apical papillae, to 6 × 3 mm. Flowers subsessile, pink, to 20 mm diam., without filamentous staminodes. Fruits unknown. July. Habitat? LB (Swellendam).•

A. Leaves with an apical diadem of coloured, sclerotinized bristles

barbatum (L.) Schwantes (= *Trichodiadema stellatum* (Mill.) Schwantes, *Ttichodiadema stelligerum* (Haw.) Schwantes)   KAREEMOER   Low succulent with tangled stems and fleshy roots. Leaves grey-green

from acute papillae, slightly recurved, with a diadem, 8–12 × 3–4 mm. Flowers solitary in axils, mauve, purple or red, to 30 mm diam. Fruits 5-locular. Apr.–Sept. Habitat?, KM, LB, SF (Little Karoo and Mossel Bay to Uitenhage).•

bulbosum (Haw.) Schwantes   Erect shrublet to 20 cm, with tuberous roots. Leaves grey-green from papillae, subcylindrical, with a diadem, 5–8 × c. 3 mm. Flowers solitary, terminal, subsessile, magenta. Fruits 5-locular. Nov.–Feb. Habitat?, SE (Port Elizabeth).•

burgeri L.Bolus   Like **T. densum** but plants less compact and fruits 5-locular. Flowering time? Habitat?, KM (Oudtshoorn).•

densum (Haw.) Schwantes   Compact, tufted shrublet with thickened fleshy roots. Leaves crowded, with acute papillae, with an apical diadem, 15–20 × 4–5 mm. Flowers carmine, 40–50 mm diam. June–Oct. Habitat?, KM, SE (Uniondale to Willowmore and Great Karoo).

emarginatum L.Bolus   Like **T. marlothii** but fruits 5-locular. July. Habitat?, KM (Little Karoo).•

fourcadei L.Bolus   Erect to spreading shrublet to 12 cm, with tuberous roots, branches with papillate internodes. Leaves brown, pubescent, with apical hairs resembling a diadem only in colour and size, 15–25 × 2–3 mm. Flowers solitary, white, with filamentous staminodes. Fruits 5-locular. Oct. Stony slopes, at c. 100 m, SE (Humansdorp).•

gracile L.Bolus   Like **T. fourcadei** but more slender and flowers pink with filamentous staminodes purple. Apr.–Aug. Dry, stony slopes, SW, LB (Caledon to Mossel Bay).•

hallii L.Bolus   Compact, closely leafy shrublet to 6 cm, with tuberous roots. Leaves imbricate, with a diadem, 11–18 × 4 mm. Flowers sessile or on short pedicels, pink with white centre, with filamentous staminodes. Fruits unknown. Flowering time? Habitat?, KM (Ladismith).•

intonsum (Haw.) Schwantes   Erect to spreading perennial. Leaves distant, inclined, slightly recurved, with a diadem, c. 13 × 4 mm. Flowers solitary on short pedicels, white or pink, to 20 mm diam. Fruits 5-locular. Oct. Habitat?, SE (Uitenhage).•

marlothii L.Bolus   Decumbent shrublet to 4 cm, branches with internodes to 12 cm long, not visible. Leaves ascending, glaucous, papillate, with a diadem, to 17 × 6 mm. Flowers pink, filamentous staminodes white with pink tip, 30–35 mm diam. Fruits 6-locular. May–June. Habitat?, LB (Swellendam).•

mirabile (N.E.Br.) Schwantes   Erect shrublet to 8 cm, branches densely white-hairy. Leaves connate basally, fresh green, papillate, subcylindrical, with a diadem, 12–26 × 4–6 mm. Flowers subsessile, white, to 40 mm diam. Fruits 6-locular. Nov.–Jan. Stony slopes, KM, SE (Witteberg to Uitenhage).•

orientale L.Bolus   Erect shrublet to 10 cm, with thickened roots, branches with papillate internodes. Leaves erect, glaucous, densely papillate, with a diadem, 10–14 × 5 mm. Flowers on pedicels to 6 mm long, pale pink becoming darker towards tips, with filamentous staminodes. Fruits 5-locular. Feb.–June. Rock outcrops, SE (Uitenhage to Ciskei).

rupicola L.Bolus   Decumbent to erect shrublet to 30 cm, with fibrous roots. Leaves imbricate, basally fused, felted-papillose, with a diadem. Flowers subsessile, whitish, with filamentous staminodes. Fruits unknown. Feb.–May. Rocky slopes, SE (Uitenhage).•

stayneri L.Bolus   Low shrublet to 6 cm, with tuberous roots, branches with internodes papillate when young. Leaves minutely papillate, with a diadem, 7–11 × 2.5 mm. Flowers on pedicels to 10 mm long, magenta, filamentous staminodes white. Fruits 4- or 5-locular. Nov.–Jan. Habitat?, SE (Uitenhage).•

[**Species excluded**   Poorly known and possibly conspecific with one of the above: **T. imitans** L.Bolus, **T. strumosum** (Haw.) L.Bolus]

VANZIJLIA   1 sp., N and W Cape Coast

annulata (A.Berger) L.Bolus (= *Vanzijlia angustipetala* (L.Bolus) N.E.Br.)   Low succulent shrublet to 13 cm, with ascending branchlets. Leaves heterophyllous, smooth. Flowers solitary, white or pale pink, 40–60 mm diam. Fruits 9- or 10-locular, with large white closing bodies. July–Sept. Sand or loam in shrubby vegetation, NW (Hondeklipbaai to Lambert's Bay).

VLOKIA•   1 sp., W Cape

ater S.A.Hammer   Dwarf succulent creeper with prostrate stems to 10 cm long. Leaves fused below, boat-shaped, keeled, 10 mm long. Flowers solitary, pink. Fruit 5–7-locular. Aug.–Sept. Shallow pans in quartzite, 1340 m, KM (Montagu).•

ZEUKTOPHYLLUM•   spookvygie   2 spp., W Cape

calycinum (L.Bolus) H.E.K.Hartmann (= *Octopoma calycinum* (L.Bolus) L.Bolus, *Ruschia calycina* L.Bolus)   Like **Z. suppositum** but leaf keels serrulate towards tips. Dec.–Feb. Gravelly sandstone slopes, KM (Little Karoo).•

suppositum (L.Bolus) N.E.Br.   Robust, closely leafy shrublet to 10 cm, with woody, ascending branches, old leaves persisting. Leaves imbricate, 4–6 per branch, ascending, united below for 2–5 mm, trigonous, obtusely keeled. Flowers solitary, yellowish pink, to 30 mm diam. Fruits 10-locular, closing bodies absent, keels ending with a small rigid wing. Jan.–Mar. Gravelly sandstone slopes, LB (Little Karoo: Phisantefontein).•

GENUS UNCERTAIN

Braunsia vanrensburgii (L.Bolus) L.Bolus   Prostrate perennial with erect flowering branches. Leaves pale-glaucous, shiny, erect, trigonous. Flowers solitary on pedicels to 7 mm long, white, to 50 mm diam. Fruits woody, 5-locular. Dec.–Apr. Limestone in shallow pockets of soil, AP (Agulhas Peninsula).•

Ruschia promontorii L.Bolus   Erect shrublet to 12 cm. Leaves spreading to erect, back surface keeled, 20–25 × 12 mm, with short mucro. Flowers 1–3 in clusters, subsessile, pink striped purplish, 30 mm diam. Fruits 5-locular, without valve wings, closing bodies large. July. Lower stony slopes, SW (Cape Peninsula).•

AMARANTHACEAE (= *CHENOPODIACEAE*)

1. Sepals dry and more or less papery or membranous:
    2. Mature flowers deflexed, falling individually; sterile flowers not reduced to spines....................**Achyranthes**
    2.' Mature flowers ascending or spreading, falling in woolly or silky clusters:
        3. Sterile flowers reduced to hooked spines; flower clusters woolly, burr-like ....**Pupalia**
        3.' Sterile flowers lacking; flower clusters densely silky, not burr-like ....**Sericocoma**
1.' Sepals herbaceous and soft:
    4. Plants more or less fleshy, often with branches articulated; leaves short and fleshy or obsolete:
        5. Branches not articulated; leaves short and fleshy...**Halopeplis**
        5.' Branches articulated; leaves much reduced or obsolete:
            6. Annuals, branches all terminating in spikes; flowers protandrous; stigmas tufted ....**Salicornia**
            6.' Perennials, branches not all terminating in spikes; flowers protogynous; stigmas 2- or 3-fid ....**Sarcocornia**
    4.' Herbs or woody shrubs with well-developed leaves:
        7. Perianth segments horizontally winged in fruit; shrublets with scale-like leaves ....**Salsola**
        7.' Perianth segments not winged:
            8. Bracts and bracteoles absent..**Chenopodium**
            8.' Bracts and bracteoles present in bisexual or female flowers:
                9. Plants glabrous; leaves linear to oblong, fleshy ....**Suaeda**
                9.' Plants silky or mealy:
                    10. Perianth segments silky with a short horn on back ....**Chenolea**
                    10.' Perianth segments not horned:
                        11. Flowers in axillary clusters, bisexual and male together..**Exomis**
                        11.' Flowers in terminal inflorescences:
                            12. Bracts enveloping fruit, flap-like or spongy, more or less connate; female flowers without perianth ....**Atriplex**
                            12.' Bracts surrounding fruit, not connate, becoming subsucculent ....**Manochlamys**

ACHYRANTHES   6 spp., Old World tropics and warm temperate

aspera L.   Shortly hairy, sprawling perennial to 50 cm. Leaves elliptic, often attenuate, pale beneath. Flowers in congested terminal and axillary spikes, nodding, white. Mainly Dec.–Mar. Forest and bush, SE (Knysna to tropical Africa and Asia).

ATRIPLEX (= *BLACKIELLA*)   saltbush   c. 250 spp., warm temperate and subtropical

bolusii C.H.Wright (= *Atriplex cinerea* Poir. in part)   Like **A. vestita** but leaves often larger and irregularly toothed, fruiting bracts rhomboid, swollen and corky at base, grey-mealy, to 10 mm long. Sept.–Oct. Coastal saline flats, seasonal streambanks, NW, SW, AP, KM, SE (Namibia to Stilbaai, ? to Uitenhage).

*__lindleyi__ Moq. (= *Blackiella inflata* (F.Muell.) Aellen)   Grey-mealy rounded annual to 30 cm. Leaves rhomboid, toothed. Flowers in axillary clusters; fruiting bracts triangular, fused into an inflated bladder, to 12 mm long. May–Oct. Dry stony flats and disturbed sites, NW, SW, KM, SE (Australian weed, widespread in southern Africa).

__patula__ L. (= *Atriplex austroafricana* Aellen)   Sprawling, grey-mealy annual to 50 cm. Leaves lanceolate to rhomboid, coarsely toothed to lobed below, glabrescent above. Flowers clustered in elongate spikes, whitish; fruiting bracts rhomboid, often with small horns, to 2 mm long. Nov.–Apr. Coastal sands, NW, SW (Namibia to Caledon, and ?N Africa).

*__semibaccata__ R.Br.   CREEPING SALTBUSH   Monoecious, grey-mealy, sprawling perennial to 30 cm. Leaves elliptic-obovate, often coarsely toothed, glabrescent above. Flowers minute in axillary clusters; fruiting bracts rhomboid, red and fleshy, 2–5 mm long. Mainly Sept.–Dec. Coastal and saline sands, NW, SW, AP, KM, LB, SE (Australian weed, widespread in southern Africa).

__vestita__ (Thunb.) Aellen (= *Atriplex halimus* C.H.Wright non L.)   Monoecious, silvery mealy subshrub to 60 cm. Leaves oblanceolate. Flowers clustered in elongate spikes; fruiting bracts, subrotund, softly papery, glabrescent and warty, 10–15 mm long. Mainly Aug.–Dec. Saline flats, streambanks, NW, SW, AP, KM, SE (Namibia to Uitenhage).

CHENOLEA   4 spp., southern Africa and Mediterranean

__diffusa__ Thunb.   Prostrate mat-forming, succulent perennial. Leaves elliptic to lanceolate, spreading to imbricate, silvery silky. Flowers axillary, inconspicuous, greenish. Feb.–Apr. Coastal salt marshes, NW, SW, AP, SE (Namibia to Mozambique).

[**CHENOPODIUM album** L., **C. ambrosioides** L., and **C. murale** L. have been recorded as weeds of disturbed places but none are considered naturalised]

EXOMIS   BRAKBOSSIE   1 sp., southern Africa

__microphylla__ (Thunb.) Aellen   Grey-mealy shrub to 1 m. Leaves ovate, oblong or sagittate. Flowers tightly clustered in upper axils, minute, yellowish. Jan.–Apr. Stony hillsides, often coastal, NW, SW, AP, SE (S Namibia to Uitenhage).

HALOPEPLIS   3 spp., warm temperate Old World

__amplexicaulis__ (Vahl) Bunge ex Ung.-Sternb.   Glaucous, red to purple annual to 25 cm. Leaves fleshy, scale-like. Flowers in axillary cones, minute. Feb.–Apr. Coastal salt marshes, NW, SW (Namaqualand to Cape Peninsula).

MANOCHLAMYS   HONDEBOSSIE   1 sp., southern Africa

__albicans__ (Aiton) Aellen   Monoecious, grey-mealy, white-stemmed shrub to 2 m, with horizontally spreading branches. Leaves rhomboid to sagittate. Flowers crowded in terminal spikes, minute, yellowish. Sept.–Jan. Dry stony slopes and flats, NW, SW, KM (S Namibia and W Karoo to Cape Peninsula and Little Karoo).

PUPALIA   4 spp., Old World tropics

__lappacea__ (L.) A.Juss. (= *Pupalia atropurpurea* (Lam.) Moq.)   Shortly hairy, sprawling perennial to 50 cm. Leaves ovate, often attenuate, pale beneath. Flowers in congested, elongate spikes elongating in fruit, whitish, fruits spiny and falling in burr-like clusters when ripe. Mainly Dec.–Apr. Forest and bush, SE (George to tropical Africa and Asia).

SALICORNIA   GLASSWORT   13 spp., cosmopolitan

__meyeriana__ Moss   Jointed, red, fleshy annual to 40 cm, with opposed branching. Leaves reduced to a small membranous collar. Flowers minute, 3 in axillary cymes, hidden by the leaves. Mar.–Apr. Sheltered saline marshes, estuaries, NW, SW, AP, SE (S Namaqualand to Cape Peninsula to KwaZulu-Natal, Madagascar).

[**Species excluded**   Probably not in our area: **S. uniflora** Toelken]

SALSOLA SALTWORT c. 150 spp., cosmopolitan

aphylla L.f. ASBOSSIE Shortly hairy, grey shrublet, to 1.2 m. Leaves scale-like, fleshy, soon becoming glabrous and wrinkled. Flowers solitary in upper axils, minute, fruits broadly winged. Dec.–Feb. Dry, often saline slopes and flats, NW, SW, KM, LB, SE (Namibia and Karoo to Uitenhage).

*kali L. (= *Salsola australis* R.Br.) ROLBOSSIE, RUSSIAN THISTLE Glabrous, rounded annual to 50 cm. Leaves sessile, linear to deltoid, acuminate, spine-tipped. Flowers 1–few in axils, whitish, fruits with a papery wing. Mainly Sept.–Nov. Widespread in dry stony places, NW, SW, AP, KM, LB, SE (Eurasian weed).

tuberculata (Moq.) Fenzl Like **S. aphylla** but leaves with dry margins, persistently hairy. Dec.–Feb. Coastal sands, NW (Namibia to Lambert's Bay).

verdoorniae Toelken Prostrate, silvery silky, mat-forming shrublet, sometimes to 15 cm. Leaves triangular, sessile, imbricate, distichous. Flowers minute. Dec. Quartzite gravels, KM (Little Karoo).•

[**Species excluded** Probably not in our area: **S. adversariifolia** Botsch.]

SARCOCORNIA 15 spp., cosmopolitan

capensis (Moss) A.J.Scott Like **S. natalensis** but flowers not hidden by the bracts. Oct.–Apr. Coastal sands hardly saline, SW, SE (Cape Peninsula, Hermanus, Port Elizabeth).•

decumbens (Toelken) A.J.Scott Sprawling, jointed perennial to 30 cm, not rooting along stems. Leaves reduced to a membranous collar. Flowers 5–7 per cluster in terminal spikes. Jan.–June. Salt marshes, AP, SE (De Hoop to S Mozambique).

littorea (Moss) A.J.Scott Like **S. pillansii** but spikes solitary, rarely more than 1 per branch, flowers becoming corky with age. Sept.–May. Rock beaches, NW, SW, AP, SE (Namaqualand to E Cape).

mossiana (Toelken) A.J.Scott Like **S. pillansii** but internodes short, to 4 mm. Mar.–May. Coastal salt marshes, SW, AP, KM, LB (Caledon to Gouritsmond).•

natalensis (Bunge ex Ung.-Sternb.) A.J.Scott SEEKORAAL Sprawling, jointed perennial to 30 cm, forming mats with erect stems. Leaves reduced to a membranous collar. Flowers in threes in terminal spikes, hidden by the bracts. Jan.–May. Coastal and inland saline habitats, NW, SW, AP, KM, SE (Angola to Mozambique, Madagascar).

perennis (Mill.) A.J.Scott Like **S. capensis** but flowers 3–5 in clusters. Jan.–June. Salt marshes, NW, SW, AP, SE (S Namibia to KwaZulu-Natal).

pillansii (Moss) A.J.Scott Sprawling, jointed shrublet to 60 cm. Leaves fleshy. Flowers in threes in terminal spikes. Feb.–Apr. Inland and coastal saline marshes, NW, SW, KM, SE (S Namibia to Mozambique).

SERICOCOMA 6 spp., southern and tropical Africa

avolans Fenzl Rounded shrublet to 50 cm. Leaves opposite below, linear to lanceolate, pungent-mucronate. Flowers in terminal spikes, silky, pink to purple. Jan.–Apr. Stony slopes and flats, KM, SE (S Namibia to Karoo and Little Karoo to Baviaanskloof Mts).

SUAEDA SEA-BLITE c. 110 spp., cosmopolitan

fruticosa (L.) Forssk. (incl. **S. caespitosa** Wolley-Dod) Softly woody sprawling to prostrate, grey-blue, glabrous shrublet, often rooting at nodes, to 1 m. Leaves fleshy, ovoid–ellipsoid. Flowers in axillary clusters, minute, with 3 styles. Aug.–Dec. Dry slopes and coastal marshes and estuaries, NW, SW, AP, KM, SE (S Namibia and Karoo to Port Elizabeth).

ANACARDIACEAE
by R.O. Moffett

HEERIA• KLIPHOUT 1 sp., W Cape

argentea (Thunb.) Meisn. Dioecious, evergreen small tree to 5 m. Leaves oblong to ovate, discolorous. Flowers in terminal or axillary panicles, cream-coloured. Jan.–July. Rocky sandstone slopes, NW, SW (Pakhuis Mts to Steenbras River mouth).•

LAUROPHYLLUS•   ystermartiens   1 sp., W Cape

capensis Thunb.   Dioecious, evergreen shrub or small tree to 6 m. Leaves elliptic to oblong, margins toothed. Flowers yellowish, male in terminal panicles, female in dense, bracteate, antler-like structures. Aug.–Jan. Streambanks and coastal scrub or forest, LB, SE (Caledon to Uitenhage).•

LOXOSTYLIS   teerhout   1 sp., South Africa

alata A.Spreng. ex Reichb.   Dioecious, evergreen tree to 6 m. Leaves pinnately compound, rachis winged. Flowers in dense terminal panicles, white. Fruits imbedded in persistent pink sepals. Sept.–Mar. Forest fringes, SE (Van Staden's Mts to KwaZulu-Natal).

RHUS   karee, korentebos, taaibos   c. 200 spp., tropical, subtropical and warm temperate regions of the world

A. Leaflets toothed (see also **R. rosmarinifolia***)*

crenata Thunb.   duinekraaibessie   Dioecious, evergreen shrub or small tree to 4 m. Leaves trifoliolate, leaflets sessile, obovate, apex blunt and crenate. Flowers cream-coloured. Drupes round, dark blue/brown, stone with lateral processes. Apr. Sandy coastal flats, SW, AP, SE (Cape Peninsula to S KwaZulu-Natal).

cuneifolia L.f.   kogelbergtaaibos   Dioecious, evergreen dwarf shrub to 60 cm. Leaves trifoliolate, leaflets sessile, stiff, obovate, coarsely toothed above. Flowers whitish. Drupes tricuspidate, smooth. Sept.–Oct. Rocky slopes, NW, SW (Clanwilliam to Caledon).•

dentata Thunb.   nanabessie   Dioecious, deciduous shrub to 2 m. Leaves trifoliolate, leaflets sessile, broadly obovate, margins prominently dentate. Flowers yellow. Drupes round, shiny. Oct.–Nov. Forest edges, SE (Humansdorp to Potgietersrus).

dissecta Thunb.   langsteel korentebossie   Dioecious, deciduous small shrub to 1.5 m. Leaves trifoliolate, leaflets sessile, obovate, discolorous, margins sharply toothed. Flowers greenish yellow. Drupes asymmetric, warty. July. Rocky slopes, NW, SW (Vanrhynsdorp to McGregor).•

incisa L.f.   rub-rub berry, baardbessie   Dioecious, deciduous shrub to 3 m. Leaves trifoliolate, leaflets sessile, ovate, discolorous, margins deeply lobed or shallowly dentate. Flowers greenish yellow. Drupes ellipsoid, hairy. June. Clay soils, NW, SW, AP, KM, LB, SE (Richtersveld to Komga).

tomentosa L.   korentebos   Dioecious, evergreen shrub or small tree to 6 m. Leaves trifoliolate, leaflets stalked, obovate, discolorous, often toothed. Flowers cream-coloured. Drupes ellipsoid, hairy. July–Aug. Rocky slopes, NW, SW, AP, KM, LB, SE (Clanwilliam to Zimbabwe).

A. Leaflets entire (see also **R. tomentosa***)*

angustifolia L.   wilgerkorentebos   Dioecious, evergreen shrub or small tree to 4 m. Leaves trifoliolate, leaflets stalked, lanceolate to narrowly elliptic, discolorous. Flowers yellowish. Drupes ellipsoid, shortly hairy. Oct.–Nov. Streambanks, NW, SW, AP, KM, LB (Bokkeveld Mts to Barrydale).•

chirindensis Baker. f. (= *Rhus legatii* Schönland)   bostaaibos   Dioecious, deciduous shrub or tree to 23 m. Leaves trifoliolate, leaflets stalked, ovate to ovate-lanceolate, weakly undulate. Flowers pale green. Drupes round, shiny. Nov.–Mar. Forest, LB, SE (Grootvadersbos to Zimbabwe).

fastigata Eckl. & Zeyh.   besemkraaibessie   Dioecious, evergreen shrub to 2 m. Leaves trifoliolate, leaflets sessile, oblanceolate. Flowers cream-coloured. Drupes round, shiny. Jan. Forest margins, SE (Humansdorp to Port Shepstone).

glauca Thunb.   bloukoeniebos   Dioecious, evergreen shrub or small tree to 4 m. Leaves trifoliolate, leaflets sessile, obovate, often bluish. Flowers greenish white. Drupes lens-shaped, shiny. June–Sept. Mostly on dunes, NW, SW, AP, KM, LB, SE (Velddrif to Kentani).

laevigata L. (= *Rhus mucronata* Thunb., *R. incana* Mill.)   duinetaaibos, umhlakothi   Dioecious, deciduous shrub to 2.5 m. Leaves trifoliolate, leaflets sessile, obovate, smooth or hairy. Flowers greenish yellow. Drupes round, shiny. Oct.–Dec. Coastal flats and slopes, NW, SW, AP, KM, SE (Lambert's Bay to East London).

lancea L.f.   karee, umhlakhotshane   Dioecious, evergreen large shrub or tree to 12 m with dark, fissured bark. Leaves trifoliolate, leaflets sessile, linear to lanceolate. Flowers greenish yellow. Drupes round, shiny. Apr.–July. Watercourses in karroid areas, NW, KM (Ceres Karoo to Zimbabwe).

longispina Eckl. & Zeyh.   DORINGTAAIBOS   Dioecious, evergreen, armed shrub to 4 m. Leaves trifoliolate, leaflets sessile, oblanceolate, shiny. Flowers pale yellow. Drupes lens-shaped, brown, shiny. May.–July. Karroid scrub, NW, SW, KM, LB, SE (Namaqualand to S KwaZulu-Natal).

lucida L. (= *Rhus schlechteri* Diels)   BLINKTAAIBOS   Dioecious, evergreen shrub to 3 m. Leaves trifoliolate, leaflets sessile, obovate to spatulate, shiny. Flowers yellow. Drupes round, shiny. Aug.–Oct. Sandy flats and slopes, NW, SW, AP, KM, LB, SE (Citrusdal to Zimbabwe).

pallens Eckl. & Zeyh.   BLEEKKOENIEBOS   Dioecious, evergreen shrub to 3 m. Leaves trifoliolate, leaflets sessile, oblanceolate. Flowers yellow. Drupes lens-shaped, shiny. Mar.–July. Karroid scrub, SW, KM, LB, SE (Robertson to KwaZulu-Natal and to Zeerust).

?*pendulina Jacq.   WITKAREE   Dioecious, evergreen shrub or tree to 12 m. Leaves trifoliolate, leaflets sessile, lanceolate. Flowers greenish yellow. Drupes round, smooth. Feb.–Mar. Banks of Berg River and Olifants River, possibly introduced from Orange River, NW (Porterville to Orange River).

pterota C.Presl   PENDORINGTAAIBOS   Dioecious, evergreen, armed shrub to 2 m with leaves in clusters. Leaves trifoliolate, leaflets sessile, obovate, margins revolute. Flowers pale yellow. Drupes ellipsoid, shiny, juice pungent. Jan.–May. Chalky and clay soils, SW, AP, LB, SE (Saldanha Bay to East London).

pyroides Burch.   GEWONETAAIBOS   Dioecious, deciduous, armed shrub or small tree to 6 m. Leaves trifoliolate, leaflets sessile, obovate, often hairy. Flowers pale yellow. Drupes round, smooth. Oct.–Jan. Near watercourses in karroid scrub, KM, LB, SE (Ladismith to Botswana and Namibia).

refracta Eckl. & Zeyh.   GROWWEBLAARTAAIBOS   Dioecious, evergreen shrub or small tree to 4 m. Leaves trifoliolate, leaflets sessile, obovate. Flowers pale yellow. Drupes round, smooth, stone with lateral processes. Jan.–Mar. Scrub, SE (Plettenberg Bay to Mount Frere).

rehmanniana Engl. (= *Rhus macowanii* Schönland)   SUURTAAIBOS   Dioecious, deciduous spreading shrub to 3 m. Leaves trifoliolate, leaflets sessile, obovate, slightly hairy. Flowers greenish yellow. Drupes round, smooth. Jan.–Apr. Clay soils, NW, SW, KM, LB, SE (Piketberg to N Province).

rimosa Eckl. & Zeyh.   SEDERBERGTAAIBOS   Dioecious, evergreen, erect, rigid shrub to 2.5 m. Leaves trifoliolate, leaflets sessile, oblanceolate, stiff. Flowers yellowish. Drupes tricuspidate, smooth. Sept.–Oct. Rocky sandstone slopes, NW (Bokkeveld Mts to Saron).•

rosmarinifolia Vahl   ROOSMARYNTAAIBOS   Dioecious, dwarf, evergreen shrub to 1 m. Leaves trifoliolate, leaflets sessile, linear, discolorous, occasionally toothed. Flowers cream-coloured. Drupes ellipsoid, mostly hairy. May–Aug. Gravelly soils, NW, SW, AP, KM, LB, SE (Clanwilliam to Port Elizabeth).•

scytophylla Eckl. & Zeyh.   ROOIBLOMTAAIBOS   Dioecious, evergreen shrub to 2 m. Leaves trifoliolate, leaflets sessile, obovate, stiff, margins revolute. Flowers red. Drupes tricuspidate, smooth. June–July. Rocky sandstone slopes, NW, SW (Bokkeveld Mts to Caledon).•

stenophylla Eckl. & Zeyh.   SMALBLAARTAAIBOS   Dioecious, dwarf, evergreen shrub to 1.2 m. Leaves trifoliolate, leaflets stalked, lanceolate, discolorous. Flowers cream-coloured. Drupes ellipsoid, hairy. July. Sandstone slopes, SW (Cape Peninsula and Paarl to Gordon's Bay).•

undulata Jacq. (= *Rhus celastroides* Eckl. & Zeyh.)   KUNI-BUSH, KOENIEBOS   Dioecious, evergreen shrub to 3 m. Leaves trifoliolate, leaflets sessile, oblanceolate, sticky, aromatic. Flowers yellowish. Drupes lens-shaped, shiny. Apr.–May. Stony slopes, NW, SW, KM (S Namibia to Ladismith).

APIACEAE

Annesorhiza by B.-E. van Wyk & P. Tilney, **Chamarea** by B.-E. van Wyk

1. Fruit without oil ducts, strongly compressed laterally (parallel to the stigmas):
    2. Leaves not peltate, petiole sheathing at base, without stipules; furrows between mericarp ribs wrinkled ....**Centella** (see **ARALIACEAE**)
    2.' Leaves peltate, with membranous stipules at base of petiole; furrows between mericarp ribs smooth ....**Hydrocotyle** (see **ARALIACEAE**)
1.' Fruit with oil ducts, not compressed or slightly compressed laterally:
    3. Leaves simple, leathery, woolly at least beneath; umbels congested and capitate, each umbel usually with a single female or bisexual flower surrounded by several male flowers ....**Hermas**
    3.' Leaves various but not as above:
        4. Plants woody, not dying down to ground level after fruiting:
            5. Fruits (and ovaries) hairy or bristly; broom-like undershrubs, leafless at flowering....................**Deverra**
            5.' Fruits (and ovaries) glabrous:
                6. Mericarps unequal, winged:
                    7. Plants leafless at flowering; margins of leaflets toothed; seed free in endocarp.......**Polemanniopsis**

7.' Plants leafy at flowering; margins of leaves or leaflets entire; seed adherent to endocarp ....**Heteromorpha**
  6.' Mericarps equal, not winged:
    8. Fruit strongly compressed dorsiventrally (at right angles to stigmas), with conspicuous vittae; leaves not or weakly dimorphic...**Peucedanum**
    8.' Fruit not compressed or slightly compressed laterally, epidermis strongly cutinised; leaves dimorphic, lower with flattened segments, upper reduced and often needle-like:
      9. Surface between mericarp ribs tuberculate or wrinkled; shrubs or shrublets with leaves often needle-like ....**Anginon**
      9.' Surface between mericarp ribs smooth; woody perennial ....**Glia**
4.' Plants herbaceous, either low creeping herbs or dying down to ground level after fruiting:
  10. Plants acaulescent, dioecious; leaves prickly, in a dense rosette; fruit 1-seeded ....**Arctopus**
  10.' Plants caulescent; leaves not prickly (at most with strong marginal cilia); fruit 2-seeded:
    11. Umbels simple, or seemingly so by forming a compact head:
      12. Involucral bracts united, conspicuous, pale green, greenish yellow or silvery white, sometimes becoming pinkish inside; fruit glabrous except for some blunt scales at top; leaves simple, coarsely toothed and ciliate ....**Alepidea**
      12.' Involucral bracts free, green; fruit with hooked bristles; leaves digitate ....**Sanicula**
    11.' Umbels manifestly compound, umbellules with conspicuous peduncles:
      13. Fruits dorsally compressed (at right angles to stigmas) :
        14. Perennials...**Peucedanum**
        14.' Annuals...**Capnophyllum**
      13.' Fruits ± round in cross section:
        15. Leaves simple:
          16. Leaves grass-like, tapering to an acute tip; flowers yellow; fruits oblong, truncate above ....**Bupleurum**
          16.' Leaves linear or spathulate, blunt; flowers white; fruits ovate, tapering into a conspicuous stylopodium; sepals sometimes filiform and horn-like ....**Itasina**
        15.' Leaves digitate, pinnate, pinnately lobed or decompound, or basal leaves simple and cauline leaves pinnately dissected:
          17. Foliage leaves all radical, those on stem reduced to a sheath with or without a vestigial blade:
            18. Fruits with oil ducts (vittae) below ridges and with a prominent stylopodium; leaves mostly green or drying at flowering..**Lichtensteinia**
            18.' Fruits with oil ducts (vittae) in furrows between ridges; leaves dry at flowering:
              19. Mericarps equal; fruits ovoid or flask-shaped with ribs well spaced, shorter than 4 mm; flowers white; involucre often absent...**Chamarea**
              19.' Mericarps equal or unequal; fruits oblong to suborbicular with thick, almost contiguous ridges, 5 mm or more long; flowers yellow; involucre always present: .....**Annesorhiza**
          17.' Foliage leaves both radical and cauline:
            20. Fruits hairy or bristly:
              21. Involucre absent; basal leaves often lacerate or simply cordate; perennials....**Pimpinella**
              21.' Involucre present or umbels leaf-opposed; leaves all finely dissected; annuals:
                22. Fruit processes barbed ....**Torilis**
                22.' Fruit processes not barbed:
                  23. Fruits oblong; leaves 2- or 3-digitate ....**Ezosciadium**
                  23.' Fruits about as broad as long; leaves pinnately dissected:
                    24. Fruit ovoid; lateral ribs marginal; umbels leaf-opposed ....**Sonderina**
                    24.' Fruit subglobose, distinctly didymous (2-lobed); lateral ribs not marginal; umbels apparently axillary ....**Stoibrax**
            20.' Fruits glabrous:
              25. Fruits at least twice as long as broad; leaf segments filiform; plant strongly aromatic; flowers yellow ....**Foeniculum**
              25.' Fruits distinctly less than twice as long as broad; plants aromatic or not; flowers mostly white:
                26. Umbels leaf-opposed, simple or compound:
                  27. Involucre absent; leaves digitate or pinnate, with broad leaflets ....**Apium**
                  27.' Involucre present; leaves dissected, with narrow leaflets ....**Sonderina**
                26.' Umbels apparently axillary, compound:
                  28. Leaves simply pinnate, robust, elongate, leaflets uniform and decreasing only slightly in size upwards; marsh plants ....**Berula**
                  28.' Leaves carrot-like or variably bi- or tripinnate:
                    29. Leaves fleshy; calyx teeth distinct; fruit with narrow but thickish wings; coastal plants ....**Dasispermum**
                    29.' Leaves soft; calyx teeth obsolete; fruit with slender ribs ....**Conium**

ALEPIDEA KATAZO, KALMOES   c. 40 spp., tropical and southern Africa

capensis (P.J.Bergius) R.A.Dyer   Slender tufted perennial to 40 cm. Leaves crowded at base, long-petiolate, oblong, with long incurving bristles. Flowers sessile in simple umbels, bracts white or pink. Mainly Nov.–Feb. Damp flats or lower slopes, KM, LB, SE (Swartberg and Langeberg Mts: Swellendam to KwaZulu-Natal).

delicatula Weim.   Like **A. capensis** but slender, to 20 cm, leaves toothed, tooth bristles spreading, intertooth bristles incurved. Jan.–Feb. High, rocky sandstone slopes, KM, SE (Swartberg and Outeniqua Mts).•

ANGINON (= *RHYTICARPUS*)   WILDESELDERY, WILDEVINKEL   12 spp., S Namibia to E Cape

A. Plants green

difforme (L.) B.L.Burtt (= *Rhyticarpus difformis* (L.) Briq.)   Rigid, stiff-leaved, few-branched shrub to 3 m. Leaves in axillary tufts, terete, pungent. Flowers in compound umbels on stout peduncles, yellowish. Fruits obovoid, ridged and warty, with a basal collar. Dec.–Apr. Rocky sandstone slopes, NW, SW, AP, KM, LB, SE (Richtersveld to Tulbagh to E Cape).

fruticosum Allison & B.-E.van Wyk   Like **A. difforme** but plants multistemmed, leaves grooved above and some shortly and unevenly 2- or 3-sect. Jan.–Apr. Clay soils along streams, NW, SW, KM, SE (W Karoo to Genadendal and to Uniondale).

pumilum Allison & B.-E.van Wyk   Slender, unbranched, rhizomatous shrublet to 70 cm. Leaves bipinnatisect, lobes reduced, cuneate to linear, toothed. Flowers in small compound umbels, yellow. Fruits subrotund, ridged and warty. July–Aug. Limestone slopes, AP (Agulhas and De Hoop).•

ternatum Allison & B.-E.van Wyk   Like **A. difforme** but leaves equally 3-sect. Dec.–Apr. Sandstone slopes, NW (Gifberg and Heerenlogementberg).•

A. Plants grey-glaucous

paniculatum (Thunb.) B.L.Burtt   Like **A. swellendamense** but leaves finely divided, ultimate segments terete, grooved above, fruits oblong, ridged, with a basal collar. Nov.–Jan. Rocky sandstone slopes, NW (Bokkeveld Mts to Citrusdal).•

rugosum (Thunb.) Raf. (= *Anginon uitenhagense* (Eckl. & Zeyh.) B.L.Burtt)   Like **A. swellendamense** but leaves grooved above, lobes linear and fruits ridged and wrinkled, with a basal collar. Dec.–Apr. Sandstone slopes, SE (Kouga Mts to Grahamstown).

swellendamense (Eckl. & Zeyh.) B.L.Burtt (= *Rhyticarpus swellendamenis* (Eckl. & Zeyh.) Briq.)   Glaucous, stiff-leaved, few-branched shrub to 2 m. Leaves in axillary clusters, simple or forked in threes, lobes terete, pungent. Flowers in small compound umbels on slender peduncles, yellowish. Fruits obovoid, ridged and wrinkled. Oct.–Jan. Stony clay slopes, NW, AP, KM, SE (Worcester to Uniondale).•

tenuior Allison & B.-E.van Wyk   Slender, few-branched, glaucous shrub to 1.5 m. Leaves bipinnatisect, lobes reduced, narrowly cuneate to linear, toothed, petioles persistent. Flowers in few-flowered, lax compound umbels on slender peduncles, yellow. Fruits subrotund, ridged and wrinkled Nov.–Dec. Stony slopes, KM (Montagu and Waboomsberg).•

ANNESORHIZA   ANYSWORTEL   13 spp., southern Africa

altiscapa (Schltr. ex) H.Wolff   BOKLAMVINKEL   Perennial to 1.5 m; roots 10 or more, slightly and evenly fleshy. Leaves dry at flowering, much dissected, segments acute. Flowers cream-coloured to yellow. Mericarps homomorphic. Aug.–Sept. Sandy and clayey slopes, NW (Namaqualand to Gifberg).

filicaulis Eckl. & Zeyh.   Small perennial to 8 cm; roots unknown. Leaves unknown. Flowers ?yellow. Mericarps linear, several times longer than broad. Flowering time? Sandy places, NW (Clanwilliam).•

grandiflora (Thunb.) Hiroe (= *Annesorhiza elata* Eckl. & Zeyh., *A. hirsuta* Eckl. & Zeyh., *A. villosa* (Thunb.) Sond.)   HARIGE-ANYSWORTEL   Perennial to 2 m; roots 6 or more, slightly or distinctly fleshy. Leaves usually present at flowering, dissected, segments rounded, sparsely to densely hairy. Flowers yellow. Mericarps homomorphic. Oct.–Feb. Flats and sandstone slopes, often damp areas, NW, SW (Bokkeveld Mts to Caledon).•

macrocarpa Eckl. & Zeyh.   WILDE-ANYSWORTEL   Perennial to 1.6 m; roots numerous, slightly or distinctly fleshy, often fluted. Leaves usually dry at flowering, finely dissected, trailing on long thin petioles and rachises, segments minute. Flowers pale yellow. Mericarps large, heteromorphic. Aug.–Jan. Coastal dunes and sandy slopes, SW, AP, LB, SE (Saldanha to East London).

nuda (Aiton) B.L.Burtt (= *Annesorhiza capensis* Cham. & Schltdl.) SOETANYSWORTEL   Perennial to 1 m; roots single (or 2, if root from previous season is still present), very fleshy. Leaves dry or present at flowering, dissected, segments rounded. Flowers whitish to yellowish. Mericarps small, heteromorphic. Dec.–Apr. Rocky, often granite slopes, NW, SW (Clanwilliam to Caledon).•

thunbergii B.L.Burtt   Perennial to 1.3 m; roots unknown. Leaves present at flowering, finely dissected, segments minute. Flowers yellow. Mericarps dissimilar, at least somewhat heteromorphic. Nov.–Feb. Sandstone slopes, SE (Humansdorp).•

sp. 1   Perennial to 70 cm; roots 15 or more, slightly fleshy, often swollen near tips. Leaves finely dissected, segments minute, densely to sparsely hairy, bases persisting to form a dense fibrous mass. Flowers yellow. Mericarps homomorphic. Nov.–Dec. Sandstone flats, NW (Bokkeveld Mts).•

sp. 2   BERG-ANYSWORTEL   Perennial to 1.6 m; roots 1–3, slightly or distinctly fleshy. Leaves usually present at flowering, finely dissected, trailing on long thin petioles and rachises, segments minute, narrow. Flowers ?yellow. Mericarps homomorphic. Oct.–Feb. Sandstone crevices at high alt., NW, SW (Worcester to Caledon).•

*APIUM   CELERY   20 spp., N and S temperate except Africa

***prostratum** Vent.   Sprawling annual or short-lived perennial to 30 cm. Leaves pinnate below, leaflets broadly toothed, upper digitate with leaflets oblanceolate. Flowers few in simple, leaf-opposed umbels, few per node. Fruits subglobose, ridged. Mainly Jan.–Mar. Vlei edges and coastal areas, SW, AP, SE (Australian weed, Cape Peninsula to Port Elizabeth).

ARCTOPUS   PLATDORING   3 spp., Namaqualand to E Cape

dregei Harv.   Like **A. echinatus** but female involucral bracts ovate, truncate above with 3–5 fine spines or bristles, lateral margins rolled inwards over fruit, somewhat enlarging and leathery in fruit. Fruits rugose. June–July. Gravelly flats, SW (Hopefield).•

echinatus L.   Dioecious, acaulescent perennial. Leaves large, prostrate, ovate to orbicular and lobed to lacerate, with bristly margins. Flowers in simple umbels, cream-coloured to pink, female involucral bracts pungent with 1 or 2 inflexed lateral spines, united below, not much enlarging in fruit. Fruits prickly. May–July. Sand and granite flats and slopes, NW, SW, AP, KM, LB, SE (S Namaqualand and W Karoo to Grahamstown).

monacanthus Carmichael ex Sond.   Like **A. echinatus** but flowers sometimes yellow and female involucral bracts suborbicular, sometimes with small lateral spines, free, much enlarging and papery in fruit. Fruits rugose. June–July. Sandstone and clay slopes and flats, NW, SW (Bokkeveld Mts to False Bay).•

BERULA   WATER PARSNIP, TANDPYNWORTEL   2 spp., cosmopolitan

erecta (Huds.) Coville (= *Berula thunbergii* (DC.) H.Wolff)   Annual to 1 m. Leaves mostly basal, pinnate, leaflets sessile elliptic to ovate, finely toothed. Flowers in compound umbels, white. Fruits subglobose, slightly ridged and rough. Jan.–Mar. Streamsides, SW, AP, KM, LB, SE (Cape Peninsula to Humansdorp, nearly cosmopolitan).

BUPLEURUM   c. 100 spp., cosmopolitan

mundii Cham. & Schltdl.   Tufted perennial to 60 cm, from woody base. Leaves mostly basal, linear to oblanceolate, grass-like. Flowers in compound umbels on branched peduncles, yellow. Fruits narrowly oblong, ridged. Aug.–Feb. Moist slopes, SW, LB, SE (Cape Peninsula and Humansdorp to Mpumalanga).

CAPNOPHYLLUM   5 spp., W Cape and Mediterranean

africanum (L.) Gaertn.   Sprawling annual to 30 cm, branching from base. Leaves finely dissected, leaflets filiform. Flowers in compound, axillary umbels, white. Fruits warty when young, elliptic, margins winged, with warty ridges on face. Sept.–Dec. Sand dunes, SW, AP (Saldanha to Gansbaai).•

leiocarpon (Sond.) J.C.Manning & Goldblatt (= *Capnophyllum jacquinii* auct., *C. africanum* var. *leiocarpon* Sond.)   Like **C. africanum** but fruits narrowly elliptic, not warty, with light ridges on face. Oct.–Nov. Deep coastal sands, NW (Port Nolloth to Rocher Pan).

CHAMAREA   CAPE CARAWAY, VINKELWORTEL   5–7 spp., southern Africa

capensis (Thunb.) Eckl. & Zeyh.   Perennial to 50 cm; roots usually single, evenly fleshy. Leaves dry at flowering, the first-formed and later ones all much-dissected, ultimate segments acute. Flowers yellow. Fruits 2.5–3.0 mm long, ovoid, mericarps homomorphic. Nov.–Apr. Sandstone slopes, SW, KM, LB, SE (Cape Peninsula to Karoo and E Cape).

esterhuyseniae B.L.Burtt   Like **C. capensis** but the first-formed leaves less finely dissected than the later ones and the fruits larger, 4 mm long, elliptic, with mericarps slightly heteromorphic. Dec.–Mar. Ledges and rocky slopes at high alt., NW (Worcester: Waaihoek Peak).•

gracillima (H.Wolff) B.L.Burtt   Like **C. capensis** but leaves less finely dissected and especially the first-formed with ultimate segments broad and rounded. Nov.–Apr. Stony slopes, NW, SW, SE (Clanwilliam to Cape Peninsula and Joubertina).

CONIUM   HEMLOCK   6 spp., southern and tropical Africa, Eurasia

chaerophylloides (Thunb.) Sond.   Like **C. sphaerocarpum** but leaves becoming dry at flowering, flowers yellowish green, and fruits ribbed, square in section. Sept.–Nov. Rocky slopes, KM, SE (Oudtshoorn to N Province).

***maculatum** L.   Like **C. sphaerocarpum** but stem speckled below and fruits ribbed. Nov.–Jan. Cultivated lands and forest clearings, SW (European weed, Stellenbosch to Bredasdorp).

sphaerocarpum Hilliard & B.L.Burtt   Robust biennial to 2 m. Leaves finely divided, leaflets soft. Flowers in flat-topped compound umbels, white. Fruits broadly ovoid, nearly smooth, elliptic to rounded in section. Aug.–Dec. Rocky slopes, NW, SW, KM, LB (Namaqualand and W Karoo to Riversdale).

DASISPERMUM   DUINESELDERY   1 sp., W Cape to KwaZulu-Natal

suffruticosum (P.J.Bergius) B.L.Burtt (= *Heteroptilis suffruticosa* (P.J.Bergius) Leute)   Sprawling to prostrate perennial to 50 cm with woody base. Leaves bipinnatisect, leathery-fleshy, leaflets narrow, toothed, often curling inward. Flowers in compound umbels on pseudoaxillary peduncles, cream-coloured. Fruits broadly ovate, broadly ridged. Mainly Aug.–Nov. Coastal sands, SW, AP, LB, SE (S Namaqualand to KwaZulu-Natal).

DEVERRA   c. 10 spp., southern and tropical Africa, Asia

denudata (Viv.) Pfisterer & Podlech (= *Deverra aphylla* (Cham. & Schltdl.) DC.)   Twiggy shrub to 2 m, leafless at flowering. Leaves bi- or tripinnatisect. Flowers in small compound umbels in diffuse panicles, yellow. Fruits ovoid, white-bristly. Nov.–Feb. Dry karroid scrub, KM (Namibia and Karoo to Little Karoo, E Cape, and N Africa).

EZOSCIADIUM   1 sp., W and E Cape

capense (Eckl. & Zeyh.) B.L.Burtt   Shortly hairy annual to 25 cm. Leaves 2- or 3-digitate, long-petiolate, leaflets oblong. Flowers few, in shortly pedunculate, leaf-opposed umbels. Fruits narrowly oblong, stiffly hairy. Sept.–Oct. Shady places, SE (Uitenhage to E Cape).

***FOENICULUM**   FENNEL 1 sp., W Europe and Mediterranean

***vulgare** Mill   Aromatic perennial to 1,5 m. Leaves mostly basal, finely dissected, segments thread-like. Flowers in compound umbels, yellow. Fruits ovoid, broadly ribbed. Sept.–April. Roadsides and waste places. NW, SW, AP, KM, LB, SE (Mediterranean weed).

GLIA•   DRONKWORTEL, MOERWORTEL   1 sp., W Cape

prolifera (Burm.f.) B.L.Burtt (= *Glia gummifera* auct.)   Sturdy, basally woody shrub 1–2 m, branching at base. Leaves dimorphic, multifid, the lower with broad lobes, the upper with narrow to subterete lobes, ascending, smaller above. Flowers in compound umbels on stout peduncles, yellow. Fruits ellipsoid, scarcely compressed, ridged. Dec.–Mar. Rocky sandstone or limestone slopes, NW, SW, AP (Tulbagh and Cape Peninsula to Agulhas).•

HERMAS• TONTELBLAAR   8 spp., W Cape

capitata L.f. (= *Hermas minima* Eckl. & Zeyh.)   Tufted perennial to 25 cm. Leaves radical, long-petiolate, ovate to elliptic, white- or brown-felted beneath, margins sometimes revolute, crenate. Flowers in small, congested compound umbels on slender peduncles, white. Fruits broadly ovate. Dec.–Mar. Damp sites on rocky sandstone slopes, SW, KM, LB, SE (Cape Peninsula to Van Staden's Mts).•

ciliata L.f.   Tufted perennial to 60 cm. Leaves radical, elliptic, tapering below, white-felted beneath, glabrous above, margins finely serrulate-ciliate. Flowers crowded in rounded compound umbels on branched peduncles, white. Fruits broadly ovate. Dec.–Mar. Sandstone slopes, NW, SW, AP, LB, SE (Waaihoek and Du Toitskloof to Van Staden's Mts).•

gigantea L.f.   Grey-felted, tufted perennial to 1 m. Leaves radical, long-petiolate, elliptic, sometimes crenulate. Flowers crowded in compound umbels on branched peduncles, purple. Fruits ovate. Dec.–Mar. Rocky sandstone slopes, NW, SW, LB (Cedarberg Mts to Albertinia).•

intermedia C.Norman   Grey-felted, tufted perennial to 60 cm. Leaves radical, long-petiolate, paler beneath, ovate-cordate, crenulate. Flowers crowded in compound umbels on woolly peduncles, maroon. Feb.–Apr. High rocky sandstone slopes, NW, SW (Cedarberg and Piketberg to Drakenstein Mts).•

pillansii C.Norman   Like **H. intermedia** but leaves shaggy-woolly and often rust-coloured. Jan.–Feb. High sandstone slopes, SW (Cape Peninsula and Jonkershoek).•

quercifolia Eckl. & Zeyh.   Tufted perennial to 45 cm. Leaves radical, shortly petiolate, white-felted but more thinly above, obovate-oblanceolate, crenate. Flowers crowded in small compound umbels on branched peduncles, white. Fruits broadly obovate. Dec.–Feb. Damp sites on rocky sandstone slopes, SW, LB (Elandskloof Mts to Houwhoek).•

quinquedentata L.f.   Tufted perennial to 45 cm. Leaves radical, shortly petiolate, white-felted beneath, glabrous above, elliptic, stiffly toothed, margins revolute. Flowers crowded in small compound umbels on mostly unbranched peduncles, white. Fruits broadly obovate. Dec.–Apr. Damp sites on rocky sandstone slopes, SW (Hottentots Holland Mts to Hermanus).•

villosa (L.) Thunb.   White velvety, single- or few-stemmed shrub, to 1 m. Leaves sessile to shortly petiolate, oblong-elliptic, leathery, glabrous above, white-felted beneath, margins revolute, toothed. Flowers in congested apparently simple umbels, cream-coloured. Fruits suborbicular, winged. Dec.–May. Rocky sandstone slopes, SW (Cape Peninsula to Hermanus).•

HETEROMORPHA   PARSLEY TREE, WILDEPIETERSELIEBOS   7 spp., sub-Saharan Africa and S Arabia

arborescens (Spreng) Cham. & Schltdl. (= *Heteromorpha trifoliata* (H.L.Wendl.) Eckl. & Zeyh.)   Shrub or small tree, 2–9 m, with smooth reddish bark peeling in waxy flakes. Leaves palmately or pinnately 1–3(–9)-foliolate, leaflets elliptic-lanceolate, crenulate. Flowers in compound umbels, greenish white, foetid. Fruits elliptic, mericarps heteromorphic, ribbed and 2- or 3-winged, sometimes pilose. Dec.–Jan. Forest margins and rocky woodland, NW, SW, KM, LB, SE (Tulbagh and Bredasdorp to S Arabia).

ITASINA•   (= *THUNBERGIELLA*)   GRASS-PARSLEY   1 sp., W Cape

filifolia (Thunb.) Raf. (= *Thunbergiella filiformis* (Koso-Pol.) H.Wolff)   Tufted perennial to 45 cm. Leaves usually dry at flowering, radical, linear to filiform, sometimes subspathulate. Flowers in compound umbels on branched peduncles, white. Fruits narrowly elliptic, ridged, sepals sometimes slender and horn-like. Mainly Nov.–Apr. Sandstone and limestone flats, NW, SW, AP, SE (Bokkeveld Mts to Plettenberg Bay).•

LICHTENSTEINIA   KALMOES   8 spp., southern Africa, St. Helena

interrupta (Thunb.) Sond.   Tufted perennial to 60 cm, sometimes to 1.2 m. Leaves radical, lacerate-pinnatisect or bipinnatisect, leaflets unevenly serrate, usually dry at flowering. Flowers in compound umbels on slender, branched peduncles, white. Fruits broadly ovoid, with prominent stylopodium. Nov.–Feb. Grassland and bush, KM, SE (Swartberg Mts and George to KwaZulu-Natal).

lacera Cham. & Schltdl.   Tufted perennial to 1.4 m. Leaves radical, large, sometimes withered at flowering, ovate or obovate, sometimes lacerate to 3-sect, margins roughly toothed and aristate. Flowers in compound umbels on long, branched, hollow peduncles, yellow. Fruits narrowly oblong, with prominent stylopodium. Dec.–Mar. Flats and lower slopes, SW (Cape Peninsula to Kleinmond).•

latifolia Eckl. & Zeyh.   Like **L. lacera** but leaves prostrate, rotund to ovate, shallowly lobed to trifid with obovate lobes, margins toothed but not aristate, flowers white. Mainly Nov.–Feb. Sandstone slopes, NW, LB, SE (Langeberg: Robertson to Uitenhage).•

obscura (Spreng.) Koso-Pol. (= *Lichtensteinia beiliana* Eckl. & Zeyh.)   Like **L. interrupta** but with 3–5 umbellules per umbel. Nov.–Jan. Flats and lower slopes, NW, SW, AP, LB (Cedarberg Mts and Saldanha to Swellendam).•

trifida Cham. & Schltdl. (incl. **L. crassijuga** E.Mey. ex Sond.)   Like **L. lacera** but leaves 3–5-sect, leaflets narrowly oblanceolate, sometimes pinnatisect, margins evenly toothed, fruits ovoid. Dec.–Jan. Sandstone slopes, NW, SW, KM, SE (Piketberg to Langkloof).•

sp. 1   Tufted perennial to 60 cm. Leaves radical, large, palmatisect, leaflets pinnatisect and lacerate, ultimate lobes irregularly toothed. Flowers in compound umbels on long, branched peduncles, white. Fruits subglobose, 2-lobed, with prominent stylopodium. Nov.–Jan. Rocky sandstone slopes, NW (Cedarberg Mts).•

PEUCEDANUM   BERGSELDERY   c. 170 spp., tropical and southern Africa, Eurasia

A. Tufted perennials with leaves basal

capillaceum Thunb.   Tufted perennial to 60 cm. Leaves in a basal tuft, divaricately dissected, long-petiolate, leaflets subterete, erect, lemon-scented. Flowers in compound umbels on simple, elongate peduncles. white to yellow. Fruits elliptic, Dec.–Feb. Lower sandstone and limestone slopes, SW, AP, LB, SE (Michell's Pass to Uitenhage).•

millefolium Sond.   Tufted perennial to 60 cm. Leaves finely divided, with primary divisions short and leaflets short, brush-like, leaf axis scabridulous. Flowers in compound umbels on elongate few-branched peduncles, ?yellow. Fruits large, obovate, broadly winged, deeply notched above. Nov.–Jan. Rocky lower slopes, NW (Twenty-Four Rivers Mts to Warm Bokkeveld).•

strictum (Spreng.) B.L.Burtt (= *Peucedanum sieberianum* Sond.)   Tufted, stiffly leafy perennial to 60 cm. Leaves in a basal tuft, deeply dissected, long-petiolate, leaflets subterete. Flowers in compound umbels on simple or few-branched, elongate peduncles, yellow. Fruits large, elliptic. Dec.–Feb. Rocky sandstone slopes, NW, SW (Cedarberg to Hottentots Holland Mts).•

triternatum Eckl. & Zeyh.   Slender tufted perennial, to 1 m. Leaves basal, tripinnate, leaflets filiform. Flowers in compound umbels on few-branched peduncles, yellow. Jan. Sandstone slopes, SW (Hottentots Holland Mts).•

typicum (Eckl. & Zeyh.) B.L.Burtt (= *Peucedanum cynorhizum* Sond.; incl. **P. olifantianum** (Koso-Pol.) Hiroe)   HONDEWORTEL   Tufted, soft-stemmed perennial to 1 m. Leaves 3-pinnate, leaflets linear to lanceolate, soft. Flowers in compound umbels on elongate peduncles, yellowish. Fruits large, obovate, broadly winged. Oct.–Jan. Coastal scrub in sand, SE (Uitenhage).•

A. Shrubs with leaves cauline
B. Leaflets lanceolate to obovate

capense (Thunb.) Sond.   LIDBOSSIE   Shrub with stiffly erect branches, to 2 m. Leaves pinnate or bipinnate, petioles winged and clasping, leaflets elliptic-lanceolate, margins sometimes revolute. Flowers in compound, orbicular umbels on unbranched peduncles, greenish. Fruits elliptic. Nov.–Mar. Rocky or sandy slopes, KM, SE (Karoo, Swartberg and Outeniqua Mts to Mpumalanga).

galbaniopse H.Wolff   Shrub to over 1 m. Leaves 3-pinnate, leaflets elliptic, glaucous, coarsely toothed and lobed. Flowers in compound umbels on few-branched, elongate peduncles, yellowish. Fruits ellipsoid. Dec.–Feb. Bushy slopes, SW (Cape Peninsula to Paarl).•

galbanum (L.) Drude   BLISTER BUSH   Robust, few-branched shrub to 3 m. Leaves pinnate, glaucous, leaflets rhomboid to obovate, toothed, sometimes 3-lobed, upper leaves not reduced to sheaths. Flowers in large, orbicular compound umbels on short, axillary peduncles, yellow. Fruits ellipsoid. July–Feb. Rocky sandstone slopes in forest and bush, NW, SW, AP, LB (Piketberg to Langeberg Mts: Riversdale).•

B. Leaflets terete to linear-oblong

dregeanum D.Dietr. (= *Peucedanum ecklonianum* Sond.)   Like **P. striatum** but leaflets oblong, broader, paler beneath and margins revolute. Dec.–Mar. Rocky slopes, SE (Humansdorp to Uitenhage).•

ferulaceum (Thunb.) Eckl. & Zeyh.   Lax erect perennial to 1.5 m. Leaves finely dissected, with winged, clasping petiole, leaflets linear. Flowers in compound umbels on simple or few-branched peduncles, pale yellow. Fruits elliptic. Dec.–Mar. Rocky slopes, NW, SW, AP, LB, SE (Namaqualand to E Cape).

polyactinum B.L.Burtt (= *Peucedanum multiradiatum* Drude)   Robust, few-branched shrub to 3 m. Leaves 3-pinnate, leaflets filiform. Flowers in large, compound umbels on short, axillary peduncles, yellow. Fruits ellipsoid. Sept.–Dec. Lower sandstone slopes, NW, SW (Cedarberg Mts to Betty's Bay).•

pungens E.Mey. ex Sond.   Robust, stiffly leafy shrub to over 1 m. Leaves 2- or 3-pinnate, leaflets terete, rigid, spine-tipped. Flowers in compound umbels on branched peduncles, yellowish. Fruits ellipsoid. Sept.–Oct. Lower slopes, SW (Botrivier).•

striatum (Thunb.) Sond.   Laxly branched to sprawling shrublet, to 40 cm. Leaves small, pinnate to tripinnate, leaflets linear with recurved apices. Flowers in small compound umbels, green. Fruits ellipsoid. May–Sept. Clay slopes in renosterveld, SW, LB (Caledon to Riversdale).•

tenuifolium Thunb. (incl. **P. hypoleucum** (Meisn.) Drude, **P. sonderi** (Hiroe) B.L.Burtt)   Shrub to 2 m. Leaves 3-pinnate, leaflets linear, often pale and glaucous beneath. Flowers in large compound umbels, often axillary, yellow. Fruits ellipsoid. Sept.–Jan. Sandstone slopes, NW, SW, KM, LB, SE (Pakhuis to Baviaanskloof Mts).•

PIMPINELLA   c. 200 spp., cosmopolitan

stadensis (Eckl. & Zeyh.) D.Dietr.   Slender, erect annual or perennial to 45 cm, finely hairy above. Leaves pinnatisect or bipinnatisect, leaflets linear to lanceolate, basal leaves long-petiolate. Flowers in compound umbels on branched peduncles, white. Fruits broadly ovate, hairy. Jan.–Mar. Grassy fynbos slopes, SE (Humansdorp to KwaZulu-Natal).

POLEMANNIOPSIS   1 sp., W Cape to Karoo

marlothii (H.Wolff) B.L.Burtt (= *Polemannia marlothii* H.Wolff)   Much-branched woody shrub to 4 m, leafless at flowering, with flexuose branches. Leaves 3–5-digitate, leaflets obovate, toothed. Flowers in compound umbels, yellow. Fruits broadly winged. Dec.–Jan. Sandstone slopes, NW (S Namibia to Pakhuis Mts).

SANICULA   SANICLE   c. 50 spp., cosmopolitan, 1 in southern Africa

elata Buch.-Ham. ex D.Don   Tufted perennial to 80 cm. Leaves 3–5-digitate, leaflets lobed and toothed. Flowers few, sessile in simple umbels, pinkish. Fruits burr-like with hooked bristles. Oct.–Jan. Shady streamsides, SW, LB, SE (Cape Peninsula to tropical Africa and Eurasia).

SONDERINA   5 spp., southern Africa

caruifolia (Sond.) H.Wolff   Annual to 45 cm, branching from below. Leaves finely dissected, leaflets linear-filiform. Flowers in compound umbels on leaf-opposed peduncles, white. Fruits ovate and ridged. Sept.–Dec. Sandy flats, NW, SW (Lambert's Bay to Cape Peninsula).•

hispida (Thunb.) H.Wolff   Like **S. caruifolia** but fruits and pedicels hispid. Mainly Sept.–Jan. Sandy or limestone flats, usually coastal, NW, SW, AP, SE (Lambert's Bay to Uitenhage).•

tenuis (Sond.) H.Wolff   Erect, soft annual to 20 cm. Leaves bipinnatisect, leaflets lanceolate, toothed above. Flowers in simple, leaf-opposed umbels, few per node, sometimes in shortly pedunculate compound umbels, pedicels scabrid, white. Fruits narrowly ellipsoid, smooth. Oct.–Nov. Low hills, NW, KM, LB (Gifberg to Swellendam).•

STOIBRAX   4 spp., W Cape and Mediterranean

capense (Lam.) B.L.Burtt (= *Stoibrax didyma* (Sond.) B.L.Burtt, *Sonderina didyma* (Sond.) Adamson)   Sprawling annual, branched from base, to 15 cm. Leaves 3-pinnatisect, lobes linear. Flowers in compound umbels, often leaf-opposed, white. Fruits bilobed, each part subglobose, tuberculate. Sept.–Nov. Sandstone and limestone flats, NW, SW, AP (Namaqualand to De Hoop).•

TORILIS   HEDGE PARSLEY   c. 15 spp., Mediterranean to Asia, also Africa

arvensis (Huds.) Link   Sprawling, soft annual to 40 cm. Leaves finely divided, leaflets toothed. Flowers few in simple, leaf opposed umbels, white. Fruits burr-like with barbed prickles. Aug.–Nov. Flats and rocky slopes, NW, SW, AP, SE (Bokkeveld Mts to Cape Peninsula, W Karoo and E Cape to Europe).

APOCYNACEAE (= ASCLEPIADACEAE)
Asclepioid genera by P.V. Bruyns

1. Flowers without a corona; filaments free; pollen grains free:
   2. Ovary entire; fruit a fleshy berry:
      3. Unarmed shrubs or trees; inflorescence axillary ....**Acokanthera**
      3.' Armed shrubs, sometimes scandent; inflorescence terminal or pseudoaxillary ....**Carissa**
   2.' Ovary of 2 separate or almost separate carpels; fruit dry:
      4. Leaves opposite or ternate; stems woody and without spines ....**Gonioma**
      4.' Leaves alternate or fascicled; stems succulent and armed with stipular spines ....**Pachypodium**
1.' Flowers with a corona; filaments united into a staminal column; pollen agglutinated into masses:
   5. Pollinarium with 4 minute pollinia attached to corpusculum; large climber with milky sap ....**Secamone**
   5.' Pollinarium with 2 pollinia attached to each corpusculum:
      6. Pollinia pendulous in anthers:
         7. Sap clear:
            8. Corona absent:
               9. Petals concave or involute, erect or inflexed and closing perianth tube; anther appendages covering style apex ....**Microloma**
               9.' Petals convex or revolute, spreading; anther appendages much shorter than the long conical style apex ....**Astephanus**
            8.' Corona present:
               10. Plant a slender climber; corona consisting of 10 lobules; corpusculum as large as pollinium ....**Oncinema**
               10.' Plant small, prostrate to shrub-like; corona consisting of 25 lobules; corpusculum much smaller than pollinium ....**Eustegia**
         7.' Sap milky:
            11. Inner and outer coronas subequal, fused or separate but together forming a more or less continuous cup around anthers ....**Cynanchum**
            11.' Outer corona much smaller than inner or absent and not forming a cup, inner dominating gynostegium:
               12. Plant a leafless succulent with cylindrical grey-green stems; outer corona enclosing gaps between bases of large inner lobes; apex of style head conical ....**Sarcostemma**
               12.' Plant leafy; outer corona minute and present as a small tooth beneath guide rails or absent; apex of style head concave:
                  13. Corona lobes entire, without a fissure, cavity or keel on upper surface...............**Xysmalobium**
                  13.' Corona lobes with a fissure, cavity or keel on upper surface:
                     14. Corona lobes folded around a deep vertical cavity, sometimes with a horn inside cavity:
                        15. Plant a robust shrub (0.5–2 m tall) arising from a small above-ground trunk..**Gomphocarpus**
                        15.' Plant with 1-several stems arising from an underground rootstock and dying back in the dry season ....**Asclepias**
                     14.' Corona lobes not concave above:
                        16. Corona lobes deeply 3-lobed with margins ascending staminal column as narrow wings ....**Woodia**
                        16.' Corona lobes not as above:
                           17. Corona lobes mostly keeled on upper surface towards base; pollinaria 3 mm or longer ....**Pachycarpus**
                           17.' Corona lobes not keeled on upper surface towards base; pollinaria shorter than 1 mm:
                              18. Inflorescences pedunculate; pollinia medially attached to caudicle with germinating mouth on the outer side..**Schizoglossum**
                              18.' Inflorescences sessile; pollinia subapically attached, usually without a germinating mouth (if this is present then it is located at point of attachment of pollinium to caudicle)..**Aspidoglossum**
      6.' Pollinia horizontal to ascending in anthers:
         19. Sap milky; pollinia joined directly to corpusculum..**Fockea**
         19.' Sap clear; pollinium joined to a slender caudicle linking it to corpusculum:
            20. Inflorescence with several leaf-like bracts; pollinium minute, joined by a long slender caudicle to corpusculum, without a germinating mouth ....**Tylophora**
            20.' Inflorescence without leaf-like bracts; pollinium joined by a caudicle to underside of wing of corpusculum, with a germinating mouth:
               21. Stems herbaceous or succulent but then not tuberculate, or if so then flowers funnel-shaped and inflated below:
                  22. Corona of 1 series of 5 lobes opposite anthers, with various processes; fruit consisting of single horn ....**Anisotoma**

22.' Corona of 2 series of 5 lobes each; fruit consisting of 2 horns:
    23. Plant a leafy climber to scrambler with large cordate leaves, arising from a cluster of brown slightly thickened roots; flowers in dense to lax many-flowered umbels ....**Riocreuxia**
    23.' Plants not as above:
        24. Corolla tube at least three times as long as broad in the middle ....**Ceropegia**
        24.' Corolla tube less than twice as long as broad in the middle:
            25. Petals reflexed against pedicel; anthers erect ....**Macropetalum**
            25.' Petals ascending or spreading; anthers more or less horizontal on top of style apex..**Brachystelma**
21.' Stems very succulent, tuberculate; leaves absent or reduced to minute rudiments at tip of tubercles; flowers fleshy:
    26. Stems with 6 or more angles:
        27. Stems with 6–8 angles..**Pectinaria**
        27.' tems with 10 or more angles ....**Hoodia**
    26.' Stems with 4 or 5 angles:
        28. Stems, pedicels and sepals pubescent ....**Stapelia**
        28.' Stems, pedicels and sepals glabrous:
            29. Young tubercles rounded, not tipped by an acute tooth or small leaf:
                30. Inner surface of corolla rough with columnar papillae (each tipped with a seta) or covered with fine crinkled hairs ....**Quaqua**
                30.' Inner surface of corolla with tiny unicellular papillae, otherwise smooth and without hairs...**Tromotriche**
        29.' Young tubercles tapering into a tooth or small acute leaf:
            31. Inflorescence(s) in upper half of stem towards apex, usually many:
                32. Outer corona forming a distinct lobe beneath or enclosing guide-rail; tubercles each tipped with a hardened sharp tooth; flowers arising in many vertically arranged small fascicles along grooves between stem angles ....**Quaqua**
                32.' Outer corona not forming a lobe beneath guide-rail and not enclosing it; tubercles without a sharp and hardened tooth; flowers arising in few inflorescences not vertically arranged...**Piaranthus**
            31.' Inflorescence arising in lower half of stem towards base, solitary:
                33. Corolla urceolate; inner corona lobes laterally flattened, touching backs of anthers only at base, rising above anthers to form a cage over them ....**Stapeliopsis**
                33.' Corolla not urceolate; inner corona lobes dorsiventrally flattened, touching backs of anthers for most of their length and not forming a cage over them:
                    34. Inner and outer coronal series clearly separated from one another at anthesis; pollinium elliptic-rectangular, with germinating mouth exactly down inner (long) side:
                      35. Corolla tube completely containing gynostegium; corolla with fine longitudinal ridges on interior, lobes concave above; wings on corpusculum much shorter than breadth of pollinium ....**Huernia**
                      35.' Corolla tube not completely containing gynostegium; corolla without fine longitudinal ridges on interior, lobes convex above, usually tightly folded into rather narrow long plates; wings on corpusculum much longer than breadth of pollinium...**Duvalia**
                  34.' Dorsal part of inner corona fused towards base with outer series; pollinium D-shaped, with germinating mouth bending up from outer side towards upper surface:
                      36. Young tubercles each bearing a small, acute, differentiated leaf, without stipular denticles ....**Tridentea**
                      36.' Young tubercles without a differentiated leaf, usually with stipular denticles..**Orbea**

ACOKANTHERA   GIFBOOM, INTLUNGUNYEMBE   15 spp., Africa

oppositifolia (Lam.) Codd   Shrub or tree to 4 m; sap milky. Leaves leathery, ovate, glossy, margins revolute. Flowers in axillary clusters, white tinged pink, fragrant. Apr.–Dec. Bush and scrub, SE (Mossel Bay to tropical Africa).

ANISOTOMA   2 spp., E Cape to KwaZulu-Natal

cordifolia Fenzl   Trailing perennial with rigid, swollen brown roots; sap clear. Leaves opposite, cordate, hairy. Flowers in umbellate cymes, brown becoming yellow to white in tube. Nov.–Feb. Rocky grassland, SE (Uitenhage to E Cape).

ASCLEPIAS MILKWEED ?150 spp., Africa, America

crispa P.J.Bergius BITTERWORTEL   Sprawling perennial to 30 cm; sap milky. Leaves opposite, narrow, hispid, with wavy margin. Flowers greenish or purple, corona lobes hollowed, greenish. Nov.–May. Flats and slopes, NW, SW, LB, SE (Bokkeveld Mts to E Cape).

expansa (E.Mey.) Schltr.   Erect, single-stemmed perennial to 70 cm; sap milky. Leaves opposite, linear, glabrous or sparsely hairy on margins. Flowers green with purple reverse, corona lobes hollowed, white. Oct.–Jan. Grassland, SE (Knysna to E Cape).

viridiflora (E.Mey.) Goyder (= *Asclepias dregeana* Schltr.)   Erect perennial to 30 cm; sap milky. Leaves opposite, lanceolate to broadly oblong, glabrous or hairy on margins. Flowers greenish. Nov.–Jan. Grassland, LB, SE (Riversdale to Mpumalanga).

ASPIDOGLOSSUM   34 spp., southern and tropical Africa

gracile (E.Mey.) Kupicha (= *Schizoglossum bolusii* Schltr., *S. burchellii* N.E.Br., *S. dregei* N.E.Br., *S. monticola* Schltr.)   Slender, single-stemmed perennial to 1 m; sap milky. Leaves opposite, linear. Flowers 4–6 mm diam., brownish green with corona usually shorter than style apex. Nov.–Dec. Grassland or fynbos on stony slopes, SW, LB, SE (Cape Peninsula to KwaZulu-Natal).

heterophyllum E.Mey.   Single-stemmed or branched perennial to 20 cm; sap milky. Leaves linear to elliptic, whorled, at least above. Flowers more than 10 mm diam., green with white corona usually much longer than style apex. Sept.–Mar. Stony slopes, SW, LB, SE (Cape Peninsula to E Cape).

ASTEPHANUS   2 spp., Namaqualand to E Cape

marginatus Decne.   Slender, thinly hairy climber to 1 m; sap clear. Leaves opposite, narrowly to broadly elliptic. Flowers white, tube twice as long as broad, hairy at base within, without corona. Apr.–Dec. Coastal bush, SE (Knysna to East London).

triflorus (L.f.) Schult.   Like **A. marginatus** but often more densely pubescent, corolla tube up to as long as broad. Apr.–Aug. Coastal or inland bush, NW, SW, AP, LB (Namaqualand to Plettenberg Bay).

BRACHYSTELMA   c. 100 spp., tropical to southern Africa, India to Australia

occidentale Schltr.   Dwarf, tuberous perennial to 10 cm; sap clear. Leaves opposite, linear, often with crisped margins, glabrescent. Flowers c. 5 mm diam., without tube, white with tips of lobes green or spotted with maroon, corona green and white. Sept.–Oct. Gravelly flats, SW (Cape Peninsula to Bredasdorp).•

thunbergii N.E.Br.   Dwarf, tuberous perennial to 10 cm; sap clear. Leaves lanceolate, pubescent. Flowers 6–15 mm diam., campanulate with cupular tube, white with green lobes, corona brilliant yellow. Sept.–Oct. Flats and lower slopes, SW, LB (Caledon to Riversdale and KwaZulu-Natal).

tuberosum (Meerb.) R.Br. ex Sims (= *Brachystelma caudatum* (Thunb.) N.E.Br.)   Dwarf, tuberous perennial to 10 cm; sap clear. Leaves opposite, lanceolate, pubescent. Flowers 20–40 mm diam., campanulate with cupular tube, cream-coloured to green with purple spots and bars, corona dark purple-brown. Sept.–Oct. Seasonally moist sandy flats, SW, AP, LB, SE (Malmesbury to E Cape).

CARISSA NUM-NUM   c. 20 spp., palaeotropics and subtropics

bispinosa (L.) Desf. ex Brenan   NUM-NUM, ISIBETHA-NKUNZI   Twiggy, shrub to 5 m armed with simple or forked thorns; sap milky. Leaves leathery, elliptic, glossy. Flowers in terminal, subumbellate cymes, white, fragrant, corolla lobes 3.5–5.0 mm. Fruits red, 16 mm long. Sept.–Dec. Coastal and karooid scrub, AP, LB, SE (Elim to tropical Africa).

haematocarpa (Eckl.) A.DC.   KAROO NUM-NUM, ISNUSA-NKUNZI   Like **C. bispinosa** but corolla lobes 2 mm and fruits black. Jan.–Apr. Karooid scrub, NW, KM, SE (Worcester to Port Elizabeth, Karoo and S Namibia).

macrocarpa (Eckl.) A.DC.   NMTHUNGULU   Like **C. bispinosa** but flowers larger, corolla lobes 15–35 mm and fruits 30–50 mm long. Nov.–Feb. Coastal bush, SE (Humansdorp to Mozambique).

CEROPEGIA   c. 160 spp., Canary Is., Africa, Madagascar, Asia, Australasia

A. Leaves lacking or vestigial and caducous

ampliata E.Mey.   BUSHMAN'S PIPE, BOESMANSPYP   Succulent climber to 1 m with finely striate, grey-green stems 2–4 mm diam.; sap clear. Leaves 3–7 mm, caducous. Flowers 50–70 mm long, tube cylindrical above and at least 10 mm di6am. throughout, cream-coloured to greenish with fine green veins outside, petals narrowing to tips, without cilia. Nov.–Apr. Karroid scrub, SE (Willowmore to Kenya, Madagascar).

fimbriata E.Mey. (= *Ceropegia connivens* R.A.Dyer, *C. geniculata* R.A.Dyer)   Succulent climber to 1 m with smooth stems; sap clear. Leaves to 5 mm, caducous. Flowers 30–60 mm long, tube flaring above, whitish heavily veined with green, petals with broad expanded apices fused together into an umbrella-like canopy, ciliate. Oct.–Mar. Karroid scrub on flats and slopes, NW, KM, SE (Worcester to E Cape).

stapeliiformis Haw.   SLANGKAMBROE   Trailing to climbing succulent with fleshy stems 5–15 mm thick; sap clear. Leaves minute, caducous. Flowers 50–70 mm long, tube funnnel-shaped above, whitish streaked and spotted with maroon, petals spreading, narrowing towards apex. Oct.–Mar. Karroid scrub, SE (Willowmore to N Province).

zeyheri Schltr.   Like **C. fimbriata** but petals fused towards tips into slender cage, with small cilia along margins. Oct.–Mar. Karroid scrub, KM, SE (Barrydale to E Cape).

A. Leaves well developed
B. Plants with 1–several depressed-globose tubers

africana R.Br.   Like **C. occidentalis** but flowers grey to green with purplish veins, outer corona lobes erect and fused into a deep cupular structure around anthers, inner lobes laterally flattened above. Oct.–Mar. Karroid scrub, NW, SW, AP, KM, LB, SE (Cape Peninsula and Worcester to KwaZulu-Natal).

linearis E.Mey. (= *Ceropegia woodii* Schltr.)   Like **C. occidentalis** but cage formed by petals nearly uniformly broad from mouth of tube to truncate apex and with blackish margins. Oct.–Mar. Rocky outcrops, often in dry forest, LB, SE (Heidelberg to KwaZulu-Natal).

occidentalis R.A.Dyer   Small succulent climber to 30 cm; sap clear. Leaves opposite, ovate-oblong. Flowers 15–30 mm long, pale green with purplish veins, petals fused at tips into a cage narrowing towards apex, outer corona spreading, shallow, inner corona lobes slender, dorsiventrally flattened. Oct.–Mar. Coastal rock outcrops, NW (S Namibia to Lambert's Bay).

occulta R.A.Dyer   Like **C. linearis** but corolla lobes broadly spathulate and forming small canopy, flowers densely greyish flecked with white and purple on petals. Oct.–Mar. Dry rocky slopes, SW, KM, LB (Caledon to Little Karoo).•

B. Plants with fusiform roots and without tubers

barbata R.A.Dyer   Slender perennial or climber to 50 cm; sap clear. Leaves opposite, linear. Flowers 25–45 mm, tube funnel-shaped above, petals slender with clavate, ciliate apices. Dec.–Jan. Stony slopes, KM, SE (Calitzdorp to Great Winterhoek Mts).•

bowkeri Harv.   Erect perennial, 8–30 cm; sap clear. Leaves opposite, linear, channelled above. Flowers 18–30 mm, tube slightly flared above, petals 15–22 mm long, pendulous. Dec.–Jan. Stony slopes, SE (Montagu Pass to E Cape).

carnosa E.Mey. (incl. **C. racemosa** N.E.Br.)   Slender climber to 1.5 m; sap clear. Leaves opposite, slightly fleshy, elliptic. Flowers 15–25 mm, tube abruptly swollen in middle and flaring above, speckled with red-brown on cream-coloured outside, petals narrowing to apex, ciliate. Dec.–May. Dry riverine scrub, SE (Langkloof to Ethiopia).

crassifolia Schltr.   Succulent climber to 1.5 m; sap clear. Leaves opposite, slightly fleshy, ovate-elliptic. Flowers 25–50 mm long, tube flaring above, pale green with purple bars and spots, petals keeled and ciliate. Oct.–Mar. Stony slopes, SE (Humansdorp to Kenya).

CYNANCHUM   BOKHORING   c. 100 spp., cosmopolitan

africanum (L.) Hoffmanns.   Climber to 60 cm, with horizontal runners, sometimes hairy; sap milky. Leaves opposite, ovate, slightly fleshy. Flowers in axillary clusters, brown sometimes green with twisted petals, corona white, twice as deep as wide, gynostegium clearly stalked, as long as corona. June–Dec. Sandy soils, mainly coastal, NW, SW, AP, LB, SE (Namaqualand to Cape Peninsula to E Cape).

ellipticum (Harv.) R.A.Dyer (= *Cynanchum capense* Thunb. non L.f.)   Climber to 3 m; sap milky. Leaves opposite, elliptic to ovate. Flowers in axillary clusters, brown or green, corona white, cup-shaped, irregularly crenate, gynostegium sessile. Jan.–Dec. Coastal bush, SE (George to Mozambique).

gerrardii (Harv.) Liede (= *Cynanchum sarcostemmatoides* K.Schum.)   Leafless succulent climber to 1 m with finely striate stems; sap milky. Leaves absent or rudimentary. Flowers on axillary branches, green, petals reflexed, corona white, shallow, gynostegium sessile. May–Dec. Dry scrub, SE (Uitenhage to Arabia, Madagascar).

natalitium Schltr.   Climber to 2 m, with woody base; sap milky. Leaves opposite, elliptic-oblong, fleshy. Flowers in axillary clusters, brown to green, corona white, cup-shaped, 5-crenate, gynostegium shortly stalked. Jan.–Dec. Coastal bush, SE (Knysna to KwaZulu-Natal).

obtusifolium L.f.   Climber with woody base to 3 m, sometimes hairy; sap milky. Leaves opposite, ovate-elliptic, often crenulate. Flowers in axillary clusters, dull green, corona white, shallow and deeply lobed, gynostegium sessile. Jan.–Dec. Coastal bush, SW, AP, SE (Cape Peninsula to Mozambique).

zeyheri Schltr.   Much-branched, mat-forming perennial to 30 cm; sap milky. Leaves opposite, ovate to suborbicular, fleshy. Flowers in axillary clusters, brown, corona white, cup-shaped, gynostegium clearly stalked, as long as corona. May–Nov. Clay and limestone flats, SW, AP (Cape Peninsula to Riversdale).•

DUVALIA   GORTJIE   16 spp., Africa, Arabia

caespitosa (Masson) Haw. (incl. **D. vestita** Meve)   Dwarf, mat-forming leafless succulent with 4- or 5-angled, tuberculate stems 1.0–1.5 cm long; sap clear. Flowers 18–25 mm diam., dark brown, with raised annulus, petals narrow with deflexed margins, finely pubescent, corona pale brown, disc-like but not completely covering the annulus. Mar.–Oct. Gravelly slopes and flats, NW, KM, LB, SE (S Namibia and Karoo to E Cape).

elegans (Masson) Haw.   Like **D. caespitosa** but flowers less than 18 mm diam., shiny blackish with coarse hairs on annulus and bases of lobes, and corona completely covering the annulus. Jan.–Oct. Karroid flats, NW, SW, KM, LB (Worcester to Riversdale).•

immaculata (C.A.Lückh.) Bayer ex L.C.Leach   Like **D. caespitosa** but stems generally somewhat rhizomatous, flowers dark red-brown, glabrescent, with bright yellow corona sunken within annulus. Apr.–Dec. Dry coastal fynbos, LB (Swellendam to Mossel Bay).•

modesta N.E.Br.   Like **D. caespitosa** but flowers pale greenish brown with paler centre, less than 18 mm diam., petal margins deflexed only in outer part. Mar.–Apr. Karroid scrub, SE (W Karoo, Uniondale to E Cape).

parviflora N.E.Br.   Dwarf, mat-forming leafless succulent with almost globose stems 1–3 cm long; sap clear. Flowers like **D. caespitosa** but less than 15 mm diam., cream-coloured, glabrous, corona cream-coloured. Feb.–Aug. Karroid scrub, KM (Ladismith to Oudtshoorn).•

pillansii N.E.Br.   Like **D. caespitosa** but flowers pale purple-brown becoming paler towards centre and corona bright yellow. Apr.–June. Karroid scrub, SE (Uitenhage to E Cape).

EUSTEGIA   1 sp., Namaqualand and W Cape

minuta (L.f.) Schult.   Slender, erect to prostrate dwarf perennial to 15 cm; sap clear. Leaves opposite, filiform to filiform-hastate. Flowers in small umbels, green, with complex white corona of 25 narrow lobules. July–Feb. Gravelly flats to lower slopes, NW, SW, LB (Namaqualand to Riversdale).

FOCKEA   KAMBROE   6 spp., southern Africa to Tanzania

capensis Endl.   (= *Fockea crispa* (Jacq.) K.Schum.)   Erect or climbing succulent to 50 cm, stems grey, 2–4 mm diam.; sap milky. Leaves opposite, ovate, grey-green, densely pubescent and crisped. Flowers yellow-green with tubular white corona, anther appendages swollen. Feb.–May. Rocky slopes and summits, KM (Little Karoo).•

comaru (E.Mey.) N.E.Br.   Slender climber to 30 cm tall, stems c. 1 mm diam.; sap milky. Leaves opposite, linear, purplish to dark grey-green, margins recurved. Flowers as in **F. capensis**. Mar.–May. Karroid scrub on rocky slopes, NW, SW, KM, LB, SE (S Namibia and Karoo to Willowmore).

edulis (Thunb.) K.Schum. (= *Fockea glabra* Decne.)   Like **F. capensis** but climbing to 2 m from a tuber up to 1 m diam. Leaves ovate, green, sparsely pubescent to glabrous and slightly undulate. Nov.–Mar. Rocky slopes in dry riverine bush, NW, LB, SE (Robertson, Riversdale to E Cape).

sinuata (E.Mey.) Druce   Like **F. comaru** but leaves crisped and margins not recurved, flowers brownish, usually with short petals, fruits warty and seeds with hairs all around margin. Jan.–May. Karroid scrub on calcareous flats, KM (Namibia and Karoo to Calitzdorp).

GOMPHOCARPUS KATOENBOS   c. 30 spp., Africa, Arabia

cancellatus (Burm.f.) Bruyns   Rigid, hairy shrub to 1.5 m; sap milky. Leaves opposite, oblong-lanceolate to elliptic, usually rounded at base, leathery. Flowers cream-coloured. Fruits inflated, ovoid-acute. Mar.–Dec. Stony slopes, NW, SW, KM, LB, SE (S Namibia to E Cape).

fruticosus (L.) W.T.Aiton   Soft shrub branching mainly from base, 1–3 m; sap milk. Leaves opposite, linear to oblanceolate, tapering gradually below. Flowers cream-coloured. Fruits inflated, ovoid-acute. Nov.–Apr. Disturbed areas, NW, SW, KM, LB, SE (Cape Peninsula and Karoo to George, widespread and almost cosmopolitan).

*****physocarpus** E.Mey.   BALBOS   Like **G. fruticosus** but branching above and fruits spherical. Nov.–Apr. Stony flats, often roadsides, NW, SW, LB, SE (tropical African weed).

GONIOMA KAMASSIEHOUT, IGALAGALA   1 sp., W and E Cape

kamassi E.Mey.   Shrub or tree to 6 m; sap clear. Leaves oblanceolate, glossy. Flowers in compact terminal cymes, white or cream-coloured, fragrant. Nov.–Apr. Coastal forest, SE (Wilderness to E Cape).

HOODIA GHAAP   13 spp., southern and tropical Africa

gordonii (Masson) Sweet ex Decne.   muishondghaap, wolweghaap   spiny, cactus-like, leafless succulent to 1 m, with 11–17-angled stems; sap clear. Flowers 40–100 mm diam., flesh-coloured, somewhat foul-smelling. Sept.–Apr. dry stony slopes and flats, NW (Namibia and Karoo to Clanwilliam).

pilifera (L.f.) Plowes   (= *Trichocaulon piliferum* (L.f.) N.E.Br.)   Spiny, cactus-like leafless succulent to 80 cm, with 21–34-angled stems; sap clear. Flowers 16–20 mm diam., dark purplish to pinkish brown, foul-smelling. May–Sept. Rocky slopes in karroid scrub, KM (Little Karoo to E Cape).

HUERNIA   c. 50 spp., southern and tropical Africa, Arabia

A. Stems more or less round in section with tubercles arranged in 6–16 rows

longii Pillans (incl. **H. echidnopsiodes** (L.C.Leach) L.C.Leach)   Leafless succulent with stems cylindrical and tesselated, teeth in 6–8, often spiralling rows; sap clear. Flowers campanulate, cream-coloured with brown-red spots. Jan.–May. Stony sandstone or conglomerate slopes, SE (Langkloof to Uitenhage).•

pillansii N.E.Br.   Leafless succulent to 10 cm, with stems cylindrical and tesselated, covered in soft bristles arranged roughly into 10–16, often spiralling rows; sap clear. Flowers like **H. longii**. Dec.–May. Stony slopes and flats in karroid scrub, KM (Montagu to Willowmore and S Karoo).

A. Stems more or less square or pentagonal in section with tubercles arranged in 4 or 5 rows

barbata (Masson) Haw. (incl. **H. campanulata** (Masson) Haw., **H. clavigera** (Jacq.) Haw.)   Leafless succulent forming dense clumps to 60 cm; sap clear. Flowers tubular-campanulate, tube entirely or irregularly and concentrically marked with maroon, with long often clavate hairs in mouth of tube and onto petals. Jan.–Aug. Karroid scrub, flats or lower slopes, NW, KM, LB, SE (Namaqualand and Karoo to Port Elizabeth).

brevirostris N.E.Br.   (incl. **H. bayeri** L.C.Leach)   Like **H. barbata** but flowers campanulate, cream-coloured, sometimes finely spotted with red to brown, smooth or covered with conical papillae each with short apical hair. Dec.–Mar. Karroid scrub, SE (Willowmore to E Cape and Karoo).

guttata (Masson) Haw. (incl. **H. reticulata** (Masson) Haw.)   Like **H. barbata** but flowers campanulate with shiny, raised annulus, cream-coloured finely spotted with maroon with spots larger and coalescing on annulus, with a few long straight hairs in throat. Dec.–Apr. Karroid scrub and stony grassland, NW, KM, SE (Namaqualand to Citrusdal, Calitzdorp to E Cape).

praestans N.E.Br.   Like **H. guttata** but annulus neither shiny nor differently coloured from rest of flower, hairs present from mouth of tube to lobes. Jan.–Mar. Stony slopes and flats in karroid scrub, KM (Montagu to Vanwyksdorp).•

[**Species excluded**   No authentic material found and probably conspecific with one of the above: **H. witzenbergensis** C.A.Lückh.]

MACROPETALUM   1 sp., South Africa and Zimbabwe

burchellii Decne.   Slender, single-stemmed perennial to 50 cm; sap clear. Leaves opposite, filiform. Flowers greenish, with reflexed, filiform petals 18–25 mm long. Oct.–Feb. Stony slopes, SE (Cockscomb to Zimbabwe).

MICROLOMA   MELKTOU, WAX CREEPER   10 spp., South Africa and Namibia

armatum (Thunb.) Schltr. ex Gilg   Somewhat rigid, spiky shrublet to 30 cm; sap clear. Leaves opposite, 2–8 mm, caducous. Flowers 2–4 mm long, greenish yellow. Mainly Sept.–Dec. Stony flats and slopes in karroid scrub, KM (Namibia and Karoo to Little Karoo).

sagittatum (L.) R.Br.   (= *Microloma gibbosum* N.E.Br., *M. glabratum* E.Mey.)   Slender climber to 1 m or more, with fibrous roots ; sap clear. Leaves opposite, 7–35 mm long, narrow. Flowers cylindrical, 5–11 mm long, at least slightly pubescent outside, pink to red, petals lanceolate. June–Oct. Stony slopes to sandy flats, NW, SW, AP, KM, LB, SE (Namaqualand to Willowmore).

tenuifolium (L.) K.Schum.   KANNETJIES   Slender deciduous climber to 1 m, with swollen roots; sap clear. Leaves opposite, 20–70 mm long, narrow. Flowers urceolate, 6–8 mm long, shiny and glabrous outside, orange to red, petals suborbicular. June–Nov. Stony slopes and flats, NW, SW, LB, SE (Gifberg to E Cape).

ONCINEMA•   1 sp., W Cape

lineare (L.f.) Bullock   Slender climber to 3 m, with fibrous roots; sap clear. Leaves opposite, linear to linear-lanceolate. Flowers cream-coloured, with long slender style head and small corona lobes. Nov.–Mar. Riverine bush or damp sandy slopes, SW, KM, LB, SE (Bainskloof to Langkloof).•

ORBEA   c. 20 spp., southern and tropical Africa

ciliata (Thunb.) L.C.Leach   Leafless succulent forming mats to 50 cm diam., stems 15–25 mm diam. (excluding teeth), with conical tubercles arranged loosely into 4 rows; sap clear. Flowers campanulate, 70–110 mm diam., cream-coloured, with funnel-shaped annulus 8–11 mm tall. Oct.–Apr. Karroid scrub on flats, NW, KM (Namaqualand to Prince Albert).

pulchella (Masson) L.C.Leach   Like **O. ciliata** but stems 5–10 mm diam. Flowers more or less rotate, 50–80 mm diam., cream-coloured to yellow spotted with purple-brown, with slightly raised, more or less pentagonal annulus, inner corona lobes much longer than anthers, rising up in centre, with small dorsal horn. Sept.–May. Coastal scrub, SE (Uitenhage to E Cape). [Probably a natural hybrid of next two species.]

variegata (L.) Haw.   CARRION FLOWER, AASBLOM   Like **O. pulchella** but annulus forming a shallowly bowl-shaped tube with spreading thinner rim, inner corona lobes much longer, tuberculate-clavate-tipped with prominent dorsal horn. Dec.–Sept. Mainly coastal, sand or granite or shale outcrops, NW, SW, LB, SE (Lambert's Bay to Humansdorp).•

verrucosa (Masson) L.C.Leach   Like **O. pulchella** but inner corona lobes just exceeding anthers and not meeting in centre, without dorsal horn. Dec.–May. Stony slopes in karroid scrub, KM, SE (Oudtshoorn to E Cape).

PACHYCARPUS   c. 30 spp., southern and tropical Africa

dealbatus E.Mey.   Stout perennial to 50 cm; sap milky. Flowers globose to globose-campanulate, lobed almost to base, green with purple-brown markings, corona lobes without keels on upper surface. Dec.–Feb. Grassy slopes, KM, LB, SE (Barrydale and Riversdale to KwaZulu-Natal).

grandiflorus (L.f.) E.Mey.   Like **P. dealbatus** but flowers globose, lobed to at most 2/3 of length, yellow with purple-brown spots, corona lobes with a pair of prominent keels on upper surface. Jan.–Apr. Mountain grassland, SE (Humansdorp to Mpumalanga).

PACHYPODIUM   DIKVOET   c. 13 spp., southern Africa and Madagascar

bispinosum (L.f.) A.DC.   Spiny succulent shrublet with swollen underground stem to 50 cm; sap clear. Leaves lanceolate, glabrescent above, hairy beneath. Flowers 1–few, pink to purple, sometimes white, broadly funnel-shaped, tube 7–8 mm diam., lobes 5–7 mm long. Aug.–Dec. Dry rocky slopes, KM, SE (Ladismith to E Cape).

succulentum (L.f.) A.DC.   Like **P. bispinosum** but flowers hairy and narrowly funnel-shaped, tube 3–4 mm diam., lobes 8–18 mm long, pink to crimson, sometimes white. Sept.–Dec. Dry rocky slopes, KM, SE (Oudtshoorn and Karoo to E Cape).

PECTINARIA   3 spp., Namaqualand and W Karoo

articulata (Aiton) Haw. (= *Pectinaria asperiflora* N.E.Br.)   GHAAP   Dwarf leafless, mat-forming succulent with tubercles arranged in 6 rows along stems, 2–7 cm; sap clear. Flowers bud-shaped with lobes remaining joined at tips, papillate, maroon to pale yellow. Sept.–Apr. Gravelly flats in karroid scrub, KM (Namaqualand and Karoo, Barrydale).

maughanii (R.A.Dyer) Bruyns   Plant like **P. articulata**. Flowers campanulate, smooth, deep yellow with pink to reddish purple centre. July–Nov. Gravelly flats in karroid scrub, NW (Bokkeveld Plateau to Calvinia).

PIARANTHUS   6 spp.,   Namibia and South Africa

comptus N.E.Br.   Like **P. geminatus** but petals 3.5–5.0 mm wide, usually cream-coloured with small red spots. Feb.–May. Flats and low hills in karroid scrub, KM (Karoo and S Karoo to Prince Albert).

geminatus (Masson) N.E.Br. (= *Piaranthus foetidus* N.E.Br., *P. pillansii* N.E.Br.; incl. **P. barrydalensis** Meve)   Dwarf leafless, mat-forming succulent, 2–5 cm, stems 4 or 5-angled and tuberculate; sap clear. Flowers shallowly campunulate, 20–26 mm diam. with petals 5–6 mm wide, white to yellow with brownish to purple banding, inner corona lobes adpressed to backs of anthers. Jan.–Apr. Gravelly lower slopes and flats in karroid scrub, KM, LB, SE (Montagu to Barrydale to E Cape).

parvulus N.E.Br.   Like **P. geminatus** but flower 5.0–13.5 mm diam., with petals to 2 mm wide, uniformly pale yellow or faintly lined with red-brown, inner corona lobes adpressed to backs of anthers at base only then erect. Feb.–May. Karroid scrub, KM (W Karoo to Ladismith).

punctatus (Masson) R.Br.   Like **P. geminatus** but flowers campanulate with tube completely enclosing corona. Feb.–June. Gravelly flats in karroid scrub, NW (Namaqualand to Klawer).

QUAQUA (= *CARALLUMA* in part)   AROENA, OURAM   15 spp., Namibia and South Africa

A. Tubercles on stems obtuse and rounded, apical tooth absent or shorter than 2 mm

linearis (N.E.Br.) Bruyns (= *Caralluma linearis* N.E.Br.)   Leafless, much-branched succulent, 5–15 cm, with 4-angled blackish stems; sap clear. Flowers few, campanulate, white in centre with red-purple tips, petals slender, without papillae, inner corona lobes much longer than anthers, erect then recurved. Jan.–Nov. Sandstone slopes and summits, KM (Touwsrivier to Witteberg).•

ramosa (Masson) Bruyns (= *Caralluma ramosa* (Masson) N.E.Br.)   Like **Q. linearis** but plant larger, to 30 cm tall, and stems purplish to greyish green. Flowers usually in dense clusters opening simultaneously, reddish in centre with shiny blackish purple tips, papillate around mouth, inner corona lobes scarcely exceeding anthers and adpressed to them. Mar.–May. Lower shale slopes, KM, LB (Little Karoo, Laingsburg).

A. Tubercles on stems not obtuse and rounded, apical tooth longer than 3 mm

arenicola (N.E.Br.) Plowes (= *Caralluma arenicola* N.E.Br.)   Leafless, clump-forming succulent with stout, erect stems, tubercles with acute teeth arranged in 4 rows; sap clear. Flowers campanulate, usually in dense clusters opening simultaneously, cream-coloured in centre with dark rings and blackish purple to reddish petals, roughly papillate in mouth of tube, foul-smelling. Aug.–May. Karroid scrub, KM (Touwsrivier to Klipplaat, Karoo).

aurea (C.A.Lückh.) Plowes (= *Caralluma aurea* C.A.Lückh.)   Small, leafless succulent forming clumps to 15 cm, tubercles on stems hardened into spreading teeth, arranged in 4(–5) angles; sap clear. Flowers campanulate, cream-coloured, petals shortly adpressed-hairy. Aug.–Oct. Karroid scrub, NW (W Karoo to Clanwilliam).

incarnata (L.f.) Bruyns (= *Caralluma hottentotorum* (N.E.Br.) N.E.Br., *C. incarnata* (L.f.) N.E.Br.)   Like **Q. aurea** but flower tube at least as long as corona and papillae restricted to mouth of tube. May–Oct. Karroid scrub, NW, SW (S Namibia to Bokbaai).

mammillaris (L.) Bruyns (= *Caralluma mammillaris* (L.) N.E.Br.)   Leafless succulent shrublet, 10–45 cm, rooting from central stem, tubercles hardened into sharp spreading teeth, obscurely arranged in 4 or 5 angles; sap clear. Flowers like **Q. arenicola**, but lower half of petals and mouth of tube roughly papillate, very foul-smelling. Mar.–June. Stony slopes and flats, NW, KM, LB (S Namibia to Little Karoo).

marlothii (N.E.Br.) Bruyns (= *Caralluma marlothii* N.E.Br.)   Like **Q. aurea** but flowers on slender pedicels 3–15 mm long, pale greenish yellow banded purple-brown with petals recurved and touching pedicel, stiffly hairy but without marginal cilia. Sept.–Mar. Stony slopes and flats in karroid scrub, KM (Montagu to Barrydale and Karoo).

parviflora (Masson) Bruyns (= *Caralluma dependens* N.E.Br.)   Like **Q. marlothii** but flowers not stiffly hairy, petals spreading and with fine, crisped marginal cilia. Mar.–July. Lower slopes and flats, NW (Namaqualand to Clanwilliam).

pillansii (N.E.Br.) Bruyns (= *Caralluma pillansii* N.E.Br.)   Leafless, much-branched succulent, 15–50 cm, with robust stems mottled with purple on grey, tubercles laterally flattened and joined into 4 angles, acute but not spine-like; sap clear. Flowers spotted with purple-brown on cream, papillate, petals oblong-ovate. Feb.–Apr. Dry sandstone slopes, NW, SW, KM, LB, SE (Cold Bokkeveld to E Cape).

[Species excluded   No authentic material found and probably conspecific with one of the above: **Q. arida** (Masson) Plowes]

RIOCREUXIA   KANDELAARTOU   c. 8 spp.,   southern to tropical Africa

torulosa Decne.   Slender, shortly hairy climber to 2 m; sap clear. Leaves opposite, soft, cordate-ovate. Flowers tubular with petals remaining fused at tips, pale yellow tinged purplish. Dec.–Mar. Forest margins and scrub, SE (George to Mpumalanga).

SARCOSTEMMA   SPANTOU   c. 10 spp., Old World tropics

viminale (L.) R.Br. (= *Cynanchum tetrapterum* (Turcz.) R.A.Dyer)   Leafless succulent scrambler or climber to 3 m, with smooth grey-green stems; sap milky. Flowers in dense umbels, yellow, corona white, inner lobes swollen and incumbent on anthers and outer ring-like, fragrant. Jan.–Dec. Arid bush on stony shale slopes, NW, SW, KM, LB, SE (Clanwilliam through Little Karoo to Arabia, India to Australia).

SCHIZOGLOSSUM   c. 20 spp., southern and tropical Africa

aschersonianum Schltr.   Dwarf, branching perennial to 15 cm; sap milky. Leaves opposite, linear. Flowers 3–4 mm diam., with recurved petals, grey-green, corona white, lobes linear arching over style-apex from broad base. Dec.–Apr. Flats and slopes, SW, LB, SE (Cape Peninsula to E Cape).

cordifolium E.Mey.   Slender, single-stemmed perennial to 60 cm tall; sap milky. Leaves opposite, usually ovate-deltoid. Flowers c. 10 mm diam., green lined with brown or yellow, corona white to green, lobes with broad erect base overtopping anthers and 2 slender lobules projecting from ventral face over top of anther. Nov.–May. Open grassland, NW, SW, LB, SE (Tulbagh and Paarl to Mpumalanga).

SECAMONE   80 spp.,   Africa, Madagascar, India to Australia

alpini Schult.   Scrambler or scandent shrub to 10 m; sap milky. Leaves opposite, lanceolate to ovate-lanceolate, 6–25 mm wide, shiny dark green. Flowers finely hairy above, corona lobes erect and incurved over anthers. Oct.–Jan. Bush and forest, to 1000 m, NW, SW, AP, LB, SE (Clanwilliam to Uganda).

filiformis (L.f.) J.H.Ross (= *Secamone frutescens* (E.Mey.) Decne.)   Slender, somewhat woody climber to 3 m; sap milky. Leaves opposite, linear-lanceolate, to 4 mm wide, pale green. Flowers glabrous above, corona lobes shorter than anthers. Nov.–Jan. Dry riverine scrub to forest margins, SE (Humansdorp to Zimbabwe).

STAPELIA   AASBLOM   c. 30 spp., mainly southern Africa

A. Stems mostly prostrate, often rhizomatous

engleriana Schltr.   Leafless, finely hairy succulent with 4-angled stems 10–22 mm diam. forming mat-like clumps to 30 cm diam. Flowers 18–22 mm diam., dark purple-brown becoming pale yellow in tube, petals tightly reflexed, shortly hairy along margins. Feb.–May. Stony slopes in karroid scrub, SW, KM (Breede River valley to Great Karroo).

A. Stems all erect, not rhizomatous

arenosa C.A.Lückh.   Leafless, finely hairy succulent with 4-angled stems 8–15 mm diam., forming diffuse clumps, 6–25 cm; sap clear. Flowers 25–35 mm diam., dark purple-brown becoming whitish in centre, transversely rugulose, petals ciliate, corona finely hairy. Jan.–June. Dry stony slopes among bushes, NW (W Karoo to Clanwilliam).

cedrimontana Frandsen (incl. **S. montana** L.C.Leach)   Leafless, finely hairy or glabrous succulent with erect, 4-angled stems 5–10 mm diam., forming clumps, 5–25 cm; sap clear. Flowers 30–50 mm diam., purple-brown usually boldly banded with cream to yellow towards centre, petals somewhat reflexed and with recurved margins, glabrous or finely hairy towards apices and ciliate. Jan.–Apr. Sandstone slopes and outcrops, NW (Cedarberg Mts and Piketberg).•

divaricata Masson   Leafless succulent with erect, 4-angled stems 4–8 mm diam., forming dense clumps, 4–13 cm; sap clear. Flowers 40–50 mm diam., pale purple to yellow, shiny and smooth with petals somewhat reflexed. Sept.–May. Stony shale slopes, LB (Swellendam to Heidelberg).•

erectiflora N.E.Br.   Like **S. cedrimontana** but stems 8–12 mm diam. Flowers 9–15 mm diam., on erect pedicels 20–120 mm long, usually button-like with strongly recurved lobes, with adpressed, often dense, transparent-white, clavate hairs especially towards centre and margins. Mar.–May. Stony slopes among bushes, NW (Botterkloof to Cedarberg Mts).•

glanduliflora Masson   Like **S. erectiflora** but pedicels pressed to ground with flowers facing upwards, flowers 25–35 mm diam., densely covered with transparent-white, clavate hairs especially towards centre and margins. Mar.–June. Stony slopes among bushes, NW (Klawer to Citrusdal).•

grandiflora Masson   (= *Stapelia desmetiana* N.E.Br., *S. flavirostris* N.E.Br.)   MAKGHAAP   Leafless, finely hairy succulent with robust, erect, 4-angled stems, 20–30 mm diam., forming large clumps, 10–30 cm; sap clear. Flowers 80–150 (–220) mm diam., purple-brown very faintly marked with cream, usually silky hairy. Mar.–May. Among bushes on lower slopes, KM (Calitzdorp to E Cape and Free State).

hirsuta L. (incl. **S. asterias** Masson, **S. obducta** L.C.Leach, **S. vetula** Masson)   Leafless, glabrous or finely hairy succulent with erect, 4-angled stems 10–20 mm diam. forming dense clumps, 10–25 cm; sap clear. Flowers 70–110 mm diam., red-purple, glabrous or softly hairy around corona and along petal margins. Mar.–Oct. Stony, often sandstone slopes, SW, KM, LB, SE (Namaqualand and Karoo, Malmesbury to Humansdorp).

paniculata Willd. (= *Stapelia nouhuysii* E.Phillips; incl. **S. immelmaniae** Pillans, **S. kougabergensis** L.C.Leach, **S. scitula** L.C.Leach)   Like **S. cedrimontana** but smaller, 3–8 cm, stems often only 3–5 mm diam. Flowers 15–45 mm diam., glabrous or finely whitish hairy, petals spreading without recurved margins. Mar.–May. Lower sandstone slopes, NW, KM, SE (Lambert's Bay to E Cape).

pillansii N.E.Br.   Like **S. hirsuta** but flowers 120–200 mm diam., uniformly dark purple or yellow, lobes ciliate, attenuated into long, slender, often twisted tails. Dec.–Apr. Stony lower slopes in karroid scrub, KM (S Karoo to Ladismith).

rufa Masson   Like **S. hirsuta** but flowers 25–40 mm diam., broadly campanulate, petals hairy towards tips and corona finely hairy. Nov.–May. Stony shale slopes in karroid scrub, KM (Montagu to Oudtshoorn and Karoo).

STAPELIOPSIS   6 spp.,   W Cape to Namibia

breviloba (R.A.Dyer) Bruyns   Leafless, dwarf, usually rhizomatous succulent to 6 cm, stems erect, 4-angled, 4–7 mm diam.; sap clear. Flowers ellipsoid, brownish outside, often partially subterranean, petals 2–3 mm long, remaining joined at tips. Apr.–May. Sand or gravel in karroid scrub, NW, SW, LB (Worcester to Swellendam).•

exasperata (Bruyns) Bruyns   Like **S. breviloba** but stems 8–15 mm diam. Flowers with cylindrical tube, pink-red to cream-coloured outside, petals 6–16 mm long, usually widely spreading. Jan.–May. Stony lower slopes in karroid scrub, KM (W Karoo to Montagu and Barrydale).

saxatilis (N.E.Br.) Bruyns (= *Pectinaria saxatilis* N.E.Br., *P. stayneri* M.B.Bayer)   Leafless, dwarf succulent with prostrate stems to 30 cm long, 4-angled, 8–25 mm diam.; sap clear. Flowers ellipsoid to obovoid or subcampanulate, pale pink to dark maroon outside, petals 4–9 mm long, usually remaining joined at tips. Jan.–May. Karroid scrub on stony N-facing slopes, NW, SW, KM, LB, SE (S Namaqualand and W Karoo, Worcester to Cockscomb).

TRIDENTEA GORTJIE   8 spp., southern Africa

gemmiflora (Masson) Haw.   Small succulent forming clumps to 1 m diam., with erect stems 10–15 mm diam., 4-angled; sap clear. Leaves minute, linear, caducous. Flowers 45–100 mm diam., purple-black often finely speckled with yellow. Mar.–May. Flats in karroid scrub, NW, SW, KM, LB, SE (Karoo, Worcester to Willowmore).

parvipuncta (N.E.Br.) L.C.Leach   Like **T. gemmiflora** but plants smaller and flowers 15–25 mm diam., cream-coloured to pale green spotted with purple-brown. Oct.–May. Karroid flats, NW (Clanwilliam and Karoo).

TROMOTRICHE   11 spp., Namibia and South Africa

baylissii (L.C.Leach) Bruyns   Leafless succulent with pendulous to creeping stems to 3 m long, 6–12 mm diam., bluntly 4-angled (square in section); sap clear. Flowers tubular-campanulate, 12–15 mm long, transversely rugulose especially in tube. Dec.–June. Sandstone cliffs and slopes, SE (Langkloof to Great Winterhoek Mts).•

choanantha (Lavranos & A.V.Hall) Bruyns   Like **T. baylissii** but stems rounded in section and flowers not rugulose within. Dec.–May. Sandstone cliffs and slopes, KM (Rooiberg to Swartberg Mts).•

revoluta (Masson) Haw.   leafless, rhizomatous succulent with erect stems 15–30 mm diam., more or less square in section; sap clear. Flowers shallowly campanulate, lobed about halfway with petals strongly recurved, 35–70 mm diam., smooth. Nov.–June. stony or sandy slopes, NW (S Namaqualand to Clanwilliam).

TYLOPHORA   c. 50 spp., Old World tropics

cordata (Thunb.) Druce   Herbaceous climber; sap clear. Leaves opposite, ovate. Flowers green to yellowish green, inflorescence bracts 1–4 mm long, not forming an involucre. Oct.–Dec. Forest, SE (George to E Cape).

lycioides (E.Mey.) Decne.   Herbaceous climber to 10 m, with woody base; sap clear. Leaves opposite, ovate-lanceolate. Flowers greenish, inflorescence bracts 2–8 mm long, forming an involucre around pedicel bases. Nov.–Mar. Coastal and riverine bush, SE (Humansdorp to Mpumalanga).

WOODIA   3 spp., southern Africa

mucronata (Thunb.) N.E.Br.   Erect perennial, 15–40 cm; sap milky. Leaves opposite, leathery, linear-lanceolate to elliptic-oblong. Flowers more or less globose, cream-coloured. Jan. Grassland, SE (Humansdorp to KwaZulu-Natal).

XYSMALOBIUM   c. 20 spp., southern and tropical Africa

gomphocarpoides (E.Mey.) D.Dietr.   Erect leafy perennial to 50 cm; sap milky. Leaves opposite, linear, acute, crisped along margins. Flowers greenish, petals glabrous. Oct.–Apr. Grassland and dry watercourses in karroid areas, NW, SW, KM, LB, SE (Karoo and Cold Bokkeveld to KwaZulu-Natal).

*****undulatum** (L.) W.T.Aiton   Sturdy, erect leafy perennial, 0.5–1.8 m; sap milky. Leaves opposite, ovate-lanceolate, obtuse, somewhat leathery, slightly undulate on margins. Flowers greenish, petals with tips recurved and densely white-pubescent inside. Oct.–Apr. Grassland, also a roadside weed, SW (Namibia, ?Cape Peninsula, KwaZulu-Natal to Mpumalanga).

AQUIFOLIACEAE

ILEX   HOLLY   c. 400 spp., cosmopolitan

mitis (L.) Radlk.   AFRICAN HOLLY, WATERBOOM, UMDUMA   Dioecious tree to 30 m. Leaves elliptic, glossy. Flowers in axillary clusters, white, fragrant. Sept.–Dec. Forest along streams, SW, LB, SE (Cape Peninsula to tropical Africa).

ARALIACEAE (= *APIACEAE* in part)
Centella by B.-E. van Wyk & M. Schubert

1. Fruit separating into 2 cocci; herbs or low perennials:
    2. Leaves not peltate, petiole sheathing at base, without stipules; furrows between fruit ribs wrinkled .....**Centella**
    2.' Leaves peltate, with membranous stipules at base of petiole; furrows between fruit ribs smooth ...**Hydrocotyle**
1.' Fruit a drupe or capsule; trees or shrubs:
    3. Flowers in panicles or racemes of small umbels ....**Schefflera**
    3.' Flowers in dense spikes or racemes, these sometimes umbellate ....**Cussonia**

CENTELLA   PENNYWORT, VARKOORTJIES   c. 50 spp., largely southern African, 1 sp. pantropical

A. Plants hermaphroditic; umbels hermaphroditic, with 3 bisexual flowers

asiatica (L.) Urban (incl. **C. coriacea** Nannfd.)   WATERNAEL   Prostrate perennial, spreading with stolons and rooting at nodes. Leaves petiolate, kidney-shaped to rounded, sparsely hairy to glabrous, crenate. Flowers bisexual, 3 per umbel, reddish. Fruits obovate, longer than bracts, wrinkled. Nov.–Mar. Marshy or damp places, SW, LB, SE (Cape Peninsula to tropical Africa, widely distributed in the tropics and S hemisphere).

ternata Schubert & B.-E.van Wyk   Tufted perennial, flowering branches spreading from a woody base. Leaves petiolate, semicircular to orbicular, serrate, petiole sparsely hairy. Flowers bisexual, 1–3 per umbel, rays single, sessile. Fruits obovate, shorter than the minutely pubescent bracts, slightly ribbed. Sept.–Dec. Sandstone slopes at high alt., NW (Cedarberg Mts: Wolfberg).•

umbellata Schubert & B.-E.van Wyk   Tufted perennial, flowering branches spreading from a woody base. Leaves petiolate, semicircular to orbicular, dentate with distinctly mucronate teeth. Flowers bisexual, 1–3 per umbel, rays 2–5 on a distinct peduncle. Fruits obovate, shorter than the glabrous bracts, slightly ribbed. Mar.–Apr. Sandstone slopes at high alt., NW (Porterville Mts).•

A. Plants andromonoecious; umbels hermaphrodandrous, with a single, sessile, central,
bisexual flower and 4 lateral male pedicellate flowers

annua Schubert & B.-E.van Wyk   Tufted annual or short-lived perennial to 10 cm. Leaves petiolate, cuneate or spathulate, densely white-hairy or rusty hairy, obtusely lobed above. Inflorescence with 4 lateral shortly pedicellate male flowers and a large central sessile bisexual flower, male flowers shorter than bracts, whitish. Fruits relatively small, obovate, shorter than bracts, ribbed, wrinkled, thinly hairy. June–Oct. Lower and upper inland slopes, NW, SW (Bokkeveld Mts to Bredasdorp).•

calcaria Schubert & B.-E.van Wyk   Tufted annual or short-lived perennial to 10 cm. Leaves petiolate, widely ovate, densely white- or brown-woolly, finely toothed above. Inflorescence pedunculate, with 4 lateral shortly pedicellate male flowers and a central sessile bisexual flower. Fruits obovate, shorter than bracts, ribbed, wrinkled, thinly hairy. June–Oct. Coastal limestones, AP (Agulhas to Gouritsmond).•

capensis (L.) Domin   Loosely tufted perennial to 10 cm, spreading from underground stems. Leaves petiolate, cuneate or spathulate, densely white-hairy or rusty hairy, obtusely lobed above. Inflorescence with an abortive central flower and 4 lateral pedicellate male flowers in functionally male plants or with a large central bisexual flower with 4 lateral abortive and sessile flowers in functionally female plants, male flowers longer than bracts, whitish, with prominent purple stylopodia. Fruits relatively large, obovate, shorter than bracts, ribbed, wrinkled, thinly hairy. June–Oct. Flats and lower slopes, often coastal, NW, SW, KM, SE (Olifants River Mts to Knysna).•

tridentata (L.f.) Drude ex Domin   Laxly branched decumbent annual or short-lived perennial, 8–20 cm. Leaves petiolate, cuneate, densely to sparsely hairy, margin mostly 3–5-toothed above. Inflorescence sessile, with 4 lateral shortly pedicellate male flowers and a central sessile bisexual flower, whitish. Fruits obovate, shorter than bracts, ribbed, wrinkled, thinly hairy. Aug.–Dec. Mainly coastal flats and lower slopes, NW, SW, AP, LB, SE (Namaqualand to Port Elizabeth).

A. Plants andromonoecious or androdioecious; inflorescence with hetero- or homosexual umbels;
male umbellules with 1–5 subsessile flowers, hermaphrodite umbellules with single subsessile flower

Group 1: Leaves widely cuneate, fleshy, distinctly petiolate; inflorescences mostly many-flowered; fruits longer than wide, relatively small, commissure not bulging or constricted

triloba (Thunb.) Drude   Sprawling, stout perennial with a woody base, 10–50 cm. Leaves cuneate, spathulate or broadly wedge-shaped, succulent, dentate with 3–9 teeth. Flowers greenish. Fruits obovate, mostly lon6ger than bracts, ribbed. Aug.–Apr. Rocky coastal flats and slopes, SW, AP (Cape Peninsula to Agulhas).•

Group 2: Leaves ovate to widely cuneate, distinctly petiolate; fruits round to ovate, slightly ribbed and wrinkled, commissure of fruits not bulging or constricted

calliodus (Cham. & Schltdl.) Drude   Sprawling, slender perennial to 50 cm. Leaves reniform to semicircular, distinctly dentate with prominent long teeth which sometimes curve at tips. Flowers cream-coloured. Fruits round, mostly longer than bracts. Aug.–Jan. Damp lower to middle slopes, NW, SW, LB, SE (Ceres to George).•

eriantha (A.Rich.) Drude   Sprawling perennial to 50 cm. Leaves reniform to widely cordate, glabrous to sparsely hairy, crenate or dentate with mucronate teeth. Flowers greenish. Fruits oblong to rounded, mostly longer than bracts. Oct.–Dec. Damp middle to upper slopes, NW, SW, LB, SE (Worcester and Cape Peninsula to E Cape).

flexuosa (Eckl. & Zeyh.) Drude   Sprawling perennial to 30 cm. Leaves widely cordate, sparsely or densely white- or brown-woolly, dentate with large triangular teeth. Flowers greenish. Fruits oblong, shorter and wider than bracts. Oct.–Apr. Sandstone rock crevices at high alt., SW (Cape Peninsula to Caledon).•

fourcadei Adamson (incl. **C. dentata** Adamson)   Sprawling perennial to 60 cm, sparsely or densely hairy on young parts. Leaves rhombic, dentate with 3–7 large teeth above. Flowers greenish. Fruits oblong to round, bract length variable. Nov.–Jan. Middle to upper slopes, LB, SE (Langeberg Mts to Humansdorp).•

lanata Compton   Sprawling perennial to 50 cm. Leaves reniform to widely cuneate, sparsely or densely white- or brown-woolly, crenate or dentate with shallow teeth. Flowers greenish. Fruits oblong, longer than bracts. Jan.–May. Rocky slopes at high alt., LB, SE (Swellendam to Uniondale).•

macrodus (Spreng.) Burtt (= *Centella hederifolia* (Burch.) Drude)   Like **C. flexuosa** but leaves somewhat smaller, less hairy, with fewer teeth (5–7) and fruits as wide as or narrower than bracts. Nov.–Apr. Sandstone crevices at medium to high alt., SW (Cape Peninsula and Paarl to Caledon).•

sp. 1   Sprawling perennial to 40 cm. Leaves distinctly rotund, sparsely hairy to glabrous, leathery, reticulately veined, brownish green, dentate. Fruits oblong to rounded, mostly slightly shorter than bracts. Sept.–May. Rocky slopes, AP (Potberg).•

sp. 2   Sprawling perennial to 30 cm. Leaves widely cordate to cordate, sparsely hairy to glabrous. Flowers greenish. Fruits oblong to rounded, longer than bracts. Oct.–Mar. Rock crevices, SW (Ceres to Franschhoek and Cape Peninsula).•

sp. 3   Loosely tufted perennial to 20 cm. Leaves fan-shaped with long petioles, sparsely hairy to glabrous, crenate above. Flowers greenish. Fruits oblong to rounded, longer than bracts. Sept.–Mar. Damp sandy flats, AP (Bredasdorp to Agulhas).•

Group 3: Leaves cuneate to linear or lanceolate, not always distinctly petiolate; fruits mostly longer than wide, longer than bracts, commissure usually bulging (visible in lateral view as 2 raised areas on either side of commissure)

affinis (Eckl. & Zeyh.) Adamson   Slender, tufted, prostrate to suberect perennial to 30 cm. Leaves variable, usually narrowly linear. Flowers yellowish. Fruits round, thin and flat, smooth, commissure sometimes slightly bulging. Aug.–Jan. Sand dunes, flats and limestone outcrops, SW, AP, LB (Malmesbury to Riversdale).•

brachycarpa Schubert & B.-E.van Wyk   Tufted, suberect perennial to 50 cm. Leaves as in **C. difformis**. Flowers whitish, only 1 (not 3–5) flower per male umbellule. Fruits small, widely elliptic, wrinkled. Aug.–Apr. Limestone flats and hills, AP (Agulhas to Gouritsmond).•

comptonii Adamson   Like **C. glabrata** but a smaller tufted, much-branched perennial to 15 cm with smaller leaves. Sept.–Jan. Flats and lower slopes, SE (Knysna to Port Elizabeth).•

difformis (Eckl. & Zeyh.) Adamson   Erect perennial, 10–40 cm. Leaves dimorphic, lanceolate, sparsely hairy to glabrous; basal leaves petiolate, entire or with 2–5 acuminate teeth in upper third, upper leaves smaller, often indistinctly petiolate, mostly entire. Flowers whitish, often in many-flowered umbels. Fruits relatively large, widely elliptic, distinctly ribbed, smooth, commissure more or less slightly bulging. Aug.–Apr. Coastal flats and lower slopes, SW, AP (Franschhoek Mts to Potberg).•

dolichocarpa Schubert & B.-E.van Wyk   Prostrate, sprawling perennial to 2 m wide. Leaves as in **C. difformis**. Fruits narrowly oblong, always much longer than bracts, slightly ribbed, smooth, borne on long, slender, often spirally curved stalks. Aug.–Apr. Rocky slopes at medium alt., SW (Riviersonderend Mts).•

fusca (Eckl. & Zeyh.) Adamson Suberect sprawling perennial, 10–40 cm. Leaves indistinctly petiolate, acicular to narrowly lanceolate, shortly and variably hairy, entire. Flowers greenish, bisexual umbellules with 4 (not 2) bracts. Fruits oblong, slightly ribbed, smooth, commissure very prominently bulging. Aug.–Jan. Middle to upper slopes, NW, SW (Gifberg to Worcester).•

glabrata L. PERSIEGRAS, SWEETKRUIE Tufted, trailing perennial. Leaves petiolate, lanceolate, sparsely hairy to glabrous. Flowers yellowish. Fruits oblong, slightly ribbed, smooth. Sept.–Jan. Slopes and mountains, NW, SW, AP, KM, LB, SE (Clanwilliam to Port Elizabeth).•

lasiophylla Adamson Like **C. villosa** but leaves larger and densely woolly. Sept.–Dec. Dry sandstone slopes, NW (Cedarberg Mts).•

pilosa Schubert & B.-E.van Wyk Tufted, sprawling annual to 25 cm. Leaves petiolate, elliptic to narrowly elliptic, pilose, reddish brown. Flowers greenish, with bisexual, female and male umbels on the same plant, only 1(not 3–5) flower per male umbellule. Fruits oblong, not ribbed, smooth. Sept.–Jan., after fires. Sandstone slopes at high alt., SW (Kogelberg).•

scabra Adamson Like **C. villosa** but leaves much larger and sparsely hairy. Aug.–Mar. Sandy slopes and rocky places, NW, SW (Gifberg to Caledon).•

villosa L. Tufted perennial to 40 cm. Leaves petiolate, widely to narrowly cordate, densely to sparsely white- or brown-hairy. Flowers greenish. Fruits distinctly angular, slightly ribbed, smooth. Sept.–Dec. Sandstone slopes in rock crevices, NW, SW (Gifberg to Caledon).•

sp. 4 Tufted perennial to 10 cm. Leaves petiolate, broadly lanceolate, sparsely hairy, entire. Fruits round to very widely ovate, slightly ribbed, wrinkled, commissure not bulging. Oct.–Dec. Shale bands at high alt., NW, SW (Cedarberg Mts to Paarl).•

Group 4: Leaves cuneate to acicular leaves, mostly not distinctly petiolate; Fruits usually wider than long, smooth and slightly ribbed and often with a constricted commissure (visible in lateral view as a groove or depression between the 2 mericarps along middle of fruit)

caespitosa Adamson Tufted, cushion-forming perennial to 8 cm. Leaves indistinctly petiolate, densely crowded, linear to acicular, subglabrous or glabrous. Flowers greenish. Fruits widely depressed-ovate, longer than bracts, mericarps separating easily. Aug.–May. Sandstone slopes at high alt., SW (Cape Peninsula to Caledon).•

cryptocarya Schubert & B.-E.van Wyk Tufted, prostrate perennial to 10 cm. Leaves petiolate, crowded, widely ovate to very widely ovate, densely hairy, entire or rarely dentate above. Flowers whitish or yellowish, male umbellules with 1–5 flowers. Fruits widely elliptic, shorter and narrower than the exceptionally large bracts. Aug.–Mar. Sandy slopes at medium alt., SW (Riviersonderend Mts: Skilpadkop).•

didymocarpa Adamson Tufted perennial to 15 cm. Leaves petiolate, crowded, lanceolate, sparsely hairy to glabrous, brownish green, entire. Flowers greenish. Fruits distinctly broader than long, longer than bracts, commissure strongly constricted. Jan. Sandstone slopes, SE (Uniondale).•

gymnocarpa Schubert & B.-E.van Wyk Tufted, prostrate perennial to 15 cm. Leaves petiolate, crowded, widely ovate to very widely ovate, densely woolly, greyish green, entire or more often shallowly toothed above. Flowers yellowish, male umbellules with 1 flower. Fruits widely depressed-ovate, distinctly ribbed, longer and wider than bracts. Aug.–Mar. Limestone outcrops, AP (Agulhas Peninsula).•

longifolia (Adamson) Schubert & B.-E.van Wyk Small tufted perennial to 10 cm. Leaves petiolate, crowded, linear-oblong, sparsely hairy to glabrous, reddish brown, entire. Flowers yellowish. Fruits widely depressed-ovate, longer and wider than bracts. Aug.–Jan. Sandy and gravelly lower slopes, SE (Knysna to Humansdorp).•

montana (Cham. & Schltdl.) Domin Tufted, sprawling perennial to 10 cm. Leaves petiolate, slightly dimorphic, elliptic to lanceolate, hairy to densely hairy, reddish brown, mostly entire. Flowers yellowish. Fruits widely depressed-ovate, shorter and wider than bracts. Aug.–Jan. Upper slopes, LB, SE (Montagu to Willowmore).•

pottebergensis Adamson Tufted, suberect perennial to 40 cm. Leaves petiolate, crowded, with long internodes between the clusters, linear-lanceolate, slightly succulent. Flowers greenish. Fruits obovate, longer than bracts. Nov.–Jan. Sandstone slopes, SW (Potberg).•

restioides Adamson   Suberect perennial to 50 cm. Leaves indistinctly petiolate, acicular to narrowly oblong, sparsely hairy to glabrous. Flowers yellowish, borne among the vegetative leaves. Fruits broadly ovate, shorter than bracts, commissure not constricted. Oct.–Dec. Rocky slopes at high alt., SW (Bainskloof to Franschhoek Mts).•

rupestris (Eckl. & Zeyh.) Adamson   Suberect woody perennial to 25 cm, stems with persistent, scale-like leaf bases. Leaves more or less sessile, linear to narrowly oblong, slightly succulent, yellowish green. Flowers yellowish. Fruits widely depressed-ovate, shorter than bracts. Aug.–Feb. Sandstone slopes, SW (Babilonstoring and Kleinrivier Mts).•

sessilis Adamson   Tufted, mat-forming perennial. Leaves indistinctly petiolate, crowded, oblanceolate to narrowly lanceolate, slightly succulent, mostly entire, sometimes 2- or 3-toothed. Flowers whitish. Fruits round to widely depressed-ovate, longer than bracts. Sept.–Jan. Moist sandstone slopes at high alt., NW, SW, AP, KM, SE (Ceres to Humansdorp).•

thesioides Schubert & B.-E.van Wyk   Erect perennial to 50 cm, stems with persistent, scale-like leaf bases. Leaves more or less sessile, narrowly linear to acicular. Flowers yellowish, borne well above the vegetative leaves, male umbellule with 3 (not 5–7) flowers and 2 (not 4) bracts. Fruits widely elliptic, shorter or as long as bracts, commissure not constricted. Aug.–Oct. Sandstone slopes at high alt., SW (Riviersonderend Mts).•

Group 5: Leaves terete to linear, entire, mostly not distinctly petiolate; fruits longer than wide, often prominently ribbed but commissure not distinctly bulging or constricted

debilis (Eckl. & Zeyh.) Drude   Slender, prostrate, trailing perennial to 15 cm. Leaves (at least the lower) petiolate, linear to narrowly lanceolate, tips acute and often curved. Flowers small, reddish, male umbellules with 1 (not 3–5) flower. Fruits round, longer than bracts, slightly ribbed, not wrinkled. Aug.–Mar. Rocky slopes, LB, SE (Mossel Bay to Uitenhage).•

glauca Schubert & B.-E.van Wyk   Robust, densely virgate perennial to 70 cm. Leaves acicular, glaucous, juvenile leaves petiolate, narrowly elliptic, with 1–3 acute teeth. Flowers yellowish. Fruits large, obovate, shorter than bracts, strongly ribbed, not wrinkled. Aug.–Apr. Dry sandstone slopes and flats at medium to high alt., NW (Cedarberg to Grootwinterhoek Mts).•

linifolia (L.f.) Drude   Erect or suberect woody shrublet to 50 cm. Leaves linear, sparsely hairy or glabrous, glaucous. Flowers yellowish. Fruits obovate, shorter than bracts, indistinctly ribbed, wrinkled. Sept.–Dec. Rocky sandstone slopes, NW, KM, LB (Worcester to Swartberg Mts and Mossel Bay).•

macrocarpa (A.Rich.) Adamson   Small, erect or suberect perennial with ascending branches from a woody base, 10–35 cm. Leaves acicular, mostly glabrous, green. Flowers yellowish. Fruits relatively large, obovate, shorter than bracts, strongly ribbed, not wrinkled. Apr.–Jan. Flats and lower slopes, NW, SW, KM (Worcester to Cape Peninsula and Swartberg Mts).

recticarpa Adamson (incl. **C. cochlearia** (Domin) Adamson)   Erect tufted perennial from a slightly woody base, to 20 cm, with persistent leaf bases. Leaves rarely somewhat petiolate, linear. Flowers purplish, with prominent purple, cone-shaped stylopodia. Fruits relatively small, oblong, longer than bracts, distinctly ribbed, not wrinkled. Flowering time? Rock crevices at high alt., NW (Bokkeveld to Cedarberg Mts).•

stenophylla Adamson   Sprawling, patch-forming perennial to 40 cm. Leaves acicular, thread-like, glabrous or very sparsely hairy. Fruits relatively small, round to obovate, longer than bracts, smooth or slightly wrinkled. Flowering time? Rocky slopes at medium to high alt., SW, LB (Bredasdorp and Langeberg Mts).•

stipitata Adamson   Sprawling perennial to 30 cm. Leaves linear to linear-lanceolate, sparsely hairy or glabrous. Flowers yellowish. Fruits obovate, slightly ribbed, variable in shape, size and surface sculpturing. Sept.–Dec. Lower to medium slopes, SE (Uniondale).•

virgata (L.f.) Drude   Erect or somewhat sprawling, sparse, virgate perennial to 60 cm. Leaves narrowly linear to acicular, densely woolly or glabrous. Flowers yellowish. Fruits borne above vegetative parts, variable in size, mostly relatively small, ovate, longer than bracts, slightly ribbed. Mostly Sept–Dec., Sandstone slopes, SW, AP, KM, LB, SE (Caledon to E Cape).

sp. 5   Robust erect to procumbent shrub to 1 m. Leaves terete below, flat in upper half, densely white-woolly when young. Flowers greenish. Fruits sessile, large, obovate, strongly ribbed. Sept.–Nov. Rocky sandstone slopes, NW (Bokkeveld Mts to Gifberg).•

CUSSONIA CABBAGE TREE, KIEPERSOL, UMSENGE c. 25 spp., Africa and Madagascar

gamtoosensis Strey GAMTOOS CABBAGE TREE Slender, several-stemmed trees to 4 m. Leaves 2-digitate, leaflets mostly simply digitate. Flower spikes umbellate, greenish yellow. May. Dry rocky slopes, SE (Gamtoos Ferry).•

paniculata Eckl. & Zeyh. BERGKIEPERSOL Thick-stemmed tree to 5 m. Leaves glaucous, digitate, leaflets sometimes toothed or lobed above. Flower spikes paniculate, green. Jan.–Apr. Dry rocky slopes, KM, SE (Swartberg Mts to Mpumalanga).

spicata Thunb. Thick-stemmed tree to 10 m. Leaves 2-digitate, leaflets vertebrate-digitate. Flower spikes umbellate, greenish yellow. Nov.–May. Rocky slopes, KM, SE (Swartberg Mts to tropical Africa).

thyrsiflora Thunb. KUSKIEPERSOL Sprawling shrub or tree to 5 m. Leaves 6–8-digitate, leaflets obovate, sometimes slightly toothed above. Flower spikes umbellate, greenish yellow. Nov.–Jan. Coastal scrub, SW, AP, LB, SE (Cape Peninsula to E Cape).

HYDROCOTYLE PENNYWORT c. 200 spp., nearly cosmopolitan

verticillata Thunb. Creeping perennial rooting at nodes, stems to 30 cm. Leaves peltate, margins crenate. Flowers whorled in interrupted spikes, white. Fruits depressed-globose. Aug.–Apr. Marshes, seeps, streamsides, SW, AP, KM, LB, SE (Cape Peninsula to Port Elizabeth, widespread in tropics and subtropics).

SCHEFFLERA BASTERKIEPERSOL, UMSENGANE c. 180 spp., pantropical and subtropical

umbellifera (Sond.) Baill. Tree to 10 m. Leaves digitate, leaflets elliptic, margins sometimes toothed or wavy. Flowers in umbellate panicles, greenish yellow. Jan. Forest margins, SE (Humansdorp to tropical Africa).

ASCLEPIADACEAE see **APOCYNACEAE**

ASTERACEAE

Amphiglossa, Elytropappus and Stoebe with M. Koekemoer, **Arctotheca, Arctotis, Eriocephalus** and **Haplocarpha** with J.B.P. Beyers, **Gymnostephium** with P. Herman

1. Style thickened or broadened near apex and often hispid at thickening, rarely not conspicuously thickened but then rays either 4-toothed or ray florets with a filiform lobe in sinus of tube; style branches convex, sometimes partially cohering, minutely downy on outer surface:
   2. Anthers long-tailed; involucral bracts many, regularly imbricate; corolla of all florets generally bilabiate or irregularly cleft:
      3. Acaulescent, tufted perennials; heads solitary, scapose, heterogamous and radiate:
         4. Rays very short; pappus hairs very long and stout, tawny ....**Perdicium**
         4.' Rays conspicuous; pappus hairs long and slender, whitish to purple ....**Gerbera**
      3.' Branched perennials or shrubs; heads not scapose, mostly homogamous, discoid, disciform or radiate:
         5. Stout, thick-stemmed shrubs; leaves thick and leathery, densely woolly below but glabrescent and glossy above; heads large, homogamous, radiate ....**Oldenburgia**
         5.' Annual or perennial herbs or shrubs; leaves not as above; heads rarely radiate, involucral bracts often pungent and recurved ....**Dicoma**
   2.' Anthers minutely tailed; corolla of all or disc florets equally toothed:
      6. Involucral bracts separate, the outer herbaceous, inner with papery margins, obtuse; rays 3-toothed:
         7. Ray florets sterile; achenes without cavities, usually woolly; pappus short, crown-like or lacking ....**Arctotheca**
         7.' Ray florets female or if sterile then achenes with cavities:
            8. Ray florets with a filiform lobe in sinus of tube; pappus of stout, scabrid, bristle-like scales in 2 rows ....**Heterolepis**
            8.' Ray florets without such a lobe:
               9. Heads sessile; mat-forming subshrub ....**Dymondia**
               9.' Heads scapose:
                  10. Achenes dorsally with 3–5 strong ribs or wings enclosing cavities, usually with a basal tuft of hairs ....**Arctotis**
                  10.' Achenes rather thinly ribbed and without cavities; scapose perennials with rosulate, discolorous leaves ....**Haplocarpha**
      6.' Involucral bracts at least partially fused and pungent; rays 4-toothed:
         11. Pappus absent; shrublets, densely leafy up to sessile heads ....**Cullumia**
         11.' Pappus present:

12. Leaves in a rosette tightly adpressed to the ground; acaulescent perennial with discoid heads crowded in a large, sessile secondary head in centre of rosette; pappus paleate ....**Platycarpha**
12.' Not as above:
  13. Receptacle with outer honeycomb cavities thick-walled and inner ones membranous; pappus scales ciliate or fimbriate:
    14. Involucral bracts in 2 rows, the outer largest and foliaceous; receptacle breaking up at maturity, outer parts adnate to outer bracts; leaves simple ....**Didelta**
    14.' Involucral bracts in 3 rows, middle row largest; receptacle not breaking up; leaves pinnatifid ....**Heterorhachis**
  13.' Receptacle uniformly honeycombed; pappus scales various:
    15. Involucral bracts shortly connate basally, pungent...**Berkheya**
    15.' Involucral bracts connate in lower half to form a smooth cup:
      16. Annuals; involucre becoming woody and enclosing achenes after anthesis:
        17. Most disc florets female-sterile; pappus of minute scales hidden among achene hairs ....**Gorteria**
        17.' All disc florets fertile; pappus scales narrow ....**Cuspidia**
      16.' Perennials; involucre not becoming woody and enclosing achenes:
        18. Outer pappus scales broad and overlapping, inner scales smaller or absent; shrublets without milky sap ....**Hirpicium**
        18.' Outer pappus scales narrow and not overlapping, inner subequal; perennial or annual herbs, usually tufted from base, with milky sap..**Gazania**
1.' Style filiform, terete or linear, not thickened above:
  19. Style branches filiform or terete, usually much exserted:
    20. Scrambling climber with opposite, sagittate leaves; anther bases obtuse ....**Mikania**
    20.' Herbs or shrubs; anther bases acute or tailed:
      21. Heads 1-flowered, corymbose, florets pink or white; tufted perennials with parallel-veined leaves ....**Corymbium**
      21.' Heads several-flowered:
        22. Spiny shrubs or shrublets; florets yellow; pappus of many, long, scabrid bristles ....**Hoplophyllum**
        22.' Unarmed herbs or shrubs; florets white to purple; pappus biseriate, outer series of short bristles or scales and inner of many, long, scabrid bristles ....**Vernonia**
  19.' Style branches linear or lanceolate:
    23. Style branches minutely and equally downy on outer surface, mostly acute:
      24. Anther bases acute or tailed:
        25. Florets unisexual, sexes on separate plants; pappus uniseriate or 0; shrubs or trees:
          26. Pappus 0; achenes woolly...**Tarconanthus**
          26.' Pappus of bristles; achenes sometimes hairy ....**Brachylaena**
        25.' Florets bisexual or male and female in the same head; pappus biseriate, of scales and bristles; perennials or shrublets:
          27. Heads heterogamous, obscurely radiate; perennial with lanceolate leaves, auriculate at base ....**Pulicaria**
          27.' Heads homogamous, discoid; shrublets with leaves tapering below or petiolate:
            28. Petals hairy at tips; pappus bristles numerous; achenes glabrous....................**Pegolettia**
            28.' Petals glabrous; pappus bristles few; achenes hairy..**Anisothrix**
      24.' Anther bases obtuse:
        29. Flower heads discoid or outer florets unilabiate (rarely radiate in **Chrysocoma** or discoid in **Felicia** but see details of these genera):
          30. Heads homogamous or outer florets ligulate:
            31. Pappus bristles free, outer series of small scales or reduced or lacking; fruits apically with 2 resin sacs in ribs; ericoid shrublets with heads often solitary on slender peduncles...**Chrysocoma**
            31.' Pappus bristles usually connate basally, of many bristles; fruits without resin sacs; shrublets, often viscid, mostly with opposite leaves with heads never solitary on slender peduncles ....**Pteronia**
          30.' Heads heterogamous with some outer florets tubular-filiform or unilabiate:
            32. Pappus 0; annual with petiolate leaves and greenish florets ....**Dicrocephala**
            32.' Pappus present; annuals or perennials with yellow or whitish florets:
              33. Heads in dense, often subglobose corymbs; florets yellow, readily visible within pappus..**Nidorella**
              33.' Heads in panicles or loose, few-headed corymbs; florets whitish, more or less concealed within pappus ....**Conyza**
        29.' Flower heads conspicuously radiate, rays mostly white, mauve or blue:
          34. Scrambling climber with petiolate, deltoid, serrate leaves; heads in panicles, with white rays ....**Microglossa**

34.' Annual or perennial herbs or shrublets with sessile leaves; heads solitary or corymbose:
    35. Pappus 0; twiggy shrublet with pinnatisect leaves ....**Garuleum**
    35.' Pappus present in some or all florets; leaves not pinnatisect:
        36. Receptacle paleate ....**Amellus**
        36.' Receptacle epaleate:
            37. Pappus bristles plumose:
                38. Softly hairy, rosette-forming herbs; leaves in a basal tuft; heads solitary and scapose; pappus persistent ....**Mairia**
                38.' Roughly hairy shrublets or woody perennials; leaves more or less cauline; heads mostly several; pappus caducous ....**Zyrphelis**
            37.' Pappus bristles scabrid or barbellate, sometimes absent or sparse in ray florets:
                39. Creeping perennials rooting from nodes; leaves succulent; anther connectives produced above into auricles; stigmas with a basal ring of long papillae or hairs........**Poecilolepis**
                39.' Not as above:
                    40. Rays discolorous, white with pink or red below; achenes glabrous, with a horny apical collar; mostly closely leafy, straggling shrublets ....**Polyarrhena**
                    40.' Rays concolorous, blue or white; achenes without a collar:
                        41. Disc florets functionally male; pappus bristles few, caducous, mostly lacking in ray florets; fertile achenes glabrous or papillate.........................**Gymnostephium**
                      41.' Disc florets perfect, rarely functionally male; pappus bristles several, persistent, rarely absent in ray florets; fertile achenes hairy, rarely glabrous:
                        42. Achenes without glands among the hairs; perennials or shrubs ....**Felicia**
                      42.' Achenes with sessile glands among tha hairs; perennials with annual stems form a persistent rootstock ....**Aster**
23.' Style branches linear and truncate, with bristly apex or tipped by a short, bristly cone; male florets with style unbranched or lacking:
    43. Pappus 0; receptacle epaleate; mostly herbaceous; heads radiate; disc florets usually sterile, with simple style; anthers acute or cuspidate:
        44. Leaves needle-like, scabrid; willowy shrublets with solitary terminal flower heads ....**Oxylaena**
        44.' Leaves lanceolate to ovate, often toothed:
            45. Style of disc florets deeply bifurcate with linear lobes, covered with papillae to well below point of bifurcation ....**Garuleum**
            45.' Style of disc florets shortly bilobed with an annular collar of hairs:
                46. Disc florets perfect; involucral bracts more or less uniseriate; disc achenes laterally flattened with thickened margins ....**Dimorphotheca**
                46.' Disc florets functionally male; involucral bracts mostly in 2 or 3 rows; fruits never flattened with 2 wings:
                    47. Fruit a drupe ....**Chrysanthemoides**
                    47.' Fruit an achene:
                      48. Achenes polymorphic (slightly curved and wingless; straight and 3-winged; or rostrate and wingless); glandular-hairy annual with entire or toothed leaves ....**Oligocarpus**
                      48.' Achenes homo- or heteromorphic:
                        49. Achenes homomorphic, kidney-shaped with a ventral cavity; shrubs with entire or toothed leaves; ovaries of ray florets obliquely lobed ....**Gibbaria**
                        49.' Achenes homomorphic or heteromorphic but not as above; ovaries not as above:
                          50. Achenes 3-winged with an apical fenestrate air chamber........................**Tripteris**
                          50.' Achenes various but not as above ....**Osteospermum**
    43.' Not as above:
        51. Involucre mostly uniseriate, with 0–several bracteoles near base:
            52. Involucres usually calyculate, with at least 1 outer small basal bract, cylindrical or cup-shaped; involucral bracts always free to base:
                53. Bracts in 2 or 3 series, the outer resembling an epicalyx; robust, woolly perennials with conspicuous, radiate heads..**Alciope**
                53.' Bracts uniseriate, sometimes with a calyculus:
                    54. Achenes flattened or winged ....**Cineraria**
                    54.' Achenes terete or angular:
                      55. Anther bases with sterile tails; climber with petiolate, toothed leaves and discoid heads in axillary corymbs..**Delairea**
                    55.' Anther bases acute:
                      56. Style branches stout, with a large sterile apical appendage and lacking sweeping hairs; shrublet with solitary, discoid heads...**Lamprocephalus**
                    56.' Style branches without a sterile apical appendage and with sweeping hairs:

57. Achenes dimorphic, in ray florets glabrous with caducous pappus, in disc florets with mucilage-producing hairs when wet and a persistent pappus; shrublet with leathery, finely toothed leaves and solitary, radiate heads with white rays ....**Phaneroglossa**
57.' Not as above:
    58. Pappus 0 in ray florets; delicate annual with lyrate to deltoid leaves and disciform or sparsely radiate heads with 3 ligulate florets ....**Stilpnogyne**
    58.' Pappus present on all florets...**Senecio**
52.' Involucres ecalyculate; bracts usually more or less fused below to form a smooth cup, rarely free but then involucre bowl-shaped:
    59. Aquatic or semi-aquatic herb with white rays; heads solitary on leaf-opposed peduncles..**Cadiscus**
    59.' Nonaquatic herbs or shrubs with yellow rays:
        60. Annual herbs:
            61. Pappus 0 ....**Steirodiscus**
            61.' Pappus present at least in ray florets:
                62. Leaves rosulate; pappus 0 in disc florets ....**Gymnodiscus**
                62.' Leaves cauline; pappus in all florets ....**Oligothrix**
        60.' Perennial herbs or shrubs:
            63. Disc florets usually bisexual; pappus caducous or 0; pappus bristles flexuous, with teeth diverging in one plane ....**Euryops**
            63.' Disc florets functionally male; pappus usually persistent, often elongating conspicuously in fruit; pappus bristles straight, with teeth diverging in several directions:
                64. Disc floret styles undivided (or minutely bifid) with a ring of sweeping hairs below apical cone ....**Othonna**
                64.' Disc floret styles divided with branches hairy outside ....**Hertia**
51.' Involucre 2–several-seriate:
    65. Anther bases usually obtuse, sometimes minutely tailed but then pappus of scales only or 0:
        66. Disc florets mostly 4-lobed and receptacle epaleate:
            67. Shrublets:
                68. Heads discoid: leaves elliptic, imbricate...**Schistostephium**
                68.' Heads radiate:
                      69. Leaves lobed or pinnatifid...**Lidbeckia**
                      69.' Leaves linear to elliptic, entire...**Thaminophyllum**
            67.' Annuals or small, rhizomatous perennials:
                70. Corolla tube of disc florets swollen and brittle, 4-ribbed ....**Oncosiphon**
                70.' Corolla tube not swollen and brittle:
                    71. Achenes flattened and often winged; ray florets usually stalked or peduncle conspicuously inflated in fruit ....**Cotula**
                    71.' Achenes not flattened ....**Foveolina**
        66.' Disc florets 5-lobed or receptacle paleate:
            72. Heads homogamous, discoid:
                73. Pappus 0, or a pseudopappus of glandular hairs present; corolla often with slender glandular hairs; glabrous or with an indumentum of stellate hairs:
                    74. Leaves opposite; heads solitary and sessile ....**Asaemia**
                    74.' Leaves alternate; heads corymbose and pedunculate...........................**Athanasia**
                73.' Pappus of scales:
                    75. Anthers tailed; leaves entire, toothed or lobed, gland-dotted ....**Inulanthera**
                    75.' Anthers obtuse or acute:
                        76. Receptacle paleate, rarely epaleate; heads narrow, few-flowered (6–10 florets), in terminal corymbs; indumentum of stellate hairs ....**Hymenolepis**
                      76.' Receptacle epaleate; hairs simple:
                        77. Pappus of 7–9 scales; shrubs with linear or tufted leaves............**Marasmodes**
                        77.' Pappus obliquely cup-shaped or divided into 3–5 scales; shrubs with simple or pinnatisect leaves ....**Pentzia**
            72.' Heads heterogamous, disciform or radiate:
                78. Heads disciform, outer female florets filiform:
                    79. Receptacle paleate; achenes densely hairy; heads solitary ....**Lasiospermum**
                    79.' Receptacle epaleate:
                      80. Heads solitary or few in corymbs; leaves simple or pinnatifid ....**Hippia**
                    80.' Heads numerous in racemes or panicles; leaves bipinnatifid ....**Artemisia**
                78.' Heads radiate, outer florets ligulate:
                  81. Receptacle paleate:
                    82. Achenes woolly; rays 2 or 3; ericoid shrublets ....**Eriocephalus**
                    82.' Achenes glabrous or with a basal tuft of hairs; rays several:

83. Inner involucral bracts broad, with membranous margins; anthers obtuse; pappus usually conspicuous, of 5 large whitish scales; rays yellow or white ....**Ursinia**
83.' Involucral bracts herbaceous; anthers tailed; pappus scales small; rays white ....**Osmitopsis**
81.' Receptacle epaleate:
84. Pappus 0; achenes with white, slime-producing hairs; twiggy shrublets with simple, adpressed leaves ....**Phymaspermum**
84.' Pappus present:
85. Pappus obliquely cup-shaped; achenes not compressed ....**Cymbopappus**
85.' Pappus of 3 scales; achenes strongly compressed........................**Leucoptera**
65.' Anthers tailed; pappus of bristles and sometimes scales; involucral bracts mostly dry, shiny (everlasting):
86. At least some heads radiate, with ligulate outer florets:
87. Ray florets white, pink or blue:
88. Leaves ericoid with involute margins, often twisted, upper surface more densely hairy than lower; ray florets few:
89. Pappus bristles plumose throughout; heads 6–15- flowered...................**Amphiglossa**
89.' Pappus bristles plumose above or barbellate; heads 1–8-flowered:
90. Cushion-forming shrublet...**Bryomorphe**
90.' Erect or sprawling shrubs ....**Disparago**
88.' Leaves sometimes ericoid but margins flat or revolute, lower surface more densely hairy than upper:
91. Involucral bracts dry, slender and aristate with recurved tips; resprouting perennials from a woody rootstock..**Athrixia**
91.' Involucral bracts green and woolly; shrublets ....**Printzia**
87.' Ray florets yellow, often coppery beneath; pappus of scales only or biseriate, of scales and bristles:
92. Plants compact, cushion-forming ....**Oreoleysera**
92.' Plants not cushion-forming:
93. Pappus bristles distinctly plumose ....**Leysera**
93.' Pappus bristles barbellate or absent:
94. Annual herbs:
95. Ovary densely hairy with long, apically coiled hairs...............**Rhynchopsidium**
95.' Ovary glabrescent ....**Nestlera**
94.' Perennial shrublets:
96. Disc florets with 1–4 pappus bristles ....**Rosenia**
96.' Disc florets without pappus bristles, scales sometimes needle-like:
97. Plants glabrous (rarely hairy); heads generally corymbose or clustered in secondary glomerules ....**Oedera**
97.' Not as above:
98. Leaves with long, glandular hairs; resprouting perennial from a woody rootstock..**Comborhiza**
98.' Leaves without long glandular hairs; shrublets ....**Relhania**
86.' Heads discoid or heterogamous and disciform but without developed ray florets:
99. Pappus biseriate, of narrow scales and a few barbellate bristles; shrublets with small, oblanceolate leaves and solitary heads...**Anisothrix**
99.' Pappus uniseriate, of barbellate or plumose bristles:
100. Leaves dimorphic, the cauline linear and involute but peduncles closely invested with imbricate, clasping, scale-like leaves; heads large (20–30 mm long), solitary, conspicuous ....**Edmondia**
100.' Leaves not dimorphic; heads mostly smaller:
101. Leaves mostly involute, ericoid, often twisted, rarely weakly involute and granular, upper surface densely white-woolly:
102. Pappus bristles plumose throughout or absent; fruits with a cupuliform apical appendage:
103. Heads 1-flowered, in spikes or glomerules ....**Stoebe**
103.' Heads 2 or more flowered:
104. Heads aggregated in glomerules or spikes; fruits without a lateral wing; ericoid shrublets but never spiny or wiry......................................**Elytropappus**
104.' Heads 1 or 2, not aggregated; fruits with a lateral wing; spiny or wiry shrublets..**Amphiglossa**
102.' Pappus bristles barbellate or plumose above:

105. Heads heterogamous; outer florets usually with pappus 0, or of 1 or 2 bristles and solitary in axils of outer bracts:
 106. Corolla cylindric throughout; styles in hermaphrodite flowers divided; annual herbs..**Ifloga**
 106.' Corolla cylindric below but campanulate above; styles in hermaphrodite flowers undivided; annual herb or woody perennials ....**Trichogyne**
105.' Heads homogamous; bracts often acute and squarrose:
 107. Petals hairy; achenes hairy with a shallow, membranous crown; heads mostly solitary:
  108. Leaves minute, granular, densely imbricate on brachyblasts; heads large and showy..**Phaenocoma**
  108.' Leaves otherwise:
   109. Apical anther appendages oblong and obtuse; apical cells of pappus bristles clavate; leaves scale-like, adpressed on short, whip-like branches ....**Dolichothrix**
   109.' Apical anther appendages lanceolate and acute; apical cells of pappus bristles not clavate ....**Lachnospermum**
 107.' Petals glabrous:
  110. Heads small (3–13-flowered), often cylindrical:
   111. Pappus bristles with flattened, more or less clavate tips ....**Metalasia**
   111.' Pappus bristles not as above ....**Planea**
  110.' Heads larger (15–15-flowered), often campanulate:
   112. Corolla cylindrical...**Calotesta**
   112.' Corolla campanulate above:
    113. Heads solitary ....**Hydroidea**
    113.' Heads corymbose ....**Atrichantha**
101.' Leaves flat or with margins revolute (rarely weakly involute), both surfaces equally woolly or the lower more densely covered:
 114. Pappus bristles plumose or barbellate, fused below into a smooth ring; style branches often rounded or truncate; silver- or grey-felted perennials with showy, everlasting heads ....**Syncarpha**
 114.' Not as above:
  115. Inner involucral bracts at least with a translucent central strip in the lower part:
   116. Involucral bracts very obtuse or truncate, usually biseriate, or if in 3 series then pappus bristles expanded towards base and fused below into a smooth ring..**Lasiopogon**
   116.' Involucral bracts usually in at least 3 series; pappus bristles never dilated below and fused into a ring:
    117. Annual or weakly perennial grey-woolly herbs; heads small (up to 4 × 3 mm), involucral bracts with opaque white tips; pappus bristles subplumose above ....**Vellereophyton**
    117.' Usually perennial herbs or shrublets, but if annual then either bracts without opaque white tips or pappus not subplumose above:
     118. Female flowers at least 5 times as many as hermaphrodite; heads campanulate with involucral bracts equalling flowers; perennial herb with lanceolate leaves expanded and clasping below and decurrent on stem ....**Pseudognaphalium**
     118.' Not as above..**Helichrysum**
  115.' Involucral bracts without a translucent strip in the lower part:
   119. Central florets hermaphrodite; inner involucral bracts without conspicuously spreading white blades; annuals, perennials or shrublets:
    120. Leaves linear to spathulate, sessile...................................**Gnaphalium**
    120.' Leaves ovate or elliptic, petiolate:
     121. Loosely tangled shrublets with heads in terminal corymbs .**Plecostachys**
     121.' Small annual or weakly perennial herbs with heads on filiform peduncles ....**Troglophyton**
   119.' Central florets functionally male; inner involucral bracts with conspicuously spreading white blades; shrublets:
    122. Leaves flat:
     123. Receptacle paleate; involucral bracts clawed; pappus bristles subplumose above; style of disc florets simple ....**Petalacte**
     123.' Receptacle epaleate; involucral bracts not clawed; pappus bristles barbellate above; style of disc florets bifid.....................**Langebergia**
    122.' Leaves revolute:

124. Flowering branches with leaves reduced above; corolla purplish; pappus reduced in female florets ....**Anaxeton**
124.' Flowering branches evenly leafy to top; corolla white or yellow; pappus well developed in all florets ....**Anderbergia**

ALCIOPE• FIRE DAISY 3 spp., W Cape

lanata (Thunb.) DC.   Robust grey-woolly perennial to 60 cm from a woody base. Leaves petiolate, ovate, discolorous, margins revolute. Flower heads radiate, few in lax terminal corymbs, golden orange, bracts densely woolly. Nov.–Jan. Shady ravines, NW (Elandskloof to Grootwinterhoek Mts).•

tabularis (Thunb.) DC.   Erect grey-felted shrublet to 50 cm from a woody base. Leaves petiolate, ovate, discolorous, margins revolute, sparsely toothed. Flower heads radiate, 1–few in loose terminal corymbs, yellow, inner bracts glabrous. Apr.–July. Sandstone slopes, mainly after fire, SW (Cape Peninsula to Betty's Bay).•

sp. 1   Thinly woolly shrublet to 1 m. Leaves obovate, not discolorous, margins revolute, sparsely toothed. Flower heads radiate, few in lax corymbs, yellow, bracts thinly woolly. Sept.–Nov. Coastal sandstone slopes, SW (Palmiet River Mts).•

AMELLUS   ASTERTJIE   12 spp., western southern Africa

A. Plants prostrate or sprawling, to 20 cm high (see also **A. asteroides**)

microglossus DC.   Roughly hairy, sprawling annual to 15 cm. Leaves oblanceolate. Flower heads with reduced rays, solitary or in lax corymbs, yellow with white rays; pappus of scales and bristles. Aug.–Dec. Sandy flats and washes, NW, KM (S Namaqualand and W Karoo to Little Karoo).

strigosus (Thunb.) Less.   Roughly hairy, sprawling annual to 10 cm. Leaves oblanceolate. Flower heads radiate, solitary, yellow with blue-mauve rays; pappus of scales and bristles. Aug.–Dec. Stony slopes, SW, AP, KM, LB, SE (Riviersonderend to E Cape and S Karoo).

tridactylus DC.   Like **A. strigosus** but pubescence spreading, not adpressed. July–Sept. Dry washes and streambeds, NW, KM (S Namibia and Karoo to Witteberg).

A. Plants erect, 20–80 cm high

alternifolius Roth   Roughly hairy annual or short-lived perennial to 60 cm. Leaves oblong or pinnately lobed. Flower heads radiate, solitary, yellow with mauve rays; pappus of scales and bristles. Sept.–Nov. Sandy flats, NW (Namaqualand to Lambert's Bay).

asteroides (L.) Druce   Silky to softly hairy, woody perennial to 40 cm. Leaves oblanceolate to spathulate. Flower heads discoid or radiate, solitary, yellow with white to mauve rays; pappus of scales and bristles. Oct.–Jan. Coastal dunes, NW, SW, AP (Lambert's Bay to Gansbaai).•

capensis (Walp.) Hutch.   Grey-silky perennial to 35 cm. Leaves oblanceolate. Flower heads discoid or radiate, solitary, yellow with mauve rays; pappus of scales only. Dec.–Jan. Coastal dunes, SW, AP (Cape Peninsula to Agulhas).•

tenuifolius Burm.   GRYSASTERTJIE   Grey-silky, much-branched perennial or shrublet to 50 cm. Leaves narrowly oblanceolate to linear. Flower heads radiate, solitary or in loose corymbs, yellow with mauve rays; pappus of scales and bristles. Sept.–Dec. Sandy flats near coast, NW, SW (S Namibia to Villiersdorp).

AMPHIGLOSSA (= *PTEROTHRIX*)   KOPSEERBOSSIE   11 spp., South Africa and Namibia, mostly W Karoo

A. Flower heads radiate

callunoides DC.   Sprawling, thinly cobwebby shrublet to 50 cm with spur shoots, layering from branch tips. Leaves narrowly lanceolate, adpressed. Flower heads radiate, small, solitary on tips of spur shoots, rays 4 or 5, white, inner bracts mucronate. Dec.–Apr. Stony riverbeds, SE (Kouga Mts to Uitenhage).•

grisea Koekemoer   Rounded, white-woolly shrublet to 30 cm. Leaves involute-ericoid, adpressed. Flower heads small, radiate, solitary at branch tips, rays 3, white. Jan.–Mar. Sandy flats and slopes, NW (Bokkeveld Mts to Clanwilliam).•

tomentosa (Thunb.) Harv. (= *Pterothrix flaccida* Schltr. ex Hutch. & E.Phillips)   Wiry-stemmed, thinly woolly shrublet to 60 cm. Leaves involute-ericoid, suberect, sometimes twisted. Flower heads radiate, few in loose terminal groups, rays 3 or 4(5), pink or white, inner bracts reddish above. Nov.–Feb. Dry karroid slopes, NW, KM (Namibia to Little Karoo).

A. *Flower heads discoid*

perotrichoides DC. (= *Pterothrix perothrichoides* (DC.) Harv.) Diffuse, stiffly branched, thinly white-woolly shrublet to 30 cm with spur shoots. Leaves involute-ericoid, suberect. Flower heads discoid, 1–few at branch tips, white, bracts acuminate-mucronate, spreading above, brown, silvery cobwebby below. Nov.–Jan. Karroid flats and slopes, NW, KM (Cedarberg Mts and Little Karoo).•

rudolphii Koekemoer Dense, brittle shrub to 1.2 m. Leaves shortly petiolate, linear-oblong with involute margins, ascending. Flower heads small, discoid, few at branch tips, white, inner bracts mucronate. Feb.–Apr. Dry karroid slopes, NW (W Karoo to Worcester).

susannae Koekemoer Erect, wiry-stemmed shrublet to 50 cm. Leaves involute-ericoid, suberect, yellowish green and sometimes tufted on young shoots. Flower heads small, discoid, white, inner bracts spathulate. Nov.–Dec. Sandy pockets on sandstone, NW (Swartruggens).•

ANAXETON• 9 spp., W Cape

angustifolium Lundgren Sparsely leafy, straggling, thinly white-woolly shrublet to 60 cm. Leaves linear, suberect, margins revolute. Flower heads discoid, small, several in small rounded clusters on long peduncles, yellow, bracts white and brown. Aug.–Sept. Sandstone slopes above 1500 m, NW (Hex River Mts: Milner Peak).•

arborescens (L.) Less. Like **A. laeve** but outer bracts glabrous. Mainly Aug.–Oct. Sandstone slopes, SW (Cape Peninsula).•

asperum (Thunb.) DC. Grey-woolly, softly woody shrublet to 30 cm. Leaves linear, often recurved and green above, margins revolute. Flower heads discoid, small, many in dense rounded clusters on elongate peduncles, yellow, bracts white and brown. Mainly Aug.–Dec. Sandstone slopes, SW, AP (Bainskloof to Bredasdorp).•

brevipes Lundgren Like **A. nycthemerum** but outer bracts obtuse. Mainly Aug.–Oct. Sandstone slopes above 1000 m, SW (Riviersonderend Mts).•

ellipticum Lundgren White-woolly, spreading shrublet to 20 cm. Leaves broadly ovate, margins revolute, green and glabrous above, reflexed. Flower heads discoid, small, many in dense rounded clusters more or less nested in the leaves, yellow, bracts white and brown. Sept.–Dec. Upper sandstone slopes above 1500 m, SW (Franschhoek Peak to Hottentots Holland Mts).•

hirsutum (Thunb.) Less. Grey-woolly shrublet to 30 cm, closely leafy below. Leaves narrowly elliptic to oblong, margins revolute, green and thinly hairy above. Flower heads discoid, small, many in dense rounded clusters on short peduncles, yellow, bracts white and brown. Mainly Oct.–Nov. Sandstone slopes and summits, 500–1500 m, SW (Riviersonderend Mts and Caledon Swartberg).•

laeve (Harv.) Lundgren White-woolly shrublet to 45 cm. Leaves linear with revolute margins, green and glabrous above. Flower heads discoid, small, many in dense rounded clusters on long peduncles, yellow, bracts whitish, the outer silky, ovaries woolly. Mainly Dec.–May. Sandstone slopes, SW (Cape Peninsula to Stanford).•

nycthemerum Less. White-woolly shrublet to 30 cm, closely leafy below. Leaves obovate with revolute margins. Flower heads discoid, small, many in dense rounded clusters, yellow, bracts white and brown. Mainly Oct.–Dec. S-facing cliffs above 600 m, SW (Cape Peninsula to Kogelberg).•

virgatum DC. Slender white-woolly shrublet to 50 cm. Leaves linear, spreading, margins revolute. Flower heads discoid, small with 2 or 3 florets, many in dense rounded clusters on elongate peduncles, yellow, bracts white and brown. July–Oct. Sandstone slopes, SW, AP (Kleinmond to Potberg).•

ANDERBERGIA• 6 spp., W Cape

elsiae B.Nord. Diffuse, grey-woolly shrublet to 50 cm. Leaves narrowly elliptic, margins revolute, glabrescent above, felted below. Flower heads in compact, rounded clusters, discoid, 4.0–4.5 mm long, yellowish, outer bracts woolly, inner petaloid bracts woolly below, male florets 6 or 7, female florets 1, ovary glabrous. Oct. Sandstone slopes, c. 1000 m, NW (Worcester: Keeromsberg).•

epaleata (Hilliard & B.L.Burtt) B.Nord. (= *Petalacte epaleata* Hilliard & B.L.Burtt) Suberect shrublet to 50 cm. Leaves oblanceolate, margins revolute, glabrescent above, felted below, c. 10 mm long. Flower heads in loose corymbs, discoid, 5 mm long, yellowish, outer bracts brownish woolly, inner petaloid bracts woolly below, male florets 4 or 5, female florets 1, ovary glabrous. Sept. Sandstone slopes, KM (Klein Swartberg).•

fallax B.Nord. Diffuse, grey-woolly shrublet to 50 cm. Leaves narrowly elliptic, cobwebby above, felted below, margins revolute, 5–10 mm long. Flower heads discoid, shorter than 5 mm, many in rounded clusters, yellowish, outer involucral bracts less than half as long as involucre, petaloid involucral bracts woolly from base, female florets (0–)1, ovary hairy, male florets 4. Oct. S-facing sandstone slopes above 1000 m, LB (Langeberg Mts: near Swellendam).•

rooibergensis B.Nord. Like **A. epaleata** but leaves mostly oblong, smaller, c. 5 mm long, outer involucral bracts white-silky and ovary shortly hairy. Oct.–Nov. Sandstone slopes, c. 1200 m, KM (Rooiberg).•

ustulata B.Nord. Like **A. elsiae** but male florets 4. Oct. Sandstone slopes above 1000 m, NW (Worcester: Saw Edge Peak).•

vlokii (Hilliard) B.Nord. (= *Petalacte vlokii* Hilliard) Like **A. elsiae** but claw of petaloid involucral bracts glabrous below and female florets usually 2. Sept.–Oct. Sandstone slopes above 1000 m, KM (Waboomsberg).•

ANISOTHRIX• 2 spp., W Cape

integra (Compton) Anderb. Twiggy shrublet to 20 cm. Leaves narrowly oblanceolate. Flower heads discoid, solitary, terminal on short peduncles, yellow. Nov.–Dec. Rocky S-facing sandstone slopes, KM (Swartberg Mts: Seweweekspoort).•

kuntzei O.Hoffm. Twiggy shrublet to 30 cm. Leaves oblanceolate, sharply toothed. Flower heads discoid, 1–few at branch tips, yellow. Mainly Nov.–May. Rocky sandstone slopes, NW (Langeberg Mts: Robertson to Montagu).•

ARCTOTHECA 5 spp., southern Africa

calendula (L.) Levyns CAPE WEED Tufted to sprawling, roughly hairy annual to 20 cm. Leaves mostly basal, lyrate to pinnatisect, sometimes bipinnatisect, lobes oblong, toothed, discolorous, woolly below, roughly hairy above. Flower heads radiate, solitary on roughly hairy scapes, black with yellow rays, sometimes with dark bases: outer bracts short, reflexed. Achenes woolly, pappus chaffy. Mainly July–Nov. Coastal areas or disturbed soil, NW, SW, AP, KM, LB, SE (Namaqualand and Karoo to Cape Peninsula and Humansdorp).

forbesiana (DC.) Lewin Tufted, thinly white-woolly, scapose perennial to 15 cm, roots thick, fusiform. Leaves in a basal tuft, lyrate to pinnatisect, sometimes bipinnatisect, lobes lanceolate, margins revolute, discolorous, glabrous above. Flower heads radiate, solitary on elongate peduncles, yellow; outer bracts long. Achenes woolly, pappus chaffy. July–Oct. Damp sandy flats, SW (Cape Peninsula to Caledon district).•

populifolia (P.J.Bergius) Norl. SEA PUMPKIN Mat-forming, white-felted perennial to 10 cm. Leaves petiolate, mostly entire, cordate, margins sparsely toothed. Flower heads radiate, solitary on woolly bracteate peduncles, yellow. Achenes woolly, pappus corona-like. Jan.–Dec. Coastal dunes, SW, AP, SE (Saldanha to Mozambique).

prostrata (Salisb.) Britten Like **A. calendula** but perennial, sprawling and rooting at nodes, softly hairy, leaves auriculate below, involucral bracts white-tipped, achenes softly hairy, pappus wanting. Mainly Sept.–Nov. Sandy slopes and flats, coastal and near streams, SW, AP, LB, SE (Cape Peninsula to E Cape).

ARCTOTIS (= *VENIDIUM*) ARCTOTIS, GOUSBLOM c. 50 spp., southern Africa to Angola

A. Plants annual

breviscapa Thunb. SANDVELDGOUSBLOM Tufted annual to 20 cm. Leaves pinnatifid, discolorous, woolly below, glabrescent above. Flower heads radiate, solitary, subscapose, blackish with orange or yellow rays red on reverse; outer bracts with slender woolly tips. Achenes with basal tuft of hairs obsolete and 2 elongate cavities, without pappus. Aug.–Nov. Rocky sandstone slopes, NW, SW (Nardouwsberg to Mamre).•

dregei Turcz. Like **A. sulcocarpa** but leaves pinnatisect with narrow lobes. Sept. Habitat?, SW (Riviersonderend).•

fastuosa Jacq. (incl. **A. gowerae** E.Phillips) NAMAKWAGOUSBLOM Like **A. hirsuta** but leaves often softly white-woolly, pinnatisect, not lyrate, flower heads radiate, solitary, orange with a dark ring. Achenes with a single elongate-undulate cavity, basal tuft of hairs obsolete. July–Sept. Sandy and gravel slopes, NW (Namaqualand and W Karoo to Biedouw Valley).

hirsuta (Harv.) Beauv.   GOUSBLOM   Slightly fleshy, often robust, caulescent annual to 45 cm. Leaves lyrate–pinnatifid, thinly hairy, often auriculate. Flower heads radiate, solitary, often several per branch, yellow, cream-coloured or orange; outer bracts shortly tailed. Achenes with 2 rounded cavities, short pappus, and basal tuft of hairs rudimentary. Aug.–Oct. Sandy slopes and flats, usually coastal, NW, SW, AP (Elandsbaai to Potberg).•

leiocarpa Harv.   Acaulescent, sometimes branched, cobwebby annual to 20 cm. Leaves oblanceolate-pinnatifid to lacerate. Flower heads radiate, solitary, usually white or pale salmon; bracts with short woolly tails. Achenes with 2 elongate cavities. Aug.–Oct. Gravel plains, KM (Namaqualand and Karoo to Little Karoo).

sulcocarpa Lewin   Thinly woolly annual to 20 cm. Leaves lyrate-pinnatifid, discolorous, grey-felted below, roughly hairy above. Flower heads radiate, solitary, pale yellow, ray florets sterile; bracts without tails. Achenes with single undulate cavity. July–Aug. Clay flats and hills, NW, KM, SE (W Karoo and Bokkeveld Mts to McGregor, Little Karoo to Gamtoos River, and Great Karoo).

sp. 1   Like **A. hirsuta** but achene cavities elongate-undulate, flower heads sometimes with a dark ring. Aug.–Sept. Sandy flats, NW (S Namaqualand to Graafwater).

A. Plants perennial, acaulescent

acaulis L. (incl. **A. fosteri** N.E.Br.)   RENOSTERGOUSBLOM   Stemless perennial to 20 cm. Leaves lyrate-pinnatifid to oblanceolate or cordate, toothed, discolorous, grey-felted below, roughly hairy above. Flower heads radiate, solitary, scapose, black with orange, yellow or cream-coloured rays; outer bracts with slender woolly tips. Achenes with 2 rounded cavities. Aug.–Oct. Clay, granitic flats and limestones, NW, SW, AP, LB (W Karoo and Bokkeveld Plateau to Langeberg Mts).

adpressa DC.   Acaulescent perennial from woody base. Leaves lyrate-pinnatifid, grey-felted. Flower heads radiate, solitary, scapose, yellow with dark reverse; outer bracts small, without tails. Achenes with 2 long, undulate cavities. Sept.–Nov. Sandstone slopes, NW (Pakhuis Mts to Cold Bokkeveld).•

undulata Jacq.   Like **A. acaulis** but leaves with long petioles, almost spathulate, not pinnatifid, flower heads orange or cream-coloured. Aug.–Oct. Sandstone slopes, NW (Bokkeveld Plateau to Gouda and Ceres).•

verbascifolia Harv.   Like **A. undulata** but leaves white-felted on both surfaces, ovate-subcordate, margins crenate, ray florets cream-coloured, pink to purple on reverse, involucral bracts with tails short or lacking. Sept.–Oct. Sandstone slopes, NW (Witzenberg and Skurweberg).•

A. Plants perennial, caulescent
*B. Pubescence entirely soft and silky or woolly (see also **A. arctotoides**)*

aenea Jacq.   Like **A. stoechadifolia** but flower heads orange and achenes with 2 elongate-undulate cavities. Sept. Rocky slopes, NW (Gifberg to Piketberg).•

argentea Thunb.   Lax, rounded, silvery felted shrub to 1 m. Leaves linear, margins revolute, sometimes toothed. Flower heads radiate, solitary on elongate pedicels, yellow; bracts not obscurely tailed. Achenes with 2 elongate-undulate cavities. Oct.–Mar. Mostly sandstone slopes, NW, KM (Swartruggens to Little Karoo and Swartberg Mts).•

pinnatifida Thunb. (incl. **A. cuneata** DC.)   Sprawling perennial to 30 cm. Leaves oblanceolate, pinnatifid, margins rolled upward, white-felted below, glabrescent above. Flower heads radiate solitary, cream-coloured or yellow, often purple on reverse; outer bracts with grey woolly tips. Achenes with 2 rounded cavities. Sept.–Nov. Rocky slopes, SW, AP, LB (Agulhas to E Cape).

stoechadifolia P.J.Bergius   KUSGOUSBLOM   Sprawling silvery woolly perennial with erect shoots, to 35 cm. Leaves oblanceolate to pinnatifid, white-felted. Flower heads radiate, solitary, cream-coloured with reddish reverse; outer bracts with woolly tails. Achenes with 2 elongate cavities. Sept.–Dec. Dunes and sandy flats, mostly coastal, SW (Yzerfontein to Cape Peninsula).•

virgata Jacq. (incl. **A. linearis** Thunb.)   Tufted grey-felted perennial to 35 cm. Leaves linear to oblanceolate, distantly toothed. Flower heads radiate, solitary, yellow to orange or pink; bracts obscurely tailed. Achenes with 2 elongate-undulate cavities. Mainly Nov.–Mar. Rocky sandstone slopes, KM, LB, SE (Tradouw Pass to Humansdorp).•

sp. 2   Tufted perennial to 40 cm, from woody base. Leaves oblanceolate to lyrate-pinnatifid, densely white-felted below. Flower heads radiate, solitary, black with deep orange rays, sometimes with dark zone; outer bracts large, obtuse, with dark edges, tailed. Achenes with elongate, sometimes undulate cavities. July–Sept. Rocky slopes in dry fynbos, NW, KM (Botterkloof to Cold Bokkeveld and Witteberg).•

B. Plants caulescent, pubescence partially rough

acuminata Lewin   Tufted, roughly hairy, thinly cobwebby perennial to 50 cm. Leaves oblanceolate to oblong, sharply toothed to pinnatifid, discolorous, margins revolute. Flower heads radiate, solitary, white to pink; bracts with sharp, leaf-like tips. Achenes with 2 elongate-undulate cavities. Nov.–Dec. Rocky sandstone slopes above 600 m, NW (Cedarberg Mts to Cold Bokkeveld and Kwadouwsberg).•

angustifolia L.   Creeping, thinly white-woolly perennial to 40 cm, with stems from diffuse underground system. Leaves broadly obovate to oblanceolate, toothed to pinnatifid, usually discolorous, margins slightly revolute. Flower heads radiate, solitary, white or yellow with reddish reverse; outer bracts obscurely tailed. Achenes with 2 rounded cavities. Mainly Sept.–Nov. Sandy slopes and flats, SW (Mamre to Franschhoek Mts and Cape Peninsula).•

arctotoides (L.f.) O.Hoffm. (incl. **A. suffruticosa** Lewin)   Sprawling soft perennial to 30 cm. Leaves lyrate-pinnatifid, petioles decurrent, discolorous, thinly woolly below. Flower heads radiate, small, solitary on slender peduncles, yellow; bracts woolly, acuminate. Achenes without pappus or basal tuft of hair, with single elongate cavity. Aug.–Mar. Mostly damp slopes and banks, SE (Knysna to KwaZulu-Natal).

aspera L. (= *Arctotis glandulosa* Thunb.)   TAAIGOUSBLOM   Softly woody, glandular-hairy, sprawling short-lived perennial to 2 m. Leaves auriculate, pinnatisect or bipinnatisect, segments narrow, sometimes grey below, margins revolute, roughly glandular-hairy. Flower heads radiate, solitary, purple or white, dark on reverse; bracts more or less tailed. Achenes with 2 elongate-undulate cavities. Mainly Aug.–Oct. Rocky slopes, NW, SW (Elandsbaai to Cape Peninsula).•

bellidifolia P.J.Bergius   Tufted to sprawling perennial from woody base, to 30 cm. Leaves oblanceolate, toothed, margins slightly revolute, white-felted below, roughly hairy above, Flower heads radiate, solitary, orange, pink or white with darker reverse; bracts tailed. Achenes with 2 elongate-undulate cavities. Sept.–Nov. Sandy slopes, NW, SW (Cedarberg Mts to Klipheuwel).•

discolor (Less.) Beauv. (incl. **A. perfoliata** (Less.) Lewin)   Sprawling perennial to 30 cm, stems roughly purple-hairy. Leaves lyrate-pinnatifid to ovate-petiolate, discolorous, margins slightly revolute, usually auriculate at base. Flower heads radiate, small, solitary on short axillary peduncles, yellow; bracts woolly, acute. Achenes without pappus or basal tuft of hairs and with single elongate cavity. Mainly Sept.–Jan. Damp rocky slopes and streambanks, SW, LB, SE (Cape Peninsula to Van Staden's Mts).•

flaccida Jacq. (incl. **A. semipapposa** (DC.) Beauv.)   Like **A. macrosperma** but leaves often auriculate below and achenes with pappus rudimentary and without a basal tuft of hair, with 2 elongate cavities. Sept.–Apr. Sandstone slopes, often wet places, NW, SW (Cedarberg to Kleinrivier Mts).•

hispidula (Less.) Beauv.   Sprawling, roughly hairy perennial to 20 cm. Leaves lyrate-pinnatisect, lobes toothed, slightly discolorous when young. Flower heads radiate, small, yellow, green on reverse; bracts scabrid, acute. Achenes without pappus, with single elongate cavity. Mainly June–Sept. Coastal dunes, SE (George to E Cape).

incisa Thunb. (incl. **A. candida** Thunb., **A. petiolata** Thunb.)   BOTTERBLOM   Sprawling perennial to 30 cm. Leaves pinnatisect to lyrate, lobes toothed, broad to narrow, discolorous, white-felted below, roughly hairy to scabrid above, margins slightly revolute. Flower heads radiate, solitary, dark with white or orange rays reddish on reverse; bracts with long woolly tails. Achenes with 2 elongate-undulate cavities. Aug.–Oct. Sandy slopes and flats, NW, SW, AP, LB, SE (Klawer to Humansdorp).•

macrosperma (DC.) Beauv.   Thinly white-woolly shrublet to 40 cm resembling **A. acuminata**. Leaves lanceolate to oblong, sessile, margins revolute, slightly toothed, discolorous, roughly hairy above. Flower heads radiate, solitary on leafy peduncles, yellow; bracts with leafy tips. Achenes with 2 elongate cavities, pappus rudimentary. Dec.–Mar. Rocky sandstone slopes, SW (Drakenstein and Du Toitskloof Mts).•

revoluta Jacq. (incl. **A. cuprea** Jacq., **A. laevis** Thunb.)   KRULBLAARGOUSBLOM   Softly woody shrub to 2 m, aromatic when crushed. Leaves usually bipinnatisect, segments linear, margins revolute, often discolorous and grey-felted below, glabrescent above. Flower heads radiate, solitary, yellow to orange; bracts tailed. Achenes with 2 cuneate cavities. Mainly Aug.–Nov. Rocky slopes, NW, SW (Namaqualand to Langebaan and Worcester).

rotundifolia Lewin   Like **A. angustifolia** but often forming dense mats, leaves white-cobwebby, leaf blades often rounded and obtuse, achene cavities elongate-undulate. Oct.–Dec. Stony sand, SW (Du Toitskloof to Franschhoek Mts).•

schlechteri Lewin   Like **A. angustifolia** but achenes with 2 elongate-undulate cavities. Feb. Sandstone slopes, AP (Agulhas Peninsula).•

tricolor Jacq.   Woolly tufted perennial to 30 cm. Leaves oblanceolate, toothed, discolorous, grey-felted below, roughly hairy above. Flower heads radiate, solitary, orange or yellow; outer bracts with slender woolly tips. Achenes with 2 round cavities. Aug.–Oct. Clay and gravelly slopes, NW, SW, LB (Cedarberg Mts to Riversdale).•

sp. 3   Softly woody, glandular-hairy, sprawling short-lived perennial near **A. aspera** to 2 m. Leaves pinnatisect, sessile above, segments pungent, roughly hairy. Flower heads radiate, solitary, yellow; bracts with leaf-like tips. Achenes with 2 elongate cavities. Mainly after fire. Mainly Aug.–Nov. Sandstone slopes, NW (Cedarberg Mts).•

sp. 4   Roughly glandular-hairy, sprawling perennial near **A. aspera**, but leaves lacerate-toothed, lobes pungent, outer bracts with long, whip-like tails, flower heads orange with darker reverse. Sept.–Nov. Sandy soils, NW, SW (S Namaqualand to False Bay).

sp. 5   Sprawling perennial to 40 cm, near **A. aspera**. Leaves oblanceolate, pinnatifid, margins revolute, roughly hairy. Flower heads radiate, solitary on leafy peduncles, white, purple below; outer bracts conspicuously tailed. Achenes with 2 elongate-undulate cavities. June–Oct. Limestone or sandy slopes, AP (De Hoop and Potberg).•

[**Species excluded**   No authentic material found and probably conspecific with one of the above: **A. caudata** Lewin, **A. decurrens** Jacq., **A. elongata** Thunb., **A. lanceolata** Harv., **A. pusilla** DC., **A. sessilifolia** Lewin, **A. setosa** Lewin (possibly = **Arctotheca**), **A. venidioides** DC.]

ARTEMISIA   WORMWOOD   c. 390 spp., mainly N hemisphere, 2 in southern Africa

afra Jacq. ex Willd.   WILDE-ALSIES   Soft aromatic shrublet to 60 cm. Leaves bipinnatifid, canescent below. Flower heads discoid, nodding, crowded at branch tips, yellowish. Mar.–May. Streamsides and damp slopes, NW, SW, AP, LB, SE (Cedarberg Mts to tropical Africa).

ASAEMIA   1 sp., W Cape, Karoo and S. Namibia

minuta (L.f.) Bremer (= *Athanasia minuta* (L.f.) Kallersjö)   VUURSIEKTEBOSSIE   Glabrous, sprawling shrublet to 30 cm. Leaves opposite, joined below, linear. Flower heads discoid, solitary and sessile at branch tips and on short shoots, yellow, epaleate, florets glandular,. Aug.–Dec. Arid slopes and flats, NW, KM (Namibia and Karoo to Worcester).

ASTER   ASTER   c. 250 spp., mostly N hemisphere

bakeranus Burtt Davy ex C.A.Sm.   Roughly hairy perennial to 70 cm. Leaves lanceolate, margins revolute, slightly toothed, 3–5-veined from base. Flower heads radiate, solitary, subscapose, yellow with blue or white rays. Mainly Nov.–May. Rocky grassland, SE (George to Tanzania).

bowiei Harv.   Glabrous, slender perennial to 25 cm. Leaves linear to filiform. Flower heads radiate, solitary, subscapose, yellow with white rays. Mar.–Apr. Grassland, LB (Swellendam Mts).•

laevigatus (Sond.) Kuntze   Glabrous perennial to 30 cm. Leaves oblanceolate, 3–5-veined from base. Flower heads radiate, solitary, yellow with ?white to mauve rays. Oct. Grassland, SE (Van Staden's Mts).•

ATHANASIA (= *STILPNOPHYTON*)   KLAASLOUWBOS   40 spp., southern Africa

A. Leaves mostly deeply lobed (see also **A. trifurcata***)*

crithmifolia (L.) L.   DRAADBLAAR-KLAASLOUWBOS   Shrub to 2 m. Leaves pinnatisect with long linear lobes. Flower heads discoid, in simple or compound terminal corymbs, yellow, paleate, florets 30–65, glandular-hairy in a basal ring. Mainly Oct.–Jan. Sandy flats and slopes, often along drainage lines, NW, SW (Piketberg to Hermanus and Riviersonderend Mts).•

elsiae Kallersjö   Mealy shrublet to 40 cm. Leaves 3-lobed, imbricate, ascending. Flower heads discoid, 1–3 in terminal corymbs, yellow, paleate, florets over 80, glandular-hairy in a basal ring. Dec.–Jan. Rocky south slopes above 1500 m, NW (Hex River Mts).•

imbricata Harv.   Densely leafy, glabrous shrub to 60 cm. Leaves obovate, adpressed, imbricate, often reflexed at tips. Flower heads discoid, large, few in simple terminal corymbs, yellow, paleate, florets more than 60, glandular-hairy in a basal ring. Sept.–Oct. Moist slopes, SW (Riviersonderend: Hassaquaskloof).•

pectinata L.f.   Erect, few-branched shrublet to 1.2 m. Leaves mostly pinnatisect with 3–5 linear lobes. Flower heads discoid, crowded in terminal compound corymbs, yellow, paleate, florets 10–15. Oct.–Dec. Damp clay soils, SW, LB (Hermanus to Gouritsmond).•

pinnata L.f.   Densely leafy, grey-velvety shrub to 2 m. Leaves pinnatisect with linear to filiform lobes. Flower heads discoid, in dense terminal compound corymbs, yellow, paleate, florets 15–20, glandular-hairy. Oct.–Dec. Rocky sandstone slopes, SE (Kouga Mts to Grahamstown).

virgata Jacq.   Erect shrub to 1 m, hairy on young parts. Leaves palmately 3–5-lobed. Flower heads discoid, in simple terminal corymbs, yellow, paleate, florets 50–60, glandular-hairy in a basal ring. Nov.–Dec. Sandstone slopes, KM, SE (Kammanassie Mts and Kouga Mts).•

A. Leaves mostly simple or toothed

adenantha (Harv.) Kallersjö   Glabrous or velvety, virgate shrub 1–2 m. Leaves linear, scattered, adpressed. Flower heads discoid, solitary or few in simple corymbs, yellow, paleate, florets 50–70, glandular-hairy in a basal ring. Flowering time? Sandstone slopes, NW (Lokenberg to Wolseley).•

alba Kallersjö   Shrublet to 50 cm. Leaves linear, sometimes with linear lobes near base. Flower heads dicoid, few in terminal clusters, cream-coloured, epaleate, florets 30–35, glandular-hairy. Nov. Rocky sandstone slopes, NW (Worcester: Audensberg).•

bremeri Kallersjö   Shrublet to 60 cm, woolly on young parts. Leaves simple, linear, imbricate, ascending. Flower heads discoid, mostly solitary and terminal, yellow, paleate, florets over 70, glandular-hairy in a basal ring. Sept.–Feb. Rocky slopes above 1000 m, NW (N Cedarberg Mts).•

calophylla Kallersjö   Erect, densely leafy, silvery velvety shrub to 1 m. Leaves cuneate, 3–12-toothed. Flower heads discoid, few in simple terminal corymbs, yellow, paleate, florets c. 50, glandular-hairy in a basal ring. Dec. Rocky slopes, NW (N Cedarberg Mts).•

capitata (L.) L.   Densely leafy shrub to 1.2 m. Leaves obovate, ascending, sometimes densely silky hairy when young, or glabrous. Flower heads discoid, few in tight clusters nested in the leaves, yellow, paleate, florets 30–60, glandular-hairy in a basal ring. Nov.–Feb. Stony flats, SW (Cape Peninsula to Klipheuwel).•

cochlearifolia Kallersjö   Shrub to 1 m, roughly mealy on young parts. Leaves subrotund, reflexed. Flower heads discoid, in simple terminal corymbs, yellow, paleate, florets over 50, glandular-hairy in a basal ring. Nov. Mostly limestone outcrops, AP (Stilbaai to Mossel Bay).•

crenata (L.) L.   Like **A. adenantha** but involucral bracts glabrous and fringed above. Oct.–Dec. Clay slopes, NW, SW (Piketberg to Drakenstein Mts).•

cuneifolia Lam. (= *Athanasia obtusa* Compton)   Densely leafy shrublet to 30 cm. Leaves obovate to rotund, recurving above. Flower heads discoid, in dense terminal compound corymbs, yellow, sparsely paleate, florets up to 10, glandular-hairy. Dec.–Jan. Rocky slopes, especially shale bands, NW (Hex River Mts to Keeromsberg).•

dentata (L.) L.   Densely leafy shrublet to 1.5 m. Leaves suborbicular, reflexed, margins serrate. Flower heads discoid, in dense terminal compound corymbs, yellow, paleate, florets 7–15. Nov.–Jan. Sandy coastal slopes, SW, AP, SE (Cape Peninsula to Struisbaai, George to Port Elizabeth).•

filiformis L.f.   klaaslouwbos   Like **A. flexuosa** but florets 15–25, glandular-hairy on tube. Oct.–Dec. Rocky slopes, NW, KM, LB, SE (Montagu to E Cape).

flexuosa Thunb. (incl. **A. glabrescens** DC.; **A. tomentella** Hutch.)   Densely leafy shrub 1–2 m, grey-mealy on young parts. Leaves linear, spreading. Flower heads discoid, mostly in simple terminal corymbs, yellow, paleate, florets 35–55, glandular-hairy in a basal ring. Mainly Sept.–Nov. Stony clay or sandstone slopes, NW, SW, KM (Namaqualand to Touwsrivier and Cape Flats).

hirsuta Thunb.   Like **A. flexuosa** but plants densely grey-velvety, heads in compound corymbs and florets 7–20, glandular-hairy on tube. Aug.–Dec. Rocky sandstone slopes, NW, KM (Michell's Pass to Montagu).•

humilis Kallersjö   Thinly white-velvety shrublet to 50 cm. Leaves narrowly oblanceolate, recurved at tips. Flower heads discoid, few in tight, simple terminal corymbs, paleate, florets 25–30, glandular-hairy in a basal ring. Nov. Dry sandstone slopes, NW, SW (Riviersonderend and Langeberg Mts: near Montagu).•

inopinata (Hutch.) Kallersjö (= *Stilpnophyton inopinatum* Hutch.)   Slender, virgate shrublet to 1 m. Leaves linear. Flower heads discoid, large, solitary to few in terminal corymbs, yellow, bracts with papery margins, epaleate, florets more than 100, glandular-hairy. Sept.–Dec. Sandstone slopes, LB (Langeberg Mts: near Tradouw Pass).•

juncea (DC.) D.Dietr.   Slender shrublet to 60 cm. Leaves simple, oblong, scattered, ascending. Flower heads discoid, in dense terminal compound corymbs, yellow, paleate, florets 12–25. Dec.–Apr. Rocky shale slopes, SW, LB (Somerset West to Great Brak River).•

leptocephala Kallersjö   Slender, mealy shrublet to 40 cm. Leaves linear. Flower heads discoid, few in simple terminal corymbs, yellow, paleate, florets 20–25, glandular-hairy in a basal ring. Nov.–Dec. Rocky sandstone slopes, NW (Gifberg).•

linifolia Burm. (= *Stilpnophyton linifolium* (L.f.) Less.)   Slender, erect shrub to 1.5 m. Leaves linear, ascending. Flower heads discoid, large, in simple terminal corymbs, yellow, epaleate, florets more than 50, glandular-hairy. Aug.–Feb. Sandstone slopes, NW, KM, SE (N Cedarberg to Tsitsikamma Mts).•

microphylla DC.   Slender, erect shrublet to 60 cm. Leaves simple, triangular, adpressed. Flower heads discoid, few in weakly compound, terminal corymbs, yellow, paleate, florets 25–50, glandular-hairy in a basal ring. Mar.–Apr. Sandy and shale slopes, NW (Cedarberg Mts).•

oocephala (DC.) Kallersjö (= *Stilpnophyton oocephalum* DC.)   Slender, erect shrub to 1 m. Leaves oblong, entire. Flower heads discoid, in dense terminal compound corymbs, yellow, epaleate, florets 15–20, glandular-hairy. Sept.–Nov. Dry, stony slopes, SW (Riviersonderend Mts).•

pachycephala DC.   Diffuse, sparsely to densely woolly shrublet to 60 cm. Leaves linear to obovate, ascending. Flower heads discoid, 1–few in simple terminal corymbs, yellow, paleate, florets 40–160, glandular-hairy in a basal ring. Dec.–May. Sandstone slopes, NW, KM (Cedarberg to Kammanassie Mts).•

pubescens (L.) L.   Densely leafy, grey-velvety shrub to 1.5 m. Leaves oblanceolate. Flower heads discoid, mostly in simple terminal corymbs, yellow, paleate, florets more than 50, glandular-hairy in a basal ring. Sept.–Nov. Rocky sandstone slopes, NW (Olifants River Mts to Piketberg).•

quinquedentata Thunb. (= *Athanasia dimorpha* DC.)   Densely leafy shrublet to 60 cm. Leaves obovate, spreading to reflexed, often 3–5-toothed. Flower heads discoid, mostly in simple terminal corymbs, yellow, paleate, florets 10–40, glandular-hairy in a basal ring. Oct.–Jan. Limestone and sandstone hills, SW, AP, SE (Stanford to Port Elizabeth).•

rugulosa E.Mey. ex DC.   Virgate, sparsely branched shrub to 60 cm, woolly on young parts. Leaves lanceolate, adpressed. Flower heads discoid, in small, simple terminal corymbs, yellow, paleate, florets 30–35, glandular-hairy in a basal ring. Oct.–Nov. Sandy flats, SW (Hopefield to Mamre).•

scabra Thunb.   Much-branched, mealy shrub to 60 cm. Leaves linear, small. Flower heads discoid, solitary at branch tips, yellow, bract margins papery, paleate, florets 50–60, glandular-hairy in a basal ring. Dec. Sandstone slopes, SW (Villiersdorp: Donkerhoek Mts).•

sertulifera DC.   Grey-felted shrub to 1 m. Leaves oblanceolate–linear. Flower heads discoid, in simple terminal corymbs, yellow, paleate, florets 30–40, glandular-hairy in a basal ring. Nov. Sandy flats, NW (Paleisheuwel).•

spathulata (DC.) D.Dietr.   Similar to **A. cuneifolia** but corymbs less crowded and on longer stalks. Leaves spathulate, ascending. Dec.–Feb. Sandstone slopes, NW (Bokkeveld Mts).•

tomentosa Thunb.   Coarsely velvety, grey shrub to 2 m. Leaves simple, oblanceolate, entire, sometimes 2- or 3-toothed. Flower heads discoid, in dense rounded, terminal compound corymbs, yellow, paleate, florets 10–15, glandular-hairy. Nov.–Dec. Rocky sandstone slopes, KM, LB, SE (Swartberg and Langeberg Mts to Antoniesberg).•

trifurcata (L.) L.   KOUTERBOS   Glabrous to grey-velvety shrub to 1.5 m. Leaves ascending cuneate, 3–5-toothed above. Flower heads discoid, in simple terminal corymbs, yellow, paleate, florets 50–100, glandular-hairy in a basal ring. Mainly Oct.–Nov. Flats and rocky slopes, NW, SW, AP, KM, LB, SE (Bokkeveld Escarpment to Port Elizabeth).•

vestita (Thunb.) Druce   Compact densely leafy shrub to 1 m. Leaves simple, linear. Flower heads discoid, small, in dense terminal compound corymbs, yellow, paleate, florets fewer than 10. Aug.–Jan. Clay slopes in karroid scrub, NW, KM, SE (Robertson to Uniondale).•

viridis Kallersjö   Densely leafy shrub to 1.5 m, mealy on young parts. Leaves subrotund, toothed, ascending. Flower heads discoid, in simple terminal corymbs, yellow, paleate, florets 60–100, glandular-hairy in a basal ring. Dec. Rocky slopes, NW, KM, LB, SE (Keeromsberg to Outeniqua Mts).•

ATHRIXIA BOESMANSTEE   14 spp., Africa and Madagascar

capensis Ker Gawl.   Thinly cobwebby perennial from woody rootstock, to 35 cm. Leaves linear with involute margins, sharply mucronate, decurrent. Flower heads radiate, solitary at branch tips, yellow with pink or purple rays. Nov.–Dec. Lower to middle slopes, SW, LB (Cape Peninsula to Humansdorp).•

crinita (L.) Druce   Thinly cobwebby perennial from woody rootstock, to 35 cm. Leaves ovoid to lanceolate with weakly involute margins, densely white-felted below. Flower heads radiate, solitary on long peduncles, yellow with pink or magenta rays. Dec.–Jan. Sandstone slopes, NW (Cedarberg to Elandskloof Mts).•

heterophylla (Thunb.) Less.   Cobwebby shrublet or perennial from woody rootstock, to 45 cm. Leaves ovate to narrowly linear, the lower much broader, with involute margins, slightly serrate, densely felted below, often scabrid above. Flower heads radiate, solitary, pink or magenta, outer pappus bristles scale-like. Aug.–Dec. Rocky slopes, SW, KM, LB, SE (Cape Peninsula to E Cape).

ATRICHANTHA•   1 sp., W Cape

gemmifera (Bolus) Hilliard & B.L.Burtt   Densely leafy, silvery woolly shrublet to 60 cm. Leaves narrowly lanceolate with involute margins. Flower heads discoid, in small terminal clusters, reddish, bracts white with maroon base. Oct.–Jan. Sandstone rocks above 1000 m, NW, SW (N Cedarberg Mts to Kogelberg).•

BERKHEYA DISSEL   c. 75 spp., southern and tropical Africa.

A. Stems partially or completely winged

carduoides (Less.) Hutch.   Perennial to 60 cm, sometimes cobwebby on young parts. Leaves oblanceolate, decurrent on stem as wings, pinnatifid, lobes rounded, toothed and spiny, margins slightly revolute. Flower heads discoid, few in terminal corymbs, yellow. Nov.–Jan. Grassy fynbos, LB, SE (Swellendam to E Cape).

decurrens (Thunb.) Willd.   White-woolly perennial to 60 cm. Leaves oblanceolate, decurrent on stem as wings, discolorous, pinnatifid, lobes rounded, toothed and spiny, margins slightly revolute. Flower heads radiate, 1–few in terminal corymbs, yellow. Mainly July–Sept. Grassy slopes, SE (Humansdorp to E Cape).

A. Stems not winged
B. Plants rosulate

armata (Vahl) Druce   GROOTDISSEL   Tufted perennial to 40 cm, with woody caudex. Leaves crowded basally, oblanceolate to obovate, glabrous above, white-felted below, margins slightly revolute, toothed and spiny. Flower heads radiate, 1–few in terminal corymbs on elongate peduncles, yellow. Sept.–Nov. Clay and granite slopes and flats, NW, SW, AP, LB (Olifants River Mts to Mossel Bay).•

carlinoides (Vahl) Willd.   BERGDISSEL   Like **B. armata** but stems glandular-hairy. Mainly Oct.–Jan. Shale and clay slopes, NW, LB, SE (Citrusdal to George).•

francisci Bolus   Densely white-woolly, tufted perennial to over 1.5 m. Leaves oblanceolate, lacerate-toothed and spiny, margins slightly revolute. Flower heads radiate, yellow, in dense clusters on elongate peduncles. Nov.–Jan. Rocky sandstone slopes, KM (Swartberg Mts).•

herbacea (L.f.) Druce   KAALDISSEL   Like **B. armata** but leaves sometimes scarcely toothed or spiny, inner involucral bracts with broad horny margins, entire. Oct.–Mar. Sandstone slopes, SW, KM (Du Toitskloof to Napier).•

B. Plants not rosulate

angusta Schltr.   Erect, densely leafy shrub to over 1 m, white-woolly on young parts. Leaves ascending, linear, margins revolute, slightly spiny. Flower heads radiate, yellow, solitary, nested in upper leaves. July–Aug. Sandstone slopes, NW, SW (Langeberg east of Worcester and Bredasdorp Mts).•

angustifolia (Houtt.) Merr.   Like **B. barbata** but leaves narrowly lanceolate. July–Sept. Rocky sandstone slopes, KM, SE (Swartberg Mts to E Cape).•

barbata (L.f.) Hutch.   White-felted shrublet to 60 cm, from woody caudex. Leaves opposite, elliptic, discolorous, leathery, white-felted below, glabrous above, margins revolute, slightly toothed and spiny. Flower heads radiate, solitary at branch tips, yellow. Mainly Aug.–Nov. Rocky sandstone slopes, NW, SW (Gifberg to Bredasdorp).•

coriacea Harv.   WITDISSEL   White-woolly shrublet to 60 cm. Leaves oblanceolate, discolorous, leathery, margins revolute, toothed and spiny. Flower heads radiate, solitary at branch tips, yellow. Mainly Sept.–Oct. Limestone and calcareous sands, AP (Agulhas to Albertinia).•

cruciata (Houtt.) Willd.   DISSELDORING   Glabrous shrub to over 1 m. Leaves spreading, rigid, striate, ovate, toothed and spiny, auriculate. Flower heads shortly radiate, yellow, in loose terminal corymbs. Nov.–Feb. Sandstone slopes, KM, SE (Swartberg Mts to Langkloof).•

cuneata (Thunb.) Willd.   VAALDISSEL   Grey-felted shrublet to 30 cm. Leaves cuneate, leathery, toothed and spiny above. Flower heads radiate, yellow, solitary at branch tips. Mar.–Oct. Sandstone slopes, KM (Montagu to Kammanassie Mts).•

dregei Harv.   Grey-woolly shrublet to 20 cm. Leaves oblong–oblanceolate, discolorous, thinly cobwebby above, white-woolly below, auriculate, margins undulate, closely toothed and spiny. Flower heads radiate, yellow, few in terminal corymbs. Dec.–Jan. Rocky sandstone slopes, NW (Pakhuis Mts).•

eriobasis (DC.) Roessler   Like **B. rigida** but densely woolly. Oct.–Dec. Rocky slopes, NW (Cold Bokkeveld and W Karoo).

fruticosa (L.) Ehrh.   VAALDISSEL   Grey-woolly shrub to 1.5 m. Leaves elliptic, discolorous, slightly toothed and spiny. Flower heads radiate, 1–few in terminal corymbs, yellow. July–Oct. Sandstone slopes and plateaus, NW (Namaqualand to Citrusdal).

heterophylla (Thunb.) O.Hoffm.   GRAWEELWORTEL   Cobwebby perennial to 1 m. Leaves oblanceolate, discolorous, woolly below, pinnatifid, lobes rhombic, margins revolute, spiny. Flower heads radiate or discoid, yellow, in terminal corymbs. Sept.–Jan. Stony waste places, NW, KM, LB, SE (Worcester to E Cape).

onobromoides (DC.) O.Hoffm. & Muschler (incl. **B. carlinifolia** (DC.) Roessler)   RUIKDISSEL   Glandular-hairy perennial to 1.5 m. Leaves oblong, pinnatifid, lobes cuneate, toothed, spiny, auriculate. Flower heads discoid, yellowish, several in terminal corymbs. Oct.–Dec. Dry watercourses, NW, SW, KM, LB, SE (Namaqualand and Karoo to Cape Peninsula and to Langkloof).

rigida (Thunb.) Adamson & T.M.Salter   Glabrescent to white-woolly perennial to 1 m. Leaves deeply pinnatifid, lobes narrow, margins revolute, spiny. Flower heads discoid, yellow, several terminal on lateral branches. Sept.–Jan. Clay and granite slopes and flats, often disturbed places, NW, SW, AP, LB (Langebaan to Riversdale).•

spinosa (L.f.) Druce   BOKDISSEL   Glabrous shrub to 1 m. Leaves obovate, leathery, margins toothed and spiny. Flower heads radiate, yellow, in lax terminal corymbs. Mainly Sept.–Nov. Dry rocky shale slopes, KM (W Karoo and Montagu to Willowmore).

tysonii Hutch.   Shortly hairy shrublet to 30 cm. Leaves obovate, margins undulate, closely toothed and spiny. Flower heads radiate, yellow, few in terminal corymbs. Nov.–Jan. Sandstone rocks, NW (Cedarberg Mts to Keeromsberg).•

viscosa (DC.) Hutch.   TAAIDISSEL   Glandular-hairy perennial to 2 m. Leaves narrowly lanceolate, pinnatifid, lobes triangular, margins undulate, spiny. Flower heads radiate, yellow, in lax terminal corymbs. Sept.–Oct. Sandstone slopes, NW (Pakhuis Mts to Piketberg).•

BRACHYLAENA   BITTERBLAAR   c. 15 spp., Africa, Madagascar, Mascarene Is

elliptica (Thunb.) DC.   BITTERBLAAR, UMPHALHLA   Dioecious shrub or tree to 15 m. Leaves aromatic, oblanceolate, broadly toothed above, white-felted below. Flower heads discoid, in short racmemes or panicles, cream-coloured. Apr.–Aug. Coastal bush, SE (Uitenhage to KwaZulu-Natal).

glabra (L.f.) Druce   MALBLAAR   Dioecious shrub or tree to 20 m, similar to **B. neriifolia** but leaves obovate to broadly oblanceolate and often toothed above with bases acute and distinctly petiolate. Sept.–Oct. Coastal forest, SE (Humansdorp to KwaZulu-Natal).

ilicifolia (Lam.) E.Phillips & Schweick.   HULSBITTERBLAAR   Dioecious, rigid, divaricately branched shrub to 3 m. Leaves on short shoots, elliptic, glossy above, densely rust- to grey-felted below. Flower heads discoid, 1–few in leaf axils, creamy yellow. July–Nov. Scrub forest, SE (Joubertina to S Mozambique).

neriifolia (L.) R.Br.   WATERWITELS   Dioecious, leafy shrub or tree to 8 m, sparsely rusty velvety. Leaves aromatic, oblanceolate, sometimes slightly toothed, leathery, margins revolute, base decurrent into a short petiole. Flower heads discoid, in short racemes or panicles, whitish. Jan.–Mar. Streamsides, NW, SW, AP, LB, SE (Gifberg to Humansdorp).•

BRYOMORPHE• 1 sp., W Cape

lycopodioides (Sch.Bip.) Levyns   Dwarf, silvery woolly, densely leafy, cushion-like shrublet to 5 cm. Leaves adpressed, linear with involute margins to granular. Flower heads radiate, in small terminal clusters, ray florets female, white or pink. Oct.–Feb. Sandstone rocks above 1000 m, NW, SW, KM (Cedarberg Mts to Kogelberg and Anysberg).•

CADISCUS•   WATER DAISY   1 sp., W Cape

aquaticus E.Mey.   Aquatic perennial with spongy, floating stems to 60 cm. Leaves narrowly oblanceolate, leathery, clasping stem. Flower heads radiate, solitary and leaf-opposed, yellow with white rays, deflexed in fruit. Aug.–Sept. Seasonal pools, NW, SW (Eendekuil to Tygerberg).•

CALOTESTA• 1 sp., W Cape

alba Karis   Densely white-woolly shrublet to 50 cm. Leaves involute ericoid, acute and often twisted. Flower heads discoid, terminal and solitary, reddish, bracts white. Flowering time? Sandstone rocks above 1800 m, KM (Klein Swartberg).•

CHRYSANTHEMOIDES   BIETOU   2 spp., southern to central Africa

incana (Burm.f.) Norl.   GRYSBIETOU   Sprawling, white-woolly, sparsely thorny shrublet to 80 cm. Leaves ovate to elliptic, toothed. Flower heads radiate, few in terminal corymbs, yellow, bracts densely woolly. Mainly Dec.–May. Coastal dunes or sandy inland slopes, NW, SW, AP, KM (Namibia and Karoo to Bredasdorp).

monilifera (L.) Norl.   BIETOU, BOSLUISBESSIE   Rounded shrub to over 1.5 m, thinly woolly on young parts. Leaves obovate to elliptic, toothed, dark green, leathery. Flower heads radiate, few in terminal corymbs on short peduncles, yellow. Mainly Mar.–July. Sandstone and limestone slopes and flats, NW, SW, AP, KM, LB, SE (Namaqualand to tropical Africa).

CHRYSOCOMA   BEESBOS   c. 20 spp., southern Africa

A. Leaves glabrous

ciliata L. (= *Chrysocoma tenuifolia* P.J.Bergius)   BITTERBOS   Glabrous, slender-stemmed, closely leafy shrublet to 60 cm. Leaves linear, ascending, 2–14 mm long. Flower heads discoid, solitary, yellow. Oct.–Jan. Rocky slopes and flats, NW, SW, AP, KM, LB, SE (Namaqualand to Mpumalanga).

coma-aurea L.   Glabrous, closely leafy shrublet to 50 cm. Leaves linear, spreading to recurved, 3–20 mm long. Flower heads discoid, solitary, yellow. Oct.–Jan. Flats and lower slopes, SW (Cape Peninsula to Hermanus).•

longifolia DC.   Densely leafy, glabrous shrub to 1 m. Leaves linear–filiform, 12–30 mm long. Flower heads discoid, in dense corymbs, yellow. Nov.–Dec. Rocky lower slopes, NW, SW (Namaqualand to Riebeek-Kasteel).

sparsifolia Hutch.   Glabrous, sparsely leafy shrublet to 60 cm. Leaves ovoid, fleshy, recurved, 2–7 mm long. Flower heads discoid, solitary, yellow. Nov.–Dec. Rocky slopes, NW (Namaqualand to Cedarberg Mts).

tridentata DC.   Glabrous, slender-stemmed shrublet to 35 cm. Leaves linear or lobed, spreading, recurving at tips, 5–23 mm long. Flower heads discoid, solitary, yellow. Aug.–Sept. Sandstone flats and lower slopes, NW, KM (Worcester to Swartberg Mts).•

valida E.Bayer   Glabrous, slender-stemmed shrublet to 35 cm. Leaves linear, ascending, 3–15 mm long. Flower heads discoid, solitary, yellow. Mainly Aug.–Sept. Sandstone flats and lower slopes, NW, KM (Lokenberg to Swartberg Mts and W Karoo).

A. Leaves hairy or scabrid, at least on margins

acicularis E.Bayer   Viscid, densely leafy shrublet to 30 cm. Leaves filiform, scabrid on margins, spreading, 7–25 mm long. Flower heads discoid, solitary, yellow. Oct.–Nov. Rocky sandstone ridges, KM (Witteberg).•

candelabrum E.Bayer   Roughly hairy shrublet to 40 cm. Leaves linear, scabrid, spreading to recurved, 4–15 mm long. Flower heads discoid, solitary, yellow. Dec.–Feb. Sandstone slopes, NW (Cedarberg Mts).•

cernua L.   Diffuse, glabrous shrublet to 35 cm. Leaves linear, ciliate on margins, 8–25(30) mm long. Flower heads discoid, solitary, yellow. July–Sept. Rocky flats and slopes, NW (Namaqualand to Cold Bokkeveld).

esterhuyseniae E.Bayer   Slender, wiry-stemmed, glabrous shrublet to 40 cm. Leaves oblong, clasping stem, imbricate, scabrid on margins, 4–9 mm long. Flower heads discoid, solitary, yellow. Aug.–Sept. Coastal sands, SW (Hopefield to Melkbosstrand).•

flava E.Bayer   Slender, glandular-hairy, wiry-stemmed shrublet to 40 cm. Leaves linear, clasping stem, imbricate, scabrid on margins, c. 5 mm long. Flower heads shortly radiate, solitary, yellow. Flowering time? Shale hillsides, LB (Riversdale).•

oblongifolia DC.   Twiggy shrublet to 30 cm. Leaves oblanceolate, ciliate on margins, 4–12(20) mm long. Flower heads discoid, solitary, yellow. Aug.–Oct. Rocky slopes, NW, KM (Namaqualand to Hex River Mts and Witteberg).

rigidula (DC.) E.Bayer   Grey-silky, much-branched shrublet to 35 cm. Leaves linear, recurved, imbricate, 3–13 mm long. Flower heads radiate, solitary, yellow with blue rays. June–Nov. Sandy flats or slopes, SE (Port Elizabeth to Alexandria).

strigosa E.Bayer   Densely leafy, dwarf shrublet to 15 cm with erect, scabrid stems. Leaves spathulate, ciliate on margins, 3–4 mm long. Flower heads discoid, solitary, yellow. Oct. Coastal limestone flats, AP (Agulhas to Stilbaai).•

CINERARIA   CINERARIA   c. 45 spp., Africa, Madagascar

A. Plants glabrous

lobata L'Hér.   Like **C. geifolia** but glabrous and flower heads smaller. Mainly Nov.–Jan. Dry rocky slopes, KM, SE (Oudtshoorn to E Cape and Karoo).

platycarpa DC.   Glabrescent annual to 20 cm. Leaves lyrate, terminal lobe cordate and crenate to toothed, petioles eared at base. Flower heads few to many in corymbs, radiate and yellow or sometimes discoid and cream-coloured. Aug.–Sept. Rocky slopes, NW, KM, SE (Cold Bokkeveld and W Karoo to E Cape).

saxifraga DC.   Glabrous diffuse perennial or soft shrub to 30 cm. Leaves palmately lobed, blades mostly cuneate at base. Flower heads radiate, in few-flowered lax corymbs or solitary. Sept.–Nov. Rocky slopes usually in shade, LB, SE (Swellendam to E Cape).

A. Plants variously hairy

alchemilloides DC.   Thinly cobwebby to glabrescent sprawling perennial to 40 cm. Leaves palmately lobed, lobes irregularly toothed. Flower heads radiate, in lax corymbs, yellow. Aug.–Oct. Damp slopes and gullies, NW, SW, KM (Velddrif and Porterville Mts to Montagu, widespread in southern Africa).

argillacea G.Kron   Thinly cobwebby annual or short-lived perennial to 50 cm. Leaves pinnatisect, lobed linear and toothed. Flower head radiate, solitary, on slender elongate peduncles, yellow. Oct.–Nov. Shale band, NW (Hex River Mts).•

canescens J.C.Wendl. ex Link   Grey to mealy perennial or shrublet, 30–60 cm. Leaves palmately lobed, lobes toothed. Flower heads radiate, in lax corymbs, yellow. Sept.–Nov. Rocky slopes, NW (Namaqualand to Ceres).

erosa (Thunb.) Harv.   Cobwebby sprawling shrublet to 50 cm. Leaves irregularly subbipinnatisect to multifid, lobes toothed. Flower heads radiate, yellow. Sept.–Oct. Lower sandstone slopes, NW, KM (Karoopoort to Montagu).•

geifolia (L.) L.   Roughly hairy perennial to 60 cm. Leaves reniform, lobed and toothed, petioles with large basal auricles. Flower heads radiate, in terminal corymbs, yellow. Aug.–Nov. Mainly coastal bush, SW, AP, LB, SE (Cape Peninsula to S KwaZulu-Natal).

pedunculosa DC.   Scapose perennial to 25 cm. Leaves mostly deltoid, margins revolute, crenate, grey-felted below. Flower heads radiate, solitary on bracteate scapes, yellow. Sept.–Oct. Sandstone slopes, especially after fire, NW, KM (Cedarberg to Swartberg Mts).•

tomentosa (DC.) Less.   Stout, white-woolly perennial to 75 cm, woody below. Leaves oblanceolate, sometimes slightly lobed, margins toothed, glabrescent above, white-woolly below. Flower heads radiate, in lax corymbs on stout peduncles, yellow. Sept.–Nov. Moist slopes and ravines, NW, SW (Grootwinterhoek Mts to Jonkershoek).•

COMBORHIZA 2 spp., W Cape to KwaZulu-Natal

longipes (Bremer) Anderb. & Bremer Sparsely branched, glandular-hairy perennial to 20 cm with woody base. Leaves narrowly lanceolate, glabrous above. Flower heads radiate, solitary on elongate, wiry peduncles, yellow. Dec.–Jan. Sandstone slopes above 1500 m, NW (Grootwinterhoek to Hex River Mts).•

CONYZA c. 60 spp., pantropical and subtropical

pinnata (L.f.) Kuntze Thinly scabrid perennial with erect stems from stolons, to 50 cm. Leaves oblanceolate, pinnatifid, lobes entire or toothed, cauline leaves sessile. Flower heads discoid, often nodding, in corymbs, yellow. Oct.–Jan. Damp areas, often marshes, SW, LB, SE (Cape Peninsula to Mpumalanga).

pinnatifida (Thunb.) Less. Softly hairy sprawling perennial to 45 cm. Leaves pinnatisect, lobes sometimes toothed, cauline leaves sessile. Flower heads discoid, in corymbs, yellow. Jan.–May. Damp sites along the coast, often seeps and marshes, SW (Cape Peninsula to Cape Hangklip).•

scabrida DC. BAKBESEMBOSSIE, OONDBOS Slender, pale-stemmed shrub with willowy branches to 2 m. Leaves petiolate, lanceolate, serrate. Flower heads discoid, in dense corymbs, cream-coloured to pale yellow. Mainly Nov.–Apr. Sandstone slopes or forest margins, often near streams, NW, SW, AP, KM, LB, SE (Clanwilliam to Zimbabwe).

ulmifolia (Burm.f.) Kuntze Erect, shortly hairy, branched perennial to 1 m. Leaves with eared petioles, ovate and toothed. Flower heads discoid, in loose corymbs, yellow. Dec.–May. Damp places, often along streams, NW, SW, KM, LB, SE (Cedarberg Mts and Stellenbosch to E Africa).

CORYMBIUM HEUNINGBOS 9 spp., W to E Cape

A. Long involucral bracts smooth

cymosum E.Mey. ex DC. Tufted perennial to 40 cm with smooth to slightly scabrid stems. Leaves filiform to linear, silky at base. Flower heads discoid, in dense corymbs, ivory, bracts smooth. Oct.–Dec. Lower sandstone slopes to 1000 m, SW, LB (Malmesbury: Perdeberg to Mossel Bay).•

enerve Markötter Tufted perennial to 40 cm with scabrid stems. Leaves narrowly lanceolate, leathery with veins inconspicuous, silky at base. Flower heads discoid, in dense corymbs, white or pink, bracts smooth. Oct.–Jan. Damp sandy flats and lower slopes, SW (Cape Peninsula to Hangklip).•

glabrum L. Tufted perennial to 60 cm with smooth stems. Leaves sword-shaped, leathery with veins prominent, silky at base. Flower heads discoid, in lax corymbs, pink or white, bracts smooth. Nov.–Jan. Rocky sandstone slopes, NW, SW, ?AP, KM, LB, SE (Cedarberg Mts to Grahamstown).

laxum Compton Tufted perennial to 35 cm with smooth stems. Leaves filiform with involute margins, smooth, silky at base. Flower heads in lax panicles, pink, bracts smooth. Dec.–Feb. Upper slopes, NW, SW (N Cedarberg Mts to Cold Bokkeveld and Franschhoek Mts).•

A. Long involucral bracts scabrid or sticky

africanum L. PLAMPERS Tufted perennial to 30 cm with scabrid stems. Leaves filiform to linear, silky at base. Flower heads discoid, in dense corymbs or panicles, purple, pink or white, bracts scabrid and sticky. Oct.–Nov. Sandy flats and slopes, NW, SW, AP, KM, LB, SE (Cedarberg Mts to Grahamstown).

congestum E.Mey. ex DC. VINGERHOED Tufted perennial to 30 cm with scabrid, viscid stems. Leaves lanceolate, scabrid, silky at base. Flower heads discoid, in panicles of small corymbs, mauve to pink, bracts scabrid and viscid. Nov.–Feb. Damp sandstone slopes, NW, SW (Pakhuis Mts to Houwhoek).•

elsiae Weitz Tufted perennial to 25 cm with scabrid stems. Leaves filiform with involute margins, smooth, silky at base. Flower heads drooping, in lax panicles, mauve or white, bracts scabrid. Dec.–Jan. upper slopes, NW, SW (Ceres to Bainskloof).•

theileri Markötter Tufted perennial to 30 cm with scabrid stems. Leaves linear to falcate, leathery with veins prominent, silky at base. Flower heads discoid, in small dense corymbs grouped in lax compound corymbs, mauve, bracts scabrid. Nov.–Jan. Rocky sandstone slopes, NW (Piketberg).•

villosum L.f. HEUNINGBOSSIE Tufted perennial to 30 cm with scabrid stems. Leaves glandular-hairy, narrowly lanceolate, silky at base. Flower heads discoid, in dense corymbs, mauve to white, bracts scabrid and sticky. Oct.–Jan. Lower sandstone slopes and flats, NW, SW, ?AP (N Cedarberg Mts to Bredasdorp).•

COTULA (= *CENIA, SPHAEROCLINIUM*)   BUTTONS, KNOPPIES   c. 50 spp., S hemisphere, mostly southern Africa

A. Flower head obviously radiate

andreae (E.Phillips) Bremer & Humphries (= *Matricaria andreae* E.Phillips)   Silky, scapose perennial to 10 cm. Leaves palmately bipinnatisect, lobes oblong. Flower heads radiate, solitary on sparsely hairy, elongate peduncles, white with purple reverse, bracts woolly. Dec.–Jan. Damp sandstone slopes above 1800 m, NW, KM (Hex River Mts to Klein Swartberg).•

duckittiae (L.Bolus) Bremer & Humphries   Like **C. turbinata** but robust, to 40 cm. Flower heads large, with conspicuous, bright orange rays. Sept. Sandy coastal slopes, SW (Yzerfontein to Bokbaai).•

macroglossa Bolus ex Schltr.   Softly hairy, scapose annual to 10 cm. Leaves bipinnatisect, lobes linear, crowded below. Flower heads conspicuously radiate, solitary on elongate, sometimes naked peduncles, yellow with blue or white rays mauve below, bracts silky. Sept.–Oct. Stony slopes, NW, KM (Cedarberg to Hex River Mts and Witteberg).•

montana Compton   Softly hairy, stoloniferous, tufted perennial to 12 cm. Leaves crowded basally, petiolate, bipinnatisect, lobes oblong. Flower heads radiate, on slender peduncles, yellow with white rays. Nov.–Jan. Sandstone slopes above 1500 m, partly shaded, NW (Cedarberg to Cold Bokkeveld Mts).•

nigellifolia (DC.) Bremer & Humphries (= *Matricaria nigellifolia* DC., *Sphaeroclinium nigellifolium* (DC.) Sch.Bip.)   RIVIERALS   Straggling to decumbent, glabrescent perennial rooting from lower nodes, to 30 cm. Leaves irregularly bipinnatisect, lobes linear to oblong. Flower heads shortly radiate, globose, solitary in upper leaf axils, yellow with white or yellow rays. Oct.–Apr. Damp areas, LB, SE (Swellendam to tropical Africa).

sericea Thunb. (= *Cenia sericea* (Thunb.) DC.)   Silky sprawling perennial 20–40 cm, spreading from woody stem. Leaves crowded below, bi- or tripinnatisect, lobes linear. Flower heads shortly radiate, solitary on elongate, naked peduncles inflated above in fruit, yellow, bracts 3-nerved. Mainly July–Dec. Stony coastal slopes, LB, SE (Mossel Bay to E Cape).

turbinata (L.) Pers. (incl. **Cenia expansa** Compton)   GANSKOS   Softly hairy annual 5–30 cm. Leaves alternate, bi- or tripinnatisect, lobes linear to filiform. Flower heads shortly radiate, solitary on slender, naked peduncles inflated above in fruit, yellow or white, bracts 3-nerved. Mainly July–Oct. Sandy or disturbed areas, NW, SW, AP (N Cedarberg Mts to Potberg).•

A. Flower heads discoid

barbata DC.   KLEINGANSKOS   Softly hairy annual to 15 cm. Leaves crowded below, opposite, clasping at base, pinnatisect or bipinnatisect. Flower heads discoid, solitary on slender, naked peduncles, yellow or white, bracts with membranous margins. July–Oct. Rocky slopes, NW (Namaqualand to Clanwilliam).

bipinnata Thunb.   KLEINKNOPPIES   Annual to 30 cm. Leaves alternate, sheathing at base, bi- or tripinnatisect. Flower heads discoid, solitary on slender leafy peduncles, yellow to white. Aug.–Oct. Near seasonal pools, NW, SW (Namaqualand to Cape Flats).

ceniifolia DC. (= *Cotula paradoxa* Schinz)   Softly hairy, slender, few-branched perennial to 15 cm, spreading on long leafy runners. Leaves opposite and crowded below, clasping at base, bipinnatisect. Flower heads discoid, solitary on slender naked peduncles, white. Aug.–Nov. Damp clay, SW, LB (Grabouw to Swellendam).•

coronopifolia L.   GANSGRAS, EENDEKOS   Erect or sprawling annual to 30 cm. Leaves alternate, sheathing at base, irregularly toothed to bipinnatisect. Flower heads discoid, solitary, on slender, minutely leafy peduncles, bright yellow. Mainly May–Oct. Seasonally wet areas, NW, SW, AP, KM, LB, SE (Namaqualand to Mpumalanga, Australia).

eckloniana (DC.) Levyns   Spreading annual branching from base, to 10 cm. Leaves linear to pinnatisect. Flower heads discoid, solitary on long, slender naked peduncles, yellow, florets broadly winged. Aug.–Oct. Sandy coastal flats, SW (Berg River to Cape Peninsula).•

filifolia Thunb.   Slender annual to 20 cm. Leaves alternate, filiform, undivided, sheathing at base. Flower heads discoid, solitary on slender naked peduncles, yellow to white, ovary winged. Aug.–Nov. Marshy ground, SW, AP (Darling to Agulhas).•

heterocarpa DC.   Softly hairy, sprawling annual to 10 cm. Leaves alternate, bi- or tripinnatisect. Flower heads discoid, solitary on slender, naked peduncles, white or yellow. Mainly Sept.–Nov. Dry rocky slopes, KM, LB, SE (Witteberg and Albertinia to E Cape and Karoo).

laxa DC.   Sparsely pilose to glabrescent, diffuse, spreading annual to 8 cm. Leaves pinnatisect, lobes linear, sometimes toothed. Flower heads discoid, small, solitary on slender naked peduncles, white. Aug.–Nov. Partly shaded rocky slopes, KM (Namaqualand to Montagu).

mariae Bremer & Humphries (= *Cenia pectinata* DC.)   Like **C. sericea** but possibly annual, leaf lobes oblong, and flower heads larger. Oct.–Nov. Coastal slopes, AP, LB, SE (Agulhas to Port Elizabeth).•

melaleuca Bolus   Like **C. barbata** but florets white with purple tips. Aug.–Oct. Sandstone slopes in sheltered sites, NW (Namaqualand to N Cedarberg Mts).

microglossa (DC.) Kuntze (= *Cenia microglossa* DC.)   KNOPPIES   Diffuse, sprawling annual branching from below, to 20 cm. Leaves tripinnatisect, lobes linear. Flower heads discoid or obscurely radiate, solitary on long, naked peduncles inflated above in fruit, tawny-haired when young, yellow or orange, bracts 1-nerved. July–Sept. Rocky flats or slopes, NW (Namaqualand to Clanwilliam and W Karoo).

myriophylloides Harv.   WATERGRAS   Prostrate aquatic ?perennial. Leaves opposite, sheathing at base, divided into many filiform segments. Flower heads discoid, solitary on slender naked peduncles in upper axils, yellow. Sept.–Dec. Coastal pools, SW (Cape Peninsula).•

nudicaulis Thunb.   WITEENDEKOS   Softly hairy annual to 15 cm. Leaves crowded below, opposite, clasping at base, bipinnatisect. Flower heads discoid, solitary on slender, naked peduncles, yellow or white, bracts large and round. Achenes winged. July–Sept. Stony flats, NW (Bokkeveld Escarpment to Worcester and W Karoo).

pedicellata Compton   Like **C. barbata** but to 30 cm. Achenes winged. July–Sept. Sandy flats, NW (S Namaqualand to Clanwilliam).

pusilla Thunb.   Dwarf, single-stemmed annual to 12 cm. Leaves alternate, linear to pinnatisect, sheathing at base. Flower heads discoid, solitary and terminal on leafy peduncles, whitish. July–Oct. Edges of pools in coastal sandy areas, SW (Saldanha to Milnerton).•

tenella E.Mey. ex DC.   Slender, glabrescent annual to 10 cm. Leaves alternate, clasping at base, linear to pinnatisect. Flower heads discoid, small, solitary on naked peduncles, yellow. Oct. Rocky slopes, NW (Olifants River valley to Hex River Pass).•

vulgaris Levyns   Possibly conspecific with **C. filifolia** but distinguished by the unwinged ovary. Mainly July–Sept. Damp ground, SW (Darling to Cape Peninsula).•

zeyheri Fenzl   Like **C. tenella** but softly hairy. May. Damp areas, SW, SE (Riviersonderend and Uitenhage).•

sp. 1 (= *Cotula sericea* auct.)   Grey-velvety sprawling perennial to 30 cm, with persistent creeping rhizome. Leaves opposite, crowded below, clasping at base with long petioles, palmately bipinnatisect. Flower heads discoid, solitary on elongate, glabrous, naked peduncles, yellow. Nov.–Jan. Sandstone slopes and ridges above 1000 m, NW (Cold Bokkeveld).•

[**Species excluded**   No authentic material found and probably conspecific with one of the above: **C. pedunculata** (Schltr.) E.Phillips]

CULLUMIA   STEEKHAARBOS   15 spp., Namaqualand, W and E Cape

A. Leaf margins unthickened or slightly revolute

aculeata (Houtt.) Roessler   Prickly, erect, densely leafy shrub to 1 m, often thinly cobwebby. Leaves ascending, oblanceolate, decurrent, margins slightly revolute, bristly in 2 unequal rows, slightly pungent. Flower heads radiate, small, pale yellow, solitary at branch tips, inner bracts unarmed. Mainly Sept.–Mar. Dry sandstone slopes, LB, SE (Langeberg Mts: Garcia's Pass to George).•

carlinoides DC.   Like **C. cirsioides** but leaves narrowly lanceolate. June–Sept. Coastal sands and limestones, AP, LB, SE (De Hoop to George).•

cirsioides DC.   Prickly, densely leafy shrublet to 60 cm. Leaves reflexed, lanceolate, margins slightly revolute, bristly, pungent. Flower heads radiate, yellow, solitary at branch tips. July–Nov. Sandstone slopes, SE (Humansdorp to Port Elizabeth).•

micracantha DC.   Like **C. aculeata** but plants glandular and sticky, weakly bristly, to 40 cm, inner bracts armed. Oct.–Jan. Sandsone slopes, NW (Olifants River valley and Piketberg).•

A. Leaf margins strongly revolute and mostly covering the underside

bisulca (Thunb.) Less.   Like **C. sulcata** but to over 1 m, leaves broader and strongly recurved, spines and bristles longer. Mainly Aug.–Nov. Dry sandstone slopes, NW, SW, KM, SE (Bokkeveld Mts to Kouga Mts and Bredasdorp).•

decurrens Less. Like **C. squarrosa** but to 90 cm, glabrous, leaves recurved, shorter, to 10 mm long. Aug.–Jan. Sandstone slopes, often near streams, KM, LB, SE (Swartberg Mts and Cloete's Pass to Port Elizabeth).•

patula (Thunb.) Less. Like **C. squarrosa** but to 30 cm, leaves spreading, acicular. Mainly Aug.–Sept. Dry sandstone slopes, NW, SW, KM, SE (Piketberg to Uniondale).•

squarrosa (L.) R.Br. GROOTSTEEKHAARBOS Prickly, robust, sprawling, densely leafy shrublet to 50 cm, cobwebby on young parts. Leaves reflexed, linear-lanceolate, (10)15–25 mm long, margins strongly revolute, bristly, pungent. Flower heads radiate, yellow, inner involucral bracts unlike the outer, unarmed. Mainly Sept.–Nov. Coastal bush, SW (Cape Peninsula to Stanford).•

sulcata (Thunb.) Less. Prickly, robust, sprawling, densely leafy shrublet to 50 cm, cobwebby on young parts. Leaves erect, linear, margins strongly revolute, bristly, pungent. Flower heads radiate, yellow, outer involucral bracts yellowish, not leaf-like. Sept.–Oct. Dry stony hillsides, NW, SW, KM (Caledon and Hex River valley to Swartberg Mts).•

A. Leaf margins thickened and hyaline

ciliaris (L.) R.Br. STEEKHAARBOS Prickly, straggling, densely leafy shrublet to 60 cm. Leaves ascending, ovate, margins thickened, bristly in 2 rows, tips pungent and reflexed. Flower heads radiate, mostly solitary at branch tips, yellow. July–Nov. Lower sandstone slopes, NW, SW (Olifants River Valley to Hottentots Holland Mts).•

floccosa E.Mey. ex DC. Prickly, densely leafy shrublet to 90 cm, cobwebby in upper axils. Leaves ascending, oblanceloate, margins thickened, bristly in 2 rows, pungent. Flower heads radiate, yellow. Nov.–Dec. Sandstone slopes, NW (Olifants River Mts and Piketberg).•

pectinata (Thunb.) Less. SLANGDISSEL Prickly, sprawling, densely leafy shrub to 1 m. Leaves incurved–adpressed, ovate, margins thick, finely bristly in 2 rows. Flower heads radiate, yellow, solitary at branch tips. Mainly Oct.–Dec. Sandstone slopes, NW (Bokkeveld Mts).•

rigida DC. Like **C. setosa** but to over 1 m. Leaves stiff and leathery, apical spine not reflexed. Sept.–Dec. Sandstone slopes to 1500 m, NW, KM (Kamiesberg, Bokkeveld Mts and Little Karoo).

selago Roessler Prickly, densely leafy shrublet to 20 cm. Leaves spreading-incurved, elliptic, margins thickened with bristles in 2 rows, tips pungent, inflexed. Flower heads radiate, yellow. Mainly Dec. Stony slopes, SW (Caledon).•

setosa (L.) R.Br. STEEKHAARBOS Prickly, densely leafy, sprawling shrublet to 60 cm, sometimes cobwebby. Leaves ovate, recurved, margins thickened and bristly sometimes in 2 rows, tips pungent and reflexed. Flower heads radiate, yellow, solitary at branch tips. Mainly Aug.–Oct. Lower mountain slopes, SW (Cape Peninsula to Stanford and Riviersonderend Mts).•

CUSPIDIA WORTELBOSSIE 1 sp., E Cape, Karoo

cernua (L.f.) B.L.Burtt Sprawling, thinly cobwebby, prickly leaved annual to 30 cm. Leaves oblanceolate, toothed and pungent. Flower heads radiate, pale yellow, bracts spreading, leafy, pungent. July–Sept. Rocky flats and lower slopes, KM, SE (Karoo to Calitzdorp and Uitenhage).

CYMBOPAPPUS 3 spp., southern Africa

adenosolen (Harv.) B.Nord. Glabrescent, twiggy shrublet to 60 cm. Leaves linear, often 3–5-lobed and hooked at tips, base often with 2 short lobes. Flower heads radiate, solitary on short terminal peduncles, often nodding, yellow with white rays. Sept.–Apr. Mostly clay slopes, renosterveld, SW, AP, LB (Caledon to Gourits River).•

[**CYPSELODONTIA** 1 sp., Uitenhage district, identity unknown, probably a synonym of another genus.]

DELAIREA 1 sp., E Cape to KwaZulu-Natal

odorata Lem. (= *Senecio mikanioides* Otto ex Harv.) Herbaceous climber. Leaves fleshy, cordate, 5–7-lobed, broadly toothed, palmately veined, petiolate, eared at base. Flower heads discoid, in dense corymbs on lateral branches, yellow. Apr.–May. Forest margins, SE (Uitenhage to KwaZulu-Natal).

DICHROCEPHALA 4 spp., Africa and Asia

integrifolia (L.f.) Kuntze Diffuse, softly hairy annual to 15 cm. Leaves ovate, sometimes with small lobes below, toothed. Flower heads discoid, in sparse panicles, greenish. Nov.–Jan. Forest floors, LB, ?SE (Swellendam to tropical Africa).

DICOMA WILDEKARMEDIK   c. 40 spp., mainly Africa, also Madagascar and India

fruticosa Compton   Twiggy shrublet to 50 cm, sticky on young parts. Leaves oblanceolate-spathulate, leathery, apiculate. Flower heads sparsely radiate, solitary at branch tips, pinkish, disc florets c. 5, bracts ovate, not spiny, margins membranous-lacerate. Sept.–Nov. Shale flats, KM (Klein Swartberg and Warmwaterberg).•

picta (Thunb.) Druce   KNOPPIESDORINGBOSSIE   Softly woody, rigid shrublet to 60 cm, branches felted-striate. Leaves spathulate, grey-mealy. Flower heads radiate, in lax corymbs, white with pink to mauve rays, bracts stiffly acuminate. Oct.–Dec. Dry stony slopes, NW, KM, SE (Cedarberg Mts to Uitenhage and Karoo).

relhanioides Less.   Cushion-like, stout, densely leafy shrublet to 15 cm. Leaves lanceolate, sessile, margins involute, sometimes woolly above, pungent. Flower heads discoid, solitary at branch tips, mauve, bracts ovate, shortly spiny. Dec.–Mar. Stony ?shale slopes, KM (Huisrivierberg).•

spinosa (L.) Druce   STEEKBLOM   Cushion-forming, usually grey-felted shrublet to 20 cm. Leaves obovate, slightly toothed and pungent, glabrescent above. Flower heads obscurely radiate, solitary at branch tips, pink or white, bracts broad below, narrow and spiny above. Flowering erratically. Dry rocky shale slopes, SW, ?AP, KM, LB, SE (Du Toitskloof to Somerset East).

DIDELTA   SLAAIBOS   2 spp., W Cape to Namibia

carnosa (L.f.) Aiton   KUSSLAAIBOS   Rounded, thinly or densely cobwebby shrublet to 1 m. Leaves fleshy, oblanceolate. Flower heads radiate, large, yellow, bracts leafy. July–Nov. Coastal dunes and sandy flats, NW, SW (S Namibia to Cape Peninsula).

spinosa (L.f.) Aiton   SLAAIBOS   Shrub or tree to 2 m, cobwebby on young parts. Leaves opposite, ovate to elliptic, cordate, margins slightly revolute, sometimes prickly. Flower heads radiate, large, solitary at branch tips, yellow, bracts leafy. July–Sept. Dry granite and sandstone slopes, NW (S Namibia to Piketberg).

DIMORPHOTHECA (= *CASTALIS*)   MARGRIET   9 spp., W Cape, and Karoo to Namibia

chrysanthemifolia (Vent.) DC.   MARGRIET   Thinly glandular-hairy perennial to 1 m. Leaves oblanceolate, toothed to pinnatifid. Flower heads radiate, large, solitary or few at branch tips, yellow or orange. Ray achenes shortly 3-winged. Oct.–Jan. Rocky sandstone slopes, NW, SW (Hex River Mts to Simonsberg).•

cuneata (Thunb.) Less.   BOSMARGRIET   Rounded, glandular-hairy, viscid shrublet to 50 cm. Leaves cuneate-oblanceolate, toothed to lobed or pinnatifid. Flower heads radiate, solitary on short, naked peduncles, yellow with white or orange rays darker on reverse. Ray achenes trigonous, glandular. Mainly Sept.–Nov. Stony and shale ridges and flats, NW, KM, SE (Namaqualand and Karoo, Ceres to Uitenhage).

montana Norl. (incl. **D. venusta** (Norl.) Norl.)   Tufted to mat-forming, glandular-hairy perennial to 30 cm, from woody caudex. Leaves crowded basally, narrowly oblanceolate, subentire to toothed. Flower heads radiate, solitary on elongate, leafy peduncles, yellow, white or pink, with dark reverse. Ray achenes 3-winged. Nov.–Jan. Sandstone slopes and ridges, NW, SW, KM, SE (Hex River Mts to Franschhoek, Swartberg to Tsitsikamma Mts).•

nudicaulis (L.) B.Nord.   WITMARGRIET   Like **D. tragus** but leaves mostly crowded basally. Flower heads white with purple to copper reverse. Ray achenes vestigial. Aug.–Oct. Sandstone slopes, NW, SW, AP, KM, LB, SE (Bokkeveld Escarpment to George).•[Possibly not distinct from **D. tragus**.]

pluvialis (L.) Moench   REËNBLOMMETJIE   Erect to sprawling, glandular-hairy annual to 30 cm. Leaves oblanceolate, lobed to toothed. Flower heads radiate, solitary at branch tips, purple with white rays purple at the base and darker on reverse, involucre shallowly cup-shaped. Ray achenes tuberculate. Mainly Aug.–Oct. Sandy and clay flats and slopes, NW, SW, AP (S Namibia to Gouritsmond).

sinuata DC.   NAMAQUALAND DAISY   Like **D. pluvialis** but often less hairy, involucre more deeply cup-shaped and flower heads yellow with yellow or orange to beige rays. Mainly Aug.–Oct. Sandy and limestone flats and ridges, NW, SW (S Namibia to Saldanha).

tragus (Aiton) B.Nord.   OX-EYE-DAISY   Glabrescent to glandular-hairy shrublet with soft annual stems to 30 cm. Leaves oblanceolate, margins entire or toothed, ciliate, sometimes glaucous. Flower heads radiate, solitary on elongate peduncles sparsely leafy below, purple with orange or yellow rays. Ray achenes vestigial. Aug.–Sept. Rocky shale slopes, NW (Namaqualand to Pakhuis Mts).

DISPARAGO BASTERSLANGBOS   9 spp., W Cape to KwaZulu-Natal

A. Ray florets sterile

anomala Schltr. ex Levyns   Like **D. kraussii** but pappus lacking. Dec.–Apr. Coastal sands and limestone, SW, AP (Cape Peninsula to Potberg).•

ericoides (P.J.Bergius) Gaertn. (= *Disparago lasiocarpa* Cass.)   Thinly cobwebby shrublet to 30 cm. Leaves linear with involute margins, spreading to recurved and twisted. Flower heads radiate, in several, shortly stalked terminal clusters, mostly 2-flowered, purple with pink rays, pappus bristles plumose, ovary of disc florets densely woolly, ray florets sterile. Nov.–Mar. Sandstone slopes, NW, SW, AP (Darling to Matroosberg and Gouritsmond).•

kraussii Sch.Bip.   Thinly cobwebby shrublet to 80 cm. Leaves linear with involute margins, spreading and twisted, suberect above, bristle-tipped. Flower heads radiate, crowded in elongate terminal clusters, 2-flowered, ?white with white or pink rays, pappus bristles plumose, ray florets sterile. Apr.–June. Sandstone, mostly coastal slopes, AP, LB, SE (Agulhas to Plettenberg Bay).•

laxifolia DC.   Spreading, thinly cobwebby shrublet to 20 cm. Leaves linear with involute margins, spreading and twisted. Flower heads radiate, in dense rounded clusters, 2-flowered, white, pappus bristles plumose, ray florets sterile. Oct.–Dec. Sandstone flats and lower slopes, SW (Cape Peninsula to Hermanus).•

A. Ray florets female

barbata Koekemoer   Trailing shrublet to 40 cm. Leaves linear with margins involute, twisted and spreading. Flower heads radiate, in dense terminal clusters, 6-flowered, brownish with white rays, pappus bristles barbed, ray florets female. Nov.–Jan. Sandstone slopes above 1200 m, NW (Matroosberg).•

gongylodes Koekemoer   Intricately branched shrublet to 35 cm. Leaves ovate to rounded, margins involute, densely white-woolly above, spreading and twisted. Flower heads radiate, in small terminal clusters, 7-flowered, white with purple rays, pappus bristles plumose, ray florets female. Sept.–Oct. Sandstone plateaus above 1200 m, NW (Grootwinterhoek Mts).•

kolbei (Bolus) Hutch. (= *Disparago rosea* Hutch.)   Densely grey-woolly shrublet to 25 cm. Leaves linear, adpressed, 4-ranked, margins involute, densely white-woolly above. Flower heads radiate, in small terminal, loosely aggregated clusters, 4–6-flowered, purple with white rays, pappus bristles plumose, ray florets female. Sept.–Apr. Rocky sandstone slopes and ridges above 1200 m, KM, SE (Witteberg and Swartberg Mts to Langkloof).•

pilosa Koekemoer   Like **D. kolbei** but florets 8 per head and bracts and ovary hairy. Sept.–Apr. Sandstone slopes above 1200 m, NW (Cold Bokkeveld to Hex River Mts).•

tortilis (DC.) Sch.Bip. (= *Disparago ericoides* auct.)   Much-branched, thinly cobwebby shrub to 90 cm. Leaves linear with involute margins, recurved and twisted, bristle-tipped. Flower heads radiate or discoid, mostly 1-flowered, grouped in dense globose clusters, pink or white, pappus bristles plumose, ray florets female. Mainly May–Sept. Sandy slopes, SW, LB, SE (Hottentots Holland to S KwaZulu-Natal).

DOLICHOTHRIX•   KLIPRENOSTERBOS   1 sp., W Cape

ericoides (Lam.) Hilliard & B.L.Burtt   Rounded, minute-leaved shrub to 70 cm, branches white-felted. Leaves adpressed to stem, scale-like. Flower heads discoid, solitary, crowded at branch tips, purplish, bracts white, reflexed at tips. Nov.–Feb. Sandstone rocks mostly above 1500 m, NW, KM, SE (Cedarberg to Great Winterhoek Mts).•

DYMONDIA•   CARPET GAZANIA   1 sp., W Cape

margaretae Compton   Mat-forming perennial to 5 cm with creeping rhizome, roots thickened, fusiform. Leaves in a basal rosette, narrowly oblanceolate, margins involute, sinuate, densely white-felted below. Flower heads radiate, sessile, yellow. Mainly Sept.–Nov. but throughout the year. Coastal flats at edge of pans and marshes, SW, AP (Agulhas to Potberg).•

EDMONDIA•   SEWEJAARTJIE   3 spp., W Cape

fasciculata (Andrews) Hilliard   Like **E. sesamoides** but tips of upper peduncular leaves becoming dry and scale-like, bracts bright yellow. Sept.–Jan. Rocky sandstone slopes, NW, SW (Cedarberg to Riviersonderend Mts).

pinifolia (Lam.) Hilliard   Sparsely branched, thinly white-woolly shrublet to 60 cm. Leaves weakly dimorphic, linear and ascending below. Flower heads discoid, solitary on long peduncles bearing short, adpressed leaves, the uppermost becoming dry and papery, yellow, bracts papery, white to pink. Sept.–Dec. Rocky sandstone slopes, NW, SW (Cedarberg Mts to Cape Peninsula and Kogelberg).•

sesamoides (L.) Hilliard   SEWEJAARTJIE   Sparsely branched, thinly white-woolly shrublet to 30 cm. Leaves dimorphic, the cauline linear, spreading, with margins rolled upward. Flower heads discoid, solitary, peduncles long, bearing short, adpressed leaves, yellow, bracts papery, white to yellow or pink. Aug.–Dec. Rocky flats and slopes, NW, SW, AP, LB (Cedarberg Mts to Mossel Bay).•

ELYTROPAPPUS   RENOSTERBOS   10 spp., W to E Cape, Karoo

adpressus Harv.   WYFIERENOSTERBOS   Like **E. rhinocerotis** but leaves linear-ericoid and flower heads several in short spikes terminating the shoots, not on lateral branches. Mar.–July. Sandstone slopes above 600 m, NW, KM (Cold Bokkeveld to Witteberg and Langkloof).•

aridus Koekemoer   Glabrescent, reddish brown shrublet to 80 cm. Leaves ericoid with margins rolled upward, widely spaced, spreading and twisted. Flower heads discoid, few in terminal glomerules, pink to red, 2–4-flowered. Mainly Feb.–Apr. Rocky sandstone slopes, KM (Swartberg Mts).•

glandulosus Less.   Glabrescent or woolly shrublet to 50 cm. Leaves linear with involute margins, glandular-hairy, adpressed or spreading and twisted. Flower heads discoid, few in axillary clusters forming slender spikes, mauve, 2(3)-flowered, florets tubular. Mar.–June. Sandstone slopes, NW, SW, LB (Cedarberg to Langeberg Mts: Swellendam).•

gnaphaloides (L.) Levyns   Like **E. glandulosus** but sparsely glandular-hairy and flower heads in open panicles. Jan.–Apr. Shale and stony slopes, NW, SW, AP, LB, SE (Bokkeveld Escarpment to Langkloof).•

hispidus (L.f.) Druce (= *Stoebe cyathiformis* DC.)   Thinly woolly shrublet to 30 cm. Leaves ericoid with involute margins, twisted, glandular-hairy. Flower heads discoid, massed in rounded terminal clusters, pinkish, 3- or 4-flowered. Achenes with horizontal ridges. Sept.–Dec. Rocky slopes, NW, KM (Bokkeveld Escarpment to Klein Swartberg Mts).•

intricata (Levyns) Koekemoer (= *Stoebe intricata* Levyns)   Intricately branched, glandular-woolly shrub to 1 m. Leaves narrowly lanceolate, often adpressed. Flower heads discoid, small, crowded in rounded clusters, purple, 1-flowered, bracts golden, acuminate. Apr.–July. Sandstone slopes, NW, KM, LB (Cold Bokkeveld to Swartberg Mts).•

longifolius (DC.) Levyns   Thinly white-woolly shrub to 1 m with stiffly erect branches. Leaves linear with involute margins, recurved and twisted, glandular-scabrid. Flower heads discoid, few in axillary clusters together forming elongate spikes or open racemes, pink, florets narrowly funnel-shaped, 3–6-flowered. Achenes warty above. Feb.–Apr. Sandstone slopes, SW (Cape Peninsula to Genadendal).•

monticola Koekemoer   Sparsely branched, single-stemmed shrublet to 40 cm. Leaves linear, densely glandular-hairy, ascending above. Flower heads many in terminal glomerules, white, 5- or 6-flowered. Oct.–Nov. Sandstone slopes, LB (Langeberg Mts: Riversdale).•

rhinocerotis (L.f.) Less.   RENOSTERBOS   Thinly grey-woolly, viscid shrub to 2 m with short whip-like branches. Leaves scale-like, adpressed. Flower heads discoid, few at tips of lateral branches, purple, mostly 3-flowered. Feb.–Apr. Dry shale and sandstone slopes and flats, NW, SW, AP, KM, LB, SE (S Namibia to E Cape and Karoo).

scaber (L.f.) Levyns   Wiry-stemmed, glandular-hairy shrub to 1 m. Leaves involute ericoid, adpressed. Flower heads discoid, few in axillary clusters scattered along upper stems, 2- or 3-flowered. Feb.–May. Sandstone slopes and plateaus, NW, SW (Bokkeveld Mts to Bredasdorp).•

ERIOCEPHALUS   KAPOKBOSSIE, WILD ROSEMARY   c. 34 spp., southern Africa

A. Flower heads inconspicuously radiate

ericoides (L.f.) Druce   GEWONE KAPOKBOSSIE   Twiggy, glabrescent shrub to 1 m. Leaves small, linear, opposite in axillary tufts. Flower heads radiate, appearing discoid, subsessile, solitary in axils of upper leaves, reddish. July–Nov. Stony clay and sandy flats, KM (widespread in dry parts of southern Africa).

racemosus L.   KAPKOPPIE   Silvery silky shrub to 1.5 m. Leaves linear, alternate, sometimes tufted. Flower heads radiate, appearing discoid, solitary in leaf axils, forming long racemes, pink. July–Sept. Coastal dunes and hills, NW, SW, AP, SE (S Namaqualand to Humansdorp).

A. Flower head conspicuously radiate

africanus L.   WILD ROSEMARY, KAPOKBOSSIE   Silvery silky, twiggy shrub to 1 m. Leaves linear or trifid, in tufts. Flower heads radiate, in small umbellate clusters at branch tips, rays conspicuous, purplish with white rays. Mainly Jan.–June. Mostly clay or granite hillsides, NW, SW, AP, LB, SE (S Namaqualand to Port Elizabeth and E Cape).

aromaticus C.A.Sm.   Like **E. brevifolius** but leaves glabrescent, linear, slightly fleshy and punctate. July–Oct. Rocky sandstone slopes, NW, KM (Swartruggens to Swartberg Mts).•

brevifolius (DC.) M.A.N.Muller   Shortly hairy shrub to 1 m. Leaves linear, punctate, crowded on spur shoots and branch tips. Flower heads radiate, in small umbellate clusters, rays conspicuous, purplish with white rays. Mainly July–Sept. Rocky slopes, NW, KM (Namaqualand and W Karoo to Swartruggens and Swartberg Mts).

capitellatus DC.   Silvery silky, slender shrub to 1.6 m. Leaves linear or trifid, in tufts. Flower heads radiate, subsessile, in small axillary glomerules arranged in spikes, rays small, white. Jan.–Apr. Dry sandstone slopes, NW, KM, SE (Cedarberg Mts to Swartberg Pass and Baviaanskloof Mts).•

grandiflorus M.A.N.Muller   Densely silvery silky, more or less thorny, rigid shrublet to 45 cm. Leaves opposite and imbricate, linear. Flower heads radiate, large, few in terminal clusters, purplish with white rays. June–Sept. Stony ground, KM (W Karoo and Witteberg to Swartberg Mts).

paniculatus Cass.   Similar to **E. africanus** but bracts and ovaries densely golden woolly. Mainly July–Sept. Sandstone slopes, NW, SW, KM, LB, SE (Cedarberg Mts to Somerset East).

punctulatus DC.   BOEGOEKAPOKBOSSIE   Like **E. brevifolius** but flower heads on longer, filiform peduncles. Aug.–Oct. Sandstone slopes, NW, SW (Namaqualand to Moorreesburg and W Karoo).

purpureus Burch.   Twiggy shrublet to 60 cm. Leaves linear, in tufts and crowded on short shoots. Flower heads radiate, solitary on short shoots, rays conspicuous, mauve. July–Aug. Dry rocky karroid slopes, NW (Namaqualand to Cedarberg Mts and W Karoo).

tenuipes C.A.Sm.   Like **E. africanus** but leaves more or less equal, punctate and often glabrous. Jan.–June. Rocky slopes, SE (Uniondale and Port Elizabeth).•

EURYOPS   HARPUISBOS   97 spp., southern and tropical Africa, S Arabia

A. Pappus lacking or vestigial

bolusii B.Nord.   Twiggy, closely leafy shrublet to 40 cm. Leaves mostly pinnatisect, 10–30 mm long, lobes short. Flower heads radiate, solitary on elongate, slender peduncles, yellow; pappus absent. Achenes with c. 10 warty ridges. Nov.–Apr. Rocky sandstone slopes, KM (Swartberg to Kammanassie Mts).•

ericifolius (Bél.) B.Nord.   Like **E. ericoides** but to 30 cm. Leaves adpressed to stems, 1.5–5 mm long. Achenes with shortly hairy ridges. Mar.–Oct. Saline or limestone flats, SE (Port Elizabeth to Sundays River).

ericoides (L.f.) B.Nord.   Twiggy, densely leafy shrublet to 60 cm. Leaves ascending, 2–5(10) mm long, linear-trigonous and apiculate. Flower heads radiate, solitary on elongate peduncles, yellow; pappus absent. Achenes with c. 10 warty ridges. Mainly Feb.–Sept. Rocky sandstone and limestone slopes, AP, LB (Swellendam and Agulhas to Ruitersbos).•

euryopoides (DC.) B.Nord.   Like **E. pinnatipartitus** but leaves mostly 3-lobed and achenes with smooth ridges. Mainly Mar.–Dec. Rocky slopes in fynbos and forest margins, SE (Humansdorp to Grahamstown).

hebecarpus (DC.) B.Nord.   Softly hairy shrublet to 35 cm. Leaves trifid, 20–50 mm long, lobes subterete. Flower heads radiate, large, mostly solitary on elongate terminal peduncles, yellow or orange; pappus vestigial. Aug.–Sept. Limestone ridges, AP (Agulhas to Potberg).•

integrifolius B.Nord.   Like **E. ericoides** but leaves obtuse and achenes minutely papillose. Nov. Rocky sandstone ridges above 1200 m, SE (Tsitsikamma Mts).•

munitus (L.f.) B.Nord.   Similar to **E. bolusii** but leaves stiff and ascending, 3–10 mm long, lobes mucronate. Achenes almost smooth. Mainly Mar.–June. Clay and sandstone slopes on dry fynbos, SE (Langkloof to Port Elizabeth).•

pinnatipartitus (DC.) B.Nord.   Densely leafy shrub to 90 cm, sometimes thinly woolly on young parts. Leaves pinnatisect, 10–30 mm long, lobes oblong to terete. Flower heads radiate, solitary, on short peduncles, yellow; pappus absent. Achenes with 10 warty ridges. Mainly July–Jan. Sandstone slopes, 500–1500 m, KM, LB, SE (Swellendam to Kammanassie Mts and Humansdorp).•

tenuilobus (DC.) B.Nord.   Like **E. bolusii** but leaf lobes longer and filiform and achenes with 5–7 faint ridges. June–Aug. Clay hills in renosterbos, SW (Caledon district).•

ursinoides B.Nord.   Like **E. bolusii** but to 2 m, thinly woolly in leaf axils, leaf lobes long and linear, pappus absent and achenes with 10 smooth ribs. July–Dec. Rocky sandstone slopes, SE (Van Staden's Mts).•

A. Pappus present
B. Disc florets widening gradually above

brevilobus Compton   Like **E. speciosissimus** but stout shrub rarely over 1 m and leaves rigid with short lobes. July–Oct. Sandstone slopes, NW (Cedarberg Mts to Swartruggens).•

decipiens Schltr.   Sprawling shrublet to 20 cm. Leaves linear, 15–60 mm long, sometimes 1- or 2-lobed. Flower heads radiate, solitary on elongate, terminal peduncles, disc florets obconic, yellow. Oct.–Jan. Steep moist sandstone slopes above 1200 m, SW (Bainskloof Mts).•

othonnoides (DC.) B.Nord.   Grey-felted shrublet to 40 cm. Leaves pinnatisect, 20–90 mm long, lobes 1–4 on each side, subterete, pungent. Flower heads radiate, solitary on elongate terminal peduncles, yellow, disc florets obconic. Aug.–Dec. Sandstone rocks and cliffs, NW, KM (Pakhuis Pass to Hex River Mts and Touwsberg).•

pectinatus (L.) Cass.   WOLHARPUISBOS   Densely grey-felted shrub to over 1.5 m. Leaves toothed to pinnatisect, 40–100 mm long, lobes linear, obtuse to acute. Flower heads radiate, large, solitary on stout elongate peduncles in upper leaf axils, with cup-shaped involucre, yellow, disc florets obconic. Sept.–Dec. Rocky sandstone slopes, NW, SW (Gifberg to Cape Peninsula).•

serra DC.   Like **E. brevilobus** but leaves with extremely short lobes, almost serrate. Aug.–Feb. Rocky sandstone slopes, NW (Grootwinterhoek Mts).•

speciosissimus DC.   PRONKHARPUISBOS   Glabrous shrub to over 2 m. Leaves pinnatisect, (40)60–200 mm long, lobes filiform, flexuose. Flower heads radiate, large, solitary, on thick terminal peduncles less than twice as long as leaves, with cup-shaped involucre, yellow, disc florets obconic. Mainly Aug.–Nov. Rocky sandstone slopes, NW (Bokkeveld Mts to Tulbagh).•

tagetoides (DC.) B.Nord.   Like **E. othonnoides** but taller, to 1 m. Leaves more finely divided, lobes 5–8 on each side. Flower heads orange, involucre cup-shaped. Aug.–Sept. Dry sandstone slopes in arid fynbos, NW (Botterkloof to N Cedarberg Mts).•

wageneri Compton   Like **E. speciosisssimus** but leaves grey. Flower heads orange or deep yellow with wider rays, peduncles more than twice as long as leaves. Aug.–Sept. Rocky sandstone slopes and plateaus, NW (Biedouw to N Cedarberg Mts).•

B. Disc florets abruptly flaring above

abrotanifolius (L.) DC.   BERGHARPUISBOS   Densely leafy shrub to 1 m. Leaves ascending, pinnatisect with linear to filiform lobes or sometimes entire, 60–90 mm long. Flower heads radiate, solitary on terminal peduncles woolly at base, yellow. Achenes glabrous, closely ribbed with terminal appendage. Mainly July–Dec. Sandstone slopes, NW, SW, LB (N Cedarberg to Langeberg Mts: Riversdale).•

algoensis DC.   KLEINHARPUISBOS   Willowy, densely leafy shrub 1.2 m. Leaves ascending, obovate, 3–5-toothed or lobed above, leathery and minutely warty, 6–25 mm long. Flower heads radiate, solitary in upper axils, yellow. Mar.–June. Mainly coastal scrub, LB, SE (Mossel Bay to Grahamstown).

brevipapposus M.D.Hend.   Shrub 2–3 m. Leaves linear, clustered at branch tips, sometimes woolly in axils, 15–90 mm long. Flower heads radiate, solitary on short peduncles clustered among leaves, yellow. Mainly Oct.–Mar. Stony slopes and riverbanks, SE (Humansdorp to N Province.)

cuneatus B.Nord.   Like **E. multifidus** but leaf lobes broad and flattened, somewhat cuneate and bracts free almost to base. July–Sept. Dry stony slopes, NW (Ceres and W Karoo).

erectus (Compton) B.Nord.   Densely leafy shrub to 1 m. Leaves needle-like, spreading to upcurved. Flower heads radiate, clustered at branch tips on filiform peduncles, bracts free. Aug.–Oct. Sandstone slopes, KM, SE (Witteberg to Swartberg Mts).•

glutinosus B.Nord.   Sparsely branched shrub to c. 1 m, woolly on young parts. Leaves linear-triquetrous, glutinous, margins prominent, 15–50 mm long. Flower heads radiate, solitary and few at branch tips, yellow. Oct.–Dec. Rocky sandstone slopes above 1500 m, KM (Klein Swartberg Mts).•

imbricatus (Thunb.) DC.   Sprawling, shrub to over 1 m. Leaves crowded above, ascending, fleshy-spathulate, rugulose, bright green, 3–17 mm long. Flower heads radiate, solitary in upper axils, yellow. May–Dec. Rocky karroid flats and slopes, NW, KM (Hex River Pass to Witteberg and W Karoo).

indecorus B.Nord. Dark-stemmed shrub to 90 cm. Leaves clustered at branch tips, leathery, oblanceolate, mostly 3-lobed at tips, 15–70 mm long. Flower heads radiate, solitary on short peduncles, yellow. Oct.–Nov. Rocky sandstone slopes and ridges, SW (Cape Hangklip).•

lasiocladus (DC.) B.Nord. Slender, white-woolly shrublet to 10 cm. Leaves linear-filiform, acuminate, loosely twisted, 15–35 mm long. Flower heads radiate, solitary on elongate peduncles, yellow. June–Aug. Sandstone slopes, 300–600 m, SW (Babilonstoring).•

lateriflorus (L.f.) DC. SOETHARPUISBOS Viscid shrub to 1 m or more, with stiffly erect branches, closely leafy above. Leaves leathery, grey, oblanceolate to obovate, often 3-veined from base, 8–30 mm long. Flower heads radiate, few, solitary on short axillary peduncles, yellow. Mainly May–Aug. Dry sandstone and shale slopes, NW, KM, SE (S Namibia and W Karoo and Ceres to Uniondale).

linearis Harv. Densely leafy shrub to over 2 m. Leaves soft, mostly entire and linear, tapering below, with distict midvein, 10–40 mm long. Flower heads radiate, on short filiform peduncles in terminal umbels, with cup-shaped involucres, yellow. Mainly Aug.–Oct. Partly shaded on limestone ridges, AP (De Hoop to Potberg).•

linifolius (L.) DC. Like **E. rehmannii** but young branches cobwebby, tips of bracts overlapping and achenes glabrous. July–Sept. Sandy flats and slopes, NW, SW (Hopefield to Stellenbosch).•

longipes DC. Like **E. rupestris** but stems more slender, leaves often entire, and peduncles longer and filiform. Nov.–Apr. Sandstone slopes and ridges, NW, SW (Witzenberg and Bainskloof to Riviersonderend Mts).•

microphyllus (Compton) B.Nord. Sparsely leafy, twiggy shrublet to 60 cm. Leaves adpressed to ascending, 1–6 mm long. Flowers heads radiate, solitary on slender peduncles, yellow. Achenes conspicuousl woolly. July–Oct. Rocky slopes, KM (W Karoo to Witteberg).

muirii C.A.Sm. STRANDHARPUISBOS Dwarf, closely leafy ericoid shrublet to 25 cm. Leaves ascending, closely imbricate, 2–5 mm long. Flower heads radiate, yellow, solitary, clustered at branch tips, bracts free. Sept. Limestone ridges, AP (near Stilbaai).•

multifidus (Thunb.) DC. HANEPOOTHARPUISBOS Shrub to 1.5 m, with stiffly erect branches. Leaves crowded on short shoots, mostly trifid with forked lateral lobes, 6–35 mm long, lobes subterete. Flower heads radiate, solitary on short peduncles in leaf axils, yellow. Achenes conspicuously woolly. June–Sept. Rocky slopes, often on outcrops, NW, SW (Namaqualand to Saldanha, Worcester and W Karoo).

oligoglossus DC. (= *Elytropappus racemosus* auct.) WATERHARPUISBOS Lax shrub to 1.5 m, sometimes sparsely woolly in axils. Leaves lanceolate to linear, spreading, 50–150 mm long. Flower heads radiate, solitary on short peduncles in upper axils, yellow. Achenes densely white-hairy, mucilaginous when wet. Dec.–June. Rocky watercourses, NW (Hex River Mts and Karoo to Lesotho).

rehmannii Compton Like **E. tenuissimus** but leaves (5–)10–50 mm long, flower heads borne above the leaves, involucres cup-shaped and achenes shortly hairy. Mainly July–Oct. Karroid scrub or arid fynbos, NW, KM, SE (N Cedarberg Mts to Worcester, Witteberg to Uitenhage and S Karoo).

rupestris Schltr. Gnarled, few-branched shrublet to 80 cm. Leaves clustered at ends of branches, 3–5-sect, 20–50 mm long, lobes linear-filiform. Flower heads radiate, solitary on long peduncles, yellow. July–Dec. Sandstone outcrops, mostly above 800 m, NW, SW (Witzenberg and Tulbagh Kloof to Riviersonderend Mts).•

spathaceus DC. HARPUISBOS Densely leafy shrub to over 1 m. Leaves terete to oblanceolate, (10)20–50 mm long. Flower heads radiate, solitary in upper axils, yellow, involucres fused when young, splitting into a few irregular segments. Mainly Jan.–Apr. Rocky karroid slopes, KM, SE (Swartberg Mts to E Cape).

subcarnosus DC. SOETHARPUISBOS Twiggy shrub to 1 m. Leaves at branch tips and on short shoots, terete, linear or with 3–5 linear lobes, 2–30 mm long. Flower heads radiate, solitary on long peduncles in upper axils, yellow. Jan.–Dec. Varied habitats, KM, SE (S Namibia and Botswana to Little Karoo).

tenuissimus (L.) DC. RESIN BUSH, GROOTHARPUISBOS Shrub to 2.5 m, often mealy on young parts. Leaves filiform, sometimes 3-lobed, 15–150 mm long. Flower heads radiate, solitary on short, filiform peduncles clustered among leaves, yellow or orange. Achenes hairy, mucilaginous when wet. Mainly Aug.–Oct. Stony karroid slopes, NW, SW, KM, LB, SE (Namaqualand and W Karoo to George).

thunbergii B.Nord. Closely leafy shrub to 1.2 m, densely cobwebby on young parts. Leaves spreading, filiform, terete or somewhat flattened, woolly on lower midline, 10–50 mm long. Flower heads radiate, on slender peduncles together forming loose umbels, yellow. Achenes white-puberulous, mucilaginous when wet. Mainly Sept.–Oct. Sandy flats and lower slopes, NW, SW (Gifberg to Tygerberg and Worcester).•

virgatus B.Nord. Twiggy shrublet to 50 cm. Leaves crowded at branch tips, small, pinnatisect, 3–15 mm long, lobes filiform. Flower heads radiate, nodding in bud, few, solitary or few on each branch, on long peduncles, yellow. July–Sept. Rocky flats, NW (Bokkeveld Escarpment).•

virgineus (L.f.) DC. RIVIERHARPUISBOS Densely leafy shrub to 3 m with stiffly erect stems. Leaves ascending, obovate, narrowly lobed to toothed above, 5–12 mm long. Flower heads radiate, small, solitary on short filiform peduncles in upper axils, yellow. Mainly July–Nov. Sandstone slopes, KM, LB, SE (Swartberg Mts and Mossel Bay to Alexandria).

FELICIA (= *CHAREIS*) ASTERTJIE 85 spp., southern and tropical Africa to Arabia

A. Involucral bracts strictly in 2 rows

aculeata Grau Roughly hairy, softly woody shrublet to 45 cm. Leaves oblong, margins revolute and harshly scabrid. Flower heads radiate, solitary, yellow with blue rays. Aug.–Sept. Coastal flats and lower slopes, SW, AP, LB, SE (Caledon to Knysna).•

aethiopica (Burm.f.) Adamson & T.M.Salter WILDE-ASTERTJIE Thinly hairy, soft shrublet to 1 m. Leaves elliptic to ovate, often deflexed. Flower heads radiate, solitary, yellow with blue rays. Jan.–Dec. Rocky flats and slopes, NW, SW, AP, KM, LB, SE (Cedarberg Mts to KwaZulu-Natal).

amelloides (L.) Voss Scabrid to almost glabrous, softly woody shrub to 1 m. Leaves ovate, spreading, 3-veined from base. Flower heads radiate, solitary on elongate peduncles, yellow with blue rays. Mainly Oct.–Feb. Coastal bush, AP, LB, SE (Stilbaai to E Cape).

amoena (Sch.Bip.) Levyns Softly hairy annual, 10–25 cm. Leaves oblanceolate, soft. Flower heads radiate, solitary, yellow with blue or white rays. Mainly June–Oct. Stony slopes, NW, SW, AP, KM, LB, SE (Cedarberg Mts to E Cape).

annectens (Harv.) Grau Annual similar to **F. bergerana** but floral bracts almost glabrous. Flowering time? Sandy soils, SW (Hopefield to Riviersonderend).• [Probably a depauperate form of **F. bergerana**.]

bellidioïdes Schltr. Softly hairy perennial to 25 cm. Leaves spathulate to narrowly oblanceolate, soft, in a basal tuft. Flower heads radiate, solitary on elongate peduncles, yellow with mauve or blue rays. Aug.–Jan. Rocks on S-facing slopes at high elevations, NW, SW, KM (Cedarberg Mts to Oudtshoorn).•

bergerana (Spreng.) O.Hoffm. Slender, hairy annual to 20 cm. Leaves obovate to oblanceolate. Flower heads radiate, solitary on elongate pedicels, yellow with blue or sometimes white rays, ray florets sometimes without pappus. Aug.–Oct. Rocky lower slopes and flats, NW, SW (Bokkeveld Mts to Botrivier).•

cymbalarioides (DC.) Grau Softly hairy, softly woody perennial to 25 cm. Leaves broadly oblanceolate, toothed, soft, in tufts at branch tips. Flower heads radiate, solitary on elongate peduncles, yellow with blue rays. Oct.–Nov. Sandstone slopes in shade of rocks at high alt., NW, SW (Cape Peninsula to Langeberg Mts: Robertson).•

diffusa (DC.) Grau Thinly hairy, soft-stemmed shrublet to 15 cm. Leaves oblong, margins revolute. Flower heads radiate, solitary on slender peduncles, yellow with blue rays, ray florets without pappus. Oct. Shady upper slopes, NW, SW (Kamiesberg to Cedarberg Mts, and Franschhoek).

ebracteata Grau Roughly hairy, closely leafy shrublet to 40 cm. Leaves oblong to lanceolate, suberect, margins revolute and spiny. Flower heads radiate, solitary, clustered at branch tips, yellow with dark blue rays. Sept.–Oct. Limestone hills, AP (De Hoop).•

elongata (Thunb.) O.Hoffm. Robust, coarsely hairy annual to 30 cm. Leaves mostly clustered at base, lanceolate, margins revolute. Flower heads radiate, solitary on elongate peduncles, yellow with white rays maroon at the base. Aug.–Oct. Coastal sands, SW (Saldanha Bay).•

heterophylla (Cass.) Grau (= *Chareis heterophylla* Cass.) BLOUBLOMASTERTJIE Roughly hairy annual with erect branches to 35 cm. Leaves oblanceolate. Flower heads radiate, solitary on elongate peduncles, blue, rarely yellow, with blue rays, ray florets without pappus. Aug.–Oct. Sandy flats and slopes, NW, SW (Clanwilliam to Cape Peninsula).•

hispida (DC.) Grau Diffuse, villous annual to 35 cm. Leaves narrowly lanceolate, margins revolute, sometimes slightly toothed. Flower heads radiate, solitary, yellow with white rays. Sept.–Dec. Rocky slopes and plateaus, NW (Cedarberg Mts to Ceres).•

joubertinae Grau   Roughly hairy, straggling shrublet to 40 cm. Leaves linear-oblong, margins revolute. Flower heads radiate, solitary, yellow with blue or white rays. Oct.–Jan. Rocky slopes often near streams, SE (Knysna to Humansdorp).•

linifolia (Harv.) Grau   Sparsely hairy shrublet to 45 cm. Leaves linear-oblong, almost glabrous, margins revolute. Flower heads radiate, solitary, yellow with blue rays. Mainly May–Oct. Rocky slopes often in damp places, KM, SE (Ladismith to Uitenhage).•

merxmuelleri Grau   Softly hairy annual to 25 cm. Leaves oblanceolate. Flower heads radiate, solitary on elongate peduncles, yellow with blue rays, ray florets without pappus. Aug.–Sept. Rocky lower slopes and plateaus, NW (Namaqualand to Ceres and W Karoo).

minima (Hutch.) Grau   Softly hairy, spreading annual, 5–20 cm. Leaves oblanceolate, obtuse. Flower heads radiate, solitary, yellow with pale blue or cream-coloured rays, ray florets without pappus. Sept.–Jan. Rocky slopes, NW, SW, KM, LB (Clanwilliam to Riversdale, Swartberg Mts and W Karoo).

namaquana (Harv.) Merxm.   Glandular-hairy, often robust annual to 30 cm. Leaves narrowly oblanceolate. Flower heads radiate, solitary on elongate peduncles, yellow with blue rays. May–Oct. Sandy flats, NW (Namibia to Worcester and W Karoo).

ovata (Thunb.) Compton   Sparsely hairy, diffuse shrublet to 60 cm. Leaves narrowly ovate, with long white hairs, tips acute and recurved. Flower heads radiate, solitary, yellow with blue rays. May–Nov. Karroid mountain slopes, NW, KM, LB, SE (Bokkeveld Escarpment to Port Elizabeth and Free State).

puberula Grau   Robust, scabrid annual to 30 cm. Leaves narrowly oblanceolate. Flower heads radiate, solitary on elongate peduncles, yellow with blue rays. July–Sept. Shale and sandy slopes, NW (S Namaqualand to Piketberg and W Karoo).

stenophylla Grau   Roughly hairy, straggling shrublet to 60 cm. Leaves linear-oblong, margins revolute. Flower heads radiate, solitary, yellow with blue to mauve rays. Oct. Mostly coastal sands, NW, SW (Namaqualand to Saldanha).

tsitsikamae Grau   Softly villous, low shrublet to 30 cm. Leaves ovate, deflexed, margins revolute. Flower heads radiate, solitary, blue. Jan.–May. Rocky slopes at high alt., SE (Tsitsikamma Mts).•

A. Involucral bracts in 3 rows
B. Annual or perennial herbs

australis (Alston) E.Phillips   SAMBREELASTERTJIE   Thinly hairy, sprawling annual, 5–25 cm. Leaves linear, sometimes slightly toothed, ciliate on margins. Flower heads radiate, solitary, yellow with blue to mauve rays. Aug.–Sept. Sand or clay flats, NW, SW (Namaqualand and W Karoo to Franschhoek).

cymbalariae (Aiton) Adamson & T.M.Salter   Diffuse or sprawling, softly hairy perennial, 10–30 cm. Leaves long-petiolate, broad and deeply toothed. Flower heads radiate, solitary, white, yellow or purple with white to mauve rays. Sept.–Feb. Sandstone slopes in shade of rocks, NW, SW (Cedarberg to Riviersonderend Mts).•

denticulata Grau   Villous perennial to 40 cm. Leaves narrowly ovate, soft, often 3-veined from base. Flower heads radiate, solitary, blue. Aug.–Sept. S-facing shale slopes, NW, KM, LB (Worcester to Bonnievale and Little Karoo).•

dubia Cass.   Diffuse, villous annual, 5–40 cm. Leaves oblanceolate, sparsely toothed. Flower heads radiate, solitary, yellow with blue rays. July–Oct. Sand or gravel flats and slopes, NW, SW (Namaqualand to Cape Peninsula and W Karoo).

fascicularis DC.   Scabrid, tufted, fine-leaved perennial to 30 cm. Leaves filiform, scabrid, often in tufts. Flower heads radiate, solitary, yellow with pale mauve rays. Feb.–June. Dry slopes, KM, SE (Montagu to N Province).

microsperma DC.   Roughly hairy annual to 50 cm. Leaves linear. Flower heads radiate, solitary, yellow with blue rays. Sept.–Jan. Sandy soil near water, NW (Namibia to Clanwilliam).

nigrescens Grau   Dwarf perennial to 6 cm. Leaves lanceolate, scabrid on margins. Flower heads radiate, solitary, yellow with blue rays. Apr. SW (Caledon).•

serrata (Thunb.) Grau   Slender, thinly hairy annual, 10–40 cm. Leaves obovate to elliptic, larger below, finely serrate. Flower heads radiate, solitary, yellow with blue rays. Dec.–Feb. Damp rocky slopes at high alt., NW (Cold Bokkeveld to Saron, ?Piketberg).•

tenella (L.) Nees   Thinly hairy annual, 5–25 cm. Leaves linear, scabrid on margins. Flower heads radiate, solitary, yellow with blue, violet or white rays. Mainly Aug.–Nov. Near water or coastal dunes, NW, SW, AP, LB (Bokkeveld Mts to Albertinia).•

sp. 1   Roughly hairy annual branching from base, to 15 cm. Leaves oblanceolate. Flower heads radiate, solitary on long peduncles, purple-blue with cream-coloured rays, ray florets without pappus. Sept.–Oct. Sandy lower slopes, NW (Graafwater to Leipoldtville).•

B.Woody shrublets

cana DC.   White-woolly, dwarf shrublet to 15 cm. Leaves narrowly oblanceolate, obtuse, clasping the stem. Flower heads radiate, solitary, yellow with mauve rays. Apr.–May. Sandstone slopes, KM, LB (Montagu to Riversdale).•

canaliculata Grau   Dwarf, roughly hairy shrublet to 20 cm. Leaves linear, canaliculate, ciliate on angles, imbricate. Flower heads radiate, solitary, yellow with blue to mauve rays. Sept. Coastal fynbos, AP (Bredasdorp to Struisbaai).•

comptonii Grau   Softly hairy shrublet to 30 cm. Leaves narrowly oblong, scabrid on margins, spreading. Flower heads radiate, solitary, yellow with blue rays. Sept.–Oct. Rocky slopes, LB (Langeberg Mts).•

dregei DC.   Sprawling, puberulous shrub to over 1 m. Leaves oblanceolate, shortly toothed, gland-dotted. Flower heads radiate, solitary, yellow with mauve rays, fragrant. July–Sept. Sandy slopes and flats, NW, SW (Namaqualand to Saldanha and W Karoo).

echinata (Thunb.) Nees   Rigid, densely leafy shrub to 1 m with white-woolly stems. Leaves ovate, glossy, margins scabrid, closely imbricate and reflexed. Flower heads radiate, solitary, crowded at ends of branches, yellow with mauve rays. Apr.–Oct. Coastal bush, LB, SE (Mossel Bay to Port Alfred).

erigeroides DC.   Sprawling shrub to 1.2 m with pale stems. Leaves obovate, hairy, in tufts, usually 3-veined. Flower heads radiate, 1–few at branch tips, yellow with blue rays. Apr.–May. Coastal bush, SE (Humansdorp to KwaZulu-Natal).

esterhuyseniae Grau   Roughly hairy shrublet to 25 cm. Leaves small, ascending and imbricate, subterete, scabrid, obtuse. Flower heads radiate, solitary, yellow with blue, white or mauve rays. Mar.–May. High sandstone slopes, KM (Kammanassie Mts).•

ferulacea Compton   Coarsely hairy, straggling shrub to 2 m. Leaves obovate, scabrid, recurved at tips. Flower heads radiate, solitary, yellow with mauve to blue or white rays. Flowering time? Damp, often marshy slopes, NW, SW, KM (Pakhuis Mts and Perdeberg to Swartberg Mts).•

filifolia (Vent.) Burtt Davy   DRAAIBOSSIE   Like **F. fruticosa** but the leaves filiform and needle-like. Mainly Oct.–Dec. Flats and slopes, NW, SW, KM, LB, SE (Namibia to Port Elizabeth to Mpumalanga).

fruticosa (L.) G.Nichols.   WILD ASTER, BOSASTERTJIE   Much-branched shrub to 1 m. Leaves oblanceolate, small, fleshy, in tufts, slightly gland-dotted. Flower heads radiate, solitary, yellow with blue to mauve rays. Sept.–Nov. Rocky lower slopes, SW (Cape Peninsula to Somerset West and N Province).

hirsuta DC.   Rounded, roughly hairy shrublet to 50 cm. Leaves narrowly linear, grey-hairy, often in tufts. Flower heads radiate, solitary, yellow with blue to mauve rays. Mainly Mar.–Oct. Dry flats and slopes, NW, KM, LB (Namibia to Swellendam to Free State).

hirta (Thunb.) Grau   KAROOASTERTJIE   Rounded, shortly hairy shrublet to 35 cm. Leaves narrowly linear-oblong. Flower heads radiate, solitary, yellow with blue rays. July–Oct. Lower slopes, NW (Namaqualand to Robertson and Karoo).

hyssopifolia (P.J.Bergius) Nees   Thinly hairy to glabrous shrublet to 60 cm. Leaves narrowly linear-oblong, acute. Flower heads radiate, solitary, yellow with blue, mauve, pink or white rays. Apr.–Jan. Sandy flats or lower slopes, NW, SW, AP, LB, SE (Kamiesberg to Port Elizabeth and Lesotho).

macrorrhiza (Thunb.) DC.   ASPOESTERTJIE   Roughly hairy, gnarled dwarf shrublet to 15 cm. Leaves linear to oblanceolate, recurved at tips, canescent. Flower heads discoid, solitary, large, yellow or orange. July–Sept. Rocky slopes, KM (Bonteberg to Witteberg and W Karoo).

microcephala Grau   Similar to **F. erigeroides** but leaves distinctly petiolate, always 1-veined. Sept.–Jan. Rocky ravines, SE (Outeniqua Mts to Humansdorp).•

muricata (Thunb.) Nees   TAAI-ASTERTJIE   Almost glabrous shrublet to 70 cm. Leaves linear, ascending. Flower heads yellow with lilac or white rays. Nov.–Apr. Flats and mountains, KM, LB, SE (Oudtshoorn and Swellendam to tropical Africa).

nordenstamii Grau   Stiffly erect, slender shrub to 60 cm. Leaves elliptic, softly hairy and scabrid below, imbricate. Flower heads radiate, solitary, nested in the leaves, large, yellow with mauve rays. Sept.–Oct. Coastal limestone hills, AP (Agulhas to Potberg).•

oleosa Grau   Shrublet to 25 cm. Leaves small, ascending and imbricate, subterete, glabrous, gland-dotted, acute, margins scabridulous. Flower heads radiate, solitary, large, yellow with blue to mauve rays. Oct. S-facing slopes below cliffs, KM (Witteberg and Swartberg Mts).•

rogersii S.Moore   Low, thinly hairy, rigid shrub to 30 cm. Leaves small, subterete, scabrid. Flower heads radiate, solitary, yellow with lilac or violet rays, fragrant. Aug.–Sept. Stony lower slopes, NW, KM (S Namaqualand and W Karoo to Hex River Valley).

scabrida (DC.) Range   Diffuse, roughly hairy shrub to over 1 m. Leaves cuneate, toothed above. Flower heads radiate, solitary, yellow with blue to violet rays. July–Sept. Rocky lower slopes, NW (Lokenberg to Karoopoort).•

venusta S.Moore   Low broom-like shrub to 40 cm. Leaves ericoid, clasping stem, scabrid on margins. Flower heads radiate, solitary, yellow with blue to mauve rays. Aug.–Oct. Rocky slopes, NW, SW, KM (Clanwilliam and Wemmershoek to Montagu).•

westae (Fourc.) Grau   Sprawling, sparsely hairy shrublet to 40 cm. Leaves linear, erect, imbricate, scabrid on margins. Flower heads radiate, solitary, nested in the leaves, yellow with white to mauve rays. Sept.–Feb. Wet places, often streambanks, SE (Knysna to Humansdorp).•

zeyheri (Less.) Nees   Roughly hairy to almost glabrous shrublet to 30 cm. Leaves linear to oblanceolate. Flower heads radiate, solitary, yellow with blue or purple rays. Mainly Oct.–Apr. Rocky slopes, KM, SE (Oudtshoorn to E Cape).

FOVEOLINA   WILDEKAMILLE   5 spp., western southern Africa

albidiformis (Thell.) Kallersjö (= *Pentzia membranacea* Hutch.)   Sprawling annual to 15 cm. Leaves bipinnatisect. Flower heads discoid, solitary on short peduncles, pale yellow, bracts with membranous margins. Sept.–Oct. Dry sandstone valleys, NW (N Cedarberg Mts to Karoo).

tenella (DC.) Kallersjö (= *Matricaria tenella* DC.)   LAZY DAISY   Sprawling, thinly hairy, aromatic annual to 25 cm. Leaves bipinnatisect. Flower heads radiate, solitary on long naked peduncles, yellow with white rays. June–Sept. Sandy slopes and flats, mostly coastal, NW, SW (S Namaqualand to Yzerfontein).

GARULEUM   8 spp., South Africa and Namibia

bipinnatum (Thunb.) Less.   Twiggy, scabrid shrublet to 30 cm. Leaves pinnatisect, segments filiform. Flower heads radiate, solitary or few in lax corymbs, yellow with white or blue rays. Achenes trigonous, warty. Mainly Sept.–Oct. Dry stony slopes, NW, KM (Bokkeveld Mts and Witteberg to E Cape and Karoo).

GAZANIA   GAZANIA   16 spp., southern and tropical Africa

A. Annuals

lichtensteinii Less.   GEELGAZANIA, KOUGOED   Annual sometimes to 30 cm. Leaves oblanceolate, serrate or toothed, woolly below, glabrescent above, margins slightly revolute. Flower heads radiate, yellow, sometimes orange; involucre smooth, collared below, inner bracts acuminate. Mainly Aug.–Sept. Gravel and sandy flats, KM (Namibia and W Karoo to Little Karoo).

pectinata (Thunb.) Spreng.   Tufted acaulescent annual to 20 cm. Leaves mostly pinnatisect, some linear-oblanceolate, leaflets linear to elliptic, white-felted below, sometimes rough above, margins revolute. Flower heads radiate, yellow or orange with dark ring; inner bracts attenuate. Aug.–Nov. Coastal flats and lower slopes, SW, AP (Saldanha to Potberg).•

tenuifolia Less.   Tufted annual to 15 cm. Leaves pinnatisect, lobes linear, white-felted below, glabrescent above, margins revolute. Flower heads radiate, small, yellow or orange with dark ring; involucre flat at base, inner bracts attenuate. Aug.–Sept. Dry flats or lower slopes, NW (Namibia to Klawer and W Karoo).

A. Perennials

ciliaris DC.   Tufted acaulescent perennial to 15 cm. Leaves linear-oblanceolate or pinnatisect, white-felted below, margins revolute, petioles white-bristly. Flower heads radiate, orange with dark ring; involucre truncate and collared below, inner bracts attenuate. July–Sept. Rocky slopes, NW, SW, LB, SE (Bokkeveld Mts to Langkloof).•

krebsiana Less.   ROOIGAZANIA   Tufted acaulescent perennial to 20 cm. Leaves linear-oblanceolate or pinnatisect, white-felted below, margins revolute. Flower heads radiate, yellow to orange; inner bracts acute. Mainly Sept.–Jan. Roadsides, flats or lower slopes, NW, SW, AP, KM, LB, SE (throughout southern Africa to Tanzania).

linearis (Thunb.) Druce   Like **G. pectinata** but perennial from a woody rootstock, decayed leaf bases accumulating as a fibrous sheath. Mainly Sept.–Dec. Grassy slopes, SE (Humansdorp to KwaZulu-Natal).

maritima Levyns   Creeping perennial rooting along rhizome, to 10 cm. Leaves pinnatisect, leaflets elliptic, thick, white-felted below, margins revolute, petioles often bristly. Flower heads radiate, yellow to orange with dark ring; inner bracts acute. Nov.–Feb. Coastal rocks and sands, NW, SW (Elandsbaai to Cape Hangklip).•

othonnites (Thunb.) Less.   Tufted, glabrous mat-forming perennial to 10 cm. Leaves oblanceolate or pinnatifid, usually glaucous, margins roughly ciliate. Flower heads radiate, yellow or orange; involucre collared below. Oct.–Dec. Sandy and shale slopes, NW (Namaqualand and W Karoo to Bokkeveld Plateau).

rigens (L.) Gaertn.   STRANDGAZANIA   Sprawling, mat-forming perennial to 20 cm. Leaves mostly simple, oblanceolate, sometimes pinnatisect, white-felted below, margins revolute, petioles smooth. Flower heads radiate, yellow; involucral bracts partly white-woolly, inner bracts acute. Mainly Oct.–Feb. Coastal dunes and sandy flats, SE (George to S Mozambique).

rigida (Burm.f.) Roessler   KAROOGAZANIA   Tufted acaulescent perennial to 25 cm. Leaves pinnatisect, lobes elliptic, sometimes linear-oblanceolate, white-woolly below, margins revolute. Flower heads radiate, yellow or orange, usually with dark ring, involucre roughly hairy; inner bracts acute. July–Nov. Flats and lower slopes, NW, SW, KM, ?LB, ?SE (S Namaqualand and W Karoo to Caledon, ?Riversdale to Humansdorp).

serrata DC.   Like **G. rigida** but leaves often linear-dentate, terminal leaf lobe largest and inner bracts attenuate. Aug.–Oct. Sandy and stony slopes and flats, NW, SW (Gifberg to Paarl).•

sp. 1   Tufted, cushion-forming perennial to 10 cm. Leaves pinnatisect, lobes broadly obovoid, white-woolly above and white-felted below. Flower heads radiate, orange with dark ring; involucre cobwebby, bracteate to base. Aug.–Oct. Dolerite ridges, NW (Robertson).•

GERBERA   AFRICAN DAISY   c. 30 spp., Africa, tropical Asia, 1 sp. South America

A. Scapes ebracteate

ambigua (Cass.) Sch.Bip. (= *Gerbera viridifolia* (DC.) Sch.Bip.)   Tufted acaulescent perennial, scapes to 35 cm. Leaves elliptic, petiolate, glabrescent or silky above, grey-felted below. Flower heads radiate, scapose, white to yellow with pink to copper reverse. Oct.–Feb. Grassland, SE (George to tropical Africa).

cordata (Thunb.) Less.   TONGBLAARBLOM   Like **G. piloselloides** but leaves elliptic, cordate at base, petiolate, white-felted below. Flower heads white. Jan.–Dec. Coastal forest and bush, SE (George to East London).

piloselloides (L.) Cass. (= *Piloselloides hirsuta* (Forssk.) C.Jeffrey)   SWARTTEE   Acaulescent perennial with scapes to 30 cm. Leaves tufted, obovate, tapering below, softly hairy or cobwebby. Flower heads radiate, white, pink, red, or yellow; bracts hairy. July–Feb. Sandstone and limestone slopes, SW, AP, LB, SE (Cape Peninsula to palaeotropics).

A. Scapes bracteate

crocea (L.) Kuntze (incl. **G. integralis** Sond. ex Harv.)   DIALSATEE   Acaulescent perennial to 40 cm. Leaves tufted, oblanceolate to elliptic, petiolate, thinly cobwebby below, margins slightly toothed. Flower heads radiate, scapose, white or pink, maroon on reverse; bracts thinly cobwebby. Sept.–May. Sandstone slopes after fire, NW, SW, ?AP, KM (Olifants River Mts to Bredasdorp and Swartberg Mts).•

linnaei Cass.   VARINGBLOM   Acaulescent perennial to 40 cm. Leaves tufted, oblanceolate-pinnatisect, lobes round and twisted, yellow-felted below, margins revolute. Flower heads radiate, scapose, cream-coloured, sometimes yellow, maroon on reverse; bracts glabrous. Oct.–Jan. Sandstone slopes after fire, SW (Cape Peninsula to Riviersonderend).•

serrata (Thunb.) Druce   Like **G. tomentosa** but bracts distinctly biseriate with inner acuminate. Aug.–Nov. Sandstone slopes in fynbos after fire, SW, LB, SE (Riviersonderend to Humansdorp).•

tomentosa DC.   TONTELBLAARBOSSIE   Tufted acaulescent perennial to 30 cm. Leaves elliptic, long-petiolate, yellow-felted below, margins retrorse-toothed and revolute. Flower heads radiate, scapose, white, maroon on reverse; bracts softly felted. Mainly Oct.–Nov. Sandstone slopes after fire, SW, KM, LB, SE (Cape Peninsula to Great Winterhoek Mts).•

wrightii Harv.   Like **G. tomentosa** but scapes to 60 cm, thinly cobwebby, and leaves softy white-felted below. Sept.–Jan. Rocky lower slopes, SW (Cape Peninsula).•

GIBBARIA• 2 spp., W and E Cape

ilicifolia (L.) Norl.   Scabrid, sprawling, densely leafy, aromatic shrub to 1 m, sometimes thinly woolly on young parts. Leaves lanceolate to ovate, margins revolute, usually sharply toothed. Flower heads radiate, solitary at branch tips, yellow. Achenes reniform, warty, c. 4.5 mm long. Aug.–Mar. Sandstone or limestone slopes, NW, SW, AP (Cape Peninsula and Ceres to Cape Infanta).•

scabra (Thunb.) Norl.   Roughly hairy, willowy shrublet to 20 cm. Leaves filiform, spreading. Flower heads radiate, solitary or few at branch tips on short wiry peduncles, yellow or white with orange reverse. Achenes reniform, smooth, c. 3 mm long. Aug.–Sept. Riverine bush and forest margins, SE (Knysna to Uitenhage).•

GNAPHALIUM   c. 50 spp., cosmopolitan

capense Hilliard   Erect or sprawling, densely leafy, white-woolly perennial to 40 cm. Leaves narrowly oblanceolate, ascending. Flower heads discoid, in small terminal corymbs, outer bracts brown and woolly, inner white-tipped. Mainly Sept.–Jan. Damp flats, SW, LB, SE (Velddrif to E Cape).

declinatum L.f.   Mat-forming, silky silvery perennial, with closely leafy stems. Leaves oblong, imbricate. Flower heads discoid, in corymbs at branch tips, bracts brown, glabrous, the inner white-tipped. Oct.–Feb. Damp flats or lower slopes, SW, AP, LB (Grabouw to Riversdale).•

gnaphalodes (DC.) Hilliard & B.L.Burtt   Nearly prostrate, stoloniferous, thinly grey-woolly perennial to 30 cm. Leaves oblong to spathulate, undulate, ascending to adpressed below, recurved above. Flower heads discoid, small, in terminal clusters, bracts white and brown. Sept.–May. Marshy areas, SE (Uitenhage to Transkei).

pauciflorum DC.   Small, erect or spreading, white-woolly annual to 10 cm. Leaves linear. Flower heads discoid, 1–few in small terminal clusters, bracts straw-coloured or golden brown tinged with purple. Sept.–Oct. Flats and slopes, NW, SW, LB (Piketberg to Riversdale).•

vestitum Thunb.   Similar to **G. capense** but inner bracts buff or dirty white. Mainly Nov.–Apr. Damp grassland, SE (Plettenberg Bay to E Cape).

GORTERIA   BEETLE DAISY   3 spp., S Cape to Namibia

diffusa Thunb.   BEETLE DAISY   Like **G. personata** but bracts acuminate, hairy to tips, mostly shorter than rays and ray florets orange with beetle-like markings. July–Oct. Clay flats or rocky lower slopes, NW, SW, ?AP, KM, ?LB (Namaqualand to Swellendam, W Karoo).

personata L.   KLITSKRUID   Roughly hairy, sprawling annual to 10 cm. Leaves oblanceolate, sometimes pinnatifid, margins revolute, roughly hairy above, white-felted below. Flower heads radiate, solitary at branch tips, yellow with greenish reverse; bracts glabrous above, pungent, longer than rays. Aug.–Oct. Rocky or sandy flats and lower slopes, NW, SW, AP, LB, SE (Bokkeveld Mts to Humansdorp).•

GYMNODISCUS   GEELKRUID   2 spp., Namaqualand to W Cape

capillaris (L.f.) DC.   Tufted, succulent annual to 20 cm. Leaves rosulate, oblanceolate to lyrate, usually lobed below. Flower heads radiate, small, in small corymbs on branched scapes, yellow. July–Oct. Sandy flats and lower slopes, NW, SW, AP, LB (Namaqualand to Mossel Bay and W Karoo).

GYMNOSTEPHIUM•   KAAPSE ASTERTJIE   7 spp., W Cape

angustifolium Harv.   Twiggy shrublet to 45 cm. Leaves linear-filiform, keeled, margins ciliate. Flower heads radiate, solitary on glandular-hairy peduncles, yellow with blue rays; bracts glabrous with ciliate margins. Apr. Sandstone slopes, SW (Riviersonderend Mts).•

ciliare (DC.) Harv.   Roughly hairy shrublet to 45 cm. Leaves linear-lanceolate, keeled, roughly hairy. Flower heads radiate, solitary on glandular-hairy peduncles, blue or mauve; bracts roughly hairy. Oct.–Jan. Sandstone slopes, SW (Houwhoek to Riviersonderend).•

corymbosum (Turcz.) Harv.   Erect, closely leafy, willowy shrub to 1.5 m. Leaves lanceolate-attenuate, overlapping, often deflexed, glossy, roughly scabrid on margins. Flower heads radiate, solitary on glandular-hairy peduncles, together forming loose corymbs, yellow with white to mauve rays; bracts scabrid. July–Sept. Wet habitats, often along streams, SW (Riviersonderend Mts).•

fruticosum DC. (incl. **G. gracile** Less)   Sprawling shrublet to 40 cm. Leaves oblong, often deflexed, margins revolute. Flower heads radiate, solitary on thinly silky peduncles, yellow with blue rays; bracts glabrous with ciliate margins. June–Sept. Steep grassy slopes, LB (Langeberg Mts: Swellendam).•

hirsutum Less.   Scabrid shrublet to 40 cm. Leaves linear, scabrid. Flower heads radiate, solitary on glandular-hairy peduncles, yellow with blue rays; bracts scabrid. Oct.–Jan. Steep S-facing mountain slopes, SW (Genadendal).•

laeve Bolus   Slender, erect, glabrous perennial with rod-like stems to 40 cm. Leaves erect, linear, margins revolute. Flower heads radiate, solitary on glabrous peduncles, yellow with mauve rays; bracts glabrous. Dec.–Apr. Damp mountain slopes, NW (Cold Bokkeveld).•

papposum Nesom (= *Mairia corymbosa* Harv.)   Very like **G. corymbosum** but involucral bracts glabrous with ciliate margins and pappus bristles conspicuously plumose not barbellate. June–Aug. Lower mountain slopes near streams, SW (Riviersonderend and Kleinrivier Mts).•

HAPLOCARPHA   BASTER GOUSBLOM   8 spp., tropical and southern Africa

lanata (Thunb.) Less.   BRANDBLOM   Tufted, densely white-woolly perennial to 15 cm. Leaves oblanceolate to elliptic, coarsely toothed, densely white-felted below, scabrid above. Flower heads radiate, solitary, scapose, yellow with red reverse. Achenes silky throughout, at base tufted with hairs longer than achene, and pappus exceeding achene. Mar.–June. Sandstone slopes, mainly after fire, SW, LB (Cape Peninsula to Swellendam).•

lyrata Harv.   Tufted, densely white-woolly perennial to 20 cm. Leaves lyrate-pinnatisect, discolorous, glabrous above. Flower heads radiate, solitary, scapose, yellow. Achenes silky throughout, at base tufted with hairs longer than achene, and pappus exceeding achene. Mainly Nov.–Jan. Grassland, SE (Langkloof to E Cape).

nervosa (Thunb.) Beauv.   Tufted, white-felted perennial to 15 cm. Leaves elliptic, shortly petiolate, with recurved teeth, discolorous, white-felted below, sometimes roughly hairy above. Flower heads radiate, solitary, on short, densely hairy scapes, yellow with greenish reverse. Achenes glabrous or ciliate, with a basal circlet of delicate hairs, sometimes almost wanting, pappus rudimentary to 1.5 mm. Sept.–Oct. Marshy grassland, SE (George to Zimbabwe).

oocephala (DC.) Beyers (= *Arctotis oocephala* DC.)   Creeping perennial rooting at nodes, to 20 cm. Leaves oblanceolate to elliptic, sinuate toothed, white-felted below, roughly hairy above, Flower heads radiate, solitary, yellow with red reverse, bracts without tails. Achenes glabrous, at base tufted with hairs longer than achene, pappus exceeding achene. Oct.–Dec. Rocky sandstone slopes, NW (Pakhuis and N Cedarberg Mts).•

parvifolia (Schltr.) Beauv.   Like **H. lanata** but leaves elliptic, petiolate, deeply sinuate, glabrescent above and achenes glabrous, at base tufted with hairs longer than achene, and pappus longer than achene. Sept.–Oct. Sandstone slopes, NW (Cold Bokkeveld).•

HELICHRYSUM   STROOIBLOM   c. 500 spp., Old World, mainly Africa

A. Heads depressed-globose, large (florets more than 300)

aureum (Houtt.) Merr.   Tufted, white-woolly perennial to 80 cm from a woody rootstock. Leaves mostly radical, oblaceolate, clasping, roughly hairy and more or less grey-woolly. Flower heads disciform, 1–6 at branch tips, large, depressed-globose, yellow, sometimes white, 15–25 mm diam., florets c. 300–1860, ovary glabrous. July–Dec. Grassland, SE (Humansdorp to Gauteng).

foetidum (L.) Moench   Robust, foetid, glandular-hairy biennial to c. 1 m. Leaves oblong-lanceolate, auriculate and clasping, roughly hairy above, grey-woolly beneath. Flower heads disciform, in leafy corymbs, depressed-globose, yellow to cream-coloured, 15–25 mm diam., florets c. 400–800, ovary glabrous. Oct.–May. Damp rocky slopes, NW, SW, KM, LB, SE (Cedarberg Mts to E Cape).

A. Heads subglobose to cylindric
B. Annuals

herniarioides DC.   Like **H. litorale** but flower heads campanulate, white sometimes pinkish, c. 4–5 × 4–6 mm, involucral bracts obtuse, abruptly spreading above. Mainly July–Sept. Sandy and stony flats, NW (Namibia and Karoo to Cold Bokkeveld).

indicum (L.) Grierson   Erect or sprawling grey-woolly annual to 30 cm, with a woolly taproot. Leaves oblanceolate-spathulate. Flower heads discoid sometimes disciform, crowded in terminal corymbs, cylindric-campanulate, white to cream-coloured, c. 3 × 2 mm, involucral bracts spreading above, florets 15–25, ovary papillate-hairy sometimes glabrous. Oct.–Feb. Sandy flats and slopes, often weedy, NW, SW (Lambert's Bay to Riviersonderend).•

leontonyx DC.   Mat-forming grey-woolly annual like **H. litorale** but heads smaller, c. 3–4 × 2.0–2.5 mm, involucral bracts acuminate and hooked, and pappus uniseriate. July–Oct. Sandy, often disturbed flats, NW, KM, SE (S Namibia and W Karoo to Little Karoo and Uniondale).

litorale Bolus   Prostrate or diffuse white-woolly annual or short-lived perennial with branches to 45 cm long. Leaves obovate-spathulate, woolly. Flower heads discoid or disciform, solitary or in glomerules enclosed by leafy bracts, cylindric, golden sometimes red, c. 5 × 2.5 mm, inner involucral bracts caducous, florets 10–30, ovary glabrous or papillate-hairy. Sept.–Dec. Coastal sands, SW, AP, SE (Saldanha to E Cape).

marmarolepis S.Moore   Like **H. indicum** but flower heads campanulate and in tightly crowded clusters. Mainly Sept.–Oct. Coastal sands, NW, SW (Namaqualand to Paternoster).•

micropoides DC.   Like **H. litorale** but heads or glomerules crowded into more elongate, racemose compound inflorescences and involucral bracts acuminate and hooked. July–Nov. Sandy flats, NW (S Namibia and W Karoo to Ceres).

moeserianum Thell.   Grey-woolly to cobwebby annual to 30 cm, branching from base. Leaves oblanceolate, woolly. Flower heads discoid or disciform, many in terminal corymbs, campanulate, yellow, 3–4 × 2–3 mm, bracts spreading above, florets 12–30, ovary papillate-hairy, sometimes glabrous. Aug.–Feb. Sandy flats and slopes, NW, SW, AP (Bokkeveld Mts to Mossel Bay).

solitarium Hilliard   Like **H. litorale** but more delicate, leaves smaller, to 7 mm long and heads smaller, c. 4 × 3 mm and always solitary. Dec. Sandy flats, NW (Bokkeveld Tafelberg).•

tinctum (Thunb.) Hilliard & B.L.Burtt   Cushion- or mat-forming annual like **H. litorale** but involucral bracts acuminate and hooked. Mainly Sept.–Dec. Sandy flats, NW, SW, AP, KM, LB, SE (Namaqualand and S Karoo to Uniondale).

versicolor O.Hoffm.& Muschl.   Like **H. moeserianum** but leaves more oblong and stereome on the outermost but one series of bracts longer, c. 2 mm long. Mainly Oct.–Nov. Coastal sands, SE (Knysna to E Cape).

zwartbergense Bolus   Like **H. litorale** but more densely branched and forming rounded cushions or mats. Mainly Dec.–Jan. Sandstone slopes, NW, SW, KM, SE (Cedarberg to Great Winterhoek Mts).•

B. Perennials
C. Stiffly erect, tufted perennials with unbranched stems;
leaves mostly basal, glabrescent above, several-veined from base

mundtii Harv.   Tufted subshrub with erect leafy stems to 1.5 m from a woody rootstock, felted above. Leaves mostly basal, oblanceolate-petiolate, decurrent, glabrous and rugose above, felted beneath, 3-net-veined from base, margins slightly revolute. Flower heads discoid, many in a large, branched corymb, campanulate, creamy white, c. 4 × 3.5 mm, florets 18–24, ovary glabrous. Feb.–Apr. Moist grassland, streamsides, and forest margins, LB, SE (Swellendam to tropical Africa).

nudifolium (L.) Less.   Subshrub with erect peduncle-like stem to 1.5 m from a woody rootstock, thinly woolly above. Leaves mostly radical, oblanceolate, glabrescent but margins and veins roughly hairy, 3–7-net-veined from base. Flower heads discoid, many in a branched corymb, campanulate, pale yellow or brownish, c. 4–5 × 3 mm, florets 15–30, ovary glabrous. Mainly Nov.–Mar. Grassland, NW, SW, KM, LB, SE (Piketberg to Middle East).

oxyphyllum DC.   Subshrub with erect peduncle-like stem to 60 cm from a woody rootstock, woolly above. Leaves mostly radical, ovate-petiolate, glabrescent above, felted beneath, 3(–5)-veined from base. Flower heads discoid, many in a compact corymb, campanulate, white and pink or red, c. 7 × 7 mm, florets 20–30, ovary glabrous. Aug.–Jan. Grassland, SW, LB, SE (Caledon to tropical Africa).

pedunculatum Hilliard & B.L.Burtt   Subshrub with erect peduncle-like stem to 50 cm from a woody rootstock, felted above. Leaves mostly radical, oblanceolate, glabrescent above, felted beneath, 5–9-net-veined from base. Flower heads discoid, many in a branched corymb, campanulate, brownish, c. 7 × 7 mm, florets 30–60, ovary glabrous. Mainly Aug.–Sept. Grassland, KM, LB, SE (Swartberg Mts and Riversdale to Lesotho).

platypterum DC. Subshrub with erect leafy stems to 1 m from a woody rootstock, thinly woolly above. Leaves mostly basal, elliptic-lanceolate, decurrent, glabrescent but roughly hairy on veins and margins, 3–5-veined from base. Flower heads discoid, many in a branched, flat-topped corymb, campanulate, silvery white, c. 5 × 5 mm, florets 30–55, ovary papillate-hairy. Mainly Feb.–Apr. Moist grassland and forest margins, SE (Tsitsikamma Mts to N Province).

> C. Plants various but if tufted then not stiffly erect and leaves not glabrescent above
> D. Flower heads large, 15–40 mm long, solitary, sometimes few; involucral bracts glossy

herbaceum (Andrews) Sweet Closely leafy, cobwebby shrublet or subshrub to 40 cm. Leaves imbricate, ascending, linear-lanceolate, cobwebby above, felted beneath, margins revolute. Flower heads solitary, sometimes few at branch tips, top-shaped, shiny golden or yellow, glabrous, c. 20 mm long, florets 100–210, ovary glabrous. Oct.–Apr. Grassland, SE (Langkloof to tropical Africa).

lancifolium (Thunb.) Thunb. Like **H. retortum** but heads on scaly peduncles distinct from the leafy shoots. Sept.–Jan. Sandstone slopes, NW, KM, SE (Bonteberg to Baviaanskloof Mts).•

retortum (L.) Willd. Straggling, closely leafy, silvery shrublet to 50 cm. Leaves imbricate, spreading or recurved above, oblong, folded, hooked, silvery silky with tissue paper-like hairs. Flower heads disciform, terminal, solitary and tips nested in leaves, top-shaped, shiny white often flushed pink and brown, glabrous, 25–40 × c. 40 mm, florets c. 100–250, ovary papillate-hairy. Aug.–Dec. Coastal sands and cliffs, SW, AP (Bloubergstrand to Stilbaai).•

stoloniferum (L.f.) Willd. Like **H. retortum** but leaves mostly broader, narrowly obovate and heads smaller, 15–20 mm long, on scaly peduncles distinct from leafy shoots. Nov.–Feb. Sandstone slopes, NW, SW (Kamiesberg to Riviersonderend Mts and Karoo).

> D. Flower heads smaller, 3–10 mm long
> E. Ericoid shrublets; leaves small, shorter than 10 mm, linear with strongly revolute margins;
> flower heads 1–4 at branch tips

asperum (Thunb.) Hilliard & B.L.Burtt Tangled, twiggy, thinly grey-woolly shrublet to 40 cm. Leaves often tufted on short shoots, linear, apiculate, woolly or glabrescent, margins strongly revolute. Flower heads discoid, 1–4 at branch tips, cylindric, brownish or white, c. 4 × 2 mm, florets c. 10, ovary papillate-hairy. Oct.–Jan. Stony slopes and flats, NW, SW, KM, LB, SE (Namibia to KwaZulu-Natal).

niveum (L.) Less. Twiggy, ericoid shrublet to 20 cm. Leaves small, spreading or reflexed, linear, glabrescent or woolly, margins strongly revolute. Flower heads discoid, in small terminal glomerules nested in leaves, cylindric, white sometimes pink, c. 4 × 1 mm, florets c. 5, ovary papillate-hairy. Mainly Dec.–Feb. Coastal sands, SW, AP (Saldanha to Stilbaai).•

> E. Plants not ericoid; leaves various, if linear and revolute then larger and flower heads numerous
> F. Closely leafy dwarf shrubs or shrublets; leaves ascending, flat;
> flower heads 1(–3) at branch tips

altigenum Schltr. & Moeser Tufted, silvery dwarf subshrub to 2 cm, stems decumbent, from a woody caudex. Leaves imbricate, obovate, some in basal rosettes, silvery with tissue paper-like hairs. Flower heads discoid or disciform, solitary, terminal and nested in leaves, campanulate, white or brownish, 8–10 × 16 mm, florets 40–60, ovary papillate-hairy. Nov.–Jan. Sandstone slopes, NW, KM, SE (Cold Bokkeveld to Great Winterhoek Mts).•

archeri Compton Like **H. saxicola** but leaves narrower, linear to oblong. Sept.–Nov. Quartzite patches and crevices, KM (Witteberg).•

saxicola Hilliard Gnarled, cushion-forming, closely leafy shrublet to 5 cm. Leaves small, imbricate, ascending, elliptic, grey-felted. Flower heads discoid, solitary, sessile at branch tips, campanulate, whitish, c. 5 × 4 mm, bracts spreading above, florets c. 25, ovary papillate-hairy. Dec.–Mar. Sandstone crevices, 1500–2000 m, KM (Swartberg Mts).•

simulans Harv. & Sond. Gnarled, closely leafy shrublet to 15 cm. Leaves small, imbricate, ascending, linear-spathulate, felted with tissue paper-like hairs. Flower heads discoid, 1–3, sessile at branch tips, cylindric, brownish, c. 5–6 × 2–3 mm, florets 10–15, ovary papillate-hairy. Nov.–Dec. Sandy and quartzitic flats, NW, KM (Vredendal to Little Karoo).•

F. Larger, mostly less densely leafy shrubs or perennials; flower heads (1–)few to many
G. Flower heads very small, cylindric, c. 3 × 2 mm, many in crowded corymbs

albanense Hilliard   Erect or sprawling, silvery felted perennial to 40 cm. Leaves elliptic, more or less hooked, silvery felted, hairs more or less skin-like. Flower heads discoid, crowded in terminal corymbs, narrowly campanulate, yellow, c. 3 × 2 mm, florets c. 10, ovary glabrous. July–Oct. Grassland, SE (Clarkson to E Cape).

cymosum (L.) D.Don.   Straggling, thinly woolly shrub to 1 m. Leaves linear to elliptic, thinly silky or hairs skin-like above, white-felted beneath, margins slightly revolute. Flower heads disciform, sometimes discoid, crowded in terminal corymbs, cylindric, yellow, c. 3 × 1 mm, florets 6–20, ovary glabrous. Sept.–Apr. Sandy slopes in damp places, SW, AP, KM, LB, SE (Mamre to Mpumalanga).

helianthemifolium (L.) D.Don   Like **H. tenuiculum** but bracts silvery to white. Dec.–Jan. Rocky slopes in damp places, NW, SW, KM, LB, SE (Cedarberg Mts to Uitenhage).•

odoratissimum (L.) Sweet   Straggling, aromatic, thinly white-woolly shrublet to 50 cm. Leaves linear to spathulate, clasping, decurrent, grey-woolly, undulate. Flower heads disciform, densely matted together in terminal cymes, narrowly campanulate, brown and yellow, c. 3 × 2 mm, florets c. 5–15, ovary papillate-hairy. Mainly Aug.–Dec. Rocky slopes, NW, SW, KM, LB, SE (Gifberg to tropical Africa).

rutilans (L.) D.Don   Stiffly branched, twiggy, grey-felted shrublet to 60 cm. Leaves linear to oblong, often folded, hooked, grey-felted. Flower heads disciform, sometimes discoid, crowded in terminal corymbs, cylindric, lemon-yellow, c. 3.5 × 2 mm, florets 3–8, ovary glabrous. All year. Rocky slopes, NW, SW, AP, KM, LB, SE (Mamre and Worcester to Free State).

simillimum DC.   Closely leafy, grey-felted shrublet to 40 cm. Leaves ascending, imbricate, lanceolate, revolute, grey-felted. Flower heads disciform, crowded in terminal corymbs, oblong-campanulate, pale yellow or whitish, c. 3.5 × 2.5 mm, florets c. 15–25, ovary glabrous. Mainly Jan.–Apr. Rocky slopes, especially forest margins, SE (George, E Cape to KwaZulu-Natal).

subglomeratum Less.   Tufted, silvery felted perennial to 60 cm from a woody rootstock. Leaves oblanceolate, silvery felted but hairs almost skin-like. Flower heads discoid, densely matted together in terminal cymes, narrowly cylindric, yellow, c. 3 × 2 mm, florets 4–13, ovary glabrous. Mar.–June. Rocky slopes, SE (Uitenhage to Okavango).

tenuiculum DC.   Straggling, thinly woolly shrub to 1 m. Leaves broadest below, linear-lanceolate to ovate, glabrous or cobwebby above, thinly woolly beneath, margins slightly revolute. Flower heads disciform, sometimes discoid, crowded in terminal corymbs, cylindric, c. 3 × 1 mm, florets 6–20, straw-coloured or tawny above, ovary glabrous. Jan.–Feb. Rocky slopes in damp places, NW, SW (Cold Bokkeveld to Paarl and Karoo).

G. Flower heads larger, 4–10 × 2–16 mm
H. Involucral bracts bright yellow

anomalum Less.   Grey-woolly shrublet to 30 cm. Leaves ascending, imbricate, linear, grey-woolly, margins revolute. Flower heads disciform, crowded in terminal cymes, cylindric-campanulate, bright yellow, c. 4 × 2.5 mm, bracts spreading above, florets 15–50, ovary glabrous, pappus reduced. Sept.–May. Stony slopes, KM, LB, SE (Outeniqua and Kammanassie Mts to Lesotho).

excisum (Thunb.) Less.   Densely twiggy, closely leafy shrublet to 45 cm, grey-felted on young parts. Leaves imbricate, spathulate, hooked, grey-felted. Flower heads discoid, in terminal corymbs, narrowly top-shaped, yellow, c. 4 × 2.5 mm, involucral bracts recurved above, florets c. 10, ovary papillate-hairy. Nov.–Feb. Sandstone slopes, SW, KM, LB, SE (Bredasdorp and Little Karoo to Langkloof).•

intricatum DC.   Wiry, grey-woolly shrublet or subshrub to 60 cm. Leaves oblanceolate, margins slightly revolute, glabrescent but roughly hairy above, woolly beneath. Flower heads discoid or disciform, many congested in terminal corymbs, campanulate, canary-yellow, 4 × 2.5 mm, florets 8–15, ovary glabrous or papillate-hairy. Dec.–Jan. Seeps on slopes, SE (Outeniqua Mts to E Cape).

splendidum (Thunb.) Less.   Grey-felted shrub to 1.5 m with rod-like branches closely leafy above. Leaves linear-oblong, often glabrescent above, felted beneath, margins revolute. Flower heads disciform, many in terminal corymbs, hemispherical, bright yellow, 4–5 × 5–8 mm, florets 50–100, ovary papillate-hairy. Oct.–Jan. Rocky slopes, KM, SE (Swartberg and Outeniqua Mts to tropical Africa).

H. Involucral bracts white to straw-coloured or pale yellow
I. Inner involucral bracts opaque white, obtuse and sharply spreading at tips; leaves soft, flat

acrophilum Bolus   Rounded or twiggy, grey-woolly shrublet to 30 cm. Leaves mostly crowded on short shoots, small, spathulate-truncate, felted. Flower heads discoid, 1–few at branch tips, campanulate, white flushed pink below, c. 7 × 12 mm, inner bracts spreading above, florets 30–60, ovary papillate-hairy. Mainly Dec.–Jan. Sandstone slopes, NW, KM (Pakhuis to Swartberg Mts).•

aureofolium Hilliard   Like **H. cylindriflorum** but flower heads broadly campanulate, florets 60–75, and tips of involucral bracts flat, acute and c. 2 mm long. Sept. Sandstone slopes, NW (Cedarberg Mts).•

cochleariforme DC.   GOLD-AND-SILVER   Rounded, grey-woolly shrublet to 20 cm. Leaves oblanceolate-spathulate, grey-woolly. Flower heads disciform, 1–few at tips of branchlets, subglobose, golden brown, c. 5–7 × 5–8 mm, involucral bracts concave, inner tipped white and minutely spreading, florets 60–120, ovary glabrous or papillate-hairy. Sept.–Dec. Coastal sands, NW, SW, AP (Aurora to Gouriqua).•

crispum (L.) D.Don   Rounded, white-woolly subshrub to 50 cm. Leaves obovate, thickly woolly. Flower heads discoid sometimes disciform, many congested in terminal clusters, campanulate, white, 3–6 × 3.5–4.0 mm, inner bracts spreading above, florets 10–50, ovary glabrous or papillate-hairy. Oct.–Dec. Mainly coastal sands, SW, AP, SE (Bloubergstrand to George).•

cylindriflorum (L.) Hilliard & B.L.Burtt (= *Helichrysum fastgiatum* Harv.)   Bushy grey-woolly shrublet to 30 cm. Leaves oblong-lanceolate, grey-woolly. Flower heads disciform sometimes discoid, clustered in branched corymbs, cylindric-campanulate, golden brown, c. 4–5 × 2.5–3.0 mm, involucral bracts sharply spreading above, inner tipped white, tips c. 1 mm long, florets c. 15–30, ovary papillate-hairy sometimes glabrous. Sept.–Dec. Rocky slopes, NW, SW, KM, LB, SE (Namaqualand to Langkloof).

diffusum DC.   Like **H. crispum** but leaves oblong with broad, clasping base and flower heads somewhat larger, c. 5–7 × 10–14 mm. Sept.–Dec. Sandstone slopes, NW, SW (Hex River Mts to Kogelberg and KwaZulu-Natal: Mtamvuna).

felinum Less.   Grey-woolly shrublet to 80 cm with rod-like branches. Leaves spreading or deflexed, ovate-lanceolate, rough and glabrescent above, felted beneath. Flower heads discoid, many in corymbs on peduncle-like stems, subglobose, white, 5–7 × 7–10 mm, inner bracts spreading above, florets 40–100, ovary papillate-hairy. Sept.–Dec. Sandstone slopes, SW, KM, LB, SE (Cape Peninsula to KwaZulu-Natal).

fruticans (L.) D.Don.   Like **H. grandiflorum** but leaves thinly woolly above, flower heads larger, c. 11 × 20 mm in a spreading corymb and involucral bracts much exceeding florets. Sept.–Dec. Sandstone slopes, SW (Cape Peninsula).•

grandiflorum (L.) D.Don.   Grey-woolly, somewhat tufted subshrub to 50 cm. Leaves obovate, densely woolly. Flower heads discoid, many congested in terminal clusters on peduncle-like stems, campanulate, white, c. 7 × 7 mm, inner bracts spreading above, florets 70–100, ovary glabrous. Dec.–Feb. Sandstone slopes, SW (Cape Peninsula).•

incarnatum DC.   Like **H. cylindriflorum** but bracts flatter and more acute, often tinged red. Sept.–Nov. Sandy flats, NW, LB (Hex River Valley to Riversdale).•

outeniquense Hilliard   Like **H. felinum** but smaller, to 30 cm, leaves more woolly and flower heads smaller, c. 4 × 3 mm. Jan. Peaty slopes, SE (Outeniqua Mts: Joubertina).•

pulchellum DC.   Like **H. cylindriflorum** but heads campanulate, with 40–70 florets and involucral bracts almost uniformly golden brown, tips not spreading. Sept.–Nov. Stony slopes, NW, KM (Namaqualand to Koo and Witteberg).

rotundifolium (Thunb.) Less.   Tufted, grey-woolly subshrub to 15 cm with decumbent stems from a woody rootstock. Leaves obovate, densely woolly, the lower in basal rosettes. Flower heads discoid, few in terminal clusters enclosed by leafy bracts, campanulate, white, 5–8 × 4–5 mm, inner bracts spreading above, florets 20–50, ovary papillate-hairy. Mainly Nov.–Jan. Sandstone slopes, NW, SW, KM, LB, SE (Grootwinterhoek Mts to Uniondale).•

sphaeroideum Moeser   Straggling, white-felted subshrub to 60 cm. Leaves oblanceolate to obovate, pseudopetiolate, glabrescent above, felted beneath. Flower heads discoid, few to many in dense terminal clusters, cylindric-campanulate, white, c. 3.5 × 3 mm, inner bracts spreading above, florets 8–30, ovary hairy. Mainly Dec.–Jan. Damp sandstone slopes, NW, SW, LB (Elandskloof to Langeberg Mts).•

stellatum (L.) Less.   Rounded, grey-woolly shrublet to 45 cm. Leaves oblanceolate-spathulate, grey-woolly. Flower heads disciform, several in terminal corymbs, broadly campanulate, c. 7 × 7 mm, golden brown and white sometimes pink, involucral bracts spreading above, florets 60–140, ovary glabrous or papillate-hairy. Mainly Sept.–Oct. Sandy flats and slopes, NW, SW (Bokkeveld Mts to Riviersonderend).•

I. Inner involucral bracts not as above or leaves firmer with margins revolute

appendiculatum (L.f.) Less.   SHEEP'S EARS   Leafy, grey-woolly subshrub to 50 cm. Leaves imbricate, oblanceolate, woolly but thinly woolly to glabrescent above. Flower heads discoid, many in congested corymbs, campanulate, creamy white flushed yellowish or red, 7–14 mm long, bracts sometimes acuminate-recurved, florets 30–60, ovary glabrous. Mainly Dec.–Feb. Rocky slopes, LB, SE (Swellendam to N Province).

bachmannii Klatt   Like **H. revolutum** and **H. rosum** but flower heads oblong-campanulate, involucral bracts oblong and tipped opaque white. Aug.–Nov. Coastal sands and granite, SW (Vredenburg).•

capense Hilliard   Diffuse, straggling shrublet to 50 cm, white-felted on young parts. Leaves obovate, spreading, often glabrescent above, white-felted beneath. Flower heads disciform, few in congested terminal clusters, campanulate, straw-coloured, c. 4 × 4 mm, bracts densely silky, florets 15–30, ovary hairy. Mainly Nov.–Dec. Sheltered sandstone slopes, SW, LB (Cape Peninsula to Robinson Pass).•

catipes (DC.) Harv.   Bushy, grey-woolly shrublet to 50 cm. Leaves oblong-ovate, grey-woolly, often glabrescent above. Flower heads disciform, few webbed together in terminal clusters enclosed by leafy bracts, campanulate, straw-coloured, c. 5 × 3 mm, florets c. 26, ovary hairy. Dec.–Mar. Sandstone slopes, NW (Cedarberg to Hex River Mts).•

dasyanthum (Willd.) Sweet   Like **H. hebelepis** but with silky, spreading, often rufous hairs as well as woolly hairs. Mainly Sept.–Nov. Sandy flats and slopes, NW, SW, AP, KM, LB, SE (Namaqualand to Baviaanskloof Mts).

dunense Hilliard   Trailing, white-woolly, short-lived perennial with slender branches to 60 cm long from a woody caudex. Leaves spathulate, grey-woolly. Flower heads discoid, several in dense terminal clusters, enclosed by leaves, cylindric-campanulate, translucent purplish, c. 4 × 3 mm, bracts apiculate, florets c. 10, ovary glabrous. Oct.–Dec. Coastal sands, NW (Namaqualand to Lambert's Bay).

fourcadei Hilliard   Twiggy, silvery shrublet. Leaves spathulate, hooked, silvery felted with tissue paper-like hairs. Flower heads discoid, 1–3 at branch tips, narrowly top-shaped, reddish brown, c. 6 × 4 mm, bracts sharply reflexed above, florets c. 20, ovary papillate-hairy. Dec. Stony slopes, SE (Uniondale and S Karoo).

hamulosum E.Mey. ex DC.   Erect, closely leafy shrublet to 60 cm. Leaves imbricate, linear, hooked, glabrous above, felted beneath, margins strongly revolute. Flower heads discoid, many in rounded corymbs, cylindric, straw-coloured, c. 6 × 2 mm, florets 5–15, ovary papillate-hairy. Dec.–Mar. Rocky slopes, NW, KM (Namaqualand to Little and Great Karoo).

hebelepis DC.   Straggling, grey-woolly shrub to 1.5 m. Leaves spreading, linear to elliptic-oblong, grey-woolly, margins often undulate and slightly revolute. Flower heads disciform, many in compact terminal corymbs, campanulate, straw-yellow, c. 4 × 3 mm, florets 15–30, ovary hairy. Mainly Aug.–Sept. Stony slopes and flats, NW, SW, KM (Namaqualand to Little Karoo).

interzonale Compton   Straggling shrublet to 50 cm, white-felted on young parts. Leaves ascending, later recurved, linear, revolute, mucronate, glabrescent above, white-woolly beneath. Flower heads disciform, sometimes discoid, 1–12 at branch tips, campanulate, tawny, c. 5 × 4 mm, florets 15–30, ovary glabrous. Sept.–Nov. Rocky slopes, NW, KM, SE (Swartruggens to Uniondale: Suurberg).•

lambertianum DC.   Erect or straggling, grey-woolly shrublet or subshrub to 60 cm. Leaves linear-lanceolate, white-felted. Flower heads discoid, several in compact terminal corymbs, campanulate, c. 5 × 5 mm, bracts narrow, entirely grey-felted, florets 20–35, ovary hairy. Oct.–Jan. Rocky slopes, NW, KM, SE (Gifberg to Uniondale).•

marifolium DC.   Like **H. capense** but leaves narrowly ovate and usually smaller, mostly 5–10 mm long. Sept.–Oct. Damp sandstone slopes, SW (Riviersonderend Mts).•

pandurifolium Schrank   Straggling, grey-woolly shrublet or shrub. Leaves ovate, pseudopetiolate and auriculate-clasping below, crisped, grey-woolly. Flower heads discoid, few to many in terminal corymbs on peduncle-like branches, campanulate, creamy, c. 5–8 × 6–10 mm, florets 12–40, ovary glabrous. Mainly Oct.–Jan. Sandy flats and slopes, NW, LB, SE (Bainskloof to Kouga Mts).•

patulum (L.) D.Don   Like **H. pandurifolium** but involucral bracts obtuse. Mainly Dec.–Jan. Sandy flats and slopes, often coastal, SW, LB, SE (Cape Peninsula to Mossel Bay).•

pentzioides Less.   Twiggy shrublet to 1 m, thinly grey-felted on young parts. Leaves obovate-spathulate, hooked, grey-felted. Flower heads discoid, few in terminal clusters, cylindric, straw-coloured, c. 5 × 3 mm, florets 8–15, ovary papillate-hairy. Mainly Dec.–Jan. Stony slopes and flats, NW, SW, KM, LB, SE (Montagu and Bredasdorp to Free State).

petiolare Hilliard & B.L.Burtt   Like **H. pandurifolium** but leaves abruptly and conspicuously petiolate and not crisped, and involucral bracts obtuse. Dec.–Jan. Sheltered slopes and forest margins, NW, SW, LB, SE (Cedarberg and Jonkershoek Mts to KwaZulu-Natal).

plebeium DC.   Like **H. hebelepis** but leaves always linear to linear-oblong with margins strongly revolute and receptacle smooth, not honeycombed nor fimbrilliferous. Aug.–Oct. Sandstone slopes, LB (Langeberg Mts).•

revolutum (Thunb.) Less.   Sprawling, thinly grey-woolly shrub to 2 m. Leaves often in axillary tufts, linear-lanceolate, slightly auriculate, revolute, glabrescent above, white-woolly beneath. Flower heads disciform, in compact terminal corymbs, campanulate, straw-coloured, c. 5 × 4 mm, florets 20–40, ovary papillate-hairy. July–Oct. Rocky or sandy flats and slopes, NW, SW, KM (S Namibia to Cape Peninsula and Witteberg).

rosum (P.J.Bergius) Less.   Like **H. revolutum** but involucral bracts tipped milk-white. Mainly Sept.–Mar. Stony slopes and flats, SW, AP, KM, LB, SE (Stellenbosch to Free State).

rotundatum Harv.   Like **H. capense** but plants coarser with larger leaves, mostly 15–20 mm long and flower heads c. 6 × 5 mm. Oct.–Dec. Damp sandstone slopes, SW (Riviersonderend Mts).•

rugulosum Less.   Thinly white-felted perennial to 30 cm, stems usually unbranched. Leaves oblong-lanceolate, slightly revolute, thinly felted above, densely so beneath. Flower heads disciform, in compact terminal corymbs, campanulate, creamy, c. 5 × 4 mm, bracts crisped, florets c. 15–40, ovary papillate-hairy. Mainly Dec.–Mar. Stony grassland, KM, LB, SE (Langeberg and Swartberg Mts to Gauteng).

scabrum Less.   Like **H. revolutum** but leaves roughly glandular-hairy and margins conspicuously crisped. Aug.–Dec. Rocky slopes, NW (Namaqualand to Ceres).

spiralepis Hilliard & B.L.Burtt   Tufted grey-woolly, short-lived perennial to 30 cm. Leaves oblanceolate, the lower in a rosette, densely woolly. Flower heads disciform sometimes discoid, congested in glomerules enclosed by leafy bracts, narrowly campanulate, creamy or reddish, c. 5 × 3 mm, bracts acuminate-hooked, florets 12–40, ovary papillate-hairy. Sept.–Feb. Sandy slopes, SW, LB, SE (Cape Peninsula to Lesotho).

teretifolium (L.) D.Don   Straggling, thinly felted, closely leafy shrublet to 30 cm. Leaves stiffly spreading, linear, revolute, hooked, glabrescent above, white-woolly beneath. Flower heads disciform, sometimes discoid, in dense terminal corymbs, campanulate, creamy white, c. 5 × 5 mm, bracts spreading above, florets 17–60, ovary glabrous or papillate-hairy. July–Nov. Sandy slopes and dunes, NW, SW, AP, KM, LB, SE (Piketberg to KwaZulu-Natal).

tricostatum (Thunb.) Less.   Straggling, grey-woolly shrublet to 1.5 m. Leaves spreading, oblanceolate to oblong, grey-woolly, margins slightly revolute. Flower heads disciform, many in compact terminal corymbs, campanulate, straw-yellow, c. 4 × 3 mm, bracts subglabrous, florets 15–30, ovary glabrous. Sept.–Dec. Coastal sands, NW, SW (Namaqualand to Bokbaai).

zeyheri Less.   Twiggy, grey-woolly shrublet mostly to 70 cm, branches often stiff. Leaves oblong to obovate, often pseudopetiolate, grey-woolly, crisped. Flower heads discoid, in compact terminal cymes, cylindric, white, c. 4 × 2 mm, florets 5, ovary glabrous. Mainly Nov.–May. Stony slopes, NW, SW, KM, LB, SE (dry parts of southern Africa).

[**Species excluded**   Not occurring in the area: **H. alsinoides** DC., **H. oxybelium** DC., **H. pilosellum** (L.f.) Less. See **Syncarpha**: **H. argenteum** Thunb., **H. chlorochrysum** DC., **H. dykei** Bolus, **H. lepidopodium** Bolus, **H. mucronatum** (L.) Willd., **H. paniculatum** (L.) Thunb., **H. recurvatum** (L.f.) Thunb., **H. sordescens** DC., **H. striatum** Thunb., **H. vestitum** (L.) Schrank]

HERTIA   SPRINGBOKBOS   c. 10 spp., South Africa to SW Asia

alata (Thunb.) Kuntze   Slender-stemmed shrub to 80 cm. Leaves decurrent in broad stem wings, oblanceolate towards base, obovoid above. Flower heads disciform, few in nodding, terminal or axillary clusters, yellow. Nov.–Dec. Rocky sandstone ridges, KM (Swartberg Mts and Karoo).

kraussii (Sch.Bip.) Fourc.   Twiggy shrublet to 50 cm. Leaves oblanceolate, leathery, sometimes slightly toothed. Flower heads disciform, small, solitary on long peduncles, terminal or axillary, yellow. June–July. Clay soils in renosterveld and karroid scrub, SE (Kammanassie Mts to Port Elizabeth).•

sp. 1   Like **H. kraussii** but leaves sometimes slightly decurrent. Flower heads few, in terminal and axillary corymbs. Apr.–Oct. Damp lower slopes, SE (Kammanassie Mts to Port Elizabeth).•

HETEROLEPIS   ROTSGOUSBLOM   4 spp., W to E Cape

aliena (L.f.) Druce   Sprawling, closely leafy, cobwebby shrublet to 30 cm, with woody rootstock. Leaves linear, margins slightly revolute, sparsely toothed, densely woolly below, (10–)15–30 mm long. Flower heads radiate, large, solitary on short, roughly hairy peduncles, yellow. Mainly Sept.–Dec. Rocky sandstone slopes and outcrops, NW, SW, KM (Cedarberg Mts to Witteberg and Hermanus).•

peduncularis DC.   Like **H. aliena** but flower heads on elongate peduncles. Sept.–Dec. Rocky slopes, NW, SW, AP, KM, LB (Hex River Mts to Langeberg Mts).•

sp. 1   Sprawling, closely leafy, thinly cobwebby shrublet to 30 cm, with woody rootstock. Leaves linear-needle-like, pungent, margins closely revolute and concealing lower surface, 10–15 mm long. Flower heads discoid, solitary on roughly hairy peduncles, yellow. Oct.–Apr. Dry N-facing sandstone slopes, LB (Langeberg Mts: Garcia's Pass).•

HETERORHACHIS•   KAAPSEKARMEDIK   2 spp., N and W Cape

aculeata (Burm.f.) Roessler   Spiny, densely leafy, cobwebby shrub to 60 cm. Leaves pinnatisect, rigid and pungent, margins usually revolute. Flower heads radiate, several at branch tips, yellow; outer bracts leaf-like, toothed and pungent, inner lanceolate. Aug.–Sept. Stony flats, NW, SW (Bokkeveld Escarpment to Koeberg).•

sp. 1   Like **H. aculeata** but flower heads discoid and involucral bracts all ovate-lanceolate. Aug.–Sept. Stony flats, NW (Graafwater).•

HIPPIA•   RANKALS   5 spp., W to E Cape

frutescens (L.) L. (= *Hippia hirsuta* DC.; incl. **H. montana** Compton)   RANKALS   Shortly hairy, straggling shrublet to 60 cm. Leaves pectinate-pinnatifid, lobes oblong to linear. Flower heads discoid, in branched corymbs, yellow. Achenes winged. Mainly Oct.–Mar. Sandstone slopes, often near streams or marshes, NW, SW, AP, KM, LB, SE (Ceres to Storms River).•

hutchinsonii Merxm.   Slender-stemmed shrublet to 25 cm. Leaves trifid, lobes linear. Flower heads discoid, in sparse terminal corymbs, yellow. Achenes ellipsoid. Nov.–Dec. Sandstone slopes, LB (Langeberg Mts: Swellendam).•

integrifolia Less.   Sprawling, slender-stemmed shrublet to 20 cm. Leaves ovate, sometimes palmately few-toothed to lobed. Flower heads discoid, few in terminal umbels, yellow. Achenes broadly winged. Mainly Sept.–Jan. Damp, shady sandstone slopes, LB (Langeberg Mts: Swellendam to Riversdale).•

pilosa (P.J.Bergius) Druce (= *Hippia bolusiae* Hutch.)   Thinly hairy, slender, erect or sprawling shrublet to 30 cm, young buds often with long russet hairs. Flower heads discoid, white or yellow. Achenes ellipsoid. Aug.–Nov. Damp sandstone slopes often in shade, NW, SW, KM, LB (Ceres to Witteberg, Cape Peninsula to Swellendam).•

sp. 1   Dwarf, tufted perennial to 10 cm, with long russet hairs on young parts. Leaves mostly in a dense basal tuft, lobes linear. Flower heads discoid, in dense terminal corymbs, yellow. Achenes ellipsoid. Nov.–Dec. Rocky sandstone slopes, SW (Worcester: Louwshoek Mts).•

HIRPICIUM   HAARBOSSIE   12 spp., tropical and southern Africa

alienatum (Thunb.) Druce   Twiggy shrublet like **H. integrifolium** but involucral bracts acuminate, shortly bristly. Aug.–Dec. Stony shale slopes, NW, KM, SE (S Namibia to Uitenhage and Karoo).•

integrifolium (Thunb.) Less.   HAARBOSSIE   Shrublet to 40 cm. Leaves linear to oblanceolate, margins revolute, white-felted below, bristly above. Flower heads radiate, solitary at branch tips, yellow; bracts filiform, plumose. Aug.–Oct. Dry rocky slopes, NW, KM (Worcester to Willowmore).•

HOPLOPHYLLUM   YLKARMEDIK   2 spp., southern Africa

spinosum DC.   Rigid spiny shrub to 1 m. Leaves linear, hard and spine-like. Flower heads discoid, sessile in small terminal clusters, yellow. Sept.–Oct. Stony karroid slopes, NW (Namaqualand to Clanwilliam and S Karoo).

HYDROIDEA• BERGBLOMBOS   1 sp., W Cape

elsiae (Hilliard) Karis   Grey-woolly shrublet to 15 cm. Leaves needle-like with involute margins, twisted, in tufts. Flower heads discoid, solitary on short peduncles, purplish, bracts white. Jan.–Mar. Sandstone rocks above 1000 m, SW (Du Toitskloof Mts).•

HYMENOLEPIS   BASTERKAROO   7 spp., W to E Cape

cynopus Bremer & Kallersjö   Grey-woolly, densely leafy shrub to 1.5 m. Leaves obovate, with revolute margins, dentate above. Flower heads discoid, small, massed in rounded corymbs, yellow; bracts densely silky. Aug.–Sept. Sandstone slopes, KM (Touwsberg).•

dentata (DC.) Kallersjö (= *Athanasia schizolepis* Harv.)   Glabrous shrub to 1.2 m. Leaves linear, entire to deeply toothed. Flower heads discoid, small, crowded in rounded corymbs, yellow. Aug.–Oct. Rocky sandstone slopes, NW (Cedarberg Mts to Matroosberg).•

gnidioides (S.Moore) Kallersjö (= *Phaeocephalus gnidioides* S.Moore)   White-woolly, slender, few-branched shrub, 1–3 m. Leaves oblong, warty below. Flower heads yellow; bracts silky. Oct.–Jan. Sandstone slopes and ridges, KM (Langeberg to Kammanassie Mts).•

incisa DC.   Glabrescent, erect-stemmed shrub, 30–90 cm. Leaves pinnatisect, segments linear with revolute margins. Flower heads discoid, small, crowded in rounded corymbs, yellow; bracts velvety. Aug.–Nov. Rocky sandstone slopes, NW, KM, LB (Ceres to Swartberg Mts).•

indivisa (Harv.) Kallersjö   Slender, few-branched shrub to 1.3 m. Leaves filiform with revolute margins. Flower heads discoid, small, crowded in rounded corymbs, yellow. Oct.–Nov. Sandstone slopes, SE (Great Winterhoek Mts to Grahamstown).

parviflora (L.) DC.   Thinly woolly shrub, 1–3 m, densely leafy above. Leaves large, pinnatisect, segments linear with revolute margins. Flower heads discoid, small, massed in dense compound corymbs, yellow. Nov.–Dec. Rocky sandstone slopes, NW, SW (Namaqualand to Villiersdorp).

speciosa (Hutch.) Kallersjö   Grey-felted shrub to 1 m, golden on young parts. Leaves pinnatisect, lobes linear with revolute margins. Flower heads discoid, small, massed in dense compound corymbs, yellow; bracts velvety. Sept.–Nov. Sandstone slopes above 1200 m, NW (Piketberg to Hex River Mts).•

IFLOGA (see also **TRICHOGYNE**)   NAALDEBOS   6 spp., southern Africa to N Africa and Middle East

anomala Hilliard   Like **I. glomerata** but ovaries of hermaphrodite flowers aborted. Mainly July–Oct. Disturbed places, AP, KM, LB (Montagu to Mossel Bay).•

glomerata (Harv.) Schltr.   Thinly white-membranous annual to 10 cm, stems stiffly erect above. Leaves linear, margins revolute. Flower heads discoid, clustered along stems, straw-coloured. Mainly Aug.–Nov. Usually disturbed sandy soil, SE (Namibia to Free State and S to Humansdorp).

thelliana Hilliard & B.L.Burtt   Erect or mat-like, thinly white-membranous annual to 15 cm. Leaves distant, linear-lanceolate, margins revolute, white-mucronate. Flower heads discoid, 1–few, nested among leaves, straw-coloured. Sept.–Oct. Sandstone flats and slopes, NW, SW (Cold Bokkeveld to Cape Peninsula and W Karoo).

INULANTHERA   10 spp., southern and tropical Africa, Madagascar

dregeana (DC.) Kallersjö (= *Athanasia dregeana* (DC.) Harv.)   Glabrescent, slender-stemmed shrub to 1.5 m. Leaves linear, crowded in tufts, slightly dentate. Flower heads discoid, in terminal corymbs, yellow. Mar.–May. Grassland and forest margins, SE (Langkloof to S KwaZulu-Natal).

[**KLEINIA**   Species from the region previously assigned to **Kleinia** are now included in **Senecio**.]

LACHNOSPERMUM•   4 spp., W Cape

fasciculatum (Thunb.) Baill.   Thinly felted shrublet to 60 cm. Leaves involute-ericoid, in tufts. Flower heads discoid, few at branch tips, yellow, bracts white-woolly, acuminate, ascending or weakly spreading. Jan.–Mar. Sandy lower slopes, NW, SW (Pakhuis Mts to Franschhoek Pass and Mamre).•

imbricatum (P.J.Bergius) Hilliard   Glabrescent shrublet to 50 cm with stiffly erect branches. Leaves ovate with involute margins, imbricate. Flower heads discoid, few in loose terminal clusters, purple, inner bracts with papery white tips. Jan.–Mar. Coastal sands and limestone outcrops, SW, AP (Mamre to Cape Flats and Elim to Agulhas).•

umbellatum (D.Don) Pillans   ROOIBLOMBOS   Thinly felted, stiffly erect shrublet to 30 cm. Leaves involute-ericoid, densely tufted. Flower heads discoid, several in corymbs at branch tips, pinkish, most bracts with papery pink tips, reflexed. Nov.–Mar. Sandy lower slopes, SW (Cape Peninsula to Hermanus).•

sp. 1 (= *L. neglectum* Schltr. ms)   Like **L. fasciculatum** but flower heads cream-coloured, bracts densely white-woolly with tips attenuate, smooth and papery, reflexed. Jan.–Feb. Dry rocky sandstone slopes, NW, SW (Hex River valley to Riviersonderend Mts).•

LAMPROCEPHALUS•   1 sp., W Cape

montanus B.Nord.   Thinly cobwebby shrublet to 50 cm. Leaves linear, keeled below, ascending, somewhat succulent, prominently apiculate. Flower heads discoid, solitary on elongate peduncles, red. Dec.–Jan. Rocky, often S-facing slopes above 1000 m, often on shale, NW (Cedarberg to Hex River Mts).•

LANGEBERGIA• (= *PETALACTE* in part)   1 sp., Langeberg Mts

canescens (DC.) Anderb. (= *Petalacte canescens* DC.)   Sprawling, silvery felted shrublet to 30 cm. Leaves oblanceolate. Flower heads discoid, small, many in dense clusters, purple, bracts white. Mainly Aug.–Dec. Rocky summits and south slopes above 1000 m, LB (Langeberg Mts: Swellendam).•

LASIOPOGON   8 spp., southern Africa, North Africa to Middle East

brachypterus O.Hoffm. ex Zahlbr.   Dwarf, white-woolly annual to 5 cm. Leaves oblanceolate. Flower heads discoid, in terminal clusters, white; pappus shortly plumose above. Mainly Sept.–Nov. Rocky sandstone slopes and outcrops, NW, SW (Kamiesberg to Hottentots Holland Mts).

glomerulatus (Harv.) Hilliard   Similar to **L. brachypterus** but pappus barbellate thoughout. May–Oct. Stony and gravelly flats, KM (S Namibia to Worcester and to Zambia).

muscoides (Desf.) DC.   Dwarf, densely grey-woolly, prostrate annual to 5 cm. Leaves spathulate. Flower heads discoid, in dense rounded woolly clusters, white; pappus plumose. Mainly Aug.–Sept. Stony flats, KM (Namaqualand to E Cape).

LASIOSPERMUM   GIFKNOPPIES   4 spp., southern Africa

bipinnatum (Thunb.) Druce   Glabrescent perennial with erect to sprawling stems from woody rootstock, to 60 cm. Leaves bipinnatisect. Flower heads radiate, globose, yellow with white rays. Aug.–Dec. Grassland or disturbed soil, NW, KM (Cedarberg Mts through Little Karoo to Gauteng).

brachyglossum DC.   Glabrescent annual with ascending stems to 40 cm. Leaves bipinnatisect. Flower heads shortly radiate, globose, yellow and red with reddish rays. July–Oct. Clay and gravel slopes, NW, KM (S Namibia to Oudtshoorn).

pedunculare Lag. (= *Lasiospermum erectum* (Lam.) Druce)   Silvery silky perennial with erect to sprawling stems from woody rootstock. Leaves bipinnatisect. Flower heads discoid, globose, yellow, fragrant. Aug.–Oct. Clay slopes, NW (Namaqualand and W Karoo to Ceres).

LEUCOPTERA   3 spp., winter-rainfall South Africa

nodosa (Thunb.) B.Nord.   Slender shrublet to 50 cm. Leaves filiform or pinnatisect with filiform lobes, bases persistent and nodular. Flower heads radiate, solitary on elongate peduncles, yellow with white rays. July–Sept. Sandveld, NW (Hondeklipbaai to Elandsbaai).

LEYSERA   TEEBOS   3 spp., southern Africa and Mediterranean to S W Asia

gnaphalodes (L.) L.   SKILPADTEEBOSSIE, TERINGTEEBOSSIE   Glabrous to cobwebby shrublet to 40 cm. Leaves linear, glandular-hairy. Flower heads radiate, solitary on slender, wiry peduncles, yellow; pappus bristles plumose from base. Mainly Sept.–Nov. Sandy flats and slopes, NW, SW, KM, AP, LB, SE (S Namibia to E Cape).

tenella DC.   VAALTEEBOSSIE   Slender, glabrescent to cobwebby annual, sometimes perennial, to 20 cm. Leaves linear, glandular-hairy. Flower heads radiate, solitary on wiry peduncles, yellow; pappus bristles plumose above. Mainly Aug.–Oct. Sandy and stony flats and slopes, NW, SW, KM, LB, SE (S Namibia to Willowmore and Karoo).

LIDBECKIA• BERGASTER   3 spp., W Cape

lobata Thunb. (= *Lidbeckia quinqueloba* Cass.)   Silky, sprawling shrublet with velvety leaves, to 60 cm. Leaves deeply palmately lobed, petiolate. Flower heads radiate, solitary on long peduncles at branch tips, yellow with white rays. Sept.–Dec. Shady sandstone slopes, NW (N Cedarberg to Olifants River Mts).•

pectinata P.J.Bergius   Glaucous, aromatic shrub to 1.5 m. Leaves pinnatifid, lobes oblong to obovate, gland-dotted below. Flower heads radiate, large, solitary on long peduncles at branch tips, yellow with white rays. Sept.–Nov. Shady sandstone slopes, NW (Grootwinterhoek Mts to Roodezandberg).•

vlokii Kallersjö ined.   Twiggy, thinly velvety shrub to 50 cm. Leaves 3–5-pinnatisect, lobes linear. Flower heads radiate, solitary on long peduncles, yellow with white rays. Mar.–May. Dry stony lower slopes, LB (Langeberg Mts to Mossel Bay).•

MAIRIA•   PLUIMASTER   5 spp., W Cape

burchellii DC. (= *Zyrphelis burchellii* (DC.) Kuntze)   Shortly hairy, tufted perennial to 15 cm. Leaves numerous, mostly in a basal tuft, linear, silky below. Flower heads radiate, 1–few, subscapose, yellow with mauve rays. Nov.–Apr. Sandy slopes after fire, NW, SW, KM (Porterville Mts and Cape Peninsula to Ladismith).•

coriacea Bolus   Hairy, tufted perennial to 12 cm. Leaves in a basal rosette, obovate, often broadly toothed above, leathery, margins thickened, silky below, becoming smooth above with age. Flower heads radiate, solitary, subscapose, yellow with purple rays. Nov.–Mar. Rocky slopes after fire, SW (Rooiels to Potberg).•

crenata (Thunb.) Nees   Tufted perennial to 15 cm. Leaves in a basal rosette, obovate, margins revolute, toothed, leathery, silky below. Flower heads radiate, solitary, subscapose, yellow with pink to mauve or white rays. Sept.–Mar. Rocky sandstone slopes and outcrops, mainly after fire, SW, KM, LB, SE (Cape Peninsula to Great Winterhoek Mts).•

hirsuta DC.   Softly woolly, tufted perennial to 30 cm. Leaves in a basal rosette, oblanceolate, margins slightly toothed and revolute, densely woolly below. Flower heads radiate, solitary, subscapose, yellow with mauve-pink rays. July–Sept. Mountain slopes after fire, NW, SW, LB (Bainskloof and Hex River Mts to Swellendam).•

sp. 1 (= *Mairia petiolaris* Z.-Wiegand ms)   Softly woolly, tufted perennial to 20 cm. Leaves in a basal rosette, long-petiolate, elliptic, margins ragged. Flower heads radiate, solitary on few-branched peduncles, yellow with pink rays. Dec.–Jan. Sandstone slopes above 1000 m, LB (Langeberg Mts: Swellendam).•

MARASMODES•   4 spp., W Cape

dummeri Bolus ex Hutch.   Divaricately branched ericoid shrublet to 30 cm. Leaves linear, adpressed. Flower heads discoid, sessile, axillary, crowded in short spikes, florets 4 or 5 per head, yellowish. May–June. Stony gravel, SW (Klipheuwel to Muldersvlei).•

oligocephalus DC.   Much-branched ericoid shrublet to 40 cm. Leaves linear, irregularly 3–5-lobed above. Flower heads discoid, few in terminal clusters or solitary and axillary, yellow. Apr.–June. Clay flats, SW (Kalbaskraal).•

polycephalus DC.   Like **M. dummeri** but leaves often in tufts, spreading and recurved above. Apr.–May. Clay flats, SW (Malmesbury to Stellenbosch).•

undulata Compton   Twiggy, ericoid shrublet to 30 cm. Leaves narrowly lanceolate, tiny, ascending. Flower heads discoid, solitary, nested in leaves, yellow, bracts papery and recurved above. Apr.–May. Gravelly flats, SW (near Paarl).•

METALASIA   BLOMBOS   52 spp., South Africa, mainly W Cape

A. Capitula 3- or 4-flowered
(see also **M. aurea***,* **M. calcicola***,* **M. densa***,* **M. luteola***,* **M. montana***,* **M. muricata** *and* **M. plicata***)*

acuta Karis   Rigid, white woolly shrub to 1.5 m. Leaves pungent, twisted, 3–15 mm long, with axillary tufts. Flower heads discoid, several in dense terminal clusters, 3- or 4-flowered, bracts erect, inner petaloid, serrate, white. Nov.–Apr. Sandy and clay flats and slopes up to 1300 m, SW, AP, KM, LB (Cape Hangklip to George).•

albescens Karis   Densely white-woolly shublet to 80 cm. Leaves reflexed, twisted, 2–6 mm long, with axillary tufts. Flower heads discoid, several, in terminal clusters, 3(4)-flowered, bracts erect, inner petaloid, pink. Dec.–Jan. Sandstone slopes, NW (S Cedarberg Mts).•

brevifolia (Lam.) Levyns (= *Metalasia intermedia* DC.)   Erect white-woolly shrub to 1 m but usually less. Leaves reflexed, twisted, 2–8 mm long, with axillary tufts. Flower heads discoid, in terminal clusters, 3-flowered, bracts erect, concave above, inner petaloid, white. Sept.–Dec. Sandstone flats and slopes, NW, SW, AP, LB, SE (Bokkeveld Mts to Cape Peninsula and to Port Elizabeth).•

capitata (Lam.) Less. (= *Metalasia bolusii* L.Bolus)   Spreading, white-woolly shrublet to 60 (–100) cm. Leaves on older branches spreading and twisted, adpressed on young shoots, imbricate and not twisted, 2–12 mm long, without axillary tufts. Flower heads discoid, in dense terminal clusters, 3- or 4(5)-flowered, bracts slightly spreading, inner petaloid, white. Aug.–Sept. Sandy coastal flats, NW, SW (Piketberg to Paarl).•

cymbifolia Harv.   Similar to **M. serrata** but leaves strongly incurved and not twisted. June–Nov. Sandstone slopes, 200–900 m, SW (Babilonstoring and Kleinrivier Mts).•

erectifolia Pillans   Erect, white-woolly shrublet to 50 cm. Leaves adpressed, imbricate, not twisted, 3–10 mm long. Flower heads discoid, several in dense terminal clusters, 3-flowered, bracts erect, inner petaloid, brown to white. Sept. Limestone hills, AP (Agulhas to Stilbaai).•

erubescens DC.   White-woolly shrub to 1 m. Leaves twisted, 2–18 mm long. Flower heads discoid, several in dense terminal clusters, 3- or 4(5)-flowered, bracts erect, inner petaloid, pink sometimes white. Sept.–Feb. Sandy flats and lower slopes, to 1000 m, SW, AP (Hottentots Holland Mts to Agulhas).•

phillipsii L.Bolus (= *Metalasia incurva* Pillans)   Glabrescent shrublet to 40 cm. Leaves slightly twisted, 2–6 mm long, with axillary tufts. Flower heads discoid, many in interlocking terminal clusters, 3-flowered, bracts erect, inner petaloid, whitish to pink. Nov.–Feb. Sandstone slopes 1500–2200 m, NW, SW (Cold Bokkeveld to Bainskloof and Hex River Mts).•

rhoderoides T.M.Salter   Spreading, coarsely hairy shrublet to 40 cm. Leaves greyish and coarsely hairy, lanceolate, smaller and fewer above, 2–12 mm long, with axillary tufts. Flower heads discoid, several, fused in terminal clusters, 3- or 4-flowered, bracts erect, inner petaloid, white. Feb.–Apr. Stony slopes, 700–1000 m, SW (Elandskloof Mts to Slanghoek Mts).•

riparia T.M.Salter   Robust, white- or green-woolly shrub to 2.5 m, young branches reddish. Leaves densely hairy, slightly twisted, 3–12 mm long. Flower heads discoid, numerous, fused in terminal clusters, 3(4)-flowered, bracts erect, inner petaloid, white. Dec.–Feb. Streambanks, SW (Hottentots Holland to Kleinrivier Mts).•

rogersii S.Moore   Slender, sprawling white-woolly shrublet to 30 cm. Leaves reflexed, twisted, 1–6 mm long, with small axillary tufts. Flower heads discoid, several in completely interlocking terminal clusters, 3-flowered, bracts erect, concave above, inner petaloid, pink sometimes white. Oct.–Jan. Sandstone slopes, NW (Cold Bokkeveld to Hex River Mts).•

seriphiifolia DC.   Slender, sparsely white-woolly shrublet to 40 cm. Leaves needle-like, sometimes hairy, twisted, mostly 2–10 mm long, with axillary tufts. Flower heads discoid, several in terminal clusters, 3-flowered, bracts erect, inner petaloid, 3-toothed, pink to whitish. Aug.–Sept. Lower sandstone slopes, to 100 m, SW (Kleinrivier Mts).•

serrata Karis   White-woolly shrublet to 60 cm. Leaves slightly twisted, 2–11 mm long, with small axillary tufts. Flower heads discoid, several in interlocking terminal clusters, 3-flowered, bracts erect, concave above, inner petaloid, keeled at tips, pink or white. July–Aug. Sandstone slopes to 450 m, SW (Kleinrivier Mts to Bredasdorp).•

serrulata Karis   Similar to **M. rogersii** but leaves 2–10 mm long and mostly without axillary tufts, heads fewer and inner bracts white. Dec.–Jan. Sandstone slopes, NW (Grootwinterhoek Mts to Skurweberg).•

tenuifolia DC.   Sprawling, slender, sparsely white-woolly shrublet to 20 cm. Leaves twisted, sometimes densely woolly, 3–14 mm long. Flower heads discoid, few in terminal clusters, 3-flowered, bracts slightly spreading, inner petaloid, whitish or cream-coloured. Aug. Sandstone slopes, 300–900 m, SW (Houwhoek to Genadendal).•

tenuis Karis   Spreading white-woolly shrublet to 20 cm. Leaves reflexed, twisted, 2–5 mm long. Flower heads discoid, many in terminal clusters, 3-flowered, bracts erect, concave above, inner petaloid, pink. Dec.–Jan. Lower, sometimes damp, sandstone slopes, SW (Riviersonderend Mts).•

trivialis Karis   Erect white-woolly shrub to 3 m. Leaves spreading, lanceolate to ovate, twisted, keeled, 4–15 mm long, with axillary tufts. Flower heads discoid, several in terminal clusters, 3–4-flowered, bracts erect, outer brown, inner petaloid, white. Sept.–Dec. Rocky slopes, KM, SE (Klein Swartberg to Cathcart).

A. Capitula with 5 or more flowers

adunca Less.   White-woolly shrub to 1.5 m. Leaves curved outwards, hooked apically, not twisted, 2–10 mm long, without axillary tufts. Flower heads discoid, several in terminal clusters, 5-flowered, inner bracts petaloid, concave, white. Mainly Aug.–Oct. Sandy flats and lower slopes, to 1000 m, NW, SW (Namaqualand to McGregor).

agathosmoides Pillans   Compact white-woolly shrublet to 60 cm. Leaves narrowly triangular, not twisted, 2–6 mm long, with axillary tufts. Flower heads discoid, several in dense, rounded terminal clusters, 5-flowered, bracts erect, slightly concave above, inner petaloid, pink to white. Oct.–Dec. Sandstone slopes, 1200–1700 m, NW, KM (Cedarberg Mts to Bonteberg).•

alfredii Pillans   Compact, white-woolly shrublet to 1 m. Leaves narrowly obovate, scarcely twisted, usually densely hairy, 2–9 mm long. Flower heads discoid, several in terminal clusters, mostly 5- or 6-flowered, bracts spreading, irregularly serrate, inner petaloid, white. Mainly Feb.–Mar. Sandstone slopes above 1600 m, SW (Riviersonderend Mts).•

aurea D.Don   GEELBLOMBOS   White-woolly shrub to 1.5 m. Leaves twisted, pungent, 4–18 mm long, with axillary tufts. Flower heads discoid, several in numerous terminal clusters, 3–5-flowered, bracts erect, inner petaloid, concave above, dull yellow. Apr.–June. Grassy flats, SW, SE (Potberg to Port Elizabeth).•

bodkinii L.Bolus   White-woolly shrublet to 20 cm. Leaves straight or twisted, woolly, 2–7 mm long. Flower heads discoid, several at branch tips, 5-flowered, innermost bracts petaloid, tips recurved, white. Sept.–Nov. Sandstone slopes, SW (Caledon Swartberg).•

calcicola Karis   Erect, densely white-woolly shrub to 1 m. Leaves strongly incurved, not twisted, 2–4 mm long, with conspicuous axillary tufts. Flower heads discoid, several in terminal clusters, 3–5-flowered, inner bracts truncate, white. May–Aug. Limestone hills and dunes, AP (Baardskeerdersbos to Stilbaai).•

cephalotes (Thunb.) Less.   Spreading, white-woolly shrub sometimes to 1 m. Leaves reflexed, scarcely twisted, 3–20 mm long, with axillary tufts. Flower heads discoid, several, fused in dense terminal clusters, 5-flowered, bracts all petaloid, usually glandular, inner spreading, pink or white. Mainly Aug.–Nov. Sandstone slopes, NW, SW (Du Toitskloof Mts to Kogelberg).•

compacta Zeyh. ex Sch.Bip.   Robust, white-woolly shrub to 1.3 m. Leaves ascending, twisted, 6–15 mm long. Flower heads discoid, several in dense terminal clusters, 6-flowered, bracts all petaloid, slightly spreading, outer reddish to pink, inner white. July–Dec. Sandy flats and slopes to 700 m, SW (Cape Peninsula).•

confusa Pillans   Spreading, white-woolly shrublet to 40 cm. Leaves lanceolate, glabrescent, slightly twisted, 3–10 mm long, with axillary tufts. Flower heads discoid, few, fused in terminal clusters, (4)5-flowered, bracts slightly spreading, inner petaloid, white or pinkish. Nov.–Mar. Damp sandstone slopes, SW (Kogelberg to Palmiet River Mts).•

densa (Lam.) Karis   BLOMBOS   Erect white-woolly shrub to 2.5(–4.0) m. Leaves often reflexed, lanceolate to ovate, twisted, 2–15 mm long, with axillary tufts. Flower heads discoid, several in terminal clusters, 3–5-flowered, bracts erect or rarely spreading, outer brown below, inner petaloid, white sometimes brown. Mainly June–Oct. Sandy or stony flats and slopes, NW, SW, KM, AP, LB, SE (Namaqualand to N Province).

distans (Schrank) DC.   Slender, weak, sparsely leafy, white-woolly shrublet to 80 cm. Leaves twisted, sparsely hairy, 10–30 mm long. Flower heads discoid, several in terminal clusters, 5-flowered, inner bracts petaloid, spreading, white to yellow sometimes with reddish tips. Mar.–June. Sandy hills and flats, SW (Darling to Kraaifontein).•

divergens (Thunb.) D.Don   Sprawling, white-woolly shrublet to 70 cm. Leaves reflexed, slightly twisted, 2–12 mm long, with axillary tufts. Flower heads discoid, several, fused, in dense terminal clusters, 5-flowered, bracts all petaloid, mostly glandular, outer brown to reddish, inner spreading, white or pink. Aug.–Jan. Sandstone slopes, SW (Cape Peninsula to Gordon's Bay).•

dregeana DC. Similar to **M. densa** but young branches densely woolly and bracts acuminate and recurved, white or cream-coloured. Mar.–Apr. Sandstone and clay slopes NW, SW (Cedarberg Mts to Helderberg).•

fastigiata (Thunb.) D.Don Densely white-woolly shublet to 1.5 m cm. Leaves reflexed, twisted, 4–20 mm long, with axillary tufts. Flower heads discoid, several, in terminal clusters, 5-flowered, bracts erect, inner petaloid, pink. Aug.–Dec. Sandy flats and slopes, NW, SW (Bokkeveld Mts to Wellington).•

galpinii L.Bolus Erect, white-woolly shrub to 1.2 m. Leaves ascending, curved inwards with reflexed tips, not twisted, 3–15 mm long. Flower heads discoid, few in terminal clusters, 5-flowered, bracts mostly spreading above, inner petaloid, white. Nov.–Dec. Stony N-facing clay slopes, 350–750 m, LB (Langeberg Mts).•

humilis Karis Similar to **M. bodkinii** but leaves glabrescent, outer involucral bracts reddish to purplish and the inner cream-coloured. Oct.–Dec. Sandstone slopes. SW (Houwhoek Mts).•

inversa Karis Like **M. pulchella** but yellowish woolly and outer bracts shorter than inner. Aug.–Dec. Stony slopes to 600 m, SW (Stellenbosch to Bredasdorp).•

juniperoides Pillans Spreading, white-woolly shrub to 3 m. Leaves glandular, 2–20 mm long, with or without axillary tufts. Flower heads discoid, few in terminal clusters exceeded by a few bract-like leaves, 5-flowered, bracts erect, inner petaloid, white. Jan.–Mar. Sandstone slopes, NW (Cold Bokkeveld).•

lichtensteinii Less. Robust white-woolly shrub to 1.5 m. Leaves twisted, 4–15 mm long, with axillary tufts. Flower heads discoid, several in terminal clusters, 8–13-flowered, bracts recurved at tips, inner petaloid, white. Aug.–Jan. Lower sandstone slopes to 600 m, SW (Gordon's Bay to Botrivier).•

luteola Karis Erect, densely white-woolly shrub to 1.5 m. Leaves curved, twisted, 4–6 mm long, with conspicuous axillary tufts. Flower heads discoid, several in terminal clusters, 4- or 5-flowered, inner bracts spreading, truncate, pale yellow. July–Aug. Limestone hills, AP (Stilbaai).•

massonii S.Moore Like **M. densa** but flower heads 5-flowered, bracts more reflexed and spreading at tips. Aug.–Oct. Mountain slopes, 300–1700 m, KM, LB, SE (Klein Swartberg Mts to Uitenhage).•

montana Karis Compact white-woolly shrublet to 30 cm. Leaves spreading, lanceolate to ovate, twisted, 4–10 mm long, with axillary tufts. Flower heads discoid, several in terminal clusters, 5-flowered, bracts erect or rarely spreading, outer brown below, inner petaloid, white sometimes brown. Dec.–May. Sandstone slopes, 1200–1700 m, SW (Stettynsberg and Wemmershoek Mts).•

muraltiifolia DC. (= *Metalasia barnardii* L.Bolus) Rigid, white-woolly shrub to 2.5 m. Leaves slightly twisted, 7–15 mm long, with small axillary tufts. Flower heads discoid, several, fused, in dense, branched terminal clusters, 5-flowered, bracts erect, inner petaloid, irregularly toothed above, pink or white. Oct.–Nov. Sandstone slopes, 600–800 m, NW, SW (Witzenberg to Wemmershoek Mts).•

muricata (L.) D.Don BLOMBOS Similar to **M. densa** but leaves hooked at tips, bracts brown, inner bracts thick and faintly keeled above. May–Sept. Coastal sands to 300 m, SW, AP, LB, SE (Yzerfontein to Transkei).

octoflora DC. Erect, white-woolly shrublet to 40 cm. Leaves twisted, 2–11 mm long, with axillary tufts. Flower heads discoid, few in terminal clusters, 5–7-flowered, bracts erect, inner bracts petaloid, dull yellow. Mar.–Apr. Clay flats in renosterveld, SW (Durbanville to Wolseley).•

oligocephala Karis Spreading white-woolly shrublet to 50 cm. Leaves twisted, hairy, 4–6 mm long. Flower heads discoid, 1 or 2 at branch tips, 5-flowered, florets slightly zygomorphic, innermost bracts petaloid, whitish, tips recurved and hairy. Nov.–Dec. Sandstone slopes, LB (Langeberg Mts).•

pallida Bolus Erect, white-woolly shrub to 2 m. Leaves curved inwards with reflexed tips, not twisted, 2–15 mm long. Flower heads discoid, several in terminal clusters, 6–15-flowered, bracts erect, inner petaloid, concave above and serrulate, light brown to white. Aug.–Nov. Rocky slopes, KM, SE (Klein Swartberg to Humansdorp and S Karoo).

plicata Karis Like **M. densa** but inner bracts often cream-coloured and apically pleated. Apr.–May. Clay slopes, SW, LB (Houwhoek to Swellendam).•

pulchella (Cass.) Karis Erect or sprawling white-woolly shrublet to 40 cm. Leaves spreading, twisted, 4–15 mm long, with axillary tufts. Flower heads discoid, several, fused in dense terminal clusters, 5-flowered, bracts all petaloid, usually glandular, inner erect, folded and irregularly toothed above, white. Oct.–Dec. Sandy flats and slopes to 200 m, SW (Cape Peninsula and Cape Flats).•

pulcherrima Less.   Erect white-woolly shrub to 1.5 m. Leaves erect, narrowly elliptic, not twisted, 2–15 mm long. Flower heads discoid, few in terminal clusters, 5(6)-flowered, bracts erect, inner petaloid, yellow to white. Mainly Sept.–Dec. Stony sandstone slopes, 600–2000 m, KM, SE (Witteberg to Langkloof).•

pungens D.Don   Like **M. densa** but leaves not twisted, outer bracts acuminate. Mainly Apr.–July. Mostly sandstone slopes, to 1800 m, AP, KM, LB, SE (Agulhas to Grahamstown).

quinqueflora DC.   Sparsely white-woolly shrublet to 50 cm. Leaves twisted, 3–18 mm long. Flower heads discoid, several in terminal clusters, 5-flowered, outer bracts recurved, inner petaloid, erect, white or pink. Aug.–Mar. Coastal sandy flats and lower slopes to 400 m, SW (Hottentots Holland Mts to Botrivier).•

strictifolia Bolus   Compact white-woolly shrublet to 50 cm. Leaves erect, scarcely twisted, 3–9 mm long. Flower heads discoid, few in dense terminal clusters, 5-flowered, inner bracts petaloid, spreading, white. Nov.–Jan. Sandstone slopes and ridges, 1200–1500 m, KM, SE (Klein Swartberg to Kouga Mts).•

tricolor Pillans   Densely white-woolly shrublet to c. 60 cm. Leaves not twisted, with axillary tufts. Flower heads discoid, few in terminal clusters, 6–8-flowered, inner bracts petaloid, spreading, pleated, pink or white. May–July. Sandstone slopes, KM (Rooiberg).•

umbelliformis Karis   Erect, white-woolly shrub to 1.2 m. Leaves slightly twisted, hooked at tips, 3–9 mm long, with axillary tufts. Flower heads discoid, several in globose terminal clusters, 5-flowered, inner bracts petaloid, speading, keeled above, white, sometimes reddish. Sept.–Oct. Limestone ridges, AP (Pearly Beach to Brandfontein).•

MICROGLOSSA   c. 10 spp., Africa, Madagascar and Asia

mespilifolia (Less.) Robins.   Scrambling, softly woody shrub, 1–2 m, sometimes to 4 m. Leaves petiolate, deltoid and coarsely toothed. Flower heads shortly radiate, several in dense corymbs, yellow with dull white rays. Jan.–Dec. Coastal bush, SE (Van Staden's Mts to KwaZulu-Natal).

MIKANIA   c. 430 spp., pantropical, mainly America, 2 in southern Africa

capensis DC.   Scrambling climber. Leaves sagittate, slightly toothed. Flower heads discoid, in corymbs, white. Aug.–Oct. Forest and bush, SE (Knysna to tropical Africa).

[**MINUROTHAMNUS** 1 sp., W Cape, identity unknown, presumably a synonym of another genus]

NESTLERA   1 sp., winter-rainfall South Africa

biennis (Jacq.) Spreng. (= *Relhania biennis* (Jacq.) Bremer)   Thinly cobwebby annual or ?biennial to 35 cm. Leaves linear, gland-dotted, with revolute margins. Flower heads radiate, solitary and terminal, yellow with yellow rays. Sept.–Oct. Sandy flats and lower slopes, NW, SW (S Namaqualand to Hopefield).

NIDORELLA   VLEIKRUID   c. 15 spp., southern and tropical Africa

auriculata DC.   Shortly hairy, single-stemmed, erect perennial to 1.5 m. Leaves oblanceolate, auriculate at base, coarsely toothed, serrate. Flower heads discoid, in dense nodding corymbs, yellow. Jan.–June. Damp places, often marshes, or forest margins, SW, LB, SE (Caledon to tropical Africa).

foetida (L.) DC.   Aromatic, scabrid, softly woody shrub to 60 cm. Leaves narrowly oblanceolate, in tufts. Flower heads discoid, in dense corymbs, often nodding, yellow. Mainly Sept.–Apr. Damp sites, often seeps and marshes, NW, SW, AP, LB, SE (Lambert's Bay to E Cape).

resedifolia DC.   Stout, roughly hairy annual to 90 cm. Leaves obovate, half-clasping at base, sometimes lobed. Flower heads discoid, in dense corymbs, yellow. Sept.–Apr. Roadsides and grassland, SW, LB, SE (Caledon to tropical Africa).

undulata (Thunb.) Sond. ex Harv.   Single-stemmed perennial to 90 cm. Leaves crowded at base, obovate, leathery, slightly toothed. Flower heads discoid, in dense corymbs, yellow. Oct.–Jan. Damp grassland and seeps, LB, SE (Riversdale to tropical Africa).

OEDERA (= *RELHANIA* in part)   PERDEKAROO   18 spp., Namaqualand to E Cape

A. Flower heads congested in false capitula

capensis (L.) Druce   Densely leafy, sprawling shrublet to 30 cm. Leaves lanceolate, spreading or recurved, glandular-scabrid on margins, sometimes silky. Flower heads radiate, several crowded in a large false head, orange or yellow. June–Sept. Dry stony flats and slopes, SW, AP, KM, LB (Cape Peninsula to Albertinia and Little Karoo).•

hirta Thunb.   Like **O. capensis** but leaves glandular-scabrid on underside, often twisted. Sept.–Dec. Sandstone slopes, NW, SW (Cold Bokkeveld and Piketberg to Wemmershoek Mts).•

imbricata Lam. (= *Oedera intermedia* DC.)   Similar to **O. capensis** but leaves broadly lanceolate to ovate. Aug.–Nov. Mountain slopes, SW, KM, AP, LB, SE (Yzerfontein to Grahamstown).

laevis DC. (= *Oedera muirii* C.A.Sm.)   Densely leafy shrublet to 45 cm. Leaves lanceolate, striate, pungent, minutely toothed. Flower heads shortly radiate, small, subsessile, crowded in terminal umbels, with 1 reduced ray floret per head, yellow. Oct. Rocky sandstone slopes, LB (Langeberg Mts: near Garcia's Pass).•

sp. 1   Densely leafy, prostrate dwarf shrublet to 10 cm. Leaves imbricate, linear-lanceolate, slightly fleshy. Flower heads shortly radiate, several crowded in a large false head, yellow, involucral bracts large, leafy, the inner ciliate, pappus scales free, needle-like. Sept.–Mar. Sandstone outcrops, NW (Cold Bokkeveld and Hex River Mts).•

A. Flower heads in umbels or solitary

foveolata (Bremer) Anderb. & Bremer   Like **O. multipunctata** but leaves smaller, flower heads campanulate and achenes densely hairy. Sept. Rocky slopes, NW (Karoopoort).•

genistifolia (L.) Anderb. & Bremer   KLEINPERDEKAROO   Twiggy shrub to 1 m. Leaves oblong, erect or spreading, often recurved at tips, often gummy above. Flower heads radiate, few in dense terminal umbels, with glandular-hairy florets, yellow. Aug.–Nov. Shale and clay slopes in renosterveld, NW, SW, AP, KM, LB, SE (Namaqualand to E Cape).

multipunctata (DC.) Anderb. & Bremer   PERDEKAROO   Twiggy shrub to 1 m, white-woolly on young parts. Leaves oblanceolate, spreading, gland-dotted. Flower heads radiate, few in slender peduncles in lax umbels, yellow. Aug.–Oct. Rocky sandstone slopes, NW (Bokkeveld Mts to Swartruggens).•

resinifera (Bremer) Anderb. & Bremer   Similar to **O. multipunctata** but leaves elliptic, densely glutinous and not gland-dotted. Aug.–Oct. Karroid lower slopes, KM (Klein Swartberg Mts and Touwsberg foothills).•

sedifolia (DC.) Anderb. & Bremer   Similar to **O. multipunctata** but leaves smaller. Flower heads subsessile and solitary. Mainly Aug.–Dec. Rocky sandstone slopes, NW, KM (Namaqualand to Witteberg and W Karoo).

squarrosa (L.) Anderb. & Bremer   VIERKANTPERDEKAROO   Similar to **O. genistifolia** but leaves obovate, recurved and mucronate at tips, young leaves 4-ranked. Mainly Sept.–Nov. Stony sand or clay slopes and flats, NW, SW, KM, AP, LB, SE (Bokkeveld Escarpment to Port Elizabeth).•

steyniae (L.Bolus) Anderb. & Bremer   Densely leafy shrublet to 30 cm, somewhat woolly on young parts. Leaves lanceolate, keeled and pungent, woolly on margins, 4-ranked. Flower heads shortly radiate, subsessile in terminal umbels, yellow. July–Sept. Limestone rocks, AP (Stilbaai to Gouritsmond).•

uniflora (L.f.) Druce   KALKSTEENPERDEKAROO   Rounded twiggy shrublet to 50 cm. Leaves oblong, channelled and recurved at tips, 4-ranked. Flower heads discoid, solitary at branch tips, yellow. Oct.–Jan. Coastal sands and limestones, SW, AP, LB (Saldanha Bay to Mossel Bay).•

viscosa (L'Hér.) Anderb. & Bremer   Like **O. genistifolia** but leaves thick, somewhat channelled and glutinous. Sept.–Nov. Clay or sandy lower slopes, NW, SW (Piketberg to Tygerberg).•

OLDENBURGIA   KREUPELBOS   4 spp., W to E Cape

intermedia Bond   Cushion-forming shrublet to 30 cm, densely woolly in axils. Leaves leathery, oblanceolate, margins revolute, woolly below. Flower heads radiate, solitary on white-woolly peduncles, white. Mainly Feb.–Mar. Sandstone rocks above 1000 m, NW, SW (Cedarberg to Hex River Mts and Hottentots Holland Mts).•

papionum DC.   Dwarf, gnarled shrublet to 1 m, woolly in axils. Leaves large, leathery, oblanceolate, margins revolute, woolly below. Flower heads radiate, 1–3 on branched, glabrescent peduncles, pinkish with white rays. Mar.–June. Sandstone slopes, to 1400 m, SW (Tulbagh Kloof to Du Toitskloof).•

paradoxa Less.   Like **O. intermedia** but flower heads sessile. Jan.–June. Sandstone rocks at high alt., SW, KM, LB, SE (Villiersdorp to Avontuur).•

OLIGOCARPUS (= *OSTEOSPERMUM* in part)   BOEGOEBOSSIE   1 sp., W to E Cape

calendulaceus (L.f.) Less   Sprawling, foetid annual to 40 cm. Leaves oblanceolate, toothed. Flower heads radiate, small, few at branch tips, yellow. Achenes mostly irregularly warty. June–Dec. Dry rocky hills and flats, NW, KM, AP, SE (Worcester to E Cape).

OLIGOTHRIX•   1 sp., W Cape

gracilis DC.   Diffuse, wiry stemmed annual to 30 cm. Leaves obovate and tapering below, coarsely toothed, becoming narrower above and clasping. Flower heads radiate, small in open panicles, yellow with yellow or white rays. Dec.–Jan. Sandstone slopes above 1000 m, NW (Cedarberg Mts).•

ONCOSIPHON   STINKKRUID   7 spp., southern Africa

africanum (P.J.Bergius) Kallersjö (= *Matricaria africana* P.J.Bergius, *M. hirta* (Thunb.) DC.)   Erect to spreading annual to 30 cm. Leaves pinnatisect, at least lower leaves bipinnatisect. Flower heads radiate, solitary on thick peduncles, yellow with white rays. Sept.–Oct. Sandy flats and lower slopes waterlogged in winter, NW, SW (Graafwater to Cape Peninsula).•

glabratum (Thunb.) Kallersjö (= *Matricaria glabrata* (Thunb.) DC.)   WILD CHAMOMILE   Like **O. africanum** but often smaller and peduncles more slender. Sept.–Oct. Seasonally waterlogged, sandy flats, SW (Darling district).•

grandiflorum (Thunb.) Kallersjö (= *Pentzia grandiflora* (Thunb.) Hutch.)   GROOTSTINKKRUID   Robust, aromatic annual to 45 cm. Leaves bipinnatisect. Flower heads discoid, solitary, yellow, bracts woolly. Aug.–Nov. Sandy and stony flats and lower slopes, NW, SW (S Namibia and W Karoo to Melkbosstrand).

intermedium (Hutch.) Kallersjö (= *Pentzia intermedia* Hutch.)   Like **O. grandiflorum** but corolla lobes elongate giving heads a fluffy appearance, peduncles thick and flowers deep yellow to orange. Sept.–Oct. Sandy lower slopes, NW (N Cedarberg Mts).•

piluliferum (L.f.) Kallersjö (= *Pentzia pilulifera* (L.f.) Fourc.)   KAROOSTINKKRUID   Similar to **O. suffruticosum** but usually to 30 cm. Leaves bipinnatisect. Flower heads solitary or few in lax corymbs on slender peduncles. Aug.–Dec. Dry stony or sandy slopes and flats, often disturbed soil, NW, KM, SE (S Namibia to Ceres and Oudtshoorn to E Cape and Karoo).

sabulosum (Wolley-Dod) Kallersjö (= *Pentzia sabulosa* (Wolley-Dod) Hutch.)   Nearly prostrate annual to 20 cm. Leaves bipinnatisect with broad lobes. Flower heads discoid, few in dense terminal corymbs, yellow. Sept.–Mar. Coastal dunes, SW, AP (Melkbosstrand to Agulhas).•

suffruticosum (L.) Kallersjö (= *Pentzia suffruticosa* (L.) Hutch. ex Merxm.)   STINKKRUID, WURMBOSSIE   Much-branched, aromatic annual to 50 cm. Leaves bi- to tripinnatisect. Flower heads discoid, many in dense corymbs, yellow. Sept.–Dec. Sandy flats and slopes, often coastal, NW, SW, AP (S Namibia and W Karoo to Gansbaai).

OREOLEYSERA•   1 sp., W Cape

montana (Bolus) Bremer   Dwarf, cushion-forming, grey-felted shrublet to 8 cm. Leaves oblanceolate, erect. Flower heads radiate, slender, solitary on wiry, pubescent peduncles, yellow. Nov.–Jan. Sandstone rocks, NW, SW (Hex River Mts to Jonkershoek).•

OSMITOPSIS•   BELSKRUIE   9 spp., W Cape

A. Pappus absent

asteriscoides (P.J.Bergius) Less.   BELSKRUIE   Sparsely branched, aromatic shrub to 2 m with erect stems densely leafy above. Leaves lanceolate, smooth or felted, ascending, 10–60(80) mm long. Flower heads radiate, in terminal corymbs, yellow with white rays; pappus absent. Mainly Aug.–Dec. Marshes and seeps on sandstone, SW (Cape Peninsula to Riviersonderend Mts).•

nana Schltr.   Perennial to 15 cm with erect stems from woody rootstock. Leaves oblanceolate, reflexed, toothed above, 5–25 mm long. Flower heads radiate, solitary, yellow with white rays; pappus absent. Oct.–Feb. Sandstone rocks mostly above 600 m, NW, SW (Grootwinterhoek Mts to Genadendal).•

parvifolia (DC.) Hofmeyr   Densely leafy shrublet to 40 cm. Leaves ovate, reflexed, toothed, 4–10 mm long. Flower heads radiate, solitary, yellow with white rays; pappus absent. Sept.–Feb. Stony sandstone slopes and rock crevices, SW (Sir Lowry's Pass to Betty's Bay).•

tenuis Bremer   Slender, sprawling, aromatic, thinly woolly shrublet to 25 cm. Leaves obovate, reflexed, sparsely toothed and apiculate, 6–15 mm long. Flower heads radiate, solitary, yellow with white rays; pappus absent. Aug.–Feb. Moist rocky slopes often in shade, SW (Bainskloof to Du Toitskloof).•

A. Pappus present at least in disc florets

afra (L.) Bremer   Densely leafy perennial with erect stems from woody rootstock, to 40 cm. Leaves oblong, the lower larger, reflexed, the upper ascending, finely toothed above, densely felted, 5–20 mm long. Flower heads radiate, solitary on long peduncles, yellow with white rays; pappus of subequal scales. Mainly Nov.–Feb. Sandstone slopes, SW (Du Toitskloof to Kogelberg).•

dentata (Thunb.) Bremer   KAAPSEBELSKRUIE   Perennial to 50 cm with erect stems from woody rootstock. Leaves oblanceolate, the lower larger, spreading, the upper acending, sharply toothed, 5–45 mm long. Flower heads radiate, large, solitary, yellow with white rays; pappus of subequal scales. Mainly Oct.–Feb. Damp rocky slopes, SW (Cape Peninsula).•

glabra Bremer   Like **O. afra** but leaves glabrous and toothed throughout. Nov.–Feb. Rocky S-facing sandstone slopes, SW (Grabouw to Palmiet River Mts).•

osmitoides (Less.) Bremer   BASTERBELSKRUIE   Softly woody shrublet 15–60 cm with slender erect branches. Leaves oblanceolate, deeply to slightly toothed, 15–80 mm long. Flower heads radiate, solitary on long peduncles, yellow with white rays; pappus of subequal scales. Aug.–Jan. Moist slopes and forest margins, SW, LB, SE (Riviersonderend Mts to Great Winterhoek Mts).•

pinnatifida (DC.) Bremer   Densely leafy shrublet to 50 cm. Leaves oblong, deeply lobed to pinnatifid, spreading to reflexed, 5–40 mm long. Flower heads radiate, solitary at branch tips, yellow with white rays; pappus of dimorphic scales. Mainly Nov.–Jan. Rocky sandstone slopes mostly above 800 m, SW (Du Toitskloof to Jonkershoek).

OSTEOSPERMUM   BIETOU   c. 70 spp., mainly Africa, Middle East (2), St Helena (1) (see also **OLIGOCARPUS, TRIPTERIS**)

A. Plants annual

acanthospermum (DC.) Norl.   ONKRUIDBIETOU   Prostrate, thinly hairy annual to 15 cm. Leaves oblanceolate, toothed. Flower heads shortly radiate, small, solitary at branch tips, pale yellow. Achenes spiny. Aug.–Oct. Gravel and sandy flats, NW (Namaqualand and Karoo to Hex River Valley).

pinnatum (Thunb.) Norl.   JAKKALSBIETOU   Sprawling, glandular-hairy annual to 20 cm. Leaves pinnatisect, lobes linear to filiform. Flower heads radiate, solitary at branch tips, disc yellow, rays white to orange or biscuit but dark at base. Achenes toothed or tubercled. July–Sept. Rocky slopes and flats, NW, KM (S Namibia and Karoo to Klawer and Little Karoo).

A. Plants perennial
B. Rays white or mauve with darker reverse

acutifolium (Hutch.) Norl. (= *Dimorphotheca acutifolia* (Hutch.) B.Nord.; incl. **O. wallianum** Norl.)   Sprawling, roughly hairy shrublet to 50 cm. Leaves linear-oblanceolate, usually toothed, sometimes sharply so. Flower heads radiate, solitary at branch tips, purple to brown with white rays purple to brown on reverse. Achenes trigonous and warty, c. 6 mm long. Aug.–Nov. Sandstone slopes, NW, SW (N Cedarberg to Shaw's Mts).•

dregei (DC.) Norl. (= *Dimorphotheca dregei* (DC.) B.Nord.)   BIETOU   Roughly hairy, tufted perennial to 20 cm, from woody base. Leaves linear, usually sharply toothed. Flower heads radiate, solitary on elongate peduncles, purple with white rays brown on reverse. Achenes trigonous and warty, c. 3 mm long. Mainly Aug.–Sept. Sandstone slopes, NW, SW, KM, LB (Bokkeveld Mts to Bredasdorp and Klein Swartberg).•

ecklonis (DC.) Norl. (= *Dimorphotheca ecklonis* (DC.) B.Nord.)   VANSTADEN'S DAISY   Sprawling, glabrescent, softly woody shrub, sometimes to 1.5 m. Leaves elliptic, mostly petiolate and toothed.

Flower heads radiate, solitary on long peduncles, violet-blue with white rays. Achenes trigonous, ridged and rugose. Apr.–Oct. Rocky slopes and flats, SE (Langkloof to Uitenhage).•

fruticosum (L.) Norl. (= *Dimorphotheca fruticosa* (L.) B.Nord.) RANKBIETOU Shortly hairy, sprawling to prostrate perennial. Leaves obovate, petiolate, fleshy, minutely toothed. Flower heads radiate, large, solitary on naked peduncles, purple with white or mauve rays. Achenes smooth, trigonous, c. 6 mm long. June–Oct. Coastal dunes and rocks, SW, AP, SE (Saldanha Bay to KwaZulu-Natal).

B. Rays yellow
C. Upright, few-branched willowy shrubs

asperulum (DC.) Norl. Roughly hairy shrub to 1 m with suberect branches, thinly woolly on young parts. Leaves linear, ascending, margins scabrid and slightly toothed and thickened. Flower heads radiate, solitary or in lax corymbs, yellow, often copper on reverse. Achenes smooth. May–Oct. Rocky sandstone ridges above 700 m, KM (Swartberg Mts).•

bolusii (Compton) Norl. Willowy perennial to 80 cm, with ribbed stems. Leaves sparse, linear, adpressed, caducous. Flower heads radiate, in lax terminal panicles, yellow. Achenes smooth and winged, c. 3 mm long. Mainly Apr.–May. Sandstone slopes, KM, ?LB, SE (Witteberg to Langkloof).•

corymbosum L. Robust, stiffly erect, single-stemmed shrub to 2.5 m. Leaves lanceolate, decreasing in size above. Flower heads radiate, in terminal, branched corymbs on roughly hairy peduncles, yellow. Achenes smooth, c. 7 mm long. Aug.–Feb. Sandstone slopes above 500 m, LB, SE (Langeberg Mts to George).•

herbaceum L.f. Willowy to straggling, glabrescent perennial to 1.2 m. Leaves opposite, ovate, clasping, toothed. Flower heads radiate, solitary on elongate peduncles, yellow. Achenes smooth, 3-angled. Mar.–Apr. Forest margins, SE (Knysna to KwaZulu-Natal).

junceum P.J.Bergius Softly woody, single-stemmed shrub to 3 m, white-woolly when young. Leaves oblanceolate, toothed, decreasing in size above. Flower heads radiate, in lax, branched corymbs on woolly peduncles, yellow. Achenes smooth, slightly lobed. July–Jan. Sandstone slopes, NW, SW, AP, KM, LB, SE (Olifants River Mts to Grahamstown).

polygaloides L. (incl. **O. imbricatum** L.) Densely leafy, glabrescent shrub to 2 m, branches stiffly erect. Leaves oblong to ovate, leathery, ascending, recurved at tips. Flower heads radiate, solitary on short, roughly hairy peduncles, yellow. Achenes ribbed and pitted, 5–7 mm long. Mainly Aug.–Dec. Rocky, mostly sandstone slopes, NW, SW, AP, KM, LB, SE (Olifants River Mts to KwaZulu-Natal).

rotundifolium (DC.) Norl. Like **O. polygaloides** but leaves broadly ovate. Flower heads in lax panicles. Achenes smooth and obscurely ribbed, c. 4 mm long. Oct.–Dec. Sandstone slopes, SW, AP, KM (Kogelberg to Franskraal, Witteberg).•

sp. 1 Like **O. junceum** but young parts and involucres densely felted. Flower heads crowded on short, densely felted peduncles. Sept.–Oct. Rocky sandstone ridges above 1000 m, SE (Great Winterhoek Mts).•

sp. 2 Like **O. polygaloides** but leaves larger, not recurved at tips, margins thickened and peduncles short and woolly. Mainly Oct.–Apr. Limestone hills, AP (Agulhas Peninsula).•

C. Straggling perennials and shrublets

aciphyllum DC. Trailing shrublet to 30 cm, thinly woolly on young parts. Leaves linear, pungent, margins revolute, minutely toothed. Flower heads radiate, solitary at branch tips, rays twice as long as bracts, yellow. Achenes smooth. Sept.–Dec. Rocky sandstone slopes, NW, SW (Piketberg to Caledon).•

bidens Thunb. Thinly woolly perennial from woody rootstock, to 60 cm. Leaves crowded at base, linear-oblanceolate, margins revolute, toothed, discolorous, densely woolly below. Flower heads radiate, in lax corymbs, yellow. Achenes smooth, oblong. Sept.–Dec. Rocky flats and slopes, NW (Namaqualand to Worcester).

ciliatum P.J.Bergius Thinly woolly, sprawling shrublet to 50 cm. Leaves elliptic, margins revolute, crenate. Flower heads radiate, small, few at branch tips, yellow. Achenes smooth. Mainly Sept.–Dec. Sandstone slopes, SW (Cape Peninsula to Houwhoek).•

elsieae Norl. Cobwebby, sprawling, tangled shrublet to 50 cm. Leaves rounded, discolorous, glabrescent above, margins revolute, undulate. Flower heads radiate, solitary on short peduncles, yellow. Achenes smooth. Sept.–Oct. Rocky southern slopes in gullies, SW (Potberg).•

glabrum N.E.Br.   Twiggy, dichotomously branched shrub to 70 cm, densely leafy above. Leaves linear-trigonous, ascending. Flower heads radiate, solitary, nested in leaves, yellow or orange with dark reverse. Achenes smooth, trigonous. Aug.–Sept. Sandstone slopes, LB, SE (Garcia's Pass to Humansdorp).•

grandidentatum DC.   Sprawling, roughly hairy perennial to 1 m. Leaves elliptic to oblanceolate, toothed, clasping at base. Flower heads radiate, solitary at branch tips, yellow. Achenes smooth, c. 8 mm long. Oct.–Jan. Grassy slopes, SE (Nature's Valley to Swaziland).

grandiflorum DC.   STINKBIETOU   Roughly hairy, foetid shrublet to 80 cm. Leaves oblanceolate to ovate, usually slightly toothed. Flower heads radiate, in lax terminal corymbs, orange or yellow. Achenes broadly winged, c. 12 mm long. July–Oct. Rocky slopes, NW, SW (Namaqualand to Saldanha Bay).

hispidum Harv.   Roughly hairy, sprawling shrublet to 30 cm. Leaves lanceolate, margins revolute, apiculate, cobwebby below. Flower heads radiate, 1–few at branch tips, yellow. Achenes smooth. Aug.–Oct. Rocky lower slopes, SW, LB (Caledon to Langeberg Mts: Heidelberg).•

pterigoideum Klatt   Like **O. ciliatum** but leaves entire and cobwebby when young. Nov.–Mar. Sandstone slopes, SE (George to Humansdorp).•

pyrifolium Norl.   Like **O. ciliatum** but leaves petiolate and white-woolly when young. Flower heads large. Oct. Sandstone slopes, LB (only Garcia's Pass, Riversdale).•

rigidum Aiton   Roughly hairy or prickly, stiffly branched shrublet to 60 cm. Leaves foetid, pinnatifid, lobes acute and pungent. Flower heads radiate, solitary at branch tips, yellow. Achenes smooth or warty and winged. Aug.–Sept. Rocky slopes or near water, NW, SW (S Namaqualand and Karoo to Paarl and to Witteberg).

spinosum L.   Glandular-hairy, aromatic, intricately branched, thorny shrub to 1.2 m. Leaves linear to pinnatisect. Flower heads radiate, solitary at branch tips, pale yellow. Achenes, smooth, c. 8 mm long, sometimes angled. Mainly May–Oct. Gravelly slopes and flats, NW, SW, KM (Pakhuis Mts to Tulbagh and Riviersonderend Mts).•

subulatum DC. (incl. **O. hafstroemii** Norl.)   Thinly white-woolly, densely leafy, prostrate shrublet to 20 cm. Leaves linear, recurved above, margins revolute. Flower heads radiate, solitary on short hairy peduncles, yellow or orange. Achenes smooth, narrowly winged, c. 5 mm long. Aug.–Apr. Limestone hills, AP (Agulhas to Potberg).•

triquetrum L.f.   Like **O. aciphyllum** but leaves linear-trigonous, acute, not pungent, margins entire and ray florets not much longer than bracts. Oct.–Dec. Sandstone slopes and gullies, KM, LB, SE (Riversdale to George).•

OTHONNA   BOBBEJAANKOOL   c. 140 spp., W Cape to KwaZulu-Natal and Namibia

A. Flower heads disciform

arbuscula (Thunb.) Sch.Bip.   TRAAP   Di- or trichotomously branched, fleshy often leafless shrub to 2 m, thinly woolly on young parts. Leaves oblanceolate. Flower heads disciform, yellow, 1 or 2 at branch tips on short peduncles; pappus of marginal florets elongated in fruit. Mainly May–July. Dry rocky slopes, NW (Namaqualand to Worcester Karoo).

digitata L.   Tuberous succulent perennial to 30 cm. Leaves oblanceolate, often lobed to pinnatisect, woolly in axils. Flower heads disciform, solitary at branch tips on elongate peduncles, mauve to purple; pappus of marginal florets much elongated in fruit. June–Oct. Sandy flats and slopes, NW, SW (Piketberg to Elim).

filicaulis Jacq.   BOBBEJAANKOOLKLIMOP   Straggling, tuberous, succulent perennial to 70 cm, stems from a woolly crown. Leaves lanceolate to rotund, clasping and cordate at base, glaucous. Flower heads disciform, solitary and terminal or axillary, white or yellow; pappus of marginal florets much elongated in fruit. May–Aug. Sandy flats and slopes, often coastal, NW, SW, KM, LB, SE (S Namibia to Uniondale).

lingua Jacq. (incl. **O. gymnodiscus** (DC.) Sch.Bip.)   Tuberous, erect or sprawling, succulent perennial to 35 cm. Leaves oblanceolate to elliptic, leathery, tapering below and weakly clasping. Flower heads disciform, solitary, terminal or axillary, yellow; pappus of marginal florets elongated in fruit. July–Aug. Sandy slopes and flats, NW, SW, KM, LB, SE (Namaqualand to Uniondale).

lobata Schltr.   Rounded, tuberous, succulent shrublet to 20 cm. Leaves pinnatifid, often woolly in axils. Flower heads disciform, 1–few on short peduncles, yellow; pappus of marginal flowers much elongated in fruit. July–Oct. Rocky sandstone slopes, NW, KM (Cedarberg Mts to Witteberg and W Karoo).

protecta Dinter   Dwarf succulent shrublet with fleshy stems to 20 cm. Leaves crowded at branch tips, fleshy, fusiform to terete, woolly in axils. Flower heads disciform, few on sparsely leafy peduncles, yellow; pappus of marginal florets much elongated in fruit. May–Oct. Karroid slopes, often under shrubs, NW, KM (S Namibia to Witteberg and W Karoo).

retrofracta Less.   Deciduous, fleshy-leaved, divaricately branched shrublet to 30 cm, with swollen fleshy caudex. Leaves oblanceolate, pinnatifid, in tufts at branch tips, woolly in axils. Flower heads disciform, 1–3 on short terminal peduncles, yellow; pappus of marginal florets elongated in fruit. May–July. Dry rocky slopes, NW, KM, SE (Worcester to Uitenhage and E Cape).•

sp. 1   Like **O. arbuscula** but branch tips densely woolly. July–Oct. Dry rocky slopes, KM (Namaqualand to Witteberg).

A. Flower heads radiate
B. Geophytic perennials with a woolly crown and with leaves mostly basal, rarely cauline

auriculifolia Licht. ex Less. (incl. **O. humilis** Schltr.)   Stemless tuberous perennial to 15 cm. Leaves radical, petiolate, oblanceolate, toothed to deeply lobed or pinnatifid, woolly in axils. Flower heads radiate, solitary on stiff, naked peduncles, yellow often with red reverse; pappus of ray florets much elongated in fruit. Apr.–Sept. Stony clay slopes and flats, NW, KM (Bokkeveld Escarpment and W Karoo to Klein Swartberg Mts).

bulbosa L.   Tuberous perennial to 30 cm. Leaves mostly basal, woolly in axils, the lower petiolate, oblanceolate to elliptic, sometimes lobed. Flower heads radiate, solitary on sparsely leafy stems, yellow. June–Oct. Sandy slopes and flats, NW, SW, AP (Cold Bokkeveld and Piketberg to Bredasdorp).•

hederifolia B.Nord.   Tuberous, tufted perennial to 15 cm. Leaves mostly basal, woolly in axils, petiolate, cuneate, broadly toothed above. Flower heads radiate, solitary on long naked peduncles, yellow. June–Aug. Sandstone pavement, NW, KM (Bokkeveld Mts to Witteberg).•

heterophylla L.f.   Similar to **O. bulbosa** but leaves usually slightly toothed to crenate and roughly hairy or woolly especially below, margins revolute. May–Sept. Sandy and clay flats and slopes, SW, LB (Cape Peninsula to Langeberg: Heidelberg).•

oleracea Compton   Like **O. bulbosa** but leaves larger. Flower heads on rigid, simple or sometimes branched peduncles. Aug.–Sept. Sandstone slopes, NW (Cedarberg Mts to Karoopoort).•

perfoliata Jacq. (incl. **O. amplexifolia** DC.)   Straggling, tuberous, succulent perennial to 70 cm, stems from a woolly crown. Leaves lanceolate to rotund, clasping and cordate at base, glaucous. Flower heads radiate, solitary and terminal or axillary, yellow. May–Sept. Sandstone slopes, NW, SW (Matsikamma to Caledon Swartberg).•

petiolaris DC.   Tuberous succulent perennial to 40 cm, stems from a woolly crown. Leaves mostly basal, large, ovate, petiolate, irregularly lobed or lacerate. Flower heads radiate, in lax corymbs on stout, elongate peduncles, yellow. June–Aug. Sandstone slopes, NW (Gifberg to Clanwilliam).•

pinnata L.f. (ncl. **O. reticulata** DC.)   Like **O. bulbosa** but leaves firm-textured with thickened margins and usually pinnatisect with elliptic lobes. June–Sept. Sandy flat and slopes, often seasonally wet, NW, SW (Olifants River Mts to Cape Peninsula and Worcester).•

sonchifolia DC.   Like **O. auriculifolia** but to 30 cm, leaves pinnatisect with lobes quadrate and flower heads 1–few on stout naked peduncles. July–Aug. Granite outcrops, NW, SW (Doringbaai to Darling).•

stenophylla Levyns   Tuberous perennial to 40 cm. Leaves mostly basal, from a woolly crown, on long wiry petioles, narrowly elliptic to linear. Flower heads radiate, solitary on terminal or axillary peduncles, yellow. June–Sept. Damp sandy flats, SW (Yzerfontein to Gordon's Bay and Worcester).•

tephrosioides Sond.   Like **O. pinnata** but flower heads several in branched corymbs. June–July. Rocky sandstone slopes, NW (Cedarberg Mts to Piketberg).•

B.' Perennials with erect wand-like stems leafy below, more or less nude and corymbose above

parviflora P.J.Bergius (= *Othonna amplexicaulis* Thunb., *O. rigens* (L.) Levyns)   BOBBEJAANKOOL   Robust shrub to 2 m. Leaves oblanceolate, weakly clasping at base, margins finely toothed, sometimes smooth. Flower heads radiate, many in dense, branched terminal corymbs, yellow. Mainly July–Nov. Sandstone slopes, NW, SW, KM, LB, SE (Bokkeveld Mts to Langkloof).•

membranifolia DC.   Like **O. parviflora** but leaves not clasping, margins with a few coarse teeth and flower heads somewhat larger (involucres c. 9 mm long). Apr.–May. Riverine bush, SE (Humansdorp to Uitenhage).•

quinquedentata DC. (= *Othonna parviflora* L.)   Shrub with several slender erect stems to 1 m. Leaves ascending, leathery, crowded below, oblanceolate to obovate, often toothed above. Flower heads radiate, many in lax terminal corymbs, yellow. Mainly Oct.–Mar. Rocky sandstone slopes, often damp places, NW, SW, LB, SE (Cape Peninsula to Langkloof).•

B."Succulents or shrublets with leaves cauline
C. Flower heads solitary

arborescens L.   BOBBEJAANKOOL   Brittle-stemmed succulent shrublet to 60 cm. Leaves oblanceolate, crowded at branch tips, margins entire or slightly lobed to toothed, woolly in axils. Flower heads radiate, mostly solitary on elongate peduncles, yellow; pappus of ray florets much elongated in fruit. May–Sept. Coastal dunes or rocks, SW (St Helena Bay to Cape Peninsula).•

cacalioides L.f. (incl. **O. minima** DC.)   Dwarf succulent to 10 cm, from tuber-like caudex. Leaves tufted, dry to green at flowering, small, from woolly crown, oblanceolate, sometimes finely toothed. Flower heads shortly radiate, mostly solitary on stiff, naked peduncles, yellow. May–Oct. Sandstone pavement, NW (Bokkeveld Mts to Gifberg).•

ciliata L.f.   Sprawling shrublet to 30 cm, woody only near base. Leaves oblanceolate, usually pinnatifid, margins conspicuously bristled. Flower heads radiate, solitary on long terminal peduncles, yellow. July–Sept. Sandy slopes, NW, SW (Ceres to Cape Peninsula and Sir Lowry's Pass).•

coronopifolia L.   SANDBOBBEJAANKOOL   Semisucculent shrub to 1.5 m. Leaves oblanceolate, often irregularly toothed, usually in tufts on short branches. Flower heads radiate, solitary on long peduncles, yellow; pappus of marginal florets much elongated in fruit. July–Nov. Rocky sandstone and granite slopes especially along coast, NW, SW (Pakhuis Mts to Cape Peninsula).•

leptodactyla Harv.   Like **O. coronopifolia** but leaves narrowly lanceolate, often sparsely and irregularly toothed. Flower heads on shorter peduncles. Aug.–Sept. Rocky sandstone slopes, NW (Namaqualand to Piketberg).

ramulosa DC. (incl. **O. mucronata** Harv.)   Succulent-leaved shrub to 1.5 m. Leaves oblanceolate, mostly in tufts on short shoots, sometimes sparsely toothed. Flower heads radiate, solitary on short axillary or terminal peduncles, yellow; pappus of marginal florets elongated in fruit. May–Sept. Rocky karroid slopes, NW, KM (Swartruggens to Worcester and Witteberg to W Karoo).

sp. 2   Sprawling, softly woody perennial to 30 cm. Leaves leathery, broadly oblanceolate, toothed above. Flower heads radiate, solitary on elongate peduncles, yellow. Apr.–Nov. Coastal limestone sands, AP (De Hoop to Stilbaai).•

C. Flower heads mostly in corymbs

capensis L.H.Bailey   Sprawling succulent to 1.5 m, rooting at nodes. Leaves fleshy, fusiform, sometimes whorled, woolly in axils. Flower heads radiate, mostly few on slender terminal peduncles, white or yellow. May–Aug. Dry rocky flats, KM, SE (Ladismith to E Cape).

cylindrica (Lam.) DC. (incl. **O. alba** Compton, **O. carnosa** Less., **O. floribunda** Schltr.)   Brittle-stemmed, succulent shrub to 1 m. Leaves fleshy, fusiform, clustered at branch tips. Flower heads radiate, few in lax, terminal corymbs on long peduncles, yellow. Mainly July–Oct. Sandy and stony flats and rocks, NW, SW, KM, SE (Namaqualand to Humansdorp, W Karoo and E Cape).

dentata L.   Succulent shrublet to 70 cm, with fleshy caudex. Leaves obovate to oblanceolate, slightly toothed, crowded at branch tips. Flower heads radiate, 1–several in lax corymbs on stout peduncles, yellow. June–Dec. Rocky outcrops, SW, AP (Cape Peninsula to Stilbaai).•

multicaulis Harv.   Tufted perennial with slender woody caudex and sprawling or suberect stems to 40 cm. Leaves crowded basally, lanceolate to terete, succulent, sometimes sparsely toothed. Flower heads radiate, few to many in terminal panicles, yellow. Mainly Apr.–July. Stony sandstone slopes, SW, LB, SE (Hottentots Holland Mts to George).•

osteospermoides DC.   Arborescent shrub to 1 m, stems often maroon. Leaves obovate, tapering below, shallowly toothed, leathery. Flower heads radiate, in lax terminal corymbs, yellow. July–Aug. Seeps on sandstone slopes, KM (Warmwaterberg and Klein Swartberg Mts).•

pygmaea Compton   Dwarf perennial to 10 cm, from tuber-like caudex. Leaves tufted, spathulate, mostly slightly toothed above, brown-woolly in axils, emerging at flowering time. Flower heads minutely radiate, several in lax corymbs on slender peduncles, yellow. Mar.–May. Sandstone pavement, NW (Bokkeveld Mts to Biedouw Valley).•

quercifolia DC. (incl. **O. macrosperma** DC., **O. obtusiloba** Harv.)   Brittle-stemmed succulent shrublet to 40 cm, from fleshy caudex. Leaves oblanceolate, variously toothed to lobed, woolly in axils. Flower heads radiate, 1–few in lax corymbs on long peduncles, yellow; pappus of ray florets much elongated in fruit. May–Sept. Sandstone and granite outcrops, NW, SW (Namaqualand to Tygerberg).

triplinervia DC.   Brittle-stemmed shrub to 1.5 m. Leaves crowded at branch tips, obovate, scarcely lobed. Flower heads radiate, many in terminal corymbs on long peduncles, yellow. Apr.–June. Rocky slopes, SE (Humansdorp to Grahamstown).

[**Species excluded**. No authentic material found and probably conspecific with one of the above: **O. umbelliformis** DC., **O. viminea** E.Mey.]

OXYLAENA•   1 sp., W Cape

acicularis (Benth.) Anderb. (= *Anaglypha acicularis* Benth.)   Slender, willowy viscid shrub. Leaves long, linear and needle-like, margin lightly revolute, glandular and scabrid, ascending but later deflexed. Flower heads radiate, solitary, yellow or ?with white rays with dark reverse. Mar. ?Fynbos or forest margins, SE (Outeniqua Mts).•

PEGOLETTIA   DRAAIBOS   9 spp., Africa to India

baccaridifolia Less.   Twiggy, aromatic shrublet to 30 cm. Leaves ovate, petiolate, coarsely toothed. Flower heads discoid, solitary and terminal, yellow. Mainly July–Nov. Dry sandstone slopes, KM, SE (Swartberg Mts to E Cape).

retrofracta (Thunb.) Kies   DRAAIBOS   Twiggy shrublet to 60 cm. Leaves obovate, shortly petiolate, margins undulate. Flower heads discoid, solitary and terminal, yellow. Mainly Sept.–Nov. Karroid scrub, NW, KM (Namibia to Oudtshoorn, E Cape and Karoo).

PENTZIA   SKAAPKAROO   c. 23 spp., southern and North Africa

dentata (L.) Kuntze   GROOTSKAAPKAROO   Silvery felted, aromatic, twiggy shrublet 60 cm. Leaves petiolate, deltoid, palmately toothed to lobed above. Flower heads discoid, in dense terminal corymbs, yellow. Mainly Aug.–Jan. Dry stony sandstone slopes, NW, KM, LB, SE (Olifants River Mts to Swartberg Mts, Riversdale to E Cape).

elegans DC.   FYNSKAAPKAROO   Grey-mealy, aromatic, twiggy shrublet to 40 cm. Leaves petiolate, palmatisect, segments lobed. Flower heads discoid, in dense terminal corymbs, yellow. Mar.–Aug. Dry rocky slopes, KM (Montagu to Swartberg Mts).•

incana (Thunb.) Kuntze   SKAAPKAROO   White-woolly stemmed, aromatic, twiggy shrub to 1 m. Leaves pinnatisect, lobes linear. Flower heads discoid, solitary on long, sometimes leafy peduncles, yellow. Mainly Nov.–Jan. Dry habitats, NW, KM, LB, SE (Namaqualand to E Cape and Karoo).

PERDICIUM   2 spp., winter-rainfall South Africa

capense L. (= *Perdicium taraxaci* Vahl)   Acaulescent tufted perennial. Leaves prostrate, lyrate-pinnatifid, lobes rounded. Flower heads obscurely radiate, solitary, scapose, white. Achenes pilose. Aug.–Oct. Sandstone slopes, NW, SW, KM (Pakhuis Mts and Witteberg to Mamre).•

leiocarpum DC.   Like **P. capense** but leaves white-woolly below and achenes glabrous. Aug.–Sept. Sandy slopes and flats, NW (Kamiesberg and Bokkeveld to Cedarberg Mts).

PETALACTE•   WILDESEWEJAARTJIE   1 sp., W Cape (see also **ANDERBERGIA**)

coronata (L.) D.Don   Erect, grey-felted, densely leafy shrublet to 50 cm, golden-hairy on young parts. Leaves oblanceolate, suberect. Flower heads discoid, in several clusters at branch tips, mauve, bracts all with white, papery appendages, receptacle paleate. Aug.–Nov. Coastal dunes and sandstone slopes, NW, SW, AP (Pakhuis Mts to Agulhas).•

PHAENOCOMA•   ROOISEWEJAARTJIE, RED EVERLASTING   1 sp., W Cape

prolifera (L.) D.Don   Divaricately branched, white-stemmed shrublet to 60 cm. Leaves densely imbricate on spur shoots, involute-granular. Flower heads discoid, large, solitary, purple, bracts papery, pink shading to red. Sept.–Mar. Sandstone slopes in fynbos, NW, SW, KM, LB (Ceres to Cape Peninsula to Robinson Pass).•

PHANEROGLOSSA•   1 sp., W Cape

bolusii (Oliv.) B.Nord.   Glabrous shrublet to 40 cm. Leaves fleshy, crowded at branch tips, woolly in axils, narrowly oblanceolate, margins revolute, slightly toothed. Flower heads radiate, large, solitary on elongate almost naked peduncles, yellow with white or yellow rays. Mainly Nov.–Jan. Sandstone slopes in fynbos, NW (Skurweberg and Hex River Mts).•

PHYMASPERMUM   BANKROTBOS   17 spp., southern Africa

appressum Bolus   Twiggy, minute-leaved shrub to 1 m, with numerous spur shoots. Leaves scale-like, closely imbricate and adpressed. Flower heads radiate, solitary at branch tips, yellow with white to pink rays. May–Dec. Sandstone slopes and ridges, KM (Anysberg to Swartberg Pass).•

leptophyllum (DC.) Benth. & Hook. ex B.D.Jacks.   Twiggy, sprawling shrub to 1 m. Leaves linear, fleshy, closely imbricate. Flower heads radiate, solitary at branch tips, yellow with white rays. Mainly May–Sept. Dry rocky slopes, KM (Barrydale to Oudtshoorn).•

PLANEA•   1 sp., W Cape

schlechteri (L.Bolus) Karis (= *Metalasia schlechteri* L.Bolus)   Rigid, densely grey-woolly shrublet 30–100 cm. Leaves involute-ericoid, in tufts. Flower heads discoid, few in dense, terminal clusters, red, bracts white. Feb. Sandstone rocks, SW (Bainskloof to Klapmuts).•

PLATYCARPHA   3 spp., E Cape to Namibia and KwaZulu-Natal

glomerata (Thunb.) Less.   Acaulescent perennial to 5 cm. Leaves prostrate, pinnatisect, lobes toothed and pungent, margins curving upward, glabrescent above, white-felted below. Flower heads discoid, crowded in a sessile compound head, purple. June–Dec. Stony or sandy soils, SE (Humansdorp to KwaZulu-Natal).

PLECOSTACHYS   VAALTEE   2 spp., southern Africa

polifolia (Thunb.) Hilliard & B.L.Burtt   Sprawling, thinly white-woolly, much-branched shrublet to 1 m. Leaves obovate. Flower heads discoid, in congested terminal clusters, yellow tinged with purple, bracts whitish. Sept.–Dec. Streambanks and forest margins, NW, SW, ?AP, KM, LB, SE (Cape Peninsula to Swaziland).

serpyllifolia (P.J.Bergius) Hilliard & B.L.Burtt   VAALTEE   Similar to **P. polifolia** but leaves suborbicular with undulate margins, floral bracts milky white. Mainly Mar.–May. Sandy coastal flats or damp slopes, often coastal, SW, AP, LB, SE (Langebaan to KwaZulu-Natal).

POECILOLEPIS   2 spp., W to E Cape

ficoidea (DC.) Grau   Sprawling perennial to 10 cm, rooting at nodes. Leaves opposite, fleshy, subterete, clasping at base. Flower heads radiate, yellow with white rays, bracts scabrid. Nov.–Apr. Saline marshes, SW, AP, LB, SE (Cape Peninsula to E Cape).

maritima (Bolus) Grau   Like **P. ficoidea** but leaves alternate and bracts glabrous with scarious margins. Jan.–June. Saline marshes, SW, AP (Cape Peninsula to Bredasdorp).•

POLYARRHENA•   4 spp., W Cape

imbricata (DC.) Grau   Erect, slightly hairy shrublet to 20 cm. Leaves closely set, narrowly oblong, ciliate on margins, ascending to spreading. Flower heads radiate, solitary, yellow with white rays pink on reverse; bracts hairy. Oct.–Jan. Sandstone rocks at high alt., NW, SW (Cedarberg Mts to Saw Edge Peak).•

prostrata Grau   Straggling, scabrid, softly woody shrublet to 40 cm. Leaves spreading, oblong, obtuse, margins revolute. Flower heads radiate, solitary, yellow with white rays reddish on reverse; bracts hairy. Sept.–Oct. Rocky slopes in fynbos, NW, SW (Bainskloof and Brandwacht to Wemmershoek Mts).•

reflexa (L.) Cass.   WILDE-ASTER   Sprawling, softly woody, tangled shrub, 30–100 cm. Leaves deltoid, reflexed, closely set, margins with prickly teeth. Flower heads radiate, solitary, yellow with white rays pink on reverse; bracts glabrous. June–Oct. Sandstone slopes in fynbos, SW (Cape Peninsula to Caledon).•

stricta Grau   Sparsely branched, hairy shrub to 30 cm. Leaves lanceolate, spreading to recurved above, margins scabrid. Flower heads radiate, several in corymbs at branch tips, yellow with white rays purplish on reverse; bracts glabrous below. July–Sept. Rocky sandstone slopes in fynbos, SW (Hottentots Holland Mts to Elim).•

PRINTZIA   6 spp., South Africa

aromatica (L.) Less.   Much-branched, densely leafy, silvery felted shrublet to 40 cm. Leaves linear with revolute margins. Flower heads radiate, in terminal corymbs or solitary, yellow with pink or white rays. Nov.–Feb. Mountain slopes, NW, SW (Clanwilliam to Riviersonderend Mts).•

polifolia (L.) Hutch.   Stiffly branched, white-woolly shrub, 1–2 m. Leaves obovate, glabrescent above, weakly toothed and undulate, decurrent. Flower heads radiate, solitary, yellow with mauve-blue rays. July–Nov. Flats and hills mostly on clay below 500 m, NW, SW, AP, KM, LB, SE (Bokkeveld Escarpment to Port Elizabeth).•

PSEUDOGNAPHALIUM   c. 80 spp., cosmopolitan

undulatum (L.) Hilliard & B.L.Burtt   Much-branched, aromatic, grey-woolly annual to 50 cm. Leaves narrowly lanceolate, clasping at base, broadly decurrent, glabrescent and glandular above. Flower heads discoid, in dense corymbs, whitish. Nov.–Apr. Damp grassy or rocky slopes, NW, SW, LB, SE (Namibia to Port Elizabeth and S Mozambique).

PTERONIA   GOMBOS   c. 80 spp., mainly southern Africa, 1 sp. in Zimbabwe

A. Leaves white-woolly

bolusii E.Phillips   WILDEBEESTGOMBOS   Straggling shrub to 60 cm, branches recurving. Leaves small, ovate, keeled below, white-woolly. Flower heads discoid, solitary at branch tips, yellow, 15 × 5 mm; outer bracts grey-woolly. Sept.–Oct. Stony shale slopes, NW (Ceres and W Karoo).

cinerea L.f.   Grey-leaved shrub to 30 cm. Leaves minute, keeled below, white-woolly. Flower heads discoid, solitary at branch tips, yellow, c. 20 × 10 mm; outer bracts partly white-woolly. Sept.–Dec. Rocky mountain slopes, NW, KM (Kamiesberg to Olifants River Mts and Witteberg).

incana (Burm.) DC.   ASBOSSIE   Divaricately branched grey-leaved shrub, 50–100 cm. Leaves small, white-woolly. Flower heads discoid, solitary at branch tips, yellow, 15 × 10 mm; bracts glabrous. Mainly Sept.–Oct. Stony slopes on sand or clay, NW, SW, KM, LB, SE (Namaqualand and W Karoo to E Cape).

ovalifolia DC.   GRYSGOMBOS   Grey-leaved shrub to 30 cm. Leaves ovate, keeled below, white-woolly. Flower heads discoid, large, solitary at branch tips, yellow, 20–30 × c. 20 mm; bracts partly white-woolly. Aug.–Oct. Rocky slopes and flats, NW, SW (S Namaqualand and W Karoo to Hex River valley).

*A.' Leaves glabrous (see also **P. camphorata**)*

ciliata Thunb.   Twiggy shrub, 30–100 cm. Leaves linear, fleshy. Flower heads discoid, solitary at branch tips, yellow, 15–20 × 6–7 mm; bracts ciliate on margins. Oct.–Dec. Dry stony hills, NW (Namibia to Clanwilliam and W Karoo).

fasciculata L.f.   Rigid shrub to 1 m, stems naked below. Leaves linear-lanceolate, rigidly coriaceous, viscid, crowded at branch tips. Flower heads discoid, 1-flowered, in tight rounded clusters, yellow, 18 × 3 mm. Oct.–Jan. Mountain fynbos, NW, SW, KM, SE (Cedarberg Mts to Witteberg and Uniondale).•

fastigiata Thunb.   Low twiggy shrublet to 40 cm. Leaves semiterete, conate below around stem, closely imbricate, viscid. Flower heads discoid, narrow, mostly solitary at branch tips, yellow, 12–15 × 4–5 mm. Nov.–Dec. Stony clay slopes, NW, KM (Bushmanland to Worcester Karoo and Witteberg).

flexicaulis L.f.   Shrub to 50 cm. Leaves subterete, viscid, connate below around stem, clustered at ends of branches. Flower heads 1–3 at branch tips, yellow, viscid, 20–25 mm long. Nov.–Dec. Karroid slopes, NW, KM (Ceres to Oudtshoorn and S Karoo).

glabrata L.f.   Spreading shrublet to 40 cm, stems brittle. Leaves oblong, fleshy. Flower heads discoid, globose, solitary at branch tips, yellow, 15–20 × 20 mm. July–Dec. Coastal dunes and saline soils, NW (S Namibia to Clanwilliam and W Karoo).

oblanceolata E.Phillips   Low spreading shrublet to 15 cm, branches sprawling. Leaves spathulate, fleshy. Flower heads discoid, solitary, white, 15–20 × 15 mm; bract margins minutely ciliate and broadly translucent. Sept.–Nov. Stony slopes, NW, KM (Vanrhynsdorp and Witteberg to Ladismith).•

pallens L.f.   AASVOËLBOSSIE, WITGATBOSSIE   Twiggy shrub to 60 cm, stems pale. Leaves subterete. Flower heads discoid, globose, 1–3 at branch tips, yellow or orange, 13 mm long; bracts obtuse with minutely ciliate margins. Sept.–Dec. Stony karroid slopes, KM (Calvinia to Kimberley and Little Karoo).

paniculata Thunb.   GOMBOSSIE   Much-branched shrub to 1 m. Leaves subterete, viscid, connate below around stem. Flower heads narrow, several in crowded corymbs at branch tips, golden yellow, 10 × 3 mm. Mainly Nov.–Jan. Karroid slopes on shale, NW, KM, LB, SE (Namibia to Port Elizabeth, E Cape and Karoo).

tenuifolia DC.   Slender shrublet to 30 cm, stems stiffly erect, arising from rhizomes. Leaves linear and channelled, warty. Flower heads discoid, solitary at branch tips, orange, 20–25 × 20 mm; bracts minutely ciliate on margins. Nov.–Dec. Sandstone slopes in fynbos, SW (Kleinmond to Elim).•

teretifolia (Thunb.) Fourc. (= *Pteronia trigona* E.Phillips)   Much-branched erect shrublet to 50 cm. Leaves subterete, keeled. Flower heads discoid, crowded at branch tips, cream-coloured or white, 10 × 3 mm. Jan.–June. Limestone outcrops or forest margins, AP, SE (Potberg to E Cape).

uncinata DC.   STRANDGOMBOS   Erect shrub to 90 cm. Leaves subterete, keeled, spreading, hooked at tips. Flower heads discoid, in loose corymbose clusters at branch tips, yellow, 10 × 3 mm. Dec.–Apr. Coastal sands, NW, SW, AP (Lambert's Bay to Stilbaai).•

*A". Leaves papillose, scabrid or ciliate (see also **P. fasciculata**)*

adenocarpa Harv.   Twiggy shrublet to 50 cm. Leaves ovate, flat, recurving at tips, scabrid on margins. Flower heads discoid, solitary at branch tips, pinkish, fragrant, 30 × 15 mm; bracts obtuse, viscid. Aug.–Dec. Karroid areas, NW, KM, SE (Tulbagh to E Cape and Karoo).

ambrariifolia Schltr.   Twiggy shrublet to 25 cm. Leaves subterete, keeled, alternate, imbricate, crowded at branch tips. Flower heads discoid, solitary at branch tips, yellow, 15 × 10 mm; bracts keeled above, margins minutely ciliate. Aug.–Sept. Sandstone slopes in fynbos, NW (Lokenberg to Cedarberg Mts).•

beckeoides DC.   Slender twiggy shrub. Leaves powdery, linear. Flower heads discoid, solitary on branch tips, yellow, 14–20 mm long; bracts with membranous margins. Flowering time? Habitat?, LB (Swellendam).•

camphorata (L.) L.   SANDGOMBOS   Slender, aromatic shrub to 1 m. Leaves linear to filiform, ciliate. Flower heads discoid, 1–few at branch tips, yellow, 15 × 15 mm; bracts shortly and closely ciliate. Aug.–Nov. Coastal to upper slopes, NW, SW, AP, KM, ?LB, SE (Kamiesberg to Uniondale).

centauroides DC.   Slender grey perennial to 30 cm with woody rootstock. Leaves oblong, powdery scabridulous. Flower heads 1 or 2 at branch tips, yellow, 25–30 × 13 mm; bracts ovate. Nov.–Feb. Granite slopes, SW (Du Toitskloof).•

diosmifolia Brusse   Rounded densely leafy shrublet to 50 cm. Leaves 4-ranked, ovate-oblong, leathery, papillose. Flower heads discoid, narrow, in dense corymbs, yellow, 8–12 × 4–6 mm. Feb.–Mar. Limestone flats, AP (De Hoop).•

divaricata (P.J.Bergius) Less.   GEELGOMBOS   Rounded leafy shrub to 2 m. Leaves broadly ovate, puberulous to scabrid. Flower heads discoid, in dense corymbs, yellow or whitish, 15 mm long; bracts ovate to linear. Aug.–Nov. Sandy and stony slopes and flats, NW, SW (S Namibia to Tygerberg Hills).

elongata Thunb.   Like **P. staehelinoides** but bracts with broad fringed margins. Oct.–Jan. Renosterveld, KM, LB, SE (Oudtshoorn and Swellendam to E Cape).

glomerata L.f.   Low twiggy shrublet to 25 cm, branches often viscid above. Leaves minute, in tight clusters, shortly ciliate on margins. Flower heads discoid, solitary at branch tips, yellow, 20 × 10 mm; bracts viscid. Aug.–Nov. Clay slopes, NW (S Namaqualand, Ceres to Witteberg and W Karoo).

hirsuta L.f.   Twiggy shrublet to 25 cm. Leaves oblong, keeled, recurving, ciliate on margins and sometimes on undersides. Flower heads discoid, cylindric, solitary at branch tips, pink, 20–30 × 10–15 mm; bracts viscid. Nov.–Jan. Sandstone slopes, NW, SW, KM, LB (Piketberg and Cape Peninsula to Mossel Bay).•

hutchinsoniana Compton   Rigid, much-branched, scabridulous shrub to 1 m. Leaves oblong-lanceolate, keeled, roughly ciliate on margins. Flower heads discoid, solitary at branch tips, cream-coloured or yellow, 25–35 × 8–12 mm; bracts brown with broad membranous, fringed margins. Mainly Sept.–Oct. Dry sandstone slopes, KM (Klein Swartberg and S Karoo).

membranacea L.f.   Shrub to 60 cm. Leaves scabrid, oblong. Flower heads discoid, large, solitary at branch tips, pale yellow, 25 × 15 mm; bracts with broad membranous margins. Oct.–Dec. Mostly dry sandstone slopes, NW, KM, SE (Montagu and Bonteberg to E Cape and Karoo).

onobromoides DC.   SAB   Rounded aromatic shrub to 1 m. Leaves oblong, dark green, margins roughly ciliate. Flower heads discoid, solitary at branch tips, yellow, 35 × 15–25 mm; bracts ovate, dry. Mainly Nov.–Dec. Sandy coastal flats, NW, SW (S Namibia to Saldanha).

oppositifolia L.   Low shrublet to 20 cm. Leaves ovate, puberulous. Flower heads discoid, solitary at branch tips, yellow, 20 × 13 mm; bracts scarious with dark tips. Nov.–Dec. Dry stony flats, KM, LB (Touwsrivier to Mossel Bay).•

scabra Harv.   Scabrid perennial to 20 cm with woody rootstock. Leaves linear, scabrid. Flower heads solitary at branch tips, large, yellow, 25–30 × 15 mm; bracts lanceolate. Dec.–Jan. Rocky slopes, SW (Houwhoek to Elim).•

staehelinoides DC.   Rigid much-branched, scabridulous shrub to 30 cm. Leaves oblong-lanceolate, keeled, roughly ciliate on margins. Flower heads discoid, solitary at branch tips, yellow, 25–30 × 20 mm; bracts viscid. Mainly Sept.–Dec. Stony slopes, KM, LB, SE (Swartberg Mts to E Cape and Karoo).

stricta Aiton   KAATJIEGERT   Densely leafy, scabrid shrub to 1.5 m, with long erect branches. Leaves in tufts, subterete, fleshy, ciliate. Flower heads discoid, 1–few at branch tips, yellow, 20 × 15 mm; bracts long and attenuate. Sept.–July. Moist upper slopes, KM, SE (Swartberg Mts to Joubertina).•

succulenta Thunb.   Twiggy shrublet to 40 cm. Leaves few, crowded at the branch tips, linear-trigonous, papillose. Flower heads discoid, 1–3 at the branch tips, pale yellow, 10–15 × 5–7 mm. Oct.–Dec. Stony slopes, KM (Namaqualand to Little Karoo).

utilis Hutch.   Spreading shrublet to 30 cm, branches prostrate, roughly hairy. Leaves oblong, scabrid, grey. Flower heads discoid, large, solitary at branch tips, yellow, 20–25 × 15 mm; bracts acute. Nov.–Dec. Sandstone mountain slopes, NW (Cedarberg and Olifants River Mts).•

PULICARIA   FLEABANE   c. 80 spp., mainly Mediterranean, North Africa and Asia

scabra (Thunb.) Druce   Erect, single-stemmed, thinly woolly perennial to 1 m. Leaves lanceolate, auriculate at base, often scabrid above. Flower heads obscurely radiate, solitary at branch tips, yellow. Mainly Dec.–Apr. Marshes and streambanks, NW, SW, AP, KM, LB, SE (Bokkeveld Mts to tropical Africa).

RELHANIA   PERDEKAROO   13 spp., southern Africa

A. Flower heads in corymbs

corymbosa (Bolus) Bremer   Felted shrublet, silvery on new parts, to 40 cm. Leaves oblanceolate with involute margins, densely white-woolly above. Flower heads radiate, small, in terminal corymbs, yellow. Oct.–Apr. Upper mountain slopes, NW, KM (Cedarberg to Witteberg Mts).•

garnotii (Less.) Bremer   Dwarf, sparsely cobwebby shrublet to 50 cm. Leaves ericoid, 4-ranked, recurving. Flower heads radiate, small, in terminal corymbs, yellow. July–Oct. Coastal sand or clay, AP, LB (Agulhas to Mossel Bay).•

tricephala (DC.) Bremer   Thinly white-cobwebby shrublet to 45 cm. Leaves linear with involute margins, white-woolly above and twisted. Flower heads discoid, small, several in rounded, terminal corymbs, yellow. Sept.–Nov. Clay or sandy flats and slopes, NW, KM (Ceres to Witteberg Mts).•

A. Flower heads solitary

calycina (L.f.) L'Hér.   Twiggy, glabrescent shrub to 1 m. Leaves ascending, lanceolate, pungent, 3–9-veined below. Flower heads radiate, solitary and terminal, yellow. Oct.–Jan. Stony shale and sandstone slopes, NW, KM, LB, SE (Cedarberg Mts to E Cape).

decussata L'Hér.   Glabrescent shrublet tp 30 cm. Leaves linear, imbricate, 4-ranked. Flower heads radiate, solitary, terminal, yellow. Sept.–Nov. Dry sandstone slopes, KM, SE (Swartberg to Kouga Mts).•

fruticosa (L.) Bremer   Thinly woolly, densely leafy shrublet to 50 cm. Leaves involute-ericoid, densely white-woolly above and recurving. Flower heads radiate, small, solitary and terminal, yellow; pappus tubular. Mainly Sept.–Nov. Clay or sandy flats in fynbos, NW, SW (Piketberg to Gordon's Bay).•

pungens L'Hér.   Thinly white-woolly shrublet to 50 cm. Leaves linear, pungent, white-woolly above. Flower heads radiate, solitary and terminal, yellow. Oct.–Feb. Stony clay and limestone flats and slopes, SW, AP, LB, SE (Caledon and Worcester to KwaZulu-Natal).

relhanioides (Schltr.) Bremer   Twiggy, white-woolly shrublet to 50 cm. Leaves narrowly oblanceolate with involute margins, spreading, densely woolly above. Flower heads radiate, solitary and terminal, yellow. Aug.–Oct. Sand or clay flats and lower slopes, NW, KM (Cold Bokkeveld Mts to Witteberg).•

rotundifolia Less.   Erect, few-branched, grey-felted shrublet to 60 cm. Leaves obovate, margins involute, densely woolly above. Flower heads radiate, solitary and terminal, yellow; pappus tubular. Sept.–Oct. Sandy hills, SW (Hopefield to Darling).•

spathulifolia Bremer   Like **R. rotundifolia** but receptacle epaleate, ovaries densely hairy and pappus lacerate. Sept.–Oct. Damp places, AP (Agulhas).•

speciosa (DC.) Harv.   Similar to **R. calycina** but leaves spreading and 5–9-veined and flower heads orange-yellow. Sept.–Dec. Rocky sandstone slopes, SW, KM, SE (Riviersonderend Mts and Witteberg to Port Elizabeth).•

RHYNCHOPSIDIUM (= *RELHANIA* in part)   GEELSNEEU   2 spp., winter-rainfall South Africa

pumilum (L.f.) DC. (= *Relhania pumila* (L.f.) Thunb.)   GEELSNEEU   Spreading, thinly cobwebby annual to 20 cm. Leaves linear, glandular-hairy. Flower heads radiate, solitary on short, slender peduncles, yellow. Aug.–Oct. Sandy and clay flats and slopes, NW, SW, KM (Namaqualand to Oudtshoorn).

sessiliflorum (L.f.) DC. (= *Relhania sessiliflora* (L.f.) Thunb.   Like **R. pumilum** but heads smaller with reduced rays, sessile. Mainly Aug.–Nov. Clay flats, NW, SW, AP, KM, LB, SE (Clanwilliam to Port Elizabeth and W Karoo).

ROSENIA   HARTEBEESKAROO   4 spp., South Africa and Namibia

humilis (Less.) Bremer   Gnarled, twiggy, cobwebby shrublet to 30 cm. Leaves linear to oblanceolate, often glandular-hairy. Flower heads radiate, solitary at branch tips, yellow. Mainly Aug.–Oct. Stony clay or sandy flats, KM, SE (Namibia and Karoo, Swartruggens to Swartberg Mts).

SCHISTOSTEPHIUM (= *PEYROUSIA*) 12 spp., southern Africa

umbellata (L.f.) Bremer & Humphries (= *Peyrousia umbellata* (L.f.) Fourc.)   Densely leafy, silvery tomentose, sparsely branched shrub to 2 m. Leaves elliptic, ascending. Flower heads discoid, solitary but grouped in lax corymbs, golden yellow. Mainly Aug.–Nov. Forest margins or fynbos to c. 1000 m, KM, LB, SE (Montagu to Humansdorp and Swartberg Mts).•

SENECIO   GROUNDSEL, HONGERBLOM, RAGWORT   c. 1200 spp., cosmopolitan

A. Annuals

abruptus Thunb. (incl. **S. diffusus** Thunb.)   BASTERGEELHONGERBLOM   Like **S. littoreus** but involucre not evidently calycled, and leaves usually more deeply pinnatifid. July–Nov. Stony slopes, NW, SW (Namaqualand to Cape Hangklip).

agapetes C.Jeffrey (= *Senecio amabilis* DC.)   Roughly hairy annual to 1 m. Leaves pinnatifid, clasping at base, often woolly below, margins sharply dentate and prickly. Flower heads discoid or radiate, in compound panicles, the lower clusters sometimes sessile, purple, involucre calycled. Mar.–May. Sandstone slopes, NW, LB (Cold Bokkeveld to Robinson Pass).•

arenarius Thunb.   HONGERBLOM   Glandular-hairy annual, 15–40 cm. Leaves toothed to pinnatisect, margins sometimes revolute. Flower heads radiate, several, terminal in branched corymbs, yellow with mauve or sometimes white rays. July–Sept. Sandy flats, NW, SW, AP (S Namibia and W Karoo to De Hoop).•

cakilefolius DC.   NAMAKWAHONGERBLOM   Like **S. arenarius** but completely glabrous and flower heads larger. Aug.–Sept. Rocky slopes mostly on clay, NW (S Namibia to Clanwilliam).

carroensis DC.   Glabrous, much-branched, sturdy annual to 30 cm. Leaves pinnatisect, lobes short and toothed, margins revolute. Flower heads radiate, few in lax corymbs, yellow, involucres calycled. Aug.–Nov. Rocky sandstone slopes, NW, KM (Namaqualand to Witteberg and Touwsberg).

elegans L.   VELD CINERARIA   Densely glandular-hairy annual to 1 m. Leaves fleshy, incised to pinnatisect, margins revolute. Flower heads radiate, numerous in dense corymbs, purple, involucre conspicuously calycled. Mainly Sept.–Nov. Coastal sands, SW, AP, SE (Saldanha to Port Elizabeth).•

glutinarius DC.   like **S. glutinosus** but involucre glabrous. Sept.–Oct. Seashore, SW (Saldanha to Cape Peninsula).•

glutinosus Thunb.   TAAIGEELHONGERBLOM   Glandular-hairy annual mostly to 40 cm. Leaves toothed to pinnatisect. Flower heads radiate, several in branched terminal corymbs, yellow. July–Sept. Rocky, mostly sandstone slopes, NW, KM (S Namibia to Piketberg and Montagu).

laevigatus Thunb.   Glabrescent to roughly hairy annual slightly woody below, to 30 cm. Leaves linear to pinnatisect. Flower heads radiate, few, terminal on slender peduncles, yellow, involucres calycled. Oct.–May. Sand over limestone, AP (Agulhas to Breede River mouth).•

laxus DC.   Slender annual to 45 cm. Leaves lanceolate, petiolate below, amplexicaul above, slightly toothed to pinnatifid. Flower heads radiate, in lax terminal corymbs, yellow, involucre calycled. July–Sept. Sandy slopes, NW (Namaqualand to Olifants River Mts). [Possibly not distinct from **S. littoreus**.]

littoreus Thunb.   GEELHONGERBLOM   Erect, glabrescent to shortly hairy annual to 40 cm. Leaves obovate to oblanceolate, toothed to slightly pinnatifid, sometimes auriculate at base. Flower heads radiate, in lax terminal corymbs, yellow, involucre calycled. Aug.–Nov. Mainly coastal sands, NW, SW (Graafwater to Cape Peninsula and Napier).•

maritimus L.   STRANDHONGERBLOM   Sprawling to prostrate annual to 30 cm. Leaves fleshy, oblong-obovate, slightly toothed, auriculate at base, obtuse to truncate. Flower heads radiate, in lax terminal corymbs, yellow, involucre calycled. Aug.–Dec. Coastal dunes and slopes, NW, SW, AP (S Namaqualand to Agulhas).

paarlensis DC.   Softly hairy annual, 15–30 cm. Leaves bipinnatisect, often with acute auricles at base. Flower heads radiate, few, terminal on slender peduncles, yellow with yellow rays blue on reverse. Sept.–Dec. Partly shaded rocky slopes and below cliffs, NW, SW (Cold Bokkeveld to Du Toitskloof Mts).•

pinnulatus Thunb.   Erect, softly woody perennial or annual to 70 cm, glabrous or hairy below. Leaves pinnatisect, lobes linear, margins revolute. Flower heads radiate, in lax terminal corymbs, yellow, involucres calycled. July–Oct. Moist sandy slopes, NW, SW, LB, SE (Bokkeveld Mts to E Cape).

*****pterophorus** DC.   Robust, thinly grey-cobwebby annual or short-lived perennial to 2 m, branching above. Leaves oblanceolate, coarsely toothed, margins revolute, glabrescent above, white-felted beneath. Flower heads radiate, in lax terminal corymbs, yellow, involucre calycled. Oct.–Jan. Pioneer of disturbed slopes, SW, KM (naturalised from E Cape and KwaZulu-Natal).

repandus Thunb.   Delicate or sprawling annual, 10–30 cm, roughly hairy on young parts and bracts. Leaves soft, deltoid to reniform, lobed to pinnatifid, lower petiolate, upper auriculate. Flower heads radiate, 1–few on short to long, slender peduncles, yellow. July–Oct. Partial shade on sandstone slopes, NW, SW (Pakhuis Mts to Cape Peninsula).•

sophioides DC.   Diffuse, minutely glandular-hairy annual to 20 cm. Leaves petiolate, pinnatisect, lobes linear, sometimes 1 or 2 toothed and with small white terminal calluses. Flower heads radiate, 1–few, terminal on slender peduncles, purple or yellow. July–Oct. Middle slopes, NW, SW (Cedarberg Mts to Caledon).•

A.' Perennials with leaves mostly basal

albanensis DC.   Stout, tufted, thinly white-woolly perennial to 80 cm, basal leaves sometimes dead at flowering, petiolate, oblanceolate, finely serrate, margins revolute. Flower heads radiate, in lax corymbs on leafy scapes, yellow, involucres conspicuously calycled. Aug.–Apr. Grassland, SE (Baviaanskloof Mts to N Province).

albifolius DC.   Silvery woolly, tufted perennial, 5–20 cm. Leaves oblanceolate, regularly pinnatisect, lobes oblong and toothed, margins revolute. Flower heads radiate, 1–3 on almost naked scapes, yellow sometimes with red reverse, involucre calycled. Aug.–Jan. Rock ledges at high alt., NW (Cedarberg to Hex River Mts).•

anthemifolius Harv.   Tufted, glabrous perennial to 30 cm. Leaves mostly basal, bipinnatisect, lobes linear. Flower heads discoid, solitary on sparsely leafy peduncles, yellow, involucres calycled. July–Aug. Clay slopes, SW, LB (Botrivier to Langeberg Mts).•

cordifolius L.f.   Like **S. hastifolius** but leaves cordate to ovate, margins revolute and finely toothed. Flower heads on slender, naked peduncles, yellow. Jan.–Apr. Sandstone slopes in sheltered sites, SW, LB (Cape Peninsula to Swellendam).•

coronatus (Thunb.) Harv.   sybossie   Tufted, thinly cobwebby perennial with woolly crown, to 40 cm. Leaves mostly basal, petiolate, obovate, leathery, margins crenulate to finely dentate. Flower heads radiate, large, in lax corymbs on sparsely leafy scapes, yellow, involucre conspicuously calycled. July–Sept. Grassland, SE (Sedgefield to tropical Africa).

crenulatus DC.   Erect, glabrous perennial to 30 cm, basal leaves long-petiolate, linear-lanceolate, margins revolute, minutely toothed. Flower heads discoid, few in corymbs on leafy scapes, yellow, involucres calycled. Apr.–May. Stony grassland, SE (Humansdorp to E Cape).

crispus Thunb.   Tufted perennial to 40 cm. Leaves petiolate, oblanceolate, sparsely toothed, margins revolute. Flower heads radiate or discoid, in lax corymbs on sparsely leafy scapes, yellow, involucre calycled. Dec.–Mar. Damp sandstone slopes above 800 m, SW, LB (Cape Peninsula with radiate heads, Babilonstoring to Swellendam with discoid heads).•

erosus L.f. (incl. **S. eriobasis** DC.)   STICKY-LEAVED GROUNDSEL   Tufted, glabrous to roughly hairy perennial with woolly crown, to 60 cm. Leaves mostly basal, petiolate, lanceolate, margins irregularly serrate to lacerate or pinnatifid. Flower heads radiate, solitary or in corymbs on sparsely leafy scapes, yellow, involucre calycled. Aug.–Sept. Rocky slopes, NW, SW, AP, KM, LB (Namaqualand to Swellendam).

gramineus Harv.   Tufted, thinly white-woolly perennial to 25 cm. Leaves linear, margins revolute, accumulating in a fibrous base. Flower heads radiate, solitary or few in lax corymbs on elongate, sparsely leafy peduncles, yellow. Nov.–Jan. Grassy slopes, SE (George to KwaZulu-Natal).

hastatus L.   GROUNDSEL   Tufted, sticky perennial from short erect rhizome, to 40 cm. Leaves oblanceolate, petiolate below, lacerate to pinnatifid. Flower heads radiate, in branched corymbs, yellow, involucre c. 8 mm long, sticky, sparsely calycled. Sept.–May. Rocky mostly karroid slopes, NW, SW, AP, KM, LB, SE (Ceres to KwaZulu-Natal and Karoo).

hastifolius (L.f.) Less. (= *Senecio cymbalariifolius* (Thunb.) Less.)   Tuberous, tufted perennial to 40 cm. Leaves often polymorphic, long-petiolate, sagittate to reniform, or lyrate and pinnatisect with terminal lobe largest, sometimes auriculate, often purple beneath, margins revolute, sometimes roughly hairy. Flower heads radiate, 1–few on scaly scapes, sometimes in lax corymbs, purple or yellow with purple or white rays, bracts broad. Sept.–Oct. Damp sandstone slopes and marshes, NW, SW, AP (Olifants River Mts to Elim).•

lanifer Mart. ex C.Jeffrey (= *Senecio erubescens* Aiton)   Glandular-hairy, rosulate perennial to 60 cm. Leaves oblanceolate, lobed to toothed to lacerate. Flower heads discoid, in lax terminal corymbs, usually magenta. July–Oct. Sandstone slopes in fynbos or grassland, NW, SW, KM, LB, SE (Cedarberg Mts and Cape Peninsula to S tropical Africa).

othonniflorus DC.   Perennial to 60 cm, with woody base, woolly in leaf axils. Leaves crowded below, linear-lanceolate, slightly toothed, margins revolute. Flower heads discoid, in lax corymbs on sparsely leafy scapes, yellow, bracts broad. Nov.–June. Grassland, SE (Knysna to Mpumalanga).

panduratus Less.   SMOOTH GROUNDSEL   Like **S. hastatus** but flower heads larger, involucre c. 12 mm long and more densely calycled. Aug.–Sept. Dry flats, NW, SW, KM (Namaqualand to Montagu and Karoo).

petiolaris DC.   Perennial to 30 cm. Leaves fleshy, basal, petiolate, obovate to round, slightly toothed. Flower heads discoid, scapose, solitary on bracteate peduncles, creamy to yellow. Sept.–Nov. Shady sandstone slopes, NW, KM (Cedarberg Mts to Witteberg).•

purpureus L. (incl. **S. odontopterus** DC.)   Like **S. lanifer** but heads crowded in compound corymbs and achenes glabrous and striate. Oct.–Feb. Moist slopes, especially after fire, SW, LB, SE (Cape Peninsula to S KwaZulu-Natal).

rhomboideus Harv.   Like **S. ruwenzoriensis** but leaves fleshy, petiolate, obovate to rhomboid, deeply toothed. Flower heads discoid. Nov.–Mar. Grassland, SE (Langkloof to Mpumalanga).

ruwenzoriensis S.Moore (= *Senecio pauciflorus* Thunb.)   Tuberous-rooted, glabrous, tufted perennial to 30 cm. Leaves lanceolate, sparsely toothed, 3-veined from base. Flower heads radiate, in lax corymbs on elongate, naked peduncles, yellow. Dec.–May. Grassy slopes, LB, SE (Swellendam to tropical Africa).

scapiflorus (L'Hér.) C.A.Sm.   PERSKOPPIE   Tufted, sparsely white-woolly perennial to 30 cm. Leaves basal, petiolate, lanceolate to cordate, toothed to pinnatifid or pinnatisect. Flower heads discoid, solitary on elongate, scaly peduncles, white or purple. Aug.–Oct. Rocky slopes and flats, NW, SW (Namaqualand to Cape Peninsula).

speciosus Willd.   Glandular-hairy, rosulate perennial to 50 cm. Leaves oblanceolate, lobed to lacerate. Flower heads radiate, several in lax terminal corymbs, magenta. July–Dec. Damp upper slopes, NW, SW, LB, SE (Cold Bokkeveld to Cape Peninsula to Mozambique).

spiraeifolius Thunb.   FERN-LEAVED GROUNDSEL   Thinly rough-woolly, tufted perennial to 30 cm. Leaves petiolate, lanceolate, regularly pinnatisect, lobes oblong and toothed. Flower heads radiate, in terminal corymbs, yellow, involucre thinly woolly and calycled. July–Sept. Rocky, mainly shale soils, NW, SW (Bokkeveld Escarpment to Riebeek-Kasteel and Tulbagh, also W Karoo).

tuberosus (DC.) Harv.   Like **S. hastiifolius** but heads discoid, white. Sept.–Oct. Sandstone slopes, NW, SW (Cold Bokkeveld to Riebeek-Kasteel).•

wittebergensis Compton   Tufted, sparsely woolly perennial to 20 cm. Leaves petiolate, lanceolate, toothed to pinnatifid, crowded at base. Flower heads discoid, mostly solitary, on elongate, sparsely leafy peduncles, yellow, involucres sparsely calycled. Oct. Sandstone slopes in arid fynbos, KM (Witteberg).•

A". Perennials with leaves mostly cauline
B. Climbers or scandent shrubs with broad leaves

angulatus L.f.   CAPE IVY   Scrambling, half-climbing shrub. Leaves petiolate, ovate to lanceolate, coarsely lobed. Flower heads radiate, in branched corymbs or panicles, yellow, involucres calycled. Apr.–Aug. Forest margins, SE (George to E Cape).

deltoideus Less.   Soft-leaved climber with zigzag branching. Leaves deltoid, petiolate, eared at base, toothed. Flower heads discoid, in corymbose panicles, orange or yellow, bracts 5. May–July. Forest margins and cliffs, LB, SE (Swellendam to tropical Africa).

macroglossus DC.   NATAL IVY   Herbaceous, somewhat succulent climber. Leaves petiolate, sagittate. Flower heads radiate, large, mostly solitary, cream-coloured to pale yellow, conspicuously calycled. Aug. Forest margins, SE (Humansdorp to tropical Africa).

quinquelobus (Thunb.) DC.   Herbaceous, semisucculent climber. Leaves ivy-like. Flower heads discoid, large, few in terminal clusters on lateral branches, conspicuously calycled, orange or yellow. Mainly Feb.–May. Forest margins, LB, SE (Swellendam to KwaZulu-Natal).

B. Erect or sprawling perennials or shrubs
C. Leaves more or less fleshy and succulent or scale-like and dry

abbreviatus S.Moore   Prostrate rhizomatous perennial. Leaves succulent, ellipsoid, glaucous, secund. Flower heads radiate, solitary and terminal on minutely bracteate, purplish peduncles, yellow, fragrant. June–Aug. Dry karroid stony slopes, NW, KM (Hex River valley to Prince Albert and Karoo).

acaulis (L.f.) Sch.Bip.   Like **S. crassulaefolius** but flower heads solitary, large, and outer florets female with reduced corolla. Oct.–Nov. Rocky karroid slopes, NW, KM, SE (Ceres Karoo and W Karoo to Kouga Mts).

aloides DC.   GROOTDIKBLAAR   Thick-stemmed, glabrous shrub to 75 cm. Leaves in terminal clusters, cylindric, fleshy. Flower heads radiate, mostly solitary on sparsely scaly, terminal peduncles, yellow, involucres calycled. July–Oct. Coastal rocks and dunes, NW, SW (S Namibia to Cape Peninsula).

articulatus (L.) Sch.Bip. (= *Kleinia articulatus* (L.) Haw.)   WORSIES   Succulent shrublet to 60 cm, with swollen jointed stems. Leaves sagittate and toothed. Flower heads discoid, few in terminal corymbs on long naked peduncles, white or yellowish. Apr.–June. Rocky slopes, KM, SE (Montagu to Uitenhage).•

bulbinifolius DC.   KRAALTJIES   Sprawling succulent to 15 cm spreading by prostrate runners. Leaves cylindric, fleshy. Flower heads radiate, solitary on sparsely scaly, slender peduncles, yellow. Aug.–Sept. Rocky slopes, NW (S Namaqualand to Klawer).

citriformis G.D.Rowley   Cushion-like succulent to 10 cm, from creeping rhizomes. Leaves compressed-fusiform, glaucous. Flower heads discoid, 1 or 2 on terminal, naked peduncles, creamy yellow. Jan.–Feb. Rocky flats, KM (Klein Swartberg foothills).•

cotyledonis DC.   Thick-stemmed, succulent shrublet to 60 cm. Leaves cylindric, fleshy, crowded apically, pungent. Flower heads radiate, in dense corymbs nested in leaves, yellow. June–Sept. Stony karroid slopes, KM (S Namibia, Namaqualand and W Karoo to Swartberg and Kammanassie Mts).

crassiusculus DC.   Loosely branched, succulent shrub to 50 cm. Leaves linear to narrowly lanceolate, fleshy, often irregularly lobed, margins slightly revolute. Flower heads radiate, in loosely branched corymbs on naked peduncles, yellow, involucres minutely calycled. Oct.–Mar. Stony slopes, LB, SE (Langeberg Mts: Cloete's Pass to Grahamstown).

crassulaefolius (DC.) Sch.Bip. (= *Kleinia archeri* Compton; incl. **K. ovoidea** Compton, **S. aizoides** (DC.) Sch.Bip)   Erect or sprawling succulent perennial to 30 cm. Leaves in terminal tufts, fusiform-terete. Flower heads discoid, 1–few on elongate, sparsely bracteate, terminal peduncles, white, sometimes yellow, involucral bracts c. 8. Mainly Aug.–Dec. Rocky outcrops and ledges, NW, SW, KM, LB, SE (Richtersveld to Uitenhage).

ficoides (L.) Sch.Bip.   Sprawling, succulent perennial to 50 cm. Leaves fleshy, compressed-fusiform glaucous. Flower heads discoid, several in 1–few-branched corymbs, white. Mar.–July. Rocky slopes, KM (Huisrivierberg and Swarteberg Mts).•

junceus (DC.) Harv.   SJAMBOKBOS   Apparently leafless succulent-stemmed shrub to 90 cm with rod-like stems from woody rootstock. Leaves scale-like, dry. Flower heads radiate, in terminal clusters, yellow. Feb.–Apr. Dry rocky lower slopes, KM, LB, SE (Namaqualand and W Karoo to Ladismith and Port Elizabeth).

muirii L.Bolus   Sprawling succulent to 40 cm, stems often rooting from nodes. Leaves obovate, glaucous, margins sometimes slightly toothed, veins 3–5 from base, transparent below. Flower heads discoid, few in lax corymbs, pale straw-coloured, involucres sparsely calycled. Mar.–Apr. Shale cliffs, KM, LB (Calitzdorp to Gourits River).•

odontophyllus C.Jeffrey (= *Senecio longifolius* L.)   Loosely branched shrub to 1 m. Leaves fleshy, filiform. Flower heads radiate, in branched corymbs on naked peduncles, yellow, involucres minutely calycled. Mainly Apr.–Nov. Stony slopes, SE (Humansdorp to E Cape).

pyramidatus DC.   Thick-stemmed, white-woolly, fleshy-leaved shrub to 40 cm. Leaves cylindric, succulent, crowded at tops of branches. Flower heads radiate, large, in small clusters on long naked peduncles, yellow, involucres calycled. July–Nov. Karroid bush, SE (Humansdorp to E Cape).

radicans (L.f.) Sch.Bip. (= *Kleinia radicans* DC.)   BOBBEJAANTOONTJIES, VINGERTJIES   Trailing, succulent perennial. Leaves fusiform, secund. Flower heads discoid, solitary or few on sparsely bracteate peduncles to 10 cm, white or mauve, fragrant. Apr.–Sept. Rock outcrops on flats and hills, NW, SW, KM, SE (S Namibia to Saldanha and E Cape).

sarcoides C.Jeffrey (= *Senecio corymbiferus* DC.)   Like **S. aloides** but flower heads smaller, several in branched corymbs. July–Oct. Rocky areas, NW, SW, KM (S Namibia to Cape Peninsula, Witteberg and W Karoo).

scaposus DC.   Silver-leaved, tufted succulent to 40 cm. Leaves lanceolate, fleshy, densely cobwebby. Flower heads radiate, large, in lax panicles on thinly white-woolly peduncles, yellow, involucres sparsely calycled. Dec.–Jan. Rocky karroid slopes and cliffs, KM, SE (Montagu to E Cape).

serpens G.D.Rowley (= *Kleinia repens* L.)   Sprawling, succulent shrublet to 20 cm. Leaves depressed-fusiform. Flower heads discoid, few on terminal corymbs, white. Feb.–May. Rock outcrops on coastal slopes, SW (Cape Peninsula to Hermanus).•

toxotis C.Jeffrey (= *Kleinia archeri* Compton)   Like **S. crassulaefolius** but leaves laterally flattened and bracts about 16. Feb.–May. Rocky karroid slopes, SW, KM, LB (Witteberg to Potberg and Cloete's Pass).•

C. Leaves not fleshy or succulent
D. Leaves or lobes mostly linear or needle-like, margins mostly revolute (see also **S. rehmannii***)*

angustifolius (Thunb.) Willd.   Twiggy, glabrous shrublet to 60 cm. Leaves linear, minutely toothed, margins revolute, often with axillary tufts. Flower heads discoid, few in small corymbs, yellow or white. Mainly Oct.–Mar. Damp slopes, NW, SW, KM (Olifants River Valley to Stormsvlei).•

burchellii DC. (= *Senecio dracunculoides* DC.)   GEELGIFBOS   Softly woody, glabrous shrublet to 40 cm, sometimes roughly hairy below. Leaves linear, margins revolute, sometimes toothed, usually with axillary tufts. Flower heads radiate, in lax corymbs, yellow, involucres conical, calycled. Mainly Apr.–July. Sandy and stony slopes, NW, SW, AP, KM, LB, SE (Namibia to Cape Peninsula to Port Elizabeth).

chrysocoma Meerb.   Willowy, softly woody perennial to 1 m, densely leafy below. Leaves filiform, margins revolute. Flower heads discoid, few in terminal umbels, yellow, sometimes white, involucres densely calycled. Nov.–Jan. Grassy slopes and forest margins, AP, LB, SE (Riversdale and Stilbaai to KwaZulu-Natal).

juniperinus L.f.   Like **S. ilicifolius** but leaves narrow, not or hardly toothed and margins stongly revolute. Sept.–Jan. Mountain slopes, NW, SW, KM, SE (Kamiesberg to Caledon and Oudtshoorn to S KwaZulu-Natal).

leptophyllus DC. (incl. **S. mucronatus** Willd.)   Like **S. angustifolius** but leaves slightly eared at base and plants usually thinly cobwebby. Mainly Aug.–Jan. Dry stony karroid slopes, LB, SE (Albertinia to Free State).

lycopodioides Schltr.   Like **S. pinifolius** but to 10 cm with stems sprawling. Leaves oblanceolate, short, reflexed. Flower heads on sparsely leafy peduncles. Mar.–Apr. Limestone rocks, AP (Agulhas to De Hoop).•

paniculatus P.J.Bergius (incl. **S. diodon** DC.)   Glabrous perennial to 80 cm, sometimes sparsely hairy below. Leaves filiform or pinnatisect with filiform lobes, margins revolute and minutely toothed. Flower heads discoid, in lax branched corymbs, white to yellow, rarely mauve, involucres calycled. Oct.–Dec. Rocky sandstone slopes, NW, SW, KM, LB (Gifberg to Langkloof).•

pillansii Levyns   Spreading, densely leafy, thinly white-woolly shrublet to 30 cm. Leaves linear, revolute-ericoid, pungent. Flower heads radiate, sometimes discoid, solitary on elongate, sparsely scaly peduncles, yellow, involucres conspicuously calycled. Nov.–Jan. Coastal slopes, SW, AP (Cape Peninsula to Elim).•

pinifolius (L.) Lam.   Sparsely branched, densely leafy shrublet to 40 cm. Leaves needle-like. Flower heads radiate, mostly solitary, subsessile, yellow, involucres calycled. Mainly Mar.–May. Sandstone slopes in fynbos, NW, SW, KM, LB, SE (Pakhuis Mts to E Cape).

retortus (DC.) Benth.   Like **S. pinifolius** and possibly not distinct, but leaves conspicuously recurved. Dec.–Mar. Sandstone slopes, SW (Riviersonderend Mts).•

rosmarinifolius L.f.   GRYSHONGERBLOM   Rounded, softly woody, thinly white-woolly shrublet to 80 cm. Leaves linear, margins revolute, sometimes toothed, usually in axillary tufts. Flower heads radiate, in dense clusters on branched corymbs, yellow, involucres calycled. Mainly Oct.–Mar. Sandy and stony slopes, NW, SW, AP, KM, LB, SE (Namaqualand to Cape Peninsula to E Cape and Karoo).

triqueter DC.   Like **S. pinifolius** but flower heads discoid, whitish, on slender peduncles bearing scale like leaves. Mar.–July. Rocky sandstone slopes, SW, AP, KM (Cape Peninsula to Ladismith and Pearly Beach).•

umbellatus L. (= *Senecio filifolius* Harv.; incl. **S. grandiflorus** P.J.Bergius, **S. leucoglossus** Sond., **S. mitophyllus** C.Jeffrey)   Glabrous perennial to 80 cm, sometimes sparsely hairy below. Leaves filiform or pinnatisect with linear to filiform lobes, margins revolute and minutely toothed. Flower heads radiate, in lax branched corymbs, yellow with magenta to pink or sometimes white rays, involucres finely calycled. Sept.–Dec. Sandstone flats and slopes, NW, SW, AP, KM, LB, SE (Cedarberg Mts to Uitenhage).•

sp. 1   Slender-stemmed perennial to 70 cm from woody base. Leaves terete, slightly fleshy. Flower heads discoid, forming lax panicles, peduncles wiry, minutely scaly below apex, cream-coloured, bracts broad, only 5. Apr.–May. Rocky sandstone slopes, KM (Klein Swartberg and Rooiberg).•

D. Leaves variously and more broadly lobed (see also **S. umbellatus***)*

arniciflorus DC.   Sprawling, white-woolly shrublet to 30 cm. Leaves oblanceolate, toothed above, margins revolute, spreading to reflexed, glabrescent above. Flower heads radiate, few to several in terminal corymbs, yellow, involucres calycled. Aug.–Oct. Sandy coastal flats and lower slopes, SW, AP (Mamre to Agulhas).•

bipinnatus (Thunb.) Less.   Like **S. pinnatifidus** but flower heads 4–6-flowered, in congested, branched corymbs. Jan.–Mar. Sandstone slopes in damp places, NW, SW, LB (Cold Bokkeveld to Cape Peninsula and Langeberg Mts: Swellendam).•

brachypodus DC.   Scandent, glabrous shrub to 2 m. Leaves ovate, shortly petiolate, fleshy, slightly toothed. Flower heads radiate, numerous in crowded, branched corymbs, yellow. Apr.–June. Forest margins, SE (Humansdorp to tropical Africa).

cinerascens Aiton   HANDJIESBOS   White-woolly shrub to 2 m. Leaves pinnatisect, lobes linear, margins revolute. Flower heads radiate, in branched terminal corymbs, yellow, involucres calycled. July–Sept. Rocky slopes, NW (S Namibia to Cedarberg Mts, and W Karoo).

coleophyllus Turcz.   Sparsely white-woolly shrublet to 50 cm. Leaves lanceolate to oblong, spreading to reflexed, discolorous and densely woolly below, stongly toothed, margins revolute. Flower heads radiate, 1–3 on elongate peduncles, pink or magenta, involucres calycled. Oct.–Nov. Sandstone slopes, SW (Riviersonderend Mts).•

crenatus Thunb. Singled-stemmed, closely leafy, thinly woolly shrub to 1 m. Leaves shortly petiolate, elliptic, margins slightly revolute, closely toothed, recurved at tips. Flower heads radiate, in dense branched corymbs, yellow, involucres calycled. Mainly Nov.–May. Sandstone slopes in fynbos, LB, SE (Swellendam to Port Elizabeth).

dissidens Fourc. (incl. **S. dumosus** Fourc.) Robust, glabrescent shrub to 2 m. Leaves lanceolate, deeply toothed, narrowed below and obscurely auriculate, margins slightly revolute. Flower heads radiate, numerous in branched terminal corymbs, yellow, involucre calycled. Nov.–Dec. Sandstone slopes often along streams, LB, SE (Langeberg Mts: near Mossel Bay to Langkloof).•

euryopoides DC. Willowy, densely leafy shrub to over 1.5 m. Leaves pinnatisect, lobes filiform. Flower heads radiate, in dense corymbs nested in upper leaves, yellow, involucres calycled. Nov.–May. Marshes and seeps, SE (Langkloof).•

foeniculoides Harv. FENNEL-LEAVED GROUNDSEL Like **S. pinnatifidus** but glabrous, densely leafy. Flower heads 12–15-flowered and more numerous in branched corymbs. Nov.–Feb. Sandy flats and lower slopes, SW (Mamre to Cape Peninsula).•

glastifolius L.f. WATERDISSEL Rigid, glabrous, softly woody perennial to 1 m. Leaves lanceolate, coarsely toothed, margins revolute, ascending, clasping at base. Flower heads radiate, large, in lax branched corymbs, yellow with mauve rays, involucres calycled. Sept.–Nov. Forest margins, SE (George to Humansdorp).•

halimifolius L. TABAKBOS Thinly white-cobwebby, glaucous shrub to 1.5 m. Leaves oblanceolate to obovate, narrow below, coarsely toothed above. Flower heads radiate, in dense terminal corymbs, yellow, involucres calycled. Nov.–Jan. Coastal sands, NW, SW (Lambert's Bay to Hermanus).•

hirtifolius DC. Densely leafy, roughly velvety shrub to 80 cm. Leaves ascending, obovate, toothed above, 3-veined from base. Flower heads radiate, yellow, involucres calycled. Dec.–Apr. SE (Uitenhage).•

hollandii Compton Densely white-felted shrub to 70 cm. Leaves ovate, petiolate, discolorous, glabrous above, sharply toothed, margins revolute. Flower heads radiate, in branched terminal corymbs, yellow, involucres calycled. Sept.–Dec. Sandstone slopes in arid fynbos, KM, SE (Witteberg to Uitenhage).

ilicifolius L. (incl. **S. aquifoliaceus** DC.) SPRINKAANBOS Thinly white-woolly, softly woody, short-lived shrub to 1 m. Leaves discolorous, oblong, sharply toothed, margins revolute, thickly white-woolly below. Flower heads radiate, in lax corymbs, yellow, involucres calycled, c. 5 mm long. June–Jan. Clay flats and slopes, KM, LB, SE (Ladismith and Langeberg Mts to E Cape).

lanceus Aiton Thinly woolly perennial to 2.5 m. Leaves lanceolate, ascending, clasping at base, margins revolute, finely toothed. Flower heads shortly radiate, crowded in small clusters on branched, terminal corymbs, yellow, involucres calycled. Jan.–Mar. Sandstone slopes near streams, NW, SW, KM, SE (Cedarberg Mts to Cape Peninsula and to KwaZulu-Natal).

lineatus (L.f.) DC. (incl. **S. quinquenervius** Sond.) Densely leafy, grey-felted shrublet to 60 cm. Leaves elliptic, 3-veined from base, usually slightly toothed above, narrowly few-lobed basally, discolorous and grey below. Flower heads radiate, in dense terminal corymbs, yellow, involucres calycled. Jan.–May. Sandstone slopes, SW, KM, LB, SE (Cape Peninsula to Grahamstown).

lyratus L.f. Similar to **S. rigidus** but softer, with lyrate to lacero-pinnatisect leaves, stems slightly woolly to softly scabrid. Nov.–Feb. Moist flats or mountain slopes, SW, LB (Cape Peninsula to Swellendam).•

mimetes Hutch. & R.A.Dyer Like **S. lanceus** but glabrous to roughly hairy below. Leaves often entire. Dec.–Feb. Sandstone slopes near water, NW, KM, SE (Hex River valley to Van Staden's Mts and E Cape).

muricatus Thunb. Scabrid, diffuse shrublet to 60 cm. Leaves pinnatisect, lobes linear and toothed, margins revolute. Flower heads radiate, few in lax corymbs, yellow, involucres calycled. Aug.–Oct. Rocky sandstone slopes, NW, KM (Gifberg to Ladismith).•

oederiifolius DC. Like **S. ilicifolius** in leaf but teeth often revolute. Flower heads few and larger with involucre c. 8 mm long. Sept.–Nov. Damp grassland, SE (Humansdorp to Van Staden's Mts).•

pauciflosculosus C. Jeffrey (= *Senecio oliganthus* DC.) Densely leafy, grey-felted shrublet to 30 cm. Leaves elliptic, discolorous, often glabrous above, entire or toothed above, usually 3-veined from base.

Flower heads discoid, in dense terminal clusters, pale yellow, involucres calycled. Feb.–May. Sandstone slopes, SW, KM, LB, SE (Du Toitskloof to E Cape).

pinnatifidus (P.J.Bergius) Less.   DILL-LEAVED GROUNDSEL   Erect, softly woody, roughly hairy perennial to 50 cm, from woody rootstock. Leaves pinnatisect to bipinnatisect, lobes narrow, margins revolute, sometimes toothed. Flower heads discoid, large, 30–40-flowered, few in lax terminal corymbs, yellow. Sept.–Feb. Sandstone slopes, NW (Pakhuis Mts to Piketberg and Ceres).•

pubigerus L. (= *Senecio expansus* Harv.; incl. **S. anapetes** C.Jeffrey, **S. incisus** Thunb.)   SKRAALBOSSIE   Diffusely branched, roughly hairy, softly woody shrub to 1 m. Leaves oblong, coarsely toothed, margins revolute. Flower heads few-rayed, few in white-woolly axillary clusters on rigid, scaly branches, yellow, involucres calycled. Mar.–July. Dry stony clay (or granite), often disturbed sites, NW, SW (Darling to Cold Bokkeveld and Riviersonderend Mts).•

rehmannii Bolus   Willowy, shortly hairy perennial to 60 cm. Leaves linear to oblanceolate, often toothed, margins revolute, thinly woolly below. Flower heads radiate, small, in lax branched corymbs, yellow, involucres calycled. Jan.–Apr. Stony flats, NW, LB (Hex River valley to Swellendam).•

rigidus L. (incl. **S. subcanescens** (DC.) Compton)   ROUGH RAGWORT   Robust, densely leafy, roughly hairy shrub to 1.5 m, with scabrid stems. Leaves oblong to oblanceolate, irregularly toothed, margins revolute, clasping at base, often woolly below with raised veins. Flower heads radiate, small, in branched corymbs, yellow, involucres calycled. Nov.–Jan. Sandstone slopes and gullies, NW, SW, LB, SE (Olifants River Valley to Uitenhage).•

serrurioides Turcz.   Erect, softly woody, densely leafy perennial to 50 cm, from woody rootstock. Leaves bipinnatisect, lobes filiform. Flower heads radiate, large, 80–100-flowered, few in lax terminal corymbs on stout peduncles, yellow. Oct.–Nov. Sandstone slopes, SE (Van Staden's Mts).•

skirrhodon DC.   Glabrous perennial to 30 cm. Leaves oblanceolate, clasping below, margins revolute, minutely toothed, curved upward above, pungent. Flower heads radiate, in lax corymbs, yellow. Mainly Aug.–Nov. Coastal sands, SE (Humansdorp to tropical Africa and Madagascar).

sociorum Bolus   Thinly woolly perennial to 45 cm from woody rootstock. Leaves elliptic, finely toothed, 5-veined from base. Flower heads radiate, few in terminal corymbs, orange or yellow, bracts broad, calycled. Mar.–May. Rocky sandstone slopes, NW, SW (Ceres to Franschhoek).•

tortuosus DC.   Twiggy, roughly hairy shrublet to 30 cm. Leaves petiolate, cuneate-obovate, toothed to pinnatifid. Flower heads radiate, solitary on short peduncles in upper leaf axils, yellow with white rays. July–Oct. Rock crevices, NW (Gifberg to Cedarberg Mts).•

verbascifolius Burm.f.   Robust, white-woolly perennial to 60 cm. Leaves ovate-cordate, petiolate and clasping at base, glabrescent above, margins revolute, minutely toothed. Flower heads radiate in compound, branched corymbs, yellow, involucre calycled, bracts thickened below. Sept.–Dec. Sheltered rocky slopes, SW (Cape Peninsula).•

vestitus P.J.Bergius   PAPERLEAF RAGWORT   Glabrous, sometimes glaucous, softly woody shrub to 1.5 m. Leaves oblanceolate, eared below, coarsely toothed. Flower heads radiate, in dense, branched terminal corymbs, yellow, involucres calycled. Sept.–Jan. Sandstone slopes, NW, KM (Namaqualand to Paarl and Montagu).•

sp. 2   Like **S. ilicifolius** but flower heads smaller, discoid. Dec.–Jan. Rocky sandstone slopes, above 1500 m, NW (Cedarberg Mts).•

sp. 3   Robust, densely leafy, sparsely white-woolly perennial to 2 m. Leaves lanceolate, ascending, margins revolute, slightly toothed. Flower heads radiate, large, in lax terminal corymbs, yellow with mauve or white rays, involucres calycled. Sept.–Nov. Sandstone slopes near water, above 800 m, SW (Bainskloof to Palmiet River Mts).•

sp. 4 (= *Senecio cinerarioides* Schltr. ms)   Delicate, shortly hairy perennial to 15 cm spreading on thin woody rhizomes. Leaves tufted, pinnatisect, lobes rhomboid, the terminal large and toothed. Flower heads radiate, 1 or 2 on slender peduncles, white. Dec.–Jan. Shady rocks above 1200 m, NW (Cedarberg Mts).•

[**Species excluded**   No authentic material found and probably conspecific with one of the above: **S. infirmus** C.Jeffrey (= *Senecio debilis* Harv.), **S. matricariifolius** DC., **S. thunbergii** Harv.; known from a single collection at Stellenbosch and probably cultivated: **S. subsinuatus** DC.; not from the area: **S. inaequidens** DC.]

STEIRODISCUS• GEELKRUID   5 spp., W Cape

capillaceus (L.f.) Less.   Annual with wiry flexuose stems, 5–20 cm. Lower leaves pinnatisect with filiform segments, upper leaves filiform. Flower heads radiate, yellow. Aug.–Sept. Sandy flats and slopes, NW (Bokkeveld Mts to Piketberg).•

gamolepis Bolus ex Schltr.   Diffuse wiry annual, 5–10 cm. Leaves pinnatisect, segments filiform. Flower heads radiate, yellow, solitary on swollen peduncles, small, bracts fused to form a cup. Aug.–Oct. Sandstone slopes, NW (Tulbagh valley).•

schlechteri Bolus ex Schltr.   Annual with wiry flexuose stems, 5–20 cm. Leaves pinnatisect with filiform segments. Flower heads radiate, yellow. July–Sept. Sandy slopes, NW (foot of Matsikamma–Gifberg).•

speciosus (Pillans) B.Nord.   Wiry-stemmed annual to 20 cm. Leaves pinnatisect, segments linear. Flower heads radiate, orange-yellow, solitary at branch tips, bracts fused to form a cup. Sept.–Oct. Sandy flats and dunes, SW (Mamre).•

tagetes (L.) Schltr.   Wiry-stemmed annual to 50 cm. Leaves pinnatisect, segments linear. Flower heads radiate, yellow, solitary at branch tips, bracts fused to form a cup. Sept.–Oct. Sandy flats, SW (Hopefield to Cape Peninsula).•

STILPNOGYNE   1 sp., winter-rainfall South Africa

bellidioides DC.   Delicate, glabrous annual, 5–20 cm. Leaves long-petiolate, lyrate to deltoid, coarsely toothed. Flower heads disciform or 3-radiate, yellow. Aug.–Oct. Shade on rocky slopes, NW, KM (S Namaqualand and W Karoo to Witteberg).

STOEBE   HARTEBEESKAROO   34 spp., Mainly W Cape, S tropical Africa, Madagascar, Reunion

A. Corolla purple, brown or yellow with small erect lobes

cinerea (L.) Thunb.   VAAL HARTEBEESKAROO   Grey-woolly, much-branched shrub to 1.5 m, with short shoots. Leaves needle-like with bulbous base, spreading and twisted. Flower heads discoid, in dense axillary glomerules together forming elongate spikes, purplish; bracts golden, acuminate. Apr.–May. Rocky slopes often shale, SW (Cape Peninsula to Riviersonderend Mts).•

incana Thunb.   Grey-woolly, ericoid shrublet to 60 cm. Leaves needle-like with involute margins, twisted and recurved. Flower heads discoid, crowded in terminal heads, brownish; bracts golden brown, acuminate. Feb.–May. Sandstone slopes, SW (Cape Peninsula to Hermanus).•

plumosa (L.) Thunb. (= *Stoebe burchellii* Levyns)   SLANGBOS   Sprawling, white-woolly, softly woody shrub to 1 m, with short shoots. Leaves granular and tufted. Flower heads discoid, in axillary glomerules forming spike-like inflorescences; bracts golden, acuminate. Mainly Apr.–June. Rocky flats and slopes, NW, SW, KM, LB, SE (throughout southern Africa).

saxatilis Levyns   Intricately branched, white-woolly shrublet 20–60 cm. Leaves linear-lanceolate, incurved, closely and obliquely set. Flower heads discoid, in terminal clusters, purple; bracts golden, acuminate. Mar.–May. Sandstone rocks above 1000 m, NW, SW (Cedarberg Mts to Bainskloof).•

spiralis Less.   Similar to **S. incana** but bracts yellow and shorter than yellow flowers. Mar.–May. Damp sandstone slopes, NW, SW, KM, LB (Elandskloof to Robinson Pass).•

sp. 1   Densely leafy, white-cobwebby shrublet to 40 cm. Leaves lanceolate, erect. Flower heads discoid, in axillary clusters aggregated in dense terminal spikes, purple?; bracts golden. Sept.–Nov. Sandstone slopes above 1000 m, SW (Kogelberg).•

A. Corolla white, pink or mauve, usually with conspicuous lobes

aethiopica L.   KNOPPIESLANGBOS   Rigid, densely leafy, glabrescent or thinly grey-woolly shrub to 1.5 m. Leaves needle-like with involute margins, usually twisted and recurved, pungent. Flower heads discoid, crowded in terminal heads, florets conspicuous, white; bracts brown. Mainly Sept.–Nov. Sandstone slopes, NW, SW, KM, SE (Bokkeveld Mts to Langkloof).•

alopecuroides (Lam.) Less.   KATSTERTSLANGBOS   Robust, thinly hairy, grey shrub to 10 cm. Leaves stiffly needle-like with involute margins, spreading, twisted. Flower heads discoid, massed in elongate spikes, florets conspicuous, white; bracts brown. July–Dec. Forest margins and fynbos, KM, LB, SE (Riversdale and Swartberg Pass to Uitenhage).•

capitata P.J.Bergius (= *Stoebe. bruniades* (Reichb.) Levyns)   Erect or spreading, thinly cobwebby shrublet to 50 cm. Leaves ericoid, spreading and twisted. Flower heads discoid, in dense, globose clusters at branch tips, florets conspicuous, mauve to pink or white; bracts brown. Dec.–Mar. Sandstone slopes and coastal sands, slopes, NW, SW, LB, SE (Piketberg to Grahamstown).

cyathuloides Schltr. (= *Stoebe sphaerocephala* Schltr., *S. humilis* Levyns)   Erect or spreading, glabrescent shrublet to 40 cm. Leaves lanceolate with slightly involute margins, twisted. Flower heads discoid, crowded in terminal heads, florets conspicuous, pink or white; bracts inconspicuous, brownish. Nov.–Feb. Coastal fynbos, SW, AP, LB (Cape Peninsula to Albertinia).•

fusca (L.) Thunb.   Densely branched, grey-woolly shrublet to 25 cm. Leaves needle-like, twisted and recurved with involute margins. Flower heads discoid, in teminal glomerules, florets conspicuous, pink or mauve; bracts golden. Mar.–May. Sandstone slopes, NW, SW (Gifberg to Bredasdorp).•

gomphrenoides P.J.Bergius   Sparsely branched, cobwebby shrublet to 30 cm. Leaves lanceolate with ciliate margins. Flower heads discoid, in dense terminal clusters, cream-coloured; bracts acuminate, cream-coloured. Mainly Nov.–Dec. Sandy hills, SW (Malmesbury to Pella).•

leucocephala DC.   Spreading, silvery woolly shrublet to 15 cm. Leaves linear with involute margins, twisted. Flower heads discoid, in terminal clusters, pink, inconspicuous among white bracts. Oct.–Jan. Sandstone slopes, NW, SW (Bokkeveld Mts to Malmesbury).•

microphylla DC.   Glabrescent slender-branched shrublet to 20 cm. Leaves scale-like, adpressed. Flower heads discoid, small, in terminal clusters, pink; bracts acuminate, white to yellow. Feb.–May. Stony slopes, KM, LB, SE (Swartberg Mts to Uitenhage).•

montana Schltr. ex Levyns   Like **S. capitata** but leaves blunt and pappus lacking. Jan. Rocky sandstone slopes, NW (Ceres: Skurweberg).•

muirii Levyns   Densely leafy, silvery woolly shrublet to 80 cm, branches often whorled. Leaves oblong with involute margins, strongly curled. Flower heads discoid, crowded in dense cylindrical spikes, florets conspicuous, pink; bracts acuminate, brown. Mar.–Apr. Coastal dunes and limestones, AP (De Hoop to Stilbaai).•

nervigera (DC.) Sch.Bip.   Stiffly branched shrublet to 30 cm. Leaves ericoid, erect or recurved, imbricate, pungent. Flower heads discoid, few in terminal clusters, florets conspicuous, cream-coloured; bracts golden, acuminate. Jan.–July. Sandy or clay slopes and flats, NW, SW, LB (Bokkeveld Escarpment to Albertinia).•

phyllostachya (DC.) Sch.Bip. (= *Stoebe copholepis* Sch.Bip., *S. ensorii* Compton)   Densely leafy, thinly white-woolly shrublet to 40 cm. Leaves linear with involute margins, often twisted, erect on peduncles but spreading elsewhere. Flower heads discoid, crowded in terminal glomerules, florets conspicuous, white; bracts brown. Feb.–Apr. Sandstone slopes, SW, LB, SE (Grabouw to Langkloof).•

prostrata L.   Sprawling, grey-woolly shrublet to 25 cm. Leaves lanceolate with involute margins, twisted and recurved, glabrescent above. Flower heads discoid, in dense terminal heads, florets conspicuous, mauve; bracts inconspicuous, brownish. Jan.–Mar. Sandstone slopes, SW (Cape Peninsula to Riviersonderend Mts).•

rosea Wolley-Dod   Stiffly branched, densely leafy shrub to 50 cm. Leaves ericoid, recuved and twisted. Flower heads discoid, crowded in dense terminal heads, florets conspicuous, pink; bracts hidden among conspicuous, feathery pappus bristles. Jan.–Mar. Sandstone slopes, SW (Cape Peninsula).•

rugulosa Harv.   Densely branched, thinly white-woolly shrublet to 40 cm. Leaves linear, adpressed, 3-veined with recurved mucro. Flower heads discoid, in terminal clusters, florets conspicuous, pink to magenta; bracts golden. Feb.–May. Coastal flats in renosterveld, AP, LB (Bredasdorp to Albertinia).•

schultzii Levyns (= *Stoebe salteri* Levyns)   Spreading, thinly white-woolly shrublet to 30 cm. Leaves diamond-shaped, adpressed. Flower heads discoid, in terminal clusters, florets conspicuous, white; pappus absent; bracts inconspicuous. Oct.–Dec. Damp coastal flats, SW, AP (Caledon to Agulhas).•

SYNCARPHA (= *HELIPTERUM* in part)   EVERLASTING, SEWEJAARTJIE   c. 25 spp., southern Africa

A. Flower heads to 15 mm diam.

dregeana (DC.) B.Nord. (= *Helipterum dregeanum* DC.)   ROOISEWEJAARTJIE   Divaricately branched, grey-felted, closely leafy shrub to 60 cm. Leaves obovate, recurved at tips, the upper often rusty on margins. Flower heads discoid, solitary, nested in leaves, hemisphaerical, 12–15 mm diam.; bracts papery above, obtuse, bright crimson above. Oct.–Jan. Dry sandstone slopes, NW, KM (Pakhuis Mts to Witteberg).•

gnaphaloides (L.) DC. (= *Helipterum gnaphaloides* (L.) DC.)   VLAKTETEE   Erect, white-felted shrublet to 30 cm. Leaves linear, ascending, margins involute. Flower heads discoid, solitary on long peduncles, fragrant, cylindrical, c. 10 mm diam.; bracts reddish brown, dry above, attenuate and sharply reflexed. Oct.–Dec. Sandstone slopes, SW, LB (Cape Peninsula and Tulbagh to Outeniqua Mts).•

marlothii (Schltr.) B.Nord. (= *Helipterum marlothii* Schltr.)   Dwarf, few-branched, densely leafy, grey-woolly shrublet to 35 cm. Leaves ovate, imbricate, ascending. Flower heads discoid, several in loose corymbs nested in leaves, subglobose, 12–15 mm diam.; bracts papery, obtuse, white. Jan.–Mar. Sandstone rocks on S slopes above 1400 m, NW, SW (Cold Bokkeveld to Villiersdorp).•

milleflora (L.f.) B.Nord. (= *Helipterum milleflorum* (L.f.) Druce)   KNOPPIESSEWEJAARTJIE   Mostly single-stemmed, robust, closely leafy, silvery felted shrub to 2 m. Leaves large, lanceolate, imbricate. Flower heads discoid, crowded in dense terminal corymbs, cylindrical, 6–8 mm diam.; bracts papery, obtuse, white to pink. July–Dec. Dry N-facing sandstone slopes, KM, LB, SE (Ladismith to Grahamstown).

paniculata (L.) B.Nord. (= *Helichrysum paniculatum* (L.) Thunb.)   SEWEJAARTJIE   Erect, densely leafy, silvery felted shrublet to 60 cm. Leaves linear, ascending, apiculate. Flower heads discoid, few to several in terminal clusters, hemisphaerical, c. 10 mm diam.; bracts papery, acute, yellow or pink in bud, ageing to white. Mainly Oct.–June. Coastal and lower slopes, NW, SW, AP, KM, LB, SE (Gifberg to Port Elizabeth).•

sordescens (DC.) B.Nord. (= *Helichrysum sordescens* DC.)   Densely leafy, grey-woolly shrublet to 30 cm. Leaves oblanceolate, recurved above, acuminate, 3-veined. Flower heads discoid, few in terminal clusters, hemisphaerical, 10–15 mm diam.; bracts papery, acute, whitish. Dec.–Jan. Dunes and sandy slopes, SE (Port Elizabeth to Alexandria).

striata (Thunb.) B.Nord. (= *Helichrysum striatum* Thunb.)   Closely leafy, hairy, softly woody shrublet to 60 cm. Leaves linear and channelled, ascending, 3-veined. Flower heads discoid, few to several in loose terminal clusters nested in leaves, hemisphaerical, 10–15 mm diam.; bracts papery, acute, white. Sept.–May. Coastal grassland, SE (Humansdorp to East London).

sp. 1 (= *Helichrysum chlorochrysum* DC.)   GOLD EVERLASTING   Erect, densely leafy, silvery felted shrub to 1.5 m. Leaves ovate, spreading to recurving, 5-veined, attenuate. Flower heads discoid, several in loose corymbs, subglobose, 10–12 mm diam.; bracts papery, acute, pale yellow. June–Nov. Limestone hills in fynbos, AP (Potberg to Stilbaai).•

sp. 2 (= *Helichrysum mucronatum* (L.) Willd.)   Silvery felted shrublet to 45 cm. Leaves linear, 1-veined, apiculate. Flower heads discoid, few in terminal clusters, subglobose, 6–10 mm diam.; bracts papery, obtuse, white or yellow. Aug.–Mar. Sandy slopes, NW, SW, AP (Gifberg to Agulhas).•

A. *Flower heads more than 15 mm diam.*
B. *Bracts white with dark brown tips*

loganiana (Compton) B.Nord. (= *Helichrysum loganianum* Compton)   Densely leafy, grey-felted shrublet to 25 cm. Leaves oblanceolate, ascending. Flower heads discoid, solitary, conical, 30–40 mm diam.; bracts papery, acute, white with chocolate-coloured tips. Oct.–Jan. Rocky sandstone ridges, NW, KM (Gydo Pass to Witteberg and Touwsberg).•

montana (B.Nord.) B.Nord. (= *Helipterum montanum* B.Nord.)   Gnarled, grey-felted shrublet to 30 cm. Leaves oblanceolate, spreading. Flower heads discoid, solitary on short peduncles, conical, 20–25 mm diam.; bracts papery, acute, white with chocolate-coloured tips. Mainly Oct.–Feb. Sandstone rocks above 1600 m, KM (Swartberg and Kammanassie Mts).•

variegata (P.J.Bergius) B.Nord. (= *Helipterum variegatum* (P.J.Bergius) DC.)   BONTSEWEJAARTJIE   Erect, closely leafy, densely grey-felted shrublet 40 cm. Leaves ovate-oblong, ascending, imbricate, upper leaves dry-tipped. Flower heads discoid, solitary, subglobose, 40–60 mm diam.; bracts papery, obtuse, white with brown tips. Sept.–Dec. Rocky sandstone slopes, often at high alt., NW, SW (N Cedarberg Mts to Bredasdorp).•

B. *Bracts unicoloured or reddish above*
C. *Flower heads few to many in corymbs (see also* **S. canescens***)*

argentea (Thunb.) B.Nord. (= *Helichrysum argenteum* Thunb.)   SILVER EVERLASTING   Sprawling, silvery felted shrublet to 60 cm. Leaves oblanceolate, 3-veined, spreading, recurved above. Flower heads discoid, 1–several in terminal clusters on long, sparsely leafy peduncles, hemispherical, 15–20 mm diam.; bracts papery, acute, white, pink above. Apr.–July. Coastal grassland and scrub, SE (Uitenhage to E Cape).

argyropsis (DC.) B.Nord. (= *Helipterum argyropsis* DC.)   WITSEWEJAARTJIE   Closely leafy, silvery felted shrublet to 70 cm. Leaves oblanceolate, ascending, imbricate. Flower heads discoid, few to several in lax terminal clusters, hemispherical, 15–20 mm diam.; bracts papery, acute, white. Aug.–Nov. Coastal slopes, sometimes on limestone, SW, ?AP, SE (Rooiels to Plettenberg Bay).•

eximia (L.) B.Nord. (= *Helipterum eximium* (L.) DC.)   STRAWBERRY EVERLASTING   Mostly single-stemmed, robust, closely leafy, silvery felted shrub to 40 cm. Leaves ovate, imbricate, ascending. Flower heads discoid, large, crowded in dense corymbs nested in leaves, hemispherical, 20–25 mm diam.; bracts papery, obtuse, bright red. Nov.–Mar. S-facing sandstone, SW, LB, SE (Riviersonderend Mts to Uitenhage).•

vestita (L.) B.Nord. (= *Helichrysum vestitum* (L.) Schrank)   CAPE SNOW   Densely leafy, grey-woolly, softly woody shrublet to 1 m. Leaves oblanceolate, ascending. Flower heads discoid, few to several in loose terminal clusters nested in leaves, conical, 35–40 mm diam.; bracts papery, acute, white. Nov.–Jan. Rocky slopes and flats, SW, LB, SE (Cape Peninsula to George).•

zeyheri (Sond.) B.Nord. (= *Helipterum zeyheri* Sond.)   Densely leafy, grey-felted shrublet to 30 cm, mostly single-stemmed below. Leaves elliptic, ascending, imbricate. Flower heads discoid, few in terminal clusters, conical, c. 30 mm diam.; bracts papery, acuminate, white to pink or purple. Aug.–Oct. Sandstone flats and slopes, SW (Cape Peninsula to Hermanus).•

C. Flower heads solitary (see also **S. argentea***)*

affinis (B.Nord.) B.Nord. (= *Helipterum affine* B.Nord.)   Straggling, grey- to rust-felted, sparsely branched shrublet to 60 cm. Leaves lanceolate, spreading, rusty on upper margins, brown-tipped. Flower heads discoid, solitary on slender peduncles, shallowly hemispherical, 20–30 mm diam.; bracts papery, obtuse, reddish or straw-coloured. July–Dec. Rocky slopes, NW (Piketberg to Cold Bokkeveld).•

canescens (L.) B.Nord. (= *Helipterum canescens* (L.) DC.)   PIENKSEWEJAARTJIE   Sparsely branched, closely leafy, grey-felted shrublet to 50 cm. Leaves small, elliptic, ascending and imbricate. Flower heads discoid, mostly solitary at branch tips, conical, 25–35 mm diam.; bracts papery, acuminate, pink to red. Jan.–Sept. Rocky sandstone slopes and flats to upper slopes, NW, SW, ?AP, KM, LB, SE (Kamiesberg, Gifberg to Humansdorp).•

dykei (Bolus) B.Nord. (= *Helichrysum dykei* Bolus)   Dwarf, grey-woolly shrublet to 10 cm. Leaves obovate, leathery. Flower heads discoid, solitary on short peduncles covered with papery scales, conical, 20–30 mm diam.; bracts papery, acute, white to pink. Dec.–Apr. Rocky sandstone slopes above 1600 m, NW (Skurweberg and Hex River Mts).•

ferruginea (Lam.) B.Nord. (= *Helipterum ferrugineum* (Lam.) DC.)   Closely leafy, grey-felted shrublet to 40 cm. Leaves oblanceolate, ascending, with long dry bristles. Flower heads solitary on short, bracteate peduncles, shallowly hemispherical, 30–50 mm diam.; bracts papery, yellow or rosy, acuminate. Aug.–Feb. Dry rocky slopes, KM, SE (Witteberg to Port Elizabeth).•

flava (Compton) B.Nord. (= *Helipterum flavum* Compton)   Closely leafy, grey-felted shrublet to 40 cm. Leaves obovate, ascending, the uppermost with long dry bristles. Flower heads solitary on short peduncles, hemispherical, 25–30 mm diam.; bracts papery, acuminate and shorter than pappus in fruit. Dec.–Mar. Dry sandstone slopes, NW, KM (Cold Bokkeveld to Swartberg Mts).•

lepidopodium (Bolus) B.Nord. (= *Helichrysum lepidopodium* Bolus)   Densely leafy, silvery felted shrublet to 60 cm. Leaves oblanceolate, ascending. Flower heads discoid, solitary on long peduncles covered with papery scales, conical-hemispherical, 40–50 mm diam.; bracts papery, acuminate, white. Oct.–Dec. Rocky slopes above 1500 m, upper slopes, SW (Franschhoek Mts).•

recurvata (L.f.) B.Nord. (= *Helichrysum recurvatum* (L.f.) Thunb.)   Spreading to prostrate, thinly woolly shrublet to 20 cm. Leaves lanceolate, channelled, strongly recurved, 3-veined, bristly on margins. Flower heads discoid, solitary on short peduncles, conical, 15–20 mm diam.; bracts acuminate, pink. Sept.–Dec. Sandstone slopes to 500 m, SE (George to Grahamstown).•

speciosissima (L.) B.Nord. (= *Helipterum speciosissimum* (L.) DC.)   CAPE EVERLASTING   Sprawling, white-woolly shrublet, 20–60 cm, with erect annual stems. Leaves oblong to linear, clasping at base. Flower heads discoid, solitary on elongate peduncles, hemispherical, 30–40 mm diam.; bracts papery, acuminate, white to cream-coloured. July–Jan. Sandstone slopes, SW, LB (Cape Peninsula to Tradouw Pass).•

virgata (P.J.Bergius) B.Nord. (= *Helipterum variegatum* (P.J.Bergius) DC.)   Erect, grey-felted shrublet to 40 cm, often rusty on peduncles. Leaves lanceolate, spreading, brown-tipped. Flower heads discoid, solitary on slender peduncles, shallowly hemispherical, 15–20 mm diam.; bracts papery, acuminate, white or yellow. Oct.–Jan. Rocky slopes, NW, SW, KM, LB (Ceres and Wemmershoek Mts to Witteberg).•

sp. 3 (= *Syncarpha scariosa* B.Nord. ms)   Closely leafy, grey-felted shrublet to 40 cm. Leaves obovate, ascending, the uppermost often with dry bristles. Flower heads solitary on long bracteate peduncles, hemispherical, 25–30 mm diam.; bracts papery, acute, yellow. Nov.–Mar. Dry upper mountain slopes, KM (Swartberg and Kammanassie Mts).•

sp. 4   Like **sp. 1** but peduncles naked. Nov. Alluvial shales on limestone, AP (Agulhas Peninsula).•

TARCHONANTHUS   CAMPHOR TREE, KANFERBOS, ISIDULI   2 spp., Africa

camphoratus L.   Dioecious, grey-felted shrub or small tree to 5 m, strongly scented of camphor. Leaves elliptic, dull green above. Flower heads discoid, in large panicles, cream-coloured. Achenes woolly. Dec.–Apr. Widespread, mainly coastal, SW, AP, ?KM, LB, SE (Namibia, Cape Peninsula to E Africa).

THAMINOPHYLLUM•   3 spp., W Cape

latifolium Bond   Sprawling, silky shrublet to 75 cm. Leaves elliptic with margins recurved, spreading or reflexed, mucronate. Flower heads radiate, 1–3 on slender peduncles at branch tips, yellow with rounded pink or white rays. Aug.–Dec. Sandstone slopes in coastal fynbos, SW (Hermanus Mts).•

multiflorum Harv.   Silvery silky shrublet to 60 cm, sometimes to 1.5 m. Leaves linear, mostly spreading. Flower heads radiate, 1–3 at branch tips, yellow with elliptic white rays. Aug.–Jan. Damp rocky sandstone slopes in fynbos, SW (Viljoen's Pass to Houwhoek).•

mundii Harv.   Similar to **T. multiflorum** but flower heads several at branch tips and ray florets only 2–4. July–Nov. Damp sandstone slopes in fynbos, SW (Kogelberg to Caledon Swartberg).•

TRICHOGYNE (= *IFLOGA* in part)   NAALDEBOS   8 spp., southern Africa

ambigua (L.) Druce (= *Ifloga ambigua* (L.) Druce)   Sprawling, grey and thinly white-hairy ericoid shrublet to 50 cm. Leaves filiform or granular, in tufts. Flower heads discoid, axillary and forming dense spike-like racemes, cream-coloured. Mainly Apr.–Nov. Sandy coastal flats and slopes, NW, SW, KM (Pakhuis Pass to Cape Peninsula and Robertson).•

decumbens (Thunb.) Less. (= *Ifloga woodii* (N.E.Br.) Burtt)   Gnarled, dwarf, white-membranous ericoid shrublet to 15 cm. Leaves filiform, suberect, imbricate. Flower heads discoid, crowded apically, white. Aug.–Mar. Stony karroid slopes, NW, KM (Cold Bokkeveld to W Lesotho).

pilulifera (Schltr.) Anderb. (= *Ifloga pilulifera* Schltr.)   GEELNAALDEBOSSIE   Similar to **T. repens** but flowering branchlets not secund and flower heads yellow. July–Sept. Sandstone slopes in fynbos, NW (Bokkeveld Mts to Cold Bokkeveld).•

polycnemoides (Fenzl) Anderb. (= *Ifloga polycnemoides* Fenzl)   Like **T. verticillata** but to 15 cm and flower heads reddish. Mainly Aug.–Sept. Sandstone slopes above 600 m, NW (Namaqualand to Worcester and W Karoo).

repens (L.) Anderb. (= *Ifloga repens* (L.) Hilliard, *Trichogyne reflexa* (L.f.) Less.)   WITNAALDEBOSSIE   Similar to **T. ambigua** but main branches prostrate, bearing secund, short erect flowering branches to 10 cm, terminating in tight rounded clusters of flower heads. July–Oct. Coastal dunes and sandy flats, SW, AP (Vredenburg to Mossel Bay).•

verticillata (L.f.) Less. (= *Ifloga verticillata* (L.f.) Fenzl)   Thinly white-membranous annual to 20 cm with stems stiffly erect above. Leaves linear, margins revolute. Flower heads discoid, in axillary clusters along stems, straw-coloured. Mainly Aug.–Oct. Coastal dunes and sandy flats, NW, SW, AP (Lambert's Bay to Stilbaai).•

TRIPTERIS (= *OSTEOSPERMUM* in part)   SKAAPBOS   c. 20 spp., southern Africa and Angola

A. Annuals

breviradiata (Norl.) B.Nord.   LEMOENBOSSIE   Glandular-hairy annual to 50 cm. Leaves soft, lanceolate, sparsely toothed, the lower petiolate, upper clasping. Flower heads radiate, small, yellow. Achenes c. 5 mm long. Aug.–Sept. Rocky hills, NW (S Namibia and W Karoo to Botterkloof).

clandestina Less.   TREKKERTJIE   Glandular-hairy, aromatic annual to 40 cm. Leaves oblanceolate, toothed to lobed. Flower heads radiate, few in branched panicles, dark purplish with pale yellow rays brown at base. Achenes 8–9 mm long. July–Sept. Sandy and rocky flats, NW, SW (S Namibia and Namaqualand to Cape Peninsula and Riviersonderend).

microcarpa Harv.   BOEGOEBOSSIE   Robust, roughly glandular-hairy annual to 80 cm. Leaves oblanceolate, toothed to pinnatifid, the upper clasping. Flower heads radiate, small, yellow. Achenes 4–5 mm long. Mainly May–Sept. Rocky slopes and flats, NW (S Angola to Pakhuis Mts).

A. Perennials or shrubs

aghillana DC (= *Osteospermum scariosum* DC.)   SKAAPBOS   Roughly hairy or bristly perennial to 30 cm, with woody base. Leaves mostly basal, oblanceolate, entire to sharply toothed. Flower heads radiate, 1–few on elongate peduncles, dark with yellow or cream-coloured rays. Achenes 9–12 mm long. July–Oct. Rocky slopes and hills, NW, SW, AP, KM, LB, SE (Worcester and Agulhas to Mpumalanga).

amplexicaulis (Thunb.) Less. (= *Osteospermum connatum* DC.)   Like **T. dentata** but lower leaves ovate and clasping. Achenes 5–6 mm long. Sept.–Nov. Sandstone slopes, NW (Gifberg to Saron).•

dentata (Burm.f.) Harv.   Sprawling, roughly hairy perennial to 60 cm. Leaves oblanceolate, toothed to pinnatifid, the lowermost opposite. Flower heads radiate, in branched corymbs, yellow. Achenes 9–12 mm long. Aug.–Oct. Sandstone slopes and flats, NW, SW (Redelinghuys to Cape Peninsula).•

oppositifolia (Aiton) B.Nord.   STINKSKAAPBOS   Rounded, glaucous shrub to 1 m. Leaves opposite, narrowly oblanceolate, succulent. Flower heads radiate, solitary, few at branch tips, dark with yellow rays. Achenes 10–15 mm long. July–Sept. Rocky sandstone or granite slopes, NW (Namaqualand to Clanwilliam).

sinuata DC.   KLEINSKAAPBOSSIE   Twiggy shrublet to 50 cm. Leaves opposite, lanceolate, toothed, fleshy. Flower heads radiate, 1–few at branch tips, yellow. Achenes 9–12 mm long. June–Sept. Rocky clay flats and slopes, NW, KM, SE (S Namibia to Uniondale and Karoo).

spathulata DC.   Like **T. aghillana** but leaves mostly entire, glabrous, spathulate and achenes with wings 1 mm wide. July–Oct. Dry karriod slopes, SE (Uitenhage to Grahamstown).

tomentosa (L.f.) Less.   White-woolly perennial with woody base, to 20 cm. Leaves spathulate, recurved at tips. Flower heads radiate, scapose, yellow. Achenes c. 9 mm long. Aug.–Oct. Sandstone slopes, NW, SW, LB (Grootwinterhoek Mts to Mossel Bay).•

TROGLOPHYTON   6 spp., southern Africa

capillaceum (Thunb.) Hilliard & B.L.Burtt   Delicate, white-woolly annual to 30 cm. Leaves petiolate, ovate, pale green and glabrescent above. Flower heads discoid, few, subracemose, white to purple, homogamous. Mainly Sept.–Oct. Damp shady slopes under shrubs, NW, SW, KM, LB, SE (S Namibia to Lesotho).

elsiae Hilliard   Delicate, white-woolly annual to 15 cm. Leaves petiolate, ovate, pale green and glabrescent above. Flower heads discoid, few, subracemose, white to purple, heterogamous. Dec.–Jan. Damp sheltered slopes above 1500 m, NW (Cold Bokkeveld to Hex River Mts).•

leptomerum Hilliard   Similar to **T. tenellum** but bracts acute. Sept.–Dec. Shade under rocks, NW (Namaqualand to Piketberg).

parvulum (Harv.) Hilliard & B.L.Burtt   Like **T. tenellum** but bracts acute and female flowers twice as many as hermaphrodite ones. Aug.–Nov. Damp, often shady slopes, NW, SW (S Namibia and W Karoo to Cape Peninsula and Riviersonderend Mts).

tenellum Hilliard   Delicate, diffuse, grey-woolly annual to 15 cm. Leaves oblanceolate, glabrescent above. Flower heads discoid, single and axillary on slender peduncles, or few at branch tips, heterogamous. Sept.–Oct. Sands in shady sites, NW (Kamiesberg to Botterkloof).

URSINIA   BERGMARGRIET   38 spp., mainly southern Africa, 1 in N Africa.

A. Paleae boat-shaped (disc shining); pappus 2-seriate

cakilefolia DC.   GANSOOGBERGMARGRIET   Sprawling annual to 45 cm. Leaves mostly bipinnatisect, 20–50 mm long, lobes linear. Flower heads radiate, solitary on long terminal peduncles, yellow or orange, c. 25–50 mm diam., paleae boat-shaped and longer than disc florets; pappus scales 5, plus 5 bristles, biseriate. July–Oct. Sandy flats and slopes, NW (Namaqualand and W Karoo to Redelinghuys).

chrysanthemoides (Less.) Harv.   Like **U. cakilefolia** but more or less woody below. Flower heads yellow or rays sometimes white or red with dark reverse, paleae shorter than disc florets. Aug.–Nov. Sandy and gravel slopes and flats, NW, SW, LB, SE (Namaqualand and Karoo to Port Elizabeth and E Cape).

nana DC.   KLEINBERGMARGRIET   Spreading annual or perennial to 20 cm. Leaves bipinnatisect, 15–50 mm long. Flower heads radiate, solitary on short peduncles, yellow, c. 10–35 mm diam., paleae boat-shaped, bracts rounded, often with dark crescent marks; pappus scales 5, plus 5 bristles, biseriate. Mainly Aug.–Oct. Gravel slopes and flats, NW, SW, KM, LB, SE (Namaquland to Mpumalanga).

speciosa DC.   Like **U. cakilefolia** but all involucral bracts rounded and papery above. Aug.–Oct. Sandy slopes and flats, NW, SW (Namaqualand to Hopefield).

A. Paleae oblong (disc dull); pappus 1-seriate or lacking

abrotanifolia (R.Br.) Spreng.   FYNKRUIE, LAMMETJIESKRUIE   Thinly grey-woolly, densely leafy shrub to 60 cm. Leaves bi- to tripinnatisect, 20–50 mm long. Flower heads radiate, solitary on elongate naked peduncles, yellow, c. 20–35 mm diam.; pappus scales 5, uniseriate. Nov.–Jan. Sandstone slopes in damp places, SW (Bainskloof to Hottentots Holland Mts).•

anethoides (DC.) N.E.Br.   Like **U. paleacea** but leaves smaller, 5–20 mm long, stems densely leafy. Flower heads on wiry peduncles, yellow. Nov.–Apr. Sandstone slopes in fynbos and grassland, SW, LB, SE (Caledon to Grahamstown).

anthemoides (L.) Poir.   MAGRIET   Annual to 50 cm. Leaves pinnatisect or bipinnatisect, 20–50 mm long. Flower heads radiate, solitary at branch tips, yellow or orange, sometimes with a dark ring and darker on reverse, 15–60 mm diam.; pappus scales 5, uniseriate. Aug.–Oct. Sandy and gravel slopes and flats, NW, SW, KM, LB, SE (S Namibia and Karoo to Port Elizabeth).

caledonica (E.Phillips) Prassler   Slender, densely leafy shrub to 1.2 m. Leaves ascending, oblanceolate, 3-lobed above, 6–35 mm long. Flower heads radiate, solitary on short peduncles nested in leaves, yellow, 15–30 mm diam.; pappus scales 5, uniseriate. Feb.–July. Damp sandstone slopes above 600 m, often seeps and marshes, SW (Groot Drakenstein to Hottentots Holland Mts).•

coronopifolia (Less.) N.E.Br.   Thinly woolly shrublet to 50 cm. Leaves oblanceolate, toothed, 20–100 mm long. Flower heads radiate, solitary on elongate, naked peduncles, yellow, 20–40 mm diam.; pappus scales 5, uniseriate. Jan.–Mar. Sandstone slopes above 600 m, near streams, NW (Grootwinterhoek Mts).•

dentata (L.) Poir.   Like **U. heterodonta** but leaves shortly bipinnatisect, lobes aristate and flower heads c. 40 mm diam. Sept.–Feb. Sandstone or limestone slopes and flats, NW, SW, AP, LB, SE (Bokkeveld Mts to Albertinia).•

discolor (Less.) N.E.Br.   Like **U. heterodonta** but grey-cobwebby, especially on young parts, leaves mostly smaller, c. 10 mm long and outer bracts also papery above. Aug.–Jan. Sandstone slopes, NW, SW, KM, LB, SE (Tulbagh to Humansdorp).•

dregeana (DC.) N.E.Br.   Like **U. nudicaulis** but stems white-woolly and leaves thinly woolly with lobes pungent. Oct.–Dec. Sandstone pavement, NW (Bokkeveld Mts to Hex River Mts).•

eckloniana (Sond.) N.E.Br.   Slender, densely leafy shrub to 1.2 m. Leaves ascending, narrowly oblong, sometimes shortly 3-toothed above, 20–80 mm long. Flower heads radiate, several densely crowded at branch tips, yellow, 30–50 mm diam.; pappus scales 5, uniseriate. July–Apr. Marshes on sandstone, SW (Groot Drakenstein to Hermanus).•

filipes (E.Mey. ex DC.) N.E.Br.   Softly woody shrublet to 40 cm. Leaves spreading, oblanceolate and pinnatifid, 10–40 mm long, lobes acute, sometimes toothed. Flower heads shortly radiate, solitary in upper axils on elongate, wiry peduncles, yellow, 6–14 mm diam.; pappus 5-lobed, uniseriate. Oct.–Mar. Sandstone slopes along streams above 600 m, NW, SW (Hex River Mts to Franschhoek).•

heterodonta (DC.) N.E.Br.   Shrublet to 50 cm. Leaves ascending, pinnatisect, 5–35 mm long, lobes linear. Flower heads radiate, solitary on elongate, wiry peduncles, yellow with dark reverse, 15–30 mm diam.; pappus scales 5, uniseriate. Mainly Aug. Sandstone slopes, NW, SW, KM, LB, SE (Hex River Mts and Caledon to Uitenhage).•

hispida (DC.) N.E.Br.   Softly hairy, slender shrublet to 40 cm. Leaves ascending, pinnatisect, 10–20 mm long, lobes linear. Flower heads radiate, solitary on short peduncles nested in leaves, yellow, c. 10–15 mm diam.; bracts narrow; pappus scales 5, uniseriate. Sept.–Dec. Sandstone slopes, LB (Langeberg Mts: Swellendam to Riversdale).•

macropoda (DC.) N.E.Br.   Like **U. nudicaulis** but leaves silvery silky and bract tips brown. Oct.–Mar. Rocky sandstone slopes, NW (Gifberg to Witzenberg).•

merxmuelleri Prassler   Sprawling, densely leafy shrublet to 60 cm. Leaves bipinnatisect, 20–40 mm long, lobes linear. Flower heads radiate, in lax corymbs on elongate peduncles, yellow, 15–30 mm diam.; pappus scales 5 plus 1–3 bristles, uniseriate. Feb.–Apr. Sandstone slopes above 1200 m, NW (Hex River Mts).•

nudicaulis (Thunb.) N.E.Br. (incl. **U. saxatilis** N.E.Br.)   Tufted shrublet to 50 cm. Leaves crowded below, pinnatisect, 15–70 mm long, lobes linear. Flower heads radiate, solitary on elongate, naked peduncles, yellow, 15–30 mm diam., outer bracts often papery above; pappus scales 5, uniseriate. Sept.–Mar. Sandstone slopes, sometimes wet places, NW, SW, ?AP, KM, LB, SE (Cedarberg Mts to Witteberg, Cape Peninsula to Humansdorp).

oreogena Schltr. ex. Prassler   Like **U. paleacea** but leaves more delicate and flower heads 20–30 mm diam. Oct.–Dec. Sandstone slopes, SW (Bainskloof to Hottentots Holland Mts).•

paleacea (L.) Moench   GEELMAGRIET   Shrub to 90 cm. Leaves pinnatisect, 20–60 mm long, lobes linear to filiform. Flower heads radiate, solitary on elongate peduncles, yellow, sometimes with greenish ring, dark on reverse, 20–50 mm diam.; pappus scales 5, uniseriate. Mainly Aug.–Dec. Sandstone slopes, NW, SW, LB, SE (Tulbagh and Cape Peninsula to Humansdorp).•

pilifera (P.J.Bergius) Poir.   GROOTBERGMARGRIET   Low, thinly woolly shrublet to 35 cm, woody below. Leaves bipinnatisect, 15–25 mm long. Flower heads radiate, large, solitary on long terminal peduncles, blackish with white or yellow rays dark on reverse, c. 30–60 mm diam.; bracts all rounded and silvery papery above; pappus scales 5, uniseriate. July–Nov. Gravel slopes in renosterveld, NW, KM, SE (Namaqualand and W Karoo to Worcester and Willowmore).

pinnata (Thunb.) Prassler   Willowy, densely leafy shrub to 1.5 m. Leaves ascending, pinnnatisect, 10–20 mm long, lobes filiform. Flower heads radiate, many in lax panicles, yellow, c. 10 mm diam.; pappus scales 5, uniseriate. Mainly Nov.–May. Marshes and streams on sandstone slopes, NW, SW (Bokkeveld Mts to Riviersonderend).•

punctata (Thunb.) N.E.Br.   Thinly hairy erect or sprawling shrublet to 50 cm. Leaves pinnatisect, 10–20 mm long, lobes oblong. Flower heads radiate, solitary on long wiry peduncles, yellow, c. 7–15 mm diam.; bracts 3- or 4-seriate; pappus scales 5, uniseriate. Sept.–Mar. Sandstone slopes, NW, KM (Bokkeveld Mts to Worcester, and Swartberg Mts).•

quinquepartita (DC.) N.E.Br.   Densely leafy shrublet to 40 cm. Leaves ascending, adpressed below, pinnatisect above, 10–20 mm long, lobes filiform. Flower heads radiate, solitary on short peduncles, sometimes nested in leaves, yellow, 10–25 mm diam.; pappus scales 10, uniseriate. Nov.–Apr. Sandstone slopes, often near water, SW (Hottentots Holland Mts to Hermanus).•

rigidula (DC.) N.E.Br.   Like **U. heterodonta** but upper leaf lobes longer and toothed and flower heads c. 45 mm diam. Sept.–Dec. Damp mountain slopes, NW, KM, LB, SE (Piketberg and Cold Bokkeveld to Riversdale and Kammanassie Mts).•

scariosa (Aiton) Poir.   Shrub to 1.5 m. Leaves pinnatisect, 20–120 mm long, lobes linear. Flower heads radiate, solitary on elongate peduncles, yellow, 20–50 mm diam.; outer bracts narrow, rounded with brown papery tips. Mainly Sept.–Feb. Forest margins and fynbos slopes, SW, KM, LB, SE (Paarl to Port Elizabeth.).•

sericea (Thunb.) N.E.Br.   Like **U. paleacea** but stems short, leaves silvery silky, 20–80 mm long, peduncles extremely long and bract tips brown. Sept.–Feb. Sandstone slopes, NW, KM (Cedarberg to Swartberg Mts).•

serrata (L.f.) Poir.   Slender, often thinly woolly, densely leafy erect shrublet to 1.5 m. Leaves ascending, narrowly oblong, sharply toothed, 15–60 mm long. Flower heads radiate, solitary on short to long axillary peduncles crowded at branch tips, yellow, c. 15–35 mm diam.; bracts all broad, golden papery; pappus scales 5, uniseriate. Aug.–Jan. Mostly marshy, sandstone slopes, SW, LB, SE (Riviersonderend to Tsitsikamma Mts).•

subflosculosa (DC.) Prassler   Densely leafy shrublet to 60 cm. Leaves ascending, pinnnatisect, 10–15 mm long, lobes filiform. Flower heads radiate, solitary on wiry peduncles clustered at branch tips, yellow, c. 15 mm diam.; bracts very narrow; pappus scales 5, uniseriate. Oct.–Dec. Sandstone slopes, NW (Gifberg to Cedarberg Mts).•

tenuifolia (L.) Poir.   Like **U. nudicaulis** but leaves linear to filiform, undivided. Sept.–Mar. Sandy flats and slopes, usually seasonally wet, SW, LB (Cape Peninsula to Albertinia).•

trifida (Thunb.) N.E.Br.   Densely leafy shrublet to 60 cm. Leaves ascending, 3-toothed above, c. 10 mm long. Flower heads radiate, solitary on wiry peduncles, yellow, c. 15 mm diam.; pappus scales 5, uniseriate or lacking. Mainly Nov.–Apr. Sandstone slopes, LB, SE (Langeberg: Swellendam to Tsitsikamma Mts).•

VELLEREOPHYTON   7 spp., W to E Cape

dealbatum (Thunb.) Hilliard & B.L.Burtt   Sprawling, white-woolly annual or ?perennial to 20 cm. Leaves oblanceolate. Flower heads discoid, crowded in dense, woolly, terminal corymbs, white, female flowers exceeding hermaphrodite ones; bracts reddish in centre. July–Mar. Damp sandstone slopes, to 750 m, NW, SW, LB, AP, SE (Namaqualand and W Karoo to Alexandria).

felinum Hilliard   Similar to **V. niveum** but grey-woolly and female flowers 1 or 2. Sept. Deep sands, NW (Cold Bokkeveld).•

gracillimum Hilliard   Sprawling, white-woolly annual to 10 cm. Leaves spathulate. Flower heads discoid, crowded in dense, woolly, terminal glomerules, white, flowers less than 20 per head; pappus tipped with white cilia. Sept.–Oct. Sandstone slopes in fynbos, NW (Bokkeveld Plateau Mts to Cold Bokkeveld).•

lasianthum (Schltr. & Moeser) Hilliard   Like **V. niveum** but heads homogamous. Jan. Sandstone flats, NW (Cold Bokkeveld).•

niveum Hilliard   Rounded, white-woolly annual to 20 cm. Leaves oblanceolate. Flower heads discoid, in glomerules arranged in corymbs. Sept.–Jan. Damp sand in streambeds or pans, AP, LB (Potberg to Albertinia and S Karoo).

pulvinatum Hilliard   Cushion-forming, grey-woolly perennial. Leaves oblanceolate. Flower heads discoid, in subcorymbose clusters. Mar. Sandy slopes, NW (near Klawer).•

vellereum (R.A.Dyer) Hilliard   Softly woody, white-felted perennial to 40 cm. Leaves oblanceolate. Flower heads discoid, in cymose clusters, white; pappus subplumose above. Nov.–Jan. Dune slacks, SE (Humansdorp to East London).

VERNONIA   BLOUTEEBOS   c. 500 spp., nearly cosmopolitan

anisochaetoides Sond.   Rampant climber. Leaves rhomboid. Flower head discoid, in divaricate panicles, yellow-beige. June–Sept. Coastal forest margins, SE (Knysna to KwaZulu-Natal).

capensis (Houtt.) Druce   BLOUNAALDETEEBOSSIE   Erect canescent perennial from woody rootstock, to 50 cm. Leaves linear, margins revolute, silvery hairy below. Flower heads discoid, in dense corymbs, purple. Oct.–Jan. Grassy fynbos, SE (George to tropical Africa).

mespilifolia Less.   Scandent or sprawling shrub to 75 cm. Leaves cuneate, deeply toothed. Flower heads discoid, in dense corymbs, white to mauve. Oct.–Mar. Forest margins, SE (George to Swaziland).

neocorymbosa Hilliard   Robust canescent shrub to 1 m. Leaves obovate, toothed above, discolorous, grey-velvety below. Flower heads discoid, in dense corymbs, mauve to white. Jan.–May. Forest margins, SE (Uitenhage to S Mozambique).

ZYRPHELIS   PLUIMASTERTJIE   13 spp., W Cape to tropical Africa

A. Flower heads on almost leafless peduncles

decumbens (Schltr.) Nesom   Sprawling or erect, shortly hairy perennial to 20 cm. Leaves oblanceolate, roughly hairy. Flower heads radiate, solitary, yellow with white to mauve rays. Oct.–Dec. Rocky sandstone slopes, SW (Bainskloof Mts).•

lasiocarpa (DC.) Kuntze   Sprawling, sparsely hairy shrublet with slender erect branches, 5–25 cm. Leaves linear, spreading, sometimes toothed. Flower heads radiate, solitary, yellow with blue, pink or white rays. Sept.–Apr. Sandstone slopes, NW, SW, KM (Grootwinterhoek to Hottentots Holland and Swartberg Mts).•

microcephala (Less.) DC.   Roughly hairy, sprawling shrublet to 30 cm. Leaves linear with recurved apiculate tips. Flower heads radiate, solitary, yellow with white or mauve rays. Mainly Sept.–Apr. Sandstone slopes, NW, KM (Ceres to Ladismith).•

montana (Schltr.) Nesom   Roughly hairy straggling subshrub, 5–25 cm. Leaves oblanceolate, sometimes with a few teeth. Flower heads radiate, solitary, yellow with mauve rays. July–Nov. Sandstone slopes, SW (Bainskloof to Wemmershoek Mts).•

monticola (Compton) Z.-Wiegand   Sparsely hairy, sprawling subshrub to 20 cm. Leaves linear, channelled. Flower heads radiate, solitary, yellow with blue rays. Sept.–Oct. Rocky sandstone slopes, KM (Witteberg).•

A. Flower heads on leafy peduncles (see also **Z. microcephala***)*

ecklonis (DC.) Sond.   Roughly hairy shrublet to 20 cm. Leaves linear, keeled. Flower heads radiate, solitary, yellow with white or yellow rays. Sept.–Dec. Rocky sandstone slopes, NW, SW (Cedarberg Mts).•

foliosa (Harv.) Kuntze   Sparsely hairy shrublet to 15 cm. Leaves linear. Flower heads radiate, 1-few in corymbs, yellow with white or mauve rays. Oct.–Feb. Sandstone slopes, SW (Cape Peninsula to Stanford).•

outeniquae (Fourc.) Z.-Wiegand   Straggling shrublet to 25 cm, thinly woolly on young parts. Leaves oblanceolate and apiculate with revolute margins. Flower heads radiate, solitary, yellow with white or mauve rays. Sept.–Jan. Sandstone slopes, SE (Outeniqua Mts).•

pilosella (Thunb.) Kuntze (= *Zyrphelis perezioides* (Nees) Nesom)   Tufted, roughly hairy perennial to 20 cm, with woody rootstock. Leaves narrowly oblanceolate to linear with revolute margins. Flower heads radiate, solitary, subscapose, yellow with white or pale pink rays. July–Oct. Sandstone flats and slopes, NW (Bokkeveld Mts to Porterville and Hex River Mts).•

spathulata Z.-Wiegand   Roughly hairy subshrub to 20 cm, base woody. Leaves oblanceolate, tips curved down. Flower heads radiate, solitary, yellow with mauve rays. Nov.–Dec. Sandstone slopes, SW (Bredasdorp district).•

taxifolia (L.) DC.   Slender, sparsely hairy, sprawling subshrub to 40 cm. Leaves linear, apiculate, margins minutely toothed. Flower heads radiate, solitary on long peduncles, yellow with blue or mauve rays. Mainly Sept.–Dec. Damp sandstone slopes, SW (Cape Peninsula to Cape Hangklip).•

BALANOPHORACEAE

MYSTROPETALON•   KAAPSEWOLWEKOS   1 sp., W Cape

thomii Harv. (= *Mystropetalon polemannii* Harv.)   Monoecious, achlorophyllous root parasite to 25 cm. Leaves scale-like, imbricate, linear-oblanceolate, ciliate. Flowers in a congested, cylindrical spike, lower female, upper male, orange or red to purple. Feb.–July. Parasitic on *Protea* and *Leucadendron*, NW, SW, LB (Hex River Mts to Cape Peninsula to Riversdale).•

BALSAMINACEAE

IMPATIENS   BALSAM   600–700 spp., worldwide

hochstetteri Warb.   Soft, brittle perennial to 60 cm. Leaves elliptic, toothed. Flowers 1–3 in axils, pink, lower sepal spurred and lower petals lobed. Oct.–Apr. Forest margins, SE (George to tropical Africa).

BIGNONIACEAE

1.   Flowers 2-lipped, tube much longer than petals; stamens arching, 4 with 1 staminode; unarmed shrub with compound leaves ....**Tecoma**
1.' Flowers nearly actinomorphic, tube shorter than petals; stamens spreading, 5 without staminodes; spinescent shrub with simple leaves ....**Rhigozum**

RHIGOZUM   WILDEGRANAAT   7 spp., Africa and Madagascar

obovatum Burch.   Rigid, spiny shrub to 4.5 m. Leaves fascicled, obovate, margins revolute. Flowers 1–3 on short shoots, funnel-shaped, yellow. Seeds papery winged. July–Dec. Dry shale slopes, NW, KM, SE (Worcester to Uitenhage and Karoo to Zimbabwe).

TECOMA (= *TECOMARIA*)   CAPE HONEYSUCKLE, ICAKATHA   14 spp., tropical and subtropical Africa and America

capensis (Thunb.) Lindl. (= *Tecomaria capensis* (Thunb.) Spach)   Scrambling shrub or small tree to 6 m. Leaves opposite, imparipinnate, leaflets toothed. Flowers in terminal racemes, trumpet-shaped, orange. Seeds papery-winged. Sept.–May. Bush and scrub, SE (Uitenhage to S Mozambique; widely cultivated).

BORAGINACEAE
by E. Retief and M.H. Buys, **Lobostemon** by M.H. Buys

1. Ovules many in each locule; petals and stamens 10–12; white-prickly shrublets .... **Codon**
1.' Ovules solitary in each locule; petals and stamens 5:
    2. Ovary entire or shallowly 4-lobed; style terminal:
        3. Style entire or bifid; fruit becoming dry, of 2–4 nutlets .... **Heliotropium**
        3.' Style deeply divided; stigmas 2 or 4; fruit a drupe:
            4. Style divided into 2 .... **Ehretia**
            4.' Style divided into 4 .... **Cordia**
    2.' Ovary deeply 4-lobed; style gynobasic; fruit dry, of 2–4 nutlets:
        5. Nutlets spiny or glochidiate:
            6. Nutlets flattened or rounded above, not or scarcely projecting above calyx .... **Cynoglossum**
            6.' Nutlets pointed above, projecting considerably above calyx .... **Lappula**
        5.' Nutlets smooth to tuberculate, sometimes winged:
            7. Filaments hairy at point of attachment to corolla:
                8. Stigma bilobed; leaves cauline .... **Lobostemon**
                8.' Stigma slightly bifid; leaves radical .... **Echiostachys**
            7.' Filaments not hairy at point of attachment to corolla:
                9. Throat of corolla closed by 5 intruding folds or pouches; nutlets white .... **Lithospermum**
                9.' Throat of corolla without intruding folds; nutlets brown to black:
                    10. Corolla without hairy scales in throat opposite corolla lobes:
                        11. Stigma bilobed; corolla actinomorphic, glabrous outside .... **Amsinckia**
                        11.' Stigma bifid; corolla 2-lipped, hairy outside .... **Echium**
                    10.' Corolla with hairy scales in throat opposite corolla lobes:
                        12. Anthers apiculate; stigma bifid; nutlets smooth and shiny .... **Myosotis**
                        12.' Anthers not apiculate; stigma bilobed; nutlets matte .... **Anchusa**

*__AMSINCKIA__ YSTERGRAS 50 spp., New World

***retrorsa** Suksd. (= *Amsinckia angustifolia* auct., *A. menziesii* (Lehm.) A.Nelson & J.F.Macbr.) Roughly hairy annual to 50 cm. Leaves ovate-lanceolate. Flowers in tightly rolled helicoid cymes, yellow to orange. July–Nov. Weed in sandy soil, disturbed areas, NW, SW, KM (American weed).

__ANCHUSA__ CAPE FORGET-ME-NOT c. 35 spp., mainly European

capensis Thunb. Softly or roughly hairy annual to 1 m. Leaves oblanceolate. Flowers in helicoid cymes lengthening in fruit, blue or dark blue. Nutlets ovoid. Sept.–Nov. Sandy flats, often disturbed places and roadsides, NW, SW, AP, KM, LB, SE (Namibia, Lesotho and drier parts of South Africa to Mpumalanga).

__CODON__ SUIKERKELK 2 spp., Namibia and South Africa

royenii L. Roughly hairy shrublet with white prickles, to 1.5 m. Leaves ovate-cuneate, distinctly petioled, white-prickly. Flowers axillary or in terminal helicoid cymes, large, cream to yellow with purple stripes. Fruit a sparsely tubercled capsule. Aug.–Apr. Dry stony slopes, NW (Namibia and W Karoo to Botterkloof).

__CORDIA__ SEPTEE TREE, UMHLOVU-HLOVU 250 spp., tropical regions

caffra Sond. Shrub or small tree to 7 m. Leaves long-petioled, lanceolate to ovate, margins irregularly serrated. Flowers creamy white. Fruits fleshy, ovoid, in a cup-like calyx, orange or red. Sept.–Oct. Dune bush, forest, woodland, SE (Baviaanskloof Mts to S Mozambique).

__CYNOGLOSSUM__ HOUND'S TONGUE, KNOPPIESKLITS 55 spp., temperate and warm regions

hispidum Thunb. (incl. **C. enerve** Turcz.) Roughly hairy annual or perennial to 90 cm. Basal leaves long-petioled, veins not prominent beneath. Flowers in scorpioid cymes, blue, pedicels deflexed in fruit, to 15 mm long, calyx lobes acute. Nutlets densely covered with short glochidia. Oct.–Nov. Disturbed areas, often along streams, NW, SE (Cold Bokkeveld and Outeniqua Mts to Zimbabwe).

lanceolatum Forssk. White-hairy annual or biennial to 1 m, mostly well branched above. Leaves with midrib and 2 lateral veins usually prominent beneath. Flowers usually in divaricately bifid inflorescences, white or pale blue, pedicels to 3 mm long in fruit. Nutlets covered with slender glochidia. Dec.–May. SE (Humansdorp to tropical Africa and Asia).

obtusicalyx Retief & A.E.van Wyk   Softly hairy perennial or biennial. Basal leaves long-petioled, winged, secondary veins not prominent. Flowers in terminal scorpioid cymes, white or blue, pedicels slightly curved, 15 mm long, calyx lobes obtuse. Nutlets densely covered with slender glochidia. Sept.–Jan. NW (W Karoo and Bokkeveld Plateau to Ceres).

ECHIOSTACHYS•   BOTTELBORSEL   3 spp., W Cape

ecklonianus (H.Buek) Levyns   Like **E. spicatus** but flowers pink, red, blue or purple, hairy on central veins, shorter than 10 mm, nutlets almost smooth. July–Sept. Gravelly or sandy slopes, SW, AP (Somerset West to Bredasdorp).•

incanus (Thunb.) Levyns   Like **E. spicatus** but leaves narrowly lanceolate and attenuate, flowers white or blue, hairy on central veins and hairs at base of filaments tufted. Aug.–Oct. Clay flats, NW, SW, LB (Clanwilliam to Swellendam).•

spicatus (Burm.f.) Levyns   White-hairy, tufted perennial from a woody caudex, to 40 cm. Leaves oblanceolate. Flowers in a pseudospike, white, more than 10 mm long, glabrous or with a few scattered hairs, hairs at base of filaments not tufted. Nutlets tuberculate. Sept.–Oct. Sandy flats, SW (St Helena Bay to Cape Flats).•

*****ECHIUM**   BLOUDISSEL   40 spp., Macronesia, Europe and W Asia

*****plantagineum** L. (= *Echium lycopsis* auct.)   Hairy annual or biennial to 65 cm. Basal leaves ovate to oblanceolate, lateral veins prominent. Flowers blue fading to pink, oblique and unequally lobed, sparsely hairy on veins and margins. Sept.–Mar. Disturbed sites, NW, SW, LB, AP, KM, SE (European weed, widespread in Africa).

EHRETIA   CAPE LILAC, DEURMEKAARBOS, UMHLELI   50 spp., tropical and warm regions

rigida (Thunb.) Druce   Many-stemmed shrub or small tree to 6 m, branches drooping and tangled. Leaves usually clustered on short shoots, obovate, margins densely ciliate. Flowers mauve to pale blue fading whitish. Fruits subglobose, orange-red. Aug.–Sept. Stony slopes, SE (Humansdorp to tropical Africa).

*****HELIOTROPIUM**   HELIOTROPE   c. 250 spp., tropical and warm regions

*****curassavicum** L.   Erect or prostrate, annual or perennial halophyte. Leaves spathulate, bluish green, succulent. Flowers in a helicoid cyme, white. Nutlets rugose. Sept.–Apr. Disturbed places, damp soil, NW, SW, KM, LB (European weed, drier parts of southern Africa).

*****supinum** L.   Prostrate annual. Leaves broadly ovate to broadly obovate, densely adpressed-hairy, veins usually sunken above and prominent beneath. Flowers in a helicoid cyme, white. Nutlets smooth. Nov.–Apr. Disturbed places in damp soil, SW, SE (European weed, Old World).

LAPPULA   50 spp., mainly temperate Eurasia

capensis (DC.) Gürke   Erect or sprawling annual to 30 cm. Leaves softly hairy, narrowly obovate with a tendency to fold. Flowers in terminal, leafy cymes elongating in fruit, pedicels very short. Nutlets ovoid-trigonous with glochidiate margins. Sept.–Oct. Disturbed places, NW, KM, LB (Namaqualand to E Cape).

LITHOSPERMUM   GROMWELL   59 spp., temperate regions, excluding Australia

papillosum Thunb.   Closely leafy perennial to 25 cm. Leaves adpressed, lanceolate, undersurface with hairs on midrib only. Flowers in terminal and axillary leafy cymes, with an annulus at base and 5 glandular invaginations in throat. Nutlets smooth. Nov.–Jan. Mountain slopes, KM, SE (Little Karoo to E southern Africa).

LOBOSTEMON   AGTDAEGENEESBOS   28 spp., South Africa, mostly W Cape

A. Flowers small (shorter than 15 mm), rotate

capitatus (L.) H.Buek (= *Lobostemon bolusii* Levyns, *L. inconspicuus* Levyns)   Shrublet, 30–60 cm. Leaves sessile, hairy, linear-lanceolate. Flowers in capitate cymes, rotate, cream-coloured with purple markings, glabrous outside, staminal scales triangular with lateral lobes, stamens exserted, style glabrous. Sept.–Oct. Shale slopes and flats, NW, SW (Porterville to Bredasdorp).•

echioides Lehm.   Shrublet, 20–80 cm. Leaves sessile, hairy (often soft, silvery), oblong-lanceolate. Flowers in cymes, rotate, blue, hairy outside, staminal scales triangular with lateral lobes, stamens exserted, style glabrous. Aug.–Oct. Stony slopes and flats, NW, SW, KM, LB, SE (Namaqualand and Karoo to E Cape).

gracilis Levyns   Shrublet, 40–70 cm. Leaves sessile, adpressed-hairy, linear. Flowers in cymes, rotate, white or pale blue, sparsely hairy outside, staminal scales large and triangular with lateral lobes, stamens exserted, style glabrous. July–Oct. Sandstone outcrops, NW (Worcester to Bonnievale).•

paniculatus (Thunb.) H.Buek (= *Lobostemon horridus* Levyns)   Shrublet, 50–80 cm. Leaves sessile, hairy, midrib and margins bristly, oblong-lanceolate. Flowers in cymes, rotate, blue, hairy outside, staminal scales triangular with lateral lobes, stamens exserted, style glabrous. Aug.–Oct. Stony slopes, NW, SW, KM, LB, SE (Namaqualand and Karoo to E Cape).

A. Flowers medium to large (longer than 15 mm), funnel-shaped or tubular
B. Flowers red or cream-coloured with red markings

belliformis M.H.Buys   Shrub, 1.0–1.5 m. Leaves sessile, hairy, margins revolute, oblanceolate. Flowers in cymes, tubular, red, hairy outside, staminal scales reduced to swellings without lateral lobes, style hairy. July–Oct. Coastal limestones, AP (Gourits River: Gouriqua).•

muirii Levyns   Shrublet, 40–90 cm. Leaves sessile, hairy, linear-oblong. Flowers in cymes, cream-coloured with red markings, glabrous outside, staminal scales ridge-like without lateral lobes, style hairy. June–Aug. Sandy northern slopes, LB (Langeberg Mts).•

regulareflorus (Ker Gawl.) M.H.Buys (= *Lobostemon grandiflorus* (Andrews) Levyns)   Shrub, 1.0–1.5 m, young branches glabrous. Leaves sessile, hairy above only, linear-lanceolate. Flowers in cymes, tubular, red, hairy outside, staminal scales reduced to swellings without lateral lobes, style hairy. Sept.–Oct. Granite slopes, SW (Stellenbosch and Du Toitskloof Mts).•

sanguineus Schltr.   Shrub, 0.5–1.2 m. Leaves sessile, sparsely hairy, leathery, ovate-elliptic. Flowers in cymes, tubular, red, hairy outside, staminal scales reduced to ridges without lateral lobes, style hairy. Feb.–Apr. Sandstone slopes, SW, AP (Bredasdorp to Potberg).•

B. Flowers pink to blue
*C. Flowers glabrous outside (see also **L. trichotomous**)*

argenteus (P.J.Bergius) H.Buek   Like **L. stachydeus** but flowers 1 per bract. July–Feb. Shale slopes, NW, SW, AP, SE (Clanwilliam to Grahamstown).

dorotheae M.H.Buys (= *Lobostemon laevigatus* (L.) H.Buek)   Like **L. glaucophyllus** but sepals unequal in width and stamens much exserted and recurved. Aug.–Nov. Rocky sandstone slopes, NW, SW (Bokkeveld Mts to Robertson).•

glaber (Vahl) H.Buek (= *Lobostemon hispidum* (Thunb.) DC.)   Shrublet, 30–60 cm. Leaves sessile, hairy, linear–lanceolate. Flowers in cymes, white sometimes pale pink, glabrous outside, staminal scales triangular with lateral lobes, style glabrescent. Aug.–Nov. Stony slopes, NW, SW, KM (Cold Bokkeveld to Cape Peninsula and Witteberg).•

glaucophyllus (Jacq.) H.Buek   Shrublet, 30–80 cm, young branches glabrous. Leaves sessile, hairy on midrib and apex, linear-lanceolate. Flowers in cymes, blue sometimes pink, glabrous outside, staminal scales rounded without lateral lobes, style hairy. July–Oct. Sandy flats and slopes, NW, SW, KM (Clanwilliam to Cape Peninsula, Worcester to Swartberg Mts).•

hottentoticus Levyns   Like **L. glaucophyllus** but leaves hairy and flowers white or pink. Aug.–Nov. Stony lower slopes, SW (Somerset West to Gordon's Bay).•

paniculiformis DC.   Shrublet, 60–90 cm. Leaves sessile, hairy but appearing glabrous above, oblong-lanceolate. Flowers in cymes, blue, glabrous outside, staminal scales triangular with lateral lobes, style sparsely hairy. Aug.–Nov. Shale and sandy slopes, NW, SW (Gifberg to Cape Peninsula).•

pearsonii Levyns   Shrublet, 60–90 cm, young branches glabrous. Leaves sessile, hairy but appearing glabrous above, oblong-lanceolate. Flowers in lax cymes, cream-coloured to blue, glabrous outside, staminal scales triangular with lateral lobes, style hairy. July–Oct. Rocky slopes, NW, KM (Namaqualand to Little Karoo).•

stachydeus DC.   Shrublet, 30–60 cm. Leaves sessile, hairy, margins revolute, linear-lanceolate. Flowers in a pseudospike with 2 flowers per bract, blue, glabrous outside except for midvein and margins, staminal scales reduced to ridges without lateral lobes, style hairy. July–Feb. Sandy slopes, KM, SE (Ladismith, Graaff-Reinet, Beaufort West).

C. Flowers hairy outside

collinus Schltr. ex C.H.Wright   Shrublet, 20–40 cm. Leaves sessile, hairy or glabrescent, leathery, oblong. Flowers in cymes, pale blue, hairy outside, staminal scales ridge-like without lateral lobes, style hairy. Sept.–Apr. Sandy coastal flats, AP (Bredasdorp).•

curvifolius H.Buek   Shrub, 30–60 cm. Leaves sessile, silvery hairy, linear to lanceolate, apex often recurved. Flowers in cymes, pink, hairy outside, staminal scales ridge-like without lateral lobes, style hairy. Aug.–Nov. Sandy flats, SW, AP (Caledon, Stanford to Cape Infanta).•

daltonii M.H.Buys   Shrublet, 30–60 cm. Leaves sessile, sparsely hairy, leathery, ovate-obovate. Flowers in cymes, blue, hairy outside, sepals tipped with conspicuous brown hairs when young, staminal scales ridge-like, style hairy. Aug.–Nov. Limestone flats, AP (Potberg to Cape Infanta).•

decorus Levyns   Shrublet, 50–100 cm. Leaves sessile, hairy near apex, linear. Flowers in cymes, blue, hairy outside, staminal scales ridge-like without lateral lobes, style hairy. Aug.–Oct. Sandstone slopes, KM (Touwsberg and Rooiberg).•

fruticosus (L.) H.Buek   DOUWURMBOS, LUIBOS   Shrublet, 50–80 cm. Leaves sessile, hairy, oblanceolate-obovate. Flowers in cymes, blue to pink, hairy outside, staminal scales ridge-like without lateral lobes, style hairy. May–Dec. Sandy flats, NW, SW (Namaqualand to Cape Peninsula and Worcester).

lindae M.H.Buys (= *Lobostemon strigosus* (Lehm.) H.Buek)   Shrublet, 40–60 cm. Leaves sessile, hairy, oblong, obtuse. Flowers in markedly one-sided cymes, blue, hairy outside, staminal scales ridge-like without lateral lobes, style hairy. Aug.–Oct. Stony clay flats, LB, AP (Swellendam to Mossel Bay).•

lucidus (Lehm.) H.Buek   Shrublet, 20–30 cm. Leaves sessile, hairy, linear-oblanceolate with persistent woody bases. Flowers in cymes, pink, hairy outside, staminal scales ridge-like without lateral lobes, style hairy. Aug.–Oct. Sandy flats and lower slopes, AP (Agulhas to Potberg).•

marlothii Levyns   Shrublet, 40–70 cm. Leaves sessile, hairy, oblong-lanceolate. Flowers in cymes, blue, hairy outside, staminal scales ridge-like without lateral lobes, style hairy. Aug.–Oct. Sandy slopes, KM, LB, SE (Witteberg and Swartberg Mts to Humansdorp).•

montanus H.Buek   Shrub, 0.8–1.2 m. Leaves sessile, silvery hairy, obovate-oblanceolate. Flowers in cymes, subtubular, blue or turquoise, hairy outside, staminal scales ridge-like without lateral lobes, stamens exserted, style hairy. July–Sept. Coastal sandstone, SW (Cape Peninsula to Onrus).•

oederiaefolius DC.   Like **L. fruticosus** but calyx hairs confined to margins and midribs. Sept.–Oct. Sandy slopes and flats, NW, KM (De Doorns to Montagu).•

trichotomus (Thunb.) DC.   Shrublet, 30–100 cm. Leaves sessile, hairy, somewhat leathery, linear-lanceolate. Flowers in cymes, white or blue, glabrous or hairy outside, staminal scales slightly triangular with lateral lobes, style glabrous or hairy. Aug.–Nov. Sandy slopes and flats, NW, SW (Bokkeveld Mts and W Karoo to Stellenbosch).

trigonus (Thunb.) H.Buek   Similar to **L. lindae** but young inflorescences condensed and without unilateral branches. Aug.–Nov. Stony slopes, SW, AP, LB, SE (Bredasdorp to Port Elizabeth).•

MYOSOTIS   FORGET-ME-NOT   50 spp., mainly Europe

***arvensis** (L.) Hill (= *Myosotis intermedia* Link)   Softly hairy annual to 25 cm. Leaves lanceolate, to 30 × 4 mm. Flowers in helicoid cymes elongating in fruit, blue fading white, calyx closed in fruit. Nutlets ovoid, smooth, shining black. Oct. Disturbed places, SW (European weed)

graminifolia DC. (= *Myosotis sylvatica* auct.)   Softly hairy, often trailing annual to 40 cm, hairs adpressed, long, white. Leaves lanceolate or obovate, sessile or petiolate. Flowers in lax helicoid cymes elongating in fruit, light blue with yellow centre, fading white, calyx open in fruit. Nutlets compressed-ovoid with a rim, dull brown. Sept.–Jan. Coastal bush, SE (Port Elizabeth to Mpumalanga).

sp. 1   Straggling, softly hairy annual to 25 cm, with long internodes. Leaves obovate to elliptic, sessile or basal ones shortly petiolate. Flowers in helicoid cymes elongating in fruit, blue, pedicels long, thin, slightly deflexed, calyx open in fruit. Nutlets light brown. Sept.–Nov. Wet places, often under shrubs in shade, NW, SW, SE (Ceres to E Cape).

TRICHODESMA   35 spp., tropical and warm regions of the Old World

africanum (L.) Lehm.   Coarsely white-hairy annual to 50 cm. Leaves ovate to lanceolate, petiolate or sessile, alternate or opposite, coarsely white-hairy. Flowers in helicoid cymes, pale pink or white, calyx as long as corolla, lobes with dark tips, margins and midrib hairy. Nutlets with glochidiate margins. July–Oct. Stony slopes and dry river beds, NW, KM (drier parts of South Africa to W Africa).

BRASSICACEAE (= *CAPPARACEAE*)

1. Stamens 5–many, not tetradynamous:
    2. Androphore elongated, longer than sepals; upper and lower sepals enclosing lateral ones; fruit cylindrical ....**Cadaba**
    2.' Androphore shorter than sepals; fruit globose:
        3. Receptacle or calyx tubular; stipules not spinescent ....**Maerua**
        3.' Receptacle or calyx not tubular; stipules spinescent ....**Capparis**
1.' Stamens 6, tetradynamous (inner 4 long, outer 2 short):
    4. Lowermost pair of 'leaves' (cotyledons) opposite, sessile, amplexicaul, broader than long; upper leaves alternate or subopposite, petiolate, cordate ....**Chamira**
    4.' Lower leaves alternate or opposite but then not as above:
        5. Fruits at least 4 times as long as broad (including length of style or beak):
            6. Hairs all branched or a mixture of simple and branched hairs; leaves pinnatisect or pinnately angular-dentate; petals generally yellow ....**Sisymbrium**
            6.' Hairs simple or absent; leaves various; petals white, mauve, purple or blue:
                7. Valves of fruit apparently nerveless; cotyledons accumbent ....**Cardamine**
                7.' Valves of fruit strongly 1–3–7-nerved; cotyledons twice transversely folded..................**Heliophila**
        5.' Fruits less than 4 times as long as broad:
            8. Fruits laterally compressed, angustiseptate:
                9. Valves falling away with seeds enclosed; all leaves entire; flowers large, usually pink to mauve (rarely white) ....**Brachycarpaea**
                9.' Valves falling away empty and seeds falling free; leaves various; flowers small, white or yellow ....**Lepidium**
            8.' Fruits latiseptate, not compressed or dorsally compressed:
                10. Valves woody with a raised keel and ridges:
                    11. Sepals 3-nerved; fruits with a long, stout beak, dehiscent ....**Cycloptychis**
                    11.' Sepals only weakly 3-nerved (sometimes apparently 1-nerved); fruits without or with a very short, persistent style but never beaked, indehiscent ....**Silicularia**
                10.' Valves membranous, sometimes with raised veins but not ridged or keeled:
                    12. Fruits pendulous, compressed, discoid, indehiscent:
                        13. Annual herbs; each of the 2 short filaments with a small appendage ....**Thlaspeocarpa**
                        13.' Perennial, woody shrublets; all filaments without appendages ....**Schlechteria**
                    12.' Fruits erect or spreading, dehiscent:
                        14. Sepals 3-nerved; anthers minutely apiculate..**Cycloptychis**
                        14.' Sepals 1-nerved (3-nerved in one species); anthers not apiculate ....**Heliophila**

BRACHYCARPAEA   WILD STOCK, BERGVIOOL   1 sp., Namaqualand and W Cape

juncea (P.J.Bergius) Marais   Willowy shrublet to 1 m. Leaves linear to narrowly oblong. Flowers white to pink to purple. Fruits subglobose, papillose. Aug.–Dec. Rocky slopes, NW, SW, AP, KM, LB, SE (Namaqualand to Langkloof Mts).

CADABA   c. 30 spp., palaeotropical and subtropical

aphylla (Thunb.) Wild   BOBBEJAANARM, SWARTSTORMBOS   Leafless often tangled shrub with purplish branches to 2 m. Flowers in corymbs or racemes on side shoots, greenish to red. Aug.–Apr. Dry bushveld or semidesert, KM, LB, SE (Montagu to Zimbabwe).

CAPPARIS   CAPER   c. 30 spp., widespread in the tropics and subtropics

sepiaria L.   CAPE CAPER, KAAPSE-KAPPERTJIE, INTSHILO   Thorny shrub or scrambler to 5 m. Leaves elliptic, margins often revolute. Flowers in terminal corymbs, hairy, white. Aug.–Jan. Coastal scrub, LB, SE (Riversdale through E Africa to Malaysia).

CARDAMINE   c. 160 spp., cosmopolitan, mainly temperate

africana L.   Soft perennial to 30 cm, sometimes coarsely hairy, rooting at lower nodes. Leaves trifoliolate, toothed. Flowers white. Fruits erect, linear. Oct.–Nov. Damp rocky places, SW, LB, SE (Cape Peninsula to Mpumalanga).

CHAMIRA•   1 sp., W Cape

circaeoides (L.f.) A.Zahlbr.   Brittle, sprawling annual to 60 cm, cotyledons persistent, opposite, depressed-ovate. Leaves rounded or cordate, coarsely toothed. Flowers few, white. Fruits compressed-lanceolate. Aug.–Oct. Sheltered sandstone slopes, NW, SW (Bokkeveld Mts to Botrivier).•

CYCLOPTYCHIS• 2 spp., W Cape

marlothii O.E.Schulz   Willowy shrub to 2 m. Leaves sparse, linear. Flowers white, pink to mauve, sepals 2–3 mm long. Fruits erect, elliptic, 8–12 mm long. Sept.–Oct. Sandstone slopes, NW (Cedarberg Mts to Swartruggens).•

virgata (Thunb.) E.Mey. ex Sond.   Willowy shrub to 1 m. Leaves linear-lanceolate. Flowers white, pink to mauve, sepals 4–5 mm long. Fruits erect, subcircular, 5–6 mm long. Sept.–Dec. Sandstone slopes, NW (Gifberg to Pakhuis Mts).•

HELIOPHILA   sporrie   c. 70 spp., Namibia and South Africa, mostly winter-rainfall areas

A. Perennials and shrublets (see also **H. subulata***)*

brachycarpa Meisn.   Brittle shrublet with striate branches to 2 m. Leaves stipulate, linear-oblanceolate. Flowers corymbose on side-branches, white or cream-coloured tinged mauve, petals with a long, papillate claw, ovules 4. Fruits elliptic-lanceolate, 20–30 mm long. Mainly July–Aug. Sandstone slopes, SE (Humansdorp to Karoo and E Cape).

callosa (L.f.) DC.   Shrublet to 1 m. Leaves stipulate, lanceolate. Flowers mauve, ovules 12–18. Fruits linear-oblong, 60–80 mm long. June–Dec. Sandy slopes, NW, SW (Tulbagh and Cape Peninsula).•

carnosa (Thunb.) Steud.   Tussock-forming shrublets to 60 cm with annual stems from a woody base. Leaves stipulate, filiform or lobed above. Flowers white, pink to violet, ovules 14–30. Fruits broadly linear, 25–80 mm long. Aug.–Oct. Dry, grassy hillsides, NW, KM, SE (Namibia to Montagu to E Cape and Gauteng).

cedarbergensis Marais   Trailing shrublet, sometimes with striate branches. Leaves stipulate, lanceolate. Flowers white or pink, ovules 10–12. Fruits erect, lanceolate, 20–40 mm long. Oct. Sandstone ledges, 1200–1650 m, NW (Cedarberg Mts).•

cinerea Marais   Densely hairy perennial to 60 cm. Leaves exstipulate, oblanceolate. Flowers blue or mauve, petals with basal appendages, ovules 32–40. Fruits linear, thinly hairy, 60–80 mm long. Oct.–Nov. Sandy coastal slopes, SW (Cape Peninsula).•

cornuta Sond.   Shrublet to 1.5 m, sometimes minutely hairy below. Leaves exstipulate, filiform to linear-oblanceolate, fleshy and subterete. Flowers white, mauve or blue, petals with basal appendages, ovules 14–38. Fruits moniliform, 30–100 mm long. July–Nov. Stony flats and slopes, NW, SW, KM, SE (S Namibia and W Karoo to Riviersonderend and Uniondale).

cuneata Marais   Thinly hairy, spindly, straggling shrub to 1.3 m. Leaves exstipulate, cuneate, toothed above, sometimes densely hairy above. Flowers blue with white centre, petals with basal appendages, ovules 6–8. Fruits linear-oblong, 60–80 mm long. Aug.–Sept. Forest fringes, SW (Jonkershoek).•

dregeana Sond.   Shrublet with rough, often striate branches to 35 cm. Leaves stipulate, elliptic-lanceolate, fleshy. Flowers white, pink or mauve, ovules 2–6. Fruits moniliform, 15–25 mm long. Aug.–Oct. Sandstone slopes, NW (Pakhuis Mts to Ceres).•

elata Sond.   Slender shrublet to 1 m, minutely hairy below. Leaves stipulate, filiform or lobed. Flowers blue to mauve, petals sometimes with a basal appendage, ovules 18–40. Fruits submoniliform, 20–80 mm long. July–Dec. Sandy flats, NW (Bokkeveld Mts to Worcester).•

elongata (Thunb.) DC.   Slender shrublet with grooved branches to 60 cm, sometimes with annual stems from a woody crown. Leaves stipulate, linear-lanceolate, sometimes toothed. Flowers white or yellowish, often tinged mauve, ovules 20–36. Fruits pendulous, linear, 50–80 mm long. Mainly May–July and Nov.–Dec. Rocky hillsides, KM, LB, SE (Langeberg and Klein Swartberg Mts to KwaZulu-Natal).

esterhuyseniae Marais   Dense shrublet with angled or narrowly winged branches to 1 m. Leaves stipulate, linear-lanceolate. Flowers white, mauve or blue, ovules 4–8. Fruits erect, pod-like, rough, 8–28 mm long. Aug.–Sept. Sandstone slopes, NW, SW (Ceres to Franschhoek Mts).•

filicaulis Marais   Slender shrublet with wiry, minutely striate branches. Leaves stipulate, linear-filiform. Flowers blue to mauve, ovules 4–8. Fruits moniliform, 15–30 mm long. Oct. Sandstone slopes, NW (Hex River Mts).•

glauca Burch. ex DC.   Willowy shrub to 2 m, sometimes with annual stems from a woody crown. Leaves exstipulate, oblanceolate, fleshy. Flowers white or mauve, ovules 4–10. Fruits oblong, 20–30 mm long. Aug.–Apr. Rocky slopes, KM, SE (Langeberg and Rooiberg Mts to Uitenhage).•

linearis (Thunb.) DC.   Perennial to 90 cm, sometimes hairy below. Leaves exstipulate, filiform to obovate, sometimes toothed, fleshy. Flowers white, mauve or purple, petals with basal appendages, ovules 20–50. Fruits linear, 25–110 mm long. Aug.–Feb. Sandy coastal flats, NW, SW, AP, LB, SE (Langebaan to E Cape).

macra Schltr.   Slender shrub to 1.3 m, often with annual stems from a woody crown. Leaves stipulate, linear. Flowers white or pinkish, ovules 18–24. Fruits linear, 25–50 mm long. Dec.–Apr. Shale slopes, SW, AP (Onrus to De Hoop).•

nubigena Schltr.   Straggling, woolly perennial to 45 cm. Leaves exstipulate, elliptic. Flowers subcorymbose, mauve to violet, ovules 2–6. Fruits elliptic, hairy, 5–13 mm long. Sept.–Oct. Sandstone crevices, 1500–2000 m, NW, SW, KM (Cold Bokkeveld to Slanghoek and Swartberg Mts).•

ramosissima O.E.Schulz   Shrublet to 25 cm. Leaves exstipulate, linear-oblanceolate, succulent. Flowers white tinged mauve, ovules 9–16. Fruits submoniliform, 25–35 mm long. Oct. Sandstone slopes, SW (Houwhoek Mts).•

rimicola Marais   Stiff shrublet with young branches striate. Leaves stipulate, linear-oblanceolate, fleshy. Flowers white to purple, sepals not saccate, ovules 4–8. Fruits lanceolate, 15–30 mm long. Dec. Moist crevices in sandstone, KM (Swartberg Mts).•

scoparia Burch. ex DC.   Leafy shrublet with striate branches to 1 m. Leaves stipulate, linear-lanceolate. Flowers white, pink to purple, few on short-shoots, ovules 10–36. Fruits erect, linear, 30–120 mm long. Apr.–Feb. Sandstone slopes, NW, SW, KM (Gifberg to Genadendal).•

suavissima Burch. ex DC.   RUIKPEPERBOSSIE   Lax shrublet to 60 cm with annual stems from a woody crown. Leaves stipulate, linear to lanceolate, fleshy. Flowers blue or purple, scented, ovules 18–26. Fruits submoniliform, 45–65 mm long. Mainly May–July. Shale slopes, KM (Montagu to Kammanassie Mts, Karoo to KwaZulu-Natal).

tricuspidata Schltr.   Slender shrublet with slightly rough stems. Leaves stipulate, cuneate and 3 lobed. Flowers mauve, ovules 2–5. Fruits ovate, 5–8 mm long. Oct. Damp sandstone crevices, SW (Hottentots Holland to Riviersonderend Mts).•

tulbaghensis Schinz   Sprawling shrublet with striate branches to 25 cm. Leaves stipulate, linear, fleshy. Flowers mauve, ovules 4–9. Fruits strongly moniliform, 15–30 mm long. Oct. Sandstone slopes, NW, SW (Skurweberg to Franschhoek Mts).•

A. Annuals

adpressa O.E.Schulz   Like **H. bulbostyla** but buds narrow and pointed. Flowers pink to blue and fruiting pedicels adpressed-erect. Sept.–Oct. Coastal sand-dunes, SW, AP (Cape Peninsula to Stilbaai).•

africana (L.) Marais   Glabrescent or hairy annual to 1.35 m. Leaves exstipulate, lanceolate, sometimes toothed. Flowers blue or mauve, petals with basal appendages, ovules 20–52. Fruits linear, 13–100 mm long. Mainly Aug.–Oct. Sandy flats, NW, SW, KM, LB (Namaqualand to Swellendam).

amplexicaulis L.f.   Annual to 45 cm, sometimes thinly hairy. Leaves exstipulate, lanceolate, clasping below. Flowers white, pink or mauve, petals with basal appendages, ovules 6–16. Fruits strongly moniliform, 25–35 mm long. Aug.–Oct. Sandy slopes, NW, SW, KM (Namaqualand and W Karoo to Saldanha and Montagu).

arenaria Sond.   Glabrescent to hairy annual to 50 cm. Leaves exstipulate, linear or lobed. Flowers blue, petals with basal appendages, ovules 16–36. Fruits moniliform, 15–55 mm long. July–Sept. Sandstone slopes, NW (Bokkeveld Mts to Piketberg).•

arenosa Schltr.   Like **H. arenaria** but flowers smaller with very narrow petals, 5–8 × 2–3 mm. Sept.–Oct. Sandstone slopes, NW (Cedarberg Mts to Swartruggens).•

bulbostyla Barnes   Annual to 35 cm, usually minutely hairy below. Leaves stipulate, divided. Flowers blue, petals with basal appendages, ovules 20–44. Fruits submoniliform, 35–55 mm long. Sept.–Oct. Rocky slopes, KM, LB (W Karoo to Swellendam).

collina O.E.Schulz   Softly hairy, sprawling annual to 20 cm. Leaves stipulate, pinnatisect, fleshy. Flowers white, sepals hairy, petals appendaged basally, ovules 4–12. Fruits moniliform, 10–20 × 2–3 mm. Aug.–Sept. Stony flats, NW (Bokkeveld Mts).•

concatenata Sond.   A complex of pubescent annuals to 45 cm. Leaves stipulate, pinnatisect. Flowers white, pink or mauve, sepals hairy, ovules 10–16. Fruits submoniliform, 10–25 mm long. Aug.–Oct. Stony slopes, SW (Kleinwinterhoek Mts to Cape Peninsula to Genadendal).•

coronopifolia L.   Annual to 60 cm, lower parts roughly hairy. Leaves sometimes stipulate, linear or variously pinnatisect. Flowers blue with white or greenish centre, petals with basal appendages, ovules 16–50. Fruits moniliform, 30–90 mm long. Aug.–Oct. Flats and slopes, NW, SW (S Namaqualand to Caledon).

crithmifolia Willd.   Thinly hairy annual to 60 cm. Leaves stipulate, pinnatifid, fleshy. Flowers white or pink to violet, petals sometimes with basal appendages, ovules 20–42. Fruits linear-oblong, 15–60 mm long, seeds broadly winged. July–Oct. Usually sandy slopes, NW, SW (S Namibia and W Karoo to Riviersonderend).

descurva Schltr.   Thinly hairy annual to 60 cm. Leaves stipulate, linear or few-lobed. Flowers blue or mauve, petals with basal appendages, ovules 22–30. Fruits on deflexed pedicels, linear, 30–75 mm long. Aug.–Sept. Sandstone slopes, NW, KM (Bokkeveld Mts to Barrydale).•

deserticola Schltr.   Annual to 30 cm. Leaves exstipulate, pinnatisect. Flowers white, mauve or blue, scented, petals sometimes with basal appendages, ovules 36–66. Fruits linear, 19–26 mm long. May–Sept. Sandy flats, NW (Namaqualand and W Karoo to Gydouw Pass).

diffusa (Thunb.) DC.   Like **H. meyeri** but ovules 1–8 and fruits narrowly oblong to elliptic, 3–14 mm long. Aug.–Nov. Sandy slopes, NW, SW (Cedarberg Mts to Caledon Swartberg).•

digitata L.f.   Glabrescent annual to 50 cm. Leaves exstipulate, pinnatisect, lobes filiform. Flowers pink or blue, petals with basal appendages, ovules 26–40. Fruits linear, 25–60 mm long. Aug.–Oct. Sandy flats and slopes, NW, SW (Clanwilliam to Caledon).•

ephemera P.A.Bean   Prostrate, cushion-like annual with papillate stems to 10 cm. Leaves oblanceolate, fleshy. Flowers mauve, ovules 4–7. Fruits ellipsoid, inflated, tuberculate, c. 2.5 mm long. Aug.–Sept. Sandstone slopes, 1750 m, KM (Swartberg Mts).•

linoides Schltr.   Usually unbranched annual to 50 cm, roughly hairy on lower nodes. Leaves stipulate, usually linear. Flowers mauve, petals with basal appendages, ovules 32–48. Fruits linear, 45–85 mm long. Aug.–Oct. Sandy slopes, NW, SW (Clanwilliam to Stellenbosch).•

macowaniana Schltr.   Thinly hairy annual to 50 cm. Leaves stipulate, mostly pinnatisect. Flowers white or blue, sepals hairy, ovules 10–24. Fruits moniliform, 10–35 mm long. Aug.–Oct. Sandy slopes, NW, SW (Olifants River Mts to Cape Peninsula).•

meyeri Sond.   Annual to 45 cm. Leaves stipulate, pinnatisect. Flowers white, petals and filaments papillate below, ovules 10–16. Fruits submoniliform, 10–25 mm long. June–Nov. Damp sandstone slopes, NW, SW, LB (Ceres to Bonnievale).•

namaquana Bolus   Glabrous or hairy annual to 30 cm. Leaves sometimes stipulate, linear or filiform. Flowers white to blue, ovules 10–24. Fruits linear, 15–37 mm long. Aug.–Oct. Sandy soils, NW (Namaqualand to Swartruggens).

patens Oliv.   Annual to 20 cm. Leaves exstipulate, filiform or linear. Flowers white, petals with basal appendages, ovules 2–4. Fruits oblong or subcircular, papillate, 3–7 mm long. July. Sandy soils, NW (Vredendal to Piketberg).•

pectinata Burch. ex DC.   Delicate, finely hairy annual. Leaves stipulate, pinnatisect. Flowers small, white, ovules 10–16. Fruits submoniliform, 8–25 mm long. Aug.–Sept. Damp, sheltered slopes, NW, KM (Namaqualand and W Karoo to Clanwilliam and Montagu).

pendula Willd.   Like **H. meyeri** but fruits larger, 15–50 × 2–3 mm with narrowly winged seeds, 2 mm diam. Aug.–Sept. Sandy flats and slopes, NW, SW, KM, LB, SE (Tulbagh to E Cape).

pinnata L.f.   Minutely hairy annual with dark, wiry stems to 20 cm. Leaves stipulate, filiform to 5-lobed. Flowers blue to mauve, ovules 4–8. Fruits moniliform, 7–30 mm long. Aug.–Oct. Sandstone slopes, NW (Bokkeveld Mts to Hex River and W Karoo).

promontorii Marais   Like **H. pinnata** but glabrous above, flowers often with a white centre and ovules 16–25. Sept.–Nov. Sandy flats, SW (Cape Peninsula).•

pubescens Burch. ex Sond.   Densely pubescent annual to 20 cm. Leaves stipulate, pinnatisect. Flowers whitish, sepals hairy, ovules 4–8. Fruits moniliform, c. 20 mm long. July–Aug. Stony slopes, NW, SW (Gifberg to Franschhoek and W Karoo).

pusilla L.f.   Slender annual to 30 cm. Leaves exstipulate, filiform to oblanceolate. Flowers white to mauve, petals sometimes with basal appendages, ovules 3–12. Fruits submoniliform, 5–18 mm long. Aug.–Oct. Clay soils, NW, SW, AP, LB (Cold Bokkeveld to De Hoop).•

refracta Sond.   Like **H. digitata** but ovules 42–60. Fruiting pedicels longer, 10–20 mm and sharply reflexed and fruiting style shorter than 2 mm. Aug.–Sept. Sandy coastal flats, NW, SW, AP (Elandsbaai to Stilbaai).•

seselifolia Burch. ex DC.   Annual to 35 cm. Leaves exstipulate, pinnatisect. Flowers white, petals usually with basal appendages, ovules 12–27. Fruits linear, 13–35 mm long. July–Sept. Clay soils, NW (Namaqualand and W Karoo to Swartruggens).

subulata Burch. ex DC.   Minutely hairy annual or perennial to 50 cm. Leaves stipulate with stipules like resinous granules, filiform to lanceolate, fleshy. Flowers blue, mauve or pink, ovules 24–42. Fruits linear, 20–60 mm long. Aug.–Sept. Mostly coastal flats and slopes, NW, SW, LB, SE (Cold Bokkeveld to Cape Peninsula to E Cape).

tabularis Wolley-Dod. (possibly not distinct from **H. concatenata**)   Annual to 12 cm. Leaves stipulate, linear or lobed. Flowers whitish, ovules 12–14. Fruits linear. Oct. Sandstone slopes, SW (Cape Peninsula).•

thunbergii Steud. (= *Heliophila latisiliqua* E.Mey. ex Sond.)   Hairy annual to 90 cm. Leaves stipulate, pinnatisect. Flowers blue to white, petals usually with basal appendages, ovules 8–18. Fruits oblong. July–Aug. Sandstone slopes, NW (Namaqualand to Ceres).

variabilis Burch. ex DC.   Bright green, finely hairy, spreading annual to 35 cm. Leaves stipulate, pinnatisect. Flowers white, ovules 16–26. Fruits linear-submoniliform, 20–40 mm long. July–Sept. Dry sandy slopes, NW (Namaqualand and W Karoo to Pakhuis Mts).

LEPIDIUM   BIRD-SEED, PEPPER WEED   c. 150 spp., cosmopolitan

africanum (Burm.f.) DC.   Pale or yellowish green biennial or short-lived perennial to 75 cm, branched only above. Leaves oblanceolate, toothed. Flowers white. Fruits ovate, emarginate, 1.8–2.7 mm long. Sept.–Mar. Often in disturbed ground, NW, SW, KM, LB, SE (widespread indigenous weed).

capense Thunb.   Shortly hairy perennial to 60 cm. Leaves crowded below, lyrate-pinnatisect. Flowers crowded, white. Fruits elliptic, obtuse, 2.3–3 mm long. Apr.–Aug. Sandy flats, SW (Cape Peninsula).•

desertorum Eckl. & Zeyh.   Sprawling, shortly hairy perennial to 25 cm. Leaves pinnatifid or laciniate. Flowers subcorymbose, petals 0, nectaries filiform. Fruits ovate, emarginate, 1.6–2.1 mm long. Aug.–Sept. Drier flats, NW (Namaqualand and W Karoo to Free State, Wuppertal and Hex River valley).

ecklonii Schrad.   Shortly hairy perennial to 60 cm. Leaves pinnatifid. Flowers apparently in leaf-opposed racemes through elongation of the lateral shoots, white. Fruits oblong-ovate, emarginate, 2.4–3.4 mm long. Apr.–Nov. Sandy flats, SW, LB, SE (Cape Peninsula to E Cape and W Karoo).

flexuosum Thunb.   Sprawling perennial with annual stems to 20 cm from a woody crown. Leaves oblanceolate or pinnatifid. Flowers subcorymbose, white, stamens 6, nectaries rounded. Fruits ovate, acute, 4.5–5.0 mm long. Oct. Sandy soils, NW, SW (St Helena Bay to Hopefield).•

pinnatum Thunb.   Like **L. capense** but fruits ovate and broadly rounded below with style scarcely projecting above, and petals larger, 0.5–0.7 mm. Aug.–Nov. Shady rock ledges, SW (Cape Peninsula to Du Toitskloof Mts).•

MAERUA   BUSH-CHERRY   c. 100 spp., Africa to Asia

cafra (DC.) Pax   WILDEBOSHOUT, UMPHUNZISA   Shrub or tree with mottled bark to 9 m. Leaves digitately 3(–5)-foliolate. Flowers in terminal corymbs, greenish. Aug.–Oct. Dune bush and forest, SE (Humansdorp to Zimbabwe).

racemulosa (A.DC.) Gilg & Ben.   FOREST BUSH-CHERRY, WITBOSHOUT   Scrambling shrub or tree to 3 m. Leaves usually simple, elliptic. Flowers in short axillary racemes, whitish. May–July. Dune forest, SE (Wilderness to Swaziland).

SCHLECHTERIA•   1 sp., W Cape

capensis Bolus   Sprawling subshrub to 30 cm. Leaves linear, striate. Flowers few-many, white or pink. Fruits pendulous, compressed-ovate, indehiscent. Sept.–Oct. Shaded sandstone slopes, 900–1800 m, NW (Cedarberg Mts).•

SILICULARIA•   1sp, W Cape

polygaloides (Schltr.) Marais   Shrublet with annual branches from a woody base. Leaves linear. Flowers white or pink. Fruits pendulous, compressed-ovate, indehiscent. Sept.–Oct. Sandstone slopes, NW (Cold Bokkeveld to Swartruggens).•

SISYMBRIUM   c. 80 spp, worldwide temperate

capense Thunb.   Grey-hairy perennial with annual stems to 1 m from a woody base. Leaves crowded below, pinnatifid. Flowers subcorymbose, yellow. Fruits erect, subterete. Oct.–Apr. Flats and slopes, SW, KM, LB, SE (Cape Peninsula to KwaZulu-Natal and Free State).

THLASPEOCARPA   2 spp., Namaqualand and W Karoo

capensis (Sond.) C.A.Sm.   Glabrous or pubescent annual to 35 cm. Leaves pinnatifid, somewhat fleshy. Flowers subumbellate, pinkish. Fruits pendulous, compressed-subcircular, indehiscent. Aug.–Sept. Shale slopes, NW (Hex River Pass and W Karoo).

BRUNIACEAE

1. Flowers solitary and terminal ....Thamnea
1.' Flowers axillary or in dense compound heads or whorls:
    2. Flowers each surrounded by 4–12 conspicuous involucral bracts:
        3. Flowers red; ovary usually 3-locular ....Audouinea
        3.' Flowers whitish or pink; ovary mostly 2-locular:
            4. Styles fused throughout their length; anthers oblong..Tittmannia
            4.' Styles free but adpressed; anthers saggitate..Linconia
    2.' Flowers each with 2 bracteoles:
        5. Ovary 1-locular:
            6. Stamens shorter than petals, curved inwards ....Mniothamnea
            6.' Stamens longer than petals, curved outwards ....Berzelia
        5.' Ovary 2-locular:
            7. Flowers longer than 5 mm; petals fused in lower quarter to half (rarely apparently free); filaments almost entirely fused to corolla tube ....Lonchostoma
            7.' Flowers shorter than 5 mm; petals free or fused at base; filaments free:
                8. Calyx constricted at and articulated with top of ovary, more or less tubercled ....Pseudobaeckia
                8.' Calyx not as above:
                    9. Styles fused for most of their length; stamens shorter than petals; flower heads surrounded by a more or less conspicuous involucre ....Staavia
                    9.' Styles free:
                        10. Sepals fused above ovary in a short cup, sepals adjacent at base and forming a V-shaped angle between one another; stamens usually shorter than petals ....Raspalia
                        10.' Sepals free above ovary, not immediately adjacent at base; stamens longer than petals:
                            11. Filaments equal in length ....Nebelia
                            11.' Filaments unequal in length ....Brunia

AUDOUINIA•   FALSE HEATH   1 sp., W Cape

capitata (L.) Brongn.   Closely leafy, laxly branched, coppicing shrub to 1.5 m. Leaves ericoid, imbricate, adpressed, margins minutely ciliate. Flowers axillary, crowded in cylindrical spikes. May–Oct. Rocky flats and slopes, SW (Cape Peninsula to Babilonstoring).•

BERZELIA   12 spp., W to E Cape

    A. Flower heads less than 10 mm diam., arranged in short racemes at the branch tips

dregeana Colozza   Like **B. squarrosa** but flower heads smaller, less than 5 mm diam., stamens short and petals spathulate. Aug.–Sept. Sandstone slopes, SW (Kogelberg to Betty's Bay).•

lanuginosa (L.) Brongn.   Finely and densely leafy shrub, 1.5–2.0 m. Leaves spreading ascending, linear-filiform. Flowers crowded in axillary heads c. 5 mm diam., arranged in short racemes and clustered terminally in loose corymbs, cream-coloured. Sept.–Dec. Damp sandstone slopes, seeps and streambanks, NW, SW (Gifberg to Bredasdorp Mts).•

rubra Schltdl.   Like **B. squarrosa** but woolly on young parts, leaves ascending, imbricate, keeled beneath and peduncles cobwebby and sparsely leafy. Mainly Feb.–Apr. Rocky sandstone slopes, SW (Kleinrivier Mts).•

squarrosa (Thunb.) Sond.   Densely leafy shrub to 1.5 m, coppicing from a woody caudex. Leaves spreading, linear-oblanceolate, furrowed beneath in lower half. Flowers crowded in small axillary heads arranged in short racemes at branch tips and aggregated in loose corymbs, white. Aug.–Oct. Rocky sandstone slopes, NW, SW (Cold Bokkeveld to Kleinrivier Mts).•

    A. Flower heads c. 10 mm diam., few arranged in corymbs at the branch tips

abrotanoides (L.) Brongn.   Densely leafy shrub to 1.5 m, coppicing from a woody caudex. Leaves ascending, oblanceolate. Flowers crowded in rounded axillary heads, c. 1 cm diam., at branch tips, aggregated in corymbs, white, peduncles red, often swollen and fleshy. Aug.–Oct. Rocky sandstone slopes, NW, SW, AP (Elandsbaai to Potberg).•

burchellii Dummer   Like **B. abrotanoides** but leaves persistently pilose. Oct.–Nov. Sandstone slopes, LB (Langeberg Mts: Riversdale).•

commutata Sond.   Like **B. abrotanoides** but flower heads less than 1 cm diam. Oct.–Jan. Sandstone slopes, SE (Langkloof to Grahamstown).

cordifolia Schltdl.   Like **B. abrotanoides** but leaves spreading, cordate, very broad below. Sept.–Nov. Sandstone and limestone slopes, AP (De Hoop to Breede River Mouth).•

ecklonii Pillans   Closely leafy, willowy shrub to 1.5 m. Leaves ascending, incurved in upper half. Flowers crowded in ovoid, mostly terminal heads aggregated into small corymbs, white. Sept.–Nov. Marshy sandstone slopes, SW (Kogelberg to Kleinmond).•

galpinii Pillans   Like **B. abrotanoides** but leaves acuminate, incurved in upper half. Sept.–Oct. Sandstone slopes, LB (Langeberg Mts: Riversdale).•

incurva Pillans   Densely leafy, twiggy shrub to 1.3 m, coppicing from a woody caudex. Leaves ascending, imbricate, linear-oblanceolate. Flowers crowded in small axillary heads in whorls near branch tips, white, peduncles densely leafy. Jan.–Feb. Sandstone rocks, SW (Babilonstoring to Stanford).•

intermedia (D.Dietr.) Schltdl.   Like **B. abrotanoides** but pilose on young parts, leaves spreading, incurved above middle. Mainly Nov.–Feb. Sandstone slopes, NW, SW, KM, LB, SE (Cedarberg Mts to E Cape).•

BRUNIA•   6 spp., W Cape

A. Leaves widest at or above the middle

laevis Thunb. (incl. **B. neglecta** Schltr.)   Closely leafy rounded shrub to 1.5 m, coppicing from a woody caudex. Leaves oblong, ascending, incurved above, puberulous above. Flowers in dense globose, terminal heads 1.5–2.0 cm diam., loosely clustered in corymbs, cream-coloured. Dec.–Feb. Rocky sandstone and limestone slopes, SW, AP, LB (Hottentots Holland Mts to Agulhas).•

macrocephala Willd.   Like **B. laevis** but leaves conspicuously villous, flower heads 1–few per branch, 2.5–3.0 cm diam. Mainly Dec.–Feb. Rocky sandstone slopes, NW, SW (Du Toitskloof to Hex River Mts and Kwadouwsberg).•

A. Leaves widest below the middle

albiflora E.Phillips   Densely leafy, sigle-stemmed shrub, 2–3 m. Leaves spreading-ascending, petiolate, linear-lanceolate, pilose, furrowed beneath. Flowers crowded in dense rounded heads crowded at branch tips in tight corymbs, white. Feb.–Apr. Marshes on peaty sandstone, SW (Hottentots Holland Mts to Hermanus).•

alopecuroides Thunb.   Finely and densely leafy shrub mostly to 1 m. Leaves spreading-ascending, linear-lanceolate. Flowers crowded in small axillary heads 3–4 mm diam., arranged in short racemes and clustered terminally in loose corymbs, cream-coloured. Sept.–Dec. Sandstone slopes, SW (Hottentots Holland to Kleinrivier Mts).•

noduliflora Goldblatt & J.C.Manning (= *Brunia nodiflora* auct.)   FONTEINBOS, STOMPIE   Closely leafy rounded shrub to 1.5 m, coppicing from a woody caudex, with minutely pubescent branches. Leaves lanceolate, adpressed, 2–3 mm long, imbricate, cuneate. Flowers in dense globose, terminal heads loosely clustered in corymbs, white, with white villous bracts. Mainly Mar.–June. Rocky sandstone slopes, NW, SW, KM, LB, SW, SE (Olifants River Mts to Van Staden's Mts).•

stokoei E.Phillips   Like **B. albiflora** but leaves broader, glabrous, with prominent midvein beneath, flowers red and ovary hairy above. Feb.–Mar. Rocky sandstone slopes, SW (Hottentots Holland Mts to Kleinmond).•

LINCONIA•   3 spp., W Cape

alopecuroidea L.   Densely leafy, sparsely branched, coppicing shrub to 1 m. Leaves linear-lanceolate, mostly 15–20 mm long, margins shortly ciliate. Flowers axillary, in subterminal whorls, urceolate, pink, c. 10 mm long. Sept.–Nov. Rocky sandstone slopes, LB, SE (Langeberg Mts: Swellendam to Tsitsikamma Mts).•

cuspidata (Thunb.) Sw. (incl. **L. deusta** (Thunb.) Pillans)   Densely leafy, coppicing shrublet to 50 cm. Leaves oblong-elliptic, mostly 5–10 mm long. Flowers axillary, in subterminal whorls, cupulate, cream-coloured, 3–5 mm long. Sept.–Jan. Rocky sandstone slopes, NW, SW (Ceres to Kleinrivier Mts).•

ericoides E.G.H.Oliv.   Densely leafy, coppicing dwarf shrublet to 10 cm. Leaves oblong-elliptic to obovate, mostly 5–8 mm long. Flowers axillary, in subterminal whorls, urceolate, pink, c. 10 mm long. Oct.–Nov. Sandstone ledges, SW (Riviersonderend Mts).•

LONCHOSTOMA•   5 spp., W Cape

esterhuyseniae Strid   Like **L. pentandrum** but spikes 4–7-flowered, and flowers deep pink and petals larger. Nov. Wet sandstone rocks, SW (Riviersonderend Mts).•

monogynum (Vahl) Pillans   Like **L. pentandrum** but flowers in terminal and lateral spikes, white, and styles entirely united. Mainly Sept.–Mar. Sandstone slopes in seeps and marshes, NW, SW (Cedarberg to Kleinrivier Mts).•

myrtoides (Vahl) Pillans   Like **L. pentandrum** but leaves broadly ovate-elliptic, flowers pink, petals fused for two-thirds their length, styles longer than corolla tube, exserted. Sept.–Oct. Marshes and seeps on sandstone slopes, NW (Witzenbergvlakte to Waaihoek).•

pentandrum (Thunb.) Druce   Densely leafy, erect, willowy shrub to 60 cm. Leaves ascending, elliptic and concave, imbricate, shortly hairy beneath. Flowers axillary, many in terminal spikes, white to pink, petals fused for up to half their length, styles free and shorter than tube. Sept.–Nov. Montane marshes, NW (Cedarberg Mts to Witzenbergvlakte).•

purpureum Pillans   Like **L. pentandrum** but leaves leathery, flowers pink to purple, petals fused for up to one-quarter their length, style longer than corolla tube and exserted. Aug.–Apr. Marshy montane slopes, SW (Bainskloof to Kogelberg).•

MNIOTHAMNEA•   2 spp., W Cape

bullata Schltr.   Like **M. callunoides** but a sprawling, mat-forming undershrub to 10 cm, leaves with spreading hairs. Sept.–Oct. Damp sandstone rocks, LB (Langeberg Mts: Swellendam).•

callunoides (Oliv.) Nied.   Thinly velvety, twiggy shrub to 1 m. Leaves minute, ovate, adpressed, imbricate, adpressed-hairy. Flowers minute, solitary in upper axils, white or pink. Mainly Dec.–Mar. Damp sandstone slopes, LB (Langeberg Mts: Swellendam to Riversdale).•

NEBELIA•   6 spp., W Cape

A. Flower heads axillary, aggregated in dense clusters

fragarioides (Willd.) Kuntze   Thinly hairy shrub to 1.5 m, coppicing from a woody caudex. Leaves linear-deltoid, ascending-incurved, imbricate. Flowers in small axillary heads crowded in globose clusters, cream-coloured. Mainly Mar.–Sept. Sandstone slopes, NW, SW (Cold Bokkeveld to Kleinrivier Mts).•

stokoei Pillans   Like **N. fragarioides** but calyx lobes smaller, c. 2 mm, hairy, petals c. 3 mm, and ovary minutely hairy above. Dec.–Jan. Sandstone slopes, NW (Cold Bokkeveld and Hex River Mts).•

A. Flower heads terminal, grouped in loose corymbs

laevis (E.Mey.) Kuntze   Like **N. paleacea** but plants glabrescent, no copicing, flowers in larger, top-shaped heads, white, involucral bracts shorter than flowers. Jan.–Feb. Rocky sandstone slopes, SW (Riviersonderend Mts).•

paleacea (P.J.Bergius) Sweet   Thinly hairy shrub to 1.5 m, resprouting from woody caudex. Leaves linear-deltoid, ascending-incurved, imbricate. Flowers in small, conspicuously involucrate heads grouped in loose corymbose clusters, cream-coloured; bracts attentuate, white, longer than flowers. Oct.–Apr. Sandstone slopes, NW, SW, LB (Cedarberg to Langeberg Mts).•

sphaerocephala (Sond.) Kuntze   Softly hairy, densely leafy shrub to 2 m. Leaves linear-deltoid, ascending-incurved, imbricate. Flowers in large, rounded, terminal heads sometimes loosely grouped in corymbs, whitish. Dec.–Apr. Sandstone slopes, SW (Bainskloof to Hottentots Holland Mts).•

[**Species excluded**   Poorly known and probably conspecific with one of the above: **N. tulbaghensis** Schltr. ex Dummer]

PSEUDOBAECKIA•   3 spp., W Cape

africana (Burm.f.) Pillans   Erect, closely leafy shrub to 3 m. Leaves linear, ascending-spreading, imbricate, Flowers in slender spikes grouped in panicles, white; bracteoles ovate or rotund. Sept.–Nov. Streambanks on sandstone slopes, NW, SW (Gifberg to Kleinrivier Mts).•

cordata (Burm.f.) Nied.   Erect, closely leafy shrub to 2 m. Leaves ovate to lanceolate, often cordate at base, imbricate, erect to spreading, glabrous or thinly hairy. Flowers in small, rounded axillary spikes grouped in panicles, white; bracteoles ovate to oblong. Nov.–Jan. Montane marshes, seeps and streams, NW, SW, KM, LB, SE (Olifants River Mts to Van Staden's Mts).•

stokoei Pillans   Like **P. cordata** but conspicuously hairy, bracteoles obovate and sepals fused for half their length. Sept.–Oct. Sandstone slopes near streams, SW (Kleinrivier Mts).•

[**Species excluded**   See under **Tittmannia: P. teres** (Oliv.) Dummer]

RASPALIA   9 spp., W Cape to KwaZulu-Natal

A. Leaves scale-like, adpressed or erect and imbricate

globosa (Lam.) Pillans   Erect, willowy shrub to 1.2 m, white-tomentose on young parts. Leaves erect to adpressed, elliptic-ovate, 4–6 mm long. Flowers few, crowded in terminal heads, pink to magenta. Mainly Mar.–June. Sandstone slopes, SW (Hottentots Holland Mts).•

microphylla (Thunb.) Brongn. (incl. **R. angulata** (Sond.) Nied., **R. variabilis** Pillans)   Sprawling softly hairy, often cushion-forming shrublet mostly to 50 cm, coppicing from a woody caudex. Leaves scale-like, closely adpressed, rotund to ovate, 2–3 mm long, hairy above, partly imbricate at first. Flowers few in small terminal heads, white, petals often pubescent. Mainly Oct.–Jan. Sandstone rocks, SW, KM, LB (Hottentots Holland to Bredasdorp and Langeberg Mts).•

phylicoides (Thunb.) Arn.   Like **R. globosa** but leaves linear-elliptic, c. 2 mm long, and flowers few in small heads, white. Feb.–Apr. Sandstone slopes, SW (Hottentots Holland Mts to Caledon Swartberg).•

virgata (Brongn.) Pillans (incl. **R. schlechteri** Dummer)   Erect willowy shrub to 1 m. Leaves linear-lanceolate, adpressed, often shortly imbricate. Flowers few in small terminal heads, white. Aug.–Oct. Moist rocky slopes and streambanks, SW, LB (Hottentots Holland to Langeberg Mts).•

A. Leaves linear to lanceolate, spreading or ascending

barnardii Pillans   Like **R. sacculata** but leaves and branchlets with bent hairs and flowers white with petals 2.75 mm long, obtuse. Oct. Sandstone slopes, LB (Langeberg Mts: Swellendam).•

dregeana (Sond.) Nied.   Silky pubescent, closely leafy, suberect shrub to 1 m. Leaves ascending, lanceolate. Flower crowded in rounded, terminal heads to 6 mm diam., white, with exserted stamens. Jan. Sandstone slopes, NW (Cedarberg to Hex River Mts).•

sacculata (Bolus ex Kirchn.) Pillans   Pilose, closely leafy, twiggy shrub to 1.5 m, coppicing from a woody caudex. Leaves elliptic, ascending and imbricate. Flowers few, in compact, raceme-like heads, white or pink. Mainly Nov.–Jan. Damp sandstone and clay slopes, NW, SW (Cold Bokkeveld to Du Toitskloof Mts).•

villosa C.Presl (incl. **R. oblongifolia** Pillans, **R. palustris** (Schltr. ex Kirch.) Pillans, **R. stokoei** Pillans)   Closely leafy shrub to 1.5 m, villous on young parts. Leaves spreading to suberect, linear-lanceolate. Flowers few in rounded, terminal heads, white, sepals glabrous. Oct.–Mar. Rocky damp to marshy sandstone slopes, NW, SW (Cedarberg Mts to Du Toitskloof).•

[**Species excluded**   See under **Staavia: R. staavioides** (Sond.) Pillans]

STAAVIA•   10 spp., W Cape

A. Flower heads mostly 8–12 mm diam. (excluding the involucre); flowers glutinous

dodii Bolus   DIAMOND-EYES   Densely leafy, willowy shrub to 1 m. Leaves elliptic, imbricate, ascending, c. 10–12 mm. Flowers in large heads surrounded by elliptic-lanceolate, white bracts, pink, sticky. Mainly Feb.–May. Rocky sandstone flats, SW (Cape Peninsula).•

glutinosa (L.) Dahl   FLY-CATCHER BUSH, VLIEËBOS   Finely leafy, willowy shrub to 1.5 m. Leaves linear, spreading, curving upward in upper half, thinly hairy to glabrous, mostly 10–15 mm long. Flowers in large heads surrounded by linear, sticky, spreading white bracts, pink. Mainly July–Sept. Cool, rocky slopes, SW (Cape Peninsula).•

A. Flower heads mostly narrower; flowers not glutinous

brownii Dummer   Like **S. zeyheri** but flower heads 1–few at branch tips, surrounded by oblanceolate bracts with broad, white, papery margins. Feb.–May. Lower sandstone slopes, SW (Kogelberg).•

capitella (Thunb.) Sond.   Finely leafy, erect shrublet to 80 cm. Leaves lanceolate, 5–8 mm long. Flowers 1–few, in small terminal heads, pink, woolly. May–Nov. Rocky sandstone slopes, SW (Hottentots Holland Mts to Bredasdorp).•

phylicoides Pillans   Erect, single-stemmed, closely leafy shrub to 1 m. Leaves linear-oblanceolate, spreading-ascending. Flowers in small heads borne in upper axils and loosely clustered in corymbs, pink, petals hairy below. Sept. Rocky sandstone flats, NW (Bokkeveld Mts).•

pinifolia Willd. (= *Staavia dregeana* C.Presl)   Like **S. capitella** but leaves spreading-recurved, narrowly oblong, and flower heads somewhat larger. June–Oct. Rocky sandstone slopes, SW (Cape Peninsula to Riviersonderend Mts).•

radiata (L.) Dahl   ALTYDBOS   Finely leafy, twiggy shrublet to 60 cm, coppicing from a woody caudex. Leaves linear-lanceolate, 4–10 mm long. Flowers few, in small heads surrounded by small white, recurving bracts, pink. Mainly Sept.–Dec. Sandy flats and plateaus, SW, AP, LB (Yzerfontein to Gouritsmond).•

verticillata (L.f.) Pillans   Like **S. capitella** but leaves oblong, 3–4 mm long. Oct.–Nov. Rocky sandstone slopes, SW (Bainskloof to Riviersonderend Mts).•

zeyheri Sond.   Densely leafy, willowy shrub to 2 m. Leaves linear-lanceolate, 13–15 mm long, long-hairy becoming glabrescent. Flowers in small heads clustered in club-like synflorescences, surrounded by short white bracts, pink, woolly. June–July. Rocky sandstone slopes, SW (Riviersonderend Mts).•

sp. 1 (Raspalia staavioides (Sond.) Pillans)   Finely leafy shrub to 70 m. Leaves linear, spreading to recurved. Flowers few in small rounded heads loosely clustered in upper axils. Mainly Jan.–Mar. Sandstone slopes, NW (Cedarberg Mts).•

THAMNEA•   6 spp., W Cape

diosmoides Oliv.   Closely leafy rounded shrublet to 50 cm, coppicing form a rootstock. Leaves minute, scale-like, linear, wider and adpressed below. Flowers solitary at branch tips, sessile, petals c. 16 mm long, white to pink with red claw. Sept.–Nov. Mainly sandstone outcrops, also shale, NW (Pakhuis Mts to Waaihoek).•

gracilis Oliv.   Like **T. uniflora** but leaves lanceolate and ovary glabrous. Jan. Rocky summits, LB (Craggy Peak, Swellendam).•

hirtella Oliv.   Like **T. diosmoides** but intricately branched shrublet to 30 cm, uppermost leaves pilose-ciliate, flowers to 7 mm long. Apr.–June. Sandstone rocks, NW (Witzenberg).•

massoniana Dummer   Like **T. diosmoides** but leaves and petals blunt, ovary hairy above. Dec.–Apr. High sandstone slopes, SW (Du Toitskloof to Hottentots Holland Mts).•

thesioides Dummer   Like **T. uniflora** but flowers with ovary, glabrous, capped with conical style base, styles up to half as long as petals. Dec.–Mar. Rocky summits, NW (Michell's Pass and Mostertshoek Twins).•

uniflora Sol. ex Brongn. (incl. **T. depressa** Oliv.)   Twiggy shrublet with slender branches, to 20 cm. Leaves, scale-like, ovate, adpressed. Flowers solitary at branch tips, white, petals c. 2 mm long, ovary hispid. Dec.–Mar. Rocky summits, NW, SW (Michell's Pass and Hex River Mts to Sir Lowry's Pass).•

TITTMANNIA•   4 spp., W Cape

esterhuyseniae Powrie   Closely leafy, erect, willowy shrub to 60 cm, coppicing from a woody caudex. Leaves linear, imbricate. Flowers axillary, in dense subterminal racemes, white, petals c. 4 mm long, calyx warty. Feb.–Mar. Rocky sandstone slopes, NW, SW (Hex River Mts to Stettynsberg).•

laevis Pillans   Like **T. laxa** but calyx longitudinally furrowed and glabrous. Mainly Sept.–Feb. High rocky slopes, NW, SW (Cold Bokkeveld Mts to Jonkershoek).•

laxa (Thunb.) C.Presl (incl. **T. hispida** Pillans)   Closely leafy, erect shrub to 60 cm. Leaves scale-like, linear, imbricate, subterete. Flowers few at branch tips, white, with petals c. 2 mm long, calyx warty. Sept.–Nov. High sandstone slopes, NW, SW, KM (Cedarberg and Du Toitskloof Mts to Klein Swartberg and Langeberg Mts at Montagu).•

sp. 1 (Pseudobaeckia teres (Oliv.) Dummer)   Sprawling, mat-forming shrublet to 10 cm, with slender, erect branchlets. Leaves granular, adpressed, with membranous, fimbriate margins. Flowers solitary in upper axils, white, petals c. 4 mm long, calyx rugose. Dec.–Feb. Rocky peaks, SW, KM (Riviersonderend Mts and Klein Swartberg).•

BUDDLEJACEAE *(= LOGANIACEAE in part)*

BUDDLEJA SAGEWOOD   c. 100 spp., pantropical and subtropical

glomerata H.L.Wendl.   Shrub to 4 m. Leaves petiolate, discolorous, ovate, margins incised and crenate. Flowers in pea-like glomerules in panicles, cup-shaped, yellow. Sept.–Mar. Dry hillsides, SE (Uniondale to Free State).

saligna Willd.   WITOLIENHOUT, UMNCEBA   Shrub or tree to 7 m. Leaves petiolate, discolorous, lanceolate. Flowers in velvety paniculate cymes, cup-shaped with anthers exserted, cream-coloured with orange throat. Aug.–Jan. Rocky slopes and scrub, SW, LB, SE (Berg River to tropical Africa).

salviifolia (L.) Lam.   IGQANGE   Willowy shrub or tree to 8 m. Leaves sessile, discolorous, lanceolate, margins crenate, base auriculate. Flowers in terminal paniculate cymes, tubular with anthers included, white to purple with orange throat. Aug.–Oct. Forest margins, along streams and rocky slopes, NW, SW, LB, SE (Clanwilliam to tropical Africa).

CAMPANULACEAE
Grammatotheca, Lobelia, Monopsis and Wimmerella by P.B. Phillipson

1. Flowers zygomorphic, 1- or 2-lipped; anthers cohering in a tube around style, very rarely free, persistent after anthesis:
    2. Petals separate, free or partly cohering by their claws above base; capsule 2-locular, many-seeded.........**Cyphia**
    2.' Petals united; capsule 1- or 2-locular, few- to many-seeded:
        3. Ovary 1-locular; capsule prismatic, elongate, 3-valved; plants usually in damp places........**Grammatotheca**
        3.' Ovary 2-locular; capsule diverse, often short, 2-valved:
            4. Corolla tube not cleft down one side; delicate herbs ....**Wimmerella**
            4.' Corolla tube cleft down one side, usually to base:
                5. Stigmatic lobes elongate, filiform; flowers not resupinate (corolla cleft beneath) ....**Monopsis**
                5.' Stigmatic lobes short, subrotund or oval; flower resupinate (corolla cleft above) ....**Lobelia**
1.' Flowers actinomorphic; anthers free, deciduous at anthesis:
    6. Capsule opening regularly by terminal valves; ovary more or less half-inferior:
        7. Stigmas and locules 5, opposite petals; annuals with flowers subsessile or on short, stout pedicels; capsules subglobose..**Microcodon**
        7.' Stigmas and locules 3–5, when 5 then opposite sepals; annuals, perennials or shrublets; capsules various ....**Wahlenbergia**
    6.' Capsule bursting irregularly at apex or operculate, rarely indehiscent; ovary inferior:
        8. Stamens epipetalous, inserted well up corolla tube, subsessile; stigmatic lobes 3:
            9. Leaves densely imbricate, broadly ovate; flowers in terminal heads ....**Rhigiophyllum**
            9.' Leaves scattered, scale-like; flowers in open cymes ....**Siphocodon**
        8.' Stamens free, inserted at base of corolla; stigmatic lobes 2:
            10. Capsule cylindrical, at length splitting longitudinally into 5 strips; flowers often in diffuse, wiry cymes..**Prismatocarpus**
            10.' Capsule obovoid, indehiscent or operculate with base of style swollen and forming a solid plug; leaves often with stout, white, marginal hairs; flowers solitary amongst leaves:
                11. Plants annual; flowers in terminal glomerules..**Treichelia**
                11.' Plants perennial or shrubby:
                    12. Ovary 1-locular (or incompletely 2-locular) with 4 basal ovules; flowers tubular, elongate, in dense cylindrical inflorescences ....**Merciera**
                    12.' Ovary 2-locular with many ovules; flowers bell-shaped, solitary or in rounded heads...........**Roella**

CYPHIA   BAROE   c. 60 spp., Africa, mainly South Africa

A. Stems erect or weakly twining

bulbosa (L.) P.J.Bergius (incl. **C. kerastes** E.Wimm., **C. stephensii** E.Wimm.)   Glabrescent tuberous perennial to 30 cm. Leaves mostly basal but grading into bracts, palmatisect, sometimes palmatifid, margins slightly revolute, paler beneath. Flowers in terminal or axillary racemes, showy, bilabiate, laterally slit, white to mauve, 8–13 mm long, stamens c. 6 mm long, 2 anthers bearded. Aug.–Nov., often after fire. Sandy and stony flats and slopes, NW, SW (Cedarberg Mts to Cape Peninsula).•

georgica E.Wimm.   Like **C. linarioides** but flowers more or less erect, on long, 10–12 mm, puberulous pedicels, slightly larger, 9–10 mm long, stamens 5 mm long. Mar. Stony slopes, SE (George).•

incisa (Thunb.) Willd.   Thinly hairy, tufted tuberous perennial to 30 cm. Leaves rosulate, oblanceolate, toothed to pinnatisect. Flowers in a scapose raceme, scape ebracteate, bilabiate, laterally slit, white to mauve, 10–14 mm long, stamens c. 6 mm long, all anthers bearded. Sept.–Oct. Sandy and stony flats and slopes, NW, SW (Bokkeveld Mts to Cape Peninsula).•

linarioides C.Presl (= *Cyphia campestris* Eckl. & Zeyh.)   Erect, sometimes weakly twining tuberous perennial to 30 cm. Leaves often reduced, linear-lanceolate, sometimes shortly lobed below. Flowers in slender, secund terminal racemes, bilabiate, laterally slit, white to mauve, 7–9 mm long, stamens 5–6 mm long, 2 anthers bearded. Apr.–Oct. Stony slopes, NW, SW, LB, SE (Hex River valley to E Cape and Free State).

phyteuma (L.) Willd.   Glabrescent tufted tuberous perennial to 40 cm. Leaves rosulate, oblanceolate, slightly toothed. Flowers in a scapose, spiralling spike-like raceme, subsessile, scape minutely bracteate, bilabiate, laterally slit, whitish to brown or mauve, 16–20 mm long, stamens c. 6 mm long, all anthers bearded. Sept.–Oct. Sandy and stony flats and slopes, NW, SW, LB (Bokkeveld Mts to Riversdale).•

triphylla E.Phillips   Erect or weakly twining tuberous perennial to 50 cm. Leaves digitately divided, lobes linear. Flowers in terminal racemes, pink to mauve, 5–10 mm long, stamens 4–6 mm long, 2 anthers bearded. Apr.–May. Stony flats, LB, SE (Riversdale and Karoo to Lesotho).

undulata Eckl.   Like **C. sylvatica** but usually erect, sometimes weakly twining, leaves sessile and flowers in terminal racemes. Apr.–Oct. Stony slopes, often on limestone, SW, AP, LB, SE (Bredasdorp to E Cape).

A. Stems twining
B. Corolla more or less equally 5-lobed, tube not completely split at sides

subtubulata E.Wimm.   Twining tuberous perennial. Leaves linear-lanceolate, toothed. Flowers in upper axils, showy, incompletely bilabiate, not completely laterally slit, white to mauve, 15–18 mm long, stamens c. 6 mm long, all anthers bearded. Aug.–Oct. Stony often clay flats, NW, SW (?Namaqualand, Ceres to Mamre and Worcester).?•

zeyheriana C.Presl ex Eckl. & Zeyh. (incl. **C. ranunculifolia** E.Wimm.)   Twining tuberous perennial. Leaves narrowly lanceolate, sometimes digitately lobed, slightly toothed. Flowers in upper axils, arcuate-hypocrateriform, showy, not laterally slit, cream-coloured to mauve, often with darker reverse, 10–18 mm long, stamens 3–5 mm long, anthers glabrous. Aug.–Oct. Sandstone slopes, NW, SW, KM (Matsikamma to Riviersonderend Mts and Little Karoo).•

B. Corolla bilabiate, tube split completely at sides

crenata (Thunb.) C.Presl   Twining tuberous perennial. Leaves linear-lanceolate to ovate, toothed, often shortly lobed below, usually ascending. Flowers 1–3 in upper axils, bilabiate, laterally slit, white to mauve, 9–11 mm long, calyx truncate below, divided almost to base, ovary more or less superior, stamens 5–6 mm long, all anthers bearded. July–Sept. Sandy flats and slopes, often coastal, NW, SW (Namaqualand to Cape Peninsula).

digitata (Thunb.) Willd. (incl. **C. dentariifolia** C.Presl)   Twining tuberous perennial. Leaves 3–7-digitate, lobes linear, sometimes linear-lanceolate and obscurely toothed. Flowers in upper axils, bilabiate, laterally slit, white to mauve, 7–14 mm long, stamens 5–9 mm, all or 2 anthers bearded. Mainly July–Oct. Sandstone and clay slopes, NW, SW, AP, KM, LB, SE (Namaqualand and W Karoo to Port Elizabeth).

heterophylla C.Presl ex Eckl. & Zeyh.   Twining tuberus perennial. Leaves ovate to sagittate, often 3-lobed below, toothed. Flowers in upper axils, bilabiate, laterally slit, white to mauve, 12–15 mm long, stamens 7–8 mm long, all anthers bearded. Sept.–Oct. Coastal bush, SE (George to E Cape).

schlechteri E.Phillips   Like **C. sylvatica** but pedicels and calyx woolly. July–Aug. Stony slopes, NW (Namaqualand and W Karoo to Clanwilliam).

sylvatica Eckl. (incl. **C. tortilis** N.E.Br.)   Twining tuberous perennial. Leaves linear to lanceolate, slightly toothed. Flowers in upper axils, bilabiate, laterally slit, white to mauve, 7–12 mm long, stamens 6–7 mm long, all anthers bearded. Mainly Sept.–Oct. Stony slopes, KM, SE (Swartberg Mts to KwaZulu-Natal).

volubilis (Burm.f.) Willd. (incl. **C. dentata** E.Wimm., **C. tricuspis** E.Wimm.)   Twining tuberous perennial. Leaves linear-lanceolate or digitately lobed, toothed. Flowers in upper axils, showy, bilabiate, laterally slit, white to purple, 10–26 mm long, stamens short, less than half as long as corolla, 3–5 mm long, all anthers bearded. Aug.–Oct. Sandy flats and slopes, NW, SW, KM, LB, SE (Namaqualand to Port Elizabeth).

[**Species excluded**   No authentic material found and probably conspecific with one of the above: **C. angustifolia** Eckl. & Zeyh., **C. longipedicellata** E.Wimm., **C. tenera** Diels]

GRAMMATOTHECA   WATER LOBELIA   2 spp., W Cape to KwaZulu-Natal

bergiana (Cham.) C.Presl   Creeping perennial, stems to 60 cm. Leaves linear-oblanceolate, sparsely toothed. Flowers sessile or pedicellate in upper axils, pink or blue. Fruits glabrous. Nov.–Apr. Marshy flats, SW, LB, SE (Bainskloof to KwaZulu-Natal).

sp. 1   Sprawling perennial with stems to 50 cm. Leaves elliptic-obovate, sharply toothed. Flowers scattered, axillary, shortly pedicellate, scabrid, blue. Fruits retrorsely hairy. Feb.–Mar. Marshy slopes, NW (Swartruggens).•

LOBELIA (= *UNIGENES*)   LOBELIA   c. 300 spp., cosmopolitan

Group 1. Anthers all with an apical brush-like tuft of straight white hairs; flowers usually in a terminal raceme of 1 to several flowers, often on a long peduncle, sometimes peduncles leaf-opposed and vegetative growth continuing from a lateral bud or flowers axillary and solitary

ardisiandroides Schltr.   Sprawling, pubescent perennial. Flowers white or mauve. Nov.–Jan. Sheltered rocks at high alt., SW, SE (Riviersonderend Mts and Langkloof).•

barkerae E.Wimm.   Erect shrublet to 30 cm, branches densely leafy below. Flowers blue. May–June. Limestone hills, AP (Agulhas coast).•

capillifolia (C.Presl) A.DC.   Broom-like perennial to 50 cm. Flowers blue and white. Sept.–May. Sandstone slopes, NW, SW, AP, LB (Cedarberg to Langeberg Mts).•

chamaepitys Lam.   Shrublet branching from base, stems to 30 cm. Flowers violet-blue. Sept.–Apr. Sandstone slopes at low to middle alt., SW, AP, KM, LB (Stellenbosch to Klein Swartberg and Bredasdorp).•

comptonii E.Wimm. (incl. **L. esterhuyseniae** E.Wimm.)   Sprawling perennial to 35 cm. Flowers blue or white. Nov.–Dec. Sheltered rocks at medium to high alt., NW (Cedarberg Mts).•

coronopifolia L.   Tufted shrublet branching from base, stems to 30 cm. Flowers large, dark blue, pink or white. Oct.–Apr. Sandy and stony flats and lower slopes, NW, SW (Gifberg to Kleinrivier Mts).•

dasyphylla E.Wimm.   Low tufted perennial to 20 cm. Flowers blue. Jan.–Feb. Middle to upper sandstone slopes, LB (Langeberg Mts: Swellendam).•

dichroma Schltr.   Shrublet to 40 cm. Leaves pubescent. Flowers large, bright blue or rose. Nov.–Jan. High rocky slopes, SE (George to Langkloof).•

linearis Thunb. (incl. **L. lasiantha** (C.Presl) A.DC., **L. spartioides** (C.Presl) D.Dietr.)   Erect, broom-like shrublet to 70 cm. Flowers blue to purple. Mainly Sept.–Mar. Dry stony lower slopes, NW, SW, AP, KM, LB, SE (Cedarberg Mts to Langkloof).•

neglecta Roem. & Schult.   Trailing shrublet to 50 cm. Flowers blue. Throughout the year. Mountain slopes and coastal grassland and bush, AP, LB, SE (Mossel Bay to Grahamstown).

patula L.f. (= *Lobelia fourcadei* Schönland ms; incl. **L. genistioides** (C.Presl) A.DC.)   Sprawling perennial with whorled branches to 40 cm. Leaves sparse or lacking. Flowers blue, pink or white. Oct.–Dec. Mountain slopes and forest, AP, KM, LB, SE (Bredasdorp to Ladismith to Uniondale).•

pinifolia L. (incl. **L. capillipes** Schltr.)   Erect shrublet to 50 cm. Flowers blue. Mainly Dec.–Apr. Rocky slopes and flats, NW, SW, AP, LB (Ceres and Cape Peninsula to Riversdale).•

setacea Thunb. (incl. **L. glaucoleuca** Schltr.)   Tufted, erect or sprawling perennial to 60 cm. Flowers blue or violet and white. Nov.–Apr. Sandstone slopes and sandy flats, SW, AP, LB (Cape Peninsula to Langeberg Mts).•

thermalis Thunb.   Perennial or annual with long trailing stems. Flowers solitary in axils, pale blue. Flowering time? Rocky flats, NW (Bokkeveld Mts to Angola and Zimbabwe).

tomentosa L.f. (incl. **L. caerulea** Hook.)   Shrublet to 40 cm, woody below and branching from base. Flowers 1–several on long peduncles, blue, violet or pink. Nov.–June. Stony lower slopes, SW, LB, AP, SE (Cape Peninsula to S Mozambique).

sp. 1 (**Lobelia eurypoda** var. **fissurarum** F.Wimm.)   Tufted annual to 15 cm. Flowers blue. Flowering time? High rocky slopes, KM (Klein Swartberg).•

Group 2. Only the lower 2 anthers with an apical brush-like tuft of straight white hairs, the upper 3 either sparsely puberulous or with single scale-like appendages; flowers solitary in axils of upper leaves, sometimes in a dense terminal many-flowered inflorescence

anceps L.f. (= *Lobelia alata* auct.) Erect or spawling perennial to 50 mm. Flowers blue, mauve, or white. Nov.–June. Damp places usually near the coast, SW, AP, LB, SE (Cape Peninsula to KwaZulu-Natal).

boivinii Sond. Prostrate perennial, usually softly pubescent, flowers blue, mauve, or white. Nov.–Feb. Coastal rocks, SW (Cape Peninsula).•

comosa L. Soft shrublet to 50 cm, branching from base. Flowers bright blue. Mainly Aug.–Jan. Sandy coastal slopes, SW, AP (Cape Peninsula to Caledon).•

cuneifolia Link & Otto Sprawling to prostrate or decumbent perennial, branches to 40 cm. Flowers pale blue or white. Oct.–Dec. Forest, LB, SE (Swellendam to Humansdorp).•

eckloniana (C.Presl) A.DC. Slender trailing annual. Flowers small, mauve. Dec.–Apr. Damp rocks on sandstone slopes, SW (Cape Peninsula).•

erinus L. (incl. **L. acutangula** (C.Presl) A.DC., **L. bicolor** Sims, **L. montaguensis** E.Wimm.) Erect or spreading annual or perennial to 10 cm. Flowers blue, violet, pink or white usually with white centre. Sept.–Dec. Lower mountain slopes and coastal flats, NW, SW, KM, LB, SE (Bokkeveld Mts to tropical Africa).

flaccida (C.Presl) A.DC. Erect or sprawling perennial to 30 cm. Flowers pale to dark blue with white markings in throat. Oct.–Apr. ?, SE (George to KwaZulu-Natal).

humifusa (A.DC.) Phillipson ined. (incl. **L. disperma** E.Wimm., **Unigenes humifusa** (A.DC.) E.Wimm.) Slender sprawling annual to 15 cm. Flowers pale blue. Nov.–Dec. Shady damp coastal and inland slopes, NW, SW, KM (Swartberg and Hex River Mts to Cape Peninsula).•

hypsibata E.Wimm. Annual to 12 cm. Flowers white. Jan. Rocky sandstone slopes, LB (Langeberg Mts: Swellendam).•

jasionoides (A.DC.) E.Wimm. Erect or sprawling perennial to 50 cm. Flowers lilac and white. Oct.–Feb. Shady upper slopes, NW, SW, AP, LB (Gifberg to Swellendam).•

laurentioides Schltr. Sprawling annual, stems to 10 cm. Flowers tiny, lilac or reddish. Dec. Rocky slopes, SW (Bainskloof to Caledon).•

limosa (Adamson) E.Wimm. (= *Lobelia depressa* auct.; incl. **L. capensis** E.Wimm.) Sprawling perennial rooting at nodes, stems to 10 cm. Flowers white or mauve. Oct.–Feb. Marshy flats, SW (Cape Peninsula to Stellenbosch).•

muscoides Cham. Minute, tufted annual to 5 cm. Flowers lilac. Nov.–Feb. Damp upper mountain slopes, LB (Langeberg Mts: Swellendam).•

nugax E.Wimm. Spreading annual, stems to 30 cm long. Flowers very small, pink. Jan. High sandstone slopes, NW (Ceres).•

pubescens Dryand. ex Aiton Spreading annual or perennial, stems to 50 cm. Flowers white. Mainly Oct.–Mar. Rocky slopes and damp rocks near the coast, SW, AP, LB, SE (Cape Peninsula to Storms River mouth).•

quadrisepala (R.Good) E.Wimm. (= *Lobelia depressa* auct.) Sprawling perennial rooting at nodes, stems to 10 cm. Flowers white or mauve. Oct.–Feb. Marshy flats and slopes, NW, SW, SE (Bokkeveld Mts to Port Elizabeth).•

stenosiphon (Adamson) E.Wimm. (incl. **L. isotomoides** E.Wimm.) Tufted to sprawling perennial to 20 cm. Flowers rose-pink to purple. Jan.–Apr. High rocky slopes, SW ((Paarl to Caledon).•

valida L.Bolus GALJOENBLOM Soft leafy shrublet to 60 cm. Flowers deep blue. Nov.–Apr. Coastal hills, AP (De Hoop to Stilbaai).•

zwartkopensis E.Wimm. Spreading annual, stems to 10 cm. Flowers minute, white or blue. Mainly Oct. Pools on limestone, SE (Uitenhage).•

MERCIERA• 3 spp., W Cape

brevifolia A.DC. (incl. **M. eckloniana** H.Buek) Like **M. tenuifolia** but leaves often shorter and flowers white or pale mauve with tube shorter, 6–8 mm long. Nov.–Mar. Shale and granite slopes, SW (Elandskloof Mts to Bredasdorp).•

leptoloba A.DC. Like **M. tenuifolia** but flowers white, tube 8–15 mm long and petals linear, almost as long as tube. Dec.–Feb. Sandy flats and lower slopes, SW, AP (Cape Peninsula to Soetanysberg).•

tenuifolia (L.f.) A.DC.   Rigid closely leafy shrublet to 30 cm. Leaves imbricate, stiffly linear, pungent, shortly hairy, margins slightly revolute and harshly ciliate, but axillary leaves glabrous. Flowers subsessile in upper axils, blue to purple, tube slender, 10–25 mm long, petals elliptic-lanceolate, much shorter than tube. Dec.–Mar. Sandstone slopes, SW (Franschhoek to Kleinrivier Mts and Caledon Swartberg).•

MICROCODON•   3 spp., W Cape

glomeratum A.DC.   Sparsely hairy annual, 5–15 cm, branching from base. Leaves linear-lanceolate, margins thickened, ciliate below. Flowers tightly clustered at branch tips, subtended by leaves, narrowly bell-shaped, pale blue with purple centre, 5–8 mm diam., calyx lobes to as long as corolla. Capsule strongly beaked. Oct.–Dec. Clay or sandy flats, NW, SW, AP (Tulbagh to De Hoop).•

hispidulum (Thunb.) Sond.   Coarsely hairy, procumbent annual, 5–8 cm. Leaves triangular-lanceolate, margins thickened, ciliate below. Flowers few at branch tips, narrowly bell-shaped, pale blue with purple centre, 6–10 mm diam., calyx lobes often longer than corolla. Capsule flat or depressed above. Sept.–Nov. Sandy flats, NW, SW, LB (Lambert's Bay to Swellendam).•

lineare (L.f.) H.Buek   Subglabrous annual, 5–15 cm, branching above. Leaves linear-lanceolate, margins thickened, ciliate below. Flowers 1 or 2 at branch tips, not subtended by leaves, narrowly bell-shaped, white to blue, 5–8 mm diam., calyx lobes to as long as corolla. Capsule strongly beaked. Sept.–Oct. Sandy slopes, NW, SW (Bokkeveld to Bainskloof Mts).

[Species excluded   No authentic material found and probably conspecific with one of the above: **M. sparsiflorum** A.DC.]

MONOPSIS   13 spp., tropical and southern Africa

A. Flowers subsessile or shortly pedicellate, yellow

flava (C.Presl) E.Wimm.   Glabrous or shortly hairy perennial, 20–50 cm,. Leaves linear-oblanceolate, sharply toothed. Flowers crowded in pseudoracemes, bilabiate, yellow, pedicels bracteate at base. Oct.–Dec. Mountain slopes, NW (Kamiesberg to Ceres).

lutea (L.) Urb. (incl. **M. arenaria** E.Wimm.)   YELLOW LOBELIA   Decumbent to prostrate perennial, stems to 60 cm. Leaves often secund, linear to elliptic, toothed. Flowers crowded at branch tips in pseudospikes, subsessile, bilabiate, yellow, pedicels bracteate at base. Nov.–Apr. Damp flats and lower slopes, NW, SW, AP, LB, SE (Grootwinterhoek Mts to Humansdorp).•

variifolia (Sims) Urb.   Closely leafy perennial to 10 cm. Leaves imbricate, linear, sharply toothed. Flowers few, terminal, nested in leaves, sessile, bilabiate, yellow, pedicels bracteate at base. Nov.–Dec. Damp flats, NW, SW (Porterville to Romans River and Wolseley).•

A. Flowers on long, slender pedicels, white, purple or yellow

acrodon E.Wimm.   Thinly silky, trailing perennial to 20 cm. Leaves elliptic to obovate, sharply toothed. Flowers bilabiate, yellow, without bracteoles at base of pedicels. Oct.–Feb. Damp and sheltered sandstone slopes, NW (Pakhuis and Cedarberg Mts).•

alba Phillipson   Prostrate to decumbent perennial, 3–15 cm but often with long trailing stems. Leaves subopposite, ovate to elliptic, sparsely toothed. Flowers bilabiate, white with purple centre, pedicels bracteate at base. Nov.–Mar. Shady rock crevices and seeps on sandstone slopes, KM, SE (Swartberg to Great Winterhoek Mts).•

debilis (L.f.) C.Presl (= *Monopsis campanulata* (Larn.) Sond., *M. simplex* auct.)   Loosely erect or tufted annual to 25 cm. Leaves elliptic-oblanceolate, slightly toothed. Flowers purple, subactinomorphic, petals subrotund, without bracteoles. Mostly Sept.–Nov. Damp sandy slopes and flats, NW, SW, AP, LB (Namaqualand to Langeberg Mts).

simplex (L.) E.Wimm. (= *Monopsis aspera* auct.)   Prostrate or ascending annual or perennial to 50 cm. Leaves narrowly elliptic to oblanceolate, sparsely toothed. Flowers bilabiate, purple with a darker centre, pedicels bracteate at base. Nov.–Apr. Damp coastal slopes and forest margins, SW, AP, LB, SE (Cape Peninsula to Grahamstown).

unidentata (Dryand. ex Aiton) E.Wimm. (= *Monopsis stricta* (C.Presl) E.Wimm., *M. scabra* auct.)   WILD VIOLET   Erect or decumbent perennial, stems to 60 cm. Leaves linear to elliptic, entire to coarsely toothed. Flowers bilabiate, purple or brown with a darker centre. Oct.–Jan. Damp sandy flats and rocky slopes at low elevations SW, LB, SE (Riviersonderend to KwaZulu-Natal).

PRISMATOCARPUS   31 spp., Africa, mainly South Africa

A. Flowers in a leafless terminal inflorescence; bracts smaller than leaves
B. Sepals fused below; flowers hypocrateriform with slender tube; style much longer than corolla

diffusus (L.f.) A.DC.   Sprawling or rounded shrublet to 45 cm, shortly hairy on young stems. Leaves needle-like, sparsely ciliate below. Flowers in leafless, divaricate terminal cymes, blue sometimes white, hypocrateriform, c. 10 mm diam., lobes linear, anthers partially or just exserted, bracts 2–3 mm long. Fruits 10–20 mm long. Nov.–Feb. Sandstone slopes, NW, SW (Kamiesberg to Riviersonderend Mts).

fastigiatus C.Presl ex A.DC.   Like **P. diffusus** but ovary hairy and bracts slightly longer, 4–6 mm long. Flowering time? Sandstone slopes, SW (Bredasdorp).•

pauciflorus Adamson   Like **P. diffusus** but leaves shorter and mostly in axillary tufts, often finely hairy, to 10 mm long and flowers with anthers well included. Jan.–Feb. Sandstone slopes, NW (N Cedarberg Mts).•

pilosus Adamson   Like **P. diffusus** but ovary and calyx finely hairy and bracts about as long as ovary, c. 5–10 mm long. Jan. Sandstone slopes, NW (Cold Bokkeveld Mts).•

B. Sepals almost free; flowers cup- or funnel-shaped; style shorter than corolla

alpinus (Bond) Adamson   Like **P. brevilobus** but mostly prostrate or mat-forming, leaves softer and broader and inflorescence few-branched or sessile. Dec.–Jan. Sandstone ledges at high alt., NW, SW (Cedarberg to Hottentots Holland Mts).•

altiflorus L'Hér.   Erect or sprawling shortly hairy shrublet to 1 m. Leaves awl-like, coarsely ciliate below, often in axillary tufts. Flowers mostly congested in subumbellate, pedunculate terminal cymes, white to blue, cup-shaped, 15–20 mm diam., ovary densely coarsely hairy. Fruits 30–45 mm long, coarsely hairy. Nov.–Dec. Sandstone slopes, NW (Cedarberg and Cold Bokkeveld Mts).•

brevilobus A.DC.   Like **P. pedunculatus** and **P. fruticosus** but sepals short and broad, ovate-oblong. Dec.–May. Rocky slopes, NW, SW, AP, KM (Cedarberg to Bredasdorp and Swartberg Mts).•

crispus L'Hér.   Robust shortly hairy annual to 50 cm. Leaves few, scattered, linear, margins slightly thickened and sparsely undulate-toothed. Flowers in leafless terminal cymes, blue or lilac, cup-shaped, (15–)20–30 mm diam., ovary glabrous or hairy. Fruits 40–60 mm long. Oct.–Dec. Dry, often sandy flats and slopes, NW, SW (Namaqualand to E Cape, mostly Bokkeveld to Riviersonderend Mts).

decurrens Adamson   Like **P. pedunculatus** but leaves decurrent, lanceolate and sparsely prickly toothed as well as ciliate below. Dec.–Mar. Sandstone slopes above 1000 m, NW (Cedarberg Mts).•

fruticosus L'Hér.   Like **P. pedunculatus** but flowers deeply cup- or funnel-shaped, white with brown or purple reverse, sepals shorter than corolla tube. Nov.–Apr. Sandy flats and slopes, NW, SW, KM, SE (Cedarberg to Riviersonderend Mts and Langkloof).•

lycopodioides A.DC.   Sprawling, closely leafy, shortly hairy shrublet to 30 cm. Leaves imbricate, spreading-incurved or reflexed, small, oblong, coarsely ciliate. Flowers 1–5 in subracemose terminal cymes on wiry peduncles, cup-shaped, white to pinkish, c. 10 mm diam. Fruits 15–20 mm long. Nov.–Jan. Sandstone slopes in sheltered places, SW (Bainskloof to Stellenbosch Mts).•

pedunculatus (P.J.Bergius) A.DC.   Erect or sprawling shortly hairy shrublet to 50 cm. Leaves awl-like, coarsely ciliate below, often in axillary tufts. Flowers in leafless terminal cymes, white to blue, dish- or bowl-shaped, 15–20 mm diam. Fruits 15–30 mm long. Sept.–Jan. Stony or shale flats and slopes, NW, SW, LB (Bokkeveld Mts to Riversdale).•

A. Flowers mostly axillary, sometimes in pseudoracemes or crowded at branch tips; bracts like leaves or larger
C. Stiff shrublets

campanuloides (L.f.) Sond.   Slender glabrescent shrublet to 80 cm. Leaves linear-lanceolate, sometimes weakly prickly toothed, coarsely ciliate below; bracts lanceolate, ciliate. Flowers sessile, more or less crowded at branch tips and upper axils, white to lilac, bowl-shaped, 20–30 mm diam. Fruits c. 20 mm long. Dec.–Apr. Sandy or limestone flats and slopes, NW, SW, AP, SE (Elandskloof Mts to E Cape).

candolleanus Cham. (incl. **P. virgatus** Fourc.)   Stiffly erect shortly hairy shrublet to 50 cm with rod-like branches. Leaves linear-lanceolate, margins revolute, entire or weakly prickly toothed; bracts broad and pinnatisect below, slightly revolute. Flowers sessile, few in upper axils, white to lilac, bell-shaped, 15–20 mm diam. Fruits c. 10 mm long. Dec.–Jan. Sandstone slopes, KM, LB, SE (Swartberg and Langeberg to Outeniqua Mts).•

cliffortioides Adamson   Stiffly erect shortly hairy shrublet to 1 m. Leaves linear-lanceolate, margins revolute and prickly toothed, often twisted and pungent; bracts broad and pinnatisect below, margins revolute. Flowers sessile, crowded in axillary glomerules, white to lilac, funnel-shaped with short lobes, 15–20 mm diam. Fruits c. 10 mm long. Dec.–Apr. Stony, often shale slopes, LB (Langeberg Mts: Cloete's Pass).•

hispidus Adamson   Sprawling, sparsely leafy roughly hairy shrublet to 60 cm. Leaves ovate, margins slightly revolute and prickly toothed, coarsely hairy; bracts pinnatisect below, coarsely hairy. Flowers in small terminal heads, white, narrowly funnel-shaped or tubular, c. 5 mm diam., ovary and calyx coarsely hairy. Fruits c. 5 mm long. Jan. Sandstone slopes, LB, SE (Langeberg: Cloete's Pass to Outeniqua Mts).•

lycioides Adamson   Thorny, stiffly branched, shortly hairy shrublet to 40 cm. Leaves oblong, margins revolute, sometimes prickly toothed below, twisted, ciliate beneath. Flowers axillary on divaricate thorny branchlets, white, bell-shaped, c. 8 mm diam. Fruits c. 5 mm long. Dec.–Apr. Dry N-facing sandstone slopes, SW (Riviersonderend Mts: Jonaskop and Hammansberg).•

rogersii Fourc.   Slender erect or sprawling shortly hairy shrublet to 50 cm. Leaves linear-lanceolate, margins slightly revolute, entire or weakly prickly toothed, coarsely ciliate below and on midrib beneath; bracts broad and pinnatisect below, ciliate. Flowers sessile, few in upper axils, white to lilac, bell-shaped, c. 10 mm diam. Fruits c. 10 mm long. Dec.–Apr. Sheltered sandstone slopes, SE (Outeniqua Mts: Outeniqua Pass).•

schlechteri Adamson   Slender shortly hairy shrublet to 40 cm. Leaves ovate-lanceolate, spreading, margins slightly revolute, weakly prickly toothed, coarsely ciliate below and on midrib beneath; bracts lanceolate, ciliate. Flowers sessile, few in upper axils, white, bell-shaped, c. 10 mm diam. Fruits c. 8 mm long. Dec.–Apr. Sandstone slopes, SW (Wemmershoek to Bredasdorp Mts).•

spinosus Adamson   Stiffly erect closely leafy coarsely hairy shrublet to 1 m. Leaves ovate, margins revolute and prickly toothed, coarsely hairy; bracts pinnatisect below, coarsely hairy. Flowers terminal and axillary, white, narrowly funnel-shaped or tubular, c. 5 mm diam., ovary and calyx coarsely hairy. Fruits c. 8 mm long. Jan. Sandstone slopes, SW (Potberg).•

C. Small or prostrate, more delicate perennials

cordifolius Adamson   Like **P. nitidus** but leaves thinly hairy and bracts distinctly toothed. Fruits c. 10 mm long. Jan. Sheltered sandstone crevices, SW (Kogelberg and Betty's Bay Mts).•

debilis Adamson (= *Prismatocarpus nitidus* var. *ovatus* Adamson)   Like **P. nitidus** but leaves often opposite, sometimes thinly hairy and upper internodes slender, wiry and peduncle-like. Fruits 10–15 mm long. Jan.–Mar. Sheltered sandstone crevices, NW, SW, LB, SE (Olifants River to Outeniqua Mts).•

implicatus Adamson   Like **P. tenellus** but flowers cup-shaped with lobes longer than tube and fruits ovoid, 5–6 mm long, not spirally twisted. Jan.–Mar. Sheltered sandstone slopes, NW (Grootwinterhoek Mts).•

lasiophyllus Adamson   Like **P. cordifolius** but sepals hairy. Jan. Sheltered sandstone crevices, LB (Langeberg Mts: Swellendam).•

nitidus L'Hér.   Prostrate, softly woody subshrub or perennial to 35 cm. Leaves alternate or subopposite, ovate to lanceolate, margins slightly revolute and toothed. Flowers 1–5 at branch tips, white to pale blue, bell-shaped, c. 8 mm diam.; bracts like leaves. Fruits 15–25 mm long. Jan.–Mar. Sheltered sandstone crevices, SW (Cape Peninsula).•

sessilis Eckl. ex A.DC.   Sprawling subshrub or shrublet to 60 cm. Leaves linear-lanceolate, margins slightly revolute and ciliate below. Flowers 1–3 in axils, sessile or pedicillate and sometimes in pseudoracemes, white or pale blue, bell-shaped, c. 5 mm diam. Fruits 4–15 mm long. Dec.–Mar. Sheltered sandstone slopes, SW, LB (Cape Peninsula to Kleinrivier Mts).•

tenellus Oliv.   Delicate, sprawling perennial with wiry stems to 50 cm. Leaves opposite, linear. Flowers on slender divaricately spreading peduncles in upper axils, white, bell-shaped, c. 5 mm diam. Fruits 5–15 mm long, spirally twisted. Jan.–Mar. Sheltered sandstone slopes, NW (Hex River Mts).•

tenerrimus H.Buek   Sprawling, minutely hairy subshrub to 30 cm. Leaves ovate-lanceolate, margins thickened and minutely prickly toothed. Flowers 1–3 in axils and in pseudoracemes, white to lilac, bell-shaped, c. 5 mm diam. Fruits 6–10 mm long. Jan.–Mar. Sandstone slopes, SW, KM, LB (Wemmershoek to Langeberg and Swartberg Mts).•

RHIGIOPHYLLUM • 1 sp., W Cape

squarrosum Hochst.   Densely leafy shrublet with erect branches to 50 cm. Leaves imbricate, 4-ranked, subrotund, leathery, shining. Flowers clustered at branch tips in bracteate heads, narrowly tubular below, deep blue. Nov.–Jan. Damp sandstone slopes, SW (Akkedisberg and Bredasdorp Mts).•

ROELLA   c. 20 spp., South Africa, mostly W Cape, 1 sp. to KwaZulu-Natal

A. Leaves ovate to elliptic

amplexicaulis Wolley-Dod   Erect, closely leafy shrublet to 50 cm. Leaves imbricate, recurved, rotund-ovate, prickly toothed, apiculate; bracts orbicular. Flowers 3–8 in terminal heads, white or pale blue, c. 10 mm diam., ovary glabrous. Nov.–Apr. Sandstone slopes, SW (Cape Peninsula).•

decurrens L'Hér.   Annual or short-lived perennial to 70 cm. Leaves scattered, spreading or recurved, ovate-lanceolate, prickly toothed, apiculate, ciliate below and decurrent; bracts broadly lanceolate. Flowers 1–5 in terminal heads, white or pale blue, c. 10–15 mm diam., ovary glabrous. Feb.–Apr. Sandy slopes, SW (Cape Peninsula).•

goodiana Adamson   Like **R. recurvata** but softer with bracts 3-toothed above and not recurved and flowers smaller. Feb.–Apr. Sandy flats, SW (Cape Peninsula: Klaver Valley).•

muscosa L.f.   Prostrate, mat-forming perennial to 5 cm. Leaves ovate-elliptic, softly prickly toothed, ciliate below; bracts lacking. Flowers solitary at branch tips, white to pale blue, c. 10 mm diam., ovary glabrous. Nov.–Feb. Sandstone rocks, SW (Cape Peninsula to Kleinrivier Mts).•

recurvata A.DC.   Closely leafy shrublet to 25 cm. Leaves imbricate, spreading or recurved, elliptic and keeled, apiculate, ciliate, slightly decurrent; bracts leaf-like, hooked. Flowers solitary at branch tips, white or blue, c. 10 mm diam., ovary minutely hairy. Jan.–Feb. Sandy flats, SW (Cape Peninsula).•

squarrosa P.J.Bergius   Straggling shrublet to 50 cm. Leaves spreading-recurved, ovate, prickly toothed, apiculate, ciliate below and decurrent; bracts rotund-ovate. Flowers 1–3 in terminal heads, white or tinged blue, c. 10 mm diam., ovary glabrous. Dec.–Mar. Sandstone slopes, SW (Cape Peninsula).•

A. Leaves linear or awl-shaped
B. Bracts like the leaves but often more crowded

arenaria Schltr.   Like **R. prostrata** but leaves spreading and sepals strongly recurved or hooked and always hairy. Dec.–Mar. Sandy flats, SW (Malmesbury to Bredasdorp).•

bryoides H.Buek   Like **R. prostrata** but leaves spreading and short, to 6 mm long and bracts always prickly toothed. Dec.–Feb. Sandy slopes, NW, SW (Clanwilliam to Caledon Swartberg).•

latiloba A.DC.   Like **R. prostrata** and **R. bryoides** but sepals broadly triangular, shortly hairy. Dec.–Feb. Sandy slopes, NW, SW (Clanwilliam, Bredasdorp).•

prostrata E.Mey. ex A.DC. (= *Roella incurva* var. *rigida* Adamson)   Erect or sprawling shrublet to 50 cm. Leaves often with axillary tufts, linear-pungent, ciliate, midrib prominent beneath; bracts leaf-like, crowded, ciliate below sometimes prickly toothed, glabrous or sparsely hairy. Flowers solitary at branch tips, white or pale blue, 10–15 mm diam., ovary hairy. Dec.–Mar. Sandy flats, SW (Hopefield to Potberg).•

B. Bracts differing from the leaves in size or form
C. Bracts broadened and clasping below; flowers usually in heads

compacta Schltr. (= *Roella cuspidata* Adamson)   Sprawling or decumbent shrublet to 30 cm. Leaves with axillary tufts, linear-pungent, coarsely ciliate, sometimes prickly toothed or roughly hairy, margins revolute; bracts ovate-acuminate, toothed above, ciliate below, sometimes roughly hairy. Flowers in terminal heads, white or pale blue, 8–10 mm diam., ovary glabrous or minutely hairy. Dec.–Feb. Rocky coastal limestones, SW, AP (Cape Peninsula to De Hoop).•

secunda H.Buek   Tangled or trailing shrublet with numerous short, often secund branchlets. Leaves with axillary tufts, small, oblong-mucronate, sometimes slightly toothed; bracts suborbicular, more or less truncate and equally 5–7-recurved-toothed above, finely hairy. Flowers 1–3 at branch tips, mauve to white, c. 8 mm diam., calyx and ovary hairy. Dec.–Mar. Sandy flats and lower slopes, NW, AP, KM, SE (Potberg and Worcester to Uitenhage).•

spicata L.f. (= *Roella lightfootioides* Schltr.)   Erect or sprawling shrublet to 40 cm. Leaves with axillary tufts, linear-pungent, coarsely ciliate below, outer bracts leaf-like but broad and clasping below, inner bracts shorter, ovate, toothed above. Flowers in terminal and lateral heads, white, tubular-campanulate, c. 10 mm diam., ovary hairy, sometimes glabrous. Jan.–Mar. Sandstone slopes, SW, KM, LB, SE (Houwhoek to Port Elizabeth).•

C. *Bracts narrow throughout; flowers not in heads, sometimes with dark markings*
D. *Ovary glabrous*

ciliata L.   Erect or sprawling shrublet to 50 cm. Leaves with axillary tufts, linear-pungent, ciliate, midrib prominent beneath; bracts larger, conspicuously white-ciliate. Flowers solitary, white or blue with a dark ring or spots on lobes, 20–25 mm diam. Aug.–Mar. Stony slopes, SW (Elandskloof Mts to Caledon Swartberg).•

incurva A.DC. (= *Roella rhodantha* Adamson)   Shrublet to 40 cm. Leaves often with axillary tufts, linear-pungent, ciliate and often prickly toothed above, midrib prominent beneath; bracts longer, ciliate and prickly toothed, glabrous or sparsely hairy. Flowers 1–3 at branch tips, white or blue, sometimes pink or red (only Potberg), mostly with dark spots on petals, 20–30 mm diam. Oct.–Jan. Sandy lower slopes, SW, AP (Elandskloof Mts to Potberg).•

D. *Ovary hairy*

dregeana A.DC. (= *Roella psammophila* Schltr.)   Erect shrublet with ascending branches to 30 cm. Leaves usually with axillary tufts, small, linear-pungent, ciliate below and usually with stiff hair-like prickles above; bracts larger, glabrous or finely hairy, margins with few to many stiff, wire-like hairs. Flowers 1–few at branch tips, pale blue or white, 10–15 mm diam., calyx and ovary hairy. Jan.–Mar. Sandstone slopes, SW (Elandskloof Mts to Kleinrivier Mts).•

dunantii A.DC.   Sprawling shrublet to 30 cm. Leaves with axillary tufts, linear-pungent, conspicuously white-ciliate, midrib prominent beneath; bracts larger. Flowers solitary, white or blue, sometimes with small spots on petals, 10–15 mm diam., ovary finely hairy. Nov.–Jan. Sandy lower slopes, SW (Mamre to Caledon Swartberg).•

maculata Adamson   Shrublet to 30 cm. Leaves with axillary tufts, linear-pungent, finely ciliate, sparsely prickly toothed above; bracts larger, finely hairy. Flowers 1–4 at branch tips, blue with large spots between petals, 20–30 mm diam., calyx and ovary finely hairy. Dec.–Feb. Sandy coastal slopes, SW, AP (Kleinmond to De Hoop).•

triflora (R.D.Good) Adamson   Erect shrublet with ascending branches to 20 cm. Leaves usually with axillary tufts, linear-pungent, ciliate below and prickly toothed above; bracts larger, finely hairy, margins with stiff, wire-like hairs. Flowers 1–3 at branch tips, pale blue with dark eye, 15–20 mm diam., calyx and ovary hairy. Dec.–May. Sandy lower slopes, SW (Cape Peninsula).•

SIPHOCODON•   2 spp., W Cape

debilis Schltr.   Like **S. spartioides** but flowers narrowly campanulate, violet to white, and filaments longer than anthers and inserted in lower half of tube. Jan.–Apr. Sandstone slopes, SW (Hottentots Holland to Kleinrivier Mts).•

spartioides Turcz.   Flexuose subshrub with slender, wand-like branches to 30 cm. Leaves minute, triangular, adpressed. Flowers in short, sometimes branched racemes, tubular, blue to purple, filaments shorter than anthers and inserted in upper part of tube. Dec.–Apr. Sandstone slopes, SW, LB (Franschhoek to Langeberg Mts).•

TREICHELIA•   1 sp., W Cape

longebracteata (H.Buek) Vatke   Coarsely hairy annual branching from base, 5–20 cm. Leaves linear-oblanceolate, slightly toothed, coarsely ciliate below. Flowers in terminal heads among long bracts, narrowly bell-shaped, pale blue to white, c. 8 mm diam., petals sparsely hairy, ovary 2-locular. Oct.–Dec. Disturbed sandy ground, SW (Cape Peninsula to Hermanus).•

WAHLENBERGIA (= *LIGHTFOOTIA, THEILERA*)   African blue-bell, blouklokkie   c. 125 spp., chiefly S temperate

A. *Petals narrowly lanceolate, more than twice as long as broad; corolla either divided to near base or hypocrateriform*
B. *Leaves channelled or concave above, ovate to lanceolate*

adamsonii Lammers (= *Lightfootia multicaulis* Adamson)   Like **W. brachyphylla** but sepals strongly keeled, flowers blue. Nov.–Jan. High sandstone slopes, NW (Pakhuis and Cedarberg Mts).•

adpressa (Thunb.) Sond. (= *Lightfootia adpressa* (Thunb.) A.DC.)   Erect or sprawling rigid shrublet to 80 cm. Leaves spreading or recurved, lanceolate, concave, toothed. Flowers in panicles, white or cream-coloured, c. 6 mm diam., tube 2–3 mm long, ovary glabrous. Dec.–Apr. Coastal sands, NW, SW (Rocher Pan to Cape Peninsula).•

asparagoides (Adamson) Lammers (= *Lightfootia asparagoides* Adamson)   Like **W. adpressa** but stems much-branched, to 1 m and inflorescence divaricately branched and becoming spinescent. Sept. Coastal sands, NW (Namaqualand to Lambert's Bay).

brachyphylla (Adamson) Lammers (= *Lightfootia brachyphylla* Adamson)   Sprawling shrublet to 30 cm. Leaves opposite, recurved, small, lanceolate, channelled, margins thickened, toothed. Flowers few in upper axils, white, c. 5 mm diam., tube c. 2 mm long, about as long as lobes, ovary half-inferior, hairy. Nov.–Dec. Sandstone slopes, NW (Cold Bokkeveld).•

cordata (Adamson) Lammers (= *Lightfootia cordata* Adamson)   Mat-forming closely leafy shrublet to 5 cm. Leaves opposite, imbricate, spreading or recurved, ovate, concave, keeled below. Flowers 1–3 in upper axils, white, c. 5 mm diam., tube to 1 mm long, ovary subglabrous. Jan.–Mar. Sheltered rock ledges at high alt., KM (Swartberg and Kammanassie Mts).•

desmantha Lammers (= *Lightfootia fasciculata* (L.f.) A.DC.)   Erect shrublet to 80 cm. Leaves recurved, ovate, margins thickened, slightly toothed below. Flowers crowded in terminal heads, white with darker reverse, 5–8 mm diam., tube c. 1 mm long, ovary hairy. Jan.–Apr. Stony and clay slopes, KM, LB, SE (Riversdale and Swartberg Mts to Uniondale).•

macrostachys (A.DC.) Lammers (= *Lightfootia spicata* H.Buek)   Slender shrublet to 50 cm. Leaves recurved, lanceolate, channelled, margins thickened, slightly toothed below. Flowers subsessile in upper axils, white, c. 5 mm diam., tube c. 1 mm long, ovary half-inferior, roughly hairy, sometimes glabrous. Sept.–Jan. Rocky sandstone slopes, NW, SW (Gifberg to Drakenstein Mts).•

microphylla (Adamson) Lammers (= *Lightfootia microphylla* Adamson)   Like **W. neorigida** but plants more delicate and leaves smaller, 1–2 mm long. Feb. Coastal sands, AP (Bredasdorp).•

neorigida Lammers (= *Lightfootia rigida* Adamson)   Erect rigid shrublet to 60 cm. Leaves recurved, ovate-lanceolate, concave, margins thickened. Flowers terminal, solitary or subumbellate, white, 5–7 mm diam., tube c. 1 mm long, ovary half-inferior, shortly hairy. Nov.–Apr. Dry rocky slopes, NW, SW, KM, LB, SE (Cold Bokkeveld to Uitenhage).•

nodosa (H.Buek) Lammers (= *Lightfootia nodosa* H.Buek)   Rigid much-branched somewhat spiny shrublet to 45 cm. Leaves recurved, ovate-lanceolate, concave, margins thickened. Flowers in divaricately branched panicles, white, 5–7 mm diam., tube c. 1 mm long, ovary half-inferior, shortly hairy. Oct.–May. Dry rocky slopes, NW, SW, KM, LB, SE (widespread in southern Africa).

oligantha Lammers (= *Lightfootia pauciflora* Adamson)   Slender wiry shrublet to 45 cm. Leaves suberect or spreading, scale-like, channelled. Flowers few in terminal axils on slender pedicels, white, c. 5 mm diam., tube obsolete, ovary half-inferior, puberulous. Dec.–Jan. N-facing sandstone slopes, LB (Langeberg Mts: Garcia's Pass).•

rubioides (A.DC.) Lammers (= *Lightfootia rubioides* A.DC.)   Sprawling to prostrate, often diffuse, shrublet to 30 cm. Leaves alternate or opposite, reflexed, lanceolate, channelled, sometimes flat, slightly toothed below. Flowers in upper axils on slender pedicels, white, c. 5 mm diam., tube c. 1 mm long, ovary half-inferior, roughly hairy. Dec.–Mar. High rocky slopes, NW, SW, LB, SE (Gifberg to Tsitsikamma Mts).•

tenella (L.f.) Lammers (= *Lightfootia diffusa* H.Buek, *L. tenella* (L.f.) A.DC.)   Erect or sprawling shrublet to 80 cm. Leaves strongly recurved, ovate, margins thickened, sometimes minutely toothed below. Flowers clustered in upper axils or paniculate, white or blue, 5–8 mm diam., tube c. 1 mm long, ovary sometimes hairy. Nov.–May. Sandy flats and slopes, often coastal, SW, AP, KM, LB, SE (Mamre to E Cape).

tenerrima (H.Buek) Lammers (= *Lightfootia tenella* Lodd.)   Sprawling diffuse shrublet to 50 cm. Leaves recurved, ovate-lanceolate, concave, margins thickened. Flowers in slender pseudoracemes, blue or white, 5–7 mm diam., tube c. 1 mm long, ovary half-inferior, shortly hairy. Oct.–May. Sandstone slopes, NW, SW, AP, KM, LB, SE (Cold Bokkeveld to E Cape).

umbellata (Adamson) Lammers (= *Lightfootia umbellata* Adamson)   Erect or sprawling shrublet to 50 cm. Leaves opposite, reflexed, linear-lanceolate, concave above. Flowers in umbellate cymes, white with darker reverse, c. 8 mm diam., tube c. 1 mm long, ovary half-inferior, roughly hairy. Flowering time? Coastal sands, NW (Lambert's Bay).•

sp. 1   Like **W. cordata** but sepals triangular and leaf and sepal margins roughly hairy below. Jan.–Mar. Sandstone slopes at high alt., NW (Cedarberg Mts).•

B. Leaves flat, linear to lanceolate
C. Leaf margins revolute

albens (Spreng. ex A.DC.) Lammers (= *Lightfootia albens* Spreng. ex A.DC.)   Twiggy shrublet to 50 cm. Leaves linear-ericoid, margins strongly revolute, greyish. Flowers in divaricate clusters, yellow or white, c. 5 mm diam., tube to 1 mm long, ovary shortly hairy. Oct.–Jan. Dry stony slopes, SW, LB, SE (Caledon and Karoo to Lesotho).

axillaris (Sond.) Lammers (= *Lightfootia axillaris* Sond.)   Like **W. unidentata** but leaves broader, oblong-lanceolate and flowers larger, 6–10 mm diam. Dec.–Apr. Sandstone slopes, SW (Kogelberg to Bredasdorp Mts).•

calcarea (Adamson) Lammers (= *Lightfootia calcarea* Adamson)   Sprawling often densely leafy shrublet to 30 cm. Leaves often tufted, linear-ericoid, margins strongly revolute. Flowers in tight terminal clusters, white with darker reverse, c. 8 mm diam., tube 2–3 mm long, ovary glabrous or shortly hairy. Oct.–Apr. Coastal limestone, AP (Pearly Beach to Stilbaai).•

cinerea (L.f.) Lammers (= *Lightfootia cinerea* (L.f.) Sond.)   Like **W. rubens** but plants distinctly cobwebby or woolly-hairy. Nov.–May. Stony slopes, SW, KM, LB, SE (Potberg and Swartberg Mts to E Cape).

guthriei L.Bolus (= *Theilera guthriei* (L.Bolus) E.Phillips)   Shortly hairy shrublet with erect or straggling branches to 30 cm. Leaves mostly in axillary tufts, narrowly triangular, rigid, margins slightly revolute. Flowers sessile in upper axils, often in pseudoracemes, hypocrateriform, tube elongate, blue to purple. Jan.–July. Sandstone slopes, KM, SE (Witteberg and Swartberg to Tsitsikamma Mts).•

levynsiae Lammers (= *Lightfootia squarrosa* Adamson)   Like **W. calcarea** but more erect with corolla tube longer than calyx, c. 3 mm long. Apr. Limestone flats, AP (Agulhas to Stilbaai).•

rubens (H.Buek) Lammers (= *Lightfootia rubens* H.Buek)   Erect shrublet to 50 cm. Leaves often tufted, spreading or suberect, linear-lanceolate, margins revolute, slightly toothed below. Flowers grouped in upper axils, white with darker reverse, c. 5 mm diam., tube to 1 mm long, ovary glabrous. Sept.–Mar. Dry lower mountain slopes, SW, AP, KM, LB, SE (Bredasdorp to E Cape).

uitenhagensis (H.Buek) Lammers (= *Lightfootia divaricata* H.Buek)   Sprawling or straggling shrublet to 50 cm. Leaves recurved, linear to lanceolate, margins slightly revolute, slightly toothed below. Flowers in upper axils, white or blue with darker reverse, 5–8 mm diam., tube 1–2 mm long, ovary glabrous. Mainly Sept.–May. Coastal sands, SW, AP, SE (De Hoop to KwaZulu-Natal).

unidentata (L.f.) Lammers (= *Lightfootia unidentata* (L.f.) A.DC)   Erect wand-like shrublet, 15–40 cm. Leaves adpressed-ascending, linear-lanceolate, margins revolute, slightly toothed. Flowers in a contracted racemose panicle, white or blue, c. 6 mm diam., tube obsolete, ovary glabrous. Dec.–Apr. Stony, often clay slopes, SW, LB, SE (Caledon Swartberg to Humansdorp).•

C. Leaf margins not revolute, sometimes thickened

capillaris (H.Buek) Lammers (= *Lightfootia oppositifolia* A.DC.)   Diffuse tangled shrublet, 10–30 cm. Leaves opposite below, spreading or reflexed, linear-lanceolate, margins slightly thickened. Flowers in upper axils on wiry pedicels, white with darker reverse, c. 5 mm diam., tube c. 2 mm long, ovary glabrous. Jan.–Mar. Sheltered sandstone slopes, NW, SW (Lambert's Bay to Riviersonderend Mts).•

effusa (Adamson) Lammers (= *Lightfootia effusa* Adamson)   Slender erect or sprawling shrublet, to 30 cm. Leaves reflexed, linear, margins thickened, minutely toothed. Flowers in diffuse, wiry panicles, white, c. 5 mm diam., ovary glabrous. Mar.–Apr. Stony slopes, LB (Swellendam).•

longifolia (A.DC.) Lammers (= *Lightfootia longifolia* A.DC.)   Erect shrublet to 40 cm. Leaves often tufted, alternate or opposite, sometimes recurved, linear, slightly toothed below. Flowers in narrow raceme-like panicles, white or cream-coloured, 5–8 mm diam., tube c. 1 mm long, ovary hairy. Dec.–Apr. Coastal sands and limestone, SW, AP (Hopefield to De Hoop).•

neostricta Lammers (= *Lightfootia stricta* Adamson)   Shrublet, 15–40 cm. Leaves often tufted, mostly opposite, spreading or reflexed, linear, margins thickened and minutely toothed. Flowers clustered in upper axils, blue, c. 5 mm diam., tube obsolete, ovary half-inferior, hairy. Flowering time? Sandy coastal plains, AP, SE (Gouritsmond to Knysna).•

parvifolia (P.J.Bergius) Lammers (= *Lightfootia parvifolia* (P.J.Bergius) Adamson)   Sprawling tangled shrublet to 45 cm with thin wiry branches. Leaves sparse, reflexed, ovate-oblong, margins thickened, sometimes slightly toothed. Flowers in lax terminal cymes, white, c. 5 mm diam., tube c. 1 mm long, ovary half-inferior, glabrous. Nov.–Mar. Sheltered rocks and streambanks, SW, KM, LB (Cape Peninsula to Langeberg and Swartberg Mts).•

polyantha Lammers (= *Lightfootia multiflora* Adamson)   Sprawling shrublet to 30 cm. Leaves often tufted, spreading or reflexed, linear, margins thickened, minutely toothed. Flowers in racemes or narrow panicles, blue, c. 5 mm diam., tube obsolete, ovary half-inferior, papillate. Dec.–Jan. Coastal sands, SW, AP, SE (Kleinmond to Knysna).•

pyrophila Lammers (= *Lightfootia tenuis* Adamson)   Like **W. capillaris** but ovary hairy. Dec.–Apr. Streamsides and marshes, SW (Cape Peninsula).•

riversdalensis Lammers (= *Lightfootia planifolia* Adamson)   Sprawling perennial to 40 cm. Leaves spreading or reflexed, ovate-oblong, margins thickened. Flowers in delicate panicles, white with darker reverse, c. 5 mm diam. Jan. Sheltered sandstone slopes, LB (Langeberg Mts: Garcia's Pass).•

subulata (L'Hér.) Lammers (= *Lightfootia subulata* L'Her.)   Erect or sprawling shrublet to 30 cm. Leaves opposite or alternate, often tufted, spreading, linear, rigid, slightly toothed. Flowers in upper axils, white to blue fading yellow, c. 6 mm diam., tube to obsolete or to 1 mm long, sepals bulbous below, ovary half-inferior, hairy, sometimes glabrous. Oct.–Feb. Stony or gravel lower slopes, NW, SW, AP (Gifberg to Potberg).•

thunbergiana (H.Buek) Lammers (= *Lightfootia thunbergiana* H.Buek)   Erect or sprawling shrublet, 15–60 cm. Leaves reflexed, lanceolate, margins thickened, sometimes minutely toothed below. Flowers in divaricate panicles becoming spinescent, white or cream-coloured with darker reverse, 6–8 mm diam., ovary half-inferior, hairy. Aug.–Nov. Stony slopes and flats in karroid scrub, NW, KM (Namaqualand and Karoo to Little Karoo).

A. Petals ovate, less than twice as long as broad; corolla bell- or bowl-shaped
D. Leaves broad, oblong to ovate
E. Leaves in a basal tuft or opposite on leafy stems

androsacea A.DC. (= *Wahlenbergia arenaria* A.DC., *W. glandulosa* Brehmer)   HARE-BELL   Tufted annual to 40 cm. Leaves mostly basal, oblanceolate, roughly hairy, margins undulate-crenate. Flowers in cymose panicles, cup-shaped, white to pale blue, 5–15 mm diam., filaments broad at base, style with 3 glands, sometimes in a continuous band below stigma, ovary glabrous. Sept.–Nov. Sandy flats, NW, SW, AP, SE (S Namibia to E Cape and tropical Africa).

annularis A.DC.   PRONKBLOUKLOKKIE   Like **W. androsacea** but flowers bowl-shaped, 15–20 mm diam., style less than half as long as corolla. Mostly Sept.–Nov. Sandy flats and lower slopes, NW (Namaqualand to Clanwilliam).

procumbens (Thunb.) A.DC. (incl. **W. saxifragoides** Brehmer)   Trailing, shortly hairy perennial forming mats and rooting at nodes. Leaves opposite, elliptic, slightly crenulate. Flowers solitary in axils, white, blue or mauve, 8–10 mm diam., ovary hairy, sometimes glabrous. Nov.–Apr. Damp sheltered slopes, NW, SW, LB, SE (Cold Bokkeveld and Cape Peninsula to E Cape).

E. Leaves alternate and not in a basal tuft

capensis (L.) A.DC.   Pubescent annual to 50 cm. Leaves oblanceolate-elliptic, sometimes clustered towards base, undulate-toothed. Flowers on elongate peduncles, bowl-shaped, blue with darker centre, ovary 5-locular, densely hairy. Sept.–Dec. Sandstone slopes and flats, NW, SW, AP, LB, SE (Clanwilliam to Knysna).•

cernua (Thunb.) A.DC. (= *Wahlenbergia ciliolata* A.DC.; incl. **W. clavatula** Brehmer, **W. maculata** Brehmer)   Roughly hairy annual to 60 cm. Leaves elliptic-obovate to ovate, incised-toothed. Flowers mostly solitary, or few on elongate peduncles, bowl-shaped, blue with dark centre or whitish, 8–15 mm diam., ovary glabrous, stigmas large and rounded. Nov.–Dec. Damp sandstone slopes, NW, SW (Witsenberg to Kleinrivier Mts).•

meyeri A.DC.   Delicate tufted annual to 20 cm, branching from base. Leaves oblanceolate, roughly hairy, undulate-crisped. Flowers in dichotomously branched paniculate cymes, bell-shaped, 5–8 mm diam., white or pale blue, style very short, ovary 3(2)-locular, glabrous. Capsules globose. Sept.–Dec. Sandstone slopes, NW (Namaqualand to Cedarberg Mts).

obovata Brehmer   Roughly hairy annual to 50 cm. Leaves ovate to elliptic, crenulate-toothed. Flower few on elongate peduncles, cup-shaped, white or blue, 12–18 mm diam., ovary glabrous, style with 3 (sometimes 6) glands below stigmas. Sept.–Feb. Moist sandstone slopes, NW, SW (Witsenberg to Kleinrivier Mts).•

undulata (L.f.) A.DC. (= *Wahlenbergia polychotoma* Brehmer)   Subglabrous perennial to 90 cm. Leaves narrowly elliptic to lanceolate, undulate-toothed, margins slightly revolute. Flowers 1–few on slender pedicels, bowl-shaped, 10–20 mm diam., blue, sometimes white or mauve, filaments broad at base, ovary 2- or 3-locular, glabrous. Oct.–Dec. Sandy slopes and flats, SE (George to tropical Africa, Madagascar).

D. Leaves narrow, linear to narrowly lanceolate
F. Ovary (4)5-locular; stigmas (4)5

decipiens A.DC. (incl. **W. longisepala** Brehmer)   Intricately branched shrublet to 15 cm. Leaves opposite below, linear-acicular, margins slightly thickened, toothed. Flowers in rigid, divaricate cymes, narrowly bell-shaped, c. 5 mm diam., white or blue, calyx lobes spreading-recurved, pungent, ovary 5-locular, scabrid. Capsules depressed, 5-lobed. Sept.–Oct. Rocky pavement and crevices near streams, NW (Gifberg to Citrusdal).•

ecklonii H.Buek (= *Wahlenbergia macra* Schltr. & Brehmer, *W. swellendamensis* H.Buek)   BOSBLOUKLOKKIE   Erect or sprawling perennial to 30 cm. Leaves opposite below, linear-lanceolate, slightly hairy, margins slightly thickened, toothed. Flowers in lax cymose panicles, narrowly bell-shaped, blue, 8–10 mm diam., calyx lobes elongate, ovary 5-locular, papillose to scabrid. Capsules hemispherical. Mainly Oct.–Dec. Rocky slopes, NW, SW, KM, LB, SE (Namaqualand to Uitenhage).

pilosa H.Buek   Roughly hairy annual to 15 cm, branching from base. Leaves lanceolate, recurved, margins thickened and slightly toothed. Flowers in paniculate cymes, narrowly bell-shaped, blue, c. 8 mm diam., ovary 5-locular, scabrid. Sept.–Oct. Sandy flats and slopes, NW (Olifants River valley).•

sphaerica Brehmer   Subglabrous annual to 30 cm, branching above. Leaves linear, margins thickened, slightly toothed. Flowers in paniculate cymes, narrowly bell-shaped, c. 10 mm diam., blue, calyx lobes elongate, ovary 5-locular, scabrid. Capsules subglobose. Oct.–Dec. Sandstone slopes, NW (Gifberg to Olifants River Mts).•

sp. 2   Minutely hairy rounded annual to 10 cm, branching from base and above. Leaves narrowly lanceolate, margins thickened and slightly toothed. Flowers in dense cymes, shortly pedicellate, large, narrowly bell-shaped, blue, 10–15 mm diam., calyx lobes long, ovary 5-locular, velvety. Oct. Sandy slopes, NW (Biedouw valley).•

sp. 3   Subglabrous annual to 8 cm, shortly dichotomously branching above. Leaves opposite, linear, margins thickened and minutely toothed, coarsely ciliate below. Flowers solitary at branch tips, sessile, bell-shaped, blue, c. 10 mm diam., ovary 5-locular, scabrid. Oct. Stony slopes, NW (Botterkloof to Pakhuis Pass).•

F. Ovary 2- or 3-locular
G. Shrublets with leaf margins revolute

capillacea (Thunb.) A.DC.   Subshrub to 50 cm, with wand-like branches. Leaves linear to filiform, often in axillary tufts, margins revolute. Flowers in lax panicles, narrowly bell-shaped, 5–10 mm diam., blue or purple, ovary glabrous. Jan.–July. Grassy slopes, SE (Langkloof and Knysna to tropical Africa).

fruticosa Brehmer   Slender shrublet to 50 cm. Leaves ascending, linear, margins revolute, slightly toothed. Flowers shortly pedicellate in upper axils, bell-shaped, 5–10 mm diam., white or pale mauve, ovary glabrous, style with 3 small glands at base of stigmas. Feb.–May. Rocky N-facing slopes (Langeberg Mts: Garcia's Pass).•

sp. 4   Sprawling shrublet, stems to 30 cm. Leaves with axillary tufts, linear-acicular, margins strongly revolute, lower margins ciliate-toothed. Flowers subsessile in upper axils, bell-shaped, blue, c. 8 mm diam., ovary glabrous. Capsules hemispherical. Apr.–June. Sandstone slopes, SE (Plettenberg Bay).•

G. Perennials or annuals with leaves plane or margins involute

brachycarpa Schltr.   Intricately branched, shortly hairy annual to 15 cm. Leaves scale-like, lanceolate, margins thickened, slightly toothed. Flowers in divaricately branched, wiry cymes, blue, bell-shaped, 5–8 mm diam., sepals stiffly spreading in fruit, ovary subglabrous. Capsules depressed. Sept.–Nov. Stony flats and slopes, NW (Pakhuis and Cedarberg Mts).•

?constricta Brehmer   Minutely hairy perennial to 20 cm, with slender stems arising from base. Leaves filiform-involute, slightly toothed. Flowers in sparse cymes, bell-shaped, blue with a pale cup and dark rings in throat and in tube, c. 8 mm diam., ovary velvety. Capsule globose. Sept.–Oct. Sandy flats, NW (Olifants River valley and Graafwater to Redelinghuys).•

exilis A.DC.   Delicate, much-branched annual to 20 cm. Leaves linear, scabrid, margins thickened, slightly toothed. Flowers in panicles on wiry pedicels, minute, c. 2 mm diam., white, ovary glabrous or scabrid. Capsules cylindrical. Oct.–Jan. Sandy and gravel slopes, NW, SW, LB, SE (Cape Peninsula and Ceres to Knysna).•

?massonii A.DC.   Slender, roughly hairy annual to 20 cm, branching from base. Leaves small, lanceolate, margins thickened. Flowers in rounded cymes, narrowly funnel-shaped and tubular below, hairy, white or blue with purple centre, 5–8 mm diam., ovary densely hairy, capsule with beak longer than wide. Sept.–Nov. Sandy and stony flats, NW (Klawer to Het Kruis).•

paniculata (Thunb.) A.DC. (incl. **W. hispidula** (Thunb.) A.DC.)   Roughly hairy annual to 15 cm, branching mainly from base. Leaves small, lanceolate, margins thickened, minutely toothed. Flowers in lax, corymbose cymes, bell-shaped, white or blue with white cup and dark ring in throat, 6–8 mm diam., ovary sparsely hairy or velvety in bands. Capsules obovoid to hemispherical Sept.–Nov. Sandy flats, mainly coastal, NW, SW, AP (Olifants River valley to Albertinia) .•

?polyclada A.DC.   Roughly hairy, rounded annual to 15 cm, brancing from base. Leaves lanceolate, more or less cordate at base, margins thickened, slightly toothed. Flower in lax cymes, narrowly bell-shaped, blue with a dark ring in throat and a dark centre, 10–12 mm diam., calyx lobes large, toothed, ovary hispid. Capsules hemispherical. Sept.–Oct. Coastal sands, NW (S Namaqualand to Klawer).

ramulosa E.Mey. (?incl. **W. debilis** H.Buek, **W. lobata** Brehmer)   Minutely hairy, delicate annual to 10 cm, branching above. Leaves linear-lanceolate, margins slightly thickened, toothed. Flowers in wiry cymes, bell-shaped, pale blue, c. 2 mm diam., ovary glabrous. Capsules hemispherical, with a short beak. Sept.–Nov. Stony flats and slopes, NW, SW, KM (Darling to Witteberg and Riviersonderend Mts).

sp. 5   Sprawling perennial to 15 cm. Leaves linear-lanceolate, margins thickened, minutely toothed. Flower 1–few at branch tips, subsessile, narrowly bell-shaped, c. 5 mm diam., white, calyx lobes leaf-like, ovary glabrous. Feb.–Mar. Rocky slopes at high alt., SE (Great Winterhoek Mts).•

[Species excluded   Incompletely known and possibly conspecific with one of the above: **W. annuliformis** Brehmer, **W. bolusiana** Schltr. & Brehmer, **W. compacta** Brehmer, **W. dichotoma** A.DC., **W. distincta** Brehmer, **W. divergens** A.DC., **W. dunantii** A.DC., **W. gracilis** E.Mey., **W. mollis** Brehmer, **W. oligotricha** Schltr. & Brehmer, **W. pseudoandrosacea** Brehmer, **W. ramifera** Brehmer, **W. schistacea** Brehmer, **W. serpentina** Brehmer, **W. sessiliflora** Brehmer, **W. sparsiflorum** A.DC., **W. subpilosa** Brehmer, **W. subtilis** Brehmer. Not from the area: **W. costata** A.DC.]

WIMMERELLA (= *LAURENTIA*)   c. 6 spp., South Africa

A. Plants sprawling, leaves subobicular, scattered along stems

frontidentata L.Serra, M.B.Crespo & Lammers (= *Laurentia frontidentata* E.Wimm.)   Sprawling to prostrate perennial, stems to 20 cm long. Leaves suborbicular, toothed at tips. Flowers axillary on short, slender pedicels, minute, white. Jan.–May. Shady rocks, 1000–1600 m, NW, KM (Hex River Mts and Anysberg).•

pygmaea (Thunb.) L.Serra, M.B.Crespo & Lammers (= *Laurentia pygmaea* (Thunb.) Sond.; incl. **W. hederacea** (Sond.) L.Serra, M.B.Crespo & Lammers, **W. hedyotidea** (Schltr.) L.Serra, M.B.Crespo & Lammers)   Thinly hairy, sprawling perennial to 5 cm. Leaves subrotund, broadly toothed or lobed. Flowers axillary on long, slender pedicels, blue or white. Sept.–Mar. Damp rocks and caves on sandstone slopes, NW, SW, KM, LB, SE (Gifberg to Grahamstown).

A. Plants tufted, leaves lanceolate or oblanceolate, basal

bifida (Thunb.) L.Serra, M.B.Crespo & Lammers (= *Laurentia bifida* (Thunb.) Sond.; incl. **W. arabidea** (C.Presl) L.Serra, M.B.Crespo & Lammers, **W. giftbergensis** (E.Phillips) L.Serra, M.B.Crespo & Lammers)   Tufted perennial to 30 cm, minutely scabrid below. Leaves mostly basal, oblanceolate-spathulate, slightly toothed. Flowers in ascending racemes in axils of linear, sometimes oblanceolate and toothed bracts, white, pink or blue with white markings. Sept.–Apr. Marshes, streambanks and seeps, NW, SW, KM, LB, SE (Bokkeveld Mts to S KwaZulu-Natal).

longitubus (E.Wimm.) L.Serra, M.B.Crespo & Lammers (= *Laurentia longituba* E.Wimm.)   Like **W. bifida** but flowers with a longer tube, c. 18 mm long (vs shorter than 10 mm). Nov.–Dec. Sandstone slopes, LB (Langeberg Mts: Riversdale).•

mariae L.Serra, M.B.Crespo & Lammers (= *Laurentia mariae* E.Wimm., *L. minuta* auct.)   Minute, acaulescent perennial to 2 cm, sometimes with weak trailing stems. Leaves spathulate. Flowers in leaf axils, blue to lilac. Dec.–Jan. Shale bands at high alt., NW (Hex River Mts).•

secunda (L.f.) L.Serra, M.B.Crespo & Lammers (= *Laurentia secunda* (L.f.) Kuntze)   Like **W. bifida** but racemes sprawling to trailing and flowers secund, in axils of leafy, slightly toothed bracts. Nov.–Apr. Near water, mainly at low elevations, NW, SW, AP (Cold Bokkeveld to Agulhas).•

CARYOPHYLLACEAE *(= ILLECEBRACEAE)*

1. Ovary with 1 or 2 ovules; fruit an indehiscent nutlet:
    2. Stigma trifid; leaves mostly alternate ....**Corrigiola**
    2.' Stigma at most bifid; leaves mostly opposite or verticillate:
        3. Stipules absent; perianth becoming crustaceous with age, pungent.....................................**Scleranthus**
        3.' Stipules present; perianth herbaceous:
            4. Perianth subsucculent, urceolate, mouth closed by a thickened, lobed disc; hairy, branched shrublet with verticillate leaves...**Pollichia**
            4.' Perianth not as above; sprawling annual or perennial herbs:
                5. Leaves aristate; sepals dorsally awned ....**Paronychia**
                5.' Leaves acute; sepals not awned ....**Herniaria**
1.' Ovary with 3 or more ovules; fruit a dehiscent capsule:
    6. Sepals united into a short or long tube:
        7. Calyx tube with commissural veins alternating with midveins of sepals; styles 3(5); leaves various, usually pubescent; petals usually with coronal scales ....**Silene**
        7.' Calyx tube without commissural veins; styles 2; leaves linear, glabrous; petals without coronal scales:
            8. Calyx tube with scarious or papery commissures; flowers in terminal heads subtended by shining, papery bracts ....**Petrorhagia**
            8.' Calyx tube green, without papery commissures; flowers not in heads subtended by papery bracts...**Dianthus**
    6.' Sepals free or very nearly so:
        9. Stipules absent; petals deeply 2-fid:
            10. Stamens 5; styles (2)3...**Stellaria**
            10.' Stamens 10; styles (4)5 ....**Cerastium**
        9.' Stipules present; petals entire or emarginate:
            11. Sepals winged up the back and sharply prominent in fruit; leaves obovate ....**Polycarpon**
            11.' Sepals not winged up the back; leaves linear:
                12. Styles 3; capsule 2- or 3-valved ....**Spergularia**
                12.' Styles (3)5; capsule (3)5-valved ....**Spergula**

CERASTIUM   MOUSE-EAR   c. 60 spp., cosmopolitan

capense Sond.   HORINGBLOM   Glandular-hairy annual to 40 cm. Leaves opposite, obovate or oblong. Flowers in open cymes, white, petals shorter than sepals, emarginate. Fruits elongating to twice as long as sepals. Sept.–Dec. Sheltered flats and slopes and waste places, NW, SW, AP, KM, LB, SE (widespread in southern Africa).

CORRIGIOLA   STRAPWORT   11 spp., cosmopolitan

capensis Willd.   Prostrate biennial or perennial to 3 cm from a woody taproot. Leaves oblanceolate, stipules 3.0–3.5 mm long, streaked brown. Flowers in compact terminal clusters, 2.0–2.5 mm long, greenish. Aug.–Sept. Sandy flats, NW (Namaqualand, Clanwilliam, Tulbagh and tropical Africa).

litoralis L.   Prostrate annual or perennial to 3 cm. Leaves oblanceolate, stipules c. 2.5 mm long, usually white. Flowers in terminal and axillary clusters, 1.25–1.75 mm long, greenish. Feb.–Sept. Weed of open ground, SW, LB, SE (European and tropical African weed, from Cape Peninsula to Port Elizabeth to Mpumalanga).

DIANTHUS   PINK   c. 300 spp., Old World

albens Aiton (incl. **D. holopetalus** Turcz.)   Loosely tufted or sprawling perennial to 40 cm. Leaves linear. Flowers several on axillary scapes, white to purple, calyx 12–18 mm long, petal limb oblanceolate, entire or toothed. Sept.–Feb. Sandy flats and slopes, often coastal, NW, SW, AP, KM, LB, SE (Cedarberg Mts to E Cape).

basuticus Burtt Davy   Densely tufted perennial to 15 cm. Leaves linear. Flowers 1–4 on axillary scapes, pink to purple, calyx 15–20 mm long, petal limbs obovate, finely toothed. Nov.–Apr. Rocky slopes, SE (Avontuur to KwaZulu-Natal).

bolusii Burtt Davy   Tufted perennial to 40 cm. Leaves linear, long. Flowers several on axillary scapes, white to purple, calyx 23–27 mm long, petal limb oblanceolate, deeply fringed. Sept.–Feb. Sandstone slopes, NW, SW, KM, LB (Cedarberg Mts to Du Toitskloof to Swartberg Mts).•

caespitosus Thunb.   Like **D. albens** but calyx 40–70 mm long and petals always toothed or fringed. Sept.–Jan. Sandstone slopes, NW, KM, SE (Botterkloof to Worcester to Uitenhage).•

thunbergii Hooper   Like **D. albens** but calyx 20–30 mm long and petals always toothed. Sept.–Feb. Sandstone slopes, AP, LB, SE (Swellendam to E Cape).

HERNIARIA   RUPTURE-WORT   47 spp., Africa, Europe and Asia

capensis Bartl.   Prostrate, mat-forming perennial with creeping runners rooting at nodes, roughly hairy with short, retrorse hairs. Leaves opposite, elliptic. Flowers in axillary clusters, greenish. Nov.–Feb. Sandy flats, usually coastal, NW, SW, LB, SE (Ceres, Cape Peninsula to George, Graaff-Reinet).

schlechteri Hermann   Prostrate, mat-forming annual with runners not rooting, softly and densely hairy with slender, erect hairs. Leaves opposite, elliptic. Flowers in axillary clusters, greenish. Aug.–Nov. Coastal limestone, AP, LB, SE (De Hoop to Port Elizabeth).•

*PARONYCHIA   50 spp., cosmopolitan

*braziliana DC.   Prostrate, mat-forming annual to 3 cm. Leaves opposite, hairy, elliptic, apiculate, stipules membranous. Flowers in axillary clusters, greenish, sepals dorsally awned. Sept.–Dec. Weed of lawns and disturbed places, NW, SW, LB, SE (South American weed).

*PETRORHAGIA   28 spp., N Temperate Old World

*prolifera (L.) P.W.Ball & Heywood   Erect or sprawling annual to 50 cm. Leaves opposite, linear-oblanceolate, sometimes roughly hairy. Flowers in terminal heads subtended by large papery bracts, pink or purple, calyx papery, c. 10 mm long, petals bifid. Sept.–Nov. Disturbed sites, NW, SW (European weed).

POLLICHIA   WAXBERRY, AARBOSSIE   1 sp., Africa and Arabia

campestris Aiton   Silky hairy shrublet to 60 cm. Leaves opposite or apparently verticillate, oblanceolate, apiculate. Flowers in axillary clusters, greenish, enclosed by fleshy, waxy white bracts in fruit. Throughout the year. Dry sandy soils, KM, LB, SE (Little Karoo through Uitenhage to Arabia).

*POLYCARPON   ALLSEED   c. 16 spp., cosmopolitan

*tetraphyllum L.f.   FOUR-LEAVED ALLSEED   Prostrate annual or perennial to 20 cm. Leaves in whorls of 4, obovate. Flowers congested in branched cymes, white, petals emarginate. Sept.–Dec. Weed of waste places, NW, SW, KM, LB, SE (cosmopolitan weed).

*SCLERANTHUS   10 spp., Old World

*annuus L.   Sprawling, thinly hairy annual to 20 cm. Leaves opposite, linear. Flowers 1–few in axillary clusters, green. Aug.–Nov. Weed of disturbed places, NW, SW, AP, KM, LB, SE (European weed, widespread throughout Africa).

SILENE   CAMPION, CATCHFLY   c. 500 spp., worldwide

A. Flowers larger, in cymes, calyx c. 20–35 mm long

bellidioides Sond. (incl. **S. ornata** Aiton)   WILD TOBACCO   Erect, shortly glandular-hairy perennial to 60 cm. Leaves oblanceolate. Flowers in cymes, white or pink to crimson, calyx 18–22 mm long, petals bifid, carpophore c. 2 mm long. Aug.–Sept. Sandy flats and slopes, NW, SW, LB, SE (Tulbagh to Cape Peninsula and Port Elizabeth to Mpumalanga).

undulata Aiton (= *Silene capensis* Otth.; incl. **S. eckloniana** Sond.)   Similar to **S. bellidioides** but sometimes sprawling. Flowers white to pink, calyx 22–35 mm long and carpophore 4–6 mm long. Aug.–Apr. Slopes and flats, NW, SW, AP, KM, LB, SE (southern and tropical Africa).

A. Flowers smaller, mostly in raceme-like monochasia, calyx to 20 mm long

crassifolia L.   Sprawling, densely hairy perennial to 50 cm. Leaves fleshy, felted beneath, obovate. Flowers in subsecund, raceme-like monochasial cymes, white to yellow, calyx c. 13 mm long, petals bifid, carpophore c. 4 mm long. Aug.–Mar. Coastal sand dunes, SW, AP (Saldanha Bay to Agulhas).•

***cretica** L. (= *Silene clandestina* Jacq., *S. dewinteri* Bocq.)   Shortly hairy annual to 45 cm. Leaves linear-oblanceolate. Flowers in raceme-like monochasial cymes, whitish, open and scented at night, calyx 9–16 mm long, petals bifid, carpophore 1–5 mm long. Aug.–Jan. Weed of sandy flats, NW, SW (European weed).

***gallica** L.   SMALL CATCHFLY   Glandular-hairy annual to 45 cm. Leaves oblanceolate, ciliate. Flowers in subsecund, spike-like monochasial cymes, white or pink, calyx coarsely hairy, 7–10 mm long, petals entire, carpophore c. 1 mm long. Sept.–Feb. Weed of disturbed places, NW, SW, AP, KM, LB, SE (European weed).

mundiana Eckl. & Zeyh.   Delicate, mat-forming, shortly hairy perennial to 10 cm. Leaves small, linear-oblanceolate. Flowers 1 or 2 in subsecund cymes, white, calyx 10–11 mm long, petals bifid, carpophore 3 mm long. Oct. Coastal and limestone flats, AP, SE (De Hoop and Plettenberg Bay).•

pilosellifolia Cham. & Schltdl. (= *Silene burchellii* Otth; incl. **S. thunbergiana** Eckl. & Zeyh.)   Erect or sprawling, shortly hairy perennial to 70 cm. Leaves linear-oblanceolate to oblanceolate. Flowers in subsecund, raceme-like monochasial cymes, white to purple, calyx 10–20 mm long, petals bifid, carpophore 6–9 mm long. Aug.–Jan. Flats and slopes, NW, SW, KM, LB, SE (throughout Africa).

primuliflora Eckl. & Zeyh. (= *Silene vlokii* D.Masson)   Sprawling, shortly hairy perennial to 50 cm. Leaves leathery, oblanceolate. Flowers in subsecund, raceme-like monochasial cymes, white, yellowish or pink, calyx 12–15 mm long, petals bifid, carpophore 4–7 mm long. Nov.–Mar. Coastal sand dunes, SE (Mossel Bay to KwaZulu-Natal).

***vulgaris** (Moench) Garcke   BLADDER CAMPION   Glaucous rhizomatous perennial to 60 cm. Leaves lanceolate or oblanceolate. Flowers in cymes, white to purple, calyx inflated, 12–15 mm long, petals deeply bifid, carpophore c. 2 mm long. Oct.–Dec. Weed of disturbed lands, SW, KM, SE (European weed).

***SPERGULA**   SPURREY   c. 5 spp., temperate parts of the world

***arvensis** L.   CORN SPURREY   Annual to 60 cm. Leaves apparently whorled, linear. Flowers in open, glandular-hairy cymes, white. June–Oct. Weed of waste places, NW, SW (cosmopolitan weed).

***SPERGULARIA**   c. 40 spp., cosmopolitan

***media** (L.) C.Presl   PERENNIAL SEA SPURREY   Glabrescent, sprawling perennial from a thick, woody rootstock, to 50 cm. Leaves opposite and in axillary tufts, linear. Flowers in open, glandular-hairy cymes, white or pink, sepals usually longer than 4 mm. Seeds winged. Oct.–Jan. Weed of coastal or inland marshes, NW, SW, AP, LB, SE (cosmopolitan weed).

***pallida** (Dumort.) Piré   LESSER CHICKWEED   Like **S. media** but sepals 2.0–3.5 mm long, petals absent or vestigial and stamens 1–3. July–Aug. Weed of waste places, SW, LB, SE (cosmopolitan weed).

***rubra** (L.) J. & C.Presl   SAND SPURREY   Sprawling annual or perennial from slender taproot, to 20 cm. Leaves opposite, linear. Flowers in open cymes, pink or lilac, sepals shorter than 4 mm. Seeds not winged. Sept.–Dec. Weed of sandy places, NW, SW (cosmopolitan weed).

***STELLARIA**   CHICKWEED   c. 100 spp., cosmopolitan

***media** (L.) Vill.   STARWORT   Diffuse annual to 30 cm with a line of hairs down each internode. Leaves opposite, ovate. Flowers in terminal cymes, white, sepals 4.5–5.0 mm long, petals bilobed, stamens mostly 5–10. June–Nov. Weed of waste places, NW, SW, AP, KM, LB, SE (cosmopolitan weed).

CELASTRACEAE
by R.H. Archer & M. Jordaan

1. Fruit a drupe or berry, indehiscent; seeds without an aril:
   2. Fruit a drupe (1-seeded); leaves alternate or opposite:
      3. Leaves opposite; endocarp thick and woody (stone) ....**Elaeodendron**
      3.' Leaves alternate; endocarp thin and crustaceous:
         4. Leaves glabrous, elastic threads present in leaves and bark (conspicuous) when pieces of leaves are pulled apart ....**Robsonodendron**
         4.' Leaves pubescent; elastic threads absent ....**Mystroxylon**
   2.' Fruit a berry (several-seeded); leaves opposite or subopposite, rarely alternate:
      5. Leaves leathery with margin revolute; ovules pendulous...**Maurocenia**
      5.' Leaves variously textured with margin plane or occasionally revolute; ovules erect:

6. Shrubs or trees; inflorescence a cyme; mature fruit pale to dark brown or purplish, succulent or fleshy ....**Cassine**
6.' Lianes, scrambling shrubs or small trees; inflorescence thyrsoid with simple or compound cymules arranged along axis; mature fruit red or purple, fleshy ....**Lauridia**
1.' Fruit a capsule or follicle-like, dehiscent when dry and mature; seeds with or without an aril:
7. Plants without spines; leaves alternate or scattered, never clustered in groups:
8. Ericoid shrublet; leaves scattered, needle-like; capsule follicle-like, dehiscing entirely along one suture ....**Empleuridium**
8.' Trees or shrubs; leaves alternate, not needle-like; capsule dehiscing along 2 or 3 sutures:
9. Capsules with horns or wing-like emergences ....**Pterocelastrus**
9.' Capsules smooth or verruculose, without horns or wing-like emergences.........................**Maytenus**
7.' Plants with spines; leaves clustered in groups on short-shoots, at least on older branches:
10. Suffrutices, shrubs or small trees; flowers always functionally unisexual with staminodes in female and pistillodes in male flowers; ovules 2 per locule; ovary (2)3(4)-locular; seeds covered completely or incompletely by an aril ....**Gymnosporia**
10.' Shrubs or woody climbers; flowers always bisexual; ovules 3–12 per locule; ovary always 3-locular; seeds covered completely by an aril:
11. Shrubs with spines always leafy and floriferous, with more than 1 node per spine; leaves with venation obsolete, margin entire; ovules usually fewer than 6 per locule ....**Gloveria**
11.' Shrubs or woody climbers with occasionally leafy and floriferous spines, when leafy not more than 1 node per spine; leaves with venation conspicuous, margin with a few teeth in upper half, occasionally subentire; ovules usually 6 or more per locule..**Putterlickia**

CASSINE (= *HARTOGIELLA*)   CAPE SAFFRON   3 spp., western and eastern South Africa

parvifolia Sond.   MOUNTAIN SAFFRON   Shrub or small tree. Leaves opposite or subopposite, narrowly elliptic, venation inconspicuous beneath, margins revolute. Flowers cream-coloured, sepals red. Berries brown to purplish. Sept.–Oct. Fynbos or forest margins, NW, SE (Porterville Mountains, Ceres and George to Clarkson).•

peragua L.   CAPE SAFFRON, IKHUKHUZI   Shrublet or tree to 15 m, bark often with yellow pigments. Leaves opposite, elliptic to orbicular, venation conspicuous beneath. Flowers cream-coloured. Berries brown to purple. Feb.–July. Coastal scrub, woodland or forest margins, NW, SW, LB, SE (Bokkeveld Mountains to Cape Peninsula to Mpumalanga).

schinoides (Spreng.) R.H.Archer (= *Hartogiella schinoides* Spreng.)   Shrub or small tree, bark occasionally with yellow pigments. Leaves opposite, narrowly elliptic to obovate, venation inconspicuous beneath. Flowers cream-coloured. Berries brown, hard. Oct.–Dec. Fynbos, woodland and forest, NW, SW, LB, SE (Cedarberg Mts to Cape Peninsula to Clarkson).•

ELAEODENDRON   SAFFRON   c. 30 spp., cosmopolitan

croceum (Thunb.) DC. (= *Cassine crocea* (Thunb.) Kuntze)   SAFFRON   Tall tree, often with yellow pigments on exposed bark. Leaves opposite, elliptic, bright green, finely toothed with prominent spines on young shoots. Flowers few, cream-coloured, ovary 2-locular. Drupe ellipsoid, cream-coloured. Aug.–Mar. Coastal and montane forest margins, LB, SE (Ladismith to E Zimbabwe).

zeyeri Spreng. ex Turcz.   SMALL-LEAVED SAFFRON, RED SAFFRONWOOD   Tree, often with yellow pigments on exposed bark. Leaves opposite or subopposite, elliptic, greyish green. Flowers greenish, ovary 4-locular. Drupe spheroidal, yellow. Oct.–Apr. Woodland. SE (Baviaanskloof Mts to Mpumalanga).

EMPLEURIDIUM•   1 sp., W Cape

juniperinum Sond.   Finely leafy shrublet. Leaves ascending, needle-like. Flowers solitary in axils, whitish. Capsule with 1 fusiform seed with a long white aril. Dec.–Apr. Sandstone slopes, SW (Wemmershoek Mts to Houwhoek).•

GLOVERIA   SPALKPENDORING   1 sp., N and W Cape

integrifolia (L.f.) M.Jordaan   Spiny shrub to 2 m. Leaves fasciculate on short shoots, cuneate, greyish, leathery. Flowers few in axillary cymes, white, sometimes tinged pink. Capsules yellow, 3-locular, seed aril pinkish. Dec.–Apr. Rocky slopes, SW, KM, LB (Namaqualand and MacGregor to Oudtshoorn).

GYMNOSPORIA (= *MAYTENUS* in part)   PENDORING   c. 40 spp., mainly Africa, also Madagascar, S Europe to Australasia

buxifolia (L.) Szyszyl. (= *Maytenus heterophylla* auct.)   STINKPENDORING, MNQUQOBA   Monoecious, spiny shrub or small tree to 7 m. Leaves in tufts, obovate, toothed above. Flowers many in axillary cymes, white, foetid. Capsules globose, warty, brown. July–Apr. Forest margins and disturbed areas, NW, SW, AP, KM, LB, SE (widespread in southern and tropical Africa).

capitata (E.Mey. ex Sond.) Loes. (= *Maytenus capitata* (E.Mey. ex Sond.) Marais)   VAALPENDORING   Monoecious, rigid spiny shrub to 1.5 m. Leaves in tufts, obovate, discolorous, leathery, sessile. Flowers in axillary glomerules, cream-coloured. Capsules yellow to bright orange-red, 3-lobed. Oct.–June. Dry scrub, AP, SE (Stilbaai to Port Alfred).

elliptica (Thunb.) Schönland   SPIKKELVRUG-DWERGPENDORING   Monoecious, spiny rhizomatous shrublet to 80 cm. Leaves in tufts, subsessile, linear-obovate, sparsely toothed above, leathery, greyish green. Flowers few in axillary cymes, whitish. Capsules globose, warty, red and brown. Mar.–May. Coastal plain, SE (Humansdorp to Port Elizabeth).•

nemorosa (Eckl. & Zeyh.) Szyszyl. (= *Maytenus nemorosa* (Eckl. & Zeyh.) Marais)   WITBOSPENDORING   Monoecious, spiny shrub or small tree to 5 m. Leaves in tufts, obovate to elliptic, toothed, dull green above, paler beneath, venation distinct. Flowers in lax axillary cymes, white. Capsules pear-shaped, brown. Dec.–June. Coastal forest margins, SE (Knysna to Mpumalanga).

szyszylowiczii (Kuntze) M.Jordaan   LEMOENDORING   Monoecious, heavily spined shrub to 2.5 m. Leaves in tufts, subsessile, obovate, leathery, glaucous. Flowers in somewhat lax axillary cymes, white, honey-scented. Capsules globose, pale yellow. Mar.–Aug. Stony slopes, KM (Oudtshoorn to Great Karoo).

LAURIDIA   2 spp., eastern southern Africa

reticulata Eckl. & Zeyh.   Shrub to small tree. Leaves opposite, elliptic. Flowers in axillary cymes, cream-coloured. Berries green, red, purple to black on same branch. Sporadic throughout the year. Scrub, SE (Humansdorp to E Cape).

tetragona (L.f.) R.H.Archer (= *Cassine tetragona* (L.f.) Loes.)   CLIMBING SAFFRON   Scrambling shrub or liana with branchlets often bent back, conspicuously 4-lined. Leaves opposite, subsessile, elliptic-ovate, toothed. Flowers in axillary cymes, cream-coloured. Berries red to purple. Sept.–Jan. Scrub, SW, AP, SE (Hermanus to N Province).

MAUROCENIA•   HOTTENTOTS CHERRY   1 sp., W Cape

frangula Mill. (= *Maurocena frangularia* (L.) Mill., sphalm.)   Often monoecious shrub to spreading tree. Leaves opposite, broadly elliptic to orbicular, leathery and rigid, margins revolute. Flowers in axillary cymes, cream-coloured, honey-scented. Berries red. Apr.–June. Coastal bush or forest, SW (Saldanha Bay to Cape Peninsula).•

MAYTENUS   c. 150 spp., cosmopolitan, mainly tropical

acuminata (L.f.) Loes.   SYBAS, UMNAMA   Shrub or small tree to 10 m. Leaves alternate, dark green and glossy above, paler beneath, ovate to lanceolate, toothed, with silky threads when broken. Flowers in axillary fascicles, white. Capsules yellow. May–Jan. Forest margins or rocky slopes, NW, SW, KM, LB, SE (Nieuwoudtville to Cape Peninsula to tropical Africa).

lucida (L.) Loes.   CAPE MAYTENUS   Much-branched shrub to 1.5 m. Leaves ovate to suborbicular, leathery, margins revolute. Flowers in axillary fascicles, greenish. Capsules light brown. June–Jan. Coastal bush, SW, AP (Saldanha Bay to Agulhas).•

oleoides (Lam.) Loes.   KLIPKERSHOUT   Shrub or tree to 4 m. Leaves leathery, obovate to lanceolate, venation obsolete, glaucescent, margins revolute. Flowers in axillary cymes, whitish. Capsules brown to orange. Apr.–Sept. Rocky slopes, NW, SW, KM, LB, SE (Richtersveld to Cape Peninsula to Great Winterhoek Mts).

peduncularis (Sond.) Loes.   KAAPSE SWARTHOUT, UMNQAYI   Tree to 8 m, hairy on young parts. Leaves ovate to lanceolate, toothed. Flowers in axillary cymes, yellowish, hairy. Capsules pendulous, green to brown. Sept.–Nov. Forest streams, SE (George to N Province).

procumbens (L.f.) Loes.   DUINEKOKOBOOM, UMPHONO-PHONO   Low scrambling shrub or small tree to 6 m. Leaves elliptic to obovate, leathery, margins revolute, coarsely toothed. Flowers in axillary sessile fascicles, greenish. Capsules bright yellow to orange. Aug.–Jan. Coastal dune forest, AP, SE (De Hoop to tropical Africa).

undata (Thunb.) Blakelock   KOKOBOOM, UMKOKANE   Shrub or tree to 10 m. Leaves ovate, toothed. Flowers in axillary fascicles, yellowish. Capsules yellow. Nov.–Mar. Coastal bush, KM, SE (Gamka and Van Staden's Mts to tropical Africa).

MYSTROXYLON (= *CASSINE* in part)   CAPE CHERRY, KUBUSBESSIE   1 sp., eastern Africa

aethiopicum (Thunb.) Loes. (= *Cassine aethiopica* Thunb.)   Scrambling shrub or tree. Leaves elliptic to circular, often densely pubescent. Flowers in axillary glomerules, yellow. Drupe light brown. Sept.–Apr. Forest margins or scrub, AP, SE (Heidelberg to tropical Africa).

PTEROCELASTRUS   CHERRYWOOD, KERSHOUT   3 spp., south-eastern Africa

rostratus (Thunb.) Walp.   RED CHERRYWOOD, ROOIKERSHOUT   Tree to 10 m. Leaves leathery, oblong-lanceolate, shining above, paler beneath, midrib raised above, petiolate. Flowers in axillary cymes, yellowish. Capsules brown with yellow decurrent horns. Apr.–June. Forest, SW, KM, LB, SE (Betty's Bay to Mpumalanga).

tricuspidatus (Lam.) Sond.   UTWINA   Shrub or small tree to 7 m. Leaves leathery, obovate, slightly emarginate, cuneate at base, margins revolute. Flowers in dense, axillary cymes, whitish. Capsules orange with 3 entire or toothed wings. May–Nov. Dune scrub or forest, NW, SW, AP, SE (Velddrif to Cape Peninsula to Port Edward).

PUTTERLICKIA   BASTERPENDORING   4 spp., southern Africa

pyracantha (L.) Szyszyl.   Rigid, straggling, spiny shrub to 3 m. Leaves fasciculate, subsessile, leathery, obovate, margins slightly revolute, sometimes toothed. Flowers in axillary cymes, white. Capsules red, seed aril orange. Feb.–Aug. Riverbanks or coastal scrub, SW, AP, SE (Velddrif to E Cape).

ROBSONODENDRON (= *CASSINE* in part)   SYBAS   2 spp., eastern southern Africa

eucleiforme (Eckl. & Zeyh.) R.H.Archer (= *Cassine eucleiformis* (Eckl. & Zeyh.) Kuntze)   WIT-SYBAS   Tree to 12 m. Leaves elliptic-oblanceolate, obtuse, margins at most slightly revolute. Flowers cream-coloured. Drupe light brown. Sporadic throughout the year. Forest margins, SW, LB, SE (Cape Peninsula, Grootvadersbos to Mpumalanga).

maritimum (Bolus) R.H.Archer (= *Cassine maritima* (Bolus) L.Bolus)   DUINE-SYBAS   Shrub to 2 m, rough on young parts. Leaves thick and leathery, margins conspicuously revolute, acute. Flowers cream-coloured. Drupe white. June–Oct. Coastal scrub, SW, AP, SE (Cape Peninsula to E Cape).

CELTIDACEAE *(= ULMACEAE in part)*

CELTIS   WHITE STINKWOOD, WITSTINKHOUT   c. 80 spp., widespread

africana Burm.f.   Monoecious, deciduous tree to 30 m, with smooth, grey bark, tawny velvety when young. Leaves obliquely ovate-lanceolate, toothed. Flowers 1–few in axillary cymes, greenish. Jan.–Apr. Forest, SW, LB, SE (Cape Peninsula to tropical Africa).

CERATOPHYLLACEAE

CERATOPHYLLUM   HORNWORT   c. 10 spp., cosmopolitan

demersum L.   Monoecious submerged aquatic perennial to 3 m. Leaves in whorls of 7–11, twice-forked, margins with small spines. Flowers 1–few in axils, green. Flowering irregularly. Estuaries, SE (George northwards and almost cosmopolitan).

CLUSIACEAE

HYPERICUM ST JOHN'S WORT c. 400 spp., widespread, especially N hemisphere

aethiopicum Thunb. Perennial to 30 cm with terete or flattened stems from a woody rootstock. Leaves spreading, ovate, margins often revolute. Flowers 1–many in a cyme, gland-dotted, yellow. Sept.–Dec. Damp clay flats and slopes, KM, LB, SE (Riversdale to Zimbabwe).

lalandii Choisy Perennial to 40 cm with 4-angled stems from a woody rootstock. Leaves ascending, linear to elliptic. Flowers 1–many in a loose cyme, yellow. Nov.–Mar. Damp clay flats and slopes, SW, LB, SE (Somerset West to N Africa).

CONVOLVULACEAE

1. Leaves minute, scale-like or absent; yellowish, twining parasites; flowers small, usually in clusters, often numerous .... **Cuscuta**
1.' Leaves well developed, green:
   2. Ovary 2- or 4-lobed or -cleft; styles 2, inserted between lobes of ovary; prostrate perennials:
      3. Ovary 2-cleft, with 2 ovules in each locule .... **Dichondra**
      3.' Ovary 4-cleft, with 1 ovule in each locule .... **Falkia**
   2.' Ovary not deeply lobed; style simple or terminal if 2; erect or twining annuals or perennials:
      4. Pollen spinose; stigma biglobose or 3-lobed .... **Ipomoea**
      4.' Pollen smooth; stigmas 2, linear or subclavate .... **Convolvulus**

CONVOLVULUS BINDWEED c. 250 spp., cosmopolitan

*****arvensis** L. LESSER BINDWEED Thinly hairy prostrate or climbing perennial with annual stems to 2 m. Leaves hastate. Flowers pink, sepals obtuse, 3.5–5.0 mm long, corolla 15–25 mm long. Dec.–Feb. Weed of disturbed places, SW, KM (European weed).

bidentatus Bernh. apud C.Krauss Like **C. sagittatus** but basal leaf lobes often bifid and sepals large, papery, obtuse, with membranous margins, 6–8 mm long, corolla 15–22 mm long. Oct.–Dec. Coastal scrub, AP, LB, SE (Bredasdorp to E Cape).

capensis Burm.f. CAPE BINDWEED Thinly hairy perennial climber to 2 m. Leaves hastate to palmatissect, often toothed. Flowers white to pink, sepals 6–10 mm long, usually silky and obtuse, corolla 15–35 mm long. Sept.–Oct. Stony slopes, NW, SW, AP, KM, LB, SE (Namaqualand and W Karoo to E Cape).

farinosus L. Thinly hairy, often silvery perennial climber to 3 m. Leaves deltoid or sagittate, often toothed. Flowers white to mauve, sepals 6–8 mm long, acute, corolla 12–15 mm long. Nov.–June. Stony slopes, NW, SW, LB, SE (Ceres to Cape Peninsula to Port Elizabeth and throughout Africa to the Mediterranean).

multifidus Thunb. Densely silky trailing perennial to 75 cm. Leaves palmatisect with linear segments. Flowers white or pink, sepals 6–7 mm long, apiculate, corolla 10–12 mm long. Nov.–Mar. Stony flats, SE (Uitenhage and Karoo to Free State).

sagittatus Thunb. Glabrescent perennial climber to 2 m. Leaves narrowly hastate. Flowers white or pink, sepals 5–6 mm long, acute, corolla c. 10 mm long. Nov.–Apr. Stony flats and slopes, NW (Namaqualand and W Karoo to Worcester and throughout Africa).

ulosepalus Hallier f. Like **C. sagittatus** but sepals very unequal, the inner round and apiculate, 5–6 mm long, corolla 7–9 mm long. Dec.–May. Stony flats, NW, SW, KM, LB, SE (widespread throughout southern Africa).

CUSCUTA DODDER c. 140 spp., cosmopolitan, mainly America

africana Willd. Leafless parasitic annual vine, stems medium. Flowers in cymes, cream-coloured, calyx shorter than corolla tube with obtuse lobes, stigmas oblong. Oct.–Dec. Stem parasite on shrubs, SW, LB, SE (Riviersonderend to Port Elizabeth).•

angulata Engelm. Leafless parasitic annual vine, stems thread-like. Flowers in cymes, white, calyx as long as corolla tube, angled and sharply protruding at sinuses, stigmas oblong. July–Dec. Stem parasite on ericoid shrubs, SW, AP (Du Toitskloof to Stilbaai).•

appendiculata Engelm. Leafless parasitic annual vine, stems thread-like. Flowers in loose paniculate cymes, white, calyx loose, warty, shorter than corolla tube, stigmas globose. Oct.–June. Stem parasite on shrubs, SW, KM, LB, SE (Cape Peninsula to Mpumalanga).

bifurcata Yunck. Leafless parasitic annual vine, stems thread-like. Flowers in umbellate cymes, greenish, 4-lobed, calyx as long as corolla tube, corolla scales bifid, stigmas globose. Jan. Stem parasite on perennials, NW, SE (Cold Bokkeveld and Port Elizabeth).•

cassytoides Nees ex Engelm. Leafless parasitic annual vine, stems c. 2 mm diam. Flowers subsessile in racemes or panicles, greenish, styles fused. Apr.–June. Stem parasite on trees and shrubs, LB, SE (Swellendam to tropical Africa and Asia).

nitida E.Mey. ex Choisy Like **C. africana** but calyx often longer than corolla tube with acute lobes and stigmas longer than styles. Aug.–Dec. Stem parasite on shrubs, NW, SW (Cedarberg to Hottentots Holland Mts).•

DICHONDRA   DAISY GRASS   c. 5 spp., worldwide

micrantha Urban (= *Dichondra repens* auct.) Hairy, creeping rhizomatous perennial to 5 cm. Leaves reniform. Flowers white. Aug.–Nov. Rock sheets or grassy flats, SW, AP, LB, SE (Cape Peninsula to tropical Africa and worldwide).

FALKIA   OORTJIES   3 spp., Africa

repens L.f. Thinly hairy, mat-forming rhizomatous perennial with short erect branches to 5 cm. Leaves cordate to reniform. Flowers white or pink to mauve. Oct.–Dec. Damp coastal flats and seeps, SW, AP, LB, SE (Darling to E Cape).

IPOMOEA   MORNING GLORY   c. 500 spp., cosmopolitan but mainly tropical

cairica (L.) Sweet   MESSINA CREEPER   Perennial climber to 3 m. Leaves palmatisect with 5–7 lanceolate segments. Flowers mauve, sepals glabrous, obtuse. Oct.–Feb. SW, SE (Uitenhage to tropical Africa and Asia but weedy on the Cape Peninsula).

ficifolia Lindl. Thinly hairy perennial climber to 2 m. Leaves 3-lobed, cobwebby beneath. Flowers pink to purple, sepals thickly hairy, acuminate. Apr.–June. Coastal bush, SE (Humansdorp to tropical Africa).

pes-caprae (L.) R.Br.   STRANDPATAT   Trailing perennial to 30 m with milky sap. Leaves leathery, suborbicular and emarginate or bilobed. Flowers pink to purple, sepals glabrous, obtuse-mucronate. Dec.–Feb. Sandy beaches, SE (Mossel Bay to tropical Africa, pantropical).

*****purpurea** (L.) Roth Thinly hairy, vigorous annual climber to 5 m. Leaves cordate. Flowers purple, sepals thickly hairy beneath, acute. Oct.–Dec. Weed of forest margins, SW (tropical American weed on the Cape Peninsula).

CORNACEAE

CURTISIA   ASSEGAAIBOS   1 sp., Africa

dentata (Burm.f.) C.A.Sm. Shrub or small tree to 13 m, rusty velvety when young. Leaves opposite, ovate, toothed, glossy above. Flowers in terminal panicles, cream-coloured. Jan.–Feb. Forest, SW, LB, SE (Cape Peninsula to Zimbabwe).

CRASSULACEAE

1. Leaves spirally arranged, often in dense clusters:
   2. Leaves soft, herbaceous, seasonal; inflorescence a single-flowered or branched thyrse .... **Tylecodon**
   2.' Leaves usually tough, perennial; inflorescence a spike-like thyrse (rarely branched) .... **Adromischus**
1.' Leaves opposite or whorled, at least on vegetative branches:
   3. Petals free or shortly united at base; stamens as many as sepals; leaves opposite and more or less fused to one another ....**Crassula**
   3.' Perianth tubular; stamens twice as many as sepals; leaves free:
      4. Flowers 5-merous; filaments fused to corolla tube in lower third ....**Cotyledon**
      4.' Flowers 4-merous; filaments fused to corolla tube at or above the middle...............................**Kalanchoe**

ADROMISCHUS   c. 26 spp., southern Africa

A. Anthers exserted; petals cuspidate, forming a frill

bicolor Hutchison Like **A. hemisphaericus** but leaves densely spotted. Dec.–Jan. Rocky slopes, SE (Baviaanskloof Mts to E Cape).

filicaulis (Eckl. & Zeyh.) C.A.Sm.   Succulent perennial to 35 cm. Leaves ovoid. Flowers in a spicate cyme, greenish and red, cylindric, not grooved, lobes short and cuspidate, forming a frill, anthers shortly exserted. Nov.–Jan. Stony slopes, NW, KM (S Namibia through Worcester to Oudtshoorn).

hemisphaericus (L.) Lem. (incl. **A. roaneanus** Uitew.)   Succulent perennial to 35 cm. Leaves obovate. Flowers in a spicate cyme, greenish and red, cylindric, not grooved, lobes short and cuspidate, forming a frill, anthers shortly exserted. Nov.–Dec. Rocky slopes, NW, SW (S Namaqualand to Cape Peninsula to Greyton).

A. Anthers included; petals ovate, spreading

caryophyllaceus (Burm.f.) Lem.   NENTABOS   Succulent perennial to 35 cm. Leaves oblanceolate. Flowers in a spicate cyme, greenish white and red, funnel-shaped, grooved above, lobes ovate, anthers included. Jan.–Apr. Sandstone slopes, KM, SW, AP, LB, SE (Hermanus and Robertson to Uniondale).•

cristatus (Haw.) Lem.   Succulent perennial to 40 cm, with tufts of fine reddish aerial roots. Leaves obovoid-cuneate, undulate at the end, often glandular-hairy. Flowers in a spicate cyme, white, cylindric, grooved, lobes ovate, anthers included. Dec.–Mar. Stony slopes, SE (Uniondale to E Cape).

inamoenus Toelken   Like **A. triflorus** but buds straight. Oct.–Jan. Rocky slopes, KM, SE (Oudtshoorn to E Cape).

leucophyllus Uitew.   Succulent perennial to 35 cm. Leaves suborbicular. Flowers in a spicate cyme, white, cylindric, grooved, lobes ovate, anthers included. Jan.–Feb. Sandstone slopes, KM (Montagu to Klein Swartberg Mts).•

maculatus (Salm-Dyck) Lem.   Like **A. triflorus** but marginal ridge of leaf continuing to base. Dec.–Jan. Sandstone slopes, NW, KM (Swartruggens through Worcester to Ladismith).•

mammillaris (L.f.) Lem.   Like **A. triflorus** but leaves ovoid. Dec. Stony slopes, KM (Calitzdorp).

marianiae (Marloth) A.Berger   Succulent perennial to 35 cm. Leaves obovoid-fusiform. Flowers in a spicate cyme, white, cylindric, grooved, lobes ovate, anthers included. Dec.–Jan. Gravelly lower slopes, NW (S Namibia to Clanwilliam and Roggeveld).

maximus Hutchison   Succulent perennial to 50 cm. Leaves oblanceolate. Flowers in a spicate cyme, greenish white and pinkish, crowded in fascicles, cylindric, not grooved, lobes elliptic, anthers included. Nov.–Dec. Sandstone slopes, NW (Gifberg and Nardouw Mts).•

sphenophyllus C.A.Sm.   Like **A. maculatus** but leaves tapering gradually below and almost auriculate and never spotted. Dec.–Jan. Rocky slopes, SE (Steytlerville to E Cape).

triflorus (L.f.) Berger   Succulent perennial to 35 cm. Leaves obovate-cuneate. Flowers in a spicate cyme, pinkish and red, cylindric, not grooved, flexed outwards at tip in bud, lobes elliptic, anthers included. Nov.–Jan. Sandstone slopes, NW, KM, LB, SE (Pakhuis Mts through Little Karoo to Mossel Bay and Steytlerville).•

COTYLEDON   9 spp., Africa to Arabia

adscendens R.A.Dyer   Scrambling succulent shrublet to 2 m. Leaves obovate, with flaking wax. Flowers several in a stout, pedunculate cyme, nodding, reddish, filaments glabrous. Oct.–Dec. Coastal dunes in scrub, SE (Port Elizabeth).•

cuneata Thunb.   Succulent shrublet to 50 cm. Leaves obovate-cuneate, rarely viscid. Flowers several in a stout, pedunculate cyme, nodding, glandular-hairy, corolla lobes about twice as long as tube, yellowish, filaments glabrous. Sept.–Dec. Lower slopes, KM, SE (Namaqualand and W Karoo to Little Karoo).

orbiculata L.   KOUTERIE, VARKOOR   Succulent shrublet to 1 m. Leaves obovate to narrowly ovoid, sometimes velvety, with a grey bloom. Flowers several in a stout, pedunculate cyme, nodding, reddish, filaments hairy below. Sept.–Dec. Sandy or stony soils in scrub, NW, SW, AP, KM, LB, SE (Namibia and South Africa).

tomentosa Harv.   Hairy succulent shrublet to 1 m. Leaves cuneate to ovoid, toothed above, grey-velvety. Flowers several in a short, pedunculate cyme, erect, corolla sometimes bulging between calyx lobes, reddish, filaments hairy below. Feb.–Aug. Sandstone slopes, KM, SE (Barrydale to Baviaanskloof Mts).•

velutina Hook.f.   Succulent shrublet to 1 m. Leaves obovate to narrowly ovoid, lower leaves often auriculate, sometimes velvety, with a grey bloom. Flowers several in a stout, pedunculate cyme, nodding, corolla bulging or pouched between calyx lobes, reddish, filaments glabrous. Oct.–Dec. Stony soils in scrub, SE (Willowmore to S KwaZulu-Natal).

woodii Schönland & Baker f.   Much-branched succulent shrublet to 1 m. Leaves obovate, with flaking wax. Flowers 1 or 2 on short peduncles to 50 mm long, tubular, spreading, reddish, filaments hairy below. Mainly Dec.–Apr. Stony slopes, KM, SE (Swartberg Mts to E Cape).

CRASSULA   STONECROP   c. 200 spp., predominantly S hemisphere, mainly southern Africa

A. Annuals or perennials with anthers 0.1–0.2(–0.3) mm long
B. Flowers cup-shaped, petals recurved and spreading above, 2–15 mm long

depressa (Eckl. & Zeyh.) Toelken   Annual with wiry stems to 5 cm. Leaves opposite, oblanceolate to obovate. Flowers in a terminal thyrse, cup-shaped, white to pinkish, petals 3–4 mm long, stigmas terminal. Sept.–Oct. Sandy coastal flats, AP, SE (Bredasdorp to Knysna).•

dichotoma L.   Annual with wiry stems to 20 cm. Leaves opposite, subterete to obovate. Flowers in a terminal thyrse, cup-shaped, yellow to orange, often marked red in throat, petals 8–15 mm long, stigmas terminal on slender styles. Sept.–Oct. Sandy flats, NW, SW, AP (Namaqualand to Agulhas).

filiformis (Eckl. & Zeyh.) Dietr.   Like **C. grammanthoides** but flowers smaller, petals 2–3 mm long and more or less contorted and stylar horn very short, about a quarter as long as ovary. Aug.–Nov. Sandy flats, NW, SW, AP, SE (Pakhuis Mts to Port Elizabeth).•

grammanthoides (Schönland) Toelken   Annual with wiry stems to 10 cm. Leaves opposite, oblanceolate to obovate. Flowers in a terminal thyrse, cup-shaped, white or yellow, petals 4–5 mm long, stigmas subterminal at base of sterile, horn-like styles. Sept.–Oct. Calcareous coastal sands, AP (De Hoop).•

sebaeoides (Eckl. & Zeyh.) Toelken   Annual with wiry stems to 8 cm. Leaves opposite, oblanceolate to obovate. Flowers in a terminal thyrse, cup-shaped, yellow, petals 4–10 mm long, stigmas subterminal, sterile style with a bilobed apical flap. Sept.–Nov. Clay flats in karroid scrub, NW, SW, KM, LB, SE (Bokkeveld Mts and W Karoo to E Cape).

B. Flowers star- or cup-shaped, petals spreading from below or incurved but then 1–2 mm long
C. Flowers in sessile axillary cymes forming an elongate, spike-like thyrse (see also C. decumbens)

bergioides Harv.   Erect, subglabrous or minutely hairy annual to 6 cm. Leaves opposite, ovate-triangular, margins papillate-ciliolate. Flowers in sessile axillary cymes, star-shaped, white or pinkish, petals 1.0–1.5 mm long, papillose, sepals 2–3 mm long, carpels tuberculate. Sept.–Oct. Gravelly slopes, NW, SW, AP (Ceres to De Hoop).•

campestris (Eckl. & Zeyh.) Endl. ex Walp.   Erect, sometimes sprawling annual to 10 cm. Leaves opposite, imbricate, ovate-triangular. Flowers in sessile axillary cymes, star-shaped, whitish, perianth 0.5–1.0 mm long, carpels smooth. Sept.–Oct. Gravelly slopes, NW, SW, AP, KM, LB, SE (S Namibia to E Cape and Lesotho).

hirsuta Schönland & Baker f.   Erect, roughly hairy annual to 6 cm. Leaves opposite, linear-triangular. Flowers in sessile axillary cymes, star-shaped, whitish, perianth 1.0–1.5 mm long, carpels echinate. July–Sept. Sandy or gravelly slopes, NW, SW, KM (Namaqualand to Malmesbury and Little Karoo).

lanceolata (Eckl. & Zeyh.) Endl. ex Walp. (= *Crassula schimperi* Fisch & C.A.Mey.)   Perennial, 5–15 cm, with wiry or fleshy stems, sometimes mat-forming. Leaves opposite, often imbricate, triangular-lanceolate. Flowers in sessile axillary cymes, star-shaped, yellowish green, sepals and petals c. 1 mm long. Dec.–July. Sheltered slopes, NW, SW, KM, LB (Cedarberg Mts to Riversdale and E Cape to tropical Africa).

muscosa L.   VETERBOS, LIZARD'S TAIL   Sprawling, closely leafy perennial with woody stems, 10–50 cm. Leaves opposite, imbricate, ovate-triangular, fleshy. Flowers in sessile axillary cymes, star-shaped, yellowish green, petals 1–2 mm long, sepals to two-thirds as long. Mainly Oct.–Feb. Rocky flats and slopes, NW, SW, AP, KM, LB, SE (S Namibia to E Cape and Free State).

thunbergiana Schult.   Sprawling annual with stems often rooting at nodes, to 8 cm long. Leaves opposite, semiterete-lanceolate, papillate. Flowers in sessile or subsessile axillary cymes, cup-shaped, white, perianth c. 1 mm long. Aug.–Nov. Sandy flats and slopes, often coastal, NW, SW, AP (S Namibia to Agulhas).

C. Flowers in terminal panicles or restricted to the uppermost axils

aphylla Schönland & Baker f.   Apparently leafless annual, 0.6–3 cm, with a larger single-noded clavate stem terminated by smaller clavate branches. Leaves fused into a fleshy ring indistinguishable from stems. Flowers terminal, star-shaped, white or pinkish, petals 1–2 mm long, ovules 1(2) per carpel. Aug.–Oct. Rock pools on sandstone, NW (Bokkeveld to Hex River Mts).•

decumbens Thunb.   Erect or sprawling annual to 12 cm. Leaves opposite, linear-lanceolate to oblanceolate. Flowers in terminal and axillary cymes, cup-shaped, white, perianth 1.5–3.0 mm long. Sept.–Nov. Moist slopes, NW, SW, AP (Kamiesberg to Stilbaai).

dodii Schönland & Baker f.   Like **C. oblanceolata** but lower leaves in clusters of 4. Sept.–Oct. Moist depressions, NW, KM (S Namaqualand and W Karoo to Montagu).

glomerata Berg.   Stiffly erect annual to 15 cm. Leaves opposite, triangular-lanceolate, papillate. Flowers sessile in glomerules in a flat-topped thyrse, cup-shaped or tubular, white, petals 1.0–1.5 mm, papillose, carpels tuberculate. Aug.–Nov. Sandy, often coastal flats and limestones, NW, SW, AP, LB, SE (Clanwilliam to Port Elizabeth).•

inanis Thunb.   Like **C. natans** but leaves broadest below, lanceolate to ovate, and flowers mostly 5–10(–20) per axil. Nov.–Mar. River margins, SE (Knysna to KwaZulu-Natal).

minuta Toelken   Like **C. decumbens** but flowers solitary in axils. Sept.–Oct. Rock flushes, NW (Hantamsberg and N Cedarberg Mts).

natans Thunb. (incl. **C. elatinoides** (Eckl. & Zeyh.) Friedr.)   Erect or floating annual, sometimes rhizomatous perennial, 2–25 cm. Leaves opposite, linear to obovate, upper floating leaves sometimes broader. Flowers 1–3 in upper axils, star-shaped, white or pinkish, petals 1.5–2.0 mm long, ovules 1 per carpel. May–Oct. Moist depressions or pools, NW, SW, AP, KM, SE (widespread in South Africa and Lesotho).

oblanceolata Schönland & Baker f.   Erect, sometimes sprawling annual to 12 cm. Leaves opposite, oblanceolate to elliptic. Flowers in branched cymes on slender pedicels, cup-shaped, white to reddish, petals 2–3 mm long. Aug.–Oct. Sheltered slopes, NW, SW, KM (S Namibia and W Karoo to Cape Peninsula and Little Karoo).

pageae Toelken   Dwarf, disciform crustose annual. Leaves partially fused to stem and rosulate, oblanceolate. Flowers sessile in centre, cup-shaped, brownish, petals 0.5–1.0 mm long. July–Oct. Sheltered clay flats, NW, KM (W Karoo to Worcester and Montagu).

strigosa L.   Like **C. oblanceolata** but leaves and sepals hairy or at least aristate. Aug.–Oct. Sheltered or moist places, NW, SW, KM (Namaqualand to Bredasdorp).

tenuipedicellata Schönland & Baker f.   Like **C. oblanceolata** but flowers smaller, petals c. 1 mm long, and ovary tuberculate. Aug.–Oct. Sheltered slopes, NW (S Namibia and W Karoo to Hex River Mts).

umbellata Thunb.   Sprawling, dwarf annual to 4 cm . Leaves opposite, more or less petiolate, rhombic or triangular. Flowers sessile in a terminal thyrse but pedicels elongating in fruit, cup-shaped, cream-coloured or pinkish, petals c. 1 mm long. July–Oct. Sandy or gravelly slopes, NW, SW, KM, SE (W Karoo, Clanwilliam to Cape Peninsula and Little Karoo).

vaillantii (Willd.) Roth   Like **C. natans** but usually more freely branching and ovules (2–)4–8 per carpel. Mainly Sept.–Jan. Moist places, often around pools, NW, SW, KM, SE (widespread in South Africa).

A. Perennials with anthers (0.4–)0.5–5.0 mm long
D. Plants with (1)2 or 3(4) pairs of flattened, thin-textured leaves; soft, deciduous geophytes with stem tubers covered with root hairs

alcicornis Schönland   Tuberous geophyte, 6–12 cm. Leaves opposite, in 1(2) pairs, palmatisect. Flowers in a rounded thyrse, tubular, white or pinkish, petals 2.5–3.5 mm long. May–June. Rocky slopes, NW (Olifants River Mts: Dasklip Pass).•

capensis (L.) Baill.   CAPE SNOWDROP   Tuberous geophyte, 5–20 cm. Leaves opposite, in 2 or 3(4) pairs, the lower subpetiolate, thin-textured, obovate to cuneate-suborbicular, crenate. Flowers in a pedunculate, subumbellate thyrse, star-shaped, white to pinkish, petals 3–8 mm long, squamae at least as broad as long. May–Nov. Damp slopes, NW, SW, AP, LB (Clanwilliam to Riversdale).•

dentata Thunb.   Erect or sprawling tuberous geophyte, 5–15 cm. Leaves opposite but the lowest in a whorl of 4, thin-textured, sometimes hairy, spathulate or petiolate, suborbicular-reniform, sometimes toothed. Flowers in terminal and axillary thyrses, star-shaped, white to cream-coloured, petals 2.5–3.5 mm long. Sept.–Nov. Sheltered and moist rocky slopes or crevices, NW, SW (Bokkeveld to Elandskloof Mts and Kasteelberg).•

nemorosa (Eckl. & Zeyh.) Endl. ex Walp.   Erect or sprawling tuberous geophyte, 6–15 cm. Leaves opposite, in 3(4) pairs, thin-textured, usually petiolate, suborbicular-reniform. Flowers in a racemose thyrse, cup-shaped, yellowish green to brown, petals 2.0–3.5 mm long. Mainly June–Aug. Sheltered rocky slopes or crevices, KM, SE (Richtersveld and W and Little Karoo to E Cape).

saxifraga Harv.   Tuberous geophyte, 5–25 cm. Leaves often emergent, opposite, in 1(2) pairs, suborbicular, thin-textured, crenate. Flowers in a pedunculate, subumbellate thyrse, tubular, white to pinkish, petals 3.5–7.5 mm long, squamae broader than long. Apr.–June. Stony slopes or crevices, NW, SW, KM, LB, SE (Richtersveld to E Cape).

simulans Schönland   Tuberous geophyte, 6–20 cm. Leaves opposite, in 3(4) pairs, thin-textured, obovate-cuneate, toothed. Flowers in a pedunculate thyrse, star-shaped, yellowish green, petals 2.5–3.5 mm long. Sept.–Oct. Sheltered rocky slopes, NW, SW (Wallekraal, Worcester to Montagu).

umbella Jacq.   Tuberous geophyte, 6–20 cm. Leaves opposite, in 1(2) pairs, thin-textured, cordate-reniform or fused and amplexicaul, crenate. Flowers in a pedunculate thyrse, star-shaped, white to yellowish green, petals 2.0–4.5 mm long. July–Sept. Sheltered rocky slopes, NW, KM, SE (Richtersveld and W Karoo through Little Karoo to Humansdorp).

D. Plants not as above
E. Flowers star- or cup-shaped with petals spreading from near base

arborescens (Mill.) Willd.   JADE PLANT   Shrub sometimes to 2 m. Leaves opposite, mostly sessile, obovate to orbicular, sometimes with horny purplish margin. Flowers in rounded, terminal clusters, stellate, cream-coloured tinged red, petals 7–10 mm long. Oct.–Dec. Rocky sandstone slopes, NW, KM, SE (S Namaqualand to E Cape).

cordata Thunb.   Perennial to 30 cm, with erect to spreading branches. Leaves opposite, broadly ovate, often speckled with red and margin reddish. Flowers in lax, rounded clusters, stellate, cream-coloured tinged pink, petals c. 4 mm long, often with leafy shoots on old flower clusters. July–Oct. Dry scrub, SE (Humansdorp to S KwaZulu-Natal).

crenulata Thunb.   Tuberous perennial to 40 cm. Leaves opposite, sessile, oblong to oblanceolate, margins entire or crenate. Flowers in flat-topped clusters, stellate, white to cream-coloured tinged pink, petals 5–8 mm long. Jan.–Apr. Sheltered grassy slopes, SE (George to KwaZulu-Natal).

expansa Dryand.   Weak sprawling perennial with herbaceous stems to 40 cm, often rooting at nodes. Leaves opposite, obovate to ellipsoid. Flowers 1–few in upper axils, cup-shaped, white tinged red, petals 2.5–4.0 mm long. Mainly July–Dec. Mainly coastal sands and limestone, NW, SW, AP, KM, LB, SE (S Namibia to Mozambique).

lactea Sol.   Perennial with spreading or scrambling branches. Leaves opposite, sessile, elliptic to oblanceolate, margin yellowish, horny. Flowers many in terminal clusters, stellate, white, tinged red at tips, petals 5–8 mm long. May–July. Rock outcrops, SE (Mossel Bay to E Cape).

ovata (Mill.) Druce   Shrublet to 2 m. Leaves opposite, mostly sessile, elliptic-oblanceolate, sometimes with reddish horny margin. Flowers in rounded terminal clusters, stellate, white, sometimes tinged pink, petals 6–10 mm long. June–Aug. Rocky slopes, KM, SE (Swartberg Mts to KwaZulu-Natal).

papillosa Schönland & Baker f.   Delicate, prostrate cushion-forming perennial to 20 cm diam., with wiry stems Leaves opposite, small, elliptic-oblanceolate. Flowers 1–few in upper axils, cup-shaped, white to pinkish, petals 2–3 mm long. Dec.–Apr. Sheltered slopes, NW, SW, AP, KM, SE (Cedarberg Mts to E Cape).

peculiaris (Toelken) Toelken & Wickens   Slender prostrate mat-forming perennial to 30 cm long. Leaves opposite, petiolate, ovate-elliptic, finely hairy. Flowers solitary in axils, bowl-shaped, white, petals 3.5–4.0 mm long. Seeds papillose. Nov.–Mar. Sheltered slopes, KM (Groot Swartberg: Swartberg Pass).•

pellucida L.   Sprawling perennial, sometimes annual. Leaves opposite, mostly sessile, elliptic to oblanceolate, margin colourless or red. Flowers in terminal clusters, sometimes solitary, stellate, white, often tinged pink, petals 3–5 mm long. Sept.–Feb. Sheltered rocky slopes, NW, SW, AP, KM, LB, SE (Bokkeveld Mts to Ethiopia).

spathulata Thunb.   Sprawling to prostrate perennial, branches often quadrangular. Leaves opposite, petiole 3–18 mm long, ovate-elliptic, serrate to crenate, margins tinged red. Flowers in terminal clusters, stellate, white often tinged red, petals 3–5 mm long. Mar.–May. Rock outcrops along forest margins, SE (Knysna to E Cape).

E. *Flowers more or less tubular or urceolate with petals erect or recurved only in upper half*
F. *Sepals hairy (sometimes glabrous) and ciliate*

ammophila Toelken   Perennial to 80 cm, with brittle branches. Leaves opposite, oblanceolate, densely recurved-adpressed-hairy. Flowers in globular clusters forming elongate to rounded panicles, tubular, cream-coloured, c. 3 mm long. Oct.–Nov. Coastal sandveld, NW, SW (Namaquland to Saldanha Bay).

atropurpurea (Haw.) Dietr.   Shrublet to 60 cm, branches fleshy to woody. Leaves opposite, oblanceolate to obovate, usually hairy and papillate, margins horny. Flowers in globular clusters forming elongate panicles, tubular, cream-coloured, c. 4 mm long. Oct.–Dec. Rocky slopes, NW, SW, KM, LB, SE (S Namibia and Karoo to Port Elizabeth).

cotyledonis Thunb.   BERGPLAKKIE   Tufted perennial to 20 cm, stems woody, few-branched. Leaves opposite, oblanceolate to rounded, densely recurved-hairy, margins with a band of cilia. Flowers in globular clusters forming elongate panicles, tubular, cream-coloured to yellow, petals c. 4 mm long. Nov.–Jan. Gravelly slopes and outcrops, KM, LB, SE (S Namibia and Karoo to Little Karoo and E Cape).

cultrata L.   Shrub to 80 cm. Leaves opposite, oblanceolate to lorate, with sharp, horny margins. Flowers in elongate clusters on short peduncles, tubular, cream-coloured, c. 4 mm long. Dec.–Jan. Dry scrub, AP, KM, LB, SE (Swellendam and Little Karoo to KwaZulu-Natal).

deceptor Schönland & Baker f.   Perennial branched from base, to 8 cm. Leaves opposite, broadly ovate, closely imbricate, forming a 4-angled column. Flowers in lax, rounded, pedunculate clusters, tubular, cream-coloured, petals c. 2.5 mm long. Jan.–Mar. Gravelly quartz outcrops, NW (S Namibia to Elandsbaai).

hirtipes Harv.   Perennial with stems spreading to prostrate. Leaves opposite, lanceolate to ovate, bent upward from base, recurved-hairy sometimes glabrous. Flowers in elongate clusters on hairy peduncles, tubular, cream-coloured to yellow, petals c. 4 mm long. Aug.–Sept. Sheltered rocky outcrops, NW (Namaqualand to Olifants River mouth).

mollis Thunb.   Shrublet to 50 cm. Leaves opposite, linear-elliptic, flat above, rounded beneath, shortly velvety. Flowers in rounded clusters on hairy peduncles, tubular, cream-coloured, petals c. 3 mm long. Dec.–Feb. Karroid scrub, KM, LB, SE (Riversdale to E Cape and Karoo).

namaquensis Schönland & Baker f.   Tufted perennial. Leaves opposite, oblong to elliptic, densely adpressed-hairy, grey- to blue-green. Flowers in dense, rounded clusters on hairy peduncles, tubular, white to yellow, 3–8 mm long. Oct.–Nov. Stony slopes, NW (S Namibia and W Karoo to Karoopoort).

nudicaulis L. Tufted perennial with fleshy branches. Leaves oblong-elliptic, to orbicular, glabrous or pubescent, margins often ciliate. Flowers in small clusters forming elongate panicles, white to cream-coloured, c. 4 mm long. Sept.–Dec. Dry stony slopes, NW, SW, AP, KM, LB, SE (Bokkeveld Mts to N Province).

perfoliata L.   Densely papillate, few-branched perennial to 1.5 m. Leaves opposite, lanceolate to triangular, green to grey sometimes with purple blotches. Flowers in flat-topped, pedunculate clusters, tubular, white, pink or red, petals 3–6 mm long. Oct.–Jan. Dry lower slopes, SE (Karoo and Uitenhage to N Province).

pubescens Thunb.   Shrublet to 30 cm. Leaves opposite, oblanceolate to obovate, hairy or smooth, margins ciliate. Flowers in rounded to elongate clusters on puberulous to smooth peduncles, tubular, cream-coloured, petals c. 2.5 mm long. Sept.–Nov. Sheltered rock crevices, NW, SW, AP, KM, LB, SE (Bokkeveld Mts to E Cape and Karoo).

rogersii Schönland   Shrublet to 30 cm, branches erect, later spreading. Leaves opposite, oblanceolate to club-shaped, densely hairy, margins often red. Flowers in globular clusters forming elongated panicles, tubular, pale yellow, petals c. 3.5 mm long. Dec.–Mar. Dry scrub, KM, SE (Little Karoo to E Cape and Karoo).

subacaulis Schönland & Baker f. (= *Crassula erosula* N.E.Br.)   Perennial or shrublet to 60 cm, branching mostly from base. Leaves opposite, oblong-elliptic, often pubescent. Flowers in small clusters forming elongated panicles on shortly hairy peduncles, tubular, cream-coloured to yellow, c. 3 mm long. Sept.–Oct. Gravelly slopes, NW (Namibia to Pakhuis Mts).

subaphylla (Eckl. & Zeyh.) Harv.   Shrublet to 80 cm, with wiry-woody branches. Leaves opposite, linear-elliptic to lanceolate, adpressed-hairy or papillate to glabrous. Flowers in globular clusters forming elongate panicles, tubular, cream-coloured, c. 4 mm long. Oct.–Nov. Dry slopes, NW, SW, KM, LB, SE (S Namibia to E Cape and Karoo).

tecta Thunb.   Tufted perennial, often much branched. Leaves opposite, oblanceolate, covered with hard papillae. Flowers in rounded clusters on papillate peduncles, tubular, whitish, petals 3–4 mm long. Apr.–June. Gravel slops, KM (S and Little Karoo).

F. Sepals glabrous or hairy but not ciliate
G. Flowers 20–60 mm long

coccinea L.   RED CRASSULA   Like **C. fascicularis** but leaves ovate-elliptic, flowers usually scarlet-red, petals 30–45 mm long. Dec.–Mar. Sandstone outcrops, SW, AP (Cape Peninsula to Stilbaai).•

fascicularis Lam.   Erect shrublet to 40 cm. Leaves opposite, linear-lanceolate, ascending, usually with recurved cilia. Flowers in subsessile, flat-topped clusters, tubular, sesssile, cream-coloured to yellow-green, petals 20–32 mm long. Sept.–Nov. Sandstone slopes, NW, SW, LB (Gifberg to Langeberg Mts).•

obtusa Haw.   Sprawling shrublet to 15 cm, branches often rooting at nodes. Leaves opposite, oblong to oblanceolate, margins ciliate. Flowers 1–5 in terminal clusters, tubular, white tinged pinkish, petals 30–40 mm long. Nov.–Jan. Sandstone ledges, NW, SW, KM, LB, SE (Bokkeveld Mts Great Winterhoek Mts).•

G. Flowers 2–15 mm long

alpestris Thunb.   Mostly unbranched, perennial, sometimes biennial to 12 cm. Leaves opposite, triangular, closely imbricate, rough and often with sand attached. Flowers in dense terminal or axillary head-like clusters, tubular, white, petals 6–9 mm long. Aug.–Nov. Rocky pavement, NW, KM (Bokkeveld to Pakhuis Mts and W Karoo to Witteberg).

ammophila Toelken   Perennial to 80 cm, with brittle branches. Leaves opposite, oblanceolate, densely recurved-adpressed-hairy. Flowers in globular clusters forming elongate to rounded panicles, tubular, cream-coloured, c. 3 mm long. Oct.–Nov. Coastal sandveld, NW, SW (Namaqualand to Saldanha Bay).

atropurpurea (Haw.) Dietr.   Shrublet to 60 cm, branches fleshy to woody. Leaves opposite, oblanceolate to obovate, usually hairy and papillate, margins horny. Flowers in globular clusters forming elongate panicles, tubular, cream-coloured, c. 4 mm long. Oct.–Dec. Rocky slopes, NW, SW, KM, LB, SE (S Namibia and Karoo to Port Elizabeth).

barbata Thunb.   Tufted biennial or annual to 30 cm. Leaves obovate to orbicular, margins ciliate. Flowers in subsessile, elongate spike-like clusters, tubular, sessile, white often tinged pink, petals 4.0–5.5 mm long. Sept.–Nov. Rocky slopes, NW, KM (Karoo and Cedarberg Mts to Little Karoo).

barklyi N.E.Br.   Like **C. columnaris** but much branched at base and leaf margins broadly membranous, with a dense fringe of cilia. June–Aug. Quartzite gravel or rock outcrops, NW (Namaqualand to Lambert's Bay).

biplanata Haw.   Perennial to 30 cm with erect to spreading, papillate branches. Leaves opposite, lanceolate, sometimes rounded in section. Flowers in dense, flat-topped clusters, tubular, whitish, petals c. 2 mm long. Feb.–Apr. Rock ledges, SW, KM, LB, SE (Franschhoek to Great Winterhoek Mts).•

brachystachya Toelken   Tufted perennial to 20 cm, with brittle branches. Leaves oblanceolate-oblong, often red on margins. Flowers in pedunculate, spike-like clusters, tubular, subsessile, white, petals c. 4.5 mm long. Nov.–Jan. Sheltered rocks, KM (Witteberg and Swartberg Mts).•

brevifolia Harvey   Much-branched shrublet to 50 cm. Leaves opposite, leaves linear-elliptic, margins horny distally, often with a thick waxy bloom. Flowers in pedunculate, rounded clusters, tubular, yellowish green or whitish tinged pink, petals 3–5 mm long. Mar.–Apr. Rocky slopes, NW (S Namibia to Cedarberg Mts).

capitella Thunb.   Tufted perennial or biennial to 40 cm. Leaves linear-lanceolate to ovate, sometimes hairy and usually with marginal cilia. Flowers in elongate, subsessile, spike-like clusters, tubular, white to pink. Nov.–Mar. Dry slopes, NW, KM, LB, SE (widespread throughout southern Africa).

ciliata L.   Perennial to 20 cm. Leaves opposite, oblong-elliptic, with recurved marginal cilia in a dense row. Flowers in rounded clusters, cream-coloured to yellow, tubular, c. 4 mm long. Nov.–Jan. Gravelly lower slopes, SW, AP, LB, SE (Cape Peninsula to Port Elizabeth).•

clavata N.E.Br.   Tufted perennial with fleshy to slightly woody branches. Leaves opposite, obovate to elliptic, mostly spreading, sometimes shortly hairy and/or margins ciliate. Flowers in globular clusters forming elongate panicles, tubular, cream-coloured to yellow, c. 3 mm long. Sept.–Oct. Sandstone pavement, NW, KM (Namaqualand to Little Karoo).

columnaris Thunb.   KHAKIBUTTON, BERGKOESNAATJIE   Perennial or biennial to 10 cm, sometimes branched at base. Leaves opposite, depressed-ovate, closely imbricate, recurved-ciliate, margins sometimes membranous. Flowers in a terminal head, tubular, white to yellow, petals 7–13 mm long. May–Aug. Quartzite gravel and rocky pavements, NW, KM (S Namibia and Karoo to Little Karoo).

congesta N.E.Br.   Biennial to 20 cm. Leaves opposite, closely imbricate, lanceolate. Flowers in a dense terminal head, tubular, cream-coloured tinged red, petals 9–13 mm long. June–July. Quartzite gravel, KM (Witteberg and Little Karoo).•

cotyledonis Thunb.   BERGPLAKKIE   Tufted perennial to 20 cm, stems woody, few-branched. Leaves opposite, oblanceolate to rounded, densely recurved-hairy, margins with a band of cilia. Flowers in globular clusters forming elongate panicles, tubular, cream-coloured to yellow, petals c. 4 mm long. Nov.–Jan. Gravelly slopes and outcrops, KM, LB, SE (S Namibia and Karoo to Little Karoo and E Cape).

cultrata L.   Shrub to 80 cm. Leaves opposite, oblanceolate to lorate, with sharp, horny margins. Flowers in elongate clusters on short peduncles, tubular, cream-coloured, c. 4 mm long. Dec.–Jan. Dry scrub, AP, KM, LB, SE (Swellendam and Little Karoo to KwaZulu-Natal).

cymosa P.J.Bergius   Shrublet to 25 cm, branching mainly from base. Leaves opposite, linear to lanceolate, margins with swollen marginal cilia. Flowers in flat-topped to rounded clusters, white to yellow, tubular, petals c. 4 mm long. Oct.–Jan. Sandy or gravelly slopes, NW, SW (S Namaqualand to Riviersonderend Mts).

deceptor Schönland & Baker f.   Perennial branched from base, to 8 cm. Leaves opposite, broadly ovate, closely imbricate, forming a 4-angled column. Flowers in lax, rounded, pedunculate clusters, tubular, cream-coloured, petals c. 2.5 mm long. Jan.–Mar. Gravelly quartz outcrops, NW (S Namibia to Elandsbaai).

dejecta Jacq.   Much-branched shrublet to 40 cm, branches recurved-hairy when young. Leaves opposite, oblong-elliptic to ovate, with rounded cilia. Flowers in subsessile, flat to rounded clusters, white tinged pinkish, tubular, petals c. 8 mm long. Nov.–Feb. Rock outcrops, NW, SW (Namaqualand to Riviersonderend Mts).

deltoidea Thunb.   KATA-KISO   Perennial with erect to spreading branches. Leaves opposite, oblanceolate to rhombic, strongly convex to keeled beneath, covered with grey, flaking layers of wax. Flowers in rounded clusters on short peduncles, urn-shaped, cream-coloured, sometimes white, urn-shaped, petals to 5 mm long. Oct.–Nov. Gravelly flats, NW, KM (S Namibia and Karoo to Little Karoo).

elsieae Toelken   Sprawling to prostrate, few-branched perennial. Leaves opposite, angular-obovate to rhombic, rough when young. Flowers 1–5 in subsessile, terminal clusters, urn-shaped, petals c. 3 mm long. Nov.–Dec. Sheltered rock outcrops, NW (N Cedarberg Mts).•

ericoides Haw.   MALMEIDBOSSIE   Erect to spreading perennial to 40 cm. Leaves opposite, lanceolate to ovate. Flowers in sessile rounded clusters, tubular, white to cream-coloured, shortly pedicellate, petals 2–4 mm long. Dec.–Mar. Dry rocky slopes, SW, KM, LB, SE (Cape Hangklip to KwaZulu-Natal).

fallax Friedrich   Shrublet to 40 cm, few-branched from base. Leaves opposite, oblong-elliptic, margins with recurved cilia. Flowers in flat-topped clusters, tubular, whitish, petals c. 8 mm long. Nov.–Mar. Sandy lower slopes and flats, NW, SW, AP (Olifants River Mts to Bredasdorp).•

flava L.   Shrublet to 40 cm, branches occasionally scabrid below flower clusters. Leaves opposite, lanceolate-triangular, with rounded marginal cilia. Flowers in rounded, subsessile clusters, white to yellow, tubular, petals c. 8 mm long. Dec.–Feb. Sandy lower slopes, NW, SW (Tulbagh to Bredasdorp Mts).•

hemisphaerica Thunb.   Tufted perennial to 15 cm. Leaves ovate, sometimes elliptic, margins ciliate. Flowers in elongate, pedunculate spike-like clusters, tubular, subsessile, white to cream-coloured, petals 2–3 mm long. Sept.–Nov. Pebbly slopes, NW, KM, SE (Kamiesberg, Bokkeveld Mts and W Karoo to Kouga Mts).

lasiantha Drège ex Harv.   Perennial with wiry, prostrate, recurved-hairy branches, rooting at nodes. Leaves opposite, obovate-elliptic, margins with recurved cilia. Flowers in subsessile, rounded clusters, whitish, tubular, petals c. 3 mm long. Dec.–Apr. Sheltered rock ledges, NW (Cedarberg to Grootwinterhoek Mts).•

macowaniana Schönland & Baker f.   Perennial or shrub to 1.2 m. Leaves opposite, linear-lanceolate, thick and fleshy. Flowers in pedunculate, rounded clusters, tubular, white, often tinged pink, petals 30, 5.5 mm long. Oct.–Dec. Rocky slopes, NW (S Namibia to Gifberg).

mesembryanthoides (Haw.) Dietr.   Shrublet to 40 cm, with spreading woody branches. Leaves opposite, linear-elliptic to triangular, densely adpressed-hairy. Flowers in rounded to flat-topped clusters on densely hairy peduncles, tubular, whitish, petals c. 4 mm long. Mar.–June. Karroid slopes, SE (Uitenhage to E Cape).

mollis Thunb.   Shrublet to 50 cm. Leaves opposite, linear-elliptic, flat above, rounded beneath, shortly velvety. Flowers in rounded clusters on hairy peduncles, tubular, cream-coloured, petals c. 3 mm long. Dec.–Feb. Karroid scrub, KM, LB, SE (Riversdale to E Cape and Karoo).

montana Thunb.   Tufted perennial to 12 cm. Leaves obovate to elliptic, margins with a dense row of cilia. Flowers in flat-topped to spike-like, pedunculate clusters, tubular, mostly sessile, white tinged pink, petals 3–5 mm long. Aug.–Oct. Sheltered sandstone outcrops, NW (Karoo and Bokkeveld Mts to Hex River valley).

multiflora Schönland & Baker f.   Shrublet to 80 cm, branching mainly above. Leaves opposite, oblong-elliptic, margins with rounded cilia. Flowers many in flat-topped to rounded clusters on short peduncles, cream-coloured, sessile, tubular, petals c. 4 mm long. Dec.–Feb. Rocky slopes, NW, SW, KM, LB (Keeromsberg to Langeberg and Swartberg Mts).•

muricata Thunb.   Much-branched shrublet to 25 cm, branches recurved-adpressed-hairy when young. Leaves opposite, lanceolate-elliptic, minutely papillate. Flowers in branched terminal clusters, sessile, white to yellow, urn-shaped, petals c. 3. mm long. Oct.–Dec. Dry sandstone slopes, NW, KM (Swartruggens and Little Karoo to S Karoo).

nudicaulis L.   Tufted perennial with fleshy branches. Leaves oblong-elliptic to orbicular, glabrous or pubescent, margins often ciliate. Flowers in small clusters forming elongate panicles, white to cream-coloured, c. 4 mm long. Sept.–Dec. Dry stony slopes, NW, SW, AP, KM, LB, SE (Bokkeveld Mts to N Province).

orbicularis L.   Perennial to 25 cm. Leaves opposite, oblanceolate to elliptic, margins with a dense row of spreading cilia. Flowers in elongate, sometimes spike-like, pedunculate clusters, tubular, white to yelow tinged pink to brown. June–Nov. Sheltered rocky slopes, NW, SW, KM, LB, SE (Montagu and Potberg to KwaZulu-Natal).

perfoliata L.   Densely papillate, few-branched perennial to 1.5 m. Leaves opposite, lanceolate to triangular, green to grey sometimes with purple blotches. Flowers in flat-topped, pedunculate clusters, tubular, white, pink or red, petals 3–6 mm long. Oct.–Jan. Dry lower slopes, SE (Karoo and Uitenhage to N Province).

perforata Thunb.   Shrublet to 60 cm. Leaves opposite, ovate to lanceolate, fused at base to opposite leaf, margins horny, red or yellow. Flowers in subsessile, elongate, rounded clusters, tubular, sessile, cream-coloured to yellow, petals c. 2.5 mm long. Nov.–Apr. Sheltered rocky slopes, SW, LB, SE (Riviersonderend Mts to KwaZulu-Natal).

pruinosa L.   Shrublet to 20 cm, branches recurved, adpressed-hairy when young. Leaves opposite, linear-lanceolate, densely adpressed-hairy. Flowers in flat-topped clusters, mostly sessile, tubular, c. 6 mm long. Nov.–Dec. Rock outcrops, SW (Cape Peninsula to Grabouw).•

pubescens Thunb.   Shrublet to 30 cm. Leaves opposite, oblanceolate to obovate, hairy or smooth, margins ciliate. Flowers in rounded to elongate clusters on puberulous to smooth peduncles, tubular, cream-coloured, petals c. 2.5 mm long. Sept.–Nov. Sheltered rock crevices, NW, SW, AP, KM, LB, SE (Bokkeveld Mts to E Cape and Karoo).

pustulata Toelken   Shrublet to 20 cm, branches recurved-adpressed-hairy when young. Leaves opposite, linear-lanceolate, densely hairy. Flowers in flat-topped clusters, whitish, tubular, petals c. 9 mm long. Nov.–Dec. Shallow sandy soils, NW (Bokkeveld to Olifants River Mts).•

pyramidalis Thunb.   RYGBOSSIE   Erect or sprawling perennial to 25 cm. Leaves opposite, angular-ovate, closely imbricate, forming 4-angled columns. Flowers in crowded, terminal heads, tubular, white or cream-coloured, petals 8–11 mm long. Aug.–Oct. Rock pavements, KM (W and Little Karoo to Great Karoo).

rogersii Schönland   Shrublet to 30 cm, branches erect, later spreading. Leaves opposite, oblanceolate to club-shaped, densely hairy, margins often red. Flowers in globular clusters forming elongated panicles, tubular, pale yellow, petals c. 3.5 mm long. Dec.–Mar. Dry scrub, KM, SE (Little Karoo to E Cape and Karoo).

rubricaulis Eckl. & Zeyh.   Much-branched shrublet to 30 cm. Leaves opposite, oblanceolate-lorate, margins often red, with recurved cilia. Flowers in rounded clusters, white, often tinged red, tubular, petals c. 5 mm long. Jan.–May. Rock outcrops, LB, SE (Langeberg Mts to Port Elizabeth).•

rupestris Thunb. CONCERTINA PLANT, SOSATIES   Much-branched shrublet to 50 cm. Leaves opposite, ovate to lanceolate, margins horny, red or yellow. Flowers in pedunculate, rounded clusters, tubular, whitish tinged pink. June–Oct. Dry stony slopes, NW, SW, KM, LB, SE (Namibia to E Cape).

scabra L.   Shrublet to 40 cm, branching mainly from base, branches recurved-adpressed-hairy when young. Leaves opposite, narrowly triangular to lanceolate, with recurved adpressed hairs. Flowers in flat-topped clusters, whitish, tubular, petals c. 9 mm long. Dec.–Feb. Dry lower slopes, NW, SW (Clanwilliam to Cape Peninsula).•

subaphylla (Eckl. & Zeyh.) Harv.   Shrublet to 80 cm, with wiry-woody branches. Leaves opposite, linear-eliptic to lanceolate, adpressed-hairy or papillate to glabrous. Flowers in globular clusters forming elongate panicles, tubular, cream-coloured, c. 4 mm long. Oct.–Nov. Dry slopes, NW, SW, KM, LB, SE (S Namibia to E Cape and Karoo).

subulata L.   Shrublet sometimes to 1 m. Leaves opposite, linear-lanceolate, mostly round in section, with stout marginal cilia or papillae. Flowers in rounded, head-like clusters, cream-coloured, tubular, c. 4 mm long. Oct.–Dec. Dry rocky slopes, NW, SW, AP, KM, LB, SE (Bokkeveld Mts to Port Alfred).

tetragona L.   KARKAI   Perennial to 1 m, with erect to spreading, sometimes papillate branches. Leaves opposite, lanceolate, sometimes rounded in section. Flowers in flat-topped clusters, tubular, cream-coloured, sometimes white, petals c. 3 mm long. Dec.–Mar. Dry slopes, NW, SW, KM, LB, SE (Namaqualand and Karoo to E Cape).

tomentosa Thunb.   Tufted perennial or biennial to 60 cm. Leaves oblong to orbicular, tomentose to rugose, margins ciliate. Flowers in elongate, subsessile, spike-like clusters, tubular, subsessile, whitish to yellow, petals 3.0–4.5 mm long. Aug.–Dec. Stony slopes, NW, SW, KM, LB (S Namibia to Gourits River).

[**Species excluded**   Not recorded from the area: **C. corallina** Thunb.]

KALANCHOE   c. 200 spp., Africa, Asia and Australasia

rotundifolia (Haw.) Haw.   NENTABOS   Succulent shrublet to 1 m. Leaves opposite, elliptic. Flowers in a rounded cyme, orange to red. Mar.–June. Bushveld, SE (Uitenhage to N Africa).

TYLECODON   27 spp., South Africa and Namibia

A. Leaves withering but not abscising cleanly

grandiflorus (Burm.f.) Toelken   ROOISUIKERBLOM   Succulent shrublet to 50 cm, often sprawling. Leaves dry at flowering, withering but not abscising, oblanceolate, margins often rolled upwards. Flowers in a narrow cyme, tubular, slightly zygomorphic, reddish, tube 30–40 mm. Jan.–Feb. Rocky outcrops, often granite, near the coast, NW, SW (S Namaqualand to Cape Peninsula).

striatus (Hutchison) Toelken   Like **T. ventricosus** but stems streaked grey and brown and flowers smaller, tube 12–15 mm. Sept.–Jan. Stony shale slopes, NW (Richtersveld and W Karoo to Clanwilliam).

ventricosus (Burm.f.) Toelken   KLIPNENTA   Succulent shrublet to 30 cm. Leaves dry at flowering, withering but not abscising, oblanceolate. Flowers in a narrow cyme, tubular, yellowish green, tube 16–19 mm. Sept.–Mar. Rocky slopes, NW, KM, SE (Namaqualand through Little Karoo to Kouga).

A. Leaves abscising cleanly

cacalioides (L.f.) Toelken   NENTA, KARKAY COTYLEDON   Like **T. wallichii** but flowers tubular, erect and sulphur-yellow, tube 17–25 mm. Dec.–Feb. Rocky sandstone slopes, KM (Ladismith to Willowmore).•

leucothrix (C.A.Sm.) Toelken   DOUBOSSIE   Succulent shrublet with peeling stems to 20 cm. Leaves dry at flowering, abscising, oblanceolate, hairy. Flowers in a narrow cyme, tubular, white or pinkish. Nov.–Feb. Rocky shale slopes, KM (Montagu to Calitzdorp).•

paniculatus (L.f.) Toelken   BOTTERBOOM   Succulent shrublet with stout, peeling stems to 1.5 m. Leaves dry at flowering, abscising, obovate. Flowers in a branched cyme, urn-shaped, nodding, yellowish to red, tube 12–16 mm. Nov.–Jan. Rocky slopes, NW, SW, KM, SE (Namibia to Cape Peninsula through Little Karoo to Willowmore).

reticulatus (L.f.) Toelken   OUKOE   Succulent shrublet with thick, peeling stems to 30 cm. Leaves dry at flowering, abscising, oblanceolate. Flowers in a twiggy, divaricately branched cyme, urn-shaped, yellowish, tube 6–8 mm. Nov.–Dec. Stony slopes, NW, KM, SE (S Namibia and W Karoo to Pakhuis Mts through Little Karoo to Willowmore).

wallichii (Harv.) Toelken   KOKERBOS, KANDELAARBOS   Succulent shrublet with warty stems to 1 m. Leaves dry at flowering, abscising, oblanceolate. Flowers in a spreading cyme, urn-shaped, nodding, greenish yellow, tube 7–12 mm. Dec.–Feb. Stony flats and slopes, NW, SW, KM (Namaqualand to Saldanha through Little Karoo to Ladismith).

CUCURBITACEAE

1. Anthers straight or curved; flowers small, less than 1 cm diam., greenish:
    2. Stamens inserted in mouth of tube; seeds pyriform or globose .... **Kedrostis**
    2.' Stamens inserted in throat or base of receptacle tube; seeds compressed .... **Zehneria**
1.' Anthers folded or sigmoid; flowers larger, more than 1 cm diam., cream-coloured or yellow:
    3. Filaments cohering or connate to form a tube; fruit scarlet, ellipsoid .... **Coccinea**
    3.' Filaments free; fruit mottled green:
        4. Male flowers solitary .... **Citrullus**
        4.' Male flowers in racemes .... **Lagenaria**

CITRULLUS   WILD MELON   4 spp., Africa and Asia

lanatus (Thunb.) Matsumura & Nakai   TSAMMA   Monoecious, hairy trailing annual to 3 m long. Leaves pinnatisect with rounded, slightly toothed lobes. Flowers solitary in axils, greenish yellow. Fruits globose, mottled green. mainly Sept.–Dec. Sandy soils, NW (Cold Bokkeveld to tropical Asia).

COCCINIA   c. 30 spp., Old World tropics

quinqueloba (Thunb.) Cogn.   BOBBEJAANKOMKOMMER   Dioecious, glabrescent perennial climber to 10 m. Leaves palmate with slightly toothed lobes. Flowers solitary in axils, yellowish. Fruits ellipsoid, scarlet. Nov.–Apr. Coastal bush, SE (Humansdorp to S KwaZulu-Natal).

KEDROSTIS   c. 23 spp., worldwide in tropics, mostly African

africana (L.) Cogn.   CAPE BRYONY, KLEINBOBBEJAANKOMKOMMER   Like **K. capensis** but male and female flowers minute and borne in the same axils. Nov.–Apr. Rocky slopes, NW, KM, SE (Namibia, Worcester and Karoo to Port Elizabeth and Mpumalanga).

capensis (Sond.) A.Meeuse   Monoecious tuberous perennial climber to 50 cm. Leaves palmatisect. Flowers often before leaves, axillary, male fascicled, female solitary, greenish. Fruits berry-like. Oct.–Apr. Rocky slopes, NW, KM (S Namibia to Lambert's Bay to Swartberg Mts and Karoo).

nana (Lam.) Cogn.   YSTERVARKPATAT   Like **K. capensis** but dioecious and leaves sometimes scarcely lobed. Feb.–Mar. Coastal scrub, SW, AP, LB, SE (Saldanha to KwaZulu-Natal).

LAGENARIA   6 spp., originally African

sphaerica (Sond.) Naudin (= *Lagenaria mascarena* Naudin)   WILDEKALABAS   Dioecious glabrescent perennial climber to 10 m. Leaves palmate with coarsely toothed lobes. Flowers axillary, female solitary, male racemose. Fruits subglobose, mottled green. Dec.–May. Coastal bush, SE (Knysna to tropical Africa).

ZEHNERIA   c. 35 spp., Old World tropics

scabra (L.f.) Sond.   Dioecious glabrescent perennial climber to 6 m. Leaves cordate, toothed, rough above and softly hairy below. Flowers in axillary umbels, cream-coloured. Fruits fleshy, scarlet. Aug.–Apr. Forest margins, SW, LB, SE (Cape Peninsula to tropical Africa).

CUNONIACEAE

1. Flowers in long-peduncled, axillary panicles; petals shorter than calyx; ovules 2 in each locule .... **Platylophus**
1.' Flowers in dense, spike-like racemes; petals longer than calyx; ovules numerous in each locule .... **Cunonia**

CUNONIA   ROOIELS   c. 17 spp., New Caledonia and 1 in southern Africa

capensis L.   BUTTERSPOON TREE   Tree to 30 m. Leaves opposite, imparipinnate, glossy, leaflets elliptic, serrate, enclosed in bud by conspicuous paddle-shaped stipules. Flowers in dense, spike-like axillary racemes, whitish, scented. Mar.–June. Streambanks and forest, NW, SW, LB, SE (Grootwinterhoek Mts to S Mozambique).

PLATYLOPHUS• WITELS   1 sp., W and E Cape

trifoliatus (Thunb.) D.Don.   Tree to 30 m. Leaves opposite, 3-foliolate, leaflets narrowly elliptic, serrate. Flowers in axillary panicles, cream-coloured or yellowish, scented. Dec.–Feb. Streambanks or forest, NW, SW, LB, SE (Piketberg to Humansdorp).•

CYTINACEAE (= *RAFFLESIACEAE* IN PART)

CYTINUS   AARDROOS   c. 10 spp., Africa, Mediterranean, Mexico

sanguineus (Thunb.) Fourc. (incl. **C. capensis** Marloth)   Dioecious root parasite to 5 cm. Leaves scale-like, obovate. Flowers in sessile clusters, often fimbriate, orange to red or purple. July–Dec. Parasitic mostly on shrubby Asteraceae, NW, SW, AP, KM, LB, SE (Namaqualand and W Karoo to Mossel Bay).

DIPSACACEAE

1. Calyx small, pappus-like; involucral bracts in several rows .... **Cephalaria**
1.' Calyx of 5 long awns; involucral bracts in 1 or 2 rows .... **Scabiosa**

CEPHALARIA   c. 65 spp., Africa and Asia

attenuata (L.f.) Roem. & Schult.   Sprawling, thinly hairy perennial to 60 cm. Leaves cauline, linear or pinnatisect below, margins revolute. Flowers in pedunculate heads, white. Dec.–Jan. Sandstone slopes, SW, LB, SE (Hottentots Holland Mts to E Cape).

decurrens (Thunb.) Roem. & Schult.   Tufted, thinly hairy perennial to 80 cm. Leaves mostly basal, soft, oblanceolate or lyrate-pinnatifid, toothed. Flowers in pedunculate heads, white. Oct. Rocky slopes, NW (Bokkeveld Mts).•

humilis (Thunb.) Roem. & Schult.   Tufted, glabrescent or roughly hairy perennial to 1 m. Leaves mostly basal, linear-oblanceolate or pinnatisect, margins revolute. Flowers in pedunculate heads. Dec.–Jan. Grassy slopes, SE (George to KwaZulu-Natal).

oblongifolia (Kuntze) Szabó   Like **C. humilis** but basal leaves obovate. Jan.–May. Rocky slopes, SE (Humansdorp to KwaZulu-Natal).

rigida (L.) Roem. & Schult.   Erect, roughly hairy perennial to 1 m. Leaves cauline, elliptic, toothed, often lobed below, margins slightly revolute. Flowers in pedunculate heads, white. Dec.–Mar. Sandstone slopes, SW (Du Toitskloof to Jonkershoek and Cape Peninsula).•

scabra (L.f.) Roem. & Schult.   Erect, glabrescent or roughly hairy perennial to 1 m. Leaves cauline, bipinnatisect, margins revolute. Flowers in pedunculate heads, white. Dec.–Feb. Sandstone slopes, NW, SW (Grootwinterhoek Mts to Palmiet River and Riviersonderend Mts).•

SCABIOSA   SCABIOUS   c. 100 spp., Africa to Asia

africana L.   CAPE SCABIOUS   Velvety hairy, straggling shrublet to 1 m. Leaves oblanceolate, toothed or incised. Flowers in pedunculate heads, mauve. July–Nov. Sheltered sandstone slopes, SW (Cape Peninsula).•

columbaria L.   JONGMANSKNOOP   Shortly hairy, tufted perennial to 80 cm. Leaves mostly basal, markedly dimorphic, lower oblanceolate to lyrate-pinnatifid, toothed or incised, upper pinnatisect. Flowers in pedunculate heads, white to mauve. Aug.–Feb. Rocky slopes, NW, SW, KM, LB, SE (widespread through Africa, Europe and Asia).

incisa Mill. (incl. **S. albanensis** R.A.Dyer)   Like **A. columbaria** but stems straggling. Leaves usually not markedly dimorphic, lyrate-pinnatisect to bipinnatisect. Sept.–Nov. Coastal sands, often on limestone, SW, AP, SE (Bokbaai to Grahamstown).

DROSERACEAE

DROSERA   SUNDEW, DOUBLOM   c. 130 spp., widespread but mainly Australia

A. Plants caulescent and woody below, with axillary scapes

capensis L.   Rhizomatous perennial to 30 cm, woody below. Leaves petiolate, linear to elliptic, stipules ovate-lacerate, brown. Flowers several, secund in an axillary helicoid cyme, scape curved outwards below, pink to magenta, stigmas swollen. Dec.–Jan. Marshy sandstone, NW, SW, KM, LB, SE (Elandsbaai and Cedarberg Mts to Langkloof).•

glabripes (Harv.) Stein   Like **D. capensis** but with long, sprawling stems, leaf blades shortly obovoid and stipules long-lacerate and stigmas shortly multifid. Dec.–Jan. Peaty sandstone slopes, SW, AP (Cape Peninsula to Potberg).•

hilaris Cham. & Schltdl.   Sprawling perennial to 40 cm. Leaves oblanceolate, densely rusty hairy beneath, stipules deeply fringed. Flowers in an axillary helicoid cyme, magenta or purple, stigmas simple or weakly divided. Sept.–Nov. Damp sandstone slopes, SW, AP (Cape Peninsula to Gansbaai).•

ramentacea Burch. ex DC.   Like **D. capensis** but with long, sprawling stems, petioles with long russet hairs and scapes often forked. Dec. Damp sandy flats and slopes, SW, LB (Cape Peninsula to Langeberg Mts).•

regia Stephens   Stout rhizomatous perennial to 40 cm, woody below. Leaves linear-attenuate, circinnate, exstipulate. Flowers in forked, subcorymbose cymes, magenta, styles 3, undivided, stigmas narrowly funnel-shaped. Jan. Wet, peaty sandstone at high alt., NW (Bainskloof Mts).•

A. Plants acaulescent or stems leafy but not woody, mostly with terminal scapes
B. Petals longer than 8 mm

cistiflora L. (incl. **D. rubripetala** Debbert)   SNOTROSIE   Slender perennial to 40 cm. Leaves dimorphic, sometimes uniform, lower radical or lacking, oblancolate, upper linear, exstipulate. Flowers large, few in subcorymbose cymes, white, yellow, mauve to purple or red, with dark centre, stigmas multifid. Aug.–Sept. Damp sandy flats, NW, SW, AP, KM, LB, SE (Namaqualand to Port Elizabeth).

cuneifolia L.f.   Tufted perennial to 20 cm. Leaves radical, cuneate, stipules lacerate. Flowers several in helicoid cymes, pink to purple, stigmas spathulate. Nov.–Jan. Peaty sandstone slopes, SW (Cape Peninsula).•

pauciflora Banks ex DC. (incl. **D. atrostyla** Debbert)   Tufted perennial to 20 cm. Leaves radical, obovate, exstipulate. Flowers few in a subcorymbose cyme, large, pink or mauve with dark centre, stigmas multifid. Aug.–Nov. Damp loamy or sandy flats, NW, SW (Piketberg to Caledon).•

B. Petals shorter than 8 mm

acaulis L.f.   Dwarf, tufted perennial to 6 cm. Leaves radical, spathulate, exstipulate. Flowers 1–few on a short scape, white to purple, stigmas multifid. Oct.–Dec. Damp sandstone slopes, NW (Cedarberg to Hex River Mts).•

alba E.Phillips   Tufted perennial to 13 cm. Leaves radical, dimorphic, lower oblanceolate and upper linear, exstipulate. Flowers few in a subcorymbose cyme, white or mauve, stigmas multifid. Aug.–Oct. Damp sandy flats and slopes, NW (Namaqualand to Cedarberg Mts).

aliciae Raym.-Hamet (incl. **D. admirabilis** Debbert, **D. venusta** Debbert)   Tufted perennial to 40 cm. Leaves radical, spathulate, hairy beneath, stipules lacerate. Flowers several on wiry scape curved below then erect, pink, stigmas 2- or 3-fid. Nov.–Jan. Peaty sandstone, SW, KM, LB, SE (Cape Peninsula to E Cape).

esterhuyseniae (T.M.Salter) Debbert   Like **D. aliciae** but leaves glabrous above in lower half and thickly woolly beneath. Dec.–Jan. Sandstone slopes, SW (Hottentots Holland Mts to Kleinmond).•

slackii M.R.Cheek   Like **D. aliciae** but leaves glabrous above in lower half and with only a few fleshy hairs below. Nov.–Dec. Damp sandstone soils, SW (Kleinrivier Mts.).•

trinervia Spreng.   Tufted perennial to 10 cm. Leaves radical, oblanceolate-cuneate, exstipulate at maturity. Flowers few in subcorymbose cymes, white to mauve or red, stigmas multifid. Aug.–Nov. Peaty sandstone slopes, NW, SW, AP, KM, LB (Namaqualand to Agulhas).

EBENACEAE

1. Calyx not accrescent; ovary on a fringed disc; fruits usually 1-seeded .... **Euclea**
1.' Calyx usually accrescent; ovary on a glabrous disc; fruits usually with 2 or more seeds .... **Diospyros**

DIOSPYROS   PERSIMMON, TOLBOS   c. 500 spp., pantropical and subtropical

austro-africana De Winter   FIRE-STICKS, KRITIKOM   Dioecious shrub to 3 m. Leaves subsessile, oblanceolate, velvety, felted beneath. Flowers solitary, axillary, cream-coloured, pink or red. Fruits shortly hairy. Aug.–Oct. Rocky flats and slopes, NW, SW, KM, LB, SE (Bokkeveld Mts to Cape Peninsula to Mpumalanga).

dichrophylla (Gand.) De Winter   POISON PEACH   Dioecious shrub or tree to 13 m. Leaves petiolate, oblanceolate, leathery, thinly hairy, margins revolute. Flowers solitary, axillary, whitish. Fruits large, velvety, calyx strongly accrescent, usually reflexed. Nov.–Mar. Coastal scrub and forest margins, AP, KM, LB, SE (Potberg and Montagu to N Province).

glabra (L.) De Winter   BLOUBESSIEBOS, KRAAIBOSSIE   Dioecious shrub to 2 m. Leaves sessile, elliptic. Flowers 1-few in axillary racemes, cream-coloured. Fruits finely glandular when young, purple or reddish, calyx not accrescent. Oct.–Dec. Sandy flats and slopes, NW, SW, KM, LB, SE (Bokkeveld Mts to Worcester, Cape Peninsula to Uniondale).•

lycioides Desf.   STAR-APPLE, SWARTBAS   Dioecious shrub or small tree to 7 m. Leaves shortly petiolate, oblanceolate, leathery, hairy when young. Flowers solitary, axillary, white to yellow, fragrant. Fruits thinly hairy, yellow or red to brown. Sept.–Jan. Rocky slopes, KM (Ladismith and drier parts of southern and tropical Africa).

pallens (Thunb.) F.White   BLOUTOLBOS   Diffuse, dioecious shrub to 2.5 m. Leaves petiolate, elliptic, glossy, margins revolute. Flowers solitary, axillary, white. Fruits thinly hairy. Aug.–Nov. Coastal scrub and dunes, SE (Knysna to E Cape).

ramulosa (E.Mey. ex A.DC.) De Winter   NAMAQUA FIRE-STICKS, NAMAKWA KRITIKOM   Rigid dioecious shrub to 5 m. Leaves subsessile, small, elliptic, shortly hairy. Flowers solitary, axillary, whitish. Fruits shortly hairy. Dec.–May. Rocky slopes, NW (S Namibia to Clanwilliam and Karoo).

whyteana (Hiern) F.White   BLADDER-NUT, BOSTOLBOS   Dioecious shrub or small tree to 6 m. Leaves shortly petiolate, elliptic, glossy above, hairy beneath, margins fimbriate. Flowers few in racemes developing into shoots, creamy. Fruits red, enclosed in inflated calyx. Aug.–Nov. Sandstone slopes, NW, SW, LB, SE (Tulbagh to Cape Peninsula to Mpumalanga).

EUCLEA   GUARRI   c. 20 spp., Africa and Arabia

A. Corolla shallowly lobed

acutifolia E.Mey. ex A.DC.   Dioecious shrublet to 1 m. Leaves elliptic, glabrescent, leathery. Flowers in hairy axillary racemes, densely grey-hairy, shallowly lobed, cream-coloured, ovary bristly. Aug.–Sept. Sandstone slopes, NW, SW, LB (Bokkeveld Mts to Riversdale).•

lancea Thunb.   Like **E. acutifolia** but leaves narrowly elliptic. Flowers in glabrous racemes, only hairy above, with glabrous anthers and long exserted styles. Sept. Rocky slopes, NW (Namaqualand to Gifberg).

polyandra (L.f.) E.Mey. ex Hiern   BAVIAANSKERS, KERSBOS   Dioecious shrub to 2.5 m, rusty velvety on young parts. Leaves elliptic to ovate, leathery, usually grey-hairy. Flowers in rusty velvety axillary racemes, hairy, shallowly lobed, white, fragrant, ovary hairy. Oct.–Nov. Coastal bush and rocky slopes, SW, KM, LB, SE (Rooiels to E Cape).

tomentosa E.Mey. ex A.DC.   KLIPKERS, HEUNINGGWARRIE   Dioecious shrub to 1.5 m, grey-velvety on young parts. Leaves obovate, leathery, usually grey-hairy. Flowers in grey-hairy axillary racemes, sometimes solitary, shallowly lobed, white, fragrant, ovary woolly. Sept.–Oct. Dry rocky slopes, NW, SW (Richtersveld to Cape Peninsula).

A. Corolla lobed halfway or more

crispa (Thunb.) Guerke   BLOUGWARRIE   Dioecious shrub or tree to 8 m, rusty granular on young parts. Leaves sometimes hairy, obovate to lanceolate. Flowers in rusty granular axillary racemes, hairy, deeply cleft, cream-coloured, fragrant, ovary bristly. Sept.–Feb. Forest margins and kloofs, LB, SE (Riversdale eastwards throughout southern Africa).

linearis Zeyh. ex Hiern   SMALBLAAR   Dioecious shrub or tree to 5 m, rusty granular on young parts. Leaves linear-oblanceolate. Flowers in rusty granular axillary racemes, glabrescent, deeply cleft, cream-coloured, ovary hairy. Aug.–Dec. Rocky slopes, NW (Bokkeveld Mts to Clanwilliam and N Province to Zimbabwe).

natalensis A.DC.   BERGGWARRIE, SWARTBASBOOM   Dioecious shrub or tree to 12 m, red-velvety on young parts. Leaves elliptic, leathery, rusty velvety. Flowers in velvety axillary panicles, thinly hairy, deeply cleft, cream-coloured, ovary hairy. May–Jan. Bush and scrub, NW, SW, SE (Pakhuis Mts to Langebaan and Humansdorp to tropical Africa).

racemosa Murray   SEEGWARRIE   Dioecious shrub or small tree to 6 m. Leaves leathery, obovate. Flowers in glabrous axillary racemes, deeply cleft, cream-coloured, fragrant, ovary shortly bristly. Dec.–June. Coastal scrub, NW, SW, AP, LB, SE (Namaqualand to E Cape).

schimperi (A.DC.) Dandy   WITSTAM   Dioecious small tree to 8 m. Leaves leathery, obovate. Flowers in glabrous axillary racemes, deeply cleft, cream-coloured, fragrant, ovary glabrous. Dec.–May. Coastal scrub and forest, SE (Humansdorp to tropical Africa).

undulata Thunb.   Dioecious shrub or tree to 7 m, rusty granular on young parts. Leaves oblanceolate, rusty granular, leathery, often undulate. Flowers in rusty granular axillary racemes, glabrescent, deeply cleft, cream-coloured, fragrant, ovary scaly. Dec.–Apr. Rocky slopes, NW, KM, SE (Worcester to Uitenhage and throughout southern Africa).

ELATINACEAE

BERGIA   c. 25 spp., cosmopolitan

glomerata L.f.   Prostrate, mat-forming, densely leafy woody perennial to 1 m diam. Leaves small, obovate, sometimes coarsely toothed above. Flowers sessile, few in axillary glomerules, white. Nov.–Feb. Damp places or temporary pools, often coastal, NW, SW, AP, KM, LB, SE (Vredendal to Grahamstown).

ERICACEAE
by E.G.H. Oliver & I.M. Oliver

ERICA (= *ACROSTEMON, ANISERICA, ANOMALANTHUS, ARACHNOCALYX, BLAERIA, COCCOSPERMA, COILOSTIGMA, EREMIA, EREMIELLA, GRISEBACHIA, NAGELOCARPUS, PHILIPPIA, PLATYCALYX, SALAXIS, SCYPHOGYNE, SIMOCHEILUS, STOKOEANTHUS, SYMPIEZA, SYNDESMANTHUS, THAMNUS, THORACOSPERMA*)   HEATHER, HEIDE   c. 860 spp., Africa, Europe and Middle East, mostly southern Africa

Flowers are described in the following size classes: *tiny,* shorter than 1 mm; *small,* 1–5 mm long; *medium,* 5–10 mm long and *large,* longer than 10 mm.

> A. Calyx lobes unequal (outer lobe usually larger than 2 side lobes); with no visible bract and bracteoles
>   B. Fruit dehiscent, capsular, 4-locular with many ovules per locule

altiphila E.G.H.Oliv. (= *Philippia alticola* E.G.H.Oliv.)   Erect bushy shrublet to 40 cm. Flowers tiny/small, cup-shaped, dull white, tinged red. Oct.–Jan. S-facing upper slopes, NW, SW (Ceres to Caledon Swartberg).•

elsieana (E.G.H.Oliv.) E.G.H.Oliv. (= *Philippia elsieana* E.G.H.Oliv.)   Erect shrublet. Flowers small, ellipsoid, cream-coloured. Sept. Moist upper slopes, LB (Langeberg Mts: Swellendam).•

esteriana E.G.H.Oliv. (= *Philippia esterhuyseniae* E.G.H.Oliv.)   Erect compact woody shrublet to 40 cm. Flowers tiny/small, cup-shaped, dull cream-coloured. Oct.–Jan. Upper rocky slopes, NW, KM (Ceres and Swartberg Mts).•

exleeana Klotzsch (= *Philippia leeana* Klotzsch)   Low wiry shrublet to 30 cm. Flowers small, cup-shaped, dull cream-coloured to reddish. Aug.–Apr. Dry slopes, SW (Paarl, Stellenbosch, Cape Peninsula to Bredasdorp).•

madida E.G.H.Oliv. (= *Philippia irrorata* E.G.H.Oliv.)   Erect sticky shrub to 2.5 m. Flowers tiny/small, cup-shaped, pale green tinged red. Nov.–Dec. Upper S slopes, KM, LB (Ladismith, Heidelberg).•

notholeeana (E.G.H.Oliv.) E.G.H.Oliv. (= *Philippia notholeeana* E.G.H.Oliv.)   Erect shrublet to 30 cm. Flowers small, cup-shaped, cream-coloured, tinged red. Nov.–Aug. Rocky slopes, SW (Franschhoek Mts to Kogelberg).•

petricola E.G.H.Oliv. (= *Philippia petrophila* E.G.H.Oliv.)   Erect small compact shrublet to 10(–30) cm. Flowers tiny to small, cream-coloured tinged red. Oct.–Jan. Rock crevices or rocky slopes, SW (Riviersonderend Mts).•

procaviana (E.G.H.Oliv.) E.G.H.Oliv. (= *Philippia procaviana* E.G.H.Oliv.)   Erect compact shrublet to 30 cm. Flowers tiny/small, urn-shaped, greenish tinged red. Sept.–Oct. Summit rocky slopes, NW (Langeberg Mts: Robertson).•

tristis Bartl. (= *Philippia chamissonis* Klotzsch)   Erect dense shrub, 1–3 m. Flowers small, cup-shaped, dull cream-coloured. Nov.–May. Rocky slopes or sandy flats, SW, AP (Langebaan to Gansbaai).•

B. Fruit indehiscent, usually 1–3-locular with 1 or 2 ovules per locule

areolata (N.E. Br). E.G.H.Oliv. (= *Coccosperma areolatum* N.E.Br.) Erect shrublet to 50 cm. Flowers tiny/small, greenish white, with very large stigma, Aug.–Nov. Stony slopes or sandy plateaus, NW, KM (Cedarberg Mts to Laingsburg).•

artemisioides (Klotzsch) E.G.H.Oliv. (= *Coccosperma micrantha* (Benth.) N.E.Br.) Dense erect shrublet to 1 m. Flowers tiny/small, dull white, with very large stigma. Oct.–Jan. Rocky upper slopes, NW, SW (Hex River to Hottentots Holland Mts).•

axillaris Thunb. (= *Salaxis axillaris* (Thunb.) Salisb. ex G.Don, *S. calyciflora* (Tausch) Druce, *S. flexuosa* Klotzsch, *S. major* N.E.Br., *S. octandra* Klotzsch, *S. puberula* Klotzsch) Erect shrublet to 80 cm. Flowers small, obovoid to cup-shaped, pale greenish, with very large stigma. May–Dec. Mountain slopes and flats SW, AP, SE (Cape Peninsula to Knysna).•

binaria E.G.H.Oliv. Compact dense shrublet to 60 cm. Flowers small, obovoid, yellow-green tinged red, with very large stigma. Feb.–May at Stellenbosch or Oct.–Dec. Rocky upper slopes, NW, SW (Stellenbosch and Riviersonderend Mts).•

boucheri E.G.H.Oliv. Compact shrublet to 50 cm. Flowers tiny/small, obovoid to cup-shaped, cream-coloured tinged red, with very large stigma. Mar.–May. Sandy slopes, SW (Riviersonderend Mts:Jonaskop).•

bredasiana E.G.H.Oliv. Low rounded shrublet to 30 cm. Flowers small, cup-shaped, creamy green tinged red, with very large stigma. Sandy calcareous flats, AP (Bredasdorp).•

burchelliana E.G.H.Oliv. (= *Coilostigma glabrum* Benth.) Compact many-stemmed resprouting shrublet to 50 cm. Flowers small, tubular to bell-shaped, white to dark pink, with 4 exserted anthers. Dec. Silcrete hills, LB (Riversdale).•

calcicola (E.G.H.Oliv.) E.G.H.Oliv. (= *Scyphogyne calcicola* E.G.H.Oliv.) Compact shrublet to 30 cm. Flowers small, cup-shaped, pale green tinged red, with very large stigma. Apr.–June. Calcareous hills AP (Gansbaai and Stilbaai).•

eglandulosa (Klotzsch) Benth. (= *Scyphogyne eglandulosa* (Klotzsch) Benth.) Dense finely branched erect shrublet to 30 cm. Flowers tiny, cup-shaped, dirty white, with very large stigma. Feb.–July. Mountain slopes, SW (Riviersonderend Mts and Caledon Swartberg).•

melanomontana E.G.H.Oliv. (= *Scyphogyne orientalis* E.G.H.Oliv.) Erect compact shrublet to 50 cm. Flowers tiny/small, obovoid, greenish, tinged red, with very large stigma. Sept.–Jan. Rocky upper slopes, KM, SE (Swartberg Mts to Uniondale).•

miniscula E.G.H.Oliv. (= *Scyphogyne tenuis* (Benth.) E.G.H.Oliv.) Erect shrublet to 50 cm. Flowers minute, obovoid, with very large stigma. Dec. Mountain slopes, LB (Langeberg Mts: Riversdale).•

muscosa (Aiton) E.G.H.Oliv. (= *Scyphogyne muscosa* (Aiton) Steud.) Erect sticky shrublet to 1 m. Flowers tiny, obconic, dirty white, with very large stigma. Jan.–Dec. Abundant on mountain slopes and flats NW, SW, AP, LB, SE (Gifberg to George).•

parviporandra E.G.H.Oliv. Erect shrublet to 1.5 m. Flowers small, cup-shaped, green tinged red, with very large stigma. Aug.–Dec. Sandy rocky lower slopes, LB (Langeberg Mts: Swellendam to Riversdale).•

perplexa E.G.H.Oliv. Erect shrub to 1.5 m. Flowers tiny/small, white, with very large stigma. Apr.–May. Moist slopes, SW (Caledon Swartberg).•

phacelanthera E.G.H.Oliv. (= *Scyphogyne capitata* (Klotzsch) Benth.) Erect sticky shrublet to 1 m. Flowers small, pear-shaped, cream-coloured, tinged red, with very large stigma. Oct.–Jan. Mountain slopes, NW, SW (Montagu and Riviersonderend Mts).•

remota (N.E.Br.) E.G.H.Oliv. (= *Scyphogyne remota* N.E.Br.) Sparse erect shrublet to 50 cm. Flowers small, cream-coloured, with very large stigma. Dec.–Jan. Sandy lower slopes, SW (Riviersonderend).•

rigidula (N.E.Br.) E.G.H.Oliv. (= *Scyphogyne brownii* Compton, *S. divaricata* (Klotzsch) Benth., *S. longistyla* N.E.Br., *S. rigidula* N.E.Br.) Erect shrublet to 50 cm. Flowers tiny/small, cup-shaped, cream-coloured tinged red, with very large stigma. July–Jan. Sandy or rocky slopes or flats, NW, SW, KM (Kamiesberg to Bredasdorp and Witteberg).

rugata E.G.H.Oliv. (= *Coccosperma rugosum* N.E.Br.) Erect shrub to 1.2 m. Flowers tiny/small, cup-shaped, greenish, with very large stigma. Sept.–Feb. in west, Mar.–June in east. Sandy flats, SW, AP (Cape Flats to Duivenhoks River mouth).•

serrata Thunb. (= *Nagelocarpus serratus* (Thunb.) Bullock)   Erect hard-leaved shrublet to 1 m. Flowers small, cup-shaped, greenish, with very large stigma. Jan.–Dec. Mountain slopes, SW (Paarl, Worcester, Caledon, Bredasdorp).•

subcapitata (N.E.Br.) E.G.H.Oliv. (= *Coccosperma hexandrum* (Klotzsch) Druce, *C. forbesianum* Klotzsch)   Erect shrublet to 80 cm. Flowers small, urn-shaped greenish, with very large stigma. Sept.–Apr. Mountain slopes, SW (Cape Peninsula).•

terniflora E.G.H.Oliv. (= *Salaxis triflora* Compton)   Erect shrublet to 50 cm. Flowers small, cup-shaped, green to reddish, with very large stigma. Jan.–May. Rocky upper slopes, NW, KM (Cedarberg Mts to Montagu and Witteberg).•

urceolata (Klotzsch) Benth. (= *Scyphogyne urceolata* (Klotzsch) Benth., *S. puberula* Klotzsch)   Erect soft shrub to 2.5 m. Flowers tiny/small, ellipsoid to urn-shaped, pale greenish, with very large stigma. Jan.–Dec., mostly Sept.–Dec. S-facing rocky upper slopes, NW, SW, LB (Cedarberg to Langeberg Mts).•

zeyheriana (Klotzsch) E.G.H.Oliv. (= *Coilostigma zeyherianum* Klotzsch)   Erect shrublet to 1 m. Flowers small, tubular to ovoid, dirty white to greenish, with 4 exserted anthers. Jan.–Dec. Sandy coastal flats, SE (Humansdorp to Alexandria).

A. Calyx more or less equally lobed; usually with 2 bracteoles on stalk
C. Fruit indehiscent or partially dehiscent, mostly 1- or 2-locular; stamens mostly 4

agglutinans E.G.H.Oliv. (= *Syndesmanthus schlechteri* N.E.Br.)   Low, small, sparse, spreading shrublet to 10 cm. Flowers in pendulous heads, small, obovoid, pink, very sticky, with 4 exserted anthers. Mar.–May. Sandy hills, SW (Elim and Napier).•

albertyniae E.G.H.Oliv.   Semispreading shrublet to 25 cm. Flowers in terminal heads, small, tubular, pink with 4 exserted anthers. Feb.–June. Sandy flats, AP (Bredasdorp).•

anguliger (N.E.Br.) E.G.H.Oliv. (= all species formerly in *Anomalanthus*)   Much-branched low spreading shrublet to 20 cm. Flowers small, tubular to bell-shaped, pink, with 4 exserted anthers, fruiting calyx enlarging and turning red or black. Apr.–Sept. Very common on lower slopes, NW, SW, AP, KM, LB (Worcester to Mossel Bay).•

angulosa E.G.H.Oliv. (= *Acrostemon fourcadei* (L.Guthrie) E.G.H.Oliv.)   Erect shrublet to 50 cm. Flowers small, tubularly cup-shaped, cream-coloured to pale pink, with 4 exserted anthers. Mar.–Apr. Dry slopes, SE (Kouga and Baviaanskloof Mts).•

arachnocalyx E.G.H.Oliv. (= *Arachnocalyx viscidus* (N.E.Br.) E.G.H.Oliv.)   Small rounded shrublet to 20 cm. Flowers in terminal heads, small, tubular, cream-coloured with exserted anthers and sticky hairy calyx. Oct.–Jan. Sandy flats and lower slopes, NW (Cold Bokkeveld).•

articulata (L.) Thunb. (= *Syndesmanthus articulatus* (L.) Klotzsch, *S. scaber* Klotzsch and *S. similis* N.E.Br.)   Erect compact shrublet to 30 cm. Flowers in terminal heads, small, tubular to narrowly funnel-shaped, pink, with 4 exserted anthers. Feb.–July. Lower slopes and flats, SW, AP, LB, SE (Cape Peninsula to George).•

atromontana E.G.H.Oliv.   Sprawling to erect shrublet to 50 cm. Flowers small, bell-shaped to ovoid, pink, with 4 exserted anthers. Aug.–Dec. Rocky upper slopes, KM (Swartberg Mts).•

bokkeveldia E.G.H.Oliv. (= *Eremia calycina* Compton)   Low compact shrublet to 30 cm. Flowers small, broadly cup-shaped, pink. Sept.–Nov. Sandy flats, NW (Cold Bokkeveld).•

bolusanthus E.G.H.Oliv. (= *Thoracosperma nanum* N.E.Br.)   Erect shrublet to 25 cm. Flowers small, cup-shaped, pink, with 4 exserted anthers. Nov.–Apr. Rocky slopes, KM, SE (Uniondale to Humansdorp).•

brownii E.G.H.Oliv. (= *Syndesmanthus breviflorus* N.E.Br.)   Erect low compact shrublet to 15 cm. Flowers in terminal heads, small, tubular, dark red or pink, with 4 exserted anthers. Aug.–Dec. Gravelly flats, SW, AP (Elim to Bredasdorp).•

caprina E.G.H.Oliv. (= *Grisebachia minutiflora* N.E.Br.)   Low compact to spreading shrublet to 20 cm. Flowers in terminal many-flowered heads, small, broadly funnel-shaped, white. Oct.–Jan. Sandy flats, NW (Ceres).•

cereris (Compton) E.G.H.Oliv. (= *Arachnocalyx cereris* Compton)   Erect shrublet to 40 cm. Flowers in terminal heads, small, ovoid, pale pink with exserted anthers and woolly calyx. Sept.–Dec. Sandy lower slopes, NW (Ceres).•

cetrata E.G.H.Oliv. (= *Eremia peltata* Compton)   Erect resprouting shrublet to 50 cm. Flowers small, broadly obconic, cream-coloured to pink. Aug.–Nov. Sand, lower slopes, NW, KM (Cold Bokkeveld to Bonteberg).•

curvistyla (N.E.Br.) E.G.H.Oliv. (= *Eremia curvistyla* (N.E.Br.) E.G.H.Oliv.)   Low spreading shrublet to 20 cm. Flowers small, obconic, white. Sept.–Dec. Rocky sandstone slopes, NW, SW (Piketberg to Ceres and Bainskloof).•

dispar N.E.Br. (= *Simocheilus dispar* N.E.Br.)   Erect to spreading shrublet to 30 cm. Flowers in terminal heads, small, tubular, pink, with 4 exserted anthers. Mar.–Sept. Sandy flats, AP, LB (Potberg to Albertinia).•

dregei E.G.H.Oliv. (= *Simocheilus oblongus* Benth.)   Erect to spreading shrublet to 30 cm. Flowers in terminal heads, small, tubular, pink, with 4 exserted anthers. June–Sept. Lower slopes, NW (Piketberg).•

ecklonii E.G.H.Oliv. (= *Sympieza eckloniana* Klotzsch)   Erect shrublet to 75 cm. Flowers in terminal heads, small, tubular, white, with 4 exserted anthers, corolla 2-lipped. July–Nov. Moist slopes, SW (Hermanus).•

eremioides (MacOwan) E.G.H.Oliv. (= *Grisebachia parviflora* (Klotzsch) Druce)   Sparse sprawling shrublet to 20 cm. Flowers small, obconic to funnel-shaped, white, slightly hairy. July–Jan. Rocky slopes, NW, SW, LB (Cedarberg to Langeberg Mts).•

erina (Klotzsch ex Benth.) E.G.H.Oliv. (= *Syndesmanthus erinus* (Klotzsch ex Benth.) N.E.Br.)   Erect low wiry shrublet to 30 cm. Flowers in terminal heads, small, obovoid, pale pink, with 4 exserted anthers. Aug.–Apr. Lower slopes, SW (Hermanus to Bredasdorp).•

eriocephala Lam. (= *Acrostemon hirsutus* Klotzsch, *A. stokoei* L.Guthrie)   Erect shrublet to 40 cm. Flowers in terminal heads, small, tubular to narrowly ovoid, bright pink, with 4 exserted anthers. Aug.–Oct. Mountain slopes, SW (Elandskloof Mts to Caledon).•

glabella Thunb. (= *Simocheilus glabellus* (Thunb.) Benth., *S. purpureus* (P.J.Bergius) Druce)   Erect to sprawling shrublet to 50 cm. Flowers small, tubular ovoid, pink, with 4 exserted anthers, slightly honey-scented. Jan.–Dec. Sandy flats and lower to middle slopes, SW, AP (Cape Peninsula to Breede River mouth).•

globiceps (N.E.Br.) E.G.H.Oliv. (= *Syndesmanthus globiceps* N.E.Br., *S. elimensis* N.E.Br., *S. gracilis* N.E.Br., *S. sympiezoides* N.E.Br., *S. zeyheri* Bolus)   Low sprawling shrublet to 30 cm. Flowers in terminal heads, small, tubular, pink, with 4 exserted anthers. Aug.–Dec. Flats to upper slopes, SW, AP (Riviersonderend Mts, Hermanus to Agulhas).•

inaequalis (Klotzsch) E.G.H.Oliver (= *Simocheilus albirameus* N.E.Br., *S. bicolor* (Klotzsch) Benth., *S. glaber* (Thunb.) Benth., *S. puberulus* (P.J.Bergius) Druce)   Low spreading shrublet to 20 cm. Flowers small, tubular ovoid, pink, with 4 exserted anthers. Sept.–Feb. Dry slopes, NW, SW (Bokkeveld Mts to Breede River valley: Worcester).•

inflaticalyx E.G.H.Oliv. Erect shrublet to 60 cm. Flowers small, obovoid, pink with 4 exserted anthers. June–Sept. Rocky middle slopes, LB (Outeniqua Mts, north side).•

innovans E.G.H.Oliv. (= *Syndesmanthus pumilus* N.E.Br.)   Erect compact shrublet to 30 cm. Flowers in terminal heads, small, tubular, pink, with 4 exserted anthers. Feb.–Apr. Sandy hills, SW (Elim).•

interrupta (N.E.Br.) E.G.H.Oliv. (= *Thoracosperma interruptum* N.E.Br.)   Low spreading shrublet to 20 cm. Flowers in terminal heads, small, narrowly ovoid, pale pink turning hard, shiny and brown in fruit, with 4 exserted anthers. Aug.–Oct. Sandy hills and flats, SW, AP (Pearly Beach to Elim).•

jonasiana E.G.H.Oliv. Low prostrate spreading shrublet to 10 cm. Flowers small, bell-shaped, fluted, bright pink. Sept.–Nov. Sandy places, SW (Jonaskop to Villiersdorp).•

kammanassiae E.G.H.Oliv.   Erect shrublet to 80 cm. Flowers small, obovoid, dark pink, with 4 exserted anthers. July–Sept. Rocky middle northern slopes, KM (Kammanassie Mts).•

karwyderi E.G.H.Oliv.   Compact shrublet to 50 cm. Flowers small, tubular ovoid, pink, with exserted anthers. May–Sept. Sandy hillslopes, SW (Botrivier, Grabouw).•

labialis Salisb. (= *Sympieza labialis* (Salisb.) Druce, *S. breviflora* N.E.Br., *S. capitellata* Licht. ex Roem. & Schult., *S. pallescens* N.E.Br., *S. vestita* N.E.Br.)   Erect compact shrublet to 50 cm. Flowers in terminal heads, small to medium, tubular to funnel-shaped, pink, with 4 exserted anthers, corolla 2-lipped. Jan.–Dec. Rocky, sandy slopes and flats, SW (Paarl to Bredasdorp).•

lateriflora E.G.H.Oliv. (= *Grisebachia secundiflora* E.G.H.Oliv.)   Low compact to sprawling shrublet to 50 cm. Flowers small, tubularly bell-shaped, pale pink to white. Sept.–Nov. Sand, NW (Cold Bokkeveld).•

macrocalyx (T.M.Salter) E.G.H.Oliv. (= *Aniserica macrocalyx* T.M.Salter, *A. gracilis* (Bartl.) N.E.Br.)   Erect shrublet to 40 cm. Flowers small, tubular, dull white to pale pink, with 4 exserted anthers, corolla 2-lipped. Aug.–Dec. Moist slopes, NW, SW, LB, SE (Piketberg to George).•

malmesburiensis E.G.H.Oliv.   Erect shrublet to 30 cm. Flowers in terminal heads, small, tubularly bell-shaped, pink, with 4 exserted anthers. Nov.–Mar. Sandy flats, SW (Malmesbury).•

montis-hominis E.G.H.Oliv.   Low semisprawling shrublet to 25 cm. Flowers small, obconic, pink, with 4 exserted anthers. Sept.–Nov. Rocky upper slopes, KM (Mannetjiesberg).•

niveniana E.G.H.Oliv.   Compact shrublet to 25 cm. Flowers in terminal fluffy heads, small, slightly inflated-tubular, pink, with 4 exserted anthers. Aug.–Nov. Sandy, stony slopes, SW (Houwhoek, Villiersdorp).•

outeniquae (Compton) E.G.H.Oliv. (= *Eremiella outeniquae* Compton)   Low compact shrublet to 15 cm. Flowers small, cup-shaped, dull red. Oct.–Jan. Upper slopes, SE (Outeniqua Mts).•

paucifolia (J.C.Wendl.) E.G.H.Oliv. (= *Syndesmanthus paucifolius* (J.C.Wendl.) Benth., *S. ciliatus* (Klotzsch) Benth., *S. squarrosus* Benth.)   Erect to spreading shrublet to 20 cm. Flowers small, narrowly obovoid, pink, with 4 exserted anthers, calyx sometimes fluffy. July–Oct. Lower slopes, SW (Houwhoek, Kleinmond, Caledon).•

phaeocarpa (= *Simocheilus quadrisulcus* N.E.Br.)   Low sprawling shrublet to 30 cm. Flowers small, tubular ovoid, pink, with 4 exserted anthers. Mar.–Dec. Rocky slopes, KM (Swartberg Mts).•

pilosiflora E.G.H.Oliv. (= *Acrostemon eriocephalus* (Klotzsch) N.E.Br.)   Erect compact greyish shrublet to 30 cm. Flowers in terminal fluffy heads, small, slightly inflated-tubular, white to pink, hairy, with 4 exserted anthers. Aug.–Nov. Sandy places, SW (Sir Lowry's Pass to Genadendal and Robertson).•

piquetbergensis (N.E.Br.) E.G.H.Oliv. (= *Simocheilus piquetbergensis* N.E.Br.)   Erect shrublet to 30 cm. Flowers in terminal heads, small, tubular, pink, with 4 exserted anthers. Aug.–Oct. Upper slopes, NW (Piketberg).•

platycalyx (= *Platycalyx pumila* N.E.Br.)   Low compact shrublet to 20 cm. Flowers small, globosely urn-shaped, pink, with 6–8 exserted anthers. Aug.–Oct. Sand on calcareous hills, AP (Riversdale, Albertinia).•

plumosa Thunb. (= *Grisebachia plumosa* (Thunb.) Klotzsch, *G. ciliaris* (L.f.) Klotzsch, *G. incana* (Bartl.) Klotzsch, *G. nivenii* N.E.Br., *G. rigida* N.E.Br.)   Erect shrublet to 50 cm. Flowers in terminal heads, small to medium, rounded at base with cup-shaped top, pink with hairy calyx and mostly 4 exserted anthers. June–Sept. Sandy flats and lower slopes, NW, SW (Bokkeveld to Langeberg Mts).•

puberuliflora E.G.H.Oliv. (= *Thoracosperma puberulum* (Klotzsch) N.E.Br.)   Erect shrublet to 50 cm. Flowers small, broadly tubular, pale to deep maroon, mostly hairy. July–Oct. Dry lower slopes, SW, LB (Hermanus to Riversdale).•

pulchelliflora E.G.H.Oliv. (= *Syndesmanthus pulchellus* N.E.Br.)   Low compact shrublet to 15 cm. Flowers in terminal heads, small, tubular, pink, with 4 exserted anthers. Dec. Lower slopes, SW (Napier).•

quadrifida (Benth.) E.G.H.Oliv. (= *Thoracosperma paniculatum* (Thunb.) Klotzsch)   Erect shrublet to 1 m. Flowers small, tubular to ovoid, cream-coloured to pale pink, with 4 exserted anthers. Apr.–Nov. Lower slopes and flats, LB, SE (Langeberg Mts to George).•

radicans (L.Guthrie) E.G.H.Oliv. (= *Thoracosperma radicans* L.Guthrie, *Acrostemon schlechteri* N.E.Br.)   Prostrate sparse shrublet rooting at nodes. Flowers small, narrowly ovoid to bell-shaped, pale pink, with 4 exserted anthers. Apr.–May. Sand on calcareous hills, SW, AP (Hermanus to Gouritsmond).•

recurvifolia E.G.H.Oliv. (= *Eremia recurvata* Klotzsch)   Low spreading grey shrublet to 30 cm. Flowers small, bell-shaped, white. Nov.–Jan. Rocky sandstone flats and slopes, NW (Clanwilliam, Ceres).•

rosacea (L.Guthrie) E.G.H.Oliv. (= *Thoracosperma bondiae* Compton, *T. fourcadei* Compton, *T. galpinii* N.E.Br., *T. marlothii* N.E.Br., *T. rosacea* L.Guthrie)   Erect shrublet, sometimes low and compact, to 50 cm. Flowers small, tubular to ellipsoid, pale to deep pink, with 4 exserted anthers. Mar.–Oct. Rocky slopes, KM, LB, SE (Witteberg and Little Karoo Mts to Hankey).•

stokoeanthus E.G.H.Oliv. (= *Stokoeanthus chionophilus* E.G.H.Oliv.)   Erect shrub to 1.2 m. Flowers small, urn-shaped, dull white. Oct.–Dec. Moist upper slopes, SW (Hottentots Holland Mts).•

thamnoides E.G.H.Oliv. (= *Thamnus multiflorus* Klotzsch)   Erect shrub to 1.5 m. Flowers small, ovoid-urn-shaped, pink, with 4 exserted anthers. Jan.–Dec. Flats and lower slopes, SE (Uniondale to Uitenhage).•

totta Thunb. (= *Eremia totta* (Thunb.) D.Don)   Low spreading grey shrublet to 30 cm. Flowers small, urn-shaped, white. Sept.–Dec. Common on rocky slopes, NW, SW (Cedarberg to Stellenbosch Mts).•

uberiflora E.G.H.Oliv. (= *Simocheilus barbiger* Klotzsch, *S. carneus* Klotzsch, *S. multiflorus* Klotzsch, *S. pubescens* Klotzsch)   Erect shrublet to 1 m. Flowers small, tubular, pale to deep pink, with 4 exserted anthers. Nov.–May. Slopes and flats, SE (George to Humansdorp).•

vallisfluminis E.G.H.Oliv.   Compact erect shrublet to 50 cm. Flowers small, bell-shaped to obconic, hairy, pink. Aug.–Nov. Sandy slopes, LB (Heidelberg to Riversdale).•

velatiflora E.G.H.Oliv. (= *Eremia brevifolia* Benth.)   Sparse erect shrublet to 30 cm. Flowers small, cup-shaped, white tinged red. Oct.–Nov. Upper slopes, LB, SE (Attaquas Kloof and Outeniqua Mts).•

venustiflora E.G.H.Oliv. (= *Syndesmanthus venustus* N.E.Br.)   Low compact shrublet to 20 cm. Flowers in terminal heads, small, tubular, pink, with 4 exserted anthers. May–Aug. Hills, SW (Bredasdorp area).•

vernicosa E.G.H.Oliv.   Prostrate mat-forming shrublet to 10 cm. Flowers small, narrowly ovoid, purple-pink, shiny/sticky, with 4 exserted stamens. Mar.–Sept. Sand over calcareous rock, AP (De Hoop to Duivenhoks River mouth).•

viscosissima (Bolus) N.E.Br. (= *Syndesmanthus viscosus* (H.Bolus) N.E.Br.) Erect compact shrublet to 40 cm. Flowers in terminal sticky heads, small, tubular ovoid, pink, with 4 exserted anthers. Nov.–Feb. Sandy flats, AP, LB (Duivenhoks River to Albertinia).•

vlokii E.G.H.Oliv.   Erect woody shrub to 1.5 m. Flowers small, ellipsoid, white to pale pink, with exserted anthers. July–Oct. Rocky slopes, KM, SE (Swartberg, Kammanassie and Outeniqua Mts).•

williamsiorum E.G.H.Oliv.   Erect shrublet to 50 cm. Flowers small, tubular, deep pink with 4 exserted anthers. Jan.–May. Rocky upper slopes, SW (Kleinrivier Mts to Napier).•

xeranthemifolia (Salisb.) E.G.H.Oliv. (= *Acrostemon xeranthemifolius* (Salisb.) E.G.H.Oliv., *Hexastemon lanatus* Klotzsch)   Low semispreading grey shrublet to 15 cm. Flowers in terminal heads, small/medium, elongate-ovoid, white, with fluffy calyx and 4 exserted anthers. Aug.–Oct. Sand flats, SW (Caledon: Shaw's Mtn).•

C. Fruit dehiscent, capsular, 4(–8)-locular
D. Stamens 4

barbigeroides E.G.H.Oliv. (= *Blaeria barbigera* (Salisb.) G.Don)   Erect soft shrublet to 30 cm. Flowers in many-flowered heads, small, tubular, pink, with 4 exserted anthers. Sept.–Feb. Lower coastal slopes and sandy flats, SW (Betty's Bay to Hermanus).•

chiroptera E.G.H.Oliv.   Erect shrublet to 50 cm. Flowers small, narrowly cup-shaped, pale pink, with 4 exserted anthers. Dec.–Mar. Sandy lower slopes, SW (Kogelberg).•

equisetifolia Salisb. (= *Blaeria equisetifolia* (Salisb.) G.Don, *B. dumosa* J.C.Wendl.)   Erect compact shrublet to 20 cm. Flowers small, tubular to narrowly cup-shaped, pink, with 4 exserted anthers. Oct.–Apr. Mountain slopes, SW, AP (Bainskloof to Agulhas coast).•

ericoides (L.) E.G.H.Oliv. (= *Blaeria ericoides* L.)   Erect compact woody shrublet to 80 cm. Flowers small, tubularly urn-shaped, pale pink, with 4 exserted anthers, honey-scented. Jan.–Apr. Lower slopes, SW (Cape Peninsula to Napier).•

fuscescens (Klotzsch) E.G.H.Oliv. (= *Blaeria fuscescens* Klotzsch)   Erect shrublet to 1 m. Flowers small, tubular with spreading lobes, pale pink to white, with 4 exserted anthers. Apr.–Jan. Lower slopes, SE (George to Uitenhage).•

hermani E.G.H.Oliv.   Erect shrublet to 50 cm. Flowers small, cup-shaped, with exserted anthers. Jan.–Feb. Sandy flats, SW (Hermanus).•

klotzschii (Alm & Fries) E.G.H.Oliv. (= *Blaeria klotzschii* Alm & Fries)   Erect slender shrublet to 30 cm. Flowers small, urn-shaped, pale to dark pink, with 4 exserted anthers. June–Oct. Dry lower slopes, SW, LB, SE (Bredasdorp and Langeberg Mts: Swellendam to Riversdale).•

longimontana E.G.H.Oliv. (= *Blaeria coccinea* Klotzsch)   Erect soft shrublet to 60 cm. Flowers small, tubular, deep pink, hairy, with 4 exserted anthers. Dec.–June. Lower to middle S-slopes, LB, SE (Langeberg and Outeniqua Mts).•

multiflexuosa E.G.H.Oliv. (= *Blaeria flexuosa* Benth.)　　Erect compact wiry shrublet to 30 cm. Flowers small, cup-shaped, pale yellow, with 4 exserted anthers. Dec.–Apr. Lower sandy slopes, SW (Hottentots Holland Mts).•

russakiana E.G.H.Oliv. (= *Blaeria kraussiana* Klotzsch ex Walp.)　　Erect shrublet to 50 cm. Flowers in terminal heads, small, narrowly urn-shaped, pink, with 4 exserted anthers. Dec.–Feb. Upper slopes, SW (Caledon Swartberg and Kleinrivier Mts).•

sagittata Klotzsch ex Benth. (= *Blaeria sagittata* (Klotzsch ex Benth.) Alm & Fries)　　Erect shrublet to 1 m. Flowers medium, narrowly urn-shaped, white, with 4 exserted anthers. Sept.–Oct. Lower slopes, SE (Van Staden's Mts).•

D. Stamens mostly 8

abelii E.G.H.Oliv. Erect gnarled woody shrublet to 60 cm. Flowers large, tubular, lime-green, finely hairy. Oct.–Nov. Rocky dry middle slopes, SE (Uitenhage).•

abietina L.　　RED HEATH, ROOIHEIDE　　Erect shrublet to 1 m. Flowers large, tubular, bright red. Jan.–Sept. Mountain plateaus, SW (Cape Peninsula).•

accommodata Klotzsch　　Erect compact shrublet to 30 cm. Flowers small, calycine, ovoid, white, viscid, with exserted anthers. Jan.–May. Upper slopes in rock crevices, SW (Riviersonderend Mts).•

acuta Andrews　　Erect shrublet to 60 cm. Flowers small/medium, calycine, ovoid, pink. Sept.–Nov. Upper rocky slopes, NW, SW (Ceres to Wemmershoek).•

adnata L.Bolus　Erect shrublet to 1 m. Flowers small, calycine, cup-shaped with spreading lobes, pink. Aug.–Nov. High alt., SW (Franschhoek to W Riviersonderend Mts).•

aemula Guthrie & Bolus　　Like **E. distorta**, erect shrublet to 50 cm. Flowers small, urn-shaped, white, hairy. Jan.–Nov. Rocky coastal slopes or inland marshy high mountain plateaus, SW (Bainskloof to Steenbras Mts).•

affinis Benth.　　Erect shrub to 1(–2) m. Flowers small, bell-shaped, pink. Sept.–Dec. Middle slopes, SE (Uniondale to Uitenhage).•

aghillana Guthrie & Bolus　　Low shrublet to 25 cm. Flowers small, narrowly cup-shaped, dark pink, with exserted anthers. Apr. Coastal flats, AP (Agulhas).•

albens L.　　Erect shrub to 1.2 m. Flowers small, ovoid with spreading lobes, white or yellowish green. Apr.–Dec. Damp upper slopes, KM, LB (Langeberg, Swartberg and Outeniqua Mts).•

albescens Klotzsch ex Benth.　　Erect shrublet to 30 cm. Flowers small, urn-shaped, white, hairy. Nov.–Feb. Upper slopes, LB (Langeberg Mts: Swellendam to Heidelberg).•

alexandri Guthrie & Bolus (= *Erica acockii* Compton)　　Erect compact shrublet to 15 cm. Flowers small, urn-shaped, pink, with slightly exserted anthers. Mar. Rare on sandy moist flats, SW (Wemmershoek, Kraaifontein).•

alfredii Guthrie & Bolus　　Erect semispreading compact shrublet to 20 cm. Flowers large, calycine, urn-shaped with spreading lobes, bright pink. Jan.–Apr. Dry stony places, high alt., SW (Riviersonderend Mts).•

alnea E.G.H.Oliv. Erect shrublet to 60 cm. Flowers small, pink. May–Oct. Damp slopes, ledges at high alt., NW, SW (Cedarberg Mts to Hex River Mts and Villiersdorp).•

altevivens H.A.Baker　　Erect dense shrublet to 50 cm. Flowers small, urn-shaped, white with musty odour. Dec.–Jan. Rocky outcrops on higher peaks NW, SW (Du Toitskloof Mts, Wemmershoek to W Riviersonderend Mts).•

amicorum E.G.H.Oliv.　　Sparse prostrate shrublet. Flowers small, saucer-shaped, deep pink, with exserted anthers. Sept.–Dec. Seeps on lower slopes, LB (Langeberg: Riversdale).•

amoena J.C.Wendl.　　Erect shrublet to 1 m. Flowers medium, bell-shaped, deep pink. Oct.–Apr. Marshy places, SW (Cape Peninsula).•

amphigena Guthrie & Bolus　　Like **E. spumosa**, erect shrublet to 60 cm. Flowers small, calycine, narrowly cup-shaped, pale pink, with exserted anthers. Sept.–Dec. Rare on upper slopes, SW (Houwhoek).•

ampullacea Curtis　BOTTLE HEATH, BOTTELHEIDE, SISSIEHEIDE　Erect shrublet to 50 cm. Flowers large, variably ampullaceous with spreading lobes, white to pink. Sept.–Jan. Slopes and hills, SW, AP (Swartberg: Caledon, Bredasdorp).•

andreaei Compton　　Like **E. zwartbergensis**, erect shrublet to 60 cm. Flowers small, urn-shaped, white. Sept.–Oct. Upper slopes, KM (E Swartberg and Baviaanskloof Mts).•

aneimena Dulfer (= *Erica hirsuta* Klotzsch ex Benth.) Erect diffuse shrublet to 60 cm. Flowers small, urn-shaped, pink, with exserted anthers. July–Oct. Moist lower slopes, SE (Outeniqua Mts: George).•

annectens Guthrie & Bolus Erect to spreading shrub, 0.6–1 m. Flowers large, tubular, orange to reddish. Jan.–Mar. Rare and local in rocky places on upper slopes, SW (Cape Peninsula).•

arcuata Compton (= *E. ostiaria* Compton) Erect shrublet to 50 cm. Flowers small, urn- to cup-shaped, white, with exserted anthers. Jan.–Dec. Upper slopes, NW, KM (Cedarberg to Laingsburg and Klein Swartberg Mts).•

ardens Andrews Erect shrublet to 70 cm. Flowers large, globose, red, sticky. Apr.–Oct. Middle slopes, LB (Langeberg Mts: Swellendam to Heidelberg).•

arenaria L.Bolus Erect shrub to 2 m. Flowers small, white. May–Aug. Coastal hills, AP (Struisbaai).•

argentea Klotzsch Erect shrublet to 40 cm. Flowers small, subcalycine, obconic, pale pink or white. Sept.–Nov. Middle to upper alt., sandy places, NW (Cedarberg to Cold Bokkeveld Mts).•

argyrea Guthrie & Bolus Erect brittle shrublet to 20 cm. Flowers small, urn-shaped, white to pale pink, hairy. Sept.–Dec. Local on shady cliffs at middle to upper alt., SW (Stellenbosch).•

aristata Andrews Erect semispreading shrublet to 50 cm. Flowers large, tubular-inflated with spreading lobes, longitudinally striped dark and light pink, very sticky. Aug.–Oct. Local on middle to upper rocky slopes, SW (Kleinrivier Mts).•

aristifolia Benth. (= *Erica sonora* Compton) Erect sparse shrublet to 1 m. Flowers small, inflated, urn-shaped, pink. Apr.–July. Sandy places, NW (Gifberg).•

armata Klotzsch ex Benth. (= *Erica umbrosa* H.A.Baker) Erect shrublet to 50 cm. Flowers medium, broadly urn-shaped, hairy, pink, with exserted anthers. Jan.–May. Rocky middle slopes, SW (Franschhoek Mts to Villiersdorp).•

articularis L. Erect resprouting shrub to 1(–2) m. Flowers in tightly packed spikes, small, calycine, urn-shaped, pink often with corolla much paler. Jan.–Dec. Coastal flats and middle to upper slopes, NW, SW, KM, LB, SE (Cedarberg Mts to Cape Peninsula, Swartberg Mts to Humansdorp).•

aspalathoides Guthrie & Bolus Spreading shrublet to 20 cm. Flowers small, narrowly cup-shaped, white to pink, hairy. Sept.–Dec. High summits, moist places, NW (Cedarberg to Cold Bokkeveld Mts).•

astroites Guthrie & Bolus Erect shrublet to 50 cm. Flowers medium, narrowly urn-shaped with spreading lobes, pink. Aug.–Oct. Rocky seeps at middle alt., KM (Swartberg Mts).•

atricha Dulfer (= *Erica carinata* Klotzsch ex Benth.) Erect shrublet to 30 cm. Flowers small, urn-shaped, pink. June–Jan. Slopes, SW (Hottentots Holland Mts).•

atropurpurea Dulfer Erect to sprawling shrublet to 30 cm. Flowers small, urn-shaped, purple-pink. Aug.–Nov. Seeps on lower to middle slopes, LB (Langeberg Mts).•

atrovinosa E.G.H.Oliv. Erect shrublet to 60 cm. Flowers medium, inflated, urn-shaped, dark red with darker lobes. Jan.–Apr. High alt., rare, NW (Hex River Mts).•

autumnalis L.Bolus Erect shrublet to 50 cm. Flowers small, bell-shaped, white to pink. Jan.–Dec. Moist slopes at middle alt., SW (Hottentots Holland Mts to Kogelberg).•

axilliflora Bartl. CRIMSON HEATH, KLOKKIESHEIDE Erect shrublet to 75 cm. Flowers medium, bell-shaped, deep pink. Feb.–Sept. Coastal flats and lower slopes, AP (Soetanysberg).•

azaleifolia Salisb. Erect shrublet to 30 cm. Flowers small, ovoid-cup-shaped, hairy, white to pink, with exserted anthers. Oct.–Feb. Middle to upper slopes, SW (Kogelberg to Potberg).•

baccans L. BERRY HEATH Erect sturdy shrub to 3 m. Flowers small, calycine, globose, rose-pink. Apr.–Aug. Mountain slopes, SW (Cape Peninsula).•

bakeri T.M.Salter Sparse but compact shrublet to 30 cm. Flowers small, urn-shaped, pink, hairy. Aug.–Oct. Rare on marshy flats, SW (Wemmershoek Mts).•

banksia Andrews Compact woody shrublet to 30 cm. Flowers large, tubular, white with green or dark pink reflexed lobes and well-exserted anthers. Apr.–Oct. Upper slopes, rocky places, SW, AP (Hottentots Holland to Bredasdorp Mts).•

barrydalensis L.Bolus Erect sparse shrub to 2 m from bushy base. Flowers medium to large, globose to conical, red. Jan.–Dec. Rocky middle slopes, KM, LB (Warmwaterberg and Langeberg Mts).•

bauera Andrews ALBERTINIA HEATH, WITHEIDE Erect shrublet to 1 m. Flowers large, tubular-inflated, white to pink. June–Oct. Local on sandy flats, LB (Albertinia).•

beatricis Compton Erect shrublet to 50 cm. Flowers medium, tubular, white, with exserted anthers. Jan. Upper slopes, SE (Tsitsikamma Mts).•

bergiana L.   Erect shrublet to 1 m. Flowers small, globosely urn-shaped, with reflexed sepals, pink. Sept.–Mar. Seeps and moist middle to upper slopes, NW, SW (Cold Bokkeveld to Riviersonderend Mts).•

berzelioides Guthrie & Bolus   Erect to sprawling shrublet to 50 cm. Flowers large, tubular, pink with white mouth. Apr.–May. Coastal calcareous flats and hills, AP (Bredasdorp).•

bibax Salisb.   Erect shrublet to 1 m. Flowers large, tubular, yellow with white tips, finely hairy. Aug.–Nov. Middle slopes, SW (Hottentots Holland Mts to Kogelberg).•

bicolor Thunb.   Erect shrub to 1.5 m. Flowers small, urn-shaped, deep pink, with exserted anthers. Aug.–Nov. Dry middle to upper alt., NW, SW (Gifberg to Stellenbosch Mts).•

blandfordia Andrews   Erect shrublet to 1 m. Flowers small, yellow, hard. Sept.–Jan. Dry middle to upper slopes, SW (Elandskloof Mts to Paarl).•

blenna Salisb.   LANTERN HEATH, RIVERSDALE HEATH, BELLETJIEHEIDE   Erect shrublet to over 1 m. Flowers large, urn-shaped to conical, orange with green tips. Apr.–Nov. Lower and middle southern slopes, LB (Langeberg Mts: Tradouw Pass to Riversdale).•

bodkinii Guthrie & Bolus   Erect shrublet to 90 cm. Flowers large, calycine, cup-shaped with long erect lobes, very finely hairy, cream-coloured. June–July. Marshy middle slopes, SW (Bredasdorp Mts).•

bolusiae T.M.Salter   Erect shrublet to 60 cm. Flowers small, urn-shaped, white to bright pink, hairy. Mar.–July. Flats, damp sandy places, SW (Yzerfontein to Kraaifontein).•

borboniifolia Salisb.   Erect dense shrublet to 40 cm. Flowers medium, calycine, conically urn-shaped with spreading lobes, deep pink. Jan.–Feb. Moist upper slopes, NW, SW (Tulbagh to Riviersonderend Mts).•

botryioides Dulfer   Erect shrublet to 50 cm. Flowers small, ovoid, dark pink, with exserted anthers. Jan.–Mar. Middle slopes, SW (Riviersonderend Mts).•

brachialis Salisb.   Erect very woody shrub to 2 m. Flowers large, tubular, green turning yellowish, finely hairy, slightly sticky. Dec.–Mar. Rocky places near sea level, SW (Cape Peninsula to Betty's Bay).•

brachycentra Benth.   Erect shrublet to 30 cm. Flowers small, cup-shaped, pink, with exserted anthers. Jan.–May. Lower to middle slopes, SE (Outeniqua Mts to Knysna).•

brachysepala Guthrie & Bolus   Low sprawling compact shrublet to 20 cm. Flowers small, urn-shaped, dull cream-coloured, with far-exserted anthers. Jan.–May. Dry gravelly flats and lower hillsides, SW, AP (Botrivier to Elim, Soetanysberg).•

bracteolaris Lam.   Erect shrublet to 50 cm. Flowers small to medium, calycine, narrowly urn-shaped with spreading lobes, dark pink. July–Feb. Upper slopes, LB (Langeberg Mts: Swellendam to Heidelberg).•

brevicaulis Guthrie & Bolus   Erect shrublet to 10 cm. Flowers small, subcalycine, bell-shaped, red. Nov.–Jan. Rock crevices at high alt., NW (Hex River Mts).•

breviflora Dulfer (= *Erica scariosa* Thunb.)   Erect shrub to 1.5 m. Flowers medium, globose, white rarely pinkish, with exserted anthers. Sept.–Jan. Middle to upper slopes, NW (Cedarberg to Cold Bokkeveld Mts).•

brevifolia Sol. ex Salisb. (= *Erica chlamydiflora* Salisb.)   Erect shrublet to 60 cm. Flowers small, calycine, urn-shaped, sticky, pale pink. Sept.–Dec. Middle to upper slopes, SW, LB, SE (Cape Peninsula to George).•

bruniades L.   KAPOKKIE   Erect shrublet to 50 cm. Flowers small, calycine, urn-shaped, woolly, with exserted anthers, corolla white hairy, calyx hairs silvery, white, pink to purple. Aug.–Dec. Sandy flats, lower slopes and plateaus, NW, SW (Piketberg to Bredasdorp).•

bruniifolia Salisb. (= *Erica incurva* J.C.Wendl.)   Erect shrublet to 50 cm. Flowers in tight heads, small, narrowly urn-shaped, white, with dark exserted anthers. July–Oct. Stony flats and lower slopes, SW, AP (Grabouw to Potberg).•

cabernetea E.G.H.Oliv.   Erect compact shrublet to 30 cm. Flowers small, urn-shaped, dark red. Oct.–Nov. Lower shale slopes in renosterveld, SW (Palmiet River valley).•

caffra L.   WATER HEATH, WATERHEIDE   Erect shrub or small tree to 4 m. Flowers medium, conical, white, finely hairy. July–Dec. Streamsides, flats and mountain slopes, NW, SW, KM, LB, SE (Bokkeveld Mts to KwaZulu-Natal).

calcareophila E.G.H.Oliv.   Erect to prostrate shrublet to 15 cm. Flowers large, urn-shaped, wax-like, white. July–Sept. Coastal limestone, AP (Pearly Beach).•

caledonica A.Spreng.   Erect shrublet to 1 m. Flowers medium, calycine, urn-shaped, pink. Feb. Rocky slopes at high alt., SW (Riviersonderend Mts).•

calycina L.   Erect rigid shrublet to 1(–2) m. Flowers small, calycine, bell-shaped with recurved lobes, white, pink or purple. Aug.–Jan. Flats to upper slopes, NW, SW (Cedarberg to Riviersonderend Mts).•

cameronii L.Bolus.   Prostrate woody shrublet to 20 cm. Flowers large, urn-shaped, crimson, hairy. Nov.–Jan.(–May). Rocky mountain summits, NW (Hex River Mts).•

campanularis Salisb.   BOTRIVER HEATH, BOTRIVIERHEIDE   Erect slender shrublet to 60 cm. Flowers small, bell-shaped, yellow. July–Oct. Marshes and streambanks, lower to middle slopes, SW (Hottentots Holland to Kleinmond Mts).•

canaliculata Andrews   Erect shrub to 2 m. Flowers small, calycine, cup-shaped, pink. Jan.–Dec. Moist flats and lower slopes, SE (George to Humansdorp).•

canescens J.C.Wendl.   Erect lax shrublet to 60 cm. Flowers small, cup-shaped, pink, hairy. Jan.–Dec. Coastal flats and lower slopes, SW, AP (Malmesbury to Bredasdorp).•

capensis T.M.Salter   Erect shrublet to 50 cm. Flowers small, cup-shaped, deep pink or white. Dec.–Apr. Local in marshes at low alt., SW (Cape Peninsula).•

capillaris Bartl.   Erect compact shrublet to 30 cm. Flowers small, tubularly bell-shaped, pale pink, with exserted anthers. Dec.–Apr. Sandy coastal flats and hills, SW (Cape Peninsula to Stanford).•

capitata L.   KAPOKKIE   Erect shrublet to 30 cm. Flowers small, calycine, urn-shaped, woolly, green becoming yellow. Aug.–Dec. Dry sandy flats and lower plateaus, SW, AP (Mamre to Baardskeerdersbos).•

carduifolia Salisb.   Erect shrublet to 40 cm. Flowers small, urn- to bell-shaped, mauve-pink, sticky. Aug.–Nov. Moist S-facing upper slopes, NW, SW, KM, SE (Cold Bokkeveld to Kouga Mts).•

casta Guthrie & Bolus   Erect shrub to 1.5 m. Flowers large, tubular, white with red calyx. May.–Oct. Rare and local on sandy flats, AP (Elim).•

caterviflora Salisb.   Erect woody shrub to 1.3 m. Flowers small, urn-shaped, pink, finely hairy. Dec.–Mar. Rocky places, upper slopes, SW (Cape Peninsula).•

cederbergensis Compton   Erect shrublet to 30 cm. Flowers small, cup-shaped, pale to deep pink. Dec.–Mar. Rocky higher peaks, NW (Cedarberg Mts).•

cedromontana E.G.H.Oliv.   Small moss-like shrublet. Flowers small, white to reddish. Sept.–Dec. Rock crevices, higher peaks, NW (Cedarberg and Cold Bokkeveld Mts).•

cerinthoides L.   FIRE HEATH, ROOIHAARTJIE   Erect resprouting shrub, mostly compact to 30 cm, sometimes sparse to 1.2 m. Flowers large, tubular-inflated, orange-red, hairy. Jan.–Dec., especially after fire. Sandy flats and slopes, NW, SW, AP, KM, LB, SE (Cedarberg Mts to Mpumalanga).

cernua Montin   Erect shrublet to 80 cm. Flowers small in heads, urn-shaped, pink. Sept.–Oct. Marshes at middle to high alt., NW (Cedarberg to Cold Bokkeveld Mts).•

chamissonis Klotzsch ex Benth.   Erect shrublet to 60 cm. Flowers small/medium, open cup-shaped, pink. Oct.–May. Flats to middle slopes, SE (Kouga Mts to Grahamstown).

chartacea Guthrie & Bolus   Erect shrublet to 50 cm. Flowers small, calycine, ovoid, white, sometimes hairy, with exserted anthers. Sept.–Dec. Middle slopes, LB (Langeberg Mts: Swellendam).•

chionophila Guthrie & Bolus   Erect shrublet to 30 cm. Flowers small, urn-shaped, pink. Mar.–Sept. Moist middle to upper slopes, SW (Bainskloof Mts).•

chloroloma Lindl.   Erect shrub to 2 m. Flowers large, tubular, bright red with green tips. Mar.–Sept. Coastal dunes and limestone, SE (Wilderness to Fish River mouth).

chlorosepala Benth.   Shrublet to 60 cm. Flowers medium, calycine, tubularly urn-shaped, bright yellow. Sept.–Mar. Middle to upper slopes, LB (Langeberg Mts: Swellendam).•

chonantha Dulfer   Erect shrublet to 30 cm. Flowers broadly bell-shaped, pink to white, with slightly exserted anthers. Dec.–May. Slopes and flats, AP (Soetanysberg).•

chrysocodon Guthrie & Bolus   Erect shrublet to 50 cm. Flowers medium, urn-shaped, hairy, golden yellow. July–Oct. Local but abundant in marshes at lower alt., SW (Franschhoek to Villiersdorp).•

cincta L.Bolus   Erect shrublet to 30 cm. Flowers small, urn-shaped, pink to white with red ring below mouth. Dec.–Feb. Local at high alt., SW (Kogelberg).•

clandestina E.G.H.Oliv.   Sparse scrambling shrublet to 20 cm. Flowers small, yellow. Jan.–Dec. Sandy flats, seeps, SW (Worcester: Slanghoek valley).•

clavisepala Guthrie & Bolus   Erect compact shrublet to 30 cm. Flowers in terminal heads, small, urn-shaped, dull red. Feb.–Apr. Local in marshes and seeps at low alt., SW (Cape Peninsula).•

coacervata H.A.Baker   Sprawling shrublet to 20 cm. Flowers small to medium, narrowly urn-shaped, pink to purple, hairy. Dec.–Mar. Higher peaks, NW (Hex River Mts).•

coarctata J.C.Wendl.   Erect to sprawling multistemmed resprouting shrublet to 30(–90) cm. Flowers tiny, reddish. Apr.–Sept. Sandy areas, flats to high alt., NW, SW, LB (Clanwilliam to Riversdale).•

coccinea L.   Erect rigid shrub to 1.2 m. Flowers large, tubular, yellow, orange or red, with far-exserted anthers. Jan.–Dec. Common on rocky flats and mountains, NW, SW, AP, LB (Clanwilliam to George).•

collina Guthrie & Bolus   Erect shrublet to 30 cm. Flowers small, urn-shaped, pink, calycine. May–Sept. Lower slopes, SW (Hermanus).•

colorans Andrews   TREGTERHEIDE   Erect shrublet to 1 m. Flowers large, tubular, white to pink, sometimes finely hairy. Jan.–Dec. Wet areas at low alt., SW (Stanford to Elim).•

columnaris E.G.H. Oliver   Erect, wand-like shrublet. Flowers in column-like inflorescences, small, urn-shaped, wine-red. Sept.–Oct. Moist sandstone slopes, SW (Riviersonderend Mts: Pilaarkop).•

comata Guthrie & Bolus   Erect pubescent shrub. Flowers small, subcalycine, obconic, red? Jan. High peaks, very rare, LB (Langeberg Mts: Swellendam).•

comptonii T.M.Salter   Erect shrublet to 60 cm. Flowers large, tubular, white with reflexed green lobes, with exserted anthers. Jan.–July. Rocky moist high alt. slopes, SW (Kogelberg area).•

condensata Benth.   Erect shrublet to 50 cm. Flowers small, cup-shaped, pink. Sept.–Feb. Middle to upper slopes, LB (Swellendam to Riversdale).•

conferta Andrews   Erect shrublet to 1 m. Flowers small, broadly urn-shaped, white, slightly sticky. Jan.–Apr. High alt. southern slopes, SW, LB, SE (Riviersonderend Mts, Langeberg, Outeniqua Mts to Kouga Mts).•

conica Lodd.   Erect shrublet to 50 cm. Flowers medium, campanulate, magenta. July–Sept. Mountain slopes, SW (Cape Peninsula).•

conspicua Sol.   Erect shrub to 2 m. Flowers large, tubular, hairy, orange to red. Oct.–Feb. Marshy or damp flats to middle slopes, NW, SW (Ceres to Paarl).•

constantia Nois. ex Benth.   Erect rigid rounded shrublet to 50 cm. Flowers small, urn-shaped, hairy, white. Nov.–June. Rocky slopes, high alt., NW, KM (Hex River Mts to Klein Swartberg).•

copiosa J.C.Wendl.   Erect shrublet to 1.5 m. Flowers small, cup-shaped, pink. Jan.–Dec. Lower to upper slopes, KM, LB, SE (Langeberg Mts to Katberg).

cordata Andrews (= *Erica arachnoidea* Klotzsch)   Erect or sprawling shrublet or shrublet to 1 m. Flowers small, urn-shaped, pink. May–Nov. Moist middle to upper slopes, LB, SE (Heidelberg to Uitenhage).•

corifolia L.   Erect shrublet to 1 m. Flowers small to medium, calycine, urn-shaped, pink soon turning brown at tips. Jan.–Dec. Common on sandy flats and middle to upper slopes, SW, AP (Malmesbury to De Hoop).•

coronanthera Compton   Like **E. copiosa**, erect shrublet to 50 cm. Flowers small, urn-shaped, pink. Apr. Middle alt., SE (Knysna).•

corydalis Salisb.   WHITE PETTICOAT HEATH, WITROKHEIDE   Erect shrublet to 50 cm. Flowers small, basally spherical, upper part cup-shaped, white. Oct.–Jan. Moist S-facing upper slopes, SW (Houwhoek to Kleinrivier Mts).•

costatisepala H.A.Baker   Erect low shrublet to 25(–50) cm. Flowers small, calycine, cup-shaped, white. Dec.–Mar. High summits, rocky places, NW, KM, LB (Robertson, Langeberg: Swellendam, Klein Swartberg to Kammanassie Mts).•

crassisepala Benth.   Erect shrublet to 20 cm. Flowers small, calycine, cup-shaped, hairy, white. Jan. High alt., LB (Langeberg Mts: Swellendam).•

crateriformis Guthrie & Bolus   Erect shrublet to 30 cm. Flowers small, cup-shaped, ?pink. Aug. ?, SW (Hermanus area).•

cremea Dulfer   Erect shrublet to 30 cm. Flowers large, tubular, cream-coloured, finely hairy. Nov. Upper mountain slopes, moist places, SW (Wellington).•

crenata E.Mey.   Erect compact shrublet to 45 cm. Flowers small, urn-shaped, pink, sticky. Jan.–May. Sandy flats and mountain plateaus, SW (Cape Peninsula and Brackenfell).•

cristata Dulfer (= *Erica pectinata* Klotzsch)   Like **E. curvifolia**, erect sparse shrublet to 60 cm. Flowers medium to large, tubularly urn-shaped, pink, sticky. Dec.–June. Middle to upper slopes, SW (Hottentots Holland Mts to Betty's Bay).•

cristiflora Salisb.   Erect shrublet to 1 m. Flowers small to medium, calycine, obconic, pink. Mar.–Nov. Lower to upper slopes, NW, SW, KM, LB (Kamiesberg, Gifberg to Antoniesberg).

cruenta Sol.   CRIMSON HEATH, ROOIHEIDE   Erect shrublet to 1 m. Flowers large, tubular, dark red. Jan.–Dec. Lower dry shale hills and slopes, SW, LB (Grabouw to Mossel Bay).•

cryptanthera Guthrie & Bolus   Erect straggling shrublet to 15 cm. Flowers small, cup-shaped, pale pink, hairy. Sept.–Dec. Middle slopes, moist places, SW (Franschhoek to Riviersonderend Mts).•

cubica L.   Erect shrublet to 45 cm. Flowers small/medium, calycine, broadly obconic, pink or reddish. July–Dec. Marshy southern slopes, LB, SE (Swellendam to KwaZulu-Natal).

cubitans E.G.H.Oliv.   Prostrate shrublet. Flowers small, cup-shaped, purplish pink, with long hairs and short glands. Dec.–Jan. Moist sandstone slopes, LB (Langeberg Mts: Lemoenshoek).•

cumuliflora Salisb.   Erect shrublet to 45 cm. Flowers in dense heads, small, narrowly urn-shaped, white with erect dark brown lobes. Sept.–Jan. Dry middle slopes, SW (Grabouw to Potberg).•

cunoniensis E.G.H.Oliv.   Erect shrublet to 50 cm. Flowers medium, narrowly cup-shaped, deep red. Sept.–Oct. Rocky southern upper slopes, SW (Kogelberg).•

curtophylla Guthrie & Bolus   Erect shrublet to 30 cm. Flowers small, urn- to cup-shaped, pink to magenta, with slightly exserted anthers. July–Oct. Limestone flats and low hills, AP (De Hoop to Stilbaai).•

curviflora L. (= *Erica sulcata* Benth.)   WATER HEATH, WATERBOS   Erect soft to stout shrub to 1.6 m. Flowers large, tubular, orange, red or yellow, hairy or glabrous. Jan.–Dec. Widespread in damp or wet areas, flats to high alt., NW, SW, AP, KM, LB, SE (Bokkeveld Mts to Grahamstown).

curvifolia Salisb.   Erect often much-branched rounded shrublet to 45 cm. Flowers medium, tubular to urn-shaped, pink, sticky. Dec.–Mar. Sandy places on upper slopes, SW (Jonkershoek to Riviersonderend Mts).•

curvirostris Salisb.   HEUNINGHEIDE   Like **E. lateralis**, erect shrublet to 60 cm. Flowers small, bell-shaped, pale to dark pink, often scented. Feb.–May. Dry stony areas, middle to upper slopes, SW (Du Toitskloof Mts to Kogelberg).•

cuscutiformis Dulfer   Slender sprawling shrublet to 25 cm. Flowers small, narrowly urn-shaped, pink. Aug.–Oct. Upper slopes, NW (Grootwinterhoek to Cold Bokkeveld Mts).•

cygnea T.M.Salter   Erect shrublet to 30 cm. Flowers medium, tubular, in nodding heads, sticky, pink. Jan.–Feb. Middle rocky slopes, SW (Kogelberg).•

cylindrica Thunb.   Erect shrub to 1.2 m. Flowers large, narrowly tubular, fragrant, pale yellow. Mar.–Oct. Dry stony slopes, SW (Elandskloof Mts).•

cymosa E.Mey.   Straggling slender shrublet against rocks. Flowers small, globular, white tinged red, hairy. Dec.–Jan. Upper slopes, rock faces, SW (Du Toitskloof).•

cyrilliflora Salisb.   Erect diffuse shrublet to 60 cm. Flowers small, narrowly urn-shaped, pink with paler tips, hairy. Feb. Marshy areas, SW (Cape Peninsula).•

daphniflora Salisb.   Erect shrublet to 1 m. Flowers medium to large, urn-shaped to tubularly urn-shaped, white, yellow, pink or red. July–Mar. Sandy flats and slopes, often beside water, NW, SW, LB (Cedarberg Mts to Heidelberg).•

deflexa Sinclair   Erect to sprawling shrublet to 30 cm. Flowers small, bell-shaped to obconic, pale pink. Nov.–Apr. Middle to upper alt., SE (Outeniqua and Tsitsikamma Mts).•

deliciosa J.C.Wendl.   Erect shrub to 1.5 m. Flowers small, ovoid, pink, with exserted anthers. June–Dec. Flats and lower mountain slopes, LB, SE (Riversdale to E Cape).

demissa Klotzsch   Like **E. simulans**, erect shrublet to 50 cm. Flowers small, urn-shaped, white, with exserted anthers. Jan.–Dec. Lower to middle mountain slopes, KM, SE (Swartberg Mts to Grahamstown).

densifolia Willd.   Erect shrub to 1.5 m. Flowers large, tubular, curved, pubescent, red with greenish yellow lobes. Sept.–May. Flats to middle slopes, KM, LB, SE (Langeberg to Kammanassie and Tsitsikamma Mts).•

denticulata L.   LEKKERRUIKHEIDE   Erect shrublet to 50 cm. Flowers medium, urn-shaped, waxy, white to cream-coloured or pinkish, distinctly fragrant. Aug.–Jan. Flats to upper slopes, NW, SW (Piketberg to Riviersonderend Mts).•

depressa Andrews   Erect dwarf shrublet to 10 cm. Flowers small, calycine, bell-shaped, white. Nov.–July. Rock crevices on upper slopes, SW (Cape Peninsula).•

desmantha Benth.   Erect shrublet to 1 m. Flowers small to medium, ovoid-cup-shaped, calycine, white, with exserted anthers. Feb.–May. Moist upper slopes, SW (Banhoek to Kleinrivier Mts).•

dianthifolia Salisb.   Erect shrublet to 70 cm. Flowers medium, calycine, white sometimes with red stripes. Aug.–Dec. Middle to upper slopes (shale bands?), SW, LB (Riviersonderend Mts to Heidelberg, Potberg).•

diaphana Spreng.   Erect shrub to 1.5 m. Flowers large, tubular, sticky, pink to dark red with pale green lobes. Jan.–Dec. Flats and rocky slopes to high alt., KM, SE (Outeniqua Mts: George to Uitenhage).•

dichrus Spreng.   Erect shrub to 2 m. Flowers large, tubular, slightly sticky, pink or purple with yellow to white lobes. Jan.–Dec. Hills and lower mountain slopes, LB, SE (Mossel Bay to George).•

diosmifolia Salisb.   Erect shrublet to 40 cm. Flowers small, calycine, open cup-shaped, white. Sept.–Dec. Rock ledges at high alt., SW (Cape Peninsula).•

diotiflora Salisb.   Erect shrublet to 50 cm. Flowers small, ovoid to cup-shaped, pink, with exserted anthers. Oct.–Jan. Moist places, lower slopes, SW (Riviersonderend Mts).•

discolor Andrews   Like **E. speciosa**, dense resprouting shrublet to 1 m. Flowers large, tubular, pink to dark red with pale tips. Jan.–Dec. Coastal flats and lower mountain slopes, SW, AP, LB, SE (Betty's Bay to Humansdorp).•

distorta Bartl.   Erect to spreading shrublet to 50 cm. Flowers small, campanulate, white to cream-coloured, hairy. Dec.–June. Sandy seeps or rock ledges, middle slopes, NW, SW (Cedarberg Mts to W Riviersonderend Mts).•

dodii Guthrie & Bolus   Erect to sprawling shrublet to 40 cm. Flowers small, narrowly bell-shaped, pink. Oct.–Dec. Moist shady upper slopes, SW, KM, LB (Cape Peninsula to Uniondale).•

doliiformis Salisb.   NINEPIN HEATH   Erect bushy shrublet to 1 m. Flowers medium to large, narrowly urn-shaped, pink, finely hairy. Sept.–Apr. Lower granitic to rocky upper slopes, SW (Elandskloof Mts to Franschhoek).•

dysantha Benth.   Low sprawling mat-forming shrublet to 30 cm. Flowers small, pubescent, white or cream-coloured with greenish tips. Nov.–Jan. Higher peaks, LB (Langeberg Mts: Riversdale).•

eburnea T.M.Salter   Erect compact shrublet to 60 cm. Flowers small, widely cup-shaped, white. Dec.–Jan. Sandy flats, SW (Cape Peninsula).•

elimensis L.   Erect shrublet to 40 cm. Flowers small, broadly cup-shaped, red to pink. Aug.–Oct. Sandy flats and hills, AP (Elim to coast).•

embothriifolia Salisb.   Sparse erect to semispreading shrublet to 45 cm. Flowers large, tubular, very narrow to broader and laterally flattened, bright pink, pedicels very sticky, with exserted anthers. Dec.–Apr. Open slopes or ledges at middle to high alt., SW (Riviersonderend Mts).•

empetrina L. (= *Erica empetrifolia* L.)   Erect shrublet to 30 cm. Flowers small, dull red to pink. Aug.–Dec. Rocky upper slopes often on ledges, SW (Cape Peninsula).•

erasmia Dulfer   Erect shrublet to 30 cm. Flowers small, calycine, urn-shaped, pink. Sept.–Oct. Upper slopes, NW (Hex River Mts).•

eriocodon Bolus   Erect shrublet to 30 cm. Flowers small, cup-shaped, pink, hairy. Oct.–Nov. Upper rocky slopes, NW (Cedarberg to Grootwinterhoek Mts).•

eriophoros Guthrie & Bolus   Prostrate spreading grey shrublet to 20 cm. Flowers in pendulous heads, small, hidden, cup-shaped, white, hairy, with exserted anthers. Sept.–Nov. Very local at high alt., NW (S Cedarberg and Gydo Mts).•

esterhuyseniae Compton   Erect shrublet to 30 cm. Flowers small, bell- to urn-shaped, white, hairy. Dec.–Mar. Rocky upper slopes, KM, SE (Swartberg to Kouga Mts).•

etheliae L.Bolus   Erect shrublet to 20 cm. Flowers tubular, pink. June. Poorly known, SE (Port Elizabeth).•

eugenea Dulfer (= *Erica nobilis* Guthrie & Bolus)   Erect shrublet to 70 cm. Flowers medium to large, calycine, urn-shaped, pink with darker lobes. Aug.–Dec. Middle to upper slopes, NW (Cedarberg and N Cold Bokkeveld Mts).•

eustacei L.Bolus   Erect shrub to 1.5 m. Flowers small, globosely urn-shaped, white. Sept.–Nov. Dry middle to upper slopes, KM (Touwsrivier to Swartberg and Gamkaberg).•

excavata L.Bolus   Like **E. propinqua**, erect shrublet to 50 cm. Flowers medium, calycine, pink. July–Oct. Coastal limestone flats and hills, SW, AP (Hermanus to Pearly Beach).•

extrusa Compton   Erect shrublet to 30 cm. Flowers small, urn-shaped, pink. Nov.–Jan. Sandy lower slopes, SW (Grabouw).•

fairii Bolus   Erect shrublet to 60 cm. Flowers medium, tubular-conical, white, sticky. Dec.–June. Localised on middle slopes, SW (Cape Peninsula).•

fascicularis L.f.   Erect sparsely branched shrub to 1.8 m. Flowers large, tubular or tubular inflated, pink with green lobes, sticky. Apr.–Dec. Sandy rocky slopes, SW (Helderberg to Riviersonderend to Bredasdorp Mts).•

fastigiata L.   FOUR SISTERS HEATH   Erect shrublet to 50 cm. Flowers medium to large, tubular with large spreading lobes, red tube with pink or white lobes with dark central ring. July–Dec. Damp to dry flats to middle alt. slopes, SW (Franschhoek to Kleinrivier Mts).•

fausta Salisb.   Erect shrublet to 50 cm. Flowers small, bell-shaped, pink, hairy. Oct.–Nov. Lower to middle slopes, SW (Stellenbosch to Kleinmond).•

feminarum E.G.H.Oliv.   Erect shrublet to 30 cm. Flowers small, cup-shaped, dark red, finely hairy. July–Sept. Seepage areas, middle slopes, SW (Stettynsberg).•

ferrea P.J.Bergius (= *Erica mucosa* L.)   Erect sparse shrublet to 1 m. Flowers small, broadly urn-shaped, magenta, sticky. Dec.–June. Sandy flats and lower slopes, SW (Malmesbury to Cape Flats).•

fervida L.Bolus   Like **E. pillansii**, erect shrublet to 1 m. Flowers medium, tubularly urn-shaped, red, finely hairy. Sept.–Nov. Moist areas on middle slopes, SW (Kogelberg).•

filamentosa Andrews   Erect shrub to 1.5 m. Flowers medium to large, pink, finely hairy, long-pedicelled. June–Nov. Dry lower slopes and hills, SW, LB (Stormsvlei to Swellendam).•

filiformis Salisb.   Erect, compact to diffuse shrublet to 30 cm. Flowers small, urn- to bell-shaped, pink, with exserted anthers. Oct.–Apr. Clay flats and lower slopes, SW (Paarl to Elgin).•

filipendula Benth.   Erect shrublet to 1 m. Flowers variable, medium to large, inflated tubular to urn-shaped, white, pink or yellow. Apr.–July. Sandy hills and flats, SW, AP (Bredasdorp).•

fimbriata Andrews   Erect shrublet to 70 cm. Flowers small, calycine, urn-shaped, white, occasionally pink. July–Dec. Middle to upper slopes, KM, SE (Swartberg and Outeniqua Mts: George).•

flacca E.Mey.   Erect shrublet to 30 cm. Flowers small, bell-shaped, magenta-pink, with exserted anthers. Oct.–Feb. Sandy flats and middle slopes, NW (Cedarberg and Cold Bokkeveld Mts).•

flavicoma Bartl. Erect shrublet to 60 cm. Flowers small, narrowly urn-shaped, pale yellow, in spikes. Sept.–Oct. Sandy flats, AP (Bredasdorp).•

flexistyla E.G.H.Oliv.   Erect compact shrublet to 30 cm. Flowers small, cup-shaped with bent style, pink. Aug.–Oct. Moist areas on flats, AP (Viljoenshof to Soetanysberg).•

floccifera Zahlbr. (= *Erica floccosa* Bartl.)   Erect robust shrublet to 70 cm. Flowers small, calycine, cup-shaped, white. Aug.–Dec. Low to middle slopes, SW (Kleinrivier and Riviersonderend Mts, Caledon: Swartberg).•

flocciflora Benth.   Erect woody shrublet to 1 m. Flowers small/medium, lime-yellow with white hairy calyx. Sept.–Dec. Dry lower slopes and rocky foothills, SE (Kouga Mts).•

florifera (Compton) E.G.H.Oliv. (= *Erica cupuliflora* Dulfer, *Eremia florifera* Compton)   Erect compact to semispreading shrub to 30 cm. Flowers small, cup-shaped, pink. Sept.–Nov. Sandy moist flats, NW (Tulbagh to Ceres).•

foliacea Andrews   Erect to slightly sprawling shrublet to 90 cm. Flowers large, tubular, very hard and waxy, green to yellow-green or orange. Apr.–Nov. Rocky mountain slopes, SW (Kogelberg to Betty's Bay).•

fontana L.Bolus   Erect shrub to 1.8 m. Flowers medium, tubular, pinkish white, hairy. Oct.–May. Marshy flats, SW (Cape Peninsula).•

formosa Thunb.   WHITE HEATH, WITHEIDE   Erect shrublet to 60 cm. Flowers small, globosely urn-shaped, often fluted, white. July–Nov. Lower slopes, SE (Mossel Bay to Humansdorp).•

fourcadei L.Bolus   Erect semispreading shrublet to 30 cm. Flowers large, tubular, red or yellowish. Apr.–Oct. Coastal rocks and lower slopes, SE (Knysna to Kareedouw).•

galgebergensis H.A.Baker   Erect to spreading shrublet to 15 cm. Flowers medium/large, calycine, bell-shaped with large spreading lobes, white. Oct.–Nov. Rock crevices at high alt., SW (Riviersonderend Mts).•

gallorum L.Bolus   Like **E. longifolia**, erect shrublet to 60 cm. Flowers medium, campanulate, white to pink. Sept.–Oct. Middle slopes, SW (Hottentots Holland Mts).•

galpinii T.M.Salter   Erect shrublet to 90 cm. Flowers large, tubular, bright yellow. Apr.–June. S-facing middle slopes, rare, SW (Kleinrivier Mts).•

garciae E.G.H.Oliv. Delicate shrublet to 30 cm. Flowers in pseudospikes, small, tubular, purplish pink. Nov.–Dec. Gravelly slopes, LB (Langeberg Mts: Garcia's Pass area).•

genistifolia Salisb. Erect delicate shrublet to 30 cm. Flowers small, narrowly urn-shaped, white with erect brown lobes. Oct.–Jan. Moist rock ledges, SW (Cape Peninsula to Napier).•

georgica Guthrie & Bolus Erect shrublet to 50 cm. Flowers small to medium, urn-shaped with spreading lobes, pink. June–Feb. Moist slopes, KM, SE (Swartberg, Outeniqua and Tsitsikamma Mts).•

gigantea Klotzsch ex Benth. Erect shrub to 2 m. Flowers small, globose, calycine, white. Aug.–Sept. Lower north slopes, LB (Langeberg Mts: Riversdale).•

gillii Benth. Erect shrublet to 1 m. Flowers small, calycine, open cup-shaped, pink. Aug.–Nov. Middle slopes, LB (Outeniqua Mts: Mossel Bay).•

gilva J.C.Wendl. GROENHEIDE Like **E. mammosa**, erect shrublet to 1 m. Flowers large, tubular, cream-coloured or pale green or tinged pink. Dec.–Jan. Slopes and flats, SW (Cape Peninsula).•

glandulifera Klotzsch Like **E. irrorata**, low viscid shrublet to 30 cm. Flowers small to medium, narrowly urn-shaped, magenta. Nov.–Jan. Sandy flats and lower slopes, NW (Cedarberg Mts to Ceres).•

glandulipila Compton (incl. **E. umbonata** Compton) Erect shrublet to 1 m. Flowers small, urn-shaped, white to pinkish. Sept.–Nov. Rocky lower slopes, KM (Witteberg to Swartberg Mts, Montagu).•

glandulosa Thunb. Erect sticky shrub to 1.5 m. Flowers medium to large, tubular, pink to orange often with darker longitudinal stripes, occasionally white. Jan.–Dec. Flats and lower slopes, LB, SE (Mossel Bay to Port Elizabeth).•

glauca Andrews CUP-AND-SAUCER HEATH, KOMMETJIE-PIERINGHEIDE Erect shrub to 2 m. Flowers large, calycine, broadly urn-shaped, dark purple with brown tips and red calyx or pale green with pink calyx. July–Dec. Lower to middle slopes, NW, SW (Ceres to Franschhoek).•

globulifera Dulfer Erect compact shrublet to 30 cm. Flowers small, ovoid, pink. Apr.–June. Sandy flats, SW, AP (Bredasdorp).•

glomiflora Salisb. Erect shrublet to 1 m. Flowers small to medium to large, urn-shaped to conical, white to deep pink, slightly sticky. Jan.–Dec. Coastal flats to middle slopes, KM, LB, SE (Langeberg: Gourits River, Swartberg and Kouga Mts).•

glumiflora Klotzsch ex Benth. Erect shrublet to 60 cm. Flowers small, calycine, urn-shaped, green to white, with exserted anthers. June–Sept. Sand dunes and lower slopes, SE (George to Humansdorp).•

glutinosa P.J.Bergius Erect compact viscid shrublet to 30 cm. Flowers medium, urn-shaped, pink or purple. Oct.–Apr. Damp sandy flat areas, middle to upper slopes, NW, SW (Ceres to Cape Peninsula to Riviersonderend Mts).•

gnaphaloides L. (= *Erica crucistigmatica* Dulfer) Erect slender shrublet to 40 cm. Flowers small, calycine, narrowly urn-shaped, pink, with distinctly cross-shaped stigma. Aug.–Dec. Marshy flats and lower slopes, SW (Wemmershoek to Kogelberg).•

goatcheriana L.Bolus Like **E. sp. 1**, erect sturdy shrub to 1.5 m. Flowers medium, calycine, pink, calyx pink or white. July–Jan. Rocky places at high alt., NW (S Cold Bokkeveld).•

gossypioides E.G.H.Oliv. Erect grey-pubescent shrublet to 45 cm. Flowers medium, calycine, cup-shaped, pink, calyx very white woolly. Dec.–Apr. Upper slopes, KM (Swartberg Mts).•

gracilipes Guthrie & Bolus Erect to spreading shrublet to 30 cm. Flowers small, calycine, urn-shaped, deep pink, with exserted anthers. Aug.–Dec. Limestone hills and ridges, AP (Pearly Beach to Agulhas).•

gracilis J.C.Wendl. Erect shrublet to 50 cm. Flowers small, narrowly urn-shaped, pink. Jan.–Dec. Flats and lower slopes, LB, SE (Heidelberg to Humansdorp).•

grandiflora L.f. Erect shrub to 1.5 m. Flowers large tubular, orange to red or yellow, sometimes with exserted anthers. Jan.–Dec., depending on area. Dry lower to middle slopes, NW, SW, KM (Bokkeveld Mts to Stellenbosch).•

granulatifolia H.A.Baker Erect shrublet to 60 cm. Flowers small, subcalycine, narrowly cup-shaped, pink. Oct. High alt., LB (Langeberg Mts: Riversdale).•

granulosa H.A.Baker Like **E. harveyana**, erect shrublet to 45 cm. Flowers small, ovoid-cup-shaped, white or pink, with exserted anthers. Sept.–Nov. S-facing upper slopes, KM (Uniondale).•

grata Guthrie & Bolus Erect or scrambling shrublet to 40 cm. Flowers small, urn-shaped, purple, with slightly exserted anthers. Aug.–Oct. Lower slopes, LB (Langeberg Mts: Heidelberg to Riversdale).•

greyi Guthrie & Bolus (= *Erica auriculata* Guthrie & Bolus)   Sparse erect shrublet to 30 cm. Flowers small, urn-shaped, white(?). Feb.–June. Middle slopes, sandy places, very rare, NW (Cold Bokkeveld).•

grisbrookii Guthrie & Bolus   Erect shrublet to 80 cm. Flowers medium, calycine, ovoid, hard and waxy, white. June–Aug. Rocky slopes, SW (Napier to Bredasdorp).•

guthriei Bolus   Sparse erect shrublet to 30 cm. Flowers small, calycine, very broadly urn-shaped, sticky, pink, with exserted anthers. Sept.–Feb. Upper slopes, NW (Cedarberg Mts and Piketberg).•

gysbertii Guthrie & Bolus   Erect shrublet to 1 m. Flowers medium to large, tubular, pink, sticky. Nov.–Mar. Sandstone middle to upper slopes, SW (Kogelberg).•

haematocodon T.M.Salter   Compact to spreading shrublet to 20 cm. Flowers small, dark red, hairy. Oct.–Jan. Rock crevices at middle alt., SW (Cape Peninsula).•

haematosiphon Guthrie & Bolus (= *Erica macropus* Guthrie & Bolus)   Erect woody shrublet to 1.2 m. Flowers large, tubular, bright dark red. Nov.–Mar. Rocky slopes and outcrops at middle to high alt., NW (Clanwilliam to Ceres).•

halicacaba L.   BLADDER HEATH   Erect woody shrublet to 1 m. Flowers large, inflated, calycine, greenish to cream-coloured. Sept.–Dec. Rock crevices and ledges at middle alt., SW (Cape Peninsula).•

hameriana L.Bolus   Like **E. sitiens**, erect shrublet to 60 cm. Flowers medium, tubular, dark red with paler or white tips. Feb.–Mar. Flats or middle slopes, SW (Kogelberg).•

hanekomii E.G.H.Oliv.   Prostrate, creeping shrublet. Flowers medium, urn-shaped, white, sepals with conspicuously long reddish hairs, anthers well exserted, with long pores. Mainly Aug.–Oct. Sandstone slopes, NW (Cold Bokkeveld Mts: Middelberg).•

hansfordii E.G.H.Oliver   Delicate straggling shrublet to 30 cm. Flowers solitary, small, cup-shaped, yellow. Mainly Nov.–Jan. Sandy seeps, SW (Du Toitskloof Mts: Slanghoek valley).•

harveyana Guthrie & Bolus (= *Erica jeppei* L.Bolus)   Like **E. granulosa**, erect shrublet to 50 cm. Flowers small, ovoid-cup-shaped, slightly sticky, white, with exserted anthers. Sept.–Nov. Upper slopes, SE (Uniondale to Uitenhage).•

hebdomadalis E.G.H.Oliv.   Erect low spreading shrublet to 20 cm. Flowers small, cup-shaped, white. Jan.–Mar. Rocky upper slopes, KM (Klein Swartberg).•

hebecalyx Benth.   Erect shrublet to 1 m. Flowers large, tubular, yellowish green, pink with greenish tips. Sept.–May. Upper rocky slopes, KM, SE (Outeniqua and Kammanassie Mts to Langklooof).•

heleogena T.M.Salter   Erect shrublet to 60 cm. Flowers medium, elongate-cup-shaped, pink, hairy. Nov.–Mar. Marshes at middle alt., SW (Cape Peninsula).•

heleophila Guthrie & Bolus   Erect shrublet to 50 cm. Flowers medium, urn-shaped with spreading lobes, pale pink. Nov.–Dec. Damp slopes, LB (Langeberg Mts).•

hendricksii H.A.Baker   Erect shrublet to 45 cm. Flowers large, tubular with spreading lobes, hairy, white and pink. Sept.–Apr. Marshy lower slopes, SW (Stanford).•

hexensis E.G.H.Oliv.   Erect shrublet to 1 m. Flowers small, dusky red to pink, with large stigma. Sept.–Nov. Rocky slopes at high alt., NW (Hex River Mts).•

hibbertia Andrews   Compact woody shrublet to 30 cm. Flowers large tubular, sticky, orange-red with greenish yellow tips. Sept.–Nov. Rock crevices and ledges at lower alt., SW (Franschhoek Pass to Villiersdorp).•

hippurus Compton   Erect shrublet to 90 cm. Flowers small, urn-shaped in dense long spike, dark pink, hairy. Apr.–July. Granite slopes, moist areas, SW (Malmesbury).•

hirta Thunb. (= *Erica sphaeroidea* Dulfer)   Erect shrublet to 80 cm. Flowers small, globosely urn-shaped, white to pink, hairy, with exserted anthers. Feb.–Aug. Dry lower slopes, SW (Malmesbury to Genadendal).•

hirtiflora Curtis   Erect shrublet to 1 m. Flowers small, ovoid, mauve-pink. Jan.–Dec. Flats and slopes, SW (?Cape Peninsula).•

hispidula L. (inc. **E. blancheana** L.Bolus, **E. inops** Bolus)   Erect shrublet to 1(–1.8) m. Flowers small, urn-shaped to bell-shaped, white, pink or red. Jan.–Dec. Widespread and common from flats to high alt., NW, SW, KM, LB, SE (Clanwilliam to Uitenhage).•

hispiduloides E.G.H.Oliv.   Erect shrub to 1.2 m. Flowers small, narrowly cup-shaped, pink to creamy yellow. Sept.–Dec. Rocky slopes at higher alt., NW, SW, KM (Cedarberg to Worcester and Swartberg Mts).•

holosericea Salisb.   Erect shrublet to 90 cm. Flowers medium, campanulate, pink with darker sepals, finely hairy. Sept.–Nov. Moist southern slopes, SW (Kogelberg to Bredasdorp).•

hottentotica E.G.H.Oliv.   Erect compact shrublet to 1 m. Flowers small to medium, red-pink, finely hairy. Sept.–Nov. Moist upper slopes, SW (Hottentots Holland Mts).•

humansdorpensis Compton   Erect shrublet to 30 cm. Flowers small, narrowly urn-shaped, pink. May–July. Coastal flats, SE (Humansdorp).•

humifusa Salisb.   Erect compact woody shrublet to 10 cm. Flowers small, pink. Sept.–Feb. Rocky upper slopes, NW, SW, KM, SE (Cedarberg to Riviersonderend to Kouga Mts).•

imbricata L.   Erect shrublet to 80 cm. Flowers small, calycine, ovoid-urn-shaped, white or pink, with exserted anthers. Jan.–Dec. Coastal sandy flats to middle alt., NW, SW, AP, KM, LB, SE (Gifberg to Kouga Mts).•

inamoena Dulfer (= *Erica blesbergensis* H.A.Baker)   Erect shrublet 80 cm. Flowers small, subcalycine, urn-shaped, pale pink. Sept.–Dec. Upper rocky slopes, KM (Anysberg, Rooiberg, Swartberg Mts).•

incarnata Thunb.   Erect shrublet to 40 cm. Flowers medium, tubularly urn-shaped, pink. Sept.–Dec. Rocky slopes at middle to high alt., NW (Cedarberg Mts to Cold Bokkeveld).•

inclusa J.C.Wendl.   Erect slender shrub to 1.2 m. Flowers medium, calycine, campanulate, pale pink with darker sepals. Aug.–Oct. Moist slopes, lower to middle slopes, LB (Langeberg Mts: Riversdale).•

inconstans Zahlbr.   Erect shrub to 2 m. Flowers small, narrowly cup- to urn-shaped, white to pink. July–Dec. Upper slopes, SE (Outeniqua Mts: Knysna).•

inflata Thunb.   Erect shrublet to 1 m. Flowers medium, urn-shaped, dark pink. Sept.–May. Sandy flats and slopes, NW (Clanwilliam to Ceres).•

infundibuliformis Andrews (= *Erica pavettiflora* Salisb.)   Erect shrublet to 90 cm. Flowers large, narrowly tubular, pink to reddish. Nov.–July. Lower to middle slopes, SW, LB (Du Toitskloof Mts to Heidelberg).•

ingeana E.G.H.Oliv.   Soft shrublet to 30 cm. Flowers in heads, small, narrowly urn-shaped with relatively large lobes, purplish pink. Jan.–Dec. Upper slopes, moist places, KM, SE (Swartberg and Outeniqua to Kouga Mts).•

inordinata H.A.Baker   Erect shrub to 1.2 m. Flowers large, tubular, very sticky, orange-red. Sept.–Feb. Local at high alt., KM (Kammanassie Mts).•

insignis E.G.H.Oliv.   Erect dwarf woody shrublet to 40 cm. Flowers large, very calycine, cream-coloured turning red, corolla very reduced around ovary, with far-exserted anthers. Sept.–Dec. Rock crevices on N-facing upper slopes, KM (Anysberg and Groot Swartberg).•

insolitanthera H.A.Baker   Erect sturdy shrublet to 1 m. Flowers medium, ovoid, sticky, red, with slightly exserted anthers. Jan.–Feb. Upper slopes, SW (Riviersonderend Mts).•

intervallaris Salisb.   Erect to sprawling shrublet to 60 cm. Flowers small, narrowly urn-shaped, pink. July–Jan. Moist places and slopes, middle to high alt., SW (Hottentots Holland to Riviersonderend Mts).•

intonsa L.   Erect compact shrublet to 30 cm. Flowers large, tubular-inflated, pink, sticky. Oct.–Feb. Rocky upper alt., SW (Kogelberg).•

intricata H.A.Baker   Erect slender diffuse shrublet to 10 cm. Flowers small, broadly cup-shaped, pink. Apr.–May. High alt. seeps on shale, SW (Du Toitskloof Mts).•

involucrata Klotzsch   Low spreading shrublet to 15 cm. Flowers in conspicuous white to pink heads, small, cup-shaped, white, hairy, with exserted anthers. Sept.–Dec. High alt., NW, SW (Cedarberg to Elandskloof Mts).•

irbyana Andrews   Erect shrublet to 50 cm. Flowers large, ampullaceous, pale pink. Sept.–Jan. Lower slopes, SW (Babilonstoring to Elim).•

irregularis Benth.   Erect shrub to 1.5 m. Flowers small, urn-shaped, pink. June–Oct. Coastal limestone flats and hills, SW, AP (Stanford to Gansbaai).•

irrorata Guthrie & Bolus   Like **E. glandulifera**, low sprawling sticky shrublet to 45 cm. Flowers large, inflated-tubular, pink. Nov.–Feb. Rocky slopes, high alt., NW (Grootwinterhoek and Cold Bokkeveld Mts).•

ixanthera Benth.   Delicate rounded shrublet to 15 cm. Flowers medium, urn-shaped, white to greenish, slightly sticky. Oct.–Dec. Shady moist overhangs, lower slopes, LB (Riversdale).•

jacksoniana H.A.Baker   Erect shrub to 1.5 m. Flowers small, cup-shaped, pink. Mar.–May. Marshes and seeps, SW (Hottentots Holland Mts).•

jasminiflora Salisb.   JASMINE HEATH, TROMPETHEIDE   Erect sparse shrublet to 60 cm. Flowers large, narrowly tubular with very broad spreading lobes, white with red-striped tube, sticky. Jan.–Feb. Clay hills, SW (Caledon).•

juniperina E.G.H.Oliv.   Erect shrublet to 50 cm. Flowers small, slightly sticky, pink, with exserted anthers. Mar.–May. Upper moist slopes, SE (Outeniqua Mts: Mossel Bay).•

junonia Bolus   Low compact spreading shrublet to 30 cm. Flowers large, tubular-ampullaceous, with spreading lobes, bright pink. Nov.–Jan. High peaks, NW (Cedarberg to Hex River Mts).•

karooica E.G.H.Oliv.   Erect shrublet to 1.2 m. Flowers small, cup-shaped, creamy brown. Mar.–Oct. Dry rocky lower slopes and hills, NW, SW, KM, LB (Robertson and Bredasdorp to Willowmore).•

keeromsbergensis H.A.Baker   Erect dense low shrub. Flowers small, calycine, cup-shaped, white. Mar. Rocks at high alt., NW (Keeromsberg).•

keetii L.Bolus   Like **E. onusta**, erect shrublet to 60 cm. Flowers small, globosely urn-shaped, pale pink. Sept. Roadside, SE (Knysna).•

kirstenii E.G.H.Oliv.   Compact shrublet to 25 cm. Flowers medium, urn-shaped, white. Sept.–Oct. Sandstone slopes, KM (Klein Swartberg).•

kogelbergensis E.G.H.Oliv.   Erect woody shrub to 1.5 m. Flowers large, tubular, yellow or orange with yellow tips, very finely hairy. Apr.–Sept. Moist upper slopes, SW (Kogelberg).•

kougabergensis H.A.Baker   Erect shrublet to 50 cm. Flowers small, ovoid-cup-shaped, pink, with exserted anthers. Sept.–Nov. S-facing upper slopes, SE (Kouga to Cockscomb Mts).•

kraussiana Klotzsch   Pubescent shrub, height? Flowers small, lilac. Flowering time? Middle slopes, SW (Genadendal).•

krugeri E.G.H.Oliv.   Low diffuse creeping shrublet. Flowers small, cup-shaped white with spreading pink sepals, with exserted anthers. Oct.–Dec. Upper moist slopes, SW (Hottentots Holland Mts).•

lachnaeifolia Salisb.   Erect woody shrublet to 30 cm. Flowers small, calycine, obconic, white. Oct.–Nov. Rock crevices at high alt., SW (W Riviersonderend Mts).•

laeta Bartl.   Erect slender shrublet to 40 cm. Flowers small, narrowly urn-shaped, pale to deep pink. Dec.–Apr. Coastal flats or lower slopes, SW (Cape Peninsula to Hermanus).•

lageniformis Salisb.   Erect to spreading sparsely branched shrublet to 70 cm. Flowers large, tubular-ampullaceous, white turning pinkish, shiny. Dec. Middle slopes, SW (Kleinrivier Mts).•

lananthera L.Bolus   Erect shrublet to 30 cm. Flowers medium, cup- to urn-shaped, pink, sticky. Oct.–Mar. Middle coastal slopes, SW (Betty's Bay to Kleinmond).•

lanata Andrews   Erect shrublet to 1 m. Flowers small, cup-shaped, woolly, white, with exserted anthers. Jan.–Dec. Mountain slopes, KM, SE (Outeniqua Mts: George, Uniondale to Humansdorp).•

langebergensis H.A.Baker   Erect shrublet to 50 cm. Flowers small, calycine, ovoid, pink. Oct. Upper slopes, LB (Langeberg Mts: Montagu).•

lanipes Guthrie & Bolus   Erect compact shrublet to 50 cm. Flowers medium/large, calycine, urn-shaped, lilac and white. Sept.–Oct. Rocky slopes, high alt., SW (Riviersonderend Mts).•

lanuginosa Andrews   Erect woody shrublet to 60 cm. Flowers large and inflated, calycine, ovoid, with long erect lobes, finely hairy, green becoming reddish. July–Aug. Rocky upper slopes, SW (Hermanus to Stanford).•

lasciva Salisb. (= *Philippia stokoei* Guthrie)   Erect shrub to 90 cm. Flowers small, calycine, cup-shaped, white to dirty yellow. Feb.–Aug. Sandy flats, SW, AP, LB (Cape Peninsula to Albertinia).•

lateralis Willd.   Erect to sprawling shrublet to 50 cm. Flowers small to medium, ovoid-urn-shaped, pink, sometimes with exserted anthers. Dec.–May. Lower to upper slopes, NW, SW (Tulbagh to Kogelberg).•

latiflora L.Bolus   Erect shrublet to 50 cm. Flowers medium, broadly funnel-shaped, pink, hairy. Sept.–Nov. Lower slopes, SW (Grabouw).•

lavandulifolia Salisb.   Erect shrub to 1.5(–2.0) m. Flowers small, calycine, open cup-shaped, lilac. Dec.–Apr. Lower to middle slopes, NW (Piketberg to Cold Bokkeveld).•

lawsonia Andrews   Erect shrublet to 50 cm. Flowers large, tubular with small spreading lobes, pink, hairy. Feb.–Apr. Middle slopes, SW (Genadendal).•

lehmannii Klotzsch   Erect shrublet to 1 m. Flowers small, cup- to bell-shaped, white, hairy. June–Jan. Moist lower to upper slopes, SE (George to Knysna).•

lepidota Rach   Erect shrublet to 70 cm. Flowers small, calycine, cup-shaped, dirty cream-coloured to greenish, with exserted anthers. July–Oct. Dry lower slopes. NW, SW (Kamiesberg to Malmesbury).

leptantha Dulfer   Erect woody shrublet to 20 cm. Flowers small, calycine, urn-shaped, rose-pink. Jan.–Mar. High alt., rocky places, NW (Grootwinterhoek to Hex River Mts).•

leptoclada Van Heurck & Müll.Arg.   Low sprawling shrublet. Flowers bell- to cup-shaped, pink, hairy. Oct.–Dec. Middle slopes, moist places, NW, SW (Piketberg to Greyton).•

leptopus Benth.   Erect semisprawling shrublet to 25 cm. Flowers small, cup-shaped, white. Nov.–Feb. Sandy flats and plateaus, NW (Clanwilliam to Ceres).•

lerouxiae Bolus   Erect shrub to 1.8 m. Flowers small, urn-shaped, white, hairy. Aug.–Sept. Middle slopes, SW (Franschhoek and Stellenbosch).•

leucantha Link   Erect shrublet to 50 cm. Flowers small, narrowly cup-shaped, white. Oct.–Feb. Lower slopes, SW (Paarl to Riviersonderend Mts).•

leucanthera L.f.   Erect shrub to 60 cm. Flowers small, calycine, yellow, with yellow exserted anthers. Sept.–Nov. Sandy flats and lower slopes, NW, SW (Cedarberg to Riviersonderend Mts).•

leucodesmia Benth.   Erect shrublet to 1 m. Flowers small, calycine, ovoid, white. Sept.–Nov. High alt., NW, KM, LB (Langeberg: Worcester to Swellendam and Waboomsberg).•

leucopelta Tausch (= *Erica maesta* Bolus)   Erect shrub to over 2 m. Flowers small, urn-shaped, cream-coloured to greenish, hairy. July–Nov. Lower to upper slopes, KM, SE (Witteberg to Free State and Mossel Bay to E Cape).

leucosiphon L.Bolus   Erect compact woody shrublet to 30 cm. Flowers large, tubular, white. Oct.–Nov. Upper slopes, rocky places, NW (Olifants River and Grootwinterhoek Mts).•

leucotrachela H.A.Baker   Erect shrub to 1.5 m. Flowers large, tubular, dark pink with white mouth, finely hairy. Apr.–Oct. Moist upper slopes, SW (Betty's Bay).•

lignosa H.A.Baker   Dwarf woody shrublet plastered against rocks. Flowers small, obconically urn-shaped, pink, finely hairy. Dec.–Feb. High peaks, KM (Groot Swartberg).•

limosa L.Bolus   Erect to semisprawling diffuse shrublet to 75 cm. Flowers small, broadly cup-shaped, white tinged red, finely hairy. Oct.–Nov. Local on marshy ground, SW (Cape Peninsula).•

lineata Benth.   Erect woody shrublet to 1(–2) m. Flowers large, shortly tubular, white to red, with exserted anthers. Mainly Dec.–Mar. Lower slopes and hills, AP (Gansbaai, Bredasdorp).•

loganii Compton   Like **E. articularis**, erect shrublet to 1 m. Flowers small, calycine, urn-shaped, white. Mar.–May. Upper slopes, NW, KM (Cedarberg to Witteberg Mts).•

longifolia Aiton (= *Erica onosmiflora* Salisb., *E. pustulata* H.A.Baker)   Erect shrublet to 1 m. Flowers large, tubular, white, yellow, orange, red, purple, greenish, sometimes bicoloured, sometimes slightly hairy, sometimes slightly sticky. Jan.–Dec. Sandy or stony slopes, SW, AP (Paarl to Bredasdorp).•

longipedunculata Lodd. (= *Erica dumosa* Andrews)   Low soft shrublet to 30 cm. Flowers small to medium, urn-shaped to conical, pink, with exserted anthers. Nov.–Apr. Dry sandy upper slopes, NW (Clanwilliam to Tulbagh).•

longistyla L.Bolus   Erect shrublet to 40 cm. Flowers small, subcalycine, cup-shaped, pink, with exserted anthers. Oct.–Apr. High alt., NW (Grootwinterhoek to Hex River Mts).•

lowryensis Bolus   Erect shrublet to 1(–2) m. Flowers small/medium, cup-shaped, white. Sept.–Nov. Middle slopes, SW (Hottentots Holland Mts to Betty's Bay).•

lucida Salisb.   Erect shrublet to 1 m. Flowers small, pink to magenta. Sept.–Nov. Rocky mountain slopes and plateaus, NW, SW (Gifberg to Wemmershoek to Montagu).•

lutea P.J.Bergius   GEELRYSHEIDE   Erect shrublet to 90 cm. Flowers small to medium, calycine, narrowly urn-shaped with spreading lobes, yellow to white. Oct.–June. Middle to upper slopes, SW (Cape Peninsula and Riviersonderend to Kleinrivier Mts).•

lycopodiastrum Lam.   Erect compact woody shrublet to 50 cm. Flowers small, calycine, obovoid, white. Sept.–Nov. Rocky places often in riverbeds, SW (Elgin to Villiersdorp).•

macilenta Guthrie & Bolus   Erect slender shrub to 1.5 m. Flowers medium, urn-shaped, pink. Dec.–Jan. Moist middle slopes, LB (Langeberg Mts).•

macowanii Cufino   Sturdy erect shrub to 1.5 m. Flowers large, tubular, yellow or red with yellow mouth, hairy. Aug.–Mar. Middle to upper slopes, SW (Kogelberg to Hermanus, Caledon: Swartberg).•

macroloma Benth.   Erect slender shrublet to 30 cm. Flowers medium, cup-shaped, pink, hairy and sticky. Aug.–Oct. Lower slopes, SW (Houwhoek to Kogelberg).•

macrophylla Klotzsch   Erect shrublet to 1 m. Flowers small, urn-shaped, white. Jan.–Aug. High alt., LB (Langeberg Mts: Swellendam to Riversdale).•

macrotrema Guthrie & Bolus   Erect shrub to 1.5(–2.0) m. Flowers small, subcalycine, cup-shaped, magenta, with exserted anthers. Sept.–Nov. Flats and lower slopes, NW (Cold Bokkeveld Mts to Witteberg).•

maderi Guthrie & Bolus   Like **E. sphaerocephala**, erect shrublet to 30 cm. Flowers small to medium in heads, ovoid, white to rose. Sept.–Feb. Middle to upper alt., NW (Cedarberg to Hex River Mts.).•

magistrati E.G.H.Oliv.   Erect woody shrub to 1.8 m. Flowers small, white. Aug.–Dec. S-facing moist slopes, SW (Hottentots Holland Mts).•

magnisylvae E.G.H.Oliv.   Erect shrub to 3.5 m. Flowers small, white. Mar.–May. Sand over calcareous rock, AP (Gansbaai).•

mammosa L.   NINEPIN HEATH, ROOIKLOSSIEHEIDE   Erect shrub to 1.5 m. Flowers large, inflated tubular with closed mouth, orange to red. Nov.–May. Sandy flats and lower mountain slopes, NW, SW, AP (Cedarberg Mts to Bredasdorp).•

margaritacea Sol.   Erect shrublet to 50 cm. Flowers small, urn-shaped, white or pinkish. Oct.–Mar. Sandy flats, SW (Cape Peninsula).•

mariae Guthrie & Bolus   Compact to straggling erect shrublet to 1(–2) m. Flowers large, tubular, dark red. Jan.–Dec. depending on rainfall. Coastal limestone hills, AP (Bredasdorp to Stilbaai).•

marifolia Sol.   Straggling slender shrublet against rocks. Leaves broad almost flat. Flowers small, urn-shaped, white, hairy, with exserted anthers. Sept.–Feb. Moist rock ledges, SW (Cape Peninsula).•

maritima Guthrie & Bolus   Erect shrublet to 30 cm. Flowers tiny, cup-shaped, dull cream-coloured to brownish. Apr.–July. Coastal and inland flats, SW (Bredasdorp and Genadendal).•

marlothii Bolus   Sprawling delicate shrublet to 30 cm. Flowers small, depressed-urn-shaped, cream-coloured to reddish, hairy or not, with exserted anthers. Oct.–Nov. Shady upper slopes and rocky outcrops, NW, SW (Cedarberg Mts to Bainskloof).•

massonii L.f.   Erect shrublet to 1 m. Flowers large, tubular, very sticky, red with green tip. Mainly Nov.–Mar. Rocky mountain slopes, SW (Hottentots Holland to Kleinrivier Mts).•

mauritanica L. (= *Erica viridipurpurea* L.)   Like **E. quadrangularis**, erect shrublet to 1 m. Flowers small, broadly urn-shaped, pink. Aug.–Nov. Lower slopes and flats, SW (Cape Peninsula to Stellenbosch).•

maximiliani Guthrie & Bolus   Erect shrub to 1.5 m. Flowers large, tubular, greenish cream-coloured to pale yellow. July–Dec. Rock crevices or rocky slopes at middle to upper alt., NW, KM (Cedarberg Mts to Klein Swartberg, Touwsberg and Rooiberg).•

melanacme Guthrie & Bolus   Erect shrublet to 45 cm. Flowers small, calycine, broadly urn-shaped, dark pink. Nov.–Dec. Coastal hills, AP (Bredasdorp).•

melanthera L.   Erect shrublet to 60 cm. Flowers small, very open cup-shaped, bright pink. June–Dec. Moist lower to upper slopes, KM, LB (Swellendam to Uitenhage, Rooiberg and Gamkaberg).•

micrandra Guthrie & Bolus   Erect shrublet to 40 cm. Flowers small, cup-shaped, whitish. Dec.–Jan. Lower slopes, NW (Ceres Mts).•

mira Klotzsch ex Benth.   Erect compact shrublet to 20 cm. Flowers small, tubular, pink. Jan.–Apr. Sandy places, KM (Witteberg).•

mitchellensis Dulfer (= *Erica saxatilis* L.Bolus)   Erect multistemmed resprouting shrublet to 20 cm. Flowers medium, tubular ovoid, slightly hairy. Jan. Upper slopes, rocky places and crevices, NW (Hex River Mts).•

modesta Salisb.   Erect shrublet to 75 cm. Flowers small, calycine, urn-shaped, pink to white, hairy. Sept.–Dec. Rock crevices, middle northern slopes, SW (Riviersonderend Mts).•

mollis Andrews   Erect shrublet to 60 cm. Flowers small, broadly urn-shaped, pink, hairy. Nov.–Dec. Marshy or moist slopes, SW (Cape Peninsula).•

monadelphia Andrews   Erect multistemmed resprouting shrublet to 50 cm. Flowers large, tubular, sticky, crimson-red, with far-exserted anthers. Jan.–Apr. or after fires. Sandy lower slopes, SW, AP (Cape Peninsula to Bredasdorp).•

monantha Compton   Erect shrublet to 10 cm. Flowers small, bell-shaped, pink, hairy. Dec.–Mar. Rocky slopes at high alt., NW (Olifants River Mts).•

monsoniana L.f.   BOKKEVELDSHEIDE   Erect sparsely branched shrub to 1.5 m. Flowers large, calycine, ovoid-conical, white with conspicuous white leaves below flowers. June–Oct. Middle to upper slopes, NW, SW (Cedarberg Mts to Potberg).•

mucronata Andrews   Erect shrublet to 50 cm. Flowers medium, calycine, cup-shaped, pink. Nov.–Jan. Middle to upper slopes, LB, SE (Riversdale to Uniondale).•

multumbellifera P.J.Bergius (= *Erica ramentacea* L.)   Erect shrublet to 40 cm. Flowers small, broadly urn-shaped, purple to red. Nov.–June. Sandy flats and mountains, NW, SW, LB (Tulbagh to Riversdale).•

mundii Guthrie & Bolus   Erect shrublet to 50 cm. Flowers small, cup-shaped, pink. Aug.–Nov. Lower mountain slopes, LB (Langeberg Mts: Swellendam).•

myriocodon Guthrie & Bolus   Erect diffuse shrublet to 20 cm. Flowers small, cup-shaped, whitish, hairy. Oct.–Dec. Marshes on sandy flats and lower slopes, SW (Wemmershoek to Riviersonderend Mts).•

nabea Guthrie & Bolus   Erect shrub to 1.5 m. Flowers large, very calycine, green and white, corolla very reduced around ovary. May–Aug. Moist flats and mountain slopes, SE (George to Uitenhage and Kouga Mts).•

nana Salisb.   Procumbent woody shrublet to 20 cm. Flowers large, tubular, bright yellow. Sept.–Oct. Cliffs and rocks at high alt., SW (Hottentots Holland Mts to Kogelberg).•

navigatoris E.G.H.Oliv.   Erect shrub to 30 cm. Flowers small, urn-shaped, white. Nov.–Jan. Middle to high alt., NW, KM (Hex River Mts to Waboomsberg).•

nematophylla Guthrie & Bolus   Erect shrublet to 50 cm. Flowers large, tubular, white. Aug.–Feb. Lower mountain slopes, LB (Langeberg: Riversdale).•

nemorosa Klotzsch ex Benth.   Erect shrublet to 1(–2) m. Flowers small, cup-shaped, dark pink. Mar. Lower slopes, SE (Uitenhage to Stutterheim).

nervata Guthrie & Bolus   Erect shrublet to 90 cm. Flowers small, calycine, cup-shaped, pink to red. Oct.–Dec. Rocky upper south slopes, KM (Swartberg to Kammanassie Mts).•

nevillei L.Bolus   Semisprawling woody shrublet to 30 cm. Flowers large, tubular, red-orange. Dec.–Mar. Upper slopes, rocky places, SW (Cape Peninsula).•

newdigateae Dulfer (= *Erica longipes* Klotzsch)   Erect shrublet to 50 cm. Flowers small/medium, open cup-shaped, pink. June–Oct. Lower to middle slopes, SE (Kouga, Baviaanskloof and Tsitsikamma Mts).•

nigrimontana Guthrie & Bolus   Like **E. corifolia**, erect shrublet to 50 cm. Flowers small, calycine, urn-shaped, pink. Oct. Upper slopes, SW (Caledon Swartberg).•

nivea Sinclair   Like **E. palliiflora**, erect shrublet to 60 cm. Flowers small, calycine, bell-shaped, white. Aug.–Oct. Middle slopes, SW (Cape Peninsula and Kogelberg).•

nubigena Bolus (= *Erica macra* Guthrie & Bolus)   Erect shrublet to 30 cm. Flowers small, urn-shaped, sticky, reddish purple. Oct.–Jan. Rocky upper slopes and summits, NW, KM, SE (Cedarberg Mts to Uniondale).•

nudiflora L.   Erect compact to sprawling shrublet to 30 cm. Flowers small, tubular to elongate-bell-shaped, pink to red, with exserted anthers. Feb.–June. Coastal flats to inland high alt., NW, SW, AP, KM (Cedarberg Mts to Witteberg and Bredasdorp).•

oakesiorum E.G.H.Oliv.   Erect shrub to 4 m. Flowers small, urn-shaped, white. Aug.–Oct. Southern slopes, SW (Riviersonderend Mts).•

obconica H.A.Baker   Erect sturdy shrublet to 1 m. Flowers small, subcalycine, open cup-shaped, rose. Oct.–Nov. High alt., LB (Langeberg Mts: Riversdale).•

obliqua Thunb.   Erect slender shrublet to 50 cm. Flowers small to medium, ovoid to urn-shaped, pink to reddish purple to lilac, hairy and sticky. Nov.–May. Lower to upper slopes, SW (Cape Peninsula to Hermanus).•

oblongiflora Benth.   Erect shrub to 1 m. Flowers medium, tubular ovoid, light green to yellowish. Dec.–Aug. Limestone hills, AP (Bredasdorp area).•

obtusata Klotzsch ex Benth.   Erect spreading or prostrate shrublet to 30 cm. Flowers small, broadly urn-shaped, white, slightly sticky, with slightly exserted anthers. Aug.–Nov. Moist rock ledges and S-facing slopes, middle to high alt., SW (Cape Peninsula to Riviersonderend Mts).•

occulta E.G.H.Oliv.   Erect compact bonsai-like shrublet to 10 cm. Flowers small, rather hidden, calycine, ovoid with long erect lobes, yellowish. Aug.–Oct. Cliffs on low limestone hills, AP (Pearly Beach).•

ocellata Guthrie & Bolus   Erect shrublet to 50 cm. Flowers small, urn-shaped, pink. Sept.–Dec. Summits and S-facing upper slopes, LB (Langeberg Mts: Swellendam).•

odorata Andrews   Erect shrublet to 90 cm. Flowers medium, very umbellate, bell-shaped, white, sticky. Aug.–Nov. Moist upper southern slopes, SW (Hottentots Holland Mts to Babilonstoring).•

oligantha Guthrie & Bolus   Prostrate spreading shrub. Flowers small, cup-shaped, white with large dark pink sepals, hairy, slightly sticky, with exserted anthers. Sept.–Nov. Marshes at middle alt., SW (Stanford, Napier to Elim).•

oliveri H.A.Baker   Erect diffuse shrublet to 15 cm. Flowers small, broadly urn-shaped, pink, slightly sticky with partially exserted anthers. Oct.–Nov. Lower to middle moist slopes, SW (Stanford to Napier).•

omninoglabra H.A.Baker   Sprawling shrublet to 10 cm. Flowers medium, narrowly urn- to cup-shaped, white. Sept.–Dec. Middle to upper slopes, wet places, LB (Langeberg Mts: Swellendam).•

onusta Guthrie & Bolus   Erect shrublet to 40 cm. Flowers small, red. Nov.–Dec. Fynbos patches between forest, SE (George to Knysna).•

oophylla Benth.   Erect woody resprouting shrublet to 10 cm. Flowers small, globose, pinkish, hairy, slightly sticky. Jan.–Feb. Rock crevices at high alt., SW (Langeberg Mts: Swellendam).•

opulenta (J.C.Wendl. ex Klotzsch) Benth.   Like **E. granulosa** and **E. harveyana**, erect shrublet to 1(–2) m. Flowers small, narrowly cup-shaped to urn-shaped, pink, with exserted anthers. Aug.–Nov. Lower slopes and coastal flats, SE (George to Humansdorp).•

orculiflora Dulfer   Like **E. oxysepala**, erect shrublet to 35 cm. Flowers small, ovoid, white to pink. Nov.–Apr. Upper slopes, NW (Ceres to Montagu).•

oreophila Guthrie & Bolus   Erect diffuse shrublet to 15 cm. Flowers small, urn-shaped, white with red tips. Sept.–Dec. Damp ledges, middle slopes, SW (Franschhoek and Stellenbosch).•

oresigena Bolus   Erect woody shrub to 2 m. Flowers small, broadly urn-shaped, mauve-pink, hairy. Oct.–Jan. Rocky summits and upper slopes, NW (Cedarberg to Hex River Mts).•

orthiocola E.G.H.Oliver   Compact shrublet. Flowers in pairs, small, calycine, shortly funnel-shaped, pink turning brownish, pedicel with substellate bracts. Feb.–Mar. Moist sandstone slopes, SW (Riviersonderend Mts: Pilaarkop).•

ovina Klotzsch ex Benth.   Erect floriferous shrublet to 9 cm. Flowers small, sometimes medium, ovoid, woolly, pink or white. July–Dec. Middle to upper rocky slopes, SW (Riviersonderend Mts).•

oxyandra Guthrie & Bolus   Erect shrublet to 50 cm. Flowers small, urn-shaped, white to pale pink, hairy. Dec.–Feb. Moist upper slopes, LB (Langeberg Mts: Swellendam).•

oxycoccifolia Salisb.   Delicate mat-forming shrublet. Flowers small, cup-shaped, white to pink, thinly hairy. Dec.–Feb. Moist rock faces, SW (Cape Peninsula).•

oxysepala Guthrie & Bolus   Erect resprouting shrublet to 35 cm. Flowers small, cup-shaped, pale yellow. Jan.–May. Middle slopes, NW (Grootwinterhoek to Cold Bokkeveld Mts).•

pageana L.Bolus   Erect shrub to 1.5m. Flowers medium, campanulate, yellow, very finely hairy. Sept.–Nov. Marshy upper slopes, SW (Kogelberg).•

palliiflora Salisb. (= *Erica nivea* Sinclair)   Erect shrublet to 60 cm. Flowers small, calycine, campanulate, pink or white. Oct.–Apr. Middle to upper slopes, NW, SW, LB, SE (Cedarberg Mts to George).•

paludicola L.Bolus   Erect diffuse to sprawling shrublet to 1 m. Flowers medium, cup-shaped, pink, minutely hairy. Jan.–Feb. Marshy mountain slopes, SW (Cape Peninsula).•

paniculata L.   Erect shrublet to 50 cm. Flowers small, bell- to urn-shaped, dark to pale pink. July–Oct. Lower slopes and hills, NW, SW, KM (Tulbagh to Caledon and Klein Swartberg).•

pannosa Salisb. (= *Erica barbata* Benth.)   Erect shrublet to 1 m. Flowers small, urn-shaped, white, hairy. Feb.–July. Lower to middle slopes, SW (Riviersonderend Mts).•

papyracea Guthrie & Bolus   PAPERY HEATH, LEMOENBLOEISELS   Erect shrublet to 1 m. Flowers medium, calycine, cup-shaped, white. Sept.–Oct. Mountain slopes, LB (Langeberg Mts: Riversdale).•

parilis Salisb. (= *Erica longisepala* Guthrie & Bolus)   Erect shrublet to 1(–2) m. Flowers small to large, urn-shaped, in spikes, bright yellow, with slightly exserted anthers. Jan.–May. Dry middle to upper slopes, NW, SW (Clanwilliam to Paarl).•

parviflora L.   Erect or spreading shrublet to 1 m. Flowers small, urn-shaped, white, pink to magenta, hairy. Apr.–Nov. Flats and slopes, often wet places, SW, AP (Du Toitskloof Mts to Cape Peninsula to Bredasdorp).•

passerinae Montin   Erect woody shrublet to 60 cm. Flowers small/medium, calycine, open cup-shaped, pink with white-woolly calyx. July–Oct. Dry rocky middle to upper slopes, KM, SE (Swartberg, Kouga and Baviaanskloof Mts).•

patersonia Andrews   MEALIE HEATH, MIELIEHEIDE   Erect sparsely branched shrublet to 1 m. Flowers large, tubular, yellow. Apr.–Nov. Marshy coastal flats, SW (Cape Peninsula to Hermanus).•

pauciovulata H.A.Baker   Diffuse sprawling shrublet, sometimes to 30 cm. Flowers small, cup-shaped, deep pink. Sept.–Nov. Marshy ground, middle slopes, SW (Stanford to Napier).•

pectinifolia Salisb.   Erect shrub, sometimes sparse, to 1.5 m. Flowers large, tubular, white, sometimes tinged with pink, hairy. Jan.–Dec. Flats to middle slopes, SE (Uniondale to Port Elizabeth).•

pellucida Sol.   Erect to spreading shrublet to 50 cm. Flowers medium, inflated tubular with spreading lobes, bright pink. Dec.–Feb. Upper slopes, SW (Riviersonderend Mts).•

peltata Andrews (= *Philippia pallida* L.Guthrie)   KER-KER, RAASHEIDE   Erect shrub to 1.2 m. Flowers small, subcalycine, cup-shaped, pink. Jan.–Dec. Lower southern slopes, SW, LB, SE (Riviersonderend Mts to George).•

penicilliformis Salisb. (= *Erica calyculata* J.C.Wendl.)   Erect shrublet to 50 cm. Flowers small, calycine, narrowly cup-shaped, white, with exserted anthers. Jan.–Dec. Lower to upper slopes, SW, LB, SE (Riviersonderend Mts to Knysna).•

perlata Sinclair   Erect shrublet to 60 cm. Flowers small, urn-shaped to bell-shaped, white, hairy with partially exserted anthers. Aug.–Dec. Middle to upper rocky slopes, SW (Riviersonderend Mts).•

permutata Dulfer (= *Erica confusa* Guthrie & Bolus)   Erect diffuse matted shrublet to 20 cm. Flowers small, ovoid, pink, hairy, with exserted anthers. Oct.–Apr. Lower slopes, SW (Riviersonderend Mts).•

perspicua J.C.Wendl. (= *Erica dulcis* L.Bolus)   PRINCE-OF-WALES HEATH, VEERHEIDE   Erect shrub to 2 m. Flowers medium to large, tubular, hairy, white to pink with white tips. Sept.–Apr. Marshy slopes and flats, SW (Betty's Bay to Hermanus).•

petiolaris Lam.   Erect shrublet to 50 cm. Flowers small, calycine, bell-shaped, white, hairy. May–Dec. Rocky damp places, upper mountain slopes, SW (Cape Peninsula and Hermanus to Napier).•

petraea Benth. (= *Erica krigeae* Compton)   Erect to semispreading shrublet to 40 cm. Flowers small, ovoid-cup-shaped, white or pink, with exserted anthers. Jan.–July. Dry rocky middle to upper slopes, KM, SE (Swartberg Mts to Uniondale).•

petrophila L.Bolus   Small mat-forming shrublet to 5 cm. Flowers small, cup-shaped, white, hairy, sticky. Sept.–Jan. Rock crevices and overhangs, middle to upper slopes, SW (Paarl, Riviersonderend and Kleinrivier Mts).•

peziza Lodd.   KAPOKKIE   Erect shrub to 2 m. Flowers small, urn-shaped, hairy, white. Aug.–Nov. Rocky lower slopes, NW, SW, LB (Robertson and Stormsvlei to Swellendam).•

philippioides Compton   Erect shrublet to 50 cm. Flowers small, cup-shaped, dull white, hairy. June–Oct. Upper slopes and summits, NW (Kamiesberg to Cedarberg Mts).

phillipsii L.Bolus   Like **E. doliiformis**, erect shrublet to 50 cm. Flowers medium, urn-shaped, reddish, hairy. Mar.–May. Upper slopes, NW, SW (Piketberg to Bainskloof Mts).•

phylicifolia Salisb. (= *Erica purpurea* Andrews)   Erect shrub to 1.5 m. Flowers large, tubular, pale magenta. Mostly Dec.–May. Sandstone slopes, SW (Cape Peninsula).•

physantha Benth.   Erect semispreading shrublet to 30 cm. Flowers small, urn-shaped, pink, with exserted anthers. Sept. Dry clay hillslopes, LB (Riversdale).•

physodes L.   Erect shrublet to 70 cm. Flowers medium, ovoid-urn-shaped, sticky, white. Feb.–Aug. Rocky upper slopes, SW (Cape Peninsula).•

physophylla Benth.   Moss-like shrublet against rocks. Flowers small, cup-shaped, reddish. Dec.–Feb. Moist rock-ledges near summits, SW (Riviersonderend Mts).•

pillansii Bolus (= *Erica pyrantha* L.Bolus)   Erect shrub to 1.5 m. Flowers large, tubular, bright orange-red, finely hairy. Jan.–July. Moist slopes and marshes, SW (Kogelberg).•

pillarkopensis H.A.Baker   Dense erect woody shrub to 2 m. Flowers small, calycine, cup-shaped, pink. Aug.–Nov. Middle to upper slopes, SW (Riviersonderend Mts).•

pilulifera L.   Erect sparsely branched shrublet to 40 cm. Flowers small, ovoid-urn-shaped, pink. May–Aug. Marshy places, upper slopes, SW (Cape Peninsula).•

pinea Thunb.   Erect shrublet to 1.5 m. Flowers large, tubular, white, yellow with white tips, or purplish pink, smooth, rarely hairy. Jan.–May. Rocky slopes and plateaus, SW (Bainskloof to Kleinrivier Mts).•

placentiflora Salisb.   KLOKKIESHEIDE   Like **E. imbricata**, erect shrublet to 60 cm. Flowers small, calycine, broadly urn-shaped, pink, with exserted anthers. June–Dec. Flats and slopes, NW, SW, AP (Cedarberg Mts to Bredasdorp).•

planifolia L.   Erect diffuse shrublet to 30 cm. Flowers small, cup-shaped, reddish, hairy. July–Feb. Mountain slopes, SW (Stellenbosch to Hottentots Holland Mts).•

plena L.Bolus   Erect soft shrublet to 30 cm. Flowers medium, narrowly cup-shaped, white turning rose, hairy or glabrous. Aug.–Oct. Marshy places on hills, SW (Napier).•

plukenetii L.   HANGERTJIE   Erect shrublet to 1.0(–1.5) m. Flowers large, inflated tubular, pink to red, sometimes white, green or yellow, with far-exserted anthers. Jan.–Dec. Widespread, NW, SW, AP, KM, LB (Namaqualand to Mossel Bay and Witteberg).

plumigera Bartl.   Erect shrublet to 70 cm. Flowers small, calycine, pink. Sept.–Nov. Middle to upper slopes, SW (Houwhoek to Napier).•

podophylla Benth.   Erect shrublet to 50 cm. Flowers small, whitish, hairy. Oct.–Dec. Rocky moist southern slopes at high alt., LB (Langeberg Mts: Swellendam).•

pogonanthera Bartl.   Erect small straggling shrublet. Flowers small, calycine, hairy, white, with exserted anthers. Sept.–Nov. Rocky slopes, SW (Caledon Swartberg).•

polifolia Salisb. ex Benth.   Erect lax spreading shrublet to 60 cm. Flowers small, broadly urn-shaped, sticky, white. Sept.–Nov. Middle to upper slopes, LB (Langeberg Mts: Swellendam).•

polycoma Benth.   Erect grey shrublet to 60 cm. Flowers small, lilac-pink. Jan.–May. Upper rocky slopes, SW (Riviersonderend Mts).•

porteri Compton   Like **E. thomae**, erect shrublet to 1 m. Flowers large, tubular, deep pink with white tips. Mar.–June. Lower rocky slopes, SW (Betty's Bay).•

praecox Klotzsch   Erect to spreading compact shrublet to 20 cm. Flowers urn-shaped to tubular urn-shaped, pink. Dec.–Feb. Upper ridges and summits, SW (Du Toitskloof Mts to Villiersdorp).•

praenitens Tausch   Erect compact shrublet to 50 cm. Flowers medium to large, tubular with spreading lobes, pink. Dec.–Feb. Rocky middle slopes, SW (Riviersonderend Mts).•

primulina Bolus   Erect shrublet to 30 cm. Flowers large, tubular, pale green to yellowish. Apr.–July. Upper slopes and summits, KM (Swartberg Mts to Willowmore).•

priorii Guthrie & Bolus   Erect shrublet to 1.2 m. Flowers small, cup-shaped, pink. Aug.–Nov. Middle slopes, SE (George).•

propendens Andrews (= *Erica dulcis* Bolus)   BELL HEATH, KLOKKIESHEIDE   Erect shrublet to 1 m. Flowers medium, campanulate, pink to dark mauve-pink sometimes with white throat, hairy. Aug.–Jan. Marshy flats and lower slopes, SW (Hottentots Holland to Kleinrivier Mts).•

propinqua Guthrie & Bolus   Erect shrublet to 50 cm. Flowers small, calycine, urn-shaped, deep pink. June–Aug. Coastal limestone hills, AP (Pearly Beach to De Hoop).•

pseudocalycina Compton   Like **E. calycina**, erect multistemmed shrublet to 50 cm. Flowers small, calycine, bell-shaped with spreading lobes, white. Sept.–Feb. Middle to upper slopes, NW, KM, SE (Cedarberg Mts to Grahamstown).

pubescens L.   Erect shrublet to 60 cm, multistemmed resprouter. Flowers small, urn-shaped, white, hairy. Oct.–Apr. Dry lower to middle slopes, NW, SW (Piketberg to Cape Peninsula).•

pubigera Salisb.   Erect shrublet to 80 cm. Flowers small, urn-shaped, hairy, white with distinctive brown lobes. Aug.–Oct. Rocky lower slopes, LB (Swellendam).•

pudens H.A.Baker   Low creeping shrublet to 15 cm. Flowers medium, clustered in pendulous heads, narrowly urn-shaped, white, sometimes hairy. July–Nov. Middle to upper slopes, NW (Kamiesberg to Grootwinterhoek Mts).

pulchella Houtt. (incl. **E. longiaristata** Benth.)   Erect shrublet to 60 cm. Flowers small, urn- to cup-shaped, sometimes narrowly so, in spikes, pink to dark red. Dec.–May. Sandy flats and lower slopes, SW, AP (Cape Peninsula to Albertinia).•

pulvinata Guthrie & Bolus   Erect shrublet to 30 cm. Flowers small, globose, white. Sept.–Oct. Limestone hills, AP (Soetanysberg).•

purgatoriensis H.A.Baker   Like **E. parviflora**, diffuse shrublet. Flowers small, urn-shaped, pink, hairy. Oct. Marshy lower slopes, SW (Franschhoek).•

pycnantha Benth.   Erect slender shrublet to 30 cm. Flowers in many-flowered heads, medium, calycine, tubularly urn-shaped with spreading lobes, pink with paler tips. Oct.–Dec. Upper moist slopes, SW (Helderberg to Betty's Bay).•

pyramidalis Sol. ex Aiton   Erect shrublet to 60 cm. Flowers medium, obconic-campanulate, rose. Oct. Marshy flats, SW (Cape Flats).•

pyxidiflora Salisb.   Erect shrublet to 60 cm. Flowers small, in dense spikes, pale pink to whitish. May–Dec. Damp upper slopes, SW (Cape Peninsula).•

quadrangularis Salisb. (= *Erica cyathiformis* Salisb.)   Like **E. mauritanica**, erect often compact floriferous shrublet to 50 cm. Flowers small, broadly bell-shaped, pink to white. July–Dec. Flats and lower slopes, NW, SW, LB, SE (Cedarberg to Hottentots Holland to Outeniqua Mts, inland to Witteberg).•

quadrisulcata L.Bolus   Erect semispreading shrublet to 30 cm. Flowers large, tubular, orange-yellow. Dec.–Feb. Rocky summits, SW (Cape Peninsula).•

racemosa Thunb.   Erect shrublet to 30 cm. Flowers small, urn-shaped, rose, with exserted anthers. Oct.–Feb. Lower mountain slopes, NW, LB (Langeberg Mts: Montagu to Riversdale).•

recta Bolus (= *Erica muirii* L.Bolus)   Like **E. speciosa**, erect shrublet with persistent rootstock to 1 m. Flowers medium, tubularly urn-shaped, white or pink, with slightly exserted anthers. Apr.–Oct. Lower dry slopes, KM, LB (Klein Swartberg and Langeberg Mts: Cloete's Pass).•

regerminans L. (= *Erica ellipticiflora* Dulfer)   Erect shrub to 1.5 m. Flowers small, globose to urn-shaped, bright pink, in long dense spikes. Aug.–Dec. Middle to upper slopes, LB (Langeberg Mts: Swellendam to Riversdale).•

regia Bartl.   ELIM HEATH   Erect shrublet to 1 m. Flowers large, tubular, striking white with purple then orange-red tips, or plain orange. Mainly Aug.–Oct. Sandy or gravelly coastal flats, SW, AP (Elim/Bredasdorp).•

rehmii Dulfer   Erect shrub to 1.5 m. Flowers small/medium, urn-shaped, yellow. Sept.–Oct. Seepage areas, middle slopes, SW (Elandsberg and Bainskloof).•

retorta Montin   BOTTLE HEATH   Erect straggling shrublet to 30 cm. Leaves reflexed. Flowers large, tubular-ampullaceous, sticky, pink. Jan.–Dec. Lower to middle slopes, SW (Viljoen's Pass to Betty's Bay).•

rhodantha Guthrie & Bolus   Erect shrublet to 30 cm. Flowers small, subcalycine, cup-shaped, rose, with exserted anthers. Sept.–Oct. Lower northern slopes, LB (Langeberg Mts: Riversdale).•

rhodopis (Bolus) Bolus   Erect rounded shrublet to 30 cm. Flowers small, ovoid, pink. Dec.–Feb. Sandy slopes and flats, SW (Botrivier).•

rhopalantha Dulfer (= *Erica nodiflora* Klotzsch)   Erect shrublet to 50 cm. Flowers small, calycine, broadly urn-shaped, dark pink. Nov.–May. Sandy flats and slopes, SW, AP (Kogelberg to Bredasdorp).•

ribisaria Guthrie & Bolus   Erect diffuse shrublet to 45 cm. Flowers small, globose to broadly urn-shaped, mauve-pink with dark tips, hairy. Jan.–Apr. Marshy places on sandy flats and lower slopes, SW (Houwhoek to Kleinmond).•

riparia H.A.Baker   Diffuse spreading shrublet to 30 cm. Flowers small, globose, dark pink to purple, hairy, with exserted anthers. Aug.–Jan. Marshy slopes, AP (Gansbaai to Soetanysberg).•

roseoloba E.G.H.Oliv.   Erect compact rounded to flattened shrublet to 30 cm. Flowers small, cup-shaped, white with pink tips. Jan.–Mar. High alt. rocky slopes, KM (Klein Swartberg).•

rubens Thunb. (= *Erica vanheurckii* Müll.Arg.)   Erect compact or spreading shrublet to 30 cm. Flowers small, urn-shaped, dark pink to almost red. Sept.–Apr. Sandy flats to upper slopes, NW, KM (Cedarberg Mts to Klein Swartberg).•

rubiginosa Dulfer (= *Erica fucata* Klotzsch ex Benth., *E. tenuipedicellata* Compton)   Erect compact shrublet to 30 cm. Flowers in dense spikes, small, broadly cup- to bell-shaped, white or pink, with dark exserted anthers. July–Oct. Flats and lower slopes, SW, AP (Botrivier to Agulhas).•

rudolfii Bolus   Erect shrub to 1.5 m. Flowers small, broadly urn-shaped, pale purplish pink. Aug.–Oct. Middle to high alt., LB (Langeberg Mts: Swellendam to Heidelberg).•

rufescens Klotzsch   Erect compact viscid shrublet to 30 cm. Flowers large, tubularly urn-shaped, magenta. Jan.–Apr. Upper rocky slopes, SW (Riviersonderend Mts).•

rupicola Klotzsch   Erect shrublet to 30 cm. Flowers small, ovoid to urn-shaped, rose. Sept.–Nov. Lower slopes, SW (Riviersonderend Mts).•

sacciflora Salisb.   SWARTBEKHEIDE   Erect shrub to 1.5 m. Flowers tubular, yellow or orange with greenish tips. July–Oct. Lower slopes, SW (Franschhoek Mts).•

salax Salisb.   Erect sprawling shrublet to 30 cm. Flowers small, broadly cyathiform, white. Oct. High alt., SW (Stellenbosch Mts).•

salteri L.Bolus   Erect compact to slender sprawling shrublet to 20 cm. Flowers small, broadly-campanulate, cream-coloured or pink, hairy. Dec.–Feb. Locally in marshes on mountains, SW (Cape Peninsula).•

saptouensis E.G.H.Oliv.   Erect shrublet to 30 cm. Flowers small, calycine, cup-shaped, pink to red, with exserted anthers. Oct.–Jan. Rocky upper slopes, SE (Kouga Mts).•

savilea Andrews   Erect shrublet to 1 m. Flowers medium, urn-shaped, bright pink. Jan.–May. High alt., SW (Hottentots Holland Mts to Villiersdorp).•

saxicola Guthrie & Bolus   Dwarf brittle shrublet to 15 cm. Flowers small, subcalycine, white. Dec.–Apr. Coastal limestone, rock ledges, AP (Pearly Beach).•

saxigena Dulfer   Erect to sprawling woody shrublet to 50 cm. Flowers medium, urn-shaped, red, finely hairy. Aug.–Nov. Rocky places at high alt., KM (Swartberg Mts).•

scabriuscula Lodd. (= *Erica gibbosa* Klotzsch)   Erect shrub to 3 m. Flowers small, urn-shaped, white to pale pink. Jan.–Dec. Lower slopes and flats, SE (Mossel Bay to Humansdorp).•

schumannii E.G.H.Oliv.   Prostrate sprawling woody shrub. Flowers medium, urn-shaped, pink. Dec.–Jan. Rocky outcrops at high alt., SW (Stettynsberg).•

scytophylla Guthrie & Bolus   Erect shrublet to 1 m. Flowers small, urn-shaped, pink, with dark exserted anthers. June–Nov. Limestone hills, AP (Bredasdorp to De Hoop).•

selaginifolia Salisb.   Erect shrublet to 1(–1.7) m. Flowers small, calycine, urn-shaped, pink. Flowering time? Rocky lower to middle slopes, NW, SW, KM (Cedarberg Mts to Willowmore).•

senilis Klotzsch   Prostrate creeping shrublet to 20 cm. Flowers in dense terminal woolly white heads, small, cup-shaped, hairy, white. Sept.–Dec. Rocky, sandy slopes and flats, NW (Cedarberg Mts to Cold Bokkeveld).•

seriphiifolia Salisb.   Erect usually compact shrublet to 30 cm. Flowers small, calycine, open cup-shaped, pink. Sept.–Jan. Moist areas, lower to middle slopes, KM, LB, SE (Rooiberg and Langeberg Mts to Uniondale).•

sessiliflora L.f.   Erect woody shrub to 2 m. Flowers large, tubular, light green, with distinctive fruiting inflorescences on older branches. Jan.–Dec. Flats and mostly lower slopes, NW, SW, LB (Piketberg to Humansdorp).•

setacea Andrews   Erect shrublet to 50 cm. Flowers small, urn-shaped, white to pink. July–Oct. Lower dry hills to middle slopes, SW (Paarl to Caledon).•

setociliata H.A.Baker   Straggling to erect shrublet to 30 cm. Flowers small, cup-shaped, white to pale pink. Sept.–Nov. Rocky middle to upper slopes, NW (Cedarberg Mts to Keeromsberg).•

setosa Bartl.   Like **E. fausta**, erect shrublet to 50 cm. Flowers small, bell-shaped, pink, hairy. Oct.–Dec. Lower slopes, SW (Stellenbosch to Somerset West).•

setulosa Benth.   Erect, sometimes sparse and spreading, shrublet to 50 cm. Flowers small, cup-shaped, white, hairy. Sept.–Nov. Rocky slopes, sometimes moist places, NW, KM (Keeromsberg to Kammanassie Mts).•

sexfaria Aiton   Like **E. spumosa**, erect shrublet to 50 cm. Flowers small, narrowly cup-shaped, white to pink, with exserted anthers. Oct.–Dec. Rocky places, upper slopes, SW, LB (Cape Peninsula to Langeberg Mts).•

shannonea Andrews   Erect semispreading shrublet to 50 cm. Flowers large, tubular-ampullaceous, white turning pink, shiny. Dec.–Feb. Lower slopes and hills, SW (Stanford).•

sicifolia Salisb.   Erect resprouting shrublet to 30 cm. Flowers small, campanulate, dark purplish red, hairy. Dec.–June. Rocky southern middle to upper slopes, SW (Riviersonderend Mts).•

simulans Dulfer   Like **E. demissa**, erect shrublet to 1(–2) m. Flowers narrowly cup-shaped to urn-shaped, white, with exserted anthers. Jan.–Dec. Drier lower to middle slopes, SE (Uniondale to Port Elizabeth and Zuurberg).

sitiens Klotzsch   Erect shrublet to 60 cm. Flowers small, elongate-urn-shaped, pale to dark pink, sometimes white. Jan.–Dec. Mountain slopes, SW (Hottentots Holland to Palmiet River Mts).•

sociorum L.Bolus   Erect shrublet to 50 cm. Flowers large, tubular-ovoid, greenish white. Feb.–Mar. Moist rock crevices and ledges, SW (Cape Peninsula).•

solandra Andrews   Erect shrublet to 40 cm. Flowers in heads, small, narrowly urn-shaped, magenta. Feb.–July. Lower slopes, SE (Outeniqua Mts: George and Kouga Mts).•

sonderiana Guthrie & Bolus   Erect compact shrublet to 45 cm. Flowers small, cup-shaped, sticky, white. Sept.–Jan. High peaks, NW, SW, KM (Cedarberg to Riviersonderend and Swartberg Mts).•

sparrmannii L.f.   Erect shrublet to 1 m. Flowers large, tubular, greenish yellow, coarsely hairy. Jan.–Dec. Flats and lower slopes, SE (Uniondale to Humansdorp).•

sparsa Lodd. (= *Erica floribunda* Lodd.) KER-KER Erect shrub to 2 m. Flowers small, calycine, cup-shaped, pink. Feb.–Aug. Lower slopes, SE (George to Humansdorp, Cockscomb Mts).•

speciosa Andrews Like **E. discolor**, erect shrublet to over 1 m. Flowers large, tubular, curved, sometimes slightly sticky, red with yellowish tip. July–Mar. Flats and mountain slopes, KM, LB, SE (Swartberg and Outeniqua Mts to Humansdorp).•

spectabilis Klotzsch ex Benth. Like **E. syngenesia**, erect stout shrublet to 70 cm. Flowers small to medium, globose to urn-shaped, white to greenish. Jan.–Dec. Coastal limestone hills, AP (Bredasdorp to Gouritsmond).•

sperata E.G.H.Oliv. Dense shrub to 2 m. Leaves small, bladder-like. Flowers small, cup-shaped, pinkish. Apr.–May. Limestone hills, AP (De Hoop).•

sphaerocephala J.C.Wendl. ex Benth. Erect shrublet to 50 cm. Flowers small, in nodding heads, urn-shaped, pink. Sept.–Jan. Middle to upper slopes, NW (Cedarberg to Hex River Mts, Piketberg).•

spumosa L. FROTHY HEATH, SWARTBEKKIE Erect compact shrublet to 45 cm. Flowers in pendulous 3-flowered heads, calycine, small, narrowly cup-shaped, pink or red, with exserted anthers. Aug.–Dec. Dry lower slopes, SW (Cape Peninsula to Riviersonderend Mts, Napier).•

squarrosa Salisb. Semispreading shrublet to 30 cm. Flowers medium, ovoid, sticky, dull red to pink. Nov.–Feb. Upper slopes, SW (Hottentots Holland Mts).•

steinbergiana H.L.Wendl. ex Klotzsch Erect shrublet to 80 cm. Flowers medium, calycine, tubularly urn-shaped with spreading lobes, deep pink. Nov.–Mar. Damp upper mountain slopes, LB, SE (Riversdale to George).•

stenantha Klotzsch ex Benth. Erect shrublet to 50 cm. Flowers small, calycine, cup-shaped, dark pink. Aug.–Dec. Upper slopes, LB (Langeberg Mts: Swellendam to Heidelberg).•

stokoei L.Bolus Erect rigid compact shrublet to 20 cm. Flowers small, cup-shaped, pale pink. Oct.–Feb. Rocks at high alt., SW (Betty's Bay Mts).•

strigilifolia Salisb. Erect shrublet to 1 m. Flowers large, tubular, carmine-red to rose, sometimes white, hairy. Jan.–Dec. Upper slopes, KM (Swartberg to Kouga Mts).•

strigosa Sol. Erect straggling shrublet to 30 cm. Flowers small, cup-shaped, reddish. July–Nov. Cliffs and rock crevices, SW (Cape Peninsula to Stellenbosch).•

stylaris Spreng. Erect shrublet to 30 cm. Flowers in heads, small, narrowly bell-shaped, white, with exserted anthers. Nov.–Dec. High alt., LB, SE (Outeniqua to Tsitsikamma Mts).•

subdivaricata P.J.Bergius Erect shrublet to 1 m. Flowers small, bell-shaped, white. Jan.–June, Nov. in east. Lower slopes and flats, SW, AP (Malmesbury to Bredasdorp).•

subulata J.C.Wendl. Erect shrublet to 1 m. Flowers small, elongate-urn-shaped, in terminal nodding heads, white or pink. Jan.–Sept. Rocky lower to middle slopes, NW, SW (Ceres to Paarl, Du Toitskloof Mts).•

suffulta J.C.Wendl. ex Benth. Like **E. pogonanthera**, erect shrublet to 45 cm. Flowers small, calycine, urn-shaped, hairy, white, with exserted anthers. Aug.–Oct. Middle slopes, SW (Houwhoek).•

syngenesia Compton Erect stout shrub to 1.7 m. Flowers small, broadly ovoid to urn-shaped, white. June–Oct. Lower to middle slopes, KM (Witteberg to Swartberg Mts).•

tarantulae E.G.H.Oliv. Low sprawling shrublet to 16 cm. Flowers small, calycine, cup-shaped, rose-pink. Aug.–Oct. Stony slopes at high alt., NW (Cold Bokkeveld Mts).•

taxifolia Bauer Like **E. corifolia**, erect shrublet to 60 cm. Flowers medium, calycine, pink. Jan.–Apr. Middle to upper slopes, NW, SW (Grootwinterhoek to Riviersonderend Mts).•

tegetiformis E.G.H.Oliv. Prostrate mat-forming shrublet. Flowers at ends of short erect branches, small, urn-shaped, white, with exserted anthers. Nov.–Jan. Middle to upper slopes, NW, KM (Ceres to Montagu).•

tegulifolia Salisb. BANKETHEIDE Erect stout shrublet to 1 m. Flowers small, calycine, cup-shaped, bright pink or red. Aug.–Dec. Middle to upper rocky slopes, SW (Hottentots Holland Mts to Villiersdorp).•

tenax L.Bolus Erect shrublet to 1 m. Flowers large tubular, very sticky, light green. Dec.–Mar. Rocky middle slopes, SW (Kleinmond).•

tenella Andrews Erect shrublet to 1 m. Flowers small, ovoid-urn-shaped, bright pink. Jan.–Dec. Middle to upper mountain slopes, SW (Houwhoek to Elim).•

tenuicaulis Klotzsch ex Benth.   Erect slender shrublet. Flowers small, bell-shaped, white to pink. Sept.–Dec. Rocks and cliffs at low to middle alt., SW, LB (Riviersonderend to Langeberg Mts).•

tenuifolia L.   Erect shrublet to 40 cm. Flowers small to medium, calycine, urn-shaped with spreading lobes, pink. Sept.–Mar. Rocky upper slopes, SW (Cape Peninsula to Riviersonderend Mts).•

tenuipes Guthrie & Bolus   Sprawling shrublet. Flowers small, narrowly cup-shaped, sticky, pink. Sept.–Jan. Local on cliffs on middle to upper slopes, SW (Du Toitskloof and Franschhoek Mts).•

tenuis Salisb.   Sprawling to erect shrublet to 1(–2) m. Flowers small, bell-shaped, white. July–Dec. Rocky wet ledges to open slopes, NW, SW, KM, LB, SE (Cedarberg Mts to Cape Peninsula to Uitenhage).•

tetragona L.f.   Erect to sprawling slender shrublet to 60 cm. Flowers small to medium, narrowly urn-shaped, yellow. July–Mar. Damp slopes from sea level to middle alt., LB, SE (Riversdale to Humansdorp).•

tetrathecoides Benth.   Like **E. cubica**, slender sprawling shrublet to 25 cm. Flowers small/medium, bell-shaped, pink. Oct.–Dec. Marshes and seeps, middle slopes, LB (Langeberg Mts: Riversdale).•

thimifolia J.C.Wendl.   Erect to scrambling shrublet. Flowers small, obconic to urn-shaped, rose-pink, with exserted anthers. Oct.–Dec. Damp lower slopes, NW, SW (Witzenberg and Cape Peninsula).•

thomae L.Bolus   Erect robust shrublet to 1 m. Flowers large, tubular, sticky, white to pink. Jan.–Apr. Rocky lower slopes, SW (Betty's Bay).•

thunbergii Montin   MALAY HEATH, GEELROKHEIDE   Erect shrublet to 60 cm. Flowers medium, spherical white tube with orange cup-shaped lobes and yellow calyx. Sept.–Dec. Sandy flats and slopes, NW (Cedarberg Mts to Cold Bokkeveld).•

tomentosa Salisb.   Erect shrublet to 50 cm. Flowers small, urn-shaped, lilac or dark pink, finely hairy. Aug.–Nov. Rocky lower southern slopes, SW (Riviersonderend Mts.).•

toringbergensis H.A.Baker   Erect semispreading shrub to 1.3 m. Flowers medium, globosely urn-shaped, dark pink. Dec.–Feb. High peaks, KM (Swartberg Mts: Ladismith).•

trachysantha Bolus   Erect shrublet to 60 cm. Flowers medium, calycine, open cup-shaped, dark pink with woolly white calyx. Aug.–Nov. Upper slopes, SE (Tsitsikamma Mts).•

tradouwensis Compton   Erect shrublet to 50 cm. Flowers small, ovoid to urn-shaped, pink, hairy, with exserted anthers. Feb.–Aug. Lower to middle slopes, LB (Langeberg Mts: Tradouw Pass).•

tragulifera Salisb.   Like **E. formosa** and **E. glomiflora**, erect shrublet to 50 cm. Flowers small, ovoid to urn-shaped, white. July–Nov. Lower slopes, SE (George to Uniondale).•

transparens P.J.Bergius   Small multistemmed resprouting shrublet to 40 cm. Flowers small, narrow-tubular, pink or white. Nov.–Mar. Moist south slopes and ledges, middle to upper slopes, SW, KM, LB (Cape Peninsula to Swartberg Mts and Uniondale).•

triceps Link (= *Erica adunca* Benth.)   Erect shrublet to 60 cm. Flowers small, calycine, urn-shaped, white, with exserted anthers. Jan.–Dec. Rocky middle slopes, LB, SE (Swellendam to Humansdorp).•

trichadenia Bolus   Erect or sprawling shrublet to 1 m. Flowers small, bell-shaped, with partially exserted anthers, sticky, white or pink. Oct.–Apr. Upper slopes, NW (Piketberg to Hex River Mts).•

trichophora Benth. (= *Erica octonaria* L.Bolus)   Erect shrublet to 1 m. Flowers medium, campanulate, pink. Sept. Marshes, lower slopes, SW (Hermanus).•

trichophylla Benth.   Erect slender shrublet to 40 cm. Flowers small, cup- to bell-shaped, white to pale pink. Oct.–Dec. Lower to middle southern slopes, SW (Riviersonderend Mts).•

trichostigma T.M.Salter   Erect shrublet to 30 cm. Flowers small, ovoid, urn-shaped, pink or white. Mar.–Aug. Sandveld, SW (Hopefield to Langebaan).•

trichroma Benth.   Like **E. tubercularis**, erect compact shrublet to 30 cm. Flowers medium, ovoid, sticky, purplish red. Feb.–May. Middle slopes, sandy places, SW (Franschhoek to Elgin).•

triflora L.   Erect robust shrub to 2 m. Flowers small, calycine, urn-shaped, white or tinged red. June–Nov. Middle to upper slopes, SW (Du Toitskloof to Riviersonderend Mts).•

truncata L.Bolus   Erect shrublet to 20 cm. Flowers small, subcalycine, cup-shaped, dull red. Sept.–Dec. Upper slopes, SW (Betty's Bay Mts).•

tubercularis Salisb.   Erect slender low shrublet to 30 cm. Flowers medium, ovoid, pink. Flowering time? Upper slopes, SW (Hottentots Holland Mts).•

tumida Ker Gawl. (= *Erica splendens* Andrews)   Erect or semisprawling shrub to 2 m. Flowers large or medium, tubular-inflated, scarlet, hairy. Oct.–Mar. Rocky upper slopes, NW (Cedarberg Mts to Matroosberg).•

turgida Salisb.   Erect compact shrublet to 20 cm. Flowers small, urn- to cup-shaped, dark pink, slightly hairy, with exserted anthers. Nov.–Dec. Very local on sand flats, SW (Cape Peninsula).•

turrisbabylonica H.A.Baker   Erect slender shrublet to 30 cm. Flowers large, narrowly tubular, green turning red, hairy. Nov.–Dec. Marshy middle to upper slopes, SW (Babilonstoring).•

umbelliflora Klotzsch ex Benth. (= *Erica ionii* H.A.Baker, *E. manifesta* Compton)   Erect compact to sprawling shrublet to 40 cm. Flowers small, urn-shaped, pink, sometimes with exserted anthers. May–Nov. Lower to middle mountain slopes, KM, SE (Groot Swartberg to Humansdorp).•

unilateralis Klotzsch ex Benth.   Erect multistemmed resprouting shrublet to 60 cm. Flowers small, subtubular, white, with exserted anthers. Feb.–Aug. Lower slopes, SE (Tsitsikamma and Baviaanskloof Mts to Port Elizabeth).•

urna-viridis Bolus   STICKY HEATH, GROENTAAIHEIDE   Erect sparse shrub to 1.5 m. Flowers medium to large, ovoid-conical, sticky, green. Dec.–July. Mountain plateaus, SW (Cape Peninsula).•

ustulescens Guthrie & Bolus   Erect shrublet to 50 cm. Flowers medium, narrowly urn-shaped, sticky, white, with exserted anthers. July–Sept. Hillsides, SW (Botrivier).•

utriculosa L.Bolus   Prostrate moss-like shrublet. Flowers small, cup-shaped, pink. Sept.–Feb. Damp rock crevices, middle to high alt., SW (Cape Peninsula to Napier).•

uysii H.A.Baker   Erect shrub to 2 m. Leaves glaucous. Flowers small, calycine, globosely urn-shaped, rose. Sept.–Oct. Limestone hills, AP (De Hoop).•

valida H.A.Baker   Erect shrublet to 50 cm. Flowers small, ovoid to urn-shaped, sticky, hairy, red or pink. Nov.–Feb. Rocky upper slopes, KM, SE (Kammanassie Mts to Uitenhage).•

vallisaranearum E.G.H.Oliv.   Erect shrublet to 1 m. Flowers large, tubular, yellow, pendulous on long red pedicels. Apr.–June. Moist upper slopes, SW (Kogelberg).•

vallisgratiae Guthrie & Bolus   Erect shrublet to 50 cm. Flowers large, subcylindrical red tube with large white spreading lobes. Aug.–Nov. High alt., SW (Riviersonderend Mts).•

velitaris Salisb.   Shrublet to 30 cm. Flowers medium, subcyathiform, pink. Flowering time? Sandy flats, SW (?Cape Peninsula).•

ventricosa Thunb.   WAX HEATH, FRANSCHHOEKHEIDE   Erect shrublet to 0.5(–1.0) m. Flowers large, ovoid to urn-shaped, pale pink, with spreading lobes. Oct.–Mar. Upper slopes, SW (Franschhoek to Hottentots Holland Mts).•

verecunda Salisb.   Erect to somewhat spreading shrub to 1.5 m. Flowers small, ovoid to urn-shaped, pale pink. Jan.–July. Sandy flats and slopes, NW, KM (Kamiesberg, Bokkeveld Mts to Witteberg).

versicolor J.C.Wendl.   Erect or sprawling shrub to over 2 m. Flowers large, tubular, pink with paler tips to red with green tips. Jan.–Dec. Sandstone slopes, NW, KM, LB (Cedarberg to Langeberg Mts).•

verticillata P.J.Bergius   Robust shrub to 1.5 m. Flowers large, tubular, mauve-pink, hairy. Jan.–May. Marshes on sandy flats, SW (Cape Flats).•

vestita Thunb.   TREMBLING HEATH, TRILHEIDE   Compact erect shrublet to 1 m. Flowers large, tubular, red, white or pink. Jan.–Dec. Mountain slopes, SW, LB (Riviersonderend to Langeberg Mts: Riversdale).•

villosa Andrews   KAPOKKIE   Like **E. bruniades**, erect rather sparse shrublet to 40 cm. Flowers small, urn-shaped, woolly, white, with exserted anthers. June–Nov. Flats and lower slopes, SW (Hermanus).•

virginalis Klotzsch ex Benth.   Erect compact shrublet to 25 cm. Flowers small, narrowly urn-shaped, pink. Dec. Sandy flats on lower slopes, NW, SW (Elandsberg: Wolseley and Cold Bokkeveld).•

viridescens Lodd. (= *Erica unicolor* J.C.Wendl.)   Shrub to 1.5 m. Flowers large, tubular, slightly sticky, green. Jan.–Dec. Hills to middle slopes, SE (Outeniqua Mts: Robinson Pass to George).•

viridiflora Andrews   Straggling erect shrublet to 1 m. Flowers large, tubular, slightly sticky, lime-green. Oct.–Jan. Middle to upper rocky slopes, SE (George to Humansdorp).•

viscaria L. (= *Erica decora* Andr.)   STICKY HEATH, KLOKKIESHEIDE   Erect shrublet to 1.0(–1.5) m. Flowers medium, tubularly bell-shaped, pink to dull red. Jan.–Dec. Flats and lower slopes, SW (Mamre to Cape Peninsula).•

vogelpoelii H.A.Baker   Like **E. corifolia**, erect shrublet to 1 m. Flowers small/medium, calycine, urn-shaped, deep rose. Dec.–Apr. Middle slopes, SW (Napier).•

walkeria Andrews   Erect shrublet to 1 m. Flowers medium, urn-shaped, pink. Aug.–Oct. Lower to upper slopes, SW (Du Toitskloof Mts).•

wendlandiana Klotzsch   Erect shrublet to 60 cm. Flowers large, tubular, brick-red. Flowering time? Habitat? NW (Tulbagh).•

winteri H.A.Baker   Erect sparse shrublet to 1 m. Flowers large, globose to urn-shaped, sticky, white to pinkish. May–Oct. Middle to high alt., LB (Langeberg Mts: Riversdale).•

wittebergensis Dulfer   Erect delicate shrublet to 20 cm. Flowers small, bell-shaped, sticky, pink, with exserted anthers. Sept.–Dec. Upper slopes, SW (Du Toitskloof Mts).•

xanthina Guthrie & Bolus (= *Erica parvulisepala* H.A.Baker)   Erect shrublet to 1 m. Flowers large, tubular, dull cream-coloured to pinkish red, very finely hairy. Sept.–Dec. Local on high peaks, SW (Riviersonderend Mts).•

xerophila Bolus   Erect, sturdy rather sparse shrub to 1.2 m. Flowers large, tubular, pale orange, hairy. Aug.–Oct. Rocky dry places on mountain slopes, KM (Witteberg to Groot Swartberg).•

zebrensis Compton   Erect shrublet to 60 cm. Flowers small, cup-shaped, magenta, with exserted anthers. Nov.–Feb. Lower to middle northern slopes, SE (Outeniqua Mts: Robinson Pass to George).•

zitzikammensis Dulfer   Like **E. newdigateae**, erect shrublet to 30 cm. Flowers medium, calycine, obconically bell-shaped, rose. Nov.–Dec. High ridges, SE (Tsitsikamma Mts).•

zwartbergensis Bolus   Like **E. andreaei**, erect shrub to 1.2 m. Flowers small, urn- to bell-shaped, slightly sticky, whitish. Sept.–Dec. Low to high alt., KM, SE (Swartberg to Outeniqua Mts).•

sp. 1 (= *Erica petrensis* (L.Bolus) E.G.H.Oliv. ms)   Like **E. goatcheriana**, erect to sprawling shrublet to 30 cm. Flowers small, calycine, ovoid, pink turning reddish. Oct.–Dec. Rocks at high alt., SW (Riviersonderend Mts).•

[**Species excluded**   Insufficiently known and possibly conspecific with one of the above: **E. schmidtii** Dulfer]

EUPHORBIACEAE
Euphorbia by P.V. Bruyns

1. Apparent flower composed of a number of stamens (male flowers) mingled with bracteoles, with or without a sessile or pedicellate ovary (female flower) in their midst, all enclosed in a more or less cup-shaped involucre; mostly succulent plants .... **Euphorbia**
1.' Flowers male, female or bisexual, with calyx, but petals present or absent:
   2. Stamens 1- or 2-seriate, the outer opposite sepals (occasionally in middle of flower); ovules 2 in each locule:
      3. Petals present in male flowers:
         4. Plants monoecious; floral disc glabrous; slender shrub with small leaves .... **Andrachne**
         4.' Plants dioecious; floral disc villous; shrubs or small trees .... **Lachnostylis**
      3.' Petals absent from male flowers:
         5. Leaves in whorls of 4, leathery; male flowers in dense, pedunculate or subsessile cymules; female flowers sessile, 1–3 in axils of leaves .... **Hyenanche**
         5.' Leaves alternate; male flowers in axillary fascicles; female flowers solitary .... **Phyllanthus**
   2.' Stamens 1- or 2-seriate, the outer alternating with sepals or more usually central; ovule 1 in each locule:
      6. Petals present in male or all flowers:
         7. Flowers in panicles or 2- or 3-branched, fasciculate cymes, female surrounded by male; stamens usually 8 in 2 series .... **Jatropha**
         7.' Flowers usually unisexual, in axillary fascicles or female often solitary; stamens in a single whorl of 5 below apex of a central column .... **Clutia**
      6.' Petals absent:
         8. Sepals of male flowers open in bud; flowers in unisexual or occasionally bisexual, axillary spikes; trees or shrubs .... **Excoecaria**
         8.' Sepals of male flowers not open in bud; inflorescence and habit various:
            9. Styles connate in a column continuous with body of carpels; stamens 20–30; filaments united into a short column; twining or climbing herbs:
               10. Flowers in dense, involucrate heads .... **Dalechampia**
               10.' Flowers in racemes or spikes .... **Ctenomeria**
            9.' Styles free or shortly united at base; stamens (2)3–10(12); filaments free or connate at base:
               11. Anther locules adnate laterally to a connective throughout their length; calyx deeply 5-lobed, imbricate .... **Adenocline**
               11.' Anther locules separate, attached to filaments by their base only; calyx 3- or 4-lobed, valvate in male flowers:
                  12. Anther locules stalked; female sepals imbricate; annual or perennial herbs .... **Acalypha**
                  12.' Anther locules sessile; female sepals valvate; soft annual herb .... **Leidesia**

ACALYPHA VALSNETEL   c. 450 spp., pantropical

capensis (L.f.) Prain & Hutch.   Monoecious, slender, sprawling shrub to 3 m, velvety on young parts. Leaves ovate-cordate, coarsely toothed, margins often slightly revolute, discolorous, grey-velvety beneath with reddish veins. Flowers unisexual, in axillary spikes, reddish. Mainly Sept.–Jan. Forest and coastal bush, LB, SE (Langeberg Mts: Riversdale to E Cape).

ecklonii Baill.   Monoecious, softly hairy annual to 50 cm. Leaves long-petioled, ovate, toothed. Flowers in unisexual axillary spikes, females crowded terminally in short spikes with leafy, laciniate bracts. Nov.–Mar. Coastal bush and forest, SE (George to KwaZulu-Natal).

glabrata Thunb.   BOSVALSNETEL   Monoecious, scrambling shrub or tall tree to 5 m, bark velvety on young parts. Leaves long-petioled, ovate, toothed, finely hairy when young. Flowers in bisexual axillary spikes, females 1 or 2 at base. Mainly Oct.–Dec. Forest and bush, often along streams, SE (Van Staden's Mts to E Zimbabwe).

peduncularis E.Mey. ex Meisn. (incl. **A. zeyheri** Baill.)   Dioecious perennial from woody caudex, to 80 cm. Leaves subsessile, coarsely hairy, ovate to elliptic, toothed. Male flowers in pedunculate axillary spikes, females in terminal bracteate clusters, with long red stigmas. Oct.–Dec. Coastal grassland, SE (Van Staden's Mts to Zimbabwe).

ADENOCLINE   4 spp., southern Africa

acuta (Thunb.) Baill.   Dioecious, scrambling shrub to 1 m. Leaves opposite, long-petioled, deltoid, toothed. Flowers in terminal racemes or panicles, cream-coloured, male flowers few per node, females solitary. Aug.–Jan. Coastal bush, SW, AP, LB, SE (Hottentots Holland Mts to Mpumalanga).

pauciflora Turcz.   (= *Adenocline humilis* Prain, *A. ovalifolia* Turcz., *A. serrata* (Sond.) Turcz., *A. sessilifolia* Turcz., *A. stricta* Prain)   Dioecious perennial from woody rootstock, 10–30 cm. Leaves ovate to linear-lanceolate, toothed. Male flowers subsessile in leaf-opposed or axillary racemes, females in leaf-opposed racemes. Sept.–Dec. Sand and limestone, SW, AP, SE (Cape Peninsula to KwaZulu-Natal).

violifolia (Kuntze) Prain   Monoecious, softly succulent annual to 10 cm. Leaves cordate, weakly toothed. Male flowers in axillary clusters, females in leaf-opposed racemes. Sept.–Oct. Sandy flats and limestone rocks, NW, SW, AP, LB, SE (Lambert's Bay to E Cape).

ANDRACHNE   BASTERBLIKSEMBOS   c. 25 spp., pantropical and warm temperate

ovalis (Sond.) Müll.Arg.   Monoecious shrub or small tree, 2–6 m. Leaves ovate-elliptic, glabrous, pale beneath, margins revolute. Flowers axillary, males few, females solitary. Nov.–Jan. Forest and bush, SE (George to Zimbabwe).

CLUTIA   BLIKSEMBOS   c. 70 spp., Africa and Arabia

A. Petals of male flowers each with 3 or more basal glands

daphnoides Lam.   VAALBLIKSEMBOS   Dioecious shrub to 2 m, white-mealy on young parts. Leaves shortly petioled, oblanceolate, 15–30 mm long, slightly discolorous. Flowers axillary, cream-coloured, males few, females solitary. June–Sept. Coastal bush, SW, AP, SE (Saldanha to E Cape).

govaertsii Radcl.-Sm. (= *Clutia vaccinioides* Prain)   Dioecious shrublet from woody base, to 50 cm. Leaves elliptic-obovate, obtuse, subsessile, leathery, pellucid-dotted and warty. Flowers axillary, cream-coloured, males clustered, females solitary. Dec.–Jan. Stony gravel slopes, AP (Albertinia to Great Brak River).•

heterophylla Thunb.   Dioecious shrublet from woody rootstock, to 30 cm. Leaves subsessile, ovate-elliptic, margins slightly revolute, pellucid-dotted, margins slightly revolute. Flowers axillary, cream-coloured, males in pairs, females solitary, ovary and fruits warty. Feb.–Apr. Grassy slopes, SE (Van Staden's Mts to E Cape).

polygonoides L.   Dioecious shrublet from woody base with stiffly erect branches, to 60 cm. Leaves subsessile, narrowly elliptic, leathery, margins revolute, often deflexed below. Flowers yellow to orange, axillary, males clustered, females solitary. Oct.–Apr. Sandstone slopes, NW, SW, KM, LB (Cedarberg to Riversdale Langeberg and Little Karoo Mts).•

thunbergii Sond.   Dioecious, grey-mealy shrublet to 60 cm. Leaves shortly petioled, obovate, small, to 10 mm long, leathery. Flowers axillary, cream-coloured, males paired, females solitary, ovary and fruits warty. Mainly Aug.–Oct. Sandstone and granite slopes, NW, KM (Namaqualand to Bokkeveld Mts and Karoo to Swartberg Mts).

sp. 1   Dioecious shrub to 1.5 m, from woody base. Leaves subsessile, ovate, leathery. Flowers axillary, males few, sometimes on short shoots, females solitary on erect pedicels. Sept.–Oct. Rocky sandstone slopes near streams, NW (Cedarberg Mts and Piketberg).•

A. Petals of male flowers with 1 or 2 basal glands
B. Ovary hairy

marginata E.Mey. ex Sond.   Like **C. tomentosa** but leaves oblanceolate, usually 15–30 mm long. June–Sept. Clay and loam in renosterveld, KM (Robertson Karoo, Montagu to Barrydale, Karoo Mts).

pubescens Thunb.   Dioecious, velvety shrub to 60 cm. Leaves shortly petiolate, narrowly elliptic to linear, margins revolute. Flowers axillary, cream-coloured, males 1–4, females solitary, ovary and fruits silky. Mainly Apr.–July. Mostly clay and granite slopes and flats, NW, SW, LB (Gifberg to Albertinia).•

tomentosa L.   WOLBLIKSEMBOS   Dioecious, twiggy shrub to 1 m. Leaves sessile, ovate to elliptic, grey-felted, ascending above, spreading below, mostly to 10 mm long. Flowers axillary, mostly solitary, cream-coloured to brownish, ovary and fruits silky. Mainly June–Sept. Clay and gravel slopes and flats in renosterveld, SW, KM, LB (Botrivier to Swellendam and Little Karoo).•

sp. 2   Like **C. ericoides** in leaf shape, but closely imbricate and spreading. Fruits subsessile, large and densely velvety. Probably June–July. Rocky slopes, SW (Bredasdorp).•

B . Ovary glabrous

affinis Sond.   OUMEISIEKNIE   Dioecious shrub to 2.5 m. Leaves lanceolate to elliptic, petiolate, mealy-pubescent, pale beneath. Flowers axillary, males 4–6, females 2 or 3. Mainly Aug.–Oct. Forest margins, SW, LB, SE (Villiersdorp to Mpumalanga).

alaternoides L. (incl. **C. rubricaulis** Eckl. ex Sond.)   Dioecious shrub, mostly 30–60 cm. Leaves oblanceolate to obovate, often very small on coppicing plants, margins minutely toothed, usually slightly revolute. Flowers axillary, males few in tight clusters, females solitary on short pedicels. Mainly Aug.–Oct. Mainly rocky sandstone or limestone slopes, NW, SW, AP, KM, LB, SE (Namaqualand to E Cape).

ericoides Thunb.   Dioecious shrublet from a woody base, to 60 cm, usually with conspicuous black axillary buds. Leaves small, subsessile, leathery, narrowly lanceolate, distinctly concave. Flowers axillary, cream-coloured, males in pairs, females solitary, ovary and fruits smooth or warty. May–July. Rocky sandstone slopes, NW, SW, AP, KM, LB, SE (Gifberg to E Cape).

laxa Eckl. ex Sond.   Dioecious, twiggy shrub to 1 m. Leaves shortly petiolate, oblanceolate to elliptic, more or less plane. Flowers axillary, cream-coloured, males 2 or 3, females solitary, ovary and fruits smooth. July–Sept. Sandstone slopes, SW, KM, AP, LB, SE (Mamre to Port Elizabeth and E Cape).

polifolia Jacq. (incl. **C. brevifolia** Sond., **C. pterogona** Müll.Arg.)   Like **C. laxa** but leaves oblanceolate to linear, usually discolorous, pale beneath, glossy above, margins revolute. July–Sept. Sandstone slopes, NW, SW, KM, LB, SE (Namaqualand to Port Elizabeth).

pulchella L.   Dioecious shrub or small tree to 2 m. Leaves ovate to broadly lanceolate, long-petiolate, thin-textured, pellucid-punctate. Flowers axillary, cream-coloured, males c. 3, females 1 or 2, ovary smooth. Capsules warty. Mainly Aug.–Sept. Sandstone and clay slopes in forest and thicket, NW, SW, LB, SE (Pakhuis Pass to N Province).

[Species excluded   No authentic material found and probably conspecific with one of the above: **C. ovalis** Sond., **C. sericea** Müll.Arg.]

CTENOMERIA   1 sp., southern Africa

capensis (Thunb.) Harv. ex Sond.   Monoecious twining perennial to 2.5 m, with stinging hairs. Leaves cordate to sagittate, slightly toothed, paler beneath. Flowers in slender, axillary bisexual spikes, greenish, females basal. Nov.–Mar. Forest and bush, SE (George to Mpumalanga).

DALECHAMPIA   c. 100 spp., pantropical

capensis Spreng.f.   Monoecious, twining, hairy perennial to 3.5 m. Leaves palmate, 5-lobed, lobes toothed, paler beneath. Flowers crowded in long-peduncled, bisexual axillary clusters subtended by large paired, yellowish, leafy bracts. Oct.–Nov. Coastal bush, SE (Baviaanskloof Mts to Tanzania).

EUPHORBIA   SPURGE   c. 1500 spp., cosmopolitan

A. Stems not succulent; leaves present at flowering
B. Plants acaulescent with leaves in a basal rosette

ecklonii (Klotzsch & Garcke) A.Hässl. (= *Euphorbia pistiifolia* Boiss.)   Like **E. tuberosa** but leaves adpressed to the ground, broadly elliptic, obtuse, narrowed abruptly below into short petioles to 20 mm long, dark green and somewhat shiny with sunken veins and cyathia 3–5 mm diam. May–Aug. Clay slopes, LB, SE (Caledon to Mossel Bay).•

mira L.C.Leach   Like **E. tuberosa** but leaves filiform and indistinguishable from petiole, acute, cyathia 2.5–3.0 mm diam. Feb.–Mar. Gravelly flats, NW, SW, LB (Tulbagh, Bredasdorp to Swellendam).•

silenifolia (Haw.) Sweet   Like **E. tuberosa** but leaves narrower, tapered below, oblanceolate or sublinear. Mainly June–Sept. Sandy and stony flats and slopes, NW, SW, AP, LB, SE (Namaqualand to E Cape).

tuberosa L. (incl. **E. crispa** (Haw.) Sweet)   MELKBOL, WILDERAMENAS   Dioecious, acaulescent, tuberous-rooted perennial to 5 cm. Leaves oblong to oblanceolate or sublinear, abruptly to gradually narrowed below into long petioles, usually minutely hairy, margins usually undulate, sometimes crisped. Flowers with several cyathia per stem, greenish, 4–6 mm diam. Mainly June–Sept. Sandy and stony flats and slopes, NW, SW, AP, LB (Namaqualand to Swellendam).

B. Plants with aerial stems and cauline leaves

epicyparissias E.Mey. ex Boiss.   PISGOED   Like **E. erythrina** but leaves deflexed, margins usually slightly revolute, 10–40 mm long. Aug.–Nov. Coastal bush and forest margins, SW, KM, LB, SE (Gansbaai to Mpumalanga).

ericoides Lam.   Like **E. genistoides** but leaves sharply deflexed, linear, dilated at base, more or less truncate-apiculate, margins revolute. Aug.–Dec. Sandstone slopes, LB, SE (Swellendam to Lesotho).

erythrina Link   Monoecious, closely leafy, softly woody perennial to 80 cm, with slender stems from woody base. Leaves elliptic to oblanceolate, ascending, imbricate, apiculate, shortly petiolate, 8–20 mm long. Flower clusters usually congested, floral glands crescent-shaped. July–Oct. Rocky sandstone slopes, NW, SW, KM, LB, SE (Cold Bokkeveld to E Cape).

foliosa N.E.Br. (incl. **E. artifolia** N.E.Br.)   Like **E. erythrina** but rounded and to 30 cm, leaves small, obovate, 2–8 mm long, usually minutely ciliate on margins. Aug.–Oct. Mostly coastal limestones and calcareous sands, SW, AP (Mamre to Stilbaai).•

genistoides P.J.Bergius   Like **E. erythrina** but leaves linear to narrowly elliptic, margins revolute, spreading, apiculate, subsessile, 7–15 mm long. Flower clusters congested in cylindrical racemes. July–Oct. Rocky sandstone slopes, NW, SW (Gifberg to Cape Peninsula).•

kraussiana Bernh.   Monoecious, erect shrub to 2 m, with slender stems. Leaves spreading or ascending, elliptic to oblancolate, petiolate, 50–120 mm long, paler beneath. Flower clusters in lax panicles, floral glands oblong. Sept.–Feb. Forest margins and coastal bush, SE (George to Mpumalanga).

striata Thunb.   MELKGRAS   Monoecious shrublet to 50 cm, with wiry stems from woody base. Leaves ascending, linear-lanceolate, sessile, acuminate, 15–40 cm long. Flower clusters in lax panicles, floral glands crescent-shaped. Mainly Sept.–Nov. Sandy slopes and flats in fynbos and grassland, SE (Humansdorp to Mpumalanga).

A. Stems succulent; leaves often absent at flowering
C. Plants with pairs of spines subtending each minute deciduous leaf and arising on hardened 'spine-shields'

coerulescens Haw.   NOORS   Monoecious spiny, decumbent rhizomatous shrub, much-branched from base, stems 0.5–1.5 m, numerous, grey-green, 30–50 mm diam., 4–6-angled, constricted at intervals, spines 6–12 mm long. Leaves shorter than 2 mm, deciduous. Cyathia in groups of 3, 5–6 mm diam., glands bright yellow. Oct.–Dec. Stony karroid slopes, KM (Calitzdorp to E Cape).

grandidens Haw.   Monoecious spiny, much-branched tree, 5–16 m, with cylindrical trunk and ascending branches bearing an apical rosette of ascending (2)3(4)-angled, green, deciduous branchlets 12–20 mm diam., spines 0.5–6.0 mm long. Leaves minute, triangular, deciduous. Cyathia 4–5 mm diam., glands entire, yellow-green. June–Aug. Steep slopes in dense valley bushveld, SE (Uitenhage to E Cape).

ledienii A.Berger   SUURNOORS   Similar to **E. coerulescens** but not rhizomatous, stems longer, 1.3–2 m, with less obvious constrictions and spines shorter, 2–6 mm long. Jan.–Apr. Arid scrub on low flats and slopes, SE (Humansdorp to E Cape).

stellata Willd.   Monoecious spiny, dwarf succulent, 2–15 cm, with clavate subterranean main stem and spreading rosette of flat branches at ground level, green mottled with purple, 5–15 mm wide, bearing marginal spines 2–4 mm long. Cyathia in groups of 3, 2.5–4.0 mm diam., glands entire, red to yellow. Sept.–Mar. Gravelly flats between dense scrub, SE (Uitenhage to E Cape).

tetragona Haw.   TREE EUPHORBIA, RIVIERNABOOM   Like **E. grandidens** but branchlets 4–6-angled, 25–50 mm diam. June–Aug. Slopes in dense valley bushveld, SE (Uitenhage to E Cape).

triangularis Desf.   TREE EUPHORBIA, RIVIERNABOOM   Monoecious spiny tree with 1–several more or less cylindrical trunks each bearing a crown of ascending 3–5-angled branches to 1.5 m long, 40–100 mm diam., deeply constricted into segments, spines 3–8 mm long. Leaves 6–7 × 5–6 mm, deciduous. Cyathia 4–5 mm diam., glands yellow. June–Aug. Steep slopes in dense valley bushveld, SE (Uitenhage to E Cape).

C. Plants sometimes spinescent but spines peduncular and never subtending leaves nor arising on hardened 'spine-shields'

arceuthobioides Boiss. (incl. **E. corymbosa** N.E.Br., **E. decussata** E.Mey. ex Boiss., **E. mundii** N.E.Br., **E. rhombifolia** Boiss., **E. tenax** Burch.)   STEENBOKBOS   Dioecious, much-branched shrublet, 5–50 cm, sometimes with tuberous rootstock; stems slender, erect, greyish, sometimes roughened, 2–6 mm diam., often slightly ridged. Leaves opposite, more or less sessile, triangular, 1–3 mm long. Cyathia 2–3 mm diam., glands more or less entire, greenish yellow. June–Nov. Karroid scrub on flats or slopes, NW, SW, KM, LB, SE (Namibia to KwaZulu-Natal).

bayeri L.C.Leach   Like **E. arceuthobioides** but plant 1–15 cm, stems c. 4 mm diam., smooth, decumbent and somewhat rhizomatous. June–Nov. Gravelly flats, LB (Mossel Bay).•

burmanii E.Mey. ex Boiss. (incl. **E. karroensis** (Boiss.) N.E.Br., **E. macella** N.E.Br.)   STEENBOKMELKBOS, SOETMELKBOS   Dioecious, much-branched shrublet, 15–100 cm; stems terete, greyish green to dark green, 3–5 mm diam., smooth. Leaves opposite, spathulate, 2–3 mm long, with subglobose, brownish stipules. Cyathia 2–4 mm diam., glands entire, pale green. June–Sept. Sandy to stony flats and slopes, NW, SW, KM, LB, SE (Namaqualand to E Cape).

caput-medusae L. (incl. **E. marlothiana** N.E.Br., **E. muirii** N.E.Br., **E. tuberculata** Jacq., **E. tuberculatoides** N.E.Br.)   MEDUSA'S HEAD, VINGERPOL   Monoecious sprawling shrublet, 20–75 × 15–100 cm with a rosette of branches from a short, thick, tuberculate stem; branches ascending, more or less clavate, tuberculate, 10–30 mm diam. Leaves on tubercules at branch tips, linear, fleshy, deciduous. Cyathia solitary, 10–18 mm diam., glands 3–6-palmate, pale yellow. May–Sept. Sandy flats and stony slopes, NW, SW, LB (Namaqualand to Mossel Bay).

clandestina Jacq.   VOLSTRUISNEK   Monoecious erect, mostly single-stemmed shrublet, 15–80 cm; stem cylindrical(–clavate), 25–50 mm diam. with stout conical tubercles in c. 10–15 spiralling rows, green often mottled purple. Leaves on tubercles in an apical tuft, linear, deciduous. Cyathia solitary, sessile in axils of tubercles, closely surrounded by several bracts, 3–5 mm diam., glands minutely toothed, pale yellow. June–Sept. Stony lower slopes and karroid scrub, SW, KM, LB (Caledon to Uniondale).•

esculenta Marloth (incl. **E. colliculina** A.C.White, R.A.Dyer & B.Sloane, **E. fortuita** A.C.White, R.A.Dyer & B.Sloane)   Like **E. caput-medusae** but branches shorter, nearly cylindrical, 4–17 mm diam., slightly tuberculate. Cyathia smaller, 4.5–5.0 mm diam., glands entire to finely notched, red to purple. Sept.–Oct. Karroid scrub on low stony hills, KM (Ladismith to Oudtshoorn).•

gamkensis Marx   Monoecious dwarf, mostly subterranean succulent with ascending or spreading branches from a stem 7–9 cm diam.; branches 10–15 mm diam., with conical tubercles. Leaves 2 mm, ovate, soon deciduous. Cyathia solitary, 7–8 mm diam., glands crenulate, pale green. Flowering time? Karroid scrub, KM (Calitzdorp).•

globosa (Haw.) Sims   EIERPOL, KNOPMELKPOL   Monoecious dwarf, cushion-forming shrublet, mostly 2.5–8.0 cm; stems many, densely packed, more or less globose but flowering stems clavate, 7–14 mm diam., with low conical tubercles arranged in 5 obscure rows. Leaves lanceolate, 2–3 mm long, deciduous. Cyathia solitary(–3), 10–20 mm diam., glands deeply dissected, segments erect, cream-coloured and pale green. Oct.–Feb. Karroid scrub in gravelly flat areas, SE (Uitenhage to Port Elizabeth).•

hallii R.A.Dyer   Monoecious mostly single-stemmed shrublet to 50 cm; stem cylindrical-clavate, 30–40 mm diam., tuberculate, pale green. Leaves to 10 mm long, deciduous. Cyathia 1–3 per peduncle in axillary clusters of 1–5, with many slender, leaf-like bracts to 100 × 2–4 mm, 8–10 mm diam., glands with 3–5 branched processes 2–4 mm long, red. Apr.–Sept. Gravelly lower slopes and flats, NW (Botterkloof and Biedouw).•

hamata (Haw.) Sweet OLIFANTSMELKBOS   Dioecious shrublet to 45 × 60 cm with many branches from a thickened main stem; branches 6–15 mm diam., with prominent conical tubercles 2–16 mm long loosely arranged in 3 angles, green to red. Leaves 8–17 mm long, more or less sessile, ovate-lanceolate, deciduous. Cyathia subtended by a whorl of 3 prominent bracts 6–10 × 4–10 mm, 5–7 mm diam., glands entire, red or yellow. Apr.–Sept. Stony slopes, NW, SW (S Namibia to Robertson).

heptagona L. (incl. **E. atrispina** N.E.Br., **E. enopla** Boiss.)   Dioecious, much-branched spinescent shrublet, 20–100 cm; branches cylindrical, with obscure tubercles arranged in 6–8 angles, 15–30 mm diam., erect, mostly covered with stout thorns, spinescent peduncles 8–60 mm long. Leaves 1–2 mm long, deciduous. Cyathia solitary, 3–5 mm diam., glands entire, green to red. June–Nov. Karroid scrub on stony N-facing slopes, KM, LB, SE (W Karoo to E Cape).

inermis Mill.   VINGERPOL   Like **E. caput-medusae** but branches thinner, 10–12 mm diam., cyathia smaller, 8–9 mm diam., with many conspicuous bracteoles among flowers giving it a woolly appearance, and glands with white processes. Sept.–Dec. Gravelly flats, SE (Port Elizabeth to E Cape).

loricata Lam.   DORINGPOL, HEDGEHOG   Monoecious, spinescent shrublet 15–150 cm; branches 8–15 mm diam., cylindrical, tuberculate, covered with stout spinescent peduncles 12–50 mm long. Leaves 25–75 × 3–10 mm, linear lanceolate, glaucous, deciduous. Cyathia solitary, glands entire, green. May–Sept. Stony sandstone slopes and flats, NW (Bokkeveld Mts to Citrusdal and Swartruggens).•

mammillaris L. (incl. **E. fimbriata** Scop.)   Like **E. heptagona** but plant smaller, to 30 cm, branches 40–60 mm diam., 7–17-angled, thorns shorter, 6–10 mm long, irregularly scattered and spreading. Mar.–Nov. Flats and stony slopes among small karroid bushes, KM, SE (Oudtshoorn and Albertinia to E Cape).

mauritanica L.   BEESMELKBOS, GEELMELKBOS   Monoecious much-branched shrub to 2 m; stems terete, 3–6 mm diam., bright green. Leaves sessile, 8–15 × 2–5 mm, lanceolate, deciduous. Cyathia usually in terminal cymes with subsessile male surrounded by 5–7 bisexual cyathia, 7–15 mm diam., glands entire, yellow. May–Oct. Flats and stony slopes, NW, SW, KM, LB, SE (widespread in southern Africa).

meloformis Aiton   BOBBEJAANKOS, ESELKOS   Dioecious, usually single-stemmed dwarf shrublet, 3–20 cm; stem spherical to cylindrical, 50–150 mm diam. with obscure tubercles joined in 8–12 angles, usually with many spinescent, simple or branched peduncles to 60 mm long. Leaves 1–3 mm long, linear, deciduous. Cyathia in cymes of 1–12, c. 4 mm diam., glands entire, pale green. Nov.–Mar. Gravelly flats, SE (Port Elizabeth to E Cape).

multiceps A.Berger   VINGERS-EN-TONE, VINGERHOEDPOL   Monoecious sparsely spiny, densely conical shrublet to 60 cm, with thick main stem to 250 mm diam., nearly as tall as plant, with many crowded spreading branches 20–75 × 15–30 mm, decreasing in length towards apex, branches with conical spreading tubercles, and a few thick spinescent peduncles towards apex of plant. Leaves 5–12 × 1–2 mm, caducous. Cyathia solitary, 5–8 mm diam., glands with 2–4 more or less linear spreading processes. May–Sept. Gravelly slopes and flats, KM (Namaqualand, Touwsrivier to Barrydale, Karoo).

multifolia A.C.White, R.A.Dyer & B.Sloane   YSTERVARKPOL   Like **E. loricata** but plant cushion-forming, 15–30 × 30–100 cm, branches 25–30 mm long, leaves 25–40 × 2.0–3.5 mm, linear-oblanceolate. May–Oct. Karroid scrub on stony slopes and summits, sandstone or shale, KM (Touwsrivier to Calitzdorp).•

nesemannii R.A.Dyer   Like **E. heptagona** but main stem mostly subterranean, branches laxly ascending, 10–30 mm diam., 6–14-angled, spines weak to absent, spreading, 10–25 mm long. Mar.–June. Karroid scrub on lower gravelly slopes, NW, LB (Breede River valley to Swellendam).•

pillansii N.E.Br.   KLEINNOORSDORING   Dioecious, much-branched spinescent shrublet, 15–30 cm; branches cylindrical-clavate, 25–60 mm diam., obtusely 7–9-angled (4 when young), with numerous spinescent peduncles 8–20 mm long often branching near apex. Leaves 1–2 mm long, deciduous. Cyathia solitary to several per peduncle, 4–5 mm diam., glands entire, green. Dec.–May. Karroid scrub on stony slopes, KM (W Karoo to Vanwyksdorp).

polygona Haw. (incl. **E. horrida** Boiss.)   Dioecious, spinescent shrub to 2 m, somewhat rhizomatous; branches cylindrical-clavate, 7–12 cm diam., with tubercles fused into 7–20 prominent, wing-like often undulating angles, with spinescent peduncles 4–10 mm long. Leaves 1–2 mm long, deciduous. Cyathia solitary to 3 per axil, 5–7 mm diam., glands dark purple. Feb.–June. Rocky sandstone slopes, KM (Swartberg Mts to E Cape).

pseudoglobosa Marloth (incl. **E. juglans** Compton)   Dioecious, dwarf succulent with subterranean main stem and rosette of slender branches just below ground; branches small, to 15 cm long, globose to cylindrical, obscurely 6–8-angled, 12–25 mm diam. Leaves bristle-like, deciduous. Cyathia solitary, 3–4 mm diam., glands entire, green to reddish brown. Apr.–July. Gravelly slopes or flats on shale, often in quartz patches, KM (Touwsrivier to Barrydale).•

pubiglans N.E.Br.   Like **E. clandestina** but to 30 cm, sometimes branching below, tubercles often rounded, leaves 20–150 × 3–6 mm, cyathia solitary on peduncles 15–60 mm long, peduncle with several bracts, 3–5 forming a cup subtending cyathium, glands entire, green. Nov.–Apr. Stony slopes in karroid bush, SE (Knysna to Port Elizabeth).•

pugniformis Boiss. (incl. **E. gorgonis** A.Berger)   Monoecious dwarf, mostly subterranean succulent with truncate stem bearing rosettes of spreading branches; branches tapering, to 3 cm long, 4–8 mm diam., covered with conical tubercles. Leaves 4–6 mm long, linear-lanceolate, deciduous. Cyathia solitary, 5–7 mm diam., glands mostly slightly toothed along edges, green to reddish. Sept.–Dec. Grassland and karroid bush, LB, SE (Mossel Bay to E Cape).

schoenlandii Pax   VOLSTRUISNEK, NOORDPOL   Monoecious spinescent, usually single-stemmed shrublet, 15–50 cm; stem 40–200 mm diam., with prominent conical tubercles arranged in many spiralling rows, with many stout spinescent peduncles, 25–50 mm long. Cyathia mostly solitary in axils of tubercles, 6–8 mm diam., glands fimbriate, cream-coloured. Apr.–July. Lower loamy slopes, NW (S Namaqualand to Gifberg).

stolonifera Marloth   KRUIPMELKBOS, RANKMELKBOS   Like **E. mauritanica** but plant 15–60 cm, rhizomatous, branching at base and near tips of stems, branches 3–10 mm diam., often tapering to both ends and somewhat glaucous. May–Oct. Stony slopes to flats, in karoo bushes, KM (S Namibia and Karoo to Ladismith).

susannae Marloth   Like **E. pseudoglobosa** but plant mostly subterranean, branch tips flush with surface, branches more or less spherical, often with flattened apex, 12–16-angled, with spreading, narrowly triangular tubercles. Apr.–June. Karroid scrub on gentle slopes, often in quartz patches, KM (east of Barrydale).•

tridentata Lam.   VINGERPOL   Like **E. globosa** but branches cylindrical to clavate and often rhizomatous, stems to 15 mm long, 6–12 mm diam., with more prominent tubercles and cyathia 12–17 mm diam., glands pale yellow to white.Oct.–Dec. Grassy flats or stony karroid slopes, KM, LB, SE (Ladismith, Riversdale to E Cape).

EXCOECARIA   c. 40 spp., Old World tropics

simii (Kuntze) Pax   Monoecious shrublet or small tree to 6 m. Leaves narrowly lanceolate, glabrous, slightly toothed. Flowers in bisexual terminal spikes, greenish, females basal. Nov.–Jan. Coastal forest and bush, SE (George to KwaZulu-Natal).

HYENANCHE•   WOLWEGIFBOOM   1 sp., W Cape

globosa (Gaertn.) Lam.   Dioecious shrub or small tree, sometimes to 5 m. Leaves leathery, in whorls of 4, lanceolate. Flowers axillary, males in crowded panicles, reddish, females 1–3. Nov.–Dec. Sandstone rocks, NW (Bokkeveld Mts to Gifberg).•

JATROPHA   c. 175 spp., pantropical and subtropical

capensis (L.f.) Sond.   Monoecious shrublet with pale corky bark, 2–3 m, sticky on young parts. Leaves oblong- to lanceolate-hastate. Flowers in loose terminal clusters, males on branched axes, females solitary, greenish. Nov.–Jan. Dry rocky slopes, SE (Baviaanskloof Mts to E Cape).

LACHNOSTYLIS•   KLIPKOOLHOUT   3 spp., W Cape

bilocularis R.A.Dyer   Dioecious shrub or small tree to 3 m, velvety on young parts. Leaves elliptic, leathery. Flowers in axillary clusters, nodding, greenish, females with ovary 2- (not 3-)locular, males with filaments free. Fruits c. 5 mm diam., on filiform pedicels c. 10 mm long, with accrescent calyx. Mainly July–Dec. Shale slopes in thicket, KM (Swartberg Mts and Rooiberg).•

hirta (L.f.) Müll.Arg.   Dioecious shrub or small tree to 3 m, velvety on young parts. Leaves elliptic to oblanceolate, leathery. Flowers in axillary clusters, nodding, greenish, males with filaments united in lower half. Fruits c. 10 mm diam., on stout pedicels c. 10 mm long, with deciduous calyx. Mainly Nov.–Feb. Coastal bush and forest, SW, LB, SE (Stanford to Port Elizabeth).•

sp. 1   Like **L. hirta** but branches reddish brown and without lenticels and fruits on short pedicels c. 5–7 mm long. Nov.–Jan. Rocky slopes in thicket, NW (Olifants River valley).•

LEIDESIA 1 sp., South Africa

procumbens (L.) Prain (= *Leidesia obtusa* (Thunb.) Müll.Arg.   Monoecious, soft, often sprawling annual to 30 cm. Leaves ovate, lobed to coarsely toothed, thinly hairy. Flowers axillary, males in spike-like cymes at branch tips, females solitary, greenish. Mainly Sept.–Dec. Forest and bush, SW, LB, SE (Cape Peninsula to Mpumalanga).

PHYLLANTHUS c. 700 spp., pantropical and subtropical

heterophyllus E.Mey. ex Müll.Arg   Like **P. incurvus** but leaves elliptic-ovate, floral disc more or less annular (not composed of separate rounded glands). Mainly Oct.–Jan. Coastal sands and limestone, AP, LB, SE (De Hoop to KwaZulu-Natal).

incurvus Thunb.   Monoecious, sometimes dioecious, twiggy perennial from woody base, with smooth reddish bark. Leaves elliptic to lanceolate. Flowers axillary, females solitary, males few, stamens 3, with filaments united, fruiting pedicels to 5 mm long. Mainly Sept.–Dec. Dry slopes and flats, LB, SE (Riversdale to Port Elizabeth and dry parts of southern Africa).

verrucosus Thunb.   Monoecious shrub to 2.5 m with grey warty bark. Leaves broadly obovate, discolorous. Flowers solitary in axils, males with 4 or 5 free stamens, fruiting pedicels c. 10 mm long. Nov.–Dec. Dry bush and thicket, SE (Humansdorp to E Cape).

[**SEIDELIA pumila** (Sond.) Baill. included in the previous edition is probably **S. firmula** (Prain) Pax & K.Hoffm., apparently an occasional exotic weed in the Cape flora.]

FABACEAE
by A.L. Schutte

Argyrolobium by T.J. Edwards & A.L. Schutte, **Indigofera** by B.D. Schrire,
Otholobium and **Psoralea** by C.H. Stirton & A.L. Schutte, **Rafnia** by G.J. Campbell,
Rhynchosia by G. Germishuizen & A.L. Schutte, **Tephrosia** by B.D. Schrire & A.L. Schutte

1. Flowers actinomorphic with petals valvate in bud, often united at base; stamens more than 10 (subfamily MIMOSOIDEAE):
    2. Stamens free or nearly so ....**Acacia**
    2.' Stamens fused into a tube ....**Paraserianthes**
1.' Flowers generally zygomorphic with petals imbricate in bud, free or some of them united; stamens 10 or fewer:
    3. Perianth weakly zygomorphic, adaxial (upper) petal overlapped by adjacent lateral petals when these are present; sepals free or fused (subfamily CAESALPINIOIDEAE):
        4. Flowers pink or red; sepals fused into a tube; anthers dehiscing by longitudinal slits......................**Schotia**
        4.' Flowers yellow; sepals free; anthers dehiscing by longitudinal slits or often by apical and/or basal pores ....**Chamaecrista**
    3.' Perianth strongly zygomorphic, pea-like, adaxial petal outside adjacent lateral petals; sepals united at base (subfamily PAPILIONOIDEAE):
        5. Anthers with an extended apical connective; biramous hairs present; petals usually caducous; upper margin of keel with a fringe or hairs ....**Indigofera**
        5.' Anthers without an extended apical connective; biramous hairs generally absent; petals usually persistent; upper margin of keel without hairs:
            6. Leaflets with conspicuous, closely parallel veins extending to margins:
                7. Leaflets with toothed margins; stipules more or less adnate to petiole; leaves (1)3(7)-foliolate ....**Trifolium**
                7.' Leaflets with smooth margins; stipules not adnate to petiole; leaves 1–many-foliolate, if 3-foliolate then lateral leaflets often slightly asymmetrical...**Tephrosia**
            6.' Leaflets with veins not extending to margins:
                8. Filaments of all or alternate stamens apically dilated; compound leaves generally with at least 3 terminal leaflets, lowermost pair sometimes stipule-like, sometimes with a glandular stipule......**Lotus**
                8.' Filaments of stamens not apically dilated:
                    9. Fruits both 1-seeded and glandular; leaves more or less glandular-punctate:
                        10. Flower pedicel subtended by a distinctive lobed cupulum....................................**Psoralea**
                        10.' Flower pedicel never subtended by a cupulum:
                            11. Fruit conspicuously black glandular-warty when mature; mat-forming shrublet ....**Cullen**
                            11.' Fruit never black glandular-warty ....**Otholobium**
                    9.' Fruits not both 1-seeded and glandular:

12. Fruits jointed and transversely septate, usually breaking up into 1-seeded segments; stipules peltate ....**Zornia**
12.' Fruits not jointed and transversely septate:
   13. Leaves pinnately compound:
      14. Leaves paripinnate; pods 4-angled or -winged ....**Sesbania**
      14.' Leaves imparipinnate; pods not 4-angled:
         15. Calyx base not intrusive, upper 2 lobes not fused higher up; pods membranous ....**Lessertia**
         15.' Calyx base intrusive, upper 2 lobes fused higher up; pods leathery:
            16. Flowers pink, purple or white; stamens free ....**Virgilia**
            16.' Flowers yellow; stamens fused ....**Calpurnia**
   13.' Leaves palmately (1)3(7)-foliolate or simple:
      17. Calyx with a trifid lower lip, upper lobes sometimes fused higher up to form an upper lip; pods often pointing upwards:
         18. Stipules semisagittate or semicordate at base, with a narrow point of attachment; glandular tubercles often present...**Melolobium**
         18.' Stipules not lobed below; glandular tubercles absent:
            19. Peduncle absent, internode below inflorescence elongated; stipules often adnate to petiole (and sometimes also connate on the leaf-opposed side); bracts and bracteoles rarely present; keel petals imbricate along lower side ....**Polhillia**
            19.' Peduncle present, internode below inflorescence not elongated; stipules rarely fused; bracts and bracteoles invariably present; keel petals rarely imbricate ....**Argyrolobium**
      17.' Calyx without a trifid lower lip; pods usually horizontal or nodding:
         20. Calyx base intrusive; stamens fused into a closed tube; flowers magenta-pink with a yellow nectar guide..**Hypocalyptus**
         20.' Calyx base intrusive or not; stamens more or less free or fused into an open tube; flowers yellow, orange or red, pink or white with a white nectar guide if present:
            21. Fruits irregularly torulose, indehiscent; leaves 1-foliolate, subtending single floriferous thorns..**Alhagi**
            21.' Fruits not as above:
               22. Leaves 1(3)-foliolate (when 3-foliolate then stamens free almost to base); seed aril conspicuous, fleshy (not fleshy in **Stirtonanthus**):
                  23. Stamens free almost to base:
                      24. Leaves digitately 3-foliolate; bracts paired ....**Cyclopia**
                      24.' Leaves simple; bracts single:
                          25. Flowers yellow, decussate (arranged in opposite pairs of 2, 4 or 6 flowers); seeds with nonfleshy rim aril ....**Stirtonanthus**
                          25.' Flowers pink, mauve or white, racemose (1–5 flowers per raceme); seeds with fleshy collar-like aril ....**Podalyria**
                  23.' Stamens diadelphous or monadelphous:
                    26. Calyx base intrusive, carinal lobe usually longer than upper 4 lobes; leaves sessile, 3- or more-veined from base; inflorescences 4–many-flowered (rarely 2-flowered); bracts often leaf-like.................................**Liparia**
                    26.' Calyx base gradually narrowing to pedicel, carinal lobe usually not longer than upper 4 lobes; leaves usually petiolate or at least with a pulvinus, single veined from base; inflorescences 1- or 2-flowered; bracts not leaf-like:
                      27. Flowers yellow, fading to brown with age; bracteoles often present; aril not extended towards lens; pods compressed between seeds ....**Xiphotheca**
                      27.' Flowers mostly pink, mauve or white (rarely yellow but then shorter than 10 mm and not fading to brown with age); bracteoles absent; aril extended towards lens; pods not compressed between seeds...**Amphithalea**
               22.' Leaves (1)3(7)-foliolate; seed aril inconspicuous, not fleshy:
                  28. Stamens fused into an open tube; lateral leaflets symmetrical, not stipellate; flowers in terminal or leaf-opposed racemes:
                      29. Plants totally glabrous, even on adaxial surface of standard; leaves simple, sessile, often drying black ....**Rafnia**
                    29.' Plants hairy, at least with some hairs on leaves or on adaxial surface of standard; leaves digitate (rarely unifoliolate or sessile but then distinctly hairy and/or not drying black):

30. Anthers 5+5; keel beaked; bracteoles present; upper part of style with 1 or 2 lines of hairs (when glabrous then with a thin upper part and a thick lower part); pods usually much inflated ....**Crotalaria**
30.' Anthers 6+4 (carinal anther often intermediate); keel beaked or not; bracteoles absent or present; upper part of style glabrous, usually not with a thin upper and thick lower part; pods flat or inflated:
   31. Ovary with 2–4 ovules; fruit 1-seeded; leaves sessile, without stipules ....**Aspalathus**
   31.' Ovary with more than 6 ovules; fruit several-seeded; leaves sessile or petiolate, with or without stipules:
      32. Bracteoles absent or vestigial; stipules sometimes present; calyx unequally lobed with lateral lobes united in pairs on either side; fruit often with upper suture verrucose ....**Lotononis**
      32.' Bracteoles present; stipules never present; calyx subequally lobed; lobes usually very short:
         33. Fruit a winged samara (if narrowly winged then 1- or 2-seeded)...**Wiborgia**
         33.' Fruit not winged (if narrowly winged then more than 2-seeded) ....**Lebeckia**
28.' Stamens diadelphous, vexillary stamen free; lateral leaflets usually markedly asymmetrical and regularly stipellate; flowers in axillary or terminal racemes:
   34. Leaflets and calyx generally with yellowish gland-dots; bracteoles absent:
      35. Ovules 3 or more ....**Bolusafra**
      35.' Ovules 2, rarely 3:
         36. Funicle of seed attached at end of linear hilum; stems usually stiff and erect ....**Eriosema**
         36.' Funicle of seed attached in middle of hilum; stems usually twining or trailing ....**Rhynchosia**
   34.' Leaflets and calyx eglandular:
      37. Style generally terete and unbearded (sometimes with a few hairs below stigma):
         38. Flowers red or pink, petals very unequal in length; trees ....**Erythrina**
         38.' Flowers yellow, petals subequal in length; climber..............**Dumasia**
      37.' Style expanded, flattened, coiled or bearded:
         39. Stigma lateral, obliquely or rarely more or less terminal ....**Vigna**
         39.' Stigma terminal:
            40. Standard appendage 1 large bilobed structure; style strongly sigmoid near base, bearded in upper half with hairs in 2 rows...**Dipogon**
            40.' Standard appendages 2–4, separate; style glabrous or hairy around stigma ....**Dolichos**

ACACIA   THORN TREE, WATTLE, DORINGBOOM   c. 1200 spp., mainly pantropical and subtropical

A. Leaves bipinnate

caffra (Thunb.) Willd.   KATDORING, GEWONE HAAKDORING   Armed shrub or tree to 14 m, spines curved. Leaves bipinnate. Flowers in cylindrical spikes, creamy yellow. Pods flat, constricted between seeds. Dec.–Mar. Karroid scrub, streambanks, NW, KM, SE (Clanwilliam, Oudtshoorn to N Province).

karroo Hayne   SWEET THORN, SOETDORING   Armed shrub or small tree to 12 m, spines straight. Leaves bipinnate. Flowers in globose heads, bright yellow. Pods flat, falcate, constricted between seeds. Nov.–Dec. Karroid scrub, sandy soil, NW, SW, KM, LB, SE (throughout southern Africa).

***mearnsii** De Wild.   BLACK WATTLE   Tree to 15 m. Leaves bipinnate, dark green, with numerous raised glands on midvein. Flowers in globose heads, yellow. Pods flat, constricted between seeds. Aug.–Nov. Watercourses, forests, roadsides, NW, SW, AP, KM, LB, SE (Australian weed).

A. Leaves reduced to simple phyllodes

***cyclops** A.Cunn. ex G.Don   REDEYE, ROOIKRANS   Shrub or tree to 4 m. Phyllodes oblong, with 3–7 longitudinal veins. Flowers in globose heads, yellow. Pods flat, undulate or twisted, not constricted between seeds. Seeds black, with prominent fleshy red aril. Oct.–May. Mountain and lowland fynbos, coastal dunes, SW, AP, LB, SE (Australian weed).

***longifolia** (Andrews) Willd.   LONG-LEAVED WATTLE   Small, resprouting tree to 6 m. Phyllodes oblong, with 3–5 longitudinal veins. Flowers in cylindrical spikes, bright yellow. Pods cylindrical, constricted between seeds. June–Nov. Mountain and lowland fynbos, streambanks, marshy areas, SW, AP, LB, SE (Australian weed).

*melanoxylon R.Br. BLACKWOOD   Resprouting shrub or tree to 35 m. Phyllodes linear-oblanceolate, with 3–7 longitudinal veins. Flowers in globose heads, yellow. Pods flat, straight or variously coiled, not constricted between seeds. Aug.–Sept. Slopes, forests and streambanks, SW, AP, LB, SE (Australian weed).

*pycnantha Benth.   GOLDEN WATTLE   Shrub or tree to 20 m. Phyllodes falcate, with 1 central longitudinal vein. Flowers in globose heads, yellow. Pods flat, slightly constricted between seeds. July–Oct. Mountain and lowland fynbos, forests, SW, AP, SE (Australian weed).

*saligna (Labill.) H.L.Wendl.   PORT JACKSON   Resprouting shrub to 6 m. Phyllodes oblong-lanceolate, with 1 central longitudinal vein. Flowers in globose heads, yellow. Pods flat, constricted between seeds. Aug.–Oct. Coastal and lowland fynbos, NW, SW, AP, LB, SE (Australian weed).

*ALHAGI   CAMEL THORN BUSH, KAMEELDORINGBOS   3 spp., Eurasia to Nepal

*maurorum Medik.   Resprouting, thorny shrub to 1.5 m. Leaves 1-foliolate, cuneate-obcordate, leathery. Flowers scattered on axillary thorns, purple. Oct.–Nov. Karroid scrub, deep soils, NW, KM (Eurasian weed).

AMPHITHALEA (= *COELIDIUM*) 42 spp., N to E Cape

A. Peduncle present

pageae (L.Bolus) A.L.Schutte (= *Coelidium pageae* L.Bolus)   Erect, reseeding shrub to 1 m. Leaves 1-foliolate, lanceolate, concave, margins strongly incurved, densely silky above, glabrescent below. Flowers in pedunculate racemes, white with keel tipped purple. July–Aug. Renosterbos-fynbos scrub, 300–450 m, KM (Montagu: Kogmanskloof).•

parvifolia (Thunb.) A.L.Schutte (= *Coelidium fourcadei* Compton, *C. parvifolium* (Thunb.) Druce) Woody, reseeding shrub to 2.5 m. Leaves 1-foliolate, ovate, more or less flat, margins slightly incurved, silky. Flowers in pedunculate racemes, rose and mauve. Sept.–Dec. Mountain fynbos, 650–1625 m, KM, SE (Swartberg and Outeniqua to Kouga Mts).•

spinosa (Harv.) A.L.Schutte (= *Coelidium spinosum* Harv.)   Erect, dense, thorny shrub to 1.2 m. Leaves 1-foliolate, elliptic-ovate, more or less flat, margins incurved, silky. Flowers in pedunculate racemes, pale rose, with dark purple keel tip. Aug.–Oct. Mountain fynbos, renosterbos-fynbos scrub, 900–1300 m, NW, KM (Hex River valley to Witteberg).•

A. Peduncle absent
B. Leaves more or less flat

axillaris Granby   Willowy, reseeding shrublet to 1 m. Leaves 1-foliolate, lanceolate, more or less flat, margins slightly recurved, silky. Flowers pale pink and purple. Apr.–Aug. Mountain fynbos, 360–1200 m, LB, SE (Langeberg and Outeniqua Mts).•

bodkinii Dummer   Decumbent shrublet. Leaves 1-foliolate, broadly ovate or elliptic, flat, with scarcely visible venation, greyish hispid. Flowers rose, mauve and dark purple. Dec.–Feb. Mountain fynbos, marshy places, 970–1300 m, SW (Wemmershoek Mts).•

cuneifolia Eckl. & Zeyh.   Robust, resprouting shrub to 2.5 m. Leaves 1-foliolate, obovate, flat, with distinct venation, silky. Flowers bright purple. Oct.–Feb. Mountain fynbos, 600–1300 m, SW (Du Toitskloof to Kleinrivier Mts).•

fourcadei Compton   Reseeding shrub to 1.5 m. Leaves 1-foliolate, ovate or elliptic, flat to weakly concave, with margins slightly thickened, silvery hairy. Flowers pink, light mauve and dark mauve. Nov.–Mar. Mountain and lowland fynbos, 225–980 m, SE (Outeniqua and Tsitsikamma Mts).•

imbricata (L.) Druce   Erect shrub to 1.8 m. Leaves 1-foliolate, ovate to obovate, flat, venation distinct, silky. Flowers deep mauve. Dec.–June. Mountain and lowland fynbos, 250–750 m, SW (Cape Peninsula to Hottentots Holland Mts).•

intermedia Eckl. & Zeyh.   Willowy, resprouting shrublet to 50 cm. Leaves 1-foliolate, elliptic, more or less flat, margins slightly recurved, silky. Flowers pale and dark violet. July–Oct. Mountain and lowland fynbos, 60–1650 m, SW, KM, SE (Hottentots Holland to Kouga Mts).•

micrantha (E.Mey.) Walp.   Dense, resprouting shrublet to 0.5 m. Leaves 1-foliolate, ovate to cordate, flat to somewhat concave, minutely black-dotted, glabrous and glossy above, hairy or glabrous beneath.

Flowers white and purple. Apr.–July. Mountain fynbos, 1625–1950 m, KM, SE (Swartberg to Van Staden's Mts).•

sericea Schltr. Erect, resprouting shrublet to 1 m. Leaves 1-foliolate, ovate-lanceolate, flat, silky. Flowers violet or purple. Apr.–July. Lowland fynbos, limestone, sandy hills, 60–500 m, AP (Pearly Beach to Stilbaai).•

tomentosa (Thunb.) Granby Willowy, resprouting shrub to 1 m. Leaves 1-foliolate, lanceolate, more or less flat, grey-silky. Flowers lemon-yellow. Apr.–Sept. Lowland fynbos, below 200 m, SW, AP, LB (Betty's Bay to Riversdale).•

violacea (E.Mey.) Benth. Erect, resprouting shrublet to 1 m. Leaves 1-foliolate, lanceolate, more or less flat, margins slightly recurved, silky. Flowers mauve and purple. Mar.–Aug. Mountain and lowland fynbos, 250–1020 m, KM, LB, SE (Gamkaberg and Langeberg to Outeniqua Mts).•

williamsonii Harv. Erect, resprouting shrub to 2 m. Leaves 1-foliolate, ovate, more or less flat to somewhat concave, venation visible, silky. Flowers pale mauve and purple. May–Aug. Mountain fynbos, 475–650 m, SE (Plettenberg Bay to E Cape).•

B. Leaves concave or convex, with margins incurved or recurved
C. Leaves convex with recurved margins

alba Granby Erect, reseeding shrub to 1.2 m. Leaves 1-foliolate, ovate, margins recurved, silky. Flowers white or creamy. May–Sept. Lowland fynbos on limestone, below 150 m, AP (Agulhas to Stilbaai).•

ericifolia (L.) Eckl. & Zeyh. Erect, resprouting shrublet to 1 m. Leaves 1-foliolate, lanceolate, margins strongly recurved, silky. Flowers pink, rose and dark pink. Apr.–Jan. Mountain and lowland fynbos below 1500 m, SW, LB (Malmesbury to Albertinia).•

phylicoides Eckl. & Zeyh. Robust, resprouting shrub to 1.5 m. Leaves 1-foliolate, ovate-lanceolate, margins recurved, densely silky above, softly hairy beneath. Flowers pale rose, white and purple. Jan.–July. Mountain fynbos, 50–1450 m, SE (Outeniqua Mts to Grahamstown).

rostrata A.L.Schutte & B.-E.van Wyk Erect, robust, resprouting shrublet to 50 cm. Leaves 1-foliolate, ovate, margins strongly recurved, hairy above, densely silky beneath. Flowers dark maroon-red, with cucullate standard apex and squarrose wing petals. June–July. Lowland fynbos, hillsides, 120 m, AP (Pearly Beach: Carruthers Hill).•

speciosa Schltr. Resprouting shrublet to 1.3 m. Leaves 1-foliolate, lanceolate to ovate-lanceolate, margins recurved, glabrous above, silky beneath. Flowers white and violet. Apr.–Sept. Lowland fynbos, 80–300 m, AP (Elim to Bredasdorp).•

stokoei L.Bolus Erect, reseeding shrublet to 60 cm. Leaves 1-foliolate, ovate-lanceolate, margins recurved, pungent, glabrous or hairy above, densely silky beneath. Flowers rose and purple. Aug.–Nov. Sandstone slopes in fynbos, 320–500 m, SW (Hottentots Holland Mts).•

virgata Eckl. & Zeyh. Erect, resprouting shrublet to 30 cm. Leaves 1-foliolate, linear-lanceolate, margins strongly recurved, thinly hairy above, silky beneath, rusty brown when dry. Flowers pinkish mauve. May–Sept. Lowland fynbos, to 450 m, SW (Caledon).•

C. Leaves concave with incurved margins
*D. Leaves twisted (see also **A. ciliaris, A. muraltoides**)*

bowiei (Benth.) A.L.Schutte (= *Coelidium bowiei* Benth.) Decumbent, resprouting shrublet to 30 cm. Leaves 1-foliolate, lanceolate-ovate, concave, margins strongly incurved, twisted, ciliate, silky above, glabrous below. Flowers yellow. June–Aug. Mountain fynbos, 300–600 m, SW (Houwhoek to Kleinmond).•

muirii (Granby) A.L.Schutte (= *Coelidium muirii* Granby) Resprouting shrublet to 70 cm. Leaves 1-foliolate, oblong, concave, margins incurved, twisted, densely silky above, glabrescent beneath. Flowers white or cream-coloured with dark purple keel tip. May–Sept. Mountain fynbos, 1000–1950 m, KM, LB (Langeberg and Little Karoo Mts).•

perplexa Eckl. & Zeyh. (= *Coelidium perplexum* (Eckl. & Zeyh.) Granby) Erect, reseeding shrub to 1.2 m. Leaves 1-foliolate, ovate to lanceolate, concave, margins slightly incurved, twisted, silky. Flowers pink, with dark violet-red keel. Sept.–Jan. Mountain fynbos, 1625 m, NW, KM (Cold Bokkeveld to Klein Swartberg).•

tortilis (E.Mey.) Steud. (= *Coelidium tortile* (E.Mey.) Druce) Prostrate, resprouting shrublet to 20 cm. Leaves 1-foliolate, narrowly elliptic, concave, margins incurved, twisted, densely silky above, glabrescent beneath. Flowers pale cream-coloured or white, with dark purple keel tip. June–Dec. Mountain fynbos, 360–1600 m, NW, SW (Grootwinterhoek to Du Toitskloof Mts).•

D. Leaves not twisted

biovulata (Bolus) Granby   Prostrate or decumbent, resprouting shrublet to 30 cm. Leaves 1-foliolate, lanceolate, concave, margins incurved, silky above, glabrous beneath. Flowers violet-blue. Sept.–Oct. Lowland fynbos, below 150 m, SW, AP (Kogelberg to De Hoop).•

bullata (Benth.) A.L.Schutte (= *Coelidium bullatum* Benth.)   Low, resprouting shrublet to 15 cm. Leaves 1-foliolate, lanceolate to broadly ovate, concave, margins strongly incurved, ciliate, silky above, thinly hairy or glabrous beneath. Flowers unknown. Flowering time? Mountain fynbos, LB (Langeberg Mts: Garcia's Pass).•

cedarbergensis (Granby) A.L.Schutte (= *Coelidium cedarbergense* Granby)   Erect, resprouting shrub to 20 cm. Leaves 1-foliolate, lanceolate to ovate-lanceolate, concave, margins slightly incurved, densely silky above, thinly hairy or glabrous beneath. Flowers pink, with purple keel tip. May–Oct. Mountain fynbos, 1300–1790 m, NW (Cedarberg to Cold Bokkeveld Mts).•

ciliaris Eckl. & Zeyh. (= *Coelidium ciliare* (Eckl. & Zeyh.) Walp.)   Erect or straggling, resprouting shrublet to 30 cm. Leaves 1-foliolate, lanceolate-ovate, concave, margins strongly incurved, sometimes twisted, silky above, glabrous beneath, bearded at apex. Flowers white with violet-brown keel. Aug.–Oct. Mountain and lowland fynbos, 160–1300 m, SW (Caledon to Bredasdorp).•

concava Granby   Low, rigid, dense, resprouting shrublet. Leaves 1-foliolate, lanceolate, concave, margins incurved, sparsely silky or glabrescent above, silky beneath. Flowers pink and dark violet. Sept.–Dec. Mountain fynbos, 1650–2000 m, SW (Wemmershoek Mts).•

cymbifolia (C.A.Sm.) A.L.Schutte (= *Coelidium cymbifolium* C.A.Sm.)   Robust, resprouting shrublet to 30 cm. Leaves 1-foliolate, narrowly cymbiform, concave, margins strongly incurved, densely silky above, glabrous beneath, venation distinct. Flowers ?pink. ?Oct. Mountain fynbos, 550 m, LB (Langeberg Mts).•

dahlgrenii (Granby) A.L.Schutte (= *Coelidium dahlgrenii* Granby)   Erect, robust, resprouting shrublet to 50 cm. Leaves 1-foliolate, linear-oblong, concave, margins incurved, densely silky above, thinly hairy or glabrous beneath. Flowers white with dark violet keel tip. Aug.–Sept. Mountain fynbos, 975–1300 m, NW (Matroosberg and Keeromsberg).•

esterhuyseniae (Granby) A.L.Schutte (= *Coelidium esterhuyseniae* Granby)   Sprawling, dense, resprouting shrublet to 35 cm. Leaves 1-foliolate, linear-oblong, concave, margins strongly incurved, densely silky above, thinly hairy or glabrous beneath. Flowers white, pink with purple keel tip. Nov.–Jan. Mountain fynbos, alpine vegetation, 1625–1950 m, NW (Hex River Mts).•

flava (Granby) A.L.Schutte (= *Coelidium flavum* Granby)   Robust, resprouting, multistemmed shrublet to 80 cm. Leaves 1-foliolate, ovate, concave, margins strongly incurved, densely silky above, glabrous beneath. Flowers yellow. Sept. Mountain fynbos, lower slopes, 470 m, SE (Outeniqua Mts).•

minima (Granby) A.L.Schutte (= *Coelidium minimum* Granby)   Erect shrublet to 80 cm. Leaves 1-foliolate, oblong, subimbricate, concave, silky or glabrous. Flowers pale rose, with purple keel tip. Sept. Fynbos, rocky outcrops, NW (Bokkeveld Mts).•

monticola A.L.Schutte (= *Coelidium humile* Schltr.)   Low, rounded, flat-topped, resprouting shrublet to 40 cm. Leaves 1-foliolate, oblong, concave, margins incurved, densely woolly. Flowers bright pink, with dark purple keel tip. Oct.–Jan. Mountain fynbos, alpine vegetation, 1625–2110 m, NW (Hex River to W Langeberg Mts).•

muraltioides (Benth.) A.L.Schutte (= *Coelidium muraltioides* Benth.)   Low, diffuse, reseeding shrublet to 40 cm. Leaves 1-foliolate, lanceolate to ovate, concave, margins strongly incurved, sometimes somewhat twisted, hairy. Flowers pale rose, with dark purple keel tip. Oct.–Mar. Mountain fynbos, 1135–2000 m, NW, KM (Cedarberg to Swartberg Mts).•

oppositifolia L.Bolus   Erect shrublet to 60 cm. Leaves opposite, 1-foliolate, linear, concave, margins incurved, silky above, glabrous beneath. Flowers mauve and dark purple. July–Sept. Mountain fynbos, lower slopes, 330–500 m, SW (Kogelberg to Betty's Bay).•

purpurea (Granby) A.L.Schutte (= *Coelidium purpureum* Granby)   Compact, reseeding shrub. Leaves 1-foliolate, oblong, concave, margins incurved, silky. Flowers bright purple or pink, with dark purple keel tip. Dec. Mountain fynbos, 1820 m, NW (Cold Bokkeveld Mts).•

villosa Schltr. (= *Coelidium villosum* (Schltr.) Granby)   Erect, robust, sometimes resprouting shrub to 1 m. Leaves 1-foliolate, linear-oblong, concave, margins incurved, densely silky above, glabrescent beneath. Flowers pink, mauve or magenta, with dark pink or mauve keel tip. Sept.–Nov. Mountain fynbos, karroid scrub, 940–1470 m, NW, KM (Cedarberg Mts to Witteberg).•

vlokii (A.L.Schutte & B.-E.van Wyk) A.L.Schutte (= *Coelidium vlokii* A.L.Schutte & B.-E.van Wyk)   Willowy, resprouting shrub to 1 m. Leaves 1-foliolate, linear to narrowly oblong, concave, margins incurved, densely silky, sometimes glabrescent beneath. Flowers creamy yellow, with dark purple keel tip. Sept. Renosterbos-fynbos scrub, 850 m, SE (Uniondale).•

ARGYROLOBIUM c. 70 spp., Africa, Mediterranean to India

A. Flowers solitary

argenteum (Jacq.) Eckl. & Zeyh.   Much-branched shrublet to 50 cm. Leaves petiolate 3-foliolate, leaflets weakly dimorphic, broadly obovate to obovate, somewhat conduplicate. Flowers solitary, sessile, yellow, fading russet. Mainly April. Fynbos, karroid scrub, lower slopes, 300–800 m, NW, SW, KM, LB, SE (widespread in dry interior).

collinum Eckl. & Zeyh.   Sparsely branched, resprouting, silvery shrublet to 3 m. Leaves subsessile 3-foliolate, leaflets sericeous narrowly to broadly obovate. Flowers solitary, sessile, yellow, fading russet. Sept.–Feb. (–July). Fynbos, grassland, 300–1500 m, SE (Joubertina to Grahamstown).

harmsianum Schltr.   Sparsely branched, resprouting shrublet to 20 cm. Leaves subsessile 3-foliolate, leaflets obovate to oblong. Flowers solitary, subsessile, yellow. Apr.–June. Fynbos, below 300 m, AP (De Hoop).•

harveyanum Oliv.   Sparsely branched, resprouting, tuberous shrublet to 30 cm. Leaves 3-foliolate, leaflets dimorphic, lower petiolate broadly obovate, elliptic to lanceolate, upper subsessile linear to oblanceolate, conduplicate. Flowers solitary, yellow. Mainly Sept.–Mar. Grassland, below 800 m, SE (Knysna to N Province).

pachyphyllum Schltr.   Well-branched shrublet, 10–20 cm. Leaves 3-foliolate, leaflets crescent-shaped. Flowers solitary, pedunculate yellow. Oct.–Apr. Fynbos below 300 m, SW (Caledon, Bredasdorp).•

polyphyllum Eckl. & Zeyh.   Virgate shrub, 0.5–2 m. Leaves 3-foliate, leaflets dimorphic, primary leaflets caducous, narrowly to broadly obovate, petiole 8–20 mm long, secondary leaflets narrowly obovate to elliptic, petiole 2–6 mm long. Flowers large, yellow. Mainly Nov.–May. Forest margins, 800–1500 m, SE (Baviaanskloof Mts to E Cape).

rarum Dummer   Trailing, sparsely branched shrublet, with stems to 40 cm long. Leaves 3-foliolate, leaflets linear, petioles long. Flowers solitary, sessile, yellow. Dec.–Mar. Fynbos, 300–800 m, KM (Rooiberg).•

A. Flowers in terminal racemes
B. Leaflets linear-oblanceolate

aciculare Dummer   Slender, resprouting shrublet to 20 cm. Leaves 3-foliolate, leaflets linear-conduplicate. Flowers yellow. Nov.–Feb. Fynbos, sandstone, below 300 m, SW (Kogelberg, Caledon).•

angustissimum (E.Mey.) T.J.Edwards   Erect, sparsely branched shrublet to 30 cm. Leaves 3-foliolate, leaflets linear. Flowers large (calyx 17–20 mm long) russet and yellow. Sept.–Dec. Fynbos, SW (Paarl Mts).•

filiforme (Thunb.) Eckl. & Zeyh.   Sparsely branched, resprouting tuberous shrublet to 15 cm. Leaves 3-foliolate, leaflets dimorphic, lower broadly obovate to lanceolate, upper linear to oblanceolate, conduplicate. Flowers subsessile, yellow. Sept.–Jan. Fynbos, below 300 m, SW, LB (Cape Peninsula to Albertinia).•

tuberosum Eckl. & Zeyh.   Sparsely branched, resprouting, tuberous shrublet to 30 cm. Leaves 3-foliolate, leaflets usually dimorphic, lower petiolate broadly obovate to narrowly oblanceolate, upper subsessile linear to oblanceolate, conduplicate. Flowers lemon-yellow and russet. Sept.–Apr. Grassland, below 1500 m, SE (Humansdorp to tropical Africa).

B. Leaflets elliptic-obovate

crassifolium (E.Mey.) Eckl. & Zeyh.   Much-branched shrublet, 0.5–1 m. Leaves 3-foliolate, leaflets broadly obovate, apiculate. Flowers yellow, fading russet. June–Nov. Grassland, below 300 m, SE (Humansdorp to Uitenhage).•

crinitum (E.Mey.) Walp.   Sparsely branched, resprouting subshrub, 30–50 cm. Leaves 3-foliolate, leaflets obovate, with very large 2–4-lobed, fused stipules. Flowers yellow. June. Renosterveld, 300–800 m, LB (Langeberg Mts: Tradouw Pass).•

incanum Eckl. & Zeyh.   Virgate shrub to 2 m. Leaves 3-foliolate, leaflets round. Flowers yellow, fading russet. July–Jan. Fynbos, grassland, 300–1500 m, SE (Baviaanskloof Mts to Grahamstown).
lunare (L.) Druce   Trailing shrublet with stems 10–40 cm long. Leaves 3-foliolate, leaflets narrowly lanceolate-elliptic, petioles very long, stipules with cordate bases. Flowers yellow. Aug.–Feb. Fynbos, below 300 m, NW, SW (Clanwilliam to Caledon).•
molle Eckl. & Zeyh.   Tufted to decumbent shrublet, 10–20 cm. Leaves 3-foliolate, leaflets slightly dimorphic, oblanceolate to obovate, weakly to strongly conduplicate, venation prominent. Flowers yellow, fading russet. Aug.–Feb. Grassland, 300–1500 m, LB, SE (Riversdale to N Province).
parviflorum T.J.Edwards   Well-branched shrublet to 1 m. Leaves 3-foliolate, leaflets broadly obovate, venation prominent beneath, reticulate. Flowers yellow. Mainly Nov.–May. Fynbos, 300–1500 m, SE (Baviaanskloof Mts).•
polyphyllum Eckl. & Zeyh.   Virgate shrub, 0.5–2 m. Leaves 3-foliolate, leaflets dimorphic, primary leaflets caducous, narrowly to broadly obovate, petiole 8–20 mm long, secondary leaflets narrowly obovate to elliptic, petiole 2–6 mm long. Flowers large yellow. Flowering time? Forest margins, 800–1500 m, SE (Baviaanskloof Mts to E Cape).
splendens (Meisn.) Walp.   Sparsely branched, silvery subshrub, 20–30 cm. Leaves 3-foliolate, leaflets oblong to obovate, margins strongly revolute, stipules obliquely cordate. Flowers yellow. Dec. Fynbos below 300 m, SW, LB (Kleinrivier Mts to Swellendam).•
tomentosum (Andrews) Druce   Well-branched, sprawling shrub, 1–1.5 m. Leaves 3-foliolate, leaflets ovate, elliptic or obovate, ciliate. Flowers yellow. Nov.–May. Forests and forest margins, below 1500 m, SE (Knysna to tropical Africa).
trifoliatum (Thunb.) Druce   Well-branched, resprouting subshrub, 2–50 cm. Leaves 3-foliolate, leaflets obovate to obcordate, emarginate to recurvo-mucronate, glabrous above, hairy below. Flowers yellow. Oct.–July. Grassland below 300 m, SE (Humansdorp).•
velutinum Eck. & Zeyh.   Sparsely branched, resprouting subshrub, 10–50 cm. Leaves petiolate, 3-foliate, leaflets broadly to narrowly obovate, sericeous to velutinous, stipules ovate. Flowers subsessile, 1-6 per peduncle, yellow. Fruits turgid. JulyNov. NW, SW, Sandveld, Elandsbaai to Langebaan.•

ASPALATHUS   CAPE GORSE   278 spp., mainly W Cape, also N Cape and KwaZulu-Natal

A. Leaves 1-foliolate
B. Leaves terete, needle-like

lebeckioides R.Dahlgren   Decumbent shrublet with branches 30–70 cm long. Leaves 1-foliolate, needle-like. Flowers yellow or standard tinted violet or purple, keel glabrous, calyx glabrous. Oct.–Nov. Lowland fynbos or renosterveld, 500–680 m, SW, AP (Mamre to De Hoop).•
linearis (Burm.f.) R.Dahlgren   ROOIBOSTEE, BUSH TEA   Erect or sprawling shrub to 2 m. Leaves 1-foliolate, needle-like. Flowers pale to bright yellow or partly purple or violet, keel hairy, calyx glabrous. Aug.–Feb. Mountain fynbos, 100–1300 m, NW, SW (Bokkeveld Mts to Cape Peninsula).•
pendula R.Dahlgren   Shrub or small tree with pendulous branches, 1.5–5 m. Leaves 1-foliolate, needle-like. Flowers on a long peduncle, pale yellow or partly violet, keel hairy, calyx glabrous. Oct.–Nov. Mountain fynbos, sometimes forming small woods, 250–600 m, NW (Cedarberg Mts and Piketberg).•

*B. Leaves more or less flat (see also **A. aemula**)*

alpestris (Benth.) R.Dahlgren   Ascending or sprawling shrublet, 50–100 cm. Leaves 1-foliolate, lanceolate, pungent, 3–9-veined from base. Flowers on a distinct peduncle, pale or bright yellow, keel glabrous, calyx glabrous. Jan.–Dec. Renosterveld and fynbos, 70–1060 m, NW, SW, KM, LB, SE (Cold Bokkeveld and Botrivier to Uniondale).•
angustifolia (Lam.) R.Dahlgren   Erect or decumbent shrublet, 15–50 cm. Leaves 1-foliolate, lanceolate, pungent, 3–5-veined from base. Flowers 1–6 at branch tips, bright yellow to partly reddish, keel hairy, calyx glabrous. Oct.–Apr. Lowland fynbos or renosterbos-fynbos scrub, 260–460 m, NW, SW, LB, SE (Bokkeveld Mts to Port Elizabeth).•
barbata (Lam.) R.Dahlgren   Rigid shrub with wand-like branches, 40–100 cm. Leaves 1-foliolate, lanceolate, pungent, subamplexicaul, white-ciliate, 7–15-veined from base. Flowers crowded at branch tips, yellow, keel hairy, calyx white-hairy. Oct.–Dec. Mountain fynbos, 160–660 m, SW (Cape Peninsula).•

commutata (Vogel) R.Dahlgren   Sprawling or prostrate, mat-forming shrublet with branches longer than 50 cm. Leaves 1-foliolate, ovate-elliptic, pungent, c. 9-veined from base, more or less reflexed. Flowers pale yellow, keel glabrous, calyx glabrous. Oct.–Apr. Mountain fynbos, upper slopes, 1000–1700 m, NW, SW (Grootwinterhoek to Hottentots Holland Mts).•

compacta R.Dahlgren   Prostrate shrub forming circular mats with slack branches to longer than 80 cm. Leaves 1-foliolate, lanceolate, pungent, 3-veined from base. Flowers on a distinct peduncle, pale yellow or standard partly purplish, keel glabrous, calyx and bracteoles sparsely and softly hairy. Oct.–Nov. Renosterbos-fynbos scrub, 1000 m, NW (Ceres: Skurweberg).•

complicata (Benth.) R.Dahlgren   Like **A. crenata** but flowers cream-coloured to pale yellow, more numerous and smaller, keel blades to 4.5 mm. Oct.–Nov. Mountain fynbos, lower slopes, NW (Piketberg).•

cordata (L.) R.Dahlgren   Rigid shrub to 1 m. Leaves 1-foliolate, ovate, subamplexicaul, pungent, 11–21-veined from base. Flowers crowded at branch tips, bright yellow, fading bright red, keel hairy, calyx white-hairy. Sept.–Dec. Mountain fynbos, lower slopes, 30–400 m, NW, SW (Piketberg to Hangklip).•

crenata (L.) R.Dahlgren   Erect or sprawling shrub, 0.2–1.5 m. Leaves 1-foliolate, ovate, subamplexicaul, pungent, margins denticulate, 7–11-veined from base. Flowers crowded at branch tips, yellow or standard partly purplish, fading red or brown, keel glabrous, calyx glabrous. Sept.–Dec. Mountain fynbos, 200–830 m, NW, SW (Olifants River Mts to Hangklip).•

elliptica (E.Phillips) R.Dahlgren   Like **A. crenata** but leaves softer, elliptic-ovate and not as pungent. Sept.–Nov. Mountain fynbos, steep, moist slopes, 660–1260 m, SW (Bainskloof to Kleinrivier Mts).•

lanceifolia R.Dahlgren   Slender decumbent to ascending shrublet to 20 cm. Leaves 1-foliolate, linear-lanceolate, pungent, sometimes softly hairy, 3-veined from base. Flowers yellow, fading reddish, keel hairy, calyx glabrous. Oct.–Jan. Mountain fynbos, marshy areas, below 500 m, NW, SW (Cedarberg to Riviersonderend Mts).•

lanifera R.Dahlgren   Procumbent or prostrate shrublet with branches to 50 cm. Leaves 1-foliolate, ovate-lanceolate, 3(–5)-veined from base, silky. Flowers on a distinct peduncle, pale yellow or violet on standard, keel glabrous, calyx sparsely and softly hairy. Oct.–Jan. Renosterbos-fynbos scrub and mountain fynbos, 30–1500 m, NW (Cedarberg and Cold Bokkeveld Mts).•

monosperma (DC.) R.Dahlgren   Sprawling shrublet. Leaves 1-foliolate, narrowly lanceolate, sometimes long-ciliate on margins, 3-veined from base. Flowers on a distinct peduncle, yellow, keel glabrous, calyx sparsely and softly hairy. Sept.–Nov. Lowland fynbos, streamsides, below 300 m, SW (Hottentots Holland Mts to Hangklip).•

nudiflora Harv.   Procumbent or decumbent shrublet to 1 m. Leaves 1-foliolate, linear-lanceolate, sometimes sparsely hairy. Flowers yellow or partly reddish, keel glabrous, calyx glabrous, stigma elongate and forward-directed. Nov.–Jan. Mountain fynbos or renosterbos-fynbos scrub, 300–1660 m, NW, SW (Pakhuis to Houwhoek Mts).•

perfoliata (Lam.) R.Dahlgren   Rigid, sparsely branched shrub, to 1.8 m. Leaves 1-foliolate, ovate-orbicular, subamplexicaul, stiff, margins with peg-like teeth, 9–15-veined from base. Flowers pale yellow or lemon-yellow, standard purplish or violet, keel glabrous, calyx sparsely and softly hairy. Sept.–Dec. Mountain fynbos, to 1700 m, NW, SW (Pakhuis to Hottentots Holland Mts).•

perforata (Thunb.) R.Dahlgren   Decumbent to prostrate shrublet with branches to 1 m. Leaves 1-foliolate, elliptic-orbicular, amplexicaul, margins sparsely tuberculate-hairy, several-veined from base. Flowers pale yellow or standard and keel partly violet, keel glabrous, calyx sometimes sparsely hairy. Sept.–Dec. Mountain fynbos, often on clay, 300–1160 m, NW, SW, LB (Olifants River to Langeberg Mts).•

A. Leaves 3-foliolate
*C. Flowers white, cream-coloured, pink, mauve or purple (see also **A. ramosissima** and **A. willdenowiana**)*

argyrella MacOwan   Decumbent to prostrate, sometimes mat-forming shrublet. Leaves 3-foliolate, leaflets linear-oblanceolate, silky. Flowers in terminal heads, mauve, keel silky, calyx silky. Sept.–Oct. Lowland fynbos, below 400 m, NW, SW (Cold Bokkeveld to Kleinrivier Mts).•

barbigera R.Dahlgren   Like **A. cephalotes** but petals more densely woolly-silky. Oct.–Nov. Renosterbos-fynbos scrub, 60 m, AP (De Hoop).•

cephalotes Thunb.   Shrub, 0.3–2 m. Leaves 3-foliolate, leaflets terete or slightly flattened, sparsely hairy. Flowers in a spike or head, pale violet or rose, sometimes almost white, keel tipped purplish violet, keel silky, calyx silky, lobes subulate. Aug.–Nov. Mountain fynbos, 160–830 m, NW, SW (Piketberg to Riviersonderend Mts).•

cerrhantha Eckl. & Zeyh.   Shrublet, 30–60 cm. Leaves 3-foliolate, leaflets terete, subglabrous. Flowers in terminal heads, white to pale violet, keel tipped purplish, keel silky, calyx sparsely silky, lobes subfiliform. Sept.–Oct. Coastal or lowland fynbos, 160–530 m, SE (George to Port Elizabeth).•

costulata Benth.   Shrub, 0.6–1.5 m. Leaves 3-foliolate, leaflets terete-ovoid. Flowers few near branch tips, rose or pale purplish, keel glabrous, calyx subglabrous. Aug.–Nov. Mountain fynbos, 500–1000 m, NW, KM (Bokkeveld Mts to Witteberg).•

forbesii Harv.   Shrub or shrublet, 0.4–2.5 m. Leaves 3-foliolate, leaflets terete, subglabrous. Flowers in terminal heads, white or cream-coloured, sometimes with pink on wings, keel glabrous, calyx shortly silky, upper lobes deeply separated. July–Apr. Coastal fynbos, limestone and marine sand, below 60 m, SW, AP (Cape Peninsula to Stilbaai).•

globosa Andrews   Erect shrub with tail-like branches, 1.5–2.5 m. Leaves 3-foliolate, leaflets subterete. Flowers in terminal heads with an involucre, pale rose or almost white, wings woolly, keel silky woolly, calyx silky. Aug.–Oct. Mountain fynbos, 400–700 m, SW (Hottentots Holland to Palmiet River Mts).•

globulosa E.Mey.   Shrublet, 0.3–0.8 m. Leaves 3-foliolate, leaflets filiform. Flowers in terminal heads with an involucre of ovate leaves, almost white, keel tipped bluish violet, keel hairy, calyx silky. Oct.–Dec. Coastal fynbos, 110–180 m, SW (Cape Peninsula to Baardskeerdersbos).•

grandiflora Benth.   Robust shrub, 0.4–1.3 m. Leaves 3-foliolate, leaflets slightly flattened, glabrous or ciliate. Flowers few at branch tips, large, cream-coloured, sometimes yellow, keel glabrous, blade 16–18 mm long, calyx silky, upper lobes larger. Aug.–Dec. Mountain fynbos, 300–1000 m, SW, LB (Breede River valley and Langeberg Mts).•

nigra L.   Shrublet to 70 cm, like **A. cephalotes** but bracts shorter than 3.5 mm, calyx lobes mostly shorter than 3 mm and flowers usually slate-blue to violet. Aug.–Nov. Renosterbos-fynbos scrub, up to 1200 m, NW, SW, KM, LB, SE (Clanwilliam to Uniondale).•

pallidiflora R.Dahlgren   Like **A. costulata** but calyx lobes subulate. Oct.–Jan. Mountain fynbos, upper slopes, 1300–1600 m, NW (Cold Bokkeveld and Hex River Mts).•

rosea Garab. ex R.Dahlgren   Decumbent shrublet to 15 cm. Leaves 3-foliolate, leaflets somewhat flattened. Flowers in terminal heads, rose to flesh-coloured, keel silky, beaked, calyx silky. Sept. Lowland fynbos, 100–200 m, SW (Botrivier to Elim).•

submissa R.Dahlgren   Spreading shrublet like **A. nigra** but calyx lobes narrowly triangular. Sept.–Nov. Renosterbos-fynbos scrub, 100–1200 m, SW, LB (Botrivier to Mossel Bay).•

C. *Flowers partly or wholly yellow, often with red or purple pattern*
D. *Leaf base spurred and spine- or thorn-like (see also **A. opaca**)*

aculeata Thunb.   Rigid shrublet to 1 m. Leaves 3-foliolate, leaflets slightly flattened, sometimes sparsely hairy, leaf bases forming woody spines. Flowers in terminal heads, bright or pale yellow, keel glabrous, calyx silky. Oct.–Jan. Lowland renosterbos-fynbos scrub, 60–330 m, NW, SW (Ceres to Cape Flats).•

calcarata Harv.   Erect or decumbent shrublet, 15–40 cm. Leaves 3-foliolate, leaflets terete, sparsely adpressed-hairy, leaf bases forming short woody spines. Flowers small, solitary on short shoots forming a pseudospike, yellow to deep red, keel glabrous, calyx silky, lobes linear-subulate. Oct.–Jan. Renosterbos-fynbos transition, 130–160 m, SW, LB (Riviersonderend and Langeberg Mts).•

dasyantha Eckl. & Zeyh.   Rigid shrub with branches knotty or almost thorn-like. Leaves 3-foliolate, leaflets oblanceolate, subglabrous, leaf base forming a short spine. Flowers 1 or 2 at branch tips but soon appearing lateral, pale yellow, fading purplish, keel silky, calyx woolly. Oct.–Dec. Coastal fynbos, 160–300 m, AP (Stilbaai to Albertinia).•

desertorum Bolus   Shrublet, often with thorn-like branches, 0.7–1 m. Leaves 3-foliolate, leaflets subterete, leaf bases forming recurved, woody spines. Flowers solitary on short shoots, pale yellow, standard sometimes purplish on back, keel glabrous, calyx silky. Oct.–Dec. Renosterbos-fynbos scrub, 330–1060 m, NW (Bokkeveld Mts).•

macrantha Harv.   Shrub or small tree, 2–4 m. Leaves 3-foliolate, leaflets subterete, sparsely hairy, leaf bases forming small, blunt spurs. Flowers large, 2–4 at branch tips, bright yellow or with reddish shades, standard with densely hairy basal calluses, keel glabrous, calyx silky. Sept.–Nov. Mountain fynbos, 200–350 m, SW (Cape Peninsula and Hottentots Holland Mts).•

macrocarpa Eckl. & Zeyh.   Lanky, sparsely branched shrub to 2 m. Leaves 3-foliolate, leaflets needle-like, leaf bases forming spurs. Flowers solitary on short shoots scattered for some distance, yellow, keel glabrous, calyx silky, anthers hairy below. Nov.–Dec. Mountain fynbos, 540 m, LB (Langeberg Mts).•

rostrata Benth.   Rigid, thorny shrub, c. 1 m. Leaves 3-foliolate, leaflets filiform, shortly hairy, leaf bases forming short spurs. Flowers solitary on short shoots, pale yellow, keel glabrous, calyx shortly silky, anthers hairy below. Sept.–Nov. Dry mountain fynbos or renosterbos-fynbos scrub, 1000 m, NW (Langeberg Mts: Robertson).•

rycroftii R.Dahlgren   Rigid, decumbent shrublet to 60 cm. Leaves 3-foliolate, leaflets subterete, leaf bases forming recurved, woody spines. Flowers solitary on short shoots, yellow or partly purple, keel glabrous, calyx shortly and sparsely hairy. Feb.–Mar. Renosterbos-fynbos scrub, 180–200 m, SW (Malmesbury).•

ternata (Thunb.) Druce   Shrub to 1 m. Leaves 3-foliolate, leaflets linear-oblanceolate, silvery silky, leaf bases forming tubercles or spurs. Flowers 1 or 2 at branch tips but soon appearing lateral, pale yellow, fading purplish or brownish purple, keel silky, calyx silky. Sept.–Nov. Coastal sands, NW, SW (Lambert's Bay to Cape Peninsula).•

tridentata L.   Shrub or shrublet, 30–90 cm. Leaves 3-foliolate, leaflets linear-oblanceolate, glabrous or silky, leaf bases usually forming a simple or ternate spine. Flowers in terminal heads with an involucre, pale yellow, keel silky, calyx silky. Oct.–Dec. Fynbos and renosterbos below 800 m, NW, SW (Bokkeveld Mts to Potberg).•

uniflora L.   Like **A. willdenowiana** but calyx glabrous or sparsely adpressed-hairy. Sept.–Jan. Mountain fynbos, lower slopes, 300–660 m, NW, SW (Cedarberg Mts to Cape Peninsula).•

willdenowiana Benth.   Shrub, 1.5–3.5 m. Leaves 3-foliolate, leaflets needle-like, subglabrous, leaf bases forming woody spurs. Flowers solitary on short shoots, irregularly clustered, pale yellow or white to rose, keel glabrous, calyx silky. June–Dec. Marshes and seeps in mountain fynbos, 100–1000 m, SW, LB (Cape Peninsula to Mossel Bay).•

D. Leaf base not spurred
E. Leaflets distinctly flat
F. Leaflets spine-tipped

acidota Garab. ex R.Dahlgren   Rigid, much-branched shrub to more than 1 m. Leaves 3-foliolate, leaflets narrowly lanceolate, pungent, flat or slightly keeled, slightly recurved, ciliate. Flowers in a terminal head, large, bright yellow, partly fading reddish, keel silky, calyx hairy, lobes elongate-spinescent. Sept.–Dec. Mountain fynbos, 500–600 m, NW (Piketberg and southern Olifants River Mts).•

borboniifolia R.Dahlgren   Like **A. acidota** but flowers smaller, standard shorter than 7 mm, and ovary and fruit hairy only along dorsal suture. Nov.–Dec. Mountain fynbos, 660 m, SW (Cape Peninsula).•

corniculata R.Dahlgren   Rigid shrublet. Leaves 3-foliolate, leaflets lanceolate, keeled, pungent. Flowers in a terminal cluster, yellow or reddish, keel hairy, calyx hairy, lobes elongate-spinescent. Dec.–Jan. Mountain fynbos, lower slopes, 830 m, NW (Witzenberg).•

erythrodes Eckl. & Zeyh.   Like **A. fusca** but flowers in an elongate raceme. Nov. Mountain fynbos, SW (Tulbagh Waterfall).•

fusca Thunb.   Rigid shrub or shrublet to 2 m. Leaves 3-foliolate, leaflets linear-subterete to obovate, usually pungent, grey-silky. Flowers in a terminal cluster, yellow, standard partly violet, keel subglabrous, calyx fleshy, glabrous. Oct.–Nov. Lowland fynbos, to 500 m, SW (Wemmershoek to Bredasdorp).•

polycephala E.Mey.   Erect or sprawling shrub, 0.2–1.5 m. Leaves 3-foliolate, leaflets lanceolate, recurved-pungent at tips. Flowers in a terminal head, pale yellow or yellow, standard more or less violet, keel silky, calyx silky. Aug.–Dec. Mountain fynbos, 700–1250 m, NW (Nardouw and Cedarberg Mts).•

rupestris R.Dahlgren   Shrublet to 80 cm. Leaves 3-foliolate, leaflets oblanceolate, pungent. Flowers in a terminal head, bright yellow, keel glabrous, calyx hairy. Nov.–Jan. Mountain fynbos, up to 1500 m, NW (Cedarberg to Hex River Mts).•

venosa E.Mey.   Dense, rigid shrublet. Leaves 3-foliolate, leaflets oblanceolate, recurved-pungent at tips. Flowers in a terminal head, yellow, fading reddish, keel hairy, calyx hairy. Oct. Mountain fynbos, 400–500 m, NW (Gifberg).•

F. *Leaflets not spine-tipped (see also* **A. bracteata, A. fusca***)*

acocksii (R.Dahlgren) R.Dahlgren   Sparsely branched, rigid shrub to 1 m. Leaves 3-foliolate, leaflets linear-oblanceolate, rather thick, glabrous or shortly hairy. Flowers crowded at branch tips, bright yellow, keel silky, calyx silky. Aug.–Dec. Mountain fynbos, middle slopes, 900–1000 m, NW (Bokkeveld to Cedarberg Mts).•

aemula E.Mey.   Sparsely branched shrub with wand-like branches, 0.6–2.5 m. Leaves 3-foliolate, median leaflet elliptic-lanceolate, thinly silky, lateral leaflets mostly reduced or lacking. Flowers crowded in terminal spikes, pale yellow, fading brownish, keel softly hairy, calyx woolly. Sept.–Nov. Mountain fynbos, 400–1250 m, NW, KM (Cedarberg and Cold Bokkeveld Mts to Witteberg).•

altissima R.Dahlgren   Sparsely branched, somewhat lanky shrub with wand-like branches to 2 m. Leaves 3-foliolate, leaflets linear-oblanceolate, densely silky. Flowers crowded in terminal spikes, pale yellow, keel silky, calyx shortly hairy. Aug.–Nov. Mountain fynbos, 660–1500 m, NW (Bokkeveld Mts to Ceres).•

aspalathoides (L.) Rothm.   Shrublet, 15–40 cm. Leaves 3-foliolate, leaflets linear-lanceolate, glabrous or thinly hairy. Flowers in a terminal head, bright yellow, often partly reddish on standard, keel hairy, calyx silky, lobes attenuate. Oct.–Jan. Lowland fynbos, 150–400 m, NW, SW, AP, LB (Cape Peninsula to Robertson and Potberg).•

bidouwensis Garab. ex R.Dahlgren   Shrublet, 30–60 cm. Leaves 3-foliolate, leaflets linear-oblanceolate, white-woolly. Flowers few in terminal heads, pale yellow, keel hairy above, calyx white-woolly, lobes large. Nov.–Dec. Mountain fynbos, lower slopes, 660–1000 m, NW (Biedouw Mts).•

bodkinii Bolus   Prostrate, more or less mat-forming shrublet. Leaves 3-foliolate, leaflets elliptic to suborbicular, silky. Flowers solitary at branch tips, pale yellow, keel silky, calyx silky. Dec.–Jan. Sandstone outcrops at high alt., 1100–1550 m, NW, SW (Cedarberg to Hex River Mts).•

caledonensis R.Dahlgren   Erect, rod-like shrub to 1.2 m. Leaves 3-foliolate, median leaflet lanceolate, silvery silky, lateral leaflets smaller. Flowers scattered along stem in a pseudo-spike, pale yellow, keel silky, calyx silky. Aug.–Nov. Lowland fynbos, 200–600 m, SW, AP (Houwhoek to Potberg).•

callosa L.   Erect shrublet, 15–60 cm. Leaves 3-foliolate, leaflets linear. Flowers crowded at branch tips, yellow, keel glabrous, calyx glabrous. Oct.–Jan. Lowland fynbos, to 200 m, SW, AP (Cape Peninsula to Bredasdorp).•

comptonii R.Dahlgren   Prostrate or procumbent, more or less mat-forming shrublet. Leaves 3-foliolate, leaflets oblanceolate, hairy to subglabrous. Flowers few on a terminal peduncle, pale yellow, keel silky, calyx silky. Nov.–Dec. Mountain fynbos, 900–1200 m, NW (Cedarberg Mts: Middelberg).•

cytisoides Lam.   Rather rigid shrublet, 20–60 cm. Leaves 3-foliolate, leaflets linear-oblanceolate, subglabrous. Flowers in a terminal head, pale to bright yellow, keel hairy, calyx hairy. Oct.–Nov. Mountain fynbos, lower slopes, below 800 m, NW, SW (Tulbagh to Riviersonderend Mts).•

diffusa Eckl. & Zeyh.   Small, much-branched shrublet with decumbent or prostrate branches to 15 cm. Leaves 3-foliolate, leaflets narrowly elliptic with incurved margins, silky. Flowers small, solitary on internodes, sessile, pale yellow, keel silky, calyx silky, lobes linear. June–Nov. Mountain fynbos, SW, LB, ?SE (Riviersonderend to Langeberg Mts, ?W Outeniqua Mts).•

dunsdoniana Alston ex R.Dahlgren   Like **A. salicifolia** but pubescence longer and more silky, flowers larger, keel 5–6 mm long, and ovules only 2. Jan.–Feb. Mountain fynbos, upper slopes, 800–1000 m, SW (Franschhoek to Hottentots Holland Mts).•

esterhuyseniae R.Dahlgren   Sprawling or ascending shrublet. Leaves 3-foliolate, leaflets linear-oblanceolate, with distinct midvein, sometimes softly hairy. Flowers 1–3 at branch tips, yellow or partly purplish, keel hairy on midline, calyx subglabrous. Dec.–Jan. Mountain fynbos, upper slopes, 1000–1400 m, NW (S Cedarberg to Grootwinterhoek Mts).•

fasciculata (Thunb.) R.Dahlgren   Rigid, much-branched shrublet, 0.5–1 m. Leaves 3-foliolate, leaflets oblanceolate, sometimes ciliate. Flowers in more or less involucrate terminal heads, bright yellow, keel glabrous, calyx hairy, lower lobe enlarged. Nov.–Jan. Mountain fynbos, lower slopes, 45–200 m, NW (Grootwinterhoek Mts).•

heterophylla L.f.   Decumbent or procumbent shrublet to 20 cm. Leaves 3-foliolate, leaflets linear-oblanceolate, subglabrous, leaves on short shoots smaller and crowded. Flowers scattered in spikes, pale yellow to yellow, keel silky, calyx thinly and softly hairy, lobes attenuate-recurved. Sept.–Dec. Lowland fynbos, 300–900 m, NW, SW (Bokkeveld Mts to Du Toitskloof).•

incana R.Dahlgren   Procumbent or decumbent shrublet with yellowish branches. Leaves 3-foliolate, leaflets oblanceolate, grey-hairy. Flowers ?in terminal groups, yellow, calyx woolly. Dec. Shale band on sandstone slopes, 1250 m, KM (Klein Swartberg).•

inops Eckl. & Zeyh.   Sprawling shrublet to 30 cm. Leaves 3-foliolate, leaflets linear-oblanceolate, leathery, subglabrous, with distinct midvein. Flowers in terminal heads, yellow to reddish yellow, keel hairy, calyx shortly hairy. Oct.–Nov. Mountain fynbos, lower slopes, below 450 m, LB (Langeberg Mts).•

intervallaris Bolus   Slender, sparsely branched sprawling shrublet. Leaves 3-foliolate, leaflets ovate-lanceolate, silky. Flowers 1 or 2 at branch tips but appearing subterminal, pale yellow, fading purplish, keel silky, calyx silky, lobes blunt. Nov.–Apr. Lowland and mountain fynbos, 750 m, SW (Houwhoek to Kleinrivier Mts).•

lanata E.Mey.   Rather large sprawling shrublet with trailing branches. Leaves 3-foliolate, leaflets lanceolate, subglabrous, with 1–3 distinct main veins. Flowers 1–4 in a shortly pedunculate spike, yellow or partly purplish, keel glabrous, calyx shortly hairy, lobes awl-like. Oct.–Dec. Mountain fynbos, lower slopes, below 600 m, NW (Bokkeveld Mts to Porterville).•

latifolia Bolus   Like **A. lanata** but leaflets broadly lanceolate, with c. 3 distinct veins, often undulate. Sept.–Nov. Mountain fynbos, 660–1330 m, NW (Piketberg).•

linearifolia (Burm.f.) DC.   Much-branched shrub to 2 m. Leaves 3-foliolate, leaflets linear-oblanceolate, silky when young. Flowers in terminal heads, pale yellow, keel silky, calyx silky. Nov. Lowland fynbos, flats and low hillsides, 150 m, NW, SW (Saron to Franschhoek).•

lotoides Thunb.   Decumbent to procumbent shrublet to 15 cm. Leaves 3-foliolate, leaflets oblanceolate, silky. Flowers in terminal spikes, pale yellow, keel silky, calyx silky. Oct.–Dec. Mainly coastal fynbos, sometimes mountain fynbos, below 300 m, SW (Saldanha Bay to Cape Peninsula).•

marginata Harv.   Compact shrublet to 30 cm. Leaves 3-foliolate, leaflets elliptic to suborbicular, mucronate, leathery, subglabrous, margins and main veins thickened. Flowers 2–4 at branch tips, yellow, fading bright or dark red, keel silky, calyx shortly hairy. Oct.–Jan. Lower slopes in fynbos, to 1300 m, SW, AP (Hangklip to Bredasdorp).•

mundtiana Eckl. & Zeyh.   Shrublet to 60 cm with erect tail-like branches. Leaves 3-foliolate, leaflets linear-sulcate, thick, incurved. Flowers scattered, yellow, standard with reddish shades, keel glabrous, calyx glabrous, lobes minute. Aug.–Apr. Lowland fynbos, 130–620 m, SW, LB, SE (Elim to Knysna).•

myrtillifolia Benth.   Like **A. marginata** but calyx glabrous, lower lobe somewhat larger. Distribution unknown.

oblongifolia R.Dahlgren   Erect, often rod-like shrublet to 1 m. Leaves 3-foliolate, leaflets oblong-oblanceolate, shortly hairy or subglabrous. Flowers scattered, pale yellow, keel silky, calyx shortly hairy. Jan.–Apr. Lowland fynbos, below 300 m, SW (Kogelberg to Elim).•

obtusata Thunb.   Rigid, stiffly branched shrublet to 40 cm, branches thorny. Leaves 3-foliolate, leaflets oblanceolate-spathulate, subglabrous. Flowers solitary in leaf axils, pale yellow, partly or entirely pale purple, keel glabrous, calyx subglabrous, lobes minute. Aug. Quartzite ridges in karroid scrub, 175–225 m, NW (Matsikamma Mts to Lambert's Bay).•

orbiculata Benth.   Rigid shrub, 0.5–1 m. Leaves 3-foliolate, leaflets suborbicular, somewhat convex, leathery, subglabrous, main veins thickened. Flowers in terminal heads, bright yellow, fading reddish, keel hairy at tips, calyx silky. Nov.–Dec. Shale bands on sandstone slopes, 1250 m, NW (Matroosberg).•

patens Garab. ex R.Dahlgren   Prostrate, mat-like shrublet. Leaves 3-foliolate, leaflets oblanceolate, subglabrous. Flowers 1(–3) on slender terminal peduncles, yellow, keel glabrous, calyx subglabrous. Dec.–Mar. Subalpine mountain fynbos, above 1500 m, KM (Swartberg and Kammanassie Mts).•

psoraleoides (C.Presl) Benth.   Shrublet to 20 cm. Leaves 3-foliolate, leaflets ovate-obovate, leathery, subglabrous, margins and main veins thickened. Flowers in terminal heads, yellow, keel hairy, calyx hairy. Oct.–Dec. Mountain fynbos, 300–500 m, SW (Cape Peninsula to Babilonstoring).•

quadrata L.Bolus   Sparsely branched shrublet to 60 cm. Leaves 3-foliolate, leaflets oblong-lanceolate, margins often involute, shortly hairy. Flowers scattered, pale yellow, keel silky, hairy. Apr.–May. Lowland fynbos, below 300 m, LB (Albertinia).•

quinquefolia L.   Erect or sprawling shrub to 1.5 m. Leaves 3-foliolate, leaflets oblong-elliptic, grey-silky or subglabrous. Flowers in terminal spikes, pale to bright yellow, keel silky, calyx silky. Aug.–Jan. Coastal, lowland and mountain fynbos, below 1000 m, NW, SW, LB (Matsikamma to Mossel Bay).•

radiata Garab. ex R.Dahlgren   Erect or sprawling shrublet to 50 cm. Leaves 3-foliolate, leaflets linear-oblanceolate, shortly hairy. Flowers in terminal, subglobose spikes, yellow to reddish brown, keel silky, calyx silky, lobes attenuate. Oct.–Dec. Sandstone slopes in mountain fynbos, 250–1250 m, SW (Hex River Mts to Riviersonderend).•

ramulosa E.Mey.   Stiff, sparsely branched shrublet with strong, knotty branches to 70 cm. Leaves 3-foliolate, leaflets ovate-oblanceolate, silvery silky. Flowers clustered at branch tips, pale lemon-yellow, keel silky, calyx silky. Sept.–Nov. Mountain fynbos, 200–350 m, SW (Houwhoek to Riviersonderend Mts).•

rugosa Thunb.   Shrub or small tree to 2 m. Leaves 3-foliolate, leaflets linear-oblanceolate, subglabrous. Flowers in short terminal racemes, bright or pale yellow, keel silky, calyx shortly hairy. Nov.–Jan. Mountain fynbos, 440–800 m, NW, SW, KM (Cold Bokkeveld to Du Toitskloof to Ladismith).•

salicifolia R.Dahlgren   Erect shrub, 1–2 m. Leaves 3-foliolate, leaflets narrowly elliptic, grey-silky. Flowers in terminal heads, pale yellow, keel silky, calyx silky, lobes long and slender. Aug.–Sept. Mountain fynbos, moist kloofs, 500–750 m, SW (Hottentots Holland to Kleinrivier Mts).•

securifolia Eckl. & Zeyh.   Shrub to 1.2 m. Leaves 3-foliolate, leaflets obovate-spathulate, leathery, glabrous. Flowers in terminal heads on long pedicels, pale yellow, keel silky, calyx shortly hairy. Jan.–Dec. Mountain fynbos, lower slopes, below 600 m, SW, LB (Bainskloof to Langeberg Mts).•

sericea P.J.Bergius   Slender shrublet with rod-like branches to 1 m like **A. aemula** but calyx silky and flowers smaller, wing blades 9–10 mm long. Sept.–Nov. Coastal fynbos, flats, below 300 m, SW, AP (Hopefield to Agulhas).•

singuliflora R.Dahlgren   Rigid shrublet with almost thorn-like branches to 20 cm. Leaves 3-foliolate, leaflets obovate, silvery silky. Flowers solitary on terminal peduncles, pale yellow, keel silky, calyx silky. Oct. Lowland fynbos, SW (Breede River valley).•

stenophylla Eckl. & Zeyh.   Erect or sprawling shrublet, 5–35 cm. Leaves 3-foliolate, leaflets linear-spathulate, leathery, with distinct midvein, subglabrous. Flowers 1–5 at branch tips, bright yellow or partly reddish brown, keel shortly hairy, calyx hairy. Sept.–Dec. Lowland fynbos, 800–1400 m, SW (Houwhoek to Bredasdorp).•

stokoei L.Bolus   Robust shrub to 3 m. Leaves 3-foliolate, leaflets linear, with prominent midvein, long-ciliate. Flowers large, in terminal heads, bright yellow, keel almost glabrous, calyx sparsely hairy. Oct.–Jan. Mountain fynbos, streamsides, 350–900 m, SW (Kogelberg).•

suaveolens Eckl. & Zeyh.   Slender, sprawling shrublet 20 cm. Leaves 3-foliolate, leaflets linear-oblanceolate, ciliate. Flowers in terminal heads, yellow, keel glabrous, calyx sparsely hairy, lower lobe enlarged. Nov.–Jan. Mountain fynbos, 700–1250 m, NW (Grootwinterhoek Mts).•

sulphurea R.Dahlgren   Sprawling shrublet. Leaves 3-foliolate, leaflets spathulate-oblanceolate, glaucous. Flowers large, strongly compressed, pale yellow, keel glabrous, calyx compressed and broad, glabrous, lobes ovate. Oct. Mountain fynbos, NW (Witzenberg and Grootwinterhoek Mts).•

taylorii R.Dahlgren   Sprawling shrublet with slender, peduncle-like branches 25–45 cm long. Leaves 3-foliolate, leaflets linear-spathulate, keeled, with long spreading hairs. Flowers small, 1–5 at branch tips, pale to bright yellow, fading red, keel shortly hairy, calyx shortly hairy, lobes awl-like. Dec. Mountain fynbos, lower slopes, 200 m, SW (Riviersonderend Mts: Spitzkop).•

truncata Eckl. & Zeyh.   Rigid shrub, 0.5–1.5 m. Leaves 3-foliolate, leaflets linear to oblanceolate, often incurved. Flowers in terminal heads, yellow, keel shortly hairy, calyx hairy below, lobes large, glabrous. Nov. Mountain fynbos, below 1000 m, SW (Elandskloof Mts to Bainskloof).•

tylodes Eckl. & Zeyh.   Like **A. callosa** but young branches shortly hairy and flowers smaller, standard blade to 6 mm long. Oct.–Nov. Coastal fynbos, 350–400 m, SW, AP (Cape Flats and Struisbaai to Albertinia).•

vacciniifolia R.Dahlgren   Shrublet to 30 cm. Leaves 3-foliolate, leaflets oblong-oblanceolate, fleshy. Flowers in terminal heads, bright yellow, keel glabrous, calyx fleshy, shortly hairy below, lobes large, glabrous. Nov.–Dec. Subalpine mountain fynbos, 1000–1300 m, SW (Hottentots Holland Mts: Somerset Sneeuwkop).•

villosa Thunb.   Slender sprawling shrublet, 10–40 cm. Leaves 3-foliolate, leaflets oblanceolate, silvery silky. Flowers 1–4 at branch tips, pale lemon-yellow, fading reddish, keel hairy, calyx silvery silky. Oct.–Dec. Mountain fynbos, below 1000 m, NW (Cedarberg Mts to Ceres).•

vulpina Garab. ex R.Dahlgren Rigid, rod-like shrublet, closely leafy. Leaves 3-foliolate, leaflets lanceolate, leathery, ciliate, 1–3 veined. Flowers scattered, yellow, keel beaked, hairy beneath, calyx campanulate, shortly hairy. Sept.–Oct. Mountain fynbos, 600 m, LB (Langeberg Mts: Barrydale to Garcia's Pass).•

*E. Leaflets subterete or angular, sometimes sulcate or somewhat flattened (see also **A. fusca**)*
*G. Leaflets mucronate or spine-tipped (see also **A. pachyloba**)*

abietina Thunb. Erect or sprawling shrublet, 10–60 cm. Leaves 3-foliolate, leaflets fascicled, subterete, recurved-pungent, subglabrous. Flowers scattered, bright yellow, fading orange, keel glabrous, calyx glabrous or sparsely adpressed-hairy, lobes pungent. Sept.–Dec. Mountain fynbos, below 500 m, NW, SW (Wolseley to Anysberg).•

acanthes Eckl. & Zeyh. Sparsely branched, densely leafy shrub, 0.5–1.5 m. Leaves 3-foliolate, leaflets terete, pungent, subglabrous. Flowers scattered, bright yellow, keel glabrous, calyx shortly hairy, lobes needle-like. Aug.–Dec. Mountain fynbos, 300–760 m, LB (Langeberg Mts).•

acanthiloba R.Dahlgren Like **A. chenopoda** but flowers smaller, standard blade 7–9 mm long. Oct.–Nov. Mountain fynbos, lower slopes, below 200 m, SW (Gordon's Bay to Cape Hangklip).•

acifera R.Dahlgren Rigid, sprawling shrublet with branches to 60 cm. Leaves 3-foliolate, leaflets needle-like, pungent. Flowers 1 or 2 at branch tips, pale yellow, partly purplish or violet, keel hairy, calyx adpressed-hairy, lobes needle-like. Aug.–Dec. Mountain fynbos, 500–1000 m, NW (Olifants River and Cedarberg Mts).•

aciphylla Harv. Rigid shrub or shrublet, 0.3–2 m. Leaves 3-foliolate, leaflets angular, pungent, bracts broadly ovate. Flowers scattered at branch tips, bright lemon-yellow, keel glabrous, calyx glabrous, lobes pungent. July–Nov. Mountain fynbos, lower to upper slopes, up to 1200 m, AP, KM, LB, SE (Elim to Humansdorp).•

acuminata Lam. Thorny shrub, 0.15–5 m. Leaves 3-foliolate, leaflets terete or angular, pungent. Flowers few on thorny branchlets, pale or bright yellow or orange, standard sometimes partly violet, keel shortly hairy, calyx adpressed-hairy, lobes pungent. Aug.–Mar. Mountain fynbos, lower slopes, 100–660 m, NW, SW, LB (Namaqualand to Albertinia).

albens L. Shrub or shrublet with spreading branches, to 1 m. Leaves 3-foliolate, leaflets terete or subterete, pungent. Flowers small, few to many in terminal spikes, pale yellow, fading brownish or rose, keel hairy, calyx large, hairy, lobes pungent. Aug.–Sept. Lowland and coastal fynbos, up to 300 m, NW, SW (Klawer to Cape Peninsula).•

alopecurus Burch. ex Benth. Erect, sometimes decumbent, sparsely branched shrub with closely leafy, tail-like branches, 20–70 cm. Leaves 3-foliolate, leaflets needle-like, pungent, sparsely hairy or glabrescent. Flowers scattered or clustered near branch ends, yellow, keel silky, calyx silky, lobes needle-like. July–Sept. Coastal fynbos, marine sand, 60–200 m, AP, SE (Bredasdorp to Knysna).•

aristata Compton Sprawling shrublet to 30 cm. Leaves 3-foliolate, leaflets recurved, subterete, pungent. Flowers solitary at branch tips, pale yellow or purplish or rose-coloured, keel glabrous, calyx glabrous, lobes needle-like and upcurved. Oct.–Dec. Mountain fynbos, upper slopes and subalpine habitats, 1000–2000 m, NW (Cedarberg to Hex River Mts).•

aristifolia R.Dahlgren Procumbent or prostrate shrublet, with branches to 1 m. Leaves 3-foliolate, leaflets angular, pungent. Flowers 1 or 2 at tips of densely leafy branchlets, lemon-yellow or violet on keel and back of standard, keel whitish hairy, calyx whitish hairy, lobes needle-like. Sept.–Nov. Mountain fynbos, 600–1160 m, NW (Piketberg and Olifants River Mts).•

astroites L. Like **A. abietina** but branch tips whitish hairy and keel tapering and beaked. Oct.–Nov. Lowland fynbos, 330–900 m, SW (Cape Peninsula to Du Toitskloof and Kleinrivier Mts).•

batodes Eckl. & Zeyh. Erect or sprawling shrublet to 50 cm. Leaves 3-foliolate, leaflets subterete, pungent. Flowers 1–3 at branch tips, pale or bright yellow, sometimes partly reddish, keel glabrous, calyx glabrous, lobes pungent. July–Oct. Lowland fynbos, 200–700 m, SW, AP (Hangklip to Bredasdorp).•

brevicarpa (R.Dahlgren) R.Dahlgren Prostrate, often mat-forming shrublet with branches to 1 m. Leaves 3-foliolate, leaflets subterete, usually incurved, apiculate or pungent. Flowers solitary at branch tips, bright yellow, keel glabrous, calyx usually shortly hairy. Oct.–Jan. Subalpine mountain fynbos, 1000–2000 m, NW (Cedarberg to Hex River Mts).•

chenopoda L.   Rigid shrub, 1–2 m. Leaves 3-foliolate, leaflets needle-like, pungent, sparsely and softly hairy. Flowers in terminal heads, bright yellow, keel glabrous, calyx densely woolly, lobes needle-like. Aug.–Dec. Mountain fynbos, 180–260 m, SW (Cape Peninsula to Caledon Swartberg).•

collina Eckl. & Zeyh.   Rigid shrub or shrublet, 0.3–1.3 m. Leaves 3-foliolate, leaflets subterete, pungent. Flowers 1 or 2 at branch tips, yellow, standard sometimes reddish or purplish, keel glabrous, calyx sometimes woolly, lobes pungent. July–Dec. Mountain or lowland fynbos, 200–1500 m, KM, SE (Swartberg to Kouga Mts).•

crassisepala R.Dahlgren   Sprawling shrublet with branches to 1 m. Leaves 3-foliolate, leaflets subterete, usually recurved, mucronate to pungent. Flowers 2–6 at branch tips, pale yellow, often violet or rose on back of standard and bottom of keel, keel glabrous, calyx subglabrous. ?Sept. Lowland fynbos, often limestone soils, 50–280 m, SW, AP, LB (Gansbaai to Mossel Bay).•

cuspidata R.Dahlgren   Like **A. acifera** but keel shortly hairy on lower margin only. Oct.–Dec. Mountain fynbos, karroid scrub, sandveld, 45–1000 m, NW (Klawer).•

divaricata Thunb.   Erect or sprawling shrublet to 1 m. Leaves 3-foliolate, leaflets subterete, mucronate or pungent. Flowers solitary or in terminal heads, pale or bright yellow or partly purplish or violet on back of standard, keel glabrous, calyx glabrous or subglabrous, lobes pungent or needle-like. Oct.–Feb. Mountain fynbos, up to 1300 m, NW, SW (Gifberg to Riviersonderend Mts).•

florifera R.Dahlgren   Like **A. glabrescens** but leaves longer, 20–35 mm, and more densely silky and calyx lobes shorter and pointed. July–Aug. Mountain fynbos, lower slopes, 100–500 m, LB, SE (Langeberg and Outeniqua Mts).•

fourcadei L.Bolus   Like **A. hirta** but keel beaked and silky and calyx lobes very short. ?Sept. Mountain fynbos, 350 m, SE (Tsitsikamma and Kouga Mts).•

glabrescens R.Dahlgren   Shrub to 2 m. Leaves 3-foliolate, leaflets needle-like, silvery silky. Flowers scattered, subsessile, pale lemon-yellow, standard partly violet, keel glabrous, calyx silky. Aug.–Oct. Renosterbos-fynbos scrub, 600–800 m, SE (Outeniqua Mts).•

glossoides R.Dahlgren   Diffuse shrublet to 1 m. Leaves 3-foliolate, leaflets subterete to angular, pungent. Flowers 1 or 2 at branch tips, soon appearing lateral, partly yellow, partly violet, keel silky, with a subapical finger-like process, calyx glabrous, lobes pungent. Oct.–Nov. Mountain fynbos, NW (Piketberg Mt).•

hirta E.Mey.   Shrub, 0.8–2 m. Leaves 3-foliolate, leaflets needle-like or slightly angular, pungent. Flowers scattered, bright yellow, keel shortly hairy below, calyx shortly hairy, lobes needle-like. Sept.–Dec. Mountain fynbos, 250–700 m, NW, SW, KM, LB, SE (Bokkeveld to Kouga Mts).•

horizontalis (R.Dahlgren) R.Dahlgren   Prostrate, mat-forming shrublet with branches to 1 m. Leaves 3-foliolate, leaflets oblong-ovoid, shortly mucronate. Flowers solitary at ends of branchlets, bright yellow or standard slightly violet on back, keel glabrous, boomerang-shaped, calyx thinly adpressed-hairy, lobes pungent. Renosterveld, 200–300 m, SW (Moorreesburg to Blouberg).•

humilis Bolus   Sprawling subshrub to 35 cm. Leaves 3-foliolate, leaflets terete, more or less incurved, mucronate. Flowers in terminal heads, pale yellow, fading red, keel glabrous, calyx thinly hairy. Oct. Mountain fynbos, 600–1100 m, SW (Cape Peninsula and Jonkershoek).•

hystrix L.f.   Like **A. hirta** but leaves grey-hairy. Aug.–Nov. Mountain fynbos or renosterbos-fynbos scrub, 700–1600 m, KM, SE (Witteberg to Willowmore). •

joubertiana Eckl. & Zeyh.   Erect shrublet with densely leafy, wand-like branches, 40–70 cm. Leaves 3-foliolate, leaflets terete, silky. Flowers scattered, bright yellow, keel glabrous, calyx hairy, lobes narrow. Sept. Renosterbos-fynbos scrub, 160–830 m, SW, AP (Bredasdorp Mts to Potberg).•

laeta Bolus   Erect to sprawling shrublet with branches to 1 m long. Leaves 3-foliolate, leaflets awl-like, pungent. Flowers 1 or 2 at branchlet tips, pale yellow, standard and keel often purplish, keel hairy, calyx thinly hairy, lobes pungent. Oct.–Dec. Mountain fynbos, 300 to more than 750 m, NW, SW, KM (Swartruggens to Witteberg).•

lamarckiana R.Dahlgren   Like **A. hystrix** but flowers larger, standard c. 16 mm broad and wing blades c. 5.5 mm broad. Sept.–Dec. Arid fynbos, 660 m, KM (Witteberg and Klein Swartberg).•

leptocoma Eckl. & Zeyh.   Like **A. divaricata** but calyx with short, spreading hairs and wings half as long as keel. Oct.–Feb. Mountain fynbos, below 900 m, NW (Grootwinterhoek to Elandskloof Mts).•

leptoptera Bolus   Prostrate mat-forming shrublet with branches to 1 m. Leaves 3-foliolate, leaflets subterete, mucronate. Flowers 1 or 2 at tips of branchlets, bright yellow, keel silky, calyx densely hairy, lobes incurved-cuspidate. Oct.–Feb. Mountain fynbos, 1500 m, NW (Skurweberg).•

longifolia Benth.   Like **A. glabrescens** but leaflets often to 40 mm long and calyx lobes much reduced. Aug. Mountain fynbos or renosterbos-fynbos scrub, 400 m, LB (Langeberg Mts: Garcia's Pass).•

neglecta T.M.Salter   Shrub, 0.7–2 m. Leaves 3-foliolate, leaflets angular, pungent, sparsely hairy. Flowers crowded at branch tips, bright yellow-orange or ferruginous, keel usually sparsely hairy, calyx sparsely silky, lobes needle-like. Sept.–Dec. Mountain fynbos, lower slopes, 240–830 m, SW (Cape Peninsula, Wemmershoek to Hottentots Holland Mts).•

opaca Eckl. & Zeyh.   Erect or sprawling shrublet to 60 cm. Leaves 3-foliolate, leaflets needle-like, acute, leaf bases sometimes spurred. Flowers scattered, bright yellow or brownish red, keel hairy, calyx shortly hairy, lobes needle-like. Jan.–Dec. Renosterbos-fynbos scrub, 300–600 m, SW, LB, SE (Caledon to Alexandria).

potbergensis R.Dahlgren   Prostrate, densely branched, matted shrublet with branches to 60 cm, like **A. batodes** but leaves glaucous and flowers smaller, keel blades 3–3.5 mm long. Dec. Mountain fynbos, 400–500 m, SW (Potberg).•

proboscidea R.Dahlgren   Like **A. glossoides** but more procumbent, to 50 cm, with mucronulate leaves and puberulous calyx. Sept.–Oct. Mountain fynbos, 700–800 m, NW (Bokkeveld Mts).•

pulicifolia R.Dahlgren   Like **A. acuminata** but wings very short, to half as long as boomerang-shaped keel. Sept.–Nov. Mountain fynbos, 160 m, NW (Namaqualand to Gifberg).

rigidifolia R.Dahlgren   Erect or sprawling shrublet to 1 m. Leaves 3-foliolate, often tufted, leaflets needle-like, pungent. Flowers 1 or 2 on short shoots, pale yellow or partly purple, keel subglabrous, calyx glabrous. Nov.–Dec. Mountain fynbos, 780–1900 m, NW, KM (Cold Bokkeveld to Swartberg Mts).•

rostripetala R.Dahlgren   Sprawling shrublet, 0.3–1 m. Leaves 3-foliolate, leaflets subterete to angular, pungent. Flowers 1–3 at branch tips, pale yellow, fading brown, keel silky, strongly beaked, calyx subglabrous. Oct.–Nov. Mountain fynbos, 200–300 m, NW (Olifants River Mts).•

sceptrum-aureum R.Dahlgren   Sparsely branched, densely leafy shrub to 2.5 m, with rod-like branches. Leaves 3-foliolate, leaflets terete, pungent. Flowers scattered, yellow, keel glabrous, calyx shortly velvety, lobes short and blunt. July–Nov. Renosterbos-fynbos scrub, 710–1500 m, KM, SE (Swartberg, Rooiberg and Outeniqua Mts).•

secunda E.Mey.   Like **A. acuminata** but standard silky on front as well as behind. Sept. Lowland fynbos, 250 m, SW (Riebeek-Kasteel).•

serpens R.Dahlgren   Procumbent shrublet with creeping branches to 1.5 m. Leaves 3-foliolate, leaflets subterete-angular, pungent. Flowers solitary on wiry peduncles from spur-shoots, pale to bright yellow, keel glabrous, calyx glabrous, lobes pungent. Sept.–Feb. Mountain fynbos, lower slopes, 150–660 m, SW (Cape Peninsula to Bredasdorp).•

setacea Eckl. & Zeyh.   Shrub, 0.25–2 m. Leaves 3-foliolate, leaflets needle-like, pungent, glabrous or hairy. Flowers several near branch tips, pale to bright yellow, often fading brownish, keel silky, calyx shortly hairy, lobes pungent. July–Mar. Mountain fynbos or grassy fynbos, lower slopes, 330–830 m, SE (Knysna to E Cape).

shawii L.Bolus   Sparsely branched shrub with densely leafy, wand-like branches, 0.5–1.2 m. Leaves 3-foliolate, leaflets needle-like, often sigmoid, pungent, usually grey-hairy. Flowers scattered or crowded near branch tips, pale to bright yellow, keel glabrous, calyx silky, lobes usually pungent. May–Dec. Mountain fynbos, 700–1500 m, NW, SW, KM (Bokkeveld Mts to Willowmore).•

spectabilis R.Dahlgren   Like **A. hirta** but flowers larger, wing and keel blades 13–17 mm long. Oct.–Nov. Mountain fynbos, 150–300 m, LB (Langeberg and Aasvogelberg Mts).•

spiculata R.Dahlgren   Like **A. collina** but flowers smaller, wing blades to 2 mm broad. Sept.–Oct. Mountain fynbos, 600–1150 m, KM (Witteberg and Waboomsberg).•

spinosissima R.Dahlgren   Erect or sprawling shrub, 0.3–1 m. Leaves 3-foliolate, leaflets subterete-angular, pungent. Flowers 1–3 on slender axillary peduncles, pale or bright yellow, sometimes reddish on back of standard, keel silky, calyx glabrous, lobes pungent. Oct.–Feb. Mountain fynbos, 500–2000 m, NW, SW (Cedarberg to Stellenbosch Mts).•

stricticlada (R.Dahlgren) R.Dahlgren   Like **A. acifera** but calyx shortly hairy. Feb. Renosterbos-fynbos scrub, SW (Malmesbury and Paardeberg).•

subulata Thunb.   Shrublet to 70 cm. Leaves 3-foliolate, leaflets needle-like, subterete, pungent. Flowers in a terminal cluster, yellow, fading orange, standard purplish on back, keel glabrous, calyx hairy. Sept.–Nov. Mountain fynbos, lower slopes, up to 600 m, SW (Groenland Mts and Caledon Swartberg).•

teres Eckl. & Zeyh.   Shrub or small tree, 1–5 m. Leaves 3-foliolate, leaflets needle-like, pungent. Flowers scattered, yellow, fading brown, keel shortly hairy, calyx shortly hairy, lobes needle-like. Oct.–June. Mountain or lowland fynbos, low alt., below 700 m, SE (Knysna to E Cape).

tulbaghensis R.Dahlgren   Like **A. acuminata** but flowers slightly larger, wing blades longer than 10 mm and keel blades longer than 9 mm. Nov. Renosterveld-fynbos scrub, SW (Wolseley).•

ulicina Eckl. & Zeyh.   Shrub, 0.2–1.5 m. Leaves 3-foliolate, leaflets subterete-angular, pungent. Flowers 1–8 in terminal racemes, pale yellow, standard often partly violet on back, keel silky, calyx glabrous, lobes pungent. Sept.–Nov. Mountain fynbos, 200–700 m, NW, SW (Olifants River to Elandskloof Mts).•

verbasciformis R.Dahlgren   Like **A. sceptrum-aureum** but calyx woolly with tapering lobes. Oct. Mountain fynbos, 800 m, NW, LB (Langeberg Mts: Montagu to Swellendam).•

vulnerans Thunb.   Shrub or small tree, 1–3 m. Leaves 3-foliolate, leaflets terete-angular, pungent. Flowers in short racemes or heads on lateral branchlets, bright yellow, keel shortly hairy, calyx glabrous, lobes pungent. Sept.–Dec. Mountain fynbos, 330–830 m, NW (Bokkeveld to Olifants River Mts).•

*G. Leaflets not spine-tipped (see also **A. grandiflora**)*

acanthoclada R.Dahlgren   Shrub to 1 m, with wand-like branches ending in thorns. Leaves 3-foliolate, leaflets ovoid-cylindric. Flowers solitary on short peduncles, pale lemon-yellow, standard violet, keel shortly silky, calyx shortly hairy, lobes more or less incurved. Jan.–Mar. Renosterbos-fynbos scrub, 300 m, SW (Breede River valley).•

acanthophylla Eckl. & Zeyh.   Like **A. spinosa** but leaves and calyx softly grey-hairy. Sept.–Feb. Renosterveld, 200–300 m, SW (Hopefield to Tygerberg).•

aciloba R.Dahlgren   Like **A. calcarea** but flowers larger, standard blade 6.5–8 mm long. ?Sept. Lowland fynbos, limestone, AP (Bredasdorp to Agulhas).•

acutiflora R.Dahlgren   Erect or sprawling shrublet to 1 m. Leaves 3-foliolate, leaflets terete. Flowers scattered, pale yellow with purple standard midrib, keel glabrous, calyx shortly hairy, lobes awl-like. Oct.–Dec. Coastal fynbos, below 50 m, AP (Stilbaai and Albertinia).•

amoena (R.Dahlgren) R.Dahlgren   Erect, closely branched shrublet, 30–70 cm. Leaves 3-foliolate, leaflets fasciculate, subterete. Flowers solitary on slender peduncles, pale yellow, keel glabrous, calyx shortly white-hairy. ?Sept. Lowland fynbos, riverbanks, 300–1000 m, NW, SW (Breede River valley: Wolseley).•

araneosa L.   Like **A. ciliaris** but calyx lobes narrower, long and linear-filiform and flowers remaining yellow. Aug.–Nov. Lowland fynbos, 200–830 m, SW (Malmesbury to Cape Peninsula).•

arenaria R.Dahlgren   Robust shrub, 1–1.7 m. Leaves 3-foliolate, leaflets subterete. Flowers scattered, bright yellow, keel beaked, glabrous, calyx glabrous. Sept.–Nov. Coastal fynbos, 20–140 m, AP (Stilbaai to Gouritsmond).•

arida E.Mey.   Erect or spreading shrub to 1.5 m. Leaves 3-foliolate, leaflets subterete or slightly depressed. Flowers scattered, yellow, or standard partly red or purple, wings hardly wrinkled, keel glabrous, calyx compressed, glabrescent . Oct.–Jan. Mountain fynbos, 30–1320 m, NW, SW (Bokkeveld Mts to Bredasdorp).•

asparagoides L.f.   Erect or spreading shrublet, 0.5–1 m. Leaves 3-foliolate, leaflets subterete-angular, glabrous to tubercled-hairy. Flowers scattered, yellow, fading rusty or dark red, keel glabrous or hairy, calyx hairy, lobes filiform. Sept.–Feb. Lowland fynbos, sometimes renosterbos-fynbos scrub, 10–650 m, SW, LB, SE (Gansbaai to Alexandria).

attenuata R.Dahlgren   Sprawling, sparsely branched shrublet with branches 30–80 cm. Leaves 3-foliolate, leaflets subterete, mucronulate. Flowers solitary at branch tips, pale yellow or standard partly violet, keel beaked, glabrous, calyx subglabrous, lobes pungent. Oct.–Feb. Mountain fynbos, lower slopes, 100–350 m, SW (Bainskloof to Houwhoek).•

aurantiaca R.Dahlgren   Like **A. biflora** but flowers larger, standard blade 9–10 mm long, bright yellow, fading orange or orange-red. Oct.–Jan. Mountain fynbos, up to 1700 m, NW, SW (Cedarberg to Du Toitskloof Mts).•

biflora E.Mey.   Sprawling shrublet. Leaves 3-foliolate, leaflets terete, mucronulate. Flowers 1–5 on terminal peduncles, pale yellow, orange or rusty, keel glabrous, sometimes hairy at tip, calyx glabrous, lobes small and triangular. Aug.–Apr. Mountain fynbos, lower slopes and hills, 1000–1200 m, SW, LB, SE (Cape Peninsula to E Cape).

bowieana (Benth.) R.Dahlgren   Erect shrub to over 1 m. Leaves 3-foliolate, leaflets terete, slender, silvery velvety. Flowers scattered, pale yellow, keel glabrous, calyx velvety. Sept. Mountain fynbos, slopes and foothills, below 850 m, SE (Outeniqua Mts).•

bracteata Thunb.   Procumbent or scrambling, diffusely branched shrublet. Leaves 3-foliolate, leaflets linear-terete, glabrous or more or less hairy. Flowers 1–3 at branch tips on a naked peduncle, pale or bright yellow or back of standard violet, keel glabrous, calyx sparsely adpressed-hairy. Sept.–Apr. Mountain fynbos, middle slopes to subalpine habitats, up to 1500 m, NW, SW (Cedarberg Mts to Caledon Swartberg).•

burchelliana Benth.   Shrublet to 80 cm. Leaves 3-foliolate, leaflets subterete to slightly flattened, slender, subglabrous. Flowers scattered, pale yellow, keel silky, calyx densely hairy. Sept.–Jan. Renosterbos-fynbos scrub, low alt., 150–300 m, LB (Riviersonderend to Swellendam).•

caespitosa R.Dahlgren   Like **A. asparagoides** but more compact, prostrate or cushion-like, wing and keel claws longer, 3.5–4 mm long. Nov. Subalpine mountain fynbos, summit and upper slopes, 1250–1500 m, SW (Du Toit's Peak and Stettynsberg).•

calcarea R.Dahlgren   Decumbent, ascending or erect shrublet to 60 cm. Leaves 3-foliolate, leaflets subterete. Flowers scattered, pale yellow, keel glabrous, calyx shortly hairy, lobes needle-like. Jan.–May. Lowland fynbos, limestone, 20–160 m, AP (Elim to Stilbaai).•

campestris R.Dahlgren   Erect or spreading shrublet, 15–35 cm. Leaves 3-foliolate, leaflets fasciculate, subterete-angular. Flowers scattered, pale yellow, fading brownish purple, keel glabrous, calyx subglabrous, lobes long. Dec.–Mar. Renosterbos-fynbos scrub, 40–50 m, NW, SW, LB (Montagu and Riviersonderend to Mossel Bay).•

candicans W.T.Aiton   Erect shrub to 1.5 m. Leaves 3-foliolate, leaflets subterete or slightly flattened, shortly white-hairy. Flowers densely scattered, pale to lemon-yellow, keel glabrous, calyx grey-hairy. Sept.–Nov. Renosterbos-fynbos scrub, 230–580 m, SW (Breede River valley: Worcester to Bonnievale).•

candidula R.Dahlgren   Like **A. calcarea** but leaves shortly silky or hairy. Nov.–Dec. Lowland fynbos, limestone, 165 m, AP (Stilbaai).•

capensis (Walp.) R.Dahlgren   Like **A. carnosa** but flowers larger, wing blades 16–18 mm long. July–Dec. Lowland fynbos, 50–400 m, SW (Cape Peninsula).•

capitata L.   Like **A. carnosa** but leaves sparsely hairy and keel rostrate. Dec.–July. Lowland and mountain fynbos, 30–950 m, SW (Cape Peninsula).•

carnosa P.J.Bergius   Much-branched shrub, 0.6–2.5 m. Leaves 3-foliolate, leaflets terete or subterete. Flowers in terminal heads, sometimes scattered, bright yellow, standard partly red, keel glabrous, calyx fleshy, glabrous or subglabrous, lobes more or less orbicular; bract adnate to pedicel. Aug.–Dec. Lowland fynbos, 10–660 m, SW (Cape Peninsula and Hangklip Mts).•

chortophila Eckl. & Zeyh.   Erect or sprawling shrublet to 1 m. Leaves 3-foliolate, leaflets subterete, glabrous or subglabrous. Flowers scattered, bright yellow or partly orange, red or purplish, keel hairy, calyx silky, lobes needle-like. Oct.–Dec. Mountain fynbos or grassy fynbos, 600–1300 m, SE (Uitenhage to KwaZulu-Natal).

chrysantha R.Dahlgren   Like **A. juniperina** but bracts larger, 3.5–5 mm long and wing blades with few folds below. Oct. Mountain fynbos, above 450 m, NW (Piketberg).•

ciliaris L.   Erect or sprawling shrub or shrublet, 0.1–1.5 m. Leaves 3-foliolate, leaflets linear-subulate, glabrous or softly tuberculate-hairy. Flowers in terminal heads, bright yellow, fading reddish or black or brown, keel glabrous or silky, calyx woolly, lobes narrow, tuberculate-hairy. Sept.–Feb. Lowland fynbos, below 800 m, NW, SW, LB, SE (Clanwilliam to Humansdorp).•

cinerascens E.Mey.   Shrub, usually 1–2 m. Leaves 3-foliolate, leaflets terete, silvery silky. Flowers scattered, pale yellow, keel glabrous, calyx silky. Jan.–Apr. Lowland fynbos, 50–200 m, SE (Uitenhage to E Cape).

citrina R.Dahlgren   Shrublet, 20–70 cm. Leaves 3-foliolate, leaflets subterete or slightly angular. Flowers scattered, bright lemon- or sulphur-yellow, keel glabrous, calyx compressed, glabrous, fleshy, lobes large and broad below. Oct.–Jan. Mountain fynbos, lower slopes, 500 m, SW, AP (Houwhoek to Agulhas).•

cliffortiifolia R.Dahlgren   Ascending, sparsely branched shrublet, 20–50 cm. Leaves 3-foliolate, leaflets terete. Flowers scattered, yellow, keel shortly hairy, calyx velvety. ?Sept. Coastal fynbos, SE (Port Elizabeth).•

cliffortioides Bolus   Like **A. ciliaris** but flowers smaller, wing blades to 7 mm long. Oct.–Jan. Mountain fynbos, lower and upper slopes, 100–1250 m, SW, LB (Stettynsberg to Langeberg Mts and Potberg).•

concava Bolus   Like **A. juniperina** but leaflets smaller, ovoid, calyx sometimes shortly hairy, lobes ovoid-globose, fleshy, c. 1 mm long. ?Sept. Mountain fynbos, 800 m, SW (Houwhoek).•

condensata R.Dahlgren   Like **A. juniperina** but standard not incurved along the sides. ?Sept. Mountain fynbos, 1000–2000 m, SW (Slanghoek Mts to Somerset Sneeuwkop).•

confusa R.Dahlgren   Decumbent or ascending, sparsely branched shrublet to 25 cm. Leaves 3-foliolate, leaflets linear-terete, sparsely long-hairy. Flowers few in a terminal head, pale yellow, keel glabrous, calyx white-silky, lobes thread-like. Oct.–Dec. Mountain fynbos, 700–800 m, NW (Clanwilliam to Ceres and Tulbagh, Ceres and Clanwilliam).•

congesta (R.Dahlgren) R.Dahlgren   Dense sprawling shrublet. Leaves 3-foliolate, leaflets subterete. Flowers scattered, yellow, keel shortly silky, calyx softly hairy, lobes needle-like. Oct.–Jan. Subalpine fynbos, 1250–1750 m, KM (Swartberg Mts).•

cordicarpa R.Dahlgren   Like **A. juniperina** but fruits cordate, shortly hairy. ?Sept. Mountain fynbos, 400–600 m, LB (Langeberg Mts: Garcia's Pass).•

corrudifolia P.J.Bergius   Slender erect shrub, 1–3 m. Leaves 3-foliolate, leaflets subterete-filiform, glabrous or subglabrous. Flowers in a terminal raceme, bright yellow, keel glabrous or thinly adpressed-hairy, calyx glabrous, lobes pungent. Oct.–Dec. Mountain fynbos, 100–1000 m, NW, SW (Grootwinterhoek Mts to Robertson).•

cymbiformis DC.   Sprawling densely branched shrublet, 15–40 cm. Leaves 3-foliolate, leaflets subterete-angular, glabrous to hairy. Flowers 1 or 2 at branch tips, bright yellow, often with violet shades on back of standard, keel glabrous, calyx hairy, lobes ovate-concave. Oct.–Nov. Mountain fynbos, 30–500 m, NW, SW (Piketberg to Bredasdorp).•

decora R.Dahlgren Shrub 1–4 m. Leaves 3-foliolate, leaflets filiform, thinly hairy. Flowers in terminal heads, pale yellow, keel glabrous, calyx hairy, lobes slender. Nov.–Dec. Mountain fynbos, 1000–1350 m, NW (Cedarberg Mts).•

densifolia Benth.   Like **A. triquetra** but leaves linear-filiform, longer, longest leaflets 6–12 mm long and standard usually more than 13.5 mm across. Sept.–Dec. Mountain fynbos, 500–1000 m, NW (Witzenberg and Waaihoek Mts).•

dianthopora E.Phillips   Erect shrub, 0.6–2 m. Leaves 3-foliolate, leaflets subterete or slightly canaliculate. Flowers 1–3 on lateral peduncles, pale or bright yellow, or partly pale purplish, keel glabrous, calyx glabrous. Sept.–Oct. Mountain fynbos or fynbos-karoo transition, 330–1100 m, NW (Bokkeveld Mts to Hex River Mts).•

digitifolia R.Dahlgren   Like **A. rubens** but keel and wings glabrous and standard subglabrous. Aug.–Sept. Mountain fynbos, 530–1000 m, SE (Outeniqua Mts near Mossel Bay).•

empetrifolia (R.Dahlgren) R.Dahlgren   Like **A. retroflexa** but leaflets c. 2.5 mm long and flowers smaller, standard blade to 5.7 mm long. ?Sept. Mountain fynbos, 1500–2000 m, NW (Cold Bokkeveld Mts).•

ericifolia L.   Erect or sprawling shrublet, 20–60 cm. Leaves 3-foliolate, leaflets terete-ovoid, glabrous or hairy. Flowers scattered, pale or bright yellow, keel hairy, sometimes glabrous, calyx shortly hairy, lobes slender-filiform. Sept.–Nov. Mountain and lowland fynbos, up to 800 m, NW, SW, AP (Porterville to Agulhas).•

excelsa R.Dahlgren   Like **A. carnosa** but bracts rhombic-ovate with slightly dentate margins, pedicels longer, 5–10 mm long, and keel rostrate. Sept.–Nov. Lowland fynbos, 50 m, SW (Hermanus).•

ferox Harv.   Rigid shrub or shrublet, thorny. Leaves 3-foliolate, leaflets subterete or slightly flattened. Flowers scattered, pale yellow, keel glabrous, calyx shortly hairy, lobes reduced. Oct. Renosterveld, SW (Breede River valley).•

filicaulis Eckl. & Zeyh.   Shrublet to 70 cm. Leaves 3-foliolate, leaflets subterete to slightly flattened. Flowers scattered or at branch tips, yellow or back of standard more or less purplish, keel glabrous, calyx shortly hairy, lobes slender. Pod linear, many-seeded. Nov.–Jan. Lowland and mountain fynbos, marshy areas, 20–1660 m, NW, SW (Cedarberg to Hottentots Holland Mts).•

flexuosa Thunb.   Like **A. ericifolia** but keel glabrous, fading orange, ferruginous or brown-purple. Aug.–Nov. Renosterbos-fynbos scrub, 80–300 m, NW, SW, LB (Nardouw Mts to Riversdale).•

florulenta R.Dahlgren   Decumbent shrublet to 50 cm. Leaves 3-foliolate, leaflets subterete, glabrous or sparsely hairy. Flowers scattered, pale yellow, standard more or less violet, keel glabrous, calyx sparsely hairy, lobes slender. Oct.–Dec. Mountain fynbos, NW (Bokkeveld to Nardouw Mts).•

galeata E.Mey.   Erect shrublet, 20–60 cm. Leaves 3-foliolate, leaflets linear-angular, subglabrous. Flowers 1–3 at branch tips, yellow, sometimes reddish on back of standard, keel glabrous, calyx sparsely silky, lobes needle-like. Sept.–Jan. Mountain fynbos, 300–900 m, NW (Cedarberg and Olifants River Mts).•

glabrata R.Dahlgren   Like **A. spinosa** but pedicels mostly glabrous and calyx always glabrous. Sept.–Dec. Renosterbos-fynbos scrub, SW (Darling).•

granulata R.Dahlgren   Erect shrublet, 0.2–1 m. Leaves 3-foliolate, leaflets succulent, ovoid-oblong. Flowers scattered, pale yellow, keel glabrous, calyx thinly woolly. Aug.–Dec. Mountain fynbos, 430–920 m, KM, LB (Langeberg, Outeniqua and Rooiberg Mts).•

grobleri R.Dahlgren   Prostrate, mat-forming shrublet with branches to 50 cm long. Leaves 3-foliolate, leaflets terete, thinly hairy, often purple. Flowers scattered, pale yellow or more or less purplish, fading dark purple, keel glabrous, calyx silky, lobes filiform. Sept. Renosterbos-fynbos scrub, LB (Swellendam).•

hispida Thunb.   Like **A. ericifolia** but bracts reduced and much smaller than leaves and flowers pale yellow or nearly white with keel apex and back of standard often dull violet. Sept.–Jan. Renosterbos-fynbos scrub or coastal fynbos, 100–1230 m, NW, SW, AP, LB, SE (Gifberg to Alexandria).

hypnoides R.Dahlgren   Like **A. juniperina** but leaflets very slender, c. 0.2 mm diam. and fruits ovate, glabrous and glossy. ?Sept. Mountain fynbos, lower slopes, LB (Langeberg Mts: Swellendam).•

incompta Thunb.   Like **A. pycnantha** but sprawling to prostrate with smaller flowers, standard blade to 4.5 mm long and wings about half as long as keel. Aug.–Sept. Renosterbos-fynbos scrub, 60–300 m, SW (Riviersonderend to Stormsvlei).•

incurva Thunb.   Like **A. linguiloba** but flowers larger, calyx lobes usually longer than 6 mm and keel blades usually longer than 8 mm. Nov.–Mar. Mountain fynbos, 600 m. SW (Cape Peninsula).•

incurvifolia Vogel ex Walp.   Shrublet with densely leafy, brush-like branches to 1 m. Leaves 3-foliolate, leaflets terete-filiform. Flowers scattered, pale yellow, keel glabrous, calyx glabrous to shortly hairy, lobes reduced. July–Sept. Lowland fynbos on limestone, 50–160 m, AP (Gansbaai to Gouritsmond).•

intermedia Eckl. & Zeyh.   Like **A. chortophila** but flowers smaller, wing and keel blades shorter than 5 mm. Dec.–Mar. Lowland fynbos, plains and lower slopes, 130–1000 m, SE (Humansdorp to East London).

intricata Compton   Like **A. wittebergensis** but branches becoming thorny and flowers mainly light yellow. Sept.–Dec. Mountain fynbos, 1000–1500 m, NW, KM (Cold Bokkeveld Mts to Witteberg).•

isolata (R.Dahlgren) R.Dahlgren   Like **A. varians** but flowers smaller, calyx lobes to 2.5 mm long and standard blade to 6.3 mm long. Nov. Mountain fynbos, 800 m, NW (Bokkeveld Mts).•

juniperina Thunb.   Sprawling or prostrate shrublet, sometimes mat-forming, with branches 0.2–2 m long. Leaves 3-foliolate, leaflets subterete. Flowers 1–3 at branch tips, pale or bright yellow, sometimes partly reddish, standard incurved on the sides, keel glabrous, calyx glabrous. Aug.–Sept. Mountain fynbos, 260–2000 m, NW, SW, AP, KM (Piketberg to Agulhas and Klein Swartberg).•

karrooensis R.Dahlgren   Shrublet, 50–80 cm. Leaves 3-foliolate, leaflets subterete, densely white-woolly. Flowers scattered, pale yellow, fading purplish, keel purplish, glabrous, calyx densely white-woolly, lobes slender. ?Sept. Mountain fynbos, KM (Rooiberg).•

keeromsbergensis R.Dahlgren   Shrublet to 30 cm. Leaves 3-foliolate, leaflets terete, shortly hairy. Flowers solitary at branchlet tips, pale yellow, keel glabrous, calyx hairy, lobes finger-like. ?Sept. Mountain fynbos, 500–740 m, NW (Keeromsberg).•

kougaensis (Garab. ex R.Dahlgren) R.Dahlgren   Like **A. chortophila** but leaves grey-hairy and flowers pale lemon-yellow. Aug.–Mar. Mountain fynbos, 200–750 m, KM, SE (Swartberg Mts to Humansdorp).•

lactea Thunb.   Erect shrublet with rigid branches becoming thorny, 0.3–1 m. Leaves 3-foliolate, leaflets subterete. Flowers scattered, pale yellow, often partly purplish, keel glabrous, calyx shortly hairy, sometimes glabrous, lobes reduced. Sept.–Mar. Renosterveld, below 1000 m, NW, SW (Worcester to Barrydale).•

lanceicarpa R.Dahlgren   Sprawling shrublet with branches to 80 cm. Leaves 3-foliolate, leaflets terete. Flowers solitary on slender, peduncle-like stalks, yellow to ferruginous or partly red, keel glabrous, calyx adpressed-hairy. Apr.–May. Mountain fynbos, SE (Van Staden's Mts).•

laricifolia P.J.Bergius   Sprawling or procumbent shrublet to 50 cm. Leaves 3-foliolate, leaflets subterete, glabrous to densely silvery hairy. Flowers scattered, pale or bright yellow, keel glabrous, calyx densely hairy. Sept.–Dec. Mountain fynbos, lower slopes, below 1000 m, SW, LB, SE (Bainskloof to George).•

lenticula Bolus   Like **A. attenuata** but leaves adpressed. Dec.–Jan. Renosterveld-fynbos scrub, 230 m, NW (Tulbagh: Saronsberg to Nieuwe Kloof).•

leucophylla R.Dahlgren   Erect shrublet, 0.5–1.5 m. Leaves 3-foliolate, leaflets ovoid-oblong, minutely grey- or white-hairy. Flowers scattered, pale yellow, keel glabrous, calyx shortly hairy. Aug.–Dec. Renosterbos-fynbos scrub, lower slopes and foothills, 330–1160 m, NW, KM, SE (Cold Bokkeveld to Kammanassie Mts).•

linguiloba R.Dahlgren   Sprawling shrublet, rarely higher than 30 cm. Leaves 3-foliolate, leaflets subterete-angular. Flowers several at branch tips, pale yellow or more or less purplish, keel glabrous, calyx hairy, lobes large. Nov.–Jan. Coastal fynbos, 30–800 m, SW, AP (Blouberg to Potberg).•

longipes Harv.   Shrublet, 30–80 cm. Leaves 3-foliolate, leaflets terete, grey- or white-hairy. Flowers 1–4 on slender peduncle-like branchlets, pale yellow, keel and standard often partly purplish, keel glabrous, calyx hairy, lobes slender. Sept.–Dec. Lowland fynbos, 35–100 m, SW, LB (Riviersonderend to Riversdale).•

marginalis Eckl. & Zeyh.   Shrublet, 15–70 cm. Leaves 3-foliolate, leaflets terete. Flowers solitary at the branch tips, yellow, keel glabrous, calyx shortly hairy, lobes with thickened, fleshy margins. ?Sept. Mountain fynbos, 600–700 m, SE (Uitenhage to Albany).

microphylla DC.   Like **A. lenticula** but keel truncate. Oct.–Feb. Lowland or coastal fynbos, 30–830 m, SW, AP, LB (Bokbaai and Cape Peninsula to Agulhas and Swellendam).•

millefolia R.Dahlgren   Like **A. ciliaris** but leaves minute, to 4 mm long. Oct.–Jan. Renosterbos-fynbos scrub, 830 m, SW, LB (Houwhoek to Riversdale).•

muraltioides Eckl. & Zeyh.   Like **A. hispida** but leaves densely white-hairy and bracteoles vestigial or lacking. Aug.–Nov. Renosterveld or mountain fynbos, below 260 m, NW, SW (Piketberg to Somerset West).•

nivea Thunb.   Much- branched shrub, 1–3 m. Leaves 3-foliolate, leaflets terete, densely silvery silky. Flowers solitary on slender peduncles, pale yellow, standard partly violet, keel hairy, calyx silvery silky. Aug.–Apr. Renosterbos-fynbos scrub, low alt., 80–430 m, SE (Uniondale to Uitenhage).•

obliqua R.Dahlgren   Erect shrub or shrublet. Leaves 3-foliolate, leaflets subterete. Flowers scattered, ?pale yellow, keel glabrous, calyx glabrous, lobes reduced. Mountain fynbos, 600–700 m, NW (Bokkeveld Mts).•

obtusifolia R.Dahlgren   Erect shrub to 1.2 m. Leaves 3-foliolate, leaflets cylindrical-ovoid. Flowers scattered, bright yellow, keel glabrous, calyx glabrous, lobes reduced. Dec.–Mar. Lowland fynbos to 130 m, AP (Stilbaai to Mossel Bay).•

odontoloba R.Dahlgren   Like **A. pinguis** but leaflets always cylindrical-ovoid and calyx lobes ovoid, fleshy. Jan.–Feb. Lowland fynbos, 10 m, AP (Albertinia).•

oliveri R.Dahlgren   Like **A. amoena** but smaller and more succulent, 1–2 mm long. ?Nov. Subalpine fynbos, 1250 m, KM (Groot Swartberg: Tierberg).•

pachyloba Benth.   Shrub or shrublet, 0.4–2 m. Leaves 3-foliolate, leaflets subterete, often oblong-ovoid, sometimes pungent. Flowers scattered, bright yellow, fading black when dried, keel glabrous, calyx glabrous or thinly hairy, lobes sometimes pungent. Oct.–Dec. Mountain fynbos, 200–1500 m, NW, SW, KM, LB (Cold Bokkeveld Mts to Caledon Swartberg, Langeberg and Swartberg Mts).•

pallescens Eckl. & Zeyh.   Erect, rigid shrub, 0.5–1.2 m. Leaves 3-foliolate, on prominent tubercles, leaflets subterete, subglabrous. Flowers scattered, pale yellow or cream-white, keel glabrous, calyx shortly hairy, lobes overlapping. Mar.–Apr. Lowland fynbos, limestone ridges, 200 m, AP (De Hoop).•

parviflora P.J.Bergius   Like **A. ericifolia** but bracts and bracteoles lacking. Sept.–Oct. Renosterbos-fynbos scrub, 30–280 m, NW, SW, AP (Piketberg to Agulhas).•

pedicellata Harv.   Sprawling or mat-forming shrublet. Leaves 3-foliolate, leaflets subterete or slightly compressed. Flowers 1–5 at branch tips, pale to bright yellow or back of standard partly purple, keel glabrous, calyx glabrous. Nov.–Jan. Subalpine mountain fynbos or renosterbos-fynbos scrub, 1000–2200 m, NW (Cold Bokkeveld to Hex River Mts).•

pedunculata Houtt.   Erect, much-branched shrub, 1–2 m. Leaves 3-foliolate, leaflets terete or very slightly flattened, densely velvety. Flowers solitary at branchlet tips, pale yellow, keel hairy, calyx shortly hairy. July–Dec. Renosterbos-fynbos, 500–1000 m, SE (Outeniqua Mts).•

petersonii R.Dahlgren   Shrub, 0.6–2 m. Leaves 3-foliolate, leaflets subterete, glabrous to thickly hairy. Flowers scattered, pale yellow, keel glabrous, calyx shortly hairy. Sept.–Feb. Dry fynbos, 200–1090 m, NW (Bokkeveld Mts to Swartruggens).•

pigmentosa R.Dahlgren   Like **A. araneosa** but keel glabrous and smaller, to 6 mm long, standard sometimes purplish. Oct.–Nov. Mountain fynbos, 250–1000 m, NW, SW (Grootwinterhoek to Groenland Mts).•

pilantha R.Dahlgren   Shrublet, 30–60 cm. Leaves 3-foliolate, leaflets terete-ovoid. Flowers solitary at branchlet tips, pale yellow, partly reddish purple, keel shortly hairy, calyx shortly white-hairy. Nov. Mountain fynbos, 700 m, NW (Matroosberg).•

pinea Thunb.   Erect shrub or shrublet with tail- or rod-like branches, 0.2–2 m. Leaves 3-foliolate, leaflets piniform, thinly hairy, on tubercle-like bases. Flowers in false spikes, pale yellow, keel rostrate, glabrous, calyx thinly hairy. Oct.–Dec. Mountain fynbos, 130–1000 m, NW, SW (Cedarberg to Stellenbosch Mts).•

pinguis Thunb.   Sparsely branched shrublet with erect, tail-like branches, 0.2–1 m. Leaves 3-foliolate, leaflets subterete to ovoid. Flowers bright scattered, yellow or partly red, keel glabrous, calyx glabrous. Aug.–Feb. Renosterbos-fynbos scrub, 100–1000 m, SW, KM, SE (Olifants River Mts to Uniondale).•

prostrata Eckl. & Zeyh.   Prostrate, mat-forming shrublet with branches to 50 cm. Leaves 3-foliolate, leaflets subterete or slightly flattened, thinly hairy. Flowers scattered, yellow, keel glabrous, calyx thinly hairy and prominently ridged, lobes ovate, overlapping. Nov.–Dec. Lowland fynbos, limestone, 50–100 m, AP (Arniston).•

puberula (Eckl. & Zeyh.) R.Dahlgren   Like **A. hispida** but flowers light bright yellow. Oct. Renosterveld, 300 m, SW (Moorreesburg to Stellenbosch).•

pycnantha R.Dahlgren   Shrublet to 50 cm. Leaves 3-foliolate, leaflets subterete, shortly hairy, sometimes glabrous. Flowers scattered, pale or bright yellow, standard partly purple, keel glabrous, calyx silky. lobes awl-like. July–Oct. Renosterbos-fynbos scrub, 10–20 m, SW, AP (Baardskeerdersbos to Agulhas).•

ramosissima R.Dahlgren   Like **A. digitifolia** but standard also entirely glabrous, flowers white or yellow. Nov. Mountain fynbos, 1300–1700 m, KM (Groot Swartberg).•

rectistyla R.Dahlgren   Rigid shrublet with almost thorn-like branches, 20–60 cm. Leaves 3-foliolate, leaflets needle-like, midrib hard and often persisting. Flowers scattered near branch tips, yellow to orange-yellow fading brown or purple, keel narrow and prolonged, glabrous, style straight and protruding, calyx adpressed-hairy, upper 2 lobes well separated. Aug.–Oct. Renosterbos-fynbos scrub, 200–850 m, NW (Piketberg and Olifants River Mts).•

recurva Benth.   Like **A. wurmbeana** but flowers larger, keel blades longer than 5.5 mm. Dec.–May. Renosterbos-fynbos scrub, 40–300 m, NW, SW, LB (Piketberg to Swellendam).•

recurvispina R.Dahlgren   Like **A. obtusifolia** but leaflets apically recurved and apiculate. Oct.–Apr. Coastal fynbos, below 100 m, SE (Port Elizabeth).•

repens R.Dahlgren   Prostrate, matted shrublet with branches to 50 cm, like **A. juniperina** but calyx lobes ovoid-fleshy. ?Oct. Lowland fynbos, limestone, 90 m, AP (Bredasdorp).•

retroflexa L.   Decumbent or sprawling shrublet with branches to 1 m long. Leaves 3-foliolate, leaflets subterete, sometimes hairy. Flowers solitary at branch tips, yellow, standard, wing and keel partly dark violet, purple or red, keel glabrous, calyx shortly hairy. Sept.–Mar. Mountain and lowland fynbos, 100–300 m, NW, SW, AP (Cedarberg Mts to Agulhas).•

rubens Thunb.   Sprawling shrublet to 25 cm. Leaves 3-foliolate, leaflets subterete, glabrous or silvery silky. Flowers scattered, yellow or orange to red or dark red, keel silky, calyx shortly hairy. Jan.–Dec. Mountain fynbos, to 2000 m, KM, LB, SE (Anysberg and Langeberg Mts to Port Elizabeth).•

rubiginosa R.Dahlgren   Like **A. ciliaris** but calyx lobes narrower, long and linear-filiform. Oct.–Dec. Mountain fynbos, lower and upper slopes, 150–1250 m, NW, SW (Ceres to Riviersonderend Mts).•

salteri L.Bolus   Sprawling shrublet to 15 cm. Leaves 3-foliolate, leaflets terete, glabrous or thinly hairy. Flowers scattered, bright or pale yellow, keel glabrous, calyx thinly hairy, ridged. Nov.–Dec. Lowland fynbos, limestone, 30–200 m, SW, AP (Cape Peninsula to Arniston).•

sanguinea Thunb.   Sprawling or erect shrub to 1 m. Leaves 3-foliolate, leaflets terete-filiform. Flowers scattered near branch tips, pale or bright yellow, partly rose or purple, keel glabrous, calyx glabrous. Jan.–Mar. Lowland fynbos, limestone, below 330 m, AP (Bredasdorp to Stilbaai).•

smithii R.Dahlgren   Rigid shrublet to 30 cm. Leaves 3-foliolate, leaflets terete. Flowers scattered, bright yellow, fading reddish, keel glabrous, calyx shortly hairy , lobes small, ovoid. Feb.–Apr. Renosterveld, 100–120 m, SW, LB (Riviersonderend to Swellendam).•

spicata Thunb.   Erect or sprawling shrub or shrublet, 0.2–2 m. Leaves 3-foliolate, leaflets subterete, rigid or weak, glabrous or tuberculate-hairy. Flowers 1 or 2 at branch tips, pale yellow to orange-yellow, sometimes rusty, keel glabrous, calyx silky, stiff or slender. Aug.–Jan. Mountain fynbos, below 1000 m, NW, SW (Bokkeveld Mts to Franschhoek).•

spinescens Thunb.   Rigid, thorny shrub or shrublet, 0.6–2 m. Leaves 3-foliolate, leaflets ovoid to cylindric. Flowers scattered, pale or bright yellow, keel silky, calyx subglabrous, lobes reduced. Sept.–Oct. Coastal fynbos, 40–560 m, NW, SW (Vredendal to Mamre).•

spinosa L.   Thorny shrub to 2 m, thorns lateral. Leaves 3-foliolate, leaflets terete to oblong. Flowers 1 or 2 in the thorn axils, yellow, partly red or purplish, keel glabrous, calyx usually adpressed-hairy, lobes reduced. Aug.–Mar. Lowland fynbos and renosterbos, 60–200 m, NW, SW, LB, SE (Olifants River Mts to KwaZulu-Natal).

steudeliana Brongn.   Erect shrublet, 0.3–1 m. Leaves 3-foliolate, leaflets ovoid-oblong or granular. Flowers scattered, pale or bright yellow, keel glabrous, calyx shortly hairy, lobes reduced. Aug.–Oct. Renosterbos-fynbos scrub, 160–1000 m, SW, LB (Robertson and Botrivier to Mossel Bay).•

subtingens Eckl. & Zeyh.   Erect or sprawling shrub or shrublet; 0.2–2 m. Leaves 3-foliolate, leaflets terete. Flowers scattered, yellow, fading orange-red, keel glabrous, calyx glabrous. Oct.–Mar. Renosterbos-fynbos scrub or marginal fynbos, 660–1300 m, KM, SE (Witteberg to Somerset East).

tenuissima R.Dahlgren   Procumbent shrublet with branches to 1 m long. Leaves 3-foliolate, leaflets subterete. Flowers solitary on filiform peduncles, yellow to orange or red-purple on back of standard, fading dark red, keel glabrous, calyx softly hairy, lobes ovoid-subglobose, fleshy. Jan.–Dec. Mountain fynbos, up to 1000 m, KM, SE (Mossel Bay to Van Staden's Mts).•

triquetra Thunb. Erect shrub or shrublet to 1.5 m. Leaves 3-foliolate, leaflets subterete-angular, glabrous or subglabrous. Flowers in terminal heads, yellow, often purplish on back of standard, keel glabrous, calyx white-silky, upper lobes broadest. Sept.–Jan. Mountain fynbos, moderate to high alt., up to 1500 m, NW, SW, KM (Cedarberg to Riviersonderend Mts and Witteberg).•

tuberculata Walp.   Erect shrublet to 1 m. Leaves 3-foliolate, leaflets subterete or slightly flattened, adpressed-hairy. Flowers scattered, pale yellow, keel glabrous, calyx shortly silky. June–Feb. Renosterbos-fynbos scrub, below 800 m, NW, SW, KM (Breede River valley to Caledon Swartberg and Anysberg to Klein Swartberg).•

varians Eckl. & Zeyh.   Like **A. ericifolia** but calyx densely white-woolly. Oct.–Nov. Renosterbos-fynbos scrub, 50 m, NW, SW (Olifants River Mts to Tygerberg).•

variegata Eckl. & Zeyh.   Slender shrublet, 15–40 cm. Leaves 3-foliolate, leaflets subterete. Flowers 1–4 in terminal heads, yellow, keel glabrous, calyx thinly hairy, bract adnate to pedicel. Nov. Lowland fynbos, 26 m, SW (Cape Flats).•

vermiculata Lam.   Shrub or shrublet, 0.4–2 m. Leaves 3-foliolate, leaflets cylindric, glabrous or thinly hairy. Flowers 1–6 in terminal clusters, bright yellow, keel shortly hairy, calyx shortly hairy. Oct.–Jan. Mountain fynbos, 600–800 m, KM, SE (Groot Swartberg to Langkloof).•

wittebergensis Compton   Erect or sprawling shrublet to 50 cm. Leaves 3-foliolate, leaflets subterete or slightly flat or angular, glabrous or grey-hairy. Flowers scattered, bright yellow or orange, keel glabrous, calyx hairy, lobes narrow. Aug.–Jan. Renosterbos-fynbos scrub, 800–1600 m, NW, KM (Bokkeveld Mts to Witteberg).•

wurmbeana E.Mey.   Shrub or shrublet to 1.5 m. Leaves 3-foliolate, leaflets terete, glabrous or thinly hairy. Flowers scattered, bright yellow, rarely purple on standard, wings and keel, keel glabrous, calyx adpressed-hairy, lobes with thickened margins. Feb. Renosterveld, 60 m, NW, SW (Gifberg to Paarl).•

zeyheri (Harv.) R.Dahlgren   Like **A. sanguinea** but wings with longer claws, c. 2.5 mm long and smaller blades, to 3.5 × 1.2 mm. Sept.–Nov. Renosterbos-fynbos scrub, 65–370 m, SW, LB (Swellendam and Potberg to Riversdale).•

BOLUSAFRA•   TAR PEA, TEER-ERTJIE   1 sp., W Cape

bituminosa (L.) Kuntze   Sprawling, resinous tar-scented shrublet. Leaves pinnately 3-foliolate. Flowers bright yellow. Aug.–Jan. Mountain fynbos, streamsides, 300–800 m, NW, SW (Tulbagh to Caledon).•

CALPURNIA   CAPE LABURNUM, GEELKEURBOOM   7 spp., southern and tropical Africa to India

intrusa (R.Br. ex W.T.Aiton) E.Mey.   Tree-like, reseeding shrub to 6 m. Leaves pinnate. Flowers yellow. Dec.–June. Renosterbos-fynbos scrub, karroid scrub, streamsides, 600–900 m, KM (Swartberg Mts: Ladismith to Oudtshoorn).•

CHAMAECRISTA (= *CASSIA* in part)   c. 250 spp., pantropical and warm temperate

capensis (Thunb.) E.Mey. (= *Cassia capensis* Thunb.)   Erect or sprawling, fine-leaved resprouting shrublet, 20–40 cm. Leaves bipinnate. Flowers few in axils, bright yellow. Dec.–Jan. Grassland, SE (Humansdorp to tropical Africa).

CROTALARIA   GEELKEURTJIE   c. 600 spp., tropics and subtropics

capensis Jacq.   Erect shrub, 2–3 m. Leaves 3-foliolate, leaflets obovate, thinly hairy, petiole usually shorter than leaflets, stipules foliaceous, sometimes lacking. Flowers in terminal racemes, yellow. May–Dec. Grassland, bushland, forest margin, below 1830 m, SE (Knysna to tropical Africa).

excisa (Thunb.) Baker f.   Decumbent, sprawling subshrub with woody rootstock, 10–15 cm. Leaves 3-foliolate, leaflets obovate, thinly hairy, petiole about as long as leaflets, stipules linear-subulate. Flowers 1 or 2 on slender peduncles, yellow. Aug.–Oct. Karroid scrub, NW, SW (Namaqualand to Cape Peninsula to Montagu).

humilis Eckl. & Zeyh.   Like **C. excisa** but annual and not developing a woody rootstock. Sept.–Oct. Karroid scrub, sandy places, 300–1050 m, NW (Richtersveld to Olifants River).

lebeckioides Bond   Rigid, wand-like shrublet to 1 m. Leaves 3-foliolate, leaflets reduced, linear-conduplicate, glaucous, petiole much longer than leaflets, stipules absent. Flowers in elongate terminal racemes, yellow. July–Dec. Karroid scrub, 500–600 m, KM (Ladismith).•

obscura DC.   Much-branched shrublet with long straggling or decumbent, shaggy hairy branches. Leaves 3-foliolate, leaflets elliptic-subrotund, long-ciliate, petiole mostly longer than leaflets, stipules linear-lanceolate or spathulate. Flowers in terminal racemes, yellow, often reddish outside. June–Nov. Grassland, 230–660 m, SE (Uitenhage to KwaZulu-Natal).

*****CULLEN**   BLUE CLOVER, BLOUKLAWER   35 spp., Old World tropics and subtropics

*****obtusifolia** (DC.) C.H.Stirt.   Spreading, mat-forming shrublet with soft stems to 1 m long. Leaves 3-foliolate, clover-like, leaflets obovate. Flowers purple, mauve. Sept.–Nov. Karroid scrub, flats, SW, KM, LB (Robertson through Little Karoo, widespread in summer-rainfall area).

CYCLOPIA•   HONEY BUSH   23 spp., W and S Cape

A. Bracts clasping the base of the calyx

bowieana Harv.   Erect, robust, resprouting or reseeding shrub to 1.8 m. Leaves 3-foliolate, leaflets linear-oblanceolate, terete, sometimes somewhat flattened, margins usually strongly revolute, softly hairy. Flowers yellow, bracts clasping base of calyx. Oct.–Dec. Mountain fynbos, upper slopes, 1220–1830 m, LB, SE (Langeberg and Outeniqua Mts).•

glabra (Hofmeyr & E.Phillips) A.L.Schutte   Robust, resprouting shrub to 1.2 m. Leaves 3-foliolate, leaflets linear, terete, margins strongly revolute, thinly hairy when young. Flowers yellow, bracts clasping base of calyx. Nov.–Dec. Subalpine mountain fynbos, 1660–2249 m, NW (Hex River Mts).•

meyeriana Walp.   WOLTEE   Erect, reseeding shrub to 2 m. Leaves 3-foliolate, leaflets linear, terete, margins strongly revolute, softly hairy or glabrescent. Flowers yellow, bracts clasping base of calyx. Sept.–Dec. Mountain fynbos, upper slopes, 1000–1800 m, NW, SW (Cedarberg to Riviersonderend Mts).•

A. Bracts not clasping the base of the calyx
*B. Leaflets terete or subterete (see also **C. buxifolia**)*

alopecuroides A.L.Schutte   Erect, resprouting or reseeding shrublet to 60 cm. Leaves 3-foliolate, leaflets linear, terete, margins strongly revolute, hirsute. Flowers yellow. Sept.–Dec. Subalpine mountain fynbos, 1500–2000 m, KM (Groot Swartberg and Kammanassie Mts).•

alpina A.L.Schutte   Sprawling, lax, resprouting shrub to 30 cm. Leaves 3-foliolate, leaflets linear-terete to oblanceolate, margins strongly revolute, sometimes slightly recurved. Flowers yellow. Nov.–Dec. Sandstone slopes at high alt., 1170–2070 m, NW, SW, KM (Hex River to Hottentots Holland to Kammanassie Mts).•

aurescens Kies   Sturdy, erect, resprouting shrub to 70 cm. Leaves 3-foliolate, leaflets linear, margins strongly revolute. Flowers yellow. Dec. Subalpine mountain fynbos, above 1800 m, KM (Klein Swartberg).•

bolusii Hofmeyr & E.Phillips   Lax, sprawling, resprouting shrublet to 30 cm. Leaves 3-foliolate, leaflets linear, margins strongly revolute, softly hairy when young. Flowers yellow. Nov.–Jan. Subalpine mountain fynbos, 1900–2270 m, KM (Groot Swartberg).•

galioides (P.J.Bergius) DC.   Robust, softly hairy, resprouting shrub to 1 m. Leaves 3-foliolate, leaflets linear-terete, margins strongly revolute, softly hairy to glabrescent. Flowers yellow. Jan.–May. Lowland fynbos, flats and slopes, 160–700 m, SW (Cape Peninsula).•

genistoides (L.) R.Br.   HEUNINGTEE, HONEYBUSH TEA   Robust, erect, resprouting shrub to 2 m. Leaves 3-foliolate, leaflets linear-terete, margins strongly revolute. Flowers yellow. Sept.–Nov. Lowland fynbos, flats slopes, 60–1170 m, SW, LB (Malmesbury to Albertinia).•

maculata (Andrews) Kies   VLEITEE   Erect, reseeding shrub to 3.5 m, rarely resprouting after fire. Leaves 3-foliolate, leaflets linear-terete, margins strongly revolute. Flowers yellow. Sept. Streamsides in lowland fynbos, 150–830 m, SW, LB (Bainskloof to Riversdale).•

pubescens Eckl. & Zeyh.   Erect, reseeding shrub to 1.7 m. Leaves 3-foliolate, leaflets linear-terete, margins strongly revolute. Flowers yellow, bracts pleated, with recurved apices. Sept. Marshes and seeps in lowland fynbos, c. 300 m, SE (Port Elizabeth).•

squamosa A.L.Schutte   ?Reseeding shrub. Leaves 3-foliolate, leaflets linear-terete, margins strongly revolute. Flowers yellow. Oct. Mountain fynbos, southern slopes, 1700 m, SW (Wemmershoek Mts).•

*B. Leaflets more or less flat (see also **C. alpina**)*

burtonii Hofmeyr & E.Phillips   SUIKERTEE, HEUNINGTEE   Robust, reseeding shrublet to 80 cm. Leaves 3-foliolate, leaflets elliptic, margins sometimes slightly recurved. Flowers yellow. Oct.–Dec. Sandstone slopes, 1600–2070 m, KM (Groot Swartberg).•

buxifolia (Burm.f.) Kies   Erect or prostrate, robust resprouting shrub to 2 m. Leaves 3-foliolate, leaflets elliptic to obovate, sometimes subterete, margins slightly recurved to revolute, minutely crisped. Flowers yellow. Sept. Sandstone slopes, 830–1670 m, NW, SW, LB, SE (Cold Bokkeveld to Outeniqua Mts).•

falcata (Harv.) Kies   Erect, robust, resprouting shrub to 1.5 m. Leaves 3-foliolate, leaflets linear to elliptic, margins slightly revolute. Flowers yellow. Sept.–Nov. Sandstone slopes, 550–1600 m, NW, SW (Cold Bokkeveld Mts to Caledon Swartberg).•

filiformis Kies   Erect, ?reseeding shrub. Leaves 3-foliolate, leaflets linear, margins slightly recurved. Flowers yellow, calyx lobes triangular, acuminate. ?Oct. Sandy flats, 100 m, SE (Van Staden's Mts).•

intermedia E.Mey.   KOUGABERGTEE, BERGTEE   Erect, robust resprouting shrub to 2 m. Leaves 3-foliolate, leaflets oblanceolate, margins slightly recurved. Flowers yellow. Sept.–Nov. Sandstone slopes, 500–1700 m, KM, LB, SE (Witteberg and Langeberg Mts to Van Staden's Mts).•

latifolia DC.   Erect, ?reseeding shrub to 1 m. Leaves 3-foliolate, leaflets cordate, margins slightly recurved, finely crisped. Flowers yellow. Sept.–Nov. Sandstone seeps, 900–1000 m, SW (Cape Peninsula).•

laxiflora Benth.   Erect, ?reseeding shrub. Leaves 3-foliolate, leaflets linear-oblanceolate, margins slightly recurved. Flowers yellow. ?Sept. Sandy flats, SE (Knysna, Plettenberg Bay).•

longifolia Vogel   Erect, reseeding shrub to 3 m. Leaves 3-foliolate, leaflets linear-oblanceolate, margins sometimes slightly recurved. Flowers bright yellow. Oct.–Nov. Sandy slopes and flats, 300–360 m, SE (Van Staden's Mts).•

sessiliflora Eckl. & Zeyh.   HEIDELBERGTEE   Erect, robust resprouting shrublet to 1 m. Leaves 3-foliolate, leaflets linear to narrowly elliptic, margins slightly recurved. Flowers pale yellow. Apr.–Sept. Sandstone slopes, 300–1500 m, KM, LB (Langeberg Mts and Warmwaterberg).•

subternata Vogel   VLEITEE   Erect, reseeding shrub to 3.5 m. Leaves 3-foliolate, leaflets obovate to oblanceolate, margins slightly recurved. Flowers yellow. Sept. Sandstone seeps, 160–1000 m, LB, SE (Langeberg to Tsitsikamma Mts).•

DIPOGON   CAPE SWEET PEA, BOSKLIMOP   1 sp., W and E Cape

lignosus (L.) Verdc.   Woody climber. Leaves pinnately 3-foliolate, leaflets rhombic, glaucous below. Flowers in pedunculate racemes, magenta or pink. Jan.–Dec. Scrub or forest, SW, KM, LB, SE (Saldanha to E Cape).

DOLICHOS   BUTTERFLY PEA   60 spp., Africa and E Asia

decumbens Thunb.   Straggling shrublet, 30–60 cm. Leaves pinnately 3-foliolate, leaflets rhombic-ovate. Flowers in pedunculate corymbs, violet, pink or blue, style terete. July–Aug. Clay flats, NW, SW (Bokkeveld Mts to Cape Peninsula).•

hastiformis E.Mey.   Straggling shrublet with long, slender stems. Leaves pinnately 3-foliolate, leaflets narrowly to broadly hastate. Flowers subumbellate on peduncles, bright pink or mauve, style channelled. Oct.–May. Coastal grassland, below 300 m, LB, SE (Riversdale to E Cape).

DUMASIA   8 spp., Africa and Asia

villosa DC.   Slender climber. Leaves pinnately 3-foliolate, leaflets ovate, glaucous and thinly adpressed-hairy beneath. Flowers in slender, pedunculate racemes, yellow, calyx truncate-cupular. Dec.–Feb. Forests, SE (Knysna to tropical Africa and Asia).

ERIOSEMA   130 spp., pantropical

squarrosum (Thunb.) Walp.   Erect or sprawling shrublet, 15–45 cm. Leaves pinnately 3-foliolate, erect, leaflets elliptic, adpressed-hairy, silvery below. Flowers in pedunculate racemes, brownish orange. Pods oblong, rusty yellow-hairy. Oct.–Nov. Grassland, SE (Knysna to E Cape).

ERYTHRINA   CORAL TREE   108 spp., pantropical and subtropical

caffra Thunb.   Deciduous, prickly tree to 20 m. Leaves pinnately 3-foliolate, leaflets rhombic-deltoid. Flowers red to pink, standard recurved, calyx bilabiate at maturity. July–Aug. Coastal forests, streamsides, SE (Humansdorp to KwaZulu-Natal).

HYPOCALYPTUS•   3 spp., W Cape

coluteoides (Lam.) R.Dahlgren   Tree-like shrub to 3 m. Leaves 3-foliolate, leaflets elliptic, paler beneath. Flowers 6–15(–25) in more or less lax terminal racemes, pink to magenta, calyx shortly hairy. Pods long-stipitate, inflated. Mainly Sept.–Nov. Sandstone slopes, 160–1670 m, SW, LB, SE (Hottentots Holland to Van Staden's Mts).•

oxalidifolius (Sims) Baill.   Sprawling shrublet to 50 cm with wiry branches. Leaves 3-foliolate, leaflets small, broadly obovate, drooping. Flowers 2–6(–15) in lax terminal racemes, pale purple to almost white, calyx glabrescent. Pods stipitate, ovate to obliquely oblong, flattened. July–Sept. Sandstone slopes, streamsides or moist places, below 700 m, SW, LB, SE (Hottentots Holland to Tsitsikamma Mts).•

sophoroides (P.J.Bergius) Baill.   Shrub or small tree to 6 m. Leaves 3-foliolate, erect, leaflets obcordate-cuneate, apiculate. Flowers 30–more in dense terminal racemes, magenta with yellow nectar guide, calyx glabrous. Pods narrowly oblong, almost segmented between the seeds. Oct.–Dec. Sandstone slopes, streamsides, 60–1670 m, NW, SW, KM, LB (Cedarberg to Langeberg and Little Karoo Mts).•

INDIGOFERA INDIGO c. 720–730 spp., worldwide tropical and subtropical

A. Leaves sparse, usually present during early stages of growth only

filifolia Thunb. (= *Indigofera juncea* DC.)   Erect, resprouting, almost leafless shrub to 3 m. Leaves mostly on younger plants or new growth, petiole 4–6 cm long, basal leaflets 6–8, elliptic to oblanceolate, thinly hairy beneath, upper leaflets scale-like. Flowers in racemes on peduncles shorter than the leaves, white to pink or purple, 9–11 mm long, back of standard petal glabrous. Pods glabrous, spreading. Oct.–Apr. Mountain and lowland fynbos, streamsides, to 1000 m, SW, LB, SE (Cape Peninsula to Tsitsikamma Mts).•

gifbergensis C.H.Stirt. & Jarvie   Stiffly branched wand-like shrub to 1 m. Leaves scarce, pinnately 3–5-foliolate, petiole c. 15 mm long, leaflets obovate, thinly hairy beneath, upper leaflets scale-like; stipules fused, adnate to the base of the petiole. Flowers in racemes on peduncles to twice as long as the leaves, brick-red, 5–7 mm long, back of standard petal minutely hairy. Aug.–Nov. Arid fynbos, 600–1650 m, NW (Bokkeveld to Cedarberg Mts).•

ionii Jarvie & C.H.Stirt. (= *Indigofera filifolia* Thunb. var. *minor* T.M.Salter)   Erect or sprawling, reseeding shrublet to 80 cm. Leaves present during early stages of growth, pinnately 3–5-foliolate, petiole 15–25 mm long, leaflets narrowly obovate or elliptic-oblanceolate, glabrous above; stipules narrowly triangular, free. Flowers in long racemes on peduncles to twice as long as the leaves, pink. Aug.–Dec. Fynbos, marshy areas, 160–700 m, SW (Cape Peninsula to Kleinrivier Mts).•

A. Leaves many, present at all stages of growth
B. Leaves 1–3-foliolate
*C. Ovary and pods glabrous (see also **I. denudata, I. sarmentosa, I. sp. 12**)*

alpina Eckl. & Zeyh. (= *Indigofera stipularis* Link)   Densely leafy, decumbent shrublet to 25 cm. Leaves digitately 3-foliolate, leaflets obovate-cuneate, coarsely long-hairy; stipules broadly cordate-ovate, as long as or longer than petioles. Flowers in racemes on robust peduncles, brick-red. Pods deflexed. Aug.–Dec. Mountain fynbos, 660–1900 m, SE (Uniondale to E Cape).

guthriei Bolus   Erect or decumbent shrublet to 25 cm. Leaves 1-foliolate, linear-oblanceolate, margins involute, glabrescent; stipules lanceolate-subulate. Flowers in racemes on filiform peduncles, pink. Nov.–Jan. Mountain fynbos, 430–1000 m, SW (Hottentots Holland to Bredasdorp Mts).•

ovata L.f.   Slender, prostrate or trailing shrublet to 20 cm. Leaves 1-foliolate, subsessile, ovate-oblong or elliptic, glabrescent, leathery; stipules subulate. Flowers in racemes on slender peduncles, pink, mauve or red. Sept.–Jan. Mountain fynbos, often high alt., 130–1660 m, SW (Franschhoek to Kleinrivier Mts).•

sp. 1   Sprawling, decumbent shrublet. Leaves 3-foliolate, leaflets narrowly elliptic-oblanceolate; stipules broadly ovate-lanceolate, about as long as petioles. Flowers in racemes on robust peduncles, pink. Aug.–Nov. Mountain fynbos in moist places, 450–1100 m, SE (Langkloof to Grahamstown).

sp. 2   Like **I. sarmentosa** but leaves 1–3-foliolate, leaflets leathery, oblong-elliptic, greyish hairy beneath with margins strongly revolute, (8–)11–26 × 3–11 mm. Nov.–Dec. Limestone fynbos, 150–600 m, AP (Potberg).•

sp. 3 (= *Indigofera culmenicola* Baker f. nom. nud.)   Like **I. capillaris** but leaves 1–3-foliolate, leaflets broadly oblong-elliptic. Flowers dark pink, purple or magenta. Dec. Mountain fynbos, c. 1650 m, KM (Swartberg Pass).•

C. Ovary and pods hairy (occasionally glabrescent when mature)
D. Petioles shorter than 1 cm
E. Rigid woody shrubs; leaflets thickened or leathery

denudata L.f. (= *Indigofera kraussiana* Meisn.)   Much-branched shrublet to 1.2 m, unarmed or branches spine-tipped. Leaves 1–3 foliolate with slender petioles, leaflets linear-oblong to obovate, fleshy to leathery, glaucous, glabrous or minutely hairy; stipules minute. Flowers in subsessile racemes, mauve. Sept.–July. Fynbos, 100–730 m, SE (Uniondale to E Cape).

nigromontana Eckl. & Zeyh. (= *Indigofera dealbata* Harv., *I. spinescens* E.Mey.)   Erect spinescent shrub to 2 m. Leaves 3-foliolate, subfasciculate, leaflets small, obovate, concave, thick, thinly hairy above; stipules obsolete. Flowers in subsessile racemes, pinkish purple. June–Mar. Karroid scrub, renosterveld, fynbos, 50–1750 m, NW, LB, KM, SE (Namaqualand and W Karoo, Little and Great Karoo to Lesotho).

obcordata Eckl. & Zeyh.   Erect or spreading shrub to 1m. Leaves 1–3-foliolate, sparse, leaflets obcordate, thinly greyish hairy; stipules obsolete. Flowers in subsessile racemes, red. July–Apr. Karroid scrub, 400–660 m, KM (Witteberg and Swartberg Mts).•

sp. 4   Like **I. denudata** but pedicels longer, 4–8 mm long (not shorter than 4 mm). Flowers magenta to deep pink. May–July. Mountain fynbos, c. 250 m, KM (Montagu).•

sp. 5   Robust, twiggy, gnarled shrublet to 50 cm tall, branches often spinescent. Leaves digitately 1–3(–7)-foliolate, often on spur shoots. Flowers in racemes on peduncles shorter than to about as long as leaves, red. June. Fynbos, 300–500 m, NW (Cedarberg Mts).•

E. Prostrate or sprawling shrublets, stems mostly herbaceous; leaflets not thickened or leathery
*F. Back of standard petal glabrous (see also **I. meyeriana**)*

gracilis Spreng.   Slender, creeping, carpet-forming shrublet. Leaves digitately 3-foliolate, rather distant, leaflets small, elliptic-oblong to obovate, glabrous or thinly hairy above; stipules small, subulate. Flowers in racemes on filiform peduncles, purple or reddish pink. Aug.–Jan. Mountain and lowland fynbos, 300–930 m, SW (Cape Peninsula, Paarl to Stellenbosch Mts).•

incana Thunb.   Spreading, prostrate to suberect shrublet, 30–60 cm. Leaves digitately 3-foliolate, leaflets obovate-oblong, coarsely long hairy; stipules ovate-lanceolate, acuminate, often longer than the petioles. Flowers in racemes on robust peduncles, rose to pink. July–Nov. Renosterbos-fynbos scrub, lower to middle slopes, below 800 m, NW, SW, KM (Piketberg to Barrydale).•

sp. 6 (= *Indigofera setacea* auct.)   Like **I. gracilis** but leaflets obovate to suborbicular, stipules absent, stamens 3.5–4 mm long (not 2–3 mm) and pods to 14 mm long with 5–7 seeds (not 5–7 mm long with 2 or 3 seeds). July–Oct. Mountain fynbos, 200–600 m, SW (Cape Peninsula).•

*F. Back of standard petal hairy (see also **I. erecta**, **I. heterophylla**, **I. meyeriana**, **I. tomentosa**)*

glomerata E.Mey.   Prostrate, much-branched closely leafy shrublet to 20 cm. Leaves imbricate, digitately 3-foliolate, leaflets obovate-oblong, softly white-hairy; stipules deltoid, long-acuminate, longer than the petioles. Flowers in subsessile racemes, pink to purple. Aug.–Nov. Lowland and mountain fynbos, 30–860 m, SW, AP (Cape Peninsula to Bredasdorp).•

leptocarpa Eckl. & Zeyh.   Slender, erect or sprawling angularly branched shrublet to 30 cm. Leaves 1–3-foliolate, leaflets linear or oblanceolate, minutely hairy; stipules minute. Flowers in racemes on short peduncles, red to purple. Nov.–Jan. Renosterbos-fynbos scrub, arid fynbos, 100–900 m, SE (George to E Cape).

thesioides Jarvie & C.H.Stirt.   Twiggy, sprawling shrublet to 40 cm. Leaves 1–3-foliolate, lateral leaflets caducous, leaflets narrowly obovate, folded, densely minutely hairy; stipules narrowly triangular. Flowers in racemes on peduncles 10–15 mm long, pink. June–Oct. Arid fynbos, karroid scrub, 700 m, KM (Groot Swartberg: Meiringspoort).•

sp. 7 (= *Indigofera depressa* Harv.)   Slender, prostrate or trailing shrublet to 50 cm. Leaves digitately 3-foliolate, leaflets cuneate-obovate, hairy; stipules lanceolate to subulate, spreading-recurved. Flowers in racemes on peduncles more or less as long as leaves, pink, rose or brick-red. Pods thinly hairy, compressed with a ridge above seeds. Nov.–Aug. Renosterveld, fynbos, 200–1000 m, SW, AP, LB, KM, SE (Riviersonderend and Little Karoo to Tsitsikamma Mts).•

D. Petioles 1–4 cm long
G. Bracts broad, ovate, more or less enclosing flower buds

amoena Aiton (= *Indigofera intermedia* Harv.)   Robust, suberect to sprawling resprouting shrub to 1 m. Leaves pinnately 3-foliolate, leaflets elliptic-oblong or obovate, thinly hairy; stipules lanceolate. Flowers in racemes on robust peduncles, pink, magenta or rose, back of standard petal glabrous, bracts ovate-acuminate to trowel-shaped. July–Oct. Karroid scrub, strandveld, 150–750 m, NW (Namaqualand to Piketberg).

venusta Eckl. & Zeyh.   Erect to prostrate resprouting shrublet, 0.2–1.5 m. Leaves digitately 3-foliolate, leaflets obovate to linear-lanceolate, shortly hairy; stipules lanceolate to attenuate. Flowers in racemes on robust peduncles, red, purple or pink, bracts ovate-cuspidate, enclosing the flower buds. June–Oct. Mountain and coastal fynbos, 350–780 m, NW (Namaqualand to St Helena Bay).

G. Bracts narrow, linear-lanceolate or subulate, not enclosing flower buds

complanata Spreng. (= *Indigofera nitida* T.M.Salter)   Straggling, suberect perennial to 40 cm. Leaves digitately 3-foliolate, leaflets lanceolate, shortly hairy, the hairs more or less perpendicular to midrib above; stipules subulate. Flowers in racemes on robust peduncles, bright red, back of standard petal hairy. July–Nov. Lowland fynbos, c. 100–300 m, SW (Cape Peninsula).•

erecta Thunb. (= *Indigofera porrecta* Eckl. & Zeyh. var. *bicolor* Harv.) Erect or spreading shrublet, 10–90 cm. Leaves digitately 3-foliolate, leaflets obovate to oblanceolate, glabrous above, thinly hairy beneath; stipules setaceous, spreading. Flowers in racemes on robust peduncles, orange-red to pink. July–Dec. Lowland and coastal fynbos, below 200 m, SE (Knysna to Port Elizabeth and E Cape).

heterophylla Thunb. (= *Indigofera adscendens* Eckl. & Zeyh.) Erect or sprawling to prostrate shrublet, 0.3–1 m, stems furrowed. Leaves pinnately or digitately 3-foliolate, leaflets sublanceolate (especially upper) to cuneate-obovate, thinly hairy; stipules subulate-setaceous, spreading to recurved. Flowers in racemes on robust peduncles, orange-pink to reddish purple; calyx thinly hairy, lobes subulate, up to 3 times as long as tube. May–Feb. Renosterveld and fynbos, 50–1500 m, NW, SW, AP, KM, LB, SE (Namaqualand and Karoo to E Cape).

meyeriana Eckl. & Zeyh. (= *Indigofera cardiophylla* Harv., *I. complicata* Eckl. & Zeyh.) Dense, erect or sprawling resprouting shrublet to 1 m, stems furrowed, often greyish. Leaves digitately 3-foliolate, leaflets linear-lanceolate to obcordate, thinly hairy; stipules minute, subulate, spreading or recurved. Flowers in racemes on robust peduncles, pink, purple or red, bracts lanceolate to subulate. June–Oct. Karroid scrub, renosterveld, strandveld, 30–1230 m, NW, SW, AP, KM, LB, SE (Namaqualand and Karoo to E Cape).

platypoda E.Mey. Prostrate or spreading shrublet to 30 cm. Leaves digitately 3-foliolate, petiole fleshy, leaflets linear-oblong, thinly hairy; stipules lanceolate-subulate, spreading or reflexed. Flowers in racemes on robust peduncles, pink. Sept.–Oct. Coastal fynbos, 100 m, SW (Saldanha Bay).•

porrecta Eckl. & Zeyh. Like **I. heterophylla** but procumbent, leaflets broadly obovate and truncate to retuse and racemes rounded at tips (vs tapering). Sept.–Apr. Coastal grassland, AP, SE (Stilbaai to E Cape).

procumbens L. (= *Indigofera discolor* E.Mey.) Spreading or trailing perennial to 10 cm, stems often running underground. Leaves digitately 3-foliolate; leaflets obovate or rhomboid, glabrous or thinly hairy above; stipules lanceolate-subulate. Flowers in racemes on fleshy peduncles, orange, copper, rose to purple, bracts subulate. June–Oct. Renosterveld, coastal fynbos, strandveld, below 160 m, NW, SW (Lambert's Bay to Cape Flats).•

psoraloides (L.) L. (= *Indigofera racemosa* L.) Erect or straggling shrublet to 80 cm. Leaves digitately 3-foliolate, leaflets lanceolate, acute, shortly hairy; stipules elongate, subulate, erect. Flowers in racemes on robust peduncles; flowers red, rose, purple. June–Dec. Lowland fynbos, 30–520 m, NW, SW (Citrusdal to Cape Peninsula).•

tomentosa Eckl. & Zeyh. Prostrate or trailing shrublet, 30–45 cm. Leaves digitately 3-foliolate, leaflets cuneate-obovate, densely white-hairy; stipules small, lanceolate-subulate. Flowers in racemes on robust peduncles, copper, orange or rose. Sept.–Nov. Coastal fynbos, 10–100 m, SE (Mossel Bay to E Cape).

triquetra E.Mey. Erect, slightly branched shrublet to 1 m, stems strongly 3-angled. Leaves digitately 3-foliolate, leaflets linear, shortly hairy; stipules subulate. Flowers in racemes on robust peduncles, purple or pink. Sept.–Jan. Renosterveld-fynbos scrub, 200–400 m, NW, SW (Piketberg to Malmesbury).•

sp. 8 Like **I. heterophylla** but calyx densely grey hairy, lobes lanceolate-acuminate, 3–4 times as long as tube and upper surface of standard petal glabrous to sparsely hairy. Sept.–Oct. Granite outcrops, 50–150 m, SW (Vredenburg to Langebaan).•

B. Leaves 5–17-foliolate (if leaflets 3 then only in a few leaves)
H. Back of standard petal with adpressed dark brown hairs (at least along midline)

stricta L.f. (= *Indigofera lateritia* Bertol., *I. notata* N.E.Br.) Erect or scrambling resprouting shrub to 1.2 m, stems angular, ribbed. Leaves pinnately 5–11-foliolate, subsessile or shortly petiolate, leaflets narrowly oblong, weakly discolorous, subglabrous above, more densely hairy beneath. Flowers in racemes on peduncles shorter than the leaves or subsessile, pink or brick-red, back of standard petal with adpressed dark brown hairs, pedicels shorter than 3 mm. Pods thinly hairy, spreading to suberect. June–Apr. Coastal bush and thicket, grassland, to 860 m, SE (Mossel Bay to E Cape).

verrucosa Eckl. & Zeyh. (= *Indigofera glabella* Fourc.) Like **I. stricta** but racemes on peduncles about as long to longer than leaves, and flowers on pedicels 3–7 mm long. Aug.–Apr. Coastal fynbos, grassland, 15–300 m, SE (Mossel Bay to E Cape).

H. Back of standard petal with white or grey hairs, or glabrous
I. Ovary and pods hairy
J. Petals persistent to fruiting stage; stamens mostly hidden in flower remnants

brachystachya (DC.) E.Mey.   Dense shrub to 1.5 m, mass flowering, stems densely grey-hairy. Leaves pinnately 5–7-foliolate, subsessile, leaflets linear-subcuneate, shortly hairy above, densely grey-hairy beneath, midrib prominent beneath, margins strongly revolute. Flowers in racemes on peduncles shorter than to twice as long as leaves, subsessile, mauve to pink, petals persistent, back of standard petal white-silky. Pods hairy. Nov.–Sept. Coastal fynbos, limestone, 20–500 m, SW, AP (Cape Peninsula to Agulhas).•

flabellata Harv.   Erect, densely hairy reseeding shrub to 1.5 m, mass flowering. Leaves subdigitately (3–)5–7-foliolate, subsessile; leaflets linear-oblanceolate, minutely hairy, midrib prominent beneath, margins revolute. Flowers in racemes on peduncles shorter than the leaves, subsessile, pink or purple, petals persistent, back of standard petal hairy, calyx densely hairy, shortly 5-toothed. Pods grey-hairy, spreading. Jan.–Sept. Mountain and lowland fynbos, 100–1330 m, LB, KM, SE (Klein Swartberg and Langeberg to Tsitsikamma Mts).•

hamulosa Schltr.   Erect or prostrate shrublet to 60 cm, mass flowering. Leaves pinnately 5–7-foliolate, subsessile, leaflets oblong-cuneate to oblanceolate, shortly hairy, midrib prominent beneath, margins revolute. Flowers in racemes on short peduncles more or less equalling the leaves, subsessile, white to pink, petals persistent, back of standard petal white-hairy. Pods densely hairy, spreading. Oct.–Mar. Lowland fynbos, limestone, 5–200 m, AP (Gansbaai to Agulhas).•

hispida Eckl. & Zeyh. (= *Indigofera rhodantha* Fourc.)   Erect shrub to 3 m, mass flowering, stems densely grey-hairy and bristly. Leaves pinnately (3–)5–7-foliolate, subsessile, leaflets oblanceolate, grey-hairy, midrib prominent beneath, margins revolute. Flowers in racemes on peduncles shorter than the leaves, subsessile, pink, purple or red, petals persistent, back of standard petal hairy, calyx softly hairy, lobes subulate-setaceous. Pods hairy, spreading. Dec.–June. Mountain fynbos, 100–1000 m, SE (Uniondale to Port Elizabeth).•

pappei Fourc.   Erect, slender leafy shrublet, 0.3–1 m, mass flowering. Leaves digitately 5–9-foliolate, subsessile, leaflets linear, folded, minutely hairy, midrib prominent beneath, margins revolute. Flowers in racemes on peduncles shorter than the leaves, subsessile, pink, petals persistent, back of standard petal minutely hairy. Pods thinly hairy, spreading. Jan.–July. Mountain fynbos, lower slopes, 260–1400 m, LB, KM, SE (Swartberg to Kareedouw Mts).•

sp. 9   Like **I. sarmentosa** but leaves pinnately (3–)5-foliolate, subsessile, leaflets leathery, to 10 mm long, 5 mm wide, obovate to oblanceolate, densely hairy or silky beneath, margins strongly revolute. Flowers in racemes on filiform peduncles about as long as leaves, pink to magenta, pedicels c. 4 mm long, petals persistent. Pods densely hairy when young. Mountain fynbos, SE (Outeniqua Mts: near George).•

*J. Petals soon caducous and stamens clearly visible after anthesis (see also **I. sp. 17**)*

cytisoides (L.) L.   Erect reseeding shrub to 3 m. Leaves pinnately (3–)5-foliolate, subsessile or petiole to 8 mm long, leaflets elliptic-obovate, minutely hairy, midrib prominent beneath, 2–5 cm long. Flowers in racemes on robust peduncles about as long as leaves, purple to pink, to 11 mm long, back of standard petal densely hairy, bracts broadly ovate. Pods greyish hairy, erect. Mar.–July. Mountain and riverine fynbos, 20–1330 m, SW (Cape Peninsula to Kleinrivier Mts).•

digitata L.f.   Sprawling shrublet, 30–60 cm. Leaves digitately 7–9-foliolate, petiole 4–7 mm long, leaflets cuneate-oblanceolate, roughly hairy. Flowers in racemes on robust peduncles more than twice as long as leaves, pink, orange, purple, back of standard petal glabrous. Pods shortly hairy, deflexed. Sept.–Dec. Renosterbos-fynbos scrub, coastal renosterveld, 130–1100 m, NW, SW (Tulbagh to Kleinmond).•

disticha Eckl. & Zeyh.   Decumbent or sprawling shrub to 80 cm or higher when clambering through other vegetation, stems greyish. Leaves pinnately 5–7-foliolate, petiole 4–8 mm long, leaflets elliptic-oblong to linear, thinly hairy. Flowers in racemes on peduncles about as long as leaves, brick-red, calyx 2–3 mm long, lobes lanceolate, more or less equalling to twice length of tube, back of standard petal grey-hairy. Pods pale, thinly hairy, apex sharply acute. Oct.–Apr. Grasslands, karroid scrub, 100–1660 m, SE (Patensie to E Cape).

exigua Eckl. & Zeyh.   Slender, prostrate or decumbent shrublet to 20 cm. Leaves pinnately 7–11-foliolate, petiole 4–10 mm, leaflets alternate, oblanceolate to obcordate, thinly hairy. Flowers in racemes on peduncles about as long as or longer than leaves, pink or mauve, 6–7 mm long, back of standard petal thinly hairy, calyx 3–4 mm long, lobes lanceolate. Pods thinly hairy, yellowish brown. July–Oct. Karroid scrub, 300–400 m, NW (Namaqualand to Bokkeveld Mts).

glaucescens Eckl. & Zeyh.   Erect or sprawling shrublet to 40 cm, branches flexuous, greyish hairy and ribbed. Leaves pinnately 3–9-foliolate, petiole 1–14 mm long; leaflets oblanceolate -cuneate, thinly hairy. Flowers in racemes on peduncles about long as leaves, orange-red to pink, back of standard petal greyish hairy, calyx 3.5–5 mm long, lobes subulate, more than twice as long as tube. Pods thinly hairy. Sept.–Apr. Grassy fynbos and dunes, 10–300 m, SE (Van Staden's Mts to E Cape).

poliotes Eckl. & Zeyh. (= *Indigofera pauciflora* Eckl. & Zeyh.)   Multistemmed shrublet to 80 cm, sometimes with reddish glands. Leaves pinnately 5–9-foliolate, subsessile or shortly petiolate; leaflets linear-oblanceolate to oblong-elliptic, thinly hairy above, more densely hairy beneath, midrib slightly sunken above, prominent beneath; stipules subulate-setaceous, spreading. Flowers few in racemes on slender peduncles scarcely longer than the leaves, pink to orange-red, back of standard petal minutely hairy. Pods thinly hairy, spreading. Nov.–Apr. Grassland, 10–700 m, SE (Knysna to E Cape).

sp. 10 (= *Indigofera pilgeriana* Schltr. nom. nud.)   Like **I. angustifolia** but leaves subdigitately (3–)5–7-foliolate (not pinnately 5–11-foliolate); leaflets linear-elliptic, yellowish grey-hairy beneath, 3–8 mm long (not to 14 mm long). Flowers rostrate, 6–9 mm long (not c. 5 mm long with a nonrostrate keel). Pods thinly hairy (not glabrous), spreading to deflexed. May–Nov. Mountain fynbos, 250–1060 m, SW (Elandskloof Mts to Riviersonderend Mts).•

sp. 11   Like **I. brachystachya** but petals shaped as normal for the genus, caducous except for standard petal (not elongated, narrowly oblong-elliptic and all persistent). May–Sept. Limestone fynbos, 20–250 m, AP (Agulhas to Stilbaai).•

sp. 12   Slender, prostrate or trailing shrublet to 50 cm. Leaves subdigitately 5–7-foliolate, leaflets cuneate-obovate, hairy; stipules lanceolate to subulate, spreading-recurved. Flowers in racemes on peduncles about as long as leaves, pink, rose or brick-red. Pods thinly hairy, compressed with a ridge above seeds. Nov.–Aug. Coastal fynbos, 20–150 m (Riversdale to Knysna).•

I. Ovary and pods glabrous
K. Leaf margins involute

concava Harv.   Erect, wiry shrublet, 20–30 cm tall. Leaves pinnately (3–)5-foliolate, subsessile, leaflets linear-oblong, glabrous, paler above, minutely hairy beneath, margins strongly involute. Flowers in racemes on peduncles about as long as leaves, pink or purple, back of standard petal hairy, calyx lobes ovate-lanceolate, keeled. Pods glabrous. June–Sept. Mountain fynbos, 600–1600 m, NW, LB (Tulbagh, Langeberg Mts).•

dillwynioides Benth. ex Harv.   Slender, erect or sprawling reseeding shrublet to 60 cm. Leaves digitately 5–7-foliolate, sessile, leaflets linear-oblanceolate, subglabrous, margins strongly involute. Flowers 6–8 in racemes on peduncles about as long as or longer than leaves, pink, mauve or red, back of standard petal minutely hairy. Pods glabrous, spreading. Sept.–Oct. Mountain and coastal fynbos, 170–600 m, NW, SW (Bokkeveld Mts to Malmesbury).•

filicaulis Eckl. & Zeyh.   Slender, decumbent, mat-like reseeding shrublet to 40 cm. Leaves digitately 3–9-foliolate, sessile, leaflets linear-oblanceolate, subglabrous above, margins strongly involute. Flowers 3 or 4 in racemes on thread-like peduncles more than twice as long as leaves, red or pink, back of standard petal minutely hairy. Pods stipitate, glabrous, spreading. Sept.–Dec. Mountain and lowland fynbos, 30–830 m, NW, SW, AP (Clanwilliam to Bredasdorp).•

quinquefolia E.Mey.   Bushy shrublet to 20 cm tall, closely leafy. Leaves digitately 5-foliolate, subsessile or shortly petiolate; leaflets cuneate-oblong, subterete with margins strongly involute, coarsely hairy. Flowers few in racemes on wiry peduncles about as long as or longer than leaves, pink or purple, back of standard petal minutely hairy. Pods glabrous, spreading. Dec.–Jan. Mountain fynbos, 1600 m, SW (Riviersonderend Mts: Galgeberg).•

sp. 13 (= *Indigofera remota* Baker f. nom. nud.; *I. pentaphylla* Burch. ex Harv.)   Like **I. concava** but a more diffuse, creeping shrublet with filiform stems. Leaves subdigitately 3–7-foliolate, subsessile, leaflets linear-oblong, paler above, minutely hairy beneath, margins strongly involute. Flowers in racemes on peduncles about as long as leaves, pink or purple, pedicels slender, to 13 mm long, back of standard petal hairy, calyx lobes ovate-lanceolate, keeled. Pods glabrous. Dec.–Jan. Mountain fynbos in moist habitats, 900–1500 m, LB (Langeberg Mts: Swellendam to Heidelberg).•

K. Leaves flat or margins revolute
*L. Peduncles shorter than to about as long as leaves (see also **I. declinata**)*

candolleana Meisn. (= *Indigofera coriacea* Aiton var. *hirta* Harv., *I. mauritanica* (L.) Thunb. var. *hirta* Harv.) Similar to **I. mauritanica** but more robust, erect or ascending, the leaves often imbricate with leaflets concolorous. Flowers on peduncles shorter than to about as long as leaves. Aug.–Mar. Mountain fynbos, 250–560 m, SW (Cape Peninsula).•

frutescens L.f. Stout shrub to 3 m. Leaves pinnately 5–9-foliolate, petiole longer than 13 mm, rachis deeply channelled; leaflets obovate to elliptic, glabrous or minutely hairy, glaucous, often fleshy, 13–25 mm long. Flowers in racemes on peduncles shorter than the leaves, mauve, red or pink, to 15 mm long, back of standard petal minutely hairy. Pods glabrous, spreading to deflexed. July–Mar. Mountain fynbos, 330–830 m, NW, SW (Bokkeveld Mts to Paarl).•

fulcrata Harv. Erect or decumbent shrublet, 3–80 cm. Leaves pinnately (3–)5–7-foliolate, petiole 8–25 mm long, leaflets elliptic to obovate-oblong, almost glabrous; stipules large, membranous, ovate-lanceolate, fused at base. Flowers in racemes on peduncles about as long as leaves, pink to purple, back of standard petal minutely hairy apically, bracts large, membranous, enclosing the buds. Pods glabrous, spreading. Nov.–Apr. Mountain fynbos, streamsides, 360 m, NW, SW (Cedarberg to Elandskloof Mts).•

grisophylla Fourc. Like **I. sulcata** but the stems more densely hairy, the leaflets broader, nonsulcate and densely grey-hairy beneath. Flowering time? Habitat? SE (Baviaanskloof and Great Winterhoek Mts).•

langebergensis L.Bolus Like **I. frutescens** but a smaller resprouting shrublet to 1 m, leaves 7–11(–13)-foliolate and leaflets broadly obovate to suborbicular. Sept.–Apr. Mountain fynbos, 330–500 m, LB (Langeberg Mts: Montagu to Heidelberg).•

sulcata DC. Erect, rigid shrublet to 80 cm, mass flowering, branches furrowed. Leaves subdigitately (3–)5-foliolate, subsessile; leaflets linear, minutely hairy, midrib prominent below, margins strongly revolute. Flowers in racemes on peduncles shorter than the leaves, subsessile, mauve, petals semipersistent but not to fruiting stage, back of standard petal mostly hairy along the midrib. Pods glabrous, spreading. Mar.–Aug. Mountain and lowland fynbos, 10–1330 m, KM, SE (Swartberg Mts to E Cape).

superba C.H.Stirt. Willowy, grey-hairy shrub, 1–3 m. Leaves pinnately (7–)9–11-foliolate, subsessile or petiole to 4 mm long; leaflets obovate to suborbicular, minutely hairy, 10–25 mm long. Flowers many in dense racemes on peduncles about as long as leaves, pink, 7–9 mm long, back of standard petal hairy. Pods glabrous, spreading. Dec.–Apr. Mountain fynbos, lower slopes, 100–400 m, SW (Kleinrivier Mts).•

sp. 14 Diffuse wiry shrublet. Leaves subdigitately 3–7-foliolate, subsessile, leaflets linear, retuse-apiculate. Flowers in subsessile racemes or on slender peduncles shorter than the leaves, the rachis sometimes much condensed, pink or red, petals persistent, pedicels thread-like, 2–8 mm long, bracts persistent. Nov.–Mar. Mountain fynbos, 300–600 m, LB, SE (Langeberg to Outeniqua Mts).•

sp. 15 Densely grey-hairy, twiggy shrublet to 40 cm. Leaves subdigitately 5–7-foliolate, subsessile; leaflets linear-oblanceolate, folded, 4–6 mm long, 1–2 mm wide. Flowers in racemes on peduncles shorter than to about as long as leaves, pink to red, back of standard petal minutely hairy. Pods glabrous or thinly hairy, oblong, spreading. Aug.–Oct. Mountain fynbos, NW (Gifberg).•

L. Peduncles longer than leaves
*M. Petiole 2 mm long or longer (see also **I. ionii**)*

capillaris Thunb. Sprawling or decumbent shrublet to 30 cm, stems reddish brown, often brittle. Leaves pinnately 5–13-foliolate, almost glabrous, petiole 1–3 cm long, leaflets linear-oblanceolate to elliptic, margins often involute; stipules subulate, ascending to erect, to 6 mm long. Flowers in racemes on slender peduncles more than twice as long as leaves, dark pink, purple, copper or magenta, back of standard petal hairy apically, stamens (4–)6–8 mm long. Pods glabrous, deflexed. Aug.–Jan. Mountain fynbos, 300–1700 m, NW, SW, KM, LB (Bokkeveld to Langeberg and Swartberg Mts).•

declinata E.Mey.   Prostrate to decumbent shrublet to 20 cm. Leaves 5–13-foliolate, petiole 2–8(–9) mm long, leaflets obovate-oblong to suborbicular or oblanceolate, glabrous or thinly hairy above, hairy beneath; stipules lanceolate to subulate, spreading to reflexed, 1–4 mm long. Flowers in racemes on slender peduncles to twice or more the length of the leaves, pink to magenta, back of standard petal hairy, stamens 4–5 mm long. Pods glabrous, spreading to suberect. Oct.–Feb. Mountain fynbos, 600–2130 m, NW, KM, LB, SE (Biedouw Mts through Little Karoo to Kouga Mts).•

humifusa Eckl. & Zeyh.   Diffuse prostrate or decumbent shrublet to 20 cm, stems pale, softly hairy or glabrous. Leaves pinnately 9–17-foliolate, petiole 4–12 mm long, leaflets obovate to oblanceolate; stipules ovate-falcate to lanceolate, spreading to reflexed, often membranous. Flowers in racemes on slender peduncles more than twice as long as leaves, mauve to purple, back of standard petal hairy, stamens 6–7 mm long. Pods glabrous, spreading. Aug.–Nov. Mountain fynbos, 380–1800 m, NW (Cedarberg to Cold Bokkeveld Mts).•

sp. 16 (= *Indigofera capillaris* auct.)   Like **I. capillaris** but an annual or short-lived perennial to 10 cm, with very narrow linear leaflets. Flowers with stamens 3–4 mm long. Sept.–Dec. Coastal fynbos, along streams or in vleis, or weed in lawns. 20–300 m, SW (Cape Peninsula, Kleinmond).•

sp. 17   Like **I. capillaris** but a robust shrub, 1.2–2 m. Sept.–Jan. Mountain fynbos. 1200–1400 m, NW (Hex River Mts).•

M. Petiole shorter than 2 mm

alopecuroides (Burm.f.) DC. (= *Indigofera coriacea* Aiton var. *alopecuroides* DC.)   Erect or creeping shrublet to 90 cm. Leaves pinnately 5–7-foliolate, subsessile, crowded; leaflets elliptic-oblanceolate to suborbicular, densely hairy, midrib prominent below, concolorous. Flowers few to many in lax or subcapitate racemes on long slender peduncles 2 or more times as long as leaves, pink, back of standard petal densely hairy. Pods glabrous, spreading. July–Feb. Mountain and lowland fynbos, 30–900 m, SW, AP, LB, SE (Stellenbosch to Humansdorp).

angustifolia L. Diffuse sprawling shrublet, 20–60 cm. Leaves pinnately 5–11-foliolate, subsessile, leaflets linear, thinly hairy above, greyish hairy beneath, midrib sunken above, prominent beneath, margins revolute. Flowers in racemes on slender peduncles 2 or more times as long as leaves, pink or red, back of standard petal greyish hairy. Pods glabrous, spreading to deflexed. May–Dec. Lowland fynbos, 15–500 m, SW, AP, LB (Cape Peninsula to Riversdale).•

filiformis L.f. (= *Indigofera wynbergensis* S.Moore)   Erect or sprawling shrublet to 60 cm, stems wiry. Leaves pinnately to subdigitately (3–)5-foliolate, subsessile, leaflets linear-oblong, thinly hairy, midrib sunken above, prominent below, margins revolute. Flowers in racemes on slender peduncles more than twice as long as leaves, pink, violet or red, back of standard petal densely grey-silky. Pods glabrous, spreading. May–Dec. Mountain fynbos, 130–800 m, SW (Cape Peninsula).•

mauritanica (L.) Thunb. (= *Indigofera coriacea* Aiton var. *cana* Harv.)   Sprawling or prostrate, hairy shrublet with stems to 30 cm long. Leaves pinnately 5-foliolate, subsessile, leaflets obovate to cuneate-obovate, discolorous, grey-hairy beneath, midrib sunken above, prominent below. Flowers in racemes on thread-like peduncles more than twice as long as leaves, mauve or pink, back of standard petal softly hairy. Pods glabrous, spreading to deflexed. Apr.–Aug. Mountain fynbos, 60–660 m, SW (Cape Peninsula).•

mundtiana Eckl. & Zeyh.   Like **I. angustifolia** but leaves 7–9-foliolate and the leaflets broader, elliptic-oblong, discolorous, silvery beneath. Sept.–Feb. Mountain and lowland fynbos, AP, LB (Langeberg Mts and De Hoop).•

sarmentosa L.f.   Diffuse, wiry, sprawling reseeding shrublet to 20 cm. Leaves pinnately 3–5-foliolate, subsessile, leaflets oblong-elliptic to obovate, thinly hairy, midrib sunken above, prominent below, margins sometimes revolute, shorter than 11 mm. Flowers in racemes on thread-like peduncles more than twice as long as leaves, pink or magenta, back of standard petal hairy. Pods glabrous, spreading. July–Mar. Mountain fynbos, 100–1500 m, SW, LB, SE (Bainskloof to Tsitsikamma Mts).•

sp. 18 (= *Indigofera mischocarpa* Schltr. nom. nud.)   Like **I. angustifolia** but more diffuse with smaller, densely arranged, subdigitately 3–7-foliolate leaves; leaflets 3–6 mm long, c. 1 mm wide. Pods stipitate. Sept.–Dec. 200–400 m, SW (Hottentots Holland Mts to Napier).•

sp. 19   Like **I. sarmentosa** but leaves subdigitately (3–)5-foliolate, crowded; leaflets leathery, oblanceolate to obcordate, dark-spotted above, greyish hairy beneath, 3–7 mm long; stipules minute. Flowers in racemes on filiform peduncles, pink to magenta, petals persistent. Oct.–Mar. Mountain fynbos, 600–1600 m, KM, SE (Swartberg and Outeniqua Mts).•

LEBECKIA GANNA c. 45 spp., southern Africa and Namibia

*A. Leaves 1-foliolate (see also **L. pungens**)*

ambigua E.Mey.   Erect, multistemmed shrublet to 1 m. Leaves 1-foliolate, terete. Flowers yellow. Oct. Karroid scrub, lowland fynbos, sandy soil, 100–300 m, NW (Namaqualand to Olifants River valley).

carnosa (E.Mey.) Druce   Sprawling shrublet to 40 cm. Leaves 1-foliolate, terete. Flowers yellow. Oct.–June. Mountain and lowland fynbos, 30–800 m, NW, SW (Cedarberg Mts to Bredasdorp).•

longipes Bolus   Sprawling shrublet with branches to 80 cm long. Leaves 1-foliolate, terete. Flowers yellow. Dec. Mountain fynbos, 500–2000 m, NW (Cedarberg Mts to Cold Bokkeveld).•

macowanii T.M.Salter   Sprawling, softly woody shrublet. Leaves 1-foliolate, terete. Flowers yellow. Sept.–Nov. Lowland fynbos, SW (Cape Peninsula).•

meyeriana Eckl. & Zeyh.   Decumbent or suberect shrublet to 40 cm. Leaves 1-foliolate, terete. Flowers yellow. Oct.–Dec. Lowland and mountain fynbos, 50–400 m, SW (Cape Peninsula to Worcester).•

pauciflora Eckl. & Zeyh.   Slender, much-branched shrublet to 80 cm. Leaves 1-foliolate, terete. Flowers yellow, the standard partly purple and the keel spirally twisted. Sept.–Jan. Mountain fynbos, 130–600 m, NW, KM (Piketberg to Outeniqua and Swartberg Mts).•

plukenetiana E.Mey.   Decumbent or ascending subshrub to 40 cm. Leaves 1-foliolate, terete. Flowers yellow. Sept.–Oct. Sandveld, 200–300 m, SW (Saldanha Bay to Cape Peninsula).•

sepiaria Thunb.   Multistemmed, diffuse or decumbent shrublet to 40 cm. Leaves 1-foliolate, terete. Flowers yellow. July–Nov. Renosterveld, strandveld, 210–700 m, SW, AP (Malmesbury to Riversdale).•

simsiana Eckl. & Zeyh.   Spreading shrublet to 50 cm. Leaves 1-foliolate, terete. Flowers yellow. Aug.–Nov. Renosterbos-fynbos scrub, strandveld, 30–600 m, NW, SW, LB (Namaqualand to Swellendam).

wrightii (Harv.) Bolus   Trailing, prostrate subshrub with branches 60 cm long. Leaves 1-foliolate, flattish. Flowers bright yellow, keel spirally twisted. June–Dec. Mountain fynbos, 400–960 m, SW (Cape Peninsula to Hermanus).•

A. Leaves 3-foliolate

bowieana Benth.   Rigid shrub to 1.5 m. Leaves 3-foliolate. Flowers bright yellow. May. Renosterbos-fynbos scrub, 250 m, LB (Bredasdorp to Mossel Bay).•

cinerea E.Mey.   Rigid shrublet to 30 cm. Leaves 3-foliolate, flattish. Flowers yellow. Aug.–Sept. Karroid scrub, 130–300 m, NW (Richtersveld to Vanrhynsdorp).

cytisoides Thunb.   Silvery shrub or small tree to 2 m. Leaves 3-foliolate, flat. Flowers bright yellow. July–Mar. Karroid scrub, fynbos, 180–1000 m, NW, SW (Bokkeveld Mts to Swellendam).•

fasciculata Benth.   Rigid shrublet to 50 cm. Leaves 3-foliolate. Flowers yellow. Sept. Lowland fynbos, 260 m, LB (Albertinia).•

inflata Bolus   Erect or prostrate shrublet to 20 cm. Leaves 3-foliolate, flat. Flowers yellow, with standard partly brown. Sept.–Nov. Mountain fynbos, lower slopes, 200–860 m, SW (Cape Peninsula to Hermanus).•

leipoldtiana Schltr. ex R.Dahlgren   Much-branched, thorny shrublet, 0.5–1 m. Leaves 3-foliolate. Flowers bright yellow. July–Oct. Renosterbos-fynbos scrub, 330–1160 m, NW (Namaqualand to Clanwilliam).

leptophylla Benth.   Shrublet to 25 cm. Leaves 3-foliolate, leaflets linear. Flowers yellow. Aug.–Nov. Mountain fynbos, high alt., LB (Langeberg: Swellendam).•

lotononoides Schltr.   Multistemmed shrub to 1 m. Leaves 3-foliolate, flattish. Flowers pale yellow, fading pink. July–Nov. Coastal fynbos, 100 m, SW (Saldanha Bay).•

melilotoides R.Dahlgren   Decumbent or suberect shrublet to 10 cm. Leaves 3-foliolate, flattish. Flowers pale yellow. Oct.–Nov. Karroid scrub, 620–700 m, KM (Touwsrivier).•

mucronata Benth.   Small, erect shrublet to 20 cm. Leaves 3-foliolate, flattish. Flowers yellow, standard brown on back. Sept.–Nov. Mountain fynbos, 600–800 m, SE (Elandsberg Mts, Port Elizabeth).•

multiflora E.Mey.   Shrub to 1.5 m. Leaves 3-foliolate, long-petiolate. Flowers yellow. Mar.–Sept. Karroid scrub, strandveld, 170–480 m, NW (S Namibia to Velddrif).

psiloloba Walp.   Rigid, thorny, almost leafless shrub to 1.5 m. Leaves 3-foliolate, flattish. Flowers yellow. Dec.–June. Grassland, streamsides, 30–260 m, SE (Uitenhage to E Cape).

pungens Thunb.   Rigid, thorny, almost leafless shrub to 1.5 m. Leaves 1–3-foliolate, flattish. Flowers bright yellow. Sept.–Oct. Karroid scrub, 700–850 m, KM, SE (Witteberg to Swartberg Mts and S. Karoo).

sericea Thunb.   Shrub to 1.5 m. Leaves 3-foliolate, flattish. Flowers yellow or cream-coloured. Aug.–Apr. Karroid scrub, 600–1200 m, NW (Namaqualand to Clanwilliam).

sessiliflora (Eckl. & Zeyh.) Benth.   Rigid shrub, 0.4–1.5 m. Leaves 3-foliolate. Flowers bright yellow. Oct.–Nov. Lowland fynbos, limestone, 50–200 m, AP (De Hoop to Stilbaai).•

spinescens Harv.   Rigid, thorny shrublet to 1 m. Leaves 3-foliolate, flattish. Flowers yellow. Oct.–Mar. Karroid scrub, 700–1160 m, NW (Namibia and Karoo to Clanwilliam).

LESSERTIA (= *SUTHERLANDIA*)   BALLOON PEA, BLAAS-ERTJIE   c. 55 spp., S and E tropical Africa

A. Flowers red, 25–35 mm long; wing petals very small, concealed in calyx; standard petal much shorter than keel

canescens Goldblatt & J.C.Manning (= *Sutherlandia tomentosa* Eckl. & Zeyh.)   EENDJIES   Shrublet to 40 cm. Leaves imparipinnate, leaflets broadly oblong-obcordate, retuse, silvery grey-hairy on both surfaces. Flowers red. Pods large, inflated, papery, adpressed-hairy. Sept.–Dec. Coastal sands, below 100 m, SW, AP (Melkbosstrand to Mossel Bay).•

frutescens (L.) Goldblatt & J.C.Manning (= *Sutherlandia frutescens* (L.) R.Br.)   KANKERBOS   Shrublet to 1 m. Leaves imparipinnate, leaflets oblong, obtuse, glabrous or thinly hairy above, greyish green. Flowers red. Pods large, inflated, papery, glabrous. July–Dec. Sandstone and shale flats and slopes, NW, SW, KM, LB, SE (Namaqualand and W Karoo to E Cape).

A. Flowers pinkish or mauve to red, to 15 mm long; wing petals conspicuous and protruding; standard petal about as long as keel
  B. Flowers congested on peduncles longer than leaves

capensis (P.J.Bergius) Druce   HARSLAGBOSSIE   Prostrate subshrub to 10 cm. Leaves imparipinnate, leaflets oblong. Flowers congested on peduncles longer than leaves, dull or dark red with pink standard. Pods oblong-elliptic, compressed, papery, glabrous. Aug.–Oct. Mountain fynbos, 100–420 m, NW, SW (Tulbagh to Hangklip).•

carnosa Eckl. & Zeyh.   Sprawling shrublet to 30 cm. Leaves imparipinnate, leaflets linear-channelled, leathery. Flowers crowded on peduncles longer than leaves. Flowers violet. Pods broadly oblong, subinflated, adpressed-hairy. Aug.–Dec. Karroid hills, SE (Uitenhage to E Cape).

diffusa R.Br.   DOU-GANSIEBOS   Softly hairy, sprawling subshrub, 20–45 cm. Leaves imparipinnate, leaflets oblong-cuneate. Flowers congested on peduncles longer than leaves, pink. Pods obliquely ovate, semilunate, compressed, prominently veined. Aug.–Sept. Fynbos, NW (Namaqualand to Piketberg).

excisa DC.   Like **L. diffusa** but pods obliquely ovoid, inflated, papery, glabrous. Aug.–Oct. Sandstone slopes and flats, 100–830 m, NW, SW (Namaqualand to Cape Peninsula).

miniata T.M.Salter   Like **L. capensis** but leaves usually narrower, linear-oblong, flowers orange-red and pods ellipsoid, inflated, leathery or woody. Sept.–Oct. Coastal fynbos, often limestone, 100 m, SW, AP (Cape Peninsula to Stilbaai).•

stenoloba E.Mey.   Like **L. capensis** but pods oblong-falcate, more than twice as long as broad. July–Oct. Coastal and arid fynbos, 100–1660 m, AP, LB, SE (Bredasdorp to E Cape).

subumbellata Harv.   Like **L. miniata** but pods not inflated. Oct. Grassy clay slopes, LB (Swellendam).•

tomentosa DC.   Like **L. excisa** but ovary hairy and pods globose and velvety. Aug.–Sept. Coastal sands and limestones, SW (Saldanha to Cape Peninsula).•

  B. Flowers scattered or congested but then peduncles shorter than leaves
    C. Pods oblong, more than twice as long as broad

annularis Burch.   Prostrate or sprawling, resprouting subshrub to 50 cm. Leaves imparipinnate, leaflets oblong-obovate. Flowers crowded on peduncles shorter than leaves, mauve to cerise. Pods narrowly oblong, strongly coiled into a semicircle, compressed, papery, adpressed-hairy. Aug.–Oct. Karroid scrub, 450–1000 m, KM (Namibia to Montagu to E Cape).

brachystachya E.Mey.   Willowy, woody annual to 60 cm. Leaves imparipinnate, leaflets oblong. Flowers more or less scattered on peduncles shorter than leaves, purple to pale pink. Pods oblong, compressed, papery, adpressed-hairy, pedicels elongating and much longer than calyx. July–Dec. Disturbed places, grassy areas, below 700 m, SE (Mossel Bay to E Cape).

lanata Harv.   Shrublet with densely white-woolly branches to 1m. Leaves imparipinnate, leaflets oblong, white-hairy. Flowers scattered on terminal peduncles, purplish pink. Pods oblong, compressed, papery. Aug.–Sept. Stony slopes, KM (Swartberg Mts).•

C. Pods broadly obovate to suborbicular, up to twice as long as broad

argentea (Thunb.) Harv.   STRAND-ERTJIESBOS   Sprawling subshrub, 30–50 cm. Leaves imparipinnate, leaflets oblong. Flowers scattered on peduncles slightly longer than leaves, dull purple. Pods broadly elliptic-oblong, obtuse at both ends, compressed, papery, adpressed-hairy. Sept.–Oct. Coastal dunes, SW (Saldanha Bay to Cape Peninsula).•

depressa Harv.   Sprawling subshrub, 15–45 cm. Leaves imparipinnate, leaflets oblong. Flowers crowded on peduncles shorter than leaves, pink. Pods obliquely ovoid, subinflated, glabrous. Aug.–Dec. Stony slopes, streamsides, SE (Uitenhage to E Cape).

globosa L.Bolus   Like **L. prostrata** but pods globose, inflated. Sept.–Oct. Sandy flats, SW (Mamre).•

prostrata DC.   Like **L. argentea** but flowers congested on peduncles shorter than leaves. Sept.–Nov. Sandy flats and slopes, NW, SW (Clanwilliam to Stellenbosch).•

rigida E.Mey. Rigid, somewhat spinescent shrublet to 50 cm. Leaves imparipinnate, leaflets oblong. Flowers scattered on rigid, at length somewhat spinescent peduncles longer than leaves, pink to purple. Pods obliquely broadly ovate, compressed around margins but subinflated in middle, mostly distinctly veined, pedicels longer than calyx and mostly blackish hairy. Aug.–Sept. Stony and sandy flats and slopes, NW, SW, KM (Bokkeveld Mts and W Karoo to Kammanassie Mts).

spinescens E.Mey. Like **L. rigida** but peduncles shorter than leaves, distinctly spinescent and pods suborbicular and not conspicuously veined. Aug.–Sept. Stony shale slopes, NW (Namaqualand to Clanwilliam).

LIPARIA• (= *PRIESTLEYA* in part)   MOUNTAIN PEA   20 spp., N and W Cape

A. Inflorescences not decussate; bracts not clasping base of calyx
B. Racemes congested and head-like

parva Vogel ex Walp.   Prostrate, resprouting shrublet to 20 cm. Leaves elliptic. Flowers 7 in congested, globose racemes, pale yellow-green, bracts longer than pedicel, not clasping base of calyx, dark reddish brown. May–Nov. Lowland and mountain fynbos, below 300 m, SW (Cape Peninsula).•

splendens (Burm.f.) Bos & De Wit   MOUNTAIN DAHLIA   Much-branched, resprouting shrub to 1 m. Leaves elliptic. Flowers 15–17 in congested, globose racemes, orange to red, bracts longer than pedicel, not clasping base of calyx, dark reddish brown. May–Jan. Mountain and lowland fynbos, 20–1200 m, SW, LB (Cape Peninsula to Hottentots Holland Mts, Albertinia).•

B. Racemes not congested and head-like

angustifolia (Eckl. & Zeyh.) A.L.Schutte (= *Priestleya angustifolia* Eckl. & Zeyh.)   Reseeding shrublet to 30 cm. Leaves linear-elliptic. Flowers 3–10 in racemes, yellow, bracts longer than pedicel, not clasping base of calyx. Oct.–Dec. Lowland fynbos, marshy places, below 130 m, SW (Cape Peninsula to Hermanus).•

confusa A.L.Schutte   Resprouting shrublet to 60 cm. Leaves narrowly elliptic-oblanceolate. Flowers 6–13 in racemes, yellow, bracts longer than pedicel, not clasping base of calyx. Nov.–Dec. Mountain fynbos, subalpine vegetation, above 1250 m, KM (Groot Swartberg).•

genistoides (Lam.) A.L.Schutte (= *Priestleya teres* (Thunb.) DC.)   Reseeding, tree-like shrub to 2.5 m. Leaves narrowly oblanceolate. Flowers 3–7 in racemes, yellow, bracts shorter than pedicel, not clasping base of calyx. Nov.–Dec. Mountain fynbos, streamsides, 1300–1660 m, SE (Kammanassie and Kouga Mts).•

graminifolia L.   Resprouting shrublet to 30 cm. Leaves linear, often twisted. Flowers 5–10 in racemes, yellow, bracts longer than pedicel, not clasping base of calyx. Lowland fynbos, SW (Cape Flats).•

hirsuta Thunb. (= *Priestleya hirsuta* (Thunb.) DC.)   Reseeding, single-stemmed, sometimes multistemmed shrub to 3 m. Leaves oblanceolate to obovate. Flowers 7–12 in racemes, yellow, bracts longer than pedicel, clasping base of calyx. Aug.–Apr. Mountain fynbos, 300–1070 m, LB, SE (Langeberg to Kareedouw Mts).•

racemosa A.L.Schutte   Reseeding, tree-like shrub to 3 m. Leaves oblanceolate to narrowly elliptic. Flowers 3–5 in racemes, yellow, bracts longer than pedicel, not clasping base of calyx. Flowers bright yellow. Dec.–Feb. Mountain fynbos, streamsides, 1600 m, KM (Groot Swartberg).•

striata A.L.Schutte   Resprouting, multistemmed shrublet to 1 m. Leaves oblanceolate. Flowers 4–6 in racemes, yellow with maroon markings on inner face of standard petal, bracts as long as pedicel, not clasping base of calyx. Dec.–Jan. Renosterveld, 200 m, AP (Heidelberg-Swellendam: Soutkloof).•

A. Inflorescences decussate
C. Bracts clasping base of calyx

boucheri (E.G.H.Oliv. & A.Fellingham) A.L.Schutte   Single-stemmed, reseeding shrublet to 50 cm. Leaves narrowly elliptic-obovate, flat. Flowers 2 in decussate racemes, pale greenish cream, bracts longer than pedicel, clasping base of calyx. Flowers pale greenish cream. Apr.–May. Mountain fynbos, 1120–1330 m, SW (Kogelberg).•

calycina (L.Bolus) A.L.Schutte (= *Priestleya calycina* L.Bolus)   Single-stemmed, reseeding shrub to 2.5 m. Leaves elliptic, concave. Flowers 4 in decussate racemes, yellow, bracts longer than pedicel, clasping base of calyx. Aug.–Nov. Mountain fynbos, streamsides, 460–1000 m, SW (Hottentots Holland to Kleinrivier Mts).•

capitata Thunb. (= *Priestleya capitata* (Thunb.) DC.)   Multistemmed, resprouting shrublet to 60 cm. Leaves linear-elliptic, concave. Flowers 4 in decussate racemes, yellow, bracts longer than pedicel, clasping base of calyx. Oct.–Dec. Mountain fynbos, crests and upper slopes, 580–2000 m, SW, KM (Hex River to Riviersonderend Mts and Klein Swartberg).•

congesta A.L.Schutte   Single-stemmed, much-branched, reseeding shrub to 2 m. Leaves obovate to elliptic. Flowers 4 in decussate racemes, yellow, bracts longer than pedicel, clasping base of calyx. Oct.–Nov. Arid mountain fynbos, 1600–1800 m, NW (Cedarberg Mts).•

umbellifera Thunb. (= *Priestleya umbellifera* (Thunb.) DC.)   Single-stemmed, reseeding shrub to 3 m. Leaves elliptic to narrowly elliptic, margins slightly incurved. Flowers 4 in decussate racemes, yellow, bracts as long as or longer than pedicel, sometimes shorter, mostly clasping base of calyx. Oct.–Feb. Marshy sandstone slopes and streamsides, 1200–2000 m, NW, SW, KM (Cedarberg Mts to Du Toitskloof and Klein Swartberg).•

vestita Thunb. (= *Priestleya vestita* (Thunb.) DC.)   Multistemmed, resprouting shrub to 2.5 m. Leaves broadly elliptic to almost circular, concave. Flowers 4 in decussate racemes, yellow, bracts longer than pedicel, clasping base of calyx. Flowers bright yellow. Mar.–Oct. Mountain fynbos, 100–1000 m, SW (Somerset West to Bredasdorp).•

*C. Bracts not clasping base of calyx (see also **L. umbellifera**)*

bonaespei A.L.Schutte   Shrublet to 30 cm. Leaves obtrullate, somewhat concave. Flowers 4 in decussate racemes, yellow, bracts shorter than pedicel, not clasping base of calyx. Mar.–Apr. Mountain fynbos, above 1400 m, SW (Hottentots Holland Mts).•

laevigata (L.) (= *Priestleya laevigata* (L.) DC.)   Single-stemmed, tree-like, reseeding shrub to 2.5 m. Leaves elliptic to narrowly elliptic. Flowers 4 in decussate racemes, yellow, bracts shorter than pedicel, not clasping base of calyx. Oct.–Nov. Mountain fynbos, marshy places, 740–1030 m, SW (Cape Peninsula).•

latifolia (Benth.) A.L.Schutte (= *Priestleya latifolia* Benth.)   Multistemmed, resprouting shrub to 1.5 m. Leaves obovate to broadly obovate. Flowers 4 in decussate racemes, yellow, bracts shorter than pedicel, not clasping base of calyx. Apr.–June. Mountain fynbos, 660–1400 m, NW, SW, KM (Cold Bokkeveld to Franschhoek Mts and Gamkaberg).•

myrtifolia Thunb. (= *Priestleya myrtifolia* (Thunb.) DC.)   Single-stemmed, tree-like, reseeding shrub to 3 m. Leaves elliptic to narrowly elliptic. Flowers 4 in decussate racemes, yellow, bracts shorter than pedicel, not clasping base of calyx. Mar.–June. Mountain fynbos, 300–1400 m, NW, SW, LB, SE (Hex River to Outeniqua Mts).•

rafnioides A.L.Schutte   Tree-like shrub to 4 m. Leaves cordate, with prominent palmate venation. Flowers 4 in decussate racemes, yellow, bracts shorter than pedicel, not clasping base of calyx. Oct.–Feb. Mountain fynbos, streamsides, 500–1600 m, SW (Groot Drakenstein and Hottentots Holland Mts).•

LOTONONIS (= *BUCHENROEDERA*) 150 spp., southern and E Africa, Eurasia and Pakistan

A. Flowers blue or blue and yellow

alpina (Eckl. & Zeyh.) B.-E.van Wyk (= *Buchenroedera multiflora* Eckl. & Zeyh.)   Much-branched shrublet to 60 cm. Leaves 3-foliolate, leaflets oblanceolate, thinly hairy, stipules paired. Flowers solitary on slender peduncles, blue, standard and keel silky. Sept.–June. Mountain fynbos, grassland, 700–1320 m, SE (Baviaanskloof Mts to E Cape).

argentea Eckl. & Zeyh.   Prostrate or procumbent shrublet. Leaves 3-foliolate, leaflets obovate, densely silvery silky above, stipules solitary or paired, unequal. Flowers solitary on slender peduncles, blue. Nov. Karroid scrub, 300–1000 m, KM (Montagu: Kogmanskloof to Waboomsberg).•

azurea (Eckl. & Zeyh.) Benth.   Prostrate or procumbent shrublet to 10 cm. Leaves 3-foliolate, stipules paired, unequal. Flowers solitary on slender peduncles, blue. Aug.–Nov. Mountain and lowland fynbos, 80–1260 m, SE (Uniondale to Port Elizabeth).•

complanata B.-E.van Wyk   Prostrate or procumbent shrublet to 10 cm. Leaves 3-foliolate, leaflets oblanceolate, ciliate, tips recurved, stipules solitary. Flowers solitary on slender peduncles, blue. June. Coastal renosterveld, 120 m, SW (Elandsberg).•

filiformis B.-E.van Wyk   Prostrate or procumbent shrublet to 10 cm. Leaves 3-foliolate, leaflets linear, thinly hairy, stipules paired, unequal. Flowers solitary on slender peduncles, blue and yellow. Oct.–Dec. Renosterbos-fynbos scrub, arid fynbos, 430–530 m, KM, LB (Oudtshoorn to Mossel Bay).•

maximiliani Schltr. ex De Wild.   Prostrate annual. Leaves 3-foliolate, obovate-obcordate, thinly hairy, leaflets elliptic-obovate, closely hairy, stipules solitary or paired. Flowers solitary on short peduncles, blue, often cleistogamous. July–Oct. Karroid scrub, 300–730 m, NW (Namaqualand to Clanwilliam).

stricta (Eckl. & Zeyh.) B.-E.van Wyk   Shrub with erect branches to 1.5 m. Leaves 3-foliolate, stipules paired. Flowers 1 or 2 on short peduncles, blue and yellow, standard glabrous. Mar.–June. Grassland, 160–1700 m, SE (Uitenhage to KwaZulu-Natal and Lesotho).

varia (E.Mey.) Steud.   Prostrate or procumbent shrublet to 10 cm. Leaves 3-foliolate, leaflets broadly obovate, glabrescent, stipules paired, unequal. Flowers often 2 or more, on slender peduncles, blue. Sept.–Apr. Mountain fynbos, 100–1000 m, SW (Franschhoek to Kleinrivier Mts).•

villosa (E.Mey.) Steud.   Prostrate or procumbent shrublet to 10 cm. Leaves 3-foliolate, leaflets elliptic, softly hairy, stipules solitary. Flowers blue. Oct. Coastal renosterveld, 180 m, NW, SW (Piketberg to Caledon).•

A. Flowers yellow, sometimes partly white, pink, purple or brown
B. Flowers sessile in dense heads or spikes

bolusii Dummer   Prostrate annual to 5 cm. Leaves 5-foliolate, leaflets obovate-cuneate, hairy below, stipules solitary. Flowers sessile in dense heads, yellow, petals shorter than calyx. Aug.–Oct. Lowland fynbos, 400 m, NW, SW (Olifants River valley to Hopefield).•

globulosa B.-E.van Wyk   Like **L.bolusii** but bracts large, cordate and leaflets 3-foliolate, keel and standard hairy at tips. Oct. Karroid scrub, 1050 m, NW, KM (Ceres to Witteberg).•

laticeps B.-E.van Wyk   Like **L. globulosa** but inflorescence discoid and standard half as long as keel. Nov. Mountain fynbos, sandy stony plateau, 1160 m, NW (Cold Bokkeveld).•

longicephala B.-E.van Wyk   Prostrate annual. Leaves 3-foliolate, leaflets obovate-cuneate, stipules solitary. Flowers sessile in dense spikes, yellow, keel hairy at tip. Oct. Karroid scrub, 900–1050 m, NW, KM (Cold Bokkeveld to Ceres and Witteberg).•

rosea Dummer   Prostrate annual to 10 cm. Leaves 3–5-foliolate, leaflets, cuneate-oblanceolate, hairy below, stipules solitary. Flowers sessile in dense heads, pink, keel hairy at tip. Oct.–Nov. Karroid scrub, 100–200 m, NW (Olifants River valley: Gifberg to Citrusdal).•

B. Flowers pedicellate on short or long peduncles
*C. Stipules paired or absent (see also **L. leptoloba, L. prostrata**)*

brevicaulis B.-E.van Wyk   Shrublet with procumbent flowering branches from a thick woody rootstock. Leaves 5–8-foliolate, leaflets elliptic, densely silky, stipules paired. Flowers 1 or 2(–4) on slender peduncles, yellow. Sept.–Oct. Karroid scrub, c. 900 m, KM (Bonteberg).•

dahlgrenii B.-F.van Wyk   Like **L. rigida** but flowers solitary, and stipules always absent. Oct. Mountain fynbos, 300–800 m, KM (Rooiberg Mts).•

densa (Thunb.) Harv.   Woody, sometimes thorny, shrub to 1.2 m. Leaves 3-foliolate, leaflets linear to obovate, thinly hairy, stipules absent. Flowers few on slender, often thorn-like peduncles, yellow. Aug.–Oct. Renosterveld, strandveld, 350–940 m, NW, SW (Namaqualand to Paarl).

dissitinodis B.-E.van Wyk   Shrublet with erect branches to 50 cm. Leaves 3-foliolate, leaflets linear-oblong, densely silky, stipules sometimes present, solitary or paired. Flowers solitary on slender peduncles, yellow. July–Sept. Karroid scrub and renosterveld, 760–900 m, KM (Swartberg Mts and Gamkaberg).•

exstipulata L.Bolus   Procumbent shrublet to 30 cm, with thick woody rootstock. Leaves 3-foliolate, leaflets oblanceolate, thinly hairy, stipules absent. Flowers few, subumbellate on slender peduncles, yellow, fading orange. Oct. Renosterveld, 1200 m, NW (Ceres).•

fastigiata (E.Mey.) B.-E.van Wyk   Shrublet with erect flowering branches from a thick woody rootstock. Leaves 3-foliolate, leaflets linear-oblong, thinly hairy, stipules paired. Flowers few, subumbellate on slender peduncles, yellow. Oct.–Nov. Lowland fynbos, 30–300 m, SW, LB (Cape Peninsula to Swellendam).•

hirsuta (Thunb.) D.Dietr.   Shrublet with slender branches from a woody rootstock. Leaves 1(–3)-foliolate, stipules absent or paired and leaf-like. Flowers few on slender peduncles, yellow, partly brown or dark purple. June–Oct. Karroid scrub, 720–2200 m, NW (Namaqualand and Karoo to Clanwilliam).

involucrata (P.J.Bergius) Benth.   Shrublet with procumbent branches from a thick woody rootstock. Leaves 3–5-foliolate, leaflets linear to oblanceolate, thinly hairy, stipules paired. Flowers few, subumbellate on slender peduncles, yellow. June–Oct. Mountain and coastal fynbos, karroid scrub, 150–750 m, NW, SW, AP (Clanwilliam to Potberg).•

lamprifolia B.-E.van Wyk   Prostrate or procumbent shrublet. Leaves 3-foliolate, leaflets lanceolate to obovate, densely silky, stipules solitary when present. Flowers solitary on slender peduncles, yellow. Aug. North slopes in fynbos, c. 300 m, LB (Langeberg Mts).•

macrocarpa Eckl. & Zeyh. Prostrate or procumbent, often mat-forming, shrublet. Leaves 3-foliolate, leaflets obovate, glabrescent, stipules paired, unequal. Flowers few on slender peduncles, bracteolate, yellow. July–Aug. Renosterveld, 130–280 m, NW (Clanwilliam to Citrusdal).•

pallens (Eckl. & Zeyh.) Benth.   Prostrate annual. Leaves 3-foliolate, leaflets obovate, thinly hairy below, stipules paired. Flowers few on slender peduncles, yellow, calyx hairs in rows on veins. Karroid scrub, NW (Clanwilliam).•

racemiflora B.-E.van Wyk   Subshrub with procumbent or erect flowering branches from a woody rootstock. Leaves 3-foliolate, leaflets oblong, thinly silky, stipules paired. Flowers several on elongate peduncles, yellow. Oct. Karroid scrub, 300 m, NW (Clanwilliam).•

*C. Stipules solitary (see also **L. dissitinodis, L. lamprifolia**)*

acocksii B.-E.van Wyk   Decumbent, sparsely branched shrublet to 50 cm. Leaves 3-foliolate, linear-oblanceolate, thinly hairy, stipules solitary. Flowers 3–5 on slender peduncles, yellow. Sept. Renosterbos-fynbos scrub, c. 500 m, KM (Rooiberg and Swartberg Mts).•

acuminata Eckl. & Zeyh.   Prostrate or procumbent shrublet to 25 cm. Leaves 3-foliolate, leaflets elliptic, silky, net-veined. stipules solitary. Flowers solitary on slender peduncles, yellow. Nov.–Dec. Disturbed renosterveld, grassy fynbos, 30–50 m, SE (Humansdorp to Port Elizabeth).•

carnea B.-E.van Wyk   Prostrate or procumbent annual, roots bright yellow. Leaves 3-foliolate, leaflets obovate, shortly silky, stipules solitary. Flowers few on vestigial peduncles, yellow, fading salmon-orange. Sept.–Oct. Karroid scrub, 800–1200 m, NW (Namaqualand to Bokkeveld Mts).

comptonii B.-E.van Wyk   Like **L. acocksii** but flowers mostly more than 5, yellow, fading orange. Oct. Karroid scrub, 1100–1200 m, KM (Swartberg Mts: Bantamskop to Seweweekspoort).•

elongata (Thunb.) D.Dietr.   Prostrate or procumbent shrublet to 20 cm. Leaves 3-foliolate, leaflets linear to oblanceolate, thinly hairy, stipules solitary. Flowers solitary on slender peduncles, yellow and purplish. Nov.–Dec. Sandstone slopes in renosterbos-fynbos scrub, 660–1800 m, KM, SE (Swartberg and Outeniqua Mts to Langkloof).•

esterhuyseana B.-E.van Wyk   Prostrate annual with sparsely leafy branches to 30 cm long. Leaves 3-foliolate, leaflets cuneate-obovate, densely silky, stipules solitary. Flowers several on short peduncles, yellow, keel silky at tip, standard half as long as keel. Nov. Sandy flats in fynbos, 1000–1330 m, NW (Swartruggens).•

falcata (E.Mey.) Benth.   Prostrate annual to 20 cm. Leaves 3-foliolate, leaflets linear-oblanceolate, shortly silky, stipules solitary. Flowers 1–few subumbellate, on short peduncles, yellow, fading orange. July–Oct. Karroid scrub, 260–900 m, NW, KM (Namibia and Karoo to Cold Bokkeveld and Swartberg Mts).

glabra (Thunb.) D.Dietr.   Prostrate shrublet. Leaves 3-foliolate, leaflets broadly ovate, shortly silky, stipules solitary. Flowers few, subumbellate on slender peduncles, yellow to white fading pink. Nov.–May. Coastal fynbos, to 250 m, LB, SE (Riversdale to E Cape).

leptoloba Bolus   Sprawling annual to 80 cm. Leaves 3-foliolate, basal leaves opposite, leaflets obovate, thinly hairy, stipules solitary or paired, dissimilar in shape and size. Flowers 1(2) on slender peduncles, yellow, often cleistogamous. Sept.–Oct. Karroid scrub, arid fynbos, 400–1250 m, NW (Namibia to Bokkeveld Mts).

monophylla Harv.   Prostrate shrublet. Leaves 1-foliolate, leaflets ovate to oblong, shortly silky below, stipules solitary. Flowers few on short peduncles, yellow. Oct.–Nov. Mountain fynbos, 280 m, SE (Van Staden's Mts).•

nutans B.-E.van Wyk   Like **L. umbellata** but flowers smaller, shorter than 7 mm. Sept. Renosterveld, SE (George to Uniondale).•

oxyptera (E.Mey.) Benth.   Prostrate or procumbent annual, with bright yellow roots. Leaves 3-foliolate, leaflets oblanceolate-obovate, shortly silky, stipules solitary. Flowers few on slender, long peduncles, yellow, fading orange. Sept.–Oct. Karroid scrub, riverbanks, 100–300 m, NW, SW (Clanwilliam to Wellington).•

parviflora (P.J.Bergius) D.Dietr.   Like **L. falcata** but softly hairy, white and pink. Sept.–Oct. Karroid scrub, 100–650 m, NW, SW (Namaqualand to Hopefield).

perplexa (E.Mey.) Eckl. & Zeyh.   Like **L. rostrata** but flowers 1 or 2 on slender peduncles. Sept.–Nov. Mountain and lowland fynbos, 100–830 m, NW, SW (Bokkeveld Mts to Cape Peninsula).•

prostrata (L.) Benth.   Prostrate or procumbent shrublet. Leaves 3-foliolate, leaflets cuneate-obovate, thinly hairy, stipules solitary or paired, unequal. Flowers solitary on slender peduncles, yellow. Aug.–Sept. Renosterveld and lowland fynbos, 100–320 m, SW (Malmesbury to Stellenbosch).•

pumila Eckl. & Zeyh.   Prostrate annual. Leaves 3-foliolate, the basal often 1-foliolate, leaflets obovate to suborbicular, shortly hairy, stipules solitary. Flowers 1–4, subumbellate on short peduncles, yellow, partly pink, with keel dark purple at tip. Sept.–Dec. Karroid scrub, renosterveld, 360–830 m, KM, SE (Karoo and Swartberg Mts to E Cape).

pungens Eckl. & Zeyh.   Prostrate or procumbent shrublet to 20 cm. Leaves 3-foliolate, leaflets linear to obovate, thinly hairy, stipules solitary. Flowers solitary on short peduncles, yellow, often cleistogamous. Aug.–Mar. Karroid scrub, 30–1100 m, NW, SW, KM, LB, SE (Klein Swartberg and Worcester to E Cape and Karoo).

purpurescens B.-E.van Wyk   Sparsely branched shrublet to 20 cm, with thick woody branches. Leaves 3-foliolate, leaflets elliptic, shortly hairy below, stipules solitary. Flowers few on slender peduncles, pale purple. June–Oct. Karroid scrub, c. 300 m, KM (Montagu).•

rigida (E.Mey.) Benth.   Rigid, widely branched, shrublet to 80 cm. Leaves 3-foliolate, leaflets linear-oblanceolate, densely silky, stipules solitary or absent. Flowers 2 or 3 on thorn-like peduncles, yellow. July–Aug. Renosterbos-fynbos scrub, 330–460 m, NW (Worcester to Robertson).•

rostrata Benth.   Like **L. stenophylla** but flowers smaller, shorter than 7 mm, on short or long peduncles. Sept.–Nov. Karroid scrub, riverbanks, 100–700 m, NW, SW (Namaqualand to Villiersdorp).

sabulosa T.M.Salter   Like **L. falcata** but flowers on long slender peduncles, keel acute. Sept.–Oct. Karroid scrub, strandveld, riverbanks, 100 m, NW, SW (Clanwilliam to Cape Peninsula).•

stenophylla (Eckl. & Zeyh.) B.-E.van Wyk   Prostrate or procumbent woody annual, with bright yellow roots. Leaves 3-foliolate, leaflets linear-oblong, shortly hairy, stipules solitary. Flowers few on short peduncles, yellow. Sept. Karroid scrub, 100–500 m, NW (Namaqualand to Piketberg).

tenella (E.Mey.) Eckl. & Zeyh.   Like **L. pungens** but standard purple-veined and keel tipped purple. Sept.–June. Karroid scrub, 600–700 m, KM, SE (Karoo and Little Karoo to E Cape).

umbellata (L.) Benth.   Prostrate shrublet with thick, woody branches to 10 cm. Leaves 3-foliolate, leaflets cuneate-obcordate, thinly hairy, stipules solitary. Flowers few, subumbellate on slender peduncles, yellow. May–Oct. Mountain and coastal fynbos, to 530 m, SW, LB, SE (Cape Peninsula to Port Elizabeth).•

venosa B.-E.van Wyk   Like **L. leptoloba** but leaflets linear-elliptic, softly hairy, stipules solitary. Flowers solitary on slender peduncles, yellow, fading blue. Sept. Karroid scrub, 1100–1460 m, KM (Karoo and Montagu to Barrydale).

viborgioides Benth.   Like **L. comptonii** but flowers smaller, shorter than 8 mm. Oct.–Nov. Renosterveld, 160–300 m, NW, LB (Montagu to Bredasdorp).•

*LOTUS   c. 100 spp., worldwide temperate

*subbiflorus Lag.   Mat-forming subshrub to 30 cm. Leaves 3-foliolate, leaflets obovate-oblanceolate, thinly hairy. Flowers subumbellate on slender peduncles, yellow. Nov.–Mar. Marshy places, SW, LB, SE (Eurasian weed).

MELOLOBIUM   c. 20 spp., southern Africa

adenodes Eckl. & Zeyh.   Glandular-hairy subshrub to 60 cm, sometimes weakly thorny. Leaves 3-foliolate, leaflets oblanceolate. Flowers in lax racemes, yellow. Sept.–Oct. Karroid scrub, 80–900 m, NW (Namaqualand, Karoo, Ceres, Piketberg, Robertson, Ladismith).

aethiopicum (L.) Druce   Softly hairy, scarcely sticky subshrub to 30 cm. Leaves 3-foliolate, leaflets oblanceolate. Flowers in lax racemes, yellow, fading orange. July–Sept. Coastal scrub and dunes, NW, SW (Lambert's Bay to Cape Peninsula).•

candicans (E.Mey.) Eckl. & Zeyh.   STROOPBOS   Rigid, thorny, scarcely glandular shrublet to 60 cm, with white-velvety stems. Leaves 3-foliolate, leaflets oblanceolate. Flowers few along the thorns, yellow, fading reddish orange. May–Jan. Karroid scrub, 400–1130 m, NW, SW, KM, SE (Namaqualand and Karoo to E Cape).

exudans Harv.   Glabrescent, sticky subshrub to 60 cm, glands not stalked. Leaves 3-foliolate, leaflets oblanceolate. Flowers in lax racemes, yellow, fading orange. June–Oct. Karroid scrub and coastal sands, 15–1330 m, NW, SW, KM (Piketberg to Little Karoo).•

humile Eckl. & Zeyh.   Like **M. adenodes** but stems with soft hairs among the glandular ones. Sept.–Oct. Karroid scrub, sandy places, NW (Namaqualand, Clanwilliam).

stipulatum (Thunb.) Harv.   Robust, glandular hairy shrublet to 40 cm. Leaves 3-foliolate, leaflets oblanceolate. Flowers imbricate in dense racemes, yellow. Oct. Karroid scrub, 900 m, NW (Hex River and Worcester).•

OTHOLOBIUM   SKAAPBOSTEE   c. 53 spp., southern and E Africa

A. Leaves 1-foliolate

accrescens C.H.Stirt.   Resprouting, rhizomatous subshrub to 20 cm. Leaves 1-foliolate, long-petiolate, broadly elliptic-obovate, densely glandular above; stipules glandular. Flowers ?purple, calyx black-ciliate on margins, accrescent. Sept.–Nov. Mountain fynbos, SE (Great Winterhoek Mts).•

dreweae C.H.Stirt.   Resprouting subshrub to 15 cm, branches dark-hairy. Leaves 1-foliolate, elliptic, hairy on veins, margins scabridulous; stipules glabrous. Flowers reddish purple, calyx shortly white- and long black-hairy. Oct.–Nov., mostly after fire. Mountain fynbos, 400 m, SW (Kleinrivier Mts).•

lanceolatum C.H.Stirt.   Resprouting, decumbent subshrub to 6 cm. Leaves 1-foliolate, subsessile, elliptic, margins scabridulous, glandular; stipules glandular. Flowers white, calyx densely white-hairy and glandular, lowest sepal larger. Nov.–Dec., after fire. Mountain fynbos, 100–200 m, SW (Shaw's Mts).•

rotundifolium (L.f.) C.H.Stirt.   Resprouting subshrub to 20 cm. Leaves 1-foliolate, rhomboid-obovate; stipules glabrous. Flowers pale lilac to white, calyx silky white-hairy and glandular, lowest sepal larger. Oct.–Jan. Mountain fynbos, 430–530 m, SW (Jonkershoek to Kleinrivier Mts).•

thomii (Harv.) C.H.Stirt.   Resprouting subshrub to 40 cm, branches white- or silver-hairy. Leaves 1-foliolate, elliptic, hairy, margins scaberulous; stipules hairy. Flowers purple, calyx densely black- and white-hairy. July–Dec., mostly after fire. Mountain fynbos, 100 m, SW (Kleinmond to Bredasdorp Mts).•

A. Leaves 3-foliolate
B. Leaves subsessile
C. Lateral leaflets symmetrical

arborescens C.H.Stirt.   Loosely branched tree to 5 m. Leaves digitately 3-foliolate, leaflets oblanceolate, adpressed-hairy; stipules glandular. Flowers white with purple keel tip, calyx grey-hairy and glandular, lower sepals recurved. Oct.–Dec. Arid fynbos, streamsides, NW, KM (Namaqualand and W Karoo to Swartberg Mts).

candicans (Eckl. & Zeyh.) C.H.Stirt.   Reseeding shrublet to 1 m. Leaves digitately 3-foliolate, leaflets oblanceolate, subglabrous; stipules ciliate. Flowers white with pink and violet keel tip, calyx glabrous to densely white-silky. Sept.–Nov. Mountain fynbos, renosterveld, 130–920 m, NW, SW, KM, LB, SE (Tulbagh to Grahamstown).

hirtum (L.) C.H.Stirt.   Erect, much-branched shrub to 2 m, white-hairy. Leaves digitately 3-foliolate, leaflets cuneate-obovate, adpressed-hairy and glandular; stipules glabrous. Flowers pale blue, mauve, calyx white-silky. Sept.–Dec. Renosterveld, 160–330 m, NW, SW (Piketberg to Gordon's Bay).•

rubicundum C.H.Stirt.   Shrub to 1.5 m. Leaves digitately 3-foliolate, leaflets oblanceolate, glabrescent, glandular; stipules thinly hairy and orange-glandular. Flowers mauve-pink and dark purple, calyx glandular with densely ciliate margins. Oct.–Dec. Mountain fynbos, 1000–1460 m, KM (Swartberg Mts).•

sabulosum C.H.Stirt. ined.   Prostrate shrublet to 5 cm. Leaves digitately 3-foliolate, leaflets obovate, emarginate; stipules glabrous. Flowers white, calyx hairy, lowest sepal larger. Sept. Coastal fynbos, limestone, 180–200 m, AP (Agulhas to Stilbaai).•

spissum C.H.Stirt. ined.   Tangled shrub to 1.5 m. Leaves digitately 3-foliolate, leaflets obovate; stipules glabrous. Flowers white and violet, calyx sparsely hairy. Oct. Renosterveld, 660 m, NW, SW, KM (Tulbagh to Barrydale).•

uncinatum (Eckl. & Zeyh.) C.H.Stirt.   Erect or spreading shrub. Leaves digitately 3-foliolate, leaflets linear-oblanceolate, glabrescent; stipules recurved, silky. Flowers purple, pink, white, calyx thickly hairy and heavily glandular, lowest sepal slightly larger. Nov.–Jan. Renosterveld, 160 m, NW, SW (Piketberg to Kleinmond).•

C. Lateral leaflets oblique

acuminatum (Lam.) C.H.Stirt.   Reseeding shrub to 2 m. Leaves digitately 3-foliolate, leaflets cuneate-emarginate, prominently glandular; stipules hairy. Flowers mauve, purple, calyx ciliate on margins and glandular, lowest sepal suborbicular-attenuate. Sept.–Mar. Fynbos at forest margins and in open areas, 160–540 m, KM, SE (Kammanassie and Outeniqua Mts to Van Staden's Mts).•

bolusii (Forbes) C.H.Stirt.   Lax, sprawling shrublet to 40 cm. Leaves digitately 3-foliolate, leaflets linear-oblanceolate, thinly hairy; stipules silky on margins. Flowers pink, calyx hairy and glandular, lowest sepal larger. Aug.–Nov. Mountain fynbos, renosterbos-fynbos scrub, strandveld, limestone, 30–1000 m, NW, SW (Cedarberg to Paarl).•

bowieanum (Harv.) C.H.Stirt.   Sprawling, resprouting subshrub to 50 cm. Leaves digitately 3-foliolate, leaflets oblanceolate, glabrescent; stipules glabrous. Flowers white with purple keel, calyx ciliate on margins and glandular, lowest sepal larger. Sept.–Dec. Mountain and lowland fynbos, 60–650 m, LB (Langeberg Mts).•

carneum (E.Mey.) C.H.Stirt.   Slender shrublet. Leaves digitately 3-foliolate, leaflets linear-oblanceolate; stipules glabrous. Flowers white, calyx thinly hairy, lowest sepal larger, ovate. Oct.–Nov. Mountain fynbos, 550–600 m, SE (Kouga and Baviaanskloof Mts).•

heterosepalum (Fourc.) C.H.Stirt.   Resprouting shrub to 2.5 m. Leaves digitately 3-foliolate, leaflets oblong, glabrescent; stipules ciliate. Flowers white and mauve, calyx subglabrous or hairy, prominently glandular, lowest sepal larger, broadly ovate. Oct.–Nov. Mountain fynbos, forest margins, 260–730 m, SE (Langkloof to Humansdorp).•

macradenium C.H.Stirt.   Erect, resprouting shrub to 2 m, aromatic. Leaves digitately 3-foliolate, leaflets cuneate, glabrescent and glandular-warty beneath; stipules hairy. Flowers creamy white and blue, calyx silky. Nov. Renosterveld, 900 m, KM, LB (Langeberg and Swartberg Mts).•

polyphyllum (Eckl. & Zeyh.) C.H.Stirt.   Erect, closely leafy shrublet to 1 m. Leaves digitately 3-foliolate, leaflets oblong, margins minutely ciliate; stipules caducous. Flowers white and blue, calyx black-hairy, accrescent, lowest sepal larger. May–Nov. Mountain fynbos, 400–600 m, SE (Baviaanskloof and Great Winterhoek Mts).•

saxosum C.H.Stirt.   Shrublet with slender branches to 20 cm. Leaves digitately 3-foliolate, leaflets obovate, glabrescent; stipules glabrous. Flowers white, calyx hairy, lowest sepal larger. Oct. Mountain fynbos, 500 m, LB (Langeberg Mts: Garcia's Pass).•

B. Leaves petiolate
D. Lateral leaflets symmetrical

bracteolatum (Eckl. & Zeyh.) C.H.Stirt.   Sprawling shrub to 2 m. Leaves digitately 3-foliolate, leaflets obovate-cuneate, glandular; stipules ciliate and glandular. Flowers blue, white, violet, calyx softly hairy and glandular, lowest sepal larger, accrescent. Nov.–Apr. Coastal sandveld, limestone hills, below 170 m, SW, AP, LB, SE (Saldanha to Grahamstown).•

flexuosum C.H.Stirt.   Dense shrub to 2.5 m. Leaves digitately 3-foliolate, leaflets obovate, sparsely silky beneath; stipules caducous. Flowers white, calyx sparsely silky and glandular. Nov. Karroid brokenveld, NW (Namaqualand to Biedouw valley).

fruticans (L.) C.H.Stirt.   Straggling subshrub to 40 cm. Leaves digitately 3-foliolate, leaflets obovate, glabrescent; stipules ciliate. Flowers purple, calyx sparsely silky and glandular, lowest sepal much larger, accrescent. Sept.–Dec. Mountain fynbos, 160–400 m, SW (Cape Peninsula).•

racemosum (Thunb.) C.H.Stirt.   Slender resprouting subshrub to 80 cm. Leaves digitately 3-foliolate, leaflets elliptic-oblanceolate, margins sparsely hairy; stipules caducous. Flowers in slender racemes, purple, calyx hispidulous and prominently glandular, lowest sepal slightly larger. Nov.–Jan. Mountain fynbos, 600–1400 m, KM, SE (Kammanassie and Outeniqua Mts).•

striatum (Thunb.) C.H.Stirt.   Willowy shrub to 2.5 m. Leaves digitately 3-foliolate, leaflets elliptic-oblanceolate, adpressed-hairy when young; stipules hairy. Flowers white to cream-coloured with dark keel tip, calyx white-hairy and glandular, accrescent. Nov.–Dec. Mountain fynbos, karoo-fynbos scrub, 300–1000 m, NW, SW, KM (Bokkeveld Mts to Montagu).•

trianthum (E.Mey.) C.H.Stirt.   Slender shrub to 2 m. Leaves digitately 3-foliolate, leaflets linear oblanceolate, glabrescent; stipules glandular. Flowers ?mauve, calyx adpressed-hairy and glandular, accrescent. Nov.–Jan. Lowland and mountain fynbos, 100–500 m, NW, SW (Clanwilliam to Franschhoek).•

virgatum (Burm.f) C.H.Stirt. (= *Otholobium decumbens* (Aiton) C.H.Stirt. ined.)   Trailing resprouting subshrub. Leaves digitately 3-foliolate, leaflets obovate-obcordate, glabrescent; stipules ciliate. Flowers pale pink, mauve-purple, calyx softly hairy and glandular, lowest sepal larger. Nov. Marshes in fynbos, to 400 m, NW, SW, KM, LB, SE (Porterville and Saldanha to E Cape).

D. Lateral leaflets oblique

argenteum (Thunb.) C.H.Stirt.   Sprawling resprouting subshrub to 1 m. Leaves digitately 3-foliolate, leaflets obovate, densely grey-hairy; stipules densely hairy and glandular below. Flowers white, yellow, cream-coloured, calyx silvery silky. Nov. Karroid scrub, NW, KM (Namaqualand to Cedarberg).•

incanum C.H.Stirt.   Sprawling shrub to 1 m. Leaves digitately 3-foliolate, leaflets obovate, shortly grey-hairy; stipules silky and glandular. Flowers pale blue to white, calyx grey-hairy. Oct.–Nov. Sandveld, coastal limestones, 60–130 m, NW (Vredendal to Lambert's Bay).•

lucens C.H.Stirt. ined.   Multistemmed, resprouting shrub to 60 cm. Leaves digitately 3-foliolate, leaflets obovate, orange glandular-warty; stipules hairy. Flowers white, pale blue, calyx black-hairy and glandular. Nov., July. Mountain fynbos, 650–1330 m, KM (Swartberg Mts).•

mundtianum (Eckl. & Zeyh.) C.H.Stirt. Densely leafy resprouting shrub to 1 m. Leaves digitately 3-foliolate, leaflets obovate-rhomboid, glabrescent and pellucid-glandular; stipules hairy. Flowers blue, calyx softly hairy and glandular, lowest sepal larger, accrescent. Dec. Mountain fynbos, 1200–1580 m, SW (Bainskloof to Riviersonderend Mts).•

nitens C.H.Stirt. ined.   Dense shrub to 1.5 m, leafy above. Leaves digitately 3-foliolate, leaflets linear-elliptic, margins finely crisped, glabrescent; stipules hairy and glandular. Flowers blue and mauve, calyx adpressed black-hairy on veins and margins, lowest sepal larger. Oct.–Dec. Mountain fynbos, SW (Bainskloof to Kogelberg).•

obliquum (E.Mey.) C.H.Stirt.   Resprouting shrub to 1 m. Leaves digitately 3-foliolate, leaflets obovate, glabrescent and glandular; stipules glabrous. Flowers blue, pink, white. calyx hairy and densely glandular. June–Dec. Mountain fynbos, 360–1300 m, NW, SW (Piketberg to Kogelberg).•

parviflorum (E.Mey.) C.H.Stirt.   Resprouting subshrub to 1 m. Leaves digitately 3-foliolate, leaflets elliptic; stipules hairy, deciduous. Flowers in elongate racemes, white, calyx hairy and glandular. Oct.–Mar. Mountain fynbos, 260–1160 m, SW (Bainskloof to Kogelberg).•

pictum C.H.Stirt.   Slender shrub to 2 m. Leaves digitately 3-foliolate, leaflets linear-oblanceolate, glabrescent; stipules orange glandular-warty. Flowers white with blue centre, calyx densely black-hairy and glandular, lowest sepal larger. Aug.–Sept. Mountain fynbos, 630–1330 m, SE (Baviaanskloof Mts).•

prodiens C.H.Stirt. ined.   Erect shrub to 2.5 m. Leaves digitately 3-foliolate, leaflets obovate; stipules caducous. Flowers white and dull blue, calyx sparsely black-hairy. July–Sept. Mountain fynbos, renosterveld, 260–700 m, LB, SE (Langeberg Mts to E Cape).

pungens C.H.Stirt. ined.   Sprawling shrublet. Leaves digitately 3-foliolate, leaflets oblanceolate to elliptic, pungent; stipules white-hairy. Flowers violet, calyx hairy, lowest sepal larger, accrescent. Sept.–Jan. Renosterbos-fynbos scrub, coastal renosterveld, 20–160 m, SW, AP (Kleinmond to Potberg).•

sericeum (Poir.) C.H.Stirt.   Sprawling or trailing shrub to 2 m, silvery hairy. Leaves digitately 3-foliolate, leaflets elliptic, prominently veined beneath, margins revolute; stipules densely hairy. Flowers crowded on long, naked peduncles, deep blue-purple, calyx silvery or black-silky, lowest sepal larger. Mainly Sept.–Jan. Lowland and mountain fynbos, riverbeds, 10–600 m, LB, SE (Albertinia to E Cape).

spicatum (L.) C.H.Stirt.   Willowy reseeding shrub to 3.5m. Leaves digitately 3-foliolate, leaflets obovate to oblong, glabrescent; stipules hairy. Flowers in elongate racemes, lilac, calyx softly hairy and glandular. Mainly Oct.–Dec. Coastal renosterveld, riverbanks, 30–400 m, SW, LB (Cape Peninsula to Riversdale).•

stachyerum (Eckl. & Zeyh.) C.H.Stirt.   Reseeding shrub to 3 m. Leaves digitately 3-foliolate, leaflets obovate-cuneate, hairy on midribs; stipules silky. Flowers in elongate racemes, violet to mauve, calyx white- or black-hairy and glandular. Sept.–Mar. Grassy fynbos, riverbanks, forest margins, 200–1450 m, SW, LB, SE (Caledon to E Cape).

swartbergense C.H.Stirt.   Sprawling shrublet to 30 cm. Leaves digitately 3-foliolate, leaflets elliptic, silky; stipules silky. Flowers crowded on long, naked peduncles, mauve-purple, calyx black-hairy, lowest sepal much larger. Nov.–Dec. Mountain fynbos, subalpine fynbos, 1560–1860 m, KM (Groot Swartberg).•

venustum C.H.Stirt.   Sprawling or trailing shrub to 1 m. Leaves digitately 3-foliolate, leaflets elliptic-oblanceolate, glabrescent; stipules finely hairy. Flowers in lax racemes, white to pale mauve, calyx black-silky. Nov. Coastal sands, 100–200 m, NW (Lambert's Bay to Langebaan).•

zeyheri (Harv.) C.H.Stirt.   Resprouting subshrub to 40 cm. Leaves digitately 3-foliolate, leaflets obcordate-oblanceolate, glabrescent, margins ciliate; stipules ciliate. Flowers crowded on slender, naked peduncles, white and mauve, calyx silky. Nov.–Apr. Fynbos, grassland, 60–800 m, SW, AP (Jonkershoek to Bredasdorp).•

*PARASERIANTHES   AUSTRALIAN ALBIZIA, STINKBOON   1 sp., Australasia

*lophantha (Willd.) Nielsen (= *Albizia lophantha* (Willd.) Benth.)   Evergreen shrub or small tree, 4–6 m. Leaves bipinnate, with elongated gland midway along petiole. Flowers in cylindrical spikes, cream-coloured. Pods flattened, oblong-apiculate. July–Aug. Lowland fynbos, grassland, streambanks, SW, LB, SE (Australian weed).

PODALYRIA   BUSH SWEET PEA, KEURTJIE, KEURBLOM   19 spp., mainly winter-rainfall South Africa

*A. Leaves shiny and glabrous or glabrescent above (see also **P. biflora, P. variabilis**)*

burchellii DC.   Willowy to dense and much-branched, resprouting shrub to 1 m. Leaves 1-foliolate, elliptic to suborbicular, thinly silky and shiny glabrescent above, felted beneath. Flowers dark pink, magenta and white, bracts broadly ovate, prominently veined, mucronate. Aug.–June. Sandstone slopes, 300–1500 m, KM, LB, SE (Swartberg and Langeberg Mts to Grahamstown).

glauca (Thunb.) DC.   Willowy, resprouting shrublet to 90 cm. Leaves 1-foliolate, elliptic-oblong, margins slightly rolled under, shiny and glabrous above, felted beneath. Flowers magenta, bracts linear-lanceolate. Aug.–Apr. Sandstone slopes, 200–1000 m, SE (Outeniqua to Kareedouw Mts).•

microphylla E.Mey.   Erect, resprouting shrublet to 60 cm. Leaves 1-foliolate, obovate, glabrescent above, silky beneath. Flowers pink and white, bracts oblanceolate. Sept. Renosterveld-fynbos scrub, rocky outcrops, 100 m, SW (Cape Flats).•

oleaefolia Salisb.   Willowy, resprouting shrub to 1 m. Leaves 1-foliolate, linear to elliptic-oblong, margins rolled under, shiny glabrous above, felted below. Flowers dark pink and white, bracts ovate. June–Dec. Sandstone lower slopes, below 500 m, SW, AP (Hottentots Holland Mts to Elim).•

orbicularis (E.Mey.) Eckl. & Zeyh.   Erect, willowy, resprouting shrub to 1.5 m. Leaves 1-foliolate, ovate to orbicular, margins slightly rolled under, hairy to shiny glabrescent above, felted beneath.

Flowers pink and white, bracts broadly obovate. Oct.–Nov. Sandstone slopes, 550 m, SW (Genadendal Mt and Caledon Swartberg).•

pearsonii E.Phillips   Willowy, resprouting shrub to 1 m. Leaves 1-foliolate, elliptic to narrowly elliptic, glabrous above except along margin, silky beneath. Flowers magenta pink and white, bracts obovate. Sept. Sandstone slopes, 830–870 m, NW (Bokkeveld Mts to Gifberg).•

A. Leaves more or less hairy above

argentea (Salisb.) Salisb.   Much-branched, resprouting shrublet to 30 cm. Leaves 1-foliolate, elliptic to broadly elliptic, silky above, silvery felted beneath. Flowers bright pink and white, bracts broadly ovate, mucronate. Sept.–Nov. Sandstone lower slopes, below 500 m, SW (Cape Peninsula and Hottentots Holland Mts).•

biflora (L.) Lam.   Prostrate, resprouting shrublet to 30 cm. Leaves 1-foliolate, elliptic to broadly elliptic, silky, sometimes glabrous above, felted beneath, margins often golden brown. Flowers bright pink and white, bracts lanceolate. Aug.–Nov. Sandstone slopes, 100–1000 m, NW, SW, LB (Clanwilliam to Mossel Bay).•

calyptrata (Retz.) Willd.   Small tree to 5 m, rarely sprouting after fire. Leaves 1-foliolate, obovate, silky. Flowers bright pink and white, bracts very broad, fused to form a calyptra. Aug.–Oct. Sandstone slopes in marshy places, below 1000 m, SW (Cape Peninsula to Potberg).•

cordata (Thunb.) R.Br.   Erect, woody, resprouting shrub to 1.7 m. Leaves 1-foliolate, ovate to suborbicular, with a cordate base, densely woolly. Flowers bright pink and white, bracts lanceolate to broadly ovate. Nov.–Feb. Sandstone slopes, 540–930 m, SW (Hottentots Holland Mts).•

hirsuta (Aiton) Willd.   Erect, woody, resprouting shrub to 3 m. Leaves 1-foliolate, ovate to suborbicular, adpressed-hairy above, felted below. Flowers bright pink and white, bracts lanceolate. June–Jan. Sandstone slopes, 400–1230 m, SW, LB (Hottentots Holland to Langeberg Mts).•

lanceolata (E.Mey.) Benth.   Small tree to 4 m. Leaves 1-foliolate, lanceolate to obovate, hairy. Flowers pink and maroon, bracts ovate. Aug. Sandstone slopes, streambanks, 350–400 m, LB (Langeberg Mts).•

leipoldtii L.Bolus ex A.L.Schutte ined.   Rigid, woody, resprouting shrub to 2 m. Leaves 1-foliolate, obovate, silky above and beneath. Flowers pink and white, bracts lanceolate. Aug.–Sept. Sandstone slopes, 180–700 m, NW (Bokkeveld to Cedarberg Mts).•

myrtillifolia (Retz.) Willd.   Erect, woody, reseeding shrub to 2 m, sometimes resprouting after fire. Leaves 1-foliolate, obovate, silky above and beneath. Flowers pink and white, bracts lanceolate. June–Nov. Sandstone, limestone or shale flats and lower slopes, 120–330 m, NW, SW, AP, LB, SE (Tulbagh to Port Elizabeth).•

reticulata Harv.   Erect, willowy, resprouting shrub up to 1.3 m. Leaves 1-foliolate, obovate to suborbicular, margins slightly rolled under, hairy above, felted below when young. Flowers pink and white, bracts broadly obovate. Aug. Sandstone slopes, SW (Caledon Swartberg).•

rotundifolia (P.J.Bergius) A.L.Schutte   Erect, woody, resprouting shrub to 3 m. Leaves 1-foliolate, ovate to suborbicular, adpressed-hairy above, felted or softly hairy beneath when young. Flowers bright pink and white, bracts depressed ovate, sheathing more than half of bud but not calyptrate. Aug.–Nov. Sandstone slopes, 260–1166 m, NW, SW, LB (Cedarberg to Langeberg Mts).•

sericea (Andrews) R.Br.   Erect, reseeding shrublet to 1 m. Leaves 1-foliolate, obovate to oblanceolate, silky above and beneath. Flowers pink and white, bracts oblanceolate. May–June. Sandstone and granite outcrops, below 500 m, SW (Saldanha to Cape Peninsula).•

variabilis A.L.Schutte   Much-branched, resprouting shrublet to 80 cm. Leaves 1-foliolate, elliptic to obovate, silky or glabrescent above, silky or felted beneath. Flowers pink and white, bracts broadly ovate to depressed ovate. Aug.–Oct. Sandstone slopes, below 1200 m, SW (Tygerberg to Caledon Swartberg).•

POLHILLIA•   7 spp., mostly W Cape, 1 sp., W Karoo

brevicalyx (C.H.Stirt.) B.-E.van Wyk & A.L.Schutte   Multistemmed shrublet to 40 cm. Leaves 3-foliolate, leaflets oblanceolate-conduplicate, recurved, thinly silky; stipules free, 3–5 mm long. Flowers 1–3 at branch tips, yellow. Oct. Renosterveld, 200–300 m, LB (Swellendam).•

canescens C.H.Stirt.   Erect, somewhat wand-like shrublet to 1.5 m. Leaves 3-foliolate, leaflets elliptic-conduplicate, silvery silky; stipules connate and fused to petiole, 7–14 mm long. Flowers 1–4 at branch tips, bright yellow. Sept.–Oct. Renosterveld, 200–300 m, SW (Bredasdorp).•

connata (Harv.) C.H.Stirt.   Like **P. canescens** but mainly tawny. Sept.–Oct. Renosterveld, 300 m, LB (Riversdale).•

obsoleta (Thunb.) B.-E.van Wyk (= *Polhillia waltersii* (C.H.Stirt.) C.H.Stirt.)   Like **P. pallens** but leaves greener, flowers brighter and pods finely hairy, conspicuously pleated. July–Sept. Renosterveld, 300–430 m, NW (Worcester, extinct at Porterville).•

pallens C.H.Stirt.   Multistemmed, resprouting shrub to 1.5 m. Leaves 3-foliolate, leaflets linear-conduplicate, recurved, silvery silky; stipules connate and fused to petiole, 1–3 mm long. Flowers 1 or 2 at branch tips, pale yellow. Pods narrowly oblong, shaggy, impressed between seeds. June–Oct. Renosterveld, 30–100 m, SW (Bredasdorp to Potberg).•

sp. 1   Like **P. obsoleta** but leaflets wider and flattened (pods unknown). Sept. Coastal granite or clay, SW (Vredenburg).•

PSORALEA   BLUE PEA, BLOUKEURTJIE   c. 50 spp., southern Africa

*A. Leaves 5–9-foliolate (see also **P. tenuifolia**)*

affinis Eckl. & Zeyh.   Slender erect shrub to 3 m. Leaves 5–9-foliolate. Flowers blue. Sept.–Nov. Mountain, lowland and coastal fynbos, marshy areas, 100–1800 m, NW, SW, LB, SE (Tulbagh to E Cape).

arborea Sims   Tree-like shrublet to 60 cm. Leaves 5–7-foliolate. Flowers mauve and blue. June–Oct. Mountain and lowland fynbos, forests, 30–330 m, AP, LB, SE (Gansbaai to Humansdorp).•

axillaris L.   Tree-like shrub to 3 m. Leaves 7-foliolate. Flowers blue. Nov.–Feb. Mountain fynbos, streambanks, 330–800 m, SW, LB, SE (Potberg, Langeberg Mts to Port Elizabeth).•

azurea C.H.Stirt. ined.   Prostrate, resprouting shrub, 20–50 cm. Leaves 5–7-foliolate. Flowers blue, mauve. June–Jan. Mountain fynbos, 260–1130 m, SW, LB, SE (Riviersonderend to Outeniqua Mts).•

elegans C.H.Stirt. ined.   Reseeding shrub to 1.8 m. Leaves 7-foliolate. Flowers pale blue. Nov.–Dec. Mountain fynbos, subalpine fynbos, gullies, 1700 m, KM (Groot Swartberg).•

filifolia Thunb.   Tree-like, resprouting shrub to 3 m. Leaves 5-foliolate. Flowers blue and white. Nov. Renosterbos-fynbos scrub, streamsides, 730 m, LB (Langeberg Mts).•

floccosa C.H.Stirt. ined.   Erect shrub. Leaves 5-foliolate. Flowers white. Sept. Mountain fynbos, 660–1000 m, LB, SE (Langeberg to Tsitsikamma Mts).•

imminens C.H.Stirt. ined.   Tree, 2–3 m. Leaves 7-foliolate. Flowers white and blue. Oct.–Dec. Mountain fynbos, streamsides, 900–1000 m, KM, SE (Swartberg and Tsitsikamma Mts to E Cape).

intonsa C.H.Stirt. ined.   Reseeding shrub to 1.5 m. Leaves 9-foliolate. Flowers white with blue striations. Nov. Mountain fynbos, 1160 m, NW, LB, SE (Grootwinterhoek, Langeberg to Outeniqua Mts).•

latifolia (Harv.) C.H.Stirt.   Erect shrub to 2 m. Leaves 7-foliolate. Flowers white and violet. Aug.–Jan. Lowland fynbos, streambanks, forest margins, 60–260 m, SE (Joubertina to E Cape).

odoratissima Jacq.   Tree-like shrub, 1–2 m. Leaves 7–9-foliolate. Flowers blue and white. Oct.–Dec. Mountain fynbos, streamsides, 300–1330 m, SW, KM (Villiersdorp to Swartberg Mts).•

pinnata L.   FONTEINBOS, PENWORTEL   Willowy tree to 4 m. Leaves 7–9-foliolate. Flowers blue. Oct.–Apr. Mountain fynbos, forest margins, riverbeds, 230–1060 m, NW, SW, LB, (Cape Peninsula to Kogelberg).•

sordida C.H.Stirt. ined.   Erect, resprouting shrub, 1.5–2.5 m. Leaves 7-foliolate. Flowers blue. Oct.–July. Mountain fynbos, 460–2000 m, KM, LB, SE (Swartberg and Langeberg to Outeniqua Mts).•

speciosa Eckl. & Zeyh.   Tree-like shrub to 2 m. Leaves 5-foliolate. Flowers deep blue. Aug.–Nov. Mountain fynbos, streamsides, 600–1200 m, LB, SE (Langeberg to Kouga Mts).•

A. Leaves 1- or 3-foliolate
B. Leaves 1-foliolate, sometimes scale-like or lacking

alata (Thunb.) T.M.Salter   Prostrate, trailing shrub to 10 cm. Leaves 1-foliolate. Flowers blue. Dec.–Jan. Lowland fynbos, 200–330 m, NW, SW (Clanwilliam to Cape Peninsula).•

aphylla L.   FONTEINBOS   Erect, reseeding or resprouting, broom-like shrub to 4 m. Leaves 1-foliolate, very small. Flowers blue with white keel petals. Sept.–May. Mountain and lowland fynbos, streambanks, 80–1460 m, NW, SW, LB (Clanwilliam to Riversdale).•

asarina (P.J.Bergius) T.M.Salter   Prostrate, trailing, mat-forming shrublet. Leaves 1-foliolate, spine-tipped. Flowers blue, purple. Oct.–Feb. Mountain fynbos, renosterbos-fynbos scrub, 20–1400 m, SW, AP, LB, SE (Cape Peninsula to Knysna).•

fleta C.H.Stirt. ined.   Tree to 6 m. Leaves 1–3-foliolate or leafless. Flowers blue, pale mauve. Oct.–Mar. Mountain fynbos, 660–1000 m, NW, SW (Elandskloof Mts to Bainskloof).•

imbricata (L.) T.M.Salter   Sprawling, resprouting shrublet to 50 cm. Leaves 1-foliolate. Flowers deep purple, blue. Oct.–Jan. Mountain and lowland fynbos, 180–800 m, NW, SW (Bokkeveld to Kleinrivier Mts).•

implexa C.H.Stirt.   Trailing, grass-like, resprouting shrublet. Leaves 1-foliolate. Flowers brick-red to salmon. Feb. Mountain fynbos, 650 m, SW (Du Toit's Peak).•

laxa T.M.Salter   Straggling, prostrate, lax shrublet to 40 cm. Leaves 1-foliolate. Flowers white or deep purple. Nov.–Mar. Mountain and coastal fynbos, streamsides, 30–1160 m, SW, AP, LB (Cape Peninsula to Albertinia).•

lucida C.H.Stirt. ined.   Densely resprouting shrub to 1.5 m. Leaves 1-foliolate but more or less leafless. Flowers purple. Oct. Mountain fynbos, 1160 m, SW (Limietberg).•

monophylla (L.) C.H.Stirt.   Trailing, decumbent, resprouting shrublet to 30 cm. Leaves 1-foliolate. Flowers blue. Nov.–Jan. Mountain fynbos, 230–1700 m, SW, LB, SE (Cape Peninsula to Humansdorp).•

peratica C.H.Stirt.   Tree to 5 m. Leaves 1-foliolate, scale-like. Flowers blue. Oct. Mountain fynbos, streamsides, 730 m, NW (Piketberg).•

plauta C.H.Stirt. ined.   Prostrate, trailing resprouting shrub. Leaves 1-foliolate. Flowers blue, dark violet. Sept.–Feb. Mountain and lowland fynbos, grassland, streamsides, 240–1500 m, SW, LB, SE (Riviersonderend to E Cape).

pullata C.H.Stirt. ined.   Erect shrub to 2 m. Leaves 1-foliolate, scale-like. Flowers purple. Aug.–Nov. Mountain fynbos, streamsides, 360–500 m, SW (Cape Peninsula to Kleinrivier Mts and Caledon Swartberg).•

ramulosa C.H.Stirt. ined.   Erect, resprouting shrub to 3 m. Leaves 1-foliolate, scale-like. Flowers purple, blue. Jan.–Apr. Mountain fynbos, streamsides, 830 m, NW, SW (Cedarberg to Hottentots Holland Mts).•

restioides Eckl. & Zeyh.   Reseeding shrublet to 50 cm. Leaves 1-foliolate. Flowers blue and white. Aug.–Dec. Mountain and lowland fynbos, marshy areas, 30–960 m, SW, AP (Cape Peninsula to Bredasdorp).•

usitata C.H.Stirt. ined.   Lax, erect shrub to 2 m. Leaves 1-foliolate, scale-like. Flowers blue and white. Nov.–May. Mountain and lowland fynbos, riverine habitats, 60–900 m, NW, SW, LB (Grootwinterhoek to Langeberg Mts).•

B. Leaves 3-foliolate (see also P. fleta)

aculeata L.   Tree-like shrub to 2 m. Leaves 3-foliolate. Flowers mauve and white. Oct.–Dec. Lowland fynbos, 30–900 m, SW (Cape Peninsula to Hermanus).•

angustifolia Jacq.   Tree-like shrub to 4 m. Leaves 3-foliolate. Flowers white and blue. Nov.–Apr. Mountain fynbos, streamsides, 330–1200 m, NW, KM, LB, SE (Bokkeveld Mts to E Cape).

crista C.H.Stirt. ined.   Low shrub. Leaves 3-foliolate. Flowers blue. Feb. Mountain fynbos, 1660 m, SE (Uitenhage).•

ensifolia (Houtt.) Merr.   Spreading shrub, 0.5–1.2 m. Leaves 3-foliolate. Flowers purple, mauve, dark blue. Dec.–Apr. Mountain and coastal fynbos, NW, SW, KM, LB, SE (Cold Bokkeveld to Kouga Mts).•

fascicularis DC.   Resprouting shrub to 2.5 m. Leaves 3-foliolate. Flowers blue. Sept.–Dec. Mountain and lowland fynbos, 100–600 m, SW (Cape Peninsula to Hermanus).•

fulcrata C.H.Stirt. ined.   Prostrate, mat-forming shrub to 1.5 m diam. Leaves 3-foliolate. Flowers dark blue and white. Dec. Mountain fynbos, subalpine fynbos, 1800 m, KM (Groot Swartberg).•

glaucescens Eckl. & Zeyh.   Lax, weeping, much-branched shrub to 2 m. Leaves 3-foliolate. Flowers pale lemon and purple. Nov.–Mar. Renosterbos-fynbos scrub, mountain fynbos, 760–1260 m, NW (Bokkeveld Mts).•

glaucina Harv.   Sprawling, straggling shrublet, 30–60 cm. Leaves 3-foliolate. Flowers white. Dec.–Feb. Lowland fynbos, SW (Cape Peninsula and Cape Flats).•

gueinzii Harv.   Erect, much-branched shrublet to 1 m. Leaves 3-foliolate. Flowers violet, mauve, purple. Oct.–Feb. Lowland fynbos, marshy places, SW (Stellenbosch).•

keetii Schönland ex Forbes   Resprouting shrub to 2 m. Leaves 3-foliolate. Flowers white and blue. Jan. Mountain fynbos, 830–1560 m, SE (Outeniqua and Tsitsikamma Mts).•

muirii C.H.Stirt. ined.   Much-branched shrub to 1 m. Leaves 3-foliolate. Flowers blue, purple. Nov. Mountain fynbos, 350–400 m, LB (Albertinia).•

nubicola C.H.Stirt. ined.   Reseeding shrub to 2 m. Leaves 3-foliolate. Flowers pale blue and purple. Dec. Mountain fynbos, streamsides, 1700–1800 m, KM, LB (Swartberg and Langeberg Mts).•

oligophylla Eckl. & Zeyh.   Resprouting, weeping shrub to 2 m. Leaves 3-foliolate. Flowers dirty white and purple. Oct.–Apr. Mountain fynbos, streamsides, NW, KM, SE (Gifberg to E Cape).

oreophila Schltr.   Sprawling shrublet to 50 cm. Leaves 3-foliolate. Flowers dark blue. Jan. Mountain fynbos, 460–1660 m, SW (Bainskloof to Franschhoek).•

oreopola C.H.Stirt. ined.   Lax shrub to 3 m. Leaves 3-foliolate. Flowers white and blue. Oct.–Apr. Mountain fynbos, riverine habitat, 550–1160 m, NW, SW (Cedarberg to Riviersonderend Mts).•

repens P.J.Bergius   Prostrate, trailing shrub. Leaves 3-foliolate, fleshy. Flowers violet. Nov.–Feb. Coastal fynbos, below 50 m, SW, AP, LB, SE (Cape Peninsula to E Cape).

tenuifolia L.   Tree-like, weeping shrub to 5 m. Leaves 3–5-foliolate. Flowers blue. Sept.–Mar. Mountain fynbos, riverbeds, 260–830 m, NW, SW (Gifberg to Franschhoek).•

tenuissima E.Mey.   Slender, trailing shrublet, 30–60 cm. Leaves 3-foliolate. Flowers blue. Dec.–Mar. Mountain fynbos, 960–1330 m, NW, SW (Cold Bokkeveld to Du Toitskloof Mts).•

triflora Thunb.   Erect, resprouting shrub to 2 m. Leaves 3-foliolate. Flowers white and purple. Nov.–Feb. Mountain fynbos, forest margins, 100–1000 m, KM, LB, SE (Swartberg and Langeberg to Baviaanskloof Mts).•

trullata C.H.Stirt. ined.   Prostrate, reseeding shrublet with branches to 1 m long. Leaves 3-foliolate. Flowers blue. Oct.–Jan. Mountain fynbos, marshy places, 400 m, SE (Outeniqua and Tsitsikamma Mts).•

verrucosa Willd.   Erect shrub to 2 m. Leaves 3-foliolate. Flowers blue, mauve and pale blue. Dec.–Mar. Mountain fynbos, streambanks, 400–1600 m, NW, KM, LB, SE (Cedarberg to Tsitsikamma Mts).•

vlokii C.H.Stirt.   Prostrate, reseeding shrublet to 60 cm diam. Leaves 3-foliolate. Flowers blue and white. Dec.–Jan. Mountain fynbos, 530–590 m, SE (Outeniqua Mts).•

RAFNIA   WIDOW PEA   19 spp., N Cape to KwaZulu-Natal

A. Calyx bilabiate with lower lobe longest; keel truncate or obtuse; wing petals usually smooth; style shorter than ovary

capensis (L.) Schinz   Erect to prostrate shrublet to 1 m. Leaves unifoliolate, alternate to opposite on flowering branches, linear to orbicular. Flowers solitary or crowded, wings smooth, keel obtuse to more or less rostrate and often subapically lobed, calyx lobes triangular to narrowly triangular, upper lobes broader than the others. Pods stipitate, oblong to broadly oblong or obliquely lanceolate. All year. Stony flats and slopes, 0–2 000 m, NW, SW, AP, KM, LB, SE (Bokkeveld Mts to Swartberg Mts).•

diffusa Thunb.   Prostrate or decumbent shrublet to 60 cm. Leaves unifoliolate, subopposite on flowering branches, linear to suborbicular, sometimes bright green. Flower solitary, small, wings sculptured, keel obtuse to more or less rostrate, calyx lobes triangular. Pods long-stipitate, more or less oblong. Aug.–Jan. Stony, sometimes limestone flats, 250–1 100 m, NW, SW (Bokkeveld Mts to Malmesbury).•

globosa G.J.Campbell & B.-E.van Wyk   Erect shrublet to 60 cm. Leaves unifoliolate, opposite on flowering branches, elliptic to orbicular. Flowers clustered in globose pseudo-racemes, wings smooth, keel more or less rostrate, slightly subapically lobed, calyx lobes often recurved in fruit. Pods more or less oblong. Sept.–Dec. Sandstone slopes, 300–700 m, NW (Cedarberg Mts: Algeria).•

schlechteriana Schinz   Erect shrublet to 60 cm. Leaves unifoliolate, opposite on flowering branches, broadly elliptic to orbicular, sometimes acuminate. Flowers clustered in large, conical pseudoracemes, wings smooth, keel truncate. Pods sessile, subtriangular, upper suture broadly winged. Sept.–Jan. Sandstone slopes, 300–1 400 m, NW (Clanwilliam to Ceres and Piketberg).•

spicata Thunb.   Procumbent, clump-forming shrublet to 30 cm. Leaves unifoliolate, subopposite to opposite on flowering branches, narrowly lanceolate to ovate, sometimes bright green. Flowers in spike-like pseudo-racemes, small, wings sometimes sculptured, keel truncate to more or less rostrate, calyx lobes triangular. Pods narrowly oblong. Sept.–Mar. Rocky slopes, 450–1 700 m, NW (Cedarberg and Swartruggens to Tulbagh).•

A. Calyx equally lobed; keel rostrate; wing petals sculptured
B. Leaves cordate and clasping

amplexicaulis (L.) Thunb.   SOETHOUTBOSSIE   Erect shrub to 2.5 m. Leaves unifoliolate, opposite on flowering branches, reniform to cordate, reticulately veined, clasping. Flowers solitary, keel long-rostrate, calyx lobes shorter than or as long as tube, upper lobes broader and falcate. Pods obliquely ovate, upper margin convex above. Sept.–Feb. Stony slopes, 300–1 550 m, NW, SW (Bokkeveld Mts to Caledon).●

inaequalis G.J.Campbell & B.-E.van Wyk   Erect shrublet to 40 cm. Leaves unifoliolate, opposite on flowering branches, oblong-cordate to orbicular-cordate, young leaves rolled over backwards. Flowers solitary, keel rostrate, calyx lobes much longer than tube, upper lobes broader and falcate, lower lobe very narrow, shorter than the others, with wing-like extension on tube between upper lobes. Pods stipitate, stipe curved, oblong. Oct.–Jan. Stony slopes, 600–700 m, NW (Piketberg).●

perfoliata (L.) Willd.   Prostrate, trailing shrublet to 30 cm and 1 m diam. Leaves unifoliolate, opposite on flowering branches, leaves cordate to oblong-cordate, reticulately veined, often clasping. Flowers solitary, keel rostrate, calyx lobes as long as or much longer than tube, upper lobes broader and falcate. Pods oblong, upper margin convex. Sept.–Apr. Stony slopes, 50–1 250 m, NW, SW, AP (Cedarberg to Bredasdorp Mts).●

B. Leaves lanceolate to orbicular but not cordate nor clasping
C. Calyx deeply cleft below (lower sinus deeper than 9 mm) and ridged or winged above

alata G.J.Campbell & B.-E.van Wyk   Procumbent, clump-forming shrublet to 40 cm. Leaves unifoliolate, subopposite to opposite on flowering, lanceolate to obovate. Flowers solitary, keel rostrate, calyx lobes longer than tube, upper lobes broader and falcate, tube winged between lobes. Pods obliquely lanceolate. Oct.–Jan. Stony slopes, 0–1 850 m, AP, KM, SE (Cape Infanta, Swartberg and Outeniqua Mts).●

crassifolia Harv.   Erect shrublet to 60 cm. Leaves unifoliolate, subopposite to opposite on flowering branches, lanceolate, bright green, margins red. Flowers solitary on elongate pedicels, keel rostrate, calyx lobes much longer than tube, upper lobes slightly broader and falcate. Pods obliquely lanceolate to oblong. Oct.–Mar. Sandstone slopes, 20–250 m, SW (Cape Peninsula to Caledon).●

elliptica Thunb.   Erect shrublet to 90 cm. Leaves unifoliolate, subopposite to opposite on flowering branches, lanceolate to elliptic. Flowers solitary, sometimes 3, large, keel rostrate, calyx lobes longer than tube, upper lobes broader, falcate. Pods obliquely lanceolate. Sept.–May. Sandy, grassy coastal fynbos and renosterveld, 0–900 m, LB, SE, KM, (Swellendam to KwaZulu-Natal).

vlokii G.J.Campbell & B.-E.van Wyk   Erect shrub to 2 m. Leaves unifoliolate, subopposite to opposite on flowering branches, lanceolate-elliptic. Flowers solitary, large, buds leaf-like, keel rostrate, calyx lobes much longer than tube, upper lobes broader and falcate, lower lobe very narrow, shorter than the others. Pods large, obliquely oblong, upper margin asymmetrically convex. Mainly June–Sept. Grassy fynbos and renosterveld, 550–750 m, SE (Outeniqua Mts: Moerasrivier).●

C. Calyx less deeply cleft below and not ridged or winged above

angulata Thunb.   Virgate or procumbent shrublet to 2.2 m. Leaves unifoliolate, alternate, sometimes opposite on flowering branches, ericoid or linear or obovate. Flowers solitary, sometimes to 6 in pseudo-racemes, keel rostrate, calyx lobes triangular to narrowly triangular, about as long as or longer than tube. Pods obliquely oblanceolate or obliquely lanceolate. Mainly Sept.–Mar. Stony slopes, 0–2 200 m, NW, SW, KM, LB (Bokkeveld to Swartberg Mts).●

crispa C.H.Stirt.   Procumbent shrublet to 40 cm. Leaves unifoliolate, subopposite to opposite on flowering branches, lanceolate, margins markedly cartilaginous. Flowers solitary, pedunculate, keel long-rostrate, calyx lobes triangular, lower lobe shorter than the others. Mainly Dec. Stony flats, 200 m, SW (Elandskloof Mts).●

lancea (Thunb.) DC.   Procumbent shrublet to 30 cm. Leaves unifoliolate, alternate on flowering branches, lanceolate. Flowers solitary, pedunculate, keel rostrate, calyx with upper and lateral lobes fused in pairs. Pods obliquely oblong, curved below. Sept.–Nov. Clay flats, 50–250 m, NW, SW (Clanwilliam to Paarl).●

ovata E.Mey.   Erect shrublet to 1.2 m. Leaves unifoliolate, subopposite to opposite on flowering branches, broadly elliptic to orbicular, acuminate. Flowers solitary, sometimes to 4, large, keel rostrate, calyx lobes triangular, shorter than tube, upper lobes broader than the others. Pods short-stipitate, obliquely lanceolate, upper suture broadly winged. Aug.–Jan. Stony slopes, 600–1 550 m, NW, SW (Bokkeveld to Stellenbosch Mts).●

racemosa Eckl. & Zeyh.   Erect or procumbent, clump-forming shrub or shrublet to 1.5 m. Leaves unifoliolate, alternate on flowering branches, lanceolate to elliptic. Flowers few to 19, wings sculptured, keel rostrate, calyx lobes triangular or narrowly triangular. Pods obliquely lanceolate. Mainly Oct.–Jan. Stony slopes, 300–1 700 m, SW, KM, SE (Worcester to Uniondale).•

rostrata G.J.Campbell & B.-E.van Wyk   Erect or procumbent shrublet to 40 cm. Leaves unifoliolate, alternate to opposite on flowering branches, lanceolate to obovate. Flowers 1–5, keel rostrate or long-rostrate, calyx lobes longer than tube, upper lobes broader and falcate. Pods obliquely lanceolate. Oct.–Mar. Stony slopes, 800–1 800 m, NW, KM (Swartruggens to Swartberg Mts).•

triflora (L.) Thunb.   Erect shrub or shrublet to 2 m. Leaves unifoliolate, subopposite to opposite on flowering branches, broadly elliptic to orbicular. Flowers solitary but in threes, keel rostrate, calyx lobes as long as or longer than tube, upper lobes broader and falcate, lower lobe shorter than the others. Pods obliquely lanceolate. Sept.–Mar. Stony slopes, 0–600 m, NW, SW, AP, LB, SE (Clanwilliam to Humansdorp).•

RHYNCHOSIA   VAAL-ERTJIE   c. 200 spp., pantropical

A. Plants climbing; flowers in axillary umbels or solitary

chrysoscias Benth. ex Harv.   Climbing, resprouting shrub to 1 m. Leaves pinnately 3-foliolate, leaflets lanceolate-oblong. Flowers yellow, large, upper calyx lobes separate almost to base. July–Dec. Sandstone slopes, 200–640 m, SW, AP, LB, SE (Caledon to E Cape).

leucoscias Benth. ex Harv.   BLINK-ERTJIE   Climbing shrub. Leaves pinnately 3-foliolate, shortly petiolate, leaflets linear to lanceolate, white-felted beneath. Flowers yellow, bracts broadly oblong, obtuse. Sept.–Dec. Sandstone slopes and flats, 15–1030 m, SW, LB, SE (Hermanus to Uitenhage).•

microscias Benth. ex Harv. (incl. **R. parviflora** (E.Mey.) Druce)   Like **R. leucoscias** but flowers smaller and bracts ovate-lanceolate, acute. July–Sept. Sandstone slopes, 160–400 m, LB, SE (Riversdale to Humansdorp).•

A. Plants erect, sprawling or climbing; flowers in racemes
B. Leaflets broad, ovate-rhomboid

argentea (Thunb.) Harv.   Woody, climbing shrub. Leaves pinnately 3-foliolate, leaflets broadly ovate, leathery, densely grey-hairy beneath. Flowers yellow. Oct.–Mar. Coastal forest margins, 360–1200 m, SE (Knysna to E Cape).

arida C.H.Stirt.   Sprawling, willowy, deciduous shrub to 1 m. Leaves pinnately 3-foliolate, leaflets ovate-elliptic, discolorous, glandular. Flowers yellow. Sept.–Oct. Karroid scrub, 260–450 m, NW (Gifberg and Matsikamma Mts).•

caribaea (Jacq.) DC.   Prostrate or climbing shrublet. Leaves pinnately 3-foliolate, leaflets ovate-rhomboid. Flowers yellow with purple venation outside of standard petal. Dec.–Apr. Coastal forests, below 600 m, SE (George to tropical Africa).

harmsiana Schltr. ex Zahlbr.   Slender, climbing shrub. Leaves pinnately 3-foliolate, leaflets ovate-rhomboid, shortly hairy, discolorous. Flowers yellow. Feb.–May. Sandstone slopes, 400–980 m, SE (Humansdorp to KwaZulu-Natal).

viscidula Steud.   Erect or straggling shrublet to 30 cm. Leaves pinnately 3-foliolate, leaflets ovate to subrotund, glandular. Flowers yellow. Aug.–Nov. Sandstone slopes, 1220 m, NW (Namaqualand to Tulbagh).

B. Leaflets narrow, linear to elliptic

angustifolia (Jacq.) DC.   Woody, climbing shrub. Leaves pinnately 3-foliolate, leaflets narrowly lanceolate, shortly stalked, tomentose below. Flowers yellow. Oct. Fynbos, SW (Caledon).•

bullata Benth. ex Harv.   Sprawling shrublet to 30 cm. Leaves pinnately 3-foliolate, leaflets small, oblong, glandular. Flowers yellow. July–Dec. Sandstone slopes, 600–1160 m, NW, SW (Namaqualand to Hex River Mts).

capensis (Burm.f.) Schinz   Decumbent, resprouting, prostrate to twining shrub. Leaves pinnately 3-foliolate, conspicuously stipulate, leaflets oblong-elliptic. Flowers yellow. Aug.–Mar. Riverbanks, below 660 m, SW, AP, KM, LB, SE (Cape Peninsula to E Cape).

ciliata (Thunb.) Schinz   Sprawling, resprouting shrublet or dwarf shrub to 30 cm. Leaves pinnately 3-foliolate, leaflets elliptic. Flowers yellow. Oct.–Mar., May–June. Sandstone slopes and flats, 230–1330, LB, SE (Heidelberg to E Cape).

ferulifolia Benth. ex Harv. (incl. **R. pinnata** Harv.)   Prostrate to sprawling woody subshrub to 10 cm. Leaves pinnately 5–9-foliolate or decompound, leaflets linear to lanceolate. Flowers yellow. July–Jan. Sandy flats and lower slopes, below 300 m, NW, SW, AP, SE (Piketberg to Humansdorp).•

totta (Thunb.) DC.   Trailing or climbing shrub. Leaves pinnately 3-foliolate, leaflets oblong-lanceolate, hispidulous. Flowers yellow. Sept.–Mar. Grassland, 240–1080 m, AP, SE (Gansbaai to N Province).

SCHOTIA   BOERBOON   c. 18 spp., tropical and southern Africa

afra (L.) Thunb.   Much-branched shrub or tree, 3–7 m. Leaves pinnate, leaflet pairs 6–18. Flowers in congested panicles, red or pink, filaments free. Aug.–Oct. Karroid scrub, often along dry watercourses, KM, LB, SE (Little Karoo to Port Elizabeth and E Cape).

latifolia Jacq.   BOSBOERBOON   Small, slender tree, 3–5 m. Leaves pinnate, leaflet pairs 3–5. Flowers in open panicles, pink to red, filaments united below. Nov.–Jan. Forest scrub, SE (Knysna to E Cape).

*SESBANIA   SESBANIA, RATTLEPOD   c. 50 spp., widespread in tropics and subtropics

*punicea (Cav.) Benth.   RED SESBANIA   Deciduous shrub to 4 m. Leaves pinnate, leaflets oblong. Flowers in axillary racemes, red and orange. Pods 4-winged, pungent. Throughout the year. Disturbed and marshy places, SW, AP, LB, SE (S American weed).

STIRTONANTHUS•   3 spp., W Cape

chrysanthus (Adamson) B.-E.van Wyk & A.L.Schutte   Single-stemmed, reseeding shrub to 4 m. Leaves 1-foliolate, subrotund, finely felted. Flowers crowded at branch tips, yellow, bracts 6–8 mm, clasping base of calyx. Pods inflated. Aug.–Sept. Mountain fynbos, streambanks, 800–1500 m, KM (Klein Swartberg).•

insignis (Compton) B.-E.van Wyk & A.L.Schutte   Multistemmed, resprouting shrub to 2 m. Leaves 1-foliolate, broadly obovate, closely felted. Flowers crowded at branch tips, yellow, bracts 10–13 mm, not clasping base of calyx. Pods compressed. Oct.–Nov. Arid fynbos, 800–1200 m, KM (Montagu: Waboomsberg).•

taylorianus (L.Bolus) B.-E.van Wyk & A.L.Schutte   Like **S. chrysanthus** but leaves obovate to elliptic and bracts 3 mm, not clasping base of calyx. Sept. Mountain fynbos, streambanks, 700–1500 m, KM (Groot Swartberg).•

TEPHROSIA   c. 400 spp., pantropical, mostly Africa

capensis (Jacq.) Pers.   Straggling subshrub, thinly hairy; bracts minute, triangular, persistent. Leaves pinnate, leaflets linear-elliptic; stipules small, triangular, 1–3-veined. Flowers scattered in slender, pedunculate racemes, pink or purple, 8–10 mm long. Aug.–June. Sandy or grassy slopes and flats, below 860 m, SW, AP, KM, LB, SE (Cape Peninsula to E Cape).

grandiflora (Aiton) Pers.   ROOI-ERTJIE   Resprouting shrub to 1.5 m, thinly hairy; bracts large, ovate, deciduous. Leaves pinnate, leaflets elliptic-obovate; stipules large, ovate, many-veined. Flowers crowded or subumbellate in pedunculate racemes, pink or magenta, 15–20 mm long. Aug.–Oct. Scrub and forest margins, 450–1000 m, SE (Humansdorp to KwaZulu-Natal).

stricta (L.f.) Pers.   Like **T. capensis** but flowers crowded on very short peduncles. Sept.–Oct. Sandy slopes, LB, SE (Swellendam to Uitenhage).•

TRIFOLIUM   WILD CLOVER   c. 250 spp., nearly worldwide but mainly N temperate

burchellianum Ser.   Prostrate perennial rooting at nodes, 30–60 cm. Leaves 3-foliolate, leaflets obcordate, finely toothed. Flowers subsessile in pedunculate heads, purple, calyx rupturing in fruit. Sept.–Mar. Grassland and forest margins, 60–1830 m, KM, LB, SE (Swellendam to tropical Africa).

stipulaceum Thunb.   Softly hairy annual, 8–15 cm. Leaves 3-foliolate, leaflets obovate, finely toothed, sparsely hairy. Flowers sessile in shortly pedunculate heads, white or red, calyx persistent and enclosing fruit, lobes awn-like. Oct.–Nov. Coastal sands and grassland, SW, SE (Saldanha Bay and Uitenhage to Port Elizabeth).•

VIGNA   WILD SWEET PEA   c. 150 spp., pantropical, mostly palaeotropical

vexillata (L.) A.Rich. (incl. **V. debilis** Fourc.)   Sprawling, usually twining, retrorsely hairy or glabrescent subshrub. Leaves pinnately 3-foliolate, leaflets linear-oblong (sometimes sublanceolate) to ovate. Flowers 2–4 on elongate peduncles, greenish white tinged magenta. Oct.–Mar. Grassland, SE (Knysna to pantropical).

VIRGILIA   KEURBOOM   2 spp., W and E Cape

divaricata Adamson   Small tree to 16 m. Leaves pinnate, leaflets glabrescent, oblong, obtuse to emarginate. Flowers in axillary racemes, violet with dark purple keel tip, bracts small, caducous before buds 5 mm long. Aug.–Nov. Forest margins, streamsides, below 1200 m, KM, SE (Klein Swartberg, George to E Cape).

oroboides (P.J.Bergius) T.M.Salter   Small tree to 20 m. Leaves pinnate, leaflets finely hairy beneath, oblong, obtuse to acute. Flowers in axillary racemes, pale pink or white with pale pink, yellowish green or dark purple keel tip, bracts large, persistent until buds 10 mm long. Mainly Jan.–Apr. Forest margins, streamsides, below 1200 m, SW, LB, SE (Cape Peninsula to George).•

WIBORGIA   PENNY POD   10 spp., N and W Cape

A. Branches glabrous

fusca Thunb.   Erect or spreading, somewhat thorny shrub, 0.6–1.5 m, branches glabrous, greyish. Leaves 3-foliolate, leaflets oblanceolate-apiculate. Flowers in terminal racemes, pale greenish yellow. Fruit with a dorsal wing, 12–30 × 9–15 mm. Aug.–Oct. Mountain and lowland fynbos, below 1400 m, NW, SW (Namaqualand to Malmesbury).

mucronata (L.f.) Druce   Like **W. fusca** but young branches usually reddish or yellow, leaflets scarcely apiculate and fruit reticulate. Aug.–Oct. Mountain fynbos or renosterbos-fynbos scrub, 160–1260 m, NW, SW, KM (Namaqualand to Anysberg).

tenuifolia E.Mey.   Like **W. tetraptera** but flowers pinkish or mauve. Sept.–Nov. Renosterbos-fynbos scrub, 160–1000 m, SW, KM, LB (Bredasdorp to Riversdale, Little Karoo).•

tetraptera E.Mey.   Erect thorny shrub, 0.4–1.5 m, branches glabrous. Leaves 3-foliolate, leaflets obovate. Flowers in terminal racemes, cream-coloured to pale greenish yellow. Fruit with conspicuous dorsal, ventral and lateral wings, papery, 15–20 × 10–14 mm. Sept.–Nov. Renosterbos-fynbos scrub, 160–1000 m, NW, SW (Bokkeveld Mts to Malmesbury and Hex River valley).•

A. Branches pubescent, at least when young

humilis (Thunb.) R.Dahlgren   Decumbent or ascending shrublet to 35 cm, branches pubescent. Leaves 3-foliolate, leaflets linear-spathulate, channelled. Flowers in terminal racemes, yellow, pale brown or pink, keel obtuse. Fruit without wings or crest, 4 × 2 mm. Sept.–Oct. Mountain fynbos near streams, 600–800 m, NW (Gifberg).•

leptoptera R.Dahlgren   Rigid, thorny shrublet, 0.3–1.2 m, branches pubescent. Leaves 3-foliolate, leaflets oblanceolate-obovate, densely grey-silky. Flowers in thorny terminal racemes, pale yellow. Fruit with a dorsal crest, reticulate, 7–12 × 6–9 mm. July–Sept. Renosterbos-fynbos scrub, 130–250 m, NW, SW (Cedarberg Mts to Darling).•

monoptera E.Mey.   Like **W. leptoptera** but leaves sparsely hairy or glabrous and fruit with a well-developed dorsal wing 4–5 mm broad. July–Sept. Karroid scrub, renosterbos-fynbos scrub, 500–1160 m, NW (Namaqualand to Clanwilliam).

obcordata (P.J.Bergius) Thunb.   Slender, stiff or willowy shrub, 1.5–3 m, branches pubescent. Leaves 3-foliolate, leaflets oblanceolate-obcordate, sparsely hairy beneath. Flowers in terminal racemes, bright yellow. Fruit with narrow dorsal crest, 7–12 × 4–7 mm. Aug.–Oct. Sandy flats and slopes, 50–800 m, NW, SW, LB (Bokkeveld Mts to Mossel Bay).•

sericea Thunb.   Like **W. leptoptera** but fruit with reticulae coalescing into a lateral ridge or wing. May–Sept. Renosterbos-fynbos scrub, arid fynbos, 660–1330 m, NW, KM (Namaqualand to Cedarberg Mts and Witteberg).

XIPHOTHECA• (= *PRIESTLEYA* in part)   SILVER PEA   9 spp., N and W Cape

A. Inflorescences shortly pedunculate

canescens (Thunb.) A.L.Schutte & B.-E.van Wyk   Single-stemmed, reseeding, tree-like shrub to 2.5 m. Leaves 1-foliolate, narrowly elliptic, flat, silky. Flowers in pedunculate pairs at tips of lateral branchlets, yellow, calyx lobes shorter than tube, ovules 5–8. Aug.–Sept. Mountain fynbos, 660–780 m, NW (Bokkeveld Mts).•

cordifolia A.L.Schutte & B.-E.van Wyk   Single-stemmed, reseeding, tree-like shrub to 2.5 m. Leaves opposite, 1-foliolate, cordate, shortly hairy beneath. Flowers in pedunculate pairs at tips of lateral branchlets, yellow, calyx intrusive, lobes slightly longer than tube, ovules 4–6. Flowering time? Mountain fynbos, streamsides, 1330–1660 m, NW (Hex River Mts).•

elliptica (DC.) A.L.Schutte & B.-E.van Wyk (= *Priestleya elliptica* DC.)   Multistemmed, resprouting shrub to 1 m like **X. canescens** but leaves elliptic with margins slightly revolute. Oct.–Dec. Mountain fynbos, 360–1400 m, SW (Wemmershoek Mts to Caledon Swartberg).•

phylicoides A.L.Schutte & B.-E.van Wyk   Multistemmed, resprouting shrub to 1.2 m. Leaves 1-foliolate, elliptic to narrowly elliptic, margins recurved, glabrescent above, densely silky hairy beneath. Flowers in pedunculate pairs crowded along branches, yellow, calyx lobes about as long as tube, ovules 2. May, after fire. Mountain fynbos, 500–800 m, SE (Outeniqua Mts).•

A. Inflorescences sessile

fruticosa (L.) A.L.Schutte & B.-E.van Wyk   Single-stemmed, reseeding, tree-like shrub to 2 m or multistemmed, resprouting shrub to 70 cm. Leaves 1-foliolate, elliptic, densely silvery silky. Flowers crowded in sessile, head-like terminal clusters nested among leaves, yellow, calyx lobes shorter than tube, ovules 2 or 3. June–Sept. Sandstone slopes in fynbos, 100–1200 m, NW, SW, KM (Hex River Mts, Cape Peninsula to Touwsberg).•

guthriei (L.Bolus) A.L.Schutte & B.-E.van Wyk (= *Priestleya guthriei* L.Bolus)   Single-stemmed, reseeding shrub to 30 cm like **X. fruticosa** but leaves narrowly elliptic and thinly velvety. July. Lowland fynbos, below 250 m, SW (Kleinrivier Mts to Potberg).•

lanceolata (E.Mey.) Eckl. & Zeyh. (= *Priestleya glauca* T.M.Salter)   Single-stemmed, reseeding shrublet to 60 cm like **X. fruticosa** but calyx lobes longer than tube. July–Sept. Lowland fynbos, granite hills, 60–200 m, SW (Cape Flats).•

reflexa (Thunb.) A.L.Schutte & B.-E.van Wyk   Multistemmed, prostrate to straggling, resprouting shrub to 50 cm. Leaves often reflexed, 1-foliolate, ovate-lanceolate, silvery silky. Flowers in sessile clusters crowded at branch tips, yellow, calyx lobes longer than tube, ovules 5–7. Aug.–Nov. Lowland fynbos, below 300 m, NW, SW, AP (Piketberg to Elim).•

tecta (Thunb.) A.L.Schutte & B.-E.van Wyk (= *Priestleya stokoei* L.Bolus)   Multistemmed, resprouting shrub to 1 m. Leaves 1-foliolate, elliptic to suborbicular, silky hairy. Flowers in sessile pairs clustered along branches, yellow, calyx lobes shorter than tube, wing petals with a pocket-like fold, ovules 5–8. May–Sept. Mountain fynbos, 500–1400 m, NW, SW (Clanwilliam to Stellenbosch).•

ZORNIA c. 80 spp., pantropical and warm temperate

capensis Pers.   Slender, wiry, glandular-punctate subshrub. Leaves 2- or 4-foliolate, leaflets elliptic; stipules leafy, peltate. Flowers in interrupted spikes, concealed by large, peltate bracts, yellow. Pods segmented and fragmenting, reticulate-tuberculate. Nov.–Feb. Grassland, 400–1000 m, SE (Knysna to Mpumalanga).

FLACOURTIACEAE

1. Petals absent; styles 2–7; much-branched spiny shrubs; fruit a berry ....**Dovyalis**
1.' Petals present (rarely absent); styles 1 or 3; plants unarmed; fruit a berry or capsule:
    2. Flowers unisexual, stamens in bundles of 3 or 4, alternating with disc glands; leaves 5–9-veined from base ....**Trimeria**
    2.' Flowers bisexual; stamens not in bundles; leaves not several-veined from base:
        3. Leaves opposite; ovary 1-locular with a single ovule on each of 2 placentas; style bilobed; fruit a 2-valved capsule ....**Pseudoscolopia**
        3.' Leaves alternate; ovary 1-locular with few ovules on 3 or 4 parietal placentas; style simple; fruit a berry...**Scolopia**

DOVYALIS   SUURBESSIE   c. 20 spp., Africa to Sri Lanka and New Guinea

lucida Sim   BLINKBLAARSUURBESSIE   Dioecious tree to 10 m. Leaves rhomboid, 3–5-veined from base. Flowers 1-few in axillary clusters, greenish. Fruits minutely papillose. July–Oct. Forests and scrub, SE (Port Elizabeth to Zimbabwe).

rhamnoides (Burch. ex DC.) Harv.   Like **D. rotundifolia** but leaves thinner in texture, narrower and usually truncate or cordate at base, sepals in fruits fringed with stalked glands and seeds hairy. Jan.–Feb. Forests and scrub, SE (George to Mpumalanga).

rotundifolia (Thunb.) Thunb. & Harv.   DUINESUURBESSIE   Dioecious, thorny tree or shrub with white, ridged bark to 6 m. Leaves leathery, rounded, 3–5-veined from base. Flowers 1-few in axillary clusters, greenish. Jan.–Mar. Coastal dune forests, SE (Humansdorp to E Cape).

PSEUDOSCOLOPIA   VALSROOIPEER   1 sp., South Africa

polyantha Gilg   Shrub or tree to 5 m. Leaves opposite, elliptic, toothed. Flowers few in axillary cymes, white. Oct. Forests, NW (Twenty Four Rivers Kloof only and E Cape to KwaZulu-Natal).

SCOLOPIA   ROOIPEER   c. 37 spp., palaeotropical and subtropical

mundii (Eckl. & Zeyh.) Warb.   Small, sometimes spiny tree to 16 m. Leaves elliptic, closely toothed. Flowers few in axillary racemes, greenish white with reduced petals. Nov.–Dec. Forests, SW, LB, SE (Cape Peninsula to Mpumalanga).

zeyheri (Nees) Harv.   DORINGROOIPEER   Like **S. mundii** but leaves narrower and entire or sparsely toothed, often waxy above. Apr.–July. Forests and scrub, SE (Knysna to Mpumalanga).

TRIMERIA   WILDEMOERBEI   5 spp., southern and tropical east Africa

grandifolia (Hochst.) Warb.   Dioecious small tree to 10 m. Leaves round, toothed, 5–7-veined from base. Flowers in axillary spikes or panicles, 4- or 5-lobed, greenish. Feb.–Apr. Forests, SE (George to Zimbabwe).

trinervis Harv.   FYNBLAARWILDEMOERBEI   Dioecious small tree to 7 m. Leaves elliptic, toothed, 3(–5)-veined from base. Flowers in axillary spikes, 3-lobed, greenish. June–Oct. Forests and scrub forests, SE (Knysna to Mpumalanga).

FRANKENIACEAE

FRANKENIA   SEA-HEATH, SANDANGELIER   c. 75 spp., cosmopolitan in saline habitats

pulverulenta L.   Spreading annual or perennial. Leaves shortly petiolate, ovate, flat or revolute. Flowers numerous, solitary in branch forks, lilac. Sept.–Jan. Salt pans or brackish places, NW, SW, AP, KM, LB, SE (widespread).

repens (P.J.Bergius) Fourc.   Prostrate perennial. Leaves subsessile, subterete with revolute margins. Flowers in congested apical cymes, pink. Sept.–Jan. Salt pans and brackish places, NW, SW, AP, LB, SE (Namaqualand to Port Elizabeth).

FUMARIACEAE

1. Ovules 2 or more in ovary; upper petal inconspicuously saccate at base; fruit ovoid or lanceolate, compressed or inflated ....**Cysticapnos**
1.' Ovules solitary in ovary; upper petal conspicuously spurred at base:
   2. Fruit asymmetrical, suboblong, much shorter than incurved pedicel.................................**Trigonocapnos**
   2.' Fruit symmetrical:
      3. Fruit turgid, more or less globose, erect ....**Fumaria**
      3.' Fruit flat, more or less discoid with a peripheral wing, nodding ....**Discocapnos**

CYSTICAPNOS (= *PHACOCAPNOS*)   AFRICAN FUMITORY   4 spp., South Africa

cracca (Cham. & Schltdl.) Lidén (= *Phacocapnos cracca* (Cham. & Schltdl.) Bernh.)   Glaucous straggling annual to 1 m. Leaves bipinnate, often tendrillous, segments cuneate, trifid. Flowers in racemes, petals connivent, upper shortly saccate at base, pink with purple tips. Fruits compressed-lanceolate, drooping. Sept.–Nov. Damp slopes, NW, SW, AP, KM, LB, SE (Bokkeveld Mts to Port Elizabeth).•

vesicaria (L.) Fedde (= *Cysticapnos grandiflora* Bernh., *Phacocapnos burmanii* (Eckl. & Zeyh.) Hutch.)   KLAPPERTJIES   Glaucous straggling annual to 1 m. Leaves bipinnate, often tendrillous, segments cuneate, trifid. Flowers in racemes, bilabiate, petals broadly winged and flaring, upper shortly saccate, pink. Fruits inflated-ovoid, nodding. Aug.–Oct. Sandy flats and slopes, NW, SW, AP, KM (Namaqualand to De Hoop).•

DISCOCAPNOS•   1 sp., W Cape

mundii Cham. & Schltdl.   Glaucous straggling annual to 1 m. Leaves bipinnate, often tendrillous, segments cuneate, incised. Flowers in racemes, petals connivent, upper spurred at base, pink with purple tips. Fruits discoid with peripheral wing, nodding. Aug.–Dec. Forest margins, SW, SE (Cape Peninsula to Jonkershoek and Witelsbos).•

*FUMARIA   FUMITORY   c. 55 spp., E Africa and Europe to India

*muralis Sond. ex Koch   DUIWELSKERWEL   Sprawling annual to 1 m. Leaves bipinnate, segments cuneate, incised. Flowers in racemes, petals connivent, upper with a globose spur at base, pink with purple tips. Fruits globose, erect. May–Oct. Weed of waste places, NW, SW, AP, KM, LB, SE (widespread European weed).

TRIGONOCAPNOS   HEKELTJIES   1 sp., W and N Cape

lichtensteinii (Cham. & Schltdl.) Lidén (= *Ttrigonocapnos curvipes* Schltr.)   Glaucous straggling annual to 1 m. Leaves bipinnate, often tendrillous, segments elliptic. Flowers in racemes, petals connivent, upper spurred at base, pink with purple tips. Fruits asymmetric, boat-shaped, nodding on incurved pedicels. Aug.–Sept. Rocky slopes, NW (Bokkeveld Mts and W Karoo to Biedouw).

GEISSOLOMATACEAE•

GEISSOLOMA• 1 sp., W Cape

marginata (L.) A.Juss.   Densely leafy shrub to 1 m, thinly woolly on young parts. Leaves decussate, subsessile, ovate, margins revolute. Flowers solitary in axils, pink. June–Sept. Sandstone slopes, 600–1200 m, LB (Langeberg Mts: Swellendam to Riversdale).•

GENTIANACEAE

1. Flowers yellow or white; ovary 2-locular ....Sebaea
1.' Flowers pink; ovary 1-locular:
    2. Calyx keeled; plants glabrous ....Chironia
    2.' Calyx rounded; plants pubescent ....Orphium

CHIRONIA   CENTAURY, BITTERWORTEL   c. 30 spp., Africa and Madagascar

A. Ovary and fruits obtuse

arenaria E.Mey.   Like **C. linoides** but typically stouter with prominent nodes, to 25 cm, and corolla lobes suborbicular. Dec.–Jan. Sandy slopes, NW, KM (Clanwilliam to Anysberg).•

baccifera L.   CHRISTMAS BERRY, AAMBEIBOSSIE   Shrublet to 1 m. Leaves linear, spreading. Flowers pink, corolla tube 3–5 mm long, constricted above ovary, ovary and fruits rounded, stigma truncate. Fruits berry-like. Nov.–Feb. Sandy flats and slopes, NW, SW, AP, LB, SE (Namaqualand to KwaZulu-Natal).

linoides L.   Shrublet to 90 cm. Leaves linear, erect or spreading. Flowers pink, corolla tube 3–5 mm long, ovary and fruits rounded, stigma truncate. Oct.–Jan. Sandy or marshy flats and slopes, NW, SW, KM, LB (Namaqualand to Cape Peninsula to Oudtshoorn and Bredasdorp).

A. Ovary and fruits acute

decumbens Levyns   Weak, decumbent perennial rooting at nodes. Leaves linear to elliptic. Flowers pink, corolla tube c. 13 mm long, ovary and fruits acuminate, stigma broadly peltate. Oct.–Jan. Coastal flats and vleis, SW, AP, LB, SE (Cape Peninsula to E Cape).

jasminoides L.   Shrublet to 90 cm. Leaves linear to ovate, erect. Flowers pink, corolla tube 9–14 mm long, calyx lobed to near base, ovary and fruits acuminate, stigma truncate. Aug.–Jan. Marshy slopes, SW, AP, LB (Bainskloof to Langeberg Mts).•

melampyrifolia Lam.   Somewhat sticky, straggling shrublet. Leaves ovate-lanceolate, cordate at base, spreading. Flowers pink, corolla tube 7–12 mm long, ovary and fruits acuminate and glutinous, style truncate. Sept.–Jan. Shady damp sandstone slopes, SW, LB, SE (Hottentots Holland Mts to E Cape).

peduncularis Lindl. Straggling perennial to 1 m. Leaves ovate-lanceolate, cordate at base. Flowers pink, corolla tube c. 16 mm long with a long narrow neck, ovary and fruits acuminate, stigma broadly peltate. Oct.–Dec. Damp, shaded places, SE (Knysna to E Cape).

serpyllifolia Lehm.   Wiry shrublet to 50 cm. Leaves ovate. Flowers pink, corolla tube constricted at neck, c. 8 mm long, ovary and fruits acuminate, stigma bifid. Dec.–Jan. Damp places, SE (Humansdorp to E Cape).

stokoei I.Verd.   Like **C. jasminoides** but flowers conspicuously bibracteate below ovary and calyx lobes acute (not acuminate). Sept.–Oct. Sandstone slopes at high alt., SW (Hottentots Holland to Kleinrivier Mts).•

tetragona L.f.   Sticky, willowy annual or biennial to 60 cm. Leaves linear to ovate. Flowers sticky, pink, corolla tube 6–10 mm long, thickened in throat, calyx lobed to halfway, ovary and fruits acuminate, stigma truncate. Oct.–Jan. Coastal sands and limestone, SW, AP, LB, SE (Cape Peninsula to E Cape).

ORPHIUM•   SEA-ROSE, TERINGBOS   1 sp., W Cape

frutescens (L.) E.Mey.   Pubescent shrublet to 80 cm. Leaves oblanceolate, margins revolute. Flowers 1 or 2 terminal and in upper axils, shining deep pink, anthers twisted, porose. Nov.–Feb. Coastal sands and pans, NW, SW, AP, LB, SE (Lambert's Bay to George).•

SEBAEA   YELLOWWORT, NAELTJIESBLOM   c. 60 spp., mainly Africa, also India and Australasia

A. Filaments inserted below corolla sinuses

exacoides (L.) Schinz   Annual to 30 cm. Leaves ovate. Flowers 5-lobed, yellow or cream-coloured with orange streaks in throat, calyx lobes strongly winged, stamens inserted below sinuses, style swelling below middle. Aug.–Oct. Sandy flats and slopes, NW, SW, AP, LB (Bokkeveld Mts to Riversdale).•

membranacea Hill   Like **S. rara** but to 15 cm, anthers with 3 glands. Dec. Sandstone slopes, NW (Namaqualand to Hex River Pass).

micrantha (Cham. & Schltdl.) Schinz   Annual to 20 cm. Leaves ovate. Flowers 5-lobed, yellow, calyx lobes strongly winged, stamens inserted below sinuses, style swelling confluent with stigma. Sept.–Nov. Flats and slopes, NW, SW, AP, LB, SE (Clanwilliam to Port Elizabeth).•

pusilla Eckl. ex Cham.   Like **S. membranacea** but style without swelling. Sept.–Dec. Sandstone slopes in seeps, NW, SW (Bokkeveld Mts to Houwhoek Mts).•

rara Wolley-Dod   Annual to 6 cm. Leaves lanceolate. Flowers 5-lobed, yellow, calyx lobes keeled, stamens inserted below sinuses, style swelling below or confluent with stigma. Oct. Sandy flats, SW, AP, LB (Cape Peninsula to Albertinia).•

A. Filaments inserted in corolla sinuses
B. Petals 4

albens (L.f.) Roem. & Schult.   Like **S. aurea** but calyx lobes rounded on back. Aug.–Nov. Damp sandy coastal flats, NW, SW, AP, LB (Piketberg to Albertinia).•

ambigua Cham.   Annual to 20 cm. Leaves ovate. Flowers 4-lobed, yellow, calyx lobes keeled at tips, corolla tube longer than lobes, 3.5–6 mm long, style with a swelling. Sept.–Dec. Sandy coastal flats, SW, AP (Cape Peninsula to Agulhas).•

aurea (L.f.) Roem. & Schult.   Annual to 30 cm. Leaves ovate. Flowers 4-lobed, yellow or white, calyx lobes keeled, corolla tube shorter than lobes, 2–6 mm long. Oct.–Dec. Sandy flats and slopes, NW, SW, AP, LB, SE (Pakhuis Mts to Cape Peninsula to Humansdorp).•

capitata Cham. & Schltdl.   Like **S. aurea** but flowers larger, calyx 6–10 mm long and petals 6.5–10 mm long. Oct.–Jan. Sandstone slopes, SW, LB, SE (Cape Peninsula to Great Winterhoek Mts).•

laxa N.E.Br.   Like **S. aurea** but delicate with numerous leaves scattered along thin, wiry stems and longer filaments, 2–2.5 mm. Sept.–Oct. Shaded sandstone slopes, LB (Langeberg Mts: Swellendam to Riversdale).•

minutiflora Schinz   Like **S. ambigua** but flowers white, smaller, petals 1–2.5 mm long and style without a swelling. Oct.–Dec. Damp sandy coastal flats, SW, AP, LB, SE (Cape Peninsula to E Cape).

schlechteri Schinz   Like **S. ambigua** but flowers yellow or white and calyx lobes strongly keeled throughout. Aug.–Nov. Damp sandy slopes, SW (Cape Peninsula to Botrivier).•

B. Petals 5 or 6

elongata E.Mey.   Tufted annual or biennial to 70 cm. Leaves in a basal rosette, ovate. Flowers 5-lobed, yellow, calyx lobes keeled, stamens inserted in sinuses, style swelling absent or below middle. Oct.–Dec. Sandstone slopes, SW, LB, SE (Riviersonderend Mts to Humansdorp).•

fourcadei Marais   Like **S. grisebachiana** but anthers with 3 glands. Oct.–Nov. Sandy slopes, LB, SE (Knysna to Humansdorp).•

grisebachiana Schinz   Annual to 15 cm. Leaves ovate. Flowers 5-lobed, yellow, calyx lobes winged, stamens inserted in sinuses, style swelling below middle. July–Jan. Sandy flats and slopes, LB, SE (Swellendam to E Cape).

macrophylla Gilg   Perennial to 75 cm. Leaves ovate. Flowers 5 or 6-lobed, yellow, calyx lobes keeled or winged, membranous, stamens inserted in sinuses, anthers with 3 glands or without glands, style swelling below middle. June–Dec. Mountain seeps, LB, SE (George to E Cape).

scabra Schinz   Annual to 20 cm. Leaves ovate. Flowers 5-lobed, yellow, calyx lobes with strongly veined scabrid wings, stamens inserted in sinuses, style swelling below middle base. Sept. Sandy flats, SW, LB (Cape Peninsula to Riversdale).•

stricta (E.Mey.) Gilg   Perennial to 60 cm. Leaves ovate, reflexed. Flowers 5-lobed, yellow, calyx lobes keeled, stamens inserted in sinuses, anthers with 2 basal glands, style swelling absent or small. Mar.–Oct. Sandstone slopes, LB, SE (Riversdale to E Cape).•

sulphurea Cham. & Schltdl.   Like **S. fourcadei** but stigmas ligulate. Sept.–Nov. Sandstone slopes, SW, LB (Cape Peninsula to Riversdale).•

zeyheri Schinz   Annual to 25 cm. Leaves ovate. Flowers 5-lobed, yellow, calyx lobes winged, stamens inserted in sinuses, style swelling confluent with stigma. Sept.–Nov. Sandy coastal flats, SW, AP, LB, SE (Cape Peninsula to KwaZulu-Natal).

GERANIACEAE
Pelargonium by P. Vorster

1.   Flowers zygomorphic; upper sepal with a nectariferous spur fused to pedicel; stamens 10, only 2–7 with anthers ....**Pelargonium**
1.'   Flowers actinomorphic; nectariferous spur absent; stamens 10–15, all fertile or alternate ones without anthers:
   2.   Stamens 10 ....**Geranium**
   2.'   Stamens 15 ....**Monsonia**

GERANIUM   GERANIUM, CRANE'S-BILL   c. 260 spp., cosmopolitan

caffrum Eckl. & Zeyh.   Diffuse perennial. Leaves digitately 3–5-lobed to base, long-petioled, finely adpressed-hairy. Flowers 1 or 2 on long axillary peduncles, 20–30 mm diam., white or sometimes pale pink. Sept.–Jan. Damp places, NW, SW, LB, SE (Cold Bokkeveld to KwaZulu-Natal).

canescens L'Hér.   Diffuse perennial, sprawling or suberect. Leaves digitately 3–5-lobed three-quarters to base, long-petioled, finely adpressed-hairy. Flowers 1 or 2 on long axillary peduncles, 15–20 mm diam., white. Sept.–Oct. Dry slopes and streambanks, NW, SW, KM, LB (Grootwinterhoek Mts and Cape Peninsula to Kammanassie Mts).•

incanum Burm.f.   Diffuse perennial with thickened taproot. Leaves digitately 3–7-lobed to base with lobes pinnatisect to bipinnatisect, long-petioled, glabrous or sparsely hairy above, densely white-hairy beneath. Flowers 1 or 2 on axillary peduncles, 15–30 mm diam., white with dark veins or pink to mauve. Mainly Aug.–Oct. Stony slopes, SW, AP, KM, LB, SE (Cape Peninsula to Port Alfred).

ornithopodon Eckl. & Zeyh.   Clump-forming perennial with thick, woody rootstock. Leaves digitately 5-lobed more than halfway to base, sparsely hairy above. Flowers usually 2, on long axillary peduncles, 15–20 mm diam., pale pink with darker veins or white. Oct.–Feb. Damp ground in scrub or forest margins, SW, LB, SE (Cape Peninsula to E Cape).

MONSONIA (= *SARCOCAULON*)   PARASOL, BOESMANSKERS   c. 40 spp., Africa, Madagascar, SW Asia

A. Stems soft or slightly woody; petioles not spinescent

emarginata (L.f.) L'Hér.   Sprawling or scrambling perennial with tuberous roots. Leaves entire, ovate to obovate, hairy. Flowers solitary on slender peduncles, cup-shaped, 15–20 mm diam., white, cream-coloured or pink. Mainly Sept.–Mar. Bushveld or scrubby grassland, SW, LB, SE (Cape Peninsula to E Cape).

speciosa L.   SAMBREELTJIE   Sprawling perennial with annual stems from a woody rhizome. Leaves subentire to digitately compound, subglabrous. Flowers solitary on long, stout peduncles, rotate, 25–65 mm diam., white to pink, deep pink beneath. Aug.–Nov. Clay and granite slopes and flats, mostly in renosterveld, NW, SW (Clanwilliam to Gordon's Bay).•

A. Stems thick and more or less succulent; petioles spinescent

crassicaule (Rehm) F.Albers (= *Sarcocaulon crassicaule* Rehm)   NOERAP   Spiny stem succulent to 50 cm, branches thicker than 10 mm, grey or greyish yellow. Leaves ovate to obovate, irregularly pinnatifid with margins crenate to dentate, usually tomentose. Flowers to 55 mm diam., pale to bright yellow, sepals with mucro longer than 2 mm. May–June. Rocky places, NW, KM (S Namibia to Cedarberg and Little Karoo).

salmoniflora (Moffett) F.Albers (= *Sarcocaulon salmoniflorum* Moffett)   PINK CANDLE BUSH   Spiny stem succulent to 40 cm, branches thinner than 4 mm, olive-green to grey. Leaves elliptic, entire. Flowers to 30 mm diam., pink to orange, petals twice as long as wide, sepals with mucro shorter than 0.5 mm. Mainly Oct.–Dec. Stony flats and slopes, often on quartz patches, NW, KM (Namibia to Swartruggens and Swartberg Mts).

spinosa L'Hér. (= *Sarcocaulon l'heritieri* Sweet)   YELLOW CANDLE BUSH   Spiny stem succulent to 80 cm, branches thinner than 10 mm, pale olive-green to grey. Leaves long- and short-petioled, obovate to round, emarginate, glaucous. Flowers to 35 mm diam., yellow, sepals with mucro longer than 2 mm. Aug.–Sept. Rocky slopes, NW (Richtersveld to Olifants River valley).

PELARGONIUM   PELARGONIUM, STORK'S-BILL, MALVA   c. 250 spp., Africa, Madagascar, Middle East, Australia

A. Annuals or apparently so

chamaedryfolium Jacq.   Sprawling annual. Leaves cordate, variously incised, c. 2.5 × 2 cm. Flowers 3–8 on short axillary peduncles, c. 7 mm diam., pink, regular; hypanthium 2–3 mm long, about as long as pedicel. Aug.–Jan. Disturbed places, mainly after fire, NW, SW (Clanwilliam to Caledon).•

columbinum Jacq.   Prostrate annual. Leaves cordate with margins variously incised, silky hairy. Flowers 2–5 on wiry axillary peduncles, c. 7 mm diam., purple; hypanthium 1.5–2.5(–5) mm long, much shorter than pedicel. Throughout the year. Cool disturbed places in mountains, NW, SW (Bokkeveld Mts to Cape Peninsula).•

grossularioides (L.) L'Hér.   Dwarf, more or less prostrate annual. Leaves cordate or somewhat digitately lobed, often purplish, 5 × 4 cm. Flowers to 50, c. 8 mm diam., purple; hypanthium c. 3 mm long, shorter than pedicel. Throughout the year. Damp places, NW, SW, AP, KM, LB, SE (Clanwilliam to KwaZulu-Natal).

minimum (Cav.) Willd.   Prostrate perennial with taproot. Leaves silvery, simple to pinnatisect with linear segments, to 2 cm diam. Flowers to 8, c. 6 mm diam., white to very pale pink or purple; hypanthium very shallow and swollen. Aug.–Nov. Dry, open places, KM (Namibia and South Africa).

nanum L'Hér.   Prostrate annual. Leaves cordate and variously lobed, 1.5–2 cm diam. Flowers to 6 on short axillary peduncles, c. 8 mm diam., typically bicoloured with upper petals pink and lower whitish; hypanthium 1–3(–8) mm long, shorter than pedicel. Aug.–Dec. Dry, open places, NW, SW, LB, SE (?Namibia, Namaqualand to E Cape).

senecioides L'Hér.   Annual to 40 cm, branching mainly from base. Leaves 2- or 3-pinnatifid, to 7 × 6 cm. Flowers 2–4 on short axillary peduncles, 8–18 mm diam., white with purple marks on upper petals and purple-veined beneath; hypanthium 6–9 mm long, longer than the very short pedicel. Sept.–Nov. Deep sand, NW, SW, KM (Bokkeveld Mts to Cape Peninsula and Witteberg).•

A. Perennials, shrubs or geophytes
 *B. Tuberous geophytes without well-developed stems; leaves mostly in a basal tuft (see also **P. minimum**)*
  C. Tubers with leathery, cracked bark; leaves in a basal tuft and on short branches, usually green at flowering; petals equal, obovate, sometimes fimbriate

anethifolium (Eckl. & Zeyh.) Steud.   Geophyte with large, woody tuber. Leaves 2- or 3-pinnatisect, to 15 × 10 cm. Flowers 6–10 on long, stout, branching peduncles, 15–20 mm diam., yellowish pink; hypanthium 25–50 mm long, much longer than pedicel. Oct.–Dec. Sandstone slopes and flats, NW, SW (Bokkeveld Mts to Cape Peninsula).•

caffrum (Eckl. & Zeyh.) Harv.   Geophyte with large, woody tuber. Leaves deeply digitate with linear segments, c. 10 cm diam. Flowers to 27 on a stout peduncle, c. 25 mm diam., yellow-green to dark maroon, petals finely dissected; hypanthium 17–32 mm long, usually slightly longer than pedicel. Oct.–Mar. Grassland, SE (Knysna to Grahamstown).

lobatum (Burm.f.) L'Hér.   KANEELBOL   Geophyte with hard, woody tuber. Leaves prostrate, ovate, more or less lobed, softly hairy, to 30 cm diam. Flowers to 20 on a stout peduncle, 15–18 mm diam., usually black with yellow margins to petals, clove-scented at night; hypanthium 25–35 mm long, much longer than pedicel. Sept.–Nov. Sandy places, NW, SW, AP, LB, SE (Graafwater to Knysna).•

luridum (Andrews) Sweet   Geophyte with large woody tuber, to 80 cm. Leaves variously subdigitately divided, usually sparsely hairy, to c. 30 cm diam. Flowers to 60 on a stout peduncle, 25–40 mm diam., white, pale yellow, or pink; hypanthium 40–80 mm long, much longer than pedicel. Sept.–Apr. Grassland, LB, SE (George to central Africa).

multiradiatum J.C.Wendl.   Geophyte with large, woody tuber. Leaves 2- or 3-pinnatisect, segments ribbon-shaped, shortly hairy, margins slightly revolute, to 45 cm diam. Flowers to 30 on stout, branching peduncles, 15–18 mm diam., petals almost black with narrow pale yellow margins, subsessile; hypanthium 25–35 mm long. Sept.–Mar. Open, usually sandy places, NW, SW (Clanwilliam to Cape Peninsula).•

pillansii T.M.Salter   Geophyte with large, woody tuber. Leaves dry at flowering, variously pinnatifid, glaucous, to 12 × 8 cm. Flowers to 17 on long subterminal peduncles, c. 20 mm diam., dirty yellow, subsessile; hypanthium to 60 mm long. Feb.–Apr. Rock crevices, NW, SW, KM, LB, SE (Gifberg to Langkloof).•

pulverulentum Sweet   Geophyte with large, woody tuber. Leaves cordate and often quite deeply incised, glaucous, leathery, sparsely softly hairy. Flowers to 14, 15–20 mm diam., black or maroon with pale yellow margins on petals, subsessile; hypanthium 20–50 mm long. Sept.–Feb. Coastal grassland, SE (Jeffreys Bay to KwaZulu-Natal).

radulifolium (Eckl. & Zeyh.) Steud.   Geophyte with large tuber. Leaves bipinnatisect, flaccid, to 30 × 15 m. Flowers to 20 on stout, branching peduncles, c. 20 mm diam., yellow to pinkish, subsessile; hypanthium 20–55 mm long. Aug.–May. Open places on well-drained soil, NW, SW, KM, LB, SE (Clanwilliam to Port Elizabeth).•

schizopetalum Sweet   MUISHONDBOSSIE   Geophyte with large, woody tuber. Leaves pinnatifid, softly hairy, to 15 × 7 cm. Flowers to 20, 30–40 mm diam., pale yellow and variously tinged with purple, petals finely dissected; hypanthium 30–70 mm long, longer than pedicel. Oct.–Feb. Grassland, SE (Humansdorp to Uitenhage).•

triste (L.) L'Hér.   KANEELTJIE   Geophyte with large, woody tuber. Leaves prostrate, 2- or 3-pinnatisect with linear segments, softly hairy, to 30 cm diam. Flowers to 20 on a stout peduncle, 15–18 mm diam., pale yellow with dark maroon to black centres, clove-scented at night; hypanthium 25–35 mm long, much longer than pedicel. Aug.–Feb. Sandy flats and slopes, NW, SW, AP (Namaqualand to Albertinia).•

C. Tubers with paper-like bark; leaves only in a basal tuft, usually dry at flowering; petals often narrow or unequal
D. Petals 2 or 4

asarifolium (Sweet) G.Don   Tuberous geophyte to 25 cm. Leaves dry at flowering, cordate, white-woolly below, to 10 × 8 cm. Flowers to 12 on branching peduncles, c. 10 mm diam., dark wine-red, subsessile, only 2 upper petals present; hypanthium 7–12 mm long. Nov.–May. Scrub, NW, SW (Piketberg to Sir Lowry's Pass).•

dipetalum L'Hér.   Tuberous geophyte to 35 cm. Leaves dry at flowering, sometimes bipinnatisect with linear segments, to c. 12 × 5 cm. Flowers to 12 on branching peduncles, c. 15 mm diam., pink, petals spoon-shaped, only upper 2 present, subsessile; hypanthium 7–18(–50) mm long. Feb.–Apr. Coastal plains, SW, AP, SE (Betty's Bay to Keurboomsrivier).•

ellaphiae E.M.Marais   Tuberous geophyte to 30 cm. Leaves dry at flowering, mostly narrowly elliptic, to 12 × 2 cm. Flowers to 15 on branching peduncles, c. 12 mm diam., dark wine-red, only upper 2 petals present, subsessile; hypanthium 9–15 mm long. Nov.–Feb. Sandy soil in fynbos, NW, SW (Ezelsbank to Cape Peninsula).•

leipoldtii R.Knuth   Tuberous geophyte to 30 cm. Leaves trifoliolate, to 20 cm diam. Flowers to 14 on branching peduncles, c. 12 mm diam., white, only 2 upper petals present, subsessile; hypanthium 7–14 mm long. Aug.–Nov. Dry outcrops in succulent vegetation, NW, KM (Namaqualand to Touwsrivier).

ternifolium Vorster   Tuberous geophyte to 25 cm. Leaves dry at flowering, trifoliolate, c. 3 cm diam. Flowers to 7 on branching peduncles, c. 15 mm diam., white to pink, only upper 2 petals present, somewhat spoon-shaped, subsessile; hypanthium 7–10 mm long. Dec.–Apr. Coastal plains, SW (Moorreesburg to Stellenbosch).•

triandrum E.M.Marais   Tuberous geophyte to 20 cm. Leaves dry at flowering, ovate, to 5 cm diam. Flowers to 30 on branching peduncles, 25–30 mm diam., pale yellow, petals 4 with upper larger and somewhat spoon-shaped, subsessile; hypanthium 23–32 mm long. Oct.–Nov. Succulent veld and dry fynbos, NW (Clanwilliam to Tanqua Karoo).

D. Petals 5

aciculatum E.M.Marais   Tuberous geophyte to 30 cm. Leaves dry at flowering, pinnatisect, to 14 × 6 cm with segments 6–12 mm wide. Flowers to 40 on branching peduncles, c. 12 mm diam., pale yellow, subsessile; hypanthium 12–16 mm long. Nov.–Dec. Clay slopes in renosterveld, NW, SW (Olifants River Mts to Langebaan).•

appendiculatum (L.f.) Willd.   Tuberous geophyte to 30 cm. Leaves 2- or 3-pinnatisect, to 10 cm long, softly hairy; stipules well-developed, ovate. Flowers to 15 on a peduncle, c. 40 mm diam., cream-coloured to pale yellow, subsessile; hypanthium 60–100 mm long. Sept.–Oct. Deep coastal sand, NW (Leipoldtville).•

aristatum (Sweet) G.Don   Tuberous geophyte to 30 cm. Leaves dry at flowering, bipinnatifid, segments linear, to 8 cm long. Flowers to 11 on branching peduncles, c. 25 mm diam., cream-yellow, subsessile; hypanthium 20–35 mm long. Oct.–Nov. Heavy soil, NW (W Karoo and Bokkeveld Mts).

attenuatum Harv.   Tuberous geophyte to 30 cm. Leaves dry at flowering, trifoliolate, to 13 cm long with segments 2–8 mm wide. Flowers to 7 on branching peduncles, c. 35 mm diam., subsessile, pale yellow, petals conspicuously ribbon-shaped; hypanthium (16–)22–31 mm long. Nov.–Jan. Sandy flats, NW (Graafwater).•

auritum (L.) Willd.   Tuberous geophyte to 25 cm. Leaves pinnatisect with elliptic pinnae, to 13 × 5 cm. Flowers to 25 on branching peduncles, c. 15 mm diam., purple-black, white to pale pink, subsessile, petals ribbon-shaped; hypanthium 8–20 mm long. Sept.–Feb. Wide range of habitats, NW, SW, KM, LB, SE (Clanwilliam to Port Elizabeth).•

caledonicum L.Bolus   Tuberous geophyte to 20 cm. Leaves simple to pinnate, to 8 × 3 cm. Flowers to 7 on branching peduncles, 15–20 mm diam., pink, subsessile; hypanthium 6–16 mm long. Dec.–Jan. Sandstone or clay soils, SW (Caledon).•

campestre (Eckl. & Zeyh.) Harv.   Tuberous geophyte to 14 cm. Leaves entire to pinnate, to 5 × 3.5 cm. Flowers to 9 on branching peduncles, c. 30 mm diam., white, subsessile; hypanthium 12–22 mm long. Oct.–Dec. Shale in grassland or scrub, SE (Port Elizabeth to Grahamstown).

carneum Jacq.   Tuberous geophyte to 40 cm. Leaves dry at flowering, pinnate or bipinnate with linear segments, to 5 × 3 cm. Flowers to 13 on branching peduncles, c. 45 mm diam., white, yellow or pinkish, subsessile; hypanthium 30–75 mm long. Nov.–Mar. Stony slopes, NW, SW, AP, LB, SE (Worcester to Gamtoos River).•

chelidonium (Houtt.) DC.   Tuberous geophyte to 18 cm. Leaves trifoliolate or 5-lobed, to 3 × 2.5 cm. Flowers to 9 on branching peduncles, c. 15 mm diam., bright pink, subsessile; hypanthium 9–16 mm long. Sept.–Oct. Clay flats in renosterveld, NW, SW (Clanwilliam to Riebeek-Kasteel).•

connivens E.M.Marais   Tuberous geophyte to 18 cm. Leaves dry at flowering, pinnate to bipinnatisect, to 13 × 6 cm. Flowers to 30 on branching peduncles, 25–30 mm diam., pale yellow to pink, subsessile, petals somewhat connivent; hypanthium 40–55 mm long. Dec.–Jan. Habitat? NW (Nieuwoudtville).•

curviandrum E.M.Marais   Tuberous geophyte to 30 cm. Leaves ovate, to 8 × 6 cm. Flowers to 30 on branching peduncles, 25–30 mm diam., white to pale yellow, subsessile, petals ribbon-shaped; hypanthium 18–30 mm long. Oct.–Nov. Sandstone or clay slopes, KM (Montagu to Oudtshoorn).•

fasciculaceum E.M.Marais   Tuberous geophyte to 50 cm. Leaves dry at flowering, bipinnate, to 27 × 12 cm with segments 4–8 mm wide. Flowers to 40 on branching peduncles, c. 15 mm diam., pale yellow, subsessile; hypanthium 40–60 mm long. Dec.–Jan. Sandy places, NW (Olifants River valley).•

fergusoniae L.Bolus   Tuberous geophyte to 22 cm. Leaves dry at flowering, palmate, to 4 cm diam, with segments c. 1.5 mm wide. Flowers to 8 on branching peduncles, 15–20 mm diam., white, pale yellow, or pale pink, subsessile; hypanthium (15–)20–42 mm long. Nov.–Jan. Loose sandy soil, NW, SW, AP, LB (Hex River valley to Riversdale).•

fissifolium (Andrews) Pers.   Tuberous geophyte to 14 cm. Leaves simple to bipinnatifid, to 7 cm long with segments 1–3 mm wide. Flowers to 14 on branching peduncles, c. 20 mm diam., cream-yellow, subsessile; hypanthium 28–65 mm long. Oct.–Nov. Succulent karoo, NW, SW, KM (W Karoo to Worcester).

fumariifolium R.Knuth   Tuberous geophyte to 23 cm. Leaves dry at flowering, pinnate to bipinnatisect, to c. 13 cm diam. Flowers to 25 on branching peduncles, 15–18 mm diam., pale yellow, subsessile, hypanthium 24–35 mm long. Oct.–Nov. Rocky karroid scrub or renosterveld, NW, KM (S Namibia to Whitehill).

gracillimum Fourc. Tuberous geophyte to 27 cm. Leaves dry at flowering, palmately compound, c. 6 cm diam. with segments 2–4 mm wide. Flowers to 16 on branching peduncles, c. 10 mm diam., white to pink, subsessile; hypanthium (15–)22–35 mm long. Jan.–Mar. Rocky sandstone slopes, KM, SE (Swartberg to Baviaanskloof Mts).•

heterophyllum Jacq. Tuberous geophyte to 17 cm. Leaves dry at flowering, simple to 3-foliolate, to 2.5 cm diam. Flowers to 11 on branching peduncles, 20–25 mm diam., white or pale pink, margins of petals conspicuously wavy, subsessile; hypanthium 9–15 mm long. Oct.–Nov. Renosterveld or coastal fynbos, SW (Darling).•

incrassatum (Andrews) Sims Tuberous geophyte to 35 cm. Leaves barely green at flowering, pinnate to pinnately lobed, to c. 11 × 7 cm. Flowers to 60 on branching peduncles, c. 25 mm diam., magenta, upper petals markedly larger; hypanthium 27–42 mm long, much longer than the very short pedicel. Aug.–Oct. Rocky slopes, NW (Namaqualand to Nardouw Mts).

leptum L.Bolus Tuberous geophyte up to 30 cm. Leaves pinnatisect to bipinnatisect, to 4 × 2 cm with segments 1–10 mm wide. Flowers to 12 on branching peduncles, c. 25 mm diam., white, pale yellow or pale pink, petals conspicuously ribbon-shaped, subsessile; hypanthium 12–30 mm long. Dec.–Feb. Sandy soil, SW (Malmesbury to Cape Flats).•

longiflorum Jacq. Tuberous geophyte to 30 cm. Leaves usually green at flowering, lanceolate, to 16 × .25 cm. Flowers to 15 on branching peduncles, c. 30 mm diam., pale yellow or pale pink, petals narrowly ribbon-shaped, subsessile; hypanthium (10–)15–44 mm long. Oct.–Nov. Sandy or stony places, NW, SW (Namaqualand to Darling and Worcester).

longifolium (Burm.f.) Jacq. Tuberous geophyte to 25 cm. Leaves sometimes bipinnatisect, to 120 × 18 mm. Flowers to 9 on branching peduncles, 20–25 mm diam., white, pale yellow, or pink, subsessile; hypanthium 8–22 mm long. Oct.–Dec. Sandy places, NW, SW (Citrusdal to Bredasdorp).•

luteolum N.E.Br. Tuberous geophyte to 30 cm. Leaves dry at flowering, palmate, c. 4 cm diam. with segments to 5 mm wide. Flowers to 16 on branching peduncles, papilionaceous, 3 lower tepals overlapping, c. 15 mm diam., pale yellow, subsessile; hypanthium 13–24 mm long. Nov.–Mar. Diverse habitats, NW, SW, KM, LB, SE (Namaqualand to Steytlerville).

luteum (Andrews) G.Don Tuberous geophyte to 20 cm. Leaves sometimes dry at flowering, pinnatisect or to bipinnatisect, to 12 × 4 cm with segments c. 2 mm wide. Flowers to 20 on branching peduncles, 15–20 mm diam., yellow, subsessile; hypanthium 20–33 mm long. Oct.–Nov. Rocky places in karoo vegetation, NW (W Karoo and Bokkeveld Mts).

moniliforme E.Mey. ex Harv. Tuberous geophyte to 40 cm. Leaves simple to tripartite, to 6 cm diam. Flowers to 50 on branching peduncles, c. 30 mm diam., cream-coloured to yellow, subsessile; hypanthium 20–77 mm long. Sept. Karoo vegetation, NW, KM (Namaqualand to Karoopoort and Matjiesfontein).

nephrophyllum E.M.Marais Tuberous geophyte to 12 cm. Leaves dry at flowering, kidney-shaped and shallowly lobed, c. 2.5 cm diam. Flowers to 7 on branching peduncles, c. 20 mm diam., pink, subsessile; hypanthium 20–30 mm long. Feb.–Apr. Stony lower slopes and sandy flats in succulent karoo, NW (Matsikamma Mts).•

nervifolium Jacq. Tuberous geophyte to 13 cm. Leaves simple to trifoliolate, to 3.5 × 2.5 cm. Flowers to 17 on branching peduncles, c. 18 mm diam., white to pale yellow, subsessile; hypanthium 30–45 mm long. Sept.–Oct. Open karoo vegetation, NW, KM (W Karoo to Karoopoort and Matjiesfontein).•

nummulifolium Salisb. Tuberous geophyte to 15 cm. Leaves ovate to cordate, obtuse, to 3 × 2 cm. Flowers to 5 on branching peduncles, c. 30 mm diam., bright pink to salmon, subsessile; hypanthium 25–40 mm long. Dec.–Mar. Rocky slopes at high alt., SW (Du Toitskloof Mts).•

parvipetalum E.M.Marais Tuberous geophyte to 30 cm. Leaves pinnatisect or bipinnatisect, to 1.5 cm long. Flowers to 18 on branching peduncles, c. 10 mm diam., white, subsessile; hypanthium 8–16 mm long. Sept. Sandy soil, NW (Namaqualand to Pakhuis Pass).

petroselinifolium G.Don Tuberous geophyte to 18 cm. Leaves pinnatisect, to 6 cm long with segments 2–3 mm wide. Flowers to 10 on branching peduncles, c. 12 mm diam., pink, subsessile; hypanthium 13–25 mm long. Oct.–Nov. Mountain fynbos or karoo vegetation, NW (Cold Bokkeveld to Karoopoort).•

pilosellifolium (Eckl. & Zeyh.) Steud. Tuberous geophyte to 32 cm. Leaves often green at flowering, rarely pinnatisect, to 10 × 2.5 cm. Flowers to 10 on branching peduncles, c. 15 mm diam., white to pale pink with conspicuous wine-red markings on each petal, subsessile; hypanthium 8–13 mm long. Oct.–Dec.(–Mar.). Fynbos, SW, LB, SE (Genadendal to Avontuur).•

pinnatum (L.) L'Hér. Tuberous geophyte to 30 cm. Leaves often green at flowering, pinnate, to 7 cm long. Flowers to 8 on branching peduncles, c. 18 mm diam., white, pale yellow, pink, or purple, subsessile; hypanthium 11–35(–45) mm long. Nov.–Mar. Clay or sandstone slopes and flats, NW, SW, AP, LB (Cedarberg Mts to Albertinia).•

proliferum (Burm.f.) Steud. Tuberous geophyte to 32 cm. Leaves often green at flowering, simple to pinnatisect, to 8 × 2 cm. Flowers to 9 on branching peduncles, c. 12 mm diam., white, pale yellow, or pink, subsessile; hypanthium 6–12 mm long. Mainly Oct.–Dec. Fynbos or renosterveld, NW, SW, AP, LB (Bokkeveld to Langeberg Mts).•

punctatum (Andrews) Willd. Tuberous geophyte to 30 cm. Leaves sometimes dry at flowering, ovate, to 9 cm diam. Flowers to 60 on branching peduncles, c. 25 mm diam., pale yellow, upper petals somewhat spoon-shaped and larger, subsessile; hypanthium 20–30 mm long. Oct.–Nov. Dry fynbos and karoo vegetation, NW (Bokkeveld to Cedarberg Mts).•

radiatum (Andrews) Pers. Tuberous geophyte to 25 cm. Leaves partly dry at flowering, mostly ovate, obtuse, to 4.5 × 3.5 cm. Flowers to 8 on branching peduncles, c. 40 mm diam., pale yellow or pink, subsessile; hypanthium 30–62 mm long. Oct.–Jan. Rocky sandstone slopes, SW (Du Toitskloof to Riviersonderend Mts).•

radicatum Vent. Tuberous geophyte to 30 cm. Leaves sometimes dry at flowering, narrowly elliptic, acute, to c. 14 × 4 cm. Flowers to 35 on branching peduncles, c. 25 mm diam., pale yellow, subsessile; hypanthium 40–60 mm long. Oct.–Dec. Sandy soil, NW, SW (Namaqualand to Hopefield).

rapaceum (L.) L'Hér. Tuberous geophyte to 40 cm. Leaves sometimes dry at flowering, bipinnatisect, to 25 cm long with segments 1–3 mm wide. Flowers to 50 on branching peduncles, c. 25 mm diam., white, yellow, or pink, papilionaceous, 3 lower petals forming a keel; hypanthium 12–55 mm long much longer than pedicel. Oct.–Feb. Stony slopes and flats, NW, SW, AP, KM, LB, SE (Namaqualand to Grahamstown).

reflexipetalum E.M.Marais Tuberous geophyte to 15 cm. Leaves pinnatisect, sometimes pinnatilobed, lobes obovate, glandular-hairy beneath. Flowers on branching peduncles, bright pink, upper petals feathered with red; hypanthium 7–12 mm long, much longer than pedicel. Sept.–Nov. Rocky sandstone slopes, NW (Pakhuis Mts).•

reflexum (Andrews) Pers. Tuberous geophyte to 15 cm. Leaves usually dry at flowering, palmate with segments 1 mm wide, c. 8 cm diam. Flowers to 5 on branching peduncles, c. 15 mm diam., white, subsessile; hypanthium 13–22 mm long. Dec.–Jan. Open places, NW (Nieuwoudtville).•

tenellum (Andrews) D.Don Tuberous geophyte to 20 cm. Leaves dry at flowering, simple to 3-foliolate, to 3.5 × 2.5 cm. Flowers to 10 on branching peduncles, c. 25 mm diam., white or pale pink, petals ribbon-shaped with margins conspicuously wavy, subsessile; hypanthium 10–19(–32) mm long. Nov.–Jan. Grassy slopes or renosterveld, SW (Malmesbury to Hottentots Holland Mts).•

trifoliolatum (Eckl. & Zeyh.) E.M.Marais Tuberous geophyte to 25 cm. Leaves dry at flowering, pinnate, to 17 cm long. Flowers to 20 on branching peduncles, c. 20 mm diam., pale yellow or pink, margins of petals conspicuously wavy, subsessile; hypanthium 10–15(–25) mm long. Oct.–Jan. Clay or sand in fynbos or renosterveld, NW, SW (Cold Bokkeveld to Cape Peninsula).•

triphyllum Jacq. Tuberous geophyte to 18 cm. Leaves mostly dry at flowering, simple or trifoliolate, c. 3 cm diam. Flowers to 10 on branching peduncles, c. 20 mm diam., pink, subsessile; hypanthium 7–19 mm long. Oct.–Dec. Shallow soil in mountain fynbos, NW, SW (Bokkeveld to Riviersonderend Mts).•

undulatum (Andrews) Pers. Tuberous geophyte to 25 cm. Leaves trifoliolate, pinnate or pinnatisect, to 12 × 10 cm with segments c. 5 mm wide. Flowers to 13 on branching peduncles, 25–30 mm diam., white, pale yellow, or pale pink, petals ribbon-shaped with wavy margins, subsessile; hypanthium 6–12(–25) mm long. Sept.–Oct. Karoo vegetation, NW, KM (Worcester to Swartberg Mts).•

viciifolium DC. Tuberous geophyte to 30 cm. Leaves dry at flowering, pinnate, to 12 cm long. Flowers to 20 on branching peduncles, c. 30 mm diam., pale yellow, petals conspicuously ribbon-shaped, subsessile; hypanthium (15–)20–38 mm long. Oct.–Jan. Shale slopes, SW (Malmesbury to Stellenbosch).•

violiflorum (Sweet) DC.   Tuberous geophyte to 25 cm. Leaves trifoliolate, pinnate or irregularly bipinnatisect, to 12 × 10 cm with segments 3–12 mm wide. Flowers 7–14 on branching peduncles, c. 10 mm diam., white, subsessile; hypanthium 8–11 mm long. Sept.–Oct. Open places, NW, SW (Worcester to Stormsvlei).•

B. Shrubs or shrublets; leaves cauline
E. Stems and branches clearly succulent
F. Leaves more or less suborbicular, often with purplish zonal marking

acetosum (L.) L'Hér.   Shrublet to 60 cm. Leaves round, leathery, glaucous to 5 cm diam. Flowers 2–7, c. 40 mm diam., salmon-pink; hypanthium 17–25 mm long, much longer than pedicel. Throughout the year. Flats, SE (Kouga Mts to Grahamstown).

articulatum (Cav.) Willd.   Semigeophyte with moniliform rhizome, stem conspicuously bracteate. Leaves round and shallowly incised, silky hairy, often with reddish zonal marking, c. 5 cm diam. Flowers 2–5 on stout peduncle, 40–50 mm diam., white to cream-coloured, subsessile, stamens exserted; hypanthium to c. 70 mm long. Oct.–Dec. Rocky slopes, often rock cracks and somewhat shady, NW, SW, KM (Namaqualand to Worcester and Witteberg).

inquinans (L.) L'Hér.   Semisucculent shrub to 2 m. Leaves round, velvety hairy, to 8 cm diam. Flowers 5–30, 25–30 mm diam., red, subsessile; hypanthium c. 40 mm long. Throughout the year. Coastal scrub and valley bushveld, SE (Gamtoos River to Umtata).

peltatum (L.) L'Hér.   Scrambling semisucculent perennial to 4 m. Leaves round, sometimes peltate, often 5–7-lobed, leathery and with reddish zonal marking. Flowers 2–9, 40–50 mm diam., pale pink to purple, subsessile; hypanthium 30–50 mm long. Mainly Sept.–Dec. Coastal or succulent scrub, NW, SW, KM, LB, SE (Bainskloof to East London).

tetragonum (L.f.) L'Hér.   Sprawling, often leafless shrublet with angular green branches c. 7 mm diam. Leaves often palmatifid, sometimes with red zonal markings, to 4 cm diam. Flowers 2, cream-coloured to pale pink with reddish veins, c. 40 mm diam., markedly asymmetric, stamens exserted; hypanthium 25–60 mm long, much longer than pedicel. Sept.–Dec. Rock outcrops, NW, SW, KM, LB, SE (Worcester to Grahamstown).

zonale (L.) L'Hér.   Shrub with somewhat succulent branches, 1–2 m. Leaves round with margins shallowly lobed and usually with a reddish zonal marking, very sparsely hairy. Flowers to 70, c. 35 mm diam., pink, subsessile; hypanthium 25–45 mm long. Mainly Sept.–Dec. Stony slopes and forest margins, NW, SW, KM, LB, SE (Piketberg to E Cape).

F. Leaves pinnate, lobed or wedge-shaped

alternans J.C.Wendl.   Woody, subsucculent shrublet to 40 cm. Leaves pinnate, conspicuously hairy, 2–6 × 1–2 cm. Flowers 1–4 on a short and nonpersistent peduncle, c. 15 mm diam., white or very pale pink, subsessile, stamens exserted; hypanthium 5–9 mm long. Apr.–Jan. Rocky slopes, NW, SW, KM (Bokkeveld Mts to Prince Albert).•

carnosum (L.) L'Hér.   Sparsely branched stem succulent to 1 m. Leaves pinnatifid to pinnate, to 20 cm long. Flowers 2–8, 10–15 mm diam., white with purple anthers; hypanthium c. 10 mm long, longer than pedicel and usually conspicuously dilated. Mar.–May and Sept.–Nov. Flats and slopes, NW, SW, KM (S Namibia and W Karoo to E Cape).

crassipes Harv.   Nearly unbranched stem succulent, c. 15 cm, stem covered with hardened remains of stipules. Leaves bipinnate, coarsely hairy, c. 4 cm long. Flowers 2–10, c. 15 mm diam., pink; hypanthium 8–10 mm long, about as long as pedicel. July–Sept. Flats, often under bushes, NW (S Namaqualand to Clanwilliam).

crithmifolium Sm.   DIKBASMALVA   Stem succulent to 1 m, with persistent, branched spine-like remains of inflorescences. Leaves bipinnatifid with almost terete segments, succulent, to 12 cm long. Flowers 4–6, c. 20 mm diam., white; hypanthium c. 5 mm long, much shorter than pedicel. Mar.–May and Sept.–Nov. Flats and rocky hills, NW, KM (S Namibia and W Karoo to N Cedarberg Mts).

dasyphyllum E.Mey. ex R.Knuth   Like **G. alternans** but smaller, to 20 cm, with larger flowers, c. 25 mm diam. on short and semipersistent peduncles; hypanthium 5–8 mm long, usually shorter than pedicel. Aug.–Dec. Rocky slopes in succulent karoo, NW (Namaqualand to Clanwilliam).

echinatum Curtis   Sparsely branched stem succulent to 40 cm, stem covered with persistent thorny stipules. Leaves often somewhat lobed, cordate, sparsely hairy above but densely white-hairy below, to

4 × 3 cm. Flowers 3–8, white or pink or purple; hypanthium 30–40 mm long, much longer than pedicel. July–Nov. Granite hills, NW (Namaqualand and W Karoo to Clanwilliam).

fulgidum (L.) L'Hér.   ROOIMALVA   Succulent-stemmed shrublet to 40 cm. Leaves pinnatifid, densely silky hairy, to 10 × 7 m, stipules large and broad. Flowers 4–9, 15–20 mm diam., red, subsessile; hypanthium 20–40 mm long. June–Nov. Rocky slopes, often coastal, NW, SW (Richtersveld to Yzerfontein).

gibbosum (L.) L'Hér.   DIKBEENMALVA   Sprawling shrublet with conspicuously swollen nodes, to 40 cm when unsupported. Leaves pinnatifid, leathery, glaucous, to 13 × 7.5 cm. Flowers 3–14, c. 15 mm diam., greenish yellow, subsessile; hypanthium 20–25 mm long. Nov.–Apr. Rock outcrops near coast, NW, SW (Namaqualand to Cape Peninsula).

grandicalcaratum R.Knuth   Shrublet to 50 cm. Leaves somewhat succulent, wedge-shaped, aromatic, 1–1.5 cm diam. Flowers to 5 on very short peduncles near branch tips, c. 14 mm diam., not opening widely, white streaked reddish; hypanthium 5–17 mm long, swollen and longer than pedicel. Mainly Oct. Rocky, granitic outcrops, NW, KM (Namibia to Matjiesfontein).

hirtum (Burm.f.) Jacq.   Somewhat succulent shrublet, often with persistent remains of petioles. Leaves bipinnate with linear segments 40 × 2 mm, thinly hairy. Flowers 3–8, c. 15 mm diam., dark pink; hypanthium c. 5 mm long, about as long as pedicel. July–Nov. Granite or sandstone slopes, SW (Velddrif to Stellenbosch).•

hystrix Harv.   Stem succulent to 20 cm, branches covered with stipular spines. Leaves bipinnatisect, to 3.5 × 2 cm. Flowers to 13 on stout, branching peduncles, c. 25 mm diam., cream-coloured, subsessile; hypanthium 30–45 mm long. Oct.–Nov. Karroid scrub, often in shelter of shrubs, KM (W Karoo to Nougaspoort).

laxum (Sweet) G.Don   Stem succulent, c. 30 cm. Leaves pinnatisect, somewhat leathery with pale margins, to 12 × 6.5 cm. Flowers to 18 on branching peduncles, white or pale pink, markedly zygomorphic with exserted stamens; hypanthium 3–18 mm long, much shorter than pedicel. July–Jan. and Mar.–Apr. Open valley bushveld or karoo, KM, SE (Little and Great Karoo to E Cape).

oreophilum Schltr.   Stem succulent armed with persistent stipules, c. 30 cm. Leaves crowded, pinnatisect with linear segments, sparsely hairy, c. 1.5 × 1 cm. Flowers 2 or 3, c. 25 mm diam., pink, on branching peduncles; hypanthium 5–12 mm long, longer than pedicel. Aug.–Oct. Rock crevices on mountains, NW (Bokkeveld to Cedarberg Mts).•

stipulaceum (L.f.) Willd.   Stem succulent to 30 cm, stem covered with persistent and broad stipules. Leaves crowded, cordate and variously incised, aromatic, c. 4 × 3 cm. Flowers 2–5 on branching peduncles, c. 25 mm diam., cream-coloured, subsessile; hypanthium 40–60 mm long. Oct. Shade in karroid bush, NW, KM (Kamieskroon to Matjiesfontein).

E. Stems and branches twig-like and somewhat woody

Group 1: *Leaves leathery, glabrous and glaucous, often with a purplish zonal marking, variously shaped, not conspicuously aromatic; flowers purple, large (more than 25 mm diam.) and showy; plants low or sprawling but sometimes erect and up to 1 m high*

fruticosum (Cav.) Willd.   Shrub to 1 m. Leaves trifoliolate with narrow segments, c. 15 mm diam., somewhat fleshy or leathery. Flowers mostly 1, on short axillary peduncles, c. 25 mm diam., pink; hypanthium 10–20 mm long, longer than pedicel. Throughout the year. Dry S-facing slopes, KM, LB, SE (Ladismith to Willowmore).•

grandiflorum (Andrews) Willd.   Straggling shrublet to 80 cm. Leaves digitately lobed, leathery, glaucous and often with a reddish zonal marking. Flowers to 5, 40–50 mm diam., pinkish purple, stamens long and exserted; hypanthium 8–10 mm long, shorter than pedicel. Oct.–Mar. Sandstone slopes, NW (Bokkeveld Mts to Tulbagh).•

incarnatum (L.) Moench (= *Erodium incarnatum* (L.) L'Hér.)   Weak-stemmed perennial to 30 cm, branching from a persistent rootstock. Leaves on long slender petioles, palmate to digitately lobed. Flowers few on elongate peduncles, petals subequal, spreading, pale pink but reddish below; hypanthium reduced to a shallow depression, much shorter than pedicel. Oct.–Nov. Rocky sandstone slopes, SW (Kogelberg to Riviersonderend Mts).•

laevigatum (L.f.) Willd.   Lax shrublet to 50 cm. Leaves trifoliolate, segments terete, somewhat succulent, c. 3 × 2 cm. Flowers solitary or rarely to 5, c. 25 mm diam., pink, on short axillary peduncles; hypanthium 13–38 mm long, much longer than pedicel. Throughout the year. Sandy slopes, KM, SE (Swartberg and Outeniqua Mts to E Cape).

lanceolatum (Cav.) Kern (= *Pelargonium glaucum* (L.f.) L'Hér.)   Shrublet to 30 cm. Leaves narrowly ovate, glaucous, to 8 × 3 cm. Flowers 1 or 2 on short lateral peduncles, c. 30 mm diam., white to pale yellow, subsessile; hypanthium 30–50 mm long. Throughout the year. Dry places, NW (Worcester to Montagu).•

patulum Jacq.   Shrublet to 30 cm. Leaves cordate to round with margin variously incised, leathery, glaucous and usually with a reddish zonal marking, c. 3 cm diam. Flowers 2 or 3, c. 25 mm diam., pink, on short, branching peduncles; hypanthium 15–20 mm long, slightly longer than pedicel. Sept.–Mar. Sandstone slopes in fynbos, NW, SW (Cedarberg to Langeberg Mts).•

praemorsum (Andrews) F.Dietr.   Woody perennial with thin branches to 2 m from a massive, partly exposed tuberous base. Leaves round and deeply 3–5 lobed, leathery, to 3 cm diam. Flowers 1 or 2, 15–50 mm diam., cream-coloured to purple, petals 4, upper much larger; hypanthium 10–40 mm long, much longer than pedicel. Aug.–Apr. Rocky slopes and flats, NW (S Namibia to Olifants River valley).

setulosum Turcz.   Tufted perennial to 15 cm. Leaves on long petioles, cordate, acute, leathery with red margins, c. 2.5 cm diam. Flowers to 5 on branching peduncles, c. 25 mm diam., white or pale pink with dark centres; hypanthium 10–15 mm long, somewhat shorter than pedicel. Sept.–Jan. Sandstone slopes, NW, SW (Montagu to Houwhoek Pass).•

tabulare (Burm.f.) L'Hér.   Shrublet to 30 cm. Leaves variously incised, almost round, leathery, glaucous. often with a reddish zonal marking, c. 4 cm diam. Flowers 2 or 3 on axillary peduncles, c. 20 mm diam., purple; hypanthium 4–8 mm long, slightly shorter than pedicel. Sept.–Jan. Cool slopes, NW, SW (Clanwilliam to Somerset West).•

ternatum (L.f.) Jacq.   Shrublet to 60 cm. Leaves dimorphic, petiolate and sessile, trifoliolate, leathery or somewhat succulent, 2–3 cm diam. Flowers solitary or to 3 on short, axillary peduncles, c. 20 mm diam., white to pink; hypanthium 10–20 mm long, longer than pedicel. Apr.–Dec. Scrub, LB (Montagu to Riversdale).•

Group 2: Leaves conspicuously aromatic, often hairy; flowers purple (rarely white), large (more than 30 mm diam.) and showy; plants often erect and up to 2 m high

betulinum (L.) L'Hér.   Shrublet to 50 cm. Leaves ovate to elliptic, somewhat leathery, c. 20 mm long. Flowers to 6 on axillary peduncles, c. 5 cm diam., white to pink; hypanthium 3–8 mm long, somewhat shorter than pedicel. Aug.–Jan. Coastal dunes, SW, AP, SE (Yzerfontein to Knysna).•

capitatum (L.) L'Hér.   KUSMALVA   Sprawling shrublet to 50 cm. Leaves cordate, margins lobed and crisped, aromatic, c. 5 cm diam. Flowers to 20 on stout axillary peduncles, 15–25 mm diam., purple, subsessile; hypanthium 3–8 mm long. Sept.–Oct. Coastal dunes and flats, NW, SW, AP, SE (Lambert's Bay to KwaZulu-Natal).

citronellum J.J.A.van der Walt   Erect shrub to 1 m. Leaves palmatifid, roughly hairy, strongly lemon-scented, c. 8 cm diam. Flowers 5–8 on branching peduncles, c. 25 mm diam., pinkish purple; hypanthium 3–8 mm long, shorter than pedicel. Aug.–Jan. Near streams on clay, KM (Ladismith).•

cordifolium (Cav.) Curtis   Shrub to 1 m. Leaves cordate, white-hairy beneath, aromatic, c. 5 cm diam. Flowers to 12 on short axillary peduncles, c. 40 mm diam., purple, lower petals much narrower than upper; hypanthium 2–12 mm long, shorter than pedicel. Aug.–Jan. Sheltered places near streams, AP, LB, SE (Bredasdorp to E Cape).

crispum (P.J.Bergius) L'Hér.   Shrublet to 70 cm. Leaves often distichous, fan-shaped, rough, margins crisped, lemon-scented, to 1 cm diam. Flowers 1–3 on short peduncles, c. 25 mm diam., pinkish purple; hypanthium 5–8 mm long, slightly longer than pedicel. Aug.–Apr. Sandy lower slopes, NW, SW, KM (Worcester to Bredasdorp).•

cucullatum (L.) L'Hér.   WILDEMALVA   Shrub to 2 m. Leaves more or less round, c. 70 mm diam. Flowers to 13, c. 4 cm diam., pinkish purple, with wide and overlapping petals; hypanthium 5–12 mm long, slightly longer than pedicel. Sept.–Feb. Sandy and granite slopes along coast, SW, AP (Saldanha Bay to Baardskeerdersbos).•

denticulatum Jacq.   Aromatic shrub to 1 m. Leaves finely 2- or 3-pinnatisect, sticky, hard, c. 8 cm diam. Flowers to 9 on short axillary peduncles, c. 20 mm diam., pinkish purple, subsessile; hypanthium 4–9 mm long. Apr.–Nov. Ravines near streams, LB (Langeberg Mts: Herbertsdale).•

englerianum R.Knuth   Shrub to 1 m. Leaves cordate, margins crisped and variously incised, roughly hairy, camphor-scented. Flowers 2–5 on short axillary peduncles, 15–20 mm diam., pale purple; hypanthium 1–10 mm long, shorter than pedicel. Aug.–Apr. Dry rocky fynbos, NW, KM (Bokkeveld to Swartberg Mts).•

glutinosum (Jacq.) L'Hér.   Erect shrub to 1.8 m. Leaves sticky, cordate but deeply 3–5-lobed, aromatic with a balm-like scent, 5–8 cm diam. Flowers to 8 on short axillary peduncles, c. 18 mm diam., pale purple, subsessile; hypanthium 3–10 mm long. Sept.–Nov. Near watercourses on sandstone slopes, SW, KM, LB, SE (Piketberg to E Cape, N Province: Soutpansberg).

graveolens L'Hér.   Shrub to c. 1 m. Leaves palmatifid, shortly hairy, strongly rose-scented, c. 4 cm diam. Flowers to 7 on short axillary peduncles, c. 30 mm diam., white to pinkish purple; hypanthium 4–15 mm long, longer than pedicel. Aug.–Jan. Moist and semishaded places, LB, SE (George to E Cape, N Province: Soutpansberg and Zimbabwe: Chimanimani Mts).

hermanniifolium (P.J.Bergius) Jacq.   Shrublet to 1 m, aromatic. Leaves distichous, wedge-shaped, 3-palmatifid, hard, rough. Flowers 1–3 on very short peduncles, c. 40 mm diam., pinkish purple; hypanthium 6–8 mm long, slightly longer than pedicel. Sept.–Apr. Upper sandstone slopes, SW (Wemmershoek to Riviersonderend Mts).•

hispidum (L.f.) Willd.   Aromatic shrub to 1 m. Leaves digitately lobed, shortly hairy, to 12 cm diam. Flowers 6–12 on a branched apical system of short peduncles, c. 20 mm diam., pinkish purple, upper petals markedly larger; hypanthium 3–4 mm long, slightly shorter than pedicel. Sept.–Apr. Shady ravines on lower slopes, NW, SW, KM (Piketberg to Bredasdorp and Swartberg Mts).•

panduriforme Eckl. & Zeyh.   Aromatic shrub to 1.2 m. Leaves cordate and deeply lobed, velvety hairy, balm-scented, c. 6 × 5 cm. Flowers to 20 on short but stout axillary peduncles, c. 40 mm diam., pink, subsessile; hypanthium 6–13 mm long. Aug.–Jan. Lower foothills and ravines, KM, SE (Willowmore to E Cape).

papilionaceum (L.) L'Hér.   Foetid shrub to 1 m. Leaves cordate with margins very shallowly lobed, 8 cm or more diam. Flowers to 20 on branching peduncles at branch tips, c. 30 mm diam., pink with conspicuous darker markings on much larger upper petals; hypanthium 2–5 mm long, much shorter than pedicel. Aug.–Jan. Forest margins near streams, SW, LB, SE (Somerset West to Humansdorp).•

pseudoglutinosum R.Knuth   Shrub to 1 m. Leaves ovate with margins deeply lobed, sticky, c. 3 cm diam. Flowers 1 or 2 on short axillary peduncles, c. 40 mm diam., pinkish purple, subsessile; hypanthium 6–10 mm long. Sept.–Jan. Along watercourses in ravines, SE (Uniondale to Keurboomsrivier).•

quercifolium (L.f.) L'Hér.   Aromatic shrub to 1.5 m. Leaves pinnatifid, roughly hairy, balm-scented, to 8 × 7 cm. Flowers to 6 on fairly short but stout axillary peduncles, c. 30 mm diam., purple, subsessile; hypanthium 6–11 mm long. Aug.–Jan. Dry rocky slopes, KM, SE (Oudtshoorn to Willowmore).•

radens H.E.Moore   Aromatic shrub to 1.5 m. Leaves bipinnatisect with ribbon-shaped segments c. 3 mm wide, rose-scented, c. 6 cm diam. Flowers to 8 on short axillary peduncles, c. 20 mm diam., purple; hypanthium 2–8 mm long, about as long as pedicel. Aug.–Jan. Damp slopes, SW, LB, SE (Elandskloof Mts to E Cape).

ribifolium Jacq.   Aromatic shrub to 1.2 m. Leaves palmate, roughly hairy, c. 5 cm diam. Flowers to 12 on short axillary peduncles, c. 20 mm diam., white, upper petals many times wider; hypanthium 6–8 mm long, about as long as pedicel. Sept.–Nov. Sandstone slopes, often along forest margins, KM, SE (Swartberg Mts to Katberg).

scabroide R.Knuth   Shrub to 70 cm. Leaves cordate and lobed, roughly hairy, c. 4 × 3 cm. Flowers to 5 on short axillary peduncles, c. 18 mm diam., white to pale pinkish purple; hypanthium 6–10 mm long, about as long as pedicel. Nov.–Feb. Sheltered places among rocks, NW, KM (Cold Bokkeveld Mts to Bonteberg).•

scabrum (L.) L'Hér.   HOENDERBOS   Aromatic shrub to 1.2 m. Leaves palmatisect, roughly hairy, lemon-scented, c. 5 cm diam. Flowers to 6 on short axillary peduncles, c. 20 mm diam., white to purplish; hypanthium 3–12 mm long, about as long as pedicel. Aug.–Jan. Rocky sandstone slopes, NW, SW, KM, LB, SE (Namaqualand to E Cape).

sublignosum R.Knuth   Shrub to 1 m. Leaves cordate and shallowly lobed with dentate margins, c. 4 cm diam. Flowers to 9 on short peduncles near branch tips, c. 25 mm diam., purplish pink; hypanthium 5–14 mm long, usually somewhat longer than pedicel. Oct.–Feb. Moist rocky places in mountain ravines, NW (Piketberg to Ceres).•

tomentosum Jacq.   Sprawling aromatic shrub to 50 cm. Leaves 3–5-palmatilobed, soft, velvety, c. 7 cm diam. Flowers to 15 on branching peduncles, c. 18 mm diam., white, upper petals several times wider; hypanthium c. 2 mm long, much shorter than pedicel. Oct.–Jan. Forest margins in mountains, SW, LB (Somerset West to Langeberg Mts).•

vitifolium (L.) L'Hér.   Shrub to 1 m. Leaves cordate and shallowly lobed, velvety hairy, aromatic, c. 8 cm diam. Flowers to 12 on short axillary peduncles, c. 18 mm diam., pinkish purple; hypanthium 2–4 mm long, usually somewhat longer than pedicel. Aug.–Jan. Shady ravines on lower slopes, SW, SE (Bainskloof to Bredasdorp, Outeniqua Mts).•

Group 3: Leaves soft-textured, often hairy, variously shaped, usually not conspicuously aromatic (but see P. odoratissimum); flowers purple, pink, yellow, or white; plants usually low or sprawling (but see P. antidysentericum)

abrotanifolium (L.f.) Jacq.   Erect, much-branched shrublet to 50 cm. Leaves small, feather-like, aromatic. Flowers 1–5, c. 20 mm diam., white to pink or purple; hypanthium c. 18 mm long, much longer than pedicel. Almost throughout the year. Flats, NW, SW, KM, LB, SE (Namaqualand to E Cape and Free State).

alchemilloides (L.) L'Hér.   Sprawling perennial to 20 cm from woody rootstock. Leaves round, usually lobed, 7 cm diam., often with reddish zonal marking, usually with sparse silky hairs. Flowers 3–6(–15), 15–20 mm diam., white or yellow or pink; hypanthium 12–35 mm long, much longer than pedicel. Sept.–Nov. Open, moist places, SW, AP, LB, SE (Saldanha Bay to N Province).

alpinum Eckl. & Zeyh.   Procumbent, softly woody perennial to 20 cm with small tubers. Leaves on long petioles, cordate with margin variously incised, c. 5 × 4 cm, softly hairy, sometimes with reddish zonal markings. Flowers usually 2 on subterminal peduncles, c. 30 mm diam., salmon-pink; hypanthium 15–70 mm long, longer than pedicel. Nov.–Jan. Near streams at high alt., NW (Grootwinterhoek to Hex River Mts).•

antidysentericum (Eckl. & Zeyh.) Kostel.   Deciduous shrub to 1.5 m, with twiggy and brittle branches from a massive and partly exposed tuber. Leaves round, margin shallowly incised, roughly hairy, 1.5–2 cm diam. Flowers 3–5 on very short peduncles, c. 20 mm diam., purple; hypanthium c. 5 mm long, shorter than pedicel. Mar.–May. Rocky places, NW (S Namibia to Tanqua Karoo).

caespitosum Turcz.   Tufted perennial to 15 cm. Leaves elliptic to almost linear, 4–5 × 0.2–1.5 cm. Flowers up to 9, c. 20 mm diam., pale pink to purple; hypanthium reduced to a shallow cavity and much shorter than pedicel. Oct.–Jan. Open places at high alt., NW (Cedarberg Mts).•

candicans Spreng.   Procumbent shrublet to 30 cm. Leaves simple or rarely 3-foliolate, cordate with margins variously lobed, silvery velvety, c. 2 cm diam. Flowers to 4 on axillary peduncles, c. 15 mm diam., white to pink, petals 4, upper larger than lower, subsessile; hypanthium 5–8 mm long, much longer than pedicel. Mainly Sept.–Nov. Open places, SW, AP, KM, LB, SE (Cape Peninsula to Plettenberg Bay).•

capillare (Cav.) Willd.   Tufted rhizomatous perennial to 20 cm. Leaves pinnatifid, c. 4 × 2 m. Flowers to 3 on branching peduncles, c. 20 mm diam., salmon-pink; hypanthium 9–22 mm long, shorter than pedicel. Aug.–Jan. Open places at high alt., NW (Cold Bokkeveld Mts).•

caucalifolium Jacq.   Sprawling shrublet to 50 cm. Leaves 1- or 2-pinnatilobed, c. 2 × 1.5 cm. Flowers 1, sometimes 2 on long erect peduncles, 25–30 mm diam., white to pinkish purple, subsessile, petals 4 with upper much larger than lower; hypanthium 10–50 mm long, much longer than pedicel. Oct.–Apr. Open places, SW, AP, KM, LB, SE (Pringle Bay to Humansdorp).•

coronopifolium Jacq.   Tufted rhizomatous perennial to 20 cm. Leaves linear to narrowly elliptic, roughly hairy, slightly toothed sometimes entire, c. 60 × 1–10 mm. Flowers to 4 on an unbranched peduncle, 20–25 mm diam., white to pink or purple; hypanthium 2–8 mm long, much shorter than pedicel. Sept.–Mar. Open places, NW (Bokkeveld Mts to Worcester).•

divisifolium Vorster   Lax scrambling perennial. Leaves well spaced, 2- or 3-pinnatisect with linear segments, c. 5 cm diam. Flowers to 5, c. 25 mm diam., pink or white; hypanthium 14–32 mm long, slightly longer than pedicel. Oct.–Jan. Lower slopes, SW (Riviersonderend Mts).•

elegans (Andrews) Willd.   Tufted, rhizomatous perennial to 45 cm. Leaves elliptic, glabrous to pubescent, somewhat leathery, c. 3 × 2.5 cm. Flowers to 7 on branching peduncles, c. 50 mm diam., pink; hypanthium 10–15 mm long, shorter than pedicel. Sept.–Jan. Coastal fynbos, SW, AP, SE (Hermanus to Stilbaai, Port Elizabeth to Grahamstown).

elongatum (Cav.) Salisb.   Soft shrublet to 25 cm. Leaves cordate, roughly hairy, often with a reddish zonal marking, 4–5 cm diam. Flowers to 6, c. 15 mm diam., cream-coloured to whitish, subsessile; hypanthium 10–15 mm long. Throughout the year. Stony slopes, NW, SW, LB, SE (Bokkeveld Mts to Uniondale).•

exstipulatum (Cav.) L'Hér.   Shrublet to c. 50 cm. Leaves variously incised, c. 1.5 cm diam. Flowers 1–5, c. 15 mm diam., pink, subsessile; hypanthium 8–9 mm long. June–Dec. Rocky places, KM (Ladismith to De Rust).•

hypoleucum Turcz.   Prostrate or scrambling perennial. Leaves cordate, aromatic, c. 4 cm diam. Flowers to 14, c. 10 mm diam., white to pink; hypanthium 6–7 mm long, longer than pedicel. Oct.–Dec. Cool slopes, SW, AP, KM, LB, SE (Hottentots Holland to Kouga Mts).•

iocastum (Eckl. & Zeyh.) Steud.   Perennial to 15 cm. Leaves cordate, sparsely hairy, to 3 cm diam. Flowers to 7, c. 15 mm diam., pale purple; hypanthium 2–8 mm long, shorter than pedicel. Sept.–Jan. Mostly high alt., SW (Piketberg to Caledon).•

karooicum Compton & P.E.Barnes   Tufted shrublet c. 40 cm. Leaves digitately compound with segments c. 1 mm wide, rarely simple, somewhat succulent, 1.5–2 cm diam. Flowers 2 on a short and apparently terminal peduncle, c. 20 mm diam., white or pale yellow or pink, subsessile; hypanthium 9–14 mm long. Jan.–Apr. Rocky slopes, NW, SW, KM (Namaqualand to Oudtshoorn).

longicaule Jacq.   Procumbent shrublet. Leaves pinnatisect, c. 4 × 3 cm. Flowers to 6, either c. 30 mm diam. and white to pale pink with red veins or c. 60 mm diam. and yellow with pinkish veins, on well developed peduncles at ends of branches, subsessile; hypanthium 30–55 mm long. Aug.–Jan. Sandy soil, NW, SW, AP (Olifants River mouth to Stilbaai).•

magenteum J.J.A.van der Walt   Rounded, twiggy shrub to 1 m. Leaves rotund-cordate, velvety, shallowly lobed, to 1.5 cm diam. Flowers to 9, c. 20 mm diam., magenta, subsessile; hypanthium c. 30 mm long. May–Oct. Rock outcrops, NW, KM (Botterkloof to Calitzdorp).•

multicaule Jacq.   Shrublet to 30 cm. Leaves pinnatisect or bipinnatisect, c. 4 × 3 cm. Flowers 3–5, c. 20 mm diam., pinkish purple, 4-petalled with upper petals markedly larger, subsessile; hypanthium 8–14 mm long. Throughout the year. Rocky places, SW, KM, LB, SE (Worcester to N Province).

myrrhifolium (L.) L'Hér.   Sprawling shrublet c. 30 cm. Leaves bipinnatisect, segments linear to ribbon-shaped, c. 5 × 3 cm. Flowers to 5, 20–25 mm diam., white to pink or pinkish purple, upper 4 petals markedly wider, subsessile; hypanthium 4–10 mm long. Aug.–Feb. Open places on stony sand, NW, SW, AP, SE (Kamiesberg to Uitenhage).

odoratissimum (L.) L'Hér.   Aromatic shrublet to 20 cm. Leaves cordate to round, velvety hairy, 3–12 cm diam. Flowers 3–10, c. 15 mm diam., white; hypanthium 5–10 mm long, about as long as pedicel. Almost throughout the year, AP, KM, LB, SE (Waboomsberg to KwaZulu-Natal).

oenothera (L.f.) Jacq.   Shrublet to 20 cm. Leaves densely crowded, oblong, tomentose, c. 4.5 × 1 cm. Flowers 1–4 on stout, branching peduncles, c. 15 mm diam., pink; hypanthium 2–7 mm long, about as long as or slightly shorter than pedicel. Sept.–Dec. Open sandy places, NW, SW (Clanwilliam to Franschhoek).•

ovale (Burm.f.) L'Hér.   Tufted rhizomatous shrublet to 30 cm. Leaves elliptic, toothed, densely hairy, c. 4 × 1–3 cm. Flowers to 7 on well-developed, branching peduncles, 25–40 mm diam., white to pink, upper petals overlapping and much larger; hypanthium 2–13 mm long, shorter than pedicel. July–Mar. Open places on mountains, SW, KM, LB, SE (Elandskloof to Great Winterhoek Mts).•

plurisectum T.M.Salter   Deciduous shrublet to 30 cm. Leaves palmatisect to pinnatisect with linear segments, c. 8 mm diam. Flowers 1 or 2 on short peduncles, c. 12 mm diam., yellowish to pinkish, subsessile; hypanthium 20–33 mm long, much longer than pedicel. Mar.–May. Renosterveld, SW (Cape Peninsula to Bredasdorp).•

reniforme Curtis   Shrublet with tuberous roots to 40 cm. Leaves reniform, usually shortly hairy, 3–7 cm diam. Flowers to 15, c. 20 mm diam., bright pink; hypanthium 15–45 mm long, much longer than pedicel. Throughout the year. Dry flats and open grassland, KM, SE (Kammanassie and Outeniqua Mts to E Cape).

sidoides DC.   Tufted perennial with swollen roots. Leaves cordate with margins somewhat lobed and crenate, silvery velvety. Flowers to 14 on limp, branching peduncles, c. 15 mm diam., almost black; hypanthium 15–37 mm long, longer than pedicel. Mainly Oct.–Jan. Dry and stony places in short grassland, SE (Uniondale to Gauteng).

suburbanum Clifford ex C.Boucher   Sprawling shrublet to 30 cm. Leaves pinnatisect to bipinnatisect, 3–7 cm diam. Flowers to 6 on stout peduncles near branch tips, 30–40 mm diam., cream-coloured to purple, 2 upper petals much wider, subsessile; hypanthium 8–50 mm long. June–Jan. Coastal dunes, SW, AP, SE (Cape Peninsula to Port Elizabeth).•

tricolor Curtis   Rhizomatous shrublet to 30 cm. Leaves crowded, elliptic, roughly hairy, c. 2.5 × 1 cm. Flowers to 3 on short, branching peduncles, c. 25 mm diam., 2- or 3-coloured with white, pink and often bright red, petals wide; hypanthium 1–2 mm long, much shorter than pedicel. Sept.–Jan. Open places on clay, KM, LB, SE (Swartberg to Langeberg and Outeniqua Mts).•

trifidum Jacq. (= *Pelargonium fragile* (Andrews) Willd.)   Sprawling, aromatic shrublet with brittle branches. Leaves almost trifoliolate, to 4 cm diam. Flowers 3–6, c. 30 mm diam., cream-yellow, subsessile; hypanthium 20–30 mm long. Sept.–Jan. Dry stony slopes and flats, NW, KM, SE (Worcester to Peddie).

[**Species excluded**   No authentic material seen and probably conspecific with one of the above: **P. sulphureum** R.Knuth]

GESNERIACEAE

STREPTOCARPUS   WILD GLOXINIA   c. 120 spp., Africa and Madagascar

meyeri B.L.Burtt   Rosulate perennial to 20 cm. Leaves elliptic, shaggy, margins toothed. Flowers several on glandular-hairy peduncles, trumpet-shaped, tube cylindric, violet with white lobes. Jan.–Apr. Shaded rocky banks, SE (Baviaanskloof Mts to E Cape and Mpumalanga).

rexii (Hook.) Lindl.   TWIN SISTERS   Rosulate perennial to 20 cm. Leaves strap-shaped, velvety, margins toothed. Flowers mostly 1 or 2 on glandular-hairy peduncles, funnel-shaped, tube flaring, whitish with purple streaks in throat. Oct.–Apr. Forest floors, LB, SE (George to S KwaZulu-Natal).

GOODENIACEAE

SCAEVOLA   SEEPLAKKIE   c. 130 spp., mainly Australia

plumieri (L.) Vahl (= *Scaevola thunbergii* Eckl. & Zeyh.)   Sprawling perennial to 40 cm. Leaves fleshy, obovate. Flowers sessile in axillary cymes, whitish, tube slit above and woolly within. Nov.–Apr. Coastal foredunes, AP, LB, SE (Indo-Pacific coasts to Agulhas).

GRUBBIACEAE•

GRUBBIA•   3 spp., W Cape

rosmarinifolia P.J.Bergius   Erect, shortly hairy ericoid shrublet to 1.5 m. Leaves spreading to reflexed, linear-lanceolate, margins revolute. Flowers 2 or 3 in axils, woolly, white. Aug.–Nov. Damp sandstone slopes, NW, SW, AP, LB, SE (Cold Bokkeveld Mts to Tsitsikamma Mts).•

rourkei Carlquist   Like **G. tomentosa** but single-stemmed at base with leaves linear, roughly hairy and somewhat spreading. Sept.–Mar. Damp sandstone slopes, SW (Kogelberg).•

tomentosa (Thunb.) Harms   Erect, shortly hairy shrublet to 1.5 m, multistemmed from a lignotuber. Leaves ascending, narrowly elliptic, margins revolute. Flowers several in cone-like axillary clusters, shortly hairy, pinkish. Dec.–Jan. Damp sandstone slopes, NW, SW, AP, LB (Wolseley to George).•

GUNNERACEAE

GUNNERA   RIVER PUMPKIN   c. 50 spp., worldwide in the tropics and subtropics

perpensa L.   Monoecious, dioecious or polygamous rhizomatous perennial to 1 m, thinly hairy. Leaves tufted, round or kidney-shaped on long petioles, toothed. Flowers sessile in spike-like racemes, greenish. Oct.–Jan. Marshes, NW, SW, KM, LB, SE (Swartruggens to Cape Peninsula and Little Karoo, to N Africa).

HALORAGACEAE

1.   Plants mainly in marshes, rooted in the soil; leaves opposite or alternate, small..........................**Laurembergia**
1.'   Plants floating aquatics; leaves in whorls of 3 or 4, with filiform segments...............................**Myriophyllum**

LAUREMBERGIA   4 spp., widespread in the tropics and subtropics

repens P.J.Bergius   Monoecious or dioecious prostrate leafy perennial to 10 cm, stems red. Leaves oblanceolate, often coarsely toothed, roughly hairy. Flowers in axillary clusters, female sessile, male on filiform pedicels. Nov.–Apr. Boggy places, NW, SW, AP, LB, SE (southern and tropical Africa).

MYRIOPHYLLUM   WATER MILFOIL   c. 45 spp., cosmopolitan

*****aquaticum** (Vell.) Verdc.   Like **M. spicatum** but leaves emergent, densely papillose with deciduous, awl-shaped stipule-like outgrowths and flowers solitary in axils. Flowering time? Weed in rivers and ponds, SW (S American weed naturalised from Berg River to Cape Peninsula and elsewhere in South Africa).

spicatum L.   Monoecious or polygamous, rhizomatous submerged aquatic perennial to 2 m. Leaves whorled, feathery pinnatisect, lobes filiform. Flowers in terminal emergent spikes, yellow. Sept.–Feb. Streams and pools, NW, SW, KM (Velddrif to Montagu, and throughout Africa and Europe).

HAMAMELIDACEAE

TRICHOCLADUS   BLACK WITCH-HAZEL   c. 6 spp., southern and tropical Africa

crinitus (Thunb.) Pers.   Monoecious or dioecious small tree to 4 m, brown-felted on young parts. Leaves opposite, elliptic, acuminate, shortly peltate at base, felted beneath. Flowers in congested terminal spikes, cream-coloured. Oct.–Feb. Coastal forests, LB, SE (Grootvadersbos and George to KwaZulu-Natal).

HYDNORACEAE

HYDNORA   BOBBEJAANKOS   12 spp., Africa and Madagascar

africana Thunb.   Leafless, achlorophyllous root parasite to 15 cm. Flower solitary, fleshy and warty or scaly, clavate with 3 or 4 lobes connate at tips, brown with orange inside. Aug.–Feb. Parasitic on *Euphorbia*, NW, SW, KM, LB, SE (S Namibia to Cape Peninsula to E Cape).

ICACINACEAE

1. Leaves opposite; inflorescence repeatedly forked; spiny shrub or tree ....**Cassinopsis**
1.' Leaves alternate; plants unarmed:
   2. Calyx and corolla present; flowers bisexual, in terminal panicles; style filiform; shrub or small tree ...**Apodytes**
   2.' Calyx absent; flowers unisexual, in spikes; stigma sessile; stems twining ....**Pyrenacantha**

APODYTES   WITPEER   c. 20 spp., Old World tropics, mainly Africa

dimidiata E.Mey. ex Arn.   UMCANDATHAMBO   Small or large tree to 20 m. Leaves ovate-elliptic, obtuse, margins undulate, glossy above. Flowers many in loose, mainly terminal panicles, white, fragrant. Fruits black with red aril. Nov.–Jan. Rocky slopes and forest, SW, AP, LB, SE (Cape Peninsula to tropical Africa).

geldenhuysii A.E.van Wyk & Potgieter   KAAPSE WITPEER   Like **A. dimidiata** but flowers few in axillary racemes, ovary hairy and aril green. Throughout the year. Sheltered sandstone slopes, SW (Jonkershoek to Kleinrivier Mts).•

CASSINOPSIS   LEMOENDORING   4 spp., Africa and Madagascar

ilicifolia (Hochst.) Kuntze   ICEGCEYA   Scrambling shrub or small tree to 5 m, often with single interpetiolar thorns. Leaves opposite, elliptic-ovate, margins revolute and sharply toothed. Flowers few in interpetiolar cymes, white. Sept.–Nov. Damp forests and kloofs, LB, SE (Riviersonderend Mts to Zimbabwe).

PYRENACANTHA   BLOUBOKTOUTJIE   c. 20 spp., Old World tropics

scandens Planch. ex Harv.   Dioecious, roughly hairy scandent shrub or climber. Leaves elliptic to ovate or lobed, toothed. Flowers in supra-axillary spikes, male elongate, female capitate, greenish. Sept.–Feb. Coastal forests, LB, SE (Swellendam to tropical Africa).

KIGGELARIACEAE

KIGGELARIA   WILD PEACH   1 sp., tropical and subtropical Africa

africana L.   Dioecious, semideciduous tree with smooth grey bark to 17 m. Leaves elliptic, usually toothed, with hairy pockets in lower vein axils, male flowers in axillary cymes, female flowers solitary in axils, yellowish. Fruits round, pale inside with red seeds. Feb.–July. Forest margins and rocky slopes, NW, SW, AP, KM, LB, SE (Namaqualand to tropical Africa).

LAMIACEAE

1. Corolla 1-lipped; stamens 4, completely exserted from between the 2 small upper lobes; ovary shortly 4-lobed; leaves 3-lobed to 3-partite ....**Teucrium**
1.' Corolla 2-lipped or nearly regularly lobed; upper lobe more or less hooded over stamens; stamens 2 or 4; ovary deeply 4-lobed; leaves seldom lobed:
   2. Stamens directed downwards upon lower side of corolla tube or lower lip, at length upcurved .....**Plectranthus**
   2.' Stamens ascending or spreading:
     3. Fertile stamens 2; anther thecae separated by a long connective; calyx 2-lipped ....**Salvia**
     3.' Fertile stamens 4; calyx not or obscurely 2-lipped:
       4. Stamens spreading; filaments straight, with 2 directed upwards and 2 downwards; corolla small, 2–5 mm long, subequally 4- or 5-lobed; perennial rhizomatous herbs..**Mentha**
       4.' Stamens all directed to upper side of tube or upper lip of corolla; flowers and habit various:
         5. Leaves 3(5)-foliolate; flowers purple, crowded ....**Cedronella**
         5.' Leaves simple; flowers various:
           6. Calyx funnel-shaped, hairy within, 10- or more-toothed, limb eventually spreading ....**Ballota**
           6.' Calyx more or less tubular, glabrous within, up to 10-toothed, limb not spreading:
             7. Calyx subequally 5-toothed; upper lip of corolla glabrous or pubescent but not with stiff brush-like hairs, usually shorter than lower lip; bracts usually reduced ....**Stachys**
             7.' Calyx 5–10-toothed; upper lip of corolla with stiff brush-like hairs, subequal or larger than lower; bracts leaf-like:
               8. Corolla usually orange (rarely cream-coloured); upper lobe 12–30 mm long, larger than lower lip..**Leonotis**
               8.' Corolla white; upper lobe shorter than 10 mm, subequal to lower lip......................**Leucas**

BALLOTA   HOREHOUND, KATTEKRUIE   c. 33 spp., mostly Mediterranean and Eurasia, also Africa

africana (L.) Benth.   Aromatic, soft, greyish shrublet to 1.2 m. Leaves softly hairy, cordate, toothed. Flowers in axillary verticils, pink to purple. May–Nov. Rocky or disturbed places, NW, SW, AP, KM, LB, SE (S Namibia to E Cape and Free State).

*****CEDRONELLA**   BALM-OF-GILEAD   1 sp., Madeira and Canary Islands

*****canariensis** (L.) Webb & Berthel.   Perennial to 2.5 m. Leaves digitately 3(–5)-foliolate, leaflets lanceolate, toothed. Flowers crowded in terminal verticils, purplish. Nov.–Jan. Weed along streams in forest clearings, SW (Canary Island weed naturalised on the Cape Peninsula).

LEONOTIS   MINARET FLOWER, WILD DAGGA   c. 10 spp., sub-Saharan Africa

leonurus (L.) R.Br.   Roughly hairy shrub to 500 cm. Leaves narrowly lanceolate, toothed. Flowers in axillary verticils, calyx subequally toothed, corolla with lower lobes reflexed, velvety orange. Nov.–July. Forest margins or rough grassland, NW, SW, AP, KM, LB, SE (Clanwilliam to Gauteng).

ocymifolia (Burm.f.) Iwarsson   Like **L. leonurus** but leaves ovate, calyx bilabiate with larger upper tooth and lower corolla lobes spreading. Mar.–May. Rocky slopes, NW, SW, AP, KM, LB, SE (Clanwilliam to tropical Africa).

LEUCAS   160+ spp., Old World, mainly Africa and Asia

capensis (Benth.) Engl.   Aromatic, greyish, twiggy shrub to 1.5 m. Leaves oblanceolate, rarely coarsely toothed above. Flowers in few-flowered verticils, white with red anthers. Oct.–Jan. Dry grassland, SE (Uitenhage to Botswana and Namibia).

MENTHA MINT c. 25 spp., cosmopolitan, mainly temperate

aquatica L. WATER MINT Trailing, mint-scented perennial to 80 cm. Leaves usually petiolate, ovate, toothed. Flowers in distant verticils, pink to purple. Feb.–May. In marshes and wet places, NW, SW, AP, LB, SE (Clanwilliam to Knysna and to Europe).

longifolia (L.) Huds. WILDEKRUISEMENT Straggling, aromatic shrublet to 150 cm, usually coarsely hairy. Leaves sessile, lanceolate, sometimes sparsely toothed. Flowers crowded in spike-like, terminal verticils, white to mauve. Nov.–Apr. Along rivers or seeps, NW, SW, AP, KM, LB, SE (southern and N Africa to Europe).

PLECTRANTHUS FOREST SAGE, SPOORSALIE c. 350 spp., Africa to Asia and Australia

ciliatus E.Mey. ex Benth. Soft, straggling, roughly hairy perennial with purplish stems to 60 cm. Leaves ovate, toothed. Flowers in verticils, calyx hairy, corolla saccate at base, whitish speckled with purple. Feb.–Apr. Forest glades and moist places, SE (Knysna to Mpumalanga).

fruticosus L'Hér. Soft, sparsely hairy shrub to 2 m. Leaves ovate, toothed. Flowers in panicles, calyx glandular-hairy, corolla saccate at base, mauve, blue or pink with darker speckling. Nov.–Apr. Forests and shaded rocky places, SW, LB, SE (Caledon to N Province).

laxiflorus Benth. Soft, citrus-scented, glandular-hairy perennial to 1.5 m. Leaves, ovate-deltoid, toothed. Flowers in racemes, calyx glandular-hairy, corolla sigmoid, whitish to mauve. Feb.–Apr. Forest margins and streambanks, SE (Humansdorp to N Province).

madagascariensis (Pers.) Benth. Sprawling, shortly hairy perennial to 1 m. Leaves leathery, broadly ovate, obscurely toothed. Flowers in verticils, bracts caducous, calyx swollen below in fruit, glandular, corolla sigmoid, white or mauve to purple. Oct.–Apr. Dry woodland, SE (Knysna to Mpumalanga).

spicatus E.Mey. ex Benth. Sprawling, succulent perennial to 60 cm. Leaves subsessile, obovate, coarsely toothed above. Flowers subspicate, bracts caducous, calyx subequally toothed in fruit, corolla sigmoid, purple. Mar.–May. Rocky woodland, SE (Humansdorp to Mpumalanga).

strigosus Benth. Sprawling, roughly hairy, semisucculent perennial to 30 cm. Leaves rounded, obscurely toothed. Flowers in verticils, calyx sparsely hairy, corolla swollen below and narrowed above, whitish to mauve. Dec.–June. Rocky outcrops in woodland, SE (Uitenhage to Mpumalanga).

verticillatus (L.f.) Druce Sprawling, semisucculent perennial to 30 cm. Leaves ovate, toothed. Flowers in verticils, calyx glabrous, corolla slightly swollen below, white to mauve. Dec.–June. Rocky outcrops in woodland, SE (Knysna to N Province).

SALVIA SAGE, SALIE 800–900 spp., cosmopolitan

A. Calyx campanulate, bilabiate, accrescent

africana-caerulea L. BLOUBLOMSALIE Grey-hairy shrub to 2 m. Leaves obovate, sometimes toothed. Flowers in verticils, calyx glandular-silky, corolla mauve to blue or pink with darker spots, 16–28 mm long, upper lip hooded. June–Jan. Sandy flats and slopes, NW, SW, KM (Namaqualand to Cape Peninsula to Montagu).•

africana-lutea L. BRUINSALIE, STRANDSALIE Aromatic grey shrub to 2 m. Leaves grey-hairy, obovate, sometimes toothed. Flowers mostly paired, calyx shortly hairy and gland-dotted, corolla golden brown, 30–50 mm long, upper lip hooded, c. 25 mm long. June–Dec. Coastal dunes and slopes, NW, SW, AP, LB, SE (Namaqualand to E Cape).

albicaulis Benth. Loose, white-velvety shrublet to 60 cm. Leaves leathery, oblanceolate, glabrescent above, coarsely toothed. Flowers in panicles, calyx densely silky, corolla purplish, 18–24 mm long. Nov.–May. Sandstone slopes, NW, SW (Clanwilliam to Paarl).•

chamelaeagnea P.J.Bergius BLOUBLOMSALIE Like **S. africana-caerulea** but leaves glabrescent and gland-dotted and calyx shortly hairy and gland-dotted. Nov.–May. Sandy slopes, NW, SW, AP (Namibia to Breede River).

dentata Aiton Like **S. africana-caerulea** and **S. chamelaeagnea** but leaves usually grey-hairy and toothed to pinnatifid and calyx shortly hairy and gland-dotted. June–Jan. Rocky hillsides, NW (Namaqualand to Pakhuis).

lanceolata Lam. Like **S. africana-lutea** but bracts caducous and corolla dull rose to grey-blue, 25–35 mm long, upper lip c. 17 mm long. Sept.–June. Mainly coastal sands and rocky outcrops, NW, SW, KM (Namaqualand to Cape Peninsula and Montagu).

A. Calyx cylindrical-campanulate, not accrescent

aurita L.f.   OOGSEERBOSSIE   Straggling, hairy perennial to 1 m. Leaves often lyrate-pinnatifid. Flowers in verticils, calyx hairy, corolla white, pinkish or mauve, 16–20 mm long. Oct.–June. Streambanks and grassland, SE (Mossel Bay to Mpumalanga).

disermas L.   GROOTBLOUSALIE   Soft, glandular-hairy shrub to 1.2 m. Leaves often crowded below, roughly hairy, ovate, toothed. Flowers in verticils, calyx glandular-hairy, corolla whitish to mauve, 15–30 mm long, upper lip deeply hooded. Aug.–May. Sandy slopes and limestone, NW, SW, LB, SE (Namibia and W Karoo to NW Province).

granitica Hochst.   Stoloniferous perennial to 60 cm. Leaves linear, hairy beneath. Flowers in pairs, calyx glandular-silky, corolla mauve-pink, c. 20 mm long, upper lip hooded. Nov.–Dec. Sandstone slopes, NW, SW (Cold Bokkeveld Mts and Babilonstoring).•

muirii L.Bolus   Shortly hairy, twiggy shrublet to 60 cm. Leaves leathery, grey-velvety, oblanceolate. Flowers mostly paired, calyx shortly hairy and gland-dotted, corolla blue with white throat, to 26 mm long, upper lip hooded. Apr.–June. Clay slopes, LB (Riversdale to Great Brak).•

namaensis Schinz   Aromatic, sticky, velvety shrub to 1.2 m. Leaves leathery, lyrate-pinnatifid, roughly hairy. Flowers in verticils, calyx roughly glandular-hairy, corolla white, mauve or blue, 8–12 mm long. Sept.–May. Rocky slopes, KM, LB, SE (Namibia, Oudtshoorn to Willowmore and Karoo).

obtusata Thunb.   Like **S. aurita** but stems and leaves almost glabrous above, corolla 20–25 mm long, blue, mauve or reddish. Flowering time? Grassland, SE (Uitenhage to E Cape).

repens Burch. ex Benth.   KRUIPSALIE   Shortly hairy perennial to 80 cm from a creeping rhizome. Leaves usually crowded below, obovate to lyrate, toothed. Flowers in verticils, calyx shortly hairy, corolla pale blue to purple, rarely white, 10–26 mm long. Oct.–May. Grassland or woodland, LB, SE (George to N Province).

runcinata L.f.   HARDESALIE, WILDESALIE   Shortly hairy perennial to 70 cm from a woody rootstock. Leaves usually lyrate. Flowers in verticils, calyx roughly hairy and gland-dotted, corolla white or mauve to purplish, 7–14 mm long. Oct.–Apr. Clay slopes, grassland or woodland, SW, LB, SE (Bredasdorp to Zimbabwe).

scabra L.f.   Like **S. aurita** but roughly hairy and corolla tubular and 25–40 mm long, mauve to purple. Throughout the year. Streamsides, SE (Humansdorp to E Cape).

stenophylla Burch. ex Benth.   Like **S. runcinata** but leaves often pinnatifid or pinnatisect with narrow segments and stems almost glabrous. Oct.–Apr. Sandy, often brackish soils, SE (Humansdorp to N Province, Botswana and Namibia).

thermara Van Jaarsv.   Stoloniferous perennial to 1 m. Leaves linear-lanceolate, toothed. Flowers in verticils in a short raceme, calyx shortly hairy, corolla reddish, 40–50 mm long. Dec.–Jan. Sandstone slopes, SW (Slanghoek Mts: Badsberg).•

triangularis Thunb.   Like **S. aurita** but leaves ovate-triangular, 20–40 mm long and calyx bilabiate in fruit. Aug.–Oct. Wooded grassland, SE (Humansdorp to E Cape).

STACHYS   WOUNDWORT, TEEBOS   c. 450 spp., cosmopolitan, mainly temperate and subtropical

A. Pubescence stellate and velvety or felted

aurea Benth.   GEELTEEBOSSIE   Aromatic, yellowish felted shrub to 1 m. Leaves subsessile, small, obovate, coarsely toothed above. Flowers in verticils, calyx yellowish woolly, corolla yellow. Sept.–Feb. Clay slopes, NW (Pakhuis and W Karoo).

linearis Burch. ex Benth.   BOESMANTEE   Grey-felted shrublet to 40 cm. Leaves sessile, linear, folded. Flowers mainly in pairs, calyx grey-felted, corolla pink to mauve. Dec.–Feb. Clay flats, NW (Pakhuis Mts and Karoo).

rugosa Aiton   VAALTEE   Grey-felted shrub to 1.2 m. Leaves sessile, lanceolate, slightly toothed. Flowers in verticils, calyx grey-felted, corolla yellow to pink or purple, often mottled. Sept.–Dec. Rocky clay slopes, NW (S Namibia and W Karoo to Piketberg to Lesotho).

A. Pubescence simple and often sparse

aethiopica L.   KATBOSSIE   Glandular-hairy, sprawling perennial to 50 cm. Leaves petiolate, ovate, toothed. Flowers in sparse verticils, calyx roughly and often glandular-hairy, corolla white, pink or mauve with darker spots. Aug.–Sept. Scrub or grassland, NW, SW, AP, KM, LB, SE Clanwilliam to Swaziland).

*arvensis L.   Hairy annual to 40 cm. Leaves petiolate, broadly ovate, toothed. Flowers in verticils, calyx softly hairy, corolla scarcely longer than calyx, mauve or white. Throughout the year. Weed of cultivation, NW, SW, SE (Eurasian weed).

bolusii Skan   Like **S. grandifolia** but upper bracts exceeding corolla. Aug.–Sept. Rock outcrops, SW, AP (Saldanha to Stanford).•

graciliflora C.Presl   Sparsely hairy, sprawling perennial to 40 cm. Leaves petiolate, ovate, toothed. Flowers in sparse verticils, calyx softly to shortly hairy, corolla white, sometimes with mauve spots. Oct.–Mar. Moist places in scrub or forest margins, SE (Knysna to S KwaZulu-Natal).

grandifolia E.Mey. ex Benth.   Softly hairy, straggling perennial to 1 m. Leaves petiolate, ovate, finely toothed. Flowers in sparse verticils, calyx softly hairy, longer than upper bracts, corolla white with mauve spots. Dec.–Apr. Forest margins and streamsides, SE (Knysna to Mpumalanga).

humifusa Burch. ex Benth.   Like **S. scabrida** but leaves sessile or subsessile and shallowly toothed. Oct.–Jan. Grassland, SE (Knysna to E Cape).

scabrida Skan   Like **S. graciliflora** but roughly hairy and leaves thicker in texture with bulbous-based hairs on upper surface. Sept.–Dec. Grassland or scrub, SE (Knysna to E Cape and Karoo).

sublobata Skan   Sparsely hairy, spreading perennial to 30 cm. Leaves shortly petiolate, narrowly triangular and often folded, coarsely lobed. Flowers mostly in pairs, calyx thinly glandular-hairy, corolla mauve. Sept.–Mar. Sandstone slopes, SW, KM, LB (Robertson and Caledon to Mossel Bay).•

thunbergii Benth.   Prickly, sprawling perennial to 2 m. Leaves petiolate, ovate, finely toothed. Flowers in sparse verticils, calyx roughly hairy, corolla tubular and twice as long as calyx, red to purple. Sept.–Mar. Bush and forest margins, SW, SE (Cape Peninsula and George to Humansdorp).•

TEUCRIUM   WOOD SAGE   c. 200 spp., worldwide, mainly Eurasia

africanum Thunb.   PADDAKLOU   Greyish shrublet to 30 cm. Leaves thinly hairy, tripartite. Flowers usually solitary in upper axils, white. Nov.–Apr. Dry areas, often among rocks, KM, LB, SE (Bredasdorp to Grahamstown and Karoo).

trifidum Retz.   AKKEDISPOOT, KAATJIEDRIEBLAAR   Soft, willowy, greyish undershrub to 1 m. Leaves thinly hairy, 3–5-partite. Flowers several in axillary cymes, white. Nov.–Apr. Dry woodlands, SE (Humansdorp to Botswana).

LENTIBULARIACEAE

UTRICULARIA   BLADDERWORT   214 spp., cosmopolitan

bisquamata Schrank (= *Utricularia capensis* Spreng.)   Annual to 12 cm. Leaves linear. Flowers in a bracteolate raceme, white to lilac (rarely yellowish) with smooth yellow palate, upper lip much smaller ovate. Fruits erect. Sept.–Jan. Boggy acid soils, NW, SW, AP, KM, LB, SE (throughout southern Africa).

gibba L.   Submerged aquatic annual or perennial to 20 cm. Leaves sparsely dissected. Flowers in an ebracteolate emergent raceme, yellow, lips subequal. Fruits erect. Nov.–Feb. Boggy or shallow water, SE (Uitenhage and throughout the tropics).

inflexa Forssk.   Free-floating, submerged aquatic annual or perennial to 1 m. Leaves finely dissected, fennel-like. Flowers in an ebracteolate emergent raceme with a basal whorl of spongy floats, white, mauve or yellow, hairy outside, lips subequal. Fruits nodding. Jan.–Mar. Shallow or deep water, SE (Uitenhage through Old World tropics).

[**Species excluded**   No authentic specimens seen from the area: **U. livida** E.Mey.]

LINACEAE

LINUM   FLAX   200 spp., pantemperate and subtropics

A. Styles partly united (see also L. esterhuyseniae)

acuticarpum C.M.Rogers   Subshrub with slender stems from persistent base, to 50 cm. Leaves opposite, sessile, linear-lanceolate, with stipular glands on lower leaves only. Flowers in panicles, yellow, styles united for more than half their length. Mainly Dec.–Mar. Sandstone slopes, SW (Kogelberg to Hermanus).•

adustum E.Mey. ex Planch.   Subshrub with slender stems from persistent base, to 50 cm. Leaves opposite, sessile, linear-lanceolate, with stipular glands throughout. Flowers in panicles, yellow, styles united almost entirely. Mainly Nov.–Jan. Sandstone slopes, NW (Gifberg to Piketberg).•

aethiopicum Thunb.   Like **L. gracile** but stems leafy to tips and flowers congested at branch tips. Sept.–Feb. Stony and grassy slopes, AP, LB, SE (Danger Point to E Cape).

africanum L.   Subshrub with slender stems from persistent base, to 50 cm. Leaves opposite, sessile, linear-lanceolate, with stipular glands throughout. Flowers in panicles, yellow, styles united for up to half their length. Mainly Nov.–Jan. Sandstone and limestone slopes and flats, SW, AP, LB, SE (Hopefield to Knysna).•

comptonii C.M.Rogers   Like **L. africanum** but flowers heterostylous. Nov.–Feb. Sandstone slopes, NW, SW (Cedarberg Mts to Hermanus).•

gracile Planch.   Like **L. adustum** but leaves elliptic-lanceolate and capsules obtuse. Mainly Sept.–Feb. Sandstone slopes, SW, LB, SE (Hermanus to Humansdorp).•

heterostylum C.M.Rogers   Like **L. acuticarpum** but flowers heterostylous, and styles united for up to half their length. Oct.–Mar. Sandstone and limestone slopes, NW, SW, AP, KM (Worcester to Little Karoo).•

A. Styles separate to base

brevistylum C.M.Rogers   Like **L. thunbergii** but outer sepals fringed and petals shorter than 10 mm. Oct.–Dec. Sandy slopes, SW (Cape Peninsula to Caledon).•

esterhuyseniae C.M.Rogers   Like **L. thunbergii** but stems velvety below, leaves somewhat crowded below, lanceolate-elliptic, styles free or basally united. Oct.–Dec. Sandstone slopes, KM, SE (Ladismith to Humansdorp).•

pungens Planch.   Like **L. thunbergii** but outer sepals conspicuously glandular-toothed. Dec.–Jan. Sandstone slopes, NW (Gifberg to Hex River Mts).•

quadrifolium L.   Slender shrublet with willowy branches, to 50 cm. Leaves opposite, sessile, elliptic to broadly ovate, in whorls of 4, with stipular glands throughout. Flowers in lax panicles, yellow, styles free. Sept.–Nov. Damp sandstone slopes, SW, LB, SE (Cape Peninsula to Knysna).•

thesioides Bartl.   Like **L. thunbergii** but stipular glands only on lower leaves, outer sepals pectinate and pedicels sparsely hairy. Oct.–Jan. Stony slopes, SW, LB (Cape Peninsula to Swellendam).•

thunbergii Eckl. & Zeyh.   Subshrub with slender stems from persistent base, to 50 cm. Leaves opposite below, sessile, linear to elliptic, sometimes ovate, with stipular glands sometimes only below. Flowers in panicles, yellow, styles free. Mainly Oct.–Jan. Mostly sandstone slopes, SW, AP, KM, LB, SE (Cape Peninsula to tropical Africa).

villosum C.M.Rogers   Subshrub to 50 cm, with softly hairy, sprawling branches leafy to tips. Leaves opposite, sessile, elliptic-obovate, stipular glands throughout. Flowers few, clustered at branch tips, yellow, styles free. Nov.–Dec. Damp sandstone slopes, SE (Outeniqua Mts).•

LOGANIACEAE

1. Anther thecae distinct, parallel throughout; leaves opposite, 3–7-veined from or near base; fruit indehiscent, large (over 10 mm diam.) with woody or leathery pericarp ....**Strychnos**
1.' Anther thecae confluent above.); leaves usually ternate, with a single midvein; fruit a capsule, small (less than 5 mm diam...**Nuxia**

NUXIA   WILD ELDER   c. 40 spp., Africa

floribunda Benth.   INGQOTA   Tree to 7 m. Leaves usually 3-foliolate, leaflets elliptic. Flowers in large terminal cymose panicles, white. May–Aug. Forests, SE (Knysna to tropical Africa).

STRYCHNOS   MONKEY-ORANGE   c. 400 spp., pantropical and subtropical

decussata (Pappe) Gilg   UMHLAMAHLAHLA   Slender tree with waxy branchlets to 12 m. Leaves glossy, obovate to elliptic, 3–5-veined from base. Flowers in axillary racemose cymes, often appearing before leaves, whitish, calyx lobes ovate, 1 mm long. Fruits turning red, 16 mm diam. Oct.–Dec. Lowland thicket, SE (Knysna to N Province).

spinosa Lam.   GREEN MONKEY-ORANGE, GROENKLAPPER, UMHLALA   Usually thorny tree with corky bark to 9 m. Leaves ovate to suborbicular, 5–7-veined from base. Flowers in terminal cymes, greenish white, calyx lobes awl-shaped, 5 mm long. Fruits turning yellow, 60–90 mm diam. Oct.–Jan. Wooded rocky slopes, SE (Knysna to tropical Africa).

LORANTHACEAE

1. Corolla 4-lobed, stellate-pubescent; filaments straight at anthesis ....**Septulina**
1.' Corolla 5-lobed, glabrous; filaments inflexed at anthesis ....**Moquiniella**

MOQUINIELLA   VUURHOUTJIES   1 sp., South Africa

rubra (A.Spreng.) Balle   Glabrescent stem parasite to 1 m. Leaves subsessile, elliptic-lanceolate. Flowers in axillary umbels, glabrous, tubular with a basal swelling, 5-lobed, mostly orange but red below and with a black tip. Mar.–June. Parasitic on various trees including *Acacia*, *Euclea*, *Rhus* and *Diospyros*, NW, KM, LB, SE (Namaqualand and Karoo to E Cape).

SEPTULINA   CANDLES, KERSIES   2 spp., southern Africa

glauca (Thunb.) Tiegh.   Stellate-pubescent stem parasite to 50 cm. Leaves grey, subsessile, elliptic. Flowers few in axils, tubular with a slight basal swelling, 4-lobed, greyish green flushed red. Feb.–Sept. Frequently parasitic on *Lycium*, NW, SW, KM (S Namibia and Karoo to Cape Peninsula and Swartberg Mts).

*LYTHRACEAE

*****LYTHRUM**   LOOSESTRIFE   c. 35 spp., cosmopolitan

*****hyssopifolium** L.   HYSSOP-LEAVED LOOSESTRIFE   Sprawling woody annual to 40 cm. Leaves elliptic. Flowers solitary in axils, pink. Nov.–Jan. Weed of disturbed, damp places especially around dams, SW, LB, SE (European weed).

MALVACEAE (= *STERCULIACEAE, TILIACEAE*)

1. Filaments free or almost so:
    2. Fruit indehiscent, without bristles; all stamens fertile; flowers mauve..**Grewia**
    2.' Fruit dehiscent and densely bristly; outer stamens sterile; flowers white................................**Sparrmannia**
1.' Filaments more or less united below into a tube or groups of 2 or 3:
    3. Anthers 2-thecous:
        4. Petals absent; flowers unisexual; stamens numerous with anthers capitate on apex of staminal column; small trees ....**Sterculia**
        4' Petals present; flowers bisexual; stamens 5 with filaments shortly fused at base; herbs or shrubs..**Hermannia**
    3.' Anthers 1-thecous:
        5. Fruit a loculicidally dehiscent capsule; style 5-branched; epicalyx of 5–12 bracts ....**Hibiscus**
        5.' Fruit of dehiscent follicles, achenes or pseudoachenes arranged around a central collumella and sometimes separating from it; style and epicalyx various:
            6. Style branches twice as many as carpels; epicalyx of 4–12 ovate to filiform bracts.....................**Pavonia**
            6.' Style branches as many as carpels:
                7. Epicalyx present, of 3(5) bracts; flowers white, pink or magenta:
                    8. Style branches with capitate or clavate stigmas; epicalyx bracts free; shrubs or subshrubs..**Anisodontea**
                    8.' Style branches with longitudinally extended stigmas; epicalyx bracts fused below into a cup; robust, single-stemmed woody annual..**Lavatera**
                7.' Epicalyx absent; flowers white, yellow or orange:
                    9. Locules containing 1 ovule each; mericarps up to 10 (rarely more), 1-seeded, not dehiscing by apical slits ....**Sida**
                    9.' Locules containing 5 or more ovules each; mericarps 10 or more (rarely fewer), (1)3(8)-seeded, dehiscing by apical slits ....**Abutilon**

ABUTILON   WILDEMALVA   c. 100 spp., cosmopolitan

sonneratianum (Cav.) Sweet   Velvety shrubby perennial to 2 m. Leaves heart-shaped, obscurely lobed, toothed or crenate, pale beneath. Flowers on long axillary pedicels, pink to yellow or orange. Nov.–Jan. Forest margins and bush, especially disturbed places, SW, AP, KM, LB, SE (Bredasdorp and Little Karoo to Zimbabwe).

ANISODONTEA AFRICAN MALLOW, BERGROOS c. 20 spp., southern Africa

A. Carpels 4–10 mm long, with 2–6 ovules

anomala (Link & Otto) D.M.Bates   Velvety felted shrub to 1 m. Leaves shallowly to deeply 3–5-palmatifid, coarsely crenate. Flowers 1–5 in upper axils, pink to magenta, epicalyx of 3 ovate lobes, adnate below to calyx. Mainly Sept.–Dec. Stony slopes, often near streams, NW, SW, KM (Namaqualand and W Karoo to Hopefield and Witteberg).

elegans (Cav.) D.M.Bates   Twiggy, white-pubescent shrublet to 1.5 m. Leaves 3–5-palmate to -palmatisect, segments variously lobed. Flowers 1–few in axils, white to deep pink, epicalyx of 3 linear to oblanceolate lobes. Mainly Sept.–Oct. Stony clay slopes, NW, SW, LB (Hex River valley to Swellendam).•

pseudocapensis D.M.Bates   Glabrescent shrublet to 50 cm. Leaves small, 3-lobed, toothed. Flowers 1 or 2 in upper axils, white to deep pink, epicalyx of 3 linear to oblanceolate lobes. Aug.–Oct. Stony clay slopes, LB (Riversdale to Gourits River).•

theronii D.M.Bates   Like **A. elegans** but leaves broadly ovate to heart-shaped and obscurely lobed, and toothed and flowers 1 or 2 in upper axils. Aug.–Sept. Stony slopes, KM (Little Karoo: Calitzdorp).•

A. Carpels 2–6 mm long, with 1 ovule
B. Staminal column glabrous, to 5 mm long

biflora (Desr.) D.M.Bates (incl. **A. alexandri** (Baker) D.M.Bates)   Thinly hairy sprawling shrublet to 1 m. Leaves 3–5-lobed to the middle, sometimes not lobed, toothed. Flowers 1–4 in upper axils, white to pink, staminal column glabrous, epicalyx of 3 linear lobes. Aug.–Nov. Stony lower slopes. NW, SW (Piketberg and Ceres to Cape Peninsula).•

dissecta (Harv.) D.M.Bates   Like **A. biflora** but leaves small, deeply bipalmatisect, flowers solitary in upper axils. Sept.–Oct. Stony lower slopes, SW, LB (Bredasdorp to Kogmanskloof and Potberg).•

B. Staminal column mostly hairy, longer than 5 mm

bryonifolia (L.) D.M.Bates   Densely, rough-hairy shrublet to 2 m. Leaves thick, deeply 3–5-lobed, central lobe largest, coarsely crenate. Flowers 1–few in tight axillary clusters, white to pink, epicalyx of 3 linear to oblanceolate lobes. Mainly Aug.–Oct. Stony slopes, NW (Namaqualand to Tulbagh).

fruticosa (P.J.Bergius) D.M.Bates   Glabrous or glabrescent, somewhat twiggy shrub to 2 m. Leaves often in fascicles, narrowly cuneate or sometimes 3-lobed, toothed. Flowers 1 or 2 in axils, sometimes nodding, white to pink, petals narrow and spreading to reflexed, epicalyx of 3 narrowly lanceolate lobes. Apr.–Nov. Stony slopes, NW, LB (Worcester to Riversdale).•

gracilis D.M.Bates   Glabrescent shrub to 1 m. Leaves deeply 3-palmatisect and coarsely toothed. Flowers 1–4 in lax axillary cymes forming loose pseudopanicles, pink, epicalyx of lanceolate lobes. Nov.–Jan. Rocky slopes, NW (E Cedarberg Mts).•

reflexa (H.L.Wendl.) D.M.Bates   Densely rough-hairy shrublet to 2 m. Leaves thick, shallowly 5-lobed, toothed. Flowers 1–few in axillary or terminal clusters, petals narrow and reflexed, epicalyx of 3 lanceolate lobes. Mainly Aug.–Nov. Stony slopes, KM, SE (Witteberg and Swartberg Mts to Uniondale).•

scabrosa (L.) D.M.Bates SANDROOS   Thinly to densely glandular-hairy shrub to 2 m. Leaves mostly obscurely 3-lobed or elliptic and toothed. Flowers 1–few in axils on slender pedicels, pink, epicalyx of narrow lobes. Mainly Sept.–Dec. Coastal sands, SW, AP, SE (Saldanha to KwaZulu-Natal).

setosa (Harv.) D.M.Bates   Shrub to 1.5 m with long stiff hairs on young branches. Leaves 3-lobed and toothed. Flowers 1–few in upper axils in loose racemes, white to pink, petals narrow, spreading, epicalyx of 3 lanceolate to elliptic lobes. Mainly Sept.–Dec. Stony clay or sandy slopes, NW, SW (Piketberg to Stellenbosch).•

triloba (Thunb.) D.M.Bates   Shortly hairy to velvety shrub to 1.5 m. Leaves often thick, broad, shallowly 3–5-lobed. Flowers 1–4 in axils, pink, epicalyx of linear to lanceolate lobes. Mainly Sept.–Oct. Rocky slopes, NW, KM (Namaqualand to Oudtshoorn).

[Species excluded   No specimens seen from the area: **A. capensis** (L.) D.M.Bates]

GREWIA CROSS-BERRY, KRUISBESSIE   400+ spp., Africa, Asia and Australia

occidentalis L.   Shrub or small tree to 3 m. Leaves glabrescent, lanceolate, toothed. Flowers usually solitary opposite leaves, purple and pink or white. Fruits 4-lobed, reddish. Oct.–Nov. Forest margins and bush, SW, LB, SE (Cape Peninsula to Zimbabwe).

robusta Burch. KAROOKRUISBESSIE   Shrub to 3 m, grey-velvety when young. Leaves discolorous, grey-velvety beneath, often clustered on short shoots, ovate, finely toothed. Flowers solitary opposite leaves, pink. Fruits 4-lobed, purplish. Sept.–Dec. Dense scrub, KM, LB, SE (George and Oudtshoorn to E Cape).

HERMANNIA   DOLL'S-ROSE, POPROSIE   c. 120 spp., dry tropics and subtropics, mostly southern Africa

> A. *Filaments cruciform, anther base not overlapping expanded portion*
> B. *Bracts connate into an amplexicaul cup (see also* **H. heterophylla**)

lacera (E.Mey. ex Harv.) Fourc.   Glandular-hairy, prostrate shrublet to 10 cm. Leaves petiolate, ovate, irregularly lacerate. Flowers 2 or 3 on axillary and terminal peduncles, orange to red, bell-shaped, bracts connate into a large cup. Capsules large, oblong and inflated. Mainly Aug.–Sept. Coastal sands on limestone, SW, AP (Cape Peninsula to De Hoop).•

saccifera (Turcz.) K.Schum.   Glabrescent and viscous sprawling shrublet to 40 cm. Leaves elliptic-oblong, regularly toothed. Flowers usually 2, on short axillary peduncles, bell-shaped, yellow, bracts connate into a small cup. Aug.–Oct. Stony clay slopes, SW, KM, LB, SE (Riviersonderend Mts and Bredasdorp to E Cape).

sp. 1   Sparsely hairy, prostrate shrublet to 10 cm. Leaves oblanceolate, irregularly lobed to toothed, stipules large, leafy, divided to base. Flowers 2, subsessile on elongate axillary peduncles, enclosed in a leafy lobed cup, pale yellow, with narrow throat and petals spreading above. July–Sept. Sandy limestone slopes, SW (Saldanha).•

> B. *Bracts not connate*

coccocarpa (Eckl. & Zeyh.) Kuntze   Glabrescent twiggy shrublet to 30 cm. Leaves linear-oblanceolate, toothed. Flowers 1 or 2 on slender axillary peduncles, purple. Capsules oblong, 2–3 times longer than calyx. Flowering time? Karroid bush, KM (dry parts of southern Africa to Swartberg Mts).

diffusa L.f.   Sprawling to prostrate, thinly glandular-hairy shrublet to 15 cm. Leaves 2-pinnatisect, with leafy lobed stipules. Flowers 1 or 2 on slender peduncles, yellow, orange or pink, bell-shaped, bracts scarcely united. Sept.–Oct. Rocky sandstone or granite slopes, NW, SW, AP, KM, LB, SE (Elandsbaai to Port Elizabeth).•

grossularifolia L.   Sprawling to prostrate, roughly hairy subshrub to 15 cm. Leaves oblanceolate-cuneate, toothed to lobed above. Flowers few, in terminal clusters, yellow, petals with stellate hairs, tightly furled. Capsules large, globose, scabrid and inflated. Sept.–Oct. Sandy flats and slopes, mostly after fire, SW, LB (Cape Peninsula to Garcia's Pass).•

heterophylla (Cav.) Thunb. (= *Hermannia humifusa* Hochr.)   Sprawling to prostrate, slightly viscid shrublet to 50 cm. Leaves oblanceolate, petiolate, toothed above, with leafy stipules. Flowers 1 or 2 on slender axillary and terminal peduncles, mauve, petals tightly furled, with bracts partly united in a cup. Sept.–Oct. Sandy, often coastal flats, NW, SW (Lambert's Bay to Cape Peninsula).•

pinnata L. (incl. **H. linifolia** Burm.f.)   Sprawling, glabrescent, mat-forming shrublet to 15 cm. Leaves often spuriously whorled, linear to 3-lobed above, with stipules divided into 2 or 3 linear lobes. Flowers on axillary peduncles, yellow, petals furled, with linear bracts. Aug.–Oct. Sandy coastal flats and dunes, SW (Velddrif to Cape Peninsula).•

sisymbriifolia (Turcz.) Hochr.   Sprawling, thinly hairy to scabrid and viscous shrublet to 40 cm, coppicing from base. Leaves sessile pinnatifid to trifid. Flowers 1 or 2 on slender axillary and terminal peduncles, white to cream-coloured fading to mauve, bell-shaped. June–Oct. Sandstone rocks, NW, KM (Gifberg to Barrydale).•

sp. 2 (= *Hermannia glabripedicellata* De Winter ined., *H. bodkinii* Pillans ms)   Glabrescent twiggy shrublet to 60 cm. Leaves oblong-lanceolate, petiolate, sparsely toothed. Flowers 1 or 2 on slender axillary peduncles, white to yellow, fragrant. Aug.–Oct. Rocky slopes and sandy flats, NW (Namaqualand to Clanwilliam).•

> A. *Filaments oblanceolate, anther base overlapping expanded portion*
> C. *Stems prostrate with flowers on erect axillary and terminal peduncles*

decumbens Willd. ex Spreng.   Sprawling shrublet with prostrate branches and short erect branchlets to 20 cm. Leaves ovate, toothed, roughly felted, with large stipules. Flowers in terminal clusters, yellow, with inflated papery calyx. Aug.–Oct. Mostly coastal sands and limestones, SW, AP, LB, SE (Mamre to Knysna).•

linifolia Burm.f.   Sprawling shrublet with prostrate branches and short erect branchlets to 10 cm. Leaves subsessile, oblanceolate and slightly lobed, with prominent leafy stipules. Flowers in terminal cymes, yellow to orange, bell-shaped. Aug.–Sept. Sandy and granite soils, NW, SW (Elandsbaai to Cape Peninsula).•

myrrhifolia Thunb.   Sprawling shrublet to 15 cm with long trailing branches from woody base. Leaves pinnatisect, subsecund. Flowers few at branch tips on nearly naked peduncles, red and yellow, tightly furled. Sept.–Oct. Granite rocks and coastal sands, NW, SW (Lambert's Bay to Mamre hills).•

prismatocarpa E.Mey. ex Harv.   Sprawling shrublet to 40 cm, with softly hairy, ascending branches. Leaves ovate, slightly toothed, discolorous, paler beneath. Flowers in terminal clusters, yellow, bell-shaped, with elongate capsules. Aug.–Nov. Sandy flats and slopes, NW, SW (Vanrhynsdorp to Tygerberg and Breede River valley).•

procumbens Cav.   Sprawling shrublet to 15 cm with long trailing branches from woody base. Leaves oblanceolate, toothed to shallowly lobed, subsecund. Flowers few at branch tips on nearly naked peduncles, tightly furled. Sept.–Oct. Coastal sands, SW (Bokbaai to Cape Peninsula).•

scordifolia Jacq.   Like **H. prismatocarpa** but shortly and roughly hairy, lower leaf surfaces soft-velvety, stipules caducous and capsules shorter and subglobose. Aug.–Oct. Rocky sandstone slopes, NW, SW (Bokkeveld Mts to Saldanha).•

C. Erect or rounded shrubs with stems erect or sprawling
D. Flowers with narrow throat and petals abruptly spreading distally

althaeoides Link   Like **H. rugosa** but leaves not crisped. Oct.–Dec. Rocky slopes, SE (Humansdorp to Queenstown).

angularis Jacq.   Like **H. rudis** but leaves oblanceolate-cuneate and calyx lobes glabrous except along margins. Sept.–Oct. Dry stony slopes, SW, KM, LB, SE (Hottentots Holland to Plettenberg Bay).•

concinnifolia I.Verd.   Erect, closely leafy shrublet to 90 cm. Leaves cuneate to oblanceolate, with recurved mucro, subsessile, with large stipules. Flowers in terminal clusters, yellow, throat narrow and petals spreading, calyx large and roughly hairy, usually reddish. Aug.–Oct. Coastal limestones, AP (Agulhas to Gouritsmond).•

diversistipula C.Presl ex Harv.   Like **H. velutina** but leaves tapering below into petiole and apparently subsessile, flowers yellow, orange or red. July–Oct. Clay slopes in renosterveld, NW, SW, AP, KM, LB, SE (Malmesbury to Uniondale).•

flammea Jacq.   Sparsely branched, often glabrescent shrublet to 80 cm. Leaves oblanceolate to cuneate, subsessile, usually sparsely toothed above. Flowers small clusters on slender peduncles, dark red, throat narrow and petals spreading, calyx lobes spreading and papery. Mainly Sept.–Oct. Mostly clay flats and slopes, SW, AP, LB, SE (Wellington to E Cape).

flammula Harv.   Like **H. flammea** but leaves smaller, narrow and folded on midline, roughly golden hairy. Mainly Aug.–Oct. Sandy, rocky clay or limestone soils, SW, AP, KM, LB, SE (Caledon to Langkloof).•

gracilis Eckl. & Zeyh.   Like **H. velutina** but flowers in slender terminal racemes and calyx not deeply lobed and flared. Mainly Aug.–Nov. Karroid slopes, KM, SE (Calitzdorp to E Cape and S Karoo).

holosericea Jacq.   Grey-mealy, twiggy shrub to 1.5 m. Leaves grey-velvety, obovate to oblanceolate, toothed above. Flowers small, in crowded, secund spreading racemose clusters, yellow. Mainly July–Oct. Clay, sandy and limestone soils, NW, SW, AP, KM, LB, SE (Stormsvlei to E Cape and S Karoo).

hyssopifolia L.   Stiffly erect, twiggy shrub to 2 m. Leaves mealy, oblanceolate to cuneate, toothed above. Flowers in dense terminal clusters, cream-coloured to pale yellow, throat narrow and petals spreading, calyx much inflated and urn-shaped. Sept.–Oct. Stony granite and clay slopes, NW, SW, KM, LB, SE (Cape Peninsula and Ceres to Grahamstown).

incana Cav.   Grey-mealy shrub to 2 m. Leaves ovate to elliptic, toothed above. Flowers in small terminal clusters, yellow. July–Oct. Mainly karroid, clay slopes, NW, AP, KM, LB (Worcester to Albertinia and De Rust).•

joubertiana Harv.   Like **H. flammea** and **H. flammula** but calyx lobes attenuate, half as long as corolla, and flowers small in tight clusters. Sept.–Oct. Limestone slopes, SW, AP (Hermanus to Mossel Bay).

lavandulifolia L.   Grey-mealy, diffusely twiggy shrub to 60 cm. Leaves grey-velvety, oblanceolate. Flowers few, in small clusters, yellow, throat narrow and petals spreading, calyx deeply lobed and flaring. Mainly Sept.–Oct. or Feb.–Apr. Clay slopes on renosterveld, NW, SW, AP, LB (Worcester to Robinson Pass).•

mucronulata Turcz.   Like **H. hyssopifolia** but leaves grey-green-velvety, oblanceolate, margins usually entire and flowers often deep yellow. Sept.–Oct. Mostly clay slopes, SE (Langkloof to E Cape).

rudis N.E.Br.   Sprawling shrublet with branches ascending to 60 cm. Leaves obovate-cuneate, in tufts. Flowers in rounded terminal clusters, yellow to red, throat narrow and petals spreading, calyx lobes large, flat, ovate and papery. Sept.–Oct. Mainly coastal sands, SW, AP (Cape Peninsula to Potberg).•

rugosa Adamson   Slightly mealy, sprawling shrublet to 60 cm. Leaves ovate, margins crisped and toothed. Flowers in terminal and axillary clusters, yellow fading red. Aug.–Oct. Granite and clay slopes, SW (Vredenburg to Cape Peninsula).•

salviifolia L.f.   Coarsely velvety, erect, twiggy shrub to 2 m. Leaves densely hairy, obovate to oblong, scarcely toothed above. Flowers in dense terminal clusters, yellow or orange, throat narrow and petals spreading, calyx inflated and tubular. Mainly Sept.–Oct. and Dec.–Apr. Stony granite and clay slopes, SW, AP, KM, LB, SE (Cape Peninsula to Grahamstown).

sulcata Harv.   Like **H. velutina** but calyx narrowly tubular and ribbed. June–Sept. Rocky slopes, often in dry bush, SE (Humansdorp to E Cape).

ternifolia C.Presl ex Harv.   Grey-mealy, closely leafy sprawling shrublet with ascending branches, to 20 cm. Leaves cuneate, usually with large leafy stipules. Flowers in terminal clusters, orange to red, throat narrow and petals spreading, calyx inflated and papery. Aug.–Sept. Coastal sands and limestones, SW, AP (Saldanha to Agulhas).•

trifoliata L.   Like **H. ternifolia** but stiffly erect, leaves often sessile, ascending and overlapping. Sept.–Oct. Coastal limestone soils, SW, AP (Hermanus to Gouritsmond).•

velutina DC.   Like **H. lavandulifolia** but usually more robust, to 3 m, with large, leafy stipules. Capsules umbonate. Aug.–Oct. Stony slopes, SE (Humansdorp to S KwaZulu-Natal).

D. Flowers bell-shaped to narrow with furled or spreading petals

alnifolia L.   Rounded, grey-mealy shrub to 1 m, with pubescent branches. Leaves cuneate to ovate, toothed above, pale mealy beneath. Flowers small, in many-flowered, elongate terminal clusters, yellow. July–Oct. Shale or rocky slopes, NW, SW, KM, LB, SE (Bokkeveld Mts to George).

althaeifolia L.   Softly hairy, mealy, grey-green shrublet to 50 cm, sometimes erect and single-stemmed. Leaves long-petiolate, ovate to elliptic, toothed and crisped, with broad, leafy stipules. Flowers in terminal and axillary clusters, yellow, calyx reddish fading to cream, inflated. Aug.–Oct. Clay, granite and limestone slopes, NW, SW, AP, KM, LB, SE (W Karoo to Langkloof).

aspera H.L.Wendl.   Stiffly hairy, erect shrublet to 1 m. Leaves oblanceolate to obovate, coarsely toothed, margins revolute. Flowers in dense terminal clusters, yellow fading reddish. Mainly Sept.–Oct. Sandstone slopes, NW, KM (Bokkeveld Mts to Montagu).•

confusa T.M.Salter   Sparsely twiggy shrublet to 60 cm. Leaves pinnatifid to pinnatisect. Flowers few at branch tips on nearly naked, branched peduncles, yellow. Aug.–Oct. Mainly clay slopes in renosterveld, NW, SW, KM, LB (Clanwilliam to Riversdale).•

conglomerata Eckl. & Zeyh.   Shrublet to 60 cm. Leaves subrotund, conspicuously petiolate, margins crenate. Flowers in congested clusters, yellow, calyx with conspicuous golden hairs. Aug.–Oct. Mainly clay soils, LB, SE (Riversdale to E Cape).

cordifolia Harv.   Robust, densely velvety shrub with stiffly erect branches to 1 m. Leaves ovate, crenulate. Flowers in tight terminal clusters, ?yellow. July–Oct. Rocky slopes, NW (Piketberg).•

cuneifolia Jacq.   Roughly scaly, twiggy shrub to 1 m. Leaves cuneate, coarsely toothed above, sometimes appearing fascicled. Flowers on subsecund racemes, yellow often fading reddish. Mainly Aug.–Oct. Clay and granitic slopes, NW, SW, KM, AP, LB, SE (Namaqualand to E Cape and Lesotho).

denudata L.f. (incl. **H. erecta** N.E.Br.)   Glabrous, stiffly erect shrub to 2 m. Leaves ascending, oblanceolate, toothed above, stipules leafy. Flowers in elongate terminal branches, few per node, pale yellow. Sept.–Oct. Sandstone slopes, often near streams, NW, LB (W Karoo to Cedarberg and Langeberg Mts near Robinson Pass).

disticha Schrad.   Prickly hairy, single-stemmed, erect shrub to 1 m. Leaves broadly ovate, sometimes slightly deflexed, shortly petiolate to subsessile, margins crenate, densely woolly beneath. Flowers in elongate, terminal clusters on lateral branches, yellow. Aug.–Sept. Rocky slopes, KM (W Little Karoo).•

filifolia L.f.   Glabrescent, twiggy shrublet to 1 m. Leaves and stipules subequal, often in tufts, linear, margins revolute. Flowers on slender elongated branches, few per node, orange to red, calyx lobes usually spreading and papery. Capsules knobbed. Aug.–Oct. Sandy or clay slopes, NW, AP, KM, LB, SE (Cold Bokkeveld to Port Elizabeth and Karoo to Free State).

helicoidea I.Verd.   Like **H. muricata** but flowers fewer and more diffuse on axillary and terminal branchlets. Aug.–Nov. Sandstone slopes, NW (Pakhuis and Olifants River Mts).•

hispidula Rchb.f.   Coarsely hairy shrub to 60 cm. Leaves rhomboid, mucronate, margins coarsely toothed, slightly revolute, discolorous, greyish beneath. Flowers in diffuse terminal panicles, yellow. Sept.–Nov. Sandstone slopes, NW (Piketberg).•

involucrata Cav. (incl. **H. decipiens** E.Mey. ex Harv.)   Coarsely velvety, erect, twiggy shrub to 1.2 m. Leaves densely hairy, oblong to cuneate, subsessile, often scarcely toothed above, stipules and bracts attenuate. Flowers in small terminal clusters, yellow or orange, calyx golden velvety. Aug.–Oct. Stony sandstone slopes, KM, SE (Swartberg to Baviaanskloof Mts).•

micrantha Adamson   Rounded shrublet to 30 cm. Leaves cuneate to ovate, toothed above. Flowers small, in elongate, few-flowered clusters, yellow. Sept.–Oct. Dry rocky slopes, SW (Cape Peninsula).•

muirii Pillans   Coarsely hairy, spreading shrublet with erect branchlets, to 20 cm. Leaves oblong, sessile, with leafy stipules as long as leaves. Flowers in terminal clusters, white to pink. Oct.–Feb. Sandy soils on limestone, AP (Stilbaai to Gouritsmond).•

multiflora Jacq.   Rounded, grey-mealy shrub to 75 cm. Leaves cuneate, toothed above, pale-mealy beneath. Flowers in small clusters, yellow, calyx brownish velvety. Aug.–Oct. Sandy and rocky flats and slopes, NW, SW (Bokkeveld Mts to Cape Peninsula).•

muricata Eckl. & Zeyh. (incl. **H. repetenda** I.Verd.)   Stiffly hairy shrublet to 50 cm, often much-branched from base. Leaves narrowly oblong, slightly toothed, white-tomentose beneath. Flowers in lax, terminal clusters on wiry branches, yellow. Dry clay and granite slopes, NW, KM (Namaqualand and S Karoo to Olifants River valley and George).

odorata Aiton   Grey-mealy, erect shrub to 1.5 m. Leaves grey-velvety, oblanceolate to elliptic, with linear stipules. Flowers in loose terminal clusters, yellow to orange, calyx prominently ribbed. Capsules umbonate. Mainly July–Oct. Rocky sandstone slopes, NW, KM, SE (Cedarberg to Baviaanskloof Mts).•

pillansii Compton   Like **H. stipulacea** but plants compact, leaves obovate, closely hairy to woolly. Sept.–Nov. Rocky sandstone slopes, KM (Witteberg).•

pulverata Andrews   Grey-mealy, gnarled shrublet from woody base, mostly to 40 cm. Leaves 1- or 2-pinnatisect. Flowers few, on slender naked peduncles, yellow to red, tightly furled. Mainly Sept.–Nov. Dry karroid slopes, NW, KM (Worcester to Port Elizabeth and Karoo).

rigida Harv.   Like **H. muricata** but flowers in racemes on short stiff pedicels. Mainly Aug.–Sept. Clay or rocky sandstone slopes, NW (Namaqualand to Ceres).

scabra Cav.   Sprawling, roughly hairy shrub to 60 cm. Leaves shortly petiolate, cuneate to linear, coarsely toothed above, terminal tooth recurving. Flowers in small clusters along elongate raceme-like terminal branches, yellow. July–Sept. Mostly sandstone slopes, sometimes granite or limestone, NW, SW (Gifberg to Durbanville).•

spinosa E.Mey. ex Harv.   Spinescent, twiggy shrublet to 40 cm, with wiry branches. Leaves obovate, slightly toothed. Flowers solitary in axils, dark pink to red, bell-shaped, peduncles stiff and becoming woody and thorn-like. Capsules horned. Mainly Aug.–Sept. but erratically after rain. Stony slopes and flats, KM (Namibia and Karoo to Little Karoo).

stipulacea Lehm. ex Eckl. & Zeyh.   Densely hairy shrublet to 40 cm. Leaves ovate to cuneate, sessile, ascending and imbricate, crenate to toothed, with large leafy stipules. Flowers mostly few in terminal clusters, yellow, calyx golden velvety. Sept.–Nov. Rocky sandstone slopes, KM, LB, SE (Ladismith and Riversdale to Port Elizabeth).•

trifurca L.   Roughly hairy shrublet to 1.5 m. Leaves oblong, 3-toothed above. Flowers solitary in upper axils, secund on horizontal racemes, bell-shaped, mauve with dark venation. Aug.–Oct. Stony soils, NW, SW (S Namibia to Darling and Breede River valley).

sp. 3   Softly hairy, erect shrub to 1 m. Leaves broadly ovate to rotund, long-petiolate, margins slightly toothed, pale beneath, with cordate, amplexicaul stipules. Flowers in slender, flexuose axillary clusters, yellow. Sept.–Oct. Rocky slopes, NW (Pakhuis Mts).•

HIBISCUS HIBISCUS, WILDESTOKROOS   c. 300 spp., cosmopolitan

aethiopicus L.   Roughly hairy subshrub to 30 cm, from woody rootstock. Leaves ovate-elliptic, toothed above, 3–5-veined from base, glabrescent above. Flowers on axillary peduncles, cream-coloured to yellow often with dark centre, epicalyx with 10–12 lanceolate to linear lobes, staminal tube c. 10 mm long. Mainly Aug.–Dec. Stony sandstone or clay slopes, SW, AP, LB, SE (Elandskloof Mts to KwaZulu-Natal).

atromarginatus Eckl. & Zeyh.   Slender glabrescent subshrub with slender erect branches to 60 cm, from woody rootstock. Leaves 3–5-palmatisect, lobes linear-oblong, usually toothed. Flowers axillary, on long peduncles, cream-coloured to yellow or pink, sometimes with dark centre, calyx lobes attenuate, epicalyx of 10–12 linear lobes. Aug.–Mar. Bush and grassland, SE (Humansdorp to N Province).

diversifolius Jacq.   Erect, prickly, perennial to 2.5 m. Leaves 3–5-palmate, coarsely toothed, on long prickly petioles. Flowers shortly pedicellate in terminal racemes, large, yellow with dark centre, epicalyx of 10–12 linear lobes, staminal tube to 20 mm long. Mainly Dec.–Apr. Forest margins and bush, SE (Plettenberg Bay to tropical Africa).

ludwigii Eckl. & Zeyh.   Stiffly hairy, shrub to 3 m, with stiffly erect branches. Leaves long-petiolate, shallowly 3–5 palmate, toothed. Flowers on short pedicels in upper axils, large, yellow with purple eye, epicalyx of 5 broad lobes, staminal tube to 30 mm long. Mainly Dec.–Feb. Coastal forests and bush, SE (George to tropical Africa).

pedunculatus L.f.   Slender, shortly hairy shrublet to 1.2 m. Leaves mostly 3–5-lobed, toothed. Flowers axillary, on long peduncles, mostly pink, epicalyx of 10–12 linear-oblanceolate lobes, staminal tube to 20 mm long. Aug.–Apr. Bush and forest margins, SE (George to tropical Africa).

pusillus Thunb.   Like *H. aethiopicus* but glabrescent, leaves rigid, leathery with prominent veins, upper leaves often 3-lobed, flowers cream-coloured, yellow or pink with dark eye, often on elongate peduncles. Mainly Jan.–May. Rocky sandstone or clay slopes, SW, AP, KM, LB, SE (Caledon and Little Karoo to tropical Africa).

*****trionum** L.   Stiffly hairy annual. Leaves mostly 3–5-palmatisect, toothed. Flowers pedunculate in upper axils, yellow with dark eye, epicalyx of 10–12 linear lobes, calyx bell-shaped, purple-veined, swollen in fruit. Sept.–Feb. Stony slopes and forest margins, SW, AP, KM, LB, SE (Cape Peninsula to Port Elizabeth, introduced from Old World tropics).

*****LAVATERA**   TREE MALLOW   c. 25 spp., Eurasia

*****arborea** L.   Glabrescent, thick-stemmed annual or short-lived perennial to 2 m. Leaves velvety, 3–5-lobed, toothed. Flowers in axillary clusters, mauve, epicalyx cup-shaped and 3 lobed, accrescent in fruit. Mainly Sept.–Nov. Waste places, dunes, NW, SW, AP, LB (Eurasian weed).

PAVONIA   c. 200 spp., pantropical and subtropical

burchellii (DC.) R.A.Dyer (= *Pavonia columella* auct.)   Softly hairy, shrubby perennial to 1 m. Leaves long-petiolate, shallowly palmately lobed, slightly toothed or crenate. Flowers axillary on long pedicels, yellow or orange, with epicalyx of 5 ovate lobes. Nov.–Apr. Forest margins and bush, SE (Humansdorp to Ethiopia).

praemorsa (L.f.) Cav.   Shortly velvety, shrubby perennial to 2 m. Leaves broadly cuneate to emarginate, toothed, rough above, pale beneath. Flowers solitary in axils, often on short shoots, yellow veined red, with epicalyx of 8–12 filiform lobes. Mainly Dec.–Mar. Bush and thicket, SE (Humansdorp to E Cape).

SIDA   MALLOW   c. 200 spp., pantemperate and tropical

ternata L.f.   Glabrescent, erect or sprawling perennial to 30 cm. Leaves shallowly to deeply palmatisect, toothed. Flowers axillary on filiform pedicels, white. Oct.–Mar. Grassy slopes, SE (George to E Cape).

SPARRMANNIA   CAPE HOLLYHOCK   c. 7 spp., Africa and Madagascar

africana L.f.   Shrub or small tree to 7 m, stiffly hairy when young. Leaves coarsely hairy, cordate, toothed. Flowers drooping in umbels, white. Capsules bristly. June–Nov. Forest margins and outcrops, LB, SE (Riversdale to E Cape).

STERCULIA   STERCULIA, STERKASTAIING   c. 200 spp., pantropical

alexandri Harv.   KAAPSE STERKASTAIING   Tree to 8 m, with smooth silvery bark, gregarious by suckering. Leaves 3–7-digitate, leaflets elliptic. Flowers in axillary cymes, mostly male by abortion, yellow with reddish cup. June–July. Coastal and riverine forests, SE (Van Staden's and Great Winterhoek Mts).•

MELIACEAE

1. Fruit a drupe or berry; leaves pinnate; filaments connate to apex .... **Ekebergia**
1.' Fruit an inflated, membranous capsule; leaves simple; filaments fused at base .... **Nymania**

EKEBERGIA   CAPE ASH, ESSENHOUT   4 spp., Africa

capensis Sparrm.   Dioecious tree to 20 m. Leaves opposite, imparipinnate, leaflets lanceolate, asymmetric at base. Flowers in axillary cymose panicles, whitish. Aug.–Sept. Forests, SE (George to tropical Africa).

NYMANIA   CHINESE LANTERNS   1 sp., South Africa and Namibia

capensis (Thunb.) Lindb.   Rigid shrub to 5 m. Leaves leathery, oblanceolate, tufted on short shoots. Flowers solitary and axillary, dull red. Capsules inflated, papery. Oct.–Dec. Karroid scrub, NW, KM, SE (Namibia to N Cape and Worcester to E Cape).

MELIANTHACEAE

MELIANTHUS   TURKEY BUSH, KALKOENTJIEBOS   6 spp., southern Africa

comosus Vahl   Like **M. elongatus** but leaves also thinly hairy above, margins not revolute and flowers solitary at nodes in pendulous racemes. Aug.–Oct. Stony slopes and stream banks, KM, SE (Namibia and W Karoo to Montagu through Little Karoo to E Cape).

elongatus Wijnands (= *Melianthus minor* L.)   Shrub to 2 m. Leaves imparipinnate, leaflets toothed, margins revolute, white-felted beneath. Flowers in axillary racemes, 2–4 at nodes, red, petals longer than sepals, ovary and fruits velvety. July–Sept. Sandstone or granite slopes and flats, NW, SW (Namaqualand to Langebaan).

major L.   KRUIDJIE-ROER-MY-NIE   Foetid shrub to 2 m. Leaves glaucous, imparipinnate, leaflets toothed. Flowers in terminal racemes, 2–4 at nodes, red, petals shorter than sepals, ovary glabrous. Aug.–Sept. Sandstone slopes, often along streams, NW, SW, AP, LB, SE (Bokkeveld Plateau to E Cape).

MENISPERMACEAE

1. Sepals and petals in female flowers 2 .... **Antizoma**
1.' Sepals and petals in female flowers 1 .... **Cissampelos**

ANTIZOMA 2 spp., southern Africa

miersiana Harv.   Dioecious greyish, rigid, twiggy shrublet to 1 m. Leaves discolorous, oblanceolate, often tufted. Flowers axillary, green, male cymose, female 1 or 2. July–Dec. Dry rocky slopes, NW (Richtersveld and W Karoo to Gifberg and Biedouw).

CISSAMPELOS   DAVIDJIES   c. 30 spp., pantropical

capensis L.f.   Dioecious glabrescent sprawling or climbing shrublet. Leaves ovate or trowel-shaped. Flowers axillary, velvety hairy, green, male cymose, female 1 or 2. Feb.–May. Sandy slopes in scrub, NW, SW, AP, KM, LB, SE (S Namibia and W Karoo to Port Elizabeth).

torulosa E.Mey. ex Harv.   Dioecious glabrescent wiry vine. Leaves glaucous, discolorous, reniform on long petioles. Flowers supra-axillary above a hairy gland, green, male cymose, female solitary or spicate. Oct.–Dec. Forest margins, SE (George to tropical Africa).

MENYANTHACEAE

1. Leaves radical, blades longer than broad; flowers in cymes; capsule 4-valvate .... **Villarsia**
1.' Leaves cauline, orbicular and deeply cordate; flowers in fascicles; capsule rupturing irregularly .... **Nymphoides**

NYMPHOIDES GEELWATERUINTJIE c. 20 spp., cosmopolitan

indica (L.) Kuntze Rhizomatous aquatic perennial. Leaves floating, orbicular-cordate. Flowers 2-several in fascicles at nodes, heterostylous, petals fringed, white or yellow. Dec.–Feb. Permanent pools, NW, SW, LB, SE (Clanwilliam to E Cape and throughout Africa, India and Australasia).

VILLARSIA YELLOW BOGBEAN c. 10 spp., mainly Australia

capensis (Houtt.) Merr. CAPE BOGBEAN Tufted aquatic perennial, mostly 15–20 cm. Leaves long-petiolate, blades ovate-oblong, attached subbasally. Flowers in paniculate cymes, heterostylous, petals fringed, yellow. Capsules with up to 4 warty seeds. Oct.–Feb. Marshy sandstone, NW, SW, LB, SE (Cold Bokkeveld Mts to Cape Peninsula to Humansdorp).•

goldblattiana Ornduff Like **V. capensis** but more robust, to 40 cm and capsules with up to 8 smooth seeds. Oct.–Dec. Marshy sandstone flats, SW (Cape Peninsula).•

MOLLUGINACEAE

1. Ovules basal, solitary in each carpel:
   2. Flowers solitary; leaves reduced, adpressed ....Polpoda
   2.' Flowers few to many in inflorescence; leaves well-developed:
      3. Leaves without stipules, sometimes narrow or in false whorls; fruit of 2 united carpels, indehiscent ....Limeum
      3.' Leaves usually stipulate, narrow and in whorls:
         4. Carpel 1; fruit indehiscent ....Adenogramma
         4.' Carpels 3–5, united; fruit dehiscent ....Psammotropha
1.' Ovules axile, many in each carpel:
   5. Calyx (or perianth) segments united; stamens attached to perianth ....Coelanthum
   5.' Calyx (or perianth) segments free or nearly so; stamens attached to receptacle:
      6. Stipules small, entire, free from leaf base, often caducous ....Mollugo
      6.' Stipules attached to leaf base, persistent:
         7. Inflorescence a simple umbel; stipules entire ....Hypertelis
         7.' Inflorescence a compound cyme; stipules lacerate:
            8. Hypogynous disc present...Pharnaceum
            8.' Hypogynous disc absent ....Suessenguthiella

ADENOGRAMMA MUGGIESGRAS 10 spp., Namibia to W Cape

A. Perennials or subshrubs

lichtensteiniana (Schult.) Druce (incl. **A. capillaris** (Eckl. & Zeyh.) Druce Spreading diffuse shrublet to 30 cm, with wiry stems. Leaves whorled, linear to oblong, apiculate. Flowers in small axillary clusters, white. Fruits subglobose, minutely pitted. Aug.–Dec. Sandstone slopes, NW, SW, AP, LB, SE (Cedarberg to Grahamstown).

rigida (Bartl.) Sond. Sprawling perennial with slender annual shoots to 60 cm. Leaves alternate and imbricate below, whorled above, linear, ascending, pungent. Flowers in small axillary clusters on thread-like peduncles, green or brown. Fruits ovoid, minutely pitted. Oct. Sandy flats, SW (Mamre to Houwhoek).•

sylvatica (Eckl. & Zeyh.) Fenzl Like **A. lichtensteinana** but leaves ovate, softly aristate, pedicels longer, often as long as leaves. Sept.–Dec. Shady, sheltered upper slopes, SW, KM, SE (Wemmershoek Mts to Uitenhage).•

teretifolia (Thunb.) Adamson Divaricately branched shrublet to 30 cm, with stiffly erect, pustulate branches from base. Leaves whorled, terete, apiculate. Flowers in small axillary clusters, white and green. Fruits top-shaped, minutely pitted. Sept.–Oct. Sandy slopes, NW (Olifants River valley).•

A. Annuals

glomerata (L.f.) Druce Diffusely branched, sprawling, wiry-stemmed annual, mostly 10–15 cm. Leaves whorled, linear, sometimes elliptic, mucronate. Flowers in small axillary clusters, white. Fruits obliquely top-shaped, minutely pitted. Aug.–Oct. Sandy slopes, NW, SW, AP, KM, LB, SE (S Namibia to Humansdorp).

littoralis Adamson Prostrate much-branched soft annual to 5 cm. Leaves closely whorled, oblanceolate, rounded at tip. Flowers in small axillary clusters, white. Fruits top-shaped, minutely pitted. Oct.–Dec. Deep sands, NW, SW (Namaqualand to False Bay).

mollugo Rchb.f. (incl. **A. congesta** Adamson)   Like **A. glomerata** but stems stiff and suberect, perianth accrescent and ultimately caducous and fruits depressed-globose, abruptly narrowed into a prominent, rough, compressed beak. Sept.–Oct. Damp sand, NW, SW, AP (Bokkeveld Mts to Stilbaai).•

physocalyx Fenzl   Like **A. glomerata** but perianth accrescent with prominent keels, fruits globose, gradually elongating into a beak as long as body. Sept.–Oct. Damp flats, NW, SW (Paarl to Kleinmond).•

COELANTHUM   2 spp., W Cape to Namibia

semiquinquefidum (Hook.f.) Druce (incl. **C. grandiflorum** E.Mey. ex Fenzl)   Delicate tufted annual mostly 8–12 cm, with wiry stems swollen at base. Leaves in a basal tuft and whorled on branches; basal leaves oblanceolate to obovate-spathulate, mucronate, upper leaves linear, in sets of 1 or 2 whorls. Flowers in lax, branched cymes, white. Oct.–Nov. Sandy flats, NW, SW (Namibia to Cape Peninsula and Worcester).

verticillatum Adamson   Like **C. semiquinquefidum** but basal leaves soon withering, upper leaves in sets of 4–8 whorls, flowers in more condensed cymes. Oct.–Mar. Coastal sands, NW, SW (S Namaqualand to Bokbaai).

HYPERTELIS   7 spp., Africa and St Helena

arenicola Sond.   Prostrate annual to 5 cm. Leaves terete, fleshy. Flowers in axillary umbels on peduncles shorter than leaves, whitish, stamens 3–5. Dec.–Apr. Saline coastal flats, SW, AP, SE (Cape Peninsula to Port Elizabeth).•

salsoloides (Burch.) Adamson   BRAKSURING   Dwarf, tufted, blue-glaucous shrublet to 30 cm, often much-grazed. Leaves terete, fleshy. Flowers in umbels on slender, glandular peduncles, white to pink, sepals reflexed, pedicels deflexed on fading, stamens 12–15. Sept.–Mar. Dry calcareous and saline flats, NW, KM, SE (Namibia to Clanwilliam, Little Karoo to Zimbabwe).

trachysperma Adamson   Like **H. arenicola** but peduncles longer than leaves. Jan.–Apr. Muddy saline flats, SW (Cape Flats).•

LIMEUM   LIZARD'S FOOT, KOGGELMANDERVOET   26 spp., W Cape to tropical Africa and India

aethiopicum Burm.   AARBOSSIE   Stunted woody shrublet with annual shoots, to 15 cm. Leaves oblong to elliptic, margins revolute. Flowers in terminal and axillary clusters on erect peduncles, green and white. Feb.–Mar. Dry, rocky slopes, NW, KM (Namibia to Worcester, Little Karoo and Swartberg Mts).

africanum L.   Sprawling annual or perennial with long trailing branches. Leaves obovate to oblanceolate, margins sometimes revolute. Flowers in crowded, terminal cymes on prostrate peduncles, green and white. Mainly Aug.–Nov. Dry rocky slopes and flats, NW, SW (Namaqualand and W Karoo to Cape Peninsula).

subnudum Friedrich   Stiffly erect, sparsely leafy shrublet to 40 cm, with rod-like branches. Leaves adpressed, subsessile, linear-elliptic. Flowers in terminal umbels on short naked peduncles soon appearing axillary, yellow. Sept.–Oct. Sandstone slopes, NW (Olifants River Mts and Piketberg).•

telephioides E.Mey. ex Fenzl   Like **L. aethiopicum** but branches often trailing, leaves broadly obovate, flat. Nov.–May. Dry stony slopes, AP, KM, LB, SE (Robertson to Oudtshoorn, Stilbaai to E Cape—probably not distinct from **L. aethiopicum**).

MOLLUGO   20 spp., cosmopolitan

cerviana (L.) Ser. ex DC.   Delicate tufted annual to 20 cm, with wiry stems. Leaves in a basal tuft and whorled on branches; basal leaves short-lived, linear to spathulate. Flowers in terminal and axillary cymes on wiry pedicels, white and green. Jan.–Mar. Sandy flats and slopes, NW, SW (Namibia and Botswana to False Bay).

namaquensis Bolus   Dwarf tufted annual to 5 cm. Leaves basal, oblanceolate. Flowers in short pseudoracemes on slender peduncles, green. Aug.–Sept. Sandy flats and slopes, NW, SW (Namaqualand to Velddrif).

pusilla (Schltr.) Adamson (incl. **M. tenella** Bolus)   Delicate tufted annual to 8 cm. Leaves oblanceolate to oblong-ovate, pseudopetiolate, prostrate. Flowers in elongate pseudoracemes on thread-like peduncles, whitish. Aug.–Sept. Sandy flats and slopes, NW, SW (S Namibia to Riviersonderend).

PHARNACEUM SNEEUVYGIE   25 spp., southern Africa

A. Leaves distantly whorled (see also P. dichotomum)

cordifolium L.   Perennial with prostrate annual stems to 50 cm. Leaves leathery, whorled, oblanceolate-spathulate, rounded at tip, with short, fringed stipules. Flowers in pedunculate cymes, white, 3–4 mm long. Mainly May–Nov. Dunes, SW (Cape Peninsula to Stanford).•

lineare L.f.   Sprawling perennial to 10 cm, with elongate internodes. Leaves in whorls, terete. Flowers in lax, branched axillary cymes on long peduncles, white, 5–8 mm long. June–Nov. Sandy slopes and flats, NW, SW (Namaqualand to Bredasdorp).

serpyllifolium L.f.   Soft, cushion-like perennial to 5 cm, with thread-like branches. Leaves 2- or 3-whorled, obovate, pseudopetiolate. Flowers 1–3 in lax axillary cymes, white, c. 1.5 mm long. Sept.–Dec. Cool shady places at high elevation, NW (Cedarberg to Cold Bokkeveld).•

thunbergii Adamson   Sprawling perennial with prostrate annual branches to 60 cm. Leaves whorled, linear to oblanceolate, acute, with long stipules cut to base. Flowers in lax axillary cymes on long peduncles, green and white, 3–4 mm long. Mainly Nov.–Jan. Coastal bush, AP, SE (Stilbaai to KwaZulu-Natal).

A. Leaves basally crowded
B. Nonwoody annuals or perennials

croceum E.Mey. ex Fenzl   Tufted annual to 10 cm. Leaves mostly basal, subterete, succulent. Flowers in long-pedunculate cymes, pink, yellow or white, sepals reflexed, 3–4 mm long. Aug.–Sept. Sandy flats, especially coastal, NW, SW, SE (Namibia to George).

exiguum Adamson   Tufted annual sometimes to 30 cm. Leaves crowded basally, linear. Flowers in lax cymes, green and white, to 2 mm long. Aug.–Sept. Sandy flats, NW, SW (Namibia to Cape Peninsula).

fluviale Eckl. & Zeyh.   Subshrub with sprawling annual branches, 5–15 cm. Leaves in a basal tuft and whorled above, linear. Flowers in lax terminal and axillary cymes, white, c. 3 mm long. Aug.–Nov. Sandy slopes, often damp sites, NW, SW, AP, SE (Grootwinterhoek Mts to Somerset East).

subtile E.Mey. ex Fenzl   Tufted annual, 3–8 cm. Leaves in a basal rosette, oblanceolate. Flowers in long, flexuose raceme-like cymes, whitish, to 1.5 mm long. Fruits shorter than perianth. Sept.–Oct. Rock outcrops, NW, KM (Worcester and Little Karoo).•

B. Perennials woody at base

albens L.f.   Like **P. aurantium** but stipules shorter, to 2 mm long, leaves often reflexed. Aug.–Oct. Rocky slopes, NW (Namibia and W Karoo to Olifants River Mts).

aurantium (DC.) Druce   Sprawling or erect shrublet to 80 cm, woody at base. Leaves scattered and alternate mostly near base, linear, stipules sheathing at base. Flowers in lax cymes on long, purplish, white-glaucous peduncles, 3–4 mm long, white. Sept.–Oct. Stony gravel slopes and flats, NW, SW, KM, LB (Namibia to Gourits River).

ciliare Adamson   Like **P. incanum** but stipules small and opaque. Sept.–Oct. High sandstone slopes, KM, LB (Montagu and Ladismith to Langeberg: Swellendam).•

confertum (DC.) Eckl. & Zeyh.   Like **P. incanum** but with sprawling branches to 1.2 m, stipules larger, floral disc red to orange. Aug.–Oct. Stony slopes and flats, NW, SW (Namibia to Malmesbury and Worcester).

dichotomum L.f.   Erect or sprawling shrublet to 30 cm, woody at base. Leaves mostly in a basal tuft, also distant along stem in whorls, linear. Flowers in terminal and axillary cymes on long peduncles, white, c. 3 mm long. July–Nov. Dry slopes, NW, SW, KM, LB, SE (Namaqualand and Karoo to E Cape).

elongatum (DC.) Adamson   Like **P. incanum** but stipules cut to base into hair-like lobes and floral disc red to yellow. Aug.–Oct. Dry slopes, NW, SW, LB, SE (Namaqualand to E Cape).

incanum L.   Sprawling or erect shrublet, woody at base with whitish branches to 40 cm. Leaves crowded near base, linear-filiform, stipules sheathing at base. Flowers in long-pedunculate cymes, 3–4 mm long, white. Aug.–Oct. Rocky slopes, NW, SW (Namaqualand to Hopefield and Worcester).

lanatum Bartl.   Like **P. aurantium** but stipular hairs curled and forming a woolly mass. Aug.–Oct. Sandy flats and slopes, NW, SW (Namaqualand and Karoo to Cape Peninsula).

lanuginosum J.C.Manning & Goldblatt (= *Pharnaceum lanatum* var. *albens* Adamson)　　Low, often tangled, closely leafy shrublet to 25 cm. Leaves erect-ascending, imbricate, ovate, stipules white-membranous with ragged margins, exceeding leaves. Flower in small umbellate cymes on elongate, terminal peduncles, white, c. 3 mm long. Sept.–Nov. Rocky sandstone slopes, NW (Cedarberg Mts to Cold Bokkeveld).•

microphyllum L.f.　DROËDASKRUIE　Tangled shrublet mostly to 20 cm, appearing grey-woolly. Leaves oblong, fleshy, stipules prominent, imbricate, curled and forming a woolly mass hiding young leaves. Flowers in umbellate cymes on short, terminal peduncles, white, 3–4 mm long, bracts woolly. Sept.–Oct. Coastal sands and limestones, NW, SW (Namaqualand to Saldanha Bay).

rubens Adamson　　Like **P. aurantium** but peduncles reddish, flowers smaller, 2–3 mm long. Flowering time? Sheltered rock ledges, NW (Cedarberg Mts).•

trigonum Eckl. & Zeyh.　　Dwarf shrublet to 10 cm, branches sprawling. Leaves crowded basally, squarrose, linear, mucronate, stipules inconspicuous, firm, opaque. Flowers in lax, umbel-like cymes on slender peduncles, white c. 3 mm long. May–Aug. Rocky slopes, NW, LB, SE (Hex River Mts to Karoo and Lesotho).

POLPODA　　2 spp., Namaqualand to W Cape

capensis C.Presl　　Silvery, lycopodium-like shrublet to 50 cm. Leaves ovate, adpressed, imbricate, with fringed, papery stipules longer than leaves. Flowers solitary in upper axils, minute, 4-merous, whitish. Mainly Apr.–July. Sandstone and gravel slopes, NW, SW, AP, LB (Namaqualand to Mossel Bay).

stipulacea (F.M.Leight.) Adamson　　Like **P. capensis** but leaves yellow-green, oblong-incurved, margins revolute, longer than stipules, flowers 5-merous, filaments fused at base. Nov.–Dec. Stony sandstone slopes, NW (Matsikamma to Wolseley).•

PSAMMOTROPHA　　11 spp., southern Africa

anguina Compton　　Like **P. quadrangularis** but leaves smaller, closely adpressed and flower clusters subsessile, with conspicuous bracts. Sept.–Oct. Rocky sandstone slopes, NW (Cedarberg and Cold Bokkeveld).•

diffusa Adamson　　Diffuse, sprawling slender shrublet 30 cm. Leaves scattered and tufted in axils, with minute papery stipules, linear-oblanceolate. Flowers in terminal, subumbellate clusters, greenish. Nov.–Dec. Shale slopes, NW (Cedarberg).•

frigida Schltr.　　Tufted, cushion-forming perennial to 2 cm, often purplish, with wiry branches. Leaves crowded at branch tips, incurved-obovate, margins whitish. Flowers in clusters on thread-like peduncles, greenish. Sept.–Dec. Sandstone rock on shallow soil, NW, KM (Cedarberg and W Karoo to Swartberg Mts).

quadrangularis (L.f.) Fenzl　　Closely leafy, gnarled shrublet to 30 cm, mostly branching from base. Leaves exstipulate, 4-ranked, linear-attenuate, spine-tipped, ascending, imbricate, margins revolute. Flowers in dense clusters on wiry peduncles, greenish cream to pinkish. Sept.–Oct. Stony slopes, NW, SW, KM (Namaqualand to Villiersdorp and Willowmore).

spicata Adamson　　Like **P. diffusa** but leaves with revolute margins, stipules long, flowers in spike-like tufts. Oct.–Nov. Sandstone slopes, NW (Gifberg).•

SUESSENGUTHIELLA　　1 sp., Namibia to W Cape

scleranthoides (Sond.) Friedrich　　Prostrate, mat-forming annual with wiry branches. Leaves whorled, linear-subterete, apiculate, fleshy. Flowers crowded at nodes, green and white. July–Sept. Sandy flats, NW (Namibia and Karoo to Clanwilliam).

MONTINIACEAE

MONTINIA　PEPERBOS　1 sp., southern Africa

caryophyllacea Thunb.　　Dioecious erect glaucous shrub to 1.5 m. Leaves oblanceolate, sometimes tufted. Flowers terminal, white, male cymose, female 1 or 2. May–Oct. Rocky slopes, NW, SW, AP, KM, LB, SE (S Angola, Namibia to Cape Peninsula to E Cape).

MORACEAE

FICUS   FIG   c. 1000 spp., pantropical and warm temperate

burtt-davyi Hutch.   VELD FIG   Monoecious shrub or liane to 5 m, bark greyish. Leaves elliptic, lateral veins c. 5. Figs axillary, pedunculate, 5–7 mm diam., yellowish. Aug.–Mar. Rock outcrops and woodland, KM, SE (Oudtshoorn to KwaZulu-Natal).

cordata Thunb.   NAMAQUA FIG   Monoecious tree to 17 m, bark ash grey. Leaves ovate. Figs axillary, sessile, 5–7 mm diam., yellowish green. June–Dec. Rock clefts, NW (Grootwinterhoek Mts northwards throughout Africa).

ilicina (Sond.) Miq.   LAUREL FIG   Monoecious shrub or small tree to 5 m, roots usually flattened and plastered over rocks, bark white. Leaves elliptic. Figs axillary, pedunculate, 5–8 mm diam. June–Dec. Rock outcrops, NW (Namibia to Clanwilliam).

ingens (Miq.) Miq.   ROCK FIG   Monoecious tree to 15 m, bark grey. Leaves ovate, lateral veins c. 8. Figs axillary, pedunculate, 9–12 mm diam., pink. June–Dec. Rock outcrops and woodland, SE (Humansdorp northwards throughout Africa).

sur Forssk.   (= *Ficus capensis* Thunb.)   CAPE FIG   Monoecious tree to 11 m, bark dark grey. Leaves elliptic to ovate, broadly toothed. Figs in panicles on trunk and main branches, 20–40 mm diam., red. Oct.–Mar. Along streams and ravines, SE (Knysna northwards throughout Africa).

MYRICACEAE

MORELLA (= *MYRICA* in part)   WASBESSIE   c. 50 spp., cosmopolitan

cordifolia (L.) Killick (= *Myrica cordifolia* L.)   CANDLE BERRY, GLASHOUT   Dioecious shrub to 3 m. Leaves imbricate, sessile, broadly ovate-cordate, toothed, margins revolute, gland-dotted. Flowers in axillary spikes. Fruits warty, 5–8 mm diam. May–Aug. Coastal sands and limestones, SW, AP, LB, SE (Yzerfontein to E Cape).

diversifolia (Adamson) Killick (= *Myrica diversifolia* Adamson) (possibly a hybrid **M. quercifolia** × **M. kraussiana**)   Dioecious shrub to 1 m. Leaves obovate, tapered below, more or less toothed, margins mostly revolute, gland-dotted. Flowers in axillary spikes. Fruits warty, c. 3 mm diam. Aug.–Sept. Sandstone slopes, SW (Cape Peninsula).•

humilis (Cham. ex Schltdl.) Killick (= *Myrica humilis* Cham. ex Schltdl.)   Like **M. kraussiana** but spikes not robust, with smaller bracts. July–Nov. Sandstone slopes, KM, LB, SE (Swellendam and Warmwaterberg to E Cape).

integra (A.Chev.) Killick (= *Myrica integra* (A.Chev.) Killick)   BASTERWATEROLIER   Dioecious shrub or tree to 3 m. Leaves narrowly elliptic, attenuate below, sometimes toothed. Flowers in axillary spikes. Fruits warty, c. 3 mm diam. Sept.–Apr. Rocky streamsides, NW, SW (Pakhuis to Kleinmond Mts).•

kraussiana (Buchinger ex Meisn.) Killick (= *Myrica kraussiana* Buchinger ex Meisn.)   Dioecious shrub to 1 m. Leaves elliptic, rounded below, sometimes toothed above, margins revolute, gland-dotted. Flowers in robust axillary spikes with large, imbricate bracts. Fruits warty, 2–3 mm diam. Aug.–Oct. Sandstone slopes, SW, AP, LB (Cape Peninsula to Swellendam).•

quercifolia (L.) Killick (= *Myrica quercifolia* L.)   MAAGPYNBOSSIE   Dioecious spreading shrub to 60 cm. Leaves obovate, attenuate below, usually pinnatifid, gland-dotted. Flowers in axillary spikes. Fruits warty, 3–4 mm diam. July–Sept. Mostly coastal sandy and limestone flats and slopes, NW, SW, AP, LB, SE (Namaqualand to E Cape).

serrata (Lam.) Killick (= *Myrica serrata* Lam.)   WATEROLIER   Like **M. integra** but leaves mostly toothed, conspicuously reticulate-veined above, usually gland-dotted beneath and with margins more or less revolute. Aug.–Dec. Rocky streamsides, NW, SW, KM, LB, SE (Bainskloof to Mpumalanga and Caprivi).

MYRSINACEAE

1. Filaments developed (sometimes short), united at base by a membrane; twigs thin and woody; leaves usually 1–4 cm long ....**Myrsine**
1.' Filaments obsolete; twigs thickish and soft; leaves up to 12 cm long ....**Rapanea**

MYRSINE CAPE MYRTLE   7 spp., Africa and Asia

africana L.   Dioecious or polygamodioecious shrub to 3 m. Leaves obovate, toothed above, margins revolute, 5–20 mm long. Flowers in axillary fascicles, cream-coloured, anthers exserted, pink to purple. Throughout the year. Sandy slopes and flats in scrub, NW, SW, AP, KM, LB, SE (Bokkeveld Mts to Cape Peninsula to tropical Africa and Azores).

pillansii Adamson   LARGE CAPE MYRTLE, GROOTMIRTING   Dioecious or polygamodioecious shrub to 4 m. Leaves elliptic, finely toothed, margins revolute, 20–50 mm long. Flowers in axillary fascicles, cream-coloured, anthers included, brown. May–Oct. Forest margins, SW, SE (Cape Peninsula to Hermanus, Knysna, KwaZulu-Natal and Mpumalanga).

RAPANEA CAPE BEECH, KAAPSEBOEKENHOUT   c. 200 spp., pantropical and subtropical

gilliana (Sond.) Mez   DWARF CAPE BEECH, DWERGBOEKENHOUT   Dioecious or polygamodioecious shrub or small tree to 4 m. Leaves leathery, subsessile, oblanceolate, margins revolute. Flowers in axillary fascicles, cream-coloured. Fruits ellipsoid, 7–9 mm diam., black. May–July. Coastal dune scrub, SE (Humansdorp to E Cape).

melanophloeos (L.) Mez   Dioecious or polygamodioecious tree to 20 m. Leaves oblong-elliptic. Flowers in axillary fascicles, cream-coloured. Fruits globose, 4–5 mm diam., purple. Oct.–Dec. Forests, SW, AP, LB, SE (Cape Peninsula to tropical Africa).

MYRTACEAE

1. Flowers in pedunculate cymes; stamens much longer than petals; leaves mostly opposite .... **Metrosideros**
1.' Flowers solitary, subsessile; stamens much shorter than petals; leaves alternate .... **Leptospermum**

*LEPTOSPERMUM AUSTRALIAN MYRTLE   79 spp., Australasia

*laevigatum (Gaertn.) F.Muell.   Shrub or small tree to 8 m with bark shedding in strips. Leaves oblanceolate, leathery, glaucous. Flowers mostly paired in axils, white. July–Oct. Weed of coastal sands, NW, SW, AP, SE (Australian weed, Elandsbaai to Port Elizabeth).

METROSIDEROS SMALBLAD   c. 50 spp., Malaysia and Pacific, 1 sp., W Cape

angustifolia (L.) Sm.   Shrub or small tree to 4(–7)m. Leaves opposite, narrowly elliptic, margins revolute. Flowers in dense axillary cymes, white. Dec.–Feb. Sandstone slopes often along watercourses, NW, SW, LB (Matsikamma to Langeberg Mts: Riversdale).•

NEURADACEAE

GRIELUM SAND-PRIMROSE, DUIKERWORTEL   4 spp., dry and winter-rainfall southern Africa

grandiflorum (L.) Druce   Sprawling white-woolly perennial, often mat-forming. Leaves pinnatisect or bipinnatisect, lobes mucronate. Flowers yellow with a green eye. Fruits depressed-pentagonal with a peripheral wing and central spines. Mainly Sept.–Oct. Sandy and stony coastal flats, NW, SW (Namaqualand to Cape Peninsula).

humifusum Thunb.   PIETSNOT   Prostrate, thinly white-woolly annual, often mat-forming. Leaves pinnatisect, glabrescent above, lobes rounded. Flowers yellow with a pale eye. Fruits depressed-pentagonal with a peripheral wing and central spines. Mainly July–Oct. Sandy lower slopes and flats, NW, SW, KM (S Namibia to Robertson and W Karoo).

OCHNACEAE

OCHNA MICKY MOUSE PLANT, ROOIHOUT   c. 86 spp., palaeotropical and subtropical

arborea Burch. ex DC.   KAAPSE ROOIHOUT   Semideciduous tree with flaking bark to 12 m. Leaves elliptic, finely toothed. Flowers several in clusters or panicles, scented, yellow. Sept.–Nov. Forest understorey, SE (George to Zimbabwe).

serrulata (Hochst.) Walp.   FYNBLAARROOIHOUT   Semideciduous shrub or small tree with warty bark to 2 m. Leaves narrowly elliptic, toothed. Flowers 1 or 2 on short shoots, yellow. Apr.–Oct. Forest margins and scrub, SE (George to Mpumalanga).

OLEACEAE

1. Fruit a membranous capsule; virgate shrublet with reduced leaves and yellow flowers ....**Menodora**
1.' Fruit a drupe or berry; flowers white:
    2. Corolla tube long and slender, lobes abruptly spreading; scramblers or bushy shrubs ....**Jasminum**
    2.' Corolla tube short or absent; shrubs or small trees:
        3. Calyx deeply lobed; inflorescence few-flowered; leaves usually with acarodomatia (pockets in axils of veins on underside); ovules basal ....**Chionanthus**
        3.' Calyx toothed or shallowly lobed; inflorescence many-flowered; leaves without acarodomatia; ovules pendulous ....**Olea**

CHIONANTHUS   IRONWOOD   c. 120 spp., pantropical and temperate

foveolatus (E.Mey.) Stearn   POKYSTERHOUT   Tree to 30 m. Leaves ovate to oblong, minutely scaly beneath, with acarodomatia. Flowers in lax axillary cymose panicles, white or tinged pink, fragrant. Mar.–June. Coastal bush and rocky slopes, SW, AP, LB, SE (Cape Peninsula to Mpumalanga).

JASMINUM   JASMINE   c. 450 spp., palaeotropical and subtropical

angulare Vahl   Scrambling shrub to 7 m, hairy on young parts. Leaves 3-foliolate, leaflets ovate, often with acarodomatia. Flowers few in terminal cymes, 5-lobed, white. Nov.–Jan. Coastal and inland bush, KM, SE (Ladismith to KwaZulu-Natal).

glaucum (L.f.) Aiton   Glaucous shrub. Leaves lanceolate-elliptic. Flowers usually 3 in cymes, 5–7-lobed, white, fragrant. Oct.–Dec. Riverbanks and rocky slopes, NW, SW (Gifberg to Riviersonderend Mts).•

multipartitum Hochst.   Shrub to 3 m, twigs shortly hairy. Leaves ovate-lanceolate. Flowers 1(–3), terminal, c. 11-lobed, white flushed red outside. Oct.–Nov. Scrub and rocky slopes, SE (Uitenhage to N Province).

tortuosum Willd.   Like **J. angulare** but leaflets linear-lanceolate, without acarodomatia. Flowers 3–5 in cymes, 6-lobed, white. Nov.–Dec. Forest margins, SW, KM, LB (Caledon, Calitzdorp, Mossel Bay).•

MENODORA   c. 25 spp., America and southern Africa

juncea Harv.   Stiffly erect shrublet with silvery hairy branches to 1.6 m. Leaves sessile, linear, adpressed. Flowers 1–few in terminal cymes, yellow. Jan. Rocky arid slopes, KM (Namaqualand and Karoo to Rooiberg Mts).

OLEA   OLIVE   c. 35 spp., Old World

capensis L.   YSTERHOUT   Shrub or tree to 12 m. Leaves elliptic-ovate. Flowers in terminal panicles, white. Feb.–Dec. Forests and scrub, NW, SW, AP, LB, SE (Olifants River Mts to E Cape and to tropical Africa).

europaea L.   WILD OLIVE   Tree to 14 m with warty branchlets. Leaves narrowly elliptic, discolorous, pale-scaled beneath. Flowers in axillary and terminal panicles, white. Oct.–Mar. Forests and rocky slopes, NW, SW, AP, KM, LB, SE (Bokkeveld Mts to E Cape and to N Africa).

exasperata Jacq.   DUNE OLIVE   Shrub or small tree to 7 m with warty branchlets. Leaves linear-oblong. Flowers in terminal panicles, white. Aug.–Oct. Coastal scrub on sand and limestone, SW, AP, LB, SE (Cape Peninsula to E Cape).

OLINIACEAE

OLINIA   HARD PEAR, HARDEPEER   c. 8 spp., southern and tropical Africa

capensis (Jacq.) Klotzsch   Like **O. ventosa** but cymes with 9 flowers at the tip of each axis, flowers smaller with hypanthium 1.5–2 mm long and bracts persistent. May–July. Coastal forests, SW, LB, SE (Vredenburg to E Cape).

ventosa (L.) Cufod.   Tree to 15 m. Leaves opposite, obovate, smelling of almonds when crushed. Flowers in crowded axillary cymes with 3 flowers at the tip of each axis, creamy white, fragrant, hypanthium 3–8 mm long; bracts caducous. May–July. Coastal forests, SW, AP, LB, SE (Cape Peninsula and Wolseley to E Cape).

ONAGRACEAE

EPILOBIUM WILLOWHERB c. 220 spp., pantemperate and tropical mountains

capense Buch.-Ham. ex Hochst. CAPE WILLOWHERB Thinly hairy perennial to 1.2 m, rhizomes with fleshy scales. Leaves ovate to lanceolate, shortly petiolate, toothed. Flowers solitary in upper axils, pink to white, stigma deeply 4-cleft. Dec.–Mar. Damp places, NW, SW, SE (Cold Bokkeveld Mts to tropical Africa).

hirsutum L. GREAT WILLOWHERB Softly velvety perennial to 2.5 m. Leaves lanceolate, sessile, finely toothed. Flowers solitary in upper axils, pink, stigma deeply 4-cleft. Damp places, NW, SW, AP, KM, LB, SE (Clanwilliam to Cape Peninsula to W Asia).

*****tetragonum** L. SQUARE-STALKED WILLOWHERB Glabrescent perennial to 1.8 m, producing leafy basal rosettes in autumn. Leaves narrowly lanceolate, sessile, finely toothed. Flowers solitary in upper axils, pink, stigma club-shaped. Nov.–Jan. Damp places, SW, KM, LB, SE (European weed, Cape Peninsula to Mpumalanga).

OROBANCHACEAE

1. Plants with reduced scale-like leaves, fully parasitic:
    2. Ovary 1-locular with parietal placentas; flowers bilabiate, lobes separated by prominent folds; filaments usually thickened or flattened near base ....**Orobanche**
    2.' Oavry 2-locular with axile placentas; flowers 3- or 5-lobed; filaments linear:
        3. Anthers 1-thecous; flowers 3-lobed ....**Hyobanche**
        3.' Anthers 2-thecous with 1 fertile and 1 barren theca; flowers 5-lobed ....**Harveya**
1.' Plants more or less leafy, not fully parasitic:
    4. Calyx 4-lobed; corolla bilabiate ....**Bartsia**
    4.' Calyx 5-lobed or 5-toothed:
        5. Anthers with 2 fertile, equal thecae; flowers yellow or orange:
            6. Calyx enlarged and inflated in fruit..**Melasma**
            6.' Calyx not enlarged in fruit ....**Alectra**
        5.' Anthers 1-thecous or with 1 fertile theca and 1 variously modified theca; flowers pink or red:
            7. Anthers 2-thecous with 1 fertile theca and 1 barren theca:
                8. Undershrubs; corolla tube funnel-shaped, longer than lobes.......................**Graderia**
                8.' Herbs; corolla tube campanulate, shorter than lobes ....**Sopubia**
            7.' Anthers 1-thecous:
                9. Corolla tube rather sharply curved at or above middle and often dilated at curve ....**Striga**
                9.' Corolla tube cylindric, straight or only slightly curved:
                    10. Petals equal or nearly equal; capsule dry, dehiscent ....**Buchnera**
                    10.' Corolla with upper lobes united and smaller than lower; capsule fleshy, indehiscent..........**Cycnium**

ALECTRA YELLOW WITCHWEED c. 40 spp., Africa, Asia

capensis Thunb. Hemiparasitic perennial to 50 cm. Leaves overlapping, entire, ascending. Flowers crowded in dense, often subglobose spikes, yellow to orange, stamens nearly equal, filaments bearded. Nov. Rocky slopes, NW, KM, LB, SE (Montagu to Mpumalanga).

lurida Harv. Hemiparasitic perennial to 30 cm. Leaves minute, slightly toothed. Flowers in spikes or racemes, cream-coloured to orange, stamens equal, filaments glabrous. Sept.–Dec. Stony slopes and flats, SW, SE (Cape Peninsula to Knysna).•

sessiliflora (Vahl) Kuntze Hemiparasitic perennial to 25 cm. Leaves coarsely toothed below, spreading, longer than flower buds. Flowers in spikes or racemes, yellow to orange, stamens unequal, filaments glabrous. Mainly Nov.–Apr. Damp flats and lower slopes, NW, SW, AP, KM, LB, SE (Gifberg to tropical Africa and Madagascar).

*****BARTSIA** c. 60 spp., N temperate and tropical mountains

*****trixago** L. Erect, glandular hemiparasitic annual to 50 cm. Leaves coarsely lobed. Flowers sessile, bilabiate, cream-coloured with a red lip. Oct.–Jan. Streamsides and damp places, NW, SW, KM, LB, SE (Eurasian weed, Namaqualand to tropical Africa).

BUCHNERA FALSE VERBENA c. 100 spp., mostly Old World tropics

dura Benth. Hemiparasitic perennial to 60 cm. Leaves mostly basal, shortly scabrid. Flowers in spikes, blue-purple, hairy. Oct.–Jan. Marshy slopes, NW, LB, SE (Ceres to tropical Africa).

glabrata Benth. Hemiparasitic perennial to 30 cm. Leaves mostly basal, shortly scabrid. Flowers crowded in dense spikes, purple or white. Dec.–Apr. Marshy mountain slopes, SW, LB, SE (Franschhoek to Mpumalanga).

CYCNIUM MUSHROOM-FLOWER c. 40 spp., southern and tropical Africa

tubulosum (L.f.) Engl. Hemiparasitic perennial to 60 cm with erect branches. Leaves narrowly oblong, minutely notched. Flowers in racemes, on long pedicels, pink to mauve turning black on fading. Oct.–Dec. Grassy slopes and flats, SE (George to S tropical Africa).

GRADERIA WILD PENTSTEMON 5 spp., Africa

scabra (L.f.) Benth. Hemiparasitic perennial to 60 cm. Leaves shortly hairy, coarsely toothed. Flowers in lax racemes, pedicels short, pink to mauve. Oct.–Mar. Grassland, LB, SE (Garcia's Pass to Mpumalanga).

HARVEYA INK FLOWER, INKBLOM c. 40 spp., Africa, Mascarene Islands

A. Flowers red to orange

bodkinii Hiern GROOT ROOI-INKBLOM Parasitic perennial to 20 cm. Flowers in dense racemes, elongate-tapering, red with yellow throat, calyx lobed for half its length, stigma oblong. Dec.–Mar. Rocky sandstone slopes, NW (Cedarberg to Hex River Mts).•

bolusii Kuntze (incl. **H. hirtiflora** Schltr.) ROOI-INKBLOM Parasitic perennial to 25 cm. Flowers in dense racemes, narrowly funnel-shaped with short petals, red to orange with yellow throat, calyx lobed for about half its length, stigma linear-oblong. Oct.–Feb. NW, SW, LB, SE (Ceres and Cape Peninsula to Humansdorp).•

squamosa (Thunb.) Steud. JAKKALSKOSINKBLOM Parasitic perennial to 15 cm. Flowers sessile in long spikes, tubular with short petals, orange with yellow throat, calyx shortly lobed, stigma subglobose. Sept.–Dec. Deep sandy soils, mostly coastal, NW, SW (Lambert's Bay to Kleinmond).•

stenosiphon Hiern SKARLAKEN-INKBLOM Parasitic perennial to 60 cm. Flowers in short racemes on long stems, narrowly funnel-shaped with broad petals, compressed in throat, red to orange with yellow throat, calyx shortly lobed, stigma subglobose. Nov.–Feb. Rocky sandstone slopes, LB, SE (Swellendam to Humansdorp).•

A. Flowers white, yellow, pink or purple

capensis Hook. WIT-INKBLOM Parasitic perennial to 40 cm. Flowers in loose racemes, narrowly funnel-shaped with large petals, compressed in throat, white or tinged pink with yellow throat, calyx shortly lobed, stigma subglobose. Nov.–Feb. Rocky sandstone slopes and flats, NW, SW, AP, KM, LB, SE (Gifberg to Port Elizabeth).•

hyobanchoides Schltr. BONT-INKBLOM Parasitic perennial to 30 cm. Flowers sessile in dense spikes on thick stems, tubular with short reflexed petals, green with yellow tube and red bracts, calyx shortly lobed, stigma subglobose. July–Sept. Deep sand, SE (Humansdorp to E Cape).

purpurea (L.f.) Harv. (incl. **H. euryantha** Schltr., **H. laxiflora** Hiern) PERS-INKBLOM Parasitic perennial to 15 cm. Flowers in short or loose racemes, broadly funnel-shaped, white to pink with yellow blotches in throat, calyx deeply lobed, stigma subglobose. Sept.–Dec. Lower slopes and flats on sandy soils, NW, SW, KM, LB, SE (Cedarberg Mts to KwaZulu-Natal).

sulphurea Hiern GEEL-INKBLOM Parasitic perennial to 9 cm. Flowers crowded in short racemes, broadly funnel-shaped, yellow, calyx lobed for half its length, stigma subglobose. Sept.–Dec. Rocky sandstone slopes, NW (Pakhuis Mts).•

tubulosa Harv. ex Hiern JAKARANDA-INKBLOM Parasitic perennial to 60 cm. Flowers in long racemes, tubular with short petals, purple with white throat, upper petals projecting forward, calyx shortly lobed, stigma slender. Nov.–Mar. Rocky mountain slopes, SW, LB, SE (Cape Peninsula to E Cape).

[**Species excluded** No authentic material found and probably conspecific with one of the above: **H. pauciflora** (Benth.) Hiern]

HYOBANCHE RED BROOMRAPE, KATNAELS, WOLWEKOS   7 spp., southern Africa

atropurpurea Bolus   Root parasite with scale-like leaves, bracteoles acuminate. Flowers dark purple to blackish, hooded, fragrant, tube inflated, stamens included. Sept.–Oct. Rocky slopes, NW, SW (Cedarberg Mts to Cape Peninsula).•

glabrata Hiern   Root parasite with scale-like leaves. Flowers scarlet red, sparsely hairy, tube subcylindrical, stamens exserted. July–Oct. Sandy slopes and flats, NW, SW, AP, KM (Namaqualand to Bredasdorp and W Karoo).

rubra N.E.Br.   Like **H. sanguinea** but flowers dark red. July–Oct. Sandy slopes and flats, LB, SE (Gourits River to E Cape).

sanguinea L.   Root parasite with scale-like leaves, bracteoles obtuse. Flowers crimson-red or pink, hooded, densely hairy, tube subcylindrical, stamens included. Aug.–Oct. Sandy slopes and flats, NW, SW, AP, LB, SE (S Namibia to Swaziland).

MELASMA   WITCH'S FOXGLOVE   5 spp., Africa, America

scabrum P.J.Bergius   Shortly hairy hemiparasitic perennial to 70 cm, sparsely branched above. Leaves narrow and slightly toothed. Flowers on long pedicels in lax racemes, cream-coloured with purple throat. Nov.–Mar. Damp mountain slopes, SW, KM, LB, SE (Cape Peninsula to Mpumalanga).

***OROBANCHE**   BROOMRAPE   c. 150 spp., more or less cosmopolitan

***minor** Sm.   Yellowish achlorophyllous, glandular-woolly root parasite to 40 cm. Leaves scale-like, imbricate at swollen base of stem. Flowers in a spike, dull brownish mauve and yellow, calyx split above and below, lobes thread-like. Aug.–Nov. Parasitic on various hosts, SW, SE (European weed, Cape Peninsula, Port Elizabeth, also tropical Africa).

***ramosa** L.   BLOUDUIWEL   Yellowish achlorophyllous, thinly glandular-hairy root parasite to 30 cm, often branched. Leaves scale-like. Flowers in a spike, bluc-mauve, calyx acutely 4-lobed, subtended by 2 thread-like bracteoles. July–Nov. Parasitic mostly on Asteraceae, NW, SW, AP, KM (European weed, Namaqualand to Riversdale, also tropical Africa).

SOPUBIA   c. 60 spp., Old World tropics and subtropics

simplex (Hochst.) Hochst.   Glabrescent hemiparasitic perennial to 60 cm. Leaves linear, denticulate. Flowers in elongate racemes, pink to mauve. Mainly Nov.–Feb. Grassland and marshes, SE (Knysna to tropical Africa).

STRIGA   WITCHWEED   c. 40 spp., Old World tropics and subtropics

bilabiata (Thunb.) Kuntze   Slender hairy hemiparasitic perennial to 50 cm. Leaves linear. Flowers pink to mauve, calyx 5-ribbed. Nov.–May. Grassland, SE (Knysna to tropical Africa).

elegans Benth.   MIELIEGIF   Slender hairy hemiparasitic perennial. Leaves linear. Flowers pink to red, lower lip enlarged, calyx 10-ribbed. Dec.–Mar. Grassland, SE (Storms River to tropical Africa).

gesnerioides (Willd.) Vatke ex Engl.   Sparsely hairy hemiparasitic perennial, black on drying. Leaves scale-like. Flowers pink to mauve, calyx 5-ribbed. July–Oct. Dry grassland, SE (Uitenhage to tropical Africa and India).

OXALIDACEAE
by B. Bayer

OXALIS   SORREL, SURING   c. 500 spp., cosmopolitan, chiefly South Africa and S America

*A. Peduncle more than 1-flowered (see also **O. anomala**)*

bowiei Lindl.   Acaulescent geophyte. Leaves trifoliolate, leaflets suborbicular-obcordate, leathery. Flowers 3–12 per peduncle, pink with greenish tube. Mar.–May. 50–300 m, SE (Port Elizabeth to E Cape).

caprina L.   BOKSPOOTJIE, BOKSURING   Small weak geophyte, stem absent or very short. Leaves terminal, trifoliolate, bilobed to the middle. Flowers 2–4 per peduncle, pale lilac or white with greenish tube. Apr.–June. 50–200 m. SW, LB, SE (frequent weed of cultivation, Cape Peninsula to Uitenhage and E Cape).

compressa L.f.   Like **O. pes-caprae** but petioles flattened and peduncles 3–6-flowered. Flowers yellow. July–Sept. Widely distributed, 100–200 m, NW, SW (Kamiesberg to Caledon).

*****corniculata** L.   RANKSURING, TUINSURING   Low branching annual with prostrate leafy stem rooting at nodes. Leaves trifoliolate, leaflets cuneate-obcordate, ciliate and hairy beneath. Flowers 1–6 per peduncle, small, yellow. Mostly warmer months. Weed of cultivation, NW, SW, LB, SE (widespread cosmopolitan weed).

dichotoma T.M.Salter   Slender, often branched geophyte to 15 cm, bulb beaked. Leaves trifoliolate, leaflets bilobed almost to base. Flowers 2–9 per peduncle, lilac. Mar.–Apr. Arid areas, 400–600 m, KM (Oudtshoorn).•

lindaviana Schltr.   Glabrous, sometimes branched geophyte, 2–20 cm, bulb beaked. Leaves trifoliolate, leaflets bilobed to middle. Flowers 3 or 4 per peduncle, white with greenish tube. June. Rocky south slopes, 250–350 m, NW (Worcester, Robertson).•

livida Jacq. (= *Oxalis dentata* Jacq., *O. lateriflora* Jacq.)   STEENTJIESURING   Caulescent geophyte with stem to 20 cm, often branched, bulb not beaked, scales with brown hairs at apex. Leaves trifoliolate, usually glabrous, leaflets deeply bilobed, purple beneath. Flowers 2–6 per peduncle, rose or lilac with yellowish tube. Apr.–May. Rocky slopes in shade, 80–350 m, NW, SW (Clanwilliam to Bredasdorp).•

pes-caprae L.   GEELSURING   Acaulescent geophyte. Leaves usually basal, trifoliolate, leaflets cuneate-obcordate, usually glabrous above, pubescent beneath. Flowers 3–20 per peduncle, yellow. June–Oct. Widespread, 50–500 m, NW, SW, AP, KM, LB, SE (Namaqualand to E Cape).

stellata Eckl. & Zeyh.   Geophyte to 9 cm, bulb beaked. Leaves terminal, trifoliolate, leaflets deeply bilobed. Flowers 3–6 per peduncle, rosy or white with yellow tube. Apr.–June. Habitat? 50–400 m, NW, SW, LB, SE (Cape Peninsula to Ceres and Port Elizabeth).•

tragopoda T.M.Salter   Acaulescent geophyte to 20 cm, bulb shortly beaked with ridged tunics. Leaves trifoliolate, leaflets obcordate. Flowers 3–7 per peduncle, lilac with greenish tube. Dec.–May. Habitat? SE (Knysna to E Cape).

A. Peduncle 1-flowered
B. Leaves unifoliolate

dregei Sond.   Acaulescent aquatic geophyte. Leaves unifoliolate, bilobed and kidney-shaped. Flowers shallowly campanulate, white with yellow cup. May–Sept. Seasonal ponds and streams, 300–500 m, NW (Kamiesberg to Worcester).

monophylla L.   Glandular-hairy acaulescent geophyte, bulb with matted, fibrous tunics. Leaves unifoliolate, linear-obovate, glandular-ciliate. Flowers white or pale lilac with yellow tube. Apr.–May. Rocky slopes, 80–450 m, NW, SW (Clanwilliam to Cape Peninsula).•

nortieri T.M.Salter   Acaulescent geophyte. Adult leaves unifoliolate, leathery, elliptic-obovate, sometimes hairy. Flowers magenta with yellow tube. June. Habitat? 550–650 m, KM, SE (Oudtshoorn, Willowmore).•

simplex T.M.Salter   Acaulescent aquatic geophyte. Leaves unifoliolate, transversely oblong, with black marginal dots beneath. Flowers shallowly campanulate, white or pale rose. July–Aug. Marshes, 400 m, NW (Pakhuis Mts: Brandewyn River).•

*B.' Leaves usually 4- or more-foliolate (see also **O. polyphylla**)*

amblyodonta T.M.Salter   Caulescent geophyte with stem to 14 cm. Leaves terminal, leaflets 5–7, oblong, densely hairy beneath. Flowers lilac with yellow tube. May–June. Renosterveld, sandier soils, 250–500 m, NW, SW (Clanwilliam to Tulbagh).•

burtoniae T.M.Salter   Caulescent geophyte with stem to 10 cm. Leaves terminal, leaflets 5–7, linear with apical calli. Flowers yellow. June. Granite and limestone outcrops, 100–200 m, SW (Paternoster to Saldanha Bay).•

engleriana Schltr. (= *Oxalis henrici* F.Bolus)   Caulescent geophyte with wiry, brown stem to 15 cm. Leaves crowded apically, leaflets 5–8, linear-oblong, ciliate and hairy beneath. Flowers rose with yellow tube. May–June. Shady southern slopes, 150–300 m, NW, SW (Hex River valley to Caledon).•

flava L.   BOBBEJAANSURING, VINGERSURING   Acaulescent geophyte. Leaves trifoliolate, leathery, glaucous, petioles articulated and with large brown stipules, leaflets 2–12, linear to obovate, conduplicate. Flowers yellow, white or lilac. May–June. Flats and lower slopes, 50–600 m, NW, SW, LB (Namaqualand to Riversdale).

tomentosa L.f. VINGERSURING   Densely silky hairy acaulescent geophyte. Leaves trifoliolate, silky hairy, leaflets 10–20, oblong-cuneate. Flowers white with yellow tube. Apr.–June. Grassy flats and slopes, 75–150 m, NW, SW (Clanwilliam to Cape Peninsula).•

variifolia Steud.   Like **O. amblyodonta** but leaflets 3–5, linear. Flowers white or lilac. May–June. Lower slopes, 100–250 m, NW (Clanwilliam to Piketberg).•

zeyheri Sond.   Shortly caulescent geophyte with stem to 3 cm, bulb tunic vertically grooved. Leaves terminal, leaflets 7–15, linear-cuneate, thinly glandular-hairy beneath. Flowers rose to lilac with yellow tube. Apr.–May. Arid flats and slopes, 300–650 m, NW, KM, LB (Ceres to Montagu).•

B". Leaves trifoliolate (see also **O. flava**, **O. variifolia**)
C. Peduncular bracts at an articulation
D. Leaves bifurcate to middle or below

bifida Thunb.   Caulescent geophyte with branched stem to 30 cm. Leaves trifoliolate, leaflets bilobed to middle. Flowers rose with greenish tube, peduncle articulated. Mar.–July. Habitat? 150–200 m, NW, SW (Ceres to Cape Peninsula and Caledon).•

heterophylla DC.   Caulescent geophyte with stem to 30 cm, often branching, bulb with hairy scales. Leaves trifoliolate, often subsessile, leaflets bilobed to middle, often silky beneath. Flowers red to purple with yellow tube, peduncle articulated. Aug.–Oct. South slopes and mountains, 150–400 m, NW, SW, KM, LB (Namaqualand to Riversdale).

orthopoda T.M.Salter   Caulescent geophyte with leafy stem to 15 cm, bulb beaked. Leaves trifoliolate, leaflets bilobed almost to base. Flowers rosy purple or white with yellowish tube, peduncle articulated. June. Lower slopes, 100–250 m, LB, AP (Swellendam to Mossel Bay).•

D. Leaves not bifurcate

duriuscula Schltr.   Caulescent geophyte with slender stem to 19 cm, bulb pointed. Leaves terminal, trifoliolate, leaflets linear-conduplicate. Flowers rosy purple with yellow tube, peduncle articulated. May. Sandy flats, 100–500 m, SW (Caledon to Bredasdorp).•

imbricata Eckl. & Zeyh.   Acaulescent geophyte with shallow, contorted bulb. Leaves trifoliolate, numerous, leaflets obcordate, densely pubescent. Flowers white with greenish tube, peduncle articulated. Apr.–June. Habitat? 50–300 m, LB, SE (Swellendam to E Cape).

incarnata L.   Caulescent geophyte with slender, branched stem, 10–30 cm. Leaves trifoliolate, leaflets obcordate. Flowers white or pale lilac with greenish tube, peduncle articulated. Jan.–Apr. Habitat? 50–300 m, SW, LB, SE (Cape Peninsula to Uitenhage).•

ioeides T.M.Salter & Exell   Small acaulescent geophyte. Leaves trifoliolate, leaflets cuneate-obcordate, purplish beneath with reddish brown marginal dots. Flowers rose-red, peduncle articulated. May–June. Upper slopes, 600 m, KM, LB (Outeniqua Mts: Robinson Pass and Kammanassie Mts).•

lanata L.f.   Caulescent geophyte with stem to 10 cm. Leaves terminal, trifoliolate, leaflets obcordate, densely hairy. Flowers white or pink with yellow tube, peduncle articulated. May–Oct. Slopes and screes, 50–300 m, SW (Cape Peninsula to Caledon).•

luteola Jacq.   Acaulescent geophyte, bulbs large and gummy. Leaves trifoliolate, leaflets cuneate-rotund. glabrous or hairy, conspicuously veined and often purple beneath. Flowers yellow, peduncle articulated. May–June. Flats and lower slopes, 100–300 m, NW, SW, KM, LB (Bokkeveld Mts to Albertinia).•

obtusa Jacq.   GEELOOGSURING   Acaulescent geophyte with deeply pitted bulb. Leaves trifoliolate, leaflets cuneate-obcordate, hairy. Flowers pink, brick-red, yellow or white, with darker veins and yellow tube, peduncle articulated. June–Oct. Mostly clay and granite, 50–600 m, NW, SW, KM, LB, SE (Namaqualand and W Karoo to Port Elizabeth).

orbicularis T.M.Salter   Acaulescent geophyte. Leaves large, trifoliolate, leaflets suborbicular, veined and often purple beneath. Flowers pale lilac with yellow tube, peduncle articulated. June. Habitat? 200–600 m, KM, LB (Caledon to Ladismith and Swellendam).•

pendulifolia T.M.Salter   Caulescent geophyte to 30 cm. Leaves terminal, trifoliolate, leaflets narrowly oblong-conduplicate, pendulous, sparsely hairy beneath, punctate. Flowers brick-red with greenish tube, viscid, peduncle articulated. Mar.–Apr. Lower slopes, grassy areas, 100–200 m, LB, SE (Swellendam to Plettenberg Bay).•

psilopoda Turcz.   Acaulescent geophyte. Leaves trifoliolate, leaflets obcordate, densely silky beneath. Flowers white or lilac with yellow tube, peduncle articulated. Mar.–June. Habitat? 150–300 m, LB, SE (Riversdale to Port Elizabeth).•

strigosa T.M.Salter   Acaulescent geophyte, bulb with long curved beak. Leaves trifoliolate, leaflets obcordate, roughly hairy. Flowers rose-red with purple margins and yellowish streaked tube, peduncle articulated. Apr.–May. Lower slopes, 250–300 m, SW (Tygerberg, Kanonkop, Hercules Pillar).•

truncatula Jacq.   Robust acaulescent geophyte. Leaves trifoliolate, leaflets cuneate-obovate, dark green, leathery, purple and densely silky beneath. Flowers lilac with yellow tube, peduncle articulated. Apr.–June. Lower and upper slopes, 75–250 m, SW (Paarl to Bredasdorp).•

zeekoevleyensis R.Knuth   Acaulescent geophyte. Leaves trifoliolate, leaflets obcordate, sparsely hairy beneath. Flowers rosy lilac with yellow tube, peduncle articulated. June–Aug. Flats, 150–250 m, SW, LB (Caledon to Riversdale).•

C. Peduncular bracts not at an articulation or lacking
E. Plants caulescent with leafy stems; lower leaves at least subsessile or sessile; peduncles often cauline

giftbergensis T.M.Salter   Caulescent geophyte with slender, leafy stem to 20 cm. Leaves trifoliolate, subsessile with scale-like petioles, leaflets linear-cuneate, glandular-ciliate, hairy beneath. Flowers pink, violet or white with yellowish tube. May–June. Upper mountain slopes, 500 m, NW (Gifberg to Clanwilliam).•

hirta L.   Softly hairy, caulescent geophyte with leafy stem, often branching, 5–30 cm, bulb scales unprotected. Leaves trifoliolate, grey-green, subsessile with scale-like petioles, leaflets linear-obovate, hairy beneath. Flowers mauve, magenta or white with sometimes elongate yellow tube. Apr.–June. Flats and lower slopes, 50–300 m, NW, SW (Bokkeveld Mts to Cape Peninsula).•

leipoldtii Schltr.   Like **O. viscosa** but pedicels slender and ebracteate and flowers purplish. June. Mountain slopes, 350 m, NW (Clanwilliam district).•

macra Schltr.   Like **O. hirta** but leaflets linear. Flowers white or pink. June. 500 m, NW, SW (Piekeniers Kloof to Piketberg).•

meisneri Sond.   Caulescent geophyte with leafy stem to 16 cm. Leaves trifoliolate, subsessile with scale-like petioles, leaflets linear-conduplicate, pubescent. Flowers yellow or lilac. Apr.–June. South slopes, 300–350 m, NW (Tulbagh to Robertson).•

multicaulis Eckl. & Zeyh.   Caulescent geophyte with leafy stem to 10 cm. Leaves trifoliolate, lower subsessile with scale-like petioles, leaflets linear-cuneate, silky hairy beneath. Flowers white or pink with yellow tube and dark reddish margins. May–Aug. Seasonally flooded areas, 150–300 m, NW, SW (Cape Peninsula to Bredasdorp).•

pardalis Sond. (= *Oxalis camelopardalis* T.M.Salter, *O. capillacea* E.Mey. ex Sond., *O. confertifolia* (Kuntze) R.Knuth, *O. grammophylla* T.M.Salter, *O. heidelbergensis* T.M.Salter, *O. leptogramma* T.M.Salter, *O. lineolata* T.M.Salter, *O. massoniana* T.M.Salter, *O. robinsonii* T.M.Salter & Exell)   Caulescent geophyte to 30 cm, bulb scales slender, many, with retrorse brown hairs. Leaves trifoliolate, linear-elliptic, glabrous or hairy, black-linear-punctate. Flowers orange, red-purple, pink, yellow, white or cream-coloured, with yellow tube. May–June. Usually in heavier soils, 150–600 m, NW, SW, AP, KM, LB (Bokkeveld Mts and W Karoo to Mossel Bay).

porphyriosiphon T.M.Salter   Glandular-hairy, caulescent geophyte to 11 cm. Leaves trifoliolate, lower sessile, leaflets cuneate-conduplicate, sparsely glandular-ciliate and hairy beneath. Flowers red to purple with white ring and dark purple tube, viscid. May–July. Damp spots, rocky sandstone slopes, c. 500 m, NW (Gifberg).•

pseudo-hirta T.M.Salter   Like **O. hirta** but slender with flexuose stem, dark, firm tunics and upper leaves shortly petiolate. Flowers lilac with yellow tube. May–June. Upper slopes, 600 m, SW (only Stettyn, Worcester).•

recticaulis Sond.   Caulescent geophyte with sparsely leafy stem to 14 cm. Leaves trifoliolate, leaflets linear-cuneate, glandular-ciliate, often hairy beneath. Flowers white or pale rose with yellow tube. June–Oct. Usually damp places, 150–300 m, NW, SW (Clanwilliam to Piketberg).•

subsessilis L.Bolus   Caulescent geophyte with polished brown, leafy stem to 25 cm. Leaves trifoliolate, subsessile with scale-like petioles, leaflets linear-cuneate, hairy beneath. Flowers pale lilac with yellow tube and dark margins. Flowering time? Habitat? 50 m, SW (Saldanha Bay).•

tenuifolia Jacq.   Hairy caulescent geophyte with leafy stem to 24 cm. Leaves trifoliolate, lower subsessile with scale-like petioles, fasciculate, leaflets linear-conduplicate, hairy beneath. Flowers white with yellow tube and purple margins. May–Aug. Slopes and flats, 50–400 m, SW (Paarl to Cape Peninsula).•

viscosa E.Mey. ex Sond.   Glandular-hairy, caulescent geophyte with leafy stem to 17 cm. Leaves, trifoliolate, lower subsessile with scale-like petioles, leaflets linear-cuneate, hairy beneath. Flowers white with yellow tube. May–July. Shady rocky slopes, 200–250 m, NW (Clanwilliam to Tulbagh).•

E. Plants acaulescent or caulescent but then leaves apically congested and distinctly petiolate; peduncles terminal
F. Leaflets linear-conduplicate

argyrophylla T.M.Salter   Grey-silky caulescent geophyte to 20 cm. Leaves terminal, trifoliolate, leaflets linear-cuneate, conduplicate, falcate, silky. Flowers white or lilac with yellow cup and darker veins beneath. May–July. 100–200 m, SW (Malmesbury to Cape Peninsula).•

burkei Sond.   Caulescent geophyte with wiry stem to 10 cm. Leaves terminal, trifoliolate, leaflets linear-conduplicate, thinly hairy beneath. Flowers rose, lilac or white with yellow tube. May–June. Habitat? 300–600 m, NW, KM (Ceres to Montagu and S Karoo).

falcatula T.M.Salter   Caulescent geophyte to 20 cm. Leaves terminal, trifoliolate, leaflets linear-falcate, thinly silky beneath. Flowers pale pink with yellow tube. Apr.–June. Lower slopes, 75–200 m, SW (Cape Peninsula to Paarl).•

fragilis T.M.Salter   Caulescent geophyte with slender stem to 10 cm. Leaves terminal, trifoliolate, leaflets linear-conduplicate. Flowers straw-coloured with yellow tube. May–July. Habitat? 170 m, SW (Moorreesburg).•

gracilipes Schltr.   Like **O. pallens** but bulb larger, 15–20 mm diam., pitted. Flowers white with yellow tube. May–June. Habitat? 100–200 m, NW (Clanwilliam, Piketberg).•

gracilis Jacq.   Caulescent geophyte with branched stem to 30 cm. Leaves mostly terminal, trifoliolate, leaflets linear-conduplicate, finely hairy beneath. Flowers apricot-pink or white with yellow tube. May–June. Sandy flats and slopes, 50–150 m, NW (Namaqualand to Clanwilliam).

involuta T.M.Salter   Shortly caulescent geophyte. Leaves terminal, trifoliolate, leaflets linear-involute with flared tips. Flowers bright yellow. May–July. Habitat? 175 m, NW (Porterville).•

levis T.M.Salter   Dwarf geophyte with stem to 3 cm. Leaves terminal, trifoliolate, leaflets linear-conduplicate, petioles dark brown. Flowers white or lilac with yellow tube. June. Clay soils, 120–160 m, SW (Malmesbury to Mamre).•

oligophylla T.M.Salter   Acaulescent geophyte. Leaves 1 or 2, trifoliolate, leaflets linear-conduplicate. Flowers on elongate peduncles, white with yellow tube, with minute bracts at calyx. May–June. Rocky sandstone slopes, c. 500 m, NW (Gifberg).•

pallens Eckl. & Zeyh.   Caulescent geophyte with slender stem to 20 cm, bulb smooth, shallow. Leaves mostly terminal, trifoliolate, leaflets linear-conduplicate, minutely hairy beneath. Flowers white with yellow tube. May–June. Flats and slopes, 150–200 m, NW (Clanwilliam to Piketberg).•

perineson T.M.Salter & Exell   Caulescent geophyte to 15 cm. Leaves terminal, trifoliolate, leaflets linear-conduplicate, with black marginal calli. Flowers mauve or white with yellow tube and dark margins. May–June. 175 m, SW (Moorreesburg).•

phloxidiflora Schltr.   Shortly caulescent geophyte to 10 cm. Leaves terminal, trifoliolate, leaflets linear-spathulate, sometimes thinly hairy beneath. Flowers purple with dull purple ring and yellow tube. May–Aug. Upper slopes, 650 m, NW (Clanwilliam).•

polyphylla Jacq.   VINGERSURING   Caulescent geophyte to 20 cm, bulb often gummy. Leaves terminal, 3(–7)-foliolate, leaflets linear-conduplicate, sometimes thinly hairy beneath. Flowers rose, lilac or white with yellow tube and often darker margins. Mar.–June. Usually on flats, light or heavy soils, 100–300 m, SW, LB, SE (Malmesbury to Port Elizabeth).•

stenopetala T.M.Salter   Caulescent geophyte with slender stem to 14 cm. Leaves terminal, trifoliolate, leaflets linear-conduplicate, often black-punctate, sometimes hairy beneath. Flowers white with yellowish tube. May–July. Habitat? 600 m, NW (Bokkeveld Mts to Clanwilliam).•

tenuipes T.M.Salter   Caulescent geophyte with slender stem to 45 cm. Leaves terminal, trifoliolate, leaflets linear-conduplicate. Flowers rosy lilac with dark purple ring and yellow tube. June. Shady slopes, 200–500 m, NW (Gifberg to Citrusdal).•

tenuis T.M.Salter   Glandular-hairy caulescent geophyte to 10 cm, rarely branching. Leaves terminal, trifoliolate, leaflets linear-conduplicate, glabrescent beneath. Flowers white with yellow tube. May–June. Shady slopes, 500 m, NW (Gifberg).•

versicolor L.   CANDYCANE SORREL   Caulescent geophyte with partly leafy stem to 20 cm, sometimes branching. Leaves mostly terminal, linear-conduplicate, often with marginal calli. Flowers white with yellow tube and reddish purple margins. May–Nov. Flats and slopes, 50–250 m, NW, SW (Clanwilliam to Hermanus).•

xantha T.M.Salter   Caulescent geophyte with slender stem to 15 cm. Leaves terminal, trifoliolate, leaflets linear-conduplicate, thinly hairy beneath. Flowers bright yellow. July. Clay soils on flats, 400 m, NW (north Pakhuis Mts).•

F. Leaflets oblong to obcordate
G. Leaflets fleshy, often small, epidermal cells large

algoensis Eckl. & Zeyh.   Acaulescent geophyte to 12 cm. Leaves trifoliolate, fleshy, leaflets oblanceolate-subrotund, ciliate and hairy beneath, with large epidermal cells. Flowers lilac or white with yellow tube. June–Sept. Habitat? KM, SE (Ladismith to Port Elizabeth).•

annae F.Bolus   Acaulescent geophyte to 6 cm. Leaves trifoliolate, leaflets rotund-obcordate, glabrous or hairy. Flowers yellow, copper-pink or white with yellow tube. June–Aug. Arid areas, 400–600 m, NW, KM (Namaqualand and W Karoo to Swartberg Mts).

attaquana T.M.Salter   Caulescent geophyte to 13 cm. Leaves congested apically, trifoliolate, fleshy, leaflets obreniform, sparsely ciliate, with large epidermal cells. Flowers rose with yellow tube. June. Habitat? 600 m, LB (Attaquas Mts).•

convexula Jacq.   Usually caulescent geophyte to 14 cm, with shallow bulb. Leaves trifoliolate, small, fleshy, forming an umbrella-like rosette, leaflets subrotund, with large epidermal cells. Flowers salmon-pink with yellow tube. June–Sept. Shale slopes, 150–350 m, NW, AP, KM, LB (Ceres to Riversdale).•

depressa Eckl. & Zeyh.   Acaulescent geophyte with shallow bulb. Leaves trifoliolate, succulent, leaflets cuneate-suborbicular, with large epidermal cells. Flowers white, lilac or pink with yellow tube. Mar.–Apr. Habitat? NW, SW, AP, KM, LB, SE (Cold Bokkeveld and Karoo to Zimbabwe).

dilatata L.Bolus   Acaulescent geophyte with shallow bulb. Leaves trifoliolate, succulent, leaflets cuneate-rotund, with large epidermal cells. Flowers yellow or reddish, sepals swollen at base. Apr.–May. Habitat? 200–400 m, NW, SW, KM, LB, SE (Ceres to Port Elizabeth).•

fergusoniae T.M.Salter   Small acaulescent geophyte with shallow bulb. Leaves trifoliolate, leaflets rotund, glandular-ciliate and with marginal black calli, with large epidermal cells. Flowers rose or white with yellow tube. Sept.–Oct. Dry flats and slopes, 300–600 m, NW, KM (Cold Bokkeveld Mts to Little Karoo).•

fourcadei T.M.Salter   Acaulescent geophyte with shallow bulb. Leaves trifoliolate, leaflets linear-oblong, ciliate and hairy, orange-punctate on margins. Flowers white with yellow tube. Oct. Dry upper slopes, KM, SE (Kammanassie to Humansdorp).•

oreithala T.M.Salter   Dwarf acaulescent geophyte. Leaves trifoliolate, fleshy, leaflets rotund, purple beneath, with large epidermal cells. Flowers white with yellow tube. July. Rocky sandstone slopes, c. 800 m, NW (Gifberg).•

pocockiae L.Bolus   Like **O. depressa** but often short-stemmed and bulb 4-angled. May–June. Varied habitats, 350–600 m, NW, SW, KM (Vanrhynsdorp to Cape Peninsula and W. Karoo).

pulchella Jacq.   Acaulescent geophyte, bulb, large, gummy, shallow. Leaves trifoliolate, fleshy, leaflets suborbicular, hairy beneath, with large epidermal cells. Flowers salmon or rose. May–June. Sandy lower slopes and flats, 250–450 m, NW, SW, LB (S Namibia to Swellendam).

punctata L.f.   Dwarf, acaulescent geophyte, bulb small, sharply angled and pitted. Leaves trifoliolate, fleshy, leaflets cuneate-rotund, with large epidermal cells. Flowers white or lilac with yellowish tube. Apr.–June. Widespread, 150–300 m, SW, KM, LB, SE (Cape Peninsula to E Cape).

G. Leaflets not as above

adspersa Eckl. & Zeyh.   Like **O. pillansiana** but flowers pale lilac, red or occasionally white with yellow tube. May–June. Renosterveld, clay flats and slopes, 200–400 m, NW, SW (Clanwilliam to Stellenbosch).•

amblyosepala Schltr.   Shortly caulescent geophyte with stem to 5 cm. Leaves terminal, trifoliolate, leaflets cuneate-obcordate, glandular-hairy beneath. Flowers white with yellow tube. Aug.–Oct. Renosterveld, 300 m, NW (Clanwilliam).•

anomala T.M.Salter   Acaulescent geophyte, 12–18 cm, bulb beaked. Leaves trifoliolate, leaflets obcordate, sparsely hairy and purple beneath. Flowers 1 or 2 per peduncle, lilac with greenish tube. May–June. Habitat? 500–600 m, KM, LB (Ladismith to Swellendam).•

argillacea F.Bolus   Like **O. pillansiana** but shorter, to 9 cm and outer filaments obtusely toothed at base. May–July. Renosterveld, hard gravelly soils, 200–500 m, NW (S Namaqualand to Clanwilliam).

aridicola T.M.Salter   Caulescent geophyte to 13 cm, thinly glandular-hairy. Leaves terminal, trifoliolate, leaflets cuneate-obovate, ciliate, sometimes hairy beneath. Flowers lilac with elongate, yellow tube. May–July. Habitat? 400–500 m, NW (Pakhuis Mts).•

aurea Schltr.   Acaulescent geophyte. Leaves trifoliolate, leaflets cuneate-rotund, often hairy beneath. Flowers golden yellow. May. Clay slopes, 200–300 m, NW (Clanwilliam to Citrusdal).•

callosa R.Knuth   Acaulescent or shortly caulescent geophyte to 10 cm. Leaves terminal, trifoliolate, leaflets elliptic-conduplicate, hairy beneath. Flowers rose-red with purple ring and yellow tube. May–June. Gravelly soils, 500 m, NW (Bokkeveld Mts and W Karoo).•

ciliaris Jacq.   Caulescent geophyte to 20 cm. Leaves mostly terminal, trifoliolate, leaflets linear-elliptic, ciliate, obscurely pustulate. Flowers pink, rose, lilac or white with yellow tube. Apr.–July. Habitat? 300–600 m, NW, SW, KM, LB, SE (Ceres to Alexandria).

commutata Sond.   Small acaulescent geophyte. Leaves trifoliolate, leaflets rotund, with large epidermal cells. Flowers rose, lilac or white with yellow tube. Apr.–June. Habitat? 50–300 m, NW, SW (Cold Bokkeveld to Cape Peninsula and Caledon).•

comptonii T.M.Salter   Caulescent geophyte with slender stem to 12 cm. Leaves terminal, trifoliolate, leaflets linear-elliptic, glabrescent beneath. Flowers white or pale rose with greenish yellow tube. May–June. Habitat? NW (Gifberg).•

disticha Jacq. (incl. **O. dines** Ornduff)   Robust glabrescent aquatic with branching, leafy stems to 40 cm. Leaves trifoliolate, floating, with large, papery stipules, leaflets obcordate. Flowers shallowly campanulate, pale yellow or white with yellow cup, scented. June–Aug. Seasonal ponds, 180–400 m, NW, SW (Bokkeveld Mts to Mamre).•

droseroides E.Mey. ex Sond.   Glandular-hairy caulescent geophyte to 15 cm. Leaves crowded terminally, trifoliolate, leaflets oblong-ovate, glandular ciliate. Flowers deep rose. May. Clay slopes, 300–500 m, NW, SW (Tulbagh to Paarl and Worcester).•

ebracteata Savign.   Glandular-hairy caulescent geophyte with slender stem to 7 cm. Leaves mostly terminal, trifoliolate, leaflets cuneate-obcordate, hairy beneath. Flowers pink or white with yellow tube. Apr.–June. Shaded places, 300–500 m, NW (Clanwilliam to Tulbagh).•

eckloniana C.Presl   Acaulescent geophyte with shallow bulb. Leaves trifoliolate, leaflets elliptic-obovate, ciliate, often purple beneath. Flowers yellow, white or pink to orange with yellow tube, anthers spreading, arrow-shaped. May–July. Usually damp situations, 50–200 m, NW, SW (Clanwilliam to Caledon).•

fibrosa F.Bolus   Acaulescent, glandular-hairy geophyte. Leaves trifoliolate, leaflets cuneate-obovate, glandular-hairy. Flowers lilac or mauve with yellow tube, anthers spreading, arrow-shaped. Apr.–June. Arid areas, 400–600 m, KM, LB (Montagu to Ladismith and S Karoo).

glabra Thunb.   TAPYTSURING   Caulescent geophyte with stem partly leafy, 5–20 cm, conspicuously stoloniferous and forming carpets. Leaves mostly terminal, trifoliolate, leaflets linear-cuneate. Flowers with long, narrow sepals, red, pink or white with yellow tube. Apr.–Nov. Sandy flats and slopes, 50–300 m, NW, SW, KM (Piketberg to Montagu).•

goniorrhiza Eckl. & Zeyh. (= *Oxalis callimarginata* Weintroub, *O. urbaniana* Schltr.)   Caulescent geophyte with slender stem to 20 cm, partly leafy below, bulb small with 3–5 vertical ridges. Leaves trifoliolate, lower sessile, leaflets linear-cuneate. Flowers white or rose with yellow tube. May–July. Marshy flats, 50–300 m, SW (Tulbagh to Caledon).•

laxiuscula R.Knuth   Lax, acaulescent geophyte. Leaves trifoliolate, pilose. Flowers white. Sept. Habitat? 140 m, NW (Clanwilliam district).•

leptocalyx Sond.   Like **O. glabra** but bulb larger, 15–25 mm diam., without stolons. Flowers rosy, sometimes white with yellow tube and darker margin. May–June. Flats and slopes, 130 m, NW, SW (Piketberg to Malmesbury).•

melanosticta Sond.   Acaulescent geophyte. Leaves trifoliolate, leaflets cuneate-obcordate, ciliate, black-dotted when dry. Flowers yellow. May–Aug. Dry mountain slopes, 500–600 m, NW, KM (Bokkeveld Escarpment and W Karoo to Montagu).

microdonta T.M.Salter   Acaulescent geophyte. Leaves trifoliolate, leaflets linear-elliptic with bristly margins. Flowers lilac with yellow tube, anthers spreading, arrow-shaped. May. Habitat? 350 m, KM (Montagu).•

minuta Thunb. Acaulescent geophyte. Leaves trifoliolate, leaflets elliptic-obovate, sometimes with long marginal hairs. Flowers white with yellow tube, anthers spreading, arrow-shaped. May–July. Usually in damp places, 50–200 m, NW, SW (Porterville Mts to Cape Peninsula).•

natans L.f. WATERKLAWER, WATERSURING Aquatic geophyte with slender, flexuose, branching stems. Leaves tufted and terminal, trifoliolate. Flowers shallowly campanulate, white with yellow tube. Sept.–Nov. Seasonal pools, 50–200 m, NW, SW (Piketberg to Cape Peninsula and Worcester).•

nidulans Eckl. & Zeyh. Acaulescent geophyte. Leaves trifoliolate, leaflets suborbicular-emarginate with widely spaced marginal hairs. Flowers white or lilac with yellow tube, anthers spreading, arrow-shaped. July–Aug. Habitat? 150–450 m, NW, SW (Clanwilliam to Caledon).•

oculifera E.G.H.Oliv. Acaulescent geophyte. Leaves trifoliolate, glaucous, leaflets oblong, obliquely peltate, petioles and peduncles glandular-hairy. Flowers violet-pink, with a white centre and dark eye. May–June. Seeps and rock flushes on sandstone pavement, c. 600 m, NW (Gifberg).•

oreophila T.M.Salter Shortly silky hairy caulescent geophyte with slender stem to 16 cm. Leaves terminal, trifoliolate, leaflets linear-conduplicate, red-punctate above, silky hairy beneath. Flowers white with yellow tube and purple margins. May–June. Rocky slopes in fynbos, c. 800 m, NW (Pakhuis Mts).•

petiolulata F.Bolus Small acaulescent geophyte. Leaves trifoliolate, leaflets broadly ovate, with dark marginal calli. Flowers deep pink with purple ring and yellow tube. Mainly Sept.–Oct. Rocky upper slopes, 1000–1500 m, NW (Cedarberg Mts).•

pillansiana T.M.Salter & Exell Caulescent geophyte to 16 cm, with soft, multicellular hairs. Leaves terminal, trifoliolate, leaflets linear-cuneate, thinly hairy beneath. Flowers yellow with purple margins. May–June. Habitat? 350 m, NW (Clanwilliam).•

purpurea L. Acaulescent geophyte, bulb large, gummy. Leaves trifoliolate, leaflets transversely obovate, ciliate, hairy and purple beneath, black-streaked when dry. Flowers purple, yellow or white with yellow tube. May–Sept. Flats and slopes, 50–500 m, NW, SW, KM, LB, SE (Namaqualand and W Karoo to Port Elizabeth).

pusilla Jacq. Like **O. glabra** but bulbils sessile, leaves terminal and sepals more obtuse. Flowers white or pale rose with yellowish tube, sometimes with purple ring. May–July. Flats and dampish places, 50–200 m, SW (Hopefield to Cape Peninsula).•

rubropunctata T.M.Salter Acaulescent geophyte. Leaves trifoliolate, leaflets transversely obovate, dark green above, purple beneath, red-dotted. Flowers white with yellow tube. May. Flats, NW (Bokkeveld Mts).•

smithiana Eckl. & Zeyh. KLAWERSURING, ROOISURING Acaulescent geophyte. Leaves trifoliolate, leaflets bilobed to near base. Flowers lilac or white with greenish tube, peduncle articulated. Jan.–Aug. Forests and bush, SE (George to Mpumalanga).

sonderiana (Kuntze) T.M.Salter Acaulescent geophyte. Leaves trifoliolate, prostrate, leaflets suborbicular, glaucous. Flowers yellow. May–June. Rocky sandstone slopes, 150 m, NW (Namaqualand to Graafwater).

stictocheila T.M.Salter Caulescent geophyte with slender stem to 12 cm. Leaves terminal, trifoliolate, leaflets oblanceolate-conduplicate, with orange marginal calli. Flowers white with yellow tube and dark margins. June–July. Habitat? 150 m, SW (Darling to Yzerfontein).•

stokoei Weintroub Dwarf acaulescent geophyte. Leaves trifoliolate, leaflets subrotund to linear, with reddish brown pubescent petioles, hairy beneath, brownish punctate. Flowers rosy purple with yellow tube and often purple eye. Apr.–June. Upper mountain slopes, 700–1000 m, NW, KM (Brandwacht Mts, Worcester).•

suavis R.Knuth Usually acaulescent geophyte to 8 cm. Leaves trifoliolate, leaflets cuneate-obcordate, glandular-ciliate, reddish brown-punctate. Flowers yellow or white with yellow tube. May–June. Lower slopes and flats, 150 m, SW (only Hopefield).•

suteroides T.M.Salter Glandular-hairy caulescent geophyte with rigid, branched stem to 25 cm, bulb long, tortuous. Leaves terminal on branches, trifoliolate, leaflets oblong-cuneate, glandular-ciliate, shortly hairy beneath. Flowers pale lilac with yellow tube streaked with purple. May–June. 600 m, NW (Bokkeveld Mts).•

tenella Jacq. Slender caulescent geophyte to 10 cm. Leaves terminal, trifoliolate, leaflets cuneate-obovate, shortly ciliate and hairy beneath. Flowers sometimes sessile, yellow, white or pale lilac with yellow tube. May–July. Clay flats, 250 m, NW (Vanrhynsdorp to Piketberg).•

uliginosa Schltr.   Acaulescent geophyte. Leaves trifoliolate, leaflets cuneate-obcordate, ciliate, reddish punctate. Flowers widely campanulate, yellow. June–Sept. Marshy places, 400 m, NW (Clanwilliam).•

viscidula Schltr.   Acaulescent geophyte, bulb tunicated. Leaves trifoliolate, leaflets obcordate, glandular-hairy. Flowers red-purple with yellow tube. Flowering time? Often associated with *Euphorbia mauritanica* patches, 250–350 m, KM, LB (Worcester-Robertson Karoo).•

PAPAVERACEAE

PAPAVER   POPPY   c. 100 spp., mainly N temperate

aculeatum Thunb.   WILD POPPY, WILDEPAPAWER   Sharply prickly tufted annual to 1 m. Leaves pinnatifid, toothed and prickly. Flowers pedunculate, orange. Oct.–Dec. Disturbed ground, SE (George to Port Elizabeth and scattered throughout southern Africa).

PENAEACEAE•

1. Style prominently 4-winged ....**Penaea**
1.' Style terete or 4–8-angled or -ridged:
   2. Anther thecae usually only half as long as connective body; flowers usually pale yellow ....**Stylapterus**
   2.' Anther thecae more than half as long as connective body; flowers variously coloured:
      3. Bracts and bracteoles partly denticulate to fimbriate, generally with an apical gland, outer bracts broader than leaves, both broadly obovate or subrotund, truncate and closely imbricate:
         4. Anthers subsessile, more or less included; ovules 2 in each locule ....**Sonderothamnus**
         4.' Anthers on prominent, flattened filaments, exserted; ovules 4 in each locule ....**Saltera**
      3.' Bracts and bracteoles entire and without an apical gland, often narrow and caducous:
         5. Ovules 2 in each locule; flowers smaller, tube to 15 mm long ....**Brachysiphon**
         5.' Ovules 4 in each locule, 2 ascending and 2 pendulous; flowers larger, tube 20–30 mm long:
            6. Bracteoles of each flower 4 or 6, often caducous; flowers yellow to red ....**Endonema**
            6.' Bracteoles of each flower 2; flowers white and red ....**Glischrocolla**

BRACHYSIPHON•   5 spp., W Cape

acutus (Thunb.) A.Juss.   Like **B. fucatus** but perianth tube 3–4 times as long as wide and flowers pale pink with reddish tube. Sept.–Dec. Rocky sandstone slopes, SW (Caledon Swartberg to Bredasdorp).•

fucatus (L.) Gilg   SISSIES   Compact, rounded shrub to 1 m. Leaves sessile, ascending, ovate-elliptic to orbicular, imbricate. Flowers crowded in upper axils, pink to reddish with darker tube up to twice as long as wide. Mainly May–Aug. Cool, rocky sandstone slopes, SW (Cape Peninsula).•

microphyllus Rourke   Compact, twiggy shrublet to 50 cm. Leaves ascending, acicular, subterete. Flowers in 2–4-flowered terminal cymes, pink. Aug.–Nov. Sandstone rocks, KM (Touwsberg and Klein Swartberg).•

mundii Sond.   Like **B. rupestris** but leaves smaller, shorter than 7 mm, ovate and broadest about the middle, flowers yellow fading red, with persistent, reddish bracts. July–Aug. Limestone rocks and cliffs, AP (De Hoop).•

rupestris Sond.   Compact rounded shrublet mostly to 20 cm. Leaves subsessile, ascending, closely set, obovate. Flowers few to several in terminal clusters, pink to red, with caducous bracts. Sept.–Oct. Sandstone rocks, SW (Kleinrivier Mts).•

ENDONEMA•   2 spp., W Cape

lateriflora (L.f.) Gilg   Densely leafy, sparsely branched shrub to 3 m. Leaves sessile, ovate-cordate, ascending and imbricate. Flowers solitary in upper axils, yellow. Mainly Jan.–Apr. Rocky sandstone slopes along streams, SW (W Riviersonderend Mts).•

retzioides Sond.   Rounded shrub to 1 m, coppicing from a woody caudex. Leaves subsessile, linear-elliptic, margins revolute. Flowers solitary in upper axils, orange-red with yellow tube. Mainly Mar.–May. Rocky southern sandstone slopes, SW (Riviersonderend Mts).•

GLISCHROCOLLA•   1 sp., W Cape

formosa (Thunb.) R.Dahlgren   Densely leafy, sparsely branched shrub to 1 m. Leaves sessile, decussate, ovate-cordate, ascending and imbricate, margins thickened and slightly revolute. Flowers few in congested terminal, head-like racemes, creamy yellow ageing pinkish, subtended by large red bracts. Mainly Jan.–Feb. Rocks and cliffs at high alt., SW (Victoria Peak to Hottentots Holland Mts).•

PENAEA• NOUGHTS-AND-CROSSES   4 spp., W Cape

acutifolia A.Juss.   Sparsely branched shrub to 1.3 m. Leaves sessile, lanceolate to elliptic. Flowers few in terminal spikes, yellow, with perianth lobes narrowly triangular and acuminate apex, bracts lanceolate-ovate. Mainly Oct.–May. Mostly damp sandstone slopes, SE (Outeniqua Mts: Robinson Pass to George).•

cneorum Meerb.   Like **P. mucronata** but young stems glabrous, leaves elliptic to cordate. Mainly Sept.–Jan. Damp sandstone slopes and streambanks, SW, LB, SE (Kogelberg and Riviersonderend Mts to Port Elizabeth).•

dahlgrenii Rourke   Densely leafy, virgate shrub, sometimes to 2 m. Leaves shortly petiolate, decussate, ascending, broadly ovate. Flowers few, in short terminal and axillary cymes, white ageing pinkish, with perianth lobes ovate, subacute. Mainly Aug.–Dec. Sandstone slopes along streams, LB (Langeberg Mts: Barrydale to Riversdale).•

mucronata L.   Ascending to erect shrub, sometimes to 1.3 m, with slightly hairy to papillate branches, coppicing from a woody caudex. Leaves sessile, cordate to lanceolate, often imbricate. Flowers in short terminal spikes, yellow, with perianth lobes triangular, subobtuse, bracts rhombic-triangular, broadened below. Mainly Oct.–Dec. Mostly rocky sandstone slopes, NW, SW, AP, LB (Robertson and Cape Peninsula to Langeberg Mts).•

SALTERA• CAPE FELLWORT, VLIEËBOS   1 sp., W Cape

sarcocolla (L.) Bullock   Closely leafy, few-branched shrub to 1.5 m, coppicing from a woody base. Leaves ascending, imbricate, opposite and decussate, obovoid to rhombic. Flowers few in congested, head-like spikes, pink. Mainly Aug.–Dec. Rocky sandstone slopes, SW, AP (Cape Peninsula to Agulhas).•

SONDEROTHAMNUS•   2 spp., W Cape

petraeus (W.F.Barker) R.Dahlgren   Sparsely branched, closely leafy shrublet to 35 cm, coppicing from a woody caudex. Leaves sessile, broadly ovate to obovate, ascending and imbricate, margins of especially upper leaves denticulate to fimbriate. Flowers several in terminal heads, pink with darker tube, tepals c. 5 mm long. Mainly Oct.–Dec. Sandstone rocks and cliffs, SW (Hottentots Holland Mts to Kleinmond).•

speciosus (Sond.) R.Dahlgren   Like **S. petraeus** but taller, to 60 cm and tepals c. 8 mm long. Aug.–Oct. Rocky sandstone slopes, SW (Babilonstoring and Kleinrivier Mts).•

STYLAPTERUS•   8 spp., W Cape

A. Ovary tapering into style, stigma lobes prominent

dubius (Stephens) R.Dahlgren   Laxly branched shrub to 50 cm. Leaves lanceolate to ovate. Flowers axillary, subsessile, in short terminal racemes, pale pink ageing white. Mainly Sept.–Oct. Sandstone slopes, LB (Langeberg Mts: Swellendam to Riversdale).•

ericifolius (A.Juss.) R.Dahlgren   Laxly branched shrub to 1 m. Leaves linear, acicular, margins rolled upward. Flowers axillary, in short terminal racemes, pale yellow. Sept.–Oct. Sandstone slopes, LB (Langeberg Mts: Swellendam).•

A. Ovary truncate and style slender, stigma lobes small

barbatus A.Juss.   Erect shrub to 60 cm. Leaves narrowly lanceolate, grooved beneath in lower half, with axillary hairs to 2 mm long. Flowers in short terminal spikes, pale yellow. Oct.–Nov. High, damp sandstone slopes, SW (Hottentots Holland Mts).•

candolleanus (Stephens) R.Dahlgren   Compact, densely branched shrublet. Leaves ascending, imbricate, ovate and acute. Flowers in short terminal spikes, ?yellow. Flowering time and distribution unknown, ?SW.•

ericoides A.Juss.   Erect, sparsely branched shrub to 1.5 m. Leaves linear, with a prominent median groove beneath. Flowers in short spikes, pale yellow, sometimes purplish, with bracts lanceolate, caducous. Sept.–Oct. Mountain streams at low elevation, SW (Elandskloof Mts to Du Toitskloof).•

fruticulosus (L.f.) A.Juss.   Closely leafy, sprawling shrublet to 60 cm, coppicing from a woody caudex. Leaves obovate to oblanceolate, often obtuse. Flowers in short terminal spikes, pale yellow, upper floral bracts broadly obovate with denticulate margins. Mainly July–Nov. Sandy flats, SW (Mamre to Cape Peninsula and Cape Flats).•

micranthus R.Dahlgren   Like **S. ericoides** but perianth tube shorter, to 3 mm long and bracts acicular, persistent. Sept.–Oct. Streambanks, SW (Hottentots Holland Mts).•

sulcatus R.Dahlgren   Like **S. fruticulosus** but plants erect with leaves ovate to lanceolate, and floral bracts with plane margins. Sept.–Dec. Rocky sandstone slopes, SW (Du Toitskloof Mts).•

PITTOSPORACEAE

PITTOSPORUM   PITTOSPORUM   c. 150 spp., palaeotropical and warm temperate

viridiflorum Sims   CHEESEWOOD, KASUUR   Tree or shrub to 20 m, thinly hairy on young parts. Leaves leathery, oblanceolate, margins revolute. Flowers in terminal panicles, creamy, fragrant. Seeds red, sticky. Nov.–Dec. Coastal bush and forests, LB, SE (Swellendam to tropical Africa).

PLANTAGINACEAE

PLANTAGO   PLANTAIN   c. 265 spp., cosmopolitan

cafra Decne.   CAPE PLANTAIN   Softly hairy annual to 15 cm, often tussock-forming. Leaves linear or filiform. Flowers in loose, oblong or subglobose spikes, whitish. Aug.–Sept. Clay slopes, NW, SW, KM (Richtersveld to Cape Peninsula through Little Karoo to Ladismith).

*****coronopus** L.   BUCK'S-HORN PLANTAIN   Thinly grey-silky annual to 18 cm. Leaves oblanceolate, pinnatifid. Flowers in dense, narrowly cylindrical spikes, whitish, corolla tube hairy. Oct.–Nov. Weed of sandy coastal flats, SW, AP (European weed, Cape Peninsula to De Hoop).

crassifolia Forssk.   FLESHY PLANTAIN   Hairy rhizomatous perennial to 30 cm. Leaves fleshy or leathery, usually woolly in axils, linear-oblanceolate, sometimes pinnatifid. Flowers in dense, narrowly cylindric spikes, whitish, corolla tube hairy. Nov.–Mar. Coastal sands and limestones, SW, AP, LB, SE (Saldanha Bay to tropical Africa).

*****lanceolata** L.   RIBWORT PLANTAIN   Thinly hairy tufted perennial to 60 cm. Leaves oblanceolate, ribbed, sometimes woolly in axils. Flowers in dense, cylindric spikes on grooved peduncles, whitish, bracts long-acuminate, lower 2 sepals fused. Mainly Oct.–Mar. Weed of waste places, NW, SW, AP, KM, LB, SE (European weed, cosmopolitan).

*****major** L.   GREAT PLANTAIN   Thinly hairy tufted perennial to 60 cm. Leaves ovate, petiolate, ribbed, sometimes woolly in axils. Flowers in elongate spikes, scattered below, whitish. Sept.–Feb. Weed of damp places, NW, SW, AP, LB, SE (European weed, cosmopolitan).

remota Lam.   Glabrescent tufted perennial to 20 cm. Leaves linear-oblanceolate, sparsely and minutely toothed, woolly in axils. Flowers scattered in elongate spikes on grooved peduncles, whitish. Aug.–Jan. Damp ground, SW, SE (Cape Peninsula, George and E Cape).

*****virginica** L.   Dioecious, softly hairy or woolly, tufted perennial to 30 cm. Leaves oblanceolate, ribbed. Flowers in elongate spikes on softly hairy peduncles, female flowers with petals connivent to tips, brownish. Oct.–Dec. Clay slopes, SW, LB (American weed, Cape Peninsula, Jeffreys Bay).

PLUMBAGINACEAE

1. Calyx with long, often glandular hairs; inflorescence a terminal spike .... **Plumbago**
1.' Calyx papery and without hairs; inflorescence racemose, cymose or paniculate .... **Limonium**

LIMONIUM   SEA-PINK, PAPIERBLOM   c. 100 spp., Old World maritime and arid regions

A. Calyx c. 10–20 mm diam. at maturity

amoenum (C.H.Wright) R.A.Dyer   Pustulate-scurfy shrublet to 30 cm, leafy above. Leaves oblanceolate, 30–50 ¥ 2–5 mm. Flowers in panicles, pink, calyx limbs 9–12 mm diam. Oct.–Dec. Shale slopes, NW (Bokkeveld Mts to Worcester and W Karoo).

capense (L.Bolus) L.Bolus   Rounded, densely leafy shrublet to 60 cm. Leaves ascending, oblanceolate, 18–25 × 3–4 mm, scurfy and minutely pitted. Flowers in spikes, pink, calyx limbs 17 mm diam. Nov.–Jan. Coastal limestone flats, SW (St Helena Bay to Saldanha Bay).•

longifolium (Thunb.) R.A.Dyer   Scurfy, tufted perennial to 60 cm. Leaves mostly basal, linear-oblanceolate, 60–200 × 5–10 mm. Flowers in scapose corymbs, pink to peach and white, calyx limbs

15–17 mm diam. Sept.–Jan. Sandy flats, NW, SW, AP, KM, LB (Namaqualand to Yzerfontein and Robertson, also Riversdale).

perigrinum (P.J.Bergius) R.A.Dyer STRANDROOS, PAPIERBLOM Shrub to 1 m, branches leafy at tips. Leaves oblanceolate, 40–80 × 10–20 mm, rough, sometimes pitted. Flowers in scapose corymbs, magenta, calyx limbs 15–17 mm diam. Aug.–Jan. Coastal dunes, NW, SW (Namaqualand to Melkbosstrand).

purpuratum (L.) Hubbard ex L.H.Bailey Like **L. perigrinum** but a tufted shrublet to 60 cm with leaves smooth on both surfaces and flowers mauve. Oct.–Feb. Sandy coastal flats, SW (Mamre to Durbanville).•

A. Calyx c. 5 mm diam. at maturity

acuminatum L.Bolus Like **L. equisetinum** but inflorescence velvety hairy. Sept.–Jan. Coastal limestone, NW, SW (Rocher Pan to Yzerfontein).•

anthericoides (Schltr.) R.A.Dyer BRAKBLOMMETJIE Roughly scurfy, tufted perennial to 30 cm. Leaves rosulate, obovate, 15–35 × 7–15 mm. Flowers in erect, open, divaricate corymbs, white to mauve, calyx limbs to 5 mm diam. with aristate lobes. Dec.–Feb. Seasonally moist flats and pans, AP (Gansbaai to Potberg).•

billardieri (Girard) Kuntze (= *Limonium depauperatum* (Boiss.) R.A.Dyer) Like **L. linifolium** but scapes developing tufts of leaves in upper axils. Aug.–Mar. Sandy coastal flats, SW, AP (Velddrif to Bredasdorp).•

dregeanum (C.Presl) Kuntze Like **L. scabrum** but older flowers in spikelets pedicellate. Dec.–May. Sandy flats, NW, KM, LB (S Namibia and Karoo to Worcester, Riversdale and Oudtshoorn).

equisetinum (Boiss.) R.A.Dyer (incl. **L. decumbens** (Boiss.) Kuntze) Roughly scurfy, often rhizomatous perennial to 10 cm. Leaves rosulate, oblanceolate to obovate 10–30 × 5–10 mm. Flowers in dense, prostrate inflorescences with numerous sterile axillary branchlets below, spikelets distichous on short branches, mauve, calyx limbs to 5 mm diam. Sept.–Jan. Coastal sand flats, NW, SW (Namaqualand to Bokbaai).

kraussianum (Buchinger ex Boiss.) Kuntze Mat-forming dwarf perennial to 10 cm. Leaves crowded, linear-subulate, 10–15 × 0.5–1 mm. Flowers in prostrate corymbs with lower branchlets sterile, mauve. Sept.–Feb. Sandy coastal flats, SW, AP (Kalbaskraal and Elim to Potberg).•

linifolium (L.f.) Kuntze Like **L. scabrum** but leaves linear, 20–35 × 1 mm. Nov.–Jan. Coastal saline pans and estuaries, AP, LB, SE (Stilbaai to E Cape).

scabrum (Thunb.) Kuntze SEA LAVENDER Roughly scurfy, tufted dwarf perennial to 25 cm. Leaves basal, oblanceolate to obovate, to 80 × 10 mm. Flowers in dense, erect or spreading corymbs with lower branchlets often sterile, mauve, scapes sometimes developing tufts of leaves in lower axils, calyx limbs to 5 mm diam. Oct.–May. Coastal dunes and estuaries, SW, AP, LB, SE (Cape Peninsula to E Cape).

PLUMBAGO PLUMBAGO c. 10 spp., more or less cosmopolitan

auriculata Lam. SYSELBOS Shrub or scrambler to 20 m. Leaves oblong, 20–25 mm long. Flowers in terminal spikes, pale blue, calyx glandular-hairy above. Dec.–May. Bush and scrub, SE (George to Mpumalanga).

tristis Aiton Roughly glandular-hairy shrublet to 30 cm. Leaves obcordate-cuneate, 5–10 mm long. Flowers in terminal spikes, dark orange-red, calyx with long dark hairs. Oct.–Mar. Dry sandy slopes, KM (Little Karoo and S Karoo).

POLYGALACEAE

1. Sepals subequal; fruit a capsule ....**Muraltia**
1.' Sepals unequal, laterals larger than the others, often coloured and wing-like; fruit a capsule or drupe:
    2. Fruit a capsule; stamens (4 or 5)8(9), rarely 7 fertile; unarmed herbs or shrubs ....**Polygala**
    2.' Fruit a drupe; stamens 7; intricate; spinescent shrubs..**Nylandtia**

MURALTIA PURPLE-GORSE, SKILPADBOS 115 spp., W Cape to Tanzania

A. Leaves solitary; flowers on distinct pedicels; capsules obtuse or shortly toothed

angulosa Turcz. Spreading shrub to 60 cm, with lax, rough, angled stems. Leaves sessile, linear, rigid, mucronate. Flowers 1–few in axils, white flushed with mauve. Aug.–Nov. Rocky sandstone slopes, NW (Cold Bokkeveld and Olifants River Mts).•

brevicornu DC. Loosely branched shrub to 90 cm. Leaves sessile, linear to linear-lanceolate, apiculate. Flowers 1–4 in axils, pink. Mainly Oct.–Dec. Rocky slopes and sandy flats, NW, SW (Piketberg to Cape Peninsula).•

carnosa E.Mey. ex Harv. Rounded shrublet to 50 cm. Leaves shortly petiolate, oblong, somewhat crowded near branch tips, fleshy, mucronate. Flowers 1 or 2 in axils, ?pink. Oct.–Nov. Rocky slopes, KM (Swartberg Mts).•

crassifolia Harv. Diffuse, erect or sprawling subshrub to 15 cm, with angled, often gummy branches. Leaves sessile, oblong to orbicular, mucronate. Flowers 1–few in axils, pale pink or cream-coloured. Dec.–Jan. Rocky sandstone slopes, NW (Cedarberg to Hex River Mts and Keeromsberg).•

elsiae Paiva Single-stemmed, closely leafy shrub to 1.5 m. Leaves shortly petiolate, linear-lanceolate, mucronate. Flowers solitary in axils, purple. Sept.–Oct. Rocky sandstone slopes, KM (Klein Swartberg).•

horrida Diels Twiggy shrublet to 60 cm, with rigid, spine-tipped branches. Leaves shortly petiolate, oblong, slightly fleshy, mucronate. Flowers 1–few in axils, pale pink. Oct.–Dec. Rocky slopes, NW (Bokkeveld to Cedarberg Mts and W Karoo).•

juniperifolia (Poir.) DC. Twiggy shrublet to 50 cm. Leaves sessile, rigid, needle-like, spine-tipped. Flowers 1 or 2 in axils, pink or white. May–Oct. Rocky slopes, KM, SE (Kammanassie Mts to Loerie and S Karoo).

leptorhiza Turcz. Loosely branched, sparsely leafy shrub to 30 cm with scabridulous stems. Leaves sessile, linear, channelled above, mucronate. Flowers 1 or 2 in axils, pink. Nov.–Dec. Rocky sandstone slopes, LB, SE (Langeberg and Outeniqua Mts).•

muraltioides (Eckl. & Zeyh.) Levyns Sprawling subshrub to 20 cm, with angled stems. Leaves sessile, linear to lanceolate, usually with reflexed mucro. Flowers solitary in axils, pale pink. Aug.–Dec. Rocky sandstone and gravel slopes, NW, SW, KM, LB, SE (Tulbagh and Riviersonderend Mts to Port Elizabeth).•

oxysepala Schltr. Prostrate subshrub to 15 cm, with winged stems. Leaves sessile, linear-elliptic, mucronate. Flowers solitary in axils, pale mauve or white. Oct.–Nov. Moist, sandstone slopes, SW (Franschhoek Mts to Hermanus).•

parvifolia N.E.Br. (= *Muraltia scoparia* sensu Levyns) Twiggy shrublet to 60 cm, with rigid, angled branches. Leaves minutely petiolate, linear to oblong, fleshy with age, mucronate. Flowers 1 or 2 in axils, pale pink. Mainly July–Sept. Dry rocky slopes, KM (Little Karoo and Witteberg and Swartberg Mts).•

pauciflora (Thunb.) DC. Slender, much-branched subshrub to 30 cm, with stems angled and rough when young. Leaves sessile, filiform, mucronate. Flowers solitary in axils, pale pink. Mainly Nov.–Jan. Moist sandstone slopes, NW, SW, LB (Elandskloof to Langeberg Mts).•

polyphylla (DC.) Levyns Low subshrub to 30 cm, with rough, angular stems. Leaves sessile, linear-lanceolate, usually scabrous, mucronate. Flowers solitary in axils, white with purple carina. Nov.–Feb. Rocky sandstone slopes, NW (Cold Bokkeveld Mts to Audensberg).•

trinervia (L.f.) DC. Sprawling subshrub to 25 cm, with ridged branches. Leaves sessile, lanceolate to ovate, mucronate. Flowers solitary in axils, pink. Aug.–Nov. Clay and granite slopes, NW, SW (Grootwinterhoek Mts to Tygerberg and Stellenbosch).•

A. Leaves fascicled (sometimes solitary), flowers sessile or on short pedicels
B. Capsules obtuse, without horns

ciliaris DC. Sprawling subshrub to 30 cm, hispid on young parts. Leaves fascicled, lanceolate, rigid and spine-tipped. Flowers sessile, axillary, pink. Apr.–Dec. Sandstone slopes, LB, SE (Swellendam to Humansdorp).•

squarrosa (L.f.) DC. Erect or spreading shrub to 1 m, tomentose on young parts. Leaves in fascicles, hairy when young, ovate to subulate, spine-tipped. Flowers sessile, axillary, pink. Capsules obtuse. Oct.–May. Sandstone slopes, SE (George to Alexandria).

B. Capsules with 4 long horns or sometimes only shortly toothed
C. Calyx small, less than half the length of corolla

acicularis Harv. Sprawling pubescent to scabrid subshrub to 25 cm. Leaves sessile, fascicled, rigid, needle-like, spine-tipped, often recurved apically. Flowers sessile, axillary, pink. Capsules ?horned. Sept.–Nov. Sandstone slopes, NW, SW (Cedarberg to Du Toitskloof Mts).•

alba Levyns   Closely leafy, pubescent shrub to 70 cm. Leaves sessile, fascicled, lanceolate-acuminate, narrowed below, hispid below, shortly spine-tipped. Flowers on short pedicels, axillary, pink. Capsules hispid, with slender, elongate horns and only 1 chamber fertile. Sept.–Feb. Rocky sandstone slopes, SW (Jonkershoek and Groot Drakenstein Mts).•

alopecuroides (L.) DC.   Pubescent or glabrescent shrub to 1 m. Leaves sessile, fascicled, subulate to ovate, usually villous, spine-tipped, straight or upcurved apically. Flowers on short pedicels, axillary, pink. Capsules often hispid, with long slender horns. Mainly Oct.–Dec. Sandstone and clay slopes and flats, NW, SW, KM, LB, SE (Bokkeveld Mts to Stutterheim).

angustiflora Levyns   Closely leafy, rigid shrub to 60 cm, pubescent on young stems. Leaves sessile, fascicled, lanceolate-attenuate, channelled, puberulous, spine-tipped. Flowers sessile, axillary, pink. Capsules hispid, with long slender horns. Oct.–Nov. Clay slopes, NW (Gydo Mts).•

aspalathoides Schltr.   Densely leafy, hispid, sprawling shrublet to 20 cm. Leaves sessile, fascicled, linear-lanceolate, long-ciliate below, spine-tipped. Flowers sessile, axillary, pink. Capsules scabrid, shortly horned. Dec.–Jan. Sandstone slopes, SW (Houwhoek and Caledon Swartberg).•

brachyceras Schltr.   Erect or sprawling, glabrescent shrublet to 30 cm. Leaves sessile, irregularly fascicled, oblong to lanceolate, puberulous becoming scabrid, shortly spine-tipped, apex often reflexed. Flowers sessile, axillary, pink. Capsules hispid, minutely horned. Sept.–Oct. Dry, sandstone slopes, NW, KM (Cedarberg Mts to Anysberg).•

capensis Levyns   Densely leafy, hispid subshrub to 20 cm, branching mainly from base. Leaves sessile, fascicled, linear, ciliate-hispid, spine-tipped. Flowers sessile, axillary, pink. Capsules ?shortly horned. Flowering time? Rocky sandstone slopes, SW (Viljoen's Pass to Houwhoek).•

chamaepitys Chodat   Slender, glabrescent subshrub to 40 cm, branching mainly from base. Leaves sessile, fascicled, linear to ovate, pilose, becoming scabrid, apiculate. Flowers on short pedicels, axillary, pink. Capsules hispid, with short slender horns. Sept.–Oct. Rocky sandstone slopes, SW (Houwhoek Mts).•

cliffortiifolia Eckl. & Zeyh.   Robust, twiggy shrub to 70 cm. Leaves usually fascicled, ovate, rigid, spike-tipped. Flowers on short pedicels, axillary, pink. Capsules with rough, tapering horns. Mar.–June. Rocky slopes, KM, LB (Rooiberg and Riversdale to Mossel Bay).•

commutata Levyns   Laxly branched, spreading shrub to 40 cm, puberulous on young parts. Leaves subsessile, fascicled, lanceolate to spathulate, pubescent below, shortly apiculate. Flowers sessile, axillary, pink. Capsules hispid, shortly horned. Sept.–Oct. Sandstone slopes, NW, KM (Cold Bokkeveld Mts to Witteberg).•

concava Levyns   Densely leafy shrublet to 30 cm, pubescent on young parts. Leaves shortly petiolate, ovate, concave, apiculate, margins long-ciliate. Flowers subsessile, axillary, pink. Capsules hispid, without horns. Flowering time? Rocky S-facing slopes, SW (Riviersonderend Mts).•

curvipetala Levyns   Sprawling or erect, glabrescent subshrub to 40 cm. Leaves sessile, fascicled, rigid, needle-like, spine-tipped. Flowers sessile, axillary, pink. Capsules scabrid, prominently horned. Aug.–Nov. Sandstone slopes, SW (Cape Peninsula).•

cuspifolia Chodat   Closely leafy, hirsute subshrub to 10 cm. Leaves sessile, fascicled, lanceolate, acuminate, fleshy, ciliate below, spine-tipped. Flowers sessile, axillary, pale pink. Capsules shortly horned. Sept.–Jan. ?Low gravel slopes, SW (Elim).•

decipiens Schltr.   Closely leafy, hispid subshrub to 20 cm, branching mainly from base. Leaves sessile, fascicled, linear-lanceolate, hispid below and on margins, spine-tipped. Flowers sessile, axillary, pink. Capsules hispid, with long slender horns. Sept.–Nov. Gravel flats and lower slopes, SW (Tygerberg to Franschhoek).•

demissa Wolley-Dod   Sprawling, glabrescent subshrub to 30 cm, with curly hairs on young parts. Leaves sessile, solitary or fascicled, linear-elliptic, shortly curly haired at least on margins, spine-tipped. Flowers on short pedicels, axillary, pink, sometimes white. Capsules sparsely hairy, shortly horned. Aug.–Nov. Sandstone slopes and flats, SW (Cape Peninsula).•

dispersa Levyns   Erect or spreading, stiffly branched, glabrescent shrublet to 40 cm. Leaves sessile, loosely or densely fascicled, mostly lanceolate, ciliate on margins, spine-tipped. Flowers sessile, axillary, pink or white. Capsules scabrid, with long horns. Sept.–Jan. Sandstone slopes, KM, LB, SE (Langeberg: Riversdale and Swartberg and Kouga Mts).•

dumosa (Poir.) DC.   Sprawling or suberect shrub to 30 cm, pilose on young parts. Leaves sessile, fascicled, mostly ovate or elliptic, apiculate, tip often reflexed, often pilose below. Flowers sessile, axillary, pink. Capsules scabrid, shortly horned. July–Sept. Mainly coastal sands, NW, SW (Olifants River mouth to Milnerton).•

ericifolia DC. Rigid, much-branched, pubescent shrub to 70 cm. Leaves sessile, fascicled, narrowly obovate, shortly spine-tipped. Flowers sessile, axillary, pink. Capsules prominently horned. Dec.–Jan. Rocky lower slopes, SW, KM, LB, SE (Robertson to Port Elizabeth).•

ferox Levyns Rigid, much-branched, glabrescent shrub to 60 cm, branch tips becoming spiny. Leaves sessile, fascicled, ovoid, thick and fleshy, spine-tipped reflexed apically. Flowers sessile, axillary, ?pink. Capsules shortly horned. Sept.–Oct. Dry rocky slopes, SW (Riviersonderend Mts).•

gillettiae Levyns Sprawling subshrub to 25 cm, branching mainly from base. Leaves sessile, fascicled, needle-like, apiculate, hairy below when young. Flowers sessile, axillary, pale pink. Capsules hispid, shortly horned. Sept.–Oct. Sandstone slopes, SW (Bredasdorp Mts near Elim).•

harveyana Levyns Stiffly branched, closely leafy, glabrescent shrub to 60 cm. Leaves solitary, oblanceolate, midrib prominent below, mucronate. Flowers on short pedicels, axillary, pink fading white. Capsules scabrid, with long slender horns. Oct.–Jan. Limestone hills, SW (Yzerfontein to Langebaan).•

heisteria (L.) DC. Erect, usually laxly branched shrub to 1(2) m, pubescent on young parts. Leaves sessile, fascicled, lanceolate-attenuate, channelled, often ciliate, spine-tipped. Flowers on short pedicels, axillary, usually purple, side petals sometimes white. Capsules with elongate, slender horns. Mainly Oct.–Dec. Rocky slopes, mainly on sandstone, NW, SW, KM, LB (Bokkeveld Mts to Riversdale).•

hirsuta Levyns Densely leafy, villous subshrub to 15 cm, branching mainly at base. Leaves sessile, fascicled, linear-lanceolate, hairy below and on margins, apiculate, Flowers sessile, axillary, white fading pink. Capsules hispid, shortly horned. Jan.–Feb. Granite slopes, SW (Hemel-en-Aarde valley).•

karroica Levyns Suberect, much-branched shrub to 60 cm, puberulous on young parts. Leaves subsessile, fascicled, mostly ovate to spathulate, puberulous to scabrid, obtuse, shortly apiculate. Flowers sessile, axillary, pink. Capsules scabrid, minutely horned. June–July. Rocky slopes, KM (Klein Swartberg to Rooiberg).•

lignosa Levyns Stiffly branched, scabrid shrub to 60 cm. Leaves shortly petiolate, fascicled, mostly obovate to oblanceolate, fleshy, pubescent, becoming scabrid, with a recurved spiny tip. Flowers sessile, axillary, pink or white. Capsules hispid, with long, slender horns. Oct.–Nov. Rocky sandstone slopes, NW (Cold Bokkeveld).•

macropetala Harv. Rigid, pubescent, divaricately branched shrublet to 45 cm. Leaves sessile, fascicled, ovate, fleshy, ciliate to hispid below, mucronate to spine-tipped. Flowers subsessile, axillary, purple with white side petals. Capsules shortly horned. July–Nov. Rocky and gravel slopes and flats, SW (Hopefield to Stellenbosch).•

mitior (P.J.Bergius) Levyns Sparsely branched, pubescent shrub to 80 cm, leaves and flowers crowded on upper branches. Leaves sessile, fascicled, linear-subulate, slightly channelled above, slightly incurved, spine-tipped. Flowers on short pedicels, axillary, pink. Capsules hispid, with slender horns. July–Feb. Sandy and limestone flats, SW, AP (Cape Peninsula to Agulhas).•

muirii F.Bolus Low subshrub to 20 cm, setose on young parts. Leaves fascicled, linear and rigid, spine-tipped. Flowers sessile, axillary, pink. Capsules hispid, shortly horned. Nov.–Apr. Rocky slopes, LB, SE (Langeberg and Outeniqua Mts).•

obovata DC. Spreading, glabrescent shrublet to 40 cm. Leaves sessile, usually solitary, obovate to oblanceolate, shortly spine-tipped, apex often reflexed. Flowers sessile, axillary, pink or white. Capsules hispid, horned. July–Sept. Dry, sandy slopes and flats, NW, SW (Olifants River mouth to Saldanha).•

occidentalis Levyns Densely leafy, hispid shrublet to 40 cm. Leaves sessile, fascicled, linear to lanceolate, ciliate on margins, spine-tipped. Flowers on short pedicels, axillary, pale pink or white. Capsules ?horned. Aug.–Oct. Sandstone slopes, SW (Hottentots Holland Mts to Kogelberg).•

origanoides C.Presl Sprawling, densely branched shrub to 20 cm, hispid on young parts, branches becoming spiny. Leaves subsessile, solitary or fascicled, obovate to elliptic, margins thickened and hispid, with a reflexed spiny tip. Flowers sessile, axillary, pink. Capsules hispid, shortly horned. July–Sept. Gravel slopes, NW, SW (Citrusdal to Tulbagh and Darling).•

rhamnoides Chodat Spreading, glabrescent shrub to 60 cm. Leaves sessile, fascicled, lanceolate-elliptic, linear to ovate, apiculate or spine-tipped. Flowers sessile, axillary, pink. Capsules hispid, prominently horned. Aug.–Nov. Sandstone slopes, NW, SW, KM, LB (Bokkeveld Mts to Little Karoo and Riversdale).•

satureioides DC.   Sprawling, divaricately branched shrub to 60 cm. Leaves sessile, fascicled, linear to oblong, shortly ciliate on margins and keel, spine-tipped. Flowers on short pedicels, axillary, pink or white. Capsules hispid, prominently horned. Mainly Sept.–Oct. Coastal calcareous sands, SW, AP, SE (Cape Peninsula to Knysna).•

serpylloides DC.   Diffuse, slightly villous shrublet to 40 cm. Leaves fascicled, shortly petiolate, broadly ovate, margins hispid and revolute. Flowers sessile, axillary, pink. Capsules prominently horned. June–Feb. Rocky sandstone slopes, SW (Cape Peninsula and Hottentots Holland Mts).•

stokoei Levyns   Densely leafy, villous subshrub to 20 cm. Leaves sessile, fascicled, lanceolate, hispid on margins and lower midrib, apiculate, Flowers sessile, axillary, pink. Capsules hispid, horned. Mar.–Apr. Rocky sandstone slopes, SW (Viljoen's Pass to Kleinmond).•

tenuifolia (Poir.) DC.   Intricately branched, glabrescent shrub to 60 cm. Leaves sessile, fascicled, rigid, needle-like, spine-tipped, recurved at tips. Flowers on short pedicels, axillary, pale pink or white. Capsules with short, slender horns. Jan.–Mar. Rocky sandstone slopes, SW (Riviersonderend Mts).•

thymifolia (Thunb.) DC.   Erect or sprawling, densely branched shrub to 60 cm, hairy on young parts. Leaves sessile, solitary or imperfectly fascicled, linear to ovate, pilose below, shortly spine-tipped. Flowers on short pedicels, axillary, pink. Capsules hispid, with long, slender horns. Mainly July–Sept. Sandy flats and lower slopes, NW, SW (Clanwilliam to Cape Peninsula).•

vulpina Chodat   Erect, stiffly branched shrub to 70 cm, villous on young parts. Leaves sessile, fascicled, linear to elliptic, long-ciliate when young, spine-tipped. Flowers on short pedicels, axillary, pink. Capsules sparsely hispid, with slender horns. Mainly Nov.–Jan. Sandstone slopes, SW (Jonkershoek to Kleinmond).•

C. Calyx at least half the length of corolla

acerosa Harv.   Sparsely hairy undershrub to 20 cm, producing slender stems from woody rootstock. Leaves sessile, fascicled or solitary, elliptic, channelled, rigid, spine-tipped. Flowers on short pedicels, axillary, pink or white. Capsules shortly horned. Oct.–Dec. Rocky sandstone slopes, LB, SE (Langeberg: Swellendam to Tsitsikamma Mts).•

acipetala Harv.   Sprawling, glabrescent subshrub to 15 cm. Leaves sessile, fascicled, linear, often channelled, rigid, spine-tipped. Flowers on short pedicels, axillary, pink or white. Capsules shortly horned. Mainly Sept.–Nov. Sandstone hills and flats, SW (Cape Peninsula).•

aciphylla Levyns   Densely leafy, hispid subshrub to 30 cm. Leaves sessile, fascicled, linear-elliptic, hispid on margins and keel, spine-tipped. Flowers sessile, axillary, usually pink. Capsules hispid, shortly horned. Sept.–Nov. High rocky slopes, SW (Jonkershoek Mts).•

arachnoidea Chodat   Sprawling shrub to 50 cm, hairy on young stems. Leaves sessile, fascicled, ovate, sparsely hairy, recurved above and spine-tipped. Flowers sessile, axillary, pink or white. Capsules hispid, horned. Sept.–Dec. Lower sandstone slopes, NW (Piketberg).•

aspalatha DC.   Densely leafy, hispid, subshrub to 20 cm, branching mainly from base. Leaves sessile, fascicled, linear-lanceolate, hispid below and on margins, spine-tipped. Flowers sessile, axillary, pink or white and yellow. Capsules ?villous, shortly horned. Aug.–Dec. Sandy and gravel slopes, SW (Hottentots Holland Mts to Botrivier and Betty's Bay).•

asparagifolia Eckl. & Zeyh.   Intricately branched shrublet to 50 cm, shortly pubescent on young parts. Leaves sessile, fascicled, needle-like, bristly on margins, spine-tipped. Flowers sessile, axillary, pink or white. Capsules with short, hispid horns. Oct.–Dec. Moist sandstone slopes, SW (Jonkershoek and Hottentots Holland Mts).•

barkerae Levyns   Sprawling woody shrublet to 50 cm, hispid on young parts. Leaves sessile, fascicled, oblong-obovate, shortly spine-tipped. Flowers sessile, axillary, pink. Capsules with short, slender horns. Aug.–Sept. Low, limestone hills, AP (Stilbaai).•

bolusii Levyns   Like **M. minuta** but flowers 3.3–4 mm long. Sept.–Jan. Sandy coastal flats, SW (Kleinmond to Hermanus).•

brachypetala Wolley-Dod   Erect, stiffly branched subshrub to 30 cm, branching mainly from base. Leaves sessile, fascicled, linear-lanceolate, acuminate, channelled, glabrescent, spine-tipped. Flowers shortly pedicellate, axillary, pink. Capsules with elongate, slender horns. July–Dec. Sandstone slopes, SW (Cape Peninsula).•

caledonensis Levyns   Sparsely branched, sprawling subshrub to 30 cm, pilose on young parts. Leaves sessile, fascicled, mostly linear or subulate, pilose below, apiculate. Flowers sessile, axillary, pink. Capsules sparsely hispid, with long, slender horns. Oct.–Nov. Gravel slopes and hills, SW (Botrivier to Caledon and Shaw's Mts).•

calycina Harv.   Twiggy, much-branched shrublet to 50 cm. Leaves fascicled, linear-oblong, villous below, mucronate. Flowers sessile, axillary, pink. Capsules horned. Feb.–May. Limestone hills, AP (Agulhas coast).•

collina Levyns   Divaricately branched, glabrescent shrublet to 35 cm. Leaves sessile, fascicled, elliptic, channelled, margins long-ciliate, spine-tipped. Flowers on short pedicels, axillary, pink. Capsules with long, slender horns. Sept.–Jan. Lower sandstone slopes, SW (Riviersonderend and Bredasdorp Mts to Potberg).•

comptonii Levyns   Like **M. stipulacea** but leaves with spiny tip often reflexed, flowers on short pedicels and with calyx about half as long as corolla. July–Sept. Sandstone slopes, SW (Cape Peninsula).•

cyclolopha Chodat   Sprawling or erect subshrub to 25 cm, branching from base and villous on young parts. Leaves sessile, fascicled, needle-like, villous or rough. Flowers on short pedicels, axillary, pale pink. Capsules hispid, shortly horned. Sept.–Jan. Gravel flats and slopes, SW, AP (Agulhas Peninsula).•

depressa DC.   Like **M. salsolacea** but leaves subulate, with curly marginal hairs, flowers sessile. Sept.–Nov. Coastal limestone slopes, AP (Stilbaai to Mossel Bay).•

diabolica Levyns   Twiggy shrublet to 30 cm, hispid on young parts. Leaves sessile, fascicled, needle-like, spine-tipped. Flowers sessile, axillary, pink or white. Capsules prominently horned. Sept.–Feb. Rocky slopes, SW (Cape Peninsula).•

divaricata Eckl. & Zeyh.   Sprawling shrublet to 50 cm, villous on young stems. Leaves sessile, fascicled, rigid, linear-subulate, apiculate. Flowers sessile, axillary, pink. Capsules hispid, prominently horned. Sept.–Dec. Rocky sandstone slopes, NW, SW (Grootwinterhoek Mts to Franschhoek and Riviersonderend).•

empleuridioides Schltr.   Sprawling, laxly branched, glabrescent shrub to 50 cm. Leaves solitary, sometimes fascicled, rigid, ovate to linear or needle-like, apiculate. Flowers sessile, axillary, pink or white. Capsules prominently horned. Aug.–Dec. Coastal calcareous sands, SW, AP (Stanford to Mossel Bay).•

ericoides (Burm.f.) Steud.   Erect or spreading, glabrescent subshrub to 40 cm. Leaves sessile, solitary or imperfectly fascicled, linear to ovate, villous below when young, apiculate. Flowers sessile, axillary, usually pink. Capsules hispid, with long, slender horns. Aug.–Apr. Low granite or sandstone slopes, SW, AP, LB, SE (Darling to Humansdorp).•

filiformis (Thunb.) DC.   Slender, glabrous subshrub to 30 cm, branching mainly at base. Leaves sessile, solitary or fascicled, needle-like to linear-lanceolate, apiculate. Flowers sessile, axillary, usually pink. Capsules with slender horns. Sept.–Dec. Sandstone slopes, often damp sites, SW, AP (Cape Peninsula to Agulhas).•

guthriei Levyns   Closely leafy, divaricately branched, pubescent shrublet to 45 cm. Leaves sessile, fascicled, linear-oblong, minutely ciliate below, spine-tipped. Flowers on short pedicels, axillary, white. Capsules with slender, elongate horns. Dec.–Jan. Rocky sandstone slopes, SW (Steenbras Mts).•

hyssopifolia Chodat   Densely leafy, sprawling or erect, glabrescent subshrub to 40 cm. Leaves sessile, fascicled, linear to elliptic, spine-tipped. Flowers on short pedicels, axillary, pink or white. Capsules with stout horns. July–Feb. Rocky sandstone slopes, SW (Franschhoek Mts to Houwhoek).•

knysnaensis Levyns   Glabrescent subshrub to 20 cm, branching mainly from base. Leaves sessile, solitary, sometimes fascicled, linear to ovate, ciliate when young, apiculate. Flowers sessile, axillary, pink. Capsules with short, slender horns. Oct.–Dec. Dry flats and hills, SE (George to Plettenberg Bay).•

langebergensis Levyns   Hispid shrublet to 50 cm. Leaves sessile, fascicled, linear-elliptic, rigid, spine-tipped. Flowers sessile, axillary, pink. Capsules with short, hispid horns. Mar.–Apr. Rocky sandstone slopes, LB (Langeberg: Swellendam).•

lewisiae Levyns   Sprawling, glabrescent, intricately branched subshrub to 20 cm. Leaves sessile, fascicled, linear-subulate, apiculate. Flowers on short pedicels, axillary, whitish. Capsules shortly horned. Dec.–Jan. Coastal limestone ridges, AP (Gansbaai).•

longicuspis Turcz.   Densely leafy, hispid subshrub to 30 cm. Leaves sessile, fascicled, elliptic, ciliate on margins, spine-tipped. Flowers sessile, axillary, pink or white. Capsules sparsely hairy, with short slender horns. Sept.–Mar. Sandstone slopes, SW (Du Toitskloof to Sir Lowry's Pass).•

macrocarpa Eckl. & Zeyh.   Spreading, much-branched shrub to 80 cm. Leaves sessile, fascicled or solitary, narrowly elliptic, hooked above and spine-tipped. Flowers sessile, axillary, white, sometimes pink. Capsules prominently horned. July–Dec. Dry, rocky slopes, KM (Little Karoo and Karoo Mts).

minuta Levyns   Erect or spreading subshrub to 20 cm, branching mainly at base. Leaves sessile, mostly fascicled, linear-lanceolate, softly hairy below, rigid, apiculate. Flowers sessile, axillary, pink, to 2.5 mm long. Capsules hispid, with slender horns. Oct.–Feb. Rocky flats, SW (Betty's Bay to Kleinmond).•

mixta (L.f.) DC.   Villous, sprawling, much-branched shrublet to 20 cm. Leaves sessile, fascicled, lanceolate-acuminate, sparsely ciliate-hispid below, spine-tipped. Flowers sessile, axillary, pale pink or white. Capsules scabrid, shortly horned. Oct.–Nov. Rocky sandstone slopes, SW (Cape Peninsula).•

montana Levyns   Sprawling shrublet to 20 cm, villous on young parts. Leaves sessile, fascicled, linear lanceolate, rigid, spine tipped. Flowers on short pedicels, axillary, deep pink. Capsules sparsely hairy, with narrow horns. Nov.–Dec. High rocky slopes, NW, SW (Grootwinterhoek and Wemmershoek Mts).•

mutabilis Levyns   Erect or spreading subshrub to 25 cm, villous on young parts. Leaves sessile, fascicled or solitary, linear, villous below, apiculate. Flowers 1 or 2 in axils, pink. Capsules hispid, prominently horned. Aug.–Jan. Low coastal sandstone slopes, SW (False Bay).•

ononidifolia Eckl. & Zeyh.   Twiggy shrublet to 50 cm. Leaves sessile, fascicled, linear-lanceolate, rigid, spine-tipped. Flowers sessile, axillary, pink or white. Capsules prominently horned. May–Sept. Rocky slopes and flats, NW, SW, AP, KM, LB (Karoopoort and Melkbos to Barrydale).•

orbicularis Hutch.   Glabrescent, sprawling shrublet to 30 cm. Leaves shortly petiolate, mostly solitary, orbicular to elliptic, sparsely curly haired, with a recurved spiny tip. Flowers on short pedicels, axillary, pink. Capsules hispid, prominently horned. May–Sept. Rocky gravel slopes, SW (Cape Peninsula).•

pageae Levyns   Low, hispid shrublet to 30 cm. Leaves sessile, fascicled, oblong, ciliate, spine-tipped. Flowers sessile, axillary, pink. Capsules with slender horns. Apr.–May. Sandy flats, SW (Cape Peninsula).•

paludosa Levyns   Erect or sprawling, glabrescent shrublet to 50 cm. Leaves solitary, distant, margins with adpressed hairs, needle-like, apiculate. Flowers sessile, axillary, pink. Capsules glabrescent, shortly horned. Aug.–Sept. Marshes on sandstone, SW (Grabouw to Kleinmond).•

pappeana Harv.   Sprawling shrublet to 25 cm, villous on young parts. Leaves sessile, fascicled, linear-oblong, channelled, recurving apically and spine tipped. Flowers sessile, axillary, white. Capsules with short, hispid horns. Apr.–July. Dry coastal hills, AP, LB (Bredasdorp to Riversdale).•

pillansii Levyns   Much-branched shrublet to 50 cm, hispid on young parts. Leaves sessile, fascicled, needle-like, spine tipped. Flowers sessile, axillary, pale pink or white. Capsules prominently horned. Sept.–Dec. Sandstone slopes, NW (Olifants River and Grootwinterhoek Mts).•

plumosa Chodat   Tufted subshrublet to 50 cm, tomentose on young parts. Leaves sessile, fascicled or solitary, slightly channelled, needle-like, spine-tipped. Flowers on short pedicels, axillary, pink or white. Capsules shortly horned. Nov.–Dec. Sandstone slopes, NW (Olifants River and Grootwinterhoek Mts).•

pottebergensis Levyns   Laxly branched shrub to 80 cm, hairy on young parts. Leaves sessile, fascicled, linear, villous below, spine-tipped. Flowers sessile, axillary, pink. Capsules hispid, with slender horns. Sept.–Nov. Sandstone slopes, SW (Potberg).•

pubescens DC.   Softly hairy spreading shrublet to 20 cm. Leaves loosely fascicled, linear-lanceolate, hairy, mucronate. Flowers shortly pedicellate, axillary, pale pink. Capsules hispid, shortly horned. Mainly Nov.–Feb. Rocky slopes, SW (Jonkershoek and Hottentots Holland Mts).•

pungens Schltr.   Densely leafy, shrublet to 50 cm, puberulous on young parts. Leaves fascicled, lanceolate, often channelled, spine-tipped. Flowers sessile, axillary, pale pink. Capsules shortly horned. Oct.–Dec. Rocky slopes and flats, SW, AP (Caledon to Agulhas).•

rosmarinifolia Levyns   Sprawling, glabrescent subshrub to 25 cm. Leaves sessile, solitary or tufted, linear to lanceolate, villous below, apiculate, often recurved above. Flowers sessile, axillary, pink. Capsules hispid, with long horns. Nov.–Jan. Sandy coastal slopes, SW (Cape Peninsula to Betty's Bay).•

rubeacea Eckl. & Zeyh.   Much-branched, villous shrub to 80 cm. Leaves sessile, fascicled, linear, hooked above and spine-tipped. Flowers sessile, axillary, usually pink. Capsules horned. June–Nov. Sandstone and limestone slopes, SW, AP (Hottentots Holland and Riviersonderend Mts to Agulhas).•

salsolacea Chodat   Sprawling, glabrescent subshrub to 25 cm. Leaves sessile, fascicled, ovate to linear, spine-tipped. Flowers on short pedicels, axillary, pink or white. Capsules with slender horns. Sept.–Feb. Coastal limestone hills, AP (Agulhas to Cape Infanta).•

schlechteri Levyns   Glabrescent shrublet to 50 cm. Leaves sessile, fascicled, lanceolate-elliptic, channelled, hispid on margins, spine-tipped. Flowers on short pedicels, axillary, pink. Capsules prominently horned. Sept.–Dec. Sandstone slopes, SW (Houwhoek and Riviersonderend Mts).•

serrata Levyns   Stiffly branched, minutely hairy shrublet to 50 cm. Leaves sessile, axillary, linear-oblanceolate, spine-tipped. Flowers sessile, axillary, pink or white. Capsules horned. Oct.–Nov. Rocky sandstone slopes, NW (Hex River Mts).•

spicata Bolus   Erect, softly hairy shrub to 60 cm. Leaves sessile, imperfectly fascicled, linear-elliptic, pilose below, mucronate. Flowers sessile, axillary, pink. Capsules villous above, shortly horned. July–Aug. Sandstone slopes, SW (Bredasdorp Mts).•

splendens Levyns   Spreading or erect subshrub to 50 cm, villous on young parts. Leaves sessile, fascicled, lanceolate, channelled, villous below, often hooked above and spine-tipped. Flowers sessile, axillary, pink or white with red keel. Capsules with short, hispid horns. June–Dec. Limestone slopes, AP (Bredasdorp to Stilbaai).•

stenophylla Levyns   Diffuse, pubescent subshrub to 40 cm, branching mainly from base. Leaves sessile, imperfectly fascicled, needle-like, ciliate when young, mucronate. Flowers sessile, axillary, pink. Capsules shortly horned. Nov.–Jan. Rocky sandstone slopes, NW, SW (Cold Bokkeveld Mts to Franschhoek Pass).•

stipulacea (Burm.f.) DC.   Sprawling, laxly branched subshrub to 30 cm, pilose on young parts. Leaves sessile, imperfectly fascicled, linear-lanceolate, pilose below, spine-tipped. Flowers sessile, axillary, pale pink, calyx more than half as long as corolla. Capsules hispid, prominently horned. June–Nov. Sandy and gravel slopes and flats, SW (Cape Peninsula).•

thunbergii Eckl. & Zeyh.   Closely leafy, hispid, erect or spreading subshrub to 40 cm. Leaves sessile, fascicled, linear-lanceolate, hispid below and on margins, spine-tipped. Flowers sessile, axillary, pink or white. Capsules hispid, with slender elongate horns. Sept.–Dec. Low sandy or clay slopes, SW (Darling to Franschhoek and Cape Peninsula).•

vulnerans Levyns   Intricately branched shrublet, mostly to 40 cm, branchlets often spiny. Leaves fascicled, sessile, linear, channelled, spine-tipped. Flowers sessile, axillary, white with purple keel. Capsules prominently horned. Aug.–Nov. Rocky slopes, KM (Witteberg, Bonteberg and SW Karoo).

NYLANDTIA   TORTOISE BERRY, SKILPADBESSIE   2 spp., Namaqualand to E Cape

scoparia (Eckl. & Zeyh.) Goldblatt & J.C.Manning   DUINEBESSIE   Erect shrub, sometimes tree-like, to 2.5 m. Leaves oblong, subsessile. Flowers solitary in upper axils, pinkish. Fruits red and fleshy. Mainly July–Aug. Sandy flats and rocky slopes, NW, SW (Bokkeveld Mts to Darling).•

spinosa (L.) Dumort.   Rounded, thorny shrub to 1 m, lateral branchlets short and pungent. Leaves oblong, subsessile. Flowers solitary in axils, purplish or pink and white. Fruits red or orange and fleshy to yellow and leathery. Mainly June–July. Sandy flats and slopes, NW, SW, AP, KM, LB, SE (Namaqualand and W Karoo to E Cape).

POLYGALA   BUTTERFLY BUSH, ERTJIEBLOM   c. 500 spp., cosmopolitan

A. Inflorescences lateral, peduncles borne at right angles to stem
B. Keel crest deeply fringed and side petals deeply lobed

lasiosepala Levyns   Like **P. teretifolia** but leaves persistently crisped-hairy, sepals often shortly hairy, and side petals with dorsal lobe shorter than sickle-shaped lower lobe. Sept.–Oct. Dry rocky slopes, NW (Namaqualand to Nardouw Mts).

pappeana Eckl. & Zeyh.   Diffuse, slender-stemmed subshrub or short-lived perennial to 30 cm. Leaves scale-like, adpressed. Flowers in short lateral racemes, purple, side petals deeply lobed with dorsal lobe larger. Sept.–Dec. Sandstone slopes, NW, SW, LB (Cedarberg to Hottentots Holland and Langeberg Mts).•

peduncularis DC.   Sprawling shrub to 1 m, often velvety on young parts. Leaves linear-oblong, margins revolute, sometimes hairy. Flowers large, few on stout axillary racemes, purple, side petals deeply subequally bilobed with a small median tooth. Sept.–Apr. Rocky slopes, NW, AP, SE (Nardouw Mts to Olifants River Mts and Agulhas to Knysna).•

refracta Burch. ex DC.   Slender subshrub to 40 cm, with slender, ridged stems. Leaves linear-lanceolate, ascending. Flowers in short axillary racemes, purple, side petals shortly bilobed. Sept.–Apr. Rocky sandstone slopes, NW, SW, LB, SE (Cedarberg Mts to E Cape).

teretifolia L.f.   Rounded shrublet to 80 cm, with stems slender and shortly velvety. Leaves linear, spreading-upcurving, often crisped-hairy when young, margins revolute. Flowers few in short lateral or sometimes terminal racemes, purple to pink, side petals deeply bilobed, dorsal lobe larger. Mainly Oct.–June. Dry stony karroid slopes, KM, LB (Little Karoo to Brak River).•

B. Either keel crest or side petals entire or shortly lobed

asbestina Burch.   Softly woolly, rounded shrublet to 20 cm. Leaves leathery, oblong to obovate, obtuse, glabrescent with age. Flowers 1–3 in short axillary racemes, blue, side petals entire. Nov.–Mar. Dry karroid slopes, KM (Little Karoo to E Cape).

brachyphylla Chodat   Rigid, stiffly branched shrublet to 35 cm, with ridged stems. Leaves sparse, elliptic, apiculate, shortly hairy. Flowers few on short, flattened axillary racemes, purple, outer sepals shortly hairy, side petals deeply lobed, dorsal lobe broadest, keel with broad, shallowly lobed crest. Sept.–Dec. Sandstone slopes, NW (Cedarberg Mts).•

illepida E.Mey. ex Harv.   Slender-stemmed, sparsely leafy shrublet to 15 cm. Leaves ascending, lanceolate-elliptic. Flowers few in lax axillary racemes, purple, side petals entire, spathulate. Mainly Oct.–Nov. Rocky slopes in grassland, SE (Van Staden's Mts to E Cape).

lehmanniana Eckl. & Zeyh.   Like **P. ludwigiana** but flowers white or pink, side petals unequally bilobed, dorsal lobe largest, keel with broad crest, barely or not fringed. Sept.–Nov. Clay and sandstone slopes, NW, SW (Cedarberg Mts to Elim).•

ludwigiana Eckl. & Zeyh.   Diffuse subshrub to 20 cm, with slender, sprawling, ridged stems. Leaves linear-elliptic. Flowers on flattened, axillary racemes, pink, side petals unequally bilobed, dorsal lobe largest, keel with minute, undivided crest. Mainly Sept.–Dec. Rocky sandstone slopes, NW, SW (Bokkeveld Mts to Tulbagh).•

nematocaulis Levyns   Diffuse, slender-stemmed subshrub or short-lived perennial to 30 cm. Leaves scale-like, adpressed. Flowers in short lateral racemes, purple, side petals shallowly lobed, keel crest about as long as body. Oct.–Apr. Marshy sandstone slopes, NW, SW (Agterwitzenberg and Cape Peninsula to Pearly Beach).•

parkeri Levyns   Like **P. ludwigiana** but side petals oblong, obtuse, entire or nearly so and carina with broad, entire, undulate crest. Sept.–Oct. Rocky slopes, SW (Franschhoek Mts to Somerset West).•

scabra L. (= *Polygala affinis* DC.)   Sprawling, sparsely leafy shrublet to 40 cm, hairy on young parts. Leaves linear-lanceolate, margins slightly revolute, shortly hairy. Flowers few in slender, axillary racemes, purple, outer sepals often shortly hairy, side petals shortly and equally lobed. July–Oct. Rocky slopes, NW, SW, KM, LB (Namaqualand and Karoo to Little Karoo and Riversdale).

A. Inflorescences terminal
*C. Side petals unequally bilobed (see also **P. teretifolia**)*

bracteolata L.   Few-branched, erect or sprawling shrub to 1 m, sometimes slightly hairy. Leaves lanceolate to elliptic, margins sometimes slightly thickened or revolute. Flowers many in terminal racemes, pink or purple, side petals unequally bilobed, lower lobe the longest, floral bracts prominent. Mainly Sept.–Nov. Sandstone slopes and flats, NW, SW, LB, SE (Gifberg to Uitenhage).•

fruticosa P.J.Bergius   Pubescent or glabrous shrub to 2 m. Leaves opposite, subsessile, lanceolate to ovate, cordate at base. Flowers in short racemes at tips of branchlets, purple, inner sepals often green, side petals deeply unequally bilobed, lower lobe often longest. Mainly Sept.–Nov. Rocky sandstone and clay slopes, NW, SW, AP, KM, LB, SE (Gydouw Pass and Houwhoek to KwaZulu-Natal).

microlopha DC. (incl. **P. levynsiana** Paiva)   Sparsely leafy shrub to 60 cm. Leaves leathery, linear to elliptic, usually shortly ciliate. Flowers in short, terminal umbel-like racemes, purple, side petals unequally bilobed with lower lobe linear and elongate, crest of keel reduced and short. Mainly Aug.–Nov. Rocky sandstone and clay slopes, KM, LB, SE (Montagu to E Cape).

myrtifolia L. (= *Polygala pinifolia* Poir.) SEPTEMBERBOS Sprawling or erect shrub to 2 m, often velvety on young parts. Leaves ascending, linear with margins slightly revolute to elliptic-obovate and flat. Flowers large, in short terminal racemes, purple, side petals bilobed, lower lobe much longer than dorsal. Mainly July–Oct. Rocky slopes, NW, SW, AP, KM, LB, SE (Bokkeveld Mts to KwaZulu-Natal).

C. Side petals entire or equally bilobed

dasyphylla Levyns   Closely leafy, hispid subshrub to 20 cm, with sprawling or ascending branches from base. Leaves linear, leathery, spreading. Flowers crowded in terminal, umbel-like racemes, pink to mauve with inner sepals white inside, side petals entire, obtuse. Aug.–Oct. Limestone hills, AP (Agulhas coast).•

ericaefolia DC.   Closely leafy shrublet to 40 cm. Leaves ascending, linear, channelled, sparsely ciliate below. Flowers in short, terminal subumbellate racemes, purple, side petals obtuse, with a short finger-like process below apex. Mainly Aug.–Nov. Sandy coastal slopes and flats, SE (George to Port Elizabeth).•

garcinii DC.   Soft subshrub to 40 cm, with slender trailing branches from woody base. Leaves linear to acicular. Flowers in long terminal racemes, purple, side petals obtuse, sparsely hairy below. Mainly Sept.–Dec. Sandy and clay slopes, NW, SW, AP, KM, LB, SE (Bokkeveld Mts to Knysna).•

hispida Burch.   Densely leafy, softly hairy subshrub to 30 cm, stems slender, sprawling or ascending. Leaves ascending, imbricate, broadly ovate to elliptic. Flowers many in congested terminal racemes, purple, side petals entire, spathulate. Oct.–Dec. Grassy slopes, SE (George to Mpumalanga).

langebergensis Levyns   Few-branched, closely leafy subshrub to 60 cm. Leaves elliptic, margins translucent, midrib prominent beneath. Flowers crowded in terminal, subumbellate racemes, mauve, side petals entire, obtuse. Sept.–Nov. Rocky sandstone slopes, LB (Langeberg Mts: Riversdale).•

leptophylla Burch.   Erect, sparsely branched shrub to 2 m. Leaves linear-elliptic, leathery. Flowers in elongate terminal racemes, mauve, side petals entire. May–Oct. Stony slopes, NW (Namaqualand and Karoo to Cedarberg Mts).

meridionalis Levyns   Closely leafy shrublet to 40 cm. Leaves ascending, linear, sparsely ciliate below. Flowers in short, terminal subumbellate racemes, purple, side petals obtuse. Mainly Sept.–Nov. Coastal sandy and limestone slopes and flats, SW, AP (Cape Peninsula to De Hoop).•

pottebergensis Levyns   Like **P. meridionalis** but outer sepals ciliate and obtuse. Mainly Sept.–Oct. Rocky sandstone slopes, AP (Potberg to Cape Infanta).•

pubiflora Burch. ex DC.   Stiffly branched, hairy shrublet to 20 cm, with sprawling to prostrate stems. Leaves ovate, cordate at base, glabrescent above. Flowers in terminal racemes, purple, side petals shortly and equally bilobed. Aug.–Mar. Limestone and stony clay, AP, LB (Cape Infanta to Mossel Bay).•

recognita Chodat   Like **P. umbellata** but keel shorter than 6 mm, with shallowly fringed crest, and stigma terminal. Sept.–Oct. Sandy slopes and flats, NW, SW (Piketberg to Cape Peninsula).•

triquetra C.Presl   Subshrub branching from base, mostly to 50 cm, with stems 3-angled, often winged. Leaves narrowly lanceolate-elliptic, rigid, margins thickened and hyaline. Flowers in terminal racemes, purple, side petals shortly, subequally bilobed. June–Oct. Rocky sandstone slopes, LB, SE (Langeberg Mts: Swellendam to Tsitsikamma Mts).•

umbellata L.   Slender subshrub to 40 cm, with sprawling branches from woody base. Leaves linear to narrowly elliptic. Flowers in dense, terminal, umbel-like clusters, purple, entire, obtuse. Aug.–Nov. Sandy and clay flats and lower slopes, NW, SW, AP, LB, SE (Ceres and Caledon to Humansdorp).•

uncinata E.Mey. ex Meisn.   Sprawling subshrub to 25 cm. Leaves ascending, linear, margins revolute, apex obtuse-apiculate and recurved. Flowers in terminal racemes, purple, side petals entire, obtuse. Dec.–Apr. Sandstone slopes in grassy fynbos, SE (Humansdorp to Zimbabwe).

virgata Thunb.   Slender-stemmed shrub, sometimes to 2 m, stems only leafy above. Leaves mostly lanceolate-elliptic. Flowers many in terminal racemes, uniformly dark purple, side petals entire, pedicels hairy. Mainly Oct.–Dec. Sandstone or clay or limestone slopes, often forest margins, AP, KM, LB, SE (Swellendam to tropical Africa).

wittebergensis Compton   Like **P. microlopha** but side petals entire, obtuse, with a shortly hairy zone below. Aug.–Oct. Rocky sandstone slopes, NW, KM, LB, SE (Cedarberg Mts to Uitenhage).•

sp. 1   Like **P. virgata** but plants glabrous including pedicels, somewhat twiggy above and flowers with crest of keel very small. Mainly Jan.–Mar. Limestone flats, AP (De Hoop).•

POLYGONACEAE

1. Perianth persistent, with short sharp spines ....**Emex**
1.' Perianth without spines:
   2. Inner perianth segments much larger than outer, enlarged in fruit ....**Rumex**
   2.' Perianth segments subequal, inner not accrescent ....**Polygonum**

EMEX   DEVIL'S THORN, DUIWELTJIE   2 spp., southern Africa, Mediterranean basin

australis Steinh.   Monoecious annual to 30 cm with sprawling to prostrate branches. Leaves long petiolate, hastate. Flowers in axillary clusters, greenish. Fruits spiny. Sept.–Oct. Sandy and stony flats and lower slopes, NW, SW, AP, KM, SE (Namaqualand and NW Province, Lambert's Bay to Uitenhage).

POLYGONUM   KNOTGRASS   c. 600 spp., cosmopolitan

*****aviculare** L.   VARKGRAS   Sprawling to prostrate annual to 30 cm, with striate stems. Leaves elliptic-oblanceolate, stipules papery, lacerate. Flowers 3–5 in axils, 5-merous, pink. Fruits shorter than perianth. Mainly Sept.–Mar. Disturbed places and saline marshes, NW, SW, AP, LB, SE (Eurasian weed).

maritimum L.   Sprawling or prostrate shrublet to 50 cm. Leaves elliptic, grey, imbricate, margins revolute, stipules conspicuous, papery, lacerate with age. Flowers 1–3 in axils, 5-merous, white to pink. Fruits longer than perianth. June–Feb. Sandy beaches, NW, SW (Namaqualand to Hermanus).

undulatum (L.) P.J.Bergius   Twiggy shrublet to 50 cm, with flaking bark. Leaves oval-oblong, punctate, margins finely crisped, stipules papery. Flowers crowded in terminal spikes, 4-merous, creamy green to pink. Fruits shorter than perianth. Nov.–Apr. Clay and sandy slopes, SW, AP, SE (Cape Peninsula to Uitenhage).•

RUMEX   SORREL   c. 200 spp., cosmopolitan

*****acetosella** L.   BOKSURING, SHEEP SORREL   Dioecious, rhizomatous perennial to 30 cm. Leaves often tufted, oblanceolate-hastate, long-petiolate, stipules conspicuous, papery. Flowers in axillary clusters on branched spikes. Fruits small, to 2 mm long, articulating at pedicel apex. Mainly Sept.–Nov. Disturbed places, NW, SW, LB, SE (cosmopolitan weed).

cordatus Desf.   TONGBLAAR   Like **R. lativalvis** but leaves ovate-cordate, often prostrate. July–Sept. Sandy flats and slopes, NW, SW, AP, LB, SE (Namaqualand and W Karoo to E Cape).

lativalvis Meisn.   Monoecious, tuberous perennial spreading on creeping rhizomes, mostly to 25 cm. Leaves mostly basal, sagittate-hastate, long-petiolate. Flowers in axillary clusters on branched spikes. Fruits enclosed in enlarged, papery sepals forming triangular-cordate wings, articulated below middle of pedicels. Aug.–Oct. Mostly clay and limestone slopes and flats, NW, SW, AP, ?SE (Clanwilliam to De Hoop, ?Uitenhage).•

sagittatus Thunb.   RANKSURING, CLIMBING SORREL   Sprawling or climbing, usually dioecious perennial. Leaves sagittate, margins undulate, often minutely crisped, with short, papery stipules. Flowers in stalked clusters forming large panicles, white. Fruits tiny, enclosed in enlarged orbicular sepals forming wings 5–8 mm diam., articulated near pedicel base. Jan.–Apr. Bush and forest margins, SW, LB, SE (Riviersonderend Mts to Port Elizabeth, widespread in southern Africa).

PORTULACACEAE

1. Leaves alternate, with papery stipules; ovules many; sepals deciduous; petals free, fugaceous ....**Anacampseros**
1.' Leaves opposite, without stipules; ovule solitary; sepals persistent, becoming somewhat rigid; petals fused into a short tube ....**Portulacaria**

ANACAMPSEROS   25 spp., Africa

albidiflora Poelln.   Like **A. retusa** but leaves truncate, clavate, woolly in axils, and seeds winged. Mainly Oct.–Jan. Stony slopes, NW, KM (Worcester to Oudtshoorn and S Karoo).

arachnoides (Haw.) Sims   Like **A. retusa** but leaves woolly in axils, ovoid with recurved tip. Nov.–Jan. Habitat? NW, KM, SE (Worcester to E Cape).

comptonii Pillans   Short-stemmed succulent to 5 cm. Leaves obovoid, shorter than 10 mm. Flowers solitary on short, filiform peduncles, white to pink, 3–5 mm diam., stamens 5. Mar.–Apr. Sandstone rock pavement, NW (Bokkeveld Mts to Clanwilliam).•

lanceolata (Haw.) Sweet (= *Anacampseros nebrownii* Poelln.)   Like **A. telephiastrum** but leaf bristles longer, often longer than leaves, flowers white or pink and seeds winged on angles. Oct.–Dec. Rock outcrops, NW, SW, AP, LB, SE (Richtersveld and W Karoo to Langkloof).

papyracea E.Mey. ex Fenzl   GANSMIS   Low perennial to 8 cm, with sprawling succulent branches. Leaves minute, hidden by white papery overlapping scales. Flowers solitary at branch tips, subsessile, whitish, stamens c. 20. Oct.–Nov. Quartz pebble patches, KM (Little Karoo to S Karoo).•

retusa Poelln.   Short-stemmed succulent to 12 cm. Leaves flattened-obovoid, shorter than 10 mm. Flowers on branched peduncles, pink, stamens 25. Mainly Sept.–Dec. Rocky slopes and flats, NW, SW, LP (S Namibia to Breede River valley and Riversdale).

telephiastrum DC.   GEMSBOKSURING   Short-stemmed succulent to 15 cm. Leaves ovate, acute, 10–30 mm long, papillate, with very short basal bristles. Flowers on stout peduncles, pink, stamens 30–45. Nov.–Dec. Rocky flats and slopes, NW, KM (Worcester to E Cape).

PORTULACARIA   SPEKBOOM   3 spp., South Africa and Namibia

afra Jacq.   Succulent shrub or small tree to 3 m. Leaves opposite, fleshy, obovate. Flowers in fascicles on terminal branches, pink. Oct.–Nov. Dry rocky slopes, KM, SE (Little Karoo to N provinces).

PRIMULACEAE

1. Plants annual; leaves opposite; flowers solitary in the axils; stamen filaments bearded ....**Anagallis**
1.' Plants perennial; leaves basal and cauline, alternate; flowers in racemes or panicles; stamen filaments hairless:
   2. Flowers white to pink; staminodes present in the mouth of the tube; stamens included; ovary half-inferior ....**Samolus**
   2.' Flowers red to purple; staminodes lacking; stamens exserted; ovary superior ....**Lysimachia**

ANAGALLIS   PIMPERNEL   c. 30 spp., cosmopolitan

*****arvensis** L.   SCARLET PIMPERNEL, BLOUSELBLOMMETJIE   Diffuse, branching annual to 20 cm. Leaves sessile, ovate. Flowers solitary in axils, often glandular-ciliate, blue or scarlet. July–Feb. Disturbed sites, NW, SW, KM, LB, SE (cosmopolitan weed).

huttonii Harv.   Sprawling, rarely branched annual to 30 cm. Leaves petiolate, orbicular. Flowers solitary in axils, white or pale pink. Oct.–Jan. Damp places, LB, SE (Riversdale to Mpumalanga).

LYSIMACHIA   LOOSESTRIFE   c. 200 spp., mostly N temperate

nutans Nees   CAPE LOOSESTRIFE   Perennial to 60 cm. Leaves lanceolate, margins slightly revolute. Flowers in a dense raceme, campanulate with exserted stamens, red to purple. Nov.–Jan. Moist places, SE (Uitenhage to E Cape).

SAMOLUS   WATER-PIMPERNEL   c. 15 spp., cosmopolitan, mostly in saline marshes

porosus (L.f.) Thunb.   Rhizomatous perennial with stiff stems to 60 cm. Leaves radical, oblanceolate. Flowers in racemes with bracteoles basal, white to pale pink. Nov.–Feb. Coastal marshes, NW, SW, AP, LB, SE (Namaqualand to KwaZulu-Natal).

*****valerandi** L.   BROOK WEED   Tufted perennial or biennial to 60 cm. Leaves basal and cauline, obovate. Flowers in racemes or panicles with bracteoles in middle of bent pedicel, minute, white. Nov.–Jan. Moist places, NW, SW, AP, LB, SE (scattered cosmopolitan weed).

PROTEACEAE

1. Flowers paired in axil of a bract; ovules 2:
   2. Leaves in whorls, oblanceolate, toothed, discolorous; fruit a velvety indehiscent achene ....**Brabejum**
   2.' Leaves alternate, usually terete but if flattened then both surfaces similar; fruit a glabrous, often warty, dehiscent follicle ....**Hakea**
1.' Flowers solitary in axil of a bract; ovule 1:
   3. Flowers unisexual:
      4. Female flowers in cones formed by woody, imbricate bracts; male flowers sessile in densely congested, globose or conical inflorescences, rarely spicate ....**Leucadendron**
      4.' Female flowers in densely involucrate, foliaceous heads, arranged racemosely on a central cone of tissue; male flowers in lax racemes, pedicellate ....**Aulax**
   3.' Flowers bisexual:
      5. Leaves pinnately or bipinnately divided, or at least divided at some stage; leaflets usually terete or subterete:

6. Inflorescence globose or cylindrical; flowers arranged in sessile groups of 4, each flower subtended by a floral bract that becomes woody in postpollination phase ....**Paranomus**
6.' Inflorescence a panicle, a capitulum or a panicle of capitula, floral bracts not enlarging in postpollination phase..**Serruria**
5.' Leaves entire or with apical teeth:
   7. Inflorescence a panicle, raceme or lax spike:
      8. Broad-leaved trees; inflorescence a lax terminal spike...**Faurea**
      8.' Terete-leaved shrubs; inflorescence a panicle, raceme or condensed raceme:
         9. Inflorescence a panicle; flowers actinomorphic, straight in bud...........................**Sorocephalus**
         9.' Inflorescence a raceme or condensed raceme; flower zygomorphic or at least abaxially curved in bud ....**Spatalla**
   7.' Inflorescence a 3–many-flowered capitulum; capitula terminal and solitary, or axillary and often numerous towards apex of a flowering shoot, or a terminal panicle of capitula:
      10. Capitula solitary and terminal or a terminal panicle of capitula:
         11. Capitula surrounded by large, highly coloured, involucral bracts; perianth separated to base into 2 parts, an adaxial sheath of 3 fused perianth segments and a free abaxial segment ....**Protea**
         11.' Capitula surrounded by small, insignificant, involucral bracts; perianth segments free or fused into a distinct tube at base:
            12. Leaves terete, acicular:
               13. Perianth limbs glabrous ....**Diastella**
               13.' Perianth limbs pubescent ....**Sorocephalus**
            12.' Leaves flat:
               14. Perianth segments free to base, perianth tube absent; hypogynous scales usually absent ....**Diastella**
               14.' Perianth segments fused below to form a distinct tube 3–8 mm long; hypogynous scales present...**Vexatorella**
      10.' Capitula axillary:
         15. Involucral bracts small, green or brown, cartilaginous, papyraceous or woody, not brightly coloured:
            16. Capitula usually pedunculate or narrowed to a peduncular region, globose or ovoid with more than 15 flowers per capitulum ....**Leucospermum**
            16.' Capitula sessile, massed among upper leaves on shoot and partly hidden by them, with 3–15 flowers per capitulum ....**Mimetes**
         15.' Involucral bracts large, conspicuous, usually brightly coloured, but occasionally brown:
            17. Perianth separated to base into 2 parts, an adaxial sheath of 3 fused perianth segments and a free, abaxial segment ....**Protea**
            17.' Perianth fused into a tube at base:
               18. Capitula sessile...**Mimetes**
               18.' Capitula pedunculate ....**Orothamnus**

AULAX• feather bush   3 spp., W Cape

cancellata (L.) Druce   Single-stemmed dioecious shrub to 2.5 m. Leaves needle-like, sometimes linear-spathulate, terete or subterete. Flowers in terminal racemes, yellow, female flowers congested, with involucre of sterile branchlets. Nov.–Feb. Sandstone slopes, SW, KM, LB, SE (Cape Peninsula to Kouga Mts).•

pallasia Stapf   Dioecious, resprouting, sparsely branched shrub to 2 m. Leaves needle-like to linear. Flowers in terminal racemes, yellow, female flowers congested, with involucre of sterile and fertile branchlets. Jan.–Apr. Sandstone slopes, NW, SW, LB (Cold Bokkeveld to Swellendam).•

umbellata (Thunb.) R.Br.   Like **A. cancellata** but leaves flat, linear-spathulate to oblanceolate. Nov.–Feb. Sandstone slopes and flats, SW, AP (Kogelberg to Stilbaai).•

BRABEJUM•   WILD ALMOND, WILDE-AMANDEL   1 sp., W Cape

stellatifolium L.   Widely spreading tree to 8 m. Leaves in whorls of (4–)6(–9), oblanceolate, toothed. Flowers in dense racemes, white. Fruits almond-shaped, velvety. Dec.–Jan. Sandstone slopes, mostly near streams, NW, SW, LB (Gifberg to Riversdale).•

DIASTELLA•   SILKY PUFF   7 spp., W Cape

A. Perianth lobes unequal; floral nectaries present

myrtifolia (Thunb.) Salisb. ex Knight   Like **D. parilis** but with sprawling branches and involucral leaves oblong-elliptic. Sept.–Jan. Seeps on sandy slopes, SW (Elandskloof Mts).•

parilis Salisb. ex Knight   Single-stemmed, rounded shrublet to 70 cm. Leaves elliptic, with 1–3 apical calluses. Flower heads 1–few at branch tips, 15–20 mm diam., pink, involucral leaves ovate with margins hairy, pink, becoming brown and papery. July–Jan. Seeps on sandstone slopes at low alt., SW (Elandskloof Mts to Slanghoek).•

A. Perianth lobes equal; floral nectaries absent

buekii (Gand.) Rourke   Mat-forming shrublet to 0.2 × 1 m, with trailing branches. Leaves needle-like. Flower heads solitary, nested among leaves, pink, 10–15 mm diam., outer involucral leaves with hairy margins. Aug.–Nov. Sandy flats, SW (Wemmershoek and Franschhoek).•

divaricata (P.J.Bergius) Rourke   Sprawling, single stemmed shrublet to 0.5 × 3 m. Leaves elliptic. Flower heads solitary, 10–15 mm diam., pink. Jan.–Dec. Sandstone flats and slopes, SW (Cape Peninsula to Kleinrivier Mts).•

fraterna Rourke   Sprawling, single-stemmed shrublet to 0.7 × 1 m. Leaves lanceolate to elliptic. Flower heads solitary, white to cream-coloured, 10–15 mm diam., involucral leaves usually becoming brown and papery. Jan.–Dec. Seeps and streambanks, SW (Kogelberg to Kleinmond).•

proteoides (L.) Druce   Sprawling shrublet, 0.5 × 3 m, with hairy stems. Leaves linear, hairy when young. Flower heads solitary, c. 10 mm diam., pink, involucral leaves hairy. Mainly July–Feb. Sandy flats, SW (Mamre and Paarl to Cape Peninsula).•

thymelaeoides (P.J.Bergius) Rourke   Erect shrub to 1.5 m. Leaves elliptic to ovate. Flower heads solitary, 10–20 mm diam., white or pink. Mainly Aug.–Nov. Sandstone slopes, SW (Hottentots Holland Mts to Betty's Bay).•

FAUREA   BEECHWOOD   18 spp., southern and tropical Africa, Madagascar

macnaughtonii E.Phillips   TERBLANS   Tree to 25 m. Leaves elliptic-lanceolate. Flowers in dense, pendent spikes, white or pale pink, scented. Dec.–Feb. Evergreen forests, SE (George to Knysna, E Cape to Mpumalanga, Madagascar).

*****HAKEA**   HAKEA   c. 100 spp., Australia

*****gibbosa** (Sm.) Cav.   ROCK HAKEA   Shrub or small tree to 4 m. Leaves needle-like, 40–80 mm long. Flowers axillary, cream-coloured. Fruits woody with knobs. June–Sept. Sandstone slopes, SW (Cape Peninsula to Caledon, E Australian weed).

*****drupacea** (C.F.Gaertn.) Roem. & Schult. (= *Hakea suaveolens* R.Br.)   SWEET HAKEA   Shrub or small tree to 4 m. Leaves pinnatisect, lobes needle-like. Flowers in axillary clusters, fragrant, cream-coloured. Fruits glossy, smooth. May–June. Sandstone slopes, SW, LB (W Australian weed).

*****sericea** Schrad.   SILKY HAKEA   Shrub or small tree to 5 m. Leaves needle-like, 30–40 mm long. Flowers in axillary clusters, white, on silky pedicels. Fruits warty. June–Oct. Sandstone slopes, NW, SW, AP, LB, SE (SE Australian weed).

LEUCADENDRON   CONE BUSH, TOLBOS   83 spp., N Cape to KwaZulu-Natal, mainly W Cape

A. Fruits retained in cones, cone scales tightly overlapping until burned
B. Fruits rounded

album (Thunb.) Fourc.   Dioecious shrub to 2 m. Leaves linear-oblanceolate, silvery appressed-hairy, 28–42 mm long (male), 45–59 mm (female), involucral leaves larger. Male flower heads c. 15 mm diam., female c. 26 mm diam., slightly scented. Nov.–Dec. Sandstone slopes, KM, LB, SE (Swartberg Mts and Langeberg to Great Winterhoek Mts).•

argenteum (L.) R.Br.   SILVER TREE, WITTEBOOM   Dioecious tree to 10 m. Leaves lanceolate, appressed silvery hairy, fringed, to 150 mm long, involucral leaves similar. Male flower heads c. 50 mm diam., female c. 40 mm diam. Sept.–Oct. Granite and clay slopes, SW (Cape Peninsula to Hottentots Holland Mts).•

cinereum (Sol. ex Aiton) R.Br.   Dioecious shrub to 1 m. Leaves linear-oblanceolate, 27 mm long (male), 47 mm long (female), glabrescent, involucral leaves similar. Male flower heads c. 17 mm long, female c. 13 mm long, slightly sweet-scented. Sept.–Oct. Sandy flats, NW, SW (Berg River mouth to Kraaifontein).•

dregei E.Mey. ex Meisn.   Sprawling, dioecious shrublet to 60 cm. Leaves linear-oblanceolate, glabrescent, 25 mm long (male), 45 mm long (female), involucral leaves similar. Male flower heads c.

16 mm diam., female c. 24 mm diam., banana-scented. Nov.–Dec. High sandstone slopes, KM (Swartberg Mts).•

galpinii E.Phillips & Hutch.   Sturdy, dioecious shrub to 3 m. Leaves linear-oblanceolate, to 40 mm long (male), to 57 mm long (female), involucral leaves similar. Male flower heads c. 15 mm long, female c. 17 mm long, with foetid-yeasty scent. Oct.–Nov. Sandy coastal flats, AP (De Hoop to Mossel Bay).•

levisanus (L.) P.J.Bergius   Much-branched, dioecious shrub to 2 m. Leaves oblanceolate-spathulate, c. 10 mm long, glabrescent, involucral leaves similar. Male flower heads c. 16 mm diam., female c. 11 mm diam., slightly sweet-scented. Sept.–Oct. Damp sandy flats, SW (Mamre to Cape Flats).•

linifolium (Jacq.) R.Br.   Densely branched, dioecious shrub to 2 m. Leaves linear-oblanceolate, 7–27 mm long (male), 15–35 mm long (female), involucral leaves similar. Male flower heads 10–12 mm long, female 12–14 mm long, yeast-scented. Sept.–Oct. Waterlogged coastal flats, SW, AP (Eersterivier to Riversdale).•

rubrum Burm.f.   Dioecious shrub to 2.5 m. Leaves oblanceolate, glabrescent, to 34 mm long (male), to 70 mm long (female), involucral leaves similar. Male flower heads clustered, c. 5 mm diam., female c. 20 mm diam. Aug.–Sept. Sandstone slopes, NW, SW, KM, LB, SE (Bokkeveld to Baviaanskloof Mts).•

B. Fruits flattened and winged
C. Cone scales flattened, not overlapping at tips; leaves needle-like in young plants

comosum (Thunb.) R.Br.   Dioecious shrub to 1.7 m. Leaves needle-like or the upper flattened, oblanceolate, 35 mm long (male), to 45 mm long (female), involucral leaves similar, pale green to yellow. Male flower heads 10–13 mm diam., female c. 16 mm diam., sweetly scented. Oct.–Dec. Sandstone slopes, SW, KM, LB, SE (Du Toitskloof to Kouga Mts and Swartberg Mts).•

muirii E.Phillips   Dioecious shrub to 2 m. Leaves dimorphic, needle-like below, spathulate-obovate above, leathery, to 30 mm long (male), to 40 mm long (female), involucral leaves similar. Male flower heads stalked, to 40 × 13 mm, female to 30 × 13 mm. Nov.–Dec. Coastal limestone slopes and flats, AP (Agulhas to Stilbaai).•

nobile I.Williams   Stout, dioecious shrub to 4 m. Leaves needle-like, c. 40 mm long (male), c. 58 mm long (female), involucral leaves similar, pale green to ivory. Male flower heads spike-like, c. 40 mm long, female narrow, c. 28 mm long, foetid. Oct.–Mar. Dry sandstone slopes, SE (Kouga and Baviaanskloof Mts to S Karoo).

osbornei Rourke   Like **L. teretifolium** but up to 4 m high, leaves terete-pungent, and male inflorescences cylindric (not globose). Oct.–Nov. Sandy or gravel slopes, KM (W Little Karoo mts).•

platyspermum R.Br.   Dioecious shrub to 1.7 m. Leaves dimorphic, the juvenile needle-like, adult oblanceolate, to 40 mm long (male), to 70 mm long (female), involucral leaves longer, yellowish. Male flower heads c. 12 mm diam., female c. 14 mm diam. Sept.–Oct. Sandy and gravel slopes, SW, AP (Villiersdorp to Agulhas coast).•

spirale (Salisb. ex Knight) I.Williams   Dioecious shrub to 1 m. Leaves terete, thinly silky, 4–9 mm long, involucral leaves similar. Male flower heads 6–8 mm long, female unknown. Nov.–Jan. Marshy flats, NW (Breede River valley: Wolseley to Botha).•

teretifolium (Andrews) I.Williams   Dioecious shrublet to 1 m. Leaves needle-like, c. 8 mm long (male), c. 22 mm long (female), involucral leaves similar. Male flower heads crowded, c. 7 mm diam., female c. 12 mm diam., slightly yeast-scented. Aug.–Sept. Sandstone and clay slopes, SW, KM, LB (Witteberg, and Kleinrivier Mts to Riversdale).•

C. Cone scales overlapping; leaves flattened in young plants

conicum (Lam.) I.Williams   Dioecious shrub or tree to 6 m. Leaves narrowly oblanceolate, 40–50 mm long, involucral leaves spreading, reddish. Male flower heads c. 15 mm diam., female c. 12 mm diam., slightly fruit-scented. Oct.–Nov. Sandstone slopes near streams, LB, SE (Langeberg Mts: Riversdale to Loerie).•

coniferum (L.) Meisn.   Dioecious shrub or small tree to 4 m. Leaves linear-oblanceolate, glabrescent, to 77 mm long (male), to 83 mm long (female), involucral leaves slightly larger, yellow. Male flower heads c. 18 mm diam., female c. 14 mm diam., bracts yellow. Aug.–Sept. Coastal sands, SW, AP (Cape Peninsula to Soetanysberg).•

cryptocephalum Guthrie   Rounded, dioecious shrub to 1 m. Leaves lanceolate, glabrescent, to 85 mm long, involucral leaves larger, yellow, concealing young flower heads. Male flower heads c. 15 mm diam., female c. 13 mm diam. Mar.–May. Clay slopes, SW (Groenland Mts to Potberg).•

diemontianum I.Williams   Spindly, dioecious shrub to 2 m, bushy at base. Leaves linear-elliptic, erect, c. 24 mm long (male), c. 36 mm long (female), involucral leaves lanceolate, yellow turning red. Male flower heads c. 12 mm diam., female c. 10 mm diam., yeast-scented. June. Sandstone slopes, NW (N Cedarberg and Grootwinterhoek Mts).•

discolor E.Phillips & Hutch.   Dioecious shrub to 3 m. Leaves obovate, 30–35 mm long (male), 40–48 mm long (female), involucral leaves larger, yellow tinged red. Male flower heads c. 25 mm diam., florets red in bud, female c. 16 mm diam., foetid. Sept.–Oct. Sandstone slopes, NW (Piketberg).•

eucalyptifolium H.Buek ex Meisn.   Dioecious shrub or tree to 5 m. Leaves linear-lanceolate, glabrescent, to 105 mm long, involucral leaves longer, yellow. Male flower heads c. 16 mm diam., female c. 12 mm diam., fruit-scented, bracts conspicuous, yellow. July–Oct. Forest margins and open sandstone slopes, SW, KM, LB, SE (Waboomsberg and Potberg to Van Staden's Mts, ?Grahamstown).•

flexuosum I.Williams   Slender, resprouting dioecious shrub to 2.5 m. Leaves linear-oblanceolate, erect, 25–27 mm long (male), 30–40 mm long (female), involucral leaves larger, yellow turning red. Male flower heads c. 14 mm diam., female c. 9 mm diam., yeast-scented. Apr.–May. Alluvial flats, NW (Breede River valley: Worcester).•

floridum R.Br.   Silvery dioecious shrub to 2 m, with pubescent branches. Leaves oblong, silvery silky, c. 25 mm long, involucral leaves larger, yellow. Flower heads c. 10 mm diam., slightly fruit-scented, bracts conspicuous, spreading, yellow. Sept.–Oct. Marshes and streamsides on sandy flats, SW (Cape Peninsula to Kuilsrivier).•

foedum I.Williams   Dioecious shrub to 2.5 m. Leaves oblanceolate, c. 27 mm long (male), c. 33 mm long (female), involucral leaves longer, pale green to ivory. Male flower heads c. 20 mm diam., female c. 12 mm diam., foetid. Sept.–Oct. Sandy flats, NW, SW (Piketberg to Hopefield).•

gandogeri Schinz ex Gand.   Rounded, dioecious shrub to 1.6 m. Leaves elliptic-oblanceolate, glabrescent, 42–85 mm long (male), 60–105 mm long (female), involucral leaves larger, yellow tinged red. Male flower heads c. 24 mm diam., female c. 18 mm diam., fruit-scented. Aug.–Oct. Rocky sandstone slopes, SW (Hottentots Holland to Bredasdorp Mts).•

lanigerum H.Buek ex Meisn.   Single-stemmed or resprouting and multistemmed dioecious shrub to 1.5 m. Leaves linear-oblanceolate, to 30 mm long, usually rough, involucral leaves linear-lanceolate, pale yellow. Male flower heads c. 11 mm diam., female 7–12 mm diam., foetid or yeast-scented. July–Sept. Clay or silt flats, SW (Darling to Strand).•

laureolum (Lam.) Fourc.   Rounded dioecious shrub to 2 m. Leaves oblong, to 75 mm long (male), to 95 mm long (female), glabrescent, involucral leaves larger, yellow, concealing young heads. Male flower heads c. 20 mm diam., female c. 14 mm diam., slightly fruit-scented. June–July. Sandstone slopes, SW, AP (Cape Peninsula to Potberg).•

loeriense I.Williams   Dioecious shrub to 2.5 m, branching from below. Leaves narrowly oblong, velvety, 30–35 mm long, often flushed red, involucral leaves slightly larger, glabrous above, whitish green. Male flower heads c. 12 mm diam., female c. 10 m diam. Dec.–Jan. Sandstone slopes, SE (Baviaanskloof Mts to Elandsberg).•

macowanii E.Phillips   Dioecious shrub to 2.5 m, with silky branches. Leaves linear-oblanceolate, thinly silky below, c. 53 mm long (male), c. 75 mm long (female), involucral leaves lacking. Male flower heads crowded, c. 14 mm diam., female c. 25 mm diam. May–July. Streamsides, SW (Cape Peninsula).•

meridianum I.Williams   Densely branched, dioecious shrub to 2 m. Leaves linear-oblanceolate, silky or glabrous, c. 40 mm long, involucral leaves longer, yellow. Flower heads c. 12 m diam., slightly scented. July–Aug. Limestone flats, AP (Gansbaai to Gouritsmond).•

microcephalum (Gand.) Gand. & Schinz   Dioecious shrub to 2 m. Leaves oblong, to 90 mm long, involucral leaves similar, yellow. Male flower heads c. 18 mm diam., female c. 11 mm diam., bracts conspicuous, obovate, brown and oily. July. Sandstone slopes, SW (Du Toitskloof to Riviersonderend Mts).•

modestum I.Williams   Slender, dioecious shrublet to 60 cm. Leaves linear-oblanceolate, rough, c. 18 mm long (male), c. 25 mm long (female), involucral leaves slightly longer, linear, pale yellow. Male flower heads c. 13 mm diam., female c. 8 mm diam. slightly foetid. Aug.–Sept. Clay and gravel flats, SW, AP (Botrivier to Potberg).•

procerum (Salisb. ex Knight) I.Williams   Like **L. foedum** but leaves larger, 26–37 mm long (male), 37–47 mm long (female), male florets red and bracts dimpled. Aug.–Sept. Sandstone slopes, NW (Gifberg to Piketberg).•

radiatum E.Phillips & Hutch.   Dioecious shrublet to 60 cm. Leaves oblanceolate, glabrescent, c. 22 mm long, involucral leaves lanceolate, softly hairy, cream-coloured. Flower heads 10–12 mm diam., slightly scented, bracts prominent, cream-coloured. Oct.–Dec. Sandstone slopes, LB (Langeberg Mts: Swellendam to Riversdale).•

rourkei I.Williams   Like **L. conicum** but leaves smaller, to 24 mm long, involucral leaves inconspicuous. Dec.–Jan. Stony or shale slopes, KM, SE (Swartberg to Kouga Mts).•

salicifolium (Salisb.) I.Williams   Dioecious shrub to 3 m. Leaves linear-falcate, to 60 mm long, involucral leaves similar, creamy yellow. Flower heads 9–10 mm diam., slightly fruit-scented, bracts spreading, yellow. July–Sept. Streams and seeps on flats and slopes, NW, SW, KM, LB (Olifants River to Langeberg Mts: Riversdale).•

salignum P.J.Bergius   Sprawling or erect, resprouting dioecious shrub to 2 m. Leaves linear-oblanceolate, 20–47 mm long (male), 48–58 mm long (female), male involucral leaves slightly longer, yellow, sometimes red, female larger, ivory or red. Male flower heads 10–14 mm diam., female 9–12 mm diam., sweet or yeast-scented. Apr.–Nov. Sandy and clay slopes and flats, NW, SW, AP, KM, LB, SE (Bokkeveld Mts to Grahamstown).

spissifolium (Salisb. ex Knight) I.Williams   Resprouting dioecious shrub to 1.3 m. Leaves linear-oblanceolate, 25–63 mm long (male), 27–80 mm long (female), involucral leaves larger, ivory or pale green. Male flower heads c. 18 mm diam., female 13–15 mm diam., lemon-scented. Aug.–Oct. Sandstone slopes, NW, SW, LB, SE (Gifberg to KwaZulu-Natal).

stelligerum I.Williams   Slender dioecious shrub to 1.3 m. Leaves erect, oblanceolate, rough, thinly silky, c. 23 mm long, involucral leaves longer, ivory. Male flower heads c. 16 m diam., female c. 10 mm diam., foetid. July–Aug. Gravel and clay flats, SW, AP (Elim to Agulhas plain).•

strobilinum (L.) Druce   Dioecious shrub to 2.6 m, branching below. Leaves elliptic, ciliate, to 67 mm long (male), to 80 mm long (female), involucral leaves larger, ivory. Male flower heads 24–36 mm diam., female 18–25 mm diam., yeast-scented. Sept.–Oct. Damp S-facing sandstone slopes, SW (Cape Peninsula).•

uliginosum R.Br.   Dioecious shrub to 4 m, bushy below. Leaves oblong, adpressed-silvery silky or glabrescent, 20–35 mm long, involucral leaves slightly larger, ivory to pale yellow. Male flower heads c. 15 mm diam., female c. 12 mm diam., slightly scented. Oct.–Dec. Sandstone slopes, LB, SE (Langeberg: Swellendam and Outeniqua to Baviaanskloof Mts).•

xanthoconus (Kuntze) K.Schum.   Dioecious shrub to 2 m. Leaves narrowly oblong-falcate, glabrescent, to 65 mm long, involucral leaves larger, yellow. Flower heads 10–11 mm diam. Aug. Sandstone slopes, SW (Cape Peninsula to Potberg).•

A. Fruits shed within a season, cone scales soon recurving to expose fruits
D. Fruits glabrous, uniformly coloured

arcuatum (Lam.) I.Williams   Sprawling or erect, resprouting dioecious shrub to 1.3 m. Leaves obovate-spathulate, margins thickened, red, 55–80 mm long, involucral leaves similar, yellow. Male flower heads 22–35 mm diam., female 14–33 mm diam., sweetly scented. Sept.–Oct. High sandstone slopes, NW (Cedarberg to Kwadouwsberg).•

barkerae I.Williams   Like **L. daphnoides** but leaves glabrescent, hairy when young, flower heads with fruity smell. Sept.–Oct. Dry sandstone slopes, NW, KM (S Cold Bokkeveld to Swartberg Mts).•

bonum I.Williams   Dioecious shrub to 1.6 m. Leaves spathulate, softly silvery hairy, to 55 mm long (males), to 62 mm (female), involucral leaves similar, pale yellow. Male flower heads c. 21 mm diam., female c. 18 mm diam., honey-scented, involucral bracts slender, softly hairy, yellow, c. 40 mm long. Oct.–Nov. High sandstone plateaus, NW (Cold Bokkeveld: Gideonskop).•

burchellii I.Williams   Dioecious shrub to 1.6 m. Leaves elliptic, margins reddish, c. 60 mm long, involucral leaves similar, yellow turning red. Male flower heads c. 42 mm diam., female c. 27 mm diam., lemon-scented, bracts recurved. Aug.–Sept. N-facing stony slopes, SW (Riviersonderend Mts).•

cadens I.Williams   Sprawling, dioecious shrublet to 50 cm diam. Leaves secund, narrowly oblanceolate, c. 22 mm long (male), c. 34 mm long (female), margins reddish, involucral leaves similar, sometimes yellowish. Male flower heads c. 15 mm diam., female c. 11 mm diam., spicy scented. Oct. Dry sandstone slopes, KM (Witteberg).•

chamelaea (Lam.) I.Williams   Dioecious shrub to 2.3 m, branching from base. Leaves linear-oblanceolate, c. 40 mm long, involucral leaves slightly longer, yellow. Flower heads c. 20 mm diam., strongly scented. Sept.–Oct. Sandstone flats, NW, SW (Cold Bokkeveld to Franschhoek valley).•

cordatum E.Phillips   Erect or sprawling, dioecious shrub to 1 m. Leaves oblong-elliptic, c. 50 mm long, involucral leaves larger, yellow to red. Flower heads nodding, male c. 25 mm diam., female c. 20 mm diam. pepper-scented, bracts recurved. June–July. Rocky sandstone slopes, NW, KM, LB (Langeberg Mts: Koo to Barrydale and Swartberg).•

daphnoides (Thunb.) Meisn.   Dioecious shrub to 1.5 m. Leaves lanceolate-elliptic, to 60 mm long, involucral leaves broader, yellow turning red. Male flower heads c. 42 mm diam., female c. 23 mm diam., with citrus scent. July–Sept. Granite slopes, SW (Du Toitskloof to Villiersdorp).•

elimense E.Phillips   Dioecious shrub to 1.5 m. Leaves elliptic-lanceolate, to 70 mm long, involucral leaves slightly larger, yellow. Male flower heads to 45 mm diam., female to 20 mm diam., strongly scented. July–Oct. Sandy flats, SW, AP (Viljoen's Pass to Gansbaai and Bredasdorp).•

glaberrimum (Schltr.) Compton   Spreading, much-branched, dioecious shrub to 1.3 m. Leaves oblanceolate, glabrous or hairy, 20–30 mm long, sometimes purplish, involucral leaves similar. Male flower heads 12–17 mm diam., female c. 10 mm diam., sweetly scented or foetid. Aug.–Oct. High sandstone slopes, NW (Cedarberg to Hex River Mts).•

globosum (J.Kennedy ex Andrews) I.Williams   Dioecious shrub to 2 m, with pubescent branches. Leaves ovate, to 60 mm long, involucral leaves slightly larger, yellow. Male flower heads c. 50 mm diam., female c. 23 mm diam., spicy scented, bracts oily, recurved above. Sept.–Oct. Loamy slopes, SW (Grabouw valley).•

grandiflorum (Salisb.) R.Br.   Like **L. globosum** but leaves pale green, twisted. Flowering time? Clay slopes, SW (Cape Peninsula).•

gydoense I.Williams   Like **L. sessile** but leaves smaller, narrowly oblanceolate, to 45 mm long, male flower heads c. 30 mm diam., female c. 12 mm diam. Oct. Sandstone slopes, NW (Cold Bokkeveld and Hex River Mts).•

loranthifolium (Salisb. ex Knight) I.Williams   Dioecious shrub to 2 m. Leaves elliptic to oblanceolate, 38–70 mm long, blue-green, involucral leaves similar. Male flower heads 20–40 mm diam., female c. 15 mm diam., foetid. July–Sept. Sandstone slopes, NW (Gifberg to Hex River Mts).•

meyerianum H.Buek ex E.Phillips & Hutch. Dioecious shrub to 2 m, with slender purplish branches. Leaves linear, c. 30 mm long, involucral leaves similar. Male flower heads c. 16 mm diam., female c. 7 mm diam., slightly scented. Aug.–Sept. Sandy plateaus, NW (Bokkeveld Mts).•

nitidum H.Buek ex Meisn.   Erect to spreading, resprouting dioecious shrub to 2 m, branching at base. Leaves linear-oblanceolate, silvery addressed-hairy, 9–15 mm long (male), 11–20 mm (female), involucral leaves twice as long, yellow. Male flower heads 15–20 mm diam., female 12–18 mm diam., sweetly scented. May–Nov. Sandstone plateaus, NW (Cedarberg Mts to Swartruggens).•

orientale I.Williams   Dioecious shrub to 1.3 m. Leaves lanceolate to oblanceolate, 90–100 mm long, involucral leaves similar, yellow. Male flower heads c. 35 mm diam., female c. 30 mm diam., sweetly scented, bracts recurved, red, oily. June–July. Sandstone slopes, SE (Cockscomb to Van Staden's Mts).•

pubescens R.Br.   Dioecious shrub to 2.5 m. Leaves oblanceolate, glabrous or silvery addressed-hairy, 16–28 mm long (male), 25–57 mm long (female). Male flower heads 9–18 mm diam., female 10–20 mm diam., sweet or yeast-scented. July–Oct. Sandstone slopes, NW, KM (Bokkeveld Mts to Witteberg and Klein Swartberg).•

pubibracteolatum I.Williams   Like **L. tinctum** but leaves shorter, to 64 mm long (male), to 75 mm long (female), bracts not oily. July–Aug. Dry sandstone slopes, KM, SE (E Swartberg and Outeniqua to Baviaanskloof Mts).•

remotum I.Williams   Dioecious shrub to 1.5 m, branching from below. Leaves linear-oblanceolate, addressed, glabrescent, c. 29 mm long (male), c. 42 mm long (female), involucral leaves similar. Male flower heads c. 16 mm diam., slightly scented. Aug.–Sept. Sandy plateaus, NW (Bokkeveld Mts).•

roodii E.Phillips   Dioecious shrub to 1.5 m. Leaves elliptic, to 44 mm long (male), to 57 mm long (female), involucral leaves similar, tinged red. Male flower heads c. 20 mm diam., female c. 15 mm diam., foetid, bracts pouched. Aug.–Sept. Sandstone rocks, NW (Gifberg).•

sericeum (Thunb.) R.Br. Slender, dioecious shrublet to 1 m. Leaves narrowly oblanceolate, adpressed-hairy, to 9 mm long (male), to 15 mm (female), involucral leaves similar. Male flower heads in small clusters, c. 11 mm diam., females c. 13 mm diam., sometimes in clusters, sweetly scented. May–Sept. Sandstone plateaus, NW (Cold Bokkeveld: Waboomsrivier).•

sessile R.Br. Dioecious shrub to 1.5 m. Leaves narrowly elliptic, to 64 mm long (male), to 80 mm long (female), involucral leaves similar, yellow turning red. Male flower heads c. 35 mm diam., female 14–18 mm diam., lemon-scented. July–Aug. Granitic slopes and flats, NW, SW (Witsenberg to Slanghoek and Jonkershoek to Kogelberg).•

sheilae I.Williams Dioecious shrub to 1 m. Leaves oblanceolate, 26–29 mm long, involucral leaves similar. Male flower heads c. 15 mm diam., female c. 6 mm diam. Aug. Sandstone plateaus, NW (Bokkeveld Mts).•

tinctum I.Williams Rounded shrub to 1.3 m. Leaves oblong-lanceolate, spreading-incurved below, to 90 mm long (male), to 115 mm (female), involucral leaves larger, yellow turning red. Male flower heads c. 35 mm diam., female c. 27 mm diam., spicy scented, bracts recurved, oily. July–Aug. Sandstone slopes, NW, SW, KM, LB (Hex River Mts to Hottentots Holland and Langeberg Mts).•

tradouwense I.Williams Dioecious shrub to 2 m. Leaves oblanceolate, c. 57 mm long (male), c. 67 mm long (female), involucral leaves larger, yellow. Male flowers heads to 25 mm diam., female c. 15 mm diam., sweetly scented, bracts red, recurved. June. Sandstone slopes, LB (Langeberg Mts: Tradouw Pass).•

D. Fruits variously hairy or mottled

brunioides Meisn. Dioecious, resprouting shrub to 2 m with many slender stems. Leaves linear to oblong, c. 23 mm long; involucral leaves similar. Male flower heads c. 17 mm diam., female c. 11 mm diam., foetid. Oct.–Nov. Sandy flats, NW, SW, KM, LB (Bokkeveld Mts to Piketberg and Bonteberg to Swellendam).•

concavum I.Williams Spindly, dioecious shrub to 1.5 m. Leaves imbricate, clasping, obovate-rotund, softly hairy, c. 9 mm long, involucral leaves similar. Male flower heads c. 20 mm diam., female c. 13 mm diam., unscented. Sept. Sandy plateaus, NW (Pakhuis Mts).•

coriaceum E.Phillips & Hutch. Resprouting, dioecious shrublet to 60 cm with many slender stems. Leaves obovate, c. 14 mm long, involucral leaves similar. Male flower heads c. 17 mm diam., female c. 8 mm diam. yeast or honey-scented. Sept. Gravel flats, SW, LB (Napier to Riversdale).•

corymbosum P.J.Bergius Slender, dioecious shrub to 2 m, with many branchlets clustered at base. Leaves stiffly subterete, erect, 15–20 mm long, involucral leaves similar. Male flower heads c. 12 mm diam., female c. 10 mm diam., foetid. Sept.–Oct. Damp clay flats, NW, SW (Michell's Pass to Paarl).•

dubium (H.Buek ex Meisn.) E.Phillips & Hutch. Like **L. concavum** but leaves smaller, 5–9 mm long. Aug.–Sept. Sandstone plateaus, NW (Cedarberg Mts).•

ericifolium R.Br. Rounded, dioecious shrub to 1.3 m (male), 3 m (female). Leaves dimorphic, lower terete, to 27 mm long, upper linear, to 8 mm long (male), to 39 mm long (female), involucral leaves similar. Male flower heads many, crowded, nodding, c. 3 mm diam., female 1-flowered, c. 8 mm diam. July–Aug. Dry rocky slopes, KM, LB, SE (Klein Swartberg and Langeberg Mts to George).•

laxum I.Williams Finely leafy, Like **L. corymbosum** but leaves narrower, c. 1 mm wide, male flower heads c. 9 mm diam. Sept.–Oct. Damp coastal flats, SW, AP (Hermanus to Agulhas).•

nervosum E.Phillips & Hutch. Dioecious shrub to 1.5 m. Leaves elliptic, glabrescent, c. 47 mm long (male), 60 mm (female), involucral leaves similar, shaggy, yellow. Male flower heads c. 22 mm diam., female c. 20 mm diam. Sept.–Oct. N-facing sandstone slopes, SW, LB (Riviersonderend Mts to Langeberg: Heidelberg).•

olens I.Williams Dioecious shrub to 1.2 m. Leaves erect, linear-lanceolate, 7–11 mm long, involucral leaves similar, yellow. Male flower heads c. 8 mm diam., female with 2 or 3 florets, c. 5 mm diam. June–July. N-facing sandstone slopes, SE (Outeniqua Mts: Grootdoring River).•

singulare I.Williams Spreading shrublet to 30 cm. Leaves linear-oblanceolate, adpressed, silvery hairy, c. 16 mm long, involucral leaves larger. Male flower heads c. 12 mm diam., female c. 18 mm diam. Oct. High rocky ridges, KM (Kammanassie Mts).•

sorocephalodes E.Phillips & Hutch. Sprawling, dioecious shrublet to 30 cm. Leaves terete, 12–18 mm long, involucral leaves similar. Male flower heads c. 20 mm diam., female c 13 mm diam., sweetly scented. Aug. High sandstone slopes, SE (Outeniqua to Baviaanskloof Mts).•

stellare (Sims) Sweet   Resprouting, multistemmed, dioecious shrub to 2 m. Leaves imbricate, clasping, oblong, 8–12 mm long, involucral bracts similar. Male flower heads 15–20 mm diam., female c. 8 mm diam., yeasty or foetid. Sept.–Oct. Silt flats, NW, SW (Aurora to Cape Flats).•

thymifolium (Salisb. ex Knight) I. Williams   Slender dioecious shrub to 2 m. Leaves imbricate, clasping, elliptic-oblong, c. 8 mm long, glabrescent, involucral bracts similar. Male flower heads c. 17 mm diam., females c. 13 mm diam., yeast-smelling. Aug.–Sept. Sandy flats, SW (Dassenberg to Klipheuwel).•

verticillatum (Thunb.) Meisn.   Slender shrub to 2 m. Leaves linear-oblanceolate, silvery adpressed-hairy, 22–25 mm long, involucral leaves similar. Male flower heads c. 11 mm diam., female c. 8 mm diam., slightly yeast-scented. Sept.–Oct. Silt flats, SW (Cape Flats around Paarl).•

LEUCOSPERMUM   PINCUSHION, LUISIESBOS   48 spp., South Africa to Zimbabwe, mainly W Cape

A. Style 10–35 mm long
B. Flower heads obconic with flat receptacle; pollen presenter scarcely thickened

gracile (Salisb. ex Knight) Rourke   Sprawling mat-forming shrublet to 0.5 × 1.5 m. Leaves secund, narrowly oblong, thinly hairy, with 1–3 apical teeth. Flower heads obconic, 25–30 mm diam., yellow, style 25–30 mm long, pollen presenter slender. July–Dec. Sandstone slopes, SW (Kleinrivier Mts to Bredasdorp).•

mundii Meisn.   Like **L. oleifolium** but leaves broadly ovate to cuneate, with 7–17 teeth and flower heads 10–20 mm diam. July–Nov. Sandstone slopes, LB (Langeberg Mts: Swellendam to Riversdale).•

oleifolium (P.J.Bergius) R.Br.   Rounded shrub to 1 m. Leaves ovate to lanceolate, with 1–5 apical teeth, glabrous or hairy. Flower heads obconic, 25–40 mm diam., yellow-green fading red, style 25–30 mm long, pollen presenter slender. Mainly Sept.–Oct. Sandstone slopes, SW (Slanghoek Mts to Caledon Swartberg and Kleinmond).•

saxatile (Salisb. ex Knight) Rourke   Sprawling shrub to 70 cm. Leaves narrowly oblong, with 1 (sometimes to 3) apical teeth, glabrescent. Flower heads obconic, 25–30 mm diam., pale lime-green to pink, style c. 30 mm long, pollen presenter slender. July–Feb. Dry rocky N-facing slopes, LB (Langeberg Mts: Garcia's Pass).•

B. Flower head subglobose with conical receptacle; pollen presenter noticeably thickened
*C. Leaves densely grey-felted or style strongly recurved (see also **L. wittebergense**)*

arenarium Rycroft   Sprawling shrublet to 75 cm, with trailing stems. Leaves secund, linear-oblong, densely grey-felted, with 1(–3) apical teeth. Flower heads 50–70 mm diam., yellow, style 30–35 mm long, incurved, pollen presenter ellipsoid; involucral bracts red. July–Oct. Sandy slopes, NW (Redelinghuys to Aurora).•

hamatum Rourke   Sprawling, mat-forming shrublet to 3 m diam. Leaves secund, narrowly oblong, glabrescent, with 3 apical teeth. Flower heads small, with 4–7 flowers, 15–20 mm diam., reddish, style strongly incurved, 18–20 mm long, pollen presenter conical. July–Nov. Dry rock outcrops, SE (Outeniqua Mts: Robinson Pass).•

harpagonatum Rourke   Like **L. hamatum** but leaves linear-oblong, channelled, with 1 apical tooth, flower heads with 7–12 flowers, style 20–25 mm long. Aug.–Nov. N-facing sandstone slopes, SW (Riviersonderend Mts: McGregor).•

hypophyllocarpodendron (L.) Druce   Sprawling to prostrate shrublet with trailing stems. Leaves secund, linear to oblanceolate, glabrous to grey-felted, with 2–4 apical teeth. Flower heads 30–40 mm diam., yellow, style 20–26 mm long, pollen presenter club-shaped. Aug.–Jan. Sandy flats, NW, SW, AP (Piketberg to Agulhas coast).•

parile (Salisb. ex Knight) Sweet   Like **L. rodolentum** but to 1.5 m, leaves linear-oblong, with 1–3 apical teeth, involucral bracts ovate, red, with only margins hairy. July–Nov. Sandy flats, SW (Dassenberg and Paardeberg).•

rodolentum (Salisb. ex Knight) Rourke   Erect or spreading shrub to 3 m. Leaves elliptic to cuneate, densely grey-velvety, with 3–6 apical teeth. Flower heads globose, 30–35 mm diam., bright yellow, style 15–25 mm long, pollen presenter ellipsoid; involucral bracts ovate. Aug.–Nov. Sandy flats and lower slopes, NW, SW (Namaqualand to Cape Peninsula).

tomentosum (Thunb.) R.Br.   Like **L. rodolentum** but resprouting shrub to 1 m, leaves linear-spathulate, channelled, 1–3 toothed and involucral bracts lanceolate. June–Nov. Coastal sands, SW (Hopefield to Bokbaai).•

*C. Leaves glabrous to thinly hairy or style strongly incurved (see also **L. hamatum**)*

bolusii Gand.   Rounded shrub to 1.5 m. Leaves ovate-elliptic, glabrescent, with 1 apical tooth. Flower heads subglobose, c. 20 mm diam., cream-coloured fading pale pink, style 15–20 mm long, pollen presenter ovoid. Sept.–Dec. Rocky sandstone slopes, SW (Gordon's Bay to Kogelbaai).•

calligerum (Salisb. ex Knight) Rourke   Shrub to 2 m. Leaves lanceolate to elliptic, grey-hairy, with 1(–3) apical teeth. Flower heads globose, 20–35 mm diam., cream-coloured fading dull red, style 21–25 mm long, pollen presenter conic-ovoid. July–Jan. Dry sandy slopes, NW, SW, KM, LB (Bokkeveld Mts to Bonteberg and Langeberg Mts: Riversdale).•

cordatum E.Phillips   Mat-forming shrublet to 2 m diam., with trailing branches. Leaves ovate-cordate, glabrescent, with 1 apical tooth. Flower heads depressed, globose, 30–40 mm diam., cream-coloured fading pink, anthers on filaments to 1.5 mm long, style incurved, 20–25 mm long, pollen presenter conical. Mainly Sept.–Nov. Shale slopes, SW (Kogelberg).•

heterophyllum (Thunb.) Rourke   Prostrate mat-forming shrublet. Leaves secund, linear to narrowly oblanceolate, glabrescent, with 3 (sometimes 1) apical teeth. Flower heads globose, 20–30 mm diam., cream-coloured fading reddish, style 18–21 mm long, pollen presenter ovoid. Mainly Sept.–Oct. Coastal slopes and flats, AP (Elim to De Hoop).•

muirii E.Phillips   Like **L. truncatum** but leaves with 3–7 apical teeth, flower heads 20–30 mm diam., perianth inflated and style 13–20 mm long. July–Oct. Coastal sands, AP (Stilbaai to Gouritsmond).•

pedunculatum Klotzsch   Prostrate, mat-forming shrublet to 30 cm, with exposed woody base. Leaves subsecund, linear, glabrescent, with 1 apical tooth. Flower heads usually solitary on lateral branchlets, globose, 25–30 mm diam., fragrant, cream-coloured fading reddish, style 17–20 mm long, pollen presenter ellipsoid. Mainly Sept.–Dec. Coastal sands, SW, AP (Danger Point to Agulhas).•

prostratum (Thunb.) Stapf   Resprouting, mat-forming shrublet to 4 m diam. Leaves secund, linear to oblong, subglabrous, with 1 apical tooth. Flower heads globose, 20–25 mm diam., scented, yellow fading orange, style 12–15 mm long, pollen presenter ellipsoid. July–Dec. Mainly coastal sands, SW, AP (Groenland Mts to Elim flats).•

royenifolium (Salisb. ex Knight) Stapf   Sprawling, mat-forming shrub to 3 m diam., with exposed rootstock. Leaves elliptic, glabrescent, with 1(–3) apical teeth. Flower heads globose, 10–20 mm diam., cream-coloured fading pink, scented, style 13–16 mm long, pollen presenter ellipsoid. July–Dec. Dry sandstone slopes, KM, SE (Swartberg to Kouga Mts).•

secundifolium Rourke   Mat-forming shrublet to 2 m diam. Leaves secund, oblanceolate, glabrescent, margins horny, with 1(–3) apical teeth. Flower heads globose, 10–15 mm diam., yellow fading red, style c. 14 mm long, pollen presenter club-shaped. Nov.–Jan. S-facing rocky upper slopes, KM (Klein Swartberg).•

truncatulum (Salisb. ex Knight) Rourke   Like **L. bolusii** but leaves elliptic-obovate, grey-felted and flower heads yellow fading red. Aug.–Dec. Sandy slopes and flats, SW (Groenland Mts to Bredasdorp).•

truncatum (H.Buek ex Meisn.) Rourke   Rounded shrub to 2 m. Leaves oblanceolate-cuneate, glabrescent, with 3 apical teeth. Flower heads globose, 30–40 mm diam., yellow fading orange, style 18–35 mm long, pollen presenter ellipsoid. Aug.–Dec. Coastal limestones, AP (Soetanysberg to Stilbaai).•

winteri Rourke   Rounded shrub to 1.3 m. Leaves broadly obovate to cuneate, glabrescent, with 5–14 apical teeth. Flower heads globose, c. 30 mm diam., yellow fading pink, style c. 20 mm long, pollen presenter club-shaped. July–Dec. Sandstone summits, LB (Langeberg Mts: Langkloof).•

wittebergense Compton   Like **L. calligerum** but pollen presenter club-shaped. Aug.–Jan. Dry rocky slopes, KM, SE (Witteberg and Swartberg to Kouga Mts).•

A. Style longer than 35 mm
D. Style 55–80 mm long, pollen presenter slender, bent at right angles to style or style sharply reflexed after anthesis

catherinae Compton   Shrub to 3 m. Leaves oblanceolate-elliptic, with 3 or 4 teeth. Flower heads discoid, c. 150 mm diam., orange, style 70–80 mm long, twisted in clockwise direction, slender, flexed outward. Sept.–Dec. Sandstone slopes along steams, NW (Cedarberg to Hex River Mts).•

formosum (Andrews) Sweet   Like **L. catherinae** but leaves grey-felted. Sept.–Oct. Moist sandstone slopes, SW, LB, SE (Riviersonderend to Outeniqua Mts).•

grandiflorum (Salisb.) R.Br. Like **L. gueinzii** but leaves ovate, with (1–)3 apical teeth, grey-felted. July–Dec. Mostly granite slopes, SW (Paardeberg to Paarl).•

gueinzii Meisn.   Stout shrub to 3 m. Leaves lanceolate, glabrescent, with 1 (sometimes to 4) apical teeth. Flower heads subglobose, 100–140 mm diam., yellow and orange, style 70–75 mm long, pollen presenter slender, flexed outward. Aug.–Dec. Clay slopes near streams, SW (Jonkershoek to Groenland Mts).•

reflexum H.Buek ex Meisn.   Single-stemmed shrub to 3 m. Leaves oblanceolate, grey-felted, with 2 or 3 apical teeth. Flower heads globose, 80–100 mm diam., scarlet, sometimes yellow, style 70–75 mm long, sharply reflexed at anthesis, pollen presenter slender. Aug.–Dec. Sandstone slopes near streams, NW (Pakhuis and N Cedarberg Mts).•

D. Style 30–60 mm long, pollen presenter swollen in line with style
E. Anthers subsessile on filaments, 1–2 mm long

cordifolium (Salisb. ex Knight) Fourc.   Rounded, spreading shrub to 1.5 m. Leaves ovate-cordate, glabrescent, with 1–6 apical teeth. Flower heads depressed-globose, 100–120 mm diam., orange-scarlet, style spreading-incurved, 45–60 mm long, pollen presenter globose and obliquely truncate. Aug.–Jan. Rocky sandstone slopes, SW, AP (Kogelberg to Soetanysberg).•

lineare R.Br.   Like **L. tottum** but leaves linear, channelled, involucral bracts hairy and cartilaginous and flower heads 60–90 mm diam., yellow to red. Mainly Sept.–Oct. Clay slopes, SW (Bainskloof to Hottentots Holland and Villiersdorp Mts).•

patersonii E.Phillips   Like **L. cordifolium** but tree to 4 m, leaves with 3–8 apical teeth, tepal claws silky. Aug.–Dec. Coastal limestone, SW, AP (Kleinmond to Agulhas).•

tottum (L.) R.Br.   Rounded shrub to 1.3 m, with spreading branches. Leaves oblong-lanceolate, with 1–3 apical teeth. Flower heads depressed globose, 90–150 mm diam., pink, style c. 50 mm long, pollen presenter ovoid. Sept.–Jan. Rocky slopes, NW, SW (Cedarberg Mts to Du Toitskloof).•

vestitum (Lam.) Rourke   Rounded shrub to 2.5 mm. Leaves ovate to elliptic, with 2–4 apical teeth. Flower heads subglobose, 70–90 mm diam., orange to scarlet, style 50–60 mm long, incurved, pollen presenter obliquely subglobose. Mainly Aug.–Nov. Rocky sandstone slopes, NW, SW (Cedarberg Mts to Botha; extinct Paarl to Cape Peninsula).•

E. Anthers sessile
F. Perianth tube cylindrical

conocarpodendron (L.) H.Buek   KREUPELHOUT   Rounded shrub or tree to 5 m, with thickly hairy branches. Leaves cuneate, subglabrous, sometimes felted, with 3–10 apical teeth. Flower heads globose to ovoid, 70–90 mm diam., yellow, style 45–55 mm long, pollen presenter conical. Aug.–Dec. Dry rocky slopes, SW (Cape Peninsula to Stanford).•

cuneiforme (Burm.f.) Rourke   Resprouting, many-stemmed shrub, to 2, sometimes 3 m, stems warty below. Leaves more or less cuneate, with 3–10 apical teeth. Flower heads ovoid, 50–90 mm diam., yellow fading to red, style 38–55 mm long, pollen presented conical. Mainly Aug.–Feb. Sandstone slopes and flats, SW, AP, KM, LB, SE (Riviersonderend and Swartberg Mts to Transkei).•

glabrum E.Phillips   Shrub to 2.5 m like **L. conocarpodendron** but mature stems glabrous, flower heads orange to red, style 50–60 mm long and involucral bracts acuminate, recurved at tips. Aug.–Oct. Moist lower slopes, SE (Outeniqua and Tsitsikamma Mts).•

pluridens Rourke   Shrub to 3 m like **L. conocarpodendron** but flower heads c. 60 mm diam., style 55–60 mm long and involucral bracts acuminate, recurved at tips. Sept.–Dec. Dry N-facing sandstone slopes, KM, LB, SE (Rooiberg to Robinson Pass).•

praemorsum (Meisn.) E.Phillips   Large shrub or tree to 5 m. Leaves petiolate, oblanceolate, with 3–5 apical teeth. Flower heads subglobose, c. 70 mm diam., orange to red, style 50–60 mm long, pollen presenter narrowly ellipsoid, involucral bracts silky. Mainly July–Sept. Dry sandy flats and plateaus, NW (Namaqualand to Cedarberg Mts).•

F. Perianth tube inflated above

erubescens Rourke   Shrub to 2 m. Leaves oblong-oblanceolate, with 3(–7) apical teeth. Flower heads subglobose, 20–30 mm diam., pale yellow fading crimson, style 40–55 mm long, pollen presenter club shaped. Aug.–Jan. Dry stony slopes and flats, KM, LB (Warmwaterberg to Langeberg: Garcia's Pass).•

fulgens Rourke   Like **L. praecox** but leaves oblong to oblanceolate, with 2–4 apical teeth, flower heads 60–80 mm diam., pink fading orange-red, style 46–53 mm long. Aug.–Jan. Coastal sands, AP (Potberg to Cape Infanta).•

praecox Rourke   Rounded shrub to 3 m. Leaves obovate-cuneate, with 5–11 apical teeth. Flower heads globose, c. 60 mm diam., yellow fading orange, style 38–48 mm long, pollen presenter narrowly conical. Apr.–Sept. Sandy flats, AP (Gouritsmond to Mossel Bay).•

profugum Rourke   Like **L. spathulatum** but leaves glabrescent, style 45–62 mm long. Sept.–Dec. Sandstone slopes, NW (Piketberg).•

spathulatum R.Br.   Mat-forming shrublet to 0.3 × 3 m. Leaves oblanceolate, secund, silvery hairy, with 1(–3) apical teeth. Flower heads globose, 50–70 mm diam., orange to red, style 30–40 mm long, incurved above, pollen presenter club-shaped. Mainly Oct. High sandstone slopes, NW (Cedarberg to Cold Bokkeveld, and Worcester: Kwadouwsberg).•

utriculosum Rourke   Like **L. erubescens** but lower branches spreading horizontally, style incurved. Mainly Sept.–Nov. Dry N-facing sandstone slopes, SW (Riviersonderend Mts and Potberg).•

MIMETES•   PAGODA BUSH, STOMPIE   13 spp., W Cape

A. Involucral bracts brightly coloured, usually longer than inflorescence leaves

capitulatus (L.) R.Br.   Rounded, single-stemmed shrub to 2 m. Leaves ovate to lanceolate, clasping stem, sometimes spreading, shaggy; inflorescence leaves similar. Flower heads in terminal spikes, white, involucral bracts orange-red, style red with yellow tip, pollen presenter club-shaped. June–Dec. Marshy sandstone slopes, SW (Kogelberg, Groenland and Kleinrivier Mts).•

hirtus (L.) Salisb. ex Knight   Single-stemmed shrub to 2 m. Leaves ovate to lanceolate, glabrescent; inflorescence leaves similar. Flower heads in terminal spikes, white, involucral bracts yellow with red tips, style red, pollen presenter needle-like, swollen at base. Mainly July–Aug. Peaty marshes, SW (Cape Peninsula to Elim).•

palustris Salisb. ex Knight   Sprawling, single-stemmed shrub sometimes to 1 m. Leaves lanceolate-elliptic, clasping stem, shaggy; inflorescence leaves similar. Flower heads in terminal spikes, white, involucral bracts yellow-green, style yellow with red tip, fading red, pollen presenter needle-like, swollen at base. Mainly July–Nov. Marshy slopes, SW (Kleinrivier Mts).•

pauciflorus R.Br.   Single-stemmed shrub, 2–4 m. Leaves ovate, ascending, hairy when young; inflorescence leaves similar. Flower heads in narrow terminal spikes, cream-coloured, involucral bracts orange-yellow, style orange with red tip, pollen presenter needle-like. Aug.–Nov. Moist S-facing slopes, SE (Outeniqua and Tsitsikamma Mts).•

A. Involucral bracts much smaller than the often brightly coloured inflorescence leaves
  B. Leaves glabrous or hairy but not silvery silky

chrysanthus Rourke   Single-stemmed shrub to 2 m. Leaves lanceolate to elliptic, glabrescent; inflorescence leaves similar. Flower heads in terminal spikes, 25–35-flowered, bright yellow, style yellow, pollen presenter needle-like. Mainly Apr.–May. Sandstone slopes, KM (Gamkaberg and Herold: Perdeberg).•

cucullatus (L.) R.Br.   ROOISTOMPIE   Many-stemmed shrub to 1.4 m. Leaves oblong-elliptic; inflorescence leaves spooned, red. Flower heads in terminal spikes, white, style red, pollen presenter needle-like. Mainly Aug.–Mar. Sandstone slopes and flats, NW, SW, AP, KM, LB, SE (Cold Bokkeveld to Outeniqua and Kouga Mts).•

fimbriifolius Salisb. ex Knight   MAANHAARSTOMPIE   Tree to 4 m. Leaves oblong to elliptic, margins hairy; inflorescence leaves spooned, dull red. Flower heads in terminal spikes, white, style yellow with red tips, pollen presenter needle-like. Mainly July–Dec. Rocky slopes, SW (Cape Peninsula).•

saxatilis E.Phillips   Robust, single-stemmed shrub to 2.5 m. Leaves elliptic to ovate, glabrescent; inflorescence leaves similar. Flower heads in terminal spikes, yellow, style yellow, pollen presenter club-shaped. July–Dec. Limestone outcrops, AP (Pearly Beach to Struisbaai).•

B. Leaves silvery silky

arboreus Rourke   Like **M. argenteus** but much-branched shrub to 6 m, leaves lanceolate. Apr.–June. Moist S-facing slopes, SW (Kogelberg).•

argenteus Salisb. ex Knight   VAALSTOMPIE   Single-stemmed shrub 3.5 m. Leaves spreading, silvery silky, elliptic; inflorescence leaves spooned, carmine to pale mauve. Flower heads in terminal spikes, pink, involucral bracts woody, style yellow, pollen presenter needle-like. Mar.–June. Moist S-facing slopes, SW (Franschhoek to Riviersonderend).•

hottentoticus E.Phillips & Hutch.   Single-stemmed shrub to 3 m. Leaves ovate, silvery silky; inflorescence leaves oblong, silvery pink. Flower heads in terminal spikes, cream-coloured, style red, pollen presenter capitate, black. Jan.–Mar. Damp upper slopes, SW (Kogelberg).•

splendidus Salisb. ex Knight   Single-stemmed shrub to 3 m. Leaves lanceolate to elliptic, silvery silky; inflorescence leaves oblong, spooned, orange-pink. Flower heads in terminal spikes, white, involucral bracts papery, style yellow, pollen presenter acute. May–July. Damp S-facing slopes, LB, SE (Langeberg to Tsitsikamma Mts).•

stokoei E.Phillips & Hutch.   Like **M. hottentoticus** but inflorescence leaves golden, involucral bracts ovate, style yellow. May–Jan. Damp S-facing sandstone slopes, SW (Palmiet River Mts).•

OROTHAMNUS•   MARSH ROSE   1 sp., W Cape

zeyheri Pappe ex Hook.f.   Slender, mostly unbranched shrub, 2–4 m. Leaves imbricate, ovate, shaggy. Flower heads 1–few at branch tips, nodding, involucral bracts large, rounded, waxy, velvety, pink to red, concealing flower heads. Mainly Sept.–Oct. Moist middle to upper slopes, SW (Kogelberg to Kleinrivier Mts).•

PARANOMUS•   DOLL'S BUSH, POPPIESBOS   18 spp., W Cape

A. Leaves distinctly dimorphic, the lower dissected, the upper more or less obovate and entire

adiantifolius Salisb. ex Knight   Sparsely branched shrub to 1.7 m. Leaves dimorphic, the lower dissected, the upper obovate, margins cartilaginous. Flower heads c. 13 mm long, in subglobose spikes, pink, style hairy below, pollen presenter top-shaped. Sept.–Nov. Sandstone slopes, SW (Groenland and Riviersonderend Mts).•

longicaulis Salisb. ex Knight   Sparsely branched shrub to 2.5 m. Leaves dimorphic, the lower dissected, the upper clasping, spathulate-truncate, margins cartilaginous. Flower heads c. 18 mm long, in terminal, subglobose spikes, pink, subtended by dry, undulate brown bracts, style ?glabrous. Mainly Sept.–Dec. Sandstone slopes, LB (Langeberg Mts: Garcia's Pass to Attaquas Kloof).•

reflexus (E.Phillips & Hutch.) Fourc.   Like **P. sceptrum-gustavianus** but flower heads 30–35 mm long, flowers reflexing at anthesis. June–Aug. Sandstone slopes, SE (Van Staden's Mts).•

roodebergensis (Compton) Levyns   Shrub to 2.5 m. Leaves dimorphic, the lower dissected, the upper obovate, margins cartilaginous. Flower heads c. 12 mm long, in cylindrical spikes, pink, involucral bracts yellow tipped red, honey-scented, style glabrous. Aug.–Oct. Sandstone slopes, KM (Touwsberg to Rooiberg).•

sceptrum-gustavianus (Sparrm.) Hyland.   Robust shrub to 1.8 m. Leaves dimorphic, the lower dissected, the upper obovate-rhomboid. Flower heads 16–22 mm long, in cylindrical, terminal spikes, cream-coloured, strongly scented, style glabrous. July–Mar. Sandstone upper slopes, SW, LB (Hottentots Holland to Langeberg Mts: Swellendam).•

spathulatus (Thunb.) Kuntze   Resprouting shrub to 2.5 m. Leaves dimorphic, glaucous, the lower dissected, the upper obovate. Flower heads 10–12 mm long, in subglobose spikes, pink, style glabrous. May–Dec. Sandstone slopes, KM, LB (Gamkaberg and Langeberg Mts: Tradouw to Garcia's Pass).•

A. Leaves all dissected, the upper sometimes less so
B. Inflorescence subglobose, to 25 mm long

abrotanifolius Salisb. ex Knight   Low much-branched shrublet to 90 cm. Leaves dissected, 15–25 mm long. Flower heads 7–9 mm long, in globose spikes, purple. May–Dec. Clay and gravel flats, SW (Grabouw, Elim flats and Potberg).•

capitatus (R.Br.) Kuntze   Shrub to 1 m. Leaves dissected, glabrescent, 10–20 mm long. Flower heads c. 9 mm long, in globose spikes, dull purple, style sparsely hairy. Oct.–Dec. Sandstone slopes, SW (Du Toitskloof and Riviersonderend Mts).•

centaureoides Levyns   Branched shrub to 1.5 m. Leaves dissected, becoming linear above, 10–60 mm long. Flower heads c. 18 mm long, in sessile, globose spikes, pink and maroon, subtended by dark brown bracts, style more or less glabrous. June–Nov. Dry upper slopes, KM (Klein Swartberg).•

dregei (H.Buek ex Meisn.) Kuntze   Shrub to 1.7 m. Leaves dissected, the upper linear, to 50 mm long. Flower heads c. 14 mm long, in globose spikes, fragrant, cream-coloured, style glabrous. May–Oct. Dry upper sandstone slopes, KM, SE (Witteberg to Kouga Mts).•

esterhuyseniae Levyns   Shrub to 70 cm. Leaves dissected or the upper entire, 20–50 mm long. Flower heads c. 8 mm long, in globose spikes, cream-coloured, style glabrous. Aug.–Nov. Dry upper slopes, SE (Outeniqua and Kouga Mts).•

B. Inflorescence cylindrical, longer than 25 mm

bolusii (Gand.) Levyns   Much-branched shrub to 1 m. Leaves dissected, to 50 mm long. Flower heads 10–12 mm long, in cylindrical spikes, purple, style thinly hairy below. June–Nov. Sandstone slopes, SW (Groenland Mts to Kleinrivier Mts).•

bracteolaris Salisb. ex Knight   Shrub to 2.5 m. Leaves dissected, to 60 mm long. Flower heads c. 14 mm long, in cylindrical spikes, purple, style thinly hairy below. Aug.–Oct. Rocky sandstone slopes and plateaus, NW (Bokkeveld Mts to Olifants River Mts).•

candicans (Thunb.) Kuntze   Much-branched shrub to 2 m. Leaves much dissected, silvery hairy when young. Flower heads 10–13 mm long, in cylindrical spikes, pale yellow, style hairy below. June–Nov. Sandstone slopes, NW, LB (Hex River Mts to Langeberg Mts: Mossel Bay).•

dispersus Levyns   Like **P. bracteolaris** but leaves sparsely dissected and lobes flattened. Mainly Aug.–Nov. Sandstone slopes, NW, SW, KM, LB, SE (Grootwinterhoek to Outeniqua Mts).•

lagopus (Thunb.) Salisb.   Shrub to 1.3 m, lower branches sprawling. Leaves dissected, 10–50 mm long. Flower heads c. 9 mm long, in cylindrical spikes, purple, style thinly hairy below. Sept.–Nov. Sandstone slopes, NW, SW (Cold Bokkeveld to Elandskloof Mts).•

spicatus (P.J.Bergius) Kuntze   PERDEBOS   Like **P. bracteolaris** but style woolly below. Sept.–Nov. Sandstone slopes, SW (Hottentots Holland Mts to Kogelberg).•

tomentosus (E.Phillips & Hutch.) N.E.Br.   Shrub to 3 m, with white-woolly branches. Leaves dissected, recurved at tips, silvery hairy. Flower heads c. 17 mm long, in cylindrical spikes, whitish, style hairy below. Aug.–Oct. High sandstone slopes, NW (Cedarberg Mts).•

PROTEA   PROTEA, SUGARBUSH   c. 115 spp., southern and tropical Africa, mainly W Cape

A. Flower heads axillary, produced close to the ground, usually clustered; involucral bracts velvety dark brown or purple

amplexicaulis (Salisb.) R.Br.   Sprawling grey-green shrublet to 40 cm. Leaves spreading, ovate and clasping at base. Flower heads axillary, borne near base of branches, cup-shaped, 60–80 mm diam., involucral bracts oblanceolate, ivory, chocolate-velvety beneath, style 25–30 mm long. June–Sept. Sandstone slopes, NW, SW (Cold Bokkeveld to Riviersonderend Mts).•

cordata Thunb.   Erect, resprouting shrublet to 50 cm. Leaves heart-shaped, glaucous blue-green. Flower heads axillary, clustered at base of branches, cup-shaped, 40–50 mm diam., involucral bracts papery, brown, style c. 30 mm long. June–July. Shale slopes, SW, AP, LB (Du Toitskloof to Langeberg Mts: Swellendam and Soetanysberg).•

decurrens E.Phillips   Resprouting shrublet to 60 cm. Leaves linear. Flower heads axillary, clustered near base, cup-shaped, 30–50 mm diam., involucral bracts pinkish brown, silvery to rusty hairy, style 30–35 mm long. Mainly July–Aug. Gravelly flats, SW, LB (Shaw's Pass to Langeberg Mts).•

humiflora Andrews   Erect to sprawling shrub to 1 m. Leaves linear-falcate, glaucous. Flower heads axillary, cup-shaped, 60–80 mm diam., involucral bracts ivory with pink tips, purple-black velvety outside, style 30–40 mm long. July–Sept. Dry, sandstone slopes, SW, KM, LB (Du Toitskloof to Gamkaberg)).•

subulifolia (Salisb. ex Knight) Rourke   Like **P. decurrens** but leaves needle-like and style 25–30 mm long. Mainly July–Sept. Sandy to clay flats, SW, AP, LB (Stettynskloof to Langeberg Mts and Agulhas plain).•

A. Flower heads terminal, usually solitary, stems aerial or subterranean
B. Stems prostrate or mostly underground

Group 1: Plants tufted and acaulescent

aspera E.Phillips   Like **P. scorzonerifolia** but leaves flat, rough, flower heads 40–60 mm diam. and style 50–70 mm long. Mainly Sept.–Oct. Clay and stony flats, SW, LB (Kleinrivier Mts to Agulhas and Langeberg Mts: Garcia's Pass to Cloete's Pass).•

lorea R.Br.   Like **P. scorzonerifolia** but leaves needle-like and style 75–90 mm long. Jan.–Feb. Stony slopes, NW, SW, LB (Ceres to Langeberg Mts).•

piscina Rourke   Like **P. scorzonerifolia** but leaf margins horny, flower heads 40–50 mm diam., pale yellow, sometimes pink-flushed and style 30–35 mm long. June–Jan. Sandstone slopes, NW, LB (Cedarberg Mts to Ceres, Langeberg Mts: Swellendam to Riversdale).•

restionifolia (Salisb. ex Knight) Rycroft   Like **P. scorzonerifolia** but leaves needle-shaped, warty, sparsely coarse white-hairy, involucral bracts brown velvety at tips and style 45–60 mm long. Aug.–Oct. Dry rocky slopes, NW, SW (Cold Bokkeveld, Wolseley to Botrivier).•

scabra R.Br.   Like **P. scorzonerifolia** but leaves needle-like to oblanceolate-channelled, rough, flower heads cup-shaped, brown and cream-coloured and style 30–35 mm long. Mainly July–Oct. Sandstone slopes, SW (Hottentots Holland to Riviersonderend Mts).•

scorzonerifolia (Salisb. ex Knight) Rycroft   Rhizomatous, mat-forming, resprouting shrublet to 1 m diam. Leaves tufted, needle-shaped to linear and channelled, margins scabrid. Flower heads obconic, involucral bracts creamy yellow to pink, silky, 50–80 mm diam., style 60–75 mm long. Aug.–Dec. Stony slopes, SW (Tygerberg and Bainskloof to Hottentots Holland Mts).•

Group 2. Involucral bracts acuminate, pinkish, white-woolly outside

cryophila Bolus   SNOW PROTEA   Like **P. pruinosa** but flower heads larger, 130–160 mm diam. and style 65–90 mm long. Jan.–Apr. Rocky ridges and summits, NW (Cedarberg Mts).•

pruinosa Rourke   Prostrate, mat-forming shrublet to 1.5 m diam. Leaves petiolate, oblanceolate, rough. Flower heads cup-shaped, 80–120 mm diam., involucral bracts acuminate, red, densely white-woolly outside, style 55–60 mm long. Jan.–Feb. Sandstone ridges and summits, KM (Swartberg Mts).•

scabriuscula E.Phillips   Like **P. pruinosa** but leaves linear, flower heads obconic, remaining more or less closed, style 30–35 mm long. Oct.–Jan., mainly Dec. High sandstone slopes, NW (Cold Bokkeveld and Hex River Mts).•

scolopendriifolia (Salisb. ex Knight) Rourke   Rhizomatous, clumped, resprouting shrublet to 1 m diam. Leaves petiolate, oblanceolate-spathulate, rough, glabrescent. Flower heads cup-shaped, 60–120 mm diam., involucral bracts acuminate, carmine, densely white-woolly outside near tips, style 30–45 mm long. Sept.–Dec. Mostly shale slopes, NW, SW, KM, LB, SE (Cedarberg to Kouga Mts).•

Group 3. Plants sprawling with prostrate stems, sometimes underground but then bracts glabrescent except on margins

acaulos (L.) Reichard   GROUND PROTEA   Mat-forming, resprouting shrublet. Leaves secund, linear-oblanceolate to obovate. Flower heads cup-shaped, 30–60 mm diam., involucral bracts green with red tips, style 25–35 mm long. June–Nov. Sandy flats and lower slopes, NW, SW, AP (Cedarberg Mts to Agulhas).•

angustata R.Br.   Mat-forming, resprouting shrublet to 30 cm with short tufted stems. Leaves linear, sometimes channelled. Flower heads cup-shaped, 30–45 mm diam., involucral bracts yellowish flushed red, style 25–30 mm long. July–Oct. Coastal flats and slopes, SW (Groenland Mts to Kleinrivier Mts).•

caespitosa Andrews   Compact, rounded shrublet to 70 cm, with tangled branches. Leaves ovate to oblanceolate, glabrescent, margins often silky. Flower heads obconic, nested among brown papery leaves, 45–60 mm diam., involucral bracts pink, silky, margins white to tawny and bearded, style 40–45 mm long. July–Nov. Shale slopes at high alt., SW (Du Toitskloof to Riviersonderend Mts).•

convexa E.Phillips   Like **P. laevis** but stems sticky to glazed, leaves broadly obovate. Aug.–Nov. Dry rocky sandstone slopes, NW, KM (N Cedarberg, Witteberg and Klein Swartberg).•

intonsa Rourke   Like **P. vogtsiae** but leaves linear to needle-like. Sept.–Nov. Dry sandstone slopes, KM, SE (Swartberg to Baviaanskloof Mts).•

laevis R.Br.   Mat-forming shrublet to 1 m diam. Leaves secund, oblong-oblanceolate, glaucous. Flower heads cup-shaped, 40–70 mm diam., involucral bracts glaucous, yellow-green, flushed red within, style 28–35 mm long. Mainly Sept.–Nov. Dry sandstone slopes, NW, KM (Cedarberg to Hex River Mts, Waboomsberg).•

montana E.Mey. ex Meisn.   Mat-forming shrublet to 4 m diam. Leaves secund, needle-like to linear. Flower heads cup-shaped, 30–50 mm diam., involucral bracts green flushed carmine, silky, the outermost with leaf-like tips, style 25–35 mm long. Feb.–June. Sandstone slopes at high alt., KM (Swartberg and Kammanassie Mts).•

pudens Rourke   Sprawling or prostrate shrublet to 1 m diam. Leaves secund, linear-spathulate, often ciliate. Flower heads cup-shaped, 30–60 mm diam., deep pink, florets tipped black, forming a woolly cone longer than bracts, style 30–45 mm long. June–Aug. Loamy clay flats, AP (Agulhas coast).•

revoluta R.Br.   Like **P. acaulos** but leaves linear, margins revolute. Nov.–Jan. Dry sandstone slopes, NW, KM (Cedarberg Mts to Witteberg).•

tenax (Salisb.) R.Br.   Trailing, resprouting shrub to 4 m diam. Leaves secund, linear-oblanceolate. Flower heads cup-shaped, 40–60 mm diam., involucral bracts green to yellow, flushed pink, margins silky, style 25–30 mm long. Mainly May–Sept. Rocky sandstone slopes, KM, SE (Swartberg and Outeniqua to Great Winterhoek Mts).•

venusta Compton   Mat-forming shrublet to 3 m diam. Leaves secund to subsecund, oblanceolate. Flower heads top-shaped, 80–100 mm diam., honey-scented, involucral bracts ivory flushed pink, style c. 55 mm long. Mainly Jan.–Feb. S-facing sandstone slopes, KM (Groot Swartberg and Kammanassie Mts).•

vogtsiae Rourke   Dwarf tufted shrublet to 25 cm. Leaves tufted, oblanceolate-spathulate, glaucous, slightly rough. Flower heads subglobose, 35–40 mm diam., involucral bracts glabrescent, green to carmine, style 25–30 mm long. Aug.–Nov. High sandstone slopes, SE (Outeniqua to Baviaanskloof Mts).•

B. Stems mainly aerial, erect to sprawling
C. Style 15–35(–40) mm long
D. Leaves needle-like

nana (P.J.Bergius) Thunb.   MOUNTAIN ROSE   Rounded shrub to 1.3 m. Leaves needle-like. Flower heads nodding, cup-shaped, 30–45 mm diam., involucral bracts red to green, ciliate, style 20–25 mm long. July–Oct. Sandstone slopes, NW, SW (Grootwinterhoek Mts to Du Toitskloof).•

pityphylla E.Phillips   Suberect to sprawling shrublet to 1 m. Leaves subsecund, needle-like. Flower heads nodding, cup-shaped, 50–80 mm diam., involucral bracts deep red, style 30–35 mm long. Mainly May–July. Sandstone slopes, NW, SW (Olifants River to Hex River Mts).•

witzenbergiana E.Phillips   Sprawling to mat-forming shrublet hairy on stems and leaves. Leaves secund, needle-like, glaucous. Flower heads nodding, cup-shaped, 40–70 mm diam., involucral bracts reddish brown, style 25–30 mm long. Mar.–June. Sandstone slopes, NW (Cedarberg to Hex River Mts).•

D. Leaves linear to obovate, flat or channelled

acuminata Sims   Rounded shrub to 2 m. Leaves linear-spathulate, glaucous. Flower heads globose, 25–35 mm diam., involucral bracts dark red, style c. 20 mm long. Mainly July–Aug. Dry sandstone slopes, NW, SW (Bokkeveld Mts to Olifants River Mts and Stettynskloof).•

canaliculata Andrews   Rounded shrub to 1.2 m. Leaves linear-spathulate, channelled above. Flower heads bowl-shaped, 35–60 mm diam., involucral bracts pink to red, style 20–30 mm long. Mar.–June. Rocky sandstone slopes, NW, KM (Hex River Mts to Waboomsberg and Swartberg Mts).•

denticulata Rourke   Densely branched, resprouting shrub to 1 m. Leaves linear, channelled, rigid and rough, softly hairy when young. Flower heads cup-shaped, 35–40 mm diam., involucral bracts reddish brown-velvety, style 25–35 mm long. Aug.–Oct. Sandstone slopes, SW (Potberg).•

effusa E.Mey. ex Meisn.   Erect or sprawling shrub to 1.5 m. Leaves oblanceolate, grey-blue. Flower heads cup-shaped, 60–100 mm diam., involucral bracts dark red, sometimes greenish yellow, style 30–35 mm long. May–Sept. Sandstone slopes, NW, SW (Cold Bokkeveld to Du Toitskloof and Naudésberg).•

foliosa Rourke   Rounded, resprouting shrub to 1.5 m. Leaves elliptic-oblanceolate, glabrescent. Flower heads subglobose, concealed by upper leaves, often clustered, 30–40 mm diam., involucral bracts greenish cream, sometimes flushed pink, style c. 25 mm long. Mainly May–June. Sandstone slopes, SE (Van Staden's Mts to Grahamstown).

mucronifolia Salisb.   Rounded shrublet to 1 m. Leaves linear, rigid, pungent. Flower heads bowl-shaped, c. 50 mm diam., involucral bracts acuminate, white and pink, style c. 16 mm long. Oct.–Dec. Gravel flats, SW (Hermon to Saron).•

odorata Thunb.   Like **P. mucronifolia** but leaves grooved below, obliquely pungent, flower heads 20–30 mm diam. Mainly Mar.–Apr. Gravel flats, SW (Kalbaskraal to Klapmuts).•

pendula R.Br.   Rounded shrub to 3 m. Leaves oblanceolate, grey-felted when young. Flower heads nodding, cup-shaped, 40–70 mm diam., involucral bracts silky, purple-red to pink, sometimes green below, style 25–30 mm long. May–Aug. Dry upper slopes, NW (Cedarberg Mts to Cold Bokkeveld).•

recondita H.Buek ex Meisn.   Sprawling shrublet to 1 m. Leaves lanceolate to broadly obovate, often flushed purple. Flower heads concealed by upper leaves, nodding, cup-shaped, 70–90 mm diam., involucral bracts green, style 30–35 mm long. Mainly May–July. Rocky sandstone slopes, NW (Cedarberg to Witsenberg).•

scolymocephala (L.) Reichard   Erect shrub to 1.5 m. Leaves linear-spathulate. Flower heads bowl-shaped, 35–35 mm diam., involucral bracts creamy green, style 12–25 mm long. Mainly Aug.–Oct. Sandy flats and lower slopes, NW, SW (Gifberg to Hermanus).•

sulphurea E.Phillips   Dense, low spreading shrub to 0.5 × 7 m. Leaves elliptic-oblanceolate. Flower heads nodding, cup-shaped, 90–130 mm diam., involucral bracts yellow inside, green with red tips outside, style 30–38 mm long. Apr.–Aug. Dry rocky slopes, NW, KM (Hex River Mts to Witteberg and Swartberg Mts).•

C. Style 40–130 mm long
E. Style longer than involucral bracts or more conspicuous

aurea (Burm.f.) Rourke   Shrub or tree to 5 m. Leaves oblong-ovate. Flower heads obconic (shuttlecock-like), 90–120 mm long, involucral bracts pink to creamy green, silky, style 85–105 mm long. Mainly Jan.–June. S-facing slopes, SW, LB, SE (Riviersonderend to Outeniqua Mts).•

glabra Thunb.   CHESTNUT SUGARBUSH   Resprouting shrub or tree to 5 m. Leaves elliptic-oblanceolate. Flower heads ovoid-subglobose, 70–120 mm diam., involucral bracts short, dull brownish, glabrous or velvety, style 40–50 mm long. Mainly Aug.–Sept. Dry sandstone slopes and plateaus, NW (Bokkeveld Mts to Cold Bokkeveld).•

inopina Rourke   Resprouting shrub to 1 m. Leaves elliptic-oblanceolate, blue-green. Flower heads ovoid, 100–120 mm diam., involucral bracts short, green below, brown velvety above, style c. 85 mm long. Sept.–Dec. Sandstone slopes, NW (S Olifants River Mts).•

lacticolor Salisb.   Like **P. punctata** but leaves cordate at base. Feb.–May. Moist sandstone slopes, SW (Bainskloof to Hottentots Holland Mts).•

lanceolata E.Mey. ex Meisn.   Shrub to 4 m. Leaves oblanceolate. Flower heads obconic, 50–70 mm diam., involucral bracts short, green with brown margins, style 55–60 m long. Mainly May–July. Calcareous sands, AP (Cape Infanta to Mossel Bay).•

mundii Klotzsch   Shrub or tree to 8(–12) m. Leaves oblanceolate, glabrescent. Flower heads oblong-obconic, 65–80 × 40–65 mm, involucral bracts white to pink, silky, fringed, style 55–65 mm long. Mainly Feb.–Apr. Moist slopes and forest margins, SW, SE (Kogelberg to Kleinrivier Mts, Outeniqua to Great Winterhoek Mts).•

nitida Mill.   WABOOM   Tree 5–10 m. Leaves elliptic, grey-glaucous. Flower heads cup-shaped, involucral bracts short, silver-grey, sometimes silky, style 60–80 mm long. Mostly May–Aug. Sandstone slopes, NW, SW, KM, LB, SE (Bokkeveld Mts to Great Winterhoek Mts).•

punctata Meisn.   Shrub to 4 m. Leaves ovate-obovate, glabrescent, glaucous. Flower heads bowl-shaped, 20–25 mm diam., involucral bracts pink or white, silky and fringed, style c. 50 mm long. Mainly Mar.–Apr. Rocky slopes, NW, SW, KM, SE (Cedarberg to Kouga Mts).•

rupicola Mund ex Meisn.   Erect or spreading shrub to 2 m. Leaves linear-oblanceolate. Flower heads ovoid, 40–100 mm diam., involucral bracts short, deep pink to brown, subglabrous, style 45–55 mm long. Sept.–Feb. Sandstone slopes at high alt., NW, SW, LB, SW (Grootwinterhoek to Great Winterhoek Mts).•

subvestita N.E.Br.   Large shrub or tree to 5 m. Leaves elliptic-oblanceolate, densely hairy when young. Flower heads oblong-obconic, 30–40 mm diam., involucral bracts cream-coloured to pink, glabrous or silky, fringed, inner recurved at tips, style c. 55 mm long. Mainly Jan.–Mar. Rocky sandstone slopes, KM (Klein Swartberg and E Cape to Mpumalanga).

E. Style shorter than involucral bracts and less conspicuous
F. Inner involucral bracts with a distinct beard

burchellii Stapf   Shrub to 2 m. Leaves linear-oblanceolate. Flower heads obconic, 50–70 mm diam., involucral bracts shiny, cream-coloured to carmine, inner fringed or bearded, style 55–65 mm long. June–Aug. Mainly clay slopes and loam, NW, SW (Piketberg to Hottentots Holland Mts).•

coronata Lam.   Erect shrub or small tree to 5 m. Leaves lanceolate. Flower heads on hairy stems, oblong, c. 100 × 60 mm diam., involucral bracts green, with a white beard, style c. 60 mm long. Mainly Apr.–June. Clay slopes, SW, LB, SE (Cape Peninsula to Zuurberg).

grandiceps Tratt.   Rounded shrub to 2 m. Leaves ovate to obovate, glaucous. Flower heads oblong, 100–140 × c. 70 mm, involucral bracts brick-red, glabrous or shortly hairy, white- to purple-bearded, style c. 65 mm long. Mainly Dec.–Jan. Sandstone slopes, SW, KM, LB, SE (Cape Peninsula and Paarl to Great Winterhoek Mts).•

holosericea (Salisb. ex Knight) Rourke   Erect or sprawling shrub to 1.2 m. Leaves oblong, glabrescent, glaucous. Flower heads on hairy branches, obconic, 60–80 mm diam., involucral bracts cream-coloured, silky and with dense blackish beard, style c. 50 mm long. Sept.–Oct. High rocky ridges, NW (Kwadouwsberg: Saw Edge Peak).•

laurifolia Thunb.   Small tree to 8 m. Leaves oblong-oblanceolate, grey-glaucous. Flower heads oblong, 100–130 × 40–60 mm, involucral bracts silky, cream-coloured to pink, inner bracts with a dense blackish beard, outer with brown horny margins, style 65–70 mm long. Mainly May–July. Sandstone slopes, NW, SW, KM (Bokkeveld Mts to Witteberg and Riviersonderend Mts).•

lepidocarpodendron (L.) L.   BLACK-BEARDED PROTEA   Like **P. neriifolia** but involucral bracts with black hairs below beard. Apr.–Aug. Mainly sandstone slopes, SW (Cape Peninsula to Kleinrivier Mts).•

lorifolia (Salisb. ex Knight) Fourc.   Rounded shrub or small tree to 3(–5) m. Leaves oblanceolate, glaucous. Flower heads oblong-obconic, 70–130 × 25–50 mm, involucral bracts pink or cream-coloured, silky with a short white to dark beard. Apr.–June. Dry sandstone slopes, NW, SW, KM, LB, SE (Cold Bokkeveld to Somerset East).

magnifica Link   QUEEN PROTEA   Sprawling grey-leaved shrub to 2.5 m. Leaves oblong-oblanceolate, undulate, glaucous. Flower heads obconic, to 140 × 150 mm, pink or cream-coloured, silky, recurved above, with a white or black beard, style 60–70 mm long. June–Jan. Sandstone slopes, NW, SW, KM, LB (Cedarberg to Klein Swartberg and Langeberg Mts).•

neriifolia R.Br.   Like **P. laurifolia** but leaves narrowly oblong, green. Feb.–Nov. Sandstone and clay slopes, SW, KM, LB, SE (Kleinwinterhoek Mts to Port Elizabeth).•

speciosa (L.) L.   Resprouting shrub to 1.2 m. Leaves oblanceolate to obovate. Flower heads on shortly hairy stems, oblong, 90–140 × c. 70 mm, involucral bracts greenish to pink, silky with a heavy brown, sometimes white beard, style 65–75 mm long. Mainly Sept.–Oct. Sandstone flats and slopes, SW, LB (Cape Peninsula to Langeberg Mts).•

stokoei E.Phillips   Shrub to 2 m. Leaves ovate-obovate. Flower heads on softly hairy stems, oblong, 90–130 × c. 60 mm, involucral bracts pink, silky and with heavy brown beard, style 65–70 mm long. Mainly May–June. Sandstone seeps, SW (Hottentots Holland Mts to Betty's Bay).•

F. Inner involucral bracts without a beard

aristata E.Phillips   Rounded shrub to 2.5 m. Leaves linear, sulphurous, black tipped. Flower heads obconic, 100–120 mm diam., involucral bracts crimson, inner narrow, velvety, style c. 70 mm long. Mainly Nov.–Dec. Sandstone slopes, KM (Klein Swartberg).•

compacta R.Br.   BOT RIVER PROTEA   Like **P. obtusifolia** but leaves oblong-ovate, cordate at base. Mainly May–June. Coastal slopes and flats, SW (Betty's Bay to Bredasdorp Mts).•

cynaroides (L.) L.   KING PROTEA   Resprouting shrub to 3 m. Leaves long-petiolate, elliptic to rhomboid. Flower heads large, obconic to cup-shaped, 120–300 mm diam., involucral bracts pale or deep pink, often silky outside, style 80–95 mm long. Jan.–Dec. Moist sandstone slopes, NW, SW, LB, SE (Gifberg to Port Elizabeth).•

eximia (Salisb. ex Knight) Fourc.   Like **P. obtusifolia** but leaves oblong-ovate, cordate at base, florets black-tipped. Mainly Aug.–Oct. Sandstone slopes, NW, KM, LB, SE (Keeromsberg to Van Staden's Mts).•

longifolia Andrews   Sprawling shrub to 1.5 m. Leaves linear. Flower heads oblong-obconic, 40–90 mm diam., involucral bracts greenish to pink, innermost linear and densely fringed, florets forming a black woolly cone longer than bracts, style 40–65 mm long. Mainly June–July. Gravel flats and lower slopes, SW, AP (Hottentots Holland Mts to Agulhas).•

obtusifolia H.Buek ex Meisn.   Large shrub to 4 m. Leaves elliptic-oblanceolate, glabrescent. Flower heads oblong-obconic, 90–120 × 50–80 mm, involucral bracts cream-coloured to red, the inner spathulate, style c. 65 mm long. Mainly June–Aug. Limestone flats and hills, SW, AP (Stanford to Stilbaai).•

repens (L.) L. SUGARBUSH, SUIKERBOS Shrub or tree to 4.5 m. Leaves linear-spathulate. Flower heads narrowly obconic, 100–160 × 70– 90 mm, involucral bracts cream-coloured to red, sticky, style 70–90 mm long. Jan.–Dec. Sandstone and clay flats and slopes, NW, SW, AP, KM, LB, SE (Bokkeveld Mts to Grahamstown).

susannae E.Phillips STINKBLAARSUIKERBOS Like **P. obtusifolia** but leaves with a sulphurous odour, and involucral bracts with brown sticky coating. Mainly May–July. Coastal limestone and sand, SW, AP (Stanford to Stilbaai).•

SERRURIA• SPIDERHEAD, SPINNEKOPBOS c. 50 spp., W Cape

*A. Flowers strongly curved in bud (see also **S. aitonii** and **S. reflexa**)*
B. Flower heads several in clusters

adscendens (Lam.) R.Br. Like **S. decipiens** but leaves 25–60 mm long, peduncle hairy and style 7–8 mm long. July–Oct. Coastal sandstone slopes, SW (Hottentots Holland to Kleinrivier Mts).•

bolusii E.Phillips & Hutch. Sprawling shrublet to 40 cm. Leaves dissected, glabrescent, 25–35 mm long. Flower heads few, adpressed-hairy, silvery pink, fragrant, style c. 6 mm long, pollen presenter club-shaped, floral bracts grooved. Sept.–Nov. Sandstone slopes, SW, AP (Elim hills and Soetanysberg).•

collina Salisb. ex Knight Sprawling, mat-forming shrublet to 3 m diam. Leaves secund, dissected, 50–150 mm long. Flower heads 8–15, clustered, adpressed-hairy, cream-coloured, fragrant, style 8–10 mm long, pollen presenter filiform, involucral bracts absent. July–Oct. Sandy flats, SW (Cape Peninsula).•

decipiens R.Br. Rounded shrub to 1 m. Leaves dissected, glabrescent, 30–45 mm long. Flower heads 5–10, clustered, adpressed-hairy, creamy white, fragrant, style 8–9 mm long, pollen presenter club-shaped, involucral bracts absent. July–Oct. Sandy flats and slopes, mainly coastal, NW, SW (Olifants River Mts to Cape Flats).•

glomerata (L.) R.Br. Compact rounded shrublet to 50 cm. Leaves dissected, 25–70 mm long. Flower heads 4–8, clustered, adpressed-hairy, cream-coloured, fragrant, style 8–10 mm long, pollen presenter club-shaped, involucral bracts few. Aug.–Oct. Sandy flats, SW (Cape Peninsula).•

nervosa Meisn. Erect shrub to 80 cm. Leaves dissected, glabrescent, 50–70 mm long. Flower heads 1–3, silvery pink, fragrant, style c. 7 mm long, pollen presenter club-shaped., involucral bracts grooved. July–Nov. Sandstone slopes, SW (Kleinrivier and Bredasdorp Mts).•

roxburghii R.Br. Sprawling shrub to 1 m. Leaves dissected, glabrescent, 10–25 mm long. Flower heads 3–7, clustered, silvery pink, fragrant, style 5–10 mm long, pollen presenter club-shaped, involucral bracts ovate. Sept.–Nov. Sandy flats, SW (Riebeek-Kasteel to Paarl).•

rubricaulis R.Br. Sprawling to prostrate, resprouting shrublet to 20 cm. Leaves secund, dissected, 35–45 mm long. Flower heads 5–25, clustered, adpressed-hairy, silvery pink, style c. 8 mm long, pollen presenter club-shaped, involucral bracts absent. Sept.–Oct. Lower sandstone slopes and flats, SW (Hottentots Holland to Kleinrivier Mts).•

*B. Flower heads solitary (see also **S. nervosa**)*
C. Style hairy below

acrocarpa R.Br. Rounded, resprouting shrublet to 50 cm. Leaves dissected, glabrescent, 20–50 mm long. Flower heads solitary, silvery pink to greenish, adpressed silky, style hairy below, c. 7 mm long, pollen presenter club-shaped, involucral bracts ovate. Mainly July–Dec. Sandy flats and slopes, NW, LB (Cold Bokkeveld to Potberg).•

balanocephala Rourke Rounded, resprouting or reseeding shrublet to 1 m. Leaves dissected, grey-hairy, 40–60 mm long. Flower heads solitary, silvery pink, style hairy below, c. 13 mm long, pollen presenter acorn-shaped, involucral bracts ovate. Aug.–Nov. Sandstone slopes, LB (Langeberg Mts: Montagu to Swellendam).•

dodii E.Phillips & Hutch. Shrublet to 1 m. Leaves dissected, silvery silky, 30–50 mm long. Flower heads solitary, silvery adpressed-hairy, fragrant, style hairy below, c. 12 mm long, pollen presenter club-shaped, involucral bracts ovate. Aug.–Nov. Sandstone slopes, NW (Hex River Mts and Keeromsberg).•

flava Meisn. Shrublet to 80 cm. Leaves dissected, silvery silky, 12–35 mm long. Flower heads solitary, yellow, silvery silky adpressed-hairy, style hairy below, c. 15 mm long, pollen presenter slender, involucral bracts ovate. Aug.–Dec. Sandstone slopes, NW (N Cedarberg).•

gremialis Rourke   Sprawling, mat-forming, resprouting shrublet to 50 cm. Leaves secund, dissected, crispy hairy when young, 10–30 mm long. Flower heads solitary, silvery pink, adpressed silky, style hairy below, 12–15 mm long, pollen presenter club-shaped. July–Feb. N-facing sandstone slopes, SW (Stettynskloof and Riviersonderend Mts, ?Potberg).•

pedunculata (Lam.) R.Br.   Erect shrublet to 1 m. Leaves dissected, grey hairy, 15–20 mm long. Flower heads solitary, silvery pink, fragrant, style hairy below, 8–12 mm long, pollen presenter club-shaped, involucral bracts ovate. Aug.–Dec. Sandstone slopes, NW, SW (Cedarberg to Riviersonderend Mts).•

C. Style glabrous

cygnea R.Br.   Mat-forming shrublet to 1 m diam. Leaves secund, dissected, glabrescent, 25–70 mm long. Flower heads solitary, silvery pink to brown, fragrant, style 9–12 mm long, pollen presenter club-shaped, involucral bracts ovate. Sept.–Nov. Sandstone and clay slopes, NW, SW (Cedarberg to Slanghoek Mts).•

effusa Rourke   Sprawling to mat-forming shrublet to 2 m diam. Leaves secund, dissected, glabrous or grey-hairy, 40–90 mm long. Flower heads solitary, adpressed-hairy, silvery pink to brown, style 12–14 mm long, pollen presenter club-shaped, involucral bracts ovate, margins sometimes hairy. Mainly Aug.–Sept. Sand or shale slopes, NW (Cedarberg to Olifants River Mts).•

fucifolia Salisb. ex Knight   Rounded shrub to 1.5 m. Leaves dissected, grey-crispy hairy, 35–60 mm long. Flower heads solitary, small, adpressed-hairy, silvery grey to purple, fragrant, style c. 9 mm long, pollen presenter slender, involucral bracts lanceolate. Aug.–Oct. Sandy flats and slopes, NW, SW (Gifberg to Hopefield).•

incrassata Meisn.   Sprawling, resprouting shrublet to 0.1 × 1 m. Leaves dissected, crispy hairy, 25–50 mm long. Flower heads solitary, adpressed silvery pink, fragrant, style 9–12 mm long, pollen presenter club-shaped, involucral bracts ovate with silky margins. July–Oct. Granite and shale slopes, SW (Paarl: Klipheuwel).•

A. Flowers more or less straight in bud
D. Flower heads sessile or subsessile and nested in leaves

brownii Meisn.   Densely leafy shrublet to 50 cm. Leaves dissected, grey-silky, 15–25 mm long. Flower heads solitary, silvery pink to brown, fragrant, style c. 10 mm long, pollen presenter club-shaped, involucral bracts lanceolate. June–Oct. Shale and granite slopes and flats, SW (Hopefield to Tygerberg).•

deluvialis Rourke   Sprawling to prostrate shrublet to 30 cm. Leaves sparsely dissected, 45–80 mm long. Flower heads solitary, sessile, silvery cream-coloured, inconspicuous, style c. 8 mm long, pollen presenter slender. Sept.–Nov. Seasonally waterlogged sandstone lower slopes, SW (Kogelberg: Palmiet River valley).•

hirsuta R.Br.   Compact rounded shrublet to 50 cm. Leaves dissected, silky hairy, 30–35 mm long. Flower heads solitary, silvery pink, fragrant, style c. 10 mm long, pollen presenter club-shaped, involucral bracts linear, with silky margins. July–Oct. Sandstone slopes, SW (Cape Peninsula).•

millefolia Salisb. ex Knight   Densely leafy shrublet to 50 cm. Leaves dissected, crispy hairy, 5–14 mm long. Flower heads solitary, silvery cream-coloured to brown, style c. 7 mm long, pollen presenter club-shaped, involucral bracts ovate. Aug.–Dec. Sandy flats and slopes, NW (Bokkeveld to Olifants River Mts).•

rostellaris Salisb. ex Knight   Compact shrublet to 50 cm. Leaves dissected, 15–25 mm long. Flower heads solitary, sessile, silvery pink, style c. 10 mm long, pollen presenter club-shaped, involucral bracts lanceolate, purple beneath. Sept.–Nov. Moist sandstone slopes, SW (Groenland to Kleinrivier Mts).•

villosa (Lam.) R.Br.   Compact rounded shrublet to 80 cm. Leaves dissected, silky hairy, 20–40 mm long. Flower heads solitary, sessile, yellow, fragrant, style c. 10 mm long, pollen presenter club-shaped, involucral bracts lanceolate. Mainly Aug.–Dec. Sandstone slopes and flats, SW (Cape Peninsula).•

reberoi Rourke   Sprawling to prostrate shrublet to 30 cm. Leaves secund, dimorphic, the lower dissected, the upper terete, glabrescent-silky, 40–55 mm long. Flower heads solitary, sessile, silvery cream-coloured, inconspicuous, style c. 5 mm long, pollen presenter slender. Sept.–Nov. Sandstone slopes, SW (Bredasdorp Mts).•

D. Flower heads on a discrete stalk
E. Involucral bracts lacking

candicans R.Br.   Shrublet to 80 cm. Leaves dissected, silvery woolly, 25–55 m long. Flower heads many, silvery pink with dark tips, style 6–9 mm long, pollen presenter club-shaped, involucral bracts absent. July–Dec. Granite and sandy slopes and flats, SW (Paardeberg to Slanghoek Mts).•

confragosa Rourke   Sparsely branched, wand-like shrub to 1 m. Leaves clustered at branch tips, dissected, 30–60 mm long. Flower heads several, silvery pink, style 10–12 mm long, pollen presenter club-shaped, involucral bracts absent. Sept.–Nov. Sandstone slopes, NW (Cold Bokkeveld and Grootwinterhoek Mts).•

decumbens (Thunb.) R.Br.   Prostrate shrublet. Leaves secund, weakly dissected. Flower heads few, adpressed-hairy, carmine, fragrant, style 12–14 mm long, pollen presenter slender. July–Oct. Rocky sandstone slopes, SW (Cape Peninsula).•

elongata (P.J.Bergius) R.Br.   Erect shrub to 1.5 m. Leaves whorled above, dissected, fleshy, 50–150 mm long. Flower heads several, on a long peduncle, adpressed-hairy, silvery pink, fragrant, style 7–11 mm long, pollen presenter club-shaped, involucral bracts absent. July–Oct. Sandy flats and slopes, SW, AP (Du Toitskloof to Agulhas).•

inconspicua L.Guthrie & T.M.Salter   Sprawling, tangled shrublet to 30 cm. Leaves dissected, crispy hairy, 15–25 mm long. Flower heads few, hidden among leaves, adpressed-hairy, cream-coloured, style c. 5 mm long, pollen presenter weakly club-shaped, involucral bracts absent. Mainly Sept.–Oct. Sandstone flats and slopes, SW (Cape Peninsula to Houwhoek).•

lacunosa Rourke   Erect shrublet to 1 m. Leaves dissected, glabrescent, 60–80 mm long. Flower heads solitary on long peduncles, adpressed-hairy, silvery pink, style 10–12 mm long, hairy below, pollen presenter club-shaped, involucral bracts absent. Sept.–Dec. Sandstone slopes, NW (Matsikamma Mts).•

leipoldtii E.Phillips & Hutch.   Multistemmed, resprouting shrub to 1 m. Leaves dissected, 65–105 mm long. Flower heads on a distinct peduncle, adpressed-hairy, silvery pink, style c. 8 mm long, pollen presenter club-shaped, involucral bracts absent. Sept.–Dec. Sandstone slopes, NW (N Cedarberg Mts).•

williamsii Rourke Like **S. elongata** but leaves 150–220 mm long, leathery. Mainly Aug.–Nov. Sandstone slopes, SW (Riviersonderend Mts).•

E. Involucral bracts present
F. Involucral bracts conspicuous, at least as long as the flowers or strongly keeled

flagellifolia Salisb. ex Knight   Prostrate, sparsely leafy shrublet with slender trailing stems. Leaves secund, needle-like or sparsely dissected. Flower heads solitary, glabrous, deep pink, style 5–8 mm long, pollen presenter club-shaped, involucral bracts ovate and keeled. June–Nov. Sandstone slopes, SW (Grabouw to Babilonstoring).•

florida (Thunb.) Salisb. ex Knight   BLUSHING BRIDE   Willowy shrub to 1.5 m. Leaves dissected, 45–100 mm long. Flower heads few, pink and white, style 8–12 mm long, pollen presenter slender, involucral bracts longer than flowers, ovate, pale pink. July–Oct. Granite slopes, SW (Franschhoek: Assegaaiboskloof).•

gracilis Salisb. ex Knight (incl. **S. pinnata** R.Br.)   Prostrate, mat-forming shrublet to 1 m diam. Leaves secund, dissected, glabrescent, 15–30 mm long. Flower heads 1 or 2, adpressed-hairy, silvery pink, style c. 11 mm long, pollen presenter club-shaped, involucral bracts lanceolate, purple. July–Oct. Sand and granite slopes, SW (Tygerberg and Du Toitskloof to Hottentots Holland Mts).•

heterophylla Meisn.   Spindly shrublet to 1 m. Leaves dissected below, becoming linear above, 25–25 mm long. Flower heads 1–3, silvery pink, fragrant, style c. 8 mm long, pollen presenter club-shaped, involucral bracts large, ovate, cream-coloured. July–Oct. Sandstone slopes, SW (Kleinmond and Kleinrivier Mts).•

meisneriana Schltr.   Slender shrublet to 50 cm. Leaves dissected, 35–90 mm long. Flower heads few, on a long peduncle, 2–7-flowered, pink with glossy maroon tips, style c. 9 mm long, pollen presenter conical, involucral bracts prominent, pink, strongly keeled. July–Oct. Shale foothills, SW (Babilonstoring).•

phylicoides (P.J.Bergius) R.Br.   Like **S. rosea** but style c. 10 mm long, involucral bracts numerous, linear-lanceolate, cream-coloured to pink. Aug.–Nov. Sandy flats and slopes, SW (Du Toitskloof to Kleinrivier Mts).•

rosea E.Phillips   Shrublet to 1.5 m. Leaves dissected, 30–60 mm long. Flower heads few, silvery pink, style c. 11 mm long, pollen presenter club-shaped, involucral bracts conspicuous, ovate, pink with silky margins. Aug.–Oct. Sandy slopes, SW (Du Toitskloof to Riviersonderend Mts).•

stellata Rourke   Prostrate shrublet to 1 m diam. Leaves secund, dissected, 50–80 mm long. Flower heads few, pink, fragrant, style 12–14 mm long, pollen presenter club-shaped, involucral bracts ovate. Sept.–Nov. Sandstone slopes, SW (Stettynsberg to W Riviersonderend Mts).•

F. Involucral bracts inconspicuous

aemula Salisb. ex Knight   Sprawling shrublet to 50 cm. Leaves dissected, glabrescent, 12–20 mm long. Flower heads solitary, adpressed-hairy, silvery pink, fragrant, style c. 8 mm long, pollen presenter club-shaped, involucral bracts linear, silky. July–Oct. Sandy flats, SW (Cape Flats).•

aitonii R.Br.   Rounded shrublet to 1 m. Leaves dissected, crispy hairy, 20–45 mm long. Flower heads 1–few, silvery grey, fragrant, style 7–10 mm long, pollen presenter club-shaped, involucral bracts ovate. July–Nov. Sandstone slopes, NW (Cedarberg to Grootwinterhoek Mts).•

altiscapa Rourke   Rounded shrublet to 30 cm with inflorescences to 2 m. Leaves whorled, dissected, 120–230 mm long. Flower heads several to many, in racemes on elongated peduncles, adpressed-hairy, silvery pink, style 12–14 mm long, pollen presenter club-shaped, involucral bracts ovate, winged. Aug.–Oct. Sandstone slopes, SW (Villiersdorp to Hottentots Holland Mts).•

cyanoides (L.) R.Br.   Densely branched, resprouting shrublet to 50 cm. Leaves dissected, glabrescent, 25–65 mm long. Flower heads solitary, adpressed or spreading hairy, silvery pink, fragrant, style 6–10 mm long, pollen presenter club-shaped, involucral bracts ovate, silky. July–Oct. Sandy flats, SW (Cape Peninsula and Cape Flats).•

fasciflora Salisb. ex Knight   Sprawling to erect shrublet to 1 m. Leaves dissected, sparsely hairy, 30–70 mm long. Flower heads many, silvery pink, sometimes glabrous, sweetly scented, style 5–7 mm long, pollen presenter slender to club-shaped, involucral bracts lanceolate. Mainly May–Dec. Sandy flats and lower slopes, SW, LB, SE (Hopefield to George).•

furcellata R.Br.   Like **S. cyanoides** but florets inflated below. Aug.–Oct. Sandy slopes and flats, SW (Elandskloof Mts and Cape Flats).•

kraussii Meisn.   Shrublet to 1 m. Leaves dissected, 50–75 mm long. Flower heads several, silvery white, style 5–9 mm long, pollen presenter knob-like, involucral bracts lanceolate, subglabrous. July–Nov. Granite and sandstone slopes, SW (Jonkershoek to Hottentots Holland Mts).•

linearis Salisb. ex Knight   Slender, resprouting shrub to 80 cm. Leaves linear, needle-like, sometimes slightly dissected, 26–60 mm long. Flower heads 1–few, silvery pink, style 6–9 mm long, pollen presenter club-shaped, involucral bracts ovate. Aug.–Nov. Sandy lower slopes and flats, SW (Mamre to Dassenberg).•

reflexa Rourke   Sparsely branched shrub to 2 m. Leaves dissected, silvery silky, 10–25 mm long. Flower heads 1–3 on spreading peduncles, white, style c. 17 mm long, pollen presenter club-shaped, involucral bracts ovate. Sept.–Nov. Sandstone slopes, NW (Cold Bokkeveld to Olifants River Mts).•

trilopha Salisb. ex Knight   Like **S. cyanoides** but leaves short, 5–25 mm long, florets glabrous below, coconut-scented. Aug.–Oct. Sandy flats, SW (Malmesbury to Cape Peninsula).•

triternata (Thunb.) R.Br.   Robust shrub to 1.5 m. Leaves dissected, 65–140 mm long, rigid. Flower heads few to many, on a shared peduncle, adpressed-hairy, cream-coloured, style 3–7 mm long, pollen presenter conical, involucral bracts ovate. Aug.–Oct. Sandstone slopes, SW (Elandskloof Mts).•

viridifolia Rourke   Prostrate, mat-forming shrublet to 2 m diam. Leaves secund, 20–50 mm long. Flower heads few, minutely hairy, pink, style c. 9 mm long, pollen presenter slender, involucral indexbracts ovate. Mainly Oct.–Nov. Deep sands on sandstone slopes, SW (Stettynsberg and W Riviersonderend Mts).•

zeyheri Meisn.   Rounded shrublet to 50 cm. Leaves dissected, glabrescent, 70–150 mm long. Flower heads many, white with glossy black tips, adpressed-hairy below, style c. 10 mm long, pollen presenter knob-like, involucral bracts lanceolate. Aug.–Nov. Peaty sandstone slopes, SW (Du Toitskloof to Riviersonderend Mts).•

SOROCEPHALUS   POWDERPUFF PROTEA   11 spp., W Cape

A. Leaves flattened above

capitatus Rourke   Slender, single-stemmed shrub to 1 m. Leaves linear, adpressed, glabrescent, 6–12 mm long, margins scabrid. Flower heads in globose clusters, 20–30 mm diam., pink or cream-coloured, inner florets with perianth tips glabrous. Sept.–Feb. Sandy flats, NW (Piketberg and Cold Bokkeveld: Onderboskloof).•

imbricatus (Thunb.) R.Br.   Slender, single-stemmed, sparsely branched shrub to 1.5 m. Leaves lanceolate, adpressed, glabrescent, c. 15 mm long, margins scabrid. Flower heads in globose, terminal clusters, c. 40 mm diam., cream-coloured, fragrant. Sept.–Dec. Sandstone and clay slopes, NW, SW (Piketberg, Grootwinterhoek and Elandskloof Mts).•

lanatus (Thunb.) R.Br.   Erect or sprawling shrublet to 80 cm. Leaves linear, adpressed, glabrous or hairy, 5–18 mm long, margins scabrid. Flower heads in globose clusters, 10–30 mm diam., pink to purple with darker tips. Sept.–Apr. Sandstone slopes at high alt, NW (Cedarberg to Hex River Mts).•

scabridus Meisn.   Erect sparsely branched shrublet to 80 cm. Leaves linear, slightly hooked at tips, glabrescent, 20–30 mm long, margins scabrid. Flower heads in globose clusters, c. 20–25 mm diam., pink with glabrous shiny black tips. Oct.–Jan. Mountain plateaus, NW (Olifants River and Grootwinterhoek Mts).•

tenuifolius R.Br.   Much-branched slender shrublet to 1 m. Leaves linear, adpressed, glabrescent, c. 10 mm long. Flower heads in globose clusters, c. 15 mm diam., cream-coloured to yellow. Jan.–Feb. Wet sandy flats, SW (Palmiet River Mts: Arieskraal and Kogelberg).•

A. Leaves channelled above

alopecurus Rourke   Single-stemmed, sparsely branched shrub to 1 m. Leaves needle-like, channelled above, 30–50 mm long. Flower heads in terminal, cylindrical clusters, c. 20 mm diam., cream-coloured with black tips, scented. July–Sept. S-facing sandstone slopes, SW (Riviersonderend Mts).•

clavigerus (Salisb. ex Knight) Hutch.   Single-stemmed shrub to 1 m. Leaves needle-like, channelled above, 7–25 mm long, glabrescent, adpressed and imbricate. Flower heads in terminal, broadly cylindrical clusters, 20–40 mm diam., cream-coloured to pale yellow with dark tips. July–Dec. Sandstone slopes, SW (Hottentots Holland to Kleinrivier Mts).•

crassifolius Hutch.   Rounded, much branched shrublet 30–50 cm, multistemmed from a woody rootstock. Leaves needle-like, channelled above, 20–25 mm long. Flower heads few, in small, solitary, terminal clusters, cream-coloured with brown tips, thinly hairy. Dec.–Feb. Rock ledges at high alt., SW (Riviersonderend Mts).•

palustris Rourke   Sprawling mat-forming shrublet to 0.2 × 1 m. Leaves needle-like, channelled above, 15–20 mm long. Flower heads in terminal, globose clusters, c. 20 mm diam., white with dark tips. Sept.–Dec. Peaty sponges at high alt., SW (Kogelberg).•

pinifolius (Salisb. ex Knight) Rourke   Single-stemmed shrub to 1 m. Leaves needle-like, channelled above, 40–60 mm long. Flower heads in terminal, subglobose clusters, c. 25 mm diam., cream-coloured with glabrous black tips. June–Oct. S-facing sandstone slopes, SW (Riviersonderend Mts).•

teretifolius (Meisn.) E.Phillips   Single-stemmed, erect or spreading shrublet to 70 cm. Leaves needle-like, channelled above, 10–14 mm long. Flower heads crowded in flattened terminal clusters, 15–20 mm diam., pink with dark tips. Nov.–Dec. Rocky sandstone slopes at high alt., SW (Du Toitskloof Mts to Blokkop).•

SPATALLA   FUZZY-BUDS   20 spp., W Cape

A. Involucres 3-flowered

argentea Rourke   Like **S. propinqua** but leaves silvery silky and style straight. Oct.–Jan. Sandstone slopes, SW (Riviersonderend Mts: Jonaskop).•

caudata (Thunb.) R.Br.   Shrub to 1 m. Leaves linear-subterete, thinly hairy above. Flower heads in cylindrical racemes, 50–60 mm long, silvery pink, sometimes cream-coloured, tepals equal, style inflexed, c. 8 mm long, pollen presenter ovoid. Aug.–Oct. Moist slopes at high alt., NW (Cedarberg and Hex River Mts).•

confusa (E.Phillips) Rourke   Sprawling to mat-forming shrublet to 1 m. Leaves linear-subterete, glabrescent. Flower heads in short racemes, 10–25 mm long, silvery pink., upper tepal larger, style inflexed, 6–7 mm long, pollen presenter ovoid. Mainly Nov.–Dec. Rocky sandstone slopes, NW, SW, KM (Cedarberg to Hottentots Holland Mts, Witteberg to Kammanassie Mts).•

incurva (Thunb.) R.Br.   Erect shrublet to 1 m. Leaves needle-like, glabrous or subglabrous. Flower heads in cylindrical racemes c. 40 mm long, silvery pink, upper tepal larger, style inflexed, c. 6 mm long, pollen presenter obovoid. Sept.–Mar. Sandstone slopes, NW, SW (Cedarberg to Du Toitskloof Mts and Kwadouwsberg).•

propinqua R.Br.   Shrublet to 1 m. Leaves needle-like, glabrescent. Flower heads in cylindrical racemes, 10–30 mm long, silvery pink, upper tepal larger, style inflexed, c. 7 mm long, pollen presenter discoid. Mainly July–Oct. Moist sandstone slopes, SW (Slanghoek to Riviersonderend Mts).•

thyrsiflora Salisb. ex Knight   Resprouting, mat-forming shrublet to 1.5 m diam. Leaves needle-like. Flower heads in cylindrical racemes, 20–40 mm long, silvery pink or whitish, lower bracteole subtending a forked shoot, tepals equal, style sigmoid, c. 10 mm long, pollen presenter ovoid, black. Mainly Oct.–Nov. Moist rocky slopes at high alt., SW (Du Toitskloof Mts and Caledon Swartberg).•

tulbaghensis (E.Phillips) Rourke   Rounded shrublet to 50 cm. Leaves linear-subterete, ciliate. Flower heads in globose clusters, silvery pink, tepals equal, style inflexed, 7–8 mm long, pollen presenter oblong-cylindrical. Sept.–Dec. Moist sandstone slopes at high alt., NW (Witsenberg and Skurweberg).•

A. Involucres 1-flowered
B. Tepals equal; pollen presenter more or less ovoid

nubicola Rourke   Rounded shrub to 1.5 m. Leaves needle-like, glabrescent. Flower heads in cylindrical racemes, 30–40 mm long, silvery pink, tepals equal, style straight, c. 9 mm long, pollen presenter more or less ovoid. Sept.–Dec. Moist S-facing sandstone slopes at high alt., LB (Langeberg Mts: Heidelberg).•

salsoloides (R.Br.) Rourke   Like **S. nubicola** but plants sprawling, leaves incurved and style inflexed. Oct.–Dec. High Sandstone slopes, SW (Du Toitskloof Mts).•

setacea (R.Br.) Rourke   Like **S. nubicola** but leaves longer, 20–30 mm long, flower heads subglobose. Sept.–Dec. Peaty sponges on S-facing slopes at high alt., SW (Slanghoek Mts to Kleinmond).•

B. Tepals unequal, the upper larger and hooded; pollen presenter discoid
C. Racemes pedunculate; leaves channelled

curvifolia Salisb. ex Knight   Rounded shrublet to 80 cm. Leaves needle-like, channelled,. Flower head in cylindrical racemes, cream-coloured, style straight, c. 7 mm long, pollen presenter discoid. Jan.–Dec. Sandstone slopes, SW, AP (Kogelberg to Agulhas coast).•

longifolia Salisb. ex Knight   Rounded shrub to 1 m. Leaves needle-like, channelled. Flower heads in cylindrical racemes, greyish pink with red markings, style incurved, c. 6 mm long, pollen presenter discoid. Oct.–Nov. Sandstone slopes, SW (Jonkershoek Mts to Kleinmond).•

racemosa (L.) Druce   Slender shrublet like **S. curvifolia** but raceme small, 10–30 mm long, style c. 5 mm long. Sept.–Mar. Sandy flats and slopes, SW (Viljoen's Pass to Kleinrivier Mts).•

, C. Racemes sessile; leaves needle-like

barbigera Salisb. ex Knight   Like **S. colorata** but leaves softly hairy, style c. 6 mm long. May–Nov. Sandstone slopes, KM, LB, SE (E Langeberg to Outeniqua Mts and Swartberg).•

colorata Meisn.   Shrublet to 80 cm. Leaves needle-like. Flower heads in short ovoid racemes, silvery pink, style incurved, c. 7 mm long, pollen presenter discoid. July–Nov. High sandstone slopes, SW, LB (Riviersonderend to Langeberg Mts: Heidelberg).•

ericoides E.Phillips   Slender shrublet to 80 cm. Leaves needle-like, fairly short, 7–12 mm long. Flower heads in cylindrical spikes, silvery pink, bracteoles fused into a prominent involucre, style incurved, c. 7 mm long, pollen presenter discoid. Aug.–Oct. Coastal sands between limestone ridges, AP (W Agulhas coast).•

mollis R.Br.   Rounded shrublet to 80 cm. Leaves needle-like. Flower heads in cylindrical racemes, white, sometimes mauve, style straight, hairy below, c. 5 mm long, pollen presenter discoid. July–Dec. Streamsides on sandstone slopes, SW (Hottentots Holland Mts to Kleinmond).•

parilis Salisb. ex Knight   Shrub to 1.5 m. Leaves needle-like. Flower heads in cylindrical racemes, silvery pink, style straight, c. 6 mm long, pollen presenter discoid. Jan.–Dec. Moist, sandstone slopes, SW, LB (Hottentots Holland to Langeberg Mts).•

prolifera (Thunb.) Salisb. ex Knight   Sparsely branched shrub to 1.5 m. Leaves needle-like. Flower heads in subglobose spikes, silvery pink, style inflexed, c. 7 mm long, pollen presenter discoid. Sept.–Dec. Marshy sandstone slopes, SW (Viljoen's Pass to Kleinmond).•

squamata Meisn.   Like **S. ericoides** but involucral bracts glabrous, accrescent, style c. 8 mm long. Aug.–Oct. Sandy flats and slopes, often between limestone ridges, SW, AP (Bredasdorp Mts and Agulhas coast).•

VEXATORELLA FALSE PINCUSHION 4 spp., Namaqualand to W Cape

amoena (Rourke) Rourke   Erect to sprawling shrub to 1.5 m. Leaves obovate, 15–30 mm long. Flower heads solitary, cream-coloured to pink, fragrant, 12–25 mm diam. Sept.–Nov. Sandstone slopes, NW (Cold Bokkeveld and Swartruggens).•

latibrosa Rourke   Single-stemmed, rounded shrub to 2 m. Leaves linear to spathulate, 50–60 mm long. Flower heads solitary, terminal, pink to carmine, fragrant, 25–30 mm diam. Oct.–Nov. Clay lower slopes, NW (Langeberg Mts: Robertson).•

obtusata (Thunb.) Rourke   Prostrate, mat-forming shrub 1–2 m diam. Leaves linear-spathulate, 9–45 mm long. Flower heads solitary, terminal, cream-coloured to pink, fragrant, 15–20 mm diam. Aug.–Dec. Sandstone slopes, NW, KM (Hex River Mts to Witteberg and Anysberg).•

RANUNCULACEAE

1. Leaves opposite; carpels with feathery tails; twining or scrambling plants, becoming woody with age ......**Clematis**
1.' Leaves radical or alternate; carpels without feathery tails; perennial herbs:
   2. Sepals petaloid, petals absent; inflorescence 1–3-flowered ....**Anemone**
   2.' Sepals and petals present; inflorescence usually several-flowered:
      3. Petals with a nectary, white or yellow; achenes dry ....**Ranunculus**
      3.' Petals without a nectary, cream-coloured to green; carpels becoming fleshy or pulpy on maturity..**Knowltonia**

ANEMONE ANEMONE, WINDFLOWER   c. 150 spp., mainly temperate

tenuifolia (L.f.) DC.   CAPE ANEMONE   Tufted perennial to 40 cm. Leaves bi- or tripinnate, segments cuneate and 3-toothed, margins revolute. Flowers solitary on woolly peduncle, pinkish white, silky beneath. June–Feb. Moist sandstone slopes, NW, SW, AP, KM, LB, SE (Bokkeveld Mts to Humansdorp).•

CLEMATIS TRAVELLER'S JOY   c. 250 spp., pantemperate

brachiata Thunb.   OLD MAN'S BEARD   Perennial climber. Leaves opposite, bi- or tripinnate, segments ovate, toothed. Flowers in axillary cymose panicles, hairy, white, fragrant. Fruits plumose. Dec.–May. Scrub and forest margins, KM, LB, SE (Montagu to Port Elizabeth, widespread in southern and tropical Africa).

KNOWLTONIA BRANDBLAAR, KATJIEDRIEBLAAR   8 spp., Africa

anemonoides H.Rasmussen   Like **K. capensis** but leaves more hairy beneath than above, scape much longer than leaves and tepals essentially alike. Sept.–Jan., after fire. Rocky slopes, NW, SW, AP, LB (Tulbagh to Cape Peninsula to Riversdale).•

capensis (L.) Huth   Silky hairy tufted perennial to 30 cm, rhizome vertical. Leaves bi- or triternate, leaflets ovate, toothed, equally hairy or more hairy above. Flowers in scapose umbels not much longer than leaves, greenish white to purple, outer tepals shorter and more hairy. Fruits hairy. June–Sept. Shaded rocky slopes, NW, SW (Porterville to Cape Peninsula).•

cordata H.Rasmussen   Thinly silky tufted perennial to 70 cm, rhizome vertical. Leaves ovate or ternate, leaflets cordate, toothed. Flowers in scapose umbels much longer than leaves, greenish to pale yellow. Fruits glabrous. Sept.–Jan. Rocky outcrops and forest margins, SE (Attaquas Mts to E Cape).

filia (L.f.) T.Durand & Schinz   Like **K. cordata** but leaves pinnate or bipinnate, leaflets coarsely toothed, never cordate. Flowers in scapose umbels, white or greenish white. Oct.–Jan. Shaded or moist slopes, KM, SE (Kammanassie Mts, Outeniqua to Tsitsikamma Mts).•

vesicatoria (L.f.) Sims   Tufted perennial to 1.2 m, rhizome horizontal. Leaves leathery, ternate or triternate, segments ovate, toothed. Flowers in scapose umbels not much longer than leaves, white to yellowish green. Fruits glabrous. Aug.–Oct. Scrub or woody ravines, NW, SW, AP, LB, SE (Bokkeveld Mts to Cape Peninsula to E Cape).

RANUNCULUS BUTTERCUP, CROWFOOT c. 400 spp., cosmopolitan, mainly temperate

multifidus Forssk. BUTTERCUP Tufted, silky perennial to 90 cm. Leaves pinnatisect or bipinnatisect, segments toothed. Flowers in open panicles, yellow. Achenes warty, 1.5–2 mm long. Aug.–Mar. Damp places, NW, SW, AP, LB, SE (Namaqualand to Cape Peninsula to Arabia).

*muricatus L. SPINY-FRUITED BUTTERCUP Annual to 30 cm. Leaves round, on long petioles, 3-lobed and toothed. Flowers in open panicles, yellow. Achenes spiny, 7–8 mm long. Sept.–Nov. Weed of damp places, SW (European weed, Cape Peninsula to Stellenbosch).

trichophyllus Chaix WATER BUTTERCUP, WATER CROWFOOT Submerged aquatic perennial. Leaves finely dissected, segments filiform. Flowers emergent, solitary, opposite leaves, white. Achenes wrinkled, c. 1 mm long. Sept.–May. Pools or streams, NW, SW (Velddrif, Cape Peninsula, Karoo, almost worldwide).

RESEDACEAE

OLIGOMERIS 9 spp., mainly N hemisphere

dipetala (Aiton) Turcz. BITTERAARBOSSIE Perennial to 70 cm, often rough. Leaves linear, often tufted. Flowers in terminal spikes, whitish, stamens mostly 8–12. Dec.–Mar. Alkaline and limestone flats, AP, KM, LB, SE (Montagu to Uitenhage to Namibia and Botswana).

RHAMNACEAE

1. Fruit a single-seeded drupe, fleshy or dry:
    2. Inflorescence a short axillary umbel; spines usually present...**Scutia**
    2.' Inflorescence an axillary raceme or cyme; spines absent ....**Rhamnus**
1.' Fruit a several-seeded capsule, fleshy or dry:
    3. Leaves elliptic-oblong, toothed; flowers in lax terminal and axillary panicles; disc inconspicuous; style shortly 3-lobed ....**Noltea**
    3.' Leaves narrower, entire; flowers in racemes, spikes or heads (rarely solitary); disc evident; style simple:
        4. Stipules present; disc densely hairy..**Trichocephalus**
        4.' Stipules absent; disc glabrous..**Phylica**

NOLTEA SEEPBLINKBLAAR 1 sp., South Africa

africana (L.) Rchb.f. Shrub or small tree to 4 m. Leaves elliptic-oblong, serrate, glossy above, paler beneath. Flowers in axillary and terminal panicles, white. Fruits globose, calyx cup persistent. Aug.–Oct. Riverine bush, SW, LB, SE (Cape Peninsula to KwaZulu-Natal).

PHYLICA PHYLICA, HARDEBOS c. 150 spp., Africa, Madagascar, S Atlantic islands

*A. Petals absent (see also **P. debilis**, **P. pubescens**)*

aemula Schltr. Closely leafy, grey-tomentose shrublet to 30 cm. Leaves linear, 8–13 mm long, rough, margins closely revolute. Flowers in large, solitary, plumose capitula surrounded by many leaves with enlarged petioles, without petals, whitish. Aug.–Jan. Sandstone plateaus, NW (Bokkeveld Mts to Cold Bokkeveld).•

alticola Pillans Closely leafy shrublet to 35 cm. Leaves linear, 5–8 mm long, smooth, margins closely revolute. Flowers in small, rounded capitula, without petals, whitish. Sept.–Oct. Upper mountain slopes, NW (Grootwinterhoek Mts).•

barbata Pillans Closely leafy, dwarf shrublet, velvety on young parts. Leaves linear, 10–14 mm long, rough, margins closely revolute. Flowers in wide, solitary, plumose capitula surrounded by many strongly ciliate leaves, without petals, whitish. Aug.–Oct. Dry rocky slopes, NW (Cedarberg Mts).•

barnardii Pillans Closely leafy shrub to 60 cm, silky on young parts. Leaves needle-like, 7–12 mm long, rough, margins closely revolute. Flowers in rounded, solitary of clustered capitula, without petals, whitish. Aug.–Oct. Rocky sandstone slopes, NW (Cedarberg to Hex River Mts).•

comptonii Pillans Compact, closely leafy shrublet to 30 cm. Leaves oblong, cordate at base, 2–5 mm long, margins closely revolute. Flowers in solitary capitula, without petals, whitish, subsessile. Dec.–Feb. Dry sandstone slopes, NW, KM (Hex River Mts to Witteberg).•

constricta Pillans Closely leafy, much-branched shrublet to 45 cm. Leaves needle-like, 5–9 mm long, smooth, margins closely revolute. Flowers in solitary, flattened capitula, whitish, without petals. Mainly Oct.–Jan. NW, SW, KM (Cedarberg to Hottentots Holland Mts and Swartberg).•

intrusa Pillans   Twiggy shrublet to 60 cm. Leaves linear, c. 5 mm long, smooth, margins closely revolute. Flowers in short spikes, whitish, without petals, tomentose outside. Nov.–Jan. High peaks, NW (Hex River Mts).•

leipoldtii Pillans   Closely leafy shrublet to 50 cm, velvety on young parts. Leaves needle-like, 10–15 mm long, rough, margins closely revolute. Flowers in wide, flattened, solitary capitula surrounded by densely pilose leaves, without petals, whitish. Aug.–Jan. Rocky slopes, NW (Cedarberg Mts to Cold Bokkeveld).•

levynsiae Pillans   Closely leafy, tomentose shrublet to 50 cm. Leaves ascending, oblong, rough, margins closely revolute. Flowers in solitary rounded capitula surrounded by many leaves, without petals, whitish. Sept.–Oct. Rocky sandstone slopes, NW (Swartruggens).•

marlothii Pillans   Closely leafy shrublet to 60 cm, velvety pubescent on young branches. Leaves needle-like, 13–15 mm long, margins closely revolute. Flowers in large, flattened, solitary capitula surrounded by many leaves and villous bracts, without petals, whitish. May–Nov. Rocky sandstone slopes, NW, KM (Cedarberg Mts to Witteberg).•

obtusifolia Pillans   Closely leafy shrublet to 35 cm. Leaves oblong, cordate at base, c. 4 mm long, margins closely revolute. Flowers in solitary capitula surrounded by leafy bracts, subsessile, without petals, whitish. Dec.–Jan. Upper slopes, NW (Cold Bokkeveld Mts).•

plumigera Pillans   Closely leafy spreading shrublet to 40 cm. Leaves needle-like, 7–15 mm long, rough, margins closely revolute. Flowers in wide, solitary, plumose capitula surrounded by many heavily ciliate leaves, without petals, white. Aug.–Dec. Rocky sandstone slopes, NW (Cedarberg Mts).•

pulchella Schltr.   Closely leafy shrublet, 40–60 cm. Leaves ascending, oblong, rough, margins closely revolute. Flowers in large, flattened, solitary capitula surrounded by villous leaves and bracts, without petals, white. Aug.–Nov. Sandstone slopes, NW (Bokkeveld Mts to Cold Bokkeveld).•

retorta Pillans   Silky haired shrublet to 40 cm. Leaves needle-like, 7–10 mm long, rough, margins closely revolute. Flowers in wide, solitary capitula surrounded by a few leaves, without petals, whitish. July–Sept. Dry sandstone slopes, KM (Bonteberg and Witteberg).•

rigida Eckl. & Zeyh.   Closely leafy, grey-tomentose shrub to 1 m. Leaves linear, 10–25 mm long, smooth, margins closely revolute. Flowers in large, flattened, solitary capitula surrounded by many pilose leaves, without petals, whitish. Sept.–Nov. Rocky slopes and outcrops, NW (Cedarberg Mts to Cold Bokkeveld).•

A. Petals present
B. Flowers in racemes or spikes

agathosmoides Pillans   Closely leafy shrub to 60 cm. Leaves ascending, needle-like, 7–10 mm long, rough, margins closely revolute. Flowers in compact spikes, lower bracts leafy, brownish green, subsessile. Aug.–Sept. Sandstone slopes, NW (Bokkeveld Mts).•

ambigua Sond.   Much-branched, densely leafy shrub to 60 cm. Leaves needle-like, 10–15 mm long, margins closely revolute, covered with silky hairs when young. Flowers axillary, forming dense racemes below branch tips, yellowish. Aug.–Nov. Rocky sandstone slopes, NW (Pakhuis Pass to Cold Bokkeveld).•

ampliata Pillans   Densely branched shrub to 1 m. Leaves ovate, 6–8 mm long, smooth, margins narrowly revolute. Flowers axillary, crowded near branch tips, whitish. Mar.–June. Mountain slopes, SW (Elandskloof Mts).•

axillaris Lam.   Much-branched, tomentose shrub to 80 cm. Leaves linear to linear-lanceolate, 5–15 mm long, smooth, margins closely revolute. Flowers axillary, well below branch tips, whitish. Mainly Dec.–Apr. Rocky slopes in coastal bush or forest margins, AP, SE (Agulhas coast to Katberg and Alexandria).

cryptandroides Sond.   Slender-branched, closely leafy shrub, to 3 m. Leaves needle-like, 10–15 mm long, rough, margins closely revolute. Flowers in dense racemes grouped in panicles, white, sometimes pinkish. Mainly Sept.–Oct. Sandstone slopes, NW (Kamiesberg and Gifberg to Piketberg).

cuspidata Eckl. & Zeyh.   Twiggy shrub to 2 m, silky on young parts. Leaves ovate-lanceolate, 4–8 mm long, recurved above middle, margins closely revolute. Flowers in rounded spikes, subtended by densely villous bracts, whitish, sessile. Aug.–Dec. Sandstone slopes, NW (Olifants River Mts to Piketberg).•

cylindrica H.L.Wendl. Virgate shrub to 1 m. Leaves linear-lanceolate, 7–12 mm long, rough, margins closely revolute. Flowers in a dense, terminal raceme, whitish. July–Sept. Sandstone slopes, NW (Olifants River Mts and Piketberg).•

elimensis Pillans Like **P. axillaris** but leaves narrowly linear, 5–12 mm long, and flowers covered with retrorse hairs. July–Aug. Lower slopes, SW, LB (Elim to Riversdale).•

hirta Pillans Densely leafy shrub to 60 cm, grey-villous on young parts. Leaves linear-lanceolate, c. 10 mm long, rough, margins closely revolute. Flowers in compact spikes, subtended by long, silky bracts, whitish, subsessile. Nov.–Jan. Sandstone slopes, NW (Nardouw and Olifants River Mts).•

mundii Pillans Closely leafy shrub to 40 cm. Leaves needle-like, 12–16 mm long, incurved above middle, margins closely revolute. Flowers in crowded racemes, pinkish. Mar.–May. Rocky sandstone slopes, SW, LB (Houwhoek to Langeberg Mts: Riversdale).•

oleaefolia Vent. BLINKHARDEBOS Rigid shrub to 2 m. Leaves ovate to broadly elliptic, 15–15 mm long, white-felted beneath, margins slightly revolute. Flowers in short, terminal racemes sometimes forming small panicles, cream-coloured. Mar.–May. Rocky slopes, NW, SW (S Namaqualand and W Karoo to Worcester).

paniculata Willd. HARDEBOS, LUISBOOM Grey-tomentose shrub or small tree mostly 1.5–3 m. Leaves ovate to lanceolate, mostly 10–15 mm long, margins revolute, white-felted beneath. Flowers usually in spikes forming compound panicles, shortly stipitate, whitish. May–June. Moist slopes and gullies, bush and forest margins, NW, KM, LB, SE (Hex River valley and Witteberg to Zimbabwe).

pinea Thunb. Closely leafy shrub to 1 m. Leaves lanceolate to linear-lanceolate, cordate at base, mostly c. 12 mm long, margins strongly revolute. Flowers in short, mostly terminal racemes, white. Dec.–Aug. Sandstone slopes, SW, KM, LB, SE (Drakenstein Mts and Klein Swartberg to Tsitsikamma Mts).•

plumosa L. VEERKOPPIE Sparsely branched shrublet to 40 cm. Leaves linear-lanceolate, 10–15 mm long, rough, margins revolute. Flowers in dense spikes, subtended by buff-plumose bracts longer than leaves, yellowish, subsessile. May–Aug. Mainly clay and granite soils, NW, SW (Piketberg to Cape Peninsula and Caledon).•

purpurea Sond. Closely leafy shrub or small tree to 3 m. Leaves lanceolate, rounded to cordate at base, 5–10 mm long, margins revolute. Flowers in dense, head-like racemes, pink to purple. Apr.–Oct. Sandstone slopes, often on forest margins, SW, KM, LB, SE (Swartberg Mts and Bredasdorp to Humansdorp).•

recurvifolia Eckl. & Zeyh. Sprawling, sparsely branched shrublet to 40 cm. Leaves linear, 12–17 mm long, recurving above middle, margins closely revolute. Flowers in lax spikes, sessile, whitish. Oct.–Nov. Sandstone slopes, LB (Langeberg Mts: Swellendam).•

rigidifolia Sond. Much-branched, closely leafy shrub to 1 m. Leaves linear to needle-like, 12–25 mm long, rough, margins closely revolute. Flowers in rounded panicles, sometimes in racemes, white. May–Oct. Sandstone slopes and plateaus, NW, KM, LB (Kamiesberg and Bokkeveld Mts to Piketberg and Witteberg to Riversdale).

spicata L.f. Closely leafy shrub to 2 m, grey-felted on young parts. Leaves ovate-lanceolate, 15–20 mm long, rough, margins slightly revolute, white-felted beneath. Flowers in solitary spikes, white, subtended by white-villous bracts, sessile. Apr.–Aug. Rocky sandstone slopes, NW, SW (Pakhuis Mts to Stellenbosch).•

strigulosa Sond. Wiry, much-branched shrublet to 30 cm. Leaves linear to needle-like, mostly 12–15 mm long, rough, margins closely revolute. Flowers in dense spikes, white, sessile. Mar.–July. Stony clay and sandstone slopes at low elevation, NW, SW (Piketberg to Stellenbosch).•

velutina Sond. Closely leafy, sparsely branched shrublet, to 30 cm. Leaves needle-like, 13–23 mm long, rough, margins closely revolute. Flowers in dense spikes, subtended by long, buff-plumose bracts, white, subsessile. Apr.–Aug. Lower slopes, LB (Langeberg Mts: Swellendam to Ruitersbos).•

villosa Thunb. NAALDHARDEBLAAR Closely leafy, much-branched shrub to 1 m. Leaves linear, mostly 7–20 mm long, tomentose when young, margins closely revolute. Flowers in dense or lax racemes subtended by leafy bracts, white. Feb.–July. Rocky sandstone slopes, NW (Pakhuis Mts and Cedarberg to Olifants River Mts).•

willdenowiana Eckl. & Zeyh. Densely leafy shrub, 30–60 cm. Leaves linear to linear-lanceolate, 5–15 mm long, margins closely revolute. Flowers in crowded racemes below branch tips, white. May–July. Sandstone slopes, SE (Outeniqua Mts to Port Elizabeth).•

B. Flowers in capitula

abietina Eckl. & Zeyh.   Closely leafy shrub to 1 m. Leaves linear-lanceolate, 4–6 mm long, smooth above, margins closely revolute. Flowers in solitary, rounded capitula surrounded by many leaves, white or pinkish, subsessile. Apr.–Aug. Dry sandstone slopes, SE (George to Uitenhage).•

acmaephylla Eckl. & Zeyh.   Spreading shrublet to 50 cm. Leaves lanceolate, 5–8 mm long, incurved, margins strongly revolute. Flowers usually in solitary capitula, surrounded by leaves and leafy bracts, whitish. Flowering time? Sandstone slopes, SW (Elandskloof Mts to Brandvlei).•

affinis Sond.   Closely leafy, grey-tomentose shrublet to 60 cm. Leaves linear, 8–13 mm long, smooth, margins closely revolute. Flowers in wide, solitary capitula surrounded by many leaves and plumose bracts, greyish white. July–Aug. Sandstone slopes, NW (Bokkeveld Mts).•

alba Pillans   Virgate shrub to 60 cm. Leaves linear-lanceolate, c. 6 mm long, cordate at base, margins closely revolute. Flowers in flattened capitula, solitary or in small clusters, whitish, subsessile. Mar.–June. Moist sandstone slopes, LB, SE (Langeberg to Tsitsikamma Mts).•

alpina Eckl. & Zeyh.   Sprawling shrublet to 20 cm. Leaves lanceolate, c. 6 mm long, spreading-incurved, margins revolute, tomentose beneath. Flowers in solitary, flattened capitula, whitish. Aug.–Sept. Sandstone slopes, NW (Olifants River Mts).•

altigena Schltr.   Much-branched shrublet to 50 cm. Leaves linear, 5–8 mm long, rough, margins closely revolute. Flowers in wide, mostly solitary capitula surrounded by many pilose leaves, whitish. Aug.–Oct. Rocky slopes, NW (Cedarberg to Grootwinterhoek Mts).•

amoena Pillans   Closely leafy shrublet to 30 cm. Leaves lanceolate, c. 8–12 mm long, margins revolute. Flowers in broad, flattened, solitary capitula surrounded by many leaves, white, sessile. Mar.–Apr. Coastal dunes, AP (Agulhas coast).•

anomala Pillans   Intricately branched, closely leafy shrublet to 40 cm. Leaves ovate-lanceolate, c. 2.5 mm long, rough, margins closely revolute. Flowers in small, solitary, flattened capitula, surrounded by several leaves, whitish. Dec. ?Sandstone slopes, SW (Caledon).•

apiculata Sond.   Erect shrub, to 60 cm, with slender tomentose branchlets. Leaves linear-lanceolate, 7–14 mm long, smooth, margins strongly revolute. Flowers in capitula surrounded by leafy bracts, sometimes grouped in loose panicles, whitish. Aug.–Sept. Mountain slopes, SW (Caledon district).•

atrata Licht. ex Roem. & Schult.   Closely leafy, spreading shrub, mostly 30–60 cm. Leaves lanceolate, 6–10 mm long, smooth, margins closely revolute. Flowers in solitary, flattened capitula, surrounded by leaves and leafy bracts, white. Oct.–Mar. Rocky slopes, SW (Elandskloof Mts to False Bay).•

bolusii Pillans   Closely leafy, grey-felted shrublet to 50 cm. Leaves needle-like, 13–20 mm long, rough, margins closely revolute. Flowers in mostly solitary capitula surrounded by buff-hairy leaves and bracts, whitish. Nov.–Jan. Rocky sandstone slopes, NW (Ceres mts).•

brachycephala Sond.   Shortly leafy shrub to 60 cm, with numerous short branches. Leaves lanceolate, 3–6 mm long, margins loosely revolute. Flowers in solitary capitula surrounded by a few small leaves, sessile, whitish. Nov. High rocky slopes, LB (Langeberg: Swellendam).•

brevifolia Eckl. & Zeyh.   Closely leafy shrublet to 40 cm. Leaves ovate-lanceolate, c. 2 mm long, incurved from middle, margins closely revolute. Flowers in rounded capitula, solitary or in small clusters, densely white woolly, whitish. July–Aug. Sandstone slopes, SW (Kleinrivier Mts).•

burchellii Pillans   Slender shrublet to 30 cm. Leaves needle-like, 10–18 mm long, margins closely revolute. Flowers in solitary, rounded capitula, whitish. Nov.–Dec. Lower mountain slopes, SW (Riviersonderend Mts).•

buxifolia L.   BUKSHARDEBLAAR   Rounded, velvety pubescent shrub or small tree, sometimes to 4 m. Leaves ovate to elliptic, 15–25 mm long, margins slightly revolute, grey-felted beneath. Flowers in small capitula grouped in panicles, shortly stipitate, white. Apr.–Aug. Lower mountain slopes, SW (Cape Peninsula to Caledon).•

calcarata Pillans   Closely leafy, slender shrublet to 35 cm. Leaves linear-lanceolate, 12–15 mm long, smooth or scabrid, margins revolute, densely felted beneath. Flowers in large, solitary capitula, surrounded by long, villous leaves, whitish. Apr.–July. Sandstone slopes, SW (Riviersonderend and Langeberg Mts).•

callosa L.f.   Closely leafy shrub to 1.2 m. Leaves ovate-lanceolate, 5–15 mm long, rough, margins revolute. Flowers in flattened, usually solitary capitula, surrounded by small leaves and leafy bracts, whitish, subsessile. Apr.–July. Lower slopes, SW (Mamre to Bredasdorp).•

cephalantha Sond.   Densely branched shrub, 40–90 cm. Leaves linear, 5–8 mm long, margins closely revolute. Flowers in solitary or panicled capitula, brownish, stipitate, bracts with silky buff hairs. Apr.–Sept. Sandy flats and lower slopes, NW, SW (Olifants River mouth and Cedarberg Mts to Cape Peninsula).•

chionocephala Schltr.   Closely leafy shrublet to 20 cm. Leaves linear, c. 6 mm long, incurved, margins closely revolute. Flowers in solitary capitula surrounded by leafy bracts with purplish hairs, whitish. Aug.–Sept. Rocky slopes, NW (Cold Bokkeveld).•

chionophila Schltr.   Densely branched shrublet to 40 cm, grey silky on young parts. Leaves linear, 10–15 mm long, margins closely revolute. Flowers in capitula surrounded by leaves and leafy bracts, whitish. Oct.–Dec. Sandstone slopes, NW (Mosterthoek Twins to Brandwacht Mt).•

comosa Steud.   Closely leafy shrublet to 40 cm. Leaves linear-lanceolate, c. 8 mm long, margins closely revolute. Flowers in solitary, rounded capitula, surrounded by leaves and leafy bracts with golden hairs, whitish. Nov.–Dec. Sandstone slopes, SW (Elandskloof Mts).•

confusa Pillans   Closely leafy shrub to 1 m. Leaves more or less linear, cordate at base, 5–15 mm long, margins closely revolute. Flowers in small capitula grouped in panicles, sessile, whitish. Mar.–May. Sandstone slopes and peaks, SE (Outeniqua and Tsitsikamma Mts).•

costata Pillans   Like **P. willdenowiana** but more densely branched, rarely exceeding 35 cm and flowers in small capitula arranged in panicles. Dec.–Jan. Rocky sandstone slopes, KM (Swartberg Mts).•

curvifolia Pillans   Closely leafy shrub to 1 m, branches usually in threes from below capitula. Leaves linear-lanceolate, 10–15 mm long, smooth, margins closely revolute. Flowers in solitary, rounded capitula surrounded by leafy bracts and leaves with enlarged petioles, whitish. Mainly Oct.–Feb. Moist sandstone slopes, SE (Outeniqua Mts).•

debilis Eckl. & Zeyh.   Twiggy shrublet, 30–60 cm, grey-tomentose on young parts. Leaves cordate-lanceolate, 2.5–5 mm long, apparently smooth, margins laxly but closely revolute. Flowers in small, flattened, solitary capitula surrounded by leaves with long petioles, whitish. Nov.–Dec. Sandstone slopes, SW, LB, SE (Caledon Swartberg to Tsitsikamma Mts).•

diffusa Pillans   Spreading shrublet to 30 cm with slender, minutely hairy branches. Leaves ovate-lanceolate, c. 6 mm long, spreading to deflexed, widely convex, margins revolute. Flowers in small, flattened capitula usually in lax panicles, whitish. Dec.–July. Sandstone slopes, SW (Riviersonderend and Langeberg Mts to Elim).•

dioica L.   Much-branched shrub to 1 m, buff-felted on young branches. Leaves ovate, obtuse, 13–18 mm long, rough, margins slightly revolute, white-felted and conspicuously veined beneath. Flowers in large, solitary capitula surrounded by many leaves. Dec.–Mar. Sandstone slopes, SW (Cape Peninsula to Jonkershoek).•

diosmoides Sond.   Closely leafy shrublet to 30 cm. Leaves lanceolate, 10–15 mm long, smooth, margins revolute. Flowers usually in solitary capitula, surrounded by leaves and leafy bracts, whitish, stipitate. Mar.–Apr. Lower stony slopes, SW (Houwhoek and Botrivier).•

disticha Eckl. & Zeyh.   Loosely branched shrub to 90 cm. Leaves linear, 4–6 mm long, margins closely revolute. Flowers in small, usually solitary, flattened capitula, whitish. Dec.–Apr. Sandstone slopes, SW (Cape Peninsula to Hermanus).•

dodii N.E.Br.   Closely leafy shrublet to 40 cm. Leaves linear, 8–20 mm long, smooth, margins closely revolute. Flowers in solitary, rounded capitula surrounded by many villous leaves with enlarged petioles, whitish, subsessile. June–Sept. Sandy or limestone flats and slopes, SW, AP, SE (Cape Peninsula to Knysna).•

ericoides L.   Closely leafy, compact shrublet, sometimes to 1 m. Leaves needle-like, mostly 5–8 mm long, often rough, margins closely revolute. Flowers in small, rounded, solitary or clustered capitula, whitish, dense white-felted. Jan.–Dec. Coastal slopes and deep sands, SW, AP, SE (Saldanha to Port Elizabeth).•

excelsa H.L.Wendl.   Closely leafy shrub to 1 m, with few willowy branches. Leaves lanceolate, 7–15 mm long, rough, margins closely revolute. Flowers in solitary, rounded capitula surrounded by leaves and bracts with prominent golden hairs, whitish, sessile, clothed with retrorse grey-silky hairs. Dec.–May. Rocky sandstone slopes, NW, SW (Grootwinterhoek Mts to Cape Peninsula and Koo).•

floccosa Pillans   Densely branched shrublet to 40 cm. Leaves linear-lanceolate, 3–6 mm long, rough, compressed near tips, margins closely revolute. Flowers in small flattened capitula arranged in panicles, whitish, densely white-woolly on outside. Jan.–Feb. High rocky slopes, KM (Kammanassie Mts).•

floribunda Pillans   Densely branched shrublet to 40 cm. Leaves linear-lanceolate, c. 5 mm long, rough, compressed near tips, deeply cordate at base, margins closely revolute. Flowers in small flattened capitula arranged in panicles, whitish, densely white-felted on outside. Sept.–Oct. Rocky slopes, SW (Bredasdorp Mts).•

fourcadei Pillans   Like **P. gnidioides** but leaves shorter, 5–7 mm long, scarcely compressed and style longer, c. 8 mm long. Mainly Jan.–Mar. Sandstone slopes, SE (Langkloof to Humansdorp).•

fruticosa Schltr.   Rigid shrub to 1.5 m. Leaves ovate to broadly lanceolate, 10–15 mm long, margins revolute, tomentose beneath. Flowers in solitary capitula surrounded by leaves, stipitate, whitish. Mar.–Apr. Rocky sandstone slopes, NW (Cedarberg Mts).•

gnidioides Eckl. & Zeyh.   Closely leafy, erect or spreading shrub, to 1 m. Leaves linear, 8–10 mm long, smooth above, margins closely revolute. Flowers in rounded capitula, grouped in small corymbs, surrounded by ciliate leaves with enlarged petioles, pink. May–July. Dunes and grassy slopes, SE (Humansdorp to Grahamstown).

gracilis (Eckl. & Zeyh.) D.Dietr.   Slender, twiggy shrublet to 40 cm. Leaves linear, 6–10 mm long, pilose on margins and midrib, margins closely revolute. Flowers in solitary, capitula surrounded by bracts with long silky hairs, whitish. Mainly Dec.–Apr. Sandy slopes and flats, SW (Malmesbury to Elim).•

greyii Pillans   Densely leafy shrublet to 30 cm, with divaricate branching. Leaves oblong-lanceolate, 4–5 mm long, smooth, cordate at base, margins closely revolute. Flowers in solitary capitula, whitish, sepals subglabrous. Flowering time? Coastal hills, SW (Saldanha Bay).•

guthriei Pillans   Closely leafy shrub to 60 cm, with wiry, tomentose branchlets. Leaves narrowly lanceolate, 10–25 mm long, margins revolute. Flowers in globose capitula surrounded by modified leaves, sometimes clustered on small branchlets or solitary, shortly stipitate, whitish. May. High sandstone slopes, SW (Franschhoek Mts).•

harveyi (Arn.) Pillans   Closely leafy shrublet to 30 cm. Leaves linear-lanceolate, 6–13 mm long, cordate at base, margins closely revolute. Flowers in rounded, solitary capitula surrounded by a few, short leaves, whitish. Sept.–Oct. Deep coastal sands, NW, SW (Piketberg to Cape Peninsula).•

humilis Sond.   Sprawling, closely leafy shrublet to 30 cm. Leaves narrowly ovate, deeply cordate at base, c. 2 mm long, margins strongly revolute. Flowers in solitary capitula surrounded by a few leaves, stipitate, whitish. Aug.–Nov. Sandstone slopes, SW (Sir Lowry's Pass to Bredasdorp).•

imberbis P.J.Bergius   Closely leafy, loosely branched shrub mostly to 50 cm. Leaves linear, 7–10 mm long, smooth except on edges, margins closely revolute. Flowers in mostly solitary, rounded capitula, whitish. Nov.–Apr. Sandstone slopes and flats, NW, SW, KM, LB, SE (Bokkeveld Mts to Knysna and Swartberg Mts).•

incurvata Pillans   Closely leafy, wiry shrublet to 30 cm. Leaves adpressed, oblong-lanceolate, c. 2.5 mm long, margins closely revolute. Flowers in small solitary, flattened capitula, whitish. Flowering time? Low slopes, SW (Elim).•

insignis Pillans   Closely leafy, stiffly branched shrub to 60 cm. Leaves linear, needle-like, 10–18 mm long, rough, margins closely revolute. Flowers in large, flattened, solitary capitula surrounded by many leaves and leafy bracts, whitish. Sept.–Oct. Rocky slopes, NW (Cedarberg and Cold Bokkeveld Mts).•

karroica Pillans   Like **P. parviflora** but leaves needle-like, 5–8 mm long, margins closely revolute, without tubercles and hairy when young, becoming smooth over entire surface. Sept.–Oct. Rocky sandstone slopes, LB, SE (Langeberg Mts to Langkloof).•

keetii Pillans   Closely leafy shrublet to 40 cm. Leaves needle-like, c. 5 mm long, smooth, margins closely revolute. Flowers in small, densely white-plumose capitula, white. Nov.–Feb. Rocky sandstone slopes, SE (Outeniqua and Tsitsikamma Mts).•

lachneaeoides Pillans   Like **P. gnidioides** but leaves short, 5–8 mm long, and flowers with cordate petals and elongate style. May–Sept. Sandstone slopes, SE (Uniondale to Humansdorp).•

laevifolia Pillans   Densely grey-pubescent shrublet to 40 cm. Leaves lanceolate, 8–10 mm long, smooth above, margins revolute. Flowers in wide, flattened, solitary capitula surrounded by many pilose leaves, whitish. Flowering itme? Lower slopes, SW (Elim).•

laevigata Pillans   Much-branched, closely leafy shrublet to 30 cm. Leaves oblong-lanceolate, cordate at base, 4–5 mm long, margins closely revolute, becoming smooth and shiny. Flowers mostly in solitary capitula, whitish, stipitate. Apr.–Aug. Limestone hills, AP (Bredasdorp to Stilbaai).•

laevis (Eckl. & Zeyh.) Steud.   Twiggy shrublet to 30 cm. Leaves ovate-lanceolate, 5–7.5 mm long, smooth or rough, margins closely revolute. Flowers in solitary capitula surrounded by many leaves. Flowering time? Sandstone slopes, SW (Houwhoek to Caledon Swartberg).•

lanata Pillans   Like **P. lasiocarpa** but leaves 4–6 mm long, entirely tubercled above, capitula usually in small clusters, and flowers subsessile and densely white-woolly. Aug.–Sept. Dry rocky sandstone slopes, KM (Witteberg, Swartberg and Little Karoo Mts).•

lasiantha Pillans   Closely leafy shrub to 60 cm. Leaves linear, 3.5–5 mm long, smooth, margins closely revolute. Flowers in small, solitary capitula, surrounded by short leaves and leafy bracts, whitish. Nov.–Feb. Sandstone slopes, LB (Langeberg: Swellendam).•

lasiocarpa Sond.   Much-branched shrub to 60 cm. Leaves linear to lanceolate, 6–12 mm long, smooth except on edges, margins strongly revolute. Flowers mostly in solitary capitula, sessile, whitish. Mainly Dec.–Apr. Clay and sandstone slopes, SW (Hottentots Holland Mts to Bredasdorp).•

linifolia Pillans   Sprawling closely leafy shrublet with slender branches. Leaves needle-like, 5–9 mm long, becoming smooth with age, margins closely revolute. Flowers in small solitary capitula surrounded by a few long leaves, whitish, subsessile. Nov.–Dec. Sandstone slopes, SW (Houwhoek Mts).•

litoralis (Eckl. & Zeyh.) D.Dietr.   Closely leafy, much-branched shrublet to 30 cm. Leaves lanceolate, 7–15 mm long, margins revolute. Flowers in solitary or clustered capitula, whitish, subsessile. Nov.–Feb. Coastal dunes, SE (Knysna to East London).

longimontana Pillans   Sprawling, closely leafy shrublet to 20 cm. Leaves lanceolate, 5–7 mm long, margins loosely revolute. Flowers in solitary capitula surrounded by short leaves, stipitate, whitish. Nov.–Dec. High rocky slopes, LB (Langeberg: Riversdale).•

lucens Pillans   Like **P. minutiflora** but leaves densely tubercled above. Oct.–Feb. Mountain slopes, SW (Riviersonderend Mts).•

lucida Pillans   Closely leafy, sparsely branched shrublet to 40 cm. Leaves lanceolate, c. 1.2 mm long, margins revolute. Flowers in solitary, flattened capitula surrounded by many villous leaves, whitish. Sept.–Oct. Coastal flats, SW, AP (Stanford to Agulhas).•

mairei Pillans   Shrub to 40 cm. Leaves lanceolate, c. 5 mm long, cordate at base, ascending and incurved, margins closely revolute. Flowers in small, usually solitary, flat capitula, whitish, subsessile. Feb.–Apr. Lower slopes, LB (Langeberg Mts).•

maximiliani Schltr.   Closely leafy shrublet to 60 cm. Leaves needle-like, 7–10 mm long, rough, margins closely revolute. Flowers in solitary, rounded capitula, whitish. Aug.–Sept. Rocky sandstone slopes, NW (Pakhuis Mts).•

meyeri Sond.   Closely leafy, sparsely branched shrublet to 40 cm. Leaves linear, 8–12 mm long, smooth, margins closely revolute. Flowers in large, solitary, rounded capitula, with long white-plumose bracts, white. Oct.–Feb. Rocky slopes, KM (Swartberg and Kammanassie Mts).•

minutiflora Schltr.   Much-branched, closely leafy shrub to 60 cm. Leaves linear to narrowly lanceolate, 5–10 mm long, almost smooth, margins closely revolute. Flowers in small, solitary capitula almost hidden by upper leaves, whitish, stipitate. Dec.–Jan. Sandstone slopes, SW (Hottentots Holland Mts to Cape Hangklip).•

nervosa Pillans   Stout, closely leafy, much-branched shrub to 1.5 m. Leaves ovate, cordate at base, 10–20 mm long, margins slightly revolute, tomentose and prominently veined beneath. Flowers in solitary capitula, subsessile, white. Sept.–Jan. Sandstone slopes, NW (Cold Bokkeveld Mts to Ceres).•

nigrita Sond.   Much-branched shrublet to 40 cm. Leaves linear-lanceolate, 5–8 mm long, rough or smooth, margins closely revolute. Flowers in solitary, flattened capitula surrounded by short bracts, whitish. July–Aug. Rocky slopes, SW, AP (Cape Peninsula to Agulhas).•

nigromontana Pillans   Closely leafy, stiffly branched shrub to 70 cm. Leaves needle-like, 10–15 mm long, smooth, margins closely revolute. Flowers in solitary, flat capitula surrounded by bracts with long golden hairs, whitish. Feb.–May. High rocky slopes, KM (Swartberg Mts).•

nodosa Pillans   Closely leafy, much-branched shrub to 1.2 m, stems swollen at positions of old capitula. Leaves lanceolate, 10–14 mm long, margins revolute. Flowers in solitary capitula, surrounded by numerous leaves, whitish. Nov.–Dec. Rocky peaks and ridges, SW (Du Toitskloof Mts).•

odorata Schltr.   Closely leafy grey-pubescent shrublet to 60 cm. Leaves linear, 6–15 mm long, rough, margins closely revolute. Flowers in solitary, flattened capitula surrounded by a few broad leaves, cream-coloured, fragrant. Sept.–Nov. Rocky sandstone slopes, NW, KM (Cedarberg to Witteberg).•

parviflora P.J.Bergius   Compact, closely leafy shrublet, mostly to 40 cm. Leaves lanceolate, c. 5 mm long, incurved, margins usually closely revolute. Flowers in small, rounded to flat capitula usually grouped in panicles, whitish, subsessile. Jan.–May. Sandy coastal flats and low hills, SW, AP, SE (Hopefield to Knysna).•

parvula Pillans   Closely leafy, sprawling shrublet to 20 cm. Leaves ovate, ascending, c. 2 mm long, cordate at base, margins closely revolute. Flowers in small, solitary capitula, surrounded by short leaves and leafy bracts, whitish, sessile. June–Sept. Coastal hills, SW, AP (Cape Peninsula to Agulhas).•

pauciflora Pillans   Closely leafy shrublet to 50 cm. Leaves linear-terete, c. 5 mm long, rough, margins closely revolute. Flowers in small, solitary, flattened capitula, surrounded by several leaves, whitish. Mainly Sept.–Nov. Rocky sandstone slopes, NW (Swartruggens).•

piquetbergensis Pillans   Twiggy shrub to 80 cm, grey-pubescent on young parts. Leaves lanceolate, 7–10 mm long, rough, margins revolute. Flowers in small, solitary or clustered capitula, white, subsessile. Sept.–Oct. Sandstone slopes, NW (Piketberg).•

propinqua Sond.   Virgate shrublet to 40 cm. Leaves lanceolate, 5–15 mm long, cordate at base, margins revolute. Flowers in solitary, flattened capitula surrounded by large leafy bracts, whitish, sessile. Apr.–Sept. Sandstone slopes, LB (Langeberg Mts).•

pubescens Aiton   FEATHERHEAD, VEERKOPPIE   Closely leafy, villous shrub 1.5 m. Leaves linear-lanceolate, 25–35 mm long, margins revolute. Flowers in large flattened, solitary capitula, surrounded by many leaves and elongate, long-villous bracts, white, sessile. May–Aug. Sandstone and limestone slopes, SW, AP (Cape Peninsula to Albertinia).•

pustulata E.Phillips   Closely leafy, slender shrub to 1 m. Leaves needle-like, 12–18 mm long, rough, margins closely revolute. Flowers in solitary, rounded capitula, whitish. May–Aug. Sandstone rocks, NW (Bokkeveld Mts to Gifberg).•

reversa Pillans   Closely leafy shrublet to 20 cm. Leaves linear, 8–10 mm long, smooth, margins closely revolute. Flowers in large, flattened, solitary capitula, surrounded by many leaves and villous bracts, white, subsessile, with petals reversed. Flowering time? Rocky sandstone slopes, NW (Hex River Mts).•

rogersii Pillans   Like **P. minutiflora** but sprawling rather than rounded, leaves mostly 5–8 mm long, conspicuously tubercled along edges and cordate at base, capitula often in panicles. June–Dec. High sandstone slopes, NW, SW (Hex River Mts to Brandvlei and Kogmanskloof).•

rubra Willd. ex Roem. & Schult.   Densely branched, closely leafy shrub to 2 m. Leaves linear to lanceolate, 5–10 mm long, smooth, margins closely revolute. Flowers in numerous small capitula forming panicles, white, sessile. Mar.–July. Lower, ?clay slopes, LB (Langeberg Mts).•

salteri Pillans   Closely leafy shrublet to 40 cm. Leaves lanceolate, c. 12 mm long, incurved above middle, margins closely revolute. Flowers in small rounded capitula in small clusters, each surrounded by many leaves, white. Apr.–May. Rocky sandstone slopes, NW (Cold Bokkeveld Mts).•

schlechteri Pillans   Sprawling, sparsely branched shrublet mostly to 40 cm, with ascending branches. Leaves linear, 8–10 mm long, buff-tomentose when young, margins closely revolute. Flowers in rounded capitula, 2 or 3 in a cluster, buff-tomentose, white, stipitate. June–July. Sandstone slopes, SW (Cape Peninsula).•

selaginoides Sond.   Closely leafy shrublet to 40 cm. Leaves linear, 3.5–6 mm long, compressed near tips, smooth, margins closely revolute. Flowers in solitary, flattened capitula surrounded by many leaves, white. Mainly Aug.–Sept. Coastal limestone flats and hills, AP (Agulhas).•

sericea Pillans   Shrublet to 60 cm, white-felted on young branches. Leaves linear, 6–20 mm long, cordate at base, rough, margins closely revolute. Flowers in solitary, flattened capitula surrounded by long-petiolate leaves, white. Nov.–Dec. Sandstone slopes, KM (Klein Swartberg).•

stenantha Pillans   Closely leafy shrublet to 25 cm. Leaves linear, 5–8 mm long, smooth, margins closely revolute. Flowers in rounded, solitary capitula surrounded by numerous leaves with enlarged petioles, white. Sept.–Oct. Sandstone slopes, SW (Riviersonderend Mts).•

stenopetala Schltr.   Closely leafy shrub to 60 cm. Leaves linear-lanceolate, 3–7 mm long, margins revolute. Flowers in small, flattened, solitary capitula, white. June. Stony flats, SW (Gouda).•

stokoei Pillans   Like **P. excelsa** but flowers stipitate, with adpressed straight golden hairs, and sepals slightly longer, c. 2 mm long. Nov.–Dec. High rocky slopes, KM (Klein Swartberg).•

strigosa P.J.Bergius   Sparsely branched shrub to 1 m, with ascending branches. Leaves lanceolate, 10–20 mm long, rough, margins revolute. Flowers in solitary, flat capitula surrounded by leaves with prominent golden hairs, yellowish. Feb.–July. Rocky sandstone slopes, NW, SW (Hopefield to Cape Peninsula and Hex River Mts).•

subulifolia Pillans   Closely leafy shrub to 50 cm. Leaves needle-like, 3–5 mm long, ultimately smooth, margins closely revolute. Flowers in small, flattened, solitary capitula, surrounded by leaves with flattened petioles, whitish, sessile. Oct.–Feb. High sandstone slopes, NW (Hex River Mts).•

thunbergiana Sond.   Closely leafy shrub, 30–50 cm. Leaves lanceolate, 2.5–6 mm long, cordate at base, smooth and shiny above, margins closely revolute. Flowers in solitary, flattened capitula surrounded by several short leaves, white. Mar.–Aug. Lower slopes, SW (Malmesbury to Stellenbosch).•

tortuosa E.Mey.   Closely leafy, tomentose shrub to 80 cm. Leaves linear, 5–7 mm long, smooth, margins closely revolute. Flowers in solitary, flattened capitula surrounded by many pilose leaves, purple. July–Oct. Rocky sandstone slopes, KM, SE (Swartberg and Tsitsikamma Mts).•

trachyphylla (Eckl. & Zeyh.) A.Dietr.   Twiggy shrub to 1 m, pilose on young branches. Leaves lanceolate, 8–12 mm long, incurved above middle, smooth, margins closely revolute. Flowers in wide, solitary capitula surrounded by many leaves, white. Oct.–Nov. Sandstone slopes, NW (Grootwinterhoek Mts).•

tuberculata Pillans   Like **P. lasiocarpa** but leaves 10–14 mm long, coarsely tubercled and rough, petals with linear claws, ovary with deciduous hairs. Dec.–Mar. Sandstone mountain slopes, NW (Cedarberg to Witzenberg).•

tubulosa Schltr.   Densely leafy, sprawling shrublet, white-felted on young parts. Leaves linear-oblong, 3–5 mm long, smooth, margins closely revolute. Flowers in solitary, rounded capitula, white, subsessile. Oct.–Feb. High sandstone slopes, SW (Riviersonderend Mts).•

variabilis Pillans   Closely leafy, densely branched shrub to 60 cm. Leaves lanceolate, 7–12 mm long, rough, margins strongly revolute, white-felted beneath. Flowers in solitary or grouped capitula, white. Jan.–Mar. High sandstone slopes, SW (Hottentots Holland Mts).•

virgata A.Dietr.   Much-branched shrub to 60 cm. Leaves linear, mostly 7–12 mm long, rough, margins revolute. Flowers in small rounded capitula in small clusters, surrounded by many small leaves with enlarged petioles, white. Mar.–Aug. Sandstone slopes, SW (Houwhoek to Hermanus).•

vulgaris Pillans   Closely leafy shrublet to 30 cm. Leaves needle-like, 6–10 mm long, rough, margins closely revolute. Flowers in flattened, solitary capitula surrounded by leaves and leafy bracts, white. June–Oct. Sandstone slopes, NW, SW, LB (Witzenberg to Witteberg and Riviersonderend Mts).•

wittebergensis Pillans   Densely leafy shrub to 60 cm, grey-tomentose on young parts. Leaves linear, c. 10 mm long, smooth, margins closely revolute. Flowers in small capitula overtopped by leaves, stipitate, white. Aug.–Sept. High rocky sandstone slopes, KM (Witteberg and Klein Swartberg).•

[**Species excluded**   Incompletely known and probably conspecific with one of the above: **P. divaricata** Vent., **P. glabrata** Thunb.]

RHAMNUS   BLINKBLAAR   9 spp., pantropical

prinoides L'Hér.   Shrub or small tree to 7 m. Leaves elliptic, serrate, glossy above. Flowers in axillary fascicles, greenish. Fruits thinly fleshy, turbinate, red to black. Aug.–Dec. Riverine scrub and forest margins, LB, SE (Riversdale to tropical Africa).

SCUTIA   KATDORING   9 spp., pantropical

myrtina (Burm.f.) Kurz   Armed scandent shrub or small tree to 8 m, with straight or recurved axillary thorns. Leaves usually opposite, ovate, leathery, glossy above. Flowers in axillary umbels, yellowish green. Fruits thinly fleshy, black. Oct.–Feb. Forest margins, SW, LB, SE (Cape Peninsula to tropical Africa).

TRICHOCEPHALUS•   DOGFACE, HONDEGESIG   1 sp., W Cape

stipularis (L.) Brongn. (= *Phylica stipularis* L.)   Rounded shrublet, sometimes to 90 cm, resprouting from persistent rootstock. Leaves with small, dry stipules, linear-lanceolate, 7–15 mm long, rough, margins closely revolute. Flowers in solitary, rounded capitula, pink, densely white-hairy on outside. May–Sept. Sandy flats and lower slopes, NW, SW, AP, SE (Cedarberg Mts to Knysna).•

RORIDULACEAE•

RORIDULA• FLYCATCHER BUSH   2 spp., W Cape

dentata L.   Shrub to 2 m. Leaves crowded on side-shoots, narrowly lanceolate, margins with narrow fine teeth and tentacles. Flowers on long pedicels in upper axils, petals persistent, pink, style shorter than ovary. Sept.–Oct. Swampy sandstone slopes, 900–1200 m, NW, SW (Pakhuis to Elandskloof Mts).•

gorgonias Planch.   Shrublet to 60 cm. Leaves crowded at branch tips, narrowly lanceolate, margins tentacled. Flowers on short pedicels in terminal racemes, petals caducous, pink, style longer than ovary, capitate. July–Oct. Swampy sandstone slopes, 100–900 m, SW (Hottentots Holland to Kleinrivier and Riviersonderend Mts).•

ROSACEAE
Cliffortia by A.C. Fellingham

1. Petals 5:
    2. Unarmed trees; leaves elliptic; fruit dry, leathery ....**Prunus**
    2.' Prickly scramblers; leaves compound; fruit succulent ....**Rubus**
1.' Petals absent:
    3. Flowers unisexual ....**Cliffortia**
    3.' Flowers bisexual:
        4. Calyx unarmed; leaves cordate..**Alchemilla**
        4.' Calyx and fruit armed with barbed bristles; leaves pinnate ....**Acaena**

ACAENA   c. 100 spp., mainly S temperate, also Hawaii, California and South Africa

latebrosa Aiton   Silky tufted perennial with woody base, stems to 30 cm. Leaves pinnate, leaflets oblong, toothed. Flowers on elongated spikes, green. Fruits woolly and barbed. Aug.–Oct. Granite and clay slopes above 1000 m, NW, KM (Namaqualand and W Karoo to Swartberg Mts).

ALCHEMILLA   LADY'S MANTLE   c. 250 spp., temperate and tropical mountains

capensis Thunb.   Silky hairy, trailing perennial. Leaves cordate, 3-lobed and toothed. Flowers in axillary spikes, green. Oct.–Jan. Moist, mainly montane sites, SW, LB, SE (Cape Peninsula to E Cape).

CLIFFORTIA   CLIMBER'S FRIEND   c. 120 spp., W Cape to tropical Africa

*A. Leaves bifoliolate, leaflets flat (see also **C. mirabilis**, **C. multiformis**)*

crenata L.f.   Virgate monoecious or dioecious shrub to 1.5 m. Leaves bifoliolate, leaflets connate around branch, oblique reniform to suborbiculate, 2–6 mm long, margins crenate. Flowers: male: stamens c. 30; female: receptacle oblong cylindrical, 4 mm long. May–Sept. Sandstone slope, NW, KM (Cold Bokkeveld to Swartberg Mts).•

crenulata Weim.   Monoecious or dioecious shrublet to 20 cm with flexuose branches. Leaves bifoliolate, leaflets small, rhomboid-obovate, 4–6-veined, 6–8 mm long, margins translucent, crenulate-denticulate. Flowers: male: unknown; female: ovoid, dull brown, glabrous. Oct. Sandstone slopes, SW (Riviersonderend Mts).•

phyllanthoides Schltr.   Monoecious or dioecious shrublet to 40 cm. Leaves bifoliolate with a rudimentary third leaflet, leaflets flat, orbicular to cordate, 3–6 mm long. Flowers: male: unknown; female: receptacle oblong, 2–2.5 mm long, 3-winged, wings small, undulate and transparent. Apr. Coastal flats, SW, AP (Botrivier to Elim).•

pulchella L.f.   Monoecious or dioecious shrublet to 60 cm. Leaves bifoliolate, leaflets flat, reniform to orbiculate, 5–6 mm long, margins shortly ciliate. Flowers: male: stamens c. 20; female: receptacle ovoid, smooth, c. 3 mm long. Apr.–May. Sandstone slopes, LB (Langeberg Mts).•

varians Weim.   Monoecious or dioecious shrublet to 60 cm. Leaves bifoliolate, occasionally trifoliolate, leaflets obliquely obovate, 4–8(–10) mm long. Flowers: male: stamens 12; female: receptacle c. 3 mm long, with brownish ribs. Sept.–Dec. Middle to upper slopes, LB (Bonnievale).•

A.' Leaves unifoliolate
*B. Leaves linear to subterete (see also **C. sp. 3**)*

aculeata Weim.   Monoecious or dioecious shrublet to 30 cm. Leaves simple, subterete, adaxially channelled, aculeate, (20–)40–60 mm long, arranged in flat imbricate fans. Flowers: male: sepals

8–10 mm long, linear-lanceolate, stamens c. 20; female: receptacle c. 4 mm long, oblong, sulcate. Nov.–Jan. Streamsides on upper slopes, KM (Swartberg Mts).•

brevifolia Weim.   Spreading monoecious or dioecious shrublet to 30 cm. Leaves simple, linear, c. 2 mm long. Flowers: male: sepals 4, stamens 4; female: sepals 4, receptacle 1.8 mm long, 4-ribbed with 2 high and 2 low ribs, style short, curved. Dec. Coastal hills and flats, sometimes on calcareous sands, SW, AP (Cape Peninsula to Bredasdorp).•

dregeana C.Presl   Monoecious or dioecious shrub to 1 m. Leaves simple, linear-lanceolate, pungent, 15–35 mm long. Flowers: male: stamens c. 40; female: receptacle c. 8 mm long, ovoid, c. 20-ribbed. July–Feb. Sandstone slopes, NW, SW (Cedarberg to Riviersonderend Mts).•

erectisepala Weim.   Monoecious or dioecious shrub, 1–2 m. Leaves simple, narrowly linear, 5–7 mm long. Flowers: male: unknown; female: sepals erect, margins coherent, receptacle c. 2.5 mm long, cylindrical, brown. Flowering time? Middle slopes, SW, KM, LB (Paarl to Swartberg Mts).•

ericifolia L.f.   Monoecious or dioecious shrublet to 60 cm. Leaves simple, linear, 4–5 mm long. Flowers: male: sepals (3)4, stamens (3)4; female: sepals 4, fugacious, receptacle 2–2.5 × 1–1.2 mm, ovoid-oblong, brownish, 4-ribbed. Feb.–June. Damp lower slopes, SW (Philadelphia to Cape Peninsula).•

neglecta Schltr.   Compact monoecious or dioecious shrublet to 30 cm. Leaves simple, needle-shaped, pungent, 5–10 mm long. Flowers: male: stamens 6(7); female: receptacle ovoid. Sept.–Oct. High rocky slopes, NW, KM (Cedarberg to Witteberg and Swartberg Mts).•

nivenioides Fellingham   Compact, spreading, monoecious or dioecious shrublet to 60 cm. Leaves simple, linear, straight to slightly falcate, bilaterally flattened, 3–33 × 1–2 mm, in distichous fans. Flowers: male: stamens (13–)16(–20); female: with thick, fleshy receptacle, faintly ridged, c. 5 mm long, style short, curved, maroon. Nov. Jan. N-facing slopes in seeps, KM (Swartberg Mts).•

pungens C.Presl   Monoecious or dioecious shrub to 1 m. Leaves simple, thick, needle-like, pungent, 4–10 mm long. Flowers: male: stamens 6; female: sepals persistent, c. 1 mm long, erect, receptacle c. 6 mm long, cylindrical, 6-angled. Sept. High sandstone slopes, SW (Riviersonderend Mts).•

uncinata Weim.   Erect or trailing, monoecious or dioecious shrublet. Leaves simple, terete, reflexed, uncinate, 10–20 mm long. Flowers: male: stamens 12–15; female: receptacle 3.5–4 mm long, subcylindrical. Sept.–Dec. Rocky sandstone slopes, SW, NW (Cedarberg to Porterville Mts).•

sp. 1   Lax monoecious or dioecious shrublet to 50 cm. Leaves simple, needle-like, bright green, c. 18 mm long. Flowers: male: stamens 6; female: sepals linear-lanceolate, erect, c. 5 mm long, receptacle turbinate, c. 1 mm long, style c. 4 mm long, hidden by erect sepals. Aug.–Sept. Streambanks, NW (Gifberg).•

*B. Leaves flat, linear to ovate (see also **C. dispar**)*

acutifolia Weim.   Like **C. phillipsii** but receptacle glabrous. Aug.–Sept. Sandstone slopes, NW (Bokkeveld Mts).•

cuneata Aiton   Monoecious or dioecious shrub to 2 m. Leaves simple, 25–30 × 5–7 mm, cuneate at base, truncate, margins toothed. Flowers: male: stamens 40–50; female: receptacle c. 7 mm long, ovoid, sulcate, glabrous. Oct.–Nov. Lower sandstone slopes, SW (Cape Peninsula to Riviersonderend Mts).•

cymbifolia Weim.   Low monoecious or dioecious shrublet. Leaves simple, boat-shaped, 6–12 × 1–2 mm. Flowers: male: stamens 7 or 8; female: receptacle 3 mm long, ovoid-oblong, brown, prominently ribbed. Sept.–Oct. High sandstone slopes, SW, KM (Franschhoek to Swartberg Mts).•

discolor Weim.   Monoecious or dioecious shrub. Leaves simple, ovate, 40–50 mm long. Flowers: male: shortly pedicellate, stamens c. 20; female: unknown. Jan. High rocky sandstone slopes, SW (Cape Peninsula).•

esterhuyseniae Weim.   Hairy monoecious or dioecious shrub to 2 m. Leaves simple, declinate, lanceolate-elliptic, 8–15 mm long. Flowers: male: stamens 9; female: sepals 0.2 mm wide, receptacle 2.5–2.8 mm long, ovoid, ribbed, brownish green. Dec. High rocky slopes, NW (Hex River Mts).•

ferruginea L.f.   GLASTEE, TERINGTEE, PYPSTEELBOS   Sprawling monoecious or dioecious shrublet, with reddish branches. Leaves simple, linear-lanceolate to obovate, margins finely dentate, apex coarsely toothed and curved to one side. Flowers male: stamens 15; female: receptacle 3 mm long, ovoid to ovoid-oblong. Nov.–July. Near water, usually on lower slopes, SW, LB, SE (Cape Peninsula to Port Elizabeth).•

graminea L.f.   VLEIROOIGRAS, WILDE-ERTJIE   Sprawling monoecious or dioecious shrub to 2 m. Leaves simple, grass-like with clasping sheath 30–60 mm long, stipules narrowly triangular, 4–10 mm long, leaf blade linear-lanceolate, 50–150 × 3–6 mm. Flowers: male: stamens c. 30; female: receptacle

4–5 mm long, oblong, greenish, sulcate. Aug.–Mar. Damp flats and slopes, NW, SW, LB, SE (Grootwinterhoek Mts to Port Elizabeth).•

grandifolia Eckl. & Zeyh.   GROOTBLAARRYSBOS   Monoecious or dioecious shrub or small tree to 5 m. Leaves simple, coarsely dentate, 50–100 × 15–30 mm. Flowers: male: stamens c. 50; female: unknown. Dec.–Jan. Wooded ravines on sandstone slopes, SW, LB (Du Toitskloof to Langeberg Mts).•

heterophylla Weim.   Slender monoecious or dioecious tree-like shrub to 2.5 m. Leaves simple, dimorphic, vegetative leaves willow-like, fertile leaves broad, amplexicaul, imbricate in female inflorescences. Flowers: male: stamens 40–50; female: receptacle 2.3–3 mm long, turbinate, fleshy, slightly striate. Fruits to 7 mm long, completely hidden between imbricate inflorescence leaves. Sept.–June. Sheltered slopes and damp ravines, SW (Kogelberg to Kleinmond).•

hirsuta Eckl. & Zeyh.   Monoecious or dioecious shrublet to 50 cm. Leaves simple, cordate, cuspidate-dentate, white-woolly below, 30–50 mm long. Flowers: male: shortly pedicellate, stamens c. 20; female shortly pedicellate, receptacle 2.5–3 mm long, ovoid, apex pointed. Apr.–Dec. Damp sites on mountain slopes, SW (Hottentots Holland to Riviersonderend Mts).•

ilicifolia L.   DORINGTEE, JANKOENSEDORING   Monoecious or dioecious shrub to 2 m. Leaves simple, ovate, 15–25 mm long, margins coarsely toothed. Flowers: male: stamens c. 40; female: sepals 1.5–2.5 mm, acute, upright, receptacle 7–9 × 4–5 mm, ovoid, with c. 20 reddish ribs. Mainly Nov.–Dec. Sandstone slopes SW, AP, KM, SE (Cape Peninsula to Port Elizabeth).•

integerrima Weim.   Monoecious or dioecious shrub to 1 m. Leaves simple, ovate to obovate, 12–16 mm long, rarely 2(3)-toothed. Flowers: male: stamens c. 50; female: receptacle 5–6 mm long, ovoid, ribbed. Jan.–May. Sandstone slopes SW, KM (Cape Peninsula to Caledon and Swartberg Mts).•

intermedia Eckl. & Zeyh.   Monoecious or dioecious shrublet to 80 cm. Leaves simple, lanceolate, acute to 2(3)-toothed, 20–30 mm long. Flowers: male: stamens 20–30; female: receptacle 5–5.5 mm long, ovoid, ribbed. July–Nov. Lower mountain slopes, SW (Cape Peninsula to Riviersonderend Mts).•

lanceolata Weim.   Monoecious or dioecious shrublet to 50 cm. Leaves simple, lanceolate, c. 20 mm long. Flowers: male: stamens c. 40, anthers with hairy connectives; female: unknown. Nov.–Jan. High rocky slopes, LB (Langeberg: Swellendam to Riversdale).•

longifolia (Eckl. & Zeyh.) Weim.   Monoecious or dioecious shrub to 1.5 m. Leaves simple, linear, to 60 mm long, minutely denticulate, cauline leaves fugacious, leaf sheath 8–13 mm long, reddish, stipules to 20 mm long, densely hairy. Flowers: male: pedicel c. 8 mm long, densely hairy, bracteoles densely hairy, situated 1.5 mm apart on pedicel, stamens c. 20; female: pedicel 1.6 mm long with half that length above upper bracteole, receptacle 2.6 mm long, ovoid, irregularly ribbed. June–Apr. Streamsides and marshes near the coast, SW, AP (Saldanha Bay to Gouritsmond).•

monophylla Weim.   Hairy monoecious or dioecious shrublet to 60 cm. Leaves simple, ovate to cordate, c. 4 mm long. Flowers: male: sepals 4, coherent for more than half their length, stamens 6; female: sepals 4, receptacle obovoid, c. 1 mm long. Oct.–Apr. Sandstone slopes, SW (Caledon Swartberg).•

multiformis Weim.   Monoecious or dioecious shrublet. Leaves simple to 2- or 3-lobed to 2- or 3-foliolate, 9–12 × 2–3 mm. Flowers: male: unknown; female: receptacle 5 mm long, oblong, 15–20-ribbed. Flowering time? Sandstone slopes, SW (Paarl to Bredasdorp).•

odorata L.f.   WILDEWINGERD   Scrambling monoecious or dioecious shrub to 1 m. Leaves simple, blade 30–60 × 20–50 mm, toothed, stipules triangular, 4–7 mm long. Flowers: male: stamens 15–20; female: receptacle 2.5 mm long, ovoid. May–Feb. Sandstone slopes, NW, SW, LB, SE (Clanwilliam to KwaZulu-Natal).

ovalis Weim.   Like **C. ilicifolia** but stems sprawling or erect and male flowers with 12 stamens; female: receptacle 3.5 mm long, ovoid, ribbed. Dec.–Jan. High rocky slopes, SW (Jonkershoek and Hottentots Holland Mts).•

phillipsii Weim.   Robust monoecious or dioecious shrub to 3 m. Leaves simple, ligulate, spiny toothed, 20–35 × 4–7 mm. Flowers: male: stamens c. 50; female: receptacle c. 10 mm long, ovoid, sulcate, densely grey-hairy. Nov.–Dec. Sandstone slopes, SW (Franschhoek and Hottentots Holland Mts).•

pilifera Bolus   Sprawling, lax, monoecious or dioecious shrublet to 60 cm. Leaves simple, ovate, 2–4 mm long, coarsely toothed. Flowers: male: stamens c. 20; female: receptacle c. 2 mm long, ovoid, smooth. Nov.–Apr. Cool middle to upper slopes, SW (Bainskloof Mts).•

reticulata Eckl. & Zeyh.   Straggling monoecious or dioecious shrublet. Leaves simple, ovate-cordate, c. 3.5 mm long, margins dentate. Flowers: male: stamens 20; female: unknown. Nov.–Feb. Moist sites on sandstone slopes, NW, SW (Grootwinterhoek Mts to Kogelberg).•

ruscifolia L. STEEKBOS Monoecious or dioecious shrub to 1.5 m. Leaves simple, pungent, vegetative leaves oblong-lanceolate, channelled, sparsely hairy, 10–12 mm long, fertile leaves shorter, tridentate, densely hairy. Flowers: male: pedicel 1–2 mm long, stamens c. 12; female: receptacle 3–4 mm long, ovoid, brownish, sulcate. Aug.–Oct. Rocky sandstone soils, NW, SW, KM, LB, SE (Bokkeveld Mts to Humansdorp).•

strigosa Weim. Straggling monoecious or dioecious shrub to 1 m. Leaves simple, cordate, hairy, 12–18 mm long. Flowers: male: stamens numerous; female: unknown. Mar. Upper slopes, SW (Paarl: Limietberg).•

strobilifera Murray KAMMIEBOS, PYPSTEELBOS, VLEIBOS Monoecious or dioecious shrub to 3 m, often bearing cone-shaped galls. Leaves simple, linear, acute, 10–15 mm long. Flowers: male: stamens 15–20; female: receptacle c. 1.5 mm long, subcylindrical, obscurely striate, apical opening as wide as receptacle. Jan.–Apr. Moist sandstone flats and lower slopes, NW, SW, KM, LB, SE (Kamiesberg and Bokkeveld Mts to KwaZulu-Natal).

theodori-friesii Weim. Like **C. ruscifolia** but leaves spine-tipped and female flowers with receptacle 4–6 mm long, ovoid, 12-ribbed. Sept.–Apr. Middle to upper slopes, SW (Cape Peninsula to Riviersonderend Mts).•

tricuspidata Harv. Sprawling monoecious or dioecious shrublet, branches to 50 cm long. Leaves simple, 2.5 × 1.5–4 mm, pubescent, apex with 3 sharp curved teeth. Flowers: male: stamens 6(7); female: receptacle 2.5 mm long, ovoid, obscurely reticulate. Dec.–Apr. Upper slopes, NW, SW (Cedarberg to Hottentots Holland Mts).•

verrucosa Weim. Monoecious or dioecious shrublet. Leaves simple, entire to irregularly dentate, hairy, 4–8 × 2–4 mm. Flowers: male: unknown; female: receptacle c. 4 mm long, oblong, verrucose. Oct. High sandstone slopes, KM (Swartberg Mts).•

virgata Weim. Sparsely branched monoecious or dioecious shrub to 2 m. Leaves simple, cordate, 20–30 mm long, apex acuminate and recurved, margins coarsely recurved-dentate. Flowers: male: stamens c. 40; female: unknown. Aug.–Oct. High sandstone slopes, NW, SW (Piketberg to Ceres and Paarl).•

viridis Weim. Spreading monoecious or dioecious shrub. Leaves simple, ovate, 30–45 mm long, margins crenate to toothed, stipules with cusps to 8 mm long. Flowers: male: stamens c. 20; female: unknown. Sept. High sandstone slopes, SW (Kogelberg Mts).•

*A". Leaves trifoliolate (see also **C. multiformis**)*
C. Leaflets flat, linear to lobed

apiculata Weim. Monoecious or dioecious shrublet to 40 cm. Leaves trifoliolate, leaflets flat, oblong-lanceolate, 5–12 mm long, apiculate. Flowers: male: stamens 7–9; female: receptacle c. 4 mm long, obovoid, apically tricostate; style 6–7 mm long, reddish purple. Oct.–Dec. SW (Palmiet River valley).•

arcuata Weim. Monoecious or dioecious shrub to 1 m. Leaves, petiolate, trifoliolate, leaflets, narrow, linear, flat, falcate, apiculate, 5–10 × 0.5–0.8 mm, margins revolute. Flowers: male: unknown; female: sepals revolute, receptacle walls thin, membranous, semitransparent. Sept.–Nov. Sandstone slopes, SE (Langkloof).•

baccans Harv. Monoecious or dioecious shrublet to 60 cm. Leaves trifoliolate, leaflets ericoid, 2–3.5 × 0.4–0.5 mm. Flowers: male: stamens 6; female: sepals 3 or 4, adpressed to receptacle, 3 mm long, spherical, deep yellow. Fruits red. Sept.–Oct. Dry rocky, upper slopes, NW, KM (Cedarberg to Witteberg Mts).•

carinata Weim. Monoecious or dioecious shrublet to 30 cm, twigs pilose. Leaves trifoliolate, leaflets lanceolate, apiculate, 5–6 × 1–1.5 mm. Flowers: male: stamens 12; female: receptacle ribbed. Jan.–Apr. Coastal slopes, SW (Cape Peninsula).•

complanata E.Mey. Spreading, monoecious or dioecious shrublet to 60 cm. Leaves trifoliolate, leaflets flat, narrowly elliptic, c. 7 mm long. Flowers: male: unknown; female: sepals 4, ovate, receptacle dorsiventrally flattened, hairy; styles 2, c. 2 mm long, rusty red. Nov.–Mar. Moist upper rocky slopes, SW (Du Toitskloof to Caledon Swartberg).•

concinna Weim. Slender monoecious or dioecious shrublet. Leaves trifoliolate, leaflets flat with midribs projecting in small hooked points, 3–5 × 1–1.5 mm. Flowers: male: stamens 6; female: receptacle prominently ribbed, style 4–5 mm long. May. High rocky slopes, KM (Rooiberg).•

conifera E.G.H.Oliv. & Fellingham Monoecious or dioecious shrub or tree to 4 m. Leaves trifoliolate, leaflets flat, narrowly elliptic to ovate, c. 15 × 5 mm, entire, sometimes 3-lobed, margins slightly revolute. Flowers: male: single or in small clusters on lower branches, with tiny pale yellowish sepals; female: in cone-like inflorescences on lateral branches near apices of main branches, hidden except for the single, linear red styles. June–Oct. E-facing sandstone summit slopes, KM (Anysberg).•

dentata Willd.   Slender, trailing monoecious or dioecious shrublet to 60 cm. Leaves trifoliolate, leaflets flat, cuneate, toothed, c. 4.5 × 3 mm. Flowers: male: sepals 4, stamens 8; female: sepals 4; receptacle dorsiventrally flattened, styles 2, 1.5–2 mm long. Mainly Sept.–Oct. Sheltered sites on sandstone slopes, SW (Cape Peninsula to Hottentots Holland Mts).•

dispar Weim.   Hairy monoecious or dioecious shrub. Leaves trifoliate sometimes unifoliolate, leaflets flat, beneath, margins revolute, 4–12 × 1–2 mm. Flowers: male: stamens 9; female: receptacle obscurely ribbed. Mainly Nov.–Jan. High sandstone slopes, LB, SE (Langeberg to Kouga Mts).•

dodecandra Weim. Monoecious or dioecious shrub to 1 m. Leaves trifoliolate, leaflets flat, linear, sharp-pointed, sparsely long white-hairy above when young, 8–12 × 0.6–1.5 mm. Flowers: male: stamens 12; female: receptacle dark reddish brown. Sept.–Dec. Lower sandstone slopes along streams, SW (Cape Peninsula to Hottentots Holland Mts).•

drepanoides Eckl. & Zeyh.   Monoecious or dioecious shrub to 1 m. Leaves trifoliolate, leaflets linear-lanceolate, somewhat falcate, subacute, 10–18 mm long. Flowers: male: stamens c. 25; female: receptacle 6-ribbed. Mar.–Apr. Sandstone slopes, SE (Langkloof to Uitenhage).•

falcata L.f.   Monoecious or dioecious shrub to 1 m. Leaves trifoliolate, leaflets 6–12 mm long, falcate, margins revolute. Flowers: male: sepals locked, pollen dispersed through round openings bordered by recurved subapical portions of sepals; female: receptacle walls thin, membranous, semitransparent. Aug.–Oct. Coastal slopes, SW, AP, LB (Cape Peninsula to Knysna).•

filicaulis Schltdl.   Sprawling or erect monoecious or dioecious shrublet. Leaves trifoliolate, stipules large and leaf-like, leaflets flat, hairy, central leaflet 2–4 × 2–3 mm, obcordate to obovate, apex trilobed, lateral leaflets 2–4 × 1–3 mm, obtuse. Flowers: male: sepals 4, stamens 8; female: sepals 4, receptacle ovoid, 1–1.5 mm long. Aug.–Dec. Lower sandstone slopes, SW, LB, SE (Cape Peninsula to Humansdorp).•

geniculata Weim.   Monoecious or dioecious shrub. Leaves trifoliolate, leaflets oblong-obovate and apiculate, 5–8 × 2–3.5 mm. Flowers: male: unknown; female: receptacle 6-ribbed, interspersed with wrinkled areas. Jan.–Feb. Lower slopes, SW (Kleinrivier Mts).•

glauca Weim.   Monoecious or dioecious shrub to 1 m. Leaves trifoliolate, leaflets broadly obovate, mucronate, 4–7 × 1–4 mm. Flowers: male: stamens 15; female: receptacle 6-ribbed, brownish. Nov.–Feb. Sandstone slopes, SW, LB (Cape Peninsula to Langeberg Mts).•

gracilis Harv.   Decumbent, much-branched, monoecious or dioecious shrublet. Leaves trifoliolate, leaflets flat, central leaflet c. 5 × 5 mm, apically with 2 notches, lateral leaflets obliquely obovate, c. 5 × 2.8 mm. Flowers: male: sepals 4, stamens 8; female: sepals 4, receptacle dorsiventrally flattened, styles 2. Oct. Shaded upper slopes, SW, LB (Du Toitskloof to Langeberg Mts).•

hantamensis Diels   Closely leafy, densely branched, hairy, monoecious or dioecious shrub to 1 m. Leaves trifoliolate, leaflets flat, oblong-ovate, obtuse, densely long-hairy, 2–3 × 1–2 mm. Flowers: male: sepals 4, c. 3 mm long, connate in lower half, apices acute recurved and hairy, stamens 8; female: sepals as for male but smaller, styles 2. Sept.–Oct. High sandstone slopes, NW, KM (W Karoo and Cedarberg to Waboomsberg).

hermaphroditica Weim.   Bisexual shrub. Leaves trifoliolate, leaflets flat, oblanceolate, c. 6.5 × 3 mm. Flowers mostly bisexual, some male, sepals 3, stamens 15, style 1, red, feathery. Nov. Sandstone slopes, SW (Jonkershoek Mts).•

hexandra Weim.   Erect, monoecious or dioecious shrub to 60 cm. Leaves trifoliolate, leaflets small, flat, oblong-lanceolate, acute, 2–2.5 × 0.5–0.7 mm. Flowers: male: stamens 6; female: receptacle fusiform, with 3 longitudinal reflexed crests. Sept.–Oct. Upper slopes, NW (Bokkeveld Mts to Ceres).•

lanata Weim.   Erect monoecious or dioecious shrublet to 30 cm. Leaves trifoliolate, leaflets flat, oblanceolate, densely grey-hairy, 3–5 × 1 mm. Flowers: male: unknown; female: fusiform, 6-winged: 3 low wings recurved and 3 high wings adpressed or patent. May–June. High rocky slopes, SW (Du Toitskloof Mts).•

lepida Weim.   Slender, sprawling monoecious or dioecious shrublet. Leaves trifoliolate, leaflets 5–12 × 2–5 mm, base cuneate, apex lobed. Flowers: male: pedicel 8–12 mm long, filiform, stamens

12; female: pedicel as for male, sepals recurved, receptacle obovoid. Dec. High sandstone slopes, NW, SW (Grootwinterhoek Mts to Du Toitskloof).•

marginata Eckl. & Zeyh.   Twiggy monoecious or dioecious shrublet to 30 cm. Leaves trifoliolate, leaflets flat, linear to narrowly lanceolate, 2–4 × 0.5–1 mm, margins transparent. Flowers: male: stamens 12; female: receptacle smooth to slightly ribbed. Mar.–Apr. Stony flats, SW (Stellenbosch to Elim).•

micrantha Weim.   Slender monoecious or dioecious shrublet to 60 cm. Leaves trifoliolate, leaflets small, broadened to apex with 3–5 obtuse teeth. Flowers: male: unknown; female: receptacle oblong, 2.5 mm long, with a few low ribs. May–June. Arid fynbos on middle to upper sandstone slopes, KM (Little Karoo Mts).•

mirabilis Weim.   Rigid monoecious or dioecious shrub to 1 m. Leaves bi- or trifoliolate, leaflets flat, obovate to oblong, 4–8 × 2.5–5 mm. Flowers: male: stamens 10–12; female: receptacle 3.5 mm long, ovoid-oblong, ribbed. July. High rocky slopes, NW, SW, KM, LB (Hex River Mts to Franschhoek, Anysberg to Langeberg Mts).•

montana Weim.   Monoecious or dioecious shrublet. Leaves trifoliolate, leaf sheaths and stipules large and prominent, leaflets flat with revolute margins, hairy, 2.5–5 × 0.5 mm. Flowers: male: stamens 6; female: receptacle smooth, brown, obscurely sulcate. Dec.–Jan. High rocky slopes, KM (Swartberg Mts).•

obcordata L.f.   Monoecious or dioecious shrub to 1.3 m. Leaves trifoliolate, leaflets flat; central leaflet obcordate, with midrib excurrent and declinate, 3–7 mm long, lateral leaflets 4–10 × 2–8 mm, obliquely rhomboid-obovate. Flowers: male: stamens 15; female: receptacle tinged with brown, shallowly 12-grooved, style 7 mm long. Mar.–Nov. Flats and lower slopes, SW, AP (Cape Peninsula to Stilbaai).•

obovata E.Mey.   Spreading monoecious or dioecious shrublet to 60 cm. Leaves trifoliolate, leaflets obovate, 2–5 mm long. Flowers: male: stamens 12; female: receptacle shallowly 6-ribbed, brownish. Apr. Middle to upper mountain slopes, NW, SW (Witzenberg to Riviersonderend Mts).•

pedunculata Schltr.   Erect to sprawling, monoecious or dioecious shrublet to 30 cm. Leaves trifoliolate, long-petiolate, leaflets cuneate-lanceolate to obovate, 2–4 mm long, irregularly toothed. Flowers on pedicels 2–4 cm long: male: sepals 7–12 mm long, stamens 50, anthers yellow; female: c. 3.5 mm long, style 1, 8–10 mm long, dull yellow. July–Jan. Damp sandstone slopes, SW (Cape Peninsula to Kleinrivier Mts).•

polygonifolia L. PADDABOS   Erect monoecious or dioecious shrub to 1.5 m. Leaves trifoliolate, leaflets flat, pubescent, oblong-lanceolate to lanceolate, obtuse to tridentate, 4–7 × 1–3 mm. Flowers: male: stamens 12; female: receptacle 3–4 mm long, fusiform, glabrous to hairy, 3-winged, wings adpressed to receptacle. Apr.–Nov. Flats and lower slopes, NW, SW (Clanwilliam to Bredasdorp).•

propinqua Eckl. & Zeyh.   Sprawling monoecious or dioecious shrublet to 60 cm. Leaves trifoliolate, leaflets 2–4 mm long, flat, pinnately veined, central leaflet with 2–4 apical teeth. Flowers: male: sepals 4, c. 2 mm long, stamens 8; female: sepals 4, receptacle dorsiventrally flattened, with long reddish hairs. Feb.–Mar. Rocky sandstone slopes, NW, SW, LB (Cedarberg to Langeberg Mts).•

ramosissima Schltr.   Much-branched monoecious or dioecious shrub to 3 m. Leaves trifoliolate, leaflets nearly flat, midvein raised abaxially, 4–6 × 0.5–1 mm. Flowers: male: stamens 6; female: receptacle c. 3 mm long, ribbed, green with a lilac tinge, walls membranous and semitransparent. Apr.–May. Flats and slopes, SW, KM, SE (Bredasdorp and Swartberg Mts to Mpumalanga).

rigida Weim.   Sprawling, monoecious or dioecious shrublet. Leaves simple and tridentate to trifoliolate, 5–7 × 1.5–2 mm. Flowers: male: stamens 6; female: receptacle 6–12-ribbed. Sept.–Oct. Sandstone slopes, SW (Du Toitskloof Mts).•

sericea Eckl. & Zeyh.   Silky monoecious or dioecious shrub to 60 cm. Leaves trifoliolate, leaflets linear-lanceolate, sericeous, 4–6 mm long. Flowers: male: stamens 9; female: receptacle c. 4 mm long, fusiform, 3-winged, wings adpressed to receptacle. Aug.–Nov. Mountain slopes, NW, SW (Tulbagh to Elim).•

serpyllifolia Cham. & Schltdl.   Monoecious or dioecious shrub to 2 m. Leaves trifoliolate, leaflets flat, margins flat or subrevolute, apex sometimes obtusely tridentate, 3–6 × 0.5–1.5 mm. Flowers: male: sepals 4, partly coherent, stamens 4–6; female: receptacle to 2 mm long, slightly ribbed. Nov.–May. Middle to upper slopes, LB, SE (Swellendam to KwaZulu-Natal).

subdura Weim.   Monoecious or dioecious shrub to 1 m. Leaves trifoliolate, leaflets narrowly lanceolate 7–10 mm long. Flowers: male: unknown; female: receptacle 4 mm long, fusiform, winged, wings recurved. May. Streambanks, SW (Du Toitskloof Mts).•

tenuis Weim.   Monoecious or dioecious shrublet to 30 cm. Leaves trifoliolate, leaflets oblanceolate, apiculate, c. 6 × 0.5 mm. Flowers: male: stamens 6; female: receptacle c. 2 mm long, 6-ribbed, ribs interspersed with fine grey-brown tubercles. Dec. Flats and lower slopes SW (Caledon, Bredasdorp).•

triloba Harv.   Sparsely branched, spreading, monoecious or dioecious shrub to 30 cm. Leaves trifoliolate, petiole 1–2 mm long, leaflets flat, cuneate-triangular, 4–8 mm long, apices 2–4-dentate. Flowers: male: stamens 20; female: sepals 3, c. 3 mm long, style 1, 8–10 mm long, red. Aug.–Sept. Upper slopes, NW, SW (Cedarberg Mts to Ceres).•

sp. 2   Densely leafy monoecious or dioecious shrub to 1 m. Leaves trifoliolate, leaflets flat, linear lanceolate, straight to twisted, c. 10 mm long. Flowers: male: stamens c. 10, maroon; female: sepals fugacious, receptacle elliptic, c. 1.5 mm long. Flowering time? Lower sandy slopes, AP (Gansbaai and Stanford).•

sp. 3   Monoecious or dioecious shrub or tree to 3 m. Primary leaves fugacious, trifoliolate, secondary leaves simple, c. 2 mm long, tongue-shaped, greyish green, grooved adaxially. Flowers: male: minute, borne singly or in small clusters in axils of vegetative leaves, stamens 3; female: hidden in compact cones c. 30 mm long, terminally on branches, styles strap-shaped, red. Flowering time? Sandstone ridges, NW (Bokkeveld Mts: Oorlogskloof).•

C. Leaflets linear to subterete

acockii Weim.   Monoecious or dioecious shrublet to 25 cm. Leaves trifoliolate, leaflets needle-shaped, 3–4 mm long, apex with distinctive red awn. Flowers: male: unknown; female: receptacle ovoid-oblong, 3.5 mm long, sulcate, dull rusty red, style 1 mm long. June–Nov. Low clay hills, SW (Paarl).•

alata N.E.Br.   Monoecious or dioecious shrublet to 50 cm. Leaves trifoliolate, leaflets linear, sparsely to densely hairy, 3–8 mm long. Flowers: male: sepals (3)4, stamens 6–8; female: pedicel c. 1 mm long, receptacle rotund-elliptic, central part grey-hairy and tubercled, 2(3)-winged, style 1 mm long, red. June–Jan. Lower rocky N slopes, LB (Langeberg Mts).•

amplexistipula Schltr.   Monoecious or dioecious shrublet to 30 cm. Leaves trifoliolate, leaflets thick, subterete, young spathulate, older elliptic, scabrid, 2–3 mm long. Flowers: male: stamens 6; female: receptacle narrowly oblong, slightly curved, 7(8)-ribbed with 2 ribs flattened and extended around base forming a continuous wing. July–Apr. Dry rocky sandstone or granite slopes, NW, KM (Kamiesberg and Bokkeveld Mts to Anysberg).•

atrata Weim.   Monoecious or dioecious shrub to 1 m. Leaves trifoliolate, leaflets needle-like, acuminate, 5–12 mm long. Flowers: male: stamens 12; female: receptacle pedicellate, dark brown to black, 12-ribbed, ribs rounded, style curled to form a dense tuft. July–Dec. Sandstone slopes, NW, SW, KM, LB (Cold Bokkeveld to Langeberg and Swartberg Mts).•

burchellii Stapf   Monoecious or dioecious shrub to 2 m. Leaves trifoliolate, leaflets linear, mucronate, margins finely toothed, 15–20 mm long. Flowers: male: stamens 12; female: sepals connate at base, fleshy, erect, receptacle 6 mm long, ovoid, shallowly ribbed to smooth, style c. 1 mm long, hidden. Apr.–May. Streamsides on sandstone slopes, SW, KM, LB, SE (Riviersonderend to Van Staden's Mts).•

burgersii E.G.H.Oliv. & Fellingham   Much-branched monoecious or dioecious shrub to 1.5 m. Leaves trifoliolate, leaflets needle-like, glaucous, 3–7 mm long. Flowers: male: pedicels 1.5 mm long, sepals 3 or 4, stamens 10–13; female: pedicels 1.5 mm long lengthening to 3.7 mm in fruit, sepals 3(4), receptacle 3(4)-winged, white-woolly, with irregular horny red spines or wing-like combs. June–Jan. Limestone flats, AP (De Hoop).•

castanea Weim.   Low monoecious or dioecious shrublet with branches densely and shortly red- or grey-hairy. Leaves trifoliolate, leaflets subterete, slightly arched, densely verruculose, 4–7 mm long. Flowers: male: unknown; female: receptacle 6-ribbed, chestnut-brown. Oct.–Nov. Dry rocky upper slopes, KM (Witteberg and Anysberg).•

cervicornu Weim.   Monoecious or dioecious shrublet. Leaves trifoliolate, leaflets 2–3 mm long, deeply 3–5-lobed. Flowers: male: unknown; female: receptacle oblong, 2.5 mm long, with a few low ribs. Oct. Rocky upper slopes, KM (Swartberg Mts).•

crassinervis Weim.   Monoecious or dioecious shrublet. Leaves trifoliolate, leaflets needle-like, shortly apiculate, bright green, 3–4 mm long, midrib broad, filling almost entire width above. Flowers: male:

unknown; female: receptacle oblong, green, sepals small, erect. Oct. High sandstone slopes, KM (Swartberg Mts: Seweweekspoort).•

cristata Weim. Monoecious or dioecious shrublet to 30 cm. Leaves trifoliolate, leaflets linear, hairy, 6–10 mm long. Flowers: male: unknown; female: receptacle ovoid-oblong, winged to crested, crests brownish red. Aug.–Nov. Upper sandstone slopes, NW, SW (Cold Bokkeveld to Cape Peninsula).•

curvifolia Weim. Spreading, densely leafy monoecious or dioecious shrub. Leaves trifoliolate, leaflets linear, curved, midrib prominent abaxially, 10–15 mm long. Flowers: male: stamens c. 25; female: receptacle fusiform, to 12 mm long. July. Low hills, SW (Bredasdorp).•

densa Weim. Densely leafy monoecious or dioecious shrub. Leaves trifoliolate, leaflets slender, curved, square in section, 6–9 mm long. Flowers: male: stamens 15; female: receptacle c. 6 mm long, fusiform, 12-ribbed. May. High sandstone slopes, LB (Langeberg Mts).•

eriocephalina Cham. Silvery hairy monoecious or dioecious shrub to 1 m. Leaves trifoliolate, leaflets linear, obtuse, sparsely to densely pilose. Flowers: male: stamens 6; female: receptacle 12-ribbed, ribs brown. Sept.–Feb. Rocky peaks, NW, SW, LB, SE (Grootwinterhoek Mts to Knysna).•

exilifolia Weim. Robust monoecious or dioecious shrub to 1.3 m. Leaves trifoliolate, leaflets needle-like, acuminate, 8–10 mm long. Flowers: male: bracteoles sparsely hairy, stamens 6; female: bracteoles as for male, sepals erect. Apr. and Sept. High rocky slopes, SW (Cape Peninsula to Kogelberg).•

filifolia L.f. Low monoecious or dioecious shrublet. Leaves trifoliolate, leaflets fine, needle-like, slightly curved, margins rough, 6–12 mm long. Flowers: male: stamens 6; female: receptacle oblong, ribs inconspicuous. May–Aug. Flats and lower slopes, SW, AP, SE (Malmesbury to Knysna).•

hirta Burm.f. Straggling monoecious or dioecious shrublet to 60 cm. Leaves trifoliolate, with sheath and stipules membranous, ciliate, leaflets linear, sparsely long-hairy, 5–12 mm long. Flowers: male: stamens c. 20; female: receptacle 3–3.5 mm long. May–Nov. Flats and lower slopes, SW (Bokbaai to Cape Peninsula).•

incana Weim. Robust monoecious or dioecious shrub. Leaves trifoliolate with sheaths and stipules densely hairy, leaflets linear, 8–14 mm long. Flowers: male: bracteoles densely grey-hairy, stamens 9–12; female: bracteoles as for male, receptacle 6-ribbed, slightly shiny. Oct. Coastal slopes, AP (Breede River Mouth).•

juniperina L.f. Closely leafy monoecious or dioecious shrub to 1 m, verruculose on young parts. Leaves trifoliolate, leaflets needle-like, verrucose, 4–10 mm long. Flowers: male: stamens c. 20; female: receptacle brownish green, rugose between the green ribs. Sept.–Mar. Granite and sandstone slopes, NW, SW (Namaqualand to Caledon).

linearifolia Eckl. & Zeyh. Monoecious or dioecious shrub to 1.5 m. Leaves trifoliolate, leaflets linear, 5–10 mm long, margins revolute, midribs prominent. Flowers: male: sepals 4, stamens 4; female: receptacle reddish brown, smooth, shiny, striate. Flowering time? Lower slopes, SE (Knysna to tropical Africa).

paucistaminea Weim. Densely leafy, monoecious or dioecious shrub to 1.2 m. Leaves trifoliolate, leaflets linear, curved, pungent, 5–13 mm long, margins minutely toothed. Flowers: male: stamens 3 to 4; female: receptacle yellowish green, shiny, oblique obovate-oblong, with 12–16 conspicuous ribs. Jan.–Apr. Lower mountain slopes, SE (George to KwaZulu-Natal).

polita Weim. Monoecious or dioecious shrublet to 60 cm. Leaves trifoliolate, leaflets linear, shiny, 3–6 mm long. Flowers: male: stamens 6; female: receptacle cylindrical, straight to slightly curved, irregularly 6-ribbed. Nov.–Jan. Dry, rocky upper slopes, KM, SE (Swartberg and Outeniqua Mts to Uitenhage).•

pterocarpa (Harv.) Weim. Monoecious or dioecious shrub to 1 m. Leaves trifoliolate, leaflets narrowly linear, adaxially flat, abaxially convex, 7–15 × 0.5–1.5 mm. Flowers: male: stamens (8)9; female: receptacle brown, 6-ridged, ridges high and sharp-edged. Aug.–Dec. Lower mountain slopes, SW, LB (Cape Peninsula to Riversdale).•

repens Schltr. Erect or spreading monoecious or dioecious shrub to 1 m. Leaves simple, linear-lanceolate, 6–15 mm long, margins scabrid. Flowers: male: stamens 6; female: receptacle 3 mm long, oblong, 9-ribbed, brownish. Mar.–Apr. Habitat? LB, SE (Swellendam to Mpumalanga).

robusta Weim. Sprawling, monoecious or dioecious shrublet to 20 cm. Leaves trifoliolate, leaflets needle-like, thick, pungent, 10–15 mm long. Flowers: male: stamens 12; female: receptacle ovoid, smooth to slightly ribbed, styles 1(2). Nov.–Feb. High rocky slopes, KM (Swartberg Mts to Uniondale).•

semiteres Weim.   Straggling monoecious or dioecious shrublet to 60 cm. Leaves trifoliolate, leaflets linear, convex below, 4–7 × 1.2–1.5 mm. Flowers: male: unknown; female: receptacle ovoid, 3-winged, wings low. June. Rocky sandstone slopes, NW, KM (Hex River valley to Klein Swartberg).•

setifolia Weim.   Sprawling monoecious or dioecious shrub to 1.5 m. Leaves trifoliolate, leaflets fine, needle-like, 4–5 mm long, apices blunt. Flowers: male: sepals 4, stamens 4; female: receptacle c. 1 mm long, smooth, style stout and curved. Dec. Rock crevices at high alt., KM (Swartberg Mts).•

stricta Weim.   Monoecious or dioecious shrub to 1.5 m. Leaves trifoliolate, sheath and stipules large and prominent, leaflets linear, hairy, 4–8 mm long. Flowers: male: stamens 6; female: receptacle 12-veined, smooth. Oct.–June. Flats and lower slopes, SW, AP, LB, SE (Cape Peninsula to Humansdorp).•

subsetacea (Eckl. & Zeyh.) Diels ex Bolus & Wolley-Dod   Monoecious or dioecious shrub to 2 m. Leaves trifoliolate, leaflets needle-like, curved, 3–5 mm long. Flowers: male: stamens 8 (1 or 2 sterile); female: receptacle with 4 high and 4 low ribs, style 1 mm long, red. Aug.–Jan. Flats and lower mountain slopes, mainly marshy areas, SW (Cape Peninsula to Kleinmond).•

teretifolia L.f.   Monoecious or dioecious shrub to 1 m. Leaves trifoliolate, leaflets terete, mucronate, 4–8 mm long. Flowers: male: sepals with long dorsal spine; female: receptacle c. 3 mm long, 3- or 4-winged. Mainly Sept.–Dec. Mountain slopes NW, KM (Namaqualand to Piketberg, Witteberg and Anysberg).

tuberculata (Harv.) Weim.   Closely leafy monoecious or dioecious shrub to 1 m. Leaves trifoliolate, leaflets linear, 5–10 mm long, margins finely scabrous. Flowers: male: stamens 12; female: receptacle reddish brown, coarsely tuberculate. Oct.–Nov. High peaks, NW, SW, KM, LB (Cedarberg to Little Karoo and Langeberg Mts).•

PRUNUS   c. 400 spp., worldwide, mainly N hemisphere

africana (Hook.f.) Kalkman   ROOISTINKHOUT   Tree to 24 m. Leaves elliptic, toothed. Flowers in axillary racemes, white, fragrant. Aug.–Sept. Evergreen forests, SE (Bloukrans River Gorge and E Cape to tropical Africa).

RUBUS   BRAMBLE   c. 250 spp., cosmopolitan, mainly N hemisphere

fruticosus L.   Sprawling thorny shrub to 2 m. Leaves 3- to 5-digitate, leaflets ovate, toothed and discolorous. Flowers in grey-velvety, terminal panicles, pink. Fruits black. Oct.–Apr. Forest margins, SW (Riebeek-Kasteel to Jonkershoek).•

pinnatus Willd.   Sprawling thorny shrub to 2 m. Leaves 5- to 7-pinnate, leaflets elliptic and toothed. Flowers in small, velvety, terminal panicles, lilac. Fruits orange. Nov.–Feb. Forest margins, SW, LB, SE (Cape Peninsula to tropical Africa).

rigidus Sm.   Like **R. pinnatus** but leaves grey-felted beneath and leaflets ovate. Oct.–Feb. Forest margins, NW, SW, LB, SE (Clanwilliam to tropical Africa).

RUBIACEAE

1. Ovules 2–many in each locule:
    2. Fruit dry, capsular; annual or perennial herbs; corolla tube shorter than 1 mm, funnel-shaped ......**Oldenlandia**
    2.' Fruit baccate, indehiscent; shrubs or small trees; corolla various:
        3. Flowers several to many, corymbose, red ....**Burchellia**
        3.' Flowers solitary, white or cream-coloured:
            4. Corolla tube cylindric; stipules connate ....**Gardenia**
            4.' Corolla tube campanulate; stipules ovate ....**Rothmannia**
1.' Ovules solitary in each locule:
    5. Style undivided, capitate or clavate; flowers conspicuous; broad-leaved trees or shrubs:
        6. Flowers yellow; cymes terminal; style included ....**Psychotria**
        6.' Flowers white; cymes axillary or lateral; style exserted:
            7. Style twice as long as corolla tube; stigmatic knob cylindrical, about twice as long as wide ....**Psydrax**
            7.' Style less than twice as long as corolla tube; stigmatic knob globose, about as long as wide ....**Canthium**
    5.' Style absent or deeply divided; flowers inconspicuous; herbs or ericoid shrubs:
        8. Stigmas capitate; stipules foliaceous, leaves thus appearing whorled; diffusely branched weak herbs with 4-angled branches:
            9. Flowers 5-merous ....**Galium**

         9.' Flowers 4-merous ....**Rubia**
      8.' Stigmas filiform; stipules unlike leaves which are thus clearly opposite:
         10. Style present; succulent-leaved coastal sand-dune runner...**Hydrophylax**
         10.' Style absent or obsolete; plants not succulent:
            11. Leaves ovate, soft, petiolate; flowers in diffuse panicles; sepals absent..............................**Galopina**
            11.' Leaves more or less ericoid, subsessile; sepals present:
               12. Sepals as long as corolla tube, sometimes only 1 or 2 developed ....**Carpococe**
               12.' Sepals very short:
                  13. Dioecious dwarf shrubs with ericoid leaves; cymes 1—flowered; stigmas usually
                     purple ....**Nenax**
                  13.' Polygamous shrubs or perennials with ericoid or oblong leaves; cymes few to many-flowered;
                     stigmas mostly white or greenish ....**Anthospermum**

ANTHOSPERMUM   39 spp., Africa and Madagascar, mostly southern Africa

aethiopicum L.   Dioecious shrub to 2 m. Leaves in whorls of 3, needle-like. Flowers in axillary clusters, yellowish. Aug.–Jan. Usually clay slopes, NW, SW, AP, KM, LB, SE (Bokkeveld Escarpment to E Cape).

bergianum Cruse   Densely hairy, mostly dioecious subshrub to 70 cm. Leaves imbricate, usually in whorls of 3, lanceolate. Flowers in axillary clusters, 5-lobed, yellowish. July–Jan. Sandstone slopes, NW, SW (Pakhuis Mts to Caledon Swartberg).•

bicorne Puff   Shrublet to 50 cm. Leaves decussate, ascending, needle-like. Flowers in axillary clusters, with only 1 fertile carpel subtended by enlarged calyx lobes, yellowish. Mainly Dec.–Apr. Sandstone slopes, NW, SW (Cedarberg Mts to Bredasdorp).•

comptonii Puff   Dioecious dwarf shrub to 30 cm. Leaves decussate, lanceolate. Flowers in axillary clusters, yellowish. Aug.–Oct. Rocky slopes, often Witteberg quartzite, NW, KM (Cold Bokkeveld to Touwsberg).•

dregei Sond.   Dwarf shrub to 40 cm. Leaves decussate, lanceolate. Flowers in axillary clusters, yellowish. Aug.–Oct. Granite or sandstone slopes, NW (Namaqualand to Tulbagh).

ericifolium (Licht. ex Roem. & Schult.) Kuntze   Shrublet to 60 cm. Leaves decussate, linear-lanceolate, margins ciliate. Flowers in axillary clusters, with only 1 fertile carpel and 1 stigma, yellowish. Oct.–Nov. Usually deep, sandy soil, SW (Mamre to Cape Peninsula to Botrivier).•

esterhuysenianum Puff   Trailing, mat-forming subshrub to 10 cm. Leaves decussate, lanceolate, sometimes hairy. Flowers solitary and terminal, 5-lobed, yellowish. Oct.–Jan. Shale bands at high alt., NW, SW (Cedarberg Mts to Jonkershoek).•

galioides Rchb.f.   Rounded or sprawling subshrub to 50 cm. Leaves decussate, linear to lanceolate, often recurved above, margins ciliate. Flowers in axillary clusters, yellowish. July–Jan. Flats and slopes, NW, SW, AP, KM, LB, SE (Bokkeveld Escarpment to E Cape).

herbaceum L.f.   Sprawling or trailing perennial to 3 m. Leaves decussate, shortly hairy, lanceolate. Flowers in elongated axillary clusters, yellowish. Throughout the year. Scrub and damp thickets, SW, AP, LB, SE (Hermanus to N Africa).

hirtum Cruse   Densely hairy shrub to 1.5 m. Leaves decussate, linear-lanceolate, often discolorous. Flowers in axillary clusters, 5-lobed, yellowish. Aug.–Dec. Damp sand or clay flats or slopes, NW, SW (Clanwilliam to Bredasdorp).•

paniculatum Cruse   Dioecious shrublet to 70 cm. Leaves decussate, linear. Flowers in terminal thyrsoid panicles, yellowish. Aug.–Apr. Habitat? LB, SE (George to E Cape).

prostratum Sond.   Dioecious, prostrate dwarf shrub to 10 cm, stems trailing and rooting at nodes. Leaves decussate, lanceolate. Flowers in axillary pairs, yellowish. Aug.–Oct. Coastal sands, SW, AP, LB, SE (Saldanha to Port Elizabeth).•

spathulatum Spreng.   Like **A. aethiopicum** but leaves decussate. June–Feb. Sandy soils, NW, SW, AP, KM, LB, SE (Namaqualand to E Cape).

BURCHELLIA   WILD POMEGRANATE, WILDEGRANAAT, UMFINCANE   1 sp., South Africa

bubalina (L.f.) Sims   Shrub or small tree to 5 m. Leaves opposite, elliptic, glossy above, margins revolute. Flowers few in terminal heads, tubular, velvety, orange. Sept.–Dec. Mainly in forests, LB, SE (Grootvadersbos to Mpumalanga).

CANTHIUM   KLIPELS   c. 40 spp., Africa to India

ciliatum (Klotzsch) Kuntze   SKAAPDROLLETJIE, DWERGBOKDROL   Thorny shrub to 4 m. Leaves opposite, small, ovate, thinly hairy and ciliate. Flowers 1 or 2 in leaf axils, cream-coloured to greenish. June–Dec. Forests and forest margins, SE (Storms River to N Province).

inerme (L.f.) Kuntze   CAPE DATE, GEWONE BOKDROL, UMNYUSHULUBE   Usually thorny shrub or tree to 10 m. Leaves opposite, elliptic. Flowers several in axillary cymes, whitish, corolla bearded. Nov.–Feb. Coastal forests, SW, AP, LB, SE (Cape Peninsula to Zimbabwe).

kuntzeanum Bridson (= *Canthium pauciflorum* (Klotzsch) Kuntze)   WATERBOKDROL   Like **C. ciliatum** but leaves glabrous, often borne on spur shoots. Oct.–Jan. Coastal forests, SE (Plettenberg Bay to N Province).

mundianum Cham. & Schltdl.   KLIPELS, ROCK ALDER, UMSANTULANE   Shrub or tree to 5 m. Leaves opposite, broadly elliptic to ovate, softly hairy, usually large. Flowers several in axillary cymes, greenish cream, calyx lobes reduced to a rim. Sept.–Nov. Coastal forest margins, SW, AP, LB, SE (Cape Peninsula to Zimbabwe).

spinosum (Klotzsch) Kuntze   DORINGKLIPELS   Thorny shrub or tree 2–9 m. Leaves opposite, ovate to elliptic, thinly hairy when young. Flowers several in axillary cymes, greenish. May–Sept. Coastal forests, SE (Humansdorp to KwaZulu-Natal).

CARPACOCE   7 spp., mostly W Cape

burchellii Puff   Dwarf shrub to 30 cm. Leaves ericoid, stipular sheath cup-shaped. Flowers 1–4 at nodes, stigma 1 and fruits 1-seeded. Nov.–Apr. Damp sandstone slopes, SW (Paarl to Riviersonderend Mts).•

curvifolia Puff   Sprawling perennial to 50 cm. Leaves rigid, shiny, recurved, lanceolate, stipular sheath with 1 bristle. Flowers 1 or 2 at nodes, greenish. Nov.–Mar. Moist spots on sandstone slopes, KM, SE (Anysberg to Uitenhage).•

gigantea Puff   Leafy shrub to 90 cm. Leaves linear-lanceolate, 40–80 mm long. Flowers solitary on lateral branchlets, greenish. Nov.–Jan. Damp sandstone slopes, LB (Langeberg Mts: Swellendam).•

heteromorpha (H.Buek) L.Bolus   Dwarf shrub to 45 cm. Leaves ericoid, spreading to recurved, margins ciliate, stipular sheath broadly cup-shaped and hairy. Flowers solitary, calyx 4-lobed, yellowish. July–Jan. Sandy flats and slopes, SW (Worcester to Bredasdorp).•

scabra (Thunb.) Sond.   Dwarf shrub to 45 cm. Leaves ascending, ericoid, stipular sheath cup-shaped. Flowers 1 or 2 at nodes, greenish. Sept.–Jan. Sandstone slopes, NW, KM (Bokkeveld Mts to Witteberg).•

spermacocea (Rchb.f.) Sond.   Foetid, lax, straggling perennial to 90 cm. Leaves lanceolate, soft, stipular sheath with several bristles. Flowers few on mostly lateral branchlets, greenish. July–Jan. Damp, sheltered places, SW, LB (Tulbagh to Port Elizabeth).•

vaginellata T.M.Salter   Like **C. scabra** but stipular sheath long and funnel-shaped. July–May. Sandy slopes and flats, NW, SW, AP, LB, SE (Bokkeveld Mts to E Cape).

GALIUM   GOOSE-GRASS   c. 400 spp., cosmopolitan

A. Inflorescences reduced, few-flowered

bredasdorpense Puff   Sprawling roughly hairy perennial to 25 cm. Leaves in whorls of 6, ericoid, almost granular (1.5–2.5 mm long). Flowers 1–3 in axils, anthers exserted, yellowish. Nov.–Dec. Limestone, AP (Agulhas to Cape Infanta).•

mucroniferum Sond.   Sprawling, glabrescent perennial to 60 cm. Leaves in whorls of 6–8, linear-lanceolate, margins ciliate. Flowers in pairs in axils, anthers well exserted, yellowish. Sept.–Dec. Sheltered sandstone slopes, SW (Du Toitskloof to Hottentots Holland Mts).•

rourkei Puff   Prostrate thinly hairy perennial to 20 cm. Leaves in whorls of 4–6, small (2–5 mm long), oblanceolate, hairy. Flowers in pairs in axils, whitish. Dec. Sheltered sandstone slopes, SW (Kogelberg).•

spurium L.   CLEAVERS, GOOSE-GRASS   Straggling prickly annual to c. 2 m. Leaves in whorls of 6–8, lanceolate-ovate, margins prickly. Flowers 1–4 in axils, anthers subsessile, ovary prickly, whitish. Sept.–Dec. Forest margins and streambanks, NW, SW, AP, KM, LB, SE (Namaqualand to tropical Africa).

A. Inflorescences many-flowered

capense Thunb. TINY TOTS Scramblig glabrescent perennial to 90 cm. Leaves in whorls of 6–10, linear-ericoid with margins revolute. Flowers many in axillary cymes, anthers exserted, yellowish. Sept.–Dec. Rocky, damp places, NW, SW, AP, KM, LB, SE (most of South Africa).

monticolum Sond. Like **G. capense** but leaves and stems shortly hairy. Oct.–Nov. Damp sandstone slopes, NW (Cedarberg Mts).•

subvillosum Sond. Scrambling glabrescent perennial to 60 cm. Leaves in whorls of 5–11, lanceolate, margins ciliate. Flowers axillary, in threes, anthers well exserted, yellowish. Sept.–Dec. Damp sandstone slopes, SW (Ceres to Kogelberg).•

tomentosum Thunb. KLEEFGRAS Dioecious, prickly, scrambling perennial to 3 m. Leaves in whorls of 6–8, ovate-lanceolate, margins prickly. Flowers many in axillary cymes, peduncles woolly, anthers subsessile, yellowish. Sept.–Nov. Scrub, NW, SW, AP, KM, LB, SE (S Namibia to E Cape and Free State).

undulatum Puff Glabrescent perennial to 1,5 m. Leaves in whorls of 6, ovate, margins prickly. Flowers 3–many in axillary cymes, anthers subsessile, whitish. Sept.–Dec. Forest margins and streambanks, LB (Langeberg Mts).•

GALOPINA 4 spp., SE Africa

circaeoides Thunb. Perennial to 1.5 m. Leaves opposite, lanceolate, soft. Flowers in a diffuse, terminal panicle, whitish or reddish. Dec.–Apr. Forests or damp scrub, LB, SE (Riviersonderend Mts to tropical Africa).

GARDENIA GARDENIA, KATJIEPIERING c. 250 spp., pantropical

thunbergia Thunb. WITKATJIEPIERING, UMKHANGAZI Grey-stemmed shrub, 2–5 m. Leaves opposite, obovate. Flowers solitary, terminal, hypocrateriform, white, fragrant. Mainly Jan.–Mar. Forest margins, SE (Witelsbos to N. KwaZulu-Natal).

HYDROPHYLAX 3 spp., Africa, Madagascar, India, Thailand

carnosa (Hochst.) Sond. Succulent creeping perennial. Leaves opposite, ovate, fleshy, stipules sheathing. Flowers solitary in axils, white. Nov.–May. Coastal sand dunes, SE (Knysna to S Mozambique).

NENAX 11 spp., southern Africa, mostly W Cape

A. Fruits dehiscent

coronata Puff Dioecious dwarf shrub to 40 cm, branching often pseudo-dichotomous. Leaves decussate, needle-like. Flowers mostly single at nodes, mostly 5-lobed, yellowish. Fruits dehiscent, glabrous or papillate, mericarps 3–5 × 2–3 mm. June–Aug. Sandy slopes, NW (Gifberg to Pakhuis Mts).•

divaricata T.M.Salter Like **N. coronata** but to 1 m, branching often divaricate, and flowers 4-lobed. July–Sept. Sandstone slopes, NW, SW (Bokkeveld Mts to Hottentots Holland Mts).•

elsieae Puff Dioecious dwarf shrub to 30 cm. Leaves decussate, needle-like. Flowers mostly single at nodes, yellowish. Fruits dehiscent, white-woolly, mericarps 3–3.5 × c. 2 mm. Sept.–Oct. Dry sandstone slopes, NW (Swartruggens to Bonteberg).•

A. Fruits indehiscent

acerosa Gaertn. Dioecious shrublet to 40 cm. Leaves decussate, needle-like, 5–12 mm long. Flowers in axillary clusters of 6–12, yellowish. Fruits indehiscent, glabrous or papillate, 3–5.5 × 2–3.5 mm. July–Nov. Sandy or gravelly flats and slopes, NW, SW (Piketberg to Riversdale).•

arenicola Puff Dioecious dwarf shrub to 10 m. Leaves decussate, linear. Flowers 1 or 2 at nodes, yellowish. Fruits indehiscent, glabrous, 5–8 × 2–3.5 mm. July–Aug. Coastal sands, NW (Namaqualand to Graafwater).

hirta (Cruse) T.M.Salter Dioecious cushion-forming shrublet to 30 cm. Leaves sometimes in whorls of 3, linear, 1.5–4 mm long. Flowers usually in axillary clusters of 3, yellowish. Fruits indehiscent, grey-hairy, 2–3 × c. 2 mm. June–Aug. Coastal sands, clays or limestone, NW, SW (Tulbagh and Hopefield to Cape Peninsula).•

sp. 1 Like **N. acerosa** but fruits smaller, c. 2–2.5 mm diam. and ribbed. July–Aug. Sandstone slopes, SW (Riviersonderend Mts).•

OLDENLANDIA   c. 300 spp., cosmopolitan in warm areas

capensis L.f.   Diffuse, sprawling or erect annual. Leaves opposite, linear to oblanceolate, thinly hairy, margins revolute. Flowers in axillary clusters, white or lilac. Jan.–Mar. Waste places, NW, SW, SE (throughout Africa).

PSYCHOTRIA   BIRD-BERRY   c. 1400 spp., pantropical

capensis (Eckl.) Vatke   BLACK BIRD-BERRY   Shrub or small tree to 7 m. Leaves opposite, leathery, obovate, margins revolute. Flowers in terminal pedunculate cymes. Flowers yellow, berries red. June–Jan. Coastal forests, SE (Knysna to Zimbabwe).

PSYDRAX   c. 40 spp., Old World tropics

obovatum (Eckl. & Zeyh.) Bridson (= *Canthium obovatum* Eckl. & Zeyh.)   KWAR, UMBOMBEMFENE   Shrub or tree to 15 m. Leaves opposite, leathery, broadly obovate to suborbicular, obtuse, margins revolute. Flowers in stoutly pedunculate axillary cymes, tube as long as lobes, white. Dec.–Jan. Coastal dunes, SE (Humansdorp to Zimbabwe).

sp. 1   Shrub or tree. Leaves opposite, mostly elliptic, glossy above. Flowers several in delicate axillary cymes, tube shorter than lobes. Dec.–Mar. Coastal and submontane forests, LB (Langeberg Mts: Grootvadersbos).•

ROTHMANNIA   GARDENIA, UMZUKUZA   c. 25 spp., sub-Saharan Africa

capensis Thunb.   CAPE GARDENIA, CANDLEWOOD   Tree to 14(–20) m. Leaves opposite, elliptic. Flowers cup-shaped, sessile, solitary, terminal, white to cream-coloured. Mainly Jan.–Feb. Forests, LB, SE (Swellendam to N Province).

RUBIA   MADDER   c. 60 spp., Old World

petiolaris DC.   Prickly, scrambling perennial to 3 m, stems 4-angled. Leaves in whorls of 6–8, ovate, petiolate, 3-veined from base. Flowers in axillary cymes, greenish. Dec.–Feb. Scrub, LB, SE (Riversdale to Free State).

RUTACEAE (= PTAEROXYLACEAE)
Agathosma by P.A. Bean

1. Leaves compound, pinnate or (1)3–5-foliolate; trees or large shrubs:
   2. Leaves (1)3(5)-foliolate; plants dioecious; stamens 8 .... **Vepris**
   2.' Leaves pinnate; stamens 4–10:
      3. Flowers bisexual; stamens 8 or 10; fruit a berry .... **Clausena**
      3.' Flowers unisexual; stamens 4; fruit a capsule:
         4. Stems and branches prickly; ovary 1-locular; fruit globose, 1-seeded .... **Zanthoxylum**
         4.' Stems and branches unarmed; ovary 2-locular; fruit compressed, 2-seeded .... **Ptaeroxylon**
1.' Leaves simple; trees or smaller, often ericoid shrubs:
   5. Tree with large pink flowers in open panicles; leaves relatively large; fruit a large (over 5 cm diam.), woody capsule covered with blunt protuberances .... **Calodendrum**
   5.' Shrubs; leaves often ericoid; fruit smaller, sometimes horned:
      6. Petals absent; flowers sometimes unisexual .... **Empleurum**
      6.' Petals 5:
         7. Style about as long as petals, longer than claw; flowers in terminal and/or axillary clusters:
            8. Staminodes present .... **Agathosma**
            8.' Staminodes absent or vestigial:
               9. Ovary 3(2 or 4)-lobed .... **Macrostylis**
               9.' Ovary 5-lobed .... **Phyllosma**
         7.' Style much shorter than petals; flowers solitary or few:
            10. Petals transversely bearded at throat .... **Euchaetis**
            10.' Petals pubescent or glabrous at throat:
               11. Staminodes hidden in a channel in petal claw, adnate below .... **Coleonema**
               11.' Staminodes free or absent:
                  12. Anthers with a stalked apical gland .... **Adenandra**

12.' Anthers with a sessile or immersed apical gland:
    13. Anthers and ovary hairy..**Sheilanthera**
    13.' Anthers glabrous and ovary at most glabrescent:
        14. Disc spreading-crenulate; staminodes absent or vestigial................................**Diosma**
        14.' Disc cylindrical or enclosing ovary; staminodes present............................**Acmadenia**

ACMADENIA   33 spp., W to E Cape

A. Petals spreading from below, exposing stamens and style

argillophila I.Williams   Like **A. matroosbergensis** but branchlets sparsely puberulous, forked at 90°, leaves glabrous and sessile, flowers lilac. July–Sept. Rocky sandstone slopes, KM (Witteberg and Bonteberg).•

matroosbergensis E.Phillips   Shrub to 60 cm. Leaves opposite and decussate, oblong, erect. Flowers in terminal pairs, sessile, pink, petals spreading, ovary 5-carpellate. Mainly Sept.–Nov. Sandstone slopes, NW, KM (Cold Bokkeveld and Hex River Mts to Waboomsberg).•

patentifolia I.Williams   Loosely branched shrub to 1 m. Leaves lanceolate, erect, opposite, margins crenulate. Flowers terminal, pedicellate, up to 6, white, ovary 5-carpellate. Fruits with long horns. Mar.–May. Marshes on sandstone slopes, NW (Cedarberg Mts).•

tenax I.Williams   Like **A. wittebergensis** but leaves glabrous and flowers larger and fruits with elongate horns. Jan.–Feb. Sandstone cliffs, NW (Hex River Mts).•

teretifolia (Link) E.Phillips   Multistemmed aromatic shrublet to 40 cm. Leaves suberect, nearly terete, channelled above, glabrescent. Flowers 2 or 3 in terminal clusters, pink, petals spreading, style at first deflexed, ovary 5-carpellate. Mar.–May. Sandstone rocks at high elevations, NW, SW, KM (Cedarberg to Jonaskop and Worcester to Swartberg Mts).•

tetracarpellata I.Williams   Loosely branched shrub to 1 m. Leaves opposite, lanceolate, erect, margins crenulate. Flowers terminal, pedicellate, up to 6, white, petals spreading, ovary 5-carpellate. Mar.–May. Marshes on sandstone slopes, NW (Cedarberg Mts).•

wittebergensis (Compton) I.Williams   Dichotomously branched shrublet to 50 cm, from a woody caudex. Leaves incurved-erect, linear-lanceolate, loosely imbricate, puberulous beneath. Flowers 2–5 in a terminal raceme, white, petals spreading, anthers red, stigma at first deflexed. July–Mar. Rocky sandstone slopes, KM (Witteberg and Little Karoo mts).•

A. Petals forming a closed throat below, concealing stamens and style
B. Anthers with a large pointed apical gland longer than 0.6 mm

baileyensis I.Williams   Like **A. sheilae** but leaves slightly folded, oblanceolate, broadly obtuse and flowers with filaments and staminodes pubescent. Oct.–Apr. Rocky sandstone slopes, KM (Rooiberg).•

flaccida Eckl. & Zeyh.   Loosely branched shrub to 1 m. Leaves lanceolate, erect, opposite, margins crenulate. Flowers terminal, pedicellate, up to 6 in opposed pairs, white, ovary 5-carpellate but only 1 or 2 carpels developing. Fruits with long horns. Mar.–May. Marshes on sandstone slopes, NW (Cedarberg Mts).•

gracilis Dummer   Like **A. rupicola** but leaves loosely imbricate or distant. Mainly Dec.–Mar. Sandstone slopes, SE (Outeniqua Mts: Robinson Pass).•

heterophylla P.E.Glover   Finely leafy, single-stemmed, aromatic shrublet to 40 cm. Leaves sometimes opposite, elliptic, loosely imbricate, obtuse, strongly keeled, silky pubescent on midribs and margins. Flowers solitary, terminal, pink, petals recurved. Mainly Sept.–Mar. Coastal limestones, AP (Agulhas to Mossel Bay).•

macropetala (P.E.Glover) Compton   Like **A. sheilae** but leaves lanceolate, ciliolate, adpressed-erect. June–Aug. Quartz outcrops on shale hills, SW, LB (Bredasdorp to Cloete's Pass).•

nivenii Sond.   Like **A. sheilae** but leaves sessile, adpressed-erect, linear-elliptic, crisped-ciliate. Sept.–Dec. Sandstone slopes, LB (Langeberg Mts: Riversdale).•

rupicola I.Williams   Like **A. sheilae** but leaves alternate. Mainly June–Nov. Local on rocky sandstone slopes, SE (Outeniqua Mts: Robinson Pass).•

sheilae I.Williams   Closely leafy, single-stemmed, aromatic shrublet to 40 cm. Leaves short-petiolate, opposite and decussate, oblong, obtuse. Flowers solitary, terminal, pink to red, anthers with a large, pointed apical gland. Mainly July–Nov. Sandstone slopes, KM (Klein Swartberg, Touwsberg and Rooiberg).•

tetragona (L.f.) Bartl. & H.L.Wendl.   PAGODA FLOWER   Like **A. sheilae** but leaves folded, suborbicular, strongly keeled. Mainly July–Nov. Sandstone slopes, LB, SE (Langeberg Mts: Cloete's Pass to Robinson Pass).•

B. Anthers with a small to minute, pointed or globose apical gland shorter than 0.6 mm
C. Ovary 2–4-carpellate

candida I.Williams   Like **A. nivea** but flowers with staminodes minute, scale-like and ovary 2-carpellate. Mar.–May. Marshes on sandstone slopes, SW (Hottentots Holland Mts: Landdroskop).•

nivea I.Williams   Slender, sparsely branched, single-stemmed shrublet to 30 cm. Leaves lanceolate, erect, margins translucent. Flowers in terminal clusters, white, ovary 3- or 4-carpellate. Mar.–Aug. High, rocky slopes, SW (Hottentots Holland Mts and Kogelberg).•

C. Ovary 5-carpellate

alternifolia Cham.   Finely leafy, single-stemmed, aromatic shrublet to 1 m. Leaves ascending, incurved above, linear to lanceolate, acute, sharply mucronate, margins scabrid. Flowers several, crowded at branch tips, pink. Fruits with long horns. Mainly June–Aug. Rocky outcrops and cliff, SE (Knysna to Keurbooms River).•

bodkinii (Schltr.) Strid   Loosely branched shrub to 1 m. Leaves obovate, adpressed below, recurved above, glandular-denticulate, sparsely pubescent. Flowers sessile, 1 or 2 at branch tips, white, ovary 5-carpellate. Mar.–May. Sandstone rocks at high elevation, NW (Cedarberg Mts).•

burchellii Dummer   Closely leafy shrublet to 30 cm. Leaves subsessile, opposite, imbricate, lanceolate, with a small, inward hooked apical mucro. Flowers terminal, solitary, bright pink. Mainly Oct.–Jan. High rocky slopes, LB (Langeberg Mts).•

densifolia Sond.   Like **A. alternifolia** but leaves adpressed-erect, margins and upper surface crisped-ciliate, flowers often solitary and fruits with short horns. Mainly Aug.–Dec. Limestone hills, AP (De Hoop to Gouriqua).•

faucitincta I.Williams   Like **A. candida** but larger in all parts, flowers with petal claws green and red-brown, and ovary 5-carpellate. Nov. High sandstone slopes, SW (Villiersdorp: Blokkop).•

fruticosa I.Williams   Closely leafy, single stemmed, aromatic shrub to 80 cm. Leaves opposite, ascending, elliptic, obtuse, margins thickened. Flowers terminal, solitary, small, pale pink or white. May–Nov. Dry sandstone slopes, KM (Klein Swartberg).•

latifolia I.Williams   Like **A. trigona** but leaves suborbicular and petal claws sparsely hairy. Mainly Oct.–Dec. N-facing sandstone slopes, LB (Langeberg Mts: Riversdale).•

laxa I.Williams   Like **A. trigona** but leaves adpressed-erect, lanceolate, obtuse and petal claws pubescent, ciliate at throat. Mainly Sept.–Nov. Low, gravel and shale slopes, LB (Swellendam).•

macradenia (Sond.) Dummer   Like **A. bodkinii** but leaves glabrescent, flowers pedicellate, white flushed with pink. Sept.–Nov. Sandstone slopes, NW (Piketberg).•

maculata I.Williams   Like **A. obtusata** but leaf margins thickened, sparsely ciliate and subscabrid, with minute black-stalked glands along upper midline. July–Aug. S-facing sandstone slopes, SE (Outeniqua Mts).•

mundiana Eckl. & Zeyh.   Like **A. alternifolia** but leaves broadly elliptic, obtuse, scabrid and ciliate, and fruits with short horns. Mainly Sept.–Feb. Limestone hills, AP (De Hoop to Potberg).•

obtusata (Thunb.) Bartl. & H.L.Wendl.   Closely leafy, single-stemmed, aromatic shrublet to 30 cm. Leaves sometimes opposite, adpressed, imbricate, linear-lanceolate, erect, margins narrow, scabrid-serrulate. Flowers solitary, terminal, often crowded on short branchlets near stem apices, bright pink. Mainly Sept.–Nov. Limestone and calcareous sands, SW, AP, LB, SE (Stanford to Alexandria).

rourkeana I.Williams   Like **A. bodkinii** but leaves with prominent midrib and distinctly grooved beneath, and flowers pedicellate. Sandstone slopes, NW (Cedarberg and Cold Bokkeveld Mts).•

trigona (Eckl. & Zeyh.) Druce   Closely leafy, sprawling aromatic shrublet to 40 cm. Leaves opposite and decussate, subimbricate or distant, linear-lanceolate, acute. Flowers solitary, terminal, pink, petal claws glabrous. Mainly Sept.–Nov. Dry N-facing sandstone slopes, LB (Langeberg Mts: Riversdale).•

ADENANDRA   CHINA FLOWER, PORSELEINBLOM   18 spp., W Cape

A. Flowers sessile or subsessile, 2–12 in involucrate heads

acuta Schltr.   Erect, aromatic shrublet sometimes to 60 cm, with slender branches. Leaves ascending, loosely imbricate, lanceolate, convex. Flowers subsessile, 1–5 in heads, white, pale pink outside.

Mainly Oct.–Dec. Rocky sandstone slopes, SW (Franschhoek to Riviersonderend and Hottentots Holland Mts).•

gracilis Eckl. & Zeyh.   Slender, sparsely branched, glandular shrublet to 40 cm. Leaves erect to spreading, sometimes imbricate, oblong to elliptic, margins thickened. Flowers subsessile, 1–few in dense heads, white, pink outside. Sept.–Oct. High sandstone slopes in damp sites, SW (Riviersonderend Mts).•

gummifera Strid   Like **A. obtusata** but sometimes to 1.5 m and ovary smooth (vs tuberculate). Mainly Sept.–Oct. Sandstone slopes, SW (Potberg).•

obtusata Sond.   Densely leafy, aromatic shrublet to 50 cm, glutinous on young parts. Leaves erect to spreading, imbricate, oblong, margins thickened, revolute. Flowers subsessile or sessile, in 1–4-flowered, glutinous heads, white, pink outside. Mainly Sept.–Nov. Limestone hills and flats, AP (Agulhas to Cape Infanta).•

rotundifolia Eckl. & Zeyh.   Densely leafy, glandular shrub to 1 m. Leaves erect to spreading, imbricate, broadly elliptic, margins thickened, revolute. Flowers subsessile, many in dense heads, white. Sept.–Oct. Limestone hills and flats, AP (Agulhas to Cape Infanta).•

viscidia Eckl. & Zeyh.   Glabrescent, aromatic shrublet to 50 cm. Leaves ascending, sometimes loosely imbricate, lanceolate to elliptic, margins thickened, revolute. Flowers subsessile, 2–10 in dense glutinous heads, white, pink outside. Aug.–Oct. Sandstone or limestone hills, SW, AP (Kleinrivier Mts to Agulhas).•

A. Flowers with short or long pedicels, solitary or in corymbs or umbels

brachyphylla Schltdl.   Closely leafy, diffuse, aromatic shrub to 80 cm. Leaves spreading above, sometimes imbricate, broadly ovate, cordate at base, margins thickened. Flowers shortly pedicellate, 2–4 in lax subumbels, white, pink to red outside. July–Nov. Rocky, sandstone slopes, SW (Sir Lowry's Pass to Kleinrivier Mts).•

coriacea Licht. ex Roem. & Schult.   Aromatic, often closely leafy shrublet to 45 cm, reddish and puberulent on young parts. Leaves ascending, ovate to elliptic, margins slightly revolute, densely glandular beneath. Flowers 2–4, in corymbs, white, pink to red outside. Mainly Sept.–Nov. Sandstone slopes, SW (Bainskloof to Kogelberg).•

dahlgrenii Strid   Densely leafy, glandular shrub to 1 m. Leaves loosely spreading, imbricate, elliptic to ovate, margins revolute. Flowers on short pedicels, 2–4 in condensed corymbs, pale pink. Mainly Oct.–Dec. Rocky sandstone slopes, KM (Anysberg).•

fragrans (Sims) Roem. & Schult.   Densely leafy shrub sometimes to 1.2 m. Leaves ascending, loosely imbricate, oblong, margins obscurely crenate, slightly thickened and revolute, heavily gland-dotted beneath. Flowers 4–18 in loose, resinous umbels, white, pink outside. Mainly Sept.–Nov. Sandstone slopes, LB (Langeberg Mts).•

lasiantha Sond.   Closely leafy, aromatic shrublet to 45 cm. Leaves ascending to spreading, loosely imbricate, elliptic to narrowly obovate, margins slightly revolute. Flowers 2–12 in dense corymbs, white, red outside, calyx villous. July–Sept. Sandstone slopes, SW (Hermanus to Bredasdorp).•

marginata (L.f.) Roem. & Schult.   Slender, laxly branched, aromatic shrublet sometimes to 1.3 m. Leaves loosely spreading, oblanceolate to narrowly cordate or elliptic, margins revolute, usually without glands. Flowers 2–15 in loose umbels or umbellate corymbs, white, pink outside. Mainly June–Nov. Sandstone slopes, NW, SW (Bokkeveld Mts to Napier).•

multiflora Strid   Aromatic, sparsely leafy shrub to 70 cm. Leaves spreading, oblong, margins thin, slightly revolute. Flowers 3–7 in loose corymbs, often 2 or more per branch tip, white or pink, darker pink outside. Aug.–Nov. Sandstone and clay slopes, SW (Kogelberg to Houwhoek).•

mundiifolia Eckl. & Zeyh.   Sparsely leafy, aromatic shrublet to 70 cm. Leaves spreading, oblong to elliptic, margins thickened, slightly revolute. Flowers 2–6, in dense, resinous umbels, white, pink outside. Aug.–Nov. Sandstone slopes, SW, LB (Riviersonderend and Langeberg Mts).•

odoratissima Strid   Densely leafy shrublet to 60 cm, branching from base. Leaves ascending, loosely imbricate, lanceolate, without glands, slightly pilose, margins slightly revolute. Flowers 2–20 in dense, head-like corymbs, often heavily gland-dotted. Sept.–Oct. Limestone slopes, AP (Agulhas Peninsula).•

schlechteri Dummer   Closely leafy shrublet to 40 cm. Leaves ascending, loosely imbricate, obovate, margins thickened, slightly revolute. Flowers 6–20 in condensed corymbs, white, pinkish outside, calyx densely villous. Apr.–July. Lower sandstone slopes, SW (Elim).•

uniflora (L.) Willd.   Sparsely branched aromatic shrublet to 50 cm. Leaves ascending to spreading, oblong to lanceolate, margins revolute. Flowers subsessile, terminal and usually solitary, white to pink. Mainly Aug.–Oct. Sandstone slopes, SW (Cape Peninsula to Kleinrivier Mts).•

villosa (P.J.Bergius) Licht. ex Roem. & Schult.   Closely leafy, aromatic shrub to 1 m. Leaves ascending, sometimes loosely imbricate, elliptic to oblong, often hairy beneath. Flowers subsessile or on short pedicels, 2–6 in condensed corymbs, white, pink to red outside, calyx usually hairy. Aug.–Nov. Sandstone slopes, NW, SW (Cold Bokkeveld and Piketberg to Langeberg and Kleinrivier Mts).•

AGATHOSMA   BUCHU, BOEGOE   c. 150 spp., southern Africa, mostly W Cape

*A. Ovary usually 4- or 5-lobed (see also **A. rosmarinifolia**)*

acocksii Pillans   Dwarf shrublet occasionally to 60 cm. Flowers in lax terminal clusters, pink. Fruits 5-chambered enclosed in calyx. Sept.–Nov. Rocky sandstone slopes, NW, KM (Worcester to Touwsrivier).•

adenandriflora Schltr.   Single-stemmed rounded shrub to 1 m, broad-leaved. Flowers large, in small lax clusters at branch tips, white to bright pink, pink beneath. Fruits 5-chambered, enclosed in calyx, lacking horns. July–Dec. Rocky slopes at middle to high alt., NW, KM (Swartruggens to Klein Swartberg).•

alligans I.Williams   Leafy shrub to 1 m, much-branched from old stumps, lemon-scented. Leaves spine-tipped. Flowers 1–5 in terminal clusters, calyx more or less woody, petals white or pink, pink-veined. Fruits 5-chambered. Nov.–Dec. Sandstone cliffs, NW (Olifants River Mts: Visgat).•

barnesiae Compton   Resprouting dwarf shrub sometimes to 1 m, lemon- or turpentine-scented. Flowers in dense heads, white. Fruits 5-chambered. Oct.–Dec. Rocky sandstone upper slopes, KM (Witteberg to Rooiberg Mts).•

bathii (Dummer) Pillans   ZEBRABUCHU   Single-stemmed, broad-leaved shrub to over 1 m. Flowers white, dark-spotted, carpels 5. July–Nov. Rocky middle to upper slopes, NW (Cedarberg Mts).•

betulina (P.J.Bergius) Pillans   BUCHU, BERGBOEGOE   Resprouting, broad-leaved shrub to over 2 m, fragrant. Flowers large, usually solitary, axillary, white to purplish pink. Fruits 5-chambered. June–Nov. Rocky sandstone slopes, NW (Bokkeveld to Grootwinterhoek Mts).•

bodkinii Dummer   Single-stemmed, slender shrublet to 40 cm, peppery or peppermint-scented. Flowers in small, dense, terminal clusters, pink to bright purple. Fruits (3)4-chambered. Sept.–Jan. Marshes on shale bands at high alt., NW (Cedarberg Mts).•

conferta Pillans   Robust, single-stemmed, densely leafy to 1,5 m shrub, peppermint-scented. Flowers in dense heads, white or mauve. Fruits 5-chambered. Dec. Sandstone slopes at high alt., NW (Cedarberg Mts).•

cordifolia Pillans   Tangled shrub to 1 m. Leaves broad, reflexed. Flowers in small dense heads, white to purple. Fruits 5-chambered. Sept.–Oct. Streambanks, NW (Skurweberg).•

craspedota E.Mey. ex Sond.   Single-stemmed, dense, rounded shrublet to 1 m. Flowers profuse, in lax terminal clusters, white. Fruits 5-chambered, more or less enclosed in calyx. Sept.–Dec. Rocky sandstone slopes, NW (Bokkeveld to Grootwinterhoek Mts).•

crenulata (L.) Pillans   BUCHU, ANYSBOEGOE   Single-stemmed shrub to 2.5 m, intensely aromatic. Flowers 1(–3) in leaf axils, relatively large, white or mauve, carpels 5. June–Nov. Middle slopes and valleys, NW, SW, LB (Ceres to Swellendam).•

decurrens Pillans   Spreading shrublet to 25 cm. Flowers axillary, solitary or in lax clusters, white to pale pink with long narrow sepals. Fruits 5-chambered. Oct.–Dec. Upper mountain slopes, SW (Slanghoek and Bainskloof Mts).•

distans Pillans   Shrub to 1 m, strongly scented. Leaves large, spine-tipped. Flowers solitary, terminal on short axillary shoots, white with purple dots. Fruits 5-chambered, horns small. Sept. Rocky places on upper slopes, NW (N Cedarberg Mts).•

divaricata Pillans   Single-stemmed, rounded shrublet to 30 cm, branching profusely at ground level, strongly unpleasantly scented. Flowers 1–4, terminal, solitary or paired, white, pink or purple with darker midline. Fruits (3)4(5)-chambered. Sept.–Nov. Shale bands, upper slopes, NW (Cedarberg Mts, Cold Bokkeveld Mts and Swartruggens).•

foetidissima (Bartl. & H.L.Wendl.) Steud.   BOKBOEGOE   Single-stemmed, much-branched shrublet to 1.2 m, strongly scented. Flowers in lax terminal clusters, white, carpels 5, processes absent. Apr.–Oct. Lower shale slopes of coastal plain, AP, LB (Bredasdorp to Riversdale).•

insignis (Compton) Pillans   Sturdy shrub to 1.4 m. Leaves relatively large. Flowers 1–3, axillary, large, white, pink-dotted, carpels 5. Sept.–Nov. Streambanks, NW (Olifants River valley).•

marlothii Dummer   Sprawling shrublet to 40 cm, lemon-scented. Flowers in lax terminal clusters with reflexed sepals, mauve to intense purple. Fruits 5-chambered, enfolded by sepals. Sept.–Jan. Rocky middle to upper slopes, NW, LB (Clanwilliam to Montagu).•

microcarpa (Sond.) Pillans   Resprouting shrublet to 40 cm. Flowers in small, lax axillary clusters below stem tips, white or pale pink. Fruits 5-chambered, processes more or less absent. June–Jan. Lower shale slopes of the coastal plain, LB, AP (Potberg to Mossel Bay).•

odoratissima (Montin) Pillans (incl. **A. hirsuta** Pillans)   BREËBLAARBOEGOE   Resprouting, dense, rounded or willowy shrub to 1 m, strongly citrus-scented. Flowers 1 or 2 in upper axils, white, mauve or purple. Fruits (3)4(5)-chambered. Aug.–Mar. Sheltered and damp rocky upper slopes, NW, SW (Cedarberg to Hottentots Holland to Langeberg Mts).•

ovata (Thunb.) Pillans   BASTERBOEGOE   Leafy shrub, usually single-stemmed to 3 m, herb-scented. Flowers axillary, white, pink or purple. Fruits 5-chambered. Jan.–Dec. Rocky sandstone and silcrete on open slopes and forest margins, KM, LB, SE (Witteberg to Lesotho).•

pattisoniae Dummer   Dwarf, gnarled shrublet to 10 cm, buchu-scented. Flowers long-stalked, in few-flowered, lax terminal clusters, white. Fruits 5-chambered. Aug.–Dec. Upper slopes, NW (Cedarberg Mts).•

pentachotoma E.Mey. ex Sond.   Usually densely leafy shrub to 60 cm, peppermint-scented. Flowers in dense terminal heads, white to purple. Fruits 5-chambered with long processes. Nov.–Mar. Damp places at high alt., NW (Ceres to Hex River Mts).•

phillipsii Dummer   Dwarf, gnarled, cushion-forming shrublet. Flowers in small terminal clusters, pale pink. Fruits 5-chambered. Nov.–Jan. Rock crevices at high alt., NW, KM (Matroosberg to Rooiberg Mts).•

purpurea Pillans   Wiry, much-branched shrublet to 60 cm, delicately lime-scented. Flowers in lax, few-flowered, terminal clusters, brilliant mauve. Fruits 4-chambered. Dec.–Jan. Rocky sandstone upper slopes, KM (Swartberg Mts).•

rubricaulis Dummer   Leafy shrub with sturdy bright mahogany single stem to over 2 m, complex diesel-herb-scented. Flowers in terminal clusters, bright mauve. Fruits (4)5-chambered. June–Sept. Damp upper slopes in shale-bands, NW (Cedarberg Mts).•

rudolphii I.Williams   Compact, leafy shrub to 1 m. Flowers axillary, solitary, bright pink or purple. Fruits 5-chambered. Nov.–Feb. Upper sandstone slopes, SW (Du Toitskloof and Drakenstein Mts).•

serratifolia (Curtis) Spreeth   Shrub to over 2 m. Leaves very long. Flowers white or pink-tinged. Fruits 5-chambered. July–Sept. Mountain slopes and damp kloofs, SW, LB (Caledon to Riversdale).•

spinescens Dummer   Single-stemmed, rounded, stiff, twiggy shrub to 1 m. Flowers in lax terminal clusters, white. Fruits 5-chambered enclosed by calyx, processes long, bifurcated. Sept.–Oct. Dry middle slopes NW (Bokkeveld Mts to Clanwilliam and Swartruggens).•

stipitata Pillans   Single-stemmed, much-branched, stiff shrub to 80 cm, lemon-scented. Flowers axillary, white. Fruits 5-chambered, stalked. Aug.–Nov. Dry rocky sandstone plateaus at middle alt., SW (Perdeberg and Riviersonderend Mts).•

subteretifolia Pillans   Single-stemmed, much-branched shrub to 40 cm. Flowers relatively large, axillary or in lax terminal clusters, white or pink. Fruits 5-chambered. Sept.–Oct. Upper slopes, NW (Montagu: Kiesiesberg and Langeberg Mts).•

tabularis Sond.   STINKBOEGOE   Slender, single-stemmed shrub to over 10 m. Leaves when crushed at first stinking, then pleasingly lemon-scented. Flowers axillary, drab pale mauve. Fruits 4-chambered. Sept.–Nov. Protected mountain slopes, forest margins, SW (Cape Peninsula to Riviersonderend Mts).•

thymifolia Schltdl.   Single-stemmed, smooth, rounded shrub to over 1 m, branching near ground level, mildly aromatic. Flowers in lax terminal clusters, pink or mauve. Fruits (3)4-chambered. Aug.–Oct. Coastal sand and dunes on limestone, SW (Vredenburg to Yzerfontein).•

zwartbergense Pillans   Single-stemmed, tangled dwarf shrublet to 20 cm, lemon-scented. Flowers 2–4 in terminal clusters, pink. Fruits 5-chambered. Nov.–Mar. Upper sandstone slopes, KM (Swartberg and Kammanassie Mts).•

sp. 1 (= *Agathosma canaliculata* P.A.Bean ms)   Densely glandular-hairy dwarf shrublet. Flowers 1 or 2 at branchlet tips, pale pink. Fruits 5-chambered. Mar. Rock cracks in sandstone cliffs at high alt., SW (Worcester: Kwadouwsberg).•

sp. 2 (= *Agathosma citriodora* P.A.Bean ms)   Single-stemmed, sparsely branched shrub to 2 m, strongly lemon-scented. Flowers in copious sessile terminal clusters forming large compound heads, pale mauve. Fruits 5-chambered. Sept.–Oct. S-facing loamy slopes and olifantsklip, KM, SE (Swartberg to Van Staden's Mts).•

sp. 3 (= *Agathosma digitata* P.A.Bean ms)   Shrublet to 50 cm. Flowers in dense terminal clusters, white or pink. Fruits 5-chambered. Oct. Sandy seeps on flats, NW (Cold Bokkeveld Mts).•

sp. 4 (= *Agathosma maculata* P.A.Bean ms)   Sturdy, bright green, leafy shrub to over 1 m. Leaves large. Flowers axillary, usually solitary, large, white, carmine-dotted. Fruits 5-chambered. Oct. Rocky upper slopes, NW (Cold Bokkeveld Mts: Hexberg).•

*A.' Ovary usually 3-lobed (see also **A. involucrata, A. sp. 14**)*

acutissima Dummer   Single-stemmed, densely tangled, sprawling shrublet to 70 cm, unpleasantly scented, branching copiously at ground level. Leaves spine-tipped. Flowers in lax terminal clusters, white with red spots, fragrant. Fruits 3-chambered. Aug.–Sept. Sandstone hills and valley bushveld ecotone, lower coastal slopes, SE (Plettenberg Bay to Uitenhage).•

aemula Schltr.   Much-branched, wiry shrublet to 40 cm. Flowers in dense terminal clusters, white. Fruits 3-chambered. Aug.–Dec. Moist sands, NW (Cedarberg Mts).•

affinis Sond.   Single-stemmed, stiff shrub to 60 cm, scarcely aromatic. Flowers in terminal clusters, pale pink or white. Fruits 3-chambered. May–Oct. Seeps on upper sandstone slopes, KM, SE (Swartberg to Kouga and Baviaanskloof Mts).•

alaris Cham.   Densely glandular shrublet to 30 cm. Flowers in terminal clusters, white. Fruits 3-chambered. Flowering time? Habitat? SE (Plettenberg Bay).•

alpina Schltr.   Resprouting shrub to 1 m, variously herb-, peppermint- or turpentine-scented. Flowers in dense heads, purple, bright pink, sometimes white. Fruits (2)3-chambered. Sept.–Dec. Sandy or rocky upper slopes and plateaus, NW, KM (Ceres to Anysberg).•

alticola Schltr. ex Dummer   Resprouting, rounded shrublet to 30 cm. Flowers in dense heads, white to mauve. Fruits 3-chambered. Dec.–Jan. Damp upper slopes and summits, shale-bands and among rocks, NW, KM (Cold Bokkeveld Mts to Witteberg).•

anomala E.Mey. ex Sond.   Robust, cushion-forming shrublet to 30 cm. Flowers in dense heads, creamy white. Fruits 3-chambered. Oct.–Apr. Rocky middle to upper slopes, SW (Worcester to Caledon).•

apiculata G.Mey.   KNOFFELBOEGOE   Densely leafy shrublet to 1.2 m, strongly sulphur-smelling, leaf tips spiny. Flowers in terminal clusters, white. Fruits 3-chambered. Apr.–Jan. Coastal dunes, clays, granite and limestone, LB, SE (Riversdale to Port Alfred).

asperifolia Eckl. & Zeyh.   Single-stemmed, harsh-textured shrub to 1 m, scarcely aromatic. Flowers white in numerous dense small heads. Fruits 3-chambered. July–Mar. Middle and lower sandstone slopes, NW, SW (Clanwilliam to Elandskloof Mts).•

bicolor Dummer   Many-stemmed, sprawling, slender shrublet to 35 cm. Flowers in lax terminal clusters, white with red buds. Fruits 3-chambered, shaggy. Aug.–Sept. Middle to upper rocky sandstone slopes, NW (Pakhuis Mts).•

bifida (Jacq.) Bartl. & H.L.Wendl.   Single-stemmed or resprouting, usually densely leafy shrublet to 80 cm. Flowers in lax terminal clusters, white to intense purple. Fruits (2)3-chambered. Jan.–Dec. Sandy mountain slopes and flats, NW, SW, AP, KM, LB (Bokkeveld to Outeniqua Mts).•

bisulca (Thunb.) Bartl. & H.L.Wendl.   STEENBOKBOEGOE   Dense, rounded shrub to over 1 m, branching at soil level. Flowers in lax terminal clusters, white. Fruits 3-chambered. June–Nov. Lower and middle slopes and flats in deep sand, NW, SW (Bokkeveld Mts to Silverstroom Strand).•

capensis (L.) Dummer   BOEGOE   Resprouting shrub to 90 cm, sweetly spice-scented. Flowers in lax terminal clusters, white, pink or purple. Fruits 3-chambered. Jan.–Dec. Slopes and flats on shale, granite or coastal sands, less often on acid sand, NW, SW, KM, LB, SE (Namaqualand to Port Elizabeth).•

capitata Sond.   Single-stemmed, rigid, harsh-textured shrublet to 25 cm. Flowers in dense heads, white. Fruits 3-chambered. Sept.–Nov. Rocky sandy plateaus, NW (Piketberg).•

cerefolium (Vent.) Bartl. & H.L.Wendl.   ANYSBOEGOE, STRANDBOEGOE   Single-stemmed shrublet to 1.4 m, strongly aniseed-scented. Flowers in lax terminal clusters, white, pink or mauve. Fruits 3-chambered. Aug.–Jan. Mostly coastal lime sands and limestone flats and hills, AP, SE (Hermanus to Humansdorp).•

ciliaris (L.) Druce   Dense, rounded shrublet to 45 cm, aniseed-scented. Flowers in terminal clusters, white or mauve. Fruits 3-chambered. May–Dec. Coastal flats to lower sandstone slopes and shale-bands, SW, AP (Cape Peninsula to Potberg).•

collina Eckl. & Zeyh.   Dense, rounded, single-stemmed yellow-green leaved shrub to over 1 m, mildly aromatic. Flowers in dense terminal clusters, white. Fruits 3-chambered. Oct.–Apr. Stabilised dunes, AP (Agulhas to Stilbaai).•

concava Pillans   Rigid, compact shrub to 50 cm, aromatic. Flowers in dense terminal clusters, mauve. Fruits 3-chambered. Dec.–Jan. Sandstone slopes, NW (Hex River Mts).•

crassifolia Sond.   Resprouting, stiff, much-branched shrublet to 1 m, scarcely aromatic. Leaves stubby. Flowers in terminal clusters, white or pale pink. Fruits 3-chambered. May–Nov. Dry rocky, sandy or loamy upper slopes at the fynbos-renosterveld ecotone, NW, KM (Cedarberg to Swartberg Mts).•

dielsiana Schltr. ex Dummer   Dense shrublet to 80 cm. Flowers in terminal clusters, white or mauve. Fruits 3-chambered. Apr.–Oct. Dunes or limestone hills, AP, SE (Bredasdorp to George).•

elata Sond.   Lax, single-stemmed shrub to 80 cm, liquorice-scented. Flowers in terminal clusters, white or pale pink. Fruits 3-chambered. Sept.–Nov. Sandy slopes, NW (Gifberg).•

eriantha (Steud.) Steud.   Single-stemmed, rigid, coarse, leafy shrub to 20 cm. Flowers in dense heads, white or purple. Fruits 3-chambered. Sept.–Nov. Coastal limestone hills, LB, AP (Bredasdorp, Swellendam to Riversdale).•

florida Sond.   Slender shrublet to 30 cm. Flowers in terminal clusters, white. Fruits 3-chambered. Sept. Coastal hills, LB (Swellendam).•

florulenta Sond.   Erect, sparsely branched shrublet to 50 cm, leaf tips greatly swollen beneath. Flowers in terminal clusters, pink. Fruits 3-chambered. Sept. Seasonally wet coastal limestone flats, AP (Agulhas).•

foleyana Dummer   Dwarf shrublet. Flowers in dense terminal clusters, magenta. Fruits 3-chambered. Jan. Stony shale-bands at high alt., NW, KM (Hex River Mts to Robertson).•

geniculata Pillans   Rigid, harsh, resprouting shrublet to 60 cm, faintly pine-scented. Flowers in dense terminal clusters, white or pale pink. Fruits 3-chambered. July–Dec. Coastal limestone, SW, AP (Stanford to Stilbaai).•

giftbergensis E.Phillips   Resprouting, rounded, glabrous shrublet to 50 cm, herb- or spice-scented. Flowers in lax terminal clusters, white or pale pink. Fruits 3-chambered. July–Nov. Sandy middle slopes, NW (Gifberg to Cedarberg Mts).•

glandulosa (Thunb.)Sond.   Sprawling, profusely gland-dotted shrub to 1 m. Flowers in lax terminal clusters, white. Fruits 3-chambered. Sept.–Oct. Low granite hills, SW (Malmesbury).•

gnidiiflora Dummer   Much-branched shrublet to 30 cm. Flowers in dense terminal clusters, white. Fruits 3-chambered. Nov. Clay flats, LB (Riversdale).•

gonaquensis Eckl. & Zeyh.   HOTTENTOTSBOEGOE   Single-stemmed, leafy shrub to over 1 m, herb-scented. Flowers in dense heads, white. Fruits 3-chambered. Jan.–Dec. Mainly coastal grassland, SE (Uitenhage to Port Elizabeth).•

hirta (Lam.) Bartl. & H.L.Wendl.   Resprouting, densely leafy shrublet to 60 cm. Flowers in terminal clusters, white. Fruits 3-chambered. Jan.–Dec. Seasonal seeps and lower slopes and dunes, SE (Humansdorp to Port Elizabeth).•

hispida (Thunb.) Bartl. & H.L.Wendl.   Resprouting, harsh-textured shrub to 80 cm, resin-scented. Flowers in dense terminal clusters, white. Fruits 3-chambered. June–Oct. Granitic hills, NW, SW (Piketberg to Paarl).•

hookeri Sond.   Resprouting shrublet to 30 cm. Flowers in dense involucrate heads, white. Fruits 3 chambered. Aug.–Nov. Coastal sandy flats and lower slopes, SW (Cape Peninsula to Betty's Bay).•

imbricata (L.) Willd.   Resprouting, leafy shrub to 1 m, sweetly or herb-scented. Flowers in dense terminal clusters, white, pink or purple. Fruits 3-chambered. June–Jan. Granite, limey or sandy well-drained or seasonally damp slopes and flats, SW, AP, SE (Tulbagh to Knysna).•

joubertiana Schltdl.   Sturdy, untidy, twisted ?resprouting shrublet to 40 cm, herb-scented. Flowers in large terminal clusters, bright mauve-pink. Fruits 3-chambered. Aug.–Oct. Gravelly ironstone flats, SW (Bredasdorp).•

krakadouwensis Dummer   ?Single-stemmed shrub to 70 cm. Flowers in dense woolly heads, white. Fruits (2)3(4)-chambered. Oct.–Dec. Moist sandy middle or upper slopes, NW (Cedarberg Mts).•

lanceolata (L.) Engl.   HEUNINGBOEGOE   Single-stemmed, tangled, spreading shrublet to 50 cm, liquorice-scented. Flowers in lax terminal clusters, pink to deep mauve. Fruits 3-chambered. Apr.–Oct. S-facing sandstone slopes, SW (Cape Peninsula).•

lancifolia Eckl. & Zeyh.   Harsh shrublet to 15 cm. Flowers in dense heads, ?pink. Fruits ?3-chambered. Sept.–Oct. Lower slopes, SW (Tulbagh Kloof).•

latipetala Sond.   Harsh, twiggy, much-branched shrublet to 30 cm. Leaves reflexed. Flowers in lax terminal clusters, white. Fruits 3-chambered. Aug.–Oct. Gravelly sand or granite lower slopes, NW, SW (Piketberg to Paarl).•

leptospermoides Sond.   Dense, sturdy, single-stemmed shrublet to 60 cm, only slightly aromatic. Flowers in dense axillary clusters well below branch tips, white. Fruits 3-chambered. Mainly Feb.–Sept. Dry sandstone upper slopes, SW (Riviersonderend Mts).•

marifolia Eckl. & Zeyh.   Single-stemmed, compact, rough shrublet to 60 cm, citronella-scented. Leaves often very reflexed. Flowers in lax terminal clusters, white. Fruits 3-chambered. June–Nov. Low sandstone slopes, NW (Clanwilliam to Piketberg).•

martiana Sond.   Single-stemmed, leafy, rounded shrub to over 3 m, acrid-smelling. Flowers in lax terminal clusters, white with maroon dots. Fruits 3-chambered. July–Sept. Lower slopes, SE (Humansdorp).•

microcalyx Dummer   Single-stemmed, rounded shrublet to 50 cm. Leaves pilose. Flowers in lax terminal clusters, white. Fruits 3-chambered. Apr.–Sept. Dry sandstone slopes and plateaus, NW (Bokkeveld Mts to Clanwilliam).•

minuta Schltdl.   Compact shrublet to 15 cm, coppicing from a woody caudex, liquorice-scented. Flowers small, in lax terminal clusters, white or mauve. Fruits 3-chambered. June–Sept. Gravelly coastal flats, AP (Bredasdorp to Agulhas).•

mirabilis Pillans   Much-branched shrublet to 40 cm, bitter lemon-scented. Flowers sessile, solitary, axillary, white. Fruits 3-chambered, long-horned. Sept.–Mar. Rock crevices, shale-bands, and sandstone ridges, NW (Cold Bokkeveld: Hansiesberg and Hex River Mts).•

mucronulata Sond.   Rounded, densely leafy shrub to 1 m, coppicing from a woody caudex, turpentine-scented. Flowers in lax terminal clusters, white with purple dots. Fruits 3-chambered. Sept.–Oct. Dry middle slopes, SE (Uniondale).•

muirii E.Phillips   Single-stemmed, sturdy shrub to 1 m. Flowers in terminal clusters, white. Fruits 3-chambered. Apr.–Aug. Coastal hills, AP, LB (Stilbaai to Mossel Bay).•

orbicularis (Thunb.) Bartl. & H.L.Wendl.   Low, tangled shrublet. Leaves minute. Flowers in lax terminal clusters, lilac. Fruits 3-chambered. July–Nov. Middle slopes, SW, LB (Caledon to Langeberg Mts).•

pallens Pillans   Single-stemmed, slender, lax shrublet to 40 cm. Flowers in terminal clusters, white. Fruits 3-chambered. Aug.–Dec. Sandy flats, LB, ?AP (Riversdale).•

propinqua Sond.   Low shrublet to 30 cm, coppicing from a woody caudex. Flowers in terminal clusters, white or purple. Fruits 3-chambered. July–Oct. Sandy flats and hills, SW (Paarl to Caledon).•

puberula (Steud.) Fourc.   Single-stemmed, densely leafy shrub to 3 m, acrid-herb- or sulphur-scented. Leaves spine-tipped. Flowers in lax terminal clusters, white, red-spotted. Fruits 3(4)-chambered. Mar.–Oct. Quartzite, sandstone or gravelly sand on middle slopes, SE (Humansdorp to Grahamstown).

pulchella (L.) Link   MUISHONDBOEGOE   Shrub to 1 m, strongly lemon-scented. Flowers axillary, magenta-pink. Fruits (2)3(4)-chambered. Nov. Damp peaty upper slopes, SW (Cape Peninsula).•

riversdalensis Dummer   Single-stemmed, slender shrublet to 60 cm, herb-scented. Flowers in terminal clusters, white or purple. Fruits 3-chambered. Oct.–Apr. Limestone flats, AP (Bredasdorp to Stilbaai).•

robusta Eckl. & Zeyh.   Laxly branched shrublet to 60 cm, sometimes rooting where buried, lemon-herb-scented. Flowers in woolly terminal clusters, mauve. Fruits 3- or 4-chambered. Oct. Calcareous coastal sands, AP (Cape Infanta to Stilbaai).•

roodebergensis Compton   Rounded, single-stemmed shrub to 1 m. Flowers in axillary clusters below branch tips, white. Fruits 3-chambered. Apr.–Nov. Middle to upper sandstone slopes, KM, SE (Rooiberg to Outeniqua Mts).•

rosmarinifolia (Bartl.) I.Williams   Single-stemmed, leafy shrub to 50 cm, unpleasantly scented. Flowers in dense terminal heads, white. Fruits 3- or 4(5)-chambered. Aug.–Feb. High sandstone slopes, SW (Perdeberg to Kleinrivier Mts).•

salina Eckl. & Zeyh.   Stiff, rounded, single-stemmed, harsh shrublet to 40 cm, scarcely aromatic. Flowers in numerous small, dense terminal clusters, bright pink. Fruits 3-chambered, without horns. Aug. Hard, sandy lower slopes, NW (Cedarberg Mts).•

scaberula Dummer   Coarse, rounded shrublet to over 1 m, coppicing from a woody caudex. Flowers in dense shaggy terminal clusters, white. Fruits 3-chambered. Mar.–June and Aug.–Dec. Calcareous coastal sands, AP (Agulhas to Stilbaai).•

sedifolia Schltdl.   BOSLUISBOEGOE   Single-stemmed, smooth shrublet to 80 cm, faintly aromatic. Flowers rather large, in lax terminal clusters, pink. Fruits 3-chambered. Mar.–Oct. Seasonally damp coastal limestone flats, AP (Agulhas).•

serpyllacea Licht. ex. Roem. & Schult.   Single-stemmed, rounded shrublet to 80 cm. Leaves narrow, swollen behind tip and slightly twisted. Flowers in many, lax terminal clusters, white, pink or purple. Fruits 3-chambered. May–Dec. Coastal or inland sand or limestone flats and slopes, NW, SW, AP, LB, SE (Piketberg to Humansdorp).•

sladeniana P.E.Glover   Single-stemmed, rounded, stiff, bristly shrublet to 40 cm. Flowers in small, dense, involucrate heads, mauve. Fruits 3-chambered. Sept.–Oct. Hard sandy ground, middle to upper slopes, NW (Citrusdal to Piketberg).•

squamosa (Roem. & Schult.) Bartl. & H.L.Wendl. (incl. **A. cedrimontana** Dummer)   Dense, single-stemmed, tussocky shrublet to 60 cm, faintly pine-scented. Leaves scale-like. Flowers in small, dense heads, white or pink. Fruits 3-chambered. Apr.–Dec. Arid rocky middle mountain slopes, NW (Cedarberg to Cold Bokkeveld Mts and Swartruggens).•

stenosepala Pillans   Single-stemmed, dense shrublet with wiry branches to 60 cm. Flowers with slender tapering sepals, usually solitary in upper axils, white. Fruits long-horned, 3-chambered. Aug.–Oct. Middle to upper rocky slopes, SW (Paarl).•

trichocarpa Holmes   Shrub to 30 cm. Leaves crowded, reflexed. Flowers in lax terminal clusters, white. Fruits 3-chambered. June–Sept. Mountain plateaus, NW (Clanwilliam to Tulbagh).•

tulbaghensis Dummer   Shrub to 40 cm. Flowers in terminal clusters, white or mauve. Fruits 2- or 3-chambered. Nov.–Jan. Sandstone summits, NW (Tulbagh, Ceres).•

sp. 5 (= *Agathosma arida* P.A.Bean ms)   Single-stemmed, rounded shrublet to 40 cm, sweetly herb-scented. Flowers in terminal clusters, pink or violet. Fruits 3-chambered. July–Nov. Gravelly loam, karoo-fynbos ecotone, KM (Little Karoo, N slopes of Langeberg and Outeniqua Mts).•

sp. 6 (= *Agathosma haelkraalensis* P.A.Bean ms)   Tough, many-stemmed, decumbent or cushion-like shrublet to 30 cm, lemon-scented, underbark yellow. Flowers in dense terminal clusters, pink or white. Fruits (2)–3-chambered, horns small. Oct.–Dec. Sheltered crevices in limestone, AP (Avila to Hagelkraal).•

sp. 7 (= *Agathosma lanata* P.A.Bean ms)   Dense, harsh, rounded shrubs to 80 cm, branching profusely at ground level, herb-scented. Flowers in dense, woolly terminal clusters, white. Fruits 3-chambered. Mainly July–Oct. Dry rocky upper slopes, KM, SE (Rooiberg and Outeniqua Mts).•

sp. 8 (= *Agathosma paralia* P.A.Bean ms)   Erect, sturdy, single-stemmed shrub to 1 m. Flowers in dense terminal clusters, white. Fruits 3-chambered. June–Aug. Fixed coastal dunes on limestone, AP (Agulhas coast).•

sp. 9 (= *Agathosma parva* P.A.Bean ms)   KLIPSPRINGERBOEGOE   Rounded, harsh, glaucous and often bronzed shrublet to 50 cm, coppicing from a woody caudex, scarcely aromatic. Flowers in terminal clusters, bright purple or pink. Fruits 3-chambered. Aug.–Oct. Rocky shallow sand on arid N-facing slopes, SW (Perdeberg and Riviersonderend Mts).•

sp. 10 (= *Agathosma parvipetala* P.A.Bean ms)   Single-stemmed, delicate shrublet to 30 cm, acrid lemon-scented. Flowers solitary in upper axils, bright mauve. Fruits 3-chambered. Oct.–Nov. Moist loam on high south slopes, SW (Hottentots Holland Mts).•

sp. 11 (= *Agathosma rotundifolia* P.A.Bean ms)   Sturdy rounded, many-stemmed shrub to 40 cm, resin-scented. Flowers in dense terminal clusters, pale pink. Fruits 3-chambered. Aug.–Sept. Stony sand over limestone, AP (De Hoop).•

sp. 12 (= *Agathosma viviersii* P.A.Bean ms)   Single-stemmed, harsh shrublet with divaricate branching to 40 cm, faintly unpleasantly scented. Flowers in multiple terminal clusters often forming large heads, white, buds red. Fruits 3-chambered. Aug.–Sept. Gravelly sand, middle slopes, NW (Cedarberg Mts).•

sp. 13 (= *Agathosma williamsii* P.A.Bean ms)   Single-stemmed shrublet to 30 cm, slightly aromatic. Flowers in lax axillary clusters well below stem tips, white. Fruits 3-chambered. Mainly Oct.–Dec. Rocky upper slopes and summit ridges, LB (Langeberg Mts).•

A". Ovary usually 1- or 2-lobed

abrupta Pillans   Single-stemmed, much-branched, tangled shrublet to 70 cm, scarcely aromatic. Flowers in axillary and terminal clusters, white or pale pink. Fruits 2-chambered. Apr.–Aug. Coastal limestone slopes, SW, AP (Grootbos to Hagelkraal).•

barosmifolia Eckl. & Zeyh.   Single-stemmed, rounded, densely leafy shrub to 2 m, branching near soil level. Flowers in terminal clusters, white. Fruits usually 2-chambered. June–Sept. Lower sandstone slopes, NW (Bokkeveld Mts to Olifants River Mts).•

blaerioides Cham. & Schltdl.   Single-stemmed shrub to 1 m. Leaves saddle-shaped, often with large stalked glands on margins. Flowers in small axillary and terminal clusters, white, sometimes pink. Fruits 1-chambered. Apr.–Jan. Moist upper sandstone slopes, SE (Mossel Bay to Uniondale).•

ciliata (L.) Link   STEENBOKBOEGOE   Slender, single-stemmed, loosely branched shrub to 60 cm, herb-scented. Flowers in lax terminal clusters, white or mauve. Fruits 2-chambered. Apr.–Sept. Sheltered sandstone slopes, SW (Bainskloof to Cape Peninsula).•

corymbosa (Montin) G.Don   Single-stemmed shrublet to 40 cm. Flowers in lax terminal clusters, white to bright purple. Fruits 2-chambered. May–Oct. Seasonally damp sandy flats, SW (Hopefield to Cape Peninsula).•

dentata Pillans   Twiggy rounded shrublet to 30 cm. Flowers in small terminal clusters, pink or white. Fruits 1-chambered. Aug.–Sept. Sandy plateaus, NW (Cedarberg Mts).•

dregeana Sond.   Slender shrub to 60 cm. Flowers in dense heads, white. Fruits 2-chambered. Sept.–Nov. Sandstone slopes, NW (Gifberg).•

elegans Cham. & Schltdl.   Willowy, single-stemmed shrub to 1.5 m, herb-scented. Flowers in terminal clusters, white. Fruits 1(2)-chambered. Jan.–Dec. Damp mountain slopes, LB, SE (Riversdale to Uniondale).•

esterhuyseniae Pillans   Wiry shrub to 1 m, strongly aromatic. Flowers in dense woolly heads, white and mauve. Fruits 2(3)-chambered. Nov.–Dec. High peaks, NW (Cedarberg Mts).•

glabrata Bartl. & H.L.Wendl.   Single-stemmed, dense, rounded shrub to 2 m, lemon-scented. Flowers in lax terminal clusters, mauve, pink or white. Fruits 2-chambered. July–Dec. Damp sandy plains and dune slacks, SW (Darling to Cape Peninsula).•

humilis Sond. (incl. **A. adnata** Pillans)   Resprouting, much-branched, cushion-like shrublet to 75 cm. Flowers in dense terminal heads, white or pale pink. Fruits 2-chambered with small horns. Oct.–Jan. Middle and upper well-drained slopes, NW (Cedarberg to Cold Bokkeveld Mts).•

involucrata Eckl. & Zeyh. (incl. **A. cephalodes** E.Mey. ex Sond., **A. sabulosa** Sond.)   Sturdy, resprouting shrub to 1 m, faintly pine-scented. Flowers in dense involucrate heads, white or bright pink. Fruits 2- or 3-chambered, without processes. Sept.–Nov. Deep sandy lower slopes, NW (Paleisheuwel).•

juniperifolia Bartl.   Willowy shrub to 2.5 m, strongly aromatic. Flowers in terminal clusters, purple, bright pink or white. Fruits 2-chambered. Apr.–Dec. Middle slopes and kloofs, NW, SW (Clanwilliam, to Riviersonderend Mts).•

kougaense Dummer   Slender leafy shrub to 20 cm. Flowers in pairs in axils, white. Fruits 1(2)-chambered. July–Nov. Rocky slopes at high alt., SE (Uniondale to Kouga Mts).•

linifolia (Roem. & Schult.) Bartl. & H.L.Wendl.   Shrub to 50 cm. Flowers in terminal clusters, white. Fruits 2-chambered. July–Feb. Sandy loam on middle S-facing slopes near streams, LB (Langeberg Mts).•

longicornu Pillans   Single-stemmed, harsh, twiggy shrublet to 25 cm, scarcely aromatic . Flowers in dense terminal clusters, white. Fruits 2-chambered, long-horned. Sept. Stony upper sandstone slopes, NW (Cedarberg Mts).•

mundtii Cham. & Schltdl.   JAKKALSPISBOS   Single-stemmed, sometimes resprouting, finely velvety, wiry shrub to 1 m, foetid. Flowers in terminal or axillary clusters, white. Fruits 2-chambered, flat-sided. June–Nov. Middle to upper dry rocky slopes, KM, SE (Witteberg to Humansdorp).•

ovalifolia Pillans   Single-stemmed, rounded shrub to 1.5 m, acrid or spice-scented. Flowers in lax terminal clusters white, red-dotted. Fruits 2-chambered. June–Oct. Rocky quartzitic upper slopes, KM, SE (Swartberg Mts to Willowmore).•

pilifera Schltdl.   Single-stemmed shrub to 60 cm, herb-scented. Flowers in terminal clusters, white with dark spots. Fruits (1)2(3)-chambered. Sept.–Oct. Upper sandstone slopes, SE (Humansdorp to Port Elizabeth).•

planifolia Sond.   Resprouting willowy shrub to 2 m, liquorice or pine-scented. Flowers in terminal clusters, white, sometimes pink. Fruits 2-chambered. Mainly Dec.–Apr. Moist loamy soils on cool upper slopes, SE (Outeniqua Mts).•

pubigera Sond.   Resprouting, glaucous, much-branched shrublet to 60 cm, slightly aromatic. Flowers in terminal clusters, white. Fruits 1-chambered. Aug.–Sept. Lower slopes NW (Cedarberg Mts).•

pungens (E.Mey. ex Sond.) Pillans   Single-stemmed, much-branched, leafy shrub to 80 cm. Leaves spine-tipped, pleasantly aromatic. Flowers axillary, usually solitary, white, pink to purple. Fruits 2-chambered. May–Nov. Upper mountain slopes, KM, SE (Swartberg and Kammanassie Mts to Uniondale).•

recurvifolia Sond.   KANFERBOEGOE   Single-stemmed, stiff, spreading shrublet to 1.5 m, turpentine-scented. Leaves recurving, with hyaline margins. Flowers in terminal clusters, white. Fruits 2-chambered. May–Nov. Dry middle to upper slopes and valley bushveld ecotone, KM, LB, SE (Rooiberg and Swartberg Mts to Uitenhage).•

spinosa Sond.   Prostrate to decumbent, resprouting shrublet, herb-scented. Leaves recurving, pungent, with hyaline margins. Flowers in terminal clusters, white. Fruits 2-chambered. Sept.–Nov. Rocky sandstone slopes, SE (Uniondale to Avontuur).•

stenopetala (Steud.) Steud.   Single-stemmed, smooth rounded shrub to 1.5 m, lemon-scented. Flowers in terminal clusters, white. Fruits 2-chambered. June–Nov. Coastal limestone hills, SE (Humansdorp to Port Elizabeth).•

stilbeoides Dummer   Single-stemmed, yellowish green leafy shrub to 1 m, liquorice-scented. Flowers in dense heads, white to mauve. Fruits 1-chambered, without horn. Oct.–Mar. Rocky upper slopes and shale-bands, NW (Cedarberg Mts).•

stokoei Pillans   Slender shrublet to 35 cm. Flowers in dense woolly heads, mauve?. Fruits 2-chambered. Nov. Upper slopes, SW (Hottentots Holland Mts to Kogelberg).•

umbonata Pillans   Single-stemmed pubescent shrublet to 60 cm. Flowers more or less sessile, solitary in leaf axils, pink or white. Fruits 1- or 2-chambered, without horns. Mainly Jan.–Mar. Peaty, pebbly, moist well-drained seeps at high alt., LB (Langeberg Mts).•

unicarpellata (Fourc.) Pillans   Single-stemmed shrublet to 45 cm, turpentine-scented. Flowers 1–3 in leaf axils, white. Fruits 1-chambered. Apr.–Dec. Rocky, grassy middle and lower slopes, SE (Kouga and Baviaanskloof Mts).•

venusta (Eckl. & Zeyh.) Pillans   GOEIEBOEGOE   Resprouting shrub with warted branches to 1 m, pleasantly liquorice-scented. Flowers in small axillary and terminal clusters, pale mauve often dark-dotted. Fruits 2(–5)-chambered, 1 or more often aborting. July–Feb. Grassy fynbos in seeps and rocky loams on upper and middle southern slopes, SE (Uniondale to E Cape).

virgata (Lam.) Bartl. & H.L.Wendl.   SKAAPBOEGOE   Resprouting, dense, rounded shrub to over 1 m, often unpleasantly scented. Flowers in lax terminal clusters, white or mauve. Fruits 2-chambered. Apr.–Dec. Middle to upper sandstone slopes NW, SW, LB (Olifants River to Langeberg Mts).•

sp. 14 (= *Agathosma anysbergensis* P.A.Bean ms)   Much-branched shrub to 1 m, slightly sweet-scented. Flowers in few-flowered terminal clusters, white. Fruits 2- or 3-chambered. May–Oct. Arid rocky slopes, KM (Anysberg).•

CALODENDRUM   CAPE-CHESTNUT, WILDEKASTAIING   1 sp., Africa

capense (L.f.) Thunb.   Tree to 20 m. Leaves elliptic, aromatic. Flowers in terminal panicles, pink dotted mauve. Sept.–Dec. Evergreen forests, SE (George to tropical Africa).

CLAUSENA   HORSEWOOD, PERDEPIS   c. 50 spp., Africa to Malaysia

anisata (Willd.) Hook.f. ex Benth.   Shrub or small tree to 5(–10) m. Leaves imparipinnate, leaflets ovate, toothed, strongly scented. Flowers in axillary cymose panicles, yellow or white. Aug.–Nov. Evergreen forests, LB, SE (Riversdale to tropical Africa).

COLEONEMA CAPE-MAY, CONFETTI BUSH   8 spp., W and E Cape

album (Thunb.) Bartl. & H.L.Wendl.   Shrub to 2 m. Leaves linear-oblong, sweet-smelling. Flowers solitary, crowded at branch tips, white, 6–7 mm diam. Aug.–Oct. Coastal sandstone or granite outcrops, SW, AP (Saldanha to Cape Infanta).•

aspalathoides Juss. ex Don   Dense shrub to 1.5 m. Leaves linear-oblong, scarcely scented. Flowers solitary, crowded, pink, petals not obviously clawed, 8–10 mm diam. Apr.–Nov. Sandstone slopes, SW, LB, SE (Potberg to Zuurberg).

calycinum (Steud.) I.Williams   Shrub to 3 m. Leaves linear, c. 15 mm long, rum-scented. Flowers solitary, white, 5.5–6.5 mm diam. Sept.–Oct. Sandstone and limestone slopes, SW, AP, KM, LB (Houwhoek to Riversdale).•

juniperinum Sond.   Shrublet to 50 cm. Leaves linear-lanceolate, ascending-adpressed, resin-scented. Flowers solitary, white, 3–4 mm diam., petals papillate. May–Nov. Sandstone slopes, NW, SW (Bokkeveld to Riviersonderend Mts).•

nubigenum Esterh.   Dense, rounded shrub to 1 m. Leaves linear-lanceolate, coconut-scented, ciliate at first. Flowers solitary, white, 7.5 mm diam., staminodes free from petals. Oct.–Feb. Sandstone slopes, SW (Du Toitskloof and Hottentots Holland Mts).•

pulchellum I.Williams   Dense shrub to 1 m. Leaves linear, pungent, sweet-smelling. Flowers solitary, often crowded, pink, 7–8 mm diam. Mar.–Oct. Coastal sands, SE (Plettenberg Bay to Port Elizabeth).•

pulchrum Hook.   Willowy shrub to 1.2 m. Leaves linear, erect, turpentine-scented, mucronate, c. 20 mm long. Flowers solitary, pink, 12–13 mm diam., petals gland-dotted. Aug.–Nov. Sandstone slopes, LB (Tradouw Pass).•

virgatum (Schltdl.) Eckl. & Zeyh.   Erect shrub to 1 m. Leaves linear, c. 30 mm. Flowers solitary, pink or white, c. 12 mm diam. June–Sept. Sandstone slopes, LB (Langeberg Mts: Swellendam to Riversdale).•

DIOSMA FALSE BUCHU   28 spp., Namaqualand to E Cape

A. Staminodes lacking
B. Leaves mostly opposite

dichotoma P.J.Bergius   Like **D. oppositifolia** but leaves adpressed, short, to 4 mm long, with apical callus and fruits without horns. Nov.–Jan. Mainly coastal sands, SW (Hopefield to False Bay).•

oppositifolia L.   Closely leafy, aromatic shrublet sometimes to 1 m, with many stems from a woody caudex. Leaves opposite and decussate, sessile, spreading-erect, loosely imbricate, linear to lanceolate, recurving towards tip. Flowers paired and aggregated in small flat-topped corymbs, white. Mainly Sept.–Jan. Sandstone, granite and limestone slopes, SW, AP (Darling to Bredasdorp and Agulhas).•

subulata J.C.Wendl.   Finely leafy, single-stemmed aromatic shrub to 1.8 m. Leaves mostly opposite, incurved-erect, linear-lanceolate, pungent. Flowers crowded terminally on 1-flowered racemes, white, petals and ovary glabrous. Mainly July–Nov. Coastal dunes, SW, AP (Hawston to Cape Agulhas).•

B. Leaves alternate

acmaeophylla Eckl. & Zeyh.   Finely leafy, single-stemmed, aromatic shrub to 2.5 m. Leaves sessile, linear-subterete, with a recurved mucro. Flowers terminal, solitary on loosely clustered, short branchlets, white, petals minutely pubescent. Mainly Aug.–Oct. Rocky sandstone slopes, NW, KM (Namaqualand to Cold Bokkeveld and Witteberg).

arenicola I.Williams   Like **D. subulata** but leaves alternate in mature plants, bracts and sepals long-crisped ciliate. Mainly Sept.–Dec. Calcareous sands on limestone, AP (Agulhas Peninsula).•

aspalathoides Lam.   Finely leafy aromatic shrub to 1 m, branching from a woody caudex. Leaves short-petiolate, ascending, linear-elliptic, mucronate. Flowers in small terminal clusters on short racemes, white. Mainly Sept.–Nov. Sandy, mostly coastal flats, SW (Hopefield to Milnerton).•

awilana I.Williams   Like **D. subulata** but leaves alternate, sessile, with recurved tip. Dec.–June. Stony calcareous sands, AP (Baardskeerdersbos).

fallax I.Williams   Finely leafy, twiggy aromatic shrublet sometimes to 40 cm. Leaves sessile, spreading, linear, pungent. Flowers few on short terminal racemes, white. Sept.–Oct. Sandstone slopes, SW (Riviersonderend Mts).•

hirsuta L.   Finely leafy aromatic shrublet sometimes to 50 cm, resprouting from a woody caudex. Leaves linear, mucronate. Flowers compact racemes grouped in lax terminal corymbs, white, petals persisting below fruit. Mainly Sept.–Nov. Sandstone and clay slopes, NW, SW, AP, LB, SE (Cedarberg Mts to Humansdorp).•

meyeriana Spreng.   Finely leafy aromatic shrub to 1 m, sometimes branched from base. Leaves sessile, ascending, linear-lanceolate, sharply mucronate. Flowers in small terminal clusters on short racemes, white. Mainly July–Dec. Rocky sandstone slopes, NW (Cedarberg and Olifants River Mts).•

parvula I. Williams   Finely leafy, single-stemmed aromatic shrublet to 40 cm. Leaves linear, mucronate. Flowers in dense terminal racemes, small, petals emarginate, disc much pitted. Mainly June–Oct. Peaty seeps on sandstone, SW (Bredasdorp and Potberg).•

pedicellata I. Williams   Like **D. hirsuta** but plants robust, to 1 m, petals soon falling, fruits pedicellate above calyx when mature and not associated with petals. Mainly Aug.–Oct. Sandy and stony slopes and flats, NW, SW (Bokkeveld Mts to False Bay).•

pilosa I. Williams   Like **D. meyeriana** but leaves adpressed, flowers larger, petals with few long weak hairs and fruit horns short and spreading (not erect). Jan.–June. Sandstone slopes, SW (Riviersonderend Mts).•

ramosissima Bartl. & H.L.Wendl.   Finely leafy aromatic shrub to 1.5 m, many-branched from base. Leaves sessile, linear-subterete, obtuse or with an apical callus. Flowers terminal, solitary or paired on loosely clustered short branchlets, white, petals minutely pubescent outside. Mainly Aug.–Nov. Rocky sandstone slopes, NW (Namaqualand to Worcester).

sabulosa I. Williams   Like **D. subulata** but leaves alternate, sessile, ovate. Mainly Aug.–Dec. Deep sands, AP, LB (De Hoop to Albertinia).•

A. Staminodes vestigial
C. Leaves shortly petiolate,

aristata I. Williams   Single-stemmed, finely leafy, aromatic shrub to 70 cm. Leaves linear-attenuate, pungent. Flowers 3–5 in short racemes, white. Jan.–May. Coastal sands, AP (Mossel Bay).

echinulata I. Williams   Like **D. guthriei** but leaves alternate, short-petiolate, sparsely setose, pungent. Mainly Dec.–Apr. Limestone hills, AP, LB (De Hoop to Albertinia).•

rourkei I. Williams   Like **D. thyrsophora** but leaves short-petiolate, linear-lanceolate, acute, flowers with sessile petals and style always erect. June–Aug. Rocky sandstone slopes, SE (Baviaanskloof Mts).•

tenella I. Williams   Finely leafy aromatic shrublet to 50 cm, single- or multistemmed. Leaves linear-oblong, spreading to suberect. Flowers few in short terminal racemes, white. June–Dec. Mainly on clays and gravel soils, AP, LB (De Hoop and Langeberg: Heidelberg to Albertinia).•

C. Leaves sessile

apetala (Dummer) I. Williams   Closely leafy sprawling, aromatic, single-stemmed shrublet. Leaves sessile, lanceolate, adpressed-erect. Flowers sessile, 1 or 2 in a terminal tuft, tiny, white, petals caducous on opening. Mainly Aug.–Nov. Sandstone slopes above 900 m, KM, SE (Swartberg and Kouga Mts).•

demissa I. Williams   Like **D. guthriei** but leaves alternate, lanceolate, 5-ranked, obtuse. Mar.–June. Coastal in sandy pockets in limestone, SW, AP (Cape Peninsula and Stanford to Hagelkraal).•

guthriei P.E.Glover   Closely leafy, spreading, aromatic shrublet to 40 cm, from a woody caudex. Leaves sessile, sometimes opposite, ovate, acute with a blunt apical callus. Flowers sessile, 2 or 3 in a terminal cluster, white. Mainly Oct.–Dec. Mainly on limestone, AP (Agulhas Peninsula).•

haelkraalensis I. Williams   Like **D. guthriei** but plants decumbent, leaves opposite, recurving, and flowers small. Apr.–Sept. Limestone hills, AP (Pearly Beach to Hagelkraal).•

passerinoides Steud.   Finely leafy, aromatic shrublet to 40 cm. Leaves sessile, elliptic to lanceolate, obtuse, minutely pubescent. Flowers solitary, terminal, white. June–Sept. Silcrete slopes, SW, KM, LB, SE (Caledon to Kouga Mts and Little Karoo).•

prama I. Williams   Like **D. passerinoides** but single-stemmed, branchlets glabrous and leaves with pustulate gland dots mainly near midrib. Mainly Sept.–Dec. Dry sandstone slopes, KM, SE (Touwsberg and Klein Swartberg to Baviaanskloof Mts).•

recurva Cham.   Finely leafy, aromatic shrub to 1 m. Leaves sessile, elliptic, obtuse, recurved. Flowers in small terminal clusters, white, petals gland-dotted. June–Dec. Rocky sandstone slopes, KM (Little Karoo Mts).•

strumosa I. Williams   Like **D. passerinoides** but style and filaments sparsely pubescent, anthers purple and fruits rough and pitted. June–Aug. Shale slopes in renosterveld, KM (Barrydale).•

thyrsophora Eckl. & Zeyh.   Single-stemmed shrub to 3 m. Leaves sessile, linear-lanceolate, obtuse. Flowers 1–3 on short terminal branchlets, white, style deflexed during anthesis. Nov.–Jan. Rocky sandstone slopes, SW (E Riviersonderend Mts).•

EMPLEURUM•   2 spp., W Cape

fragrans P.E.Glover   Monoecious or polygamous wand-like shrub to 1.7 m. Leaves oblong, 6–8 mm long. Flowers 1 or 2 in axils, greenish. Oct. Streamside seeps, LB (Langeberg Mts).•

unicapsulare (L.f.) Skeels   Monoecious or polygamous willowy shrub or small tree to 4 m. Leaves linear-lanceolate, finely serrate, resin-scented, 20–60 mm long. Flowers 1 or 2 in axils, greenish. Mainly Apr.–Sept. Streambanks or seeps, NW, SW, KM, LB, SE (Cedarberg Mts to Port Elizabeth).•

EUCHAETIS•   23 spp., W and E Cape

A. Disc spreading-crenulate below ovary

diosmoides (Schltr.) I.Williams   Like **E. schlechteri** but leaf apices incurved-apiculate and flowers many. July–Jan. Rocky shale slopes, SW (Elim to Potberg).•

elsieae I.Williams   Shrublet to 1(–1.5) m. Leaves ascending, oblong, fleshy, obtuse. Flowers 2–4 at branch tips, white. Throughout the year. Sandstone slopes, NW (Cold Bokkeveld to Hex River Mts).•

linearis Sond.   Slender willowy shrublets to 40 cm. Leaves adpressed, linear. Flowers 4 in terminal clusters, white. Feb.–July. Sandstone slopes, NW, SW (Grootwinterhoek Mts to Caledon Swartberg).•

longicornis I.Williams   Foetid shrublet like **E. linearis** but leaves pungent and fruit horns longer, 7 mm long. May–June. Clay slopes, LB (Riversdale).•

schlechteri Schinz   Diffuse shrublet to 40 cm. Leaves ascending, subsessile, lanceolate, finely hairy. Flowers to 9 in terminal clusters, white. May–Aug. Clay slopes, SW (Sir Lowry's Pass to Bredasdorp).•

A'. Disc collar-like around ovary

avisylvana I.Williams   Diffuse shrub to 1 m. Leaves ascending, subsessile, linear-lanceolate, long-ciliate. Flowers several in terminal glomerules, white, petals acute. Apr.–June. Sandstone slopes, LB (Swellendam).•

cristagalli I.Williams   Like **E. avisylvana** but petals obtuse and larger, 5.5 mm long. Feb.–Mar. Sandstone slopes, SE (Great Winterhoek Mts).•

elata Eckl. & Zeyh.   Diffuse shrublet to 40 cm. Leaves adpressed, lanceolate, sessile, minutely ciliate. Flowers several in terminal glomerules, white. Throughout the year. Sandstone slopes, SW (Slanghoek to Palmiet River Mts).•

esterhuyseniae I.Williams   Twiggy shrublet to 60 cm. Leaves spreading, fleshy, ovate. Flowers few in terminal glomerules, white. Dec.–May. Sandstone crevices, NW (Cold Bokkeveld Mts).•

flexilis Eckl. & Zeyh.   Like **E. elata** but leaves blunt and like the prow of a boat at tips and margins membranous. Feb.–June. Sandstone slopes, SW (Du Toitskloof to Riviersonderend Mts).•

glabra I.Williams   Shrublet to 50 cm. Leaves ascending, shortly petiolate, elliptic. Flowers c. 12 in terminal glomerules, white. Mar.–May. Sandstone slopes, SW (Jonkershoek to Kogelberg).•

glomerata Bartl. & H.L.Wendl.   Like **E. elata** but fruit horns longer, c. 3 mm. Throughout the year. Sandstone slopes, NW (Pakhuis Mts to Cold Bokkeveld Mts).•

longibracteata Schltr.   Shrublet to 80 cm. Leaves ascending, subsessile, ovate to lanceolate, c. 10 mm long, sometimes minutely hairy, upper whitish and forming a conspicuous involucre below flowers. Flowers 1-several in terminal clusters, pink. Dec.–Apr. Limestone hills, AP (Agulhas to Cape Infanta).•

A". Disc arching over and partly enclosing ovary

albertiniana I.Williams   Like **E. burchellii** but single-stemmed at base and sparser. Throughout the year. Coastal sands and limestone, AP (Albertinia).•

burchellii Dummer   Dense shrub to 1 m. Leaves opposite, ascending-recurved, sessile, ovate. Flowers 2 at branch tips, white or pink. Throughout the year. Coastal sands, AP, SE (Gansbaai to George).•

ericoides Dummer   Diffuse shrublet to 70 cm. Leaves alternate or opposite, ascending, sessile, lanceolate. Flowers 3 or 4 at branch tips, white. Mar.–Sept. Sandstone slopes, NW (Cold Bokkeveld Mts).•

intonsa I.Williams   Like **E. meridionalis** but leaves, bracts and sepals shortly hairy and ciliolate. Apr.–Nov. Coastal limestone, AP (De Hoop).•

laevigata Turcz.   Dense shrublet to 80 cm. Leaves opposite, fleshy, ascending-recurved, subsessile, suborbicular. Flowers 2–4 at branch tips, white. July–Nov. Limestone ridges, AP (Bredasdorp).•

meridionalis I.Williams   Dense shrublet to 1.5 m. Leaves imbricate, decussate, ascending, subsessile, ovate, folded, ciliolate. Flowers 4–6 at branch tips, white or pink. Apr.–Nov. Coastal limestone, AP (Agulhas to Cape Infanta).•

pungens (Bartl. & H.L.Wendl.) I.Williams   Like **E. burchellii** but leaves somewhat pungent and fruit horns short and bifid. Mar.–Nov. Sandstone slopes, NW (Breede River valley).•

scabricosta I.Williams   Like **E. meridionalis** but leaves lanceolate. July–Nov. Coastal sands, AP (Gansbaai to Potberg).•

tricarpellata I.Williams   Diffuse shrublet to 40 cm. Leaves adpressed, subsessile, linear. Flowers 2 or 3 at branch tips, white, ovary 3-carpellate. Sept.–Nov. Sandstone slopes, NW (Piketberg).•

vallis-simiae I.Williams   Shrub to 1.2 m. Leaves ascending, elliptic, finely hairy, apex knobbed. Flowers 1–4 at branch tips, white. Aug.–Oct. Sandstone slopes, SE (Kouga and Baviaanskloof Mts).•

MACROSTYLIS•   10 spp., W Cape

barbigera (L.f.) Bartl. & H.L.Wendl.   Shrublet to 30 cm. Leaves opposite, sessile, spreading-reflexed, ovate-cordate, 6–12 mm long. Flowers to 8 in terminal, pendulous umbels, white. July–Jan. Sandstone slopes, NW (Cold Bokkeveld Mts).•

cassiopoides (Turcz.) I.Williams   Wand-like shrublet to 1.5 m. Leaves sessile, adpressed, imbricate, ovate, 5 mm long. Flowers 1–5 at branch tips, white. Aug.–Nov. Sandy flats and slopes, NW, SW (Nardouw Mts to Bellville).•

cauliflora I.Williams   Like **M. crassifolia** but leaves gland-dotted on midrib and margins only. Feb.–July. Rocky hills, SW (Elim to Bredasdorp).•

crassifolia Sond.   Like **M. tenuis** but leaves thicker with prominent margins and midrib. June–Dec. Coastal sandy flats, NW, SW (Lambert's Bay to Hopefield).•

decipiens E.Mey. ex Sond.   Like **M. villosa** but leaves smaller, 4–6 mm and flowers up to 8. Jan.–Sept. Sandstone slopes, NW (Gifberg to Ceres).•

hirta E.Mey. ex Sond.   Twiggy shrublet to 1 m. Leaves sessile, spreading, ovate-cordate, 2–5 mm long. Flowers to 7 at branch tips, white. Aug.–Sept. Sandy slopes, NW (Nardouw Mts).•

ramulosa I.Williams   Sprawling shrublet to 30 cm. Leaves imbricate, sessile, ascending, awl-shaped, softly hairy, 6–8 mm long. Flowers 4–6 at branch tips, white. Mar.–July. Sandstone slopes, NW (Cold Bokkeveld Mts).•

squarrosa Bartl. & H.L.Wendl.   Twiggy shrublet to 30 cm. Leaves sessile, recurved, elliptic, finely hairy, 2–3 mm. Flowers 5–7 at branch tips, white. Aug.–Feb. Sandstone slopes, NW (Bokkeveld to Cedarberg Mts).•

tenuis E.Mey. ex Sond.   Twiggy shrublet to 90 cm. Leaves sessile, ascending, imbricate, ovate, 2–3 mm long. Flowers to 6 at branch tips, white. Sept.–Mar. Sandstone slopes, NW (Gifberg to Worcester).•

villosa (Thunb.) Sond.   Rounded shrublet to 30 cm. Leaves subsessile, ascending, imbricate, lanceolate, 4–14 mm long. Flowers 11–14 at branch tips, white. Dec.–July. Sandy coastal flats, SW (Mamre to Cape Peninsula).•

PHYLLOSMA•   2 spp., W Cape

barosmoides (Dummer) I.Williams   Shrublet to 30 cm. Leaves obovate, ciliate, conspicuously gland-dotted. Flowers crowded at branch tips, pink. Jan.–Sept. Sandstone crevices, NW (Hex River Mts).•

capensis Bolus   Shrublet to 40 cm. Leaves linear-oblong, leathery, margins revolute, foetid. Flowers crowded at branch tips, pale pink. July–Sept. Sandstone crevices, NW (Pakhuis to Cold Bokkeveld Mts).•

PTAEROXYLON   SNEEZEWOOD, NIESHOUT   1 sp., southern Africa

obliquum (Thunb.) Radlk.   Dioecious, deciduous shrub or tree rarely to 20 m. Leaves opposite, paripinnate, leaflets in 3–7 pairs, oblique. Flowers in axillary thyrses, pale yellow. Apr.–Sept. Forests and bush, SE (Humansdorp to tropical Africa).

SHEILANTHERA• 1 sp., W Cape

pubens I.Williams   Shrublet to 40 cm. Leaves oblong, leathery, margins revolute, thinly hairy when young. Flowers crowded at branch tips, pink to white. Oct. Sandstone crevices, NW (Cold Bokkeveld Mts).•

VEPRIS   WHITE IRONWOOD, WITYSTERHOUT   c. 15 spp., Africa and Mascarene Islands

lanceolata (Lam.) G.Don   Dioecious shrub or small tree to 5 m. Leaves trifoliolate, leaflets narrowly elliptic, undulate. Flowers unisexual, in terminal panicles, yellowish. Dec.–Mar. Evergreen forests, LB, SE (Swellendam to tropical Africa).

ZANTHOXYLUM   KNOBWOOD, PERDEPRAM   c. 250 spp., Africa, E Indies, America

capense (Thunb.) Harv.   SMALL KNOBWOOD, KLEINPERDEPRAM   Dioecious, small prickly tree to 7 m. Leaves imparipinnate, leaflets obovate, lower smaller, toothed, with 4–8 pairs of veins. Flowers unisexual, in terminal or axillary panicles, whitish. Oct.–Dec. Evergreen forests, SE (George to tropical Africa).

davyi (I.Verd.) P.G.Waterman   Like **Z. capense** but leaflets lanceolate, acute, with 16–20 pairs of veins. Oct.–Jan. Evergreen forests, SE (Knysna to tropical Africa).

SALICACEAE

SALIX   WILLOW   c. 400 spp., mainly N hemisphere

mucronata Thunb.   CAPE WILLOW   Monoecious shrub or small tree to 12 m, with rough scaly bark, branches sometimes drooping. Leaves lanceolate, silvery hairy, finely toothed, paler beneath. Flowers in spikes on axillary shoots; seeds woolly. Sept.–Oct. Along rivers and streams, NW, SW, AP, KM, LB, SE (throughout southern Africa).

SALVADORACEAE

AZIMA   NEEDLE BUSH, SPELDEDORING   c. 4 spp., palaeotropics and subtropics

tetracantha Lam.   Dioecious shrub or scrambler to 9 m with 4 straight thorns at each node. Leaves more or less opposite, leathery, elliptic to suborbicular. Flowers 1–few in axillary glomerules or cymes, greenish to yellow. Oct.–Feb. Lowland scrub and bush, LB, SE (Cape Infanta to tropical Africa).

SANTALACEAE

1. Leaves opposite, well developed; flowers bisexual; fruit fleshy:
    2. Leaves subsessile, rounded or cordate at base; style relatively long; stigma 5-lobed....................**Rhoiacarpos**
    2.' Leaves shortly petiolate, cuneate at base; style relatively short; stigma 4-lobed ....**Osyris**
1.' Leaves alternate, rarely subopposite, reduced; fruit dry or fleshy:
    3. Flowers unisexual; perianth lobes and stamens 4 ....**Thesidium**
    3.' Flowers bisexual; perianth lobes and stamens 5 ....**Thesium**

OSYRIS (= *COLPOON*)   CAPE SUMACH, PRUIMBAS   6 or 7 spp., Africa to India

compressa (P.J.Bergius) A.DC. (= *Colpoon compressum* P.J.Bergius)   Glaucous hemiparasitic shrub or small tree to 5 m. Leaves mostly opposite, ovate-elliptic, leathery, margins thickened. Flowers in delicate terminal cymose panicles, greenish, buds 1.7–2.4 mm diam. Fruits a red to black drupe, to 17 × 12 mm. Dec.–June. Sandstone slopes, NW, SW, AP, KM, LB, SE (Cedarberg Mts to tropical Africa).

speciosa (A.W.Hill) J.C.Manning & Goldblatt (= *Colpoon speciosum* (A.W.Hill) P.A.Bean)   Many-stemmed resprouting shrublet to 2 m. Leaves ovate-elliptic, leathery, margins thickened. Flowers in robust terminal cymose panicles, greenish, buds 2.7–3.1 mm diam. Fruits a red to black drupe, to 25 × 21 mm. Sept.–May, after fire. Coastal sandstone and limestone flats and slopes, SW, AP (Houwhoek to Agulhas).•

RHOIACARPOS 1 sp., South Africa

capensis (Harv.) A.DC.   Straggling shrublet to 60 cm. Leaves opposite, ovate-elliptic, margins revolute and minutely crisped. Flowers in terminal cymose panicles or racemes, greenish. Fruit a red drupe. Mar.–July. Coastal bush, SE (Mossel Bay to KwaZulu-Natal).

THESIDIUM 4 or 5 spp., W and E Cape

fragile (Thunb.) Sond. (incl. **T. microcarpum** (A.DC.) A.DC. & **T. podocarpum** (A.DC.) A.DC.)   Dioecious brittle hemiparasitic shrublet to 50 cm, often yellowish. Leaves and bracts adpressed, scale-like, shorter than flowers, lower leaves sometimes longer and spreading. Flowers in spikes, greenish. Fruits whitish with orange calyx. Throughout the year. Sandy flats and slopes, SW, AP, KM, LB, SE (Little Karoo, Saldanha Bay to E Cape).

fruticulosum A.W.Hill (incl. **T. minus** A.W.Hill)   Dioecious, sexually dimorphic hemiparasitic shrublet to 40 cm. Leaves and bracts very much larger in female, spreading or ascending, lanceolate, keeled, longer than flowers. Flowers in spikes, greenish. Fruits greenish. Throughout the year. Sandstone and limestone flats and slopes, SW, AP, SE (?Namaqualand, Cape Peninsula to Humansdorp).?•

hirtum Sond.   Dioecious, sexually dimorphic roughly hairy, closely leafy hemiparasitic shrublet to 20 cm. Leaves and bracts imbricate, incurved, lanceolate, keeled, margins and midrib roughly hairy, much longer than flowers. Flowers in spikes, greenish. Fruits pale with orange calyx. Throughout the year. Coastal sandstone slopes, SW (Cape Peninsula to Potberg).•

leptostachyum (A.DC.) Sond.   Dioecious, sexually dimorphic diffuse, slender hemiparasitic shrublet to 50 cm. Leaves spreading, linear, margins revolute, bracts as long as flowers, ciliolate. Flowers in lax spikes, greenish. Throughout the year. Coastal forests, SE (Knysna).•

THESIUM   more than 300 spp., Old World

A. Tepals glabrous or fringed with minute papillae (without an apical beard)
*B. Flowers solitary in bract axil and arranged in simple racemes or spikes (see also **T. squarrosum**)*

acutissimum A.DC.   Sprawling, hemiparasitic shrublet to 30 cm, with flexuose, often grey-glaucous branches. Leaves linear, mucronate. Flowers in simple racemes, minute, white. Sept.–Jan. Marshy streambanks, SE (Great Winterhoek Mts to KwaZulu-Natal).

foliosum A.DC.   Densely leafy, hemiparasitic shrub to 90 cm with ribbed stems. Leaves linear, obtuse. Flowers in crowded spikes, whitish, with floral disc, tepals glabrous. May–Aug. Sandstone slopes, SE (Knysna to E Cape).

lineatum L.f.   Rigid, nearly leafless, hemiparasitic shrub to 2 m, with stems grooved, rigid, spine-tipped. Leaves linear, soon deciduous. Flowers in short racemes on spinescent branchlets, whitish, bracteoles like bracts and as long as or longer than flowers. Aug.–Feb. Dry stony slopes and flats, NW, KM (Namibia and Karoo to Cedarberg and Swartberg Mts).

sedifolium A.DC. ex Levyns (= *Thesium crassifolium* A.DC.)   Hemiparasitic shrublet to 20 cm, with angular branches. Leaves succulent, linear, recurved, convex beneath. Flowers sessile, in leafy spikes, bracts and bracteoles exceeding flowers, concealed by bracts, whitish. Feb.–Apr. Sandstone slopes, SW (Cape Peninsula to Riviersonderend Mts).•

spinosum L.f.   Densely branched, hemiparasitic shrub to 1 m, with spiny, often grey-glaucous branches. Leaves triangular-terete, spine-tipped. Flowers in simple racemes, whitish. Aug.–Jan. Coastal sands, NW, SW (Lambert's Bay to Yzerfontein).•

spinulosum A.DC.   Slender, hemiparasitic shrublet to 15 cm, with slender, angular, spine-tipped branches. Leaves needle-like, upper spine-tipped. Flowers in axillary, flexuose racemes, whitish. Oct.–Feb. Sandy slopes, NW, SW (Namaqualand to Caledon).

whitehillensis Compton   Sprawling, sparsely leafy, hemiparasitic shrublet to 20 cm. Leaves linear, adpressed, mucronate. Flowers in simple racemes, whitish. Aug.–Sept. Stony slopes, NW (Namaqualand to Witteberg).

B. Flowers in cymules variously arranged in loose heads, racemes or panicles

albomontanum Compton   Robust, twiggy shrublet to 30 cm. Leaves reduced, scale-like. Flowers in small globose heads, minute, white. Oct.–Jan. Rocky sandstone slopes, NW, KM (Cold Bokkeveld to Witteberg).•

archeri Compton   Robust, thorny, nearly leafless shrub to 1.5 m, with grooved stems. Leaves obsolete. Flowers in rounded, head-like clusters on short peduncles, whitish. Aug.–Sept. Stony slopes and flats, KM (Witteberg and Gamkaberg).•

asperifolium A.W.Hill   Sprawling, scabrid, hemiparasitic shrublet to 25 cm. Leaves linear to lanceolate, subacute, keeled beneath. Flowers in axillary cymules arranged in loose panicles, whitish. Flowering time? Sandstone slopes, SE (George to Transkei).

commutatum Sond.   Densely branched, nearly leafless hemiparasitic shrub, to 30 cm. Leaves of 2 kinds, few and terete below, small and subulate-lanceolate above, decurrent. Flowers sessile, few, crowded in terminal clusters, whitish. Nov.–Mar. Flats and slopes, NW, SW, LB, SE (Cold Bokkeveld to Uitenhage).•

dissitiflorum Schltr.   Sprawling, twiggy, hemiparasitic shrublet to 1 m, branchlets spine-tipped. Leaves sparse, linear-terete. Flowers in axillary cymules arranged in head-like clusters, whitish, bracts caducous. Sept.–Jan. Sandstone slopes, NW (Bokkeveld Mts to Cold Bokkeveld).•

ericifolium A.DC.   Densely leafy, much-branched, hemiparasitic shrublet to 20 cm. Leaves linear, keeled. Flowers sessile, in small terminal cymose clusters, bracts exceeding flowers, whitish, with conspicuous external glands. Sept.–Jan. Rocky slopes, NW, SW, LB, SE (Cedarberg to Uitenhage).•

euphorbioides L.   Single-stemmed, slender, hemiparasitic shrub to 2 m, with angular branchlets. Leaves ovate to suborbicular, cordate at base, glaucous. Flowers in cymules arranged in terminal, head-like clusters, cream-coloured, bracts broad, yellowish. Aug.–Jan. Sandstone slopes, SW, LB, SE (Elandskloof Mts to Uitenhage).•

galioides A.DC.   Sprawling hemiparasitic shrublet to 30 cm, with numerous short branchlets. Leaves small. Flowers in axillary cymules arranged dichotomously branched panicles, whitish. Aug.–Oct. Sandstone slopes, SW, LB, SE (Sir Lowry's Pass to Graaff-Reinet).

glomeruliflorum Sond.   Low, hemiparasitic shrublet to 40 cm with angular to winged branches. Leaves linear, flat, rounded beneath. Flowers in axillary cymules arranged in determinate racemes, whitish. May–June. Sandstone slopes, LB, SE (Swellendam to Humansdorp).•

juncifolium A.DC.   Slender, rush-like, nearly leafless, hemiparasitic shrublet to 60 cm with minutely pustulate branchlets. Leaves mostly minute, terete. Flowers in lax corymbs, whitish. Oct.–Jan. Sandstone slopes, NW, SW (Cedarberg to Bainskloof and Hex River Mts).•

leptocaule Sond.   Like **T. nigromontanum** but flowers without visible external glands. July–Aug. Sandy slopes, SE (Port Elizabeth to Transkei).

nigromontanum Sond.   Heath-like, hemiparasitic shrublet to 25 cm, with slender suberect branches. Leaves linear, fleshy, subobtuse below, sparsely leafy above and appressed, terete-lanceolate, acute. Flowers sessile, in axillary cymules aggregated in terminal panicles, whitish, with conspicuous external glands. Sept.–Nov. Sandstone slopes and flats, SW, LB, SE (Cape Flats to Sundays River).•

nudicaule A.W.Hill   Nearly leafless, divaricately branched, hemiparasitic shrublet. Leaves scale-like, ovate-triangular, margins fimbriate-ciliate. Flowers in small terminal clusters, whitish, with conspicuous external glands. Flowering time? Sandy slopes and flats, NW, SW (Olifants River valley to Hopefield).•

oresigenum Compton   Densely leafy shrublet to 45 cm. Leaves linear, upper surface flat. Flowers in axillary clusters, whitish. Feb.–Oct. Rocky upper slopes, NW (Cedarberg and Piketberg to Hex River Mts).•

pinifolium A.DC.   Leafy, hemiparasitic shrub to 2 m, with ribbed branches. Leaves terete, obtuse, slightly decurrent. Flowers in cymules arranged in terminal, head-like clusters, whitish. Oct.–Dec. Rocky sandstone slopes, SW, LB (Cape Peninsula to Riversdale).•

pseudovirgatum Levyns   Sparsely leafy, much-branched shrublet to 30 cm. Leaves minute, adpressed or weakly spreading. Flowers in small terminal clusters, whitish. Sept.–Feb. Lower slopes, SW (Cape Peninsula).•

quinqueflorum Sond.   Densely leafy, hemiparasitic shrub to 1 m, with brittle, pustulate branches. Leaves linear, with brackish subacute tips, finely pustulate. Flowers in small terminal cymose clusters, whitish. June–Sept. Sandstone slopes, SW (Sir Lowry's Pass to Houwhoek).•

scandens E.Mey.   Scrambling, hemiparasitic undershrub with flexuose trailing branches. Leaves fleshy, terete. Flowers in axillary cymules arranged in branched racemes, whitish. June–Dec. Rocky slopes in bush, SE (Uitenhage to Fort Beaufort).

schumannianum Schltr.   Sparsely leafy, hemiparasitic shrublet to 40 cm with slightly angled, purplish branches. Leaves oblong-lanceolate, obtuse to subacute, keeled or convex beneath, with thickened hyaline margins. Flowers in small terminal clusters, whitish. Aug.–Feb. Lower sandstone slopes, SW (Cape Peninsula to Houwhoek).•

squarrosum L.f.   Densely branched, hemiparasitic shrublet to 20 cm, branches angular when young. Leaves fleshy, linear, subacute, recurved at tips. Flowers in shortly pedunculate racemes, whitish, with prominent disc. Nov.–Jan. Flats or lower slopes, SE (Knysna to E Cape).

strictum P.J.Bergius   TERINGBOS   Sparsely leafy, broom-like shrub to 2 m. Leaves lanceolate to needle-like, adpressed. Flowers crowded in dense, terminal corymbs, whitish. Sept.–Feb. Sandstone slopes, NW, SW, KM, LB, SE (S Namaqualand to Grahamstown).

susannae A.W.Hill   Leafy hemiparasitic shrublet to 2.5 m, with willowy, angled branches. Leaves linear-terete, trigonous. Flowers in axillary cymules arranged in determinate racemes, whitish. Sept.–Nov. Sandstone slopes, LB (Langeberg Mts: Heidelberg).•

triflorum Thunb. ex L.f.   Scrambling, hemiparasitic undershrub with trailing branches angular and sulcate when young. Leaves linear, flat or subterete. Flowers in axillary cymules arranged in branched racemes, whitish. Nov.–Jan. Forests and coastal bush, SE (Humansdorp to N Province).

virgatum Lam. (incl. **T. corymbuligerum** A.DC.)   Much-branched, hemiparasitic, sparsely leafy shrublet to 60 cm with stems angled and minutely pustulate. Leaves small, terete. Flowers in lax corymbs, whitish. Aug.–Feb. Stony flats and lower slopes, NW, SW, LB, SE (Cederberg to Port Elizabeth).•

A. Tepals with an apical beard of stiff or woolly hairs
C. Tepals with tufts of hairs at throat, between filament bases
D. Flowers in elongated racemes or spikes

funale L.   Slender, nearly leafless, hemiparasitic shrublet to 40 cm. Leaves few, or lacking, terete, acute. Flowers in lax, elongated, terminal spikes, whitish. Jan.–Dec. Dry to marshy lower slopes, NW, SW, KM, LB, SE (Tulbagh and Cape Peninsula to Uniondale).•

macrostachyum A.DC.   Erect, nearly leafless, hemiparasitic shrub to 1 m. Leaves lanceolate to linear, flat and fleshy. Flowers 1–3 in bract axils, in elongate racemes, whitish. Sept.–Dec. Lower slopes and flats, NW, SW (Citrusdal to Cape Peninsula).•

micropogon A.DC.   Sprawling, hemiparasitic shrublet to 15 cm. Leaves subulate, with recurved subacute tips. Flowers in dense, slender spikes, whitish. Aug.–Sept. Clay slopes, SW (Caledon Swartberg).•

patulum A.W.Hill   Nearly leafless, hemiparasitic shrublet to 30 mm. Leaves lanceolate to linear, acute, often pungent. Flowers in lax, flexuose spikes, whitish. Sept.–Dec. Stony flats and lower slopes, NW, SW (Piketberg to Cape Peninsula).•

urceolatum A.W.Hill   Sprawling, glaucous, hemiparasitic shrublet to 40 cm. Leaves linear, fleshy, rounded beneath. Flowers in spikes, dense at first, whitish. Dec.–Apr. Rocky slopes, NW (Namaqualand to Bokkeveld Mts).

D. Flowers in heads or spikes
E. Bracts imbricate, conspicuous

aggregatum A.W.Hill   Hemiparasitic, nearly leafless shrublet to 50 cm. Leaves minute, terete, ascending. Flowers in oblong heads, whitish, with toothed bract margins. Aug.–Jan. Sandstone flats and slopes, NW, SW, SE (Namaqualand to Bredasdorp, Humansdorp).

bathyschistum Schltr.   Erect, hemiparasitic shrublet to 30 cm. Leaves fleshy, subterete, pungent. Flowers in short, dense, oblong capitate spikes. Nov.–Apr. Sandstone slopes, SW (Kleinrivier Mts).•

diversifolium Sond.   Densely leafy, hemiparasitic shrublet to 80 cm, with finely ribbed branches. Leaves linear-lanceolate, recurved above, sometimes glandular-hairy on margins. Flowers in crowded, ovoid spikes, whitish, with toothed bract margins. Flowering time? Sandstone slopes, SW (Elandskloof Mts).•

spicatum L.   LIDJIESTEE   Sparsely leafy, hemiparasitic shrublet to 50 cm, with angled branches. Leaves terete, acute, adpressed with spreading tips. Flowers in dense, oblong spikes, whitish. July–Dec. Sandstone slopes, SW (Cape Peninsula to Riviersonderend Mts).•

subnudum Sond.   Broom-like, hemiparasitic shrub to 60 cm, only leafy below. Leaves sparse, needle-like. Flowers in slender, terminal spikes, whitish. July–Jan. Sandstone slopes, NW, SW, KM, LB, SE (Olifants River Mts to Port Elizabeth).•

E. Bracts inconspicuous

annulatum A.W.Hill   Densely leafy, hemiparasitic shrublet to 10 cm. Leaves adpressed, linear to lanceolate, acute, rounded beneath. Flowers in small, dense, globose heads, minute, white. Flowering time? High rocky slopes, NW (Hex River Mts).•

brachygyne Schltr.   Sparsely leafy shrublet to 15 cm with slender, angular branches. Leaves scattered, spreading, subterete, acute. Flowers in small, globose, terminal clusters, whitish. Oct.–Nov. High sandstone slopes, SW (Franschhoek Mts).•

elatius A.DC.   Twiggy, hemiparasitic shrublet to 30 cm. Leaves leathery, oblong, channelled above, keeled beneath, recurved at tips. Flowers in short spikes or heads, whitish. Aug.–Oct. Sandy slopes and coastal flats, NW (Namaqualand to Olifants River valley and Bokkeveld Mts).

frisea L.   Sprawling, hemiparasitic shrublet with trailing stems rooting at nodes. Leaves linear, pungent. Flowers in cymules arranged in dense, terminal spikes, whitish, bracts with fringed margins. Aug.–Oct. Sandstone slopes and flats, NW, SW, LB, SE (Elandsbaai to Uitenhage).•

patersoniae A.W.Hill   Sparsely leafy, hemiparasitic shrublet to 25 cm. Leaves spreading, linear-subterete, acute. Flowers many in crowded ovoid heads, whitish. Flowering time? Sandstone slopes, SE (Port Elizabeth).•

C. Perianth without tufts of hairs between filaments but anthers attached to perianth by a tuft of hair
F. Flowers solitary or up to 5 at branch tips

capituliflorum Sond.   Sprawling, much-branched, hemiparasitic shrublet to 15 cm. Leaves lanceolate, terete, acute and recurved above. Flowers in small terminal clusters, whitish. Aug.–Jan. Flats and lower slopes, NW, SW (Cape Peninsula to Ceres and Riviersonderend Mts).

cuspidatum A.W.Hill   Hemiparasitic shrublet to 20 cm, with slightly ribbed branchlets. Leaves scattered, of 2 kinds, the lower terete, spreading, the upper almost scale-like, adpressed, triangular-lanceolate. Flowers in small terminal clusters, whitish. Nov.–Jan. Lower slopes, SW (Cape Peninsula to Elim).•

euphrasioides A.DC.   Stout hemiparasitic shrublet to 20 cm. Leaves of 2 kinds, the lower linear, the upper scale-like, adpressed below, recurved above. Flowers solitary or few at branch tips, whitish. Sept.–Nov. Sandstone slopes, NW, SW (Tulbagh to Cape Peninsula and Caledon).•

litoreum Brenan   Broom-like hemiparasitic shrublet to 30 cm. Leaves reduced, scale-like. Flowers 1–4 at branch tips, whitish. Dec.–Jan. Sandy coastal flats, SW (Melkbos).•

micromeria A.DC.   Slender, hemiparasitic shrublet. Leaves almost scale-like, scattered, lanceolate, recurved above. Flowers subsessile, in small terminal clusters, whitish, with exserted anthers. Flowering time? Lower slopes, SW, LB (Paarl to Swellendam).•

paniculatum L.   Slender, much-branched, hemiparasitic shrublet to 40 cm. Leaves usually few, linear, grooved above. Flowers in lax, dichotomous cymes, whitish. Oct.–Mar. Sandstone slopes, SW (Cape Peninsula to Paarl).•

rariflorum Sond.   Slender, sparsely branched, hemiparasitic shrublet. Leaves scattered, fleshy, terete, subacute. Flowers 1–3 at tips of lateral branchlets, whitish. Nov.–Mar. Marshy places, SW (Cape Peninsula to Hermanus).•

sertulariastrum A.W.Hill   Minutely puberulous, hemiparasitic shrublet with minutely puberulous branchlets. Leaves of 2 kinds, the lower terete, scattered, subobtuse, the upper subulate-lanceolate, with black, acute tips. Flowers sessile, 1–3 at branch tips, white. Flowering time? Sandy slopes, SW (Stanford to Bredasdorp).•

F. Flowers in heads, spikes or racemes

acuminatum A.W.Hill   Sprawling, much-branched, hemiparasitic shrublet to 20 cm, much-branched from base. Leaves subterete, acute. Flowers in small terminal clusters, minute, white. Mostly Sept.–Nov. Flats and lower slopes, NW, SW (Hex River Mts to Hermanus).•

capitatum L.   Densely leafy shrublet to 30 cm. Leaves adpressed to spreading, linear, with thickened pungent tip, strongly keeled, margins minutely translucent and scabrid. Flowers minute, in dense, terminal, bracteate heads, whitish. Jan.–Dec. Sandstone slopes, SW, LB, SE (Hopefield to Humansdorp).•

capitellatum A.DC.   Sparsely leafy, hemiparasitic shrublet to 30 cm with angular branchlets. Leaves linear-lanceolate, fleshy, grooved above. Flowers few in small terminal heads, whitish. Aug.–Oct. Sandstone slopes, SW (Franschhoek to Elim).•

carinatum A.DC.   Like **T. capitatum** but flowers with short style or stigma sessile. Jan.–Dec. Sandstone slopes, NW, SW, KM, LB, SE (Pakhuis to Great Winterhoek Mts).•

densiflorum A.DC.   Erect, sparsely leafy, hemiparasitic shrublet to 40 cm. Leaves linear-terete, rigid, subacute. Flowers in small terminal clusters, whitish. Oct.–Mar. Sandstone slopes, NW, SW (Clanwilliam to Bredasdorp).•

ecklonianum Sond.   Sprawling, hemiparasitic shrublet. Leaves large, linear to lanceolate, often falcate, convex beneath. Flowers few, in small terminal clusters, whitish. Sept.–Apr. Sandy flats, SW, AP (Cape Peninsula to Bredasdorp).•

fallax Schltr.   Robust hemiparasitic shrublet. Leaves flat, linear-lanceolate, with thickened hyaline margins. Flowers in head-like terminal clusters, whitish. Nov.–Dec. Sandstone slopes, SW (Bredasdorp Mts).•

fimbriatum A.W.Hill   Hemiparasitic shrublet to 30 cm. Leaves linear, subacute, adpressed, becoming spreading towards branch tips. Flowers in dense terminal clusters, whitish. Flowering time? Lower slopes, NW (mts near Tulbagh).•

flexuosum A.DC.   Straggling, hemiparasitic shrublet to 30 cm, with whip-like, grooved branches. Leaves sparse, subterete, acute. Flowers crowded in stout spikes, whitish. May–Sept. Dry lower slopes, KM, SE (Swartberg Mts to Graaff-Reinet).

glaucescens A.W.Hill   Sprawling, hemiparasitic shrub to 60 cm. Leaves mostly scale-like, linear, rigid, obtuse. Flowers in small terminal clusters, whitish, with exserted anthers. Flowering time? Dry flats, SW (Riviersonderend).•

glomeratum A.W.Hill   Slender, hemiparasitic shrublet. Leaves linear, acute, flat or concave above, sometimes scabrous on margins. Flowers in short racemes of small terminal heads, whitish. Oct.–Nov. Rocky slopes, SE (George and Langkloof).•

gnidiaceum A.DC.   Closely leafy, hemiparasitic shrublet to 50 cm. Leaves ascending, terete, with recurved, white tips. Flowers in dense terminal spikes, white. Flowering time? Sandstone slopes, SW, SE (Caledon Swartberg to Transkei).

hillianum Compton   Sprawling, hemiparasitic shrublet to 20 cm. Leaves oblanceolate. Flowers in lax racemes, whitish. Aug.–Sept. Stony slopes and flats, KM (Witteberg).•

hollandii Compton   Like **T. scabrum** but flowers in oblong spikes. Aug.–Oct. Sandstone slopes, SE (Uniondale to Port Elizabeth).•

junceum Bernh.   Sparsely leafy, hemiparasitic shrublet to 60 cm with slender whip-like stems. Leaves few, terete. Flowers in slender, lax to dense spikes, whitish. Oct.–Feb. Moist lower slopes, NW, LB, SE (Clanwilliam to Transkei).

karooicum Compton   Much-branched, pilose to scabrid, densely leafy, hemiparasitic shrublet to 35 cm. Leaves squarrose, fleshy, terete. Flowers whitish. July–Sept. Rocky sandstone slopes, NW, KM (Hex River Mts to Witteberg).•

litoreum Brenan   Broom-like hemiparasitic shrublet to 30 cm. Leaves reduced, scale-like. Flowers 1–4 at branch tips, whitish. Dec.–Jan. Sandy coastal flats, SW (Melkbos).•

microcephalum A.W.Hill   Densely leafy, hemiparasitic shrublet. Leaves adpressed, narrowly lanceolate, subacute. Flowers few, in compact terminal clusters. Flowering time? High mountain slopes, NW (Hex River Mts).•

paronychioides Sond.   Slender, hemiparasitic annual to 12 cm. Leaves few, linear, obtuse to subacute. Flowers in lax spikes, whitish. Mar.–Apr. Lower slopes, LB, SE (Swellendam to George).•

penicillatum A.W.Hill (= *Thesium helichrysoides* A.W.Hill)   Stout hemiparasitic shrub to 1 m, branches purple-angled. Leaves fleshy, linear, with slightly thickened hyaline margins. Flowers in dense terminal corymbs, whitish. Sept.–Feb. Sandstone slopes, SW, LB, SE (Caledon to Tsitsikamma Mts).•

phyllostachyum Sond.   Leafy, hemiparasitic shrublet to 25 cm. Leaves linear, recurved above, acute. Flowers in lax spikes, whitish. Aug.–Oct. Low hills, KM, LB, SE (Swartberg Mts and Riversdale to Port Elizabeth).•

prostratum A.W.Hill   Slender, prostrate shrublet. Leaves fleshy, terete. Flowers in small terminal clusters, whitish, with exserted anthers. Sept.–Oct. High sandstone slopes, NW (Cedarberg to Ceres).•

pubescens A.DC.   Sprawling, densely leafy, hemiparasitic shrublet to 30 cm, with reflexed hairs on branchlets. Leaves linear, keeled beneath, triangular in section, shortly pubescent on margins or throughout. Flowers in small, crowded leafy clusters, whitish. Aug.–Nov. Sandstone slopes, NW, SW, ?KM (Olifants River Mts to Cape Peninsula, ?Witteberg).•

pycnanthum Schltr.   Sprawling, hemiparasitic shrublet to 30 cm. Leaves linear, with a prominent midrib, rounded beneath. Flowers in small terminal heads, white. Sept.–May. Sandstone slopes, NW, SW, LB (Hex River Mts and Paarl to Swellendam).•

repandum A.W.Hill   Sprawling, hemiparasitic shrublet to 15 cm, with ridged branches. Leaves linear, acute, recurved. Flowers sessile, in small terminal heads, whitish. Aug.–Sept. Flats and lower slopes, SW (Malmesbury to Paarl).•

rufescens A.W.Hill   Sprawling, densely leafy, hemiparasitic shrublet, branches red-brown, with short reflexed hairs. Leaves spreading, linear, shortly pubescent or glabrous. Flowers in dense oblong spikes, whitish. Oct.–Nov. Stony slopes, SW, LB (Bredasdorp to Riversdale).•

scabrum L.   Densely leafy, hemiparasitic shrub to 1 m. Leaves linear, triquetrous, margins scabrid-serrulate. Flowers in dense globose heads, whitish. June–Jan. Sandstone slopes, NW, SW, AP (Hex River Mts to Cape Peninsula and Agulhas).•

selagineum A.DC.   Slender, much-branched, hemiparasitic shrublet to 20 cm, with angular to subterete branchlets. Leaves linear-lanceolate, recurved, acute. Flowers sessile, in small clusters at branch tips, whitish. Sept.–Nov. Sandstone slopes, NW (Piketberg and Cold Bokkeveld to Hex River Mts).•

sonderianum Schltr.   Densely leafy, hemiparasitic shrub to 1 m. Leaves linear, pungent, recurved, scabridulous on margins. Flowers in crowded, oblong spikes, whitish. Aug.–Sept. Sandstone slopes, SE (George to Grahamstown).

translucens A.W.Hill   Leafy shrublet to 45 cm, with angular branches. Leaves ascending, linear-terete, pungent, keeled beneath. Flowers in small, dense, bracteate heads, whitish. Apr.–July. Sandstone slopes, SW, LB (Houwhoek Mts to Riversdale).•

umbelliferum A.W.Hill   Robust hemiparasitic shrub to 2 m. Leaves fleshy, subterete, obtuse. Flowers in dense terminal corymbs, whitish. Dec.–Jan. Rocky sandstone slopes, KM, SE (Swartberg Mts and George to Humansdorp).•

viridifolium Levyns   Densely leafy, hemiparasitic shrublet to 50 cm. Leaves linear, spreading, margins scabrid. Flowers in dense heads, white, bracts nearly as long as flowers. July–Dec. Sandy flats, SW (Cape Peninsula to Caledon).•

SAPINDACEAE

1. Leaves simple:
    2. Petals (4)5(6); ovary 2- or 3-locular with a single ovule in each locule; fruit globose .... **Pappea**
    2.' Petals absent; ovary 2–6-locular with 2 ovules in each locule; fruit angled or winged .... **Dodonaea**
1.' Leaves compound (rarely 1-foliolate and apparently simple):
    3. Leaves (1)3(5)-foliolate .... **Allophyllus**
    3.' Leaves paripinnate:
        4. Fruit a samara with a broad wing .... **Atalaya**
        4.' Fruit a drupe or berry, not winged:
            5. Fruit globose, obscurely 3-lobed; leaf rachis more or less winged; sepals unequal, fimbriate or ciliate; petals simple .... **Hippobromus**
            5.' Fruit conspicuously lobed; sepals small, ovate; petals with a conspicuous appendage .... **Smelophyllum**

ALLOPHYLUS   c. 180 spp., pantropics and subtropics

decipiens (Sond.) Radlk.   BASTER-TAAIBOS   Dioecious shrub or small tree to 4 m, velvety on young parts. Leaves trifoliolate, discolorous, leaflets elliptic, crenate, vein axils furry beneath. Flowers in axillary spikes, white. Fruits globose. Nov.–Jan. Coastal bush, LB, SE (Gourits River to Mpumalanga).•

ATALAYA   WING-NUT, KRANS-ESSEBOOM   c. 9 spp., Old World tropics

capensis R.A.Dyer   CAPE WING-NUT, KAAPSE KRANS-ESSEBOOM   Small tree to 5(–10)m. Leaves paripinnate, leaflets lanceolate, undulate. Flowers in terminal panicles, white. Fruits winged. Dec.–Jan. Coastal forests, SE (Humansdorp to Port Elizabeth).•

DODONAEA   c. 60 spp., widespread, mostly Australia

angustifolia L.f. (= *Dodonaea viscosa* Jacq.)   SAND OLIVE, SANDOLIEN   Dioecious small tree to 5(–10) m. Leaves linear-oblanceolate, resinous. Flowers in small, rounded axillary and terminal panicles, greenish yellow. Fruits winged. July–Oct. Riverine thicket and rocky outcrops, NW, SW, KM, LB (Namaqualand to Stellenbosch through Little Karoo to tropical Africa).

HIPPOBROMUS BASTERPERDEPIS   1 sp., South Africa

pauciflorus (L.f.) Radlk.   Monoecious or polygamous, aromatic small tree to 5 m, velvety on young parts. Leaves paripinnate, rachis winged, leaflets obovate, toothed above, margins slightly revolute. Flowers in axillary panicles, golden velvety. Fruits globose. July–Nov. Riverine thickets and forest margins, SE (Humansdorp to Mpumalanga).

PAPPEA DOPPRUIM   1 sp., Africa

capensis Eckl. & Zeyh.   Spreading tree to 7(–13) m, velvety on young parts. Leaves oblong, minutely toothed, crowded at branch tips. Flowers in axillary racemes, yellowish. Fruits globose, velvety. Nov.–Mar. Rocky slopes and open woodland, KM, SE (S Namibia, and W Karoo, Little Karoo from Montagu to Willowmore to E tropical Africa).

SMELOPHYLLUM• BUIG-MY-NIE   1 sp., W and E Cape

capense (Sond.) Radlk.   Shrub or small tree to 4 m. Leaves imparipinnate, leaflets obliquely elliptic, deeply scalloped. Flowers in small axillary panicles, greenish. Fruits deeply 2-lobed. Dec.–Jan. Evergreen forests, SE (Baviaanskloof Mts to Port Elizabeth).•

SAPOTACEAE

SIDEROXYLON   c. 100 spp., pantropical and subtropical

inerme L.   WITMELKHOUT, UMQWASHU   Shrub or small tree to 10 m. Leaves dark green, leathery, elliptic, obtuse. Flowers 1-few in axillary clusters or cauliflorous, greenish white. Fruits fleshy, black. Dec.–June. Sand dunes and coastal bush, SW, AP, LB, SE (Cape Peninsula to tropical Africa).

SCROPHULARIACEAE

Colpias, Diascia and **Hemimeris** by K.E. Steiner, **Selago** with O.M. Hillard

1. Ovary with 1 apical, pendulous ovule in each fertile locule:
    2. Calyx 1- or 2-lobed; corolla split down the front for about half the length of tube, lacking a lower lip and expanded above into a 4-lobed upper lip; stamens 4:
        3. Calyx spathe-like, subhyaline, adnate to supporting bract ....**Hebenstretia**
        3.' Calyx 2-lobed, free from supporting bract..**Dischisma**
    2.' Calyx 3–5-lobed; corolla usually with a lower lip but if not then stamens 2:
        4. Ovary with one locule aborted or barren:
            5. Calyx adnate to supporting bract...**Microdon**
            5.' Calyx free from supporting bract:
                6. Calyx very shortly toothed; corolla limb herbaceous ....**Globulariopsis**
                6.' Calyx lobed at least halfway to base; corolla limb leathery................................**Gosela**
        4.' Ovary bilocular with both locules fertile:
            7. Leaves opposite, at least below, never on short shoots; annual or perennial herbs:
                8. Lower lip of calyx shallowly bifid; lower lip of corolla 1-lobed, minute or absent; fruit hard-walled; seed fusiform ....**Chenopodiopsis**
                8.' Calyx deeply 5-lobed; lower lip of corolla 1–3-lobed; fruit soft-walled; seed elliptic .....**Pseudoselago**
            7.' Leaves alternate, usually on short shoots; perennial herbs or shrublets:
                9. Upper corolla lip bearded ....**Cromidon**
                9.' Upper corolla lip glabrous: ....**Selago**
1.' Ovary with 4–many ovules in each locule:
    10. Leaves alternate (rarely opposite); corolla without a basal inflation or spur; flowers blue or purple:
        11. Anthers all perfect; capsule ovoid-conical ....**Peliostomum**
        11.' Anthers of posterior pair of stamens smaller than others, often empty; capsule short, obcordate ....**Aptosimum**
    10.' Leaves opposite, at least near base, or rosulate or verticillate; flowers variously coloured:
        12. Corolla tube pocketed, sacculate or spurred near base:
            13. Corolla tube very short:
                14. Stamens 2; flowers yellow ....**Hemimeris**
                14.' Stamens 4; flowers usually pink to orange:
                    15. Corolla split to base between the 2 smaller lobes; pedicels resupinate, largest corolla lobe upper-most ....**Alonsoa**
                    15.' Corolla not split; pedicels not resupinate, largest corolla lobe lowermost.......................**Diascia**
            13.' Corolla tube somewhat elongate:

16. Corolla funnel-shaped, with 2 pockets at base, yellow .... **Colpias**
16.' Corolla often personate, with 1 pocket or spur at base:
   17. Anther thecae separate, distinct; capsule loculicidal and partly septicidal; corolla pale blue, lightly pouched below .... **Charadrophila**
   17.' Anther thecae confluent; corolla usually spurred:
      18. Capsule septicidal; plants never creeping; flowers various colours .... **Nemesia**
      18.' Capsule loculicidal; plants creeping with suborbicular leaves; flowers white to mauve .... **Diclis**
12.' Corolla tube without pockets or spurs:
   19. Fruit fleshy:
      20. Ovules usually 4 in each locule; flowers solitary .... **Oftia**
      20.' Ovules many in each locule; flowers in cymes or subfasciculate:
         21. Calyx deeply 5-lobed .... **Teedia**
         21.' Calyx shallowly 3–5-lobed .... **Halleria**
   19.' Fruit capsular:
      22. Anther thecae parallel, separate; shrubs .... **Freylinia**
      22.' Anther thecae diverging, usually confluent:
         23. Anther thecae distinct, not confluent; corolla glabrous bilabiate .... **Ilysanthes**
         23.' Anther thecae confluent; corolla usually partially hairy:
            24. Calyx 3-parted; leaves densely whorled, linear, serrated in upper half; erect shrub ...... **Ixianthes**
            24.' Calyx 5-partite; plants not as above:
               25. Anticous filaments geniculate, crossing the posticous filaments at right angles; flowers solitary in the axils; marsh plants or aquatics .... **Limosella**
               25.' Anticous and posticous filaments straight, not crossing at right angles:
                  26. Posticous filaments not decurrent down corolla tube:
                     27. Posticous stamens included, inserted halfway up corolla tube or higher, anticous stamens either included or anthers just visible in mouth .... **Manulea**
                     27.' One or both pairs of stamens exserted at anthesis:
                        28. Inside of corolla tube either glabrous or with longitudinal bands of clavate hairs, sometimes coalescing or extending onto base of petals; seeds patterned with transversely elongated pits arranged like a chequer board .... **Sutera**
                        28.' Clavate hairs confined to an orange/yellow patch at base of posticous lip; seeds sinuously wrinkled in longitudinal bands .... **Trieenia**
                  26.' Posticous filaments decurrent down corolla tube, often to base:
                     29. Stigma very short, stigmatic surface more or less terminal; corolla tube cylindrical, abruptly dilated at throat:
                        30. Leaves opposite and decurrent to form ridges or narrow wings down stem; seeds black .... **Lyperia**
                        30.' Leaves alternate or if opposite then not decurrent; seeds brown .... **Jamesbrittenia**
                     29.' Stigma tongue-like, stigmatic surface comprising 2 marginal bands:
                        31. Bract adnate to pedicel and at most base of calyx tube; corolla with an orange/yellow patch at base of anticous lip; seeds pallid, greenish or amber, sinuously wrinkled .... **Trieenia**
                        31.' Bract adnate at least halfway up calyx tube (rarely less):
                           32. Calyx distinctly bilabiate, anticous lip more or less 2-toothed, posticous lip 3-toothed, strongly 5-ribbed and plicate in flower; staminodes absent; corolla actinomorphic, hypocrateriform, unmarked or symmetrically marked .... **Zaluzianskya**
                           32.' Calyx bilabiate or not, rarely plicate and then 2 staminodes present, lobing various; corolla various, often zygomorphic, markings asymmetric:
                              33. Hairs on stems always spreading, gland-tipped; seeds 3-angled or 3-winged, testa translucent and loose .... **Polycarena**
                              33.' Hairs on stems either eglandular or mixed, nearly always downward facing; seeds never winged, testa opaque and tight .... **Phyllopodium**

ALONSOA MASKFLOWER   c. 12 spp., mostly S and central America, South Africa

peduncularis (Kunze) Wettst.   Erect or decumbent perennial to 50 cm. Leaves ovate, incised-serrate. Flowers in racemes, resupinate, 2-saccate, split to base above, yellow with pink lobes, 10–12.5 mm long, filaments straight or more or less curved, anthers yellow and glabrous. Oct.–Mar. Stony slopes, KM, ?SE (Touwsberg, Rooiberg and ?Uitenhage).•

unilabiata (L.f.) Steud. (= *Diascia unilabiata* (L.f.) Benth.)   Erect annual to 40 cm. Leaves lanceolate to ovate, dentate to pinnatifid. Flowers in racemes, resupinate, split to base above, pink to orange,

12–24 mm long, filaments strongly curved and thickened below, anthers blue and ciliate. July–Sept. Coastal and inland sands or clay, NW (Elandsbaai, Bokkeveld Plateau and W Karoo).

APTOSIMUM   KAROO VIOLET   c. 20 spp., southern and tropical Africa

indivisum Burch. ex Benth.   Dwarf tufted shrublet. Leaves oblong and spine-tipped, exceeding flowers, narrowed below into a petiole. Flowers blue and violet, whitish on outside, tube filiform below. Aug.–Dec. Dry stony flats, KM (Swartberg Mts and Little Karoo, Namaqualand to Botswana).

procumbens (Lehm.) Steud.   Gnarled, woody, prostrate shrublet to 1 m diam. Leaves small, rounded, narrowed below into a petiole. Flowers blue and violet, whitish on outside, tube filiform below. Aug.–Dec. Dry stony flats, KM, SE (Swartberg Mts to Uitenhage, Namaqualand, Karoo and Botswana).

spinescens (Thunb.) F.E.Weber   Spreading shrublet with erect branches to 25 cm. Leaves linear, rigid becoming spiny with age, sessile. Flowers blue and purple, fawn on outside. Mainly Oct.–Dec. Rocky karroid slopes and flats, NW, KM (Namibia to Clanwilliam and S Karoo).

CHARADROPHILA•   CAPE GLOXINIA   1 sp., W Cape

capensis Marloth   Rosulate, hairy perennial to 10 cm. Leaves elliptic, coarsely toothed. Flowers several in a terminal raceme, nodding on long pedicels, mauve. Nov. Damp, mossy rocks, SW (Jonkershoek and Betty's Bay Mts).•

CHENOPODIOPSIS•   3 spp., W Cape

chenopodioides (Diels) Hilliard   Hairy annual to 12 cm. Leaves narrowly elliptic, closely toothed. Flowers in spikes, 4–20 mm long in fruit, white, 4-lobed, corolla tube 0.8 mm long, stamens 2. Aug.–Sept. Stony slopes, NW (Botterkloof).•

hirta (L.f.) Hilliard (= *Selago hirta* L.f.)   Hairy annual to 25 cm. Leaves obovate. Flowers crowded in spikes, 35–120 mm long in fruit, white, 5-lobed, corolla tube c. 1.5 mm long, stamens usually 2. Sept.–Nov. Moist sandy slopes, NW, SW (Matsikamma to Riviersonderend Mts).•

retrorsa Hilliard   Retrorsely hairy annual to 15 cm. Leaves oblanceolate. Flowers crowded in spikes, 10–40 mm long in fruit, white, 5-lobed, corolla tube 2–2.5 mm long, stamens 4. Oct. Sandstone slopes, SW (Kogelberg, Kleinrivier Mts and Caledon Swartberg).•

COLPIAS   KLIPBLOM   1 sp., N Cape

mollis E.Mey. ex Benth.   Softly hairy, sometimes glabrous, tufted shrublet to 20 cm. Leaves ovate, dentate. Flowers axillary on slender pedicels, funnel-shaped, 2-saccate below, yellow to white, scented, 20–30 mm diam. July–Sept.(–Dec.). Rock crevices, mostly in granite, NW (Namaqualand to Bokkeveld Mts).

CROMIDON   12 spp., South Africa and Namibia

gracile Hilliard   Sprawling, glandular-hairy, much-branched annual or perennial to 10 cm. Leaves elliptic, glandular-hairy, sometimes coarsely toothed. Flowers in small racemes, 4-lobed, white with orange patch. Dec.–Apr. Sandstone slopes, 2000–2200 m, NW, KM (Hex River and Swartberg Mts).•

microechinos Hilliard   Sprawling, hairy annual to 10 cm. Leaves elliptic, hairy, sometimes obscurely toothed. Flowers in heads, sometimes these in panicles, 4-lobed, white. Sept.–Nov. Rocky slopes, NW, KM (Namaqualand to Hex River Mts to Swartberg Mts).

plantaginis (L.f.) Hilliard (= *Polycarena plantaginea* (L.f.) Benth.)   Sprawling, shortly hairy annual to 10 cm. Leaves mostly radical, ovate. Flowers in heads, 1–few in corymbs or panicles, 5-lobed, white, bracts densely hairy above. Aug.–Nov. Shale slopes, NW (Swartruggens to Cold Bokkeveld and W Karoo).

varicalyx Hilliard   Shortly hairy annual to 10 cm. Leaves elliptic, glandular-hairy, sometimes coarsely toothed. Flowers in heads in corymbs, 5-lobed, white with orange patch, bracts glabrescent above. Sept.–Oct. Clay flats, KM (Hex River Pass and W Karoo).

DIASCIA   TWINSPURS, HORINKIES   c. 70 spp., southern Africa

In *Diascia* the morphologically lower (*anterior*) stamens are twisted to lie above the morphologically upper (*posterior*) ones. The stamens are described here in relation to their morphological position and not apparent disposition.

A. *Flowers in racemes*

parviflora Benth. (incl. **D. burchellii** Benth.)   Annual to 40 cm. Leaves ovate, serrate. Flowers in racemes, 2-saccate, greyish to reddish violet with round yellow windows, 6–12 mm long, sacs c. 2.5–3.5 mm long, posterior stamens strongly recurved just below anthers. Capsules oblong-ovate, 4–12 × 2–4.5 mm. Mainly Aug.–Oct. Renosterveld and karroid flats in loam, NW, KM, LB, SE (W Karoo and Worcester to Kouga Mts).

patens (Thunb.) Fourc. (incl. **D. dielsiana** Hiern)   Perennial to 1 m. Leaves linear to ovate, entire or with a few sharp teeth. Flowers in racemes, 2-spurred, pink to red with a yellow window, 12–17 mm long, spurs projecting backwards and down or diverging at nearly right angles, c. 4–7 mm long. Capsules ovate, 4.5–15 × 4–6 mm. (Mar.–)July–Sept. Fynbos and renosterveld in sand or clay, AP, KM, LB, SE (Bredasdorp and Huisrivier Pass to Joubertina).•

veronicoides Schltr.   Annual to 75 cm. Leaves ovate, serrate. Flowers in racemes, 2-spurred, deep violet with yellow windows, 8–13 mm long, spurs c. 3–4.5 mm long, projecting backwards and diverging, posterior stamens straight. Capsules linear, more or less straight or curved. Aug.–Sept. Renosterveld and karroid flats in clay loam, NW (Namaqualand to Porterville).

A. *Flowers axillary*
B. *Flowers spurred, spurs 3–22 mm long*

albicornis K.E.Steiner   Annual to 30 cm. Leaves ovate, obovate or elliptic, sinuate or sparsely toothed. Flowers axillary, 2-spurred, lilac with dark violet centre and no yellowish spots, limb 10–15 mm long, spurs obtuse, white, c. 6–8 mm, long. Capsules lanceolate, 10–15 × 1.5–2 mm. July–Oct. Renosterveld and streambanks in loam, LB, SE (Garcia's Pass to Joubertina).•

albiloba K.E.Steiner   Annual to 22 cm. Leaves oblong or elliptic, sinuate or dentate, sometimes pinnatifid. Flowers axillary, 2-spurred, white to reddish lilac with maroon centre and pale yellow windows, limb 9–13.5 mm long, spurs acute, c. 4–6 mm long. Capsules oblong-ovate, 7–10 × 2–2.5 mm. Aug.–Sept. Renosterveld, LB, SE (Garcia's Pass to Grahamstown).

bicornuta K.E.Steiner   Annual to 43 cm. Leaves obovate to elliptic, pinnatifid to pinnatisect. Flowers axillary, 2-spurred, greyish magenta with darker centre and 0–5 yellow spots below each upper lobe, limb 13–19 mm long, spurs mostly brown mottled purple, 14–17 mm long. Capsules ovate and falcate, 5.5–7 × 2.5–3 mm. Aug.–Sept. Karroid flats in clay, NW (Botterkloof Pass to Cedarberg Mts, ?Ceres Mts: Slab Peak).•

dimorpha K.E.Steiner   Annual to 15 cm. Leaves obovate to elliptic, dentate. Flowers axillary, 2-saccate or 2-spurred, reddish lilac with dark violet centre and 0–5 small yellow dots below each upper lobe, limb 11–12 mm long, spurs dark violet, 3–7 mm long. Capsules lanceolate, 10–14 × c. 2.5 mm. Sept.–Oct. Renosterveld, KM (Montagu to Calitzdorp).•

hexensis K.E.Steiner   Annual to 22 cm. Leaves ovate or obovate to elliptic, pinnatifid to pinnatisect. Flowers axillary, 2-spurred, light violet with greyish magenta centre and with 2 or 3 small yellow dots in an oblique line below each upper lobe, limb 11–15 mm long, spurs c. 9.5–11.5 mm long. Capsules lanceolate and falcate, 9–10.5 × c. 2.5 mm. June–Oct. Renosterveld and karroid flats in loam, KM (W and Little Karoo).

longicornis (Thunb.) Druce   Annual to 32 cm. Leaves obovate to elliptic, sinuate to pinnatisect. Flowers axillary, 2-spurred, reddish or white with deep magenta centre and a large and small yellow spot below each upper lobe, limb 9–19 mm long, spurs turned upwards or projecting downwards, short or long, 4–18 mm. Capsules linear to lanceolate and falcate, 8–10 × 1.5–2 mm. Aug.–Oct. Renosterveld in loam, NW, SW, KM (Olifants River Mts to Montagu).•

minutiflora Hiern   Annual to 35 cm. Leaves obovate to elliptic, incised to pinnatisect. Flowers axillary, 2-spurred, greyish magenta to violet with darker centre and 1 or 2 yellow patches below upper lobes, limb 7–14 mm long, spurs 1–1.5 mm long. Capsules ovate and falcate, 4–5.5 × 2–2.5 mm. July–Sept. Renosterveld or karroid flats in sandy loam, KM (Namaqualand and Swartberg Mts).

sacculata Benth.   Annual to 30 cm. Leaves ovate to elliptic, dentate, sometimes entire. Flowers axillary, 2-spurred, greyish magenta with darker centre and 2–4 yellow spots in an oblique line below each upper lobe, limb 9–11 mm long, spurs upturned and more or less straight, c. 3–3.5 mm long. Capsules linear to oblong and falcate, 10–12 × c. 2 mm. July–Oct. Renosterveld and karroid flats in clay, NW, KM, SE (Namaqualand to Worcester to Joubertina).

stenocarpa K.E.Steiner   Annual to 38 cm. Leaves ovate, elliptic or obovate, dentate, sometimes pinnatifid to laciniate. Flowers like those of **D. sacculata** but larger, limb 10–16.5 mm long, spurs 3–6 mm long, capsules slightly longer and narrower. Aug.–Oct. Fynbos in sandy loam, NW (Lambert's Bay to Clanwilliam).•

variabilis K.E.Steiner.   Annual to 26 cm. Leaves elliptic, oblong or obovate, sinuate to pinnatisect. Flowers like those of **D. sacculata** but with a yellow spot in tube below sinus of lateral and lower lobes and sometimes with much longer attenuate spurs, spurs 3.5–11 mm long. July–Sept. Fynbos in sandy loam, NW (Citrusdal to Piketberg).•

whiteheadii K.E.Steiner   Annual to 45 cm. Leaves elliptic, oblong or obovate, sinuate to pinnatisect, sometimes entire. Flowers axillary, 2-spurred, pink with deep magenta centre and 1–4 small yellow spots below each upper lobe, limb 13–20 mm long, spurs projecting backwards and diverging, 18–22 mm long. Mainly July–Sept. Fynbos in loam, NW (Bokkeveld Mts to Citrusdal).•

B. Flowers saccate, sacs to 3 mm long
C. Flowers yellow with violet centre

bicolor K.E.Steiner   Erect or decumbent annual to 30 cm. Leaves ovate or elliptic to obovate, sinuate or pinnatifid. Flowers axillary, 2-saccate, light yellow with red or violet centre and greenish yellow sacs, limb 13–23 mm long, sacs c. 1.5–3 mm long, stamens projecting forward, filaments greyish ruby with yellow hairs above. Capsules ovate-falcate. May–Oct. Succulent karroid flats in clay and silt, KM (Little Karoo: Barrydale to Zebra).•

cuneata E.Mey. ex Benth.   Like **D. bicolor** but flowers smaller, limb 6.5–10 mm long, with sacs absent or vestigial, upper half of centre greenish yellow and filaments with hairs absent or purple. Mar.–Oct., after rains. Succulent karroid flats in clay or silt, KM (Little and Great Karoo to western Free State).

decipiens K.E.Steiner   Like **D. bicolor** but sacs deep violet, stamens erect and filaments with purple hairs. May–Oct. Succulent karroid flats in clay and silt, KM (Little Karoo).•

C. Flowers not yellow
D. Fruiting pedicels S-shaped

elongata Benth.   Annual to 40 cm. Leaves lanceolate to ovate or elliptic, pinnatifid to pinnatisect. Flowers axillary, 2-saccate, greyish pink, apricot or white with red centre and yellow patch below sinus of each lobe, limb 9–21 mm long, sacs yellow, 1–1.6 mm long, posterior filaments geniculate and swollen. Capsules broadly ovate, 5–10 × 4–7 mm, fruiting pedicels S-shaped. Aug.–Oct. Renosterveld and fynbos in loam or sand, NW, SW, LB (Bokkeveld Mts to Riversdale).•

grantiana K.E.Steiner   Annual to 23 cm. Leaves ovate to elliptic, sinuate to pinnatisect. Flowers axillary, 2-saccate, reddish lilac to greyish magenta with darker centre and 2 or 4 yellow spots, limb 8–15 mm long, sacs bulbous, yellow, 1–2 mm long, tube with a horn-shaped callus behind stamens. Capsules ovate, fruiting pedicels S-shaped. Aug.–Sept. Renosterveld and fynbos in sand and loam, NW, SW, LB (Piketberg to Swellendam).•

speciosa K.E.Steiner   Like **D. grantiana** but flowers often larger, limb 11–24 mm long and greyish magenta. Sacs compressed-oblong and usually longer, c. 1.6–3 mm long, and anterior filaments strongly downcurved. Leaves pinnatifid to pinnatisect. Aug.–Sept. Renosterveld in seasonally moist loam, NW, SW (Piketberg to Bredasdorp).•

*D. Fruiting pedicels not S-shaped (see also **D. dimorpha**)*

appendiculata K.E.Steiner   Decumbent annual to 22 cm. Leaves oblong, mostly pinnatifid to pinnatisect. Flowers axillary, 2-saccate, reddish lilac, with dark red or purplish centre with yellow sacs and stamen-bearing boss, limb 7.5–14 mm long, sacs very shallow, c. 0.5–0.7 mm long, stamens erect, glabrous, posterior filaments geniculate with a sterile branch to c. 1 mm long from knee. Capsules ovate-falcate. Aug.–Sept., especially after fire. Fynbos in alluvial sand, NW (Citrusdal to Piketberg).•

arenicola K.E.Steiner   Erect or decumbent annual to 35 cm. Leaves pinnatifid to pinnatisect. Flowers axillary, 2-saccate, greyish violet with dark magenta centre, yellow sacs and a yellow stamen-bearing boss, limb 12–23 mm long, sacs c. 2–5.5 mm long, posticous stamens straight or upcurved, anterior stamens downcurved, style upcurved. Capsules ovate-falcate. Aug.–Oct. Fynbos and sandveld in sand or loam, NW (Elandsbaai to Piketberg).•

capensis (L.) Britten (incl. **D. bergiana** Link & Otto)   Like **D. arenicola** but stamens and style strongly downcurved away from upper lip. Aug.–Oct. Mainly coastal sandveld, NW, SW, AP (Piketberg to Stilbaai).•

catherineae K.E.Steiner   Erect or decumbent annual to 13 cm. Leaves ovate to elliptic, sinuate to pinnatisect. Flowers axillary, 2-saccate, pastel red with dark magenta centre and yellow sacs, stamen-bearing boss and base of lateral lobes, with a small callus-like outgrowth behind stamens, limb 8.5–10 mm long, sacs c. 1 mm long, stamens erect, posterior filament thickened and bilobed, pollen orange. Capsules ovate-falcate. Aug.–Oct. Renosterveld in clay loam, LB, SE (Barrydale to Langkloof).•

collina K.E.Steiner   Decumbent annual to 20 cm. Leaves elliptic to oblong, pinnatifid to pinnatisect. Flowers axillary, 2-saccate, greyish magenta with dark magenta centre and yellow sacs and stamen-bearing boss, limb 13–20 mm long, sacs c. 4–5 mm long, stamens yellow, posterior filaments with a prominent protuberance at bend, pollen orange. Capsules ovate-falcate. Aug.–Sept. Strandveld in sandy loam, SW (Langebaan and Postberg).•

diffusa (Thunb.) Benth.   Annual to 30 cm. Leaves ovate, elliptic, oblong to obovate, pinnatifid to pinnatisect, sometimes entire to dentate. Flowers like those of **D. appendiculata** but mostly larger, limb 12–24 mm long, with larger sacs, c. 1–2 mm long, stamens pubescent, projecting forwards and downwards and posterior filaments with a sterile branch that is expanded, rounded and flattened near tip. Aug.–Nov. Fynbos and renosterveld in sand or loam, NW, SW (Piketberg to Cape Peninsula).•

dilatata K.E.Steiner   Like **D. catherinae** but flowers without callus or reddish purple patches below upper lobes and with a sterile lobe on posterior filament that is thicker and more bulbous. Aug.–Oct. Renosterveld in loam, NW, KM (Hex River Pass to Montagu).•

ellaphieae K.E.Steiner   Like **D. pachyceras** but leaves elliptic to oblong, sinuate to dentate (not pinnatifid to pinnatisect), stamen-bearing boss reddish purple, capsules narrower, linear to lanceolate, to 2 mm wide and seeds semicircular (not nearly circular). July–Aug. Fynbos and karroid shrubland in sand or loam, NW (Matsikamma to Nardouw Mts).•

gracilis Schltr.   Annual to 30 cm. Leaves elliptic to oblong, sinuate or dentate. Flowers like those of **D. maculata** but stamens erect and capsules linear-oblong, 5–10 × 1.5–2 mm. Aug.–Sept. Fynbos in loam, NW (Bokkeveld to Olifants River Mts).•

humilis K.E.Steiner   Erect or decumbent annual to 25 cm. Leaves ovate, elliptic, oblong or obovate, dentate to pinnatifid, upper leaves pinnatisect. Flowers axillary, 2-saccate, pastel red to greyish magenta with yellow below lateral lobes, limb 6–7.5 mm long, sacs c. 1.5–2 mm long, yellow, upper filaments swollen or lobed at knee, pollen orange. Capsules ovate-falcate, 5–6.5 × 2.5–3 mm. Aug.–Oct., especially after fire. Renosterveld or fynbos in loam, NW (Namaqualand and W Karoo to Worcester).

maculata K.E.Steiner   Annual to 32 cm. Leaves ovate, elliptic or oblong, pinnatifid to pinnatisect, sometimes entire, upper leaves pinnatisect. Flowers axillary, obscurely 2-saccate, greyish magenta to peachy orange with dark magenta centre and yellow patches below upper and lower corolla lobes and stamen-bearing boss, limb 6.5–11 mm long, sacs to 1 mm long, pale greyish magenta, stamens projecting forward. Capsules ovate-falcate, 5–6 × c. 3 mm. Aug.–Oct. Fynbos or renosterveld in loam, NW (Namaqualand to Hex River Pass).

occidentalis K.E.Steiner   Like **D. arenicola** but flowers smaller, limb 9.5–13 mm long with smaller sacs, c. 2–3 mm long. Aug.–Sept. Sandveld in loose sand, NW (Lambert's Bay to Elandsbaai).•

pachyceras E.Mey. ex Benth.   Annual to 32 cm. Leaves elliptic to obovate, pinnatifid or pinnatisect. Flowers axillary, 2-saccate, greyish magenta with dark centre, yellow stamen-bearing boss and a small yellow spot below each upper corolla lobe and 3 larger spots at base of lower corolla lobe, limb 12.5–26.5 mm long, sacs c. 2.5–4 mm long, greyish magenta. Capsules 6–8.5 × 2–3 mm, ovate to oblong-ovate. July–Oct. Coastal sands, NW (S Namaqualand to Leipoldtville).•

pusilla K.E.Steiner   Like **D. collina** but flowers smaller, limb 9–13 mm long, sacs shorter and not strongly divergent, c. 2 mm long and posterior filaments thickened. Aug.–Sept. Sandveld in loose sand, NW, SW (Lambert's Bay to Langebaanweg).•

DICLIS   DWARF SNAPDRAGON   c. 10 spp., Africa and Madagascar

reptans Benth.   Spreading to prostrate, slightly hairy perennial rooting at nodes. Leaves rounded, toothed. Flowers on long pedicels in axils of upper leaves, bilabiate and spurred, white to violet. Aug.–Jan. Forests and shady places, SE (George to Mpumalanga).

DISCHISMA   FALSE SLUGWORT, BASTERSLAKBLOM   11 spp., Namibia to W Cape

*A. Shrublets (see also **D. leptostachyum**)*

ciliatum (P.J.Bergius) Choisy   Densely hairy shrublet to 40 cm. Leaves narrow, spreading, bracts densely hairy. Flowers crowded in elongate spikes, white, calyx lobes narrow, hairy on margins. Mainly Aug.–Nov. Rocky slopes and flats, NW, SW, AP, LB, SE (Lokenberg to Port Elizabeth).•

crassum Rolfe   Shrublet 30–50 cm. Leaves broadly ovate and overlapping, fleshy, crenulate. Flowers in dense spikes, white, fragrant, calyx lobes narrow and sparsely ciliate. Sept.–Oct. Sandy coastal dunes, NW (Lambert's Bay to St Helena Bay).•

fruticosum (L.f.) Rolfe   Glabrous shrublet to 50 cm. Leaves narrow, deeply toothed. Flowers in dense spikes, white, calyx lobes broadly ovate, folded in midline. Flowering time? Sandstone slopes, NW (Lambert's Bay to Piketberg).•

squarrosum Schltr.   Minutely hairy shrublet to 60 cm. Leaves broadly ovate and overlapping, sometimes minutely toothed. Flowers in compact spikes, white, calyx lobes ciliate. Aug.–Oct. Sandy flats, NW (S Namaqualand to Clanwilliam).

tomentosum Schltr.   White-woolly shrublet with sprawling to ascending branches. Leaves narrow, ascending, woolly, bracts densely woolly. Flowers in dense oblong spikes, white, calyx lobes hairy. Sept.–Oct. Rocky sandstone flats, NW (Cold Bokkeveld and Swartruggens).•

A. Annuals

arenarium E.Mey.   Annual to 10 cm, branched near base. Leaves narrow, spreading, minutely toothed, bracts ciliate below. Flowers in rounded spikes elongating in fruit, white, calyx narrow and ciliate. Aug.–Oct. Sandy flats NW, SW, AP (Clanwilliam to Bredasdorp).•

capitatum (Thunb.) Choisy   Annual to 10 cm, branched near base. Leaves linear, spreading, minutely toothed, bracts ciliate below, drawn into long leaf-like tips. Flowers in dense rounded spikes, white, calyx narrow and ciliate. Aug.–Sept. Sandy flats NW, SW (Bokkeveld Escarpment to Villiersdorp).•

clandestinum E.Mey.   Erect hairy annual, 10–20 cm. Leaves narrow, spreading, bracts broad and hairy below, drawn into leaf-like tips. Flowers in elongate spikes, tiny, white, calyx lobes narrow, hairy on margins. Aug.–Nov. Rocky slopes and flats, NW (Namaqualand to N Cedarberg Mts).

leptostachyum E.Mey.   Erect hairy annual or subshrub to 30 cm, with ascending cobwebby branches. Leaves narrow and spreading, sparsely toothed, bracts hairy. Flowers in elongate spikes, white, calyx lobes narrow, hairy on margins. Sept.–Oct. Coastal dunes, NW (Namaqualand to Lambert's Bay).

spicatum (Thunb.) Choisy   Erect hairy annual to 30 cm, with ascending branches. Leaves narrow and spreading with sparse teeth, bracts hairy. Flowers in elongate spikes, white, calyx lobes narrow, hairy on margins. July–Oct. Sandy flats, NW (S Namibia to Piketberg).

FREYLINIA   BELL BUSH, KLOKKIESBOS   8 spp., southern and central Africa

A. Leaves lanceolate

densiflora Benth. (= *Freylinia decurrens* Levyns ex Van Jaarsv.)   Erect, slender shrub to 2 m. Leaves spreading or recurved, lanceolate, channelled, 15–25 mm long. Flowers in short narrow panicles, spreading, white to purple, c. 16 mm long, petals recurving, stamens unequal, style 12–14 mm long. June–Sept. Rocky slopes, KM, SE (Montagu to Grahamstown).

lanceolata (L.f.) G.Don   HEUNINGKLOKKIESBOS   Small tree, 2–4(–6) m. Leaves linear-lanceolate, glabrous or shortly hairy, ascending, margins slightly revolute, 40–120 mm long. Flowers in densely branched panicles, cream-coloured to yellow fading orange to brown, honey-scented, 10–15 mm long, stamens unequal, style 6–10 mm long. Feb.–July. Streambanks, often on sandstone, NW, SW, AP, KM, LB, SE (Namaqualand and W Karoo to E Cape).

visseri Van Jaarsv.   Erect wand-like shrub, 1.5–3 m. Leaves oblanceolate, margins slightly revolute, 15–30 mm long. Flowers crowded in short racemes, spreading, purple, 20–30 mm long, stamens unequal, style 15 mm long. Sept.–Nov. Strandveld, NW (Aurora).•

A. Leaves elliptic to ovate

crispa Van Jaarsv.   Erect, slender shrub to 2.5 m. Leaves ovate, margins thickened and strongly crisped, 8–15 mm long. Flowers in short racemes, subpendent, purple, 22–25 mm long, stamens subequal, style exserted, 22–25 mm long. Apr.–Oct. Rocky sandstone scree, SE (Kouga Mts).•

longiflora Benth.   Erect slender shrub to 1.5 m. Leaves elliptic, 15–30 mm long. Flowers in elongate racemes, spreading, narrowly tubular, white to mauve, 15–20 mm long, stamens unequal, style half as long, 5–7 mm long. Sept.–Oct. Stony slopes, SW (Elgin: Arieskraal).•

undulata (L.f.) Benth.   Erect, rigid shrub to 2 m. Leaves ovate, glabrous or sparsely hairy, often minutely crisped or undulate, 5–15 mm long. Flowers in lax, narrow racemes, subpendent, white to purple, 15–20 mm long, stamens unequal, style 10–15 mm long. June–Dec. Shale slopes, SW, AP, KM, LB, SE (Grabouw to Port Elizabeth).•

vlokii Van Jaarsv.   Erect, slender shrub, 2–3 m. Leaves elliptic, ascending, velvety hairy, 10–20 mm long. Flowers in short racemes or panicles, subpendent, purple, 20 mm long, stamens subequal, style 17–18 mm long. July–Aug. Dry sandstone slopes in succulent karoo, KM (Rooiberg Mts).•.

GLOBULARIOPSIS•   7 spp., W Cape

adpressa (Choisy) Hilliard (= *Selago adpressa* Choisy)   ?Perennial, 15–45 cm. Leaves fascicled, linear, margins thinly hairy. Flowers in small glomerules arranged in long narrow panicles, white, back of throat yellow to orange, bracts hairy. Mainly Nov. Sandy or stony slopes, NW, SW (Piketberg to Somerset West).•

montana Hilliard (= *Selago wittebergensis* Compton)   Dwarf shrublet. Leaves opposite, fascicled, elliptic, minutely hairy. Flowers in racemes, 1–4 cm long, arranged in panicles, white, back of throat orange, bracts hairy. Nov. Rock ledges and scree, KM (Witteberg).•

obtusiloba Hilliard   Perennial to 40 cm. Leaves fascicled, linear, subglabrous. Flowers in small glomerules arranged in corymbose panicles, white, back of throat yellow, calyx lobes obtuse, bracts hairy on margins. Nov.–Jan. Habitat? NW (Grootwinterhoek Mts).•

pumila Hilliard   Annual to c. 22 cm. Leaves opposite below, weakly fascicled, narrowly elliptic, hairy beneath. Flowers in terminal glomerules, white, back of throat yellow, bracts hairy. Oct. Habitat? NW (Piketberg).•

stricta (P.J.Bergius) Hilliard (= *Selago stricta* P.J.Bergius)   Annual to 45 cm. Leaves fascicled, linear-elliptic, margins and midrib hairy. Flowers in small glomerules arranged in a corymbose panicle, white, back of throat yellow to orange, bracts hairy. Sept.–Oct. Stony or sandy slopes and flats, NW, SW (Bokkeveld Mts to Paarl).•

tephrodes (E.Mey.) Hilliard (= *Selago tephrodes* E.Mey., *S. laxiflora* Choisy)   Perennial to 1 m. Leaves fascicled, linear, subglabrous. Flowers in small racemes arranged in elongated panicles, white, back of throat yellow to orange, bracts hairy. Sept.–Oct. Sandstone slopes and flats, NW, SW (Matsikamma to Stellenbosch).•

wittebergensis Compton   Dwarf shrublet, 6–15 cm. Leaves mostly opposite and decussate, minute, elliptic-oblong, toothed, leathery, subglabrous. Flowers in terminal glomerules, white, ?back of throat yellow, bracts hairy beneath. Oct.–Dec. Rocky sandstone slopes, NW, KM (Bonteberg to Witteberg).•

GOSELA•   1 sp., W Cape

eckloniana Choisy   Shrublet to 80 cm, with slender ascending branches. Leaves narrow, in tufts. Flowers in compact spikes, corolla tube long and slender, light brown. Sept.–Dec. Sandstone slopes, NW (Olifants River and Cold Bokkeveld Mts).•

HALLERIA   TREE FUCHSIA, NOTSUNG   c. 4 spp., W Cape to tropical Africa, Madagascar

elliptica Thunb.   Shrub to 1.2 m. Leaves elliptic, sharply toothed, margins revolute. Flowers orange, corolla tubular, 4-lobed. Sept.–Apr. Rocky slopes, often near streams, NW, SW, LB, SE (Clanwilliam to Swellendam, also tropical Africa).

lucida L.   UMBINZA   Shrub or small tree to 12 m. Leaves ovate, shiny, toothed. Flowers orange or greenish yellow, corolla tubular, curved, 5-lobed. July–Feb. Inland or coastal bush or forests, NW, SW, AP, KM, LB, SE (Gifberg to tropical Africa).

ovata Benth.   Shrub to 3.5 m. Leaves ovate, bluntly serrate. Flowers orange, corolla flared, straight, 5-lobed. Aug.–Sept. Streambanks, NW (Olifants River Mts).•

HEBENSTRETIA   SLUGWORT, SLAKBLOM   c. 40 spp., southern and tropical Africa

A. Perennials or shrublets

cordata L.   Shrublet to 30 cm, with closely leafy, suberect branches. Leaves heart-shaped. Flowers crowded in short dense spikes, white, anthers subsessile, calyx large. Mainly Sept.–Feb. Coastal sands, NW, SW, LB, AP, SE (Namaqualand to Port Alfred).

dregei Rolfe   Subshrub, 25–45 cm, with slender branches. Leaves lanceolate, with prominent teeth. Flowers cream-coloured, papillate below lobes, anthers subsessile, calyx toothed above. Sept.–Dec. Stony mountain slopes, SW, LB (Genadendal to Swellendam).•

lanceolata (E.Mey.) Rolfe   Hairy shrub to 60 cm. Leaves broad and toothed, bracts hairy. Flowers on elongate spikes, white with orange marks, papillate below lobes, calyx toothed at tip. Sept.–Nov. Rocky sandstone slopes, NW, SW (Cedarberg Mts to Stellenbosch).•

paarlensis Roessler   Shrublet to 40 cm, branches erect, densely leafy. Leaves linear, slightly toothed, deflexed. Flowers in elongate spikes, white with orange marks, anthers subsessile, calyx glabrous. Sept.–Dec. Rocky sandstone slopes, NW, SW (S Cedarberg Mts to Somerset West).•

robusta E.Mey.   Shrublet with erect branches. Leaves linear, slightly toothed, spreading. Flowers in elongate spikes, white with orange to red marks, anthers subsessile, calyx glabrous. Aug.–Oct. Rocky sandstone soils, NW, SW, KM, LB, SE (Namaqualand to Uniondale and E Cape).

sp. 1 (aff. **H. paarlensis**)   Shrublet to 40 cm, branches erect, densely leafy. Leaves short, needle-like, slightly toothed, deflexed. Flowers in elongate spikes, white with orange marks, anthers stalked, calyx glabrous. Sept.–Oct. Rocky slopes, KM (Bonteberg).•

A. Annuals

dentata L.   Erect, sparsely hairy annual to 40 cm, with ascending branches. Leaves linear and toothed. Flowers white with orange marks, anthers subsessile, calyx glabrous. July–Oct. Rocky sandstone soils, NW, SW (Namaqualand to Cape Peninsula).

fastigiosa Jaroscz   Annual to 45 cm, with suberect branches. Leaves narrow, slightly toothed. Flowers in elongate spikes, white, anthers sessile, calyx glabrous. Aug.–Oct. Rocky slopes, NW, SW (Pakhuis to Cape Peninsula).•

integrifolia L.   More or less glabrous erect annual to 60 cm, with ascending branches. Leaves linear and slightly toothed. Flowers white with red or orange marks, anthers subsessile, calyx glabrous. Oct.–Jan. Rocky soils, often in grassland, LB, SE (Namibia to E Cape).

neglecta Roessler   Erect annual to 30 cm, with ascending branches. Leaves narrow, slightly toothed. Flowers in elongate spikes, white, papillate at base of lobes, anthers stalked, calyx slightly hairy on margins. July–Sept. Sandstone slopes and plateaus, NW (Bokkeveld Mts to Piketberg).•

parviflora E.Mey.   Erect, sparsely hairy annual to 30 cm, with ascending branches. Leaves linear and toothed. Flowers white with orange marks, anthers subsessile, calyx glabrous. July–Oct. Rocky sandstone soils, NW, KM (Namibia to Oudtshoorn and Karoo).

ramosissima Jaroscz   Slender annual branching from base, to 15 cm. Leaves narrow, sparsely toothed, glabrous. Flowers in elongate spikes, white, small, anthers subsessile, calyx broad and hairy. Aug.–Oct. Rocky sandstone slopes, NW, SW (Piekenierskloof to Riviersonderend).•

repens Jaroscz   Sprawling branched annual to 45 cm, branches erect terminally. Leaves narrow, sparsely toothed. Flowers in elongate spikes, white, anthers stalked, calyx glabrous. July–Oct. Sand flats and slopes, NW, SW, AP, KM, LB (Namaqualand to Albertinia).•

HEMIMERIS   YELLOW-FACES, GEELGESIGGIE   6 spp., N and W Cape

A. Flowers spurred, spurs longer than lower corolla lip

centrodes Hiern   Annual, branching from base, 2–30 cm. Leaves ovate, entire or toothed. Flowers axillary, umbelliform, 2-spurred, pale yellow with paired brown spots on upper lip, 8–14 mm long, spurs projecting backwards, c. 3–6 mm long. Aug.–Oct. Karroid flats in clay, NW, KM, SE (Bokkeveld Mts and W Karoo to Baviaanskloof Mts).

gracilis Schltr.   Annual, 2–30 cm. Leaves ovate, toothed to crenate. Flowers axillary, umbelliform, 2-spurred, yellow, 5–8 mm long, spurs divergent, c. 4–5 mm. July–Oct. Moist shaded spots around rocks or small streams, NW, KM, SE (Bokkeveld to Baviaanskloof Mts).•

A. *Flowers saccate or minutely spurred, sacs much shorter than lower corolla lip*

racemosa (Houtt.) Merrill (incl. **H. montana** L.f.)   Annual, 3.5–44 cm high. Leaves ovate, toothed, sometimes pinnatifid. Flowers enantiostylous, axillary, umbelliform, 2-spurred, yellow, c. 7.5–13 mm long, spurs c. 1.5–3 mm long. July–Oct. Coastal and inland sands and clay, NW, SW, AP, KM, LB, SE (Richtersveld to Port Elizabeth).

sabulosa L.f.   Annual, 3–50 cm high. Leaves pinnatifid, sometimes toothed. Flowers axillary, umbelliform, 2-saccate with invaginations over stamens, yellow, 9–11.5 mm long, sacs c. 1–2 mm long. July–Oct. Sandy coastal flats, NW, SW, AP (Namaqualand to Stilbaai).

sp. 1   Like **H. sabulosa** but plants clammy glandular-hairy, leaves ovate, toothed and flowers pale yellow with fruiting pedicels much longer, c. 5–6 cm long. Aug.–Oct. Granite and coastal sands, SW (Saldanha Bay).•

sp. 2   Like **H. sabulosa** but leaves ovate, toothed and flowers with 2 tail-like extensions over stamens. Aug.–Oct. Sandstone slopes, NW (Piketberg).•

ILYSANTHES   c. 50 spp., tropics and subtropics

dubia (L.) Bernh. (= *Ilysanthes gratioloides* (L.) Benth.)   Diffuse annual to 20 cm. Leaves opposite, elliptic, somewhat glossy. Flowers in axils of upper leaves, white or pale blue. Capsules deflexed. Dec.–Apr. Marshes, seeps and streamsides, SW, LB, SE (Cape Peninsula to Mpumalanga).

IXIANTHES•   WATERBOSSIE   1 sp., W Cape

retzioides Benth.   Glabrescent shrub to 2 m. Leaves linear-oblanceolate, toothed above. Flowers crowded among leaves, bilabiate, yellow. Sept.–Nov. Mountain streams with roots in water, NW, SW (Cedarberg to Elandskloof Mts).•

JAMESBRITTENIA   83 spp., mainly central and southern Africa, 1 sp. extending to India

A. *Flowers green to brown, yellow or orange*

albomarginata Hilliard   Dwarf glandular-hairy shrublet to 40 cm. Leaves small, often toothed above. Flowers axillary, tube long and inflated above, lobes orange to maroon with white margins. Jan.–Dec. Coastal limestone flats and dunes in scrub, AP (Gansbaai to Stilbaai).•

atropurpurea (Benth.) Hilliard (= *Sutera atropurpurea* (Benth.) Hiern   SAFFRAANBOSSIE   Wiry glandular shrublet to 1 m. Leaves mostly entire. Flowers axillary, long-tubed and 2-lipped with narrow lobes, yellowish to brown. Dec.–Apr. Stony or rocky slopes in karroid scrub, KM, SE (Botswana through W Karoo to Joubertina).

A. *Flowers white, pink, blue or mauve*

albanensis Hilliard   Sprawling glandular-hairy shrublet to 45 cm. Leaves lobed. Flowers axillary, pink to mauve with orange throat. Jan.–Dec. Karroid scrub, SE (Uniondale to Alice).

argentea (L.f.) Hilliard (= *Sutera argentea* (L.f.) Hiern   Erect or straggling glandular shrub to 1 m. Leaves mostly toothed. Flowers axillary, tube inflated above, white with yellow throat. Jan.–Dec. Rock outcrops or forest margins in moist, shaded places, KM, SE (Montagu to Port Elizabeth).•

aspalathoides (Benth.) Hilliard (= *Sutera aspalathoides* (Benth.) Hiern)   Spreading glandular-hairy and often glistening dwarf shrublet to 50 cm. Leaves crowded, very small, mostly entire, narrow. Flowers axillary, tube inflated above, pink to blue, rarely white, with yellow throat. Jan.–Dec. Scrub, SW, KM, LB, SE (Elim to Baviaanskloof Mts).•

calciphila Hilliard   Gnarled, very twiggy, glandular-hairy shrublet to 45 cm. Leaves minute, mostly entire, rotund. Flowers axillary, tube inflated above, pink to blue, rarely white, with yellow throat. Jan.–Dec. Coastal limestone rocks and cliffs, AP (Pearly Beach to Stilbaai).•

foliolosa (Benth.) Hilliard (= *Sutera foliolosa* (Benth.) Hiern)   Densely leafy, glandular shrublet to 40 cm. Leaves lobed, leathery. Flowers axillary among reduced leaves, purple to mauve, rarely white, with yellow throat. Jan.–Dec. Karroid scrub, KM, SE (Karoo to Alexandria).

microphylla (L.f.) Hilliard (= *Sutera microphylla* (L.f.) Hiern)   Glandular-hairy shrublet to 50 cm. Leaves small, imbricate, leathery, entire. Flowers axillary, tube inflated above, purple to mauve with red median streaks and yellow throat. Jan.–Dec. Coastal scrub or grassland, SE (Knysna to Port Alfred).

pinnatifida (L.f.) Hilliard   Straggling glandular-hairy ?perennial to 45 cm. Leaves lobed. Flowers axillary, tube inflated above, white with yellow throat. Jan.–Dec. Scrubby grassland, SE (Kouga Mts to Grahamstown).

stellata Hilliard (= *Sutera pedunculata* auct.)   Glandular-hairy shrublet to 40 cm. Leaves small, often toothed above. Flowers axillary, tube inflated above, white or pink with wedge-shaped yellow to red patch at base of each lobe. Jan.–Dec. Sheltered places on coastal limestone cliffs, SW, AP (Cape Peninsula, Struisbaai to Gouritsmond).•

tenuifolia (Bernh.) Hilliard (= *Sutera atrocaerulea* Fourc.)   Glandular-hairy shrublet to 60 cm. Leaves sparsely toothed. Flowers axillary tending to racemose, tube inflated above, purple to blue, rarely white to pink. Jan.–Dec. Sandy and rocky slopes and dunes, SE (Mossel Bay to Humansdorp).•

tortuosa (Benth.) Hilliard (= *Sutera tortuosa* (Benth.) Hiern)   Glandular-hairy shrublet to 20 cm. Leaves mostly entire, small and tufted. Flowers in racemes, white to mauve with dark median streaks. Sept.–Dec. Stony and shaly slopes or flats in karroid scrub, KM, SE (Prince Albert to E Cape).

LIMOSELLA   MUDWORT   c. 15 spp., cosmopolitan

africana Glück   Tufted or submerged annual with floating branches. Leaves long-petioled, ovate-oblong. Flowers small, petals shorter than calyx, white or lilac. Aug.–Nov. Pools and marshes, NW, SW, LB, SE (Bokkeveld Plateau to tropical Africa).

grandiflora Benth.   Like **L. africana** but flowers larger with petals longer than calyx, white or lilac with yellow throat. Aug.–Nov. Pools and marshes, NW, SW, LB, SE (Clanwilliam to Mpumalanga).

LYPERIA   WIDOW'S-PHLOX, TRAANBLOMMETJIE   6 spp., Namibia to W Cape

antirrhinoides (L.f.) Hilliard (= *Sutera antirrhinoides* (L.f.) Hiern, *S. ochracea* Hiern)   Glandular-hairy annual to 30 cm. Leaves coarsely toothed. Flowers laxly racemose, tube funnel-shaped, 14–18 mm long, white to cream-coloured with dark patch at base of each lobe, fertile stamens 2. July–Oct. Stony ground, NW, SW, KM (Piketberg to Oudtshoorn).•

formosa Hilliard   Glandular-hairy annual to 40 cm. Leaves serrate. Flowers laxly racemose, tube long and funnel-shaped, c. 24 mm long, white with orange patch at base of each lobe, fertile stamens 2. Aug. Shallow, rocky soils, NW (Langeberg Mts: Robertson).•

lychnidea (L.) Druce (= *Sutera lychnidea* (L.) Hiern)   Leafy, glandular-hairy, glistening perennial to 1 m. Leaves mostly in axillary tufts, shortly toothed above. Flowers in racemes, tube long and inflated above, 23–28 mm long, lobes narrow, greenish to yellow, clove-scented at night, fertile stamens 4. Aug.–Nov. Coastal sands in scrub, SW, AP (Saldanha Bay to Stilbaai).•

tenuiflora Benth. (= *Sutera tenuiflora* (Benth.) Hiern)   Glandular-hairy annual to 20 cm. Leaves sometimes toothed. Flowers in racemes, tube long and inflated above, 24–32 mm long, pink to mauve with yellow star in throat, fertile stamens 2. June–Oct. Sandy or gravelly flats, NW, KM (W Karoo and Swartruggens to Groot Swartberg).

tristis (L.f.) Benth. (= *Sutera tristis* (L.f.) Hiern)   Glandular-hairy annual to 60 cm. Leaves sometimes toothed. Flowers in racemes, tube long and inflated above, 20–29 mm long, lobes narrow, whitish to yellow or brown, clove-scented at night, fertile stamens 4. Mainly July–Oct. Sandy, gravelly or stony ground, often in scrub, NW, SW, AP, KM, LB (Namibia through W Cape and Karoo to Willowmore).

violacea (Jaroscz) Benth.   Sparsely glandular-hairy annual to 45 cm. Leaves coarsely toothed. Flowers laxly racemose, tube funnel-shaped, 9–15 mm long, pink to blue with dark streaks at base of each lobe, fertile stamens 2. June–Sept. Sandy or stony ground, including limestone, SW, LB, SE (Cape Peninsula, Swellendam to Uniondale).•

MANULEA   FINGER-PHLOX, VINGERTJIES   74 spp., southern and S tropical Africa, mainly W Cape

A. Stigma usually shortly exserted, much longer than style

adenocalyx Hilliard   Glandular-hairy annual to 45 cm. Leaves crowded basally, deeply toothed. Flowers crowded in capitate racemes, tube dilated above, lobes rounded, creamy white with orange centre. Aug.–Oct. Sandy slopes, NW (Piketberg to Karoopoort).•

annua (Hiern) Hilliard   Glandular-hairy annual to 30 cm. Leaves mostly toothed. Flowers in racemes, tube short and narrowly funnel-shaped, white with yellow on tube. July–Oct. Damp, sandy places, NW (Pakhuis Pass to Porterville).•

arabidea Schltr. ex Hiern   Minutely glandular-hairy annual to 21 cm. Leaves sometimes toothed. Flowers in racemes, tube shortly funnel-shaped, white with yellow tube. Aug. Sandy soil, NW (Clanwilliam).•

augei (Hiern) Hilliard   Minutely glandular-hairy annual to 15 cm. Leaves toothed. Flowers in racemes, tube cylindric, lobes broad, white to mauve with yellow tube. Aug.–Oct. Sandy lowlands, SW (Saldanha to Hopefield).•

calciphila Hilliard   Glandular-hairy annual to 15 cm. Leaves mostly basal, usually toothed. Flowers in racemes, tube shortly funnel-shaped, white with yellow tube. July–Nov. Damp sandy soil on coastal limestone outcrops, AP (Namaqualand and Bredasdorp to Stilbaai).

corymbosa L.f.   Glandular-hairy annual to 45 cm. Leaves crowded basally, toothed. Flowers crowded in capitate racemes, tube dilated above, lobes rounded, creamy white with orange centre. July–Nov. Sandy soils near the coast, SW (Velddrif to Cape Peninsula).•

derustiana Hilliard   Glandular-hairy annual to 20 cm. Leaves crowded basally. Flowers in racemes, tube shortly funnel-shaped, lobes rounded, white with yellow tube. July–Aug. Deep sands in karroid scrub, KM (De Rust).•

latiloba Hilliard   Minutely glandular-hairy annual to 12 cm. Leaves crowded basally, sometimes shallowly toothed. Flowers in racemes, tube shortly funnel-shaped, white with yellow tube. July–Oct. Damp sand or clay soils, KM (Tanqua Karoo to Ladismith).

paucibarbata Hilliard   Glandular-hairy annual to 40 cm. Leaves toothed. Flowers laxly racemose, tube narrowly funnel-shaped, white to lilac with yellow tube. Aug.–Oct. Sandy or stony slopes, NW (Botterkloof to Cedarberg Mts).•

psilostoma Hilliard   Glandular-hairy annual to 23 cm. Leaves crowded basally, toothed. Flowers in cymules crowded above, tube inflated above, lobes rounded, white with yellow centre. Aug. Sandy flats, NW (Graafwater).•

    A. *Stigma well included, shorter than to longer than style*
    B. *Stamens inserted in middle of corolla tube; tube 6.5–15 mm long*

altissima L.f.   Glandular-hairy, foetid, short-lived perennial to 1 m. Leaves crowded basally, obscurely toothed. Flowers crowded in capitate racemes, tube inflated above, lobes rounded, white with yellowish centre, scented. July–Sept. Deep sandy soils, NW, SW (Namaqualand to Malmesbury).

cephalotes Thunb.   Coarse, glandular-hairy ?perennial to 1 m, stems ridged. Leaves coarsely toothed. Flowers in panicles, tube inflated above, lobes rounded, yellow to brown. Dec.–Apr. Rocky sandstone slopes, NW (Bokkeveld Mts to Ceres).•

juncea Benth.   Glandular-hairy short-lived perennial to 75 cm, stems ridged. Leaves coarsely toothed. Flowers in moderately dense racemes, tube inflated above, lobes rounded, dull yellow to red. Jan.–May. Rocky sandstone slopes, NW (N Cedarberg Mts to Ceres).•

linearifolia Hilliard   Coarse glabrous ?perennial to 1 m. Leaves very narrow and usually entire. Flowers in panicles, tube inflated above, lobes rounded, orange. Jan.–Dec. Rocky sandstone slopes, NW (Cedarberg to Hex River Mts).•

multispicata Hilliard   Shortly glandular-hairy short-lived perennial to 70 cm, stems ridged. Leaves deeply toothed. Flowers laxly racemose, tube short and inflated above, lobes rounded, dull yellow to brown. June–Jan. Rocky sandstone slopes, NW, KM (Cold Bokkeveld to Montagu).•

rigida Benth.   Coarse, glandular-hairy, softly woody shrub to 1 m, with ridged, spreading branches. Leaves sharply toothed. Flowers in panicles, tube inflated above, lobes rounded, white with yellow tube. June–Dec. Rocky sandstone slopes in wet places, NW (Cedarberg Mts to Citrusdal).•

    B. *Stamens inserted in upper third of corolla tube, rarely in the middle but then tube to 6 mm long*

adenodes Hilliard   Glandular-hairy annual to 15 cm. Leaves lobed or toothed. Flowers laxly racemose, bilabiate, tube short and expanded above, lobes very narrow, orange. Sept.–Oct. Rocky slopes, NW (N Cedarberg Mts).•

caledonica Hilliard   Hairy somewhat bushy perennial to 75 cm. Leaves sometimes crowded below, usually toothed. Flowers in cymules, tube inflated above, reddish brown to orange. Apr.–Jan. Sandy calcareous soils, SW, AP (Stanford to Stilbaai).•

cheiranthus (L.) L.   Glandular-hairy annual to 30 cm. Leaves coarsely toothed. Flowers in racemes, bilabiate, tube short and inflated above, lobes very narrow, ochre to brown. July–Nov. Sandy and rocky slopes and flats, SW, AP, LB, SE (Malmesbury to Knysna).•

chrysantha Hilliard   Minutely glandular-hairy annual to 15 cm. Leaves crowded basally, more or less toothed. Flowers in racemes, sometimes bilabiate, tube short and inflated above, bright yellow with orange patch down back of throat. Apr.–Oct. Riverbeds and gravel patches, KM, SE (S Karoo and Swartberg Mts to Uitenhage).

decipiens Hilliard   Glandular-hairy annual to 40 cm. Leaves toothed. Flowers in cymules, bilabiate, tube short and inflated above, lobes very narrow, cream-coloured or yellow to brown or maroon. Aug.–Oct. Sandy and stony slopes, NW (Namaqualand to Clanwilliam).

exigua Hilliard   Annual to 37 cm, shortly glandular-hairy below. Leaves toothed. Flowers in cymules, bilabiate, tube short and inflated above, white with back of tube orange. July–Nov. Mainly sandy flats and slopes, SW, AP (Betty's Bay to Gansbaai).•

glandulosa E.Phillips   Perennial to 55 cm, shortly hairy below. Leaves narrow, toothed. Flowers in laxly arranged cymules, tube inflated above, lobes narrow, orange to brown. Sept.–Oct. Dry sandy flats, NW (Gifberg to N Cedarberg Mts).•

laxa Schltr.   Erect perennial to 80 cm, hairy below. Leaves greyish, mostly entire. Flowers in racemes or lax cymules, tube inflated above, lobes narrow, brown. Sept.–Oct. Mountain slopes, NW, KM, LB, SE (Klawer to George).•

leiostachys Benth.   Perennial to 1.2 m, hairy below. Leaves toothed. Flowers in laxly arranged cymules, tube inflated above, brownish. Sept.–Nov. Rocky mountain slopes, NW, SW (Cedarberg Mts to Du Toitskloof).•

minor Diels   Minutely glandular-hairy annual to 30 cm. Leaves crowded basally, usually toothed. Flowers in racemes or cymules, somewhat bilabiate, tube short and inflated above, lobes rounded, white with a yellow centre. July–Oct. Stony slopes, NW (Hex River valley).•

montana Hilliard   Perennial to 45 cm, shortly hairy below. Leaves narrow, sharply toothed. Flowers in laxly arranged cymules, tube inflated above, lobes narrow, brown. Sept.–Dec. Mountain slopes, NW (N Cedarberg Mts).•

obovata Benth.   Annual or short-lived perennial to 90 cm, hairy below. Leaves toothed. Flowers in cymules, bilabiate, tube inflated above, yellow to orange with yellow patch in back of throat, turning brown. Aug.–Dec. Sand dunes or coastal scrub, SE (Humansdorp to Port Alfred).

obtusa Hiern   Bushy, hairy perennial. Leaves mostly entire. Flowers in crowded cymes, tube inflated above, dull orange. Known only from the type, possibly a hybrid, no other information.•

ovatifolia Hilliard   Slender, hairy shrublet to 75 cm. Leaves toothed. Flowers in laxly arranged cymules, tube inflated above, yellowish brown. Oct.–Nov. Rocky sites, NW (Piketberg).•

pillansii Hilliard   Minutely hairy shrub to 1 m, stems reddish. Leaves sometimes with blunt teeth. Flowers in racemes or laxly arranged cymules, tube inflated above, lobes narrow, orange. Mar.–Sept. Sandy soils, NW (Lambert's Bay to Clanwilliam).•

praeterita Hilliard   Glandular-hairy annual to 30 cm. Leaves crowded basally, obscurely toothed. Flowers in racemes, tube short and inflated above, white to mauve with a yellow centre. May–Dec. Sandy slopes, NW (Bokkeveld Mts to Clanwilliam).•

pusilla E.Mey. ex Benth.   Minutely glandular-hairy annual to 12 cm. Leaves crowded basally, rounded and mostly entire. Flowers in racemes, bilabiate, tube short and inflated above, lobes very narrow, yellow to brown or maroon. July–Sept. Dry rocky places in sand, NW, KM (S Namaqualand to W Karoo).

rubra (P.J.Bergius) L.f.   Glabrescent perennial to 70 cm. Leaves crowded basally. Flowers in laxly arranged cymes, tube inflated above, reddish brown. May–Nov. Sandy flats near the coast, SW (Velddrif to Somerset West).•

stellata Benth.   Similar to **M. virgata** and possibly not distinct. NW (Lambert's Bay and Elandsbaai).•

thyrsiflora L.f.   Densely hairy shrub to 1 m. Leaves toothed. Flowers in bracteate cymules, bilabiate, tube inflated above, greenish to golden yellow with yellow patch in back of throat, turning brown. Aug.–Oct. Coastal dunes in scrub, SW, AP (Velddrif to Blouberg and De Hoop to Stilbaai).•

tomentosa (L.) L.   Grey-hairy perennial to 60 cm. Leaves toothed. Flowers in crowded cymes, tube inflated above, orange to brown. Aug.–Dec. Coastal sands, SW, AP (Saldanha Bay to Pearly Beach).•

turritis Benth.   Softly hairy perennial to 1 m. Leaves doubly toothed. Flowers in laxly arranged cymules, tube inflated above, brownish yellow to reddish brown. Sept.–Jan. Rocky slopes, NW, SW (Piketberg to Breede River valley).•

virgata Thunb.   Densely hairy, erect shrublet to 45 cm. Leaves toothed. Flowers in lax, delicately arranged cymules, bilabiate, tube inflated above, lobes narrow, yellow with orange patch in back of throat, turning brown. Aug.–Oct. Rocky slopes in scrub, NW (N Cedarberg and Olifants River Mts).•

MICRODON (= *AGATHELPIS*)   CAT'S-TAIL BUSH, KATSTERTBOS   7 spp., N and W Cape

A. Stamens 2

dubius (L.) Hilliard (= *Agathelpis dubia* (L.) Hutch., *A. angustifolia* Choisy)   Densely leafy shrublet to 70 cm. Leaves oblong-linear, bracts ovate, keeled, less than half as long as corolla tube. Flowers in a long spike, long-tubed, yellow often with maroon to brown lobes. Mainly Sept.–Jan. Rocky sandstone slopes, NW, SW, AP, KM, LB (Kamiesberg to Ladismith).

nitidus (E.Mey.) Hilliard (= *Agathelpis nitida* E.Mey.)   Densely leafy shrublet to 80 cm with pubescent stems. Leaves long and linear-oblanceolate, bracts ovate, keeled, shiny, more than half as long as corolla tube. Flowers in a compact spike, long-tubed, creamy white. Aug.–Sept. Rocky sandstone slopes, SW (Cape Peninsula).•

A. Stamens 4

capitatus (P.J.Bergius) Levyns (= *Microdon linearis* Choisy)   Ericoid shrublet to 60 cm. Leaves fascicled, short, needle-like, bracts large, heart-shaped. Flowers in dense ovoid heads, white with orange throat. Oct.–Dec. Rocky sandstone slopes, NW, SW (Clanwilliam to Cape Peninsula).•

orbicularis Choisy   Twiggy shrublet to 30 cm. Leaves narrowly elliptic, bracts large, rounded, papery. Flowers in dense ovoid heads, white. July–Nov. Rocky sandstone and clay slopes, NW (Bokkeveld Mts to Olifants River valley).•

parviflorus (P.J.Bergius) Hilliard (= *Agathelpis parviflora* (P.J.Bergius) Choisy, *Microdon bracteatus* (Thunb.) Hartley, *M. lucidus* (Vent.) Choisy)   Densely leafy shrub to 60 cm. Leaves ovate, imbricate, bracts heart-shaped. Flowers in elongate spikes, white. Sept.–Jan. Shale-bands at middle to upper elevations, NW, SW (Cedarberg to Hex River Mts and Du Toitskloof).•

polygaloides (L.) Druce (= *Microdon cylindricus* E.Mey.)   Ericoid shrublet to 45 cm. Leaves needle-like, ascending, bracts ovate. Flowers in elongate spikes, white. Sept.–Oct. Rocky sandstone slopes and middle to upper elevations, NW, SW, LB (Gifberg to Swellendam).•

NEMESIA   CAPE SNAPDRAGON, LEEUBEKKIES   c. 60 spp., southern Africa

A. Corolla pouched or saccate

acornis K.E.Steiner (= *Nemesia latifrons* A.L.Grant ms)   Softly hairy, short-lived perennial to 60 cm. Leaves opposite, ovate, sharply toothed, 5-veined from base. Flowers in racemes, saccate, white, lower lip with a yellow palate, upper lobes rounded. Capsules slightly longer than wide. Mainly Aug.–Dec. Sheltered rocky outcrops, NW (Piketberg).•

bodkinii Bolus   Short-lived perennial to 20 cm. Leaves opposite, elliptic-lanceolate, margins revolute, toothed. Flowers few in racemes, strongly saccate, purple to blackish, lobes rounded, velvety in throat. Capsules longer than wide. Mainly Sept.–Nov., mainly after fire. Sandstone slopes, NW, SW (Bokkeveld Mts to Hex River valley).•

glandulosa E.Phillips   Like **N. strumosa** but flowers blue, sometimes white, lower lip with inflated black velvety palate. Sept. Sandstone slopes, NW (Gifberg and Nardouw Mts).•

leipoldtii Hiern   Annual to 30 cm. Leaves opposite, ovate, toothed. Flowers in racemes, saccate, sac sometimes acute, white to mauve with a raised yellow palate, upper lobes oblong, sepals broadly ovate. Capsules as long as wide. Aug.–Sept. Clay flats, NW (W Karroo and Bokkeveld Mts to Hex River Pass).

picta Schltr.   Short-lived perennial to 30 cm. Leaves opposite, lanceolate, toothed. Flowers in lax racemes, deeply and narrowly saccate, red with a raised dark red palate, throat yellow, hairy, upper lobes oblong. Capsules as long as wide. Sept.–Nov., especially after fire. Rocky sandstone slopes, SW (Bainskloof Mts).•

strumosa Benth.   Annual to 40 cm. Leaves opposite, linear-oblanceolate, slightly toothed. Flowers in subumbellate racemes, saccate, throat coarsely hairy, white, cream-coloured, pink, mauve, sometimes red, mottled brown in throat, lobes rounded. Capsules longer than wide. Aug.–Sept. Sandy flats, often in sandveld, SW (Hopefield to Melkbos).•

sp. 1   Annual to 30 cm. Leaves opposite, ovate, toothed. Flowers in racemes, acutely saccate or minutely spurred, purple-blue with a raised yellow palate, upper lobes oblong. Capsules as long as wide. Aug.–Sept., mainly after fire. Clay slopes, NW (W Karoo to Hex River Pass).

A. *Corolla spurred*
B. *Short-lived perennials or soft shrublets; leaves mostly 3–5-veined from base*

acuminata Benth.   Diffuse, sprawling perennial to 40 cm. Leaves opposite, ovate, 3–5-veined from base, sharply toothed. Flowers in terminal and axillary flexuose racemes, pedicels filiform, white to lilac, lower lip with raised velvety orange palate, hairy within, upper lobes rounded, spur c. 2 mm long. Capsules triangular. Mainly Oct.–Feb. Sheltered sandstone slopes often near streams, NW, SW (Cold Bokkeveld to Hottentots Holland Mts).•

diffusa Benth. (incl. **N. anfracta** Hiern, **N. brevicalcarata** Schltr.)   Short-lived perennial to 30 cm. Leaves opposite, lanceolate, toothed, 3–5-veined from base. Flowers often in flexuose racemes, lilac with darker veins and a raised velvety yellow palate, upper lobes oblong, spur c. 2.5 mm long. Capsules longer than wide. Mainly Sept.–Feb. Sandstone slopes, NW, SW, LB, SE (Cedarberg Mts to Humansdorp).•

fruticans (Thunb.) Benth. (= *Nemesia capensis* (Spreng.) Kuntze, *N. foetens* Vent.)   Shrublet to 40 cm. Leaves opposite, linear-lanceolate, slightly toothed, margins slightly revolute. Flowers in racemes, pink or lilac with a raised yellow palate, hairy in throat, upper lobes oblong, spur c. 4 mm long. Capsules as long as or longer than wide. Mainly Sept.–Nov. Stony slopes, KM, LB, SE (widespread in southern Africa).

macrocarpa (Aiton) Druce   Short-lived perennial to 40 cm. Leaves opposite, ovate, 3–5-veined from base, sharply toothed. Flowers crowded in upper axils, white to pink, lips subequal, lower lip with a raised velvety palate, upper lobes rounded, spur c. 2 mm long, sepals leaf-like. Capsules slightly longer than wide. Flowering time? Sheltered slopes and forest margins, SW, LB (Cape Peninsula to Langeberg Mts).•

melissifolia Benth.   Like **N. petiolina** but flowers 4 at some nodes. Sept.–Dec. Forest margins, and sheltered slopes, LB, SE (Langeberg Mts: Swellendam to KwaZulu-Natal).

petiolina Hiern   Short-lived perennial to 50 cm. Leaves opposite, distinctly petiolate, ovate-rhombic, sharply serrate, several-veined from base. Flowers in racemes, white with grey veins, lower lip with a raised velvety palate, upper lobes small, rounded, spur c. 3 mm long. Capsules longer than wide. Mainly Nov.–Dec. Sheltered slopes and forest margins, NW, SW, LB, (Cold Bokkeveld to E Cape).

sp. 2 (= *Nemesia elata* A.L.Grant ms)   Like **N. diffusa** but more robust, to 60 cm, leaves ovate, sharply toothed. Flowering time? Moist rocky slopes, SE (Outeniqua Mts).•

B. *Annuals; leaves mostly with a single main vein from base*
C. *Upper corolla lobes linear to narrowly oblong, often white with lower lip yellow*

anisocarpa E.Mey. ex Benth.   Annual to 40 cm. Leaves opposite, elliptic-lanceolate, slightly toothed. Flowers in racemes, white or pink, sometimes orange, with yellow lower lip and raised palate, velvety within, upper lobes narrowly oblong, spur c. 3 mm long. Capsules longer than wide, oblique at base. Aug.–Sept. Stony flats, NW, KM (Namaqualand to Bokkeveld Mts and W Little Karoo).

bicornis (L.) Pers.   Diffuse annual to 80 cm, branching above. Leaves narrowly elliptic-lanceolate, toothed to pinnatifid. Flowers in branched racemes, white to pale lilac with grey veins, lower lip with 4 velvety bosses, upper lobes linear-oblong, spur swollen, c. 4 mm long. Capsules triangular. Mostly July–Sept. Coastal sands, NW, SW, AP (Namaqualand to Stilbaai).

cheiranthus E.Mey. ex Benth. (incl. **N. pulchella** Schltr. ex Hiern)   Annual to 40 cm. Leaves opposite, elliptic-lanceolate, slightly toothed. Flowers in racemes, white, sometimes marked purple, with yellow lower lip with a slightly raised palate with 2 velvety bosses, upper lobes linear, elongate, spur 3–5 mm long. Capsules as long as wide. Aug.–Sept. Mainly sandy slopes and flats, NW (Bokkeveld Mts to Piketberg).•

ligulata E.Mey. ex Benth. (incl. **N. calcarata** E.Mey. ex Benth., **N. macroceras** Schltr.)   Annual to 40 cm. Leaves opposite, elliptic-lanceolate, slightly toothed. Flowers in racemes, white and yellow to orange or blue, sometimes entirely yellow, palate with 2 velvety bosses, upper lobes linear to narrowly oblong, spur curved and often blunt or slightly swollen, 6–12 mm long. Aug.–Oct. Sandy slopes and flats, NW, SW, KM (Namaqualand to Hex River valley and Bonteberg).

sp. 3   Sprawling annual to 20 cm. Leaves opposite, ovate to orbicular, distinctly petiolate below, toothed. Flowers axillary and in short racemes, white with a yellow palate and orange spur, lower lip with 2 velvety bosses, upper lobes narrowly oblong, spur c. 3 mm long. Capsules longer than wide. Sept.–Nov. Limestone slopes, AP (Agulhas coast).•

C. Upper corolla lobes oblong to rounded

affinis Benth. (= *Nemesia versicolor* auct.; incl. **N. floribunda** Lehm., **N. psammophila** Schltr.)   Annual to 30 cm. Leaves opposite, elliptic-lanceolate, toothed. Flowers in racemes, white, blue, yellow, sometimes red with a raised cream-coloured to yellow palate with 2 velvety bosses, upper lobes oblong, spur 3–5 mm long. Capsules as long as to slightly longer than wide. Aug.–Nov. Sandy and granite slopes and flats, NW, SW, AP, KM, SE (S Namibia to E Cape).

barbata (Thunb.) Benth. (= *Nemesia guthriei* Hiern; incl. **N. grandiflora** Diels)   Annual to 30 cm. Leaves opposite, ovate, toothed. Flowers in compact racemes, white to cream-coloured with blue to blackish lower lip and a raised hairy palate, upper lobes small, rounded, spur short, obtuse, to 2 mm long. Capsules longer than wide. Aug.–Oct., often after fire. Sandy flats and slopes, NW, SW, AP, KM, LB (Kamiesberg to Riversdale).

deflexa K.E.Steiner   Softly glandular-hairy annual or short-lived perennial to 20 cm. Leaves opposite, broadly ovate, toothed, the lower petiolate. Flowers in racemes, white with red at base of upper lobes, lower lip with a raised yellow rugose palate, spur c. 4.5 mm long. Capsules on deflexed pedicels, slightly longer than wide. Oct.–Mar. Sheltered slopes and streambanks, KM, LB, SE (Swartberg and Langeberg Mts to Langkloof).•

gracilis Benth.   Slender annual to 25 cm. Leaves opposite, ovate-elliptic, slightly toothed. Flowers in racemes, tiny, orange, lower lip with 2 raised velvety bosses, upper lobes minute, oblong, spur c. 2 mm long. Capsules small, triangular. July–Sept. Sandy and gravel flats, NW, SW (Bokkeveld Mts to Paarl).•

lucida Benth.   Annual to 30 cm. Leaves opposite, ovate, toothed. Flowers in lax racemes, tiny, white with black streaks, palate with 2 papillose bosses, upper lobes rounded, spur 0.5–1.5 mm long. Capsules oblong, twice as long as wide. Mainly Aug.–Oct., mainly after fire. Mainly clay and gravel slopes, NW, SW (Michell's Pass to Cape Peninsula).•

micrantha Hiern   Leafy annual to 15 cm. Leaves opposite, lanceolate, slightly toothed. Flowers in racemes, white with grey veins, upper lobes oblong, with spur c. 1 mm long. Capsules unknown. June–Aug. Wet rocks, SW (Cape Peninsula).•

pageae L.Bolus   Annual to 15 cm, sometimes to 40 cm. Leaves opposite, elliptic-lanceolate, toothed. Flowers in racemes, orange with yellow palate, lower lip narrowed proximally, with inflated, velvety boss, upper lobes oblong, spur blunt, 1–2 mm long. Capsules longer than wide. Aug.–Sept. Clay flats and slopes, NW, KM (Worcester and Robertson to Ladismith).•

pinnata (L.f.) E.Mey. ex Benth.   Slender annual to 25 cm. Leaves opposite, narrowly pinnatisect. Flowers in racemes, orange-yellow, sometimes white, lower lip with a velvety palate with 2 raised bosses, upper lobes minute, rounded, spur c. 1 mm long. Capsules small, triangular. Aug.–Sept., often after fire. Sandstone slopes, SW, LB (Cape Peninsula to Riversdale).•

[**Excluded species**   Incompletely known and probably conspecific with one of the above: **N. parviflora** Benth.; outside our area: **N. pubescens** Benth.]

OFTIA   LAZY BUSH, SUKKELBOSSIE   3 spp., Namaqualand to E Cape

africana (L.) Bocq.   Sprawling, roughly hairy shrublet with trailing branches to 1 m. Leaves large, 10–40 mm long, toothed. Flowers in axils of upper leaves, white, fragrant. Mainly Sept.–Dec. Rocky sandstone and granite slopes, NW, SW, AP, KM, LB, SE (Bokkeveld Mts to Uitenhage).•

glabra Compton   Sprawling, glabrous shrublet with ascending branches to 50 cm. Leaves small, 7–12 mm long, toothed, spreading. Flowers in axils of upper leaves, white to pale mauve. Sept.–Oct. Rocky sandstone slopes, KM (Witteberg and Warmwaterberg).•

PELIOSTOMUM   c. 7 spp., Africa

leucorrhizum E.Mey. ex Benth.   Laxly branched shrublet to 25 cm. Leaves linear to narrowly obovate, nearly glabrous. Flowers purplish, anthers long-hairy. Sept.–Mar. Karroid flats, KM (Namaqualand to Botswana and S tropical Africa).

virgatum E.Mey. ex Benth.   Laxly branched shrublet to 30 cm. Leaves narrowly ovate, glandular-hairy. Flowers violet, anthers shortly hairy. Aug.–Nov. Stony slopes and flats, NW (S Namibia and W Karoo to Clanwilliam).

PHYLLOPODIUM CAPEWORT, OPSLAG 26 spp., Namibia and South Africa, mostly W Cape

A. Flowers in rounded heads

alpinum N.E.Br. (= *Polycarena alpina* (N.E.Br.) Levyns) Hairy perennial. Leaves elliptic-spathulate, thick-textured, glandular-punctate, sometimes toothed. Flowers massed in crowded heads on leafy twiglets, pink, mauve or white. Oct.–Dec. Sandstone slopes, 600–1750 m, NW, SW (Skurweberg to Riviersonderend Mts).•

capillare (L.f.) Hilliard (= *Polycarena capillaris* (L.f.) Benth., *P. parvula* Schltr.) Glandular-hairy annual to 26 cm. Leaves elliptic-oblanceolate, slightly toothed. Flowers in compact heads on nude peduncles, tube very short, white. July–Oct. Sandy or stony places below 120 m, NW, SW, AP (Lambert's Bay to Albertinia).•

cephalophorum (Thunb.) Hilliard (= *Polycarena cephalophora* (Thunb.) Levyns) Hairy annual to 30 cm. Leaves oblanceolate, slightly toothed. Flowers many in crowded heads arranged in corymbs, mauve, pink or white. Sept.–Oct. Sandy flats, below 300 m, NW, SW (S Namaqualand to Cape Peninsula).

cordatum (Thunb.) Hilliard Hairy annual to 30 cm. Leaves mostly basal, ovate-elliptic, toothed. Flowers in small heads elongating into racemes on nude scapes, white or cream-coloured. July–Oct. Sandy or stony places up to 750 m, NW, SW, LB (Bokkeveld Mts to Albertinia).•

heterophyllum (L.f.) Benth. (= *Polycarena heterophylla* (L.f.) Levyns, *P. capitata* (L.f.) Levyns) Annual to 30 cm. Leaves elliptic, toothed. Flowers in small heads elongating into racemes, cream-coloured to pale yellow with patches of orange at base of upper lobes and inside tube. Aug.–Sept. Sandy flats and slopes, NW, SW, AP (S Namaqualand to Port Beaufort).

mimetes Hilliard Glandular-hairy annual to 20 cm. Leaves basal, ovate-elliptic, slightly toothed. Flowers in rounded heads on long, bare peduncles, mauve. Sept.–Oct. Sandy soils, NW, SW (Aurora to Mamre).•

phyllopodioides (Schltr.) Hilliard (= *Polycarena selaginoides* Schltr. ex Hiern) Hairy annual to 25 cm. Leaves oblanceolate, slightly toothed. Flowers few to several in crowded heads arranged in corymbs, mauve or rarely white. July–Sept. Sandy flats, below 300 m, NW, SW (S Namaqualand to Saldanha Bay).

pubiflorum Hilliard Hairy annual to 15 cm. Leaves basal, elliptic-oblanceolate, slightly toothed. Flowers in crowded solitary heads on long nude peduncles, finely hairy on outside of tube, yellow. Sept.–Oct. Sandstone slopes, NW (Pakhuis Mts).•

A. Flowers in elongated racemes

anomalum Hilliard (= *Polycarena plantaginea* auct.) Minutely glandular-hairy annual to 25 cm. Leaves ovate-elliptic, slightly toothed, grading into floral bracts. Flowers in leafy racemes, white, stamens only 2. July–Sept. Sandy and gravelly places, NW, KM (Namaqualand and W Karoo to Montagu).

bracteatum Benth. (= *Polycarena bracteata* (Benth.) Levyns) Hairy annual to 30 cm. Leaves elliptic-ovate, toothed, grading into floral bracts. Flowers in lax leafy racemes, white with orange patch at base of upper lobes. Jan.–Dec. Damp, sandy places, SE (Mossel Bay to KwaZulu-Natal).

caespitosum Hilliard Hairy tufted, cushion-like annual to 5 cm. Leaves elliptic, slightly toothed. Flowers in heads rapidly elongating into racemes, tube very short, white possibly with yellow patch at base of upper lobes. Sept.–Oct. Rock ledges and shallow basins, 1200–2075 m, NW (Cedarberg and Hex River Mts).•

cuneifolium (L.f.) Benth. (= *Polycarena cuneifolia* (L.f.) Levyns) Hairy annual to 40 cm. Leaves elliptic-ovate, toothed, sharply differentiated from bracts. Flowers in rounded heads rapidly elongating into long racemes, mauve with orange patch at base of upper lobes. Jan.–Dec. Damp sandy places, SE (Humansdorp to Kentani).

diffusum Benth. (= *Polycarena diffusa* (Benth.) Levyns) Hairy annual to 35 cm. Leaves elliptic-ovate, toothed. Flowers in crowded heads rapidly elongating into lax racemes, white with orange patch at base of upper lobes and violet patch in tube. July–Jan. Sandy grassland, SE (Uitenhage to King William's Town).

dolomiticum Hilliard Twiggy glandular-hairy perennial to 30 cm. Leaves elliptic-ovate, slightly toothed. Flowers few in lax racemes, white or light mauve with orange patch at base of upper lobes. Aug. Dolomite hills in low scrub, KM (Swartberg Mts).•

elegans (Choisy) Hilliard (= *Polycarena linearifolia* (Bolus) Levyns, *Selago elegans* Choisy) Glandular-hairy annual or short-lived perennial to 45 cm. Leaves linear-oblanceolate, slightly toothed, often seemingly fascicled below. Flowers in crowded heads rapidly elongating into long spikes, white ?with yellow patch at base of upper lobes. Jan.–Dec. Sandstone slopes, KM, LB, SE (Napier and Montagu to Humansdorp).•

micranthum (Schltr.) Hilliard   Glandular-hairy annual to 15 cm. Leaves elliptic-oblanceolate, toothed. Flowers in slender racemes on bare peduncles, tube very short, white or cream-coloured. Sept.–Oct. Rocky sandstone slopes or cliffs, NW, KM (Cedarberg Mts to Anysberg).•

multifolium Hiern (= *Polycarena multifolia* (Hiern) Levyns)   Glandular-hairy annual to 37 cm. Leaves ovate-oblanceolate, often seemingly fascicled below. Flowers in crowded heads rapidly elongating into spikes, blue or mauve with yellow patch at base of upper lobes. Sept.–Nov. Mountain scrub, SW, LB, SE (Potberg to George).•

rustii (Rolfe) Hilliard (= *Selago rustii* Rolfe)   Glandular-hairy annual to 30 cm. Leaves elliptic, toothed, upper leaves progressively narrower. Flowers in crowded heads rapidly elongating into somewhat lax racemes, white or mauve with orange patch at base of upper lobes. Aug.–Nov. Loamy or deep sand mostly below 250m, LB SE (Swellendam to Hankey).•

tweedense Hilliard   Like **P. micranthum** but flowers larger, with prominent dark anthers. Sept.–Oct. Sandy slopes, NW, KM (Swartruggens to Tweedside).•

viscidissimum Hilliard   Like **P. micranthum** but to 6 cm, bracts glandular-pubescent. Oct. Sandstone ledges, NW (Swartruggens).•

POLYCARENA   CAPE-PHLOX   17 spp., N and W Cape

A. Corolla tube pubescent

aemulans Hilliard   Glandular-hairy annual to 13 cm. Leaves broad and coarsely toothed. Flowers few in lax racemes, often small and autogamous, tube hairy, cream-coloured with yellow patch at base of upper lobes, flushed purple on back and on tube. Aug.–Oct. Sandy places among rocks, NW (Bokkeveld Mts to Cold Bokkeveld).•

capensis (L.) Benth.   Glandular-hairy annual to 28 cm. Leaves narrow and mostly toothed. Flowers in corymbosely arranged heads, tube hairy, cream-coloured to yellow, darker around throat. Aug.–Oct. Sandy soils, SW (Hopefield to Cape Peninsula).•

gilioides Benth.   Glandular-hairy annual to 20 cm. Leaves sometimes coarsely toothed. Flowers in small, terminal clusters, tube hairy, white or cream-coloured with orange patch below upper lobes, flushed purple on back. Aug.–Oct. Sandy soils, NW, SW (Gifberg to Paarl).•

lilacina Hilliard   Shortly glandular-hairy annual to 28 cm. Leaves narrow and often toothed. Flowers in corymbosely arranged heads, tube hairy, white or pale mauve with yellow patch at base of upper lip. Sept.–Oct. Sandy places below 200 m, NW, SW (Piketberg to Bokbaai).•

nardouwensis Hilliard   Glandular-hairy annual to 16 cm. Leaves narrow and often toothed. Flowers in small heads, tube long and hairy, cream-coloured flushed orange in throat, purplish on back. Aug.–Sept. Sandstone slopes, NW (Nardouw Mts).•

silenoides Harv. ex Benth.   Glandular-hairy annual to 14 cm. Leaves often coarsely toothed. Flowers in spikes, tube short and minutely hairy, cream-coloured. Aug.–Sept. Granite slopes, SW (Cape Peninsula).•

A. Corolla tube glabrous

aurea Benth.   Shortly glandular-hairy annual to 20 cm. Leaves narrow and sparsely toothed. Flowers in small terminal heads, tube short and glabrous, yellow or upper lip white tipped yellow. Sept. Sandy or clay flats and slopes, NW (Bokkeveld Plateau to Ceres and W Karoo).•

batteniana Hilliard   Glandular-hairy annual to 18 cm. Leaves obscurely toothed. Flowers in terminal heads, tube mostly glabrous, cream-coloured to white with yellow patch at base of upper lip. Sept. Sandy slopes, NW (Garies to Wuppertal).

exigua Hilliard   Glandular-hairy annual to 13 cm. Leaves broad and mostly coarsely toothed. Flowers few in lax racemes, tube short and mostly glabrous, white or cream-coloured with yellow patch at base of upper lobes, reddish on back. Aug.–Oct. Damp silty places, NW (Cedarberg Mts).•

formosa Hilliard   Glandular-hairy annual to 20 cm. Leaves narrow and mostly toothed. Flowers in terminal heads, tube mostly glabrous, cream-coloured to yellow, upper lobes tipped orange and with orange patch at base. Aug.–Sept. Sandy or clay slopes or flats, NW (Bokkeveld to Pakhuis Mts).•

gracilis Hilliard   Glandular-hairy annual to 23 cm. Leaves narrow and coarsely toothed. Flowers in lax racemes, tube glabrous, white to cream-coloured with orange patch at base of upper lobes. Aug.–Nov. Sandstone slopes, NW (Pakhuis Pass to Piketberg).•

pubescens Benth.   Glandular-hairy annual to 28 cm. Leaves narrow and obscurely toothed. Flowers in small terminal heads, tube short and glabrous, white with yellow patch in throat. July–Sept. Moist rocky sites, NW, KM (Namaqualand and W Karoo to Montagu).

rariflora Benth.   Glandular-hairy annual to 25 cm. Leaves sparsely toothed. Flowers in lax spikes, often cleistogamous, tube short and glabrous, white or cream-coloured with orange patch at base of upper lobes and lobes often tipped orange. July–Oct. Moist sandy patches, NW, KM (Namaqualand to Malmesbury, W Karoo to Outeniqua Mts).

subtilis Hilliard   Glandular-hairy annual to 24 cm. Leaves narrow, often toothed. Flowers in small heads, tube mostly glabrous, yellow with orange throat. July–Sept. Sandy flats, NW (Clanwilliam to Piketberg).•

PSEUDOSELAGO•   POWDERPUFF   28 spp., N and W Cape

A. Corolla tube narrowly cylindric, abruptly expanded above

arguta (E.Mey.) Hilliard (= *Selago arguta* E.Mey.)   Perennial to 45 cm, stems narrowly winged. Leaves elliptic, toothed, bracts minutely glandular-hairy beneath. Flowers in loose corymbs, tube narrowly cylindric, upper lip 2-lobed, white or mauve, anthers included. Oct.–Nov. Sandy slopes, NW (Nieuwoudtville to Cold Bokkeveld).•

burmannii (Choisy) Hilliard (= *Selago burmannii* Choisy)   Annual to 80 cm. Leaves short, elliptic, adpressed, bracts glabrous. Flowers in compact corymbs, tube narrowly cylindric, upper lip 2-lobed, white or mauve with orange patch. Nov.–Dec. Rocky slopes, NW (N Cedarberg Mts and Olifants River valley).•

candida Hilliard   Annual to 75 cm, hairy below. Leaves narrowly oblanceolate, toothed, subpetiolate, bracts glabrous. Flowers in loose corymbs, tube narrowly cylindric, upper lip 2-lobed, white with orange patch. Sept.–Dec. Stony slopes, LB (Swellendam to Riversdale).•

densifolia (Hochst.) Hilliard   Like **P. subglabra** but densely leafy. Oct.–Dec. Damp slopes, NW, SW (Pakhuis Mts to Du Toitskloof).•

rapunculoides (L.) Hilliard   Like **P. subglabra** but completely glabrous. Oct.–Feb. Sandy flats, SW, AP (Cape Peninsula to Pearly Beach).•

subglabra Hilliard   Annual to 80 cm. Leaves linear, sometimes toothed, margins often sparsely hairy, bracts glabrous. Flowers in compact corymbs, tube narrowly cylindric, upper lip 2-lobed, white or mauve with orange patch. Oct.–Dec. Sandy slopes, NW, SW (Piketberg and Grootwinterhoek to Hottentots Holland Mts).•

sp. 1   Like **P. burmannii** but leaves deeply toothed. Oct.–Feb. Rocky slopes, NW, SW (Piketberg, Cold Bokkeveld to Hex River Mts).•

A. Corolla tube flared, gradually widening above
B. Upper lip 4-lobed

bella Hilliard   Prostrate deflexed-hairy perennial to 15 cm. Leaves obovate, toothed above, bracts hairy. Flowers crowded in spikes, tube funnel-shaped, upper lip 4-lobed, mauve with orange patch. Jan. Stony slopes, SW, KM (Wemmershoek to Riviersonderend and Swartberg Mts).•

diplotricha Hilliard   Glandular-hairy perennial to 75 cm. Leaves ascending, elliptic, toothed, bracts glandular-hairy. Flowers crowded in spikes or panicles, tube funnel-shaped, upper lip 4-lobed, mauve with orange patch. Dec.–Jan. Stony slopes, NW (Hex River Mts and Kwadouwsberg).•

humilis (Rolfe) Hilliard (= *Selago rudolphii* (Hiern) Levyns)   Spreading perennial to 20 cm, hairy below. Leaves obovate, toothed, bracts glabrescent. Flowers crowded in spikes, tube funnel-shaped, upper lip 4-lobed, white with orange patch. Dec.–Feb. Stony slopes, NW (Cedarberg to Cold Bokkeveld Mts).•

prolixa Hilliard   Like **P. humilis** but plants hairy. Leaves sparsely toothed and flowers mauve. Dec.–Apr. Stony slopes, NW (Hex River Mts).•

prostrata Hilliard   Like **P. bella** but smaller with leaves entire or sparsely toothed and flowers white, with corolla tube to 4 mm. Dec.–May. Stony slopes, KM (Klein Swartberg).•

similis Hilliard   Glabrescent, sprawling perennial to 50 cm. Leaves ascending, narrowly elliptic, sparsely toothed, bracts sparsely hairy. Flowers crowded in spikes or panicles, tube funnel-shaped, upper lip 4-lobed, mauve with white patch. Dec.–Jan.(–Apr.). Stony slopes, NW (Cedarberg to Hex River Mts).•

B. Upper lip 2-lobed

ascendens (E.Mey.) Hilliard (= *Selago ascendens* E.Mey., *S. incisa* Hochst.) Spreading, glabrescent, short-lived perennial to 20 cm, stems narrowly winged. Leaves oblanceolate, toothed, bracts glabrous. Flowers crowded in corymbs, tube funnel-shaped, upper lip 2-lobed, white with orange patch. Mainly Nov.–Jan. Rocky slopes, NW, SW (Cold Bokkeveld to Riviersonderend Mts).•

caerulescens Hilliard Sprawling hairy annual to 60 cm, branches narrowly winged. Leaves oblanceolate, toothed, often tufted in axils, bracts glabrous. Flowers in small corymbs in panicles, tube funnel-shaped, upper lip 2-lobed, blue with orange patch. Nov.–Mar. Rocky slopes, SW, LB (Riviersonderend to Langeberg Mts: Swellendam).•

gracilis Hilliard Like **P. spuria** but stems usually branched. Leaves ascending and not crowded below and flowers typically white. Sept.–Dec. Rocky slopes, NW, SW, AP, LB (Bokkeveld Mts to Cape Peninsula to Agulhas).•

guttata (E.Mey.) Hilliard (= *Selago guttata* E.Mey.) Like **P. ascendens** but stems erect almost from base with leaves ascending and bracts glandular beneath. Nov.–Jan. Rocky slopes, NW (Pakhuis Mts to Cold Bokkeveld).•

langebergensis Hilliard Shrublet with narrowly winged, rod-like branches to 50 cm. Leaves narrowly oblanceolate, toothed, often tufted in axils, bracts minutely glandular at base. Flowers in corymbs, tube funnel-shaped, upper lip 2-lobed, mauve with orange patch. Dec.–Mar. Scrub on flats and slopes, KM, LB (Swartberg, Langeberg and Outeniqua Mts).•

outeniquensis Hilliard Glabrescent annual to 75 cm, stems narrowly winged. Leaves ascending, linear, toothed, often tufted in axils, bracts glabrous. Flowers in loose corymbs, tube funnel-shaped, upper lip 2-lobed, white with orange patch. Jan.–Dec. Flats and slopes, SW, LB, SE (Riviersonderend Mts to Tsitsikamma Mts).•

parvifolia Hilliard Glabrous perennial to 90 cm, stems wand-like, narrowly winged. Leaves overlapping, ascending, small, oblong, toothed, bracts glabrous. Flowers in spikes or sparse corymbs, tube funnel-shaped, upper lip 2-lobed, white with orange patch. Nov.–Dec. Stony slopes, NW, KM (Bokkeveld Mts to Montagu).•

peninsulae Hilliard Like **P. violacea** but leaves toothed only in upper part and flowers white. Sept.–Jan. Rocky slopes, SW (Cape Peninsula).•

pulchra Hilliard Stout leafy perennial to 40 cm, stems with tapering wings. Leaves imbricate, ovate-oblong, toothed, bracts glabrous. Flowers crowded in corymbs, tube funnel-shaped, upper lip 2-lobed, mauve ?with orange patch. Dec.–Jan. Stony slopes, SW (Kleinrivier Mts to Bredasdorp).•

quadrangularis (Choisy) Hilliard (= *Selago quadrangularis* Choisy) Sparsely hairy ?perennial to 75 cm, stems with tapering wings. Leaves oblanceolate, toothed, bracts glandular-hairy. Flowers crowded in corymbs, tube funnel-shaped, upper lip 2-lobed, white to mauve ?with orange patch. Dec.–Jan. Stony slopes, NW (Ceres).•

recurvifolia Hilliard Tufted perennial to 50 cm, stems rod-like with narrow wings. Leaves recurved, folded, toothed, bracts glandular. Flowers crowded in corymbs, tube funnel-shaped, upper lip 2-lobed, white to mauve with orange patch. Dec.–Feb. Stony slopes, NW, SW (Pakhuis Mts to Wolseley).•

serrata (P.J.Bergius) Hilliard (= *Selago serrata* P.J.Bergius) Like **P. pulchra** but leaves recurved at ends and sparsely toothed and floral axis glandular-hairy. Sept.–Mar. Stony slopes, SW (Cape Peninsula and Bainskloof to Palmiet River Mts).•

spuria (L.) Hilliard (= *Selago spuria* L.) Sparsely hairy perennial to 75 cm, stems rod-like, narrowly winged. Leaves linear-oblanceolate, toothed above, often tufted in axils, crowded below, bracts glabrous. Flowers in loose corymbs, tube funnel-shaped, upper lip 2-lobed, mauve with orange patch. Aug.–Feb. Slopes and flats, NW, SW (Tulbagh to Hangklip).•

verbenacea (L.f.) Hilliard (= *Selago verbenacea* L.f.) Glabrescent, short-lived perennial to 2 m. Leaves opposite throughout, oblanceolate, toothed, bracts minutely glandular-hairy. Flowers in corymbs, tube funnel-shaped, upper lip 2-lobed, mauve with orange patch. Sept.–Feb. Seeps or streamsides, NW, SW, AP (Hex River Mts to Agulhas).•

violacea Hilliard Sprawling, glandular-hairy, short-lived perennial to 40 cm, stems narrowly winged. Leaves oblanceolate, toothed, bracts glandular-hairy. Flowers crowded in corymbs, tube funnel-shaped, upper lip 2-lobed, mauve with orange patch. Nov.–Feb. Shaded rocky slopes, NW (Piketberg, Olifants River Mts to Ceres).•

SELAGO (= *WALAFRIDA*)   BITTER BUSH, AARBOSSIE   c. 190 spp., southern Africa

Group 1: Dwarf shrublets; leaves not tufted flowers in spikes or racemes, either solitary or in loose panicles, white (mauve in **S. pinea**); *calyx 5-fid halfway or more; anthers all well exserted or posterior in mouth*

albomontana Hilliard   Dwarf, densely branched shrublet. Leaves obtuse, without a mucro. Flowers in racemes, white, posterior anthers included. Aug.–Sept. Rocky slopes, KM (Voetpadsberg and Witteberg).•

aspera Choisy   Densely leafy, glandular-hairy, dwarf shrublet. Leaves not tufted, linear, with midrib raised beneath. Flowers in dense, oblong spikes, white. Mostly Sept.–Dec. Dry flats and slopes, NW, AP, KM, LB (Brandvlei to Gouriqua and Gamkaberg).•

bilacunosa Hilliard   Like **S. mundii** but cocci with 2 large lacunae. Oct.–Dec. Dry sandstone slopes, KM (Voetpadsberg to Swartberg Mts).•

cryptadenia Hilliard   Like **S. ramosissima** but leaves spreading to recurved, spikes larger (15–50 × 8–10 mm vs 5–12 × 6–7 mm) and cocci with seed occupying whole loculus. Aug.–Sept. Sandstone and granite slopes, NW, SW (Pakhuis Mts to Malmesbury).•

cupressoides Hilliard   Like **S. triquetra** but leaves small, adpressed, bracts broader and glandular on adaxial surface, often sticking to calyx. Sept.–Oct. Dry, stony slopes, NW (Gydo Pass to Hex River Mts).•

curvifolia Rolfe   Densely leafy, dwarf shrublet to 30 cm. Leaves not tufted, linear, spreading to recurved. Flowers in dense oblong spikes, white. Mostly Aug.–Nov. Dry sandstone slopes, NW, KM (Hex River Mts to Anysberg).•

diffusa Thunb. (= *Selago diosmoides* Rolfe)   Dwarf, densely branched shrublet. Leaves not tufted, linear, imbricate. Flowers in loose panicles or spikes, white, calyx baggy and strongly keeled, pair of cocci cohering, bony, subglobose. Oct.–Nov. Coastal slopes, SW, AP (Gansbaai to Stilbaai).•

dregeana Hilliard   Like **S. glutinosa** but stamens all well exserted. Sept.–Oct. Sandstone slopes, NW (Pakhuis and N Cedarberg Mts).•

eckloniana Choisy (= *Selago elata* Choisy)   Dwarf shrublet to 30 cm, branches with hairs mostly in longitudinal bands and often longitudinally channelled. Leaves not tufted, crowded, linear, glabrous except for sunken glands. Flowers in racemes, distinctly pedicellate, white, bracts nipped below to clasp pedicel, calyx glandular and sticky. Mostly Nov.–Feb. Rocky slopes, NW, SW, KM, LB (Hex River valley and Hessaquaskloof to Uniondale and W Karoo).

fruticosa L. (= *Selago fruticulosa* Rolfe in part)   Dwarf, divaricately branched shrublet with many short, lateral branchlets. Leaves not tufted, linear to oblong, often grey-hairy. Flowers in terminal, oblong to rounded spikes congested even in fruit, white, calyx hairs reflexed. Mostly Sept.–Dec. Stony slopes, NW, SW, AP (Olifants River Mts to De Hoop).•

gliodes Hilliard   Closely leafy shrublet to 40 cm. Leaves not tufted, linear, spreading,. Flowers in crowded spikes, white. Aug.–Oct. Stony slopes, NW, KM (Hex River Mts to Witteberg Mts and W Karoo).•

glutinosa E.Mey.   Dwarf shrublet with adpressed-hairy stems. Leaves not tufted, crowded, linear to terete. Flowers in dense, oblong spikes, white, posterior stamens in mouth or included in tube. Mostly Aug.–Nov. Sandy and stony slopes, NW, KM (Namaqualand to Montagu).

grandiceps Hilliard   Like **S. aspera** but spikes very broad (15–40 × 15–18 mm) and bracts and calyx strongly glandular. Aug.–Oct. Sandstone slopes, SW, LB (Riviersonderend Mts to Garcia's Pass).•

heterotricha Hilliard   Like **S. fruticosa** but branches virgate and lax, spikes elongate, leaves slender with midrib raised beneath, and acute and obtuse patent hairs on calyx. July–Aug. Sandy flats, NW (Graafwater).•

hispida L.f.   Densely leafy shrublet to 30 cm. Leaves not tufted, linear, spreading or recurved, strongly hispid. Flowers in dense, oblong spikes, white. Sept.–Nov. Sandstone slopes, NW (Olifants River Mts to Tulbagh).•

lamprocarpa Rolfe   Dwarf shrublet. Leaves not tufted, linear-lanceolate, spreading, puberulous. Flowers in dense, short spikes, white. Oct.–Jan. Sandstone slopes, NW, KM (Cedarberg to Porterville Mts and Witteberg).•

morrisii Rolfe   Like **S. fruticosa** but branches virgate and lax, and spikes more elongate. Aug.–Sept. Rocky slopes, NW (Namaqualand to Bokkeveld Mts).

mucronata Hilliard   Dwarf, densely branched shrublet. Leaves terete, almost clavate, tipped with a conspicuous, recurved mucro. Aug.–Sept. Dry rocky slopes, KM (Witteberg to Klein Swartberg).•

neglecta Hilliard   Like **S. ramosissima** but style pubescent and cocci with seed occupying entire loculus. Oct.–Nov. Rocky slopes, SW (Potberg).•

nigrescens Rolfe   Like **S. aspera** but more loosely branched, leaves narrower, and bracts and corolla tube shorter. Oct.–May. Sandstone slopes, SW, LB, SE (Riviersonderend Mts to Plettenberg Bay).•

oresigena Compton   Like **S. lamprocarpa** but stems minutely globular-glandular, leaves larger (1.4–2.8 mm vs 0.7–1.2 mm wide) and posterior anthers in mouth of tube or shortly exserted. Mostly Dec.–Feb. High rocky slopes, NW (Cedarberg to Grootwinterhoek Mts).•

perplexa Hilliard   Like **S. lamprocarpa** but posterior anthers included or barely exserted, leaves short, spreading. Mostly Sept.–Dec. Stony slopes, NW, KM (Grootwinterhoek to Witteberg and Langeberg Mts).•

pinea Link (= *Selago spinea* orth. var.)   Densely leafy shrublet to 45 cm. Leaves not tufted, crowded, linear, hispidulous. Flowers in dense, oblong spikes, mauve. Nov.–Dec. Coastal slopes and flats, SW, AP (Kleinmond to Arniston).•

pustulosa Hilliard   Like **S. glutinosa** but leaves pustulate. Mainly Sept.–Nov. Sandstone slopes, NW (Cedarberg and Cold Bokkeveld Mts).•

ramosissima Rolfe   Dwarf, closely branched shrublet. Leaves not tufted, crowded, adpressed, linear, 2–4 mm long. Flowers on short spikes crowded together, white, calyx shaggy, cocci broad, with bony tissue filling one-third of the volume. Sept.–Dec. Clay flats, SW, LB (Riviersonderend to Great Brak River).•

scabribractea Hilliard   Like **S. glutinosa** but stems with spreading glandular hairs, bracts and calyx with broad-based gland-tipped and acute hairs on outside. Sept.–Oct. Stony slopes, NW, SW (Namaqualand and W Karoo to Pakhuis Mts and St. Helena Bay).•

thomii Rolfe   Like **S. ramosissima** but bracts and calyx less hairy and cocci with seed occupying entire loculus. Mostly Dec.–June. Stony slopes, NW, SW, AP, KM, LB, SE (Hermanus and Hex River valley to Avontuur).•

trichophylla Hilliard   Like **S. ecklonii** but leaves with long hairs and bracts with long coarse hairs on outside (vs hairs mostly on margins). Flowering time? Stony slopes, NW, KM (Grootwinterhoek to Witteberg Mts).•

triquetra L.f.   Dwarf shrublet with virgate branches. Leaves not tufted, linear-oblong, spreading, usually shortly grey-hairy. Flowers in long, narrow spikes, solitary at tips of lateral branchlets, white. July–Oct. Sandstone slopes, NW, SW, KM (Piketberg to Montagu).•

valliscitri Hilliard   Like **S. triquetra** but leaves terete, to 0.3 mm diam., flowers distinctly pedicellate and bracts narrower. Aug.–Sept. Rocky sandstone slopes, NW (Cold Bokkeveld Mts).•

Group 2: Dwarf shrublets or subshrubs; leaves various; inflorescence a dense spike or raceme, often solitary, or in corymbs or panicles; calyx divided at least halfway into 5 (or 3) narrow, acute, subequal lobes; corolla white; anthers shortly exserted, posterior ones often included

cedrimontana Hilliard   Like **S. mundii** but branches with spreading glandular hairs, primary leaves mostly 1.5–2.2 mm long (vs 0.8–1.2 mm). Jan.–Feb. Sandstone slopes at high alt., NW (Cedarberg and Cold Bokkeveld Mts).•

glandulosa Choisy   Dwarf shrublet to 40 cm. Leaves tufted, linear, margins closely revolute. Flowers in dense spikes, often several at branch tips, white. Mainly Oct.–Apr. Limestone and clay slopes, AP, LB (Arniston to Albertinia).•

impedita Hilliard   Like **S. scabrida** but loosely branched with weak stems, leaves glabrous or hairy only on margins, not in tufts, and spikes usually solitary. Nov.–Feb. Sandstone slopes, SW (Wemmershoek Mts to Kogelberg).•

levynsiae Hilliard (= *Selago scabrida* auct.)   Like **S. scabrida** but branches divaricate, leaves loosely tufted and blades mostly wider (1.2–2 mm vs 1–1.3 mm). Mainly Jan.–Mar. Stony slopes, SW (Cape Peninsula to Shaw's Mts).•

micradenia Hilliard   Like **S. mundii** but branching divaricate, leaves spreading, bracts and calyx minutely glandular and corolla lobes shorter and minutely puberulous. Nov.–Dec. Stony slopes, NW (Nardouwsberg to Cold Bokkeveld).•

mundii Rolfe   Slender, closely leafy shrublet to 1 m, branches with retrorse, often grey hairs in decurrent lines. Leaves not tufted, ascending, more or less spreading above, oblanceolate to linear, glaucous or minutely grey-hairy. Flowers in dense, capitate to oblong spikes, white, corolla hairy, bracts and calyx minutely glandular. Sept.–Nov. Stony slopes, NW, SW (Grootwinterhoek to Elandskloof Mts).•

oppositifolia Hilliard   Dwarf shrublet. Leaves tufted, opposite and decussate. Flowers in dense spikes, white. Nov.–Dec. Rocky slopes, KM (Swartberg Mts: Seweweekspoort).•

polystachya L.   Densely leafy shrublet to 60 cm, young stems with recurved-hairs but leaf bases decurrent into more or less glabrous wings. Leaves tufted, linear-oblanceolate. Flowers in short oblong spikes, white. Oct.–Mar. Rocky slopes, SW (Saldanha to Gansbaai).•

praetermissa Hilliard   Like **S. psammophila** but flowering spikes only to 25 × 15 mm (vs 30–40 × c. 12 mm) and glandular hairs few, only at base of outside of bracts. Sept.–Oct. Rocky slopes, NW (Bokkeveld Mts to N Cedarberg).•

prostrata Hilliard   Like **S. setulosa** but stems divaricately branched, primary leaves reflexed and with weakly developed white scales and flowers in spikes 30–35 mm long (vs 10–15 mm long). Feb.–Mar. Limestone slopes, AP (Bredasdorp Poort).•

psammophila Hilliard   Like **S. polystachya** but stems hairy all around, leaves not or scarcely tufted, bracts with coarse hairs covering minute glands and corolla often densely hairy. Nov.–Dec. Stony slopes, NW, SW (Aurora to Blouberg).•

scabrida Thunb. (incl. **S. cylindrica** Levyns)   Densely leafy, virgately branched shrublet to 45 cm, with adpressed-hairy stems. Leaves tufted, oblong. Flowers in narrow spikes mostly clustered in corymbose panicles, white. Mostly Oct.–Mar. Rocky slopes, SW, AP, LB (Cape Peninsula to Swellendam).•

seticaulis Hilliard   Like **S. levynsiae** but with long, coarse, spreading hairs on stems, leaves, bracts and calyx, and corolla minutely glandular on reverse. Mainly Sept.–Oct. Stony slopes, SW, AP (McGregor to Agulhas).•

setulosa Rolfe   Dwarf shrublet, stems with long, patent hairs. Leaves tufted, ovate, hispid mainly on margins, often with numerous white scales. Flowers in spikes 10–15 mm long, white. Oct.–Nov. Limestone hills, AP (Hagelkraal to Mossel Bay).•

Group 3: Dwarf shrublets; leaves tufted, glabrous, ericoid; flowers in dense spikes or racemes, solitary or arranged in corymbose or elongate panicles, white; calyx divided more than halfway into 5 subequal lobes; anthers mostly all well exserted (posterior ones in mouth or shortly exserted in **S. brevifolia**, **S. karooica**, **S. nigromontana**)

brevifolia Rolfe   Densely leafy shrublet to 30 cm, branches with coarse patent hairs. Leaves tufted, oblong, fleshy, glossy, apiculate. Flowers in short, oblong spikes, white, bracts and calyx hairy, cocci deeply concave on inner face. Mostly Oct.–Nov. Sandstone slopes, LB, KM (Langeberg Mts: Riversdale to Kammanassie Mts).•

dolichonema Hilliard   Like **S. venosa** but filaments very long in relation to corolla tube, thus anthers strikingly exserted, bracts narrower (0.6–1 mm vs 1.2–2 mm), hairy towards adaxial base and veins not raised. Nov.–Dec. Rocky sandstone slopes, NW (S Cedarberg Mts).•

esterhuysenii Hilliard   Like **S. nigromontana** but primary leaves 0.8–1.1 mm wide (vs 1.5–2 mm) and minutely glandular-hairy, corolla tube 2.2–3 mm long (vs 4–6 mm), bracts minutely glandular, longer than calyx. Mar.–Apr. Sandstone slopes at high alt., KM (Klein Swartberg: Towerkop).•

glomerata Thunb.   Densely leafy shrub to 1 m, with shortly pubescent branches, hairs more or less recurved. Leaves tufted, linear-oblong, glossy, glabrescent. Flowers in spikes aggregated in corymbose panicles, white, bracts glabrous, with sunken veins, cocci deeply concave on adaxial surface. Mostly Sept.–Mar. Sandstone slopes, SE (Outeniqua to Great Winterhoek Mts).•

karooica Hilliard   Like **S. glomerata** but habit lax, spikes solitary or few in lax clusters, bracts longer and with veins on outside raised, and anthers barely exserted or included. Nov.–Dec. Rocky slopes, KM, SE (Witteberg to Kouga Mts).•

luxurians Choisy (= *Selago dregei* Rolfe)   Densely leafy shrublet to 40 cm, branches with coarse, patent hairs. Leaves tufted, linear-oblong, thick, apiculate. Flowers crowded in short spikes arranged in well branched, corymbose panicles, white, inner face of cocci plane with a median groove. Nov.–Feb. Rocky slopes, KM, LB, SE (Swartberg Mts and Swellendam to Port Elizabeth).•

mediocris Hilliard   Like **S. glomerata** but leaves narrower, the primary up to 0.6–1 mm wide (vs 0.8–2 mm), bracts usually longer than calyx (not shorter) and cocci with interior face plane. Mainly Apr.–May. Rocky slopes, SE (Baviaanskloof Mts to Zuurberg).

myriophylla Hilliard   Like **S. venosa** but flowering branchlets closely leafy, bracts c. 0.9 mm wide (vs 1.2–2 mm) and corolla and filaments slightly shorter. Oct.–Nov. Sandstone slopes, KM, LB, SE (Witteberg and Langeberg to Kouga Mts).•

nigromontana Hilliard   Like **S. luxurians** but primary leaves mostly longer (3.5–9.5 mm vs 2.4–4.5 mm), and bracts larger and glabrous abaxially. Dec.–Jan. Rocky sandstone slopes, KM (Groot Swartberg).•

parvibractea Hilliard   Like **S. glomerata** but bracts mostly smaller, to 1 mm wide and calyx shorter and lobes obtuse. Jan.–Feb. Sandstone slopes, SW, AP, KM (W Little Karoo to Napier and Mossel Bay).•

rubromontana Hilliard   Like **S. brevifolia** but leaves obtuse to subacute, thinner, bracts glabrous on outside or minutely hairy at base, cocci plane on inner surface. Aug.–Sept. Stony lower slopes, KM (Rooiberg).•

venosa Hilliard   Like **S. glomerata** but leaves and bracts with raised veins beneath (not sunken), bracts sticky and cocci almost flat on inner surface. Jan.–Feb. Sandstone slopes, NW, SW (Bokkeveld Mts to Riviersonderend).•

Group 4: Dwarf shrublets; leaves in loose to tight tufts, primary leaves often fairly broad; flowers in congested or lax racemes, solitary or in panicles, white or shades of blue to violet or pink; calyx often lobed to about halfway, mostly 3-lobed or irregularly 3–5-lobed; anthers all well exserted

chalarantha Hilliard   Twiggy shrublet to 30 cm. Leaves tufted, linear. Flowers in spikes, usually on distinct peduncles, often only female, white, bracts long and recurved, staminodes well exserted, calyx irregularly 3–5-lobed. Aug.–Sept. Stony flats in renosterveld, NW (Bokkeveld Plateau).•

linearifolia Rolfe   Loosely branched shrublet to 30 cm. Leaves tufted, linear, puberulous. Flowers in small, compact panicles, white. July–Sept. Stony slopes, NW (Olifants River valley: Bulshoek).•

pinguicula E.Mey.   Twiggy shrublet to 40 cm, with glabrescent branches. Leaves in loose tufts, oblanceolate, margins slightly revolute. Flowers in short, compact, rounded racemes, white, calyx inflated, including at least half the corolla tube, bracts pouched, wrinkled at base. Aug.–Oct. Stony clay soils in renosterveld, NW (Namaqualand and Karoo to Hex River valley).

polygala S.Moore   Glabrescent shrublet to 40 cm. Leaves tufted, lanceolate, primary ones 2–5 mm wide. Flowers in dense racemes, white, calyx with scattered, minute glands on outside. July–Sept. Stony karroid slopes, NW (W Karoo to Hex River valley).

subspinosa Hilliard   Twiggy, white-hairy shrublet to 30 cm, flowering branches persisting as spines. Leaves tufted, linear, margins strongly revolute. Flowers in small, mainly single clusters, mauve or pink, stamens well exserted, bracts and calyx with stalked glands. Aug.–Sept. Stony flats, KM (W Little Karoo).•

Group 5: Dwarf shrublets or subshrubs; leaves not tufted, relatively broad; flowers in congested spikes sometimes arranged in loose panicles, white, violet or pink; calyx deeply 3-lobed, one or both lateral lobes bifid; anthers all well exserted or posterior ones included or shortly exserted

ciliata L.f. (= *Walafrida ciliata* (L.f.) Rolfe)   Densely leafy shrublet to 60 cm. Leaves not tufted, crowded, ovate-lanceolate, prominently ciliate. Flowers in crowded, oblong spikes, mauve or white. Mostly Apr.–Sept. Gravel flats and lower slopes, SW, LB (Caledon to Little Brak River).•

elsiae Hilliard   Like **S. myrtifolia** but stems hairy all around, leaves with upcurved hairs on margins and midline, flowers mauve and ovary glabrous in upper half. Mar.–Apr. Rocky slopes, SE (Great Winterhoek Mts).•

myrtifolia Rchb. (= *Walafrida myrtifolia* (Rchb.) Rolfe, *Selago nitida* E.Mey.)   Twiggy shrublet to 70 cm, branches with hairs in longitudinal bands. Leaves not tufted, ovate, leathery, glandular-punctate, sometimes hairy on margins. Flowers in dense ovoid spikes, magenta-pink, corolla tube unusually long, coccus with 2 large spurious cells. Mainly July–Nov. Stony slopes, SE (Langkloof to Port Elizabeth).•

recurva E.Mey. (= *Walafrida recurva* (E.Mey.) Rolfe)   Divaricately branched shrublet to 60 cm. Leaves not tufted, linear-lanceolate, recurved apically, margins hispid. Flowers in ovoid spikes, mauve or white. Apr.–Sept. Gravel flats and lower slopes, SE (Langkloof to Alexandria).

zeyheri Choisy (= *Walafrida zeyheri* (Choisy) Rolfe)   Densely leafy shrublet to 35 cm, branches pubescent. Leaves not tufted, linear-oblong, fleshy, recurved at tip. Flowers in short dense spikes, mauve. Nov.–Mar. Dry flats, SE (Port Elizabeth to E Cape).

Group 6: Subshrubs; leaves tufted; flowers in large, congested spikes either solitary or in loose panicles, white or violet; calyx deeply 3-lobed, 2-lobed in **S. congesta***; posterior anthers included or shortly exserted, anterior anthers well exserted*

cinerea L.f. (= *Walafrida cinerea* (L.f.) Rolfe)   Densely leafy shrublet to 75 cm, branches puberulous. Leaves tufted, linear-oblong, glaucous or grey-hairy. Flowers in dense corymbose panicles, mauve or white. Aug.–Jan. Gravel or limestone slopes and flats, AP, KM, LB, SE (Swartberg Mts and De Hoop to East London).

congesta Rolfe   Dwarf shrublet. Leaves tufted, elliptic-lanceolate, hairy only on margins of clasping base. Flowers in congested, mostly solitary spikes, white, calyx deeply 2-lobed, bracts lanceolate, concave, with stiff hairs on margins and base. Flowering time? Habitat? SE (Langkloof to Port Elizabeth).•

decipiens E.Mey. (= *Walafrida decipiens* (E.Mey.) Rolfe)   Sprawling, finely leafy shrublet to 60 cm, branches minutely puberulous. Leaves tufted, linear-oblong. Flowers in short spikes aggregated in corymbose panicles, mauve or white. Aug.–Dec. Gravel flats, SE (Uitenhage to Grahamstown).

polycephala Otto   Subshrub to 30 cm. Leaves tufted, linear. Flowers in dense spikes, ?white, bracts linear-lanceolate with recurved tips. Flowering time? Habitat? SE (Uitenhage to Port Elizabeth).•

rotundifolia L.f. (= *Walafrida rotundifolia* (L.f.) Rolfe)   Densely leafy shrublet to 60 cm, branches puberulous. Leaves tufted, oblong to rounded, glaucous. Flowers in short spikes arranged in dense panicles, white or mauve. Aug.–Mar. Grassy flats, SE (Knysna to Port Elizabeth).•

*Group 7: Dwarf shrublets or subshrubs; leaves usually tufted; flowers in racemes arranged in elongate or corymbose panicles, usually mauve, sometimes white (***S. pulchra***); calyx either 5-lobed in upper half, tending to be 3-lobed, or 3-toothed; lateral lobes sometimes bifid; anthers all well exserted*

adenodes Hilliard   Like **S. pulchra** but lowermost bracts 3–4 × 1.4–2.3 mm (vs 2.3–3 × 1–1.3 mm), obtuse, with minute glandular hairs on both surfaces, and cocci without spurious cells. Sept.–Nov. Sandstone slopes, KM (Swartberg Mts).•

burchellii Rolfe (= *Selago pubescens* Rolfe)   Densely leafy shrub to 50 cm, with branches pubescent, mainly in decurrent lines. Leaves tufted, lanceolate, minutely puberulous. Flowers in small heads forming compact, narrow panicles, mauve or white. Mostly Aug.–Jan. Coastal slopes and flats, SE (George to Plettenberg Bay).•

canescens L.f. (= *Selago forbesii* Rolfe, *S. ramulosa* E.Mey., *S. thunbergii* Choisy)   Densely leafy shrublet to 50 cm, with puberulous branches. Leaves tufted, linear-lanceolate, minutely puberulous. Flowers in small round heads forming narrow compact panicles, mauve. Mostly July–Sept. Dry, mostly clay slopes, SW, AP, LB, SE (Bellville to Port Elizabeth).•

lilacina Hilliard   Like **S. canescens** but leaves broader, and flowers few in small heads arranged in lax panicles, lilac. Sept.–Apr. Rocky slopes, SE (Kammanassie Mts and Langkloof).•

pulchra Hilliard   Like **S. canescens** but leaves broader, panicles shorter, flowers white, calyx 3-lobed, bracts glabrous except for minute hairs on margins and cocci with 2 spurious cells. Aug.–Apr. Rocky sandstone slopes, KM, SE (Swartberg, Rooiberg and Outeniqua Mts).•

villicaulis Rolfe   Densely leafy shrublet to 45 cm, with villous branches. Leaves tufted, linear, puberulous. Flowers in compact racemes arranged in rounded panicles, mauve. Apr.–Oct. Limestone and sandy slopes, AP, SE (Stilbaai to Knysna).•

Group 8: Dwarf shrublets; leaves usually tufted; flowers white or mauve, in spikes or racemes either solitary or in loose panicles; calyx mostly obliquely 5-lobed in upper half, or irregularly 3–5-lobed; anthers all well to far exserted

acocksii Hilliard (= *Selago zeyheri* Rolfe)   Densely leafy, minutely puberulous shrublet to 30 cm. Leaves tufted, oblong, minutely glandular throughout. Flowers in oblong spikes, white, bracts minutely glandular and with few to many acute hairs. Feb.–Oct. Dry stony slopes, NW, KM (W Karoo and Swartberg Mts to Karoo and Free State).

articulata Thunb. (= *Walafrida loganii* Hutch.)   Finely leafy, dwarf shrublet to 20 cm. Leaves tufted, needle-like, spreading. Flowers in short spikes, white to pale mauve. Sept.–Oct. Stony flats and slopes, NW (Bokkeveld Mts and W Karoo).

distans E.Mey. (= *Walafrida distans* (E.Mey.) Rolfe)   Densely leafy shrublet to 30 cm. Leaves in loose tufts, linear, glandular-punctate, often becoming viscid, margins thickened. Flowers in short spikes, white, calyx with minute globular glands. Apr.–Sept. Sandstone slopes, KM (Witteberg and Anysberg).•

exigua Hilliard   Minutely glandular-pubescent shrublet to 30 cm. Leaves tufted, linear, puberulous. Flowers in narrow, mostly solitary spikes, ?white. Jan.–Feb. Sandstone slopes, KM (Swartberg Mts).•

ferruginea Rolfe   Finely leafy, densely glandular shrublet to 30 cm. Leaves tufted, oblong, hispidulous. Flowers in oblong to elongate spikes, mauve. Sept.–Oct. Gravel flats, KM, SE (Little Karoo to George).•

fourcadei Hilliard   Densely leafy shrublet to 30 cm. Leaves in loose tufts, linear, minutely glandular-punctate, margins thickened. Flowers in short spikes, purple or whitish flushed mauve. Sept.–Nov. Stony slopes, SE (Witteberg and Swartberg Mts to Avontuur and Karoo).

geniculata L.f. (= *Walafrida geniculata* (L.f.) Rolfe)   Grey-hairy, finely leafy shrublet to 60 cm. Leaves tufted, linear. Flowers in elongate, narrow spikes, violet, sometimes white, calyx 3–5-lobed, bracts with abruptly acute to shortly acuminate, recurved tips. Nov.–Apr. Stony slopes and flats, NW, KM, LB, SE (Hex River valley to E Cape and Free State).

thermalis Hilliard   Finely leafy shrublet to 40 cm, with short divaricate branchlets each ending in a broad spike, stem hairs in longitudinal bands. Leaves not or scarcely tufted, linear, glabrescent except for globular or sunken glands. Flowers in short subcapitate spikes on leafy divaricate branches, white. June–July. Rocky flats, SW, KM (Worcester Karoo to Warmwaterberg).•

Group 9: Dwarf shrublets or subshrubs; leaves tufted, ericoid; flowers in small spikes or racemes either solitary or in loose panicles, white or mauve; calyx obliquely 3- or 5-lobed, sometimes irregularly 3–5-lobed; anthers all exserted

confusa Hilliard   Like **S. gracilis** but primary leaves longer and broader, spikes arranged in narrow panicles. Nov.–Feb. Stony slopes, SE (Hankey to Fort Beaufort) .

gracilis (Rolfe) Hilliard (= *Walafrida gracilis* Rolfe, *W. pubescens* Rolfe, *W. squarrosa* Rolfe)   Sprawling, finely leafy shrublet to 30 cm. Leaves tufted, linear, puberulous. Flowers in short oblong spikes, either solitary at tips of leafy branchlets or branchlets forming panicles, white. Mostly June–Nov. Gravel slopes, NW, SW, KM, LB, SE (Caledon and Hex River valley to Port Elizabeth).•

linearis Rolfe   Densely leafy shrub to 30 cm. Leaves tufted, linear, glabrous. Flowers in short, terminal spikes, mauve, bracts sparsely hairy on lower margins. Mostly Nov.–Jan. Along streams, LB, SE (Langeberg: Attaquas Kloof to Port Elizabeth).•

Group 10: Dwarf shrublets or subshrubs; leaves tufted, often ericoid; flowers in a long, narrow panicle of small, globular spikes or racemes, white or shades of violet; calyx mostly 5-lobed, rarely irregularly 3–5-lobed; anthers all well exserted

albida Choisy   Finely pubescent to velvety shrublet to 40 cm. Leaves in tufts, linear. Flowers in notably pedunculate racemes forming loose panicles, usually shades of violet, bracts narrow. Mainly May–Sept. Stony karroid slopes, NW, KM (Worcester and Little Karoo to Free State and Great Karoo).

capituliflora Rolfe (= *Selago albida* sensu Rolfe in part)   Like **S. albida** but leaves, bracts and calyx much less hairy and flowers in large, lax panicles. Aug.–Sept. Rocky slopes, NW (Bokkeveld Mts to Piketberg).•

divaricata L.f. (= *Selago albida* sensu Rolfe in part, *S. minutissima* Choisy)   Twiggy, divaricately branched, sometimes lax, minutely puberulous shrublet to 60 cm, with pale bark. Leaves tufted, linear-oblong, sometimes glabrescent, primary leaf much larger and strongly reflexed. Flowers in small rounded clusters, axes becoming spiny, white to violet. Mostly Sept.–Dec. Stony slopes and flats, KM (S Namibia to Little Karoo and E Cape).

glabrata Choisy   Densely leafy, virgate shrublet to 35 cm. Leaves tufted, linear, calyx 3–5(6)-lobed. Flowers in dense spikes arranged in narrow, lax panicles, white, calyx 5-lobed. Sept.–Oct. Stony slopes in renosterveld and karroid scrub, NW, SW, KM (Namaqualand to Worcester and Little Karoo).

inaequifolia Hilliard   Like **S. multiflora** but branching virgate, hairs on stem spreading and primary leaves 8–20 mm long (vs 2.5–8 mm). Aug.–Oct. Rocky slopes, NW (Gifberg slopes).•

michelliae Hilliard   Like **S. singularis** but leaves, stems, bracts and calyx with shaggy, simple hairs, leaf margins plane, flowers white and calyx irregularly 3–5-lobed. Sept.–Oct. Rocky slopes, NW (Hex River Mts). •

multiflora Hilliard   Like **S. divaricata** but bark dark, leaves markedly pilose and glandular-punctate, larger bracts 2.2–3.2 mm long (vs 1.4–2.4 mm) and corolla tube longer (2.3–3.8 mm vs 1.3–2.4 mm). Sept.–Nov. Sandstone slopes, NW (Namaqualand to Hex River valley and Bonteberg).

singularis Hilliard   Like **S. albida** but stems with branched (not simple) hairs on young branches, primary leaves 2–7 mm long (vs 6–15 mm), velvety, margins strongly revolute, and corolla tube shorter than 3 mm. Sept.–Oct. Rocky slopes, NW (Cedarberg to Karoo Poort).•

stenostachya Hilliard   Like **S. inaequifolia** but with branches with retrorse hairs, leaves glandular-punctate and sparsely hairy. Aug.–Sept. Stony slopes, NW (S Namaqualand to Klawer).

Group 11: Shrublets; leaves tufted and ericoid; flowers in corymbose panicles, white; calyx shallowly and subequally 5-toothed (irregularly 3–5-lobed in **S. variicalyx***); anthers mostly well exserted*

corymbosa L.   Densely leafy shrub to 60 cm, with pubescent branches. Leaves tufted, linear, the primary 5–15 × 0.5–1 mm, margins revolute, puberulous to subglabrous. Flowers in rounded corymbose panicles, white, calyx subequally and obtusely 5-toothed, bracts narrow, obtuse, the lowermost 0.4–0.7 mm wide. Mostly Feb.–Apr. Stony slopes and flats, NW, SW, KM, SE (Cape Peninsula and Gydo Pass to Hankey).•

dolosa Hilliard   Like **S. corymbosa** but leaves glabrous, shorter, relatively broader, the primary 2–7 × 5–1.2 mm and broader bracts, the lowermost 0.7–1.25 mm wide. Mainly Sept.–Mar. Stony and gravelly slopes, SW, AP, LB, SE (Hottentots Holland Mts to Karoo and Transkei).

variicalyx Hilliard   Like **S. corymbosa** but major bracts larger, acute, 0.8–1.1 mm wide (vs 0.4–0.7 mm) and calyx acutely and irregularly 3–5-lobed. Sept.–Dec. Stony slopes, SE (Van Staden's Mts to Grahamstown).

SUTERA   SKUNK BUSH, STINKBOSSIE   49 spp., mainly South Africa but also tropical Africa

A. All 4 stamens exserted

affinis (Bernh.) Kuntze   Minutely hairy perennial, 15–30 cm. Leaves narrow, more or less entire. Flowers in racemes or loosely paniculate, tube broadly funnel-shaped, mauve with a yellow tube. Oct.–Feb. Dry, stony ground, KM, SE (Oudtshoorn to Langkloof).•

caerulea (L.f.) Hiern   Erect, glandular-hairy perennial to 1 m. Leaves sometimes coarsely toothed. Flowers in long racemes, tube shortly funnel-shaped, mauve or violet with a yellow tube, style glandular-hairy. July–Oct. NW, SW, KM, LB, SE (S Namaqualand to Humansdorp).

calciphila Hilliard   Sprawling or creeping, glandular-hairy perennial to 30 cm. Leaves often coarsely toothed. Flowers shortly racemose, tube broadly funnel-shaped, mauve to violet with orange tube. Oct. Limestone in scrub, AP (Bredasdorp to Stilbaai).•

campanulata (Benth.) Kuntze   Glandular-hairy perennial to 50 cm. Leaves coarsely toothed. Flowers in racemes, tube broadly funnel-shaped, mauve with a yellow tube. Jan.–Dec. Sandy places in scrub or grassland, AP, SE (Stilbaai to Port Alfred).

cinerea Hilliard   Sprawling cobwebby woolly shrublet with slender stems. Leaves narrow. Flowers axillary forming short racemes, tube broadly funnel-shaped, mauve with a yellow tube. Sept.–Feb. Sandy slopes in scrub, SE (Baviaanskloof Mts).•

foetida Roth   Minutely glandular annual to 60 cm, foetid, leafy throughout. Leaves coarsely serrate. Flowers in racemose or paniculate cymules, tube narrowly funnel-shaped, white, pink or violet with orange throat. July–Dec. Damp, sheltered spots, often below rocks, NW, SW, KM, LB, SE (Kamiesberg to Malmesbury and east to Kouga Mts).

halimifolia (Benth.) Kuntze   Gnarled glandular shrublet to 35 cm, hairs granular. Leaves sometimes coarsely toothed. Flowers in long racemes, tube shortly funnel-shaped, pink or mauve with a yellow tube. Nov.–Jan. Rocky slopes, SE (Humansdorp, S Namibia to E Cape).

pauciflora (Benth.) Kuntze   Twiggy glandular-hairy shrublet to 45 cm. Leaves suborbicular and toothed. Flowers axillary, tube broadly funnel-shaped, mauve with a yellow tube. Sept.–Apr. Rocky mountain slopes, NW, KM, SE (W Karoo to Humansdorp).

placida Hilliard   Sprawling glandular-hairy perennial to 30 cm, stems narrowly winged. Leaves often with a few teeth above. Flowers in racemes or panicles, tube broadly funnel-shaped, mauve with yellow throat. Mar.–Aug. Limestone hills, AP (Stilbaai).•

polyantha (Benth.) Kuntze   Soft, glandular-hairy short-lived perennial to 30 cm. Leaves mostly toothed. Flowers in leafy racemes, tube shortly funnel-shaped, white or mauve with a yellow tube. Sept.–May. Mostly coastal dunes, SE (Knysna to Peddie).

subnuda (N.E.Br.) Hiern   Glabrescent perennial to 45 cm. Leaves narrow. Flowers usually paniculate, tube short and broadly funnel-shaped, pink to mauve with a yellow tube. Nov.–May. Stony mountain slopes, KM, LB (S Karoo, Ladismith to De Rust).

A. Upper 2 stamens included

aethiopica (L.) Kuntze   Twiggy minutely glandular-hairy shrublet to 30 cm. Leaves crowded above, mostly coarsely toothed above. Flowers in short racemes, tube long, narrowly funnel-shaped, white, pink to violet with yellow throat. Aug.–Sept. SW, LB (Caledon to Heidelberg, ?Mossel Bay).•

comptonii Hilliard   Like **S. paniculata** but main leaves shorter than 20 mm, flower tube 10.5–12.5 mm long. Oct.–Nov. Sandstone slopes, KM (W Karoo and Witteberg).

cordata (Thunb.) Kuntze   Sprawling softly hairy perennial. Leaves bluntly toothed, tube funnel-shaped, white with a yellow tube. Jan.–Dec. Forests and forest margins, SE (George to East London).

decipiens Hilliard   Tufted or twiggy, softly and glandular-hairy perennial. Leaves usually coarsely toothed. Flowers axillary, forming racemes, white, pink or mauve with a yellow tube. (Mar.–)Aug.–Dec. Sheltered places in kloofs or below rocks, NW, SW, KM (S Namaqualand to Franschhoek Mts to Avontuur).

denudata (Benth.) Kuntze   Glabrescent perennial to 45 cm. Leaves narrow. Flowers in racemes or loose panicles, tube funnel-shaped, pink to purple with a yellow tube. Jan.–Dec. Rocky hills, SE (Uniondale to Humansdorp).•

glabrata (Benth.) Kuntze   Minutely glandular-hairy shrublet to 45 cm, branches slender. Leaves linear and obtuse. Flowers in axillary racemes or narrow panicles, tube funnel-shaped, white, pink or mauve with a yellow tube. July–Dec. Stony shale slopes and cliffs, NW, KM (Worcester to Barrydale).•

hispida (Thunb.) Druce   Roughly glandular-hairy bushy shrublet to 50 cm. Leaves coarsely toothed. Flowers axillary, sometimes in racemes or narrow panicles, tube narrowly funnel-shaped, pink to mauve with yellow throat. Jan.–Dec. Rocky sandstone or limestone, SW (Cape Peninsula to Bredasdorp).•

integrifolia (L.f.) Kuntze   Twiggy shortly glandular-hairy shrublet to 60 cm. Leaves sometimes toothed. Flowers axillary forming leafy racemes or panicles, tube narrowly funnel-shaped, white or rarely mauve with yellow throat. Jan.–Dec. Coastal scrub and forest margins, LB, SE (Albertinia to Humansdorp).•

langebergensis Hilliard   Tangled sparsely glandular-hairy perennial to 30 cm. Leaves coarsely toothed. Flowers in laxly arranged racemes or panicles, tube narrowly funnel-shaped, white or mauve with yellow throat. May–Sept. Rocky sandstone slopes, LB (Langeberg Mts: Tradouw Pass to Garcia's Pass).•

longipedicellata Hilliard   Like **S. uncinata** but longest pedicels to 20 mm. Mainly June–Sept. Rocky sandstone slopes, NW (Bokkeveld escarpment).•

marifolia (Benth.) Kuntze   Cobwebby woolly shrublet with slender stems. Leaves mostly toothed. Flowers axillary eventually forming leafy panicles, tube narrowly funnel-shaped, pink to mauve with a yellow tube. Jan.–Dec. Rocky slopes, SE (Avontuur to Port Elizabeth).•

paniculata Hilliard   Like **S. revoluta** but hairs on stem 0.3–1 mm long, flower tube 6.5–9.5 mm. Mostly May–Sept. Sandy streambanks or rocky slopes in scrub, NW (Kamiesberg to Clanwilliam).

revoluta (Thunb.) Kuntze   Glandular-hairy shrublet to 60 cm. Leaves narrow. Flowers axillary forming racemes or narrow panicles, tube funnel-shaped, 5–13 mm, white, pink or mauve with yellow throat. Mar.–Nov. Stony shale slopes, SW, AP, KM, LB, SE (W Karoo to Botrivier to Baviaanskloof Mts).

subsessilis Hilliard   Tufted softly and glandular-hairy perennial to 35 cm. Leaves mostly entire, soft. Flowers axillary forming leafy racemes or narrow panicles, tube funnel-shaped, mauve with yellow throat. Oct.–Dec. Sheltered places, often below rocks, NW (Cedarberg Mts).•

subspicata (Benth.) Kuntze   Gnarled dwarf shrublet to 30 cm, hairs confined to narrow bands. Leaves leathery, coarsely toothed. Flowers axillary forming short, crowded racemes or panicles, tube narrowly funnel-shaped, white, pink or mauve with yellow throat. Apr.–Oct. Coastal dune scrub, AP (Stanford to Cape Infanta).•

tenuicaulis Hilliard   Glabrescent perennial, 15–45 cm. Leaves narrow. Flowers laxly racemose or paniculate, tube narrowly funnel-shaped, pink with yellow throat. Oct.–May. Rocky slopes in scrub, KM (Swartberg Mts).•

titanophila Hilliard   Closely leafy, glandular-hairy gnarled shrublet to 20 cm. Leaves mostly coarsely toothed. Flowers in short racemes, tube narrowly funnel-shaped, white, pink to mauve with yellow throat. Oct. Cracks on limestone cliffs, AP (De Hoop).•

uncinata (Desr.) Hilliard   Glandular-hairy shrublet to 60 cm with narrowly winged stems, leafy throughout. Leaves mostly entire. Flowers in racemes or narrow panicles, tube narrowly funnel-shaped, pink to purple with yellow throat. May–Oct. Sandy or rocky places in scrub, NW, SW (Gifberg to Cape Peninsula to Bonnievale).•

violacea (Schltr.) Hiern   Glandular-hairy shrublet, 50–150 cm. Leaves linear. Flowers axillary forming simple racemes, mauve to purple with orange throat. Apr.–Nov. Rocky slopes, NW, KM (W Karoo and Bokkeveld Plateau to De Rust).

TEEDIA   LILAC BERRY   2 spp., Namaqualand to E Cape

lucida (Aiton) Rudolphi   Sprawling shrublet to 1.2 m. Leaves with winged petioles, finely toothed, shiny. Flowers mauve. Sept.–Jan. Rocky outcrops at middle to upper elevations, NW, SW, AP, KM, LB, SE (Namaqualand to Swaziland).

pubescens Burch.   Shrublet to 60 cm, stems densely hairy. Leaves toothed. Flowers pink or mauve. Sept.–Nov. Rock outcrops at middle to upper elevations, NW, SW, KM, LB, SE (Gifberg to Kouga Mts).•

TRIEENEA•   9 spp., W Cape, mostly Cedarberg Mts

elsiae Hilliard   Glandular-hairy ?annual. Leaves opposite below becoming alternate above. Flowers in rounded heads on nude peduncles, tube short and glandular-hairy, white with orange patch at base of upper lobes and down back of tube. Oct.–Dec. Sandstone slopes, NW (S Cedarberg and Cold Bokkeveld Mts).•

frigida Hilliard   Glandular-hairy ?annual. Leaves opposite. Flowers in lax racemes, tube short and minutely glandular-hairy, colour unknown. Jan.–Feb. Sandstone slopes, NW (Cold Bokkeveld Mts).•

glutinosa (Schltr.) Hilliard (= *Polycarena glutinosa* (Schltr.) Levyns)   Glandular-hairy annual or short-lived perennial. Leaves opposite becoming alternate above and passing into bracts. Flowers in heads elongating into racemes, tube glandular-hairy, whitish to pale mauve with 2 orange patches at base of upper lobes. Mainly Sept.–Dec. Sandy peaty slopes from 1600–2100 m, NW, KM, SE (Gifberg to Great Winterhoek Mts).•

lanciloba Hilliard   Glabrous shrublet to 20 cm. Leaves opposite. Flowers in round heads, white or cream-coloured with orange patch at base of upper lobes and down back of tube. Dec.–Jan. Sandstone slopes, NW (N Cedarberg Mts).•

lasiocephala Hilliard   Hairy shrublet to 30 cm. Leaves opposite. Flowers in rounded heads, minutely glandular-hairy, white or cream-coloured with orange at base of upper petals and down back of tube. Dec.–Jan. Sandstone slopes, NW (N Cedarberg Mts).•

laxiflora Hilliard   Shrublet to 45 cm, glandular-hairy on inflorescence axes. Leaves opposite. Flowers in lax heads arranged in open corymbose panicles, mauve with orange at base of upper petals and on back of tube. Dec.–Jan. Sandstone slopes, NW (N Cedarberg Mts).•

longipedicellata Hilliard   Glandular-hairy perennial with sprawling rooting branches. Leaves opposite becoming alternate above and passing into bracts. Flowers solitary in upper axils on long pedicels, tube short and glandular-hairy, white with orange patch at base of lower lobes and down back of tube. Mainly Nov.–Feb. Damp places under sandstone overhangs, SW (Paarl to Genadendal).•

schlechteri (Hiern) Hilliard (= *Polycarena schlechteri* (Hiern) Levyns)   Glandular-hairy perennial. Leaves mostly opposite. Flowers in lax racemes, often autogamous, tube short and glandular-hairy, white with orange patch at base of lower lobes and down back of tube. Nov.–Dec. Under sandstone overhangs, NW (Cedarberg and Cold Bokkeveld Mts).•

taylorii Hilliard   Hairy perennial to 30 cm. Leaves opposite, sometimes alternate above. Flowers in rounded heads elongating in fruit, tube short, minutely glandular-hairy, upper lip dark violet with orange patch running down tube, lower lip white or cream-coloured. Oct.–Dec. Sheltered sandstone slopes, NW (Cedarberg Mts).•

ZALUZIANSKYA DRUMSTICKS, VERFBLOMMETJIE 55 spp., southern Africa, annuals mainly W Cape, perennials mainly E southern Africa

A. Petals entire

acrobareia Hilliard   Like **Z. glandulosa** but flower tube 22–30 mm long. Aug. Moist sandstone pavement, NW (Bokkeveld Mts to Gifberg).•

benthamiana Walp. (= *Z. ramosa* Hiern)   Annual to 33 cm, stems with spreading hairs. Leaves glandular-hairy. Flowers in a crowded spike, tube glandular-hairy, 10–20 mm long, lobes rounded, white to yellow inside with star-shaped orange centre and maroon outside, stamens 2. June–Aug. Sandy or gravelly flats and slopes, KM (S Namibia and W Karoo to Oudtshoorn).

divaricata (Thunb.) Walp.   Annual to 25 cm, stems with retrorse hairs. Leaves thinly hairy. Flowers in a crowded spike, tube long, 18–20 mm long, lobes rounded, yellow inside with a red star-shaped centre and brown outside, stamens 4. July–Oct. Stony or gravelly slopes, below 750 m, NW, SW, LB (Pakhuis Mts to Albertinia).•

glandulosa Hilliard   Dwarf annual to 4 cm, stems with spreading, glandular hairs. Leaves glandular-hairy. Flowers few in a crowded spike, tube minutely glandular-hairy, c. 10 mm long, lobes rounded, yellow inside and brown outside, stamens 4. Aug.–Sept. Sandy soil, 900 m, NW (Pakhuis Mts).•

isanthera Hilliard   Dwarf annual to 4 cm, stems with spreading hairs. Leaves hairy. Flowers in a dense spike, tube minutely glandular-hairy, 9–11 mm long, lobes rounded, creamy white inside with a yellow centre and mauve outside, stamens 4. Sept.–Oct. Sand on rock sheets, 900–1300 m, NW (Cold Bokkeveld Mts).•

peduncularis (Benth.) Walp. (= *Zaluzianskya gilioides* Schltr.)   Annual to 25 cm, stems with retrorse hairs. Leaves thinly hairy. Flowers in a terminal head, tube long and glandular-hairy, 17–25 mm long, lobes rounded, cream-coloured to lemon-yellow inside with a yellow or red centre and maroon outside, opening at dusk, stamens 4. June–Sept. Sandy or stony places, 100–1900m, KM (S Namibia and W Karoo to Matroosberg to Lesotho).

pusilla (Benth.) Walp. (= *Zaluzianskya collina* Hiern)   Annual to 15 cm, stems with retrorse hairs. Leaves thinly hairy. Flowers initially in a crowded spike, tube sometimes sparsely glandular-hairy, 6–17 mm long, lobes rounded, white or cream-coloured inside with star-shaped orange centre and maroon outside, stamens 4. July–Oct. Rocky or sandy slopes and flats, 150–950 m, NW, LB (Namaqualand and W Karoo to Swellendam).

A. Petals bifid or notched
B. Stamens 2

affinis Hilliard   Like **Z. villosa** but leaves and bracts lanceolate, acute, mostly entire and glabrescent. June–Nov. Sandy flats and slopes, to 900 m, NW (Richtersveld to Vredenburg).

gracilis Hilliard   Like **Z. parviflora** but spikes not floriferous from near base. Aug.–Oct. Calcareous sands near the coast, AP (De Hoop to Stilbaai).•

parviflora Hilliard   Annual to 30 cm, stems with spreading hairs. Leaves thinly hairy. Flowers in long, leafy spikes, very small and shorter than bracts, tube 8–10 mm long, lobes deeply bifid, cream-coloured with a yellow centre, stamens 2. Aug.–Sept. Limestone and granite outcrops, SW (Paternoster to Saldanha Bay).•

villosa F.W.Schmidt   Annual to 30 cm, stems with retrorse hairs. Leaves more or less oblanceolate, densely hairy. Flowers in crowded spikes later elongating, tube 10–25 mm long, lobes deeply bifid, white to mauve with yellow or red star in centre, stamens 2. July–Nov. Sandy flats along the coast, to 200 m, SW, AP (Langebaan to Pearly Beach).•

B. Stamens 4

capensis (L.) Walp.   Annual or short-lived perennial, stems with retrorse hairs. Leaves usually hairy. Flowers in a spike, tube 25–40 mm long, lobes deeply notched, white inside and red outside, opening at dusk and then scented, stamens 4. Mainly July–Oct. Sandy places, NW, SW, LB, SE (Namaqualand to E Cape).

maritima (L.f.) Walp.   Annual or short-lived perennial, stems with retrorse hairs. Leaves often fleshy, nearly glabrous. Flowers in a crowded head, tube 25–50 mm long, lobes deeply notched, white inside and crimson outside, opening at dusk and then scented, stamens 4. Jan.–Dec. Coastal dunes along foreshore, SE (George to E Cape).

muirii Hilliard & B.L.Burtt   Annual or short-lived perennial, stems with retrorse hairs. Leaves narrow and thinly hairy. Flowers in a spike, tube 15–20 mm long, glabrous in mouth, lobes deeply notched, white or pink inside and red outside, probably opening at dusk, stamens 4. June–Sept. Sandy soil in scrub, LB, AP (Potberg to Stilbaai).•

ovata (Benth.) Walp.   Twiggy shrublet to 45 cm, usually coarsely hairy. Leaves shaggy. Flowers few in spikes, tube 30–60 mm long, lobes deeply notched, white inside (sometimes orange in centre) and red outside, opening at dusk, stamens 4. Nov.–Dec. NW, KM (W Karoo to Tulbagh, Little Karoo to KwaZulu-Natal).

synaptica Hilliard   Annual to 24 cm, stems with spreading hairs. Leaves thinly glandular-hairy. Flowers in dense spikes, tube glandular-hairy, 17–26 mm long, lobes notched or almost entire, white or orange inside with star-shaped orange centre and coppery orange outside, stamens 4. Aug.–Oct. Sandy or shale slopes, 600–1200 m, NW, KM, SE (Hex River valley to Graaff-Reinet).

venusta Hilliard   Annual to 15 cm, stems with spreading hairs. Leaves glandular-hairy. Flowers crowded in a head, tube glandular-hairy, 15–26 mm long, lobes bifid, mauve to pink inside with star-shaped orange centre and orange outside, stamens 4. June–Nov. Sandy or gravelly soils, 350–1500 m, KM (W Karoo and Great Karoo to Free State).

SOLANACEAE
Solanum by W.G. Welman

1. Fruit a capsule; slender shrub with drooping, tubular yellow flowers ....**Nicotiana**
1.' Fruit a berry:
   2. Stamens inserted in mouth or throat of corolla; anthers dehiscing by pores ....**Solanum**
   2.' Stamens inserted near base of corolla tube; anthers dehiscing by longitudinal slits ....**Lycium**

LYCIUM   HONEY THORN   c. 200 spp., cosmopolitan in warm and temperate regions

afrum L.   KRAAL HONEY THORN, KRAALKRIEKDORING   Stiffly branched, thorny shrub or small tree to 3 m. Leaves leathery, linear, in tufts on short shoots. Flowers 12–20 mm long, tubular with petals about a quarter as long as tube, purple, stamens included, inserted half way up tube. Mainly July–Sept. Mostly dry stony slopes and flats, NW, SW, LB, SE (Lambert's Bay to Uitenhage).•

cinereum Thunb. (incl. **L. horridum** Thunb., **L. tetrandrum** Thunb.)   BOKSDORING   Stiffly branched, thorny shrub to 2 m. Leaves succulent to leathery, obovoid, in tufts on short shoots. Flowers to 12 mm long, with tube abruptly constricted below, petals half to one-third as long as tube, white to pale mauve, anthers and style exserted. Mainly July–Sept. Sandy or karroid flats, NW, SW, AP, KM, LB, SE (Namaqualand and Karoo to Gauteng).

ferocissimum Miers   SLANGBESSIE   Stiffly branched, thorny shrub to 2 m. Leaves leathery, oblanceolate to obovate, in tufts on short shoots. Flowers to 12 mm long, bell-shaped, petals half to one-third as long as tube, white to mauve, anthers and style exserted. Mainly July–Sept. Dry stony flats, NW, SW, AP, KM, LB, SE (Namaqualand and W Karoo to E Cape).

oxycarpum Dunal   WOLWEDORING   Like **L. afrum** but leaves usually oblanceolate, flowers cream-coloured with mauve petals, and anthers and style often exserted, stamens inserted in lower third of tube. Mainly July–Sept. Dry karroid slopes and flats, NW, KM, SE (S Namibia and Karoo to E Cape).

*****NICOTIANA**   WILD TOBACCO   67 spp., America, Australasia, 1 sp., Namibia

*****glauca** Graham   TABAKBOOM   Slender shrublet or small tree to 3 m. Leaves ovate-elliptic, long-petioled. Flowers in loose terminal cymes, tubular, 35–40 mm long, yellow. Aug.–Mar. Stony slopes and dry river courses, NW, SW, KM, LB, SE (American weed, widespread in dry areas).

SOLANUM   NIGHTSHADE   c. 1400 spp., cosmopolitan, mainly tropical

A. Plants unarmed

africanum Mill. (= ?*Solanum aggerum* Dunal, *S. quadrangulare* Thunb. ex L.f.)   DRONKBESSIE   Scrambling or prostrate semisucculent shrub to 3 m, stems squared when young. Leaves lanceolate to ovate, lower often lobed, to 6 cm long. Flowers c. 30 in terminal panicles, white, mauve or purple, 10 mm diam. Berries purplish black, to 15 mm diam. Mainly Jan.–Oct. Coastal dunes in bush, SW, AP, LB, SE (Cape Peninsula to KwaZulu-Natal).

crassifolium Lam.   Sprawling perennial with bristly stems to 30 cm, rooting at nodes. Leaves leathery, softly hairy, ovate or sinuately angled, to 4 cm long. Flowers in terminal cymes, blue-mauve, 5 mm diam. Berries ?black, 5 mm diam. Feb.–Apr. Coastal dunes in scrub, SW, AP (Gordon's Bay to Bredasdorp).•

guineense L. (= *Solanum dasypus* E.Mey.)   Erect or sprawling shrub to 1.5 m. Leaves ovate to elliptic, softly leathery, to 7 cm long. Flowers 1–few in axils, mauve to light blue, to 18 mm diam. Berries yellow, orange or red, to 15 mm diam. Mar.–Aug. Coastal dunes, slopes and river banks, NW, SW, AP, LB, SE (Namaqualand to E Cape).

*****mauritianum** Scop.   BUGWEED, LUISBOOM   Whitish felted small tree to 5 m. Leaves ovate to elliptic, discolorous, to 25 cm long. Flowers crowded in terminal corymbs, purple, to 10 mm diam. Berries yellow, 10 mm diam., velvety. Mainly May–July. Weed of disturbed places, scrub and forests, SW, LB, SE (S American weed).

nigrum L.   NASTERGAL   Glabrous to thinly hairy annual to 1 m. Leaves soft, ovate-lanceolate, often toothed, to 10 cm long. Flowers 5–10 in clusters, white, to 5 mm diam. Berries black, to 10 mm diam. Mainly Oct. Sheltered sites and disturbed places, SW, AP, LB, SE (Cape Peninsula to Eurasia).

*****pseudocapsicum** L.   JERUSALEM CHERRY, BOSGIFAPPEL   Softly woody shrublet to 1 m. Leaves bright green, elliptic to lanceolate, to 12 cm long. Flowers 1–few in clusters, white, 5 mm diam. Berries orange-red, 10–15 mm diam. Oct.–Jan. Weed of disturbed areas and forest margins, NW, SW, LB, SE (S American weed, widely cultivated).

retroflexum Dunal   Like **S. nigrum** but leaves deeply serrate, to 8 cm long, flowers 3–7 in clusters, white with purple keels, to 6 mm diam. Jan.–Dec. Usually forest margins or clearings, also weedy, SW, KM, LB, SE (Cape Peninsula to Arabia).

A. Plants armed

aculeastrum Dunal   GOAT APPLE, BOKAPPEL   Shrub or small tree to 5 m with young stems whitish woolly; spines to 15 mm long, recurved, brownish. Leaves deeply lobed, dark green glabrescent above, whitish tomentose beneath, to 15 cm long. Flowers in clusters of 5–10, whitish, 10 mm diam. Berries yellow, to 50 mm diam. Mar.–Oct. Grassy slopes and forest margins, also roadsides, LB, SE (Riversdale to tropical Africa).

aculeatissimum Jacq.   Softly hairy shrublet to 1 m with purple and green stems; spines to 15 mm long, straight. Leaves lobed, spiny, to 12 cm long. Flowers in clusters of 3–5, white, sometimes purplish, 10 mm diam. Berries yellow, to 25 mm diam. Nov.–Feb. Weedy on forest margins and roadsides, SE (George to tropical Africa).

capense L.   Sprawling or prostrate shrublet to 60 cm; spines to 5 mm long, recurved, yellow. Leaves elliptic, often deeply lobed, midrib spiny beneath, to 5 cm long. Flowers few in clusters, white, to 5 mm diam. Berries orange to red, to 10 mm diam. Mainly Dec. Rocky slopes and disturbed areas, KM, SE (Little Karoo to E southern Africa).

coccineum Jacq.   KLEINGRYSBITTERAPPEL   Softly hairy shrublet to 60 cm; spines to 5 mm long, straight. Leaves ovate, sinuate, rarely spiny, to 5 cm long. Flowers few in clusters, mauve to violet, to 5 mm diam. Berries orange-red, to 10 mm diam. Mainly Mar.–Apr. Sandy and rocky slopes and flats and waste places, NW, KM, LB, SE (Worcester to Mpumalanga and Botswana).

giftbergense Dunal   Shrub to 1.5 m; spines to 12 mm long, slender, straight, yellow to reddish brown. Leaves ovate, obtusely lobed, spiny on veins, to 4 cm long. Flowers 1–4 in clusters, mauve to purple, to 10 mm diam. Berries orange to red, 10 mm diam. May–Sept. Rocky sandstone slopes, NW (Namaqualand to Pakhuis Mts).

giganteum Jacq.   GENEESBLAARBOOM   Shrub or small tree to 3 m; white-woolly on young parts; spines to 4 mm long, stout. Leaves elliptic, white-woolly beneath, to 15 cm long. Flowers crowded in terminal corymbs, mauve to purple, 6 mm diam. Berries red, to 8 mm diam. Dec.–Apr. Forest margins, river banks and disturbed places, SW, LB, SE (Cape Peninsula to tropical Africa).

linnaeanum Hepper & E.Jaeger (= *Solanum hermannii* Dunal)   BITTER APPLE, BITTERAPPEL   Shrub to 1 m; spines to 12 mm long, straight, yellow. Leaves deeply lobed, usually spiny, to 15 cm long. Flowers 1–few in axils, mauve to purple, 15 mm diam. Berries yellow, 25 mm diam. Mainly June–Sept. Rocky slopes and flats and roadsides, NW, SW, AP, KM, LB, SE (Worcester and Darling to KwaZulu-Natal).

macowanii Fourc.   Shrub to 1.8 m; spines long, straight. Leaves deeply lobed, to 18 cm long. Flowers paired, white, to 12 mm diam. Berries red, 25 mm diam. Apr. Forests, SE (Humansdorp to E Cape: perhaps not distinct from **S. aculeatissimum** Jacq. or the exotic **S. casicoides** Allioni).

rigescens Jacq. WILDE-LEMOENTJIE   Hairy erect or scrambling shrub to 1 m; spines to 8 mm long, straight. Leaves lobed, spiny on main veins, to 8 cm long. Flowers few in clusters, mauve, blue or purple, to 12 mm diam. Berries yellow to orange, to 15 mm diam. May–Sept. Coastal scrub and forests, SW, AP, LB, SE (Caledon to N Province).

tomentosum L.   SLANGAPPELBOS   Yellowish green, densely felted shrub to 1 m; spines to 12 mm, straight or slightly curved. Leaves ovate, sinuate, rarely spiny, to 8 cm. Flowers few in clusters, mauve or purple, to 15 mm long. Berries orange, to 20 mm. Mainly Aug.–Oct. Stony slopes and flats, NW, SW, KM, LB, SE (Namaqualand to E Cape).

STILBACEAE• (= *RETZIACEAE*)

1. Corolla apparently 4-lobed, upper 2 lobes more or less completely fused:
   2. Sepals free; corolla lobes glabrous..**Campylostachys**
   2.' Sepals fused below into a tube; corolla lobes pubescent ....**Thesmophora**
1.' Corolla 5-lobed:
   3. Corolla bilabiate, with 2 larger erect posterior lobes and 3 narrow anterior lobes ....**Stilbe**
   3.' Corolla actinomorphic, lobes equal or almost equal:
      4. Corolla lobes glabrous ....**Euthystachys**
      4.' Corolla lobes pubescent:
         5. Flowers 45–55 mm long, orange-red with black tips; leaves 30–70 mm long............................**Retzia**
         5.' Flowers shorter than 10 mm, white to mauve; leaves shorter than 20 mm ....**Kogelbergia**

CAMPYLOSTACHYS•   2 spp., W Cape

cernua (L.f.) Kunth   Closely leafy, resprouting shrublet to 80 cm, with velvety branches. Leaves oblong-lanceolate, in whorls of 4, 2-grooved below, spreading to recurved. Flowers in nodding, subglobose spikes, white, corolla hairy in throat. Nov.–Mar. Sandstone slopes, SW (Cape Peninsula and Du Toitskloof to Bredasdorp Mts).•

sp. 1   Compact, resprouting shrublet to 50 cm. Leaves oblong-elliptic, recurved at tips, 2-grooved below. Flowers 2 or 3 at branch tips, white, corolla hairy in throat. Oct.–Nov. Rocky slopes, SW (Riviersonderend Mts).•

EUTHYSTACHYS•   1 sp., W Cape

abbreviata (E.Mey.) A.DC.   Closely leafy, resprouting shrublet to 80 cm. Leaves linear, in whorls of 4, 2-grooved below, spreading to recurved. Flowers minute, in compact, nodding, subglobose spikes, concealed by long leafy bracts, white, calyx bilabiate, corolla hairy in throat. Aug.–Oct. Sandstone slopes, at high alt., SW (Bainskloof to Franschhoek Mts).•

KOGELBERGIA•   2 spp., W Cape

phylicoides (A.DC.) Rourke (= *Stilbe phylicoides* A.DC.)   Like **K. verticillata** but leaves erect to spreading, in whorls of 6 or 7. Oct.–Nov. Sandstone slopes, LB (Langeberg Mts: Swellendam to Riversdale).•

verticillata (Eckl. & Zeyh.) Rourke (= *Stilbe mucronata* N.E.Br., *S. verticillata* Eckl. & Zeyh.) Moldenke) Closely leafy shrub to 1 m, branches velvety. Leaves lanceolate, spreading to reflexed, in whorls of 4 or 5, 2-grooved below, glabrescent. Flowers in subglobose spikes, petals densely silky, grey-mauve, corolla hairy in throat. May–Oct. Moist slopes at high alt., SW, AP (Hottentots Holland Mts to Agulhas coast).•

RETZIA•   HEUNINGBLOM   1 sp., W Cape

capensis Thunb.   Stiffly erect, densely leafy shrublet to 1 m, branches velvety. Leaves in whorls of 4, ascending, imbricate, linear-lanceolate, 2-grooved beneath, glabrescent. Flowers few in axils, tubular, silky hairy, red with black lobes tipped with a tuft of white hairs. Sept.–Mar. Sandstone slopes, SW (Hottentots Holland to Bredasdorp Mts).•

STILBE• (= *EURYLOBIUM, XEROPLANA*)   BOTTLEBRUSH   7 spp., W Cape

albiflora E.Mey.   Erect, resprouting shrublet to 1.2 m, with velvety branches. Leaves imbricate, ascending to spreading, linear, in whorls of 4–6, 2-grooved below, recurved at tips, mucronate. Flowers in ovoid spikes, white, calyx cartilaginous, corolla hairy in throat. Nov.–Feb. Sandstone slopes, NW, SW, LB (Cedarberg to Langeberg Mts: Swellendam).•

ericoides L.   Erect or straggling, resprouting shrublet to 80 cm, with velvety branches. Leaves imbricate, ascending, linear, in whorls of 4, 2-grooved below. Flowers in subglobose spikes, pink, calyx membranous, corolla hairy in throat. Apr.–Sept. Sandy flats or limestone hills, SW, AP (Hopefield to De Hoop).•

gymnopharyngia (Rourke) Rourke (= *Xeroplana gymnopharyngia* Rourke)   Sprawling, densely branched shrublet to 30 cm. Leaves imbricate, sometimes spreading, in whorls of 3 or 4, linear, 2-grooved below. Flowers in dense spikes, pink, corolla weakly bilabiate, throat glabrous. Oct.–Nov. Sandstone slopes, LB (Langeberg Mts: Riversdale).•

overbergensis Rourke (= *Xeroplana zeyheri* Briq.)   Like **S. ericoides** but single-stemmed, leaves mainly in whorls of 3. Apr.–July. Lower sandstone and limestone slopes, SW, AP (Riviersonderend Mts to Agulhas coast and De Hoop).•

rupestris Compton   Sprawling shrublet to 20 cm. Leaves becoming spreading or reflexed, linear, in whorls of 2–4, 2-grooved below, glabrescent. Flowers in crowded spikes, pink to white, corolla hairy in throat. Dec.–Apr. Rock crevices on sandstone slopes, SW (Hottentots Holland to Kleinrivier Mts).•

serrulata (Hochst.) Rourke (= *Eurylobium serrulatum* Hochst.)   Closely leafy shrub to over 1 m. Leaves imbricate, ascending, linear-elliptic, in whorls of 4, minutely serrate, punctate above, 2-grooved below. Flowers in crowded spikes, white, calyx membranous, corolla weakly bilabiate, hairy in throat. Jan.–Feb. Sandstone slopes, SW (Riviersonderend Mts).•

vestita P.J.Bergius   Like **S. albiflora** but petals more or less silky hairy along margins, whitish. Mostly July–Dec. Sandstone slopes, SW (Cape Peninsula to Houwhoek).•

THESMOPHORA•   1 sp., W Cape

scopulosa Rourke   Sprawling, closely leafy shrublet to 30 cm. Leaves imbricate, lanceolate, in whorls of 4, minutely serrate, punctate above, 2-grooved beneath. Flowers in spikes, pinkish mauve, lobes woolly at tips. Sept.–Nov. High sandstone slopes, NW (Ceres).•

THYMELAEACEAE

Gnidia by A. Beaumont, **Lachnaea** and **Struthiola** by J.B.P. Beyers, **Passerina** by C. Bredenkamp

1. Flowers without petaloid scales; stamens 8, well exserted...**Passerina**
1.' Flowers usually with petaloid scales, rarely lacking but then stamens included or shortly exserted:
   2. Scales 8, arising in hypanthium below level of stamen insertion ....**Lachnaea**
   2.' Scales 4–12, arising at mouth of hypanthium above level of stamen insertion:
      3. Stamens 4 in a single whorl; anthers usually with an apical appendage; flowers solitary in axils in elongate racemes ....**Struthiola**
      3.' Stamens 8 in 2 whorls, rarely upper whorl lacking; flowers usually terminal in clusters, rarely solitary in upper axils..**Gnidia**

GNIDIA (= *ARTHROSOLEN, LASIOSIPHON*)   SAFFRON BUSH, SAFFRAAN   c. 150 spp., mainly Africa, also Madagascar and India

A. Sepals 5 (rarely 4 or 6)

anthylloides (L.f.) Gilg   Slender shrub to 1.5 m. Leaves alternate, oblong-elliptic to elliptic, silky hairy beneath; involucral leaves slightly smaller with robust bases. Flowers c. 10–20 in terminal heads, bright yellow, hypanthium silky hairy, petaloid floral scales 5, tiny. Jan.–Dec. Coastal scrub and grassland, SE (Mossel Bay to N provinces).

capitata L.f.   KERRIEBLOM   Many-stemmed shrublet to 70 cm from a fleshy rootstock. Leaves linear-lanceolate, bluish green; involucral leaves similar but with robust bases. Flowers 10–15 in terminal heads, bright yellow, petaloid floral scales 0–5, membranous. Oct.–Feb. Grassland, SE (Humansdorp to N provinces).

cuneata Meisn. (= *Lasiosiphon meisnerianus* Endl.)   KOORBOSSIE   Shrub to 1 m. Leaves alternate, obovate-oblong to lanceolate, smoothly silky hairy; involucral leaves shorter, with robust bases. Flowers 3–12 in terminal clusters, ochre, hypanthium densely silky hairy, petaloid floral scales 5, membranous. Mar.–July. Lower slopes, SE (Humansdorp to KwaZulu-Natal).

A. Sepals 4
B. Petaloid floral scales absent

inconspicua Meisn.   Shrublet to 25 cm. Leaves opposite or whorled, oblong-ovate to lanceolate, moderately silky hairy (sparkle in direct light); involucral leaves similar. Flowers solitary or paired at branch tips, pale yellow, hypanthium silky hairy. Sept.–Dec. Flats and lower slopes, NW, SW (Ceres to Cape Peninsula).•

laxa (L.f.) Gilg   Shrub to 1 m. Leaves subopposite, linear-oblong, pilose when young; involucral leaves similar. Flowers 4–8 at branch tips, greenish yellow, sometimes tinged dark pink, hypanthium funnel-shaped, hairy. Jan.–Dec. Lower slopes, SW, LB (Cape Peninsula to Swellendam).•

ornata (Meisn.) Gilg (= *Gnidia vesiculosa* Eckl. & Zeyh. ex Drège)   Shrublet to 30 cm or more. Leaves opposite, sometimes alternate, oblong-lanceolate, sparsely hairy beneath; involucral leaves similar but pilose on margins and beneath. Flowers 1–few at branch tips, white, hypanthium silky villous. July–Oct. Marshy flats and lower slopes, SW (Kleinrivier Mts to Bredasdorp).•

spicata (L.f.) Gilg   Shrublet to 80 cm. Leaves scattered, ovate to broadly lanceolate, hairy beneath. Flowers in racemes, cream-coloured to pale pink, hypanthium funnel-shaped, hairy, bracteoles 2. Aug.–Oct. Marshy flats and dune slacks, SW, AP (Malmesbury to Bredasdorp).•

B. Petaloid floral scales present
C. Floral scales 8

anomala Meisn.   Lanky shrub up to 2 m. Leaves opposite, ovate-oval to oval-oblong, densely silky hairy; involucral leaves similar. Flowers few at branch tips, pale yellow, hypanthium densely silky hairy, petaloid floral scales 8, upper stamens reduced or lacking. Sept.–Dec. Mountain slopes, SW, AP, LB, SE (Cape Peninsula to Knysna).•

caniflora Meisn.   Shrublet to 50 cm. Leaves alternate or subopposite, oblong or lanceolate; involucral leaves similar. Flowers few at branch tips, cream-coloured; hypanthium densely villous, petaloid floral scales 8, anther-like. Apr.–July. Lower slopes, NW, SW, LB (Bokkeveld Mts to Riversdale).•

denudata Lindl.   Graceful shrub to 4 m. Leaves opposite, oblong-elliptic, densely silky hairy, becoming glabrous beneath; involucral leaves similar. Flowers 4–9 at branch tips, pale yellow, hypanthium densely pilose, petaloid floral scales 8, small. Aug.–Jan. Coastal forest margins, LB, SE (Langeberg to Tsitsikamma Mts).•

ericoides C.H.Wright   Dwarf shrub. Leaves opposite, ericoid, glabrescent, slightly pubescent when young; involucral leaves similar. Flowers 6–9 at branch tips, greenish yellow; hypanthium funnel-shaped, silky hairy, petaloid floral scales 8, anther-like. June–Sept. Lower slopes, LB (Swellendam to Riversdale).•

francisci Bolus   Shrublet to 60 cm. Leaves opposite, subulate, acuminate; involucral leaves shorter and wider, silky. Flowers few–5 at branch tips, cream-coloured, hypanthium silky hairy, petaloid floral scales 8, slightly fleshy. Nov.–Apr. Mountain slopes, KM, SE (Swartberg to Outeniqua Mts).•

imbricata L.f.   Shrublet to 30 cm., silvery hairy. Leaves opposite to subopposite, elliptic-lanceolate, densely silky hairy; involucral leaves similar. Flowers 2–4 at branch tips, cream-coloured, hypanthium densely silky hairy, petaloid floral scales 8, anther-like. June–Jan. Lower and middle slopes, NW, SW (Namaqualand to Cape Peninsula).

leipoldtii C.H.Wright   Spreading slender shrublet to 60 cm. Leaves opposite, ovate, densely silky hairy; involucral leaves similar. Flowers 2–6 at branch tips, cream-coloured, hypanthium densely silky hairy, petaloid floral scales 8, anther-like. Mainly July–Jan. Sandstone slopes, NW (Bokkeveld Mts).•

meyeri Meisn.   Spreading slender shrublet to 80 cm. Leaves alternate, linear-lanceolate; involucral leaves similar. Flowers few–8 at branch tips, pale yellow to yellow-green, hypanthium minutely hairy, petaloid floral scales 8, membranous. Aug.–Oct. Dry slopes, NW, SW (Namaqualand and Karoo to Malmesbury).

nitida Bolus ex C.H.Wright   Small shrub. Leaves opposite, crowded towards branch tips, elliptic-subacute, smoothly silky, becoming glabrous. Flowers paired at branch tips, cream-coloured to yellow, hypanthium densely silky hairy, petaloid floral scales 8, anther-like. Mar.–Oct. Rocky sandstone slopes, KM (Namaqualand and W Karoo to Ladismith).

nodiflora Meisn.   Shrublet to 30 cm. Leaves alternate to verticillate, linear-oblong, hairy becoming subglabrous. Flowers 3–6 at branch tips, surrounded by a whorl of leaves, pale blue and white, hypanthium silky hairy, petaloid floral scales 8, linear. July–Dec. Coastal bush and slopes, SE (Mossel Bay to KwaZulu-Natal).

obtusissima Meisn.   Shrub to 30 cm. Leaves opposite, oblong-subulate; involucral leaves elliptic-lanceolate. Flowers 2 or 3 at branch tips, cream-coloured, hypanthium slightly funnel-shaped, silky hairy, petaloid floral scales 8, slightly fleshy, oblong. Sept.–Oct. Mountain slopes, NW, SW, KM, SE (Hex River Mts to Uitenhage).•

parvula Wolley-Dod   Shrublet to 30 cm. Leaves scattered, narrowly lanceolate; involucral leaves similar. Flowers 6–8 at branch tips, yellowish, hypanthium sparsely hairy, petaloid floral scales 8, oblong, as long as sepals. Aug.–Nov. Granite slopes, SW (Cape Peninsula, Durbanville).•

scabra Thunb.   Shrub to 1 m. Leaves alternate or subopposite, oblong-lanceolate to linear-lanceolate. Flowers up to 3 at branch tips, cream-yellow, hypanthium puberulous, petaloid floral scales 8. Sept.–Apr. Lower slopes, watercourses, NW, KM (Cold Bokkeveld and W Karoo to Little Karoo).

scabrida Thunb.   Shrub to 1.3 m. Leaves linear to linear-lanceolate, long-ciliate when young, later sparsely hairy; involucral leaves more hairy. Flowers few at branch tips, cream-yellow, hypanthium densely villous, petaloid floral scales 8, lanceolate-acuminate. July. Middle slopes, SW, LB, SE (Du Toitskloof to George).•

sericea L.   Willowy shrub to 1.5 m. Leaves opposite, elliptic, adpressed-silky hairy, sometimes velvety above; involucral leaves similar. Flowers 5–9 at branch tips, cream, hypanthium adpressed-silky hairy, petaloid floral scales 8, small. July–Sept. Scrub, often along streams, NW, SW, AP, LB, SE (Worcester and Caledon to E Cape).

setosa (Thunb.) Wikstr.   Shrublet to 45 cm. Leaves alternate, lanceolate. Flowers in condensed spikes elongating in fruit, yellow, hypanthium silky hairy, sepals very small, petaloid floral scales 8, subulate. Fruits teardrop-shaped, hairy. Sept. Lower slopes, SW, LB (Malmesbury to Riversdale).

squarrosa (L.) Druce (incl. **G. polystachya** P.J.Bergius)   Lax, much-branched willowy shrub to 2 m. Leaves alternate, linear-lanceolate; involucral leaves whorled, slightly broader, margins sometimes ciliate. Flowers 6–30 at branch tips, pale cream-green, ovary portion and sepal tips often pink, hypanthium pilose, petaloid floral scales 8, finger-like, variable. June–Oct. Coastal limestone, sandy slopes, SW, AP, SE (Cape Peninsula to E Cape).

strigillosa Meisn.   Shrublet to 15 cm. Leaves opposite, linear-oblong, glabrescent. Flowers paired at branch tips, cream-coloured, hypanthium densely silky hairy, petaloid floral scales 8, awl-shaped. June. Lower slopes, LB (Swellendam).•

C. Floral scales 4, sometimes divided or fringed
D. Floral scales fleshy

chrysophylla Meisn.   Shrublet to 60 cm. Leaves opposite, elliptic-obovate, densely and smoothly silky hairy; involucral leaves similar. Flowers 6–9 at branch tips, cream-coloured, hypanthium densely silky hairy, petaloid floral scales 4, fleshy. June–Sept. Coastal flats, SW, AP, LB (Bredasdorp to Riversdale).•

geminiflora E.Mey. ex Meisn.   Shrub to 60 cm. Leaves opposite, linear-lanceolate, glabrous or silky hairy above; involucral leaves similar. Flowers paired at branch tips, cream-yellow, hypanthium silky hairy, petaloid floral scales 4, fleshy, bifid. June–Dec. Sandy flats and slopes, NW, SW (Namaqualand to Langebaan).

insignis Compton   Large-leaved shrub to 1m. Leaves opposite, broadly ovate to orbicular; involucral leaves slightly smaller. Flowers 6–15 at branch tips, yellow, hypanthium minutely hairy, ribbed, petaloid floral scales 4, fleshy. Sept.–Oct. S-facing sandstone slopes, SW (Du Toitskloof: Molenaar's Peak).•

nana (L.f.) Wikstr.   Shrub to 3 m. Leaves scattered, narrowly lanceolate, minutely warty, pilose on margins and beneath when young; involucral leaves similar. Flowers 3–5 at branch tips, grey-lilac to dull purple, sometimes cream-coloured, hypanthium densely hairy above, sepals reflexed, hairy above and beneath, petaloid floral scales in 4 groups of c. 3 slender fleshy segments forming a continuous ring interspersed with hairs, yellow, upper row of stamens abortive. Jan.–Dec. Sandstone slopes, SW (Elandskloof to Bredasdorp Mts).•

oppositifolia L.   Erect, willowy shrub to 3 m. Leaves opposite-decussate, oblong-lanceolate to elliptic-lanceolate; involucral leaves similar but crimson-edged. Flowers 4–6 at branch tips, pale yellow, hypanthium silky villous, ribbed, petaloid floral scales 4, fleshy, pale yellow (brown when dry). Jan.–Dec. Sandstone slopes, NW, SW, LB, SE (Clanwilliam to E Cape).

orbiculata C.H.Wright   Like **G. oppositifolia** but leaves broadly elliptic to orbicular. Jan.–Aug. Sandy flats, SE (George to Humansdorp).•

penicillata Licht. ex Meisn.   Shrublet to 40 cm. Leaves opposite, linear, margins densely ciliate, becoming glabrous; involucral leaves similar. Flowers 2–6 at branch tips, bright blue or pink, hypanthium slightly funnel-shaped, silky hairy, petaloid floral scales 4, divided into 4 slender fleshy segments each, hairy, upper stamens occasionally reduced. Aug.–May. Marshy flats and lower slopes, SW (Cape Peninsula to Caledon Swartberg).•

pinifolia L.   Shrub to 1 m. Leaves crowded, alternate, needle-like, narrowly linear-oblong, pungent-acuminate; involucral leaves wider. Flowers c. 10 at branch tips, white, hypanthium villous, petaloid floral scales 4, anther-like, fleshy, densely villous. Jan.–Dec. Flats to middle slopes, NW, SW, LB, SE (Piketberg to E Cape).

racemosa Thunb.   Shrub to 60 cm. Leaves scattered, obovate to lanceolate. Flowers racemose, pale yellow-green, hypanthium puberulous, petaloid floral scales 4, deeply bifid, fleshy. Flowering time? Riverbanks, LB, SE (Riversdale to E Cape).

tomentosa L.   Shrub to 1 m. Leaves alternate, ovate to lanceolate, warty beneath; involucral leaves similar. Flowers c. 6 at branch tips, white, hypanthium densely silky hairy, petaloid floral scales 4, fleshy, yellow, hairy. Jan.–Dec. Marshy sandstone slopes, SW (Du Toitskloof to Palmiet River Mts).•

sp. 1 (= *Gnidia struthioloides* Moss ms)   Shrublet to 30 cm. Leaves alternate, obovate-lanceolate; involucral leaves shorter, linear-oblong. Flowers 1–4 at branch tips, green-yellow, hypanthium silky hairy, petaloid floral scales 4. June–Aug. Coastal flats, SE (Humansdorp).•

D. Floral scales membranous

coriacea Meisn.   Much-branched shrublet to 30 cm, blackening when dried. Leaves opposite, ovate-oblong; involucral leaves slightly larger. Flowers 2–4 at branch tips, yellow, hypanthium glabrous, petaloid floral scales 4, pale and membranous. Oct.–Apr. Middle slopes, SW, SE (Kogelberg and Outeniqua Mts to KwaZulu-Natal).

deserticola Gilg   SAFFRAAN   Twiggy shrub to 30 cm. Leaves clustered at branch tips, alternate to subopposite, oblong-ovate, hispidulous; involucral leaves velvety, with swollen bases. Flowers 5–7 at branch tips, ochre-yellow, hypanthium densely silky hairy, base with long, spreading hairs persisting in fruit, petaloid floral scales 4, small, membranous. Jan.–Dec. Dry flats and lower slopes, NW, KM (W Karoo and Bokkeveld Mts to Worcester, Little Karoo to E Cape).

galpinii C.H.Wright   Slender-branched shrublet to 60 cm. Leaves opposite, oblong-lanceolate; involucral leaves slightly wider. Flowers paired at branch tips, yellow; hypanthium funnel-shaped, glabrous, petaloid floral scales 4, membranous, pale. Sept.–May. Sandstone slopes, SW, KM, LB (Somerset West to Joubertina).•

humilis Meisn.   Shrublet to 30 cm, branches slender, hairy. Leaves opposite, oblong-ovate, pilose; involucral leaves similar. Flowers solitary or paired at branch tips, yellow, hypanthium funnel-shaped, pilose, petaloid floral scales 4, membranous. Oct.–Mar. Damp sandstone slopes and bogs, SW (Cape Peninsula to Houwhoek Mts).•

juniperifolia Lam.   Erect or spreading shrublet to 50 cm. Leaves scattered, linear-subulate; involucral leaves slightly wider. Flowers paired at branch tips, yellow, hypanthium funnel-shaped, glabrous; petaloid floral scales 4, membranous, pale yellow. Jan.–Dec. Mountain slopes, SW, LB (Cape Peninsula to Riversdale).•

linearifolia (Wikstr.) B.Peterson   Shrublet to 30 cm. Leaves alternate to subopposite, lanceolate-ovate, white-dotted, pilose on margins and beneath when young; involucral leaves similar. Flowers few at branch tips, rose-pink to magenta, hypanthium hairy, petaloid floral scales 4, membranous, deeply and narrowly laciniate, hairy, upper stamens often abortive. Sept.–Dec. Mountains, SW, AP (Stellenbosch to Bredasdorp Mts).•

linoides Wikstr.   Delicate shrublet to 20 cm. Leaves opposite, linear-subulate. Flowers solitary in upper axils, white, hypanthium funnel-shaped, hairy, petaloid floral scales 4, membranous, bilobed. Oct.–Mar. Mountains, NW, SW, LB, SE (Worcester to N provinces).

parviflora Meisn.   Shrublet to 20 cm. Leaves opposite, linear-lanceolate; involucral leaves similar. Flowers usually solitary at branch tips, yellow, hypanthium glabrous, petaloid floral scales 4, small, membranous. Sept.–Oct. Mountain slopes, SW (Piketberg to Riviersonderend Mts).•

simplex L.   Shrublet to 30 cm. Leaves linear-lanceolate to awl-shaped, leathery; involucral leaves similar. Flowers 2–4 at branch tips, yellow, hypanthium funnel-shaped, glabrous, petaloid floral scales 4, membranous. Nov.–Apr. Mountain slopes, SW, LB (Cape Peninsula to Langeberg Mts).•

sonderiana Meisn.   Slender shrublet. Leaves opposite, leathery, ovate-lanceolate, hairy on margins and beneath. Flowers axillary or paired at branch tips, hypanthium funnel-shaped, sparsely hairy, petaloid floral scales 4, membranous. June–Aug. Rocky slopes, SW (Babilonstoring).•

styphelioides Meisn. (incl. **G. quadrifaria** C.H.Wright)   Spreading or erect shrublet to 30 cm. Leaves opposite, lanceolate-pungent; involucral leaves slightly larger. Flowers 1–3 at branch tips, yellow, hypanthium funnel-shaped, glabrous to sparsely hairy, petaloid floral scales 4, membranous. Aug.–Apr. Lower and middle slopes, SE (Humansdorp to E Cape).

tenella (Meisn.) Meisn. (= *Gnidia albicans* var. *tenella* Meisn.) Lax, silvery shrub to 2 m. Leaves opposite, oblong-oval, densely silky hairy. Flowers 2–4 at branch tips, cream-coloured, hypanthium densely silky hairy, petaloid floral scales 4, membranous to slightly fleshy. May–Nov. Mountain slopes, NW, SW (Ceres to Bredasdorp).•

sp. 2 (= *Gnidia fourcadei* Moss ms) Shrublet to 20 cm. Flowers yellow, petaloid floral scales 4, membranous. Sept.–Oct. Lower slopes, SW (Cape Peninsula to Caledon).•

LACHNAEA• MOUNTAIN-CARNATION, BERGANGELIER, LETJIESBOS 40 spp., W and E Cape

A. Flowers solitary, terminal (may appear axillary when much-reduced lateral branches comprise 1 pair of leaves); leaves usually decussate, floral scales usually included

axillaris Meisn. (= *Lachnaea micrantha* Schltr.) Compact, flexuose shrublet to 75 cm. Leaves ascending, narrowly ovate to lanceolate. Flowers cream-coloured to dark pink, glabrous outside, stigma capitate, papillate. Jan.–Dec. Sandy flats, SW, AP, LB (Hopefield and Darling, Elim to Gouritsmond).•

filicaulis (Meisn.) Beyers Erect, flexuosely branched shrublet to 40 cm. Leaves ascending to patent, lanceolate. Flowers cream-coloured or pink, silky outside, stigma conical, papillate. Jan.–Dec. Flats and lower mountain slopes SW, AP, LB (Betty's Bay to Riversdale).•

gracilis Meisn. Erect, laxly branched shrub to 2 m, compact and decumbent when grazed. Leaves adpressed to ascending, narrowly elliptic to obovate. Flowers cream-coloured, usually turning pink with age, silky outside, stigma capitate, papillate. Mostly July–Jan. Sandy flats and slopes, 300–1660 m, NW, SW, KM (Cedarberg Mts to Touwsrivier and Elandskloof Mts).•

grandiflora (L.f.) Baill. (= *Cryptadenia breviflora* Meisn.) Erect, usually compact, rounded, resprouting shrub, 20–60 cm, sometimes to 1 m. Leaves subadpressed to ascending, narrowly ovate to obovate or elliptic. Flowers large, pink or white, silky outside, stigma capitate, long papillate. Aug.–June. Sandy flats and lower slopes, 15–1000 m, NW, SW, AP (Cedarberg to Agulhas).•

laxa (C.H.Wright) Beyers Erect to sprawling shrublet to 30 cm. Leaves adpressed to ascending, narrowly elliptic to lanceolate. Flowers cream-coloured to pink, silky outside, scales exserted, stigma brush-like. Oct.–Dec. Damp, sandstone slopes, 600–2150 m, SW (Jonkershoek Mts to Caledon Swartberg and Elim).•

pudens Beyers Erect to sprawling, much-branched shrublet to 60 cm. Leaves ascending-incurved, elliptic. Flowers on declinate branches, dark red, silky outside, stigma brush-like. Aug.–May. Rocky slopes, 330–1530 m, SW (Riviersonderend Mts).•

ruscifolia Compton Moderately branched, resprouting shrub to 80 cm, multistemmed at base. Leaves elliptic to orbicular, palmately ribbed beneath. Flowers creamy white, hairy outside, scales exserted, stigma capitate, long papillate. Jan.–Dec. Sandstone slopes, 650–1830 m, KM, LB (Little Karoo mts).•

uniflora (L.) Beyers Like **L. grandiflora** but to 45 cm, single- or multistemmed at base, branches flexuose, leaves narrowly elliptic. Flowers pink, sometimes white, silky outside but hairs on lower hypanthium not acicular but blunt, stigma narrowly conical to ellipsoid, long papillate. July–Mar. Sandy flats and rocky sandstone slopes, 15–860 m, NW, SW (Porterville to Yzerfontein and Hottentots Holland Mts).•

sp. 1 (= *Lachnaea leipoldtii* Beyers ms) Erect to sprawling, much-branched shrublet to 40 cm. Leaves adpressed, narrowly elliptic to ovate. Flowers white, silky outside, stigma conical, papillate. Dec.–Feb. Sandstone slopes, c. 1000 m, NW (Cedarberg Mts).•

sp. 2 (= *Lachnaea pusilla* Beyers ms) Like **L. sp. 1** but leaves narrowly elliptic, flowers white or tinged pink, lower part of hypanthium with ascending, obtuse hairs, stigma linear-conical, papillate. Oct.–Mar. Sandstone slopes, 70–1000 m, SW (Elandskloof Mts).•

A. Flowers several in terminal umbels or heads

Group 1: Inflorescence a terminal, pedunculate, ebracteate capitulum; leaves alternate, floral scales usually exserted, stigma brush-like

alpina Meisn. Woody, much-branched shrublet to 1m. Leaves ascending, narrowly elliptic to spathulate. Flowers in capitula, 15–25 mm diam., pale blue or cream-coloured, honey-scented. Nov.–Jan. Sandstone slopes, 1600–2100 m, NW (Tulbagh, Ceres).•

buxifolia Lam.   Erect, moderately branched shrub to 1.7 m. Leaves ascending to patent, shape variable, narrowly ovate to obovate. Flowers many in capitula, 27–55 mm diam., cream-coloured, sometimes tinged blue, sweetly scented. Aug.–Jan. Sandstone slopes, 500–2160 m, SW, NW, KM (Olifants River Mts to Du Toitskloof and to Witteberg and Klein Swartberg).•

capitata (L.) Crantz   Slender, erect shrub to 1.8 m. Leaves ascending, linear to narrowly elliptic. Flowers in capitula, 5–16 mm diam., few open at a time, cream-coloured, receptacle hemispherical, becoming conical, floral scales included. June–Mar. Sandy flats and lower slopes, NW, SW (Clanwilliam to Cape Peninsula).•

densiflora Meisn.   Like **L. capitata** but to 50 cm, branching corymbose, flowers many and cream-coloured to dark pink, scales exserted. Aug.–Mar. Flats and lower slopes, SW, AP (Cape Peninsula to Bredasdorp).•

filamentosa Meisn.   BERGASTER   Erect, moderately branched, shrub to 1.5 m. Leaves ascending, shape variable, narrowly ovate to obovate. Flowers many in globose capitula, 3–10 cm diam., cream-coloured, lilac-blue or blue, sepals unequal, anterior sepal 2–3 times the size of others. July–Feb. Sandy or stony slopes in seasonally damp areas, 660–1900 m, NW (Cedarberg to Waaihoek Mts).•

macrantha Meisn.   Like **L. filamentosa** but leaves obovate, capitula daisy-like, 35–60 mm diam., and flowers white, sometimes tinged pink or purplish. Sept.–Jan. Rock outcrops, 1330–2250 m, SW, LB (Slanghoek Mts to Langeberg).•

Group 2: Inflorescence a terminal, sessile, bracteate, capitulum-like umbel, floral scales usually exserted, stigma various

aurea Meisn.   Slender, erect shrub to 1 m. Leaves alternate, adpressed to ascending, sometimes patent to reflexed, narrowly elliptic or narrowly obovate. Flowers many in capitula, 20–55 mm diam., yellow, sepals unequal, posterior sepal much smaller than others, stigma brush-like. July–Sept., sometimes Mar.–June. Coastal slopes below 900 m, SW, AP (Hermanus to Agulhas).•

funicaulis Schinz   Erect, felted shrublet to 45 cm. Leaves decussate, adpressed, narrowly elliptic or narrowly obovate. Flowers in capitula, 4–8 mm diam., cream-coloured, scales included, stigma capitate, papillate to long papillate. July–Mar. Sandy or stony slopes, 500–1000 m, NW (mountains south of Ceres).•

globulifera Meisn.   Slender, adpressed-hairy or felted shrublet to 60 cm, sometimes to 1 m. Leaves decussate, adpressed or ascending, narrowly elliptic. Flowers in capitula, 5–20 mm diam., pale mauve or cream-coloured, stigma brush-like. Jan.–Dec. Sandy flats and slopes, 200–1660 m, NW, SW (Cedarberg to Stettynsberg).•

penicillata Meisn. (= *Lachnaea passerinoides* N.E.Br.)   Felted, compact, rounded shrublet to 40 cm. Leaves decussate, adpressed, narrowly elliptic or lanceolate. Flowers apparently solitary in capitula, 3–4 mm diam., cream-coloured, stigma capitate, long papillate. Jan.–Dec. Sandstone slopes, 300–1330 m. LB (Montagu to Langeberg Mts and Gourits River).•

sp. 3 (= *Lachnaea montana* Beyers ms)   Erect, moderately branched shrublet to 60 cm. Leaves decussate, linear, narrowly elliptic or narrowly ovate. Flowers in capitula to 30 mm diam., cream-mauve or dirty violet, stigma brush-like. Oct.–Feb. Rocky slopes, 1330–2330 m, NW (Piketberg, Hex River Mts and Keeromsberg).•

sp. 4 (= *Lachnaea oliverorum* Beyers ms)   Like **L. funicaulis** but leaves adpressed to ascending, narrowly elliptic or lanceolate and inflorescence and flowers larger, stigma subcapitate, long papillate. June–Aug. Sandy soil, 1280–1400 m, KM (Waboomsberg, Montagu).•

sp. 5 (= *Lachnaea pedicellata* Beyers ms)   Like **L. sp. 3** but capitula to 15 mm diam., flowers cream-coloured, shorter and almost enclosed by bracts and flattened pedicels narrowly obovate to obovate usually with emarginate apex. Oct.–Jan. Sandy, loam soil probably on underlying shale, 1760–1830 m. NW (Cold Bokkeveld).•

sp. 6 (= *Lachnaea pendula* Beyers ms)   Erect to decumbent, felted to silky shrublet to 45 cm. Leaves decussate, adpressed, narrowly ovate to elliptic. Flowers in nodding capitula, 3–5 mm diam., cream-coloured or pale yellow, stigma capitate, long papillate. Sept.–Dec. Sandstone slopes, 1330–2000 m, NW (Witsenberg to Hex River Mts).•

Group 3: Inflorescence a terminal, sessile, pseudobracteate, capitulum-like umbel, floral scales exserted, stigma various

eriocephala L. (= *Lachnaea purpurea* Andrews)   BERGANGELIER   Erect, sparsely to moderately branched shrublet to 60 cm. Leaves decussate, ascending or subadpressed, lanceolate or narrowly elliptic. Flowers many in a capitulum, 25–55 mm diam., cream-coloured or mauve, sepals unequal, the posterior the smallest, stigma brush-like. July–Nov. Sandstone slopes, 100–500 m, NW, SW (Tulbagh and Malmesbury to Betty's Bay).•

laniflora (C.H.Wright) Bond   Rounded, much-branched shrub to 1 m. Leaves decussate, ascending, slightly incurved, narrowly oblong or narrowly elliptic. Flowers in capitula, 8–15 mm diam., pink, white, or white tinged violet-blue or pink, stigma brush-like. July–Jan. Rock outcrops, 1000–2200 m, NW (Cedarberg to Brandwag Peak, Worcester).•

marlothii Schltr.   Slender, erect to spreading, thinly tomentose to felted shrublet to 28 cm. Leaves decussate, adpressed or ascending, narrowly elliptic. Flowers in capitula, 5–10 mm diam., cream-coloured or creamy pink, stigma capitate, long papillate. Dec. Rocky slopes, 1800–2330 m., NW (Hex River Mts).•

naviculifolia Compton   Erect, sparsely to moderately branched shrub to 1.5 m. Leaves decussate, ascending, sometimes patent, narrowly elliptic to elliptic, sometimes narrowly ovate to ovate. Flowers in capitula, 18–35 mm, white, pale blue, mauve or pale yellow, scented, stigma brush-like. Aug.–Nov. Sandy flats, rocky slopes, 1000–1860 m, NW (Cedarberg and E of Cold Bokkeveld).•

sp. 7 (= *Lachnaea elsiae* Beyers ms)   Like **L. penicillata** but capitula pseudobracteate and with up to 4 open flowers at once. Jan.–Dec. Shale slopes, 1260–2300 m, NW, KM (Cedarberg Mts to Seweweekspoort).•

sp. 8 (= *Lachnaea greytonensis* Beyers ms)   Erect, felted shrublet to 45 cm. Leaves decussate, adpressed, narrowly elliptic or narrowly obovate. Flowers in capitula, 4–8 mm diam., cream-coloured, stigma capitate, papillate to long papillate. Sept.–Jan. Sandstone slopes, 640–830 m, SW (Riviersonderend Mts: Genadendal to Greyton).•

sp. 9 (= *Lachnaea rupestris* Beyers ms)   Like **L. marlothii** but branches covered with adpressed hairs mixed with short crisped ones, flowers mauve, scales narrowly obovoid to ellipsoid, not capitate, stigma brush-like. July–Oct. rocky slopes above 1500 m, SW (Villiersdorp: Stettynsberg to Riviersonderend Mts: Genadendal).•

sp. 10 (= *Lachnaea villosa* Beyers ms)   Erect, felted shrublet to 45 cm. Leaves decussate, adpressed, narrowly elliptic or narrowly obovate. Flowers in capitula, 4–8 mm diam., cream-coloured, skunk-scented, stigma capitate, papillate to long papillate. Sept.–July. Sandstone slopes and flats, 480–1500 m, NW (Cold Bokkeveld).•

Group 4: Inflorescence a terminal, ebracteate umbel, floral scales usually exserted, stigma various

burchellii Meisn.   Erect to sprawling shrublet to 30 cm. Leaves decussate occasionally subopposite, ascending to spreading, lanceolate to narrowly elliptic. Flowers in ebracteate umbels, apparently bracteate on short lateral branches, cream-coloured to pale pink, stigma brush-like. June–Mar. Mountain slopes, 30–1660 m, LB, SE (Langeberg: Albertinia to Port Elizabeth).•

diosmoides Meisn.   Moderately branched, flexuose shrub to 1.5 m. Leaves alternate, ascending to spreading, linear elliptic to narrowly elliptic. Flowers in ebracteate umbels, cream-coloured, stigma capitate, papillate. Jan.–Dec. Lower and middle slopes, 250–1660 m, LB, SE (Attaquas Mts to Tsitsikamma).•

ericoides Meisn.   Erect to straggling, much-branched shrublet to 50 cm. Leaves alternate, subadpressed to ascending, narrowly elliptic to obovate, tufted at tip. Flowers in ebracteate umbels, cream-coloured, outside sericeo-villous but basal portion of hypanthium glabrous, stigma capitate, long papillate. July–Dec. Lower slopes, 330–660 m, LB (Swellendam to Riversdale).•

glomerata Fourc.   Erect shrublet, to 45 cm. Leaves decussate, ascending to spreading, linear-elliptic to elliptic or obovate. Flowers in ebracteate umbels, apparently bracteate on short, reduced, lateral branches, cream-coloured or pale pink, tomentose outside, sweetly skunk-scented, stigma capitate, long papillate. Aug.–Oct. Sandstone slopes, 330–1130 m, KM, SE (Rooiberg to Humansdorp).•

nervosa (Thunb.) Meisn. (= *Lachnaea ambigua* Meisn.)   Erect, moderately to much-branched shrublet, 20–60 cm, sometimes to 1 m. Leaves alternate, ascending, rarely spreading, narrowly elliptic. Flowers in ebracteate umbels, cream-coloured, pink or cream-coloured with mauve tinge, silky outside, stigma brush-like. July–Dec. Rocky slopes, summit ridges, 830–2000 m, NW, SW, LB (Elandskloof to Langeberg Mts: Swellendam).•

sociorum Beyers   Like **L. ericoides** but a resprouter, leaf apex glabrous and flowers white or white tinged pink, skunk-scented, stigma brush-like. Aug.–Jan. Stony sandstone slopes, 300–1100 m, LB (E Langeberg to Attaquas Mts).•

striata (Lam.) Meisn. (= *Lachnaea elegans* Compton)   Corymbosely branched shrublet to 60 cm. Leaves alternate, ascending, narrowly elliptic to elliptic or ovate, 1–5-ribbed abaxially. Flowers in ebracteate umbels, cream-coloured, purple or pale rose, stigma brush-like. Sept.–Jan. Seasonally damp areas, 800–1160 m, NW (N Cedarberg to Witsenberg).•

sp. 11 (= *Lachnaea stokoei* Beyers ms)   Like **L. ericoides** but hypanthium obconic and scales included. Flowering time? Habitat? LB (Lemoenshoek near Barrydale).•

PASSERINA   GANNA BUSH   c. 17 spp., southern Africa

A. Fruits fleshy

ericoides L.   CHRISTMAS BERRY, DRONKBESSIE   Willowy shrublet to 1 m. Leaves 4-ranked, oblong, hairy beneath, 2–3 mm long, bracts leaf-like but dilated below. Flowers in spikes, tube initially 3 mm long but enlarging rapidly in fruit, neck absent. Fruits fleshy, red. Oct.–Nov. Coastal sands, SW (Blouberg to Hermanus).•

rigida Wikstr.   SEEKOPPIESGANNA   Robust shrub with branches nodding at tips and pendulous branchlets, to 2 m. Leaves ovate-lanceolate, compressed, hairy beneath, 2.5–4 mm long, bracts acuminate, furrowed. Flowers in spikes, tube c. 3 mm long, neck c. 1 mm. Fruits fleshy, orange. Nov. Coastal dunes, SW, AP, SE (Cape Peninsula to KwaZulu-Natal).

A. Fruits dry

burchellii Thoday   Perennial to 30 cm. Leaves ovate to obovate, softly hairy beneath with apical tuft, 2–3 mm long, bracts leaf-like, larger. Flowers in spikes, tube 3 mm long. Fruits dry. Oct.–Nov. Sandstone crevices and outcrops, SW (Riviersonderend Mts: Baviaanskloof).•

comosa C.H.Wright   Shrublet. Leaves linear, softly hairy beneath with apical tuft, 3 mm long, bracts broadly ovate, hairy beneath, larger. Flowers in spikes, tube 3 mm long, neck 1 mm. Fruits dry. Flowering time? Sandstone slopes, KM (Kamiesberg, Swartberg Mts).

falcifolia C.H.Wright   Shrub or small tree, 1.5–3 m. Leaves falcate, hairy beneath, 4–10 mm long, bracts ovate-acuminate, wings prominently veined. Flowers in spikes, tube 5–6 mm long, neck slender and exserted. Fruits dry. Oct.–Nov. Rocky slopes and river banks, LB, SE (Outeniqua to Van Staden's Mts).•

filiformis L.   BRUINGANNA   Shrub, 1.5–2 m. Leaves concave, linear, hairy beneath, 7–15 mm long, bracts membranous-winged. Flowers in spikes, tube c. 4 mm long, neck slender and exserted. Fruits dry. Flowering time? Stony slopes, riverbanks and grassland, NW, SW, AP, LB, SE (Clanwilliam to tropical Africa).

galpinii C.H.Wright   Shrub or shrublet. Leaves linear, incurved, hairy beneath, 3–5 mm long, bracts membranous-winged. Flowers in dense spikes, tube c. 3 mm long, neck c. 1 mm. Fruits dry. Oct.–Nov. Coastal limestones, AP (Arniston to Mossel Bay).•

glomerata Thunb.   Shrub or shrublet. Leaves truncate or hump-backed, hairy beneath, 2–4 mm long, bracts obovate, ribbed, larger. Flowers few in dense spikes, tube c. 3.5 mm long, neck very short. Fruits dry. Oct.–Nov. Sandy and stony flats, NW, SW (Bokkeveld Mts to Worcester).•

nivicola Bredenkamp & A.E.van Wyk   Shrub to 2 m. Leaves subterete, with a hairy groove beneath and tufted at apex when young, 2.5–4.5 mm long, bracts winged with a bulla on each side. Flowers in spikes, tube c. 3 mm long, neck c. 1.5 mm, hairy. Fruits dry. Oct.–Nov. High mountain slopes, NW (Cold Bokkeveld).•

obtusifolia Thoday   Shrub, 1–1.5 m. Leaves linear-oblong, hairy beneath, 4–8 mm long, bracts dilated and ribbed below. Flowers in dense spikes, tube. c. 3.5 mm long, neck bent outwards. Fruits dry. Oct.–Nov. Rocky slopes and flats, KM, SE (W and Little Karoo to E Cape).

paleacea Wikstr.   Shrublet to 1 m. Leaves small, compressed above, 1.5–2.5 mm long, bracts suborbicular, keeled above. Flowers in dense spikes, tube c. 2 mm long, neck absent. Fruits dry. Oct.–Nov. Coastal sands and estuaries, SW, AP (Saldanha Bay to Agulhas).•

paludosa Thoday   Shrub to 2 m. Leaves adpressed, lanceolate, hairy beneath and tufted at apex, 6–10 mm long, bracts ovate-acuminate, wings membranous and faintly veined. Flowers in spikes, tube c. 5 mm long, neck slender and longer. Fruits dry. Nov. Coastal marshes and seeps, SW (Cape Peninsula and Cape Flats).•

pendula Eckl. & Zeyh.   Shrub to 1.5 m. Leaves adpressed, ovate-lanceolate, with a hairy groove beneath, 2.5–3 mm long, bracts rhombic, margins somewhat dry and ciliate, larger. Flowers in more or less pendulous spikes, tube c. 3 mm long, neck c. 0.5 mm. Fruits dry. Oct.–Nov. Sandstone slopes and river banks, SE (Uniondale to Port Elizabeth).•

rubra C.H.Wright   Shrublet to 1 m. Leaves lanceolate, hairy beneath, 2.5–5 mm long, bracts spreading, ovate-acuminate. Flowers in long, lax spikes, tube c. 3 mm long, neck c. 1 mm. Fruits dry. Oct.–Nov. Stony and sandy flats, LB, SE (Riversdale to E Cape).

vulgaris Thoday   Shrub or small tree to 2 m. Leaves linear, with a hairy groove beneath, 3.5–10 mm long, bracts ovate-acuminate, wings longitudinally folded. Flowers in spikes, tube c. 4 mm long, neck slender, c. 2 mm. Fruits dry. Oct.–Nov. Sandy, often disturbed flats and slopes, NW, SW, AP, KM, LB, SE (Tulbagh to E Cape).

STRUTHIOLA   FEATHERHEAD, KATSTERTJIE, ROEMENAGGIE, VEERTJIE   c. 40 spp., tropical and southern Africa, mainly W Cape

A. Floral scales 4

striata Lam.   Shrub to 1 m, hairy on young branches. Leaves opposite, imbricate, ovate-oblong, subacute, strongly striate below, ciliate at first. Flowers axillary, cream-coloured, yellow or pinkish. Sept.–June. Flats and lower slopes, SW, AP, LB, SE (Yzerfontein to Mossel Bay, Uitenhage).•

tetralepis Schltr. Shrublet to 30 cm, hairy on young branches. Leaves opposite, imbricate, lanceolate, acuminate, striate below when dry, ciliate. Flowers axillary, reddish, sepals lanceolate, acuminate. Oct.–Feb. Lower to middle slopes, SW (Paarl to Caledon).•

A.' Floral scales 8
B. Hypanthium hairy outside

ciliata (L.) Lam. (incl. **S. angustifolia** Lam., **S. flavescens** Gilg, **S. longiflora** Lam., **S. lucens** Lam., **S. pillansii** Hutch., **S. rustiana** Gilg, **S. schlechteri** Gilg, **S. virgata** L.)   Shrub to 1.5 m with adpressed-hairy to white-woolly, tetragonal branches. Leaves opposite, linear-lanceolate to narrowly elliptic, faintly ribbed beneath, ciliate. Flowers axillary, cream-coloured, pink or reddish, hypanthium sometimes glabrescent, scales oblong, shorter to longer than perigonal hairs. Jan.–Dec. Flats and slopes, NW, SW, AP, KM, LB (Namaqualand to Albertinia and Little Karoo).

confusa C.H.Wright   Slender shrublet to 60 cm with tetragonal branches. Leaves opposite, linear-lanceolate to narrowly elliptic, subacute, sparsely ciliate. Flowers axillary, cream-coloured or pink, hypanthium sparsely adpressed-hairy, scales oblong, as long as perigonal hairs. Aug.–Dec., sometimes Feb. Lower slopes, NW, SW, KM (Citrusdal to Little Karoo).•

leptantha Bolus   Shrub to 2 m with densely to sparsely hairy, slightly tetragonal branches. Leaves opposite, narrowly elliptic to narrowly obovate, obtuse to subacute, at first hairy below, later glabrous and pustulate. Flowers axillary, cream-coloured, sometimes reddish, scales oblong, as long as or shorter than perigonal hairs. June–Oct. Sandy flats and mountain slopes, NW, SW, KM (Namaqualand to Malmesbury and Little Karoo).

lineariloba Meisn.   Slender shrublet to 30 cm with tetragonal branches. Leaves opposite, lanceolate, ciliate. Flowers axillary, red, hypanthium adpressed-hairy only in lower half, sepals lanceolate, acute, scales oblong, as long as perigonal hairs. Sept.–Oct. Upper slopes, NW (Pakhuis Mts).

B. Hypanthium glabrous outside

dodecandra (L.) Druce (= *Struthiola erecta* Lam.)   Erect shrub to 80 cm with lax, tetragonal branches. Leaves opposite, lanceolate to narrowly elliptic, acute, smooth or faintly ribbed below. Flowers axillary, white or pink, scales narrowly ovoid, as long as or longer than perigonal hairs. Jan.–Dec. Flats and lower slopes, SW, AP, SE (Cape Peninsula to Bredasdorp, Knysna).•

eckloniana Meisn.   Shrub to over 2 m with tetragonal, glabrous or sparsely adpressed-hairy branches. Leaves opposite, narrowly lanceolate to narrowly elliptic, acute. Flowers axillary, twice as long as leaves, white, cream-coloured, pink or sometimes yellow, scales linear, longer or shorter than perigonal hairs. Jan.–Dec. Mountain slopes, LB, KM, SE (Langeberg Mts and Little Karoo to Uniondale).•

ericoides C.H.Wright   Erect shrub to 80 cm with stiffly hairy, tetragonal branches. Leaves opposite, narrowly elliptic, at first ciliate, faintly striate below. Flowers axillary, deep yellow to red, scales ovoid, as long as the white to cream-coloured perigonal hairs. Oct.–Mar. Sand over limestone hills and mountain slopes, AP, SE (Cape Infanta to George).

hirsuta Wikstr. (incl. **S. pentheri** Moore, **S. tuberculosa** Lam.)   Shrub to 2 m with stiffly hairy branches. Leaves alternate, densely adpressed-hairy to glabrous beneath, ciliate. Flowers axillary, white, scented, scales ovoid, longer than perigonal hairs. Oct.–Aug. Coastal hills and lower slopes, AP, LB, SE (Gansbaai to Uitenhage).•

macowanii C.H.Wright   Shrub to 1.3 m, hairy on young branches. Leaves opposite or ternate. Flowers axillary, white, pale yellow to reddish, scales ovoid, as long as or shorter than perigonal hairs. Jan.–Dec. Mountain slopes and coastal hills, AP, LB, SE (Stilbaai to E Cape).

myrsinites Lam. (= *Struthiola ovata* Thunb.)   Erect, flexuose shrub to 2 m with tetragonal branches. Leaves opposite, narrowly elliptic, acute. Flowers axillary, twice as long as leaves, white or pale pink, scales narrowly ellipsoid, longer than perigonal hairs. Jan.–Dec. Sandy soils, NW, SW, LB, SE (Bokkeveld Mts to E Cape).

parviflora Bartl. ex Meisn.   Straggling shrublet to 60 cm with tetragonal branches, stiffly or woolly hairy on ridges. Leaves opposite, narrowly elliptic to lanceolate, obtuse, ciliate. Flowers axillary, deep cream-coloured, greenish yellow or sometimes red, scales oblong, shorter than the golden yellow perigonal hairs. Jan.–Dec. Flats or lower slopes, LB, SE (Riversdale to E Cape).

salteri Levyns   Erect to spreading shrublet to 80 cm with hairy branches. Leaves in whorls of 3, sometimes 4, narrowly ovate to narrowly elliptic, acute or acuminate. Flowers axillary, cream-coloured, pinkish, sometimes yellow, scales narrowly ellipsoid, slightly shorter than perigonal hairs. Jan.–Dec. Limestone hills, SW, AP (Cape Peninsula to Agulhas).•

A". Floral scales 12
C. Hypanthium glabrous outside

mundii Eckl. ex Meisn. (incl. **S. ramosa** C.H.Wright)   Shrublet to 60 cm with stiffly hairy branches. Leaves alternate, imbricate, narrowly elliptic, acute, 3–5-ribbed and sometimes adpressed-hairy below, densely white-ciliate. Flowers axillary, crowded at apex of branches, yellow, scales ovoid shorter to longer than perigonal hairs. Aug.–Oct. Lower slopes, SW (Gouda to Swellendam).•

rigida Meisn.   Shrublet to 50 cm with crisped-hairy branches. Leaves alternate, imbricate, narrowly elliptic to narrowly ovate, subacute, ciliate. Flowers axillary, yellow, scales as long as or shorter than perigonal hairs. June–Jan. In sand on limestone, SW, AP (Potberg to Gouritsmond).•

C. Hypanthium hairy outside

argentea Lehm.   AANDGONNA   Shrub to 2 m with adpressed-hairy branches. Leaves opposite, imbricate, elliptic to suborbicular, sometimes narrowly elliptic, faintly ribbed below, white-ciliate. Flowers axillary, yellow, sometimes reddish orange, scales ovoid, longer than perigonal hairs. Mar.–Dec. Coastal flats or slopes, SW, AP, KM, LB, SE (Hottentots Holland Mts and Montagu to E Cape).

fasciata C.H.Wright   Slender shrublet to 60 cm with crisped-woolly branches. Leaves opposite, lanceolate to narrowly elliptic, blunt, slightly keeled below, white-ciliate. Flowers axillary, cream-coloured or yellow, scales linear, slightly longer than perigonal hairs. Jan.–Aug. Flats or slopes, LB, SE (Swellendam to Knysna).•

garciana C.H.Wright   Shrub to 1 m with adpressed-hairy branches. Leaves opposite, lanceolate, acute, 5- or 7-ribbed below. Flowers yellow, cream-coloured or pinkish, scales narrowly ellipsoid, as long as or shorter than perigonal hairs. Mar.–Nov. Forest margins and slopes, KM, LB, SE (Swellendam and Little Karoo to Humansdorp).•

martiana Meisn. (incl. **S. fourcadei** Compton, **S. leiosiphon** Gilg)   Shrub to 2 m with silky branches. Leaves alternate, narrowly elliptic, striate below, silky when young, later sparsely hairy and pustulate. Flowers axillary, white or pinkish; hypanthium softly hairy to almost glabrous, scales linear, as long as or shorter than perigonal hairs. May–Jan. Upper slopes, SW, LB, SE (Stellenbosch to Humansdorp).•

tomentosa Andrews   Slender shrublet to 60 cm with silky woolly branches. Leaves opposite, imbricate, narrowly elliptic to obovate, acute to rounded, striate below, white-silky when young. Flowers axillary, yellow to pale orange, sometimes cream-coloured, hypanthium woolly, scales linear, shorter than the yellow perigonal hairs. July–Dec., sometimes Mar. Lower slopes, sometimes high, SW, LB, SE (Stellenbosch to George).•

[**Species excluded**   Poorly known and probably conspecific with one of the above: **S. bachmanniana** Gilg, **S. cicatricosa** C.H.Wright, **S. concava** Moore, **S. epacridioides** C.H.Wright, **S. floribunda** C.H.Wright, **S. galpinii** C.H.Wright, **S. longifolia** C.H.Wright, **S. recta** C.H.Wright]

URTICACEAE

1. Plants with stinging hairs; perianth segments free; stamens 4 or 5 .... **Laportea**
1.' Plants without stinging hairs; perianth of female flowers absent or apparently so; stamen 1:
   2. Flowers bracteate but not enclosed by a common involucre; annual herbs .... **Didymodoxa**
   2.' Flowers almost completely enclosed by a common involucre of free or fused bracts:
      3. Involucral bracts mostly free, accrescent; leaves with hooked hairs .... **Forsskaolea**
      3.' Involucral bracts fused to tips in a cup; leaves without hooked hairs .... **Droguetia**

DIDYMODOXA   2 spp., southern Africa to Ethiopia

caffra (Thunb.) Friis & Wilmot-Dear (= *Australina acuminata* Wedd.)   Like **D. capensis** but leaves lanceolate to ovate, sharply toothed with apical tooth longer than wide and stipules lanceolate. Nov.–Feb. Forest margins and among rocks, SE (Port Elizabeth to Ethiopia).

capensis (L.f.) Friis & Wilmot-Dear (= *Australina integrifolia* Wedd., *A. lanceolata* (Thunb.) N.E.Br., *A. procumbens* N.E.Br.)   Monoecious sprawling to upright, glabrescent or hairy annual to 30 cm. Leaves soft, ovate, entire or crenate to bluntly toothed, with ovate-lanceolate stipules. Flowers in axillary clusters, green. Aug.–Nov. Sheltered sites, forest margins and clearings, NW, SW, LB, SE (Namibia and Namaqualand to Knysna).

DROGUETIA   7 spp., southern Africa to Indonesia

iners (Forssk.) Schweinf. (= *Droguetia thunbergii* N.E.Br.)   Monoecious softly woody perennial, mostly to 50 cm. Leaves mostly opposite, lanceolate, toothed, apical tooth longer than wide, with tailed apices. Flowers in small axillary clusters, greenish. July–Nov. Coastal forest, scrub, and among rocks, SW, LB, SE (Cape Peninsula to Indonesia).

FORSSKAOLEA   6 spp., southern Africa to Indonesia

candida L.f.   Monoecious, softly woody, roughly hairy shrublet to 80 cm, often with red stems. Leaves ovate, discolorous, usually grey-felted beneath, coarsely toothed, margins revolute. Flowers in axillary, involucrate clusters, greenish, involucre lobes enlarging in fruit. Mainly Aug.–Oct. Dry rocky slopes, river beds, karroid flats, KM (Namibia to Karoo and Little Karoo).

LAPORTEA   c. 15 spp., cosmopolitan

grossa (Wedd.) Chew   Monoecious annual to 1 m, with deflexed stinging hairs on raised base. Leaves triangular, coarsely toothed to lobed, sparsely armed, often spotted white. Flowers in axillary panicles, small, greenish. Dec.–Mar. Coastal forests, SE (George to Kenya).

peduncularis (Wedd.) Chew   Monoecious scrambling to creeping annual to 1.5 m, woody at base, young plants with small stinging hairs. Leaves ovate, thin-textured, finely serrate. Flowers in axillary panicles, small, greenish. Mainly Jan.–Mar. Forests and coastal bush, SE (George to E Africa).

VAHLIACEAE

VAHLIA   VERKLEURMANNETJIEKRUID   5 spp., Africa, Madagascar and India

capensis (L.f.) Thunb.   Shortly hairy shrublet from a woody caudex, mostly to 30 cm. Leaves opposite, linear-oblanceolate. Flowers paired, axillary, yellow fading maroon or brown. Mainly Sept.–Nov. Sandstone and granite slopes, NW, SW (Namibia to Mamre, widespread in the drier parts of southern Africa).

VALERIANACEAE

VALERIANA   VALERIAN   c. 200 spp., N temperate and tropical mountains

capensis Thunb.   WILDEBALDERJAN   Tufted rhizomatous perennial to 1 m. Leaves pinnatisect, leaflets smaller towards base, upper leaves opposite and sessile. Flowers in corymbose panicles, cream-coloured to pinkish. Fruits with a plumose apical pappus. Nov.–Feb. Sandstone slopes in moist sites, SW, KM, LB, SE (Cape Peninsula to S tropical Africa).

[**VALERIANELLA eriocarpa** Desr. included in previous edition as endemic to Langeberg Mts and collected there in the early 19th century is almost certainly a casual introduction from Europe. Not recorded since 1850, it cannot be considered even a naturalised exotic.]

VERBENACEAE

CHASCANUM (= *PLEXIPUS*)   c. 25 spp., Africa

cernuum (L.) E.Mey. (= *Chascanum integrifolium* (H.Pearson) Moldenke)   Twiggy shrublet, mostly to 30 cm. Leaves sessile, mostly ternate, cuneate, 3–5-toothed above. Flowers in compact terminal spikes, white or tinged mauve often with yellow throat. All year but mainly Aug.–Nov. Rocky slopes and flats in karroid scrub and fynbos, often on limestone, SW, AP, KM, LB, SE (Cape Peninsula to KwaZulu-Natal).

cuneifolium (L.f.) E.Mey. (= *Chascanum dehiscens* (L.f.) Moldenke)   Twiggy shrublet to 50 cm. Leaves opposite, obovate to round or flabellate, sharply narrowed below, mostly 5–7-toothed. Flowers in compact spikes, white or mauve. July–Nov. Shale and limestone slopes, KM (Little Karoo to coastal KwaZulu-Natal).

VIOLACEAE

1.   Sepals without basal lobes; stipules small, narrow ....**Hybanthus**
1.' Sepals with ear-like basal lobes; stipules foliaceous ....**Viola**

HYBANTHUS   LADY'S-SLIPPER   c. 100 spp., tropical and subtropical, mostly New World

capensis (Thunb.) Engl.   Roughly hairy perennial from a woody rootstock to 15 cm. Leaves obovate, shallowly toothed. Flowers solitary in axils, nodding, violet, mauve or white. Nov.–May. Grassland, SE (Uitenhage to Mpumalanga).

VIOLA   VIOLET   c. 300 spp., cosmopolitan, mainly N temperate

decumbens L.f.   CAPE VIOLET   Shrublet to 25 cm. Leaves linear. Flowers solitary in axils, nodding, faintly scented, violet to purple. July–Dec. Damp sandstone slopes, SW, AP, LB (Bainskloof to Riversdale).•

VISCACEAE

VISCUM   MISTLETOE, LIDJIESTEE, VOËLENT   c. 100, Old World, mainly tropical

capense L.f.   Dioecious stem parasite to 50 cm. Leaves scale-like. Inflorescences unisexual. Berries sessile, smooth, white. July–Oct. Parasitic on various shrubs including *Chrysanthemoides*, *Euclea*, *Maytenus* and *Pterocelastrus*, NW, SW (S Namibia to Caledon).

continuum E.Mey. ex Sprague   Dioecious stem parasite to 1 m, leafless. Inflorescences unisexual. Berries pedicellate, smooth, pale yellow. July–Aug. Parasitic on *Acacia* only, NW, SW, KM, LB (Worcester to E Cape).

crassulae Eckl. & Zeyh.   Dioecious stem parasite to 50 cm. Leaves orbicular, fleshy. Inflorescences unisexual. Berries pedicellate, smooth, bright orange. July–Aug. Parasitic mainly on *Portulacaria afra* but also *Euphorbia*, SE (Patensie to Port Elizabeth and Steytlerville).

hoolei (Wiens) Polhill & Wiens   Monoecious stem parasite to 50 cm. Leaves scale-like or lacking. Inflorescences bisexual. Berries sessile, smooth, white. June–July. Parasitic mainly on *Rhus*, SW, AP, KM, LB, SE (Caledon to Lesotho).

minimum Harv.   Minute monoecious stem parasite to 3 mm, leafless. Inflorescences bisexual. Berries pedicellate, smooth, bright orange. June–July. Parasitic on *Euphorbia polygona* and *E. horrida*, KM, SE (Rooiberg Mts to E Cape).

obscurum Thunb.   Dioecious stem parasite to 1 m. Leaves oblanceolate, 3-veined from base. Inflorescences unisexual. Berries pedicellate, smooth, cream-coloured to pale pink. June–July. Parasitic on various trees including *Acacia*, *Maytenus*, *Olea* and *Rhus*, KM, LB, SE (Touwsrivier to KwaZulu-Natal).

pauciflorum L.f.   Monoecious stem parasite to 75 cm. Leaves large, elliptic, leathery, often 3-veined from base. Inflorescences bisexual. Berries pedicellate, smooth, orange. Apr.–Oct. Parasitic on *Euclea*, *Maytenus* and *Rhus*, NW, SW (Bokkeveld Plateau to Hangklip).•

rotundifolium L.f.   Like **V. pauciflorum** but leaves smaller, up to 12 mm long, and usually obtuse at base. Feb.–May. Parasitic on various trees including *Acacia, Euclea, Olea* and *Rhus*, NW, SW, KM, LB, SE (southern Africa).

VITACEAE
by E. Retief

1. Petals 5 or 6; stamens with anthers bending over gynoecium .... **Rhoicissus**
1.' Petals 4; stamens not bending noticeably over gynoecium .... **Cyphostemma**

CYPHOSTEMMA   WILDEDRUIF   c. 150 spp., warm regions

cirrhosum (Thunb.) Desc. ex Wild & R.B.Drumm.   DROOG-MY-KEEL   Scrambling shrub or canopy climber. Leaves digitately 5-foliolate, leaflets sessile, margins broadly dentate, succulent. Flowers dull yellow. Fruits subglobose, c. 10 mm diam. Nov.–Feb. Coastal bush, SE (Port Elizabeth to KwaZulu-Natal).

RHOICISSUS   BOSDRUIF   10 spp., southern and tropical Africa

digitata (L.f.) Gilg & M.Brandt   BOBBEJAANDRUIF, WILDEPATAT   Scrambling vine. Leaves digitately 3-foliolate, leaflets subsessile, lateral ones usually sessile, margins entire. Flowers yellowish green. Fruits subglobose, reddish brown to purple. Jan.–May. Coastal dunes, SW, AP, SE (Betty's Bay to Mozambique).

kougabergensis Retief & Van Jaarsv.   Sprawling shrub. Leaves secund, simple, narrowly obovate. Flowers in leaf-opposed, reddish brown-hairy cymes, greenish yellow. Fruits globose, 8–10 mm diam. Oct.–Nov. Subtropical thicket, sandstone slopes, SE (Kouga Mts).•

tomentosa (Lam.) Wild & R.B.Drumm.   MONKEYROPE   Scrambling shrub or canopy climber. Leaves simple, usually reniform, 3-veined from base, margins lobed with shallow serrations, dark green above, rusty velvety beneath. Flowers in dense reddish brown-hairy cymes, greenish. Fruits large, to 20 mm diam., red to dark purple. Oct.–Feb. Forests, SE (Knysna to Tanzania).

tridentata (L.f.) Wild & R.B.Drumm.   DROOG-MY-KEEL, WILDERUIF   Shrub with branches scandent or erect, to 1 m. Leaves 3-foliolate, leaflets with margins more or less entire or terminal leaflet 1–4-toothed or -crenate, glabrescent or grey-hairy. Flowers yellowish green. Fruits subglobose, red to purplish black. Sept.–June. Scrub, LB, SE (Riversdale to E Cape).

ZYGOPHYLLACEAE
Zygophyllum by L. van Zyl

1. Petals absent; sepals with a horseshoe-shaped ridge on inner face with deflexed, glandular hairs from upper portion of ridge; stamens 5(7 or 8); prostrate, woody annual with woolly nodes .... **Seetzenia**
1.' Petals 4 or 5; sepals simple or outer 2 pouched at base; stamens 8–10 with ornamented filaments:
    2. Leaves simple, clavate; petals 3-fid; succulent annual or short-lived perennial .... **Augea**
    2.' Leaves compound; petals simple; woody perennials or annual:
        3. Shrublets; leaves (1)2- or 3-foliolate; ovary and fruit glabrous .... **Zygophyllum**
        3.' Annual; leaves pinnate; ovary bristly and fruit spiny .... **Tribulus**

AUGEA   1 sp., W Karoo and Namaqualand to Little Karoo

capensis Thunb.   Brittle, succulent annual or short-lived perennial to 40 cm. Leaves opposite, clavate. Flowers few at nodes, whitish, petals 3-toothed. Fruits large, ellipsoid, splitting irregularly; seeds woolly. Aug.–Oct. Dry sandy flats, NW, KM (S Namibia and W Karoo to Little Karoo).

SEETZENIA   KNIETJIESBOS, LIDJIESBOS   1 sp., southern Africa to India

lanata (Willd.) Bullock   Glandular, prostrate annual with jointed stems woolly at nodes. Leaves opposite, trifoliolate, leaflets obovate, succulent. Flowers solitary, terminal, erect, sepals greenish yellow, with long filaments and styles. Sept.–Dec. Rocky sandstone slopes, NW (Bokkeveld Mts to Twenty Four Rivers Mts and N Africa to India).

TRIBULUS DUBBELTJIEDORING   c. 20 spp., cosmopoloitan in drier parts

terrestris L.   Glabrescent or hairy, prostrate annual with stems radiating from a crown. Leaves opposite, unequal, pinnate, leaflets oblong, silky. Flowers solitary in axils, yellow. Fruits spiny, fragmenting into segments. Oct.–Apr. Sandy flats and roadsides, NW, SW, AP, KM, LB, SE (throughout southern Africa).

ZYGOPHYLLUM   TWINLEAF, SPEKBOS   c. 100 spp., Africa, Middle East and Australia, mostly southern Africa

A. Leaves simple

cordifolium L.f.   Erect or sprawling shrublet to 60 cm. Leaves obovate to suborbicular, glaucous, succulent. Flowers lime-yellow with red or brown markings. Fruits oblong, 5-angled and 5-winged, wings 2 mm wide. Apr.–Oct. Sandy coastal dunes and rocky slopes, NW, SW (S Namibia to Saldanha).

A. Leaves bifoliolate
B. Leaves petiolate
C. Fruits angled and winged

debile Cham. & Schltdl.   Spreading, soft shrublet to 20 cm. Leaves bifoliolate, leaflets elliptic, glaucous. Flowers pale yellow. Fruits oblong, 5-angled with wings, wings 2 mm wide, red-brown. May–Oct. Dry shale slopes, KM, SE (Swartberg Mts to Uniondale).•

lichtensteinianum Cham. & Schltdl.   VAALSPEKBOS   Erect shrublet to 80 cm. Leaves bifoliolate, leaflets obovate, glaucous and waxy. Flowers lime-yellow. Fruits oblong, 5-angled with wings, wings 2 mm wide. Mainly Sept. Shale flats, KM (W and Little Karoo to Steytlerville).

morgsana L.   SLAAIBOS   Erect shrub or shrublet to 1.5 m. Leaves bifoliolate, leaflets asymmetric, obovate. Flowers yellow, with only 4 petals. Fruits large, 4-angled with wings, wings 15 mm wide. Apr.–Oct. Sandy and stony slopes and flats, mostly coastal, NW, SW, AP, KM, LB, SE (S Namibia to Grahamstown).

retrofractum Thunb.   Erect, rounded shrub or shrublet to 1.5 m. Leaves bifoliolate, leaflets oblong, succulent. Flowers minute, white or cream-coloured. Fruits small, elliptic, 5-angled with wings. Sept.–Dec. Sandy and shale flats, KM, SE (S Namibia to Steytlerville).

C. Fruits lobed and ribbed

foetidum Schrad. & J.C.Wendl. (incl. **Z. meyeri** Sond.)   SLYMBOS   Foetid, sprawling and scandent shrub to 2 m or more. Leaves bifoliolate, leaflets asymmetric, obovate. Flowers deep yellow with red markings. Fruits roundish when fresh, 5-lobed with prominent bony ribs when dry. July–Oct. Slopes, flats and streambanks, NW, KM, SE (S Namibia to Grahamstown).

maculatum Aiton   Erect shrublet to 80 cm, stems striate. Leaves bifoliolate, leaflets linear. Flowers deep yellow with red markings. Fruits oblong, 5-lobed with ribs. Mar.–July. Shale flats and rocky ridges, KM (Tanqua Karoo to Witteberg).

uitenhagense Sond.   Trailing shrublet with branches to 2 m long. Leaves bifoliolate, leaflets obovate to elliptic. Flowers yellow with red or brown markings. Fruits round or oblong and 5-sutured when fresh, 5-angled with heavy ribs when dry. Mainly Sept. Coastal sands and limestone among scrub, AP, LB, SE (Bredasdorp to Grahamstown).

B. Leaves sessile
D. Fruits rounded and not lobed when fresh

flexuosum Eckl. & Zeyh.   Sprawling shrublet to 70 cm. Leaves bifoliolate, leaflets obovate, often succulent. Flowers golden yellow with red markings, tepals reflexed. Fruits subrotund, wider than long, 5-sutured, seeds sticky. June–Oct. Coastal sands and limestone, SW, AP, SE (Velddrif to Knysna).•

fulvum L.   Erect or straggling shrublet to 1 m. Leaves bifoliolate, leaflets obovate with smooth, bony margins. Flowers cream-coloured to yellow with red markings. Fruits oblong, 5-sutured when fresh but 5-angled when dry. July–Oct. Sandy flats and rocky slopes, NW, SW, KM, LB, SE (Gifberg to Port Elizabeth).•

procumbens Adamson   Sprawling soft shrublet to 20 cm. Leaves bifoliolate, leaflets linear with margins flat or revolute. Flowers yellow. Fruits round and 5-sutured when fresh but 5-angled and slightly ribbed when dry. June–Aug. Sandy flats and rocky ridges, SW (Cape Peninsula).•

spinosum L.   Erect shrublet to 1 m. Leaves bifoliolate, leaflets cylindrical, grooved beneath. Flowers yellow with or without red markings. Fruits round, acute and 5-sutured when fresh, 5-angled and slightly ribbed when dry. June–Sept. Coastal sands, NW, SW (Lambert's Bay to Cape Peninsula).•

D. Fruits lobed or angled when fresh

cuneifolium Eckl. & Zeyh.   Erect shrublet to 60 cm. Leaves bifoliolate, leaflets wedge-shaped, succulent. Flowers yellow with khaki markings. Fruits oblong, 5-angled with ribs. June–Sept. Red sands and shale slopes, NW (Namaqualand to Klawer).

fuscatum Van Zyl   Erect shrublet to 1 m. Leaves bifoliolate, leaflets elliptic to obovate. Flowers deep yellow with or without red markings. Fruits oblong, 5-angled with ribs. May–Aug. Sandy flats on coastal limestone, SW, AP (Betty's Bay to De Hoop).•

pygmaeum Eckl. & Zeyh.   Erect shrublet to 60 cm. Leaves bifoliolate, leaflets cylindrical, grooved beneath. Flowers yellow with red markings. Fruits oblong, 5-lobed and slightly ribbed. June–Sept. Sandstone and shale slopes, NW, KM, SE (Bokkeveld Mts and Karoo to Uniondale).

rogersii Compton   Soft shrublet to 20 cm. Leaves bifoliolate, leaflets cylindrical, grooved beneath. Flowers dark pink. Fruits round and 5-sutured when fresh, 5-angled and slightly ribbed when dry. Aug.–Oct. Sandstone and shale slopes, NW, KM (Swartruggens and Witteberg).•

sessilifolium L.   WITSPEKBOS   Sprawling or prostrate shrublet to 10 cm. Leaves bifoliolate, leaflets obovate with rough, bony margins. Flowers cupped, whitish with red veins and markings. Fruits round to oblong, 5-angled when fresh, 5-lobed and ribbed when dry. July–Sept. Shale and sandy slopes, often under bushes, SW (Moorreesburg to Cape Peninsula).•

sp. 1   Like **Z. sessilifolium** but erect, to 1.2 m. June–Sept. Sandy and stony slopes, NW, SW (Graafwater to Villiersdorp).•

sp. 2 (= *Zygophyllum calcaricum* Van Zyl ined.)   Erect shrublet to 1 m. Leaves bifoliolate, leaflets obovate. Flowers cupped, whitish with red veins. Fruits oblong, 5-lobed and ribbed. Apr.–Sept. Coastal sands on limestone, AP (Gansbaai to Stilbaai).•

sp. 3   Erect shrub to 1.5 m. Leaves bifoliolate, leaflets elliptic. Flowers deep yellow with red or brown markings. Fruits oblong, 5-angled and ribbed. May–Sept. Rocky slopes, KM (Swartberg Mts).•

sp. 4   Large shrub to 1.5 m, young stems red-brown. Leaves bifoliolate, leaflets obovate, cuneate below, dark green. Flowers yellow with or without red markings. Fruits oblong, 5-lobed and ribbed. June–Oct. Mostly shale slopes, NW, KM (Hex River Pass and Little Karoo to Garcia's Pass).•

KEYS TO THE FAMILIES

A. FERNS & FERN ALLIES

1. Leaves simple with an unbranched central vein:
   2. Sporangia borne on peltate sporangiophores arranged in a terminal strobilus .... **EQUISETACEAE**
   2.' Sporangia borne in axils of sporophylls or adaxially near base:
      3. Plants homosporous; leaves without ligules .... **LYCOPODIACEAE**
      3.' Plants heterosporous; leaves ligulate:
         4. Plants wholly or partially aquatic; stems short .... **ISOETACEAE**
         4.' Plants not aquatic; stems long, erect or creeping .... **SELAGINELLACEAE**
1.' Fronds (leaves) simple or compound with a branched vascular system:
   5. Sporangia fused to form a synangium:
      6. Synangia borne abaxially on lamina; stem massive .... **MARATTIACEAE**
      6.' Synangia 2, elongate, borne at apex of spike; stems small and subterranean............ **OPHIOGLOSSACEAE**
   5.' Sporangia borne singly or in sori but never fused:
      7. Plants heterosporous; sporangia heteromorphic:
         8. Plants rooted; fronds with 4 leaflets borne apically .... **MARSILEACEAE**
         8.' Plants free-floating:
            9. Leaves borne in whorls, 2 floating, others submerged and root-like; true roots absent .... **SALVINIACEAE**
            9.' Leaves not in whorls, bilobed; true roots present.. **AZOLLACEAE**
      7.' Plants homosporous; sporangia isomorphic:
         10. Sporangia sessile or subsessile, or with a stalk 4 or more cells thick:
            11. Sporangia with a poorly developed lateral annulus .... **OSMUNDACEAE**
            11.' Sporangia with a well developed apical or subapical annulus:
               12. Sori indusiate, indusium inferior... **CYATHEACEAE**
               12.' Sori exindusiate:
                  13. Sporangia borne on small fertile pinnae arranged at apex of a narrow, linear frond.. **SCHIZAEACEAE**
                  13.' Sporangia borne at or near vein apices of differentiated or undifferentiated fertile part of lamina .... **ANEMIACEAE**
         10.' Sporangia borne on a short or long slender stalk up to 3 cells thick:
            14. Fronds pseudodichotomously branched, with an arrested bud between branchlets.. **GLEICHENIACEAE**
            14.' Fronds not pseudodichotomously branched:
               15. Lamina uniseriate, without stomata... **HYMENOPHYLLACEAE**
               15.' Lamina multiseriate, with stomata:
                  16. Fronds articulated at stipe base:
                     17. Sporangia acrostichoid; fronds dimorphic................................. **LOMARIOPSIDACEAE**
                     17.' Sporangia arranged in distinct sori; fronds monomorphic .... **POLYPODIACEAE**
                  16.' Fronds not articulated at stipe base:
                     18. Pinnae articulated along rachis.. **NEPHROLEPIDACEAE**
                     18.' Pinnae not articulated along rachis:
                        19. Rhizome paleate, paleae clathrate (reticulately thickened); stipe with 2 outward-facing vascular bundles:
                           20. Lamina simple, entire; sporangia in 2 submerged parallel grooves, exindusiate.. **VITTARIACEAE**
                           20.' Lamina 1–4-pinnatifid; sori linear along veins, indusiate .... **ASPLENIACEAE**
                        19.' Rhizome hairy or paleate, paleae sometimes clathrate but then stipe with 2 or more inward-facing vascular bundles:
                           21. Sori marginal:
                              22. Rhizome paleate; an abaxial indusium absent............................. **PTERIDACEAE**
                              22.' Rhizome hairy; sori indusiate or recurved margin serving as a pseudo-indusium, or rhizome paleate and a superficial indusium absent.............. **DENNSTAEDTIACEAE**
                           21.' Sori superficial:
                              23. Sori elongate, adjacent and mostly parallel to segment axis .... **BLECHNACEAE**
                              23.' Sori circular, or if elliptic then not parallel to segment axis:
                                 24. Stipe with 2 strap-shaped vascular bundles .... **THELYPTERIDACEAE**
                                 24.' Stipe with 3 or more approximately circular vascular bundles:
                                    25. Sori indusiate or if exindusiate then without receptacular trichomes .... **DRYOPTERIDACEAE**
                                    25.' Sori exindusiate, receptacle often with a few receptacular hairs .... **GRAMMITIDACEAE**

B. GYMNOSPERMS

1. Palm-like plants with pinnate leaves and with leaf scars girdling stem ....**ZAMIACEAE**
1.' Trees or shrubs with hardwood stems and simple leaves along branches:
    2. Leaves small and scale-like; seeds winged, in an indehiscent roundish fruit or in a woody dehiscent cone..**CUPRESSACEAE**
    2.' Leaves well developed, needle-like or strap-shaped:
        3. Leaves needle-like, 1 or few united at base by a membranous sheath; seeds many, borne on spirally arranged woody bracts in a cone..**PINACEAE**
        3.' Leaves strap-shaped, mostly alternate; seeds 1 or 2, large and drupe-like, borne on a receptacle (sometimes fleshy) ....**PODOCARPACEAE**

C. ANGIOSPERMS

NOTE: In checking placentation it is often easier to examine ovaries after fertilization when they have slightly swollen. When only a single perianth whorl is present it is taken to represent the calyx.

1. Leaves usually parallel-veined and without a petiole; flowers usually with parts in multiples of 3; vascular bundles scattered in stem; embryo with 1 cotyledon; herbs or climbers, rarely shrubby.................**MONOCOTYLEDONS**
1.' Leaves usually reticulate-veined and often petiolate; flowers usually with parts in multiples of 4 or 5; vascular bundles usually arranged in a cylinder in stem; embryo with 2 (rarely more) cotyledons; herbs, climbers, shrubs or trees ....**DICOTYLEDONS**

MONOCOTYLEDONS

1. Gynoecium composed of 2 or more free carpels with separate styles and stigmas ....**KEY 1**
1.' Gynoecium composed of 1 carpel or of 2 or more united carpels with free or united styles:
    2. Perianth absent or reduced to bristles or to 1–3 scales:
        3. Flowers unisexual ....**KEY 2**
        3.' Flowers (florets) bisexual or unisexual, arranged in small spikes (spikelets) with scale-like bracts (glumes or lemmas) ....**KEY 3**
    2.' Perianth present, not reduced to bristles:
        4. Ovary more or less inferior...**KEY 4**
        4.' Ovary superior:
            5. Perianth composed of separate calyx and corolla, calyx often herbaceous, corolla usually petaloid or otherwise different from calyx ....**KEY 5**
            5.' Perianth composed of similar or subsimilar segments, these either all petaloid or all herbaceous or dry ....**KEY 6**

KEY 1

1. Perianth absent or cupular; stamens 1 or 2:
    2. Flowers bisexual, in spikes, emergent above water at anthesis, spikes consisting of 2 naked flowers facing in opposite directions and inserted at unequal heights; stamens 2, free; carpels 4–8, becoming stipitate in fruit and appearing umbellate ....**RUPPIACEAE**
    2.' Flowers unisexual, axillary, cymose or solitary, submerged; stamens 1 or 2, fused; carpels 1–9, stipitate ....**ZANNICHELLIACEAE**
1.' Perianth present, composed of 1–4 free segments; stamens 4–16:
    3. Leaves radical; spikes simple or 2-branched on elongate peduncles, at first enclosed in a spathe; tepals 1–3; stamens 6(–16), with elongate filaments; ovules 2 or more in each carpel..................**APONOGETONACEAE**
    3.' Leaves cauline; spikes simple on axillary peduncles, without a spathe; tepals 4; stamens 4, sessile; ovules 1 in each carpel...**POTAMOGETONACEAE**

KEY 2

1. Terrestrial plants or minute, floating aquatics:
    2. Leaves either absent or with a broad blade and pinnate venation; flowers either embedded in a minute, floating body or in a spike subtended by a large, petaloid spathe.....................................**ARACEAE**
    2.' Leaves strap-shaped; flowers in a dense, cylindrical spike ....**TYPHACEAE**
1.' Submerged aquatic herbs with elongate stems and linear leaves:
    3. Leaves opposite or verticillate; flowers solitary or few together in leaf axils; fresh-water plants ..**NAJADACEAE**
    3.' Leaves alternate; flowers several in a spike of alternating male and female flowers, enclosed in a spathe; marine plants ....**ZOSTERACEAE**

KEY 3

1. Florets each enclosed by a bract (lemma) on outside and a bracteole (palea) on inside, spikelet usually with 2 empty bracts (glumes) at its base; perianth represented by 2 minute scales (lodicules); stems usually with hollow

internodes, terete or compressed; leaf sheaths usually open, with a hairy or papery fringed (ligule) at mouth ....**POACEAE**

1.' Florets each enclosed by a single bract (glume) on outside, without a bracteole or female florets sometimes each surrounded by a closed bracteole (utricle); perianth absent or represented by bristles or scales; stems usually solid, often triquetrous; leaf sheaths usually closed, without a ligule...**CYPERACEAE**

KEY 4

1. Submerged aquatics rooting from a rhizome; flowers mainly unisexual from a tubular or 2-lobed spathe; stamens 3 ....**HYDROCHARITACEAE**
1.' Terrestrial plants, or emergent aquatics but then rootstock cormous or bulbous and stamens 6:
   2. Stamen 1; pollen agglutinated into masses (pollinia); ovary 1-locular with numerous ovules on parietal placentas; flowers highly zygomorphic, upper or lower median tepal often spurred ....**ORCHIDACEAE**
   2.' Stamens 3–6; pollen free:
      3. Flowers unisexual; plants twining; leaves with reticulate venation; fruit a 3-winged or 3-angled capsule ....**DIOSCOREACEAE**
      3.' Flowers bisexual:
         4. Fertile stamens 5; flowers in axil of a leathery spathe, outer tepals united into a calyx-like tube, lateral inner tepals fused, very showy; leaves large, stiff, petiolate ....**STRELITZIACEAE**
         4.' Fertile stamens 3(4) or 6; leaves seldom petiolate:
            5. Stamens 3 opposite outer tepals; leaves usually unifacial; rootstock a corm or rhizome; capsules loculicidally dehiscent...**IRIDACEAE**
            5.' Stamens 6 or 3 but then opposite inner tepals:
               6. Flowers in umbels (sometimes 1-flowered) subtended by 1 or more spathaceous bracts and borne on naked peduncles; herbs with radical leaves; rootstock a bulb or rhizome ....**AMARYLLIDACEAE**
               6.' Flowers in racemes or corymbs or solitary but not subtended by spathaceous bracts; rootstock a rhizome or corm:
                  7. Flowers solitary or few on leafless scapes or pedicels, subumbellate or in racemes; ovules many in each locule...**HYPOXIDACEAE**
                  7.' Flowers numerous on bracteate stems, in helicoid cymes; ovules 1 or 2 in each locule:
                     8. Stamens 6; inflorescence hairs plumose; leaves bifacial, channelled; rhizome not red ....**LANARIACEAE**
                     8.' Stamens 3; inflorescence hairs simple; leaves unifacial, ensiform (African genera); stem tissue bright orange-red..**HAEMODORACEAE**

KEY 5

1. Ovary 1-locular; leaves mostly radical; flowers in spikes or heads on elongated peduncles; style 3-branched...**XYRIDACEAE**
1.' Ovary 2- or 3-locular; leaves cauline; flowers in congested axillary and terminal clusters; style simple ....**COMMELINACEAE**

KEY 6

1. Tepals dry and glumaceous or the inner hyaline; leaves narrow or reduced to sheaths; rush-like plants:
   2. Flowers unisexual, sexes on separate plants; ovule 1 in each locule, pendulous....................**RESTIONACEAE**
   2.' Flowers bisexual; ovules 3 or more in each locule:
      3. Plants caulescent, fibrous, with apical leaf rosette; leaves serrate; ovary 3-locular, ovules few; stigmas sessile or subsessile ....**PRIONIACEAE**
      3.' Plants rhizomatous, with basal leaves; leaves entire; ovary 1–3-locular; ovules numerous; style short or long...**JUNCACEAE**
1.' Tepals petaloid or herbaceous:
   4. Tepals herbaceous; fruit separating into 3 cocci; anthers sessile; flowers without bracts, in racemes on elongated peduncles; leaves radical, linear or filiform ....**JUNCAGINACEAE**
   4.' Tepals petaloid or sepaloid; fruit a capsule or berry; flowers bracteate:
      5. Stamens 3; leaves unifacial, pleated; stem tissue bright orange-red ....**HAEMODORACEAE**
      5.' Stamens 6, outer whorl of filaments sometimes without anthers; leaves bifacial, flat or channelled:
         6. Anthers dehiscing by an apical pore, often unequal; filaments short, inserted at mouth of perianth tube; rootstock a corm ....**TECOPHILAEACEAE**
         6.' Anthers dehiscing by longitudinal slits (these rarely very short but then plants leafless); rootstock a rhizome or bulb or tuber:
            7. Flowers in umbels subtended by 2 spathaceous bracts and borne on naked peduncles:
               8. Flowers perfectly actinomorphic with stamens spreading or anthers sessile; capsule ovoid; plants usually smelling of onion or garlic; rootstock a bulb or rhizome ....**ALLIACEAE**
               8.' Flowers slightly zygomorphic with stamens declinate and recurved at tips; capsule 3-angled or -winged; plants not smelling of onion or garlic; rootstock a rhizome ....**AGAPANTHACEAE**
            7.' Flowers in spikes, racemes or panicles:
               9. Rootstock a bulb, rarely with loose scales..**HYACINTHACEAE**
               9.' Rootstock a corm, rhizome or tuber:

10. Fruit a fleshy berry or drupe; plants shrubby or scandent:
   11. Flowers longer than 20 mm, tufted in dense racemes or panicles, tubular below with exserted stamens; leaves large, tough and crowded; plants rhizomatous or tree-like ....**CONVALLARIACEAE**
   11.' Flowers shorter than 10 mm, few in axillary racemes or cymes; plants shrubby or scandent:
      12. Climbers; leaves well developed, ovate, with distinct cross-connections between the numerous prominent parallel veins; flowers in lax, axillary cymes; anthers included ....**BEHNIACEAE**
      12.' Climbers or shrubs, often spiny; leaves small or rudimentary, with a spinous or soft spur, often replaced by cladodes which may be ovate and leaf-like but never with cross-connections between veins; flowers in fascicles or racemes; anthers exserted ....**ASPARAGACEAE**
10.' Fruit a capsule:
   13. Seeds woolly; flowering stems leafless except for a clasping basal leafy bract; leaves absent at flowering, petiolate; rootstock a tuber ....**CONVALLARIACEAE**
   13.' Seeds glabrous; flowering stems leafy or leafless but then without a basal bract:
      14. Rootstock a corm; seeds brown; flowers usually opposite bracts ....**COLCHICACEAE**
      14.' Rootstock a rhizome, sometimes short and with swollen roots; seeds black; flowers in axils of bracts:
         15. Flowers solitary in axils of bracts ....**ASPHODELACEAE**
         15.' Flowers more than 1 in axils of bracts:
            16. Leaves without a pseudopetiole; ovules 6–30 in each locule; perianth persistent, white; flowers erect or spreading, with white filaments; seeds flattened or angular ....**ANTHERICACEAE**
            16.' Leaves with a flattened pseudopetiole; ovules 2 in each locule; perianth caducous, blue; flowers often pendulous, with banded filaments; seeds pear-shaped with a white aril ....**HEMEROCALLIDACEAE**

DICOTYLEDONS

1. Gynoecium composed of 2 or more free carpels with separate styles and stigmas ....**KEY 1**
1.' Gynoecium composed of 1 carpel or 2 or more united carpels with free or united styles, or if carpels free below then style or stigmas united:
   2. Ovules attached to outer wall of ovary, 2 or more (placentation parietal):
      3. Ovary superior...**KEY 2**
      3.' Ovary more or less inferior..**KEY 3**
   2.' Ovules attached to central axis, base or apex of ovary, 1 or more (placentation axile):
      4. Ovary superior:
         5. Petals absent ....**KEY 4**
         5.' Petals present:
            6. Petals free ....**KEY 5**
            6.' Petals more or less fused ....**KEY 6**
      4.' Ovary more or less inferior:
         7. Petals absent ....**KEY 7**
         7.' Petals present:
            8. Petals free ....**KEY 8**
            8.' Petals more or less fused ....**KEY 9**

KEY 1
1. Leaves opposite, without stipules; carpels as many as petals; plants often succulent ....**CRASSULACEAE**
1.' Leaves alternate, sometimes all radical:
   2. Aquatic herbs with floating leaves on long petioles, blade more or less peltate; flowers solitary, on long peduncles, many-petalled and conspicuous; carpels immersed in an expanded torus ....**NYMPHAEACEAE**
   2.' Terrestrial plants; leaf blades not peltate; carpels free:
      3. Sepals fused below; stamens perigynous, inserted at mouth of calyx tube; stipules herbaceous, conspicuous ....**ROSACEAE**
      3.' Sepals free; stamens hypogynous, not inserted on calyx; stipules rudimentary or absent ....**RANUNCULACEAE**

KEY 2
1. Gynoecium composed of 1 carpel (only 1 placenta in ovary):
   2. Stipules absent; leaflets dotted with pellucid glands at least at margin ....**RUTACEAE**

2.' Stipules present; leaves without pellucid dots .... **FABACEAE**
1.' Gynoecium composed of 2 or more united carpels (2 or more placentas in ovary):
   3. Petals more or less fused:
      4. Flowers zygomorphic:
         5. Leaves pinnate; shrubs, trees or woody climbers; stamens 4 or 5 .... **BIGNONIACEAE**
         5.' Leaves simple, sometimes reduced to scales; herbs:
            6. Stamens 2; leaf or leaves well developed; not parasitic .... **GESNERIACEAE**
            6.' Stamens 4; leaves reduced to scales; root parasites .... **OROBANCHACEAE**
      4.' Flowers actinomorphic:
         7. Flowers unisexual .... **ACHARIACEAE**
         7.' Flowers bisexual:
            8. Petals and stamens 10–12; style deeply 2-cleft; white-prickly shrublets .... **BORAGINACEAE**
            8.' Petals and stamens 5; herbs, style entire or 2-lobed; shrubs or trees, sometimes spiny but not prickly:
               9. Leaves alternate or radical; petals induplicate-valvate, fimbriate, yellow; aquatic or marsh plants .... **MENYANTHACEAE**
               9.' Leaves opposite; petals contorted, entire, yellow, white or pink:
                  10. Shrubs or woody climbers with latex; fruit a large berry .... **APOCYNACEAE**
                  10.' Herbs without latex; fruit a capsule, rarely berry-like .... **GENTIANACEAE**
   3.' Petals free or absent:
      11. Flowers more or less zygomorphic; ovary sessile:
         12. Connective of anthers produced into a membranous appendage; flowers mostly solitary, violet; fruit a 3-valved capsule; leaves simple .... **VIOLACEAE**
         12.' Connective without an appendage; flowers in spikes or racemes, whitish; fruit a capsule gaping at apex; leaves often dissected .... **RESEDACEAE**
      11.' Flowers actinomorphic or rarely zygomorphic but then ovary stipitate on a distinct gynophore:
         13. Leaves opposite or whorled:
            14. Leaves dotted with pellucid glands; flowers yellow; stamens more than twice as many as petals .... **CLUSIACEAE**
            14.' Leaves not gland-dotted; flowers pink to lilac; stamens (4 or 5)6 .... **FRANKENIACEAE**
         13.' Leaves alternate or all radical:
            15. Calyx absent; flowers unisexual, in catkins; fruit a 2-valved capsule; seeds with a basal tuft of long, fine hairs .... **SALICACEAE**
            15.' Calyx present:
               16. Stamens as many as and alternating with petals:
                  17. Styles 2–5; leaves with stalked glands; insectivorous herbs .... **DROSERACEAE**
                  17.' Style 1; leaves not glandular-hairy:
                     18. Stamens accompanied by 2 series of staminodes, the outer filiform, the inner petaloid; stipules present .... **OCHNACEAE**
                     18.' Stamens not accompanied by staminodes; stipules absent .... **PITTOSPORACEAE**
               16.' Stamens more numerous than petals or as many as petals and opposite them:
                 19. Prickly herbs; leaves pinnately lobed; sepals 2 or 3, caducous; fruit a capsule dehiscing by 4–6 short apical valves .... **PAPAVERACEAE**
                 19.' Unarmed herbs, shrubs or trees; leaves simple or digitate; sepals persistent:
                     20. Sepals free or fused below to receptacle but then ovary on a distinct gynophore .... **BRASSICACEAE**
                     20.' Sepals fused below, tube sometimes adnate to ovary:
                        21. Petals or inner perianth segments with a fleshy gland on inner face; fruit a globose, leathery capsule with several orange seeds .... **KIGGELARIACEAE**
                        21.' Petals or petaloid stamens without fleshy glands or petals absent .... **FLACOURTIACEAE**

KEY 3
1. Petals absent; plant parasitic on roots; flowers unisexual, solitary, red .... **CYTINACEAE**
1.' Petals present; plants not parasitic:
   2. Flowers unisexual; plants usually trailing or climbing with tendrils; leaves often palmately lobed or deeply divided .... **CUCURBITACEAE**
   2.' Flowers bisexual; plants without tendrils:
      3. Petals more or less fused; stipules interpetiolar .... **RUBIACEAE**
      3.' Petals free; stipules not interpetiolar or absent:
         4. Trees or shrubs; leaves alternate; petals 5–9 .... **FLACOURTIACEAE**
         4.' Succulent herbs; leaves opposite; petals numerous .... **AIZOACEAE**

KEY 4
1. Ovary with 2 or more ovules in each locule:
   2. Leaves opposite or whorled or all radical:

3. Flowers unisexual; leaves in whorls of 4; stamens numerous, spirally arranged on a prolonged torus ....**EUPHORBIACEAE**
3.' Flowers bisexual:
   4. Sepals free or almost so; stamens hypogynous:
      5. Ovary 3–5-locular with axile placentas ....**MOLLUGINACEAE**
      5.' Ovary 4-locular, with (1)2 pendulous ovules in each locule ....**GEISSOLOMATACEAE**
   4.' Sepals fused below; stamens perigynous, inserted on calyx tube:
      6. Styles 2–5, or if 1 then ovary 1-locular with few ovules:
         7. Ovules many in each locule, axile..**MOLLUGINACEAE**
         7.' Ovules 1–few in each locule, pendulous ....**AIZOACEAE**
      6.' Style 1:
         8. Ovary 1–5-locular with numerous ovules in each locule ....**LYTHRACEAE**
         8.' Ovary 4-locular with 2 or 4 basal or apical ovules in each locule ....**PENAEACEAE**
2.' Leaves alternate, not all radical:
   9. Ovary 1-locular:
      10. Flowers bisexual; calyx scarious; fruit a circumscissile capsule; herbs..................**AMARANTHACEAE**
      10.' Flowers unisexual; calyx not scarious; fruit a drupe; woody plants:
         11. Leaves with stipules; calyx imbricate; stamens opposite sepals; styles 3 ....**EUPHORBIACEAE**
         11.' Leaves without stipules; calyx valvate; stamens alternating with sepals; stigmas sessile and multiradiate ....**ICACINACEAE**
   9.' Ovary 2- or more-locular:
      12. Gynoecium of 3 or more loosely united carpels, in fruit separating into follicles; trees with unisexual flowers; leaves simple or digitate; stamens united into a column ....**MALVACEAE**
      12.' Gynoecium of completely united carpels, in fruit forming a capsule or indehiscent or separating into winged cocci:
         13. Ovary stipitate on a distinct gynophore ....**BRASSICACEAE**
         13.' Ovary sessile or subsessile:
            14. Leaves with stipules, simple ....**EUPHORBIACEAE**
            14.' Leaves without stipules, simple or pinnate..**SAPINDACEAE**
1.' Ovary with 1 ovule in each locule:
   15. Ovary 2- or more-locular:
      16. Leaves pinnate; trees or shrubs; fruit a drupe or berry ....**SAPINDACEAE**
      16.' Leaves simple, sometimes lobed or divided or reduced to scales or stipular spines:
         17. Flowers unisexual or polygamous:
            18. Ovary 5-locular, carpels loosely united and separating in fruit; calyx present...............**MALVACEAE**
            18.' Ovary 2–4-locular, carpels completely united; calyx present or absent ....**EUPHORBIACEAE**
         17.' Flowers bisexual:
            19. Style 1 ....**BRASSICACEAE**
            19.' Styles 2–5, free or fused at base:
               20. Ovary 7–10-locular; fruit a berry; flowers in spike-like racemes ....**PHYTOLACCACEAE**
               20.' Ovary 2-locular; fruit a capsule; flowers in cymes ....**AIZOACEAE**
   15.' Ovary 1-locular:
      21. Stipules absent:
         22. Submerged, aquatic herbs; leaves whorled, bifurcately dissected with linear or filiform segments; flowers unisexual, solitary and sessile in axils of leaves..**CERATOPHYLLACEAE**
         22.' Plants not aquatic; leaves alternate or opposite, simple or pinnate:
            23. Calyx absent; flowers in spikes:
               24. Flowers unisexual; stamens 3–12; styles 2, free or shortly fused below; trees or shrubs..**MYRICACEAE**
               24.' Flowers bisexual; stamens 2; stigma sessile; herbs ....**PIPERACEAE**
            23.' Calyx present, at least in male flowers:
               25. Anthers opening by valves; calyx 6-lobed; stamens 6–12, often accompanied by staminodes ....**LAURACEAE**
               25.' Anthers opening by longitudinal slits; calyx 3–5-lobed:
                  26. Stamens twice as many as sepals, inserted on an elongate calyx tube ....**THYMELAEACEAE**
                  26.' Stamens fewer than twice as many as sepals, sometimes accompanied by staminodes:
                     27. Calyx with 2–4 valvate segments; trees or shrubs; flowers bisexual in large bracteate heads or racemes; stamens 4, inserted on sepals ....**PROTEACEAE**
                     27.' Calyx with 3–5 imbricate segments, or almost completely tubular; herbs or shrubs...**AMARANTHACEAE**
      21.' Stipules present, sometimes forming a sheath surrounding stem:
         28. Leaves 3- or 4-pinnate; calyx petaloid; fruit a stipitate achene borne on a slender pedicel; herbs...**RANUNCULACEAE**
         28.' Leaves simple or digitate:

29. Ovule pendulous from apex or near apex of ovary; flowers unisexual or polygamous:
  30. Flowers crowded inside a hollow, almost closed receptacle (fig) .... **MORACEAE**
  30.' Flowers in open inflorescences or solitary:
    31. Stamens as many as sepals, 4 or 5; style 2-branched; fruit a drupe; trees or shrubs .... **CELTIDACEAE**
    31.' Stamens more numerous than sepals; style unbranched:
      32. Flowers solitary or paired in axils of leaves; leaves 1–3-foliolate with narrow leaflets; heath-like shrubs .... **ROSACEAE**
      32.' Flowers in axillary racemes or panicles; leaves simple; trees .... **EUPHORBIACEAE**
29.' Ovule arising from base or near base of ovary:
  33. Calyx absent; flowers minute, in dense spikes; shrubs, sometimes climbing .... **PIPERACEAE**
  33.' Calyx present:
    34. Styles 2 or 3, free or fused below; stamens 4–8; fruit a small nut; stipules often forming a sheath..**POLYGONACEAE**
    34.' Style 1 or absent; stamens 5 or fewer:
      35. Flowers unisexual; sepals free, without an epicalyx; style terminal; leaves often marked with cystoliths...**URTICACEAE**
      35.' Flowers bisexual:
        36. Leaves alternate, palmate; sepals fused below, free portion alternating with lobes of an epicalyx; style arising from near base of ovary; herbs with palmate leaves........**ROSACEAE**
        36.' Leaves whorled, simple; sepals free; style arising terminally or subterminally..**MOLLUGINACEAE**

KEY 5

1. Ovary 1-locular, sometimes septate towards base:
  2. Sepals 1 or 2, free, sometimes caducous:
    3. Flowers unisexual; leaf blade peltate or subpeltate; woody climbers........................**MENISPERMACEAE**
    3.' Flowers bisexual; leaf blade not peltate; herbs:
      4. Leaves much divided; flowers zygomorphic; ovary with 1 ovule; fruit a nut ....**FUMARIACEAE**
      4.' Leaves simple, fleshy; flowers actinomorphic; ovary with numerous ovules; fruit a capsule ....**PORTULACACEAE**
  2.' Sepals 3 or more:
    5. Leaves opposite or verticillate, not all radical:
      6. Petals and stamens perigynous, inserted on calyx tube; style 1:
        7. Ovary with 1 apical, pendulous ovule; heath-like shrubs ....**THYMELAEACEAE**
        7.' Ovary with numerous ovules on a free-basal or free-central placenta; herbs ....**LYTHRACEAE**
      6.' Petals and stamens hypogynous or slightly perigynous, not inserted on calyx:
        8. Herbs; ovary with numerous ovules on a free-basal or free-central placenta, rarely with 1 or 2 basal ovules but then stipules present; fruit a capsule or a nut enclosed by calyx:
          9. Ovary with 1 or 2 basal ovules; petals absent or minute; fruit indehiscent, enclosed by calyx; stipules present ....**ILLECEBRACEAE**
          9.' Ovary with 3 or more ovules; petals rarely absent; fruit a 3–5-valved capsule; stipules present or absent...**CARYOPHYLLACEAE**
        8.' Trees or shrubs; ovary with 1 or 2 basal ovules; fruit a drupe:
          10. Filaments united at base into a tube or cup..**SALVADORACEAE**
          10.' Filaments free:
            11. Ovule 1; styles 3, free or united at base ....**ANACARDIACEAE**
            11.' Ovules 2–8; style 1 or stigma sessile:
              12. Ovules 2, pendulous from apex of ovary...**ICACINACEAE**
              12.' Ovules 2–8, erect from base of ovary ....**CELASTRACEAE**
    5.' Leaves alternate or all radical:
      13. Stipules present:
        14. Stamens more numerous than petals; petals and stamens perigynous, inserted at mouth of calyx tube ....**ROSACEAE**
        14.' Stamens as many as petals or sepals; petals and stamens not perigynous ....**ILLECEBRACEAE**
      13.' Stipules absent:
        15. Ovules 2 or more:
          16. Petals imbricate; stamens opposite petals; ovules on a free-basal placenta ....**MYRSINACEAE**
          16.' Petals valvate; stamens alternating with petals; ovules 2, pendulous from apex of a central placenta ....**ICACINACEAE**
        15.' Ovule 1:
          17. Flowers zygomorphic, inner 2 sepals larger than others, lowest petal forming a keel and upper 2 petals vestigial or absent ....**POLYGALACEAE**
          17.' Flowers actinomorphic:

18. Sepals united into an elongate tube; stamens up to twice as many as petals, inserted on calyx tube; heath-like shrubs .... **THYMELAEACEAE**
18.' Sepals free or almost so:
19. Stamens united; leaf blade peltate or subpeltate; woody climbers with unisexual flowers .... **MENISPERMACEAE**
19.' Stamens free or almost so; leaf blade not peltate; trees or shrubs: .... **ANACARDIACEAE**
1.' Ovary 2- or more-locular:
20. Stamens as many as and opposite petals:
21. Filaments more or less united, sometimes only at base; sepals valvate; herbs or shrubs with stellate hairs .... **MALVACEAE**
21.' Filaments free:
22. Plants viscid; leaves with gland-tipped tentacles which trap insects; undershrubs .... **RORIDULACEAE**
22.' Plants not viscid:
23. Ovules numerous in each locule; fruit a loculicidal capsule; inflorescences terminal; trees with gland-dotted leaves .... **MYRTACEAE**
23.' Ovules 1 or 2 in each locule; fruit a drupe or berry:
24. Inflorescences axillary; trees or shrubs, often spiny, without tendrils; ovule 1 in each locule; fruit a drupe .... **RHAMNACEAE**
24.' Inflorescences leaf-opposed; herbaceous or woody plants, often climbing with tendrils; ovules 2 in each locule; fruit a berry .... **VITACEAE**
20.' Stamens as many as and alternating with petals, or more numerous or fewer:
25. Leaves compound, with 2 or more leaflets:
26. Leaves opposite or subopposite:
27. Stipules absent .... **RUTACEAE**
27.' Stipules present; stamens twice as many as petals:
28. Flowers solitary; ovary 4- or 5-chambered (rarely less); petals larger than sepals; herbs or shrubs .... **ZYGOPHYLLACEAE**
28.' Flowers in racemes or panicles; ovary 2-chambered; petals usually smaller than sepals; trees or shrubs .... **CUNONIACEAE**
26.' Leaves alternate:
29. Stipules present, lateral or intrapetiolar; flowers more or less zygomorphic, in racemes; stamens 4 or 5 .... **MELIANTHACEAE**
29.' Stipules absent:
30. Filaments united into a tube; leaves pinnate or bipinnate .... **MELIACEAE**
30.' Filaments free or only shortly united at base:
31. Ovules 2 or more in each locule:
32. Geophytic herbs or annuals; leaves mostly digitate; plants tristylous with stamens in 2 series; ovules numerous per locule .... **OXALIDACEAE**
32.' Shrubs or trees; leaves simple or pinnate; ovules 1 or 2 per locule .... **RUTACEAE**
31.' Ovule 1 in each locule:
33. Styles 3–5, separated at base .... **ANACARDIACEAE**
33.' Styles 1 or 2, terminal:
34. Styles 2; fruit with carpels separating from a persistent central column .... **RUTACEAE**
34.' Style 1; fruit a drupe or capsule not as above .... **SAPINDACEAE**
25.' Leaves simple or 1-foliolate (sometimes deeply divided):
35. Leaves opposite or verticillate, not all radical:
36. Sepals united below into a tube:
37. Petals hypogynous; ovules 1 or 2 in each locule; spiny trees or shrubs .... **SALVADORACEAE**
37.' Petals perigynous, inserted at mouth of calyx tube; ovules many in each locule; herbs .... **LYTHRACEAE**
36.' Sepals free or almost so:
38. Style 1; trees or shrubs:
39. Stamens as many as petals and alternating with petaloid staminodes; leaves dotted with pellucid glands .... **RUTACEAE**
39.' Stamens not alternating with staminodes; leaves without pellucid glands .... **CELASTRACEAE**
38.' Styles 3–5; herbs or shrubs:
40. Stipules present:
41. Ovules 2 in each locule; ovary beaked; capsule dehiscing elastically .... **GERANIACEAE**
41.' Ovules numerous in each locule; ovary not beaked; capsule not dehiscing elastically .... **ELATINACEAE**
40.' Stipules absent:
42. Stamens 5; sepals toothed or lobed at apex; ovules 2 in each locule; leaves without glands .... **LINACEAE**

42.' Stamens indefinite; sepals entire; ovules few to many in each locule; leaves usually dotted or streaked with pellucid or opaque glands ....**CLUSIACEAE**
35.' Leaves alternate or all radical:
 43. Ovule 1 in each fertile locule:
  44. Flowers unisexual or polygamous:
   45. Stipules absent; trees or shrubs; flowers small, in racemes or narrow raceme-like panicles...**SAPINDACEAE**
   45.' Stipules present ....**EUPHORBIACEAE**
  44.' Flowers bisexual:
   46. Anthers 1-thecous; stamens numerous, filaments more or less united into a tube; sepals valvate; plants often with stellate hairs..**MALVACEAE**
   46.' Anthers 2-thecous:
    47. Sepals 4; style 1; leaves entire or variously toothed or lobed ....**BRASSICACEAE**
    47.' Sepals 5:
     48. Style 1; stamens many; shrubs with yellow flowers ....**OCHNACEAE**
     48.' Styles 2–10:
      49. Ovule basal, erect; styles 2 ....**MOLLUGINACEAE**
      49.' Ovule pendulous; styles 10..**NEURADACEAE**
43.' Ovules 2 or more in each locule:
 50. Stamens as many as or fewer than petals:
  51. Ovary 5-locular; ovules 2 in each locule:
   52. Petals imbricate; ovary and fruit beaked; leaves often lobed or divided; stipules conspicuous ....**GERANIACEAE**
   52.' Petals contorted; ovary and fruit unbeaked; leaves simple; stipules 0 or gland-like ....**LINACEAE**
  51.' Ovary 2–4-locular:
   53. Flowers unisexual; styles 2 or 3 ....**EUPHORBIACEAE**
   53.' Flowers bisexual; style 1 ....**CELASTRACEAE**
 50.' Stamens more numerous than petals:
  54. Stipules absent:
   55. Filaments free; leaves dotted with pellucid glands ....**RUTACEAE**
   55.' Filaments more or less united; leaves not dotted with glands........................**MELIACEAE**
  54.' Stipules present:
   56. Sepals valvate; plants often with stellate hairs...**MALVACEAE**
   56.' Sepals imbricate; plants without stellate hairs ....**GERANIACEAE**

KEY 6

1. Ovary 1-locular, sometimes septate towards base:
 2. Flowers unisexual, males with united petals, females with 1 or 2 free petals; stamens united; leaf blade peltate or subpeltate; woody climbers ....**MENISPERMACEAE**
 2.' Flowers bisexual:
  3. Ovary with 1 ovule:
   4. Flowers zygomorphic, pea-like; stamens 10..**FABACEAE**
   4.' Flowers actinomorphic; stamens 4 or 5:
    5. Leaves opposite; stamens alternating with petals ....**SALVADORACEAE**
    5.' Leaves alternate; stamens opposite petals ....**PLUMBAGINACEAE**
  3.' Ovary with 2 or more ovules:
   6. Stamens fewer than petals, 2; flowers strongly zygomorphic, spurred; leaves often bearing insectivorous bladders; herbs of wet places ....**LENTIBULARIACEAE**
   6.' Stamens as many as petals; flowers actinomorphic or almost so:
    7. Stamens opposite petals:
     8. Trees or shrubs; fruit indehiscent, usually 1-seeded; leaves alternate ....**MYRSINACEAE**
     8.' Herbs; fruit a many-seeded circumscissile or 5-valved capsule (rarely indehiscent); leaves opposite or alternate..**PRIMULACEAE**
    7.' Stamens alternating with petals:
     9. Leaves sessile, whorled; ovules erect ....**STILBACEAE**
     9.' Leaves opposite or alternate; ovules pendulous ....**ICACINACEAE**
1.' Ovary 2- or more-locular:
 10. Petals numerous, more than 10; stamens indefinite; styles 5; fruit a 5-valved capsule; herbs with fleshy leaves...**AIZOACEAE**
 10.' Petals fewer than 10 (if more, then stamens only 2):
  11. Stamens more numerous than petals:
   12. Stipules present; flowers unisexual; leaves often lobed ....**EUPHORBIACEAE**
   12.' Stipules absent:

13. Flowers zygomorphic, lowest petal forming a keel; filaments united into a split sheath; ovary 2-locular with 1 ovule in each locule ....**POLYGALACEAE**
13.' Flowers actinomorphic; filaments not united into a sheath:
  14. Ovules 1 or 2 in each locule; fruit an indehiscent berry ....**EBENACEAE**
  14.' Ovules several in each locule; fruit a loculicidal capsule; anthers often opening by apical pores ....**ERICACEAE**
11.' Stamens as many as or fewer than petals:
  15. Stamens fewer than petals, 2–4:
    16. Perianth actinomorphic; stamens 2; trees or shrubs ....**OLEACEAE**
    16.' Perianth zygomorphic:
      17. Ovules more than 4 in each locule:
        18. Leaves pinnate or trifoliolate; stamens 4; fruit a loculicidal capsule with winged seeds ....**BIGNONIACEAE**
        18.' Leaves simple:
          19. Ovules arranged in 1 or 2 series on each placenta; seeds often borne on hardened, hook-like funicles ....**ACANTHACEAE**
          19.' Ovules arranged in more than 2 series on each placenta:
            20. Posticous (upper) petals entirely or partially interior in bud; parasitic or hemi-parasitic perennials or shrublets ....**OROBANCHACEAE**
            20.' Posticous (upper) petals exterior in bud; autotrophic annuals, perennials or shrubs ....**SCROPHULARIACEAE**
      17.' Ovules 1–4 in each locule:
        21. Ovary more or less deeply 4-lobed, style arising from between lobes; fruit separating into nutlets; plants often aromatic ....**LAMIACEAE**
        21.' Ovary not deeply 4-lobed, style terminal:
          22. Anthers 1-thecous; leaves mostly alternate, narrow; ovary 2-locular with 1 apical, pendulous ovule in each locule ....**SCROPHULARIACEAE**
          22.' Anthers 2-thecous, thecae sometimes confluent; leaves opposite or whorled:
            23. Herbs; leaves opposite ....**VERBENACEAE**
            23.' Ericoid undershrubs; leaves whorled ....**STILBACEAE**
  15.' Stamens as many as petals, 4 or more:
    24. Stamens opposite petals; trees or shrubs with alternate leaves ....**SAPOTACEAE**
    24.' Stamens alternating with petals:
      25. Leaves opposite or verticillate, not all radical:
        26. Stamens hypogynous, not inserted on corolla; anthers opening by apical slits; heath-like shrubs ....**ERICACEAE**
        26.' Stamens inserted on corolla tube:
          27. Petals imbricate:
            28. Ovules 1 or 2 in each locule; petals and stamens 5; plants often hispid ...**BORAGINACEAE**
            28.' Ovules numerous in each locule:
              29. Trees or shrubs; flowers in cymes, racemes or panicles ....**BUDDLEJACEAE**
              29.' Herbs; flowers solitary or paired in leaf axils.......................**SCROPHULARIACEAE**
          27.' Petals contorted or valvate:
            30. Petals valvate; leaves often with 3 or more longitudinal veins ....**LOGANIACEAE**
            30.' Petals contorted:
              31. Fruit a septicidal capsule; herbs...**GENTIANACEAE**
              31.' carpels separating in fruit; trees, shrubs or herbs but then flowers with a corona...**APOCYNACEAE**
      25.' Leaves (at least lower ones) alternate, or all radical, sometimes rudimentary:
        32. Leaves all radical; flowers small, in pedunculate spikes; fruit a circumscissile capsule ....**PLANTAGINACEAE**
        32.' Leaves not all radical:
          33. Ovules more than 2 in each locule:
            34. Petals contorted; fruit of 2 separated carpels; stipules spiny ....**APOCYNACEAE**
            34.' Petals not contorted; fruit a capsule or berry; stipules not spiny:
              35. Fruit a capsule with winged seeds; plants never prickly ....**BIGNONIACEAE**
              35.' Fruit a berry or capsule but seeds unwinged; plants often prickly
                36. Petals 5; style simple..**SOLANACEAE**
                36.' Petals 10–12; style deeply 2-cleft ....**BORAGINACEAE**
          33.' Ovules 1 or 2 in each locule:
            37. Flowers strongly zygomorphic, lowest petal forming a keel; filaments united into a split sheath ....**POLYGALACEAE**
            37.' Flowers not as above; filaments free:

                38. Corolla tube split down the front, with 4 lobes; anthers 1-thecous..**SCROPHULARIACEAE**
                38.' Corolla tube not split; anthers 2-thecous:
                    39. Style absent, stigma sessile; stipules present; flowers unisexual ....**AQUIFOLIACEAE**
                    39.' Style/s present; stipules absent; flowers bisexual:
                        40. Fruit separating into nutlets; petals imbricate or contorted .....**BORAGINACEAE**
                        40.' Fruit a capsule or indehiscent; petals plicate or valvate; plants often twining or trailing..**CONVOLVULACEAE**

KEY 7

1. Ovary 2- or more-locular:
    2. Flowers bisexual; leaves without stipules ....**AIZOACEAE**
    2.' Flowers unisexual; leaves with stipules:
        3. Flowers in heads; stamens as many as sepals; ovary with 1 ovule in each locule; trees or shrubs with entire leaves ....**HAMAMELIDACEAE**
        3.' Flowers in cymes or panicles; stamens numerous; ovary with very numerous ovules in each locule; herbs often with oblique leaves ....**BEGONIACEAE**
1.' Ovary 1-locular:
    4. Ovule 1:
        5. Fleshy herbs, parasitic on roots; leaves reduced to scales; flowers unisexual in involucrate heads ....**BALANOPHORACEAE**
        5.' Plants not parasitic; leaves radical, on long petioles, reniform, toothed; flowers in paniculate-spicate heads ....**GUNNERACEAE**
    4.' Ovules 2 or more:
        6. Plants leafless, parasitic on roots; flowers solitary, subsessile on rhizome; ovules very numerous on apical, pendulous placentas...**HYDNORACEAE**
        6.' Plants with leaves, these sometimes reduced and small; ovules 2–4:
            7. Stamens twice as many as sepals; flowers usually in threes surrounded by an involucre or in miniature cone-like aggregations ....**GRUBBIACEAE**
            7.' Stamens as many as sepals:
                8. Ovules 2–4, pendulous from a free-basal placenta; style 1; plants often parasitic on roots and with reduced leaves..**SANTALACEAE**
                8.' Ovules 4, pendulous from apex of ovary; styles 4; herbs, plants usually associated with water, not parasitic ....**HALORAGIDACEAE**

KEY 8

1. Ovary 1-locular:
    2. Ovules numerous on a free-basal placenta; sepals 2, often deciduous; fruit a circumscissile capsule...**PORTULACACEAE**
    2.' Ovules pendulous from apex of ovary or on pendulous placentas or erect from base of ovary; sepals 2–5; fruit not circumscissile:
        3. Trees or shrubs; style 1; leaves mostly ericoid:
            4. Petals 5; ovules pendulous from apex of ovary; flowers in spikes, panicles or congested heads..**BRUNIACEAE**
            4.' Petals 4; ovules erect from base of ovary; flowers solitary, axillary ....**CELASTRACEAE**
        3.' Herbs; style or style branches 2–6:
            5. Ovules numerous on pendulous placentas; fruit an apically dehiscent capsule; flowers in axillary pairs; petals and stamens 5 ....**VAHLIACEAE**
            5.' Ovules 1–4, pendulous from apex of ovary; fruit indehiscent:
                6. Leaves small, sessile; flowers in axillary fascicles...**HALORAGIDACEAE**
                6.' Leaves large, on long petioles; flowers in terminal heads or spikes ....**GUNNERACEAE**
1.' Ovary 2- or more locular:
    7. Ovules 2 or more in each locule:
        8. Leaves usually small, ericoid, densely spirally imbricate..**BRUNIACEAE**
        8.' Leaves alternate or opposite:
            9. Stipules present; leaves often oblique; flowers unisexual; stamens numerous; ovules numerous in each locule ....**BEGONIACEAE**
            9.' Stipules absent:
                10. Leaves fleshy and succulent...**AIZOACEAE**
                10.' Leaves not succulent:
                    11. Stamens numerous, more than twice as many as petals; shrubs or small trees with gland-dotted leaves ....**MYRTACEAE**
                    11.' Stamens as many or twice as many as petals:
                        12. Flowers unisexual; male flowers corymbose, female solitary ....**MONTINIACEAE**
                        12.' Flowers bisexual:

```
         13.  Petals valvate, alternating with incurved scales; trees or shrubs ..................... OLINIACEAE
         13.' Petals twisted, not alternating with scales; herbs ................................... ONAGRACEAE
   7.' Ovule 1 in each locule:
      14. Stamens twice as many as petals:
         15. Leaves verticillate, pinnatisect with filiform segments; aquatic herbs ................... HALORAGIDACEAE
         15.' Leaves alternate; plants not aquatic ................................................... ROSACEAE
      14.' Stamens as many as petals:
         16. Stamens opposite petals .............................................................. RHAMNACEAE
         16.' Stamens alternating with petals:
            17. Fruit with oil ducts, separating into 2 cocci; flowers umbellate ..................... APIACEAE
            17.' Fruit without oil ducts, a drupe or capsule or separating into 2 cocci:
               18. Petals and stamens 4; ovary 4-locular ......................................... CORNACEAE
               18.' Petals and stamens 5; ovary 2- or 3-locular:
                  19. Flowers unisexual ............................................................ HAMAMELIDACEAE
                  19.' Flowers bisexual:
                     20. Leaves variously divided or simple, alternate or opposite ........................... ARALIACEAE
                     20.' Leaves simple, ericoid, densely spirally imbricate ................................ BRUNIACEAE
```

KEY 9

1. Parasitic shrubs growing on other shrubs or trees; ovule scarcely distinguishable from ovary tissue; calyx truncate or shortly lobed; stamens epipetalous .... LORANTHACEAE
1.' Free-growing plants; ovules clearly distinguishable:
   2. Ovary 1-locular:
      3. Ovule 1; flowers in involucrate heads; fruit indehiscent, often crowned by persistent calyx forming a pappus of bristles or scales:
         4. Anthers free; ovule pendulous from apex of ovary .... DIPSACACEAE
         4.' Anthers united into a tube surrounding style; ovule erect from base of ovary .... ASTERACEAE
      3.' Ovules 2–many (rarely 1); flowers solitary, in open inflorescences or in heads but these not involucrate:
         5. Calyx of 2 often deciduous sepals; ovules on a free basal placenta; fruit a circumscissile capsule; herbs .... PORTULACACEAE
         5.' Calyx 4–10-lobed:
            6. Stamens opposite petals; capsule opening above by 5 valves; soft or wiry herbs with rosulate leaves .... PRIMULACEAE
            6.' Stamens alternating with petals:
               7. Ovules pendulous, usually 2 in each locule; pollen not shed onto style at anthesis; ericoid shrubs or shrublets with densely spirally imbricate leaves .... BRUNIACEAE
               7.' Ovules axile or basal, rarely pendulous; pollen shed onto style at anthesis; herbs or shrublets with alternate, opposite or whorled leaves .... CAMPANULACEAE
   2.' Ovary 2- or more-locular (sometimes 3-locular with 1 fertile locule and 2 empty locules):
      8. Stamens very numerous, indefinite in number:
         9. Petals fused into a deciduous mass (calyptra); trees or shrubs with gland-dotted leaves .... MYRTACEAE
         9.' Petals more or less fused into a tube; herbs or shrubs with fleshy leaves .... AIZOACEAE
      8.' Stamens never more than 10, definite in number:
         10. Ovary 3-locular with 2 empty locules, fertile locule with 1 apical, pendulous ovule; stamens 3; fruit indehiscent, often crowned by persistent calyx forming a feathery pappus; herbs with opposite leaves .... VALERIANACEAE
         10.' Ovary 2- or more-locular without empty locules, or not as above:
            11. Trailing or climbing plants with tendrils; flowers unisexual; stamens 3–5; leaves often lobed or divided .... CUCURBITACEAE
            11.' Plants without tendrils:
               12. Leaves opposite or whorled, with interpetiolar stipules (sometimes leaf-like) .... RUBIACEAE
               12.' Leaves alternate, opposite or spirally imbricate, without stipules:
                  13. Stamens twice as many as petals; anthers opening by apical pores .... ERICACEAE
                  13.' Stamens as many as petals; anthers opening by longitudinal slits:
                     14. Fruit a drupe; ovule 1 in each locule; erect or ascending; flowers zygomorphic, corolla tube split down the back; maritime shrubs or undershrubs .... GOODENIACEAE
                     14.' Fruit a capsule; ovules 2 or more in each locule (rarely 1 but then pendulous):
                        15. Ovules pendulous, usually 2 in each locule; pollen not shed onto style at anthesis; ericoid shrubs or shrublets with densely spirally imbricate leaves .... BRUNIACEAE
                        15.' Ovules axile or basal, rarely pendulous; pollen shed onto style at anthesis; herbs or shrublets with alternate, opposite or whorled leaves .... CAMPANULACEAE

TAXONOMIC NOTES

Families recognized and generic realignments

This volume incorporates name changes and new species and genera described since the completion of the first edition and has a major addition, species of pteridophyta, not treated in the first edition. Some generic alignments have also been changed. Most notably *Retzia*, only genus of Retziaceae, is now included in **Stilbaceae** where DNA sequence analysis has shown its affinities lie (Bremer *et al.* 1994). The genus *Agapanthus* is assigned to **Agapanthaceae** (not Alliaceae), again following DNA analysis which shows the genus to lie between the two large families Amaryllidaceae and Alliaceae (Chase *et al.* 1995; Fay & Chase 1996). Similarly, DNA analysis has indicated that *Lanaria* is related to the Hypoxidaceae (and not nested in Tecophilaeaceae) and *Caesia* belongs in **Hemerocallidaceae** and not Anthericaceae where it had been placed by Dahlgren *et al.* (1985). We accordingly recognize **Lanariaceae** for *Lanaria* and add Hemerocallidaceae, with one genus, *Caesia*, to the Cape flora. *Dracaena* and *Sanseviera*, assigned to **Dracaenaceae** in Bond & Goldblatt (1984), and *Eriospermum* to the monogeneric **Eriospermaceae**, are now considered to belong in the otherwise largely northern hemisphere family **Convallariaceae** (Chase *et al.* 1995; Angiosperm Phylogeny Group 1998). The taxonomically isolated *Behnia* has no close relatives (e.g. Chase *et al.* 1995) and molecular data suggest that it is best placed in its own family, **Behniaceae** (Conran *et al.* 1997). *Prionium*, traditionally included in Juncaceae, is sister to the clade including Cyperaceae and Juncaceae (Munro & Linder 1998), and is now regarded as comprising the monotypic **Prioniaceae**.

DNA sequence analysis has provided an answer to the old question of the relationships of *Grielum* and its allies, often included in Rosaceae. Alverson *et al.* (1998) have shown that these genera belong in the Malvales and should be treated as a separate family, **Neuradaceae**. Among the dicots, we follow classification of the Angiosperm Phylogeny Group (1998) in recognizing **Buddlejaceae, Celtidaceae** and **Orobanchaceae**. We thus remove *Buddleja* from **Scrophulariaceae**, *Celtis* from **Ulmaceae** and the parasitic genera from **Scrophulariaceae**. **Ulmaceae** are no longer represented in the Cape flora. Family circumscriptions in Malvales are unresolved, and both **Sterculiaceae** and **Tiliaceae** appear to be polyphyletic with **Malvaceae**. Sterculiaceae and Tiliaceae are included in Malvaceae in this account.

Generic changes are legion and cannot be listed here in detail. Major revision in the understanding of generic concepts in **Iridaceae** has left the family in southern Africa with fewer genera. Many of those reduced are now understood to have been recognized on the basis of adaptations for a particular pollination syndrome, often bird pollination, and some were even polyphyletic. Thus *Homoglossum* and *Anomalesia* have been included in *Gladiolus*, *Anapalina* in *Tritoniopsis*, *Antholyza* in *Babiana* (Goldblatt & De Vos 1989; Goldblatt 1990), and *Galaxia, Gynandriris, Hexaglottis* and *Homeria* in *Moraea* (Goldblatt 1998). In Amaryllidaceae we follow Snijman's (1994) generic concepts which are phylogenetically founded, rather than the radical restructuring proposed by D. Müller-Doblies & U. Müller-Doblies (1985, 1996). *Hessea* and *Strumaria* are thus broadly circumscribed. *Crossyne* has been revived to accommodate two misplaced species of *Boophone*.

In **Hyacinthaceae** several new species have been described by Müller-Doblies & Müller-Doblies (1996, 1997) and Müller-Doblies (1994, 1995) some of which we do not recognize and synonymize below. We have also revised the generic circumscription of *Daubenya* to include two monotypic genera that are evidently nested in the genus together with two species of *Massonia*. The subtribes Disinae and Coryciinae of **Orchidaceae** have been extensively revised (Linder 1981) and they now comprise fewer genera, *Corycium* now including *Anochilus*, and *Disa* including *Evota, Herschelia, Monadenia* and *Penthea*.

Among the dicots, generic changes in **Apiaceae** (Burtt 1991) leave that family much changed since the first edition of *Plants of the Cape flora*. Included among these changes is the transfer of *Centella*, previously the largest genus of that family (49 species in the Cape flora). It is now treated as a member of **Araliaceae** following Plunkett *et al.* (1997). **Mesembryanthemaceae**, third-largest family in the flora in the first edition with 660 species, is now widely acknowledged as nested in **Aizoaceae**. With the two families combined, Aizoaceae now includes some **658** species. Critical revisions of some key genera are responsible for the decrease in species numbers. For example, *Cephalophyllum* has been reduced from 33 to 10 species, and *Conophytum* from 32 to 12 species in the Cape Region (and from over 200 species in total to just 84). Such stark adjustments to genera make it seem likely that revisions of *Antimima*

(34 species), *Drosanthemum* (c. 70 species), *Lampranthus* (c. 120 species), and *Ruschia* (c. 90 species) will result in comparable reductions in species numbers. Nevertheless, Aizoaceae remain an important family in the Cape flora. Among the five largest families in terms of species numbers, it is well represented and diverse in all major vegetation types except forest, not only in karroid scrub and semidesert.

The face of **Asteraceae** has been changed radically. We reluctantly accept the many small genera recently described for the family, many of them monotypic and segregated apparently on the most trivial grounds. As a result preparation of a generic key to that family has been extremely difficult. We question the rigid application of the results of a particular cladistic analysis to generic classification unless clades are strongly supported. The treatments of the *Relhania* group of genera by Bremer (1976) and Anderberg & Bremer (1991), for example, differ substantially. Both are based on cladistic analysis and the genera recognized as a result of the later account are for the most part weakly supported. For a more useful and stable classification based on Anderberg & Bremer's results *Comborhiza*, *Leysera*, *Oedera*, *Oreoleysera*, *Relhania* and *Rhynchopsidium* should be united, an action that would yield a single strongly supported genus. Another analysis done with a new perspective on the characters employed, or the addition of new characters or taxa, will yield other trees and another classification. Other generic constellations will likewise see new rounds of restructuring after cladistic re-analysis. This instability has unfortunate repercussions for systematics. In similar vein we question the grounds for separating *Tripteris* from *Osteospermum* and cannot accept the transfer of section *Blaxium* of *Osteospermum* to *Dimorphotheca*. This makes preparation of a key to the genera of Calenduleae impossible without recourse to chemical analysis and microscopic investigation of epidermal papillae.

Scrophulariaceae have been extensively revised over the past 20 years and, in particular, the generic limits of species previously placed in *Selago*, *Sutera* and *Walafrida* have been changed (Hilliard 1994a, 1999). *Sutera* is more narrowly circumscribed, and *Jamesbrittenia* and *Lyperia* are recognized while *Walafrida* is included in *Selago*, a few species of which have been transferred to *Microdon*. Detailed study has also resulted in a substantial increase in the number of species in several genera of the family.

Although the Cape flora is well known in general, new species remain to be discovered, and there are many more known but not yet formally described. Regrettably, there are still genera without revision since their treatment in *Flora capensis* (1866–1932), most of these completed more than a century ago. Even genera revised no more than 40 years ago now require re-evaluation as a result of the spectacular increase in the numbers of specimens available for study and the movement toward more natural (phylogenetic) classification. Most seriously in need of study are large genera with important centres within the Cape Region, including *Drosanthemum*, *Lampranthus* and *Ruschia* (Aizoaceae), *Arctotis*, *Pteronia* and *Senecio* (Asteraceae), *Thesium* (Santalaceae), and *Wahlenbergia* (Campanulaceae), to name some prominent examples. The accounts of several large genera currently under revision, including *Centella* (Araliaceae), *Indigofera* (Fabaceae), *Lobelia* (Campanulaceae) and *Selago* (Scrophulariaceae), were contributed by their authors so that treatments of these genera incorporate unpublished information.

Taxonomic and nomenclatural changes

We continue the practice in the first edition of *Plants of the Cape flora* of including undescribed species, either as 'sp.' or with an unpublished name if one is available. This reflects our aim to most accurately record the flora, both at genus and species levels. Sometimes this results in a disregard for nomenclatural considerations which we hope is justified. Wherever possible, however, we have followed current nomenclatural practice. We have also taken the opportunity to make some taxonomic changes that seemed necessary without additional study. These are listed below.

AIZOACEAE

1. Morphologically the monotpyic genus *Muiria* shares most of the diagnostic features of *Gibbaeum*, including habit, floral characters and some fruit characters. In particular, the epidermis of the leaves of *Muiria* are covered by long velvety hairs, which are also found in some species of *Gibbaeum*, e.g., *G. cryptopodium*. *Muiria* differs mainly by its completely connate leaves, capsules that lack covering membranes and the expanding keels are closely contiguous. These characters may be considered autapomorphies. DNA sequencing analysis shows that *Muiria* and *Gibbaeum* are a clade within Aizoaceae and thus comprise a monophyletic group. We take this opportunity to reduce *Muiria* to synonymy and provide the new combination in *Gibbaeum* below.

Gibbaeum Haw., Revisiones plantarum succulentarum 104 (1821). Type: G. pubescens ((Haw.) N.E. Br.

Muiria N.E.Br., Gardeners Chronicle ser. 3, 81: 116 (1927), syn. nov. Type: M. *hortenseae* N.E.Br.

Gibbaeum hortenseae (N.E.Br.) Thiede & Klak, comb. nov. Basionym: *Muiria hortenseae* N.E.Br., Gardeners Chronicle, ser. 3, 81: 116 (1927).

AMARYLLIDACEAE

2. Phylogenetic analysis of Amaryllidaceae: Strumariinae (Snijman 1994) indicates that the two species of the genus *Dewinterella* are nested in *Hessea*. There seems no merit in the genus and the synonymy of the two species is outlined below.

Hessea *Herb.*, Amaryllidaceae: 289 (1837). Type: *H. stellaris* (Jacq.) Herb.
Dewinterella D. & U.Müll.-Doblies: 341 (1994), syn. nov. Type: *D. pulcherrima* (D. & U.Müll.-Doblies) D. & U.Müll.-Doblies (= *H. pulcherrima* (D. & U.Müll.-Doblies) Snijman).

Hessea mathewsii *W.F.Barker* in The Flowering Plants of South Africa 11: t. 404 (1931).
Dewinterella mathewsii (W.F.Barker) D. & U.Müll.-Doblies: 342 (1994), syn. nov.

Hessea pulcherrima *(D. & U.Müll.-Doblies) Snijman* in Contributions from the Bolus Herbarium 16: 81 (1994).
Gemmaria pulcherrima D. & U.Müll.-Doblies: 35 (1985). *Dewinterella pulcherrima* (D. & U.Müll.-Doblies) D. & U.Müll.-Doblies: 342 (1994), syn. nov.

APIACEAE

3. The genus *Capnophyllum* was considered by Burtt (1991) to be endemic to South Africa and to contain a single species, *C. africanum*. Within the species, var. *leiocarpon* was recognized on the basis of differences in mericarp sculpturing. Examination of several recent collections leaves us in no doubt that the two are specifically distinct. Broadly resembling *C. africanum*, *C. leiocarpon* has fruits that are narrowly elliptic, not warty, with light ridges on the face. This is in marked contrast to the warty, prominently ridged fruits of *C. africanum*. Records of *C. leiocarpon* extend from Port Nolloth southward along the coast in sandveld to Elandsbaai and Rocher Pan. The type collection, said to be from 'near Cape Town,' *Drège 6843*, may be an error as there are no other records of the plant this far south. Records indicate that *C. leiocarpon* is a strandveld species, favouring lime-enriched habitats.

Capnophyllum leiocarpon *(Sond.) J.C.Manning & Goldblatt*, comb. et stat. nov.
Capnophyllum africanum var. *leiocarpon* Sond., Flora capensis 2: 562 (1862); B.L.Burtt: 189 (1991). Type: South Africa, 'near Cape Town', *Drège 6843* (G, S, not seen, SAM!).

ASTERACEAE

4. The genus **Asaemia** has had a chequered history over the past decade and has twice been included in *Athanasia* (Kallersjö 1991). Provisionally, we maintain it separate from the morphologically very different *Athanasia*.

5. Although two species of **Edmondia** were recorded by T.P. Stokoe from the Kamiesberg (Hilliard 1983), neither has been re-collected there despite a fair amount of collecting in the Kamiesberg during the past 20 years. We conclude that Stokoe's specimens are mislabelled. *Edmondia* is almost certainly endemic to the Cape Region where no species have been recorded further north than the Cedarberg Mountains.

6. Examination of plants matching the type collection of *Arctotis oocephala*, and from the same general area, shows the species to have glabrous achenes without cavities, tufted at the base, and a well-developed pappus longer than the achene, thus characteristic of *Haplocarpha* and unlike those of *Arctotis*. These achenes closely match those of *H. parviflora* (Schltr.) P.Beauv. to which *H. oocephala* appears most closely allied. The latter is a creeping perennial, rooting at the nodes, whereas *H. parviflora* is a tufted, acaulescent perennial.

Haplocarpha oocephala *(DC.) Beyers*, comb. nov.
Arctotis oocephala DC., Prodromus systematis naturalis regni vegetabilis 6: 486 (1838a); Harv.: 452 (1865); Lewin: 60 (1922). Type: South Africa, Western Cape, Cedarberg, *Drège s.n.* (G, holotype, not seen; PRE photo!).

7. The genus **Minurothamnus** DC. (1838b: 286) is not currently recognized in herbaria and is known from fragmentary material from the Caledon district. The description matches *Heterolepis* Cass., described in 1820 and *Minurothamnus* is provisionally regarded as a later synonym of that genus.

8. In **Othonna** there has been confusion over the the names and circumscirptions of three tuberous species that have amplexicaul leaves. The earliest name for the only radiate species is *O. perfoliata* Jacq. (dating from 1797), which is *O. amplexifolia* DC., a later synonym for the species used in *Flora capensis*. The morphologically similar species with discoid (or prodominantly discoid florets) is *O.filicaulis,* which is readily distinguished by its very long pappus in the fruiting state. The third species is *O. lingua* (including *O. gymnodiscus* DC.), which has a shorter erect stem and leaves tapering to the base, and it also has a very long pappus in fruit.

BORAGINACEAE

9. The genus **Codon** has traditionally been assigned to the Hydrophyllaceae, in which it has always seemed misplaced geographically, as the family is otherwise largely restricted to the New World. Molecular sequence data show that *Codon* does not fall within the core Hydrophyllaceae, which is itself nested within Boraginaceae (Ferguson, 1999). We follow the implications of this new information and treat *Codon* as a member of the Boraginaceae in this account.

BRUNIACEAE

10. *Brunia nodiflora* was described in *Species plantarum* (1753), where Linnaeus assigned it to *Pentandria monogynia* (plants with five stamens and one carpel). One of the type elements of the protologue, however, was a sterile specimen of the cupressoid gymnosperm *Widdringtonia* in the Clifford Herbarium, with which the descriptive phrase closely accords. This specimen was chosen as the lectotype by Powrie (1972), who also made the combination *Widdringtonia nodiflora* to replace the later name *W. cupressoides*. Although this lectotypification remains highly controversial, and has not been universally recognized (e.g. Bond & Goldblatt 1984), we accept it here in the interest of nomenclatural stability. The species of Bruniaceae therefore currently called *B. nodiflora* is without a name. We describe it here as a new species.

Brunia noduliflora *Goldblatt & J.C.Manning*, sp. nov. Type: South Africa, Western Cape, Dutoitskloof, *Esterhuysen 9690* (BOL, holotype).

Frutex ad 1 m altus, caudice lignoso, foliis adpressis sessilibus 2–3 mm longis infra medium quam supra latioribus, floribus in capitula globosa dispositis c.10 mm diam., petalis albis c. 3 mm longis, staminibus exsertis, filamentis inaequalibus, ovario infero villoso.

11. We have adopted unpublished conclusions of E. Powrie for the account of the **Bruniaceae** in this work. Thus fewer species are recognized than were admitted by Pillans (1947) in the most current monograph of the family.

CAMPANULACEAE

12. In his revision of *Prismatocarpus,* Adamson (1951) treated *P. nitidus* as comprising two varieties, var. *nitidus* restricted to the Cape Peninsula and var. *ovatus* from the mountains to the east, extending as far as George. Var. *ovatus* was distinguished by having broader leaves and flowers on extended axes whereas in var. *nitidus* the flowers are crowded at the stem apices and the leaves are lanceolate. Var. *ovatus*, however, resembles far more closely *P. debilis* in its smaller, ovate leaves of soft texture, slender sprawling stems and elongate floral axes. We cannot distinguish the latter two taxa and after examining the material available to Adamson and additional collections made more recently, we conclude that they represent a single species distinct from *P. nitidus*. *Prismatocarpus nitidus* as now circumscribed is restricted to the Cape Peninsula.

Prismatocarpus debilis *Adamson* in Journal of South African Botany 12: 37 (1946).
Prismatocarpus nitidus var. *ovatus* Adamson: 120 (1951), syn. nov.

13. In *Roella* we also depart from Adamson's (1951) account of the genus. *Roella rhodantha* was based on limited material of pink-flowered plants from the Potberg, west of Bredasdorp. The plants otherwise match *R. incurva* closely, having the distinctive crowded bracts forming a conspicuous bulge below the inflorescence, which distinguishes *R. incurva*. Blue-flowered plants from the Hermanus area, to the west of

Bredasdorp, are virtually identical to specimens of so-called *R. rhodantha* and their flowers dry pink. *Roella rhodantha* should be regarded as no more than a striking colour form of *R. incurva*. Plants that Adamson assigned to *R. incurva* var. *rigida* from the Potberg appear to have confused him, leading to his treatment of the pink-flowered plants as a different species. However, after examining specimens of var. *rigida* we are convinced that it is identical to another species, *R. prostrata*, the latter belonging to a different section. These specimens have the narrow and leaflike bracts that characterize *R. prostrata* and we have no hesitation in uniting *R. incurva* var. *rigida* with *R. prostrata*.

Roella dregeana and *R. psammophila* are unusual in the genus in that the bract margins are fringed with stiff, wiry hairs. The two were distinguished by the relative lengths of the bracts and calyx lobes and by trivial differences in degree of hairiness. The distinction does not appear to hold when more recent collections are examined and the two must be united. Among the species of series Spicatae, *R. lightfootioides* was recognized on the basis of the somewhat leafy outer bracts, densely hairy calyx lobes and style longer than the corolla, but it otherwise closely resembles *R. spicta*. Examination of the two cited specimens of *R. lightfootioides* shows that the style and corolla length differences are misleading. Greater shrinkage of the corolla of some flowers results in the style appearing longer, but in other flowers of the same collection the corolla is less shrunken and is as long as the style. Other differences specified by Adamson fall within the range of variation of the widespread and somewhat variable *R. spicata*, the calyx lobes of which are usually hairy. We therefore include *R. lightfootioides* in *R. spicata*. We also unite *R. compacta* and *R. cuspidata*, species distinguished by the greater or lesser degree of crowding of the leaves, a feature which appears to be associated with habitat.

Roella compacta *Schltr.* in Botanische Jahrbücher 27: 193 (1900).
Roella cuspidata Adamson: 151 (1951), syn. nov.
Roella cuspidata var. *hispida* Adamson: 151 (1951), syn. nov.

Roella dregeana *DC.*, Prodromus systematis naturalis regni vegetabilis 7: 446 (1839).
Roella psammophila Schltr.: 446 (1898), syn. nov.

Roella incurva *A.DC.*, Monographie des Campanulées: 172 (1830).
Roella rhodantha Adamson: 135 (1951), syn. nov.

Roella prostrata *E.Mey. ex DC.*, Prodromus systematis naturalis regni vegetabilis 7: 447 (1839).
Roella incurva var. *rigida* Adamson: 135 (1951), syn. nov.

Roella spicata *L.f.*, Supplementum plantarum systematis vegetabilium: 143 (1782).
Roella lightfootioides Schltr.: 445 (1898), syn. nov.

14. In the *Wahlenbergia* alliance, *Theilera guthriei*, only species of *Theilera*, appears to be nested in section *Laricifolia* (Adamson 1955) of *Lightfootia*, the latter now included in *Wahlenbergia*. Section *Laricifolia* includes species such as *W. polyantha*, which also have leaves in fascicles, the revolute margins and flowers aggregated in small clusters with a 3-locular ovary. The distinguishing feature of *Theilera* is the long, cylindrical floral tube, although the corolla lobes themselves are deeply cut and comparable to other species in the *W. polyantha* alliance. The relatively elongate corolla tube serves to distinguish the species within the section but does not, in our estimation, signal a distinct genus.

Wahlenbergia glandulosa, described by Von Brehmer in 1915, was based on plants otherwise resembling the common *W. androsacea*, but with the glands below the stigma in a continuous ring rather than as three discrete structures. Glands vary considerably in size in *W. androsacea*, even within a population, and are occasionally continuous or discrete within the same collection, e.g. *Burger & Louw 276* (NBG), *Walters 16* (NBG). There seems no ground for maintaining *W. glandulosa* and we reduce the species to synonymy. The type collections of *W. macra*, *Schlechter 2505*, and *W. swellendamensis*, *Ecklon & Zeyher 2383*, correspond closely to the range of material now available for *W. ecklonii*, notably in the elongate calyx lobes, and must be considered conspecific. In the *W. undulata* complex, Von Brehmer recognized *W. polychotoma* on the basis of its broadly cylindric, elongate capsules. Among the syntypes cited we have examined *Ecklon & Zeyher 251* and *2373* and conclude that the capsules fall entirely within the range of variation for *W. undulata*. Specimens reputedly collected by Pappe at Riviersonderend are unlikely to be from this far to the west and the species is otherwise recorded only east of Mossel Bay.

Wahlenbergia Schrad. *ex Roth*, Novae plantarum species: 399 (1821).
Theilera E.Phillips: 369 (1932), syn. nov.

Synonymy or new combinations for species transferred to *Wahlenbergia*

Wahlenbergia androsacea *A.DC.*, Monographie des Campanulées: 150, t. 19, f. 1 (1830).
Wahlenbergia glandulosa Brehmer: 140 (1915), syn. nov.

Wahlenbergia ecklonii *H.Buek*, in Eckl. & Zeyh., Enumeratio plantarum africae australis extratropicae: 380 (1837).
Wahlenbergia swellendamensis H.Buek in Eckl. & Zeyh.: 381 (1837), syn. nov.
Wahlenbergia macra Schltr. & Brehmer: 91 (1915), syn. nov.

Wahlenbergia guthriei *L.Bolus* in Annals of the Bolus Herbarium 1: 193 (1915).
Theilera guthriei (L.Bolus) E.Phillips: 369 (1932).

Wahlenbergia undulata *(L.f.) A.DC.*, Monographie des Campanulées: 148 (1830).
Campanula undulata L.f.: 142 (1782).
Wahlenbergia polychotoma Brehmer: 123 (1915), syn. nov.

CARYOPHYLLACEAE

15. *Silene vlokii* D.Masson was distinguished from *S. primuliflora* by its smaller size and leaves clustered toward the base (Masson 1989). The plants of the type collection barely differ from several collections of the fairly widespread, mostly sand dune species *S. primuliflora*, and we cannot justify recognition of *S. vlokii* which falls into synonymy here.

Silene primuliflora *Eckl. & Zeyh.*, Enumeratio plantarum africae australis extratropicae: 32 (1834).
Silene vlokii D.Masson: 485 (1989). Type: South Africa, Western Cape, cliffs near Herold's Bay, *Masson 1225* (G, holotype), syn. nov.

FABACEAE

16. As with *Astroloba* (Asphodelaceae), we strongly recommend withholding recognition of a genus based solely on adaptations for a particular pollination syndrome when other morphological characters indicate that it is nested in another genus. We therefore reduce *Sutherlandia* to synonymy in *Lessertia*. The two share derived, large, papery (sometimes inflated) capsules, similar compound leaves with small oblong, sessile leaflets, and an unusual foveolate-papillate seed testa. There seems no sound reason, except for tradition, to continue to recognize the genus *Sutherlandia*, which is based on large, red flowers with prominent keels and inflated, bladder-like pods. An analogous situation in Fabaceae is the Australian bird-pollinated *Clianthus formosus*, which is now included in the larger genus *Swainsona* (Thompson 1990). Both *S. formosus* and *C. puniceus* exhibit specializations for bird-pollination that are strikingly similar to those in *Sutherlandia*–large red flowers with an elongated, strongly beaked keel and reduced wing petals.

Lessertia canescens *Goldblatt & J.C.Manning*, nom. nov., non *Lessertia tomentosa* DC.
Sutherlandia tomentosa Eckl. & Zeyh., Enumeratio plantarum africae australis extratropicae: 251 (1834).

Lessertia frutescens *(L.) Goldblatt & J.C.Manning*, comb. nov.
Colutea frutescens L., Species plantarum: 723 (1753). *Sutherlandia frutescens* (L.) R.Br.: 327 (1812).

Lessertia humilis *(E.Phillips & R.A.Dyer) Goldblatt & J.C.Manning*, comb. nov.
Sutherlandia humilis E.Phillips & R.A.Dyer in Revista Sudamericana de Botánica 1: 78 (1934).

Lessertia microphylla *(Burch. ex DC.) Goldblatt & J.C.Manning*, comb. nov.
Sutherlandia microphylla Burch. ex DC., Prodromus systematis naturalis regni vegetabilis: 273 (1825).

Lessertia montana *(E.Phillips & R.A.Dyer) Goldblatt & J.C.Manning*, comb. nov.
Sutherlandia montana E.Phillips & R.A.Dyer in Revista Sudamericana de Botánica 1: 78 (1934).

Lessertia speciosa *(E.Phillips & R.A.Dyer) Goldblatt & J.C.Manning*, comb. nov.
Sutherlandia speciosa E.Phillips & R.A.Dyer in Revista Sudamericana de Botánica 1: 75 (1934).

FUMARIACEAE

17. '*Cysticapnos grandiflora* Bernh.' was recognized by Lidén (1986) in his monograph of Fumarieae with the homotypic synonym, *Phacocapnos burmannii* (Eckl. & Zeyh. ex Harv.) Hutch. (basionym: *Corydalis burmannii* Eckl. & Zeyh. ex Harv.). In fact, *Cysticapnos grandiflora* was not described by Bernhardi and is merely a nomen nudum listed by Meyer (1843). *Corydalis burmanni* Eckl. & Zeyh. ex Harv. is based on a mixed gathering (*Ecklon & Zeyher 23*) consisting of flowering material of *Cysticapnos vesicaria* and fruits of *C. cracca*. Duplicates of the type collection at MO and SAM also consist of the same mixture. Although Harvey (1859), describing *Corydalis burmannii* for the first time, noted that the specimen he examined 'has the foliage and petals of *Cysticapnos africana* (a later synonym of *C. vesicaria*) but the fruit is that of a *Corydalis* (referring to *Cysticapnos cracca*)', he did not actually realize that the type collection was a mixture of two species. The species does not exist except in name. In order to dispose of it effectively, we designate the flowering material as the lectotype, and thus *Corydalis burmannii* falls into the synonymy of *C. vesicaria*.

Cysticapnos vesicaria *(L.) Fedde*, Repertorium specierum novarum regni vegetabilis 19: 287 (1924).
Cysticapnos grandiflora E. Mey.: 95 (1843), nom. nud. *Cysticapnos grandiflora* [Bernh.] sensu Lidén: 106 (1986), nom. nud. (as *C. grandiflora* Bernh. in Linnaea 12 (as '13' by Lidén): 664 (1838), but not mentioned therein).
Corydalis burmannii Eckl. & Zeyh. ex Harv.: 17 (1859). *Phacocapnos burmannii* (Eckl. & Zeyh. ex Harv.) Hutch.: 110 (1921). Type: South Africa, *Ecklon & Zeyher 23* (SAM, lectotype here designated, MO, isolectotype).
Cysticapnos africana Gaertner: 161 (1791).

HAEMODORACEAE

18. Barker (1940) recognized five species of *Dilatris,* all endemic to the Cape Region, one of them, *D. paniculata,* known only from the protologue. Later, when she was able to examine the type of the species in the Thunberg Herbarium, she noted that it represented the northern, paniculate form of *D. viscosa*. We have examined the type and concur with her conclusion.

Dilatris viscosa *L.f.*, Supplementum plantarum systematis vegetabilium: 101 (1782).
D. paniculata L.f.: 101 (1782), syn. nov.

19. Nordenstam (1970b) lectotypified *Babiana multiflora* Klatt from an isotype in Klatt's herbarium at Stockholm, correctly pointing out that it actually represented a species of *Wachendorfia* (Haemodoraceae). Unfortunately he was mistaken in referring it to *W. paniculata* L. when it clearly falls within the circumscription of *W. parviflora* W.F.Barker and Klatt's name for the species takes priority.

Wachendorfia multiflora *(Klatt) J.C.Manning & Goldblatt*, comb. nov.
Babiana multiflora Klatt in Abhandlungen der Naturforschenden Gesellschaft zu Halle 15: 351 (1882).
Wachendorfia parviflora W.F.Barker: 29 (1949). Type: South Africa, Cape Peninsula, Camps Bay, *Salter 7457* (NBG, holotype), syn. nov.

HYACINTHACEAE

20. In *Albuca* we describe one new species and propose changes to the nomenclature of several others.

Albuca papyracea *J.C.Manning & Goldblatt*, sp. nov. Type: South Africa, Western Cape, Touwsberg, 7 Oct. 1993, *Snijman 1397* (NBG, holotype).

Haec species ab aliis speciebus sectionis *Falconerae* foliis duobus linearibus, cataphyllis basalibus longevaginantibus membranaceis, aetate leviter fibrosis papyraceisque, saepe verrucosis, longe supra terram extensis, inflorescentis juvenibus nutantibus, floribus staminibus omnibus fertilibus, antheris curvatis, stylo omnino gracili distinguitur.

Other material seen: Western Cape: Oudtshoorn District, Farm Doornkraal, 28 Sept. 1971, *Dåhlstrand 2077* (NBG); Boomplaas, Cango Valley, 5 Nov. 1974, *Moffett 452* (NBG); Ouberg Pass, Montagu, Sept. 1997, *Goldblatt & Manning 10764* (NBG).

Typical of section *Falconera* in its unspecialized inner tepals, six fertile stamens and filiform style, *Albuca papyracea* is largely distinguished by its two long leaves enclosed in firmly papery to more or less fibrous sheaths and the inflorescence drooping in bud. The sheaths are usually lightly warty and extend up to 15 cm above the ground, giving the species a distinctive appearance. *A. papyracea* appears to most closely resemble *A. fragrans* Jacq. which also has drooping racemes, but short sheaths and several leaves, usually withered at flowering. *A. robertsoniae* U. & D.Müll.-Doblies, known to us only from the protologue (the type is not available at the herbaria cited), appears similar except that it is described as having a single leaf.

Some new species of *Albuca* described by U. Müller-Doblies (1994, 1995) appear to represent minor variants of well-known species, possibly atypical in cultivation. The following species occurring in our area are formally synonymized:

Albuca cooperi *Baker* in Journal of Botany, British and Foreign 12: 366 (1874).
Albuca karooica U.Müll.-Doblies: 367 (1994). Type: South Africa, Northern Cape, *Müller-Doblies & Müller-Doblies 86045e* (PRE, holotype), syn. nov.

Albuca flaccida *Jacq.*, Collectanea 4: 201 (1791).
Albuca materfamilias U.Müll.-Doblies: 367 (1994). Type: South Africa, Western Cape, *Müller-Doblies & Müller-Doblies 78052a* (PRE, holotype), syn. nov.

Albuca hallii *U.Müll.-Doblies* in Feddes Repertorium 105: 366 (1994).
Albuca brucebayeri U.Müll.-Doblies: 366 (1994). Type: South Africa, Western Cape, *Bayer 1756* (PRE, holotype), syn. nov.

Albuca viscosa *L.f.*, Supplementum plantarum systematis vegetabilium: 196 (1782).
Albuca aspera U.Müll.-Doblies: 355 (1995). Type: South Africa, Western Cape, *Müller-Doblies 82161f* (PRE, holotype), syn. nov.
Albuca bontebokensis U.Müll.-Doblies: 358 (1995). Type: South Africa, Western Cape, *Müller-Doblies 78102b* (PRE, holotype), syn. nov.
Albuca jacquinii U.Müll.-Doblies: 366 (1995). Type: South Africa, Western Cape, *Müller-Doblies 80109ab* (PRE, holotype), syn. nov.
Albuca viscosella U.Müll.-Doblies: 369 (1995). Type: South Africa, Western Cape, *Müller-Doblies 82184e* (PRE, holotype), syn. nov.

21. The genera of Hyacinthaceae with spurred bracts form a natural, monophyletic group comprising the subfamily Urgineoideae. Within the group, however, and particularly within the *Drimia* alliance, the genera are poorly defined and several species blur the boundaries traditionally used to distinguish them. The two core genera, *Drimia* and *Urginea*, are traditionally separated by the degree of the fusion of the perianth and the orientation of the tepals and stamens. The distinction is not absolute and both Jessop (1977), working with southern African material, and Stedje (1997), working with tropical African material, concurred that the two genera could not be maintained as distinct genera. Jessop (1977) also included the two small genera *Tenicroa* and *Thuranthos* within *Drimia*, but retained *Bowiea*, *Litanthus*, *Rhadamanthus* and *Schizobasis*. Although Obermeyer (1980) believed that the bulb of *Tenicroa* differed from that in related genera in being derived solely from the swollen bases of the cataphylls, she was mistaken. Dissected bulbs of *T. filifolia* consist of the swollen bases of both cataphylls and leaves, although the latter are smaller in size. Bulbs of *Tenicroa* thus differ from other species groups in the *Drimia* alliance only in having those bulb scales derived from leaf bases smaller than those derived from the cataphyll bases.

Rhadamanthus was established for species with porose or at least tardily longitudinally dehiscent anthers and often a nodding, urceolate perianth, but even Nordenstam (1970a) in his revision of the genus, was not convinced that it was entirely monophyletic. Since the description of two further species with longitudinally dehiscent anthers (Obermeyer 1980) that are clearly allied to more typical members of the group, the genus is now very weakly circumscribed by short filaments. In addition, an odd character thought to be restricted to some species of *Rhadamanthus*, the scapes longitudinally scabrid below, is actually well represented in the smaller species of *Drimia-Urginea* and it is evident that the genus is nested within the group. The small genera *Litanthus* (1 species) and *Schizobasis* (c. 5 species) share the character of scabrid scapes and thus also appear to be nested in this alliance. Recognizing any of these segregates deprives the remaining species of any synapomorphy and it is evident that they represent clusters of derived species defined by ecological strategies, mainly of pollination. Regarded as a single genus, the *Drimia* alliance is well defined by the apomorphic short-lived flowers with the tepals mostly more or less fused below and the perianth caducous, abscising at the base and withering as a cap on the developing fruit. It constitutes a sister clade (M. Fay *et al.*, pers. comm.) to *Bowiea*, the remaining genus in the subfamily, which has plesiomorphic long-lived flowers with a persistent perianth, the tepals entirely free and remaining attached to the base of the developing fruit. *Bowiea* is defined by its twining, somewhat fleshy inflorescence, quite unlike the wiry inflorescence of *Schizobasis* with which it has been allied. The Madagascan *Rhodocodon* has the spurred bracts of the subtribe and is clearly related to *Drimia*, and has been reduced to synonymy in *Rhadamanthus* by Speta (1998). The nodding, urceolate flowers recall those of *Rhadamanthus* but the tepals are fused for most of their length and although abscising below, are not as deciduous as those of *Drimia*, instead persisting around the developing fruit until the capsule is nearly mature. Study of living plants will be necessary to determine the floral longevity. The pedicels of several species of *Rhodocodon* have a well-defined abscission layer below the flower not encountered in the mainland species of the *Drimia* alliance. We provide new combinations below for the species of the *Drimia* complex occurring in the Cape flora.

The additional generic segregates of *Drimia* (or *Urginea*) described by Speta (1998), obviously fall within our broader concept of the genus. Consequently, *Ebertia*, *Rhadamanthopsis*, *Urginavia* as well as *Fusifilum*, *Thuranthos* and *Urgineopsis*, all of which Speta recognized, are included in the synonymy below.

Drimia Jacq. ex Willd., Species plantarum 2: 162 (1799).
Urginea Steinh.: 321 (1834). Type: *U. fugax* Steinh.
Fusifilum Raf.: 27 (1836). Type: *F. physodes* (Jacq.) Speta, syn. nov.
Tenicroa Raf.: 52 (1836). Type: *T. fragrans* (Jacq.) Raf., syn. nov.
Litanthus Harv.: 315, t. 9 (1844). Type: *L. pusillus* Harv., syn. nov.
Rhadamanthus Salisb.: 37 (1866). Type: *R. convallarioides* (L.f.) Baker, syn. nov.
Schizobasis Baker: 105 (1873). Type: *S. macowanii* Baker (= *S. intricata* (Baker) Baker), syn. nov.
Thuranthos C.H.Wright: 233 (1916). Type: *T. macranthum* (Baker) C.H.Wright, syn. nov.
Urgineopsis Compton: 107 (1930). Type: *U. salteri* Compton, syn. nov.
Ebertia Speta: 65 (1998). Type: *E. nana* (Oyewole) Speta, syn. nov.
Rhadamanthopsis (Oberm.) Speta: 74 (1998). Type: *R. namibensis* (Oberm.) Speta, syn. nov.
Urginavia Speta: 86 (1998). Type: *U. micrantha* (A.Rich.) Speta, syn. nov.

Drimia albiflora *(B.Nord.) J.C.Manning & Goldblatt*, comb. nov.
Rhadmanthus albiflora B.Nord. in Botaniska Notiser 123: 177 (1970a).

Drimia arenicola *(B.Nord.) J.C.Manning & Goldblatt*, comb. nov.
Rhadamanthus arenicola B.Nord. in Botaniska Notiser 123: 166 (1970a).

Drimia convallarioides *(L.f.) J.C.Manning & Goldblatt*, comb. nov.
Hyacinthus convallarioides L.f., Supplementum plantarum systematis vegetabilium: 204 (1782). *Rhadamanthus convallarioides* (L.f.) Baker: 434 (1871).
R. montanus B.Nord.: 162 (1970a), syn. nov.

Drimia cuscutoides *(Burch. ex Baker) J.C.Manning & Goldblatt*, comb. nov.
Asparagus cuscutoides Burch. ex Baker in Journal of the Linnean Society. Botany 14 : 606 (1875). *Schizobasis cuscutoides* (Burch. ex Baker) Benth.: 786 (1883).

Drimia cyanelloides *(Baker) J.C.Manning & Goldblatt*, comb. nov.
Rhadamanthus cyanelloides Baker in Thiselton-Dyer, Flora capensis 6: 444 (1897).

Drimia dregei *(Baker) J.C.Manning & Goldblatt*, comb. nov.
Urginea dregei Baker in Thiselton-Dyer, Flora capensis 6: 467 (1897).

Drimia fasciata *(B.Nord.) J.C.Manning & Goldblatt*, comb. nov.
Rhadamanthus fasciata B.Nord. in Botaniska Notiser 123: 174 (1970a).

Drimia filifolia (Jacq.) J.C.Manning & Goldblatt, comb. nov.
Anthericum filifolium Jacq., Icones plantarum rariorum 2, 15: 18, t. 414 (1794). *Tenicroa filifolia* (Jacq.) Oberm.: 577 (1981).

Drimia fragrans *(Jacq.) J.C.Manning & Goldblatt*, comb. nov.
Anthericum fragrans Jacq., Plantarum rariorum horti caesarei schoenbrunnensis 1: 45, t. 86 (1797). *Tenicroa fragrans* (Jacq.) Raf.: 52 (1836).

Drimia intricata *(Baker) J.C.Manning & Goldblatt*, comb. nov.
Anthericum intricatum Baker in Journal of Botany, British and Foreign 10: 140 (1872).

Drimia involuta *(J.C.Manning & Snijman) J.C.Manning & Goldblatt*, comb. nov.
Rhadamanthus involutus J.C.Manning & Snijman in Snjiman et al., Novon 8: 112 (1999).

Drimia karooica *(Oberm.) J.C.Manning & Goldblatt*, comb. nov.
Rhadamanthus karooica Oberm. in Bothalia 13: 138 (1980).

Drimia namibensis *(Oberm.) J.C.Manning & Goldblatt*, comb. nov.
Rhadamanthus namibensis Oberm. in Bothalia 13: 137 (1980).

Drimia platyphylla *(B.Nord.) J.C.Manning & Goldblatt*, comb. nov.
Rhadamanthus platyphyllus B.Nord. in Botaniska Notiser 123: 172 (1970a).

Drimia revoluta *(A.V.Duthie) J.C.Manning & Goldblatt*, comb. nov.
Urginea revoluta A.V.Duthie in Annale van die Universiteit van Stellenbosch 6A, 2: 9 (1928).

Drimia salteri *(Compton) J.C.Manning & Goldblatt*, comb. nov.
Urgineopsis salteri Compton in Journal of Botany, British and Foreign 68: 107 (1930).

Drimia sclerophylla *J.C.Manning & Goldblatt*, nom. nov. pro *Urginea rigidifolia* Baker: 323 (1878), non *D. rigidifolia* Baker: 420 (1871).

Drimia secunda *(B.Nord.) J.C.Manning & Goldblatt*, comb. nov.
Rhadamanthus secundus B.Nord. in Botaniska Notiser 123: 168 (1970a).

Drimia uniflora *J.C.Manning & Goldblatt*, nom. nov. pro *Litanthus pusillus* Harv.: 315 (1844), non *D. pusilla* Jacq.

Drimia uranthera *(R.A.Dyer) J.C.Manning & Goldblatt*, comb. nov.
Rhadamanthus urantherus R.A.Dyer in Hooker's icones plantarum 1934: t. 3247 (1934).

Drimia virens *(Schltr.) J.C.Manning & Goldblatt*, comb. nov.
Urginea virens Schltr. in Journal of Botany, British and Foreign 35: 433 (1897).

22. *Amphisiphon, Androsiphon, Daubenya* and *Massonia* are morphologically similar genera of subfamily Hyacinthoideae distinguished by their highly contracted peduncles, prostrate leaves, flowers with a narrow perianth tube and subglobose, glossy seeds. A particularly close relationship between *Amphisiphon, Androsiphon, Daubenya* and *M. angustifolia* is supported by DNA sequence data, but other *Massonia* species are not immediately related to them. The monotypic *Androsiphon, Amphisiphon* and *Daubenya* appear to have been recognised on the basis of substantial floral specialisations, which at least in some of them are clearly related to their pollination strategies. These highly adaptive specializations alone are not sufficient basis for generic segregation despite the exaggerated effect they have on the appearance of the flowers. *Amphisiphon* and *Androsiphon* are distinguished by the fusion of the filaments into a narrow tube, accompanied in *Amphisiphon* by the almost complete fusion of the perianth. Although occuring outside our area on the Roggeveld Escarpment, *Daubenya* is part of this complex and must be dealt with here as well. *Daubenya* is distinguished by having the outer (lower) flowers of the inflorescence zygomorphic, the three outer tepals being grossly enlarged and conspicuous. The central flowers, however, have short tepals and are quite unexceptional among species in the group.

We suggest uniting *Amphisiphon, Androsiphon, Daubenya* and *Massonia angustifolia* into a single genus on the basis of their close molecular relationship. The alternative of several monotypic genera is less informative. Morphologically these species are much more heterogeneous than the remaining species of *Massonia*, and differ from one another in several floral, bract and fruit characters. They do, however, differ from all other species of *Massonia* in their yellow (or red) flowers. True *Massonia* species have white to pink flowers subtended by large floral bracts and prominently winged capsules. Evidently the prostrate leaves and acaulescent habit evolved independently in these two lineages. The only other member of the *Massonia* group that appears to belong here is *Neobakeria namaquensis*. Included by Jessop (1976) in *M. angustifolia*, it has small bracts and yellow flowers in an elongate inflorescence like that of *M. angustifolia*. Although it occurs outside the Cape region, we take this opportunity to transfer it to *Daubenya* as well.

Daubenya Lindl.: 21, pl. 1813 (1835). Type: *D. aurea* Lindl.
Neobakeria Schltr.: 150 (1924), syn. nov. Type: *N. namaquensis* Schltr.
Amphisiphon W.F.Barker: 19 (1936), syn. nov. Type: *A. stylosus* W.F.Barker.
Androsiphon Schltr.: 148 (1924), syn. nov. Type: *A. capense* Schltr.

Daubenya angustifolia *(L.f.) A.M. van der Merwe & J.C.Manning,* comb. nov.
Massonia angustifolia L.f.: 193 (1782), syn. nov.

Daubenya capensis *(Schltr.) A.M. van der Merwe & J.C.Manning,* comb. nov.
Androsiphon capense Schltr. in Notizblatt des Botanischen Gartens und Museums zu Berlin-Dahlem 9: 148 (1924), syn. nov.

Daubenya namaquana *(Schltr.) J.C.Manning & Goldblatt,* comb. nov.
Neobakeria namaquensis Schltr.: 150 (1924), syn. nov.

Daubenya stylosa *(W.F.Barker) A.M. van der Merwe & J.C.Manning,* comb. nov.
Amphisiphon stylosus W.F.Barker in Journal of South African Botany 2: 19 (1936), syn. nov.

In addition to the above, *Massonia tenella* was omitted by Jessop (1977) from his account of *Massonia*, although it is clear from his discussion of *M. echinata* that he regarded the two as conspecific. *Massonia tenella* is here reduced to synonymy.

Massonia echinata *L.f.*, Supplementum plantarum systematis vegetabilium: 193 (1782).
Massonia tenella Sol. ex Baker: 389 (1871), syn. nov.

23. For *Ornithogalum*, revised by U. & D. Müller-Doblies (1996), we have examined available material and prefer to follow Obermeyer (1978) in including *O. ovatum* Thunb. in *O. unifolium* Retz. The arguments advanced by U. & D. Müller-Doblies about differences in flower size, leaf number, vestiture and marginal structure are unconvincing. Likewise, we are not convinced that *O. semipedale* (Baker) U. & D.Müll.-Doblies, included by Obermeyer in *O. polyphyllum,* is specifically distinct. Some new species recognized by U. & D. Müller-Doblies appear to be no more than minor variants of other species and those that occur in our area are synonymized below.

Ornithogalum ciliiferum *U. & D.Müll.-Doblies* in Feddes Repertorium 107: 414 (1996). Type: South Africa, Northern Cape, Calvinia, *Müller-Doblies 82166c* (PRE, holotype).
Ornithogalum gifbergense U. & D.Müll.-Doblies: 418 (1996). Type: South Africa, Western Cape, *Müller-Doblies & Müller-Doblies 80105k* (PRE, holotype), syn. nov.

Ornithogalum nannodes *F.M.Leight.* in Journal of South African Botany 9: 113 (1943).
Ornithogalum hesperanthum U. & D.Müll.-Doblies: 443 (1996). Type: South Africa, Western Cape, *Müller-Doblies & Müller-Doblies 80113a* (PRE, holotype), syn. nov.

Ornithogalum suaveolens *Jacq.*, Icones plantarum rariorum 2: 19, t. 431 (1789).
Ornithogalum namaquanum U. & D.Müll.-Doblies: 500 (1996). Type: South Africa, Northern Cape, *Müller-Doblies & Müller-Doblies 79192c* (PRE, holotype), syn. nov.

Ornithogalum tortuosum *Baker* in Thiselton-Dyer, Flora capensis 6: 510 (1897).
Ornithogalum thunbergianulum U. & D.Müll.-Doblies: 440 (1996), as nom. nov. pro *Anthericum filifolium* Thunb. Type: South Africa, Western Cape, *Thunberg s.n.* (UPS-Thunb. 8372, holotype), syn. nov.

In the past, *Ornithogalum thyrsoides* has been much confused with *O. conicum*, for example by Obermeyer (1978). The difference, inner filaments filiform or at most with an ovate expansion below in the latter versus filaments broad below and auriculate, clasping the ovary and abruptly narrowed above, separates the two. *O. conicum* is largely a Karoo species, but also occurs along the Bokkeveld Escarpment and in Botterkloof and the lower Olifants River valley, whereas *O. thyrsoides* occurs from Namaqualand south to the Agulhas coast, mostly to the west of the main Cape mountain ranges.

24. Traditionally included in *Polyxena* Kunth (e.g. by Baker 1896, in *Flora capensis*), *Periboeia* Kunth has been revived and expanded by Müller-Doblies & Müller-Doblies (1997). We see no merit in the genus, distinguished largely by leaf shape (linear to narrowly lanceolate versus lanceolate to suborbicular), plus some overlapping qualitative features (perianth tube shorter than the tepals versus as long or longer, scape usually aerial at flowering versus subterranean but later aerial). These authors also recognized one new

species of *Periboeia*, *P. oliveri*, distinguished from *P. paucifolia* by wider outer tepals, shorter inner whorl of stamens, and a shorter style. Populations at the type localities of *P. paucifolia* and *P. oliveri*, both near Saldanha Bay, and an additional population some distance to the south examined in May 1998, showed that the distinguishing characters of *P. oliveri* overlap completely with the corresponding features of *P. paucifolia*. There are therefore essentially no differences of any taxonomic significance between the two species and *P. oliveri* cannot be upheld. We also provide the combination *Polyxena paucifolia*, the species having been originally assigned to *Hyacinthus*.

Polyxena paucifolia *(W.F.Barker) A.M.van der Merwe & J.C.Manning*, comb. nov.

Hyacinthus paucifolius W.F.Barker in Journal of South African Botany 7: 198 (1941).

Periboeia oliveri U. & D.Müll.-Doblies: 84 (1997), syn. nov.

HYPOXIDACEAE

25. The genus *Forbesia* is a synonym of *Empodium* but some taxa referred to it have not yet been transferred to that genus, among them *F. flexilis* Nel. Field studies indicate that this taxon is quite distinct and the necessary combination in *Empodium* is made below.

Empodium flexile (Nel) *M.F.Thomps.ex Snijman,* comb. nov. Type: South Africa, Western Cape, Cango, Mund & Maire s.n. (B). Basionym: *Forbesia flexilis* Nel, Botanische Jahrbucher 51: 288 (1914).

IRIDACEAE

26. Described by L. Bolus in 1933, *Freesia elimensis* was recognized by Goldblatt (1982) in a revision of the genus as a local endemic of limestone outcrops in the southern Cape. We find that it cannot be distinguished from the fairly widespread *F. caryophyllacea* which typically grows on clay soils. Both flowering early in the season, mainly in May and June, they have been separated by minor differences in the length of the perianth tube (20–25 mm long with the lower part 6–8 mm long and widening gradually in *F. caryophyllacea* versus 25–35 mm long and 8–10 mm long and widening abruptly in the lower part in *F. elimensis*) and the markings on the lower tepals. These characters vary across the range of *F. caryophyllacea* and there seems no merit in maintaining the limestone populations as taxonomically separate. The possibility that *F. elimensis* is a hybrid between *F. caryophyllacea* and *F. alba*, as suggested by Goldblatt (1982), still exists but this is no argument for maintaining the species.

Freesia caryophyllacea *(Burm.f) N.E.Br.* in Kew Bulletin 1929: 134 (1929). Goldblatt: 64 (1982).

Ixia caryophyllacea Burm.f.: 1 (1768).

Freesia elimensis L.Bolus: 167 (1933); Goldblatt: 68 (1982), syn. nov.

IRIDACEAE

27. With the genus *Galaxia* now included in *Moraea* (Goldblatt, 1998), the new name *M. ovalifolia* was made for *G. ovata,* that epithet being preoccupied in *Moraea*. An earlier, valid synonym *Ixia galaxia* was unfortunately overlooked and the name M. ovalifolia is a later synonym. We provide the necessary combination in *Moraea* here.

Moraea galaxia (L.f.) *Goldblatt & J.C.Manning,* comb. nov.

Ixia galaxia L.f., Supplementum plantarum 93 (1782).

Galaxia ovata Thunb., Nova Genera Plantarum 50 (1782), non M. ovata Thunb. 1800.

Moraea ovalifolia Goldblatt, Novon 8: 374 (1998).

28. As a member of the genus *Anapalina*, *Tritoniopsis longituba* was distinguished by Lewis (1960) from *T. antholyza* (= *A. nervosa*) by the upper cauline leaves bract-like rather than vestigial, the lower part of the perianth tube slightly longer and the anther apiculi more well developed. In practice we find the distinction has no merit. The ranges and flowering times of specimens assigned by Lewis to each species also overlap and they appear to us to be identical.

Tritoniopsis antholyza *(Poir.) Goldblatt* in South African Journal of Botany 57: 226 (1991).

Anapalina longituba Fourc.: 76 (1932); G.J.Lewis: 61 (1960). *Tritoniopsis longituba* (Fourc.) Goldblatt: 580 (1990), syn. nov. Type: South Africa, Western Cape, *Fourcade 2003* (BOL, holotype).

MOLLUGINACEAE

29. *Pharnaceum microphyllum* var. *albens* was described by Adamson (1958) as an unusual plant from the Cold Bokkeveld which differed from *P. microphyllum* in being a closely leafy, densely white-woolly shrublet with stipules exceeding the leaves and elongate peduncles. The variety seems to us so different from *P. microphyllum* which has short peduncles and longer leaves that it must be regarded as a separate species.

Pharnaceum lanuginosum J.C.Manning & Goldblatt, nom. et stat. nov. pro *P. microphyllum* var. *albens* Adamson: 23 (1958), non *P. albens* L.f. (1782). Type: South Africa, Western Cape, Cold Bokkeveld, *Schlechter 10114* (SAM, holotype).

OXALIDACEAE

30. Dreyer & Van Wyk (1998) provide convincing evidence that *Oxalis henrici* is a synonym of *O. engleriana*. We concur with their decision and formally reduce *O. henrici* to synonymy.

Oxalis engleriana Schltr. in Botanische Jahrbücher 24: 437 (1898).
Oxalis henrici F.Bolus: 24 (1916), syn. nov.

POLYGALACEAE

31. Although Johnson & Weitz (1991) admit only one species in *Nylandtia*, versus four recognized by Ecklon & Zeyher (1834), we are convinced that there are at least two species in the genus. One, a low, twiggy, spinescent, early-flowering shrub, is *N. spinosa*. A taller, weakly spinescent, several-stemmed shrub currently recognized as *N. spinosa* var. *scoparia* (Eckl. & Zeyh.) C.T.Johnson & F.M.Weitz, is more than an ecological race and represents a second species. The last, recognized as *Mundtia scoparia* by Ecklon & Zeyher (1834), has almost the same distribution as *N. spinosa* in the western Cape, and the two sometimes grow side by side. In such situations *N. spinosa* is mostly in late flower and already in leaf when *N. scoparia* is in full bloom, and without leaves.

Populations of *Nylandtia* with grey leaves and spine-tipped branches from the Cape west coast on limestone and calciferous sands are unusual in having fairly dry, yellow fruits with leathery exocarps (unlike the fleshy, reddish berries of *N. spinosa* and *N. scoparia*) and may represent a third species.

Nylandtia scoparia *(Eckl. & Zeyh.) Goldblatt & J.C. Manning*, comb. nov.
Mundtia scoparia Eckl. & Zeyh., Enumeratio plantarum africae australis extratropicae: 30 (1834–35). Type: South Africa, Western Cape, Langevallei, *Ecklon & Zeyher 235* (S, lectotype designated by Johnson & Weitz 1991).

Nylandtia scoparia poses a second problem. Its basionym, *Mundtia scoparia*, is also the basionym of *Muraltia scoparia* (Eckl. & Zeyh.) Levyns, as neotypified by Levyns (1954: 41–42) (*Zeyher 67* at K) which Levyns maintained had toothed capsules. The protologue of *Mundtia scoparia* states that the fruits are red drupes and the epithet must remain in *Nylandtia* which is defined by having fleshy fruits and has been lectotypified by Johnson & Weitz. The species that Levyns regarded as belonging in *Muraltia* has a synonym, **Muraltia parvifolia** N.E.Br., and that name must be used for *Muraltia scoparia* sensu Levyns.

RUBIACEAE

32. Specimens of *Psydrax* from Swellendam do not accord with the current circumscription of *P. obovatum* (Bridson 1985) and appear to represent an undescribed species. *P. obovatum* and *P. pyrifolium*, both based on types from the 'Uitenhage District' in the southern Cape, were treated as conspecific by Bridson but the specimens included under the name appear to represent two separate species. One has flowers with a short tube and short peduncles (*P. obovatum*) whereas specimens matching isotypes of *Canthium pyrifolium*, the basionym of *P. pyrifolium*, at SAM differ in their broader leaves, larger inflorescences with longer peduncles and larger, longer-tubed flowers. These may be the Swellendam plant. Provisionally we include the latter as *Psydrax* sp. in our account of the flora, although it seems to resemble *P. pyrifolium* closely.

SANTALACEAE

33. The genus *Colpoon* was merged with *Osyris* by Hilliard (1994b) but the transfer of *C. speciosum* (Bean 1990) has not been made. We provide the new combination for this narrow endemic of the Western Cape Province.

Osyris speciosa *(A.W.Hill) J.C.Manning & Goldblatt*, comb. nov.
Osyris abyssinica var. *speciosa* A.W.Hill in Thiselton-Dyer, Flora capensis 5,2: 209 (1925). *Colpoon speciosum* (A.W.Hill) P.A.Bean: 667 (1990).

SCROPHULARIACEAE

34. Additional collections of *Freylinia densiflora* show that *F. decurrens* cannot be maintained as a separate species and we formally unite the two.

Freylinia densiflora *Benth.* in Hook., Companion to the Botanical Magazine 2: 55 (1836a).
Freylinia decurrens Levyns ex Van Jaarsv.: 61 (1983), syn. nov.

35. Bergius, in his description of *Eranthemum parviflorum,* the basionym of *Agathelpis parviflorum,* overlooked the two lower anthers which are included in the tube and incorrectly described it as having only two anthers. This error was confirmed by dissection of the type by Dr. U.-M. Hultgaard (pers. comm.) at the Stockholm Herbarium. Dr. O.M. Hilliard (pers. comm.) confirms that the relationships of the species are thus with *Microdon bracteatus* and not with those previously assigned to *Agathelpis*, a genus now included in *Microdon* (Hilliard 1999), and where it was previously included in *A. dubia* (Hartley & Balkwill 1990). A comparison between the type of *E. parviflorum* and collections of *M. bracteatus* reveals no differences between the two in any taxonomic detail and we have no hesitation in regarding them as conspecific. *M. bracteatus* thus falls into synonymy under *M. parviflorus.*

Microdon parviflorus *(P.J.Bergius) Hilliard*, The tribe Selagineae (Scrophulariaceae) :12 (1999).
Eranthemum parviflorum P.J.Bergius: 2 (1767). *Agathelpis parviflora* (P.J.Bergius) Choisy: 95 (1823), syn. nov.
Microdon bracteatus (Thunb.) I.H.Hartley: 57 (1991), syn. nov.

36. Known only from the single specimen that comprises the type collection, *Nemesia guthriei* has remained enigmatic since it was described by Hiern in *Flora capensis* in 1804. No natural vegetation remains at the type locality, Raapenberg, Cape Town, and the type population cannot be examined. The type specimen at BOL *(Guthrie 1221)* has no mature open flowers, but the presence of a short spur is evident in the buds; the flowers are not saccate as stated by Hiern. The species is clearly an annual and the general appearance of the plant matches exactly the widespread *N. barbata* which has also been recorded at the northern end of the Cape Peninsula. In particular, the rather deeply toothed leaves with lightly revolute margins, the crowded, subcorymbose raceme, shortly spurred corolla and capsule longer than wide and narrowed just below the apex of *N. guthriei* match *N. barbata* closely. Because the flowers are immature, however, it is not possible to see the fully developed lower corolla lobe (described as pale by Hiern), which in *N. barbata* is enlarged and darkly coloured either blue or purple to black, but the palate is coarsely hairy, as in *N. barbata*. All available information suggests that *N. guthriei* is a synonym of *N. barbata* and they are treated as a single species here.

Nemesia barbata *(Thunb.) Benth.* in Hook., Companion to the Botanical Magazine 2: 19 (1836b).
Antirrhinum barbatum Thunb.: 105 (1800).
Nemesia guthriei Hiern: 173 (1904), syn. nov.

VERBENACEAE

37. In his monograph of *Chascanum* Moldenke (1933a) commented that *C. integrifolium* differed minimally from *C. cernuum*, and 'may not prove specifically distinct.' The important distinguishing feature of *C. integrifolium* was opposite leaves compared with ternate leaves in *C. cernuum*. We have examined a range of specimens and find that the leaves of *C. cernuum* are usually ternate in the southwestern part of its range (sometimes opposite on parts of the same plant) whereas they are generally opposite toward the east of the range, which encompasses the type locality of *C. integrifolium* near Knysna. We therefore unite the two species.

Chascanum cernuum *(L.) E.Mey.*, Commentariorum de plantis Africae australioris: 276 (1836).
Bouchea cernuum L.: 251 (1771).
Bouchea integrifolium H.Pearson: 179 (1905). *Chascanum integrifolium* (H.Pearson) Moldenke: 18 (1933b), syn. nov.

BIBLIOGRAPHY

ADAMSON, R.S. 1946. *Prismatocarpus debilis* Adamson sp. nov. (Campanulaceae). *Journal of South African Botany* 12: 37.
ADAMSON, R.S. 1951. A revision of the genera *Prismatocarpus* and *Roella*. *Journal of South African Botany* 17: 93–166.
ADAMSON, R.S. 1955. The South African species of *Lightfootia*. *Journal of South African Botany* 21: 155–221.
ADAMSON, R.S. 1958. The South African species of Aizoaceae. IV. *Mollugo*, *Pharnaceum*, *Coelanthus*, and *Hypertelis*. *Journal of South African Botany* 24: 11–65.
ALVERSON, W.S., KAROL, K.G., BAUM, D.A., CHASE, M.W., SWENSON, S.M., McCOURT, R. & SYTSMA, K.J. 1998. Circumscription of the Malvales and relationships to other Rosidae. *American Journal of Botany* 98: 876–887.
ANDERBERG, A.A. & BREMER, K. 1991. Parsimony analysis and cladistic reclassification of the *Relhania* generic group (Asteraceae–Gnaphalieae). *Annals of the Missouri Botanical Garden* 78: 1061–1072.
ANGIOSPERM PHYLOGENY GROUP. 1998. An ordinal classification for the families of flowering plants. *Annals of the Missouri Botanical Garden* 85: 531–553.
ARNOLD, T.H. & DE WET, B.C. (eds) 1993. Plants of southern Africa: names and distribution. *Memoirs of the Botanical Survey of South Africa* No. 62. National Botanical Institute, Pretoria.
ARROYO, M.T.K. & CAVIERES, L. 1997. The Mediterranean-type climate flora of central Chile–what do we know and how can we assure its protection. *Noticias Biologicas* 5: 48–56.
ARROYO, M.T.K., CAVIERES, L., MARTICORENA, C. & MUNOZ-SCHICK, M. 1994. Convergence in the Mediterranean floras in central Chile and California: insights from comparative biology. In M.T.K. Arroyo, P.H. Zedler & M.D. Fox, *Ecology and biogeography of Mediterranean ecosystems in Chile, California and Australia*: 43–88. Ecological Studies 108. Springer-Verlag, New York.
AXELROD, D.I. & RAVEN, P.H. 1978. Late Cretaceous and Tertiary vegetation history of Africa. In M.J.A. Werger, *Biogeography and ecology of southern Africa*: 77–130. Junk, The Hague.
BAKER, J.G. 1871. A revision of the genera and species of herbaceous capsular gamophyllous Liliaceae. *Journal of the Linnean Society. Botany* 11: 349–436.
BAKER, J.G. 1872. Revision of the nomenclature and arrangement of the Cape species of *Anthericum*. *Journal of Botany, British and Foreign* 10: 135–141.
BAKER, J.G. 1873. On *Schizobasis*, a new genus of Liliaceae from Cape Colony. *Journal of Botany, British and Foreign* 11: 105.
BAKER, J.G. 1874. Description of new species of Scilleae and other Liliaceae. *Journal of Botany, British and Foreign* 12: 363–368.
BAKER, J.G. 1875. Revision of the genera and species of Asparagaceae. *Journal of the Linnean Society. Botany* 14: 508–632.
BAKER, J.G. 1878. Descriptions of new and little known Liliaceae. *Journal of Botany, British and Foreign* 16: 321–326.
BAKER, J.G. 1896–97. Liliaceae. In W.T. Thiselton-Dyer, *Flora capensis* 6: 253–528. Reeve, London.
BARKER, W.F. 1931. *Hessea mathewsii*. *The Flowering Plants of South Africa* 11: t. 404.
BARKER, W.F. 1936. *Amphisiphon*, a new genus of Liliaceae. *Journal of South African Botany* 2: 19–23.
BARKER, W.F. 1940. The genus *Dilatris* Berg. *Journal of South African Botany* 6: 147–164.
BARKER, W.F. 1941. *Hyacinthus paucifolius* Barker. *Journal of South African Botany* 7: 198–200.
BARKER, W.F. 1949. *Plantae novae africanae*. *Journal of South African Botany* 15: 39–42.
BEAN, P.A. 1990. The identity of *Osyris abyssinica* var. *speciosa* (Santalaceae). *South African Journal of Botany* 56: 665–669.
BEARD, J.S. (ed.) 1970. *An annotated checklist of the plants of Western Australia*. Kings Park Board, Perth.
BENTHAM, G. 1836a. Observations of some new or little known genera and species of Scrophulariaceae. In W.J. Hooker, *Companion to the Botanical Magazine* 2: 53–60. Couchman, London.
BENTHAM, G. 1836b. A synopsis of the Hemimerideae, a tribe of Scrophulariaceae. In W.J. Hooker, *Companion to the Botanical Magazine* 2: 13–23. Couchman, London.
BENTHAM, G. 1883. In G. Bentham & J.D. Hooker, *Genera plantarum*, Vol. 3. Reeve, London.
BERG, R.Y. 1975. Myrmecochorous plants in Australia and their dispersal by ants. *Australian Journal of Botany* 23: 475–508.
BERGIUS, P.J. 1767. *Descriptiones plantarum ex Capite Bonae Spei*. Stockholm.
BERNHARDI, J.J. 1838. Nachträgliche Bemerkungen über Papaveracéen und Fumariacéen. *Linnaea* 12: 651–668.
BOLUS, F. 1916. Novitates Africanae *Contributions from the Bolus Herbarium* 2,1: 19–32.
BOLUS, L. 1915. *Wahlenbergia guthriei* L.Bolus (Campanulaceae–Campanuleae). *Annals of the Bolus Herbarium* 1: 193.
BOLUS, L. 1933. Plants—new and noteworthy. *South African Gardening* 23: 167, 168.
BOND, P. & GOLDBLATT, P. 1984. Plants of the Cape flora: a descriptive catalogue. *Journal of South African Botany*, Suppl. Vol. 13.
BOND, W.J. 1983. On alpha diversity and the richness of the Cape flora: a study in southern Cape fynbos. In F.J.

Kruger, D.T. Mitchell & J.U.M. Jarvis, *Mediterranean-type ecosystems. The role of nutrients*: 225–243. Springer-Verlag, Berlin.
BOND, W.J. & SLINGSBY, P. 1983. Seed dispersal by ants in shrublands of the Cape Province and its evolutionary implications. *South African Journal of Science* 79: 231–233.
BRAKO, L. & ZARUCCHI, J.L. 1993. Catalogue of the flowering plants and gymnosperms of Peru. *Monographs in Systematic Botany from the Missouri Botanical Garden* 45.
BREHMER, W.G.B.A. VON. 1915. Über die systematische Gliederung und Entwiklung der Gattung *Wahlenbergia* in Afrika. *Botanische Jahrbücher für Systematik* 53: 9–143.
BREMER, B., OLMSTEAD, R.G., STRUWE, L. & SWEERE, J.A. 1994. *Rbc*L sequences support the exclusion of *Retzia*, *Desfontainia*, and *Nicodenia* from the Gentianales. *Plant Systematics and Evolution* 190: 213–230.
BREMER, K. 1976. The genus *Relhania* (Compositae). *Opera Botanica* 40.
BRENAN, J.P.M. 1978. Some aspects of the phytogeography of tropical Africa. *Annals of the Missouri Botanical Garden* 65: 437–478.
BRIDSON, D. 1985. The reinstatement of *Psydrax* (Rubiaceae, subfam. Cinchonoideae tribe Vanguerieae) and a revision of the African species. *Kew Bulletin* 40: 687–725.
BROWN, N.E. 1929. The Iridaceae of Burmann's *Florae capensis prodromus*. *Kew Bulletin* 1929: 129–139.
BROWN, R. 1812. *Hortus kewensis*, edn 2. London.
BURMAN, N.L. 1768. *Prodromus florae capensis*. Amsterdam.
BURTT, B.L. 1991. Umbelliferae of southern Africa: an introduction and annotated check-list. *Edinburgh Journal of Botany* 48: 133–282.
CHASE, M.W., DUVALL, M.R., HILLS, H.G., CONRAN, J.G., COX, A.V., EGUIARTE, L.E., HARTWELL, J., FAY, M.F., CADDICK, L.R., CAMERON, K.M. & HOOT, S. 1995. Molecular phylogenetics of Lilianae. In P.J. Rudall, P.J. Cribb, D.F. Cutler & C.J. Humphries, *Monocotyledons: systematics and evolution*: 101–110. Royal Botanic Gardens, Kew.
CHOISY, J.D. 1823. Mémoire sur la famille de Sélaginées. *Mémoires de la Société de Physique et d'Histoire Naturelle de Genève*, le janvier 1822. Paschoud, Geneva.
COETZEE, J.A. 1993. African flora since the terminal Jurassic. In P. Goldblatt, *Biological relationships between Africa and South America*: 37–61. Yale University Press, Newhaven.
COETZEE, J.A. & MULLER, J. 1984. The phytogeographic significance of some extinct Gondwana pollen types from the Tertiary of the southwestern Cape (South Africa). *Annals of the Missouri Botanical Garden* 71: 1088–1099.
COETZEE, J.A. & PRAGLOWSKI, J. 1984. Pollen evidence for the occurrence of *Casuarina* and *Myrica* in the Tertiary of South Africa. *Grana* 23: 23–41.
COETZEE, J.A. & ROGERS, J. 1982. Palynological and lithological evidence for the Miocene palaeoenvironment in the Saldanha region (South Africa). *Palaeogeography, Palaeoclimatology and Palaeoecology* 39: 71–85.
COMPTON, R.H. 1930. Novitates Africanae. *Journal of Botany, British and Foreign* 68: 107.
CONRAN, J.G., CHASE, M.W. & RUDALL, P.J. 1997. Two new monocotyledon families: Anemarrhenaceae and Behniaceae (Lilianae: Asparagales). *Kew Bulletin* 52: 995–999.
COWLING, R.M. 1990. Diversity components in a species-rich area of the Cape Floristic Region. *Journal of Vegetation Sciences* 1: 699–710.
COWLING, R.M. 1992. *The ecology of fynbos*. Oxford University Press, Cape Town.
COWLING, R.M. & HOLMES, P.M. 1992a. Flora and vegetation. In R.M. Cowling, *The ecology of fynbos*: 23–61. Oxford University Press, Cape Town.
COWLING, R.M. & HOLMES, P.M. 1992b. Endemism and speciation in a lowland flora from the Cape Floristic Region. *Biological Journal of the Linnean Society* 47: 367–383.
COWLING, R.M., HOLMES, P.M. & REBELO, A.G. 1992. Plant diversity and endemism. In R.M. Cowling, *The ecology of fynbos*: 62–112. Oxford University Press, Cape Town.
COWLING, R.M., RUNDELL, P.W., LAMONT, B.B., ARROYO, M.T.K. & ARIANOUTSOU, M. 1996. Plant diversity in Mediterranean-climate regions. *Trends in Ecology and Evolution* 11: 362–366.
D'ARCY, W.D. 1987. *Flora of Panama. Checklist and index*. Missouri Botanical Garden, St. Louis.
DAHLGREN, R., CLIFFORD, H.T. & YEO, P.F. 1985. *The families of the monocotyledons*. Springer-Verlag, Berlin.
DAHLGREN, R., NIELSEN, B.J., GOLDBLATT, P. & ROURKE, J.P. 1979. Further notes on Retziaceae: its chemical contents and affinities. *Annals of the Missouri Botanical Garden* 66: 545–556.
DEACON, H.J. 1979. Palaeoecology. In J. Day, W.R. Siegfried, G.N. Louw & M.L. Jarman, *Fynbos ecology: a preliminary synthesis*. *South African National Scientific Programmes Report* 40: 58–66. CSIR, Pretoria.
DE CANDOLLE, A.L.P.P. 1830. *Monographie des Campanulées*. Paris.
DE CANDOLLE, A.P. 1825. *Prodromus systematis naturalis regni vegetabilis*, Vol. 2. Treuttel & Würtz, Paris.
DE CANDOLLE, A.P. 1826. *Minurothamnus. Prodromus systematis naturalis regni vegetabilis* Vol. 5.
DE CANDOLLE, A.P. 1838a. *Prodromus systematis naturalis regni vegetabilis*, Vol. 6. Treuttel & Würtz, Paris.
DE CANDOLLE, A.P. 1838b. *Prodromus systematis naturalis regni vegetabilis*, Vol. 7, Suppl. Treuttel & Würtz, Paris.
DE CANDOLLE, A.P. 1839. *Prodromus systematis naturalis regni vegetabilis*, Vol. 7. Treuttel & Würtz, Paris.

DREYER, L.L. & VAN WYK, A.E. 1998. Oxalidaceae. Taxonomic delimitation of *Oxalis engleriana*. *Bothalia* 28: 65.
DUTHIE, A.V. 1928. The species of *Urginea* of the Stellenbosch Flats. *Annale van die Universiteit van Stellenbosch* 6A, 2: 3–16.
DYER, R.A. 1934. *Rhadamanthus urantherus*. *Hooker's icones plantarum* 1934: t. 3247.
ECKLON, C.F. & ZEYHER, C.L. 1834–1837. *Enumeratio plantarum africae australis extratropicae*. Hamburg.
ESLER, K.J. & RUNDEL, P.W. 1998. Unusual geophytes of the succulent karoo. *Veld & Flora* 84: 6–9.
ESLER, K.J., RUNDEL, P.W. & VORSTER, P. 1999. Biogeography of prostrate-leaved geophytes in semi-arid South Africa: hypotheses on functionality. *Plant Ecology* 142: 105–120.
FAY, M.F. & CHASE, M.W. 1996. Resurrection of Themidaceae for the *Brodiaea* alliance, and recircumscription of Alliaceae, Amaryllidaceae and Agapanthoideae. *Taxon* 45: 441–451.
FEDDE, F.K.G. 1924. *Cysticapnos vesicarius*. *Repertorium specierum novarum regni vegetabilis* 19: 285–288.
FENNER, M., LEE, W.G. & WILLIAMS, J.B. 1997. A comparative study of the distribution of genus size in twenty angiosperm floras. *Biological Journal of the Linnean Society* 62: 225–237.
FERGUSON, D. M. 1999. Phylogenetic analysis and relationships in Hydrophyllaceae based on *ndhF* sequence data. *Systematic Botany* 23: 253—268.
FOURCADE, G.H. 1932. Contributions to the flora of the Knysna and neighbouring divisions. *Transactions of the Royal Society of South Africa* 21: 75–102.
GAERTNER, J. 1791. *De fructibus et seminibus plantarum*. Stuttgart, Tübingen.
GENTRY, A.H. 1988a. Tree species richness of upper Amazonian species. *Proceedings of the National Academy of Sciences of the United States of America*. 85: 156–159.
GENTRY, A.H. 1988b. Changes in plant community diversity and floristic composition on environmental and geographical gradients. *Annals of the Missouri Botanical Garden* 75: 1–34.
GENTRY, A.H. & DODSON, C. 1987. Diversity and phytogeography of neotropical vascular epiphytes. *Annals of the Missouri Botanical Garden* 74: 205–233.
GERBAULET, M. 1998. *Aptenia*, *Mesembryanthemum*. In H.E.K. Hartmann, *IOS lexicon of succulent plants*. Fischer, Jena.
GIBBS RUSSELL, G.E. 1985. Analysis of the size and composition of the southern African flora. *Bothalia* 15: 613–629.
GIBBS RUSSELL, G.E. 1991. An overview of the systematics, phylogeny and biology of the African Iridaceae. *Contributions from the Bolus Herbarium* 13: 1–74.
GOLDBLATT, P. 1978. An analysis of the flora of southern Africa: its characteristics, relationships, and origins. *Annals of the Missouri Botanical Garden* 65: 369–436.
GOLDBLATT, P. 1979. Miscellaneous chromosome counts in angiosperms. II, including new family and generic records. *Annals of the Missouri Botanical Garden*. 66: 856–861.
GOLDBLATT, P. 1982. Systematics of *Freesia* Klatt (Iridaceae). *Journal of South African Botany* 48: 39–91.
GOLDBLATT, P. 1990. Status of the southern African *Anapalina* and *Antholyza* (Iridaceae) genera, based solely on characters for bird pollination, and a new species of *Tritoniopsis*. *South African Journal of Botany* 56: 577–582.
GOLDBLATT, P. 1991. Nomenclatural notes on African Iridaceae. *South African Journal of Botany* 57: 226.
GOLDBLATT, P. 1992. Phylogenetic analysis of the South African genus *Sparaxis* (including *Synnotia*) (Iridaceae: Ixioideae), with two new species and a review of the genus. *Annals of the Missouri Botanical Garden* 79: 143–159.
GOLDBLATT, P. 1997. Floristic diversity in the Cape flora of South Africa. *Biodiversity & Conservation* 6: 359–377.
GOLDBLATT, P. 1998. Reduction of *Barnardiella*, *Galaxia*, *Gynandriris*, *Hexaglottis* and *Homeria* in *Moraea* (Iridaceae: Irideae). *Novon* 8: 371–377.
GOLDBLATT, P. & DE VOS, M.P. 1989. The reduction of *Oenostachys*, *Homoglossum* and *Anomalesia*, putative sunbird pollinated genera, in *Gladiolus* L. (Iridaceae: Ixioideae). *Bulletin du Muséum National de l'Histoire Naturelle, 4 Série, Sect. B, Adansonia* 11: 417–428.
GOLDBLATT, P. & MANNING, J.C. 1996. Phylogeny and speciation in *Lapeirousia* subgenus *Lapeirousia* (Iridaceae: Ixioideae). *Annals of the Missouri Botanical Garden* 83: 346–361.
GOLDBLATT, P. & MANNING, J.C. 1998. *Gladiolus in southern Africa*. Fernwood Press, Vlaeberg.
GOOD, R. 1974. *The geography of the flowering plants*, edn 4. Longmans Green, London.
GROVES, R.H., BEARD, J.S., DEACON, H.J., LAMBRECHTS, J.J.N., RABINOVITCH-VIN, A., SPECHT, R.L. & STOCK, W.D. 1983. Introduction: the origins and characteristics of Mediterranean ecosystems. In A. Day, *Mineral nutrients in Mediterranean ecosystems*. *South African National Scientific Porgrammes Report* 71: 1–17. CSIR, Pretoria.
HARTLEY, I.H. 1991. Selaginaceae. *Microdon bracteatus*—the correct name for *M. lucidus*. *Bothalia* 21: 57–59.
HARTLEY, I.H. & BALKWILL, K. 1990. A taxonomic account of *Agathelpis*, *Globulariopsis* and *Gosela* (Selaginaceae). *South African Journal of Botany* 56: 471—481.
HARVEY, W.H. 1844. *Litanthus*, a new genus of Asphodeleae, from South Africa. In W.J. Hooker in *The London Journal of Botany* 3: 314, 315, t. 9.
HARVEY, W.H. 1859. Fumariaceae. In W.H. Harvey & O.W. Sonder, *Flora capensis* 1: 15–19. Hodges, Smith & Co.,

Dublin.
HARVEY, W.H. 1865. Compositae. In W.H. Harvey & O.W. Sonder, *Flora capensis* 3: 44–530. Reeve, London.
HENDEY, Q.B. 1982. *Langebaanweg. A record of past life*. South African Museum, Cape Town.
HERBERT, W. 1837. *Amaryllidaceae*. Ridgway, London.
HIERN, W.P. 1904. Scrophulariaceae. In W.T. Thiselton-Dyer, *Flora capensis* 4,2: 121–420. Reeve, London.
HILL, A.W. 1925. Santalaceae. In W.T. Thiselton-Dyer, *Flora capensis* 5,2: 315–212. Reeve, London.
HILLIARD, O.M. 1983. Asteraceae: Inuleae: Gnaphaliinae. *Flora of southern Africa* 33,7,2 (first part). Botanical Research Institute, Pretoria.
HILLIARD, O.M. 1994a. *The Manuleae: a tribe of the Scrophulariaceae*. Edinburgh University Press, Edinburgh.
HILLIARD, O.M. 1994b. A note on *Colpoon. Edinburgh Journal of Botany* 51: 391–392.
HILLIARD, O.M. 1999. *The tribe Selagineae (Scrophulariaceae)*. Royal Botanic Garden, Edinburgh.
HOPPER, S.D. 1992. Patterns of plant diversity at the population and species levels in south-west Australian Mediterranean ecosystems. In R.J. Hobbs, *Biodiversity in Mediterranean ecosystems in Australia*: 27–46. Surrey Beatty & Sons, Chipping Norton.
HUTCHINSON, J. 1921. The genera of Fumariaceae and their distribution. *Kew Bulletin* 1921: 97–115.
JACQUIN, N.J. VON. 1789. *Icones plantarum rariorum*, Vol. 2, Fasc. 4, t. 431 Vienna.
JACQUIN, N.J. VON. 1791. *Collectanea*, Vol. 4. Vienna.
JACQUIN, N.J. VON. 1794. *Icones plantarum rariorum*, Vol. 2, Fasc. 15: t. 414. Vienna.
JACQUIN, N.J. VON. 1797. *Plantarum rariorum horti caesarei schoenbrunnensis* Vol. 1. Vienna.
JESSOP, J. 1976. Studies in the bulbous Liliaceae in South Africa: 6. The taxonomy of *Massonia* and allied genera. *Journal of South African Botany* 42: 401–437.
JESSOP, J. 1977. Studies in the bulbous Liliaceae in South Africa: 7. The taxonomy of *Drimia* and certain allied genera. *Journal of South African Botany* 43: 265–319.
JOHNSON, C.T. & WEITZ, F.M. 1991. A re-evaluation of *Nylandtia* (Polygalaceae). *South African Journal of Botany* 57: 229–233.
KALLERSJÖ, M. 1991. The genus *Athanasia* (Compositae: Anthemidae). *Opera Botanica* 106.
KLEIN, R.G. 1997. The ecology of early man in southern Africa. *Science* 197: 115–126.
KRUGER, F.J. & TAYLOR, H.C. 1979. Plant species diversity in Cape fynbos: gamma and delta diversity. *Vegetatio* 41: 85–93.
LAMONT, B.B., HOPKINS, A.J.M. & KNATIUK, R.J. 1984. The flora—composition, diversity and origin. In J.S. Pate & J.S. Beard, *Kwongan: plant life of the sandplain*: 27–50. University of Western Australia Press, Nedlands.
LEBRUN, J.-P. & STORK, A.L. 1997. *Énumération des plantes à fleurs d'Afrique tropicale*. Vol. 5. Conservatoire et Jardin Botanique de la Ville de Genève, Geneva.
LEIGHTON, F.M. 1943. *Ornithogalum nannodes* Leighton nom. nov. *Journal of South African Botany* 9: 113.
LEVIN, D.A. 1993. Local speciation in plants: the rule not the exception. *Systematic Botany* 18: 197–208.
LEVYNS, M.L. 1954. The genus *Muraltia. Journal of South African Botany*, Suppl. Vol. 2.
LEWIN, K. 1922. Systematische Gliederung und geographische Verbreitung der Arctotideae-Arctotidinae. *Feddes Repertorium* 11: 1–75.
LEWIS, G.J. 1960. South African Iridaceae. The genus *Anapalina. Journal of South African Botany* 26: 51–71.
LIDÉN, M. 1986. Synopsis of Fumarioideae (Papaveraceae) with a monograph of tribe Fumarieae. *Opera Botanica* 88: 1–133.
LINDER, H.P. 1981. Taxonomic studies on the Disinae. III. A revision of *Disa* Berg. excluding sect. *Macranthae* Lindl. *Contributions from the Bolus Herbarium* 9.
LINDER, H.P. 1985. Gene flow, speciation and species diversity patterns in a species-rich area: the Cape flora. In E.S. Vrba, *Species and speciation*: 53–57. Transvaal Museum Monograph 4. Transvaal Museum, Pretoria.
LINDER, H.P. & VLOK, J.H.J. 1991. The morphology, taxonomy and evolution of *Rhodocoma* (Restionaceae). *Plant Systematics and Evolution* 175: 139–160.
LINDLEY, J. 1835. *Daubenya aurea. Botanical Register* 21: t. 1813. Ridgway, London.
LINNAEUS, C. 1753. *Species plantarum*. Salvius, Stockholm.
LINNAEUS, C. 1771. *Mantissa plantarum*. Salvius, Stockholm.
LINNAEUS, C. (fil.) 1782. ('1781'). *Supplementum plantarum systematis vegetabilium*. Orphanotrophei, Brunsvigae.
LOW, A.B. & A.G. REBELO. 1996. *Vegetation of South Africa, Lesotho and Swaziland*. Department of Environmental Affairs and Tourism, Pretoria.
MAGGS, G.L., CRAVEN, P. & KOLBERG, H. 1998. Plant species richness, endemism and genetic resources in Namibia. *Biodiversity & Conservation* 7: 435–446.
MASSON, D. 1989. *Silene vlokii* D. Masson sp. nov., new species of Caryophyllaceae from South Africa. *Candollea* 44: 485–491.
MEADOWS, M.E. & SUGDEN, J.M. 1991. A vegetation history of the last 14 000 years on the Cedarberg, south-western Cape Province. *South African Journal of Science* 87: 34–43.
MÉDAIL, F. & QUÉZEL, P. 1997. Hot spots analysis for conservation of plant biodiversity in the Mediterranean

Basin. *Annals of the Missouri Botanical Garden* 84: 112–127.
MEYER, E.H.F. 1836. *Commentariorum de plantis Africae australioris*. Leipzig.
MEYER, E.H.F. 1843. Zwei pflanzengeographische Documente. *Flora* 26.
MOLDENKE, H.N. 1933a. A monograph of the genus *Chascanum*. XXV. *Repertorium Specierum Novarum Regni Vegetabilis* 46: 300–319.
MOLDENKE, H.N. 1933b. Studies of new and noteworthy tropical American plants 1. *Phytologia* 1: 5–18
MÜLLER-DOBLIES, D. & MÜLLER-DOBLIES, U. 1985. De Liliifloris notulae 2. De taxonomia subtribus Strumariinae (Amaryllidaceae). *Botanische Jahrbücher für Systematik* 107: 17–47.
MÜLLER-DOBLIES, D. & MÜLLER-DOBLIES, U. 1994. De Liliifloris notulae. 5. Some new taxa and combinations in the Amaryllidaceae tribe Amaryllideae from arid southern Africa. *Feddes Repertorium* 105: 331–363.
MÜLLER-DOBLIES, D. & MÜLLER-DOBLIES, U. 1996. Tribes and subtribes and some species combinations in Amaryllidaceae. *Feddes Repertorium* 107: S.c. 1–9.
MÜLLER-DOBLIES, U. 1994. Enumeratio Albucarum (Hyacinthaceae) Austro-Africanarum adhuc cognitarum. 1. Subgenus *Albuca*. *Feddes Repertorium* 105: 365–368.
MÜLLER-DOBLIES, U. 1995. Enumeratio Albucarum (Hyacinthaceae) Austro-Africanarum adhuc cognitarum. 2. Subgenus *Falconera* (Salisb.) Baker emend. U.M.-D. 1987. *Feddes Repertorium* 106: 353–370.
MÜLLER-DOBLIES, U. & MÜLLER-DOBLIES, D. 1996. Revisionula incompleta Ornithogalorum Austro-Africanorum (Hyacinthaceae). *Feddes Repertorium* 107: 361–548.
MÜLLER-DOBLIES, U. & MÜLLER-DOBLIES, D. 1997. A partial revision of the tribe Massonieae (Hyacinthaceae). *Feddes Repertorium* 108: 49–96.
MUNRO, S.L. & LINDER, H.P. 1998. The phylogenetic position of *Prionium* (Juncaceae) within the order Juncales based on morphological and *rbc*L sequence data. *Systematic Botany* 23: 43–55.
NANDI, O.I., CHASE, M.W. & ENDRESS, P.K. 1998. A combined cladistic analysis of angiosperms using *rbc*L and non-molecular data sets. *Annals of the Missouri Botanical Garden* 85: 137–212.
NORDENSTAM, B. 1970a. Studies in South African Liliaceae III. The genus *Rhadamanthus*. *Botaniska Notiser* 123: 155–182.
NORDENSTAM, B. 1970b. Notes on South African Iridaceae: *Lapeirousia* and *Babiana*. *Botaniska Notiser* 123: 433–443.
OBERMEYER, A.A. 1978. *Ornithogalum*: a revision of the southern African species. *Bothalia* 12: 323–376.
OBERMEYER, A.A. 1980. A new subgenus *Rhadamanthopsis* and two new species of *Rhadamanthus*. *Bothalia* 13: 137–139.
OBERMEYER, A.A. 1981. Liliaceae: some name changes in the *Urgineae* complex. *Journal of South African Botany* 47: 577.
PATE, J.S. & DIXON, K.W. 1982. *Tuberous, cormous and bulbous plants: biology of an adaptive strategy in Western Australia*. University of Western Australia Press, Perth.
PEARSON, H.H.W. 1905. South African Verbenaceae. *Transactions of the South African Philosophical Society* 15: 175–182.
PENNINGTON, T.D. & STYLES, B.T. 1975. A generic monograph of the Meliaceae. *Blumea* 22: 419–540.
PHILLIPS, E.P. 1932. Description of three new South African plants. *Bothalia* 2: 368, 369.
PHILLIPS, E.P. & DYER, R.A. 1934. The genus *Sutherlandia*. *Revista Sudamericana de Botánica* 1: 69–80.
PILLANS, N. 1947. A revision of Bruniaceae. *Journal of South African Botany* 13: 121–206.
PLUNKETT, G.M., SOLTIS, D.E. & SOLTIS, P.S. 1997. Classification of the relationship between Apiaceae and Araliaceae based on *mat*K and *rbc*L sequence data. *American Journal of Botany* 84: 565–580.
POWRIE, E. 1972. The typification of *Brunia nodiflora* L. *Journal of South African Botany* 38: 301–304.
RAFINESQUE-SCHMALTZ, C.S. 1836. *Flora telluriana*. Philadelphia.
RAVEN, P.H. & AXELROD, D.I. 1978. Origin and relationships of the California flora. *University of California Pubications in Botany* 72.
REBELO, A.G. 1987. Bird pollination in the Cape flora. *A preliminary synthesis of pollination biology in the Cape flora*: 83–108. *South African National Scientific Programmes Report* 141. CSIR, Pretoria.
ROTH, A.W. 1821. Novae plantarum species. Halberstadt.
RUTHERFORD, M.C. & WESTFALL, R.H. 1994. Biomes of southern Africa: an objective categorization. *Memoirs of the Botanical Survey of South Africa* No. 63. National Botanical Institute, Pretoria.
SALISBURY, R.A. 1866. *The genera of plants*. Van Voorst, London.
SCHATZ, G.E., LOWRY, P.P., LESCOT, M., WOLF, A.E., ANDRIAMBOLOLONERA, S., RANDRINASOLO, V. & RAHARIMAMPIONONA, J. 1996. Conspectus of the vascular plants of Madagascar: a taxonomic and conservation electronic data base. In L.J.G. van der Maesen *et al.*, *The biodiversity of African plants*: 10–17. Proceedings of the XIV AETFAT Congress, Wageningen, The Netherlands. Kluwer Academic Publications, Dordrecht.
SCHLECHTER, R. 1897. Decades plantarum novarum austro-africanarum. *Journal of Botany, British and Foreign* 35: 428–433.
SCHLECHTER, R. 1898. Plantae Schlechterianae novae vel minus cognitae describuntur. I. *Botanische Jahrbücher für*

Systematik 24: 434–459.
SCHLECHTER, R. 1900. Plantae Schlechterianae novae vel minus cognitae describuntur. II *Botanische Jahrbücher für Systematik* 27: 86–220.
SCHLECHTER, R. 1924. Drei neue Gattungen der Liliaceen aus Südafrika. *Notizblatt des Botanischen Gartens und Museums zu Berlin-Dahlem* 9: 145–151.
SCHLECHTER, R. & BREHMER, W. VON. 1915. *Wahlenbergia macra* Schltr. et v. Brehmer n. sp. *Botanische Jahrbücher für Systematik* 53: 91.
SCHOLZ, A. 1985. The palynology of the upper lacustrine sediments of the Arnot pipe, Banke, Namaqualand. *Annals of the South African Museum* 95: 1–109.
SNIJMAN, D. 1994. Systematics of *Hessea*, *Strumaria* and *Carpolyza* (Amaryllideae: Amaryllidaceae). *Contributions from the Bolus Herbarium* 16.
SNIJMAN, D.A., J.C. MANNING & P. GOLDBLATT. 199. A new *Rhadamanthus* species (Hyacinthaceae) from the northwestern Cape, South Africa. Novon 9: 111-113.
SONDER, O.W. 1862. Umbelliferae. *Flora capensis* 2: 524–567.
SPETA, F. 1998. Systematische Analyse der Gattung *Scilla* L. s.l. (Hyacinthaceae). *Phyton* (Austria) 38: 1–141.
STEDJE, B. 1996. Hyacinthaceae. In R.M. Polhill, *Flora of tropical East Africa*. Balkema, Rotterdam.
STEINHEIL, A. 1834. Nore sur le genre *Urginea*, nouvellement formé dans la famille des Liliacées. *Annales des Sciences Naturelles, Paris*, Ser. 2, 1: 321–332.
TAKHTAJAN, A. 1986. *Floristic regions of the world*. University of California Press, Berkeley (translated by T. Crovello).
THOMPSON, J. 1990. New species and combinations in the genus *Swainsona* (Fabaceae) in New South Wales. *Telopea* 4: 1–5.
THUNBERG, C.P. 1800. *Prodromus plantarum capensium*. Edman, Uppsala.
TILMAN, D. 1982. *Resource competition and community structure*. Princeton University Press, Princeton.
TILMAN, D. 1983. Some thoughts on resource competition and diversity in plant communities. In F.J. Kruger, D.T. Mitchell & J.U.M. Jarvis, *Mediterranean-type ecosystems. The role of nutrients*: 322–336. Springer-Verlag, Berlin.
TILMAN, D., BOND, W.J., CAMPBELL, B.M., KRUGER, F.J., LINDER, H.P., SCHOLZ, A., TAYLOR, H.C. & WITTER, M. 1983. Origin and maintenance of plant species diversity. In J.A. Day, *Mineral nutrients in Mediterranean ecosystems*: 125–135. South African National Scientific Programmes Report 71. CSIR, Pretoria.
VAN JAARSVELD, E.J. 1983. *Freylinia visseri* en *Freylinia decurrens* (Scrophulariaceae): twee nuwe spesies van die Suidwes-Kaapland. *Journal of South African Botany* 49: 57–64.
VILLAGRÁN, C.M. 1994. Quarternary history of the Mediterranean vegetation of Chile. In M.T. Kalin Arroyo, P.H. Zedler & M.D. Fox, *Ecological and biogeography of Mediterranean ecosystems in Chile, California and Australia*: 3–20. Ecological Studies 108. Springer-Verlag, New York.
VOGEL, S. 1954. Blütenbiologische Typen als Elemente der Sippengliederung. *Botanische Studien* 1: 1–338.
WAGNER, W.L. 1991. Evolution of waif floras: a comparison of the Hawaiian and Marquesan Archipelagoes. In E.C. Dudley, *The unity of evolutionary biology* Vol. 1: 267–284. Dioscorides Press, Portland, Oregon.
WAGNER, W.L., HERBST, D.R. & SOHMER, S.H. 1990. *Manual of the flowering plants of Hawaii*. University of Hawaii Press, Honolulu.
WEIMARCK, H. 1941. Phytogeographical groups, centres and intervals within the Cape flora. *Lunds Universitets Arsskrift N.F. Avd.* 2, 37(5): 1–143.
WILLDENOW, C.L. 1799. *Species plantarum*, Vol. 2, edn 4. Nauk, Berlin.
WRIGHT, C.H. 1916. Diagnoses Africanae LXIX. *Bulletin of Miscellaneous Information. Kew* 1916: 229–235.
XIANG, Q.-Y., SOLTIS, D.E., MORGAN, D.R. & SOLTIS, P.S. 1993. Phylogenetic relationships of *Cornus* L. sensu lato and putative relatives inferred from *rbc*L sequence data. *Annals of the Missouri Botanical Garden* 80: 723–734.

Appendix: Statistics for families of the Cape flora

Peridophytes	Genera	Endemic	Species	Endemic	NW spp.	Endemic	SW spp.
Anemiaceae	1	0	2	1	2	1	1
Aspleniaceae	2	0	14	0	5	0	6
Blechnaceae	1	0	6	0	5	0	6
Cyatheaceae	1	0	2	0	0	0	1
Dennstaedtiaceae	4	0	5	0	2	0	3
Dryopteridaceae	4	0	8	0	3	0	7
Equisetaceae	1	0	1	0	0	0	0
Gleicheniaceae	1	0	1	0	1	0	1
Grammitidaceae	1	0	1	0	0	0	1
Hymenophyllaceae	3	0	5	0	3	0	4
Isoetaceae	1	0	2	2	0	0	2
Lomariopsidaceae	1	0	3	0	1	0	3
Lycopodiaceae	3	0	6	0	4	0	6
Marattiaceae	1	0	1	0	0	0	0
Marsileaceae	1	0	4	0	2	0	3
Ophioglossaceae	1	0	4	3	3	0	3
Osmundaceae	2	0	2	0	2	0	2
Polypodiaceae	4	0	5	0	1	0	3
Pteridaceae	5	0	28	2	13	0	18
Schizaeaceae	1	0	2	0	2	0	2
Selaginellaceae	1	0	3	1	1	0	1
Thelypteridaceae	5	0	6	1	0	0	4
Vittariaceae	1	0	1	0	0	0	1
Total	46	0	112	10	50	1	78

Seed Plants							
Acanthaceae	13	0	25	4	1	0	0
Achariaceae	2	0	2	0	0	0	0
Agapanthaceae	1	0	3	2	0	0	2
Aizoaceae	76	18	661	526	316	146	211
Alliaceae	2	0	5	1	3	0	3
Amaranthaceae	12	0	22	3	16	0	17
Amaryllidaceae	16	1	93	56	48	14	39
Anacardiaceae	4	2	28	9	18	1	18
Anthericaceae	1	0	12	5	8	3	5
Apiaceae	23	3	72	49	39	10	45
Apocynaceae	36	1	112	21	43	2	32
Aponogetonaceae	1	0	5	2	2	0	2
Aquifoliaceae	1	0	1	0	0	0	1
Araceae	4	0	5	0	1	0	4
Araliaceae	4	0	55	46	19	6	31
Asparagaceae	1	0	36	5	18	1	20
Asphodelaceae	8	0	157	81	69	7	60
Asteraceae	121	32	1035	654	537	107	506
Balanophoraceae	1	1	1	1	1	0	1
Balsaminaceae	1	0	1	0	0	0	0
Behniaceaeae	1	0	1	0	0	0	0
Bignoniaceae	2	0	2	0	1	0	0
Boraginaceae	9	1	41	24	19	1	22
Brassicaceae	13	4	72	40	47	12	37
Bruniaceae	11	9	64	63	23	8	44
Buddlejaceae	1	0	3	0	1	0	2
Campanulaceae	13	6	184	140	86	22	116
Caryophyllaceae	6	0	17	5	9	0	9
Celastraceae	12	2	26	6	7	0	15
Celtidaceae	1	0	1	0	0	0	1
Ceratophyllaceae	1	0	1	0	0	0	0
Clusiaceae	1	0	2	0	0	0	1
Colchicaceae	6	2	32	18	23	4	17
Commelinaceae	2	0	3	0	0	0	1
Convallariaceae	3	0	49	24	30	7	18

Appendix: Statistics for families of the Cape flora

Endemic	AP spp.	Endemic	KM spp.	Endemic	LB spp.	Endemic	SE spp.	Endemic
0	0	0	1	0	1	0	1	0
0	1	0	5	0	10	0	14	0
0	0	0	4	0	4	0	6	0
0	0	0	1	0	2	0	1	0
0	0	0	1	0	3	0	5	0
0	0	0	4	0	6	0	5	0
0	1	0	1	0	0	0	1	0
0	0	0	1	0	1	0	1	0
0	0	0	0	0	1	0	0	0
0	0	0	1	0	5	0	5	0
2	0	0	0	0	0	0	0	0
0	0	0	0	0	2	0	2	0
0	0	0	5	0	5	0	4	0
0	0	0	0	0	0	0	1	0
0	2	0	1	0	1	0	2	0
0	1	0	2	0	2	0	2	0
0	0	0	1	0	1	0	2	0
0	0	0	1	0	4	0	5	0
0	2	0	15	0	13	0	17	0
0	1	0	2	0	2	0	2	0
0	1	0	1	0	0	0	2	0
0	0	0	0	0	2	0	6	0
0	0	0	0	0	0	0	1	1
2	9	0	47	0	65	0	85	1
0	1	0	8	2	9	1	21	1
0	0	0	0	0	0	0	2	0
1	1	0	0	0	2	0	1	0
97	46	16	171	69	132	43	112	41
0	1	0	1	0	3	0	4	0
0	12	0	6	1	3	0	17	0
10	23	0	15	0	15	2	30	8
1	9	0	12	0	13	0	18	0
0	0	0	0	0	2	0	3	0
9	22	1	18	1	26	0	31	3
1	8	0	47	4	40	3	64	1
0	2	0	1	0	2	0	2	0
0	0	0	0	0	1	0	1	0
0	1	0	0	0	2	0	4	0
12	10	5	9	0	15	0	22	2
0	14	0	16	1	20	1	21	0
7	15	1	51	5	40	6	54	11
128	140	26	327	31	268	19	349	26
0	0	0	0	0	1	0	0	0
0	0	0	0	0	0	0	1	0
0	0	0	1	0	0	0	1	0
4	10	4	12	0	7	1	13	0
7	6	0	18	2	14	0	16	0
23	4	1	5	0	15	6	4	0
0	1	0	0	0	2	0	3	0
31	42	5	33	1	59	9	71	5
0	7	0	7	0	10	0	14	0
2	9	0	7	0	8	0	22	0
0	1	0	0	0	1	0	1	0
0	0	0	0	0	0	0	1	0
0	0	0	1	0	2	0	2	0
1	5	0	12	1	10	0	8	0
0	1	0	1	0	2	0	2	0
1	7	0	20	4	11	1	18	1

Appendix: Statistics for families of the Cape flora

Seed Plants	Genera	Endemic	Species	Endemic	NW spp.	Endemic	SW spp.
Convolvulaceae	5	0	17	4	6	0	9
Cornaceae	1	0	1	0	1	0	1
Crassulaceae	5	0	123	35	81	6	53
Cucurbitaceae	5	0	7	0	3	0	3
Cunoniaceae	2	1	2	1	2	0	2
Cupressaceae	1	0	3	2	2	1	1
Cyperaceae	29	3	206	101	81	5	161
Cytinaceae	1	0	1	0	1	0	1
Dioscoreaceae	1	0	6	2	1	0	0
Dipsacaceae	2	0	9	4	3	1	7
Droseraceae	1	0	14	10	7	2	11
Ebenaceae	2	0	17	2	11	0	8
Elatinaceae	1	0	1	0	1	0	1
Ericaceae	1	0	658	636	152	55	375
Euphorbiaceae	13	2	80	25	25	5	29
Fabaceae	37	6	760	627	293	92	374
Flacourtiaceae	4	0	8	0	1	0	1
Frankeniaceae	1	0	2	0	2	0	2
Fumariaceae	3	1	4	2	3	0	3
Geissolomaceae	1	1	1	1	0	0	0
Gentianaceae	3	1	31	18	11	0	20
Geraniaceae	3	0	155	91	98	17	88
Gesneriaceae	1	0	2	0	0	0	0
Goodeniaceae	1	0	1	0	0	0	0
Grubbiaceae	1	1	3	3	2	0	3
Gunneraceae	1	0	1	0	1	0	1
Haemodoraceae	2	1	8	7	7	0	8
Haloragidaceae	2	0	2	0	2	0	2
Hamamelidaceae	1	0	1	0	0	0	0
Hemerocallidaceae	1	0	3	2	2	0	3
Hyacinthaceae	14	0	192	87	126	22	93
Hydnoraceae	1	0	1	0	1	0	1
Hydrocharitaceae	2	0	2	0	0	0	0
Hypoxidaceae	4	1	31	16	15	2	16
Icacinaceae	3	0	5	1	0	0	3
Iridaceae	28	6	663	521	356	123	379
Juncaceae	1	0	14	4	11	1	10
Juncaginaceae	1	0	2	0	2	0	2
Kiggelariaceae	1	0	1	0	1	0	1
Lamiaceae	8	0	43	6	17	0	18
Lanariaceae	1	0	1	0	0	0	1
Lauraceae	3	0	3	1	2	0	3
Lentibulariaceae	1	0	3	0	1	0	1
Linaceae	1	0	14	12	4	2	9
Loganiaceae	2	0	3	0	0	0	0
Loranthaceae	2	0	2	0	2	0	1
Malvaceae	8	0	88	47	39	5	43
Meliaceae	2	0	2	0	1	0	0
Melianthaceae	1	0	3	0	2	0	2
Menispermaceae	2	0	3	0	1	0	1
Menyanthaceae	2	0	3	2	2	0	3
Molluginaceae	9	0	53	17	40	9	29
Montiniaceae	1	0	1	0	1	0	1
Moraceae	1	0	5	0	2	0	0
Myricaceae	1	0	7	3	3	0	6
Myrsinaceae	2	0	4	0	1	0	3
Myrtaceae	1	0	1	1	1	0	1
Najadaceae	1	0	1	0	0	0	0
Neuradaceae	1	0	2	0	2	0	2
Nymphaeaceae	1	0	1	0	1	0	1
Ochnaceae	1	0	2	0	0	0	0

Appendix: Statistics for families of the Cape flora

Endemic	AP spp.	Endemic	KM spp.	Endemic	LB spp.	Endemic	SE spp.	Endemic
0	5	0	3	0	8	0	13	0
0	0	0	0	0	1	0	1	0
1	24	1	66	5	32	0	60	0
0	1	0	2	0	2	0	5	0
0	1	0	0	0	2	0	2	0
0	0	0	0	0	1	0	2	1
29	66	2	22	1	73	1	117	2
0	1	0	1	0	1	0	1	0
0	0	0	1	0	0	0	5	2
2	1	0	1	0	2	0	5	0
3	4	0	4	0	5	0	3	0
0	1	0	5	0	8	0	11	0
0	1	0	1	0	1	0	1	0
248	75	33	93	22	121	47	109	42
1	12	1	27	6	33	1	48	2
141	83	26	148	39	164	28	196	42
0	0	0	0	0	1	0	7	0
0	2	0	1	0	2	0	2	0
0	2	0	2	0	1	0	2	0
0	0	0	0	0	1	1	0	0
1	13	0	3	0	23	1	17	0
13	23	0	56	3	41	2	58	2
0	0	0	0	0	1	0	2	0
0	1	0	0	0	1	0	1	0
1	2	0	0	0	2	0	2	0
0	0	0	1	0	1	0	1	0
1	4	0	1	0	5	0	3	0
0	1	0	1	0	1	0	1	0
0	0	0	0	0	1	0	1	0
1	1	0	2	0	1	0	1	0
17	41	3	62	5	40	1	56	2
0	0	0	1	0	1	0	1	0
0	0	0	0	0	0	0	2	0
3	10	0	3	0	7	0	10	0
1	1	0	1	0	4	0	4	0
132	103	3	117	24	148	11	114	9
0	6	0	3	0	9	0	9	0
0	1	0	1	0	2	0	2	0
0	1	0	1	0	1	0	1	0
0	9	0	10	0	15	1	31	0
0	0	0	0	0	1	0	1	0
0	1	0	1	0	3	0	3	0
0	1	0	1	0	1	0	3	0
2	4	0	3	0	6	0	7	1
0	0	0	0	0	0	0	3	0
0	0	0	2	0	1	0	1	0
5	22	2	32	3	30	1	38	1
0	0	0	1	0	0	0	2	0
0	0	0	1	0	1	0	1	0
0	1	0	1	0	1	0	2	0
1	0	0	0	0	2	0	2	0
3	7	0	10	0	9	0	12	0
0	1	0	1	0	1	0	1	0
0	0	0	1	0	0	0	3	0
1	3	0	2	0	5	0	4	0
0	2	0	1	0	2	0	4	0
0	0	0	0	0	1	0	0	0
0	0	0	0	0	0	0	1	0
0	0	0	1	0	0	0	0	0
0	1	0	0	0	1	0	0	0
0	0	0	0	0	0	0	2	0

Appendix: Statistics for families of the Cape flora

Seed Plants	Genera	Endemic	Species	Endemic	NW spp.	Endemic	SW spp.
Oleaceae	4	0	7	2	3	0	6
Oliniaceae	1	0	1	0	0	0	1
Onagraceae	1	0	2	0	2	0	2
Orchidaceae	25	2	227	138	98	6	144
Orobanchaceae	8	0	21	6	9	2	9
Oxalidaceae	1	0	118	95	78	35	56
Papaveraceae	1	0	1	0	0	0	0
Penaeaceae	7	7	23	23	1	0	17
Piperaceae	2	0	3	0	0	0	1
Pittosporaceae	1	0	1	0	0	0	0
Plantaginaceae	1	0	3	0	1	0	3
Plumbaginaceae	2	0	15	6	6	0	9
Poaceae	61	3	207	80	124	9	146
Podocarpaceae	2	0	3	1	1	0	2
Polygalaceae	3	0	141	122	46	10	89
Polygonaceae	3	0	6	2	4	0	6
Portulacaceae	2	0	7	2	6	1	2
Potamogetonaceae	1	0	5	0	2	0	2
Primulaceae	3	0	3	0	1	0	1
Prioniaceae	1	0	1	0	1	0	1
Proteaceae	14	9	330	319	104	44	181
Ranunculaceae	4	0	9	3	6	0	6
Resedaceae	1	0	1	0	0	0	0
Restionaceae	19	10	318	294	132	28	224
Rhamnaceae	5	1	137	126	52	32	61
Roridulaceae	1	1	2	2	1	1	1
Rosaceae	5	0	120	105	35	5	78
Rubiaceae	16	0	50	26	22	3	28
Ruppiaceae	1	0	2	0	1	0	2
Rutaceae	15	6	273	258	95	65	87
Salicaceae	1	0	1	0	1	0	1
Salvadoraceae	1	0	1	0	0	0	0
Santalaceae	4	0	94	69	37	7	58
Sapindaceae	6	1	6	1	1	0	1
Sapotaceae	1	0	1	0	0	0	1
Scrophulariaceae	33	7	418	297	241	92	153
Solanaceae	2	0	18	2	9	0	12
Stilbaceae	6	6	14	14	2	1	11
Strelitziaceae	1	0	3	2	0	0	0
Tecophilaeaceae	2	0	5	0	5	0	2
Thymelaeaceae	4	1	124	94	44	15	70
Typhaceae	1	0	1	0	1	0	1
Urticaceae	4	0	6	0	1	0	2
Vahliaceae	1	0	1	0	1	0	1
Valerianaceae	1	0	1	0	0	0	1
Verbenaceae	1	0	2	0	0	0	1
Violaceae	2	0	2	1	0	0	1
Viscaceae	1	0	8	1	4	0	5
Vitaceae	2	0	5	1	0	0	1
Xyridaceae	1	0	1	0	1	0	1
Zamiaceae	1	0	4	0	0	0	0
Zannichelliaceae	2	0	3	1	2	0	3
Zosteraceae	1	0	1	0	0	0	0
Zygophyllaceae	3	0	23	12	11	0	8
Seed Pl. Totals	942	160	8992	6182	4010	1055	45768
Seed Pl. Center %		17.0%		69.5%	45.1%	26.3%	51.5%
Pteridophytes	46	0	112	10	50	1	78
Vasc. Pl. Totals	988	160	9004	6192	4060	1056	4656
Vasc. Pl. %		16.2%		68.8%	45.1%	26.0%	51.7%

Appendix: Statistics for families of the Cape flora

Endemic	AP spp.	Endemic	KM spp.	Endemic	LB spp.	Endemic	SE spp.	Endemic
0	4	0	4	0	5	0	8	0
0	1	0	0	0	1	0	1	0
0	1	0	1	0	1	0	2	0
21	49	0	50	1	92	4	120	4
0	2	0	5	0	15	0	18	0
37	5	1	29	9	29	13	20	0
0	0	0	0	0	0	0	1	0
14	3	1	1	1	5	3	2	0
0	0	0	1	0	3	0	3	0
0	0	0	0	0	1	0	1	0
0	1	0	1	0	1	0	2	0
2	5	1	3	0	4	0	3	0
16	60	3	62	2	95	1	113	1
0	0	0	1	0	1	0	2	1
46	28	9	27	4	32	2	29	2
0	4	0	1	0	2	0	4	0
0	1	0	5	1	2	0	3	0
0	1	0	1	0	1	0	5	0
0	1	0	0	0	2	0	3	0
0	1	0	1	0	1	0	1	0
99	29	8	51	11	55	5	42	10
0	4	0	3	0	5	0	6	0
0	1	0	1	0	1	0	1	0
101	27	6	59	9	85	16	57	3
34	12	3	18	8	22	8	24	6
1	0	0	0	0	0	0	0	0
35	11	3	32	12	29	5	24	2
8	10	1	9	0	21	2	27	0
0	1	0	0	0	0	0	2	0
49	51	27	39	13	47	16	54	21
0	1	0	1	0	1	0	1	0
0	0	0	0	0	1	0	1	0
21	6	0	14	2	25	1	37	4
0	0	0	2	0	2	0	5	0
0	1	0	0	0	1	0	1	0
28	60	9	131	26	62	5	98	10
0	8	0	9	0	13	0	17	0
7	3	0	0	0	3	2	0	0
0	0	0	0	0	0	0	3	1
0	1	0	2	0	2	0	1	0
20	25	2	21	1	42	6	41	3
0	1	0	1	0	1	0	1	0
0	0	0	1	0	2	0	5	0
0	0	0	0	0	0	0	0	0
0	0	0	1	0	1	0	1	0
0	0	0	2	0	1	0	1	0
0	1	0	0	0	1	0	1	0
0	1	0	5	0	4	0	5	0
0	1	0	0	0	1	0	4	1
0	0	0	1	0	1	0	1	0
0	0	0	0	0	0	0	4	0
0	1	0	0	0	0	0	1	0
0	0	0	0	0	0	0	1	0
2	5	1	12	1	4	0	8	0
1485	1365	205	2102	331	2299	276	2747	274
32.4%	15.4%	15.0%	23.6%	15.7%	25.9%	12.0%	30.9%	10.0%
2	9	0	47	0	65	0	85	1
1487	1374	205	2149	331	2364	276	2832	275
31.9%	15.3%	14.9%	23.9%	15.4%	26.3%	11.7%	31.5%	9.7%

Index

Abutilon	535	Amphithalea	462
Acacia	461	Amsinckia	374
Acaena	608	Anacampseros	573
Acalypha	453	ANACARDIACEAE	270
ACANTHACEAE	219	Anagallis	574
Acanthopsis	220	*Anapalina* see Tritoniopsis	
Acharia	222	Anaxeton	302
ACHARIACEAE	222	Anchusa	374
Achyranthes	268	Anderbergia	302
Acmadenia	621	Andrachne	453
Acokanthera	281	Androcymbium	75
Acrodon	224	Andropogon	179
Acrolophia	156	*Androsiphon* see Daubenya	
Acrosanthes	225	ANEMIACEAE	37
Acrostemon see Erica		Anemone	598
Adenandra	622	Anginon	274
Adenocline	453	Angraecum	157
Adenogramma	543	*Aniserica* see Erica	
Adiantum	46	Anisodontea	536
Adromischus	409	Anisothrix	303
Aethephyllum	225	Anisotoma	281
Afrocarpus	50	Annesorhiza	274
AGAPANTHACEAE	52	*Anochilus* see Pterygodium	
Agapanthus	52	Anogramma	46
Agathelpis see Microdon		*Anomalanthus* see Erica	
Agathosma	624	*Anomalesia* see Gladiolus	
Agrostis	179	*Anomatheca* see Freesia	
Aira	179	Antegibbaeum	226
AIZOACEAE	222	ANTHERICEAE	60
Aizoon	225	*Anthericum* see Chlorophytum	
Albuca	94	Anthochortus	200
Alchemilla	608	*Antholyza* see Babiana	
Alciope	301	Anthospermum	617
Alectra	550	Anthoxanthum	179
Alepidia	274	Antimima	226
Alhagi	462	Antizoma	542
ALLIACEAE	52	Apatesia	228
Allium	52	APIACEAE	272
Allophylus	642	Apium	275
Aloe	64	APOCYNACEAE	280
Alonsoa	644	Apodolirion	54
Althenia	219	Apodytes	529
AMARANTHACEAE	268	Aponogeton	61
AMARYLLIDACEAE	53	APONOGETONACEAE	61
Amaryllis	54	Aptenia	228
Amauropelta	49	Aptosimum	645
Amellus	301	AQUIFOLIACEAE	290
Ammocharis	54	ARACEAE	61
Ammophila	179	*Arachnocalyx* see Erica	
Amphibolia	226	ARALIACEAE	291
Amphiglossa	301	Arctopus	275
Amphisiphon see Daubenya		Arctotheca	303

Arctotis	303	BLECHNACEAE	39
Arenifera	228	Blechnum	39
Argyrolobium	465	Blepharis	220
Aridaria	229	Blotiella	40
Aristea	112	Bobartia	118
Aristida	180	Bolusafra	483
Artemisia	306	Bonatea	157
Arthrosolen see Gnidia		Boophone	54
Arundinella	180	BORAGINACEAE	374
Arundo	180	Bowiea	96
Asaemia	306	Brabejum	575
ASCLEPIADACEAE see APOCYNACEAE		Brachiaria	180
Asclepias	282	Brachycarpaea	378
Askidiosperma	201	Brachycorythis	157
Aspalathus	466	Brachylaena	310
ASPARAGACEAE	62	Brachypodium	180
Asparagus	62	Brachysiphon	560
ASPHODELACEAE	64	Brachystelma	282
Aspidoglossum	282	BRASSICACEAE	378
ASPLENIACEAE	37	Braunsia	229
Asplenium	37	Briza	181
Aster	306	Bromus	181
ASTERACEAE	295	Brownanthus	229
Astephanus	282	Brownleea	158
Astroloba	66	Brunia	384
Atalaya	642	BRUNIACEAE	383
Athanasia	306	Brunsvigia	54
Athrixia	309	Bryomorphe	311
Atrichantha	309	*Buchenroedera* see Lotononis	
Atriplex	268	Buchnera	551
Audouinia	383	Buddleja	388
Augea	688	BUDDLEJACEAE	388
Aulax	575	Bulbine	67
Avena	180	Bulbinella	68
Azima	636	Bulboschoenus	82
Azolla	38	Bulbostylis	82
AZOLLACEAE	38	Bupleurum	275
Babiana	115	Burchellia	617
Baeometra	76	Cadaba	378
BALANOPHORACEAE	373	Cadiscus	311
Ballota	530	Caesia	93
BALSAMINACEAE	373	Calanthe	158
Barleria	220	Calodendrum	631
Bartholina	157	Calopsis	201
Bartsia	550	Calotesta	311
BEHNIACEAE	75	Calpurnia	483
Behnia	75	CAMPANULACEAE	388
Bergia	423	Campylostachys	675
Berkheya	309	Cannomois	203
Berula	275	Canthium	618
Berzelia	383	Capeobolus	82
BIGNONIACEAE	373	Capnophyllum	275
Bijlia	229	*CAPPARACEAE* see BRASSICACEAE	
Blackiella see Atriplex		Capparis	378
Blaeria see Erica		Cardamine	378

Carex	82	Chrysanthemoides	311
Carissa	282	Chrysitrix	83
Carpanthea	229	Chrysosoma	311
Carpha	82	Cineraria	312
Carpobrotus	229	Circandra	231
Carpococe	618	Cissampelos	542
Carpolyza	55	Citrullus	419
Carruanthus	230	Cladium	83
Caryotophora	230	Cladoraphis	182
CARYOPHYLLACEAE	402	Clausena	631
Cassia see Chamaecrista		Clematis	598
Cassine	405	Cleretum	231
Cassinopsis	529	Cliffortia	608
Cassytha	51	CLUSIACEAE	408
Castalis see Dimorphotheca		Clutia	453
Catapodium	181	Coccinia	419
Cedronella	530	*Coccosperma* see Erica	
CELASTRACEAE	404	Codon	374
CELTIDACEAE	407	Coelanthum	544
Celtis	407	*Coelidium* see Amphithalea	
Cenchrus	181	*Coilostigma* see Erica	
Cenia see Cotula		COLCHICACEAE	75
Centella	291	Coleonema	632
Cephalaria	420	Colpias	645
Cephalophyllum	230	*Colpoon* see Osyris	
Cerastium	402	Comborhiza	313
Ceratandra	158	Commelina	77
Ceratiosicyos	222	COMMELINACEAE	77
Ceratocaryum	203	Conicosia	231
CERATOPHYLLACEAE	407	Conium	276
Ceratophyllum	407	Conophytum	232
Cerochlamys	231	CONVALLARIACEAE	78
Ceropegia	282	CONVOLVULACEAE	408
Ceterach	38	Convolvulus	408
Chaetacanthus	220	Conyza	313
Chaetobromus	181	Cordia	374
Chamaecrista	483	CORNACEAE	409
Chamarea	276	Corpuscularia	233
Chamira	378	Corrigiola	402
Charadrophila	645	Cortaderia	182
Chareis see Felicia		Corycium	158
Chascanum	687	Corymbium	313
Chasmanthe	119	Corynephorus	182
Cheilanthes	46	*Costularia* see Capeobolus	
Cheirodopsis	231	Cotula	314
Chenolea	269	Cotyledon	410
CHENOPODIACEAE see AMARANTHACEAE		Crassula	411
Chenopodiopsis	645	CRASSULACEAE	409
Chenopodium	269	Crinum	55
Chionanthus	549	Crocosmia	119
Chironia	514	Cromidon	645
Chloris	181	Crossyne	55
Chlorophytum	60	Crotalaria	483
Chondropetalum	203	Cryptocarya	51
Christella	49	Ctenomeria	454

CUCURBITACEAE	419	Didelta	317
Cullen	483	Didymodoxa	686
Cullumia	315	Dierama	119
Cunonia	419	Dietes	119
CUNONIACEAE	419	Digitaria	182
CUPRESSACEAE	50	Dilatris	92
Curtisia	409	Dimorphotheca	317
Cuscuta	408	Dioscorea	92
Cuspidia	316	DIOSCOREACEAE	92
Cussonia	295	Diosma	632
Cyanella	218	Diospyros	421
Cyanotis	77	Dipcadi	97
CYATHACEAE	39	Diplachne	183
Cyathea	39	Diplosoma	235
Cyathocoma	83	Dipogon	485
Cybistetes	55	DIPSACACEAE	420
Cyclopia	483	Disa	159
Cycloptychis	379	Dischisma	649
Cyclosorus	49	Discocapnos	513
Cycnium	551	Disparago	318
Cylindrophyllum	233	Disperis	166
Cymbopappus	316	Disphyma	235
Cymbopogon	182	Dodonaea	642
Cynanchum	283	Dolichos	485
Cynodon	182	Dolichothrix	318
Cynoglossum	374	Dorotheanthus	235
Cynosurus	182	Dovea	204
CYPERACEAE	81	Dovyalis	512
Cyperus	83	Dracaena	78
Cyphia	388	*DRACAENACEAE* see CONVALLARIACEAE	
Cyphostemma	688	Drimia	97
Cypselodontia	316	Droguetia	686
Cyrtanthus	55	Drosanthemum	235
Cyrtorchis	159	Drosera	420
Cysticapnos	513	DROSERACEAE	420
Cystopteris	40	DRYOPTERIDACEAE	40
CYTINACEAE	420	Dryopteris	40
Cytinus	420	Dumasia	485
Dalechampia	454	Duvalia	284
Dasispermum	276	Dymondia	318
Daubenya	96	EBENACEAE	421
Deilanthe	233	Echinochloa	183
Delairea	316	Echiostachys	375
Delosperma	233	Echium	375
DENNSTAEDTIACEAE	39	Edmondia	318
Deverra	276	Ehretia	375
Dianthus	402	Ehrharta	183
Diascia	645	Ekebergia	542
Diastella	575	Elaeodendron	405
Dichondra	409	Elaphoglossum	42
Dichrocaulon	235	ELATINACEAE	423
Dichrocephala	316	Elegia	204
Dicliptera	220	Eleocharis	84
Diclis	648	Eleusine	185
Dicoma	317	Elionurus	185

Elytropappus	319	Foveolina	326
Emex	573	Frankenia	513
Empleuridium	405	FRANKENIACEAE	513
Empleurum	634	Freesia	120
Empodium	108	Freylinia	649
Encephalartos	51	Fuirena	87
Endonema	560	Fumaria	514
Enneapogon	185	FUMARIACEAE	513
Epilobium	550	*Galaxia* see Moraea	
Epischoenus	84	Galenia	242
EQUISETACEAE	41	Galium	618
Equisetum	41	Galopina	619
Eragrostis	185	Gardenia	619
Eremia see Erica		Garuleum	326
Eremiella see Erica		Gasteria	69
Erepsia	239	Gastridium	186
Erica	423	Gastrodia	168
ERICACEAE	423	Gazania	326
Eriocephalus	319	Geissoloma	514
Eriosema	485	GEISSOLOMATACEAE	514
ERIOSPERMACEAE see CONVALLARIACEAE		Geissorhiza	120
Eriospermum	78	GENTIANACEAE	514
Erythrina	485	GERANIACEAE	516
Esterhuysenia	241	Geranium	516
Euchaetis	634	Gerbera	327
Euclea	422	GESNERIACEAE	528
Eucomis	99	Gethyllis	57
Eulalia	186	Gibbaeum	243
Eulopia	167	Gibbaria	328
Euphorbia	455	Gladiolus	125
EUPHORBIACEAE	452	Glechenia	41
Eurylobium see Stilbe		GLEICHENIACEAE	41
Euryops	320	Glia	276
Eustachys	186	Glischrocolla	560
Eustegia	284	Globulariopsis	650
Euthystachys	675	Glottiphyllum	244
Evotella	168	Gloveria	405
Excoecaria	458	Gnaphalium	328
Exomis	269	Gnidia	676
Ezosciadium	276	Gomphocarpus	285
FABACEAE	459	Gonioma	285
Falkia	409	GOODENIACEAE	528
Faucaria	242	Gorteria	328
Faurea	576	Gosela	650
Felicia	323	Graderia	551
Ferraria	119	Grammatotheca	390
Festuca	186	GRAMMITIDACEAE	41
Ficinia	84	Grammitis	41
Ficus	547	Grewia	536
Fimbristylis	87	Grielum	548
Fingerhuthia	186	*Grisebachia* see Erica	
FLACOURTIACEAE	512	Grubbia	528
Fockea	284	GRUBBIACEAE	528
Foeniculum	276	Gunnera	528
Forsskaolea	686	GUNNERACEAE	528

Gymnodiscus	328	Hordeum	187
Gymnosporia	406	Huernia	285
Gymnostephium	328	Huperzia	43
Gynandriris see Moraea		HYACINTHACEAE	93
Habenaria	168	Hybanthus	687
Haemanthus	58	Hydnora	529
HAEMODORACEAE	92	HYDNORACEAE	529
Hainardia	187	HYDROCHARITACEAE	108
Hakea	576	Hydrocotyle	295
Halleria	650	Hydroidea	337
Hallianthus	245	Hydrophilus	205
Halopeplis	269	Hydrophylax	619
Halophila	108	Hyenanche	458
HALORAGACEAE	528	Hymenogyne	245
HAMAMELIDACEAE	529	Hymenolepis	337
Haplocarpha	329	HYMENOPHYLLACEAE	42
Harpochloa	187	Hymenophyllum	42
Hartogiella see Cassine		Hyobanche	552
Harveya	551	Hyparrhenia	188
Haworthia	70	Hypericum	408
Hebenstretia	651	Hypertelis	544
Heeria	270	Hypocalyptus	485
Helichrysum	329	Hypodiscus	206
Helictotrichon	187	Hypoestes	220
Heliophila	379	Hypolepis	40
Heliotropium	375	HYPOXIDACEAE	108
Helipterum see Syncarpha		Hypoxis	108
Hellmuthia	87	ICACINACEAE	529
Hemarthria	187	Ifloga	337
HEMEROCALLIDACEAE	93	Ilex	290
Hemimeris	651	*ILLECEBRACEAE* see CARYOPHYLLACEAE	
Herschelia see Disa		Ilysanthes	652
Hereroa	245	Impatiens	373
Hermannia	537	Imperata	188
Hermas	277	Indigofera	486
Herniaria	403	Inulanthera	337
Hertia	335	Ipomoea	409
Hesperantha	131	IRIDACEAE	110
Hessea	58	Ischyrolepis	206
Heterolepis	336	ISOETACEAE	42
Heteromorpha	277	Isoetes	42
Heteropogon	187	Isoglossa	221
Heterorhachis	336	Isolepis	87
Hexaglottis see Moraea		Itasina	277
Hibiscus	541	Ixia	133
Hippia	336	Ixianthes	652
Hippobromus	643	Jamesbrittenia	652
Hirpicium	336	Jasminum	549
Histiopteris	40	Jatropha	458
Holcus	187	Jordaaniella	246
Holothrix	168	JUNCACEAE	154
Homeria see Moraea		JUNCAGINACEAE	155
Homoglossum see Gladiolus		Juncus	154
Hoodia	285	Justicia	221
Hoplophyllum	336	Kalanchoe	418

Karroochloa	188	Leysera	338
Kedrostis	419	Lichtensteinia	277
Kensitia see Erepsia		Lidbeckia	339
Kiggelaria	530	*Lightfootia* see Wahlenbergia	
KIGGELARIACEAE	530	Limeum	544
Klattia	135	Limonium	562
Kleinia	337	Limosella	653
Kniphofia	73	LINACEAE	533
Knowltonia	598	Linconia	384
Koeleria	188	Linum	533
Kogelbergia	675	Liparia	495
Kyllinga	88	Liparis	169
Lachenalia	99	*Litanthus* see Drimia	
Lachnaea	680	Lithops	253
Lachnospermum	337	Lithospermum	375
Lachnostylis	458	Lobelia	390
Lagarosiphon	108	Lobostemon	375
Lagenaria	419	LOGANIACEAE	534
Lagurus	188	Lolium	188
LAMIACEAE	530	LOMARIOPSIDACEAE	42
Lampranthus	246	Lonchostoma	385
Lamprocephalus	338	Lophochloa	189
Lanaria	155	LORANTHACEAE	535
LANARIACEAE	155	Lotononis	496
Langebergia	338	Lotus	500
Lapeirousia	135	Loxostylis	271
Laportea	686	Lycium	673
Lappula	375	LYCOPODIACEAE	43
Lasiochloa see Tribolium		Lycopodiella	43
Lasiopogon	338	Lycopodium	43
Lasiosiphon see Gnidia		Lyperia	653
Lasiospermum	338	Lysimachia	574
LAURACEAE	51	LYTHRACEAE	535
Laurembergia	529	Lythrum	535
Laurentia see Wimmerella		Machairophyllum	253
Lauridia	406	*Macrochaetum* see Cyathocoma	
Laurophyllus	271	Macropetalum	286
Lavatera	541	Macrostylis	635
Lebeckia	493	Maerua	382
Ledebouria	104	Mairia	339
Leersia	188	Malephora	253
Leidesia	459	MALVACEAE	535
Leipoldtia	253	Manochlamys	269
Lemna	61	Manulea	653
LEMNACEAE see ARACEAE		Marasmodes	339
LENTIBULARIACEAE	533	Marattia	43
Leonotis	530	MARATTIACEAE	43
Lepidium	382	Mariscus	88
Lepisorus	45	Marlothistella	254
Leptospermum	548	Marsilea	44
Lessertia	494	MARSILEACEAE	44
Leucadendron	576	Massonia	104
Leucas	530	Mastersiella	209
Leucoptera	338	Maurocenia	406
Leucospermum	582	Maytenus	406

Melasma	552	MYRTACEAE	548
Melasphaerula	136	Mystacidium	170
MELIACEAE	542	Mystropetalon	373
MELIANTHACEAE	542	Mystroxylon	407
Melianthus	542	*Nagelocarpus* see Erica	
Melica	189	NAJADACEAE	155
Melinis	189	Najas	155
Melolobium	500	Nassella	190
MENISPERMACEAE	542	Nebelia	385
Menodora	549	Neesenbeckia	89
Mentha	531	Nemesia	656
MENYANTHACEAE	542	Nenax	619
Merciera	391	Neodregea	76
Merxmuellera	189	Neopatersonia	104
MESEMBRYANTHEMACEAE see AIZOACEAE		NEPHROLEPIDACEAE	44
Mesembryanthemum	254	Nephrolepis	44
Mestoklema	255	Nerine	59
Metalasia	339	Nestlera	343
Metrosideros	548	NEURADACEAE	548
Micranthus	136	Nevillea	209
Microcodon	392	Nicotiana	673
Microdon	656	Nidorella	343
Microglossa	343	Nivenia	144
Microloma	286	Noltea	599
Micropterum see Cleretum		Nuxia	534
Microstegium	190	Nylandtia	570
Mikania	343	Nymania	542
Mimetes	585	Nymphaea	51
Minurothamnus	343	NYMPHAEACEAE	51
Miscanthidium see Miscanthus		Nymphoides	543
Miscanthus	190	Ochna	548
Mniothamnea	385	OCHNACEAE	548
Mohria	37	Ocotea	51
MOLLUGINACEAE	543	Octopoma	255
Mollugo	544	Odyssea	190
Monadenia see Disa		Oedera	344
Monechma	221	Oftia	658
Monilaria	255	Oldenburgia	344
Monopsis	392	Oldenlandia	620
Monsonia	516	Olea	549
Montinia	546	OLEACEAE	549
MONTINIACEAE	546	Oligocarpus	345
Moquiniella	535	Oligomeris	599
MORACEAE	547	Oligothrix	345
Moraea	137	Olinia	549
Morella	547	OLINIACEAE	549
Muiria see Gibbaeum		ONAGRACEAE	550
Muraltia	563	Oncinema	286
Myosotis	377	Oncosiphon	345
Myrica see Morella		Onixotis	76
MYRICACEAE	547	OPHIOGLOSSACEAE	44
Myriophyllum	529	Ophioglossum	44
MYRSINACEAE	547	Oplismenus	190
Myrsine	548	Orbea	286
Myrsiphyllum see Asparagus		ORCHIDACEAE	155

Oreoleysera	345	Phaneroglossa	352
Ornithogalum	104	Pharnaceum	545
Ornithoglossum	76	*Philippia* see Erica	
OROBANCHACEAE	550	Phragmites	194
Orobanche	552	Phylica	599
Orothamnus	586	Phyllanthus	459
Orphium	515	Phyllobolus	256
Oscularia	255	Phyllopodium	659
Osmitopsis	345	Phyllosma	635
Osmunda	44	Phymaspermum	352
OSMUNDACEAE	44	Piaranthus	287
Osteospermum	346	Pillansia	145
Osyris	636	Pimpinella	279
Othonna	348	PINACEAE	50
Otholobium	500	Pinus	50
OXALIDACEAE	552	Piper	52
Oxalis	552	PIPERACEAE	52
Oxylaena	351	PITTOSPORACEAE	562
Pachites	170	Pittosporum	562
Pachycarpus	286	*Plagiochloa* see Tribolium	
Pachypodium	286	Planea	352
Panicum	190	PLANTAGINACEAE	562
Papaver	560	Plantago	562
PAPAVERACEAE	560	*Platycalyx* see Erica	
Pappea	643	Platycarpha	352
Paranomus	586	Platycaulos	209
Parapholis	190	Platylophus	420
Paraserianthes	503	*Platythyra* see Aptenia	
Paronychia	403	Plecostachys	352
Paspalum	190	Plectranthus	531
Passerina	683	Pleiospilos	257
Pauridia	109	Pleopeltis	45
Pavonia	541	×Pleopodium	45
Pectinaria	287	*Plexipus* see Chascanum	
Pegolettia	351	PLUMBAGINACEAE	562
Pelargonium	517	Plumbago	563
Peliostomum	658	Poa	195
Pellaea	47	Polypogon	195
Penaea	561	POACEAE	174
PENAEACEAE	560	Podalyria	503
Pennisetum	191	PODOCARPACEAE	50
Pentameris	191	Podocarpus	50
Pentaschistis	192	Poecilolepis	352
Pentzia	351	*Poellnitzia* see Astroloba	
Peperomia	52	Polemanniopsis	279
Perdicium	351	Polhillia	504
Periboea see Polyxena		Pollichia	403
Peristrophe	221	Polpoda	546
Petalacte	351	Polyarrhena	352
Petrorhagia	403	Polycarena	660
Peucedanum	278	Polycarpon	403
Peyrousia see Schistostephium		Polygala	570
Phacocapnos see Cysticapnos		POLYGALACEAE	563
Phaenocoma	351	POLYGONACEAE	573
Phalaris	194	Polygonum	573

POLYPODIACEAE	45	Relhania	355
Polypodium	45	RESEDACEAE	599
Polypogon	195	Restio	210
Polystachya	170	RESTIONACEAE	199
Polystichum	41	Retzia	675
Polyxena	107	*RETZIACEAE* see STILBACEAE	
PORTULACACEAE	573	*Rhadamanthus* see Drimia	
Portulacaria	574	RHAMNACEAE	599
Potamogeton	199	Rhamnus	607
POTAMOGETONACEAE	199	*Rheome* see Moraea	
Prenia	257	Rhigiophyllum	395
Priestleya see Liparia, Xiphotheca		Rhigozum	373
PRIMULACEAE	574	Rhinephyllum	258
Printzia	353	Rhoicissus	688
Prionanthium	195	Rhodocoma	214
PRIONIACEAE	199	Rhoiacarpos	637
Prionium	199	Rhombophyllum	258
Prismatocarpus	393	Rhus	271
Protasparagus see Asparagus		*Rhynchelytrum* see Melinis	
Protea	587	Rhynchopsidium	356
PROTEACEAE	574	Rhynchosia	509
Prunus	616	Rhynchospora	89
Psammotropha	546	*Rhyticarpus* see Anginon	
Pseudobaeckia	385	Riocreuxia	288
Pseudognaphalium	353	Robsonodendron	407
Pseudopentameris	195	Roella	395
Pseudoschoenus	89	*Roggeveldia* see Moraea	
Pseudoscolopia	513	Romulea	145
Pseudoselago	661	Roridula	608
Psilocaulon	257	RORIDULACEAE	608
Psoralea	505	ROSACEAE	608
Psychotria	620	Rosenia	356
Psydrax	620	Rothmannia	620
PTAEROXYLACEAE see RUTACEAE		Rubia	620
Ptaeroxylon	635	RUBIACEAE	616
PTERIDACEAE	45	Rubus	616
Pteridium	40	Ruellia	221
Pteris	48	Rumex	573
Pterocelastrus	407	Rumohra	41
Pteronia	353	Ruppia	217
Pterothrix see Amphiglossa		RUPPIACEAE	217
Pterygodium	170	Ruschia	258
Puccinella	195	RUTACEAE	620
Pulicaria	355	*Salaxis* see Erica	
Pupalia	269	SALICACEAE	636
Putterlickia	407	Salicornia	269
Pycreus	89	Salix	636
Pyrenocantha	529	Salsola	269
Quaqua	287	Saltera	561
RAFFLESIACEAE see CYTINACEAE		SALVADORACEAE	636
Rafnia	507	Salvia	531
RANUNCULACEAE	598	Salvinia	48
Ranunculus	599	SALVINIACEAE	48
Rapanea	548	Samolus	574
Raspalia	386	Sanicula	279

Sansevieria	80	Silene	403
SANTALACEAE	636	Silicularia	382
Saphesia	263	*Simocheilus* see Erica	
SAPINDACEAE	642	Siphocodon	396
SAPOTACEAE	643	Siphonoglossa	221
Sarcocaulon see Monsonia		Sisymbrium	382
Sarcocornia	270	Skiatophytum	264
Sarcostemma	288	Smelophyllum	643
Satyridium see Satyrium		Smicrostigma	264
Satyrium	171	SOLANACEAE	673
Scabiosa	420	Solanum	673
Scadoxus	59	Sonderina	279
Scaevola	528	Sonderothamnus	561
Sceletium	263	Sopubia	552
Schefflera	295	Sorocephalu	595
Schismus	196	Sparaxis	148
Schistostephium	356	Sparrmannia	541
Schizaea	48	Spartina	196
SCHIZAEACEAE	48	Spatalla	596
Schizobasis see Drimia		Spergula	404
Schizodium	173	Spergularia	404
Schizoglossum	288	Sphaerocionium	42
Schlechteria	382	*Sphaeroclinium* see Cotula	
Schoenoplectus	89	*Sphalmanthus* see Phyllobolus	
Schoenoxiphium	89	Sphenopus	196
Schoenus	90	Spiloxene	109
Schotia	510	Spirodella	62
Scilla	107	Sporobolus	196
Scirpoides	90	Staavia	386
Scirpus	90	Staberoha	215
Scleranthus	403	Stachys	532
Scleria	90	Stapelia	288
Scolopia	513	Stapeliopsis	289
Scopelogena	264	Stayneria	264
SCROPHULARIACEAE	643	Stegnogramma	49
Scutia	607	Steirodiscus	364
Scyphogyne see Erica		Stellaria	404
Sebaea	515	Stenoglottis	174
Secamone	288	Stenotaphrum	197
Seetzenia	688	Sterculia	542
Seidelia	459	*STERCULIACEAE* see MALVACEAE	
Selaginella	49	STILBACEAE	675
SELAGINELLACEAE	49	Stilbe	675
Selago	663	Stilpnogyne	364
Semnanthe see Erepsia		*Stilpnophyton* see Athanasia	
Senecio	356	Stipa	197
Septulina	535	Stipagrostis	197
Sericocoma	270	Stirtonanthus	510
Serruria	592	Stoebe	364
Sesbania	510	Stoeberia	264
Sessilistigma see Moraea		Stoibrax	279
Setaria	196	*Stokoeanthus* see Erica	
Sheilanthera	636	Stomatium	264
Sida	541	Strelitzia	218
Sideroxylon	643	STRELITZIACEAE	218

Streptocarpus	528	Trichocephalus	607
Striga	552	Trichocladus	529
Strumaria	59	Trichodesma	377
Struthiola	684	Trichodiadema	267
Strychnos	534	Trichogyne	368
Stylapterus	561	Trichomanes	42
Suaeda	270	Tridactyle	174
Suessenguthiella	546	Tridentea	290
Sutera	669	Trieenia	671
Sutherlandia see Lessertia		Trifolium	510
Sympieza see Erica		Triglochin	155
Syncarpha	365	Trigonocapnos	514
Syndesmanthus see Erica		Trimeria	513
Synnotia see Sparaxis		Tripteris	368
Syringodea	149	Triraphis	198
Tanquana	265	Tristachya	198
Tarchonanthus	368	Tritonia	150
Tecoma	373	Tritoniopsis	151
Tecomaria see Tecoma		Troglophyton	369
TECOPHILAEACEAE	218	Tromotriche	290
Teedia	671	Tulbaghia	52
Tenicroa see Drimia		Tylecodon	418
Tephrosia	510	Tylophora	290
Tetragonia	265	Typha	218
Tetraria	90	TYPHACEAE	218
Teucrium	533	*ULMACEAE* see CELTIDACEAE	
Thaminophyllum	368	*Unigenes* see Lobelia	
Thamnea	387	*Urginea* see Drimia	
Thamnochortus	215	*Urochlaena* see Tribolium	
Thamnus see Erica		Ursinia	369
Theilera see Wahlenbergia		URTICAEAE	686
THELYPTERIDACEAE	49	Utricularia	533
Thelypteris	50	Vahlia	686
Themeda	197	VAHLIACEAE	686
Thereianthus	149	Valeriana	686
Thesidium	637	VALERIANACEAE	686
Thesium	637	Valerianella	687
Thesmophora	676	Vanzijlia	267
Thinopyrum	197	Vellereophyton	372
Thlaspeocarpa	383	Veltheimia	107
Thoracosperma see Erica		*Venidium* see Arctotis	
Thunbergia	221	Vepris	636
Thunbergiella see Itasina		VERBENACEAE	687
THYMELAEACEAE	676	Vernonia	372
TILIACEAE see MALVACEAE		Vexatorella	598
Tittmannia	387	Vigna	510
Todea	45	Villarsia	543
Torilis	279	Viola	687
Trachyandra	73	VIOLACEAE	687
Trachypogon	197	Virgilia	511
Tragus	197	VISCACEAE	687
Treichelia	396	Viscum	687
Trianoptiles	92	VITACEAE	688
Tribolium	198	Vittaria	50
Tribulus	689	VITTARIACEAE	50

Vlokia	267	Xiphotheca	511
Vulpia	199	XYRIDACEAE	218
Wachendorfia	93	Xyris	218
Wahlenbergia	396	Xysmalobium	290
Walafrida see Selago		Zaluzianskya	672
Walleria	218	ZAMIACEAE	51
Watsonia	152	Zannichellia	219
Whiteheadia	108	ZANNICHELLIACEAE	218
Wiborgia	511	Zantedeschia	62
Widdringtonia	50	Zehyneria	419
Willdenowia	217	Zeuktophyllum	268
Wimmerella	401	Zornia	512
Witsenia	154	Zostera	219
Wolffia	62	ZOSTERACEAE	219
Woodia	290	ZYGOPHYLLACEAE	688
Wurmbea	76	Zygophyllum	689
Xanthoxylum	636	Zyrphelis	372
Xenoscapa	154		
Xeroplana see Stilbe			

STRELITZIA

1. Botanical diversity in southern Africa. 1994. B.J. Huntley (ed.). ISBN 1-874907-25-0.
2. Cyperaceae in Natal. 1995. K.D. Gordon-Gray. ISBN 1-874907-04-8.
3. Cederberg vegetation and flora. 1996. H.C. Taylor. ISBN 1-874907-28-5.
4. Red Data List of southern African plants. 1996. Craig Hilton-Taylor. ISBN 1-874907-29-3.
5. Taxonomic literature of southern African plants. 1997. N.L. Meyer, M. Mössmer & G.F. Smith (eds). ISBN 1-874907-35-8.
6. Plants of the northern provinces of South Africa: keys and diagnostic characters. 1997. E. Retief & P.P.J. Herman. ISBN 1-874907-30-7.
7. Preparing herbarium specimens. 1999. Lyn Fish. ISBN 1-919795-38-3
8. Bulbinella in South Africa. 1999. Pauline L. Perry. ISBN 1-919795-46-4.
9. Cape plants. A conspectus of the Cape flora of South Africa. 2000. P. Goldblatt & J.C. Manning. ISBN 0-620-26236-2.
10. Seed plants of southern Africa: families and genera. 2000. O.A. Leistner (ed.). ISBN 1-919795-51-0.

MEMOIRS OF THE BOTANICAL SURVEY OF SOUTH AFRICA
(discontinued after No. 63)

Still available are:

2. Botanical survey of Natal and Zululand. 1921. R.D. Aitken & G.W. Gale.
8. Researches on the vegetation of Natal. Series II. 1925. J.W. Bews & R.D. Aitken.
17. The vegetation of the Divisions of Albany and Bathurst. 1937. R.A. Dyer.
29. The wheel-point method of survey and measurement of semi-open grasslands and karoo vegetation in South Africa. 1955. C.E.M. Tidmarsh & C.M. Havenga.
31. Studies of the vegetation of parts of the Bloemfontein and Brandfort Districts. 1958. J.W.C. Mostert.
33. The vegetation of the Districts of East London and King William's Town, Cape Province. 1962. D.M. Comins.
39. Flora of Natal. 1973. J.H. Ross. ISBN 0-621-00327-1.
41. The biostratigraphy of the Permian and Triassic. Part 3. A review of Gondwana Permian palynology with particular reference to the northern Karoo Basin, South Africa. 1977. J.M. Anderson. ISBN 0-621-03834-2.
42. Vegetation of Westfalia Estate on the north-eastern Transvaal escarpment. 1977. J.C. Scheepers. ISBN 0-621-03844-X.
43. The bryophytes of southern Africa. An annotated checklist. 1979. R.E. Magill & E.A. Schelpe. ISBN 0-621-04718-X.
44. A conspectus of the African Acacia species. 1979. J.H. Ross. ISBN 0-621-05309-0.
45. The plant ecology of the Isipingo Beach area, Natal, South Africa. 1980. C.J. Ward. ISBN 0-621-05307-4.
46. A phytosociological study of the Upper Orange River Valley. 1980. M.J.A. Werger. ISBN 0-621-05308-2.
47. A catalogue of South African green, brown and red algae. 1984. S.C. Seagrief. ISBN 0-621-07971-5.
49. Pattern analysis in savanna-woodlands at Nylsvley, South Africa. 1984. R.H. Whittaker, J.W. Morris & D. Goodman. ISBN 0-621-08265-1.
50. A classification of the mountain vegetation of the Fynbos Biome. 1985. B.M. Campbell. ISBN 0-621-08862-5.
52. A plant ecological bibliography and thesaurus for southern Africa up to 1975. 1986. A.P. Backer, D.J.B. Killick & D. Edwards. ISBN 0-621-08871-4.
53. A catalogue of problem plants in southern Africa, incorporating the National Weed List of South Africa. 1986. M.J. Wells, A.A. Balsinhas, H. Joffe, V.M. Engelbrecht, G. Harding & C.H. Stirton. ISBN 0-621-09688-1.
55. Barrier plants of southern Africa. 1987. L. Henderson. ISBN 0-621-10338-1.
57. Veld types of South Africa 3rd edn. 1988. J.P.H. Acocks. With separate wall map. ISBN 0-621-11394-8.
58. Grasses of southern Africa. 1990. G.E. Gibbs Russell, L. Watson, M. Koekemoer, L. Smook, N.P. Barker, H.M. Anderson & M.J. Dallwitz. ISBN 0-620-14846-2.
59. Tannin-like substances in grass leaves. 1990. R.P. Ellis. ISBN 0-620-15151-X.
61. The marine red algae of Natal, South Africa: Order Gelidiales (Rhodophyta). 1992. Richard E. Norris. ISBN 1-874907-01-3.
63. Biomes of southern Africa: an objective categorization. 2nd edn. 1994. M.C. Rutherford & R.H. Westfall. ISBN 1-874907-24-2.

ANNALS OF KIRSTENBOSCH BOTANIC GARDENS
(discontinued after Vol. 19)

The following volumes are available:

14. The moraeas of southern Africa. 1986. P. Goldblatt. ISSN 0-258-3305. ISBN 0-620-09974-7.
15. The botany of the southern Natal Drakensberg. 1987. O.M. Hilliard & B.L. Burtt. ISSN 0-258-3305. ISBN 0-620-10625-5.
17. The Lachenalia handbook. 1988. G. Duncan. ISSN 0-258-3305. ISBN 0-620-11953-5.
18. The way to Kirstenbosch. 1988. D.P. McCracken & E.M. McCracken. ISSN 0-258-3305. ISBN 0-620-11648-X.
19. The genus Watsonia. 1989. P. Goldblatt. ISSN 0-258-3305. ISBN 0-620-12517-9.

ENQUIRIES:

Bookshop, National Botanical Institute, Private Bag X101, Pretoria, 0001 South Africa.
Tel. (012) 804-3200 Fax (012) 804-3211 E-mail bookshop@nbipre.nbi.ac.za

Topography & Mountain Ranges

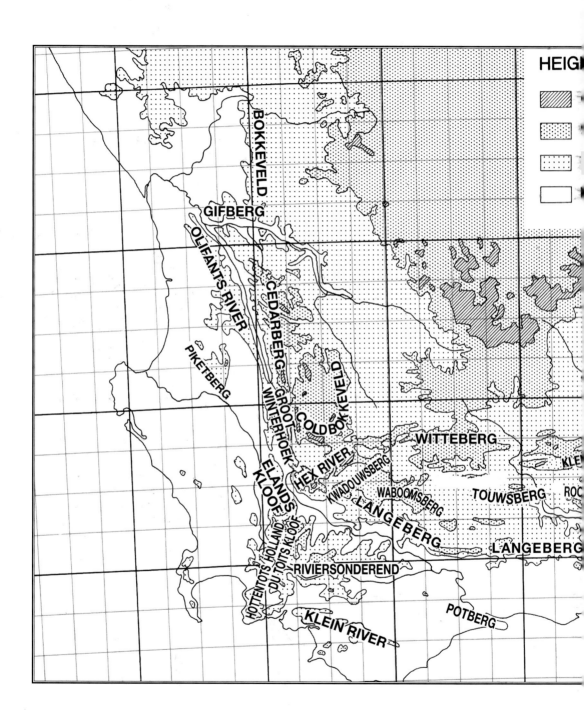